PRINCIPLES OF GENETICS

SECOND EDITION

D. Peter Snustad
University of Minnesota

Michael J. Simmons
University of Minnesota

Epilogue by James F. Crow, University of Wisconsin, Madison
Problems Edited by H. James Price, Texas A & M University

John Wiley & Sons, Inc.

New York • Chichester • Weinheim • Brisbane • Singapore • Toronto

EXECUTIVE EDITOR	David Harris
SENIOR DEVELOPMENT EDITOR	Nancy Perry
MARKETING MANAGER	Clay Stone
PRODUCTION EDITOR	Barbara Russiello
SENIOR DESIGNER	Harry Nolan
PHOTO EDITOR	Jill Hilycord
PHOTO ASSISTANTS	Jennifer Atkins, Ramon Rivera Moret
ILLUSTRATION EDITOR	Edward Starr
ILLUSTRATION	Radiant/Poole VCG

Cover Photos: (*From left to right, top to bottom*)
Tunneling micrograph of DNA—Lawrence Berkeley Laboratory/SPL/Photo Researchers.
Honey bees—Brownie Harris/The Stock Market.
Albino frog—Joe & Carol McDonald/Bruce Coleman.
Sweet pea—Bill Ivy/Tony Stone Images.
HIV virus—Scott Camazine/Photo Researchers.
Drosophila—Oliver Meckes/Photo Researchers.
Dividing breast cancer cell—Quest/SPL/Photo Researchers.
Immunofluorescent light micrograph of HeLa cancer cells—Nancy Kedersha/Immunogen/SPL/Photo Researchers.
SEM of human metaphase chromosomes—Biophoto Associates/SS/Photo Researchers.
Cell division, mitosis—Conly L. Rider/BPS/Tony Stone Images.

This book was set in 10/12 Palatino by York Graphic Services, Inc. and printed and bound by Von Hoffmann Press, Inc. The cover was printed by Phoenix Color.

This book is printed on acid free paper. ∞

Library of Congress Cataloging-in-Publication Data
Snustad, D. Peter.
Principles of genetics / D. Peter Snustad, Michael J. Simmons. —2nd ed.
p. cm.
Includes bibliographical references and index.
ISBN 0-471-29800-X (cloth: alk. paper)
1. Genetics. I. Simmons, Michael J.7001- . II. Title.
QH430.S68 1999 99-15482
576.5—dc21 CIP

Printed in the United States of America

10 9 8 7 6 5 4 3 2

DEDICATIONS

To my fiancée Judy, my son Eric, and my lifelong friend Oak.
D.P.S.

To my family, especially to Benjamin John.
M.J.S.

About the Authors

D. Peter Snustad is a Professor in the Department of Genetics and Cell Biology at the University of Minnesota, Twin Cities. He received his B.S. degree in science specialization from the University of Minnesota and his M.S. and Ph.D. degrees in genetics from the University of California, Davis. During his 34 years as a member of the faculty at Minnesota, he has taught courses at all levels from general biology to advanced biochemical genetics. For 20 years, his research focused on bacteriophage T4 morphogenesis and the interaction between T4 and its host, *Escherichia coli.* For the past 14 years, his research group has studied the genetic control of the cytoskeleton in *Arabidopsis thaliana* and the glutamine synthetase gene family in *Zea mays.* He has served on the National Institutes of Health Molecular Cytology Study Section and as Program Chairperson for the Annual Meeting of the Genetics Society of America. His honors include the Morse-Amoco and Stanley Dagley Memorial teaching awards. A lifelong love of the Canadian wilderness has kept him in nearby Minnesota.

Michael J. Simmons is a Professor in the Department of Genetics and Cell Biology at the University of Minnesota, Twin Cities. He received his B.A. degree in biology from St. Vincent College in Latrobe, Pennsylvania, and his M.S. and Ph.D. degrees in genetics from the University of Wisconsin, Madison. Dr. Simmons has taught courses in general biology, genetics, population biology, population genetics, and molecular biology. His research activities are focused on the genetic significance of transposable genetic elements, especially those present in the genome of *Drosophila melanogaster.* He has served on advisory committees at the National Institutes of Health and is currently a member of the Editorial Board of *Genetics,* published by the Genetics Society of America. In 1986, Dr. Simmons received the Morse-Amoco teaching award from the University of Minnesota in recognition of his contributions to undergraduate education. One of his favorite activities, figure skating, is especially compatible with the Minnesota climate.

Preface

Since the first edition of *Principles of Genetics* was published, the science of genetics has advanced on many fronts: the sheep Dolly was cloned from a differentiated adult cell; the genomic DNA sequences of several organisms, including two eukaryotes, were determined; microchip technologies were developed to analyze gene expression; new insights were obtained into the genetic basis of cancer; and minute quantities of DNA isolated from a Neanderthal fossil were successfully amplified and sequenced. These and many other developments have prompted us to prepare a second edition of *Principles of Genetics.* Our intention has been to build on the strengths of the first edition, to update its content selectively, and to incorporate helpful suggestions from the students and teachers who have used it.

We are gratified by how warmly the first edition was received. Students and teachers have found it to be a useful introduction to genetics—one that stresses fundamental concepts and that shows how genetic analysis is carried out. Genetic approaches have proven invaluable in the investigation of all sorts of biological processes, ranging from biochemical and developmental pathways to the etiology of human diseases, and genetic techniques are now being used in fields as diverse as medicine, agriculture, and forensics. *Principles of Genetics* is the culmination of our efforts to present the exciting new developments in genetics along with the fundamental principles. The second edition of this book has the same overall goals as the first edition:

- **To focus on the basis principles of genetics** by presenting the important concepts of classical, molecular, and population genetics. The rapid growth of genetics is rooted in basic principles about gene replication, transmission, expression, and mutation. This textbook strives to develop these principles carefully and thoroughly. We believe that an understanding of current advances in genetics, and an appreciation for their practical significance, must be based on a strong foundation. Furthermore, we believe that the breadth and depth of coverage in the different areas of genetics—classical, molecular, and populational—must be balanced.

- **To focus on the scientific process** by showing how scientific concepts develop from observation and experimentation. No one should lose sight of the fact that genetics, like every other science, is a human endeavor. Thousands of individuals have labored to document the facts and develop the concepts that make up the science of genetics. Our book provides numerous examples to show how genetic principles have emerged through the work of different scientists. We emphasize that genetics is an ongoing process of observation, experimentation, and discovery, and we include *Conversations* with well-known geneticists to show the personal element in this process.

- **To focus on human genetics** by incorporating human examples and showing the relevance of genetics to societal issues. Experience has shown us that our students are keenly interested in human genetics, and that because of this interest, they find it easier to comprehend complex concepts that are illustrated with human examples. Consequently, we have used human examples to illustrate genetic principles wherever possible. We have also included discussions of the Human Genome Project, human gene mapping, human gene therapy, human genetic disorders, and genetic counseling throughout the text. Special *Human Genetics Sidelights* that detail these discussions are found in most of the chapters. In addition, we have dedicated an entire chapter to behavior genetics (Chapter 26), and therein examine some of the genetic influences on human traits such as intelligence and personality.

Genetics, perhaps more than any other science, has sparked numerous social, legal, and ethical debates, which is one reason why students find the subject so fascinating. Many of the controversial issues—genetic screening and its potential for misuse, DNA fingerprinting, genetic engineering, and gene therapy—have been highlighted by advances in the Human Genome Project. We believe that it is important to involve students in intelligent discussions about these issues, and we hope that this textbook will provide them with the background to address these issues thoughtfully.

- **To focus on developing critical thinking skills** by emphasizing the analysis of experimental data and problems. Genetics has always been a bit different from other disciplines in biology because of its heavy emphasis on problem solving. In this text, we have fleshed out the analytical nature of genetics in many ways—

in the development of principles in classical genetics, in the discussion of experiments in molecular genetics, and in the presentation of calculations in population and quantitative genetics. Throughout the book we have emphasized the integration of observational and experimental evidence with logical analysis in the development of key concepts. In the *Testing Your Knowledge* section at the end of each chapter we have provided special worked-out problems to help students develop their analytical skills. In addition, we have provided a set of *Questions and Problems*—significantly expanded in this edition—with each chapter.

Content and Organization of the Second Edition

As in the first edition, the most difficult decisions in preparing this text related to organization and content. Should the text begin with classical genetics? Should it begin with DNA and present classical transmission genetics afterward? Should population and quantitative genetics follow classical genetics, or should they be placed at the very end of the text? Should the different areas of genetics be interwoven throughout the text? Obviously, there is no single "correct" way to teach the basic principles of genetics. Thus, we have tried to create a text that covers all the core topics and that has an adaptable organization. We believe that *Principles of Genetics* can be adapted to a variety of class formats. For example, instructors who prefer to start with molecular genetics can begin with Chapters 10–15 and then return to Chapters 3–9.

We have also tried to keep the length of the text reasonable. Parts of the first edition had to be trimmed to accommodate new material in the second edition. Because every topic is someone's favorite topic, it was often very difficult to decide what to trim. We hope that we have made reasonable decisions about what to include and exclude in the second edition.

The text comprises 29 chapters, one more than the first edition, and an epilogue. Chapters 1–2 introduce the science of genetics and the basic features of living organisms; Chapters 3–9 present the concepts of classical genetics; Chapters 10–15 present the core concepts of molecular genetics; Chapters 16–19 cover special genetic systems: viruses, bacteria, transposable elements, and eukaryotic organelles; Chapters 20–22 contain more advanced topics in molecular genetics; Chapters 23–28 deal with the regulation of gene expression and the genetic bases of development, behavior, immunity, and cancer; Chapter 29 is a capstone chapter that presents population and evolutionary genetics, including a section on human evolution. The epilogue, by James F. Crow, provides a historical perspective on genetics and considers some prospects for the future.

Here are some of the new features of the second edition:

- **Chapter 1. The Science of Genetics.** In this chapter, we provide an overview of the basic phenomena of molecular genetics–the structure, replication, expression, and mutation of DNA. This overview prepares students for the detailed analyses of these phenomena in Chapters 10–14, and it gives them a molecular perspective for the classical genetics discussed in Chapters 3–9.

- **Chapter 5. The Inheritance of Complex Traits.** On the recommendation of readers, we brought the inheritance of complex traits into the unit on classical genetics. This chapter follows up the principles discussed in Chapter 4 by introducing the analysis of quantitative and threshold traits, and it extends the analysis of pedigrees by investigating the effects of consanguineous mating. We think its placement early in the book will help students see the connection between simple Mendelian heredity and multifactorial heredity. However, some instructors may prefer to cover this chapter near the end of a genetics course, perhaps in conjunction with population and evolutionary genetics (Chapter 29).

- **Chapter 21. Genomics.** In response to emerging trends in genetics, we have included a new chapter on genomics. This chapter treats both the structural and functional aspects of genomics and discusses the impact of the Human Genome Project.

- **Chapter 28. The Genetic Basis of Cancer.** The Human Genome Project has helped to elucidate the genetic mechanisms that underlie cancer. In a new chapter devoted to expanded coverage of cancer genetics, we present some of the important discoveries that have been made in this rapidly advancing field.

- **Chapter 29. Population and Evolutionary Genetics.** To connect the mathematical side of genetics with the elements of evolutionary theory, we have integrated population and evolutionary genetics into a single chapter. We hope that this amalgamation will help students see the interrelationships between these disciplines.

- **Other Updated Content.** Although it was impossible to include all the latest developments in genetics in this book, we thought it important to update several areas: the analysis of chromosome structure (Chapter 10), molecular genetic techniques (Chapters 20 and 22), eukaryotic gene expression (Chapter 24), developmental genetics (Chapter 25), and behavior genetics (Chapter 26). In addition, we have adjusted other chapters to provide clearer, more effective coverage, and to limit the size of the book.

- **Art Program.** Well-designed illustrations are an essential component of any science textbook. As in the first edition, we have strived to create illustrations that are clear, attractive, and pedagogically effective. Approximately 700 figures and 300 photographs have been included to illustrate basic concepts, experimental procedures, and various genetic phenomena. More of the figures have been stepped out with succinct labels to help students break down difficult processes into manageable parts. We have also incorporated more micrographs with explanatory diagrams to show students how cellular and molecular images are interpreted.

- **Conversations with Geneticists.** Readers of the first edition indicated how much the conversations with prominent geneticists enhanced their appreciation for the science of genetics. The second edition contains nine conversations to carry on this tradition. We are indebted to each of the following for sharing their insights about science and education:

  Deborah and Brian Charlesworth, *University of Edinburgh*
  James F. Crow, *University of Wisconsin, Madison*
  Joseph G. Gall, *Carnegie Institution of Washington*
  Mitzi Kuroda, *Baylor University*
  Edward B. Lewis, *California Institute of Technology*
  Gisela Mosig, *Vanderbilt University*
  Valery N. Soyfer, *George Mason University*
  Nancy Wexler, *Hereditary Disease Foundation, Santa Monica, California*
  Eric Wieschaus, *Princeton University*

- **Epilogue.** As in the first edition, *Genetics: Yesterday, Today, and Tomorrow,* by James F. Crow, presents a personal view of genetics from one who has lived through much of its history. Dr. Crow's wit, charm, and wisdom are greatly appreciated. Many thanks, Jim, for a truly special ending to this book.

Pedagogy of the Second Edition

This text includes special features designed to emphasize the relevance of the topics discussed, to facilitate the comprehension of important concepts, and to assist students in evaluating their grasp of these concepts. Users of the first edition have helped us make the following features even more useful to students.

- **Sidelights.** Throughout the text, special topics are presented in separate *Sidelight* sections: *Technical Sidelights* describe important experimental techniques; *Historical Sidelights* provide insights into the history of genetics; and *Human Genetics Sidelights* examine important aspects of human genetics. Some of the new *Sidelights* discuss the topical areas of cloning mammals from differentiated cells, analyzing fossilized DNA, and genetic counseling for inherited cancers.

- **Chapter-Opening Vignettes.** Each chapter opens with a vignette or brief story—usually related to human genetics or to historical developments—that emphasizes the relevance of the topics discussed in the chapter.

- **Key Points.** These learning aids are in-text summaries that appear at the end of each major section of the text. They are designed to help students review for exams and focus on the major concepts covered in each section.

- **Testing Your Knowledge.** At the end of each chapter, we have provided worked-out problems to help students hone their analytical and problem-solving skills. The answers walk the students step by step through the solutions to the problems. New worked problems have been added in this edition.

- **Questions and Problems.** Each chapter ends with 20 to 40 questions and problems, more than in the first edition. Approximately 20 percent are new in this edition, some contributed by teachers who have used the text. The questions and problems have been organized to correspond to the sequence of topics in the chapter, and the more difficult ones have been marked by red numbers. The range of questions and problems will provide students with the opportunity to enhance their understanding of the concepts covered in the chapter and to develop their analytical skills further.

- **Answers.** Answers to the odd-numbered problems are given at the back of the text, and answers to all problems are included in *The Problems Workbook and Study Guide,* by H. James Price of Texas A & M University.

Supplements

The Problems Workbook and Study Guide, by H. J. Price of Texas A & M University, is a hands-on workbook designed to improve problem solving skills and to reinforce terminology and concepts from the text. Features included are important concepts, terms, and names; additional problems for self-testing; key figures from the text; thought challenging exercises; answers to all text questions and approaches to problem solving.

Take Note! is a collection of figures, diagrams, and art which clearly illustrate key concepts featured in the text. This convenient notebook provides students with plenty of space to take notes beside key figures, in class and during lectures. We thank Stuart Brody at UC-San Diego for helping us develop *Take Note!*

Instructor's Resource CD-ROM, designed as a lecture enhancer and database of instructor's materials, is highly functional and easy to use. This cross-platform CD-ROM contains the following components and is free to adopters of the text: (1) A database of all the illustrations and selected photographs from the text from which the instructor can create presentations, download to the desktop, and/or print to acetates (2) *Instructor's Manual and Test Bank,* written by Bob Ivarie of the University of Georgia, which contains sample syllabi, lecture outlines, key concepts, teaching strategies for difficult material, and approximately 40 test questions per chapter. The complete set of answers to problems in the text are included here as well. The material can be printed from the CD or exported to word-processing programs for creation of handouts or lecture notes. The *Instructor's Manual and Test Bank* will also be available in a print format. (3) The *Transparency Set* which contains full-color figures from the text. (4) *World Wide Web.* Materials relevant to *Principles of Genetics* which are available to faculty and students.

GenLink, http://www.wiley.com/college/genlink, integrates the Table of Contents to the vast resources of the World Wide Web and provides relevant links to genetic sites for you and your students. Linked sites include the latest research findings, forums, and simulations. In addition, we have added *Problem Assist!* to the Web site. *Problem Assist!* provides additional practice problems for students. It includes model problems with step-by-step solutions. We invite you to make comments and suggestions at the GenLink homepage so that we can make this program an even more effective learning tool.

Of Related Interest: Drlica, *Understanding DNA & Gene Cloning: A Guide For the Curious,* 3/e, 1997.

Acknowledgments

This book has been greatly influenced by the genetics courses we have taught over the last 30 years. We must, therefore, acknowledge the many contributions of our students at the University of Minnesota. We must also acknowledge the help and suggestions of our colleagues who have contributed to our ongoing education. We are especially grateful to Jeffrey Simon and Judith Berman of the University of Minnesota and to Johng K. Lim of the University of Wisconsin-Eau Claire, all of whom made suggestions on the text.

We are grateful to the many people—teachers, students, and reviewers—who suggested ways of improving this book. Their constructive comments helped to shape the organization and content. We deeply appreciate their efforts. In particular, we acknowledge the help of the following reviewers:

Karen Anderson, *Hofstra University*
Paul J. Bottino, *University of Maryland-College Park*
Brian P. Bradley, *University of Maryland-Baltimore County*
LaJoyce H. Debro, *Jacksonville State University*
Anne Stewart Gerber, *University of North Dakota*
Elliott Goldstein, *Arizona State University*
Shubha Govind, *City College of New York*
Allen J. Moore, *University of Kentucky*
Jay Phelan, *University of California-Los Angeles*
Gregory C. Phillips, *New Mexico State University*
John P. Phillips, *University of Guelph*
Robert Podolsky, *University of Michigan-Flint*
Douglas Ruden, *University of Kansas*
Len Seligman, *Pomona College*
Jeffrey B. Stavenhagen, *University of Dayton*
Christine Tachibana, *Penn State University*
Randall Terry, *University of Montana-Missoula*

Student Focus Group Coordinators
Texas A & M University

Marian Beremand
John Ellison
Spencer Johnston

Many people contributed to the development and production of this edition. David Harris, executive editor at John Wiley & Sons, catalyzed our efforts to revise the first edition and provided expert guidance throughout the project. We are grateful for David's enthusiastic encouragement at every turn. Cathy Donovan, David's editorial program assistant, provided help on many of the details, especially in the project's early phases. We thank Nancy Perry, senior development editor, for steering the project so skillfully and for providing input on innumerable issues during the creation of the text. Nancy's dedication to the success of the project and her excellent professional judgment are especially appreciated. We also thank her for carrying out the time-consuming task of obtaining permission to reproduce illustrations and photographs from diverse sources. As with the first edition, Barbara Conover, developmental consultant, was responsible for editing our early drafts and for helping to convert our ideas and crude diagrams into stepped-out illustrations in the text. Her constructive criticisms and advice have been invaluable. Many thanks, Barbara. Edward Starr, art coordinator at John Wiley, added significant improvements to the illustrations carried over from the first edition and executed many new figures. We appreciate all his efforts. We thank Jill Hilycord, photo editor, for obtaining all the photographs that we requested, and Harry Nolan, designer, for creating an appealing cover and for developing the text layout. Barbara Russiello, production editor, superbly coordinated the production of the book. We are grateful for her tireless attention to detail and her unflagging ability to keep the project on schedule. Betty Pessagno, copy editor, did a wonderful job of preparing the manuscript for publication. We are also grateful to Catherine Beckham and Clay Stone, marketing managers, for helping to get this edition into the hands of prospective users, and Jennifer Yee, supplements editor, for coordinating the creation of the *Problems Workbook and Study Guide* by H. James Price and the *Instructor's Manual and Test Questions* by Bob Ivarie. Our thanks also go to the following people, who have contributed questions and problems used in this edition:

John Ellison, *Texas A & M University*
Robert Fowler, *San Jose State University*
David Francis, *University of Delaware*
Margaret J. Hollingsworth, *State University of New York-Buffalo*
Bruce McDonald, *Texas A & M University*
Jay Phelan, *University of California-Los Angeles*
Robert Podolsky, *University of Michigan-Flint*
Therese M. Poole, *Georgia State University*
H. James Price, *Texas A & M University*
Beat Suter, *McGill University*

We are especially grateful to Jim Price for collecting, creating, and organizing the questions and problems used in this edition. Through Jim's efforts, students will be in a better position to develop the critical thinking skills that are an important part of genetics.

Lastly, we would like to encourage everyone—students, teaching assistants, instructors, and other readers—to send us their comments on the text in care of D. Peter Snustad at John Wiley & Sons, Inc., 605 Third Avenue, New York, NY, 10158-0012, or at PSnustad@Wiley.com.

Brief Contents

An Introduction to the Science of Genetics
Chapter 1 The Science of Genetics 3
Chapter 2 Reproduction as the Basis of Heredity 22
Classical Genetic Analysis
Chapter 3 Mendelism: The Basic Principles of Inheritance 52
Chapter 4 Extensions of Mendelism 72
Chapter 5 Inheritance of Complex Traits 91
Chapter 6 The Chromosomal Basis of Mendelism 116
Chapter 7 Variation in Chromosome Number and Structure 138
Chapter 8 Linkage, Crossing Over, and Chromosome Mapping in Eukaryotes 167
Chapter 9 Advanced Linkage Analysis 190
The Molecular Biology of Genes
Chapter 10 DNA and the Molecular Structure of Chromosomes 219
Chapter 11 Replication of DNA and Chromosomes 255
Chapter 12 Transcription and RNA Processing 292
Chapter 13 Translation and the Genetic Code 326
Chapter 14 Mutation, DNA Repair, and Recombination 358
Chapter 15 Definitions of the Gene 397
The Genetics of Viruses, Bacteria, Transposons, and Eukaryotic Organelles
Chapter 16 The Genetics of Viruses 429
Chapter 17 The Genetics of Bacteria 453
Chapter 18 Transposable Genetic Elements 476
Chapter 19 The Genetics of Mitochondria and Chloroplasts 496
Molecular Genetic Analysis
Chapter 20 The Techniques of Molecular Genetics 517
Chapter 21 Genomics 553
Chapter 22 Molecular Analysis of Genes and Gene Products 585
The Regulation of Gene Expression
Chapter 23 Regulation of Gene Expression in Prokaryotes and Their Viruses 615
Chapter 24 Regulation of Gene Expression in Eukaryotes 644
The Genetic Control of Biological Processes
Chapter 25 The Genetic Control of Animal Development 675
Chapter 26 The Genetic Control of Behavior 710
Chapter 27 The Genetic Control of the Vertebrate Immune System 728
Chapter 28 The Genetic Basis of Cancer 750
The Genetics of the Evolutionary Process
Chapter 29 Population and Evolutionary Genetics 774
Epilogue by James F. Crow 807

Contents

An Introduction to the Science of Genetics

A Conversation with Valery N. Soyfer 1

1 THE SCIENCE OF GENETICS 3
A Passion for Experimenting 3
Classical and Molecular Genetics 4
Genetics in the News 5
Genetics and Medicine 7
Genetics and Modern Agriculture 11
Genetics and Society 15
The Principles of Genetics: An Overview 16
DNA: The Genetic Material 17
DNA Replication: Transmission of Genetic Information 17
Gene Expression: Control of Growth and Development 18
Mutation: Changes in the Genetic Material over Time 20
Human Genetics Sidelight: Hemophilia: Successful Treatment of a Once Deadly Disorder 10

2 REPRODUCTION AS THE BASIS OF HEREDITY 22
An Accident in Cell Division 22
The Cell as the Basic Unit of Life 23
The Prokaryotic Cell 23
The Eukaryotic Cell 24
The Chromosome: An Overview 27
The Cell Cycle 28
Cell Division: Mitosis 28
Cell Division: Meiosis 32
Meiotic Abnormalities 37
The Evolutionary Significance of Meiosis 37
The Formation and Union of Gametes 37
Oogenesis: The Formation of the Egg 38
Spermatogenesis: The Formation of Sperm 39
Gamete Formation in Plants 40
The Life Cycles of Some Genetically Important Organisms 40
Neurospora crassa: *The Simple Bread Mold* 41
Saccharomyces cerevisiae: *Baker's Yeast* 43
Zea mays: *Corn* 44
Drosophila melanogaster: *The Fruit Fly* 45
Humans 46
Technical Sidelight: Dolly: To Clone or Not to Clone 42

Classical Genetic Analysis

A Conversation with Joseph G. Gall 50

3 MENDELISM: THE BASIC PRINCIPLES OF INHERITANCE 52
The Birth of Genetics: A Scientific Revolution 52
Mendel's Study of Heredity 53
Mendel's Experimental Organism, the Garden Pea 53
Monohybrid Crosses: The Principles of Dominance and Segregation 54
Dihybrid Crosses: The Principle of Independent Assortment 56
Applications of Mendel's Principles 58
The Punnett Square Method 58
The Forked-Line Method 59
The Probability Method 59
Formulating and Testing Genetic Hypotheses 61
The Chi-Square Test 62
Mendelian Principles in Human Genetics 64
Pedigrees 64
Mendelian Segregation in Human Families 65
Genetic Counseling 66
Technical Sidelight: The Multiplicative and Additive Rules of Probability 61
Technical Sidelight: Binomial Probabilities 67

4 EXTENSIONS OF MENDELISM 72
Genetics Grows Beyond Mendel's Monastery Garden 72
Allelic Variation and Gene Function 73
Incomplete Dominance and Codominance 73
Multiple Alleles 74
Allelic Series 75
Testing Gene Mutations for Allelism 75
Variation Among the Effects of Mutations 76
Genes Function to Produce Polypeptides 77
Why Are Some Mutations Dominant and Others Recessive? 79
Gene Action: From Genotype to Phenotype 80
Influence of the Environment 80
Environmental Effects on the Expression of Human Genes 81
Penetrance and Expressivity 81
Gene Interactions 82

Epistasis 84
Pleiotropy 86
The Genetic Basis of Continuous Phenotypic Variation 86
Historical Sidelight: Genetic Symbols: Evolution in a Dynamic Science 78
Human Genetics Sidelight: The Hapsburg Jaw: A Dominant Trait with Incomplete Penetrance and Variable Expressivity in the European Nobility 83

5 INHERITANCE OF COMPLEX TRAITS 91
Cardiovascular Disease: A Combination of Genetic and Environmental Factors 91
Complex Patterns of Inheritance 92
Quantitative Traits 92
Threshold Traits 95
Statistics of Quantitative Genetics 97
Frequency Distributions 97
The Mean and the Modal Class 98
The Variance and the Standard Deviation 98
Analysis of Quantitative Traits 99
The Multiple Factor Hypothesis 99
Partitioning the Phenotypic Variance 100
Broad-Sense Heritability 101
Narrow-Sense Heritability 102
Predicting Phenotypes 103
Artificial Selection 104
Quantitative Trait Loci 105
Inbreeding and the Resemblance Between Relatives 107
The Effects of Inbreeding 107
Genetic Analysis of Inbreeding 109
Correlations Between Relatives 112
Human Genetics Sidelight: Genetic Counseling and Multifactoral Traits 96
Technical Sidelight: Artificial Selection for Body Size 106

6 THE CHROMOSOMAL BASIS OF MENDELISM 116
Sex, Chromosomes, and Genes 116
Chromosomes 117
Chromosome Number 117
Sex Chromosomes 117
The Chromosome Theory of Heredity 119
Experimental Evidence Linking the Inheritance of Genes to Chromosomes 119
Chromosomes as Arrays of Genes 120
Nondisjunction as Proof of the Chromosome Theory 121
The Chromosomal Basis of Mendel's Principles of Segregation and Independent Assortment 123
Sex-Linked Genes in Human Beings 126
Hemophilia, an X-Linked Blood-Clotting Disorder 126
Color Blindness, an X-Linked Vision Disorder 127
The Fragile X Syndrome and Mental Retardation 127
Genes on the Human Y Chromosome 128
Genes on Both the X and Y Chromosomes 128
Sex Chromosomes and Sex Determination 128
Sex Determination in Human Beings 129
Sex Determination in Drosophila *130*
Sex Determination in Other Animals 131
Dosage Compensation of X-Linked Genes 132
Hyperactivation of X-Linked Genes in Male Drosophila 133
Inactivation of X-Linked Genes in Female Mammals 133
Historical Sidelight: Drosophila, T. H. Morgan, and "The Fly Room" 122

7 VARIATION IN CHROMOSOME NUMBER AND STRUCTURE 138
Chromosomes, Agriculture, and Civilization 138
Cytological Techniques 139
Analysis of Mitotic Chromosomes 139
The Human Karyotype 141
Cytogenetic Variation: An Overview 142
Polyploidy 143
Sterile Polyploids 144
Fertile Polyploids 145
Chromosome Doubling and the Origin of Polyploids 146
Experimental Production of Polyploids 147
Tissue-Specific Polyploidy and Polyteny 148
Aneuploidy 149
Trisomy in Human Beings 150
Monosomy 152
Deletions and Duplications of Chromosome Segments 153
Rearrangements of Chromosome Structure: An Overview 156
Inversions 156
Translocations 157
Compound Chromosomes and Robertsonian Translocations 159
Phenotypic Effects of Chromosome Rearrangements 162
Human Genetics Sidelight: Amniocentesis and Chorionic Biopsy: Procedures to Detect Aneuploidy in Human Fetuses 154
Technical Sidelight: Oenothera: *Chromosome Rings and Other Things 160*

8 LINKAGE, CROSSING OVER, AND CHROMOSOME MAPPING IN EUKARYOTES 167
The World's First Chromosome Map 167
Linkage, Recombination, and Crossing Over 168

Exceptions to the Mendelian Principle of Independent Assortment 168
Frequency of Recombination as a Measure of Linkage Intensity 169
Crossing Over as the Physical Basis of Recombination 171
Evidence That Crossing Over Causes Recombination 172
Chiasmata and the Time of Crossing Over 173
Chromosome Mapping 174
Crossing Over as a Measure of Genetic Distance 174
Recombination Mapping with a Two-Point Testcross 175
Recombination Mapping with a Three-Point Testcross 176
Recombination Frequency and Genetic Map Distance 180
Chiasma Frequency and Genetic Map Distance 182
Genetic Distance and Physical Distance 182
Recombination and Evolution 183
Evolutionary Significance of Recombination 183
Suppression of Recombination by Inversions 184
Genetic Control of Recombination 185
Human Genetics Sidelight: Evidence for Linkage Between the Genes for Hemophilia and Color Blindness 178

9 ADVANCED LINKAGE ANALYSIS 190
Huntington's Disease: A Challenge in Gene Mapping 190
Detection of Linkage in Experimental Organisms 191
Tetrad Analysis to Detect Linkage in Fungi 191
Balancer Chromosome Technique to Assign a Gene to a Chromosome in Drosophila 194
Specialized Mapping Techniques 196
Centromere Mapping with Ordered Tetrads in Neurospora 196
Cytogenetic Mapping with Deletions and Duplications in Drosophila 198
Linkage Analysis in Humans 201
Detection of Linked Loci by Pedigree Analysis 201
Somatic-Cell Genetics: An Alternative Approach to Gene Mapping 204
The Human Gene Map 211
Technical Sidelight: The Relationship Between Genes and Bands in the Drosophila *Polytene X Chromosome* 202

The Molecular Biology of Genes

***A Conversation with Edward B. Lewis* 217**

10 DNA AND THE MOLECULAR STRUCTURE OF CHROMOSOMES 219
Discovery of Nuclein 219
Functions of the Genetic Material 220
Proof That Genetic Information Is Stored in DNA 220
Discovery of Transformation in Bacteria 220
Proof That DNA Mediates Transformation 222
Proof That DNA Carries the Genetic Information in Bacteriophage T2 222
Proof That RNA Stores the Genetic Information in Some Viruses 224
The Structures of DNA and RNA 225
Nature of the Chemical Subunits in DNA and RNA 225
DNA Structure: The Double Helix 225
DNA Structure: Alternate Forms of the Double Helix 230
DNA Structure: Negative Supercoils in Vivo 232
Chromosome Structure in Prokaryotes and Viruses 233
Chromosome Structure in Eukaryotes 234
Chemical Composition of Eukaryotic Chromosomes 235
One Large DNA Molecule per Chromosome 236
Three Levels of DNA Packaging in Eukaryotic Chromosomes 239
Centromeres and Telomeres 242
Eukaryotic Genomes: Repeated DNA Sequences 246
Detection of Repeated Sequences: DNA Renaturation Kinetics 246
Localization of Repeated DNA Sequences in Chromosomes 249
Repeated Sequences in the Human Genome 251
Technical Sidelight: DNA Renaturation Kinetics 248

11 REPLICATION OF DNA AND CHROMOSOMES 255
Monozygotic Twins: Are They Identical? 255
Basic Features of DNA Replication *In Vivo* 256
Semiconservative Replication 256
Visualization of Replication Forks by Autoradiography 259
Unique Origins of Replication 262
Bidirectional Replication 263
DNA Polymerases and DNA Synthesis *In Vitro* 264
Discovery of DNA Polymerase I in Escherichia coli 266
Multiple DNA Polymerases 267
DNA Polymerase III: The Replicase in Escherichia coli 269
Proofreading Activities of DNA Polymerases 270
The Complex Replication Apparatus 271
Continuous Synthesis of One Strand; Discontinuous Synthesis of the Other Strand 272

Covalent Closure of Nicks in DNA by DNA Ligase 273
Initiation of DNA Chains with RNA Primers 273
Unwinding DNA with Helicases, DNA-Binding Proteins, and Topoisomerases 275
The Replication Apparatus: Prepriming Proteins, Primosomes, and Replisomes 277
Rolling-Circle Replication 280
Unique Aspects of Eukaryotic Chromosome Replication 280
The Cell Cycle 281
Multiple Replicons per Chromosome 281
Two DNA Polymerases at a Single Replication Fork 283
Duplication of Nucleosomes at Replication Forks 284
Telomerase: Replication of Chromosome Termini 284
Telomere Length and Aging in Humans 285
Technical Sidelight: Cesium Chloride Equilibrium Density-Gradient Centrifugation and Sucrose Velocity Density-Gradient Centrifugation: Two Important and Distinct Tools 260

12 TRANSCRIPTION AND RNA PROCESSING 292
Storage and Transmission of Information with Simple Codes 292
The Genetic Control of Metabolism: An Overview 293
Transfer of Genetic Information: The Central Dogma 294
Transcription and Translation 294
Four Types of RNA Molecules 295
The Process of Gene Expression 298
An mRNA Intermediary 298
General Features of RNA Synthesis 300
Transcription in Prokaryotes 302
RNA Polymerase: A Complex Enzyme 302
Initiation of RNA Chains 303
Elongation of RNA Chains 303
Termination of RNA Chains 304
Concurrent Transcription, Translation, and mRNA Degradation 304
Transcription and RNA Processing in Eukaryotes 305
Three RNA Polymerases/Three Sets of Genes 307
Initiation of RNA Chains 307
RNA Chain Elongation and the Addition of 5′ Methyl Guanosine Caps 308
Termination by Chain Cleavage and the Addition of 3′ Poly(A) Tails 310
RNA Editing: Altering the Information Content of mRNA Molecules 310
Interrupted Genes in Eukaryotes: Exons and Introns 312
Early Evidence for Noncoding Sequences in Eukaryotic Genes 313
Some Very Large Eukaryotic Genes 315
Introns: Biological Significance? 316
Removal of Intron Sequences by RNA Splicing 317
tRNA Precursor Splicing: Unique Nuclease and Ligase Activities 318
Autocatalytic Splicing 319
Pre-mRNA Splicing: snRNAs, snRNPs, and the Spliceosome 320
Technical Sidelight: An mRNA Intermediary: Evidence from Phage-Infected E. coli 299

13 TRANSLATION AND THE GENETIC CODE 326
Sickle-Cell Anemia: Devastating Effects of a Single Base-Pair Change 326
Protein Structure 327
Polypeptides: Twenty Different Amino Acid Subunits 327
Proteins: Complex Three-Dimensional Structures 329
Protein Synthesis: Translation 331
Overview of Protein Synthesis 331
Components Required for Protein Synthesis: Ribosomes and Transfer RNAs 332
Translation: The Synthesis of Polypeptides Using mRNA Templates 337
The Genetic Code 344
Properties of the Genetic Code: An Overview 344
Three Nucleotides per Codon 344
Deciphering the Code 346
Initiation and Termination Codons 349
A Degenerate and Ordered Code 350
A Nearly Universal Code 351
Codon-tRNA Interactions 352
Recognition of Codons by tRNAs: The Wobble Hypothesis 352
Suppressor Mutations That Produce tRNAs with Altered Codon Recognition 352
In Vivo Confirmation of the Nature of the Genetic Code 354
Technical Sidelight: Cracking the Genetic Code: Synthetic mRNAs 348
Historical Sidelight: Hannah Bernstein and the Amber Codon 351

14 MUTATION, DNA REPAIR, AND RECOMBINATION 358
Xeroderma Pigmentosum: Defective Repair of Damaged DNA in Humans 358
Mutation: Source of the Genetic Variability Required for Evolution 359

Mutation: Basic Features of the Process 360
Mutation: Somatic or Germinal 360
Mutation: Spontaneous or Induced 360
Mutation: Usually a Random, Nonadaptive Process 361
Mutation: A Reversible Process 362
Mutation: Phenotypic Effects 363
Mutations with Phenotypic Effects: Usually Deleterious and Recessive 364
Effects of Mutations in Human Globin Genes 364
Mutation in Humans: Blocks in Metabolic Pathways 366
Conditional Lethal Mutations: Powerful Tools for Genetic Studies 369
Muller's Demonstration That X Rays Are Mutagenic 370
The Molecular Basis of Mutation 371
Mutations Induced by Chemicals 373
Mutations Induced by Radiation 378
Mutations Induced by Transposable Genetic Elements 380
Expanding Trinucleotide Repeats and Inherited Human Diseases 381
Screening Chemicals for Mutagenicity: The Ames Test 381
DNA Repair Mechanisms 383
Light-Dependent Repair 383
Excision Repair 383
Mismatch Repair 385
Postreplication Repair 386
The Error-Prone Repair System 387
Inherited Human Diseases with Defects in DNA Repair 387
DNA Recombination Mechanisms 388
Recombination: Cleavage and Rejoining of DNA Molecules 388
Gene Conversion: DNA Repair Synthesis Associated with Recombination 390
Human Genetics Sidelight: Tay-Sachs Disease, A Childhood Tragedy 367

15 DEFINITIONS OF THE GENE 397
Sir Archibald Garrod and Human Inborn Errors of Metabolism 397
Evolution of the Concept of the Gene: Summary 398
Evolution of the Concept of the Gene: Function 401
Mendel: Constant Factors Controlling Phenotypic Traits 401
Garrod: One Mutant Gene–One Metabolic Block 402
Early Evidence That Enzymes Are Controlled by Genes 402
Beadle and Tatum: One Gene–One Enzyme 404
One Gene–One Polypeptide 406
Evolution of the Concept of the Gene: Structure 406
The Pre-1940 Beads-on-a-String Concept 406
Discovery of Recombination Within the Gene 407
Recombination Between Adjacent Nucleotide Pairs 407
Colinearity Between the Coding Sequence of a Gene and Its Polypeptide Product 408
A Genetic Definition of the Gene 410
The Complementation Test as an Operational Definition of the Gene 411
Intragenic Complementation 415
Limitations on the Use of the Complementation Test 418
Complex Gene–Protein Relationships 419
Alternate Pathways of Transcript Splicing: Protein Isoforms 420
Assembly of Genes During Development: Human Antibody Chains 420
Human Genetics Sidelight: Human Inborn Errors of Metabolism 399

The Genetics of Viruses, Bacteria, Transposons, and Eukaryotic Organelles

A Conversation with Gisela Mosig 427

16 THE GENETICS OF VIRUSES 429
A Killer Unleashed 429
The Discovery of Viruses 430
The Origin of Viruses 431
The Structure and Life Cycle of a Bacterial Virus 432
Mapping the Bacteriophage Genome 434
Phage Phenotypes 434
Genetic Recombination in Phage 435
Genetic Fine Structure 436
Deletion Mapping 437
T4: A Circular Genetic Map but a Linear Chromosome 441
Genes-Within-Genes: Bacteriophage ϕX174 443
HIV: A Eukaryotic Virus 445
What Is AIDS? 445
The Structure of HIV 445
The HIV Life Cycle 445
The HIV Genome 449
The Course of HIV Infection 449

17 THE GENETICS OF BACTERIA 453
Drug Resistance and Gene Transfer in Bacteria 453
Genetic Exchange in Bacteria: An Overview 454
Mutant Phenotypes in Bacteria 455
Basic Test for Transformation, Conjugation, and Transduction 456
Transformation 457
The Process of Transformation 457
Transformation and Gene Mapping 458
Conjugation 460

The Discovery of Conjugation 460
$F^+ \times F^-$ Mating 461
Hfr Conjugation 461
F′ Conjugation or Sexduction 463
Conjugation and Gene Mapping 464
Mapping Closely Linked Genes 465
On the Origin of Plasmids 466
Transduction 467
The Discovery of Transduction 467
Generalized Transduction 467
Specialized Transduction 467
Transduction and Gene Mapping 470
The Evolutionary Significance of Sexuality in Bacteria 472

18 TRANSPOSABLE GENETIC ELEMENTS 476
Maize: A Staple Crop with a Cultural Heritage 476
Transposable Elements in Bacteria 477
IS Elements 477
Composite Transposons 479
Tn3 Elements 480
The Medical Significance of Bacterial Transposons 481
Transposable Elements in Eukaryotes 481
Ac *and* Ds *Elements in Maize* 482
P *Elements and Hybrid Dysgenesis in* Drosophila 484
mariner, *an Ancient and Widespread Transposon* 487
Retrotransposons 487
Retroviruslike Elements 487
Retroposons 490
The Genetic and Evolutionary Significance of Transposable Elements 491
Transposons and Genome Organization 491
Transposons and Mutation 492
Evolutionary Issues Concerning Transposable Elements 493
Historical Sidelight: Barbara McClintock, The Discoverer of Transposable Elements 478
Technical Sidelight: Genetic Transformation of Drosophila *with* P *Elements* 488

19 THE GENETICS OF MITOCHONDRIA AND CHLOROPLASTS 496
Mitochondria, Chloroplasts, and the Biological Energy Wheel 496
The Classical Genetics of Organelles 497
Leaf Variegation in Plants 497
Cytoplasmic Male Sterility in Maize 499
Antibiotic Resistance in Chlamydomonas 500
Metabolic Defects in Yeast 502
The Molecular Genetics of Mitochondria 503
Mitochondrial DNA 503
Expression of Mitochondrial Genes 505
Interplay Between Mitochondrial and Nuclear Gene Products 506
Mitochondrial DNA and Human Disease 506
The Molecular Genetics of Chloroplasts 507
Chloroplast DNA 507
Chloroplast Biogenesis 507
The Origin and Evolution of Mitochondria and Chloroplasts 509
Eukaryotic Organelles as Endosymbionts 509
The Evolution of Mitochondria and Chloroplasts 512
Human Genetics Sidelight: Using Mitochondrial DNA to Study Human Evolution 510

Molecular Genetic Analysis

***A Conversation with Nancy Wexler* 515**

20 THE TECHNIQUES OF MOLECULAR GENETICS 517
Treatment of Pituitary Dwarfism with Human Growth Hormone 517
Basic Techniques Used to Clone Genes 518
The Discovery of Restriction Endonucleases 518
The Production of Recombinant DNA Molecules In Vitro 521
Amplification of Recombinant DNA Molecules in Cloning Vectors 522
Construction and Screening of DNA Libraries 525
Construction of Genomic Libraries 526
Construction of cDNA Libraries 528
Screening DNA Libraries for Genes of Interest 528
Biological and Physical Containment of Recombinant DNA Molecules 529
The Manipulation of Cloned DNA Sequences *In Vitro* 531
Phagemids: The Biological Purification of DNA Single Strands 531
In Vitro *Site-Specific Mutagenesis* 534
The Molecular Analysis of DNA, RNA, and Protein 535
Analysis of DNAs by Southern Blot Hybridizations 536
Analysis of RNAs by Northern Blot Hybridizations 539
Analysis of Proteins by Western Blot Techniques 540
The Molecular Analysis of Genes and Chromosomes 541
Amplification of DNAs by the Polymerase Chain Reaction (PCR) 541
Physical Maps of DNA Molecules Based on Restriction Enzyme Cleavage Sites 543
Nucleotide Sequences: The Ultimate Fine Structure Maps 543

Human Genetics Sidelight: Treatment of Diabetes Mellitus with Human Insulin Produced in Bacteria 520
Technical Sidelight: Detection of a Mutant Gene Causing Cystic Fibrosis by Southern Blot Analysis 538

21 GENOMICS 553
Human Gene Prospecting in Iceland 553
Genomics: An Overview 555
Correlated Genetic, Cytological, and Physical Maps of Chromosomes 555
Restriction Fragment-Length Polymorphism (RFLP) and Microsatellite Maps 558
Cytogenetic Maps 562
Physical Maps and Clone Banks 562
Map Position-Based Cloning of Genes 564
Chromosome Walks 564
Chromosome Jumps 566
The Human Genome Project 567
Bacterial Genomes 567
The Saccharomyces cerevisiae *(Yeast) Genome 569*
The Human Genome 571
RNA and Protein Assays of Genome Function 573
Expressed Sequences 573
Array Hybridizations and Gene Chips 573
Use of Green Fluorescent Protein as a Reporter of Protein Synthesis 574
Evolution of Genomes 577
Genome Evolution in the Cereal Grasses 577
Genome Evolution in Mammals 579
Human Genetics Sidelight: The Race to Sequence the Human Genome 556

22 MOLECULAR ANALYSIS OF GENES AND GENE PRODUCTS 585
Detection of the Tay-Sachs Mutation in Eight-Cell Pre-embryos 585
Use of Recombinant DNA Technology to Identify Human Genes 586
Huntington's Disease 586
Cystic Fibrosis 590
Molecular Diagnosis of Human Diseases 593
DNA Fingerprints 594
Paternity Tests 595
Forensic Applications 595
Human Gene Therapy 597
Production of Eukaryotic Proteins in Bacteria 601
Human Growth Hormone 601
Proteins with Industrial Applications 602
Transgenic Plants and Animals 603
Transgenic Animals: Microinjection of DNA into Fertilized Eggs 603
Transgenic Plants: The Ti Plasmid of Agrobacterium tumefaciens *605*
Blocking Gene Expression with Antisense RNA 608

Technical Sidelight: DNA Tests and the Mystery of the Duchess Anastasia 596

The Regulation of Gene Expression

A Conversation with Mitzi I. Kuroda 613

23 REGULATION OF GENE EXPRESSION IN PROKARYOTES AND THEIR VIRUSES 615
D'Hérelle's Dream of Treating Dysentery in Humans by Phage Therapy 615
Constitutive, Inducible, and Repressible Gene Expression 617
Positive and Negative Control of Gene Expression 618
Operons: Coordinately Regulated Units of Gene Expression 620
The Lactose Operon in *E. coli*: Induction and Catabolite Repression 622
Induction 622
Catabolite Repression 625
The Tryptophan Operon in *E. coli:* Repression and Attenuation 626
Repression 626
Attenuation 627
Bacteriophage Lambda: Lysogeny or Lysis 630
Repression of Lambda Lytic Pathway Genes During Lysogeny 631
The Lambda Lytic Regulatory Cascade 634
The Lambda Switch: Lytic Development or Lysogeny 635
Temporal Sequences of Gene Expression During Phage Infection 637
Translational Control of Gene Expression 638
Posttranslational Regulatory Mechanisms 639

24 REGULATION OF GENE EXPRESSION IN EUKARYOTES 644
African Trypanosomes: A Wardrobe of Molecular Disguises 644
Spatial and Temporal Regulation of Eukaryotic Genes 645
Spatial Regulation of Tubulin Genes in Plants 645
Temporal Regulation of Globin Genes in Animals 646
Ways of Regulating Eukaryotic Gene Expression 647
Controlled Transcription of DNA 647
Alternate Splicing of RNA 648
Cytoplasmic Control of Messenger RNA Stability 649
Induction of Transcriptional Activity by Environmental and Biological Factors 649
Temperature: The Heat-shock Genes 650
Light: The Ribulose 1,5-bisphosphate Carboxylase Genes in Plants 650

Signal Molecules: Genes That Respond to Hormones 651
Molecular Control of Transcription in Eukaryotes 653
DNA Sequences Involved in the Control of Transcription 653
Proteins Involved in the Control of Transcription: Transcription Factors 655
Gene Expression and Chromosome Organization 658
Transcription in Lampbrush Chromosome Loops 658
Transcription in Polytene Chromosome Puffs 659
Molecular Organization of Transcriptionally Active DNA 659
Euchromatin and Heterochromatin 660
Gene Silencing 661
DNA Methylation and Imprinting 665
Gene Amplification 666
Activation and Inactivation of Whole Chromosomes 668
Inactivation of X Chromosomes in Mammals 668
Hyperactivation of X Chromosomes in Drosophila *669*
Hypoactivation of X Chromosomes in Caenorhabditis *670*
Technical Sidelight: GAL4, a Transcription Factor That Regulates the Genes Involved in Galactose Metabolism in Yeast 656

The Genetic Control of Biological Processes

A Conversation with Eric Wieschaus 673

25 THE GENETIC CONTROL OF ANIMAL DEVELOPMENT 675
Insect Wings: Evidence for the Genetic Control of Development 675
The Process of Development in Animals 676
Oogenesis and Fertilization 676
The Embryonic Cleavage Divisions and Blastula Formation 678
Gastrulation and Morphogenesis 678
Genetic Analysis of Development in Model Organisms 679
Drosophila *as a Model Organism 679*
Caenorhabditis *as a Model Organism 679*
Genetic Analysis of Developmental Pathways 681
Sex Determination in Drosophila *682*
Sex Determination in Caenorhabditis *685*
The Use of Genetic Mosaics to Study Development 686
X Chromosome Loss in Drosophila*: Gynandromorphs 687*
Somatic Recombination in Drosophila*: Twin Spots 687*
Molecular Analysis of Genes Involved in Development 689
Maternal Gene Activity in Development 690
Maternal-Effect Genes 690
Determination of the Dorsal-Ventral and Anterior-Posterior Axes in Drosophila *Embryos 690*
Zygotic Gene Activity in Development 695
Body Segmentation 695
Specification of Cell Types 699
Organ Formation 702
Genetic Analysis of Development in Vertebrates 703
Vertebrate Homologs of Invertebrate Genes 703
The Mouse: Insertional Mutagenesis, Transgenics, and Knockout Mutations 704
Technical Sidelight: The Development of Butterfly Wings: Insights from Drosophila *Genetics 700*

26 THE GENETIC CONTROL OF BEHAVIOR 710
Twins and Personality 710
Genetic Analysis of Behavior in Experimental Animals 711
Nest-Cleaning Behavior in Honeybees 711
Biological Rhythms in Drosophila *and Mice 712*
Sexual Preference in Drosophila *716*
Nurturing Behavior in Mice 719
Chromosome Abnormalities and Insights into Human Behavior 720
Single-Gene Mutations and Human Behavior 721
Phenylketonuria 721
Lesch-Nyhan Syndrome 722
Huntington's Disease 722
Complex Human Behaviors 723
Quantitative Human Genetics: The Twin Correlation Method 723
Intelligence 724
Personality 725
Human Genetics Sidelight: A Genetic Basis for Male Homosexuality? 718

27 THE GENETIC CONTROL OF THE VERTEBRATE IMMUNE SYSTEM 728
AIDS Emphasizes the Vital Role of the Immune System 728
The Immune Response: An Overview 729
Components of the Mammalian Immune System 730
Specialized Cells Mediate the Immune Response 730
Specialized Proteins Provide Immunological Specificity 732
Major Histocompatibility Antigens: Distinguishing Self from Nonself 735

The Antibody-Mediated (Humoral) Immune Response 736
The T Cell-Mediated (Cellular) Immune Response 738
Immunological Memory 739
Genome Rearrangements During B Lymphocyte Differentiation 740
Lambda Light Chain Genes Assembled from Two Gene Segments 740
Kappa Light Chain Genes Assembled from Three Gene Segments 740
Heavy Chain Genes Assembled from Four Gene Segments 741
Signal Sequences Control Somatic Recombination Events 742
Additional Diversity: Variable Joining Sites and Somatic Hypermutation 743
Antibody Class Switching 745
Assembly of T Cell Receptor Genes by Somatic Recombination 746
Regulation of Immunoglobin Gene Expression 746
Allelic Exclusion: Only One Functional Rearrangement per Cell 746
Heavy Chain Gene Transcription: A Tissue-Specific Enhancer 746
Technical Sidelight: Monoclonal Antibodies: Powerful Probes for Detecting Proteins and Other Antigens 747

28 THE GENETIC BASIS OF CANCER 750
A Common Killer 750
Cancer: A Genetic Disease 751
The Many Forms of Cancer 751
Cancer and the Cell Cycle 752
A Genetic Basis for Cancer 753
Oncogenes 753
Tumor-Inducing Retroviruses and Viral Oncogenes 753
Cellular Homologs of Viral Oncogenes: The Proto-Oncogenes 755
Mutant Cellular Oncogenes and Cancer 756
Chromosome Rearrangements and Cancer 758
Tumor Suppressor Genes 759
Inherited Cancers and Knudson's Two-Hit Hypothesis 759
Cellular Roles of Tumor Suppressor Proteins 761
Genetic Pathways to Cancer 767
Human Genetics Sidelight: Cancer and Genetic Counseling 766

The Genetics of the Evolutionary Process

A Conversation with Deborah and Brian Charlesworth 772

29 POPULATION AND EVOLUTIONARY GENETICS 774
D'ou venons nous? Que sommes nous? Ou allons nous? 774
The Emergence of Evolutionary Theory and Population Genetics 775
Darwin's Theory of Evolution 775
Population Genetics 776
The Theory of Allele Frequencies 777
Estimating Allele Frequencies 777
Relating Genotype Frequencies to Allele Frequencies: The Hardy-Weinberg Principle 778
Applications of the Hardy-Weinberg Principle 778
Exceptions to the Hardy-Weinberg Principle 780
Using Allelic Frequencies in Genetic Counseling 781
Natural Selection 782
Natural Selection at the Level of the Gene 782
Natural Selection at the Level of the Phenotype 785
Random Genetic Drift 786
Random Changes in Allele Frequencies 786
The Effects of Population Size 787
Populations in Genetic Equilibrium 788
Balancing Selection 788
Mutation-Selection Balance 789
Mutation-Drift Balance 791
Genetic Variation in Natural Populations 793
Phenotypic Variation 793
Polymorphism of Chromosome Structure 794
Genetic Variation at the Molecular Level 794
Speciation 795
Definition of Species 795
Modes of Speciation 796
Human Evolution 799
Humans and the Great Apes 799
Human Evolution in the Fossil Record 799
DNA Sequence Variation and Human Origins 800
Technical Sidelight: Molecular Evolution and the Neutral Theory 790
A Conversation with James F. Crow 805

Epilogue: Genetics Yesterday, Today, and Tomorrow, by James F. Crow: A Personal View 807
Glossary 814
Answers to Odd-Numbered Questions and Problems 832
Photo and Illustration Credits 850
Index 855

PRINCIPLES OF GENETICS

SECOND EDITION

A CONVERSATION WITH

VALERY N. SOYFER

When did you first become interested in science?

As an eighth-grade student, I was invited by Professor Peter A. Suvorov from Gorky State University to participate in his research project on *in vitro* propagation of forest fungi. It was natural that after high school graduation I enrolled in the Moscow Timiryazev Academy for undergraduate studies in biology and agronomy. There I won a scientific award for research by undergraduate students, and later this work became a key part of my Ph.D. thesis in biology. After four years of undergraduate education in biology, I applied for undergraduate study at the Department of Physics at Moscow State University and ultimately received doctoral degrees in both biology and physics.

Did any specific event during your undergraduate years have a major effect on your future career as a scientist?

When I was an undergraduate, the science of genetics was officially forbidden in the USSR as a "bourgeois perversion." But forbidden fruit is always sweeter. I began to work—sometimes past midnight—in the first laboratory of genetics that was reopened in Moscow after almost 20 years of T. D. Lysenko's dictatorship over all aspects of biology and agriculture. It was a fascinating time because a new sense of purpose appeared in my life. During school vacations I returned to my native city of Gorky and spent practically all my time with the founder of population genetics, Professor Sergei S. Chetverikov. During the academic year, I enjoyed evening work with the famous Russian geneticists Vladimir V. Sakharov and Nikolai P. Dubinin. Dubinin introduced me to Igor Ye Tamm, a physicist-theoretician, who two years later received the Nobel Prize. Our frequent discussions with Professor Tamm changed my life—I shall never forget my feelings of inspiration.

Valery Nikolayevich Soyfer was born and raised in the USSR. He received his Ph.D. in biology and Doctor of Science degrees in genetics and biophysics. A leading Soviet geneticist, he organized the USSR Scientific Research Institute of Molecular Biology and Genetics in Moscow. However, in 1978, Dr. Soyfer appealed to Communist leaders to allow more open expression of ideas. Immediately punished, he and his wife were jobless for eight long years. Despite offers from many American universities, Soviet authorities refused to grant the Soyfers exit visas. Ronald Reagan asked Mikhail Gorbachev to release them three times; similar requests were made by many U.S. legislators. In 1988, the Soyfers at last came to the United States. Since 1990, after two years at The Ohio State University, Dr. Soyfer has been Distinguished University Professor and Director of the Laboratory of Molecular Genetics at George Mason University. His recently published books are *Lysenko and the Tragedy of Soviet Science* (Rutgers University Press, 1994) and *Triple Helical Nucleic Acids* (Springer Verlag, 1996).

Your scientific career spans two very different cultures: the Soviet Union during Communist rule and the United States. Is pure science the same in both cultures?

Pure science is always pure science, and no scientific laws are appropriate for one society and wrong for another. Unfortunately, however, this simple rule is rejected by totalitarian rulers, who seek to create "useful" science that will better serve their purposes. Lenin and Stalin repeated many times that the sciences are divided into two conflicting entities: socialist versus bourgeois sciences. The Soviet authorities condemned and forbade research in genetics, cybernetics, and several branches of economics. In reality, their ideology led several scientific disciplines in the Soviet Union to disaster. The consequences of this politicization of science are now painful for the new Russia.

When you wrote to Leonid Brezhnev on behalf of Andrei Sakharov, did you think that there might be repercussions?

I understood that my letter would provoke the powers. In that quiet letter I stated that Sakharov's concerns about the USSR's development were justified. I didn't think I would be dismissed from the Institute since it had been based on my laboratory. I was not an enemy of the country where I was born, for I wasn't doing anything to endanger her. Everything that happened—my dismissal, the revocation of my scientific degrees, and my unemployment (and my wife's)—was the result of paranoid ideological dictates in the Soviet Union.

My wife and I lost our official jobs and were trying to raise two children. Soon after being dismissed, we began to repair private apartments in Moscow—painting, hanging wallpaper. Fortunately, after we received permission to emigrate, my wife and I were able to resume our work as scientists in the United States.

Emotionally, however, the years of scientific unemployment in the USSR were personally rewarding for my wife and me because we "grew" in our power to survive in extreme circumstances. We also understood the importance of real friendships and of independent and free thinking. Several scientists who were our friends had also been dismissed from their jobs; I organized an ongoing scientific seminar for them. Every week a couple dozen of us gathered at my home to discuss subjects from different disciplines. Later, many of the reports from my seminar were published in *The Annals of the New York Academy of Sciences* (vol. 452) and as a separate volume, *Science Behind Iron Curtains* (Henry Stewart Publications, 1990).

Your book *Lysenko and the Tragedy of Soviet Science* has received glowing reviews. What was the most rewarding aspect of bringing the book to fruition?

I worked on this project for eight years when I was jobless in the USSR. Writing the book sustained me as a person and as a scientist. However, every day I was aware that the KGB might seize the manuscripts. After several versions had been deposited in the University of Wisconsin library, I felt free to give copies to my colleagues; an enjoyable result was that they shared with me new documents from their private archives. After publication of the book, I was honored to be awarded two Gregor Mendel Medals, one from the Czech Academy of Sciences and another, international one from the Society of the History of Science.

During your prestigious career, what has been your greatest challenge?

In 1969, scientists from the Oak Ridge National Laboratory in the United States published a series of articles in which the authors concluded that higher plants either did not acquire excision and post-replicative repair enzymes in evolution or lost these enzymes during evolution. The more I pondered these findings, the more I understood that plants ought to have repair enzyme systems—because every day they encounter unpleasant effects of environmental stress that should induce mutations. Thus, I began our own experiments to recheck their findings. In 1973, we published data confirming excision repair in higher plants, and the next year we found post-replicative repair. In parallel, we studied the physical structure of plant DNAs before and after the repair reactions and established a parallelism between the efficiency of repair and the rate of mutagenesis in plants.

What is your favorite story about one of your famous colleagues?

There are two stories that always amaze me. In 1946, a 21-year-old student, Joshua Lederberg, together with his teacher Edward Tatum, discovered the process of genetic recombination, and in 1958, they were awarded the Nobel Prize for this discovery. In 1946, nobody understood DNA structure, and Lederberg's work is a great example of the power of human genius. Another example is the 1953 discovery of DNA structure by James Watson and Francis Crick. In the same year, Linus Pauling, who was already a Nobel Prize winner, published in collaboration with Robert Corey their hypothesis that the DNA structure should be a triple helix. The very fact that Watson and Crick went against the opinion of Nobelist Pauling and created a model of double helical DNA was one of the brightest events in the history of science.

One of your responsibilities at George Mason University is teaching genetics to undergraduates. What advice do you give undergraduates interested in careers in genetics and molecular biology?

I try to convey to the students a very important message—that the scientific search is never completed. Year after year new frontiers are opening to inquisitive minds, and there is no more personal satisfaction in life than to find out that you have created, through your own experimentation, knowledge that did not exist before. Surely nothing can surpass this feeling of satisfaction.

What are your current research projects?

We recently discovered a substantial increase of mutability in humans inhabiting one of the most contaminated regions on our planet—near the city of Chelyabinsk in the southern Ural Mountains where the Soviets had released huge quantities of radioactive waste. I also serve as Chairman of the Board of the International Soros Science Education Program, which operates in several countries, including Russia.

A photo taken in 1881 of Charles Darwin at the age of 72. This was one of the last pictures ever taken of Darwin, who died a year later.

1

The Science of Genetics

CHAPTER OUTLINE

Classical and Molecular Genetics
Genetics in the News
Genetics and Medicine
Genetics and Modern Agriculture
Genetics and Society
The Principles of Genetics: An Overview

A Passion for Experimenting

If you happened to be out walking in the English countryside just beyond the small village of Downe in Kent on a summer evening 125 years ago, you might have come across a remarkable sight. In a greenhouse down a path about 100 yards from a large, elegant home, a tall man in his sixties stooped over some small potted plants. Next to him sat a younger man playing a bassoon. The older man was Charles Darwin; the younger man was his son, Frank. They were conducting an experiment on insect-eating plants called sundews. Darwin wanted to know what caused the sundew to close its leaves when a fly lands on it and becomes trapped in its sticky liquid. He was exploring various factors that might trigger the leaves' closing. He had already experimented with numerous substances including sand, milk, urine, water, roast beef, and bits of hard-boiled egg. None of them worked. Now he was experimenting with sound, even though he did not think sound waves would trigger the plant's response. Darwin was conducting an experiment, methodically working through a list of potential stimulants.

Darwin never did find out what triggered the sundew's response, but he was not discouraged by this failure. He was content to leave the sundew problem for others to solve. Darwin in his sixties was a world-renowned scientist who had changed our whole understanding and perspective of the natural world. He remained intensely curious about nature, content to perform experiments that would bear fruit in the future.

Darwin was an explorer with a passion to uncover the mysteries of the natural world. In a way, that is a goal of this book. It asks questions; it encourages exploration; and it details discovery.

CLASSICAL AND MOLECULAR GENETICS

As sciences go, genetics is young—not yet 150 years old. However, for thousands of years people have sought answers to questions about inheritance. Scientifically sound answers became possible only about 130 years ago. In 1865, Gregor Johann Mendel, an Austrian monk, reported his discoveries of the fundamental laws of inheritance. He suggested that every cell contained pairs of "factors" and that each pair determined a specific trait. The members of each pair segregated from each other during the process of sex-cell formation, so that a gamete contained one member of each pair. Furthermore, the segregation of each pair of factors was independent of the segregation of other pairs of factors. These deceptively simple ideas form the foundation of the modern science of genetics and are the core of classical genetics.

As profound and as far reaching as Mendel's observations and interpretations were, they were not recognized as such for 35 years. Part of the reason for this delay was the absence of an understanding of cell structure and of the process by which cells divide. However, in 1900, with the description of cell structure and of cell division, there at last existed a solid cellular framework within which Mendel's principles could be properly interpreted. The year 1900 is important, for it marks the beginning of the modern era of genetics.

The spectacular unfolding of modern genetic concepts has taken less than a century. We have moved from the obscure units that Mendel called "factors," segregating and assorting independently of each other in the nuclei of sex cells, to the identification of DNA (deoxyribonucleic acid) as the chemical basis of inheritance. We have moved from a vague understanding of the relationship between the units of inheritance and the physical appearance of an organism to an understanding of the elegantly organized sequence of events by which Mendel's factors, now called *genes*, express their encoded information in cells. We have moved from exclusive reliance on classical genetic analysis to the merger of classical genetic techniques with modern molecular techniques. Today, molecular technology that allows us to manipulate genetic material has opened entirely new vistas to us, vistas of how genes work, how they are regulated, and how genetic defects can be detected, modified, or corrected.

This unfolding of genetic concepts is much more than the erection of a structure within which we can interpret various aspects of heredity. It is a model of scientific methodology, of human genius and creativity, and, above all, of human potential. The thousands of participants who have contributed and continue to contribute to our understanding of modern genetics reflect human qualities we should always treasure and support. As we explore the many aspects of genetics, we encounter individuals who discover or observe something very important about the natural world, but fail to interpret it correctly; others, who though incorrect in their thinking, stimulate others to make new and important discoveries; and those few who integrate diverse and seemingly unrelated observations to arrive at unique and important insights that have a major influence on the thinking of others. Although the basic concepts of inheritance are well established, genetics remains a rapidly expanding discipline that promises to challenge future generations of scientists. As you read this text, we believe that you will discover that genetics is a vital and dynamic science with relevance to many aspects of your life and to humankind in general.

In the Epilogue of this text, James F. Crow, who has lived through much of the history of genetics and participated in its development, captures the excitement of those who contributed to the rapid expansion of knowledge in this field. As authors, we hope that you will enjoy his view of the birth, growth, and development of this science and his personal insights as much as we do.

Key Points: **The science of genetics began in 1865 with the work of Gregor Johann Mendel, an Austrian monk who discovered the fundamental laws of inheritance. Genetics has grown at an explosive rate since 1900, when Mendel's work was discovered by the scientific world. During the last half century, geneticists have determined the molecular structure of Mendel's "factors" and the mechanisms by which they control the characteristics of an organism.**

GENETICS IN THE NEWS

Barely a week passes without the appearance of some item in the news media about a new genetic discovery or a new way that genetics applies to our daily lives (Figure 1.1). As we will see in the next section, some of the most dramatic discoveries have been in the field of medical genetics. Geneticists now understand the metabolic basis for several hundred inherited disorders. Researchers have isolated defective genes that result in inherited disorders, including hemophilia, cystic fibrosis, Duchenne muscular dystrophy, Huntington's disease, the fragile X syndrome, neurofibromatosis, Alzheimer's disease, familial hypercholesterolemia, adult polycystic kidney disease, and breast cancer, to name only a few. Beyond medical applications, however, researchers study which aspects of our behavior and personality are controlled by our genetic constitution. This question has been debated for hundreds of years. Important ongoing studies centered at the University of Minnesota using identical twins raised apart show that many aspects of our personality and behavior are strongly influenced by genes. Indeed, evidence is accumulating on the significant role genes may play in such behaviors as alcoholism and sexuality. Geneticists have known for some time that defective genes lie at the root of mental disorders such as schizophrenia and manic-depressive illness. Today, armed with powerful tools of molecular genetic analysis, researchers are poised to identify the genes that, when defective, make individuals more susceptible to these maladies.

Molecular genetic research has given us powerful tools to study genes. In 1985 Kary Mullis developed a technique called the *polymerase chain reaction (PCR)* that allows researchers to take a minute sample of DNA, perhaps only a single cell's worth, and produce millions of copies of the DNA for use in molecular analyses. One application of this technique has been in forensic science, the use of scientific techniques in legal settings. With the use of PCR, DNA isolated from a small sample of tissue—perhaps a single sperm, white blood cell, or hair follicle—recovered at a crime scene can be subjected to detailed molecular analyses. Forensic scientists use a type of DNA analysis called *DNA fingerprinting* to examine variation in the sequence of the subunits that compose DNA. A tissue sample not belonging to the victim, such as a hair follicle, a drop of blood, or a semen sample, can be analyzed and used to identify a primary suspect or to exclude a person as a suspect.

A man was recently arrested in Camden, New Jersey, for the 1985 murder of a young Florida girl. The man smoked an unusual brand of cigarette, a brand found at the murder scene. The investigating detective, who had stayed with this case over the years, suspected the man but could not arrest him based only on the cigarettes. The cigarette butts were saved, however, and cells were isolated from them. In 1985, DNA testing was in its infancy. However, in 1993, those cells were analyzed by more sophisticated DNA fingerprinting techniques, and a link was established between the sample and the suspect, who was then prosecuted and found guilty. In other cases, DNA fingerprint analyses have exonerated individuals convicted of crimes on the basis of less definitive evidence.

Genetic research may also be seen in the corner grocery store. To eat or not to eat genetically engineered food products was the question that many people asked as the Flavr Savr™ tomato hit the supermarkets. This tomato was created by manipulating the tomato's genetic material. This new variety of tomato contains a gene that causes the fruit to remain firm longer during the ripening process. As a result, Flavr Savr™ tomatoes can be harvested later and marketed with a "vine-ripened" flavor. There are those who openly fear food products that have been genetically manipulated and thus avoid them. Is the fear justified? Probably not, at least not in the case of the Flavr Savr™ tomato. However, as with any new technology, unexpected risks may be associated with the use of genetic engineering to enhance food quality and yield.

A list of recent news items can easily fill a book. Only a few have been mentioned here. Other topics that have received major attention by the news media include the cloning of a sheep named Dolly (see Technical Sidelight in Chapter 2) and the race to characterize all the genetic information in the human species (see Human Genetics Sidelight in Chapter 21). One objective of this book is to help you gain a deeper understanding of the science behind the headlines and in the process to think more critically about issues at the interface of science and society.

Privatizing the Human Genome?

Principals behind joint-venture proposal and public effort seek to define relationships

By Paul Smaglik

A private effort to sequence the human genome four years ahead of the Human Genome Project's 2005 goal could either compete directly with the federal project or meld seamlessly with it. Before any rel... ship between the...

J. Craig Venter, president and director of The Institute for Genomic Research (TIGR) in Rockville, Md., would be embodied by a new company formed to sequence the entire human genome by 2001, and at an estimated cost of $150 million to $300 million. Faster, more automated sequencing machines recently unveiled by Perkin-Elmer, and a different sequencing strategy espoused by Venter make that target feasible. The federally sponsored Human Genome Project, which has a total estim... f $3 billion, is on schedule ... g the approximately 3 bil... DNA that make up the ... 005. ... director of the National ... Institute, cautions ...

tain each gene's function. Also, the federal project's larger budget has paid for much of the research—including gene mapping—that the private venture will use as its foundation. Collins told *The Scientist* that only about $100 million has been spent on human genome sequencing thus far. He and others are unsure whether the private product will be better—only that it will be different, since the public and private efforts will use different sequencing strategies.

That di... whether th... compatibl... will indee... Collins says, may be the sequencing of *Drosophila*, which the venture intends to

complete as a test. Even the success of that ...

The Scientist
June 8, 1998

Human Or Chimp? 50 Genes Are the Key

By NICHOLAS WADE

Theologians may ponder the difference between God and the creature ... but biologists have ... ler question: How ... m other animals? ... r active discussion ... an answer of possi... ecision. ... ntify the genes that ... ans by sequencing ... DNA, of the chim... ring it with the hu...

... nzees are so closely ... s, those genes that ... people than in chim... all that is needed to ... mode great ape into a ...

... that is exploring ... edical reasons. ... elcoming pursuit of ... foresee certain haz... company it, like the ... er more human hu... ced versions of the ... ies, as well as the ... erent in trying to ... genes by inserting ... es. ... have an ancestry ... earliest animals ... mmon with other ... distant from hu... or roundworm ... A sequence is ... o their human ...

The New York Times
October 20, 1998

Cloned mice hold promise for medical research

ROBERT S. BOYD WASHINGTON BUREAU — WASHING...

Dolly the clone is no longer alone. An international team of scientists announced Wednesday they have cloned three generations of genetically identical ... from adult mouse cells, surpassing the scientific achie... first accomplished with a sheep two years ago. The more than 50 carbon-copy mice born at the Univ... ... in Honolulu — including clones of ... research animal and human concept... h, scientists said. ... used to help preserve ... , panthers or pandas ... them. ... British journal Nat... A series of reports in ... from a 6-year-old ...

St. Paul Pioneer Press
July 23, 1998

Flies Invade Human Genetics

By Ricki Lewis

Recent issues of *The American Journal of Human Genetics* have featured a newcomer: *Drosophila melanogaster*. The fruit fly is a frequent star of a series of review articles called "Insights From Model Systems." The insect's appearance in a human genetics journal is a telling sign that this model among model organisms, long used to decipher general principles of in... a new status in helping ... human genome informa... Researchers of the ... that alter flies in noticeable ways: eye col... wing shape, or mixed-up parts ... people" take a more ... sequencin...

plex organism. Such discoveries increasingly point the way to identifying related functions in humans. And the sequence similarities, called homologies, between many fly and human genes are "stunning and striking," says **Thomas C. Kaufman**, professor of biology at Indiana University in Bloomington. "Whole pathways in flies are maintained in humans, using the same molecules. Who would have ... d 20 years ago that that would be the ... hen, we were just working on ... fruit flies," he add...

The Scientist
June 22, 1998

In the Hunt for Useful Genes, a Lot Depends on 'Snips'

By NICHOLAS WADE

A NEW word is edging into the study of human genetics: the snip.

Snips, variations in the DNA that make each individual unique, are the genetic determinants of health and disease. Because of machines that rapidly decode DNA, the chemical chain that embodies the genetic instructions, snips are at last becoming widely accessible.

Snips promise to yield two troves of information, one past and one present.

They will help population geneticists reconstruct the size and timing of the early human migrations that peopled the globe. And they should enable medical geneticists to trace elusive links between genes and disease, particularly the common diseases ... genes contribute. With the

A Special Chip to Hunt for Snips

To locate and identify the genetic variations known as Single Nucleotide Polymorphisms (SNPs or snips), researchers have developed the "snip chip." Unlike the silicon chips used in computers, the snip chip uses human DNA to search for its target.

IDENTIFYING SNIPS
Normal Sequence
NUCLEOTIDES
Snip Sequence
VARIATION
Snips are areas on the human genome where variations in the ... nucleotide pairs that compose a ...

HOW THE CHIP WORKS
GENE CHIP WITH ARTIFICIAL DNA
GLASS
0.5 INCHES
CHIP DNA SEQUENCES
SAMPLE DNA SEQUENCES
1 The chip is composed of multiple strands of DNA repre... ... nucleotide ...
2 Single stra... samples der... man tissue and containing ... are then ...
... are present in the ... the bonded sequences are ... by their location.

The New York Times
August 11, 1998

FILE PHOTO/THE WASHINGTON POST

J. Craig Venter with his standard poodle, Shadow.

One Man's Race to Map Genetic Code

High-Stakes Md. Scientist Has Long History of Ignoring Naysayers

By Justin Gillis
Washington Post Staff Writer

When a Rockville scientist named J. Craig Venter announced a few months ago that he would decode the entire sequence of genetic instructions that make up a human being—and do it in three years—some in the scientific world scoffed.

The federal government was already embarked on such a project, they pointed out, employing thousands of scientists and technicians around the world and spending billions of dollars on a grand endeavor called the Human Genome Project. Even so, it wasn't supposed to be finished until 2005.

And here was Venter, head of a lone laboratory off Interstate 270, maintaining that with a mere $250 million—all of it private—he could beat the government and harness the power of human genes.

The business of biotechnology is moving along at a breathtaking pace, but even so, Venter's audacity was surprising. His announcement turn... human genes into a... quickly, and under ... tists will develop ...

picture of the instructions that describe the growth and functioning of the human body. Such a map is universally agreed to be a Holy Grail of modern science, a key to 21st century advances that could turn many common diseases into bad memories.

Less than four months after Venter's announcement, he seems to be pulling his venture together at a brisk clip. At two sites along the I-270 corridor, construc... ...tion have begun on labora... ...e Venter's new company.

A6, Col. 1

The Washington Post
August 22, 1998

Figure 1.1 Genetics in the news; a sampling of some recent headlines.

Key Points: **The relevance of genetics to many aspects of human life and society, including health, behavior, food production, forensics, and even politics, is chronicled almost daily in the news media.**

GENETICS AND MEDICINE

Nature is nowhere accustomed to more openly display her secret mysteries than in cases where she shows traces of her workings apart from the beaten path; nor is there any better way to advance the proper practice of medicine than to give our minds to the discovery of the usual law of Nature by careful investigation of cases of rarer forms. For it has been found, in almost all things, that what they contain of useful or applicable nature is hardly perceived unless we are deprived of them, or they become deranged in some way.

A letter written by William Harvey in 1657 (quoted by A. Garrod in 1928 in an article about rare maladies. *Lancet, i*, pp. 1055–1060, 1928)

Figure 1.2 Sir Archibald E. Garrod, biochemist/physician, whose one mutant gene–one metabolic block concept was described in his book *Inborn Errors of Metabolism*, published in 1909.

Modern genetics has had a profound impact on medicine. Although connections between certain diseases and inheritance were made centuries ago, one of the first and most important links between the newly discovered Mendelian principles and disease was published by Sir Archibald Garrod in 1902. This British physician was studying rare metabolic disorders that ran in families and thus seemed to be inherited. One of these disorders was *alkaptonuria,* a metabolic disorder in which a substance called *alkapton* accumulates in cells and tissues and is excreted in the urine. Alkapton, known properly as *homogentisic acid,* is an intermediary metabolic product formed from the breakdown of the amino acid phenylalanine. Normally, homogentisic acid is converted into another compound (maleylacetoacetate), but in this case the enzyme that makes the conversion possible is defective. Homogentisic acid thus accumulates and is oxidized to a black product that accumulates in areas of the body rich in cartilage, turning ears, nose, joints, and other tissues black. The urine, too, turns black when exposed to oxygen. Although this accumulation of homogentisic acid is a relatively benign clinical condition, at its worst causing moderate arthritis, it persists throughout a patient's life.

Garrod, after discussions with William Bateson, one of the early champions of Mendel's ideas, concluded that alkaptonuria was inherited according to Mendelian principles and that the Mendelian factors were controlling cellular chemistry. Garrod was the first to associate a defective or *mutant* gene with a block in a metabolic pathway causing an inherited disorder. In recognition of his contributions, Garrod (Figure 1.2) is often referred to as "the father of biochemical genetics."

Garrod's insightful analysis of alkaptonuria stands today as the basis of modern genetic medicine. Even though many inherited disorders are rare, understanding them provides insight as to how genes control normal cellular processes. For example, studies of alkaptonuria, where a normal metabolic process is disrupted, led to important discoveries of how the amino acid phenylalanine is normally metabolized. Individuals with an inherited disorder called phenylketonuria, or simply PKU, are unable to convert phenylalanine to another amino acid, tyrosine, because they lack the enzyme needed for this reaction. As a result, derivatives of phenylalanine that are toxic to the central nervous system accumulate in the bloodstream and cause severe mental retardation. However, once scientists had determined the molecular basis of the mental retardation in individuals with PKU, an effective treatment was soon established. When PKU is diagnosed at childbirth, the infant is placed on a diet with carefully controlled levels of phenylalanine—high enough to support normal growth but low enough to avoid toxicity to nerve cells. (See Chapter 26 for more information about PKU.)

Since Garrod's time, researchers have made enormous strides in establishing links between defective genes and disease. We will discuss a few of the major linkages here and come back to many of them throughout the text. The genes in normal individuals are called *wild-type genes;* those with defects that result in the synthesis of nonfunctional products are called *mutant genes.* They result in abnormal or *mutant* conditions.

The defects in mutant genes are called *mutations*. We discuss the nature of these mutations in Chapter 14.

Hemophilia, or bleeder's disease, was one of the first human disorders to be linked to a mutant gene and its nonfunctional product. In the past, individuals with this inherited defect in blood clotting usually died during childhood. Today, they often live full and normal lives thanks to effective treatment with clotting factor produced by genetically engineered mammalian cells growing in culture (see Human Genetics Sidelight: Hemophilia: Successful Treatment of a Once Deadly Disorder).

The mutant gene that is responsible for Huntington's disease, a fatal neurological disorder, was isolated in 1993 after an intense 10-year search (Chapter 22). The mutant gene responsible for cystic fibrosis was also isolated and its gene product identified (Chapter 22). With the understanding of disease mechanisms comes the opportunity for treatment.

Gene therapy—the introduction of normal genes into the cells of patients with defective genes—offers great promise as an effective treatment of inherited diseases. Unfortunately, to date, gene therapy has not lived up to its promise. Gene therapy has been used in combination with other treatments on children with a devastating immune system disorder called severe combined immunodeficiency disease, but the results have been discouraging. The introduced genes were expressed for only a short period of time. Clinical gene therapy trials are currently being performed on patients with various types of cancer, Gaucher's disease, familial hypercholesterolemia, hemophilia, Fanconi's anemia, Hunter's syndrome, and acquired immune deficiency syndrome (AIDS). Although there is every reason to expect that gene therapy will eventually be successful in the treatment of these and other disorders, current results indicate that more research is needed to determine how and why genes are turned off shortly after they enter new host cells.

Some forms of cancer are familial (i.e., they occur frequently within families) or hereditary; others are nonhereditary and occur sporadically among all members of a population. However, all cancers are genetic diseases in the sense that they are caused by changes in the genetic information that are transmitted to progeny cells. The available evidence indicates that all cancers result from the accumulation of damage to genes that control or influence cell division, cell differentiation, and related processes. Although there are hundreds of different types of cancer, they all have one thing in common—loss of the normal control of cell division (see Chapter 28).

Mutant genes that cause an increased risk of cancer are being identified and studied intensely. One such gene, designated *BRCA1* for *br*east *ca*ncer gene *1*, results in increased susceptibility to breast and ovarian cancer in certain families. The *BRCA1* gene was isolated, characterized, and shown to be located on chromosome 17 (Figure 1.3). A second breast cancer gene, *BRCA2*, was subsequently identified. They join a long list of genes that cause familial cancer, including Wilm's tumor, a kidney cancer in children, and Gardner syndrome, a type of colon cancer. As we learn more about how these cancer genes function in normal and abnormal settings, we move that much closer to effective therapeutic treatments.

However, as is the case for other genes identified by this strategy, it is a long way from isolating a gene to identifying its product and determining what the product does in the cell. Even with information about the function of a gene, designing a therapy is another problem of major proportions. For example, the gene for Huntington's disease was isolated in 1993, and its protein product, *huntingtin*, was identified shortly thereafter. Yet, over five years later, the function of *huntingtin* is still unknown. Nancy Wexler, who was instrumental in the discovery of the Huntington's disease gene, shares her thoughts with us about the scientific, personal, and social aspects of Huntington's disease in a conversation at the beginning of Chapter 20. Once the function of the Huntington's disease gene is fully understood, it may still be a long time before

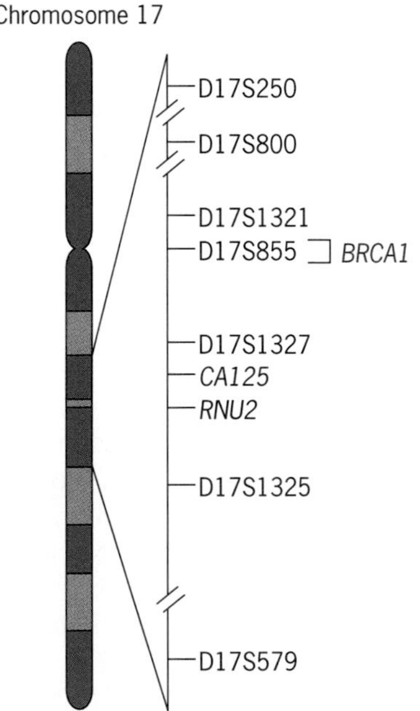

Figure 1.3 Schematic of human chromosome 17 showing the location of the breast cancer gene, *BRCA1*. Genes (*RNU2* and *CA125*) and molecular markers (everything that begins with D) in the vicinity of *BRCA1* are also shown.

an effective therapy for this devastating disease is established.

In contrast to Huntington's disease, we understand a great deal about the function of the cystic fibrosis gene product, called CFTR (*c*ystic *f*ibrosis *t*ransmembrane conductance *r*egulator). This knowledge suggests that it may someday be possible to treat cystic fibrosis by gene therapy, that is, by introducing functional copies of the *CFTR* gene into cells of individuals with defective versions of the gene. In one strategy, a normal form of the *CFTR* gene is inserted into a common cold virus that has been genetically altered so that it cannot complete its normal life cycle. This engineered virus is then applied to the nasal passages of CF patients. The envelope that encapsulates the cold virus functions as the vehicle for transferring the normal *CFTR* gene into nasal epithelial cells. Once in the cells, the *CFTR* gene should direct the synthesis of a normal gene product, and the cells—previously crippled by the lack of CFTR—should regain some of their normal function. Unfortunately, to date, this type of gene therapy has not proven effective. Too few cells actually acquire functional copies of the gene, and the levels of CFTR synthesized in cells of treated CF patients are too low to eliminate the symptoms of the disease. Furthermore, this type of gene therapy does not correct the pancreatic defect of this disease, nor does it correct CF-caused defects in other cells. Although the initial gene therapy experiments have been unsuccessful, it seems safe to predict that in the future—the near future, we hope—gene therapy will provide an effective treatment of cystic fibrosis.

The isolation and characterization of the mutant *BRCA1* and *BRCA2* genes that increase the risk of breast cancer are creating important social and ethical issues. For example, *BRCA1* is a very large gene, and over 100 distinct *BRCA1* mutations have been identified in humans. It is an expensive and labor-intensive job to screen women for all possible variants of this gene on a routine basis. To complicate matters further, some variants of *BRCA1* are benign and do not predispose women to breast cancer. Therefore, it appears that the testing will be done primarily on women with a family history of breast cancer. Even in these cases, however, there is the risk that laboratory tests will produce false positive or false negative results. In cases where a diagnostic test indicates a woman is carrying a mutant *BRCA1* or *BRCA2* gene that may predispose her to develop breast cancer, what treatment options can be offered? Is it always in the best interest of the woman to be told that she is carrying a mutation that predisposes her to develop breast cancer? Prophylactic mastectomy is a recommended treatment for carriers of mutant *BRCA* genes, but this treatment may not always be successful. Furthermore, since mutant *BRCA* genes also increase the risk of ovarian cancer, should genetic counselors suggest that women with *BRCA* mutations also have their ovaries removed? Many vexing problems are associated with the identification and isolation of breast cancer genes, problems that will have to be addressed.

Recently, scientists have combined genetic techniques with computer applications to develop powerful new *gene-chip technologies* that can be used to diagnose inherited diseases quickly, the presence of viruses, or changes in patterns of gene expression in tissues of patients with various diseases (Figure 1.4). Gene chips can be used to compare quickly the levels of expres-

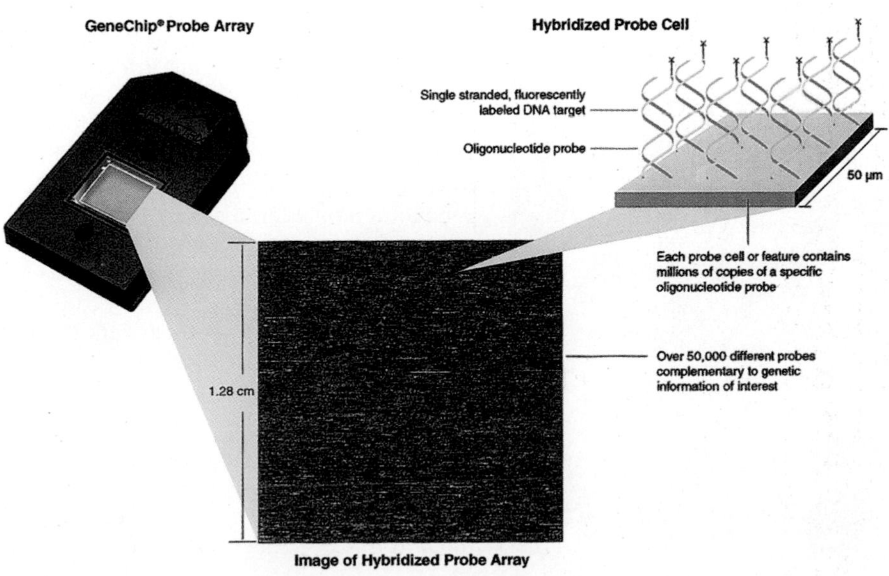

Figure 1.4 Photograph of a gene chip (top left) and an illustration of the use of the gene chip to detect alterations in the human *p53* tumor-suppressor gene. Alterations in this gene are associated with several types of human cancer (see Chapter 28). Other applications of the gene-chip technology are discussed in Chapter 21.

HUMAN GENETICS SIDELIGHT

Hemophilia: Successful Treatment of a Once Deadly Disorder

A small defect in an important gene can cause a fatal human disease. In the past, hemophilia, excess bleeding caused by a defect in blood clotting, was such a disease—often fatal early in life. Before the 1960s, when scientist-physicians developed the first effective treatment, the life expectancy of individuals with hemophilia was about 20 years. Today, hemophiliacs in most of the world have a nearly normal life expectancy. An understanding of the molecular basis of the disease resulted in the development of an effective treatment.

There are two major types of hemophilia. About 80 percent of the individuals with this disease have hemophilia A (classical hemophilia), and about 20 percent have hemophilia B (also called Christmas disease because it was first detected in a patient named Stephen Christmas). Both types of hemophilia are caused by defective genes on the X chromosome, the human chromosome that is present in two copies in females and one copy in males (Chapter 6). Most hemophiliacs are males, because they only need one copy of the defective gene to have the disease. Hemophilia is rare in females, because they need two copies of the defective gene, one on each X chromosome, to have the disorder.

Hemophilia A is sometimes called "royal hemophilia" because of its prevalence in the royal families of Europe. England's Queen Victoria (Figure 1) did not have hemophilia, although she carried the defective gene that causes hemophilia A on one of her X chromosomes. However, she passed the defective gene to two of her daughters—Alice, who transmitted the gene to the imperial families of Russia (see Figure 6.9) and Germany, and Beatrice, who passed the gene to the royal family of Spain—and to her son Prince Leopold, who died at age 31 from hemorrhages after a fall. Several of the queen's grandsons and great-grandsons died early in life because of excess bleeding or hemorrhages after surgery or accidents.

The mode of transmission of hemophilia was probably recognized in ancient civilizations. The Jewish Talmud, which dates to about 400 B.C. and was compiled into a single document in the 4th and 5th centuries A.D., decreed that boys whose older brothers or male cousins had died from excessive bleeding after circumcision were exempt from this procedure.

Hemophilia A and hemophilia B both result from defects in blood coagulation—the cascade of reactions that causes blood to clot at the site of a wound. A simplified version of part of this pathway is shown in Figure 2. Individuals with hemophilia A are deficient in a gene product called factor VIII; those with hemophilia B are lacking factor IX. In the absence of either of these blood-clotting factors, an individual can bleed to death after suffering a small cut or can die from internal hemorrhages after an otherwise minor bruise.

When scientists discovered that hemophilia was caused by the absence of specific blood-clotting factors, they realized that the disease could be treated with transfusions of concentrates of the missing factor. Initially, beginning in the 1960s, the proteins were purified from blood obtained from large numbers of donors. This process was expensive, and the concentrates were either unavailable or were too expensive for use by hemophiliacs in many countries. Fortunately, the advent of genetic engineering brought positive changes. The genes that encode factor VIII and factor IX were both isolated, and each gene was introduced into mammalian cells growing in culture. By this procedure, cell culture lines were produced that synthesize large quantities of either factor VIII or factor IX. The clotting factors are now purified from these cells and used to prepare concentrates for use in transfusions. As a result, both clotting factors are now available in essentially unlimited quantity to treat people suffering from hemophilia.

Figure 1 A portrait of Great Britain's Queen Victoria, her husband Prince Albert, and five of their nine children. Queen Victoria passed the defective gene that is responsible for hemophilia to at least three of her children. They, in turn, passed the gene to the royal families of Germany, Russia, and Spain (see Figure 6.9). The present British royal family is free of hemophilia. They are descendants of Victoria's son King Edward VII, who did not inherit the hemophilia gene from his mother.

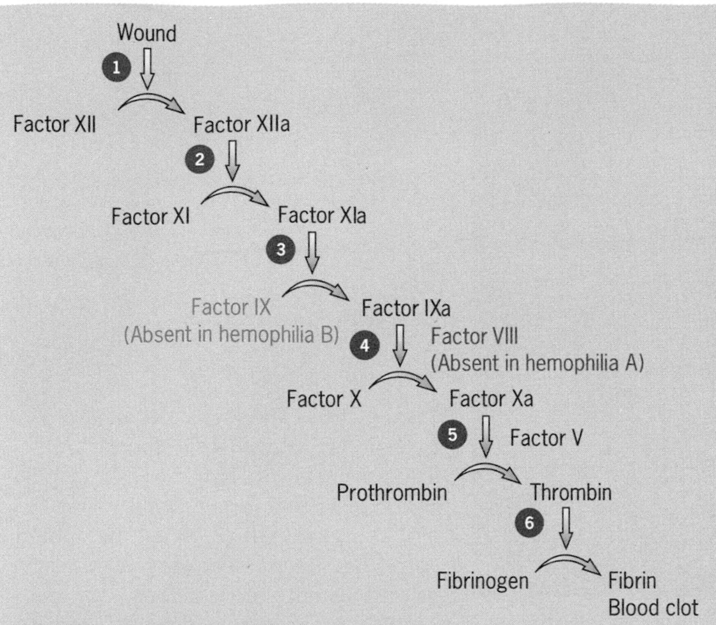

Figure 2 Simplified pathway of blood coagulation in humans. The cascade is activated by cell damage at a wound site. Each factor is the product of a specific gene. At each step in the pathway, a factor is converted to its active form, which then activates the next factor in the pathway. Some factors, such as factor VIII, participate in the activation of other factors without prior activation. In the last two steps in the pathway, the factors prothrombin and fibrinogen are converted to their active forms thrombin and fibrin, respectively. Long strands of fibrin stabilize blood clots by trapping cells called platelets (photograph above). The absence of factors VIII and IX in individuals with hemophilia A and B, respectively, blocks the clotting process prior to the formation of fibrin, which results in excessive bleeding or hemorrhaging from a cut or bruise.

sion of 10,000 or more genes in selected tissues of individuals with a given disease and of normal individuals. The first gene chips to become available commercially were designed to detect HIV (the virus that causes AIDS), to detect alterations in the structure of the human *p53* tumor-suppressor gene (altered in several types of cancer; see Chapter 28), and to quantitate the RNA products of 7700 human genes (Chapter 21). Other gene chips will soon be available. The gene-chip technologies promise to expand our understanding of the relationships between altered gene expression and human diseases. This knowledge should help scientists develop effective methods for alleviating the pain and suffering associated with these diseases.

Key Points: **Genetics has had and continues to have a profound impact in the field of medicine. Defective genes place humans at increased risk for developing many diseases. Treatment of inherited disorders by gene therapy has great potential but has not yet proven effective.**

GENETICS AND MODERN AGRICULTURE

In addition to its impact on medicine, modern genetics has had a tremendous impact on agriculture. One of the greatest achievements in modern agriculture was the application of Mendelian principles to the development of hybrid corn (Figure 1.5). Hybrid corn is produced by crossing different inbred strains, each of which has valuable properties, such as disease resistance, high protein or sugar content, and drought tolerance. During the period from 1940 to 1980, the average yield for corn increased by over 250 percent, in large part because of the development and introduction of hybrid corn varieties. In the United States, hybridization has resulted in dramatic yield increases in nearly all the important food crops, such as barley, beans, oats, rice, and wheat, though none as great as in hybrid corn.

From the 1950s through the 1970s, Norman Borlaug and his team of researchers used classic genetic

(a)

(b)

Figure 1.5 The increased yield of hybrid corn is one of the spectacular achievements resulting from the application of genetic principles to plant and animal breeding. When two inbred corn lines are crossed, the hybrid progeny are usually more vigorous (*a*) and produce larger ears with more kernels (*b*), resulting in higher yields.

principles to develop new strains of wheat in Mexico that would do well under the stressed conditions often found in developing countries (Figure 1.6). His success in developing this wheat as well as other crop plants earned him a Nobel Peace Prize in 1970. His enormous contributions to agricultural productivity, and thus to nutrition, in developing countries helped launch the much-publicized "green revolution."

The modern tomato has benefited tremendously from the application of genetic principles. Genes that confer resistance to various pathogens, such as fungi and nematodes, have been bred into modern varieties. Most of these resistance genes came from tomato relatives growing in the wild. The growth pattern of the tomato plant has also been genetically altered so that the plant is bushier and more compact—characteristics that help keep the fruits off the ground. Breeders have developed a wide variety of morphological tomato types. These include round, medium-sized lemon boys; round, large beefsteaks; small, round cherry; pear-shaped yellow and red; and others (Figure 1.7).

Selective breeding programs have produced chickens that are meatier, grow faster, are more disease resistant, and lay more eggs (Figure 1.8). Selective breed-

Figure 1.6 Norman E. Borlaug and his prizewinning semidwarf wheat at the International Maize and Wheat Improvement Center, Londres, Mexico.

Figure 1.7 Some of the many varieties of tomatoes produced by hybridization and selection.

Gold Laced Wyandotte *Silver Laced Wyandotte* *Black Wyandotte Bantam* *Silver Wyber*

Black Belgian Bantam *White Crested Polands* *Ancona Cockerel*

Figure 1.8 Some of the many varieties of chickens produced by hybridization and selection.

ing programs have also produced cattle, sheep, and pigs that are meatier, grow faster, are more efficient at converting feed into meat, and are better adapted to regional environments (Figure 1.9). Milk production per cow and butterfat content have also increased dramatically as a consequence of selective breeding.

One of the most effective techniques used to improve livestock quality is artificial insemination. A bull known to sire cows that produce large quantities of milk or offspring that are outstanding meat producers has his sperm collected at regular intervals and stored. This sperm is used to fertilize thousands of females all over the world. The favorable genes of this bull are thus transmitted to a large number of progeny, resulting in dramatic increases in milk or meat production.

In addition to the genetic crosses and selection for desired traits that have been used by plant and animal breeders for centuries, new genetic technologies already have had and continue to have a major impact on agriculture. Inserting genes for resistance to insects or pathogens into crop plants is becoming a major weapon in the war against organisms that destroy so much of the world's food supply. The corn borer is an insect that burrows through corn stalks (Figure 1.10*a*), interrupting the flow of nutrients to the ears and decreasing yield. The tunnels produced by the corn borer also weaken the stalks, causing the plants to blow over in high winds. In combination, the effects of the corn borer cause major decreases in yield and cost the corn growers about $1 billion per year in the United States alone. In the past, farmers have attempted to control corn borer damage by spraying their fields with insecticides. They now have another option. The bacterium *Bacillus thuringiensis* contains a gene encoding a protein that is toxic to many insects. Different subspecies of this bacterium produce toxins that kill different insects. One subspecies produces a toxin that kills the corn borer. The gene for this protein toxin has been isolated from the bacterium and inserted into corn plants. When corn borers feed on plants that carry the toxin gene, they die within a few days, whereas corn plants lacking this gene are unprotected (Figure 1.10*b*).

Another application of molecular genetics to plant and animal improvement is the use of molecular markers to determine the locations of genes or groups of genes that control important traits such as growth rate and yield. These traits usually exhibit continuous vari-

Figure 1.9 Some of the many varieties of beef cattle and sheep produced by hybridization and selection.

ation, for example, from the shortest corn plant to the tallest one (see Chapter 5). Therefore, they are called *quantitative traits* (QTs), and the genes or groups of genes that control them are called *quantitative trait loci*. A *locus* (plural *loci*) is simply the site or location of a gene or a mutation on a chromosome. Once QT loci have been mapped to regions of individual chromosomes, geneticists can design crosses to introduce them into important crop varieties or animal breeds. Thus, QT loci are being used extensively to speed up plant and animal breeding projects.

While the application of the principles of genetics to plant and animal breeding was producing dramatic increases in agricultural productivity in most of the world, agricultural production in the Soviet Union was stagnant. From 1937 to 1964, biology and agriculture in the Soviet Union were dominated by one person, Trofim D. Lysenko, a young plant breeder from the Ukraine. Lysenko believed that plant development and agricultural productivity could be vastly improved by manipulating the environment. Environmentally induced changes in plant growth, according to Lysenko, would be assimilated into the genetic material and passed on to the next generation. His ideas were similar to those formalized by Lamarck in the early nineteenth century. Lysenko's ideas were eventually discredited, but only after agricultural productivity in the Soviet Union had fallen far behind that in other developed countries of the world. The Lysenko era in the Soviet Union provides a vivid illustration of the danger of placing too much power in the hands of a single person.

Although molecular and classical genetic principles have been applied to agriculture for less than 100 years, we should not lose sight of the fact that the human species has been using genetics in agriculture for centuries without formal knowledge of the laws of inheritance. Early humans carried out the first "genetic selection" experiments in wheat between 7000 and 10,000 years ago. Evidence indicates that almost all of

(a)

(b)

Figure 1.10 (*a*) A European corn borer eating away the stalk of a corn plant. (*b*) Side-by-side comparison of corn stalks from plants that are resistant (top) and susceptible (bottom) to the corn borer. The resistant plant is expressing a gene for an insecticidal protein derived from *Bacillus thuringiensis*.

our present food crops were domesticated during this early Neolithic period, coincident with the development of stone tools. The largest, most vigorous individuals in the population were simply selected as parents for subsequent generations, a practice that is still the mainstay of modern plant and animal breeding. However, this practice is now being supplemented with the powerful new tools of molecular genetics and genetic engineering. As a result, we can expect rapid progress and major breakthroughs in agriculture in the future, just as in the past.

Key Points: **Genetics has had a huge impact on agriculture. Today, both selective breeding and new molecular genetic approaches are being applied in the challenge to feed the world's rapidly expanding human population.**

GENETICS AND SOCIETY

The goals of the **Human Genome Project,** launched in 1990, are to map and sequence all the genetic information of humans and a few other important genetic organisms by the year 2005. Progress toward these goals is ahead of schedule (see Chapter 21), and a newly formed, private company has announced that it will sequence the human genome in just three years—by 2001. Another private company is sequencing the expressed portion—the genes—of the human genome (see the Human Genetics Sidelight: The Race to Sequence the Human Genome, in Chapter 21). The complete sequences of the genomes of 18 bacteria and the yeast *Saccharomyces cerevisiae* have already been determined; the sequence of the genome of the worm *Caenorhabditis elegans* is 99.9 percent complete; and the sequences of the genomes of the plant *Arabidopsis thaliana* and the fruit fly *Drosophila melanogaster* will soon be available. Knowledge of the structure of all the genetic information of humans and other important genetic organisms will have profound effects on society. This information will have dramatic effects on scientists' ability to diagnose and devise effective treatments for human diseases. Thus, this information should have a very positive impact on human health. However, it will also raise complex moral, ethical, and legal questions that must be dealt with by individuals and by society. Let's consider a few of these issues.

Information about the human genome is accumulating at a rapid pace. Given this new information, scientists may soon be able to make important predictions about the future health of individuals: for example, the probability that a person will develop cancer, mental illness, or some other genetically based disorder. A complex problem that lies at the core of this issue is access to the information being gathered. Most of the research on the human genome is publicly funded. Should the public have unlimited access to the information? What about insurance companies and employers? They might want to deny a person employment or insurance based on genetic information. How will society protect the privacy of the individual given the availability of this new genetic information?

A young couple has a daughter with cystic fibrosis. To have this devastating genetic disease, the child must inherit two copies of the defective gene. Thus, each of her parents had to carry one copy of the defective gene and pass that copy on to her. The mother became pregnant again, even though she and her husband knew there was a 25 percent risk they would produce another affected child. Their insurance company requested that they obtain a prenatal diagnosis of the fetus to assess its cystic fibrosis status. The test results indicated that the fetus had cystic fibrosis. Representatives of the insurance company then requested that they terminate the pregnancy or risk losing insurance coverage. The couple refused to terminate the pregnancy, the insurance company terminated their insurance, and the couple sued. Do insurance companies have the right to cancel health insurance policies or insist on pregnancy termination when confronted with the birth a child with a medically expensive, usually fatal genetic disorder? The courts said no. However, such an issue does not end with a legal decision. In fact, families that are at risk for developing certain inherited diseases are being denied health insurance.

It is not fair, however, to argue that the insurance companies are insensitive and interested only in the bottom line. Genetic disorders such as cystic fibrosis are expensive to treat, often costing hundreds of thousands, even millions, of dollars. Is it appropriate to ask the insurance company's policyholders to cover the entire costs for treatments? What rights or obligations do insurance companies have in the prevention of genetically defective births? What responsibilities do parents who are at risk have in the prevention of genetic disorders? As you use this text to lay a genetic foundation for a thoughtful analysis of these and other issues, you will be in a stronger position to evaluate them.

Genetics has the potential to enhance the quality of human life. Throughout this book we highlight how our knowledge of genetics has grown and how our lives have been made better by that knowledge. But we should bear in mind some unfortunate, even tragic, misuses of genetics.

Darwin's theory of evolution by natural selection stated, among other things, that organisms with traits that are beneficial produce larger numbers of progeny than organisms with less beneficial traits. As a result, these beneficial traits become the norm in the population. This seminal idea has been inappropriately applied to the human species. Francis Galton, Charles Darwin's cousin, believed that many human mental and physical qualities were inherited and thus subject to the forces of selection. But Galton took this idea a step further. He suggested that the genetic constitution of the human species could be improved through the use of artificial selection, an idea he called **eugenics:** Parents expressing favorable traits would be encouraged to have larger families (*positive eugenics*), and parents with undesirable traits would be encouraged not to have children (*negative eugenics*). Traits that Galton thought were favorable included high intelligence, high levels of achievement, artistic creativity, and excellent health. Traits that he believed should be selected against included low intelligence, mental illness, criminal behavior, and alcoholism.

In the United States, the eugenics movement gained strength during the early part of the twentieth century, especially negative eugenics. In 1907, Indiana passed laws that mandated sterilization of individuals who were "imbeciles, idiots, convicted rapists, and habitual criminals." By 1931 nearly half of the states had adopted such laws, and mandatory sterilization was extended to include such things as sexual perversion, drug addiction, alcoholism, and epilepsy. Implicit in these sterilization laws was the connection between inheritance and behavior. Sterilization of the mentally retarded may sound barbarous to us; however, during the 1920s and 1930s, sterilization procedures were advocated because, it was argued, mentally retarded people were poor parents and provided poor environments for their children. Because there were no good drug treatments enabling individuals with mental disorders to live normal lives and care for children, many more were institutionalized than are now. Sterilization was thought to be kinder than incarceration.

Another outgrowth of the eugenics movement during the 1920s was the immigration restriction laws. These laws, motivated largely by economics, favored "genetically desirable" ethnic groups (mainly northern European) and placed severe restrictions on the less "genetically desirable" groups (the Mediterranean area, Central Europe, China). These laws were founded on flawed, unsubstantiated "data" and on bigotry. Many of them remained on the books until the 1960s.

The eugenics movement took on its most twisted and perverted form in Nazi Germany. Between 1930 and 1945, the Hitler regime systematically exterminated millions of Jews, Gypsies, and others in an attempt to "cleanse" Germany of "inferior" genetic material.

In the United States, England, and elsewhere, most geneticists were appalled at the way the science of genetics, still so young, was being abused. There was no solid evidence that genetics played a role in most of the human traits deemed desirable or undesirable. People were taking anecdotal information and elevating it to the level of scientific fact (e.g., Mr. Smith was a wife and child abuser; his son was also a wife and child abuser; therefore, abusing one's wife and child is an inherited behavior). This perversion of genetics and society's willingness to accept it uncritically prompted many geneticists to avoid the study of human genetics for fear of being grouped together with the eugenicists. The field of human genetics suffered greatly because of these abuses.

Today, all universities that accept federal research grants have established committees that must approve proposed research involving human subjects prior to initiation of the work. In addition, the National Institutes of Health of the United States Department of Public Health have appointed committees that must evaluate and approve proposed genetics research on humans before that work can begin. Safeguards of this type should help us avoid inappropriate genetics research. Society should heed the lessons learned by past mistakes and proceed with optimism into the twenty-first century. The Human Genome Project is producing a wealth of new information about the human genome, and genes that influence the risk of human disease are rapidly being identified. With a better understanding of genes and genomes, scientists should be able to devise effective methods of treating many human diseases, just as they have for phenylketonuria and hemophilia (see Human Genetics Sidelight in the preceding section) in the past.

Key Points: **The Human Genome Project is producing a wealth of new information about the genetics of humans. If used wisely, the new information promises to enhance our quality of life. However, the availability of this information also raises complex moral, ethical, and legal questions that must be carefully evaluated by society.**

THE PRINCIPLES OF GENETICS: AN OVERVIEW

Genetics has had and continues to have a significant impact on our daily lives. At various times in this text, we will discuss the complex personal and societal implications of some of the recent discoveries in genetics. Mainly, however, this book will address four major questions, though not necessarily in this order: *What is the chemical nature of the genetic material? How is the genetic material transmitted? What does it do? And how does it change or evolve over time?* These questions form the core of modern genetics.

In this section, we will briefly discuss the answers to these four questions. We will examine (1) the molecular structure of the genetic material, (2) the mechanism by which the genetic material is duplicated so that it can be transmitted to progeny, (3) the process by which genetic information is expressed to control the growth and development of organisms, and (4) the way the genetic material changes over time, allowing evolution to occur. Then, after considering the structure of cells and the transmission of genes and chromosomes in Chapters 2 through 9, we will return to these four topics and discuss them in detail in Chapters 10 through 14. For most of you, this section will be a review of topics covered in high school or college biology. However, if this is your first exposure to biology, this overview will introduce you to some of the most important concepts and processes of genetics.

DNA: The Genetic Material

The genetic information of living organisms is stored in large macromolecules called nucleic acids. These nucleic acids are of two types: *DNA,* which contains the sugar 2′-deoxyribose, and *RNA,* which contains the sugar ribose. In some viruses, the genetic information is encoded in RNA molecules. In all cellular organisms, the genetic material is DNA. In these organisms, the DNA is organized into structures called chromosomes, the number depending on the species. We will discuss the structure of DNA and chromosomes in more detail in Chapter 10.

The genetic information in nucleic acids is specified by the sequence of four nitrogen-containing compounds called *bases.* Thus, the genetic blueprints of all living organisms are encoded using a four-letter alphabet: the four bases adenine (A), guanine (G), cytosine (C), and thymine (T) in DNA. In RNA, uracil (U) occurs in place of T. In DNA, which has a double-stranded structure, the bases are present in pairs (Figure 1.11). A is always paired with T, and G is always paired with C. Although there are only four letters in the genetic alphabet, a vast amount of information can be stored because of the large sizes of nucleic acid molecules. For example, one complete copy of the human genome (all the genetic information in one complete set of the human chromosomes) contains 3 billion (3×10^9) base pairs of DNA. Since the number of possible sequences of four letters using n at a time is 4^n, we can see that, with $n = 3 \times 10^9$, the human genome has the capacity to store a huge amount of information. The large information storage capacity of the human genome is reassuring given the enormous variability in features and traits in the human species.

DNA Replication: Transmission of Genetic Information

The genetic information of an organism is transmitted from cell to cell during development and from generation to generation during reproduction by the accurate replication of the sequence of bases in nucleic acids due to the precise base-pairing in double-stranded nucleic acids: A with T and G with C. Two properly base-paired strands of DNA are complementary. Thus, in terms of transmission of genetic information to progeny DNA molecules, the two strands of a DNA double helix contain the same genetic information. When the two strands of a parental double-helix of DNA separate, the base sequence of each parental strand can serve as a template for the synthesis of a new complementary strand (Figure 1.12). As a result, the two progeny DNA molecules will be identical to the parental DNA molecule. Thus, the structure of DNA is perfect for its role in transmitting genetic information from cell to cell and from generation to generation.

Although the use of parental strands of DNA as templates for the synthesis of new complementary strands is an elegant and simple way to duplicate genetic information, the mechanism by which DNA replicates is quite complex. We will discuss the complexities of DNA replication in Chapter 11.

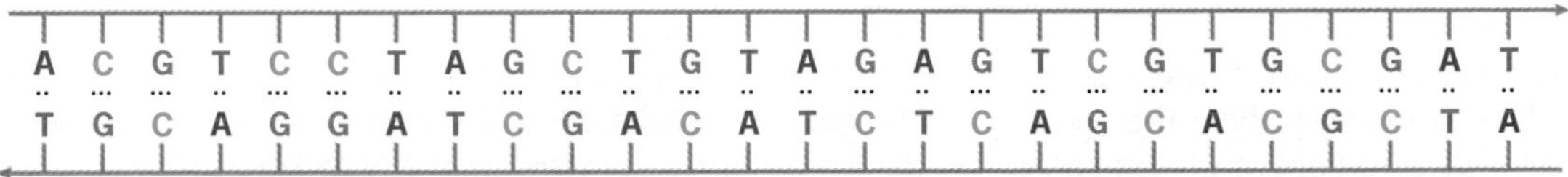

Figure 1.11 Two-dimensional view of DNA structure. Genetic information is stored in the sequences of base-pairs: A:T, T:A, G:C, and C:G, in giant DNA molecules present in the chromosomes of living organisms.

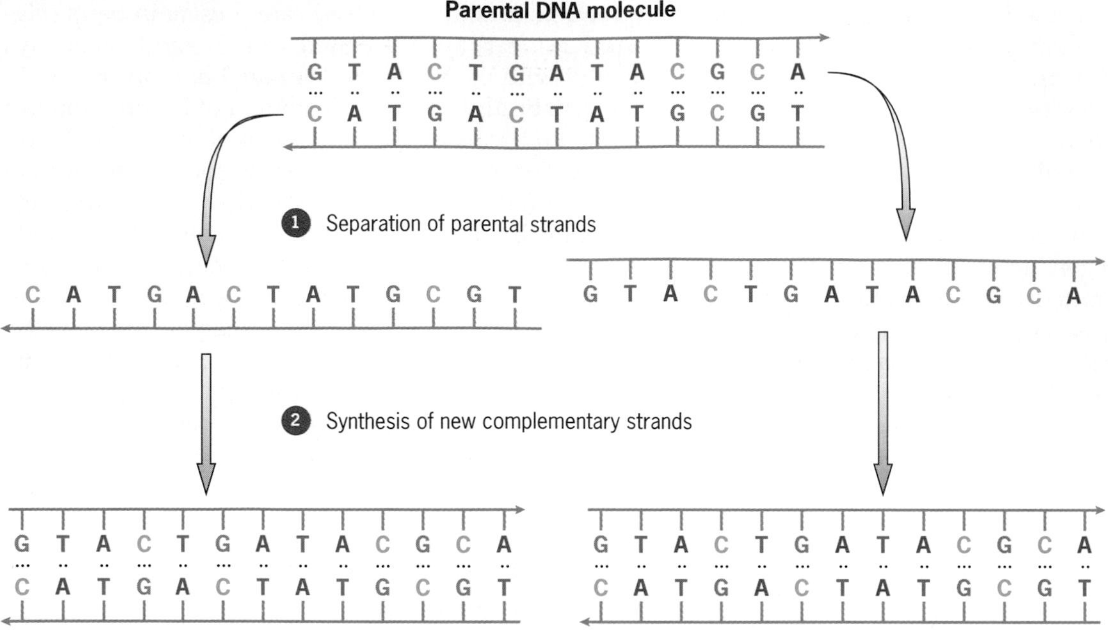

Figure 1.12 DNA replication. Strand separation (top) and the synthesis of new complementary strands (bottom) occur in a concerted, processive manner during the replication of the DNA molecules in the chromosomes of living organisms. They are separated here to emphasize the two steps in the process.

Gene Expression: Control of Growth and Development

The genetic information controls the morphogenesis of the organism, be it a virus, a bacterium, a plant, or an animal. This genetic information must be expressed accurately both spatially and temporally to produce the appropriate three-dimensional form of the organism. In multicellular organisms, the genetic information must control the growth and differentiation of the organism from the single-celled zygote to the mature adult. To accomplish this complex process, each gene of an organism must be expressed at the proper time and in the proper cells during development. The initial steps in the pathways of gene expression are called *transcription* and *translation;* these steps are illustrated for the expression of the human β-globin gene in Figure 1.13. This gene specifies a major component of the oxygen-transporting macromolecule, hemoglobin, an essential macromolecule present in red blood cells.

Transcription and translation result in the synthesis of *proteins,* macromolecules that catalyze the metabolic reactions essential to life and contribute much of the structure of living organisms. Proteins are large, complex molecules containing from one to many *polypeptides*. Polypeptides, in turn, are long chains of small, repeating subunits called *amino acids*. Twenty different amino acids—alanine, arginine, asparagine, aspartic acid, cysteine, glutamic acid, glutamine, glycine, histidine, isoleucine, leucine, lysine, methionine, phenylalanine, proline, serine, threonine, tryptophan, tyrosine, and valine—are incorporated into polypeptides in response to the genetic information stored in DNA molecules. Polypeptides are often several hundred amino acids long. Given 20 different amino acids, living organisms can produce a vast number of distinct proteins. There are 20^{100} different sequences, for example, of a polypeptide 100 amino acids long. This enormous variation in protein structure is largely responsible for the vast variability of living organisms.

The first step in gene expression, transcription, involves converting genetic information stored in the form of base-pairs in double-stranded DNA into the sequence of bases in a single-stranded molecule of messenger RNA (mRNA). Catalyzed by enzymes called RNA polymerases, this process occurs when one strand of the DNA acts as a template for the synthesis of a complementary strand of RNA. Transcription uses the same base-pairing rules as DNA replication except that the base uracil is incorporated into RNA at positions where thymine would be present in DNA (Fig-

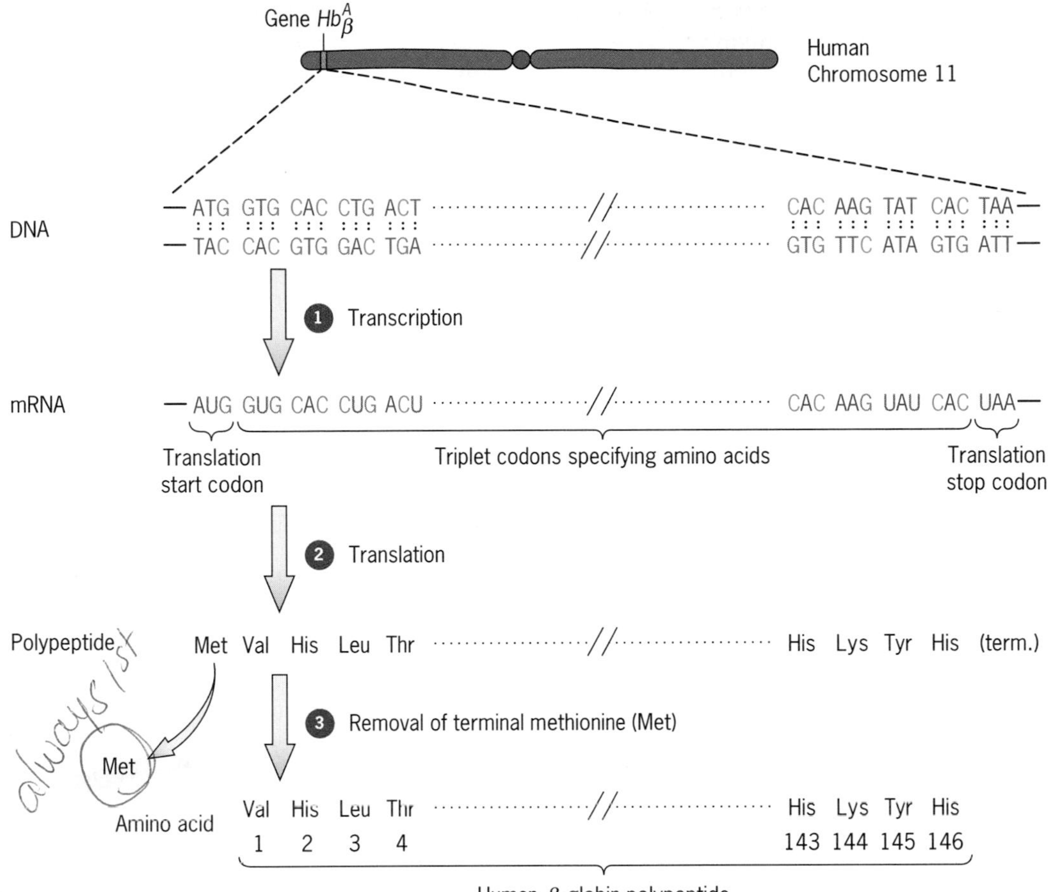

Figure 1.13 Transcription and translation, the initial steps in gene expression, are illustrated by using the synthesis of human β-globin as an example. During transcription (step 1), the two strands of DNA are locally unwound, and one strand—the bottom strand of the DNA molecule shown—serves as a template for the synthesis of a complementary strand of RNA. After undergoing modifications that we will discuss in Chapter 12, the resulting mRNA (*messenger RNA*) is transported to the cytoplasm. In the cytoplasm, the sequence of bases in the mRNA molecule is "translated" into the sequence of amino acids in the polypeptide gene product according to the rules of the genetic code. Translation (step 2) takes place on complex structures called ribosomes; it is discussed in detail in Chapter 13. The synthesis of mature β-globin requires the posttranslational removal of the terminal methionine residue (step 3).

ure 1.13, top). The sites at which transcription starts and terminates are determined by specific base sequences that are recognized by some of the protein factors that participate in the transcription process. We will discuss the signals and the proteins involved in transcription in Chapter 12.

During translation, the sequence of bases in an mRNA molecule is converted ("translated") into the specified sequence of amino acids in a polypeptide according to the rules of the *genetic code* (Figure 1.13, center). Each amino acid is specified by one *codon*, a triplet of three adjacent bases in the mRNA molecule. Translation is a complex process occurring on cytoplasmic macromolecular structures called ribosomes and requiring the participation of many other macromolecules. We will discuss translation in detail in Chapter 13.

Subsequent steps in the pathway through which a gene exerts its effect on the morphogenesis of an organism are usually complex, especially in multicellular eukaryotes. Pathways of gene expression frequently involve macromolecular interactions, cell–cell interactions and intercellular communication by hormones and other signal molecules, tissue and organ interactions, and restrictions imposed by environmental factors. Current studies on the genetic control of the growth and development of plants and animals are yielding novel and exciting results. Many genes that control the growth and development of plants and animals have recently been identified and characterized, and detailed pathways of morphogenesis are being worked out for several model organisms such as the worm *Caenorhabditis elegans*, the fly *Drosophila melanogaster*, the plant *Arabidopsis thaliana*, and even the mouse. Some of the exciting breakthroughs in our understanding of the genetic control of development are discussed in Chapter 25.

Mutation: Changes in the Genetic Material over Time

Although genetic information must be transmitted from generation to generation with considerable accuracy, it is not static. Rather, it undergoes occasional changes, or *mutations,* to produce new genetic variability that provides the raw material for evolution. The new variant genes produced by mutation, called mutant genes, often result in organisms with abnormal traits. We will take a quick look at mutation and its effects here but will examine this important process in detail in Chapter 14.

Single-site or point mutations within individual genes may be either base-pair substitutions or the insertion or deletion of one or a few contiguous base pairs. The insertion or deletion of one or two base pairs within the coding sequence of a gene will alter the codon reading frame—the series of base triplets that specify amino acids during translation—in the mRNA (see Figure 1.13). Thus, such mutations are referred to as frameshift mutations. Frameshift mutations usually result in nonfunctional gene products. In contrast, base-pair substitutions usually result in the replacement of a single amino acid with another amino acid because each base-pair substitution changes only one codon.

The relationship between base-pair substitutions and amino acid replacements can be best illustrated with an example. Sickle-cell anemia occurs in humans who carry two copies of a β-globin gene that differs from the normal adult β-globin gene by a single base-pair substitution. The one base-pair substitution in the Hb_{β}^{S} gene changes the sixth amino acid of the β-hemoglobin polypeptide from glutamic acid in normal individuals to valine in individuals with sickle-cell anemia (Figure 1.14). The single amino acid change in the human β-globin chain results in sickle-shaped red blood cells and in sickle-cell anemia in individu-

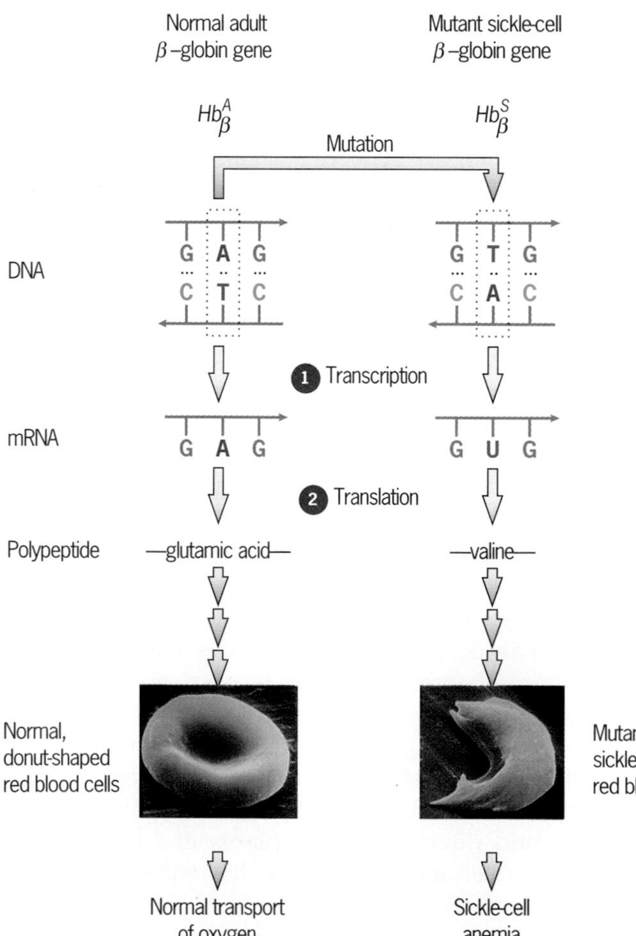

Figure 1.14 Mutation: change in the sequence of base pairs in DNA. The mutant gene (Hb_{β}^{S}; top right) responsible for sickle-cell anemia resulted from a single base pair substitution in the human adult β-globin gene (Hb_{β}^{A}; top left). Transcription and translation of the mutant gene produce a β-globin polypeptide containing the amino acid valine (center right) at the position where normal β-globin contains glutamic acid (center left). This single amino acid substitution results in the formation of sickle-shaped red blood cells (bottom right) rather than the normal donut-shaped cells (bottom left). The sickle-shaped red blood cells, in turn, cause severe anemia in individuals with this inherited disorder.

als with two copies of the Hb^S_β gene. This alteration of the red blood cells not only results in the reduced oxygen-carrying capacity of the blood characteristic of sickle-cell anemia but also causes a large number of secondary effects, such as impaired growth, general fatigue, enlarged spleen, recurrent pain, and increased susceptibility to microbial and viral infections (see the opening section of Chapter 13 for additional details). Thus, a single base-pair substitution in DNA can have a very large effect on the phenotype of the organism harboring the mutation.

Key Points: **The genetic information of living organisms is stored in the sequence of four bases in DNA or, in some viruses, in RNA. DNA is a double-stranded molecule. During DNA replication, the two strands separate, and each strand serves as a template for the synthesis of a new complementary strand, producing two identical progeny DNA molecules. The genetic information of an organism controls its growth and development through complex pathways of gene expression. Occasional changes in the genetic material provide new genetic variability that is essential for the process of evolution.**

QUESTIONS AND PROBLEMS

1.1 Select one or two recent newspaper or magazine articles about genetics and write a brief report about them. Why are these articles interesting, and how do they relate to the issues raised in this chapter?

1.2 Several participants in the Human Genome Project maintain Web sites that provide up-to-date results. Consult the following Web sites and discuss with classmates which of the sites are the most informative:

Whitehead Institute/MIT
http://www.genome.wi.mit.edu/

Stanford Human Genome Center
http://www.shgc.stanford.edu/

The Sanger Centre, United Kingdom
http://www.sanger.ac.uk/

Wellcome Trust Centre, United Kingdom
http://www.well.ox.ac.uk/

Généthon, France
http://www.genethon.fr/

1.3 As of September 3, 1998, The Sanger Centre in the United Kingdom had determined the sequence of 107,865,862 of the approximately 3 billion nucleotides in the human genome. Consult their Web site at http://www.sanger.ac.uk/ to find out how much progress has been made at The Sanger Centre since that date.

BIBLIOGRAPHY

Bearn, A. E. 1993. *Archibald Garrod and the Individuality of Man.* Clarendon Press, London.

Borlaug, N. E. 1983. Contributions of conventional plant breeding to feed production. *Science* 219:689–693.

Horgan, J. 1993. Eugenics revisited. *Sci. Amer.* 268(6):123–131.

Kay, M. A., D. Liu, and P. M. Hoogerbrugge. 1997. Gene therapy. *Proc. Natl. Acad. Sci. USA* 94:12744–12746.

Koonin, S. E. 1997. An independent perspective on the Human Genome Project. *Science* 279:36–37.

Lawn, R. M., and G. A. Vehar. 1986. The molecular genetics of hemophilia. *Sci. Amer.* 254(3):48–54.

Lewis, R. 1999. *Human Genetics, Concepts and Applications,* 3rd ed. WCB/McGraw-Hill, Dubuque, IA.

Marshall, A., and J. Hodgson. 1998. DNA chips: An array of possibilities. *Nature Biotechnology* 16:27–31.

Ramsay, G. 1998. DNA chips: State-of-the-art. *Nature Biotechnology* 16:40–44.

Scazzocchio, C. 1997. Alkaptonuria: From humans to moulds and back. *Trends in Genetics* 13:125–127.

Scientific American Special Report. 1997. *Making Gene Therapy Work.* (A series of five papers on gene therapy.) *Sci. Amer.* 276(6):96–123.

Soyfer, V. 1994. *Lysenko and the Tragedy of Soviet Science.* Rutgers University Press, New Brunswick, NJ.

2

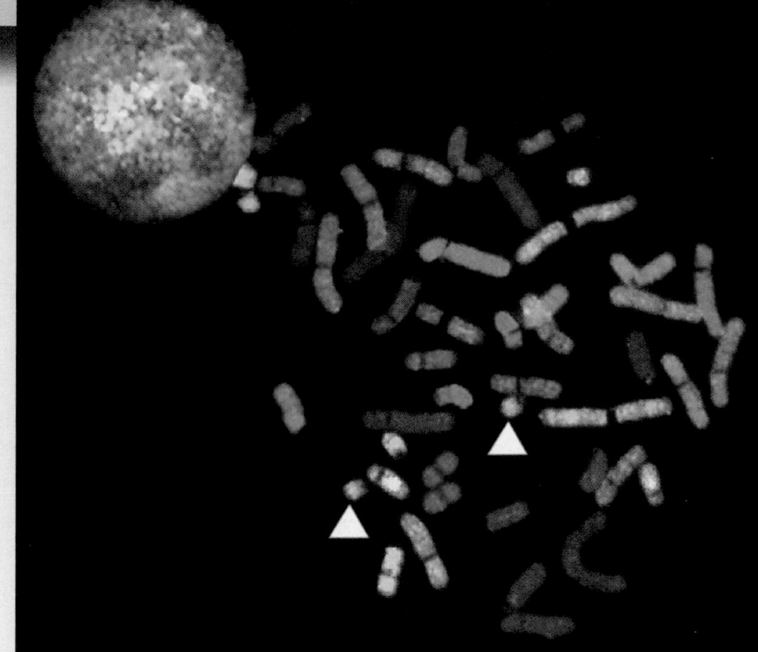

The chromosomes of a normal woman. Each of the 23 distinct human chromosomes has been "painted" a different color by the binding of differentially labeled chromosome-specific DNA sequences and subsequent spectral imaging. Individuals with Down syndrome have three copies of chromosome 21 (white arrows).

Reproduction as the Basis of Heredity

CHAPTER OUTLINE

The Cell as the Basic Unit of Life
The Chromosome: An Overview
The Cell Cycle
Cell Division: Mitosis
Cell Division: Meiosis
The Evolutionary Significance of Meiosis
The Formation and Union of Gametes
The Life Cycles of Some Genetically Important Organisms

An Accident in Cell Division

After five years of marriage, Michael and Carole decided that they were ready to begin their family. Carole soon became pregnant, and she and Michael eagerly awaited the birth of their baby. It was a normal pregnancy and a normal delivery. When their baby boy was born, however, they knew immediately that something was wrong. The child was lethargic and had poor muscle tone. His head was short and very flat at the back; the eyes had a slanted appearance, and the irises had speckles around the edges; the nose had a low bridge and the tongue tended to protrude. These and other abnormalities all pointed to one conclusion: The baby boy had Down syndrome. Examination of the baby's cells showed that he had 47 instead of 46 chromosomes. The extra chromosome was the smallest chromosome, chromosome number 21. The diagnosis was confirmed.

The birth of a baby with Down syndrome occurs at a frequency of about 0.15 percent in the human population but happens about 2 percent of the time in 45-year-old women. Down syndrome illustrates several important genetic concepts. It emphasizes the central role of chromosomes in inheritance and the necessity for cell division processes to be error free. In this case, an error occurring in the cell division process that produced the mother's egg resulted in an extra chromosome 21 in that egg. After fertilization there were three copies of chromosome 21, two from the mother and one from the father. Thus, the child carried three doses of chromosome 21 genes instead of the normal two. This dosage imbalance caused the broad spectrum of abnormalities that are called Down syndrome. In addition, the boy will grow up to be seriously retarded and will have a 1500 percent increased risk for developing leukemia. If he survives to adulthood, he will almost certainly develop early-onset Alzheimer's disease. All of these problems develop because of the extra dose of genetic material he carries.

Chapter 1 has provided us with an introduction to the societal relevance of genetics and an overview of major questions that genetics addresses. In this chapter, as further preparation for our journey, we review the structure of cells and the behavior of chromosomes during cell division. We examine how sex cells, or gametes, are formed. We also look at the various strategies that have evolved in the natural world for transmitting the genetic material to the next generation, strategies known as life cycles. The cell is the basic unit of life, and the chromosome is the carrier of the genetic information. The behavior of chromosomes during cell division accounts for the principles of inheritance first described by Mendel.

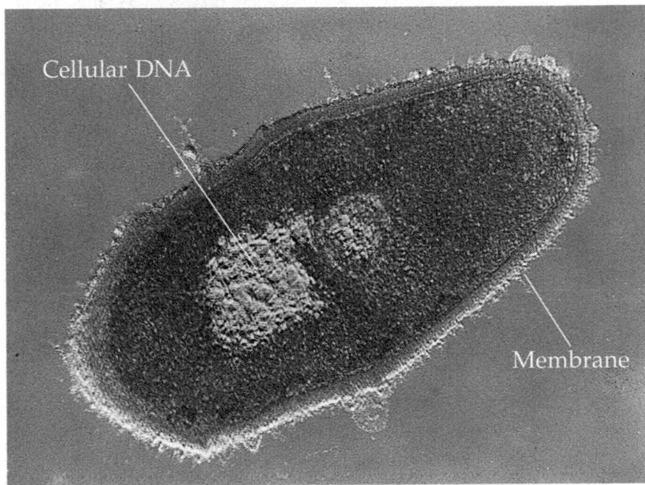

Figure 2.1 An electron micrograph of a longitudinal section through a bacterium, *Escherichia coli.* The cell's DNA is concentrated in a special region of the cytoplasm called the nucleoid.

THE CELL AS THE BASIC UNIT OF LIFE

All living things, with the exception of viruses, are made up of cells—small membrane-bounded structures filled with a variety of chemicals in an aqueous solution that interact to give life its basic properties. These properties include the acquisition and use of energy, as well as the capacity to reproduce, to respond to the environment, to carry out a variety of controlled chemical reactions, and to maintain a relatively constant internal environment despite fluctuations in the external environment.

There are two major types of cells in the living world, distinguishable by the type of internal structures they contain: *prokaryotes* and *eukaryotes*. Although both contain the same kinds of molecules, there are fundamental differences in the ways these cells are organized.

The Prokaryotic Cell

The simpler of the two cell types is the prokaryotic cell (*pro* = before; *karyote* = nucleus). Prokaryotic cells include the various types of bacteria and their relatives, all members of the kingdom Monera (Figure 2.1). They possess a **cell wall** constructed of peptidoglycan, a substance unique to prokaryotes. Beneath the cell wall is a **plasma membrane** that encloses a **cytoplasm** containing DNA, RNA, proteins, and various small molecules. Under the highest magnification, the cytoplasm does not appear to have any distinct internal organization, although the DNA is sequestered in a poorly demarcated region of the cytoplasm called a **nucleoid.** The nucleoid is not surrounded by a membrane, nor are there any membrane-bound cytoplasmic structures in prokaryotes. We discuss the structure of nucleoids, which contain the genetic material of prokaryotes, in Chapter 10.

Bacteria are microscopic organisms that reproduce quickly by binary fission—simple, asexual division. A single cell in an environment with plentiful food resources divides about once every 20 minutes. At this rate, a single cell would produce 5 billion progeny cells in just under 11 hours, a number equal to the approximate human population of the earth! Bacteria are the most abundant cells in the living world.

Bacteria may be simple in terms of their structure, but they are biochemically diverse. The ancestral prokaryote, a bacterium that appeared on earth over 3 billion years ago, evolved into a myriad of bacterial types that have adapted to and inhabit virtually all niches in the world, from the human gut to garden soil, from the ocean depths to mountain peaks, from swamps to hot acid springs. There are bacterial species that can utilize virtually any organic molecule as a food source, including some that use carbon from CO_2 and nitrogen from N_2 to make all of their necessary molecules. Bacteria are ancestral to all life-forms on earth and have occupied this planet longer than any other life-form. We discuss the genetics of bacteria and the viruses that infect them in Chapters 16 and 17.

The Eukaryotic Cell

The eukaryotic cell (*eu* = true; *karyote* = nucleus) is generally much larger than the prokaryotic cell and far more complex in terms of its structure and functions. The single most distinctive feature of a eukaryotic cell is a membrane-bound **nucleus** that houses the genetic material. In addition, eukaryotic cells contain various **organelles,** specialized structures that perform important functions such as the generation of energy for growth and movement. Between the nucleus and the plasma membrane is the cytoplasm, an aqueous matrix containing various membranous and nonmembranous organelles that perform a variety of complex metabolic functions. Sometimes a cell wall lies just exterior to the plasma membrane. The eukaryotic cell wall is made of cellulose and other molecules, but *not* peptidoglycan. The vast majority of living species, including animals, plants, fungi, and protists, are eukaryotic. Let's look more closely at eukaryotic cellular structure and functions (Figure 2.2).

Surrounding the cell is a plasma membrane constructed of a fluid phospholipid bilayer penetrated by a variety of proteins. Sometimes, membrane-bound microvilli extend outward from the cell surface making contact with neighboring cells or other components of its microenvironment. The plasma membrane has two main functions:

1. It forms a protective barrier between the extracellular and intracellular matrices.
2. It regulates the flow of molecules into and out of the cell.

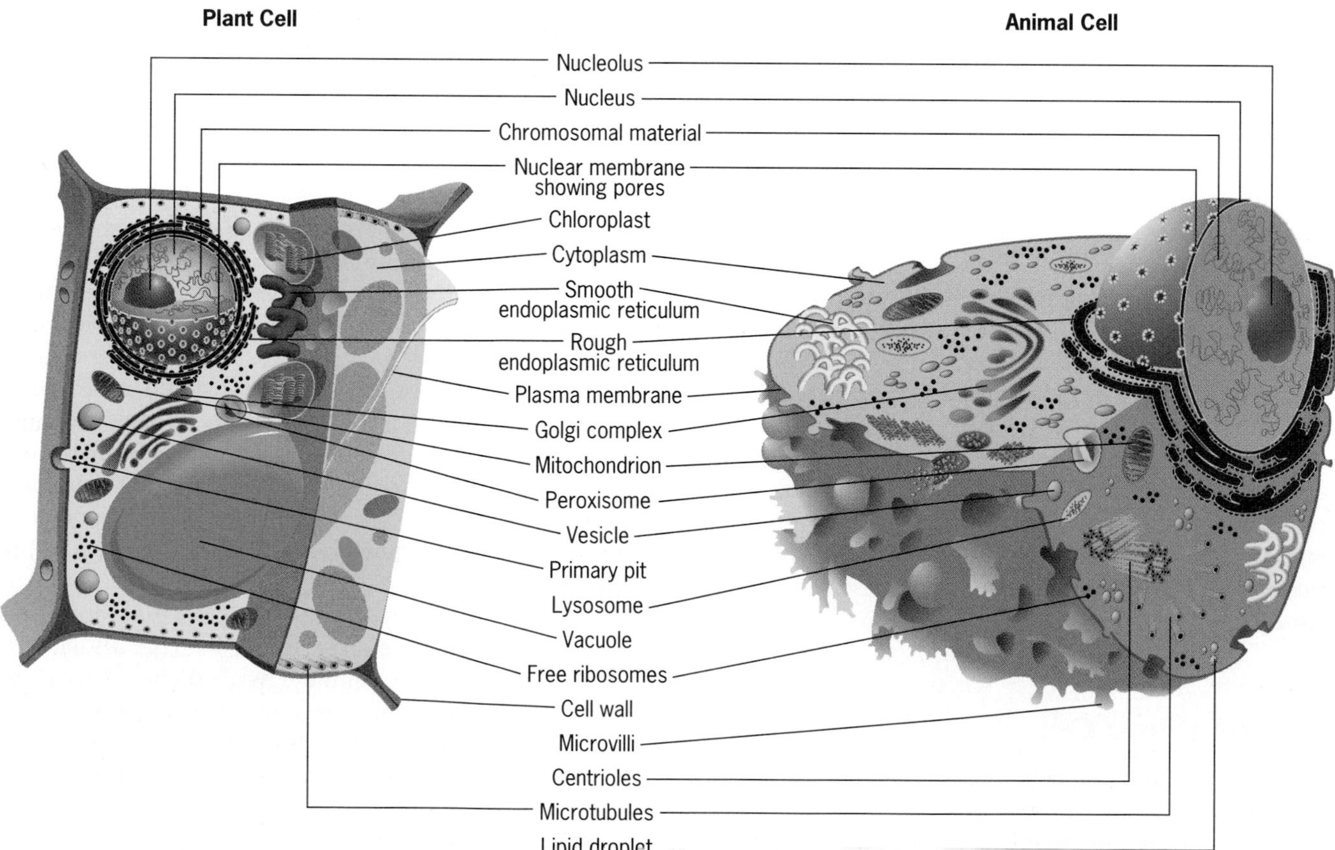

Figure 2.2 The generalized structure of eukaryotic cells. Both plant and animal cells have a plasma membrane, a membrane-bound nucleus, and cytoplasmic organelles. The two kinds of cells differ in the types of organelles they contain and in the presence or absence of a cell wall. Plant cells have cell walls; animal cells do not.

The membrane-bound proteins are commonly complexed with carbohydrates; thus, they are glycoproteins. They usually interact with other molecules in the extracellular world. Those proteins and glycoproteins that receive signals from the external world and communicate with the cell's interior are called receptors. Receptors have particular importance to humans because mutations in the genes that code for them can lead to diseases such as cancer and cystic fibrosis.

A variety of organelles are distributed throughout the cytoplasm, some of which are surrounded by membranes. Though cell structure and function vary widely from one cell type to another, all eukaryotic cells have the same basic types of organelles. The cell differences arise from the kinds of gene products being synthesized in the cells, the nature of their distribution in the cell, and the distribution and type of organelle systems in cells. We might think of organelles as compartments or structures in cells that carry on specific types of metabolic activity, chemical processes by which energy is provided and used.

The membrane-bound nucleus is the largest structure in a eukaryotic cell (Figure 2.2). Inside the nucleus are the **chromosomes,** made up of DNA, RNA, and various proteins. Chromosomes contain the genetic information of eukaryotes. Also inside the nucleus and associated with the chromosomes are often found one or more **nucleoli** (sing. = nucleolus), structures that function in RNA synthesis. The chromosomes and nucleoli are contained in a nuclear matrix called the **nucleoplasm.** The nuclear membrane separating the cytoplasm from the nucleoplasm has numerous pores in it so that molecules can traverse it. The RNA synthesized in the nucleus is transported to the cytoplasm through the nuclear pores.

The cellular cytoplasm is filled with a vast interconnected network of tubular membranes called the **endoplasmic reticulum,** or **ER** (Figure 2.2). Some of the ER is rough in appearance because **ribosomes,** cytoplasmic structures made of RNA and protein involved in protein synthesis, are bound to it. Ribosomes are assembled in the nucleolus. Rough ER is the site of protein synthesis. Proteins synthesized here are transported across the ER membrane where they are directed to different cellular locations. ER that is not complexed with ribosomes is called smooth ER. This type of ER functions in the synthesis of certain types of hormones and in other cellular reactions.

The **Golgi complex** is an organized pile of flattened membranous sacs. It receives, stores, and often modifies proteins synthesized on the rough ER. For example, the insulin protein that regulates blood sugar levels is synthesized on rough ER as a large precursor protein called proinsulin that is subsequently cleaved to its functional form, insulin, in the Golgi complex and membranous vesicles derived from it. Inherited defects in this Golgi-based cleavage process lead to diabetes. The proteins stored in the Golgi complex also function in the synthesis of other important cellular molecules, such as the complex carbohydrates that make up the plant cell wall. Sometimes Golgi vesicles move to the plasma membrane, fuse with it, and release their contents to the external environment, a process called exocytosis.

Small vesicles called lysosomes pinch off from the Golgi complex, forming membranous sacs containing powerful digestive enzymes capable of degrading nearly any molecule inside the cell. The membrane surrounding the lysosome keeps these powerful enzymes harmlessly sequestered inside the cell. Mutations in genes coding for lysosomal enzymes cause such devastating conditions as Tay-Sachs disease. In the case of Tay-Sachs, a defective lysosomal enzyme is unable to degrade a ganglioside, a type of lipid molecule found in abundance in nerve cell membranes. Intermediate breakdown products accumulate in the lysosome to toxic levels and destroy brain cell functions, causing death before the age of three.

Similar to lysosomes are the peroxisomes. They are membrane-bound vesicles containing a different battery of enzymes that use molecular oxygen to oxidize organic molecules, especially fatty acids. Some of the enzymes in peroxisomes generate dangerously reactive hydrogen peroxide (H_2O_2), a powerful oxidizing agent, during chemical reactions; other enzymes degrade hydrogen peroxide.

In plants, membrane-bound, liquid-filled vesicles called vacuoles may occupy as much as 90 percent of a cell's total volume (Figure 2.2). The plant vacuole may contain a variety of dissolved molecules such as salts, sugars, toxic waste, and pigments that account for the color of flowers and leaves. Vacuoles also function to maintain high internal water pressure in plant cells, which aids in the physical support of plant tissues.

Among the most prominent organelles in eukaryotic cells are the mitochondria and chloroplasts (Figure 2.3). Mitochondria are found in virtually all eukaryotic cells, whereas chloroplasts are found only in photosynthesizing plant cells. Animal and fungal cells do not contain chloroplasts. Mitochondria and chloroplasts probably originated from small prokaryotic cells that invaded or somehow became incorporated into larger cells. Supporting this hypothesis is the fact that both of these organelles contain their own genetic material: small circular molecules of DNA, like those found in modern-day bacteria. They also synthesize many of their own proteins.

Mitochondria are cellular "power plants." They convert energy stored in energy-rich macromolecules (lipids and carbohydrates) into adenosine triphosphate (ATP), which is used to power virtually all the

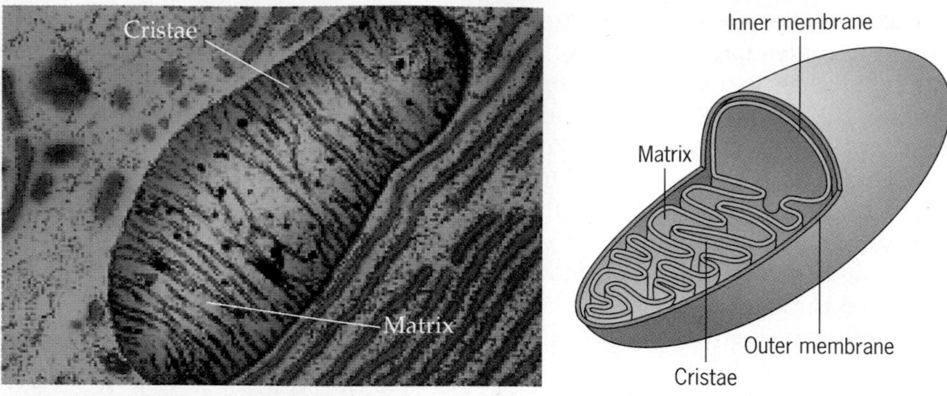

Figure 2.3*a* Electron micrograph and drawing of mitochondria showing the outer and inner membranes. The highly folded inner membrane forms the cristae, which project into the matrix containing DNA, RNA, ribosomes, and various proteins.

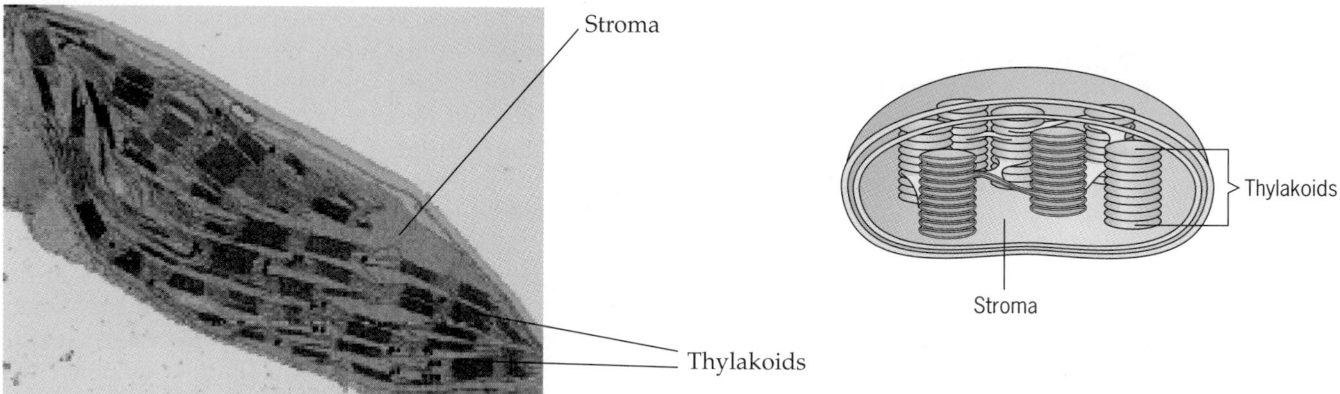

Figure 2.3*b* Electron micrograph and drawing of a chloroplast showing the inner and outer membranes as well as the stacks of membranous thylakoids contained in the central stroma. The stroma contains DNA, RNA, ribosomes, and various proteins.

activities of the cell. As might be expected, the most energetically active cells have the largest number of mitochondria. The mitochondrion has a double membrane. Proteins embedded in the inner membrane function in energy transfer reactions. The inner membrane has a specific structure and organization so that reactions are carried out in a precise and orderly fashion. It is folded into cristae that project into a matrix, thus increasing the surface area for these reactions.

Mutations in mitochondrial genes cause a number of disorders. For example, a type of eye disease called Leber's hereditary optic neuropathy (LHON) is caused by mutations in mitochondrial genes. Aging is associated with degenerative changes in mitochondrial DNA. Mutations in mitochondrial genes produce slow-growing colonies of the bread mold, *Neurospora crassa,* and strains of budding yeast, *Saccharomyces cerevisiae,* that cannot grow at all on certain media.

The **chloroplasts,** which are found only in plant cells, are the sites of photosynthesis, a complex process by which plants (and some bacteria) capture and use energy from the sun to drive the synthesis of complex organic molecules from carbon dioxide and water. Like mitochondria, chloroplasts have a double membrane system. The interior of the chloroplast contains a system of flattened membranous discs called thylakoids. Piles of thylakoids are located in the central region called the stroma. Chlorophyll, the green, sunlight-capturing pigment, is contained in the thylakoids.

Finally, all eukaryotic cells have a **cytoskeleton.** The cytoskeleton is a network of protein filaments, giving the cell its shape, its ability to move, and its ability to organize and position its organelles within its cytoplasm. Two of the most important filaments are the microfilaments and microtubules. The microfilaments are involved primarily in cell movement, like the flowing movements of the amoeba. Microtubules are the main structural elements in cilia and flagella—cell surface structures that propel cells through their environment.

Important inherited disorders are caused by defects in the cytoskeleton. For example, the gene causing Duchenne muscular dystrophy codes for a protein called *dystrophin,* which is a member of a family of cytoskeletal proteins. Although geneticists understand a great deal about dystrophin's structure and cellular location, its function remains elusive.

Key Points: **The prokaryotic cell has no membrane-bound nucleus. The prokaryote equivalent of the true nucleus is the nucleoid, composed of DNA, RNA, and proteins. The eukaryotic cell has a membrane-bound nucleus that houses the chromosomes. The eukaryotic cell has localized many of its metabolic processes into membrane-bound compartments or organelles, such as lysosomes, Golgi apparatus, mitochondria, and chloroplasts. Protein synthesis in eukaryotes takes place on ribosomes in the cytoplasm, in mitochondria, and in chloroplasts.**

THE CHROMOSOME: AN OVERVIEW

The key event in cell division is the proper distribution of genetic material to daughter cells. This genetic material is organized into structures called **chromosomes** (*chromo* = color; *soma* = body). One of the most important developments in the history of modern genetics was the research that associated genes with chromosomes, a topic that we will discuss in depth in Chapter 6.

Chromosomes are involved in two main activities:

1. The transmission of genetic information from cell to cell and from generation to generation.
2. The orderly release of this information to control cellular function and development.

In prokaryotes, the chromosome is composed of a molecule of double-stranded DNA packaged into the nucleoid by interactions with RNA and proteins. The RNA and proteins maintain the DNA molecule in a highly folded and coiled structure (Chapter 10).

The eukaryotic chromosome is considerably more complex. Prior to replication, each eukaryotic chromosome contains a huge DNA molecule packaged in a protein matrix and extending from one end of the chromosome to the other. After replication, the chromosome contains two giant DNA molecules. Chromosomal DNA is compacted by coiling and supercoiling (Chapter 10). If the DNA in a single human chromosome were uncoiled so that it was a perfectly linear molecule, it would measure between 2.5 and 5 centimeters in length. If all the DNA molecules in the nucleus of a human cell were uncoiled and lined up end to end, they would measure more than a meter in length. To imagine how this much DNA is packaged into chromosomes and then packed into a nucleus, we might visualize twisting a rubber band over and over: it coils and then forms supercoils as it compacts.

In nondividing eukaryotic cells, the chromosome is extended and its structure is difficult to study. In dividing cells, the chromosomal material condenses, forming relatively thick, dense, rodlike structures. As seen under a light microscope, the chromosome has some distinct morphological features. The replicated chromosome is made up of two **sister chromatids**, two identical copies of the parental chromosome, joined together by a **centromere** (Figure 2.4). The **kinetochore,**

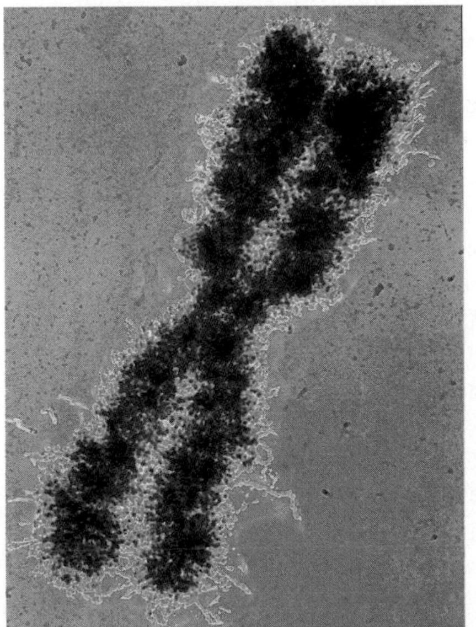

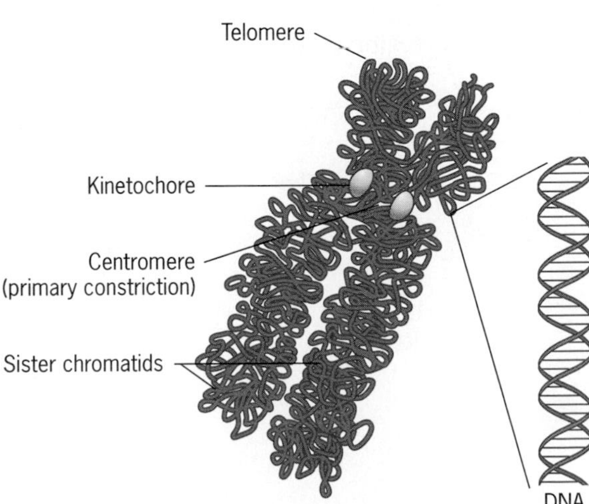

Figure 2.4 The structure of a replicated chromosome.

a protein structure at the centromere, functions in chromosome movement during cell division. The position of the centromere varies among different chromosomes. The ends of the chromosomes are called **telomeres.**

Key Points: **Chromosomes are threadlike structures consisting mainly of a complex of DNA (the genetic material), RNA, and proteins. Chromosomes have key morphological features such as centromeres, telomeres, and chromatids.**

THE CELL CYCLE

In order for an organism to grow, whether it is single-celled or multicellular, three events must take place: (1) the cell mass must increase; (2) there must be a duplication of the genetic material; and (3) a division process must occur assuring that each daughter cell receives an equal and identical complement of the genetic material to ensure perpetuation of the cell line. In eukaryotes, these occurrences take place in an ordered progression of events during the cell's life span, or **cell cycle.**

The initial event in the cell cycle (Figure 2.5) is the growth and increase in cell mass that occurs following cell division. This is called the **G_1 phase** (G = gap). For a mammalian cell taking 24 hours to complete the cell cycle, the G_1 phase might last about 10 hours. G_1 is devoted to the metabolic activities associated with cell growth and preparation for DNA replication. Following G_1, DNA replication occurs. This is the synthesis, or **S phase,** which lasts about nine hours. During the S phase, the genetic material of each and every chromosome is replicated. Thus, each chromosome consists of two sister chromatids joined at the centromere. After completion of DNA replication, the cell enters another growth phase called the **G_2 phase.** During this post–DNA replication phase, which typically lasts about four hours, preparations for mitotic cell division are taking place. The **M phase,** or division phase, is the final part of the cell cycle. The division process takes about one hour to complete. The two sister chromatids separate from each other, one going to each of the two daughter cells. The two daughter cells are genetically identical to each other and to the parent cell.

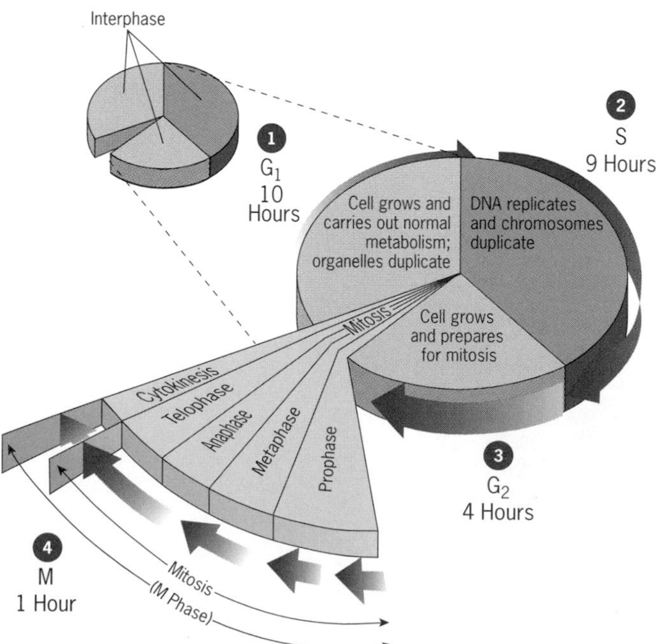

Figure 2.5 The cell cycle. Times indicated are for mammalian cells in tissue culture.

Key Points: **The cell cycle is a sequence of events involving the periodic replication of DNA and the segregation of this replicated DNA with cellular constituents to daughter cells. There are four phases of the cell cycle: G_1, S, G_2, and M or division. The two G phases are called "gaps"; the S (synthesis) phase is the period of DNA replication; and the M (mitotic) phase signals the actual division of the cell.**

CELL DIVISION: MITOSIS

The capacity of a cell to reproduce itself is perhaps the most fundamental property of life. There are two main processes by which eukaryotic cells divide: mitosis and meiosis. Mitotic cell division, which occurs in virtually all eukaryotic cells, is the process by which single cells reproduce themselves and multicellular organisms grow. The key feature of mitotic cell division is that the daughter cells are identical to each other and to the parent cell. The parent and daughter cells are **diploid (2n):** there are two copies of each type of chromosome. This is not true in meiotic cell division. Meiotic cell division is the basis of sexual reproduction in all higher plants and animals. It occurs during gamete formation in animals and spore formation in plants and fungi. The parent cell is diploid (2n), but following meiosis, the daughter cells (gametes or spores) are **haploid (n):** each gamete or spore has one member of each chromosome pair.

Mitosis, or mitotic cell division, was first described in 1879 by Walter Flemming. It consists of two interrelated processes: (1) mitosis, the division of the nucleus (*mitos* = a thread, a term that refers to the threadlike appearance of the chromosomes); and (2) **cytokinesis,** the changes in the cytoplasm that include division of the cell proper. Figure 2.6 illustrates these processes as

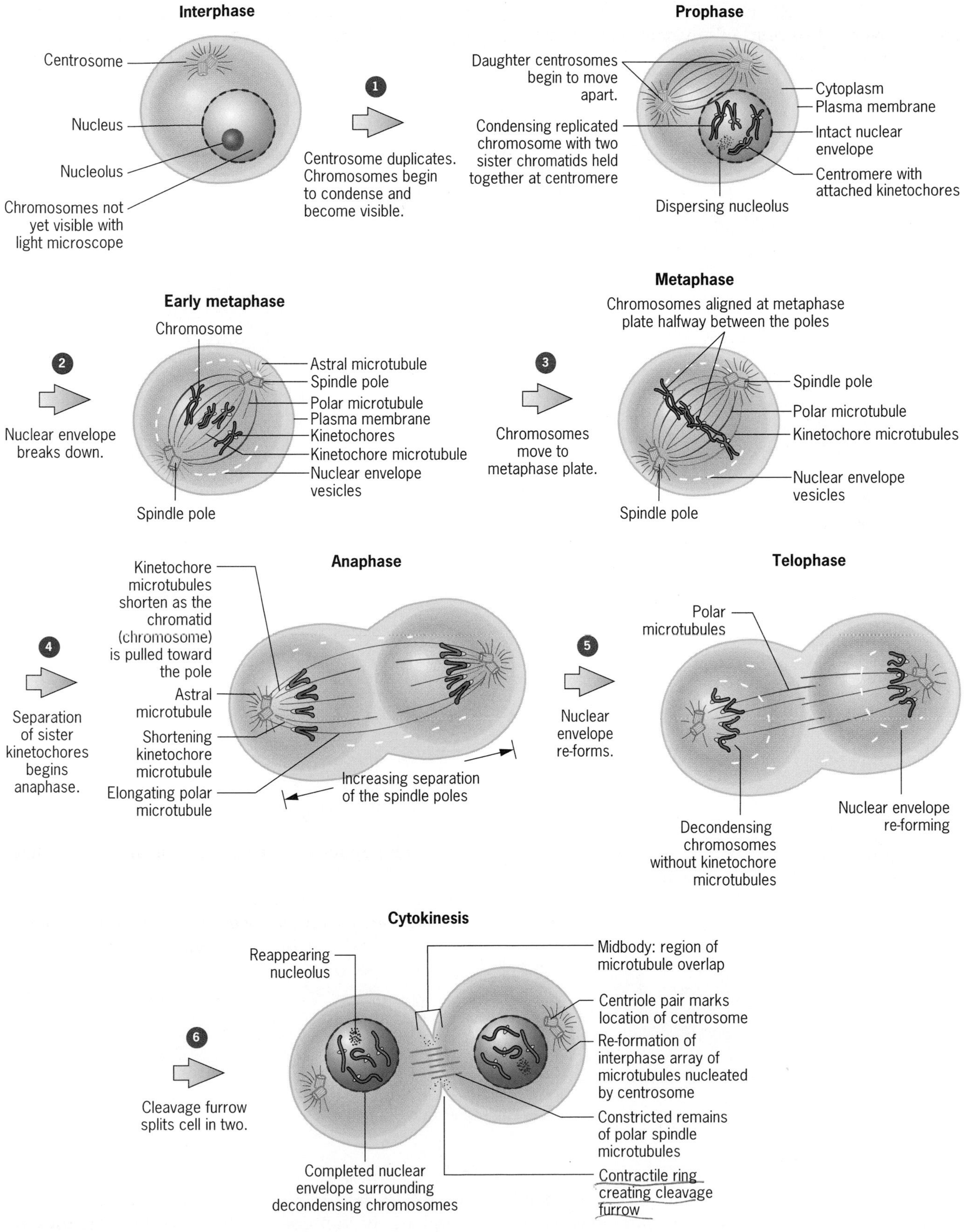

Figure 2.6 Mitosis in animal cells.

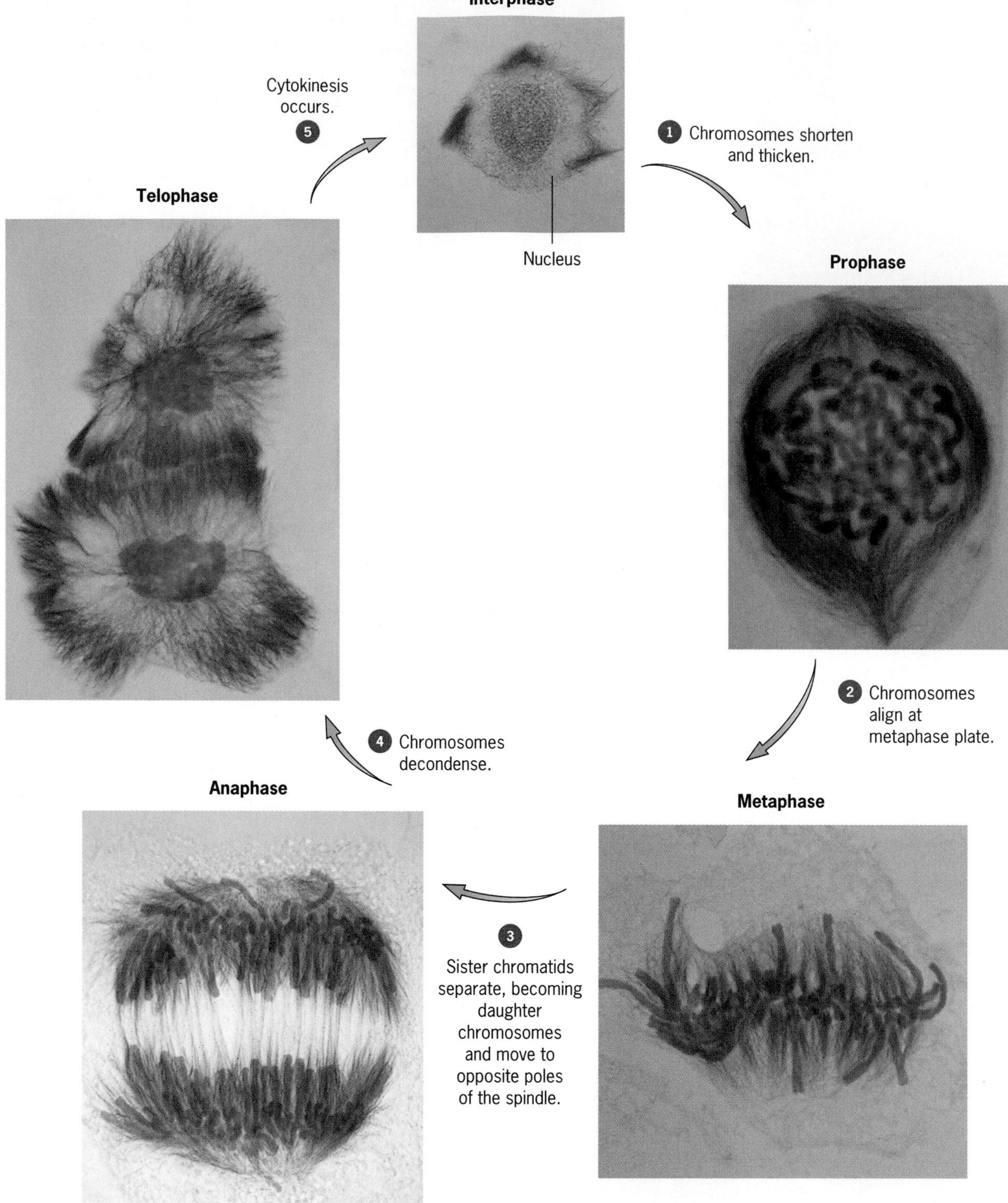

Figure 2.7 Photographs of the major stages of mitosis in the blood lily, *Haemanthus.*

they occur in animal cells. The process is virtually identical in plants (Figure 2.7), with a few variations. Mitosis is a continuous process, but for discussion purposes it is divided into five sequential phases: interphase (between divisions—consisting of G_1, S, and G_2 stages of the cell cycle), prophase, metaphase, anaphase, and telophase. Each phase is defined by the structure and behavior of the chromosomes. Prophase and telophase are usually long and involved, whereas metaphase and anaphase are commonly brief.

The first indications of approaching mitosis in animal cells are observed in the cytoplasm of the cell at **interphase.** The **centrosome** (Figure 2.8), a centrally located organelle, is the primary microtubule organizing center and acts as the spindle pole during mitosis in animals. The microtubules, cytoskeletal structures that radiate from the centrosome, organize and coordinate the movement of chromosomes during mitotic cell division. The centrosome is duplicated by the cell during interphase so that each daughter cell eventually receives one. The centrosome in most animal cells has a pair of **centrioles** found in its center, and they, too, are replicated prior to division. As mitosis begins, the centrosome begins splitting into two, and the microtubules radiate out forming an **aster.**

The centrosome and its associated centrioles have perplexed and fascinated biologists for over a hundred years. How exactly do they replicate? What are they constructed of? What is their evolutionary history? Answers to these questions have so far eluded biologists. Indeed, centrioles are not present in plants. Instead, plants have less distinct spindle pole organizing centers called **MTOCs** (*m*icro*t*ubule *o*rganizing *c*enters), whose structure is still being investigated.

In **early prophase** the two centrosomes separate and move toward opposite poles of the cell. The microtubules extending from them form the "rails" for chromosome movement. The chromosomes, which were replicated during the S phase of the cell cycle, are highly extended. They begin condensing through coiling and now appear as discrete entities.

By **late prophase** the chromosomes are highly condensed. The two chromatids, which will become the daughter chromosomes when mitosis is complete, of each chromosome are held together at the centromere, or primary constriction. The kinetochores complex with the centromere and function as the sites to which microtubules attach and guide chromosome movement. Centrosomes have now moved to opposite poles of the cell. A continuous microtubule network extends around the nuclear membrane and interconnects the poles. The nuclear membrane and nucleolus, the nuclear structure where ribosomes are assembled, become fragmented and dispersed in the cytoplasm. Microtubules now invade the nuclear region and become attached to each chromatid at the kinetochore in the centromeric region. Each of the two sister chromatids becomes attached to a different pole of the spindle, but the centromeres remain together.

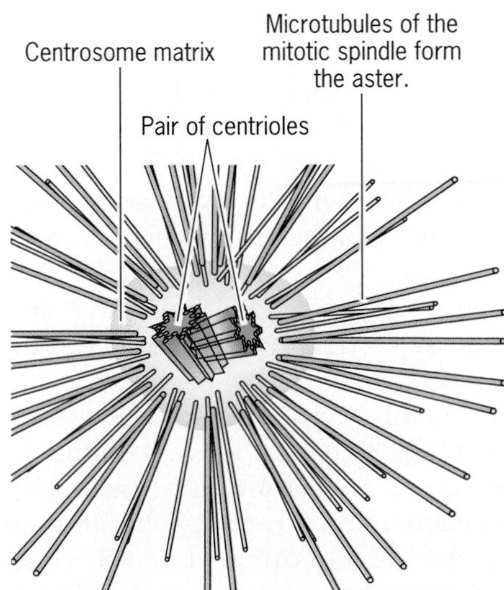

Figure 2.8 The centrosome.

At **metaphase** condensed chromosomes composed of sister chromatids assume positions in the cell's center, or **equatorial plate,** between the two poles. Metaphase chromatids are tightly coiled and discrete, thus facilitating accurate chromosome counts and gross structural analysis. The diagnosis of disorders caused by the structural changes in chromosomes is commonly made on these metaphase chromosomes. Arms of sister chromatids are extended from the centromere region, but the chromatids are held together at the centromeres until the beginning of the next phase.

Anaphase begins when the centromeres separate and the sister chromatids of each chromosome disengage and move toward opposite poles of the cell. Movement is the result of microtubule shortening. After anaphase separation occurs, each chromatid has its own centromere and is now considered to be a chromosome. Anaphase chromosomes elongate somewhat by relaxation of the tight metaphase coiling and move to respective poles of the spindle.

The last mitotic phase, **telophase,** occurs when chromosome movement is completed and the microtubules disassemble. By the completion of telophase, a nuclear membrane is reconstructed around each daughter nucleus, and the nucleolus begins to reappear. The chromosomes uncoil and become more extended. Mitosis is completed. Mitosis ensures that each daughter cell has the same genetic information as the parent cell.

The cytoplasmic part of the cell now divides by cytokinesis (Figures 2.6, 2.7, and 2.9). Animal cells, with their flexible outer layers, accomplish this by a midline constriction that eventually separates the two daughter cells. The surface around the equatorial region of the cell pushes in toward the center and pinches the cell into two parts. Plant cells, with their rigid cell walls, form a partition, or **cell plate,** between the daughter cells. After the cell plate is formed, walls of cellulose are deposited on either side.

As noted earlier, each cell division (nuclear and cytoplasmic phases) is a continuous process from the time

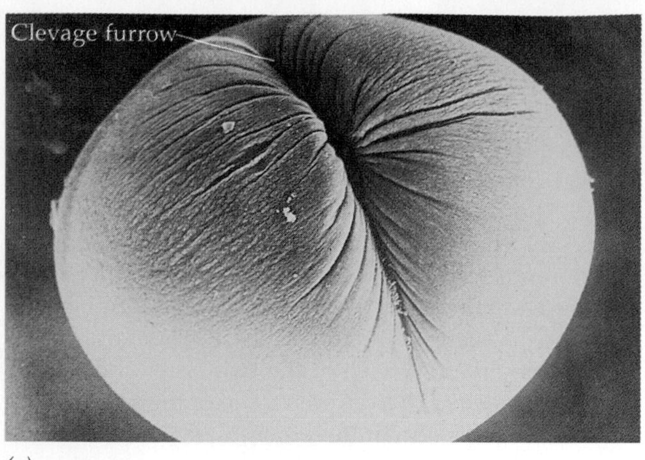

(a)

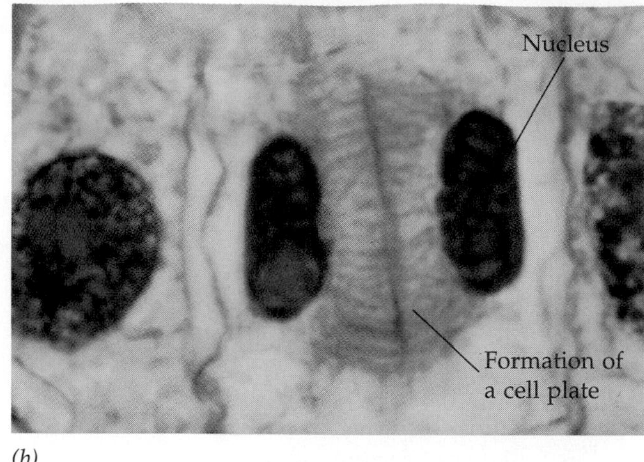

(b)

Figure 2.9 Cytokinesis in animal (*a*) and plant (*b*) cells. The animal cells constrict; the plant cells build a cell plate.

a cell first shows evidence of beginning to divide until the two daughter cells are completely formed. Nuclear and cytoplasmic phases are distinct but coordinated processes. The entire process ordinarily requires a few hours to several days, with variation dependent on the type of organism and environmental conditions. The process of mitosis described above occurs in animals and most plants. Variations of the process described here occur in fungi and unicellular eukaryotes.

Key Points: **Mitosis, a mode of nuclear and cytoplasmic division, provides for the production of two daughter cells that contain identical chromosome sets and are genetically identical to each other and to the parent cell from which they arose. Biologists divide mitosis into five phases: interphase, prophase, metaphase, anaphase, and telophase. Cytokinesis, the division of the cytoplasm, follows telophase.**

CELL DIVISION: MEIOSIS

In 1883, four years after mitosis was first described, Edouard van Beneden discovered that the egg of the roundworm, *Ascaris*, contains only half as many chromosomes as somatic cells (nonsex cells) and that the chromosome number characteristic of the species is reestablished at fertilization. He accurately described the process of **meiosis** (from the Greek meaning diminution), or meiotic cell division, whereby the diploid chromosome number ($2n$) is reduced by one-half to the haploid state (n) during gamete formation. But van Beneden erroneously concluded that all maternal chromosomes (those chromosomes contributed to the individual by the mother) go to one nucleus and all paternal chromosomes to the other. In fact, maternal and paternal chromosomes are shuffled and randomly distributed to daughter nuclei during meiosis. The behavior of chromosomes during meiosis was thus recorded, but no one at this time related this behavior to heredity.

In its broadest sense, meiosis is a lengthy process in which the chromosomal material replicates once and the cell divides twice, thus reducing the chromosome number by one-half. In pre-meiotic diploid cells, each chromosome has a pairing mate, or **homolog**. For each homologous pair, one member is contributed by the sperm (paternal chromosome), and the other is contributed by the egg (maternal chromosome). At the conclusion of meiosis, each cell has one member of a homologous pair, either a maternal or a paternal chromosome, and is thus haploid.

In the absence of meiosis, the fusion of two diploid sex cells would produce a zygote with twice the number of chromosomes as the parents. In humans, where the diploid chromosome number is 46 (23 pairs), the fusion of two diploid cells would produce progeny with 92 chromosomes. The next generation would have 184 and the next 368. This type of chromosome doubling is not compatible with life. Meiosis ensures a constant chromosome number from generation to generation.

Meiosis has two successive divisions. The first (meiosis I), called the **reduction division** because it reduces the chromosome number by one-half, is the more complex. During this division, homologous chromosomes pair, then segregate and move to different nuclei. During the second meiotic division (meiosis II), the **equational division,** the centromere splits and the sister chromatids separate into different nuclei. Thus, four haploid nuclei result from the meiotic division of one diploid nucleus. Figure 2.10 details the two divi-

Prophase I

Leptonema

Replicated chromosomes become visible.

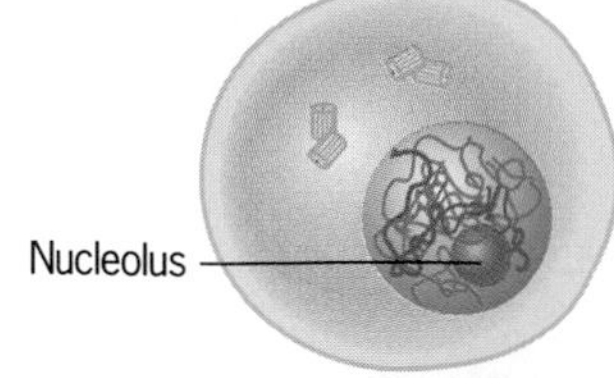

Zygonema

Homologous chromosomes pair.

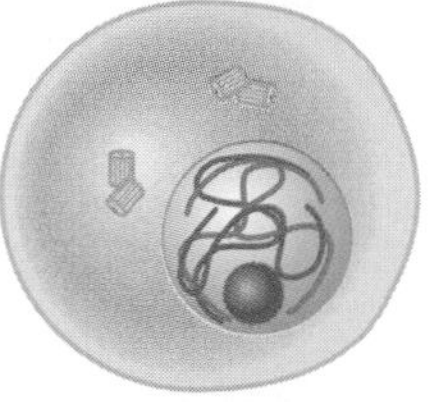

Pachynema

Homologous chromosomes fully paired. Crossing over occurs.

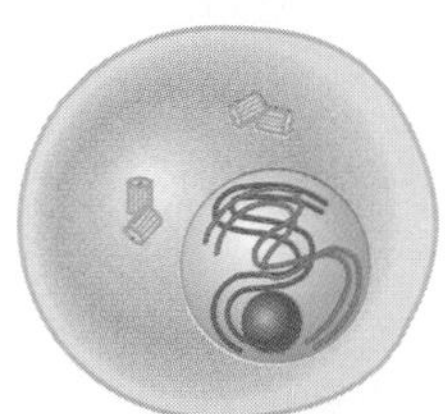

Diplonema

Homologous chromosomes begin to repel each other. Chromatids become fully visible. Chiasmata become visible.

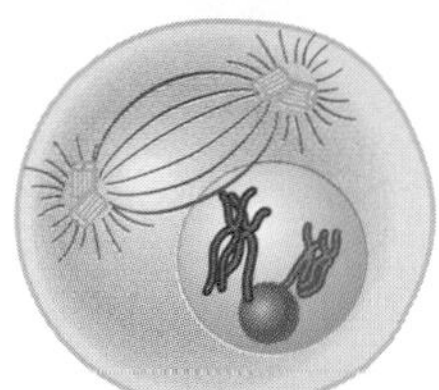

Diakinesis

Chromosomes continue to shorten and thicken. Nucleolus and nuclear membrane disappear. Microtubules attach to kinetochores.

Metaphase I

Assembly of spindle is completed. Each chromosome pair aligns across the metaphase plate of the spindle.

Anaphase I

Homologous chromosome pairs separate and migrate toward opposite poles.

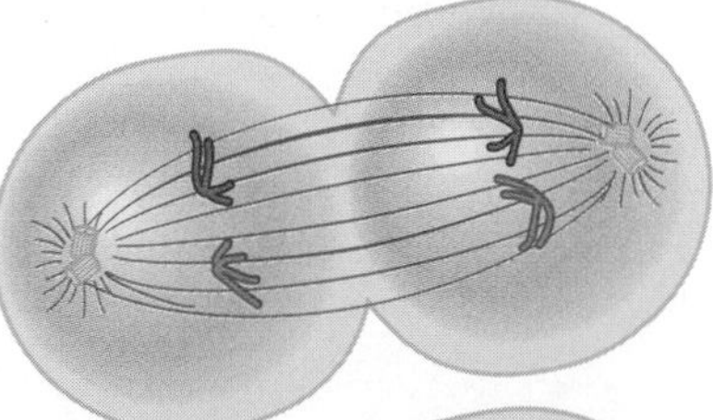

Telophase I

Chromosomes (each with two sister chromatids) complete migration to the poles, and new nuclear membranes may form.

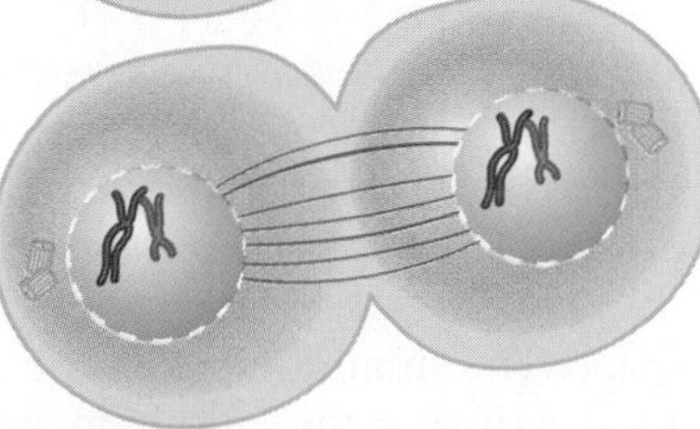

Cytokinesis

In most species, cytokinesis produces two daughter cells. Chromosomes do not replicate before meiosis II.

Prophase II

Chromosomes condense and move to metaphase plate.

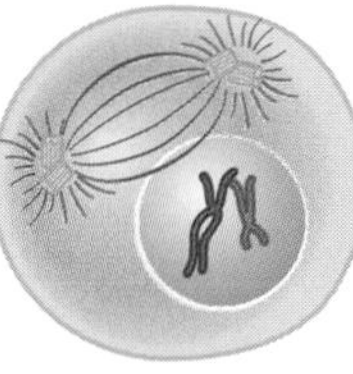

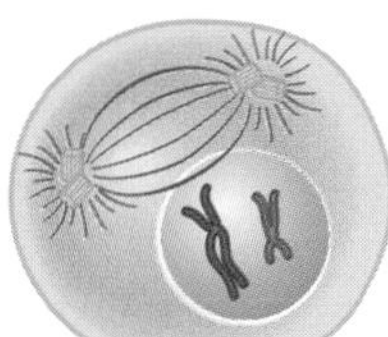

Metaphase II

Kinetochores attach to spindle fibers. Chromosomes line up on metaphase plate.

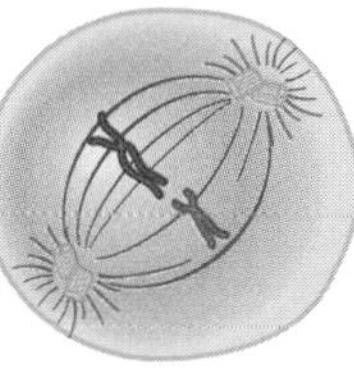

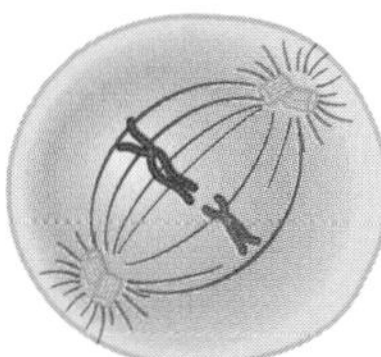

Anaphase II

Sister chromatids separate and move to opposite poles as separate chromosomes.

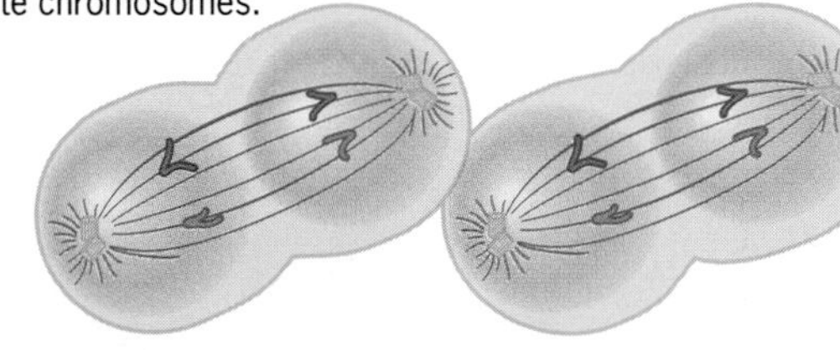

Telophase II

Nuclear membrane forms around chromosomes and chromosomes uncoil. Nucleolus reforms.

Reforming nucleolus

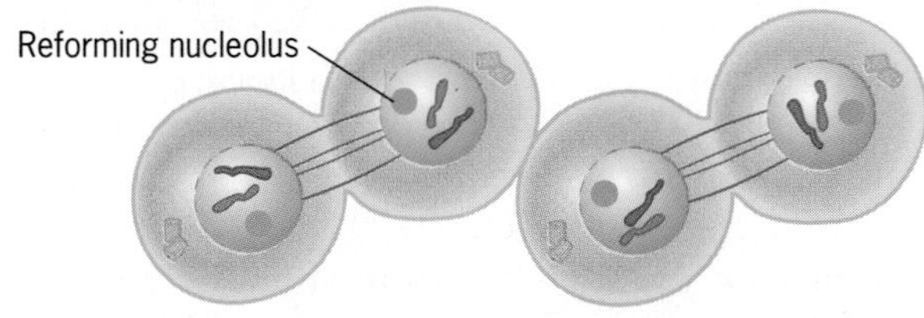

Four haploid cells form after cytokinesis.

Nucleolus

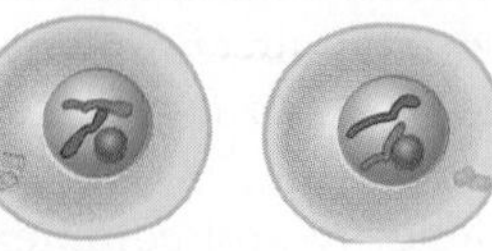

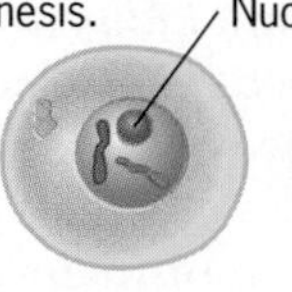

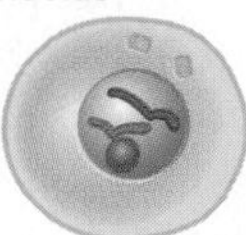

Figure 2.10 Meiosis in animals.

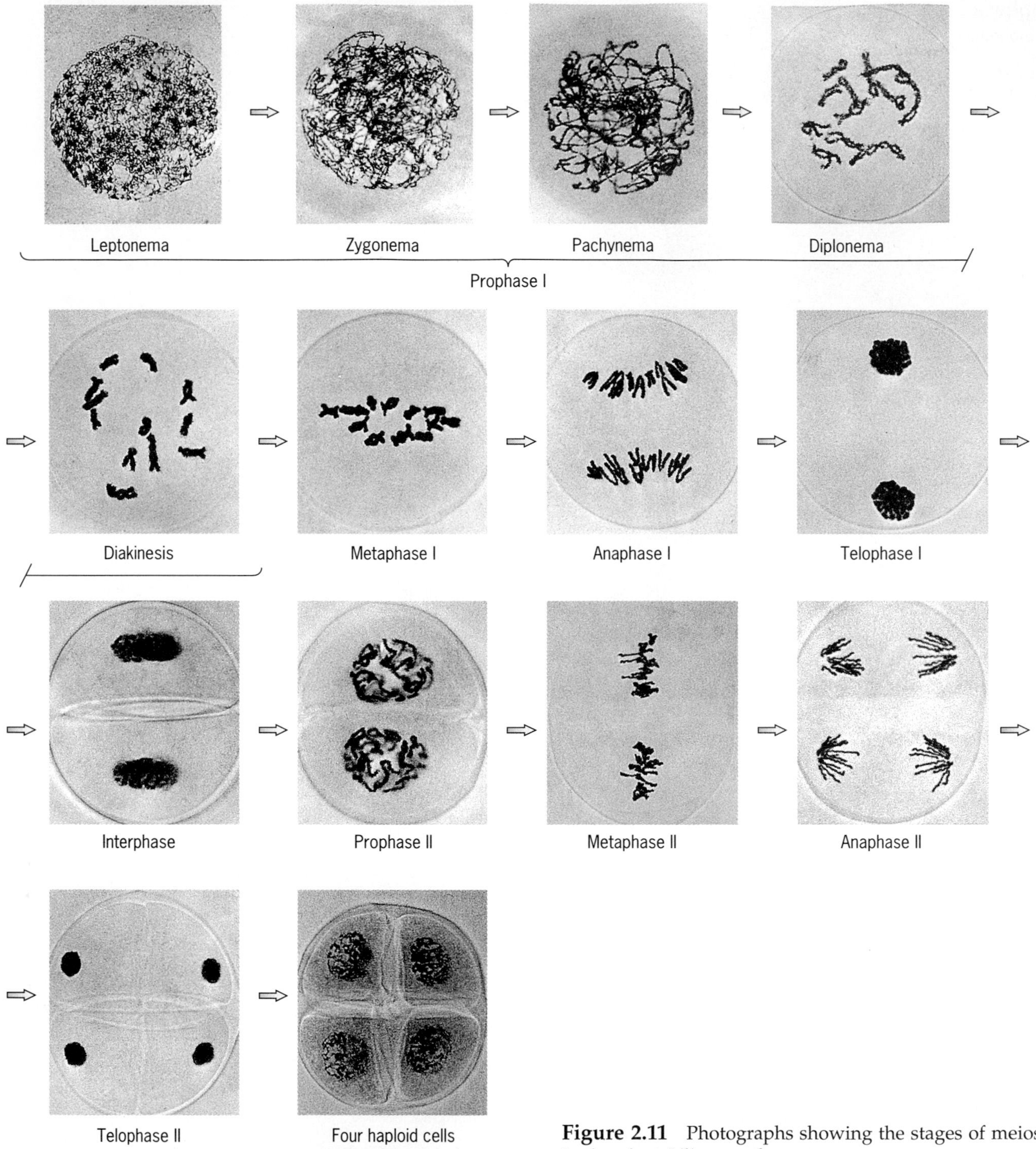

Figure 2.11 Photographs showing the stages of meiosis in the plant *Lilium regale.*

sions of meiosis and their stages in animals, and Figure 2.11 shows photomicrographs of the stages of meiosis in the plant *Lilium regale*. As in mitosis, the major differences between meiosis in animals and plants involve the processes of spindle formation and cytokinesis.

The first stage of meiosis, a lengthy and complex *prophase I*, is made up of five sequential substages: leptonema, zygonema, pachynema, diplonema (*-tene* in adjective form; e.g., leptotene stage), and diakinesis. The first substage that can be recognized as distinct from interphase is *leptonema*. During this substage, the chromosomes become visible with the light microscope. The small regions of the chromosomes that first appear during the leptotene stage are called **chromomeres**. As leptonema proceeds, the chromomeres

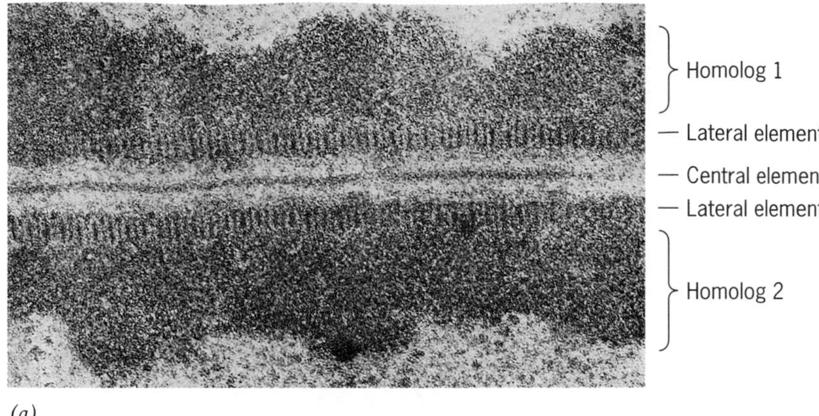

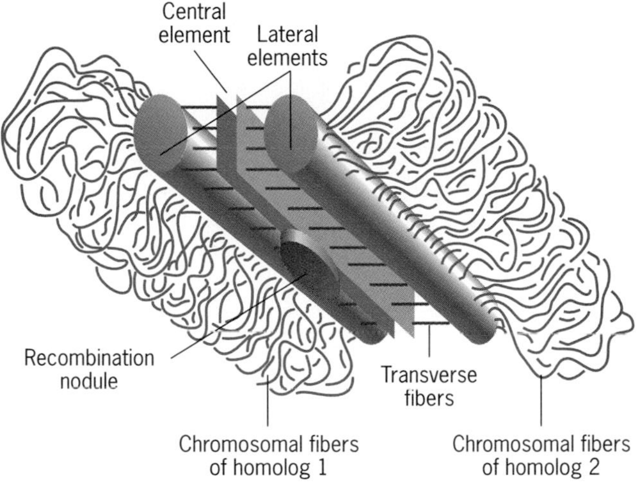

Figure 2.12 The synaptinemal complex. Electron micrograph (*a*) and diagram (*b*) showing the structure of the synaptinemal complex that forms between homologous chromosomes during the zygotene stage of prophase I of meiosis. The recombination nodules are thought to contain the enzymatic machinery required for genetic recombination.

thicken and join to form visible threadlike structures. The telomeres of leptotene chromosomes are attached to the nuclear membrane, and this attachment appears to play an important role in the subsequent pairing of homologous chromosomes. The chromosomes have already replicated (during interphase) at this substage, but the sister chromatids cannot be seen with the light microscope.

During the next substage of prophase I, *zygonema,* the homologous chromosomes align side by side (point for point) to form **bivalents,** pairs of tightly associated homologous chromosomes. Because the chromosomes have already replicated, each bivalent contains four DNA molecules, one in each chromatid. Therefore, a pair of homologous chromosomes is a four-stranded complex, or **tetrad.** The pairing process, which occurs continuously through zygonema at multiple points along the chromosome, is called **synapsis.** How do homologous chromosomes find one another to begin the pairing process? One possibility is that the telomeres of homologous chromosomes are attached to adjacent sites on the nuclear membrane and that synapsis begins in telomeric regions of homologous chromosomes. Regardless of the mechanism by which synapsis occurs, we know that it involves a ladderlike structure called the **synaptinemal complex** (Figure 2.12). This intricate structure, composed of DNA and protein, participates in both synapsis and the process of genetic exchange called **crossing over,** which results in new combinations of genes on the participating chromosomes. The chromosomes continue to shorten and thicken throughout zygonema, but the two chromatids of each homolog are usually not visible at this stage.

The third substage of prophase I is called *pachynema*. For the first time, the chromosomes can be seen as replicated structures—that is, as seen under the light microscope, each chromosome consists of two chromatids. The chromosomes continue to shorten and condense during pachynema. Crossing over occurs during pachynema, but the results of this exchange process do not become visible until the next stage of prophase I. During crossing over, nonsister chromatids within each homologous chromosome pair exchange segments of genetic material with each other. The synaptinemal complex mediates this process. The central element of the synaptinemal complex (Figure 2.12) contains loops of DNA from each chromosome, and it appears likely that crossing over occurs between these DNA loops, coordinated by nodules of recombination enzymes. The genetic significance of crossing over, especially as it pertains to gene mapping, will be explored in more detail in Chapter 8.

During the fourth substage of prophase I, *diplonema*, homologous chromosomes seem to repel each other, or desynapse, and X-shaped structures called **chiasmata** (sing. = chiasma) become visible (Figure 2.13). Chiasmata are points of contact between homologous chromosomes, where chromatids exchange pairing partners. The chiasmata hold the homologous chromosomes of bivalents together and provide visual evidence that crossing over has occurred. The chiasmata begin to terminalize or move toward the ends of the chromosomes, and the synaptinemal complexes disassemble during diplonema.

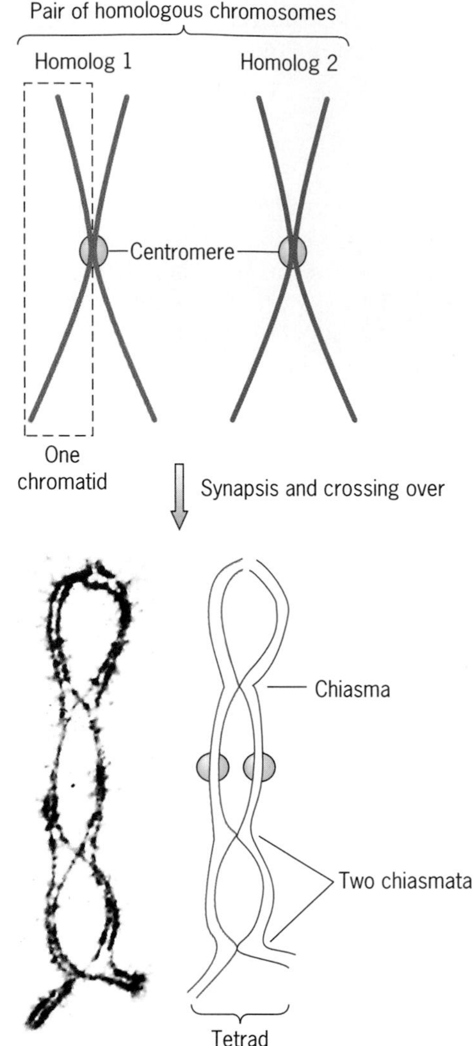

Figure 2.13 Chiasmata are cytologically visible X-shaped regions where crossing over has occurred in a synapsed pair of homologous chromosomes.

The diplotene stage is very long in many animals. In human females, for example, the diplotene stage of prophase I is initiated in the ovary during fetal life. About 400,000 immature eggs (oocytes) attain this stage, and then meiosis I is arrested. The eggs remain in this arrested state until the onset of puberty. After puberty and the onset of the menstrual cycle, one egg a month resumes meiosis in the fallopian tube. Some eggs remain in the suspended diplotene state for several decades. This can have profound genetic consequences. Older eggs have an increasingly difficult time completing a normal meiosis because of changes that accumulate in the proteins controlling chromosome movement. The result is eggs with abnormal chromosome numbers—extra or missing chromosomes.

During the last substage of prophase I, *diakinesis*, chromosome pairs continue to thicken and move toward the center of the nucleus, and the chiasmata move toward the ends of the chromosomes. The nucleolus and nuclear membrane disperse, and microtubules emanating from the centrosomes attach to the kinetochore. Prophase I is now completed.

At *metaphase I* the chromosome pairs are highly condensed and arranged on opposite sides of the equatorial plate. Microtubules are attached to the kinetochores in the centromeric regions of chromosomes. The chiasmata have moved to the ends of each chromosome pair.

During *anaphase I* the members of a homologous chromosome pair move toward opposite poles. The centromeres have not yet divided; thus, each chromosome still consists of two chromatids. *Telophase I* occurs when the chromosomes reach the opposite poles of the cell. Often this stage is accompanied by the reformation of nuclear membranes around each complement of chromosomes, cytokinesis, and then by a brief interphase (but an interphase in which the DNA is *not* replicated). Sometimes, however, membrane formation, cytokinesis, and interphase do not occur, and cells move directly into the second meiotic division.

Meiotic division II, the equational division, resembles a normal mitosis, except there is only one member of a chromosome pair in each nucleus. At *prophase II*, the chromosomes condense and move to the equatorial region. The centromere divides at *metaphase II*, and the former chromatids, now daughter chromosomes, move toward the opposite poles in *anaphase II*. When the chromosomes have reached the poles and stop their movement (*telophase II*), nuclear membranes form around each set of haploid chromosomes, and cytokinesis follows. The result of meiotic divisions I and II is four haploid cells each containing one member of each homologous chromosome pair.

Meiotic Abnormalities

On occasion, a problem occurs during meiosis such that homologous chromosomes fail to pair properly and separate from each other. This phenomenon is called **nondisjunction.** A result of nondisjunction, which can occur in either meiosis I or meiosis II, is that both members of a homologous chromosome pair (or both sister chromatids) may go to the same pole instead of segregating to opposite poles of the cell. Studies of sperm cells from normal fertile men indicate that almost 5 percent of these cells carry abnormal chromosome numbers; thus, meiotic mistakes are surprisingly common.

The causes of nondisjunction are many and varied. A number of environmental agents cause nondisjunction. For example, heat and cold shock interfere with chromosome pairing in many species and lead to irregular meiosis. X rays produce meiotic abnormalities by disrupting either synapsis or the integrity of the spindle fiber apparatus. Chemicals, such as colchicine, an alkaloid commonly used to treat gout, cause the rapid disintegration of the spindle apparatus and thus disrupt normal cell division.

In humans, there is a striking relationship between advanced maternal age and nondisjunction. The frequency of Down syndrome, the disorder described in the opening of this chapter, is 4 per 10,000 births in 20-year-old women but increases dramatically to about 200 per 10,000 in women who are 45 or older. Mothers over the age of 35 give birth to 35 to 50 percent of the infants with abnormal numbers of chromosomes. The underlying cause of nondisjunction in older women is probably the deteriorating state of the egg, which has been held in suspended meiosis for 30 to 40 years or more. While maternal age is a major factor in meiotic mistakes, advanced paternal age is also important. It is estimated that about 20 percent of all children born with abnormal chromosome numbers are the result of paternal nondisjunction, mainly in older fathers.

Genes have also been implicated in meiotic mistakes, although it is difficult to identify and characterize specific genes. High frequencies of meiotic mistakes and children born with multiple meiotic abnormalities have been found in some families. These observations suggest that mutant genes in some families disrupt the meiotic process. Some of these genes may control or influence the structure of the synaptinemal complex or perhaps the attachment of microtubules at the centromeric region. We will consider the consequences of meiotic mistakes in detail in Chapter 7.

Key Points: **Meiosis, a fundamental process of cell division in sexually reproducing eukaryotes, involves three main events: the pairing of homologous chromosomes; the exchange of genetic material by crossing over; and the segregation of the members of a homologous pair of chromosomes into different daughter nuclei. Meiosis involves one round of DNA replication followed by two separate cell divisions. The result of meiosis in a diploid cell is four haploid cells. Meiotic nondisjunction is the failure of sister chromatids or homologous chromosomes to disjoin.**

THE EVOLUTIONARY SIGNIFICANCE OF MEIOSIS

The behavior of chromosomes during meiotic cell division accounts for the basic principles of Mendelian inheritance (Chapter 3). Mendel's principles of gene segregation and assortment describe perfectly the behavior of chromosomes during meiosis. But meiosis is important for other reasons too. Van Beneden, in his original description of meiosis, had suggested that at metaphase all the maternal chromosomes lined up on one side of the equatorial plate and all paternal chromosomes on the other. Such an arrangement would produce gametes carrying all maternal or all paternal chromosomes. This is not the case, however. Chromosomes align randomly along the equatorial plate so that the gametes carry a mixture of maternal and paternal chromosomes. For example, the human has 23 pairs of chromosomes. For each pair, the probability that a gamete will get a maternal chromosome of any pair is 1/2. The probability that a gamete will get *all* maternal chromosomes is $(1/2)^{23}$! Stated another way, there are 2^{23} possible combinations of chromosomes. When crossing over is factored into this equation, the number of possible different genetic combinations becomes astronomical. This is the true significance of meiosis and sexual reproduction: It is a mechanism for generating tremendous amounts of genetic variation in sexually reproducing populations.

Key Points: **The most important consequence of meiosis is that it produces different combinations of chromosomes.**

THE FORMATION AND UNION OF GAMETES

The completion of meiosis does not necessarily mean that gametes have formed. Meiosis produces four haploid cells. These cells must now mature into functional gametes, eggs and sperm. The meiotic process that produces haploid cells and the subsequent maturation of these cells into functioning gametes is called **gametogenesis.** Let's now look at the relationship between meiosis and gamete formation in males and females.

Oogenesis: The Formation of the Egg

In the early stages of embryonic animal development, groups of cells become committed to differentiate into various cell types, such as liver cells, nerve cells, and muscle cells. One group of cells becomes committed to form the germ cell line, cells that will eventually develop into eggs if the embryo is female or sperm if the embryo is male. These are the only cells that undergo meiosis. These committed cells, primordial germ cells, increase in number through mitotic cell division. Later, they divide meiotically to produce mature sperm or eggs.

The process of forming mature eggs is called **oogenesis.** There are different stages of oogenesis, and although there is variation from species to species, the general stages of egg formation are similar (Figure 2.14). In this section we describe oogenesis in mammals. The primordial germ cells that migrate to the developing ovary during early embryogenesis become **oogonia** (sing. = oogonium). They multiply rapidly by undergoing several rounds of mitotic cell division, and eventually differentiate into **primary oocytes.**

The primary oocytes begin meiotic cell division. They enter the diplotene stage of prophase I and then division stops. They remain in this suspended prophase for a period of time that varies from species to species. During this period, the oocyte undergoes many changes that prepare it for the completion of meiosis and for fertilization. For example, it acquires a special coat that protects the developing egg from mechanical damage and in many cases acts as a barrier to sperm from other species. Just beneath the membrane, **cortical granules** develop that alter the egg coat so that only a single sperm can fertilize the egg. In addition, the primary oocyte accumulates large quantities of nutrients and other molecules that will nourish the early embryo and coordinate and direct its development.

With the attainment of sexual maturity, the next stage of oocyte development occurs, triggered by hormones. The oocyte completes meiosis I (the reductional division of meiosis) and produces two haploid nuclei, each containing one member of each chromosome pair in a replicated state. But cytokinesis is very asymmetrical. One cell, called the **secondary oocyte,** gets virtually all the cytoplasm and will be ancestral to the mature egg. The other cell, called a **polar body,** gets very little cytoplasm. Both of these cells, the secondary oocyte and the polar body, undergo meiosis II to produce a total of four haploid nuclei. Again, cytoki-

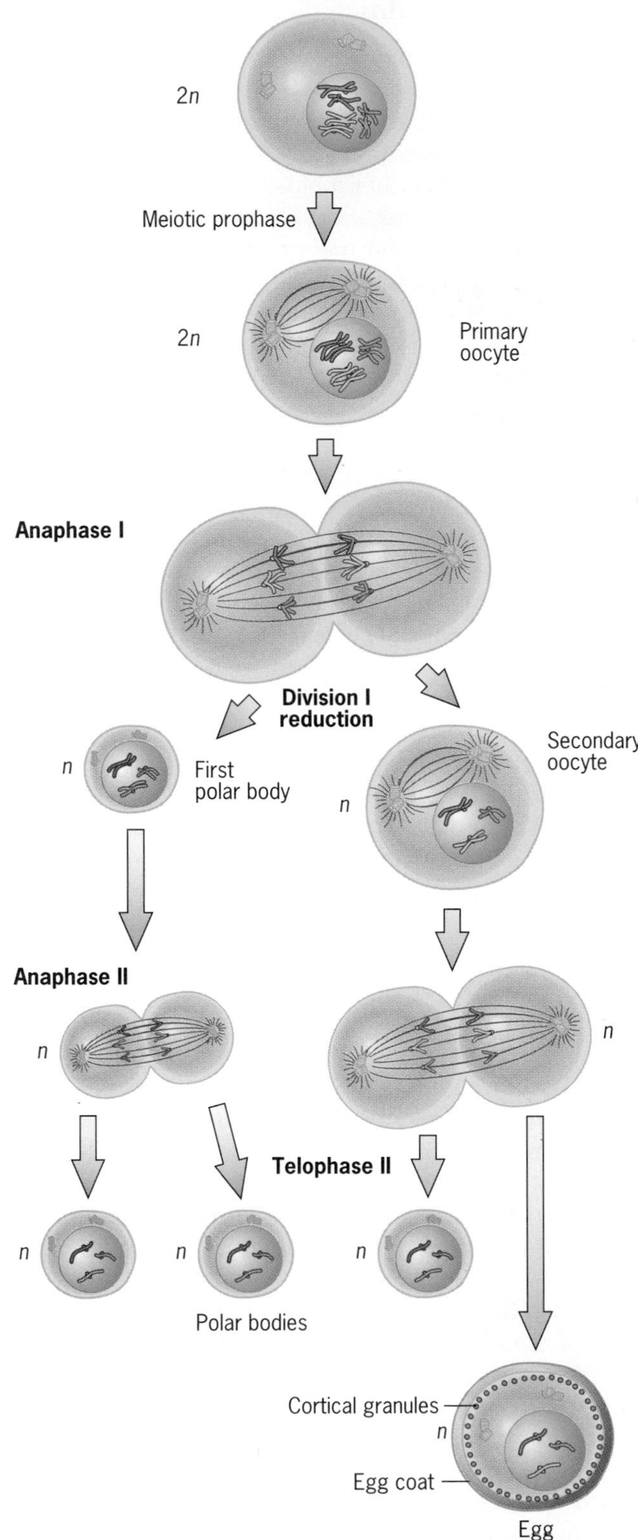

Figure 2.14 Oogenesis in the female results in the formation of one egg and three polar bodies.

nesis is asymmetrical: Meiosis II in the secondary oocyte produces one large cell, the **ovum,** or egg, which has virtually all the cytoplasm, and a small polar body with very little cytoplasm. Thus, of the four meiotic products, only one forms the mature egg. The polar bodies, which are small with little cytoplasm to support their metabolism, eventually degenerate.

Oogenesis occurs in the ovaries of the female. Each primary oocyte is surrounded by a spherical cluster of cells in a cavity or sac called the primary or Graafian follicle. In response to hormone signals, the primary oocyte completes meiosis I to become a secondary oocyte. The follicle then ruptures, releasing the secondary oocyte into the oviduct, where it begins the second meiotic division. In humans, the oviduct is called the fallopian tube. The release of the secondary oocyte from the Graafian follicle is called **ovulation.**

As the secondary oocyte travels down the oviduct, it commences the second meiotic division. If it is fertilized, meiosis II is completed. If it is not fertilized, meiosis II is not completed and the egg degenerates. Upon fertilization, both male and female nuclei occupy the egg cytoplasm. The haploid nuclei fuse, restoring the diploid chromosome number. The **zygote**, or fertilized egg, passes down the oviduct into the uterus and attaches there. We discuss fertilization in more detail shortly.

Spermatogenesis: The Formation of Sperm

The egg is among the largest cells in an organism; the sperm is among the smallest. This size difference is related to the function of the two cell types. The egg is nonmotile and stores large quantities of material needed to support the growth and development of the early embryo. The sperm is highly motile and is constructed to move through the female reproductive tract to fertilize the egg. In many respects, the sperm is simply a haploid nucleus with a tail. There is neither an endoplasmic reticulum nor a Golgi apparatus, nor are there any ribosomes. The mitochondria coalesce around the tail and serve as an energy source for tail movement. The head of the sperm contains an acrosomal vesicle with several enzymes that help the sperm penetrate the egg.

Spermatogenesis begins in the human male at puberty. (Recall that in the human female, meiosis begins during early embryogenesis.) It takes place in the testes in coiled tubes called seminiferous tubules. The immature premeiotic germ cells, called **spermatogonia,** are located around the border of the seminiferous tubules. The spermatogonia proliferate mitotically. At various points, a subset of the spermatogonia stops dividing, enlarges, and differentiates into **primary spermatocytes.** Primary spermatocytes undergo meiosis I to produce **secondary spermatocytes** and then meiosis II to produce four haploid **spermatids.** The spermatids differentiate into mature **spermatozoa,** which pass into the lumen of the seminiferous tubule, and from there into the epididymis, a tube overlying the testis, where they undergo further maturation. Figure 2.15 summarizes the stages of spermatogenesis.

The developing sperm cells remain connected to each other by a common cytoplasm during the entire process. With the differentiation of spermatids into spermatozoa, the individual sperm are released. The common cytoplasm provides the developing spermatids with all the cytoplasmic products they require to differentiate into spermatozoa. In contrast to the mature egg that has sequestered virtually all the cytoplasm from the meiotic products, the sperm cells lose cytoplasm during spermatogenesis, producing spermatids that contain very little cytoplasm.

In mammals, **fertilization** occurs with the union of sperm and egg in the oviduct. The sperm head binds to the **zona pellucida,** or egg coat. The zona pellucida has two main functions: (1) it contains various glycoproteins that bind the sperm and serve as recognition

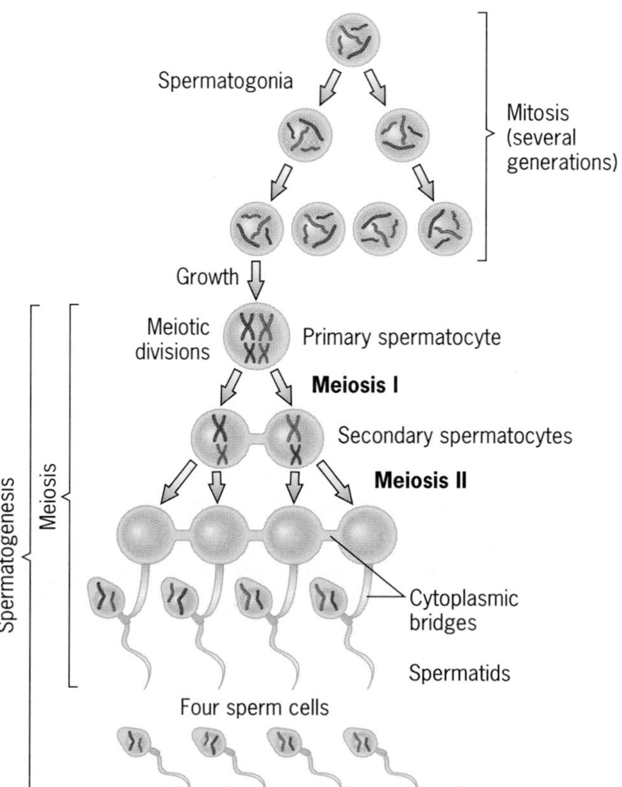

Figure 2.15 Spermatogenesis in the male results in the formation of four sperm cells. During the process, sperm remain connected via a common cytoplasm until the formation of mature sperm.

signals for sperm of the same species; and (2) it functions as a barrier to sperm from other species. On binding, the contents of the sperm's acrosomal vesicle are released. This vesicle contains enzymes that aid the sperm in its penetration of the zona pellucida and facilitate the fusion of the sperm and egg membranes. After membrane fusion, the cortical granules just inside the egg membrane release a variety of molecules, including enzymes that alter the structure of the zona pellucida in such a way that additional sperm are prevented from fusing with the egg membrane. This assures that only a single sperm can fertilize an egg. Once the sperm and egg nuclei fuse into a single diploid nucleus, the zygote begins development.

Gamete Formation in Plants

The major difference between the reproductive cycles in plants and animals is the mode of gamete formation. In animals, meiosis produces haploid cells that then mature into gametes. In plants, gametes rarely arise directly from meiosis. Meiosis takes place in plants, but the resultant haploid cells do not become gametes. The haploid products of meiosis in plants are called **spores.**

Before gametes form and fertilization occurs, haploid spores divide mitotically to produce a fully developed, multicellular haploid organism called a **gametophyte,** or gamete-producing plant. The gametophyte, which develops from the spore, produces gametes by mitosis. When two haploid gametes fuse, a diploid zygote is produced. The zygote divides mitotically and develops into a multicellular diploid organism called the **sporophyte,** or spore-producing plant. Later, specialized cells within the sporophyte, analogous to the germ cells in animals, undergo meiosis and produce haploid spores. The cycle then starts anew. Thus plants, unlike animals, exist as fully developed haploid and diploid organisms during their life cycle. Alternating sporophyte and gametophyte phases, referred to as the **alternation of generations,** vary in length and importance in different plants (Figure 2.16). In most flowering plants, the gametophytic phase is limited to a few cell divisions that produce the embryo sac (female gametophyte) and the germinating pollen grain (male gametophyte). However, some plants such as ferns, mosses, and liverworts have more prominent gametophytic phases.

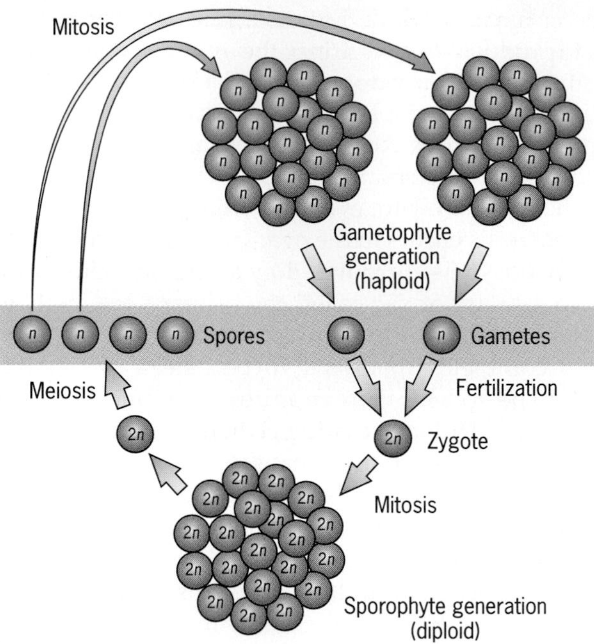

Figure 2.16 In plants there is an alternation of haploid (gametophytic) and diploid (sporophytic) generations.

Key Points: **Gametogenesis is the formation of male and female gametes or sex cells. Spermatogenesis and oogenesis are the formation of male and female gametes, respectively, by the processes of mitosis and meiosis. Gametes unite in the process of fertilization. In plants, meiosis in the sporophyte produces spores that develop into haploid gametophytes. Gametophytes produce haploid gametes that fuse to produce a diploid zygote, which divides mitotically to produce a diploid sporophyte.**

THE LIFE CYCLES OF SOME GENETICALLY IMPORTANT ORGANISMS

The life cycle of an organism is the sequence of events that occur from its origin as a zygote to its death. In this section, we will examine the life cycles of some species that have played and continue to play important roles in advancing the science of genetics. Unfortunately, space limitations dictate that some important model organisms must be left out. However, many of the species that are omitted here are discussed in subsequent chapters. The worm *Caenorhabidtis elegans*, which is contributing immensely in the field of developmental genetics, for example, is discussed in detail in Chapter 25. The life cycles of some nonsexually reproducing organisms, such as bacteria and viruses, are examined in Chapters 16 and 17.

In eukaryotes, meiosis generates haploid cells in which the parental genetic material has been reshuffled by chromosome assortment and crossing over. The fusion of haploid cells produces an almost infinite variety of new genetic combinations that the evolutionary processes can act upon. Thus, the life cycles of organisms present opportunities for the reshuffling of the genetic material to produce new genetic combinations. Note that these natural life cycles have the opposite effect of "cloning," the artificial production of genetically identical organisms (see Technical Sidelight: Dolly: To Clone or Not to Clone).

Neurospora crassa: The Simple Bread Mold

An organism that has made enormous contributions to our understanding of genetic concepts, especially to issues relating to recombination and the relationship between genes and gene products, is the salmon-colored bread mold, *Neurospora crassa* (Figure 2.17). Molds or fungi are predominantly terrestrial and haploid through most of their life cycle. Although some fungi are unicellular, most are multicellular and filamentous. Collectively, these vegetative filaments are called the **mycelium** (pl. = mycelia). *Neurospora* is a haploid filamentous fungus. The filaments, or hyphae, are divided into segments by perforated cross-walls called septae. The perforations allow cytoplasm to flow freely between the hyphal segments. *Neurospora* reproduces asexually by the formation of specialized multinucleated spores called **conidia** (sing. = conidium). These asexually produced spores germinate into hyphal segments. Asexual reproduction is the predominant mode of reproduction in *Neurospora* and related fungi.

Neurospora also has a sexual mode of reproduction. There are two mating types of *Neurospora*, *A* and *a*. Under certain conditions, a conidium from one mating type lands on a large, specialized hypha called a protoperitheicium, or fruiting body, from an organism that is of the opposite mating type. The conidium fuses with a special mating structure, the trychogyne. The two nuclei now share a common cytoplasm, but they do not immediately fuse. Rather, each divides mitotically to form a multinucleated cell. Nuclei of opposite mating types fuse to form a diploid nucleus, the only diploid phase in the entire *Neurospora* life cycle. Each diploid nucleus becomes walled off in an **ascus** (pl. = asci), a cell with a thick outer wall. Meiosis produces

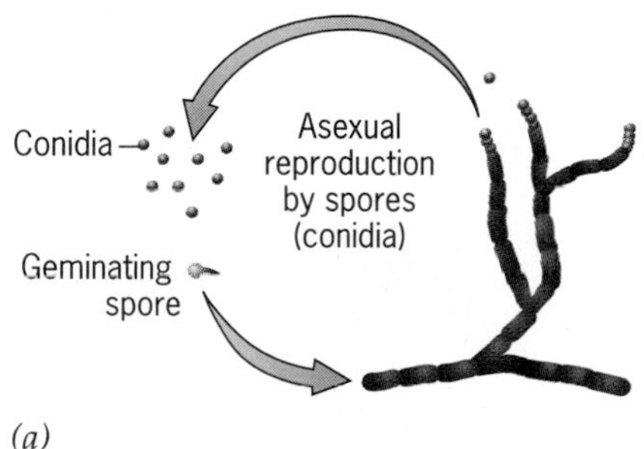

(*a*)

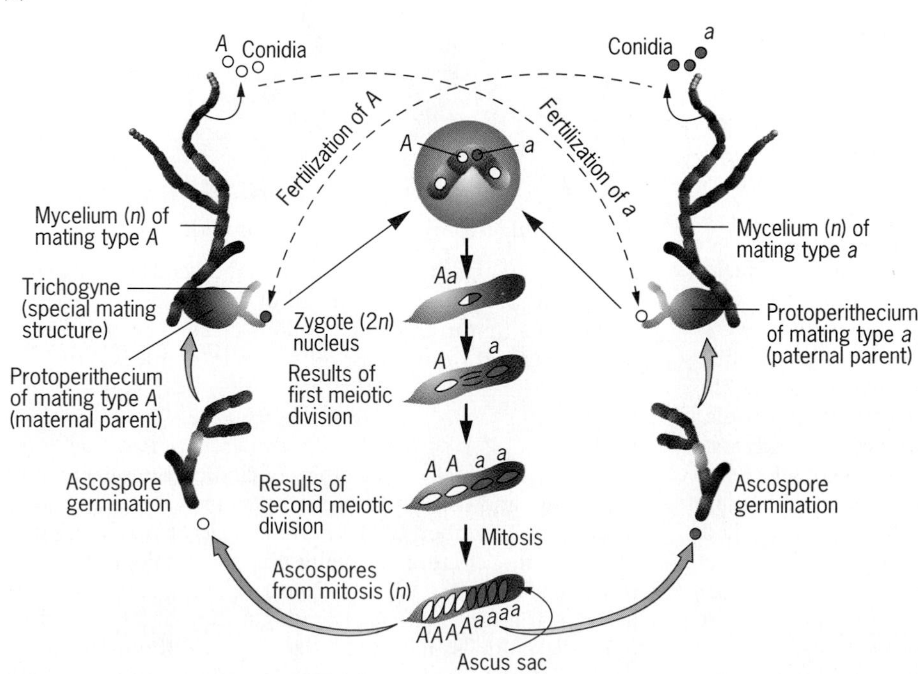

(*b*)

Figure 2.17 The life cycle of bread mold, *Neurospora crassa*. *Neurospora* reproduces (*a*) asexually or (*b*) sexually.

TECHNICAL SIDELIGHT

Dolly: To Clone or Not to Clone

On February 27, 1997, a star was born. On that day, the world was introduced to Dolly (Figure 1*a*), a Dorset lamb born to a Scottish Blackface ewe at the Roslin Institute near Edinburgh, Scotland, seven months earlier. To no one's surprise, the white-faced Dolly looked nothing like her black-faced mother, because Dolly shared no nuclear genes with her surrogate mother—although she did share mitochondrial genes. Instead, Dolly was a clone produced by removing the nucleus from an egg obtained from a Scottish Blackface ewe, fusing the enucleated egg with a mammary cell from a six-year-old Finn Dorset ewe, and implanting the embryo produced by the fusion egg in the uterus of a Scottish Blackface ewe (Figure 2*a–c*).

The announced cloning of a mammal was not the reason for the excitement. Indeed, lambs had previously been produced by introducing nuclei from embryonic cells into enucleated eggs and implanting them in surrogate mothers. But this was the first time that scientists had been successful in using the nucleus from an apparently differentiated cell of an adult mammal to produce a clone. Initially, some scientists were skeptical of Dolly's origin. Perhaps Dolly had been produced by a rare contaminating cell, not by a mature mammary cell as claimed. Moreover, Dolly was a single clone; could the process be repeated?

Subsequent comparisons of Dolly's DNA with the DNA of the Dorset cell donor have provided strong support for Dolly's clonal origin. Moreover, Dolly has now been joined by Cumulina (Figure 1*b*), the first of three generations of cloned mice produced by injecting diploid nuclei from cells of adult mice into enucleated eggs (Figure 2*d*), and implanting embryos produced by these eggs in surrogate mothers. These new results have largely eliminated skepticism about Dolly's origin. However, they have also heightened concerns about the possible misuse of cloning technology.

Whether justified or not, the cloning of Dolly has received more media attention than any other scientific breakthrough of this decade. Almost immediately after Dolly was introduced to the world, members of the news media were discussing the likelihood and the ethics of cloning humans. To many members of society, Dolly was startling, giving life to the monsters of science fiction horror stories. Would cloning be used to produce spare body parts for the rich and famous? Would it dehumanize our species? Even before the birth of Dolly, the British Parliament had banned human cloning. In June 1997, the president of the United States called for a federal law banning the cloning of humans and asked a task force to make recommendations regarding the regulation of cloning techniques. The governments of China and Switzerland called for similar bans. However, by early 1998, a physicist in Chicago announced that he would initiate experiments designed to develop methods for cloning humans.

(*a*) Dolly

(*b*) Cumulina (front) and her foster mother

Figure 1 (*a*) Dolly, the first mammal cloned by using the nucleus from a differentiated cell of an adult. (*b*) Cumulina, the first in three generations of mice cloned using nuclei obtained from donor cells of mature mice.

Clearly, the debates on the ethics of cloning humans are just beginning, and their outcome remains uncertain.

In its December 19, 1997, issue, *Science* magazine pronounced the cloning of Dolly the "technological breakthrough of the year." Whereas human cloning seems unlikely in the near future, the use of cloning technologies to produce domestic animals that synthesize valuable human proteins such as the blood-clotting factor IX—absent in individuals with one type of hemophilia—is already underway, and

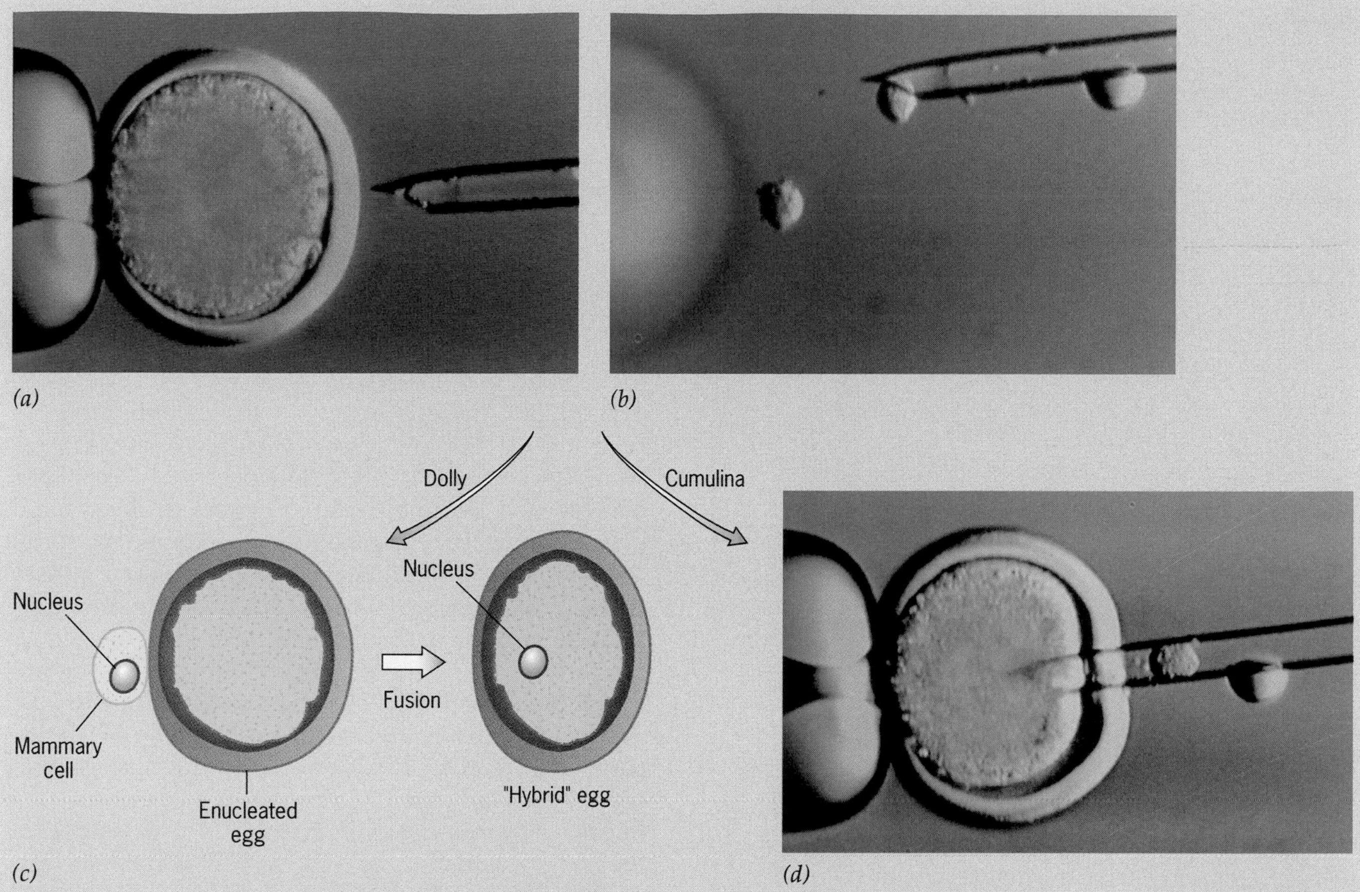

Figure 2 The nuclear transfer techniques used to produce Dolly and Cumulina. In both cases, an egg was enucleated by (*a*) holding it on the tip of a pipette with light suction and (*b*) withdrawing its nucleus with a glass needle. In Dolly's case, (*c*) a short pulse of electricity was used to fuse the enucleated egg with a mature mammary cell. To produce Cumulina, (*d*) scientists used a glass needle to inject a nucleus removed from a cell of an adult mouse into the enucleated egg. In both cases, the nucleated eggs were grown briefly in embryo culture before being implanted in surrogate mothers.

many other potential applications are on the drawing board. Thus, discussions about the potential applications of cloning and the ethics of various types of cloning can be expected to continue well into the future.

When scientists first learned how to produce recombinant DNA molecules, they called for a moratorium on some types of research because of questions about potential misuses of this new technology. Perhaps a similar moratorium on human cloning experiments would be appropriate until the ethical questions that these experiments raise have been discussed at length. These discussions, we hope, would then lead to the wise use of this new scientific breakthrough.

four haploid products, which in turn divide mitotically to produce eight haploid products contained in a linear array. Each of the four meiotic products is represented twice. The eight haploid products mature into **ascospores** and are released when the ascus ruptures. Each ascospore gives rise to a haploid hyphal segment that continues to divide mitotically. The ordered arrangement of the ascospores in the ascus has important genetic significance in gene mapping and the mechanisms of recombination, which we will discuss in Chapters 9 and 14.

Saccharomyces cerevisiae: Baker's Yeast

Today, another fungus, the yeast *Saccharomyces cerevisiae,* is the most popular unicellular eukaryotic model system for genetic studies. Like *Neurospora, Saccharomyces* reproduces both sexually and asexually (Fig-

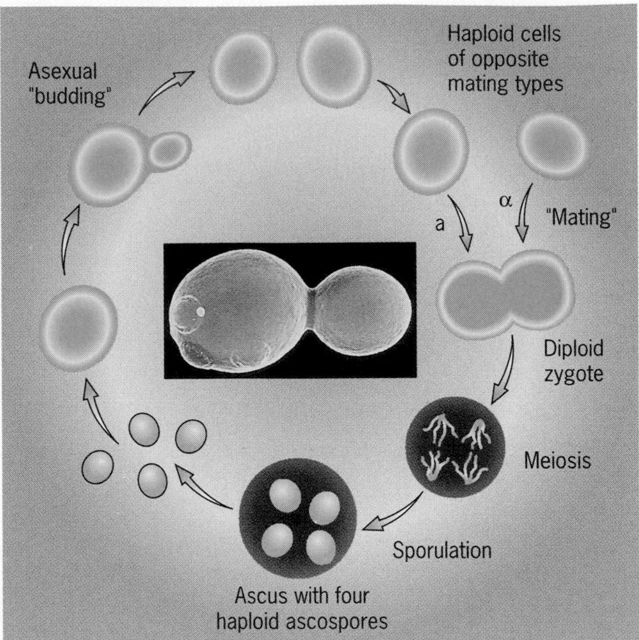

Figure 2.18 The life cycle of baker's yeast, *Saccharomyces cerevisiae.*

ure 2.18). Asexual reproduction occurs by the process of budding, which simply involves a mitotic division of the haploid nucleus, during which one daughter nucleus moves into a small "bud" or progeny cell. Sexual reproduction occurs as in *Neurospora* except the mating types are denoted "a" and α (alpha) and the ascospores are not ordered within the bag-shaped ascus. The fusion of haploid cells of opposite mating type creates a diploid cell, which promptly undergoes meiosis to produce four haploid ascospores inside the ascus. By dissecting this sac, a researcher can isolate each product and place it in a culture dish to start a new yeast colony. Many mutations have been identified in *S. cerevisiae,* including some that alter colony shape and color. These mutations have been useful in studies of chromosome pairing, genetic recombination, the cell cycle, cytokinesis, chromosome segregation, the cytoskeleton, and other components of cell growth and division.

Because of its relatively simple structure and the ease with which it can be grown in culture, *S. cerevisiae* is used extensively for genetic research. The complete sequence of the over 12 million base pairs in the *S. cerevisiae* genome was reported in 1996, and techniques are available by which researchers can produce mutations or "gene disruptions" in essentially any of its approximately 6000 genes. In the near future, scientists hope to determine the functions of these genes and look for the presence of genes with similar functions in other eukaryotes, including humans.

Zea mays: Corn

Genetic studies using corn have led to many important discoveries. Several of the early Mendelian geneticists used corn as their experimental material to explore the basic laws of inheritance as well as variations of these laws. Barbara McClintock was awarded the Nobel Prize in 1983 for her pioneering research into genetic elements that move around the corn genome. These mobile genetic elements have been discovered in many species and will be discussed in Chapter 18.

In *Neurospora* and in most lower plants, the haploid condition is the dominant phase of the life cycle. The diploid phase is fleeting, existing only after the fusion of two haploid spores. Meiosis follows this fusion to produce the haploid state once again. As plants evolved and became more complicated, the diploid phase of the life cycle became the dominant form. Thus, we can see in the more ancient ferns and mosses a gradual lengthening of the diploid phase and a diminishing of the haploid phase.

In most flowering plants, including the pea plant which Mendel studied, the flower contains both male and female reproductive structures (Figure 2.19). In the corn plant, which represents the sporophyte generation, male and female reproductive structures are housed in separate flowers (Figure 2.20). The tassel at the top of the plant is an aggregation of **stamens** containing the diploid male spore mother cells, or **microspore mother cells.** The silks on the side of the plant are an aggregation of **pistils.** The diploid female spore mother cells **(megaspore mother cells)** are present in the ovaries at the base of the pistils.

In the stamens, microspore mother cells divide meiotically to produce four haploid **microspores,** the first cells of the gametophyte generation. Each microspore differentiates into a **pollen grain,** the male gametophyte. The haploid pollen nucleus divides mitotically, forming two nuclei. One becomes a tube nu-

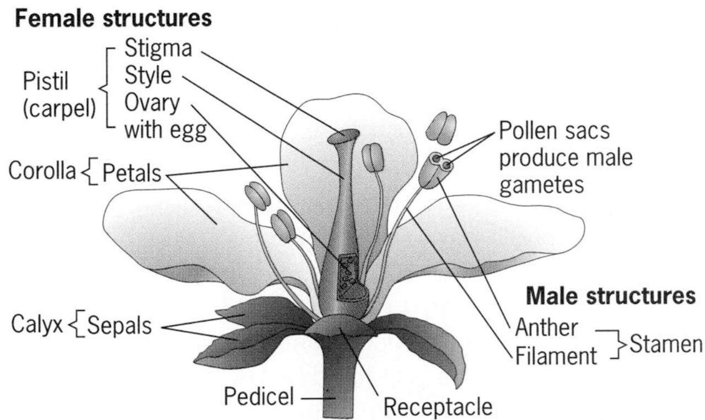

Figure 2.19 The structures of a typical flower, showing the male and female reproductive systems.

cleus, which controls the growth of the pollen tube during fertilization, and the other a generative nucleus, which gives rise to the nuclei involved in fertilization. The generative nucleus divides mitotically again to produce two sperm nuclei. The pollen grain now has three haploid nuclei—two sperm nuclei and one tube nucleus.

In the pistils, megaspore mother cells undergo meiosis to form four haploid nuclei. As in oogenesis, only one of the four meiotic products becomes functional and occupies the embryo sac or female gametophyte. The other three meiotic products degenerate. The remaining nucleus divides three times mitotically to produce eight identical haploid nuclei arranged linearly in the embryo sac. Two nuclei in the center fuse to produce a diploid endosperm nucleus. The three nuclei farthest away from the point in the sac where the pollen tube enters become the antipodal nuclei. The three closest to the pollen tube's point of entry form an egg nucleus and two synergid nuclei.

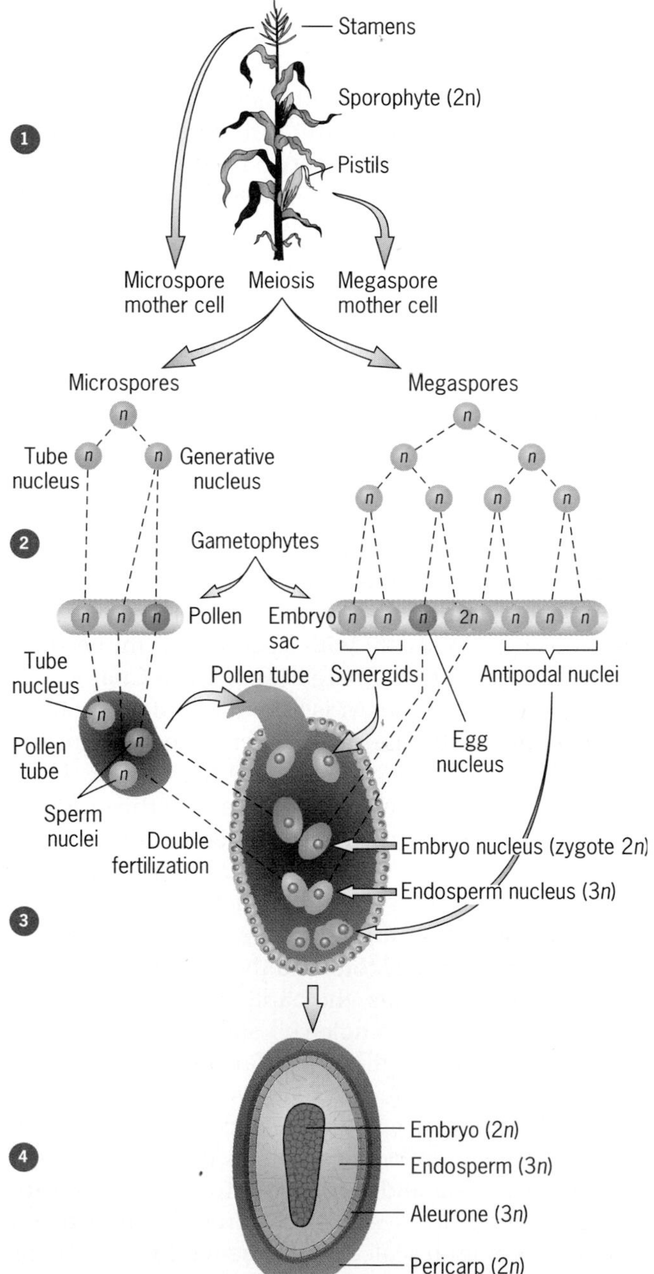

Figure 2.20 The life cycle of corn, *Zea mays*.

Pollen grains land on the stigma (silk) of the pistil and produce long pollen tubes that grow down the style. The development and growth of the tube are under the control and direction of the tube nucleus. The sperm nuclei enter the ovary (embryo sac). One sperm nucleus fertilizes the diploid endosperm nucleus, and the other sperm nucleus fertilizes the egg nucleus. The fusion of one sperm nucleus with the diploid endosperm nucleus produces the triploid **endosperm,** a tissue with three copies of each chromosome that supplies nourishment to the developing embryo until it is able to photosynthesize. The other sperm nucleus fertilizes the egg nucleus to produce the diploid nucleus of the zygote, which divides to form the **embryo.** The synergid and antipodal nuclei degenerate after fertilization; in some plants, they produce a few cells that provide nourishment to the egg and zygote. It is not known if the synergid and antipodal nuclei have any other function.

The outer layer of the triploid endosperm differentiates into the aleurone. The aleurone can accumulate pigments, giving corn kernels a range of colors, from the more familiar market varieties of white and yellow if *no pigment* is deposited in the aleurone (white and yellow are properties of the endosperm, visible through the colorless aleurone) to purples and reds if pigments are deposited. Nuclear genes control pigment deposition in the aleurone and the endosperm color. The pericarp, which originates from the maternal plant tissue, surrounds the developing kernel, which becomes the seed that germinates into the diploid sporophyte generation.

Drosophila melanogaster: The Fruit Fly

Perhaps no organism has proven so valuable to genetic research as the little fruit fly, *Drosophila melanogaster*. The development of the entire chromosome theory of inheritance, which established experimental links between Mendelian principles of inheritance and chromosomes, has been based almost entirely on research using *Drosophila*. Today, *Drosophila* is giving researchers valuable insights into the genetic control of developmental processes.

Drosophila, a small organism about 2 mm in length, can complete its life cycle in about 10 days at 25° C

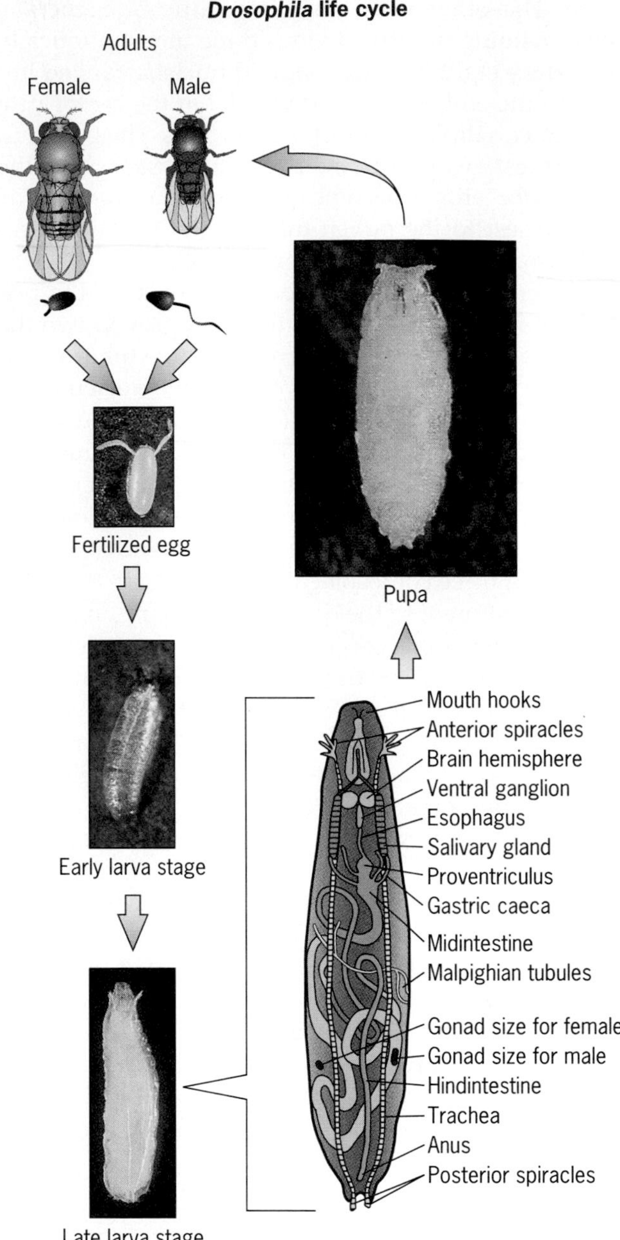

Figure 2.21 The life cycle of the fruit fly, *Drosophila melanogaster*.

(Figure 2.21). The fertilized egg becomes an embryo that hatches into a small, segmented first instar **larva** that is very motile and that feeds rapidly. Specialized jaw hooks in the head region function like rakes as the larva burrows through the food material (rotting and fermenting fruit in the natural world; corn meal, agar, molasses, and yeast in the lab). As the larva grows, it eventually outgrows its chitinous cuticle, shedding it and growing a new one. This is the second instar stage of development. Further growth and development result in another cuticle molting to produce the large third instar larva. This stage appears about five days after fertilization.

Near the end of the third instar stage, the larva slows down and crawls out of the food onto a relatively dry place. The larval body shortens, and the cuticle thickens and accumulates pigment as it becomes the chitinous pupal case. Major developmental changes are now taking place as the **pupa** forms inside the pupal case. Adult structures, such as head, wings, and legs, developing in the pupal case, were present in the larva as small pre-adult clusters of determined tissue called **imaginal discs.** These discs are genetically committed to form specific adult structures, but they do not differentiate into those structures until the pupal stage of development.

As adult structures develop, larval structures break down. About 10 days after fertilization, the adult emerges from the pupal case. If it is male, spermatogenesis produces mature sperm; if female, oogenesis produces mature eggs. Fertilization produces the zygote and the cycle begins anew.

Humans

Some of the most exciting genetic research today is being carried out in humans. We mentioned the Human Genome Project, the diagnosis of inherited diseases, and the possible treatment of these disorders by somatic-cell gene therapy in Chapter 1. We discuss these topics and other applications of human genetics in more detail in Chapters 21 and 22. The genetic control of the immune system in humans and its malfunction in individuals with acquired immune deficiency syndrome (AIDS) are examined in Chapter 27. The genetic basis of cancer is the subject of Chapter 28.

In terms of its essential elements, the human life cycle is basically the same as that of the fruit fly: Haploid gametes (sperm and egg) unite in fertilization to produce a diploid zygote that develops into an embryo and, finally, a mature multicellular organism. Specialized cells in the gonad (testicle or ovary) of the mature organism produce gametes that unite in fertilization to produce a zygote, and the cycle begins anew.

As in most animals, the haploid phase of the human is reduced to a single-cell stage, sperm or egg. There are significant differences, however, between the larval instar developmental stages in *Drosophila* and human development. Six to eight days after fertilization, the human embryo completes its journey down the fallopian tube and implants into the uterus, rupturing uterine blood vessels in the process. These blood vessels become resealed, but the embryo remains bathed in a pool of maternal blood that is continuously replenished by fresh blood from maternal blood vessels. A **placenta,** which functions as the connection between mother and embryo, is constructed from embryonic and maternal membranes. It embeds in the uterine wall and is the site for the exchange of respiratory gases, nutrients, and waste matter. Connecting blood vessels in the umbilical cord link the embryonic circulatory system to the highly vascularized placenta.

During the first two months of development, all of the major organ systems appear in the embryo. Most of them are functioning by the end of this period. Beginning with the third month of development, the embryo is referred to as a **fetus.** During the last seven months of development, the fetus grows and refinements occur in its organ systems. The fetus is preparing for life outside the uterus.

From conception to birth, approximately 266 days pass. These 266 days are divided into three periods, or trimesters. The most dramatic developmental changes occur during the first trimester (Figure 2.22). During this period, the single-celled zygote is transformed into a fetus with functioning organ systems that are only slightly less complex than those found in a full-term baby.

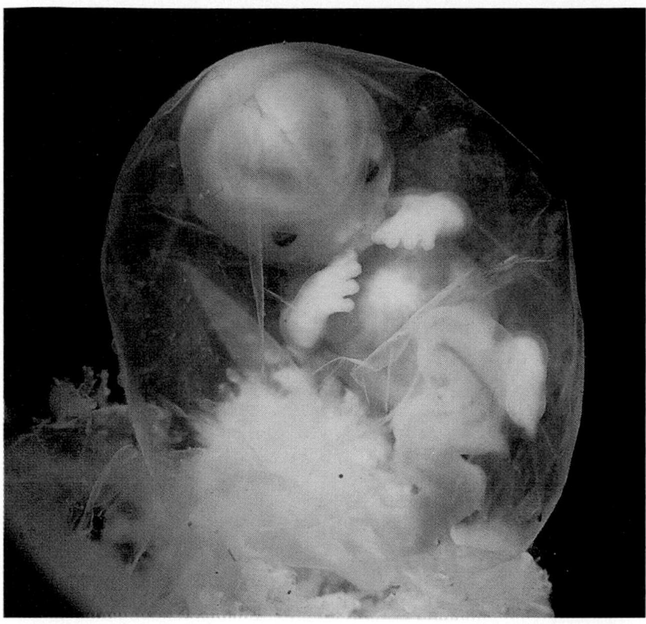

Figure 2.22 A two-month human embryo. During this period, the single-celled zygote is transformed into a fetus with functioning organ systems that are only slightly less complex than those found in a full-term baby.

During the final trimester, the fetus prepares to live on its own. It increases in body weight 500 to 600 percent. The brain and peripheral nervous system mature at a rapid rate. As the end of the third trimester approaches, the fetus develops the ability to regulate its own temperature and to control its own breathing. Fat deposits form under the skin, giving the fetus a chubby appearance. Most fetuses change position in the uterus, becoming aligned for a head-first delivery through the birth canal.

One system does not mature until after birth. Human babies have no functioning immune system. During the last month of fetal life, antibodies from the mother cross the placental barrier and protect the baby from infectious diseases until its own immune system begins functioning. If the baby is breast-fed, maternal milk provides additional antibodies for the baby, augmenting its developing immune system.

Key Points: **The life cycle is the sequence of events from the individual's origin as a zygote to its ultimate death. It can also be described as the stages through which an organism passes between the production of gametes by one generation and the next. In many plants and some animals, the life cycle involves a regular alternation of generations between haploid and diploid phases.**

TESTING YOUR KNOWLEDGE

1. Study the diagrams below of dividing cells. Is either of these cells haploid? What stages of mitosis are represented in these cells?

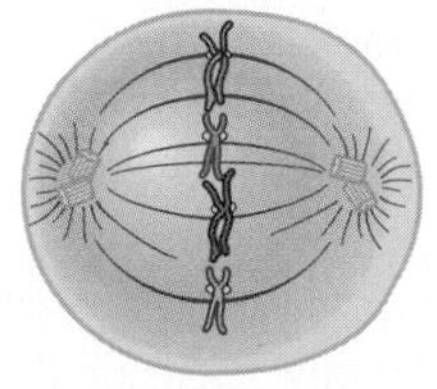

(a)

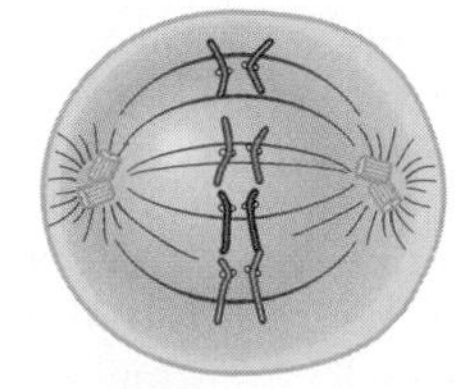

(b)

ANSWER

Careful examination of these two cells shows that they are both haploid because the four chromosomes are all different and in an unpaired state. Cell (*a*) is in metaphase because the chromosomes have replicated and lined up at the equatorial region, but the centromere has not yet divided. Cell (*b*) is in anaphase because the sister chromatids have separated and are beginning their journey to opposite poles of the cell.

2. What meiotic stages are represented by the following cells?

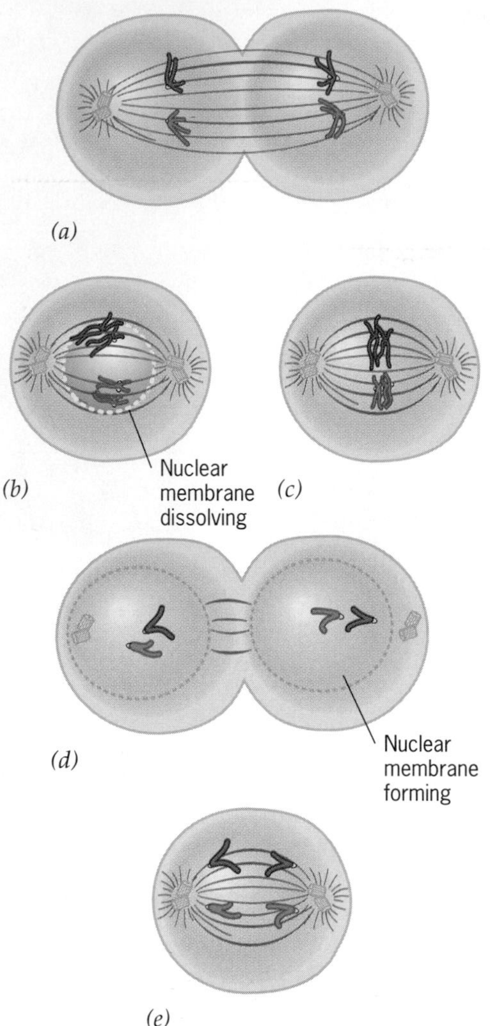

ANSWER

Cell (*a*) is in anaphase I because the pairs of replicated chromosomes are segregating from each other; the centromere has not divided. Cell (*b*) is in prophase I because the replicated chromosomes have paired but not yet aligned at the equatorial plate; the dissolving nuclear membrane is indicative of the approaching metaphase. Cell (*c*) is in metaphase I because the replicated chromosomes have paired at the equatorial plate and the nuclear membrane has dissolved. Cell (*d*) is in telophase II because the chromosomes in each of the haploid nuclei are unreplicated and a nuclear membrane is forming around them. Cell (*e*) is in anaphase II because the former sister chromatids are migrating to opposite poles of the cell as they approach telophase II.

QUESTIONS AND PROBLEMS

These questions and problems represent varying degrees of difficulty. The more difficult questions and problems are indicated by red numbers. This designation will be used in all subsequent chapters.

2.1 Mark the true statements with a (+) and the false with a zero (0). (a) Skin cells and gametes of the same animal contain the same number of chromosomes. (b) Any chromosome may pair with any other chromosome in the same cell in meiosis. (c) The gametes of an animal may contain more maternal chromosomes than its body cells contain. (d) Of 10 chromosomes in a mature sperm cell, 5 are always maternal. (e) Of 22 chromosomes in a primary oocyte, 15 may be paternal. (f) The homologous parts of two chromosomes lie opposite one another in pairing. (g) A sperm has half as many postmitotic chromosomes as a spermatogonium of the same animal.

2.2 In each somatic cell of a particular animal species, there are 46 postmitotic chromosomes. How many should there be in a (a) mature egg, (b) first polar body, (c) sperm, (d) spermatid, (e) primary spermatocyte, (f) brain cell, (g) secondary oocyte, (h) spermatogonium?

2.3 If spermatogenesis is normal and all cells survive, how many sperm will result from (a) 50 primary spermatocytes and (b) 50 spermatids?

2.4 A diploid, eukaryotic organism has $2n = 8$ chromosomes. (a) How many bivalents are present in a nucleus at diplonema? (b) How many chromatids are present in a pachytene nucleus? (c) How many DNA molecules are present in a nucleus at zygonema? (d) How many chromosomes will be present at telophase II of meiosis?

2.5 In humans, a type of myopia (an eye abnormality) is dependent on a gene (M). The normal gene is m. Assuming that genes are located on chromosomes, diagram (placing the genes on homologous chromosomes) a cross between a woman with myopia who is genetically Mm and a normal man (*mm*). Show the kinds of gametes that each parent could produce and summarize the expected results from the cross.

2.6 Beginning with the myopic woman in Problem 2.5 and again assuming that the genes are present on chromosomes, diagram the process of oogenesis that produces the egg involved in the production of an *Mm* child. Label all stages.

2.7 At three points during meiosis, events occur that will lead to gametes that differ from each other. Describe these three events.

2.8 In what ways is cell division similar and different in animals and plants?

2.9 How does meiosis differ from mitosis? Consider differences in mechanism as well as end results.

2.10 Organize the following processes of meiosis in the order in which they occur: (I) centromeres divide; (II) cells first become haploid; (III) homologous chromosomes segregate; (IV) chromosomes pair; (V) gamete nuclei are formed; (VI) DNA replicates; (VII) crossing over occurs.

2.11 How does gamete formation in higher plants differ from that in higher animals with reference to (a) gross mechanism and (b) chromosome mechanism?

2.12 How is double fertilization accomplished in plants, and what is the fate of the egg and the endosperm nuclei?

2.13 In humans, an abnormality of the large intestine called intestinal polyposis is dependent on gene *A*, and a neurological disorder called Huntington's disease is determined by gene *H*. The normal genes are *a* and *h*, respectively. A man who is *Aa hh* married a woman who is *aa Hh*. Assume the *A* and *H* are on nonhomologous chromosomes. Diagram the chromosomal constitution of the two parents and the gametes that each would produce. Which gametes would form a zygote that has an *A and* an *H*?

2.14 Beginning with the oogonium in the woman described in Problem 2.13 and again assuming that genes are located on chromosomes, diagram the steps in the process of oogenesis necessary for formation of the egg that carries the chromosome with the *H* gene. Label all stages.

2.15 Once again assuming that genes are located on chromosomes, diagram completely the process of spermatogenesis involved in producing the sperm in Problem 2.13 necessary for production of a child carrying a chromosome with the *A* gene. Label all stages.

2.16 A man produces the following kinds of sperm in equal proportions: *AB*, *Ab*, *aB*, and *ab*. Assuming that the genes *A* (or *a*) and *B* (or *b*) are present on chromosomes, diagram his chromosomal constitution.

2.17 Would greater genetic variability be expected among asexually reproducing organisms, self-fertilizing organisms, or bisexual organisms? Explain.

2.18 If biopsies were taken from follicle tissues of the human ovary at the following development periods, what stages in the process of oogenesis might be observed: (a) in the fifth month of intrauterine development, (b) at birth, (c) at 10 years of age, (d) at 17 years of age?

2.19 A cell has four pairs of homologous chromosomes designated *Aa*, *Bb*, *Cc*, and *Dd*, where *A* and *a* represent a pair of homologs. How many different kinds of gametes can this cell produce?

2.20 The imaginary jelly bean tree has $2n = 32$ chromosomes. Plant breeders developed a triploid (triploids have three sets of chromosomes), seedless jelly bean for the American consumer. (a) What would be the maximum and minimum number of chromosomes present in the meiotic end products? (b) To what reason do you attribute the sterility of the triploid jelly bean plant?

2.21 A fruit fly has eight chromosomes in its somatic cells. How many of those chromosomes does it receive from its father?

2.22 The garden pea has 14 chromosomes. How many chromosomally different kinds of gametes can it form?

2.23 How is it possible for a species that is basically haploid to undergo meiosis?

2.24 Why is meiosis so important in evolution?

BIBLIOGRAPHY

ALBERTS, B., D. BRAY, J. LEWIS, M. RAFF, K. ROBERTS, AND J. D. WATSON. 1994. *Molecular Biology of the Cell*, 3rd ed. Garland Publishing, New York.

The Cell Cycle. 1991. *Cold Spring Harbor Symp. Quant. Biol.* 56.

GILBERT, S. F. 1997. *Developmental Biology*, 5th ed. Sinauer, Sunderland, MA.

KARP, G., 1999. *Cell and Molecular Biology: Concepts and Experiments*, 2nd ed. John Wiley and Sons, NY.

LEWIS, R. 1999. *Human Genetics, Concepts and Applications*, 3rd ed. WCB/McGraw-Hill, Dubuque, IA.

LODISH, H., D. BALTIMORE, A. BERK, S. L. ZIPURSKY, P. MATSUDAIRA, AND J. DARNELL. 1995. *Molecular Cell Biology*, 3rd ed. W. H. Freeman, New York.

MIYAZAKI, W. Y., AND T. ORR-WEAVER. 1994. Sister chromatid cohesion in mitosis and meiosis. *Ann. Rev. Genet.* 28:167–188.

MURRAY, A., AND T. HUNT. 1993. *The Cell Cycle*. Oxford University Press, Oxford, UK.

THERMAN, E., AND M. SUSMAN. 1992. *Human Chromosomes: Structure, Behavior, and Effects*. 3rd ed. Springer-Verlag, New York.

A CONVERSATION WITH

JOSEPH G. GALL

When did you first become interested in science?

I can't remember a time when I was not interested in living creatures, particularly insects and wild animals. One of my earliest memories—when I was about five years old—is of catching toads and frogs and raising tadpoles from their eggs. Undoubtedly my mother was the greatest influence, not only because she let me bring in anything I wanted to, but also because she herself was keenly interested in whatever I found. She saw to it that I had plenty of equipment, as well as books on natural history—not just children's books but technical ones as well. Although I didn't know it at the time, she taught me that curiosity about the natural world was an important activity.

Did you have mentors, and, if so, how did they influence your career?

My interest in natural history was pretty much developed before I went to college. As an undergraduate at Yale I found genetics particularly fascinating, and I was strongly influenced by Donald Poulson, who was professor of genetics in the Zoology Department. He encouraged me to enter the graduate program at Yale. I wanted to work on *Drosophila,* particularly because of its giant polytene chromosomes. The best thing Don Poulson did was to leave me alone to find my own thesis project, on amphibian chromosomes. I learned from him to give my students the widest possible latitude in choosing their own problems and approaches.

Joseph G. Gall is a cell biologist whose major interest is in chromosomes and the cell nucleus. His earlier work focused on the giant lampbrush chromosomes of amphibian oocytes and amplification of the genes coding for ribosomal RNA in amphibians, insects, and protozoa. With his student Mary Lou Pardue, he developed the technique of *in situ* hybridization, widely used to localize DNA and RNA sequences at the cellular level. A study with postdoctoral fellow Elizabeth Blackburn defined the six-nucleotide sequence that characterizes the telomeres of most chromosomes. His current research deals with RNA processing in the nucleus. Dr. Gall received his Ph.D. at Yale University and taught at the University of Minnesota and Yale. Currently he is a Staff Member in the Department of Embryology of the Carnegie Institution in Baltimore.

During your long and prestigious career, what has been your greatest challenge, and how did you overcome it?

When I got my first position—Instructor in the Zoology Department at the University of Minnesota—I was unprepared for the fact that much of my time would be involved in teaching. Not only did I not realize how much time it took to prepare lectures, I was painfully shy in front of any audience. I can't say I've overcome this, but it gradually got better. Most professors, when they are honest, admit to a greater or lesser degree of stage fright.

You have made many contributions to molecular genetics. When you look back over your career, do you recall any one time as being the most exciting?

Probably the most exciting time was when we worked out the molecular details of gene amplification in amphibian oocytes, developed the *in situ* hybridization technique, and applied that technique to the demonstration of satellite DNA sequences in the heterochromatic parts of chromosomes, all within two to three years. Gene amplification came first. Although it was not widely known, several investigators (particularly Brachet in Belgium and, independently, Painter in the United States) had claimed that the multiple nucleoli in frog oocytes contain a small amount of extrachromosomal DNA. This was largely ignored—in part I am sure because scientists at the time were trying to establish that DNA was the hereditary material, and it was embarrassing to have DNA outside the chromosomes. Nevertheless, Brachet and Painter were correct, as proven by our biochemical and cytological experiments and those of Donald Brown and Igor Dawid made at the same time.

I realized from this work that the immature amphibian oocyte has a most unusual nucleus: Some two-thirds of all its DNA is a single sequence—thousands of copies of the gene coding for rRNA. I had been trying to work out a cytological technique for visualizing specific DNA sequences in nuclei by hybridizing with a radioactive DNA probe. Nothing had worked, however. With the discovery that the early oocyte nucleus was essentially a bag of one kind of DNA, I felt that I had the perfect test object. If I couldn't get radioactive DNA to hybridize to the rRNA genes in this nucleus, I obviously couldn't get it to hybridize to the much less abundant genes in normal nuclei. Fortunately, some of the first attempts were a success—photographic emulsion was exposed only over the part of the nucleus that contained the rRNA genes.

At this point my graduate student Mary Lou Pardue made improvements in the sensitivity of the technique. She then turned her attention to the so-called satellite DNA of the mouse. About 10 percent of the DNA isolated from mouse cells bands in a CsCl gradient at a density lighter than the bulk of the DNA—hence the name *satellite.* Mary Lou made a radioactive probe from the satellite DNA and showed that it hybridized specifically in the heterochromatic segment of each mouse chromosome. This was the first demonstration that heterochromatin has a special DNA composition, consisting of a simple repeated sequence that contains essentially no ordinary genes.

Students find lampbrush chromosomes fascinating. When did you first become aware of these huge chromosomes?

My first introduction to lampbrush chromosomes was a photo in a 1939 genetics textbook by C. H. Waddington, which I happened upon as a graduate student at Yale. The photo was taken by an American investigator named Duryee, who had only a few years earlier discovered that these giant chromosomes could be isolated by hand from the living oocyte of a frog or newt. I could hardly believe the photo and had to see for myself if such giant chromosomes really existed—which I did by repeating Duryee's observations. I soon found that since the 1880s others had studied these chromosomes, but mostly as sectioned material in which most of the detail is uninterpretable. Another critical factor was that the Zoology Department at Yale had just acquired a phase contrast microscope. With this newfangled instrument, I was able to make the first phase contrast observations on freshly isolated chromosomes. To this day I have not tired of looking at these wonderful structures that have revealed so much about chromosomes in general.

What advice would you give undergraduate students interested in the study of genetics and molecular biology?

My advice to undergraduates, regardless of their particular interests, is to expose themselves to as many areas of science and mathematics as possible. Truly novel insight almost always involves bringing together ideas and approaches—not simply facts—from disparate fields.

What are your current research projects?

Currently we have two projects going in the lab. The first is to understand an organelle in the cell nucleus called the *coiled body* (from its appearance in sections prepared for the electron microscope). This body contains factors that are involved in three separate pathways of RNA processing: mRNA splicing, pre-rRNA processing, and histone pre-mRNA processing. We and others believe this organelle plays a critical role in nuclear metabolism, but to date precise functions have been difficult to demonstrate.

Our second project stems from a totally unexpected observation we made recently. We injected sperm heads from a frog into the nucleus of a frog oocyte. We found that the sperm heads swell up enormously and eventually transform into giant lampbrush chromosomes of exactly the same type as are already present in the nucleus. The same thing happened when sperm heads of other organisms were injected into the frog nucleus. Our hope is to produce giant chromosomes from organisms that do not have lampbrush chromosomes, or whose nuclei are too small for easy analysis. We would especially like to produce human lampbrush chromosomes, which would have considerable value for the study of human genes and their activity.

What do you see as the major breakthroughs in genetics during the next few decades?

I always shy away from predictions about the future—the record of such predictions is abysmal! Genetics is already having a major impact on our understanding of development and will continue to do so. Similarly, many aspects of cell biology will be illuminated by the combination of genetics and molecular biology. Beyond these, there will be major contributions to the understanding and treatment of human hereditary diseases. And then there will be the unforeseen consequences, some of which will be the most exciting of all.

3

Front page of Mendel's paper superimposed on a portrait of Mendel.

Mendelism: The Basic Principles of Inheritance

CHAPTER OUTLINE

Mendel's Study of Heredity
Applications of Mendel's Principles
Formulating and Testing Genetic Hypotheses
Mendelian Principles in Human Genetics

The Birth of Genetics: A Scientific Revolution

Science is a complex endeavor involving the careful observation of natural phenomena, reflective thinking about these phenomena, and the formulation of testable ideas about their causes and effects. Progress in science often depends on the work of a single insightful individual. Consider, for example, the effect that Nicolaus Copernicus had on astronomy, that Isaac Newton had on physics, or that Charles Darwin had on biology. Each of these individuals altered the course of their scientific disciplines by introducing radically new ideas. In effect, they began scientific revolutions.

In the middle of the nineteenth century, the Austrian monk Gregor Mendel, a contemporary of Darwin, laid the foundation for another revolution in biology, one that eventually produced an entirely new science—genetics. Mendel's ideas, published in 1866 under the title "Experiments in Plant Hybridization," endeavored to explain how the characteristics of organisms are inherited. Many people had attempted such an explanation previously but without much success. Indeed, Mendel commented on their failures in the opening paragraphs of his article:

To this object, numerous careful observers such as Kölreuter, Gärtner, Herbert, Lecoq, Wichura and others, have devoted a part of their lives with inexhaustible perseverance. . . . [However,] Those who survey the work in this department will arrive at the conviction that among all the numerous experiments made, not one has been carried out to such an extent and in such a way as to make it possible to determine the number of different forms under which the offspring of hybrids appear, or to arrange these forms with certainty according to their separate generations, or definitely to ascertain their statistical relations.[1]

He then described his own efforts to elucidate the mechanism of heredity:

It requires indeed some courage to undertake a labor of such far-reaching extent; this appears, however, to be the only right way by which we can finally reach the solution of a question the importance of which cannot be overestimated in connection with the history of the evolution of organic forms.

The paper now presented records the results of such a detailed experiment. This experiment was practically confined to a small plant group, and is now, after eight years' pursuit, concluded in all essentials. Whether the plan upon which the separate experiments were conducted and carried out was the best suited to attain the desired end is left to the friendly decision of the reader.[2]

This chapter describes Mendel's discoveries and their fundamental implications for the study of heredity. It also introduces the basic terms and techniques that are used in modern genetic analysis.

MENDEL'S STUDY OF HEREDITY

The life of Gregor Johann Mendel (1822–1884) spanned the middle of the nineteenth century. His parents were farmers in Moravia, then a part of the Hapsburg Empire in Central Europe. A rural upbringing taught him plant and animal husbandry and inspired an interest in nature. At the age of 21, Mendel left the farm and entered a Catholic monastery in the city of Brünn (today, Brno in the Czech Republic). In 1847 he was ordained a priest, adopting the clerical name Gregor. He subsequently taught at the local high school, taking time out between 1851 and 1853 to study at the University of Vienna. After returning to Brünn, he resumed his life as a teaching monk and began the genetic experiments that eventually made him famous.

Mendel performed experiments with several species of garden plants, and he even tried some experiments with honeybees. His greatest success, however, was with peas. He completed his experiments with peas in 1863 and spent the next two years analyzing and summarizing the data. In 1865, Mendel presented the results before the local Natural History Society, and the following year, he published a detailed report in the society's proceedings. Unfortunately, this paper languished in obscurity until 1900, when it was rediscovered by three botanists—Hugo de Vries in Holland, Carl Correns in Germany, and Eric von Tschermak-Seysenegg in Austria. As these men searched the scientific literature for data supporting their own theories of heredity, each found that Mendel had performed a detailed and careful analysis 35 years earlier. Mendel's ideas quickly gained acceptance, especially through the promotional efforts of a British biologist, William Bateson. This champion of Mendel's discoveries coined a new term to describe the study of heredity: genetics, from the Greek word meaning "to generate."

Mendel's Experimental Organism, the Garden Pea

One reason for Mendel's success is that he chose his experimental material astutely. The garden pea, *Pisum sativum*, is a dicot, a type of plant that sprouts two

[1, 2] Peters, J. A., ed. 1959. *Classic Papers in Genetics*. Prentice-Hall, Englewood Cliffs, NJ.

leaves, or cotyledons, from a germinating seed. Peas are easily grown in experimental gardens or in pots in a greenhouse.

One peculiarity of pea reproduction is that the petals of the flower close down tightly, preventing pollen grains from entering or leaving. This enforces a system of self-fertilization, in which sperm and eggs from a particular flower unite with each other to produce seeds. As a result, individual pea strains are highly inbred, displaying little if any genetic variation from one generation to the next. Because of this uniformity, we say that such strains are **true-breeding.**

At the outset, Mendel obtained many different true-breeding varieties of peas, each distinguished by a particular characteristic. In one strain, the plants were 6 to 7 feet high, whereas in another they measured only 9 to 18 inches. Another variety produced green seeds, and still another produced yellow seeds. Mendel took advantage of these contrasting traits to determine how the characteristics of pea plants are inherited. His focus on these singular differences between pea strains allowed him to study the inheritance of a trait one at a time—for example, plant height. Other biologists had attempted to follow the inheritance of many traits simultaneously, but because the results of such experiments were complex, they were unable to discover any fundamental principles about heredity. Mendel succeeded where these biologists had failed because he focused his attention on contrasting differences between plants that were otherwise the same—tall versus short, green seeds versus yellow seeds, and so forth. In addition, he kept careful records of the experiments that he performed.

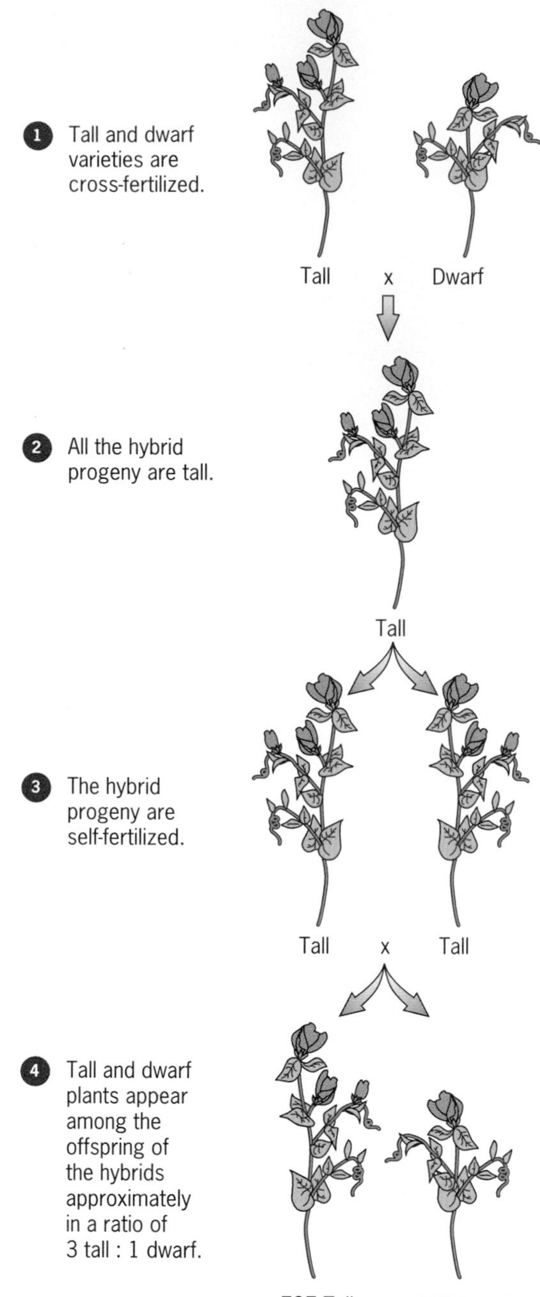

Figure 3.1 Mendel's crosses involving tall and dwarf varieties of peas.

Monohybrid Crosses: The Principles of Dominance and Segregation

In one experiment, Mendel **cross-fertilized**—or, simply, crossed—tall and dwarf pea plants to investigate how height was inherited (Figure 3.1). He carefully removed the anthers from one variety before its pollen had matured and then applied pollen from the other variety to the stigma, a sticky organ on top of the pistil that leads to the ovary. The seeds that resulted from these cross-fertilizations were sown the next year, yielding hybrids that were uniformly tall. Mendel obtained tall plants regardless of the way he performed the cross (tall male with dwarf female or dwarf male with tall female); thus, the two reciprocal crosses gave the same results. Even more significantly, however, Mendel noted that the dwarf characteristic seemed to have disappeared in the progeny of the cross, for all the hybrid plants were tall. To explore the hereditary makeup of these tall hybrids, Mendel allowed them to undergo self-fertilization—the natural course of events in peas. When he examined the progeny, he found that they consisted of both tall and dwarf plants. In fact, among 1064 progeny that Mendel cultivated in his garden, 787 were tall and 277 were dwarf—a ratio of approximately 3:1.

Mendel was struck by the reappearance of the dwarf characteristic. Clearly, the hybrids that he had made by crossing tall and dwarf varieties had the ability to produce dwarf progeny even though they themselves were tall. Mendel inferred that these hybrids carried a latent genetic factor for dwarfness, one that was masked by the expression of another factor for tallness. He said that the latent factor was **recessive** and that the expressed factor was **dominant.** He also inferred that these recessive and dominant factors separated from each other when the hybrid plants reproduced. How else could he explain the reappearance of the dwarf characteristic in the next generation?

Mendel performed similar experiments to study the inheritance of six other traits: seed texture, seed color, pod shape, pod color, flower color, and flower position (Table 3.1). In each experiment—called a **monohybrid cross** because a single trait was being studied—Mendel observed that only one of the two contrasting characteristics appeared in the hybrids and that when these hybrids were self-fertilized, they produced two types of progeny, each resembling one of the plants in the original crosses. Furthermore, he found that these progeny consistently appeared in a ratio of 3:1. Thus, each trait that Mendel studied seemed to be controlled by a heritable factor that existed in two forms, one dominant, the other recessive. These factors are now called **genes**, a word coined by the Danish plant breeder Wilhelm Johannsen in 1909; their dominant and recessive forms are called **alleles**—from the Greek word meaning "of one another."

The regular numerical relationships that Mendel observed in these crosses led him to another important conclusion: that genes come in pairs. Mendel proposed that each of the parental strains that he used in his experiments carried two identical copies of a gene—in modern terminology, they are diploid and **homozygous.** However, during the production of gametes, Mendel proposed that these two copies are reduced to one; that is, the gametes that emerge from meiosis carry a single copy of a gene—in modern terminology, they are haploid.

Mendel recognized that the diploid gene number would be restored when sperm and egg unite to form a zygote. Furthermore, he understood that if the pollen and egg came from genetically different plants—as they did in his crosses—the hybrid zygote would inherit two different alleles, one from the mother and one from the father. Such an offspring is said to be **heterozygous.** Mendel realized that the different alleles that are present in a heterozygote must coexist even though one is dominant and the other recessive, and that each of these alleles would have an equal chance of entering a gamete when the heterozygote reproduces. Furthermore, he realized that random fertilizations with a mixed population of gametes—half carrying the dominant allele and half carrying the recessive allele—would produce some zygotes in which both alleles were recessive. Thus, he could explain the reappearance of the recessive characteristic in the progeny of the hybrid plants.

Mendel used symbols to represent the hereditary factors that he postulated—a methodological breakthrough. With symbols, he could describe hereditary phenomena clearly and concisely, and he could analyze the results of crosses mathematically. He could even make predictions about the outcome of future crosses. Although the practice of using symbols to analyze genetic problems has been much refined since Mendel's time, the basic principles remain the same. The symbols stand for genes (or, more precisely, for their alleles), and they are manipulated according to the rules of inheritance that Mendel discovered. These manipulations are the essence of formal genetic analysis. As an introduction to this subject, let us consider the symbolic representation of the cross between tall and dwarf peas (Figure 3.2).

The two true-breeding varieties, tall and dwarf, are homozygous for different alleles of a gene controlling plant height. The allele for dwarfness, being recessive, is symbolized by a lower case letter *d*; the allele for tallness, being dominant, is symbolized by the corresponding upper case letter *D*. In genetics, the letter that is chosen to denote the alleles of a gene is usually taken from the word that describes the recessive trait (*d*, for *d*warfness). Thus, the tall and dwarf pea strains are symbolized by *DD* and *dd*, respectively. The allelic constitution of each strain is said to be its **genotype.**

TABLE 3.1
Results of Mendel's Monohybrid Crosses

Parental Strains	F_2 *Progeny*	*Ratio*
Tall plants × dwarf plants	787 tall, 277 dwarf	2.84:1
Round seeds × wrinkled seeds	5474 round, 1850 wrinkled	2.96:1
Yellow seeds × green seeds	6022 yellow, 2001 green	3.01:1
Violet flowers × white flowers	705 violet, 224 white	3.15:1
Inflated pods × constricted pods	882 inflated, 299 constricted	2.95:1
Green pods × yellow pods	428 green, 152 yellow	2.82:1
Axial flowers × terminal flowers	651 axial, 207 terminal	3.14:1

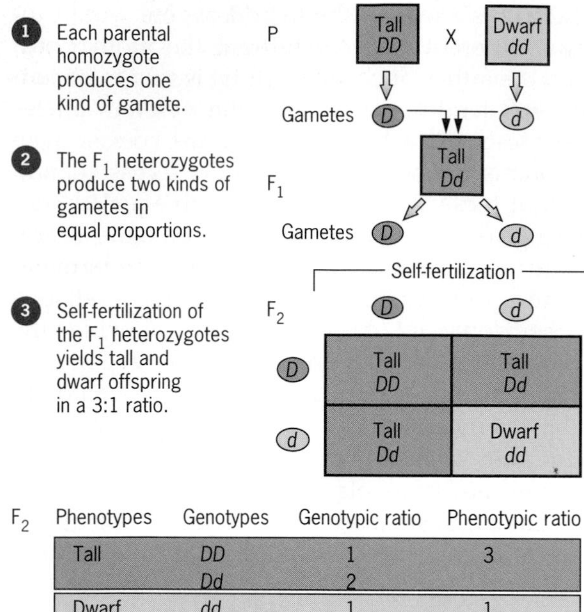

Figure 3.2 Symbolic representation of the cross between tall and dwarf peas and a summary of the phenotypic and genotypic results.

By contrast, the appearance of each strain—that is, the tall or dwarf characteristic—is said to be its **phenotype.**

As the **parental** strains, the tall and dwarf pea plants form the **P** generation of the experiment. Their hybrid progeny are referred to as the first **filial** generation, or $\mathbf{F_1}$, from a Latin word meaning "son" or "daughter." Because each parent contributes equally to its offspring, the genotype of the F_1 plants must be *Dd;* that is, they are heterozygous for the alleles of the gene that controls plant height. Their phenotype, however, is the same as that of the *DD* parental strain because *D* is dominant over *d*. During meiosis, these F_1 plants produce two kinds of gametes, *D* and *d*, in equal proportions. Neither allele is changed by having coexisted with the other in a heterozygous genotype; rather, they separate, or **segregate,** from each other during gamete formation. This process of allele segregation is perhaps the most important discovery that Mendel made.

Upon self-fertilization, the two kinds of gametes produced by heterozygotes can unite in all possible ways. Thus, they produce four kinds of zygotes: *DD* (the contribution of the egg is usually written first), *Dd*, *dD*, and *dd*. However, because of dominance, three of these genotypes have the same phenotype. Thus, in the next generation, called the $\mathbf{F_2}$, the plants are either tall or dwarf, in a ratio of 3:1.

Mendel took this analysis one step further. The F_2 plants were self-fertilized to produce an F_3. All the dwarf F_2 plants produced only dwarf offspring, demonstrating that they were homozygous for the *d* allele, but the tall F_2 plants comprised two categories. Approximately one-third of them produced only tall offspring, whereas the other two-thirds produced a mixture of tall and dwarf offspring. Mendel concluded—correctly—that the third that were true-breeding were *DD* homozygotes and that the two-thirds that were segregating were *Dd* heterozygotes. These proportions, 1/3 and 2/3, were exactly what his analysis predicted because, among the tall F_2 plants, the *DD* and *Dd* genotypes occur in a ratio of 1:2.

We summarize Mendel's analysis of this and other monohybrid crosses by stating two key principles that he discovered:

1. **The Principle of Dominance:** *In a heterozygote, one allele may conceal the presence of another*. This principle is a statement about genetic function. Some alleles evidently control the phenotype even when they are present in a single copy. We consider the physiological explanation for this phenomenon in later chapters.

2. **The Principle of Segregation:** *In a heterozygote, two different alleles segregate from each other during the formation of gametes*. This principle is a statement about genetic transmission. An allele is transmitted faithfully to the next generation, even if it was present with a different allele in a heterozygote. The biological basis for this phenomenon is the pairing and subsequent separation of homologous chromosomes during meiosis, a process we discussed in Chapter 2. We will consider the experiments that led to this chromosome theory of heredity in Chapter 6.

Dihybrid Crosses: The Principle of Independent Assortment

Mendel also performed experiments with plants that differed in two traits (Figure 3.3). He crossed plants that produced yellow, round seeds with plants that produced green, wrinkled seeds. The purpose of the experiments was to see if the two seed traits, color and texture, were inherited independently. Because the F_1 seeds were all yellow and round, the alleles for these two characteristics were dominant. Mendel grew plants from these seeds and allowed them to self-fertilize. He then classified the F_2 seeds and counted them by phenotype.

The four phenotypic classes in the F_2 represented all possible combinations of the color and texture traits. Two classes—yellow, round and green, wrinkled—resembled the parental strains. The other two—green, round and yellow, wrinkled—showed new combinations of traits. Mendel noticed that the four classes appeared in an approximate ratio of 9 yellow, round: 3

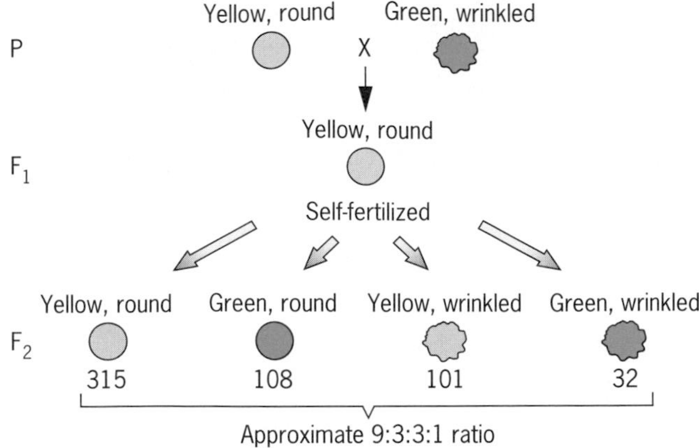

Figure 3.3 Mendel's crosses between peas that produced yellow, round seeds and peas that produced green, wrinkled seeds.

green, round: 3 yellow, wrinkled: 1 green, wrinkled (Figure 3.3). To his insightful mind, these numerical relationships suggested a simple explanation: Each trait was controlled by a different gene segregating two alleles, and the two genes were inherited independently.

Let us analyze the results of this two-factor, or **dihybrid cross,** using Mendel's methods. We denote each gene with a letter, using lower case for the recessive allele and upper case for the dominant (Figure 3.4). For the seed color gene, the two alleles are *g* (for green) and *G* (for yellow), and for the seed texture gene, they are *w* (for wrinkled) and *W* (for round). The parental strains, which were true-breeding, must have been doubly homozygous; the yellow, round plants were *GG WW* and the green, wrinkled plants were *gg ww*. Such two-gene genotypes are customarily written by separating pairs of alleles with a space.

The haploid gametes produced by a diploid plant contain one copy of each gene. Gametes from *GG WW* plants therefore contain one copy of the seed color gene (the *G* allele) and one copy of the seed texture gene (the *W* allele). Such gametes are symbolized by *G W*. By similar reasoning, the gametes from *gg ww* plants are written *g w*. Cross-fertilization of these two types

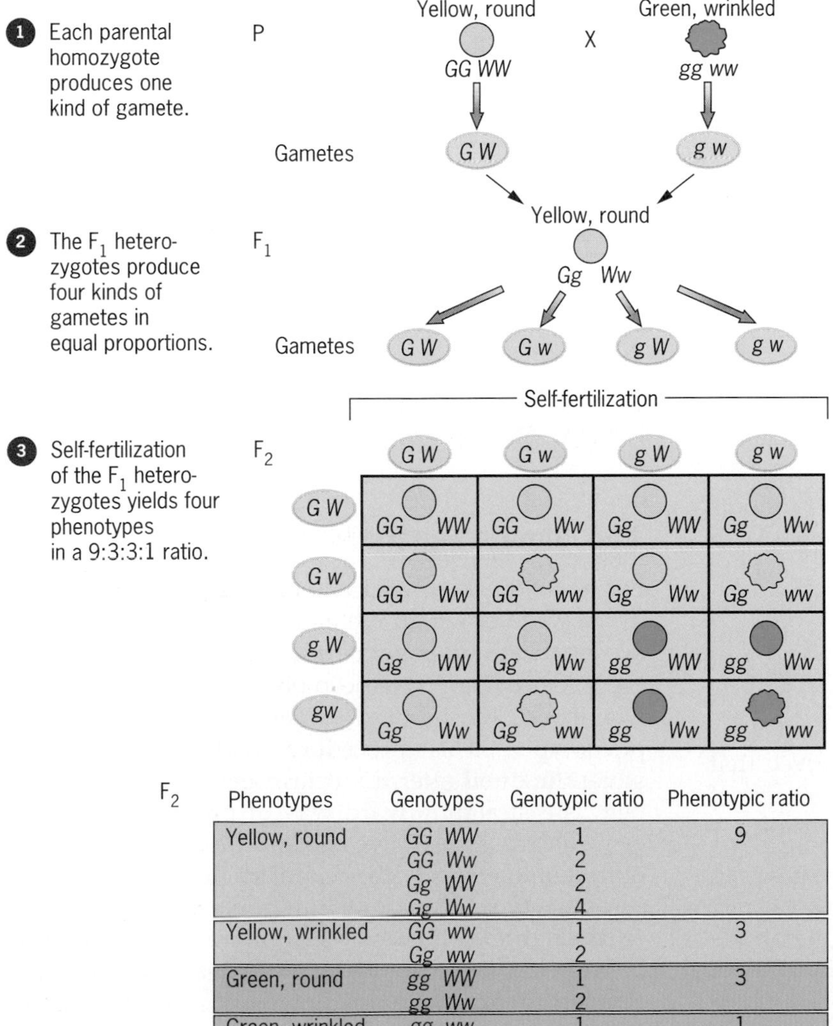

Phenotypes	Genotypes	Genotypic ratio	Phenotypic ratio
Yellow, round	GG WW	1	9
	GG Ww	2	
	Gg WW	2	
	Gg Ww	4	
Yellow, wrinkled	GG ww	1	3
	Gg ww	2	
Green, round	gg WW	1	3
	gg Ww	2	
Green, wrinkled	gg ww	1	1

Figure 3.4 Symbolic representation of the results of a cross between a variety of peas with yellow, round seeds and a variety with green, wrinkled seeds.

of gametes produces F_1 hybrids that are doubly heterozygous, symbolized by *Gg Ww*, and their yellow, round phenotype indicates that the *G* and *W* alleles are dominant.

The Principle of Segregation predicts that the F_1 hybrids will produce four different gametic genotypes: (1) *G W*, (2) *G w*, (3) *g W*, and (4) *g w*. If each gene segregates its alleles independently, these four types will be equally frequent; that is, each will be 25 percent of the total. On this assumption, self-fertilization in the F_1 will produce an array of 16 equally frequent zygotic genotypes. We obtain the zygotic array by systematically combining the gametes, as shown in Figure 3.4. We then obtain the phenotypes of these F_2 genotypes by noting that *G* and *W* are the dominant alleles. Altogether, there are four distinguishable phenotypes, with relative frequencies indicated by the number of positions occupied in the array. For absolute frequencies, we divide each number by the total, 16:

yellow, round	9/16
yellow, wrinkled	3/16
green, round	3/16
green, wrinkled	1/16

This analysis is predicated on two assumptions: (1) that each gene segregates its alleles and (2) that these segregations are independent of each other. The second assumption implies that there is no connection or linkage between the segregation events of the two genes. For example, a gamete that receives *W* through the segregation of the texture gene is just as likely to receive *G* as it is to receive *g* through the segregation of the color gene.

Do the experimental data fit with the predictions of our analysis? Figure 3.5 compares the predicted and observed frequencies of the four F_2 phenotypes in two ways—by proportions and by numerical frequencies. For the numerical frequencies, we calculate the predicted numbers by multiplying the predicted proportion by the total number of F_2 seeds examined. With either method, there is obviously good agreement between the observations and the predictions. Thus, the assumptions on which we have built our analysis—independent segregation of the seed color and seed texture genes—are consistent with the observed data.

Mendel conducted similar experiments with other combinations of traits and in each case observed that the genes segregated independently. The results of these experiments led him to a third key principle:

3. **The Principle of Independent Assortment:** *The alleles of different genes segregate, or as we sometimes say, assort, independently of each other.* This principle is another rule of genetic transmission, based, as we will see in Chapter 6, on the behavior of different pairs of chromosomes during meiosis. However, not all genes abide by the Principle of Independent Assortment. In Chapters 8 and 9 we consider some important exceptions.

	Observed		Expected	
F_2 phenotypes	Number	Proportion	Number	Proportion
Yellow, round	315	0.567	313	0.563
Green, round	108	0.194	104	0.187
Yellow, wrinkled	101	0.182	104	0.187
Green, wrinkled	32	0.057	35	0.063
Total	**556**	**1.000**	**556**	**1.000**

Figure 3.5 Comparison of observed and expected results in the F_2 of Mendel's experiment involving the genes for seed color and seed texture in peas.

Key Points: **Mendel's experiments established three basic genetic principles: (1) Some alleles are dominant, others recessive. (2) During gamete formation, different alleles segregate from each other. (3) Different genes assort independently.**

APPLICATIONS OF MENDEL'S PRINCIPLES

If the genetic basis of a trait is known, Mendel's principles can be used to predict the outcome of crosses. There are three general procedures, two relying on the systematic enumeration of all the zygotic genotypes or phenotypes and one relying on mathematical insight.

The Punnett Square Method

For situations involving one or two genes, it is possible to write down all the gametes and combine them systematically to generate the array of zygotic genotypes. Once these have been obtained, the Principle of Dominance can be used to determine the associated phenotypes. This procedure, called the **Punnett square method** after the British geneticist R. C. Punnett, is a straightforward way of predicting the outcome of crosses. We have used it to analyze the zygotic output of the cross with Mendel's yellow, round F_1 hybrids—a type of mating commonly called an **intercross** (Figure 3.4). However, in more complicated situations, like those involving more than two genes, the Punnett square method is unwieldy.

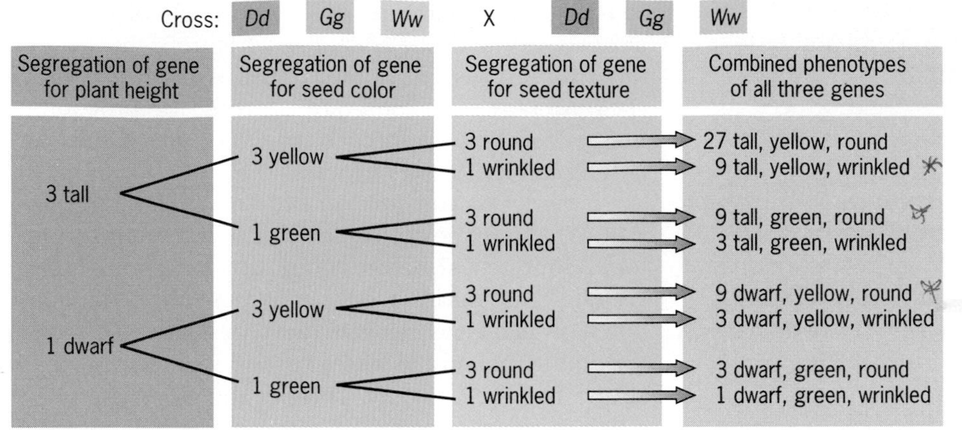

Figure 3.6 The forked-line method for predicting the outcome of an intercross involving three independently assorting genes in peas.

The Forked-Line Method

Another procedure for predicting the outcome of a cross involving two or more genes is the **forked-line method.** However, instead of enumerating the progeny in a square by combining the gametes systematically, we tally them in a diagram of branching lines.

As an example, let us consider an intercross between peas that are heterozygous for three independently assorting genes—one controlling plant height, one controlling seed color, and one controlling seed texture. This is a trihybrid cross—*Dd Gg Ww* × *Dd Gg Ww*—that can be partitioned into three monohybrid crosses—*Dd* × *Dd*, *Gg* × *Gg*, and *Ww* × *Ww*—because all the genes assort independently. For each gene, we expect the phenotypes to appear in a 3:1 ratio. Thus, for example, *Dd* × *Dd* will produce a ratio of 3 tall plants:1 dwarf plant. Using the forked-line method (Figure 3.6), we can combine these separate ratios into an overall phenotypic ratio for the offspring of the cross.

We can also use this method to analyze the results of a cross between multiply heterozygous individuals and multiply homozygous individuals. This type of cross is called a **testcross.** For example, if *Dd Gg Ww* pea plants are crossed with *dd gg ww* pea plants, we can predict the phenotypes of the progeny by noting that each of the three genes in the heterozygous parent segregates dominant and recessive alleles in a 1:1 ratio, and that the homozygous parent transmits only recessive alleles of these genes. Thus, the genotypes—and ultimately the phenotypes—of the offspring of this cross depend on which alleles the heterozygous parent transmits (Figure 3.7).

The Probability Method

An alternative method to the Punnett Square and forked-line methods—and a quicker one—is based on the principle of **probability** (see Technical Sidelight: The Multiplicative and Additive Rules of Probability). Mendelian segregation is like a coin toss; when a heterozygote produces gametes, half contain one allele and half contain the other. If two segregating heterozygotes are crossed, their gametes are combined randomly, producing the zygotic genotypes (Figure

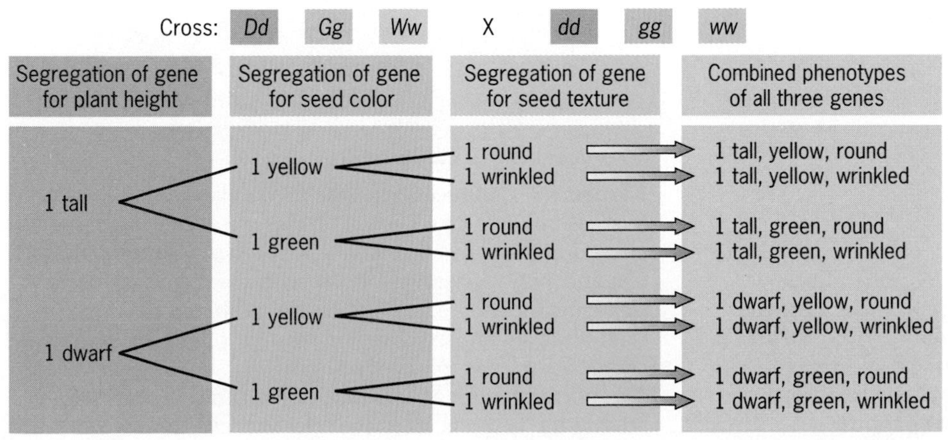

Figure 3.7 The forked-line method for predicting the outcome of a testcross involving three independently assorting genes in peas.

3.8). Let us suppose the cross is $Aa \times Aa$. The chance that a zygote will be AA is simply the probability that each of the uniting gametes contains A, or $(1/2) \times (1/2) = (1/4)$, since the two gametes are produced independently. The chance for an aa homozygote is also 1/4. However, the chance for an Aa heterozygote is 1/2 because there are two ways of creating a heterozygote—A may come from the egg and a from the sperm, or vice versa. Because each of these events has a one-quarter chance of occurring, the total probability that an offspring is heterozygous is $(1/4) + (1/4) = (1/2)$. We therefore obtain the following **probability distribution** of the genotypes from the mating $Aa \times Aa$:

AA	1/4
Aa	1/2
aa	1/4

We conclude that $(1/4) + (1/2) = (3/4)$ of the progeny will have the dominant phenotype and 1/4 will have the recessive.

For such a simple situation, use of the probability method may seem unnecessary. However, in more complicated situations, it is clearly the most practical approach to predict the outcome of crosses. Consider, for example, a cross between plants heterozygous for four different genes, each assorting independently. What fraction of the progeny will be homozyous for all four recessive alleles? To answer this question, we consider the genes one at a time. For the first gene, the fraction of offspring that will be recessive homozygotes is 1/4, as it will be for the second, third, and fourth genes. Therefore, by the Principle of Independent Assortment, the fraction of offspring that will be quadruple recessive homozygotes is $(1/4) \times (1/4) \times (1/4) \times (1/4) = (1/256)$. Surely, using the probability method is a better approach than diagramming a Punnett Square with 256 entries!

Now let us consider an even more difficult question. What fraction of the offspring will be homozygous for all four genes? Before computing any probabilities, we must first decide what genotypes satisfy the question. For each gene there are two types of homozygotes, the dominant and the recessive, and together they constitute half the progeny. The fraction of progeny that will be homozygous for all four genes will therefore be $(1/2) \times (1/2) \times (1/2) \times (1/2) = (1/16)$.

To see the full power of the probability method, we need to consider one more question. Suppose the cross is $Aa\ Bb \times Aa\ Bb$ and we want to know what fraction of the progeny will show the recessive phenotype for at least one gene (Figure 3.9). Three kinds of genotypes would satisfy this condition: (1) *A- bb* (the dash stands for either *A* or *a*), (2) *aa B-*, and (3) *aa bb*. The answer to the question must therefore be the sum of the probabilities corresponding to each of these genotypes. The probability for *A- bb* is $(3/4) \times (1/4) = (3/16)$, that for *aa B-* is $(1/4) \times (3/4) = (3/16)$, and that for *aa bb* is $(1/4) \times (1/4) = (1/16)$. Adding these together, we find that the answer is 7/16.

Cross: *Aa* X *Aa*

Male gametes

	A (1/2)	*a* (1/2)
Female gametes *A* (1/2)	*AA* (1/4)	*Aa* (1/4)
Female gametes *a* (1/2)	*aA* (1/4)	*aa* (1/4)

Progeny:

Genotype	Frequency	Phenotype	Frequency
AA	1/4	Dominant	3/4
Aa	1/2		
aa	1/4	Recessive	1/4

Figure 3.8 An intercross showing the probability method in the context of a Punnett Square. The frequency of each genotype from the cross is obtained from the frequencies in the Punnett Square, which are, in turn, obtained by multiplying the frequencies of the two types of gametes produced by the heterozygous parents.

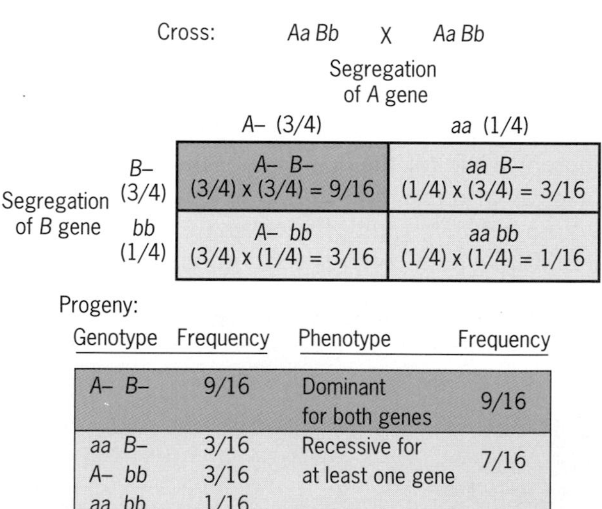

Figure 3.9 Application of the probability method to an intercross involving two genes. In this cross, each gene segregates dominant and recessive phenotypes, with probabilities 3/4 and 1/4, respectively. Because the segregations occur independently, the frequencies of the combined phenotypes within the square are obtained by multiplying the marginal probabilities. The frequency of progeny showing the recessive phenotype for at least one of the genes is obtained by adding the frequencies in the relevant cells (yellow).

TECHNICAL SIDELIGHT

The Multiplicative and Additive Rules of Probability

Probability theory accounts for the frequency of events—for example, the chance of getting a head on a coin toss, drawing an ace from a deck of cards, or obtaining a dominant homozygote from a mating between two heterozygotes. In each case, the event is the outcome of a process—tossing a coin, drawing a card, producing an offspring. To determine the probability of a particular event, we must consider all possible outcomes of the process. The collection of all events is called the *sample space.* For a coin toss, the sample space contains two events, head and tail; for drawing a card, it contains 52, one for each card; and for heterozygotes producing an offspring, it contains three, *GG*, *Gg*, and *gg*. *The probability of an event is the frequency of that event in the sample space.* For example, the probabilities associated with each of the progeny from a mating between two heterozygotes are 1/4 (for *GG*), 1/2 (for *Gg*), and 1/4 (for *gg*).

Two kinds of questions often arise in problems involving probabilities: (1) What is the probability that two events, A and B, will occur together? (2) What is the probability that at least one of two events, A or B, will occur at all? The first question specifies the joint occurrence of two events—A *and* B must occur together to satisfy this question. The second question is less stringent—if *either* A *or* B occurs, the question will be satisfied. A simple diagram can help to explain the different meanings of these two questions.

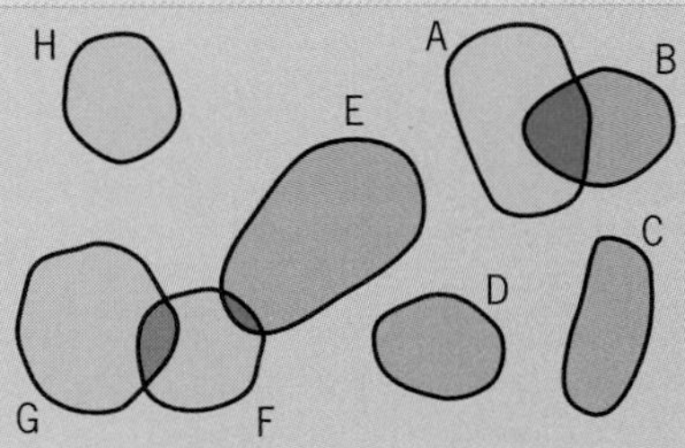

The shapes in the diagram represent events in the sample space, and the sizes of the shapes reflect their relative frequencies. Overlaps between shapes indicate the joint occurrence of two events. If the events do not overlap, then they can never occur together. The first question seeks the probability that both A and B will occur; this probability is represented by the size of the overlap between the two events. The second question seeks the probability that either A or B will occur; this probability is represented by the combined shapes of the two events, including, of course, the overlap between them.

The Multiplicative Rule: If the events A and B are independent, the probability that they occur together, denoted P(A and B), is P(A) × P(B).

Here P(A) and P(B) are the probabilities of the individual events. Note that independent does not mean that they do not overlap in the sample space. In fact, nonoverlapping, or disjoint, events are not independent, for if one occurs, then the other cannot. In probability theory, independent means that one event provides no information about the other. For example, if a card drawn from a deck turns out to be an ace, we have no clue about the card's suit. Thus, drawing the ace of hearts represents the joint occurrence of two independent events—the card is an ace (A) and it is a heart (H). According to the Multiplicative Rule, P(A and H) = P(A) × P(H), and because P(A) = 4/52 and P(H) = 1/4, P(A and H) = (4/52) × (1/4) = 1/52.

The Additive Rule: If the events A and B are independent, the probability that at least one of them occurs, denoted P(A or B), is P(A) + P(B) − [P(A) × P(B)].

Here the term P(A) × P(B), which is the probability that A and B occur together, is subtracted from the sum of the probabilities, P(A) + P(B), because the straight sum includes this term twice. As an example, suppose we seek the probability that a card drawn from a deck is either an ace or a heart. According to the Additive Rule, P(A or H) = P(A) + P(H) − [P(A) × P(H)] = (4/52) + (1/4) − [(4/52) × (1/4)] = 16/52.

If the two events do not overlap in the sample space, the Additive Rule reduces to a simpler expression: P(A or B) = P(A) + P(B). For example, suppose we seek the probability that a card drawn from a deck is either an ace or a king (K). These two events do not overlap in the sample space; they are said to be mutually exclusive. Thus, P(A or K) = P(A) + P(K) = (4/52) + 4(52) = 8/52.

Key Points: **The outcome of a cross can be predicted by the systematic enumeration of the genotypes in a Punnett square. However, when more than two genes are involved, the forked-line or probability methods are used to predict the outcome of a cross.**

FORMULATING AND TESTING GENETIC HYPOTHESES

A scientific investigation always begins with observations of a natural phenomenon. The observations lead to ideas or questions about the phenomenon, and these ideas or questions are explored more fully by conducting further observations or by performing experiments. A well-formulated scientific idea is called an **hypothesis.** Data collected from observations or from experimentation enable scientists to test hypotheses—that is, to determine if a particular hypothesis should be accepted or rejected.

As an example of hypothesis testing in genetics, let us consider the inheritance of flower color in the snapdragon, *Antirrhinum majus*, a popular garden plant (Figure 3.10*a*). Two true-breeding strains were obtained from nursery stock, one with deep red flowers and one with pure white flowers. Do these color differences have a genetic basis? To answer this question, the two strains were crossed (Figure 3.10*b*) to produce F_1 hybrids, which all had pink flowers. When the F_1 hybrids were intercrossed, they produced three kinds of F_2 plants: red (62), pink (131), and white (57), with the actual numbers shown in parentheses. How might we explain the data?

We could hypothesize that flower color is controlled by a single gene with two alleles, *W* (for red) and *w* (for white), and that the flowers of *Ww* heterozygotes are pink because *W* is only partially dominant over *w*. According to this hypothesis, the P generation would be *WW* (red) × *ww* (white), producing F_1 hybrids that would be *Ww* (pink), which, when intercrossed, would yield *WW* (red), *Ww* (pink), and *ww* (white) F_2 progeny in a 1:2:1 ratio. The actual numbers seem to bear this out, giving credence to the hypothesis.

The Chi-Square Test

We may ask whether data really do support a particular hypothesis. This question is critical, since the value of a hypothesis depends on its ability to explain the data. A hypothesis that does not fit needs to be modified or discarded altogether in favor of something better. One procedure for testing the fit between the predictions of a hypothesis and actual data uses a statistic called **chi-square (χ^2).** A statistic is a number calculated from data—for example, the mean of a set of examination scores. The χ^2 statistic allows a researcher to compare data, such as the numbers we get from a breeding experiment, with their predicted values. If the comparison is unfavorable—that is, the data are not in line with the predicted values—the χ^2 statistic will exceed a critical number and the genetic hypothesis will be rejected. If the χ^2 statistic is below this number, the hypothesis will be accepted. The χ^2 statistic therefore reduces hypothesis testing to a simple, objective procedure.

As an example, let us consider the snapdragon crosses described earlier. The F_2 data seemed to be consistent with the hypothesis that a single gene is segregating two alleles. However, to evaluate this hypothesis objectively we need to compare the data with their predicted values. Figure 3.11 illustrates the calculations.

The procedure is straightforward. For each phenotypic class, we compute the expected number of off-

Figure 3.10 (*a*) Snapdragons, *Antirrhinum majus.* (*b*) Results of crosses between red and white varieties of snapdragons.

F_2 Phenotype	Observed number	Expected number
Red	62	(1/4) x 250 = 62.5
Pink	131	(1/2) x 250 = 125
White	57	(1/4) x 250 = 62.5
Total	**250**	**250**

Calculation of chi-square statistic to test for agreement between observed and expected numbers:

$$\chi^2 = \sum \frac{(\text{Observed} - \text{Expected})^2}{\text{Expected}}$$

$$= \frac{(62 - 62.5)^2}{62.5} + \frac{(131 - 125)^2}{125} + \frac{(57 - 62.5)^2}{62.5}$$

$$= 0.776$$

Figure 3.11 Comparison of the observed and expected results and calculation of χ^2 for an intercross with hybrid snapdragons.

spring by multiplying the Mendelian proportion and the total sample size. We then compute the difference between the observed and expected numbers and square these differences to eliminate the canceling effects of positive and negative values. After dividing each squared difference by the corresponding expected number of offspring, we sum all the terms and refer the resulting χ^2 statistic to a distribution of χ^2 values (Figure 3.12).

The distribution of χ^2 values, which is established by statistical theory, shows how often χ^2 will exceed a particular value solely by chance. Statisticians recommend focusing on the value that cuts off the upper 5 percent of the distribution. If the hypothesis is correct, the χ^2 statistic will exceed this **critical value** 5 percent of the time. However, if the hypothesis is incorrect, there is a much greater chance that χ^2 will exceed the critical value. An incorrect hypothesis is more likely to produce large differences between observations and expectations. When such large differences are squared, they will produce a large value of the χ^2 statistic, one situated to the right on the theoretical scale. It is customary to reject the hypothesis if the χ^2 statistic exceeds the critical value. Thus, if the hypothesis is true, there is a 5 percent chance of erroneously rejecting it.

To return to the example: The computed χ^2, 0.776, must be compared to the critical value from a theoretical distribution. It turns out, however, that there are many such distributions, and to select the appropriate one, we need to know the **degrees of freedom** associ-

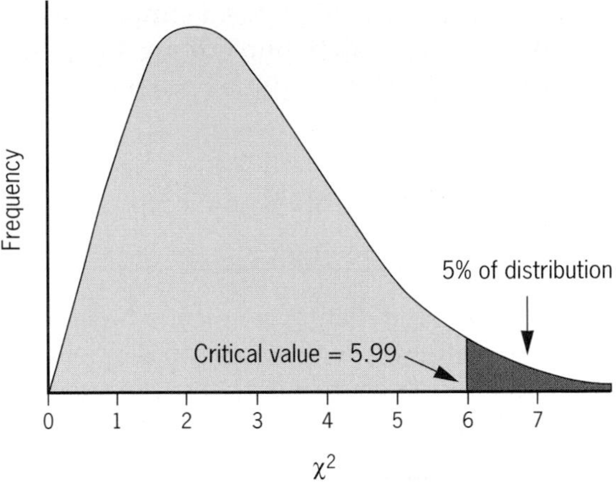

Figure 3.12 Distribution of a χ^2 statistic.

ated with the χ^2 statistic. This index to the set of χ^2 distributions is determined by subtracting one from the number of phenotypic classes; in this example, the number of degrees of freedom is $3-1 = 2$. We can now compare the computed χ^2 statistic with the critical value from the theoretical distribution with 2 degrees of freedom (see Table 3.2 for a list of critical values). Because the computed statistic, 0.776, is less than the critical value, 5.991, the hypothesis of one gene segregating two alleles is *not* rejected. We conclude that this hypothesis is an adequate explanation for the data.

A worked problem at the end of the chapter shows what happens when the χ^2 statistic is greater than the critical value. Other problems provide opportunities to use the χ^2 statistic.

TABLE 3.2
Table of Chi-Square (χ^2) 5% Critical Values[a]

Degrees of Freedom	*5% Critical Value*
1	3.841
2	5.991
3	7.815
4	9.488
5	11.070
6	12.592
7	14.067
8	15.507
9	16.919
10	18.307
15	24.996
20	31.410
25	37.652
30	43.773

[a]Selected entries from R. A. Fisher and Yates, 1943, *Statistical Tables for Biological, Agricultural and Medical Research.* Oliver and Boyd, London.

Key Points: **The chi-square test is a simple way of evaluating whether the predictions of a genetic hypothesis agree with data from an experiment.**

MENDELIAN PRINCIPLES IN HUMAN GENETICS

The application of Mendelian principles to human genetics began soon after the rediscovery of Mendel's paper in 1900. However, because it is not possible to make controlled crosses with human beings, progress was obviously slow. The genetic analysis of human heredity depends on family records, which are often incomplete. In addition, human beings—unlike experimental organisms—do not produce many progeny, making it difficult to discern Mendelian ratios. Mistaken paternity is another problem in human genetics, introducing an element of confusion into the data. Time is also a factor, because some genetic conditions do not manifest themselves until a person reaches middle age. For all these reasons, human genetic analysis has been a difficult endeavor. Nonetheless, the drive to understand human heredity has been very strong, and today, despite all the obstacles, we have learned about thousands of human genes. Table 3.3 lists some of the conditions they control. We discuss many of these conditions in later chapters of this book.

TABLE 3.3
Inherited Conditions in Human Beings

Dominant Traits
Achondroplasia (dwarfism)
Brachydactyly (short fingers)
Congenital night blindness
Ehler-Danlos syndrome (a connective tissue disorder)
Huntington's disease (a neurological disorder)
Marfan syndrome (tall, gangly stature)
Neurofibromatosis (tumorlike growths on the body)
Phenylthiocarbamide (PTC) tasting
Widow's peak
Woolly hair
Recessive Traits
Albinism (lack of pigment)
Alkaptonuria (a disorder of amino acid metabolism)
Ataxia telangiectasia (a neurological disorder)
Cystic fibrosis (a respiratory disorder)
Duchenne muscular dystrophy
Galactosemia (a disorder of carbohydrate metabolism)
Glycogen storage disease
Phenylketonuria (a disorder of amino acid metabolism)
Sickle-cell anemia (a hemoglobin disorder)
Tay-Sachs disease (a lipid storage disorder)

Pedigrees

Pedigrees are diagrams that show the relationships among the members of a family (Figure 3.13*a*). It is customary to represent males as squares and females as circles. A horizontal line connecting a circle and a square represents a mating. The offspring of the mating are shown beneath the mates, starting with the first born at the left and proceeding through the birth order to the right. Individuals that have a genetic condition are indicated by coloring or shading. The generations in a pedigree are usually denoted by Roman

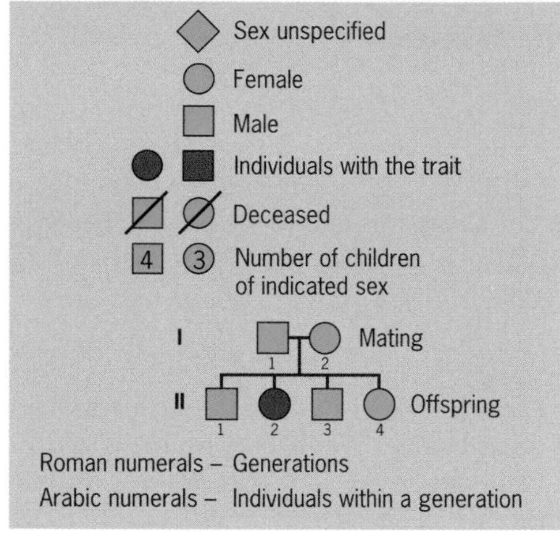

(*a*) Pedigree conventions

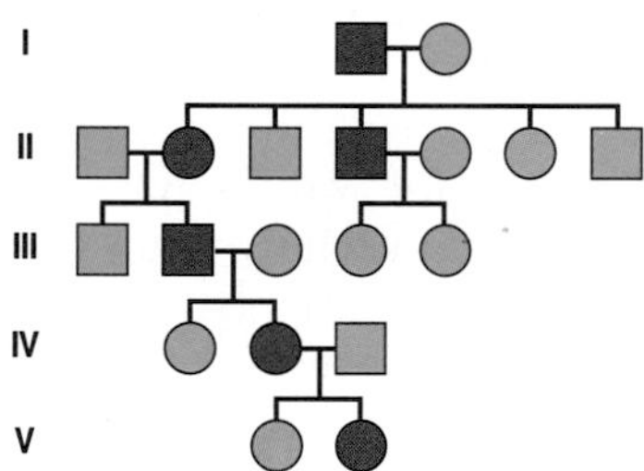

(*b*) Dominant trait

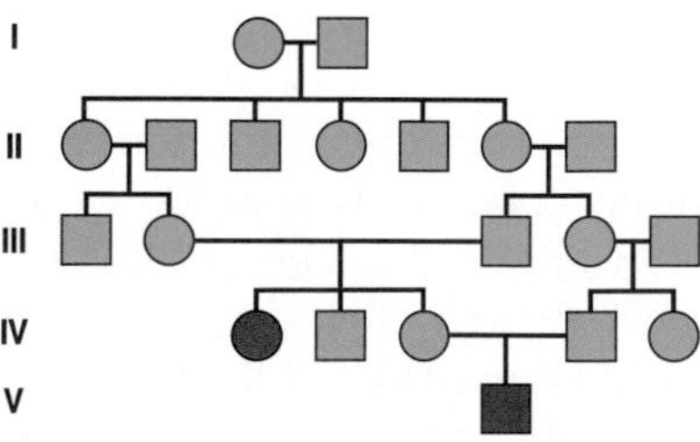

(*c*) Recessive trait

Figure 3.13 (*a*) Pedigree conventions. (*b*) Inheritance of a dominant trait. The trait appears in each generation. (*c*) Inheritance of a recessive trait. The two affected individuals are the offspring of relatives.

numerals, and particular individuals within a generation are referred to by Arabic numerals following the Roman numeral.

Traits caused by dominant alleles are the easiest to identify. Usually, every individual who carries the dominant allele manifests the trait, making it possible to trace the transmission of the dominant allele through the pedigree (Figure 3.13*b*). Every affected individual is expected to have at least one affected parent, unless, of course, the dominant allele has just appeared in the family as a result of a new mutation—a change in the gene itself. However, the frequency of new mutants is very low—on the order of one in a million; consequently, the spontaneous appearance of a dominant condition is an extremely rare event. Dominant traits that are associated with reduced viability or fertility never become frequent in a population. Thus most of the people who show such traits are heterozygous for the dominant allele. If their spouses do not have the trait, half their children should inherit the condition.

Recessive traits are not so easy to identify because they may occur in individuals whose parents are not affected. Sometimes several generations of pedigree data are needed to trace the transmission of a recessive allele (Figure 3.13*c*). Nevertheless, a large number of recessive traits have been observed in human beings—at last count, over 4000. Rare recessive traits are more likely to appear in a pedigree when spouses are related to each other—for example, when they are first cousins. This increased incidence occurs because relatives share alleles by virtue of their common ancestry. Siblings share one-half their alleles, half siblings one-fourth their alleles, and first cousins one-eighth their alleles. Thus, when such relatives mate, they have a greater chance of producing a child who is homozygous for a particular recessive allele than do unrelated parents. Many of the classical studies in human genetics have relied on the analysis of matings between relatives, principally first cousins. We will consider this subject in more detail in Chapter 5.

Mendelian Segregation in Human Families

In human beings, the number of children produced by a couple is typically small. Today in the United States, the average is around two. In developing countries, it is six to seven. Such numbers provide nothing close to the statistical power that Mendel had in his experiments with peas. Consequently, phenotypic ratios in human families often deviate significantly from their Mendelian expectations.

As an example, let us consider a couple who are each heterozygous for a recessive allele that, in homozygous condition, causes cystic fibrosis. If the couple were to have four children, would we expect exactly three to be normal and one to be affected by cystic fibrosis? The answer is no. Although this is a possible outcome, it is not the only one. There are, in fact, five distinct possibilities:

1. Four normal, none affected.
2. Three normal, one affected.
3. Two normal, two affected.
4. One normal, three affected.
5. None normal, four affected.

Intuitively, the second outcome seems to be the most likely, since it conforms to Mendel's 3:1 ratio. We can calculate the probability of this outcome, and of each of the others, by using Mendel's principles and by treating each birth as an independent event (Figure 3.14).

For a particular birth, the chance that the child will be normal is 3/4. The probability that all four children will be normal is therefore $(3/4) \times (3/4) \times (3/4) \times (3/4) = (3/4)^4 = 81/256$. Similarly, the chance that a particular child will be affected is 1/4; thus, the probability that all four will be affected is $(1/4)^4 = 1/256$. To find the probabilities for the three other outcomes,

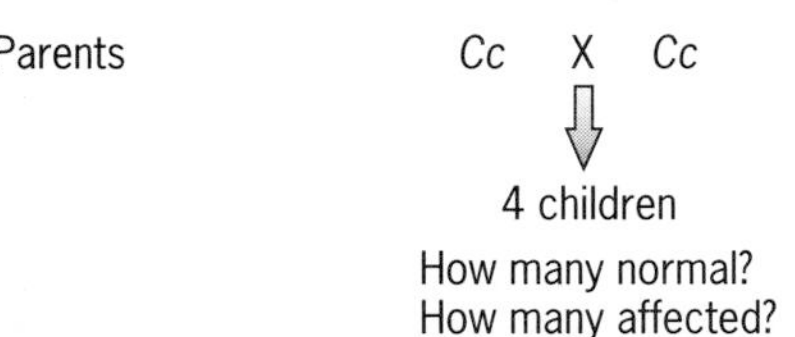

Number of children that are:

Normal	Affected	Probability
4	0	1 x (3/4) x (3/4) x (3/4) x (3/4) = 81/256
3	1	4 x (3/4) x (3/4) x (3/4) x (1/4) = 108/256
2	2	6 x (3/4) x (3/4) x (1/4) x (1/4) = 54/256
1	3	4 x (3/4) x (1/4) x (1/4) x (1/4) = 12/256
0	4	1 x (1/4) x (1/4) x (1/4) x (1/4) = 1/256

Probability distribution

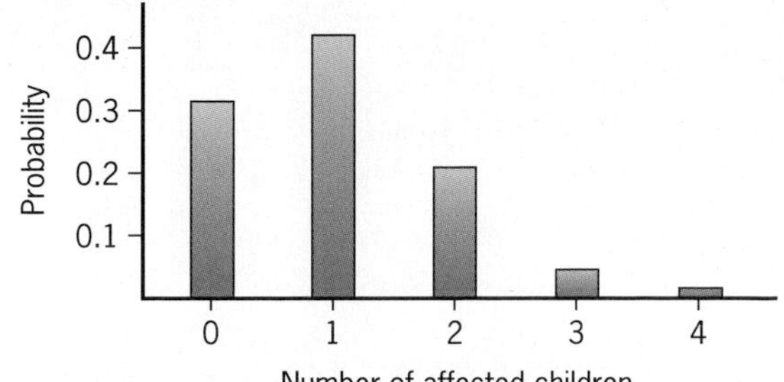

Figure 3.14 Probability distribution of families with four children segregating a recessive trait.

we need to recognize that each actually represents a collection of distinct events. The outcome of three normal children and one affected child, for instance, comprises four distinct events; if we let N symbolize a normal child and A an affected child, and if we write the children in their order of birth, we can represent these events as

NNNA, NNAN, NANN, and ANNN

Because each has probability $(3/4)^3 \times (1/4)$, the total probability for three normal children and one affected, regardless of birth order, is $4 \times (3/4)^3 \times (1/4)$. The coefficient 4 is the number of ways in which three children could be normal and one could be affected in a family with four children. Similarly, the probability for two normal children and two affected is $6 \times (3/4)^2 \times (1/4)^2$, since in this case there are six distinct events. The probability for one normal child and three affected is $4 \times (3/4) \times (1/4)^3$, since in this case there are four distinct events. Figure 3.14 summarizes the calculations in the form of a probability distribution. As expected, three normal children and one affected child is the most probable outcome (probability 108/256). However, this outcome is not expected a majority of the time, because the other four outcomes have a combined probability of 148/256. The Technical Sidelight: Binomial Probabilities generalizes this procedure to other situations in which the children fall into two possible phenotypic classes. Because there are only two classes, the probabilities associated with the outcomes are called **binomial probabilities.**

Genetic Counseling

The diagnosis of genetic conditions is often a difficult process. Typically, diagnoses are made by physicians who have been trained in genetics. The study of these conditions requires a great deal of careful research, including examining patients, interviewing relatives, and sifting through vital statistics on births, deaths, and marriages. The accumulated data provide the basis for defining the condition clinically and for determining its mode of inheritance.

Prospective parents may want to know whether their children are at risk to inherit a particular condition, especially if other family members have been affected. It is the responsibility of the genetic counselor to assess such risks and to explain them to the prospective parents. Risk assessment requires familiarity with probability and statistics, as well as a thorough knowledge of genetics.

As an example, let us consider a pedigree showing the inheritance of a rare form of **dwarfism** (Figure 3.15). The condition is manifested in all but the first generation, and every affected individual except the very first has an affected parent. This pattern strongly argues that the condition is caused by a dominant allele. The dwarfed woman in generation II probably represents a rare new mutation in the population.

The counseling issue arises in generation V. What is the chance that the dwarfed man will produce a dwarfed child? To answer this question we need to know the genotypes of the prospective parents. Because the condition is caused by a dominant allele, the

(a)

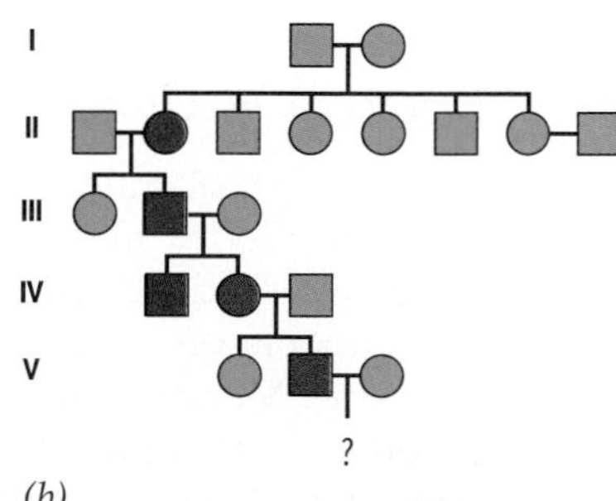

(b)

Figure 3.15 (*a*) Dwarfism illustrated by a family of seven siblings. The trait is caused by a dominant allele. (*b*) Pedigree showing the inheritance of dwarfism through five generations. The affected woman in generation II appears to be a new mutant because neither of her parents was affected.

TECHNICAL SIDELIGHT

Binomial Probabilities

The progeny of crosses sometimes segregate into two distinct classes—for example, male or female, healthy or diseased, normal or mutant, dominant phenotype or recessive phenotype. To be general, we can refer to these two kinds of progeny as P and Q, and note that for any individual offspring, the probability of being P is p and the probability of being Q is q. Because there are only two classes, $q = 1 - p$. Suppose that the total number of progeny is n and that each one is produced independently. We can calculate the **binomial probability** that exactly x of the progeny will fall into one class and y into the other:

Probability of x in class P and y in class Q =

$$\left[\frac{n!}{x!\ y!}\right] p^x q^y$$

The bracketed term contains three factorial functions ($n!$, $x!$, and $y!$), each of which is computed as a descending series of products. For example, $n! = n\ (n-1)\ (n-2)\ (n-3) \ldots (3)\ (2)\ (1)$. If 0! is needed, it is defined as one. In the formula, the bracketed term enumerates the different ways, or orders, in which n offspring can be segregated so that x fall in the P class and y fall in the Q class. The other term, $p^x\ q^y$, gives the probability of obtaining a particular way or order. Because each of the orders is equally likely, multiplying this term by the bracketed term gives the probability of obtaining x progeny in the P class and y in the Q class, regardless of the order of occurrence.

If, for fixed values of n, p, and q, we systematically vary x and y, we can calculate a whole set of probabilities. This set constitutes a binomial probability distribution. With the distribution, we can answer questions such as "What is the probability that x will exceed a particular value?" or "What is the probability that x will lie between two particular values?" For example, let us consider a family with six children. What is the probability that at least four will be girls? To answer this question, we note that for any given child, the probability that it will be a girl (p) is 1/2 and the probability that it will be a boy (q) is also 1/2. The probability that exactly four children in a family will be girls (and two will be boys) is therefore $[(6!)/(4!\ 2!)](1/2)^4\ (1/2)^2 = 15/64$, which is one of the terms in the binomial distribution. However, the probability that at least four will be girls (and that no more than two will be boys) is the sum of three terms from this distribution:

Event	Binomial Formula	Probability
4 girls and 2 boys	$[(6!)/(4!\ 2!)] \times (1/2)^4\ (1/2)^2 =$	15/64
5 girls and 1 boy	$[(6!)/(5!\ 1!)] \times (1/2)^5\ (1/2)^1 =$	6/64
6 girls and 0 boys	$[(6!)/(6!\ 0!)] \times (1/2)^6\ (1/2)^0 =$	1/64

Therefore, the answer is $(15/64) + (6/64) + (1/64) = 22/64$.

The binomial distribution also provides answers to other kinds of questions. For example, what is the probability that at least one but no more than four of the children will be girls? Here the answer is the sum of four terms:

Event	Binomial Formula	Probability
1 girl and 5 boys	$[(6!)/(1!\ 5!)] \times (1/2)^1\ (1/2)^5 =$	6/64
2 girls and 4 boys	$[(6!)/(2!\ 4!)] \times (1/2)^2\ (1/2)^4 =$	15/64
3 girls and 3 boys	$[(6!)/(3!\ 3!)] \times (1/2)^3\ (1/2)^3 =$	20/64
4 girls and 2 boys	$[(6!)/(4!\ 2!)] \times (1/2)^4\ (1/2)^2 =$	15/64

Summing up, we find that the answer is 56/64.

mother must be homozygous for the normal (recessive) allele (*dd*) and the father must be heterozygous for the dwarfism allele (*Dd*). The chance that the couple will have a dwarfed child is therefore 1/2.

As another example, consider the situation shown in Figure 3.16. A couple, denoted R and S, is concerned about the possibility that they will have a child (T) with **albinism**, an autosomal recessive condition characterized by a complete absence of melanin pigment in the skin, eyes, and hair. S, the prospective father, is an albino, and R, the prospective mother, has two albino siblings. It would therefore seem that the child has some risk of being born albino.

This risk depends on two factors: (1) the probability that R is a heterozygous carrier of the albinism allele (*a*), and (2) the probability that she will transmit this allele to T if she actually is a carrier. S, who is obviously homozygous for the albinism allele, must transmit this allele to his offspring.

To determine the first probability, we need to consider the possible genotypes for R. One of these, that she is homozygous for the recessive allele (*aa*), is excluded because we know that she is not an albino. However, the other two genotypes, *AA* and *Aa*, remain distinct possibilities. To calculate the probabilities associated with each of these, we note that both of R's

(a)

R S
T

(b)

Normal Aa X Normal Aa

	A	a
A	*AA* Normal	*Aa* Normal
a	*aA* Normal	*aa* Albino

Among nonalbinos, 2/3 are heterozygotes.

(c)

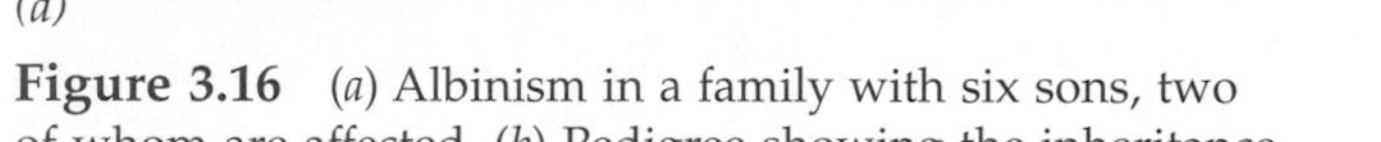

Figure 3.16 (*a*) Albinism in a family with six sons, two of whom are affected. (*b*) Pedigree showing the inheritance of albinism. (*c*) Punnett Square showing that among nonalbinos, the frequency of heterozygotes is 2/3.

parents must be heterozygotes, because they have had two albino children. The mating that produced R was therefore $Aa \times Aa$, and from such a mating we would expect 2/3 of the *nonalbino* offspring to be *Aa* and 1/3 to be *AA*. Thus, the probability that R is a heterozygous carrier of the albinism allele is 2/3. To determine the probability that she will transmit this allele to her child, we simply note that *a* will be present in half of her gametes.

In summary,

Risk that T is *aa*

= [Probability that R is *Aa*] × [Probability that R transmits *a*, assuming that R is *Aa*]

$= (2/3) \times (1/2) = (1/3)$

The examples in Figures 3.15 and 3.16 illustrate simple counseling situations in which the risk can be determined precisely. Often the circumstances are much more complicated, making the task of risk assessment quite difficult. The genetic counselor's responsibility is to analyze the pedigree information and determine the risk as precisely as possible.

Key Points: **Pedigrees are used to identify dominant and recessive traits in human families. The analysis of pedigrees allows genetic counselors to determine the probability that an individual will inherit a particular trait.**

TESTING YOUR KNOWLEDGE

1. Phenylketonuria, a metabolic disease in humans, is caused by a recessive allele, *k*. If two heterozygous carriers of the allele marry and plan a family of five children: (a) What is the chance that all their children will be normal? (b) What is the chance that four children will be normal and one affected with phenylketonuria? (c) What is the chance that at least three children will be normal? (d) What is the chance that the first child will be a normal girl?

ANSWER

Before answering each of the questions, note that from a mating between two heterozygotes, the probability that a particular child will be normal is 3/4, and the probability that a particular child will be affected is 1/4. Furthermore, for any one child born, the chance that it will be a boy is 1/2 and the chance that it will be a girl is 1/2.

(a) To calculate the chance that all five children will be normal, use the Multiplicative Rule. For each child, the chance that it will be normal is 3/4, and all five children are independent. Consequently, the probability of five normal children is $(3/4)^5 = 0.237$. This is the first term of the binomial probability distribution with $p = 3/4$ and $q = 1/4$.

(b) To calculate the chance that four children will be normal and one affected, compute the second term of the binomial distribution:

$$[5!/(4!\ 1!)] \times (3/4)^4 \times (1/4)^1 = 5 \times (81/1024) = 0.399$$

(c) To find the probability that at least three children will be normal, sum the first three terms of the binomial distribution:

Event	Binomial Formula	Probability
5 normal, 0 affected	$[(5!)/(5!\ 0!)] \times (3/4)^5\ (1/4)^0 =$	0.237
4 normal, 1 affected	$[(5!)/(4!\ 1!)] \times (3/4)^4\ (1/4)^1 =$	0.399
3 normal, 2 affected	$[(5!)/(3!\ 2!)] \times (3/4)^3\ (1/4)^2 =$	0.264
	Total	0.900

(d) To determine the probability that the first child will be a normal girl, use the Multiplicative Rule: P(normal child and girl) = P(normal child) × P(girl) = (3/4) × (1/2) = 3/8.

2. Mice from wild populations typically have gray-brown (or *agouti*) fur, but in one laboratory strain, some of the mice have yellow fur. A single yellow male is mated to several agouti females. Altogether, the matings produce 40 progeny, 22 with agouti fur and 18 with yellow fur. The agouti F_1 animals are then intercrossed with each other to produce an F_2, all of which are agouti. Similarly, the yellow F_1 animals are intercrossed with each other, but their F_2 progeny segregate into two classes; 30 are agouti and 54 are yellow. Subsequent crosses between yellow F_2 animals also segregate yellow and agouti progeny. What is the genetic basis of these coat color differences?

ANSWER

We note that the cross agouti × agouti produces only agouti animals and that the cross yellow × yellow produces a mixture of yellow and agouti. Thus, a reasonable hypothesis is that yellow fur is caused by a dominant allele, *Y*, and that agouti fur is caused by a recessive allele, *y*. According to this hypothesis, the agouti females used in the initial cross would be *yy* and their yellow mate would be *Yy*. We hypothesize that the male was heterozygous because he produced approximately equal numbers of agouti and yellow F_1 offspring. Among these, the agouti animals should be *yy* and the yellow animals *Yy*. These genotypic assignments are borne out by the F_2 data, which show that the F_1 agouti mice have bred true and the F_1 yellow mice have segregated. However, the segregation ratio of yellow to agouti (54:30) seems to be out of line with the Mendelian expectation of 3:1. Is this lack of fit serious enough to reject the hypothesis?

We can use the χ^2 procedure to test for disagreement between the data and the predictions of the hypothesis. According to the hypothesis, 3/4 of the F_2 progeny from the yellow × yellow intercross should be yellow and 1/4 should be agouti. Using these proportions, we can calculate the expected numbers of progeny in each class and then calculate a χ^2 statistic with 2 − 1 = 1 degree of freedom.

F_2 Phenotype	Obs	Exp	$(\text{Obs-Exp})^2/\text{Exp}$
yellow (*YY* and *Yy*)	54	(3/4) × 84 = 63	1.286
agouti (*yy*)	30	(1/4) × 84 = 21	3.857
Total	84	84	5.143

The χ^2 statistic (5.143) is much greater than the critical value (3.841) for a χ^2 distribution with 1 degree of freedom. Consequently, we reject the hypothesis that the coat colors are segregating in a 3:1 Mendelian fashion.

What might account for the failure of the coat colors to segregate as hypothesized? We obtain a clue by noting that subsequent yellow × yellow crosses failed to establish a true-breeding yellow strain. This suggests that the yellow animals are all *Yy* heterozygotes and that the *YY* homozygotes produced by matings between heterozygotes do not survive to the adult stage. Embryonic death is, in fact, why the yellow mice are underrepresented in the F_2 data. Examination of the uteruses of pregnant females reveals that about 1/4 of the embryos are dead. These dead embryos must be genotypically *YY*. Thus, a single copy of the *Y* allele produces a visible phenotypic effect (yellow fur), but two copies cause death. Taking this embryonic mortality into account, we can modify the hypothesis and predict that 2/3 of the live-born F_2 progeny should be yellow (*Yy*) and 1/3 should be agouti (*yy*). We can then use the χ^2 procedure to test this modified hypothesis for consistency with the data.

F_2 Phenotype	Obs	Exp	$(\text{Obs-Exp})^2/\text{Exp}$
yellow (*Yy*)	54	(2/3) × 84 = 56	0.071
agouti (*yy*)	30	(1/3) × 84 = 28	0.143
Total	84	84	0.214

This χ^2 statistic is less than the critical value for a χ^2 distribution with 1 degree of freedom. Thus, the data are in agreement with the predictions of the modified hypothesis.

QUESTIONS AND PROBLEMS

3.1 On the basis of Mendel's observations, predict the results from the following crosses with peas: (a) a tall (dominant and homozygous) variety crossed with a dwarf variety; (b) the progeny of (a) self-fertilized; (c) the progeny from (a) crossed with the original tall parent; (d) the progeny of (a) crossed with the original dwarf parent.

3.2 Mendel crossed pea plants that produced round seeds with those that produced wrinkled seeds and self-fertilized the progeny. In the F_2, he observed 5474 round seeds and 1850 wrinkled seeds. Using the letters *W* and *w* for the seed texture alleles, diagram Mendel's crosses, showing the genotypes of the plants in each generation. Are the results consistent with the Principle of Segregation?

3.3 A geneticist crossed wild, gray-colored mice with white (albino) mice. All the progeny were gray. These progeny were intercrossed to produce an F_2, which consisted of 198 gray and 72 white mice. Propose an hypothesis to explain these results, diagram the crosses, and compare the results with the predictions of the hypothesis.

3.4 A woman has a rare abnormality of the eyelids called ptosis, which prevents her from opening her eyes completely.

This condition is caused by a dominant allele, *P*. The woman's father had ptosis, but her mother had normal eyelids. Her father's mother had normal eyelids. (a) What are the genotypes of the woman, her father, and her mother? (b) What proportion of the woman's children will have ptosis if she marries a man with normal eyelids?

3.5 In pigeons, a dominant allele *C* causes a checkered pattern in the feathers; its recessive allele *c* produces a plain pattern. Feather coloration is controlled by an independently assorting gene; the dominant allele *B* produces red feathers, and the recessive allele *b* produces brown feathers. Birds from a true-breeding checkered, red variety are crossed to birds from a true-breeding plain, brown variety. (a) Predict the phenotype of their progeny. (b) If these progeny are intercrossed, what phenotypes will appear in the F_2, and in what proportions?

3.6 In mice, the allele *C* for colored fur is dominant over the allele *c* for white fur, and the allele *V* for normal behavior is dominant over the allele *v* for waltzing behavior, a form of discoordination. Give the genotypes of the parents in each of the following crosses: (a) colored, normal mice mated with white, normal mice produced 29 colored, normal and 10 colored, waltzing progeny; (b) colored, normal mice mated with colored, normal mice produced 38 colored, normal, 15 colored, waltzing, 11 white, normal, and 4 white, waltzing progeny; (c) colored, normal mice mated with white, waltzing mice produced 8 colored, normal, 7 colored, waltzing, 9 white, normal, and 6 white, waltzing progeny.

3.7 In rabbits, the dominant allele *B* causes black fur and the recessive allele *b* causes brown fur; for an independently assorting gene, the dominant allele *R* causes long fur and the recessive allele *r* (for *rex*) causes short fur. A homozygous rabbit with long, black fur is crossed with a rabbit with short, brown fur, and the offspring are intercrossed. In the F_2, what proportion of the rabbits with long, black fur will be homozygous for both genes?

3.8 In shorthorn cattle, the genotype *RR* causes a red coat, the genotype *rr* causes a white coat, and the genotype *Rr* causes a roan coat. A breeder has red, white, and roan cows and bulls. What phenotypes might be expected from the following matings, and in what proportions? (a) red × red; (b) red × roan; (c) red × white; (d) roan × roan.

3.9 How many different kinds of F_1 gametes, F_2 genotypes, and F_2 phenotypes would be expected from the following crosses: (a) *AA* × *aa*; (b) *AA BB* × *aa bb*; (c) *AA BB CC* × *aa bb cc*? (d) What general formulas are suggested by these answers?

3.10 A researcher studied six independently assorting genes in a plant. Each gene has a dominant and a recessive allele: *R* black stem, *r* red stem; *D* tall plant, *d* dwarf plant; *C* full pods, *c* constricted pods; *O* round fruit, *o* oval fruit; *H* hairless leaves, *h* hairy leaves; *W* purple flower, *w* white flower. From the cross (P1) *Rr Dd cc Oo Hh Ww* × (P2) *Rr dd Cc oo Hh ww*, (a) how many kinds of gametes can be formed by P1? (b) How many genotypes are possible among the progeny of this cross? (c) How many phenotypes are possible among the progeny? (d) What is the probability of obtaining the *Rr Dd cc Oo hh ww* genotype in the progeny? (e) What is the probability of obtaining a black, dwarf, constricted, oval, hairy, purple phenotype in the progeny?

3.11 Mendel testcrossed pea plants grown from yellow, round F_1 seeds to plants grown from green, wrinkled seeds and obtained the following results: 31 yellow, round; 26 green, round; 27 yellow, wrinkled; and 26 green, wrinkled. Are these results consistent with the hypothesis that seed color and seed texture are controlled by independently assorting genes, each segregating two alleles?

3.12 For each of the situations below, determine the degrees of freedom associated with the χ^2 statistic and decide whether or not the observed χ^2 value warrants acceptance or rejection of the hypothesized genetic ratio.

	Hypothesized Ratio	Observed χ^2
(a)	3:1	7.0
(b)	1:2:1	7.0
(c)	1:1:1:1	7.0
(d)	9:3:3:1	5.0

3.13 Perform a chi-square test to determine if an observed ratio of 30 tall: 20 dwarf pea plants is consistent with an expected ratio of 1:1 from the cross *Dd* × *dd*.

3.14 Seed capsules of the Shepherd's purse are either triangular or ovoid. A cross between a plant with triangular seed capsules and a plant with ovoid seed capsules yielded F_1 hybrids that all had triangular seed capsules. When these F_1 hybrids were intercrossed, they produced 80 F_2 plants, 72 of which had triangular seed capsules and 8 of which had ovoid seed capsules. Are these results consistent with the hypothesis that capsule shape is determined by a single gene with two alleles?

3.15 Albinism in humans is caused by a recessive allele *a*. From marriages between normally pigmented people known to be carriers (*Aa*) and albinos (*aa*), what proportion of the children would be expected to be albinos? Among three children what is the chance of one normal and two albino?

3.16 If both husband and wife are known to be carriers of the allele for albinism, what is the chance of the following combinations in a family of four children: (a) all four normal; (b) three normal and one albino; (c) two normal and two albino; (d) one normal and three albino?

3.17 In humans, cataracts in the eyes and fragility of the bones are caused by dominant alleles that assort independently. A man with cataracts and normal bones marries a woman without cataracts but with fragile bones. The man's father had normal eyes, and the woman's father had normal bones. What is the probability that the first child of this couple will (a) be free from both abnormalities; (b) have cataracts but not have fragile bones; (c) have fragile bones but not have cataracts; (d) have both cataracts and fragile bones?

3.18 Are the traits shown in the pedigrees on the following page due to a dominant or a recessive allele?

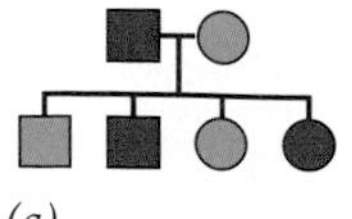

(a)

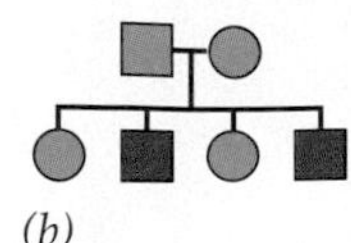

(b)

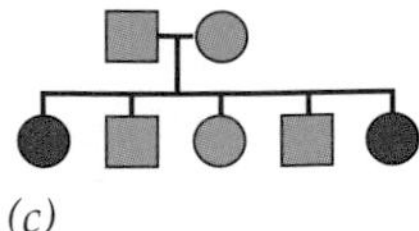

(c)

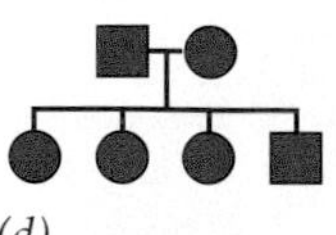

(d)

3.19 If a man and a woman are heterozygous for a gene, and if they have three children, what is the chance that all three will also be heterozygous?

3.20 If four babies are born on a given day: (a) What is the chance that two will be boys and two girls? (b) What is the chance that all four will be girls? (c) What combination of boys and girls among four babies is most likely? (d) What is the chance that at least one baby will be a girl?

3.21 In a family of six children, what is the chance that at least three are girls?

3.22 The pedigree below shows the inheritance of a dominant trait. What is the chance that the offspring of the following matings will show the trait: (a) III-1 × III-3; (b) III-2 × III-4?

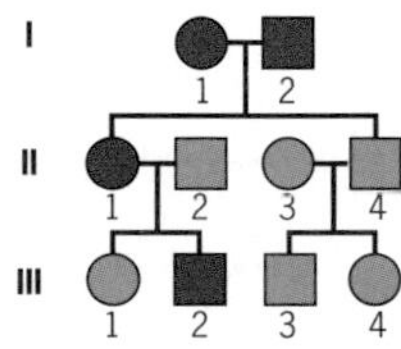

3.23 The pedigree below shows the inheritance of a recessive trait. Unless there is evidence to the contrary, assume that the individuals who have married into the family do not carry the recessive allele. What is the chance that the offspring of the following matings will show the trait: (a) III-1 × III-12; (b) III-4 × III-14; (c) III-6 × III-13; (d) IV-1 × IV-2?

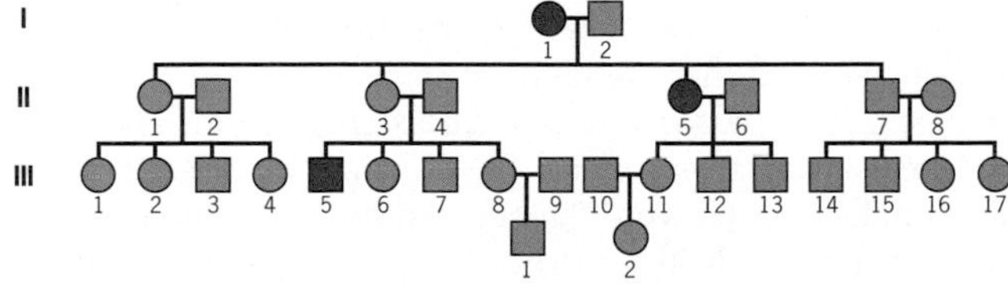

3.24 In the pedigrees below, determine whether the trait is more likely to be due to a dominant or a recessive allele.

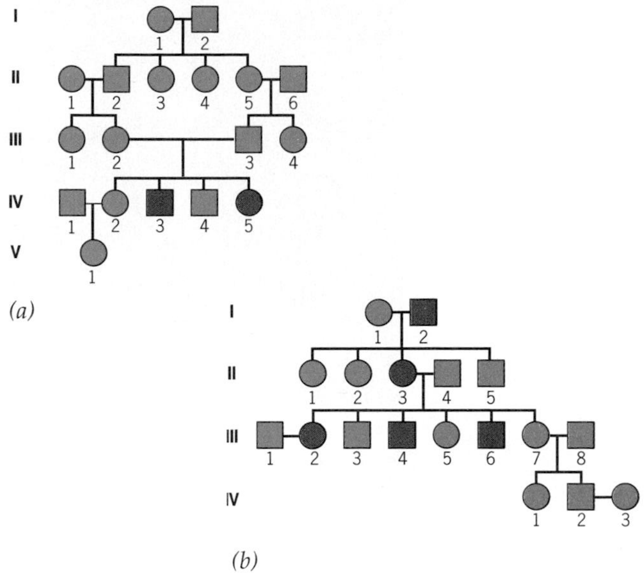

(a)

(b)

3.25 In pedigree (*b*) above, what is the chance that the couple III-1 and III-2 will have an affected child? What is the chance that the couple IV-2 and IV-3 will have an affected child?

3.26 The pedigree below shows the inheritance of a recessive trait. What is the chance that the couple III-3 and III-4 will have an affected child?

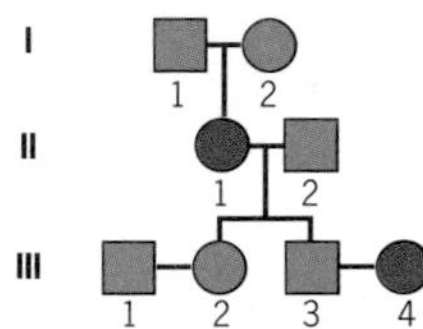

3.27 Peas heterozygous for three independently assorting genes were intercrossed. (a) What proportion of the offspring will be homozygous for all three recessive alleles? (b) What proportion of the offspring will be homozygous for all three genes? (c) What proportion of the offspring will be homozygous for one gene and heterozygous for the other two? (d) What proportion of the offspring will be homozygous for the recessive allele of at least one gene?

BIBLIOGRAPHY

CARLSON, E. A. 1966. *The Gene: A Critical History*. W. B. Saunders, Philadelphia.

DUNN, L. C. 1965. *A Short History of Genetics*. McGraw-Hill, New York.

ILTIS, H. 1932. *Life of Mendel*. (Translation by E. and C. Paul.) W. W. Norton, New York.

PETERS, J. A., ED. 1959. *Classic Papers in Genetics*. Prentice-Hall, Engelwood Cliffs, NJ.

STERN, C. 1973. *Principles of Human Genetics*. W. H. Freeman, San Francisco.

Diverse species of plants growing in a garden. Experiments with many different plants extended Mendel's Principles of Dominance, Segregation, and Independent Assortment.

4 Extensions of Mendelism

CHAPTER OUTLINE

Allelic Variation and Gene Function
Gene Action: From Genotype to Phenotype

Genetics Grows Beyond Mendel's Monastery Garden

In 1902, enthused by what he read in Mendel's paper, the British biologist William Bateson published an English translation of Mendel's German text and appended to it a brief account of what he called "Mendelism"—the Principles of Dominance, Segregation, and Independent Assortment. Later, in 1909, he published *Mendel's Principles of Heredity*, in which he summarized all the evidence then available to support Mendel's findings. This book was remarkable for two reasons. First, it examined the results of breeding experiments with many different plants and animals and in each case demonstrated that Mendel's principles applied. Second, it considered the implications of these experiments and raised questions about the fundamental nature of genes, or, as Bateson called them, "unit-characters." At the time Bateson's book was published, the word "gene" had not yet been invented.

Bateson's book played a crucial role in spreading the principles of Mendelism to the scientific world. Botanists, zoologists, naturalists, horticulturalists, and animal breeders all got the message in plain and simple language: Mendel's principles—tested by experiments with peas, beans, sunflowers, cotton, wheat, barley, tomatoes, maize, and assorted ornamental plants, as well as with cattle, sheep, cats, mice, rabbits, guinea pigs, chickens, pigeons, canaries, and moths—were universal. In the preface to his book, Bateson remarked that "The study of heredity thus becomes an organized branch of physiological science, already abundant in results, and in promise unsurpassed."[1]

[1]Bateson, W. 1909. *Mendel's Principles of Heredity*. University Press, Cambridge, England.

In this chapter we consider some of the early extensions of Mendelian analysis: that a gene can exist in many different allelic states, that a particular gene can affect several different traits, and that a particular trait can be affected by several different genes. We also consider how genetic and environmental factors influence phenotypic variation.

Phenotype	Genotype	Amount of gene product
Red	*WW*	2x
Pink	*Ww*	x
White	*ww*	0

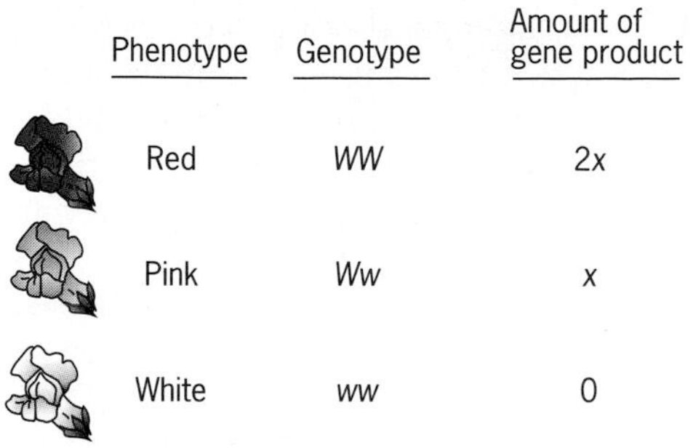

Figure 4.1 Genetic basis of flower color in snapdragons. The allele *W* is incompletely dominant over *w*. Differences among the phenotypes could be due to differences in the amount of the product specified by the *W* allele.

ALLELIC VARIATION AND GENE FUNCTION

Mendel's experiments established that genes can exist in alternate forms. For each of the seven traits that he studied—seed color, seed texture, plant height, flower color, flower position, pod shape, and pod color—Mendel identified two alleles, one dominant, the other recessive. This discovery suggested a simple functional dichotomy between alleles, as if one allele did nothing and the other did everything to determine the phenotype. However, research early in the twentieth century demonstrated this to be an oversimplification. Genes can exist in more than two allelic states, and each allele can have a different effect on the phenotype.

Incomplete Dominance and Codominance

An allele is dominant if it has the same phenotypic effect in heterozygotes as in homozygotes—that is, the genotypes *Aa* and *AA* are phenotypically indistinguishable. Sometimes, however, a heterozygote has a phenotype different from that of either of its associated homozygotes. Flower color in the snapdragon, *Antirrhinum majus*, is an example. White and red varieties are homozygous for different alleles of a color-determining gene; when crossed, they produce heterozygotes that have pink flowers. The allele for red color (*W*) is therefore said to be **incompletely,** or **partially, dominant** over the allele for white color (*w*). The most likely explanation is that the intensity of pigmentation in this species depends on the amount of a product made by the color gene (Figure 4.1). If the *W* allele makes this product and the *w* allele does not, *WW* homozygotes will have twice as much of the product as *Ww* heterozygotes do and will therefore show deeper color. When the heterozygote's phenotype is midway between the phenotypes of the two homozygotes, as it is here, the partially dominant allele is sometimes said to be **semidominant** (from the Latin word for "half"—thus half-dominant).

Another exception to the principle of simple dominance arises when a heterozygote shows characteristics found in each of the associated homozygotes. This occurs with human blood types, which are identified by testing for special cellular products called *antigens*. An antigen is detected by its ability to react with factors obtained from the serum portion of the blood. These factors, which are produced by the immune system, recognize antigens quite specifically. Thus, for example, one serum, called anti-M, recognizes only the M antigen on human blood cells; another serum, called anti-N, recognizes only the N antigen on these cells (Figure 4.2). When one of these sera detects its specific antigen in a blood-typing test, the cells clump together, causing a reaction called *agglutination*. Thus, by testing cells for agglutination with different sera, a medical technologist can identify which antigens are present and thereby determine the blood type.

The ability to produce the M and N antigens is determined by a gene with two alleles. One allele allows the M antigen to be produced; the other allows the N antigen to be produced. Homozygotes for the M allele

Genotype	Blood type (antigen present)	Reactions with anti-sera: Anti-M serum	Anti-N serum
$L^M L^M$	M		
$L^M L^N$	M N		
$L^N L^N$	N		

Figure 4.2 Detection of the M and N antigens on blood cells by agglutination with specific anti-sera. With the anti-M and anti-N sera, three blood types can be identified.

produce only the M antigen, and homozygotes for the N allele produce only the N antigen. However, heterozygotes for these two alleles produce both kinds of antigens. Because the two alleles appear to contribute equally to the phenotype of the heterozygotes, they are said to be **codominant.** Codominance implies that there is an independence of allele function. Neither allele is dominant, or even partially dominant, over the other. It would therefore be inappropriate to distinguish the alleles by upper and lower case letters, as we have in all previous examples. Instead, codominant alleles are represented by superscripts on the symbol for the gene, which in this case is the letter *L*—a tribute to Karl Landsteiner, the discoverer of blood-typing. Thus, the M allele is L^M and the N allele is L^N. Figure 4.2 shows the three possible genotypes formed by the L^M and L^N alleles, and their associated phenotypes.

Multiple Alleles

The Mendelian concept that genes exist in no more than two allelic states had to be modified when genes with three, four, or more alleles were discovered. A classic example of a gene with **multiple alleles** is the one that controls coat color in rabbits (Figure 4.3). The color-determining gene, denoted by the lower case letter *c*, has four alleles, three of which are distinguished by a superscript: *c* (*albino*), c^h (*himalayan*), c^{ch} (*chinchilla*), and c^+ (wild-type). In homozygous condition, each allele has a characteristic effect on the coat color: *cc*—white hairs over the entire body; c^hc^h—black hairs on the extremities, white hairs everywhere else; $c^{ch}c^{ch}$—white hair with black tips on the body; c^+c^+—colored

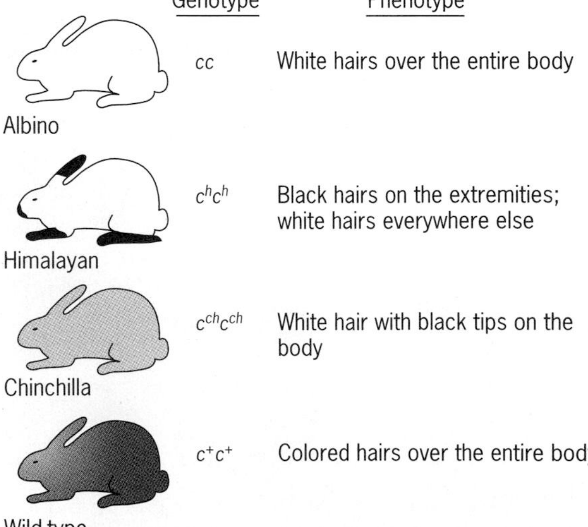

Figure 4.3 Coat colors in rabbits. The different phenotypes are caused by four different alleles of the *c* gene.

hairs over the entire body. Because most rabbits in wild populations are homozygous for the c^+ allele, this allele is called the **wild-type.** In genetics it is customary to represent wild-type alleles by a superscript plus sign after the letter for the gene. When the context is clear, the letter is sometimes omitted and only the plus sign is used; thus, c^+ may be abbreviated simply as +.

The other alleles of the *c* gene are **mutants**—altered forms of the wild-type allele that must have arisen sometime during the evolution of the rabbit. The *himalayan* and *chinchilla* alleles are denoted by superscripts, but the *albino* allele is denoted simply by the letter *c* (for *c*olorless, another word for the albino condition). This notation reflects another custom in genetics nomenclature: genes are often named for a mutant allele, usually the allele associated with the most abnormal phenotype. The convention of naming a gene for a mutant allele is generally consistent with the convention we discussed in Chapter 3—that of naming genes for a recessive allele—because most mutant alleles are recessive. However, sometimes a mutant allele is dominant, in which case the gene is named after its associated phenotype. For example, a gene in mice controls the length of the tail. The first mutant allele of this gene that was discovered caused a shortening of the tail in heterozygotes. This dominant mutant was therefore symbolized by *T*, for *t*ailless. All other alleles of this gene—and there are many—have been denoted by an upper or lower case letter, depending on whether they are dominant or recessive; different alleles are distinguished from each other by superscripts.

Another example of multiple alleles comes from the study of human blood types. The A, B, AB, and O blood types, like the M, N, and MN blood types discussed previously, are identified by testing a blood sample with different sera. One serum detects the A antigen, another the B antigen. When only the A antigen is present on the cells, the blood is type A; when only the B antigen is present, the blood is type B. When both antigens are present, the blood is type AB, and when neither antigen is present, it is type O. Blood typing for the A and B antigens is completely independent of blood typing for the M and N antigens.

The gene responsible for producing the A and B antigens is denoted by the letter *I*. It has three alleles: I^A, I^B, and I^O. The I^A allele specifies the production of the A antigen, and the I^B allele specifies the production of the B antigen. However, the I^O allele does not specify an antigen. Among the six possible genotypes, there are four distinguishable phenotypes—the A, B, AB, and O blood types (Table 4.1). In this system, the I^A and I^B alleles are codominant, since each is expressed equally in the I^A I^B heterozygotes, and the I^O allele is recessive to both the I^A and I^B alleles. All three alleles are found at appreciable frequencies in human popu-

TABLE 4.1
Genotypes, Phenotypes, and Frequencies in the ABO Blood-Typing System

Genotype	*Blood Type*	*A Antigen Present*	*B Antigen Present*	*Frequency in U.S. White Population (%)*
I^AI^A or I^AI^O	A	+	−	41
I^BI^B or I^BI^O	B	−	+	11
I^AI^B	AB	+	+	4
I^OI^O	O	−	−	44

lations; thus, the *I* gene is said to be **polymorphic,** from the Greek words for "having many forms." We consider the population and evolutionary significance of genetic polymorphisms in Chapter 29.

Allelic Series

The functional relationships among the members of a series of multiple alleles can be studied by making heterozygous combinations through crosses between homozygotes. For example, the four alleles of the *c* gene in rabbits can be combined with each other to make six different kinds of heterozygotes: cc^h, cc^{ch}, cc^+, c^hc^{ch}, c^hc^+, and $c^{ch}c^+$. These heterozygotes allow the dominance relations among the alleles to be studied (Figure 4.4). The wild-type allele is completely dominant over all the other alleles in the series; the *chinchilla* allele is partially dominant over the *himalayan* and *albino* alleles, and the himalayan allele is completely dominant over the *albino* allele. These dominance relations can be summarized as $c^+ > c^{ch} > c^h > c$.

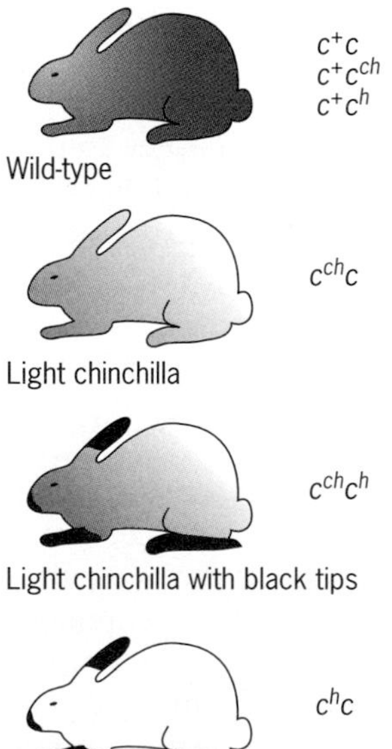

Figure 4.4 Phenotypes of different combinations of *c* alleles in rabbits. The alleles form a series, with the wild-type allele, c^+, dominant over all the other alleles and the null allele, *c* (albino), recessive to all the other alleles; one hypomorphic allele, c^{ch} (chinchilla), is partially dominant over the other, c^h (himalayan).

Notice that the dominance hierarchy parallels the effects that the alleles have on coat color. A plausible explanation is that the *c* gene controls a step in the formation of black pigment in the fur. The wild-type allele is fully functional in this process, producing colored hairs throughout the body. The *chinchilla* and *himalayan* alleles are only partially functional, producing some colored hairs, and the *albino* allele is not functional at all. Nonfunctional alleles are said to be **null** or **amorphic** (from the Greek words for "without form"); they are almost always completely recessive. Partially functional alleles are said to be **hypomorphic** (from the Greek words for "beneath form"); they are recessive to alleles that are more functional, including (usually) the wild-type allele. Later in this chapter we consider the biochemical basis for these differences.

Testing Gene Mutations for Allelism

A mutant allele is created when an existing allele is changed to a new genetic state—a process called **mutation.** This event always involves a change in the physical composition of the gene (see Chapter 14) and sometimes produces an allele that has a detectable phenotypic effect. If, for example, the c^+ allele mutated to a null allele, a rabbit homozygous for this mutation would have the albino phenotype. However, it is not always possible to assign a new mutation to a gene on the basis of its phenotypic effect. In rabbits, for example, several genes determine coat color, and a mutation in any one of them could reduce, alter, or abolish pigmentation in the hairs. Thus, if a new coat color appears in a population of rabbits, it is not immediately clear which gene has been mutated.

A simple test can be used to determine the allelic identity of a new mutation as long as the new mutation is recessive. The procedure involves crosses to combine the new recessive mutation with recessive mutations of known genes (Figure 4.5). If the hybrid progeny show a mutant phenotype, then the new mutation and the tester mutation *are* alleles of the same gene. If the hybrid progeny show a wild phenotype, then the new mutation and the tester mutation *are not* alleles of the same gene. This test is based on the principle that mutations of the same gene impair the same genetic function. If two such mutations are combined, the organism will be abnormal for this function and will show a mutant phenotype, even if the two mutations had an independent origin.

New recessive mutation		Tester genotype		Hybrid phenotype	Conclusion
c^*c^*	X	*a a*	⇨	Wild-type	*a* and c^* not alleles
		b b	⇨	Wild-type	*b* and c^* not alleles
		c c	⇨	Mutant	*c* and c^* alleles
		d d	⇨	Wild-type	*d* and c^* not alleles

Figure 4.5 A general scheme to test recessive mutations for allelism. Two mutations are alleles if a hybrid that contains both of them has the mutant phenotype.

It is important to remember that this test applies only to recessive mutations. Dominant mutations cannot be tested in this way because they exert their effects even if a wild-type copy of the gene is present.

As an example, let us consider the analysis of two recessive mutations affecting eye color in the fruit fly, *Drosophila melanogaster* (Figure 4.6). This tiny organism has been investigated by geneticists for nearly a century, and many different mutations have been identified. Two independently isolated mutations, called *cinnabar* and *scarlet*, are phenotypically indistinguishable, each causing the eyes to be bright red. In wild-type flies, the eyes are dark red. We wish to know whether the *cinnabar* and *scarlet* mutations are alleles of a single color-determining gene or if they are mutations in two different genes. To find the answer, we must cross the homozygous mutant strains with each other to produce hybrid progeny. If the hybrids have bright red eyes, we will conclude that *cinnabar* and *scarlet* are alleles of the same gene. If they have dark red eyes, we will conclude that they are mutations in different genes.

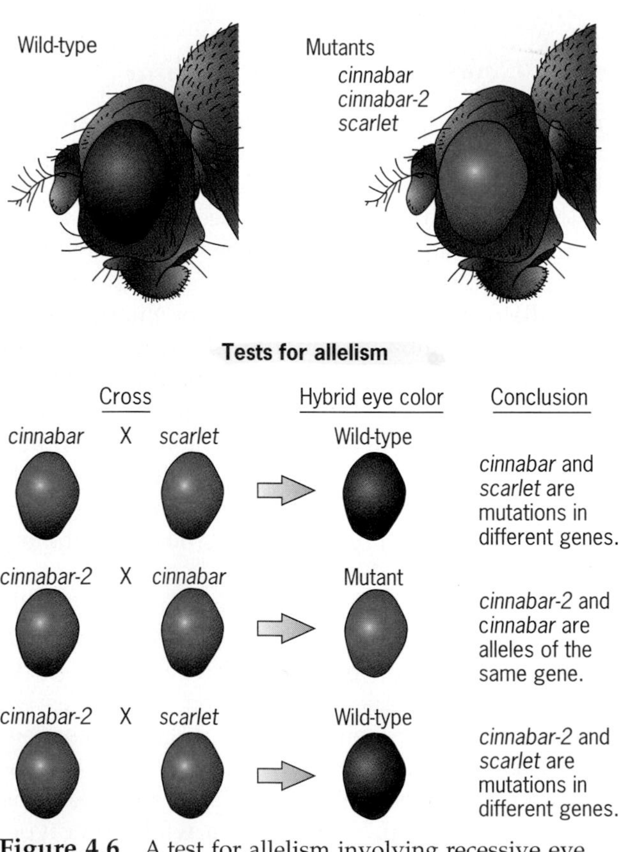

Figure 4.6 A test for allelism involving recessive eye color mutations in *Drosophila*. Three phenotypically identical mutations, *cinnabar*, *scarlet*, and *cinnabar-2*, are tested for allelism by making pairwise crosses between flies homozygous for different mutations. The phenotypes of the hybrids show that the *cinnabar* and *cinnabar-2* mutations are alleles of a single gene and that the *scarlet* mutation is not an allele of this gene.

The hybrid progeny turn out to have dark red eyes; that is, they are wild-type rather than mutant. Thus, *cinnabar* and *scarlet* are not alleles of the same gene but, rather, mutations in two different genes, each apparently involved in the control of eye pigmentation. When we test a third mutation, called *cinnabar-2*, for allelism with the *cinnabar* and *scarlet* mutations, we find that the hybrid combination of *cinnabar-2* and *cinnabar* has the mutant phenotype (bright red eyes) and that the hybrid combination of *cinnabar-2* and *scarlet* has the wild phenotype (dark red eyes). These results tell us that the mutations *cinnabar* and *cinnabar-2* are alleles of one color-determining gene and that the *scarlet* mutation is not an allele of this gene. Rather, the *scarlet* mutation defines another color-determining gene.

The test to determine whether mutations are alleles of a particular gene is based on the phenotypic effect of combining the mutations in the same individual. If the hybrid combination is mutant, we conclude that the mutations are alleles; if it is wild-type, we conclude that they are not alleles. Chapter 15 discusses how this test—called the *complementation test* in modern terminology—enables geneticists to define the functions of individual genes.

Variation Among the Effects of Mutations

Genes are identified by mutations that alter the phenotype in some conspicuous way. For instance, a mutation may change the color or shape of the eyes, alter a behavior, or cause sterility or even death. The tremendous variation among the effects of individual mutations suggests that each organism carries many different kinds of genes and that each of these can mutate in different ways. In nature, mutations provide the raw material for evolution (see Chapter 29).

Mutations that alter some aspect of morphology, such as seed texture or color, are called **visible mutations.** Most visible mutations are recessive, but a few are dominant. Geneticists have learned much about genes by analyzing the properties of these mutations.

We will encounter many examples of this analysis throughout this textbook.

Mutations that limit reproduction are called **sterile mutations.** Some sterile mutations affect both sexes, but most affect either males *or* females. As with visible mutations, steriles can be either dominant or recessive. Some steriles completely prevent reproduction, whereas others only impair it slightly.

Mutations that interfere with necessary vital functions are called **lethal mutations.** Their phenotypic effect is death. We know that many—perhaps a majority of—genes are capable of mutating to the lethal state. Thus, a large fraction of all genes are absolutely essential for life.

Dominant lethals are lost one generation after they occur because the individuals that carry them die; however, recessive lethals may linger a long time in a population because they can be hidden in heterozygous condition by a wild-type allele. Recessive lethal mutations are detected by observing unusual segregation ratios in the progeny of heterozygous carriers. An example is the *yellow-lethal* mutation, Y^l, in the mouse (Figure 4.7). This mutation is a dominant visible, causing the fur to be yellow instead of gray-brown (the wild-type color, also known as *agouti*). In addition, however, the Y^l mutation is a recessive lethal, killing Y^lY^l homozygotes early in their development. A cross between Y^ly heterozygotes produces two kinds of viable progeny, yellow (Y^ly) and gray-brown (yy), in a ratio of 2:1. Given a sufficiently large number of progeny, this ratio can easily be distinguished from the 3:1 ratio that would be obtained if Y^l were simply a dominant visible mutation.

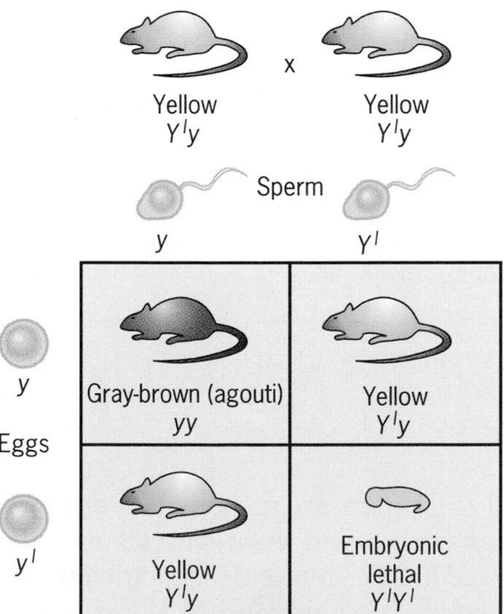

Figure 4.7 Y^l, the yellow-lethal mutation in mice: a dominant visible that is also a recessive lethal. A cross between carriers of this mutation produces yellow heterozygotes and gray-brown (agouti) homozygotes in a ratio of 2:1. The yellow homozygotes die as embryos.

Geneticists have used different conventions to symbolize genes and their mutations. Mendel began the practice of using letters to denote genes. However, he simply started with the letter *A* and proceeded through the alphabet as symbols were needed to represent genes in his crosses. William Bateson was the first person to use letters mnemonically to symbolize genes. For the symbol, Bateson chose the first letter of the word that described the gene's phenotypic effect–thus, *B* for a gene causing *b*lue flowers, *L* for a gene causing *l*ong pollen grains. As the number of known genes grew, it became necessary to use two or more letters to represent newly discovered genes. Variations in the procedures for naming and symbolizing genes are discussed in the Historical Sidelight, Genetic Symbols: Evolution in a Dynamic Science.

Genes Function to Produce Polypeptides

The extensive variation revealed by mutations indicates that organisms contain many different genes and that these genes can exist in multiple allelic states. However, it does not tell us how genes actually affect the phenotype. What is it about a gene that enables it to control a trait such as eye color, seed texture, or plant height?

The early geneticists had no answer to this question. However, today it is clear that most genes make a product that subsequently affects the phenotype. This idea, which was discussed in Bateson's book and which was supported by the research of many scientists, including, most notably, the British physician Sir Archibald Garrod, was forcefully brought out in the middle of the twentieth century when George Beadle and Edward Tatum discovered that the products of genes are *polypeptides* (Figure 4.8).

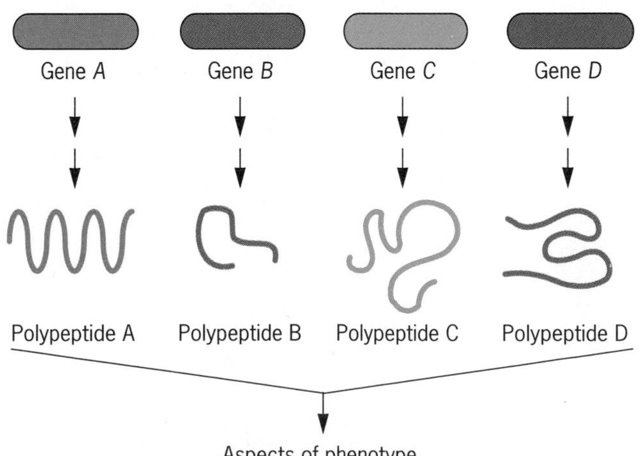

Figure 4.8 Relationship between genes and polypeptides. Each gene specifies a different polypeptide. These polypeptides then function to determine the organism's phenotype.

HISTORICAL SIDELIGHT

Genetic Symbols: Evolution in a Dynamic Science

William Bateson started the practice of choosing gene symbols mnemonically. In discussing Mendel's work, for example, he symbolized the dominant allele for tall pea plants as *T* and the recessive allele for short plants as *t*. Later, when it became customary to choose allele symbols based on the mutant trait, these symbols were changed to *D* (for tall) and *d* (for *d*warf). This convention provided a simple and consistent notation in which the dominant and recessive alleles of a particular gene were represented by a single letter, and that letter was mnemonic for the trait influenced by the gene. Bateson also coined the words *genetics*, *alleleomorph* (which was later shortened to *allele*), *homozygote*, and *heterozygote*, and he introduced the practice of denoting the generations in a breeding scheme as P, F_1, F_2, and so forth.

The gene-naming system that Bateson developed worked well until the number of genes that had been identified exceeded the capacity of the English alphabet; thereupon it became necessary to use two or more letters to symbolize a gene. For example, a particular mutant allele in *Drosophila* causes the eyes to be carmine instead of red. When this allele was discovered, it was given the symbol *cm* because the single letter *c* had already been used to represent a mutant allele that causes the wings to be curved instead of straight. Today, with thousands of genes identified, it is often necessary to use three or four letters, or combinations of letters and numbers, to symbolize genes. For example, mutations in the *cmp* gene of *Drosophila* cause the wings to be *crumpled*, and mutations in the *Sh1* and *Sh2* genes of maize cause the kernels to be *sh*runken.

The discovery of multiple alleles made genetic notation even more complicated; because upper and lower case letters were no longer adequate to distinguish among alleles, geneticists began to combine a basic gene symbol with an identification symbol. *Drosophila* geneticists were the first to apply this procedure. They made the identification symbol a superscript on the basic gene symbol. Usually, both the gene symbol and the superscript had some mnemonic significance. Thus, for example, cn^2 was used to symbolize the *second* cin*n*abar eye color allele that was discovered in *Drosophila*; and ey^D was used to symbolize a *d*ominant allele that causes *Drosophila* to be *ey*eless. This convention was extended to other experimental animals, such as rabbits and mice. For example, c^{ch} was used to symbolize the *ch*inchilla allele of the gene that determines whether a rabbit's fur is colored or *c*olorless. Plant geneticists adopted a variation of this practice. They use hyphenated symbols to identify mutant alleles; for example, *sh2–6801* represents a mutant allele of the *Sh2* gene that was discovered in *1968*.

As genetic nomenclature developed, it became necessary to use a special symbol to represent the wild-type allele. The early *Drosophila* geneticists proposed using a plus sign (+), sometimes written as a superscript on the basic gene symbol (for example, c^+). This simple notation conveys the idea that the wild-type allele is the standard, or normal, allele of the gene, and is widely used today. However other gene-naming practices persist. Plant geneticists tend to use the gene symbol itself to represent the wild-type allele, but to make it stand out, they capitalize the first letter. Thus, *Sh2* is the wild-type allele of the second shrunken gene discovered in maize, whereas *sh2* is a mutant allele.

Genetic nomenclature has been further complicated by the discovery of genes through the polypeptides they specify. These discoveries have introduced gene symbols that are mnemonic for polypeptide gene products. For example, the human gene that specifies the polypeptide *h*ypoxanthine-guanine *p*hosphoribosyl *t*ransferase is symbolized by *HPRT*, and the plant gene that specifies the polypeptide *a*lcohol *de*-*h*ydrogenase is symbolized by *Adh*. Whether upper case letters are used throughout the gene symbol or only for the first letter depends on the organism.

Today there are many specialized systems for symbolizing genes and alleles. Researchers who work with different organisms—*Drosophila*, mice, plants, and humans—speak slightly different languages. Later, we will see that still other genetic dialects have been created to describe the genes of viruses, bacteria, and fungi. These different systems of nomenclature indicate that the symbols in genetics have evolved in response to new discoveries—visible evidence of growth in a dynamic, young science.

Polypeptides are macromolecules built of a linear chain of *amino acids*. Every organism makes thousands of different polypeptides, each characterized by a specific amino acid sequence. These polypeptides are the fundamental constituents of *proteins*. Two or more polypeptides may combine to form a protein. Some proteins, called *enzymes*, function as catalysts in biochemical reactions; others form the structural components of cells; and still others are responsible for transporting substances within and between cells. Beadle and Tatum proposed that each gene is responsible for the synthesis of a particular polypeptide. When a gene is mutated, its polypeptide product either is not made or is altered in such a way that its role in the organism is changed. Mutations that eliminate or alter a polypeptide are often associated with a phenotypic effect. Whether this effect is dominant or recessive depends on the nature of the mutation. In Chapter 13 we consider the details of how genes produce polypeptides, and in Chapter 14 we discuss the molecular basis of mutation.

Why Are Some Mutations Dominant and Others Recessive?

The discovery that genes produce polypeptides provides insight into the nature of dominant and recessive mutations. Dominant mutations have phenotypic effects in heterozygotes as well as in homozygotes, whereas recessive mutations have these effects only in homozygotes. What accounts for this striking difference in expression?

Recessive mutations occur when a gene loses its function, that is, when it no longer specifies a polypeptide or when it specifies a defective polypeptide (Figure 4.9). Recessive mutations are therefore typically **loss-of-function** alleles. In contrast, dominant mutations occur when a gene acquires a new function—that is, when it specifies a polypeptide that behaves in a new and different way. Dominant mutations are therefore typically **gain-of-function** alleles. According to this view, a recessive phenotype results from a partial or total loss of polypeptide function, whereas a dominant phenotype results from the gain of a new polypeptide function—a function that is expressed even in the presence of the polypeptide made by the wild-type allele. A recessive mutation that causes a total loss of polypeptide function is a null or amorphic allele, and one that causes a partial loss of function is a hypomorphic allele. A dominant mutation that creates a new polypeptide function is called a **neomorphic** allele, from the Greek words for "new form."

Why is a loss-of-function mutation recessive to a wild-type allele? In a heterozygote, such a mutation either will not produce a polypeptide, or it will produce a defective polypeptide that has no effect on the phenotype (Figure 4.9). However, the wild-type allele will produce a fully functional polypeptide that will carry out its normal role in the organism. The phenotype of a mutant/wild heterozygote will therefore be the same, or essentially the same, as that of a wild-type homozygote. The only possible difference might be that the homozygote, with its two wild-type alleles, will have a more intense phenotype (for example, a deeper eye color) than the heterozygote, with its one such allele. In most instances, however, a single wild-type allele is sufficient to produce the wild phenotype.

We see this, for example, with the *cinnabar* gene in *Drosophila*. This gene specifies a polypeptide that functions as an enzyme in the synthesis of the brown pigment that is deposited in *Drosophila* eyes. Flies that are homozygous for a loss-of-function mutation in the *cinnabar* gene cannot produce this enzyme; conse-

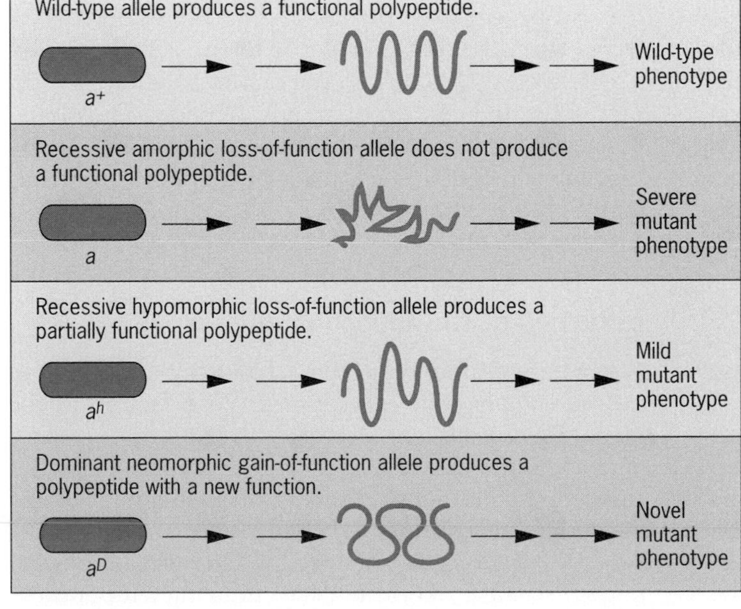

(a)

Genotype	Polypeptides present	Phenotype	Nature of mutant allele
$a^+ a$		Wild-type	Recessive
$a^+ a^h$		Wild-type	Recessive
$a^+ a^D$		Mutant	Dominant

(b)

Figure 4.9 Differences between recessive loss-of-function mutations and dominant gain-of-function mutations. (*a*) Polypeptide products of recessive and dominant mutations. (*b*) Phenotypes of heterozygotes carrying a wild-type allele and different types of mutant alleles.

quently, they do not synthesize any brown pigment in their eyes. The phenotype of homozygous *cinnabar* mutants is bright red—the color of the mineral cinnabar, for which the gene is named. However, flies that are heterozygous for the *cinnabar* mutation and its wild-type allele have dark red eyes; that is, they are phenotypically identical to wild-type. In these flies, the loss-of-function allele is recessive to the wild-type allele because the latter produces enough enzyme to synthesize brown eye pigment.

Why is a gain-of-function mutation dominant to a wild-type allele? Such a mutation will produce a polypeptide with a novel function. In a mutant/wild heterozygote, this polypeptide can replace or interfere with the function of the normal polypeptide and thereby cause a mutant phenotype.

As an example, let us consider dominant mutations in the *T* gene of the mouse. We have already seen that in heterozygous condition, these mutations cause a shortening of the tail. In homozygous condition, they are lethal. This lethality is due to the complete failure of the trunk of the animal to develop. The *T* gene is therefore critical for the formation of a substantial portion of the mouse's body. The first dominant *T* mutation was discovered in 1927 by N. Dobrovolskaia-Zavadskaia, who worked at the Pasteur Institute in Paris. Many other dominant *T* mutations have since been identified, and recently several have been analyzed at the molecular level. From this analysis, mouse geneticists have learned that the wild-type *T* gene specifies a polypeptide of 436 amino acids.

Biochemical studies have indicated that this polypeptide is able to bind to DNA. It may therefore act as a master regulator of other genes whose products are needed for normal mouse development. Dominant mutations of the *T* gene result in the production of polypeptides that are slightly shorter than the wild-type polypeptide and also somewhat different in structure. The available evidence suggests that these mutant polypeptides counteract the function of the wild-type polypeptide, perhaps by interfering with its ability to bind DNA. This interference then alters the expression of the genes that are regulated by the T polypeptide, and thereby upsets the course of mouse embryological development. If only one dominant *T* mutation is present in the genotype, the animal is born with a short tail, or perhaps with no tail at all. If two dominant *T* mutations are present in the genotype, the animal dies as an embryo in the uterus.

In later chapters, we encounter other examples of dominant mutations. Some, like the *T* mutations in the mouse, are genuine gain-of-function mutations; however, others are not. We will see that the view presented here—that recessive alleles are typically loss-of-function mutations and that dominant alleles are typically gain-of-function mutations—is an oversimplification. The phenotypes associated with some dominant alleles, for example, result from a loss of gene function. A discussion of these atypical situations must, however, be postponed until after we have considered the molecular biology of genes.

Key Points: **Genes often have multiple alleles. Different alleles are created by the mutation of a wild-type allele. Mutant alleles may be dominant, recessive, incompletely dominant, or codominant. The allelism of different recessive mutations can be tested by combining them in the same individual. If the individual has a mutant phenotype, then the mutations are alleles; if it has a wild phenotype, then they are not alleles. The function of most genes is to produce a polypeptide. Recessive mutations can cause a partial or complete loss of polypeptide activity. Dominant mutations can endow a polypeptide with a new activity that may replace or interfere with the activity of the wild-type polypeptide.**

GENE ACTION: FROM GENOTYPE TO PHENOTYPE

At the beginning of the twentieth century, geneticists had imprecise ideas about how genes evoke particular phenotypes. They knew nothing about the chemistry of gene structure or function, nor had they developed the techniques to study it. Everything that they proposed about the nature of gene action was inferred from the analysis of phenotypes. These analyses showed that genes do not act in isolation. Rather, they act in the context of an environment, and sometimes they act in concert with other genes. These analyses also showed that a particular gene can influence many different traits.

Influence of the Environment

A gene must function in the context of both a biological and a physical environment. The factors in the physical environment are easier to study, for particular genotypes can be reared in the laboratory under controlled conditions, allowing an assessment of the effects of temperature, light, nutrition, and humidity. As an example, let's consider the *Drosophila* mutation known as *shibire*. At the normal culturing temperature, 25°C, *shibire* flies are viable and fertile, but are extremely sensitive to a sudden shock. When a *shibire* culture is shaken, the flies—temporarily paralyzed—fall to the bottom of the culture. Indeed, *shibire* is the word for "paralysis" in Japanese. However, if a culture of *shibire* flies is placed at a slightly higher temperature, 29° C, all the flies fall to the bottom and die, even without a shock. Thus, the phenotype of the *shibire* mutation is temperature-sensitive. At 25° C, the mutation is viable, but at 29° C it is lethal. A plausible explanation is that at 25° C the mutant gene makes a partially functional protein, but at 29° C this protein is totally nonfunctional.

Environmental Effects on the Expression of Human Genes

Human genetic research provides an example of how the physical environment can influence a phenotype. **Phenylketonuria** (PKU) is a recessive disorder of amino acid metabolism. Infants homozygous for the mutant allele accumulate semitoxic substances in their brains; though not lethal, these substances can impair mental ability by affecting the brain's development. The harmful aspects of PKU are traceable to a particular amino acid, phenylalanine, which is ingested in the diet. Infants who are fed normal diets ingest enough phenylalanine to bring out the worst manifestations of the disease. However, infants who are fed low-phenylalanine diets usually mature without serious mental impairment. Because PKU can be diagnosed in newborn babies, the clinical impact of this disease can be reduced if infants that are PKU homozygotes are placed on a low-phenylalanine diet shortly after birth. This example illustrates how an environmental factor—diet—can be manipulated to modify a phenotype that would otherwise become a personal tragedy.

The biological environment can also influence the phenotypic expression of genes. **Pattern baldness** in humans is a well-known example (Figure 4.10). Here the relevant biological factor is gender. Premature pattern baldness is due to an allele that is expressed differently in the two sexes. In males, both homozygotes and heterozygotes for this allele develop bald patches, whereas in females, only the homozygotes show a tendency to become bald, and this is usually limited to general thinning of the hair. The expression of this allele is probably triggered by the male hormone **testosterone.** Females produce much less of this hormone and are therefore seldom at risk to develop bald patches. The sex-influenced nature of pattern baldness shows that biological factors can control the expression of genes.

Figure 4.10 Man with premature balding beginning on the crown of his head.

Penetrance and Expressivity

When individuals do not show a trait even though they have the appropriate genotype, the trait is said to exhibit **incomplete penetrance.** An example of incomplete penetrance in humans is **polydactyly**—the presence of extra fingers and toes (Figure 4.11*a*). This condition is due to a dominant mutation, *P*, that is manifested in some of its carriers. In the pedigree in Figure 4.11*b*, the individual denoted III-2 must be a carrier even though he does not have extra fingers or toes. The reason is that both his mother and three of his chil-

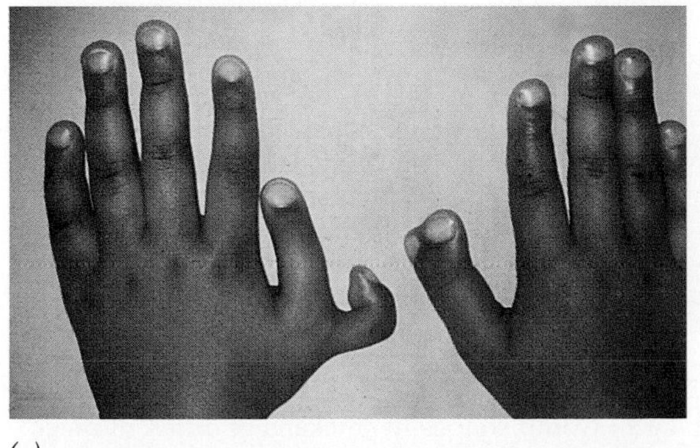

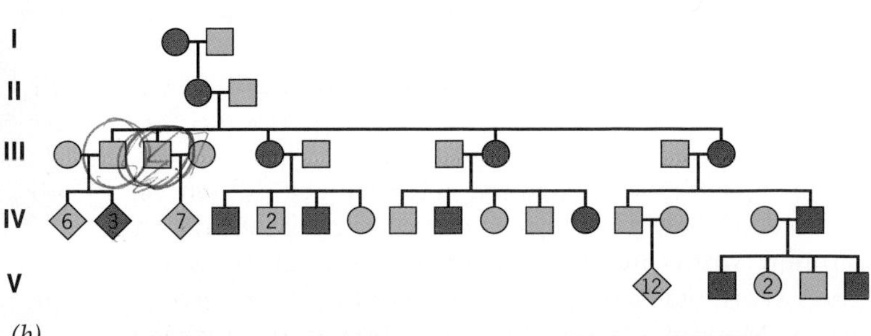

(*a*) (*b*)

Figure 4.11 Polydactyly in human beings. (*a*) Phenotype showing extra fingers. (*b*) Pedigree showing the inheritance of this dominant trait.

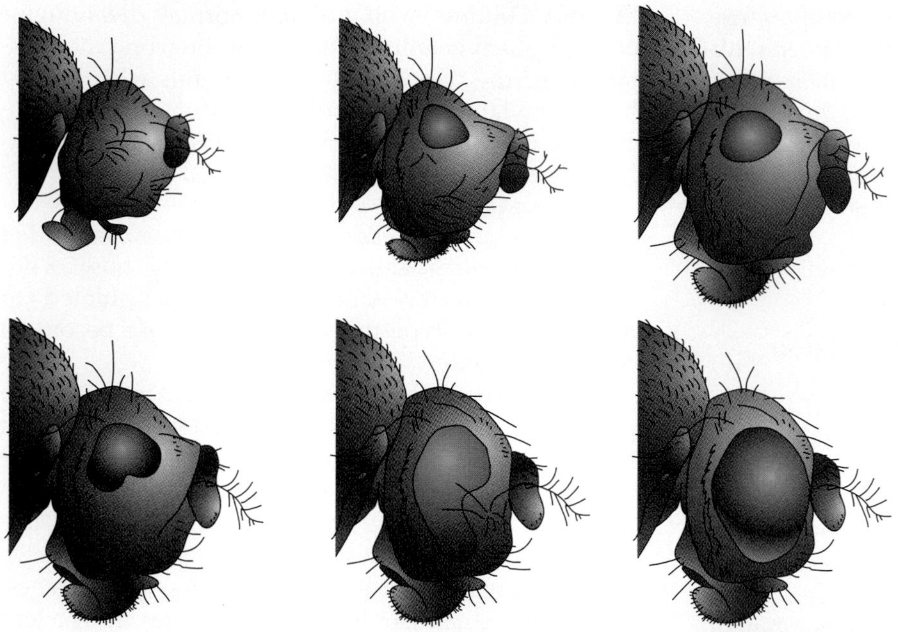

Figure 4.12 Variable expressivity of the *Lobe* mutation in *Drosophila*. Each fly is heterozygous for this dominant mutation; however, the phenotypes vary from complete absence of the eye to a nearly wild-type eye.

dren are polydactylous—an indication of the transmission of the mutation through III-2. Incomplete penetrance can be a serious problem in pedigree analysis because it can lead to the incorrect assignment of genotypes.

The term **expressivity** is used if a trait is not manifested uniformly among the individuals that show it. The dominant *Lobe* eye mutation (Figure 4.12) in *Drosophila* is an example. The phenotype associated with this mutation is extremely variable. Some heterozygous flies have tiny compound eyes, whereas others have large, lobulated eyes; between these extremes, there is a full range of phenotypes. The *Lobe* mutation is therefore said to have **variable expressivity.**

Incomplete penetrance and variable expressivity indicate that the pathway between a genotype and its phenotypes is subject to considerable modulation. (For an interesting example from European history, see the Human Genetics Sidelight: The Hapsburg Jaw.) Geneticists know that some of this modulation is due to environmental factors, but some is also due to factors in the genetic background. Clearcut evidence for such factors comes from breeding experiments showing that two or more genes can control a particular trait.

Gene Interactions

Some of the first evidence that a trait can be influenced by more than one gene was obtained by Bateson and Punnett from breeding experiments with chickens. Their work was carried out shortly after the rediscovery of Mendel's paper. Domestic breeds of chickens have different comb shapes (Figure 4.13): Wyandottes have "rose" combs, Brahmas have "pea" combs, and

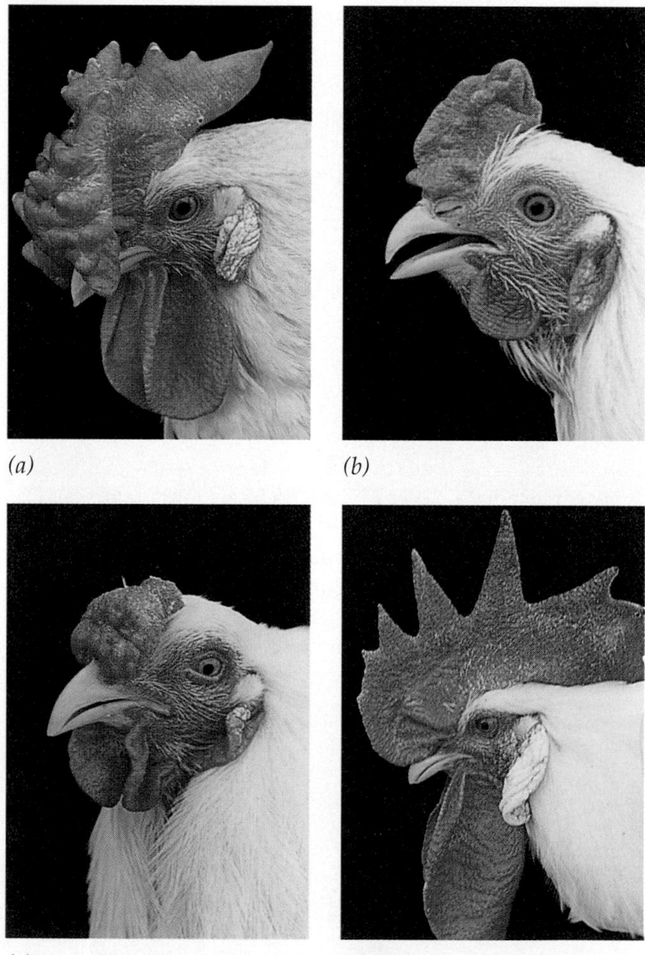

Figure 4.13 Comb shapes in chickens of different breeds. (*a*) rose, Wyandottes; (*b*) pea, Brahmas; (*c*) walnut, hybrid from cross between chickens with rose and pea combs; (*d*) single, Leghorns.

HUMAN GENETICS SIDELIGHT

The Hapsburg Jaw: A Dominant Trait with Incomplete Penetrance and Variable Expressivity in the European Nobility[1]

From the Middle Ages to the beginning of the twentieth century, the House of Hapsburg ruled much of Europe. The members of this family included kings and queens, emperors and empresses. Although Hapsburg power was concentrated in Central Europe—especially in Austria—at various times it encompassed Spain, the Low Countries, and parts of Italy. The extension of Hapsburg power in Europe was due, in part, to the family's uncanny ability to arrange politically beneficial marriages. This skill was celebrated in a line of Latin verse: *Bella gerant allii, tu, felix Austria, nube!*—"Let others wage wars; you, happy Austria, marry!" Thus, through a foreign policy that was based on matrimonial vows, the Hapsburgs spread their genes among the European nobility. Marie Antoinette, the wife of the ill-fated French king Louis XVI, was, for example, a Hapsburg.

As the Hapsburgs spread their genes, so too did they spread the phenotypes associated with those genes. The most famous of these phenotypes is the Hapsburg jaw, a protrusion of the mandible often associated with difficulty in chewing. This condition is more properly known as **mandibular prognathism**. Its presence is recorded in the portraits of the Hapsburgs—in paintings, on coins, in sculpture, and, more recently, in photographs.

The first documented case of prognathism among the Hapsburgs appears in portraits of the Emperor Maximilian I of Austria (Figure 1*a*), who lived from 1459 to 1519. Maximilian's son, Philip I, married Joanna ("The Mad"), the daughter of King Ferdinand and Queen Isabella of Spain. In this way, Philip I became king of Spain and established a Spanish line of Hapsburgs. Although it is not known whether Philip I manifested the Hapsburg jaw, his son, Charles V (1500–1558), clearly did (Figure 1*b*). Charles's severe prognathism caused his mouth to hang open. He grew up in Flanders, another Hapsburg dominion, and when he first came to Spain, it is said that a peasant shouted to him: "Your Majesty, shut your mouth; the flies in this country are very insolent."[2] In spite of Charles' condition, he proved to be an excellent leader, and during his reign, Spain became the most powerful country in Europe.

When poor health forced Charles V to abdicate in 1556, his son, Philip II, became king. Portraits show that Philip II also had the Hapsburg jaw. It is not known whether Philip II's son, Philip III, manifested this trait. However, his son, Philip IV, clearly did. Diego Rodriguez de Silva y Velazquez (1599–1660)—regarded by some as Spain's greatest painter—produced a magnificent portrait of Philip IV when Philip was 18 years old (Figure 1*c*). The slightly empty expression and the sense of shyness in this portrait suggest that Philip IV was less than comfortable in the political realm. In fact, he turned the government of Spain over to professional politicians and focused his attention on the arts; Philip IV was a poet and a dramatist, and he esteemed Velazquez as one of his closest friends.

The Spanish line of Hapsburgs ended with the death of Philip IV's son, Carlos II, in 1700. Before his death, Carlos II designated Philip of Anjou as his successor. This Philip was from the House of Bourbon, the French royal family. However, because Bourbons had married Hapsburgs in previous generations, some of the Spanish Bourbons inherited the Hapsburg jaw. One example is Alfonso XIII (Figure 1*d*), who was king of Spain in the early part of the twentieth century.

Five hundred years of Hapsburg portraiture indicate that mandibular prognathism is caused by a dominant allele with incomplete penetrance and variable expressivity. This condition has been found to varying degrees in the portraits of both men and women. Although the mode of inheritance is typical for that of a dominant allele, it sometimes skips a generation, presumably because the dominant allele is not fully penetrant. The physiological basis of this allele's effects is currently unknown.

[1]Hodge, G. P. 1977. A medical history of the Spanish Habsburgs. *J. Am. Med. Assn.* 238: 1169–1174.

[2]Langdon-Davies, J. 1963. *Carlos, the King Who Would Not Die.* Prentice-Hall, Englewood Cliffs, NJ.

(*a*)

(*b*)

(*c*)

(*d*)

Figure 1 (*a*) Emperor Maximilian I of Austria, (*b*) King Charles V of Spain, (*c*) King Philip IV of Spain, and (*d*) King Alfonso XIII of Spain.

Leghorns have "single" combs. Crosses between Wyandottes and Brahmas produce chickens that have yet another type of comb called "walnut." Bateson and Punnett discovered that comb type is controlled by two independently assorting genes, *R* and *P*, each with two alleles (Figure 4.14). Wyandottes (with rose combs) have the genotype *RR pp*, and Brahmas (with pea combs) have the genotype *rr PP*. The F_1 hybrids between these two varieties are therefore *Rr Pp*, and phenotypically they have walnut combs. If these hybrids are intercrossed with each other, all four types of combs appear in the progeny: 9/16 walnut (*R- P-*), 3/16 rose (*R- pp*), 3/16 pea (*rr P-*), and 1/16 single (*rr pp*). The Leghorn breed, which has the single-comb type, must therefore be homozygous for both of the recessive alleles.

The work of Bateson and Punnett demonstrated that two independently assorting genes can control a trait. Different combinations of alleles from the two genes resulted in different phenotypes, presumably because of interactions between their products at the biochemical or cellular level.

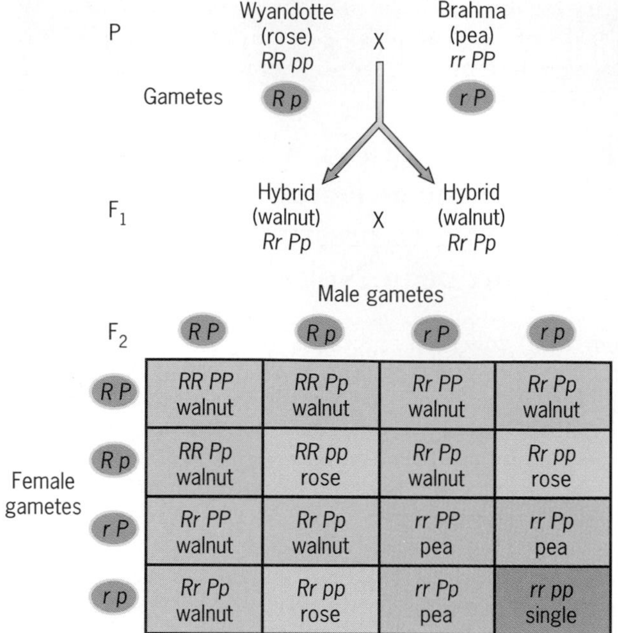

Figure 4.14 Bateson and Punnett's experiment on comb shape in chickens. The intercross in the F_1 produces four phenotypes, each highlighted by a different color in the Punnett Square, in a 9:3:3:1 ratio.

Epistasis

When two or more genes influence a trait, an allele of one of them may have an overriding effect on the phenotype. When an allele has such an overriding effect, it is said to be epistatic to the other genes that are involved; the term **epistasis** comes from a Greek word meaning to "stand above." For example, we know that eye pigmentation in *Drosophila* involves a large number of genes. If a fly is homozygous for a null allele in any one of these genes, the pigment-synthesizing pathway can be blocked, and an abnormal eye color will be produced. This allele essentially nullifies the work of all the other genes, masking their contributions to the phenotype.

A mutant allele is said to be epistatic when it conceals the presence of another mutant allele in the genotype. We have already seen that a recessive mutation in the *cinnabar* gene of *Drosophila* causes the eyes of the fly to be bright red. A recessive mutation in a different gene causes the eyes to be white. When both of these mutations are made homozygous in the same fly, the eye color is white. Thus the *white* mutation is epistatic to the *cinnabar* mutation.

What physiological mechanism makes the *white* mutation epistatic to the *cinnabar* mutation? For many years, the answer to this question was not known. However, recent molecular analysis has shown that the polypeptide product of the *white* gene transports pigment into the *Drosophila* eye. When this gene is mutated, the transporter polypeptide is not made and the eyes remain colorless, even though red pigment is synthesized in other tissues of the fly. A fly that is homozygous for the *cinnabar* and *white* mutations therefore has white eyes.

The analysis of epistatic relationships such as the one between *cinnabar* and *white* can suggest ways in which genes control a phenotype. A classic example of this analysis is again from the work of Bateson and Punnett, who studied the genetic control of flower color in the sweet pea, *Lathyrus odoratus* (Figure 4.15*a*). The flowers in this plant are either purple or white—purple if they contain anthocyanin pigment and white if they do not. Bateson and Punnett crossed two different varieties with white flowers to obtain F_1 hybrids, which all had purple flowers. When these hybrids were intercrossed, Bateson and Punnett obtained a ratio of 9 purple: 7 white plants in the F_2. They explained the results by proposing that two independently assorting genes, *C* and *P*, are involved in anthocyanin synthesis and that each gene has a recessive allele that abolishes pigment production (Figure 4.15*b*).

Given this hypothesis, the parental varieties must have had complementary genotypes: *cc PP* and *CC pp*. When the two varieties were crossed, they produced *Cc Pp* double heterozygotes that had purple flowers. In this system, a dominant allele from each gene is necessary for the synthesis of anthocyanin pigment. In the F_2, 9/16 of the plants are *C- P-* and have purple flowers; the remaining 7/16 are homozygous for at least one of the recessive alleles and have white flowers. Notice that the double recessive homozygotes, *cc pp*, are not phenotypically different from either of the single recessive homozygotes. Bateson and Punnett's work established that each of the recessive alleles is epistatic over the dominant allele of the other gene. A plausible explanation is that each dominant allele produces an enzyme that controls a step in the synthesis of anthocyanin from a biochemical precursor. If a dominant allele is not present, its step in the biosynthetic pathway is blocked and anthocyanin is not produced:

Gene		*C*		*P*	
	Precursor	⟶	Intermediate	⟶	Anthocyanin
Genotype					
C- P-	+		+		+
cc P-	+		−		−
C- pp	+		+		−

Another classic study of epistasis was performed by George Shull using a weedy plant called the shepherd's purse, *Bursa bursa-pastoris* (Figure 4.16*a*). The seed capsules of this plant are either triangular or

(*a*)

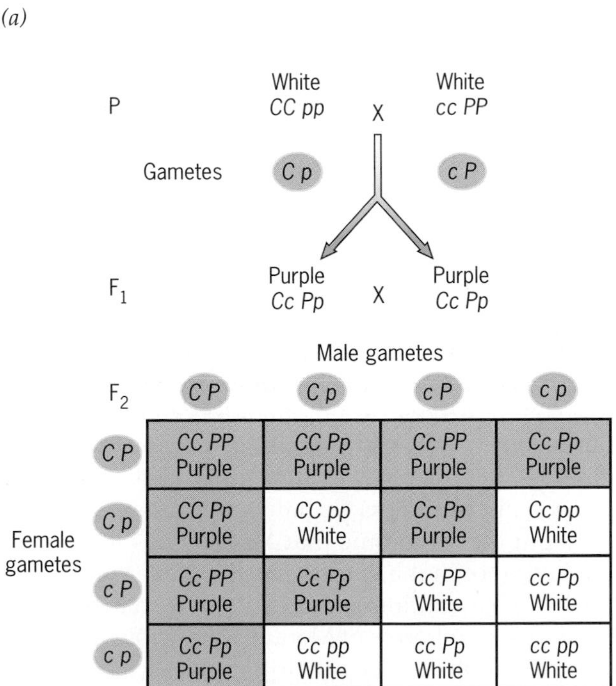

	C P	C p	c P	c p
C P	*CC PP* Purple	*CC Pp* Purple	*Cc PP* Purple	*Cc Pp* Purple
C p	*CC Pp* Purple	*CC pp* White	*Cc Pp* Purple	*Cc pp* White
c P	*Cc PP* Purple	*Cc Pp* Purple	*cc PP* White	*cc Pp* White
c p	*Cc Pp* Purple	*Cc pp* White	*cc Pp* White	*cc pp* White

Summary: 9/16 purple, 7/16 white

(*b*)

Figure 4.15 (*a*) White and purple flowers of the sweet pea. (*b*) Bateson and Punnett's experiment on the genetic control of flower color in sweet peas.

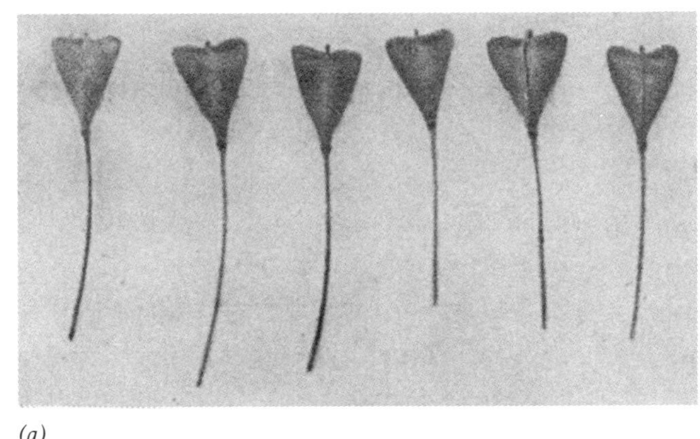

(*a*)

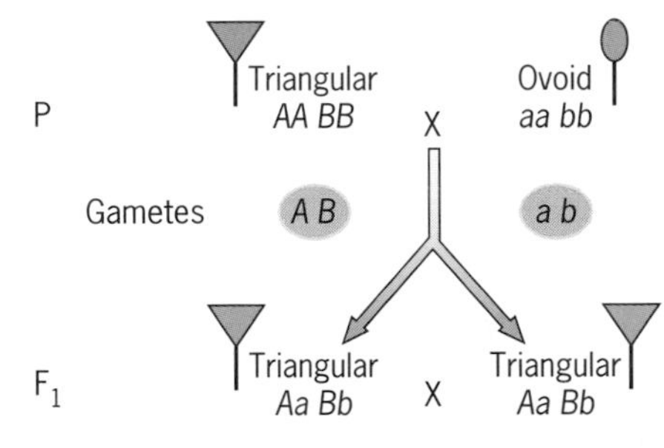

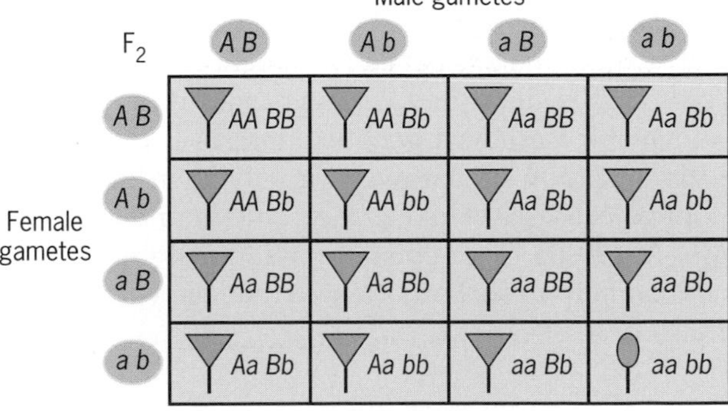

	A B	A b	a B	a b
A B	*AA BB*	*AA Bb*	*Aa BB*	*Aa Bb*
A b	*AA Bb*	*AA bb*	*Aa Bb*	*Aa bb*
a B	*Aa BB*	*Aa Bb*	*aa BB*	*aa Bb*
a b	*Aa Bb*	*Aa bb*	*aa Bb*	*aa bb*

Summary: 15/16 triangular, 1/16 ovoid

(*b*)

Figure 4.16 (*a*) Triangular seed capsules of the shepherd's purse, *Bursa bursa-pastoris*. (*b*) Crosses showing duplicate gene control of seed capsule shape in the shepherd's purse.

ovoid in shape. Ovoid capsules are produced only if a plant is homozygous for the recessive alleles of two genes—that is, if it has the genotype *aa bb*. If the dominant allele of either gene is present, the plant produces triangular capsules. The evidence for this conclusion comes from crosses between doubly heterozygous plants (Figure 4.16*b*). Such crosses produce progeny in a ratio of 15 triangular: 1 ovoid, indicating that the dominant allele of one gene is epistatic over the recessive allele of the other. The data suggest that capsule shape is determined by duplicate developmental pathways, either of which can produce a triangular capsule. One pathway involves the dominant allele of the *A* gene, and the other the dominant allele of the *B* gene. A precursor substance can be converted into a product that leads to a triangular seed capsule through either of these pathways. Only when both pathways are blocked by homozygous recessive alleles is the triangular phenotype suppressed and an ovoid capsule produced:

Genotype

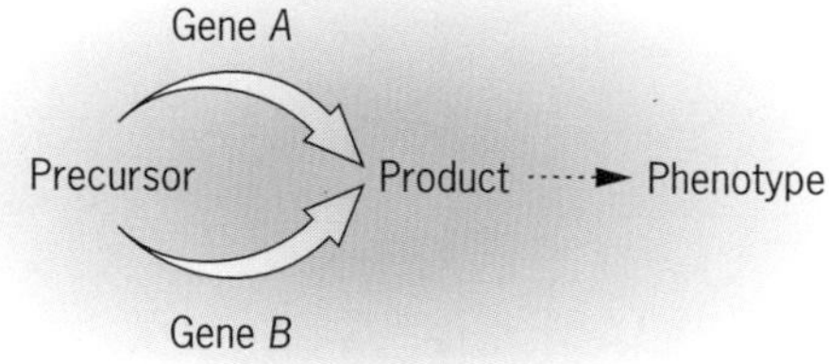

Genotype			
A– B–	+	+	triangular
aa B–	+	+	triangular
A– bb	+	+	triangular
aa bb	+	–	ovoid

These examples indicate that a particular phenotype is often the result of a process controlled by more than one gene. Each gene governs a step in a pathway that is part of the process. When a gene is mutated to a nonfunctional or partially functional state, the process can be disrupted, leading to a mutant phenotype. Much of modern genetic analysis is devoted to the investigation of pathways involved in important biological processes such as metabolism and development. Studying the epistatic relationships among genes can help to sort out the role that each gene plays in these processes.

Pleiotropy

Not only is it true that a phenotype can be influenced by many genes, but it is also true that a gene can influence many phenotypes. When a gene affects many aspects of the phenotype, it is said to be **pleiotropic,** from the Greek words for "to take many turns." The gene for phenylketonuria in human beings is an example. The primary effect of recessive mutations in this gene is to cause toxic substances to accumulate in the brain, leading to mental impairment. However, these mutations also interfere with the synthesis of melanin pigment, lightening the color of the hair; therefore, individuals with PKU frequently have light brown or blond hair. Biochemical tests also reveal that the blood and urine of PKU patients contain compounds that are rare or absent in normal individuals. This manifold of phenotypic effects is typical of most genes and probably results from interconnections between the biochemical and cellular pathways that the genes control.

Another example of pleiotropy comes from the study of mutations affecting the formation of bristles in *Drosophila*. Wild-type flies have long, smoothly curved bristles on the head and thorax. Flies homozygous for the *singed* bristle muation have short, twisted bristles on these body parts—as if they had been scorched. Thus, the wild-type *singed* gene product is needed for the proper formation of bristles. It is also needed for the production of healthy, fertile eggs. We know this fact because females that are homozygous for certain *singed* mutations are completely sterile; they lay flimsy, ill-formed eggs that never hatch. However, these mutations have no adverse effect on male fertility. Thus, the *singed* gene pleiotropically controls the formation of both bristles and eggs in females and the formation of bristles in males.

The Genetic Basis of Continuous Phenotypic Variation

Mendel's principles met with rapid acceptance because they were able to explain the inheritance of discrete traits such as flower color and seed shape. However, more complex traits such as height, weight, and agricultural yield seemed to defy Mendelian analysis. These complex traits vary more or less continuously in a population, making it difficult, if not impossible, to define distinct phenotypes. Attempts to study the inheritance of such traits began in the nineteenth century about the same time that Mendel was investigating inheritance in the garden pea. However, it was not until 1918 that continuous phenotypic variation could be understood in terms of Mendel's principles. The key paper was published by the British geneticist and statistician Ronald A. Fisher.

Fisher proposed that continuously varying traits are influenced by a large number of genes, each segregating different alleles. He further proposed that these traits are influenced by a host of environmental factors, each having a small effect. This combination of genetic and environmental factors creates a wide range of phenotypes distributed more or less continuously in a population of individuals. Thus, for example, the height of adult human males in the United States ranges from scarcely 90 centimeters to over 200 centimeters. However, very few individuals are found at the extremes. Most men in the U.S. population are between 160 and 180 centimeters tall (Figure 4.17). The distribution is approximately bell-shaped, with a peak corresponding to the mean height of 175.2 centimeters (5 feet 9 inches). Fisher developed a theory to explain the inheritance of continuously varying traits such as height by extending Mendel's principles to a large number of genes. We discuss this theory, which has become the cornerstone of modern quantitative genetics, in Chapter 5.

Key Points: **Gene action is affected by biological and physical factors in the environment. Two or more genes may determine a trait, and one of them may exert an overriding effect on it. A gene may affect many different phenotypes. Continuous variation in a trait can be explained by the combined action of many genes and environmental factors.**

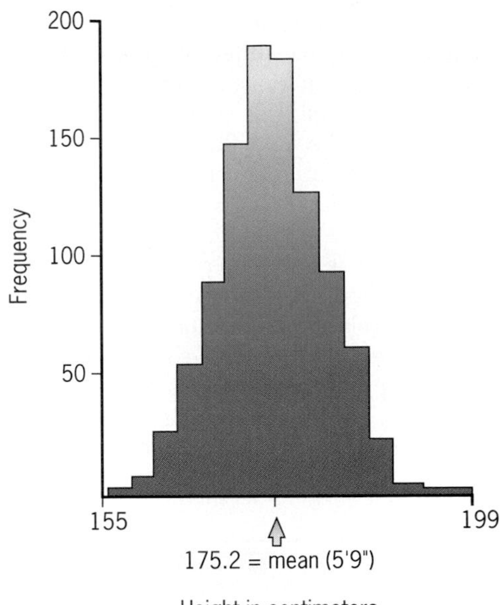

Figure 4.17 Distribution of height among 1000 male students at Harvard University early in the twentieth century.

TESTING YOUR KNOWLEDGE

1. A geneticist has obtained two true-breeding strains of mice, each homozygous for an independently discovered recessive mutation that prevents the formation of hair on the body. One mutant strain is called *naked*, and the other is called *hairless*. To determine whether the two mutations are alleles, the geneticist crosses *naked* and *hairless* mice with each other. All the offspring are phenotypically wild-type; that is, they have hairs all over their bodies. After intercrossing these F_1 mice, the geneticist observes 115 wild-type mice and 85 mutant mice in the F_2. Are the *naked* and *hairless* mutations alleles? How would you explain the segregation of wild-type and mutant mice in the F_2?

ANSWER

The *naked* and *hairless* mutations are not alleles because the F_1 hybrids are phenotypically wild-type. Thus, *naked* and *hairless* are mutations of two different genes. To explain the phenotypic ratio in the F_2, let us first adopt symbols for these mutations and their dominant wild-type alleles:

n = *naked* mutation, N = wild-type allele

h = *hairless* mutation, H = wild-type allele

With these symbols, the genotypes of the true-breeding parental strains are *nn HH* (*naked*) and *NN hh* (*hairless*). The F_1 hybrids produced by crossing these strains are therefore *Nn Hh*. When these hybrids are intercrossed, we expect many different genotypes to appear in the offspring. However, each recessive allele, when homozygous, prevents the formation of hair on the body. Thus, only mice that are genotypically *N- H-* will develop hair; all the others—homozygous *nn* or homozygous *hh*, or homozygous for both recessive alleles—will fail to develop body hair. We can predict the frequencies of the wild and mutant phenotypes if we assume that the naked and hairless genes assort independently. The frequency of mice that will be *N- H-* is $(3/4) \times (3/4) = 9/16 = 0.56$ (by the Multiplicative Rule of probability), and the frequency of mice that will be either *nn* or *hh* (or both) is $(1/4) + (1/4) - [(1/4) \times (1/4)] = 7/16 = 0.44$ (by the Additive Rule of probability). Thus, in a sample of 200 F_2 progeny, we expect $200 \times 0.56 = 112$ to be wild-type and $200 \times 0.44 = 88$ to be mutant. The observed frequencies of 115 wild-type and 85 mutant mice are close to these expected numbers, suggesting that the hypothesis of two independently assorting genes for body hair is, indeed, correct.

2. In fruit flies a recessive mutation, *w*, causes the eyes to be white. Another recessive mutation, *v*, causes them to be vermilion, and a third recessive mutation, *bw*, causes them to be brown. The wild-type eye color is reddish-brown. Hybrids produced by crossing any two homozygous mutants have reddish-brown eyes, and all the doubly homozygous mutant combinations have white eyes. How many genes do these three mutations define? If the reddish-brown color of wild-type eyes is due to the accumulation of two different pigments, one red and the other brown, which gene controls the expression of which pigment? Can the genes be ordered into a pathway for pigment accumulation?

ANSWER

The three mutations define three different genes because when any two homozygous mutations are crossed, the offspring have wild-type eye color. The *w* mutation prevents the expression of all pigment because flies homozygous for it have neither red nor brown pigment in their eyes; the *v* mutation prevents the expression of brown pigment because flies homozygous for it have vermilion (bright red) eyes; and the *bw* mutation prevents the expression of red pigment because flies that are homozygous for it have brown eyes. Thus, the wild-type *v* gene controls the expression of brown pigment, the wild-type *bw* gene controls the expression of red pigment, and the wild-type *w* gene is necessary for the expression of both pigments. We can summarize these findings by proposing that each pigment is expressed in a different pathway and that the functioning of these pathways depends on the wild-type *w* gene.

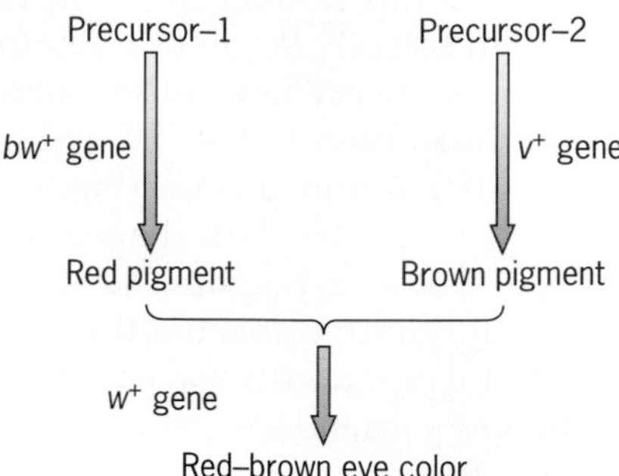

QUESTIONS AND PROBLEMS

4.1 What blood types could be observed in children born to a woman who has blood type M and a man who has blood type MN?

4.2 In rabbits, coloration of the fur depends on alleles of the gene *c*. From information given in the chapter, what phenotypes and proportions would be expected from the following crosses: (a) $c^+c^+ \times cc$; (b) $c^+c \times c^+c$; (c) $c^+c^h \times c^+ c^{ch}$; (d) $cc^{ch} \times cc$; (e) $c^+c^h \times c^+c$; (f) $c^hc \times cc$?

4.3 The flower colors of plants in a particular population may be blue, purple, turquoise, light-blue, or white. A series of crosses between different members of the population produced the following results:

Cross	Parents	Progeny
1	purple × blue	all purple
2	purple × purple	76 purple, 25 turquoise
3	blue × blue	86 blue, 29 turquoise
4	purple × turquoise	49 purple, 52 turquoise
5	purple × purple	69 purple, 22 blue
6	purple × blue	50 purple, 51 blue
7	purple × blue	54 purple, 26 blue, 25 turquoise
8	turquoise × turquoise	all turquoise
9	purple × blue	49 purple, 25 blue, 23 light-blue
10	light-blue × light-blue	60 light-blue, 29 turquoise, 31 white
11	turquoise × white	all light-blue
12	white × white	all white
13	purple × white	all purple

How many genes and alleles are involved in the inheritance of flower color? Indicate all possible genotypes for the following phenotypes: (a) purple; (b) blue; (c) turquoise; (d) light-blue; (e) white.

4.4 In mice, a series of five alleles determines fur color. In order of dominance, these alleles are: A^Y, yellow fur but homozygous lethal; A^L, agouti with light belly; A^+, agouti (wild-type); a^t, black and tan; and *a*, black. For each of the following crosses, give the coat color of the parents and the phenotypic ratios expected among the progeny: (a) $A^YA^L \times A^YA^L$; (b) $A^Ya \times A^La^t$; (c) $a^ta \times A^Ya$; (d) $A^La^t \times A^LA^L$; (e) $A^LA^L \times A^YA^+$; (f) $A^+a^t \times a^ta$; (g) $a^ta \times aa$; (h) $A^YA^L \times A^+a^t$, and (i) $A^Ya^L \times A^YA^+$.

4.5 In several plants, such as tobacco, primrose, and red clover, combinations of alleles in eggs and pollen have been found to influence the reproductive compatibility of the plants. Homozygous combinations, such as S^1S^1, do not develop because S^1 pollen is not effective on S^1- stigmas. However, S^1 pollen is effective on S^2S^3 stigmas. What progeny might be expected from the following crosses (seed parent written first): (a) $S^1S^2 \times S^2S^3$; (b) $S^1S^2 \times S^3S^4$; (c) $S^4S^5 \times S^4S^5$; and (d) $S^3S^4 \times S^5S^6$?

4.6 From information in the chapter about the ABO blood types, what phenotypes and ratios are expected from the following matings: (a) $I^AI^A \times I^BI^B$; (b) $I^AI^B \times I^OI^O$; (c) $I^AI^O \times I^BI^O$; and (d) $I^AI^O \times I^OI^O$?

4.7 A woman with type O blood gave birth to a baby, also with type O blood. The woman alleged that a man with type AB blood was the father of the baby. Is there any merit to her allegation?

4.8 Another woman with type AB blood gave birth to a baby with type B blood. Two different men claim to be the father. One has type A blood, the other type B blood. Can the genetic evidence decide in favor of either?

4.9 A woman who has blood type O and blood type M marries a man who has blood type AB and blood type MN. If we assume that the genes for the A-B-O and M-N blood-typing systems assort independently, what blood types might the children of this couple have, and in what proportions?

4.10 A Japanese strain of mice has a peculiar uncoordinated gait called waltzing, which is due to a recessive allele, *v*. The dominant allele *V* causes mice to move in a coordinated fashion. A mouse geneticist has recently isolated another recessive mutation that causes uncoordinated movement. This mutation, called *tango*, could be an allele of the *waltzing* gene, or it could be a mutation in an entirely different gene. Propose a test to determine whether the *waltzing* and *tango* mutations are alleles, and if they are, propose symbols to denote them.

4.11 Congenital deafness in human beings is inherited as a recessive condition. In the pedigree below, two deaf individuals, each presumably homozygous for a recessive mutation, have married and produced four children with normal hearing. Propose an explanation.

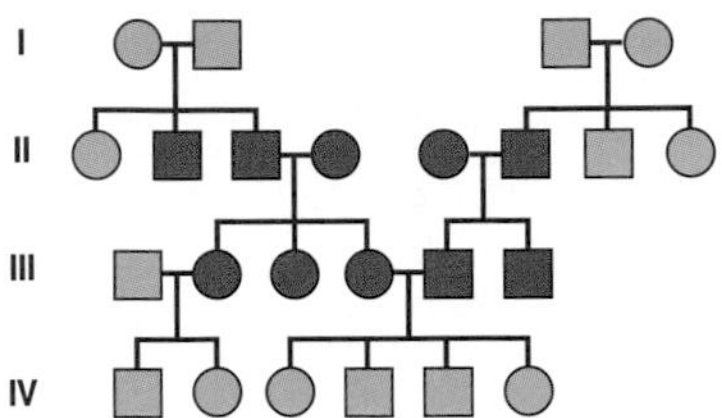

4.12 In the fruit fly, recessive mutations in either of two independently assorting genes, *brown* and *purple*, prevent the synthesis of red pigment in the eyes. Thus, homozygotes for either of these mutations have brownish-purple eyes. However, heterozygotes for both of these mutations have dark red, that is, wild-type eyes. If such double heterozygotes are intercrossed, what kinds of progeny will be produced, and in what proportions?

4.13 The dominant mutation *Plum* in the fruit fly also causes brownish-purple eyes. Is it possible to determine whether *Plum* is an allele of the *brown* or *purple* genes?

4.14 From information given in the chapter, explain why mice with yellow coat color are not true-breeding.

4.15 A couple has four children. Neither the father nor the mother is bald; one of the two sons is bald, but neither of the daughters is bald. (a) If one of the daughters marries a nonbald man and they have a son, what is the chance that the son will become bald as an adult? (b) If the couple has a daughter, what is the chance that she will become bald as an adult?

4.16 The pedigree below shows the inheritance of ataxia, a rare neurological disorder characterized by uncoordinated movements. Is ataxia caused by a dominant or a recessive allele? Explain.

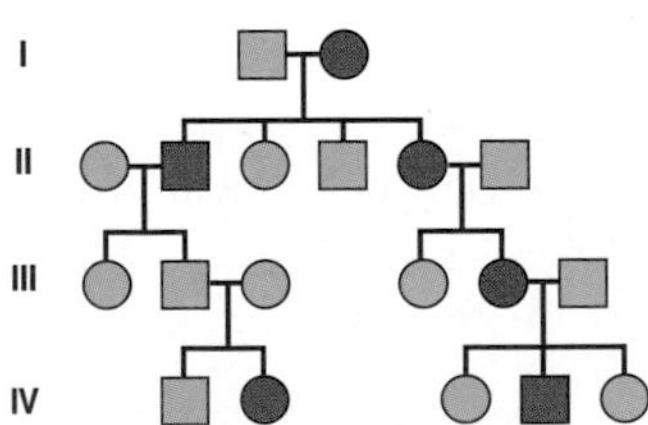

4.17 Chickens that carry both the alleles for rose comb (*R*) and pea comb (*P*) have walnut combs, whereas chickens that lack both of these alleles (that is, they are genotypically *rr pp*) have single combs. From the information about interactions between these two genes given in the chapter, determine the phenotypes and proportions expected from the following crosses: (a) *RR Pp* × *rr Pp*; (b) *rr PP* × *Rr Pp*; (c) *Rr Pp* × *Rr pp*; (d) *Rr pp* × *rr pp*.

4.18 Rose-comb chickens mated with walnut-comb chickens produced 15 walnut, 14 rose, 5 pea, and 6 single-comb chicks. Determine the genotypes of the parents.

4.19 Summer squash plants with the dominant allele *C* bear white fruit, whereas plants homozygous for the recessive allele *c* bear colored fruit. When the fruit is colored, the dominant allele *G* causes it to be yellow; in the absence of this allele (that is, with genotype *gg*), the fruit color is green. What are the F_2 phenotypes and proportions expected from intercrossing the progeny of *CC GG* and *cc gg* plants? Assume that the *C* and *G* genes assort independently.

4.20 The white Leghorn breed of chickens is homozygous for the dominant allele *C*, which produces colored feathers. However, this breed is also homozygous for the dominant allele *I* of an independently assorting gene that inhibits coloration of the feathers. Consequently, Leghorn chickens have white feathers. The white Wyandotte breed of chickens has neither the allele for color nor the inhibitor of color; it is therefore genotypically *cc ii*. What are the F_2 phenotypes and proportions expected from intercrossing the progeny of a white Leghorn hen and a white Wyandotte rooster?

4.21 Fruit flies homozygous for the recessive mutation *scarlet* have bright red eyes because they cannot synthesize brown pigment. Fruit flies homozygous for the recessive mutation *brown* have brownish-purple eyes because they cannot synthesize red pigment. Fruit flies homozygous for both of these mutations have white eyes because they cannot synthesize either type of pigment. The *brown* and *scarlet* mutations assort independently. If fruit flies that are heterozygous for both of these mutations are intercrossed, what kinds of progeny will they produce, and in what proportions?

4.22 Consider the following hypothetical scheme of determination of coat color in a mammal. Gene *A* controls the conversion of a white pigment P_0 into a gray pigment P_1; the dominant allele *A* produces the enzyme necessary for this conversion, and the recessive allele *a* produces an enzyme without biochemical activity. Gene *B* controls the conversion of the gray pigment P_1 into a black pigment P_2; the dominant allele *B* produces the active enzyme for this conversion, and the recessive allele *b* produces an enzyme without activity. The dominant allele *C* of a third gene produces a polypeptide that completely inhibits the activity of the enzyme produced by gene *A*; that is, it prevents the reaction $P_0 \longrightarrow P_1$. Allele *c* of this gene produces a defective polypeptide that does not inhibit the reaction $P_0 \longrightarrow P_1$. Genes *A*, *B*, and *C* assort independently, and no other genes are involved. In the F_2 of the cross *AA bb CC* × *aa BB cc*, what is the expected phenotypic segregation ratio?

4.23 What F_2 phenotypic segregation ratio would be expected for the cross described in the preceding problem if the dominant allele, *C*, of the third gene produced a product that completely inhibited the activity of the enzyme produced by gene *B*—that is, prevented the reaction $P_1 \longrightarrow P_2$, rather than inhibiting the activity of the enzyme produced by gene *A*?

4.24 The Micronesian Kingfisher, *Halcyon cinnamomina*, has a cinnamon-colored face. In some birds, the color continues onto the chest, producing one of three patterns: a circle, a shield, or a triangle; in other birds, there is no color on the chest. A male with a colored triangle was crossed with a female that had no color on her chest, and all their offspring had a colored shield on the chest. When these offspring were intercrossed, they produced an F_2 with a phenotypic ratio of 3 circle: 6 shield: 3 triangle: 4 no color. (a) Determine the mode of inheritance for this trait and indicate the genotypes of the birds in all three generations. (b) If a male without color on his chest is mated to a female with a colored shield on her chest and the F_1 segregate in the ratio of 1 circle: 2 shield: 1 triangle, what are the genotypes of the parents and their progeny?

4.25 In a species of tree, seed color is determined by four independently assorting genes: *A*, *B*, *C*, and *D*. The recessive alleles of each of these genes (*a*, *b*, *c*, and *d*) produce abnormal enzymes that cannot catalyze a reaction in the biosynthetic pathway for seed pigment. This pathway is diagrammed as follows:

White precursor $\xrightarrow{A}$ Yellow $\xrightarrow{B}$ Orange $\xrightarrow{C}$ Red

Orange $\xrightarrow{D}$ Blue

When both red and blue pigments are present, the seeds are purple. Trees with the genotypes *Aa Bb Cc Dd* and *Aa Bb Cc dd* were crossed. (a) What color are the seeds in these two parental genotypes? (b) What proportion of the offspring from the cross will have white seeds? (c) Determine the relative proportions of red, white, and blue offspring from the cross.

4.26 Multiple crosses were made between true-breeding lines of black and yellow Labrador retrievers. All the F_1 progeny were black. When these progeny were intercrossed, they produced an F_2 consisting of 91 black, 39 yellow and 30 chocolate. (a) Propose an explanation for the inheritance of coat color in Labrador retrievers. (b) Propose a biochemical pathway for coat color determination and indicate how the relevant genes control coat coloration.

4.27 Two plants with white flowers, each from true-breeding strains, were crossed. All the F_1 plants had red flowers. When these F_1 plants were intercrossed, they produced an F_2 consisting of 177 plants with red flowers and 142 with white flowers. (a) Propose an explanation for the inheritance of flower color in this plant species. (b) Propose a biochemical pathway for flower pigmentation and indicate which genes control which steps in this pathway.

4.28 Consider the following genetically controlled biosynthetic pathway for pigments in the flowers of a hypothetical plant:

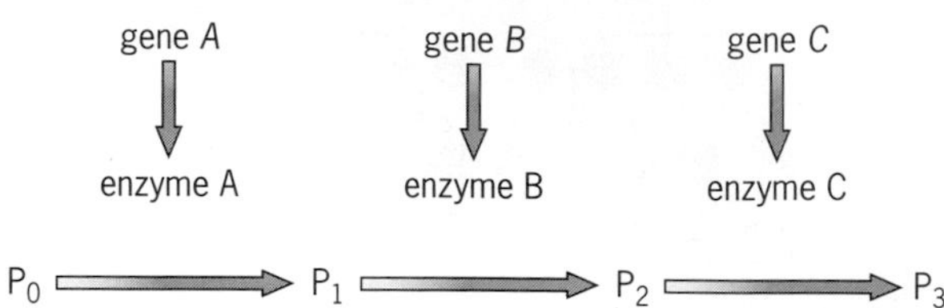

Assume that gene *A* controls the conversion of a white pigment, P_0, into another white pigment, P_1; the dominant allele *A* specifies an enzyme necessary for this conversion, and the recessive allele *a* specifies a defective enzyme without biochemical function. Gene *B* controls the conversion of the white pigment, P_1, into a pink pigment, P_2; the dominant allele, *B*, produces the enzyme necessary for this conversion, and the recessive allele, *b*, produces a defective enzyme. The dominant allele, *C*, of the third gene specifies an enzyme that converts the pink pigment, P_2, into a red pigment, P_3; its recessive allele, *c*, produces an altered enzyme that cannot carry out this conversion. The dominant allele, *D*, of a fourth gene produces a polypeptide that completely inhibits the function of enzyme C, that is, it blocks the reaction $P_2 \longrightarrow P_3$. Its recessive allele, *d*, produces a defective polypeptide that does not block this reaction. Assume that flower color is determined solely by these four genes and that they assort independently. In the F_2 of a cross between plants of the genotype *AA bb CC DD* and plants of the genotype *aa BB cc dd*, what proportion of the plants will have (a) red flowers? (b) pink flowers? (c) white flowers?

BIBLIOGRAPHY

BATESON, W. 1909. *Mendel's Principles of Heredity.* University Press, Cambridge, England.

CARLSON, E. A. 1966. *The Gene: A Critical History.* W. B. Saunders, Philadelphia.

DUNN, L. C. 1965. *A Short History of Genetics.* McGraw-Hill, New York.

HERRMANN, B. G., AND A. KISPERT. 1994. The *T* genes in embryogenesis. *Trends in Genet.* 10: 280–286.

PETERS, J. A., ED. 1959. *Classic Papers in Genetics.* Prentice-Hall, Englewood Cliffs, NJ.

STEWART, A. ED. 1994. Genetic nomenclature guide. *Trends in Genet.* (supp.)

STURTEVANT, A. H. 1965. *A History of Genetics.* Harper and Row, New York.

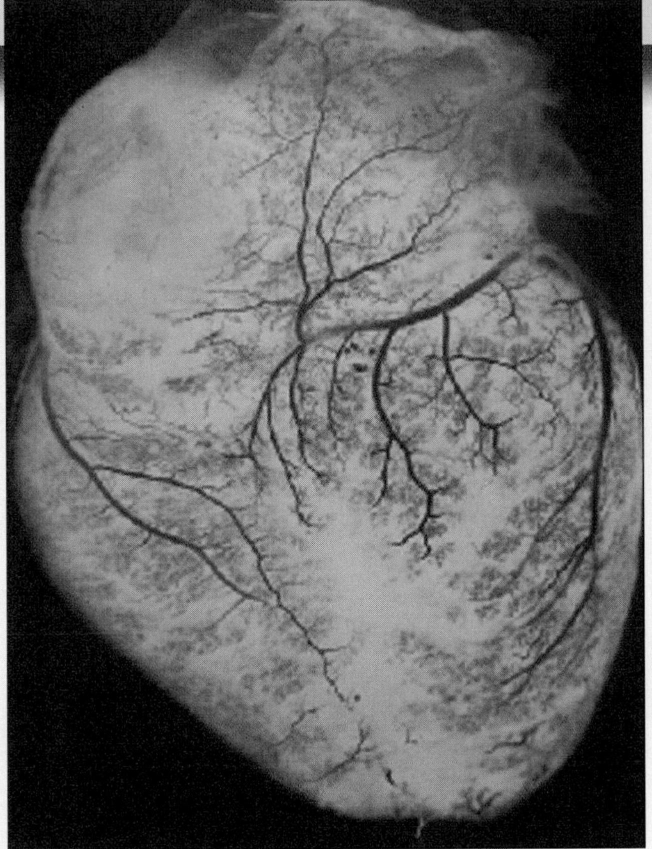

A human heart showing the arteries and veins that supply blood to the cardiac muscles.

5

Inheritance of Complex Traits

CHAPTER OUTLINE

Complex Patterns of Inheritance
Statistics of Quantitative Genetics
Analysis of Quantitative Traits
Inbreeding and the Resemblance Between Relatives

Cardiovascular Disease: A Combination of Genetic and Environmental Factors

Near the end of December 1996, Paul Reston, a 47-year-old biology teacher in a suburban high school outside Pittsburgh, Pennsylvania, was spending his Saturday morning grading examinations. He was somewhat tired that day and felt a bit of stomach distress. He also had a slight pain in his left arm and shoulder. These symptoms had persisted for a few days. At first, Mr. Reston thought he had a mild case of the flu, but the arm and shoulder pain suggested another possibility: that he was having a heart attack. This possibility seemed more real when he remembered that his father had died from a sudden heart attack many years earlier at the relatively young age of 45. After a telephone conversation with a nurse in his health care clinic, Mr. Reston had his son drive him to a nearby hospital, where he spent two hours in the emergency room. The attending physician gave Mr. Reston a battery of tests to evaluate his condition. His heartbeat was regular, his blood pressure was normal, and an electrocardiogram revealed no abnormalities. Biochemical tests for telltale signs of heart damage were also negative. In addition, except for a family history of heart disease, Mr. Reston did not present other major risk factors. He was not overweight, he did not smoke, and he exercised regularly. The physician released Mr. Reston but advised him to return to the hospital the following week for a cardiac stress test. The following Monday, he was tested for heart function while running on a treadmill. The test results were good. Based on his performance, the supervising cardiologist concluded that Mr. Reston had less than a 1 percent chance of suffering a fatal heart attack.

Cardiovascular disease is a serious health problem in the United States. Along with cancer, it is a major cause of death. Many factors contribute to the risk of heart disease: poor diet, lack of physical activity, smoking, and, not least of all, a family history of the disease. However, this disease does not show a simple pattern of inheritance; rather, it seems to involve the interplay of many different genetic and environmental factors.

In this chapter we will consider traits that show complex patterns of inheritance. For human beings, these traits are important in medicine and agriculture, and in nature, they play a key role in the evolution of species.

COMPLEX PATTERNS OF INHERITANCE

Many interesting and important traits, such as disease susceptibility, body size, nutritional content, and various aspects of behavior, do not show simple patterns of inheritance. Nonetheless, we know that variation in these types of traits is under genetic control. One indication is that genetically related individuals resemble each other. We see these resemblances between siblings, between parents and offspring, and sometimes between more distant relatives. The extreme case is monozygotic twins—twins that have developed from a single fertilized egg; these twins often look remarkably alike and sometimes even share mannerisms. The resemblance between relatives is a strong indication that anatomical, physiological, and behavioral traits that do not show simple inheritance patterns are genetically influenced. Another indication of genetic influence is that these types of traits can be altered by selective breeding. In agriculture, plants and livestock have been shaped by propagating individuals with desirable features—greater protein content, reduced body fat, greater productivity, resistance to disease, and so forth. This ability to change phenotypes by selective breeding indicates that variation in the traits has a genetic basis. However, we know from experimentation that the genetic basis is complex. In the following sections, we consider two types of genetically complex traits: (1) traits that vary continuously in a population, and (2) traits that are manifested when an underlying genetic threshold is crossed.

Quantitative Traits

Nature is full of examples of traits that vary continuously—body size, height, weight, the activity of a particular enzyme, reproductive ability, and general vigor and fitness, to name just a few. These are among the most interesting traits in biology, especially from an evolutionary standpoint. One phenotype seems to blend imperceptibly into the next; however, if we examine many individuals in a population, we usually find significant differences among them. Often this type of variation can be quantified by measuring the trait in a sample of individuals from the population. We might, for example, capture mice in a barn and weigh them or collect corncobs from a field and count the number of kernels on each. With such a quantitative approach, the phenotype of every individual in the sample is reduced to a number. These numbers can be analyzed with a variety of statistical techniques, enabling us to study the trait and, in particular, to investigate its genetic basis. Traits that are amenable to this kind of treatment are called **quantitative traits.** Their essential characteristic is that they can be measured.

Charles Darwin thought that such quantitative variation was crucial for the evolution of a species. He hypothesized that this variation provided the raw material for species to change gradually over time. The mechanism for change that Darwin proposed was the differential reproduction of individuals according to phenotype—the process he called natural selection. For example, the largest individuals in a species might produce more offspring than other individuals. To the extent that size is inherited, we would expect the next generation to be a little larger, on average, than the parental generation. Over time, size changes would accumulate and the members of the species would be consistently larger than their ancestors were.

When Darwin proposed his ideas in 1859, there was no mechanistic understanding of heredity; genes, chromosomes, and the Principles of Segregation and Independent Assortment had not yet been discovered—or, more correctly, rediscovered. Mendel was a contemporary of Darwin, but his research on the mechanism of inheritance apparently never came to Darwin's attention. After Mendel's discoveries were brought to light in 1900, biologists eagerly sought to link them to Darwin's evolutionary theory. The difficulty was that the traits that were important for evolution—quantitative traits such as body size, vigor, and reproductive ability—did not "mendelize"; that is, they did not segregate as simple Mendelian factors. Instead, their inheritance was complex. For nearly 20 years, biologists struggled to reconcile the complexity of quantitative traits with the simplicity of Mendel's theory. Some scientists, such as the British statistician Karl Pearson, maintained that they could not be reconciled. These scientists asserted that Mendel's principles simply did not apply to the analysis of complex inheritance patterns. However, this view was eventually overturned. A growing body of evidence suggested that the complex inheritance of quantitative traits could be explained by Mendelian theory. As we will see, two insights made this possible:

1. Quantitative traits are influenced by many genes.
2. Quantitative traits are influenced by many factors in the environment.

It is the combination of genetic and environmental influences that makes quantitative traits difficult to study. These traits depend on a complex mix of factors; that is, they are **multifactorial.**

The Danish biologist Wilhelm Johannsen was one of the first people to show that a portion of the observed variation in a quantiative trait is due to environmental factors. Johannsen worked with the broad bean, *Phaseolus vulgaris*. Under normal field conditions where extensive intercrossing occurs, bean size varies considerably; the smallest beans weigh about 150 mg and the largest weigh about 750 mg. Johannsen established "pure" lines of beans by selecting either the largest or smallest beans from the population and inbreeding for several generations. Although the small and large bean strains were phenotypically distinct from each other, within each strain there was still significant phenotypic variation. Johannsen recognized that these inbred strains should be homozygous for most of their genes. Thus, the residual variation that he observed could not be due to segregating alleles; rather, it must have been due to uncontrolled factors in the environment. Johannsen's work, published in 1903 and 1909, led to the realization that phenotypic variation has two components—one genetic, the other environmental.

Another Scandinavian, Herman Nilsson-Ehle, provided evidence that the genetic component of this variation could involve the contributions of several different genes. Nilsson-Ehle studied color variation in wheat grains. When he crossed a white-grained variety with a dark red-grained variety, he obtained an F_1 with an intermediate red phenotype (Figure 5.1). Self-fertilization of the F_1 produced an F_2 with seven distinct classes, ranging from white to dark red. The num-

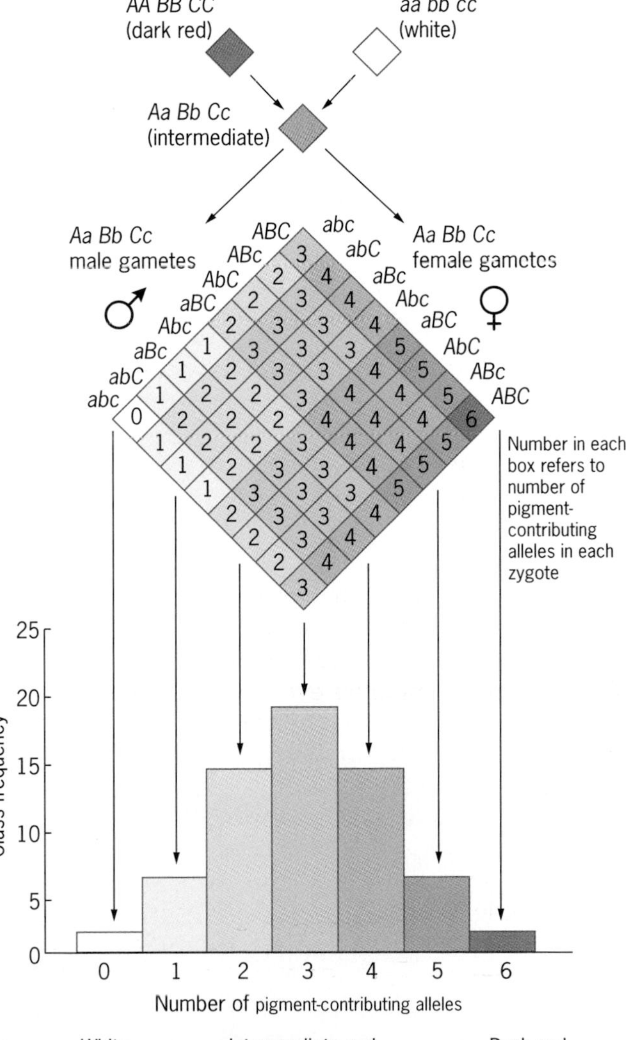

Figure 5.1 Nilsson-Ehle's experiment with kernel color in wheat indicated that three pairs of independently assorting alleles and incomplete dominance accounted for kernel color variation.

ber of F_2 classes and the phenotypic ratio that Nilsson-Ehle observed suggested that three independently assorting genes were involved in the determination of grain color. Nilsson-Ehle hypothesized that each gene had two alleles, one causing red grain color and the other white grain color, and that the allele for red grain color was semidominant over the allele for white grain color. Based on this hypothesis, the genotype of the white-grained parent could be represented as *aa bb cc*, and the genotype of the red-grained parent could be represented as *AA BB CC*. The F_1 genotype would be *Aa Bb Cc*, and the F_2 would contain an array of genotypes that would differ in the number of semidominant, that is, pigment-contributing, alleles present. Each phenotypic class in the F_2 would carry a different number of pigment-contributing alleles. The white class, for example, would carry zero such alleles, the intermediate red class would carry three, and the dark red class would carry six. Nilsson-Ehle's work, published in 1909, showed that a complex inheritance pattern could be explained by the segregation and assortment of multiple genes.

The American geneticist Edward M. East extended Nilsson-Ehle's studies to a trait that did not show Mendelian ratios in the F_2. East studied the length of the corolla in tobacco flowers (Figure 5.2). In one pure line, the corolla length averaged 41 mm; in another, it averaged 93 mm. Within each pure line, East observed some phenotypic variation—presumably the result of environmental influences. By crossing the two lines, East obtained an F_1 that had intermediate corolla length and approximately the same amount of variation that he had seen within each of the parental strains. When East intercrossed the F_1 plants, he obtained an F_2 with about the same corolla length, on average, that he saw in the F_1; however, the F_2 plants were much more variable than the F_1. This variability was due to two sources: (1) the segregation and independent assortment of different pairs of alleles controlling corolla length, and (2) environmental factors. East inbred some of the F_2 plants to produce an F_3 and observed less variation within the different F_3 lines than in the F_2. The reduced amount of variation within the F_3 lines was presumably due to the segregation of fewer allelic differences. Thus, the complex inheritance pattern that East observed with corolla length could be explained by a combination of genetic segregation and environmental influences.

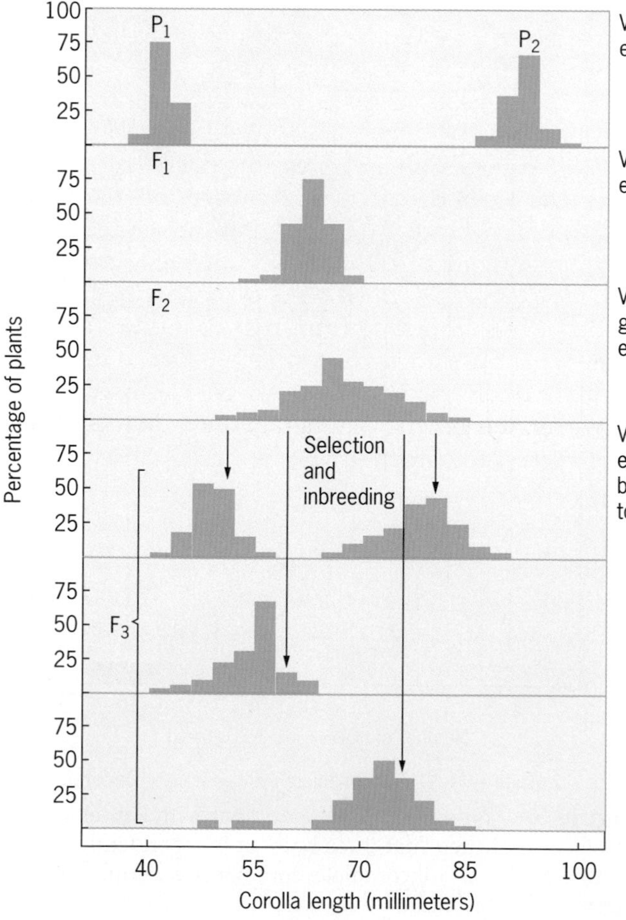

Figure 5.2 E. M. East studied the inheritance of corolla length in tobacco (*Nicotiana longiflora*). The figure shows the frequencies with which individuals fall into various corolla-length classes. The P_1 and P_2 populations are pure lines; the F_1 is uniformly heterozygous. The F_2 was derived by intercrossing the F_1 plants. The four F_3 populations were obtained by inbreeding particular classes of F_2 plants.

How many genes were involved in determining corolla length in East's strains of tobacco? We can make a crude guess by comparing the F_2 plants with each of the inbred parental strains. Let's suppose that the strain with the shorter corollas was homozygous for one set of alleles and that the strain with the longer corollas was homozygous for another set of alleles. Furthermore, let's suppose that the long-corolla alleles are semidominant, that all length-controlling genes assort independently, and that each gene makes an equal contribution to the phenotype. If corolla length were determined by one gene, with alleles *a* (for short corolla) and *A* (for long corolla), we would expect 1/4 of the F_2 plants to have short corollas (like the short parental strain) and 1/4 to have long corollas (like the long parental strain). If two genes determined corolla length, we would expect 1/16 of the F_2 plants to resemble the short corolla parent and 1/16 to resemble the long-corolla parent. If three genes were involved, the frequency of each parental type in the F_2 would be 1/64, and if four genes were involved, it would be 1/256. With five genes, the parental frequencies in the F_2 would each be 1/1024. East studied 444 F_2 plants and failed to find even one with either of the parental phenotypes. This failure would seem to rule out the hypothesis of four or fewer genes controlling corolla length. Thus, we can conclude that at least five genes are responsible for the difference in corolla length between East's two inbred strains.

Threshold Traits

Continuously varying traits such as corolla length, grain color, and bean size are controlled by multiple factors, both genetic and environmental. Geneticists have found that some discontinuous traits also appear to be multifactorial. An example of such a trait in humans is cleft lip with or without cleft palate, a developmental error in which there is a failure of midline fusion during embryogenesis. Twin and family studies have suggested this trait is influenced by multiple genetic and environmental factors. However, cleft lip is not a quantitative trait in the usual sense; individuals either express it or they do not. Nonetheless, the genetic and environmental factors that predispose an individual to show it appear to vary quantitatively. Thus, although cleft lip with or without cleft palate is manifested discontinuously, its underlying variation seems to be continuous. Geneticists theorize that when the underlying variable exceeds a certain level, the trait appears. This type of trait is therefore called a **threshold trait** (Figure 5.3). The underlying variable is a function of factors in the genotype as well as factors in the environment. Some individuals carry alleles that predispose them to develop cleft lip, with or without cleft palate. There is, in effect, a genetic "liability" to develop the condition. However, only those individuals whose liability crosses the threshold are actually at risk.

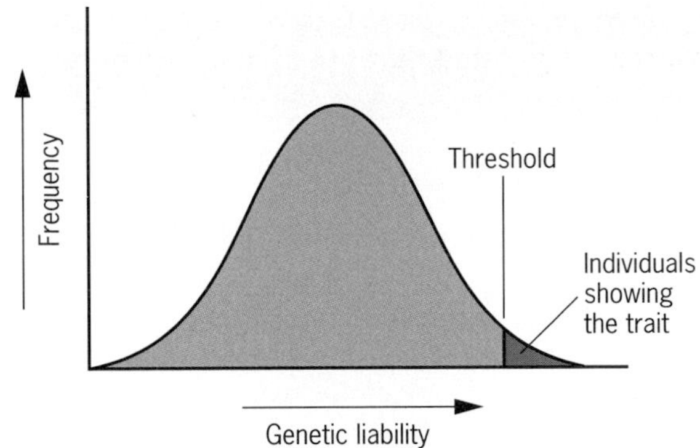

Figure 5.3 A model for expression of a threshold trait. When the underlying variable, the "liability," reaches a threshold value, the trait is expressed. This variable is assumed to be continuously distributed in the population.

How do geneticists know that cleft lip with or without cleft palate is inherited? Two types of evidence suggest that genetic factors are important. The first type comes from twin studies (Table 5.1). Monozygotic twins are genetically identical. When one twin expresses cleft lip, the other twin also expresses it 40 percent of the time—we say that the *concordance rate* is 40 percent. Among dizygotic or fraternal twins, who share 50 percent of their genes, the concordance rate is 4 percent. These data suggest that genes predispose an individual to express this trait, but that environmental factors trigger its ultimate development. If only the genotype controlled the expression of cleft lip, the concordance rate for monozygotic twins would be 100 percent. Table 5.1 shows the concordance rates for other human conditions that seem to be threshold traits. However, the genetic factors that underlie these traits are poorly understood.

TABLE 5.1
Concordance Among Monozygotic (MZ) and Dizygotic (DZ) Twins for Common Malformations and Diseases

	Concordance (%)	
Trait	***MZ***	***DZ***
Cleft lip ± cleft palate	40	4
Clubfoot	30	2
Pyloric stenosis	22	2
Congenital dislocation of hip	33	3
Schizophrenia	60	10
Insulin-dependent diabetes mellitus	50	10

HUMAN GENETICS SIDELIGHT

Genetic Counseling and Multifactorial Traits

For traits that do not follow single-gene modes of inheritance, genetic counseling can be difficult. Nevertheless, there are ways to test for inheritance and ways to counsel families that express multifactorial traits. C. O. Carter suggests five features of multifactorial traits that are important in counseling situations.[1]

1. For multifactorial traits, the risk of recurrence is greater in relatives of an affected person than the incidence in the general population. The magnitude of the risk difference decreases as the incidence of the trait in the general population increases. For example, cleft lip/palate has an incidence of about 0.001 in the general population, and siblings of a person with cleft lip/palate have a 50 times higher risk. Spina bifida, in contrast, has a higher general population incidence of about 0.005, but for the siblings of an affected person, the risk is increased only about sevenfold.
2. For multifactorial traits, the risk of recurrence increases with increasing severity of the malformation. For example, if a couple has a child with mild cleft lip but no cleft palate, the risk that their next child will have the trait is 2.5 percent. However, if the first child is severely affected with cleft lip and palate, the risk that their next child will have the trait more than doubles to about 6 percent.
3. In single-gene inheritance, the risks do not increase with the number of affected family members. For example, in cystic fibrosis, if a couple has an affected child, the risk is 25 percent for each of their subsequent children. That risk will not change no matter how many affected children they have. For multifactorial traits, however, the risk of recurrence increases with the number of affected family members. For example, if a couple has a child with spina bifida, the risk for the next child is about 5 percent. However, if the couple has two children with spina bifida, the risk for their next child doubles to about 10 percent. The risk increases because two cases of spina bifida imply more strongly than one that the parents of the affected children carry genes for this condition.
4. For multifactorial traits, if the trait is expressed more in one sex than in the other, the risk of recurrence is greater in the offspring of the affected person belonging to the less frequently affected sex. For example, pyloric stenosis (narrowing of the pylorus, the band of muscle that functions like a valve at the junction of the stomach and small intestine) is five times more common in males than in females. The risk to a child is greater if the mother had pyloric stenosis than if the father had it.
5. The incidence of multifactorial traits is higher in the offspring from matings between relatives than in matings between unrelated persons.

Before considering these five principles, genetic counselors must be absolutely certain that they are dealing with a multifactorial trait. Although many conditions are multifactorial, some are the result of a pleiotropic effect of a single gene. For example, cleft palate without cleft lip is most likely the result of a single dominant allele. Cleft lip alone or cleft lip with cleft palate is usually multifactorial.

[1]Carter, C. O. 1969. Genetics of common disorders. *Br. Med. Bull.* 25:52–57.

The second type of evidence supporting a genetic basis for cleft lip comes from family studies (Table 5.2). First-degree relatives of a person with the trait are far more likely to express it than are second- and third-degree relatives. This finding is consistent with a genetic basis because first-degree relatives share 50 percent of their genes, whereas second- and third-degree relatives share only 25 and 12.5 percent of their genes, respectively. Thus, for threshold traits such as cleft lip, the risk of showing the condition decreases with the degree of separation between relatives. The Human Genetics Sidelight: Genetic Counseling and Multifactorial Traits discusses how genetics counselors deal with threshold traits.

TABLE 5.2
Family Studies of Cleft Lip, With or Without Cleft Palate

Relatives	*Percentage of Relatives Affected*	*Incidence Relative to General Population*
First degree		
Sibs	4.1	× 40
Children	3.5	× 35
Second degree		
Aunts and uncles	0.7	× 7
Nephews and nieces	0.8	× 8
Third degree		
First cousins	0.3	× 3

Source: C. O. Carter. 1969. Genetics of common disorders. *Br. Med. Bull.* 25:52–57.

Key Points: **Quantitative traits vary continuously on some scale of measurement. Breeding experi-**

ments indicate that these traits are influenced by many genetic and environmental factors. Threshold traits are manifested discontinuously; however, the underlying liability for them varies continuously. An individual whose liability crosses a critical threshold exhibits the trait. In humans, evidence of a genetic basis for threshold traits comes from twin and family studies.

STATISTICS OF QUANTITATIVE GENETICS

In nature, variation in many important traits forms a continuum rather than discrete phenotypic classes. This variation is quantitative, not qualitative, and it poses a formidable problem for the geneticist. Segregation ratios are difficult, if not impossible, to discern because the number of phenotypes is large and one phenotype blends imperceptibly into the next. For quantitatively varying traits, routine genetic analyses of the sort that we have done with seed color in peas and with human disorders such as albinism are out of the question. For these types of traits we must resort to a different kind of analysis, one that is based on statistical descriptions of the phenotype in a population. In the sections that follow, we introduce the basic statistical concepts that are needed for this type of analysis.

Frequency Distributions

The first step in the study of any quantitative trait is to collect measurements of the trait from individuals in a population. Usually, only a small fraction of all the individuals in the population can be measured. We call this group the **sample.** The data from the sample can be presented graphically as a **frequency distribution.** In the graph the horizontal or x-axis measures values of the trait. This axis is divided into regular intervals that allow each individual in the population to be categorized for the trait. Thus, each observation in the sample can be placed into one of the intervals on the x-axis. The vertical or y-axis measures the frequency of the observations within each interval.

Figure 5.4 shows frequency distributions that were obtained in a genetic study of wheat. The investigators measured the time that wheat takes to mature. Four different populations of wheat were grown in test plots in the same season, and 40 plants from each population were monitored until the heads of grain matured. The time to maturity for each plant was recorded in days. Two of the populations (A and B) were inbred strains, and one was an F_1 produced by crossing these two strains. The fourth population was an F_2 produced by intercrossing the F_1 plants.

The two parental strains A and B were highly inbred varieties that were completely or almost completely homozygous. As the frequency distributions indicate, strain A matured quickly and strain B matured slowly. The lack of phenotypic overlap between the samples from these two strains demonstrates their genetic distinctiveness. Apparently, strains A and B were homozygous for different alleles of genes controlling maturation time. Within each strain, however, there was still some phenotypic variation, presumably the

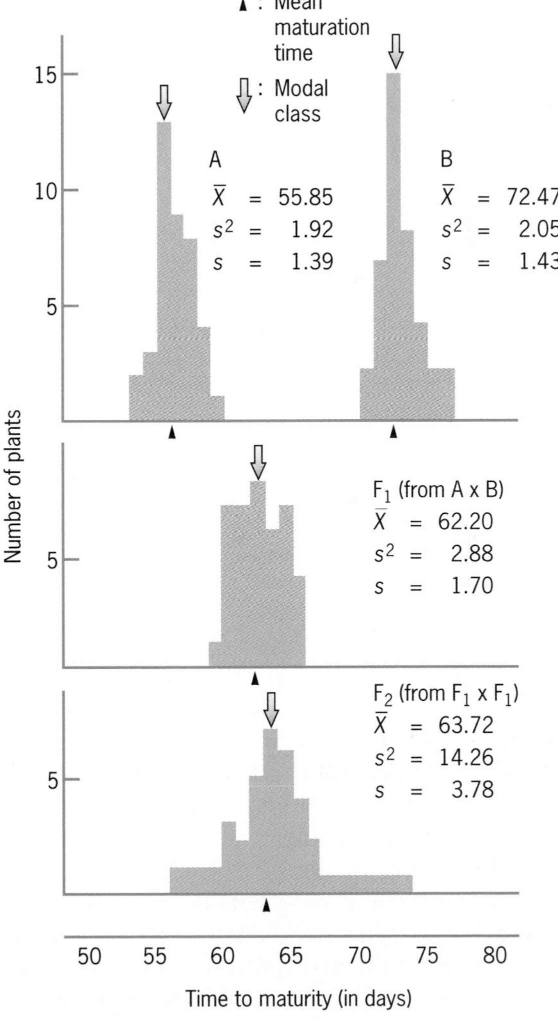

Figure 5.4 Frequency distributions and descriptive statistics of time to maturity in four populations of wheat. A and B are inbred strains that were crossed to produce F_1 hybrids. The F_1 plants were intercrossed to produce an F_2. Seed from all four populations was planted in the same season to determine the time to maturity. In each case, data were obtained from 40 plants. The mean, mode, variance, and standard deviation are given.

result of microenvironmental differences within the test plots.

The distributions of the F_1 and F_2 samples indicate that these populations had intermediate maturation times. Their intermediate position on the x-axis suggests that the alleles controlling maturation time are semidominant. Notice that the distribution of the F_2 sample is considerably broader than that of the F_1. The additional variability seen in the F_2 population reflects the genetic segregation that occurred when the F_1 plants reproduced.

This type of analysis accomplishes two main objectives. First, it describes the trait quantitatively, and second, it tells us that genetic and environmental factors influence the variability of the trait. We are now in a position to consider ways of summarizing the features of a frequency distribution.

The Mean and the Modal Class

The essential characteristics of a frequency distribution can be summarized by simple statistics calculated from the data. One of these summary statistics is called the **mean** or average. It gives us the "center" of the distribution—the "typical" value. We calculate the sample mean ($\overline{X}$) by summing all the data in the sample and dividing by the total number of observations (n). In mathematical notation, the mean is:

$$\overline{X} = (\Sigma X_i)/n$$

The Greek letter Σ in this formula is a mathematical shorthand for the sum of all the individual measurements in the sample; thus, $\Sigma X_i = (X_1 + X_2 + X_3 + \ldots X_n)$, where X_i represents the ith of the n individual observations. In Figure 5.4 the positions of the sample means are indicated by triangles beneath the distributions; the numerical values of these means are given on the right. The means of the F_1 and F_2 samples are 62.20 and 63.72 days, respectively; both are a little less than the average of the means of the two inbred parental strains (64.16 days).

The **modal class** in a sample is the class that contains the most observations. Like the mean, it also captures the "center" of the distribution. In Figure 5.4 the modal classes are indicated by short arrows. We see that in each of the distributions the mean is within or very close to the modal class. This coincidence reflects the symmetry of the distributions; in each case, roughly equal numbers of observations are above and below the mean and the modal class. Not all distributions have this feature. Some are skewed, with most of the observations clustered at one end and only a few at the other end forming a long tail. Statisticians have

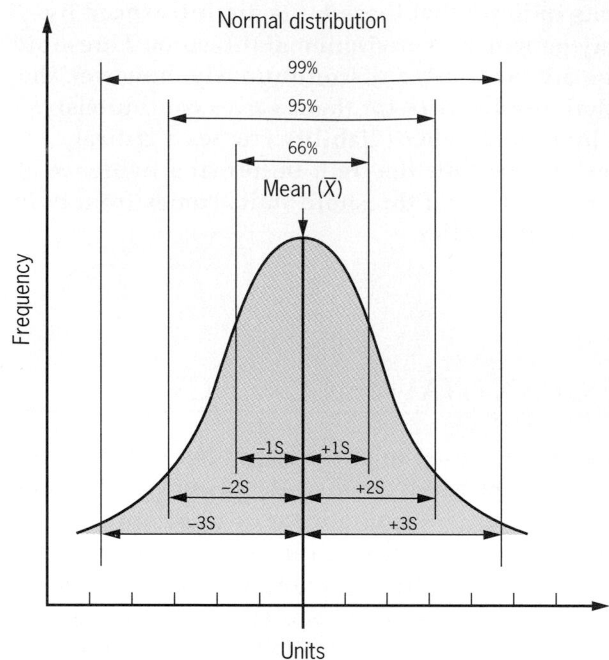

Figure 5.5 A normal frequency distribution showing the proportions of the data included within 1, 2, and 3 standard deviations from the mean.

developed an extensive theory about a particular type of symmetrical distribution called a **normal distribution** (Figure 5.5). In this bell-shaped distribution, the mean and the modal class are located exactly in the center. Often distributions of sample data approximate the shape of a normal distribution. Thus, we can apply the extensive theory about normal distributions to analyze such data.

The Variance and the Standard Deviation

The mean tells us nothing about how data are scattered in a sample distribution. The data could be dispersed or they could be clustered. To measure the scatter, we use a statistic called the **variance.** Data that are widely dispersed produce a large value for the variance, whereas data that are tightly clustered produce a small value. The sample variance, denoted s^2, is calculated from the formula

$$s^2 = \Sigma(X_i - \overline{X})^2/(n - 1)$$

In this formula, $(X_i - \overline{X})^2$ is the squared difference between the ith observation and the sample mean (often called the *squared deviation from the mean*), and the Greek letter Σ indicates that all such squared deviations are summed. The sum of the squared deviations is averaged by dividing by $n - 1$. (For technical reasons, the divisor is one less than the sample size.)

We should note two features of the variance. First, it measures the dispersion of the data around the *mean*. When we calculate the variance, we take the mean to be the central value of the distribution and find the difference between it and each of the observations in the sample. Second, the variance is always positive. When we calculate the variance, we square the difference between each observation and the mean, and then sum the squared differences. Because each of the squared differences is positive, the variance, calculated by summing these squared differences, is also positive.

Although the variance has desirable mathematical properties, it is difficult to interpret because the units of measurement are squared (for example, $s^2 = 2.88$ days^2). Consequently, another statistic, called the **standard deviation,** is often used to describe the variability of a sample. The standard deviation (s) is the square root of the sample variance

$$s = \sqrt{s^2}$$

This statistic is easier to interpret than the variance because it is expressed in the same units as the original measurements.

The variances and standard deviations of the four wheat populations are given in Figure 5.4. The F_2 population has the greatest variance and standard deviation, no doubt because it is segregating for genes that control maturation time. In the F_2 plants, both genetic and environmental differences produce the observed variability. In the other populations, most if not all of the observed variation is due to environmental factors alone. Each of the two parental strains is highly inbred and is therefore expected to be homozygous for most of its genes. The F_1 plants are heterozygous for the alleles that are different in the two parental strains, but they all have the same genotype. Thus, in neither the parental strains nor the F_1 do we expect to find much genetic variation among plants. In a later section we will see how to estimate that part of the variance in a quantitative trait that is due to genetic differences among individuals in a population.

As mentioned above, the distribution of a quantitative trait often looks like a normal distribution. The shape and the position of a normal distribution are completely specified by its mean and standard deviation. Thus, if we know only the mean and standard deviation of a quantitative trait and assume that the trait is normally distributed, we can construct the approximate shape of the trait's distribution: 66 percent of all the individuals will be within one standard deviation of the mean, 95 percent will be within two standard deviations of the mean, and 99 percent will be within three standard deviations of the mean (Figure 5.5).

Key Points: **The mean and the mode are statistics that point to the center of a frequency distribution. The variance and the standard deviation are statistics that indicate the extent to which data are scattered around the mean.**

ANALYSIS OF QUANTITATIVE TRAITS

In this section we will see how statistics are used in the genetic analysis of quantitative traits. The thrust of the analysis is to partition the observed variation in the trait into genetic and environmental components, and then to use the genetic component to make predictions about the phenotypes of the offspring of particular crosses.

The Multiple Factor Hypothesis

The key idea in quantitative genetics is that traits are controlled by many different factors in the environment and in the genotype. This **Multiple Factor Hypothesis** emerged in the second decade of the twentieth century through the experimental investigations of E. M. East, W. Johannsen, H. Nilsson-Ehle, and others. However, it was a theoretician, R. A. Fisher, who crystallized the Multiple Factor Hypothesis into its modern form. Fisher did this work during World War I while he was teaching school in Great Britain. His theoretical analysis was published in 1918, the year the war ended.

Fisher hypothesized that a particular value of a quantitative trait, T, is the result of the combined influence of genetic and environmental factors. He represented the effects of these factors as deviations from the overall population mean:

$$T = \mu + g + e$$

In this equation, the Greek letter μ represents the population mean, g represents the deviation from the mean that is due to genetic factors, and e represents the deviation from the mean that is due to environmental factors. In Fisher's scheme, the position of a particular value of the trait, T, in the population depends on the genetic and environmental factors that have affected it (Figure 5.6). Some factors produce large values of T, and some produce small values of T. For each individual, these factors are different. Furthermore, Fisher emphasized that a multitude of factors are involved. He hypothesized that many genes contribute to a quantitative trait, and he assumed that many aspects of the environment also make contributions. Today we

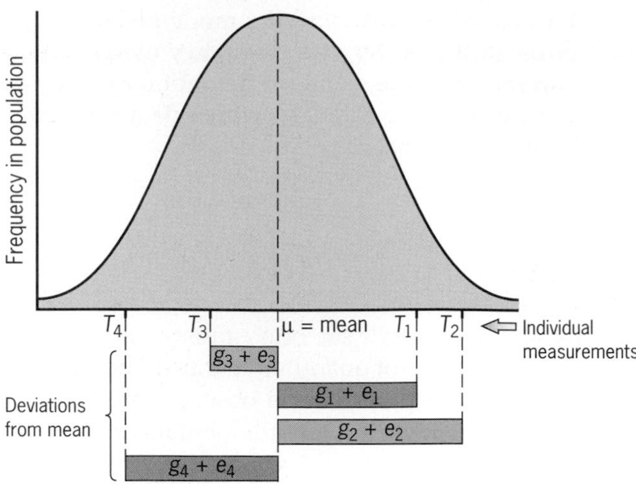

Figure 5.6 Quantitative phenotypes and the deviations of individual measurements from the population mean. Each individual's deviation is hypothesized to consist of a deviation due to its genotype (g) and a deviation due to its environment (e).

say that a trait that is controlled by many genes is **polygenic.**

With these simple ideas, Fisher was able to develop a procedure to analyze the variability of a quantitative trait in terms of the contributing genetic and environmental factors. In essence, he addressed the following question:

To what extent is the variation observed in a population due to genetic differences among individuals, and to what extent is it due to environmental differences?

Fisher approached this question by studying the variance as a measure of scatter in the distribution of a quantitative trait. Specifically, he discovered how to split the overall variance of a trait into two component variances, one measuring the effects of genetic differences and the other measuring the effects of environmental differences. Thus, in Fisher's analysis, the variance of a quantitative trait, symbolized V_T, was equal to the sum of a *genetic variance*, symbolized V_g, and an *environmental variance*, symbolized V_e:

$$V_T = V_g + V_e$$

Often, the total variance of the quantitative trait, V_T, is referred to as the *phenotypic variance.*

A discussion of Fisher's method of splitting the total phenotypic variance into its genetic and environmental components is beyond the scope of this book. However, this method has since been used in many different contexts and has given rise to a general statistical technique called *analysis of variance.* We will consider an example of how the genetic and environmental variances are estimated after we have introduced another important concept—the notion of heritability.

Heritability is the proportion of a population's phenotypic variance that is attributable to genetic differences. Heritability reflects the relative contributions of genetic and environmental differences to the observed variance in a quantitative trait within a particular population. Among populations, the value of the heritability may differ. Why is heritability so important to understand? In agricultural programs, many quantitative traits—body weight in cattle, the number of eggs laid by chickens, or the amount of milk produced by dairy cows—are influenced by the genotype. Knowledge of the heritability of these traits allows agriculturalists to develop goal-oriented selective breeding programs. Many ecologically important traits such as body size, fertility, and disease resistance also depend on genetic factors. Knowing the heritability of these traits helps us to understand how natural populations evolve. In addition, human traits such as blood pressure and plasma cholesterol levels are influenced by multiple genetic factors. Knowing about the underlying genetic variation in these traits can be useful in medicine and health care.

Partitioning the Phenotypic Variance

To estimate heritability, researchers must determine the total variance in a trait (V_T) and then partition this variance into the variance due to genetic differences (V_g) and the variance due to environmental differences (V_e). A simple example with three different scenarios will help to elucidate this idea.

The trait we will focus on is the biochemical activity of an enzyme extracted from the tissues of a plant species. The activity of this enzyme varies quantitatively and depends, as we will see, on both the genotype of the plants and the environment in which they were grown. In the first hypothetical scenario, we examine the enzyme activity of four individual plants representing two different genotypes (A and B) and two different environments (1 and 2):

Individual	Genotype	Environment	Enzyme Activity
1	A	1	32
2	A	2	28
3	B	1	24
4	B	2	20

The mean enzyme activity in this sample is 26, and the variance is 26.6. (We can verify these values using the formulas given earlier.) The value 26.6 represents the total phenotypic variance, V_T, and it reflects both genetic and environmental differences among the four individuals that were assayed for enzyme activity.

In the second hypothetical scenario, we eliminate all the variance due to environmental differences by growing the two different genotypes in the same environment.

Individual	Genotype	Environment	Enzyme Activity
1	A	1	32
2	A	1	32
3	B	1	24
4	B	1	24

The mean in this population is 28, and the variance is 21.3. Because there is no environmental variance (V_e = 0), the entire phenotypic variance is due to genetic differences. Thus, $V_T = V_g = 21.3$.

In the third hypothetical scenario, the genotypes are identical, but the environments vary:

Individual	Genotype	Environment	Enzyme Activity
1	A	1	32
2	A	1	32
3	A	2	28
4	A	2	28

The mean in this sample is 30, and the variance is 5.3. Because the genotypes of the four individuals are the same, all the variation in this case is due to environmental differences; therefore, $V_g = 0$, and $V_T = V_e = 5.3$.

Using this simple example, we can partition the total variance in the first scenario into a variance due to genetic differences and a variance due to environmental differences:

$$V_T = V_g + V_e$$
$$26.6 = 21.3 + 5.3$$

Most of the observed variation in enzyme activity is therefore due to genotypic differences. The remainder is due to the effects of different environments.

Broad-Sense Heritability

Broad-sense heritability, symbolized H^2, is that fraction of the total variance (V_T) that is due to genetic differences among individuals in a population (V_g). In mathematical notation,

$$H^2 = V_g/V_T$$

The symbol H^2 is used to remind us that the broad-sense heritability is calculated from variances, which are squared quantities. In our enzyme activity example, the broad-sense heritability is

$$H^2 = 21.3/26.6 = 0.80$$

Broad-sense heritability ranges from 0 to 1. A 0 means that *none* of the phenotypic variability is due to genetic differences; a 1 means that *all* the phenotypic variability is due to genetic differences.

Let us return to the earlier discussion of corolla length in tobacco flowers and use East's data (Figure 5.2) to estimate the broad-sense heritability of this quantitative trait. The P_1, P_2, and F_1 are all genetically homogeneous, or nearly so ($V_g = 0$); the two parental classes are highly inbred and homozygous, and the F_1 progeny are all identically heterozygous. Because there is essentially no genetic variation in these three populations, all the observed variation in each of them must be due to environmental effects. Thus, using East's published data, we can estimate V_e for corolla length by calculating the average of the variances of the P_1, P_2, and F_1 samples:

$$V_e \text{ (estimated)} = (V_{P1} + V_{P2} + V_{F1})/3 = 7 \text{ mm}^2$$

The F_2 generation is genetically and environmentally heterogeneous. The total variance in this population, 43 mm^2, is due to genetic and environmental differences. However, because we have already estimated V_e, we can estimate V_g by subtraction from the total variance:

$$V_T = V_g + V_e$$
$$V_g = V_T - V_e$$
$$V_g = 43 \text{ mm}^2 - 7 \text{ mm}^2$$
$$= 36 \text{ mm}^2$$

The estimated heritability for corolla length is therefore

$$H^2 = 36/43 = 0.84$$

Broad-sense heritability includes all types of genetic variation. It includes variation due to the effects of individual alleles, to the dominance relationships between alleles, and to the epistatic interactions between alleles of different genes. In the next two sections we will see that by breaking out these components of the broad-sense heritability and by focusing on the component that involves the effects of individual alleles, we can predict the phenotypes of offspring from information about the phenotypes of their parents.

Narrow-Sense Heritability

Estimates of genetic variance are important in predicting the phenotypes of offspring. However, the ability to make these predictions depends on the amount of genetic variation that is due to the additive effects of individual alleles. Genetic variation that is due to the effects of dominance and epistasis has little predictive power.

To see how dominance limits the ability to make predictions, consider the ABO blood types in humans (Table 4.1 in Chapter 4). This trait is determined strictly by the genotype; environmental variation has essentially no effect on the phenotype. However, because of dominance, two individuals with the same phenotype can have different genotypes. For example, a person with type A blood could be either I^AI^A or I^AI^O. If two people with type A blood produce a child, we cannot predict precisely what phenotype the child will have. It could be either type A or type O, depending on the genotypes of the parents; however, we know that it will not have type B or type AB blood. Thus, although we can make some kind of prediction about the child's phenotype, dominance prevents us from making a precise prediction.

Our ability to make predictions about an offspring's phenotype is improved in situations where the genotypes are not confused by dominance. Consider, for example, the inheritance of flower color in the snapdragon, *Antirrhinum majus*. Flowers in this plant are white, red, or pink, depending on the genotype (Figure 4.1 in Chapter 4). As with the ABO blood types, variation in flower color has essentially no environmental component; all the variance is the result of genetic differences. However, for the flower color trait, the phenotype of an individual is determined without the complications of dominance. A plant with two *w* alleles is white, a plant with one *w* allele and one *W* allele is pink, and a plant with two *W* alleles is red. In this system, the phenotype depends simply on the number of *W* alleles present; each *W* allele intensifies the color by a fixed amount. Thus, we can say that the color-determining alleles contribute to the phenotype in a strictly additive fashion. This additivity improves our ability to make predictions in crosses between different plants. A red × red mating predictably produces all red offspring; a white × white mating produces all white offspring; and a red × white mating produces all pink offspring. The only uncertainty is in a cross between two heterozygotes, and in this case the uncertainty is due to Mendelian segregation, not to dominance.

Quantitative geneticists distinguish between genetic variance that is due to alleles that act additively (such as those in the flower color example just discussed) and genetic variance that is due to dominance. These different variance components are symbolized as:

$$V_a = \text{additive genetic variance}$$

$$V_d = \text{dominance variance}$$

In addition, geneticists define a third variance component that measures variation due to epistatic interactions between alleles of different genes:

$$V_i = \text{epistatic variance}$$

Epistatic interactions, like dominance, are of little help in predicting phenotypes.

Altogether, these three variance components constitute the total genetic variance:

$$V_g = V_a + V_d + V_i$$

If we recall that $V_T = V_g + V_e$, we can express the total phenotypic variance as the sum of four components:

$$V_T = V_a + V_d + V_i + V_e$$

Of these four variance components, only the additive genetic variance, V_a, is useful in predicting the phenotypes of offspring from the phenotypes of their parents. The **narrow-sense heritability** (symbolized h^2) is based on the additive genetic variance. Specifically, h^2 is the proportion of the total phenotypic variance that is additive genetic variance:

$$h^2 = V_a/V_T$$

Like the broad-sense heritability, h^2 lies between 0 and 1. The closer it is to one, the greater is the proportion of the total phenotypic variance that is additive genetic variance, and the greater is our ability to predict an offspring's phenotype. Table 5.3 gives some estimates

TABLE 5.3
Estimates of Narrow-Sense Heritability (h^2) for Quantitative Traits

Trait	h^2
Stature in human beings	0.65
Milk yield in dairy cattle	0.35
Litter size in pigs	0.05
Egg production in poultry	0.10
Tail length in mice	0.40
Body size in *Drosophila*	0.40

Source: D. S. Falconer. 1981. *Introduction to Quantitative Genetics*, 2nd ed., p. 51. Longman, London.

for the narrow-sense heritability for several traits. Human stature is highly heritable, but litter size in pigs is not. Thus, if we knew the parental phenotypes, we would be better able to predict the height of a human's offspring than the litter size of a pig's offspring.

Predicting Phenotypes

To gain insight into the meaning of the narrow-sense heritability, let's consider the situation diagrammed in Figure 5.7. Michael (M) and Frances (F) have taken a standardized intelligence test, and their Intelligence Quotients (IQs) have been determined. Michael's score is 110 and Frances' score is 120. The mean IQ score in the population is 100. Michael and Frances had an infant son Oswald (O), who was given up for adoption when he was born, and the adoptive parents wish to predict Oswald's IQ. If IQ had no genetic component, our best estimate for Oswald's IQ would be 100, the mean of the population. We have no way of predicting what kind of home environment Oswald will receive and therefore cannot predict what kind of nongenetic factors will influence his mental development. Nor can we use the IQs of Michael and Frances to predict anything about Oswald's IQ, since, by assumption, the genes they gave to him would have nothing to do with mental development. However, several

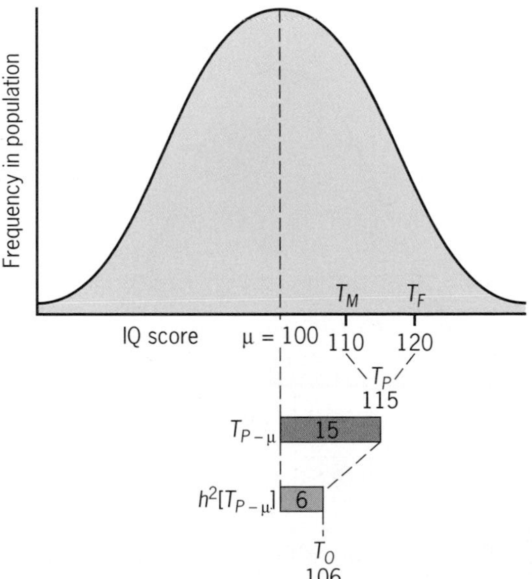

Figure 5.7 Predicting an offspring's phenotype based on the phenotypes of its parents and the narrow-sense heritability of the trait. Only a portion of the deviation of the midparent value from the population mean is heritable. The magnitude of this portion is determined by the narrow-sense heritability.

studies have indicated that variation in IQ scores does have a genetic component. In fact, the narrow-sense heritability of IQ has been estimated to be about 0.4—that is, about 40 percent of the observed variation in IQ scores is due to the additive effects of alleles. Can we use this statistic along with the parental IQs to predict Oswald's IQ score?

Let's symbolize the IQs of Oswald, Michael, and Frances as T_O, T_M, and T_F, respectively, and let's symbolize the population mean as μ. The best prediction for Oswald's IQ is

$$T_O = \mu + h^2\,[(T_M + T_F)/2 - \mu]$$

The expression with parentheses, $(T_M + T_F)/2$, is usually called the *midparent value*. It is the average of the phenotypes of the two parents. If we denote the midparent value with the symbol T_P, the prediction equation for Oswald's phenotype simplifies to

$$T_O = \mu + h^2[T_P - \mu]$$

The expression in brackets, $[T_P - \mu]$, is the difference between the midparent value and the mean of the population. The product of this difference and the narrow-sense heritability is the predicted deviation of the offspring's phenotype from the mean of the population. In effect, the narrow-sense heritability translates the difference between the midparent value and the mean of the population into a "heritable" difference that we can expect to see in the offspring. By adding this heritable difference to the mean, we can predict the offspring's phenotype.

Let's now substitute the known quantities for each of the terms in the prediction equation: $\mu = 100$, $T_P = (110 + 120)/2 = 115$, and $h^2 = 0.4$. Thus, the predicted value of T_O is

$$\begin{aligned} T_O &= 100 + (0.4)[115 - 100] \\ &= 106 \end{aligned}$$

This result tells us that Oswald's IQ is expected to be between the midparent value (115) and the mean of the population (100). In fact, it is at a point 40 percent of the distance between the population mean and the midparent value. This 40 percent corresponds to the narrow-sense heritability (0.4). If the narrow-sense heritability of IQ were greater than 0.4, the predicted value of Oswald's IQ would be closer to the midparent value. For a perfectly heritable trait, $h^2 = 1$ and the predicted value of the offspring's phenotype would equal the average of the two parents' phenotypes. Thus, the narrow-sense heritability is a critical statistic. It tells us how closely the offspring will resemble the average of their parents. We should emphasize, however, that the IQ score we have calculated for Oswald is a predicted value—not one we know for certain. If we were to look at thousands of couples, each having a midparent IQ value of 115, the IQs of their children would be expected to form a frequency distribution. The mean IQ of this distribution would be 106; however, most children would have higher or lower IQs—some even higher than the IQs of either parent and some even lower than the population mean of 100. The variability in this distribution comes from Mendelian segregation of the alleles that influence IQ and from factors in the environment. If, for example, Oswald were raised in a home with little or no intellectual stimulation, with a poor diet, and with other unfavorable conditions, his IQ might turn out to be considerably lower than 106. Conversely, in a nurturing home environment, Oswald's IQ might turn out to be much greater than 106. We have predicted Oswald's IQ to be 106; however, we should keep in mind that this number is a predicted value, not a mathematically set value.

Artificial Selection

In addition to predicting an offspring's phenotype, the narrow-sense heritability has another use: to predict the outcome of a program of selective breeding in a population. Let's suppose that a trait has been measured in a population and that the resulting data have been compiled into a frequency distribution (Figure 5.8). The mean of the distribution is 40 units. To form the next generation, we select the individuals in the upper tail of the distribution to be parents; let's suppose that the mean of these selected individuals is 46 units. Can we predict the mean value of the trait in the offspring of these selected parents? The answer is yes, providing we know the narrow-sense heritability of the trait. The prediction equation is

$$T_O = \mu + h^2 [T_S - \mu]$$

where T_O is the mean of the offspring, μ is the mean of the overall population, T_S is the mean of the selected parents, and h^2 is the narrow-sense heritability. Notice that this equation is the same as the prediction equation for the phenotype of a single offspring, except that T_S has been substituted for T_P. In effect, we have adapted the single-offspring prediction equation to a situation in which many parents (albeit *selected* parents) produce a whole group of offspring, which then forms the population in the next generation. Thus, the new equation allows us to predict how the mean of the population will change by selecting the individuals that will be parents. We call this process **artificial selection.** It is a practice common in plant and animal breeding, and to a large extent, it is responsible for the highly productive strains of crop and farm species that are used in agriculture today.

We can see more clearly how selection changes the mean of a quantitative trait in a population by rearranging the terms in the selection equation. After subtracting μ from both sides of the equation and introducing brackets around the term on the left, we have

$$[T_O - \mu] = h^2 [T_S - \mu]$$

The bracketed term on the right, $[T_S - \mu]$, is called the *selection differential*; it is the difference between the

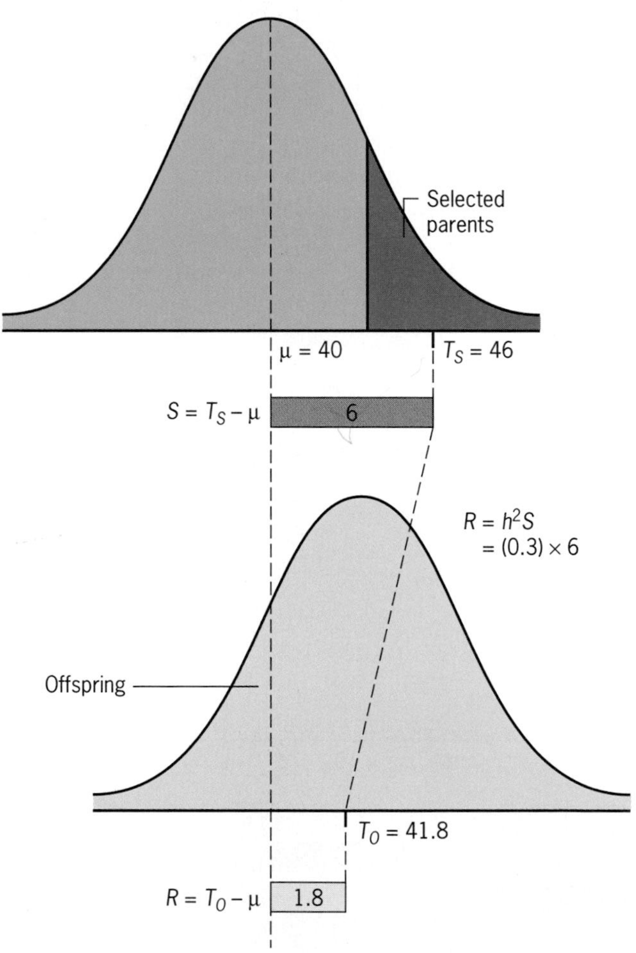

Figure 5.8 The process of artificial selection. The selection differential (S) is the difference between the mean of the selected parents and the mean of the population. The response to selection (R) is the difference between the mean of the offspring and the mean of the overall population that included their parents. The ratio R/S equals the narrow-sense heritability.

mean of the selected parents and the mean of the population from which they were selected. The selection differential measures the intensity of artificial selection. The bracketed term on the left, $[T_O - \mu]$, is called the *response to selection;* it is the difference between the mean of the offspring and the mean of the entire population in the previous generation. Thus, the response to selection measures how much the mean of the trait has changed in one generation. We can put this in even simpler terms if we denote the response to selection by R and the selection differential by S; then

$$R = h^2 S$$

Thus, the response to selection is the product of the selection differential and the narrow-sense heritability. Let's now return to our example; $\mu = 40$, $T_S = 46$, and let's suppose that $h^2 = 0.3$. With these values, $S = 6$ and $R = (0.3) \times 6 = 1.8$; thus, $T_O = 40 + 1.8 = 41.8$. If the selection process were repeated generation after generation, we would expect the mean of the population to increase incrementally. The Technical Sidelight: Artificial Selection for Body Size shows how this is accomplished in practice.

Now let's suppose that we select for a change in another trait whose narrow-sense heritability is unknown. For this trait, the mean of the population is 100 and the mean of the selected parents is 120. Among the offspring of these parents, we find that the mean is 104. What is the narrow-sense heritability? From the equation for the response to selection, we see that $R/S = h^2$, and in this example, $R = 104 - 100 = 4$, and $S = 120 - 100 = 20$. Thus, $R/S = 4/20 = 0.2 = h^2$, the narrow-sense heritabilty. From this example, we see that the response to an artificial selection experiment can be used to estimate the narrow-sense heritability.

Quantitative Trait Loci

Statistical analysis has been a mainstay of quantitative genetics since Fisher's 1918 paper. With this type of analysis, quantitative geneticists have studied many different traits in many different organisms, and recently they have developed techniques to identify individual genes that influence complex traits. Each gene occupies a specific position in the set of DNA molecules that constitutes the genome of an organism. These DNA molecules are organized into discrete structures called chromosomes. A gene's position in a chromosome is called a locus (plural, loci), and a gene that influences a quantitative trait is called a quantitative trait locus—abbreviated QT locus, or more simply QTL.

The identification of QT loci has become feasible with the advent of techniques to detect and analyze genetic variation at the molecular level. We will consider some of these techniques in Chapters 20, 21, and 22. In addition, the Human Genome Project has provided a wealth of information that has led to the identification of particular genes that influence complex traits. QT loci have been identified and mapped on specific chromosomes in model laboratory organisms such as the fruit fly and the mouse, in agriculturally significant plants such as corn and rice, in livestock such as pigs and cows, and in our own species. The traits that have been studied include bristle number in the fruit fly, obesity in the mouse, crop yield in rice and corn, milk production in dairy cattle, fatness and growth rate in pigs, and susceptibility to illnesses such as diabetes, cancer, cardiovascular disease, and schizophrenia in human beings.

We began this chapter with a story about cardiovascular disease, which is a major cause of death among people in postindustrial societies. It has long been known that susceptibility to this disease is influenced by genetic factors. For example, first-degree relatives of people who have had coronary heart disease are seven times more likely to develop this disease themselves than are the first-degree relatives of unaffected people. Furthermore, the relative hazard of a monozygotic twin dying of coronary heart disease when its co-twin died of this disease before age 65 is three to seven times greater than the relative hazard for dizygotic twins. These and other statistical data indicate that susceptibility to cardiovascular disease is under genetic control. Current research is focusing on efforts to identify specific genes that contribute to variation in the factors that put people at risk to develop cardiovascular disease. These factors include plasma cholesterol level, obesity, blood pressure, high- and low-density lipoprotein levels, and triglyceride level. Table 5.4 lists some of the QT loci that have been identified in these efforts.

Key Points: **The phenotypic variance is equal to the variance due to genetic differences as well as the variance due to environmental differences. Broad-sense heritability is the proportion of the phenotypic variance due to genetic differences in the population. Narrow-sense heritability is the proportion of the phenotypic variance due to the additive effects of alleles in the population. The narrow-sense heritability of a quantitative trait can be used to predict the outcome of a selective breeding program. Recent advances have allowed quantitative geneticists to identify specific genes that are involved in the inheritance of complex traits.**

TECHNICAL SIDELIGHT

Artificial Selection for Body Size

Artificial selection is a standard practice to improve crop plants and livestock. However, improvement is usually slow because the generation time of agriculturally significant species is typically measured in years rather than weeks or months. To study the efficacy of artificial selection, Franklin Enfield and his colleagues carried out extensive experiments with a laboratory animal, the flour beetle, *Tribolium castaneum* (Figure 1). In these experiments, Enfield selected for increased body size. He measured the weight of the animals at the pupal stage and selected the heaviest pupae to be the parents of the next generation. This process was continued for 125 generations. At the start of the experiment, the weight of the individual pupae ranged from 1800 to 3000 μg, the mean was 2400 μg, and the variance was 40,000 μg^2. After 125 generations of selection, the mean pupa weight had increased to 5800 μg, more than twice the mean of the starting population (Figure 2*a*). Moreover, none of the individuals in the selected population was as small as the largest individuals in the original starting population (Figure 2*b*). This complete lack of overlap in the frequency distributions indicates that the genetic makeup of the population had been radically altered.

Figure 1 Contrasting sizes of flour beetles. The smaller beetle is from a standard laboratory stock. The larger beetle shows the result of over 120 generations of selection for increased body size.

To achieve this stunning result, Enfield used a selection differential of 200 μg in each generation. Initially, the narrow-sense heritability for pupa weight was estimated to be about 0.3; thus, the predicted response to selection was 0.3 × 200 μg = 60 μg per generation (see Figure 5.8). For the first 40 generations, this was approximately what Enfield observed. However, the cumulative response during this time was 2000 μg, a little less than the 2400 μg that was expected (60 μg/generation × 40 generations). This discrepancy was due to factors that reduced the selection efficiency, including such things as infertility among the selected individuals. Thus, although the narrow-sense heritability is a reasonably good predictor of the response to selection over a few generations, in the long term it tends to overestimate this response.

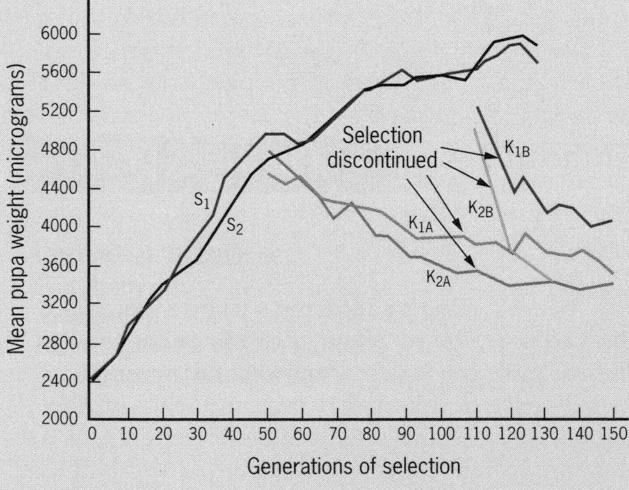

(*a*)

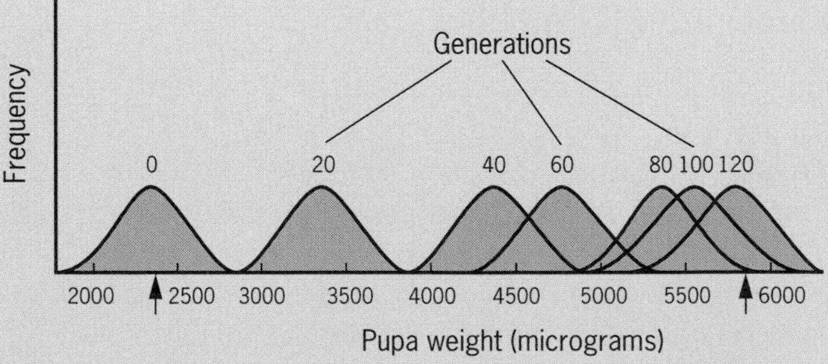

(*b*)

Figure 2 (*a*) Artificial selection for increased size in the flour beetle, *Tribolium*. Size was measured by pupa weight. The curves S_1 and S_2 show the response to selection in two replicate populations. The curves K_{1A}, K_{2A}, K_{1B}, and K_{2B} show what happened when artificial selection was discontinued in subpopulations that were established from the selected populations. (*b*) Frequency distributions of pupa weight in *Tribolium* populations selected for increased size. The shape of the distributions is only approximate. The means at generations 0 and 120 are indicated by arrows.

The later generations of Enfield's project dramatically demonstrate this point. Between generations 40 and 125, the cumulative response was 1400 μg, which, though impressive, is much less than the expected response of 5100 μg (60 μg/generation × 85 generations). Enfield checked narrow-sense heritability in these later generations to see if any of the additive genetic variance had been lost during the long selective process. To his surprise, he found that h^2 was only slightly changed, indicating that the population still retained selectable genetic variability. A more detailed analysis demonstrated that during these generations, the efficiency of selection was severely reduced by a negative correlation between size and reproductive ability. (After a certain point, the larger the beetle, the less reproductively successful it is.) This reduced the effective selection differential and made it difficult to select for further increases in size. In fact, when selection was relaxed in generation 50 and again in generation 110, mean pupa weight began dropping back to a nearly normal value. This result suggests that a force of natural selection strongly opposed the artificial selection that was practiced in the main experiment. Enfield's attempts to increase pupa weight beyond 5800 μg failed, not for any lack of genetic variability, but simply because artificial selection had exceeded the natural limits.

TABLE 5.4
Quantitative Trait Loci That Contribute to Variation in Risk Factors for Cardiovascular Disease

Locus	*Gene Product*	*Chromosome*	*Risk Factor*
AGT	Angiotensin	1	Blood pressure
APOA-1	Apolipoprotein A1	11	HDL[a] cholesterol
APOA-2	Apolipoprotein A2	1	HDL cholesterol
APOA-4	Apolipoprotein A4	11	HDL cholesterol, triglycerides
APOB	Apolipoprotein B	2	LDL[b] cholesterol
APOC-3	Apolipoprotein C3	11	Triglycerides
APOE	Apolipoprotein E	19	LDL cholesterol, triglycerides
CETP	Cholesterol ester transfer protein	16	HDL cholesterol
DCP	Dipeptidyl carboxypeptidase	17	HDL cholesterol, blood pressure
FGA/B	Fibrinogen A and B	4	Fibrinogen
HRG	Histidine-rich glycoprotein	3	Histidine-rich glycoprotein
LDLR	Low-density lipoprotein receptor	19	LDL cholesterol
LPA	Lipoprotein (a)	6	HDL cholesterol, triglycerides
LPL	Lipoprotein lipase	8	Triglycerides
PLAT	Plasminogen activator tissue-type	8	Tissue plasminogen activator level
PLANH1	Plasminogen activator inhibitor-1	7	PAI-1 level

Source: G. P. Vogler, *et al.* 1997. Genetics and behavioral medicine: risk factors for cardiovascular disease. *Behavioral Medicine* 22:141–149.
[a]High-density lipoprotein.
[b]Low-density lipoprotein.

INBREEDING AND THE RESEMBLANCE BETWEEN RELATIVES

Geneticists have always been interested in the phenomenon of inbreeding, whether to make true-breeding strains or to reveal the homozygous effects of recessive alleles. In addition, when inbreeding occurs in nature, it can affect the character of plant and animal populations. In this section we consider ways to analyze the effects of inbreeding. We also introduce the techniques needed to study common ancestry in pedigrees.

The Effects of Inbreeding

Inbreeding occurs when mates are related to each other by virtue of common ancestry. A mating between relatives is often referred to as a **consanguineous mating**, from Latin words meaning "of the same blood." In human populations, these types of matings are rare, with the incidence depending on cultural and ethnic traditions and on geography. One classic study, for example, demonstrated that in the United States only 0.08 percent of 133,228 Roman Catholic marriages recorded in 1959–1960 were between first cousins and only 0.11 percent were between second cousins. In Fouta-Djallon, Guinea, by comparison, 19.08 percent of 739 marriages surveyed in 1955 were between first cousins and 6.22 percent were between second cousins. In many cultures, marriages between close relatives—for example, between siblings or half siblings—are expressly forbidden, and marriages between more distant relatives, though allowed, must be approved by civil or religous authorities before they can occur. The reason for these restrictions is that inbreeding brings out recessive alleles that cause disorders such as mental retardation, phenylketonuria, and cystic fibrosis. Even though people had no understanding of genetic principles during most of human history, they did recognize that consanguineous matings tend to produce more diseased and debilitated children than matings between unrelated individuals. In some cultures, however, consanguineous matings have been accepted and even encouraged. In ancient Egypt, for example, the royal line was perpetuated by brother–sister marriages, presumably to preserve the "purity" of the royal blood. Similar practices existed in Polynesia until relatively recent times.

The occurrence of consanguineous matings in human populations has helped in the analysis of genetic conditions caused by recessive alleles. Some of the classic studies have focused on the Amish, a religious group scattered in small communities in the eastern and midwestern United States. Because marriages are usually within the group, it is not uncommon to find families in which the parents are related to each other. Figure 5.9 shows an Amish pedigree in which 10 individuals have albinism. The affected individuals, who lived in rural Indiana in the United States, were descendants of two people (I-1 and I-2) who lived in the Alsace area of Europe. The consanguineous matings in the pedigree are indicated by double lines connecting the mates. All the affected individuals come from such matings. Thus, this pedigree shows how inbreeding brings out a recessive condition.

The effects of inbreeding are also evident in experimental species where it is possible to arrange matings between relatives. For example, animals such as rats, mice, and guinea pigs can be mated brother to sister, generation after generation, to create an **inbred line.** Although these lines are genetically quite pure—that is, they do not segregate different alleles of particular genes—they often are less vigorous than lines maintained by matings between unrelated individuals. We refer to this loss of vigor as **inbreeding depression.** In plants where self-fertilization is possible, very highly inbred lines can be created by repeated self-fertilization over several generations. Figure 1.5 shows the result of this process in maize. The inbred plant is short and produces small ears with a few kernels. By comparison, the noninbred plant is tall and produces robust ears with many kernels.

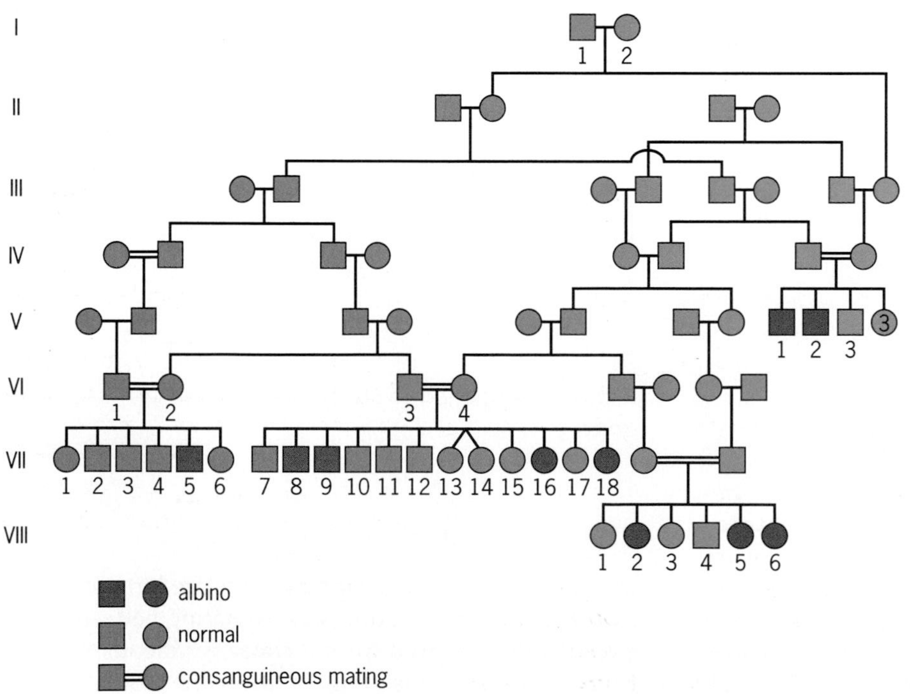

Figure 5.9 The effects of inbreeding in a pedigree. The albinos, which are homozygous for a recessive allele, all come from consanguineous matings.

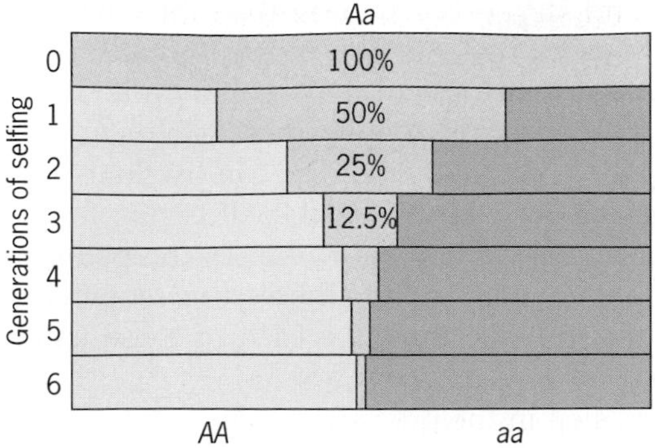

Figure 5.10 Increase in frequency of homozygotes with repeated self-fertilization.

Inbreeding creates genotypes that are homozygous for many of their genes. We can see this result by considering the most extreme case of inbreeding—self-fertilization (Figure 5.10). Suppose that we begin with a population of individuals that are all *Aa* heterozygotes. If each individual reproduces by self-fertilization, half its progeny will be heterozygous and half will be homozygous (either *AA* or *aa* with equal probability). If these progeny also reproduce by self-fertilization, the homozygotes will produce only homozygous offspring, and the heterozygotes will produce equal numbers of heterozygotes and homozygotes. Thus, with repeated self-fertilization, the frequency of homozygotes increases by 50 percent each generation. Ultimately, we would expect the population to consist entirely of homozygous individuals. Another way of describing the effect of inbreeding is to say that the frequency of heterozygotes decreases. With repeated self-fertilization, the frequency of heterozygotes is reduced by 50 percent each generation. We will return to this phenomenon in Chapter 29 when we consider the inbreeding effect of small populations and the evolutionary process of random genetic drift.

Inbred strains that have been produced by repeated self-fertilization would be expected to be homozygous for different alleles of the genes that were segregating in the original population. One strain might be *AA bb CC dd*, and another might be *aa BB CC DD*. A cross between two such strains would produce offspring heterozygous for many different genes. Such multiple-gene heterozygotes often exhibit much greater vigor than their inbred parents. Geneticists call this phenomenon of increased vigor in the hybrid offspring of inbred parents **heterosis.** The term was introduced in 1914 by George Shull, a pioneering plant breeder who began the practice of crossing inbred strains to produce uniformly high-yielding heterozygous offspring. Shull's technique has since become standard in the plant breeding industry.

Genetic Analysis of Inbreeding

Matings between full siblings, between half siblings, and between first cousins are all examples of inbreeding. When such matings occur, we speak of the offspring as being *inbred*. Inbred individuals differ from the offspring of unrelated parents in one important way: the two copies of a gene they carry may be identical to each other by virtue of common ancestry— that is, because they have descended from a gene that was present in an ancestor of the inbred individual. To understand this concept, let's consider a simple pedigree that illustrates a mating between half siblings.

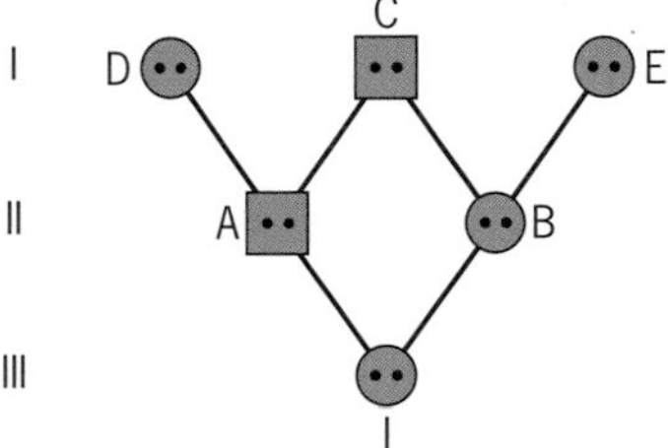

The two dots in each individual represent the two copies of a particular gene, and the lines that connect individuals show how genes have passed from parent to offspring. This way of drawing a pedigree is different from the one we have used previously. It clarifies how each parent contributes genes to its offspring, and it allows us to trace the descent of a particular gene through multiple generations.

The two individuals in Generation II, labeled A and B, are half siblings. These individuals had a common father, C, but different mothers (D and E). The mating between A and B produced an offspring, I, who is inbred. Notice that I inherits one gene copy from A and one copy from B. However, both of these copies might have originated in C, the common father of A and B. Thus, the two gene copies in I might be identical to each other by descent from one of the gene copies that was present in C. This possibility of **identity by descent** is the important consequence of inbreeding. Any individual whose gene copies are identical by descent must be homozygous for a particular allele of that gene. Thus, consanguineous matings are expected to produce relatively more homozygotes than matings between unrelated individuals, which, as we have seen, is one of the conspicuous effects of inbreeding.

In the pedigree we are considering, C is referred to as the *common ancestor* of I because two paths of descent from C converge in I, the inbred individual. The two paths are C $\rightarrow$ A $\rightarrow$ I and C $\rightarrow$ B $\rightarrow$ I, and together they form what geneticists call an *inbreeding loop*. This loop shows how a particular gene copy in C can be passed down both sides of the pedigree to produce two identical gene copies in I.

The fundamental determination in any analysis of inbreeding is to calculate the probability that two gene copies in an individual are identical by descent. Intuitively, this probability should increase with the intensity of inbreeding. Thus, the offspring of a mating between full siblings should have a greater probability of identity by descent than the offspring of a mating between half siblings. The effort to measure inbreeding intensity began with the pioneering work of the American geneticist Sewall Wright. In 1921 Wright discovered a mathematical quantity he called the **inbreeding coefficient.** Wright's investigations—too complicated to be discussed here—involved an analysis of correlations between the individuals in a pedigree. In these investigations, he discovered how to calculate the inbreeding coefficient and used it to measure the intensity of inbreeding. Then, in the 1940s, another American, Charles Cotterman, showed that Wright's inbreeding coefficient was equivalent to the probability of identity by descent. Thus, we can define the inbreeding coefficient, symbolized by the letter F, as the probability that two gene copies in an individual are identical by descent from a common ancestor.

To calculate the inbreeding coefficient, we follow the procedures developed by Wright and Cotterman. First, we identify the common ancestor(s) of the inbred individual. A common ancestor is connected to the inbred individual through both of that individual's parents. In the pedigree we are considering, I has only one common ancestor; however, in other types of pedigrees, an inbred individual might have more than one common ancestor. For example, the offspring of a mating between full siblings has two common ancestors:

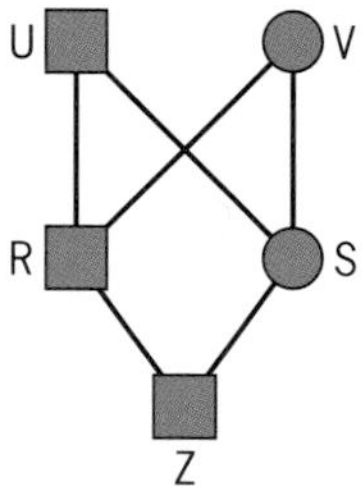

In this case, both of Z's grandparents (U and V) are common ancestors. Two genetic paths descend from each grandparent and converge in Z. Thus, the pedigree for full-sib mating has two distinct inbreeding loops:

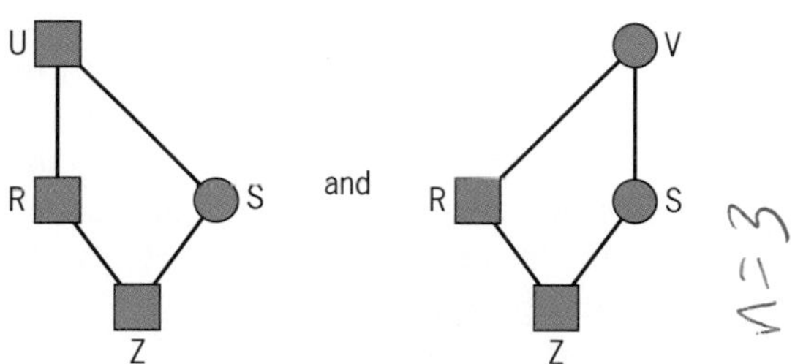

The second step in calculating the inbreeding coefficient is to count the number of individuals (n) in each inbreeding loop defined by a common ancestor. In the pedigree for mating between half siblings, there is one inbreeding loop and it has three individuals. (We do not count the inbred individual itself.) Thus, for the pedigree with half-sib mating, $n = 3$. In the pedigree for full-sib mating there are two inbreeding loops, each with three individuals; thus, for each of these loops, $n = 3$.

The third step in the procedure to calculate the inbreeding coefficient is to compute the quantity $(1/2)^n$ for each inbreeding loop and then sum the results. The sum we obtain is the inbreeding coefficient, F, of the inbred individual—that is, the probability that its two gene copies are identical to each other by descent from a common ancestor. For the offspring of a mating between half siblings, we obtain $F = (1/2)^3 = 1/8$. For the offspring of a mating between full siblings, we obtain $F = (1/2)^3 + (1/2)^3 = 1/4$. Thus, the inbreeding coefficient of the offspring of full-sib mating is greater than the inbreeding coefficient of the offspring of half-sib mating, as expected.

The factor $(1/2)^n$ that we compute for each inbreeding loop is the probability that *either* of the two gene copies in the common ancestor of that loop produces two identical gene copies in the inbred individual. To understand this probability, let's focus on the mating between half siblings. We must consider two cases, labeled 1 and 2 in the following illustration.

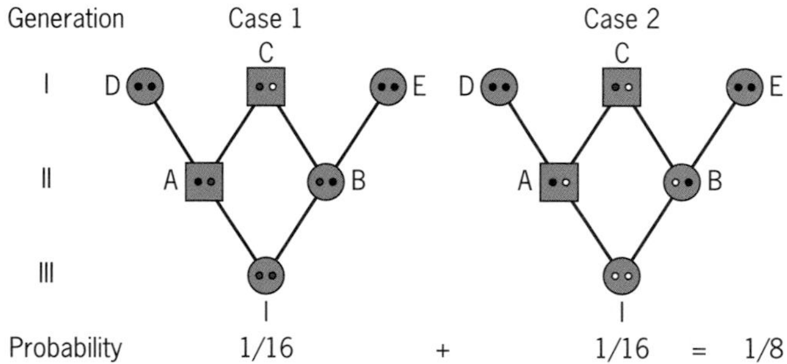

In Case 1, the chance that the gene copy on the left (shown in red) in the common ancestor C is transmitted to the daughter A is 1/2; once in A, the chance that this gene copy is transmitted to I is 1/2. Thus, the probability that the "left" gene copy in C makes its way down to I through A is $(1/2) \times (1/2) = 1/4$. Similarly, the chance that the "left" gene copy makes its way down to I through B is $(1/2) \times (1/2) = 1/4$. Altogether, then, the probability that the "left" gene copy in C produces two identical gene copies in I, one transmitted through A and the other through B, is $(1/4) \times (1/4) = 1/16$. By similar reasoning in Case 2, we find the probability that the "right" gene copy (shown in white) in C produces two identical gene copies in I to be 1/16. Thus, the probability that *either* the "left" or the "right"

gene copies in C will produce two identical gene copies in I is $(1/16) + (1/16) = 1/8$, which, as we have seen, is $(1/2)^3$. The procedure of calculating the factor $(1/2)^n$ is therefore a shortcut to find the probability that either of the gene copies in a particular common ancestor will give rise to two identical gene copies in the inbred individual.

This method of calculating inbreeding coefficients works for most pedigrees. However, when a common ancestor is itself inbred, the method needs to be modified. We multiply the factor $(1/2)^n$ for the common ancestor by the term $[1 + F_{CA}]$, where F_{CA} is the inbreeding coefficient of the common ancestor. For example, in the pedigree

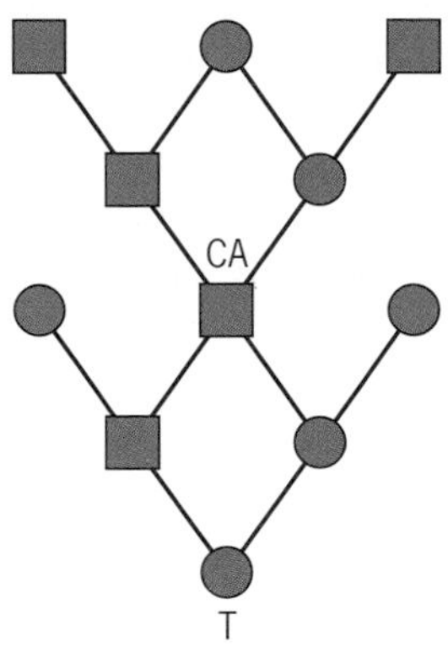

the inbreeding coefficient of T is $F_T = (1/2)^3 \times [1 + F_{CA}]$, and because $F_{CA} = (1/2)^3 = 1/8$, we conclude that $F_T = (1/8) \times [1 + (1/8)] = 9/64$. The modifying term $[1 + F_{CA}]$ accounts for the possibility that the "left" and "right" gene copies in CA are already identical by descent.

The inbreeding coefficient as defined by Wright and Cotterman is an accurate measure of inbreeding intensity. Figure 5.11 presents values of this coefficient for the offspring of different types of consanguineous matings.

One use of the inbreeding coefficient is to explain the increased frequency of recessive disorders among the offspring of consanguineous matings. In the human population, for example, the incidence of phenylketonuria (PKU) among the offspring of unrelated parents is about 1/10,000; among the offspring of first-cousin marriages, it is about 7/10,000. The difference between these frequencies, 6/10,000, is the effect of inbreeding with $F = 1/16$. For the offspring of closer relatives, we would expect a greater difference in the frequencies of PKU. For example, the offspring of half siblings have an inbreeding coefficient of 1/8, twice that of the offspring of first cousins. Because the effect of inbreeding is proportional to F, we would expect the incidence of PKU among the offspring of half siblings to be twice the inbreeding effect seen with the offspring of first cousins, plus the incidence of PKU in the general population. Thus, the predicted frequency of PKU among the offspring of half siblings is $2 \times (0.0006) + 0.0001 = 0.0013$. Among the offspring of full siblings,

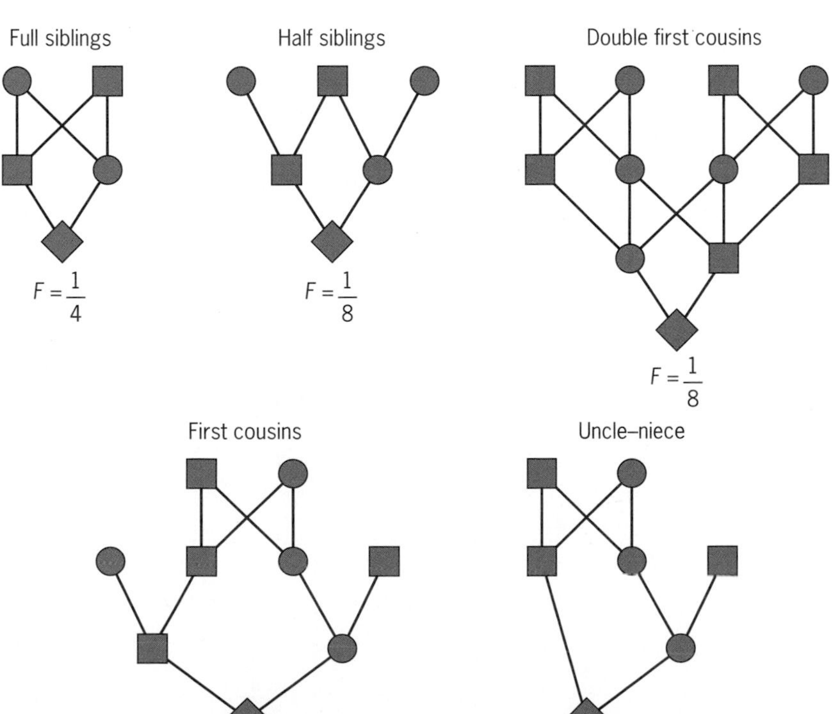

Figure 5.11 Values of the inbreeding coefficient, F, for different pedigrees.

the predicted frequency is $4 \times (0.0006) + 0.0001 = 0.0017$ (because they have an inbreeding coefficient four times that of the offspring of first cousins).

This line of reasoning can be extended to all types of recessive disorders. A study by Jan Böök indicated that in Sweden about 16 percent of the children of first-cousin marriages manifested some type of genetic disease, whereas only about 4 percent of the children of unrelated parents did. The difference between these percentages, 12 percent, must reflect the inbreeding effect of first-cousin marriage, an effect we know must be proportional to $F = 1/16$. If we imagine complete inbreeding in the human population so that $F = 1$, the inbreeding effect will be 16 times that seen in the offspring of first-cousin marriages, or an increased incidence of genetic disease of $16 \times 12\% = 192\%$. This calculation therefore suggests that if each of the genes in a human gamete were doubled to make every pair of gene copies identical by descent, the resulting zygote would, on average, suffer from two genetic diseases. Thus, inbreeding reveals the extent to which individuals would suffer if they were homozygous for all their genes. Some geneticists have called this effect our "genetic load."

Another use of the inbreeding coefficient is to measure the decline in the mean value of a quantitative trait. Figure 5.12 shows data collected from inbred strains of maize that were obtained through a program of repeated self-fertilization. Seed was saved at each stage of the inbreeding process, and at the end, maize plants were grown from the seed in test plots to study two traits, plant height and crop yield. As Figure 5.12 shows, both of these traits declined linearly as a function of the inbreeding coefficient. The simplest explanation for this linear decline is that recessive alleles of different genes were made homozygous as the inbreeding proceeded—that is, in proportion to the value of F—and that these homozygotes manifested lower values for the traits. Thus, an increase in the incidence of deleterious recessive homozygotes is the basis for inbreeding depression.

Correlations Between Relatives

The inbreeding coefficient can also be used to measure the closeness of genetic relationships. Full siblings are obviously more closely related than half siblings. Are aunt and nephew more closely related than half siblings? Half siblings are more closely related than first cousins. Are they more closely related than double first cousins? To answer these questions, we must determine the fraction of genes that two relatives share by virtue of common ancestry. As we will see, the inbreeding coefficient plays a key role in this determination.

For regular relatives—that is, relatives that are not themselves inbred—we can calculate the fraction of genes that are shared by imagining that the relatives have mated and produced an offspring. Obviously, because this offspring is inbred, we can calculate its inbreeding coefficient according to the usual procedure. Then, to determine the fraction of genes that the two relatives share, we simply multiply the offspring's inbreeding coefficient by 2. The result is sometimes called the **coefficient of relationship.** For full siblings, the inbreeding coefficient of an imaginary offspring is 1/4; thus, the coefficient of relationship of full siblings (or the fraction of genes they share) is $2 \times (1/4) = 1/2$. By similar reasoning, the coefficient of relationship of half siblings is 1/4, that of first cousins is 1/8, and that of double first cousins is 1/4. For aunt and nephew, the coefficient of relationship is 1/4. Thus, half siblings, double first cousins, and aunt and nephew are equivalently related because each shares the same fraction of their genes, 1/4. Siblings, by comparison, are more closely related because they share half their genes, and

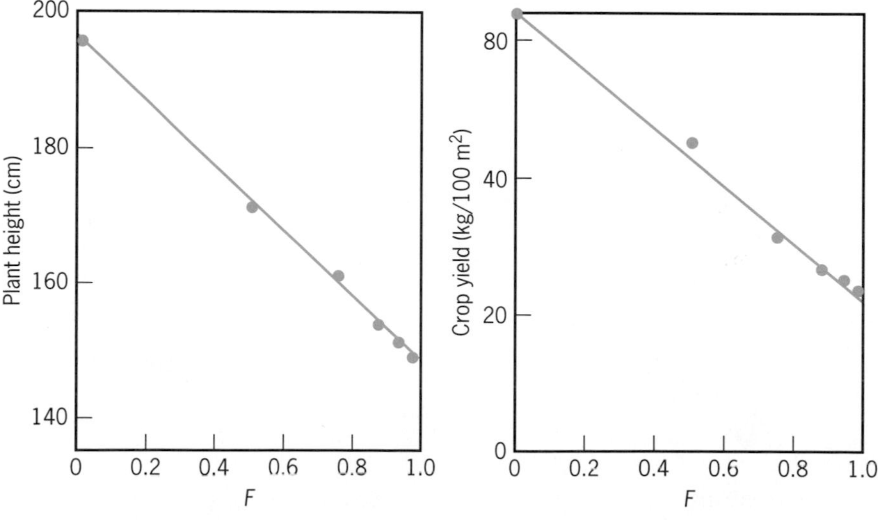

Figure 5.12 Inbreeding decline in plant height and crop yield in maize. The intensity of inbreeding is measured by the inbreeding coefficient, F.

single first cousins are less closely related because they share only one-eighth of their genes.

Key Points: **Inbreeding increases the frequency of homozygotes and decreases the frequency of heterozygotes. The effects of inbreeding are proportional to the inbreeding coefficient, which is the probability that two gene copies in an individual are identical by descent from a common ancestor. The coefficient of relationship is the fraction of genes that two individuals share by virtue of common ancestry.**

TESTING YOUR KNOWLEDGE

1. Two inbred strains of *Drosophila* differ from each other in the number of abdominal bristles they have. These two strains are intercrossed, and the F_1 and F_2 are analyzed. The mean number of bristles in the F_1 is 25, with a standard deviation of 2. The F_2 mean is also 25, but the standard deviation is 3. Calculate V_g, V_e, and H^2 for this quantitative trait.

ANSWER

The F_1 are identically heterozygous because each parental strain was inbred. Thus, V_g for the F_1 is 0. The total variance (V_T) in the F_1 is s^2 or 4 because s is the square root of the variance. Since $V_g = 0$, all the variance must be due to V_e; thus, $V_e = 4$.

The F_2 is genetically heterogeneous because of independent assortment and segregation of the alleles affecting abdominal bristle number. V_T in the F_2 is 3^2 or 9, and because we have already estimated that $V_e = 4$, V_g can be obtained by subtraction; thus, $V_g = 9-4 = 5$. H^2 is therefore $5/9 = 0.56$.

2. The mean value of a trait is 100 units, and the narrow-sense heritability is 0.3. A male and a female measuring 130 and 90 units, respectively, mate and produce a large number of offspring, which are reared in randomized environments. What is the expected value of the trait among these offspring?

ANSWER

The midparent value (the average of the two parents) is $(130 + 90)/2 = 110$. This value deviates from the population mean (100) by 10 units. If the narrow-sense heritability for the trait is 0.3, 30 percent of this deviation should be heritable. Consequently, the predicted value of the trait for the offspring of these two parents is $100 + (0.3) \times (10) = 103$.

3. In the pedigree below, calculate the inbreeding coefficient of M.

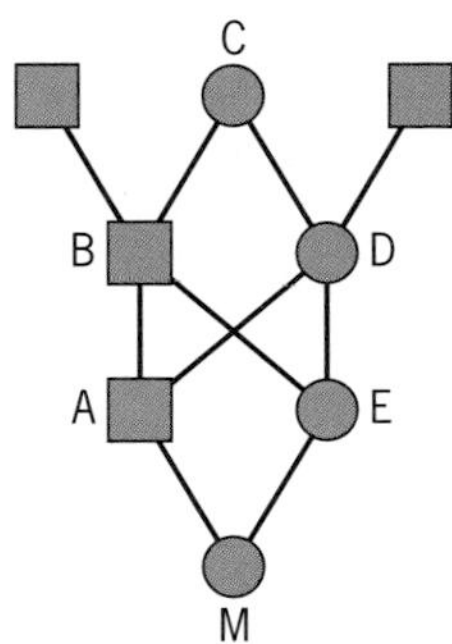

ANSWER

M has three common ancestors, B, C, and D, because two lines of descent from each of these individuals ultimately converge in M. There are four distinct inbreeding loops (common ancestor underlined):

(1) A B <u>C</u> D E ($n = 5$)
(2) A D <u>C</u> B E ($n = 5$)
(3) A <u>B</u> E ($n = 3$)
(4) A <u>D</u> E ($n = 3$)

To calculate the inbreeding coefficient of M, F_M, we raise 1/2 to the power n for each of the loops and sum the results:

$$F_M = (1/2)^5 + (1/2)^5 + (1/2)^3 + (1/2)^3 = 5/16$$

QUESTIONS AND PROBLEMS

5.1 A wheat variety with red kernels (genotype $A'A'$ $B'B'$) was crossed with a variety with white kernels (genotype AA BB). The F_1 were intercrossed to produce an F_2. If each primed allele increases the amount of pigment in the kernel by an equal amount, what phenotypes will be expected in the F_2? Assuming that the A and B loci assort independently, what will the phenotypic frequencies be?

5.2 How would you distinguish between a phenotype controlled by polygenic inheritance and one controlled by a single gene with two alleles and variable expressivity?

5.3 Nilsson-Ehle studied the inheritance of kernel color in wheat. In one experiment, he crossed a true-breeding red variety with a true-breeding white variety. The F_1 was red, as was the F_2. When he self-fertilized the 78 F_2 plants, he obtained the following results:

Number of F_2 Plants	Progeny from Self-Fertilized Plants
50	all red
15	15 red:1 white
8	3 red:1 white
5	63 red:1 white

How many genes are segregating in this series of crosses? What are the parental and F_2 phenotypes?

5.4 Assume that size in rabbits is determined by genes with equal and additive effects. From a total of 2012 F_2 progeny from crosses between true-breeding large and small varieties, eight rabbits were as small as the small variety and eight were as large as the large variety. How many size-determining genes were segregating in these crosses?

5.5 The height of the seed head in wheat at maturity is determined by several genes. In one variety, the head is just 9 inches above the ground; in another, it is 33 inches above the ground. Plants from the 9-inch variety were crossed to plants from the 33-inch variety. Among the F_1, the seed head was 21 inches above the ground. After self-fertilization, the F_1 plants produced an F_2 population in which 9-inch and 33-inch plants each appeared with a frequency of 1/256. (a) How many genes are involved in the determination of seed head height in these strains of wheat? (b) How much does each allele of these genes contribute to seed head height? (c) If a 21-inch F_1 plant were crossed to a 9-inch plant, how often would you expect 18-inch wheat to occur in the progeny?

5.6 A sample of 20 plants from a population was measured in inches as follows: 18, 21, 20, 23, 20, 21, 20, 22, 19, 20, 17, 21, 20, 22, 20, 21, 20, 22, 19, and 23. Calculate (a) the mean, (b) the variance, and (c) the standard deviation.

5.7 Quantitative geneticists use the variance as a measure of scatter in a sample of data; they calculate this statistic by averaging the squared deviations between each measurement and the sample mean. Why don't they simply measure the scatter by computing the average of the deviations without bothering to square them?

5.8 Two inbred strains of corn were crossed to produce an F_1, which was then intercrossed to produce an F_2. Data on ear length from a sample of F_1 and F_2 individuals gave phenotypic variances of 15.2 cm^2 and 27.6 cm^2, respectively. Why was the phenotypic variance greater for the F_2 than for the F_1?

5.9 A study of quantitative variation for abdominal bristle number in female *Drosophila* yielded estimates of $V_T = 6.08$, $V_g = 3.17$, and $V_e = 2.91$. What was the broad-sense heritability?

5.10 The following listing presents data on the variances of two phenotypic traits in sparrows (wing span and beak length):

Wing Span	**Beak Length**
$V_T = 271.4$	$V_T = 627.8$
$V_e = 71.2$	$V_e = 107.3$

Estimate the broad-sense heritability for each trait.

5.11 Measurements on ear length were obtained from three populations of corn—two inbred varieties and a randomly pollinated population derived from a cross between the two inbred strains. The phenotypic variances were 9.2 cm^2 and 9.6 cm^2 for the two inbred varieties and 26.4 cm^2 for the randomly pollinated population. Estimate the broad-sense heritability of ear length for these populations.

5.12 Suppose that for a particular threshold trait, the incidence in males is four times the incidence in females. In family A, the mother shows the trait and the father does not. In family B, the father shows the trait and the mother does not. Which of these two families has the greater risk for an affected child? Why?

5.13 A quantitative geneticist claims that the narrow-sense heritability for body mass in human beings is 0.7, while the broad-sense heritability is only 0.3. Why must there be an error?

5.14 The mean value of a trait is 100 units, and the narrow-sense heritability is 0.4. A male and a female measuring 124 and 126 units, respectively, mate and produce a large number of offspring, which are reared in an average environment. What is the expected value of the trait among these offspring?

5.15 The narrow-sense heritability for abdominal bristle number in a population of *Drosophila* is 0.3. The mean bristle number is 12. A male with 10 bristles is mated to a female with 20 bristles, and a large number of progeny are scored for bristle number. What is the expected number of bristles among these progeny?

5.16 A breeder is trying to decrease the maturation time in a population of sunflowers. In this population, the mean time to flowering is 100 days. Plants with a mean flowering time of only 90 days were used to produce the next generation. If the narrow-sense heritability for flowering time is 0.2, what will the average time to flowering be in the next generation?

5.17 A fish breeder wishes to increase the rate of growth in a stock by selecting for increased length at six weeks after hatching. The mean length of six-week-old fingerlings is currently 10 cm. Adult fish that had a mean length of 15 cm at six weeks of age were used to produce a new generation of fingerlings. Among these, the mean length was 12.5 cm. Estimate the narrow-sense heritability of fingerling length at six weeks of age and advise the breeder about the feasibility of the plan to increase growth rate.

5.18 Leo's IQ is 86 and Julie's IQ is 110. The mean IQ in the population is 100. Assume that the narrow-sense heritability for IQ is 0.4. What is the expected IQ of Leo and Julie's first child?

5.19 A selection differential of 40 μg per generation was used in an experiment to select for increased pupa weight in *Tribolium*. The narrow-sense heritability for pupa weight was estimated to be 0.3. If the mean pupa weight was initially 2000 μg and selection was practiced for 10 generations, what was the mean pupa weight expected to become?

5.20 In the pedigrees on the following page, what are the inbreeding coefficients of A, B, and C?

Questions due 3/13/01

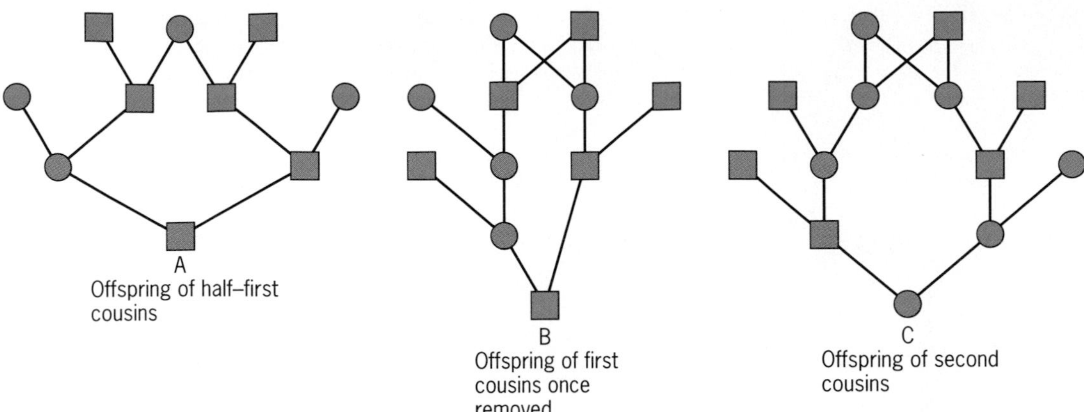

5.21 Suppose that the inbreeding coefficient of I in the following pedigree is 0.25. What is the inbreeding coefficient of I's common ancestor, C?

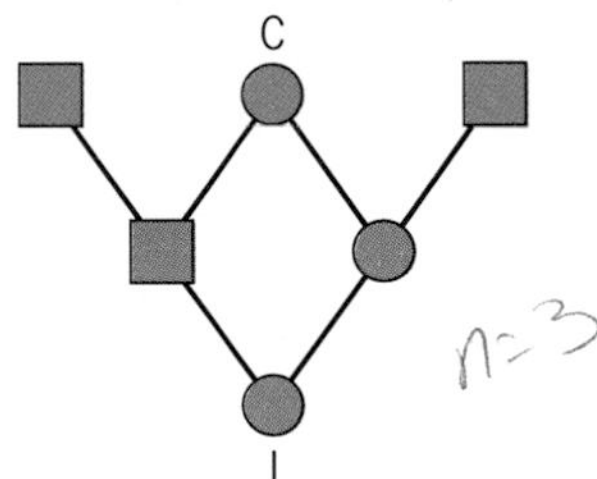

5.22 Mabel and Frank are half siblings, as are Tina and Tim. However, these two pairs of half sibs do not have any common ancestors. If Mabel marries Tim and Frank marries Tina and each couple has a child, what fraction of their genes will these children share by virtue of common ancestry? Will the children be more or less closely related than first cousins?

5.23 A, B, and C are inbred strains of mice, assumed to be completely homozygous. A is mated to B and B to C. Then the A × B hybrids are mated to C, and the offspring of this mating are mated to the B × C hybrids. What is the inbreeding coefficient of the offspring of this last mating?

5.24 A randomly pollinated strain of maize produces ears that are 24 cm long, on average. After one generation of self-fertilization, the ear length is reduced to 20 cm. Predict the ear length if self-fertilization is continued for one more generation.

BIBLIOGRAPHY

BODMER, W. F., AND L. L. CAVALLI-SFORZA. 1976. *Genetics, Evolution, and Man.* W. H. Freeman, San Francisco.

BÖÖK, J. A. 1957. Genetical investigations in a north Swedish population: the offspring of first-cousin marriages. *Ann. Human. Genet.* 21:191–221.

CROW, J. F. 1998. 90 years ago: the beginning of hybrid maize. *Genetics* 148:923–928.

EAST, E. M. 1916. Studies on size inheritance in *Nicotiana*. *Genetics* 1:164–176.

FALCONER, D. 1989. *Introduction to Quantitative Genetics*, 3rd ed. Longman Scientific and Technical Publications, London.

LYNCH, M., AND B. WALSH. 1998. *Genetics and Analysis of Quantitative Traits*. Sinauer Associates, Sunderland, MA.

MARX, J. 1990. Dissecting the complex diseases. *Science* 247:1540–1542.

6

Male and female Drosophila. *The male has white eyes because it carries a mutation on its X chromosome.*

The Chromosomal Basis of Mendelism

CHAPTER OUTLINE

Chromosomes
The Chromosome Theory of Heredity
Sex-Linked Genes in Human Beings
Sex Chromosomes and Sex Determination
Dosage Compensation of X-Linked Genes

Sex, Chromosomes, and Genes

What causes organisms to develop as males or females? Why are there only two sexual phenotypes? Is the sex of an organism determined by its genes? These and related questions have intrigued geneticists since the rediscovery of Mendel's work at the beginning of the twentieth century.

The discovery that genes play a role in the determination of sex emerged from a fusion between two previously distinct scientific disciplines, genetics—the study of heredity—and cytology—the study of cells. Early in the twentieth century, these disciplines were brought together through a friendship between two remarkable American scientists, Thomas Hunt Morgan and Edmund Beecher Wilson. Morgan was the geneticist and Wilson the cytologist.

As the cytologist, Wilson was interested in the behavior of chromosomes. These structures would prove to be important for sex determination in many species, including our own. Wilson was one of the first to investigate differences in the chromosomes of the two sexes. Through careful study, he and his colleagues showed that these differences were confined to a special pair of chromosomes called sex chromosomes. Wilson found that the behavior of these chromosomes during meiosis could account for the inheritance of sex.

As the geneticist, Morgan was interested in the identification of genes. He focused his research on the fruit fly, *Drosophila melanogaster*, and rather quickly discovered a gene that gave different phenotypic ratios in males and females. Morgan hypothesized that this gene was located on one of the sex chromosomes, and one of his students, Calvin Bridges, eventually proved this hypothesis to be correct. Morgan's discovery that genes reside on chromosomes was a great achievement. The abstract genetic factors postulated by Mendel were finally localized on visible structures within cells. Geneticists could now explain the Principles of Segregation and Independent Assortment in terms of meiotic chromosome behavior.

The discovery that specific genes determine the sex of an organism came much later, only after another scientific discipline, molecular biology, had joined forces with genetics and cytology. Through their combined efforts, cytologists, geneticists, and molecular biologists identified specific sex-determining genes by studying rare individuals in which the sexual phenotype was inconsistent with the sex chromosomes that were present. Today, researchers in all three fields are earnestly trying to figure out how these genes control sexual development.

In this chapter we will see how cytological studies of differences between the sexes led to the localization of genes on chromosomes. We also discuss the genetic basis of sex determination and the ways in which the two sexes compensate for different numbers of sex chromosomes.

CHROMOSOMES

Chromosomes were discovered in the second half of the nineteenth century by a German cytologist, W. Waldeyer. Subsequent investigations with many different organisms established that chromosomes are characteristic of the nuclei of all cells. They are best seen by applying dyes to dividing cells; at this time, the material in a chromosome is packed into a small volume, giving it the appearance of a tightly organized cylinder. During the interphase between cell divisions, chromosomes are not so easily seen, even with the best of dyes. Interphase chromosomes are loosely organized, forming thin threads that are distributed throughout the nucleus. Consequently, when dyes are applied, the whole nucleus is stained and individual chromosomes cannot be identified. This diffuse network of threads is called **chromatin.** Some regions of the chromatin stain more darkly than others, suggesting an underlying difference in organization. The light and dark regions, respectively called the **euchromatin** (eu = true chromatin) and the **heterochromatin** (heter = different chromatin), have different densities of chromosomal threads.

Chromosome Number

Within a species, the number of chromosomes is almost always an even multiple of a basic number. In human beings, for example, the basic number is 23; mature eggs and sperm have this number of chromosomes. Most other types of human cells have twice as many (46), although a few kinds, such as some liver cells, have four times (92) the basic number.

The **haploid,** or basic, chromosome number **(n)** defines a set of chromosomes called the **haploid genome.** Most somatic cells contain two of each of the chromosomes in this set and are therefore **diploid (2n).** Cells with four of each chromosome are **tetraploid (4n),** those with eight of each are **octaploid (8n),** and so on.

The basic number of chromosomes varies among species. Chromosome number is unrelated to the size or biological complexity of an organism, with most species containing between 10 and 40 chromosomes in their genomes (Table 6.1). The muntjac, a tiny Asian deer, has only three chromosomes in its genome, whereas some species of ferns have many hundreds.

Sex Chromosomes

In some animal species—for example, grasshoppers—females have one more chromosome than males (Figure 6.1*a*). This extra chromosome, originally observed in other insects, is called the **X chromosome.** Females of these species have two X chromosomes, and males have only one; thus females are cytologically XX and males are XO, where the "O" denotes the absence of a chromosome. During meiosis in the female, the two X

TABLE 6.1
Chromosome Number in Different Organisms

Organism	*Haploid Chromosome Number*
Simple Eukaryotes	
Baker's yeast (*Saccharomyces cerevisiae*)	16
Bread mold (*Neurospora crassa*)	7
Unicellular green alga (*Chlamydomonas reinhardtii*)	17
Plants	
Maize (*Zea mays*)	10
Bread wheat (*Triticum aestivum*)	21
Tomato (*Lycopersicon esculentum*)	12
Broad bean (*Vicia faba*)	6
Giant sequoia (*Sequoia sempivirens*)	11
Crucifer (*Arabidopsis thaliana*)	5
Invertebrate Animals	
Fruit fly (*Drosophila melanogaster*)	4
Mosquito (*Anopheles culicifacies*)	3
Starfish (*Asterias forbesi*)	18
Nematode (*Caenorhabditis elegans*)	6
Mussel (*Mytilus edulis*)	14
Vertebrate Animals	
Human being (*Homo sapiens*)	23
Chimpanzee (*Pan tryglodites*)	24
Cat (*Felis domesticus*)	36
Mouse (*Mus musculus*)	20
Chicken (*Gallus domesticus*)	39
Toad (*Xenopus laevis*)	17
Fish (*Esox lucius*)	25

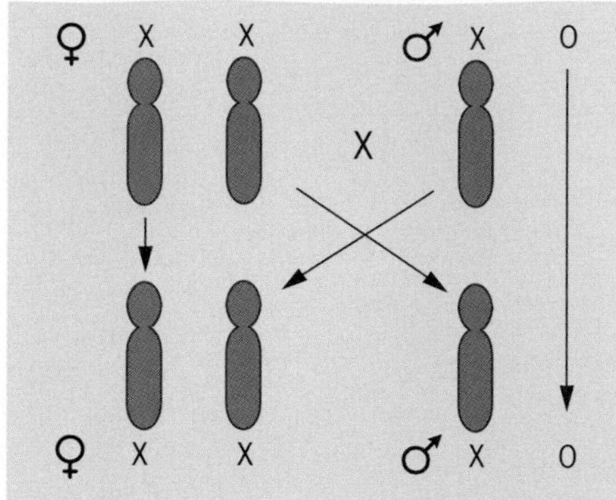

(*a*) Inheritance of sex chromosomes in animals with XX females and XO males.

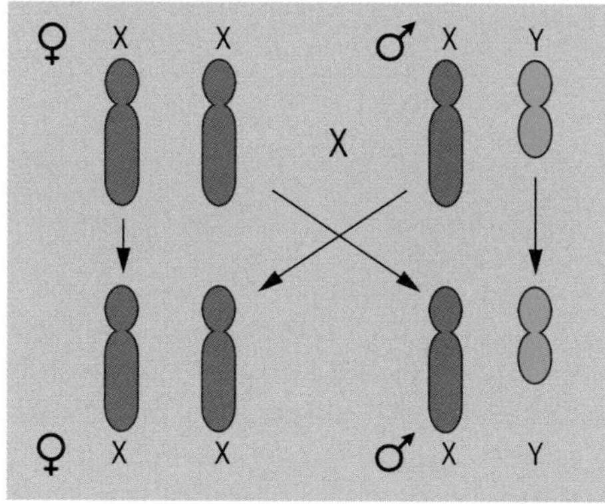

(*b*) Inheritance of sex chromosomes in animals with XX females and XY males.

Figure 6.1 Inheritance of sex chromosomes in animals.

chromosomes pair and then separate, producing eggs that contain a single X chromosome. During meiosis in the male, the solitary X chromosome moves independently of all the other chromosomes and is incorporated into half the sperm; the other half receive no X chromosome. Thus, when sperms and eggs unite, two kinds of zygotes are produced: XX, which develop into females, and XO, which develop into males. Because each of these types is equally likely, the reproductive mechanism preserves a 1:1 ratio of males to females in these species.

In many other animals, including human beings, males and females have the same number of chromosomes (Figure 6.1*b*). This numerical equality is due to the presence of a chromosome in the male, called the **Y chromosome,** which pairs with the X during meiosis. The Y chromosome is morphologically distinguishable from the X chromosome. In humans, for example, the Y is much shorter than the X, and its centromere is located closer to one of the ends (Figure 6.2). The material common to the human X and Y chromosomes is limited, consisting mainly of short terminal segments. During meiosis in the male, the X and Y chromosomes separate from each other, producing two kinds of sperm, X-bearing and Y-bearing; the frequencies of the two types are approximately equal. XX females produce only one kind of egg, which is X-bearing. If fertilization were to occur randomly, approximately half the zygotes would be XX and the other half would be XY, leading to a 1:1 sex ratio at conception. However, in human beings, Y-bearing sperm have a fertilization advantage, and the zygotic

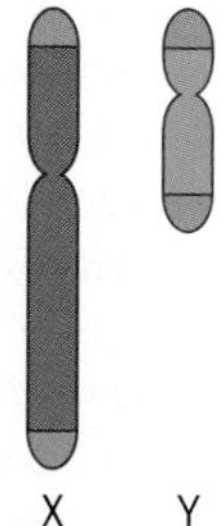

Figure 6.2 Human X and Y chromosomes. The terminal regions are common to both sex chromosomes.

sex ratio is about 1.3:1. During development, the excess of males is diminished by differential viability of XX and XY embryos, and at birth, males are only slightly more numerous than females (sex ratio 1.07:1). By the age of reproduction, the excess of males is essentially eliminated and the sex ratio is very close to 1:1.

The X and Y chromosomes are called **sex chromosomes.** All the other chromosomes in the genome are called **autosomes.** Sex chromosomes were discovered in the first few years of the twentieth century through the work of the American cytologists C. E. McClung, N. M. Stevens, W. S. Sutton, and E. B. Wilson. This discovery coincided closely with the emergence of Mendelism and stimulated research on the possible relationships between Mendel's principles and the meiotic behavior of chromosomes.

Key Points: **Chromosomes emerge from a diffuse network of chromatin fibers during cell division. Diploid somatic cells have twice as many chromosomes as haploid gametes. In some species, the X and Y sex chromosomes distinguish the cells of males and females—XY in males and XX in females. In other species, the female has two X chromosomes and the male has a single X and no Y.**

THE CHROMOSOME THEORY OF HEREDITY

By 1910 many biologists suspected that genes were situated on chromosomes, but they did not have definitive proof. Researchers needed to find a gene that could be unambiguously linked to a chromosome. This goal required that the gene be defined by a mutant allele and that the chromosome be morphologically distinguishable. Furthermore, the pattern of gene transmission had to reflect the chromosome's behavior during reproduction. All these requirements were fulfilled when the American biologist Thomas H. Morgan discovered a particular eye color mutation in the fruit fly, *Drosophila melanogaster*. Morgan began experimentation with this species of fly about 1909. It was ideally suited for genetics research because it reproduced quickly and prolifically and was inexpensive to rear in the laboratory. In addition, it had only four pairs of chromosomes, one being a pair of sex chromosomes—XX in the female and XY in the male. The X and Y chromosomes were morphologically distinguishable from each other and from each of the autosomes. Through careful experiments, Morgan was able to show that the eye color mutation was inherited along with the X chromosome, suggesting that a gene for eye color was physically situated on that chromosome. Later, one of his students, Calvin B. Bridges, obtained definitive proof for this Chromosome Theory of Heredity.

Experimental Evidence Linking the Inheritance of Genes to Chromosomes

Morgan's experiments commenced with his discovery of a mutant male fly that had white eyes instead of the red eyes of wild-type flies. When this male was crossed to wild-type females, all the progeny had red eyes, indicating that white was recessive to red. When these progeny were intercrossed with each other, Morgan observed a peculiar segregation pattern: all of the daughters, but only half of the sons, had red eyes; the other half of the sons had white eyes. This pattern suggested that the inheritance of eye color was linked to the sex chromosomes. Morgan proposed that a gene for eye color was present on the X chromosome, but not on the Y, and that the white and red phenotypes were due to two different alleles, a mutant allele denoted w and a wild-type allele denoted w^+.

Morgan's hypothesis is diagrammed in Figure 6.3. The wild-type females in the first cross are assumed to be homozygous for the w^+ allele. Their mate is as-

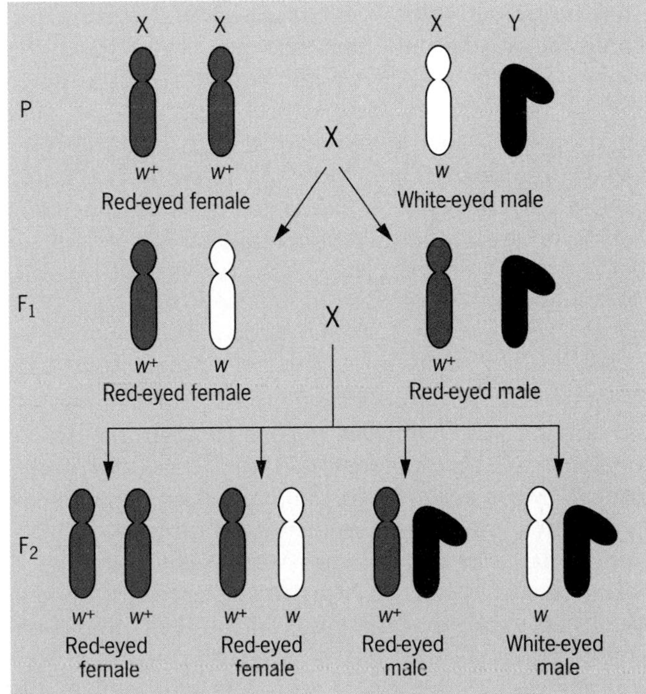

Figure 6.3 Morgan's experiment studying the inheritance of white eyes in *Drosophila*. The transmission of the mutant condition in association with sex suggested that the gene for eye color was present on the X chromosome but not on the Y chromosome.

sumed to carry the mutant *w* allele on its X chromosome and neither of the alleles on its Y chromosome. An organism that has only one copy of a gene is called a **hemizygote.** Among the progeny from the cross, the sons inherit an X chromosome from their mother and a Y chromosome from their father; because the maternally inherited X carries the w^+ allele, these sons have red eyes. The daughters, in contrast, inherit an X chromosome from each parent—an X with w^+ from the mother and an X with *w* from the father. However, because w^+ is dominant to *w*, these heterozygous F_1 females also have red eyes.

When the F_1 males and females are intercrossed, four genotypic classes of progeny are produced, each representing a different combination of sex chromosomes. The XX flies, which are female, have red eyes because at least one w^+ allele is present. The XY flies, which are male, have either red or white eyes, depending on which X chromosome is inherited from the heterozygous F_1 females. Segregation of the *w* and w^+ alleles in these females is therefore the reason why half the F_2 males have white eyes.

Morgan carried out additional experiments to confirm the elements of his hypothesis. In one (Figure 6.4*a*), he crossed females assumed to be heterozygous for the eye color gene to mutant white males. As he expected, half the progeny of each sex had white eyes, and the other half had red eyes. In another experiment (Figure 6.4*b*), he crossed white-eyed females to red-eyed males. This time, all the daughters had red eyes, and all the sons had white eyes. When he intercrossed these progeny, Morgan observed the expected segregation: half the progeny of each sex had white eyes, and the other half had red eyes. Thus, Morgan's hypothesis that the gene for eye color was linked to the X chromosome withstood additional experimental testing.

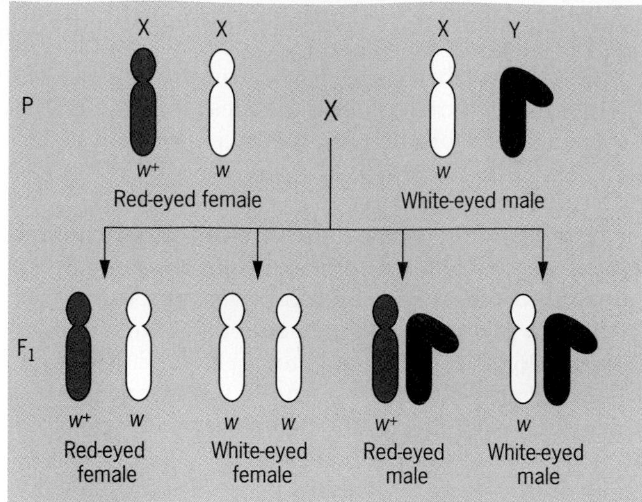

(*a*) Cross between a heterozygous female and a hemizygous mutant male.

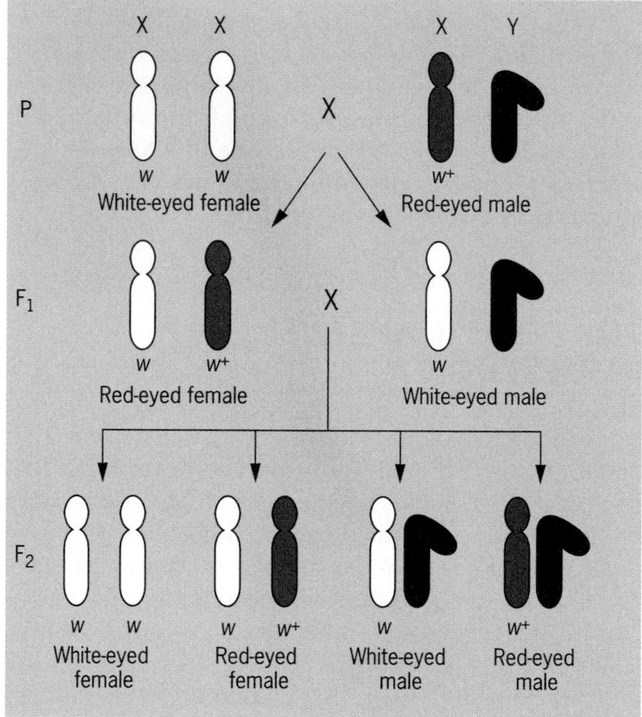

(*b*) Cross between a homozygous mutant female and a hemizygous wild-type male.

Figure 6.4 Experimental tests of Morgan's hypothesis that the gene for eye color in *Drosophila* is X-linked. In each experiment, eye color is inherited along with the X chromosome. Thus, the results of these crosses supported Morgan's hypothesis that the gene for eye color is X-linked.

Chromosomes as Arrays of Genes

Morgan and his students soon identified other X-linked genes in *Drosophila*. In each case, simple breeding experiments demonstrated that recessive mutations of these genes were transmitted along with the X chromosome. As the evidence accumulated, it became clear that many genes were located on the X chromosome. However, Morgan's research group also identified genes that were not on the X chromosome. These genes followed the Mendelian Principle of Segregation, but they did not segregate with sex, as the gene for eye color did. Morgan correctly concluded that such genes were located on one of the three autosomes in the *Drosophila* genome. Thus, each *Drosophila* chromosome appeared to contain a different set of genes.

Morgan's laboratory then attempted to determine the relationships among the genes on a particular chromosome. They proceeded on the assumption that the genes were arranged in a linear array—an idea inspired by cytological evidence that the chromosome

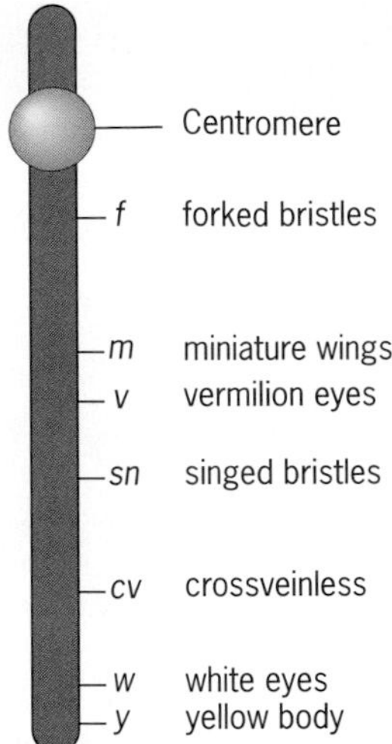

Figure 6.5 A map of genes on the X chromosome of *Drosophila.*

was a long, thin thread. In just a few years, Morgan's students were able to show that genes were indeed situated at different sites, or loci (from the Latin word for "place"; singular: locus), on a linear structure. This analysis, which we will discuss in Chapter 8, produced the world's first genetic maps—diagrams showing the positions of genes and the relative distances between them (Figure 6.5). Morgan's laboratory pioneered the methods for genetic mapmaking and laid the foundation for subsequent research on the physical structure of chromosomes. Eventually, the linearity of chromosomes was connected to the linear structure of DNA (see Chapter 10).

These early studies with *Drosophila*—primarily the work of Morgan and his students (Historical Sidelight: *Drosophila,* T. H. Morgan, and "The Fly Room")—greatly strengthened the view that all genes were located on chromosomes and that Mendel's principles could be explained by the transmissional properties of chromosomes during reproduction. This idea, called the **Chromosome Theory of Heredity,** stands as one of the most important achievements in biology. Since its formulation in the early part of the twentieth century, the Chromosome Theory of Heredity has provided a unifying framework for all studies of inheritance.

Nondisjunction as Proof of the Chromosome Theory

Morgan showed that a gene for eye color was on the X chromosome of *Drosophila* by correlating the inheritance of that gene with the transmission of the X chromosome during reproduction. However, as noted earlier, it was one of his students, C. B. Bridges, who secured proof of the chromosome theory by showing that exceptions to the rules of inheritance could also be explained by chromosome behavior.

Bridges performed one of Morgan's experiments on a larger scale. He crossed white-eyed female *Drosophila* to red-eyed males and examined many F_1 progeny. Although as expected, nearly all the F_1 flies were either red-eyed females or white-eyed males, Bridges found a few exceptional flies—white-eyed females and red-eyed males. He crossed these exceptions to determine how they might have arisen. The exceptional males all proved to be sterile; however, the exceptional females were fertile, and when crossed to normal red-eyed males, they produced many progeny, including large numbers of white-eyed daughters and red-eyed sons. Thus, the exceptional F_1 females, though rare in their own right, were prone to produce many exceptional progeny.

Bridges explained these results by proposing that the exceptional F_1 flies were the result of abnormal X chromosome behavior during meiosis in the females of the P generation. Ordinarily, the X chromosomes in these females should **disjoin,** or separate, from each other during meiosis. Occasionally, however, they might fail to separate, producing an egg with two X chromosomes or an egg with no X chromosome at all. Fertilization of such abnormal eggs by normal sperm would produce zygotes with an abnormal number of sex chromosomes. Figure 6.6 illustrates the possibilities.

If an egg with two X chromosomes (usually called a diplo-X egg; genotype X^wX^w) is fertilized by a Y-bearing sperm, the zygote will be X^wX^wY. Since each of the X chromosomes in this zygote carries a mutant w allele, the resulting fly will have white eyes. If an egg without an X chromosome (usually called a nullo-X egg) is fertilized by an X-bearing sperm (X^+), the zygote will be X^+O. (Once again, "O" denotes the absence of a chromosome.) Because the single X in this zygote carries a w^+ allele, it will develop into a red-eyed fly. Bridges inferred that XXY flies were female and that XO flies were male. The exceptional white-eyed females that he observed were therefore X^wX^wY, and the exceptional red-eyed males were X^+O. Bridges confirmed the chromosome constitutions of these exceptional flies by direct cytological observation. Because the XO animals were male, Bridges concluded that in *Drosophila* the Y chromosome has nothing to do with the determination of the sexual phenotype. However, because the XO males were always sterile, he realized that this chromosome must be important for male sexual function.

Bridges recognized that the fertilization of abnormal eggs by normal sperm could produce two additional kinds of zygotes: $X^wX^wX^+$, arising from the

HISTORICAL SIDELIGHT

Drosophila, T. H. Morgan, and "The Fly Room"

Many people regard fruit flies as pests; they buzz around orchards and vineyards, especially in late summer, and make their way into houses when fruit is harvested. Because of this annoying behavior, geneticists often have difficulty explaining why these tiny creatures are among their favorite animals.

The genus *Drosophila* (from the Greek words meaning "lover of dew") comprises a large number of species, the most intensively studied being *D. melanogaster*. This species was described in the middle of the nineteenth century under the name *D. ampelophilia*, a name meaning "lover of grape vines." After being established as an experimental animal, *D. ampelophilia* was renamed *D. melanogaster*, which means "black belly."

C. W. Woodworth seems to have been the first person to culture *Drosophila* in the laboratory. It is from Woodworth that W. E. Castle, a professor of zoology at Harvard University, learned of the advantages of using this animal as an experimental organism. Castle then recommended it to T. H. Morgan, who began to culture *Drosophila* in 1909.

From its intensive work with *Drosophila*, Morgan's laboratory at Columbia University became known as "The Fly Room" (see Figure 1). Numerous students worked in this laboratory from 1910 until 1926, the year Morgan moved his research to the California Institute of Technology. The most famous of Morgan's students were Calvin Bridges, Alfred Sturtevant, and Hermann Muller. Bridges provided the proof for the Chromosome Theory of Heredity, Sturtevant produced the world's first chromosome map, and Muller discovered that mutations could be induced by X-irradiation. These and other "*Drosophila* workers" were instrumental in developing much of classical genetic analysis. Indeed, William Bateson commented that "not even the most skeptical of readers can go through the *Drosophila* work unmoved by a sense of admiration for the zeal and penetration with which it has been conducted, and for the great extension of genetic knowledge to which it has led—greater far than has been made in any one line of work since Mendel's own experiments."[1]

Figure 1 T. H. Morgan in his laboratory at Columbia University.

In 1939 Morgan reminisced about the early days in the Fly Room:

> *It was not unusual for the six of us to carry on in this small room, the only space at our disposal. These were the days when bananas were used as fly food and in one corner of the room a bunch of bananas was generally on hand,—an adjunct to our researches which interested other members of the laboratory in a different way. As there were no incubators, a bookcase and a wallcase were rigged up with electric bulbs and a cheap thermostat, which behaved badly at times, with consequent loss of cultures. The use of milk bottles came into the program at an early date, but where they came from was not known, or at least not mentioned. . . . Our proximity to each other led to cooperation in everything that went on.*[2]

Later, Morgan's student, Alfred Sturtevant, noted that "There was a give-and-take atmosphere in the fly room. As each new result or new idea came along, it was discussed freely by the group."[3] The excitement of using *Drosophila* as an experimental organism, and the intellectual camaraderie created by the close quarters of the Fly Room combined to stimulate research. The rapid isolation and analysis of new mutations turned *Drosophila* into the premier organism for experimental genetics.

Since the days of the Fly Room, research on *Drosophila* has grown into a worldwide endeavor. Hundreds of different laboratories are currently investigating the genes and chromosomes of this animal. In the United States, one large stock center maintains cultures of mutant flies for distribution to interested researchers, and every year *Drosophila* workers gather for an international five-day meeting to discuss the results of their investigations; typically, more than a thousand people attend this annual meeting. In 1992, at the meeting in Philadelphia, a compendium of 80 years of *Drosophila* research was unveiled. This 1133-page volume, entitled *The Genome of* Drosophila melanogaster,[4] describes more than 4000 genes. However, even as it went to press, the book was out of date. Information about *Drosophila* genes is now available from a regularly updated electronic database, called FlyBase, which is accessible on the Internet through servers such as Gopher (flybase.bio.indiana.edu) or through the World Wide Web (http://morgan.harvard.edu/). At the last major update, FlyBase listed more than 38,000 alleles and more than 11,000 genes—a legacy that would certainly make T. H. Morgan happy.

[1]Bateson, W. 1916. The mechanism of Mendelian heredity (a review). *Science* 44:536–543.

[2]Morgan, T. H. 1939. "Personal recollections of Calvin B. Bridges." *Journal of Heredity* 30: 355.

[3]Sturtevant, A. H. 1965. *A History of Genetics*. Harper and Row, New York, p. 49.

[4]Lindsley, D. L., and G. G. Zimm. 1992. *The Genome of* Drosophila melanogaster. Academic Press, New York.

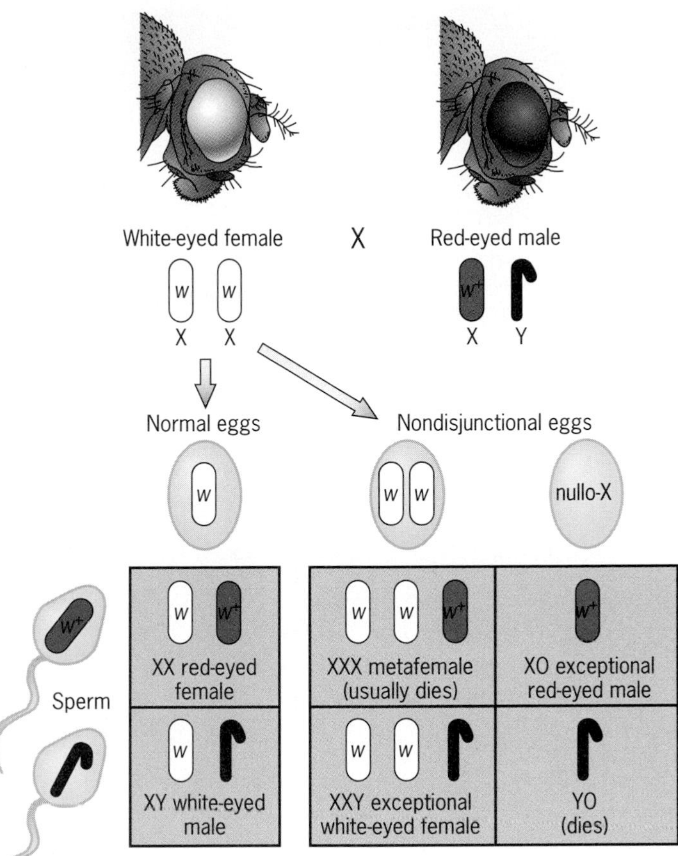

Figure 6.6 X chromosome nondisjunction is responsible for the exceptional progeny that appeared in Bridges' experiment. Nondisjunctional eggs that contain either two X chromosomes or no X chromosome unite with normal sperm that contain either an X chromosome or a Y chromosome to produce four types of zygotes. The XXY zygotes develop into white-eyed females, the XO zygotes develop into red-eyed, sterile males, and the XXX and YO zygotes die.

union of a diplo-X egg and an X-bearing sperm, and *YO*, arising from the union of a nullo-X egg and a Y-bearing sperm. The $X^wX^wX^+$ zygotes develop into females that are red-eyed, but weak and sickly. These "metafemales" can be distinguished from XX females by a syndrome of anatomical abnormalities, including ragged wings and etched abdomens. Generations of geneticists have inappropriately called them "superfemales"—a term coined by Bridges—even though there is nothing super about them. The *YO* zygotes turn out to be completely inviable; that is, they die. In *Drosophila*, as in most other organisms with sex chromosomes, at least one X chromosome is needed for viability.

Bridges' ability to explain the exceptional progeny that came from these crosses showed the power of the chromosome theory. Each of the exceptions was due to anomalous chromosome behavior during meiosis. Bridges called the anomaly **nondisjunction,** because it involved a failure of the chromosomes to disjoin during one of the meiotic divisions. This failure could result from faulty chromosome movement, imprecise or incomplete pairing, or centromere malfunction. From Bridges' data, it is impossible to specify the exact cause. However, Bridges did note that the exceptional XXY females go on to produce a high frequency of exceptional progeny, presumably because their sex chromosomes can disjoin in different ways: the X chromosomes can disjoin from each other, or either X can disjoin from the Y. In the latter case, a diplo- or nullo-X egg is produced because the X that does not disjoin from the Y is free to move to either pole during the first meiotic division. When fertilized by normal sperm, these abnormal eggs will produce exceptional zygotes.

The Chromosomal Basis of Mendel's Principles of Segregation and Independent Assortment

Mendel established two principles of genetic transmission: (1) the alleles of a gene segregate from each other, and (2) the alleles of different genes assort independently. The finding that genes are located on chromosomes made it possible to explain these principles (as well as exceptions to them) in terms of the meiotic behavior of chromosomes.

The Principle of Segregation (Figure 6.7). During the first meiotic division, homologous chromosomes pair. One of the homologs comes from the mother, the other from the father. If the mother was homozygous for an

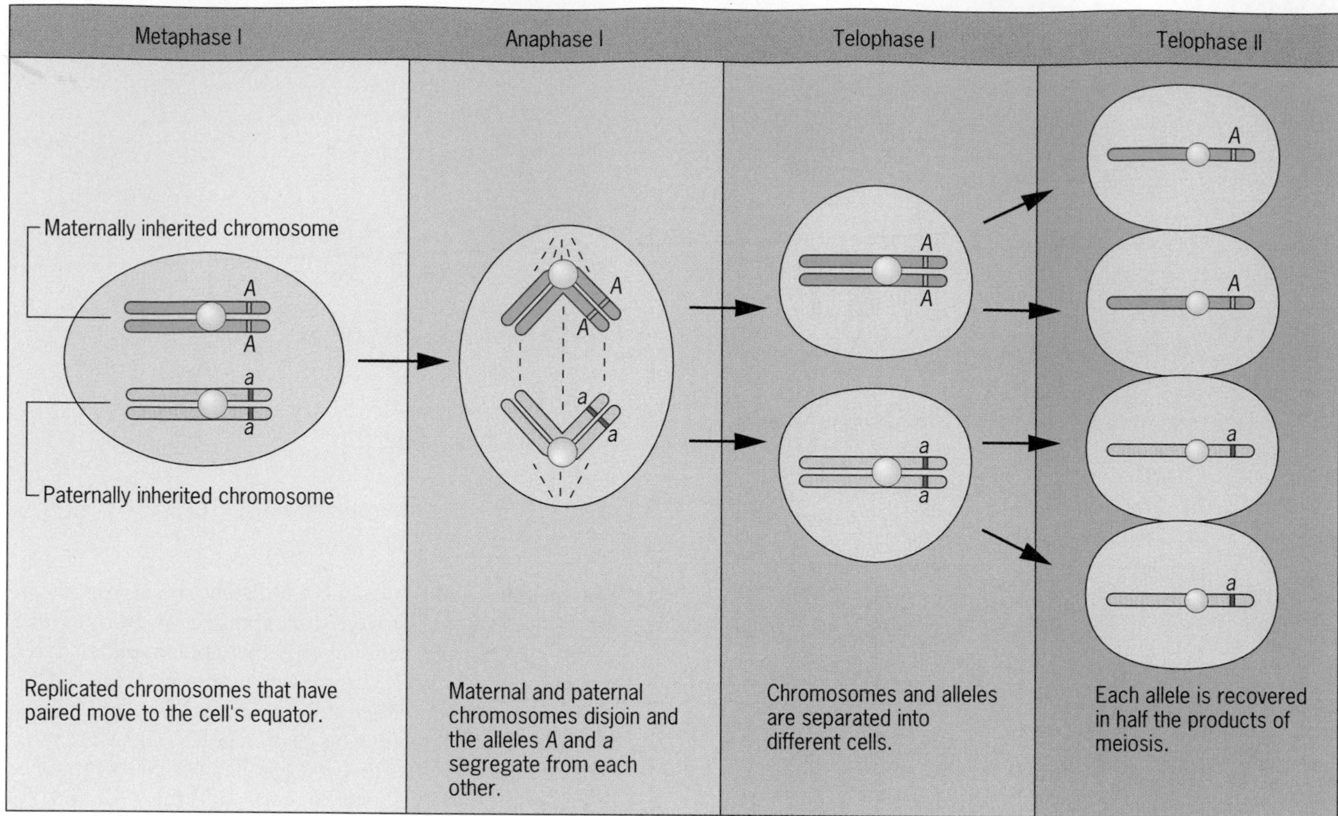

Figure 6.7 Mendel's Principle of Segregation and meiotic chromosome behavior. The segregation of alleles corresponds to the disjunction of paired chromosomes in the anaphase of the first meiotic division.

allele, *A*, of a gene on this chromosome, and the father was homozygous for a different allele, *a*, of the same gene, the offspring must be heterozygous, that is, *Aa*. In the anaphase of the first meiotic division, the paired chromosomes separate and move to opposite poles of the cell. One carries allele *A* and the other allele *a*. This physical separation of the two chromosomes segregates the alleles from each other; eventually, they will reside in different daughter cells. Mendel's Principle of Segregation is therefore based on the separation of homologous chromosomes during the anaphase of the first meiotic division.

The Principle of Independent Assortment (Figure 6.8). The Principle of Independent Assortment is also based on this anaphase separation. To understand the relationship, we need to consider genes on two different pairs of chromosomes. Suppose that a heterozygote *Aa Bb* was produced by mating an *AA BB* female to an *aa bb* male; also, suppose that the two genes are on different chromosomes. During the prophase of meiosis I, the chromosomes with the *A* and *a* alleles will pair, as will the chromosomes with the *B* and *b* alleles. At metaphase, the two pairs will take up positions on the meiotic spindle in preparation for the upcoming anaphase separation. Because there are two pairs of chromosomes, there are two distinguishable metaphase alignments:

$$\frac{A}{a}\;\frac{B}{b} \quad \text{or} \quad \frac{A}{a}\;\frac{b}{B}$$

Each of these alignments is equally likely. Here the space separates different pairs of chromosomes, and the bar separates the homologous members of each pair. During anaphase, the alleles above the bars will move to one pole, and the alleles below them will move to the other. When disjunction occurs, there is therefore a 50 percent chance that the *A* and *B* alleles will move together to the same pole and a 50 percent chance that they will move to opposite poles. Similarly, there is a 50 percent chance that the *a* and *b* alleles will move to the same pole and a 50 percent chance that they will move to opposite poles. At the end of meiosis, when the chromosome number is finally reduced, half the gametes should contain a parental combination of alleles (*A B* or *a b*), and half should contain a new combination (*A b* or *a B*). Altogether, there will be four types of gametes, each one-fourth of the total. This equality of gamete frequencies is a result of the independent behavior of the two pairs of chromosomes

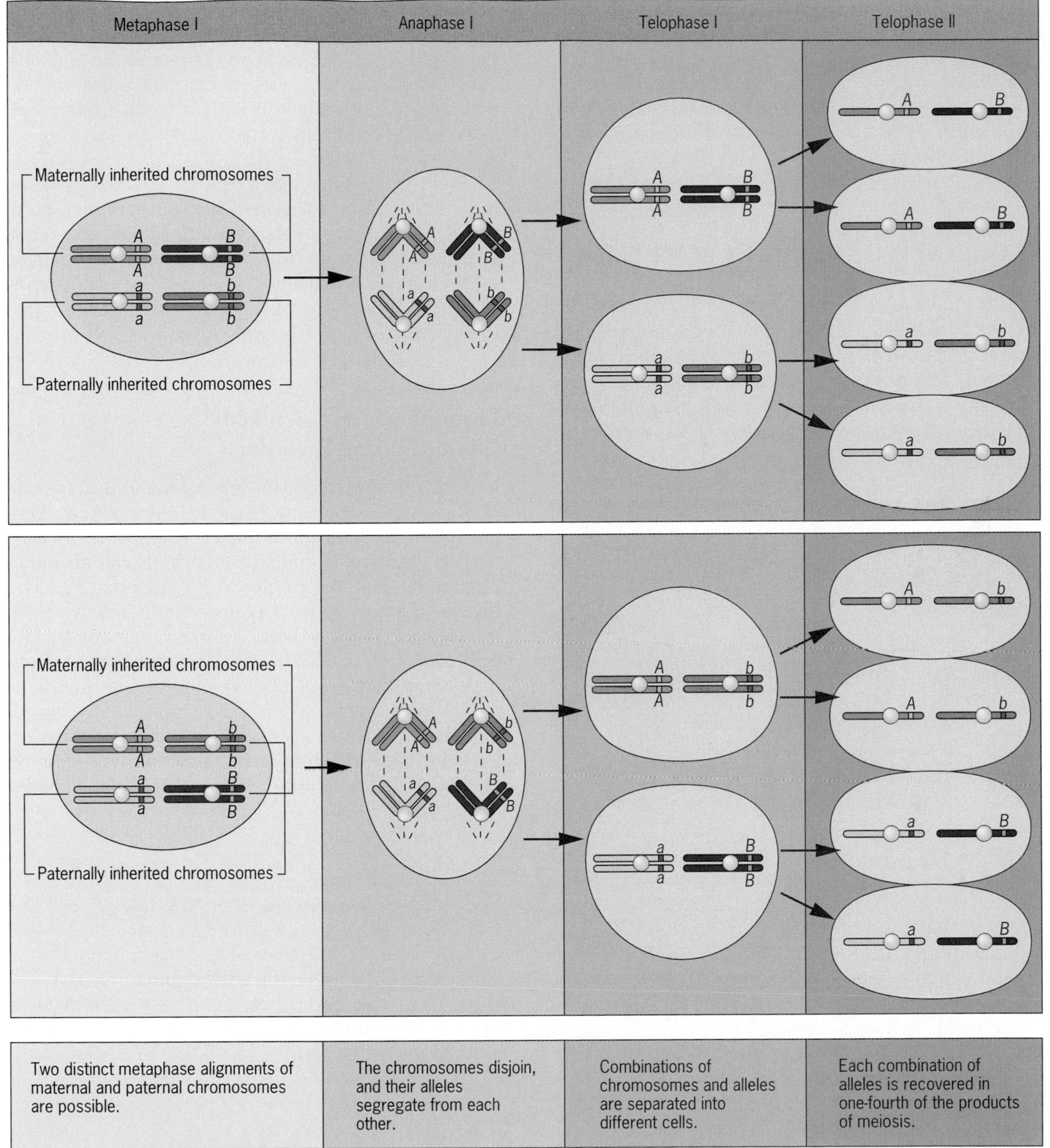

Figure 6.8 Mendel's Principle of Independent Assortment and meiotic chromosome behavior. Alleles on different pairs of chromosomes assort independently in the anaphase of the first meiotic division because maternally and paternally inherited chromosomes have aligned randomly on the cell's equator.

during the first meiotic division. Mendel's Principle of Independent Assortment is therefore a statement about the anaphase separation of genes on different pairs of chromosomes. Later, we will see that genes on the same pair of chromosomes do not assort independently. Instead, because they are physically linked to each other, they tend to travel together through meiosis, violating the Principle of Independent Assortment.

Key Points: **Genes are located on chromosomes. The disjunction of chromosomes during meiosis is responsible for the segregation and independent assortment of genes.**

SEX-LINKED GENES IN HUMAN BEINGS

The development of the chromosome theory depended on the discovery of the *white* eye mutation in *Drosophila*. Subsequent analysis demonstrated that this mutation was a recessive allele of an X-linked gene. Although some of us might credit this important episode in the history of genetics to extraordinarily good luck, Morgan's discovery of the *white* eye mutation was not so remarkable. Such mutations are among the easiest to detect because they show up immediately in hemizygous males. In contrast, autosomal recessive mutations show up only after two mutant alleles have been brought together in a homozygote—a much more unlikely event.

In human beings too, recessive X-linked traits are much more easily identified than are recessive autosomal traits. A male needs only to inherit one recessive allele to show an X-linked trait; however, a female needs to inherit two—one from each of her parents. Thus, the preponderance of people who show X-linked traits are male.

Hemophilia, an X-Linked Blood-Clotting Disorder

In human beings, a certain type of **hemophilia** is one of the best known examples of an X-linked trait. People with this disease are unable to produce a factor needed for blood clotting; the cuts and wounds of hemophiliacs continue to bleed and, if not stopped by therapeutic treatment, can cause death. Nearly all the individuals affected with X-linked hemophilia are male. Other blood-clotting disorders are found in both males and females because they are due to mutations in autosomal genes.

The most famous case of X-linked hemophilia occurred in the Russian imperial family at the beginning of the twentieth century (Figure 6.9). Czar Nicholas

(*a*)

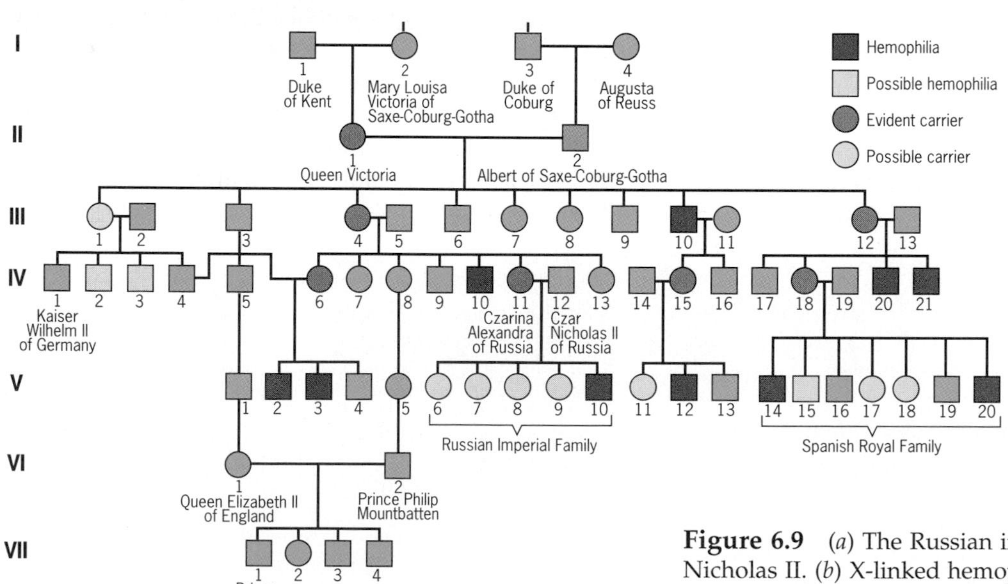

(*b*)

Figure 6.9 (*a*) The Russian imperial family of Czar Nicholas II. (*b*) X-linked hemophilia in the royal families of Europe. Through intermarriage, the mutant allele for hemophilia was transmitted from the British royal family to the German, Russian, and Spanish royal families.

and Czarina Alexandra had four daughters and one son. The son, Alexis, suffered from hemophilia. The X-linked mutation responsible for Alexis' disease was transmitted to him by his mother, who was a heterozygous carrier. Czarina Alexandra was a granddaughter of Queen Victoria of Great Britain, who was also a carrier. Pedigree records show that Victoria transmitted the mutant allele to three of her nine children: Alice, who was Alexandra's mother, Beatrice, who had two sons with the disease, and Leopold, who had the disease himself. The allele that Victoria carried evidently arose as a new mutation in the germ cells of one of her parents, or in those of a more distant maternal ancestor. Some scholars have even argued that this mutation changed the course of history by contributing to the downfall of the Romanov dynasty in Russia and the subsequent rise of the Soviet state.

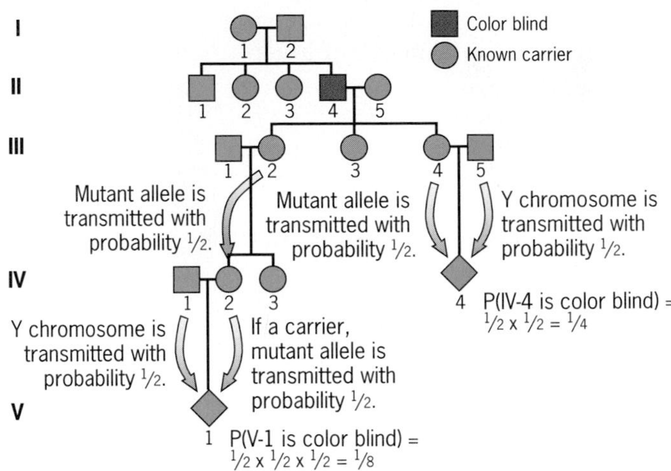

Figure 6.10 Analysis of a pedigree showing the segregation of X-linked color blindness.

Color Blindness, an X-Linked Vision Disorder

In human beings, color perception is mediated by light-absorbing proteins in the specialized cone cells of the retina in the eye. Three such proteins have been identified—one to absorb blue light, one to absorb green light, and one to absorb red light. Color blindness may be caused by an abnormality in any of these receptor proteins. The classic type of color blindness, involving faulty perception of red and green light, follows an X-linked pattern of inheritance. About 5 to 10 percent of human males are red-green color blind; however, a much smaller fraction of females, less than 1 percent, has this disability, suggesting that the mutant alleles are recessive. Molecular studies have shown that there are two distinct genes for color perception on the X chromosome; one encodes the receptor for green light, and the other encodes the receptor for red light. Detailed analyses have demonstrated that these two receptors are structurally very similar, probably because the genes encoding them evolved from an ancient color-receptor gene. A third gene for color perception, the one encoding the receptor for blue light, is located on an autosome.

In Figure 6.10 color blindness is used to illustrate the procedures for calculating the risk of inheriting a recessive X-linked condition. A heterozygous carrier, such as III-4 in the figure, has a 1/2 chance of transmitting the mutant allele to her children. However, the risk that a particular child will be color blind is only 1/4 since the child must be a male in order to manifest the trait. The female labeled IV-2 in the pedigree could be a carrier of the mutant allele for color blindness, because her mother was. This uncertainty about the genotype of IV-2 introduces another factor of 1/2 in the risk of having a color-blind child; thus, the risk for her child is $1/4 \times 1/2 = 1/8$.

The Fragile X Syndrome and Mental Retardation

In human beings, many cases of mental retardation appear to follow an X-linked pattern of inheritance. Most of these are associated with a cytological anomaly that is detectable in cells that have been cultured in the absence of certain nucleotides, the building blocks of DNA. This anomaly—a constriction near the tip of the long arm of the X chromosome—gives the impression that the tip is ready to detach from the rest of the chromosome (Figure 6.11*a*); hence the name fragile X chromosome. The clinical features of the fragile X syndrome vary considerably, making diagnosis difficult. Most patients show significant mental impairment, and some exhibit facial and behavioral abnormalities; both males and females can be affected. Among children, the incidence of the fragile X syndrome is about one in 2000.

The fragile X syndrome has been described as an X-linked dominant disorder with incomplete penetrance. Affected females are heterozygous for the fragile X chromosome, and affected males are hemizygous for this chromosome. However, some carriers of the chromosome, both male and female, are asymptomatic for the disorder (Figure 6.11*b*). This lack of full penetrance complicates the analysis of pedigrees and makes genetic counseling difficult.

Molecular techniques have been used to isolate and analyze the fragile X site. It consists of a short repeating unit in the DNA adjacent to *FMR1* (*F*ragile X *M*ental *R*etardation *1*), a gene of unknown function

(a)

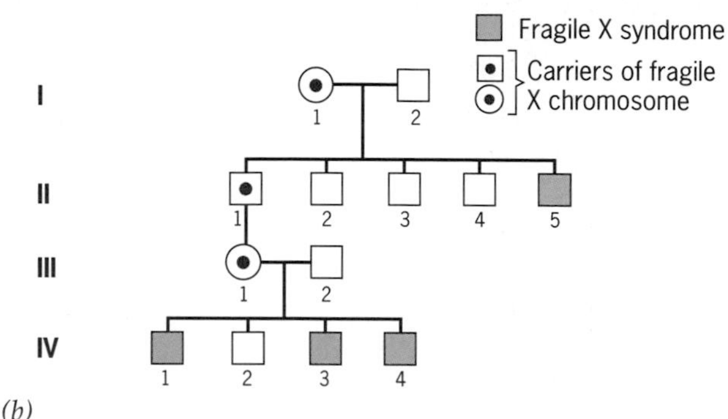

(b)

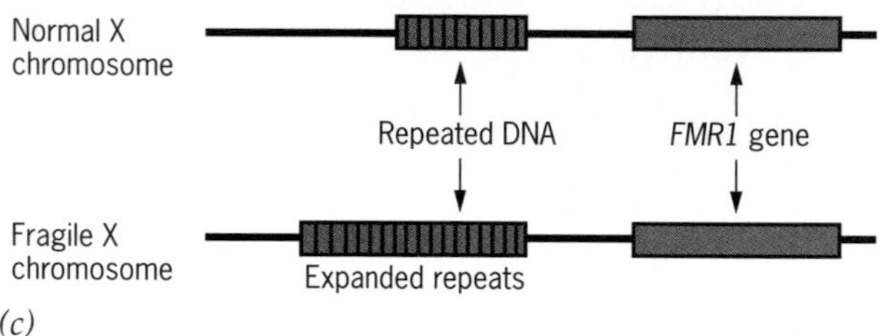

(c)

Figure 6.11 The fragile X chromosome. (*a*) A female (left) showing the fragile X and a normal X chromosome, and a male (right) showing the fragile X and a normal Y chromosome. (*b*) A pedigree showing the inheritance of the fragile X syndrome. The asymptomatic male II-1 is a carrier, indicating that the condition has incomplete penetrance. (*c*) Molecular basis of the fragile X syndrome. The mutation in the fragile X chromosome is due to an expansion of a region of repeats in the DNA flanking the *FMR1* gene. Chemical modification of the DNA around these repeats adversely affects the expression of the *FMR1* gene.

(Figure 6.11*c*). This repeating unit varies in length, being much larger in individuals who show the disorder. A combination of genetic and molecular analyses has demonstrated that the repeating unit tends to increase in size, probably as a result of faulty chromosome replication. This instability explains why individuals who do not show the disorder may have children who do; the repeating unit may expand in the mother's germ line and be passed on to the children, who will have it in all their somatic cells. Although the physiological details are not known, an expanded unit appears to be associated with chemical modification of the surrounding DNA. This modification has an adverse effect on the expression of the *FMR1* gene and is probably the actual cause of the fragile X syndrome.

Genes on the Human Y Chromosome

Very few genes have been localized to the human Y chromosome. This failure to find Y-linked genes is somewhat surprising, since a mutation in one of these genes should have an immediate phenotypic effect on the man who carries it. Furthermore, such a mutation should be passed on to all the man's sons but to none of his daughters. A Y-linked gene should therefore be the easiest kind of gene to identify in conventional pedigree analysis. To date, however, only a few Y-linked genes have been discovered. One is responsible for the synthesis of a male-specific substance called the H-Y antigen, which is found on cell surfaces. Another is involved in the production of a factor that is critical for the differentiation of the testes and the subsequent acquisition of male sexual characteristics (see below). Advances in molecular genetics have provided new techniques to identify other Y-linked genes, but even with these, the view that the Y chromosome has fewer genes than any other human chromosome probably will not change.

Genes on Both the X and Y Chromosomes

Some genes are present on both the X and Y chromosomes, mostly near the ends of the short arms (see Figure 6.2). Alleles of these genes do not follow a distinct X- or Y-linked pattern of inheritance. Instead, they are transmitted from mothers and fathers to sons and daughters alike, mimicking the inheritance of an autosomal gene. Such genes are therefore called **pseudoautosomal genes.** In males, the regions that contain these genes seem to mediate pairing between the X and Y chromosomes.

Key Points: **Disorders caused by recessive X-linked mutations such as hemophilia and color blindness are more common in males than in females. In humans, the Y chromosome carries few genes. Some of these genes are also carried by the X chromosome.**

SEX CHROMOSOMES AND SEX DETERMINATION

In the animal kingdom, sex is perhaps the most conspicuous phenotype. Animals with distinct males and females are sexually dimorphic. Sometimes this dimorphism is established by environmental factors. In one species of turtles, for example, sex is determined by temperature. Eggs that have been incubated above 30°C hatch into females, whereas eggs that have been incubated at a lower temperature hatch into males. In many other species, sexual dimorphism is established by genetic factors, often involving a pair of sex chromosomes.

Sex Determination in Human Beings

The discovery that human females are XX and that human males are XY suggested that sex might be determined by the number of X chromosomes or by the presence or absence of a Y chromosome. As we now know, the second hypothesis is correct. In humans and other placental mammals, maleness is due to a dominant effect of the Y chromosome (Figure 6.12). The evidence for this fact comes from the study of individuals with an abnormal number of sex chromosomes. XO animals develop as females, and XXY animals develop as males. The dominant effect of the Y chromosome is manifested early in development, when it directs the primordial gonads to develop into testes. Once the testes have formed, they secrete testosterone, a hormone that stimulates the development of male secondary sexual characteristics.

Recent research has pinpointed the **testis-determining factor *(TDF)*** on the human Y chromosome. This factor appears to correspond to a gene called ***SRY*** (for **sex-determining *r*egion *Y***), which is located just outside the pseudoautosomal region in the chromosome's short arm. The discovery of *SRY* was made possible by the identification of unusual individuals whose sex was inconsistent with their chromosome constitution—XX males and XY females (Figure 6.13). Some of the XX males were found to carry a small piece of the Y chromosome inserted into one of the X chromosomes. This piece evidently carried a gene responsible for maleness. Some of the XY females were found to carry an incomplete Y chromosome. The part of the Y chromosome that was missing corresponded to the piece that was present in the XX males; its absence in

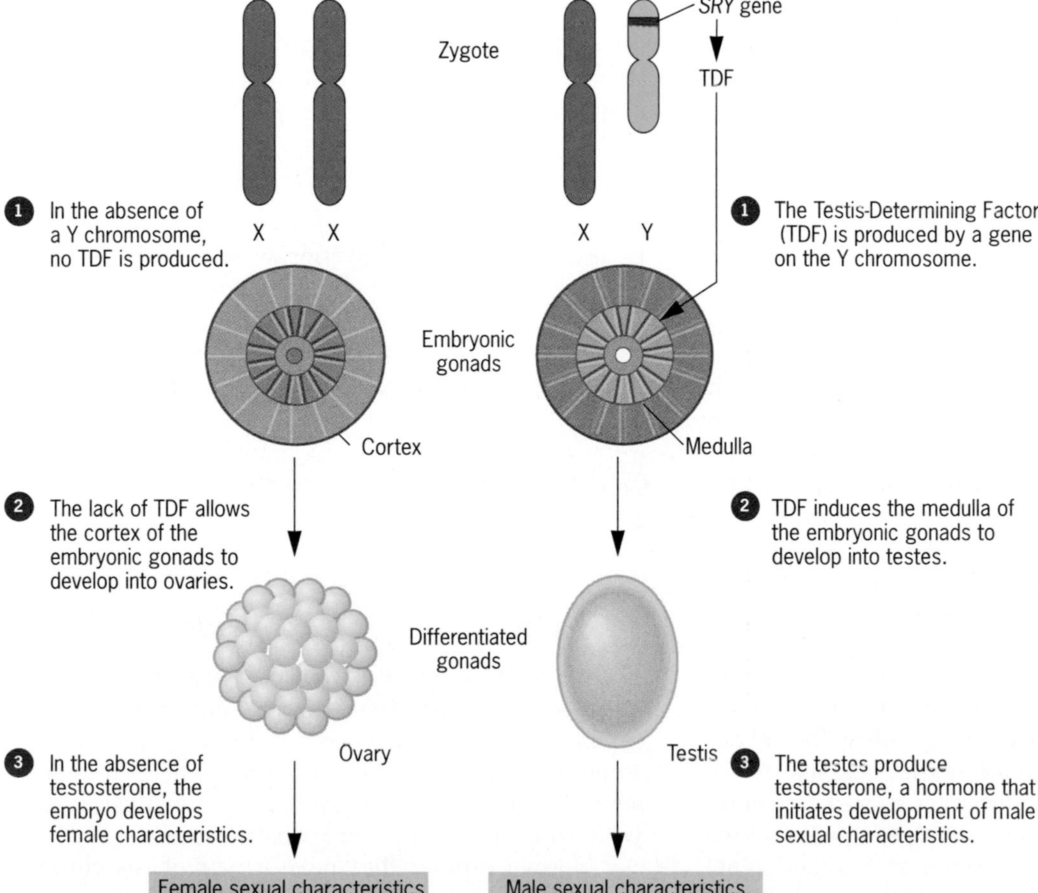

Figure 6.12 The process of sex determination in human beings. Male sexual development depends on the production of the testis-determining factor (TDF) by a gene on the Y chromosome. In the absence of this factor, the embryo develops as a female.

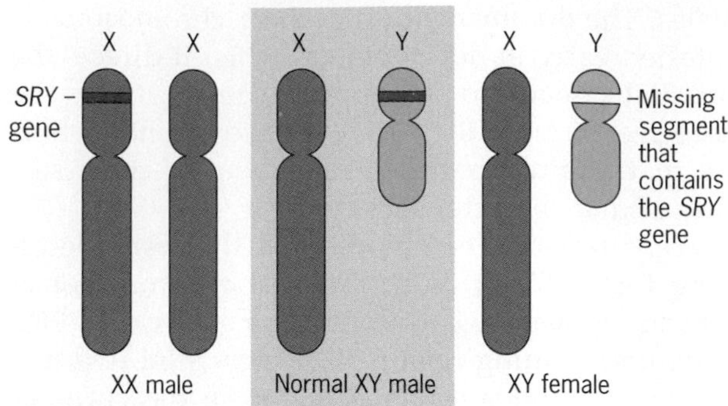

Figure 6.13 Evidence localizing the gene for the testis-determining factor (TDF) to the short arm of the Y chromosome in normal males. The TDF is the product of the *SRY* gene. In XX males, a small region containing this gene has been inserted into one of the X chromosomes, and in XY females, it has been deleted from the Y chromosome.

the XY females apparently prevented them from developing testes. These complementary lines of evidence showed that a particular segment of the Y chromosome was needed for male development. Molecular analyses subsequently identified the *SRY* gene in this male-determining segment. Additional research has shown that an *SRY* gene is present on the Y chromosome of the mouse, and that—like the human *SRY* gene—it specifies male development.

After the testes have formed, testosterone secretion initiates the development of male sexual characteristics. Testosterone is a hormone that binds to receptors on many kinds of cells. Once bound, the hormone-receptor complex transmits a signal to the cell, instructing it in how to differentiate. The concerted differentiation of many types of cells leads to the development of distinctly male characteristics such as heavy musculature, beard, and deep voice. If the testosterone signaling system fails, these characteristics do not appear and the individual develops as a female. One reason for failure is an inability to make the testosterone receptor (Figure 6.14). XY individuals with this biochemical deficiency initially develop as males—testes are formed and testosterone is produced. However, the testosterone has no effect because it cannot reach its target cells to transmit the developmental signal. Individuals lacking the testosterone receptor therefore switch sexes during embryological development and acquire female sexual characteristics. They do not, however, develop ovaries and are therefore sterile. This syndrome, called **testicular feminization,** results from a mutation in an X-linked gene, *tfm,* which encodes the testosterone receptor. The *tfm* mutation is transmitted from mothers to sons (who are phenotypically female) in a typical X-linked pattern.

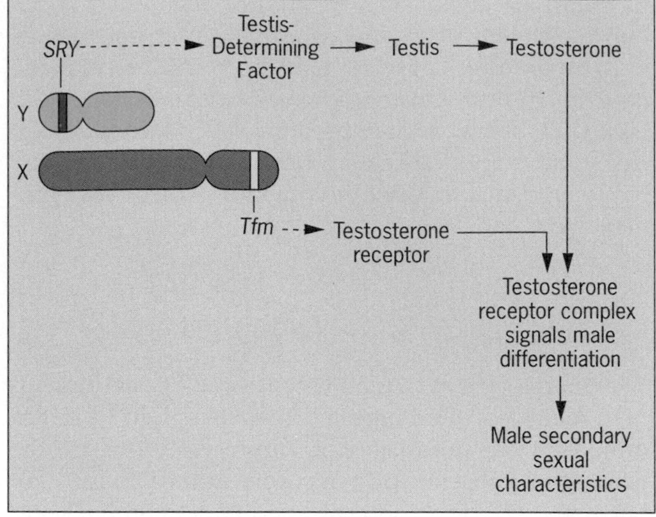

(*a*) Normal male with the wild-type *Tfm* gene.

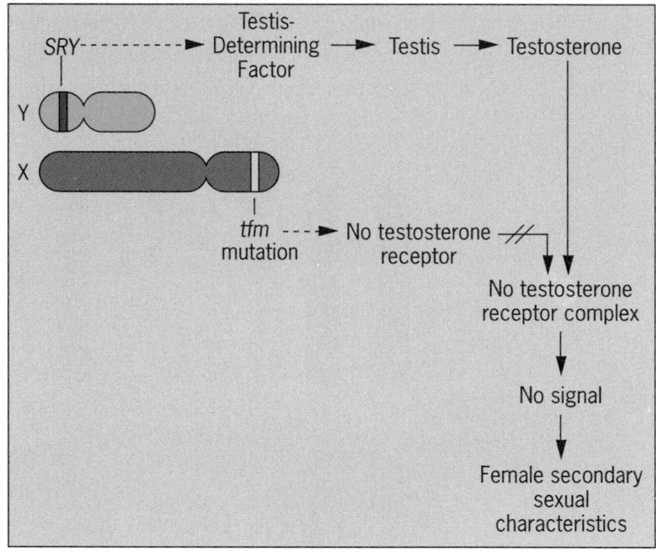

(*b*) Male with the *tfm* mutation and testicular feminization.

Figure 6.14 Testicular feminization, a condition caused by an X-linked mutation, *tfm,* that prevents the production of the testosterone receptor.

Sex Determination in *Drosophila*

The Y chromosome in *Drosophila*—unlike that in humans—plays no role in sex determination. Instead, the sex of the fly is determined by the ratio of X chromosomes to autosomes. This mechanism was first demonstrated by Bridges in 1921 through an analysis of flies with unusual chromosome constitutions.

Normal diploid flies have a pair of sex chromosomes, either XX or XY, and three pairs of autosomes, usually denoted AA; here, each A represents one haploid set of autosomes. Through genetic trickery,

TABLE 6.2
Ratio of X Chromosomes to Autosomes and the Corresponding Phenotype in *Drosophila*

X Chromosomes (X) and Sets of Autosomes (A)	*X:A Ratio*	*Phenotype*
1X 2A	0.5	Male
2X 2A	1.0	Female
3X 2A	1.5	Metafemale
4X 3A	1.33	Metafemale
4X 4A	1.0	Tetraploid female
3X 3A	1.0	Triploid female
3X 4A	0.75	Intersex
2X 3A	0.67	Intersex
2X 4A	0.5	Tetraploid male
1X 3A	0.33	Metamale

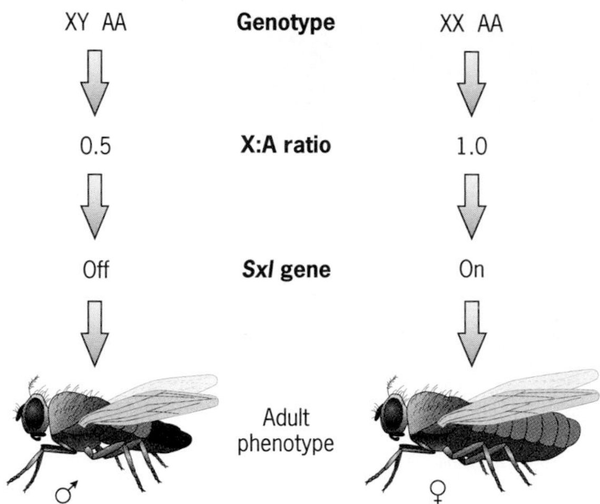

Figure 6.15 The *Sxl* sex determination switch in *Drosophila*. The *Sxl* gene is turned on in zygotes in which the X:A ratio is 1.0, and off in zygotes in which the X:A ratio is 0.5. Expression of the *Sxl* gene causes the zygote to develop into a female, whereas nonexpression of this gene causes the zygote to develop into a male.

Bridges contrived flies with abnormal numbers of chromosomes (Table 6.2). He observed that whenever the ratio of X's to A's was 1.0 or greater, the fly was female, and whenever it was 0.5 or less, the fly was male. Flies with an X:A ratio between 0.5 and 1.0 developed characteristics of both sexes; thus Bridges called them intersexes. In none of these flies did the Y chromosome have any effect on the sexual phenotype. It was, however, required for male fertility.

We now know that an X-linked gene called *Sex-lethal (Sxl)* is a key player in the *Drosophila* sex-determination system (Figure 6.15). An elaborate network of other X-linked genes, working with factors already present in the *Drosophila* egg, sets the level of *Sxl* activity in a zygote. If the X:A ratio is greater than or equal to one, the *Sxl* gene is activated and the zygote develops as a female; if it is less than or equal to 0.5, the *Sxl* gene is inactivated and the zygote develops as a male. A ratio between 0.5 and 1.0 causes a mix of signals, and the zygote develops both male and female characteristics. This system has the remarkable ability to count chromosomes, compute the ratio of X's to A's, and then flip the *Sxl* switch either on or off. Inappropriate expression of *Sxl*—off in females or on in males—leads to embryonic death—hence the name of the gene, *Sex-lethal*. The molecular details of sex determination in *Drosophila*, and the nature of the lethality associated with inappropriate expression of the *Sxl* gene, are discussed in Chapters 24 and 25.

Sex Determination in Other Animals

In *Drosophila* and human beings, males produce two kinds of gametes, X-bearing and Y-bearing. For this reason, they are referred to as the **heterogametic** sex; in these species females are the **homogametic** sex. In birds, butterflies, and some reptiles, this situation is reversed (Figure 6.16). Males are homogametic (usually denoted ZZ) and females are heterogametic (ZW). However, little is known about the mechanism of sex determination in the Z-W sex chromosome system.

In honeybees, sex is determined by whether the animal is haploid or diploid (Figure 6.17). Diploid embryos, which develop from fertilized eggs, become females; haploid embryos, which develop from unfertilized eggs, become males. Whether or not a given female will mature into a reproductive form (queen)

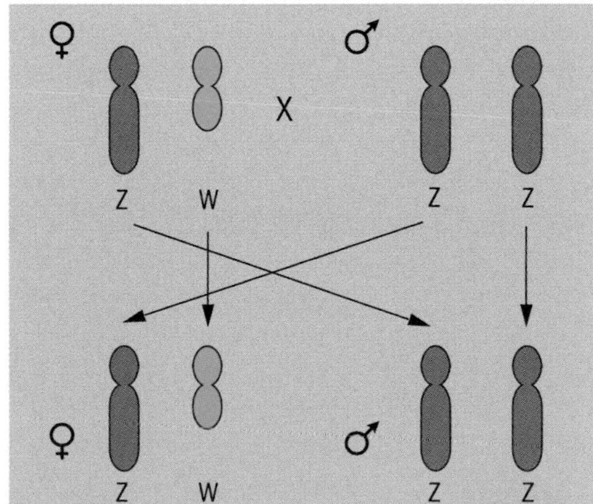

Figure 6.16 Sex determination in birds. The female is heterogametic (ZW), and the male is homogametic (ZZ). The sex of the offspring is determined by which of the sex chromosomes, Z or W, is transmitted by the female.

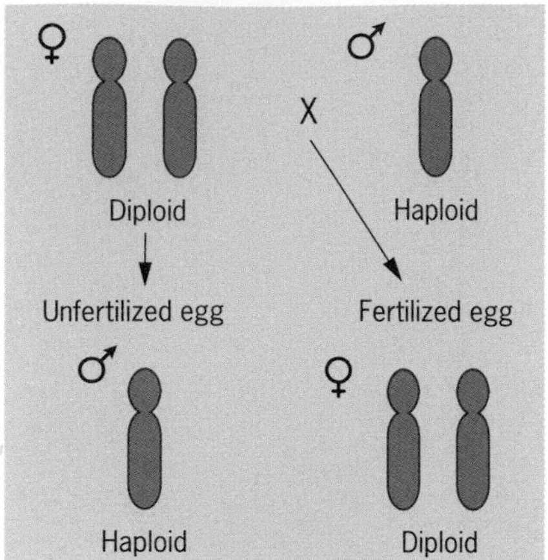

Figure 6.17 Sex determination in honeybees. Females, which are derived from fertilized eggs, are diploid, and males, which are derived from unfertilized eggs, are haploid.

depends on how she was nourished as a larva. In this system, a queen can control the ratio of males to females by regulating the proportion of unfertilized eggs that she lays. Because this number is small, most of the progeny are female, albeit sterile, and serve as workers for the hive. In a haplo-diplo system of sex determination, eggs are produced through meiosis in the queen, and sperm are produced through mitosis in the male. This system ensures that fertilized eggs will have the diploid chromosome number and that unfertilized eggs will have the haploid number.

Some wasps also have a haplo-diplo method of sex determination. In these species diploid males are sometimes produced, but they are always sterile. Detailed genetic analysis in one species, *Bracon hebetor,* has indicated that the diploid males are homozygous for a sex-determining locus, called *X*; diploid females are always heterozygous for this locus. Evidently, the sex locus in *Bracon* has many alleles; crosses between unrelated males and females therefore almost always produce heterozygous diploid females. However, when the mates are related, there is an appreciable chance that their offspring will be homozygous for the sex locus, in which case they develop into sterile males.

Key Points: **The mechanisms of sex determination vary among different types of organisms. In humans the testis-determining factor (*TDF*) on the Y chromosome causes an embryo to develop into a male. Without this factor, an embryo develops into a female. In *Drosophila,* sex is determined by the ratio of X chromosomes to sets of autosomes (X:A). For X:A $\leq$ 0.5 , the fly develops as a male; for X:A $\geq$ 1.0, it develops as a female; and for 0.5 $<$ X:A $<$ 1.0, it develops as an intersex. In honeybees, sex is determined by the number of chromosome sets.**

DOSAGE COMPENSATION OF X-LINKED GENES

Animal development is usually sensitive to an imbalance in the number of genes. Normally, each gene is present in two copies. Departures from this condition, either up or down, can cause abnormal phenotypes, and sometimes even death. It is therefore puzzling that so many species should have a sex-determination system based on females with two X chromosomes and males with only one. In these species, how is the numerical difference of X-linked genes accommodated? *A priori,* two mechanisms may compensate for this difference: (1) each X-linked gene works twice as hard in males as it does in females, or (2) one copy of each X-linked gene is inactivated in females. Extensive research has shown that both mechanisms are utilized, the first in *Drosophila* and the second in mammals.

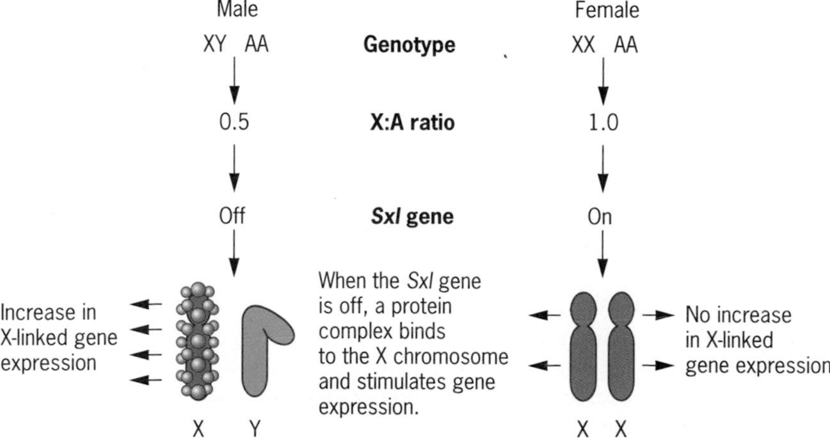

Figure 6.18 Dosage compensation in *Drosophila.* The single X chromosome in males is hyperactivated owing to a protein complex that binds to the chromosome. In females, where the *Sxl* gene is expressed, this complex does not bind to the X chromosomes, and these chromosomes are not hyperactivated.

Hyperactivation of X-linked Genes in Male *Drosophila*

In *Drosophila,* dosage compensation of X-linked genes is achieved by an increase in the activity of these genes in males. This phenomenon, called **hyperactivation,** involves the *Sex-lethal* gene, which also plays a key role in sex determination (Figure 6.18). The *Sxl* gene is turned on in females and off in males. When the *Sxl* gene product is absent (as it is in males), a complex of different proteins binds to many sites on the X chromosome and triggers a doubling of gene activity (see Chapter 24). When the *Sxl* gene product is present (as it is in females), this protein complex does not bind, and hyperactivation of X-linked genes does not occur. In this way, total X-linked gene activity in males and females is approximately equalized.

Inactivation of X-linked Genes in Female Mammals

In placental mammals, dosage compensation of X-linked genes is achieved by the **inactivation** of one of the female's X chromosomes (Figure 6.19). This mechanism was first proposed in 1961 by the British geneticist Mary Lyon, who inferred it from studies on mice. Subsequent research by Lyon and others has shown that the inactivation event occurs when the mouse embryo consists of a few thousand cells. At this time, each cell makes an independent decision to silence one of its X chromosomes. The chromosome to be inactivated is chosen at random; once chosen, however, it remains inactivated in all the descendants of that cell. Thus, female mammals are **genetic mosaics** containing two types of cell clones; the maternally inherited X chromosome is inactivated in roughly half of these clones, and the paternally inherited X is inactivated in the other half. A female that is heterozygous for an X-linked gene is therefore able to show two different phenotypes.

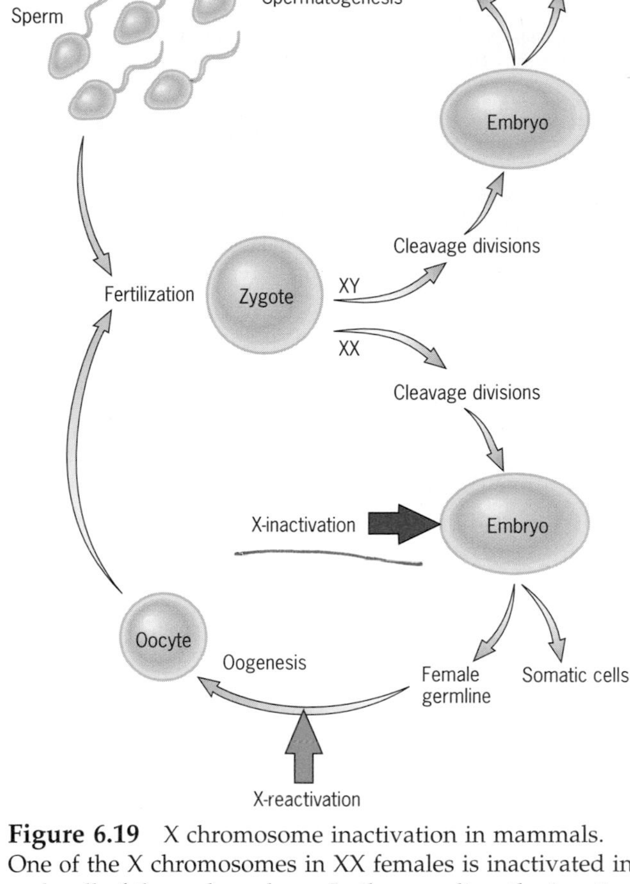

Figure 6.19 X chromosome inactivation in mammals. One of the X chromosomes in XX females is inactivated in each cell of the early embryo. In the germline, the inactivated X chromosomes are subsequently reactivated during oogenesis.

One of the best examples of this phenotypic mosaicism comes from the study of fur coloration in cats and mice (Figure 6.20). In both of these species, the X chromosome carries a gene for pigmentation of the fur. Females heterozygous for different alleles of

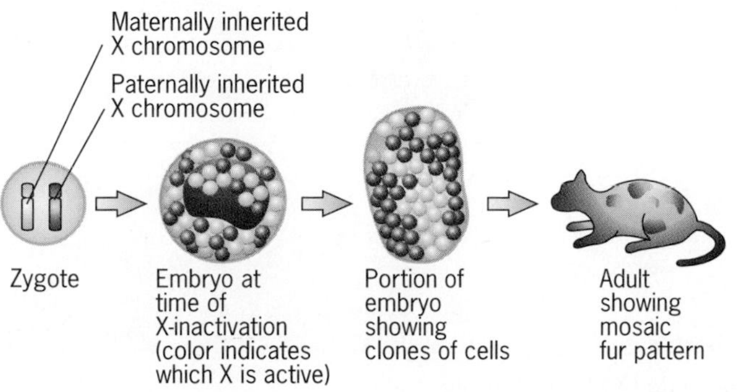

(a)

(b)

Figure 6.20 Color mosaics resulting from X chromosome inactivation in female mammals. (*a*) Formation of clones of cells in a cat embryo that produce different patches of fur in the adult. (*b*) A tortoiseshell cat. This female is heterozygous for an X-linked coat color gene. The orange and black patches are due to inactivation of different alleles in the pigment-producing cells of the body.

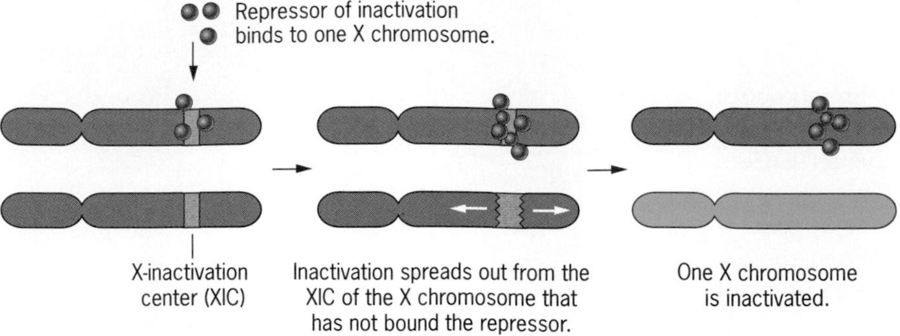

Figure 6.21 Mechanism of X chromosome inactivation.

this gene show patches of light and dark fur. The light patches express one allele, and the dark patches express the other. In cats, where one allele produces black pigment and the other produces orange pigment, this patchy phenotype is called tortoiseshell. Each patch of fur defines a clone of pigment-producing cells, or melanocytes, that were derived by mitosis from a precursor cell present at the time of X chromosome inactivation.

Many aspects of the mechanism for X chromosome inactivation are still a mystery. Genetic analyses have shown that in both humans and mice it begins at a site in the long arm of the X chromosome, and that it spreads in both directions to the chromosome's ends (Figure 6.21). The initiating site is called the **X-inactivation center (XIC).**

An X chromosome that has been inactivated does not look or act like other chromosomes. Chemical analyses show that its DNA is modified by the addition of numerous methyl groups. In addition, it condenses into a darkly staining structure called a **Barr body** (Figure 6.22), after the Canadian geneticist Murray Barr, who first observed it. This structure becomes attached to the inner surface of the nuclear membrane, where it replicates out of step with the other chromosomes in the cell. The inactivated X chromosome remains in this altered state in all the somatic tissues. However, in the germ tissues it is reactivated, perhaps because two copies of some X-linked genes are needed for the successful completion of oogenesis.

Cytological studies have identified human beings with more than two X chromosomes (see Chapter 7). For the most part, these people are phenotypically normal females, apparently because all but one of their X chromosomes is inactivated. Often all the inactivated X's congeal into a single Barr body. These observations suggest that cells may have a limited amount of some factor needed to prevent X-inactivation. Once this factor has been used to keep one X chromosome active, all the others quietly succumb to the inactivation process.

Key Points: **In *Drosophila,* dosage compensation for X-linked genes is achieved by hyperactivating the single X chromosome in males. In mammals, dosage compensation is achieved by inactivating one of the two X chromosomes in females.**

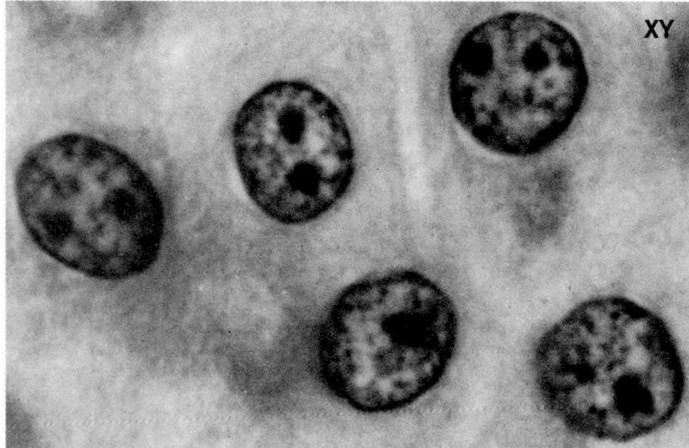

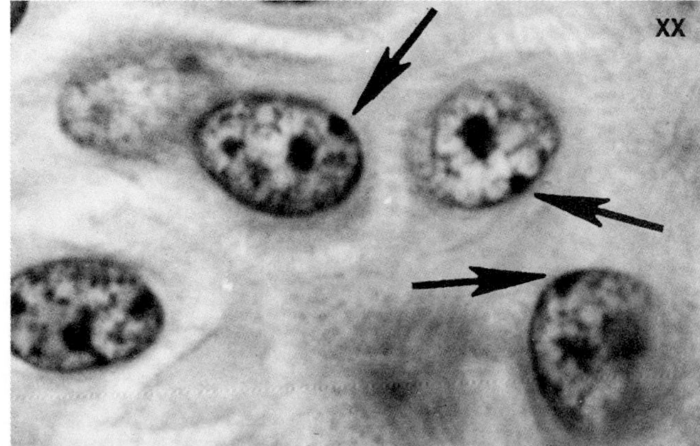

Figure 6.22 Barr bodies. Photograph of nuclei of normal human male (XY) and female (XX) cells. The Barr bodies (arrows) are visible only in the female nuclei.

TESTING YOUR KNOWLEDGE

1. The Lesch–Nyhan syndrome is a serious metabolic disorder affecting about one in 50,000 males in the population of the United States. A class of molecules called purines, which are biochemical precursors of DNA, accumulate in the nervous tissues and joints of people with the Lesch–Nyhan syndrome. This biochemical abnormality is caused by a deficiency for the enzyme hypoxanthine-guanine phosphoribosyltransferase (HPRT), which is encoded by a gene located on the X chromosome. Individuals deficient for this enzyme are unable to control their movements and unwillingly engage in self-destructive behavior such as biting and scratching themselves. The males labeled IV-5 and IV-6 in the pedigree below have the Lesch–Nyhan syndrome. What are the risks that V-1 and V-2 will inherit this disorder?

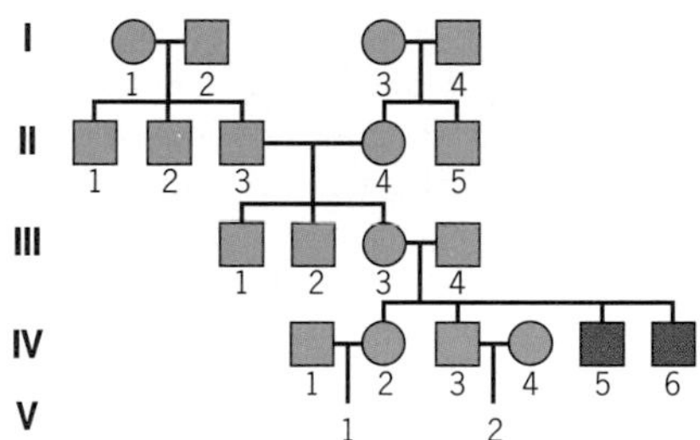

ANSWER

We know that III-3 must be a heterozygous carrier of the mutant allele *(h)* because two of her sons are affected. However, because she herself does not show the mutant phenotype, we know that her other X chromosome must carry the wild-type allele *(H)*. Given that III-3 is genotypically *H/h*, there is a one-half chance that she passed the mutant allele to her daughter (IV-2). If she did, there is a one-half chance that IV-2 will transmit this allele to her child (V-1), and there is a one-half chance that this child will be a male. Thus, the risk that V-1 will have the Lesch–Nyhan syndrome is $(1/2) \times (1/2) \times (1/2) = 1/8$. For V-2, the risk of inheriting the Lesch–Nyhan syndrome is essentially zero. This child's father (IV-3) is not a carrier, and even if he were, he would not transmit the mutant allele to a son. The child's mother comes from outside the family and is very unlikely to be a carrier because the trait is rare in the general population. Thus, V-2 has virtually no chance of suffering from the Lesch–Nyhan syndrome.

2. A geneticist crossed female *Drosophila* that had white eyes and ebony bodies to wild-type males, which had red eyes and gray bodies. Among the F_1, all the daughters had red eyes and gray bodies, and all the sons had white eyes and gray bodies. These flies were intercrossed to produce F_2 progeny, which were classified for eye and body color and then counted. Among 384 total progeny, the geneticist obtained the following results:

Combination of traits			
Eye Color	**Body Color**	**Males**	**Females**
white	ebony	20	21
white	gray	70	73
red	ebony	28	25
red	gray	76	71

How would you explain the inheritance of eye color and body color?

ANSWER

The results in the F_1 tell us that both mutant phenotypes are caused by recessive alleles. Furthermore, because the males and females have different eye color phenotypes, we know that the eye color gene is X-linked and that the body color gene is autosomal. In the F_2, the two genes assort independently, as we would expect for genes located on different chromosomes. Below, we show the genotypes of the different classes of flies in this experiment, using *w* for the *white* mutation and *e* for the *ebony* mutation; the wild-type alleles are denoted by plus signs. Following the convention of *Drosophila* geneticists, we write the sex chromosomes (X and Y) on the left and the autosomes on the right. A question mark in a genotype indicates that either the wild-type or mutant alleles could be present.

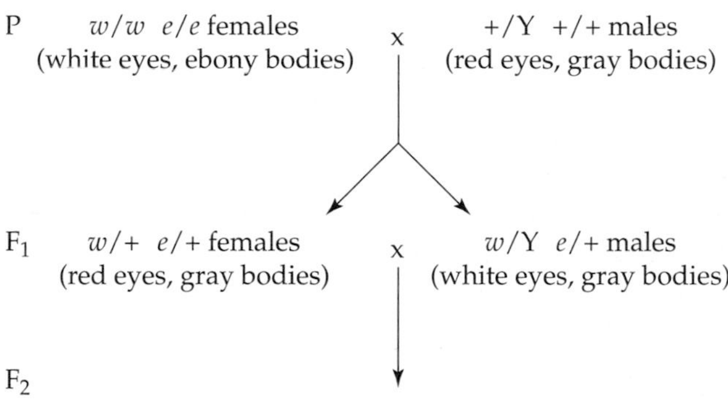

Phenotypes		Genotypes	
Eye Color	**Body Color**	**Males**	**Females**
white	ebony	*w*/Y *e/e*	*w/w e/e*
white	gray	*w*/Y +/?	*w/w* +/?
red	ebony	+/Y *e/e*	+/*w e/e*
red	gray	+/Y +/?	+/*w* +/?

QUESTIONS AND PROBLEMS

6.1 What are the genetic differences between male- and female-determining sperm in animals with heterogametic males?

6.2 A male with singed bristles appeared in a culture of *Drosophila*. How would you determine if this unusual phenotype was due to an X-linked mutation?

6.3 In grasshoppers, rosy body color is caused by a recessive mutation; the wild-type body color is green. If the gene for body color is on the X chromosome, what kind of progeny would be obtained from a mating between a homozygous rosy female and a hemizygous wild-type male? (In grasshoppers, females are XX and males are XO.)

6.4 In the mosquito *Anopheles culicifacies*, *golden* body *(go)* is a recessive X-linked mutation, and *brown* eyes *(bw)* is a recessive autosomal mutation. A homozygous XX female with golden body is mated to a homozygous XY male with brown eyes. Predict the phenotypes of their F_1 offspring. If the F_1 progeny are intercrossed, what kinds of progeny will appear in the F_2, and in what proportions?

6.5 What are the sexual phenotypes of the following genotypes in *Drosophila*: XX, XY, XXY, XXX, XO?

6.6 In human beings, a recessive X-linked mutation, *g*, causes green-defective color vision; the wild-type allele, *G*, causes normal color vision. A man (a) and a woman (b), both with normal vision, have three children, all married to people with normal vision: a color-defective son (c), who has a daughter with normal vision (f); a daughter with normal vision (d), who has one color-defective son (g) and two normal sons (h); and a daughter with normal vision (e), who has six normal sons (i). Give the most likely genotypes for the individuals (a to i) in this family.

6.7 If a father and son both have defective color vision, is it likely that the son inherited the trait from his father?

6.8 A normal woman, whose father had hemophilia, marries a normal man. What is the chance that their first child will have hemophilia?

6.9 A man with X-linked color blindness marries a woman with no history of color blindness in her family. The daughter of this couple marries a normal man, and their daughter also marries a normal man. What is the chance that this last couple will have a child with color blindness? If this couple has already had a child with color blindness, what is the chance that their next child will be color blind?

6.10 A man who has color blindness and type O blood has children with a woman who has normal color vision and type AB blood. The woman's father had color blindness. Color blindness is determined by an X-linked gene, and blood type is determined by an autosomal gene. (a) What are the genotypes of the man and the woman? (b) What proportion of their children will have color blindness and type B blood? (c) What proportion of their children will have color blindness and type A blood? (d) What proportion of their children will be color blind and have type AB blood?

6.11 A *Drosophila* female homozygous for a recessive X-linked mutation that causes vermilion eyes is mated to a wild-type male with red eyes. Among their progeny, all the sons have vermilion eyes, and nearly all the daughters have red eyes; however, a few daughters have vermilion eyes. Explain the origin of these vermilion-eyed daughters.

6.12 A *Drosophila* female heterozygous for the recessive X-linked mutation *w* (for *white* eyes) and its wild-type allele w^+ is mated to a wild-type male with red eyes. Among the sons, half have white eyes and half have red eyes. Among the daughters, nearly all have red eyes; however, a few have white eyes. Explain the origin of these white-eyed daughters.

6.13 In *Drosophila,* vermilion eye color is due to a recessive allele *(v)* located on the X chromosome. Curved wings is due to a recessive allele *(cu)* located on one autosome, and ebony body is due to a recessive allele *(e)* located on another autosome. A vermilion male is mated to a curved, ebony female, and the F_1 males are phenotypically wild-type. If these males were backcrossed to curved, ebony females, what proportion of the F_2 offspring will be wild-type males?

6.14 In *Drosophila,* a recessive mutation called *chocolate (c)* causes the eyes to be darkly pigmented. The mutant phenotype is indistinguishable from that of an autosomal recessive mutation called *brown (bw)*. A cross of chocolate-eyed females to homozygous brown males yielded wild-type F_1 females and darkly pigmented F_1 males. If the F_1 flies are intercrossed, what types of progeny are expected, and in what proportions? (Assume the double mutant combination has the same phenotype as either of the single mutants alone.)

6.15 Suppose that a mutation occurred in the *SRY* gene on the human Y chromosome, knocking out its ability to produce the testis-determining factor. Predict the phenotype of an individual who carried this mutation and a normal X chromosome.

6.16 A woman carries the testicular feminization mutation *(tfm)* on one of her X chromosomes; the other X carries the wild-type allele *(Tfm)*. If the woman marries a normal man, what fraction of her children will be phenotypically female? Of these, what fraction will be fertile?

6.17 Would a human with two X chromosomes and a Y chromosome be male or female?

6.18 Predict the sex of *Drosophila* with the following chromosome compositions (A = haploid set of autosomes): (a) 4X 4A; (b) 3X 4A; (c) 2X 3A; (d) 1X 3A; (e) 2X 2A; (f) 1X 2A.

6.19 In *Drosophila,* the gene for *bobbed* bristles (recessive allele *bb*, bobbed bristles; wild-type allele +, normal bristles) is located on the X chromosome and on a homologous segment of the Y chromosome. Give the genotypes and phenotypes of the offspring from the following crosses: (a) $X^{bb} X^{bb} \times X^{bb} Y^+$; (b) $X^{bb} X^{bb} \times X^+ Y^{bb}$; (c) $X^+ X^{bb} \times X^+ Y^{bb}$; (d) $X^+ X^{bb} \times X^{bb} Y^+$.

6.20 In chickens, the absence of barred feathers is due to a recessive allele. A barred rooster was mated with a nonbarred hen, and all the offspring were barred. These F_1 chickens were intercrossed to produce F_2 progeny, among which all the males were barred; half the females were barred and half were nonbarred. Are these results consistent with the hypothesis that the gene for barred feathers is located on one of the sex chromosomes?

6.21 A *Drosophila* male carrying a recessive X-linked mutation for yellow body is mated to a homozygous wild-type female with gray body. The daughters of this mating all have uniformly gray bodies. Why aren't their bodies a mosaic of yellow and gray patches?

6.22 How many Barr bodies would be visible in the nuclei of human cells with the following chromosome compositions: (a) XY; (b) XX; (c) XXY; (d) XXX; (e) XXXX; (f) XYY?

6.23 Males in a certain species of deer have two nonhomologous X chromosomes, denoted X_1 and X_2, and a Y chromosome. Each X chromosome is about half as large as the Y chromosome, and its centromere is located near one of the ends; the centromere of the Y chromosome is located in the middle. Females in this species have two copies of each of the X chromosomes and lack a Y chromosome. How would you predict the X and Y chromosomes to pair and disjoin during spermatogenesis to produce equal numbers of male- and female-determining sperm?

6.24 A breeder of sun conures (a type of bird) has obtained two true-breeding strains, A and B, which have red eyes instead of the normal brown found in natural populations. In Cross 1, a male from strain A was mated to a female from strain B, and the male and female offspring all had brown eyes. In Cross 2, a female from strain A was mated to a male from strain B, and the male offspring had brown eyes and the female offspring had red eyes. When the F_1 birds from each cross were mated brother to sister, the breeder obtained the following results:

Phenotype	**Proportion in F_2 of Cross 1**	**Proportion in F_2 of Cross 2**
Brown male	6/16	3/16
Red male	2/16	5/16
Brown female	3/16	3/16
Red female	5/16	5/16

Provide a genetic explanation for these results.

BIBLIOGRAPHY

CLINE, T. W. 1993. The *Drosophila* sex determination signal: How do flies count to two? *Trends in Genet.* 9:385–390.

GOULD, L. 1996. *Cats Are Not Peas: A Calico History of Genetics.* Springer-Verlag, New York.

HEARD, E., P. CLERC, AND P. AVNER. 1997. X-chromosome inactivation in mammals. *Ann. Rev. Genetics.* 31:571–610.

KOHLER, R. E. 1994. *Lords of the Fly: Drosophila Genetics and the Experimental Life.* University of Chicago Press, Chicago.

LYON, M. F. 1993. Epigenetic inheritance in mammals. *Trends in Genet.* 9:123–128.

7

Wheat field in central North America.

Variation in Chromosome Number and Structure

CHAPTER OUTLINE

Cytological Techniques
Polyploidy
Aneuploidy
Rearrangements of Chromosome Structure: An Overview

Chromosomes, Agriculture, and Civilization

The cultivation of wheat originated some 10,000 years ago in the Middle East. Today, wheat is the principal food crop for more than a billion people. Wheat is grown in diverse environments, from Norway to Argentina. More than 17,000 varieties have been developed, each adapted to a different locality. The total wheat production of the world is 50 million metric tons annually, accounting for more than 20 percent of the food calories consumed by the entire human population. Wheat is clearly an important agricultural crop and, some would argue, a mainstay of civilization.

Modern cultivated wheat, *Triticum aestivum*, is a hybrid of at least three different species. Its progenitors were low-yielding grasses that grew in Syria, Iran, Iraq, and Turkey. Some of these grasses appear to have been cultivated by the ancient peoples of this region. Although we do not know the exact course of events, two of the grasses apparently interbred, producing a species that excelled as a crop plant. Through human cultivation, this hybrid species was selectively improved, and then it, too, interbred with a third species, yielding a triple-hybrid that was even better suited for agriculture. Modern wheat is descended from these triple-hybrid plants.

What made the triple-hybrid wheats so superior to their ancestors? They had larger grains, they were more easily harvested, and they grew in a wider range of conditions. We now understand the chromosomal basis for these improvements. Triple-hybrid wheat contains the chromosomes of each of its progenitors. Genetically, it is an amalgamation of the genes of three different species.

Such chromosomal mixtures are common in the plant kingdom, especially among plants with an agricultural significance. Variations in chromosome number are also found in animals, including our own species, and some of these variations are associated with conspicuous phenotypes. In this chapter we consider different kinds of changes in chromosome number. We also consider several types of changes in chromosome structure.

CYTOLOGICAL TECHNIQUES

Geneticists study chromosome number and structure by staining dividing cells with certain dyes and then examining them with a microscope. The analysis of stained chromosomes is the main activity of the discipline called **cytogenetics.**

Cytogenetics had its roots in the research of several nineteenth-century European biologists who discovered chromosomes and observed their behavior during mitosis, meiosis, and fertilization. This research blossomed during the twentieth century, as microscopes improved and better procedures for preparing and staining chromosomes were developed. The demonstration that genes reside on chromosomes boosted interest in this research and led to important studies on chromosome number and structure. Today, cytogenetics has significant applied aspects, especially in medicine, where it is used to determine whether disease conditions are associated with chromosome abnormalities.

Analysis of Mitotic Chromosomes

Researchers perform most cytological analyses on dividing cells, usually cells in the middle of mitosis. To enrich for cells at this stage, they have traditionally used rapidly growing tissues such as animal embryos and plant root tips. However, the development of cell-culturing techniques has made it possible to study chromosomes in other types of cells (Figure 7.1). For example, human white blood cells can be collected from peripheral blood, separated from the nondividing red blood cells, and put into culture. The white cells are then stimulated to divide by chemical treatment, and midway through division a sample of the cells is prepared for cytological analysis. The usual procedure is to treat the dividing cells with a chemical that

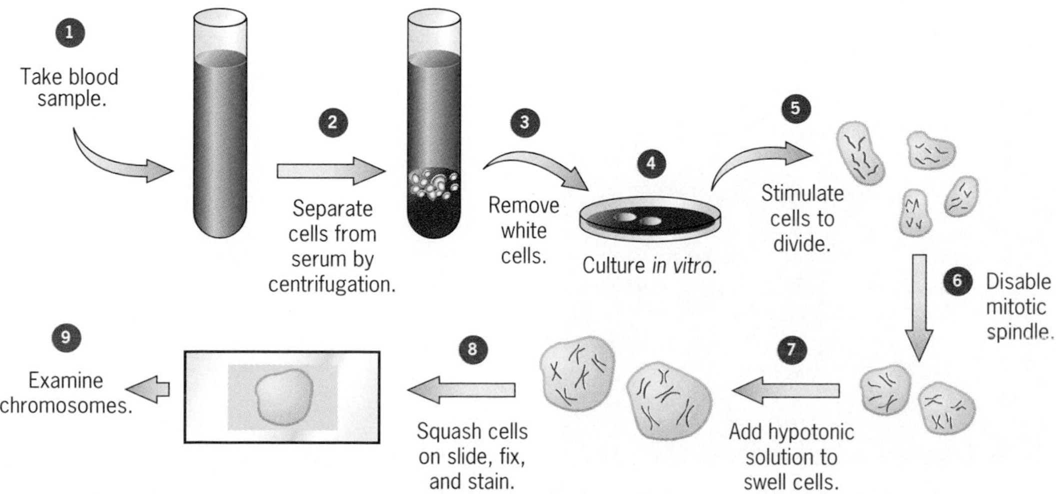

Figure 7.1 Preparation of cells for cytological analysis.

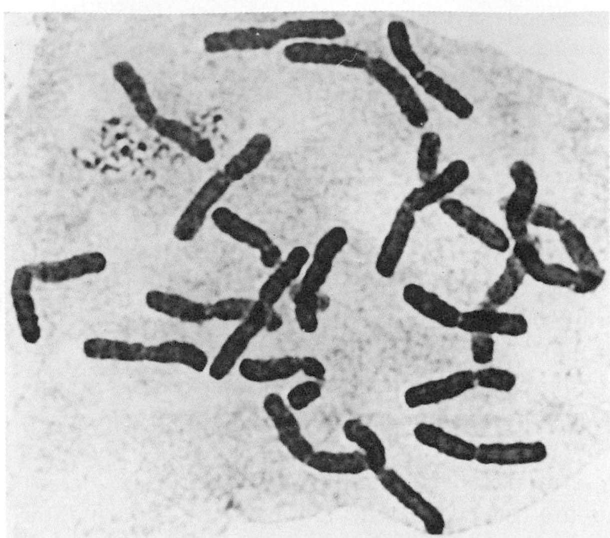

Figure 7.2 Chromosomes of the incense cedar (*Calocedrus decurrens*) stained with aceto-carmine.

disables the spindle apparatus. The effect of this interference is to trap the chromosomes in mitosis, when they are most easily seen. Mitotically arrested cells are then swollen by immersion in a hypotonic solution that causes the cells to take up water by osmosis. The contents of each cell are diluted by the additional water, so that when the cells are squashed on a microscope slide, the chromosomes are spread out in an uncluttered fashion. This technique greatly facilitates subsequent analysis, especially if the chromosome number is large. For many years it was erroneously thought that human cells contained 48 chromosomes. The correct number, 46, was determined only after the swelling technique was used to separate the chromosomes within individual mitotic cells.

Until the late 1960s and early 1970s, chromosome spreads were stained with Feulgen's reagent, a purple dye that reacts with the sugar molecules in DNA, or with aceto-carmine, a deep red dye (Figure 7.2). Although these stains are still used for routine cytological analyses, more detailed studies now employ staining compounds that insert between the base-pairs of DNA. A molecule with this insertional property is called an **intercalating agent.** Quinacrine, a chemical relative of the antimalarial drug quinine, is one such molecule. Chromosomes that have been stained with quinacrine show a characteristic pattern of bands. However, because quinacrine is a **fluorescent** compound, the bands appear only when the chromosomes are exposed to ultraviolet (UV) light. Ultraviolet irradiation causes some of the quinacrine molecules that have inserted into the chromosome to emit energy. Parts of the chromosome shine brightly, whereas other parts remain dark. This differential fluorescence creates a pattern of bright and dark bands that is highly reproducible. Using these bands, cytogeneticists can identify particular chromosomes in a cell, and they can also analyze these chromosomes for structural abnormalities. This staining procedure is called **Q banding,** and the bands that it produces are called Q bands (Figure 7.3).

Excellent nonfluorescent staining techniques have also been developed. The most popular of these uses **Giemsa** stain, a mixture of dyes named after its inventor, Gustav Giemsa. Like quinacrine, Giemsa creates a reproducible pattern of bands on each chromosome. However, the nature of the banding pattern

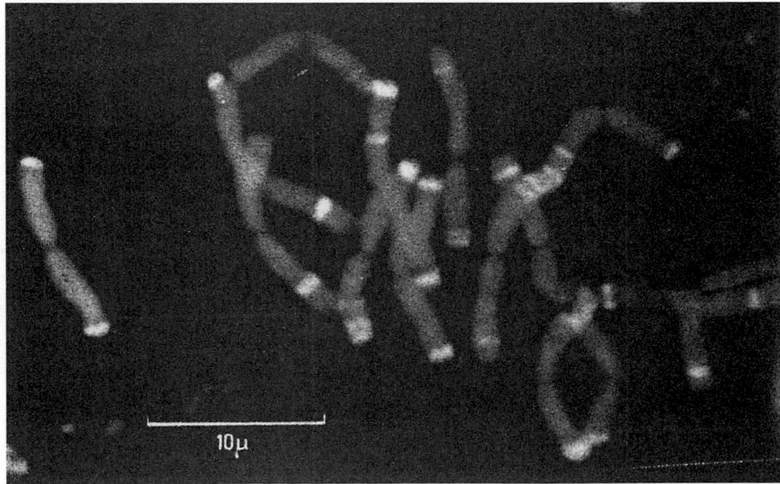

Figure 7.3 Metaphase chromosomes of the plant *Allium carinatum*, stained with quinacrine to reveal Q bands.

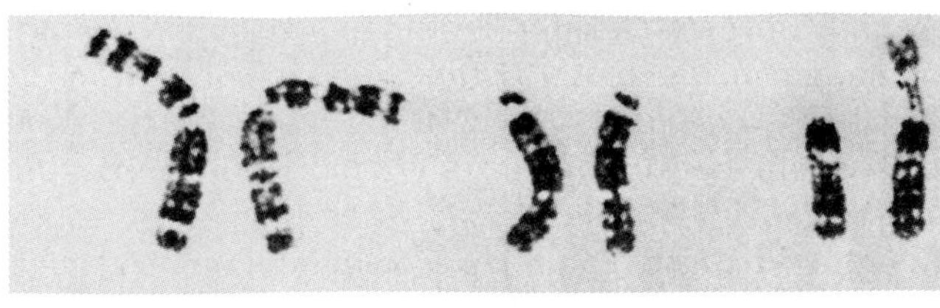

(a)

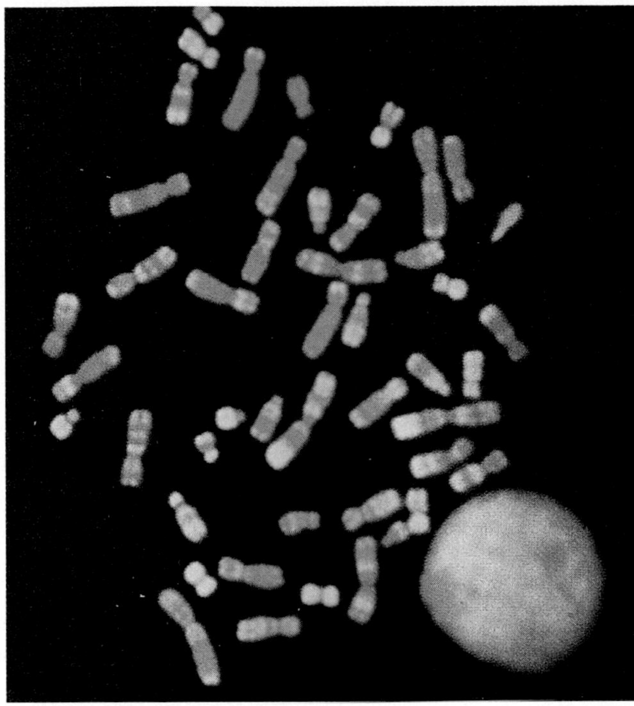

(b)

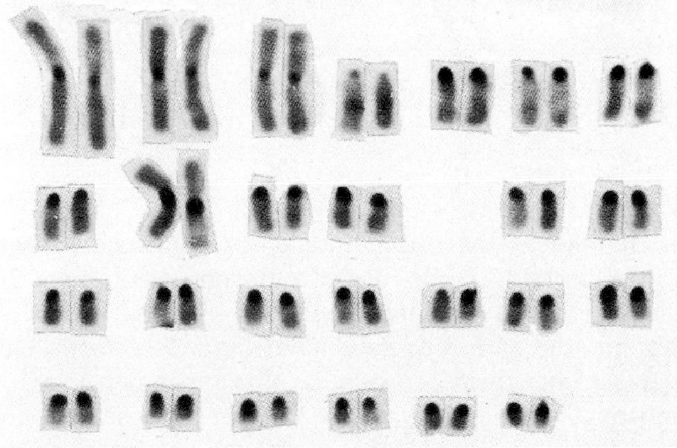

(c)

Figure 7.4 (*a*) Metaphase chromosomes of the deerlike Asian muntjak stained to show G banding. (*b*) Metaphase human chromosomes stained with acridine orange to show R banding. (*c*) Metaphase chromosomes of the domestic sheep stained to show C banding.

depends on how the chromosomes were prepared prior to staining. One procedure, called **G banding,** gives dark bands that correspond roughly to the bright bands obtained with quinacrine; another procedure, called **R banding,** gives the reverse pattern—dark bands that correspond roughly to light G bands. A third procedure, called **C banding,** stains the region around the centromere of each chromosome. These different banding techniques provide cytogeneticists with the means to analyze fine details of chromosome structure (Figure 7.4).

The Human Karyotype

Diploid human cells contain 46 chromosomes—44 autosomes and two sex chromosomes, which are XX in females and XY in males. At mitotic metaphase, each of the chromosomes can be recognized by its size, shape, and banding pattern. For cytological analysis, well-stained metaphase spreads are photographed, and then each of the chromosome images is cut out of the picture, matched with its partner, and arranged from largest to smallest on a chart (Figure 7.5). The largest autosome is number 1, and the smallest is number 21. (For historical reasons, the second smallest chromosome has been designated number 22.) The X chromosome is intermediate in size, and the Y chromosome is about the same size as chromosome 22. This chart of chromosome cutouts is called a **karyotype** (from the Greek word meaning "kernel," a reference to the contents of the nucleus). A skilled researcher can use a karyotype to identify abnormalities in chromosome number and structure.

Before the banding techniques were available, it was difficult to distinguish one human chromosome from another. Cytogeneticists could only arrange the chromosomes into groups according to size, classifying the largest as group A, the next largest as group B, and so forth. Although they could recognize seven different groups, within these groups it was nearly impossible to identify a particular chromosome. Today—as a result of the banding techniques—we can routinely identify each of the chromosomes. The banding techniques also make it possible to distinguish each arm of a chromosome. The length of the arms is determined by the position of the centromere, which is the place where the spindle fibers attach to move the chromosome during cell division. The centromere divides the chromosome into long and short arms. The short arm is denoted by the letter **p** (from the French word *petite,* meaning "small") and the long arm by the letter **q** (because it follows "p" in the alphabet). Thus, for example, a cytogeneticist can refer specifically to the short arm of chromosome 5 simply by writing "5p."

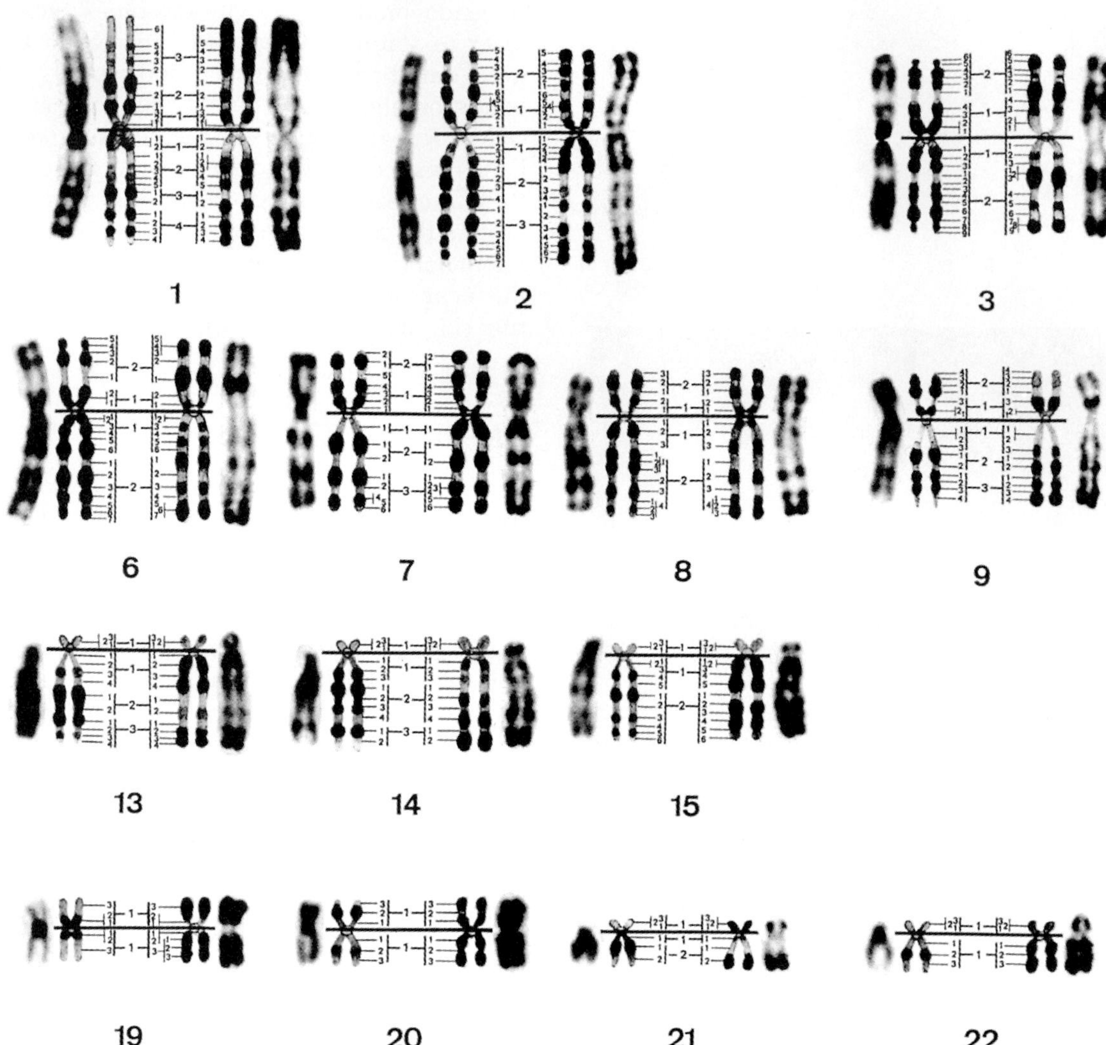

Figure 7.5 The human karyotype. Photographs of chromosomes stained to show G banding (left) and R banding (right). Autosomes are numbered from 1 to 22. The short arm (p) and long arm (q) of each chromosome are subdivided into numbered regions, starting at the centromere.

Cytogenetic Variation: An Overview

The phenotypes of many organisms are affected by changes in the number of chromosomes in their cells; sometimes even changes in part of a chromosome can be significant. These numerical changes are usually described as variations in the **ploidy** of the organism (from the Greek word meaning "fold," as in "twofold"). Organisms with complete, or normal, sets of chromosomes are said to be *euploid* (from the Greek words meaning "good" and "fold"). Organisms that carry extra sets of chromosomes are said to be *polyploid* (from the Greek words meaning "many" and "fold"), and the level of polyploidy is described by referring to a basic chromosome number, usually denoted n; n is defined as the number of chromosomes in a set. Thus, diploids, with two chromosome sets, have $2n$ chromosomes; triploids, with three sets, have $3n$; tetraploids, with four sets, have $4n$; and so forth. Organisms in which a particular chromosome, or chromosome segment, is under- or overrepresented are said to be *aneuploid* (from the Greek words meaning "not," "good," and "fold"). These organisms therefore suffer from a specific genetic imbalance. The distinction between aneuploidy and polyploidy is that aneuploidy refers to a numerical change in part of the genome, usually just a single chromosome, whereas polyploidy refers to a numerical change in a whole set

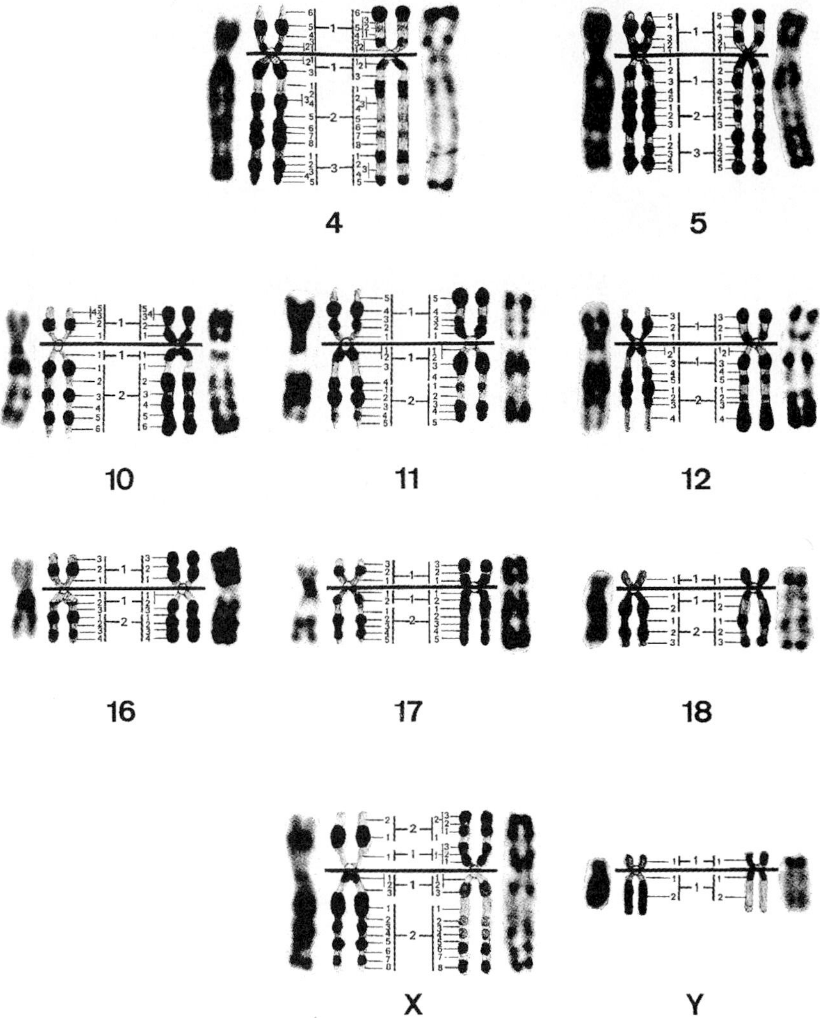

of chromosomes. Aneuploidy implies a genetic imbalance, but polyploidy does not.

Cytogeneticists have also catalogued various types of structural changes in the chromosomes of organisms. For example, a piece of one chromosome may be fused to another chromosome, or a segment within a chromosome may be inverted with respect to the rest of that chromosome. These structural changes are called rearrangements. Because rearrangements segregate irregularly during meiosis, they are often associated with aneuploidy. In the sections that follow, we consider all these cytogenetic variations—polyploidy, aneuploidy, and chromosome rearrangements.

Key Points: **Chromosomes are most easily analyzed in dividing cells. Among the various dyes used to stain chromosomes, quinacrine and Giemsa produce banding patterns that are useful in identifying individual chromosomes within a karyotype.**

POLYPLOIDY

Polyploidy, the presence of extra chromosome sets, is fairly common in plants but very rare in animals. One-half of all known plant genera contain polyploid species, and about two-thirds of all grasses are polyploids. Many of these species reproduce asexually. In animals, where reproduction is primarily by sexual means, polyploidy is rare, probably because it interferes with the sex-determination mechanism.

One general effect of polyploidy is that cell size is increased, presumably because there are more chro-

Figure 7.6 Polyploid plants with agricultural or horticultural significance.

mosomes in the nucleus. Often this increase in size is correlated with an overall increase in the size of the organism. Polyploid species tend to be larger and more robust than their diploid counterparts. These characteristics have a practical significance for human beings, who depend on many polyploid plant species for food. These species tend to produce larger seeds and fruits, and therefore provide greater yields in agriculture. Wheat, coffee, potatoes, bananas, strawberries, and cotton are all polyploid crop plants. Many ornamental garden plants, including roses, chrysanthemums, and tulips, are also polyploid (Figure 7.6).

Sterile Polyploids

In spite of their robust physical appearance, many polyploid species are sterile. Extra sets of chromosomes segregate irregularly in meiosis, leading to grossly unbalanced (that is, aneuploid) gametes. If such gametes unite in fertilization, the resulting zygotes almost always die. This inviability among the zygotes explains why many polyploid species are sterile.

As an example, let us consider a triploid species with three identical sets of n chromosomes. The total number of chromosomes is therefore $3n$. When meiosis occurs, each chromosome will try to pair with its homologs (Figure 7.7). One possibility is that two homologs will pair completely along their length, leaving the third without a partner; this solitary chromosome is called a **univalent.** Another possibility is that all three homologs will synapse, forming a **trivalent** in which each member is partially paired with each of the others. In either case, it is difficult to predict how the chromosomes will move during anaphase of the first meiotic division. The more likely event is that two of the homologs will move to one pole, and one homolog will move to the other, yielding gametes with one or two copies of the chromosome. However, all three homologs might move to one pole, producing gametes with zero or three copies of the chromosome. Because

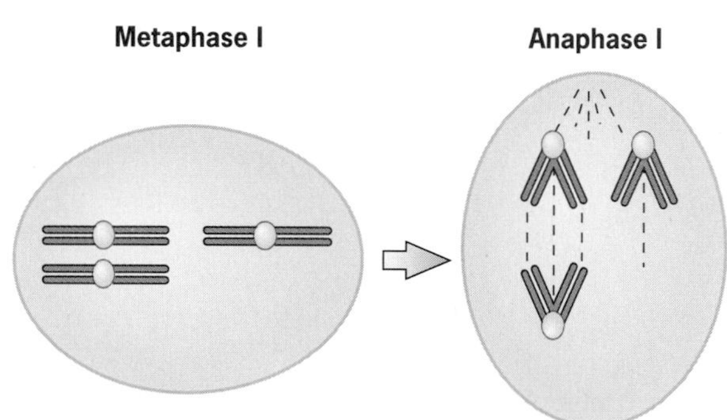

(*a*) Two of the three homologs synapse, leaving a univalent free to move to either pole during anaphase.

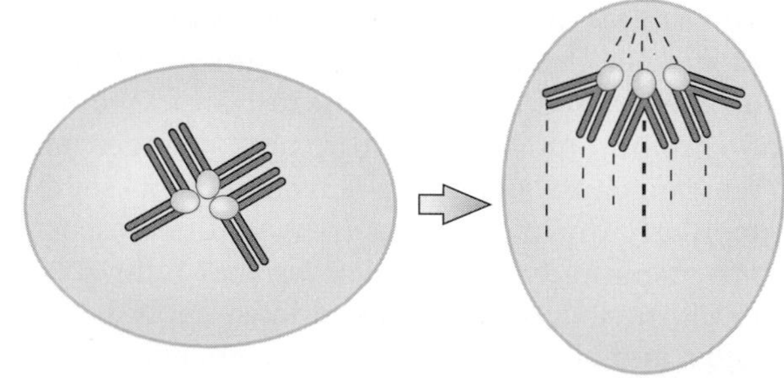

(*b*) All three homologs synapse, forming a trivalent, which may move as a unit to one pole during anaphase. However, other anaphase disjunctions are possible.

Figure 7.7 Meiosis in a triploid. (*a*) Univalent formation. (*b*) Trivalent formation.

this segregational uncertainty applies to each trio of chromosomes in the cell, the total number of chromosomes in a gamete can vary from zero to $3n$.

Zygotes formed by fertilization with such gametes are almost certain to die; thus, most triploids are completely sterile. In agriculture and horticulture, this sterility is circumvented by propagating the species asexually. The many methods of asexual propagation include cultivation from cuttings (bananas), grafts (Winesap, Gravenstein, and Baldwin apples), and bulbs (tulips). In nature, polyploid plants can also reproduce asexually. One mechanism is **apomixis**, which involves a modified meiosis that produces unreduced eggs; these eggs then form seeds that germinate into new plants. A common plant that reproduces in this way is the dandelion, *Taraxacum officinale*, a highly successful polyploid weed.

Fertile Polyploids

The meiotic uncertainties that occur in triploids also occur in tetraploids with four identical chromosome sets. Such tetraploids are therefore also sterile. However, some tetraploids are able to produce viable progeny. Close examination shows that these species contain two distinct sets of chromosomes and that each set has been duplicated. Thus, fertile tetraploids seem to have arisen by chromosome duplication in a hybrid that was produced by a cross of two different, but related, diploid species; most often these species have the same or very similar chromosome numbers. Figure 7.8 shows a plausible mechanism for the origin of such a tetraploid. Two diploids, denoted A and B, are crossed to produce a hybrid that receives one set of chromosomes from each of the parental species. Such a hybrid will probably be sterile because the A and B chromosomes cannot pair with each other. However, if the chromosomes in this hybrid are duplicated, meiosis will proceed in reasonably good order. Each of the A and B chromosomes will be able to pair with a perfectly homologous partner. Meiotic segregation can therefore produce gametes with a complete set of A and B chromosomes. In fertilization, these "diploid" gametes will unite to form tetraploid zygotes, which will survive because each of the parental sets of chromosomes will be balanced.

This scenario of hybridization between different but related species followed by chromosome doubling has evidently occurred many times during plant evolution. In some cases, the process has occurred repeatedly, generating complex polyploids with distinct chromosome sets. One of the best examples is modern bread wheat, *Triticum aestivum* (Figure 7.9). This important crop species is a hexaploid containing three different chromosome sets, each of which has been duplicated. There are seven chromosomes in each set, for a total of 21 in the gametes and 42 in the somatic cells. Thus, as we noted at the beginning of this chapter,

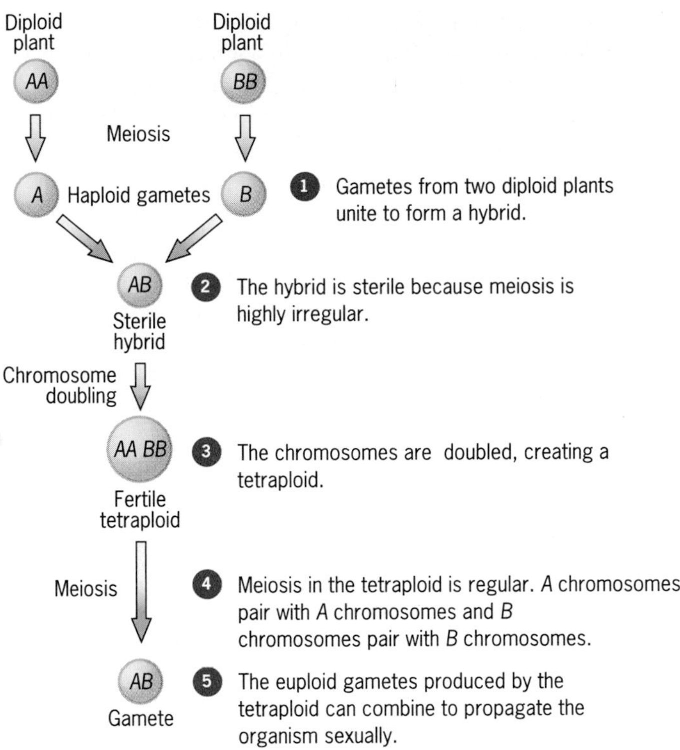

Figure 7.8 Origin of a fertile tetraploid by hybridization between two diploids and subsequent doubling of the chromosomes.

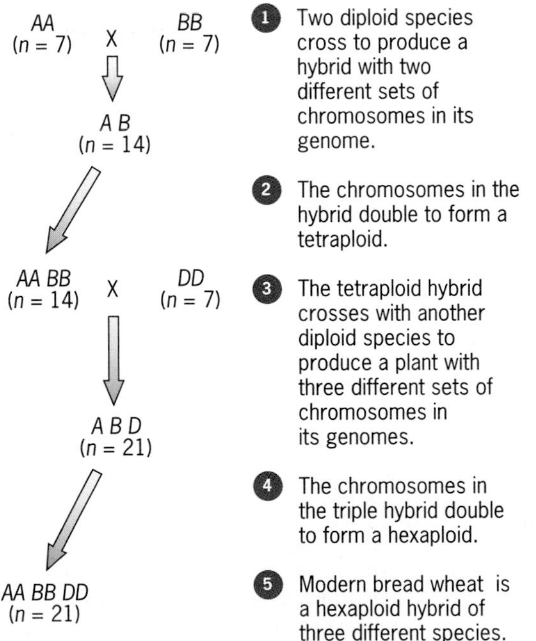

Figure 7.9 Origin of hexaploid wheat by sequential hybridization of different species. Each hybridization event is followed by doubling of the chromosomes.

modern wheat seems to have been formed by two hybridization events. The first involved two diploid species that combined to form a tetraploid, and the second involved a combination between this tetraploid and another diploid, to produce a hexaploid. Cytogeneticists have identified primitive cereal plants in the Middle East that may have participated in this evolutionary process.

Because chromosomes from different species are less likely to interfere with each other's segregation during meiosis, polyploids arising from hybridizations between different species have a much greater chance of being fertile than do polyploids arising from the duplication of chromosomes in a single species. Polyploids created by hybridization between different species are called ***allo*polyploids** (from the Greek prefix for "other"); in these polyploids, the contributing genomes are qualitatively different. Polyploids created by chromosome duplication within a species are called ***auto*polyploids** (from the Greek prefix for "self"); in these polyploids, a single genome has been multiplied to create extra chromosome sets.

Chromosome Doubling and the Origin of Polyploids

Where does the chromosome doubling process occur in the evolution of a polyploid plant species (Figure 7.10)? There are two possibilities: (1) in the cells that divide to produce the different tissues of plants (cells that are called meristematic cells), or (2) in the repro-

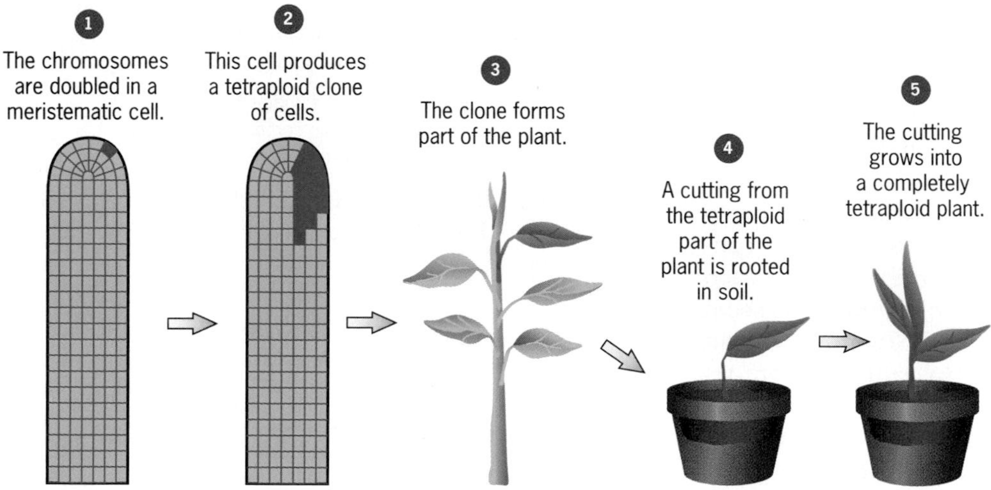

(*a*) Chromosome doubling in meristematic tissue.

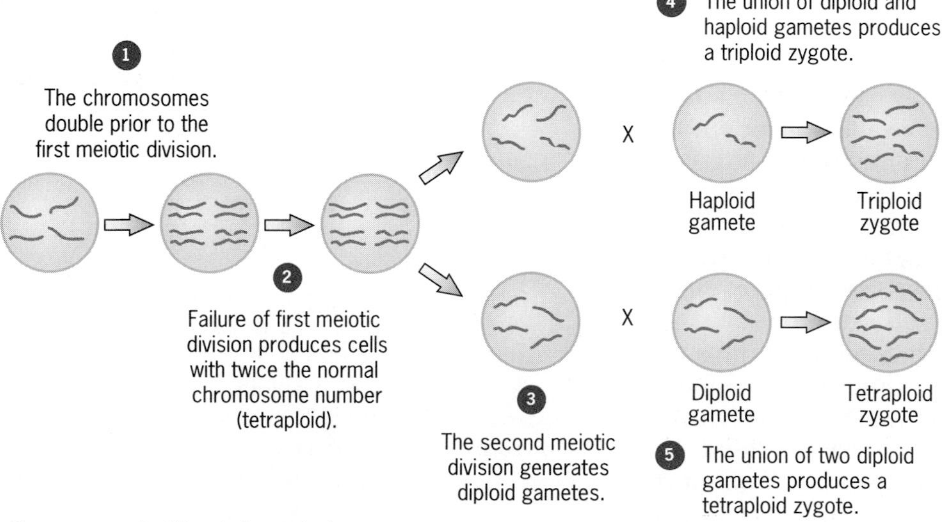

(*b*) Chromosome doubling during meiosis.

Figure 7.10 Chromosome doubling and the origin of polyploids.

ductive cells as they go through meiosis. In a diploid meristematic cell, the sister chromatids may fail to separate during mitosis. This failure of the normal mitotic mechanism will produce a tetraploid cell, which will divide to form a tetraploid clone of cells within an otherwise diploid organism. If this clone of tissue is detached from the plant and rooted, the tetraploid condition can be propagated in a new organism. Cuttings from this tetraploid plant can then be rooted to establish many additional tetraploid plants. If the original chromosome doubling occurs in a hybrid between two different species, the resulting tetraploids may be fertile, permitting subsequent sexual propagation. Allotetraploids of this sort are sometimes called **amphidiploids** (from the Greek words meaning "of two kinds"), in recognition of their status as the sum of two diploid species.

Chromosome doubling could also take place in a cell undergoing meiosis. The chromosomes might fail to separate in either of the meiotic divisions, leading to gametes with twice the normal chromosome number. If such a diploid gamete unites with a normal haploid gamete, a triploid zygote is produced. Alternatively, two diploid gametes might unite with each other, creating a tetraploid zygote.

Experimental Production of Polyploids

Numerous polyploids have been created in the laboratory, partly for scientific interest and partly for agricultural benefit. One of the earliest experiments involved hybrids between the radish, *Raphanus sativus*, and the cabbage, *Brassica oleracea*. Although these plants belong to different genera, they can be crossed to make hybrids, which are usually sterile. Both species have nine pairs of chromosomes and the hybrids have one complete set of chromosomes from each parent, that is, 18 chromosomes altogether.

In the 1920s, the Russian cytologist G. D. Karpechenko found some unusual radish–cabbage hybrids that made unreduced gametes. Combining these gametes to form progeny that had 36 chromosomes—18 from each of the parental species—he established a strain of the new tetraploid and named it *Raphanobrassica*. With this work, Karpechenko demonstrated that cytogeneticists could create fertile interspecific hybrids in the laboratory and that they could combine traits from different species into a single polyploid plant. Unfortunately, his hope of creating a plant with the root of the radish and the foliage of the cabbage was not realized; both the foliage and the root of *Raphanobrassica* were inedible.

Polyploids have been induced experimentally in other plants, mainly by using the spindle-poisoning drug **colchicine,** an alkaloid extracted from the tissues of the autumn crocus, *Colchicum autumnale*. The root tips are placed in a solution of colchicine to inhibit spindle formation, thereby arresting mitosis and doubling the chromosome number. When the treated roots are subsequently propagated, tetraploid plants can be obtained. The feasibility of this procedure was demonstrated in the 1940s by J. O. Beasley, who studied hybrids between two species of cotton in the genus *Gossypium* (Figure 7.11). Old World cotton has 13 pairs of large chromosomes, whereas American cotton, which originated in Central or South America, has 13 pairs of small chromosomes. New World cotton has 26 pairs, 13 large and 13 small. These numbers suggest that New World cotton arose as an allotetraploid when

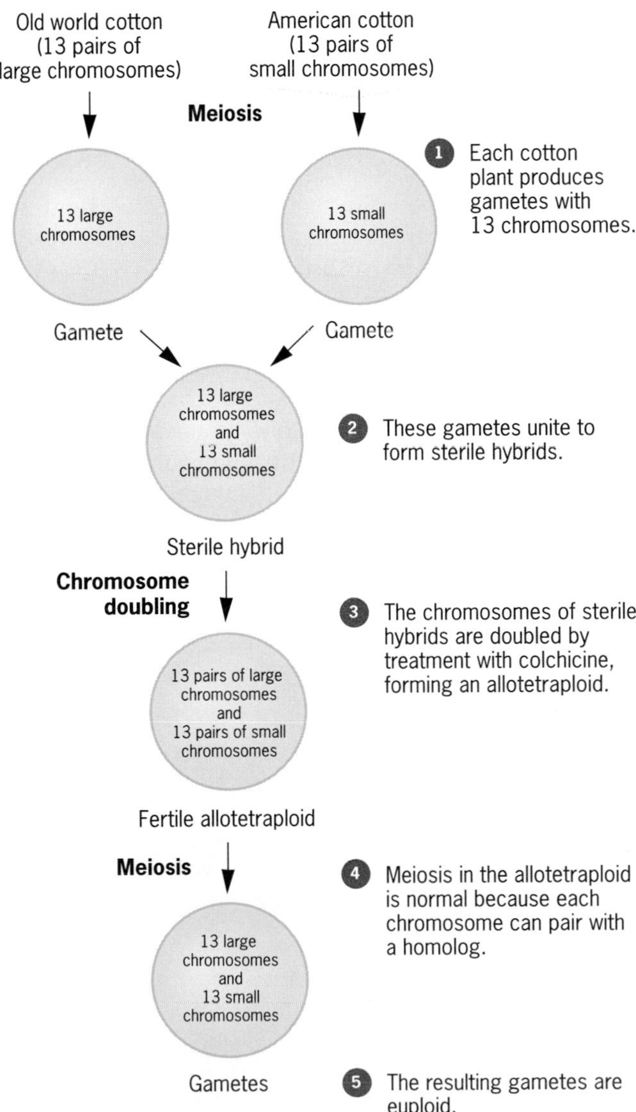

Figure 7.11 Experimental induction of polyploidy in cotton.

the Old World and American species hybridized. Beasley confirmed this hypothesis by creating hybrids between Old World and American cotton and then doubling their chromosomes with colchicine. The resulting plants, which had 26 pairs of chromosomes, were fertile and resembled New World cotton. Thus, Beasley duplicated in the laboratory the process by which polyploid cotton may have originated in nature.

Tissue-Specific Polyploidy and Polyteny

In some organisms, certain tissues become polyploid during development. This polyploidization is probably a response to the need for multiple copies of each chromosome and the genes it carries. The process that produces such polyploid cells, called **endomitosis,** involves chromosome duplication, followed by separation of the resulting sister chromatids. However, because there is no accompanying cell division, extra chromosome sets accumulate within a single nucleus. In human beings, for example, one round of endomitosis produces tetraploid cells in the liver and kidney.

Sometimes polyploidization occurs without the separation of sister chromatids. In these cases, the duplicated chromosomes pile up next to each other, forming a bundle of strands that are aligned in parallel. The resulting chromosomes are said to be **polytene,** from the Greek words meaning "many threads." The most spectacular examples of polytene chromosomes are found in the salivary glands of *Drosophila* larvae. Each chromosome undergoes about nine rounds of replication, producing a total of about 500 copies in each cell. All the copies pair tightly, forming a thick bundle of chromatin fibers. This bundle is so large that it can be seen under low magnification with a dissecting microscope. Differential coiling along the length of the bundle causes variation in the density of the chromatin. When dyes are applied to these chromosomes, the denser chromatin stains more deeply, creating a pattern of dark and light bands (Figure 7.12). This pattern is highly reproducible, permitting detailed analysis of chromosome structure.

The polytene chromosomes of *Drosophila* show two additional features:

1. *Homologous polytene chromosomes pair.* Ordinarily, we think of pairing as a property of meiotic chromosomes; however, in many insect species the somatic chromosomes also pair—probably as a way of organizing the chromosomes within the nucleus. When *Drosophila* polytene chromosomes pair, the large chromatin bundles become even larger. Because this pairing is precise—point-for-point along the length of the chromosome—the two homologs come into perfect alignment. Thus, the banding patterns of each are exactly in register, so much so that it is almost impossible to distinguish the individual members of a pair.
2. *All the centromeres of* Drosophila *polytene chromosomes congeal into a darkly staining body called the* ***chromocenter.*** Material flanking the centromeres is also drawn into this dark mass. The result is that the chromosome arms seem to emanate out of the chromocenter. These arms, which are banded, consist of euchromatin, that portion of the chromosome that contains most of the genes; the chromocenter, which is uniformly stained, consists of heterochromatin, a gene-poor material that surrounds the centromere. The distinction between these different materials is that heterochromatin

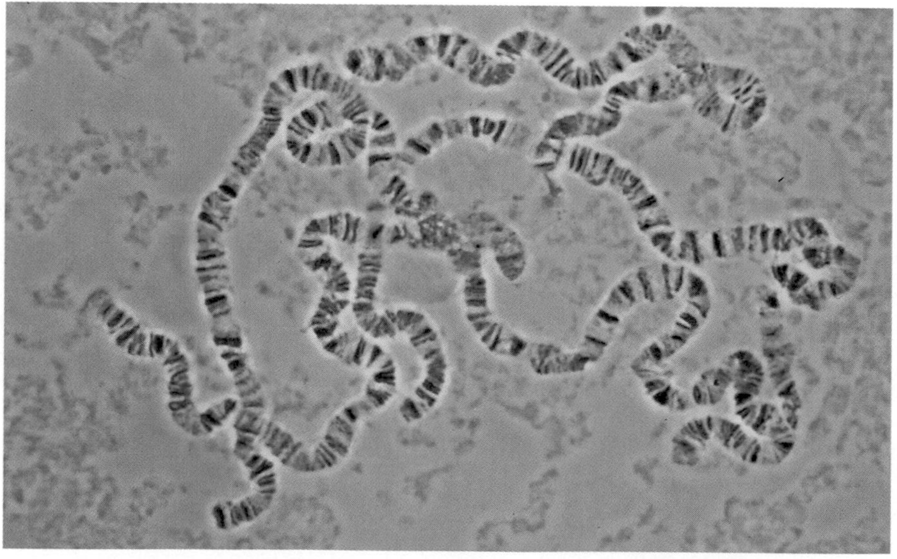

Figure 7.12 Polytene chromosomes of *Drosophila.*

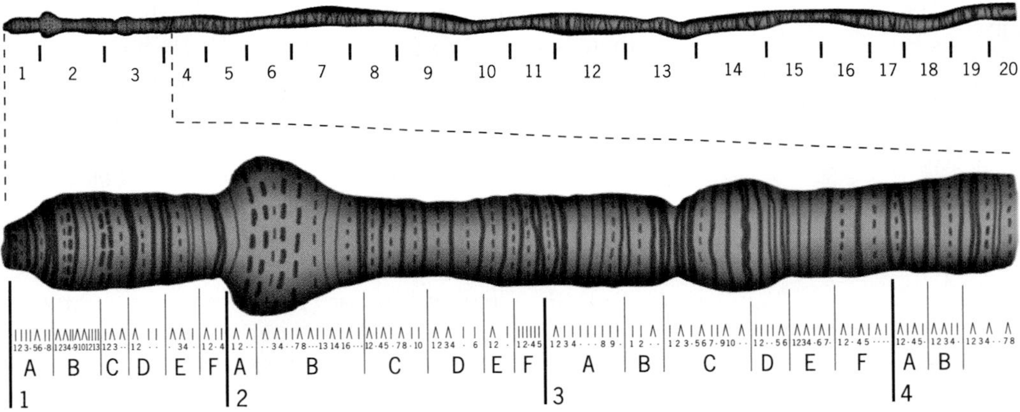

Figure 7.13 Bridges' polytene chromosome maps. (*Top*) Banding pattern of the polytene X chromosome. The chromosome is divided into 20 numbered sections. (*Bottom*) Detailed view of the left end of the polytene X chromosome showing Bridges' system for denoting individual bands.

always stains darkly with dyes such as Feulgen and aceto-carmine.

Polytene chromosomes were discovered in the nineteenth century by the Italian cytologist E. G. Balbiani. It was not until the 1930s, however, that Emil Heitz and Hans Bauer found them in garden midges, and Theophilus Painter found them in *Drosophila* larvae. Painter's discovery was extremely important because *Drosophila* had become the favorite organism for genetic analysis. After Painter's discovery, it was possible to combine sophisticated genetic work with detailed cytology to study the organization and structure of chromosomes. This combination of genetic and cytological analysis eventually culminated in maps showing the location of individual genes on these chromosomes. C. B. Bridges laid the foundation for this genetic cartography by publishing detailed drawings of the polytene chromosomes (Figure 7.13). Bridges arbitrarily divided each of the chromosomes into sections, which he numbered; each section was then divided into subsections, which were designated by the letters A to F. Within each subsection, Bridges enumerated all the dark bands, creating an alphanumeric directory of sites along the length of each chromosome. Bridges' alphanumeric system is still used today to describe the features of these remarkable chromosomes.

The polytene chromosomes of *Drosophila* are trapped in the interphase of the cell cycle. Thus, although most cytological analyses are performed on mitotic chromosomes, the most thorough and detailed analyses are performed on polytenized interphase chromosomes. Such chromosomes are found in many species within the insect order Diptera, including flies and mosquitoes. Unfortunately, human beings do not have polytene chromosomes; thus, the high-resolution structural analysis that is possible for *Drosophila* is not possible for our own species.

Key Points: **Polyploidy involves the presence of extra sets of chromosomes. Many polyploids are sterile because their multiple sets of chromosomes segregate irregularly in meiosis. However, polyploids produced by chromosome doubling in interspecific hybrids may be fertile if their constituent genomes segregate independently. In some polyploid tissues, sister chromatids remain together, forming a large, polytene chromosome.**

ANEUPLOIDY

Aneuploidy describes a numerical change in part of the genome, usually a change in the dosage of a single chromosome. Individuals that have an extra chromosome, that are missing a chromosome, or that have a combination of these anomalies are aneuploid. This definition also includes pieces of chromosomes. Thus, an individual in which a chromosome arm has been deleted is also considered to be aneuploid.

Aneuploidy was originally studied in plants, where it was shown that a chromosome imbalance usually has a phenotypic effect. The classic study was one by Albert Blakeslee and John Belling, who analyzed chromosome anomalies in Jimson weed, *Datura stramonium*. This diploid species has 12 pairs of chromosomes, for a somatic total of 24. Blakeslee collected plants with altered phenotypes and discovered that in some cases the phenotypes were inherited in an irregular way. These peculiar mutants were apparently caused by dominant factors that were transmitted primarily through the female. By examining the chromosomes of the mutant plants, Belling found that in every case an extra chromosome was present. Detailed analysis established that the extra chromosome was differ-

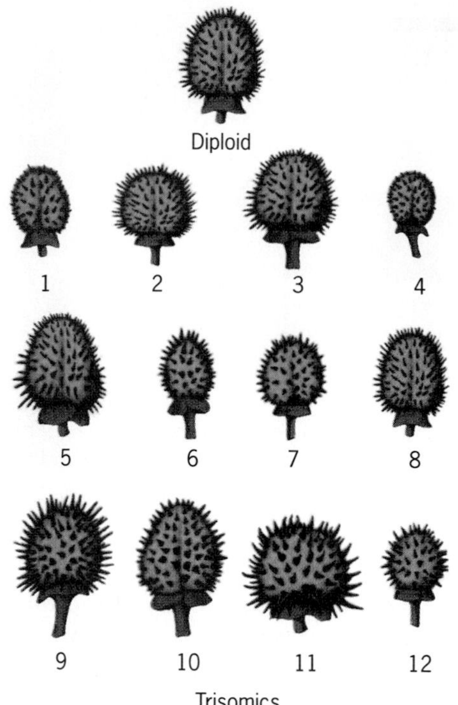

Figure 7.14 Seed capsules of normal and trisomic *Datura stramonium*. Each of the 12 trisomies is shown.

ent in each mutant strain. Altogether there were 12 different mutants, each corresponding to a triplication of one of the *Datura* chromosomes (Figure 7.14). Such triplications are called **trisomies**. The transmissional irregularities of these mutants were due to anomalous chromosome behavior during meiosis.

Belling also discovered the reason for the preferential transmission of the trisomic phenotypes through the female. During pollen tube growth, aneuploid pollen—in particular, pollen with $n+1$ chromosomes—does not compete well with euploid pollen. Consequently, trisomic plants almost always inherit their extra chromosome from the female parent. Belling's work with *Datura* demonstrated that each chromosome must be present in the proper dosage for normal growth and development.

Since Belling's work, aneuploids have been identified in many species, including our own. An organism in which a chromosome, or a piece of a chromosome, is underrepresented is referred to as a ***hypo*ploid** (from the Greek prefix for "under"). When a chromosome or chromosome segment is overrepresented, the organism is said to be ***hyper*ploid** (from the Greek prefix for "over"). Each of these terms covers a wide range of abnormalities.

Trisomy in Human Beings

The best-known and most common chromosome abnormality in humans is **Down syndrome,** a condition associated with an extra chromosome 21 (Figure 7.15*a*). This syndrome was first described in 1866 by a British physician, Langdon Down, but its chromosomal basis was not clearly understood until 1959. People with Down syndrome are typically short in stature and loose-jointed, particularly in the ankles; they have broad skulls, wide nostrils, large tongues with a distinctive furrowing, and stubby hands with a crease on the palm. Impaired mental abilities require that they be given special training and care.

The extra chromosome 21 in Down syndrome is an example of a trisomy. Figure 7.15*b* shows the karyotype of a female Down patient. Altogether, there are 47 chromosomes, including two X chromosomes, as well as the extra chromosome 21. The karyotype of this individual is therefore written **47, XX,+21.**

Trisomy 21 can be caused by chromosome nondisjunction in one of the meiotic cell divisions (Figure

(*a*)

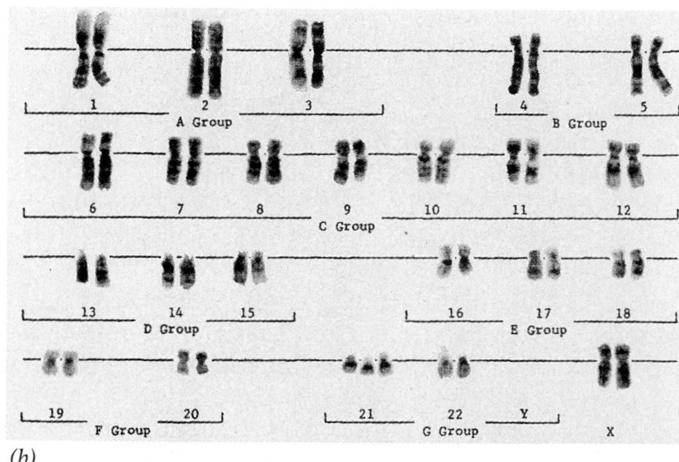

(*b*)

Figure 7.15 Down syndrome. (*a*) Facial features of a child with Down syndrome. (*b*) Karyotype of a child with Down syndrome, showing trisomy for chromosome 21 (47,XX, +21).

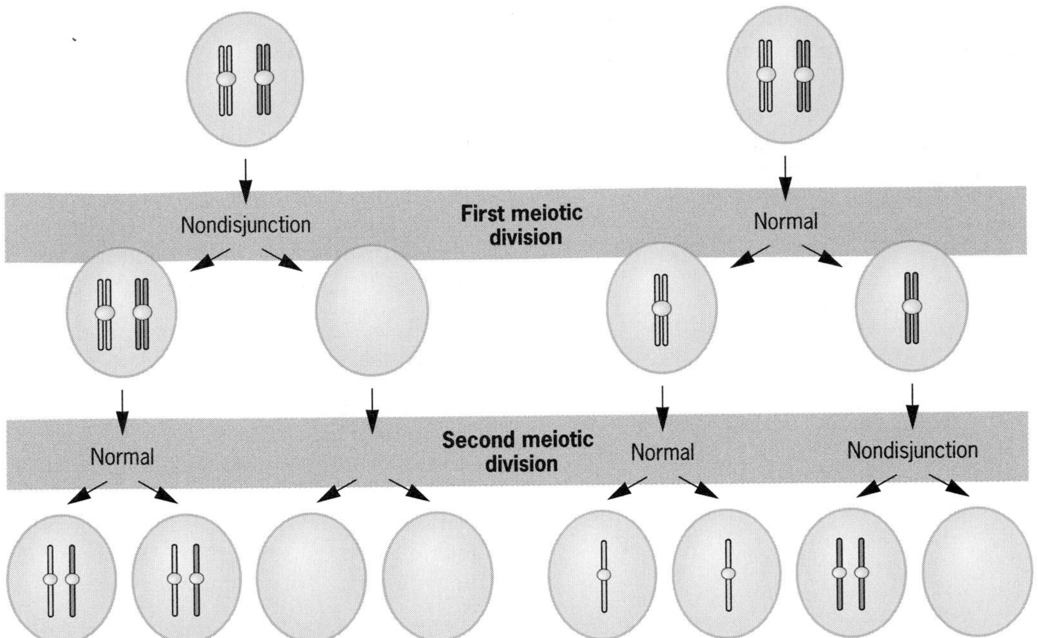

Figure 7.16 Meiotic nondisjunction and the origin of Down syndrome. Nondisjunction at meiosis I produces no normal gametes. Nondisjunction at meiosis II produces a gamete with two identical sister chromosomes, a gamete lacking chromosome 21, and two normal gametes.

7.16). The nondisjunction event can occur in either parent, but it seems to be more likely in females. In addition, the frequency of nondisjunction increases with maternal age. Thus, among mothers younger than 25 years old, the risk of having a child with Down syndrome is about 1 in 1500, whereas among mothers 40 years old, it is 1 in 100. This increased risk is due to factors that adversely affect meiotic chromosome behavior as a woman ages. In human females, meiosis begins in the fetus, but it is not completed until after the egg is fertilized. During the long time prior to fertilization, the meiotic cells are arrested in the prophase of the first division. In this suspended state, the chromosomes may become unpaired. The longer the time in prophase, the greater the chance for unpairing and subsequent chromosome nondisjunction. Older females are therefore more likely than younger females to produce aneuploid eggs.

Trisomies for chromosomes 13 and 18 have also been reported. However, these are rare, and the affected individuals show serious phenotypic abnormalities and are short-lived, usually dying within the first few weeks after birth. Another viable trisomy that has been observed in human beings is the triplo-X karyotype, **47, XXX.** These individuals survive because two of the three X chromosomes are inactivated, effectively reducing the dosage of the X chromosome to the normal level of one. Triplo-X individuals are female and are phenotypically normal, or nearly so; sometimes they exhibit a slight mental impairment and reduced fertility.

The **47, XXY** karyotype is also a viable trisomy in human beings. These individuals have three sex chromosomes, two X's and one Y. Phenotypically, they are male, but they can show some female secondary sexual characteristics and are usually sterile. In 1942 H. F. Klinefelter described the abnormalities associated with this condition, now called **Klinefelter syndrome;** these include small testes, enlarged breasts, long limbs, knock-knees, and underdeveloped body hair. The XXY karyotype can originate by fertilization of an exceptional XX egg with a Y-bearing sperm or by fertilization of an X-bearing egg with an exceptional XY sperm. The XXY karyotype accounts for about three-fourths of all cases of Klinefelter syndrome. Other cases involve more complex karyotypes such as XXYY, XXXY, XXXYY, XXXXY, XXXXYY, and XXXXXY. All individuals with Klinefelter syndrome have one or more Barr bodies in their cells, and those with more than two X chromosomes usually have some degree of mental impairment.

The **47, XYY** karyotype is another viable trisomy in human beings. These individuals are male, and except for a tendency to be taller than 46, XY men, they do not show a consistent syndrome of characteristics. A possible connection between the XYY karyotype and antisocial behavior has not been studied thoroughly enough to draw a firm conclusion. All the other trisomies in human beings are embryonic lethals, demonstrating the importance of correct gene dosage. Unlike *Datura*, in which each of the possible trisomies is vi-

TABLE 7.1
Aneuploidy Resulting from Nondisjunction in Human Beings

Karyotype	*Chromosome Formula*	*Clinical Syndrome*	*Estimated Frequency at Birth*	*Phenotype*
47,+21	$2n+1$	Down	1/700	Short, broad hands with palmar crease, short stature, hyperflexibility of joints, mental retardation, broad head with round face, open mouth with large tongue, epicanthal fold.
47,+13	$2n+1$	Patau	1/20,000	Mental deficiency and deafness, minor muscle seizures, cleft lip and/or palate, cardiac anomalies, posterior heel prominence.
47,+18	$2n+1$	Edward	1/8000	Congenital malformation of many organs, low-set, malformed ears, receding mandible, small mouth and nose with general elfin appearance, mental deficiency, horseshoe or double kidney, short sternum, 90 percent die within first six months after birth.
45,X	$2n-1$	Turner	1/2500 female births	Female with retarded sexual development, usually sterile, short stature, webbing of skin in neck region, cardiovascular abnormalities, hearing impairment.
47,XXY 48,XXXY 48,XXYY 49,XXXXY 50,XXXXXY	$2n+1$ $2n+2$ $2n+2$ $2n+3$ $2n+4$	Klinefelter	1/500 male births	Male, subfertile with small testes, developed breasts, feminine-pitched voice, knock knees, long limbs.
47,XXX	$2n+1$	Triplo-X	1/700	Female with usually normal genitalia and limited fertility, slight mental retardation.

able, human beings do not tolerate many types of chromosomal imbalance (see Table 7.1).

Monosomy

Monosomy occurs when one chromosome is missing in an otherwise diploid individual. In human beings, there is only one viable monosomy, the **45, X** karyotype. These individuals have a single X chromosome as well as a diploid complement of autosomes. Phenotypically, they are female, but because their ovaries are rudimentary, they are almost always sterile. 45, X individuals are usually short in stature; they have webbed necks, hearing deficiencies, and significant cardiovascular abnormalities. Henry H. Turner first described the condition in 1938; thus, it is now called **Turner syndrome.**

45, X individuals can originate from eggs or sperm that lack a sex chromosome or from the loss of a sex chromosome in mitosis sometime after fertilization (Figure 7.17). This latter possibility is supported by the finding that many Turner individuals are **somatic mosaics.** These people have two types of cells in their bodies; some are 45, X and others are 46, XX. This karyotypic mosaicism evidently arises when an X chromosome is lost during the development of a 46, XX zygote. All the descendants of the cell in which the loss occurred are 45, X. If the loss occurs early in develop-

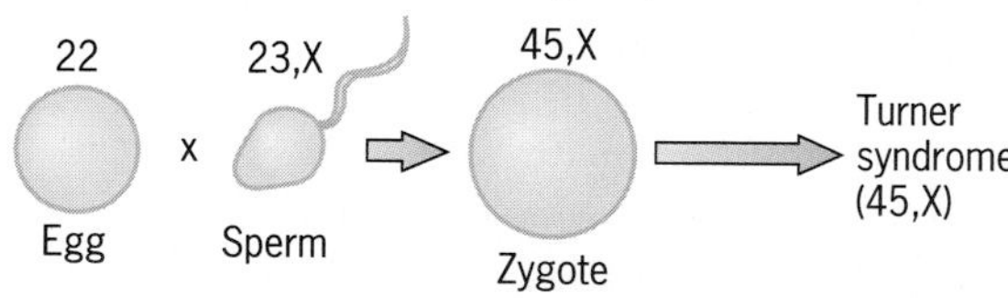

(a) Origin of monosomy at fertilization.

(b) Origin of monosomy in the cleavage division following fertilization.

Figure 7.17 Origin of the Turner syndrome karyotype.

ment, an appreciable fraction of the body's cells will be aneuploid and the individual will show the features of Turner syndrome. If the loss occurs later, the aneuploid cell population will be smaller and the severity of the syndrome is likely to be reduced. (For a discussion of procedures used to detect aneuploidy in human fetuses, see Human Genetics Sidelight: Amniocentesis and Chorionic Biopsy.)

XX/XO chromosome mosaics also occur in *Drosophila*, where they produce a curious phenotype. Because sex in this species is determined by the ratio of X chromosomes to autosomes, such flies are part female and part male. XX cells develop in the female direction, and XO cells develop in the male direction. Flies with both male and female structures are called **gynandromorphs** (from Greek words meaning "a combination of the two sexes").

People with the 45, X karyotype have no Barr bodies in their cells, indicating that the single X chromosome that is present is not inactivated. Why, then, should Turner patients, who have the same number of active X chromosomes as normal XX females, show any phenotypic abnormalities at all? The answer probably involves a small number of genes that remain active on both of the X chromosomes in normal 46, XX females. These noninactivated genes are apparently needed in double dose for proper growth and development. The finding that at least some of these special X-linked genes are also present on the Y chromosome would explain why XY males grow and develop normally. In addition, the X chromosome that has been inactivated in 46, XX females is reactivated during oogenesis, presumably because two copies of some X-linked genes are required for normal ovarian function. 45, X individuals, who have only one copy of these genes, cannot meet this quantitative requirement and are therefore sterile.

Curiously, the cognate of the XO Turner karyotype in the mouse is fully fertile and exhibits no anatomical abnormalities. This finding implies that the mouse homologs of the human genes that are involved in Turner syndrome need only be present in one copy for normal growth and development.

Deletions and Duplications of Chromosome Segments

A missing chromosome segment is referred to either as a **deletion** or as a **deficiency.** Large deletions can be detected cytologically, but small ones cannot. In a diploid organism, the deletion of a chromosome segment makes part of the genome hypoploid. This hypoploidy may be associated with a phenotypic effect, especially if the deletion is large. A classic example is the ***cri-du-chat* syndrome** (from the French words for "cry of the cat") in human beings (Figure 7.18). This condition is caused by a conspicuous deletion in the short arm of chromosome 5; about half the arm appears to be missing. Individuals heterozygous for this deletion and a normal chromosome have the karyotype 46 (5p-), where the term in parentheses indicates

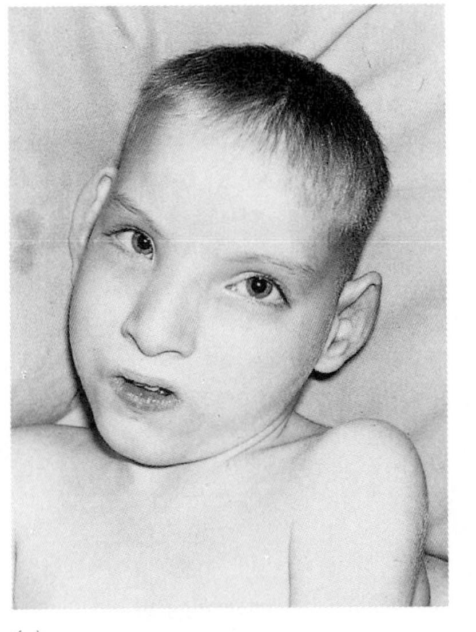

(a)

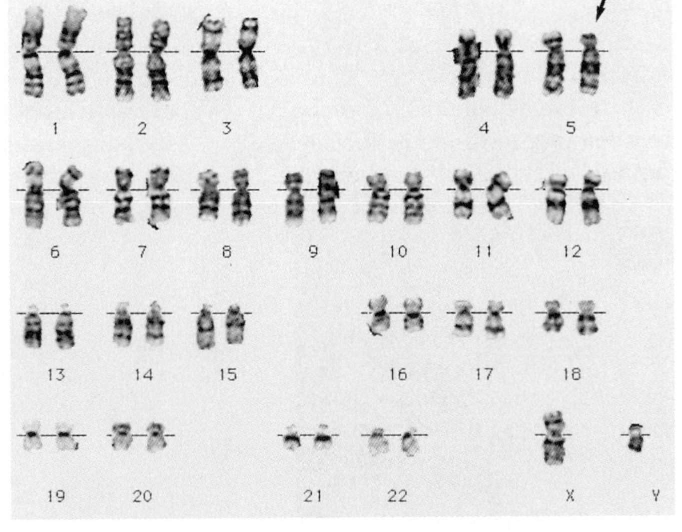

(b)

Figure 7.18 *Cri-du-chat* syndrome. (*a*) Patient with *cri-du-chat* syndrome. (*b*) Karyotype of infant with *cri-du-chat* syndrome, 46, XY(5p-). There is a deletion in the short arm of chromosome 5 (arrow).

HUMAN GENETICS SIDELIGHT

Amniocentesis and Chorionic Biopsy: Procedures to Detect Aneuploidy in Human Fetuses

The Andersons, a couple living in Minneapolis, were expecting their first baby. Neither Donald nor Laura Anderson knew of any genetic abnormalities in their families, but because of Laura's age—38—they decided to have the fetus checked for aneuploidy.

Laura's physician performed a procedure called **amniocentesis.** A small amount of fluid was removed from the cavity surrounding the developing fetus by inserting a needle into Laura's abdomen (Figure 1). This cavity, called the amnionic sac, is enclosed by a membrane. To prevent discomfort during the procedure, Laura was given a local anesthetic. The needle was guided into position by following an ultrasound scan, and some of the amnionic fluid was drawn out. Because this fluid contains nucleated cells sloughed off from the fetus, it is possible to determine the fetus's karyotype (Figure 2). Usually the fetal cells are purified from the amnionic fluid by centrifugation, and then the cells are cultured for several days to a few weeks. Cytological analysis of these cells will reveal if the fetus is aneuploid. Additional tests may be performed on the fluid recovered from the amnionic sac to detect other sorts of abnormalities, including neural tube defects and some kinds of mutations. The results of all these tests may take up to three weeks. In Laura's case, no abnormalities of any sort were detected, and 20 weeks after the amniocentesis, she gave birth to a healthy baby girl.

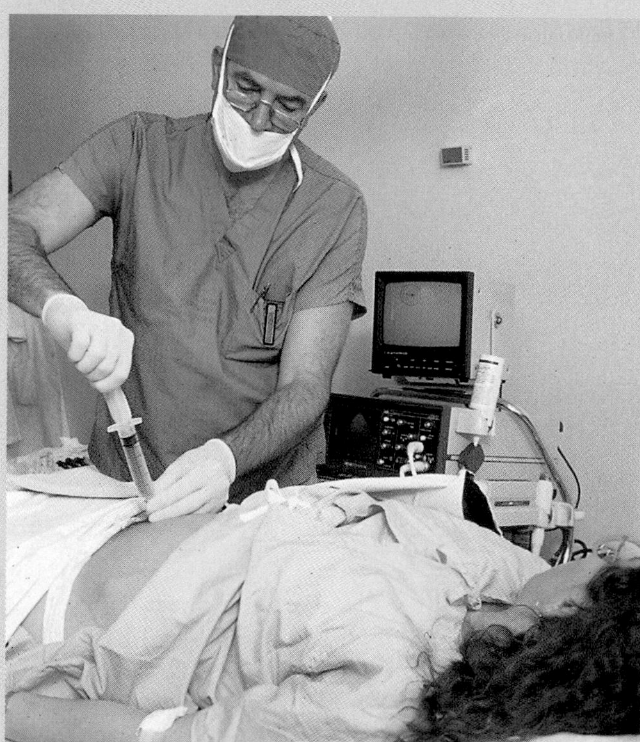

Figure 1 A physician taking a sample of fluid from the amniotic sac of a pregnant woman for prenatal diagnosis of a chromosomal or biochemical abnormality.

Chorionic biopsy provides another way of detecting chromosomal abnormalities in the fetus. The chorion is a fetal membrane that interdigitates with the uterine wall, eventually forming the placenta. The minute chorionic projections into the uterine tissue are called *villi* (singular, villus). At 10–11 weeks of gestation, before the placenta has developed, a sample of chorionic villi can be obtained by passing a hollow plastic tube into the uterus through the cervix. This tube can be guided by an ultrasound scan, and when it is in place, a tiny bit of material can be drawn up into the tube by aspiration. The recovered material usually consists of a mixture of maternal and fetal tissue. After these tissues are separated by dissection, the fetal cells can be analyzed for chromosome abnormalities.

Chorionic biopsy can be performed earlier than amniocentesis (10–11 weeks gestation versus 14–16 weeks), but it is not as reliable. In addition, it seems to be associated with a slightly greater chance of miscarriage than amniocentesis, perhaps 2 to 3 percent. For these reasons, it tends to be used only in pregnancies where there is a strong reason to expect a genetic abnormality. In routine pregnancies, such as Laura Anderson's, amniocentesis is the preferred procedure.

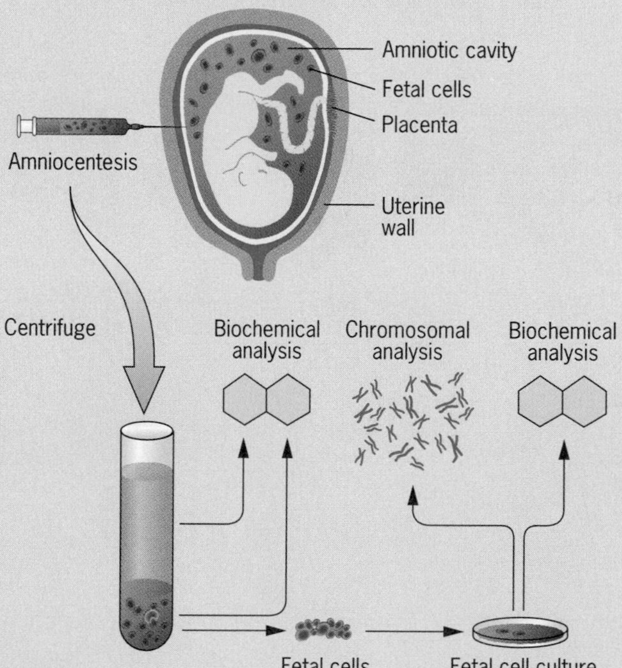

Figure 2 Amniocentesis and procedures for prenatal diagnosis of chromosomal and biochemical abnormalities.

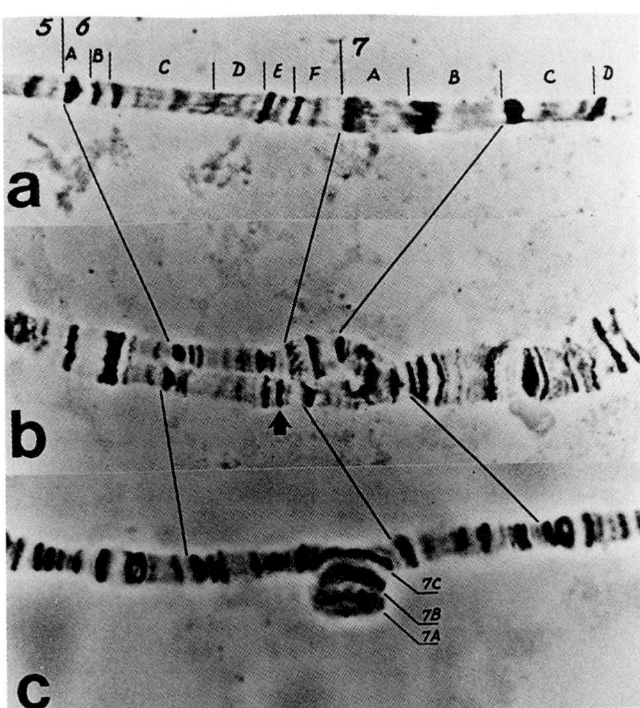

Figure 7.19 Polytene chromosomes showing (*a*) the normal structure of regions 6 and 7 in the middle of the *Drosophila* X chromosome, (*b*) a heterozygote with a deletion of region 6F-7C in one of the chromosomes (arrow), and (*c*) an X chromosome showing a reverse tandem duplication of region 6F-7C. In (*b*) the prominent bands in regions 7A and 7C are present in the upper chromosome but absent in the lower one, indicating that the lower chromosome has undergone a deletion.

that part of the short arm (p) of one of the chromosomes 5 is missing. These individuals are severely impaired, mentally as well as physically; their plaintive, catlike crying gives the syndrome its name.

An extra chromosome segment is referred to as a **duplication.** The extra segment can be attached to one of the chromosomes, or it can exist as a new and separate chromosome, that is, as a "free duplication." In either case, the effect is the same: The organism is hyperploid for part of its genome. As with deletions, this hyperploidy can be associated with a phenotypic effect.

Deletions and duplications are two types of aberrations in chromosome structure. Large aberrations can be detected by examination of mitotic chromosomes that have been stained with banding agents such as quinacrine or Giemsa. However, small aberrations are difficult to detect in this way and are usually identified by other genetic and molecular techniques. The best organism for studying deletions and duplications is *Drosophila,* where the polytene chromosomes afford an unparalleled opportunity for detailed cytological analysis. Figure 7.19*b* shows a deletion in one of two paired homologous chromosomes in a *Drosophila* salivary gland. Because the two chromosomes have separated slightly, we can see that a small region is missing in the lower one.

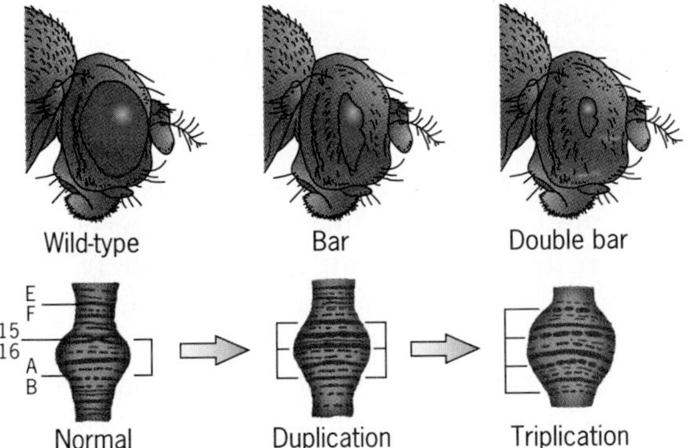

Figure 7.20 Effects of duplications for region 16A of the X chromosome on the size of the eyes in *Drosophila.*

Duplicated segments can also be recognized in polytene chromosomes. Figure 7.19*c* shows a tandem duplication of a segment in the middle of the X chromosome of *Drosophila.* Because tandem copies of this segment pair with each other, the chromosome appears to have a knot in its middle. The *Bar* eye mutation in *Drosophila* is associated with a tandem duplication (Figure 7.20). This dominant X-linked mutation alters the size and shape of the compound eyes, transforming them from large, spherical structures into narrow bars. In the 1930s C. B. Bridges analyzed X chromosomes carrying the *Bar* mutation and found that the 16A region, which apparently contains a gene for eye shape, had been tandemly duplicated. Tandem triplications of 16A were also observed, and in these cases the compound eye was extremely small—a phenotype referred to as double-bar. Many other tandem duplications have been found in *Drosophila,* where polytene chromosome analysis makes their detection relatively easy. Today, molecular techniques have made it possible to detect very small tandem duplications in a wide variety of organisms. For example, the genes that encode the hemoglobin proteins have been tandemly duplicated in mammals (Chapter 24). Gene duplications appear to be relatively common and provide a significant source of variation for evolution.

Key Points: **Aneuploidy involves the under- or overrepresentation of a chromosome or chromosome segment. In a trisomy, such as Down syndrome in human beings, a chromosome is overrepresented; in a monosomy, such as Turner syndrome, it is underrepresented. Deletions and duplications of particular chromosome segments also cause aneuploidy—hypoploidy in the case of a deletion and hyperploidy in the case of a duplication.**

REARRANGEMENTS OF CHROMOSOME STRUCTURE: AN OVERVIEW

In nature there is considerable variation in the number and structure of chromosomes, even among closely related organisms. For example, *Drosophila melanogaster* has four pairs of chromosomes, including a pair of sex chromosomes, two pairs of large, metacentric autosomes (chromosomes with the centromere in the middle), and a pair of small, dotlike autosomes. *Drosophila virilis*, which is not too distantly related, has a pair of sex chromosomes, four pairs of acrocentric autosomes (chromosomes with the centromere near one end), and a pair of dotlike autosomes. Thus, even in the same genus, species can have different chromosome arrangements. These differences imply that over evolutionary time, segments of the genome are rearranged. In fact, the observation that chromosome rearrangements can be found as variants within a single species suggests that the genome is continuously being reshaped. These rearrangements may change the position of a segment within a chromosome, or they may bring together segments from different chromosomes. In either case, the order of the genes is altered. Cytogeneticists have identified many kinds of chromosome rearrangements. Here we consider two types, inversions, which involve a switch in the orientation of a segment within a chromosome, and translocations, which involve the fusion of segments from different chromosomes.

Inversions

Inversions occur when a chromosome segment is detached, flipped around 180°, and reattached to the rest of the chromosome; as a result, the order of the segment's genes is reversed (Figure 7.21). Such rearrangements can be induced in the laboratory by X-irradiation, which breaks chromosomes into pieces. Sometimes the pieces reattach, but in the process a segment gets turned around and an inversion occurs. There is also evidence that inversions are produced naturally through the activity of transposable elements, unusual DNA sequences capable of moving from one chromosomal position to another (Chapter 18). Sometimes, in the course of moving, these elements break a chromosome into pieces and the pieces reattach in an aberrant way, producing an inversion. Inversions may also be created by the reattachment of chromosome fragments generated by mechanical shear, perhaps as a result of chromosome entanglement within the nucleus. No one really knows what fraction of naturally occurring inversions is caused by each of these mechanisms.

Cytogeneticists distinguish between two types of inversions based on whether or not the inverted segment includes the chromosome's centromere (Figure 7.22). **Pericentric** inversions include the centromere, whereas **paracentric** inversions do not. The significance is that a pericentric inversion may change the relative lengths of the two arms of the chromosome, whereas a paracentric inversion has no such effect. Thus, if an acrocentric chromosome acquires an inversion with a breakpoint in each of the chromosome's arms (that is, a pericentric inversion), it can be transformed into a metacentric chromosome. However, if an acrocentric chromosome acquires an inversion in which both of the breaks are in the chromosome's long arm (that is, a paracentric inversion), the morphology of the chromosome will not be changed. Hence, with the use of stan-

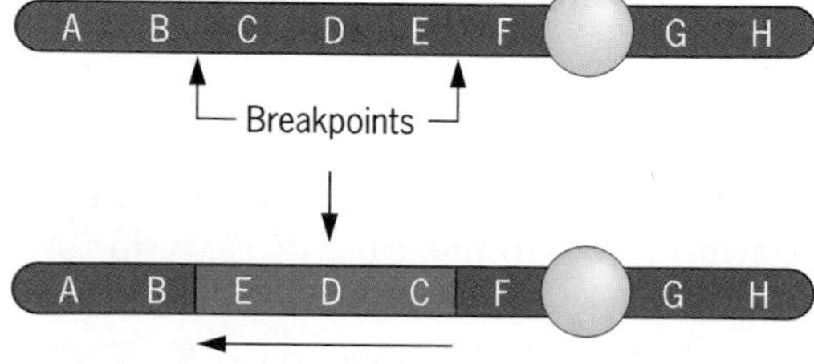

Figure 7.21 Structure of an inversion. The chromosome has been broken at two points, and the segment between them (containing regions C, D, and E) has been inverted.

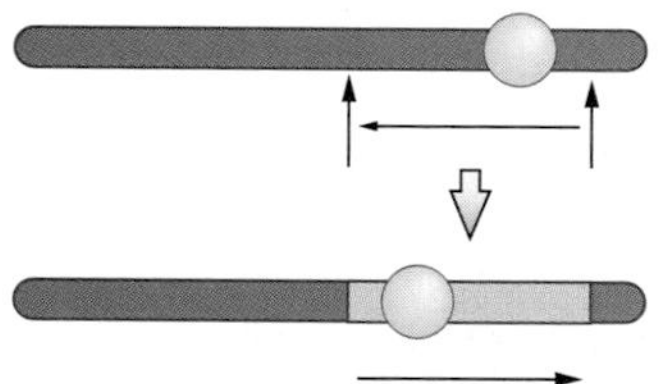

(a) Pericentric inversion—includes centromere

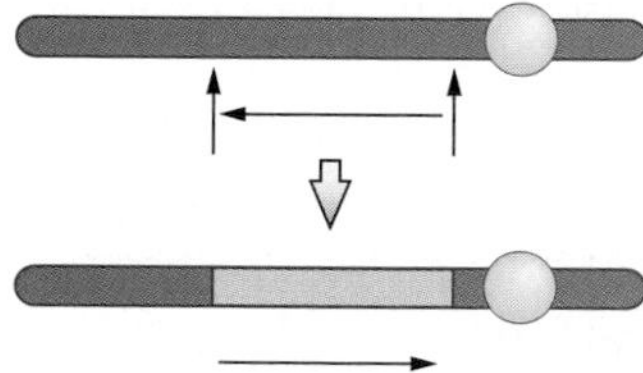

(b) Paracentric inversion—excludes centromere

Figure 7.22 Pericentric and paracentric inversions. A pericentric inversion changes the size of the chromosome arms because the centromere is included within the inversion.

dard cytological methods, pericentric inversions are much easier to detect than paracentric inversions.

An individual in which one chromosome is inverted but its homolog is not is said to be an inversion heterozygote. During meiosis, the inverted and noninverted chromosomes pair point-for-point along their length. However, because of the inversion, the chromosomes must form a loop to allow for pairing in the region where their genes are in reversed order. Figure 7.23*a* shows this pairing configuration; only one of the chromosomes is looped, and the other conforms around it. In practice, either the inverted or noninverted chromosome can form the loop to maximize pairing between them. However, near the ends of the inversion, the chromosomes are stretched, and there is a tendency for some de-synapsis. We consider the genetic consequences of inversion heterozygosity in Chapter 8.

In most organisms, meiotic inversion loops are difficult to observe because the chromosomes are so small. Detailed analysis is practically impossible. However, the loop configuration is easily seen in the somatically paired polytene chromosomes of *Drosophila* (Figure 7.23*b*). Because of the banding pattern, the looped region can be easily identified, and sometimes even the endpoints of the inversion can be precisely determined. Inversion loops are found only in heterozygotes. When inversion chromosomes are homozygous, they pair smoothly along their length without any loop formation.

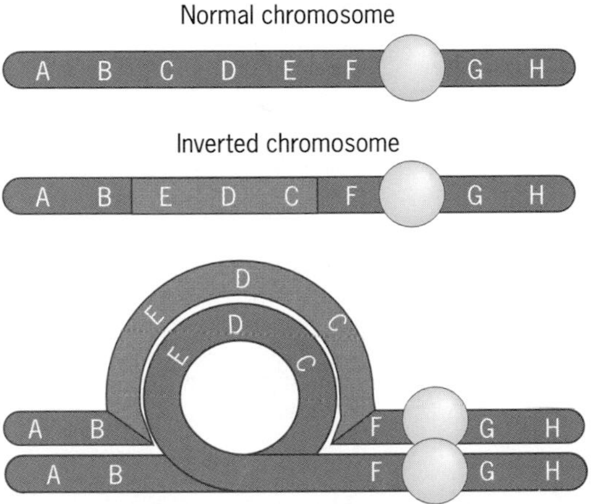

(*a*) Pairing between normal and inverted chromosomes.

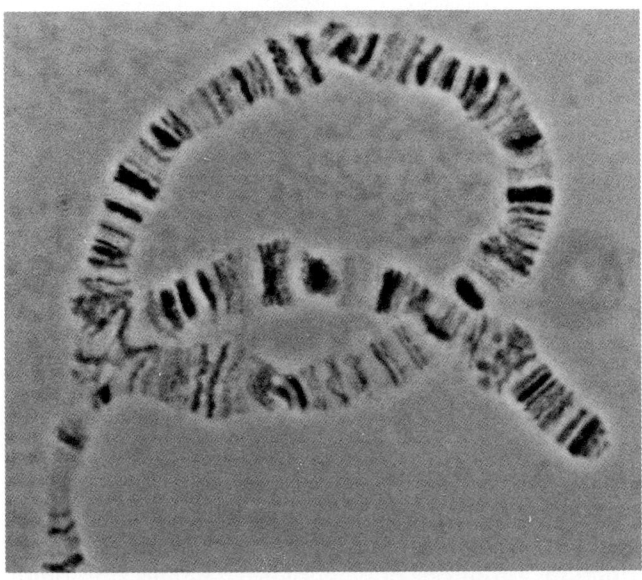

(*b*)

Figure 7.23 Pairing in inversion heterozygotes. (*a*) Diagram showing pairing between normal and inverted chromosomes. (*b*) Polytene chromosomes from an inversion heterozygote.

Translocations

Translocations occur when a segment from one chromosome is detached and reattached to a different (that is, nonhomologous) chromosome. The genetic significance is that genes from one chromosome are transferred to another and their linkage relationships are altered. Translocations can be induced by X-irradiation, which can break two chromosomes simultaneously. Occasionally, the broken fragments are reattached in a novel way, and segments from the different chromosomes are fused. In nature, other factors such as transposable elements and mechanical shear may be involved in producing translocations; however, little is currently known about the relative importance of these factors.

When pieces of two nonhomologous chromosomes are interchanged without any net loss of genetic material, the event is referred to as a **reciprocal translocation.** Figure 7.24*a* shows a reciprocal translocation between the two large autosomes of *Drosophila*. These chromosomes have interchanged pieces of their right arms. During meiosis (and also in the polytene cells of the larval salivary glands), these translocated chromosomes would be expected to pair with their untranslocated homologs in a cruciform, or crosslike, pattern (Figure 7.24*b*). The two translocated chromosomes face each other opposite the center of the cross, and the two untranslocated chromosomes do likewise; to maximize pairing, the translocated and untranslocated chromosomes alternate with each other, forming the arms of the cross. This pairing configuration is diagnostic of a translocation heterozygote. Cells in which the translocated chromosomes are homozygous do not form a cruciform pattern. Instead, each of the translocated chromosomes pairs smoothly with its structurally identical partner.

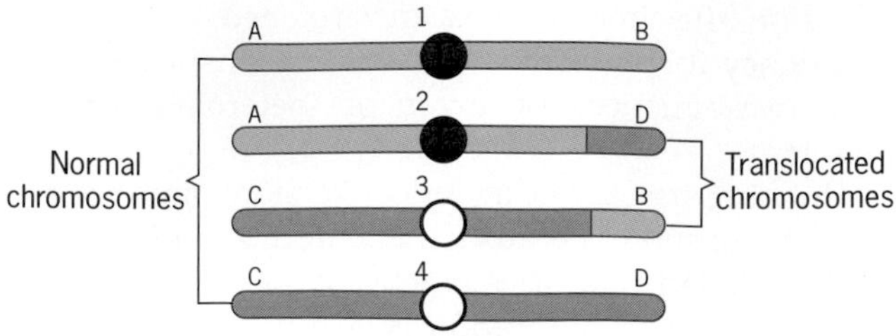

(*a*) Structure of chromosomes in translocation heterozygote.

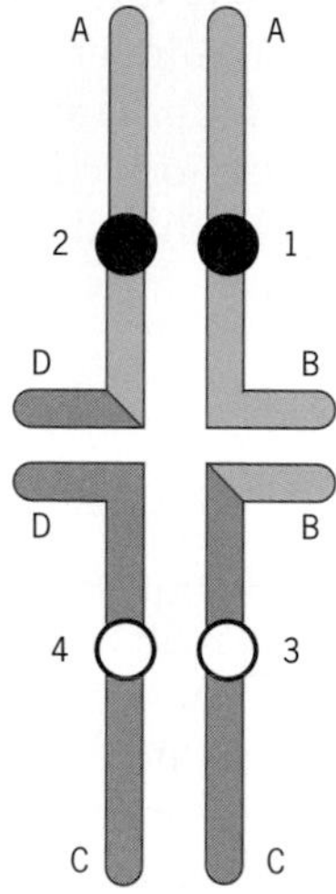

(*b*) Pairing of chromosomes in translocation heterozygote.

Figure 7.24 Structure and pairing behavior of a reciprocal translocation between chromosomes.

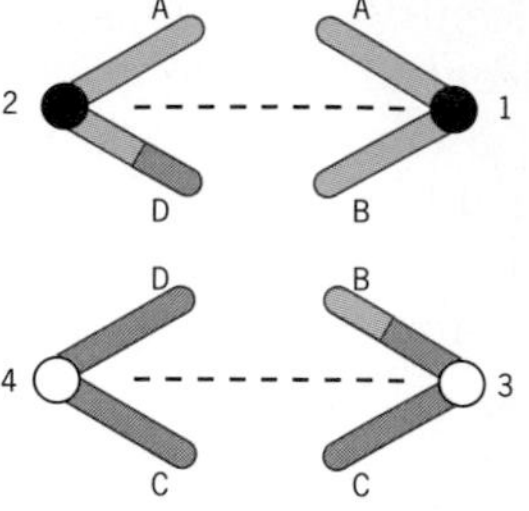

(*a*) Adjacent disjunction I.

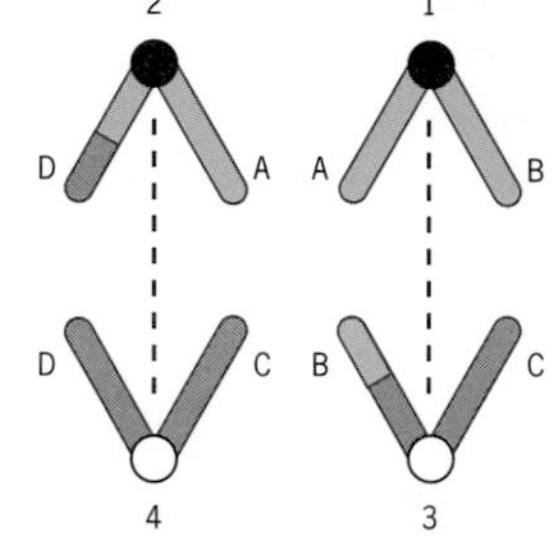

(*b*) Adjacent disjunction II.

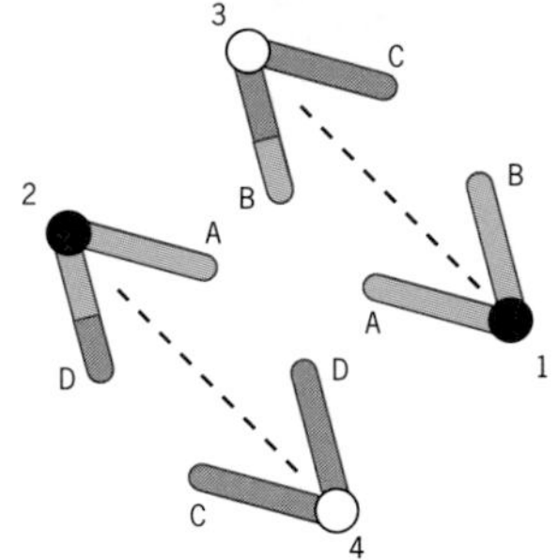

(*c*) Alternate disjunction.

Figure 7.25 Types of disjunction in a translocation heterozygote. (*a*) One form of adjacent disjunction in which homologous centromeres go to opposite poles during anaphase. (*b*) Another form of adjacent disjunction in which homologous centromeres go to the same pole during anaphase. (*c*) Alternate disjunction in which homologous centromeres go to opposite poles during anaphase.

Because cruciform pairing involves four centromeres, which may or may not be coordinately distributed to opposite poles in the first meiotic division, chromosome disjunction in translocation heterozygotes is a somewhat uncertain process, prone to produce aneuploid gametes. Altogether there are three possible disjunctional events, illustrated in Figure 7.25. This simplified figure shows only one of the two sister chromatids of each chromosome. In addition, each of the centromeres is labeled to keep track of chromosome movements; the two white centromeres are homologous (that is, derived from the same chromosome pair), as are the two black centromeres.

If centromeres 2 and 4 move to the same pole, forcing 1 and 3 to the opposite pole, all the resulting gametes will be aneuploid—because some chromosome segments will be deficient for genes, and others will be duplicated (Figure 7.25*a*). Similarly, if centromeres 1 and 2 move to one pole and 3 and 4 to the other, only aneuploid gametes will be produced (Figure 7.25*b*). Each of these cases is referred to as **adjacent disjunction** because centromeres that were next to each other in the cruciform pattern moved to the same pole. When the centromeres that move to the same pole are from different chromosomes (that is, they are *heterologous*), the disjunction is referred to as adjacent I (Figure 7.25*a*); when the centromeres that move to the same pole are from the same chromosome (that is, they are *homologous*), the disjunction is referred to as adjacent II (Figure 7.25*b*). Another possibility is that centromeres 1 and 4 move to the same pole, forcing 2 and 3 to the opposite pole. This case, called **alternate disjunction,** produces only euploid gametes, although half of them will carry only translocated chromosomes (Figure 7.25*c*).

The production of aneuploid gametes by adjacent disjunction explains why translocation heterozygotes have reduced fertility. When such gametes fertilize a euploid gamete, the resulting zygote will be genetically unbalanced and therefore will be unlikely to survive. In plants, aneuploid gametes are themselves often inviable, especially on the male side, and fewer zygotes are produced. Translocation heterozygotes are therefore characterized by low fertility.

Sometimes the aneuploid gametes derived from adjacent disjunction do produce a euploid, viable zygote. Aneuploid gametes that have complementary deficiencies and duplications can unite in fertilization (Figure 7.26). A deficiency in one gamete is compensated for by a duplication in the other, and vice versa. The resulting zygote therefore has exactly two copies of all the chromosome segments. In this way, crosses between translocation heterozygotes can produce viable offspring. However, there are usually far fewer offspring from such crosses than from crosses between chromosomally normal individuals because the frequency of euploid zygotes is so low.

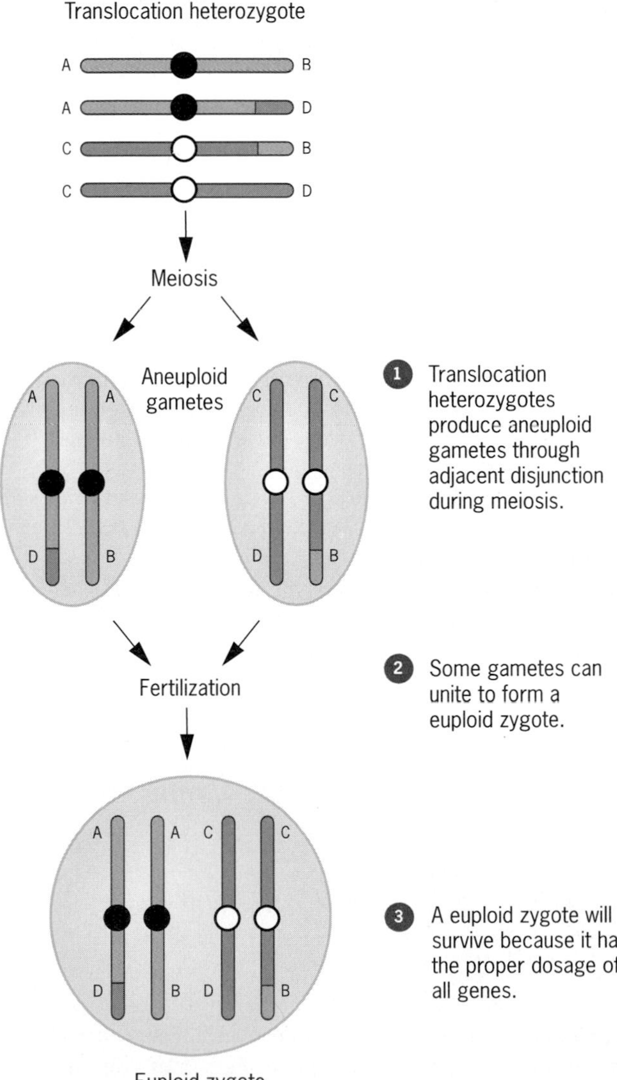

Figure 7.26 Euploid zygotes formed by the union of aneuploid gametes derived from a translocation heterozygote.

In some species, such as the evening primroses in the genus *Oenothera*, the chromosomes are serially translocated with each other: chromosome 1 is translocated with chromosome 2, chromosome 2 with chromosome 3, chromosome 3 with chromosome 4, and so forth (see Technical Sidelight: *Oenothera*). During meiosis, this complicated situation is handled in a simple way: the chromosomes congregate to form a ring, with homologous segments synapsed along their lengths. Then they disjoin in an alternate pattern, producing two types of euploid gametes. Self-fertilization with these gametes produces three kinds of zygotes; however, only the zygotes that are genotypically like their parents survive. Thus, alternate disjunction during meiosis and selective survival among the offspring perpetuate the system of serially translocated chromosomes.

Compound Chromosomes and Robertsonian Translocations

Sometimes one chromosome fuses with its homolog, or two sister chromatids become attached to each other, forming a single genetic unit. A **compound chromosome** can exist stably in a cell as long as it has a single functional centromere; if there are two centromeres, each may move to a different pole during division, pulling the compound chromosome apart. A compound chromosome may also be formed by the union of homologous chromosome segments. For example, the right arms of the two second chromosomes in *Drosophila* might detach from their left arms and fuse at the centromere, creating a compound half-chromosome. Cytogeneticists sometimes call this structure an **isochromosome** (from the Greek prefix for "equal"), because its two arms are equivalent. Compound chromosomes differ from translocations in that they involve fusions of homologous chromosome segments. Translocations, by contrast, always involve fusions between nonhomologous chromosomes.

The first compound chromosome was discovered in 1922 by Lillian Morgan, the wife of T. H. Morgan. This compound was apparently formed by fusion of the two X chromosomes in *Drosophila*, creating double-X or **attached-X chromosomes.** The discovery was made through genetic experimentation rather than cytological analysis. Lillian Morgan crossed females homozygous for a recessive X-linked mutation to wild-type males. From such a cross, we would ordinarily expect all the daughters to be wild-type and all the sons to be mutant. However, Morgan observed just the opposite: all the daughters were mutant and all the sons were wild-type. Further work established that the X chromosomes in the mutant females had

TECHNICAL SIDELIGHT

Oenothera: Chromosome Rings and Other Things

Hugo de Vries, one of the botanists who rediscovered Mendel's principles, was the first scientist to study the genetic properties of the evening primroses (Figure 1). In 1886 he discovered phenotypically different varieties of *Oenothera lamarckiana* growing in a waste field near Amsterdam. Intrigued by these differences, he brought specimens into his experimental garden, where he cultivated them under controlled conditions for several years. During this time, de Vries observed that the predominant form of *O. lamarckiana* occasionally produced offspring with unusual phenotypes. He gave each of these "sports" Latin names—*scintillans* (smooth, narrow leaves), *lata* (bubbly, broad leaves), *nanella* (dwarf), *gigas* (giant), and *rubrinervis* (red veins in the leaves)—and noted that each was heritable.

De Vries regarded these sports as new species of *Oenothera*. His extensive research led him to propose that "species are derived from other species by means of sudden small changes."[1] This view has come to be known as the mutation theory of evolution. Subsequent research has shown that de Vries reached the right conclusion for the wrong reason. The phenotypic changes that he observed were not caused by mutations in the true sense of the word; that is, they were not due to changes in genes. Rather, they were caused by chromosome aberrations—polyploidy, trisomy, and unusual segregations in complex translocation heterozygotes.

Most species of *Oenothera*, including *O. lamarckiana*, have 14 metacentric chromosomes in their somatic cells. During meiosis, these chromosomes associate in an unusual way. In *O. lamarckiana*, for example, two of the chromosomes pair, but the other 12 congregate to form a ring. In this ring, the arms of each chromosome appear to be paired with the arms of two adjacent chromosomes. This association suggested to the early cytologists—John Belling, in particular—that the chromosomes of *Oenothera* had been serially translocated. This hypothesis was subsequently verified by extensive cytological observations on a wide array of *Oenothera* species and on hybrids between them. This cytological analysis was mainly the work of the American botanist Ralph Cleland. Each species of *Oenothera* proved to carry different translocation complexes—called Renner complexes after the German cytologist Oscar Renner, whose genetic analyses had first suggested their existence.

Figure 1 Photograph of one of the evening primroses in the genus *Oenothera.*

In *O. lamarckiana*, the ring of 12 chromosomes is produced by two sets of serial translocations. One set of translocated chromosomes constitutes the *gaudens* Renner complex; the

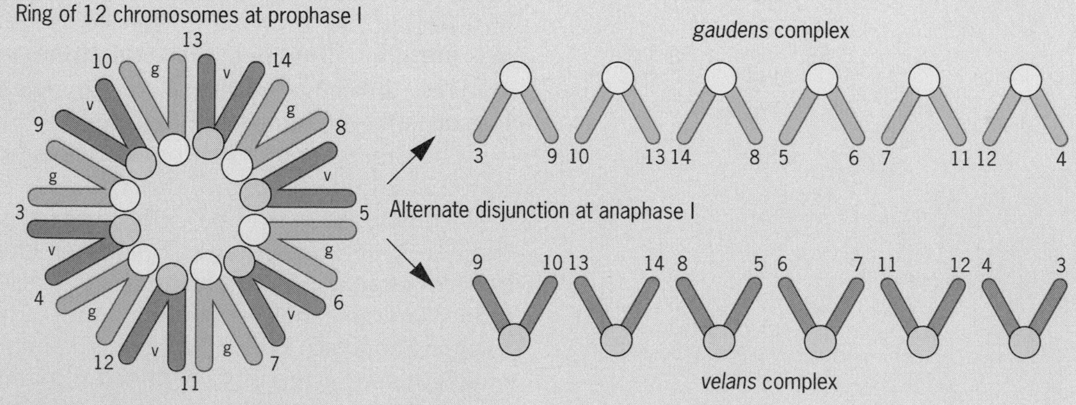

Figure 2 Formation of a ring of 12 chromosomes during prophase I in *O. lamarckiana*, and disjunction of the *gaudens* and *velans* complexes from that ring during anaphase I. The end of each chromosome arm is labeled with a number to indicate homology between the members of the *gaudens* and *velans* complexes. The chromosomes with arms 1 and 2 (not shown) form a bivalent during prophase I.

other set constitutes the *velans* Renner complex. In the prophase of meiosis I, these two complexes associate to form a ring of 12 chromosomes (Figure 2). During anaphase I, the chromosomes disjoin alternately so that, at the conclusion of meiosis, two types of gametes are formed; one contains the *gaudens* complex, and the other the *velans* complex. Fertilization with these types of gametes produces three types of zygotes: *gaudens/gaudens*, *velans/velans*, and *gaudens/velans*; however, only the *gaudens/velans* zygotes survive. The other two genotypes die because they are homozygous for different recessive lethal mutations. This system therefore perpetuates the *gaudens/velans* genotype generation after generation.

The lethality of the *gaudens* and *velans* homozygotes explains why most of the offspring of *O. lamarckiana* are phenotypically like their parents; only the *gaudens/velans* heterozygotes survive, and they resemble their parents. At a frequency of about 1 percent, however, different offspring are produced. After all, it was these different phenotypes that attracted de Vries to the study of *Oenothera*. Some of the phenotypes that he observed were due to an exchange of alleles between the *gaudens* and *velans* chromosomes. (See Chapter 8 for a discussion of the nature and significance of genetic exchange between chromosomes.) For example, *nanella*, the dwarf type that he discovered, was caused by a reciprocal exchange between one chromosome in the *gaudens* complex and another in the *velans* complex. This exchange made a recessive allele for dwarfism homozygous in the otherwise heterozygous *gaudens/velans* genotype. Other phenotypes, such as *lata* and *scintillans*, were due to trisomy—undoubtedly caused by aberrant chromosome segregation from the meiotic ring—and the extremely rare *gigas* was due to polyploidy. Thus, the remarkable mutability that de Vries observed in *O. lamarckiana* was the result of occasional slipups and breakdowns in a rather peculiar genetic system.

[1]de Vries, H. 1919. *Plant breeding*. Open Court Publishing Company, Chicago, p. 9.

become attached to each other. Figure 7.27 illustrates the genetic significance of this attachment. The attached-X females produce two kinds of eggs, diplo-X and nullo-X, and their mates produce two kinds of sperm, X-bearing and Y-bearing. The union of these gametes in all possible ways produces two kinds of viable progeny: mutant XXY females, which inherit the attached-X chromosomes from their mothers and a Y chromosome from their fathers; and phenotypically wild-type XO males, which inherit an X chromosome from their fathers and no sex chromosome from their mothers. Because the Y chromosome is needed for fertility, these XO males are sterile. Lillian Morgan was able to propagate the attached-X chromosomes by backcrossing XXY females to wild-type XY males from another stock. Because the sons of this cross inherited a Y chromosome from their mothers, they were fertile and could be crossed to their XXY sisters to establish a stock in which the attached-X chromosomes were permanently maintained in the female line.

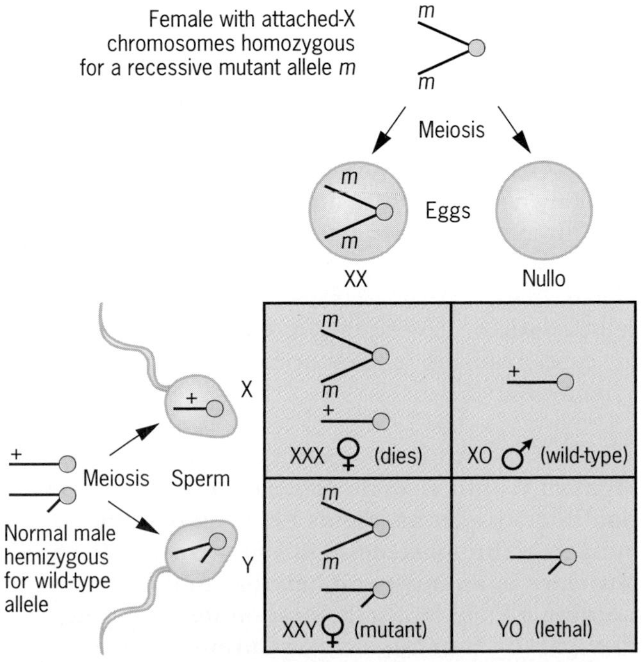

Figure 7.27 Results of a cross between a normal male and a female with attached-X chromosomes.

Nonhomologous chromosomes can also fuse at their centromeres, creating a structure called a **Robertsonian translocation** (Figure 7.28), named for the cytologist F. W. Robertson. For example, if two acrocentric chromosomes fuse, they will produce a metacentric chromosome; the tiny short arms of the participating

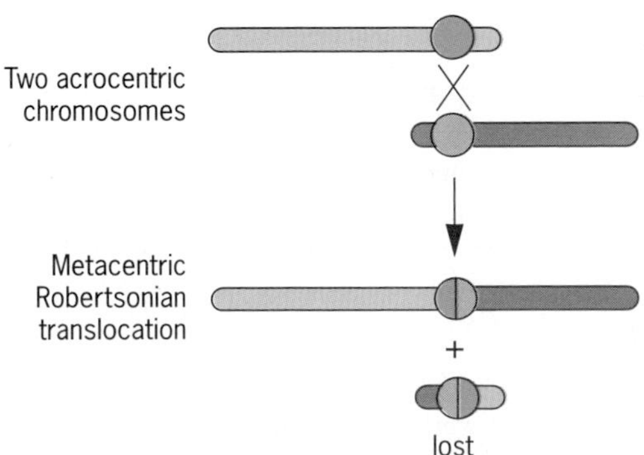

Figure 7.28 Formation of a metacentric Robertsonian translocation by exchange between two nonhomologous acrocentric chromosomes.

chromosomes are simply lost in this process. Apparently, such chromosome fusions have occurred quite often in the course of evolution.

Chromosomes can also fuse end-to-end to form a structure with two centromeres. If one of these is inactivated, the chromosome fusion will be stable. Such a fusion evidently occurred in the evolution of our own species. Human chromosome 2, which is a metacentric, has arms that correspond to two different acrocentric chromosomes in the genomes of the great apes. Detailed cytological analysis has shown that the ends of the short arms of these two chromosomes apparently fused to create human chromosome 2.

Phenotypic Effects of Chromosome Rearrangements

In homozygous condition, deletions that remove several genes are almost always lethal because some of the missing genes are likely to be essential for life. Duplications, in contrast, may be viable in homozygous condition, provided that they are not too large.

In heterozygous condition, deletions and duplications can affect the phenotype by altering the dosage of groups of genes. Usually, the larger the chromosome segment involved, the greater the phenotypic effect. In fact, aneuploidy for very large chromosome segments is typically lethal. However, sometimes even small deletions or duplications can have a lethal effect in heterozygous condition, indicating that the aneuploid region contains at least one gene with a strict requirement for proper dosage. In a diploid organism, a gene that is unable to sustain life when present in a single dose is said to be **haplo-lethal,** and a gene that kills the organism when it is duplicated is said to be **triplo-lethal**. Such genes have been identified in *Drosophila*, where researchers have analyzed thousands of different deficiencies and duplications scattered around the genome.

Inversions and translocations may also affect the phenotype. Sometimes the breakpoints of these rearrangements disrupt genes, rendering them mutant. When the rearrangements are then made homozygous, the mutant phenotype appears. In other cases, the breakpoints are not themselves disruptive, but the genes near them are put into a different chromosomal environment, where they may not function normally. A classic example involves the *white-mottled* (w^m) allele of *Drosophila* (Figure 7.29). This allele, discovered by Hermann J. Muller, is associated with an inversion on the X chromosome. One breakpoint is in the euchromatin near the *white* gene, and the other is in the distant heterochromatin flanking the centromere. Although the inversion does not physically disrupt the *white* gene, it adversely affects its function. Instead of having pigment evenly distributed throughout their eyes, flies with the inversion have a spotty distribution of pigment—a mosaic suggesting that the *white* gene is functioning properly in some eye cells but not in others. Muller reasoned that this effect was due to the juxtaposition of the *white* gene to centric heterochromatin, which tends to remain condensed during the cell cycle. In its new chromosomal environment, the *white* gene does not function normally. It is said to be influenced by a chromosomal **position effect.** Such effects typically occur when euchromatic genes are juxtaposed near heterochromatin. We will consider this phenomenon more fully in Chapter 24.

Chromosome rearrangements may also change the way in which the expression of a particular gene is regulated. Inappropriate expression of a gene may cause cells to become cancerous. We will consider examples of cancer-causing chromosome rearrangements in Chapter 28.

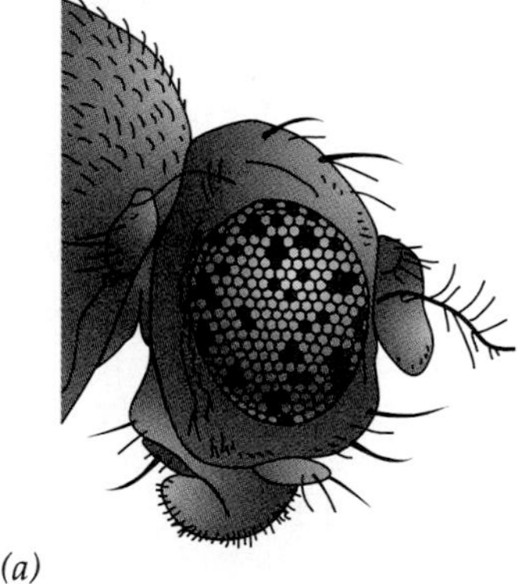

(a)

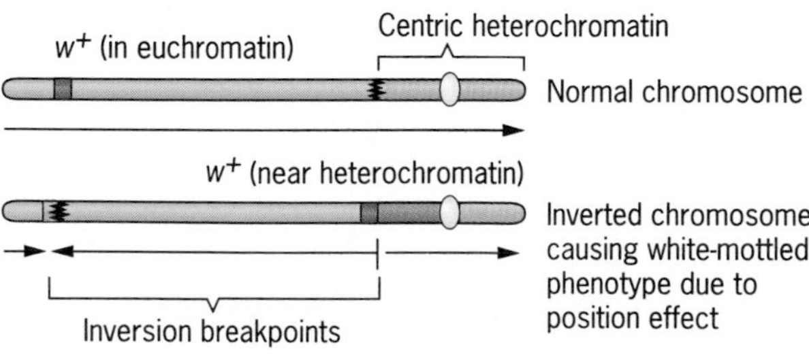

(b)

Figure 7.29 The *white-mottled* (w^m) allele in *Drosophila*. (*a*) Phenotype of the eye. (*b*) Origin of the phenotype through an inversion that moves the wild-type white gene (w^+) near the centric heterochromatin, creating a position effect.

Key Points: **An inversion reverses the order of a segment within a single chromosome; a translocation interchanges segments between two nonhomologous chromosomes. During meiosis, the chromosomes in an inversion heterozygote pair by forming a loop; in a translocation heterozygote, they pair by forming a cross. Chromosome rearrangements may alter the expression of genes by moving them to new chromosomal positions.**

TESTING YOUR KNOWLEDGE

1. A *Drosophila* geneticist has obtained females that carry attached-X chromosomes homozygous for a recessive mutation (*y*) that causes the body to be yellow instead of gray. In one experiment, she crosses some of these females to ordinary wild-type males, and in another, she crosses these females to wild-type males that have their X and Y chromosomes attached to each other; that is, they carry a compound XY chromosome. Predict the phenotypes of the progeny from these two crosses and indicate which, if any, will be sterile.

ANSWER

To predict the phenotypes of the progeny, we need to know their genotypes. The easiest way to determine these genotypes is to diagram the kinds of zygotes produced by each cross.

First, we consider the cross between the yellow-bodied attached-X females and the ordinary wild-type males. The females produce two kinds of gametes, XX and nullo. The males also produce two kinds of gametes, X and Y. When these are combined in all possible ways, four types of zygotes are produced; however, only two types are viable. The XXY zygotes will develop into yellow-bodied females—like their mothers except that they carry a Y chromosome—and the XO zygotes will develop into gray-bodied males—like their fathers except that they lack a Y chromosome. The extra Y chromosome in the females will have no effect on fertility, but the missing Y chromosome in the males will cause them to be sterile.

		Eggs	
		$\widehat{X^yX^y}$	0
Sperm	X^+	$\widehat{X^yX^y}\,X^+$ (die)	X^+0 gray males
	Y	$\widehat{X^yX^y}\,Y$ yellow females	Y0 (die)

Now we consider the cross between the yellow-bodied attached-X females and the males with a compound XY chromosome. Both sexes produce two kinds of gametes—for the females, the same as above, and for the males, either XY or nullo. When these are united in all possible ways, we find that two types of zygotes will be viable: yellow-bodied females with attached-X chromosomes and gray-bodied males with a compound XY chromosome. Both types of progeny will be fertile.

		Eggs	
		$\widehat{X^y\,X^y}$	0
Sperm	$\widehat{X^+Y}$	$\widehat{X^y\,X^y}\,\widehat{X^+Y}$ (die)	$\widehat{X^+Y}\,0$ gray males
	0	$\widehat{X^y\,X^y}\,0$ yellow females	0 0 (die)

2. A phenotypically normal man carries a translocated chromosome that contains the entire long arm of chromosome 14, part of the short arm of chromosome 14, and most of the long arm of chromosome 21:

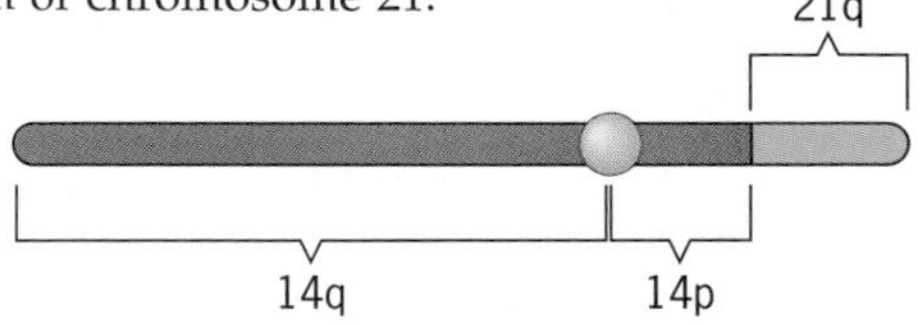

The man also carries a normal chromosome 14 and a normal chromosome 21. If he marries a cytologically (and phenotypically) normal woman, is there any chance that the couple will produce phenotypically abnormal children?

ANSWER

Yes, the couple could produce children with Down syndrome as a result of meiotic segregation in the cytologically abnormal man. During meiosis in this man, the translocated chromosome, T(14, 21), will synapse with the normal chromosomes 14 and 21, forming a trivalent. Disjunction from this trivalent will produce six different types of sperm, four of which are aneuploid.

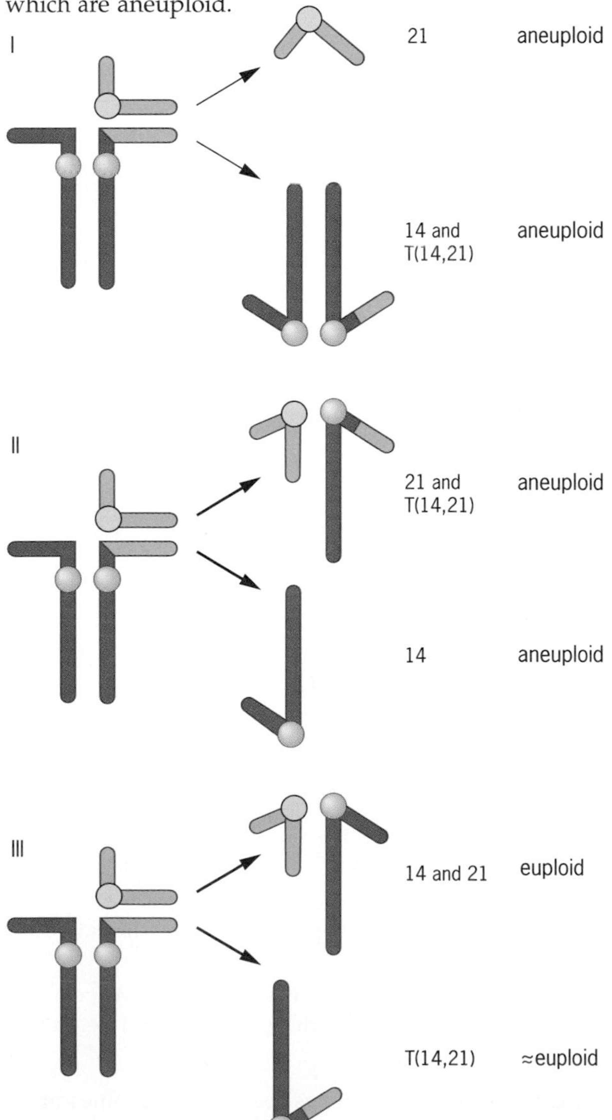

Fertilization of an egg containing one chromosome 14 and one chromosome 21 by any of the aneuploid sperm will produce an aneuploid zygote.

Although trisomy or monosomy for chromosome 14 and monosomy for chromosome 21 are all lethal conditions, trisomy for chromosome 21 is not. Thus, it is possible for the couple to give birth to a child with Down syndrome.

Disjunction	Sperm	Zygote	Condition	Outcome
I	21	14, 21, 21	monosomy 14	dies
	14, T(14, 21)	14, 14, T(14, 21), 21	trisomy 14	dies
II	14	14, 14, 21	monosomy 21	dies
	T(14, 21), 21	14, T(14, 21), 21, 21	trisomy 21	Down
III	14, 21	14, 14, 21, 21	euploid	normal
	T(14, 21)	14, T(14, 21), 21	euploid	normal

QUESTIONS AND PROBLEMS

7.1 In the human karyotype, the X chromosome is approximately the same size as seven of the autosomes (the so-called C group of chromosomes). What procedure could be used to distinguish the X chromosome from the other members of this group?

7.2 In human beings, a cytologically abnormal chromosome 22, called the "Philadelphia" chromosome because of the city in which it was discovered, is associated with chronic leukemia. This chromosome is missing part of its long arm. How would you denote the karyotype of an individual who had 46 chromosomes in his somatic cells, including one normal 22 and one Philadelphia chromosome?

7.3 During meiosis, why do some tetraploids behave more regularly than triploids?

7.4 The following table presents chromosome data on four species of plants and their F_1 hybrids:

Species or F_1 Hybrid	Root Tip Chromosome Number	Meiosis I Metaphase	
		Number of Bivalents	Number of Univalents
A	20	10	0
B	20	10	0
C	10	5	0
D	10	5	0
A × B	20	0	20
A × C	15	5	10
A × D	15	5	10
C × D	10	0	10

(a) Deduce the chromosomal origin of species A.

(b) How many bivalents and univalents would you expect to observe at meiotic metaphase I in a hybrid between species C and species B?

(c) How many bivalents and univalents would you expect to observe at meiotic metaphase I in a hybrid between species D and species B?

7.5 A plant species A, which has seven chromosomes in its gametes, was crossed with a related species B, which has nine. The hybrids were sterile, and microscopic observation of their pollen mother cells showed no chromosome pairing. A section from one of the hybrids that grew vigorously was propagated vegetatively, producing a plant with 32 chromosomes in its somatic cells. This plant was fertile. Explain.

7.6 A plant species X with $n = 5$ was crossed with a related species Y with $n = 7$. The F_1 hybrid produced only a few pollen grains, which were used to fertilize the ovules of species Y. A few plants were produced from this cross, and all had 19 chromosomes. Following self-fertilization, the F_1 hybrids produced a few F_2 plants, each with 24 chromosomes. These plants were phenotypically different from either of the original species and were highly fertile. Explain the sequence of events that produced these fertile F_2 hybrids.

7.7 Identify the sexual phenotypes of the following genotypes in human beings: XX, XY, XO, XXX, XXY, XY

7.8 If nondisjunction of chromosome 21 occurs in the division of a secondary oocyte in a human female, what is the chance that a mature egg derived from this division will receive two number 21 chromosomes?

7.9 A *Drosophila* female homozygous for a recessive X-linked mutation causing yellow body was crossed to a wild-type male. Among the progeny, one fly had sectors of yellow pigment in an otherwise gray body. These yellow sectors were distinctly male, whereas the gray areas were female. Explain the peculiar phenotype of this fly.

7.10 The *Drosophila* fourth chromosome is so small that flies monosomic or trisomic for it survive and are fertile. Several genes, including *eyeless* (*ey*), have been located on this chromosome. If a cytologically normal fly homozygous for a recessive eyeless mutation is crossed to a fly monosomic for a wild-type fourth chromosome, what kinds of progeny will be produced, and in what proportions?

7.11 In humans, Hunter syndrome is known to be an X-linked trait with complete penetrance. In family A, two phenotypically normal parents have produced a normal son, a *daughter* with Hunter and Turner syndromes, and a son with Hunter syndrome. In family B, two phenotypically normal parents have produced two phenotypically normal daughters and a *son* with Hunter and Klinefelter syndromes. In family C, two phenotypically normal parents have produced a phenotypically normal daughter, a *daughter* with Hunter syndrome, and a son with Hunter syndrome. For each family, explain the origin of the child indicated in italics.

7.12 A woman with X-linked color blindness and Turner syndrome had a color-blind father and a normal mother. In which of her parents did nondisjunction of the sex chromosomes occur?

7.13 Although XYY men are phenotypically normal, would they be expected to produce more children with sex chromosome abnormalities than XY men? Explain.

7.14 In a *Drosophila* salivary chromosome, the bands have a sequence of 1 2 3 4 5 6 7 8. The homolog with which this chromosome is synapsed has a sequence of 1 2 3 6 5 4 7 8. What kind of chromosome change has occurred? Draw the synapsed chromosomes.

7.15 Other chromosomes have sequences as follows: (a) 1 2 5 6 7 8; (b) 1 2 3 4 4 5 6 7 8; (c) 1 2 3 4 5 8 7 6. What kind of chromosome change is present in each? Illustrate how these chromosomes would pair with a chromosome whose sequence is 1 2 3 4 5 6 7 8.

7.16 Both inversion and translocation heterozygotes display pollen abortion in plants. The extent of pollen abortion varies greatly among different inversion heterozygotes, but it is generally about 50 percent for different translocation heterozygotes. Explain this difference.

7.17 One chromosome in a plant has the sequence A B C D E F, and another has the sequence M N O P Q R. A reciprocal translocation between these chromosomes produced the following arrangement: A B C P Q R on one chromosome and M N O D E F on the other. Illustrate how these translocated chromosomes would pair with their normal counterparts in a heterozygous individual during meiosis.

7.18 In *Drosophila*, the genes *bw* and *st* are located on chromosomes 2 and 3, respectively. Flies homozygous for *bw* mutations have brown eyes, flies homozygous for *st* mutations have scarlet eyes, and flies homozygous for *bw* and *st* mutations have white eyes. Doubly heterozygous males were mated individually to homozygous *bw; st* females. All but one of the matings produced four classes of progeny: wild-type, brown-, scarlet- and white-eyed flies. The single exception produced only wild-type and white-eyed progeny. Explain the nature of this exception.

7.19 A phenotypically normal boy has 45 chromosomes, but his sister, who has Down syndrome, has 46. Suggest an explanation for this paradox.

7.20 How do chromosomal position effects originate, and how can they be explained?

7.21 A yellow-bodied *Drosophila* female with attached-X chromosomes was crossed to a white-eyed male. Both of the parental phenotypes are caused by X-linked recessive mutations. Predict the phenotypes of the progeny.

7.22 A man has attached chromosomes 21. If his wife is cytologically normal, what is the chance their first child will have Down syndrome?

7.23 Analysis of the polytene chromosomes of three populations of *Drosophila* has revealed three different banding sequences in a region of the second chromosome:

Population	Banding Sequence
P1	1 2 3 4 5 6 7 8 9 10
P2	1 2 3 9 8 7 6 5 4 10
P3	1 2 3 9 8 5 6 7 4 10

Explain the evolutionary relationships among these populations.

7.24 Each of six populations of *Drosophila* in different geographic regions had a specific arrangement of bands in one of the large autosomes:

(a) 12345678
(b) 12263478
(c) 15432678
(d) 14322678
(e) 16223478
(f) 154322678

Assume that arrangement (a) is the original one. In what order did the other arrangements most likely arise, and what type of chromosomal aberration is responsible for each change?

7.25 The following diagram shows two pairs of chromosomes in the karyotypes of a man, a woman, and their child. The man and the woman are phenotypically normal, but the child (a boy) suffers from a syndrome of abnormalities, including poor motor control and severe mental impairment. What is the genetic basis of the child's abnormal phenotype? Is the child hyperploid or hypoploid for a segment in one of his chromosomes?

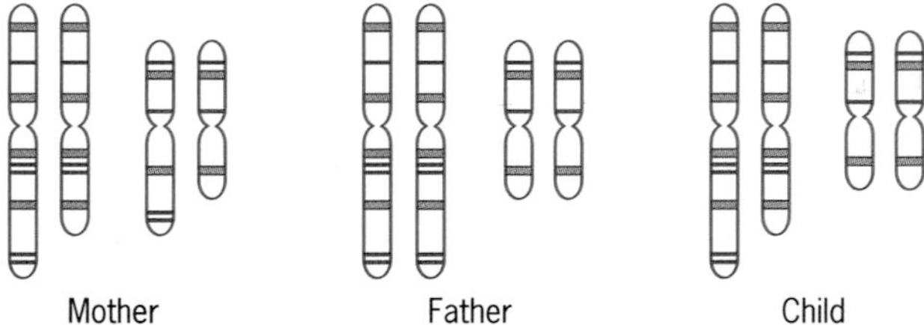

7.26 A male mouse that is heterozygous for a reciprocal translocation between the X chromosome and an autosome is crossed to a female mouse with a normal karyotype. The autosome involved in the translocation carries a gene responsible for coloration of the fur. The allele on the male's translocated autosome is wild-type, and the allele on its nontranslocated autosome is mutant; however, because the wild-type allele is dominant to the mutant allele, the male's fur is wild-type (dark in color). The female mouse has light color in her fur because she is homozygous for the mutant allele of the color-determining gene. When the offspring of the cross are examined, all the males have light fur and all the females have patches of light and dark fur. Explain these peculiar results.

7.27 In *Drosophila*, the autosomal genes *cinnabar* (*cn*) and *brown* (*bw*) control the production of brown and red eye pigments, respectively. Flies homozygous for *cinnabar* mutations have bright red eyes, flies homozygous for *brown* mutations have brown eyes, and flies homozygous for mutations in both of these genes have white eyes. A male homozygous for mutations in the *cn* and *bw* genes has bright red eyes because a small duplication that carries the wild-type allele of *bw* (bw^+) is attached to the Y chromosome. If this male is mated to a karyotypically normal female that is homozygous for the *cn* and *bw* mutations, what types of progeny will be produced?

7.28 In *Drosophila*, vestigial wing (*vg*), hairy body (*h*), and eyeless (*ey*) are recessive mutations on chromosomes 2, 3, and 4, respectively. Wild-type males that had been irradiated with X rays were crossed to triply homozygous recessive females. The F_1 females (all phenotypically wild-type) were then testcrossed to triply homozygous recessive males. Most of the F_1 females produced eight classes of progeny in approximately equal proportions, as would be expected if the *vg*, *h*, and *ey* genes assort independently. However, one F_1 female produced only four classes of offspring, each approximately one-fourth of the total: (1) wild-type, (2) eyeless, (3) vestigial, hairy, and (4) vestigial, hairy, eyeless. What kind of chromosome aberration did the exceptional F_1 female carry, and which chromosomes were involved?

7.29 Cytological examination of the sex chromosomes in a man has revealed that he carries an insertional translocation. A small segment has been deleted from the Y chromosome and inserted into the short arm of the X chromosome; this segment contains the gene responsible for male differentiation (*SRY*). If this man marries a karyotypically normal woman, what types of progeny will the couple produce?

BIBLIOGRAPHY

Bridges, C. B. 1935. Salivary chromosome maps with a key to the banding of the chromosomes of *Drosophila melanogaster*. *J. Hered*. 26:60–64.

Burnham, C. R. 1962. *Discussions in Cytogenetics*. Burgess, Minneapolis.

DeGrouchy, J., and C. Turleau. 1984. *Clinical Atlas of Human Chromosomes*. 2nd ed. Wiley, New York.

Hsu, T. C. 1973. The longitudinal differentiation of chromosomes. *Ann. Rev. Genet*. 7:153–176.

Simmonds, N. W., ed. 1976. *The Evolution of Crop Plants*. Longman, New York.

Stebbins, G. L. 1950. *Variation and Evolution in Plants*. Columbia University Press, New York.

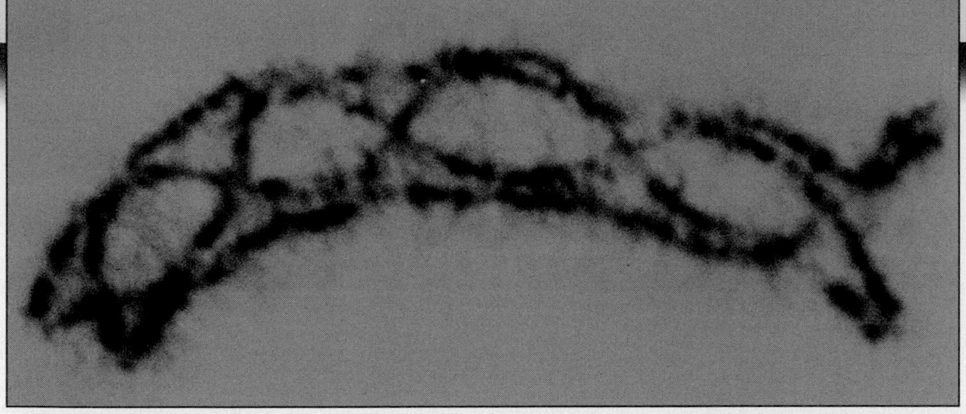

Chiasmata in the late prophase of the first meiotic divison. These cross-shaped figures are the result of exchanges between paired chromosomes.

8

Linkage, Crossing Over, and Chromosome Mapping in Eukaryotes

CHAPTER OUTLINE

Linkage, Recombination, and Crossing Over
Chromosome Mapping
Recombination and Evolution

The World's First Chromosome Map

The modern picture of chromosome organization emerged from a combination of genetic and cytological studies. T. H. Morgan laid the foundation for these studies when he demonstrated that the gene for *white* eyes in *Drosophila* was located on the X chromosome. Soon afterward Morgan's students showed that other genes were X-linked, and eventually they were able to locate each of these genes on a map of the chromosome. This map was a straight line, and each gene was situated at a particular point, or **locus**, on it (Figure 8.1). The structure of the map

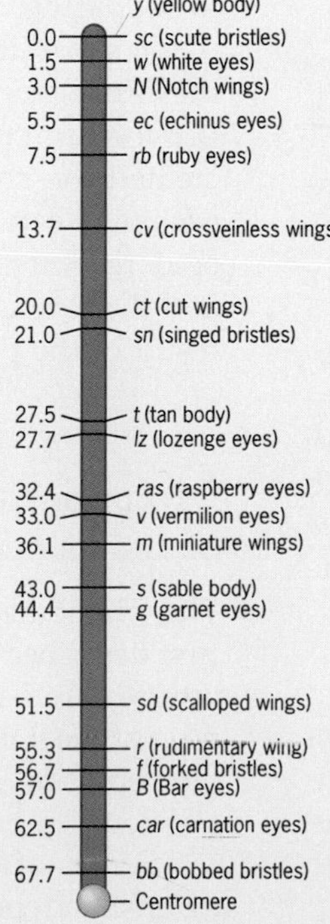

Figure 8.1 A map of genes on the X chromosome of *Drosophila melanogaster*.

therefore implied that a chromosome was simply a linear array of genes. All subsequent research has supported this view. Genes are arranged in a linear order along the length of a chromosome. There are no branches or discontinuities. The basic geometric figure in genetics is a line, reflecting, as we now know, the linear structure of DNA.

The procedure for mapping chromosomes was invented by Alfred H. Sturtevant (Figure 8.2), an undergraduate working in Morgan's laboratory. One night in 1911 Sturtevant put aside his algebra homework in order to evaluate some experimental data. Before the sun rose the next day, he had constructed the world's first chromosome map.

The cytological appearance of chromosomes inspired Sturtevant to make his map a straight line. How was Sturtevant able to determine the map locations of individual genes? No microscope was powerful enough to see genes, nor was any measuring device accurate enough to obtain the distances between them. In fact, Sturtevant did not use any sophisticated instruments in his work. Instead, he relied completely on the analysis of data from experimental crosses with *Drosophila*.

Figure 8.2 Alfred H. Sturtevant.

In this chapter we will consider the procedures for mapping genes in eukaryotic organisms. Understanding these procedures will require an examination of the process of genetic exchange between homologous chromosomes which occurs when these chromosomes are paired during meiosis.

LINKAGE, RECOMBINATION, AND CROSSING OVER

Sturtevant based his mapping procedure on the principle that genes on the same chromosome should be inherited together. Because such genes are physically attached to the same structure, they should travel as a unit through meiosis. This phenomenon is called **linkage.** The early geneticists were unsure about the nature of linkage, but some of them, including Morgan and his students, thought that genes were attached to one another much like beads on a string. Thus, these researchers clearly had a linear model of chromosome organization in mind.

The early geneticists also knew that linkage was not absolute. Their experimental data demonstrated that genes on the same chromosome could be separated as they went through meiosis, and that new combinations of genes could be formed. However, this phenomenon, called **recombination**, was difficult to explain by simple genetic theory.

One hypothesis was that during meiosis, when homologous chromosomes paired, a physical exchange of material separated and recombined genes. This idea was inspired by the cytological observation that chromosomes could be seen in pairing configurations that suggested they had switched pieces with each other. At the switch points, the two homologs were crossed over, as if each had been broken and then reattached to its partner (see opening photograph). A crossover point was called a **chiasma** (plural, **chiasmata**), from the Greek word meaning "cross." Geneticists began to use the term *crossing over* to describe the process that created the chiasmata—that is, the actual process of exchange between paired chromosomes. They considered recombination—the separation of linked genes and the formation of new gene combinations—to be a result of the physical event of crossing over.

Exceptions to the Mendelian Principle of Independent Assortment

The phenomena of linkage and recombination were first described by W. Bateson and R. C. Punnett shortly after the rediscovery of Mendel's work at the beginning of the twentieth century. Initially, this linkage was viewed as an exception to Mendel's Principle of Independent Assortment.

Some of the first evidence for linkage—and against independent assortment—came from experi-

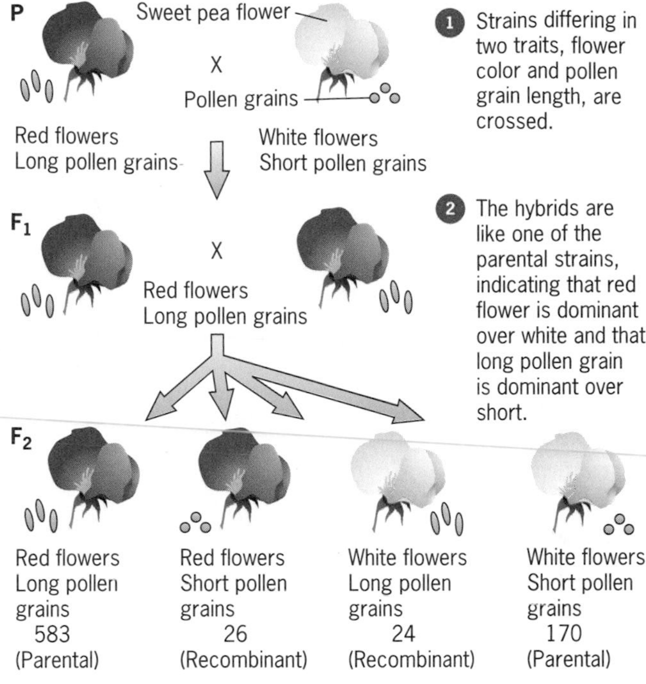

3 The frequencies of the four classes of F_2 progeny deviate significantly from the Mendelian expectation of 9/16, 3/16, 3/16, and 1/16, indicating that the genes controlling the trait are linked.

Figure 8.3 Bateson and Punnett's experiment with sweet peas. The results in the F_2 indicate that the genes for flower color and pollen length do not assort independently.

ments with sweet peas (Figure 8.3). Bateson and Punnett crossed varieties that differed in two traits, flower color and pollen length. They crossed plants with red flowers and long pollen grains to plants with white flowers and short pollen grains. All the F_1 plants had red flowers and long pollen grains, thus indicating that the alleles for these two phenotypes were dominant. When the F_1 plants were self-fertilized, Bateson and Punnett observed a peculiar distribution of phenotypes among the offspring. Instead of the 9:3:3:1 ratio expected for two independently assorting genes, they obtained a ratio of 23.3:1:1:6.8. Obviously, the two parental classes were significantly overrepresented in the progeny. This departure from the expected Mendelian results was due to linkage between the gene for flower color and that for pollen length.

Although Bateson and Punnett devised a complicated explanation for their results, it turned out to be wrong. The correct explanation is that the genes for flower color and pollen length are located on the same chromosome; consequently, they tend to travel through meiosis together. This explanation is diagrammed in Figure 8.4. The alleles of the flower color gene are *R* (red) and *r* (white), and the alleles of the pollen length gene are *L* (long) and *l* (short); the *R* and *L* alleles are dominant. (Note here that for historical reasons, the allele symbols are derived from the dominant rather than the recessive phenotypes.) Because the flower color and pollen length genes are linked, the parental combinations of alleles (*R* and *L*, and *r* and *l*) are more prevalent than the nonparental combinations (*R* and *l*, and *r* and *L*) in the gametes of the F_1 plants. Self-fertilization of the F_1 therefore produces a preponderance of parental phenotypes in the F_2. However, the genes for flower color and pollen length are not inextricably linked. Some nonparental progeny do appear, although at low frequency. Because these progeny indicate that the alleles of the two genes were recombined in the F_1, they are called **recombinants.**

Frequency of Recombination as a Measure of Linkage Intensity

We can use the frequency of recombination to measure the intensity of linkage between genes. Genes that are tightly linked seldom recombine, whereas genes that are loosely linked recombine often. The most direct way of estimating the frequency of recombination is to use a testcross, in which the individual to be analyzed genetically is crossed to a homozygote carrying the recessive alleles. For example, the heterozygous F_1 sweet peas discussed above could have been crossed to homozygotes carrying the recessive alleles of the genes for flower color and pollen length. The progeny of this testcross would therefore reveal what kinds of gametes the F_1 plants produced and in what proportions.

Figure 8.5 presents the analysis of testcross data. Among 1000 progeny scored, 920 resemble one or the other of the parental strains and the remaining 80 are recombinant. The frequency of recombinant gametes produced by the heterozygous F_1 plants is therefore $80/1000 = 0.08$. We can use this frequency as a measure of how often the flower color and pollen length genes recombine during meiosis. Notice that it represents the summation of two approximately equal frequencies, one for each of the two types of recombinant gametes.

For any two genes, the frequency of recombinant gametes never exceeds 50 percent. This upper limit is reached when genes are very far apart, perhaps at opposite ends of a chromosome. It is also reached when genes are on different chromosomes; 50 percent recombination is, in fact, what is meant when it is said that the genes assort independently. For example, let's assume that genes *A* and *B* are on different chromosomes and that an *AA BB* individual is crossed to an *aa bb* individual. From this cross, the *Aa Bb* offspring are then testcrossed to the double recessive parent. Be-

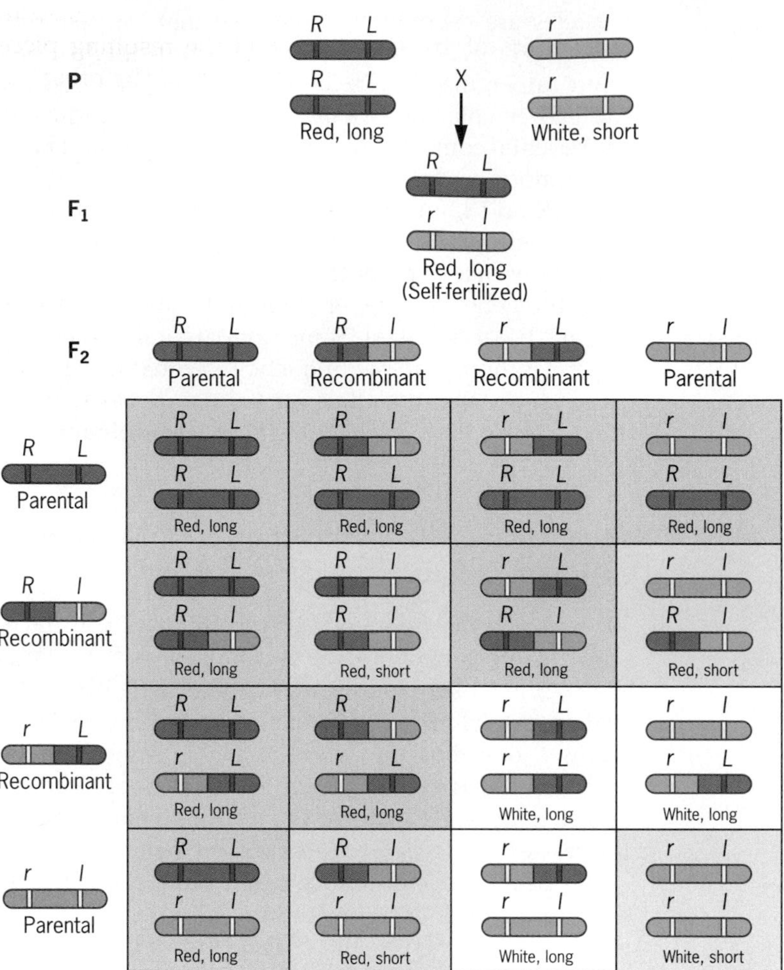

Figure 8.4 Genetic basis of the results of Bateson and Punnett's experiment. Each phenotype is highlighted by a different color in the Punnett Square. These phenotypes do not appear in a 9:3:3:1 ratio because the genes for flower color and pollen length are located on the same chromosome. In the gametes of the F_1 this linkage causes the parental combinations of flower color and pollen length alleles to be more frequent than the nonparental (recombinant) combinations.

cause the *A* and *B* genes assort independently, the F_2 will consist of two classes (*Aa Bb* and *aa bb*) that are phenotypically like the parents in the original cross, and two classes (*Aa bb* and *aa Bb*) that are phenotypically recombinant. Furthermore, each F_2 class will occur with a frequency of 25 percent. Thus, the total frequency of recombinant progeny from a testcross involving two genes on different chromosomes will be 50 percent. A frequency of recombination less than 50 percent implies that the genes are linked on the same chromosome.

Crosses involving linked genes are usually diagrammed to show the **linkage phase**—the way in which the alleles are arranged in heterozygous individuals (Figure 8.6). In Bateson and Punnett's sweet pea experiment, the heterozygous F_1 plants received

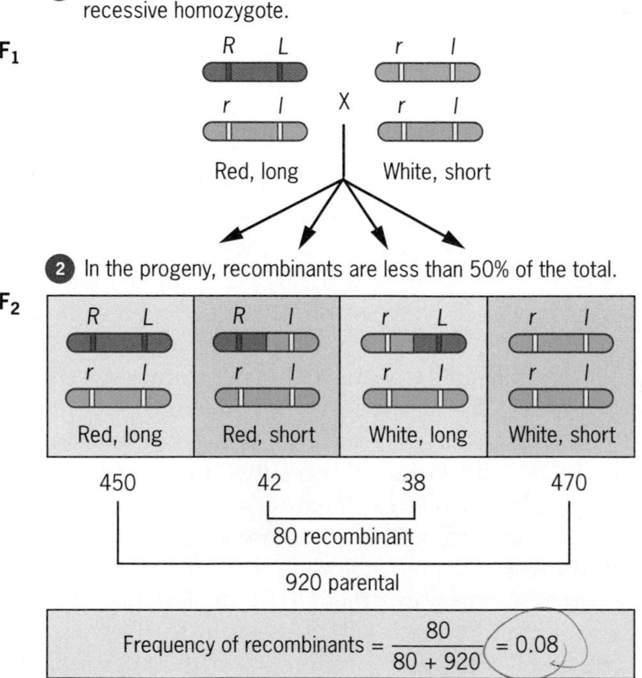

Figure 8.5 A testcross for linkage between genes in sweet peas. Because the recombinant progeny in the F_2 are 8 percent of the total, the genes for flower color and pollen length are tightly linked.

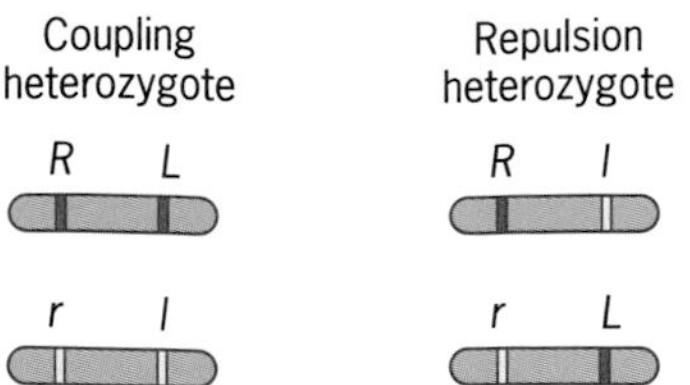

Figure 8.6 Coupling and repulsion linkage phases in double heterozygotes.

two dominant alleles, *R* and *L*, from one parent and two recessive alleles, *r* and *l*, from the other. Thus, we write the genotype of these plants *R L*/*r l*, where the slash (/) separates alleles inherited from different parents. Another way of interpreting this symbolism is to say that the alleles on the left and right of the slash entered the genotype on different homologous chromosomes, one from each parent. Whenever the dominant alleles are all on one side of the slash, as in this example, the genotype has the *coupling* linkage phase. When the dominant and recessive alleles are split on both sides of the slash, as in *R l*/*r L*, the genotype has the *repulsion* linkage phase. These terms provide us with a way of distinguishing between the two kinds of double heterozygotes.

Crossing Over as the Physical Basis of Recombination

Recombinant gametes are produced as a result of crossing over between homologous chromosomes. This process involves a physical exchange between the chromosomes, as diagrammed in Figure 8.7. The exchange event occurs during the prophase of the first meiotic division, when duplicated chromosomes have paired. Although four homologous chromatids are present, forming what is called a **tetrad**, only two cross over at any one point. Each of these chromatids breaks at the site of the crossover, and the resulting pieces reattach to produce the recombinants. The other two chromatids are not recombinant at this site. Each crossover event therefore produces two recombinant chromatids among a total of four.

The evidence for crossing over between two chromatids within a tetrad comes from the study of certain fungi in the class Ascomycetes, the most important of which is baker's yeast, *Saccharomyces cerevisiae*. This single-celled haploid organism ordinarily reproduces asexually, but it also has a sexual phase that is initiated by the fusion of haploid cells of opposite mating types, denoted "a" and α (*alpha*). This fusion creates a diploid cell, which promptly undergoes meiosis to produce four haploid cells, or ascospores, inside a small sac called the ascus. Each ascus therefore contains all four products of a single meiosis. By dissecting this sac, a researcher can isolate each product and place it in a culture dish to start a new yeast colony. Many mutations have been identified in *S. cerevisiae*, including some that alter colony shape and color. These mutations have been useful in the study of genetic recombination.

One important finding from studies with *S. cerevisiae* is that crossing over occurs after homologous chromosomes have duplicated. An ascus may contain two recombinant and two nonrecombinant ascospores, each of which has a different genotype (Figure 8.7). This observation can only be explained if crossing over occurs after chromosome duplication. If it occurred before duplication, an ascus could never contain more than two kinds of ascospores.

A second finding is that only two chromatids are involved in an exchange at any one point. However, the other two chromatids may cross over at a different point. Thus, there is a possibility for multiple exchanges in a tetrad of chromatids (Figure 8.8). There may, for example, be two, three, or even four separate

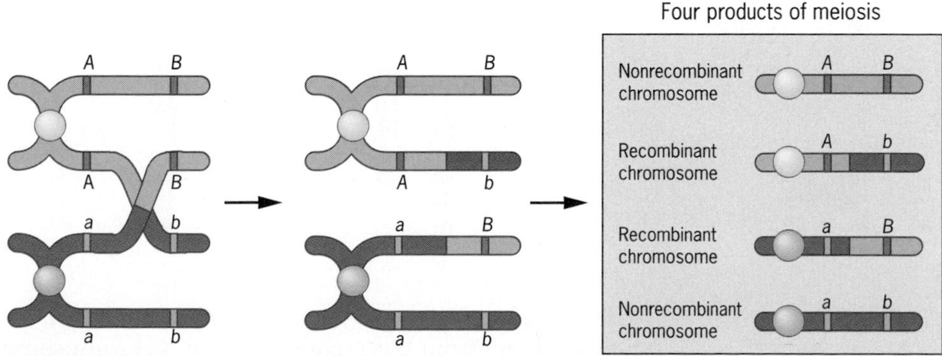

Figure 8.7 Crossing over as the basis of recombination between genes. An exchange between paired chromosomes during meiosis produces recombinant chromosomes at the end of meiosis.

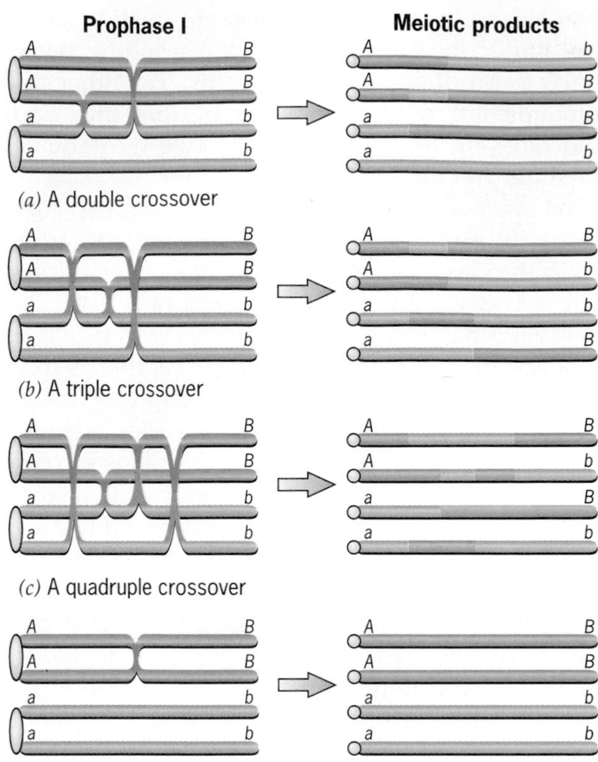

Figure 8.8 Consequences of multiple exchanges between chromosomes and exchange between sister chromatids during prophase I of meiosis.

exchanges—customarily called double, triple, or quadruple crossovers. (We consider the genetic significance of these in a later section.) Note, however, that an exchange between sister chromatids does not produce genetic recombinants because the sister chromatids are identical.

What is responsible for the breakage of chromatids during crossing over? One hypothesis is that the chromatid fibers become entangled with each other and torsional stress causes them to break. Although such tangles may sometimes occur, it seems more likely that the breaks are induced by enzymes acting on the DNA within the chromatids. Enzymes are also responsible for repairing these breaks, that is, for reattaching chromatid fragments to each other. We consider the molecular details of this process in Chapter 14.

Evidence That Crossing Over Causes Recombination

In 1931 Curt Stern, Harriet Creighton, and Barbara McClintock obtained the first evidence that genetic recombination was associated with a material exchange between chromosomes. Stern worked with *Drosophila*, and Creighton and McClintock worked with maize.

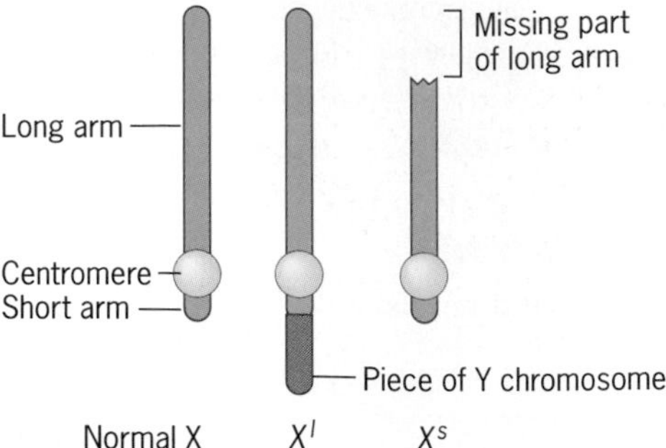

Figure 8.9 Structure of the X chromosomes used in Stern's experiment.

Each set of experiments depended on the availability of homologous chromosomes that were morphologically distinguishable. The goal was to determine whether physical exchange between these homologs was correlated with recombination between some of the genes they carried.

Stern used two different X chromosomes, one shorter and one longer than normal (Figure 8.9). The shorter X, X^s, had lost a piece from its long arm through a translocation with the tiny fourth chromosome, and the longer X, X^l, had gained a piece on its short arm through a translocation with the Y chromosome. Thus, the long X chromosome had a normal long arm attached to an abnormal short arm, and the short X chromosome had an abnormal long arm attached to a normal short arm. The short X chromosome also carried mutant genes to allow genetic recombination to be detected. One of these was the dominant mutation for *Bar* eyes (symbol *B*), and the other was the recessive mutation for *carnation* eyes (symbol *car*). The long X chromosome carried the wild-type alleles of these two genes (B^+ and car^+).

Stern crossed females that were heterozygous for the long and short X chromosomes to males that carried a structurally normal X marked with B^+ and *car*.

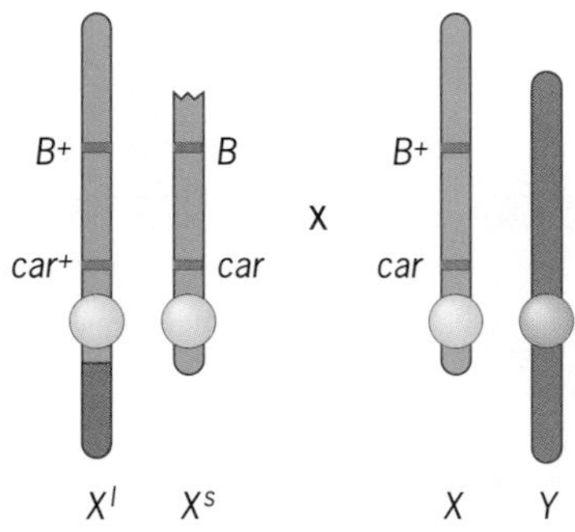

The critical question was whether or not the recombinant progeny from this cross carried an X chromosome that was produced by crossing over between the two abnormal X chromosomes. Stern showed that the answer was yes by demonstrating that the *B* car^+ recombinants carried an X chromosome in which both arms were abnormal:

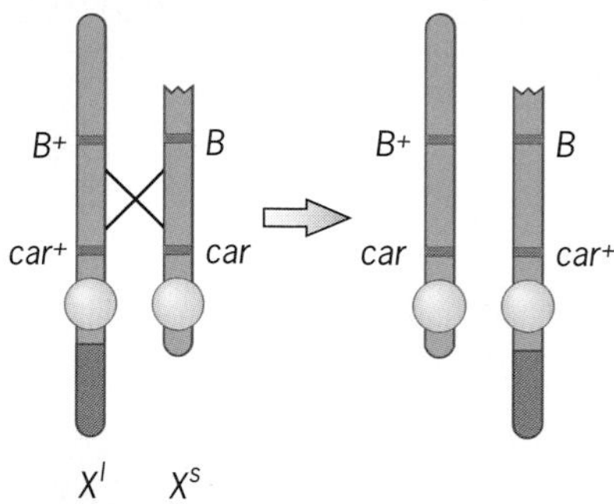

Such a chromosome could only have been produced by exchange between the two structurally abnormal X chromosomes that were used in the experiment.

This demonstration that recombination was correlated with chromosome exchange was independently confirmed by Creighton and McClintock in a study with maize. Two forms of chromosome 9 were available for analysis; one was normal, and the other had cytological aberrations at each end—a heterochromatic knob at one end and a piece of a different chromosome at the other. These two forms of chromosome 9 were also genetically marked to detect recombination. One marker gene controlled kernel color (*C*, colored; *c*, colorless), and the other controlled kernel texture (*Wx*, starchy; *wx*, waxy). Creighton and McClintock performed the following cross:

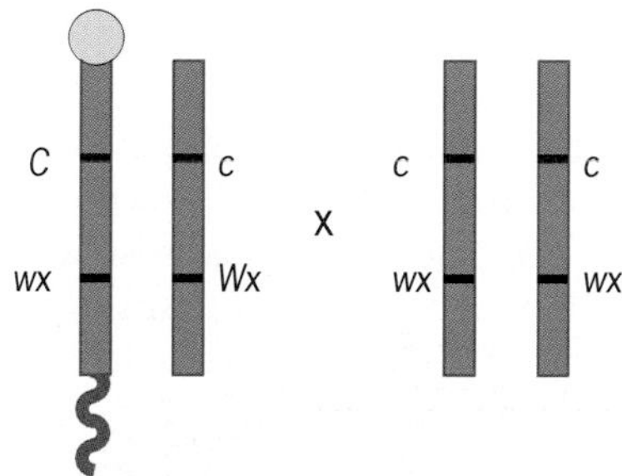

They then examined the recombinant progeny for evidence of exchange between the two different forms of chromosome 9. Their results showed that the *C Wx* and *c wx* recombinants carried a chromosome with only one of the abnormal cytological markers; the other abnormal marker had evidently been lost through an exchange with the normal chromosome 9 in the previous generation:

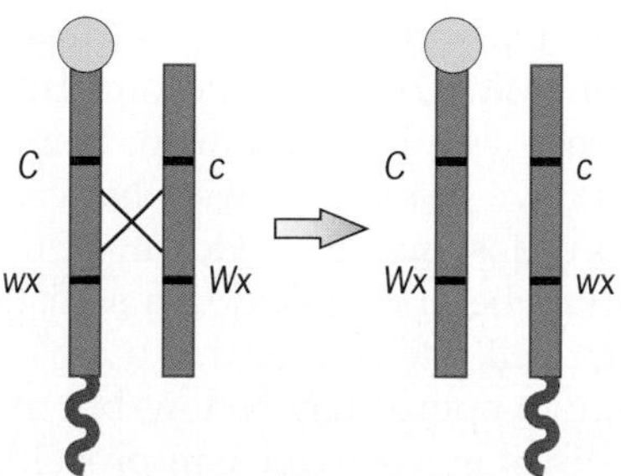

These findings strongly argued that recombination was caused by a physical exchange between paired chromosomes.

Chiasmata and the Time of Crossing Over

The cytological evidence for crossing over can be seen during late prophase of the first meiotic division when the chiasmata become clearly visible. At this time paired chromosomes repel each other slightly, maintaining close contact only at the centromere and at each chiasma (Figure 8.10). This partial separation makes it possible to count the chiasmata accurately. As we might expect, large chromosomes typically have more chiasmata than small chromosomes. Thus, the number of chiasmata is roughly proportional to chromosome length.

The appearance of chiasmata late in the first meiotic prophase might imply that it is then that crossing over occurs. However, evidence from several different experiments suggests that it occurs earlier. Some of these experiments used heat shocks to alter the frequency of recombination. When the heat shocks were administered late in prophase, there was little effect, but when they were given earlier, the recombination frequency was changed. Thus, the event responsible

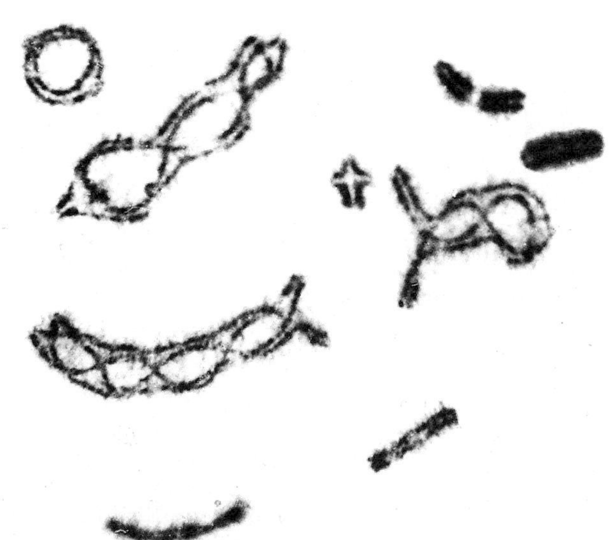

Figure 8.10 Diplonema of male meiosis in the grasshopper *Chorthippus parallelus*. There are eight autosomal bivalents and an X-chromosome univalent. One of the bivalents and the X-univalent are heteropycnotic and stain more densely than the remaining autosomes. The four smaller bivalents have single chiasma that are terminalized in three of them. The remaining bivalents carry two, three, four, and five chiasmata respectively.

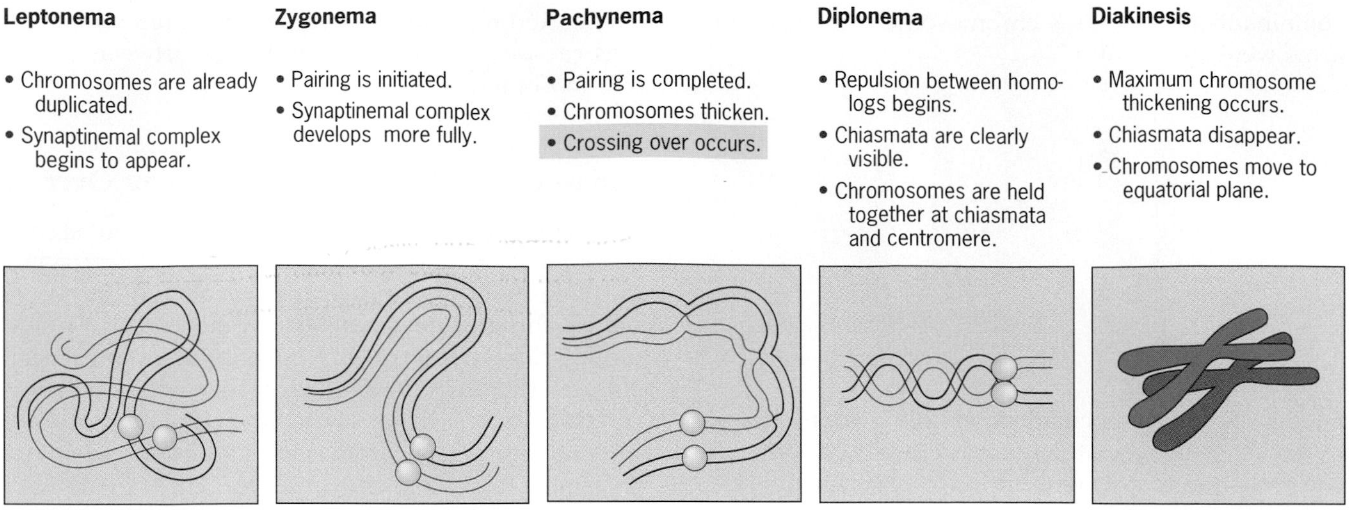

Leptonema	**Zygonema**	**Pachynema**	**Diplonema**	**Diakinesis**
• Chromosomes are already duplicated. • Synaptinemal complex begins to appear.	• Pairing is initiated. • Synaptinemal complex develops more fully.	• Pairing is completed. • Chromosomes thicken. • Crossing over occurs.	• Repulsion between homologs begins. • Chiasmata are clearly visible. • Chromosomes are held together at chiasmata and centromere.	• Maximum chromosome thickening occurs. • Chiasmata disappear. • Chromosomes move to equatorial plane.

Figure 8.11 Five stages of prophase of meiosis I. Crossing over occurs in pachynema. Chiasmata become visible in diplonema and disappear during diakinesis.

for recombination, namely, crossing over, occurs rather early in the meiotic prophase. Additional evidence comes from molecular studies on the time of DNA synthesis. Although almost all the DNA is synthesized during the interphase that precedes the onset of meiosis, a small amount is made during the first meiotic prophase. This limited DNA synthesis has been interpreted as part of a process to repair broken chromatids, which, as we have discussed, is thought to be associated with crossing over. Careful timing experiments have shown that this DNA synthesis occurs in early to mid-prophase, but not later. The accumulated evidence therefore suggests that crossing over occurs in early to mid-prophase, long before the chiasmata can be seen.

What, then, are chiasmata, and what do they mean? Most geneticists believe that the chiasmata are merely vestiges of the actual exchange process. Chromatids that have experienced an exchange probably remain entangled with each other during most of prophase. Eventually, these entanglements are resolved, and the chromatids are separated by the meiotic spindle apparatus to opposite poles of the cell. Therefore, each chiasma probably represents an entanglement that was created by a crossover event earlier in prophase, probably in pachynema (Figure 8.11).

Key Points: **The frequency of recombination is a measure of linkage between genes on the same chromosome. Recombination is caused by a physical exchange between paired homologous chromosomes early in prophase of the first meiotic division, after the chromosomes have duplicated. At any one point, the process of exchange—crossing over—involves only two of the four chromatids in a meiotic tetrad. Later in prophase I, these crossovers become visible as chiasmata.**

CHROMOSOME MAPPING

Crossing over during the prophase of the first meiotic division has two observable outcomes:

1. Formation of chiasmata in late prophase.
2. Recombination between genes on opposite sides of the crossover point.

However, the second outcome can only be seen in the next generation, when the genes on the recombinant chromosomes are expressed.

Geneticists construct chromosome maps by counting the number of crossovers that occur during meiosis. However, because the actual crossover events cannot be seen, they cannot count them directly. Instead, they must estimate how many crossovers have taken place by counting either chiasmata or recombinant chromosomes. Chiasmata are counted through cytological analysis, whereas recombinant chromosomes are counted through genetic analysis. Before we consider either of these procedures, we must define what we mean by distance on a chromosome map.

Crossing Over as a Measure of Genetic Distance

Sturtevant's fundamental insight was to estimate the distance between points on a chromosome by counting the number of crossovers between them. Points that are far apart should have more crossovers between them than points that are close together. However, the number of crossovers must be understood in a statistical sense. In any particular cell, the chance that a crossover will occur between two points may be low, but in a large population of cells, this crossover will probably occur sev-

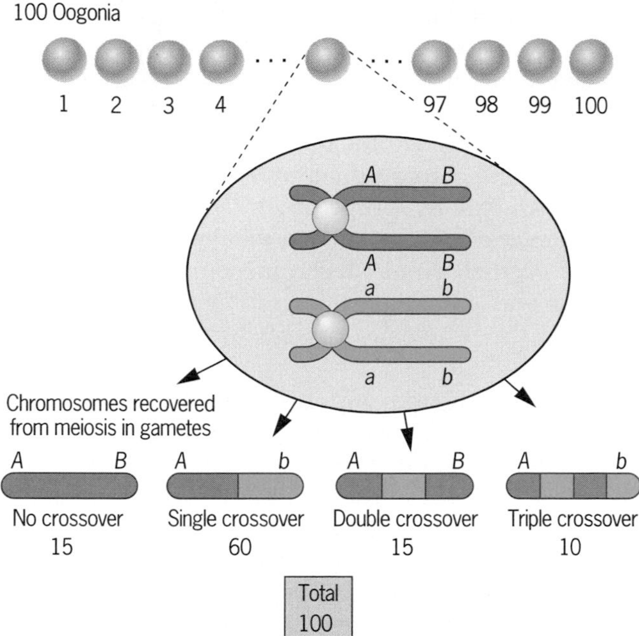

Average number of crossovers between A and B =

$$0\times\left(\frac{15}{100}\right) + 1\times\left(\frac{60}{100}\right) + 2\times\left(\frac{15}{100}\right) + 3\times\left(\frac{10}{100}\right) = 1.2$$

Figure 8.12 Calculating the average number of crossovers between genes on chromosomes recovered from meiosis.

eral times simply because there are so many independent opportunities for it. Thus, the quantity that we really need to measure is the *average* number of crossovers in a particular chromosome region. Genetic map distances are, in fact, based on such averages. This idea is sufficiently important to justify a formal definition: *The distance between two points on the genetic map of a chromosome is the average number of crossovers between them.*

One way for us to understand this definition is to consider 100 oogonia going through meiosis (Figure 8.12). In some cells, no crossovers will occur between sites A and B; in others, one, two, or more crossovers will occur between these loci. At the end of meiosis, there will be 100 gametes, each containing a chromosome with either zero, one, two, or more crossovers between A and B. We estimate the genetic map distance between these loci by calculating the average number of crossovers in this sample of chromosomes:

$$0 \times (15/100) + 1 \times (60/100) + 2 \times (15/100) + 3 \times (10/100) = 1.2$$

In practice, we cannot "see" each of the exchange points on the chromosomes coming out of meiosis. Instead, we infer their existence by observing the recombination of the alleles that flank them. A chromosome in which alleles have recombined must have arisen by crossing over. Counting recombinant chromosomes therefore provides a way of counting crossover exchange points.

Recombination Mapping with a Two-Point Testcross

To illustrate the mapping procedure, let's consider the two-point testcross in Figure 8.13. Wild-type *Drosophila* females were mated to males homozygous for two autosomal mutations—*vestigial* (*vg*), which produces short wings, and *black* (*b*), which produces a black body. All the F_1 flies had long wings and gray bodies; thus the wild-type alleles (vg^+ and b^+) are dominant.

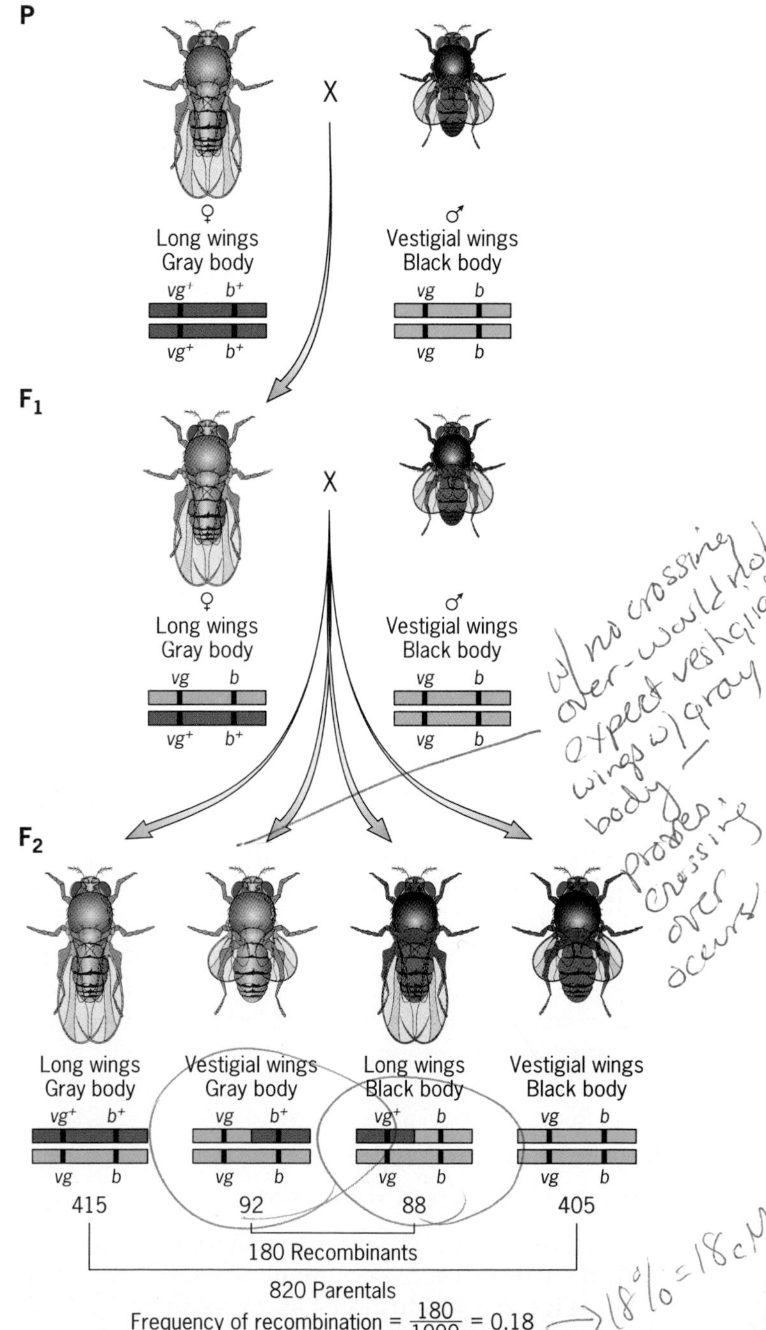

Figure 8.13 An experiment involving two linked genes, *vg* (*vestigial* wings) and *b* (*black* body), in *Drosophila*.

The F_1 females were then testcrossed to vestigial, black males, and the F_2 progeny were sorted by phenotype and counted. As the data show, there were four phenotypic classes, two abundant and two rare. The abundant classes had the same phenotypes as the original parents, and the rare classes had recombinant phenotypes. We know that the vestigial and black genes are linked because the recombinants are much fewer than 50 percent of the total progeny counted. These genes must therefore be on the same chromosome. To determine the distance between them, we must estimate the average number of crossovers in the gametes of the doubly heterozygous F_1 females. We can do this by calculating the frequency of recombinant F_2 flies and noting that each such fly inherited a chromosome that had crossed over once between *vg* and *b*. The average number of crossovers in the whole sample of progeny is therefore

$$\underset{(0) \times 0.82}{\text{nonrecombinants}} + \underset{(1) \times 0.18}{\text{recombinants}} = 0.18$$

In this expression, the number of crossovers for each class of flies is placed in parentheses; the other number is the frequency of that class. The nonrecombinant progeny obviously do not add any crossover chromosomes to the data, but we include them in the calculation to emphasize that we must calculate the average number of crossovers by using all the data, not just those from the recombinants.

This simple analysis indicates that, on average, 18 out of 100 chromosomes recovered from meiosis had a crossover between *vg* and *b*. Thus, *vg* and *b* are separated by 18 **units** on the genetic map. Sometimes geneticists call a map unit a **centiMorgan**, abbreviated cM, in honor of T. H. Morgan; 100 centiMorgans equal one Morgan (M). We can therefore say that *vg* and *b* are 18 cM (or 0.18 M) apart. Notice that the map distance is equal to the frequency of recombination, written as a percentage. Later we will see that when the frequency of recombination approaches 0.5, it underestimates the map distance. The Human Genetics Sidelight: Evidence for Linkage Between the Genes for Hemophilia and Color Blindness shows how pedigrees can be analyzed for evidence of linkage between loci on a human chromosome.

Recombination Mapping with a Three-Point Testcross

We can also use the recombination mapping procedure with data from testcrosses involving more than two genes. Figure 8.14 illustrates an experiment by C. B. Bridges and T. M. Olbrycht, who crossed wild-type *Drosophila* males to females homozygous for three recessive X-linked mutations—*scute* (*sc*) bristles, *echinus* (*ec*) eyes, and *crossveinless* (*cv*) wings. They then intercrossed the F_1 progeny to produce F_2 flies, which they classified and counted. We note that the F_1 females in this intercross carried the three recessive mutations on one of their X chromosomes and the wild-type alleles of these mutations on the other X chromosome. Furthermore, the F_1 males carried the three recessive mutations on their single X chromosome. Thus, this intercross was equivalent to a testcross with all three genes in the F_1 females present in the coupling configuration.

The F_2 flies from the intercross comprised eight phenotypically distinct classes, two of them parental and six recombinant. The parental classes were by far the most numerous. The less numerous recombinant classes each represented a different kind of crossover chromosome. To figure out which crossovers were involved in producing each type of recombinant, we must first determine how the genes are ordered on the chromosome.

Determination of the Gene Order There are three possible gene orders:

(i) *sc ec cv*
(ii) *ec sc cv*
(iii) *ec cv sc*

Other possibilities, such as *cv* – *ec* – *sc*, are the same as one of these because the left and right ends of the chromosome cannot be distinguished. Which of the orders is correct?

To answer this question, we must take a careful look at the six recombinant classes. Four of them must have come from a single crossover in one of the two regions delimited by the genes. The other two must have come from double crossing over—one exchange in each of the two regions. Because a double crossover switches the gene in the middle with respect to the genetic markers on either side of it, we have, in principle, a way of determining the gene order. Intuitively, we also know that a double crossover should occur much less frequently than a single crossover. Consequently, among the six recombinant classes, the two rare ones must represent the double crossover chromosomes.

In our data, the rare, double crossover classes are 7 ($sc\ ec^+\ cv$) and 8 ($sc^+\ ec\ cv^+$), each containing a single fly (Figure 8.14). Comparing these to parental classes 1 ($sc\ ec\ cv$) and 2 ($sc^+\ ec^+\ cv^+$), we see that the *echinus* allele has been switched with respect to *scute* and *crossveinless*. Consequently, the *echinus* gene must be located between the other two. The correct gene order is therefore (i) *sc* – *ec* – *cv*.

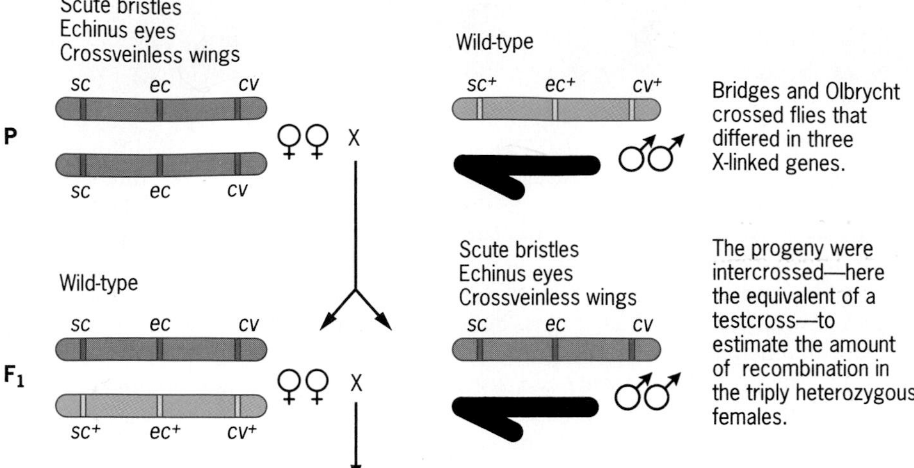

F2

Class	Phenotype	Genotype of maternally inherited X chromosome			Number observed
1	Scute, echinus, crossveinless	sc	ec	cv	1158
2	Wild-type	sc^+	ec^+	cv^+	1455
3	Scute	sc	ec^+	cv^+	163
4	Echinus, crossveinless	sc^+	ec	cv	130
5	Scute, echinus	sc	ec	cv^+	192
6	Crossveinless	sc^+	ec^+	cv	148
7	Scute, crossveinless	sc	ec^+	cv	1
8	Echinus	sc^+	ec	cv^+	1
				Total:	3248

Figure 8.14 Bridges and Olbrycht's three-point cross with the X-linked genes *sc* (*scute* bristles), *ec* (*echinus* eyes), and *cv* (*crossveinless* wings) in *Drosophila*.

Calculation of the Distances Between Genes Having established the gene order, we can now determine the distances between adjacent genes. Again, the procedure is to compute the average number of crossovers in each chromosomal region (Figure 8.15).

We can obtain the length of the region between *sc* and *ec* by identifying the recombinant classes that involved a crossover between these genes. There are four such classes: 3 ($sc\ ec^+\ cv^+$), 4 ($sc^+\ ec\ cv$), 7 ($sc\ ec^+\ cv$), and 8 ($sc^+\ ec\ cv^+$). Classes 3 and 4 involved a single

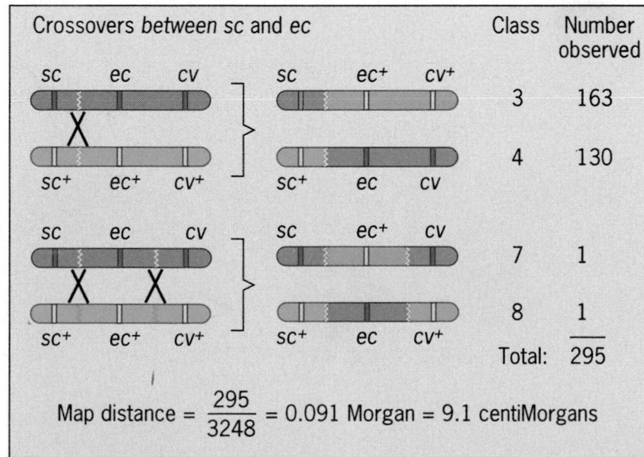

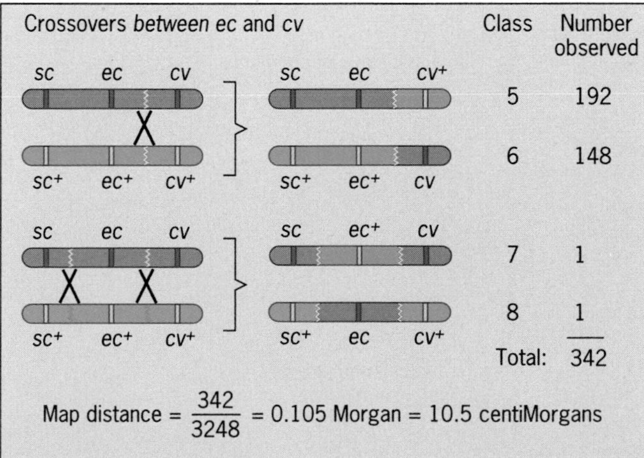

Figure 8.15 Calculation of genetic map distances from Bridges and Olbrycht's data. The distance between each pair of genes is obtained by estimating the average number of crossovers.

HUMAN GENETICS SIDELIGHT

Evidence for Linkage Between the Genes for Hemophilia and Color Blindness

Detection and analysis of linkage in human beings have never been easy. A geneticist cannot arrange matings between men and women with particular genotypes in order to obtain linkage data. Rather, he or she must collect the data from pedigrees, which are often incomplete, and then analyze them for evidence of linkage. The procedures used in this analysis are considerably more complex than those used in the analysis of data from experimental organisms (see Chapter 9).

Some of the first evidence for linkage between human genes was obtained by studying the inheritance of red-green color blindness and hemophilia. Because both of these traits are caused by mutant genes on the X chromosome, they could be linked. However, if the genes controlling them were sufficiently far apart on the X chromosome, the traits would appear to be unlinked. Only pedigree analysis could distinguish between these two possibilities.

In the first half of the twentieth century, geneticists in England, the United States, Germany, and the Netherlands searched for linkage between color blindness and hemophilia. Their strategy was to study families in which both traits were segregating. Figure 1 shows three of several pedigrees that were discovered. To analyze these pedigrees, we note that both traits are caused by recessive mutations: c = color blindness, h = hemophilia; their normal alleles are denoted by the letters C and H, respectively.

Pedigree (*a*) shows four generations of a family described by M. Madlener in 1928. Because the great-grandfather, I-1, has both color blindness and hemophilia, we know that his genotype is $c\ h$. His daughter, II-1, is phenotypically normal and must therefore carry the normal alleles, C and H. Moreover, because II-1 inherited c and h on the X chromosome transmitted by I-1, the two normal alleles must be carried by the other X chromosome—that is, the one she inherited from her mother. II-1's genotype is therefore $C\ H/c\ h$—that is, she is a coupling heterozygote for the two loci. III-2, the daughter of II-1, is also a coupling heterozygote. We infer that she has this genotype because her son has both color blindness and hemophilia ($c\ h$) and her father is phenotypically normal ($C\ H$). Evidently, III-2 inherited the $c\ h$ chromosome from her mother.

To detect linkage between the two loci, we must examine the genotypes of the children produced by II-1 and III-2. Among the three sons of II-1, one is normal ($C\ H$) and two are doubly mutant ($c\ h$). Thus, each of the sons must have inherited a nonrecombinant X chromosome from II-1. As discussed above, II-1's first daughter (III-2) also inherited a nonrecombinant X chromosome; her second daughter (III-4) cannot be

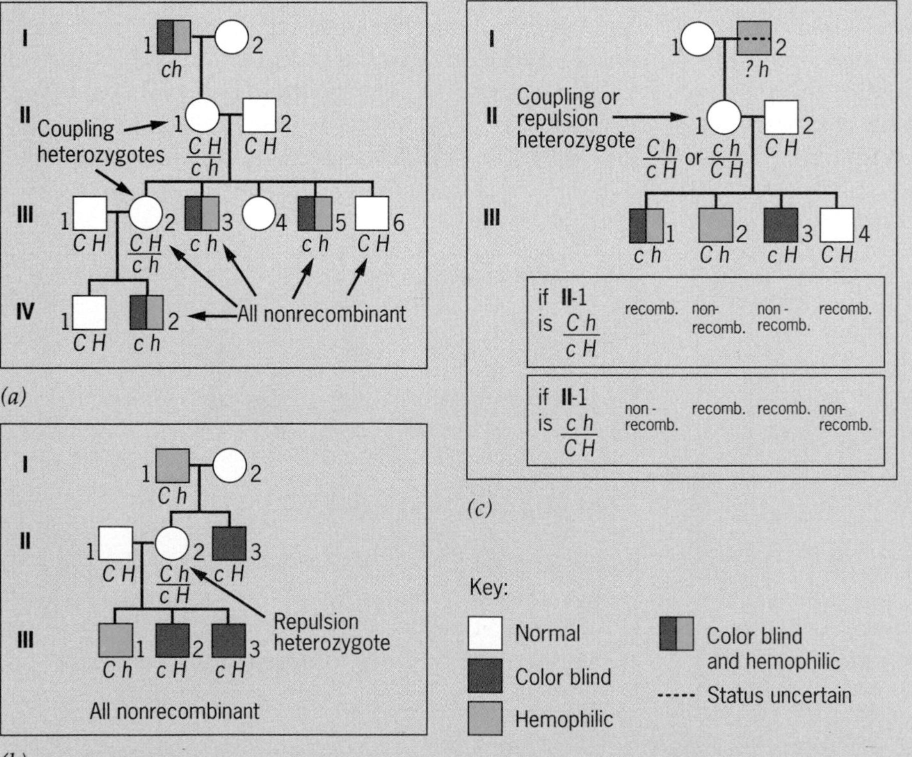

Figure 1 Pedigrees showing the inheritance of two X-linked traits, color blindness and hemophilia.

assigned a genotype. The other coupling heterozygote, III-2, produced two sons, one normal (*C H*) and one doubly mutant (*c h*). Both of these sons must also have inherited nonrecombinant X chromosomes from their mother. Thus, we can ascertain the genotypes of six of the seven X chromosomes that were transmitted by II-1 and III-2, and we find that all six are nonrecombinant. This result argues rather persuasively that the color blindness and hemophilia loci are linked.

Pedigree (*b*), described by C. L. Birch in 1937, provides additional evidence for linkage. I-1 is genotypically *C h*. His wife has an uncertain genotype; however, because the couple has a color-blind son, we can infer that she carries the mutant allele *c*. The couple's phenotypically normal daughter (II-2) must carry both mutant alleles, *c* and *h*, because she has both color-blind and hemophilic sons. Moreover, her genotype must be *C h*/*c H* because she inherited a *C h* chromosome from her father. Thus, II-2 is a repulsion heterozygote for the two loci. Among II-2's sons, we find that all three inherited nonrecombinant X chromosomes—an observation that suggests the two loci are linked.

Pedigree (*c*), described by B. Rath in 1938, provides evidence for recombination between the color blindness and hemophilia loci—the first ever observed. The key individual in this pedigree is II-1. This woman has four sons, each with a different genotype: one is color blind and hemophilic (*c h*), one is hemophilic (*C h*), one is color blind (*c H*), and one is normal (*C H*). Thus, II-1 must carry both *c* and *h*. However, we cannot determine whether she is a coupling or a repulsion heterozygote because we do not know her father's complete genotype. (He died of hemophilia before he could be tested for color blindness.) All we can say is that II-1 is either *C H*/*c h* or *C h*/*c H*. In either case, half her sons have recombinant genotypes, and half have nonrecombinant genotypes. Thus, we can infer that recombination occurs between these two X-linked loci. The frequency of recombination (50 percent) suggests that the loci are far apart on the X chromosome.

More detailed analyses of other pedigrees have shown that there are two forms of hemophilia, A and B, each involving the deficiency of a different blood clotting factor (VIII in A and IX in B). These two factors are produced by genes at different positions on the X chromosome (Figure 2). Factor VIII is encoded by a gene near the end of the chromosome's long arm, and factor IX is encoded by a gene in the middle of the long arm. The color blindness gene is tightly linked to the factor VIII gene but loosely linked to the factor IX gene. These linkage relationships might explain why there is no recombination between color blindness and hemophilia in pedigrees (*a*) and (*b*), and 50 percent recombination in pedigree (*c*). The hemophilia observed in pedigrees (*a*) and (*b*) may have been caused by a factor VIII deficiency, which is the more common type; in pedigree (*c*), it may have been caused by a factor IX deficiency.

There are also two genes for color blindness on the X chromosome (see Chapter 6). However, these genes are so close together that they behave as one locus. In fact, the two genes appear to have been created by the duplication of a single gene sometime during the course of human evolution.

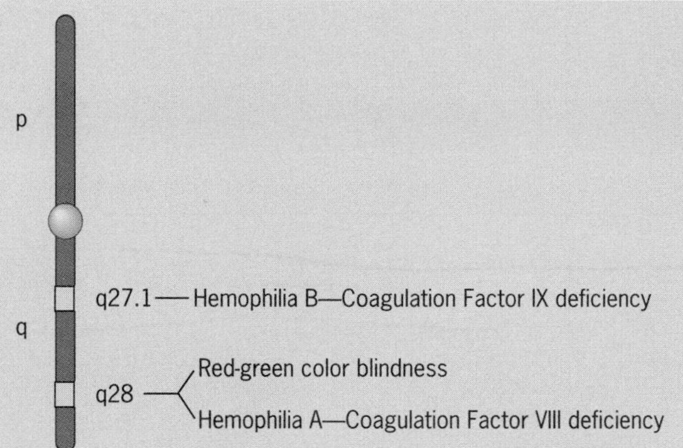

Figure 2 Map of the human X chromosome showing the location of the genes for hemophilia A and B and for red-green color blindness.

crossover between *sc* and *ec*, and classes 7 and 8 involved two crossovers, one between *sc* and *ec* and the other between *ec* and *cv*. We can therefore use the frequencies of these four classes to estimate the average number of crossovers between *sc* and *ec*:

$$\frac{\overset{\text{Class 3}}{163} + \overset{\text{Class 4}}{130} + \overset{\text{Class 7}}{1} + \overset{\text{Class 8}}{1}}{\text{Total}} = \frac{295}{3248} = 0.091$$

Thus, in every 100 chromosomes coming from meiosis in the F_1 females, 9.1 had a crossover between *sc* and *ec*. The distance between these genes is therefore 9.1 map units (or, if you prefer, 9.1 centiMorgans).

In a similar way, we can obtain the distance between *ec* and *cv*. Four recombinant classes involved a crossover in this region: 5 (*sc ec* cv^+), 6 (sc^+ ec^+ *cv*), 7, and 8. The double recombinants are also included here because one of their two crossovers was between *ec* and *cv*. The combined frequency of these four classes is:

$$\frac{\overset{\text{Class 5}}{192} + \overset{\text{Class 6}}{148} + \overset{\text{Class 7}}{1} + \overset{\text{Class 8}}{1}}{\text{Total}} = \frac{342}{3248} = 0.105$$

Consequently, *ec* and *cv* are 10.5 map units apart.

Combining the data for the two regions, we obtain the map

sc—9.1—*ec*—10.5—*cv*

Map distances computed in this way are additive. Thus, we can estimate the distance between *sc* and *cv*

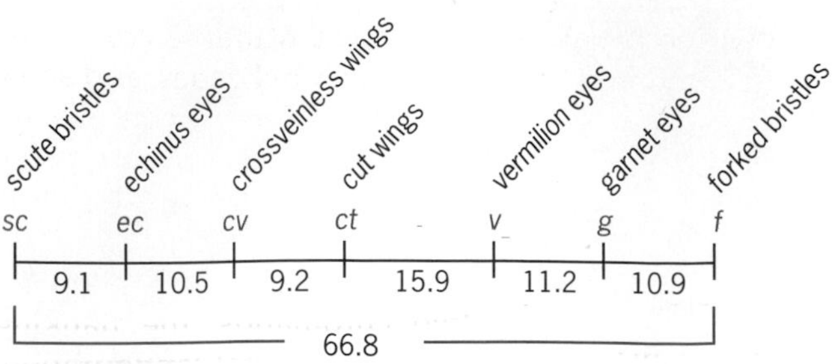

Figure 8.16 Bridges and Olbrycht's map of seven X-linked genes in *Drosophila*. Distances are given in centiMorgans.

by summing the lengths of the two map intervals between them:

$$9.1 \text{ cM} + 10.5 \text{ cM} = 19.6 \text{ cM}$$

We can easily verify this estimate by directly calculating the average number of crossovers between these genes:

Non-crossover classes	Single crossover classes	Double crossover classes	
(1 and 2)	(3, 4, 5, and 6)	(7 and 8)	
(0) × 0.805 +	(1) × 0.195	+ (2) × 0.0006	= 0.196

Here the number of crossovers is given in parentheses, and its multiplier is the combined frequency of the classes with that many crossovers. In other words, each recombinant class contributes to the map distance according to the product of its frequency and the number of crossovers it represents.

Bridges and Olbrycht studied seven X-linked genes in their recombination experiment: *sc*, *ec*, *cv*, *ct* (*cut* wings), *v* (*vermilion* eyes), *g* (*garnet* eyes), and *f* (*forked* bristles). By calculating recombination frequencies between each pair of adjacent genes, they were able to construct a map of a large segment of the X chromosome (Figure 8.16); *sc* was at one end, and *f* was at the other. Each of the seven genes that Bridges and Olbrycht studied was, in effect, a **marker** for a particular site on the X chromosome. Summing all the map intervals between these markers, they estimated the total length of the mapped segment to be 66.8 cM. Thus, the average number of crossovers in this segment was 0.668.

Interference and the Coefficient of Coincidence A three-point cross has an important advantage over a two-point cross: it allows the detection of double crossovers, permitting us to determine if exchanges in adjacent regions are independent of each other. For example, does a crossover in the region between *sc* and *ec* (region I on the map of the X chromosome) occur independently of a crossover in the region between *ec* and *cv* (region II)? Or does one crossover inhibit the occurrence of another nearby?

To answer these questions, we must calculate the expected frequency of double crossovers, based on the idea of independence. We can do this by multiplying the crossover frequencies for two adjacent chromosome regions. For example, in region I on Bridges' and Olbrycht's map, the crossover frequency was (163 + 130 + 1 + 1)/3248 = 0.091, and in region II, it was (192 + 148 + 1 + 1)/3248 = 0.105. If we assume independence, the expected frequency of double crossovers in the interval between *sc* and *cv* would therefore be (0.091) × (0.105) = 0.0095. We can now compare this frequency with the observed frequency, which was 2/3248 = 0.0006. Double crossovers between *sc* and *cv* were much less frequent than expected. This result suggests that one crossover inhibited the occurrence of another nearby, a phenomenon called **interference**. The extent of the interference is customarily measured by the **coefficient of coincidence**, *c*, which is the ratio of the observed frequency of double crossovers to the expected frequency:

$$c = \frac{\text{observed frequency of double crossovers}}{\text{expected frequency of double crossovers}}$$

$$= \frac{0.0006}{0.0095} = 0.063$$

The level of interference, symbolized *I*, is calculated as $I = 1 - c = 0.937$.

Because in this example the coefficient of coincidence is close to zero, its lowest possible value, interference was very strong (*I* is close to 1.). At the other extreme, a coefficient of coincidence equal to one would imply no interference at all; that is, it would imply that the crossovers occurred independently of each other.

Many studies have shown that interference is strong over map distances less than 20 cM; thus, double crossovers seldom occur in short chromosomal regions. However, over long regions, interference weakens to the point that crossovers occur more or less independently. The strength of interference is therefore a function of map distance.

Recombination Frequency and Genetic Map Distance

In the preceding sections, we have considered how to construct chromosome maps from data on the recombination of genetic markers. These data allow us to infer where crossovers have occurred in a sample of chromosomes. By localizing and counting these crossovers, we can estimate the distances between genes and then place the genes on a chromosome map.

This method works well as long as the genes are fairly close together. However, when they are far apart, the frequency of recombination may not reflect the true map distance (Figure 8.17). As an example, let us consider the genes at the ends of Bridges' and Olbrycht's map of the X chromosome; *sc*, at the left end, was 66.8 cM away from *f*, at the right end. However, the frequency of recombination between *sc* and *f* was 50 percent—the maximum possible value. Using this frequency to estimate map distance, we would conclude that *sc* and *f* were 50 map units apart. Of course, the distance obtained by summing the lengths of the intervening regions on the map, 66.8 cM, is much greater. This example shows that the true genetic distance, which depends on the average number of crossovers on a chromosome, may be much greater than the observed recombination frequency. Multiple crossovers may occur between widely separated genes, and some of these crossovers may not produce genetically recombinant chromosomes (Figure 8.18). To see this, let us assume that a single crossover occurs between two chromatids in a tetrad, causing recombination of the flanking genetic markers. If another crossover occurs between these same two chromatids, the flanking markers will be restored to their original configuration; the second crossover essentially cancels the effect of the first, converting the recombinant chromatids back into nonrecombinants. Thus, even though two crossovers have occurred in this tetrad, none of the chromatids that come from it will be recombinant for the flanking markers.

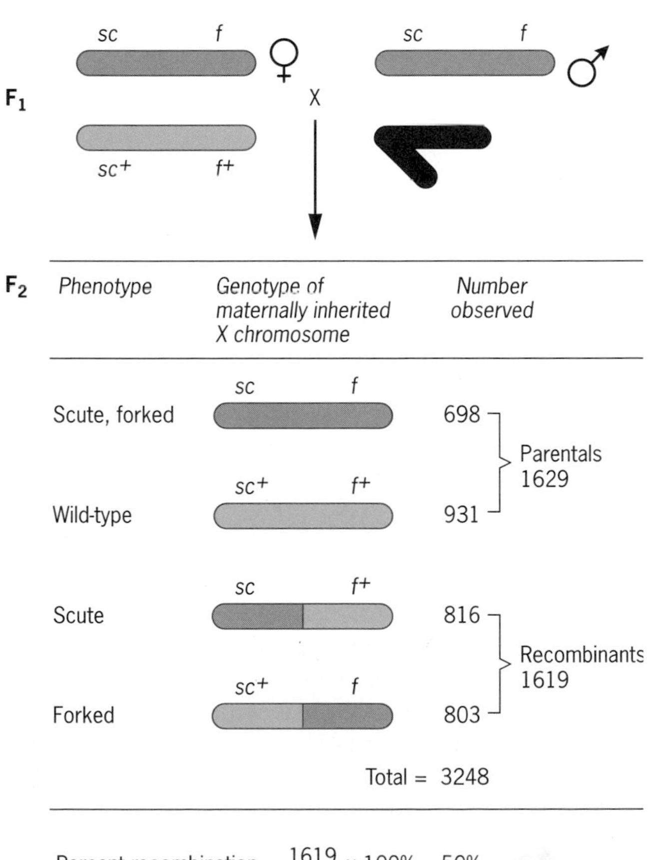

Figure 8.17 A discrepancy between map distance and percent recombination. The map distance between the genes *sc* and *f* is greater than the observed percent recombination between them.

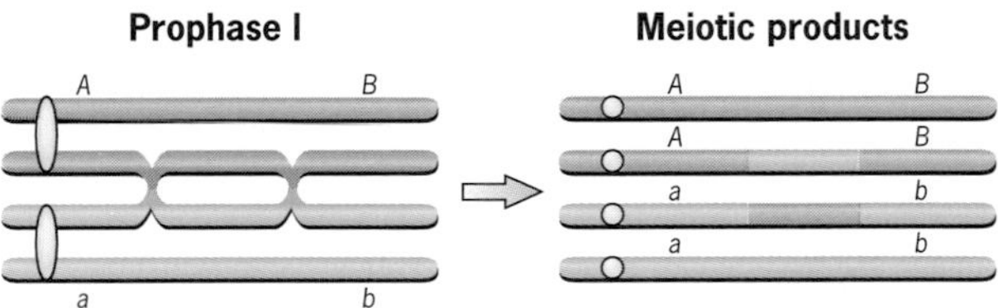

(*a*) Two-strand double crossovers produce all parental chromosomes.

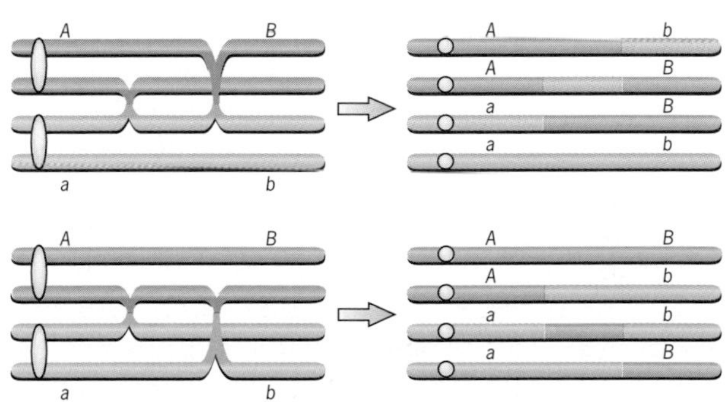

(*b*) Three-strand double crossovers produce half parental, half recombinant chromosomes.

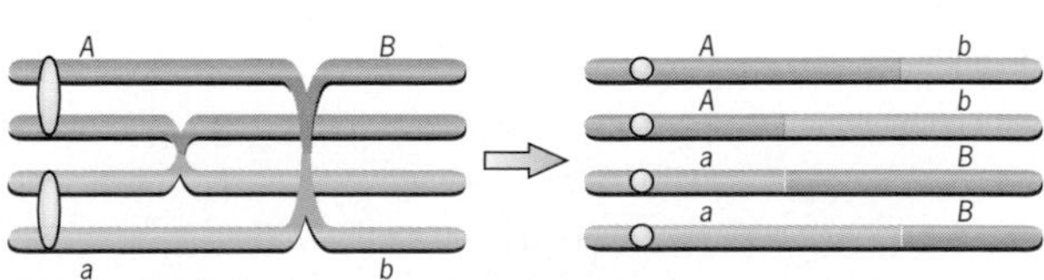

(*c*) Four-strand double crossovers produce all recombinant chromosomes.

Figure 8.18 Consequences of double crossing over between two loci. Two-strand double crossovers produce only nonrecombinant chromosomes, and four-strand double crossovers produce only recombinant chromosomes. Three-strand double crossovers produce half recombinant and half nonrecombinant chromosomes.

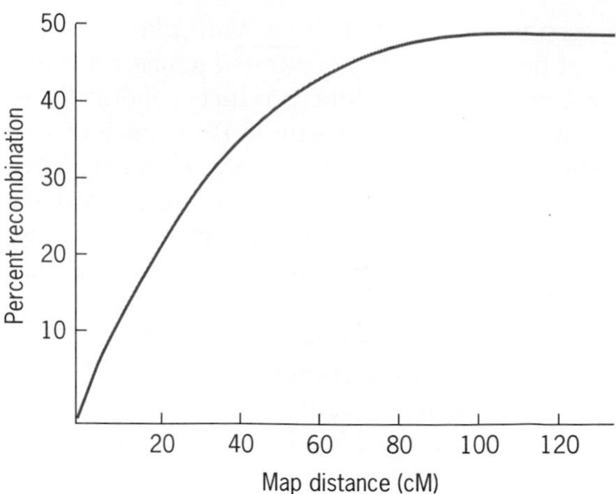

Figure 8.19 Relationship between frequency of recombination and genetic map distance. For values less than 25 cM, there is approximately a linear relationship between percent recombination and map distance; for values greater than 25 cM, the percent recombination underestimates the map distance.

This second example shows that a double crossover may not contribute to the frequency of recombination, even though it contributes to the average number of exchanges on a chromosome. A quadruple crossover would have the same effect. These and other multiple exchanges are responsible for the discrepancy between recombination frequency and genetic map distance. In practice, this discrepancy is small for distances less than 20–25 cM. Over such distances, interference is strong enough to suppress almost all multiple exchanges, and the recombination frequency is a good estimator of the true genetic distance. For values greater than 25 cM, these two quantities diverge, principally because multiple exchanges become much more likely. Figure 8.19 shows the mathematical relationship between recombination frequency and genetic map distance.

Chiasma Frequency and Genetic Map Distance

Recombination between genetic markers is one outcome of crossing over. The other is chiasma formation during prophase of the first meiotic division. Each chiasma is thought to represent the resolution of a crossover that occurred earlier in prophase. Thus, by counting chiasmata, we should be able to estimate the average number of crossovers occurring on a chromosome, which we can then use as an estimate of genetic map length.

Let's suppose, for example, that in a group of 100 cells going through meiosis, 5 cells show five chiasmata in a particular pair of chromosomes, 15 show four chiasmata in this pair, 15 show three, 30 show two, 25 show one, and 10 show none (Figure 8.20). Per cell, the average number of chiasmata in this pair of chromosomes is 2.15. This translates into an average of 1.07 chiasmata per chromatid because each chiasma affects only two of the four chromatids in the chromosome pair. Thus, at the completion of meiosis, the average number of crossovers per chromatid will be 1.07. This implies that the genetic length of this chromosome is 1.07 M, or 107 cM. If 2.15 chiasmata are equivalent to 107 cM, then one chiasma is equivalent to 107/2.15 = 50 cM. In general, we can conclude that each chiasma corresponds to 50 cM on the genetic map.

Meiotic chromosomes	Number chiasmata (A)	Number of cells observed (B)	Product (A × B)
	5	5	25
	4	15	60
	3	15	45
	2	30	60
	1	25	25
	0	10	0
	Total:	100	215

Average number of chiasmata per cell $= \frac{215}{100} = 2.15$

Average number of chiasmata per chromatid $= \frac{2.15}{2} = 1.07$

Genetic length of chromosome = 1.07 Morgans = 107 centiMorgans

Relationship between genetic length and average number of chiasmata $= \frac{107 \text{ cM}}{2.15 \text{ chiasmata}} = 50$ cM/chiasma

Figure 8.20 Computing map length from chiasmata frequencies in meiosis. The average number of chiasmata per chromatid estimates the genetic length of the chromosome.

Genetic Distance and Physical Distance

The procedures for measuring genetic distance are based on the incidence of crossing over between paired chromosomes. Intuitively, we expect that long chro-

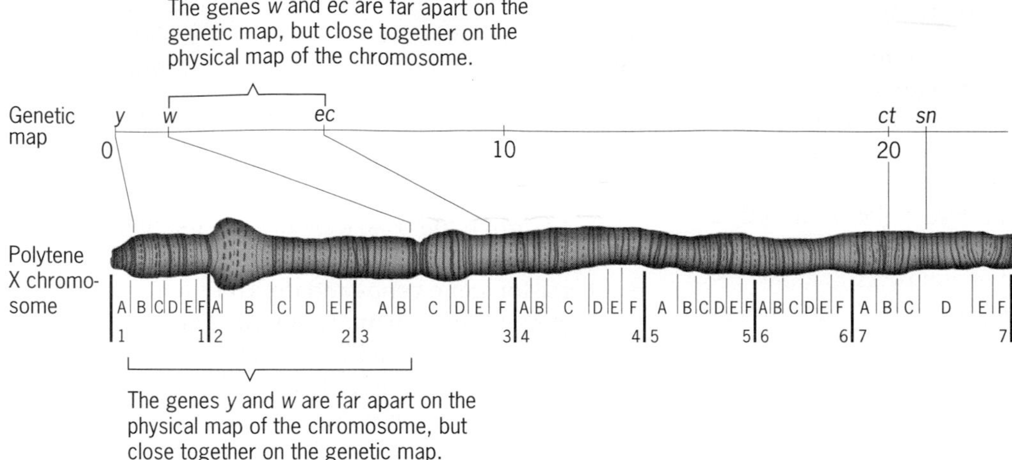

Figure 8.21 Left end of the polytene X chromosome of *Drosophila* and the corresponding portion of the genetic map showing the genes for *yellow* body (*y*), *white* eyes (*w*), *echinus* eyes (*ec*), *cut* wings (*ct*), and *singed* bristles (*sn*).

mosomes should have more crossovers than short ones and that this relationship will be reflected in the lengths of their genetic maps. For the most part, our assumption is true; however, within a chromosome some regions are more prone to crossing over than others. Thus, distances on the genetic map do not correspond exactly to physical distances along the chromosome (Figure 8.21). Crossing over is less likely to occur near the ends of a chromosome and also around the centromere; consequently, these regions are condensed on the genetic map. Other regions, in which crossovers occur more frequently, are expanded.

Even though there is not a uniform relationship between genetic and physical distance, the genetic and physical maps of a chromosome are colinear; that is, particular sites have the same order. Recombination mapping therefore reveals the actual order of the genes along a chromosome. However, it does not tell us the actual physical distances between them.

Key Points: **The genetic maps of chromosomes are based on the average number of crossovers that occur during meiosis. Map distances are estimated by calculating the frequency of recombination between genes in experimental crosses or by counting chiasmata in cytological preparations. One chiasma is equivalent to a distance of 50 cM. Recombination frequencies less than 20 to 25 percent estimate map distance directly. However, frequencies greater than 20 to 25 percent usually underestimate map distance because some of the multiple crossovers that occur do not contribute to the recombination frequency.**

RECOMBINATION AND EVOLUTION

Recombination is an essential feature of sexual reproduction. During meiosis, when chromosomes come together and cross over, there is an opportunity to create new combinations of alleles. Some of these may benefit the organism by enhancing survival or reproductive ability. Over time, such beneficial combinations would be expected to spread through a population and become standard features of the genetic makeup of the species. Meiotic recombination is therefore a way of shuffling genetic variation to potentiate evolutionary change.

Evolutionary Significance of Recombination

We can appreciate the evolutionary advantage of recombination by comparing two species, one capable of reproducing sexually and the other not. Let's suppose that a beneficial mutation has arisen in each species. Over time, we would expect these mutations to spread. Let's also suppose that while they are spreading, another beneficial mutation occurs in a nonmutant individual within each species. In the asexual organism, there is no possibility that this second mutation will be recombined with the first, but in the sexual organism, the two mutations can be recombined to produce a strain that is better than either of the single mutants by itself. This recombinant strain will be able to spread through the whole species population. In evolutionary terms, recombination can allow favorable alleles of different genes to come together in the same organism.

Suppression of Recombination by Inversions

The gene-shuffling effect of recombination can be thwarted by chromosome rearrangements. Crossing over is usually inhibited near the breakpoints of a rearrangement in heterozygous condition, probably because the rearrangement disrupts chromosome pairing. Many rearrangements are therefore associated with a reduction in the frequency of recombination. This effect is most pronounced in inversion heterozygotes because the inhibition of crossing over that occurs near the breakpoints of the inversion is compounded by the selective loss of chromosomes that have undergone crossing over within the inverted region.

To see this recombination-suppressing effect, we consider an inversion in the long arm of a chromosome (Figure 8.22). If a crossover occurs between inverted and noninverted chromatids within the tetrad, it will produce two recombinant chromatids; however, both of these chromatids are likely to be lost during or after meiosis. One of the chromatids lacks a centromere—it is an *acentric* fragment—and will therefore be unable to move to its proper place during anaphase of the first meiotic division. The other chromatid has two centromeres and will therefore be pulled in opposite directions, forming a *dicentric chromatid bridge.* Eventually, this bridge will break and split the chromatid into pieces. Even if the acentric and dicentric chromatids produced by crossing over within the inversion survive meiosis, they are not likely to form viable zygotes. Both of these chromatids are aneuploid—duplicate for some genes and deficient for others—and such aneuploidy is usually lethal. These chromatids will therefore be eliminated by natural selection in the next generation. The net effect of this chromatid loss is to suppress recombination between inverted and noninverted chromosomes in heterozygotes.

Sometimes, however, euploid products result from crossing over between inverted and noninverted chromatids, for example, when two crossovers occur within the inverted region (Figure 8.23). Both of the crossovers must involve the same two chromatids—a so-called two-strand double exchange. If they involve different chromatids, the products of the exchanges will be aneuploid.

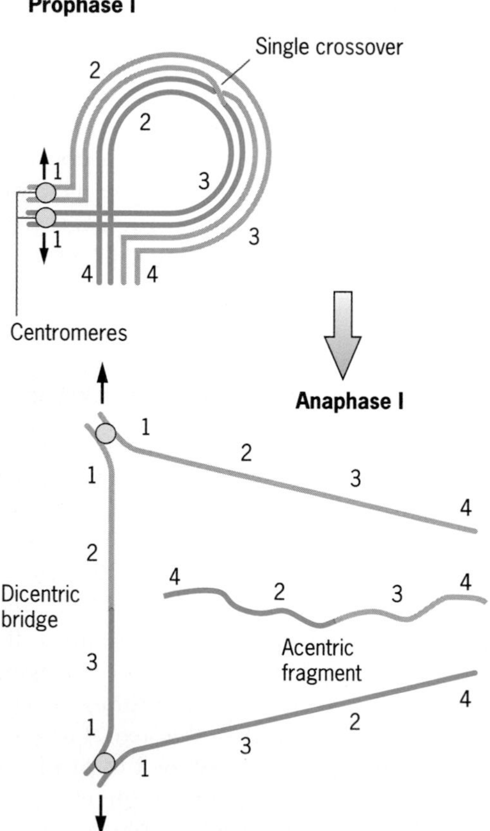

Figure 8.22 Suppression of recombination in an inversion heterozygote. The dicentric (1 2 3 1) and acentric (4 3 2 4) chromosomes formed from the crossover chromatids are aneuploid and will cause inviability in the next generation.

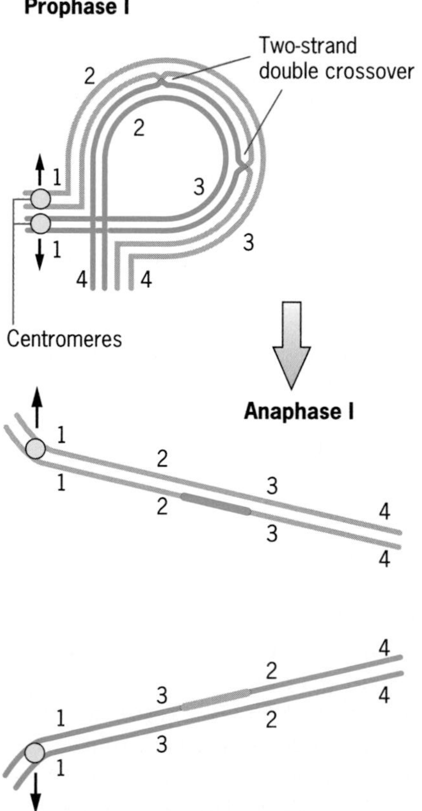

Figure 8.23 Double crossing over in an inversion heterozygote. None of the chromatids from a two-strand double crossover tetrad is aneuploid; consequently, they will be recovered in the next generation.

Geneticists have exploited the recombination-suppressing properties of inversions to keep alleles of different genes together on the same chromosome. Let us assume, for example, that a chromosome that is structurally normal carries the recessive alleles *a*, *b*, *c*, *d* and *e*. If this chromosome is paired with another structurally normal chromosome that carries the corresponding wild-type alleles a^+, b^+, c^+, d^+, and e^+, the recessive and wild-type alleles will be scrambled by recombination. To prevent this scrambling, the chromosome with the recessive alleles can be paired with a wild-type chromosome that has an inversion. Unless double crossovers occur within the inverted region, this structural heterozygosity will suppress recombination. The multiply mutant chromosome can then be transmitted to the progeny as an intact genetic unit.

This recombination-suppressing technique has often been used in experiments with *Drosophila*, where the inverted chromosome usually carries a dominant mutation that permits it to be tracked through a whole series of crosses without cytological examination. Such marked inversion chromosomes are called **balancers** because they allow a mutant chromosome to be kept in heterozygous condition over the inversion.

Genetic Control of Recombination

It is not surprising that a process as important as recombination should be under genetic control. Studies with several organisms, including yeast and *Drosophila*, have demonstrated that recombination involves the products of many genes. Some of these gene products play a role in chromosome pairing, others catalyze the process of exchange, and still others help to rejoin broken chromatid segments. We will consider some of these activities in greater detail in Chapter 14.

One curious phenomenon, which no one has yet explained, is that there is no crossing over in *Drosophila* males. In this regard, *Drosophila* is different from most species, including our own, where crossing over occurs in both sexes. In addition, we know that the amount of recombination varies among species. Perhaps the events that lead to recombination are themselves subject to evolutionary change.

Key Points: **Recombination can bring favorable mutations together in a population. Chromosome rearrangements, especially inversions, can suppress recombination.**

TESTING YOUR KNOWLEDGE

1. R. K. Sakai, K. Akhtar, and C. J. Dubash (1985, *J. of Heredity* 76:140–141) reported data from a set of testcrosses with the mosquito *Anopheles culicifacies*, a vector for malaria in southern Asia. The data involved three mutations: *bw* (*brown* eyes), *c* (*colorless* eyes), and *Blk* (*black* body). In each cross, repulsion heterozygotes were mated to mosquitoes homozygous for the recessive alleles of the genes, and the progeny were scored as having either a parental or a recombinant genotype. Are any of the three genes studied in these crosses linked? If so, construct a map of the linkage relationships.

Cross	Repulsion Heterozygote	Progeny Parental	Progeny Recombinant	Percent Recombination
1	*bw* +/+ *c*	850	503	37.2
2	*bw* +/+ *Blk*	750	237	24.0
3	*c* +/+ *Blk*	629	183	22.5

ANSWER

In each cross, the frequency of recombination is less than 50 percent, so all three loci are linked. To place them on a linkage map, we estimate the distances between each pair of genes from the observed recombination frequencies:

bw —- 24.0 —- *Blk* —- 22.5 —- *c*

37.2

Notice that the recombination frequency between *bw* and *c* (37.2 percent, from Cross 1) is substantially less than the actual distance between these genes (46.5). This shows that for widely separated genes, the recombination frequency underestimates the true map distance.

2. Singed bristles (*sn*), crossveinless wings (*cv*), and vermilion eye color (*v*) are due to recessive mutant alleles of three X-linked genes in *Drosophila melanogaster*. When a female heterozygous for each of the three genes was testcrossed with a singed, crossveinless, vermilion male, the following progeny were obtained:

Class	Phenotype	Number
1	singed, crossveinless, vermilion	3
2	crossveinless, vermilion	392
3	vermilion	34
4	crossveinless	61
5	singed, crossveinless	32
6	singed, vermilion	65
7	singed	410
8	wild-type	3
		Total = 1000

What is the correct order of these three genes on the X chromosome? What are the genetic map distances between *sn* and *cv*, *sn* and *v*, and *cv* and *v*? What is the coefficient of coincidence?

ANSWER

Before attempting to analyze these data, we must establish the genotype of the heterozygous female that produced the eight classes of offspring. We do this by identifying the two parental classes (2 and 7), which are the most numerous in the data. These classes tell us that the heterozygous female had the *cv* and *v* mutations on one of her X chromosomes and the *sn* mutation on the other. Her genotype was therefore (*cv* + *v*)/(+ *sn* +), with the parentheses indicating uncertainty about the gene order.

To determine the gene order, we must identify the double crossover classes among the six types of recombinant progeny. These are classes 1 and 8—the least numerous. They tell us that the *singed* gene is between *crossveinless* and *vermilion*. We can verify this by investigating the effect of a double crossover in a female with the genotype

$$\frac{cv + v}{+ sn +}$$

Two exchanges in this genotype will produce gametes that are either *cv sn v* or + + +, which correspond to classes 1 and 8, the observed double crossovers. Thus, the proposed gene order—*cv sn v*—is correct.

Having established the gene order, we can now determine which recombinant classes represent crossovers between *cv* and *sn* and which represent crossovers between *sn* and *v*.

Crossovers between *cv* and *sn*:

Class:	3		5		1		8		
Number:	34	+	32	+	3	+	3	=	72

Crossovers between *sn* and *v*:

Class:	4		6		1		8		
Number:	61	+	65	+	3	+	3	=	132

We determine the distances between these pairs of genes by calculating the average number of crossovers. Between *cv* and *sn*, the distance is 72/1000 = 7.2 cM, and between *sn* and *v* it is 132/1000 = 13.2 cM. We can estimate the distance between *cv* and *v* as the sum of these values: 7.2 + 13.2 = 20.4 cM. The linkage map of these three genes is therefore:

$$cv\text{—}7.2\text{—}sn\text{—}13.2\text{—}v$$

To calculate the coefficient of coincidence, we use the observed and expected frequencies of double crossovers:

$$c = \frac{\text{observed frequency of double crossovers}}{\text{expected frequency of double crossovers}} = \frac{(0.006)}{(0.072) \times (0.132)} = 0.63$$

which indicates only moderate interference.

3. A woman has two dominant traits, each caused by a mutation in a different gene: cataract (an eye abnormality), which she inherited from her father, and polydactyly (an extra finger), which she inherited from her mother. Her husband has neither trait. If the genes for these two traits are 15 cM apart on the same chromosome, what is the chance that the first child of this couple will have both cataract and polydactyly?

ANSWER

To calculate the chance that the child will have both traits, we first need to determine the linkage phase of the mutant alleles in the woman's genotype. Because she inherited the cataract mutation from her father and the polydactyly mutation from her mother, the mutant alleles must be on opposite chromosomes, that is, in the repulsion linkage phase:

$$\frac{C\ +}{+\ P}$$

For a child to inherit both mutant alleles, the woman would have to produce an egg that carried a recombinant chromosome, *C P*. We can estimate the probability of this event from the distance between the two genes, 15 cM, which, because of interference, should be equivalent to 15 percent recombination. However, only half the recombinants will be *C P*. Thus, the chance that the child will inherit both mutant alleles is (15/2) percent = 7.5 percent.

QUESTIONS AND PROBLEMS

8.1 Mendel did not know of the existence of chromosomes. Had he known, what change might he have made in his Principle of Independent Assortment?

8.2 From a cross between individuals with the genotypes *Cc Dd Ee* × *cc dd ee*, 1000 offspring were produced. The class that was *C- D- ee* included 351 individuals. Are the genes *c*, *d*, and *e* on the same or different chromosomes? Explain.

8.3 If *a* is linked to *b*, and *b* to *c*, and *c* to *d*, does it follow that a recombination experiment would detect linkage between *a* and *d*? Explain.

8.4 Two yeast strains differing in three linked genes were crossed: *A B C* × *a b c*. Among the tetrads that were analyzed, one contained the following spores: *A B C*, *A b C*, *a B c*, and *a b c*. How did this tetrad originate?

8.5 Mice have 19 autosomes in their genome, each about the same size. If two autosomal genes are chosen randomly, what is the chance that they will be on the same chromosome?

8.6 If two loci are 10 cM apart, what proportion of the cells in prophase of the first meiotic division will contain a single crossover in the region between them?

8.7 Genes *a* and *b* are 20 cM apart. An $a^+\ b\ +/\ a\ +\ b^+$ individual was mated with an *a b*/ *a b* individual. (a) Diagram the cross and show the gametes produced by each parent and the genotype of the F_1. (b) What gametes can the F_1 produce, and in what proportions? (c) If the F_1 was crossed to *a b*/ *a b* individuals, what offspring would be expected, and in what proportions? (d) Is this an example of the coupling or repulsion linkage phase? (e) If the F_1 were intercrossed, what offspring would be expected, and in what proportions?

8.8 Answer questions (a)–(e) in the preceding problem under the assumption that the original cross was $a^+\ b/\ a^+\ b \times a\ b^+/\ a\ b^+$.

8.9 If the recombination frequency in the previous two problems were 40 percent instead of 20 percent, what change would occur in the proportions of gametes and testcross progeny?

8.10 A homozygous variety of maize with red leaves and normal seeds was crossed with another homozygous variety with green leaves and tassel seeds. The hybrids were then backcrossed to the green, tassel-seeded variety, and the following offspring were obtained: red, normal 124; red, tassel 126; green, normal 125; green, tassel 123. Are the genes for plant color and seed type linked? Explain.

8.11 A phenotypically wild-type female fruit fly that was heterozygous for genes controlling body color and wing length was crossed to a homozygous mutant male with black body (allele *b*) and vestigial wings (allele *vg*). The cross produced the following progeny: gray body, normal wings 126; gray body, vestigial wings 24; black body, normal wings 26; black body, vestigial wings 124. Do these data indicate linkage between the genes for body color and wing length? What is the frequency of recombination? Diagram the cross, showing the arrangement of the genetic markers on the chromosomes.

8.12 Another phenotypically wild-type female fruit fly heterozygous for the two genes mentioned in the previous problem was crossed to a homozygous black, vestigial male. The cross produced the following progeny: gray body, normal wings 23; gray body, vestigial wings 127; black body, normal wings 124; black body, vestigial wings 26. Do these data indicate linkage? What is the frequency of recombination? Diagram the cross, showing the arrangement of the genetic markers on the chromosomes.

8.13 In rabbits, the dominant allele *C* is required for colored fur; the recessive allele *c* makes the fur colorless (albino). In the presence of at least one *C* allele, another gene determines whether the fur is black (*B*, dominant) or brown (*b*, recessive). A homozygous strain of brown rabbits was crossed with a homozygous strain of albinos. The F_1 were then crossed to homozygous double recessive rabbits, yielding the following results: black 34; brown 66; albino 100. Are the genes *b* and *c* linked? What is the frequency of recombination? Diagram the crosses, showing the arrangement of the genetic markers on the chromosomes.

8.14 In tomatoes, tall vine (*D*) is dominant over dwarf (*d*), and spherical fruit shape (*P*) is dominant over pear shape (*p*). The genes for vine height and fruit shape are linked with 20 percent recombination between them. One tall plant (I) with spherical fruit was crossed with a dwarf, pear-fruited plant. The cross produced the following results: tall, spherical 81; dwarf, pear 79; tall, pear 22; dwarf spherical 17. Another tall plant with spherical fruit (II) was crossed with the dwarf, pear-fruited plant, and the following results were obtained: tall, pear 21; dwarf, spherical 18; tall, spherical 5; dwarf, pear 4. Diagram these two crosses, showing the genetic markers on the chromosomes. If the two tall plants with spherical fruit were crossed with each other, that is, I × II, what phenotypic classes would you expect from the cross, and in what proportions?

8.15 In *Drosophila*, genes *a* and *b* are located at positions 22.0 and 42.0 on chromosome 2, and genes *c* and *d* are located at positions 10.0 and 25.0 on chromosome 3. A fly homozygous for the wild-type alleles of these four genes was crossed with a fly homozygous for the recessive alleles, and the F_1 daughters were backcrossed to their quadruply recessive fathers. What offspring would you expect from this backcross, and in what proportions?

8.16 In *Drosophila*, the genes *sr* (*stripe* thorax) and *e* (*ebony* body) are located at 62 and 70 cM, respectively, from the left end of chromosome 3. A striped female homozygous for e^+ was mated with an ebony male homozygous for sr^+. All the offspring were phenotypically wild-type (gray body and unstriped). (a) What kind of gametes will be produced by the F_1 females, and in what proportions? (b) What kind of gametes will be produced by the F_1 males, and in what proportions? (c) If the F_1 females are mated with striped, ebony males, what offspring are expected, and in what proportions? (d) If the F_1 males and females are intercrossed, what offspring would you expect from this intercross, and in what proportions?

8.17 The *Drosophila* genes *vg* (*vestigial* wings) and *cn* (*cinnabar* eyes) are located at 67.0 and 57.0, respectively, on chromosome 2. A female from a homozygous strain of vestigial flies was crossed with a male from a homozygous strain of cinnabar flies. The F_1 hybrids were phenotypically wild-type (long wings and reddish-brown eyes). (a) How many different kinds of gametes could the F_1 females produce, and in what proportions? (b) If these females are mated with cinnabar, vestigial males, what kinds of progeny would you expect, and in what proportions?

8.18 In *Drosophila*, the genes *st* (*scarlet* eyes), *ss* (*spineless* bristles), and *e* (*ebony* body) are located on chromosome 3, with map positions as indicated:

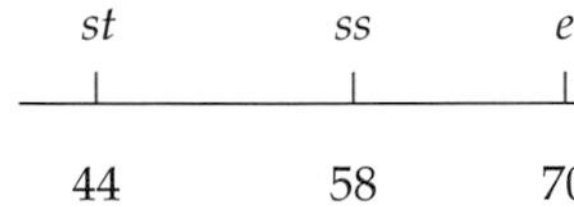

Each of these mutations is recessive to its wild-type allele (st^+, reddish-brown eyes; ss^+, smooth bristles; e^+, gray body). Phenotypically wild-type females with the genotype $st\ ss\ e^+/\ st^+\ ss^+\ e$ were crossed with triply recessive males. Predict the phenotypes of the progeny and the frequencies with which they will occur assuming (a) no interference and (b) complete interference.

8.19 In maize, the genes *Pl* for purple leaves (dominant over *pl* for green leaves), *sm* for salmon silk (recessive to *Sm* for yellow silk), and *py* for pigmy plant (recessive to *Py* for normal size plant) are on chromosome 6, with map positions as shown:

Pl *sm* *py*

45 55 65

Hybrids from the cross *Pl sm py/Pl sm py* × *pl Sm Py/ pl Sm Py* were testcrossed with *pl sm py/pl sm py* plants. Predict the phenotypes of the offspring and their frequencies assuming (a) no interference and (b) complete interference.

8.20 In maize, the genes *Tu*, *j2*, and *gl3* are located on chromosome 4 at map positions 101, 106, and 112, respectively. If plants homozygous for the recessive alleles of these genes are crossed with plants homozygous for the dominant alleles, and the F_1 plants are testcrossed to triply recessive plants, what genotypes would you expect, and in what proportions? Assume that interference is complete over this map interval.

8.21 A *Drosophila* geneticist made a cross between females homozygous for three X-linked recessive mutations (*y*, *yellow* body; *ec*, *echinus* eye shape; *w*, *white* eye color) and wild-type males. He then mated the F_1 females to triply mutant males and obtained the following results:

Females	Males	Number
+ + +/*y ec w*	+ + +	475
y ec w/*y ec w*	*y ec w*	469
y + +/*y ec w*	*y* + +	8
+ *ec w*/*y ec w*	+ *ec w*	7
y + *w*/*y ec w*	*y* + *w*	18
+ *ec* +/*y ec w*	+ *ec* +	23
+ + *w*/*y ec w*	+ + *w*	0
y ec +/*y ec w*	*y ec* +	0

Determine the order of the three loci *y*, *ec*, and *w*, and estimate the distances between them on the linkage map of the X chromosome.

8.22 A *Drosophila* geneticist crossed females homozygous for three X-linked mutations (*y*, *yellow* body; *B*, *bar* eye shape; *v*, *vermilion* eye color) to wild-type males. The F_1 females, which had gray bodies and bar eyes with reddish-brown pigment, were then crossed to *y* B^+ *v* males, yielding the following results:

Phenotype	Number
yellow, bar, vermilion wild-type	546
yellow bar, vermilion	244
yellow, vermilion bar	160
yellow, bar vermilion	25

Determine the order of these three loci on the X chromosome and estimate the distances between them.

8.23 Female *Drosophila* heterozygous for three recessive mutations *e* (*ebony* body), *st* (*scarlet* eyes), and *ss* (*spineless* bristles) were testcrossed, and the following progeny were obtained:

Phenotype	Number
wild-type	67
ebony	8
ebony, scarlet	68
ebony, spineless	347
ebony, scarlet, spineless	78
scarlet	368
scarlet, spineless	10
spineless	54

(a) What indicates that the genes are linked? (b) What was the genotype of the original heterozygous females? (c) What is the order of the genes? (d) What is the map distance between *e* and *st*? (e) Between *e* and *ss*? (f) What is the coefficient of coincidence? (g) Diagram the crosses in this experiment.

8.24 Consider a female *Drosophila* with the following X chromosome genotype:

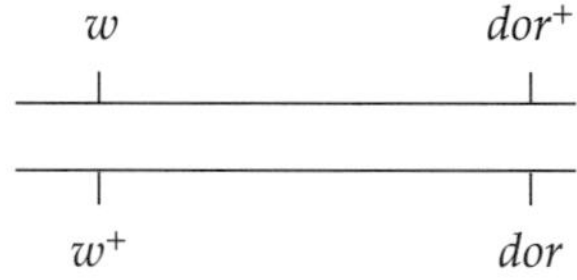

The recessive alleles *w* and *dor* cause mutant eye colors (white and deep orange, respectively). However, *w* is epistatic over *dor*; that is, the genotypes *w dor*/*Y* and *w dor*/*w dor* have white eyes. If there is 40 percent recombination between *w* and *dor*, what proportion of the sons from this heterozygous female will show a mutant phenotype? What proportion will have either red or deep orange eyes?

8.25 Assume that in *Drosophila* there are three genes *x*, *y*, and *z*, with each mutant allele recessive to the wild-type allele. A cross between females heterozygous for these three loci and wild-type males yielded the following progeny:

Females	+ + +	1010
Males	+ + +	39
	+ + *z*	430
	+ *y z*	32
	x + +	27
	x y +	441
	x y z	31
	Total =	2010

Using these data, construct a linkage map of the three genes and calculate the coefficient of coincidence.

8.26 A normal woman with a color blind father married a normal man, and their first child, a boy, had hemophilia. Both color blindness and hemophilia are due to X-linked recessive mutations, and the relevant genes are separated by 10 cM. This couple plans to have a second child. What is the probability that it will have hemophilia? Color blindness?

Both hemophilia and color blindness? Neither hemophilia nor color blindness?

8.27 In the nematode *Caenorhabditis elegans*, the linked genes *dpy* (*dumpy* body) and *unc* (*uncoordinated* behavior) recombine with a frequency *P*. If a repulsion heterozygote carrying recessive mutations in these genes is self-fertilized, what fraction of the offspring will be both dumpy and uncoordinated?

8.28 In the following testcross, genes *a* and *b* are 20 cM apart, and genes *b* and *c* are 10 cM apart: $a + c / + b + \times a\ b\ c / a\ b\ c$. If the coefficient of coincidence is 0.5 over this interval on the linkage map, how many triply homozygous recessive individuals are expected among 1000 progeny?

8.29 *Drosophila* females heterozygous for three recessive mutations, *a*, *b*, and *c*, were crossed to males homozygous for all three mutations. The cross yielded the following results:

Phenotype	Number
+ + +	75
+ + *c*	348
+ *b* *c*	96
a + +	110
a *b* +	306
a *b* *c*	65

Construct a linkage map showing the correct order of these genes and estimate the distances between them.

8.30 A chromosome is 120 cM long. On the average, how many chiasmata will occur when it goes through meiosis?

8.31 The total map length of the *Drosophila* genome is about 270 cM. On the average, how many chiasmata will occur in an oocyte going through meiosis?

8.32 One of the metacentric chromosomes in a species of bean typically has one chiasma in each of its arms during meiosis. What is the genetic map length of this chromosome?

8.33 Two strains of maize, M1 and M2, are homozygous for four recessive mutations, *a*, *b*, *c*, and *d*, on one of the large chromosomes in the genome. Strain W1 is homozygous for the dominant alleles of these mutations. Hybrids produced by crossing M1 and W1 yield many different classes of recombinants, whereas hybrids produced by crossing M2 and W1 do not yield any recombinants at all. What is the difference between M1 and M2?

8.34 A *Drosophila* geneticist has identified a strain of flies with a large inversion in the left arm of chromosome 3. This inversion includes two mutations, *e* (*ebony* body) and *cd* (*cardinal* eyes), and is flanked by two other mutations, *sr* (*stripe* thorax) on the right and *ro* (*rough* eyes) on the left. The geneticist wishes to replace the *e* and *cd* mutations inside the inversion with their wild-type alleles; he plans to accomplish this by recombining the multiply mutant, inverted chromosome with a wild-type, inversion-free chromosome. What event is the geneticist counting on to achieve his objective? Explain.

BIBLIOGRAPHY

Bridges, C. B., and T. M. Olbrycht. 1926. The multiple stock "XPLE" and its use. *Genetics* 11:41–56.

Carlson, E. A. 1966. *The Gene: A Critical History*. W. B. Saunders, Philadelphia.

Creighton, H. B., and B. McClintock. 1931. A correlation of cytological and genetical crossing over in *Zea mays*. *Proc. Natl. Acad. Sci. U.S.A.* 17:492–497.

Lindsley, D. L., and G. Zimm. 1992. *The Genome of* Drosophila melanogaster. Academic Press, New York.

Peters, J. A., ed. 1959. *Classic Papers in Genetics*. Prentice-Hall, Englewood Cliffs, NJ.

Sturtevant, A. H., and G. W. Beadle. 1939. *An Introduction to Genetics*. W. B. Saunders, Philadelphia. (Republished in 1962 by Dover, New York).

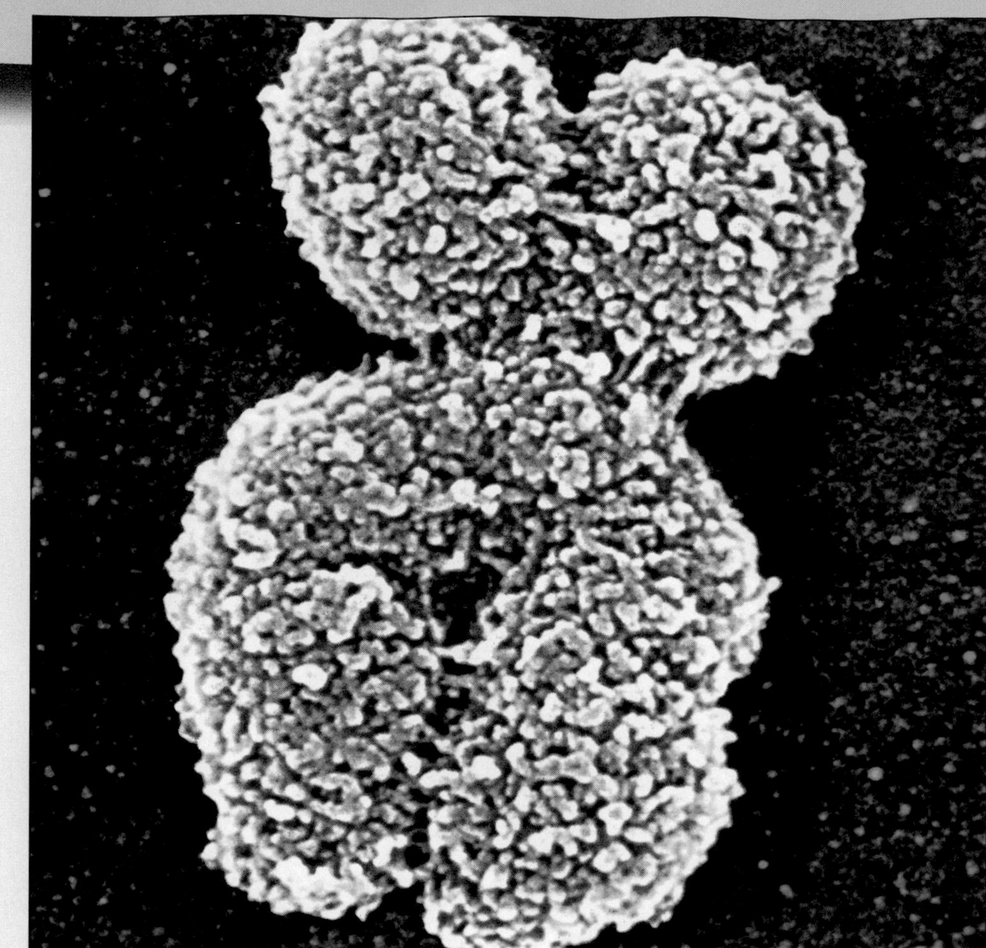

Highly magnified image of a human mitotic metaphase chromosome.

9

Advanced Linkage Analysis

CHAPTER OUTLINE

Detection of Linkage in Experimental Organisms
Specialized Mapping Techniques
Linkage Analysis in Humans

Huntington's Disease: A Challenge in Gene Mapping

In 1872 a young boy accompanied his physician father as he made his rounds among the isolated villages of rural Long Island. As they passed through the countryside, they chanced upon a sight so mesmerizing that the young, impressionable boy never forgot it. Two women moved through the tall grass blanketing the clearing. Both were thin to the point of emaciation, and both were twisting, bowing, and grimacing. The father paused, spoke to the women, and left. It was a disturbing sight, but it made an enormous impact on the youngster.

The boy was George Huntington, and so fascinated was he by this encounter that he pondered it for many years. He wanted to understand this bizarre and terrifying malady. Young Huntington followed in his father's footsteps and entered into a career in medicine. His perseverance paid off, and the disease he observed in those two women today bears his name: **Huntington's disease.** Huntington concluded that the disease is inherited and that it involves a gradual deterioration of the central nervous system. This is a late-onset disease, with symptoms first appearing in approximately the third or fourth decade of life. Limb movement becomes spastic and uncontrolled. Mental functions deteriorate. Patients become bedridden and helpless. Death follows about 10 to 15 years after the symptoms make their first appearance. There is no cure, and little relief is possible once the disease begins its course.

Though rare, this disease has fascinated researchers for decades. Because its biochemical basis is unknown, researchers first have tried to locate the gene, then to isolate it, and finally to determine what cell functions the gene controls. This strategy is working. In 1983 the gene for Huntington's disease was located on chromosome 4. Ten years later it was isolated and the protein that it encodes was determined. The next step is to elucidate this protein's function. Clearly, the key advance in the attack on Huntington's disease was to map the gene.

In this chapter we consider some advanced techniques in gene mapping, including procedures to detect linkage in experimental organisms and to position genes with respect to chromosomal structures such as centromeres and bands. We also examine various strategies for mapping genes in human beings.

DETECTION OF LINKAGE IN EXPERIMENTAL ORGANISMS

During the hundred or so years that genetics has been a formal science, researchers have studied linkage in numerous organisms including peas, fruit flies, bread mold, mice, corn, yeast, mosquitoes, dogs, chickens, nematodes, tomatoes, and human beings. The most intensively studied organisms have been fungi—especially yeast—and fruit flies. In this section we consider some of the techniques that have been developed to detect linkage in these organisms.

Tetrad Analysis to Detect Linkage in Fungi

Fungi in the class Ascomycetes are unusual because all four products of a single meiosis are retained in the ascus sac. Geneticists have exploited this property to study the events of meiosis, including crossing over and recombination. Because these fungi are ordinarily haploid, each meiotic product, or ascospore, germinates into a new organism. Phenotypic analysis of these organisms can therefore provide information about the genotypes of the ascospores from which they were derived.

Ascospore formation in *Saccharomyces cerevisiae*, or baker's yeast, reflects the meiotic process (Figure 9.1). Each ascospore receives one of the chromatids that was

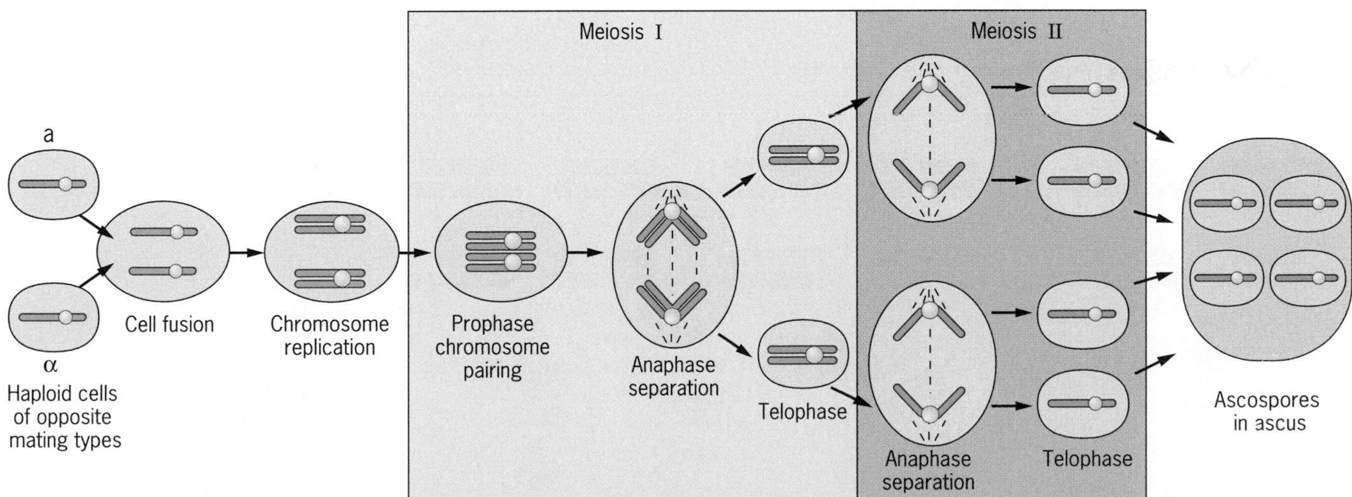

Figure 9.1 Formation of tetrads in *Saccharomyces cerevisiae*, or baker's yeast. All four products of meiosis are recovered in the ascus.

present in a tetrad during prophase of the first meiotic division. The ascospores in an ascus therefore contain all four of the chromatids from a tetrad.

We detect linkage in yeast by analyzing tetrads of ascospores. A cross between two yeast strains with mutations on different chromosomes produces different classes of tetrads (Figure 9.2). We distinguish between alleles by using upper and lower case letters; the lower case letters represent mutants, and the upper case letters represent wild-type. The segregation and independent assortment of mutations on different chromosomes produce two main types of asci. In one type of ascus, two of the ascospores are genotypically like one of the parents and two are like the other parent. An ascus with this pattern is called a **parental ditype** ascus. In the other type, all the spores are recombinant; two of the spores show one genotype, and two show the reciprocal genotype. Neither of the two genotypes is like the parental genotypes; thus, the ascus is called a **nonparental ditype** ascus. Because different chromosomes assort independently during the first meiotic division, a cross involving unlinked genes produces approximately equal frequencies of parental and nonparental ditype asci.

The segregation and independent assortment of unlinked genes can also produce a third type of ascus, called a **tetratype.** In this type there are four kinds of ascospores, one that is like each of the parents and two that are recombinant. Such asci occur whenever there is crossing over between one of the genes and its centromere (Figure 9.2). The existence of tetratype asci does not, however, break the rule that for unlinked genes, parental and nonparental ditype asci occur with equal frequency.

A cross between two yeast strains with mutations on the same chromosome produces different results (Figure 9.3). If the mutations are tightly linked, we expect most of the asci to show the parental ditype pattern. The only exceptions arise from crossing over between the mutant loci. For example, a single crossover between these loci will produce an ascus with four kinds of ascospores: one spore like one parent, one like the other, and two with recombinant genotypes. Such an ascus shows the tetratype pattern. If a double crossover occurs between the two loci, the resulting ascospore pattern will be either ditype or tetratype, depending on which chromatids are involved in the exchanges. If the exchanges are confined to only two of

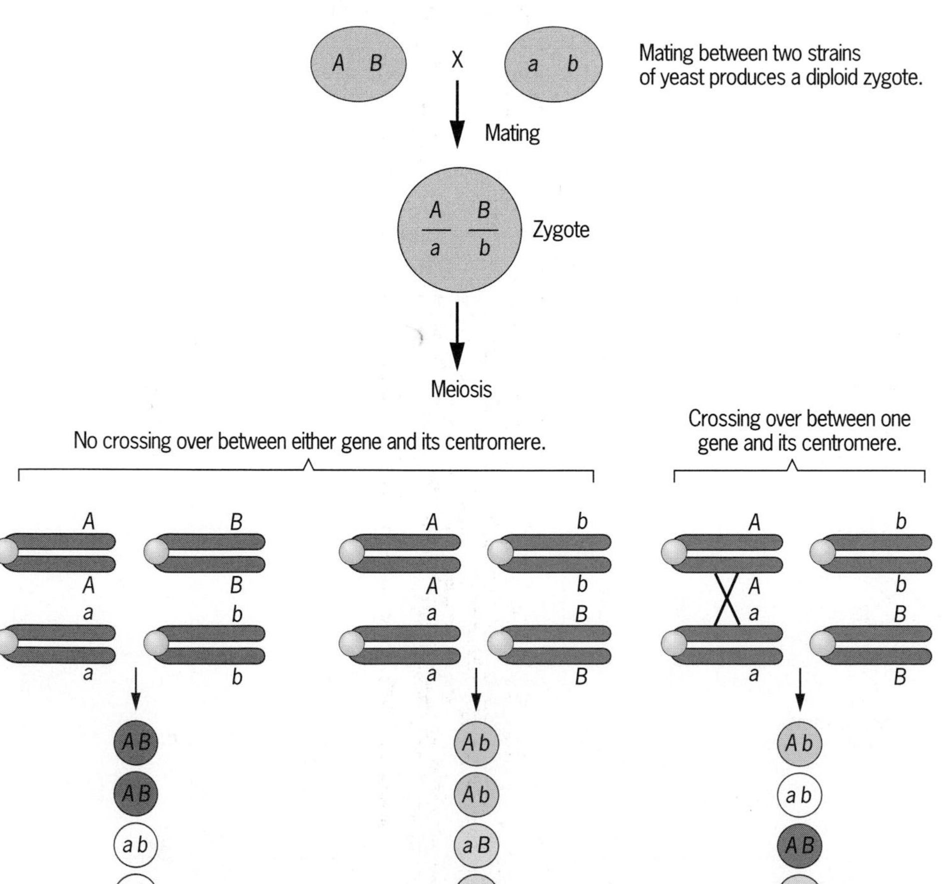

Figure 9.2 Cross with unlinked yeast mutations. The frequencies of parental and nonparental ditype asci are expected to be equal.

the chromatids in the tetrad (a two-strand double crossover), the resulting ascus will have the parental ditype pattern. If the exchanges affect three of the chromatids (a three-strand double crossover), the ascus will have a tetratype pattern, and if they affect four of the chromatids (a four-strand double crossover), it will have a nonparental ditype pattern. Nonparental ditype asci can be produced only by a four-strand double crossover between the mutant loci. Even for loosely linked genes, this is a rare event. Consequently, a low frequency of nonparental ditype asci is considered strong evidence for genetic linkage.

Once we establish linkage, we can use the tetrad data to estimate the distance between the genes. A cross between two yeast strains illustrates how genetic distances are estimated (Figure 9.4). One strain carries two recessive mutations, *py* and *th*, which prevent growth unless the vitamins pyridoxine and thiamin are added to the medium; the *py* mutation causes the pyridoxine requirement, and the *th* mutation causes the thiamin requirement. The other strain, which can grow without any vitamin supplements, carries the wild-

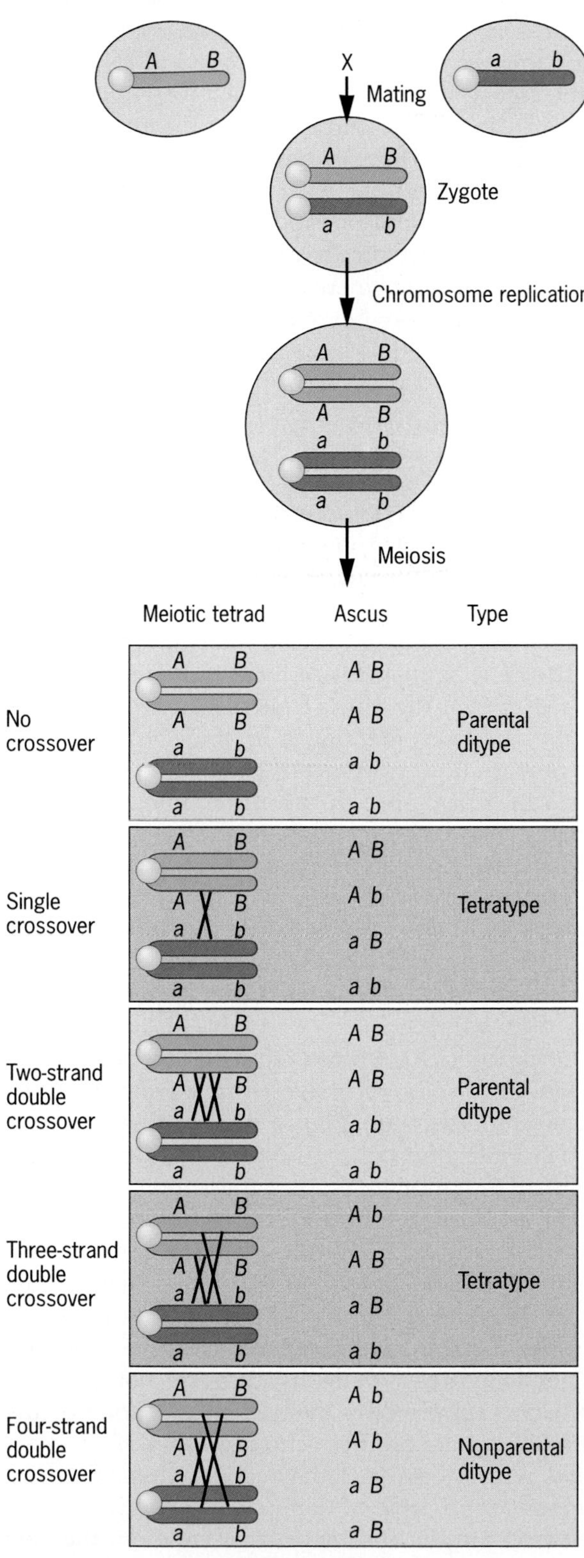

Figure 9.3 Cross with linked yeast mutations. The frequencies of parental and nonparental ditype asci are expected to be unequal. The nonparental ditype asci, which are produced by four-strand double crossovers, should be much less numerous than the parental ditype asci.

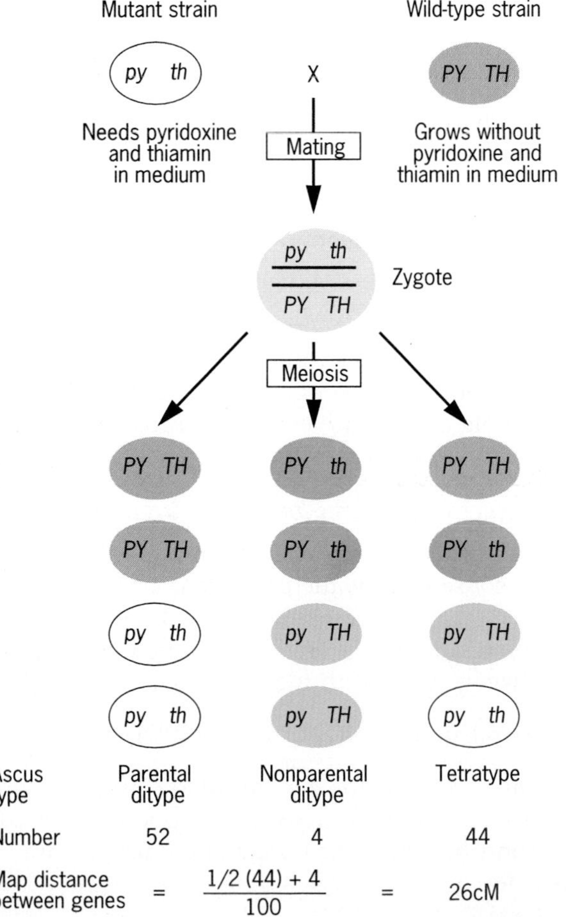

Figure 9.4 Tetrad analysis with two linked yeast mutations, *py* (pyridoxine requirement) and *th* (thiamin requirement).

type alleles of these two mutations, *PY* and *TH*. Diploid cells formed by crossing the two haploid strains were induced to go through meiosis, and the ascospores from 100 of the resulting tetrads were carefully isolated and grown into independent yeast colonies. Cells from each colony were then separately tested for their ability to grow in the absence of pyridoxine or thiamin, thereby revealing the genotype of the spore from which they came. When all the tests were finished, each tetrad of spores could be classified into one of the three possible types.

The data indicate that the two genes are linked because nonparental ditype tetrads occur much less frequently than do parental ditype tetrads. To calculate the distance between the genes, we must estimate the average number of crossovers between them on a chromosome. A simple estimate is the frequency of recombination, which we obtain from the formula

$$\text{Recombination frequency} = [(1/2)\ T + \text{NPD}]/\text{total}$$

Here T and NPD are the observed numbers of tetratype and nonparental ditype tetrads. We multiply T by one-half because only half the chromatids in tetratype tetrads are genetically recombinant. However, all the chromatids in a nonparental ditype tetrad are recombinant; thus, we count these tetrads fully in the formula. We do not count the parental ditype tetrads because none of their chromatids is genetically recombinant.

Applying the formula, we find that

$$\text{Recombination frequency} = [(1/2)\ (44) + 4]/\text{total} = 0.26\ \text{Morgan}$$

or a distance of 26 centiMorgans.

This mapping procedure usually underestimates the true distance between genes, albeit slightly. Two-strand double crossovers, which produce parental ditype tetrads, are not counted at all, and three-strand double crossovers, which produce tetratype tetrads, are not counted correctly. In these latter tetrads, the average number of exchanges per chromatid is 3/4 rather than 1/2; consequently, the coefficient 1/2 in the formula is too small. To compensate for these shortcomings, some researchers prefer to estimate the map distance between genes with the formula

$$\text{Map distance} = [(1/2)\ T + 3\ \text{NPD}]/\text{total}$$

This formula corrects for the failure to count the two- and three-strand double crossovers properly by inflating the contribution of the nonparental ditype tetrads. In this case,

$$\text{Map distance} = [(1/2)\ (44) + 3\ (4)]/100 = 0.26\ \text{Morgan} = 34\ \text{centiMorgans}$$

Tetrad analysis is also possible for several other experimental organisms, including *Chlamydomonas reinhardii*, a unicellular green alga. In a later section, we consider a special application of tetrad analysis involving the Ascomycete *Neurospora crassa*. In this fungus, the ascospores are arranged in a way that reflects the progression of the chromosomes through the two meiotic divisions. These **ordered tetrads** permit mapping of the loci responsible for chromosome movement, that is, the centromeres. In yeast and *Chlamydomonas*, where the tetrads are unordered, it is also possible to map centromeres, but the statistical procedures for doing so are beyond the scope of this book.

Balancer Chromosome Technique to Assign a Gene to a Chromosome in *Drosophila*

Nearly a century of intensive work has made *Drosophila melanogaster* one of the best understood organisms for genetics research. Hundreds of genes have been mapped on its four chromosomes, and many structural aberrations in the chromosomes have been identified. The availability of these mutants and aberrations in *Drosophila* stock centers has greatly facilitated the isolation, identification, and localization of many new mutations in this organism.

One analytical procedure for mapping genes takes advantage of the recombination-suppressing effects of inversions (see Chapter 8). These aberrations are present in balancer chromosomes, which also carry dominant markers to allow them to be followed through a series of crosses. A new mutation can be localized to a chromosome on the basis of how it segregates in crosses with balancers.

The procedure involves crossing the new mutation—assumed to be recessive—to flies that carry the balancer chromosomes (Figure 9.5). In this example, the balancer second chromosome is marked with *Cy*, a dominant mutation for Curl*y* wings, and the balancer third chromosome is marked with *Tb*, a dominant mutation for Tu*b*by body. The homologs of these chromosomes also carry dominant markers—*Pm* for Plu*m* eyes on the second chromosome and *Sb* for Stu*b*ble bristles on the third chromosome. However, because they lack extensive inversions, they are not good recombination-suppressors themselves. All four of the dominant markers on these chromosomes are also associated with recessive lethal effects; thus, flies homozygous for any one of them cannot survive.

To ascertain the chromosomal location of the new mutation, we cross females homozygous for the new mutation to males carrying the balancer chromosomes. If all the sons from this cross show the mutant phenotype, we immediately know that the mutation is X-linked; otherwise, it must be autosomal. Should it turn

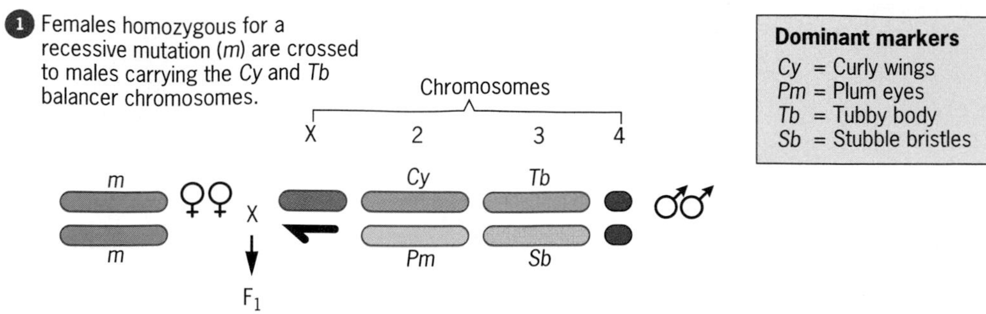

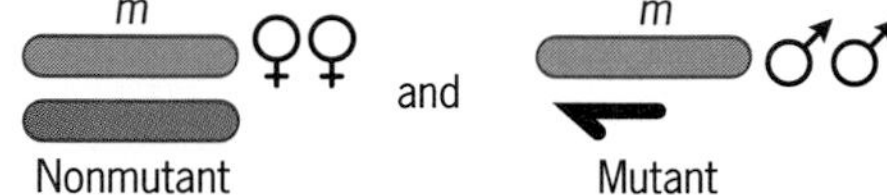

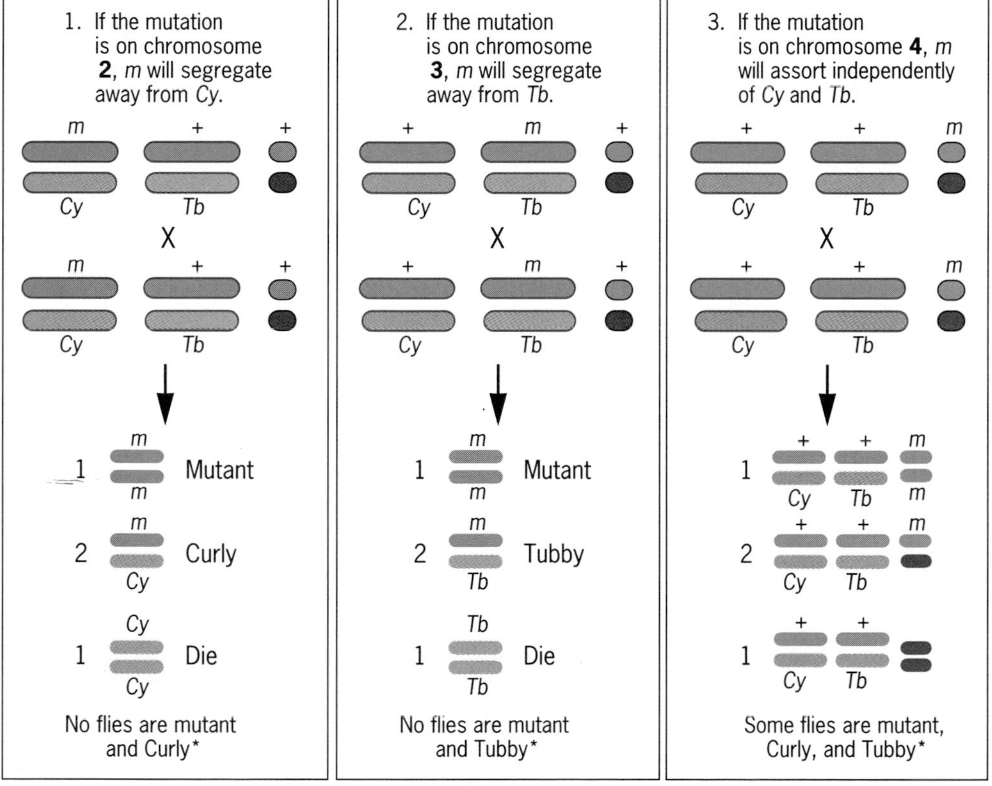

Figure 9.5 Balancer chromosome technique to assign a gene marked by the recessive mutation *m* to a chromosome in *Drosophila*. The balancer chromosomes are marked by the dominant mutations *Cy* (Curl*y* wings) and *Tb* (*T*u*b*by body), on chromosomes 2 and 3, respectively.

out to be autosomal, we determine which autosome carries the mutation by intercrossing Curly, Tubby F_1 flies. A mutation on the second chromosome will segregate away from *Cy* because the *Cy* balancer suppresses recombination. The result is that none of the F_2 progeny will show the mutant *and* Curly phenotypes simultaneously. Segregation of a mutation on the third chromosome away from *Tb* prevents any F_2 flies from showing the mutant *and* Tubby phenotypes simultaneously. If the mutation is on the fourth chromosome, it will assort independently of both balancers and yield some flies that are simultaneously mutant, Curly, and Tubby. Analysis of the F_2 phentoypes can therefore reveal which of the four *Drosophila* chromosomes carries the new mutation. Testcrosses can then be designed to map its position precisely.

Key Points: **In organisms amenable to tetrad analysis, linkage is indicated when parental ditype tetrads are more frequent than nonparental ditype tetrads. In *Drosophila*, genes can be assigned to chromosomes by analyzing crosses involving specially marked balancer chromosomes.**

SPECIALIZED MAPPING TECHNIQUES

In some fungi, the ascospores are arranged in an ascus sac in a precise sequence. This enables us to map the centromere of the chromosome. In this section, we discuss how to map centromeres by analyzing ordered tetrads in the fungus *Neurospora crassa*. We also discuss procedures for aligning the genetic and cytological maps of *Drosophila* chromosomes.

Centromere Mapping with Ordered Tetrads in *Neurospora*

The bread mold *Neurospora crassa* (Figure 9.6) grows as a branching mass of haploid filaments called a *mycelium*. Cells from different *Neurospora* strains, if they belong to different mating types, can fuse to form a diploid cell. This diploid cell undergoes meiosis to produce four haploid nuclei within a narrow ascus. Constraints of space force these nuclei to separate in a linear fashion according to the way in which they were created on the meiotic spindles. Each of the four nuclei divides mitotically to produce two twin nuclei, which remain next to each other in the ascus. The end result is a set of eight nuclei representing the four meiotic products; each product is present twice as a result of the mitotic doubling that occurred after meiosis. These eight nuclei are then partitioned into separate cells, which develop into ascospores capable of germinating into new organisms. The genetic significance of this system is that the order of the ascospores in the ascus reflects the way in which the chromatids segregated during meiosis. Through careful dissection, each of the ascospores can be removed from the ascus and grown to determine its genotype.

The ordered array of ascospores makes it possible to map the loci responsible for chromosome movement during meiosis. These loci, or centromeres, are the

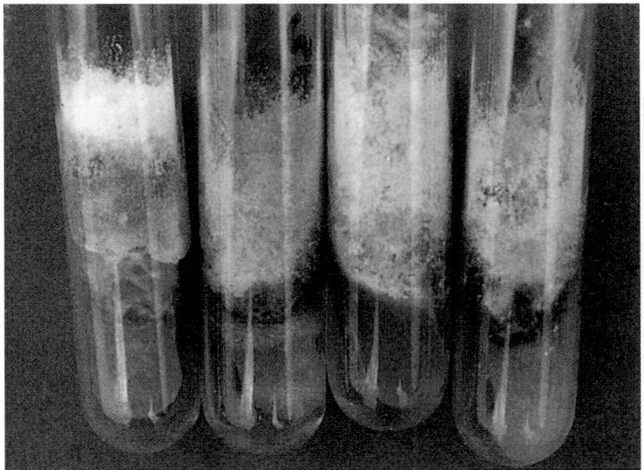

(*a*)

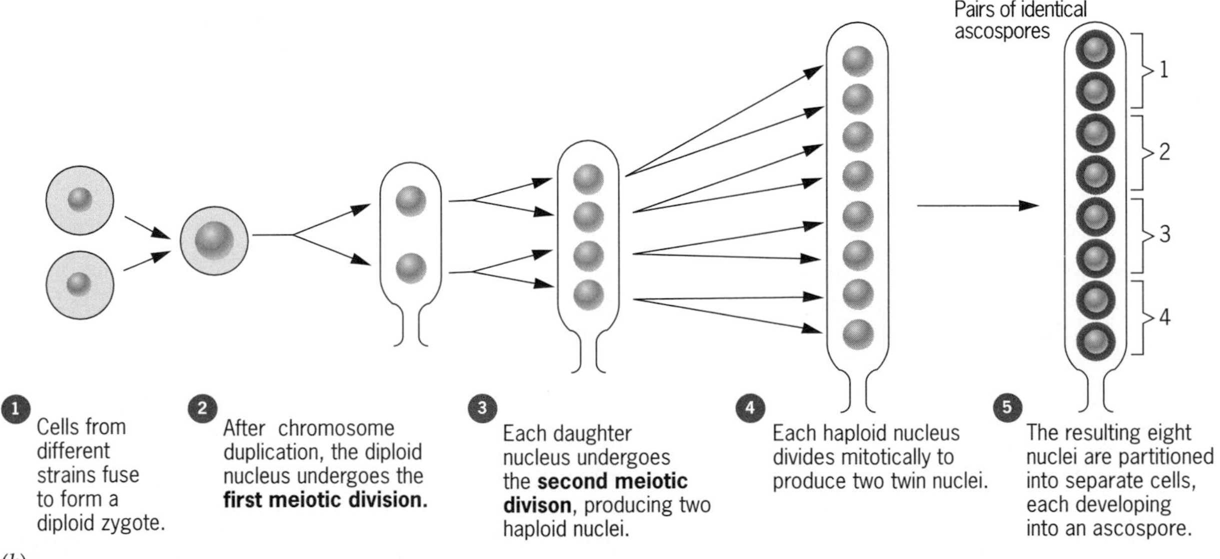

(*b*)

Figure 9.6 (*a*) Cultures of *Neurospora crassa*. (*b*) Production of ordered tetrads of ascospores in *Neurospora*.

points at which spindle fibers attach to the chromosomes. Although the centromeres have no phenotype themselves, we can map them by studying the segregation of nearby genes. Figure 9.7 illustrates the principles.

To map the centromere, let's assume that the *A* locus is a short distance away from the centromere and that we can distinguish its two alleles, *A* and *a*, in haploid *Neurospora*. (They might, for example, produce different pigments in the ascospores; see Figure 9.8.) A heterozygous *Neurospora* cell going through meiosis can segregate these alleles in two different ways. One type of segregation produces an ascus in which the ascospores with the *A* allele are all at one end and those with the *a* allele are at the other; this is called the **first division segregation** pattern. The other type produces an alternating array of ascospores; this is called the **second division segregation** pattern.

The genetic basis for these two patterns is simple. The first division segregation pattern occurs when the *A* and *a* alleles separate in the first meiotic division, that is, there is no crossing over. The sister chromatids with the *A* allele move to one pole, and the sister chromatids with the *a* allele move to the other. Because this positioning takes place within the narrow confines of the developing ascus, the alleles become "stacked" in a linear array. This array, which is preserved through the second meiotic division and the subsequent mitotic division, generates a total of eight ascospores. The four that carry *A* wind up at one end, and the four that carry *a* wind up at the other. The set of four that goes to the top of the ascus will depend on how the paired *A* and *a* alleles were oriented on the spindle of the first meiotic division.

The second division segregation pattern occurs when there is a crossover between the *A* locus and its centromere in the first meiotic division. Such a crossover puts both *A* and *a* into each of the daughter nuclei generated by that division. Then, in the second meiotic division, these alleles segregate from each other, creating an alternating pattern that persists through the ensuing mitotic division.

Because second division segregation patterns are produced by crossing over between the gene and its centromere, their frequency allows us to estimate the distance between these two loci. The gene-to-

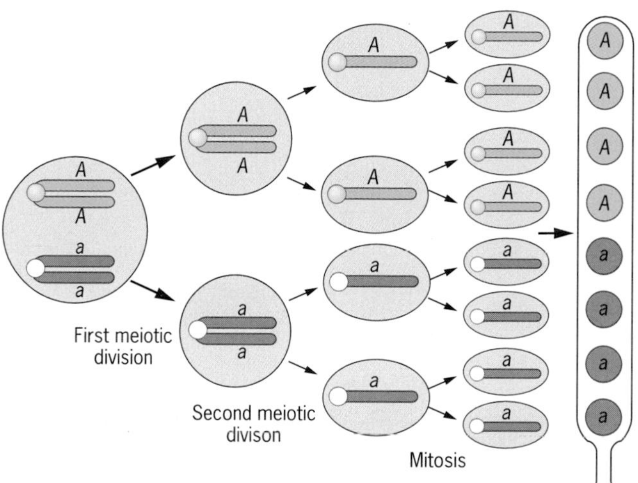

(a) No crossover between gene and centromere.

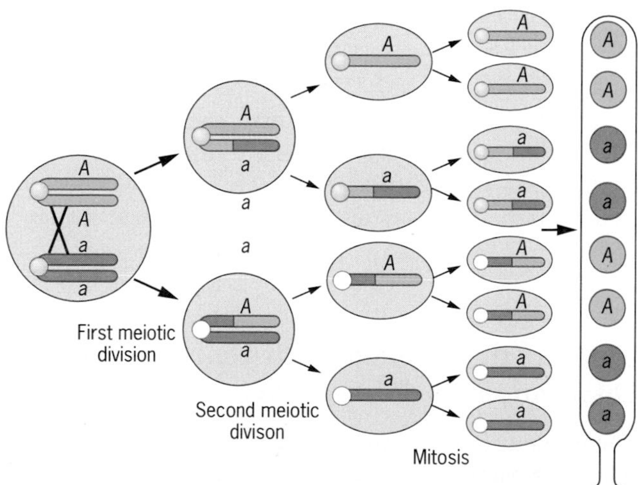

(b) Crossover between gene and centromere.

Figure 9.7 Segregation patterns in *Neurospora* asci. (*a*) First division segregation, resulting from the absence of a crossover between a gene and its centromere. (*b*) Second division segregation, resulting from a crossover between a gene and its centromere.

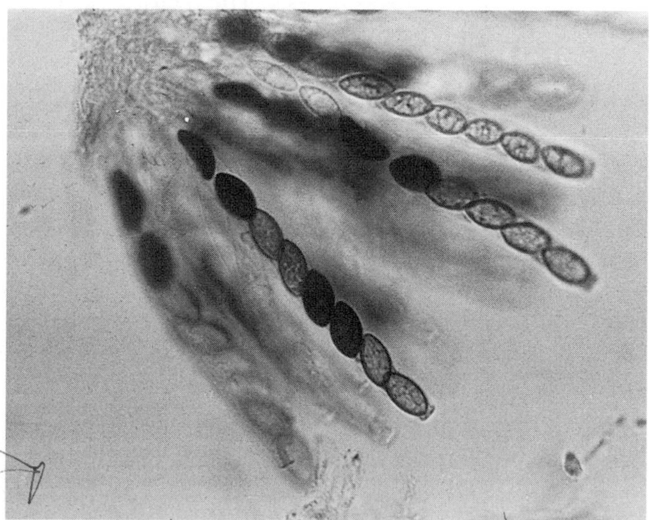

Figure 9.8 Photomicrograph showing the segregation of dark and light ascospores in *Neurospora*.

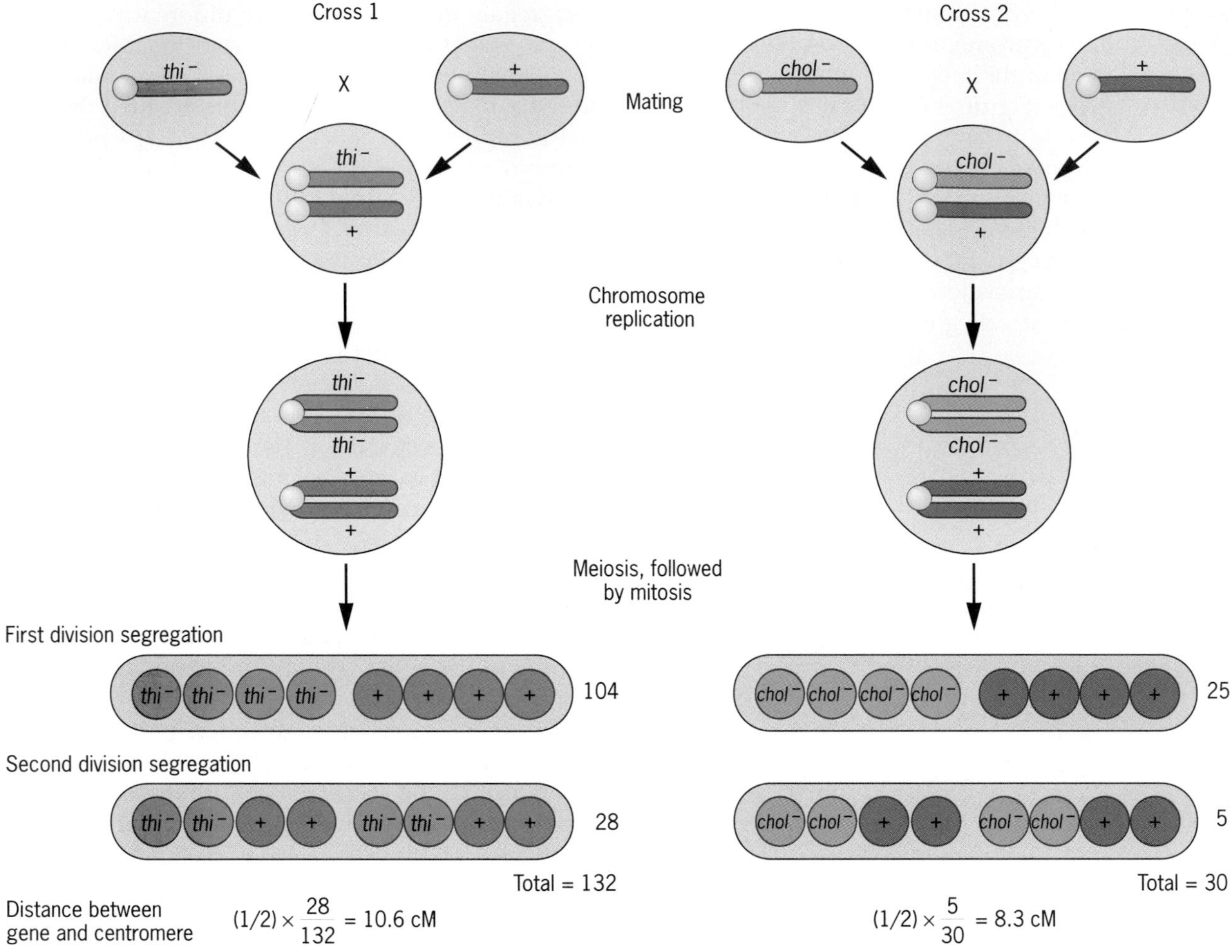

Figure 9.9 Centromere mapping in *Neurospora*. In Cross 1, the mutant strain, *thi*$^-$, requires thiamin; in Cross 2, the mutant strain, *chol*$^-$, requires choline.

centromere distance is simply (1/2) × frequency of second division segregation asci. The factor 1/2 enters in this expression because only half the chromatids in a second division segregation tetrad have crossed over between the gene and the centromere.

As an example of ordered tetrad analysis, let us consider data from two crosses performed with different strains of *Neurospora* by Mary Houlahan, George Beadle, and Hermione Calhoun (Figure 9.9). These researchers mapped a large number of *Neurospora* genes. In one experiment, they crossed a mutant strain (*thi*$^-$) that could not grow in the absence of the vitamin thiamin with a wild-type strain (*thi*$^+$); 132 asci were analyzed for first and second division segregation patterns. From the results, Houlahan, Beadle, and Calhoun inferred that the *thi* gene was (1/2) × (28/132) = 10.6 cM from its centromere. In another experiment, they crossed a mutant strain (*chol*$^-$) that could not grow in the absence of choline with a wild-type strain (*chol*$^+$). From the 30 asci that were analyzed, they estimated the gene-to-centromere distance to be (1/2) × (5/30) = 8.3 cM. Of course, these results do not tell us if the *thi* and *chol* genes are linked to the same centromere or to different ones. A cross between the two mutant strains would be needed to test for such linkage.

Cytogenetic Mapping with Deletions and Duplications in *Drosophila*

Recombination mapping allows us to determine the relative positions of genes by using the frequency of crossing over as a measure of distance. However, it does not allow us to localize genes with respect to cytological landmarks, such as bands, on chromosomes. This kind of localization requires a different procedure based on the phenotypic effects of combining a recessive mutation with a cytologically recognizable deletion or duplication. This approach has been most thoroughly developed in *Drosophila* genetics, where the large, banded polytene chromosomes make it possible to define deletions and duplications accurately. Furthermore, many such aberrations have been collected and maintained in *Drosophila* research laboratories around the world.

As an example, let us consider the cytogenetic mapping of the X-linked *white* gene of *Drosophila*, a wild-type copy of which is required for pigmentation in the eyes. This gene is situated at map position 1.5 near one end of the X chromosome. But which of the two ends is it near, and how far is it, in physical terms, from that end? To answer these questions, we need to find the position of the *white* gene on the cytological map of the polytene X chromosome.

w/Df Genotype
Phenotype
w
white eyes
Df removes the w+ gene.
w
w+
red eyes
Df does not remove the w+ gene.

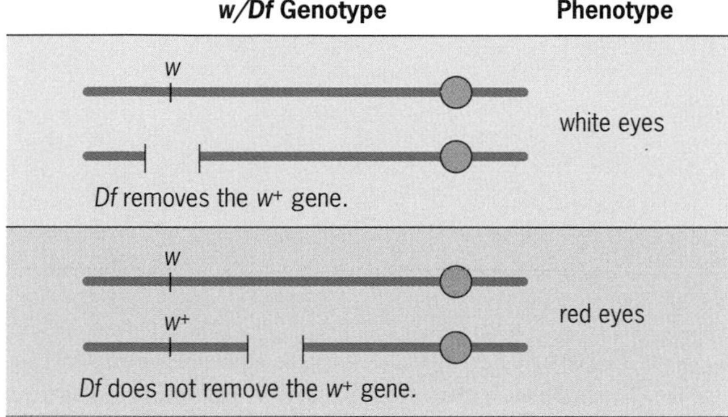

Figure 9.10 Principles of deletion mapping to localize a gene within a *Drosophila* chromosome. The *white* gene on the X chromosome, defined by the recessive mutation *w* which causes white eyes, is used as an example.

The procedure is to produce flies that are heterozygous for a recessive null mutation of the *white* gene (*w*) and a cytologically defined deletion (or deficiency, usually symbolized *Df*) for part of the X chromosome (Figure 9.10). These *w/Df* heterozygotes provide a functional test for the location of *white* relative to the deficiency. If the *white* gene has been deleted from the *Df* chromosome, the *w/Df* heterozygotes will not be able to make eye pigment because they will not have a functional copy of the *white* gene on either of their X chromosomes. The eyes of the *w/Df* heterozygotes will therefore be white (mutant). If, however, the *white* gene has not been deleted from the *Df* chromosome, the *w/Df* heterozygotes will have a functional *white* gene somewhere on that chromosome, and their eyes will be red (wild-type). By looking at the eyes of the *w/Df* heterozygotes, we can therefore determine whether or not a specific deficiency has deleted the *white* gene. If it has, *white* must be located within the boundaries of that deficiency.

Different X-chromosome deficiencies allowed researchers to locate the *white* gene to a position near the left end of the X chromosome (Figure 9.11). Each deficiency was combined with a recessive *white* mutation, but only one of them, *Df*(1)w^{rJ1}, produced white eyes. Thus, we know that the *white* gene must be located within the chromosome segment that is missing in

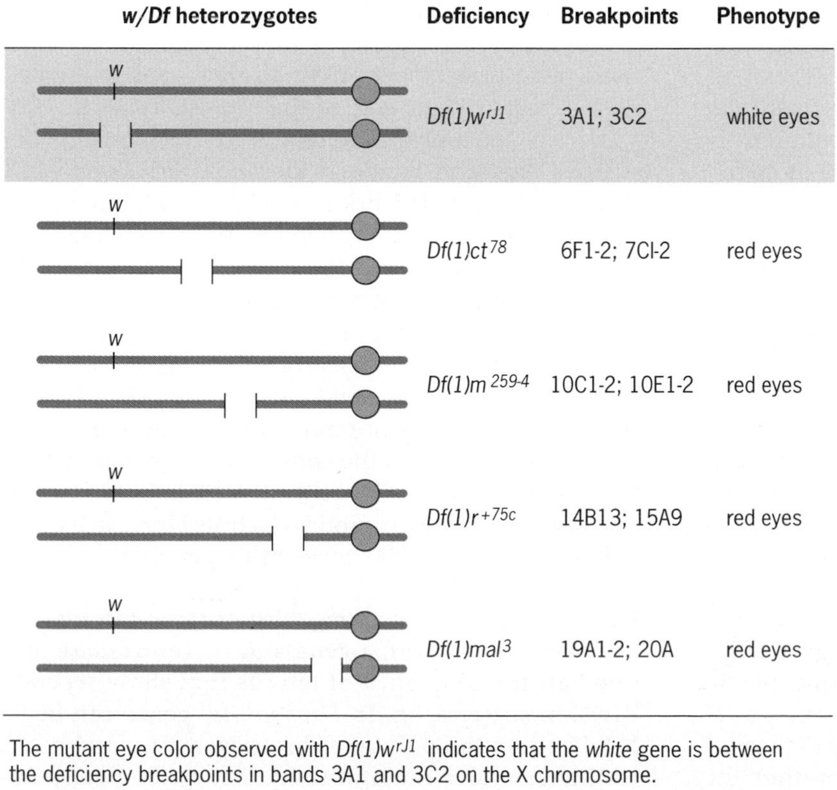

w/Df heterozygotes
Deficiency
Breakpoints
Phenotype
w
Df(1)w^rJ1
3A1; 3C2
white eyes
w
Df(1)ct^78
6F1-2; 7Cl-2
red eyes
w
Df(1)m^259-4
10C1-2; 10E1-2
red eyes
w
Df(1)r^+75c
14B13; 15A9
red eyes
w
Df(1)mal^3
19A1-2; 20A
red eyes
The mutant eye color observed with Df(1)w^rJ1 indicates that the white gene is between the deficiency breakpoints in bands 3A1 and 3C2 on the X chromosome.

Figure 9.11 Localization of the *white* gene in the *Drosophila* X chromosome by deletion mapping.

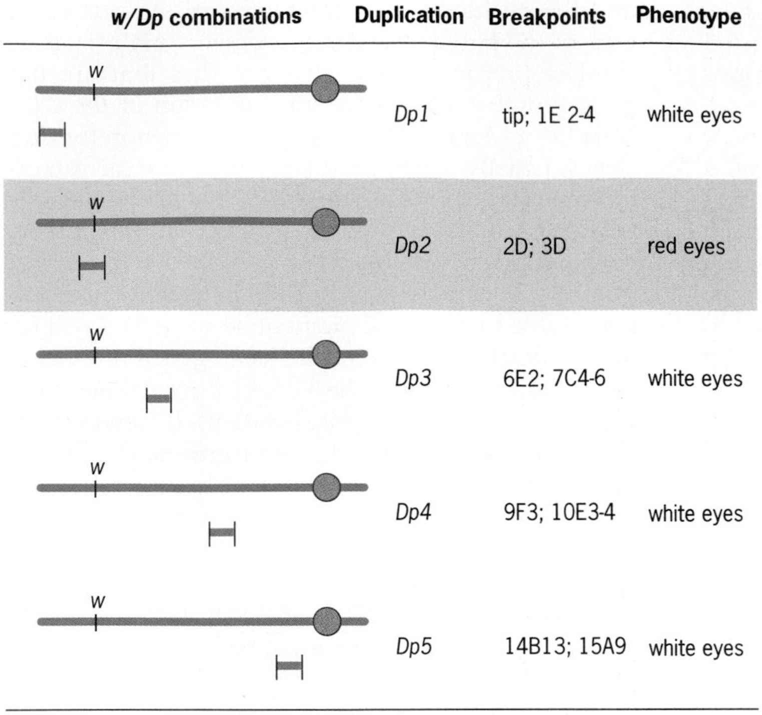

w/Dp combinations	Duplication	Breakpoints	Phenotype
w	*Dp1*	tip; 1E 2-4	white eyes
w	*Dp2*	2D; 3D	red eyes
w	*Dp3*	6E2; 7C4-6	white eyes
w	*Dp4*	9F3; 10E3-4	white eyes
w	*Dp5*	14B13; 15A9	white eyes

The wild-type eye color observed with *Dp2* indicates that the *white* gene is between the duplication breakpoints in regions 2D and 3D on the X chromosome.

Figure 9.12 Localization of the *white* gene in the *Drosophila* X chromosome by duplication mapping.

$Df(1)w^{rJ1}$, that is, somewhere between polytene chromosome bands 3A1 and 3C2. With smaller deficiencies, the *white* gene has been localized to polytene chromosome band 3C2, near the right boundary of $Df(1)w^{rJ1}$.

Many other X-linked genes have been localized in this fashion. When we compare the resulting cytological map with the genetic map (see Figure 8.21), we see that the mapped sites are colinear but not identically spaced. Thus, crossing over does not occur uniformly along the chromosome. Rather, it takes place more frequently in the middle than at the two ends.

We can also use duplications to determine the cytological locations of genes. The procedure is similar to that using deletions, except that we look for a duplication that masks the phenotype of a recessive mutation. Figure 9.12 shows an example utilizing duplications for small segments of the X chromosome. Only one of these duplications, *Dp2*, masked the *white* mutation; thus, a wild-type copy of *white* must be present within it. This localizes the *white* gene somewhere between sections 2D and 3D on the polytene X chromosome, which is consistent with the results of the deletion tests discussed above.

Deletions and duplications have been extraordinarily useful in locating genes on the cytological maps of *Drosophila* chromosomes. The basic principle in **deletion mapping** is that a deletion that *uncovers* a recessive mutation must lack a wild-type copy of the mutant gene. This fact localizes that gene within the boundaries of the deletion. The basic principle in **duplication mapping** is that a duplication that *covers* a recessive mutation must contain a wild-type copy of the mutant gene. This fact localizes that gene within the boundaries of the duplication.

Burke Judd, Margaret Shen, and Thomas Kaufman used these principles to map all the genes in a small region near the left end of the *Drosophila* X chromosome (see Technical Sidelight: The Relationship Between Genes and Bands in the *Drosophila* Polytene X Chromosome). This work was so thorough that nearly all the genes could be localized to individual bands on the cytological map, suggesting a one gene–one band relationship. From subsequent work, we now know that this is an oversimplification; some bands in the polytene chromosomes clearly contain more than one gene. However, because the average number of genes per band is probably not much more than two, the total number of genes in the entire genome is somewhere between one and two times the total number of bands. The band number is around 5000; thus *Drosophila* probably has 5000 to 10,000 genes in its genome.

Key Points: **In organisms with ordered tetrads, the distance between a gene and its centromere is one-half the frequency of tetrads that show second division segregation. In *Drosophila*, genes can be localized on the polytene chromosome maps by testing recessive mutations against cytologically de-**

fined deletions and duplications. A deletion will reveal the phenotype of a recessive mutation located between its endpoints, whereas a duplication will conceal the mutant phenotype.

LINKAGE ANALYSIS IN HUMANS

Mapping human genes is critically important for a number of reasons. Knowledge of genes and their location gives us insight into the evolutionary relationship of humans to other primates and other vertebrate species. Equally important, mapping disease genes creates an opportunity for researchers to isolate the gene and understand how it causes a disease. In case after case, researchers have mapped a gene to a specific chromosome and then proceeded to isolate the gene for further study. Duchenne muscular dystrophy, cystic fibrosis, Huntington's disease, and breast cancer are examples of the success of mapping disease genes to specific regions of specific chromosomes and subsequently isolating those genes.

Until the advent of advanced techniques associated with the Human Genome Project, mapping genes in humans has been difficult. It is not possible to make controlled crosses as it is with *Drosophila* or *Neurospora.* Although informative matings that help to locate genes do sometimes occur by chance, even in these instances, the number of offspring produced is usually small. To circumvent this problem, human geneticists must collect data from many pedigrees and analyze them with sophisticated statistical techniques. In addition, they can use cultured somatic cells and a variety of biochemical and molecular techniques to help them locate genes on chromosomes. In this chapter we consider classical approaches to human gene mapping. These approaches are fundamental to the techniques employed in the Human Genome Project, which is discussed in Chapter 21.

Detection of Linked Loci by Pedigree Analysis

The linkage relationships that are easiest to study in human beings are those between genes on the X chromosome. Such genes follow a pattern of inheritance that is readily identified. If two genes show this pattern, they must be linked. (See, for example, the discussion of the genes for red-green color blindness and hemophilia in the Human Genetics Sidelight in Chapter 8.) Determining linkage between autosomal genes is, however, much more difficult. The human genome has 22 different autosomes, and a gene that does not show X-linkage could be on any one of them. Which autosome is the gene on, what other genes are linked to it, and what are the map positions of these genes? These are challenging questions for the human geneticist. Often, painstaking pedigree analysis is required to answer them.

One of the first instances of linkage between autosomal genes was reported in 1955 by J. H. Renwick and S. D. Lawler. These researchers studied an association between the gene controlling the ABO blood groups (see Chapter 4) and a dominant mutation responsible for a rare, autosomal disorder called the nail-patella syndrome. People with this disorder have malformed nails and kneecaps; often they also have elbow and kidney abnormalities. A pedigree showing the inheritance of the nail-patella syndrome and the ABO blood types is given in Figure 9.13*a*. If we examine the pedigree carefully, we see that with one exception

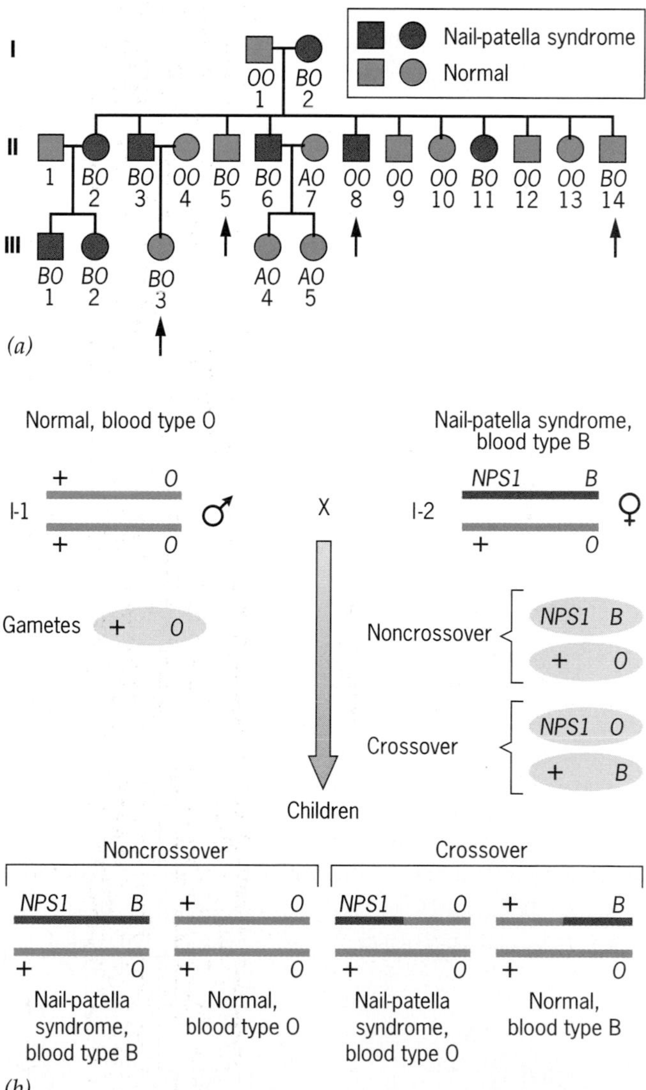

Figure 9.13 (*a*) A pedigree for the nail-patella syndrome (NPS1), with alleles of the ABO blood-type locus noted. An arrow indicates an individual produced by a recombination event. (*b*) A genetic interpretation of the pedigree.

TECHNICAL SIDELIGHT

The Relationship Between Genes and Bands in the *Drosophila* Polytene X Chromosome

Since their discovery by Theophilus Painter, the polytene chromosomes of *Drosophila* have afforded an unparalleled opportunity to study the fine details of chromosome organization. One idea that emerged from these studies is that polytene chromosome bands might correspond to individual genes. Over the years, this idea has been investigated by many researchers, but none have been as thorough as Burke Judd, Margaret Shen, and Thomas Kaufman. These researchers focused their attention on a small region near the left end of the X chromosome that was known to contain two genes that affect eye color: *zeste* (*z*) and *white* (*w*) (Figure 1*a*). Because these genes appeared to bracket the region, the segment between them has come to be called the *zeste-white* region. Cytological analysis had indicated that this region contained 15 polytene chromosome bands. Did this finding imply that it contained 15 genes? Judd, Shen, and Kaufman attempted to answer this question by collecting mutations in the *zeste-white* region and then mapping each of them cytologically.

The mutations were collected by screening for unusual phenotypes in flies that were heterozygous for an X chromosome that had been treated with a mutation-inducing agent, such as X rays, and an X chromosome that was missing the *zeste-white* region. The missing region was the result of a deletion called $Df(1)w^{rJ1}$. These flies therefore had the genotype

Mutagenized X chromosome: ——*——

Deletion X chromosome: ——⌋ ⌊—— $Df(1)w^{rJ1}$

where the gap represents the deletion and the asterisk represents a newly induced recessive mutation.

If the induced mutation occurred in a gene in the *zeste-white* region, it would be expressed phenotypically because the $Df(1)w^{rJ1}$ chromosome has no wild-type gene to "cover" it. In the parlance of *Drosophila* genetics, we would say that the mutation was "uncovered" by the deletion. However, if the mutation occurred in a gene outside the *zeste-white* region, it would not be expressed because the $Df(1)w^{rJ1}$ chromosome would carry a wild-type copy of such a gene.

Mutation in *zeste-white* region	Mutation outside *zeste-white* region
——*——	——*——
——⌋ ⌊——	——⌋ ⌊——
$Df(1)w^{rJ1}$	$Df(1)w^{rJ1}$
Mutant phenotype	Wild phenotype

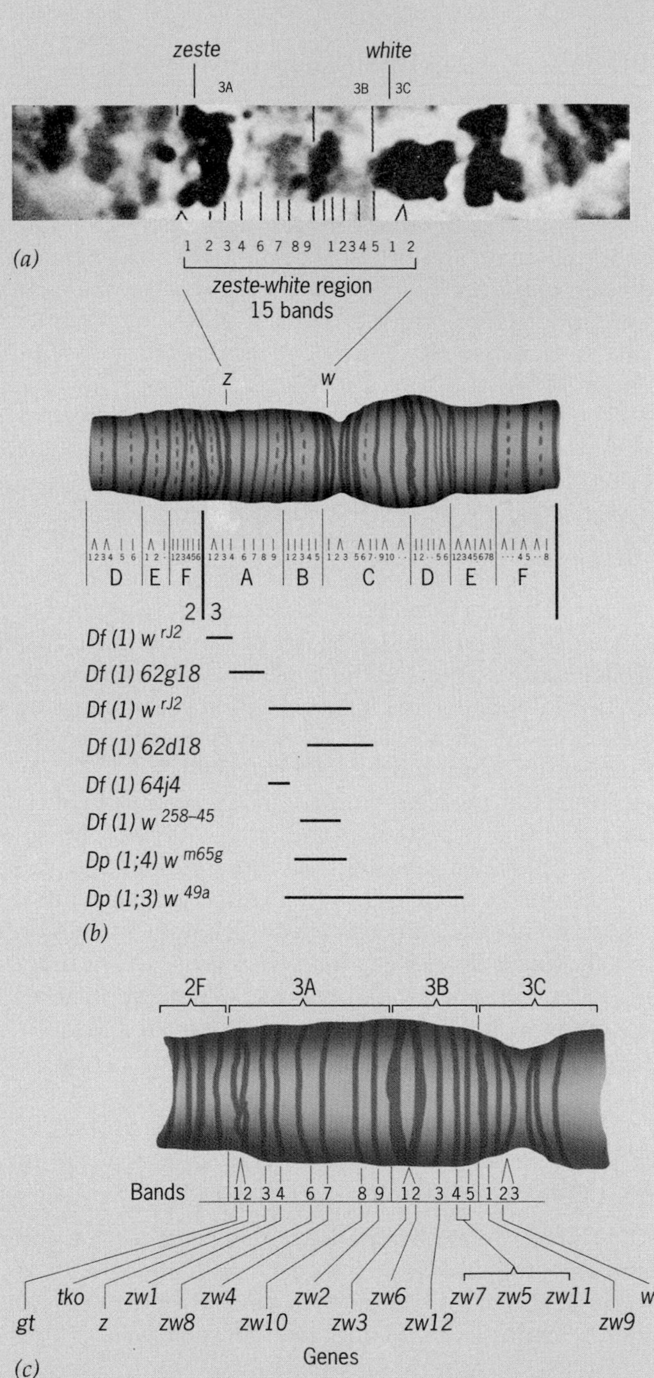

Figure 1 (*a*) Photomicrograph of the *zeste-white* region of the polytene X chromosome. (*b*) Some deletions and duplications that were used to localize mutations in the *zeste-white* region. (*c*) Cytogenetic map of the *zeste-white* region showing the locations of 16 different genes.

Judd, Shen, and Kaufman found many flies with a mutant phenotype, indicating that a mutation had occurred in the *zeste-white* region. Most of these mutations were recessive lethals, manifested by the failure of the */*Df(1)w*^{rJ1} flies to appear in the culture. Fortunately, these lethals could be saved for further analysis by putting them over a deletion-free, balancer X chromosome.

Judd, Shen, and Kaufman then tested each of the mutations they collected with a battery of cytologically defined deletions and duplications in and around the *zeste-white* region (Figure 1*b*). If a mutation was uncovered by a deletion or covered by a duplication, it could be localized within the deleted or duplicated segment. For example, let's consider tests with a mutation, *m*, and three different deletions, *Df(1)62g18*, *Df(1)64j4*, and *Df(1)62d18*:

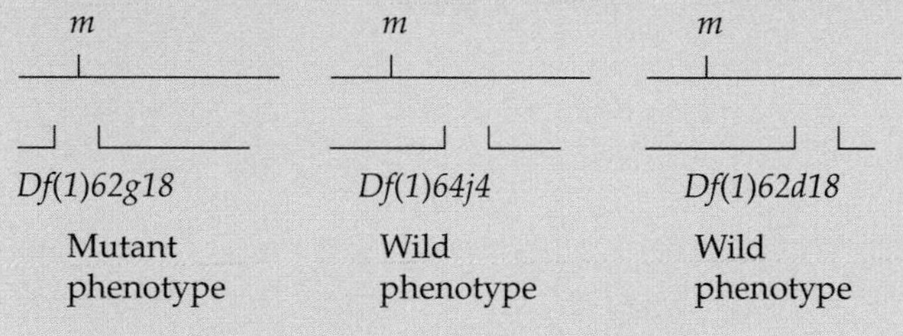

Df(1)62g18 uncovers the mutation, but the other two deletions do not. This result tells us that the mutation resides in the region that is missing in *Df(1)62g18*, a segment that consists of four bands. By such analysis, Judd, Shen, and Kaufman were eventually able to localize most of the mutations they collected to individual bands. In the end, they were able to sort these mutations into 16 different genes, one more than the number of bands (Figure 1*c*). Thirteen of the genes each unambiguously corresponded to a single band. The other three, which were mapped to a pair of bands next to each other, could not be localized precisely because of technical limitations. Judd, Shen, and Kaufman therefore showed that the one gene–one band idea was *almost* correct.

Subsequent genetic studies have uncovered at least four more genes in the *zeste-white* region, bringing the number of known genes to 20. At the same time, electron microscopy has revealed that this region contains several more bands, perhaps a total of 23. The relationship between genes and bands is therefore still unclear. However, as Judd, Shen, and Kaufman suggested, it is probably not far from one to one.

(II-8), each person with the nail-patella syndrome also has blood type B. This finding rather persuasively argues that in this pedigree, the nail-patella mutation, denoted *NPS1*, is linked to the *B* allele of the *ABO* blood group locus. If we assume this inference to be correct, the woman in generation I has the genotype *NPS1 B*/+ *O*; that is, she is a repulsion heterozygote. Her husband's genotype is clearly + *O*/+ *O*.

Figure 9.13*b* illustrates the genetic phenomena underlying this pedigree and suggests a strategy to estimate, albeit crudely, the distance between the *NPS1* and *ABO* loci. The mating indicated in Figure 9.13*b* is essentially a testcross. The woman I-2 can produce four different kinds of gametes, two carrying recombinant chromosomes and two carrying nonrecombinant chromosomes. When these gametes are combined with the single type of gamete (+ *O*) produced by the man I-1, four different genotypes can result. As the pedigree in Figure 9.13*a* shows, I-1 and I-2 produced all four types of children. However, only three (II-5, II-8, and II-14, indicated by arrows in Figure 9.13*a*) of their 11 children were recombinants; the other 8 were nonrecombinants. Thus, we can estimate the frequency of recombination between the *NPS1* and *ABO* loci as 3/11 = 27%. However, this estimate does not use all the information in the pedigree. To refine it, we can incorporate the information from the couple's five grandchildren, only one (III-3) of which was a recombinant. Altogether, then, 3 + 1 = 4 of the 11 + 5 = 16 offspring in the pedigree were recombinants. Thus, we conclude that the frequency of recombination between the *NPS1* and *ABO* loci is 4/16 = 25%, or, in terms of a linkage map, we estimate that the distance between these genes is about 25 cM. Since the original study by Renwick and Lawler, this estimate has been revised downward. The *NPS1* and *ABO* genes are now known to be about 10 cM apart, near the tip of the long arm of chromosome 9.

Renwick and Lawler's study of the *NPS1* and *ABO* loci established that these two genes are linked, but it could not identify the specific autosome that carried them. The first localization of a gene to a specific autosome came in 1968, when R. P. Donahue and coworkers demonstrated that the Duffy blood group locus, denoted *FY*, is on chromosome 1. This demonstration hinged on the discovery of a variant of chromosome 1 that was longer than normal. The longer chromosome turned out to have an abnormally large region of heterochromatin around its centromere. Pedigree analysis showed that in a particular family, this long chromosome segregated specific *FY* alleles. Thus, the *FY* locus was assigned to chromosome 1. Subsequent research has placed this locus at region 1p31 on that chromosome.

Until the early 1980s, progress in human gene mapping was extremely slow because it was difficult to find pedigrees that were segregating linked markers, say, for example, two different genetic diseases. In the 1980s, however, it became possible to identify genetic variants in the DNA itself. These variants result from differences in the DNA sequence in parts of chromosomes. For example, in one individual a particular sequence might be GAATTC on one of the DNA strands, and in another individual the corresponding DNA sequence might be GATTTC—a difference of just one nucleotide. Although we must defer to later chapters a discussion of the techniques that are used to reveal such molecular differences, here we can explore how they have helped to map human genes, including many that are involved in serious inherited diseases.

Small differences in DNA sequences are common in human populations, and often they can serve as molecular genetic markers in pedigrees that are segregating for a particular gene. If, in addition to the usual phenotypic analysis, the members of a pedigree are analyzed for the presence or absence of these markers, a researcher can look for linkage between each marker and the gene. Then, with appropriate statistical techniques, he or she can estimate the distances between the gene and the markers that are linked to it. In this fashion, genetic maps can be built up from completely independent analyses. If gene *A* has been shown to be linked to marker *x*, and gene *B* has also been shown to be linked to marker *x*, then gene *A* and gene *B* are obviously linked to each other. Thus, this procedure allows human geneticists to determine linkage relationships between genes that are not segregating in the same pedigrees.

This strategy for compiling human linkage data into maps has worked exceedingly well. A very large number of molecular markers based on DNA sequence differences and hundreds of different genes have been positioned. Figure 9.14 shows a map of molecular markers on chromosome 4, the one introduced at the beginning of this chapter as the chromosome that carries the Huntington's disease (*HD*) gene. When James Gusella, Nancy Wexler, and colleagues analyzed large pedigrees from Venezuela, they found that the *HD* gene was tightly linked to one of the molecular markers known to reside on chromosome 4. Recombination between the gene and the marker was only 4 percent. Subsequent analysis, described in Chapter 22, pinpointed the *HD* gene and paved the way for studies of its function.

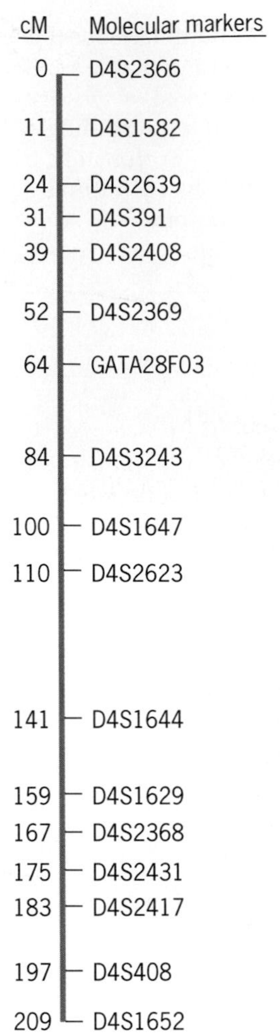

Figure 9.14 Linkage map of molecular markers on human chromosome 4.

Somatic-Cell Genetics: An Alternative Approach to Gene Mapping

We can often tell from the analysis of pedigrees whether genes are sex-linked or autosomal. Through the application of direct or indirect methods, we can also usually ascertain whether or not particular genes are linked to each other. However, we usually wish to answer more pointed questions. On *which* specific autosome is a particular gene located? *Where* on that chromosome is the gene?

With *Drosophila* and other experimental organisms, we are able to carry out carefully designed crosses that provide answers to these questions, as we have discussed in this and the previous chapter. With human beings, however, different procedures must be used to localize genes on particular chromosomes.

Somatic-Cell Hybridization In the early 1960s, G. Barski and B. Ephrussi developed techniques for fusing different cells either from the same species or different species. The fused cells are called hybrids. These **cell hybridization** or **cell fusion** techniques have had a profound impact on human gene mapping. To form interspecific cell hybrids, human cells and cells from another species, usually a mouse, are mixed together in culture with a fusing agent, such as inactivated Sendai viruses or polyethylene glycol. The fusing agent first causes fusion of cell membranes followed by fusion of nuclear membranes (Figure 9.15). Then, as the hybrid cells divide, human chromosomes are randomly lost; however, the hybrids retain all the mouse chromosomes. Eventually, after several rounds of division, the hybrid stabilizes, retaining only one or a very few human chromosomes. We do not yet understand the mechanism that causes human chromosomes to be lost and mouse chromosomes to be retained. As we will see below, hybrid cells with just a few human chromosomes can be used to locate a human gene on a specific chromosome.

Even under ideal research conditions, hybrid cells are not produced in large numbers. Some cells do not fuse, and some fusions involve same-species cells (mouse–mouse or human–human). To eliminate all but the true hybrids, the cell mixtures are placed on a medium that selects against all but the hybrid cells. The most commonly used selective medium is called the **HAT medium** because it contains three chemical additives—*h*ypoxanthine, *a*minopterin, and *t*hymidine.

This medium prevents the growth of cells with certain kinds of biochemical defects. In particular, cells deficient for either of two enzymes, thymidine kinase (TK) or hypoxanthine phosphoribosyl transferase (HPRT), cannot grow in HAT medium. Thus, if human cells that are TK^- (that is, deficient in thymidine kinase) and mouse cells that are $HPRT^-$ (that is, deficient in hypoxanthine phosphoribosyl transferase) are placed in HAT medium, neither type will grow. However, if the two types of cells fuse, the resulting hybrids will grow in HAT medium because the fused cells compensate for each other's defects. Though deficient for the TK enzyme, human cells can produce HPRT (that is, they are $HPRT^+$), and though deficient for HPRT, mouse cells can produce TK (that is, they are TK^+). Thus, hybrids formed by fusing human and mouse cells produce both enzymes and are able to survive in HAT medium. Other types of cells, including unfused human and mouse cells, mouse–mouse hybrids, and human–human hybrids, die (Figure 9.16).

Once hybrid cells have been selected in HAT medium, they are used to establish different lines. Each line carries a sample of human chromosomes. These lines can be characterized by cytological analysis and assayed for particular gene products. The presence of a particular product can then be correlated with the presence of a specific human chromosome.

In 1971, Oscar Miller and his colleagues used somatic-cell hybridization to determine the chromosomal location of the *TK* and *HPRT* genes (Figure 9.17). TK^- mouse cells (which were $HPRT^+$) were fused to $HPRT^-$ human cells (which were TK^+). After selection on HAT medium, the surviving hybrid cells ($HPRT^+$ TK^+) were analyzed for the human chromosomes that were present. This analysis was possible because human and mouse chromosomes are cytologically distinct from each other. One of the hybrid lines retained only one human chromosome, 17, and still synthesized TK. Miller and his colleagues concluded that the *TK* gene must be located on chromosome 17. In a reciprocal experiment they fused $HPRT^-$ mouse cells with TK^- human cells and found that all the surviving hybrid cells carried the human X chromosome. Thus, they confirmed that the human *HPRT* gene is X-linked.

Somatic-cell hybrids can be used to map many human genes. The only requirement is that the hybrid cells express the human gene product and furthermore that the product be detectable and distinguishable

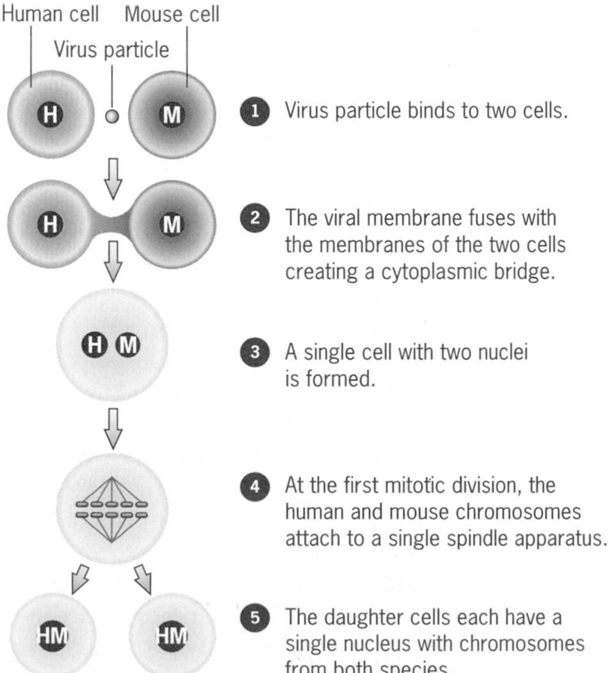

Figure 9.15 Fusion of human and mouse cells mediated by an inactivated Sendai virus.

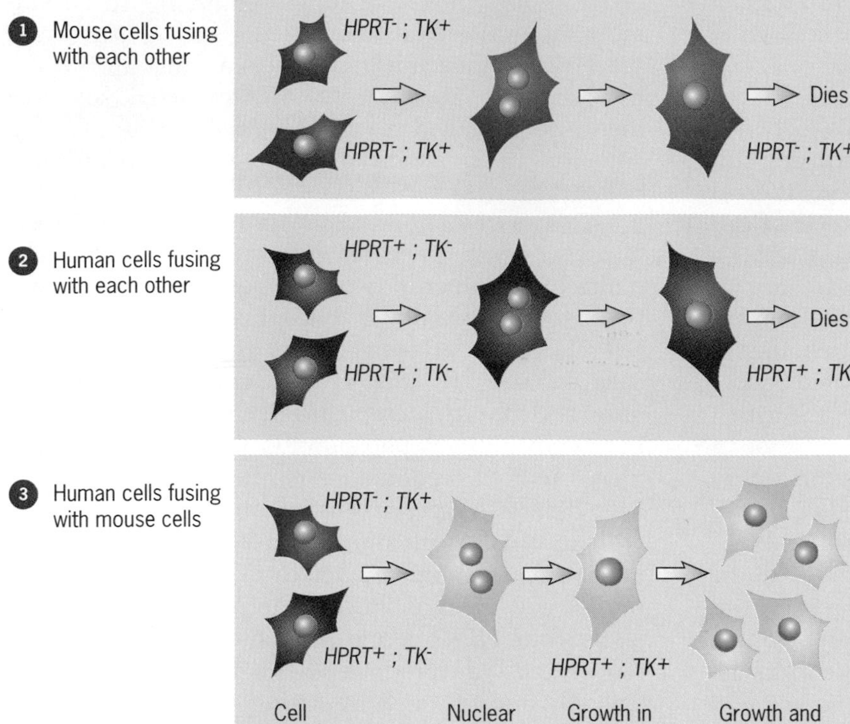

Figure 9.16 Selection of hybrid cells in HAT medium. Only cells that are $HPRT^+$ and TK^+ are able to grow in this medium. Parental cells and fused cells containing only one parental type do not survive in HAT medium; the true hybrid cells, however, survive and divide.

from the mouse product being expressed in the hybrids. For example, the gene for UMP kinase has been mapped to chromosome 1 using somatic-cell hybridization techniques (Table 9.1). The enzyme is expressed and detectable in hybrid cells. In addition, mouse UMP kinase is distinct from human UMP kinase because of a few amino acid differences. These differences give the two forms of the enzyme different electrical properties that can be detected by gel electrophoresis, a technique that separates molecules by forcing them through a porous gel in an electrical field. The human and mouse UMP kinases migrate to different regions in the gel, where they can be identified by biochemical assays.

Until recently, somatic-cell genetics could be used to map genes only if the hybrid cells expressed detectable gene products. If a gene was not expressed in the hybrid cells, it could not be mapped. However, with the development of molecular techniques, it is now possible to detect the genes themselves. For example, the genes for the α and β globins, the proteins that make up adult hemoglobin, are not expressed in hybrid cells. Nevertheless, we can detect each of the globin genes using gene-specific DNA probes. The techniques for this type of analysis are discussed in detail in Chapter 20.

Gene Mapping Using Chromosomal Rearrangements When particular types of chromosome rearrangements are available, the somatic-cell hybridization technique can be used to map a gene to a specific region of a chromosome. One type of rearrangement that is valuable in localizing a gene is a translocation (see Chapter 7). To see the usefulness of such rearrangements, let's consider an example in which hybrid cell lines were developed from a patient with a reciprocal translocation between the X chromosome and chromosome 14 (Figure 9.18). In this rearrangement, most of the long arm of the X chromosome was translocated to the tip of chromosome 14, and a small piece of chromosome 14 was translocated to the X chromosome. After human–mouse hybrid cells were selected in HAT medium, the only human chromosome in one of the resulting cell lines was the 14 with the long arm of the X attached to it. The normal X, normal 14, and the X carrying the small piece of chromosome 14 were not present in this line. Because this cell line survived in HAT medium, it must have been $HPRT^+$, and because the *HPRT* gene is X-linked, we infer that the *HPRT* gene is located in the long arm of the X chromosome. Two other human X-linked genes were also expressed in this particular hybrid cell line: *PGK* (encoding phosphoglycerate kinase) and *G6PD* (encoding glucose-6-phosphate dehydrogenase). Obviously, these genes are also located in the long arm of the X chromosome. In addition, this same cell line expressed nucleoside phosphorylase, an enzyme encoded by *NP*, an autosomal human gene. Because this hybrid line carried no other

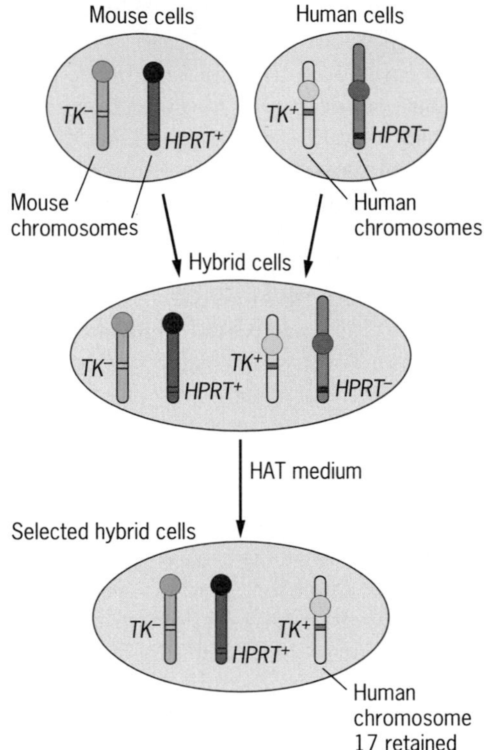

1. Human and mouse cells are hybridized in culture.

2. The hybrid cells are grown on HAT medium to select for the TK^+ and $HPRT^+$ genes. During this time, human chromosomes are randomly lost from the hybrid cells.

3A. In panel A, the selected cells carry the mouse $HPRT^+$ gene and the human TK^+ gene. Because human chromosome 17 is common to all selected cell lines, the human TK^+ gene must be located on this chromosome.

3B. In panel B, the selected cells carry the mouse TK^+ gene and the human $HPRT^+$ gene. Because the human X chromosome is common to all the selected cell lines, the human $HPRT^+$ gene must be located on this chromosome.

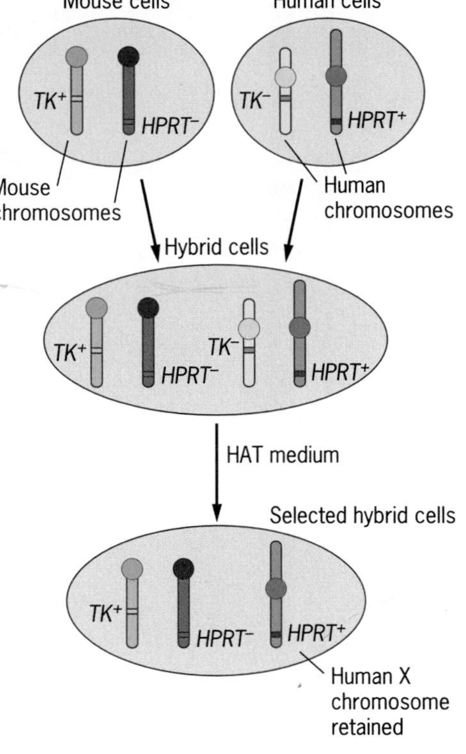

Figure 9.17 Somatic-cell hybridization experiments to locate the human TK^+ and $HPRT^+$ genes on specific chromosomes. Cultured mouse and human cells were fused to produce hybrid cells, which were selected for retention of either the human TK^+ (panel A) or $HPRT^+$ (panel B) genes during growth in HAT medium. For simplicity, only one member of the relevant mouse and human chromosome pairs is shown.

human autosomes, the *NP* gene must be located on the part of chromosome 14 that was present in the hybrid cells.

By using other X-autosome translocations, it has been possible to establish the order of the genes *PGK*, *HPRT*, and *G6PD* on the long arm of the X chromosome

TABLE 9.1
Assignment of the UMP-Kinase (*UMPK*) Gene to Chromosome 1 Using a Hybrid Cell Line Panel

Cell Line	*UMPK Activity*	*Human Chromosomes Present*[a]
WA-lla	–	2, 3, 4, 11, 12, 14, 16, X
JFA-14b	+	1, 2, 5, 8, 13, 16, 17, 18, X
WA-la	+	1, 2, 4, 7, 8, 10, 12, 13, 15, 16, 18, 19, 20, 21, 22, X
J10-H12	+	1, 3, 6, 7, 10, 11, 12, 14, 15, 16, 17, 18, 21, X
AIM-3a	–	2, 4, 5, 7, 8, 10, 11, 12, 13, 14, 17, 18, 20, X
AIM-8a	–	6, 7, 9, 11, 12, 13, 14, 15, 16, 17, 18, 19, 20, 21, 22, X
AIM-11a	–	2, 3, 7, 10, 12, X
AIM-23a	+	1, 2, 3, 4, 5, 7, 8, 10, 12, 14, 15, 16, 18, 21, 22, X

[a]Every time chromosome 1 is present, so is the enzyme; every time it is absent, so is the enzyme.

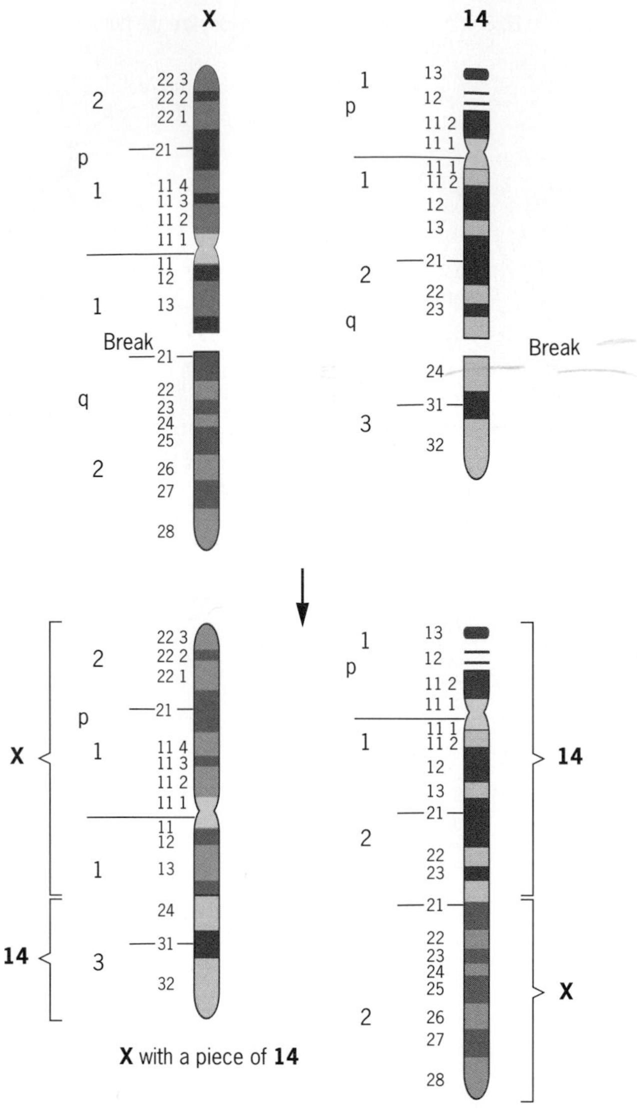

Figure 9.18 A reciprocal translocation between the X chromosome and chromosome 14. A translocation of this type established that the *HPRT* gene is located in the long arm of the X chromosome.

(Figure 9.19). Only translocated segments containing Xq13 expressed PGK; thus, this locus must be in the q13 region of the long arm, near the centromere. Only translocated segments carrying Xq26 expressed HPRT, and only segments carrying Xq28 expressed G6PD. Therefore, the gene order in the long arm of the X chromosome must be centromere–*PGK–HPRT–G6PD.*

One of the most exciting stories in modern medical genetics is the discovery and eventual isolation of the X-linked gene causing **Duchenne muscular dystrophy (DMD).** DMD is a fatal neuromuscular disease. Symptoms usually begin by age 6; the victim is chairbound by age 12 and dead by age 20. Mapping the *DMD* gene began in 1984 with the discovery by Ron Worton that rare females with DMD carry an X-autosome translocation. Because DMD is an X-linked recessive disorder and males do not survive long enough to pass the defective allele on to their daughters, mutant homozygotes are never produced. How, then, is it possible that a female should ever develop DMD? The answer to this puzzle involves the phenomenon of X-chromosome inactivation (Chapter 6). Early in embryonic development, one of a female's two X chromosomes is randomly inactivated. In some cells the maternally inherited X is inactivated, and in others the paternally inherited X is inactivated. However, in DMD females that have X-autosome translocations, inactivation is *not* random. The structurally intact, nontranslocated X is preferentially inactivated, and the translocated X remains as the active chromosome (Figure 9.20). Although the autosome involved in this type of translocation varies from patient to patient, the breakpoint in the X chromosome is always in the same region: Xp21. This break must therefore be directly responsible for the DMD mutation, presumably because it interferes with the expression of the *DMD* gene itself. Analysis of these rare X-autosome translocations therefore revealed the location of the *DMD* gene.

Deletion Mapping Translocations are not the only type of chromosome abnormality used in gene localization. Sometimes a segment of a chromosome and the genes it carries is simply deleted. If we are able to associate missing genes or gene products with a specific chromosome deletion, we can use this information to construct a chromosome map. Let's consider the case of a

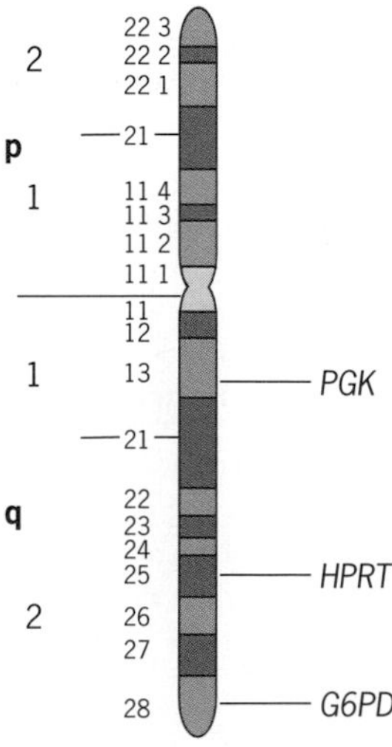

Figure 9.19 A schematic of the X chromosome showing the locations of the *PGK, HPRT,* and *G6PD* genes. These genes were physically mapped using translocations involving different X-chromosome breakpoints.

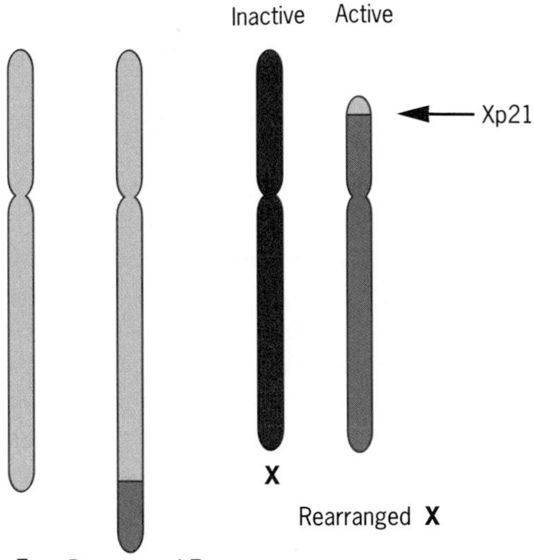

Figure 9.20 A reciprocal translocation between the X chromosome and chromosome 5. The translocation breakpoint in band p21 of the X chromosome disrupted the function of the *DMD* gene. The untranslocated X chromosome was inactivated, and the active X chromosome carried a disrupted *DMD* gene; consequently, the woman who had this reciprocal translocation suffered from Duchenne muscular dystrophy.

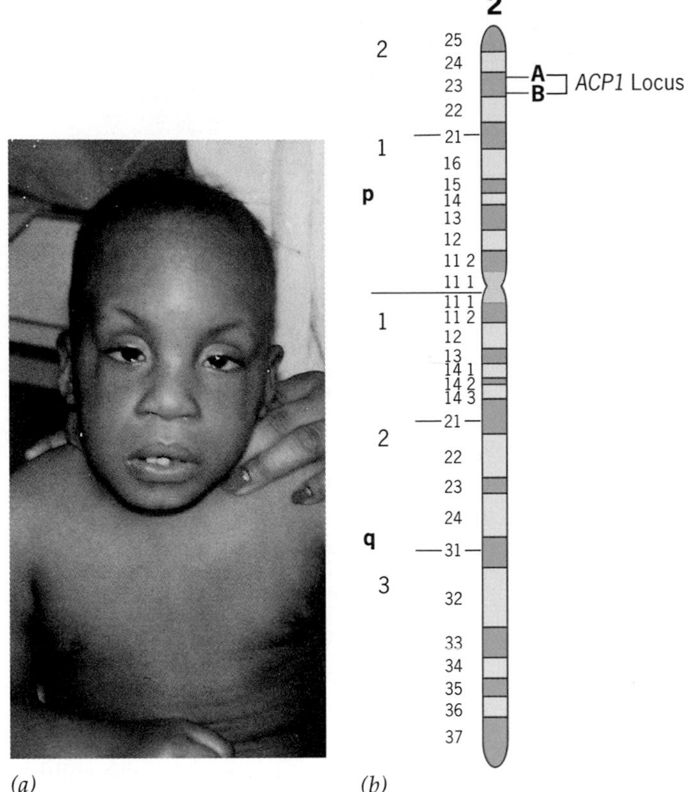

Figure 9.21 Genes can be located on chromosomes using deletions. (*a*) A young male who is missing the tip of the short arm of one of his chromosomes 2. (*b*) The breakpoints in two different patients with 2p deletions. The patient with the **A** breakpoint retains the *ACP1* gene; the patient with the **B** breakpoint has lost the *ACP1* gene. Thus, the *ACP1* gene must lie between the two breakpoints.

young boy who had multiple anomalies at birth (Figure 9.21*a*) and who was suspected of carrying a chromosome abnormality. When his karyotype was prepared, it was discovered that he was missing the tip of one of his chromosomes 2 (breakpoint **A** in Figure 9.21*b*). Other studies had already suggested that the gene for acid phosphatase (*ACP1*) mapped in this area of chromosome 2. Was the boy missing one of his two *ACP1* alleles? The *ACP1* gene has three distinct alleles: *A*, *B*, and *C*. Each allele encodes a form of *ACP1* that has slightly different electrical charges, and these differences are detectable using gel electrophoresis. Because the boy carried two alleles, *A* and *B*, we can conclude that the missing segment of chromosome 2 did *not* contain the *ACP1* locus. A second abnormal child was discovered to have a deletion of the tip of one of his chromosomes 2 (breakpoint **B** in Figure 9.21*b*); however, this child carried only one allele of the *ACP1* gene. Because pedigree records indicated that the child could not be homozygous for the *ACP1* allele he carried, the deleted segment of chromosome 2 must have contained the *ACP1* locus. Thus, the *ACP1* gene could be localized to the small segment between the deletion breakpoints observed in the two different children.

Deletion mapping also helped researchers confirm the location of the Duchenne muscular dystrophy locus (*DMD*) on the X chromosome. Recall that the translocation DMD females suggested that the *DMD* gene is located in the short arm of the X chromosome. In 1985 Uta Francke and her colleagues examined a young boy with several rare clinical conditions, all known to be X-linked recessive Mendelian traits: chronic granulomatous disease (CGD), an immune disorder associated with a deficiency of cytochrome b; the McLeod red blood cell disorder; retinitis pigmentosa (RP), a vision disorder; and Duchenne muscular dystrophy. Careful analysis of his X chromosome revealed that he was missing a very small segment: Xp21. All four of these genes must therefore be located in this segment of the X chromosome. Later that same year, Louis Kunkel and his colleagues in Boston and Ron Worton and his colleagues in Toronto, using DNA from this young boy and from one of the translocation DMD females, isolated the *DMD* gene.

Duplication Mapping Chromosome rearrangements sometimes result in extra copies of genes. If we can determine how many copies of a gene are present and then correlate gene duplication events with a dupli-

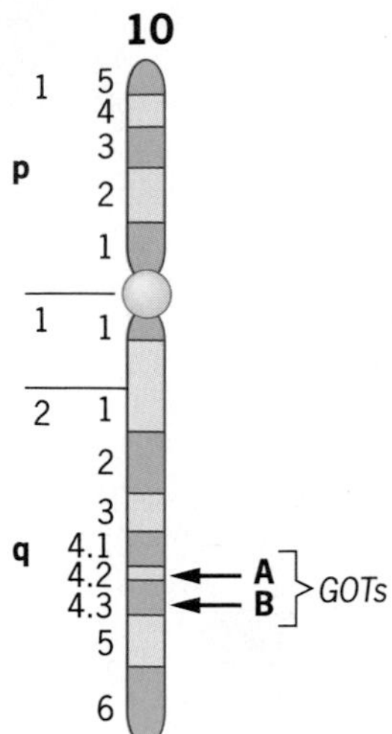

Figure 9.22 Two different translocation breakpoints, **A** and **B,** in chromosome 10 were used to locate the *GOTs* gene.

cated chromosome segment, we can assign a gene to that specific duplicated region of a chromosome.

We can see how **duplication mapping** works by looking at the methodology used to map the soluble (cytoplasmic) glutamic oxaloacetic transaminase (GOTs) gene. The *GOTs* gene was first assigned to the distal segment of the long arm of chromosome 10 by means of the somatic-cell techniques we have already discussed. Then, using hybrid cell lines with rearrangements that resulted in duplicated segments of chromosome 10, researchers were able to locate the gene in a specific region on the long arm of that chromosome (10q24). The researchers used three different hybrid cell lines to map the *GOTs* locus. Cell line 1 carried a translocation between chromosomes 10 and 17, and cell line 2 carried a translocation between chromosomes 10 and 21. The breakpoints in chromosome 10 were different in both cases (Figure 9.22). Cell line 1, with breakpoint **A,** carried two copies of chromosome 10 and a small extra piece of chromosome 10 attached to chromosome 17; cell line 2, with breakpoint **B,** carried two copies of chromosome 10 and a smaller extra piece of chromosome 10 attached to chromosome 21. A third cell line with two copies each of the normal chromosomes 10, 17, and 21 served as a control. The researchers measured the GOTs enzyme activity in each cell line (Figure 9.23). The control line, with two copies of the *GOTs* gene, produced approximately 15 units of GOTs activity. Cell line 2, with the 10/21 translocation, produced about the same amount of GOTs activity. Thus, because cell line 2 appeared to have only two copies of the *GOTs* gene, the researchers concluded that the translocated segment of chromosome 10 in cell line 2 must *not* carry a copy of the *GOTs* locus. This finding implies that the *GOTs* gene is more proximal (that is, closer to the centromere) than breakpoint **B.** Cell line 1, with the 10/17 translocation, produced about 30 units of GOTs activity. This is about 100 percent more than the activity seen in the control

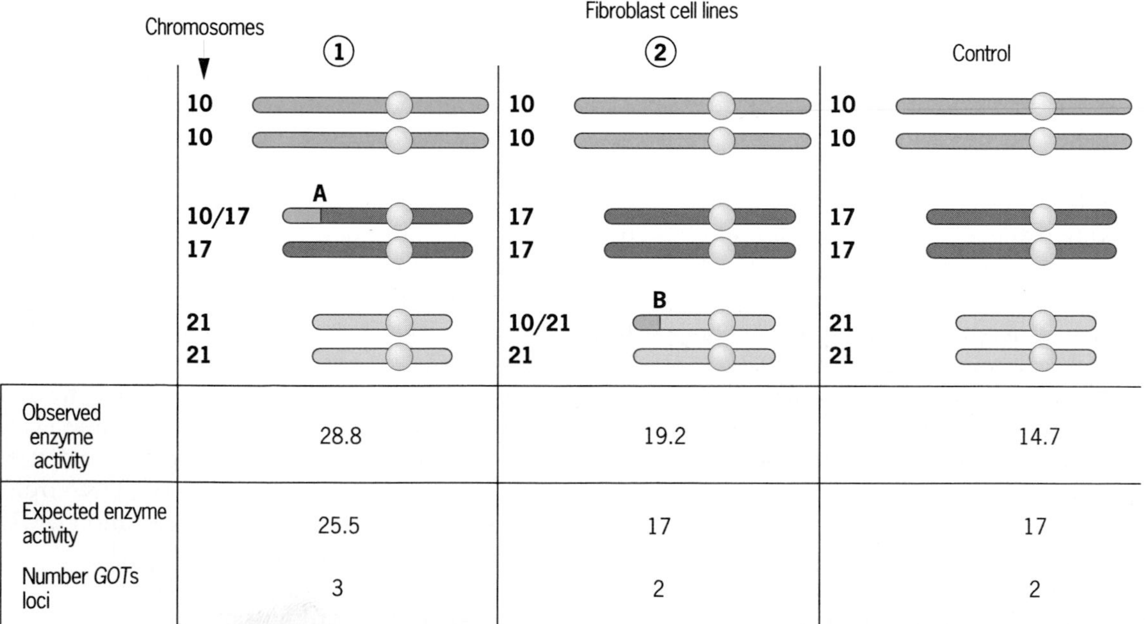

	1	2	Control
Observed enzyme activity	28.8	19.2	14.7
Expected enzyme activity	25.5	17	17
Number *GOTs* loci	3	2	2

Figure 9.23 Duplication mapping. Three cell lines were used, each with a normal pair of chromosome 10s. Cell line 1, with translocational breakpoint **A,** had three copies of the *GOTs* gene. Cell line 2, with translocation breakpoint **B,** had two copies. Thus, the *GOTs* gene must lie between the two breakpoints.

line and is consistent with the expression of *three* copies of the *GOTs* locus. Consequently, the researchers concluded that the translocated segment of chromosome 10 present in this cell line did carry the *GOTs* locus. This finding implies that the *GOTs* gene is distal (that is, farther away from the centromere) than breakpoint **A.** By comparing the results from the two cell lines, the researchers inferred that the *GOTs* gene lies between the two translocation breakpoints, **A** and **B,** in the q24 region of chromosome 10.

The Human Gene Map

Somatic-cell hybridization techniques have provided a major impetus for mapping human genes. However, more recent molecular techniques have greatly accelerated the process of human gene mapping. A major goal of the much publicized **Human Genome Project** is the construction of high-density maps for each human chromosome. The Human Genome Organization (HUGO) was established to coordinate this research throughout the world. Once this mapping is accomplished, the next step will be to sequence the DNA in each chromosome. The ultimate goal of the several research laboratories involved in HUGO is to identify the approximately 70,000 to 100,000 human genes and to sequence each one of them. This is an enormous undertaking if we consider that humans have about 3 billion DNA base-pairs.

The rapidly developing map of the human genome describes the location and sequence of genes and molecular markers and the spacing between them on each chromosome. These maps are being constructed at several levels. The coarsest level of resolution is the genetic linkage map, which consists of the location and order of genes and DNA markers. More refined physical genetic maps describe the chemical characteristics of the DNA molecule itself at each chromosomal location. We discuss the techniques involved in mapping and sequencing the genomes of human beings and other organisms—bacteria, yeast, *Drosophila*, and plants—in Chapter 21.

At this time, approximately 50,000 human genes have been identified by sequence data; about 10 percent of these genes correspond to previously known human genes. Thus, the entire project still has a long way to go. Many of the genes that have been identified are involved in some kind of genetic disorder or disease state. We are just beginning to identify loci that influence normal, nondisease types of variation, such as height, weight, skin pigmentation, hair color, and disease resistance/susceptibility.

Key Points: **Pedigree analysis can provide evidence for linkage between human genes. A gene can be localized to a particular human chromosome by correlating it with the human chromosomes that are retained in interspecific somatic-cell hybrids. The location can be refined by analyzing somatic-cell hybrids that carry rearranged human chromosomes. The Human Genome Project is an international effort to map and sequence the entire human genome.**

TESTING YOUR KNOWLEDGE

1. Mary Houlahan, George Beadle, and Hermione Calhoun (*Genetics* 34:493–507, 1949) studied two mutant strains of *Neurospora crassa*. One strain (*pdx*) could not grow without pyridoxine, and the other (*pan*) could not grow without pantothenic acid. Tetrads of ascospores from a cross between these strains were dissected and analyzed. From the results (shown at the right), determine the linkage relationships between the *pdx* and *pan* genes, and between each gene and its centromere. (A + represents the wild-type allele of a gene; thus, for example, *pdx* + = *pdx* pan^+ is the genotype of a spore that cannot grow without pyridoxine but that can grow without pantothenic acid.)

Tetrad Classes

1	2	3	4	5	6
pdx +	*pdx pan*	*pdx* +	*pdx* +	*pdx* +	*pdx* +
pdx +	*pdx pan*	*pdx pan*	+ +	+ *pan*	+ *pan*
+ *pan*	+ +	+ +	*pdx pan*	*pdx* +	*pdx pan*
+ *pan*	+ +	+ *pan*	+ *pan*	+ *pan*	+ +

Number observed

1	2	3	4	5	6
15	1	17	1	13	2

ANSWER

Before analyzing the data, we note that the cross was between strains with the genotypes *pdx* + and + *pan*; these genotypes are indicative of parental-type ascospores in the progeny. Nonparental (that is, recombinant) ascospores are either *pdx pan* or + +. Using this nomenclature, we can classify each tetrad as parental ditype (PD), nonparental ditype (NPD), or tetratype (T). Furthermore, because *Neurospora* tetrads are ordered, we can classify them as having first (F) or second (S) division segregation patterns with respect to each marker:

Tetrad class	1	2	3	4	5	6
Number observed	15	1	17	1	13	2
Type	PD	NPD	T	T	PD	T
Segregation with respect to *pdx*	F	F	F	S	S	S
Segregation with respect to *pan*	F	F	S	F	S	S

The analysis of the data involves several steps. First, we note that the NPD tetrads (1) are much less numerous than the PD tetrads (15 + 13 = 28). This observation indicates that *pdx* and *pan* are linked. Second, we use the standard mapping formula to estimate the distance between *pdx* and *pan*:

$$\frac{(1/2)\text{T} + \text{NPD}}{\text{total}} = \frac{(1/2)(17 + 1 + 2) + 1}{49} = 22.2 \text{ cM}$$

Third, we estimate the distance between *pdx* and its centromere by calculating the frequency of second division segregation patterns:

$$\frac{(1/2)\text{S}}{\text{total}} = \frac{(1/2)(1 + 13 + 2)}{49} = 16 \text{ cM}$$

Similarly, we estimate the distance between *pan* and its centromere:

$$\frac{(1/2)\text{S}}{\text{total}} = \frac{(1/2)(17 + 13 + 2)}{49} = 32.7 \text{ cM}$$

The combined results tell us that *pdx* and *pan* are on the same side of the centromere and that *pan* is farther away. We summarize our analysis in a linkage map:

centromere—16 cM—*pdx*—22.2 cM—*pan*

Notice that the frequency of second division segregation patterns underestimates the distance between *pan* and the centromere because it overlooks a few double exchanges. We can obtain a better estimate by summing the distances between the centromere and *pdx* and between *pdx* and *pan*: 16 + 22.2 = 38.2 cM.

2. The pedigree below is for an X-linked recessive trait, ocular albinism (OA). Beneath some members of this family is information about a molecular marker located about 5 centiMorgans from the gene. The marker has two codominant alleles, *1* and *2*.

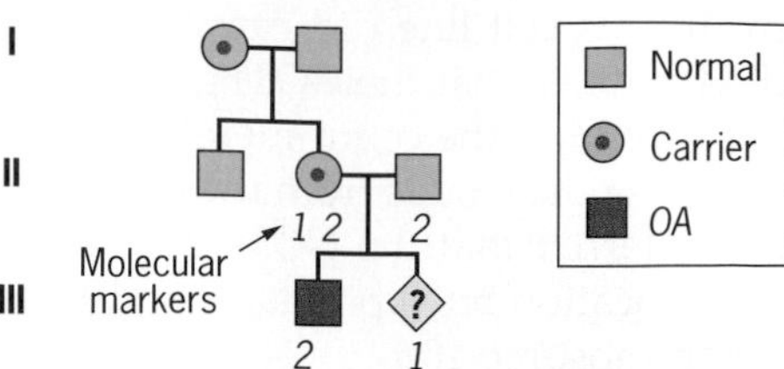

(a) Based on the information in this pedigree, what is the most likely arrangement of the alleles on the mother's (II-2) X chromosomes?

(b) The mother wishes to know if her fetus (III-2) inherited the trait. What is the most likely diagnosis of the fetus based on data for the molecular marker and the knowledge that the fetus is a male?

(c) At a later point, person I-2, the maternal grandfather, comes in for testing and is found to carry molecular marker allele *2*. How does this information affect your answer to (a)?

(d) Based on this additional information, how do you explain the affected male (III-1)?

(e) What further studies are needed to confirm the diagnosis of III-1?

(f) Does this additional information alter your diagnosis of the fetus (III-2)? Explain.

ANSWER

(a) The mother passed the *OA* allele and marker *2* to her affected son. Since the two points are about 5 cM apart, there is a 95 percent chance she is:

———*2*——— *OA* —

———*1*——— + —

(b) The fetus received marker *1* from his mother. There is a 95 percent chance the mother has the genotype diagrammed in (a) and a 5 percent chance she has the *1* allele linked to *OA*. If she is as shown in (a), there is a 95 percent chance she will pass *1* and + to the fetus and a 5 percent chance she will pass *1* and *OA* to the fetus. If she has the *1* allele linked to *OA*, then there is a 95 percent chance she will pass *1* and *OA* to the fetus. Thus, in total, there is a $(0.95 \times 0.05) + (0.05 \times 0.95) = 9.5$ percent chance that the fetus will be affected.

(c) Since the grandfather was —*2*—+—, the mother must be:

———*1*——— *OA* —

———*2*——— + —

(d) The affected male is —*2*—*OA*—. He must have received a recombinant X chromosome from the mother.

(e) By looking at other markers to the left and right of the *OA* gene in the mother and her offspring, we can confirm a crossover event.

(f) Since we now know with certainty what the arrangement of the alleles is in the mother, there is a 95 percent chance that the fetus (III-2) will have *OA* and a 5 percent chance he will not.

QUESTIONS AND PROBLEMS

9.1 In yeast, the *ad* mutation requires adenine for growth, and the *in* mutation requires inositol for growth. The wild-type alleles of these two mutations are *AD* and *IN*. C. C. Lindegren analyzed 48 tetrads from the cross *AD IN* × *ad in*; among these tetrads, 22 were parental ditype, 3 were nonparental ditype, and 23 were tetratype. Are the *AD* and *IN* genes linked? If so, what is the distance between them?

9.2 *Drosophila* females homozygous for a recessive mutation causing purple eyes (*pr*) were crossed to *Cy/Pm; Tb/Sb* males (see Figure 8.5 for an explanation of these genetic symbols). None of the F_1 flies had purple eyes, but when Curly, Tubby F_1 flies were intercrossed, they produced some purple offspring, and none of these had a Tubby body. Which chromosome carries the *pr* gene?

9.3 A geneticist obtained the following ordered tetrad data from a cross with *Neurospora:*

Spore Pairs				
Top of Ascus			Bottom of Ascus	Number of Tetrads
(1,2)	(3,4)	(5,6)	(7,8)	
A	*A*	*a*	*a*	61
a	*a*	*A*	*A*	55
a	*A*	*a*	*A*	40
A	*a*	*A*	*a*	44
				Total = 200

What is the distance between the *A* locus and its centromere?

9.4 The following tetrad data were obtained from the cross *A B* × *a b* in *Neurospora:*

Spore Pairs				
Top of Ascus			Bottom of Ascus	Number of Tetrads
(1,2)	(3,4)	(5,6)	(7,8)	
AB	*AB*	*ab*	*ab*	1766
AB	*aB*	*Ab*	*ab*	220
AB	*Ab*	*aB*	*ab*	14
				Total = 2000

Use these data to construct a linkage map of the two genes and their centromere.

9.5 Analysis of unordered yeast tetrads from the cross + + + × *a b c* yielded the following data:

Tetrad Class	Spores				Number of Asci
1	*a b c*	*a b c*	+ + +	+ + +	36
2	*a b c*	*a* + *c*	+ *b* +	+ + +	14
3	*a* + +	*a* + +	+ *b c*	+ *b c*	32
4	*a b* +	*a* + +	+ *b c*	+ + *c*	16
5	*a b* +	*a b* +	+ + *c*	+ + *c*	2

Which of these genes is linked, and what is the map distance between them?

9.6 In *Neurospora*, the mutations *arg*, *thi*, and *leu* block the synthesis of arginine, thiamine, and leucine, respectively. The following ordered tetrad data come from two crosses with these mutations:

Cross	Spore Pairs				Number of Tetrads
	(1,2)	(3,4)	(5,6)	(7,8)	
arg × *thi*	*arg* +	*arg* +	+ *thi*	+ *thi*	46
	arg thi	*arg thi*	+ +	+ +	56
					Total = 100
arg × *leu*	*arg* +	*arg* +	+ *leu*	+ *leu*	155
	arg +	*arg leu*	+ +	+ *leu*	44
	arg +	+ *leu*	*arg* +	+ *leu*	1
					Total = 200

Using these data, construct the linkage map(s) for these genes. Show the position of the centromere(s).

9.7 The order of three genes and the centromere on one *Neurospora* chromosome is

centromere *x y z*

A cross between + + + and *x y z* produced one ascus with the following ordered array of ascospores (only one member of each spore pair is shown):

(+ + *z*) (+ *y z*) (*x* + +) (*x y* +)

(a) Is this ascus most likely the result of a meiotic event in which 0, 1, 2, or 3 crossovers occurred?

(b) In what interval(s) did the crossover(s) most likely occur?

(c) If double or triple crossovers were involved, were they two-strand, three-strand, or four-strand multiple crossovers?

9.8 From a cross of a wild-type *Neurospora* strain with a strain that has nutritional requirements for the amino acids histidine (*his*) and tyrosine (*tyr*), the following ordered tetrads were obtained:

Spore Pairs	Tetrads 1	2	3	4	5	6	7	8	
1–2	*his tyr*	*his tyr*	*his tyr*	*his* +	*his* +	+ *tyr*	+ *tyr*	*his* +	
3–4	*his tyr*	*his* +	+ *tyr*	*his* +	*his tyr*	*his tyr*	+ +	+ *tyr*	
5–6	+ +	+ *tyr*	+ +	+ *tyr*	+ *tyr*	*his* +	*his tyr*	*his tyr*	
7–8	+ +	+ +	*his* +	+ *tyr*	+ +	+ +	*his* +	+ +	
Number	30	8	5	30	10	7	8	2	Total = 100

(a) What is the map distance between the *his* locus and the centromere?

(b) What is the linkage relationship between the *his* and *tyr* loci?

9.9 The following tetrads were produced by a cross of a *Neurospora* strain that had white spores (*w*) and a nutritional requirement for the amino acid arginine (*arg*) with a strain that had dark spores and no arginine requirement:

Spore Pairs	Tetrads 1	2	3	4	5	6
1–2	*w arg*	*w arg*	*w arg*	*w arg*	*w* +	*w* +
3–4	*w arg*	*w* +	+ *arg*	+ +	*w* +	+ +
5–6	+ +	+ *arg*	+ +	+ *arg*	+ *arg*	*w arg*
7–8	+ +	+ +	*w* +	*w* +	+ *arg*	+ *arg*
No.	58	14	15	2	1	10
						Total = 100

(a) What is the map distance between the *arg* locus and the centromere?

(b) What is the map distance between the *w* and *arg* loci?

(c) Are the *arg* and *w* loci found on the same or different arms of the chromosome?

9.10 *Drosophila* females with X chromosomes attached at the centromere carried the recessive mutations *y* (*yellow* body) and *sn* (*singed* bristles) on one of their X chromosomes and the wild-type alleles of these genes on the other. When these females were crossed to wild-type males, they produced mostly wild-type progeny. However, a few of their daughters had yellow bodies, and fewer still had yellow bodies *and* singed bristles. How would you explain the occurrence of these exceptional females? Can you infer anything about the location of *y* and *sn* with respect to the centromere?

9.11 A *Drosophila* second chromosome that carried a recessive lethal mutation was balanced over the *Cy* chromosome. *l*/*Cy* females were systematically crossed to males carrying different second chromosome deletions (all homozygous lethal) balanced over the *Cy* chromosome. Each cross was scored for the presence or absence of non-Curly progeny.

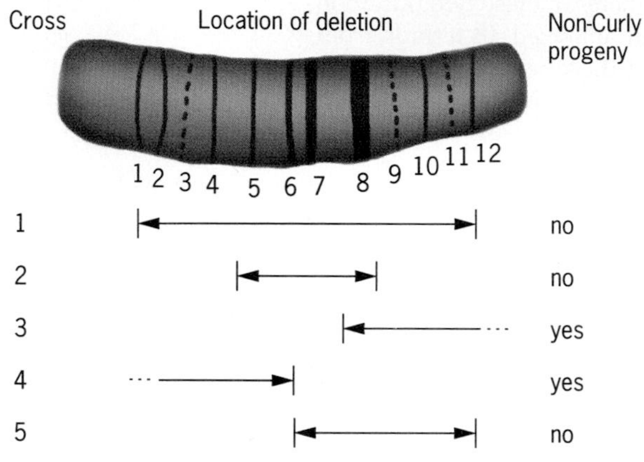

In which band is the lethal mutation located?

9.12 A woman gives birth to six sons. Two are normal, three are both color blind and hemophilic, and one is just color blind. These are X-linked traits. Draw the most likely linkage relationship of the genes on the two woman's X chromosomes.

9.13 A woman gives birth to seven sons. Three are hemophilic, three are color blind, and one is normal. Draw the most likely linkage relationship of the genes on the woman's two X chromosomes.

9.14 A woman with the genetic constitution

A ———— *b*
a ———— *B*

marries a man with the genetic constitution

a ———— *b*
a ———— *b*

A and *B* are dominant to *a* and *b*, and the genes are 20 map units apart. What classes of offspring would this couple produce, and in what approximate proportions?

9.15 A woman is heterozygous for three X-linked genes: *Aa*, *Bb*, and *Cc*. She has seven sons with the following genotypes:

2 *A B c*
3 *a b C*
1 *A B C*
1 *a B c*

Draw the mother's two X chromosomes with the appropriate alleles on each. Which of the sons are the result of recombination?

9.16 A human cell line carries a gene that codes for enzyme E. Mouse cells do not carry this gene. Human–mouse hybrids are formed, and the following stable lines are studied for the human chromosomes present and the presence or absence of enzyme E.

Hybrid Cell Population	Human Chromosomes Present	E Activity Present
1	21, 18, 16, 7, 5, 2	Yes
2	18, 16, 8, 3	Yes
3	21, 18, 1	Yes
4	23, 19, 17, 15, 14, 11, 9, 8, 3, 1	No
5	18, 6, 2	Yes
6	21, 5	No

Using these data, determine where the gene for E is located.

9.17 The following translocation occurred between the X chromosome and chromosome 14:

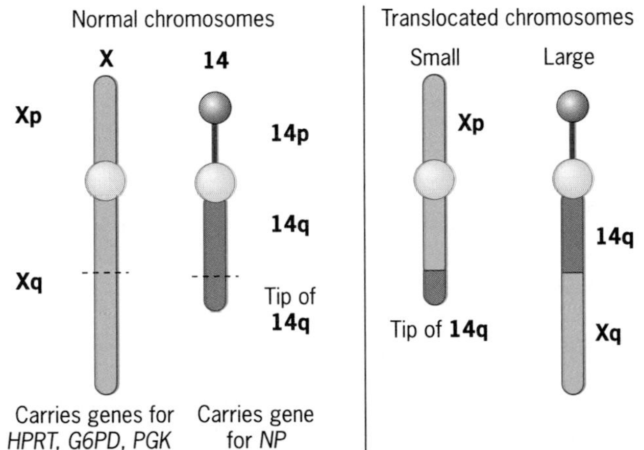

A cell line carrying this translocation was fused with a mouse cell line, and the following hybrid cell lines were isolated:

Translocated chromosomes in hybrid cell lines

Enzymes contained in hybrid cell lines	Both	Large one	Small one	Neither
G6PD	+	+	–	–
HPRT	+	+	–	–
PGK	+	+	–	–
NP	+	+	–	–

You know that the genes for *G6PD*, *HPRT*, and *PGK* are X-linked and that the gene for *NP* is autosomal. What else can you deduce from these data?

9.18 A child was born with the karyotype shown below:

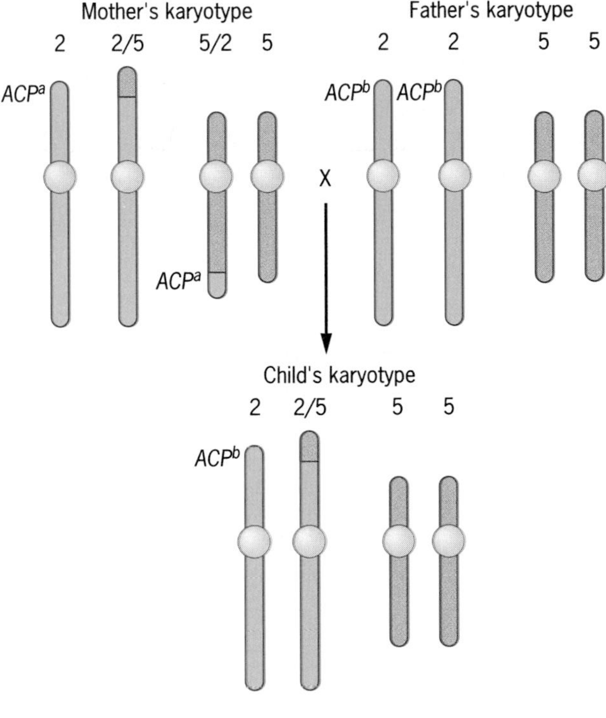

The mother carried a reciprocal translocation and was homozygous for an acid phosphatase allele (ACP^a). The father's karyotype was normal, and he was homozygous for a different *ACP* allele, ACP^b. The *ACP* alleles are codominant, but the child expressed only the ACP^b allele. What can you conclude about the location of the *ACP* gene?

9.19 There is a heritable fragile site on the long arm of chromosome 16. Somatic-cell hybrids show that the *HPA* gene (encoding a polypeptide of the haptoglobin protein) is also on chromosome 16. Among the offspring produced by persons heterozygous for both the fragile site and the *HPA* gene, about 8 percent were recombinant. What do you conclude from this observation?

9.20 The X-linked genes *A* and *B* show 10 percent recombination. Genes *A* and *C* show 16 percent recombination. Construct the gene maps that are compatible with these data.

9.21 Which of the answers to Problem 9.20 would be correct if *B* and *C* showed about 6 percent recombination?

9.22 Fujibayashi and his colleagues (*Am. J. Hum. Genet.* 37:741, 1985) mapped the *SAP2* gene (encoding sphingolipid activator protein-2) using somatic-cell hybrids. Their results are presented in the accompanying table. On which chromosome is the human *SAP2* gene?

Hybrid	Human Chromosomes Present	SAP-2
CP3-1	4, 5, 11, 12, 14, 16, 17, 18, 19, 20, 21, X	−
CP4-1	4, 5, 8, 11, 14, 22, X	−
CP5-1	1, 5, 8, 9, 12, 14, 15, 17, 19, 21, 22	−
CP6-1	4, 12, 14, 17, 18, 19, 21, 22	−
CP12.3.2	2, 8, 9, 11, 12, 21, 22, X	−
CP12-7	2, 4, 8, 9, 10, 11, 12, 13, 21, 22, X	+
CP14-1	4, 5, 17, 21, 22, X	−
CP15-1	4, 5, 11, 12, 15, 16, 17	−
CP16-1	5, 14, 17, 19, 20, 21, 22	−
CP17-1	1, 4, 5, 12, 17	−
CP18-1	1, 8, 11, 14, 15, 17, 18, 19	−
CP20-1	2, 5, 9, 17, 21	−
CP26-1	1, 4, 5, 6, 7, 9, 10, 11, 12, 13, 17, 22	+
CP28-1	1, 4, 5, 8, 9, 18, 19, 20	−
153-E4A-1	4, 5, 8, 9, 11, 13, 15	−
822-19B3	1, 2, 3, 4, 5, 6, 10, 11, 12, 14, 15, 19, 21	+
Q72-18	4, 5, 6, 8, 9, 10, 11, 12, 14, 15, 16, 19, 20, 21, 22, X	+
640-53B	3, 4, 9, 10, 13, 14, 15, 21, 22	+
640-34	4, 6, 9, 10, 12, 18, 19, 21	+
Q21-10A	1, 3, 4, 5, 6, 11, 12, 14, 15, 19, 20, 21, X	−
Q21-3A	1, 3, 4, 5, 6, 10, 11, 14, 15, 19, 20, 21	+
Q68-20-1A	3, 5, 10, 11, 12, 14, 18, 19, 21, X	+

9.23 The following table shows which human chromosomes or chromosome arms were retained in seven clones of human–mouse hybrid cells (+ means the entire chromosome was present, p means only the short arm was present, and q means only the long arm was present) and which of four enzyme activities were detected in these clones:

	Human–Mouse Hybrid Cell Clones						
	A	B	C	D	E	F	G
Human chromosomes:							
2	+	−	+	−	+	−	p
5	−	+	+	q	+	+	−
8	−	+	−	+	−	+	−
17	−	+	+	p	−	q	+
X	+	p	p	+	q	+	+
Enzymes:							
E1	−	+	−	+	−	+	−
E2	+	−	+	−	+	−	+
E3	−	+	+	−	+	+	−
E4	−	+	+	−	−	+	+

Locate the genes that encode these enzymes. If possible, determine on which chromosome arm each gene is located.

BIBLIOGRAPHY

Ashburner, M. 1989. *Drosophila: A Laboratory Handbook.* Cold Spring Harbor Laboratory Press, Cold Spring Harbor, NY.

Houlahan, M. G., Beadle, and H. Calhoun. 1949. Linkage studies with biochemical mutants of *Neurospora crassa. Genetics* 34:493–507.

Judd, B. H. 1998. Genes and chromomeres: a puzzle in three dimensions. *Genetics* 150:1–9.

Judd, B. H., M. W. Shen, and T. C. Kaufman. 1972. The anatomy and function of a segment of the *X* chromosome of *Drosophila melanogaster. Genetics* 71:139–156.

Kao, F., C. Jones, and T. T. Puck. 1976. Genetics of somatic mammalian cells: genetic, immunologic, and biochemical analysis with Chinese hamster cell hybrids containing selected human chromosomes. *Proc. Natl. Acad. Sci. USA* 73:193–197.

Lindegren, C. C. 1949. *The Yeast Cell, Its Genetics and Cytology.* Educational Publishers, St. Louis.

Ruddle, F. H., and R. S. Kucherlapati. 1974. Hybrid cells and human genes. *Sci. Amer.* 231(1):36–44.

Thompson, M. W., R. R. McInnes, and H. F. Willard. 1991. *Genetics in Medicine*, 5th ed. W. B. Saunders, Philadelphia.

A CONVERSATION WITH
EDWARD B. LEWIS

What inspired you to become a biologist? Did you have any mentors or role models who were especially inspiring to you?

An interest in animals, especially toads, turtles, and snakes, probably inspired me to become a biologist. In high school I was most influenced by a book, *The Biological Basis of Human Nature*, by H. S. Jennings in which there were figures of the results of *Drosophila* matings that demonstrated sex linkage of the *white* eye gene. I had no mentors in high school, but a fellow student, Edward Novitski (now Emeritus Professor of Genetics at the University of Oregon), and I teamed up in high school to carry out experiments with *Drosophila*. Novitski corresponded with Calvin Bridges of Caltech, who provided us with cultures and instructions. As an undergraduate at the University of Minnesota, I was greatly indebted to Professor C. P. Oliver (a student of H. J. Muller) who gave me a desk in his laboratory so I could carry on my *Drosophila* experiments. He was studying lozenge eye mutants which he showed undergo mutation associated with crossing over—a result that was a forerunner of pseudoallelism.

You concentrate most of your research on *Drosophila* genetics. What is so enticing about *Drosophila* genetics?

Drosophila genetics combines virtually all of the requirements needed to deduce the laws of inheritance for higher organisms: a short life cycle of 10 days at room temperature; high fertility; only 4 pairs of chromosomes; giant salivary gland chromosomes. For a long time it did not seem to be a good organism for studying embryology; but new tools and a host of mutant genes now make *Drosophila* a leading organism for studying development.

Edward B. Lewis is currently Professor Emeritus at the California Institute of Technology in Pasadena. He received his Ph.D. at the California Institute of Technology in 1942. He served as Captain in the United States Army Air Force from 1942 to 1945, as a meteorologist and oceanographer. He received an honorary Doctor of Science degree from the University of Umea in Sweden in 1982, and an honorary Doctor of Science degree from the University of Minnesota in 1993. His research has focused on developmental genetics and on the somatic effects of radiation. He has received numerous awards and prizes including the Nobel Prize in Physiology or Medicine in 1995. Dr. Lewis, Christiane Nüsslein-Volhard, and Eric Wieschaus were awarded this Nobel Prize for their research on the genetic control of development in *Drosophila*. He was a member of the National Advisory Committee on Radiation from 1958 to 1961 and served on several Advisory Committees of the National Academy of Sciences on the Biological Effects of Ionizing Radiation.

Do studies of *Drosophila* genes help us to understand how human genes function?

Now there are many examples in which genes first found in *Drosophila* turn out to have homologues in human beings; for example, the homeotic genes that control the head, thorax and abdomen of the fly and of the human being are highly conserved between these organisms in the homebox portion of these genes.

In your career as a scientist, what was one of your greatest challenges, and how did you overcome it?

One of my greatest challenges as a scientist was how to alert the public to the potential dangers of fallout from atom-bomb testing by the United States and other countries. This led to a publication [E. B. Lewis, Leukemia and Ionizing Radiation, *Science* 125:965–972 (1957)] in which I analyzed several groups of persons who had been exposed to ionizing radiation (including A-bomb survivors) and showed that the risk of leukemia in these groups was compatible with a linear relation between radiation dose and number of cases of leukemia with no indication of a threshold dose below which no cases would be expected. By extrapolation, the leukemia risk was estimated as one or two cases per million persons per year (0.1 rad being the approximate annual dose to the body from natural background radiation). A threshold dose of many rads for cancer induction had been assumed by those in charge of testing atomic weapons. Two years later I pointed out [E. B. Lewis, Thyroid Radiation Doses from Fallout, *Proc. Natl. Acad. Sci.* 45:894–897 (1959)] that aboveground testing of such weapons in Nevada produced levels of radioiodine in the thyroid of infants consuming fresh milk that probably averaged 0.1 rad per year for several years (with some regions of the United States receiving doses that could have averaged one rad). Infants receive a dose ten to twenty times that which the adult thyroid receives for the same amount taken up by the gland, owing to the small size of the infant thyroid, and this had been overlooked by those in charge of testing. The public had been assured that fallout levels would result in exposures that were well below the natural background level of 0.1 rad. I also carried out an epidemiological study [E. B. Lewis, Leukemia, Multiple Myeloma, and Aplastic Anemia in American Radiologists, *Science* 142:1492–1494 (1963)] that showed that American radiologists, occupationally exposed to radiation, experienced during the years 1948 to 1961, inclusive, a statistically highly significant increased death rate from leukemia, multiple myeloma, and aplastic anemia.

That cancers result from somatic mutation had been proposed by H. J. Muller [*Science* 66:844 (1927)] in his famous paper showing that X rays can induce genetic mutations in *Drosophila*. The absence of a detectable threshold for certain cancers as well as for germline mutations after exposure to ionizing radiation lent further support to Muller's hypothesis. Molecular evidence for such mutations is now well established for certain human cancers.

What do you see as the most challenging and exciting problems to be explored in genetics today?

The most exciting challenges are to determine how genes are able to control the development and behavior of organisms, including human beings, and to determine how gene duplication has been able to accomplish the evolution of higher organisms. Computer analysis of DNA sequence data will be increasingly helpful in identifying increasing numbers of genes involved in producing birth defects and other heritable diseases as well as those involved in cancer.

What advice would you give to undergraduate students interested in the study of genetics?

One, obtain a firm foundation in all branches of genetics and in the related disciplines of cytology, statistics, biochemistry, and molecular biology and two, retain a keen interest in the biology of living organisms, whether they be plants or animals.

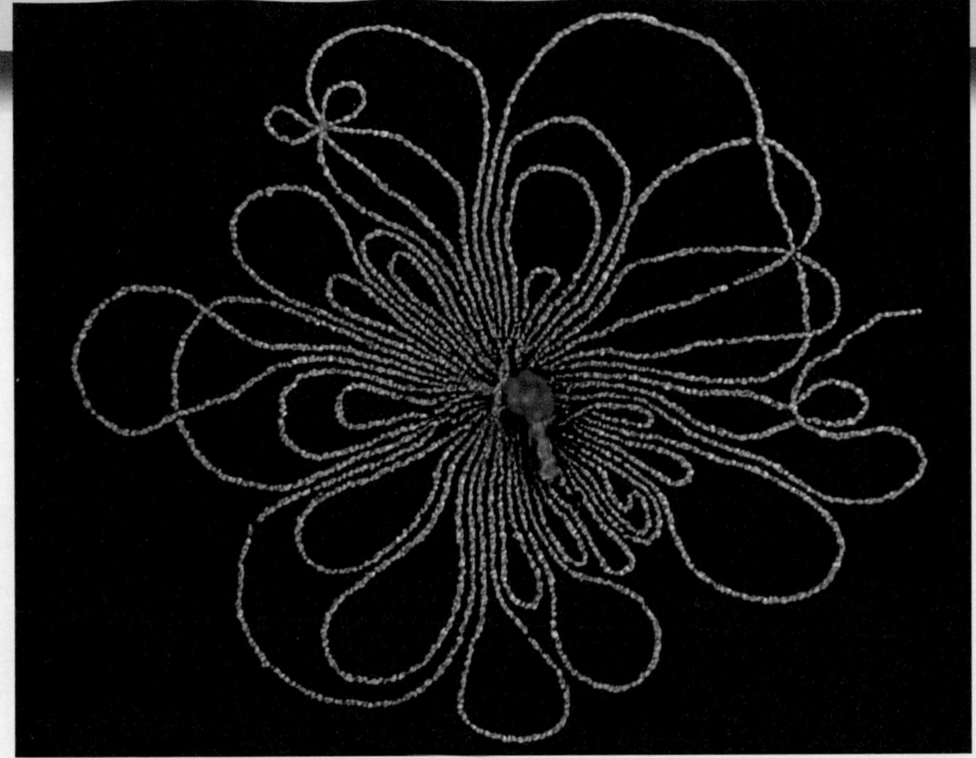

Electron micrograph of the DNA released from a bacteriophage.

DNA and the Molecular Structure of Chromosomes

CHAPTER OUTLINE

Functions of the Genetic Material
Proof That Genetic Information Is Stored in DNA
The Structures of DNA and RNA
Chromosome Structure in Prokaryotes and Viruses
Chromosome Structure in Eukaryotes
Eukaryotic Genomes: Repeated DNA Sequences

Discovery of Nuclein

In 1868, Johann Friedrich Miescher, a young Swiss medical student, became fascinated with an acidic substance that he isolated from pus cells obtained from bandages used to dress human wounds. To obtain larger quantities of pus for his investigations, Miescher began purchasing used bandages from a surgical clinic in Tübingen, Germany. He first separated the pus cells from the bandages and associated debris, and then treated the cells with pepsin, a proteolytic enzyme that he isolated from the stomachs of pigs. After the pepsin treatment, he recovered an acidic substance that he called "nuclein." Miescher's nuclein was unusual in that it contained large amounts of both nitrogen and phosphorus, two elements known at the time to coexist only in certain types of fat. Miescher wrote a paper describing his discovery of nuclein in human pus cells and submitted it for publication in 1869. However, the editor of the journal to which the paper was sent was skeptical of the results and decided to repeat the experiments himself. As a result, Miescher's paper describing nuclein was not published until 1871, two entire years after its submission.

At the time, the importance of the substance that Miescher called nuclein could not have been anticipated. The existence of polynucleotide chains, the key component of the acidic material in Miescher's nuclein, was not documented until the 1940s. The role of nucleic acids in storing and transmitting genetic information was not established until 1944, and the double-helix structure of DNA was not discovered until 1953. Even in 1953, many geneticists were reluctant to accept the idea that nucleic acids, rather than proteins, carried the genetic information because nucleic acids exhibited less structural variability than proteins.

FUNCTIONS OF THE GENETIC MATERIAL

In 1865, Mendel showed that genes transmitted genetic information, and in the first few decades of the twentieth century, their patterns of transmission from generation to generation were studied extensively. Although these classical genetic studies provided little insight into the molecular nature of genes, they did demonstrate that, whatever its chemical nature, the genetic material must perform three essential functions.

1. The genotypic function, **replication.** The genetic material must store genetic information and accurately transmit that information from parents to offspring, generation after generation.
2. The phenotypic function, **gene expression.** The genetic material must control the development of the phenotype of the organism. That is, the genetic material must dictate the growth and differentiation of the organism from the single-celled zygote to the mature adult.
3. The evolutionary function, **mutation.** The genetic material must undergo change so that organisms can adapt to modifications in the environment. Without such changes, evolution could not occur.

Other early genetic studies established a precise correlation between the patterns of transmission of genes and the behavior of chromosomes during sexual reproduction, providing strong evidence that genes are usually located on chromosomes. Thus further attempts to discover the chemical basis of heredity focused on the molecules present in chromosomes.

Chromosomes are composed of two types of large organic molecules (macromolecules) called **proteins** and **nucleic acids.** The nucleic acids are of two types: **deoxyribonucleic acid (DNA)** and **ribonucleic acid (RNA).** During the 1940s and early 1950s, the results of elegant experiments clearly established that the genetic information is stored in nucleic acids, not in proteins. These experiments are discussed in the next section of this chapter. In most organisms, the genetic information is encoded in the structure of DNA. However, in some small viruses, the genetic information is encoded in RNA.

Key Points: **The genetic material must provide the genotypic, phenotypic, and evolutionary functions of replication, gene expression, and mutation, respectively.**

PROOF THAT GENETIC INFORMATION IS STORED IN DNA

Several lines of indirect evidence suggested that DNA harbors the genetic information of living organisms.

1. Most of the DNA of cells is located in the chromosomes, whereas RNA and proteins are also abundant in the cytoplasm.
2. A precise correlation exists between the amount of DNA per cell and the number of sets of chromosomes per cell. For example, most somatic cells of diploid organisms contain twice the amount of DNA as the haploid germ cells (gametes) of the same species.
3. The molecular composition of the DNA is the same (with rare exceptions) in all the different cells of an organism, whereas the composition of both RNA and proteins is highly variable from one cell type to another.
4. DNA is more stable than RNA or proteins, which are synthesized and degraded quite rapidly in living organisms. Since the genetic material must store and transmit information from parents to offspring, we might expect it to be stable, like DNA.

Although these correlations strongly suggest that DNA is the genetic material, they by no means prove it.

Discovery of Transformation in Bacteria

Bacterial transformation, a type of recombination occurring in bacteria, was discovered by Frederick Grif-

TABLE 10.1
Characteristics of *Streptococcus pneumoniae* Strains When Grown on Blood Agar Medium

	Colony Morphology				*Reaction with Antiserum Prepared Against*	
Type	*Appearance*	*Size*	*Capsule*	*Virulence*	*Type IIS*	*Type IIIS*
IIR[a]	Rough	Small	Absent	Avirulent	None	None
IIS	Smooth	Large	Present	Virulent	Agglutination	None
IIIR[a]	Rough	Small	Absent	Avirulent	**None**	**None**
IIIS	Smooth	Large	Present	Virulent	**None**	**Agglutination**

[a]Although Type R cells are nonencapsulated, they carry genes that would direct the synthesis of a specific kind (antigenic Type II or III) of capsule if the block in capsule formation were not present. When Type R cells mutate back to encapsulated Type S cells, the capsule Type (II or III) is determined by these genes. Thus, R cells derived from Type IIS cells are designated Type IIR. When these Type IIR cells mutate back to encapsulated Type S cells, the capsules are of Type II.

fith in 1928. Although his experiments on *Streptococcus pneumoniae* (pneumococcus) provided no evidence that DNA was involved in the transformation process, they set the stage for later discoveries.

Pneumococci, like all other living organisms, exhibit genetic variability that can be recognized by the existence of different phenotypes (Table 10.1). The two phenotypic characteristics of importance in Griffith's demonstration of transformation are (1) the presence or absence of a polysaccharide (complex sugar polymer) capsule surrounding the bacterial cells, and (2) the type of capsule—that is, the specific molecular composition of the polysaccharides present in the capsule. When grown on blood agar medium in petri dishes, pneumococci with capsules form large, smooth colonies (Figure 10.1) and are thus designated Type S. Encapsulated pneumococci are virulent (pathogenic), causing pneumonia in mammals such as mice and humans. The virulent Type S pneumococci mutate to an avirulent (nonpathogenic) form that has no polysaccharide capsule at a frequency of about once per 10^7 cells. When grown on blood agar medium, such nonencapsulated, avirulent pneumococci produce small, rough-surfaced colonies (Figure 10.1) and are thus designated Type R. The polysaccharide capsule is required for virulence because it protects the bacterial cell from destruction by white blood cells. When a capsule is present, it may be of several different antigenic types (Type I, II, III, and so forth), depending on the specific molecular composition of the polysaccharides and, of course, ultimately depending on the genotype of the cell.

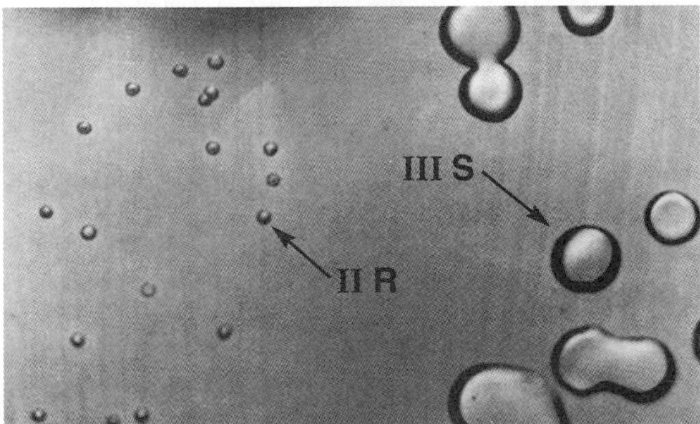

Figure 10.1 Colony phenotypes of the two strains of *Streptococcus pneumoniae* studied by Griffith in 1928.

The different capsule types can be identified immunologically. If Type II cells are injected into the bloodstream of rabbits, the immune system of the rabbits will produce antibodies that react specifically with Type II cells. Such Type II antibodies will agglutinate Type II pneumococci but not Type I or Type III pneumococci.

Griffith's unexpected discovery was that if he injected heat-killed Type IIIS pneumococci (virulent when alive) plus live Type IIR pneumococci (avirulent) into mice, many of the mice succumbed to pneumonia, and live Type IIIS cells were recovered from the carcasses (Figure 10.2). When mice were injected with heat-killed Type IIIS pneumococci alone, none of the mice died. The observed virulence was therefore not due to a few Type IIIS cells that survived the heat treatment. The live pathogenic pneumococci recovered from the carcasses had Type III polysaccharide capsules. This result is important because nonencapsulated Type R cells can mutate back to encapsulated Type S cells. However, when such a mutation occurs in a Type IIR cell, the resulting cell will become Type IIS, not Type IIIS. Thus, the transformation of avirulent Type IIR cells to virulent Type IIIS cells cannot be explained by mutation. Instead, some component of the dead Type IIIS cells (the "transforming principle") must have converted living Type IIR cells to Type IIIS.

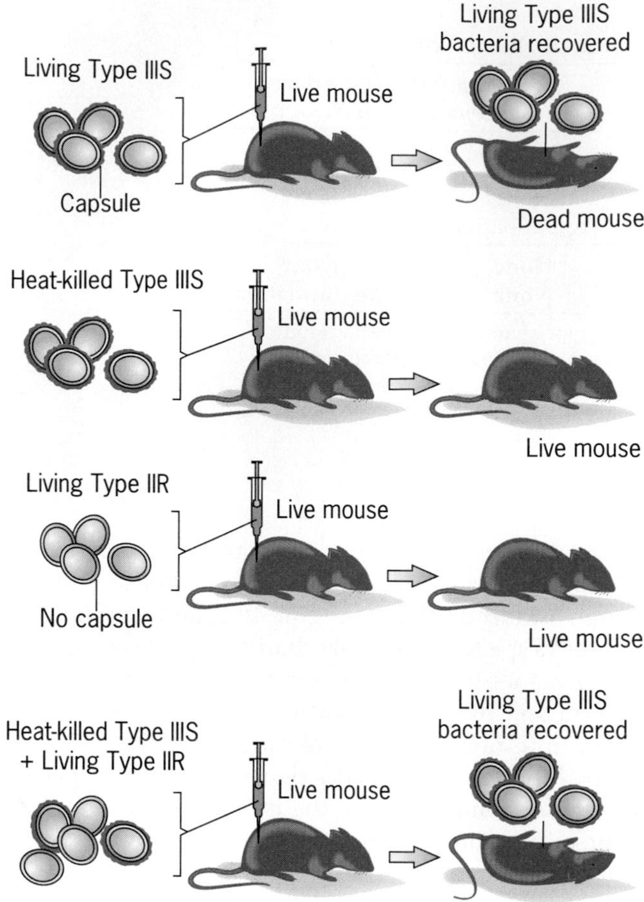

Figure 10.2 Griffith's discovery of transformation in *Streptococcus pneumoniae.*

Subsequent experiments showed that the phenomenon described by Griffith, now called **transformation,** was not mediated in any way by a living host. The same phenomenon occurred in the test tube when live Type IIR cells were grown in the presence of dead Type IIIS cells or extracts of Type IIIS cells. Since Griffith's experiments demonstrated that the Type IIIS phenotype of the transformed cells was passed on to progeny cells—that is, was due to a permanent inherited change in the genotype of the cells—the demonstration of transformation set the stage for determining the chemical basis of heredity in pneumococcus.

Proof That DNA Mediates Transformation

The first direct evidence showing that the genetic material is DNA rather than protein or RNA was published by Oswald Avery, Colin MacLeod, and Maclyn McCarty in 1944. They showed that the component of the cell responsible for transformation (the "transforming principle") in *Streptococcus pneumoniae* is DNA. Avery and colleagues demonstrated that if highly purified DNA from Type IIIS pneumococci was present with Type IIR pneumococci, some of the pneumococci were transformed to Type IIIS (Figure 10.3). But how could they be sure that the DNA was really pure? Proving the complete purity of any macromolecular substance is extremely difficult. Maybe the DNA preparation contained a few molecules of protein, and these contaminating proteins were responsible for the observed transformation. The most definitive experiments in Avery, MacLeod, and McCarty's proof that DNA was the transforming principle involved the use of enzymes that degrade DNA, RNA, or protein. In separate experiments, highly purified DNA from Type IIIS cells was treated with the enzymes (1) **deoxyribonuclease (DNase),** which degrades DNA, (2) **ribonuclease (RNase),** which degrades RNA, or (3) **proteases,** which degrade proteins; the DNA was then tested for its ability to transform Type IIR cells to Type IIIS. Only DNase treatment had any effect on the transforming activity of the DNA preparation—it eliminated all transforming activity (Figure 10.3).

Although the molecular mechanism by which transformation occurs remained unknown for many years, the results of Avery and coworkers clearly established that the genetic information in pneumococcus is present in DNA. Geneticists now know that the segment of DNA in the chromosome of pneumococcus that carries the genetic information specifying the synthesis of a Type III capsule is physically inserted into the chromosome of the Type IIR recipient cell during the transformation process.

Proof That DNA Carries the Genetic Information in Bacteriophage T2

Additional evidence demonstrating that DNA is the genetic material was published in 1952 by Alfred Hershey (1969 Nobel Prize winner) and Martha Chase. The results of their experiments showed that the genetic information of a particular bacterial virus (bacteriophage T2) was present in DNA. Their results had a major impact on scientists' acceptance of DNA as the genetic material. This impact was the result of the simplicity of the Hershey–Chase experiment.

Viruses are the smallest living organisms; they are living, at least in the sense that their reproduction is controlled by genetic information stored in nucleic acids via the same processes as in cellular organisms (Chapter 16). However, viruses are acellular parasites that can reproduce only in appropriate host cells. Their reproduction is totally dependent on the metabolic machinery (ribosomes, energy-generating systems, and

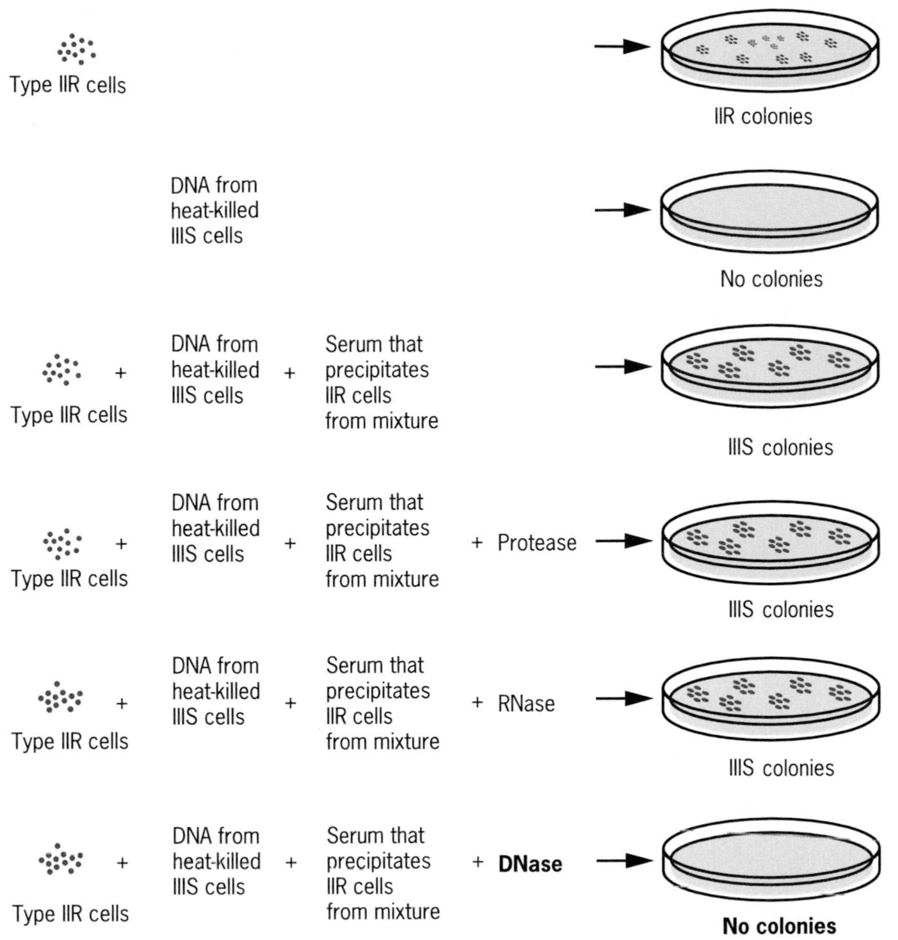

Figure 10.3 Avery, MacLeod, and McCarty's proof that the "transforming principle" is DNA.

other components) of the host. Viruses have been extremely useful in the study of many genetic processes because of their simple structure and chemical composition (many contain only proteins and nucleic acids) and their very rapid reproduction (15 to 20 minutes for some bacterial viruses under optimal conditions).

Bacteriophage T2, which infects the common colon bacillus *Escherichia coli*, is composed of about 50 percent DNA and about 50 percent protein (Figure 10.4). Experiments prior to 1952 had shown that all bacteriophage T2 reproduction takes place within *E. coli* cells. Therefore, when Hershey and Chase showed that the DNA of the virus particle entered the cell, whereas most of the protein of the virus remained adsorbed to the outside of the cell, the implication was that the genetic information necessary for viral reproduction was present in DNA. The basis for the Hershey–Chase experiment is that DNA contains phosphorus but no sulfur, whereas proteins contain sulfur but virtually no phosphorus. Thus, Hershey and Chase were able to label specifically either (1) the phage DNA by growth in a medium containing the radioactive isotope of phosphorus, ^{32}P, in place of the normal isotope, ^{31}P; or (2) the phage protein coats by growth in a medium containing radioactive sulfur, ^{35}S, in place of the normal isotope, ^{32}S (Figure 10.4).

When T2 phage particles labeled with ^{35}S were mixed with *E. coli* cells for a few minutes and the phage-infected cells were then subjected to shearing forces in a Waring blender, most of the radioactivity (and thus the proteins) could be removed from the cells without affecting progeny phage production. When T2 particles in which the DNA was labeled with ^{32}P were used, however, essentially all the radioactivity was found inside the cells; that is, the DNA was not subject to removal by shearing in a blender. The sheared-off phage coats were separated from the infected cells by low-speed centrifugation, which pellets (sediments) cells while leaving phage particles suspended. These results indicated that the DNA of the virus enters the host cell, whereas the protein coat remains outside the cell. Since progeny viruses are produced inside the cell, Hershey and Chase's results indicated that the genetic information directing the synthesis of both the DNA molecules and the protein coats of the progeny viruses must be present in the parental DNA. Moreover, the progeny particles were shown to contain some of the ^{32}P, but none of the ^{35}S of the parental phage.

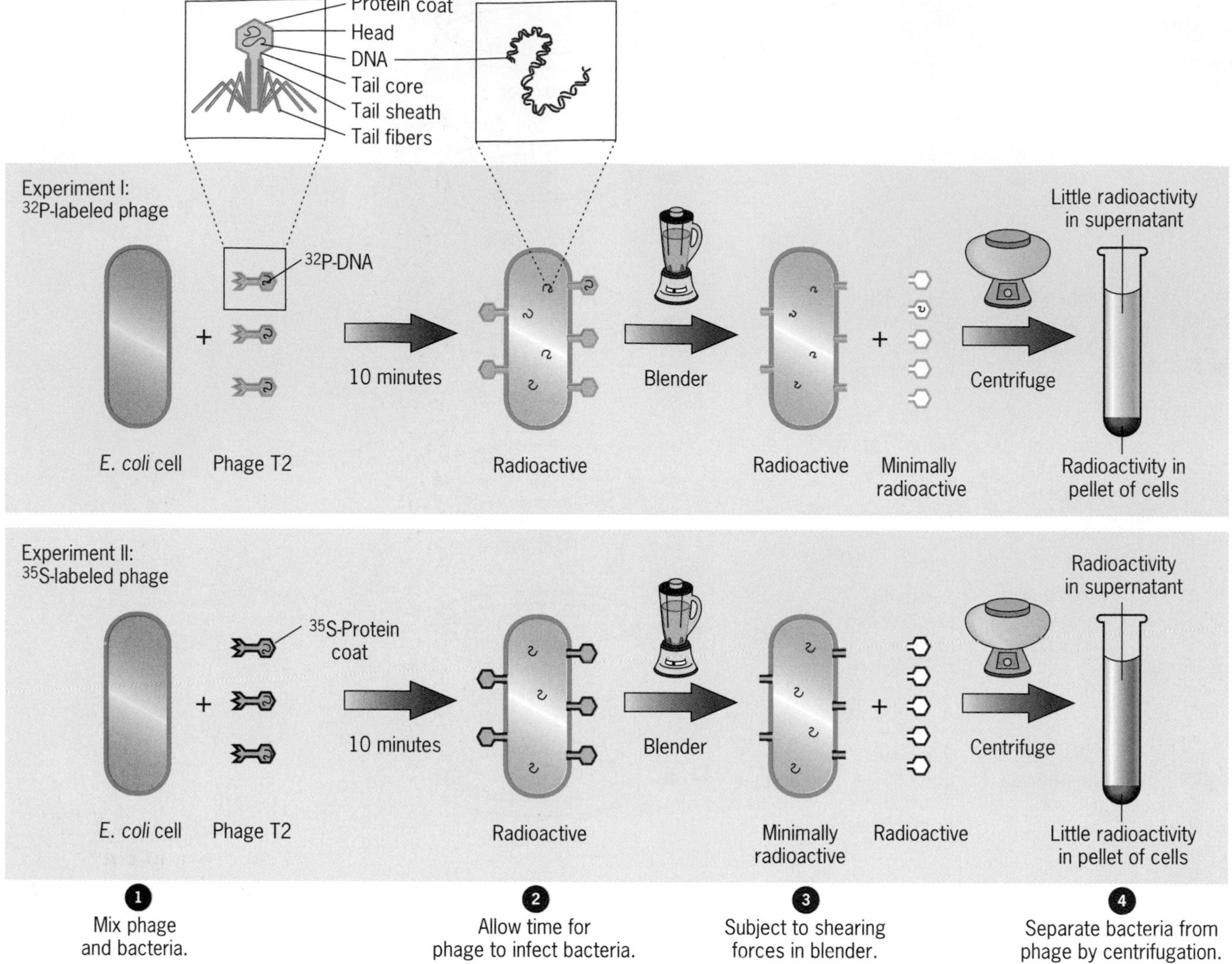

Figure 10.4 Demonstration by Hershey and Chase that the genetic information of bacteriophage T2 resides in its DNA.

There was one flaw in Hershey and Chase's proof that the genetic material of phage T2 is DNA. A significant amount of ^{35}S (and thus protein) was found to be injected into the host cells with the DNA. Thus, it could be argued that this small fraction of the phage proteins contained the genetic information. More recently, scientists have developed procedures by which protoplasts (cells with the walls removed) of *E. coli* can be infected with pure phage DNA. Normal infective progeny phage are produced in these experiments, called **transfection** experiments, proving that the genetic material of such bacterial viruses is DNA.

Proof That RNA Stores the Genetic Information in Some Viruses

As more and more viruses were identified and studied, it became apparent that many of them contain RNA and proteins, but no DNA. In all cases studied to date, it is clear that these RNA viruses—like all other organisms—store their genetic information in nucleic acids rather than in proteins, although in these viruses the nucleic acid is RNA. One of the first experiments that established RNA as the genetic material in RNA viruses was the so-called reconstitution experiment of Heinz Fraenkel-Conrat and coworkers, published in 1957. Their simple, but definitive, experiment was done with tobacco mosaic virus (TMV), a small virus composed of a single molecule of RNA encapsulated in a protein coat. Different strains of TMV can be identified on the basis of differences in the chemical composition of their protein coats.

Fraenkel-Conrat and colleagues treated TMV particles of two different strains with chemicals that dissociate the protein coats of the viruses from the RNA molecules and separated the proteins from the RNA. Then they mixed the proteins from one strain with the

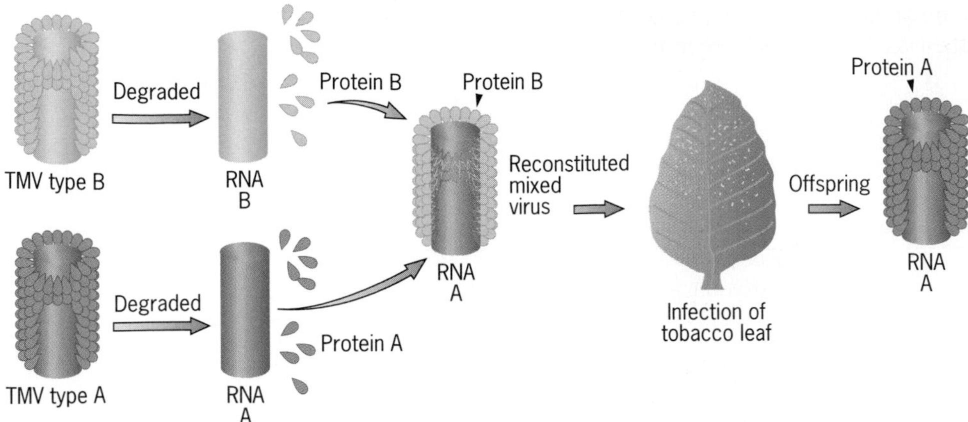

Figure 10.5 Demonstration that the genetic material of tobacco mosaic virus (TMV) is RNA, not protein. TMV contains no DNA; it is composed of just RNA and protein.

RNA molecules from the other strain under conditions that result in the reconstitution of complete, infective viruses composed of proteins from one strain and RNA from the other strain. When tobacco leaves were infected with these reconstituted mixed viruses, the progeny viruses were always phenotypically and genotypically identical to the parent strain from which the RNA had been obtained (Figure 10.5). Thus, the genetic information of TMV is stored in RNA, not in protein.

Key Points: **The genetic information of living organisms is stored in deoxyribonucleic acid (DNA). In some viruses, ribonucleic acid (RNA) is the genetic material.**

THE STRUCTURES OF DNA AND RNA

The genetic information of all living organisms, except the RNA viruses, is stored in DNA. What is the structure of DNA, and in what form is the genetic information stored? What features of the structure of DNA facilitate the accurate transmission of genetic information from generation to generation? The answers to these questions are without doubt three of the most important facets of our understanding of the nature of life.

Nature of the Chemical Subunits in DNA and RNA

Nucleic acids, the major components of Miescher's nuclein, are macromolecules composed of repeating subunits called **nucleotides**. Each nucleotide is composed of (1) a phosphate group, (2) a five-carbon sugar (or pentose), and (3) a cyclic nitrogen-containing compound called a base (Figure 10.6). In DNA, the sugar is 2-deoxyribose (thus the name deoxyribonucleic acid); in RNA, the sugar is ribose (thus ribonucleic acid). Four different bases commonly are found in DNA: **adenine (A)**, **guanine (G)**, **thymine (T)**, and **cytosine (C).** RNA also usually contains adenine, guanine, and cytosine but has a different base, **uracil (U)**, in place of thymine. Adenine and guanine are double-ring bases called **purines;** cytosine, thymine, and uracil are single-ring bases called **pyrimidines.** Both DNA and RNA, therefore, contain four different subunits or nucleotides, two purine nucleotides, and two pyrimidine nucleotides (Figure 10.7). In polynucleotides such as DNA and RNA, these subunits are joined together in long chains (Figure 10.8). RNA usually exists as a single-stranded polymer that is composed of a long sequence of nucleotides. DNA has one additional—and very important—level of organization; it is usually a double-stranded molecule.

DNA Structure: The Double Helix

One of the most exciting breakthroughs in the history of biology occurred in 1953 when James Watson and Francis Crick (Figure 10.9) deduced the correct structure of DNA. Their double-helix model of the DNA molecule immediately suggested an elegant mechanism for the transmission of genetic information (see below and Chapter 11). Watson and Crick's double-helix structure was based on two major kinds of evidence.

1. When Erwin Chargaff and colleagues analyzed the composition of DNA from many different organisms, they found that the concentration of thymine was always equal to the concentration of adenine and the concentration of cytosine was always equal to the concentration of guanine. Their results strongly suggested that thymine and ade-

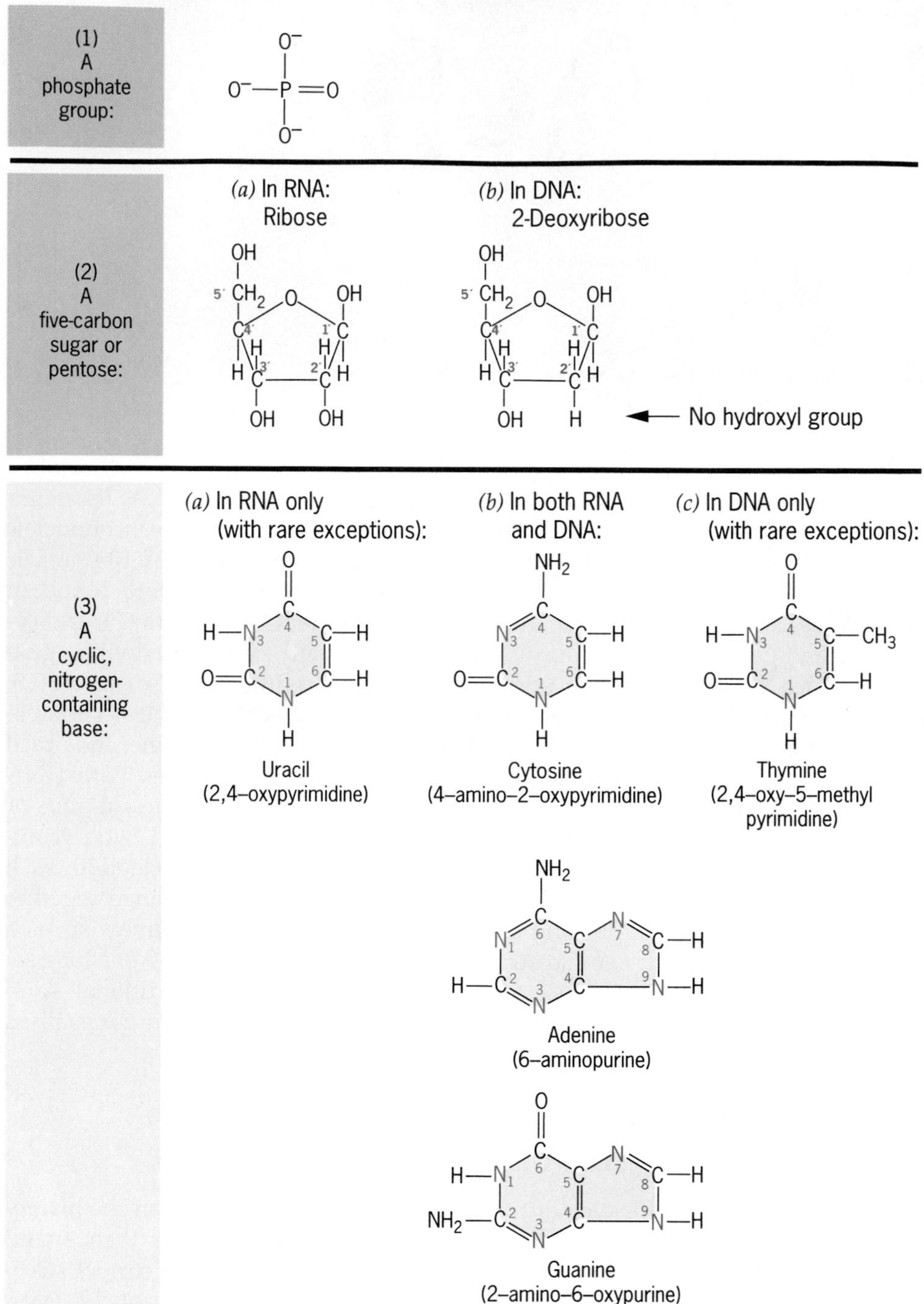

Figure 10.6 Structural components of nucleic acids.

nine as well as cytosine and guanine were present in DNA in some fixed interrelationship. Their data also showed that the total concentration of pyrimidines (thymine plus cytosine) was always equal to the total concentration of purines (adenine plus guanine; see Table 10.2). In contrast, the [thymine + adenine]/[cytosine + guanine] ratio varied widely in DNAs of different species.

2. When X rays are focused through crystals of purified molecules, the rays are deflected by the atoms of the molecules in specific patterns, called diffraction patterns, which provide information

Pyrimidine nucleotides

Deoxythymidine monophosphate, dTMP

Deoxycytidine monophosphate, dCMP

Purine nucleotides

Deoxyadenosine monophosphate, dAMP

Deoxyguanosine monophosphate, dGMP

Figure 10.7 Structures of the four common deoxyribonucleotides present in DNA.

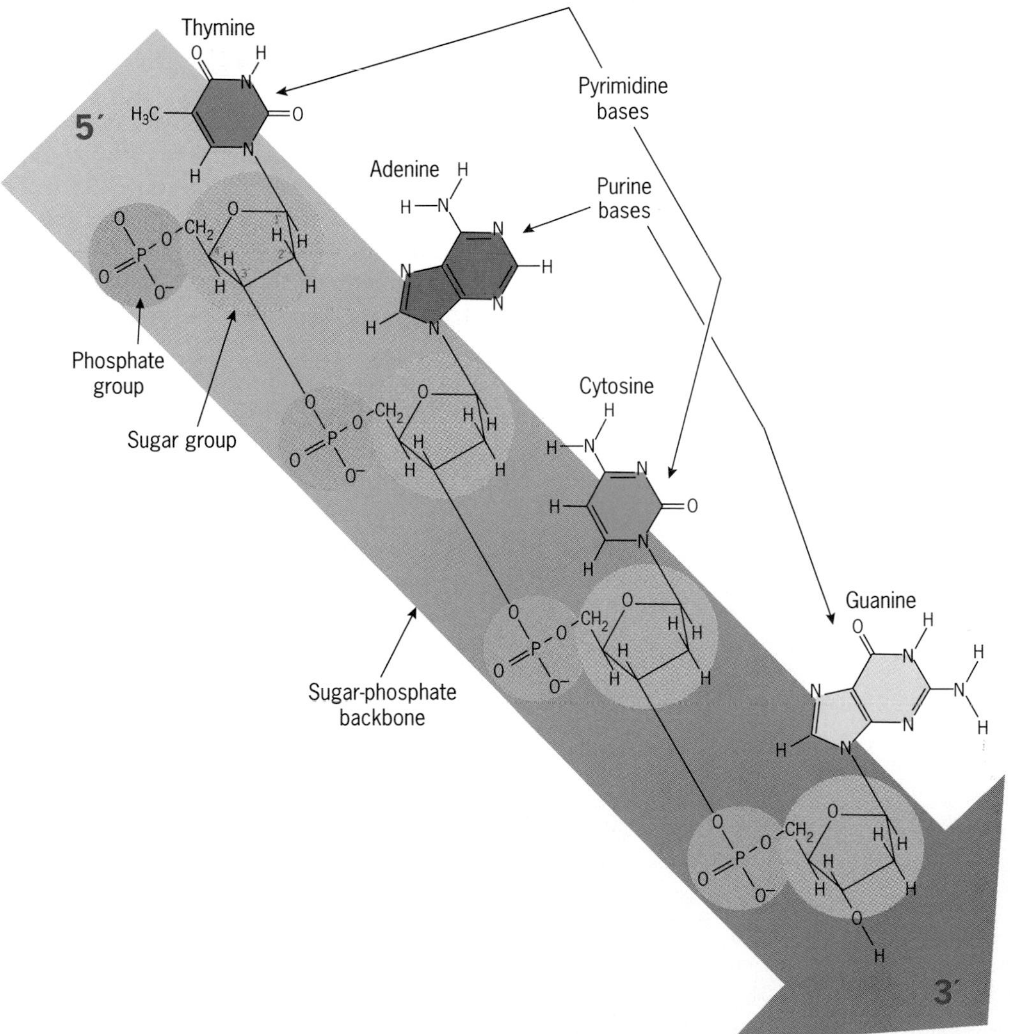

Figure 10.8 Formation of a polynucleotide chain by joining nucleotides with phosphodiester linkages. Note that the polynucleotide has a 5′ to 3′ chemical polarity (top to bottom); each phosphodiester linkage joins the 5′ carbon of 2′-deoxyribose in one nucleotide to the 3′ carbon of 2′-deoxyribose in the adjacent nucleotide, resulting in the chemical polarity of the polynucleotide chain.

Figure 10.9 James D. Watson (left) and Francis H. C. Crick.

about the organization of the components of the molecules. These **X-ray diffraction patterns** can be recorded on X-ray-sensitive film just as patterns of light can be recorded with a camera and light-sensitive film. Watson and Crick used X-ray crystallographic data on DNA structure (Figure 10.10) provided by Maurice Wilkins, Rosalind Franklin, and their coworkers. These data indicated that DNA was a highly ordered, two-stranded structure with repeating substructures spaced every 0.34 nanometer (1 nm = 10^{-9} meter) along the axis of the molecule.

On the basis of Chargaff's chemical data, Wilkins' and Franklin's X-ray diffraction data, and inferences from model building, Watson and Crick proposed that DNA exists as a right-handed **double helix** in which the two polynucleotide chains are coiled about one another in a spiral (Figure 10.11). Each polynucleotide chain consists of a sequence of nucleotides linked together by phosphodiester bonds, joining adjacent deoxyribose moieties (Table 10.3). The two polynucleotide strands are held together in their helical configuration by hydrogen bonding (Table 10.3) between bases in opposing strands; the resulting base pairs are stacked between the two chains perpendicular to the axis of the molecule like the steps of a spiral staircase (Figure 10.11). The base-pairing is specific: adenine is always paired with thymine, and guanine is always paired with cytosine. Thus, all base pairs consist of one purine and one pyrimidine. The specificity of base-pairing results from the hydrogen-bonding capacities of the bases in their normal configurations (Figure 10.12). In their common structural configurations, adenine and thymine form two hydrogen bonds, and guanine and cytosine form three hydrogen bonds.

TABLE 10.2
Base Composition of DNA from Various Organisms

					Molar Ratios	
Species	% *Adenine*	% *Guanine*	% *Cytosine*	% *Thymine*	$\frac{A+G}{T+C}$	$\frac{A+T}{G+C}$
I. **Viruses**						
Bacteriophage λ	26.0	23.8	24.3	25.8	0.99	1.08
Bacteriophage T2	32.6	18.1	16.6	32.6	1.03	1.88
Herpes simplex	13.8	37.7	35.6	12.8	1.06	0.36
Vaccinia	31.5	18.0	19.0	31.5	0.98	1.70
II. **Bacteria**						
Escherichia coli	26.0	24.9	25.2	23.9	1.04	1.00
Micrococcus lysodeikticus	14.4	37.3	34.6	13.7	1.07	0.39
Ramibacterium ramosum	35.1	14.9	15.2	34.8	1.00	2.32
III. **Fungi**						
Neurospora crassa	23.0	27.1	26.6	23.3	1.00	0.86
Aspergillus niger	25.0	25.1	25.0	24.9	1.00	1.00
Saccharomyces cerevisiae	31.7	18.3	17.4	32.6	1.00	1.80
IV. **Higher Eukaryotes**						
Zea mays (corn)	25.6	24.5	24.6	25.3	1.00	1.04
Nicotiana tabacum (tobacco)	29.3	23.5	16.5	30.7	1.12	1.50
Drosophila melanogaster	30.7	19.6	20.2	29.4	1.01	1.51
Homo sapiens (human)	30.2	19.9	19.6	30.3	1.01	1.53

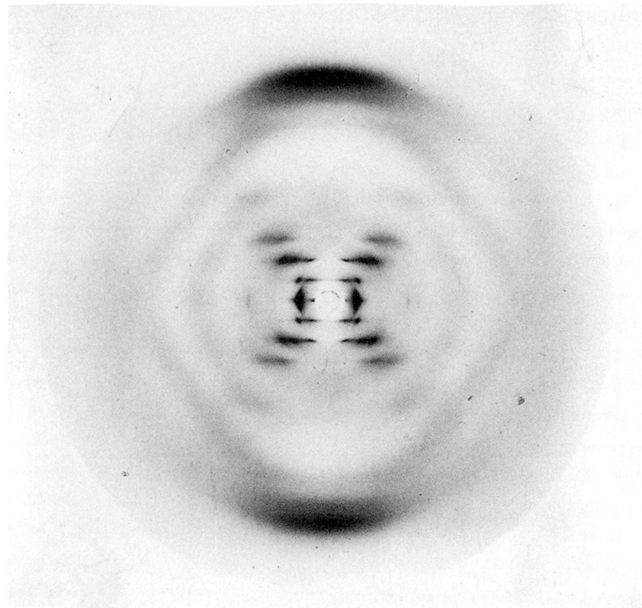

Figure 10.10 Photograph of the X-ray diffraction pattern obtained with DNA. The central cross-shaped pattern indicates that the DNA molecule has a helical structure, and the dark bands at the top and bottom indicate that the bases are stacked perpendicular to the axis of the molecule with a periodicity of 0.34 nm.

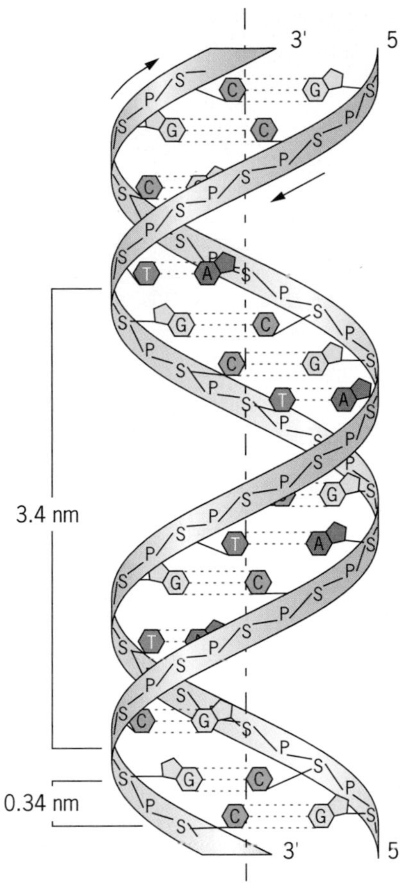

Figure 10.11 Diagram of the double-helix structure of DNA.

TABLE 10.3
Chemical Bonds Important in DNA Structure

(*a*) *Covalent bonds*

Strong chemical bonds formed by sharing of electrons between atoms.

(1) In bases and sugars

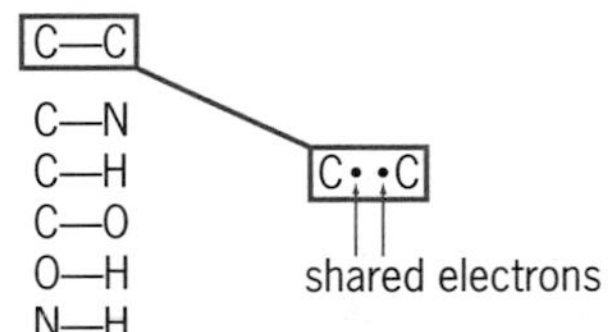

(2) In phosphodiester linkages

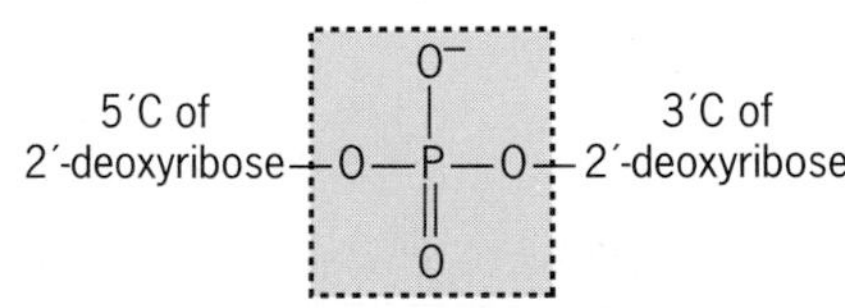

(*b*) *Hydrogen bonds*

A weak bond between an electronegative atom and a hydrogen atom (electropositive) that is covalently linked to a second electronegative atom.

δ^+ δ^-
N — H• • • • •O —
N — H• • • • •N

(*c*) *Hydrophobic "bonds"*

The association of nonpolar groups with each other when present in aqueous solutions because of their insolubility in water.

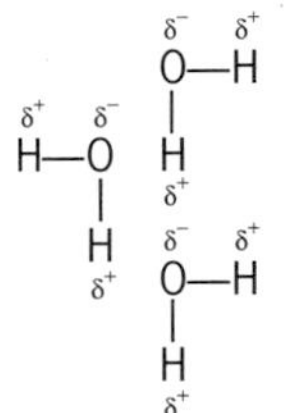

Water molecules are very polar (δ^- O and δ^+ Hs)
Compounds that are similarly polar are very soluble in water ("hydrophilic").
Compounds that are nonpolar (no charged groups) are very insoluble in water ("hydrophobic").

The stacked base pairs provide a hydrophobic core.

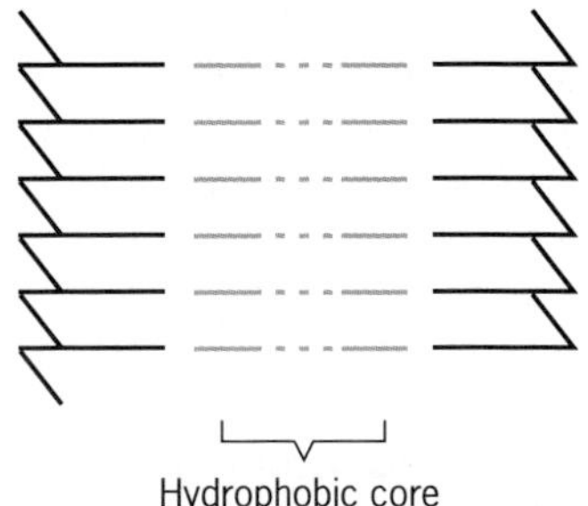

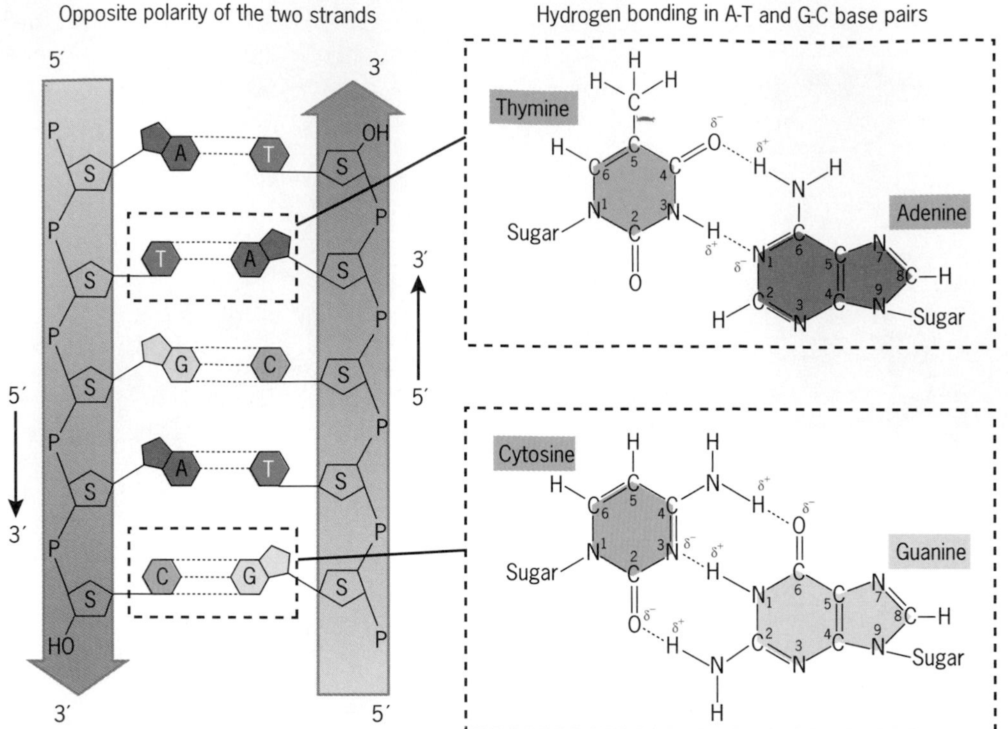

Figure 10.12 Diagram of a DNA double helix, illustrating the opposite chemical polarity (see Fig. 10.8) of the two strands and the hydrogen bonding between thymine (T) and adenine (A) and between cytosine (C) and guanine (G). The base-pairing in DNA, T with A and C with G, is governed by the hydrogen-bonding potential of the bases. S = the sugar 2-deoxyribose; P = a phosphate group.

Hydrogen bonding is not possible between cytosine and adenine or thymine and guanine when they exist in their common structural states.

Once the sequence of bases in one strand of a DNA double helix is known, the sequence of bases in the other strand is also known because of the specific base-pairing. The two strands of a DNA double helix are thus said to be complementary. *This property, the* **complementarity** *of the two strands of the double helix, makes DNA uniquely suited to store and transmit genetic information from generation to generation.*

The base pairs in DNA are stacked about 0.34 nm apart, with 10 base pairs per turn (360°) of the double helix (Figure 10.11). The sugar-phosphate backbones of the two complementary strands are **antiparallel** (Figure 10.12). Unidirectionally along a DNA double helix, the phosphodiester bonds in one strand go from a 3′ carbon of one nucleotide to a 5′ carbon of the adjacent nucleotide, whereas those in the complementary strand go from a 5′ carbon to a 3′ carbon. This "opposite polarity" of the complementary strands of a DNA double helix plays an important role in DNA replication, transcription, and recombination.

The stability of DNA double helices results in part from the large number of hydrogen bonds between the base pairs (even though each hydrogen bond by itself is weak, much weaker than a covalent bond) and in part from the hydrophobic bonding (or stacking forces) between adjacent base pairs (Table 10.3). The stacked nature of the base pairs is best illustrated with a space-filling diagram of DNA structure (Figure 10.13). The planar sides of the base pairs are relatively nonpolar and thus tend to be hydrophobic (water insoluble). Because of this insolubility in water, the hydrophobic core of stacked base pairs contributes considerable stability to DNA molecules present in the aqueous protoplasms of living cells. The space-filling drawing also shows that the two grooves of a DNA double helix are not identical; one, the major groove, is much wider than the other, the minor groove.

DNA Structure: Alternate Forms of the Double Helix

The Watson–Crick double-helix structure just described is called **B-DNA.** B-DNA is the conformation that DNA takes under physiological conditions (in aqueous solutions containing low concentrations of salts). The vast majority, if not all, of the DNA molecules present in the aqueous protoplasms of living cells exist in the B conformation. However, DNA is not a

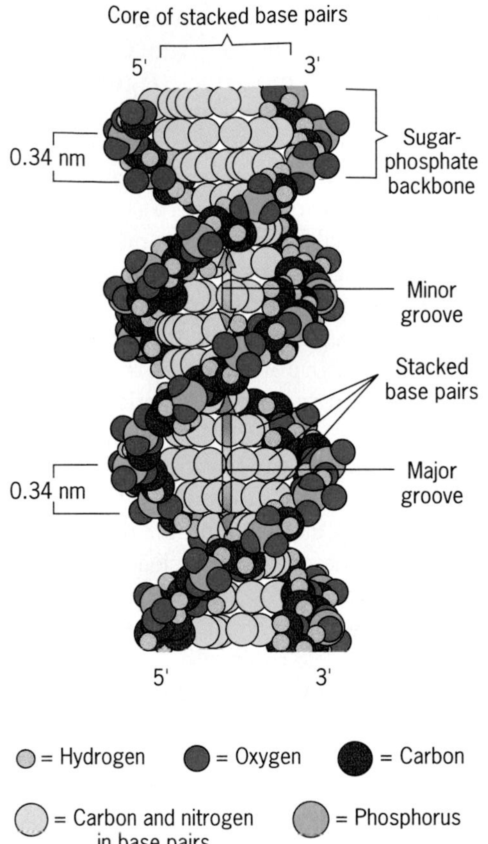

Figure 10.13 Space-filling diagram of a DNA double helix.

static, invariant molecule. On the contrary, DNA molecules exhibit a considerable amount of conformational flexibility.

The structures of DNA molecules change as a function of their environment. The exact conformation of a given DNA molecule or segment of DNA molecule will depend on the nature of the molecules with which it is interacting. In fact, intracellular B-DNA appears to have an average of 10.4 nucleotide pairs per turn, rather than precisely 10 as shown in Figure 10.11. In high concentrations of salts or in a partially dehydrated state, DNA exists as **A-DNA,** which is a right-handed helix like B-DNA, but with 11 nucleotide pairs per turn (Table 10.4). A-DNA is a shorter, thicker double helix with a diameter of 0.23 nm. It has a narrow, deep major groove and a broad, shallow minor groove. DNA molecules almost certainly never exist as A-DNA *in vivo*. However, the A-DNA conformation is important because DNA-RNA heteroduplexes (double helices containing a DNA strand base-paired with a complementary RNA strand) or RNA-RNA duplexes exist in a very similar structure *in vivo*. Additional right-handed helical forms of DNA, designated C, D, and E, have been shown to exist under specific environmental conditions (dehydration, etc.). However, since these forms are not believed to exist in living cells, they will not be discussed here.

Certain DNA sequences have been shown to exist in a left-handed, double-helical form called **Z-DNA** (Z for the zigzagged path of the sugar-phosphate backbones of the structure). Z-DNA was discovered by X-ray diffraction analysis of crystals formed by DNA oligomers containing alternating G:C and C:G base pairs. Z-DNA occurs in double helices that are G:C-rich and contain alternating purine and pyrimidine residues. In addition to its unique left-handed helical structure, Z-DNA differs from B-DNA in having 12 base-pairs per turn, a diameter of 0.18 nm, and a single deep groove (Table 10.4; Figure 10.14). Whether Z-DNA exists in living cells is still uncertain.

TABLE 10.4
Alternate Forms of DNA

Helix Form	*Helix Direction*	*Base Pairs per Turn*	*Helix Diameter*
A	Right-handed	11	2.3 nm
B	Right-handed	10	1.9 nm
Z	Left-handed	12	1.8 nm

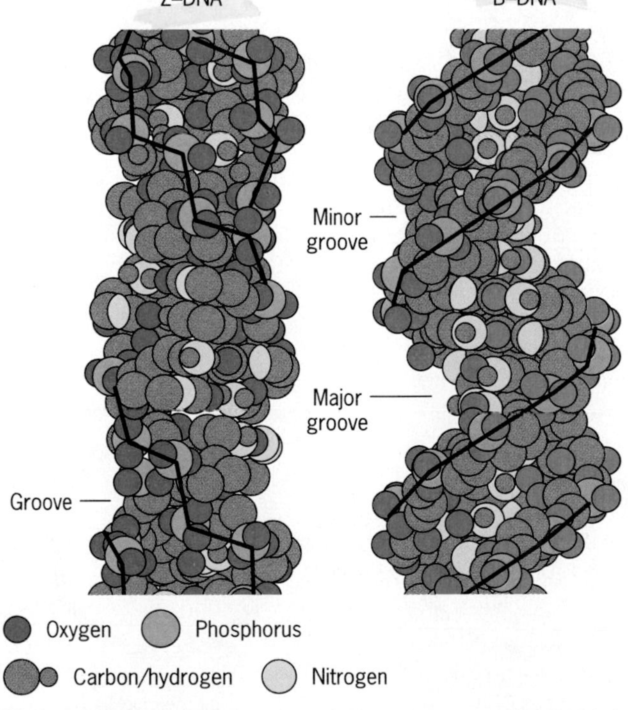

Figure 10.14 Comparison of the structures of Z-DNA and B-DNA. Note the left-handed winding and the zigzagged paths of the sugar-phosphate backbones (solid lines) of the Z-DNA double helix in contrast to the smooth paths and right-handed winding of the backbones in the B-DNA helix.

DNA Structure: Negative Supercoils *In Vivo*

All the functional DNA molecules present in living cells display one other very important level of organization—they are supercoiled. **Supercoils** are introduced into a DNA molecule when one or both strands are cleaved and when the complementary strands at one end are rotated or twisted around each other with the other end held fixed in space—and thus not allowed to spin. This supercoiling causes a DNA molecule to collapse into a tightly coiled structure similar to a coiled telephone cord or twisted rubber band (Figure 10.15, lower right). Supercoils are introduced into and removed from DNA molecules by enzymes that play essential roles in DNA replication (Chapter 11) and other processes.

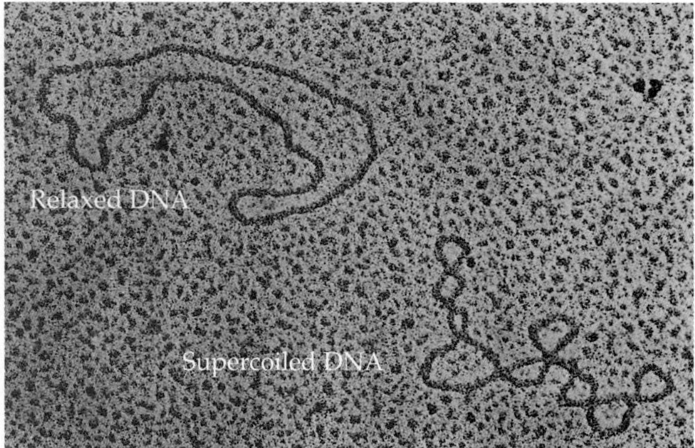

Figure 10.15 Comparison of the relaxed and negatively supercoiled structures of DNA. The relaxed structure is B-DNA with 10.4 base pairs per turn of the helix. The negatively supercoiled structure results when B-DNA is underwound, with less than one turn of the helix for every 10.4 base pairs.

Supercoiling occurs only in DNA molecules with fixed ends, ends that are not free to rotate. Obviously, the ends of the circular DNA molecules (Figure 10.15) present in most prokaryotic chromosomes and in the chromosomes of eukaryotic organelles such as mitochondria are fixed. The large linear DNA molecules present in eukaryotic chromosomes are also fixed by their attachment at intervals and at the ends to non-DNA components of the chromosomes. These attachments allow enzymes to introduce supercoils into the linear DNA molecules present in eukaryotic chromosomes, just as they are incorporated into the circular DNA molecules present in most prokaryotic chromosomes.

We can perhaps visualize supercoiling most easily by considering a circular DNA molecule. If we cleave one strand of a covalently closed, circular double helix of DNA, and rotate one end of the cleaved strand a complete turn (360°) around the complementary strand while holding the other end fixed, we will introduce one supercoil into the molecule (Figure 10.16). If we rotate the free end in the same direction as the DNA double helix is wound (right-handed), a positive supercoil (overwound DNA) will be produced. If we rotate the free end in the opposite direction (left-handed), a negative supercoil (underwound DNA) will result. Although this is the simplest way to define supercoiling in DNA, it is not the mechanism by which supercoils are produced in DNA *in vivo*. That mechanism is discussed in Chapter 11.

The DNA molecules of all organisms, from the smallest viruses to the largest eukaryotes, exhibit **negative supercoiling** (secondary coiling produced by left-handed rotations in DNA resulting in underwound double helices) *in vivo*, and many of the biological functions of chromosomes can be carried out only when the participating DNA molecules are neg-

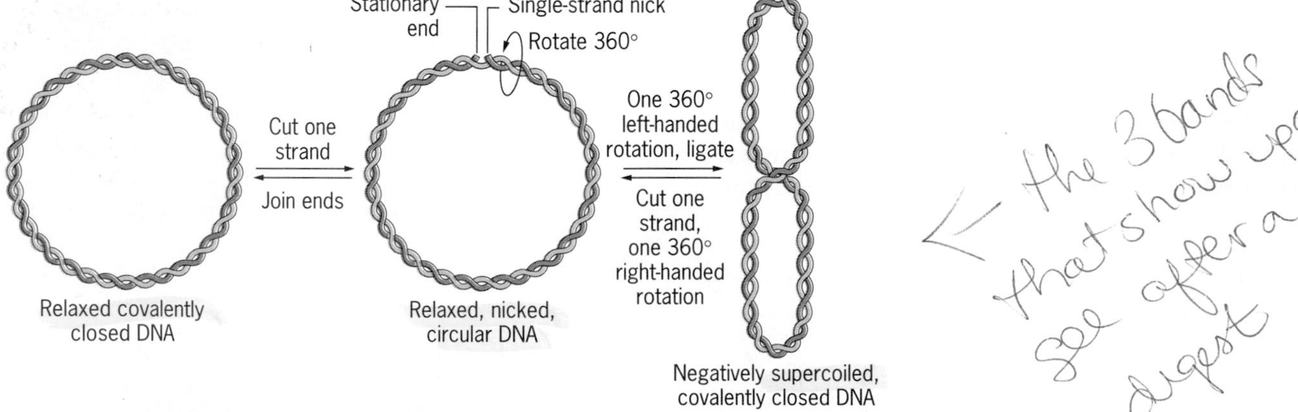

Figure 10.16 A visual definition of negatively supercoiled DNA. Although the structure of DNA supercoils is most clearly illustrated by the mechanism shown here, DNA supercoils are produced by a different mechanism *in vivo* (see Chapter 11).

atively supercoiled. Considerable evidence indicates that negative supercoiling is involved in replication (Chapter 11), recombination, gene expression, and the regulation of gene expression. Supercoils are introduced into DNA by enzymes called **topoisomerases,** enzymes that catalyze changes in the topology of DNA molecules. The topoisomerase **DNA gyrase,** which catalyzes the formation of negative supercoils in DNA during its replication, has been isolated from several species, both prokaryotic and eukaryotic. The DNA gyrase of *E. coli* is inhibited by novobiocin and nalidixic acid, two potent inhibitors of DNA synthesis in bacteria; this result indicates that DNA gyrase activity is required for DNA replication. Similar amounts of negative supercoiling exist in the DNA molecules present in bacterial chromosomes and eukaryotic chromosomes.

Key Points: **DNA usually exists as a double helix, with the two strands held together by hydrogen bonds between the complementary bases: adenine paired with thymine and guanine paired with cytosine. The complementarity of the two strands of a double helix makes DNA uniquely suited to store and transmit genetic information. The two strands of a DNA double helix have opposite chemical polarity. RNA usually exists as a single-stranded molecule containing uracil instead of thymine. The DNA molecules present in chromosomes are negatively supercoiled.**

CHROMOSOME STRUCTURE IN PROKARYOTES AND VIRUSES

Much of the information about the structure of DNA has come from studies of prokaryotes, primarily because they are less complex, both genetically and biochemically, than eukaryotes. Prokaryotes are monoploid (*mono* = one); they have only one set of genes (one copy of the genome). In most viruses and prokaryotes, the single set of genes is stored in a single chromosome, which in turn contains a single molecule of nucleic acid (either RNA or DNA).

The smallest known RNA viruses have only three genes, and the complete nucleotide sequences of the genomes of many viruses are known. For example, the single RNA molecule in the genome of bacteriophage MS2 consists of 3569 nucleotides and contains 4 genes. The smallest known DNA viruses have only 9 to 11 genes. Again, the complete nucleotide sequences are known in several cases. For example, the genome of bacteriophage ϕX174 is a single DNA molecule 5386 nucleotides in length that contains 11 genes. The largest DNA viruses, like bacteriophage T2 and the animal pox viruses, contain about 150 genes. Bacteria like *E. coli* have 2500 to 3500 genes, most of which are present in a single molecule of DNA.

Until recently, prokaryotic chromosomes were often characterized as "naked molecules of DNA," in contrast to eukaryotic chromosomes with their associated proteins and complex morphology. This misconception resulted in part because (1) most of the published pictures of prokaryotic "chromosomes" were electron micrographs of isolated DNA molecules, not metabolically active or functional chromosomes, and (2) most of the published photographs of eukaryotic chromosomes were of highly condensed meiotic or mitotic chromosomes—again, metabolically inactive chromosomal states. Functional prokaryotic chromosomes, or nucleoids (nucleoids rather than nuclei because they are not enclosed in a nuclear membrane), are now known to bear little resemblance to the isolated viral and bacterial DNA molecules seen in electron micrographs, just as the metabolically active interphase chromosomes of eukaryotes have little morphological resemblance to mitotic or meiotic metaphase chromosomes.

The contour length of the circular DNA molecule present in the chromosome of the bacterium *Escherichia coli* is about 1100 μm. Because an *E. coli* cell has a diameter of only 1 to 2 μm, the large DNA molecule present in each bacterium must exist in a highly condensed (folded or coiled) configuration. When *E. coli* chromosomes are isolated by gentle procedures in the absence of ionic detergents (commonly used to lyse cells) and are kept in the presence of a high concentration of cations such as polyamines (small basic or positively charged proteins) or 1*M* salt to neutralize the negatively charged phosphate groups of DNA, the chromosomes remain in a highly condensed state comparable in size to the nucleoid *in vivo*. This structure, called the **folded genome,** is the functional state of a bacterial chromosome. Though smaller, the functional intracellular chromosomes of bacterial viruses are very similar to the folded genomes of bacteria.

Within the folded genome, the large DNA molecule in an *E. coli* chromosome is organized into 50 to 100 **domains** or loops, each of which is independently negatively supercoiled (Figure 10.17). Both RNA and protein are components of the folded genome, which can be partially relaxed by treatment with either deoxyribonuclease (DNase) or ribonuclease (RNase). Because each domain of the chromosome is independently supercoiled, the introduction of single-strand "nicks" in DNA by treatment of the chromosomes with a DNase that cleaves DNA at internal sites will relax the DNA only in the nicked domains, and all unnicked loops will remain supercoiled. Destruction of the RNA

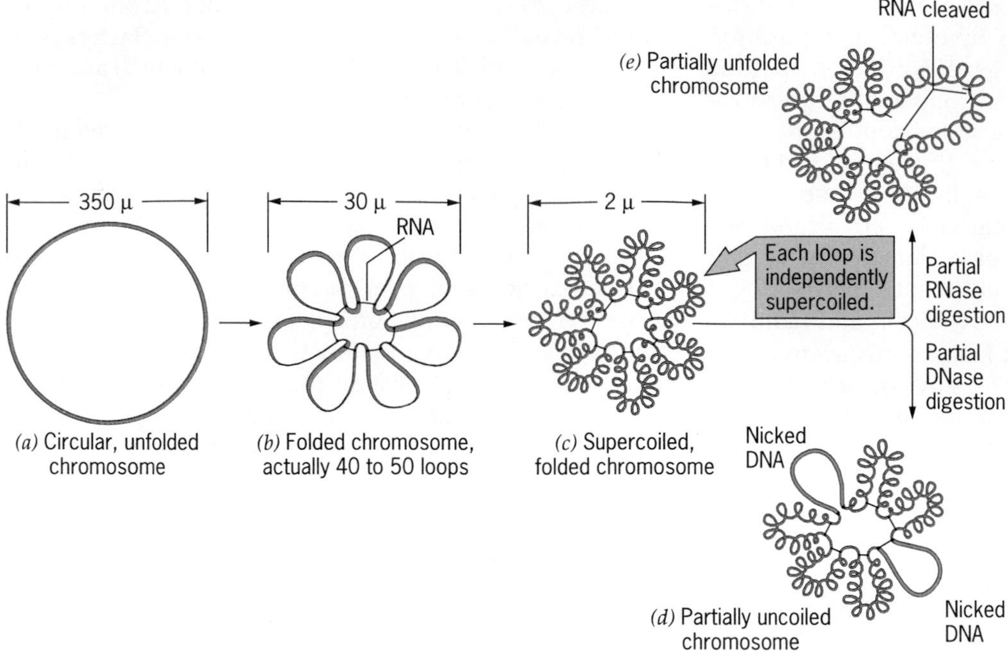

Figure 10.17 Diagram of the structure of the functional state of the *E. coli* chromosome.

connectors by RNase will unfold the folded genome partially by eliminating the organization of the DNA molecule into 50 to 100 loops. However, RNase treatment will not affect the supercoiling of the domains of the chromosome.

Key Points: **The DNA molecules in prokaryotic chromosomes are segregated into negatively supercoiled domains. Bacterial chromosomes contain circular molecules of DNA segregated into about 50 such domains.**

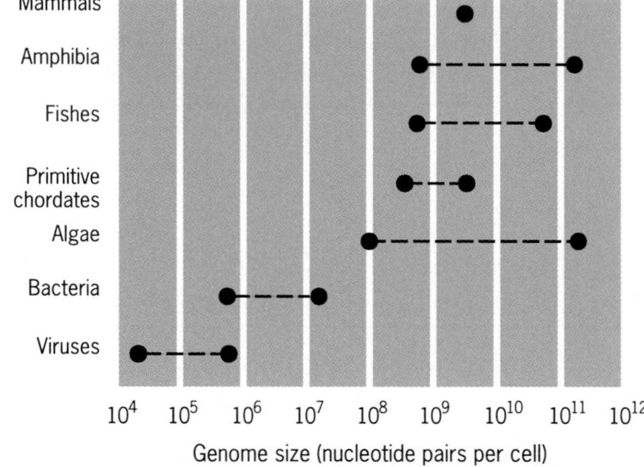

Figure 10.18 Increased genome size in organisms with increased developmental complexity.

CHROMOSOME STRUCTURE IN EUKARYOTES

Eukaryotic genomes contain levels of complexity that are not encountered in prokaryotes. In contrast to prokaryotes, most eukaryotes are diploid, having two complete sets of genes, one from each parent. As we discussed in Chapter 7, some flowering plants are polyploid; that is, they carry several copies of the genome. Moreover, although eukaryotes have only about 2 to 25 times as many genes as *E. coli*, they have orders of magnitude more DNA (Figure 10.18). Moreover, much of this DNA does not contain genes, at least not genes encoding proteins or RNA molecules.

Not only do most eukaryotes contain many times the amount of DNA in prokaryotes, but also this DNA is packaged into several chromosomes, and each chromosome is present in two (diploids) or more (polyploids) copies. Recall that the chromosome of *E. coli* has a contour length of 1100 μm, or about 1 mm. Now consider that the haploid chromosome complement, or genome, of the human contains about 1000 mm of DNA (or about 2000 mm per diploid cell). Moreover, this meter of DNA is subdivided among 23 chromosomes of variable size and shape, each chromosome containing 15 to 85 mm of DNA. Until recently, geneticists had little information as to how this DNA was arranged in the chromosomes. Is there one molecule of DNA per chromosome as in prokaryotes, or are there many? If many, how are the molecules arranged relative to each other? How does the 85 mm (85,000 μm) of DNA in the largest human chromosome get condensed into a mitotic metaphase structure that is about 0.5 μm in diameter and 10 μm long? What are the structures of the metabolically active interphase chromosomes? We consider the answers to some of these questions in the following sections.

Chemical Composition of Eukaryotic Chromosomes

Interphase chromosomes are usually not visible with the light microscope. However, chemical analysis, electron microscopy, and X-ray diffraction studies of isolated **chromatin** (the complex of the DNA, chromosomal proteins, and other chromosome constituents isolated from nuclei) have provided valuable information about the structure of eukaryotic chromosomes.

When chromatin is isolated from interphase nuclei, the individual chromosomes are not recognizable. Instead, one observes an irregular aggregate of nucleoprotein. Chemical analysis of isolated chromatin shows that it consists primarily of DNA and proteins with lesser amounts of RNA (Figure 10.19). The proteins are of two major classes: (1) basic proteins (positively charged at neutral pH) called **histones** and (2) a heterogeneous, largely acidic (negatively charged at neutral pH) group of proteins collectively referred to as **nonhistone chromosomal proteins.**

Histones play a major structural role in chromatin. They are present in the chromatin of all higher eukaryotes in amounts equivalent to the amounts of DNA. This relationship suggests that an interaction occurs between histones and DNA that is conserved in eukaryotes. The histones of all plants and animals consist of five classes of proteins. These five major histone types, called **H1, H2a, H2b, H3,** and **H4,** are present in almost all cell types. A few exceptions exist, most notably some sperm, where the histones are replaced by another class of small basic proteins called **protamines.**

The five histone types are present in molar ratios of approximately 1 H1: 2 H2a: 2 H2b: 2 H3: 2 H4. They are specifically complexed with DNA to produce the basic structural subunits of chromatin, small (approximately 11 nm in diameter by 6 nm high) ellipsoidal beads called **nucleosomes.** The histones have been highly conserved during evolution—four of the five types of histone are similar in all eukaryotes.

Most of the 20 amino acids in proteins are neutral in charge; that is, they have no charge at pH 7. However, a few are basic and a few are acidic. The histones are basic because they contain 20 to 30 percent arginine and lysine, two positively charged amino acids (Figure 10.20). The exposed $-NH_3^+$ groups of arginine and lysine allow histones to act as polycations. The positively charged side groups on histones are important in their interaction with DNA, which is polyanionic because of the negatively charged phosphate groups.

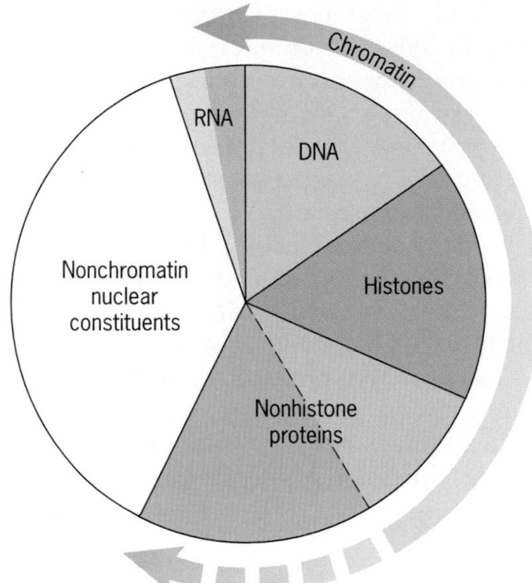

Figure 10.19 The chemical composition of chromatin as a function of the total nuclear content. The DNA and histone contents of chromatin are relatively constant, but the amount of nonhistone proteins present depends on the procedure used to isolate the chromatin (dashed arrow).

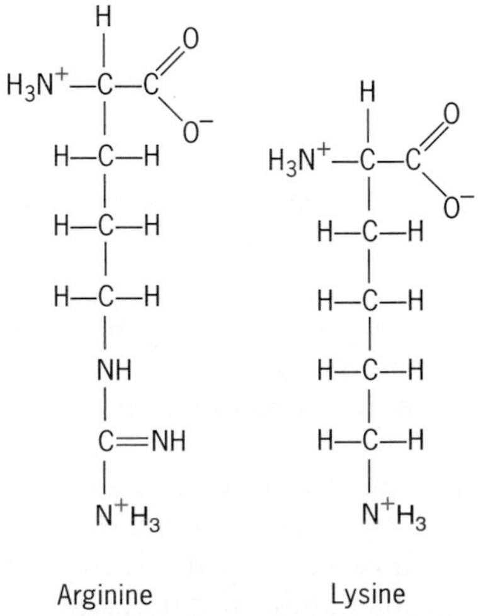

Arginine: $H_3N^+-CH(COO^-)-CH_2-CH_2-CH_2-NH-C(=NH)-N^+H_3$

Lysine: $H_3N^+-CH(COO^-)-CH_2-CH_2-CH_2-CH_2-N^+H_3$

Arginine Lysine

Figure 10.20 Structures of the amino acids arginine and lysine (at pH 7), which together account for 20 to 30 percent of the amino acids in histones.

The remarkable constancy of histones H2a, H2b, H3, and H4 in all cell types of an organism and even among widely divergent species is consistent with the idea that they are important in chromatin structure (DNA packaging) and are only nonspecifically involved in the regulation of gene expression.

In contrast, the nonhistone protein fraction of chromatin consists of a large number of heterogeneous proteins. Moreover, the composition of the nonhistone chromosomal protein fraction varies widely among different cell types of the same organism. Thus, the nonhistone chromosomal proteins probably do not play central roles in the packaging of DNA into chromosomes. Instead, they are likely candidates for roles in regulating the expression of specific genes or sets of genes.

One Large DNA Molecule per Chromosome

A typical eukaryotic chromosome contains 1 to 20 cm (10^4 to 2×10^5 μm) of DNA. During metaphase of meiosis and mitosis, this DNA is packaged in a chromosome with a length of only 1 to 10 μm. How is all of this DNA condensed into the compact chromosomes that are present during mitosis and meiosis? Do many DNA molecules run parallel throughout the chromosome—the **multineme** or "multistrand" model—or is there just one DNA double helix extending from one end of the chromosome to the other—the **unineme** or "single-strand" model? (Note that strand here refers to the DNA double helix, not the individual polynucleotide chains of DNA.) Are there many DNA molecules joined end-to-end or arranged in some other fashion in the chromosome, or does one giant, continuous molecule of DNA extend from one end to the other in a highly coiled and folded form? The evidence supporting the unineme model of chromosome structure is now conclusive. In addition, solid evidence supports the concept of chromosome-size DNA molecules; that is, each chromosome appears to contain a single, giant molecule of DNA that extends from one end through the centromere all the way to the other end of the chromosome.

Some of the strongest evidence supporting the unineme model of chromosome structure has come from studies of **lampbrush chromosomes** (so-named because they resemble the brushes used to clean the mantles of kerosene lamps) present during prophase I of oogenesis in many vertebrates, particularly amphibians. Lampbrush chromosomes are up to 800 μm long, and their large size has allowed cytologists to carry out microscopic studies of chromosome structure that are not feasible with smaller chromosomes. The homologous chromosomes are paired, and each has duplicated to produce two chromatids. Each lampbrush chromosome contains a central axial region, where the two chromatids are highly condensed, and numerous pairs of lateral loops (Figure 10.21). The loops are transcriptionally active regions of single chromatids. The integrity of both the central axis and the lateral loops depends on DNA. Treatment with DNase fragments both the axis and the loops. Treatment with RNase or proteases removes surrounding matrix material, but does not destroy the continuity of either the axis or the loops. Electron microscopy of RNase- and protease-treated lampbrush chromosomes reveals a central filament of about 2 nm in diameter in the lateral loops. Since each loop is a segment of one chromatid, and since the diameter of a DNA double helix is 1.9 nm, these lampbrush chromosomes must be unineme structures (Figure 10.21). This conclusion is supported by studies on the kinetics of nuclease digestion of lampbrush chromosomes; that is, the kinetics observed are those expected if the central filament of each loop is one DNA double helix. Thus, the axial region contains two DNA molecules, one from each of the two tightly paired chromatids.

Because lampbrush chromosomes are germ-line chromosomes, their structure is particularly relevant to an understanding of genetic phenomena. Chromosomes of somatic cells may have different structures. Although most are unineme, some—such as the giant polytene chromosomes in the salivary glands of *Drosophila* (Chapter 7)—are known to be multineme structures.

The question of whether the unineme chromosomes of eukaryotes contain a single large molecule of DNA or many smaller molecules linked end-to-end has proven difficult to answer with rigorous experimental evidence. A centimeter-long molecule of DNA has a length-to-width (diameter) ratio of 5 million to 1. Such a structure is extremely sensitive to shearing. If such a DNA molecule is in solution in a test tube, the slightest vibration will break the molecule into many fragments. For this and other reasons, accurate estimates of the sizes of eukaryotic DNAs cannot be obtained with the procedures used to analyze prokaryotic DNA molecules. However, by modifying old techniques and developing some new ones and applying these to the problem, scientists have obtained solid evidence indicating that each eukaryotic chromosome, no matter how large, contains one giant DNA double helix.

Some lower eukaryotic organisms, such as the mold *Neurospora crassa* and the yeast *Saccharomyces cerevisiae*, have relatively small chromosomes. With these organisms, a procedure called **pulsed-field gel electrophoresis** has been used to demonstrate that each chromosome contains a single molecule of DNA. The technique of gel electrophoresis is a powerful tool

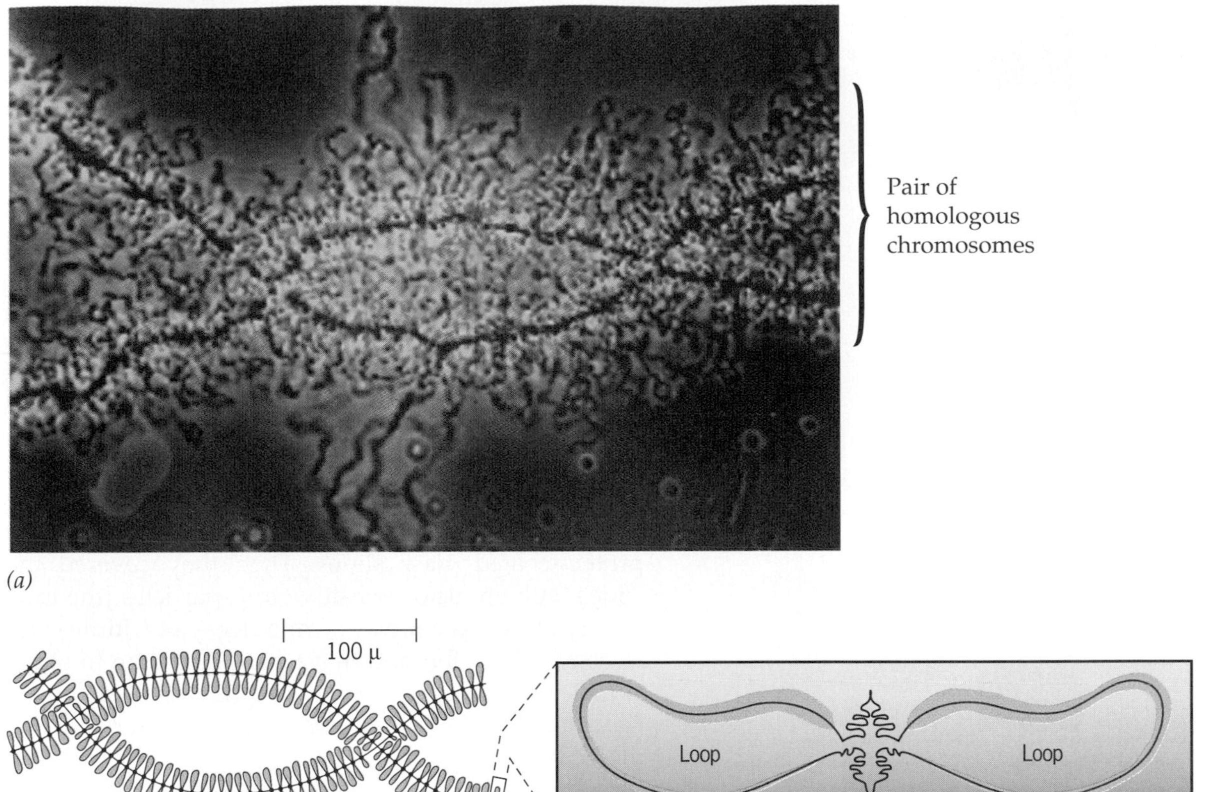

(b) Diagram of the chromosome pair shown above.

(c) Diagram of the two chromatids of one homolog.

Figure 10.21 Phase contrast micrograph (*a*) and diagram (*b*) of a pair of lampbrush chromosomes in an oocyte of the newt, *Triturus viridescens*. The structures of axial and loop regions of the two chromatids of a single lampbrush chromosome are shown in (*c*). The central element in each chromatid (both axial regions and lateral loops) is a single molecule of DNA. The material surrounding the DNA molecule is primarily nascent RNA that is synthesized on the extended DNA in the loop regions.

for separating macromolecules such as proteins and nucleic acids based on their size and charge (see Chapter 20). A semisolid gel (usually polyacrylamide or agarose) provides an inert matrix with pores in a given size range through which the macromolecules migrate when placed in an electric field. Positively charged molecules migrate toward the cathode (the negative electrode), and negatively charged molecules move toward the anode (the positive electrode). Proteins may be either positively or negatively charged, depending on their amino acid composition. Nucleic acids are negatively charged with one phosphate group per nucleotide. Thus, nucleic acids have an approximately constant charge per unit of mass, and all would migrate at the same rate in the absence of sieving. However, polyacrylamide gels have relatively small pores, and agarose gels have somewhat larger pores. These gels act as molecular sieves such that small molecules migrate faster than larger molecules with the same charge per unit of mass. As a result, the rate of migration of a nucleic acid during gel electrophoresis is almost exclusively a function of its size. Sometimes, conformation is a factor; for example, supercoiled DNAs migrate faster than relaxed molecules of the same size.

Pulsed-field gel electrophoresis differs from standard gel electrophoresis in that instead of a single (one-dimensional), constant electric field, two electric fields offset by about 90° are applied across the gel in an alternating or pulsed manner. In standard gel electrophoresis, the DNA molecules pass through the gel in an end-first or snakelike fashion. In pulsed-field gel electrophoresis, the application of intermittent and alternating electric fields requires the molecules to reorient themselves before continuing to migrate through the gel. Larger molecules take longer to undergo these reorientation events and move more slowly. As a result, pulsed-field gel electrophoresis yields superior separation of very large DNA molecules. When this technique was used to separate intact

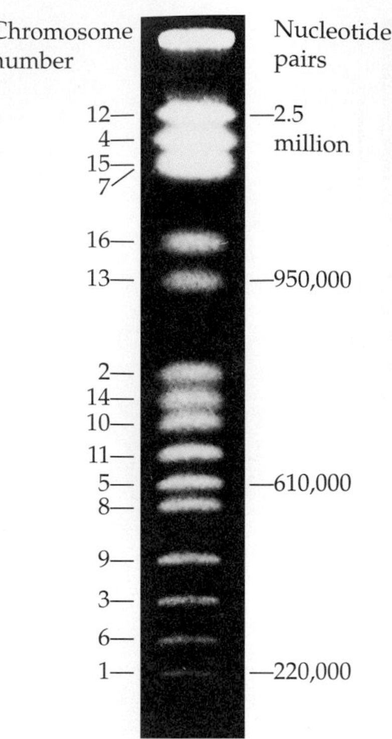

Figure 10.22 Separation of the chromosome-size DNA molecules of the yeast *Saccharomyces cerevisiae* by pulse-field agarose gel electrophoresis. The large DNA molecules present in all 16 of the yeast chromosomes can be resolved by this procedure.

DNA molecules from the fungi *N. crassa* and *S. cerevisiae,* the results showed that the number of different sized DNA molecules was equal to the number of nonhomologous chromosomes in these species (Figure 10.22).

Unfortunately, the very large DNA molecules present in the chromosomes of higher eukaryotes such as *Drosophila* and humans cannot be separated even by pulsed-field gel electrophoresis. Researchers have used additional approaches in attempts to demonstrate that the large chromosomes of higher animals and plants each contain one molecule of DNA. Autoradiography and viscoelastometry are two approaches that have yielded important results.

Autoradiography is a method for detecting and localizing radioactive isotopes in cytological preparations or macromolecules by exposure to a photographic emulsion that is sensitive to low-energy radiation. The emulsion contains silver halides that produce tiny black spots — often called silver grains — when they are exposed to the charged particles emitted during the decay of radioactive isotopes. Autoradiography permits a researcher to prepare an image of the localization of radioactivity in macromolecules, cells, or tissues, just as photography permits us to make a picture of what we see. The difference is that the film used for autoradiography is sensitive to radioactivity, whereas the film we use in a camera is sensitive to visible light. Autoradiography is particularly useful in studying DNA metabolism because DNA can be specifically labeled by growing cells on ^{3}H-thymidine, a deoxyribonucleoside of thymine that contains a radioactive isotope of hydrogen (tritium). Thymidine is incorporated almost exclusively into DNA; it is not present in any other major component of the cell.

In the early 1960s, Ruth Kavenoff, Lynn Klotz, and Bruno Zimm grew *Drosophila* cells in culture medium containing ^{3}H-thymidine for 24 hours, lysed the cells gently so as not to break the chromosomal DNA molecules, and carefully collected the DNA molecules on protein-coated glass slides. Then they covered the slides with emulsion sensitive to β-particles (the low-energy electrons emitted during decay of tritium) and stored them in the dark for a period of time to allow sufficient radioactive decays. The greatest challenge to Kavenoff and her coworkers was to spread out the molecules with no tangles or overlaps on the slides so that the entire length of a molecule would be visible. Their best autoradiographs showed DNA molecules with contour lengths of up to 1.2 cm (Figure 10.23). DNA molecules of this length would have a mass of about 3×10^{10} daltons (one dalton is the mass of one hydrogen atom) and would contain about two-thirds of the DNA known to be present in the largest chromosomes of *D. melanogaster.* Thus, these results provide support for the concept of chromosome-size DNA molecules in *Drosophila.*

Kavenoff and colleagues also used a technique called **viscoelastometry,** a procedure for analyzing the viscosity of molecules in solution, to determine the sizes of the DNA molecules in the largest *Drosophila* chromosomes. They measured the sizes of the largest DNA molecules present in *Drosophila* cells growing in suspension cultures. Kavenoff and coworkers' viscoelastometric data indicate that the largest DNA molecules in *Drosophila* have a mass of 4.1×10^{10} daltons. Because the largest chromosomes of *Drosophila* have been shown to contain about 4.3×10^{10} daltons of DNA (total, whether one molecule or many) by direct biochemical analysis, the viscoelastometric estimate of the size of the largest DNA molecules in *Drosophila* nuclei correlates almost exactly with the total amount of DNA present in the largest chromosome.

These and other results have provided convincing evidence that each eukaryotic chromosome contains one long double helix of DNA extending from one end of the chromosome through the centromere all the way to the other end. However, as we will discuss in the

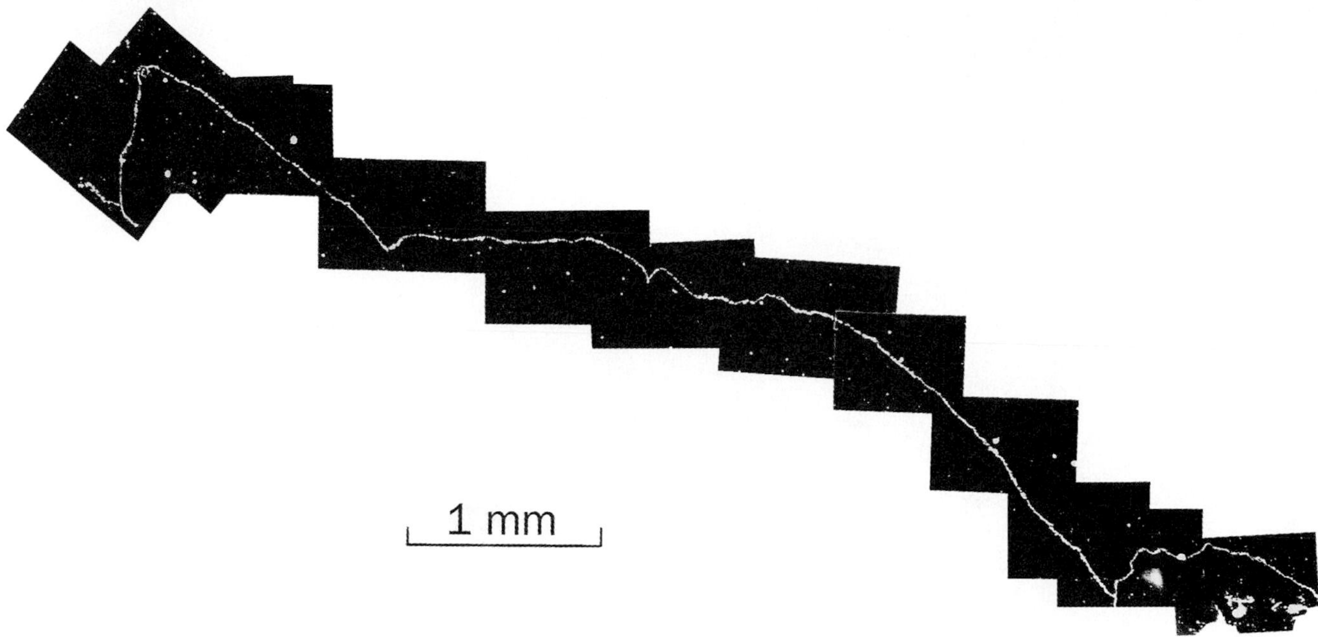

Figure 10.23 Autoradiograph of a 1.2-cm-long DNA molecule from *Drosophila melanogaster*.

following section, this giant DNA molecule is highly condensed (coiled and folded) within the chromosome.

Three Levels of DNA Packaging in Eukaryotic Chromosomes

The largest chromosome in the human genome contains about 85 mm (85,000 μm, or 8.5×10^7 nm) of DNA that is believed to exist as one giant molecule. This DNA molecule somehow gets packaged into a metaphase structure that is about 0.5 μm in diameter and about 10 μm in length—a condensation of almost 10^4-fold in length from the naked DNA molecule to the metaphase chromosome. How does this condensation occur? What components of the chromosomes are involved in the packaging processes? Are DNA molecules packaged in different chromosomes in different ways, or is there a universal packaging scheme? Are there different levels of packaging? Clearly, meiotic and mitotic chromosomes are more extensively condensed than are interphase chromosomes. What additional levels of condensation occur in these special structures that are designed to assure the proper segregation of the genetic material during cell divisions? Are DNA sequences of genes that are being expressed packaged differently from those of genes that are not being expressed? Let us investigate some of the evidence that establishes the existence of three different levels of packaging of DNA into chromosomes.

When isolated chromatin is examined by electron microscopy, it is found to consist of a series of ellipsoidal beads (about 11 nm in diameter and 6 nm high) joined by thin threads (Figure 10.24). Further evidence for a regular, periodic packaging of DNA has come from studies on the digestion of chromatin with various nucleases. These studies have shown that segments of DNA 146 nucleotide pairs in length were somehow protected from degradation by nucleases. Moreover, partial digestion of chromatin with these nucleases yielded fragments of DNA in a set of discrete sizes that were integral multiples of the smallest size fragment. These results are nicely explained if chromatin has a repeating structure, supposedly the bead seen by electron microscopy (Figure 10.24), within which the DNA is packaged in a nuclease-resistant form (Figure 10.25). This "bead" or chromatin subunit is called the **nucleosome.** According to the present concept of chromatin structure, the **linkers,** or interbead threads of DNA, are susceptible to nuclease attack.

After partial digestion of the DNA in chromatin with an endonuclease (an enzyme that cleaves DNA internally), DNA approximately 200 nucleotide pairs in length is associated with each nucleosome (produced by a cleavage in each linker region). After extensive nuclease digestion, a 146-nucleotide-pair-long segment of DNA remains present in each nucleosome. This nuclease-resistant structure is called the **nucleosome core.** Its structure—essentially invariant in all eukaryotes—consists of a 146-nucleotide-pair length

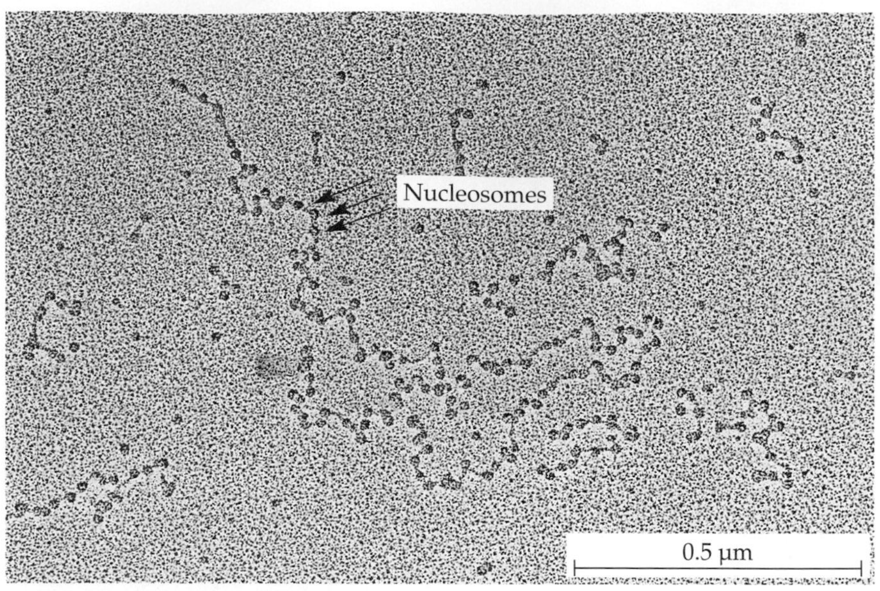

Figure 10.24 Electron micrograph of rat liver chromatin showing the beads-on-a-string nucleosome substructure.

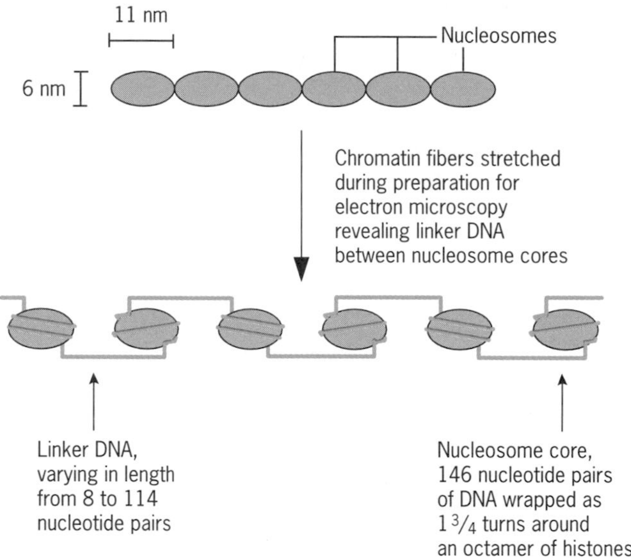

Figure 10.25 Diagram of the nucleosome structure of chromatin.

of DNA and two molecules each of histones H2a, H2b, H3, and H4. The histones protect the segment of DNA in the nucleosome core from cleavage by endonucleases. Physical studies (X-ray diffraction and similar analyses) of nucleosome-core crystals have shown that the DNA is wound as $1\frac{3}{4}$ turns of a superhelix around the outside of the histone octamer (Figure 10.26*a*).

The structure of the nucleosome core has been determined with resolution to 0.28 nm by X-ray diffraction studies. The resulting high-resolution map of the nucleosome core shows the precise location of all eight histone molecules and the 146 nucleotide pairs of negatively supercoiled DNA (Figure 10.26*b* and *c*). Some of the terminal segments of the histones pass over and between the turns of the DNA superhelix to add stability to the nucleosome. The interactions between the various histone molecules and between the histones and DNA are seen most clearly in the structure of one-half of the nucleosome core (Figure 10.26*d*), which contains only 73 nucleotide pairs of supercoiled DNA.

The complete chromatin subunit consists of the nucleosome core, the linker DNA, an average of one molecule of histone H1, and the associated nonhistone chromosomal proteins. However, it has not been firmly established that histone H1 is evenly distributed, one molecule per nucleosome or linker, in chromatin. The size of the linker DNA varies from species to species and from one cell type to another. Linkers as short as eight nucleotide pairs and as long as 114 nucleotide pairs have been reported. Evidence suggests that the complete nucleosome (as opposed to the nucleosome core) contains one molecule of histone H1, which stabilizes two full turns of DNA superhelix (a 166-nucleotide-pair length of DNA) on the surface of the histone octamer (Figure 10.26*e*).

The basic structural component of eukaryotic chromatin clearly is the nucleosome. But are the structures of all nucleosomes the same? What role(s), if any, does nucleosome structure play in gene expression and the regulation of gene expression? The structure of nucleosomes in genetically active regions of chromatin is known to differ from that of nucleosomes in genetically inactive regions. But what are the details of this structure–function relationship? Present and future studies on the fine structure of nucleosomes will un-

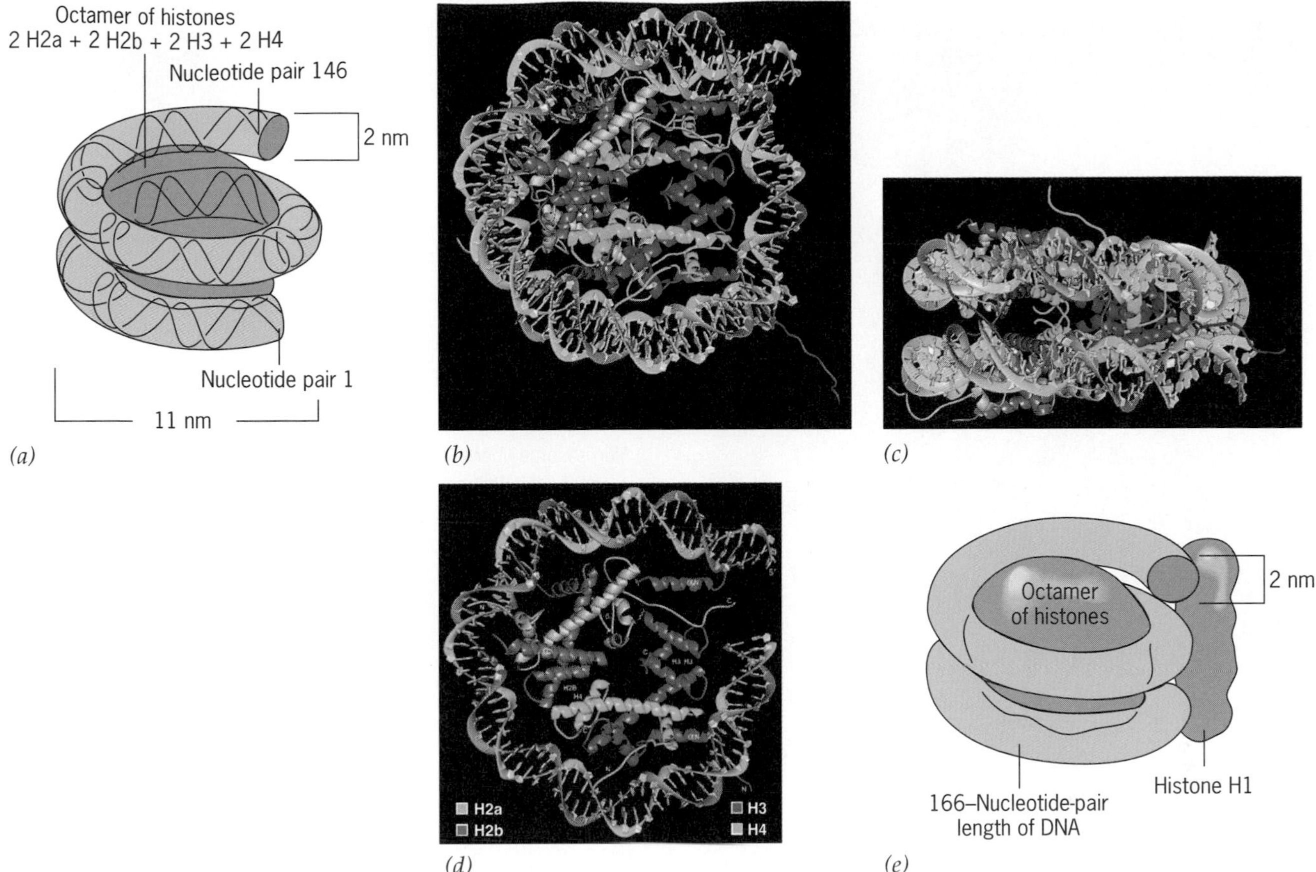

Figure 10.26 Diagrams of (*a*) the gross structure of the nucleosome core, the macromolecular composition of the nucleosome core looking down (*b*) or perpendicular (*c*) to the superhelix axis, (*d*) the macromolecular structure of a half-nucleosome, which shows the relative positions of the DNA superhelix and histones more clearly, and (*e*) the gross structure of the complete nucleosome. In (*b*), (*c*), and (*d*), the complementary strands of DNA are shown in brown and green, and histones H2a, H2b, H3, and H4 are shown in yellow, red, blue, and green, respectively.

doubtedly prove informative with regard to these and other questions.

Electron micrographs of isolated metaphase chromosomes show masses of tightly coiled or folded lumpy fibers (Figure 10.27). These **chromatin fibers** have an average diameter of 30 nm. When the structures seen by light and electron microscopy during earlier stages of meiosis are compared, it becomes clear that the light microscope simply permits one to see those regions where these 30-nm fibers are tightly packed or condensed.

What is the substructure of the 30-nm fiber seen in mitotic and meiotic chromosomes? Although biologists do not have a definitive answer to this question, they do know that the DNA is wound as a supercoil about a histone octamer to yield the roughly 10 nm in diameter nucleosome. *In vivo*, the nucleosomes are probably in direct juxtaposition with each other without detectable linker regions; if so, they will form a 10-nm nucleosome fiber (see Figure 10.25, top). If this 10-nm fiber, in turn, is wound in a higher-order supercoil (a solenoid), a 30-nm fiber can be generated. Although scientists still do not understand all the details of the structure of this 30-nm chromatin fiber, there is good evidence that it represents a solenoidlike structure such as that shown in Figure 10.28.

Metaphase chromosomes are the most condensed of normal eukaryotic chromosomes. Clearly, the role of these highly condensed chromosomes is to organize and package the giant DNA molecules of eukaryotic chromosomes into structures that will facilitate their segregation to daughter nuclei without the DNA molecules of different chromosomes becoming entangled and, as a result, being broken during the anaphase separation of the daughter chromosomes. As we noted in the preceding section, the basic structural unit of the metaphase chromosome is the 30-nm chromatin fiber. However, how are these 30-nm fibers further condensed into the observed metaphase structure? Unfortunately, there is still no clear answer to this

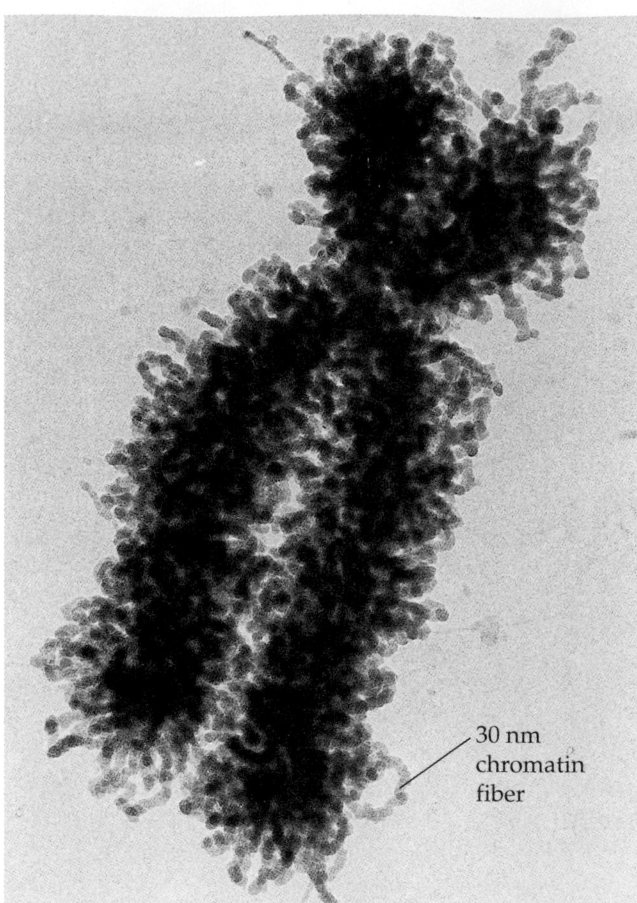

Figure 10.27 Electron micrograph of a human metaphase chromosome showing the presence of 30-nm chromatin fibers. The available evidence indicates that each chromatid contains one large, highly coiled or folded 30-nm fiber.

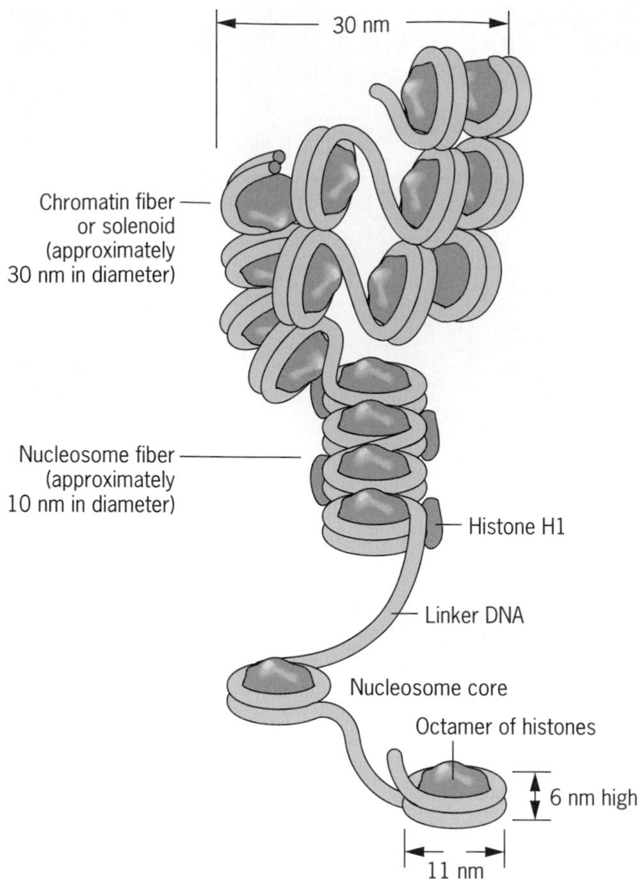

Figure 10.28 Diagram of the solenoid model of the 30-nm chromatin fiber. Histone H1 appears to stabilize the 10-nm nucleosome fiber and contribute to the formation of the 30-nm chromatin fiber.

question. There is evidence that the gross structure of metaphase chromosomes is not dependent on histones. Electron micrographs of isolated metaphase chromosomes from which the histones have been removed reveal a **scaffold,** or central core, which is surrounded by a huge pool or halo of DNA (Figure 10.29). This chromosome scaffold must be composed of nonhistone chromosomal proteins. Note the absence of any apparent ends of DNA molecules in the micrograph shown in Figure 10.29; this finding again supports the concept of one giant DNA molecule per chromosome.

In summary, at least three levels of condensation are required to package the 10^3 to 10^5 μm of DNA in a eukaryotic chromosome into a metaphase structure a few microns long. (1) The first level of condensation involves packaging DNA as a supercoil into nucleosomes, to produce the 10-nm-diameter interphase chromatin fiber. This clearly involves an octamer of histone molecules, two each of histones H2a, H2b, H3, and H4. (2) The second level of condensation involves an additional folding or supercoiling of the 10-nm nucleosome fiber, to produce the 30-nm chromatin fiber characteristic of mitotic and meiotic chromosomes. Histone H1 is involved in this supercoiling of the 10-nm nucleosome fiber to produce the 30-nm chromatin fiber. (3) Finally, nonhistone chromosomal proteins form a scaffold that is involved in condensing the 30-nm chromatin fiber into the tightly packed metaphase chromosomes. This third level of condensation appears to involve the separation of segments of the giant DNA molecules present in eukaryotic chromosomes into independently supercoiled domains or loops (Figure 10.29). The mechanism by which this third level of condensation occurs is not known.

Centromeres and Telomeres

As we discussed in Chapter 2, the two homologous chromosomes (each containing two sister chromatids) of each chromosome pair separate to opposite poles of the meiotic spindle during anaphase I of meiosis. Similarly, during anaphase II of meiosis and the single

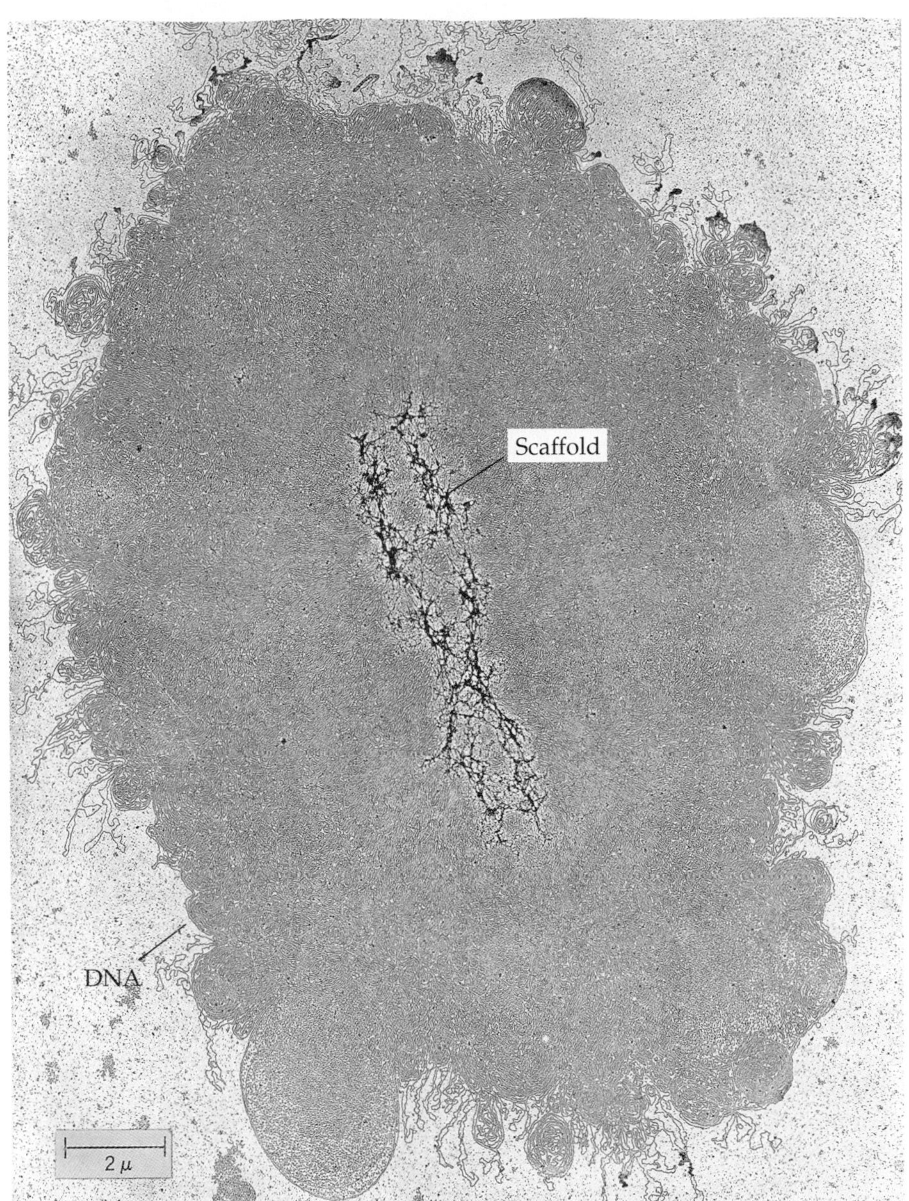

Figure 10.29 Electron micrograph of a human metaphase chromosome from which the histones have been removed. A huge pool of DNA surrounds a central "scaffold" composed of nonhistone chromosomal proteins. Note that the scaffold has roughly the same shape as the metaphase chromosome prior to removal of the histones.

anaphase of mitosis, the sister chromatids of each chromosome move to opposite spindle poles and become daughter chromosomes. These anaphase movements depend on the attachment of spindle microtubules to specific regions of the chromosomes, the centromeres. Because all centromeres perform the same basic function, it is not surprising that the centromeres of different chromosomes of a species contain similar structural components.

The centromere of a metaphase chromosome can usually be recognized as a constricted region (see Figure 10.27) where the chromosome does not appear to have duplicated (Figure 10.30). In fact, the production of two functional centromeres from one parental centromere is a key step in the transition from metaphase to anaphase, and a functional centromere must be present on each daughter chromosome to avoid the dele-

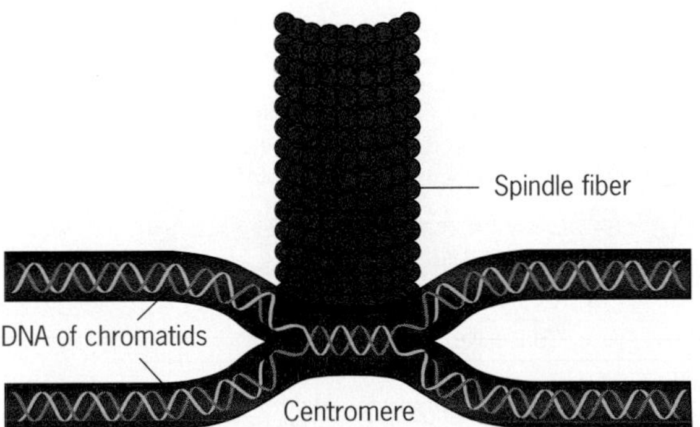

Figure 10.30 Model of centromere structure in a metaphase chromosome. The spindle fibers, which attach to centromeres, are reponsible for the separation of homologous chromosomes during anaphase I of meiosis and progeny chromosomes (derived from chromatids) during anaphase II of meiosis and anaphase of mitosis (Chapter 2).

terious effects of nondisjunction. Acentric chromosomal fragments are usually lost during mitotic and meiotic divisions.

The centromeres (*CEN* regions) of all the chromosomes of baker's yeast, *S. cerevisiae*, have been isolated and characterized. The *CEN* regions of different chromosomes are interchangeable: replacing the *CEN* region of one chromosome with the *CEN* region of another chromosome has no detectable effect on the host cell or its capacity to undergo a normal cell division. Molecular studies have shown that a functional *S. cerevisiae* centromere is 110 to 120 nucleotide pairs in length and has three essential regions (Figure 10.31). Regions I and III are short, conserved boundary sequences, and region II is an A:T-rich (>90 percent A:T) central segment about 90 nucleotide pairs long. The length and A:T-rich nature of region II are probably more important than its actual nucleotide sequence, whereas regions I and III contain specific sequences that serve as binding sites for proteins thought to be involved in spindle-fiber attachment. The precise macromolecular structure of the centromere spindle-fiber attachment complex has not yet been determined for any eukaryotic chromosome.

It has been known for several decades that the **telomeres** (from the Greek terms *telos* and *meros* meaning "end" and "part") or ends of eukaryotic chromosomes have unique properties. Herman J. Muller, who introduced the term *telomere* in 1938, demonstrated that *Drosophila* chromosomes without natural ends—produced by breaking chromosomes with X rays—were not transmitted to progeny. In a classical study of maize chromosomes, Barbara McClintock (see Historical Sidelight: Barbara McClintock, the Discoverer of Transposable Elements in Chapter 18) demonstrated that new ends of broken chromosomes are sticky and tend to fuse with each other. In contrast, the natural

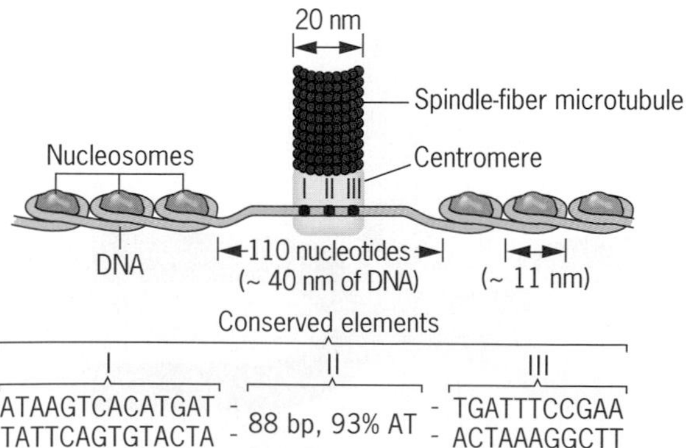

Figure 10.31 Diagram of the conserved structure of the centromeres in *Saccharomyces cerevisiae* (top) and the sequence of the *CEN* region of chromosome 3 of this species (bottom).

ends of normal (unbroken) chromosomes are stable and show no tendency to fuse with other broken or native ends. McClintock's results indicated that telomeres must have special structures different from the ends produced by breakage of chromosomes.

Another reason for postulating that telomeres have unique structures is that the known mechanisms of replication of linear DNA molecules do not permit duplication of both strands of DNA all the way to the ends of the molecules (Chapter 11). Thus, telomeres must have unique structures that facilitate their replication, or there must be some special replication enzyme that resolves this enigma. Whatever their structure, telomeres must provide at least three important functions. They must (1) prevent deoxyribonucleases from degrading the ends of the linear DNA molecules, (2) prevent fusion of the ends with other DNA molecules, and (3) facilitate replication of the ends of the linear DNA molecules without loss of material.

The telomeres of eukaryotic chromosomes have unique structures that include short nucleotide sequences present as tandem repeats. Although the sequences vary somewhat in different species, the basic repeat unit has the pattern $5'\text{-}T_{1-4}A_{0-1}G_{1-8}\text{-}3'$ in all but a few species. For example, the repeat sequence in humans and other vertebrates is TTAGGG, that of the protozoan *Tetrahymena thermophila* is TTGGGG, and that of the plant *Arabidopsis thaliana* is TTTAGGG. In most species, additional repetitive DNA sequences are present adjacent to telomeres; these are referred to as telomere-associated sequences.

In vertebrates, the TTAGGG repeat is highly conserved; it has been identified in more than 100 species, including mammals, birds, reptiles, amphibians, and fishes. The number of copies of this basic repeat unit in telomeres varies from species to species, from chromosome to chromosome within a species, and even on the same chromosome in different cell types. In normal (noncancerous) human somatic cells, telomeres usually contain 500 to 3000 TTAGGG repeats and gradually shorten with age. In contrast, the telomeres of germ-line cells and cancer cells do not shorten with age (see "Telomere Length and Aging in Humans" in Chapter 11).

The telomeres of a few species are not composed of short tandem repeats of the type described earlier. In *D. melanogaster*, for example, telomeres are composed of two specialized DNA sequences that can move from one location in the genome to other locations. Because of their mobility, such sequences are called transposable genetic elements (see Chapter 18).

Most telomeres terminate with a G-rich single-stranded region of the DNA strand with the 3′ end (a so-called 3′ overhang). These overhangs are short (12 to 16 bases) in ciliates such as *Tetrahymena*, but they

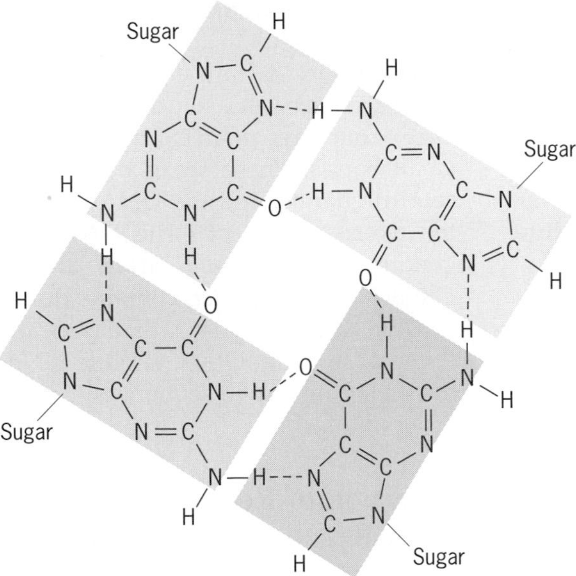

(*a*) Structure of a G-quartet

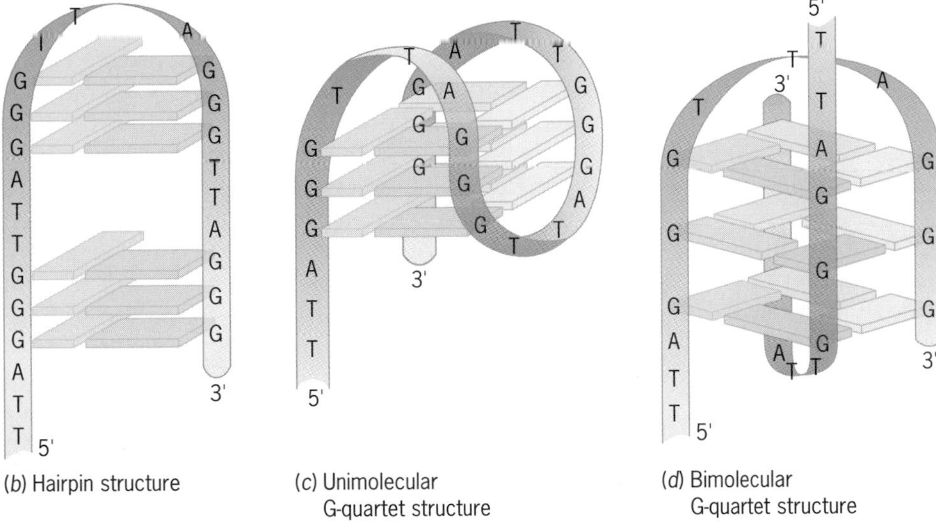

(*b*) Hairpin structure (*c*) Unimolecular G-quartet structure (*d*) Bimolecular G-quartet structure

Figure 10.32 (*a*) Structure of the G-quartet formed in solution by telomerelike G-rich oligonucleotides. (*b–d*) Three models of human telomere structure: (*b*) a unimolecular hairpin structure, (*c*) a unimolecular G-quartet structure, and (*d*) a bimolecular G-quartet structure. For brevity, the structures shown here contain only a few TTAGGG repeat units, whereas this sequence is repeated 250 to 1500 times in the telomeres of normal human chromosomes.

are quite long (125 to 275 bases) in humans. Terminal bases of these single-stranded ends often exhibit unique patterns of methylation (covalently attached methyl groups) that undoubtedly contribute to telomere structure. The guanine-rich repeat sequences of telomeres have the ability to form numerous hydrogen-bonded structures distinct from those produced by Watson–Crick base-pairing in DNA. For example, four guanine residues can form a G-quartet by a special type of hydrogen bonding called Hoogstein base-pairing (Figure 10.32*a*). Oligonucleotides that contain tandem telomere repeat sequences form G-quartets in solution, but whether these sequences form G-quartets *in vivo* remains unknown. Three models of human telomere structure are shown in Figure 10.32, *b-d*. One proposal is that bimolecular G-quartets (Figure 10.32*d*) may play a role in initiating chromosome pairing during meiosis.

Although the actual three-dimensional structures of telomeres are still uncertain, telomeres clearly do possess special features that are responsible for their unique properties. In addition, telomere sequences are known to be added to chromosomes by special RNA-containing enzymes called telomerases (Chapter 11). Lastly, specific telomere-binding proteins have been identified in several species, and these proteins are thought to cap the telomeric DNA and provide further stability to the ends of chromosomes. Indeed, a protein called **TRF-2** (**T**elomere **R**epeat-binding **F**actor **2**) has been shown to prevent end-to-end fusion of human chromosomes.

Key Points: **Each eukaryotic chromosome contains one giant molecule of DNA packaged into 10-nm ellipsoidal beads called nucleosomes. In the condensed chromosomes present during meiosis and mitosis, the 10-nm nucleosome fibers are further coiled into chromatin fibers about 30 nm in diameter. At metaphase, these 30-nm fibers are organized into domains by scaffolds composed of nonhistone chromosomal proteins. The spindle-fiber-attachment regions (centromeres) and ends (telomeres) of chromosomes have unique structures that facilitate their respective functions.**

EUKARYOTIC GENOMES: REPEATED DNA SEQUENCES

The chromosomes of prokaryotes almost exclusively contain DNA molecules with unique (nonrepeated) base-pair sequences; that is, each gene (with a few exceptions) is present only once in the genome. If the DNA molecules in prokaryotic chromosomes are broken into many short fragments, each fragment will contain a different sequence of base pairs. The chromosomes of eukaryotes are much more complex in this respect. Certain base sequences are repeated many times in the haploid chromosome complement, sometimes as many as a million times. DNA containing such repeated sequences, called **repetitive DNA,** often represents a major component (20 to 50 percent) of the eukaryotic genome.

The first evidence for repetitive DNA came from centrifugation studies of eukaryotic DNA. When the DNA of a prokaryote, such as *E. coli*, is isolated, fragmented, and centrifuged at high speeds for long periods of time in a 6*M* cesium chloride (CsCl) solution, the DNA will form a single band in the centrifuge tube at the position where its density is equal to the density of the CsCl solution (Chapter 11). For *E. coli*, this band will form at a position where the CsCl density is equal to the density of DNA containing about 50 percent A:T and 50 percent G:C base pairs. DNA density increases with increasing G:C content. The extra hydrogen bond in a G:C base pair is believed to result in a tighter association between the bases and thus a higher density than for A:T base pairs. The centrifugation of DNAs from eukaryotes to equilibrium conditions in such CsCl solutions usually reveals the presence of one large main band of DNA and one to several small bands. These small bands of DNA are called **satellite bands,** and the DNAs in these bands are often referred to as **satellite DNAs.** For example, the genome of *Drosophila virilis*, a distant relative of *Drosophila melanogaster*, contains three distinct satellite DNAs, each composed of a repeating sequence of seven base pairs. Other satellite DNAs in eukaryotes have long repetitive sequences.

Detection of Repeated Sequences: DNA Renaturation Kinetics

A more complete picture of the frequency and complexity of repetitive DNA sequences in eukaryotes has resulted from studies of DNA renaturation rates. The two strands of a DNA double helix are held together by a large number of relatively weak hydrogen bonds between complementary bases. When DNA molecules in aqueous solution are heated to near 100° C, these bonds are broken and the complementary strands of DNA separate. This process is called **denaturation.** If the complementary single strands of DNA are cooled slowly under the right conditions, the complementary base sequences will find each other and will re-form base-paired double helices. This re-formation of double helices from the complementary single strands of DNA is called **renaturation.** Analyses of the kinetics of renaturation of DNAs of eukaryotic organisms have yielded a wealth of information about the types of repetitive DNA sequences present in the genomes of eukaryotes.

Let's consider a long DNA molecule with no repeated sequences (for example, a DNA molecule from a prokaryotic chromosome). If such a molecule is sheared into fragments of a particular length, say 400 nucleotide pairs, is denatured, and is then allowed to renature under appropriate conditions, the rate of renaturation will depend on (1) the concentration of DNA in solution and (2) the complexity of the DNA, that is, the number of different 400 base-pair fragments. These two factors will determine the probability that any two complementary single strands of DNA will undergo a collision that can lead to renaturation. The effect of DNA concentration is quite obvious, because the higher the concentration of single strands in the solution, the greater the chance of a collision between any two of them.

The complexity of DNA is the total size of the nonrepeated nucleotide-pair sequences in the genome. In

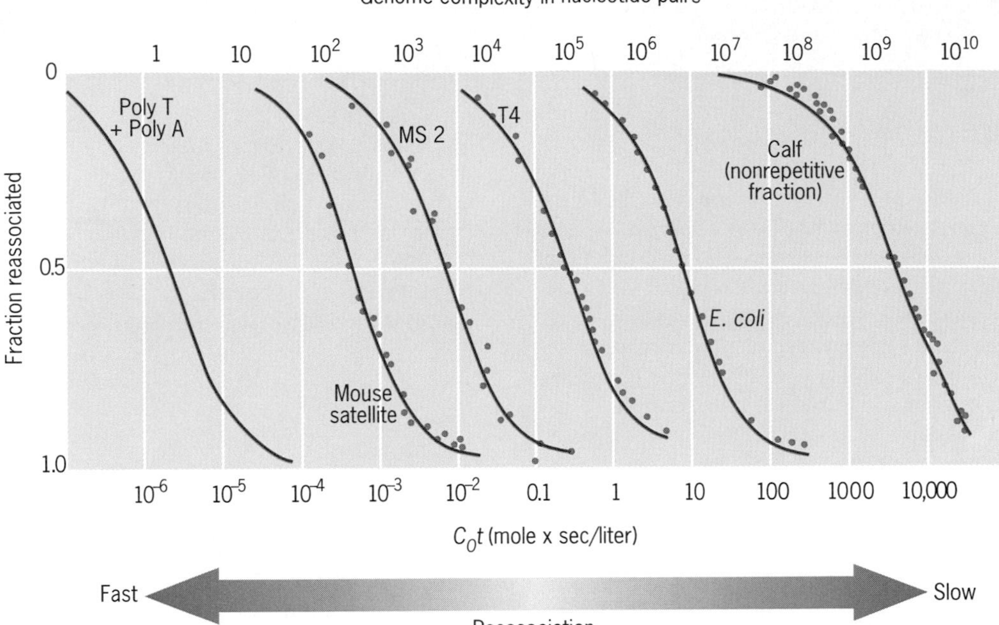

Figure 10.33 Illustration of the effect of genome complexity on DNA renaturation rates. In poly T/poly A duplexes, all fragments produced by shearing will have the same complementary strands, and renaturation will occur rapidly (left). In complex genomes with lots of different sequences, for example, the nonrepeated sequences of calf DNA (right), renaturation will occur less rapidly.

a double helix with poly A in one strand and poly T in the other stand, the complementary strands of every fragment will be the same. Thus, all collisions between DNA single strands derived from complementary strands will lead to renaturation. As the complexity of the DNA increases, the proportion of the random collisions between single strands of DNA that are between complementary single strands will decrease and the rate of renaturation will decrease accordingly. Consider genomes that contain 4, 400, 4,000, and 40,000 different genes. In a solution containing 100 μg of DNA per ml, the concentration of any one gene will be 25, 0.25, 0.025, and 0.0025 μg/ml, respectively, for the four genomes. As a result, random collisions between the complementary strands derived from any one gene will decrease in frequency as the complexity of the genome (the number of different genes) increases. This effect of genome complexity on renaturation kinetics is illustrated in Figure 10.33, which shows the very rapid renaturation of poly A:T and the very slow renaturation of the nonrepeated sequences in the genome of *Bos taurus* (domestic cattle).

Mathematical analyses (see Technical Sidelight: DNA Renaturation Kinetics) of the kinetics of renaturation of eukaryotic DNAs have been extremely informative. The results of DNA renaturation kinetics experiments have demonstrated that eukaryotic genomes contain many different kinds of repetitive DNA sequences. The DNA sequences present in eukaryotes are commonly divided into three classes:

1. **Unique** or **single-copy DNA sequences**—1 to 10 copies per genome.
2. **Moderately repetitive DNA sequences**—10 to 10^5 copies per genome.
3. **Highly repetitive DNA sequences**—more than 10^5 copies per genome.

Figure 10.34 shows a DNA renaturation curve demonstrating the presence of these three classes of DNA sequences in the human genome. The proportions of these three classes of DNA sequences vary from species to species. Single-copy sequences make up about 40 to 70 percent of the genome in most plants and animals. The moderately repetitive class of DNA sequences appears to be very heterogeneous in most eukaryotes; it contains many different sequences with different degrees of reiteration. For example, in *Drosophila melanogaster*, 12 percent of the DNA contains moderately repetitive sequences, with an average reiteration frequency of 70. The highly repetitive DNAs of eukaryotes frequently contain both satellite and nonsatellite DNA sequences.

When the sequence organization of eukaryotic genomes is studied in greater detail—by combining the techniques of density-gradient centrifugation, renaturation kinetic analysis, biochemical analyses, elec-

TECHNICAL SIDELIGHT

DNA Renaturation Kinetics

Definitive evidence for the presence of repeated DNA sequences in eukaryotic genomes has been obtained by analyzing the kinetics of DNA renaturation. Sequences that are present in multiple copies per genome will renature more rapidly than sequences that are present only once. Indeed, as we discuss below, the rate at which a specific sequence renatures is inversely proportional to its copy number—the higher the copy number, the less time required for renaturation.

Consider a particular 400 base-pair fragment composed of two complementary strands, *a* and *a*′.

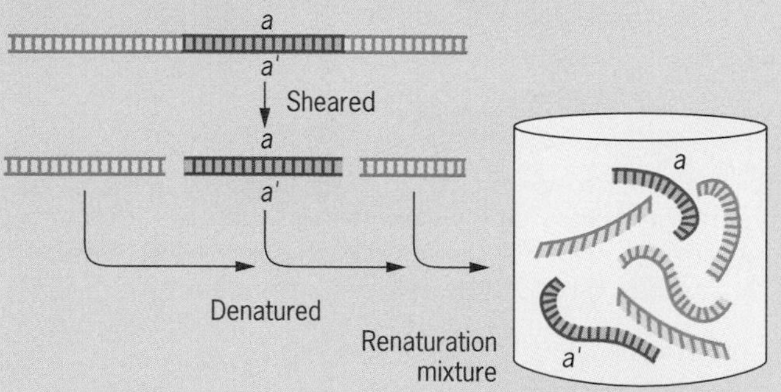

Reassociation of *a* and *a*′ will require a specific collision between these two single strands. Collisions between *a* or *a*′ and any other single strand will not lead to renaturation. For a given concentration, the larger the DNA molecule, and thus the more nonidentical 400 base-pair fragments present, the slower the reassociation reaction will be. With a larger number of different fragments, a smaller proportion of the random collisions between molecules will occur between complementary single strands such as *a* and *a*′.

Note that every reassociation event, like *a* with *a*′, will require a collision between two single strands that are present in the renaturation mixture in equal concentration. Because the reaction requires the interaction of two equally frequent molecules, the rate of renaturation will be a function of the square of the concentration of single strands (so-called second-order or bimolecular reaction kinetics), or

$$\frac{-dC}{dt} = kC^2$$

$$\left(\text{or } \frac{-dC}{C^2} = kdt, \text{ rearranged for integration}\right)$$

where

C = the concentration of single-stranded DNA in moles of nucleotides per liter

t = time in seconds

k = a second-order rate constant in liters per mole seconds

This equation states that the instantaneous change (decrease) in concentration of single-stranded DNA ($-dC$) with time (dt) is equal to the proportionality constant (k) times the square of the concentration of single-stranded DNA.

Integration of the preceding equation from the initial conditions ($t = 0$ seconds and $C = C_0$, where C_0 equals the concentration of single-stranded DNA at $t = 0$) yields

$$\frac{1}{C} - \frac{1}{C_0} = kt$$

or, rearranged,

$$\frac{C}{C_0} = \frac{1}{1 + kC_0t}$$

This equation states that the fraction of input single-stranded DNA remaining in a renaturation reaction mixture (C/C_0) at any given time (t) is a function of the initial concentration (C_0) times elapsed time (t), or C_0t. It is thus convenient to present data on renaturation kinetics in a plot of C/C_0 versus C_0t. These so-called **C_0t** (pronounced "caught") **curves** (see Figures 10.33 and 10.34) have provided a great deal of information about the types of repetitive DNA in eukaryotic genomes.

Consider a DNA molecule containing a 400 base-pair sequence that is repeated (present twice).

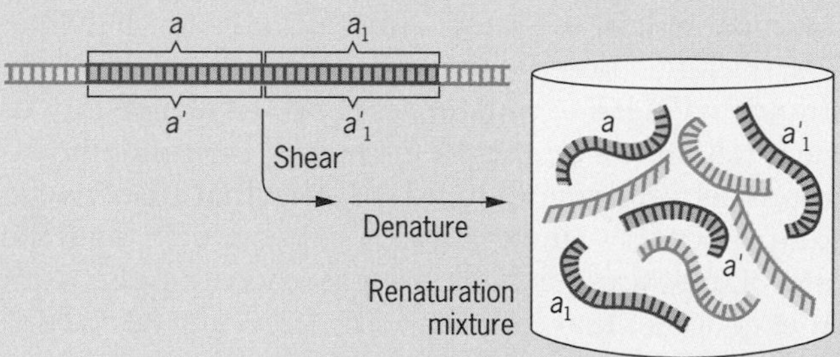

Now a and a_1 are identical single strands, and a' and a_1' are identical complementary strands. Reassociation of each repeated sequence will take only half as long as in the previous example, where each fragment contained a unique base sequence. Now reassociation will result from a collision of a with either a' or a_1' (also for a_1 and either a' or a_1'). Thus, the time required for reassociation of a particular DNA sequence is inversely proportional to the number of times that sequence is present in the genome. Clearly, highly repetitive DNA sequences will renature very rapidly.

The proportion of the DNA that has renatured at any time can be quantified in several ways. One method is simply to treat samples taken at various times with a nuclease that is specific for single-stranded nucleic acids. The DNA remaining in the samples after digestion (all double-stranded) can then be quantified by direct chemical analysis.

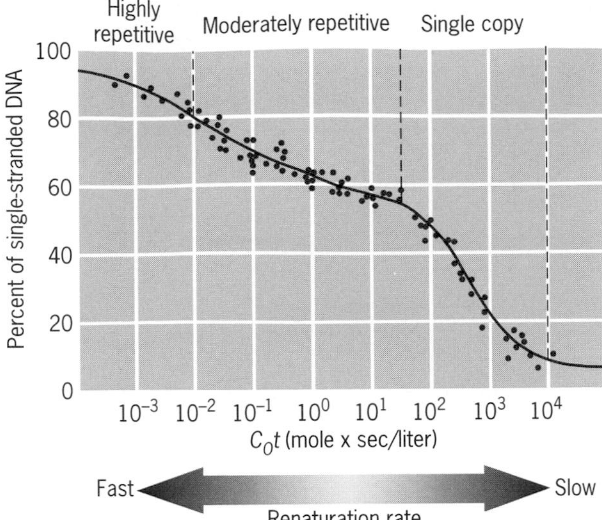

Figure 10.34 The renaturation kinetics of human DNA. Sequences that are highly repetitive (>10^5 copies per genome) renature rapidly (left); moderately repeated sequences (from 10 to 10^5 copies per genome) renature at intermediate rates; and single-copy sequences (one to ten copies per genome) renature slowly.

tron microscopy, and cloning and sequencing of substantial segments of chromosomal DNA—a pattern emerges. Much of the genome consists of moderately repetitive sequences interspersed with single-copy sequences. In toads, sea urchins, and humans, these sequences are quite short. The moderately repetitive sequences average 300 nucleotide pairs in length, and the single-copy sequences are about 800 to 1200 nucleotide pairs long. *Drosophila melanogaster* DNA also exhibits interspersion of moderately repetitive and single-copy sequences, but the sequences are much longer (5,600 and 13,000 nucleotide pairs, respectively).

What are the functions of the different kinds of DNA sequences in eukaryotes? Most of the structural genes (genes coding for proteins or RNA molecules) are single-copy sequences. The genes for histones, rRNA molecules, and ribosomal proteins, all gene products that are present in the cell in large quantities, are exceptions; these are redundant (present in moderately repetitive DNA) genes. Many geneticists have postulated that most other moderately repetitive sequences may be involved in the regulation of gene expression. Their interspersion with single-copy sequences, and thus their location adjacent to structural genes, is consistent with a regulatory role. Another important group of moderately repetitive DNA sequences are the **transposable genetic elements** (Chapter 18) which can move from one location in a chromosome to another or even to a different chromosome. The transposability of these moderately repetitive sequences has led to much speculation about their possible involvement in the regulation of gene expression during development and their roles in the evolution of eukaryotic genomes. These transposable elements are responsible for a surprisingly large number of the naturally occurring mutations in eukaryotic organisms.

The function(s) of highly repetitive DNA—most, if not all, of which is located in genetically inactive heterochromatic regions of chromosomes—is completely unknown. Postulated functions for highly repetitive DNA include (1) structural or organizational roles in chromosomes; (2) involvement in chromosome pairing during meiosis; (3) involvement in crossing over or recombination (Chapter 14); (4) protection of important structural genes, like histone, rRNA, or ribosomal protein genes; (5) a repository of unessential DNA sequences for use in the future evolution of the species; and (6) no function at all—just junk DNA that is carried along by the processes of replication and segregation of chromosomes. The validity of any of these postulated roles remains to be established.

Localization of Repeated DNA Sequences in Chromosomes

In 1969, Mary Lou Pardue and Joseph Gall developed a procedure by which they could anneal radioactive single strands of DNA with complementary strands of DNA in chromosomes on glass slides. By using this procedure, called ***in situ* hybridization,** Pardue and Gall were able to determine the chromosomal locations of repetitive DNA sequences. (The Latin term *in situ* means "in its original place"; hybridization is the formation of "hybrid" duplex molecules by the annealing of complementary or partially complementary strands of DNA or RNA.) *In situ* hybridization involves spreading mitotic or meiotic chromosomes on a glass slide (see Figure 7.1), denaturing the DNA in the chromosomes by exposure to alkali (0.07 *N* NaOH for a few minutes), rinsing with buffer to remove the alkaline solution, incubating the slide in hybridization solution containing radioactive copies of the nucleotide sequence of interest, washing off the radioactive strands that have not hybridized with complementary sequences in the chromosomes, exposing the slide to a photographic emulsion that is sensitive to low-energy radioactivity, developing the autoradiograph, and superimposing the autoradiograph and a photograph of the chromosomes (Figure 10.35*a* and *b*).

One of the first *in situ* hybridization experiments that Pardue and Gall performed demonstrated that the satellite DNA sequence of the mouse is located in heterochromatic regions that flank the centromeres of the

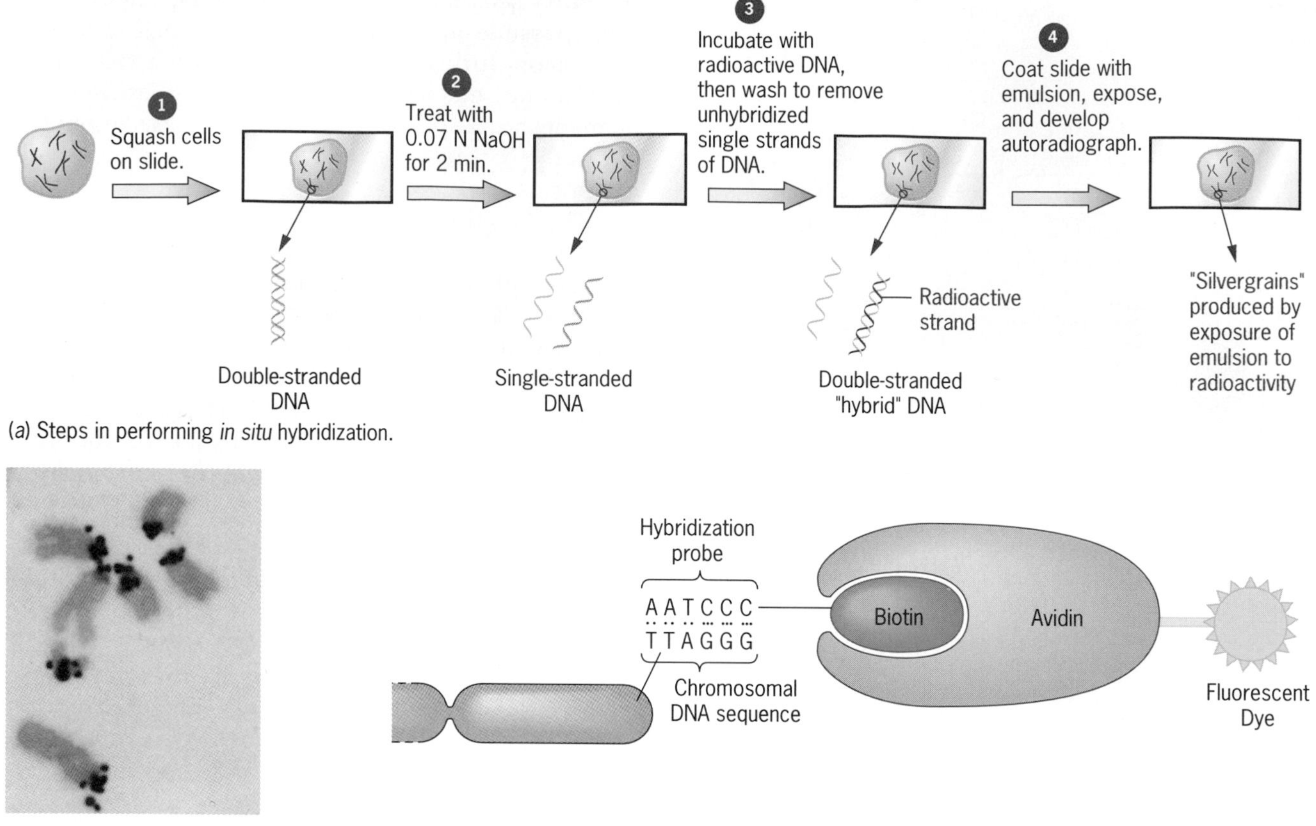

(*a*) Steps in performing *in situ* hybridization.

(*b*) Autoradiograph showing chromosomal locations of mouse satellite DNA sequences.

(*c*) Visualization of human telomeres by using fluorescent dyes and *in situ* hybridization.

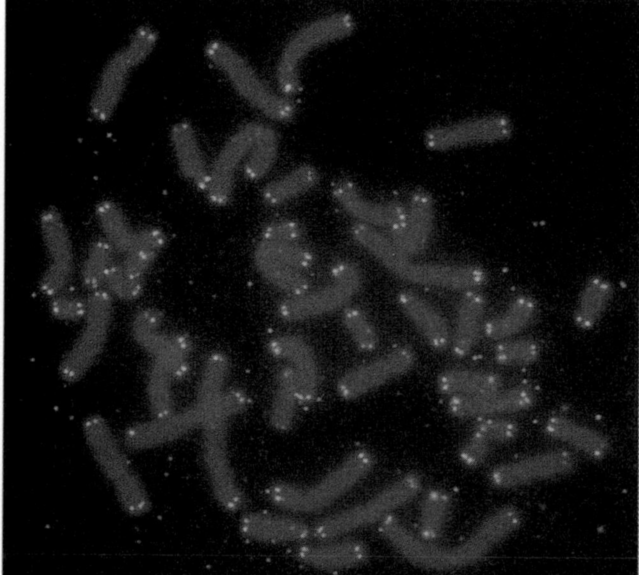

(*d*) Human telomeres visualized using fluorescent probes and *in situ* hybridization

Figure 10.35 Localization of repeated DNA sequences in chromosomes by *in situ* hybridization performed with radioactive probes (*a* and *b*) or fluorescent probes (*c* and *d*). The *in situ* hybridization procedure developed by Pardue and Gall is shown in (*a*), and one of their autoradiographs demonstrating the presence of the mouse satellite DNA sequence in centromeric heterochromatin is shown in (*b*). Use of fluorescent dyes to localize the TTAGGG repeat sequence to the telomeres of human chromosomes is illustrated in (*c*), and a photomicrograph demonstrating its telomeric location is shown in (*d*).

mouse chromosomes (Figure 10.35*b*). The mouse genome contains about 10^6 copies of this satellite DNA sequence, which is about 400 nucleotide pairs long and makes up about 10 percent of the mouse genome. Similar studies have subsequently been done with the satellite DNAs of several other species, and these repetitive DNA sequences are usually located in centromeric heterochromatin or adjacent to telomeres. A repetitive DNA sequence can be identified as satellite DNA only if the sequence has a base composition sufficiently different from that of main-band DNA to produce a distinct band during density-gradient centrifugation. Therefore, centrifugation cannot be used to identify all repetitive DNA sequences. Satellite DNA sequences usually are not expressed; that is, they do not encode RNA or protein products.

Today, *in situ* hybridization experiments are often done by using hybridization probes that are linked to fluorescent dyes or antibodies tagged with fluorescent compounds (Figure 10.35*c* and *d*). In one protocol, DNA or RNA hybridization probes are linked to the vitamin biotin, which is bound with high affinity by the egg protein avidin (Figure 10.35*c*). By using avidin covalently linked to a fluorescent dye, the chromosomal location of the hybridized probe can be detected by the fluorescence of the dye. This procedure, called **FISH** (**F**luorescent ***I**n **S**itu* **H**ybridization), has been used to demonstrate the presence of the repetitive sequence TTAGGG in the telomeres of human chromosomes (Figure 10.35*d*). The FISH procedure is very sensitive and can be used to detect the locations of single-copy sequences in human mitotic chromosomes (see Figure 21.6*b*).

Repeated Sequences in the Human Genome

In humans, as in other eukaryotes, much of the dispersed moderately repetitive DNA consists of families of transposable elements. Two important groups of moderately repetitive transposable elements are called the **LINE** (**L**ong **I**nterspersed **N**uclear **E**lement) and the **SINE** (**S**hort **I**nterspersed **N**uclear **E**lement) families.

LINE transposable elements have an average length of 6500 nucleotide pairs and are dispersed throughout the human genome. The best known member of the human LINE family of sequences is the transposable element *L1* (Chapter 18), which is present in 20,000 to 50,000 copies per genome and accounts for 5 percent of all human DNA.

SINE elements are only 150 to 300 base pairs in length. The most extensively studied SINE sequences are the **Alu sequences** in humans. (The designation Alu is derived from the name of an enzyme that cleaves DNA at a specific sequence located within the Alu elements.) There are approximately 500,000 Alu sequences, each about 300 nucleotide pairs long, in the haploid human genome. They are related, but not identical, in nucleotide sequence and are widely dispersed on all the human chromosomes. Collectively, they constitute about 5 percent of the human genome, with an Alu sequence present on average once every 5000 nucleotide pairs. Despite speculation that the Alu elements might contain sequences involved in the initiation of DNA replication and the regulation of gene expression, their biological function(s) is still unknown.

As is observed with other transposable genetic elements (Chapter 18), the transposition of an Alu sequence into genes, and DNA rearrangements caused by recombination between two Alu sequences can produce mutant phenotypes. Mutant alleles that cause **acholinesterasemia** (a deficiency of the enzyme cholinesterase resulting in transient paralysis upon exposure to muscle relaxants) and **neurofibromatosis** (tumors of nerve tissues just under the skin and the central nervous system) have been shown to result from the insertion of Alu sequences into genes. In addition, numerous cases of **hypercholesterolemia** (excessive amounts of cholesterol in the blood due to a deficiency of cholesterol receptors on liver cells) and **thalassemia** (hemoglobin deficiencies) have been shown to result from gene rearrangements produced by recombination between closely linked Alu sequences. These and other results have clearly demonstrated that Alu sequences have played, and continue to play, a major role in the evolution of the human genome.

Key Points: **Eukaryotic genomes contain repeated DNA sequences, with some sequences present a million times or more. Eukaryotic DNA sequences are commonly grouped into three classes: (1) unique or single-copy sequences present in one to a few copies per genome, (2) moderately repetitive sequences present in 10 to 10^5 copies, and (3) highly repetitive sequences present in over 10^5 copies per genome. Two families of moderately repetitive transposable elements make up 10 percent of the human genome and play an important role in its evolution.**

TESTING YOUR KNOWLEDGE

1. The red alga *Polyides rotundus* stores its genetic information in double-stranded DNA. When DNA was extracted from *P. rotundus* cells and analyzed, 32 percent of the bases were found to be guanine residues. From this information, can you determine what percentage of the bases in this DNA were thymine residues? If so, what percentage? If not, why not?

ANSWER

The two strands of a DNA double helix are complementary to each other, with guanine (G) in one strand always paired with cytosine (C) in the other strand and, similarly, adenine (A) always paired with thymine (T). Therefore, the concentrations of G and C are always equal, as are the concentrations of A and T. If 32 percent of the bases in double-stranded DNA are G residues, then another 32 percent are C residues. Together, G and C comprise 64 percent of the bases in *P. rotundus* DNA; thus, 36 percent of the bases are A's and T's. Since the concentration of A must equal the concentration of T, 18 percent (36% × 1/2) of the bases must be T residues.

2. The *E. coli* virus ϕX174 stores its genetic information in single-stranded DNA. When DNA was extracted from ϕX174 virus particles and analyzed, 21 percent of the bases were found to be G residues. From this information, can you determine what percentage of the bases in this DNA were thymine residues? If so, what percentage? If not, why not?

ANSWER

No! The A = T and G = C relationships occur only in double-stranded DNA molecules, because of their complementary strands. Since base-pairing doesn't occur or occurs only as limited intrastrand pairing in single-stranded nucleic acids, you cannot determine the percentage of any of the other three bases from the G content of the DNA.

3. If each G_1-stage human chromosome contains a single molecule of DNA, how many DNA molecules would be present in the chromosomes of the nucleus of (a) a human egg, (b) a human sperm, (c) a human diploid somatic cell in stage G_1, (d) a human diploid somatic cell in stage G_2, (e) a human primary oocyte?

ANSWER

A normal human haploid cell contains 23 chromosomes, and a normal human diploid cell contains 46 chromosomes, or 23 pairs of homologs. If prereplication chromosomes contain a single DNA molecule, postreplication chromosomes will contain two DNA molecules, one in each of the two chromatids. Thus, normal human eggs and sperm contain 23 chromosomal DNA molecules; diploid somatic cells contain 46 and 92 chromosomal DNA molecules at stages G_1 and G_2, respectively; and a primary oocyte contains 92 such DNA molecules.

QUESTIONS AND PROBLEMS

10.1 (a) How did the transformation experiments of Griffith differ from those of Avery and his associates? (b) What was the significant contribution of each? (c) Why was Griffith's work not evidence for DNA as the genetic material, whereas the experiments of Avery and coworkers provided direct proof that DNA carried the genetic information?

10.2 A cell-free extract is prepared from Type IIIS pneumococcal cells. What effect will treatment of this extract with (a) protease, (b) RNase, and (c) DNase have on its subsequent capacity to transform recipient Type IIR cells to Type IIIS? Why?

10.3 How could it be demonstrated that the mixing of heat-killed Type III pneumococcus with live Type II resulted in a transfer of genetic material from Type III to Type II rather than a restoration of viability to Type III by Type II?

10.4 What is the macromolecular composition of a bacterial virus or bacteriophage such as phage T2?

10.5 What chemical properties do DNA and proteins possess that allow researchers to label specifically one or the other of these macromolecules with a radioactive isotope?

10.6 (a) What was the objective of the experiment carried out by Hershey and Chase? (b) How was the objective accomplished? (c) What is the significance of this experiment?

10.7 (a) What background material did Watson and Crick have available for developing a model of DNA? (b) What was their contribution to building the model?

10.8 (a) Why did Watson and Crick choose a double helix for their model of DNA structure? (b) Why were hydrogen bonds placed in the model to connect the bases?

10.9 (a) If a virus particle contained double-stranded DNA with 200,000 base pairs, how many nucleotides would be present? (b) How many complete spirals would occur on each strand? (c) How many atoms of phosphorus would be present? (d) What would be the length of the DNA configuration in the virus?

10.10 If one strand of DNA in the Watson–Crick double helix has a base sequence of 5′-GTCATGAC-3′, what is the base sequence of the complementary strand?

10.11 What are the differences between DNA and RNA?

10.12 DNA was extracted from cells of *Staphylococcus afermentans* and analyzed for base composition. It was found that 37 percent of the bases are cytosine. With this information, is it possible to predict what percentage of the bases are adenine? If so, what percentage? If not, why not?

10.13 RNA was extracted from TMV (tobacco mosaic virus) particles and found to contain 20 percent cytosine (20 percent of the bases were cytosine). With this information, is it possible to predict what percentage of the bases in TMV are adenine? If so, what percentage? If not, why not?

10.14 Indicate whether each of the following statements about the structure of DNA is true or false. (Each letter is used to refer to the concentration of that base in DNA.) (a) $A + T = G + C$. (b) $A = G$; $C - T$. (c) $A/T = C/G$. (d) $T/A = C/G$. (e) $A + G = C + T$. (f) $G/C = 1$. (g) $A = T$ within each single strand. (h) Hydrogen bonding provides stability to the double helix in aqueous cytoplasms. (i) Hydrophobic bonding provides stability to the double helix in aqueous cytoplasms. (j) When separated, the two strands of a double helix are identical. (k) Once the base sequence of one strand of a DNA double helix is known, the base sequence of the second strand can be deduced. (l) The structure of a DNA double helix is invariant. (m) Each nucleotide pair contains two phosphate groups, two deoxyribose molecules, and two bases.

10.15 The nucleic acids from various viruses were extracted and examined to determine their base composition. Given the results outlined below, what can you hypothesize about the physical nature of the nucleic acids from these viruses? (a) 35% A, 35% T, 15% G, and 15% C. (b) 35% A, 15% T, 25% G, and 25% C. (c) 35% A, 30% U, 30% G, and 5% C.

10.16 Compare and contrast the structures of the A, B, and Z forms of DNA?

10.17 The temperature at which one-half of a double-stranded DNA molecule has been denatured is called the melting temperature, T_m. Why does T_m depend directly on the GC content of the DNA?

10.18 The relationship between the melting T_m and GC content can be expressed, in its much simplified form, by the formula $T_m = 69 + 0.41$ (% GC). (a) Calculate the melting temperature of *E. coli* DNA that has about 50% GC. (b) Estimate the % GC of DNA from a human kidney cell where $T_m = 85°C$.

10.19 The available evidence indicates that each eukaryotic chromosome (excluding polytene chromosomes) contains a single giant molecule of DNA. What different levels of organization of this DNA molecule are apparent in chromosomes of eukaryotes at various times during the cell cycle?

10.20 A diploid rye plant, *Secale cereale*, has $2n = 14$ chromosomes and approximately 1.6×10^{10} bp of DNA. How much DNA is in a nucleus of a rye cell at (a) mitotic metaphase, (b) meiotic metaphase I, (c) mitotic telophase, and (d) meiotic telophase II?

10.21 A diploid nucleus of *Drosophila melanogaster* contains about 3.4×10^8 nucleotide pairs. Assume (1) that all the nuclear DNA is packaged in nucleosomes and (2) that an average internucleosome linker size is 60 nucleotide pairs. How many nucleosomes would be present in a diploid nucleus of *D. melanogaster*? How many molecules of histone H2a, H2b, H3, and H4 would be required?

10.22 The satellite DNAs of *Drosophila virilis* can be isolated, essentially free of main-band DNA, by density-gradient centrifugation. If these satellite DNAs are sheared into approximately 40-nucleotide-pair-long fragments and are analyzed in denaturation–renaturation experiments, how would you expect their hybridization kinetics to compare with the renaturation kinetics observed using similarly sheared main-band DNA under the same conditions? Why?

10.23 Experimental evidence indicates that most highly repetitive DNA sequences in the chromosomes of eukaryotes do not produce any RNA or protein products. What does this indicate about the function of highly repetitive DNA?

10.24 (a) Are (1) single-copy DNA sequences, (2) moderately repetitive DNA sequences, (3) highly repetitive DNA sequences more prevalent in euchromatin or heterochromatin? (b) Are most of the single-copy DNA sequences in chromosomes transcribed? Are the highly repetitive DNA sequences in chromatin transcribed? (c) What are the presumed functions of most (1) single-copy DNA sequences, (2) highly repetitive DNA sequences? (d) What roles are the moderately repetitive DNA sequences believed to play in higher eukaryotes?

10.25 Are eukaryotic chromosomes metabolically most active during prophase, metaphase, anaphase, telophase, or interphase?

10.26 (a) What functions do (1) centromeres, (2) telomeres provide? (b) Do telomeres have any unique structural features? (c) What is the function of telomerase? (d) When chromosomes are broken by exposure to high-energy radiation such as X rays, the resulting broken ends exhibit a pronounced tendency to stick to each other and fuse. Why might this occur?

10.27 Of what special interest is the biophysical technique of viscoelastometry to geneticists?

10.28 How many DNA molecules are present in (a) the axial regions, (b) the lateral loops of lampbrush chromosomes?

10.29 Are the scaffolds of eukaryotic chromosomes composed of histone or nonhistone chromosomal proteins? How has this been determined experimentally?

10.30 (a) Which class of chromosomal proteins, histones or nonhistones, is the more highly conserved in different eukaryotic species? Why might this difference be expected? (b) If one compares the histone and nonhistone chromosomal proteins of chromatin isolated from different tissues or cell types of a given eukaryotic organism, which class of proteins will exhibit the greater heterogeneity? Why are both classes of proteins not expected to be equally homogeneous in chromosomes from different tissues or cell types?

10.31 During DNA renaturation experiments, why is the rate of renaturation ($-dC/dt$) proportional to the square of the concentration of single strands (kC^2) rather than simply proportional to the concentration of single strands (kC)?

10.32 Studies of the renaturation kinetics of DNA from the human genome reveal that at $t = 0$ (the earliest point in time at which the proportion of double-stranded DNA can be measured in standard renaturation experiments) approximately 6 percent of the DNA is already double stranded (see Figure 10.34). Assuming that all of the DNA was initially denatured in these experiments, how can this result (6 percent renatured at $t = 0$) be explained?

10.33 (a) If the haploid human genome contains 3×10^9 nucleotide pairs and the average molecular weight of a nucleotide pair is 660, how many copies of the human genome are present, on average, in 1 mg of human DNA? (b) What is the weight of one copy of the human genome? (c) If the haploid genome of the small plant *Arabidopsis thaliana* contains 7.7×10^7 nucleotide pairs, how many copies of the *A. thaliana* genome are present, on average, in 1 mg of *A. thaliana* DNA? (d) What is the weight of one copy of the

A. thaliana genome? (e) Of what importance are calculations of the above type to geneticists?

10.34 The relationship between DNA sequence length N (in base pairs) and $C_0t_{1/2}$ (the C_0t at which one-half of the DNA has renatured) can be expressed by the equation $N = lC_0t_{1/2}$ where l, the length constant, is 5×10^5 under certain standard conditions. Calculate the length of genomic DNA in the following organisms: (a) lambda phage with $C_0t_{1/2} = 1 \times 10^{-1}$, and (b) *E. coli* with $C_0t_{1/2} = 8$.

10.35 Analysis of the C_0t curve of a particular eukaryotic organism identified three renaturing components, whose respective $C_0t_{1/2}$ values were 5×10^{-3}, 2×10^{-1}, and 1×10^3 mol-sec/l. If only one copy of the unique sequences is present in the genome of this eukaryote, how many copies are there of (a) the moderately repetitive sequences, and (b) the highly repetitive sequences?

BIBLIOGRAPHY

AVERY, O. T., C. M. MACLEOD, AND M. MCCARTY. 1944. Studies on the chemical nature of the substance inducing transformation in pneumococcal types. *J. Expl. Med.* 79:137–158.

BLACKBURN, E. H., AND C. W. GREIDER, EDS. 1995. *Telomeres.* Cold Spring Harbor Laboratory Press, Cold Spring Harbor, NY.

BRITTEN, R. J., AND D. E. KOHNE. 1970. Repeated segments of DNA. *Sci. Amer.* 222(4):24–31.

HERSHEY, A. D., AND M. CHASE. 1952. Independent functions of viral protein and nucleic acid in growth of bacteriophage. *J. Gen. Physiol.* 36:39–56. (Reprinted in G. S. Stent. 1960. *Papers on Bacterial Viruses*, 2nd ed. Little, Brown, Boston.)

KORNBERG, R. D., AND A. KLUG. 1981. The nucleosome. *Sci. Amer.* 244(2):52–64.

LUGER, K., A. W. MÄDER, R. K. RICHMOND, D. F. SARGENT, AND T. J. RICHMOND. 1997. Crystal structure of the nucleosome core particle at 2.8 Å resolution. *Nature* 389:251–260.

TRUN, N. J., AND J. F. MARKO. 1998. Architecture of a bacterial chromosome. *ASM News* 64: 276–283.

VAN STEENSEL, B., A. SMOGORZEWSKA, AND T. DE LANGE. 1998. TRF-2 protects human telomeres from end-to-end fusions. *Cell* 92:401–413.

WATSON, J. D. 1968. *The Double Helix.* Atheneum, New York.

WATSON, J. D., AND F. H. C. CRICK. 1953. A structure for deoxyribose nucleic acid. *Nature* 171:737–738.

Seven pairs of identical twins. Although the twins were given no instructions regarding how to pose for the photograph, note the similar posture, hand placement, and facial expressions of both members of each pair of twins.

Replication of DNA and Chromosomes

CHAPTER OUTLINE

Basic Features of DNA Replication *In Vivo*

DNA Polymerases and DNA Synthesis *In Vitro*

The Complex Replication Apparatus

Unique Aspects of Eukaryotic Chromosome Replication

Monozygotic Twins: Are They Identical?

From the day of their birth, through childhood, adolescence, and adulthood, Merry and Sherry have been mistaken for one another. When they are apart, Merry is called Sherry about half of the time, and Sherry is misidentified as Merry with equal frequency. Even their parents have trouble distinguishing them. Merry and Sherry are monozygotic ("identical") twins; they both developed from a single fertilized egg. At an early cleavage stage, the embryo split into two cell masses, and both groups of cells developed into complete embryos. Both embryos developed normally, and on April 7, 1955, one newborn was named Merry, the other Sherry.

People often explain nearly identical phenotypes of monozygotic twins like Merry and Sherry by stating that "they contain the same genes." Of course, that is not true. To be accurate, the statement should be that identical twins contain progeny replicas of the same parental genes. But this simple colloquialism suggests that most people do, indeed, believe that the progeny replicas of a gene actually are identical. If the human genome contains 70,000 to 100,000 genes, are the progeny replicas of all these genes exactly the same in identical twins?

A human life emerges from a single cell, a tiny sphere about 0.1 mm in diameter. That cell gives rise to hundreds of billions of cells during fetal deveiopment. At maturity, a human of average size contains about 65 trillion (65,000,000,000,000) cells. With some exceptions, each of these 65 trillion cells contains a progeny replica of each of the 70,000 to 100,000 genes. Moreover, the cells of the body are not static; in some tissues, old cells are continuously being replaced by new cells. For example, in a healthy individual, the bone marrow cells produce about 2 million red blood cells per minute. Although not all of the progeny replicas of genes in the human body are identical, the process by which these genes are duplicated is very accurate. The human haploid genome contains about 3×10^9 nucleotide pairs of DNA, all of which must be duplicated during each cell division. In this chapter, we examine how DNA replicates and we focus on the mechanisms that ensure the fidelity of this process.

BASIC FEATURES OF DNA REPLICATION *IN VIVO*

In humans, the synthesis of a new strand of DNA occurs at the rate of about 3000 nucleotides per minute. In bacteria, about 30,000 nucleotides are added to a nascent DNA chain per minute. Clearly, the cellular machinery responsible for DNA replication must work very fast, but, even more importantly, it must work with great precision. Indeed, the fidelity of DNA replication is amazing, with an average of only one mistake per billion nucleotides incorporated. Thus, the majority of the genes of identical twins are indeed identical, but some will have changed owing to replication errors and other types of mutations (Chapter 14). Most of the key features of the mechanism by which the rapid and accurate replication of DNA occurs are now known, although many molecular details remain to be elucidated.

The synthesis of DNA, like the synthesis of RNA (Chapter 12) and proteins (Chapter 13), involves three steps: (1) chain initiation, (2) chain extension or elongation, and (3) chain termination. In this and the following two chapters, we examine the mechanisms by which cells carry out each of the three steps in the synthesis of these important macromolecules. First, however, we consider some key features of DNA replication.

Semiconservative Replication

When Watson and Crick deduced the double-helix structure of DNA with its complementary base-pairing, they immediately recognized that the base-pairing specificity could provide the basis for a simple mechanism for DNA duplication. If the two complementary strands of a double helix separated, by breaking the hydrogen bonds of each base pair, each parental strand could direct the synthesis of a new complementary strand because of the specific base-pairing requirements (Figure 11.1). That is, each parental strand could function as a template, a single strand of DNA that specifies the nucleotide sequence of a new complementary strand. Adenine, for example, in the parent strand would serve as a template via its hydrogen-bonding potential for the incorporation of thymine in the nascent complementary strand. This mechanism of DNA replication is called **semiconservative replication,** because each of the complementary strands of the parental double helix is conserved (or the double helix is "half-conserved") during the process.

Hypothetically, DNA replication could occur by any of three different mechanisms (Figure 11.2). In addition to (1) the semiconservative mechanism proposed by Watson and Crick, replication could occur by either (2) a conservative mechanism, with the parental double helix being conserved and directing the synthesis of a new progeny double helix, or (3) a dispersive mechanism, with segments of parental and progeny strands interspersed as a result of the synthesis and rejoining of short segments of DNA. The first critical test of Watson and Crick's proposal that DNA replicates by a semiconservative mechanism was performed by Matthew Meselson and Franklin Stahl in 1958. Their results showed that the chromosome (now known to contain a single double helix of DNA) of the common colon bacillus *E. coli*, a prokaryote, replicates semiconservatively.

Meselson and Stahl grew *E. coli* cells for many generations in a medium in which the heavy isotope of nitrogen, ^{15}N, had been substituted for the normal, light isotope, ^{14}N. The purine and pyrimidine bases in DNA contain nitrogen; thus, the DNA of cells grown on medium containing ^{15}N will have a greater density (mass per unit volume) than the DNA of cells grown on medium containing ^{14}N. Because molecules of different densities can be separated by a procedure called **equilibrium density-gradient centrifugation** (see Technical Sidelight: Cesium Chloride Equilibrium Density-Gradient Centrifugation . . .), Meselson and Stahl were able to distinguish between the three pos-

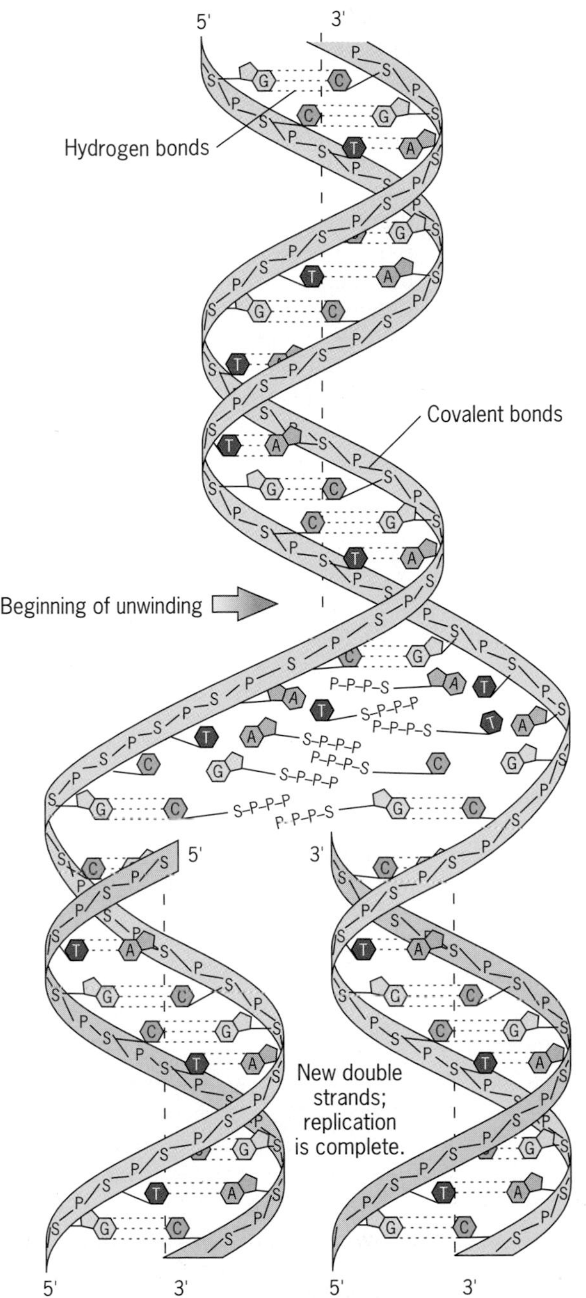

Figure 11.1 Semiconservative model of DNA replication based on complementary base-pairing as proposed by Watson and Crick. Note that each of the parental strands is conserved and serves as a template for the synthesis of a new complementary strand; that is, the base sequence in each progeny strand is determined by the hydrogen-bonding potentials of the bases in the parental strand.

sible modes of DNA replication by following the changes in the density of DNA from cells grown on ^{15}N medium and then transferred to ^{14}N medium for various periods of time—so-called density-transfer experiments.

The density of most DNAs is about the same as the density of concentrated solutions of heavy salts such as cesium chloride (CsCl). For example, the density of 6*M* CsCl is about 1.7 g/cm^3. *Escherichia coli* DNA containing ^{14}N has a density of 1.710 g/cm^3. Substitution of ^{15}N for ^{14}N increases the density of *E. coli* DNA to 1.724 g/cm^3. When a 6*M* CsCl solution is centrifuged at very high speeds for long periods of time, an equilibrium density gradient is formed (see Technical Sidelight). If DNA is present in such a gradient, it will move to a position where the density of the CsCl solution is equal to its own density.

Meselson and Stahl took cells that had been growing in medium containing ^{15}N for several generations (and thus contained "heavy" DNA), washed them to remove the medium containing ^{15}N, and transferred them to medium containing ^{14}N. After the cells were allowed to grow in the presence of ^{14}N for varying periods of time, the DNAs were extracted and analyzed in CsCl equilibrium density gradients. The results of their experiment (Figure 11.3) are consistent only with semiconservative replication, excluding both conservative and dispersive models of DNA synthesis. All the DNA isolated from cells after one generation of growth in medium containing ^{14}N had a density halfway between the densities of "heavy" DNA and "light" DNA. This intermediate density is usually referred to as "hybrid" density. After two generations of growth in medium containing ^{14}N, half of the DNA was of hybrid density and half was light. These results are precisely those predicted by the Watson and Crick semiconservative mode of replication (Figure 11.1). One generation of semiconservative replication of a parental double helix containing ^{15}N in medium containing only ^{14}N would produce two progeny double helices, both of which had ^{15}N in one strand (the "old" strand) and ^{14}N in the other strand (the "new" strand). Such molecules would be of hybrid density.

Conservative replication would not produce any DNA molecules with hybrid density; after one generation of conservative replication of heavy DNA in light medium, half of the DNA still would be heavy and the other half would be light. If replication were dispersive, Meselson and Stahl would have observed a shift of the DNA from heavy toward light in each generation (that is, "half heavy" or hybrid after one generation, "quarter heavy" after two generations, and so forth). These possibilities are clearly inconsistent with the results of Meselson and Stahl's experiment.

The semiconservative replication of eukaryotic chromosomes was first demonstrated in 1957 by the results of experiments carried out by J. Herbert Taylor, Philip Woods, and Walter Hughes on root-tip cells of the broad bean, *Vicia faba*. Taylor and colleagues labeled *Vicia faba* chromosomes by growing root tips for

Semiconservative
Conservative
Dispersive
Parental DNA
First-generation progeny DNA
Second-generation progeny DNA

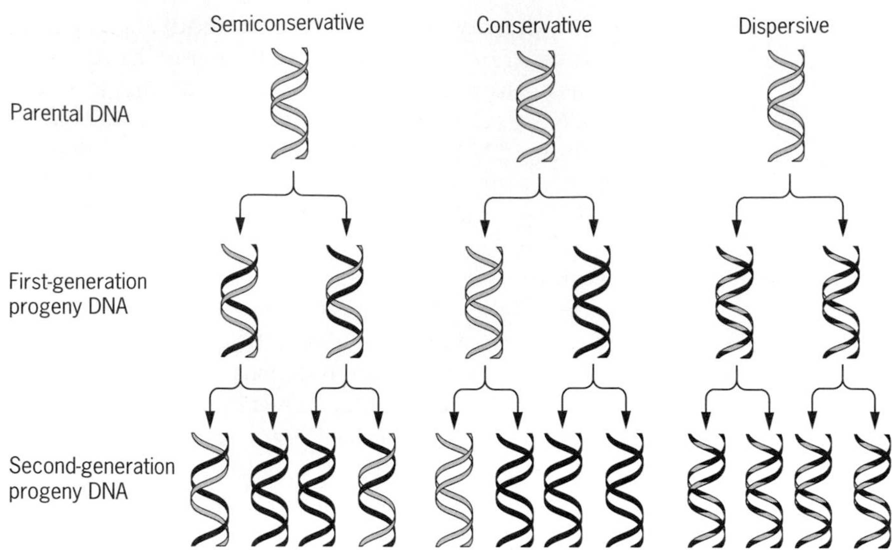

Figure 11.2 Three possible modes of DNA replication: (1) semiconservative, in which each strand of the parental double helix is conserved and directs the synthesis of a new complementary progeny strand; (2) conservative, in which the parental double helix is conserved and directs the synthesis of a new progeny double helix; and (3) dispersive, in which segments of each parental strand are conserved and direct the synthesis of new complementary strand segments that are subsequently joined to produce new progeny strands.

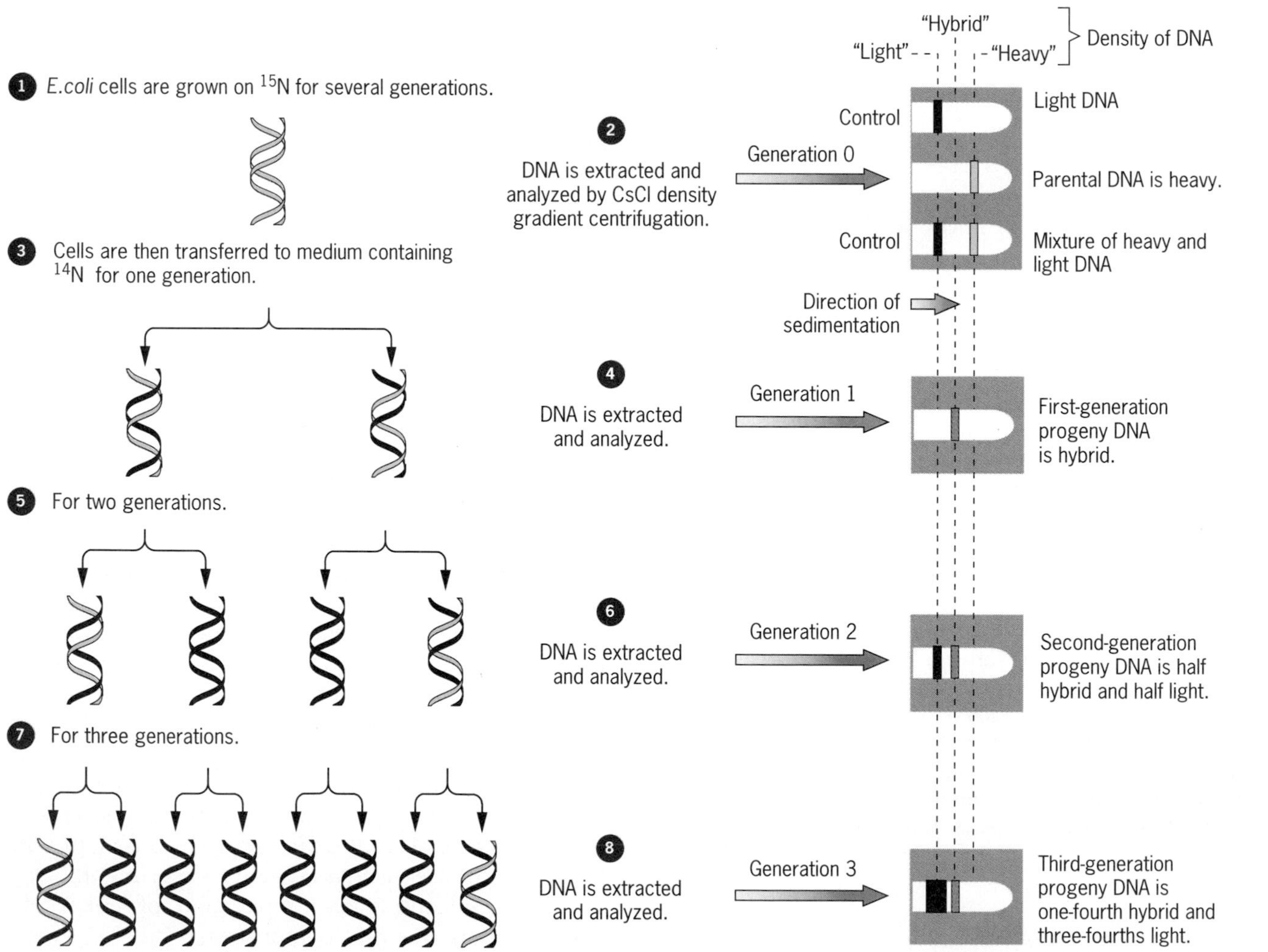

"Hybrid"
"Light"
"Heavy"
Density of DNA
1 E.coli cells are grown on 15N for several generations.
2 DNA is extracted and analyzed by CsCl density gradient centrifugation.
Control
Light DNA
Generation 0
Parental DNA is heavy.
Control
Mixture of heavy and light DNA
3 Cells are then transferred to medium containing 14N for one generation.
Direction of sedimentation
4 DNA is extracted and analyzed.
Generation 1
First-generation progeny DNA is hybrid.
5 For two generations.
6 DNA is extracted and analyzed.
Generation 2
Second-generation progeny DNA is half hybrid and half light.
7 For three generations.
8 DNA is extracted and analyzed.
Generation 3
Third-generation progeny DNA is one-fourth hybrid and three-fourths light.

Figure 11.3 Meselson and Stahl's experiment demonstrating that DNA replicates by a semiconservative mechanism in *E. coli*. The diagram shows that the results of their experiment are those expected if the *E. coli* chromosome replicates semiconservatively. Different results would have been obtained if DNA replication in *E. coli* were either conservative or dispersive (see Fig. 11.2).

Autoradiographs of *Vicia faba* chromosomes

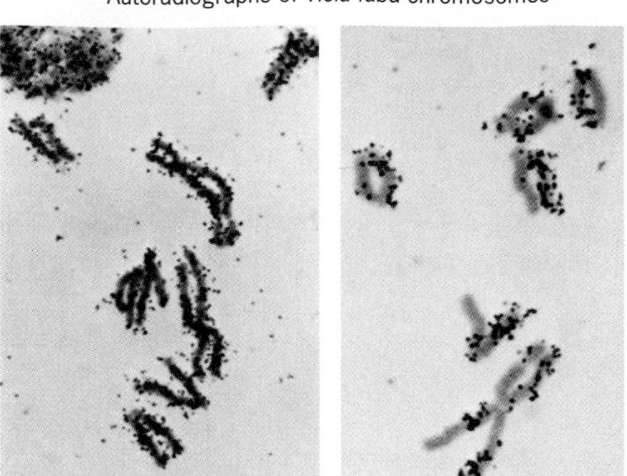

(*a*) First metaphase after replication in ^{3}H-thymidine. (*b*) Second metaphase after an additional replication in ^{1}H-thymidine.

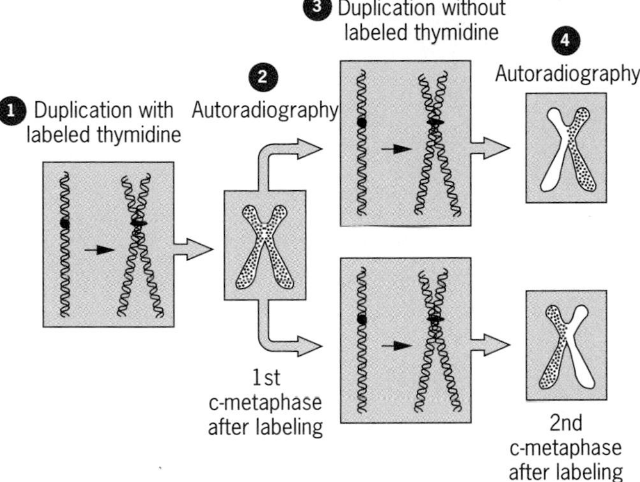

(*c*) Interpretation of the autoradiographs above in terms of semiconservative replication.

Figure 11.4 Results and interpretation of the experiment by Taylor and colleagues showing that the chromosomal DNA of the broad bean, *Vicia faba*, replicates by a semiconservative mechanism.

eight hours (less than one cell generation) in medium containing radioactive ^{3}H-thymidine. The root tips were then removed from the radioactive medium, washed, and transferred to nonradioactive medium containing the alkaloid colchicine. As we discussed in Chapter 7, colchicine binds to microtubules and prevents the formation of functional spindle fibers. As a result, daughter chromosomes do not undergo their normal anaphase separation. Thus, the number of chromosomes per nucleus will double once per cell cycle in the presence of colchicine. This doubling of the chromosome number each cell generation allowed Taylor and his colleagues to determine how many DNA duplications each cell had undergone subsequent to the incorporation of radioactive thymidine. At the first metaphase in colchicine (c-metaphase), nuclei will contain 12 pairs of chromatids (still joined at the centromeres). At the second c-metaphase, nuclei will contain 24 pairs, and so on.

When the distribution of radioactivity in the *Vicia faba* chromosomes was examined by autoradiography (Chapter 10), both chromatids of each pair were similarly labeled at the first c-metaphase (Figure 11.4*a*). However, at the second c-metaphase, only one of the chromatids of each pair was radioactive (Figure 11.4*b*). These are precisely the results expected if DNA replication is semiconservative, given one DNA molecule per chromosome (Figure 11.4*c*). In 1957, Taylor and his colleagues were able to conclude that chromosomal DNA in *Vicia faba* segregated in a semiconservative manner during each cell division. The conclusion that the double helix replicated semiconservatively in the broad bean had to await subsequent evidence indicating that each chromosome contains a single molecule of DNA. Analogous experiments have subsequently been carried out with several other eukaryotes, and, in all cases, the results indicate that replication is semiconservative.

Visualization of Replication Forks by Autoradiography

The gross structure of replicating bacterial chromosomes was first determined by John Cairns in 1963, again by means of autoradiography. Cairns grew *E. coli* cells in medium containing [^{3}H]thymidine for varying periods of time, lysed the cells gently so as not to break the chromosomes (long DNA molecules are sensitive to shearing), and carefully collected the chromosomes on membrane filters. These filters were affixed to glass slides, coated with emulsion sensitive to β-particles (the low-energy electrons emitted during decay of tritium), and stored in the dark for a period of time to allow sufficient radioactive decay. When the films were developed, the autoradiographs (Figure 11.5*a*) showed that the chromosomes of *E. coli* are circular structures that exist as θ-shaped intermediates during replication. The autoradiographs further indicated that the unwinding of the two complementary parental strands (which is necessary for their separation) and their semiconservative replication occur simultaneously or are closely coupled. Since the parental double helix must rotate 360° to unwind each gyre of the helix, some kind of "swivel" must exist. Geneticists now know that the required swivel is a transient single-strand break (cleavage of one phosphodiester bond in one strand of the double helix) produced by the action of enzymes called topoisomerases.

TECHNICAL SIDELIGHT

Cesium Chloride Equilibrium Density-Gradient Centrifugation and Sucrose Velocity Density-Gradient Centrifugation: Two Important and Distinct Tools

Although cesium chloride (CsCl) and sucrose density gradients are both used to study nucleic acids, these two techniques are used for totally different purposes.

CsCl EQUILIBRIUM DENSITY-GRADIENT CENTRIFUGATION

The CsCl density gradients that are used to analyze nucleic acids are **equilibrium density gradients.** They are used to separate nucleic acid molecules based on their densities, which determine the positions at which the molecules will band in the linear density gradient produced in the centrifuge tube at equilibrium conditions.

When a heavy salt solution such as 6*M* CsCl is centrifuged at very high speeds (30,000 to 50,000 revolutions per minute) for 48 to 72 hours, an equilibrium density gradient is formed (Figure 1). The centrifugal force caused by spinning the solution at high speeds sediments the salt toward the bottom of the tube. However, diffusion, a countervailing force, results in movement of salt molecules back toward the top (low salt concentration) of the tube. After a sufficient period of high-speed centrifugation, an equilibrium between sedimentation and diffusion is reached, at which time a linear gradient of increasing density exists from the top to the bottom of the tube.

The density of 6*M* CsCl is about 1.7 g/cm^3; at equilibrium, the linear density gradient produced by high-speed centrifugation of a 6*M* CsCl solution will range from about 1.65 g/cm^3 at the top of the tube to about 1.75 g/cm^3 at the bottom of the tube. The densities of most naturally occurring nucleic acids fall within this range. If DNA is present in such a gradient, it will move to a position where the density of the salt solution is equal to its own density. Thus, if a mixture of *E. coli* DNA containing the heavy isotope of nitrogen, ^{15}N, and *E. coli* DNA containing the normal light nitrogen isotope, ^{14}N, is subjected to CsCl equilibrium density-gradient centrifugation, the DNA molecules will separate into two "bands," one consisting of "heavy" (^{15}N-containing) DNA and the other consisting of "light" (^{14}N-containing) DNA.

SUCROSE VELOCITY DENSITY-GRADIENT CENTRIFUGATION

The sucrose density gradients that are used to analyze nucleic acids are **velocity density gradients.** They are used to

1 Prepare 6*M* CsCl solution and add mixture of DNAs containing ^{14}N and ^{15}N.

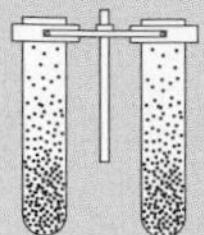

2 Centrifuge at 50,000 rpm for 48 to 72 hours.

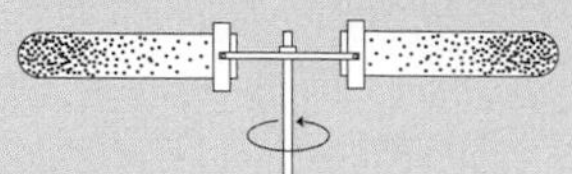

An equilibrium is established between

centrifugal force and diffusion

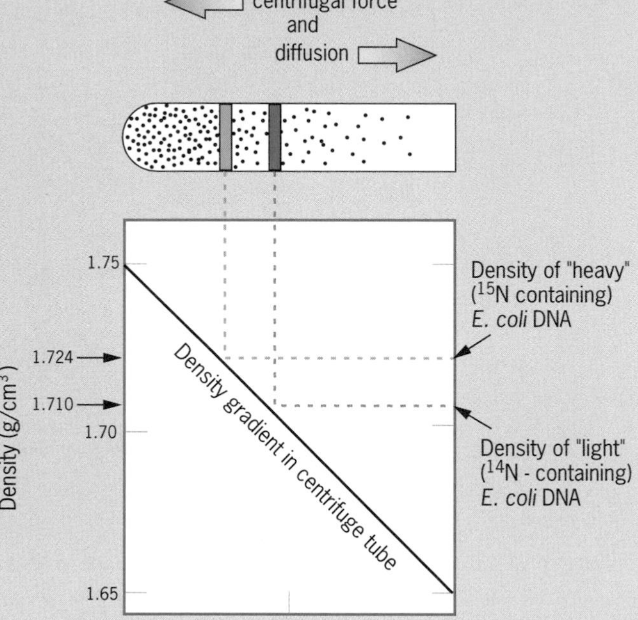

3 Punch hole in centrifuge tube and collect fractions.

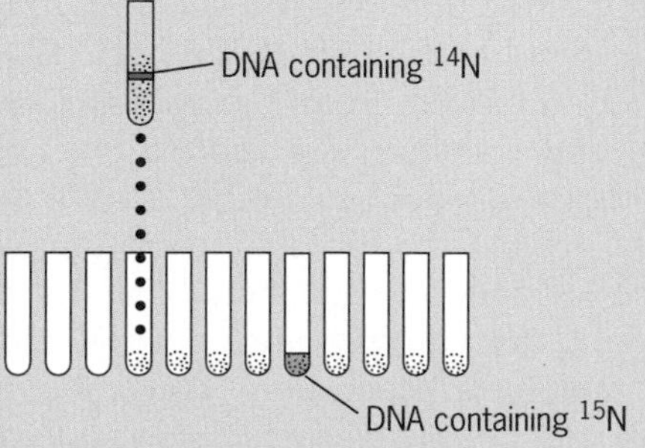

Figure 1 CsCI Equilibrium Density Gradient.

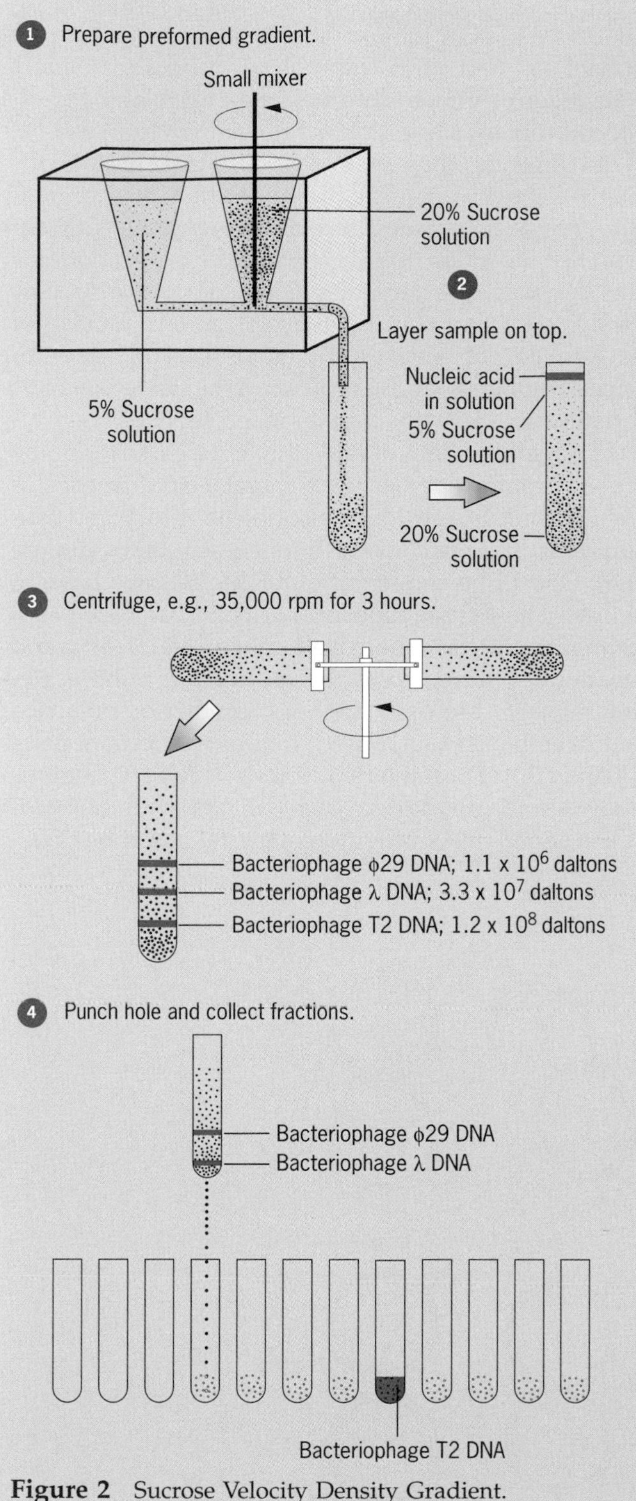

Figure 2 Sucrose Velocity Density Gradient.

estimate the sizes of nucleic acid molecules based on their rates or velocities of movement through a preformed sucrose density gradient when exposed to a centrifugal force.

A gradient maker is used to prepare a preformed density gradient in a centrifuge tube (Figure 2). The gradient maker contains two chambers for sucrose solutions. One chamber is filled with a sucrose solution of the concentration desired at the bottom of the centrifuge tube; the other is filled with a sucrose solution of the concentration desired at the top of the tube. As the tube is filled, the high concentration sucrose solution is progressively diluted with the low concentration solution, producing a linear gradient in the tube.

Although any range of sucrose concentrations can be used, 5 to 20 percent and 10 to 40 percent sucrose gradients are widely used to separate nucleic acids. The solution containing the molecules to be analyzed is layered on the top of the sucrose gradient prior to centrifugation. The density of this solution must be lower than the density of the sucrose at the top of the centrifuge tube so that it will stay on the surface of the sucrose gradient. The tube is then placed in a swinging bucket rotor and centrifuged at high speed (20,000 to 30,000 revolutions per minute) for one to a few hours. The centrifugal force causes the macromolecules to migrate through the gradient, with larger molecules moving faster than smaller molecules with the same shape. If centrifugation is continued too long, all the macromolecules will accumulate at the bottom of the tube. If the dye ethidium bromide, which binds to DNA and fluoresces when exposed to ultraviolet (UV) light, is present, the bands of DNA in the gradient can be seen under UV illumination. After centrifugation, a small hole is punched through the bottom of the centrifuge tube, and drops of solution are collected in test tubes for subsequent analysis. The collected fractions represent successive layers of the sucrose gradient and any molecules therein at the time that centrifugation was terminated.

The velocity at which a macromolecule will move through such a gradient is determined primarily by its molecular weight and shape. A molecule's rate of movement through a solution of lower density during sedimentation is measured by its sedimentation coefficient (s), which equals velocity/centrifugal force. Most macromolecules have s values between 10^{-13} and 10^{-11} seconds. Thus, a sedimentation coefficient of 10^{-13} seconds has been designated one Svedberg (S) unit. It was named in honor of The Svedberg, the scientist who invented the ultracentrifuge. The Svedberg unit is the commonly used unit of sedimentation velocity in dealing with macromolecules or macromolecular complexes. For example, nucleic acids are sometimes referred to as 5S or 10S molecules, and ribosomal subunits of prokaryotes are frequently called 30S and 50S subunits.

Cairns' interpretation of the autoradiographs was that semiconservative replication of the *E. coli* chromosome started at a specific site, which he called the *origin,* and proceeded sequentially and unidirectionally around the circular structure (Figure 11.5*b*). Subsequent evidence has shown his original interpretation to be incorrect on one point: replication of the *E. coli* chromosome actually proceeds bidirectionally, not unidirectionally. Each Y-shaped structure is a **replication fork,** and the two replication forks move in opposite directions sequentially around the circular chromosome (Figure 11.5*c*).

Unique Origins of Replication

Cairns' results established the existence of a site of initiation or **origin** of replication on the circular chromosome of *E. coli,* but provided no hint as to whether the origin was a unique site or occurred at randomly located sites in a population of replicating chromosomes. In bacterial and viral chromosomes, there is usually one unique origin per chromosome, and this single origin controls the replication of the entire chromosome. In the large chromosomes of eukaryotes, multiple origins collectively control the replication of the giant DNA molecule present in each chromosome. Current evidence indicates that these multiple replication origins in eukaryotic chromosomes also occur at specific sites. Each origin controls the replication of a unit of DNA called a *replicon;* thus, most prokaryotic chromosomes contain a single replicon, whereas eukaryotic chromosomes usually contain many replicons.

The single origin of replication, called ***OriC,*** in the *E. coli* chromosome has been characterized in considerable detail. *OriC* is 245 nucleotide pairs long and contains two different conserved repeat sequences (Figure 11.6). One 13-bp sequence is present as three tandem repeats. These three repeats are rich in A:T base pairs, facilitating the formation of a localized region of strand separation referred to as the **replication bubble.** Recall that A:T base pairs are held together by only two hydrogen bonds as opposed to three in G:C base pairs (Chapter 10). Thus, the two strands of A:T-rich regions of DNA come apart more easily, that is, with the input of less energy. The formation of a localized zone of de-

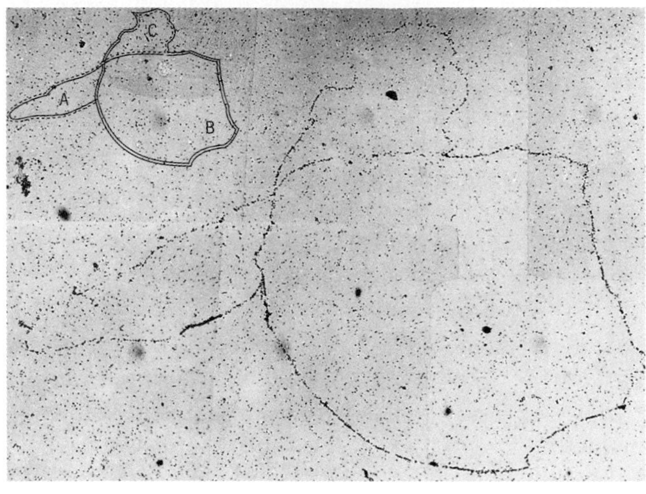

(*a*)

Swivel

Replication fork

(*b*) Original interpretation: unidirectional replication.

Origin

Replication fork

Replication fork

(*c*) Correct interpretation: bidirectional replication.

Figure 11.5 Visualization of the replication of the *E. coli* chromosome by autoradiography. (*a*) One of Cairns' autoradiographs of a θ-shaped replicating chromosome from a cell that had been grown for two generations in the presence of ^{3}H-thymidine, with his interpretative diagram shown at the upper left. Radioactive strands of DNA are shown as solid lines and nonradioactive strands as dashed lines. Loops A and B have completed a second replication in ^{3}H-thymidine; section C remains to be replicated the second time. The two possible interpretations of Cairns' results are shown in (*b*) and (*c*). Cairns originally interpreted his results in terms of unidirectional replication (*b*). DNA replication in *E. coli* was subsequently shown to be bidirectional (*c*).

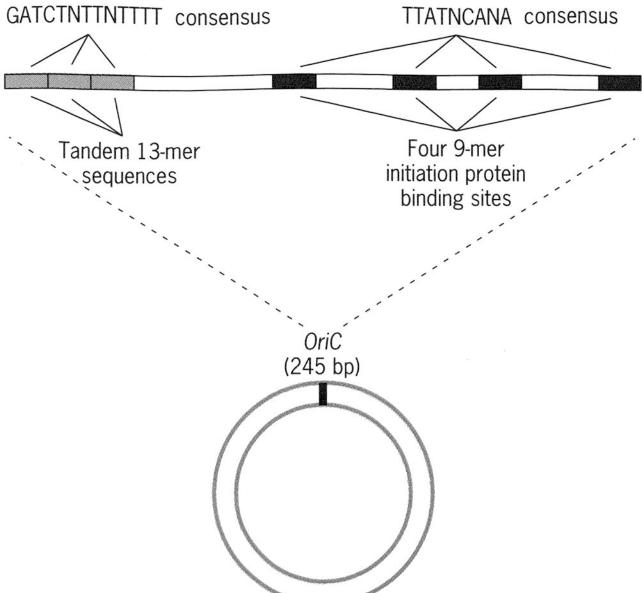

Figure 11.6 Structure of *OriC*, the single origin of replication in the *E. coli* chromosome.

naturation is an essential first step in the replication of all double-stranded DNAs. Another conserved component of *OriC* is a 9-bp sequence that is repeated four times and is interspersed with other sequences. These four sequences are binding sites for a protein that plays a key role in the formation of the replication bubble. Later in this chapter we discuss additional details of the process of initiation of DNA synthesis at origins and the proteins that are involved.

The multiple origins of replication in eukaryotic chromosomes also appear to be specific DNA sequences. In the yeast *Saccharomyces cerevisiae*, segments of chromosomal DNA that allow a fragment of circularized DNA to replicate as an independent unit (autonomously), that is, as an extrachromosomal self-replicating unit, have been identified and characterized. These sequences are called **ARS** (for **A**utonomously **R**eplicating **S**equences) **elements.** Their frequency in the yeast genome corresponds well with the number of origins of replication, and some have been shown experimentally to function as origins. ARS elements are about 50 base pairs in length and include a core 11-bp A:T-rich sequence,

AT TT AT PuT T T A
T AAAT APyAAAT

(where Pu is either of the two purines and Py is either of the two pyrimidines) and additional imperfect copies of this sequence. The ability of ARS elements to function as origins of replication is abolished by base-pair changes within this conserved core sequence. Although the molecular details remain to be worked out, the A:T-rich cores of ARS elements presumably are the sites of the initial strand separations leading to DNA replication.

Bidirectional Replication

Cairns interpreted his autoradiograms of replicating *E. coli* chromosomes based on the prevailing view that replication started at a single origin and that a single replication fork moved unidirectionally around the circular chromosome (Figure 11.5*b*). However, his results can be explained equally well by the formation of two forks at the single origin, with the two forks moving in opposite directions or bidirectionally around the circular chromosome (Figure 11.5*c*). We now have definitive evidence showing that replication in *E. coli* and many other prokaryotes proceeds bidirectionally from a unique origin.

Bidirectional replication was first convincingly demonstrated by experiments with some of the small bacterial viruses that infect *E. coli.* Bacteriophage lambda (phage λ) contains a single linear molecule of DNA only 17.5 μm long. The phage λ chromosome is somewhat unique in that it has a single-stranded region, 12 nucleotides long, at the 5′ end of each complementary strand (Figure 11.7). These single-stranded ends, called "cohesive" or "sticky" ends, are complementary to each other. The cohesive ends of a lambda chromosome can thus base-pair to form a hydrogen-bonded circular structure. One of the first events to occur after a lambda chromosome is injected into a host cell is its conversion to a covalently closed circular molecule (Figure 11.7). This conversion from the hydrogen-bonded circular form to the covalently closed circular form is catalyzed by *DNA ligase*, an important enzyme that seals single-stranded breaks in DNA double helices. DNA ligase is required in all organisms for DNA replication, DNA repair, and recombination between DNA molecules. Like the *E. coli* chromosome, the lambda chromosome replicates in its circular form via θ-shaped intermediates.

The feature of the lambda chromosome that facilitated the demonstration of bidirectional replication is its differentiation into regions containing high concentrations of adenine and thymine (A:T-rich regions) and regions with large amounts of guanine and cytosine (G:C-rich regions). In particular, it contains a few segments with high A:T content (A:T-rich clusters). In the late 1960s, Maria Schnös and Ross Inman used these A:T-rich clusters as physical markers to demonstrate, by means of a technique called denaturation mapping, that replication of the lambda chromosome is initiated at a unique origin and proceeds bidirectionally rather than unidirectionally.

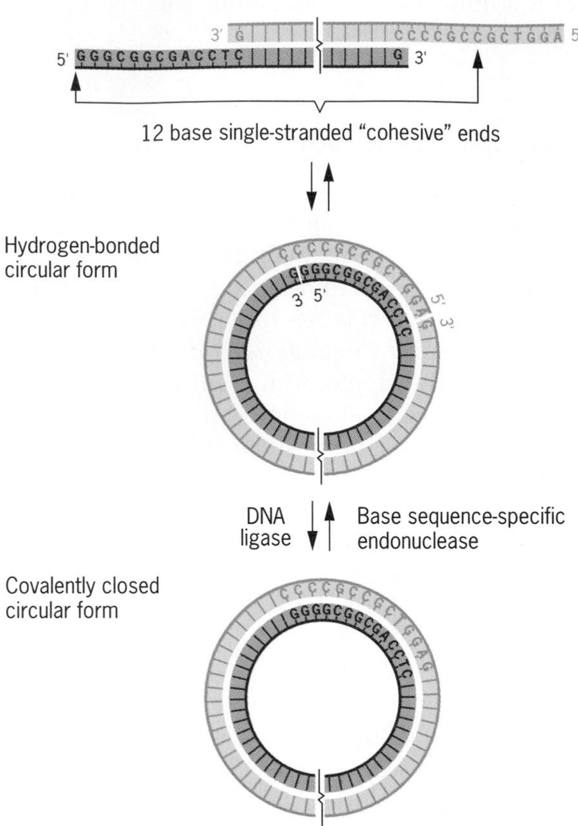

Figure 11.7 Interconversion of the linear lambda chromosome with its complementary cohesive ends, the hydrogen-bonded circular lambda chromosome, and the covalently closed circular lambda chromosome. The linear form of the chromosome appears to be an adaptation to facilitate its injection from the phage head through the small opening in the phage tail into the host cell during infection. Prior to replicating in the host cell, the chromosome is converted to the covalently closed circular form. Only the ends of the chromosome of the mature phage are shown; the jagged vertical line indicates that the central portion of the chromosome is not shown. The entire lambda chromosome is about 4.8×10^4 nucleotide pairs long.

When DNA molecules are exposed to high temperature (100°C) or high pH (11.4), the hydrogen and hydrophobic bonds that hold the complementary strands together in the double-helix configuration are broken, and the two strands separate—a process called denaturation. Because A:T base pairs are held together by only two hydrogen bonds, compared with three hydrogen bonds in G:C base pairs, A:T-rich molecules denature more easily (at lower pH or temperature) than G:C-rich molecules. When lambda chromosomes are exposed to pH 11.05 for 10 minutes under the appropriate conditions, the A:T-rich clusters denature to form denaturation bubbles, which are detectable by electron microscopy, whereas the G:C-rich regions remain in the duplex state (Figure 11.8). These denaturation bubbles can be used as physical markers whether the lambda chromosome is in its mature linear form, its circular form, or its θ-shaped replicative intermediates. By examining the positions of the branch points (Y-shaped structures) relative to the positions of the denaturation bubbles in a large number of θ-shaped replicative intermediates, Schnös and Inman demonstrated that both branch points are replication forks that move in opposite directions around the circular chromosome. Figure 11.9 shows the results expected in Schnös and Inman's experiment if replication is (*a*) unidirectional or (*b*) bidirectional. The results clearly demonstrated that replication of the lambda chromosome is bidirectional.

Bidirectional replication from a fixed origin has also been demonstrated for several organisms with chromosomes that replicate as linear structures. Replication of the chromosome of phage T7, another small bacteriophage, begins at a unique site near one end to form an "eye" structure (Figure 11.10*a*) and then proceeds bidirectionally until one fork reaches the nearest end. Replication of the Y-shaped structure (Figure 11.10*b*) continues until the second fork reaches the other end of the molecule, producing two progeny chromosomes.

Replication of chromosomal DNA in eukaryotes is also bidirectional in those cases where it has been investigated. However, bidirectional replication is not universal. The chromosome of coliphage P2, which replicates as a θ-shaped structure like the lambda chromosome, replicates unidirectionally from a unique origin.

Key Points: **DNA replication is semiconservative. As the two complementary strands of a parental double helix unwind and separate, each serves as a template for the synthesis of a new complementary strand. The hydrogen-bonding potentials of the bases in the template strands specify complementary base sequences in the newly synthesized strands. Replication is initiated at unique origins and usually proceeds bidirectionally from each origin.**

DNA POLYMERASES AND DNA SYNTHESIS *IN VITRO*

Much has been learned about the molecular mechanisms involved in biological processes by disrupting cells, separating the various organelles, macromolecules, and other components, and then reconstituting

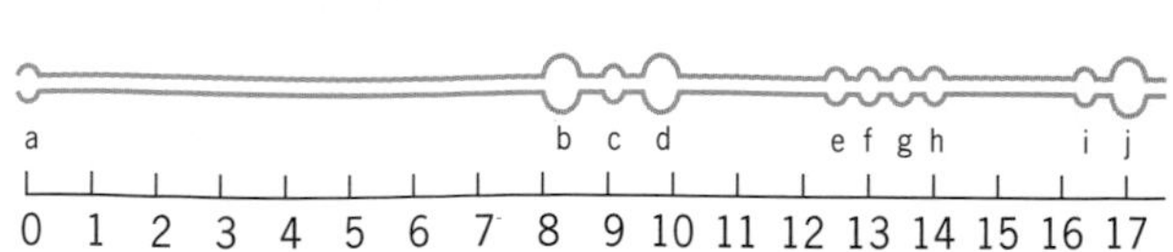

(*a*) Diagram showing the locations of AT-rich denaturation sites in the linear λ chromosome.

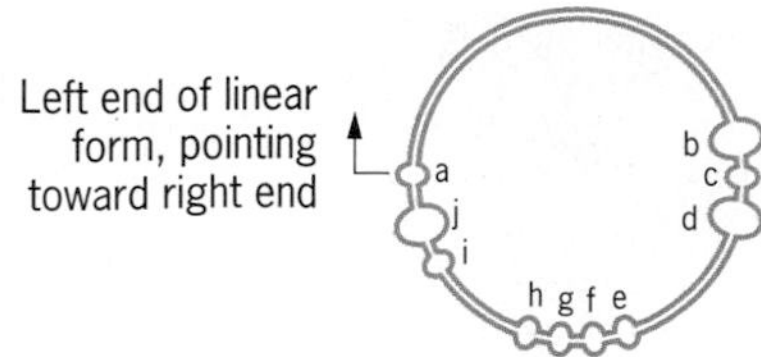

(*b*) Diagram showing the positions of AT-rich denaturation sites in the circular form of the λ chromosome.

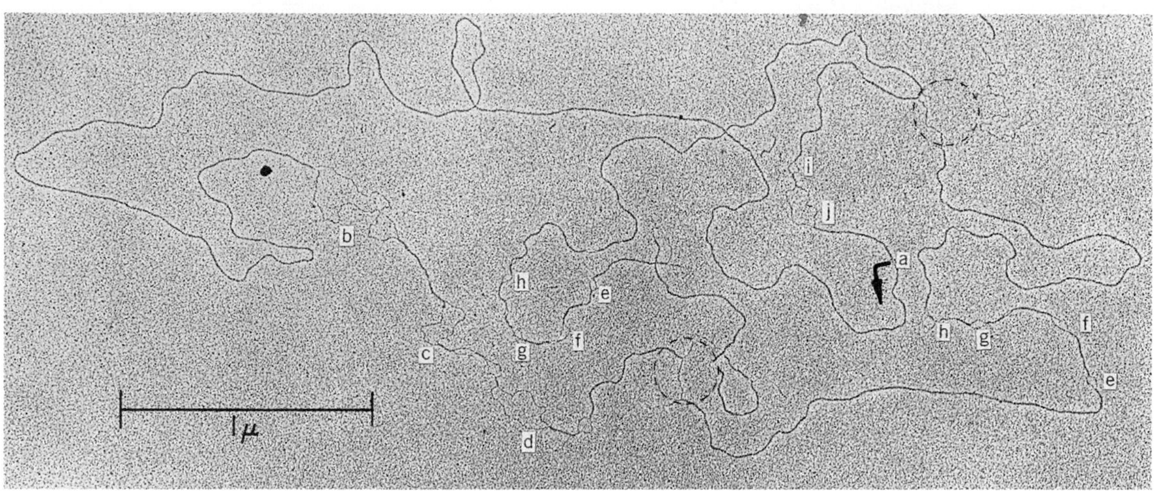

(*c*) Electron micrograph showing the locations of the AT-rich denaturation bubbles in a θ -shaped replicating λ chromosome.

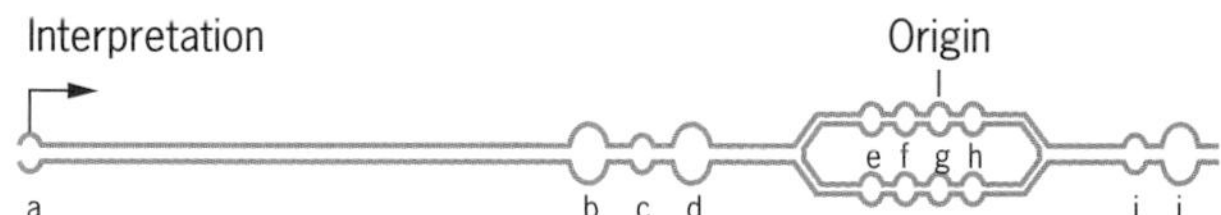

(*d*) Diagram in linear form of the λ replicative intermediate in (*c*) showing the positions of denaturation bubbles relative to the replication forks.

Figure 11.8 Illustration of the use of AT-rich denaturation sites as physical markers to prove that the phage λ chromosome replicates bidirectionally rather than unidirectionally.

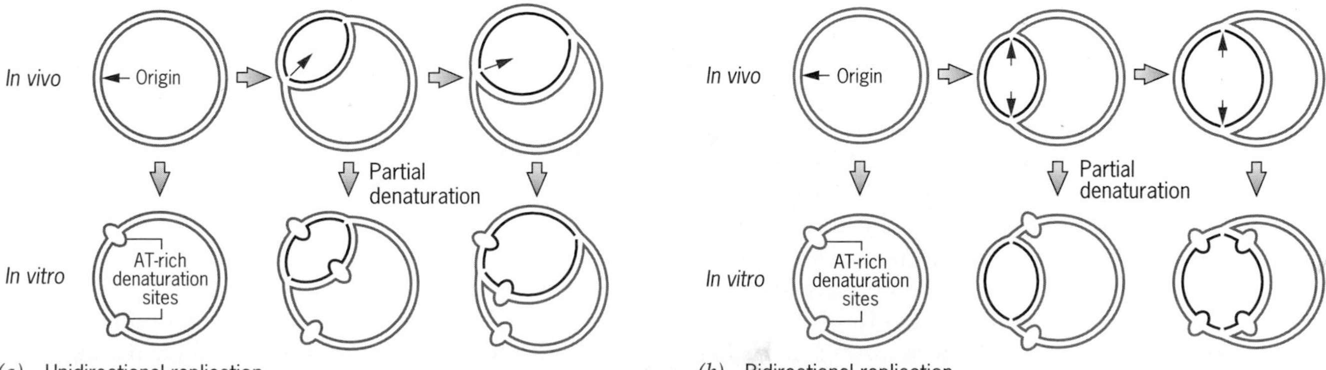

(*a*) Unidirectional replication.

(*b*) Bidirectional replication.

Figure 11.9 Rationale of the denaturation mapping procedure used by Schnös and Inman to distinguish between (*a*) unidirectional and (*b*) bidirectional modes of chromosome replication.

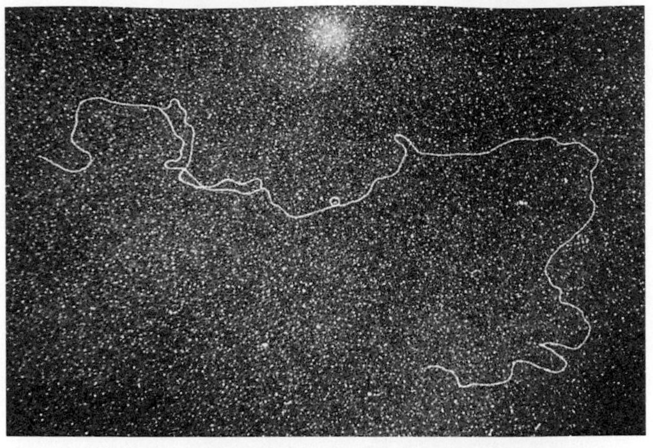

(a)

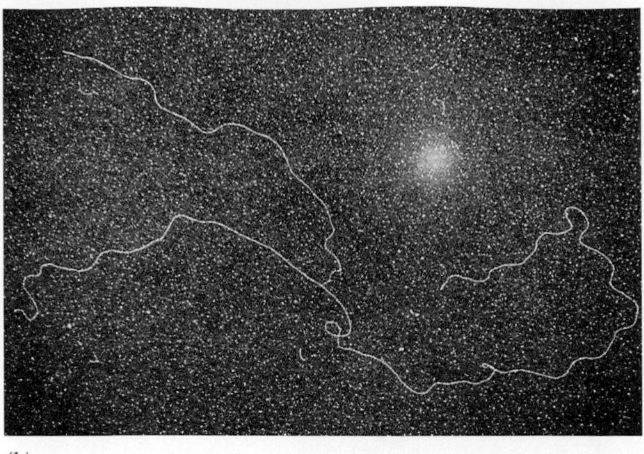

(b)

Figure 11.10 Electron micrographs of replicating bacteriophage T7 chromosomes. The phage T7 chromosome, unlike the *E. coli* and phage λ chromosomes, replicates as a linear structure. Its origin of replication is located 17 percent of the length of the chromosome from one end (the left end of the chromosomes shown). The chromosome in (*a*) illustrates the "eye" form (–O–) characteristic of early stages in the replication of linear DNA molecules. Replication proceeds bidirectionally from the origin until the fork moving in a leftward direction reaches the left end of the molecule, yielding a Y-shaped structure such as that shown in (*b*).

systems in the test tube, so-called *in vitro* systems that are capable of carrying out particular metabolic events. Such *in vitro* systems can be dissected biochemically much more easily than *in vivo* systems. Clearly, the information obtained from studies on *in vitro* systems has been invaluable. However, we should never assume that a phenomenon demonstrated *in vitro* occurs *in vivo*. Such an extrapolation should be made only when independent evidence from *in vivo* studies validates the *in vitro* studies.

Discovery of DNA Polymerase I in *Escherichia coli*

The *in vitro* synthesis of DNA was first accomplished by Arthur Kornberg and his coworkers in 1957. Kornberg, who received a Nobel Prize in 1959 for this work, isolated an enzyme from *E. coli* that catalyzes the covalent addition of nucleotides to preexisting DNA chains. Initially called DNA polymerase or "Kornberg's enzyme," it is now known as **DNA polymerase I.** The enzyme requires the 5′-triphosphates of each of the four deoxyribonucleosides—deoxyadenosine triphosphate (dATP), deoxythymidine triphosphate (dTTP), deoxyguanosine triphosphate (dGTP), and deoxycytidine triphosphate (dCTP)—and is active only in the presence of Mg^{2+} ions and preexisting DNA. This DNA must provide two essential components, one serving a primer function and the other a template function (Figure 11.11).

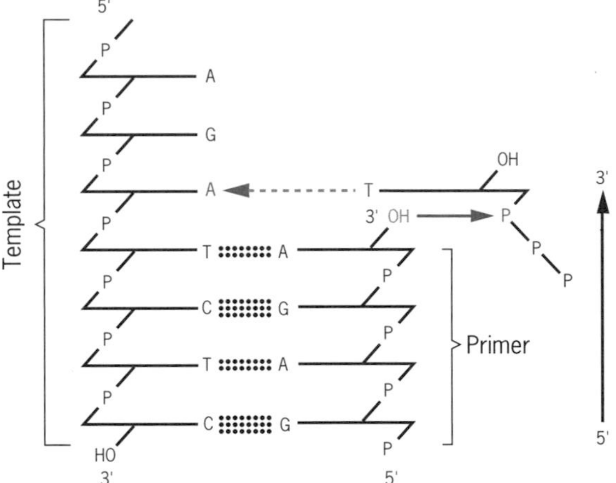

Figure 11.11 Template and primer requirements of DNA polymerases. All DNA polymerases require a primer strand (shown on the right) with a free 3′-hydroxyl. The primer strand is covalently extended by the addition of nucleotides (such as dTMP, derived from the incoming precursor dTTP shown). In addition, DNA polymerases require a template strand (shown on the left), which determines the base sequence of the strand being synthesized. The new strand will be complementary to the template strand.

1. The **primer DNA** provides a terminus with a free 3′-OH to which nucleotides are added during DNA synthesis. DNA polymerase I cannot initiate the synthesis of DNA chains *de novo*. It has an absolute requirement for a free 3′-hydroxyl on a preexisting

DNA chain. DNA polymerase I catalyzes the formation of a phosphodiester bridge between the 3′-OH at the end of the primer DNA chain and the 5′-phosphate of the incoming deoxyribonucleotide.

2. The **template DNA** provides the nucleotide sequence that specifies the complementary sequence of the nascent DNA chain. DNA polymerase I requires a DNA template whose base sequence dictates, by its base-pairing potential, the synthesis of a complementary base sequence in the strand being synthesized.

The reaction catalyzed by DNA polymerase I is a nucleophilic attack by the 3′-OH at the terminus of the primer strand on the nucleotidyl or interior phosphorus atom of the nucleoside triphosphate precursor with the elimination of pyrophosphate. This reaction mechanism explains the absolute requirement of DNA polymerase I for a free 3′-OH group on the primer DNA strand that is being covalently extended and dictates that *the direction of synthesis is always 5′ → 3′* (Figure 11.12).

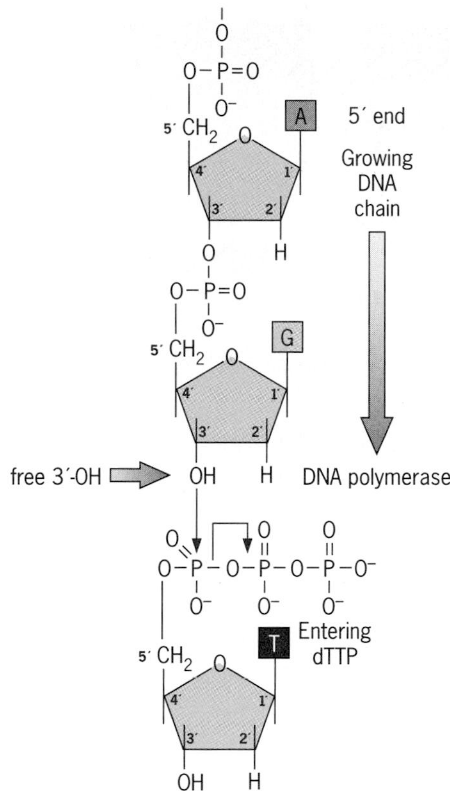

Figure 11.12 Mechanism of action of DNA polymerase I: covalent extension of a DNA primer strand in the 5′ → 3′ direction. The existing chain terminates at the 3′ end with the nucleotide deoxyguanylate (or deoxyguanosine-5′-phosphate). The diagram shows the DNA polymerase-catalyzed addition of deoxythymidine monophosphate (from the precursor deoxythymidine triphosphate, dTTP) to the 3′ end of the chain with the release of pyrophosphate (P_2O_7).

DNA polymerase I is a single polypeptide with a molecular weight of 109,000 encoded by a gene called *polA*. However, subsequent research has shown that DNA polymerase I is not the true "DNA replicase" in *E. coli*. It does not catalyze the semiconservative replication of the *E. coli* chromosome; that function is performed by another enzyme. Nevertheless, DNA polymerase I does perform important functions in the *E. coli* cell, including playing a key role in chromosome replication and a central role in repairing damaged DNA. To understand how DNA polymerase I performs these functions, we must get to know this enzyme better.

In addition to the polymerase activity illustrated in Figure 11.12, DNA polymerase I has two other enzymatic activities, both exonuclease activities. A **nuclease** is an enzyme that degrades nucleic acids. An **exonuclease** degrades nucleic acids starting at one or both ends, whereas an **endonuclease** cleaves nucleic acids at internal sites. DNA polymerase I contains both **5′ → 3′ exonuclease activity,** which cuts back DNA strands starting at 5′ termini, and **3′ → 5′ exonuclease activity,** which cleaves off mononucleotides from the 3′ termini of DNA strands. The 5′ → 3′ exonuclease activity of DNA polymerase I usually excises small oligomers containing up to 10 nucleotides. Thus, DNA polymerase I has three different enzymatic activities: (1) a 5′ → 3′ polymerase activity, (2) a 5′ → 3′ exonuclease activity, and (3) a 3′ → 5′ exonuclease activity. These three activities are illustrated in Figure 11.13.

The first evidence that DNA polymerase I was not the true DNA replicase was published in 1969 by Paula DeLucia and John Cairns. They reported that DNA replication occurred in an *E. coli* strain lacking this enzyme owing to a mutation in the *polA* gene. Their results demonstrated that DNA replication in *E. coli* does not require DNA polymerase I activity, at least not the 5′ → 3′ polymerase activity of the enzyme. However, DeLucia and Cairns also discovered that this *polA1* mutant was extremely sensitive to ultraviolet light (UV). All three enzymatic activities of DNA polymerase I play important roles in the cell. The major function of DNA polymerase I in *E. coli* is to repair defects in DNA, such as those induced by UV (Chapter 14). However, as we will see later in this chapter, the 5′ → 3′ exonuclease activity of DNA polymerase I is also involved at one stage of chromosome replication.

Multiple DNA Polymerases

If DNA polymerase I does not catalyze the semiconservative replication of the *E. coli* chromosome, another

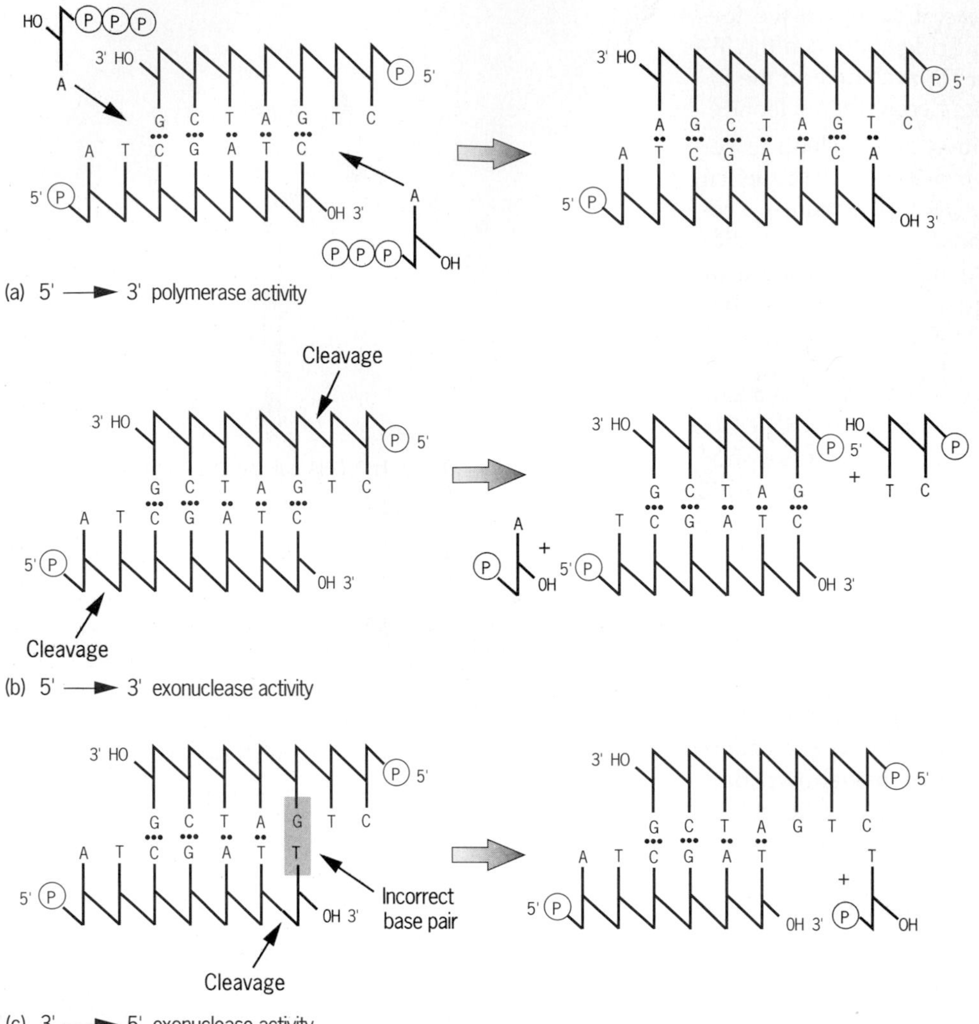

Figure 11.13 The three activities of DNA polymerase I of *E. coli.* As is discussed in the text, all three activities play important roles in the replication of the *E. coli* chromosome.

polymerase must carry out this function. In fact, there are two other DNA polymerases, **DNA polymerase II** and **DNA polymerase III,** in *E. coli.* Like DNA polymerase I, DNA polymerase II is a DNA repair enzyme; but it represents a small proportion of the polymerase activity in an *E. coli* cell. DNA polymerase II is a single polypeptide with $5' \rightarrow 3'$ polymerase and $3' \rightarrow 5'$ exonuclease activities. However, it has no $5' \rightarrow 3'$ exonuclease activity. In contrast to DNA polymerases I and II, DNA polymerase III is a complex enzyme composed of many different subunits. Like DNA polymerase II, DNA polymerase III has $5' \rightarrow 3'$ polymerase and $3' \rightarrow 5'$ exonuclease activities, but no $5' \rightarrow 3'$ exonuclease activity.

Eukaryotic organisms are even more complex—with up to five different DNA polymerases having been identified. The five DNA polymerases characterized in mammals have been named α (sometimes I), β, γ, δ (sometimes III), and ϵ (sometimes II). Two of the DNA polymerases (α and δ) work together to carry out the semiconservative replication of nuclear DNA. DNA polymerase γ is responsible for the replication of DNA in mitochondria, and DNA polymerases β and ϵ are nuclear DNA repair enzymes. Some of the eukaryotic DNA polymerases lack the $3' \rightarrow 5'$ exonuclease activity that is present in prokaryotic DNA polymerases.

All of the DNA polymerases studied to date, prokaryotic and eukaryotic, catalyze the same basic reaction: a nucleophilic attack by the free 3'-OH at the primer strand terminus on the nucleotidyl phosphorus of the nucleoside triphosphate precursor. Thus, all DNA polymerases have an absolute requirement for a free 3'-hydroxyl group on a preexisting primer strand;

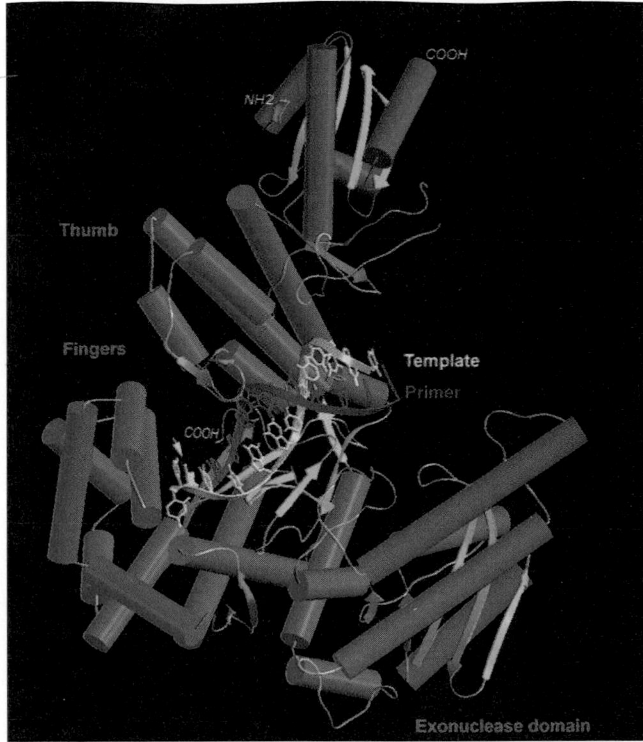

(a)

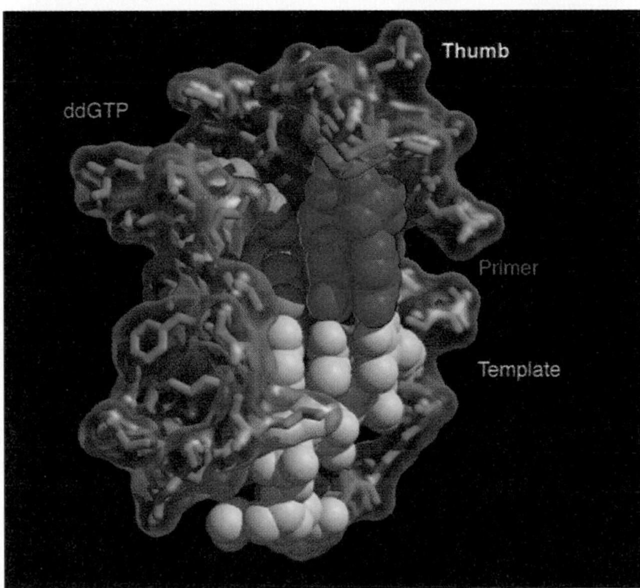

(b)

Figure 11.14 Schematic diagram (*a*) and space-filling model (*b*) of the structure of the complex between the phage T7 DNA polymerase, template-primer DNA, and a nucleoside triphosphate (ddGTP) precursor molecule. The template strand, primer strand, and nucleoside triphosphate are shown in yellow, magenta, and cyan, respectively. Protein components are shown in purple, green, orange, and gray. Note the tight juxtaposition between the nucleoside triphosphate, the primer terminus, and the template strand in (*b*).

none of these DNA polymerases can initiate new DNA chains *de novo*, and all DNA synthesis occurs in the $5' \rightarrow 3'$ direction.

As previously mentioned, DNA polymerases are amazingly accurate, with error frequencies of 10^{-5} to 10^{-6}. Recent studies of the crystal structure of the complex formed by a monomeric DNA polymerase, a nucleoside triphosphate precursor, and a template-primer DNA have contributed to our understanding of the high fidelity of DNA synthesis. In these studies, published in 1998, Sylvie Doublié and colleagues determined the structure on the phage T7 polymerase, which is similar to DNA polymerase I of *E. coli*, with resolution to 0.22 nm. The results show that the polymerase is shaped like a little hand, with the incoming nucleoside triphosphate, the template, and the primer terminus all tightly grasped between the thumb, the fingers, and the palm (Figure 11.14). The enzyme positions the incoming nucleoside triphosphate in juxtaposition with the terminus of the primer strand in a position to form hydrogen bonds with the first unpaired base in the template strand. Thus, the structure of this polymerase complex provides a simple explanation for the template-directed selection of incoming nucleotides during DNA synthesis.

DNA Polymerase III: The Replicase in *Escherichia coli*

Evidence indicating that DNA polymerase III is the true DNA replicase responsible for the semiconservative replication of DNA in *E. coli* was first provided by isolating and characterizing a mutant strain with a mutation in a gene called *polC*, now renamed *dnaE*. This mutant strain produced active DNA polymerase III when grown at 25°C, but totally inactive polymerase III when grown at 43°C. When *dnaE* mutant cells growing at 25°C were shifted to 43°C, DNA replication stopped, indicating that the product of the *dnaE* gene is required for DNA synthesis. The *dnaE* gene was subsequently shown to encode the catalytic α subunit, the subunit with $5' \rightarrow 3'$ polymerase activity, of DNA polymerase III.

DNA polymerase III is a multimeric enzyme (an enzyme with many subunits) with a molecular mass of about 900,000 daltons in its complete or **holoenzyme** form. The minimal core that has catalytic activity *in vitro* contains three subunits: α (the *dnaE* gene product), ϵ (the *dnaQ* product), and θ (the *holE* product). Addition of the τ subunit (the *dnaX* product) results in dimerization of the catalytic core and increased activity. The catalytic core synthesizes rather short DNA strands because of its tendency to fall off the DNA template. In order to synthesize the long DNA molecules present in chromosomes, this frequent dissociation of the polymerase from the template must be eliminated. The β subunit (the *dnaN* gene product) of DNA polymerase III forms a dimeric clamp that keeps the polymerase from falling off the template DNA (Figure 11.15). The β-dimer forms a ring that encircles the replicating DNA molecule and allows DNA polymerase III to slide along the DNA while remaining tethered to it. The DNA polymerase III holoenzyme, which is responsible for the synthesis of both nascent DNA strands at a replication fork, contains at least 20 polypeptides. The structural complexity of the DNA polymerase III holoenzyme is illustrated in Figure 11.16; the diagram shows 16 of the best-characterized polypeptides encoded by seven different genes. We will consider the functions of some of the subunits in subsequent sections of this chapter.

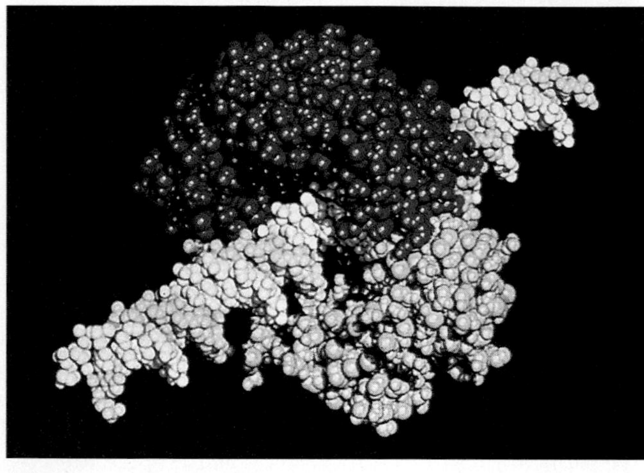

(*a*)

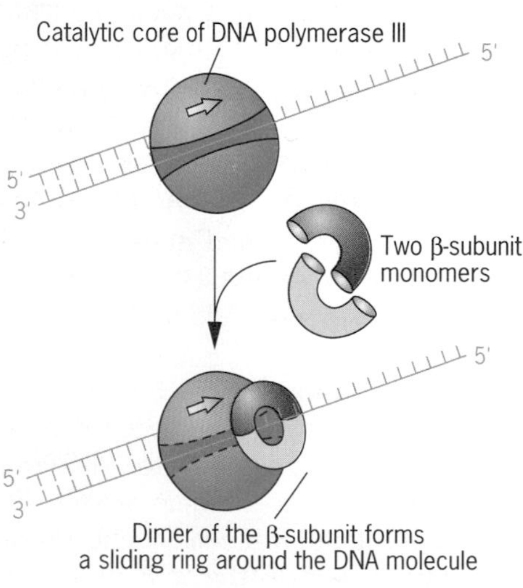

(*b*)

Figure 11.15 Space-filling model (*a*) and diagram (*b*) showing how two β-subunits (red and yellow) of DNA polymerase III clamp the enzyme to the DNA molecule (blue).

Proofreading Activities of DNA Polymerases

As we discussed earlier, the fidelity of DNA replication is amazing—with only about one error in every billion base-pair duplications. This high fidelity is necessary to keep the mutation load at a tolerable level, especially in large genomes such as those of mammals, which contain 3×10^9 nucleotide pairs. Without the high fidelity of DNA replication, the monozygotic twins discussed at the beginning of this chapter would be less similar in phenotype. Indeed, based on the dynamic structures of the four nucleotides in DNA, the observed fidelity of DNA replication is much higher than expected. The thermodynamic changes in nucleotides that allow the formation of hydrogen-bonded base pairs other than A:T and G:C predict error rates of 10^{-5} to 10^{-4}, or one error per 10,000 to 100,000 incorporated nucleotides. The predicted error rate of 10,000 times the observed error rate raises the question of how this high fidelity of DNA replication can be achieved.

Living organisms have solved the potential problem of insufficient fidelity during DNA replication by evolving a mechanism for **proofreading** the nascent DNA chain as it is being synthesized. The proofreading process involves scanning the termini of nascent DNA chains for errors and correcting them. This process is carried out by the 3′ → 5′ exonuclease activities of DNA polymerases. When a template-primer DNA has a terminal mismatch (an unpaired or incorrectly paired base or sequence of bases at the 3′ end of the primer), the 3′ → 5′ exonuclease activity of the DNA polymerase clips off the unpaired base or bases (Figure 11.17). When an appropriately base-paired terminus is produced, the 5′ → 3′ polymerase activity of the enzyme begins resynthesis by adding nucleotides to the 3′ end of the primer strand. In monomeric enzymes like DNA polymerase I of *E. coli*, this activity is built in. In multimeric enzymes, the 3′ → 5′ proofreading exonuclease activity is often present on a separate subunit. In the case of DNA polymerase III of

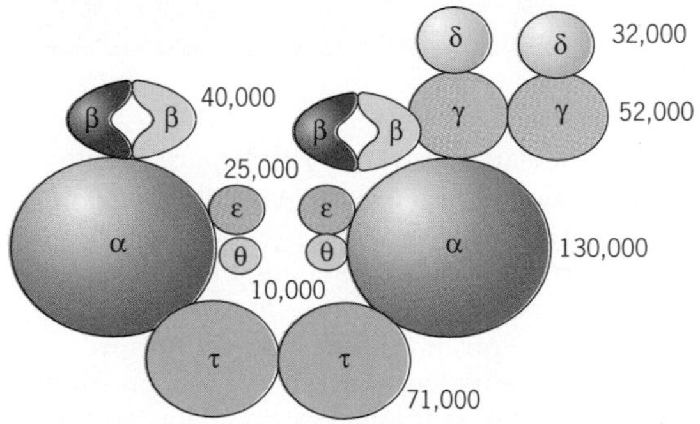

Figure 11.16 Structure of the *E. coli* DNA polymerase III holoenzyme. The numbers give mass in daltons.

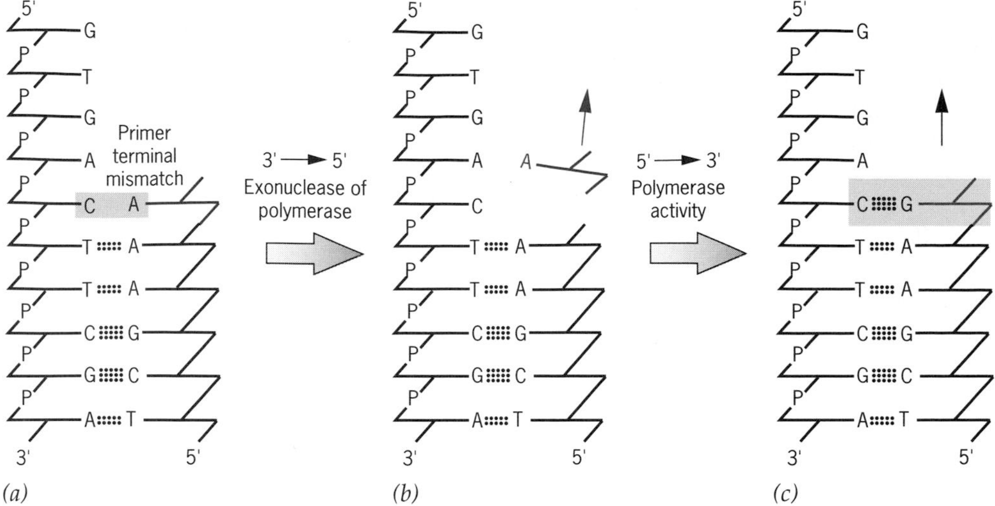

Figure 11.17 Proofreading by the 3′→ 5′ exonuclease activity of DNA polymerases during DNA replication. If DNA polymerase is presented with a template and primer containing a 3′ primer terminal mismatch (*a*), it will not catalyze covalent extension (polymerization). Instead, the 3′ → 5′ exonuclease activity, an integral part of many DNA polymerases, will cleave off the mismatched terminal nucleotide (*b*). Then, presented with a correctly base-paired primer terminus, DNA polymerase will catalyze 5′ → 3′ covalent extension of the primer strand (*c*).

E. coli, this proofreading function is carried out by the ϵ subunit. In eukaryotes, DNA polymerases γ, δ, and ϵ contain 3′ → 5′ proofreading exonuclease activities, but polymerases α and β lack this activity. Given the importance of proofreading, we might speculate that accessory proteins must carry out the proofreading function for DNA polymerases α and β.

Without proofreading during DNA replication, Merry and Sherry, the twins discussed at the beginning of this chapter, would be less identical in appearance. Without proofreading, changes would have accumulated in their genes during the billions of cell divisions that occurred during their growth from small embryos to adults. Indeed, the identity of the genotypes of identical twins depends both on DNA proofreading during replication and on the activity of an army of DNA repair enzymes (Chapter 14). These enzymes continually scan DNA for various types of damage and make repairs before the alterations cause inherited genetic changes.

Key Points: **DNA synthesis is catalyzed by enzymes called DNA polymerases. All DNA polymerases have an absolute requirement for a primer strand, which is extended, and a template strand, which is copied. All DNA polymerases have an absolute requirement for a free 3′-OH on the primer strand, and all DNA synthesis occurs in the 5′ to 3′ direction. The 3′ → 5′ exonuclease activities of DNA polymerases proofread nascent strands as they are synthesized, removing any mispaired nucleotides at the 3′ termini of primer strands.**

THE COMPLEX REPLICATION APPARATUS

The results of studies of DNA replication by autoradiography and electron microscopy indicate that the two progeny strands being synthesized at each replicating fork are being extended in the same overall direction at the macromolecular level. Because the complementary strands of a double helix have opposite polarity, synthesis is occurring at the 5′ end of one strand (3′ → 5′ extension) and the 3′ end of the other strand (5′ → 3′ extension). However, as we have previously discussed, all known polymerases have an absolute requirement for a free 3′-hydroxyl; they only carry out 5′ → 3′ synthesis. These apparently contradictory results created an interesting paradox. For many years biochemists searched for new polymerases that could catalyze 3′ → 5′ synthesis. No such polymerase was ever found. Instead, experimental evidence has shown that all DNA synthesis occurs in the 5′ → 3′ direction.

Clearly, the mechanism of DNA replication must be more complex than researchers originally thought (see Figure 11.1). In addition, given the absolute requirement of DNA polymerase for a free 3′-OH on the primer strand, this enzyme cannot begin the synthesis of a new strand *de novo*. How is the synthesis of a new DNA strand initiated? Since the two parental strands of DNA must be unwound, we have to deal with the need for a swivel or axis of rotation, especially for circular DNA molecules like the one present in the *E. coli* chromosome. Finally, how does the localized zone of strand separation or replication bubble form at the origin? These considerations and others indicate that DNA replication is more complicated than scientists thought when the semiconservative mechanism of replication was proposed by Watson and Crick in 1953.

Continuous Synthesis of One Strand; Discontinuous Synthesis of the Other Strand

As discussed in the preceding section, the two nascent DNA strands being synthesized at each replicating fork are being extended in the same direction at the macromolecular level. Because the complementary strands of a DNA double helix have opposite chemical polarity, one strand is being extended in an overall 5′ → 3′ direction and the other strand is being extended in an overall 3′ → 5′ direction (Figure 11.18, top). But DNA polymerases can only catalyze synthesis in the 5′ → 3′ direction. This paradox was resolved with the demonstration that the synthesis of one strand of DNA is **continuous,** whereas synthesis of the other strand is **discontinuous.** At the molecular level, synthesis of the complementary strands of DNA is occurring in opposite physical directions (Figure 11.18, bottom), but both new strands are extended in the same 5′ → 3′ chemical direction. The synthesis of the strand being extended in the overall 5′ → 3′ direction, called the **leading strand,** is **continuous.** The strand being extended in the overall 3′ → 5′ direction, called the **lagging strand,** grows by the synthesis of short fragments (synthesized 5′ → 3′) and the subsequent covalent joining of these short fragments. Thus, the synthesis of the lagging strand occurs by a discontinuous mechanism.

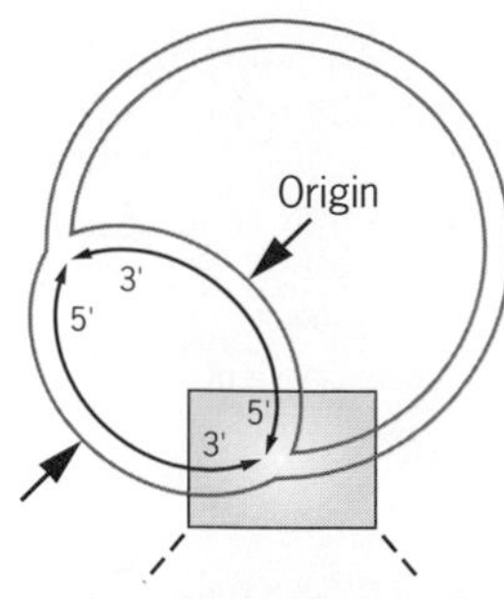

(*a*) Relatively low-resolution techniques such as autoradiography and electron microscopy show that at the macromolecular level both nascent DNA chains are extended in the same overall direction at each replication fork.

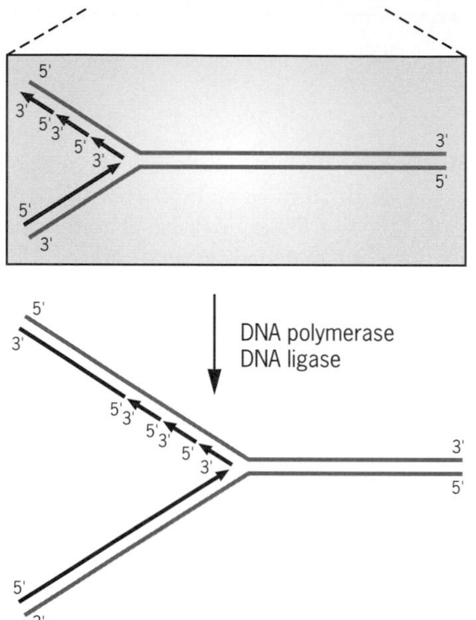

(*b*) High-resolution biochemical techniques such as pulse-labeling and density gradient analysis show that replication of the lagging strand is discontinuous—short fragments are synthesized in the 5' → 3' direction and subsequently joined by DNA ligase.

Figure 11.18 Continuous synthesis of the leading strand and discontinuous synthesis of the lagging strand at a DNA replication fork.

The first evidence for this discontinuous mode of DNA replication came from studies in which intermediates in DNA synthesis were radioactively labeled by growing *E. coli* cells and bacteriophage T4-infected *E. coli* cells for very short periods of time in medium containing ^{3}H-thymidine (pulse-labeling experiments). The labeled DNAs were then isolated, denatured, and characterized by measuring their velocity of sedimentation through sucrose gradients during high-speed centrifugation (see Technical Sidelight: Cesium Chloride . . .). When *E. coli* cells were pulse-labeled for 15 seconds, for example, much of the label was found in small fragments of DNA, 1000 to 2000 nucleotides long. These small fragments of DNA have been named Okazaki fragments after Reiji and Tuneko Okazaki, the scientists who discovered them in the late 1960s. In eukaryotes, the Okazaki fragments are only 100 to 200 nucleotides in length. When longer pulse-labeling periods are used, more of the label is recovered in large DNA molecules, presumably the size of *E. coli* or phage T4 chromosomes. If cells are pulse-labeled with ^{3}H-thymidine for a short period and then are transferred to nonradioactive medium for an extended period of growth (pulse-chase experiments), the labeled thymidine is present in chromosome-size DNA molecules. The results of these pulse-chase experiments are important because they indicate that the Okazaki fragments are true intermediates in DNA replication and not some type of metabolic by-product.

Covalent Closure of Nicks in DNA by DNA Ligase

If the lagging strand of DNA is synthesized discontinuously as described in the preceding section, a mechanism is needed to link the Okazaki fragments together to produce the large DNA strands present in mature chromosomes. This mechanism is provided by the enzyme **DNA ligase.** DNA ligase catalyzes the covalent closure of nicks (missing phosphodiester linkages; no missing bases) in DNA molecules by using energy from nicotinamide adenine dinucleotide (NAD) or adenosine triphosphate (ATP). The *E. coli* DNA ligase uses NAD as a cofactor, but some DNA ligases use ATP. The reaction catalyzed by DNA ligase is shown in Figure 11.19. First, AMP of the ligase-AMP intermediate forms a phosphoester linkage with the 5′-phosphate at the nick, and then a nucleophilic attack by the 3′-OH at the nick on the DNA-proximal phosphorus atom produces a phosphodiester linkage between the adjacent nucleotides at the site of the nick. DNA ligase has no activity at breaks in DNA where one or more nucleotides are missing—so-called gaps. Gaps can be filled in and sealed only by the combined action of a DNA polymerase and DNA ligase. DNA ligase plays an essential role not only in DNA replication, but also in DNA repair and in the pathways of recombination, as we will see in Chapter 14.

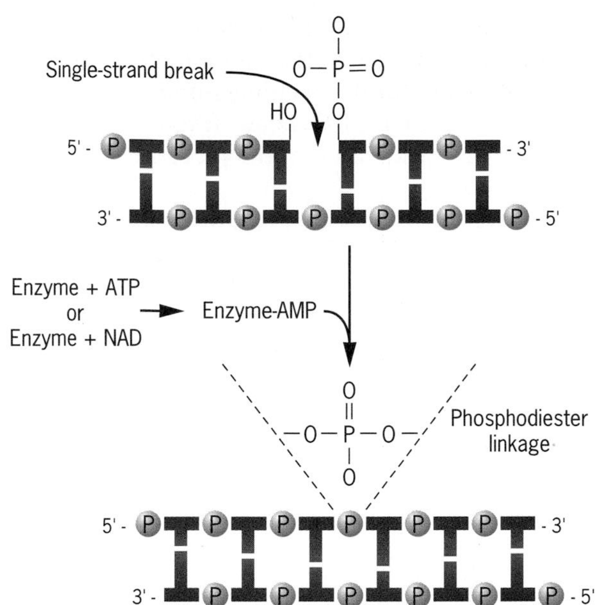

Figure 11.19 DNA ligase catalyzes the covalent closure of nicks in DNA. The energy required to form the ester linkage is provided by either adenosine triphosphate (ATP) or nicotinamide-adenine dinucleotide (NAD), depending on the species.

Initiation of DNA Chains with RNA Primers

As discussed earlier, all known DNA polymerases have absolute requirements for a free 3′-OH on a DNA primer strand and an appropriate DNA template strand for activity. No known DNA polymerase can initiate the synthesis of a new strand of DNA. Thus, some special mechanism must exist to initiate or prime new DNA chains. Whereas the continuous synthesis of the leading strand requires the priming function only at the origin of replication, a priming event is required to initiate each Okazaki fragment during the discontinuous synthesis of the lagging strand. RNA polymerase, a complex enzyme that catalyzes the synthesis of RNA molecules from DNA templates, has long been known to be capable of initiating the synthesis of new RNA chains at specific sites on the DNA. When this occurs, an RNA-DNA hybrid is formed in which the nascent RNA is hydrogen-bonded to the DNA template. Because DNA polymerases are capable of extending polynucleotide chains containing an RNA primer with a free 3′-OH, scientists began testing the idea that DNA synthesis is initiated by RNA primers. Their results proved that this idea is correct.

Subsequent research has shown that each new DNA chain is initiated by a short **RNA primer** synthesized by **DNA primase** (Figure 11.20). The *E. coli* DNA primase is the product of the *dnaG* gene. In prokaryotes, these RNA primers are 10 to 60 nucleotides long, whereas in eukaryotes they are shorter, only about 10 nucleotides long. The RNA primers provide the free 3′-OHs required for covalent extension of polynucleotide chains by DNA polymerases. In *E. coli*, deoxyribonucleotides are added to the RNA primers

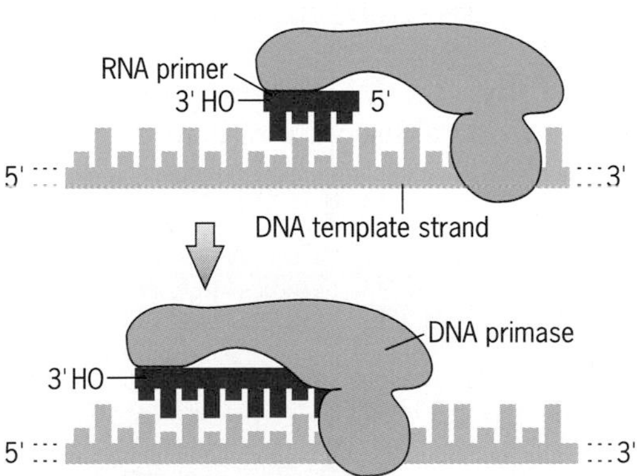

Figure 11.20 The initiation of DNA strands with RNA primers. The enzyme DNA primase catalyzes the synthesis of short (10 to 60 nucleotides long) RNA strands that are complementary to the template strands. DNA polymerase III then uses the free 3′-hydroxyls of the RNA primers to extend the chains by the addition of deoxyribonucleotides (see Figure 11.21).

by DNA polymerase III, either continuously on the leading strand or discontinuously by the synthesis of Okazaki fragments on the lagging stand. DNA polymerase III terminates an Okazaki fragment when it bumps into the RNA primer of the preceding Okazaki fragment.

The RNA primers subsequently are excised and replaced with DNA chains. This step is accomplished by DNA polymerase I in *E. coli*. Recall that of the three DNA polymerases in *E. coli*, only DNA polymerase I possesses 5′ → 3′ exonuclease activity. The 5′ → 3′ exonuclease activity of DNA polymerase I excises the RNA primer, and, at the same time, the 5′ → 3′ polymerase activity of the enzyme replaces the RNA with a DNA chain by using the adjacent Okazaki fragment with its free 3′-OH as a primer. As we might expect based on this mechanism of primer replacement, *E. coli polA* mutants that lack the 5′ → 3′ exonuclease activity of DNA polymerase I are defective in the excision of RNA primers and the joining of Okazaki fragments. After DNA polymerase I has replaced the RNA primer with a DNA chain, the 3′-OH of one Okazaki fragment is next to the 5′-phosphate group of the preceding Okazaki fragment. This product is an appropriate substrate for DNA ligase, which catalyzes the formation of a phosphodiester linkage between the adjacent Okazaki fragments. The steps involved in the synthesis and replacement of RNA primers during the discontinuous replication of the lagging strand are illustrated in Figure 11.21.

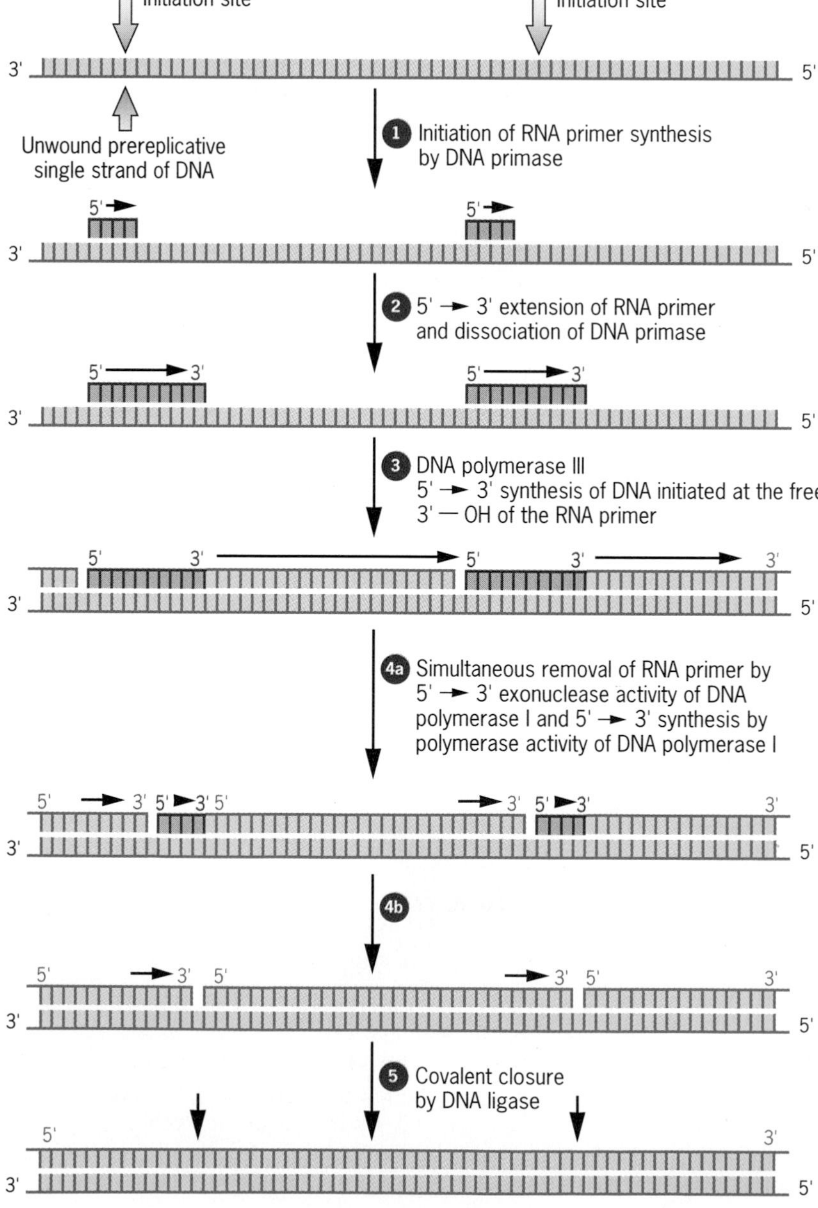

Figure 11.21 Synthesis and replacement of RNA primers during replication of the lagging strand of DNA. A short RNA strand is synthesized to provide a 3′-OH primer for DNA synthesis (see Figure 11.20). The RNA primer is subsequently removed and replaced with DNA by the dual 5′ → 3′ exonuclease and 5′ → 3′ polymerase activities built into DNA polymerase I. DNA ligase then covalently closes the nascent DNA chain, catalyzing the formation of phosphodiester linkages between adjacent 3′-hydroxyls and 5′-phosphates (see Figure 11.19).

Unwinding DNA with Helicases, DNA-Binding Proteins, and Topoisomerases

Semiconservative replication requires that the two strands of a parental DNA molecule be separated during the synthesis of new complementary strands. Since a DNA double helix is a plectonemic coil with two strands that cannot be separated without untwisting them turn by turn, DNA replication requires an unwinding mechanism. Given that each gyre, or turn, is about 10 nucleotide pairs long, a DNA molecule must be rotated 360° once for each 10 replicated base pairs. In *E. coli*, DNA replicates at a rate of about 30,000 nucleotides per minute. Thus, a replicating DNA molecule must spin at 3000 revolutions per minute to facilitate the unwinding of the parental DNA strands. The unwinding process (Figure 11.22*a*) is catalyzed by enzymes called **DNA helicases.** The major DNA helicase in *E. coli* is the product of the *dnaB* gene. DNA helicases unwind DNA molecules using energy derived from ATP.

Once the DNA strands are unwound by DNA helicase, they must be kept in an extended single-stranded form for replication. They are maintained in this state by a coating of **single-strand DNA-binding proteins** (SSB proteins) (Figure 11.22*b*). The binding of SSB proteins to single-stranded DNA is cooperative; that is, the binding of the first SSB monomer stimulates the binding of additional monomers at contiguous sites on the DNA chain. Because of the cooperativity of SSB protein binding, an entire single-stranded region of DNA is rapidly coated with SSB protein. Without the SSB protein coating, the complementary strands could renature or form intrastrand hairpin structures by hydrogen bonding between short segments of complementary or partially complementary nucleotide sequences. Such hairpin structures are known to impede the activity of DNA polymerases. In *E. coli*, the SSB protein is encoded by the *ssb* gene.

Recall that the *E. coli* chromosome contains a circular molecule of DNA. With the *E. coli* DNA spinning at 3000 revolutions per minute to allow the unwinding of the parental strands during replication (Figure 11.23), what provides the swivel or axis of rotation that prevents the DNA from becoming tangled (positively supercoiled) ahead of the replication fork? The required axes of rotation during the replication of circular DNA molecules are provided by enzymes called **DNA topoisomerases.** The topoisomerases catalyze transient breaks in DNA molecules but use covalent linkages to themselves to hold on to the cleaved molecules. The topoisomerases are of two types: (1) DNA

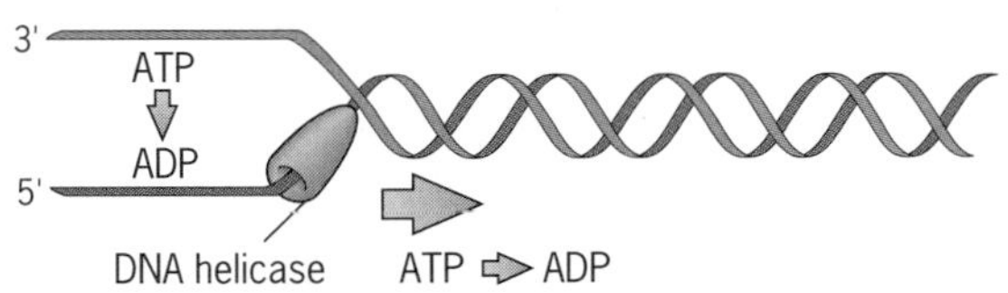

(*a*) DNA helicase catalyzes the unwinding of the parental double helix.

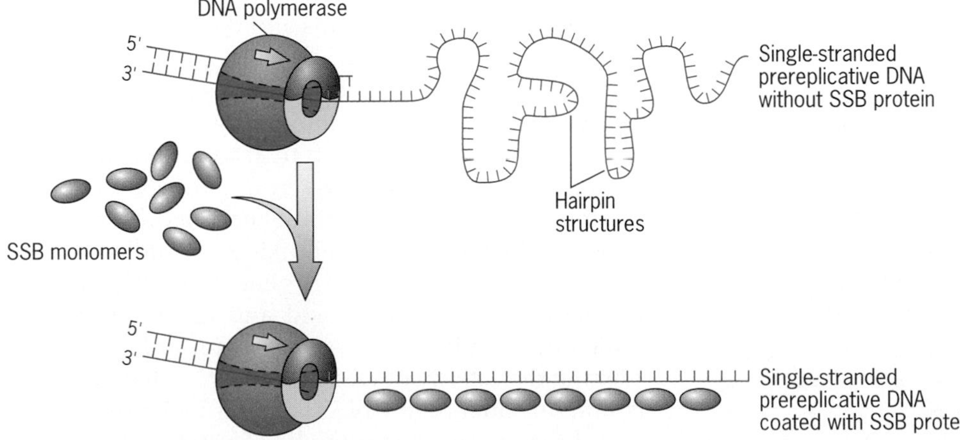

(*b*) Single-stranded DNA-binding (SSB) proteins keep the unwound strands in an extended form for replication.

Figure 11.22 The formation of functional template DNA requires (*a*) DNA helicase, which unwinds the parental double helix, and (*b*) single-strand DNA-binding (SSB) protein, which keeps the unwound DNA strands in an extended form. In the absence of SSB protein, DNA single strands can form hairpin structures by intrastrand base-pairing (*b*, top), and the hairpin structures will retard or arrest DNA synthesis.

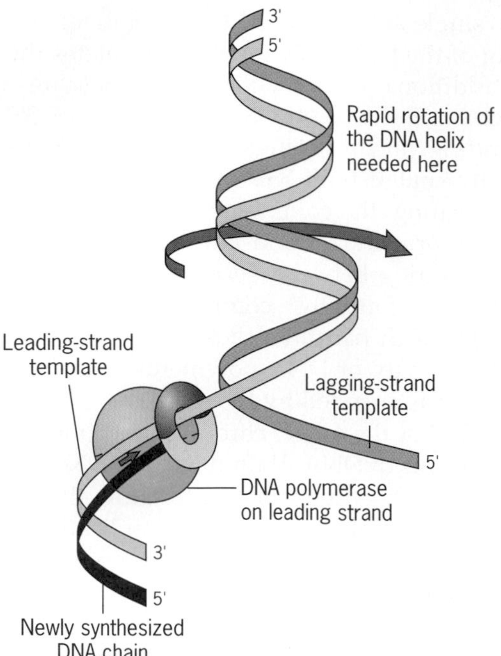

(*a*) To unwind the template strands in *E. coli*, the DNA helix in front of the replication fork must spin at 3000 rpm.

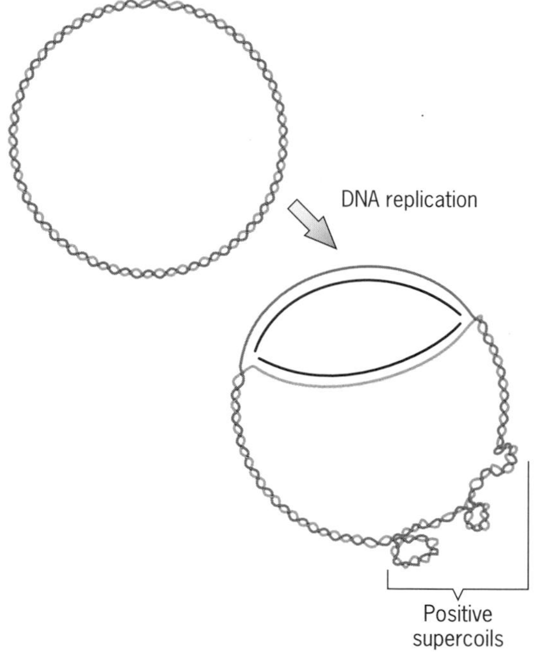

(*b*) Without a swivel or axis of rotation, the unwinding process would produce positive supercoils in front of the replication forks.

Figure 11.23 A swivel or axis of rotation is required during the replication of circular molecules of DNA like those in the *E. coli* or phage λ chromosomes.

topoisomerase I enzymes produce temporary single-strand breaks or nicks in DNA, and (2) DNA topoisomerase II enzymes produce transient double-strand breaks in DNA. An important result of this difference is that topoisomerase I activities remove supercoils from DNA one at a time, whereas topoisomerase II enzymes remove and introduce supercoils two at a time.

The transient single-strand break produced by the activity of topoisomerase I provides an axis of rotation that allows the segments of DNA on opposite sides of the break to spin independently, with the phosphodiester bond in the intact strand serving as a swivel (Figure 11.24*a*). Thus, during DNA replication, only a short segment of DNA in front of the replication fork needs to spin—the segment up to the closest transient nick by topoisomerase I. Topoisomerase I enzymes are energy efficient. They conserve the energy of the cleaved phosphodiester linkages by storing it in covalent linkages between themselves and the phosphate groups at the cleavage sites; they then reuse this energy to reseal the breaks. Some topoisomerase I enzymes can relax both negatively and positively supercoiled DNAs; others, such as the *E. coli topA* gene product, can act only on negatively supercoiled DNAs. The *E. coli* topoisomerase I removes one negative supercoil from DNA by breaking one strand, passing the intact strand through the break, and resealing the transient break (Figure 11.24*b*).

DNA topoisomerase II enzymes induce transient double-strand breaks and add or remove supercoils two at a time by an energy (ATP)-requiring mechanism. They carry out this process by cutting both strands of DNA, holding on to the ends at the cleavage site via covalent bonds, passing the intact double helix through the cut, and resealing the break (Figure 11.25). In addition to relaxing supercoiled DNA and introducing negative supercoils into DNA, topoisomerase II enzymes can separate interlocking circular molecules of DNA. Given the high concentration of DNA in nuclei and nucleoids and the large sizes of the DNA molecules involved, we can appreciate the important role of type II topoisomerases in keeping these DNA molecules from becoming irreversibly entangled.

The best-characterized type II topoisomerase is an enzyme named **DNA gyrase** in *E. coli*. DNA gyrase is a tetramer with two α subunits encoded by the *gyrA* gene (originally *nalA*, for nalidixic acid) and two β subunits specified by the *gyrB* gene (formerly *cou*, for coumermycin). Nalidixic acid and coumermycin are antibiotics that block DNA replication in *E. coli* by inhibiting the activity of DNA gyrase. Nalidixic acid and coumermycin inhibit DNA synthesis by binding to the α and β subunits, respectively, of DNA gyrase. Thus, DNA gyrase activity is required for DNA replication to occur in *E. coli*.

Recall that chromosomal DNA is negatively supercoiled in *E. coli* (Chapter 10). The negative supercoils in bacterial chromosomes are introduced by DNA

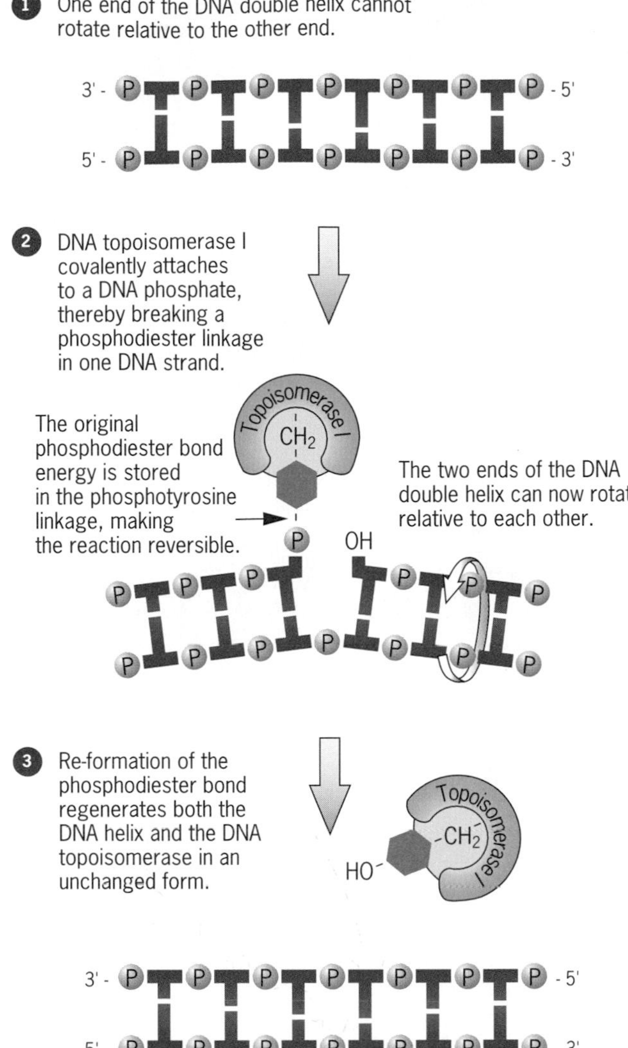

(*a*) DNA topoisomerase I provides transient single-strand breaks that serve as axes of rotation or swivels in DNA molecules.

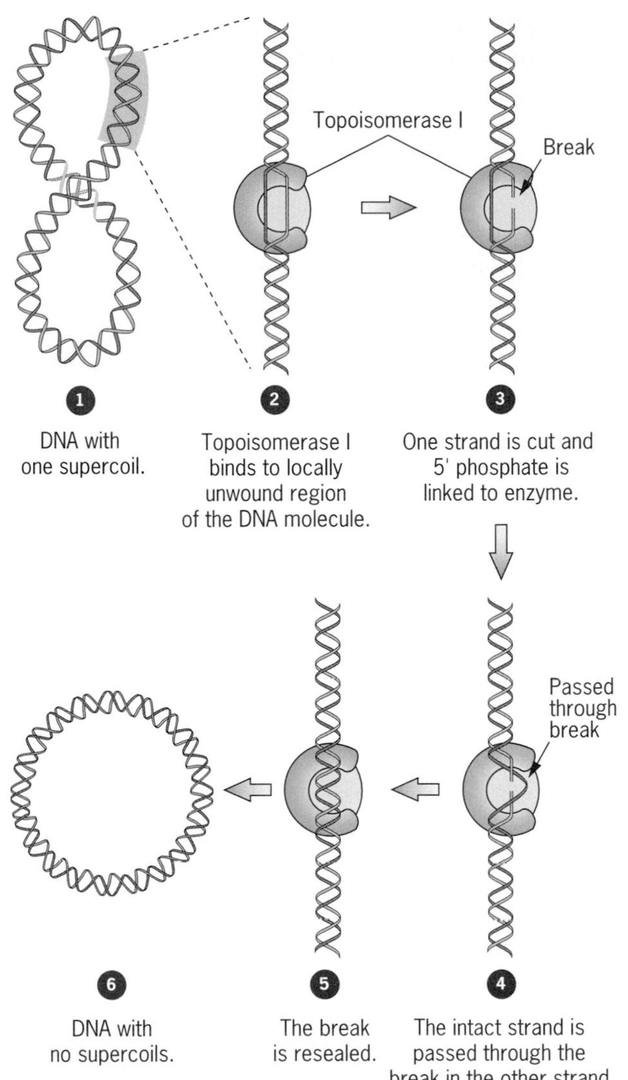

(*b*) DNA topoisomerase I removes supercoils from DNA one at a time.

Figure 11.24 DNA topoisomerase I produces transient single-strand breaks in DNA (*a*) that act as axes of rotation or swivels during DNA replication. The reactions catalyzed by DNA topoisomerase I remove supercoils from DNA one at a time (*b*), whereas those catalyzed by DNA topoisomerase II remove supercoils two at a time (see Figure 11.25).

gyrase, with energy supplied by ATP. This activity of DNA gyrase provides another solution to the unwinding problem. Instead of creating positive supercoils ahead of the replication fork by unwinding the complementary strands of relaxed DNA, replication may produce relaxed DNA ahead of the fork by unwinding negatively supercoiled DNA. Because superhelical tension is reduced during unwinding—that is, strand separation is energetically favored—the negative supercoiling behind the fork may drive the unwinding process. If so, this mechanism nicely explains why DNA gyrase activity is required for DNA replication in bacteria.

The Replication Apparatus: Prepriming Proteins, Primosomes, and Replisomes

The basic features of DNA replication and most of the important components of the replication apparatus have been introduced in the preceding sections of this chapter. Now we will put the pieces together and look at the coordinated activities of these components during the replication of chromosomal DNA. To do so, we will consider the sequence of events that occur during the replication of the circular molecule of DNA in the *E. coli* chromosome. Figure 11.26 shows the most important components of the replication apparatus in *E.*

Figure 11.25 Mechanism of action of DNA gyrase, an *E. coli* DNA topoisomerase II required for DNA replication.

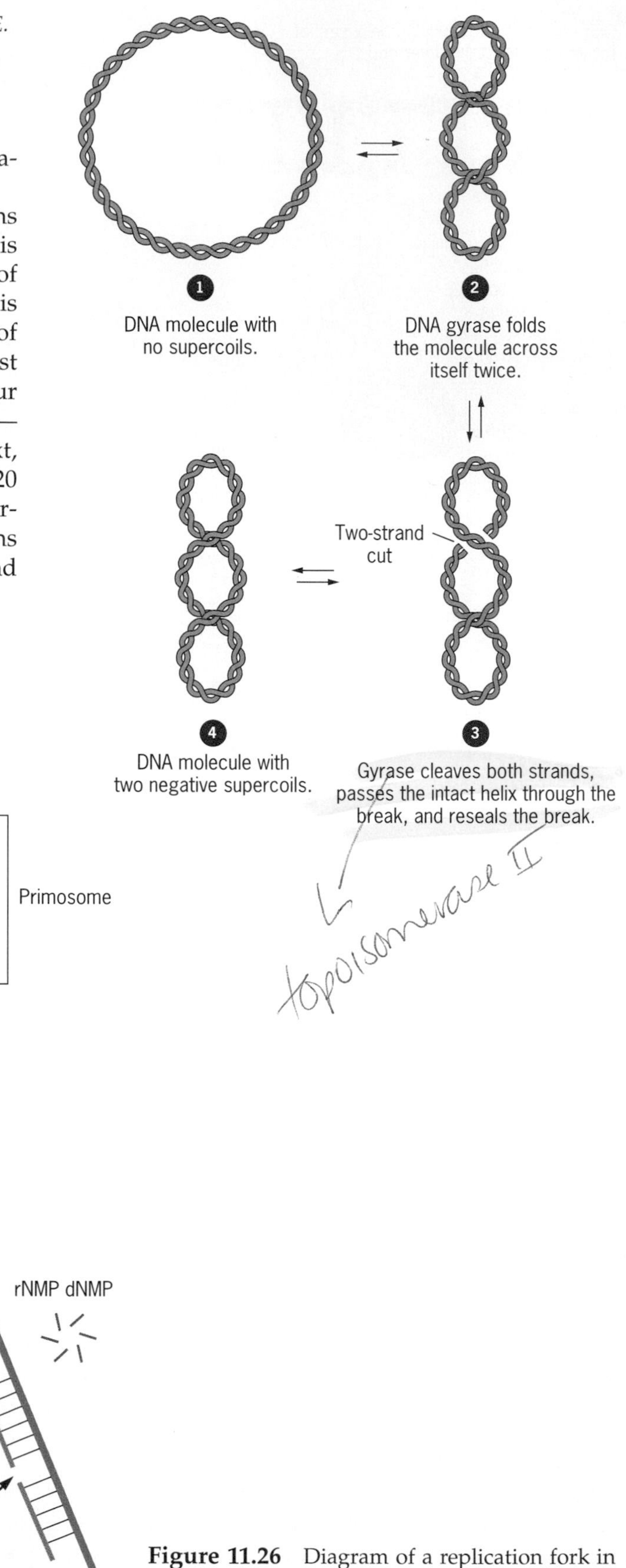

coli and indicates where they are located on a replication fork.

The replication of the *E. coli* chromosome begins at *OriC,* the unique sequence at which replication is initiated, with the formation of a localized region of strand separation called the *replication bubble.* This replication bubble is formed by the interaction of **prepriming proteins** with *OriC* (Figure 11.27). The first step in prepriming appears to be the binding of four molecules of the *dnaA* gene product—**DnaA protein**—to the four 9-base-pair (bp) repeats in *OriC*. Next, DnaA proteins bind cooperatively to form a core of 20 to 40 polypeptides with *OriC* DNA wound on the surface of the protein complex. Strand separation begins within the three tandem 13-bp repeats in *OriC* and

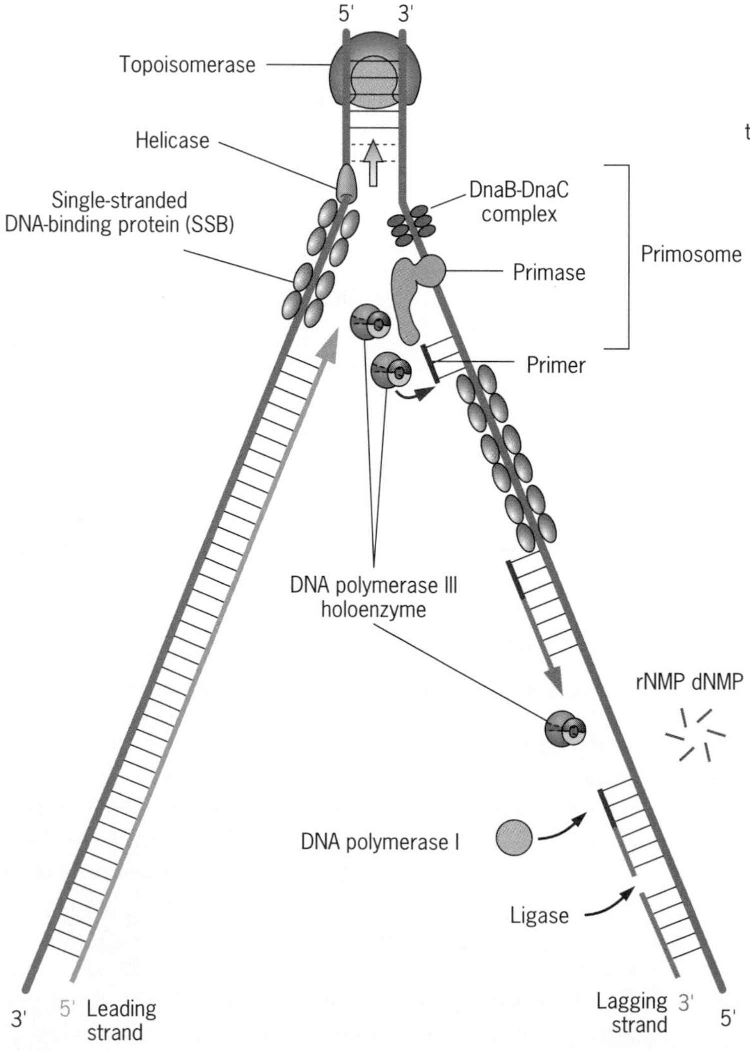

Figure 11.26 Diagram of a replication fork in *E. coli* showing the major components of the replication apparatus.

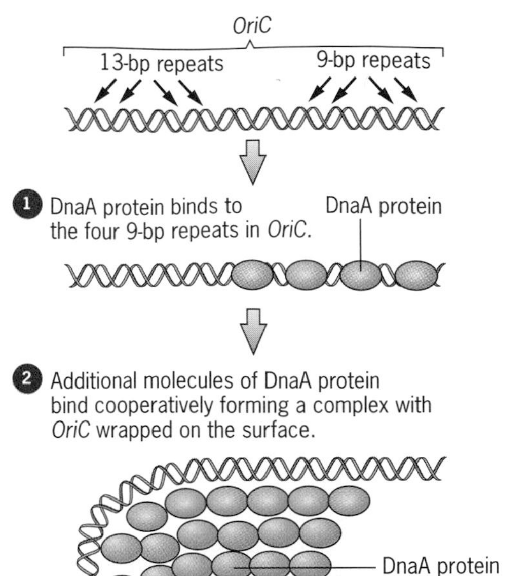

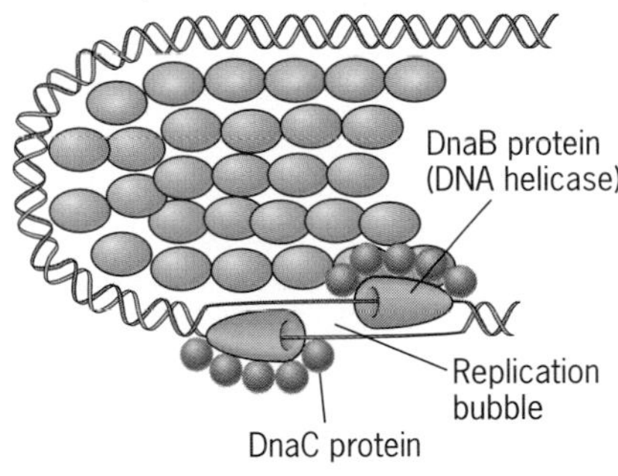

Figure 11.27 Prepriming of DNA replication at *OriC* in the *E. coli* chromosome.

spreads until the replication bubble is created. A complex of DnaB protein (the hexameric DNA helicase) and **DnaC protein** (six molecules) joins the initiation complex and contributes to the formation of two bidirectional replication forks. The DnaT protein also is present in the prepriming protein complex, but its function is unknown. Other proteins associated with the initiation complex at *OriC* are DnaJ protein, DnaK protein, PriA protein, PriB protein, PriC protein, DNA-binding protein HU, DNA gyrase, and single-strand DNA-binding (SSB) protein. In some cases, however, their functional involvement in the prepriming process has not been established; in other cases, they are known to be involved, but their roles are unknown. The DnaA protein appears to be largely responsible for the localized strand separation at *OriC* during the initiation process.

Once a replication fork has formed, the synthesis of new DNA strands is initiated by RNA primers synthesized by DNA primase. A single RNA primer is sufficient for the continuous replication of the leading strand, but the discontinuous replication of the lagging strand requires an RNA primer to start the synthesis of each Okazaki fragment. The initiation of Okazaki fragments on the lagging strand is carried out by the **primosome,** a protein complex containing DNA primase and DNA helicase. The primosome moves along a DNA molecule, powered by energy from ATP; DNA helicase unwinds the double helix, and DNA primase synthesizes the RNA primers for successive Okazaki fragments. The RNA primers are covalently extended with deoxyribonucleotides by DNA polymerase III. DNA topoisomerases provide transient breaks in DNA that serve as swivels for DNA unwinding and keep the DNA untangled. Single-strand DNA-binding protein coats the unwound prereplicative DNA and keeps it in an extended state for DNA polymerase III. The RNA primers are replaced with DNA by DNA polymerase I, and the single-strand breaks left by polymerase are sealed by DNA ligase. The DNA is then condensed into the nucleoid or folded genome of *E. coli,* in part through negative supercoiling introduced by DNA gyrase. All of these enzymes and DNA-binding proteins function in concert at each replication fork.

As a replication fork moves along a parental double helix, two DNA strands (the leading strand and the lagging strand) are replicated in the highly coordinated series of reactions described above. The complete replication apparatus moving along the DNA molecule at a replication fork is called the **replisome** (Figure 11.28). The replisome contains the DNA polymerase III holoenzyme; one catalytic core replicates the leading strand, the second catalytic core replicates the lagging strand, and the primosome unwinds the parental DNA molecule and synthesizes the RNA primers needed for the discontinuous synthesis of the lagging strand. In order for the two catalytic cores of the polymerase III holoenzyme to synthesize both the nascent leading and lagging strands, the lagging strand is thought to form a loop from the primosome to the second catalytic core of DNA polymerase III (Figure 11.28).

At the beginning of this chapter, we noted the striking fidelity of DNA replication. Now that we have examined the cellular machinery responsible for DNA replication in living organisms, this fidelity seems less amazing. A very sophisticated apparatus, with built-in safeguards against malfunctions, has evolved to assure that the genetic information of *E. coli* is transmitted accurately from generation to generation.

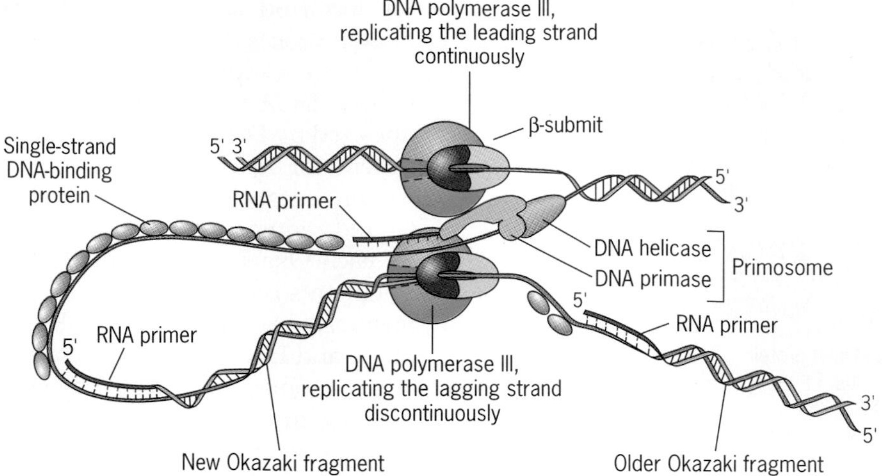

Figure 11.28 Diagram of the *E. coli* replisome, showing the two catalytic cores of DNA polymerase III replicating the leading and lagging strands and the primosome unwinding the parental double helix and initiating the synthesis of new chains with RNA primers. The entire replisome moves along the parental double helix, with each component performing its respective function in a concerted manner. Actually, the replication complex probably does not move. Instead, the DNA is pulled through the replisome. Replication is proceeding from left to right.

Rolling-Circle Replication

In the preceding sections of this chapter, we have considered θ-shaped, eye-shaped, and Y-shaped replicating DNAs. We will now examine another important type of DNA replication called **rolling-circle replication.** Rolling-circle replication is used (1) by many viruses to duplicate their genomes (Chapter 16), (2) to transfer DNA from donor cells to recipient cells during one type of genetic exchange in bacteria (Chapter 17), and (3) to amplify extrachromosomal DNAs carrying clusters of ribosomal RNA genes during oogenesis in amphibians (Chapter 24).

As the name implies, rolling-circle replication is a mechanism for replicating circular DNA molecules. The unique aspect of rolling-circle replication is that one parental circular DNA strand remains intact and rolls (thus the name rolling circle) or spins while serving as a template for the synthesis of a new complementary strand (Figure 11.29). Replication is initiated when a sequence-specific endonuclease cleaves one strand at the origin, producing 3′-OH and 5′-phosphate termini. The 5′ terminus is displaced from the circle as the intact template strand turns about its axis. Covalent extension occurs at the 3′-OH of the cleaved strand. Since the circular template DNA may turn 360° many times, with the synthesis of one complete or unit-length DNA strand during each turn, rolling-circle replication generates single-stranded tails longer than the contour length of the circular chromosome (Figure 11.29). Rolling-circle replication can produce either single-stranded or double-stranded progeny DNAs. Circular single-stranded progeny molecules are produced by site-specific cleavage of the single-stranded tails at the origins of replication and recircularization of the resulting unit-length molecules. To produce double-stranded progeny molecules, the single-stranded tails are used as templates for the discontinuous synthesis of complementary strands prior to cleavage and circularization. The enzymes involved in rolling-circle replication and the reactions catalyzed by these enzymes are basically the same as those responsible for DNA replication involving θ-type intermediates.

Key Points: **DNA replication is complex, requiring the participation of a large number of proteins. DNA synthesis is continuous on the progeny strand that is being extended in the overall 5′ → 3′ direction, but is discontinuous on the strand growing in the overall 3′ → 5′ direction. New DNA chains are initiated by short RNA primers synthesized by DNA primase. The enzymes and DNA-binding proteins involved in replication assemble into a replisome at each replication fork and act in concert as the fork moves along the parental DNA molecule.**

UNIQUE ASPECTS OF EUKARYOTIC CHROMOSOME REPLICATION

Most of the information about DNA replication has resulted from studies of *E. coli* and some of its viruses. Less information is available about DNA replication in eukaryotic organisms. However, enough information

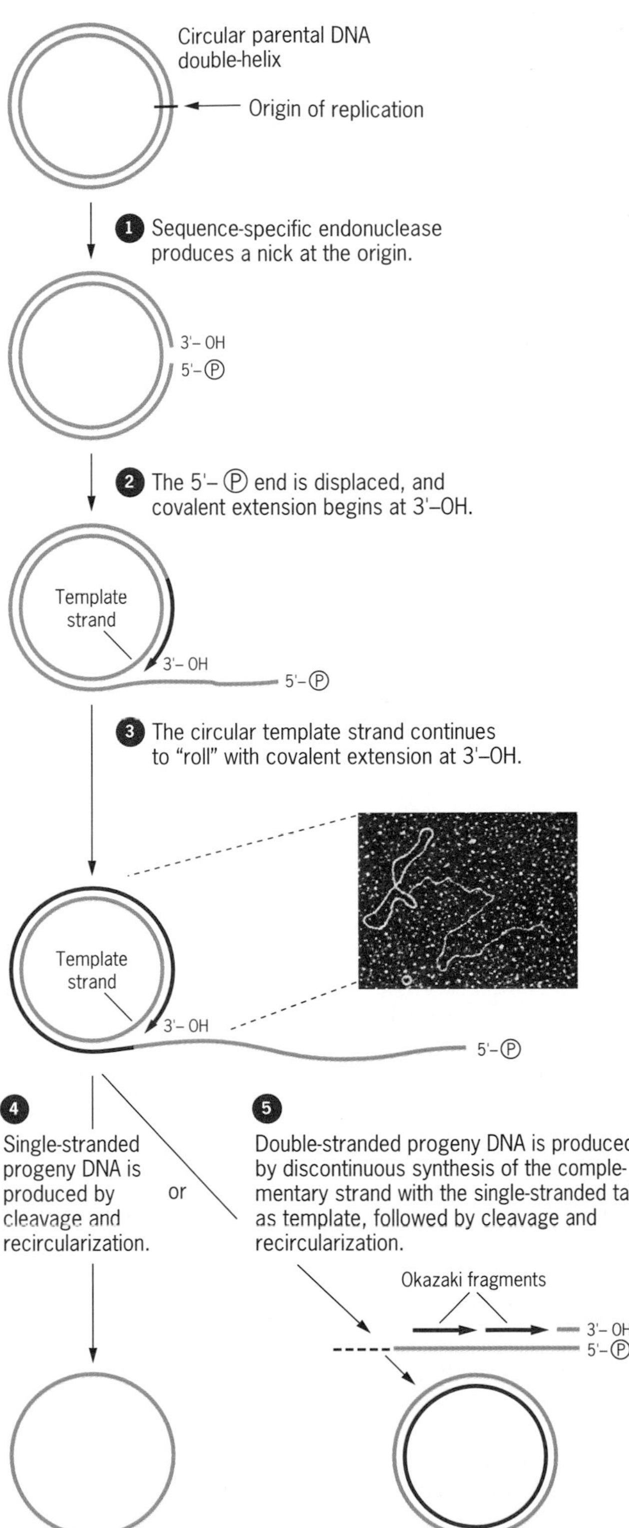

Figure 11.29 The rolling-circle mechanism of DNA replication. The inset shows an electron micrograph of a bacteriophage ΦX174 DNA molecule replicating by the rolling-circle mechanism. A single-stranded tail extends from a double-stranded, circular, replicative DNA.

is available to conclude that most aspects of DNA replication are basically the same in both prokaryotes and eukaryotes, including humans. RNA primers and Okazaki fragments are shorter in eukaryotes than in prokaryotes, but the leading and lagging strands replicate by continuous and discontinuous mechanisms, respectively, in eukaryotes just as in prokaryotes. Nevertheless, a few aspects of eukaryotic DNA replication are unique to these structurally more complex species. For example, DNA synthesis takes place within a small portion of the cell cycle in eukaryotes, not continuously as in prokaryotes. The giant DNA molecules present in eukaryotic chromosomes would take much too long to replicate if each chromosome contained a single origin. Thus, eukaryotic chromosomes contain multiple origins of replication. Rather than using two catalytic complexes of one DNA polymerase to replicate the leading and lagging strands at each replication fork, eukaryotic organisms utilize two different polymerases.

As we discussed in Chapter 10, eukaryotic DNA is sequestered in histone-containing structures called nucleosomes. Do these nucleosomes impede the movement of replication forks? If not, how does a replisome move past a nucleosome? Is the nucleosome completely or partially disassembled, or does the fork somehow slide past the nucleosome as the replisome duplicates the DNA molecule while it is still present on the surface of the nucleosome? Lastly, eukaryotic chromosomes contain linear DNA molecules, and the discontinuous replication of the ends of linear DNA molecules creates a special problem. We will address these aspects of chromatin replication in eukaryotes in the final sections of this chapter.

The Cell Cycle

When bacteria are growing on rich media, DNA replication occurs nonstop throughout the cell cycle. However, in eukaryotes, DNA replication is restricted to the S phase (for synthesis; Chapter 2). Recall that a normal eukaryotic cell cycle consists of G_1 phase (immediately following the completion of mitosis; G for gap), S phase, G_2 phase (preparation for mitosis), and M phase (mitosis). In rapidly dividing embryonic cells, G_1 and G_2 are very short or nonexistent. In all cells, decisions to continue on through the cell cycle occur at two points: (1) entry into S phase and (2) entry into mitosis (see Chapter 2 for details).

Multiple Replicons per Chromosome

The giant DNA molecules in the largest chromosomes of *Drosophila melanogaster* contain about 6.5×10^7 nu-

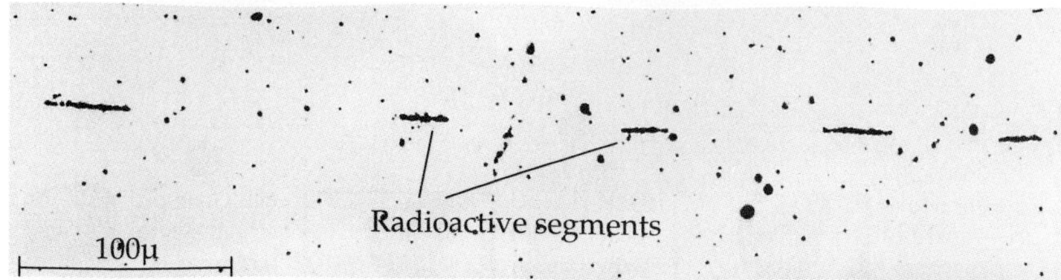

(a) Autoradiograph of a portion of a DNA molecule from a Chinese hamster cell that had been pulse-labeled with ^{3}H-thymidine.

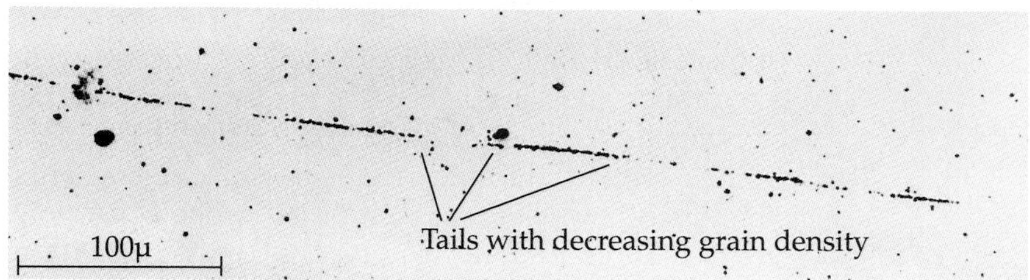

(b) Autoradiograph of a segment of a DNA molecule from a Chinese hamster cell that was pulse-labeled with ^{3}H-thymidine and then transferred to nonradioactive medium for an additional growth period.

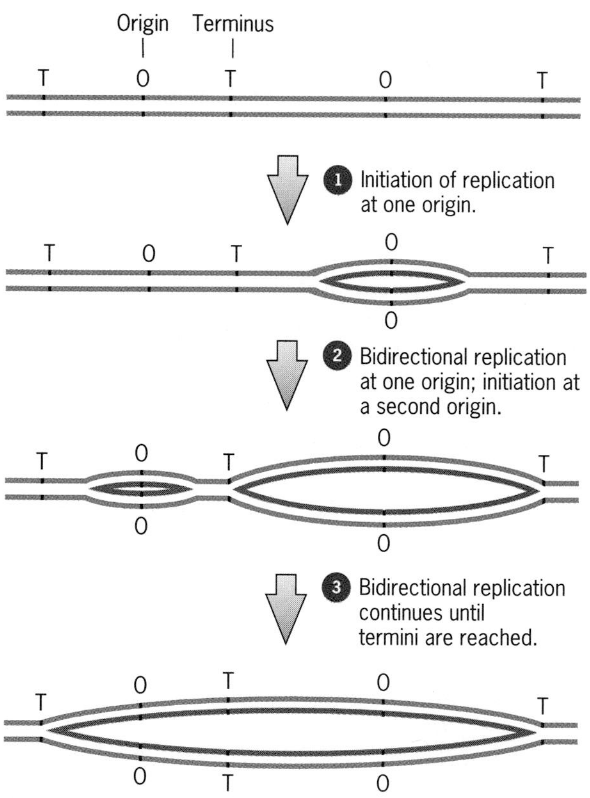

(c) Diagrammatic interpretation of the replication of the DNA molecules visualized above.

Figure 11.30 Evidence for bidirectional replication of the multiple replicons in the giant DNA molecules of eukaryotes. The tandem arrays of radioactivity in (*a*) indicate that replication occurs at multiple origins; tails with decreasing grain density observed in (*b*) indicate that replication occurs bidirectionally from each origin (*c*).

cleotide pairs. The rate of DNA replication in *Drosophila* is about 2600 nucleotide pairs per minute at 25°C. A single replication fork would therefore take about 17.5 days to replicate one of these giant DNA molecules. With two replication forks moving bidirectionally from a central origin, such a DNA molecule could be replicated in just over 8.5 days. Given that the chromosomes of *Drosophila* embryos replicate within 3 to 4 minutes and the nuclei divide once every 9 to 10 minutes during the early cleavage divisions, it is clear that each giant DNA molecule must contain many origins of replication. Indeed, the complete replication of the DNA of the largest *Drosophila* chromosomes within 3.5 minutes would require over 7000 replication forks distributed at equal intervals along the molecules. Thus, multiple origins of replication are required to allow the very large DNA molecules in eukaryotic chromosomes to replicate within the observed cell division times.

The first evidence for multiple origins in eukaryotic chromosomes resulted from pulse-labeling experiments with Chinese hamster cells growing in culture. In 1968, when Joel Huberman and Arthur Riggs pulse-labeled cells with [^{3}H]thymidine for a few minutes, extracted the DNA, and performed autoradiographic analysis of the labeled DNA, they observed tandem arrays of exposed silver grains. The simplest interpretation of their results is that individual macromolecules of DNA contain multiple origins of replication (Figure 11.30*a*). When the pulse-labeling period was followed by a short interval of growth in nonradioactive medium (pulse-chase experiments), the tandem arrays contained central regions of high-grain density with tails of decreasing grain density at *both* ends (Figure 11.30*b*). This result indicates that replication in eukaryotes is bidirectional just as it is in most prokaryotes. The tails of decreasing grain density result from the gradual dilution of the intracellular pools of ^{3}H-thymidine by ^{1}H-thymidine as the replication forks move bidirectionally from central origins toward replication termini (Figure 11.30*c*).

A segment of DNA whose replication is under the control of one origin and two termini is called a **replicon.** In prokaryotes, the entire chromosome is usually one replicon. The existence of multiple replicons per eukaryotic chromosome has been verified directly by electron microscopy (Figure 11.31). Clearly, the number of replicons per chromosome is not fixed throughout the growth and development of a multicellular eukaryote. Replication is initiated at more sites during the very rapid cell divisions of embryogenesis than during later stages of development. Unfortunately, geneticists don't know what factors determine which origins are operational at any given time or in a particular type of cell.

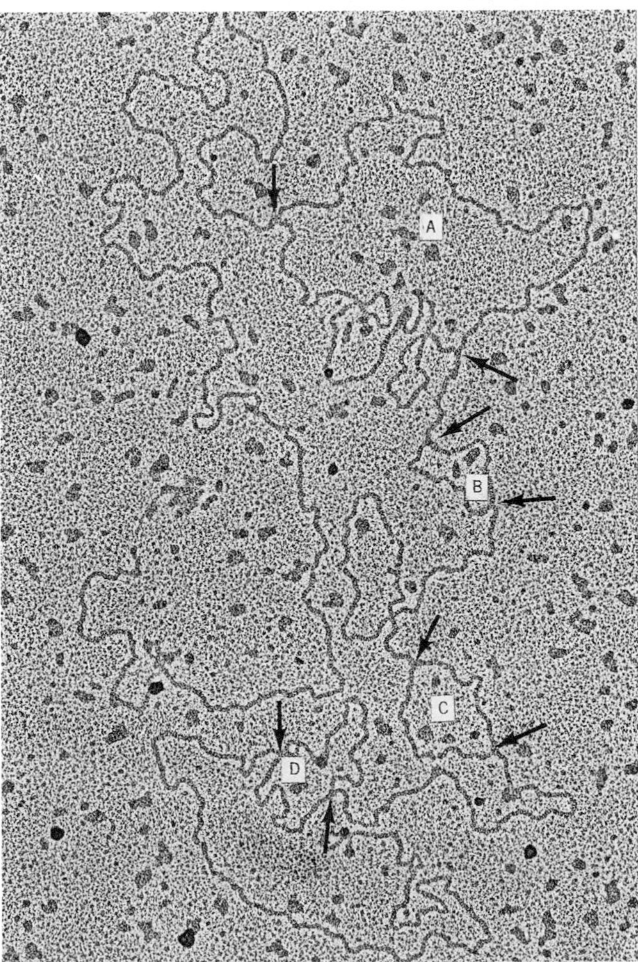

Figure 11.31 Electron micrograph of a DNA molecule in *D. melanogaster* showing multiple sites of replication. Four eye-shaped (—⬭—) replication structures (labeled A-D) are present in the segment of the DNA molecule shown. The arrows show the positions of replication forks.

Two DNA Polymerases at a Single Replication Fork

Given the complexity of the replisome in the simple bacterium *E. coli* (see Figures 11.26 and 11.28), it seems likely that the replication apparatus is even more complex in eukaryotes. Although knowledge of the structure of the replicative machinery in eukaryotes is still somewhat limited, many features of DNA replication are the same in both eukaryotes and prokaryotes. One difference is that two different DNA polymerases function at each replication fork in eukaryotes.

As in the case of prokaryotes, much of the information about DNA synthesis in eukaryotes has come from the development and dissection of *in vitro* DNA replication systems. Studies of the replication of DNA viruses of eukaryotes have proven informative, and, of these viruses, Simian virus 40 (SV40) has proven particularly useful. The replication of SV40 is carried out almost entirely by the host cell's replication apparatus. Only one viral protein, the so-called T antigen, is required for replication of the SV40 chromosome. Two of the host cell's DNA polymerases, α and δ, are required to replicate the SV40 DNA molecule *in vitro* (Figure 11.32). DNA polymerase α appears to be the most similar to DNA polymerase III of *E. coli.* **DNA polymerase α** is multimeric in structure, and two of its subunits interact to provide DNA primase activity. Thus, polymerase α contains the enzymatic activities required to carry out the discontinuous replication of the lagging strand. **DNA polymerase δ** lacks primase activity, and evidence suggests that it catalyzes the replication of the leading strand. As in prokaryotes, a helicase is required to unwind the double helix ahead of the replication fork.

Several other proteins have been identified as replication factors (for example, RF-A, RF-B, RF-C, and so forth); their functions remain to be discovered. A protein called PCNA (proliferating cell nuclear antigen) appears to be a cofactor of polymerase δ; its presence results in a much increased rate of replication. Other accessory proteins undoubtedly will be identified in the future. Given the utility of the *in vitro* DNA replication systems now available, we can anticipate a more detailed picture of the eukaryotic replisome in the near future. However, the real challenges are (1) to identify the signals that regulate DNA replication during growth and differentiation in multicellu-

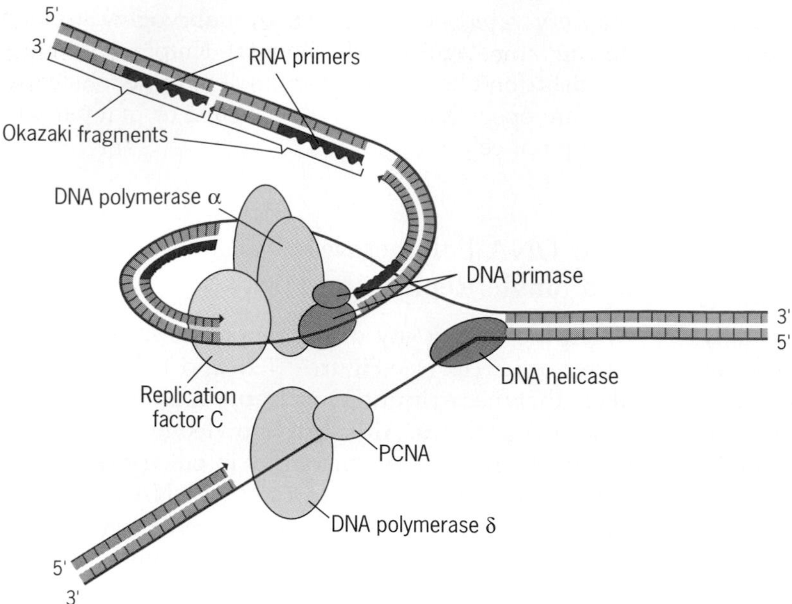

Figure 11.32 Two DNA polymerases, α and δ, function at the replication fork in eukaryotes. DNA polymerase δ replicates the leading strand, and DNA polymerase α replicates the lagging strand. PCNA = proliferating cell nuclear antigen.

lar organisms such as humans, (2) to determine which signal pathways are malfunctioning in the various types of cancer (Chapter 28), and (3) to figure out how to restore normal function to the various types of cancer cells.

Duplication of Nucleosomes at Replication Forks

As we discussed in Chapter 10, the DNA in eukaryotic interphase chromosomes is packaged in approximately 10-nm beads called nucleosomes. Each nucleosome contains 166 nucleotide pairs of DNA wound in two turns around an octamer of histone molecules. Given the size of nucleosomes and the large size of DNA replisomes, it seems unlikely that a replication fork can move past an intact nucleosome. Yet, electron micrographs of replicating chromatin in *Drosophila* clearly show nucleosomes with apparently normal structure and spacing on both sides of replication forks (Figure 11.33*a*); that is, nucleosomes appear to have the same structure and spacing immediately behind a replication fork (postreplicative DNA) as they do in front of a replication fork (prereplicative DNA). Thus, various models have been proposed for transient alterations in nucleosome structure that would allow replisomes to synthesize DNA as they move past nucleosomes. In one popular model, the nucleosome splits into two half-nucleosomes while the replisome moves past, and the halves then reassemble into intact nucleosomes behind the replication fork (Figure 11.33*b*).

Regardless of the mechanism by which the replication fork moves past nucleosomes, DNA replication and nucleosome assembly are tightly coupled in eukaryotes. Since the mass of the histones in nucleosomes is equivalent to that of the DNA, large quantities of histones must be synthesized during each cell generation in order for the nucleosomes to duplicate. Although histone synthesis occurs throughout the cell cycle, there is a burst of histone biosynthesis during S phase that generates enough histones for chromatin duplication. When density-transfer experiments were performed to examine the mode of nucleosome duplication, the nucleosomes on both progeny DNA molecules were found to contain both old or prereplicative histone complexes and new or postreplicative complexes. Thus, at the protein level, nucleosome duplication appears to occur by a dispersive mechanism.

Telomerase: Replication of Chromosome Termini

We discussed the unique structures of telomeres, or chromosome ends, in Chapter 10. An early reason for thinking that telomeres must have special structures was that DNA polymerases cannot replicate the ter-

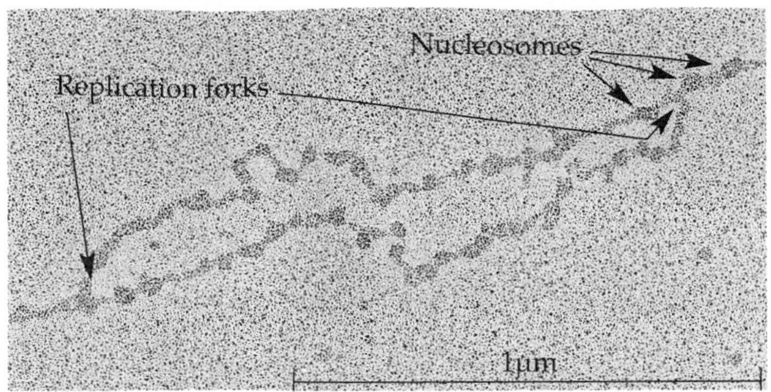

(*a*) Electron micrograph showing nucleosomes on both sides of each of two replication forks in *Drosophila*.

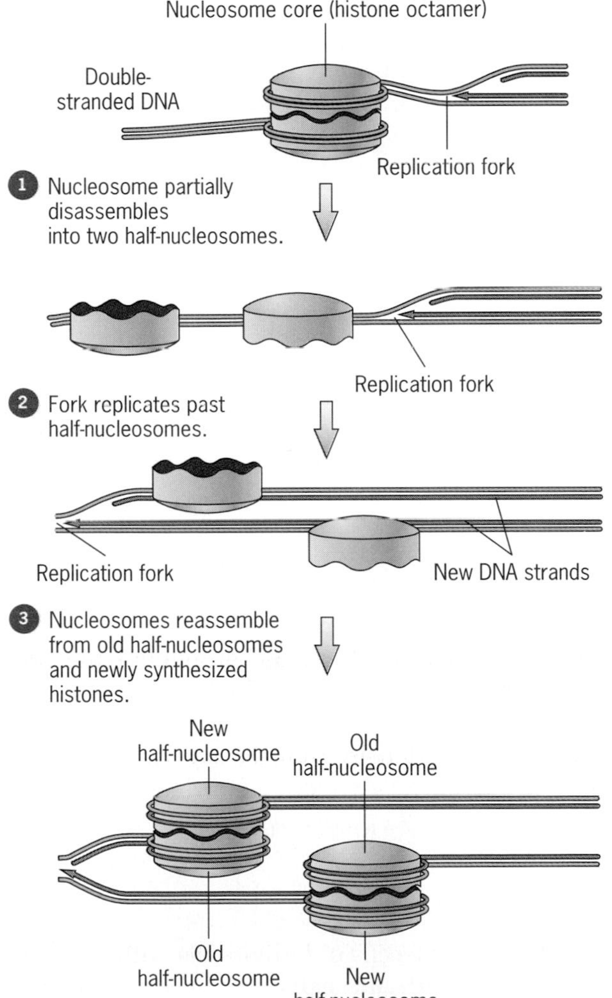

(*b*) Model of the movement of replisomes past nucleosomes.

Figure 11.33 Movement of a replication fork past nucleosomes. (*a*) Electron micrograph showing nucleosomes on both sides of two replication forks in *Drosophila*. Recall that DNA replication is bidirectional in eukaryotes; thus, each branch point is a replication fork. (*b*) One model for the movement of replication forks past nucleosomes.

minal DNA segment of the lagging strand of a linear chromosome. At the end of the DNA molecule being replicated discontinuously, there would be no DNA strand to provide a free 3′-OH (primer) for polymerization of deoxyribonucleotides after the RNA primer of the terminal Okazaki fragment has been excised (Figure 11.34*a*). Either (1) the telomere must have a unique structure that facilitates its replication or (2) there must be a special enzyme that resolves this enigma of replicating the terminus of the lagging strand. Indeed, evidence has shown that both are correct. The special structure of telomeres provides a neat mechanism for the addition of telomeres by an RNA-containing enzyme called **telomerase.**

The telomeres of humans, which contain the tandemly repeated sequence TTAGGG, will be used to illustrate how telomerase adds ends to chromosomes. Telomerase recognizes the G-rich telomere sequence on the 3′ overhang and extends it 5′ → 3′ one repeat unit at a time. The unique feature of telomerase is that it contains a built-in RNA template (Figure 11.34*b*). After several telomere repeat units are added by telomerase, DNA polymerase catalyzes the synthesis of the complementary strand. Without telomerase activity, linear chromosomes would become progressively shorter. If the resulting terminal deletions extended into an essential gene or genes, this chromosome shortening would be lethal.

Telomere Length and Aging in Humans

Unlike germ-line cells, most human somatic cells lack telomerase activity. When human somatic cells are grown in culture, they divide only a limited number of times (usually only 20 to 70 cell generations) before senescence and death occur. When telomere lengths are measured in various somatic cell cultures, a correlation is observed between telomere length and the number of cell divisions preceding senescence and death. Cells with longer telomeres survive longer—go through more cell divisions—than cells with shorter telomeres. As would be expected in the absence of telomerase activity, telomere length decreases as the age of the cell culture increases. Occasionally, somatic cells are observed to acquire the ability to proliferate in culture indefinitely, and these immortal cells have been shown to contain telomerase activity, unlike their progenitors. Since the one common feature of all cancers is uncontrolled cell division or immortality, scientists have proposed that one way to combat human cancers would be to inhibit the telomerase activity in cancer cells.

Further evidence of a relationship between telomere length and aging in humans has come from stud-

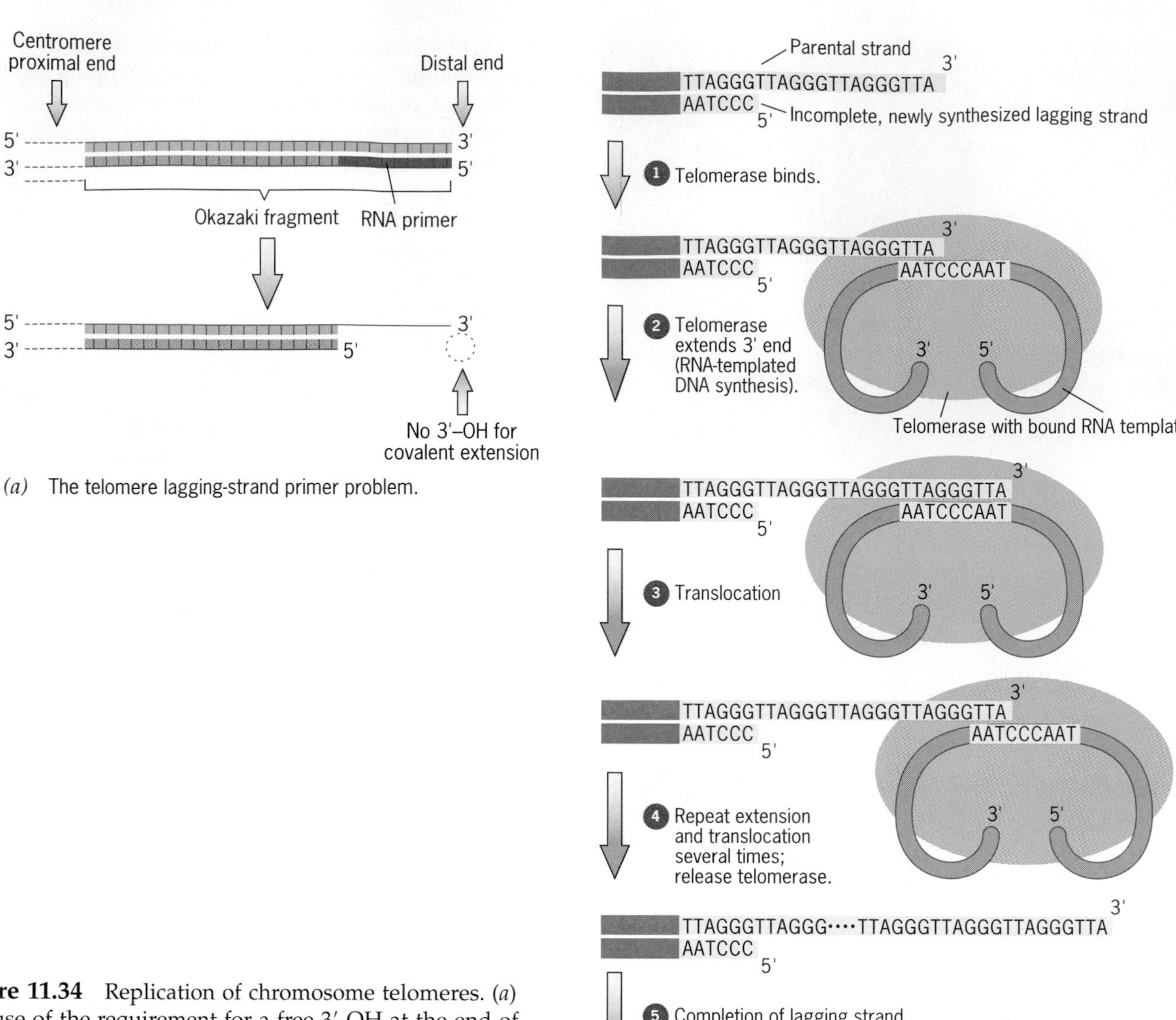

(*a*) The telomere lagging-strand primer problem.

(*b*) Telomerase resolves the terminal primer problem.

Figure 11.34 Replication of chromosome telomeres. (*a*) Because of the requirement for a free 3′-OH at the end of the primer strand, DNA polymerases cannot replace the RNA primer that initiates DNA synthesis at the terminus of the lagging strand. (*b*) These termini of chromosomes are replicated by a special enzyme called telomerase, which prevents the ends of chromosomes from becoming shorter during each replication. The nucleotide sequence at the terminus of the lagging strand is specified by a short RNA molecule present as an essential component of telomerase. The telomere sequence shown is that of humans.

ies of individuals with disorders called **progerias,** inherited diseases characterized by premature aging. In the most severe form of progeria, Hutchinson–Gilford syndrome (Figure 11.35), senescence—wrinkles, baldness, and other symptoms of aging—begins immediately after birth and death usually occurs in the teens. In a less severe form, Werner syndrome, senescence begins in the teenage years, with death usually occurring in the 40s. Consistent with the hypothesis that decreasing telomere length contributes to the aging process, the somatic cells of individuals with progeria have short telomeres and exhibit decreased proliferative capacity when grown in culture.

At present, the relationship between telomere length and cell senescence is entirely correlative. There is no direct evidence indicating that telomere shortening causes aging. Nevertheless, the correlation is striking, and the hypothesis that telomere shortening contributes to the aging process in humans warrants further study.

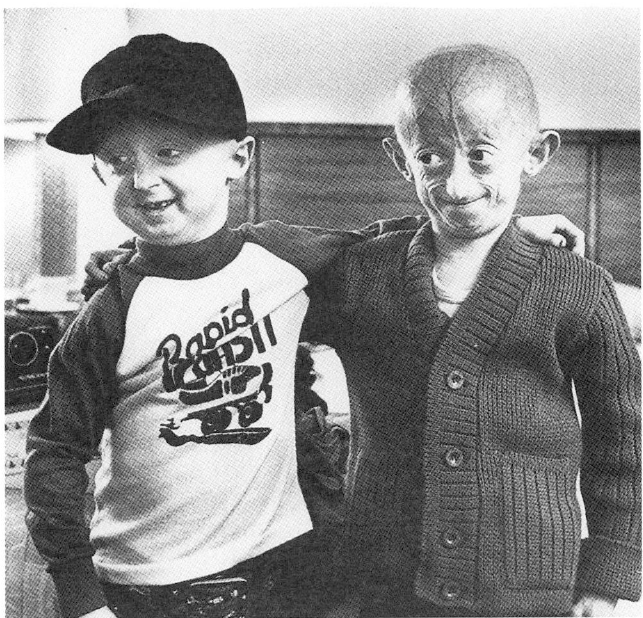

Figure 11.35 Photograph of children with Hutchinson–Gilford syndrome, showing the premature aging characteristic of this disorder. The children are eight (right) and nine years old, respectively.

Key Points: **DNA replication is basically the same in both prokaryotes and eukaryotes, including humans. Replication of the giant DNA molecules in eukaryotic chromosomes occurs bidirectionally from multiple origins. Two DNA polymerases (α and δ) are present at each replication fork. Telomeres, the unique sequences at the ends of chromosomes, are added to chromosomes by a special RNA-containing enzyme called telomerase.**

TESTING YOUR KNOWLEDGE

1. *Escherichia coli* cells were grown for many generations in a medium in which the only available nitrogen was the heavy isotope ^{15}N. The cells were then collected by centrifugation, washed with a buffer, and transferred to a medium containing ^{14}N (the normal light nitrogen isotope). After two generations of growth in the ^{14}N-containing medium, the cells were transferred back to ^{15}N-containing medium for one final generation of growth. After this final generation of growth in the presence of ^{15}N, the cells were collected by centrifugation. The DNA was then extracted from these cells and analyzed by CsCl equilibrium density-gradient centrifugation. How would you expect the DNA from these cells to be distributed in the gradient?

ANSWER

Meselson and Stahl demonstrated that DNA replication in *E. coli* is semiconservative. Their control experiments showed that DNA double helices with (1) ^{14}N in both strands, (2) ^{14}N in one strand and ^{15}N in the other strand, and (3) ^{15}N in both strands separated into three distinct bands in the gradient, called (1) the light band, (2) the hybrid band, and (3) the heavy band, respectively. If you start with a DNA double helix with ^{15}N in both strands, and replicate it semiconservatively for two generations in the presence of ^{14}N and then for one generation in the presence of ^{15}N, you will end up with eight DNA molecules, two with ^{15}N in both strands and six with ^{14}N in one strand and ^{15}N in the other strand, as shown below. Therefore, 75 percent (6/8) of the DNA will appear in the hybrid band, and 25 percent (2/8) will appear in the heavy band.

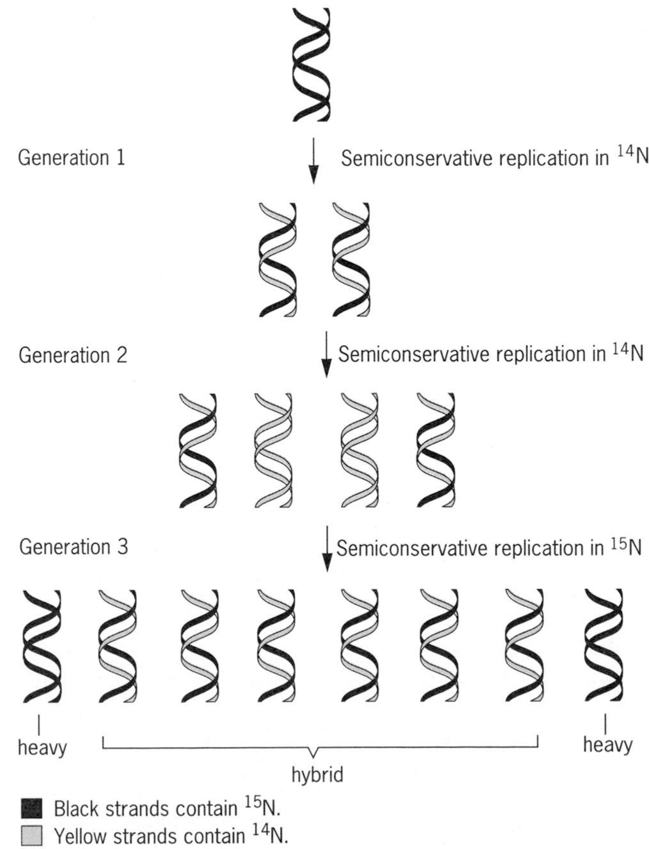

2. *Haplopappus gracilis* is a diploid plant with only two pairs of chromosomes. Assume that a G_1-stage cell of this plant, not previously exposed to radioactivity of any kind, was growing in culture medium containing ^{3}H-thymidine. After one generation of growth in this medium, the two progeny cells were transferred to medium containing ^{1}H-thymidine and colchicine. They were allowed to grow in this medium for about one and a half generations. Then, the chromosomes from each cell were spread on a microscope slide, stained, photographed, and exposed to an emulsion sensitive to low-energy radiation. One of the daughter cells exhibited a metaphase plate with eight chromosomes, each with two daughter chromatids. Draw this metaphase plate showing the predicted distribution of radioactivity on the autoradiograph.

ANSWER

Each chromosome contains one giant double helix that replicates semiconservatively. At the start of the experiment, both strands of this double helix contained nonradioactive ^{1}H-thymidine. Since all four chromosomes will go through the same processes of replication, you need to follow only one of them and then multiply your result by four. Since colchicine was not present during the first replication, the chromosome number per cell will remain unchanged, but the newly synthesized strands will be radioactive (contain ^{3}H-thymidine). Because the second and third replications occurred in the presence of ^{1}H-thymidine, strands synthesized during these replications will be nonradioactive. However, colchicine was present; thus, the chromosome number will double during each duplication. These replications are as follows, with radioactive strands of DNA shown in red and nonradioactive strands in blue.

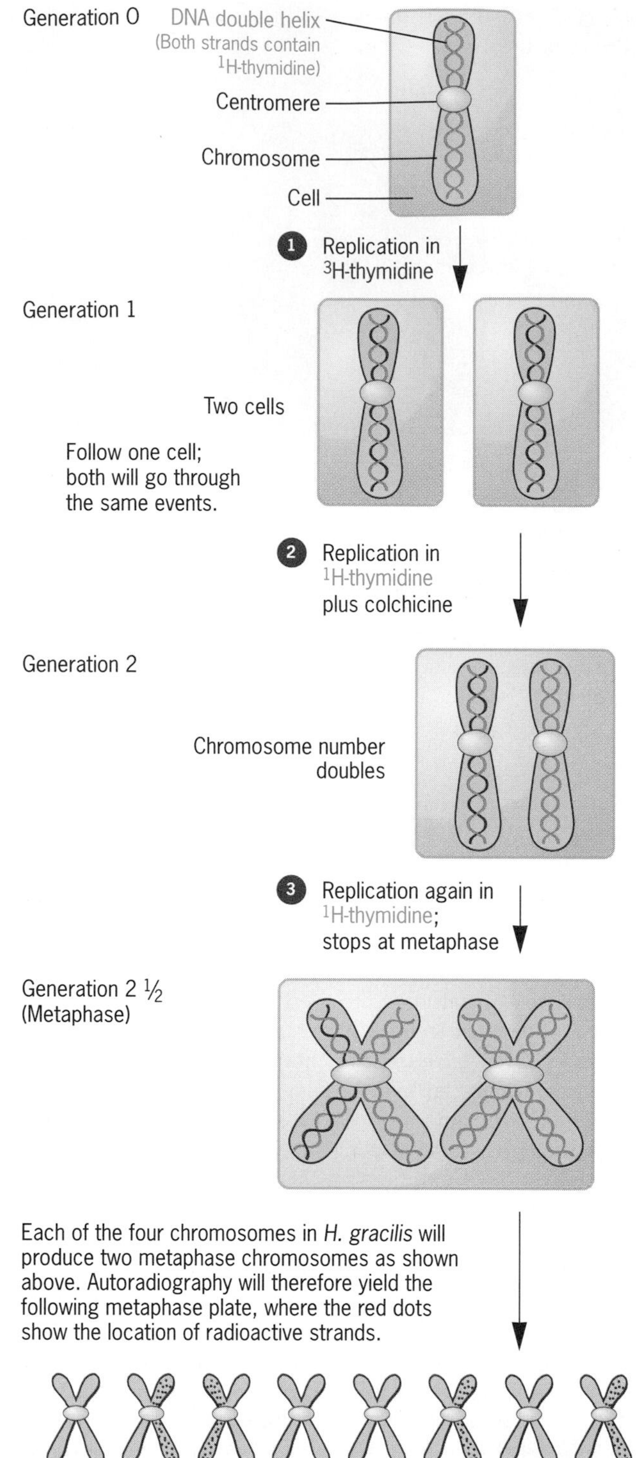

QUESTIONS AND PROBLEMS

11.1 *Escherichia coli* cells are grown for many generations in a medium in which the only available nitrogen is the heavy isotope ^{15}N. They are then transferred to a medium containing ^{14}N as the only source of nitrogen. (a) What distribution of ^{15}N and ^{14}N would be expected in the DNA molecules of cells that had grown for one generation in the ^{14}N-containing medium assuming that DNA replication was (i) conservative, (ii) semiconservative, or (iii) dispersive? (b) What distribution would be expected after two generations of growth in the ^{14}N-containing medium assuming (i) conservative, (ii) semiconservative, or (iii) dispersive replication?

11.2 A culture of bacteria is grown for many generations in a medium in which the only available nitrogen is the heavy isotope (^{15}N). The culture is then switched to a medium containing only ^{14}N for one generation of growth; it is then returned to a ^{15}N-containing medium for one final generation of growth. If the DNA from these bacteria is isolated and centrifuged to equilibrium in a CsCl density gradient, how would you predict the DNA to band in the gradient?

11.3 DNA polymerase I of *E. coli* is a single polypeptide of molecular weight 109,000. (a) What enzymatic activities other than polymerase activity does this polypeptide possess? (b) What are the *in vivo* functions of these activities? (c) Are these activities of major importance to an *E. coli* cell? Why?

11.4 A DNA template plus primer with the structure

3′ Ⓟ—TGCGAATTAGCGACAT—Ⓟ 5′
5′ Ⓟ—ATCGGTACGACGCT TAAC—OH 3′

(where P = a phosphate group) is placed in an *in vitro* DNA synthesis system (Mg^{2+}, an excess of the four deoxyribonucleoside triphosphates, etc.) containing a mutant form of *E. coli* DNA polymerase I that lacks 5′ → 3′ exonuclease activity. The 5′ → 3′ polymerase and 3′ → 5′ exonuclease activities of this aberrant enzyme are identical to those of normal *E. coli* DNA polymerase I. It simply has no 5′ → 3′ exonuclease activity. (a) What will be the structure of the final product? (b) What will be the first step in the reaction sequence?

11.5 How might continuous and discontinuous modes of DNA replication be distinguished experimentally?

11.6 Identify the proteins that are involved in DNA replication in *E. coli* and list their known or putative functions.

11.7 The Boston straggler is an imaginary plant with a diploid chromosome number of 4. Boston straggler cells are easily grown in suspended cell cultures. ^3H-Thymidine was added to the culture medium in which a G_1-stage cell of this plant was growing. After one cell generation of growth in ^3H-thymidine-containing medium, colchicine was added to the culture medium. The medium now contained both ^3H-thymidine and colchicine. After two "generations" of growth in ^3H-thymidine-containing medium (the second "generation" occurring in the presence of colchicine as well), the two progeny cells (each now containing eight chromosomes) were transferred to culture medium containing nonradioactive thymidine (^1H-thymidine) plus colchicine. Note that a "generation" in the presence of colchicine consists of a normal cell cycle's chromosomal duplication, but no cell division. The two progeny cells were allowed to continue to grow, proceeding through the "cell cycle," until each cell contained a set of metaphase chromosomes that looked like the following.

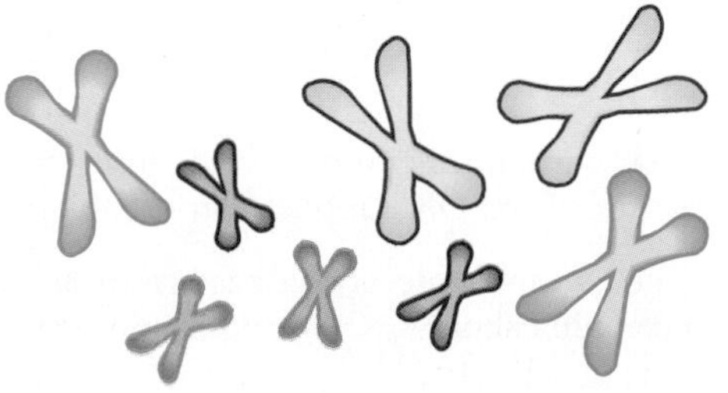

If autoradiography were carried out on these metaphase chromosomes (four large plus four small), what pattern of radioactivity (as indicated by silver grains on the autoradiograph) would be expected? (Assume no recombination between DNA molecules.)

11.8 Suppose that the experiment described in Problem 11.7 was carried out again, except this time replacing the ^3H-thymidine with nonradioactive thymidine at the same time that the colchicine was added (after one cell generation of growth in ^3H-thymidine-containing medium). The cells were then maintained in colchicine plus nonradioactive thymidine until the metaphase shown in Problem 11.7 occurred. What would the autoradiographs of these chromosomes look like?

11.9 Suppose that the DNA of cells (growing in a cell culture) in a eukaryotic species was labeled for a short period of time by the addition of ^3H-thymidine to the medium. Next assume that the label was removed and the cells were resuspended in nonradioactive medium. After a short period of growth in nonradioactive medium, the DNA was extracted from these cells, diluted, gently layered on filters, and autoradiographed. If autoradiographs of the type

———.... ———.... ———.....

were observed, what would this indicate about the nature of DNA replication in these cells? Why?

11.10 Arrange the following enzymes in the order of their action during DNA replication in *E. coli*: (1) DNA polymerase I, (2) DNA polymerase III, (3) DNA primase, (4) DNA gyrase, and (5) DNA helicase.

11.11 Five distinct DNA polymerases—α, β, γ, δ, and ε—have been characterized in mammals. What are the intracellular locations and functions of these polymerases?

11.12 The *E. coli* chromosome contains approximately 4×10^6 nucleotide pairs and replicates as a single bidirectional replicon in approximately 40 minutes under a wide variety of growth conditions. The largest chromosome of *D. melanogaster* contains about 6×10^7 nucleotide pairs. (a) If this chromosome contains one giant molecule of DNA that replicates bidirectionally from a single origin located precisely in the middle of the DNA molecule, how long would it take to replicate the entire chromosome if replication in *Drosophila* occurred at the same rate as replication in *E. coli*? (b) Actually, replication rates are slower in eukaryotes than in prokaryotes. If each replication bubble grows at a rate of 5000 nucleotide pairs per minute in *Drosophila* and 100,000 nucleotide pairs per minute in *E. coli*, how long will it take to replicate the largest *Drosophila* chromosome if it contains a single bidirectional replicon as described in (a) above? (c) During the early cleavage divisions in *Drosophila* embryos, the nuclei divide every 9 to 10 minutes. Based on your calculations in (a) and (b) above, what do these rapid nuclear divisions indicate about the number of replicons per chromosome in *Drosophila*?

11.13 What experimental techniques can be used to separate DNA molecules of mass 3×10^7 daltons isolated from lambda bacteriophage and DNA molecules of mass 1.3×10^8 daltons isolated from T2 bacteriophage?

11.14 The bacteriophage lambda chromosome has several AT-rich segments that denature when exposed to pH 11.05 for 10 minutes. After such partial denaturation, the linear packaged form of the lambda DNA molecule has the structure shown in Figure 11.9*a*. Following its injection into an *E. coli* cell, the lambda DNA molecule is converted to a covalently closed circular molecule by hydrogen bonding between its complementary single-stranded termini and the action of DNA ligase. It then replicates as a θ-shaped structure. The entire lambda chromosome is 17.5 μm long. It has a unique origin of replication located 14.3 μm from the left end of the linear form shown in Figure 11.9*a*. Draw the structure that would be observed by electron microscopy after both (1) replication of an approximately 6 μm-long segment of the lambda chromosomal DNA molecule (*in vivo*) and (2) exposure of this partially replicated DNA molecule to pH 11.05 for 10 minutes (*in vitro*), (a) *if* replication had proceeded bidirectionally from the origin, and (b) *if* replication had proceeded unidirectionally from the origin.

11.15 What enzyme activity catalyzes each of the following steps in the semiconservative replication of DNA in prokaryotes: (a) the formation of negative supercoils in progeny DNA molecules, (b) the synthesis of RNA primers, (c) the removal of RNA primers, (d) the covalent extension of DNA chains at the 3′-OH termini of primer strands, and (e) proofreading of the nucleotides at the 3′-OH termini of DNA primer strands?

11.16 One species of tree has a very large genome consisting of 2.0×10^{10} base pairs of DNA. (a) If this DNA was organized into a single linear molecule, how long (meters) would this molecule be? (b) If the DNA is evenly distributed among 10 chromosomes and each chromosome has one origin of DNA replication, how long would it take to complete the S phase of the cell cycle, assuming that DNA polymerase can synthesize 2×10^4 bp of DNA per minute? (c) An actively growing cell can complete the S phase of the cell cycle in approximately 300 minutes. Assuming that the origins of replication are evenly distributed, how many origins of replication are present on each chromosome? (d) What is the average number of base pairs between adjacent origins of replication?

11.17 Why must each of the giant DNA molecules in eukaryotic chromosomes contain multiple origins of replication?

11.18 In *E. coli*, *polA* mutants have been isolated that produce a defective gene product with little or no 5′ → 3′ polymerase activity, but normal 5′ → 3′ exonuclease activity. However, no *polA* mutant has been identified that is completely deficient in the 5′ → 3′ exonuclease activity, while retaining 5′ → 3′ polymerase activity, of DNA polymerase I. How can these results be explained?

11.19 Other *polA* mutants of *E. coli* lack the 3′ → 5′ exonuclease activity of DNA polymerase I. Will the rate of DNA synthesis be altered in these mutants? What effect(s) will these *polA* mutations have on the phenotype of the organism?

11.20 Many of the origins of replication that have been characterized contain A:T-rich core sequences. Are these A:T-rich cores of any functional significance? If so, what?

11.21 (a) Why isn't DNA primase activity required to initiate rolling-circle replication? (b) DNA primase is required for the discontinuous synthesis of the lagging strand, which occurs on the single-stranded tail of the rolling circle. Why?

11.22 DNA polymerase I is needed to remove RNA primers during chromosome replication in *E. coli*. However, DNA polymerase III is the true replicase in *E. coli*. Why doesn't DNA polymerase III remove the RNA primers?

11.23 In *E. coli*, three different proteins are required to unwind the parental double helix and keep the unwound strands in an extended template form. What are these proteins, and what are their respective functions?

11.24 How similar are the structures of DNA polymerase I and DNA polymerase III in *E. coli*? What is the structure of the DNA polymerase III holoenzyme? What is the function of the *dnaN* gene product in *E. coli*?

11.25 The *dnaA* gene product of *E. coli* is required for the initiation of DNA synthesis at *OriC*. What is its function? How do we know that the DnaA protein is essential to the initiation process?

11.26 What is a primosome, and what are its functions? What essential enzymes are present in the primosome? What are the major components of the *E. coli* replisome? How can geneticists determine whether these components are required for DNA replication?

11.27 The chromosomal DNA of eukaryotes is packaged into nucleosomes during the S phase of the cell cycle. What obstacles do the size and complexity of both the replisome and the nucleosome present during the semiconservative replication of eukaryotic DNA? How might these obstacles be overcome?

11.28 Design an experiment with *E. coli* that will allow you to distinguish between (1) continuous synthesis of one DNA strand and discontinuous synthesis of the other DNA strand and (2) discontinuous synthesis of both progeny DNA strands.

11.29 Two mutant strains of *E. coli* each have a temperature-sensitive mutation in a gene that encodes a product required for chromosome duplication. Both strains replicate their DNA and divide normally at 25°C, but are unable to replicate their DNA or divide at 42°C. When cells of one strain are shifted from growth at 25°C to growth at 42°C, DNA synthesis stops immediately. When cells of the other strain are subjected to the same temperature shift, DNA synthesis continues, albeit at a decreasing rate, for about a half hour. What can you conclude about the functions of the products of these two genes?

11.30 In what ways does chromosomal DNA replication in eukaryotes differ from DNA replication in prokaryotes?

11.31 (a) The chromosome of the bacterium *Salmonella typhimurium* contains about 4×10^6 nucleotide pairs. Approximately how many Okazaki fragments are produced during one complete replication of the *S. typhimurium* chromosome? (b) The largest chromosome of *D. melanogaster* contains approximately 6×10^7 nucleotide pairs. About how many Okazaki fragments are produced during the replication of this chromosome?

11.32 In the yeast *S. cerevisiae*, haploid cells carrying a mutation called *est1* (for *e*ver-*s*horter *t*elomeres) lose distal telomere sequences during each cell division. Predict the ultimate phenotypic effect of this mutation on the progeny of these cells.

11.33 The sequence of a double-stranded DNA is as follows.

5′-GATTCCCCGGGAAGCTTGGGGGGCCCATCTTCGTACGTCTTTGCA-3′
3′-CTAAGGGGCCCTTCGAACCCCCCGGGTAGAAGCATGCAGAAACGT-5′

Assuming a eukaryotic polymerase could replicate this small piece of DNA, what would you hypothesize that the replication products would look like?

BIBLIOGRAPHY

ALBERTS, B., D. BRAY, J. LEWIS, M. RAFF, K. ROBERTS, AND J. D. WATSON. 1994. *Molecular Biology of the Cell*, 3rd ed. Garland, New York.

CAIRNS, J. 1966. The bacterial chromosome. *Sci. Amer.* 214(1):36–44.

DE LANGE, T. 1994. Activation of telomerase in a human tumor. *Proc. Natl. Acad. Sci. U.S.A.* 91:2882–2885.

HUBERMAN, J. A., AND A. D. RIGGS. 1968. On the mechanism of DNA replication in mammalian chromosomes. *J. Mol. Biol.* 32:327–341.

KORNBERG, A., AND T. A. BAKER. 1992. *DNA Replication,* 2nd ed. Freeman, San Francisco.

LEE, H-W., M. A. BLASCO, G. J. GOTTLIEB, J. W. HORNER II, C. W. GREIDER, AND R. A. DEPINHO. 1998. Essential role of mouse telomerase in highly proliferative organs. *Nature* 392:569–574.

LEWIN, B. 1994. *Genes V.* Cell Press, Cambridge, MA/Oxford University Press, Oxford.

MESELSON, M. S., AND F. W. STAHL. 1958. The replication of DNA in *Escherichia coli*. *Proc. Natl. Acad. Sci. U.S.A.* 44:671–682.

SNYDER, L., AND W. CHAMPNESS. 1997. *Molecular Genetics of Bacteria.* ASM Press, Washington, DC.

TAYLOR, J. H., P. S. WOODS, AND W. L. HUGHES. 1957. The organization and duplication of chromosomes as revealed by autoradiographic studies using tritium-labeled thymidine. *Proc. Natl. Acad. Sci., U.S.A.* 43:122–128.

12

Spliceosome processing a gene transcript.

Transcription and RNA Processing

CHAPTER OUTLINE

The Genetic Control of Metabolism: An Overview
Transfer of Genetic Information: The Central Dogma
The Process of Gene Expression
Transcription in Prokaryotes
Transcription and RNA Processing in Eukaryotes
Interrupted Genes in Eukaryotes: Exons and Introns
Removal of Intron Sequences by RNA Splicing

Storage and Transmission of Information with Simple Codes

We live in the age of the computer. It has an impact on virtually all aspects of our lives, from driving to work to watching spaceships land on the moon. These electronic wizards can store, retrieve, and analyze data with lightninglike speed. The "brain" of the computer is a small chip of silicon, the microprocessor, which contains a sophisticated and integrated array of electronic circuits capable of responding almost instantaneously to coded bursts of electrical energy. In carrying out its amazing feats, the computer uses a binary code, a language based on 0's and 1's. Thus, the alphabet used by computers is like that of the Morse code (dots and dashes) used in telegraphy. Both consist of only two symbols—in marked contrast to the 26 letters of the English alphabet. Obviously, if the computer can perform its wizardry with a binary alphabet, vast amounts of information can be stored and retrieved without using complex codes or lengthy alphabets. In this and the following chapter, we examine (1) how the genetic information of living creatures is written in an alphabet with just four letters, the four base pairs in DNA, and (2) how this genetic information is expressed during the growth and development of an organism. We shall see that RNA plays a key role in the process of gene expression.

THE GENETIC CONTROL OF METABOLISM: AN OVERVIEW

The ability of living organisms to grow and reproduce depends on a vast number of chemical reactions. Organisms must synthesize the many different molecules of which they are composed and degrade other molecules to obtain energy for growth. Plants have the ability to use energy from sunlight to synthesize macromolecules, but animals must obtain energy from the food they eat. This energy is obtained by breaking large molecules into smaller molecules and converting the energy derived from this process to stored chemical energy. All of the reactions that occur in living organisms collectively are called *metabolism,* and an organic molecule that is synthesized or degraded is called a *metabolite.* In the next four chapters, we examine a detailed picture of the genetic control of metabolism and present evidence documenting the most important features of this picture. We begin with a brief overview of the genetic control of metabolism, which should help us integrate the various pieces of the picture as each component is added.

The recognition of reactions specific to living creatures dates back to at least the early 1800s, when scientists observed the digestion of meat by secretions from the stomachs of animals and the conversion of starch to sugar by extracts of plant tissues. In the 1850s, Louis Pasteur studied the conversion of sugar into alcohol by "ferments" in yeasts and concluded that the fermentation process required the intact yeast cell. Pasteur's ferments were later called *enzymes,* and, in 1897, Eduard Buchner first extracted these enzymes from yeast cells, allowing their activities to be studied *in vitro.*

A catalyst is a substance that causes a reaction to occur without its being modified in the process. Enzymes catalyze metabolic reactions by binding tightly to specific molecules, called *substrates,* and holding them in close juxtaposition so that the energy required for the reaction, the *activation energy,* is reduced. The net result is that covalent bonds can be rearranged at biological temperatures, the body temperatures of living organisms.

Although cell lysates were used extensively to study enzyme-catalyzed reactions throughout the first quarter of the twentieth century, the chemical nature of enzymes remained open to debate. Then, in 1926, James Sumner extracted the enzyme urease from jack beans, purified it to crystalline form, and demonstrated that it consisted entirely of a type of macromolecule called *protein.* Proteins were known to contain constituents called *amino acids* in the early 1900s, but it was not until the 1950s, when the structure of the small protein insulin was determined, that the complexity of each protein was appreciated. Amino acids are small organic molecules with a range of chemical structures, but all contain an amino ($-NH_2$) group and a carboxyl ($-COOH$) group. The 20 different amino acids present in proteins provide them with great structural diversity. A protein may contain a single chain of amino acids, called a *polypeptide,* or two or more chains (Chapter 13).

Thousands of enzymes have now been purified. Most are proteins with catalytic activity, but some are more complex, with additional components. Enzymes exhibit striking specificity—each enzyme catalyzes one reaction or a few very similar reactions. They range in size from about 12,000 to over 1 million daltons. The specificity of each enzyme results from its unique sequence of the 20 amino acids. Indeed, it was this essentially unlimited variability of protein structure that led many early geneticists to believe that proteins, not nucleic acids, stored genetic information.

The amino acid sequence of each polypeptide is controlled by one gene, and each polypeptide has either a structural, a regulatory, or an enzymatic role in the cell. Moreover, each enzyme catalyzes a specific metabolic reaction. However, the picture gets more complex at this point, because metabolic processes seldom involve just one enzyme-catalyzed reaction. Instead, the synthesis or degradation of a particular metabolite usually occurs by a series of enzyme-catalyzed reactions. The complete set of enzyme-catalyzed reactions required to synthesize or degrade a given metabolite is called a **metabolic pathway.** For example, consider the biosynthesis of arginine, one of the 20 amino acids present in most proteins. In *E. coli,* the synthesis of arginine occurs in eight steps, each catalyzed by a specific enzyme (Figure 12.1). The eight enzymes are composed of nine gene products; one enzyme, ornithine carbamoyltransferase, contains two different gene products (ArgF and ArgI).

In general, each step in a metabolic pathway requires the activity of an enzyme, and each enzyme contains one or more polypeptides. Since each polypeptide is specified by one gene, each of the steps in a metabolic pathway will require the product of at least one gene. A generalized picture of the genetic control of metabolic pathways is presented in Figure 12.2.

Whereas many genes specify polypeptide products that are components of enzymes, other genes encode polypeptides with structural or regulatory roles. Some structural and regulatory proteins are components of cell membranes; others are present in the cytoskeleton, an intracellular network of fibers that controls cell shape and participates in the intracellular transport of macromolecules, organelles, and other macromolecular assemblages. Although most genes encode proteins, the final products of some genes are

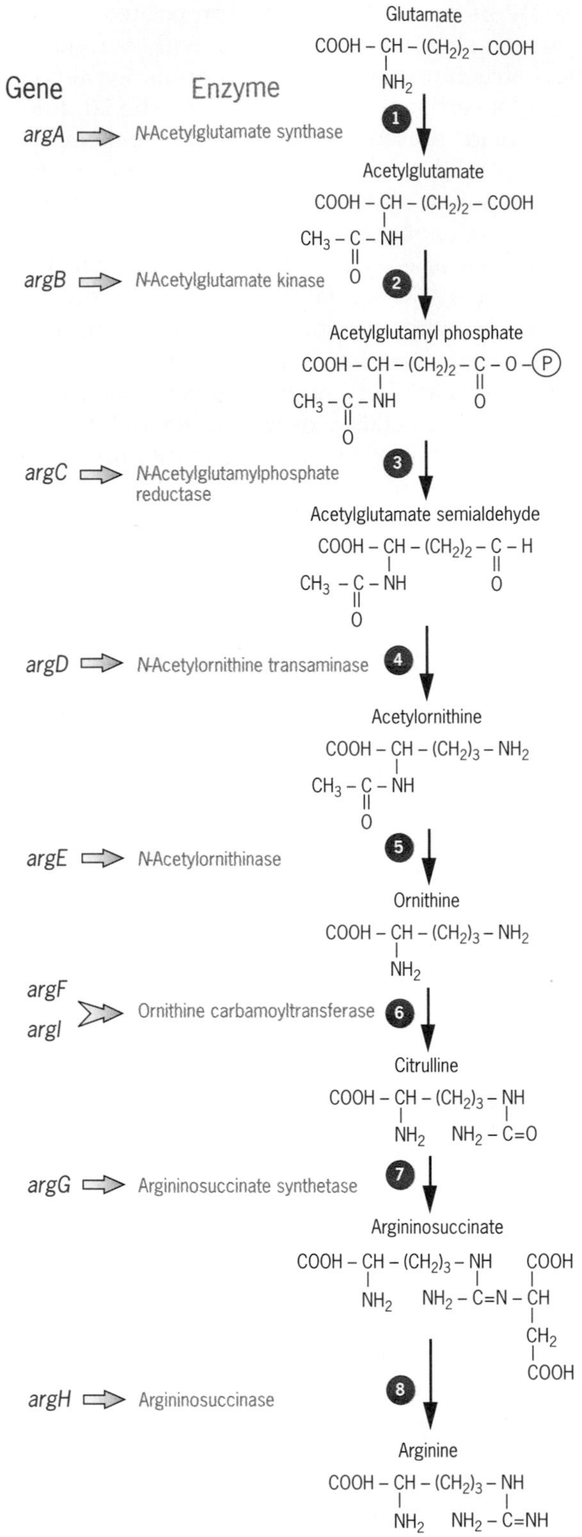

Gene	Enzyme	Step	Compound
			Glutamate: COOH – CH – $(CH_2)_2$ – COOH, with NH_2 on CH
argA	N-Acetylglutamate synthase	1	Acetylglutamate: COOH – CH – $(CH_2)_2$ – COOH, with CH_3 – C(=O) – NH on CH
argB	N-Acetylglutamate kinase	2	Acetylglutamyl phosphate: COOH – CH – $(CH_2)_2$ – C(=O) – O – (P), with CH_3 – C(=O) – NH on CH
argC	N-Acetylglutamylphosphate reductase	3	Acetylglutamate semialdehyde: COOH – CH – $(CH_2)_2$ – C(=O) – H, with CH_3 – C(=O) – NH on CH
argD	N-Acetylornithine transaminase	4	Acetylornithine: COOH – CH – $(CH_2)_3$ – NH_2, with CH_3 – C(=O) – NH on CH
argE	N-Acetylornithinase	5	Ornithine: COOH – CH – $(CH_2)_3$ – NH_2, with NH_2 on CH
argF, *argI*	Ornithine carbamoyltransferase	6	Citrulline: COOH – CH – $(CH_2)_3$ – NH, with NH_2 on CH and NH_2 – C=O on NH
argG	Argininosuccinate synthetase	7	Argininosuccinate: COOH – CH – $(CH_2)_3$ – NH, with NH_2 on CH; NH_2 – C=N – CH(COOH) – CH_2 – COOH on NH
argH	Argininosuccinase	8	Arginine: COOH – CH – $(CH_2)_3$ – NH, with NH_2 on CH and NH_2 – C=NH on NH

Figure 12.1 The genetic control of the arginine biosynthetic pathway in *E. coli*. Nine genes encode eight enzymes that catalyze the eight steps in this metabolic pathway.

RNA molecules. Several of these RNA molecules play essential roles in protein synthesis; we consider their functions in detail in Chapter 13. Because genes control the structures of RNAs and proteins, we should next ask how sequences of nucleotide pairs in DNA molecules specify the sequences of nucleotides in RNA and amino acids in protein molecules.

Key Points: Metabolism occurs by sequences of enzyme-catalyzed reactions, with each enzyme specified by one or more genes. Therefore, each step in a metabolic pathway is under genetic control.

TRANSFER OF GENETIC INFORMATION: THE CENTRAL DOGMA

According to the central dogma of molecular biology, genetic information flows (1) from DNA to DNA during its transmission from generation to generation and (2) from DNA to protein during its phenotypic expression in an organism (Figure 12.3). The transfer of genetic information from DNA to protein involves two steps: (1) **transcription,** the transfer of the genetic information from DNA to RNA, and (2) **translation,** the transfer of information from RNA to protein. In addition, genetic information flows from RNA to DNA during the conversion of the genomes of RNA tumor viruses to their DNA proviral forms (Chapter 16). Thus, the transfer of genetic information from DNA to RNA is sometimes reversible, whereas the transfer of information from RNA to protein is always irreversible.

Transcription and Translation

As we discussed above, the expression of genetic information occurs in two steps: transcription and translation (Figure 12.3). During transcription, one strand of DNA of a gene is used as a template to synthesize a complementary strand of RNA, called the gene **transcript.** For example, in Figure 12.3, the DNA strand containing the nucleotide sequence AAA is used as a template to produce the complementary sequence UUU in the RNA transcript. During translation, the sequence of nucleotides in the RNA transcript is converted into the sequence of amino acids in the polypeptide gene product. This conversion is governed by the **genetic code,** the specification of amino acids by nucleotide triplets called **codons** in the gene transcript. For example, the UUU triplet in the RNA transcript shown in Figure 12.3 specifies the amino acid phenylalanine (Phe) in the polypeptide gene product. Translation takes place on intricate macromolecular machines called **ribosomes,** which are composed of three to five RNA molecules and 50 to 90 different proteins. However, the process of translation also requires the

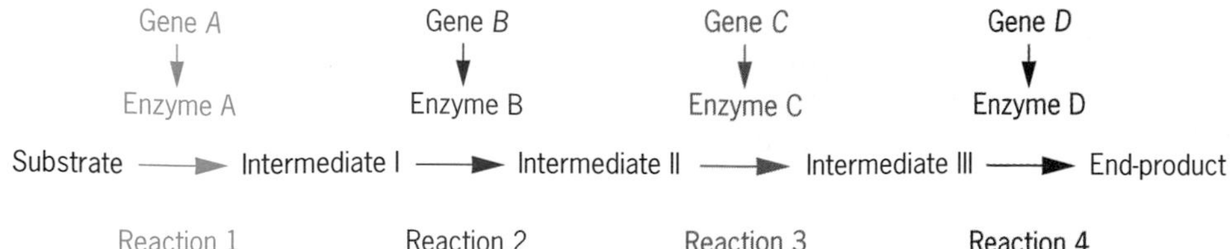

Figure 12.2 The genetic control of metabolism. The metabolic pathway shown contains four steps; some pathways are shorter and others are much longer. In addition, some enzymes contain the products of two or more different genes.

participation of many other macromolecules. This chapter focuses on transcription; translation is the subject of Chapter 13.

The RNA molecules that are translated on ribosomes are called **messenger RNAs (mRNAs).** In prokaryotes, the product of transcription, the **primary transcript,** usually is equivalent to the mRNA molecule (Figure 12.4*a*). In eukaryotes, primary transcripts often must be processed by the excision of specific sequences and the modification of both termini before they can be translated (Figure 12.4*b*). Thus, in eukaryotes, primary transcripts usually are precursors to mRNAs and, as such, are called **pre-mRNAs.** Most of the nuclear genes in higher eukaryotes and some in lower eukaryotes contain noncoding sequences called *introns* that separate the coding sequences or *exons* of these genes. The entire sequences of these *split genes* are transcribed into pre-mRNAs, and the noncoding intron sequences are subsequently removed by *splicing reactions* carried out on macromolecular structures called *spliceosomes*.

Four Types of RNA Molecules

Four different classes of RNA molecules play essential roles in gene expression. We have already discussed messenger RNAs, the intermediaries that carry genetic

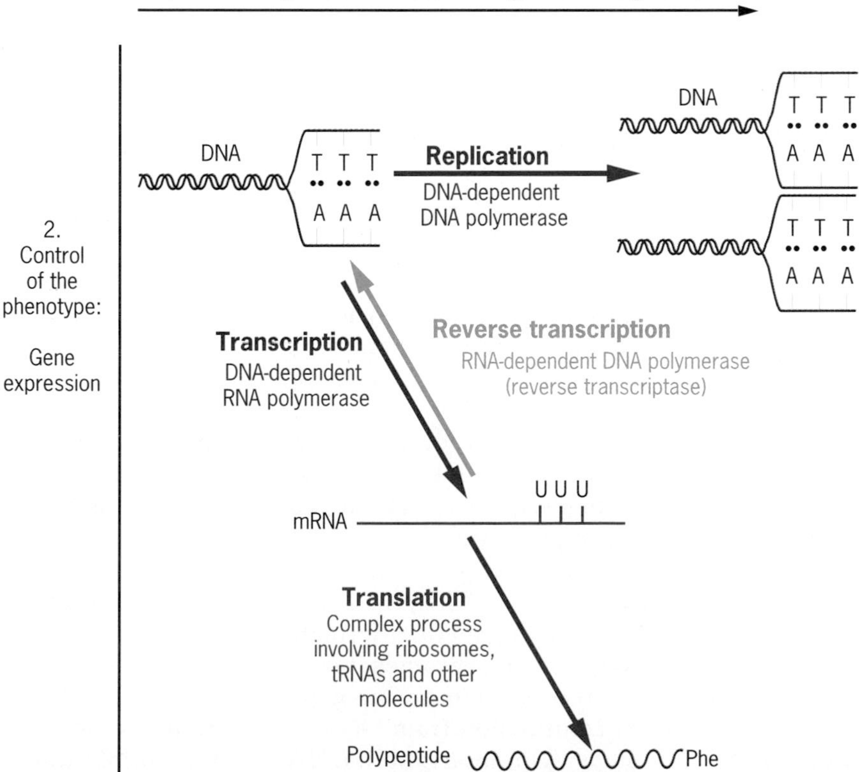

Figure 12.3 The flow of genetic information according to the central dogma of molecular biology. Replication, transcription, and translation occur in all organisms; reverse transcription occurs in cells infected with certain RNA viruses.

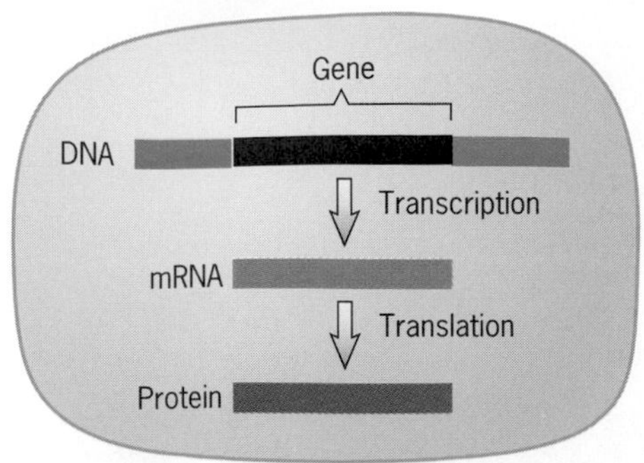

(*a*) Prokaryotes.

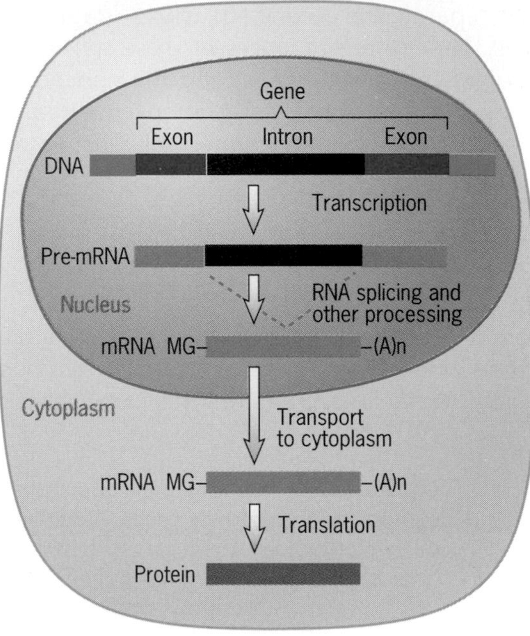

(*b*) Eukaryotes.

Figure 12.4 Protein synthesis involves two steps: transcription and translation in both prokaryotes (*a*) and eukaryotes (*b*). In addition, in eukaryotes, the primary transcripts or pre-mRNAs often must be processed by the excision of introns and the addition of 5′ 7-methyl guanosine caps (MG) and 3′ poly(A) tails $[(A)_n]$.

information from DNA to the ribosomes where proteins are synthesized. **Transfer RNAs (tRNAs)** are small RNA molecules that function as adaptors between amino acids and the codons in mRNA during translation. **Ribosomal RNAs (rRNAs)** are structural components of the ribosomes, the intricate machines that translate nucleotide sequences of mRNAs into amino acid sequences of polypeptides. **Small nuclear RNAs (snRNAs)** are structural components of spliceosomes, the nuclear structures that excise introns from nuclear genes. The roles of mRNAs and snRNAs are discussed in this chapter. Because we examine the structures and functions of tRNAs and rRNAs in detail in Chapter 13, the discussion here will be brief.

There are one to four tRNAs for each of the 20 amino acids specified by the genetic code. The tRNA molecules are 70 to 90 nucleotides long and fold into cloverleaf-shaped structures by intramolecular base-pairing. In its folded form, each tRNA has a specific amino acid attached at one end and a triplet nucleotide sequence called the **anticodon** at the other end. The recognition of nucleotide sequences in mRNAs by tRNAs is mediated by base-pairing between the anticodon in the tRNAs and the codon in the mRNA.

Ribosomal RNAs are essential components of ribosomes. The ribosomal RNAs interact with over 50 different ribosomal proteins to produce the complex three-dimensional structure of the ribosome. Prokaryotic ribosomes contain three rRNAs, about 120, 1540, and 2900 nucleotides in length, and about 50 proteins. Prokaryotic ribosomal RNA molecules are designated 5S, 16S, and 23S rRNAs based on their rates of sedimentation during sucrose gradient centrifugation (see Technical Sidelight in Chapter 11). Eukaryotic ribosomes are more complex than those of prokaryotes; they contain four rRNA molecules and about 80 different proteins. The four eukaryotic rRNAs, designated 5S, 5.8S, 18S, and 28S rRNAs, are approximately 120, 160, 1900, and 4700 nucleotides in length, respectively.

Five different small nuclear RNAs are present in spliceosomes. As the name implies, they are small in size, ranging from 100 to 215 nucleotides long in mammals. Unlike the other RNAs discussed here, snRNAs never leave the nucleus. Instead, they interact with about 40 nuclear proteins to form spliceosomes, and they play key roles in the excision of noncoding sequences from the transcripts of nuclear genes.

All four types of RNA—mRNA, tRNA, rRNA, and snRNA—are produced by transcription. Unlike mRNAs, which specify polypeptides, the final products of tRNA, rRNA, and snRNA genes are RNA molecules. Transfer RNA, ribosomal RNA, and snRNA molecules are not translated. Figure 12.5 shows an overview of protein synthesis in eukaryotes, emphasizing the transcriptional origin and functions of the four types of RNA molecules. The process is similar in prokaryotes. However, in prokaryotes, the DNA is not separated from the ribosomes by a nuclear envelope. In addition, prokaryotic genes seldom contain noncoding sequences that are removed during RNA transcript processing.

Key Points: **The central dogma of molecular biology is that genetic information flows from DNA to DNA during chromosome replication and from DNA to protein during gene expression. The flow of information from DNA to protein occurs in two steps: (1) transcription, DNA to RNA, and (2) translation, RNA to protein. Transcription involves the synthesis of an RNA transcript complementary to**

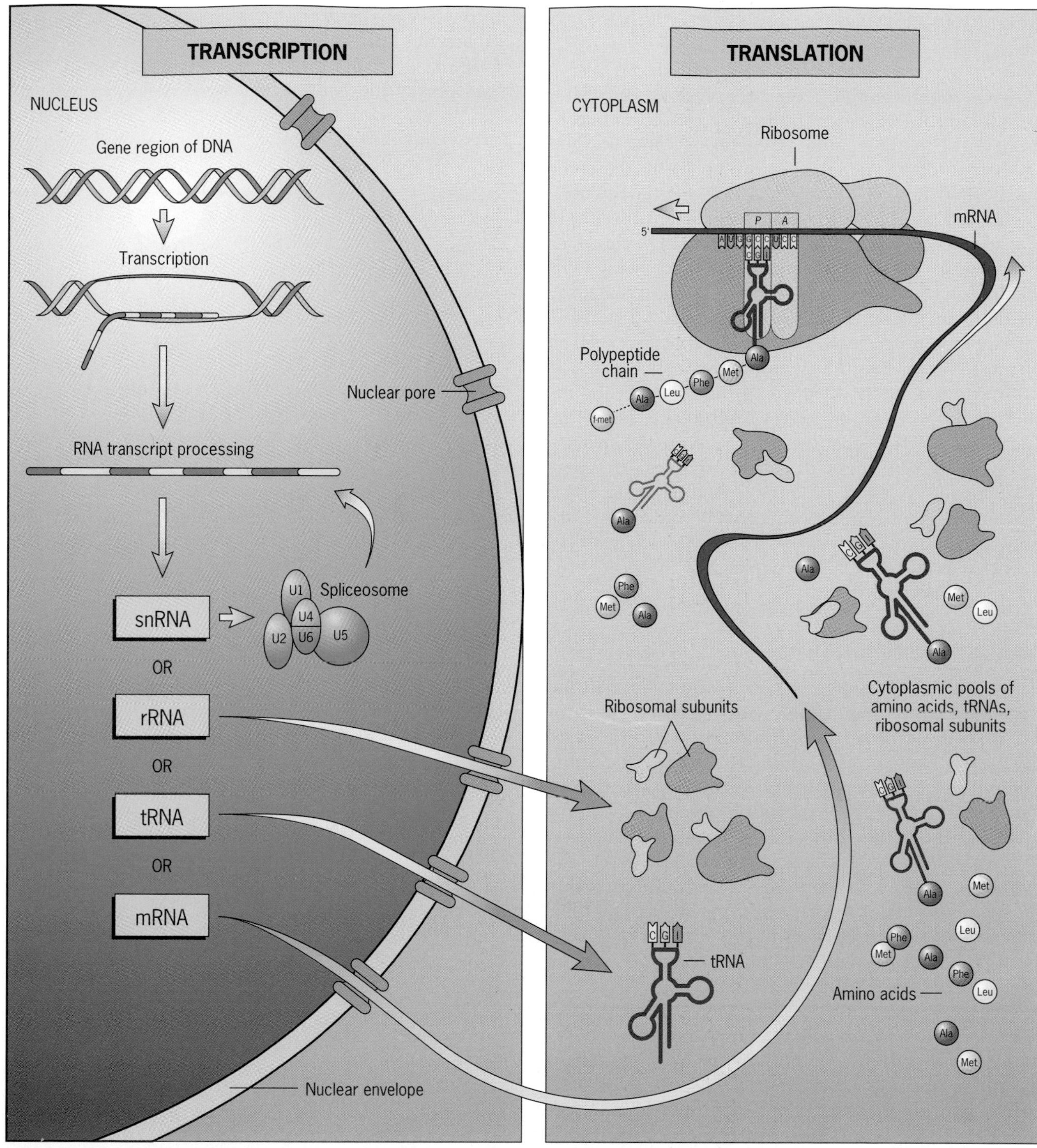

(*a*) Transcription and RNA processing occur in the nucleus.

(*b*) Translation occurs in the cytoplasm.

Figure 12.5 An overview of protein synthesis, emphasizing the transcriptional origin of snRNA, tRNA, rRNA, and mRNA, the splicing function of snRNA, and the translational roles of tRNA, rRNA, mRNA, and ribosomes.

one strand of DNA of a gene. Translation is the conversion of information stored in the sequence of nucleotides in the RNA transcript into the sequence of amino acids in the polypeptide gene product, according to the specifications of the genetic code.

THE PROCESS OF GENE EXPRESSION

How do genes control the phenotype of an organism? How do the nucleotide sequences of genes direct the growth and development of a cell, a tissue, an organ, or an entire living creature? Geneticists know that the phenotype of an organism is produced by the combined effects of all its genes, acting within the constraints imposed by the environment. They also know that the number of genes in an organism varies over an enormous range, with gene number increasing with the developmental complexity of the species. The RNA genomes of the smallest viruses such as phage MS2 contain only four genes, whereas large viruses such as phage T4 have about 200 genes. Bacteria such as *E. coli* have approximately 4000 genes, and mammals, including humans, have 70,000 to 100,000 genes. In this and the following chapter, we focus on the mechanisms by which genes direct the synthesis of their products, RNAs and proteins. The mechanisms by which these gene products collectively control the phenotypes of mature organisms are discussed in subsequent chapters, especially Chapter 25.

An mRNA Intermediary

If most of the genes of a eukaryote are located in the nucleus, and if proteins are synthesized in the cytoplasm, how do these genes control the amino acid sequences of their protein products? The genetic information stored in the sequences of nucleotide pairs in genes must somehow be transferred to the sites of protein synthesis in the cytoplasm. Messengers are needed to transfer genetic information from the nucleus to the cytoplasm. Although the need for such messengers is most obvious in eukaryotes, the first evidence for their existence came from studies of prokaryotes (see Technical Sidelight: An mRNA Intermediary: Evidence from Phage-Infected *E. coli*).

The synthesis of messenger RNAs or precursors to mRNAs in the nuclei of eukaryotes and their subsequent transport to the cytoplasm can be documented by pulse-labeling experiments, pulse-chase experiments, and autoradiography. If a cell growing in culture is exposed to a radioactive RNA precursor such as ^{3}H-uridine or ^{3}H-cytidine for a few minutes, and the intracellular location of the incorporated radioactivity is determined by autoradiography, the labeled RNA is present almost exclusively in the nucleus (Figure 12.6*a*). However, if the short exposure to the radioactive precursor is followed by a period of growth in nonradioactive medium, most of the incorporated radioactivity is present in the cytoplasm (Figure 12.6*b*). Thus, the unstable RNA intermediaries are synthesized in the nucleus and are transported to the cytoplasm, where they direct the synthesis of proteins. In this chapter, we focus on the synthesis, processing, and transport of these messenger RNA molecules.

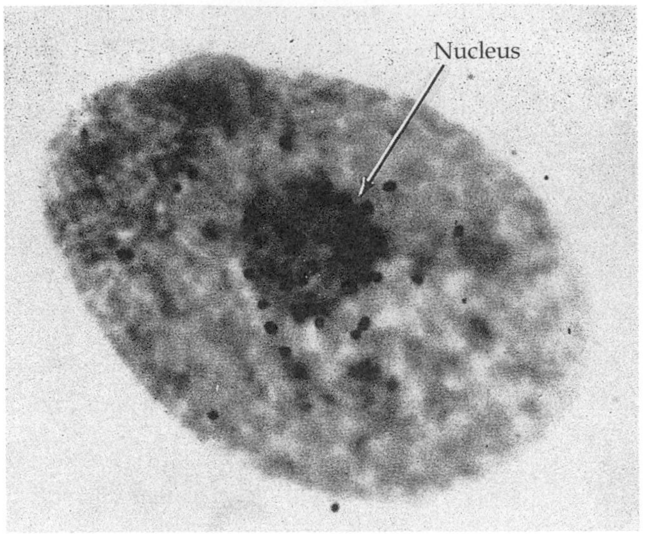

(*a*)

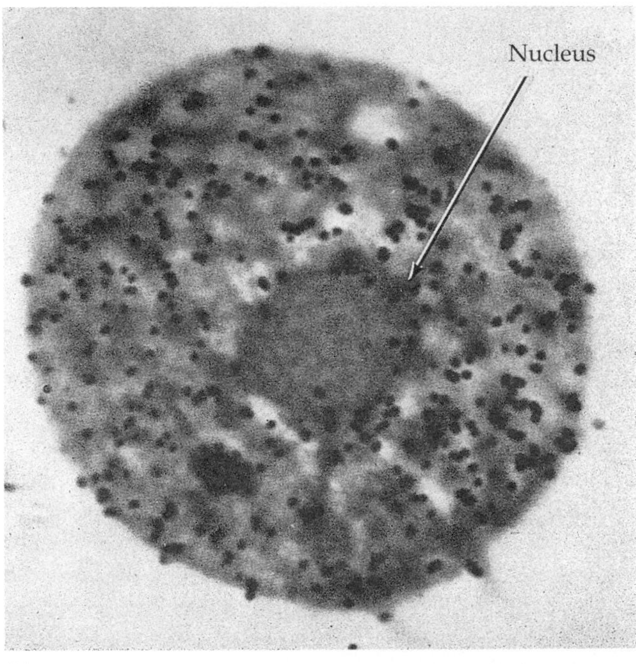

(*b*)

Figure 12.6 Autoradiographs demonstrating (1) the synthesis of RNA in the nucleus and (2) its subsequent transport to the cytoplasm. Each autoradiograph is superimposed on a photograph of a thin section of the cell. The black dots represent silver granules in the autoradiographic emulsion that have reacted with electrons emitted by the decay of ^{3}H atoms. (*a*) A *Tetrahymena* cell labeled with ^{3}H-cytidine for 15 minutes. (*b*) A *Tetrahymena* cell that was grown on nonradioactive medium for 88 minutes after exposure to ^{3}H-cytidine for 12 minutes.

TECHNICAL SIDELIGHT

An mRNA Intermediary: Evidence from Phage-Infected *E. coli*

The first evidence for the existence of an RNA intermediary in protein synthesis came from studies by Elliot Volkin and Lawrence Astrachan on bacteria infected with bacterial viruses. Their results, published in 1956, suggested that the synthesis of viral proteins in infected bacteria involved unstable RNA molecules specified by viral DNA. Volkin and Astrachan observed a burst of RNA synthesis after infecting *E. coli* cells with bacteriophage T2. By labeling RNA with the radioactive isotope ^{32}P, they demonstrated that the newly synthesized RNA molecules were unstable, turning over with half-lives of only a few minutes. In addition, they showed that the nucleotide composition of the unstable RNAs was similar to the composition of T2 DNA and unlike that of *E. coli* DNA. Their results were soon extended by studies in other laboratories.

In 1961, Sol Spiegelman and coworkers reported that the unstable RNAs synthesized in phage T4-infected cells could form RNA–DNA duplexes with denatured T4 DNA, but not with denatured *E. coli* DNA. They pulse-labeled bacteria with ^3H-uridine at various times after infection with T4 phage, isolated total RNA from these cells, and determined whether the radioactive RNA molecules hybridized with *E. coli* DNA or phage T4 DNA. Their experiment is diagrammed in Figure 1.

(*Sidelight continues on next page.*)

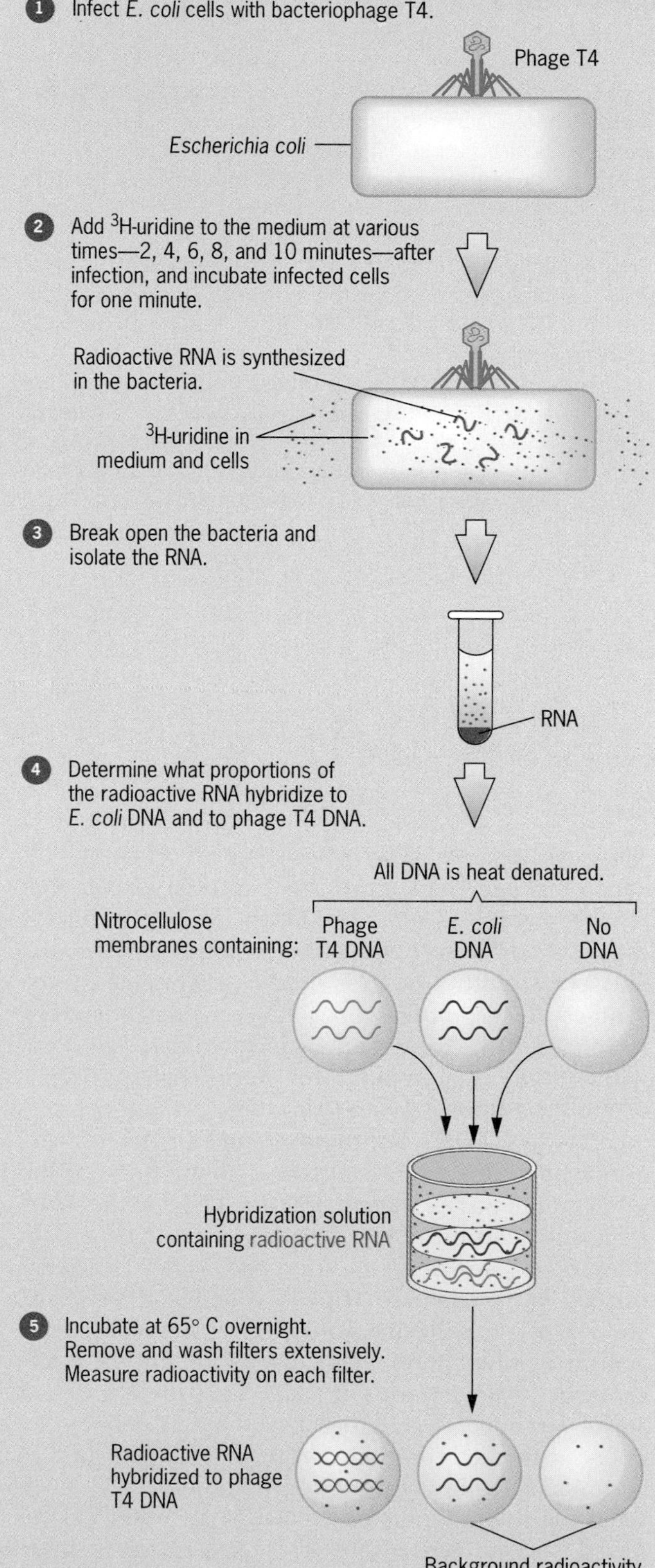

Figure 1 Spiegelman's experiment.

Their results (Figure 2) demonstrated that most of the short-lived RNA molecules synthesized after infection were complementary to single strands of phage T4 DNA and noncomplementary to single strands of *E. coli* DNA, indicating that they were produced from phage T4 DNA templates, not from *E. coli* DNA templates.

In the same year that Spiegelman and colleagues published their results, Sydney Brenner, François Jacob, and Matthew Meselson demonstrated that phage T4 proteins were synthesized on *E. coli* ribosomes. Thus, the amino acid sequences of T4 proteins were not controlled by components of the ribosomes. Instead, the ribosomes provided the workbenches on which protein synthesis occurred, but did not provide the specifications for individual proteins. These results strengthened the idea, first formally proposed by François Jacob and Jacques Monod in 1961, that unstable RNA molecules carried the specifications for the amino acid sequences of individual gene products from the genes to the ribosomes. Subsequent research firmly established the role of these unstable RNAs, now called messenger RNAs or mRNAs, in the transfer of genetic information from genes to the sites of protein synthesis in the cytoplasm.

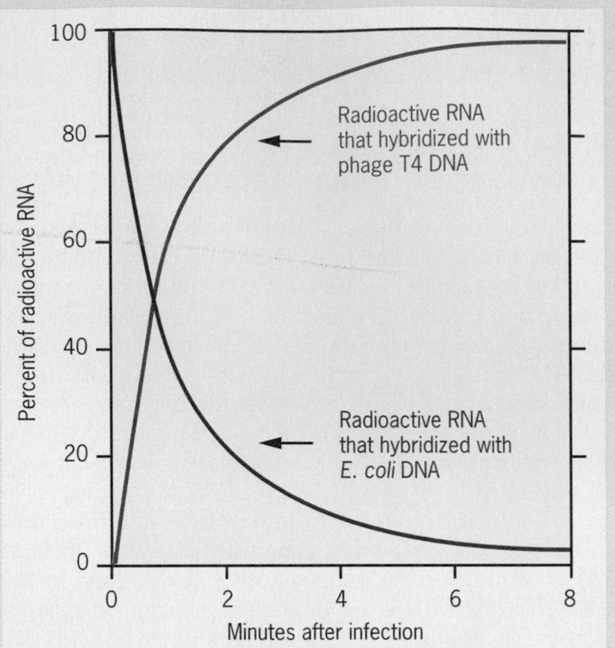

Figure 2 Rapid switch from the transcription of *E. coli* genes to phage T4 genes in T4-infected bacteria.

General Features of RNA Synthesis

RNA synthesis occurs by a mechanism that is very similar to that of DNA synthesis (Chapter 11) except that (1) the precursors are **ribonucleoside triphosphates** rather than deoxyribonucleoside triphosphates, (2) only one strand of DNA is used as a template for the synthesis of a complementary RNA chain in any given region, and (3) RNA chains can be initiated *de novo,* without any requirement for a preexisting primer strand. The RNA molecule produced will be complementary to the DNA **template strand** and identical, except that uridine residues replace thymidines, to the DNA **nontemplate strand** (Figure 12.7). If the RNA molecule is an mRNA, it will specify amino acids in the protein gene product. This specification is accomplished by nucleotide triplets called codons. Each codon specifies the incorporation of a single amino acid in a polypeptide chain. Therefore, mRNA molecules are coding strands of RNA. They are also called **sense strands** of RNA because their nucleotide sequences "make sense" in that they specify sequences of amino acids in the protein gene products. An RNA molecule that is complementary to an mRNA is referred to as **antisense RNA.** This terminology is sometimes extended to the two strands of DNA. However, usage of the terms *sense* and *antisense* to denote DNA strands has been inconsistent. Thus, we will use *template strand* and *nontemplate strand* to refer to the transcribed and nontranscribed strands, respectively, of a gene.

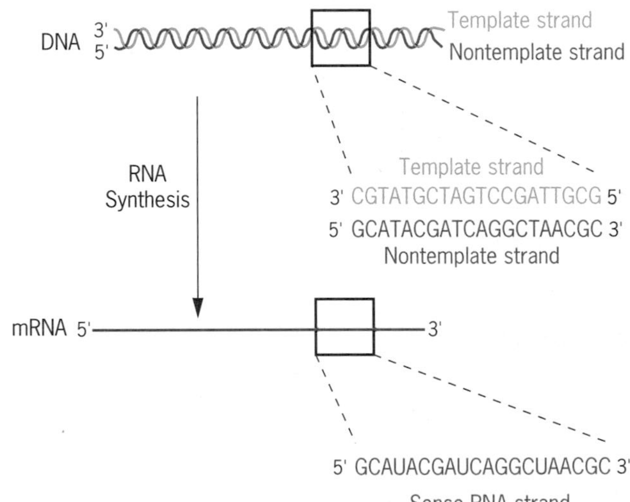

Figure 12.7 RNA synthesis utilizes only one DNA strand of a gene as template.

The synthesis of RNA chains, like DNA chains, occurs in the $5' \rightarrow 3'$ direction, with the addition of ribonucleotides to the 3′-hydroxyl group at the end of the chain (Figure 12.8). The reaction involves a nucleophilic attack by the 3′-OH on the nucleotidyl (interior) phosphorus atom of the ribonucleoside triphosphate precursor with the elimination of pyrophosphate, just as in DNA synthesis. This reaction is catalyzed by enzymes called **RNA polymerases.** The overall reaction is as follows:

$$n(\text{RTP}) \xrightarrow[\text{RNA polymerase}]{\text{DNA template}} (\text{RMP})_n + n(\text{PP})$$

where n is the number of moles of ribonucleotide triphosphate (RTP) consumed, ribonucleotide monophosphate (RMP) incorporated into RNA, and pyrophosphate (PP) produced.

RNA polymerases initiate transcription at specific nucleotide sequences called **promoters,** which are different in prokaryotes and eukaryotes. A single RNA polymerase carries out all transcription in most prokaryotes, whereas three different RNA polymerases are present in eukaryotes, with each polymerase responsible for the synthesis of a distinct class of RNAs. RNA synthesis takes place within a locally unwound segment of DNA, sometimes called a **transcription**

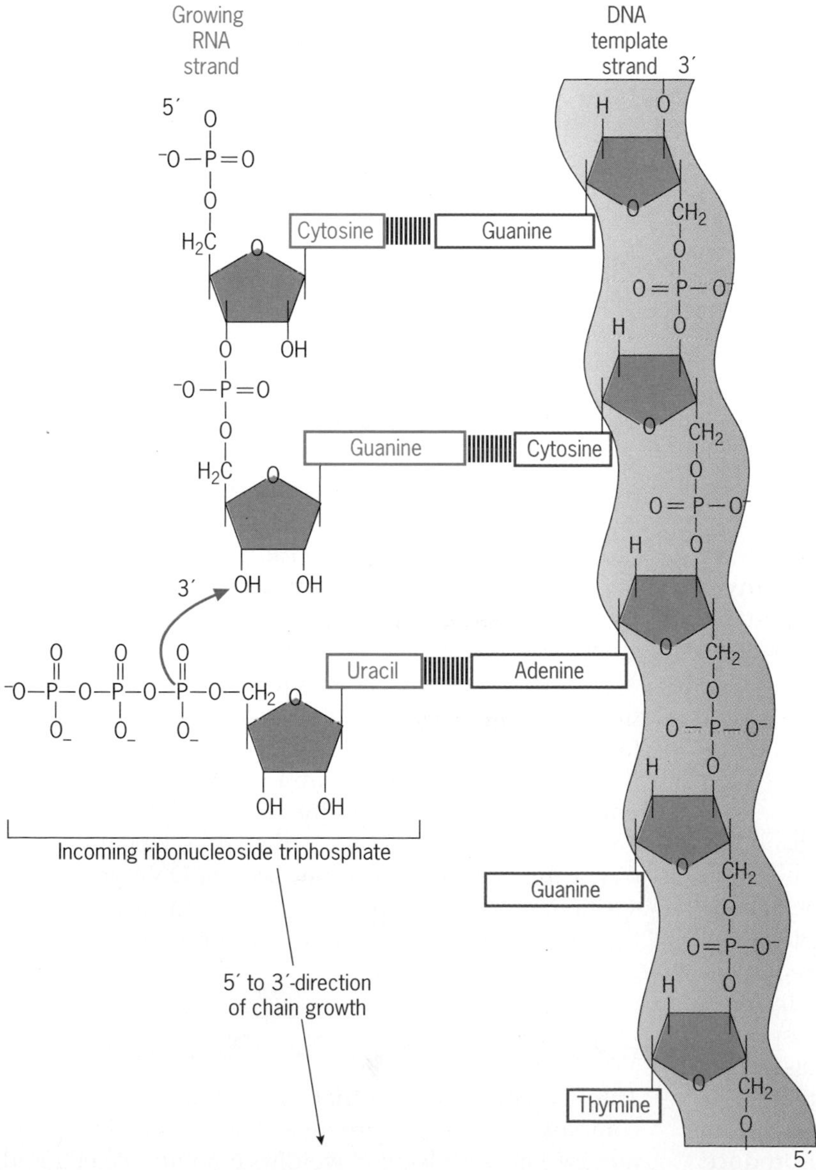

Figure 12.8 The RNA chain elongation reaction catalyzed by RNA polymerase.

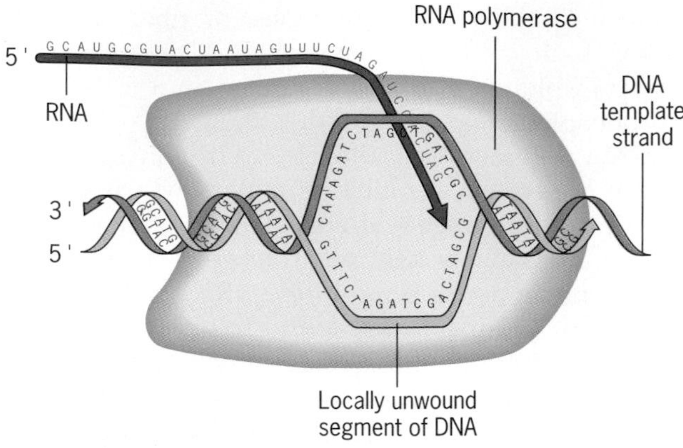

Figure 12.9 RNA synthesis occurs within a locally unwound segment of DNA. This *transcription bubble* allows a few nucleotides in the template strand to base-pair with the growing end of the RNA chain. The unwinding and rewinding of the DNA molecule are catalyzed by RNA polymerase.

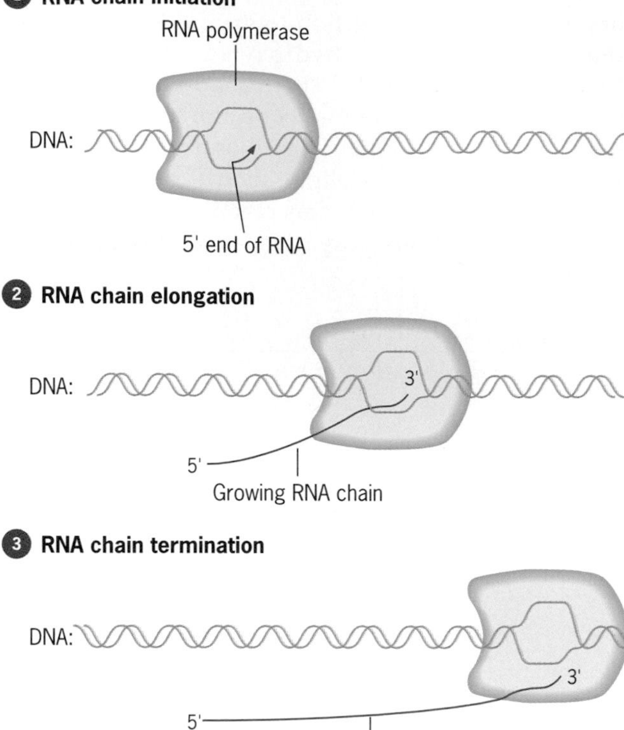

Figure 12.10 The three stages of transcription: initiation, elongation, and termination.

bubble, which is produced by RNA polymerase (Figure 12.9). The nucleotide sequence of an RNA molecule is complementary to that of its DNA template strand, and RNA synthesis is governed by the same base-pairing rules as DNA synthesis, but uracil replaces thymine.

Key Points: **In eukaryotes, genes are present in the nucleus, whereas polypeptides are synthesized in the cytoplasm. Messenger RNA molecules function as intermediaries that carry genetic information from DNA to the ribosomes, where proteins are synthesized. RNA synthesis, catalyzed by RNA polymerases, is similar to DNA synthesis in many respects. However, RNA synthesis occurs within a localized region of strand separation, and only one strand of DNA functions as a template for RNA synthesis.**

TRANSCRIPTION IN PROKARYOTES

The basic features of transcription are the same in both prokaryotes and eukaryotes, but many of the details—such as the promoter sequences—are different. The RNA polymerase of *E. coli* has been studied in great detail. It catalyzes all RNA synthesis in this species. There are about 7000 copies of RNA polymerase in an *E. coli* cell, and the majority of them are engaged in RNA synthesis.

A segment of DNA that is transcribed to produce one RNA molecule is called a **transcription unit.** Transcription units may be equivalent to individual genes, or they may include several contiguous genes. Large transcripts that carry the coding sequences of several genes are common in bacteria. The process of transcription can be divided into three stages: (1) initiation of a new RNA chain, (2) elongation of the chain, and (3) termination of transcription and release of the nascent RNA molecule (Figure 12.10).

When discussing transcription, biologists often use the terms *upstream* and *downstream* to refer to regions located toward the 5′ end or the 3′ end, respectively, of the transcript from some site in the mRNA molecule. These terms are based on the fact that RNA synthesis always occurs in the 5′ to 3′ direction. Upstream and downstream regions of genes are the DNA sequences specifying the corresponding 5′ and 3′ segments of their transcripts relative to a specific reference point.

RNA Polymerase: A Complex Enzyme

The RNA polymerases that catalyze transcription are complex, multimeric proteins. The *E. coli* RNA polymerase has a molecular weight of about 480,000 and consists of five polypeptides. Two of these are identical; thus, the enzyme contains four distinct polypeptides. The complete RNA polymerase molecule, the **holoenzyme,** has the composition $\alpha_2\ \beta\ \beta'\sigma$. The α subunits are involved in the assembly of the **tetrameric core** ($\alpha_2\ \beta\ \beta'$) of RNA polymerase. The β subunit contains the ribonucleoside triphosphate binding site, and the β' subunit harbors the DNA template binding region.

One subunit, the **sigma (σ) factor,** is involved only in the initiation of transcription; it plays no role in chain elongation. After RNA chain initiation has occurred, the σ factor is released, and chain elongation (see Figure 12.8) is catalyzed by the core enzyme ($\alpha_2\ \beta\beta'$). The function of sigma is to recognize and bind RNA polymerase to the transcription initiation or **promoter sites** in DNA. The core enzyme (with no σ) will catalyze RNA synthesis from DNA templates *in vitro,* but, in so doing, it will initiate RNA chains at random sites on both strands of DNA. In contrast, the holoenzyme (σ present) initiates RNA chains *in vitro* only at sites used *in vivo.*

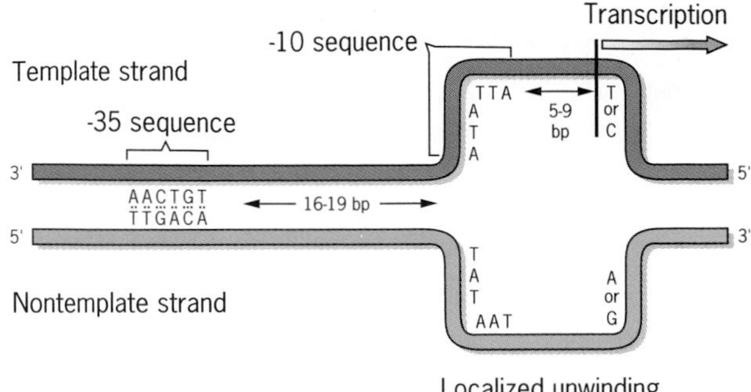

Figure 12.11 Structure of a typical promoter in *E. coli.* RNA polymerase binds to the –35 sequence of the promoter and initiates unwinding of the DNA strands at the AT-rich –10 sequence. Transcription begins within the transcription bubble at a site five to nine base pairs beyond the –10 sequence.

Initiation of RNA Chains

Initiation of RNA chains involves three steps: (1) binding of the RNA polymerase holoenzyme to a promoter region in DNA; (2) the localized unwinding of the two strands of DNA by RNA polymerase, providing a template strand free to base-pair with incoming ribonucleotides; and (3) the formation of phosphodiester bonds between the first few ribonucleotides in the nascent RNA chain. The holoenzyme remains bound at the promoter region during the synthesis of the first eight or nine bonds; then the sigma factor is released, and the core enzyme begins the elongation phase of RNA synthesis. During initiation, short chains of two to nine ribonucleotides are synthesized and released. This abortive synthesis stops once chains of ten or more ribonucleotides have been synthesized and RNA polymerase has begun to move downstream from the promoter.

As mentioned earlier, the sigma subunit of RNA polymerase mediates its binding to promoters in DNA. Hundreds of *E. coli* promoters have been sequenced and found to have surprisingly little in common. Two short sequences within these promoters are sufficiently conserved to be recognized, but even these are seldom identical in two different promoters. The midpoints of the two conserved sequences occur at about 10 and 35 nucleotide pairs, respectively, before the transcription-initiation site (Figure 12.11). Thus they are called the **−10 sequence** and the **−35 sequence,** respectively. Although these sequences vary slightly from gene to gene, some nucleotides are highly conserved. The nucleotide sequences that are present in such conserved genetic elements most often are called **consensus sequences.** The −10 consensus sequence in the nontemplate strand is TATAAT; the −35 consensus sequence is TTGACA. The sigma subunit initially recognizes and binds to the −35 sequence; thus, this sequence is sometimes called the **recognition sequence.** The AT-rich −10 sequence facilitates the localized unwinding of DNA, which is an essential prerequisite to the synthesis of a new RNA chain. The distance between the −35 and −10 sequences is highly conserved in *E. coli* promoters, never being less than 15 or more than 20 nucleotide pairs in length. In addition, the first or 5′ base in *E. coli* RNAs is usually (>90%) a purine. Except for these short conserved sequences, different *E. coli* promoters have little in common.

Elongation of RNA Chains

Elongation of RNA chains is catalyzed by the RNA polymerase core enzyme, after the release of the σ subunit. The covalent extension of RNA chains (see Figure 12.8) takes place within the transcription bubble, a locally unwound segment of DNA. The RNA polymerase molecule contains both DNA unwinding and DNA rewinding activities. RNA polymerase continuously unwinds the DNA double helix ahead of the polymerization site and rewinds the complementary DNA strands behind the polymerization site as it moves along the double helix (Figure 12.12). In *E. coli,* the average length of a transcription bubble is 18 nucleotide pairs, and about 40 ribonucleotides are incorporated into the growing RNA chain per second. The nascent RNA chain is displaced from the DNA tem-

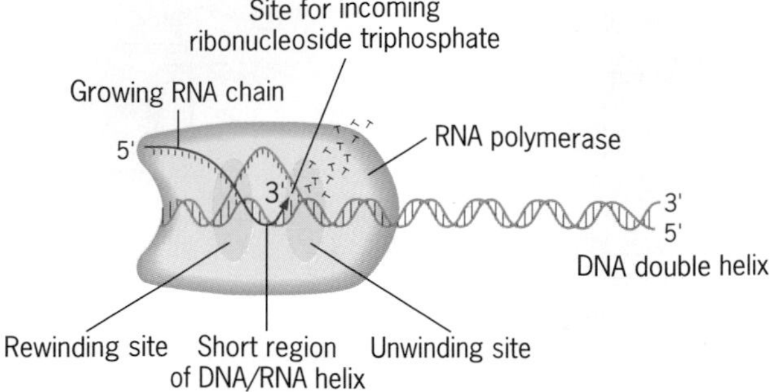

(*a*) RNA polymerase is bound to DNA and is covalently extending the RNA chain.

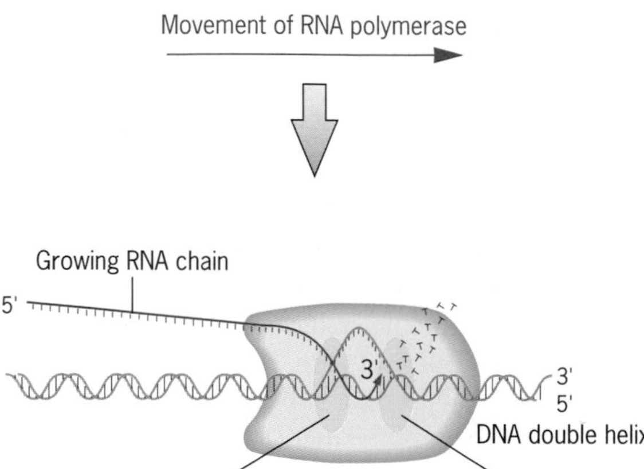

(*b*) RNA polymerase has moved downstream from its position in (*a*) processively extending the nascent RNA chain.

Figure 12.12 Elongation of an RNA chain catalyzed by RNA polymerase in *E. coli*.

plate strand as RNA polymerase moves along the DNA molecule. The region of transient base-pairing between the growing chain and the DNA template strand is very short, perhaps only three base pairs in length. The stability of the transcription complex is maintained primarily by the binding of the DNA and the growing RNA chain to RNA polymerase, rather than by the base-pairing between the template strand of DNA and the nascent RNA.

Termination of RNA Chains

Termination of RNA chains occurs when RNA polymerase encounters a **termination signal.** When it does, the transcription complex dissociates, releasing the nascent RNA molecule. There are two types of transcription terminators in *E. coli*. One type results in termination only in the presence of a protein called *rho* (ρ); therefore, such termination sequences are called *rho-dependent terminators*. The other type results in the termination of transcription without the involvement of rho; such sequences are called *rho-independent terminators*.

Rho-independent terminators contain a G:C-rich region followed by six or more A:T base pairs, with the A's present in the template strand (Figure 12.13, top). The nucleotide sequence of the G:C-rich region is such that regions of the single-stranded RNA can base-pair and form hairpinlike structures (Figure 12.13, bottom). The RNA hairpin structures form immediately after the synthesis of the participating regions of the RNA chain and retard the movement of RNA polymerase molecules along the DNA, causing pauses in chain extension. Since A:U base-pairing is weak, requiring less energy to separate the bases than any of the other standard base pairs, the run of U's after the hairpin region is thought to facilitate the release of the newly synthesized RNA chains from the DNA template when the hairpin structure causes RNA polymerase to pause at this site.

The mechanism by which rho-dependent termination of transcription occurs is still uncertain. Rho-dependent termination sequences are 50 to 90 base pairs long and specify RNA transcripts that are rich in C residues and largely devoid of G's. Beyond that, different rho-dependent termination signals have little in common. The rho protein binds to the growing RNA chain and moves 5′ to 3′ along the RNA, seeming to pursue the RNA polymerase molecule catalyzing the synthesis of the chain. When RNA polymerase slows down or pauses at the rho-dependent termination sequence, rho catches up with the polymerase and pulls the nascent RNA chain from the transcription bubble.

Concurrent Transcription, Translation, and mRNA Degradation

In prokaryotes, the translation and degradation of an mRNA molecule often begin before its synthesis (transcription) is complete. Since mRNA molecules are synthesized, translated, and degraded in the 5′ to 3′ direction, all three processes can occur simultaneously on the same RNA molecule. In prokaryotes, the polypeptide-synthesizing machinery is not separated by a nuclear envelope from the site of mRNA synthesis. Therefore, once the 5′ end of an mRNA has been synthesized, it can immediately be used as a template for polypeptide synthesis. Indeed, transcription and translation often are tightly coupled in prokaryotes. Oscar Miller, Barbara Hamkalo, and colleagues developed electron microscopy techniques that allowed them to visualize this coupling between transcription and translation in bacteria directly. One of their pho-

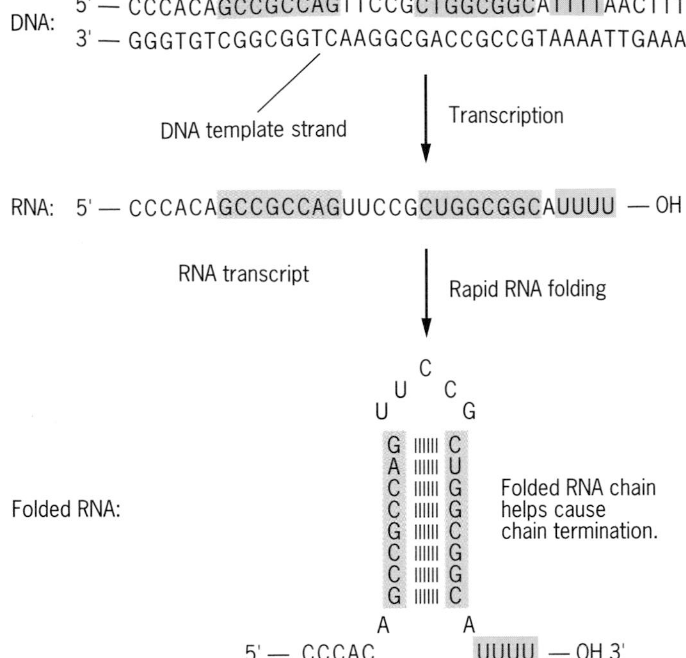

Figure 12.13 Structure of a rho-independent transcription terminator. Rho-independent terminator sequences contain a G:C-rich region followed by at least six A:T base pairs. Transcription of such a terminator sequence produces an RNA chain with G:C-rich segments that base-pair with each other immediately after synthesis to form a hairpinlike structure. This structure retards the movement of RNA polymerase along the DNA molecule, which results in the termination of transcription in the adjacent A:T tract and the release of the nascent RNA chain.

tographs showing the coupled transcription of a gene and translation of its mRNA product in *E. coli* is reproduced in Figure 12.14.

Key Points: **RNA synthesis occurs in three stages: (1) initiation, (2) elongation, and (3) termination. The RNA polymerases that catalyze transcription are complex, multimeric proteins. Covalent extension of RNA chains occurs within transcription bubbles, locally unwound segments of DNA. Chain elongation stops when RNA polymerase encounters transcription–termination signals in DNA. Transcription, translation, and degradation of a specific mRNA molecule often occur simultaneously in prokaryotes.**

TRANSCRIPTION AND RNA PROCESSING IN EUKARYOTES

Although the overall process of RNA synthesis is similar in prokaryotes and eukaryotes, the process is considerably more complex in eukaryotes. In eukaryotes, RNA is synthesized in the nucleus, and RNAs that encode proteins must be transported to the cytoplasm for

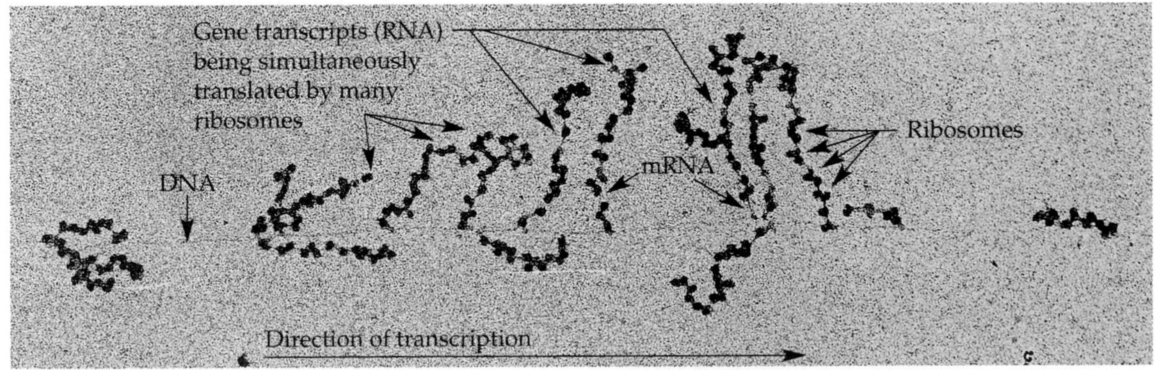

Figure 12.14 Electron micrograph prepared by Oscar Miller and Barbara Hamkalo showing the coupled transcription and translation of a gene in *E. coli*. DNA, mRNAs, and the ribosomes translating individual mRNA molecules are visible. The nascent polypeptide chains being synthesized on the ribosomes are not visible as they fold into their three-dimensional configuration during synthesis.

translation on ribosomes. Prokaryotic mRNAs often contain the coding regions of two or more genes; such mRNAs are said to be multigenic. In contrast, many of the eukaryotic transcripts that have been characterized contain the coding region of a single gene (are monogenic). However, recent evidence indicates that up to one-fourth of the transcription units in the small worm *Caenorhabditis elegans* may be multigenic. Clearly, eukaryotic mRNAs may be either monogenic or multigenic. Three different RNA polymerases are present in eukaryotes; each enzyme catalyzes the transcription of a specific class of genes. Moreover, in eukaryotes, the majority of the primary transcripts of genes that encode polypeptides undergo three major modifications prior to their transport to the cytoplasm for translation (Figure 12.15).

1. 7-Methyl guanosine caps are added to the 5′ ends of the primary transcripts.
2. Poly(A) tails are added to the 3′ ends of the transcripts, which are generated by cleavage rather than by termination of chain extension.
3. When present, noncoding intron sequences are spliced out of transcripts.

The **5′ cap** on most eukaryotic mRNAs is a 7-methyl guanosine residue joined to the initial nucleoside of the transcript by a 5′-5′ phosphate linkage. The 3′

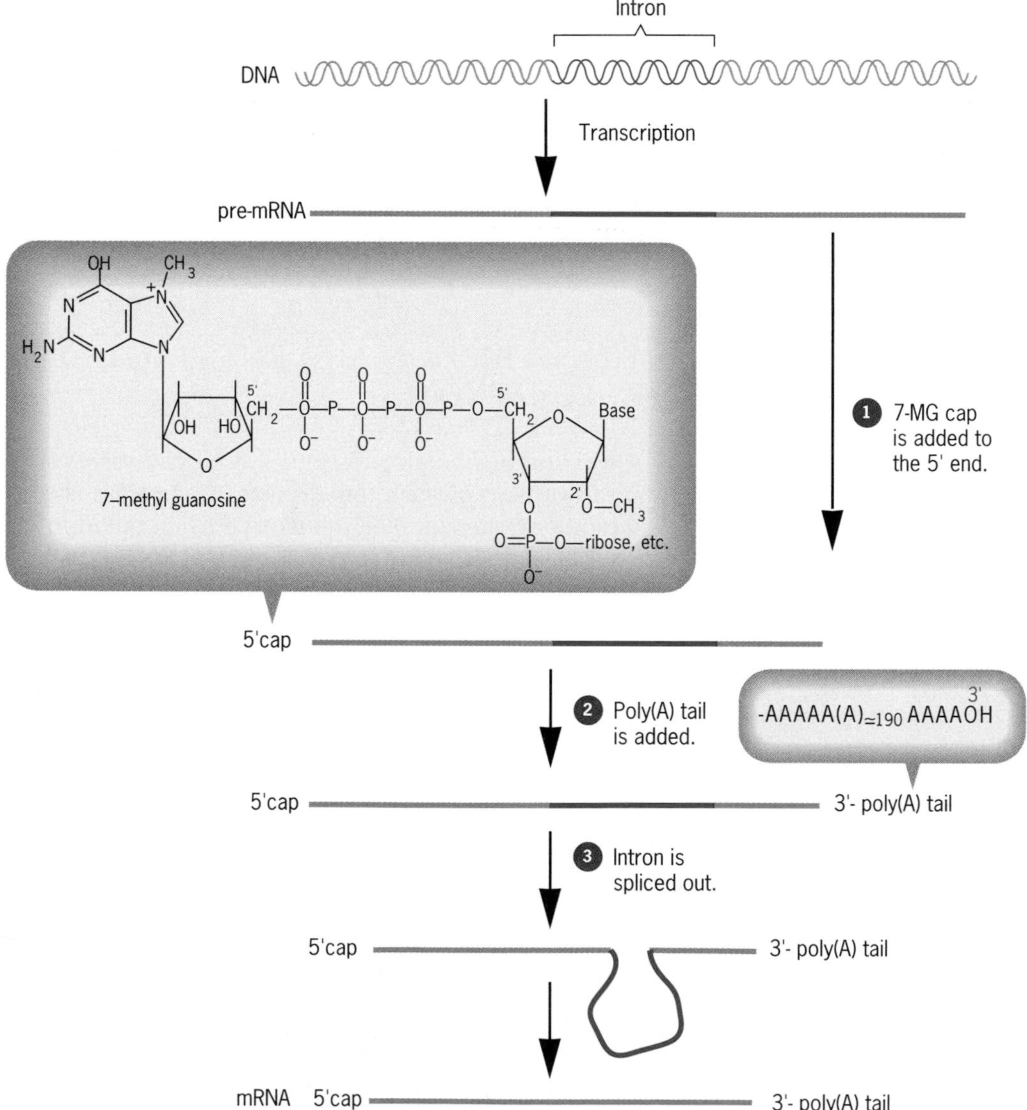

Figure 12.15 In eukaryotes, most gene transcripts undergo three different types of post-transcriptional processing.

poly(A) tail is a polyadenosine tract 20 to 200 nucleotides long.

In eukaryotes, the population of primary transcripts in a nucleus is called **heterogeneous nuclear RNA (hnRNA)** because of the large variation in the sizes of the RNA molecules present. Major portions of these hnRNAs are noncoding intron sequences, which are excised from the primary transcripts and degraded in the nucleus. Thus, much of the hnRNA actually consists of pre-mRNA molecules undergoing various processing events before leaving the nucleus. Also, in eukaryotes, RNA transcripts are coated with RNA-binding proteins during or immediately after their synthesis. These proteins protect gene transcripts from degradation by ribonucleases, enzymes that degrade RNA molecules, during processing and transport to the cytoplasm. The average half-life of a gene transcript in eukaryotes is about five hours, in contrast to an average half-life of less than five minutes in *E. coli*. This enhanced stability of gene transcripts in eukaryotes is provided, at least in part, by their presence in complexes with RNA-binding proteins.

Three RNA Polymerases/ Three Sets of Genes

Whereas a single RNA polymerase catalyzes all transcription in *E. coli*, eukaryotes ranging in complexity from the single-celled yeasts to humans contain three different RNA polymerases. All three eukaryotic enzymes, designated **RNA polymerases I, II,** and **III,** are more complex, with 10 or more subunits, than the *E. coli* RNA polymerase. Moreover, unlike the *E. coli* enzyme, all three eukaryotic RNA polymerases require the assistance of other proteins called **transcription factors** in order to initiate the synthesis of RNA chains.

The key features of the three eukaryotic RNA polymerases are summarized in Table 12.1. RNA polymerase I is located in the nucleolus, a distinct region of the nucleus where rRNAs are synthesized and combined with ribosomal proteins. RNA polymerase I catalyzes the synthesis of all ribosomal RNAs except the small 5S rRNA. RNA polymerase II transcribes nuclear genes that encode proteins and perhaps other genes specifying hnRNAs. RNA polymerase III catalyzes the synthesis of the transfer RNA molecules, the 5S rRNA molecules, and small nuclear RNAs.

The three RNA polymerases exhibit very different sensitivities to inhibition by α-amanitin, a metabolic poison produced by the mushroom *Amanita phalloides*. Whereas RNA polymerase I activity is insensitive to α-amanitin, RNA polymerase II activity is inhibited completely by low concentrations of α-amanitin, and RNA polymerase III exhibits an intermediate level of sensitivity to this drug. Thus, α-amanitin can be used to determine which RNA polymerase catalyzes the transcription of a particular gene.

Initiation of RNA Chains

Unlike their prokaryotic counterparts, eukaryotic RNA polymerases cannot initiate transcription by themselves. All three eukaryotic RNA polymerases require the assistance of protein transcription factors to start the synthesis of an RNA chain. Indeed, these transcription factors must bind to a promoter region in DNA and form an appropriate initiation complex before RNA polymerase will bind and initiate transcription. Different promoters and transcription factors are utilized by the RNA polymerases I, II, and III. In this section, we focus on the initiation of pre-mRNA synthesis by RNA polymerase II, which transcribes the vast majority of eukaryotic genes.

In all cases, the initiation of transcription involves the formation of a locally unwound segment of DNA, providing a DNA strand that is free to function as a template for the synthesis of a complementary strand of RNA (see Figure 12.9). The formation of the locally unwound segment of DNA required to initiate transcription involves the interaction of several transcription factors with specific sequences in the promoter for the transcription unit. The promoters recognized by RNA polymerase II consist of short conserved elements, or modules, located upstream from the transcription startpoint (Figure 12.16). The conserved element closest to the transcription start site (position +1) is called the **TATA box;** it has the consensus sequence TATAAAA (reading 5′ to 3′ on the nontemplate strand) and is centered at about position −30. The TATA box plays an important role in positioning the transcrip-

TABLE 12.1
Characteristics of the Three RNA Polymerases of Eukaryotes

Enzyme	*Location*	*Products*	*Sensitivity to α-Amanitin*
RNA polymerase I	Nucleolus	Ribosomal RNAs, excluding 5S rRNA	No sensitivity
RNA polymerase II	Nucleus	Nuclear Pre-mRNAs	Complete sensitivity
RNA polymerase III	Nucleus	tRNAs, 5S rRNA, and other small nuclear RNAs	Intermediate sensitivity

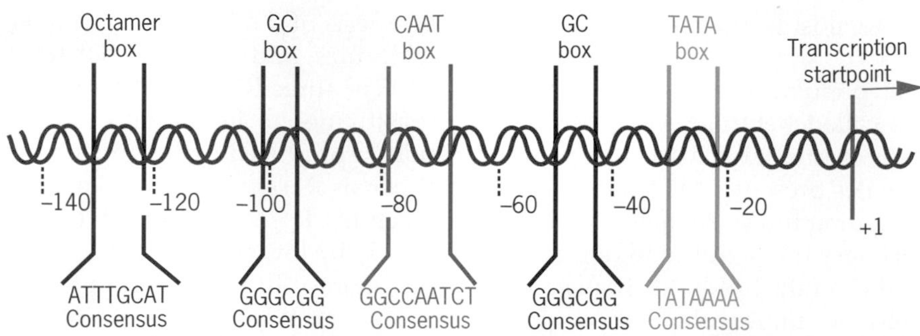

Figure 12.16 Structure of a promoter recognized by RNA polymerase II. The TATA and CAAT boxes are located at about the same positions in the promoters of most nuclear genes encoding proteins. The GC and octamer boxes may be present or absent; when present, they occur at many different locations, either singly or in multiple copies. The sequences shown here are the consensus sequences for each of the promoter elements. The conserved promoter elements are shown at their locations in the mouse thymidine kinase gene.

tion startpoint. The second conserved element is called the **CAAT box;** it usually occurs near position −80 and has the consensus sequence GGCCAATCT. Two other conserved elements, the **GC box,** consensus GGGCGG, and the **octamer box,** consensus ATTTGCAT, often are present in RNA polymerase II promoters; they influence the efficiency of a promoter in initiating transcription.

The initiation of transcription by RNA polymerase II requires the assistance of several **basal transcription factors.** Still other transcription factors modulate the efficiency of initiation (Chapter 24). The basal transcription factors must interact with promoters in the correct sequence to initiate transcription effectively (Figure 12.17). Each basal transcription factor is denoted **TFIIX** (**T**ranscription **F**actor for polymerase **II,** where **X** is a letter identifying the individual factor). The positions at which these transcription factors bind to the promoter DNA have been established in part by determining which segments of DNA are protected from degradation by nucleases when a specific factor has been added to the initiation complex.

TFIID is the first to interact with the promoter; it contains a TATA-binding protein (TBP) and several small TBP-associated proteins (Figure 12.17). Next, TFIIA joins the complex, followed by TFIIB. TFIIF first associates with RNA polymerase II, and then TFIIF and RNA polymerase II join the transcription initiation complex together. TFIIF contains two subunits, one of which has DNA-unwinding activity. Thus, TFIIF probably catalyzes the localized unwinding of the DNA double helix required to initiate transcription. TFIIE then joins the initiation complex, binding to the DNA downstream from the transcription startpoint. Two other factors, TFIIH and TFIIJ, join the complex after TFIIE, but their locations in the complex are unknown.

Similar transcription factor/promoter complexes are required to initiate transcription by RNA polymerases I and III. Polymerases I and III use some of the same transcription factors as polymerase II and some factors specific to each enzyme. The promoters of genes transcribed by polymerases I and III are quite different from the promoters utilized by RNA polymerase II. However, these promoters often contain some of the same *cis*-acting elements as the promoters recognized by RNA polymerase II. Interestingly, the promoters of some of the genes transcribed by RNA polymerase III are located within the transcription units, downstream from the transcription startpoints, rather than upstream as in units transcribed by RNA polymerase II.

RNA Chain Elongation and the Addition of 5′ Methyl Guanosine Caps

Once eukaryotic RNA polymerases have been released from their initiation complexes, they catalyze RNA chain elongation by the same mechanism as the RNA polymerases of prokaryotes (see Figures 12.8 and 12.9). Other details of the elongation process in eukaryotes are still uncertain. The exact structures of eukaryotic RNA polymerases are still unknown, but it is known that they are more complex than their prokaryotic counterparts. RNA polymerase II of the yeast *S. cerevisiae* contains 13 or more polypeptides, at least 10 of which are different gene products. Structural analysis of this enzyme reveals the presence of surface grooves thought to be DNA and RNA binding sites (Figure 12.18).

Early in the elongation process, the 5′ ends of eukaryotic pre-mRNAs are modified by the addition of 7-methyl guanosine (7-MG) caps. These 7-MG caps are added when the growing RNA chains are only about 30 nucleotides long (Figure 12.19). The 7-MG cap contains an unusual 5′-5′ triphosphate linkage (see Figure 12.15) and two or more methyl groups. These 5′ caps are added post-transcriptionally by the biosynthetic

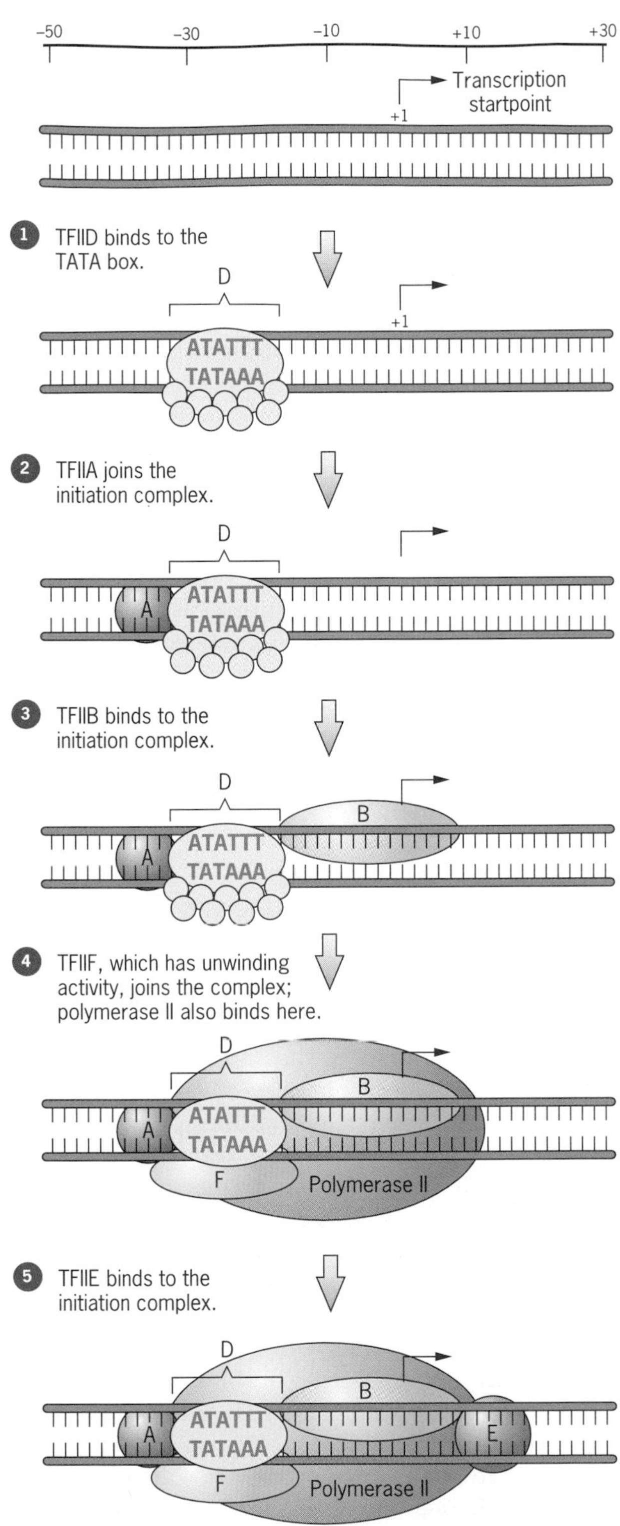

Figure 12.17 The initiation of transcription by RNA polymerase II requires the formation of a basal transcription initiation complex at the promoter region. The assembly of this complex begins when TFIID, which contains the TATA-binding protein (TBP), binds to the TATA box. The other transcription factors and RNA polymerase II join the complex in the sequence shown.

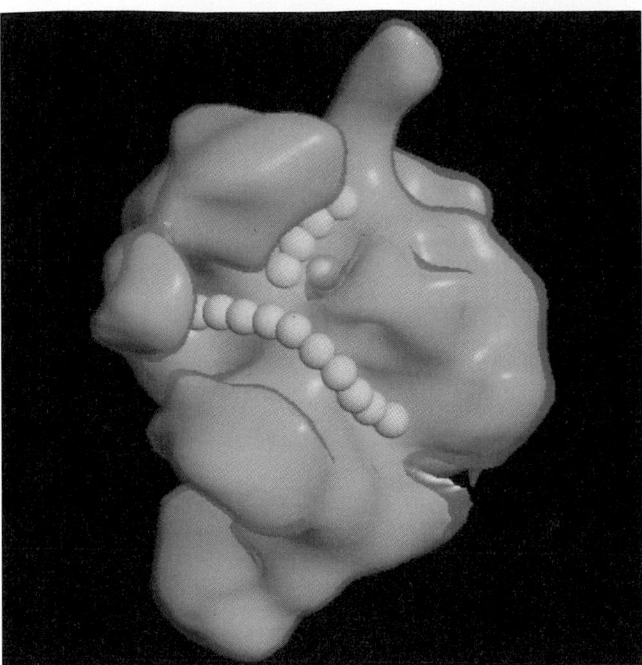

Figure 12.18 Model of the structure of RNA polymerase II in the yeast *S. cerevisiae*. Surface grooves on the enzyme are thought to be binding sites for DNA and the growing RNA chain. One groove about 2.5 nm wide and 1 nm deep is the putative DNA (the ten-beads-long chain) binding site; another groove about 1.5 nm wide and 2 nm deep could bind the growing RNA chain (the four-beads-long chain).

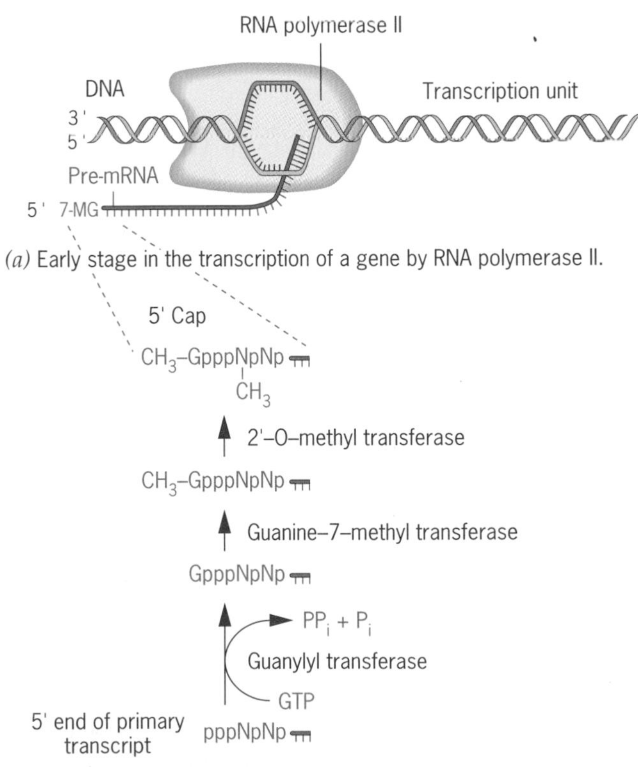

Figure 12.19 7-Methyl guanosine (7-MG) caps are added to the 5′ ends of pre-mRNAs shortly after the elongation process begins.

pathway shown in Figure 12.19. The 7-MG caps are recognized by protein factors involved in the initiation of translation (Chapter 13) and also help protect the growing RNA chains from degradation by nucleases.

Termination by Chain Cleavage and the Addition of 3′ Poly(A) Tails

The 3′ ends of RNA transcripts synthesized by RNA polymerase II are produced by endonucleolytic cleavage of the primary transcripts rather than by the termination of transcription (Figure 12.20). The actual transcription termination events often occur at multiple sites that are located 1000 to 2000 nucleotides downstream from the site that will become the 3′ end of the mature transcript. That is, transcription proceeds beyond the site that will become the 3′ terminus, and the distal segment is removed by endonucleolytic cleavage. The cleavage event that produces the 3′ end of a transcript usually occurs at a site 11 to 30 nucleotides downstream from a conserved sequence, consensus AAUAAA, located near the end of the transcription unit. After cleavage, the enzyme **poly(A) polymerase** adds poly(A) tails, tracts of adenosine monophosphate residues about 200 nucleotides long, to the 3′ ends of the transcripts (Figure 12.20). The addition of poly(A) tails to eukaryotic mRNAs is called **polyadenylation.**

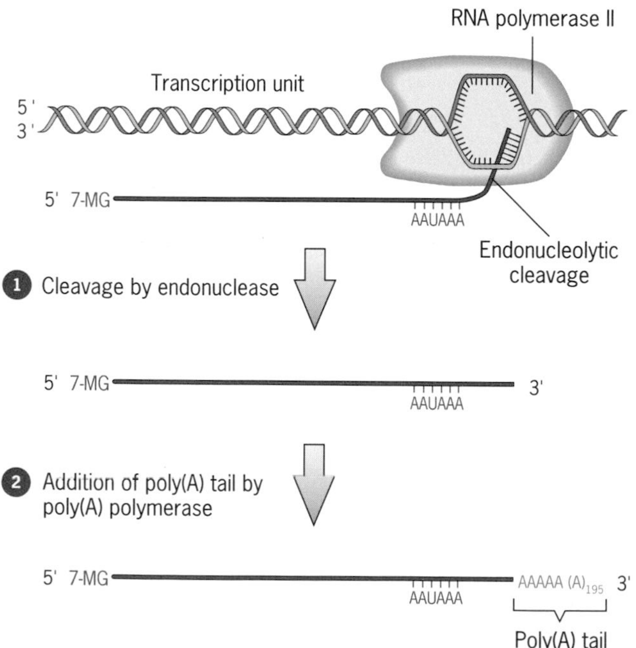

Figure 12.20 Poly(A) tails are added to the 3′ ends of transcripts by the enzyme poly(A) polymerase. The 3′ end substrates for poly(A) polymerase are produced by endonucleolytic cleavage of the transcript downstream from a polyadenylation signal, which has the consensus sequence AAUAAA.

The formation of poly(A) tails on transcripts requires a **specificity component** that recognizes the AAUAAA sequences in RNAs, binds to them, and directs both the cleavage and the polyadenylation reactions. The specificity component, the endonuclease, and the poly(A) polymerase are present in a multimeric complex that catalyzes both cleavage and polyadenylation in tightly coupled reactions. The poly(A) tails of eukaryotic mRNAs enhance their stability and play an important role in their transport from the nucleus to the cytoplasm.

In contrast to RNA polymerase II, both RNA polymerase I and III respond to discrete termination signals. RNA polymerase I terminates transcription in response to an 18-nucleotide-long sequence that is recognized by an associated terminator protein. RNA polymerase III reponds to a termination signal that is similar to the rho-independent terminator in *E. coli* (see Figure 12.13). However, many details of the mechanism of termination of polymerase III-catalyzed transcription are still unknown.

RNA Editing: Altering the Information Content of mRNA Molecules

According to the central dogma of molecular biology, genetic information flows from DNA to RNA to protein during gene expression. Normally, the genetic information is not altered in the mRNA intermediary. However, the discovery of **RNA editing** has shown that exceptions do occur. RNA editing processes alter the information content of gene transcripts in two ways: (1) by changing the structures of individual bases, and (2) by inserting or deleting uridine monophosphate residues.

The first type of RNA editing, which results in the substitution of one base for another base, is rare. This type of editing was discovered in studies of the apolipoprotein-B *(apo-B)* genes and mRNAs in rabbits and humans. Apolipoproteins are blood proteins that transport certain types of fat molecules in the circulatory system. In the liver, the *apo-B* mRNA encodes a large protein 4563 amino acids long. In the intestine, the *apo-B* mRNA directs the synthesis of a protein only 2153 amino acids long. Here, a C residue in the pre-mRNA is converted to a U, generating an internal UAA translation–termination codon, which results in the truncated apolipoprotein (Figure 12.21). UAA is one of three codons that terminates polypeptide chains during translation. If a UAA codon is produced within the coding region of an mRNA, it will prematurely termi-

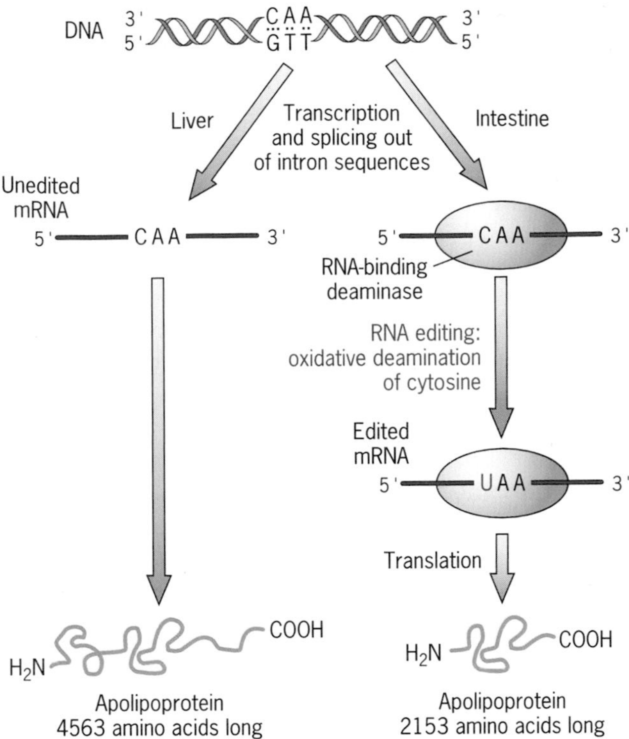

Figure 12.21 Editing of the apolipoprotein-B mRNA in the intestines of mammals.

nate the polypeptide during translation, yielding an incomplete gene product. The C → U conversion is catalyzed by a sequence-specific RNA-binding protein with an activity that removes amino groups from cytosine residues. A similar example of RNA editing has been documented for an mRNA specifying a protein (the glutamate receptor) present in rat brain cells. More extensive mRNA editing of the C → U type occurs in the mitochondria of plants, where most of the gene transcripts are edited to some degree. Mitochondria have their own DNA genomes and protein-synthesizing machinery (Chapter 19). In some transcripts present in plant mitochondria, most of the C's are converted to U residues.

A second, more complex type of RNA editing occurs in the mitochondria of trypanosomes (a group of flagellated protozoa that causes sleeping sickness in humans). In this case, uridine monophosphate residues are inserted (occasionally deleted) into gene transcripts, causing major changes in the polypeptides specified by the mRNA molecules. The trypanosome mitochondrial type of editing is illustrated in Figure 12.22, which shows the sequences of the DNA coding strand, the pre-edited RNA transcript, and the edited mRNA of the cytochrome b gene of *Leishmania taren-*

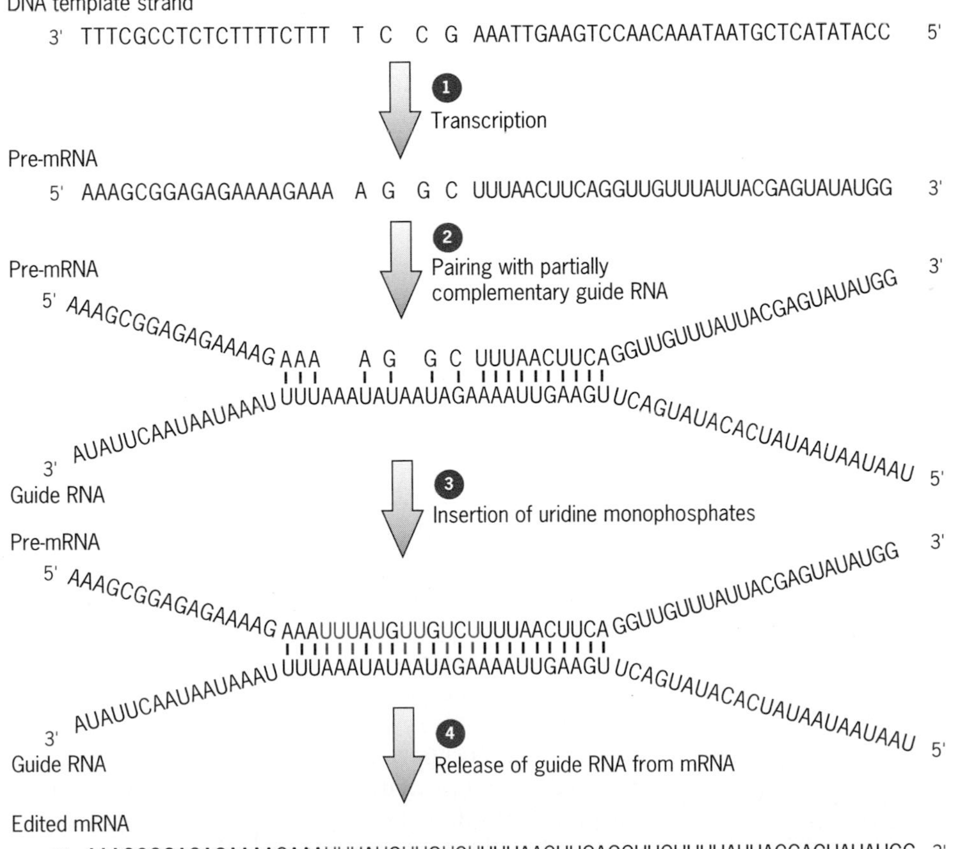

Figure 12.22 Editing of the mitochondrial cytochrome b pre-mRNA in the trypanosome *Leishmania tarentolae.* The uridine monophosphate residues (U) that are inserted in the gaps in the pre-mRNA during the editing process are highlighted by orange shading. Base-pairing between the pre-mRNA and the guide RNA is represented by vertical lines between the two molecules.

tolae. (Cytochrome b is a protein involved in the production of chemical energy.) This RNA editing process is mediated by **guide RNAs** transcribed from distinct mitochondrial genes. The guide RNAs contain sequences that are partially complementary to the pre-mRNAs to be edited. Pairing between the guide RNAs and the pre-mRNAs results in gaps with unpaired A residues in the guide RNAs. The guide RNAs serve as templates for editing, as U's are inserted in the gaps in pre-mRNA molecules opposite the A's in the guide RNAs (Figure 12.22).

In some cases, two or more different guide RNAs participate in the editing of a single pre-mRNA. The guide RNAs not only provide the templates for the editing process, but also donate the uridine monophosphates that are inserted into the pre-mRNAs. The guide RNAs have oligo(U) 3′ tails, 5 to 24 nucleotides long. These U residues are inserted into pre-mRNAs during the editing process by a complex two-step mechanism (Figure 12.23). In the first step, the free 3′-OH on the guide RNA attacks and breaks an internal ester bond in the pre-mRNA, forming a covalent linkage with the 3′ portion of the cleaved pre-mRNA. In the second step, the free 3′-OH on the 5′ portion of the cleaved pre-mRNA attacks an ester linkage between two U residues at the terminus of the oligo(U) tail of the guide RNA, re-forming the pre-mRNA molecule, but now with a U residue inserted at the site of the initial attack.

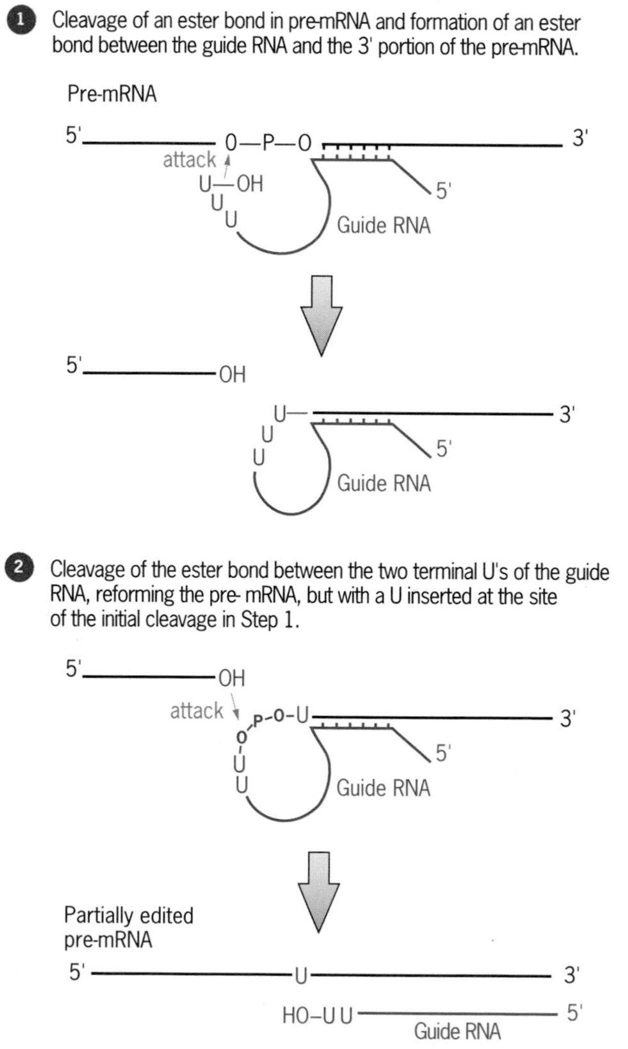

Why do these RNA editing processes occur? Why are the final nucleotide sequences of these mRNAs not specified by the sequences of the mitochondrial genes as they are in most nuclear genes? As yet, answers to these interesting questions are purely speculative. Trypanosomes are primitive single-celled eukaryotes that diverged from other eukaryotes early in evolution. Some evolutionists have speculated that RNA editing was common in ancient cells, where many reactions are thought to have been catalyzed by RNA molecules instead of proteins. Another view is that RNA editing is a primitive mechanism for altering patterns of gene expression. For whatever reason, RNA editing plays a major role in the expresson of genes in the mitochondria of trypanosomes and plants.

Figure 12.23 The two-step mechanism by which uridine monophosphate residues are inserted into pre-mRNA molecules during their editing in mitochondria of trypanosomes.

Key Points: **Three different RNA polymerases are present in eukaryotes, and each polymerase transcribes a distinct set of genes. Eukaryotic gene transcripts usually undergo three major modifications: (1) the addition of 7-methyl guanosine caps to 5′ termini, (2) the addition of poly(A) tails to 3′ ends, and (3) the excision of noncoding intron sequences. The information content of some eukaryotic transcripts is altered by RNA editing, which changes the nucleotide sequences of transcripts prior to their translation.**

INTERRUPTED GENES IN EUKARYOTES: EXONS AND INTRONS

Most of the well-characterized genes of prokaryotes consist of continuous sequences of nucleotide pairs, which specify colinear sequences of amino acids in the polypeptide gene products. However, in 1977, molecular analyses of three eukaryotic genes yielded a major surprise. Studies of mouse and rabbit β-globin (one of two different proteins in hemoglobin) genes and the chicken ovalbumin (an egg storage protein) gene revealed that they contain noncoding sequences intervening between coding sequences. Soon thereafter, noncoding sequences were shown to occur within many genes in a number of eukaryotic species. The coding sequences of these interrupted genes are called **exons** (for **ex**pressed sequences); the noncoding sequences between the coding sequences are called **introns** (for **int**ervening sequences). Geneticists now know that introns interrupt most, but not all, eukaryotic genes.

Early Evidence for Noncoding Sequences in Eukaryotic Genes

In the mid-1970s, researchers began to isolate or "clone" and characterize individual genes (Chapter 20). Comparisons of the structures of eukaryotic genes and the mRNA molecules produced by these genes yielded a totally unexpected result. The genes were shown to contain nucleotide sequences that were not present in the mRNA products of these genes. In October through December 1977, evidence for interrupted genes was reported from three laboratories. Philip Leder and colleagues reported evidence for a nucleotide sequence in a mouse β-globin gene that was not present in the corresponding β-globin mRNA molecule. One month later, Pierre Chambon and coworkers published evidence for noncoding sequences in the chicken ovalbumin gene, and, the following month, Alan Jeffries and Richard Flavell reported the presence of two introns in a rabbit β-globin gene. These initial reports were rapidly followed by numerous publications from these and other laboratories documenting the presence of introns in a large number of eukaryotic genes.

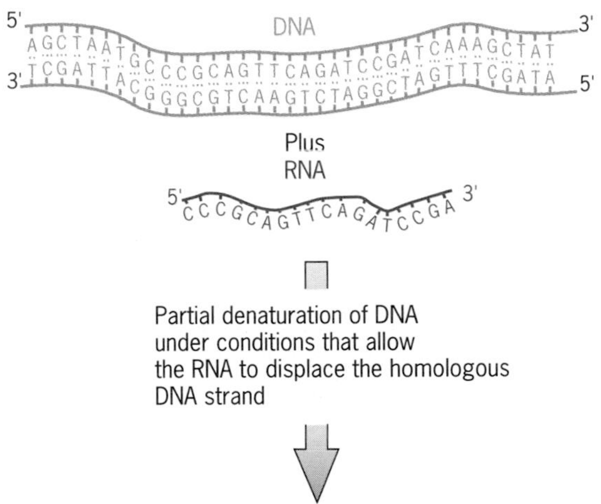

Partial denaturation of DNA under conditions that allow the RNA to displace the homologous DNA strand

The RNA pairs with the complementary strand of DNA forming a DNA–RNA duplex, leaving a single-stranded region of DNA called an R–loop.

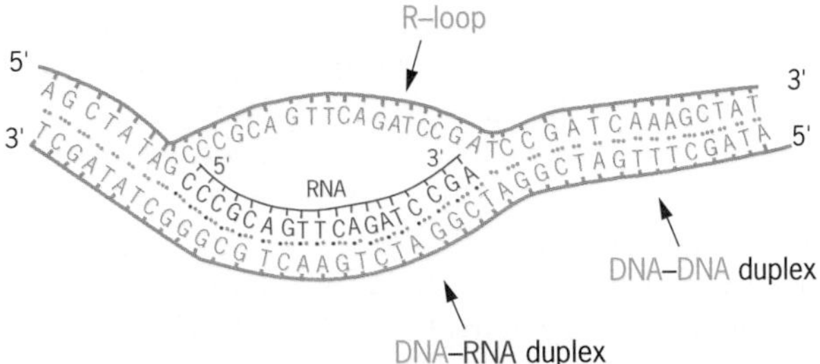

Figure 12.24 The technique of R-loop hybridization.

Some of the earliest evidence for introns in mammalian β-globin genes resulted from the visualization of genomic DNA-mRNA hybrids by electron microscopy. DNA-RNA duplexes are more stable than DNA double helices. Thus, if DNA double helices are partially denatured and are incubated with homologous RNA molecules under the appropriate conditions, the RNA strands will hybridize with the complementary DNA strands, displacing the equivalent DNA strands (Figure 12.24). The resulting DNA-RNA hybrid structures will contain single-stranded regions of DNA called **R-loops,** where RNA molecules have displaced DNA strands to form DNA-RNA duplex regions. These R-loop structures can be visualized directly by electron microscopy. Thus, R-loop hybridization and electron microscopy provide a powerful tool that has been used extensively in studies of eukaryotic gene structure.

When Philip Leder and colleagues hybridized purified mouse β-globin mRNA to a DNA molecule that contained the mouse β-globin gene, they observed two R-loops separated by a loop of double-stranded DNA (Figure 12.25*a*). Their result demonstrated the presence of a sequence of nucleotide pairs in the middle of the β-globin gene that is not present in β-globin mRNA and, therefore, does not code for amino acids in the β-globin polypeptide. When Leder and coworkers repeated the R-loop experiments using purified β-globin gene transcripts isolated from nuclei and believed to be primary gene transcripts or pre-mRNA molecules, in place of cytoplasmic β-globin mRNA, they observed only one R-loop (Figure 12.25*b*). This result indicated that the primary transcript contains the complete structural gene sequence, including both exons and introns. Together, the R-loop results obtained with cytoplasmic mRNA and nuclear pre-mRNA demonstrate that the intron sequence is excised and the exons sequences are spliced together during processing events that convert the primary transcript to the mature mRNA.

Leder and colleagues quickly verified their interpretation of the R-loop results by sequencing a segment of the mouse β-globin gene that spanned the junction between this large intron and the exon preceding it. When they used the established codon as-

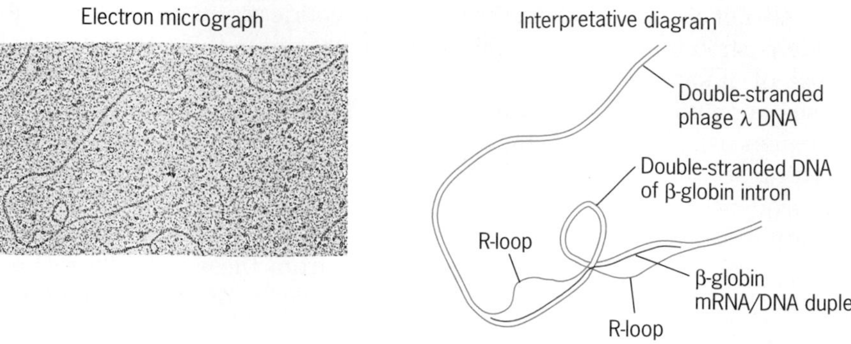

(a) R-loops formed by β-globin mRNA.

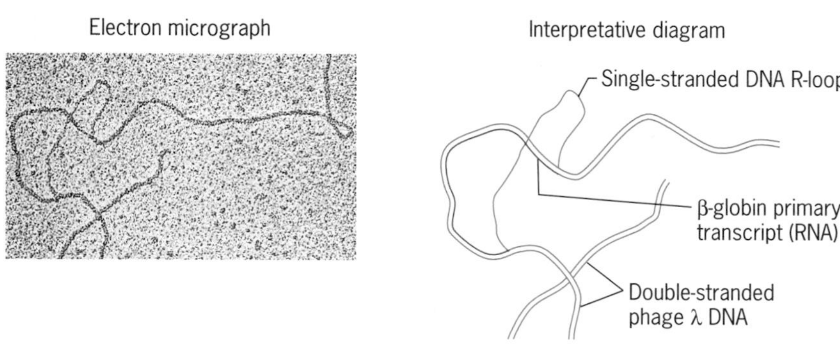

(b) R-loop formed by β-globin primary transcript (pre-mRNA).

Figure 12.25 R-loop evidence for an intron in the mouse β-globin gene. *(a)* When mouse β-globin genes and mRNAs were hybridized under R-loop conditions, two R-loops were observed in the resulting DNA-RNA hybrids. *(b)* When primary transcripts or pre-mRNAs of mouse β-globin genes were used in the R-loop experiments, a single R-loop was observed. These results demonstrate that the intron sequence is present in the primary transcript, but is removed during the processing of the primary transcript to produce the mature mRNA.

signments to compare the nucleotide sequence of the mouse β-globin gene with the known amino acid sequence of mouse β-globin, the presence of the noncoding sequence was clear (Figure 12.26). If there were no intron at this position in the mouse β-globin gene, the nucleotide sequence predicts that amino acid residues 105 through 110 would be Val-Ser-Leu-Met-Gly-Thr. However, the amino acid sequence of mouse β-globin has been determined experimentally, and amino acids 105 through 110 are Leu-Leu-Gly-Asn-Met-Ile. In a later paper, Leder and coworkers sequenced the rest of the mouse β-globin gene and showed that amino acids 105 through 110 are encoded by the first six nucleotide-pair triplets in the exon distal to this intron. The large β-globin gene intron characterized in these early studies is 653 nucleotide pairs long. Shortly after the discovery of the large intron, Leder and colleagues found that the mouse β-globin gene contains a second, smaller intron, 116 nucleotide pairs long (Figure 12.26).

Studies by Tom Maniatis, Richard Flavell, and many others have shown that the structure of the human β-globin gene is very similar to that of the mouse β-globin gene; both contain two introns that separate three exons. Each of the two human α-globin genes also contains two introns. In fact, the human genes encoding the embryonic α-like (ζ) and β-like (ε) globin chains, the fetal β-like (γ) globin chain, and the adult minor β-like (δ) globin chain, all contain two introns and three exons. Moreover, all of the human genes encoding α-like globins have the introns at the same positions (separating codon 31 from 32 and codon 99 from 100), as do all the human genes encoding β-like globins (separating codon 30 from 31 and codon 104 from 105). Thus, intron positions have been highly conserved in the human globin genes.

While Philip Leder, Richard Flavell, and their coworkers were demonstrating the presence of two introns in mammalian β-globin genes, Pierre Chambon, Bert W. O'Malley, and their colleagues were perform-

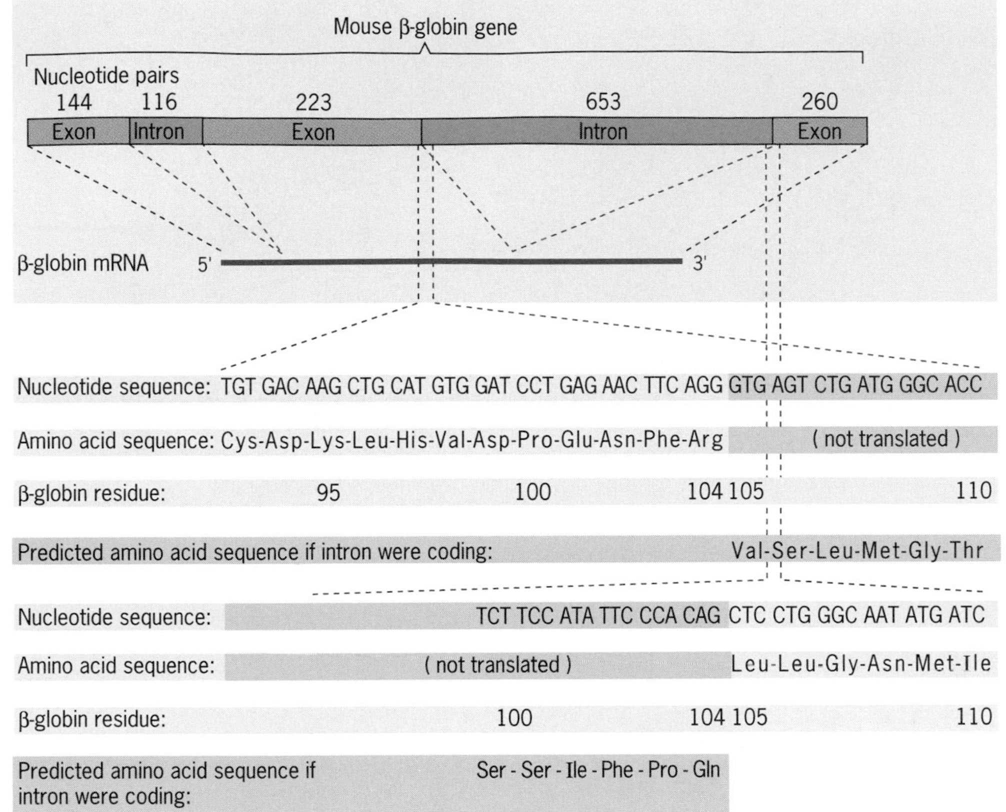

Figure 12.26 Structure of the mouse major β-globin gene. The three exons (coding sequences) are shown in purple, and the two introns (noncoding sequences) are shown in red. The relationship of the exon sequences in the gene to mRNA sequences is indicated by the dashed diagonal lines. The nucleotide sequences of the nontranscribed DNA strand are shown for two exon–intron junctions. These nucleotide sequences are the same as the mRNA sequences, but T replaces U.

ing similar experiments on the chicken ovalbumin gene. When the transcribed strand of the chicken ovalbumin gene was hybridized to the ovalbumin mRNA and visualized by electron microscopy, seven single-stranded DNA loops were observed (Figure 12.27), suggesting that the ovalbumin gene contains seven introns. Subsequent experiments, including direct comparisons of the sequences of the gene and the mRNA, showed conclusively that the chicken ovalbumin gene contains seven introns separating eight exons. In the case of the ovalbumin gene, one intron is present in the 5′ transcribed but untranslated region of the gene. Thus, introns are not restricted to the protein coding regions of genes.

Some Very Large Eukaryotic Genes

Subsequent to the pioneering studies on the mammalian globin genes and the chicken ovalbumin gene, noncoding introns have been demonstrated in a large number of eukaryotic genes. In fact, interrupted genes are much more common than uninterrupted genes in higher animals and plants. For example, the rat serum albumin gene contains 13 introns, and the *Xenopus laevis* gene that encodes vitellogenin A2 (which ends up as egg yolk protein) contains 33 introns. The chicken 1α2 collagen gene contains at least 50 introns, and many of these introns are only 45 or 54 nucleotide pairs long. This gene spans 37,000 nucleotide pairs, but gives rise to an mRNA molecule only about 4600 nucleotides long. Other genes contain relatively few introns, but some of the introns are very large. For example, the *Ultrabithorax* (*Ubx*) gene of *Drosophila* contains an intron that is approximately 70,000 nucleotide pairs in length. The largest gene characterized to date is the human *DMD* gene, which causes Duchenne muscular dystrophy when rendered nonfunctional by mutation. The *DMD* gene spans 2.5 million nucleotide pairs and contains 78 introns.

The first evidence for introns in plant genes was the demonstration that the gene that encodes phaseolin (a major storage protein) in the French bean contains three introns. Subsequent studies have shown

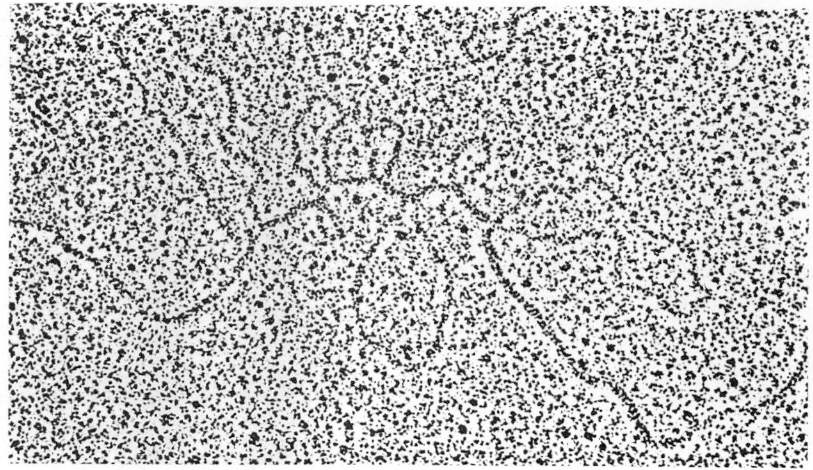

(*a*) Electron micrograph of ovalbumin DNA–mRNA heteroduplex.

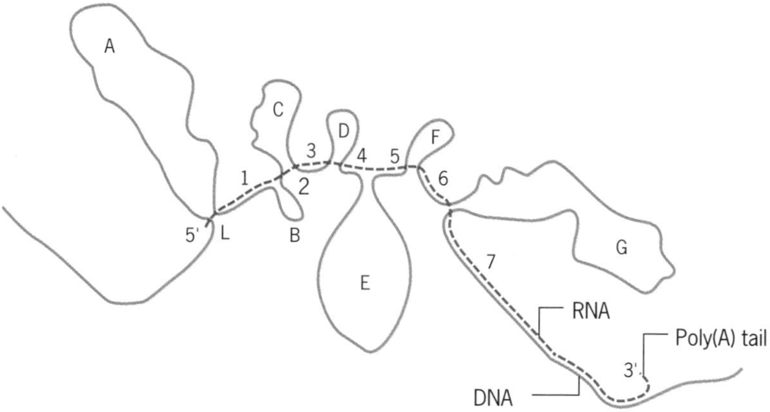

(*b*) Interpretative diagram of the DNA–mRNA heteroduplex.

Figure 12.27 Evidence for seven introns in the chicken ovalbumin gene. The mRNA molecule is represented by a red line; the DNA strand is shown as a green line. Intron sequences are single-stranded DNA loops labeled A through G.

that most of the genes of higher plants—like those of higher animals—contain multiple introns.

The only structural features that seem to be shared by different introns are the dinucleotide sequences at their ends. The primary transcripts of genes almost always begin introns with the sequence GU (5′) and end them with the sequence AG (3′). These consensus sequences at intron–exon junctions are important in the mechanism by which introns are spliced out of primary transcripts.

Although introns are present in most genes of higher animals and plants, they are not essential because not all such genes contain introns. The sea urchin histone genes and four *Drosophila* heat shock genes were among the first animal genes shown to lack introns. We now know that many genes of higher animals and plants lack introns.

Introns: Biological Significance?

At present, scientists know relatively little about the biological significance of the exon–intron structure of eukaryotic genes. Introns are highly variable in size, ranging from a few nucleotide pairs to thousands of nucleotide pairs in length. This fact has led to speculation that introns may play a role in regulating gene expression. The transcription of genes with large introns will take longer than the transcription of genes with small introns or no introns. Thus, the presence of large introns in genes would be expected to decrease the rate of transcript accumulation in a cell. The fact that introns accumulate new mutations much more rapidly than exons indicates that the specific nucleotide-pair sequences of introns, excluding the ends, are not very important.

In some cases, the different exons of genes encode different functional domains of the protein gene products. This is most apparent in the case of the genes encoding heavy and light antibody chains (Chapter 27). In the case of the mammalian globin genes, the middle exon encodes the heme-binding domain of the protein. There has been considerable speculation that the exon–intron structure of eukaryotic genes has resulted from the evolution of new genes by the fusion of uninterrupted (single exon) ancestral genes. If this hypothesis is correct, introns may merely be relics of the evolutionary process.

Alternatively, introns may provide a selective advantage by increasing the rate at which coding sequences in different exons of a gene can reassort by recombination, thus speeding up the process of evolution. In some cases, alternate ways of splicing a transcript produce a family of related proteins (see Figure 15.21). In these cases, introns result in multiple products from a single gene. The alternate splicing of the rat troponin T transcript is illustrated in Figure 24.4. In the case of the mitochondrial gene of yeast encoding cytochrome *b*, the introns contain exons of genes encoding enzymes involved in processing the primary transcript of the gene. Thus, different introns may indeed play different roles, and many introns may have no biological significance. Since many eukaryotic genes contain no introns, these noncoding regions are not required for normal gene expression.

Key Points: **Most, but not all, eukaryotic genes are split into coding sequences called exons and noncoding sequences called introns. Some genes contain very large introns; others harbor large numbers of small introns. The biological significance of introns is still open to debate.**

REMOVAL OF INTRON SEQUENCES BY RNA SPLICING

Most nuclear genes that encode proteins in multicellular eukaryotes contain introns. Fewer, but still many, of the genes of unicellular eukaryotes such as the yeasts contain noncoding introns. Rare genes of a few viruses of prokaryotes and of an archebacterium (a primitive bacterium) also contain introns. In the case of these "split" genes, with coding sequences interrupted by noncoding sequences, the primary transcript contains the entire sequence of the gene and the noncoding sequences are spliced out during RNA processing (see Figure 12.15).

For genes that encode proteins, the splicing mechanism must be precise; it must join exon sequences

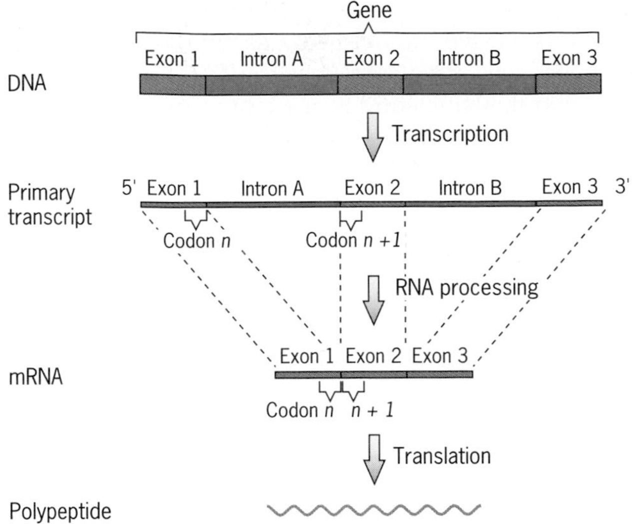

Figure 12.28 The excision of intron sequences from primary transcripts by RNA splicing. The splicing mechanism must be accurate to the single nucleotide to assure that codons in downstream exons are translated correctly to produce the right amino acid sequence in the polypeptide product.

with accuracy to the single nucleotide to assure that codons in exons distal to introns are read correctly (Figure 12.28). Accuracy to this degree would seem to require precise splicing signals, presumably nucleotide sequences within introns and at the exon–intron junctions. However, in the primary transcripts of nuclear genes, the only completely conserved sequences of different introns are the dinucleotide sequences at the ends of introns, namely,

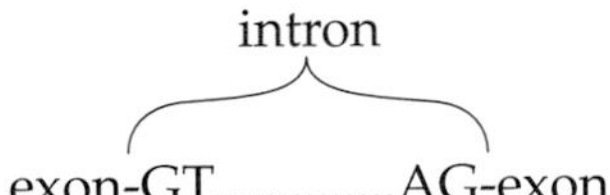

The sequences shown here are for the DNA nontemplate strand (equivalent to the RNA transcript, but with T rather than U). In addition, there are short consensus sequences at the exon–intron junctions. For nuclear genes, the consensus junctions are

exon intron exon

$A_{64}G_{73}G_{100}T_{100}A_{68}A_{68}G_{84}T_{63}$.............$6Py_{74\text{-}87}N\ C_{65}A_{100}G_{100}\ N$

The numerical subscripts indicate the percentage frequencies of the consensus bases at each position; thus, a 100 subscript indicates that a base is always present at that position. N indicates that any of the four standard nucleotides may be present at the indicated position. The exon–intron junctions are different for tRNA genes and structural genes in mitochondria and chloroplasts, which utilize different RNA splicing mechanisms (see the following section). There is only one short conserved sequence, the **TACTAAC box,** located about 30 nucleotides upstream from the 3′ splice site of introns in nuclear genes, and it is rather poorly conserved. The TACTAAC box does exhibit a strong preference for either a purine or a pyrimidine at each site as follows:

$$\mathrm{Py_{80}\ N\ Py_{80}\ Py_{87}\ Pu_{75}\ A_{100}\ Py_{95}}$$

The adenine residue at position six in the TACTAAC box is completely conserved and is known to play a key role in the splicing reaction. With the exception of the terminal dinucleotides and the TACTAAC box, the intron sequences of nuclear genes are highly divergent, apparently random sequences. The introns of genes of mitochondria and chloroplasts also contain conserved sequences, but they are different from those of nuclear genes.

The highly conserved nature of the 5′ and 3′ splice sites and the TACTAAC box indicates that they play an important role in the process of gene expression. Direct evidence for their importance has been provided by mutations at these sites that cause mutant phenotypes in many different eukaryotes. Indeed, such mutations are sometimes responsible for inherited diseases in humans, such as hemoglobin disorders.

The discovery of noncoding introns in genes stimulated intense interest in the mechanism(s) by which intron sequences are removed during gene expression. The early demonstration that the intron sequences in eukaryotic genes were transcribed along with the exon sequences focused research on the processing of primary gene transcripts. Just as *in vitro* systems provided important information about the mechanisms of transcription and translation, the key to understanding RNA splicing events was the development of *in vitro* splicing systems. By using these systems, researchers have shown that there are three distinct types of intron excision from RNA transcripts, presented here in the order of increasing complexity, not in the order of importance.

1. The introns of tRNA precursors are excised by precise endonucleolytic cleavage and ligation reactions catalyzed by special splicing endonuclease and ligase activities.
2. The introns of some rRNA precursors are removed autocatalytically in a unique reaction mediated by the RNA molecule itself. (No protein enzymatic activity is involved.)
3. The introns of nuclear pre-mRNA (hnRNA) transcripts are spliced out in two-step reactions carried out by complex ribonucleoprotein particles called spliceosomes.

Many genes contain large numbers of introns. For example, the chicken 1α2 collagen (a structural protein in cartilage) gene contains over 50 introns, which leads to the question of the order in which multiple introns are removed. For certain genes that have been studied, the introns are excised in a preferred, but not fixed, order. Other introns have been found to undergo alternative pathways of splicing leading to mRNAs that produce families of related proteins. An intron in the mitochondrial cytochrome *b* gene of yeast includes part of the coding sequence for a protein, an RNA maturase, which is responsible for excising the second intron from the transcript of that gene. This discovery suggests interesting mechanisms for regulating the expression of genes at the level of intron processing. Clearly, variations in the use, structure, and excision of intron sequences occur in various oganisms, and novel intron structures may be discovered in the future. We will consider the three major mechanisms of intron excision in the following three sections.

tRNA Precursor Splicing: Unique Nuclease and Ligase Activities

The tRNA precursor splicing reaction has been worked out in detail in the yeast *Saccharomyces cerevisiae*. Both *in vitro* splicing systems and temperature-sensitive splicing mutants have been used in dissecting the tRNA splicing mechanism in *S. cerevisiae*. The excision of introns from yeast tRNA precursors occurs in two stages (Figure 12.29). In stage I, a nuclear membrane-bound **splicing endonuclease** makes two cuts precisely at the ends of the intron. Then, in stage II, a **splicing ligase** joins the two halves of the tRNA to produce the mature form of the tRNA molecule. The specificity for these reactions resides in conserved three-dimensional features of the tRNA precursors, not in the nucleotide sequences per se.

Cleavage of the tRNA precursor yields 5′-OH termini and 2′-3′ cyclic phosphate groups at the 3′ termini. The stage II ligation process involves four separate reactions. (1) The first reaction is the addition of a phosphate group to the 5′-OH terminus; this reaction requires kinase activity and a phosphate donor (ATP). (Kinases are enzymes that add phosphate groups to molecules.) (2) Then, the 5′ phosphate group is activated by the transfer of an AMP group to the terminus from an AMP-ligase intermediate. (3) The 2′-3′ cyclic phosphate is opened by a cyclic phosphodiesterase that produces a 2′ phosphate and a free 3′ hydroxyl. (4) The final ligation reaction occurs via a nucleophilic attack of the free 3′-OH on the interior 5′ phosphate with the release of AMP. All four of these reactions are catalyzed by the splicing ligase. Finally,

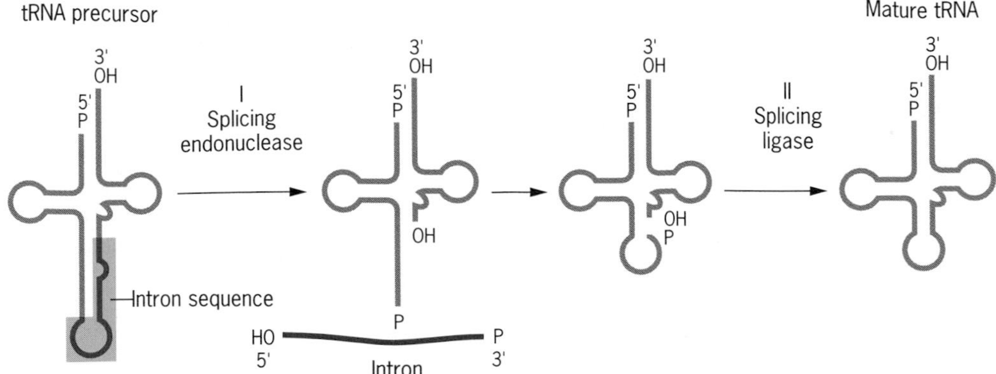

Figure 12.29 The excision of introns from tRNA precursors is a two-stage process: (I) endonucleolytic cleavages at the ends of the intron, and (II) ligation of the newly formed ends of the two exons. The ligation process actually involves three distinct reactions, all catalyzed by the splicing ligase. This unique enzyme contains phosphodiesterase, kinase, and ligase activities.

the 2′ phosphate group (remaining from the 2′-3′ cyclic phosphate produced by the original cleavage reaction) is removed by a phosphatase activity to yield the mature tRNA molecule.

The overall two-stage mode of tRNA intron excision appears to occur in other organisms as well. In fact, the mechanism may involve the same reactions in plants. However, in mammals, the reactions are not the same. Splicing still occurs in two stages, but the ligation reaction directly joins the 2′-3′ cyclic phosphate terminus to the 5′-OH terminus. The details of this process of tRNA precursor splicing in mammalian cells are not yet as clearly established as they are in yeast.

Autocatalytic Splicing

As we reviewed in the first section of this chapter, a general theme in biology is that metabolism occurs via sequences of enzyme-catalyzed reactions. These all-important enzymes are generally proteins, sometimes single polypeptides and sometimes complex heteromultimers. Occasionally, enzymes require nonprotein cofactors to perform their functions. When covalent bonds are being altered, it is usually assumed that the reaction is being catalyzed by an enzyme. Thus, the discovery in *Tetrahymena thermophila* that the intron in the rRNA precursor was excised without the involvement of any protein was quite surprising to most biologists. However, it is now clearly established that the splicing activity that excises the intron from this rRNA precursor is intrinsic to the RNA molecule itself. Moreover, such **self-splicing** or **autocatalytic activity** has been shown to occur in rRNA precursors of several lower eukaryotes and in a large number of rRNA, tRNA, and mRNA precursors in mitochondria and chloroplasts of many different species. In the case of many of these introns, the self-splicing mechanism is the same as or very similar to that utilized by the *Tetrahymena* rRNA precursors (see Figure 12.30). For others, the self-splicing mechanism is similar to the splicing mechanism observed with nuclear mRNA precursors, but without the involvement of the spliceosome (see Figure 12.31).

The autocatalytic excision of the intron in the *Tetrahymena* rRNA precursor and certain other introns requires no external energy source and no protein. Instead, the splicing mechanism involves a series of phosphoester bond transfers, with no bonds lost or gained in the process. The reaction requires a guanine nucleoside or nucleotide with a free 3′-OH group (GTP, GDP, GMP, or guanosine all work) as a cofactor plus a monovalent cation and a divalent cation. The requirement for the G-3′-OH is absolute; no other base can be substituted in the nucleoside or nucleotide cofactor. The intron is excised by means of two phosphoester bond transfers, and the excised intron can subsequently circularize by means of another phosphoester bond transfer. These reactions are diagrammed in Figure 12.30.

The autocatalytic circularization of the excised intron suggests that the self-splicing of these rRNA precursors resides primarily, if not entirely, within the intron structure itself. Presumably, the autocatalytic activity is dependent on the secondary structure of the intron or at least the secondary structure of the RNA precursor molecule. The secondary structures of these self-splicing RNAs must bring the reactive groups into close juxtaposition to allow the phosphoester bond transfers to occur. Since the self-splicing phosphoester bond transfers are potentially reversible reactions, rapid degradation of the excised introns or export of

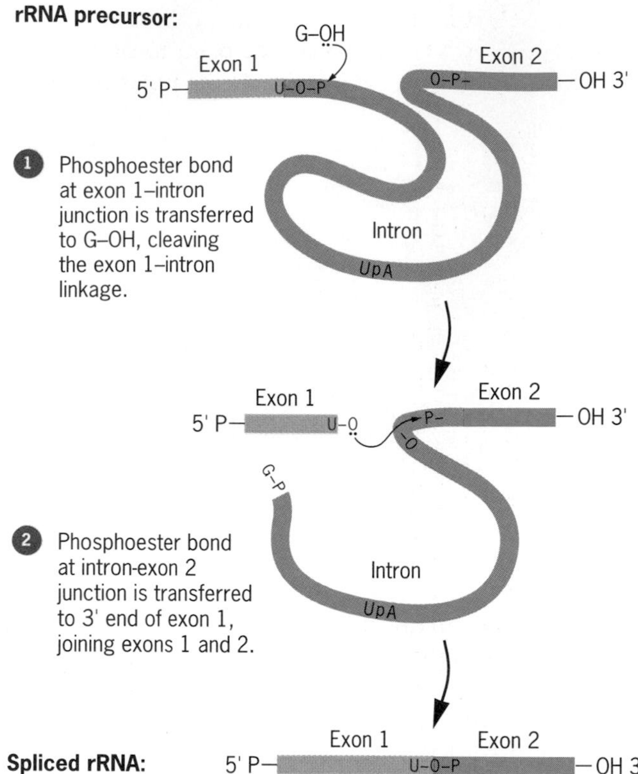

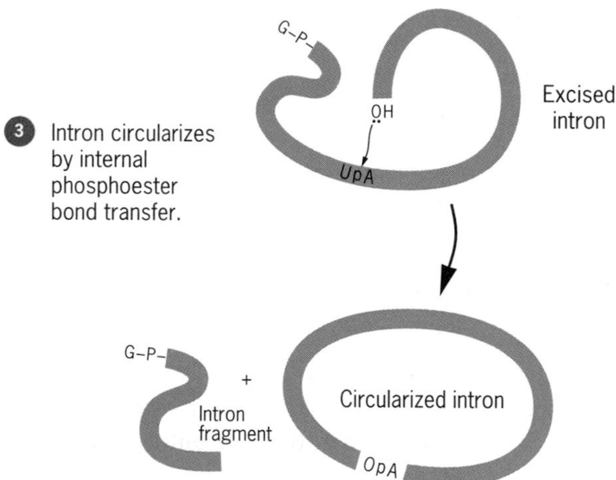

Figure 12.30 Diagram of the mechanism of self-splicing of the *Tetrahymena thermophila* rRNA precursor and the subsequent circularization of the excised intron.

the spliced rRNAs to the cytoplasm may drive splicing in the forward direction.

Note that the autocatalytic splicing reactions are intramolecular in nature and thus are not dependent on concentration. Moreover, the RNA precursors are capable of forming an active center in which the guanosine-3′-OH cofactor binds. The autocatalytic splicing of these rRNA precursors demonstrates that catalytic sites are not restricted to proteins; however, there is no *trans* catalytic activity as for enzymes, only *cis* catalytic activity.

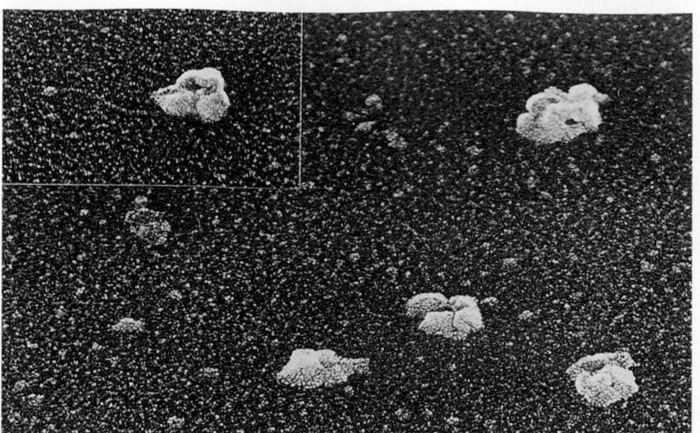

Figure 12.31 Electron microscope photographs of purified spliceosomes. Note the striking substructure of the particles, which have dimensions of 40 to 60 nanometers. The inset (top left) shows a particle with a thin filament and a smaller particle at the end of the filament.

Pre-mRNA Splicing: snRNAs, snRNPs, and the Spliceosome

The introns in nuclear pre-mRNAs are excised in two steps like the introns in yeast tRNA precursors and *Tetrahymena* rRNA precursors that were discussed in the preceding two sections. However, the introns are not excised by simple splicing nucleases and ligases or autocatalytically. Instead, nuclear pre-mRNA splicing is carried out by complex RNA/protein structures called **spliceosomes** (Figure 12.31). These structures are in many ways like small ribosomes. They contain a set of small RNA molecules called snRNAs (small nuclear RNAs) and a set of proteins that are still not completely defined. The two steps in nuclear pre-mRNA splicing are known (Figure 12.32); however, some of the details of the splicing process are still uncertain.

Five snRNAs, called U1, U2, U4, U5, and U6, are involved in nuclear pre-mRNA splicing as components of the spliceosome. (snRNA U3 is localized in the nucleolus and probably is involved in the formation of ribosomes.) In mammals, these snRNAs range in size from 100 nucleotides (U6) to 215 nucleotides (U3). Some of the snRNAs in the yeast *S. cerevisiae* are much larger. These snRNAs do not exist as free RNA molecules. Instead, they are present in small nuclear RNA-protein complexes called **snRNPs** (small *n*uclear *r*ibo*n*ucleo*p*roteins). Spliceosomes are assembled from four different snRNPs and protein splicing factors dur-

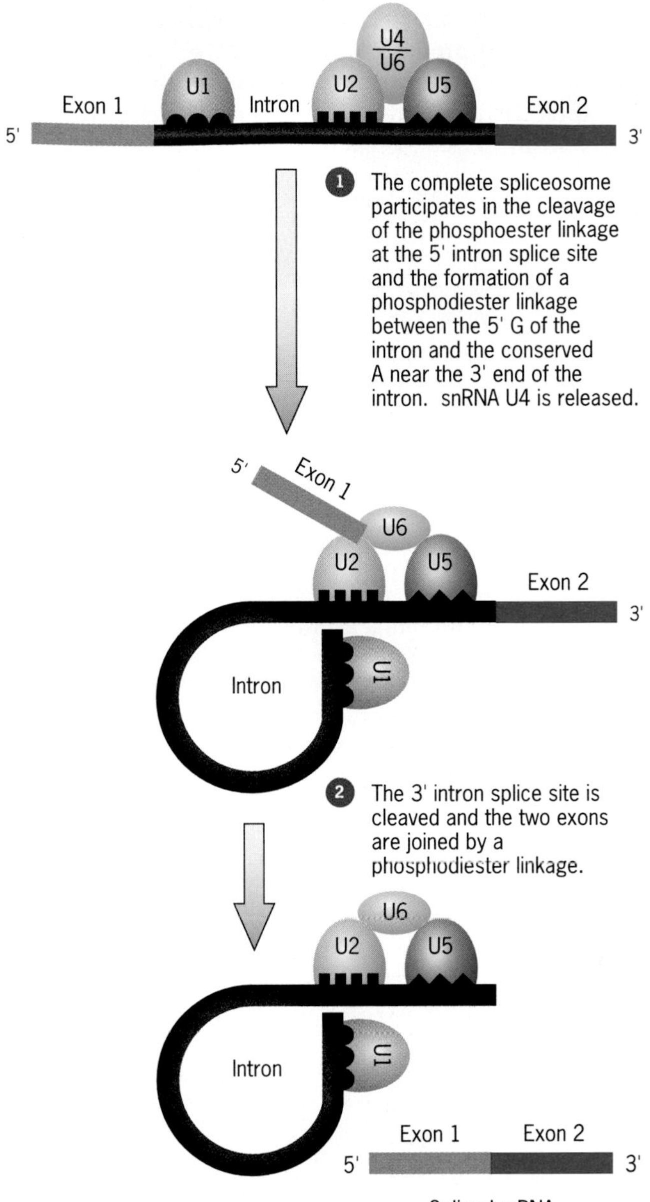

Figure 12.32 The postulated roles of the snRNA-containing snRNPs in nuclear pre-mRNA splicing.

ing the splicing process. Characterization of snRNPs has been facilitated by the discovery that some patients with an often fatal autoimmune disease called systemic lupus erythematosus produce antibodies that react with many of their own cellular components including snRNP proteins. These antibodies are called autoantibodies because they react with the patient's own proteins; normally, the human immune system will produce only antibodies that react with foreign proteins (Chapter 27). In lupus patients, the autoantibodies cause inflammations of tissues and organs that often result in heart, kidney, or liver malfunction and even death. The autoantibodies from patients with systemic lupus erythematosus can be used to precipitate snRNPs; thus, they greatly facilitate the purification of snRNPs for structural and functional studies.

Each of the snRNAs U1, U2, and U5 is present by itself in a specific snRNP particle. snRNAs U4 and U6 are present together in a fourth snRNP; U4 and U6 snRNAs contain two regions of intermolecular complementarity that probably are base-paired in the U4/U6 snRNP. Each of the four types of snRNP particles contains a subset of seven well-characterized snRNP proteins plus one or more proteins unique to the particular type of snRNP particle. All four snRNP complexes are present in the isolated spliceosomes shown in Figure 12.31. However, the exact protein composition of intact spliceosomes still is not established.

The first step in nuclear pre-mRNA splicing involves cleavage at the 5′ intron splice site (↓GU-intron) and the formation of an intramolecular phosphodiester linkage between the 5′ carbon of the G at the cleavage site and the 2′ carbon of a conserved A residue near the 3′ end of the intron. This step occurs on complete spliceosomes (Figure 12.32) and requires the hydrolysis of ATP. Evidence indicates that the U1 snRNP must bind at the 5′ splice site prior to the initial cleavage reaction. Recognition of the cleavage site at the 5′ end of the intron probably involves base-pairing between the consensus sequence at this site and a complementary sequence near the 5′ terminus of snRNA U1. However, the specificity of the binding of at least some of the snRNPs to intron consensus sequences involves both the snRNAs and specific snRNP proteins. Thus, the base-pairing between the intron 5′ consensus sequence and the complementary sequence in the snRNA U1 may provide only part of the specificity for the functional binding of the U1 snRNP to the pre-mRNA molecule.

The second snRNP to be added to the splicing complex appears to be the U2 snRNP; it binds at the consensus sequence that contains the conserved A residue that forms the branch point in the lariat structure of the spliced intron. Thereafter, the U5 snRNP binds at the 3′ splice site, and the U4/U6 snRNP is added to the complex to yield the complete spliceosome (Figures 12.31 and 12.32). When the 5′ intron splice site is cleaved in step 1, the U4 snRNA is released from the spliceosome. In step 2 of the splicing reaction, the 3′ splice site of the intron is cleaved, and the two exons are joined by a normal 5′ to 3′ phosphodiester linkage (Figure 12.32). The spliced mRNA is now ready for export to the cytoplasm and translation on ribosomes.

Key Points: **Noncoding intron sequences are excised from RNA transcripts in the nucleus prior to**

their transport to the cytoplasm. Introns in tRNA precursors are removed by the concerted action of a splicing endonuclease and ligase, whereas introns in some rRNA precursors are spliced out autocatalytically—with no protein involved. The introns in nuclear pre-mRNAs are excised on complex ribonucleoprotein structures called spliceosomes. This excision process must be precise, with accuracy to the nucleotide level, to assure that codons in exons distal to introns are read correctly during translation.

TESTING YOUR KNOWLEDGE

1. Certain medically important human proteins such as insulin and growth hormone are now being produced in bacteria. By using the tools of genetic engineering, DNA sequences encoding these proteins have been introduced into bacteria. You wish to introduce a human gene into *E. coli* and have that gene produce large amounts of the human gene product in the bacterial cells. Assuming that the human gene of interest can be isolated and introduced into *E. coli*, what problems might you encounter in attempting to achieve your goal?

ANSWER

The promoter sequences that are required to initiate transcription are very different in mammals and bacteria. Therefore, your gene will not be expressed in *E. coli* unless you first fuse its coding region to a bacterial promoter. In addition, your human gene probably will contain introns. Since *E. coli* cells do not contain spliceosomes or equivalent machinery with which to excise introns from RNA transcripts, your human gene will not be expressed correctly if it contains introns. As you can see, expressing eukaryotic genes in prokaryotic cells is not a trivial task.

2. A human β-globin gene has been purified and inserted into a linear bacteriophage lambda chromosome, producing the following DNA molecule.

λ DNA Exon 1 Intron 1 Exon 2 Intron 2 Exon 3 λ DNA

If this DNA molecule is hybridized to human β-globin mRNA using conditions that favor DNA:RNA duplexes over DNA:DNA duplexes (R-loop mapping conditions) and the product is visualized by electron microscopy, what nucleic acid structure would you expect to see?

ANSWER

The primary transcript of this human β-globin gene will contain both introns and all three exons. However, prior to its export to the cytoplasm, the intron sequences will be spliced out of the transcript. Thus, the mature mRNA molecule will contain the three exon sequences spliced together with no intron sequences present. Under R-loop conditions, the mRNA will anneal with the complementary strand of DNA, displacing the equivalent DNA strand. However, since the mRNA contains no intron sequences, the introns will remain as regions of double-stranded DNA as shown in the following diagram.

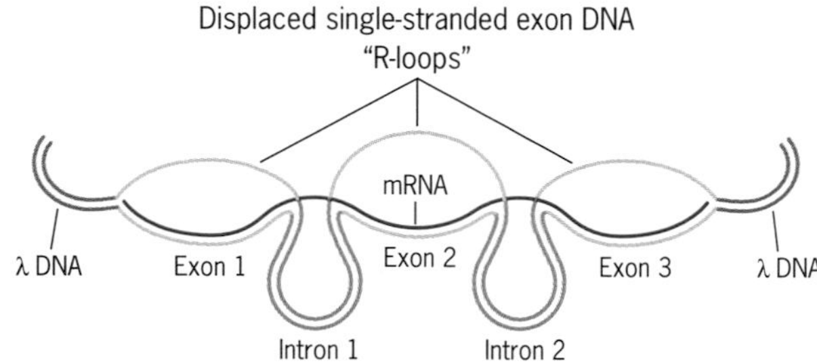

QUESTIONS AND PROBLEMS

12.1 Distinguish between DNA and RNA (a) chemically, (b) functionally, and (c) by location in the cell.

12.2 What bases in the mRNA transcript would represent the following DNA template sequence: 5′-TGCAGACA-3′?

12.3 What bases in the transcribed strand of DNA would give rise to the following mRNA base sequence: 5′-CUGAU-3′?

12.4 On the basis of what evidence was the messenger RNA hypothesis established?

12.5 At what locations in the cell does protein synthesis occur?

12.6 List three ways in which the mRNAs of eukaryotes differ from the mRNAs of prokaryotes.

12.7 What different types of RNA molecules are present in prokaryotic cells? In eukaryotic cells? What roles do these different classes of RNA molecules play in the cell?

12.8 Many eukaryotic genes contain noncoding introns that separate the coding sequences or exons of these genes. At what stage during the expression of these mosaic genes are the noncoding intron sequences removed?

12.9 For several decades, the dogma in biology has been that molecular reactions in living cells are catalyzed by enzymes composed of polypeptides. We now know that the introns of some precursor RNA molecules such as the rRNA precursors in *Tetrahymena* are removed autocatalytically ("self-spliced") with no involvement of any protein. What does the demonstration of autocatalytic splicing indicate about the dogma that biological reactions are always catalyzed by proteinaceous enzymes?

12.10 What role(s) do spliceosomes play in pathways of gene expression? What is their macromolecular structure?

12.11 What components of the introns of nuclear genes that encode proteins in higher eukaryotes are conserved and required for the correct excision of intron sequences from primary transcripts by spliceosomes?

12.12 Match one of the following terms with each of the descriptions given below.

Terms: (1) sigma (σ) factor; (2) poly(A) tail; (3) TATAAT; (4) exons; (5) TATAAAA; (6) RNA polymerase III; (7) intron; (8) RNA polymerase II; (9) heterogeneous nuclear RNA (hnRNA); (10) snRNA; (11) RNA polymerase I; (12) TTGACA; (13) GGCCAATCT (CAAT box).

Descriptions:

(a) Intervening sequence found in many eukaryotic genes.

(b) A conserved nucleotide sequence (−30) in eukaryotic promoters involved in the initiation of transcription.

(c) Small RNA molecules that are located in the nuclei of eukaryotic cells, most as components of the spliceosome, that participate in the excision of introns from nuclear gene transcripts.

(d) A sequence (−10) in the nontemplate strand of the promoter of *E. coli* that facilitates the localized unwinding of DNA when complexed with RNA polymerase.

(e) The RNA polymerase in the nucleus that catalyzes the synthesis of all rRNAs except for the small 5S rRNA.

(f) The subunit of prokaryotic RNA polymerase that is responsible for the initiation of transcription at promoters.

(g) An *E. coli* promoter sequence located 35 nucleotides upstream from the transcription-initiation site; it serves as a recognition site for the sigma factor.

(h) The RNA polymerase in the nucleus that catalyzes the synthesis of the transfer RNA molecules and small nuclear RNAs.

(i) A polyadenosine tract 20 to 200 nucleotides long that is added to the 3′ end of most eukaryotic messenger RNAs.

(j) The RNA polymerase that transcribes nuclear genes that encode proteins.

(k) A conserved sequence in the nontemplate strand of eukaryotic promoters that is located about 80 nucleotides upstream from the transcription start site.

(l) Segments of a eukaryotic gene that correspond to the sequences in the final processed RNA transcript of the gene.

(m) The population of primary transcripts in the nucleus of a eukaryotic cell.

12.13 (a) Which of the following nuclear pre-mRNA nucleotide sequences potentially contains an intron?

(1) 5′ — UGACCAUGGCGCUAACACUGCCAAUU-GGCAAUACUGACCUGAUAGCAUCAGCCAA — 3′

(2) 5′ — UAGUCUCAUCUGUCCAUUGACUUCGA-AACUGAAUCGUAACUCCUACGUCUAUGGA — 3′

(3) 5′ — UAGCUGUUUGUCAUGACUGACUGGUC-ACUAUCGUACUAACCUGUCAUGCAAUGUC — 3′

(4) 5′ — UAGCAGUUCUGUCGCCUCGUGGUGCU-GCUGGCCCUUCGUCGCUCGGGCUUAGCUA — 3′

(5) 5′ — UAGGUUCGCAUUGACGUACUUCUGAGA-CUACUAACUACUAACGCAUCGAGUCUCAA — 3′

(b) One of the five pre-mRNAs shown in (a) may undergo RNA splicing to excise an intron sequence. What mRNA nucleotide sequence would be expected to result from this splicing event?

12.14 What is the function of the introns in eukaryotic genes?

12.15 A particular gene is inserted into the phage lambda chromosome and is shown to contain three introns. (*a*) The primary transcript of this gene is purified from isolated nuclei. When this primary transcript is hybridized under R-loop conditions with the recombinant lambda chromosome carrying the gene, what will the R-loop structure(s) look like? Label your diagram. (*b*) The mRNA produced from the primary transcript of this gene is then isolated from cytoplasmic polyribosomes and similarly examined by the R-loop hybridization procedure using the recombinant lambda chromosome carrying the gene. Diagram what the R-loop structure(s) will look like when the cytoplasmic mRNA is used. Again, label the components of your diagram.

12.16 A segment of DNA in *E. coli* has the following sequence of nucleotide pairs:

When this segment of DNA is transcribed by RNA polymerase, what will be the sequence of nucleotides in the RNA transcript if the promoter is located to the left of the sequence shown?

12.17 A segment of DNA in *E. coli* has the following sequence of nucleotide pairs:

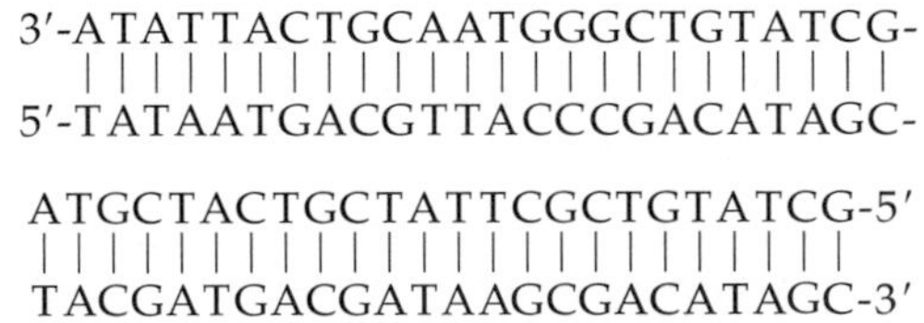

When this segment of DNA is transcribed by RNA polymerase, what will be the sequence of nucleotides in the RNA transcript?

12.18 A segment of DNA in *E. coli* has the following sequence of nucleotide pairs:

```
3′-AACTGTACGTGCTACCTTGCTGATATTACT-
   ||||||||||||||||||||||||||||||
5′-TTGACATGCACGATGGAACGACTATAATGA-
```

```
GCAATGGGCTGTATCGATGCTACTGCTAT-5′
|||||||||||||||||||||||||||||
CGTTACCCGACATAGCTACGATGACGATA-3′
```

When this segment of DNA is transcribed by RNA polymerase, what will be the sequence of nucleotides in the RNA transcript?

12.19 A segment of human DNA has the following sequence of nucleotide pairs:

```
3′-ATATTTACGTGCTACCTTGCTGATAGGACT-
   ||||||||||||||||||||||||||||||
5′-TATAAATGCACGATGGAACGACTATCCTGA-

GCAATGGGCTGTATCGATGCTACTGCTAT-5′
|||||||||||||||||||||||||||||
CGTTACCCGACATAGCTACGATGACGATA-3′
```

When this segment of DNA is transcribed by RNA polymerase, what will be the sequence of nucleotides in the RNA transcript?

12.20 The genome of a human must store a tremendous amount of information using the four nucleotide pairs present in DNA. What do the Morse code and the language of computers tell us about the feasibility of storing large amounts of information using an alphabet composed of just four letters?

12.21 The biosynthesis of metabolite X occurs via six steps catalyzed by six different enzymes. What is the minimal number of genes required for the genetic control of this metabolic pathway? Might more genes be involved? Why?

12.22 What is the central dogma of molecular genetics? What impact did the discovery of RNA tumor viruses have on the central dogma?

12.23 What do the processes of DNA synthesis, RNA synthesis, and polypeptide synthesis have in common?

12.24 What are the two stages of gene expression? Where do they occur in the cell?

12.25 Compare the structures of primary transcripts with those of mRNAs in prokaryotes and eukaryotes. On average, in which group of organisms do they differ the most?

12.26 What four types of RNA molecules participate in the process of gene expression? What are the functions of each type of RNA? Which types of RNA perform their function(s) in (a) the nucleus and (b) the cytoplasm?

12.27 Two eukaryotic genes encode two different polypeptides, each of which is 335 amino acids long. One gene contains a single exon; the other gene contains an intron 41,324 nucleotide pairs long. Which gene would you expect to be transcribed in the least amount of time? Why? When the mRNAs specified by these genes are translated, which mRNA would you expect to be translated in the least time? Why?

12.28 Why was the need for an RNA intermediary in protein synthesis most obvious in eukaryotes? How did researchers first demonstrate that RNA synthesis occurred in the nucleus and that protein synthesis occurred in the cytoplasm?

12.29 Design an experiment to demonstrate that RNA transcripts are synthesized in the nucleus of eukaryotes and are subsequently transported to the cytoplasm.

12.30 What was the first evidence for an unstable mRNA intermediary in prokaryotes? How was this initial evidence substantiated?

12.31 DNA and RNA synthesis occur by very similar mechanisms. In what ways do they differ?

12.32 Total RNA was isolated from human cells growing in culture. This RNA was mixed with nontemplate strands (single strands) of the human gene encoding the enzyme thymidine kinase, and the RNA–DNA mixture was incubated for 12 hours under renaturation conditions. Would you expect any RNA–DNA duplexes to be formed during the incubation? If so, why? If not, why not? The same experiment was then performed using the template strand of the thymidine kinase gene. Would you expect any RNA–DNA duplexes to be formed in this second experiment? If so, why? If not, why not?

12.33 Two preparations of RNA polymerase from *E. coli* are used in separate experiments to catalyze RNA synthesis *in vitro* using a purified fragment of DNA carrying the *argH* gene (Figure 12.1) as template DNA. One preparation catalyzes the synthesis of RNA chains that are highly heterogeneous in size. The other preparation catalyzes the synthesis of RNA chains that are all the same length. What is the most likely difference in the composition of the RNA polymerases in the two preparations?

12.34 Transcription and translation are coupled in prokaryotes. Why is this not the case in eukaryotes?

12.35 You are studying the expression of a recently identified gene in *Drosophila*, and you have shown that this gene is expressed in *Drosophila* cells growing in culture. How can you determine whether this gene is transcribed by RNA polymerase I, RNA polymerase II, or RNA polymerase III?

12.36 What two elements are almost always present in the promoters of eukaryotic genes that are transcribed by RNA polymerase II? Where are these elements located relative to the transcription start site? What are their functions?

12.37 In what ways are most eukaryotic gene transcripts modified? What are the functions of these post-transcriptional modifications?

12.38 How does RNA editing contribute to protein diversity in eukaryotes? What role(s) do guide RNAs play in RNA editing?

12.39 How do the mechanisms by which the introns of tRNA precursors, *Tetrahymena* rRNA precursors, and nuclear pre-mRNAs are excised differ? In which process are snRNAs involved? What role(s) do these snRNAs play?

12.40 What role did the human disease systemic lupus erythematosus play in the characterization of human snRNPs?

12.41 A mutation in an essential human gene changes the 5′ splice site of a large intron from GT to CC. Predict the phenotype of an individual homozygous for this mutation.

12.42 Total RNA was isolated from nuclei of human cells growing in culture. This RNA was mixed with a purified, denatured DNA fragment that carried a large intron of a housekeeping gene (a gene expressed in essentially all cells), and the RNA–DNA mixture was incubated for 12 hours under renaturation conditions. Would you expect any RNA–DNA duplexes to be formed during the incubation? If so, why? If not, why not? The same experiment was then performed using total cytoplasmic RNA from these cells. Would you expect any RNA–DNA duplexes to be formed in this second experiment? If so, why? If not, why not?

BIBLIOGRAPHY

Blum, B., N. R. Sturm, A. M. Simpson, and L. Simpson. 1991. Chimeric gRNA-mRNA molecules with oligo(U) tails covalently linked at sites of RNA editing suggest that U addition occurs by transesterification. *Cell* 65: 543–550.

Cech, T. R. 1986. RNA as an enzyme. *Sci. Amer.* 255(5):64–75.

Chambon, P. 1981. Split genes. *Sci. Amer.* 244(5):60–71.

Conaway, R. C., and J. W. Conaway. 1993. General initiation factors for RNA polymerase II. *Annu. Rev. Biochem.* 62:161–190.

Guthrie, C., and B. Patterson. 1988. Spliceosomal snRNAs. *Annu. Rev. Genet.* 22:387–419.

Henkin, T. M. 1996. Control of transcription termination in prokaryotes. *Annu. Rev. Genet.* 30:35–57.

Reed, R., J. Griffith, and T. Maniatis. 1988. Purification and visualization of native spliceosomes. *Cell* 53:949–961.

Sharp, P. 1987. Splicing of messenger RNA precursors. *Science* 238:729–730.

Steitz, J. A. 1988. Snurps. *Sci. Amer.* 258(6):56–63.

Tilghman, S. M., D. C. Tiemeier, J. G. Seidman, B. M. Peterlin, M. Sullivan, J. V. Maizel, and P. Leder. 1978. Intervening sequence of DNA identified in the structural portion of a mouse β-globin gene. *Proc. Natl. Acad. Sci. USA* 75:725–729.

Uptain, S. M., C. M. Kane, and M. J. Chamberlin. 1997. Basic mechanisms of transcript elongation and its regulation. *Annu. Rev. Biochem.* 66:117–172.

13

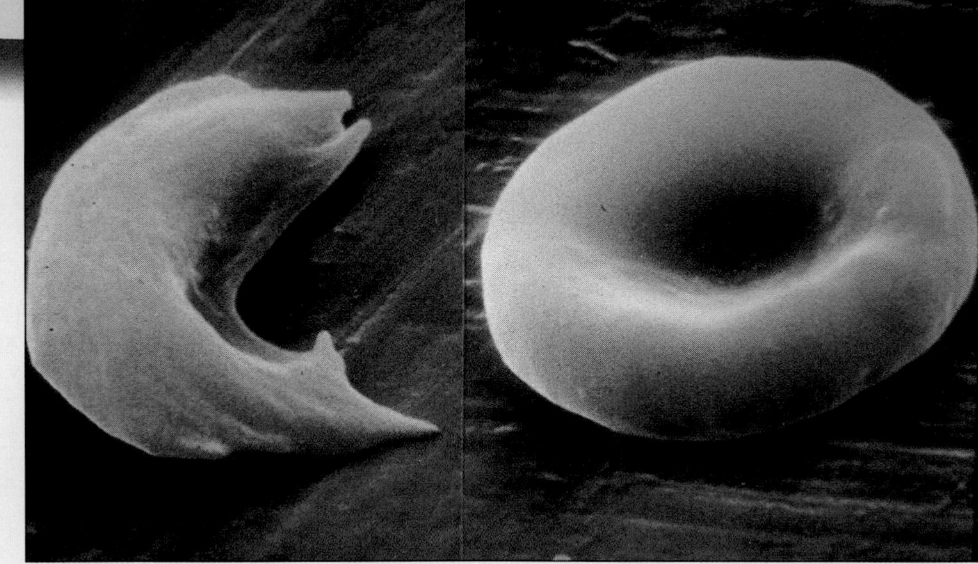

Scanning electron micrograph of sickle-shaped (left) and normal red blood cells.

Translation and the Genetic Code

CHAPTER OUTLINE

Protein Structure
Protein Synthesis: Translation
The Genetic Code
Codon-tRNA Interactions
In Vivo Confirmation of the Nature of the Genetic Code

Sickle-Cell Anemia: Devastating Effects of a Single Base-Pair Change

In 1904, James Herrick, a Chicago physician, and Ernest Irons, a medical intern working under Herrick's supervision, examined the blood cells of one of their patients. They noticed that many of the red blood cells of the young man were thin and elongated, in striking contrast to the round, donutlike red cells of their other patients. Herrick and Irons had examined the blood cells of many other patients but had never seen red cells like these. They obtained fresh blood samples and repeated their microscopic examinations several times, always with the same result. The blood of this patient always contained cells shaped like the sickles that farmers used to harvest grain at that time.

The patient was a 20-year-old college student who was experiencing periods of weakness and dizziness. In many respects, the patient seemed normal, both physically and mentally. His major problem was fatigue. However, a physical exam showed an enlarged heart and enlarged lymph nodes. His heart always seemed to be working too hard, even when he was resting. Blood tests showed that the patient was anemic; the hemoglobin content of his blood was about half the normal level. Hemoglobin is the complex protein that carries oxygen from the lungs to other tissues. Herrick charted this patient's symptoms for six years before publishing his observations in 1910. In his paper, Herrick emphasized the chronic nature of the anemia and the presence of the sickle-shaped red cells. In 1916, at age 32, the patient died from severe anemia and kidney damage.

James Herrick was the first to publish a description of sickle-cell anemia, the first inherited human disease to be understood at the molecular level. In 1949, Linus Pauling and coworkers documented a difference between the hemoglobin of healthy individuals and those with sickle-cell anemia. Hemoglobin contains four polypeptides, two α-globin chains and two β-globin chains, and an iron-containing heme group. In 1957, Vernon Ingram and colleagues demonstrated that the sixth amino acid of the β-chain of sickle-cell hemoglobin was glutamic acid, whereas valine was present at this position in normal adult human hemoglobin. This single amino acid change in a single polypeptide chain is responsible for all the symptoms of sickle-cell anemia. Otherwise, the hemoglobin molecules in sickled and normal red blood cells are identical.

How does the genetic information of an organism, stored in the sequence of nucleotide pairs in DNA, control the phenotype of the organism? How does a nucleotide-pair change in a gene—like the mutation that causes sickle-cell anemia—alter the structure of a protein, the emissary through which the gene acts? In Chapter 12, we discussed the transfer of genetic information stored in the sequences of nucleotide pairs in DNA to the sequences of nucleotides in mRNA molecules, which carry that information from the nucleus to the sites of protein synthesis in the cytoplasm. The transfer of information from DNA to RNA (transcription) and RNA processing occur in the nucleus. In this chapter, we examine the process by which genetic information stored in sequences of nucleotides in mRNAs is used to specify the sequences of amino acids in polypeptide gene products. This process, **translation**, takes place in the cytoplasm on complex workbenches called ribosomes and requires the participation of many macromolecules.

PROTEIN STRUCTURE

Collectively, the proteins constitute about 15 percent of the wet weight of cells. Water molecules account for 70 percent of the total weight of living cells. With the exception of water, proteins are by far the most prevalent component of living organisms in terms of total mass. Not only are proteins major components in terms of cell mass, but they also play many roles vital to the lives of all cells. Before discussing the synthesis of proteins, we need to become more familiar with their structure.

Polypeptides: Twenty Different Amino Acid Subunits

As we know from Chapter 12, proteins are composed of polypeptides, and every polypeptide is encoded by a gene. Each polypeptide consists of a long sequence of amino acids linked together by covalent bonds. Twenty different amino acids are present in most proteins. Occasionally, one or more of the amino acids are chemically modified after a polypeptide is synthesized, yielding a novel amino acid in the mature protein. The structures of the 20 common amino acids are shown in Figure 13.1. All the amino acids except proline contain a **free amino group** and a **free carboxyl group.**

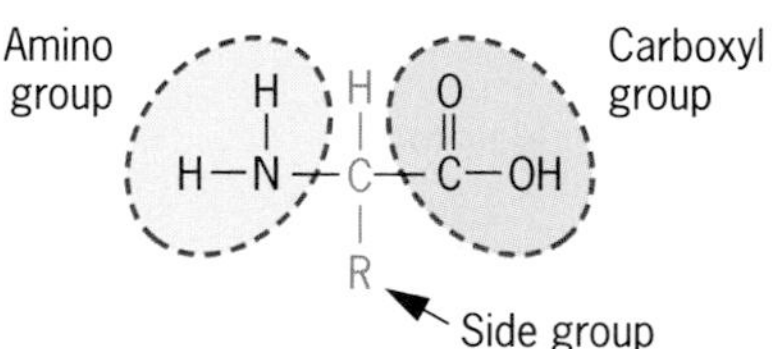

The amino acids differ from each other by the **side groups** (designated **R** for *r*adical) that are present. The highly varied side groups provide the structural diversity of proteins. These side chains are of four types: (1) hydrophobic or nonpolar groups, (2) hydrophilic or polar groups, (3) acidic or negatively charged groups, and (4) basic or positively charged groups (Figure 13.1). The chemical diversity of the side groups of the amino acids is responsible for the enormous structural and functional versatility of proteins.

A **peptide** is a compound composed of two or more amino acids. Polypeptides are long sequences of amino acids, ranging in length from 51 amino acids in insulin to over 1000 amino acids in the silk protein fibroin. Given the 20 different amino acids commonly found in polypeptides, the number of different polypeptides that are possible is truly enormous. For example, the number of different amino acid sequences that can occur in a polypeptide containing 100 amino acids is 20^{100}. Since 20^{100} is too large to comprehend, let's consider a short peptide. There are 1.28 billion (20^7) different amino acid sequences possible in a peptide seven amino acids long. The amino acids in polypeptides are covalently joined by linkages called **peptide bonds.** Each peptide bond is formed by a reaction between the amino group of one amino acid and the carboxyl group of a second amino acid with the elimination of a water molecule (Figure 13.2).

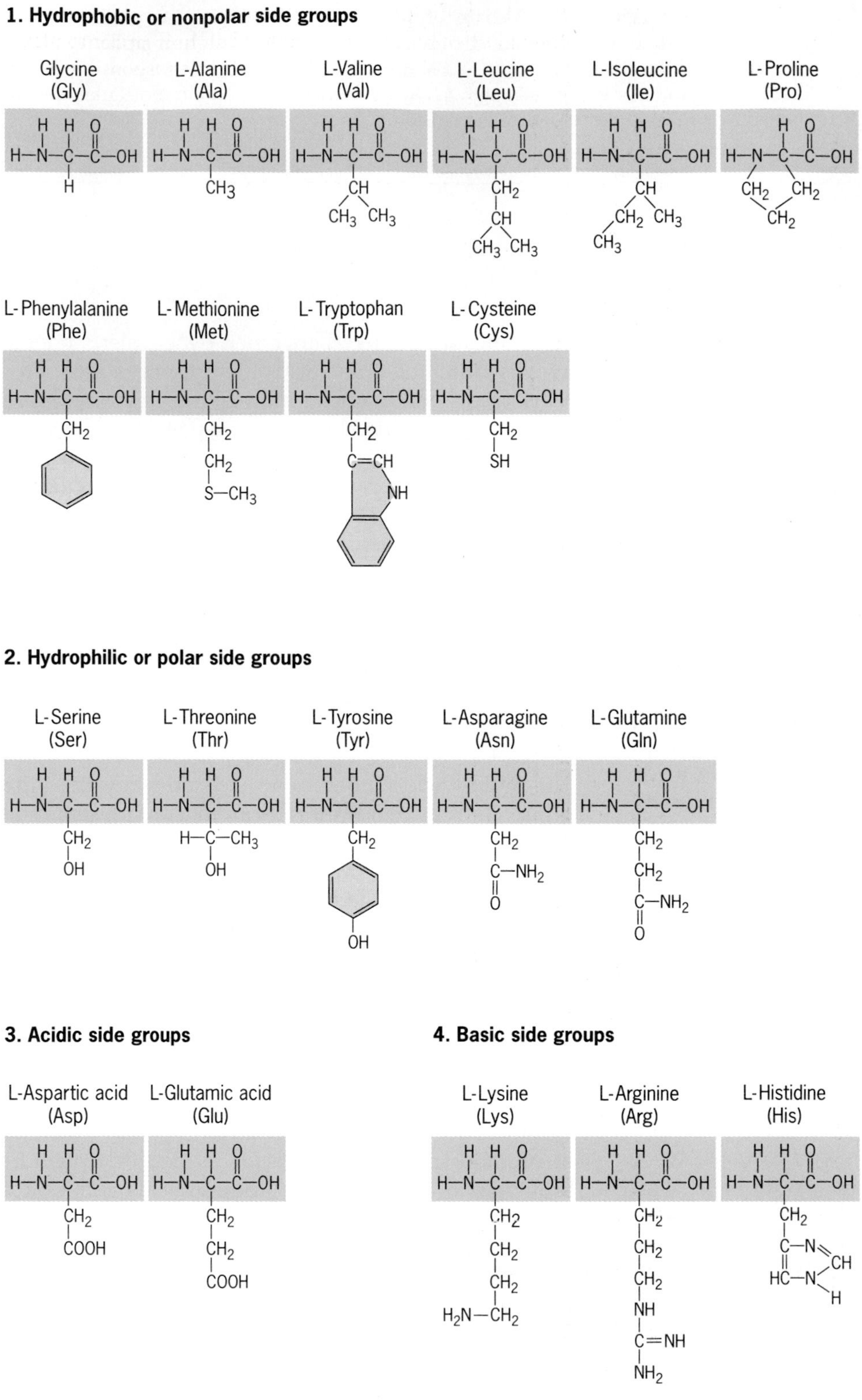

1. Hydrophobic or nonpolar side groups

Glycine (Gly) L-Alanine (Ala) L-Valine (Val) L-Leucine (Leu) L-Isoleucine (Ile) L-Proline (Pro)

L-Phenylalanine (Phe) L-Methionine (Met) L-Tryptophan (Trp) L-Cysteine (Cys)

2. Hydrophilic or polar side groups

L-Serine (Ser) L-Threonine (Thr) L-Tyrosine (Tyr) L-Asparagine (Asn) L-Glutamine (Gln)

3. Acidic side groups

L-Aspartic acid (Asp) L-Glutamic acid (Glu)

4. Basic side groups

L-Lysine (Lys) L-Arginine (Arg) L-Histidine (His)

Figure 13.1 Structures of the 20 amino acids commonly found in proteins. The amino and carboxyl groups, which participate in peptide bond formation during protein synthesis, are shown in the shaded areas. The side groups, which are different for each amino acid, are shown below the shaded areas. The standard three-letter abbreviations are shown in parentheses.

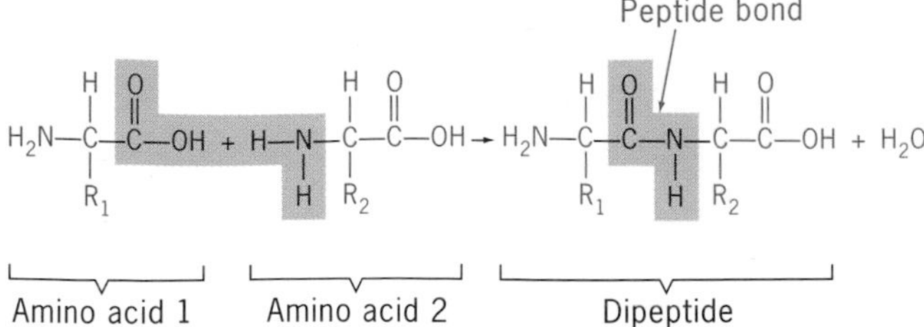

Figure 13.2 The formation of a peptide bond between two amino acids by the removal of water. Each peptide bond connects the amino group of one amino acid and the carboxyl group of the adjacent amino acid.

Proteins: Complex Three-Dimensional Structures

Four different levels of organization—primary, secondary, tertiary, and quaternary—are distinguished in the complex three-dimensional structures of proteins. The **primary structure** of a polypeptide is its amino acid sequence, which is specified by the nucleotide sequence of a gene. The **secondary structure** of a polypeptide refers to the spatial interrelationships of the amino acids in segments of the polypeptide. The **tertiary structure** of a polypeptide refers to its overall folding in three-dimensional space, and the **quaternary structure** refers to the association of two or more polypeptides in a multimeric protein. Hemoglobin provides an excellent example of the complexity of proteins, exhibiting all four levels of structural organization (Figure 13.3).

Most polypeptides will fold spontaneously into specific conformations dictated by their primary structures. If denatured (unfolded) by treatment with appropriate solvents, most proteins will re-form their original conformations when the denaturing agent is removed. Thus, in most cases, all of the information required for shape determination resides in the primary structure of the protein. In some cases, protein folding involves interactions with proteins called **chaperones** that help nascent polypeptides form the proper three-dimensional structure.

The two most common types of secondary structure in proteins are **α helices** and **β sheets** (Figure 13.4). Both structures are maintained by hydrogen bonding between peptide bonds located in close proximity to one another. The α helix is a rigid cylinder in which every peptide bond is hydrogen bonded to another nearby peptide bond (Figure 13.4*a*). A β sheet occurs when a polypeptide folds back onto itself, sometimes repeatedly, and the parallel segments are held in place by hydrogen bonding between neighboring peptide bonds (Figure 13.4*b*).

Whereas the spatial organization of adjacent amino acids and segments of a polypeptide determine its secondary structure, the overall folding of the complete polypeptide defines its tertiary structure or **conformation.** In general, amino acids with hydrophilic side chains are located on the surfaces of proteins (in contact with the aqueous cytoplasm), whereas those with hydrophobic side chains interact with each other in the interior regions. The tertiary structure of a protein is maintained primarily by a large number of rel-

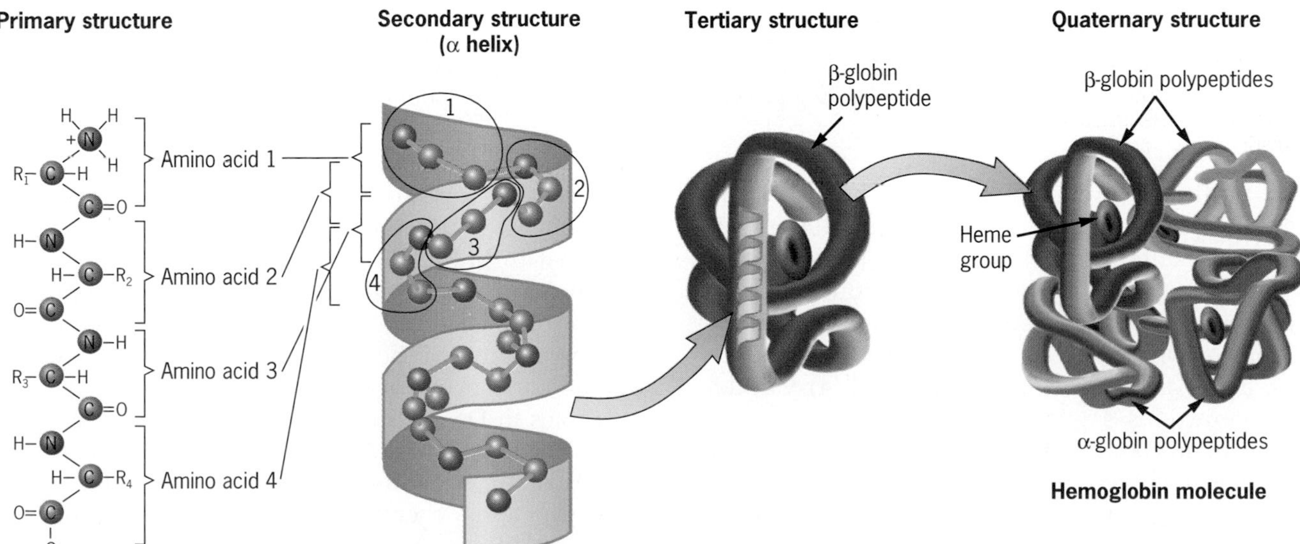

Figure 13.3 The four levels of organization in proteins—(1) primary, (2) secondary, (3) tertiary, and (4) quaternary structure—are illustrated using human hemoglobin as an example.

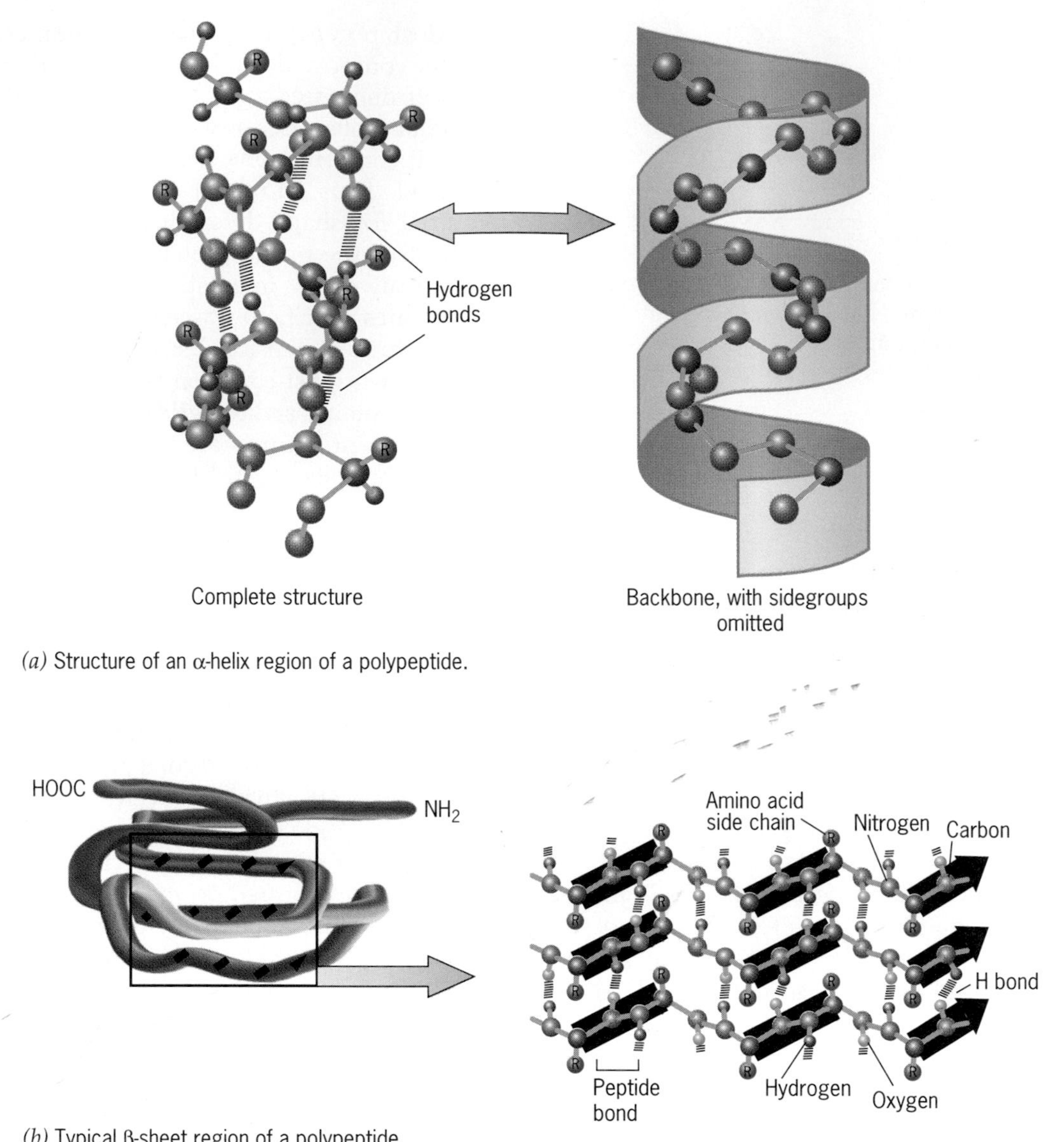

(*a*) Structure of an α-helix region of a polypeptide.

(*b*) Typical β-sheet region of a polypeptide.

Figure 13.4 Secondary structure in proteins. The α helix shown is from myoglobin, the oxygen-transporting protein in vertebrate muscle; the β sheet illustrated here is from a human immunoglobulin (antibody) molecule.

atively weak noncovalent bonds. The only covalent bonds that play a significant role in protein conformation are disulfide (S—S) bridges that form between appropriately positioned cysteine residues (Figure 13.5). However, four different types of noncovalent interactions are involved: (1) ionic bonds, (2) hydrogen bonds, (3) hydrophobic interactions, and (4) Van der Waals interactions (Figure 13.5).

Ionic bonds occur between amino acid side chains with opposite charges—for example, the side groups of lysine and glutamic acid (see Figure 13.1). **Ionic bonds** are strong forces under some conditions, but they are relatively weak interactions in the aqueous interiors of living cells because the polar water molecules partially neutralize or shield the charged groups. **Hydrogen bonds** are weak interactions between electronegative atoms (which have a partial negative charge) and hydrogen atoms (which are electropositive) that are linked to other electronegative atoms. **Hydrophobic interactions** are associations of nonpolar groups with each other when present in aqueous solutions because of their insolubility in water. Hydrogen bonds and hydrophobic interactions play important roles in DNA structure; thus, we have discussed them in some detail in Chapter 10 (see Table 10.3). **Van der Waals interactions** are weak attractions that occur between atoms when they are placed in close proximity to one another. Van der Waals forces are very weak, with about one one-thousandth of the strength of a covalent bond, but they play an important role in maintaining the conformations of closely aligned regions of macromolecules.

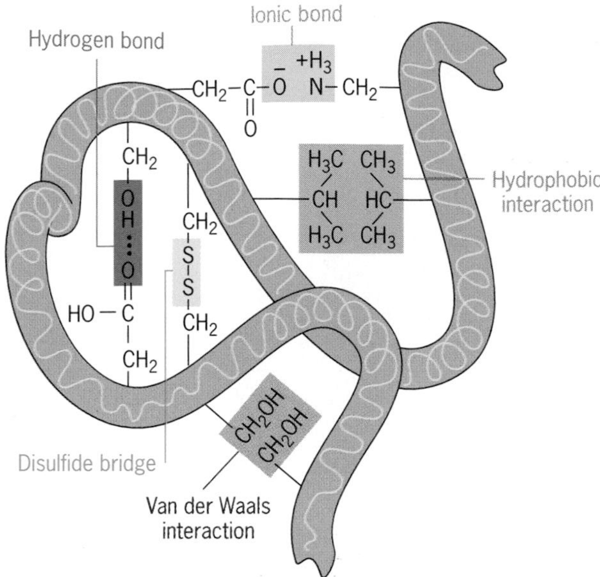

Figure 13.5 The five types of molecular interactions that determine the tertiary structure or three-dimensional conformation of a polypeptide. The disulfide bridge is a covalent bond; all other interactions are noncovalent.

Quaternary structure exists only in proteins that contain more than one polypeptide. Hemoglobin provides a good illustration of quaternary structure, being a tetrameric molecule composed of two α-globin chains and two β-globin chains, plus four iron-containing heme groups (see Figure 13.3).

In a few cases, the primary translation products contain short amino acid sequences, called **inteins,** that excise themselves from the nascent polypeptides. Inteins occur in both eukaryotes and prokaryotes. For example, one of the first inteins discovered is in the RecA protein, which is involved in recombination and DNA repair (Chapter 14), in *Mycobacterium tuberculosis*, the bacterium that causes tuberculosis.

Since the secondary, tertiary, and quaternary structures of proteins and intein excision usually are determined by the primary structure(s) of the polypeptide(s) involved, we will focus in the rest of this chapter on the mechanisms by which genes control the primary structures of polypeptides.

Key Points: **Most genes exert their effect(s) on the phenotype of an organism through proteins, which are large macromolecules composed of polypeptides. Each polypeptide is a chainlike polymer assembled from different amino acids. The amino acid sequence of each polypeptide is specified by the nucleotide sequence of a gene. The vast functional diversity of proteins results in part from their complex three-dimensional structures.**

PROTEIN SYNTHESIS: TRANSLATION

The process by which the genetic information stored in the sequence of nucleotides in an mRNA is translated, according to the specifications of the genetic code, into the sequence of amino acids in the polypeptide gene product is complex, requiring the functions of a large number of macromolecules. These include (1) over 50 polypeptides and 3 to 5 RNA molecules present in each ribosome (the exact composition varies from species to species), (2) at least 20 amino acid-activating enzymes, (3) 40 to 60 different tRNA molecules, and (4) numerous soluble proteins involved in polypeptide chain initiation, elongation, and termination. Because many of these macromolecules, particularly the components of the ribosome, are present in large quantities in each cell, the translation system makes up a major portion of the metabolic machinery of each cell.

Overview of Protein Synthesis

Before focusing on the details of the translation process, we should preview the process of protein synthesis in its entirety. An overview of protein synthesis, illustrating its complexity and the major macromolecules involved, is presented in Figure 13.6. The first step in gene expression, transcription, involves the transfer of information stored in genes to messenger RNA (mRNA) intermediaries, which carry that information to the sites of polypeptide synthesis in the cytoplasm. Transcription is discussed in detail in Chapter 12. The second step, translation, involves the transfer of the information in mRNA molecules into the sequences of amino acids in polypeptide gene products.

Translation occurs on ribosomes, which are complex macromolecular structures located in the cytoplasm. Translation involves three types of RNA, all of which are transcribed from DNA templates (chromosomal genes). In addition to mRNAs, three to five RNA molecules (rRNA molecules) are present as part of the structure of each ribosome, and 40 to 60 small RNA molecules (tRNA molecules) function as adaptors by mediating the incorporation of the proper amino acids into polypeptides in response to specific nucleotide sequences in mRNAs. The amino acids are attached to the correct tRNA molecules by a set of activating enzymes called **aminoacyl-tRNA synthetases.**

The nucleotide sequence of an mRNA molecule is translated into the appropriate amino acid sequence according to the dictations of the genetic code. Each amino acid is specified by one or more codons, and each codon contains three nucleotides. Of the 64 possible nucleotide triplets, 61 specify amino acids and 3 specify polypeptide chain termination. The tRNA molecules contain nucleotide triplets called anticodons,

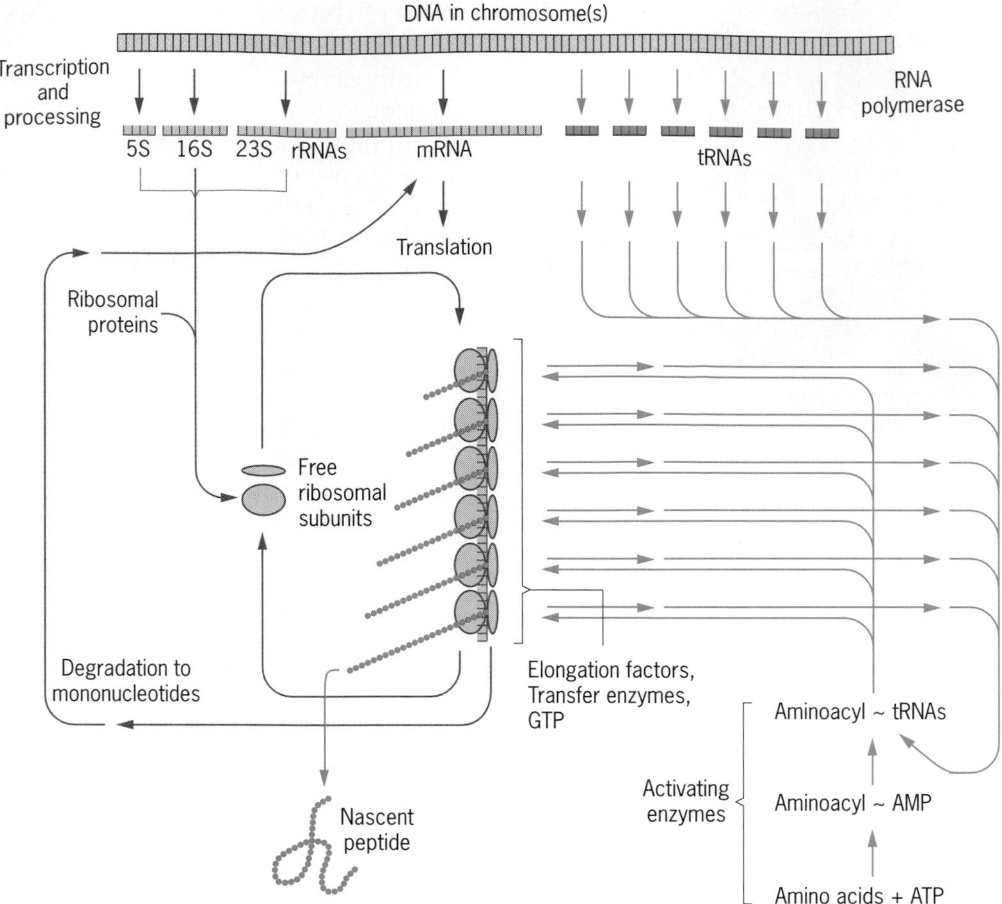

Figure 13.6 Overview of protein synthesis. The sizes of the rRNA molecules shown are correct for bacteria; larger rRNAs are present in eukaryotes. For simplicity, all RNA species have been transcribed from contiguous segments of a single DNA molecule. In reality, the various RNAs are transcripts of genes located at different positions on from one to many chromosomes. Details of the various stages of protein synthesis are discussed in subsequent sections of this chapter.

which base-pair with the codons in mRNA during the translation process.

The ribosomes may be thought of as workbenches, complete with machines and tools needed to make a polypeptide. They are nonspecific in the sense that they can synthesize any polypeptide (any amino acid sequence) encoded by a particular mRNA molecule, even an mRNA from a different species. Each mRNA molecule is simultaneously translated by several ribosomes, resulting in the formation of polyribosomes. Given this brief overview of protein synthesis, we will now examine some of the more important components of the translation machinery more closely.

Components Required for Protein Synthesis: Ribosomes and Transfer RNAs

Living cells devote more energy to the synthesis of proteins than to any other aspect of metabolism. About one-third of the total dry mass of most cells consists of molecules that participate directly in the biosynthesis of proteins. In *E. coli*, the approximately 200,000 ribosomes account for 25 percent of the dry weight of each cell. This commitment of a major proportion of the metabolic machinery of cells to the process of protein synthesis documents its importance in the life forms that exist on our planet.

When the sites of protein synthesis were labeled in cells grown for short intervals in the presence of radioactive amino acids and were visualized by autoradiography, the results showed that proteins are synthesized on the ribosomes. In prokaryotes, ribosomes are distributed throughout cells; in eukaryotes, they are located in the cytoplasm, frequently on the extensive intracellular membrane network of the endoplasmic reticulum.

Ribosomes are approximately half protein and half RNA (Figure 13.7). They are composed of two subunits, one large and one small, which dissociate when the translation of an mRNA molecule is completed and reassociate during the initiation of translation. Ribosome sizes are most frequently expressed in terms of their rates of sedimentation during centrifugation, in Svedberg units (see Technical Sidelight in Chapter 11). The *E. coli* ribosome, like the ribosomes of other prokaryotes, has a molecular weight of 2.5×10^6, a size of 70S, and dimensions of about 20 nm × 25 nm. The ribosomes of eukaryotes are larger (usually about 80S); however, size varies from species to species. The ribosomes present in the mitochondria and chloroplasts of eukaryotic cells are smaller (usually about 60S).

Although the size and macromolecular composition of ribosomes vary, the overall three-dimensional structure of the ribosome is basically the same in all organisms. In *E. coli*, the small (30S) ribosomal subunit contains a 16S (molecular weight about 6×10^5) RNA molecule plus 21 different polypeptides, and the large (50S) subunit contains two RNA molecules (5S, molecular weight about 4×10^4, and 23S, molecular weight about 1.2×10^6) plus 31 polypeptides. In mammalian ribosomes, the small subunit contains an 18S RNA molecule plus 33 polypeptides, and the large subunit contains three RNA molecules of sizes 5S, 5.8S, and 28S plus 49 polypeptides. In organelles, the corresponding rRNA sizes are 5S, 13S, and 21S.

Masayasu Nomura and his colleagues were able to disassemble the 30S ribosomal subunit of *E. coli* into the individual macromolecules and then reconstitute functional 30S subunits from the components. In this way, they studied the functions of individual rRNA and ribosomal protein molecules.

The ribosomal RNA molecules, like mRNA molecules, are transcribed from a DNA template. In eukaryotes, rRNA synthesis occurs in the nucleolus (Figure 13.8) and is catalyzed by RNA polymerase I. The nucleolus is a highly specialized component of the nucleus devoted exclusively to the synthesis of rRNAs and their assembly into ribosomes. The ribosomal

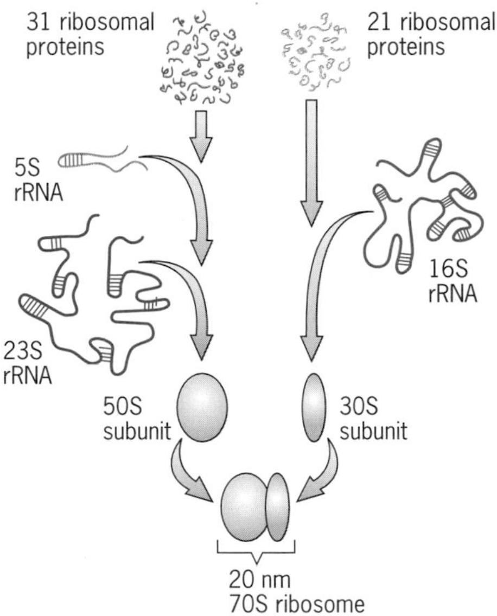

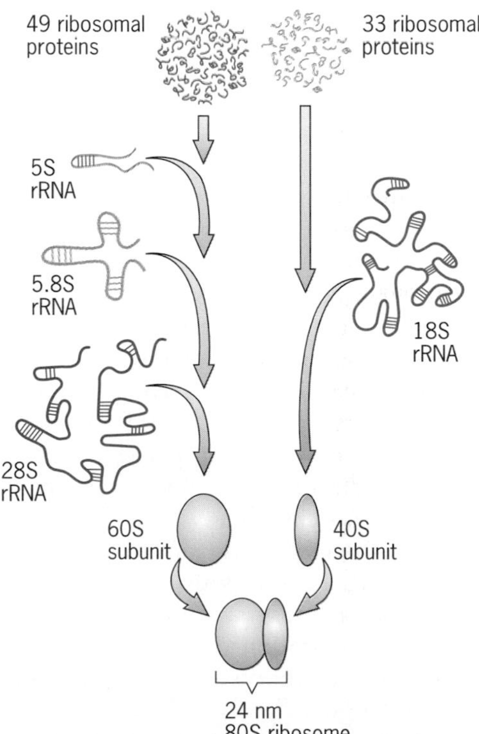

Figure 13.7 Macromolecular composition of prokaryotic and eukaryotic ribosomes.

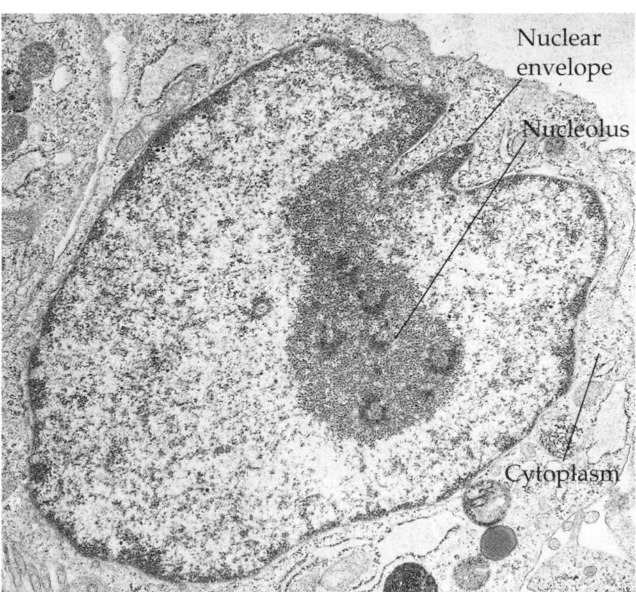

Figure 13.8 Electron micrograph of a human fibroblast, showing the nucleolus, the site of rRNA synthesis and ribosome assembly.

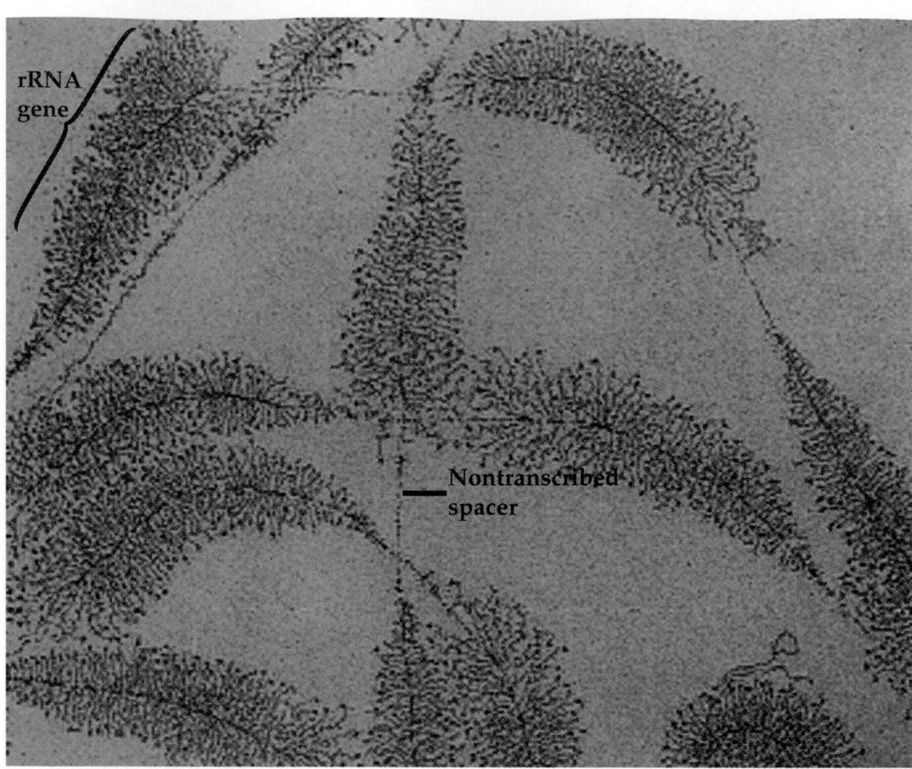

Figure 13.9 Electron micrograph showing the transcription of tandemly repeated rRNA genes in the nucleolus of the newt *Triturus viridescens*. A gradient of fibrils of increasing length is observed for each rRNA gene, and nontranscribed spacer regions separate the genes.

RNA genes are present in tandemly duplicated arrays separated by intergenic spacer regions. The transcription of these tandem sets of rRNA genes can be visualized directly by electron microscopy (Figure 13.9).

The transcription of the rRNA genes produces precursors that are much larger than the RNA molecules found in ribosomes. These rRNA precursors undergo post-transcriptional processing to produce the mature rRNA molecules. In *E. coli*, the rRNA gene transcript is a 30S precursor, which undergoes endonucleolytic cleavages to produce the 5S, 16S, and 23S rRNAs plus one 4S transfer RNA molecule (Figure 13.10*a*). In mammals, the 5.8S, 18S, and 28S rRNAs are cleaved from a 45S precursor (Figure 13.10*b*), whereas the 5S rRNA is produced by post-transcriptional processing of a separate gene transcript. In addition to the post-transcriptional cleavages of rRNA precursors, many of the nucleotides in rRNAs are post-transcriptionally methylated. The methylation is thought to protect rRNA molecules from degradation by ribonucleases.

Multiple copies of the genes for rRNA are present in the genomes of all organisms that have been studied to date. This redundancy of rRNA genes is not surprising considering the large number of ribosomes present per cell. In *E. coli*, seven rRNA genes (*rrnA-rrnE, rrnG, rrnH*) are distributed among three distinct sites on the chromosome. In eukaryotes, the rRNA genes are present in hundreds to thousands of copies. The 5.8S-18S-28S rRNA genes of eukaryotes are present in tandem arrays in the **nucleolar organizer regions** of the chromosomes. In some eukaryotes, such as maize, there is a single pair of nucleolar organizers (on chromosome 6 in maize). In *Drosophila* and the South African clawed toad, *Xenopus laevis*, the sex chromosomes carry the nucleolar organizers. Humans have five pairs of nucleolar organizers located on the short arms of chromosomes 13, 14, 15, 21, and 22. The 5S rRNA genes in eukaryotes are not located in the nucleolar organizer regions. Instead, they are distributed over several chromosomes. However, the 5S rRNA genes are highly redundant, just as are the 5.8S-18S-28S rRNA genes.

The nucleolar organizer region in *Xenopus laevis* contains about 500 copies of the 5.8S-18S-28S rRNA gene. The estimates of rRNA gene redundancy are similar for other animals. Plants exhibit a greater variation in rRNA gene redundancy, with several thousand copies present in some genomes. Intraspecies variation in the amount of rRNA gene redundancy has also been documented in several species.

Although the ribosomes provide many of the components required for protein synthesis, and the specifications for each polypeptide are encoded in an mRNA molecule, the translation of a coded mRNA message into a sequence of amino acids in a polypeptide requires one additional class of RNA molecules, the transfer RNA (tRNA) molecules. Chemical considerations suggested that direct interactions between the

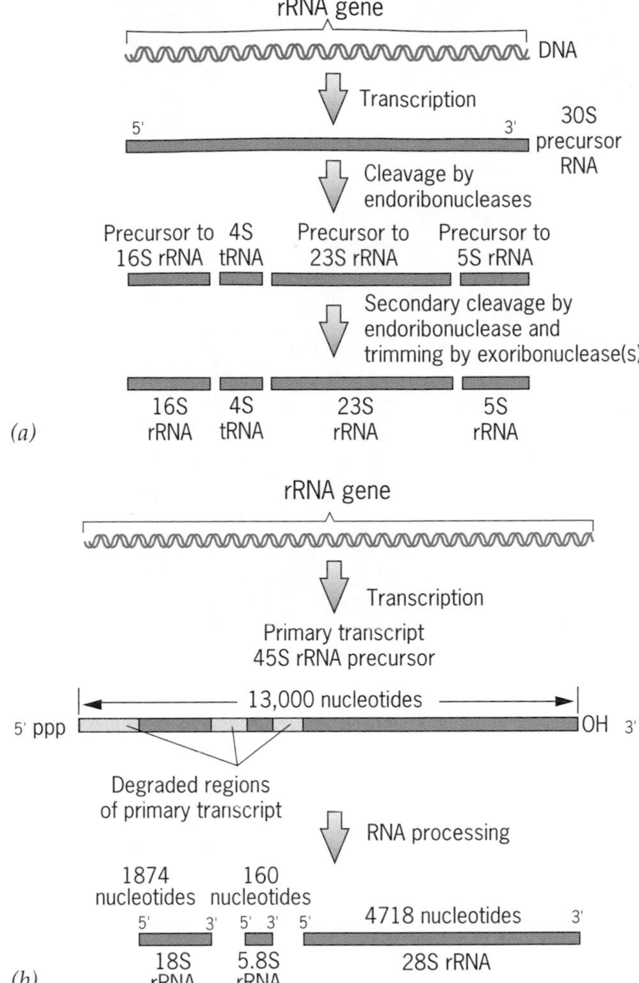

Figure 13.10 Synthesis and processing of *(a)* the 30S rRNA precursor in *E. coli* and *(b)* the 45S rRNA precursor in mammals.

amino acids and the nucleotide triplets or codons in mRNA were unlikely. Thus, in 1958, Francis Crick proposed that some kind of an adaptor molecule must mediate the specification of amino acids by codons in mRNAs during protein synthesis. The adaptor molecules were soon identified by other researchers and shown to be small (4S, 70–90 nucleotides long) RNA molecules. These molecules, first called soluble RNA (sRNA) molecules and subsequently transfer RNA (tRNA) molecules, contain a triplet nucleotide sequence, the anticodon, which is complementary to and base-pairs with the codon sequence in mRNA during translation. There are one to four tRNAs for each of the 20 amino acids.

The amino acids are attached to the tRNAs by high-energy (very reactive) bonds between the carboxyl groups of the amino acids and the 3′-hydroxyl termini of the tRNAs. The tRNAs are activated or charged with amino acids in a two-step process, with both reactions catalyzed by the same enzyme, aminoacyl-tRNA synthetase. There is at least one aminoacyl-tRNA synthetase for each of the 20 amino acids. The first step in aminoacyl-tRNA synthesis involves the activation of the amino acid using energy from adenosine triphosphate (ATP):

$$\text{amino acid} + \text{ATP} \xrightarrow{\text{aminoacyl-tRNA synthetase}} \text{amino acid} \sim \text{AMP} + \textcircled{P} \sim \textcircled{P}$$

The amino acid~AMP intermediate is not normally released from the enzyme before undergoing the second step in aminoacyl-tRNA synthesis, namely, the reaction with the appropriate tRNA:

$$\text{amino acid} \sim \text{AMP} + \text{tRNA} \xrightarrow{\text{aminoacyl-tRNA synthetase}} \text{amino acid} \sim \text{tRNA} + \text{AMP}$$

The aminoacyl~tRNAs are the substrates for polypeptide synthesis on ribosomes, with each activated tRNA recognizing the correct mRNA codon and presenting the amino acid in a steric configuration (three-dimensional structure) that facilitates peptide bond formation.

The tRNAs are transcribed from genes. As in the case of rRNAs, the tRNAs are transcribed in the form of larger precursor molecules that undergo post-transcriptional processing (cleavage, trimming, methylation, and so forth). The mature tRNA molecules contain several nucleosides that are not present in the primary tRNA gene transcripts. These unusual nucleosides, such as inosine, pseudouridine, dihydrouridine, 1-methylguanosine, and several others, are produced by post-transcriptional, enzyme-catalyzed modifications of the four nucleosides incorporated into RNA during transcription.

Because of their small size (70 to 80 nucleotides long), tRNAs have been more amenable to structural analysis than the other, larger molecules of RNA involved in protein synthesis. The complete nucleotide sequence and proposed cloverleaf structure of the alanine tRNA of yeast (Figure 13.11) were published by Robert W. Holley and colleagues in 1965; Holley shared the 1968 Nobel Prize in physiology or medicine for this work. Since then, many tRNAs have been sequenced, and the yeast alanine tRNA gene has been synthesized *in vitro* from mononucleotides by H. Ghobind Khorana and coworkers. The three-dimensional structure of the phenylalanine tRNA of yeast was determined by X-ray diffraction studies in 1974

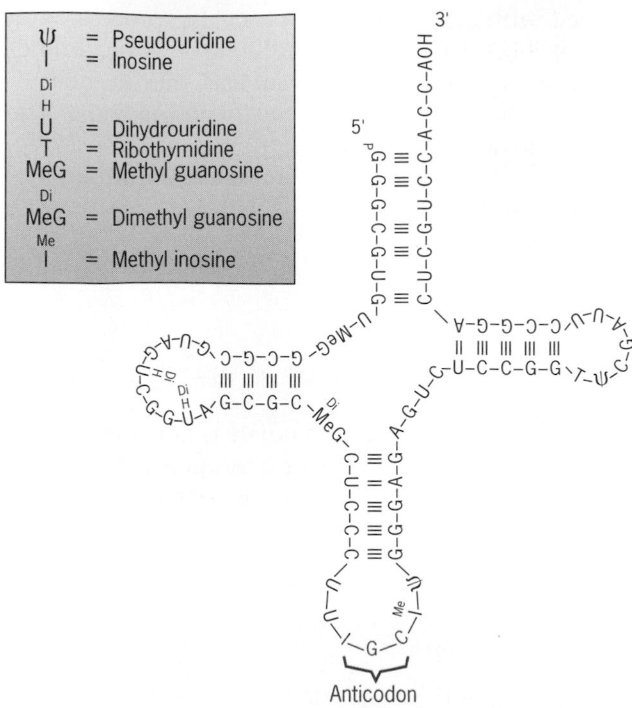

Figure 13.11 Nucleotide sequence and cloverleaf configuration of the alanine tRNA of *S. cerevisiae*. The names of the modified nucleosides present in the tRNA are shown in the inset.

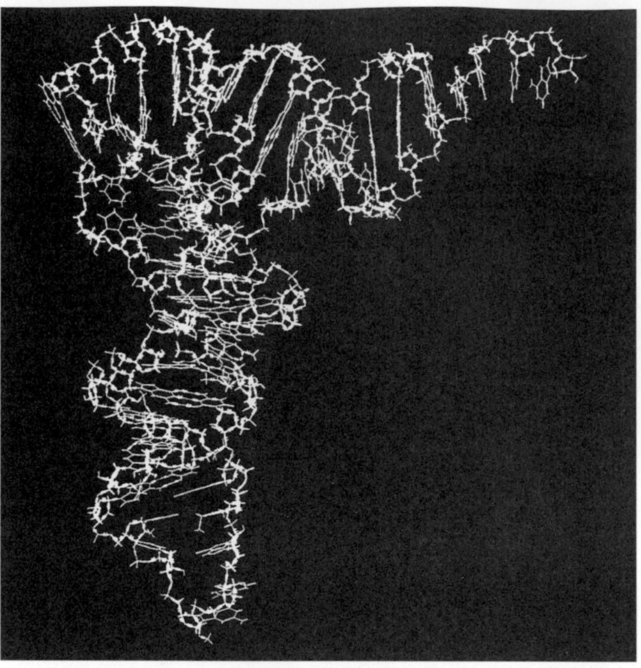

(a)

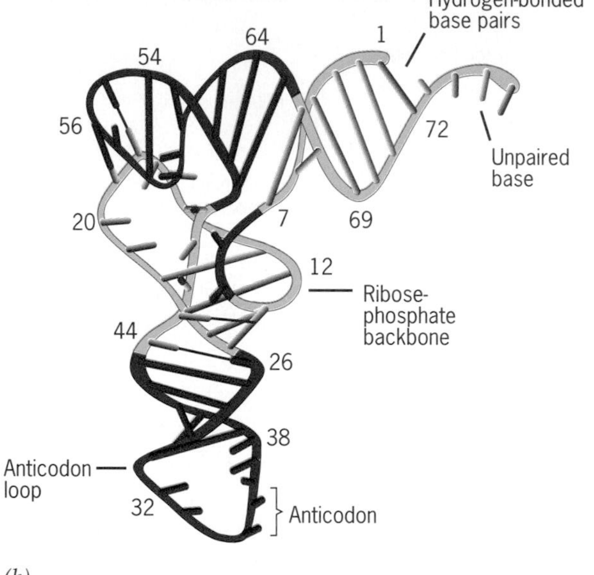

(b)

Figure 13.12 Photograph (*a*) and interpretative drawing (*b*) of a molecular model of the yeast phenylalanine tRNA based on X-ray diffraction data.

(Figure 13.12). The anticodon of each tRNA occurs within a loop (nonhydrogen-bonded region) near the middle of the molecule.

It should be apparent that tRNA molecules must contain a great deal of specificity despite their small size. Not only must they (1) have the correct anticodon sequences, so as to respond to the right codons, but they also must (2) be recognized by the correct aminoacyl-tRNA synthetases, so that they are activated with the correct amino acids, and (3) bind to the appropriate sites on the ribosomes to carry out their adaptor functions.

François Chapeville and Günter von Ehrenstein and colleagues have proven, by means of a simple and direct experiment (Figure 13.13), that the specificity for codon recognition resides in the tRNA portion of an aminoacyl-tRNA, rather than in the amino acid. They treated cysteyl-$tRNA^{Cys}$ (the cysteine tRNA activated with cysteine) with a strongly reducing nickel powder (Raney nickel), which converted (reduced) the cysteine to alanine while it was still attached to the cysteine tRNA. When this hybrid aminoacyl-tRNA, alanyl-$tRNA^{Cys}$, was used in *in vitro* protein-synthesizing systems, alanine was incorporated into polypeptides at positions normally occupied by cysteine. Thus, tRNAs really do function as the adaptor molecules that Crick proposed must mediate the interaction between the codons in mRNAs and the amino acids that the codons specify during the translation process.

There are three tRNA binding sites on each ribosome (Figure 13.14*a*). The ***A*** or **aminoacyl site** binds the incoming aminoacyl-tRNA, the tRNA carrying the next amino acid to be added to the growing polypeptide chain. The ***P*** or **peptidyl site** binds the tRNA to which the growing polypeptide is attached. The ***E*** or **exit site** binds the departing uncharged tRNA.

Although the aminoacyl-tRNA binding sites are located largely on the 50S subunit and the mRNA mol-

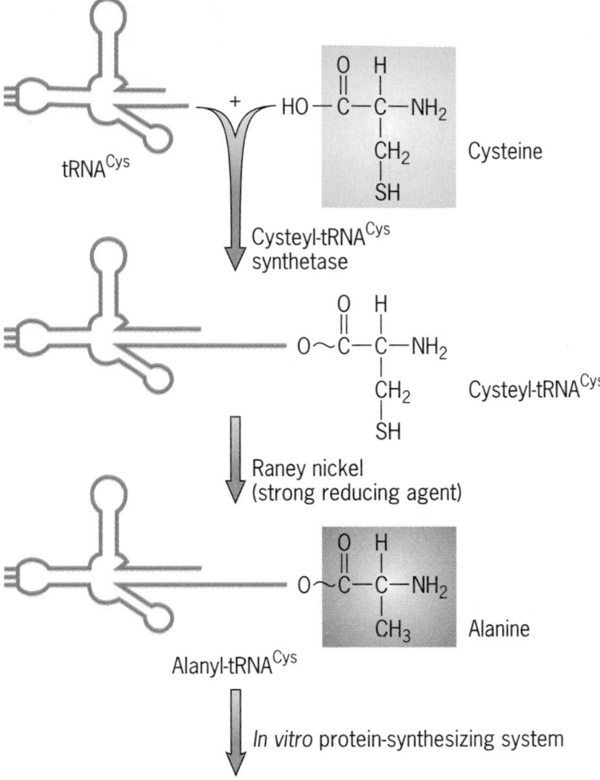

Figure 13.13 Proof that the codon-recognizing specificity of an aminoacyl-tRNA complex resides in the tRNA rather than in the amino acid.

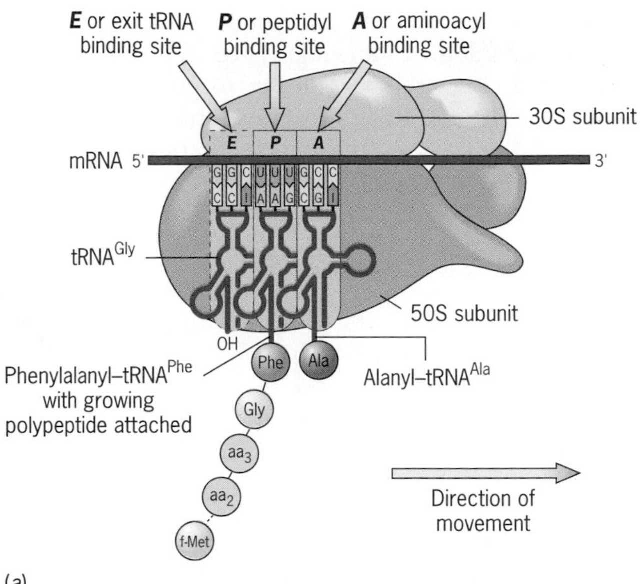

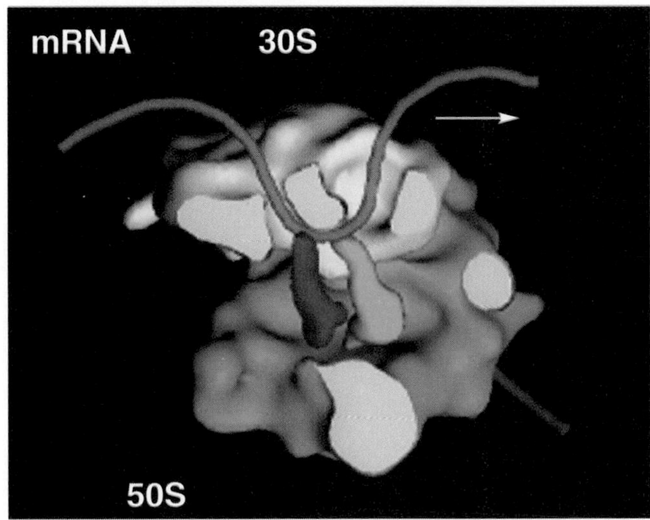

(*b*)

Figure 13.14 Diagram (*a*) and cutaway view (*b*) of a model of the 70S ribosome of *E. coli*. (*a*) Each ribosome-mRNA complex contains three aminoacyl-tRNA binding sites. The *A* or aminoacyl-tRNA site is occupied by an alanyl-tRNA^{Ala} complex. The *P* or peptidyl site is occupied by a phenylalanyl-tRNA^{Phe} complex, with the growing polypeptide chain covalently linked to the phenylalanine tRNA. The *E* or exit site is occupied by the tRNA^{Gly} prior to its release from the ribosome. (*b*) An mRNA molecule (orange), which is attached to the 30S subunit (light green) of the ribosome, contributes specificity to the tRNA-binding sites, which are located largely on the 50S subunit (blue) of the ribosome. The aminoacyl-tRNAs located in the *P* and *A* sites are shown in red and dark green, respectively. The *E* site is unoccupied.

ecule is bound by the 30S subunit, the specificity for aminoacyl-tRNA binding in each site is provided by the mRNA codon that makes up part of the binding site (Figure 13.14*b*). As the ribosome moves along an mRNA (or as the mRNA is shuttled across the ribosome), the specificity for the aminoacyl-tRNA binding in the *A*, *P*, and *E* sites changes as different mRNA codons move into register in the binding sites. The ribosomal binding sites by themselves (minus mRNA) are thus capable of binding any aminoacyl-tRNA.

Translation: The Synthesis of Polypeptides Using mRNA Templates

We now have reviewed all the major components of the protein-synthesizing system. The mRNA molecules provide the specifications for the amino acid sequences of the polypeptide gene products. The ribosomes provide many of the macromolecular components required for the translation process. The tRNAs provide the adaptor molecules needed to incorporate amino acids into polypeptides in response to codons in mRNAs. In addition, several soluble proteins participate in the process. The translation of the sequence of nucleotides in an mRNA molecule into the sequence of amino acids in its polypeptide product can be divided into three stages: (1) polypeptide chain initiation, (2) chain elongation, and (3) chain termination.

The **initiation** of translation includes all events that precede the formation of a peptide bond between the first two amino acids of the new polypeptide chain.

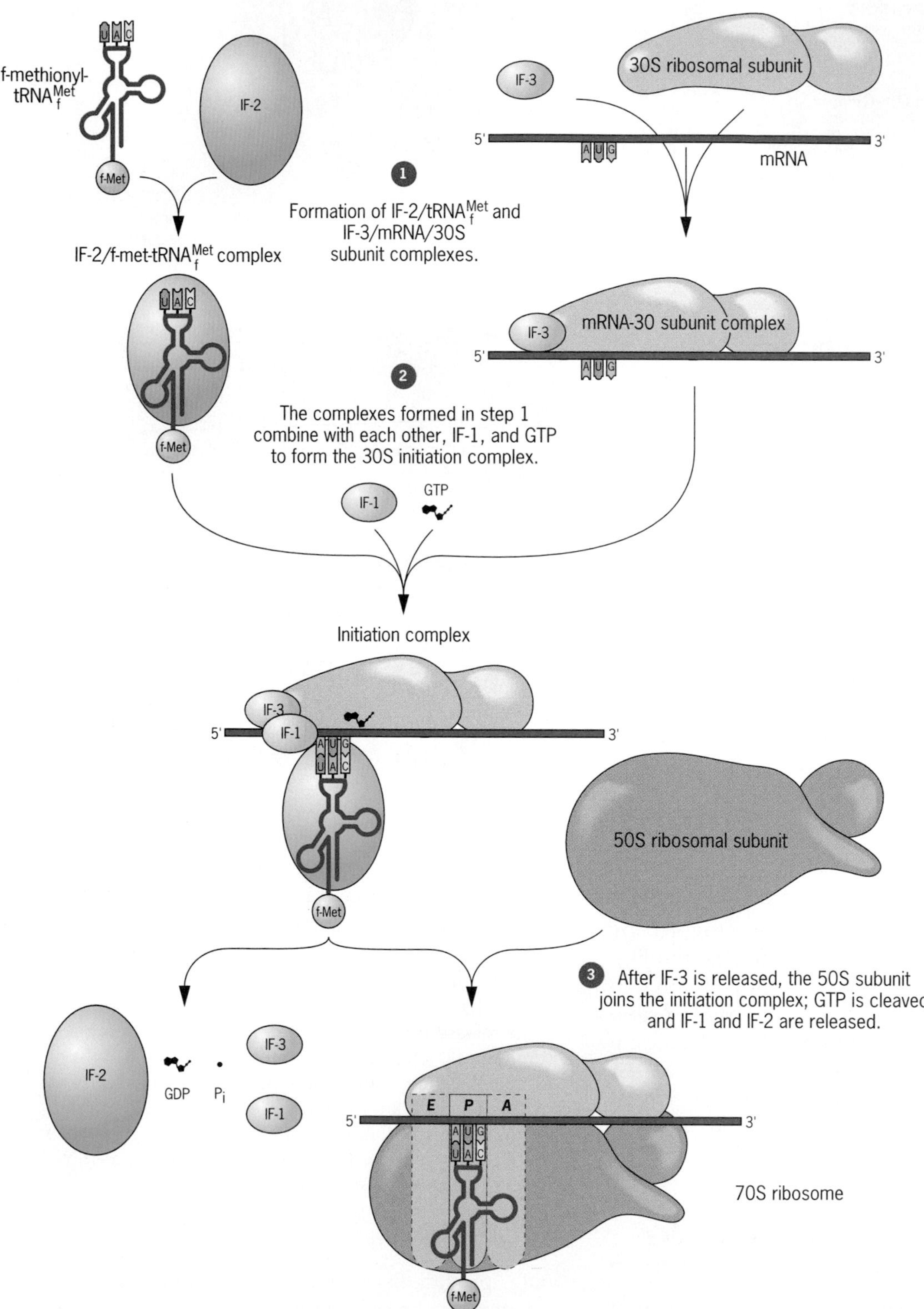

Figure 13.15 The initiation of translation in *E. coli*.

Although several aspects of the initiation process are the same in prokaryotes and eukaryotes, some are different. Accordingly, we will first examine the initiation of polypeptide chains in *E. coli*, and we will then look at the unique aspects of translational initiation in eukaryotes.

In *E. coli*, the initiation process involves the 30S subunit of the ribosome, a special initiator tRNA, an mRNA molecule, three soluble protein **initiation factors: IF-1, IF-2,** and **IF-3,** and one molecule of GTP (Figure 13.15). Translation occurs on 70S ribosomes, but the ribosomes dissociate into their 30S and 50S subunits each time they complete the synthesis of a polypeptide chain. In the first stage of the initiation of translation, a free 30S subunit interacts with an mRNA molecule and the initiation factors. The 50S subunit joins the complex to form the 70S ribosome in the final step of the initiation process.

The synthesis of polypeptides is initiated by a special tRNA, designated **$tRNA_f^{Met}$**. Therefore, all polypeptides begin with methionine during synthesis. The amino-terminal methionine is subsequently cleaved from many polypeptides. Thus, functional proteins need not have an amino-terminal methionine. The methionine on the initiator $tRNA_f^{Met}$ has the amino group blocked with a formyl $(-\overset{O}{\overset{\|}{C}}-H)$ group. A distinct methionine tRNA, **$tRNA^{Met}$**, responds to internal methionine codons. Both methionine tRNAs have the same anticodon, and both respond to the same codon (AUG) for methionine. However, the formylated amino group on methionyl-$tRNA_f^{Met}$ prevents the formation of a peptide bond between the amino group and the carboxyl group of the amino acid at the end of the growing polypeptide chain. In addition, only methionyl-$tRNA_f^{Met}$ interacts with protein initiation factor IF-2 to begin the initiation process (Figure 13.15). Thus, only methionyl-tRNA binds to the ribosome in response to AUG initiation codons in mRNAs, leaving methionyl-$tRNA^{Met}$ to bind in response to internal AUG codons. Methionyl-$tRNA_f^{Met}$ also binds to ribosomes in response to an alternate initiator codon, GUG (a valine codon when present at internal positions), that occurs in some mRNA molecules.

Polypeptide chain initiation begins with the formation of two complexes: (1) one contains initiation factor IF-2 and methionyl-$tRNA_f^{Met}$, and (2) the other contains an mRNA molecule, a 30S ribosomal subunit and initiation factor IF-3 (Figure 13.15). The 30S subunit/mRNA complex will form only in the presence of IF-3; thus, IF-3 controls the ability of the 30S subunit to begin the initiation process. The formation of the 30S subunit/mRNA complex depends in part on base-pairing between a nucleotide sequence near the 3′ end of the 16S rRNA and a sequence near the 5′ end of the mRNA molecule (Figure 13.16). Prokaryotic mRNAs contain a conserved polypurine tract, consensus AGGAGG, located about seven nucleotides upstream from the AUG initiation codon. This conserved hexamer, called the **Shine-Dalgarno sequence** after the scientists who discovered it, is complementary to a sequence near the 5′ terminus of the 16S ribosomal RNA. When the Shine-Dalgarno sequences of mRNAs are experimentally modified so that they can no longer base-pair with the 16S rRNA, the modified mRNAs either are not translated or are translated very inefficiently, indicating that this base-pairing plays an important role in translation.

The IF-2/methionyl-$tRNA_f^{Met}$ complex and the mRNA/30S subunit/IF-3 complex then combine with each other and with initiation factor IF-1 and one molecule of GTP to form the complete 30S initiation complex. The final step in the initiation of translation is the addition of the 50S subunit to the 30S initiation complex to produce the complete 70S ribosome. Initiation factor IF-3 must be released from the complex before the 50S subunit can join the complex; IF-3 and the 50S subunit are never found to be associated with the 30S subunit at the same time. The addition of the 50S subunit requires energy from GTP and the release of initiation factors IF-1 and IF-2. The results of early experiments suggested that one molecule of GTP was required for polypeptide chain initiation. More recent data indicate that two molecules of GTP are required to form the initiation complex; however, there is still not complete agreement as to whether one or two molecules of GTP are required for initiation.

The addition of the 50S ribosomal subunit to the complex positions the initiator tRNA, methionyl-$tRNA_f^{Met}$, in the peptidyl (*P*) site with the anticodon of the tRNA aligned with the AUG initiation codon of the mRNA. Methionyl-$tRNA_f^{Met}$ is the only aminoacyl-tRNA that can enter the *P* site directly, without first passing through the aminoacyl (*A*) site. With the initiator AUG positioned in the *P* site, the second codon of the mRNA is in register with the *A* site, dictating

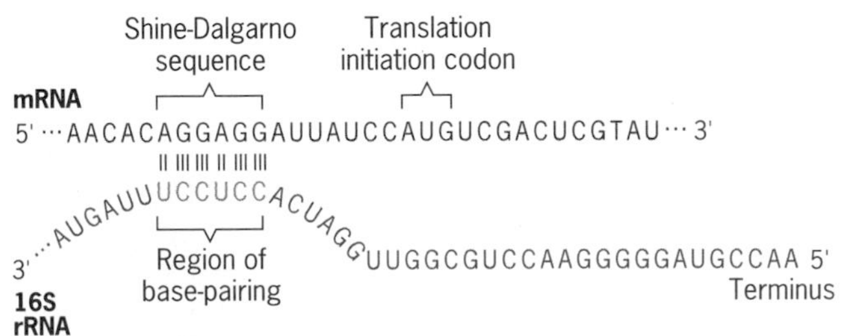

Figure 13.16 Base-pairing between the Shine-Dalgarno sequence in a prokaryotic mRNA and a complementary sequence near the 5′ terminus of the 16S rRNA is involved in the formation of the mRNA/30S ribosomal subunit initiation complex.

the aminoacyl-tRNA binding specificity at that site and setting the stage for the second phase in polypeptide synthesis, chain elongation.

The initiation of translation is more complex in eukaryotes, involving several soluble initiation factors. Nevertheless, the overall process is similar except for two features. (1) The amino group of the methionine on the initiator tRNA is not formylated as in prokaryotes. (2) The initiation complex forms at the 5′ terminus of the mRNA, not at the Shine-Dalgarno/AUG translation start site as in *E. coli*. In eukaryotes, the initiation complex scans the mRNA, starting at the 5′ end,

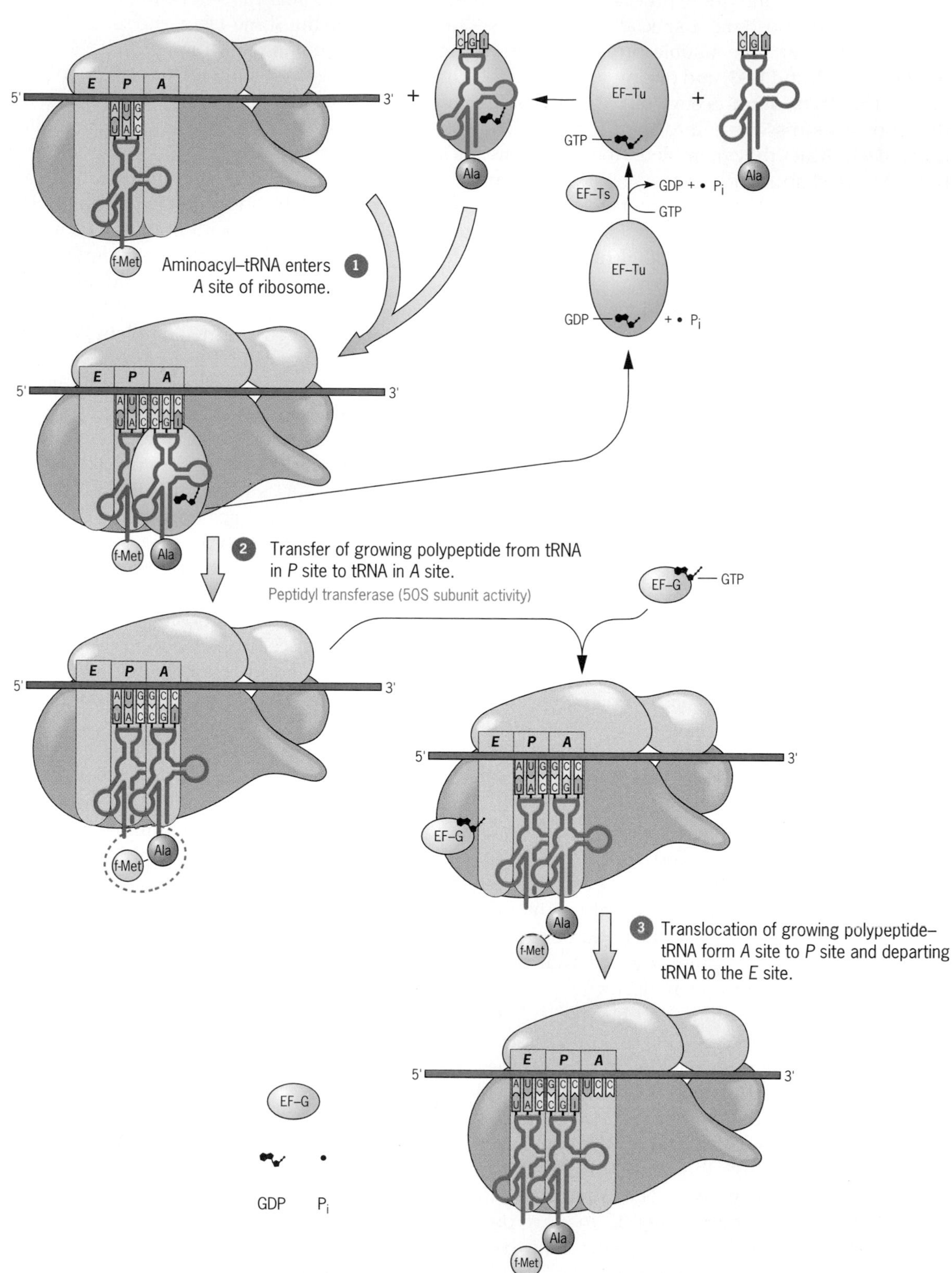

Figure 13.17 Polypeptide chain elongation in *E. coli*.

searching for an AUG translation-initiation codon. Thus, in eukaryotes, translation usually begins at the AUG closest to the 5′ terminus of the mRNA molecule, although the efficiency with which a given AUG is used to initiate translation depends on the contiguous nucleotide sequence.

Like prokaryotes, eukaryotes contain a special initiator tRNA, **tRNA$_i^{Met}$**, but the amino group of the methionyl-tRNA$_i^{Met}$ is not formylated. The initiator methionyl-tRNA$_i^{Met}$ interacts with a soluble initiation factor and enters the *P* site directly during the initiation process, just as in *E. coli*.

In eukaryotes, a cap-binding protein (CBP) binds to the 7-methyl guanosine cap at the 5′ terminus of the mRNA. Then, other initiation factors bind to the CBP-mRNA complex, followed by the small (40S) subunit of the ribosome. The entire initiation complex moves 5′ → 3′ along the mRNA molecule, searching for an AUG codon. When an AUG triplet is found, the initiation factors dissociate from the complex, and the large (60S) subunit binds to the methionyl-tRNA/mRNA/40S subunit complex, forming the complete (80S) ribosome. The 80S ribosome/mRNA/tRNA complex is ready to begin the second phase of translation, chain elongation.

The process of polypeptide chain **elongation** is basically the same in both prokaryotes and eukaryotes. The addition of each amino acid to the growing polypeptide occurs in three steps: (1) binding of an aminoacyl-tRNA to the *A* site of the ribosome, (2) transfer of the growing polypeptide chain from the tRNA in the *P* site to the tRNA in the *A* site by the formation of a new peptide bond, and (3) translocation of the ribosome along the mRNA to position the next codon in the *A* site (Figure 13.17). During step 3, the nascent polypeptide-tRNA and the uncharged tRNA

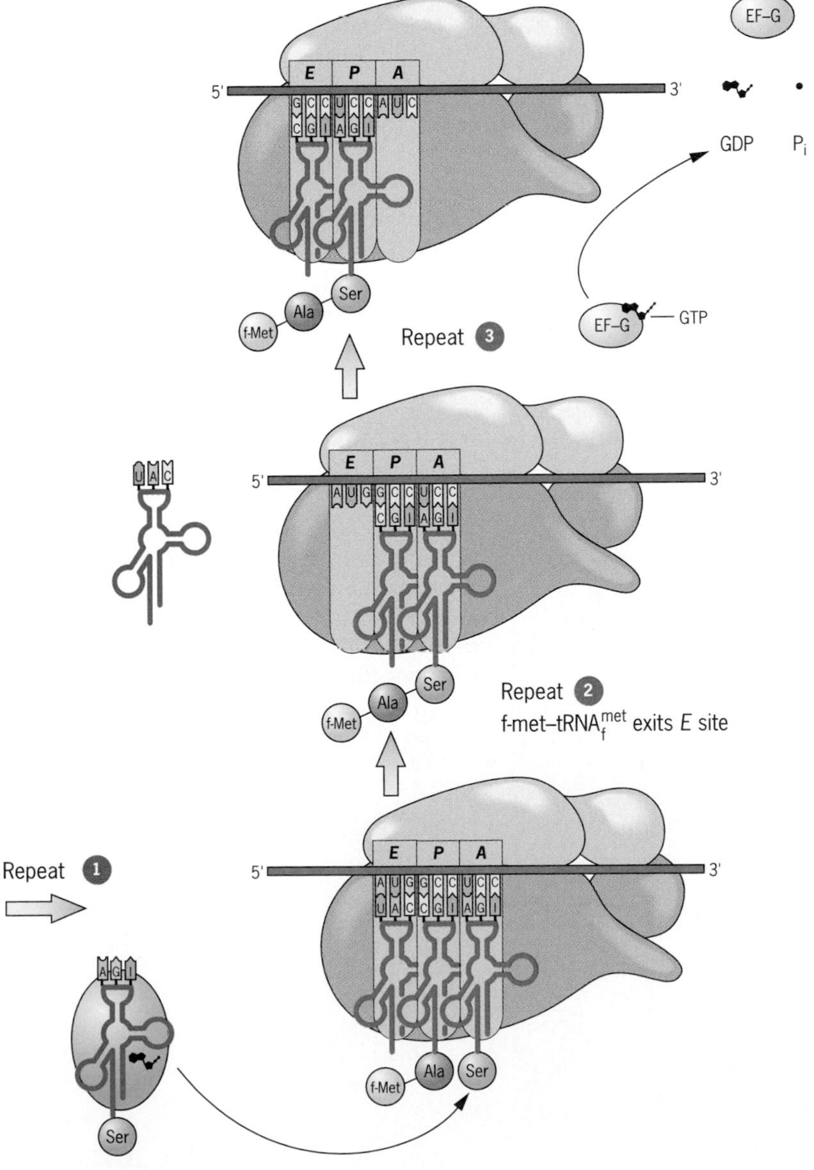

Figure 13.17 *(Continued)*

are translocated from the *A* and *P* sites to the *P* and *E* sites, respectively. These three steps are repeated in a cyclic manner throughout the elongation process. The soluble factors involved in chain elongation in *E. coli* are described here. Similar factors participate in chain elongation in eukaryotes.

In the first step, an aminoacyl-tRNA enters and becomes bound to the *A* site of the ribosome, with the specificity provided by the mRNA codon in register with the *A* site (Figure 13.17). The three nucleotides in the anticodon of the incoming aminoacyl-tRNA must pair with the nucleotides of the mRNA codon present at the *A* site. This step requires **elongation factor Tu** carrying a molecule of GTP **(EF-TuGTP).** The GTP is required for aminoacyl-tRNA binding at the *A* site, but is not cleaved until the peptide bond is formed. After the cleavage of GTP, EF-TuGDP is released from the ribosome. EF-TuGDP is inactive and will not bind to aminoacyl-tRNAs. EF-TuGDP is converted to the active EF-TuGTP form by **elongation factor Ts (EF-Ts),** which hydrolyzes one molecule of GTP in the process. EF-Tu interacts with all of the aminoacyl-tRNAs except methionyl-tRNA.

The second step in chain elongation is the formation of a peptide bond between the amino group of the aminoacyl-tRNA in the *A* site and the carboxyl terminus of the growing polypeptide chain attached to the tRNA in the *P* site. This uncouples the growing chain from the tRNA in the *P* site and covalently joins the chain to the tRNA in the *A* site (Figure 13.17). This key reaction is catalyzed by **peptidyl transferase,** an enzymatic activity built into the 50S subunit of the ribosome. We should note that the peptidyl transferase activity resides in the 23S rRNA molecule rather than in a ribosomal protein. Peptide bond formation requires the hydrolysis of the molecule of GTP brought to the ribosome by EF-Tu in step 1.

During the third step in chain elongation, the peptidyl-tRNA present in the *A* site of the ribosome is translocated to the *P* site, and the uncharged tRNA in the *P* site is translocated to the *E* site, as the ribosome moves three nucleotides toward the 3′ end of the mRNA molecule. The translocation step requires GTP and **elongation factor G (EF-G).** The ribosome undergoes changes in conformation during the translocation process, suggesting that it may shuttle along the mRNA molecule. The energy for the movement of the ribosome is provided by the hydrolysis of GTP. The translocation of the peptidyl-tRNA from the *A* site to the *P* site leaves the *A* site unoccupied and the ribosome ready to begin the next cycle of chain elongation.

The elongation of one eukaryotic polypeptide, the silk protein fibroin, can be visualized with the electron microscope by using techniques developed by Oscar Miller, Barbara Hamkalo, and colleagues. Most proteins fold up on the surface of the ribosome during their synthesis. However, fibroin remains extended from the surface of the ribosome under the conditions used by Miller and coworkers. As a result, nascent polypeptide chains of increasing length can be seen attached to the ribosomes as they are scanned from the 5′ end of the mRNA to the 3′ end (Figure 13.18). Fibroin is a large protein with a mass of over 200,000 daltons; it is synthesized on large polyribosomes containing 50 to 80 ribosomes.

Polypeptide chain elongation proceeds rapidly. In *E. coli*, all three steps required to add one amino acid to the growing polypeptide chain occur in about 0.05

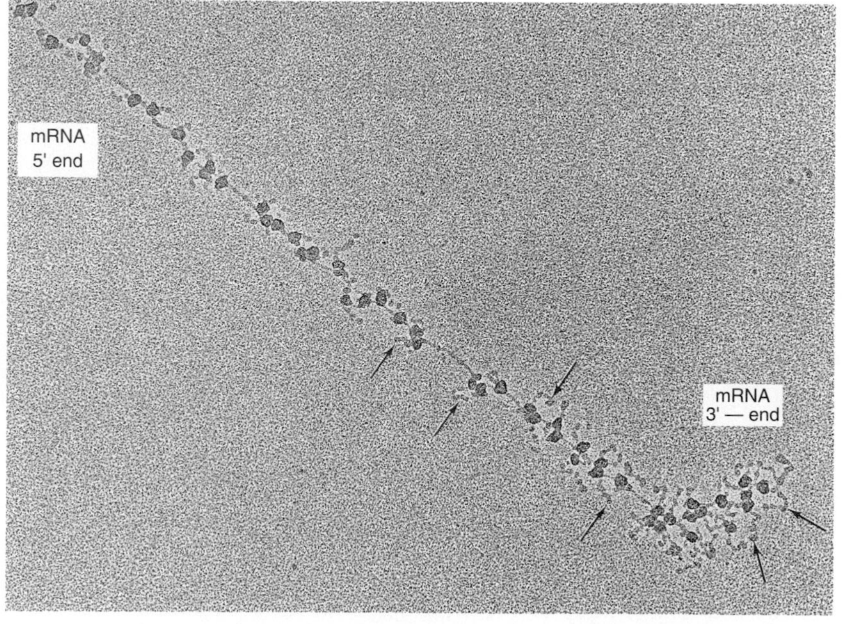

Figure 13.18 Visualization of the elongation of fibroin polypeptides in the posterior silk gland of the silkworm *Bombyx mori*. The arrows point to growing fibroin polypeptides. Note their increasing length as one approaches the 3′ end of the mRNA molecule.

second. Thus, the synthesis of a polypeptide containing 300 amino acids takes only about 15 seconds. Given its complexity, the accuracy and efficiency of the translational apparatus are indeed amazing.

Polypeptide chain elongation undergoes **termination** when any of three **chain-termination codons** (UAA, UAG, or UGA) enters the *A* site on the ribosome (Figure 13.19). These three stop codons are rec-

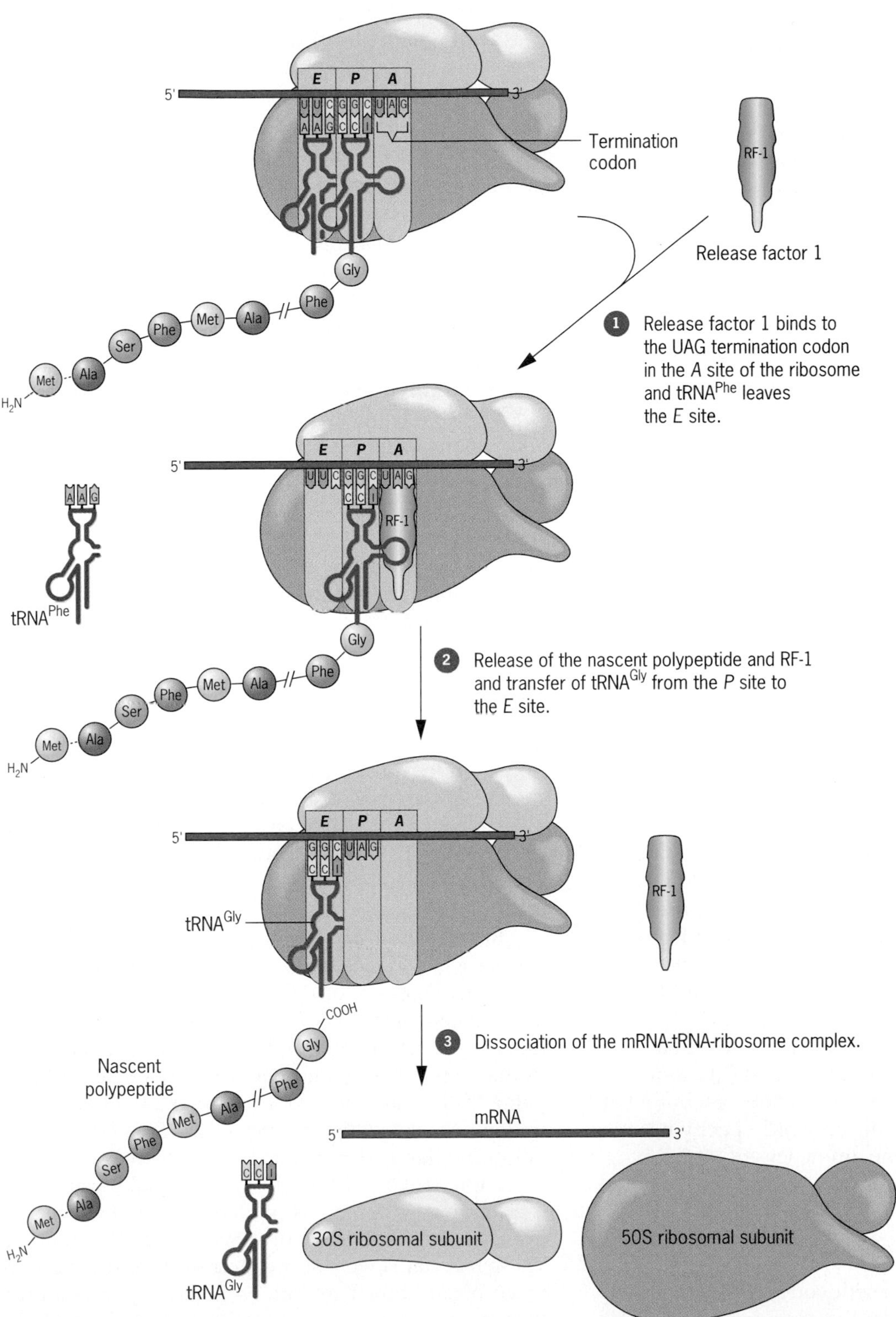

Figure 13.19 Polypeptide chain termination in *E. coli*. The formyl group of formylmethionine is removed during translation.

ognized by soluble proteins called **release factors (RFs).** In *E. coli*, there are two release factors, RF-1 and RF-2. RF-1 recognizes termination codons UAA and UAG; RF-2 recognizes UAA and UGA. In eukaryotes, a single release factor **(eRF)** recognizes all three termination codons. The presence of a release factor in the *A* site alters the activity of peptidyl transferase such that it adds a water molecule to the carboxyl terminus of the nascent polypeptide. This reaction releases the polypeptide from the tRNA molecule in the *P* site and triggers the translocation of the free tRNA to the *E* site. Termination is completed by the release of the mRNA molecule from the ribosome and the dissociation of the ribosome into its subunits. The ribosomal subunits are then ready to initiate another round of protein synthesis, as previously described.

Key Points: **Genetic information carried in the sequence of nucleotides in an mRNA molecule is translated into a sequence of amino acids in a polypeptide gene product by intricate macromolecular machines called ribosomes. The translation process is complex, requiring the participation of many different RNA and protein molecules. Transfer RNA molecules serve as adaptors, mediating the interaction between amino acids and codons in mRNA. The process of translation involves the initiation, elongation, and termination of polypeptide chains and is governed by the specifications of the genetic code.**

THE GENETIC CODE

As it became evident that genes controlled the structure of polypeptides, attention focused on how the sequence of the four different nucleotides in DNA could control the sequence of the 20 amino acids present in proteins. With the discovery of the mRNA intermediary, the question became one of how the sequence of the four bases present in mRNA molecules could specify the amino acid sequence of a polypeptide. What is the nature of the genetic code relating mRNA base sequences to amino acid sequences? Clearly, the symbols or letters used in the code must be the bases; but what comprises a codon, the unit or word specifying one amino acid or, actually, one aminoacyl-tRNA?

Properties of the Genetic Code: An Overview

The main features of the genetic code were worked out during the 1960s. Cracking the code was one of the most exciting events in the history of science, with new information reported almost daily. By the mid-1960s, the genetic code was largely solved. Before focusing on specific features of the code, let us consider its most important properties.

1. *The genetic code is composed of nucleotide triplets.* Three nucleotides in mRNA specify one amino acid in the polypeptide product; thus, each codon contains three nucleotides.
2. *The genetic code is nonoverlapping.* Each nucleotide in mRNA belongs to just one codon except in rare cases where genes overlap.
3. *The genetic code is comma-free.* There are no commas or other forms of punctuation within the coding regions of mRNA molecules. During translation, the codons are read consecutively.
4. *The genetic code is degenerate.* All but two of the amino acids are specified by more than one codon.
5. *The genetic code is ordered.* Multiple codons for a given amino acid and codons for amino acids with similar chemical properties are closely related, usually differing by a single nucleotide.
6. *The genetic code contains start and stop codons.* Specific codons are used to initiate and to terminate polypeptide chains.
7. *The genetic code is nearly universal.* With minor exceptions, the codons have the same meaning in all living organisms, from viruses to humans.

Three Nucleotides per Codon

Twenty different amino acids are incorporated into polypeptides during translation. Thus, at least 20 different codons must be formed with the four bases available in mRNA. Two bases per codon would result in only 4^2 or 16 possible codons—clearly not enough. Three bases per codon yield 4^3 or 64 possible codons—an apparent excess.

In 1961, Francis Crick and colleagues published the first strong evidence in support of a **triplet code** (three nucleotides per codon). Crick and coworkers carried out a genetic analysis of mutations induced at the *rII* locus of bacteriophage T4 by the chemical proflavin. Proflavin is a mutagenic agent that causes single base-pair additions and deletions (Chapter 14). Phage T4 *rII* mutants are unable to grow in cells of *E. coli* strain K12, but grow like wild-type phage in cells of *E. coli* strain B (Chapter 16). Wild-type T4 grows equally well on either strain. Crick and coworkers isolated proflavin-induced revertants of a proflavin-induced mutation. These revertants were shown to result from the occurrence of additional mutations at nearby sites rather than reversion of the original mutation. Second-site mutations that restore the wild-type phenotype in a mutant organism are called **suppressor mutations** because they cancel, or suppress, the effect(s) of the original mutation.

Crick and colleagues reasoned that if the original mutation was a single base-pair addition or deletion, then the suppressor mutations must be single base-pair deletions or additions, respectively, occurring at a site or sites near the original mutation. If sequential nucleotide triplets in an mRNA specify amino acids, then every nucleotide sequence can be recognized or read during translation in three different ways. For example, the sequence AAAGGGCCCTTT can be read (1) AAA, GGG, CCC, TTT, (2) A, AAG, GGC, CCT, TT, or (3) AA, AGG, GCC, CTT, T. The **reading frame** of an mRNA is the series of nucleotide triplets that are read (positioned in the *A* site of the ribosome) during translation. A single base-pair addition or deletion will alter the reading frame of the gene and mRNA for that portion of the gene distal to the mutation. This effect is illustrated in Figure 13.20*a*. The suppressor mutations were then isolated as single mutants by screening progeny of backcrosses to wild-type. Like the original mutation, the suppressor mutations were found to produce mutant phenotypes. Crick and colleagues next isolated proflavin-induced suppressor mutations of the original suppressor mutations, and so on.

Crick and colleagues then classified all the isolated mutations into two groups, plus (+) and minus (−) (for additions and deletions, although they had no idea which group was which), based on the reasoning that a (+) mutation would suppress a (−) mutation, but not another (+) mutation, and vice versa (Figure 13.20). Then, Crick and coworkers constructed recombinants that carried various combinations of the (+) and the (−) mutations. Like the single mutants, recombinants with two (+) mutations or two (−) mutations always had the mutant phenotype. The critical result was that recombinants with three (+) mutations (Figure 13.20*b*) or three (−) mutations often exhibited the wild-type phenotype. This indicated that the addition of three base pairs or the deletion of three base pairs left the distal portion of the gene with the wild-type reading

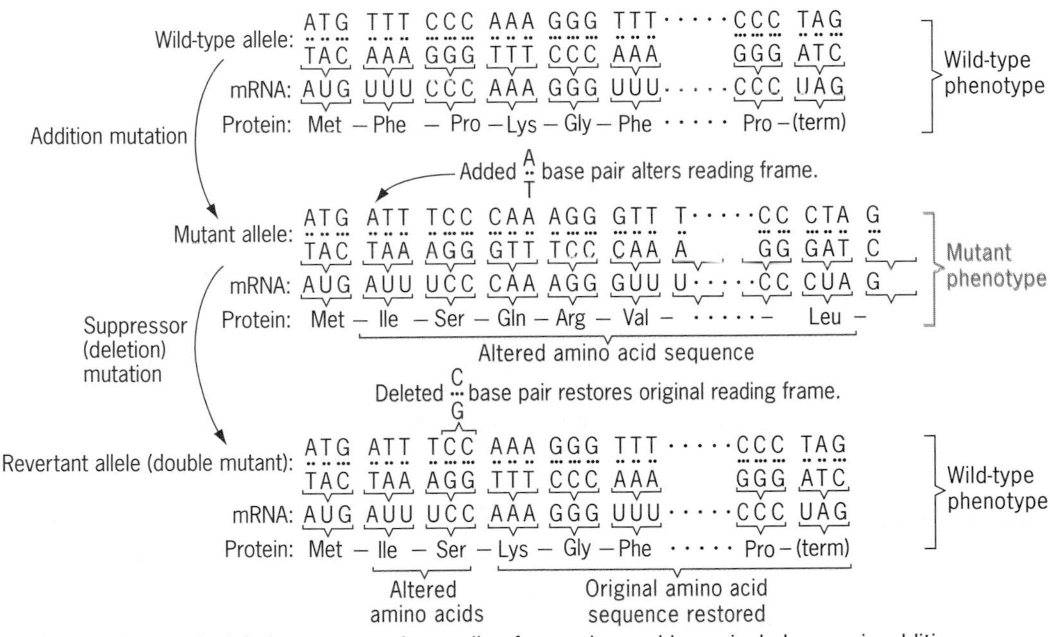

(*a*) A single base-pair deletion restores the reading frame changed by a single base-pair addition.

(*b*) Recombinant containing three single base-pair additions has the wild-type reading frame.

Figure 13.20 Early evidence that the genetic code is a triplet code. See the text for details.

frame. This result would be expected only if each codon contained three nucleotides.

Evidence from *in vitro* translation studies soon supported the results of Crick and colleagues and firmly established the triplet nature of the code. Some of the more important results follow: (1) Trinucleotides were sufficient to stimulate specific binding of aminoacyl-tRNAs to ribosomes. For example, 5′-UUC-3′ stimulated the binding of phenylalanyl-$tRNA^{Phe}$ to ribosomes. (2) Chemically synthesized mRNA molecules that contained repeating dinucleotide sequences directed the synthesis of copolymers (large chainlike molecules composed of two different subunits) with alternating amino acid sequences. For example, when $poly(UG)_n$ was used as an artificial mRNA in an *in vitro* translation system, the repeating copolymer $(cys\text{-}val)_m$ was synthesized. (The subscripts n and m refer to the number of nucleotides and amino acids in the respective polymers.) (3) In contrast, mRNAs with repeating trinucleotide sequences directed the synthesis of a mixture of three homopolymers (initiation being at random on such mRNAs in the *in vitro* systems). For example, $poly(UUG)_n$ directed the synthesis of a mixture of polyleucine, polycysteine, and polyvaline. These results are consistent only with a triplet code, with its three different reading frames. When $poly(UUG)_n$ is translated in reading frame 1, UUG, UUG, polyleucine is produced, whereas translation in reading frame 2, UGU, UGU, yields polycysteine, and translation in reading frame 3, GUU, GUU, produces polyvaline. Ultimately, the triplet nature of the code was definitively established by comparing the nucleotide sequences of genes and mRNAs with the amino acid sequences of their polypeptide products.

Deciphering the Code

The cracking of the genetic code in the 1960s took several years and involved intense competition between many different research laboratories. New information accumulated rapidly but sometimes was inconsistent with earlier data. Indeed, cracking the code proved to be a major challenge.

Deciphering the genetic code required scientists to obtain answers to several questions. (1) Which codons specify each of the 20 amino acids? (2) How many of the 64 possible triplet codons are utilized? (3) How is the code punctuated? (4) Do the codons have the same meaning in viruses, bacteria, plants, and animals? The answers to these questions were obtained primarily from the results of two types of experiments, both of which were performed with cell-free systems. The first type of experiment involved translating artificial mRNA molecules *in vitro* and determining which of the 20 amino acids were incorporated into proteins. In the second type of experiment, ribosomes were activated with mini-mRNAs just three nucleotides long. Then, researchers determined which aminoacyl-tRNAs were stimulated to bind to ribosomes activated with each of the trinucleotide messages.

The first major breakthrough came in 1961 when Marshall Nirenberg (1968 Nobel Prize recipient) and J. Heinrich Matthaei demonstrated that synthetic RNA molecules could be used as artificial mRNAs to direct *in vitro* protein synthesis. When ribosomes, aminoacyl-tRNAs, and the soluble factors required for translation are purified free of natural mRNAs, these components can be combined *in vitro* and stimulated to synthesize polypeptides by the addition of chemically synthesized RNA molecules. If these synthetic mRNA molecules are of known nucleotide content, the amino acid composition of the resulting polypeptides can be used to deduce which codons specify which amino acids.

The first codon assignment (UUU for phenylalanine) was made when Nirenberg and Matthaei demonstrated that polyuridylic acid [poly(U) = $(U)_n$] directed the synthesis of polyphenylalanine [$(phenylalanine)_m$]. Nirenberg and Matthaei used radioactively labeled phenylalanine (^{14}C-phenylalanine) as a substrate in an *in vitro* translation system with poly(U), poly(C), and poly(A) as artificial messenger RNAs. The ^{14}C-phenylalanine was incorporated into polyphenylalanine only when poly(U) was used as the mRNA (Table 13.1). When each of 17 other labeled amino acids was used in the poly(U)-stimulated translation system, no radioactivity was incorporated into polypeptides. Given the triplet nature of the genetic code, Nirenberg and Matthaei concluded that UUU must be a codon for phenylalanine. Shortly thereafter, poly(A) and poly(C) were shown to encode polylysine and polyproline, re-

TABLE 13.1

Incorporation of ^{14}C-phenylalanine into Polyphenylalanine in an *In Vitro* Translation System Activated with Synthetic RNA Homopolymers[a]

Synthetic mRNA	*Radioactivity Incorporated (counts per minute)*
None	44
Poly (U)	39,800
Poly (A)	50
Poly (C)	38
Poly (I)[b]	57

[a]Data are from Nirenberg and Matthaei, *Proc. Natl. Acad. Sci. USA* 47:1588–1602, 1961.

[b]Poly (I) is polyinosinic acid, which contains the purine hypoxanthine. Hypoxanthine is like guanine in that it base-pairs with cytosine.

spectively, allowing Nirenberg and colleagues to assign codon AAA to lysine and CCC to proline. Poly(G) does not function as an mRNA in the *in vitro* translation systems because of base-pairing between the guanine residues, resulting in the formation of complex three-stranded structures.

Researchers next began using synthetic mRNAs that contained two or more different nucleotides—first in random order and later in known order—to probe the nature of the code. The results of some of these experiments are discussed in Technical Sidelight: Cracking the Genetic Code. By using RNAs with both random and fixed nucleotide sequences as mRNAs in *in vitro* translation systems, researchers were able to establish the meaning—which trinucleotide sequences specified which amino acids—of most of the 64 possible codons.

Additional information on the nature of the genetic code was obtained by assaying the binding of aminoacyl-tRNAs to ribosomes activated with small RNA oligomers. When an *in vitro* translation system is activated with poly(U), only one aminoacyl-tRNA, phenylalanyl-tRNA^{Phe}, binds to the ribosome. This specificity exists because the mRNA codon is part of the binding site, and binding involves base-pairing between the codon and the anticodon of the tRNA. In 1964, Nirenberg and Philip Leder developed an assay for aminoacyl-tRNA binding to ribosomes activated with *trinucleotides*, mini-mRNAs only three nucleotides long. Nirenberg and Leder synthesized trinucleotides of known sequence and tested their ability to stimulate the binding of specific aminoacyl-tRNAs to ribosomes. They assayed the ability of each trinucleotide to serve as a mini-mRNA by using labeled amino acids to detect the formation of trinucleotide-aminoacyl-tRNA-ribosome complexes (Figure 13.21). For example, the trinucleotides 5′-UUU-3′ and 5′-UUC-3′ both stimulated the binding of phenylalanyl-tRNA^{Phe} to ribosomes, indicating that UUU and UUC are both phenylalanine codons.

This simple binding assay allowed Nirenberg, Leder, and others to test the ability of each of the 64 possible trinucleotides to activate ribosomes so that specific aminoacyl-tRNA would be bound. Not all of the trinucleotides worked; in some cases, the level of binding was very low, yielding ambiguous results. However, in most instances, a given trinucleotide caused one specific aminoacyl-tRNA to be bound to the activated ribosomes. By combining the results of trinucleotide binding assays and *in vitro* translation experiments performed with synthetic mRNAs, Nirenberg, Severo Ochoa, Khorana, and others were able to

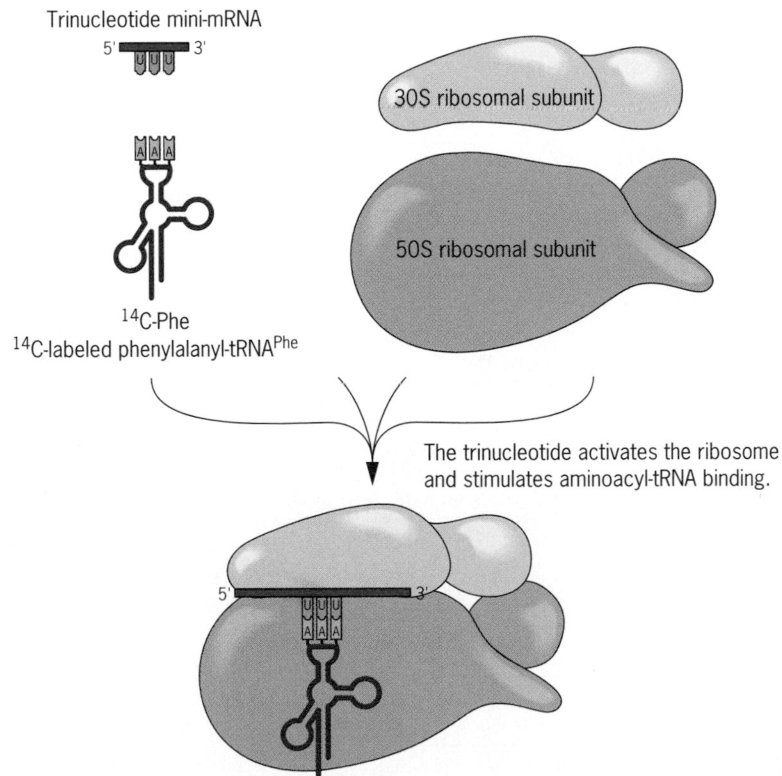

Figure 13.21 Stimulation of aminoacyl-tRNA binding to ribosomes by synthetic trinucleotide mini-mRNAs. The results of these trinucleotide-activated ribosome binding assays helped scientists crack the genetic code.

TECHNICAL SIDELIGHT

Cracking the Genetic Code: Synthetic mRNAs

Marshall Nirenberg and J. Heinrich Matthaei's 1961 demonstration that poly(U) directed the synthesis of polyphenylalanine in an *in vitro* translation system created excitement about the possibility of using such artificial mRNAs to probe the nature of the genetic code and stimulated a flurry of activity in several laboratories. Researchers in the laboratories of Nirenberg and Severo Ochoa (1959 Nobel Prize recipient) quickly extended the use of artificial mRNAs to synthetic copolymers with random sequences. For example, Nirenberg and coworkers synthesized a random copolymer containing approximately equal amounts of adenine and cytosine and used it as an artificial mRNA in their *in vitro* translation system. A random AC copolymer composed of equal amounts of A and C will contain 12.5 percent ($1/2 \times 1/2 \times 1/2 = 1/8$) of each of the eight possible codons: AAA, AAC, ACA, CAA, CCA, CAC, ACC, and CCC. Nirenberg and colleagues observed that poly(AC) directs the incorporation of six amino acids—asparagine, glutamine, histidine, lysine, proline, and threonine—into polypeptides. Because they already knew that AAA and CCC were lysine and proline codons, their results indicated that codons composed of two A's plus one C and two C's plus one A specified asparagine, glutamine, histidine, and threonine. Approximately equal amounts of asparagine, glutamine, histidine, and lysine were incorporated, but threonine and proline were incorporated in about double the amount of each of the other four amino acids. This result indicated that random AC copolymers might contain twice as many threonine and proline codons as asparagine codons. Indeed, there are two threonine codons (ACC and ACA) and two proline codons (CCC and CCA), but only one asparagine codon (AAC), in random AC copolymers.

The use of random copolymers as synthetic mRNAs provided lots of information about the nucleotide composition of codons but relatively little information about the specific nucleotide sequences of codons. By varying the nucleotide composition of random copolymers, Nirenberg, Ochoa, and their colleagues could alter the relative frequencies of the eight codons and look for correlations with the relative frequencies of the amino acids in the polypeptides synthesized in response to the copolymers. For example, Ochoa and coworkers synthesized random copolymers containing A and C in 5:1 and 1:5 ratios. When these copolymers were used as mRNAs *in vitro*, the same six amino acids were incorporated as in Nirenberg's experiments with a 1A:1C random copolymer, but their relative frequencies in the polypeptide products were very different. The results of Ochoa and colleagues are summarized in the following table.

(*a*) Predicted Codon Frequencies in Random Copolymers

		Frequency of Codon in Random Copolymer of Composition			
		5A:1C		*1A:5C*	
Codon Composition	*Possible Codons*	*Each Codon*	*Total*	*Each Codon*	*Total*
3A	AAA	$(5/6)^3 = 57.9\%$	57.9%	$(1/6)^3 = 0.4\%$	0.4%
2A + 1C	AAC, ACA, CAA	$(5/6)^2(1/6) = 11.6\% \times 3 =$	34.8%	$(1/6)^2(5/6) = 2.3\% \times 3 =$	6.9%
1A + 2C	CCA, CAC, ACC	$(5/6)(1/6)^2 = 2.3\% \times 3 =$	6.9%	$(1/6)(5/6)^2 = 11.6\% \times 3 =$	34.8%
3C	CCC	$(1/6)^3 = 0.4\%$	0.4%	$(5/6)^3 = 57.9\%$	57.9%

(*b*) Amino Acids Incorporated into Proteins *In Vitro*

	Proportion of Total Amino Acid Incorporation with Copolymer of Composition	
Amino Acid	*5A:1C*	*1A:5C*
Lysine	51.0%	1.2%
Glutamine	12.3%	4.6%
Asparagine	12.7%	4.4%
Threonine	13.8%	13.6%
Histidine	6.4%	15.4%
Proline	3.8%	60.8%

[a]Data are from Speyer et al., *Cold Spring Harbor Symp. Quant. Biol.* 28:559–567, 1963.

With the 1A:5C copolymer, proline and lysine represented 60.8 percent and 1.2 percent, respectively, of the amino acid residues incorporated, in agreement with the earlier demonstration that CCC and AAA were proline and lysine codons. The values were reversed when the 5A:1C copolymer was used; 51 percent and 3.8 percent of the incorporated residues were lysine and proline, respectively.

Ochoa's results indicated that the glutamine and asparagine codons contain 2 A's and 1 C, whereas the histidine codon in poly(AC) contains 1 A and 2 C's. Approximately the same amount of threonine was incorporated in response to both the 5A:1C and the 1A:5C copolymers. At first, the threonine incorporation data did not seem logical. However, recall that Nirenberg's results with the 1A:1C copolymer indicated that poly(AC) contains two threonine codons. Taken together with Nirenberg's results, Ochoa's results with the 5A:1C and 1A:5C copolymers suggested that one of the threonine codons contains 1 A and 2C's and the other contains 2 A's and 1 C. Also, given that one proline codon is CCC, the combined results of the two studies indicated that the second proline codon in poly(AC) probably contains 1 A and 2 C's. In the 5A:1C copolymer, 0.4 percent of the codons will be CCC and 6.9 percent will contain 1 A and 2 C's (2.3 percent of each ACC, CAC, and CCA). If one of the 1 A and 2 C codons specifies proline, the expected frequency of proline in the 5A:1C product will be 2.3 percent (ACC or CAC or CCA) plus 0.4 percent (CCC) or 2.7 percent. This value agrees well with the experimental value of 3.8 percent.

A major breakthrough in deciphering the genetic code occurred when H. Ghobind Khorana (1968 Nobel Prize recipient) and colleagues developed a procedure by which copolymers with known repeating di-, tri- and tetranucleotide sequences could be synthesized. Because the codon sequences in these repeating copolymers were fixed, their use as synthetic mRNAs yielded results that were more easily interpreted than those obtained with random copolymers. For example, an RNA molecule with a repeating UG dinucleotide sequence directed the synthesis of polypeptides containing alternating cysteine and valine residues.

5′-UGU GUG UGU GUG UGU GUG-3′

H_2N—Cys—Val—Cys—Val—Cys—Val—COOH

Because the triplets UGU and GUG alternate in poly$(UG)_n$, these two codons must specify cysteine and valine, but the result doesn't tell us which codon specifies which amino acid. In contrast, a repeating UUG trinucleotide polymer directed the synthesis of a mixture of polyleucine, polycysteine, and polyvaline.

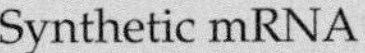
Synthetic mRNA

5′-UUG UUG UUG UUG UUG UUG UUG

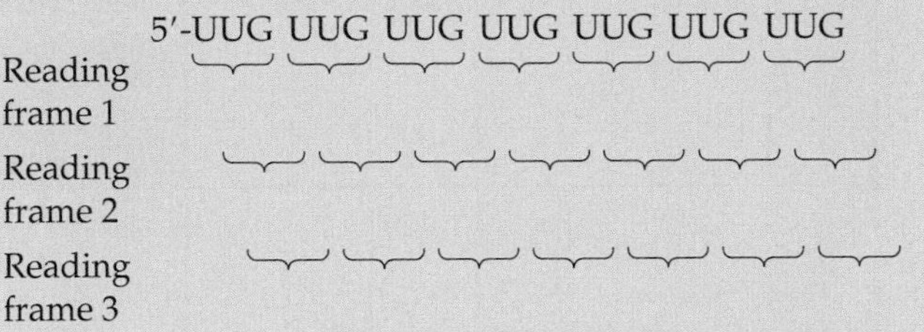

Polypeptide Products

Polyleucine Leu—Leu—Leu—Leu—Leu—Leu—Leu
Polycysteine Cys—Cys—Cys—Cys—Cys—Cys—Cys
Polyvaline Val—Val—Val—Val—Val—Val—Val

Because the initiation of translation occurs at random in these *in vitro* systems, some ribosomes will translate these polymers as UUG, UUG, UUG, and so on, whereas others will translate them as UGU, UGU, UGU. . . , and still others as GUU, GUU, GUU. . . . Thus, these three codons must specify leucine, cysteine, and valine. Because the poly$(UG)_n$ product was a cysteine-valine copolymer, one of the new codons in poly$(UUG)_n$, either UUG or GUU, must be a leucine codon. By analyzing the amino acids incorporated in response to many different repeating di-, tri-, and tetranucleotide copolymers, Khorana and coworkers were able to assign many of the codons to specific amino acids. For example, UGU specifies cysteine, UUG leucine, and GUU valine.

By using synthetic RNAs of varied composition and sequence to activate *in vitro* translation systems, Nirenberg, Ochoa, and Khorana, their coworkers, and other researchers were able to determine the amino acids specified by most of the 64 possible trinucleotides. Their results were soon supported by the results of trinucleotide-binding experiments (see text) and were subsequently verified by direct comparisons of the nucleotide sequences of genes and the amino acid sequences of their polypeptide products.

decipher the meaning of all 64 triplet codons (Table 13.2). These codon assignments are now firmly established, supported by definitive data from both *in vitro* and *in vivo* studies.

Initiation and Termination Codons

The genetic code also provides for punctuation of genetic information at the level of translation. In both prokaryotes and eukaryotes, the codon AUG is used to initiate polypeptide chains (Table 13.2). In rare instances, GUG is used as an initiation codon. In both cases, the initiation codon is recognized by an initiator tRNA, $tRNA_f^{Met}$ in prokaryotes and $tRNA_i^{Met}$ in eukaryotes. In prokaryotes, an AUG codon must follow an appropriate nucleotide sequence in the 5′ nontranslated segment of the mRNA molecule in order to serve as translation initiation codon. In eukaryotes, the codon must be the first AUG encountered by the ribosome as it scans from the 5′ end of the mRNA molecule. At internal positions, AUG is recognized by $tRNA^{Met}$, and GUG is recognized by a valine tRNA.

TABLE 13.2
The Genetic Code[a]

First (5′) letter	Second letter: U	Second letter: C	Second letter: A	Second letter: G	Third (3′) letter
U	UUU, UUC: Phe UUA, UUG: Leu	UCU, UCC, UCA, UCG: Ser	UAU, UAC: Tyr UAA: *Ochre* (terminator) UAG: *Amber* (terminator)	UGU, UGC: Cys UGA: *Opal* (terminator) UGG: Trp	U C A G
C	CUU, CUC, CUA, CUG: Leu	CCU, CCC, CCA, CCG: Pro	CAU, CAC: His CAA, CAG: Gln	CGU, CGC, CGA, CGG: Arg	U C A G
A	AUU, AUC, AUA: Ileu AUG: *Met* (initiator)	ACU, ACC, ACA, ACG: Thr	AAU, AAC: Asn AAA, AAG: Lys	AGU, AGC: Ser AGA, AGG: Arg	U C A G
G	GUU, GUC, GUA, GUG: Val	GCU, GCC, GCA, GCG: Ala	GAU, GAC: Asp GAA, GAG: Glu	GGU, GGC, GGA, GGG: Gly	U C A G

(light shading) = Polypeptide chain initiation codon

(dark shading) = Polypeptide chain termination codon

[a]Each triplet nucleotide sequence or codon refers to the nucleotide sequence in **mRNA** (not DNA) that specifies the incorporation of the indicated amino acid or polypeptide chain termination.

Three codons—UAG, UAA, and UGA—specify polypeptide chain termination (Table 13.2). Because of a purely frivolous historical event, the UAG, UAA, and UGA codons are sometimes called *amber*, *ochre*, and *opal* terminators, respectively (see Historical Sidelight: Hannah Bernstein and the *Amber* Codon). These codons are recognized by protein release factors, rather than by tRNAs. Prokaryotes contain two release factors, RF-1 and RF-2. RF-1 terminates polypeptides in response to codons UAA and UAG, whereas RF-2 causes termination at UAA and UGA codons. Eukaryotes contain a single release factor that recognizes all three termination codons.

A Degenerate and Ordered Code

All the amino acids except methionine and tryptophan are specified by more than one codon (Table 13.2). Three amino acids—leucine, serine, and arginine—are each specified by six different codons. Isoleucine has three codons. The other amino acids each has either two or four codons. The occurrence of more than one codon per amino acid is called **degeneracy** (although the usual connotations of the term are hardly appropriate). The degeneracy in the genetic code is not at random; instead, it is highly ordered. In most cases, the multiple codons specifying a given amino acid differ by only one base, the third or 3′ base of the codon. The degeneracy is primarily of two types. (1) Partial degeneracy occurs when the third base may be either of the two pyrimidines (U or C) or, alternatively, either of the two purines (A or G). With partial degeneracy, changing the third base from a purine to a pyrimidine, or vice versa, will change the amino acid specified by the codon. (2) In the case of complete degeneracy, any of the four bases may be present at the third position in the codon, and the codon will still specify the same amino acid. For example, valine is encoded by GUU, GUC, GUA, and GUG (Table 13.2).

Scientists have speculated that the **order** in the genetic code has evolved as a way of minimizing mutational lethality. Many base substitutions at the third position of codons do not change the amino acid specified by the codon. Moreover, amino acids with similar

HISTORICAL SIDELIGHT

Hannah Bernstein and the *Amber* Codon

As we discussed in the text, mutations that produce in-frame UAG codons, which specify polypeptide chain termination, within the coding regions of mRNAs are called *amber* mutations. They produce truncated gene products—polypeptides lacking their normal carboxyl termini. *Amber* mutants—bacteria or bacteriophages that carry *amber* mutations—have proven extremely valuable in analyzing gene structure and function. But why are they called *amber* mutants? Are all *amber* mutants amber-colored?

Most scientific terms are descriptive of the component, process, or event that they represent. However, sometimes a frivolous term is introduced and becomes established in science. The use of the term *amber* to refer to the UAG termination codon is one such example.

In 1960, Harris Bernstein was a graduate student working on the fungus *N. crassa* in the laboratory of Herschel Mitchell at the California Institute of Technology in Pasadena. One evening, after working all day in the laboratory, Harris walked down the hallway to the "phage lab"—as Max Delbrück's research laboratory was called—to invite his friends Charley Steinberg and Dick Epstein to go to a movie with him. Unbeknown to Harris, Charley and Dick had just started a large, laborious experiment. They were attempting to isolate mutants of bacteriophage T4 that could grow on *E. coli* strain C but could not grow on *E. coli* strain B.

Screening for such mutants involves repeated transfers of phage from one plate to two other plates containing different host strains. Each potential phage mutant must be individually transferred. At the time, the transfers were done with inoculating needles that were flame-sterilized between transfers—an extremely laborious process. Thus, Steinberg and Epstein urged Bernstein to forget about the movie and stay and help them with their experiment. Harris agreed to stay and help with the mutant hunt—but only if Steinberg and Epstein agreed to name the desired mutants after his mother, Hannah Bernstein, if the experiment were successful.

When Bernstein arrived at the laboratory the next morning, the desired mutants had indeed been identified. In addition, they had already been named *amber* mutants—amber being the English translation of the German word *Bernstein*. However, there is disagreement on one point. Dick Epstein, who actually named the *amber* mutations, has stated that he named them after Harris Bernstein rather than Harris' mother.

When these mutations were shown to result in UAG chain-termination codons within the coding regions of mRNAs, the term *amber* was extrapolated to the termination codon as well. Later, the UAA and UGA terminator codons were named *ochre* and *opal* to have analogous terms for all three termination codons. Thus, although the term *amber* was introduced in a frivolous manner, it is now an established component of the vocabulary of geneticists.

chemical properties (such as leucine, isoleucine, and valine) have codons that differ from each other by only one base. Thus, many single base-pair substitutions will result in the substitution of one amino acid for another amino acid with very similar chemical properties (for example, valine for isoleucine). In most cases, conservative substitutions of this type will yield active gene products, which minimizes the effects of mutations.

A Nearly Universal Code

Vast quantities of information are now available from *in vitro* studies, from amino acid replacements due to mutations, and from correlated nucleic acid and polypeptide sequencing, which allow a comparison of the meaning of the 64 codons in different species. These data all indicate that the genetic code is nearly **universal;** that is, the codons have the same meaning, with minor exceptions, in all species.

The most important exceptions to the universality of the code occur in mitochondria of mammals, yeast, and several other species. Mitochondria have their own chromosomes and protein-synthesizing machinery (Chapter 19). Although the mitochondrial and cytoplasmic systems are similar, there are some differences. In the mitochondria of humans and other mammals, (1) UGA specifies tryptophan rather than chain termination, (2) AUA is a methionine codon, not an isoleucine codon, and (3) AGA and AGG are chain-termination codons rather than arginine codons. The other 60 codons have the same meaning in mammalian mitochondria as in nuclear mRNAs (Table 13.2). There are also rare differences in codon meaning in the mitochondria of other species and in nuclear transcripts of some protozoa. However, since these exceptions are rare, the genetic code should be considered nearly universal.

Key Points: **Each of the 20 amino acids in proteins is specified by one or more nucleotide triplets in mRNA. Of the 64 possible triplets, given the four bases in mRNA, 61 specify amino acids and 3 signal chain termination. The code is nonoverlapping,**

with each nucleotide part of a single codon, degenerate, with most amino acids specified by two or four codons, and ordered, with similar amino acids specified by related codons. The genetic code is nearly universal; with minor exceptions, the 64 triplets have the same meaning in all organisms.

CODON-tRNA INTERACTIONS

The translation of a sequence of nucleotides in mRNA into the correct sequence of amino acids in the polypeptide product requires the accurate recognition of codons by aminoacyl-tRNAs. Because of the degeneracy of the genetic code, either several different tRNAs must recognize the different codons specifying a given amino acid or the anticodon of a given tRNA must be able to base-pair with several different codons. Actually, both of these phenomena occur. Several tRNAs exist for certain amino acids, and some tRNAs recognize more than one codon.

Recognition of Codons by tRNAs: The Wobble Hypothesis

The hydrogen bonding between the bases in the anticodons of tRNAs and the codons of mRNAs appears to follow strict base-pairing rules only for the first two bases of the codon. The base-pairing involving the third base of the codon is apparently less stringent, allowing what Crick has called **wobble** at this site. On the basis of molecular distances and steric (three-dimensional structure) considerations, Crick proposed that wobble would allow several types, but not all types, of base-pairing at the third codon base in the codon–anticodon interaction. His proposal has since been strongly supported by experimental data. Table 13.3 shows the base-pairing predicted by Crick's wobble hypothesis.

The **wobble hypothesis** predicted the existence of at least two tRNAs for each amino acid with codons that exhibit complete degeneracy, and this has proven to be true. The wobble hypothesis also predicted the occurrence of three tRNAs for the six serine codons. Three serine tRNAs have been characterized: (1) $tRNA^{Ser1}$ (anticodon AGG) binds to codons UCU and UCC, (2) $tRNA^{Ser2}$ (anticodon AGU) binds to codons UCA and UCG, and (3) $tRNA^{Ser3}$ (anticodon UCG) binds to codons AGU and AGC. These specificities were verified by the trinucleotide-stimulated binding of purified aminoacyl-tRNAs to ribosomes *in vitro*.

Finally, several tRNAs contain the base inosine, which is made from the purine hypoxanthine. Inosine is produced by a post-transcriptional modification of adenosine. Crick's wobble hypothesis predicted that when inosine is present at the 5′ end of an anticodon (the wobble position), it would base-pair with adenine, uracil, or cytosine in the codon. In fact, purified alanyl-tRNA containing inosine (I) at the 5′ position of the anticodon (see Figure 13.11) binds to ribosomes activated with GCU, GCC, or GCA trinucleotides (Figure 13.22). The same result has been obtained with other purified tRNAs with inosine at the 5′ position of the anticodon. Thus, Crick's wobble hypothesis nicely explains the relationships between tRNAs and codons given the degenerate, but ordered, genetic code.

TABLE 13.3
Base-Pairing Between the 5′ Base of the Anticodons of tRNAs and the 3′ Base of Codons of mRNAs According to the Wobble Hypothesis

Base in Anticodon	*Base in Codon*
G	U or C
C	G
A	U
U	A or G
I	A, U, or C

Suppressor Mutations That Produce tRNAs with Altered Codon Recognition

Even if we exclude the mitochondria, the genetic code is not absolutely universal. Minor variations in codon recognition and translation are well documented. In *E. coli* and yeast, for example, some mutations in tRNA genes alter the anticodons and thus the codons recog-

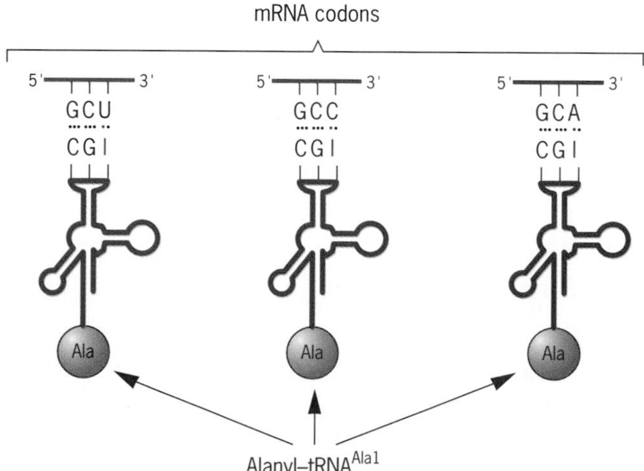

Figure 13.22 Base-pairing between the anticodon of alanyl-$tRNA^{Ala1}$ and mRNA codons GCU, GCC, and GCA according to Crick's wobble hypothesis. Trinucleotide-activated ribosome binding assays have shown that alanyl-$tRNA^{Ala1}$ does indeed base-pair with all three codons.

nized by the mutant tRNAs. These mutations were initially detected as suppressor mutations, nucleotide substitutions that suppressed the effects of other mutations. The suppressor mutations were subsequently shown to occur in tRNA genes. Many of these suppressor mutations changed the anticodons of the altered tRNAs.

The best-known examples of suppressor mutations that alter tRNA specificity are those that suppress UAG chain-termination mutations within the coding sequences of genes. Such mutations, called *amber* mutations (see Historical Sidelight: Hannah Bernstein and the *Amber* Codon), result in the synthesis of truncated polypeptides. Mutations that produce chain-termination triplets within genes have come to be known as **nonsense mutations,** in contrast to **missense mutations,** which change a triplet so that it specifies a different amino acid. A gene that contains a missense mutation encodes a complete polypeptide, but with an amino acid substitution in the polypeptide gene product. A nonsense mutation results in a truncated polypeptide, with the length of the chain depending on the position of the mutation within the gene. Nonsense mutations frequently result from single base-pair substitutions, as illustrated in Figure 13.23*a*.

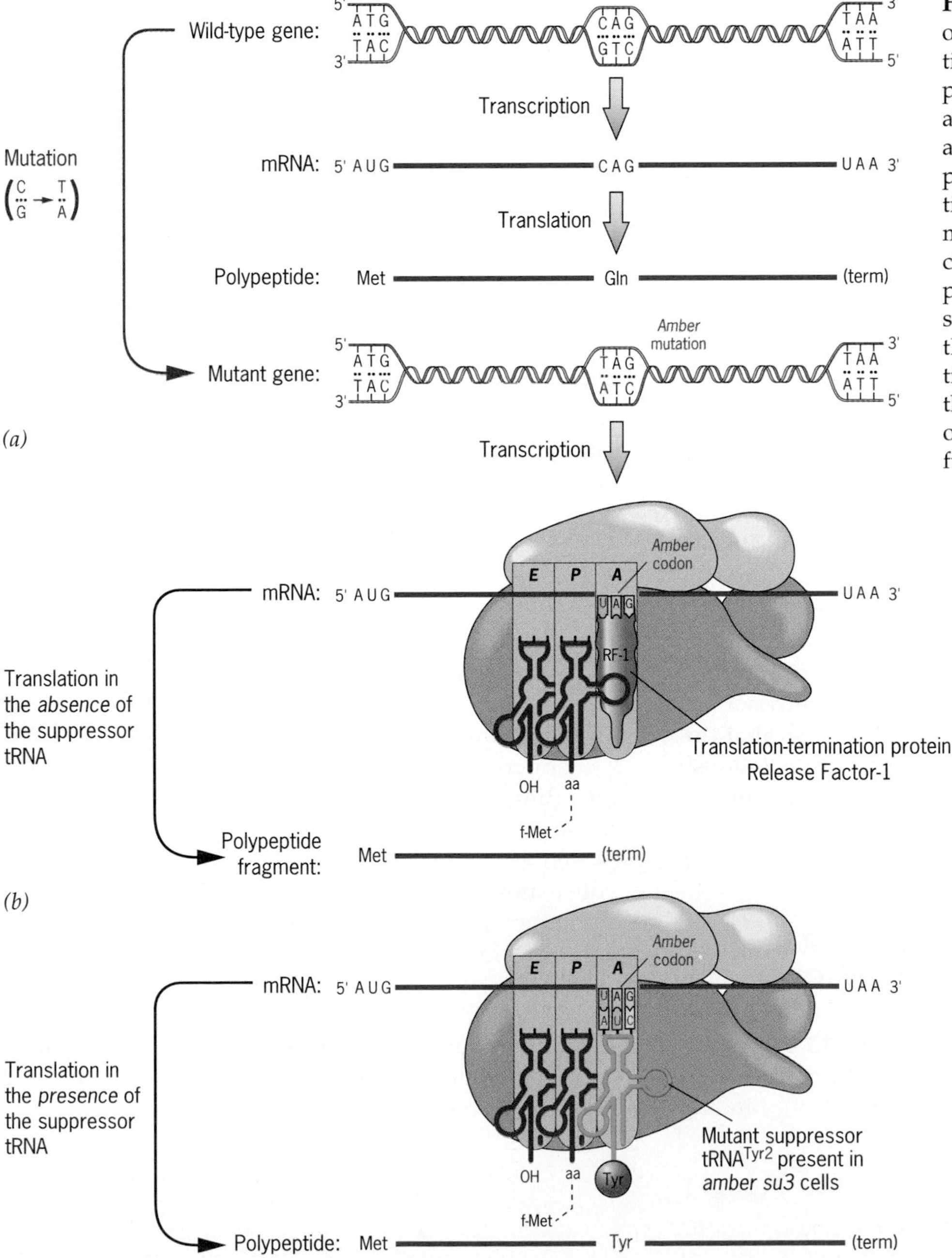

Figure 13.23 (*a*) The formation of an *amber* (UAG) chain-termination mutation. (*b*) Its effect on the polypeptide gene product in the absence of a suppressor tRNA, and (*c*) in the presence of a suppressor tRNA. The *amber* mutation shown changes a CAG glutamine (Gln) codon to a UAG chain-termination codon. The polypeptide containing the tyrosine inserted by the suppressor tRNA may or may not be functional; however, suppression of the mutant phenotype will occur only when the polypeptide is functional.

The polypeptide fragments produced from genes containing nonsense mutations (Figure 13.23*b*) often are completely nonfunctional.

Suppression of nonsense mutations has been shown to result from mutations in tRNA genes that cause the mutant tRNAs to recognize the nonsense (UAG, UAA, or UGA) codons, albeit with varying efficiencies. These mutant tRNAs are referred to as suppressor tRNAs. When the *amber* (UAG) suppressor tRNA produced by the *amber su3* mutation in *E. coli* was sequenced, it was found to have an altered anticodon. This particular *amber* suppressor mutation occurs in the $tRNA^{Tyr2}$ gene (one of two tyrosine tRNA genes in *E. coli*). The anticodon of the wild-type (nonsuppressor) $tRNA^{Tyr2}$ was shown to be 5′-G′UA-3′ (where G′ is a derivative of guanine). The anticodon of the mutant (suppressor) $tRNA^{Tyr2}$ is 5′-CUA-3′. Because of the single-base substitution, the anticodon of the suppressor $tRNA^{Tyr2}$ base-pairs with the 5′-UAG-3′ *amber* codon (recall that base-pairing always involves strands of opposite polarity); that is,

mRNA: 5′-UAG-3′ (codon)
tRNA: 3′-AUC-5′ (anticodon)

Thus, suppressor tRNAs allow complete polypeptides to be synthesized from mRNAs containing nonsense codons (Figure 13.23*c*). Such polypeptides will be functional if the amino acid inserted by the suppressor tRNA does not significantly alter the protein's chemical properties.

Key Points: **The wobble hypothesis explains how a single tRNA can respond to two or more degenerate codons. According to this hypothesis, the pairing between the third base of a codon and the first base of an anticodon is less stringent than normal, permitting wobble at this site. Some suppressor mutations alter the anticodons of tRNAs so that the mutant tRNAs recognize chain-termination codons and insert amino acids in response to their presence in mRNA molecules.**

IN VIVO CONFIRMATION OF THE NATURE OF THE GENETIC CODE

The codon assignments shown in Table 13.2 were initially based on the results of *in vitro* translation and trinucleotide-stimulated aminoacyl-tRNA binding studies. The source of these assignments raised an obvious question. Are the assignments based on *in vitro* experiments valid *in vivo*? Several lines of evidence now indicate that these codon assignments are correct for protein synthesis *in vivo* for most, if not all, species. When the amino acid substitutions that result from mutations induced by chemical mutagens with specific effects (Chapter 14) are determined by amino acid sequencing, the substitutions are almost always consistent with the codon assignments given in Table 13.2 and the known effect of the mutagen.

More convincingly, when the nucleotide sequences of genes or mRNAs are determined and compared with the amino acid sequences of the polypeptides encoded by those genes or mRNAs, the observed correlations are those predicted from the codon assignments shown in Table 13.2. This relationship was first demonstrated for bacteriophage MS2 by Walter Fiers and colleagues in 1972. Fiers and coworkers determined the amino acid sequence of the phage MS2 coat protein and the nucleotide sequence of the MS2 gene that encodes the coat protein (see Figure 15.12). Phage MS2 stores its genetic information in single-stranded RNA; its chromosome is equivalent to an mRNA molecule in organisms with DNA genomes. Fiers and colleagues then compared the nucleotide sequence of the coat protein gene with the amino acid sequence of the coat polypeptide. The amino acid sequence of the coat protein was exactly the sequence predicted from the nucleotide sequence of the coat protein gene and the codon assignments shown in Table 13.2.

In the last two decades, similar comparisons of the nucleotide sequences of genes and the amino acid sequences of their polypeptide products have clearly established that the codon assignments shown in Table 13.2 are indeed valid *in vivo*. When the nucleotide sequence of the normal human β-globin gene was determined, the sequence predicted the sequence of the 146 amino acids in the human β-globin polypeptide, including the valine at position six. When the sequence of the sickle-cell allele was determined, it predicted the presence of glutamic acid at position six. Indeed, scientists now understand the molecular basis of all the symptoms of sickle-cell anemia described by Herrick and Irons in 1910.

Key Points: **Comparisons of the nucleotide sequences of genes with the amino acid sequences of their polypeptide products verify the codon assignments deduced from *in vitro* studies.**

TESTING YOUR KNOWLEDGE

1. The average mass of the 20 common amino acids is about 137 daltons. Estimate the approximate length of an mRNA molecule that encodes a polypeptide with a mass of 65,760 daltons. Assume that the polypeptide contains equal amounts of all 20 amino acids.

ANSWER

Based on this assumption, the polypeptide would contain about 480 amino acids (65,760 daltons/137 daltons per amino acid). Since each codon contains three nucleotides, the coding region of the mRNA would have to be 1440 nucleotides long (480 amino acids × 3 nucleotides per amino acid).

2. What sequence of nucleotide pairs in a gene in *Drosophila* will encode the amino acid sequence methionine-tryptophan (reading from the amino terminus to the carboxyl terminus)?

ANSWER

The codons for methionine and tryptophan are AUG and UGG, respectively. Thus, the nucleotide sequence in the mRNA specifying the dipeptide sequence methionine-tryptophan must be 5′-AUG-UGG-3′. The template DNA strand must be complementary and antiparallel to the mRNA sequence (3′-TAC-ACC-5′), and the other strand of DNA must be complementary to the template strand. Therefore, the sequence of base pairs in the gene must be

5′-ATG-TGG-3′
3′-TAC-ACC-5′

3. The antibiotic streptomycin kills sensitive *E. coli* by inhibiting the binding of $tRNA_f^{Met}$ to the *P* site of the ribosome and by causing misreading of codons in mRNA. In sensitive bacteria, streptomycin is bound by protein S12 in the 30S subunit of the ribosome. Resistance to streptomycin can result from a mutation in the gene-encoding protein S12 so that the altered protein will no longer bind the antibiotic. In 1964, Luigi Gorini and Eva Kataja isolated mutants of *E. coli* that grew on minimal medium supplemented with either the amino acid arginine or streptomycin. That is, in the absence of streptomycin, the mutants behaved like typical arginine-requiring bacteria. However, in the absence of arginine, they were streptomycin-dependent conditional-lethal mutants. That is, they grew in the presence of streptomycin but not in the absence of streptomycin. Explain the results obtained by Gorini and Kataja.

ANSWER

The streptomycin-dependent conditional-lethal mutants isolated by Gorini and Kataja contained missense mutations in genes encoding arginine biosynthetic enzymes. If arginine was present in the medium, these enzymes were unessential. However, these enzymes were required for growth in the absence of arginine (one of the 20 amino acids required for protein synthesis).

Streptomycin causes misreading of mRNA codons in bacteria. This misreading allowed the codons that contained the missense mutations to be translated ambiguously—with the wrong amino acids incorporated—when the antibiotic was present. When streptomycin was present in the mutant bacteria, an amino acid occasionally would be inserted (at the site of the mutation) that resulted in an active enzyme, which, in turn, allowed the cells to grow, albeit slowly. In the absence of streptomycin, no misreading occurred, and all of the mutant polypeptides were inactive.

QUESTIONS AND PROBLEMS

13.1 In a general way, describe the molecular organization of proteins and distinguish proteins from DNA, chemically and functionally. Why is the synthesis of proteins of particular interest to geneticists?

13.2 At what locations in the cell does protein synthesis occur?

13.3 Characterize ribosomes in general as to size, location, function, and macromolecular composition.

13.4 (a) Where in the cells of higher organisms do ribosomes originate? (b) Where in the cells are ribosomes most active in protein synthesis?

13.5 Identify three different types of RNA that are involved in translation and list the characteristics and functions of each.

13.6 (a) How is messenger RNA related to polysome formation? (b) How does rRNA differ from mRNA and tRNA in specificity? (c) How does the tRNA molecule differ from that of DNA and mRNA in size and helical arrangement?

13.7 Outline the process of aminoacyl-tRNA formation.

13.8 What types of experimental evidence were used to decipher the genetic code?

13.9 In what sense and to what extent is the genetic code (a) degenerate, (b) ordered, and (c) universal?

13.10 Draw an analogy between the processes of transcription and translation and the process of building a house.

13.11 The thymine analog 5-bromouracil is a chemical mutagen that induces single base-pair substitutions in DNA called transitions (substitutions of one purine for another purine and one pyrimidine for another pyrimidine). Using the known nature of the genetic code (Table 13.2), which of the following amino acid substitutions should you expect to be induced by 5-bromouracil with the highest frequency: (a)

Met → Val; (b) Met → Leu; (c) Lys → Thr; (d) Lys → Gln; (e) Pro → Arg; or (f) Pro → Gln? Why?

13.12 Using the information given in Problem 13.11, would you expect 5-bromouracil to induce a higher frequency of His → Arg or His → Pro substitutions? Why?

13.13 How is translation (a) initiated and (b) terminated?

13.14 If the average molecular mass of an amino acid is assumed to be 100 daltons, about how many nucleotides will be present in an mRNA coding sequence specifying a single polypeptide with a molecular mass of 27,000 daltons?

13.15 Of what significance is the wobble hypothesis?

13.16 The bases A, G, U, C, I (inosine) all occur at the 5′ positions of anticodons in tRNAs. (a) Which base can pair with three different bases at the 3′ positions of codons in mRNA? (b) What is the minimum number of tRNAs required to recognize all codons of amino acids specified by codons with complete degeneracy?

13.17 What is the minimum number of tRNAs required to recognize the six codons specifying the amino acid leucine?

13.18 Assume that in the year 2025, the first expedition of humans to Mars discovers several Martian life-forms thriving in hydrothermal vents that exist below the planet's surface. Several teams of molecular biologists extract proteins and nucleic acids from these organisms and make some momentous discoveries. Their first discovery is that the proteins in Martian life-forms contain only 14 different amino acids instead of the 20 present in life-forms on Earth. Their second discovery is that the DNA and RNA in these organisms have only two different nucleotides instead of the four nucleotides present in living organisms on Earth. (a) Assuming that transcription and translation work similarly in Martians and Earthlings, what is the minimum number of nucleotides that must be present in the Martian codon to specify all the amino acids in Martians? (b) Assuming that the Martian code proposed above has translational start-and-stop signals, would you expect the Martian genetic code to be degenerate like the genetic code used on Earth?

13.19 (a) Why is the genetic code a triplet code instead of a singlet or doublet code? (b) How many different amino acids are specified by the genetic code? (c) How many different amino acid sequences are possible in a polypeptide 146 amino acids long?

13.20 What are the basic differences between translation in prokaryotes and eukaryotes?

13.21 What is the function of each of the following components of the protein-synthesizing apparatus: (a) aminoacyl-tRNA synthetase, (b) release factor 1, (c) peptidyl transferase, (d) initiation factors, (e) elongation factor G?

13.22 An *E. coli* gene has been isolated and shown to be 68 nm long. What is the maximum number of amino acids that this gene could encode?

13.23 (a) What is the difference between a nonsense mutation and a missense mutation? (b) Are nonsense or missense mutations more frequent in living organisms? (c) Why?

13.24 The human α-globin chain is 141 amino acids long. How many nucleotides in mRNA are required to encode human α-globin?

13.25 What are the functions of the *A*, *P*, and *E* aminoacyl-tRNA binding sites on the ribosome?

13.26 (a) In what ways does the order in the genetic code minimize mutational lethality? (b) Why do base-pair changes that cause the substitution of a leucine for a valine in the polypeptide gene product seldom produce a mutant phenotype?

13.27 (a) What is the function of the Shine-Dalgarno sequence in prokaryotic mRNAs? (b) What effect does the deletion of the Shine-Dalgarno sequence from a mRNA have on its translation?

13.28 (a) In what ways are ribosomes and spliceosomes similar? (b) In what ways are they different?

13.29 If you were to (1) purify cysteine transfer RNA and charge it with labeled cysteine (that is, activate it by attaching ^{3}H-labeled cysteine), (2) use Raney nickel (a highly reducing nickel powder) to convert the cysteine to alanine still attached to the cysteyl-specific transfer RNA, and (3) place the alanine-charged cysteyl transfer RNA into an *in vitro* protein-synthesizing system activated with poly UG templates that normally stimulate the incorporation of cysteine, but not alanine, into polypeptide chains (that is, when you use the normal alanine-charged alanyl and cysteine-charged cysteyl transfer RNAs), what result would you expect?

13.30 A partial (5′ subterminal) nucleotide sequence of a prokaryotic mRNA is as follows:

5′-.....AGGAGGCUCGAACAUGUCAAUAUGCUUGUUC-CAAUCGUUAGCUGCGCAGGACCGUCCCGGA......3′.

When this mRNA is translated, what amino acid sequence will be specified by this portion of the mRNA?

13.31 The 5′ terminus of a human mRNA has the following sequence:

5′ cap-GAAGAGACAAGGTCAUGGCCAUAUGCUUGU-UCCAAUCGUUAGCUGCGCAGGAUCGCCCUGGG......3′.

When this mRNA is translated, what amino acid sequence will be specified by this portion of the mRNA?

13.32 Alan Garen extensively studied a particular nonsense (chain-termination) mutation in the alkaline phosphatase gene of *E. coli*. This mutation resulted in the termination of the alkaline phosphatase polypeptide chain at a position where the amino acid tryptophan occurred in the wild-type polypeptide. Garen induced revertants (in this case, mutations altering the same codon) of this mutant with chemical mutagens that induced single base-pair substitutions and sequenced the polypeptides in the revertants. Seven different types of revertants were found, each with a different amino acid at the tryptophan position of the wild-type polypeptide (termination position of the mutant polypeptide fragment). The amino acids present at this position in the various revertants included tryptophan, serine, tyrosine, leucine, glutamic acid, glutamine, and lysine. Was the nonsense mutation studied by Garen an *amber* (UAG), an *ochre* (UAA), or an *opal* (UGA) nonsense mutation? Explain the basis of your deduction.

13.33 The following DNA sequence occurs in a bacterium (the promoter sequence is located to the left but is not shown).

↓
5′-CAATCATGGACTGCCATGCTTCATATGAATAGTTGACAT-3′
3′-GTTAGTACCTGACGGTACGAAGTATACTTATCAACTGTA-5′

(a) What is the ribonucleotide sequence of the mRNA molecule that is transcribed from the template strand of this piece of DNA? Assume that both translational start and termination codons are present. (b) What is the amino acid sequence of the polypetide encoded by this mRNA? (c) If the nucleotide indicated by the arrow undergoes a mutation that causes this C:G base pair to be deleted, what will be the polypeptide encoded by the mutant gene?

13.34 The following DNA sequence occurs in the template strand of a structural gene in a bacterium (the promoter sequence is located to the left but is not shown):

↓
5′-GAATGTCAGAACTGCCATGCTTCATATGAATAGACCTCTAG-3′

(a) What is the ribonucleotide sequence of the mRNA molecule that is transcribed from this piece of DNA? (b) What is the amino acid sequence of the polypeptide encoded by this mRNA? (c) If the nucleotide indicated by the arrow undergoes a mutation that changes T to A, what will be the resulting amino acid sequence following transcription and translation?

BIBLIOGRAPHY

Crick, F. H. C. 1962. The genetic code. *Sci. Amer.* 207(4):66–77.

Crick, F. H. C. 1966. Codon-anticodon pairing: The wobble hypothesis. *J. Mol. Biol.* 19:548–555.

Crick, F. H. C. 1966. The genetic code: III. *Sci. Amer.* 215(4):55–62.

The Genetic Code. 1966. *Cold Spring Harbor Symp. Quant. Biol.*, Vol. 31. Cold Spring Harbor Laboratory Press, Cold Spring Harbor, NY.

Green, R., and H. F. Noller. 1997. Ribosomes and translation. *Annu. Rev. Biochem.* 66:679–716.

Lake, J. A. 1981. The ribosome. *Sci. Amer.* 245(2):84–97.

The Mechanism of Protein Synthesis. 1970. *Cold Spring Harbor Symp. Quant. Biol.*, Vol. 34. Cold Spring Harbor Laboratory Press, Cold Spring Harbor, NY.

Nirenberg, M. W. 1963. The genetic code: II. *Sci. Amer.* 208(3):80–94.

Nomura, M. 1984. The control of ribosome synthesis. *Sci. Amer.* 250(1):102–114.

Yanofsky, C. 1967. Gene structure and protein structure. *Sci. Amer.* 216(5):80–94.

14

Scanning electron micrograph of Edward Lewis' four-winged fruit fly produced by combining three mutations at the bithorax *locus.*

Mutation, DNA Repair, and Recombination

CHAPTER OUTLINE

Mutation: Source of the Genetic Variability Required for Evolution
Mutation: Basic Features of the Process
Mutation: Phenotypic Effects
Muller's Demonstration That X Rays Are Mutagenic
The Molecular Basis of Mutation
Screening Chemicals for Mutagenicity: The Ames Test
DNA Repair Mechanisms
Inherited Human Diseases with Defects in DNA Repair
DNA Recombination Mechanisms

Xeroderma Pigmentosum: Defective Repair of Damaged DNA in Humans

The sun shone brightly on a mid-summer day—a perfect day for most children to spend at the beach. All of Nathan's friends were dressed in shorts or swimsuits. As Nathan prepared to join his friends, he pulled on full-length sweatpants and a long-sleeved shirt. Then he put on a wide-brimmed hat and applied a thick layer of sunscreen to his hands and face. Whereas his friends enjoy sunshine and the tan that results, Nathan lives in constant fear of the effects of sunlight. Nathan was born with the inherited disorder xeroderma pigmentosum, an autosomal recessive trait that affects about one out of 250,000 children. Nathan's skin cells are extremely sensitive to ultraviolet radiation—the high-energy rays of sunlight. Ultraviolet light causes chemical changes in the DNA in Nathan's skin cells, changes that lead not only to intense freckling but also to skin cancer (Figure 14.1).

Nathan's friends gave little thought to playing in the sun; sunburn was their only major concern. Their skin cells contain enzymes that correct the changes in DNA resulting from exposure to ultraviolet light. However, Nathan's skin cells are lacking one of the enzymes required to repair ultraviolet light-induced alterations in the structure of DNA. Xeroderma pigmentosum can result from inher-

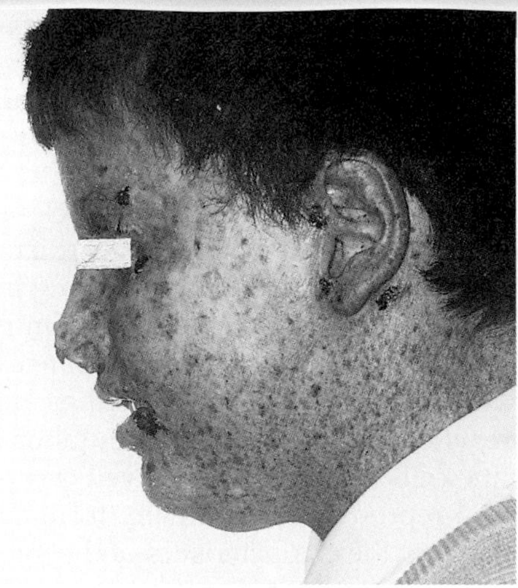

ited defects in any of nine different human genes. Moreover, other inherited disorders are known to result from the failure to repair DNA damaged by other physical and chemical agents. The life-threatening consequences of these inherited defects in the DNA repair enzymes dramatically emphasize their importance.

Figure 14.1 Phenotypic effects of the inherited disease xeroderma pigmentosum. Individuals with this malignant disease develop extensive skin tumors after exposure to sunlight.

Given the key role that DNA plays in the growth and development of living organisms, the evolution of mechanisms to protect its integrity would seem inevitable. Indeed, as we discuss below, living cells contain numerous enzymes that constantly scan DNA to search for damaged or incorrectly paired nucleotides. When detected, these defects are corrected by a small army of **DNA repair enzymes,** each having evolved to combat a particular type of damage. In some cases, incorrect nucleotide pairs are produced during recombination between homologous DNA molecules. The repair of these mismatched base pairs results in a type of nonreciprocal recombination called **gene conversion,** a process during which one allele of a gene is changed or converted to another allele. In this chapter, we examine the types of changes that occur in DNA, the processes by which these alterations are corrected, and the related processes of recombination between homologous DNA molecules.

MUTATION: SOURCE OF THE GENETIC VARIABILITY REQUIRED FOR EVOLUTION

We know from preceding chapters that inheritance is based on genes that are transmitted from parents to offspring during reproduction and that the genes store genetic information encoded in the sequences of nucleotide pairs in DNA or nucleotides in RNA. We have examined how this genetic information is accurately duplicated during the semiconservative replication of DNA. This accurate replication was shown to depend in part on proofreading activities built into the DNA polymerases that catalyze DNA synthesis. Thus, mechanisms have evolved to facilitate the faithful transmission of genetic information from generation to generation. Nevertheless, mistakes in the genetic material do occur. Such heritable changes in the genetic material are called mutations.

The term **mutation** refers to both (1) the change in the genetic material and (2) the process by which the change occurs. An organism that exhibits a novel phenotype resulting from a mutation is called a **mutant.** Used in its broad historical sense, mutation refers to any sudden, heritable change in the genotype of an organism. However, changes in the genotype, and thus in the phenotype, of an organism that result from recombination events that produce new combinations of preexisting genetic variation must be carefully distinguished from changes caused by new mutations. Both events sometimes give rise to new phenotypes at very low frequencies. Mutational changes in the genotype of an organism include changes in chromosome number and structure (Chapter 7), as well as changes in the structures of individual genes. Today, the term *mutation* often is used in a narrow sense to refer only to changes occurring within genes. In this chapter, we explore the process of mutation as defined in the narrow sense. Mutations that involve changes at specific sites in a gene are referred to as **point mutations.** They include the substitution of one base pair for another or the insertion or deletion of one or a few nucleotide pairs at a specific site in a gene.

Mutation is the ultimate source of all genetic variation; it provides the raw material for evolution. Recombination mechanisms rearrange genetic variability into new combinations, and natural or artificial selection preserves the combinations best adapted to the existing environmental conditions or desired by the plant or animal breeder. Without mutation, all genes would exist in only one form. Alleles would not exist, and classical genetic analysis would not be possible. Most important, organisms would not be able to evolve and adapt to environmental changes. Some level of mutation is essential to provide new genetic variability and allow organisms to adapt to new environments. At the same time, if mutations occurred too frequently, they would disrupt the faithful transfer of genetic information from generation to generation. Moreover, most mutations with easily detected phenotypic effects are deleterious to the organisms in which they occur. As we would expect, the rate of mutation is under genetic control, and mechanisms have evolved that regulate the level of mutation that occurs under various environmental conditions.

Key Points: **Mutations are heritable changes in the genetic material that provide the raw material for evolution. Many mutations have deleterious effects.**

MUTATION: BASIC FEATURES OF THE PROCESS

Mutations occur in all genes of all living organisms. These mutations provide new genetic variability that allows organisms to adapt to environmental changes. Thus, mutations have been, and continue to be, essential to the evolutionary process. Before we discuss specific phenotypic effects of mutations and the mechanisms by which various types of mutation occur, we will consider some of the basic features of this important process.

Mutation: Somatic or Germinal

A mutation may occur in any cell and at any stage in the development of a multicellular organism. The immediate effects of the mutation and its ability to produce a phenotypic change are determined by its dominance, the type of cell in which it occurs, and the time at which it takes place during the life cycle of the organism. In higher animals, the germ-line cells that give rise to the gametes separate from other cell lineages early in development (Chapter 2). All nongerm-line cells are somatic cells. **Germinal mutations** are those that occur in germ-line cells, whereas **somatic mutations** occur in somatic cells.

If a mutation occurs in a somatic cell, the resulting mutant phenotype will occur only in the descendants of that cell. The mutation will not be transmitted through the gametes to the progeny. The Delicious apple and the navel orange are examples of mutant phenotypes that resulted from mutations occurring in somatic cells. The fruit trees in which the original mutations occurred were somatic mosaics. The changes that give these two fruits their desirable qualities are believed to have resulted from spontaneous mutations occurring in somatic cells. In each case, the cell carrying the mutant gene reproduced, eventually forming an entire branch that had the characteristics of the mutant type. Fortunately, vegetative propagation was feasible for both the Delicious apple and the navel orange, and today numerous progeny from grafts and buds have perpetuated the original mutations.

If dominant mutations occur in germ-line cells, their effects may be expressed immediately in progeny. If the mutations are recessive, their effects are often obscured in diploids. Germinal mutations may occur at any stage in the reproductive cycle of the organism, but they are more common during some stages than others. If the mutation arises in a gamete, only a single member of the progeny is likely to have the mutant gene. If a mutation occurs in a primordial germ-line cell of the testis or ovary, several gametes may receive the mutant gene, enhancing its potential for perpetuation. Thus, the dominance of a mutant allele and the stage in the reproductive cycle at which a mutation occurs are major factors in determining the likelihood that the mutant allele will be manifest in an organism in a population.

The earliest recorded dominant germinal mutation in domestic animals was that observed by Seth Wright in 1791 on his farm by the Charles River in Dover, Massachusetts. Among his flock of sheep, Wright noticed a peculiar male lamb with unusually short legs (Figure 14.2). It occurred to him that it would be an advantage to have a whole flock of these short-legged sheep, which could not jump over the low stone fences in his New England neighborhood. Wright used the new short-legged ram to breed his 15 ewes in the next season. Two of the 15 lambs had short legs. Short-legged sheep were then bred together, and a line was developed in which the new trait was expressed in all individuals.

Mutation: Spontaneous or Induced

When a new mutation—such as the one that produced Wright's short-legged sheep—occurs, is it caused by some agent in the environment or does it result from

(a)

(b)

Figure 14.2 Effect of a dominant mutation. (*a*) Short-legged sheep of the Ancon breed; (*b*) sheep with normal length legs.

an inherent process in living organisms? **Spontaneous mutations** are those that occur without a known cause. They may truly be spontaneous, resulting from a low level of inherent metabolic errors, or they may actually be caused by unknown agents present in the environment. **Induced mutations** are those resulting from exposure of organisms to physical and chemical agents that cause changes in DNA (or RNA in some viruses). Such agents are called **mutagens;** they include ionizing irradiation, ultraviolet light, and a wide variety of chemicals.

Operationally, it is impossible to prove that a particular mutation occurred spontaneously or was induced by a mutagenic agent. Geneticists must restrict such distinctions to the population level. If the mutation rate is increased a hundredfold by treatment of a population with a mutagen, an average of 99 of every 100 mutations present in the population will have been induced by the mutagen. Researchers can thus make valid comparisons between spontaneous and induced mutations statistically by comparing populations exposed to a mutagenic agent with control populations that have not been exposed to the mutagen.

Spontaneous mutations occur infrequently, although the observed frequencies vary from gene to gene and from organism to organism. Measurements of spontaneous mutation frequencies for various genes of phage and bacteria range from about 10^{-8} to 10^{-10} detectable mutations per nucleotide pair per generation. For eukaryotes, estimates of mutation rates range from about 10^{-7} to 10^{-9} detectable mutations per nucleotide pair per generation (considering only those genes for which extensive data are available). In comparing mutation rates per nucleotide with mutation rates per gene, the average gene is usually assumed to be 1000 nucleotide pairs in length. Thus, the mutation rate per gene varies from about 10^{-4} to 10^{-7} per generation.

Treatment with mutagenic agents can increase mutation frequencies by orders of magnitude. The mutation frequency per gene in bacteria and viruses can be increased to over 1 percent by treatment with potent chemical mutagens. That is, over 1 percent of the genes of the treated organisms will contain a mutation, or, stated differently, over 1 percent of the phage or bacteria in the population will have a mutation in a given gene.

Mutation: Usually a Random, Nonadaptive Process

The rats in many cities are no longer affected by the anticoagulants that have traditionally been used as rodent poisons. Many cockroach populations are insensitive to chlordane, the poison used to control them in the 1950s. Housefly populations often exhibit high levels of resistance to many insecticides. More and more pathogenic microorganisms are becoming resistant to antibiotics such as penicillin and streptomycin developed to control them. The introduction of these pesticides and antibiotics by humans produced new environments for these organisms. Mutations producing resistance to these pesticides and antibiotics occurred, and the mutant organisms were at a large selective advantage in environments where the agents were present. The sensitive organisms were killed, and the mutants multiplied to produce new resistant populations. Many such cases of evolution via mutation and natural selection are well documented.

These examples raise a basic question about the nature of mutation. Is mutation a purely random event in which the environmental stress merely preserves preexisting mutations? Or is mutation directed by the environmental stress? For example, if you cut off the tails of mice for many generations, will you eventually produce a strain of tailless mice? Despite the beliefs of Jean Lamarck and his follower Trofim Lysenko, whom we discussed in Chapter 1, the answer is no; the mice will continue to be born with tails.

Today, it is hard to understand how Lysenko could have sold his belief in Lamarckism—the inheritance of acquired traits—to those in power in the Soviet Union from 1937 through 1964. However, disproving Lamarckism was not an easy task, especially in the case of microorganisms, where even small cultures often contain billions of organisms. As an example, let us consider a population of bacteria such as *E. coli* growing in a streptomycin-free environment. When exposed to streptomycin, most of the bacteria will be killed by the antibiotic. However, if the population is large enough, it will soon give rise to a streptomycin-resistant culture in which all the cells are resistant to the antibiotic. Does streptomycin simply select rare, randomly occurring mutants that preexist in the population, or do all of the cells have some low probability of developing resistance in response to the presence of streptomycin? How can geneticists distinguish between these two possibilities? They can detect resistance to streptomycin only by treating the culture with the antibiotic. How, then, can they determine whether resistant bacteria are present prior to exposure to streptomycin or are induced by the antibiotic?

In 1952, Joshua and Esther Lederberg developed an important new technique called **replica plating.** This technique allowed them to demonstrate the presence of antibiotic-resistant mutants in bacterial cultures prior to their exposure to the antibiotic (Figure 14.3). They first diluted the bacterial cultures, spread the bacteria on the surface of semisolid nutrient agar medium in petri dishes, and incubated the plates until each bacterium had produced a visible colony on the surface of the agar. They next inverted each plate and pressed it onto sterile velvet placed over a wood block. Some of the cells from each colony stuck to the velvet. They then gently pressed a sterile plate of nutrient agar medium containing streptomycin onto the velvet. They repeated this replica-plating procedure with many plates, each containing about 200 bacterial colonies. After they incubated the selective plates (those containing streptomycin) overnight, rare streptomycin-resistant colonies had formed.

The Lederbergs subsequently tested the colonies on the nonselective plates (those not containing streptomycin) for their ability to grow on medium containing streptomycin. Their results were definitive. The colonies that grew on the selective replica plates almost always contained streptomycin-resistant cells, whereas those that did not grow on the selective medium seldom contained any resistant cells (Figure 14.3). If a mutation that makes a bacterium resistant to streptomycin occurs at an early stage in the growth of a colony, the resistant cell will divide and produce two, then four, then eight, and eventually a large number of resistant bacteria. Thus, if mutation is a randomly occurring, nonadaptive process, many of the colonies that form on the nonselective plates will contain more than one antibiotic-resistant bacterium and will give rise to resistant cultures when tested for growth on selective media. However, if mutation is adaptive and the mutations to streptomycin resistance occur only after exposure to the antibiotic, then the colonies on the nonselective plates that gave rise to resistant colonies on the selective plates after replica plating would be no more likely to contain streptomycin-resistant cells than the other colonies on the nonselective plates. Thus, by using their replica-plating technique, the Lederbergs demonstrated the existence of streptomycin-resistant mutants in a population of bacteria prior to their exposure to the antibiotic. Their results and the results of many other experiments have shown that environmental stress does not direct or cause genetic changes as Lysenko believed; it simply selects rare preexisting mutations that result in phenotypes better adapted to the new environment.

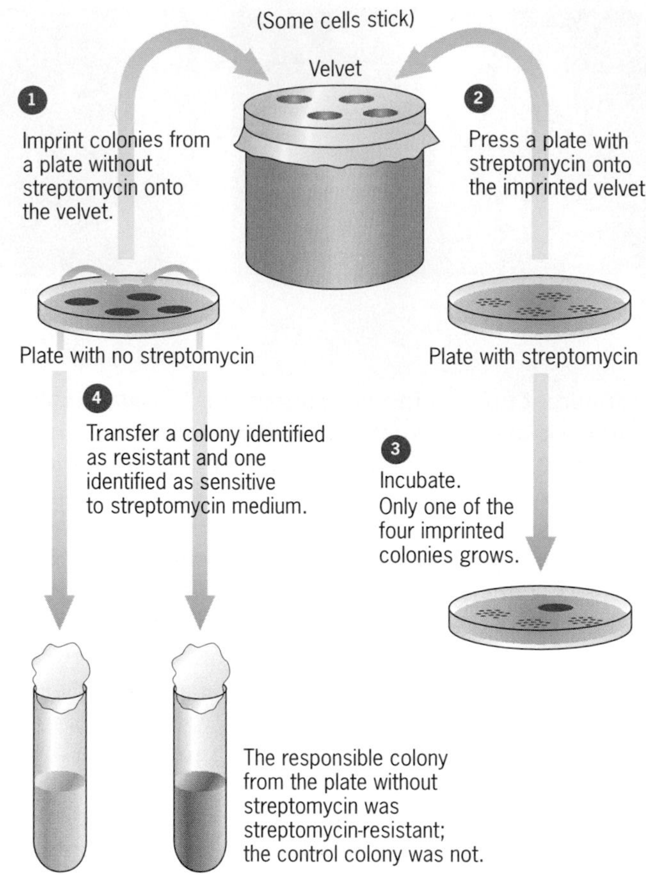

Figure 14.3 Joshua and Esther Lederberg's use of replica plating to demonstrate the random or nondirected nature of mutation. For simplicity, only four colonies are shown on each plate, and only two are tested for streptomycin resistance in step 4. Actually, each plate contains about 200 colonies, and many plates must be used to find an adequate number of mutant colonies.

Mutation: A Reversible Process

As we discussed earlier, a mutation in a wild-type gene can produce a mutant allele that results in an abnormal phenotype. However, the mutant allele can also

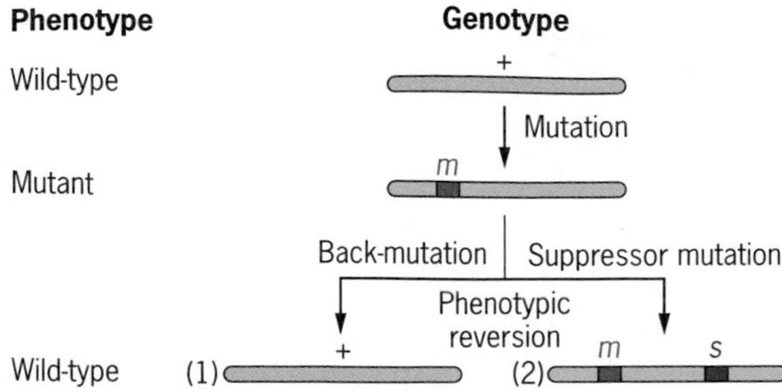

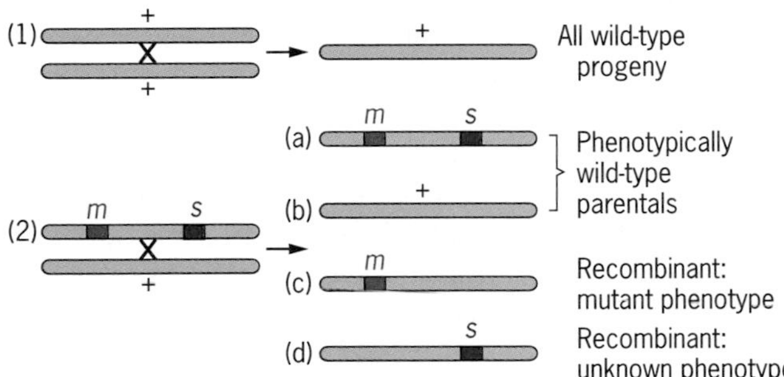

Figure 14.4 Restoration of the original wild-type phenotype of an organism may occur by (1) back mutation or (2) suppressor mutation. Some mutants can revert to the wild-type phenotype by both mechanisms. Revertants of the two types can be distinguished by backcrosses to the original wild-type. If back mutation has occurred, all backcross progeny will be wild-type. If a suppressor mutation is responsible, some of the backcross progeny will have the mutant phenotype (2c).

mutate back to a form that restores the wild-type phenotype. That is, mutation is a reversible process.

The mutation of a wild-type gene to a form that results in a mutant phenotype is referred to as **forward mutation.** However, sometimes the designation of the wild-type and mutant phenotypes is quite arbitrary. They may simply represent two different, but normal, phenotypes. For example, geneticists consider the alleles for brown and blue eye color in humans both to be wild-type. However, in a population composed almost entirely of brown-eyed individuals, the allele for blue eyes might be thought of as a mutant allele.

When a second mutation restores the original phenotype lost because of an earlier mutation, the process is called **reversion** or **reverse mutation.** Reversion may occur in two different ways: (1) by **back mutation,** a second mutation at the same site in the gene as the original mutation, restoring the wild-type nucleotide sequence, or (2) by the occurrence of a **suppressor mutation,** a second mutation at a different location in the genome, which compensates for the effects of the first mutation (Figure 14.4). Back mutation restores the original wild-type nucleotide sequence of the gene, whereas a suppressor mutation does not. Suppressor mutations may occur at distinct sites in the same gene as the original mutation or in different genes, even on different chromosomes.

Some mutations revert primarily by back mutation, whereas others do so almost exclusively through the occurrence of suppressor mutations. Thus, in genetic studies, researchers often must distinguish between these two possibilities by backcrossing the phenotypic revertant with the original wild-type organism. If the wild-type phenotype is restored by a suppressor mutation, the original mutation will still be present and can be separated from the suppressor mutation by recombination (Figure 14.4). If the wild-type phenotype is restored by back mutation, all of the progeny of the backcross will be wild-type.

Key Points: **Mutations occur in both germ-line and somatic cells, but only germ-line mutations are transmitted to progeny. Mutations may occur spontaneously or be induced by mutagenic agents in the environment. Mutation usually is a nonadaptive process in which an environmental stress simply selects organisms with preexisting, randomly occurring mutations. Restoration of the wild-type phenotype in a mutant organism may occur by either back mutation or the occurrence of a suppressor mutation.**

MUTATION: PHENOTYPIC EFFECTS

To be recognized, mutations must result in some detectable phenotypic change. The effects of mutations on phenotype range from alterations so minor that they can be detected only by special genetic or biochemical techniques, to gross modifications of morphology, to lethals. A gene is a sequence of nucleotide pairs that usually encodes a specific polypeptide. Any mutation occurring within a given gene will thus produce a new allele of that gene. Genes containing mutations with small effects that can be recognized only by special techniques are called **isoalleles.** Other mutations produce **null alleles** that result in totally nonfunctional gene products. If mutations of the latter type occur in genes that are required for the growth of the organism, individuals that are homozygous for the mutation will not survive. Such mutations are called **recessive lethals.**

Mutations can be either recessive or dominant. In monoploid organisms such as viruses and bacteria, both recessive and dominant mutations can be recognized by their effect on the phenotype of the organism in which they occur. In diploid organisms such as fruit

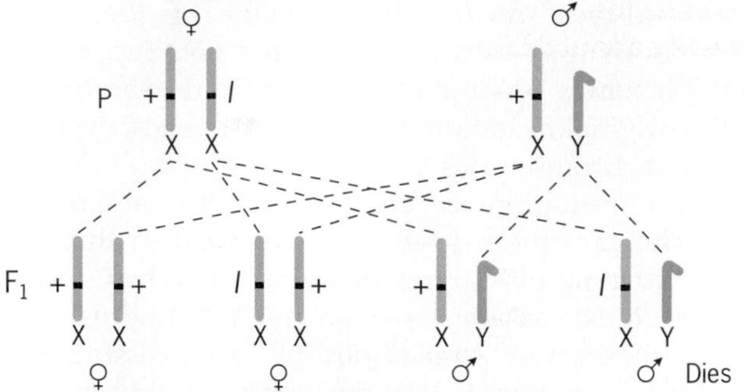

Figure 14.5 Alteration of the sex ratio by an X-linked recessive lethal mutation. Females heterozygous for an X-linked recessive lethal will produce female and male progeny in a 2:1 ratio.

flies and humans, recessive mutations will alter the phenotype only when present in the homozygous condition. Thus, in diploids, most recessive mutations will not be recognized at the time of their occurrence because they will be present in the heterozygous state. X-linked recessive mutations are an exception; they will be expressed in the hemizygous state in the heterogametic sex (males in humans and fruit flies; females in birds). X-linked recessive lethal mutations will alter the sex ratio of offspring because hemizygous individuals that carry the lethal will not survive (Figure 14.5).

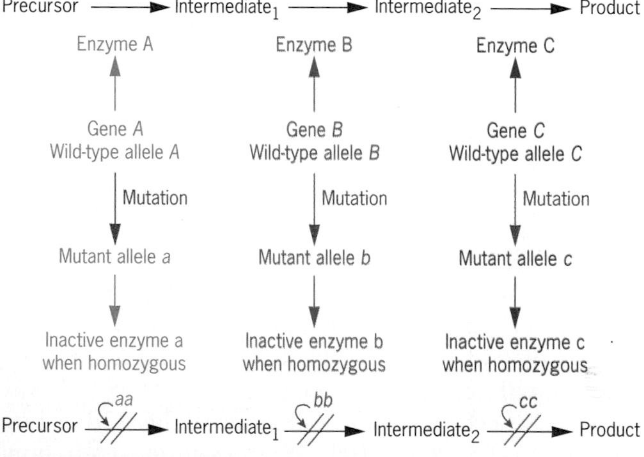

Figure 14.6 Recessive mutant alleles often result in blocks in metabolic pathways. The pathways can be only a few steps long, as diagrammed here, or many steps long. The wild-type allele of each gene usually encodes a functional enzyme that catalyzes the appropriate reaction. Most mutations that occur in wild-type genes result in altered forms of the enzyme with reduced or no activity. In the homozygous state, mutant alleles that produce inactive products cause metabolic blocks (⇸) owing to the lack of the required enzyme activity.

Mutations with Phenotypic Effects: Usually Deleterious and Recessive

Most of the thousands of mutations that have been identified and studied by geneticists are deleterious and recessive. This result is to be expected if we consider what is known about the genetic control of metabolism and the techniques available for identifying mutations. As we discussed in Chapter 12, metabolism occurs by sequences of chemical reactions, with each step catalyzed by a specific enzyme encoded by one or more genes. Mutations in these genes frequently produce blocks in metabolic pathways (Figure 14.6). These blocks occur because alterations in the base-pair sequences of genes often cause changes in the amino acid sequences of polypeptides (Figure 14.7), which may result in nonfunctional products (Figure 14.6). Indeed, this is the most commonly observed effect of easily detected mutations. Given a wild-type allele encoding an active enzyme and mutant alleles encoding less active or totally inactive enzymes, it is apparent why most of the observed mutations would be recessive. If a cell contains both active and inactive forms of a given enzyme, the active form usually will catalyze the reaction in question. Therefore, the allele specifying the active product usually will be dominant, and the allele encoding the inactive product will be recessive (Chapter 4).

Because of the degeneracy and order in the genetic code (Chapter 13), many mutations have no effect on the phenotype of the organism; they are called **neutral mutations.** But why should most mutations with phenotypically recognizable effects result in decreased gene-product activity or no gene-product activity? A wild-type allele of a gene encoding a wild-type enzyme or structural protein will have been selected for optimal activity during the course of evolution. Thus mutations, which cause random changes in the highly adapted amino acid sequences, usually will produce less active or totally inactive products. You can make an analogy with any complex, carefully engineered machine such as a watch or an automobile. If you randomly modify an essential component, the machine seldom performs as well as it did prior to the change. This view of mutation and the interaction between mutant and wild-type alleles fits with the observation that most mutations with recognizable phenotypic effects are recessive and deleterious.

Effects of Mutations in Human Globin Genes

Mutant human hemoglobins provide good illustrations of the deleterious effects of mutation. In Chapter 13, we discussed the structure of hemoglobin and the

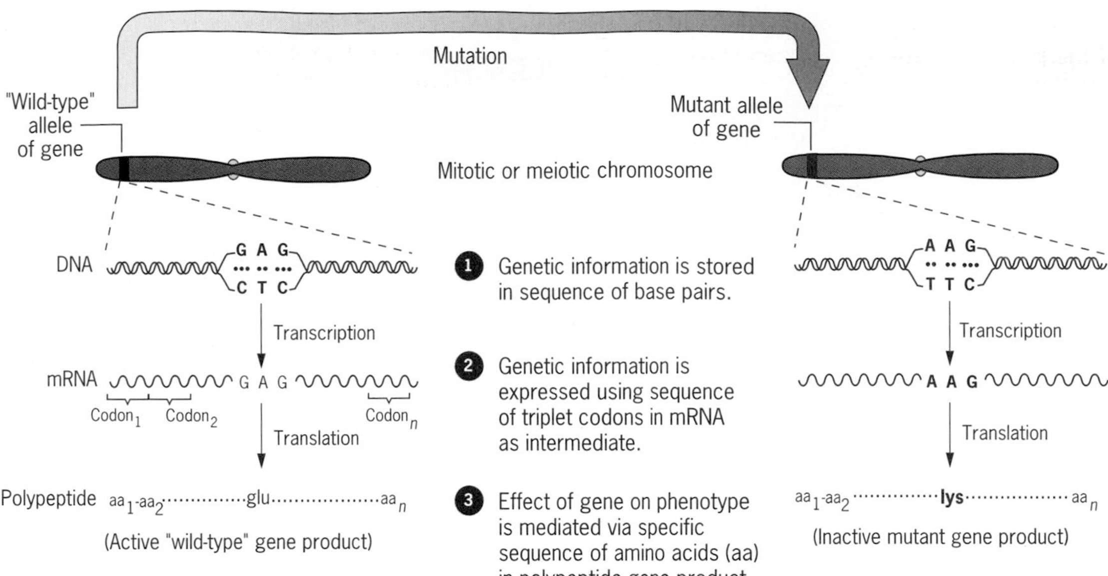

Figure 14.7 Overview of the mutation process and the expression of wild-type and mutant alleles. Mutations alter the sequences of nucleotide pairs in genes, which, in turn, cause changes in the amino acid sequences of the polypeptides encoded by these genes. A G:C base pair (top, left) has mutated to an A:T base pair (top, right). This mutation changes one mRNA codon from GAG to AAG and one amino acid in the polypeptide product from glutamic acid (glu) to lysine (lys). Such changes often yield nonfunctional gene products.

traumatic effects of one hemoglobin variant, sickle-cell hemoglobin. Recall that the major form of hemoglobin in adults (hemoglobin A) contains two identical **alpha (α) chains** and two identical **beta (β) chains**. Each α polypeptide consists of a specific sequence of 141 amino acids, whereas each β chain is 146 amino acids long. Because of similarities in their amino acid sequences, all the globin chains (and, thus, their structural genes) are believed to have evolved from a common progenitor.

Many different variants of adult hemoglobin have been identified in human populations, and several of them have severe phenotypic effects. Many of the variants were initially detected by their altered electrophoretic behavior (movement in an electric field due to charge differences). The hemoglobin variants provide an excellent illustration of the effects of mutation on the structures and functions of gene products and, ultimately, on the phenotypes of the affected individuals.

When the amino acid sequences of the β chains of hemoglobin A and the hemoglobin in patients with sickle-cell anemia (hemoglobin S) were determined and compared, hemoglobin S was found to differ from hemoglobin A at only one position. The sixth amino acid from the amino terminus of the β chain of hemoglobin A is glutamic acid (a negatively charged amino acid). The β chain of hemoglobin S contains valine (no charge at neutral pH) at that position. The α chains of hemoglobin A and hemoglobin S are identical. Thus, the change of a single amino acid in one polypeptide can have severe effects on the phenotype. A large number of similar effects of mutation on protein structure, and ultimately on the phenotype, are now well documented.

In the case of hemoglobin S, the substitution of valine for glutamic acid at the sixth position in the β chain allows a new bond to form, which changes the conformation of the protein and leads to aggregation of hemoglobin molecules. This change results in the grossly abnormal shape of the red blood cells. The mutational change in the Hb_β^A allele that gave rise to Hb_β^S was a substitution of an A:T base pair for a T:A base pair, with a T in the transcribed strand in the first case and an A in the transcribed strand in the second case (Figure 14.8). This T:A → A:T base-pair substitution was first predicted from protein sequence data and the known codon assignments, and was later verified by sequencing the Hb_β^A and Hb_β^S alleles.

Over 100 hemoglobin variants with amino acid changes in the β chain are known. Most of them differ from the normal β chain of hemoglobin A by a single amino acid substitution. A few examples are illustrated in Figure 14.9. Numerous variants of the α polypeptide also have been identified.

The hemoglobin examples show that mutation is a process in which changes in gene structure, often changes in one or a few base pairs, cause changes in

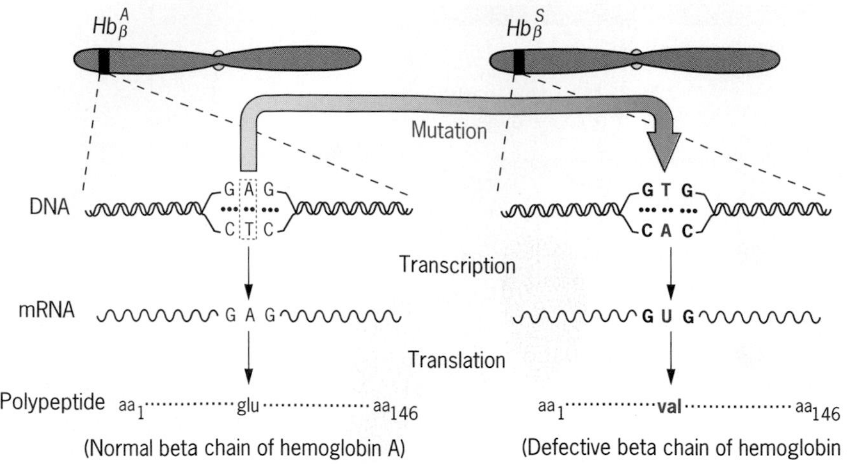

Figure 14.8 The mutational origin of sickle-cell hemoglobin (hemoglobin S). The mutation that produced the Hb_β^S allele from the Hb_β^A gene was the substitution of a $\overset{T}{A}$ base pair for an $\overset{A}{T}$ base pair, in which the bottom strand of DNA is the transcribed strand. The result is the substitution of the amino acid valine in the β chain of hemoglobin S for the glutamic acid residue present at the same position in hemoglobin A.

the amino acid sequences of the polypeptide gene products. These alterations in protein structure, in turn, cause changes in the phenotype that are recognized as mutant.

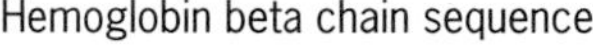

Mutation in Humans: Blocks in Metabolic Pathways

In Chapter 12, we discussed the genetic control of metabolic pathways, in which each step in a pathway is catalyzed by an enzyme encoded by one or more genes. When mutations occur in such genes, they often cause metabolic blocks (see Figure 14.6) that lead to abnormal phenotypes. This picture of the genetic control of metabolism is valid for all living organisms, including humans (see Human Genetics Sidelight: Tay-Sachs Disease).

We can illustrate the effects of mutations on human metabolism by considering virtually any metabolic pathway. However, the metabolism of the aromatic amino acids phenylalanine and tyrosine provides an especially good example because some of the early studies of mutations in humans revealed

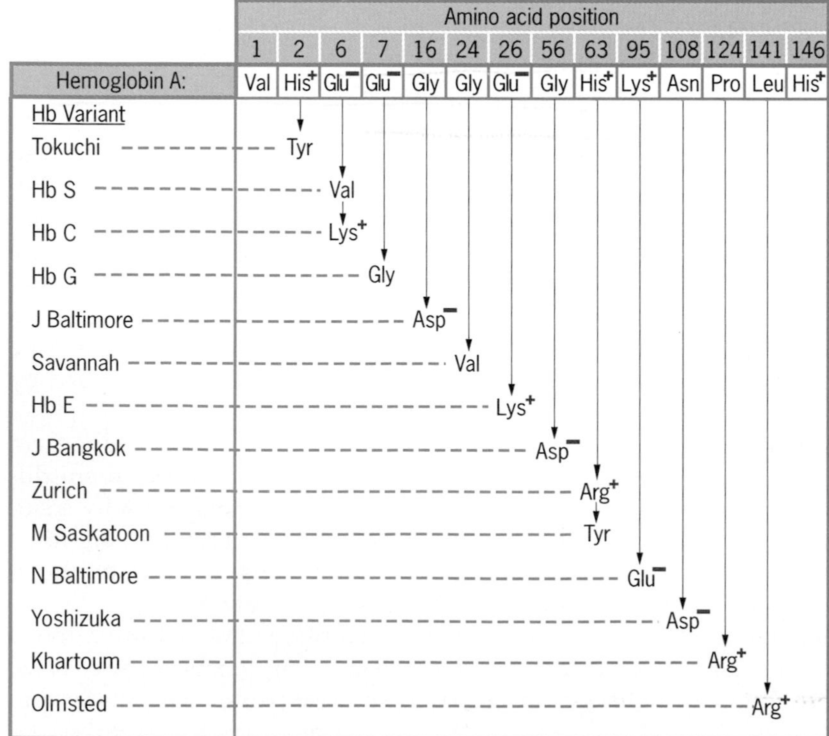

Figure 14.9 A few of the amino acid substitutions that have been documented in human β-globin chains. All the examples shown resulted from single base-pair substitutions in the Hb_β^A gene (analogous to that shown for hemoglobin S in Figure 14.8). Note that many of the amino acid substitutions involve changes in charge, even though only 5 of the 20 common amino acids have a net charge at neutral pH. Substitutions involving changes in charge alter protein conformation more often than do other amino acid substitutions. Thus, such substitutions are more likely to cause changes in protein function.

HUMAN GENETICS SIDELIGHT

Tay-Sachs Disease, a Childhood Tragedy

Of all the inherited human disorders, one of the most tragic is Tay-Sachs disease. Infants homozygous for the mutant gene that causes Tay-Sachs disease are normal at birth. However, within a few months, they become hypersensitive to loud noises and develop a cherry-red spot on the retina of the eye. These early symptoms of the disease often go undetected by parents and physicians. At six months to one year after birth, Tay-Sachs children begin to undergo progressive neurological degeneration that rapidly leads to mental retardation, blindness, deafness, and general loss of control of body functions. By two years of age, they are usually totally paralyzed and develop chronic respiratory infections. Death commonly occurs at three to four years of age.

Although the molecular defect responsible for Tay-Sachs disease is known, there is no effective treatment of the disorder. The only positive aspect of Tay-Sachs disease is that it is rare in most populations. However, this is of little comfort to the Ashkenazi Jewish people of Central Europe and their descendants. Tay-Sachs disease occurs in about 1 of 3600 of their children, and about 1 of 30 adults in these Jewish populations carries the mutant gene in the heterozygous state. If two individuals from these populations marry, the chance that both will carry the mutant gene is about 1 in 1000 (0.033×0.033); if both are carriers, on average, one-fourth of their children will be homozygous for the mutant gene and develop Tay-Sachs disease.

The mutation that causes Tay-Sachs disease is located in the *HEXA* gene, which encodes the enzyme hexosaminidase A. This enzyme acts on a complex lipid called ganglioside G_{M2}, cleaving it into a smaller ganglioside (G_{M3}) and *N*-acetyl-*D*-galactosamine, as shown in Figure 1.

The function of ganglioside G_{M2} is to coat nerve cells, insulating them from events occurring in neighboring cells and thus speeding up the transmission of nerve impulses. In the absence of the enzyme that breaks it down, ganglioside G_{M2} accumulates and literally smothers nerve cells. This buildup of complex lipids on neurons blocks their action, leading to deterioration of the nervous system and eventually paralysis.

Although Tay-Sachs disease was described by Warren Tay in 1881 and the biochemical basis has been known for over 20 years, there is still no effective treatment of this tragic disorder. Whereas some inherited disorders can be treated by enzyme therapy—by supplying the missing enzyme to patients—this approach is not feasible with Tay-Sachs disease because the enzyme will not penetrate the barrier separating brain cells from the circulatory system. Moreover, somatic-cell gene therapy—providing functional copies of the defective gene to somatic cells (Chapter 22)—is not possible at present because there is no established procedure for introducing genes into neurons. Indeed, scientists still don't know which nerve cells are responsible for the neurological degeneration that occurs in children with the disease.

Tay-Sachs disease can be detected prenatally by amniocentesis (Chapter 7), and this procedure has been used extensively to diagnose the disorder. Recently, a sensitive DNA test has been developed that allows scientists to detect the mutant gene that causes Tay-Sachs disease in DNA isolated from a single cell (Chapter 22). This DNA test has been used to screen eight-cell pre-embryos produced by *in vitro* fertilization for the Tay-Sachs mutation. One cell is used for the DNA test, and the other seven cells retain the capacity to develop into a normal embryo when implanted into the uterus of the mother. Only embryos that test normal—those not homozygous for the deadly Tay-Sachs gene—are implanted. This procedure allows parents who are both carriers of the mutant gene to have children without worrying about the birth of a child with Tay-Sachs disease.

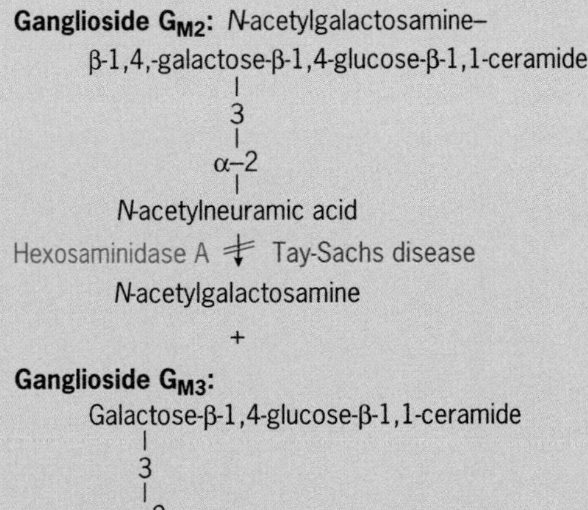

Figure 1 The metabolic defect in humans with Tay-Sachs disease.

blocks in this pathway (Figure 14.10). Phenylalanine and tyrosine are essential amino acids required for protein synthesis; they are not synthesized *de novo* in humans as they are in microorganisms. Thus, both amino acids must be obtained from dietary proteins.

The best-known inherited defect in phenylalanine-tyrosine metabolism is phenylketonuria, which is caused by the absence of phenylalanine hydroxylase, the enzyme that converts phenylalanine to tyrosine. Newborns with phenylketonuria, an autosomal reces-

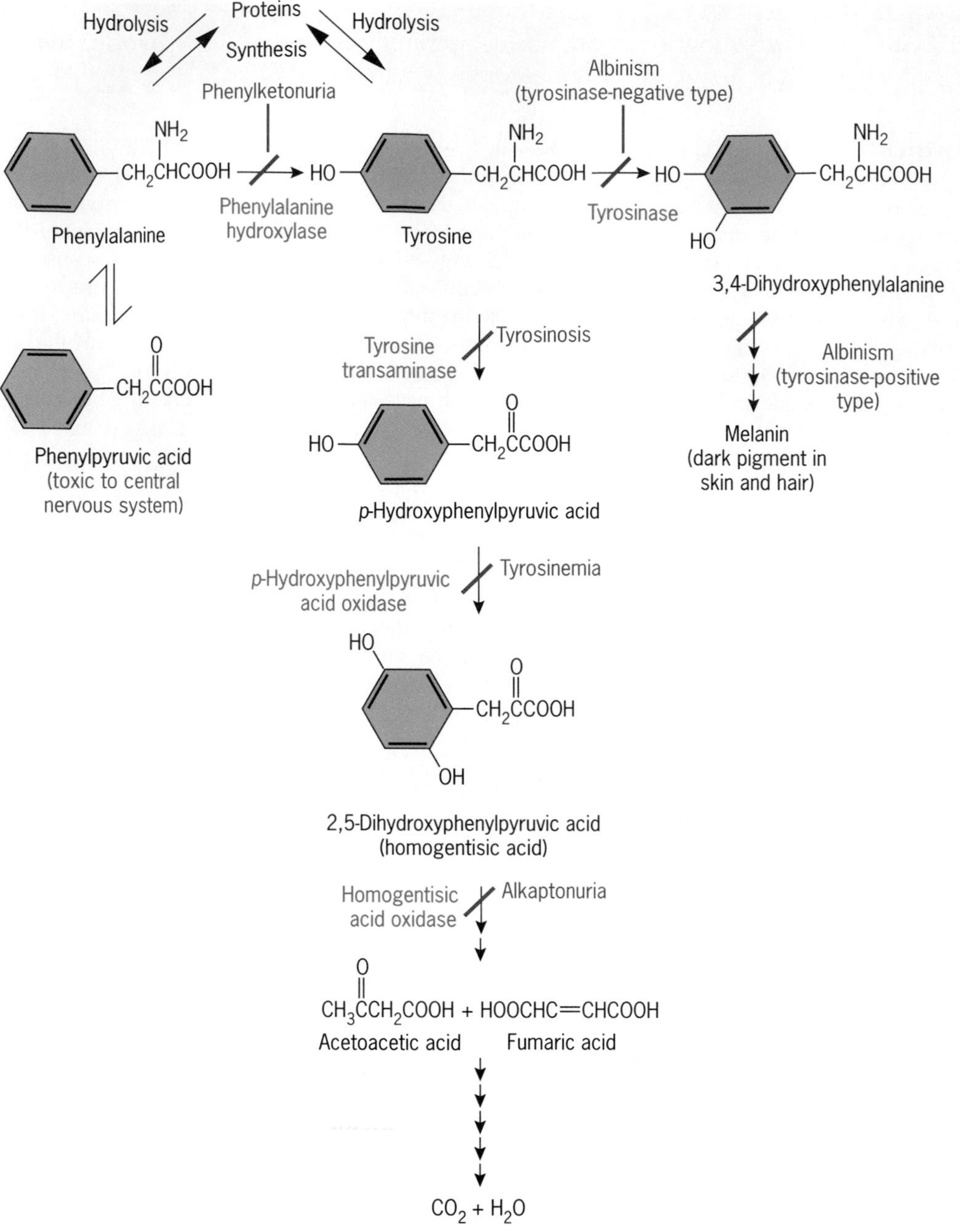

Figure 14.10 Inherited human disorders with defects in phenylalanine-tyrosine metabolism: phenylketonuria, tyrosinosis, tyrosinemia, alkaptonuria, and albinism. All five disorders are caused by autosomal recessive mutations. The mutations, which result in the synthesis of inactive enzymes, block phenylalanine-tyrosine metabolism at the steps indicated.

sive disease, develop severe mental retardation if not placed on a diet low in phenylalanine. (This disorder is discussed in detail in Chapter 26.) The first inherited disorder in the phenylalanine-tyrosine metabolic pathway to be studied in humans was alkaptonuria, which is caused by autosomal recessive mutations that inactivate the enzyme homogentisic acid oxidase. Alkaptonuria, which played an important role in the evolution of the concept of the gene, is discussed in Chapter 15.

Two other inherited disorders are caused by mutations in genes encoding enzymes required for the catabolism of tyrosine; both are inherited as autosomal recessives. Tyrosinosis and tyrosinemia result from the lack of the enzymes tyrosine transaminase and *p*-hydroxyphenylpyruvic acid oxidase, respectively

(Figure 14.10). Tyrosinosis is very rare; only a few cases have been studied. Individuals with tyrosinosis show pronounced increases in tyrosine levels in their blood and urine and have various congenital abnormalities. Individuals with tyrosinemia have elevated levels of both tyrosine and *p*-hydroxyphenylpyruvic acid in their blood and urine. Most newborns with tyrosinemia die within six months after birth because of liver failure.

Albinism, the absence of pigmentation in the skin, hair, and eyes, results from a mutational block in the conversion of tyrosine to the dark pigment melanin. One type of albinism is caused by the absence of tyrosinase, the enzyme that catalyzes the first step in the synthesis of melanin from tyrosine. Other types of albinism result from blocks in subsequent steps in the conversion of tyrosine to melanin. Albinism is inherited as an autosomal recessive trait; heterozygotes usually have normal levels of pigmentation. Therefore, two albinos who have mutations in different genes can produce normally pigmented children.

Thus, studies of a single metabolic pathway, phenylalanine-tyrosine metabolism, have revealed five different inherited disorders, all caused by mutations in genes that control steps in this pathway. The same picture of the genetic control of metabolism can be obtained by examining essentially any other metabolic pathway in humans.

Conditional Lethal Mutations: Powerful Tools for Genetic Studies

Of all the mutations—from isoalleles to lethals, **conditional lethal mutations** are the most useful for genetic studies. These are mutations that are (1) lethal in one environment, the **restrictive conditions,** but are (2) viable in a second environment, the **permissive conditions.** Conditional lethal mutations allow geneticists to identify and study mutations in essential genes that result in complete loss of gene-product activity even in haploid organisms. Mutants carrying conditional lethals can be propagated under permissive conditions, and information about the functions of the gene products can be inferred by studying the consequences of their absence under the restrictive conditions. Conditional lethal mutations have been used to investigate a vast array of biological processes from development to photosynthesis. They have also been used to construct detailed genetic maps of chromosomes (Chapters 16 and 17).

The three major classes of mutants with conditional lethal phenotypes are (1) auxotrophic mutants, (2) temperature-sensitive mutants, and (3) suppressor-sensitive mutants. **Auxotrophs** are mutants that are unable to synthesize an essential metabolite (amino acid, purine, pyrimidine, vitamin, and so forth) that is synthesized by wild-type or *prototrophic* organisms of the same species. The auxotrophs will grow and reproduce when the metabolite is supplied in the medium (the permissive conditions); they will not grow when the essential metabolite is absent (the restrictive conditions). **Temperature-sensitive mutants** will grow at one temperature but not at another. Most temperature-sensitive mutants are heat-sensitive; however, some are cold-sensitive. The temperature sensitivity usually results from increased heat or cold lability of the mutant gene product, for example, an enzyme that is active at low temperature but partially or totally inactive at higher temperatures. Occasionally, only the synthesis of the gene product is sensitive to temperature, and, once synthesized, the mutant gene product may be as stable as the wild-type gene product. **Suppressor-sensitive mutants** are viable when a second genetic factor, a suppressor, is present, but they are nonviable in the absence of the suppressor. The suppressor gene may correct or compensate for the defect in phenotype that is caused by the suppressor-sensitive mutation, or it may render the gene product altered by the mutation nonessential. We have discussed one class of suppressor-sensitive mutations, the *amber* mutations, in Chapter 13.

Now, let's briefly consider how conditional lethal mutations can be used to investigate biological processes—to dissect biological processes into their individual parts or steps. Let's begin with a simple biosynthetic pathway:

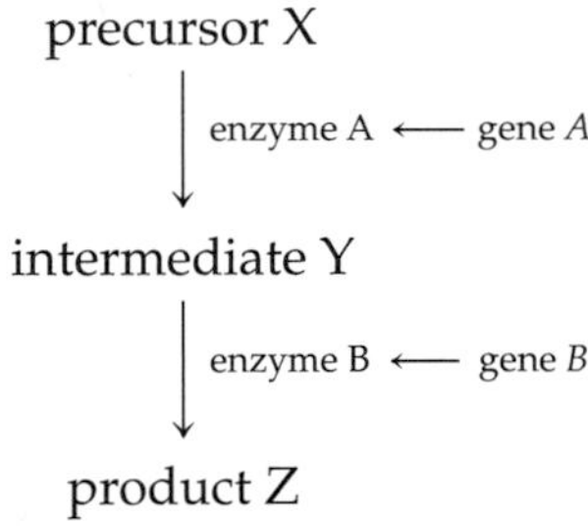

Intermediate Y is produced from precursor X by the action of enzyme A, the product of gene *A*, but intermediate Y may be rapidly converted to product Z by enzyme B, the product of gene *B*. If so, intermediate Y may be present in minute quantities and be difficult to isolate and characterize. However, in a mutant organism that has a mutation in gene *B*, resulting in the synthesis of either an inactive form of enzyme B or no enzyme B, intermediate Y may accumulate to much higher concentrations, facilitating its isolation and characterization. Similarly, a mutation in gene *A* may aid in the identification of precursor X.

In this way, the sequence of steps in a given metabolic pathway can usually be determined. Morphogenesis in living organisms occurs in part by the sequential addition of proteins to macromolecular structures to produce the final three-dimensional conformations, and the sequence of protein additions can often be determined by isolating and studying mutant organisms with defects in the genes encoding the proteins involved. Because an appropriate mutation will eliminate the activity of a single polypeptide, mutations provide a powerful tool with which to dissect biological processes—to break the processes down into individual steps.

The resolving power of mutational dissection of biological processes has been elegantly documented by the research of Robert Edgar, Jonathan King, William Wood, and colleagues, who worked out the complete pathway of morphogenesis of bacteriophage T4. This complex process involves the products of about 50 of the roughly 200 genes in the T4 genome. Each gene encodes a structural protein of the virus or an enzyme that catalyzes one or more steps in the morphogenetic pathway. By (1) isolating mutant strains of phage T4 with temperature-sensitive and suppressor-sensitive conditional lethal mutations in each of the approximately 50 genes, and (2) analyzing the structures that accumulate when these mutant strains are grown under the restrictive conditions by electron microscopy and biochemical techniques, Edgar, King, Wood, and coworkers have established the complete pathway of phage T4 morphogenesis (Figure 14.11).

Many other biological processes also have been successfully dissected by mutational studies. Examples include the photosynthetic electron transport chains in plants and pathways of nitrogen fixation in bacteria. Currently, mutational dissection is yielding new insights into the processes of differentiation and development in higher plants and animals (Chapter 25). Researchers are also using mutations to dissect behavior and learning in *Drosophila*. In principle, scientists should be able to use mutations to dissect any process that is under genetic control. Every gene can mutate to a nonfunctional state. Thus, mutational dissection of biological processes is limited only by the ingenuity of researchers in identifying mutations of the desired types.

Key Points: **The effects of mutations on the phenotypes of living organisms range from minor to lethal changes. Most mutations exert their effects on the phenotype by altering the amino acid sequences of polypeptides, the primary gene products. The mutant polypeptides, in turn, cause blocks in metabolic pathways. Conditional lethal mutations provide powerful tools with which to dissect biological processes such as morphogenesis.**

MULLER'S DEMONSTRATION THAT X RAYS ARE MUTAGENIC

In 1927, Hermann J. Muller first demonstrated that mutation could be induced by an external factor. Muller showed that X-ray treatment markedly increased the frequency of sex-linked recessive lethal mutations in *D. melanogaster*. X rays are a form of electromagnetic radiation with shorter wavelengths and higher energy than visible light; we discuss their effects on DNA later in this chapter. Muller's unambiguous demonstration of the mutagenicity of X rays became possible by his development of a technique facilitating the simple and

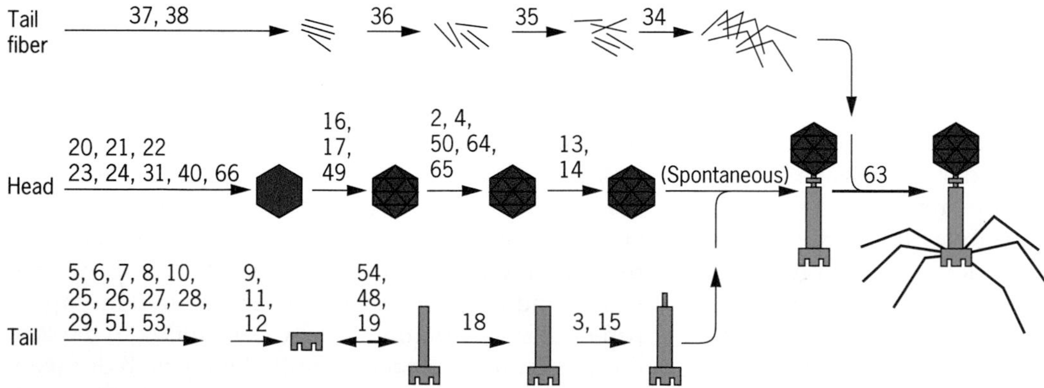

Figure 14.11 Abbreviated map of morphogenesis in bacteriophage T4. The pathway of morphogenesis is based on the work of Edgar, Wood, King, and colleagues. The head, the tail, and the tail fibers are produced via separate branches of the pathway and are then joined in the final stages of morphogenesis. The numbers identify the T4 genes whose products are required at each step in the pathway. The sequences of early steps in head and tail formation are known, but are omitted here to keep the diagram concise.

accurate identification of lethal mutations on the X chromosome of *Drosophila*. This technique, called the ***ClB* method,** involves the use of females heterozygous for a normal X chromosome and an X chromosome—the ***ClB* chromosome**—specifically constructed for Muller's experiment.

The *ClB* chromosome has three essential components. (1) C, for crossover suppressor, refers to a long inversion that suppresses recombination between the *ClB* chromosome and the structurally normal X chromosome in heterozygous females. The inversion does not prevent crossing over between the two chromosomes, but causes progeny carrying recombinant X chromosomes produced by crossing over between the two chromosomes to abort because of duplications and deficiencies (see Chapter 8). (2) *l* refers to a recessive lethal mutation on the *ClB* chromosome. Homozygous females and hemizygous males carrying this X-linked lethal mutation are nonviable. (3) *B* refers to a mutation that causes the bar eye phenotype, which is a narrow, bar-shaped eye. Because it is partially dominant, the *B* mutation allows females heterozygous for the *ClB* chromosome to be identified readily. Both the recessive lethal (*l*) and the bar eye mutation (*B*) are located within the inverted segment of the *ClB* chromosome.

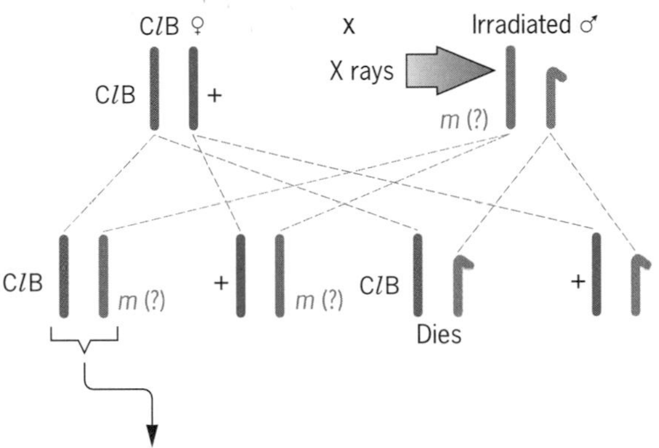

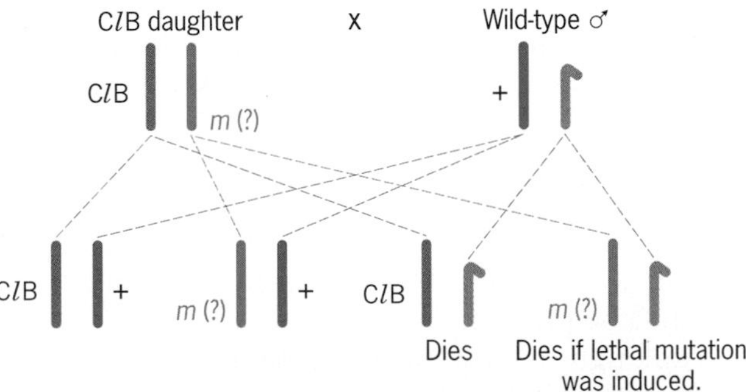

Figure 14.12 The *ClB* technique used by Muller to detect X-linked recessive lethal mutations in *Drosophila*. The mating shown in cross II will produce only female progeny if an X-linked recessive lethal is present on the irradiated X chromosome. One-third of the progeny produced from cross II will be males if there is no recessive lethal on the irradiated X chromosome. Thus, scoring for lethal mutations simply involves screening the progeny of cross II for the presence or absence of males.

Muller irradiated male flies and mated them with *ClB* females (Figure 14.12). The bar-eyed daughters of this mating all carried the *ClB* chromosome of the female parent and the irradiated X chromosome of the male parent. Because the entire population of reproductive cells of the males was irradiated, each bar-eyed daughter carried a potentially mutated X chromosome. The bar-eyed daughters were then mated individually (in separate cultures) with wild-type males. If the irradiated X chromosome carried by a bar-eyed daughter acquired an X-linked lethal, all the progeny of the mating will be female. Males hemizygous for the *ClB* chromosome will die because of the recessive lethal (*l*) that the chromosome carries. Males hemizygous for the irradiated X chromosome will also die if a recessive lethal has been induced on it. Matings of bar-eyed daughters carrying an irradiated X chromosome in which no lethal mutation has been induced will produce female and male progeny in a ratio of 2:1 (only the males with the *ClB* chromosome will die). With the use of the *ClB* technique, detecting newly induced recessive, X-linked lethals is unambiguous and error free—simply scoring for the presence or absence of male progeny. By this procedure, Muller was able to demonstrate a 150-fold increase in the frequency of X-linked lethals after treatment of male flies with X rays.

After Muller's pioneering experiments with *Drosophila*, other researchers soon demonstrated that X rays are mutagenic to other organisms, including plants, other animals, and microbes. Moreover, other types of high-energy electromagnetic radiation and many chemicals were soon shown to be potent mutagens. We examine the mechanisms by which some of these agents induce mutations in the following sections of this chapter.

Key Points: **Hermann J. Muller first demonstrated that an external agent can be mutagenic when, in 1927, he showed that X rays induce recessive lethals on the X chromosome of *Drosophila*.**

THE MOLECULAR BASIS OF MUTATION

When Watson and Crick described the double-helix structure of DNA and proposed its semiconservative

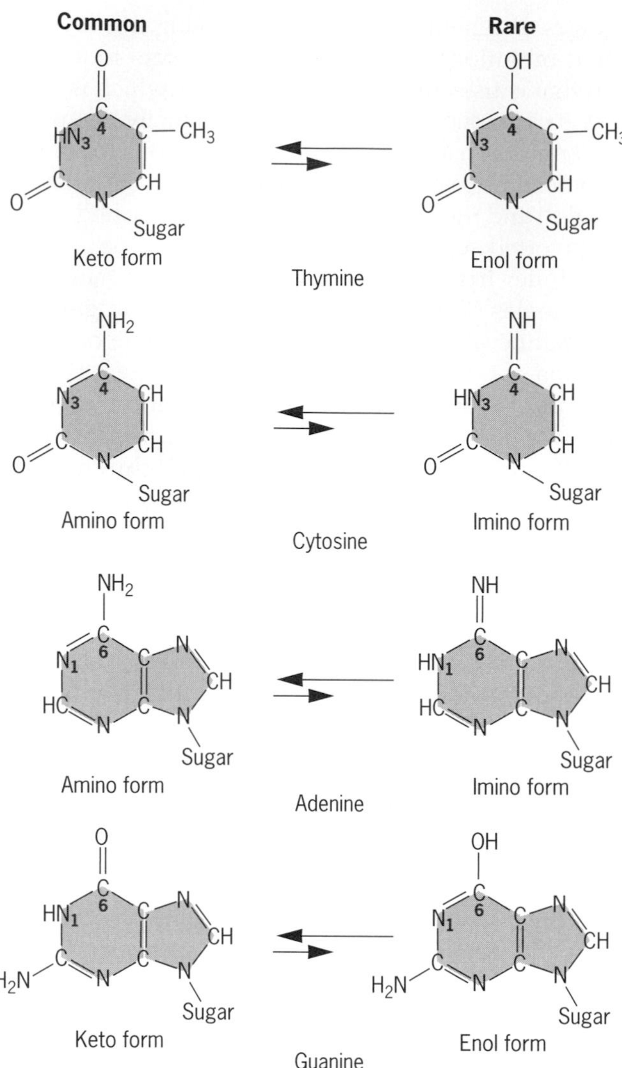

Figure 14.13 Tautomeric forms of the four common bases in DNA. The shifts of hydrogen atoms between the number 3 and number 4 positions of the pyrimidines and between the number 1 and number 6 positions of the purines change their base-pairing potential.

replication based on specific base-pairing to account for the accurate transmission of genetic information from generation to generation, they also proposed a mechanism to explain spontaneous mutation. Watson and Crick pointed out that the structures of the bases in DNA are not static. Hydrogen atoms can move from one position in a purine or pyrimidine to another position, for example, from an amino group to a ring nitrogen. Such chemical fluctuations are called **tautomeric shifts.** Although tautomeric shifts are rare, they may be of considerable importance in DNA metabolism because some alter the pairing potential of the bases. The nucleotide structures that we discussed in Chapter 10 are the common, more stable forms, in which adenine always pairs with thymine and guanine always pairs with cytosine. The more stable keto forms of thymine and guanine and the amino forms of adenine and cytosine may infrequently undergo tautomeric shifts to less stable enol and imino forms, respectively (Figure 14.13). The bases would be expected to exist in their less stable tautomeric forms for only short periods of time. However, if a base existed in the rare form at the moment that it was being replicated or being incorporated into a nascent DNA chain, a mutation would result. When the bases are present in their rare imino or enol states, they can form adenine-cytosine and guanine-thymine base pairs (Figure 14.14*a*). The net effect of such an event, and the subsequent replication required to segregate the mismatched base pair, is an A:T to G:C or a G:C to A:T base-pair substitution (Figure 14.14*b*).

Mutations resulting from tautomeric shifts in the bases of DNA involve the replacement of a purine in one strand of DNA with the other purine and the replacement of a pyrimidine in the complementary strand with the other pyrimidine. Such base-pair substitutions are called **transitions.** Base-pair substitutions involving the replacement of a purine with a pyrimidine and vice versa are called **transversions.** Four different transitions and eight different transversions are possible (Figure 14.15*a*). A third type of point mutation involves the addition or deletion of one or a few base pairs. Base-pair additions and deletions are collectively referred to as **frameshift mutations** because they alter the reading frame of all base-pair triplets (specifying codons in mRNA and amino acids in the polypeptide gene product) in the gene that are distal to the site at which the mutation occurs (Figure 14.15*b*).

All three types of point mutations—transitions, transversions, and frameshift mutations—are present among spontaneously occurring mutations. A surprisingly large proportion of the spontaneous mutations that have been studied in prokaryotes are found to be single base-pair additions and deletions rather than base-pair substitutions. These frameshift mutations almost always result in the synthesis of nonfunctional protein gene products.

Although much remains to be learned about the causes, molecular mechanisms, and frequency of spontaneously occurring mutations, three major factors are (1) the accuracy of the DNA replication machinery, (2) the efficiency of the mechanisms that have evolved for the repair of damaged DNA, and (3) the degree of exposure to mutagenic agents present in the environment. Perturbations of the DNA replication apparatus or DNA repair systems have been shown to cause large increases in mutation rates.

Rare A:C base pair

Cytosine (rare imino form)

Adenine (common amino form)

Rare G:T base pair

Thymine (common keto form)

Guanine (rare enol form)

(*a*) Hydrogen-bonded A:C and G:T base pairs that form when cytosine and guanine are in their rare imino and enol tautomeric forms.

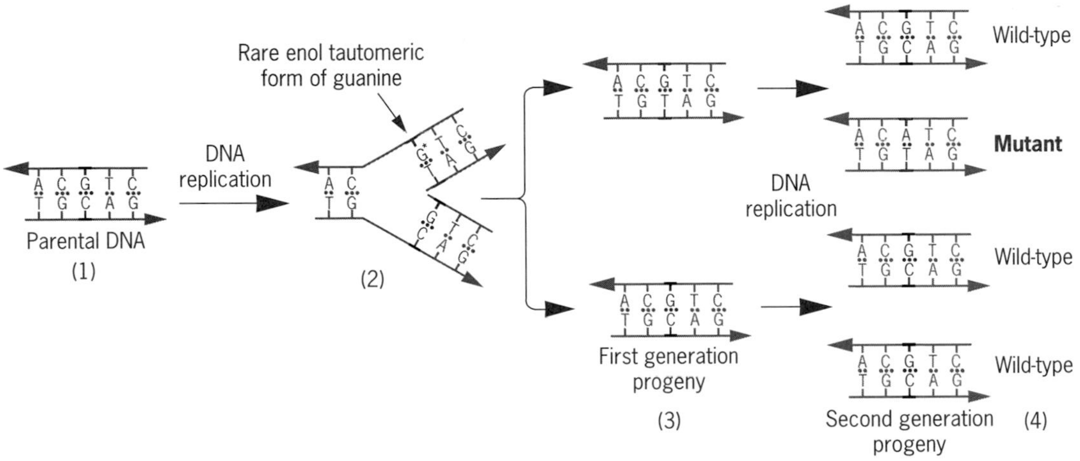

(*b*) Mechanism by which tautomeric shifts in the bases in DNA cause mutations.

Figure 14.14 The effects of tautomeric shifts in the nucleotides in DNA on (*a*) base-pairing and (*b*) mutation. Rare A:C and G:T base pairs like those shown in (*a*) also form when thymine and adenine are in their rare enol and imino forms, respectively. (*b*) A guanine (1) undergoes a tautomeric shift to its rare enol form (G^*) at the time of replication (2). In its enol form, guanine pairs with thymine (2). During the subsequent replication (3 to 4), the guanine shifts back to its more stable keto form. The thymine incorporated opposite the enol form of guanine (2) directs the incorporation of adenine during the next replication (3 to 4). The net result is a G:C to A:T base-pair substitution.

Mutations Induced by Chemicals

Muller's discovery of the mutagenic effects of X rays in 1927 provided a method by which researchers could induce large numbers of mutations. However, because X rays have many effects on living tissues, X-ray-induced mutations provided little information about the molecular mechanisms that are involved. The discovery of chemical mutagens with known effects on DNA led to a better understanding of mutation at the molecular level. Many chemicals are mutagenic; our discussion here focuses on a few chemicals that have specific or potent mutagenic effects (Table 14.1).

Mustard gas (sulfur mustard) was the first chemical shown to be mutagenic. Charlotte Auerbach and her associates discovered the mutagenic effects of mustard gas and related compounds during World War II. However, because of the potential use of mustard gas in chemical warfare, the British government placed their results on the classified list. Thus, Auerbach and coworkers could neither publish their results nor discuss them with other geneticists until the war ended.

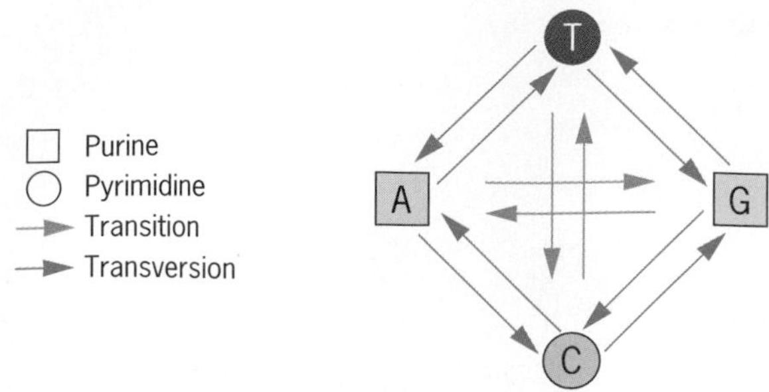

(a) Twelve different base substitutions can occur in DNA.

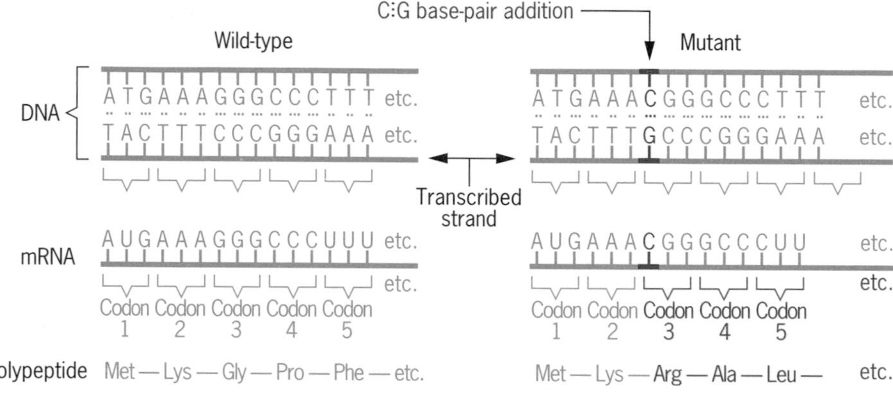

(b) Insertions or deletions of one or two base pairs alter the reading frame of the gene distal to the site of the mutation.

Figure 14.15 Types of point mutations that occur in DNA: (a) base substitutions, and (b) frameshift mutations. (*a*) The base substitutions include four transitions (purine for purine and pyrimidine for pyrimidine; green arrows) and eight transversions (purine for pyrimidine and pyrimidine for purine; blue arrows). In general, transitions are more common than transversions. (*b*) A mutant gene (top, right) was produced by the insertion of a C:G base pair between the sixth and seventh base pairs of the wild-type gene (top, left). This insertion alters the reading frame of that portion of the gene distal to the mutation, relative to the direction of transcription and translation (left to right, as diagrammed). The shift in reading frame, in turn, changes all of the codons in the mRNA and all of the amino acids in the polypeptide specified by base-pair triplets distal to the mutation.

The compounds that they studied are examples of a large class of chemical mutagens that transfer alkyl (CH_3–, CH_3CH_2–, and so forth) groups to the bases in DNA; thus, they are called alkylating agents.

Chemical mutagens can be divided into two groups: (1) those that are mutagenic to both replicating and nonreplicating DNA, such as the alkylating agents and nitrous acid; and (2) those that are mutagenic only to replicating DNA, such as base analogs—purines, and pyrimidines with structures similar to the normal bases in DNA. The base analogs must be incorporated into DNA chains in the place of normal bases during replication in order to exert their mutagenic effects. The first group of mutagens also includes the acridine dyes, which intercalate into DNA and increase the probability of mistakes during replication.

The mutagenic **base analogs** have structures similar to the normal bases and are incorporated into DNA during replication. However, their structures are sufficiently different from the normal bases in DNA that they increase the frequency of mispairing and thus mutation during replication. The two most commonly used base analogs are 5-bromouracil and 2-aminopurine. The pyrimidine 5-bromouracil is a thymine analog; the bromine at the 5 position is similar in several respects to the methyl (-CH_3) group at the 5 position

TABLE 14.1
Some Potent Chemical Mutagens

Chemical Name	*Common Name or Abbreviation*	*Structure*
I. Alkylating agents		
Di-(2-chloroethyl) sulfide	Mustard gas or sulfur mustard	$Cl—CH_2—CH_2—S—CH_2—CH_2—Cl$
Di-(2-chloroethyl) methylamine	Nitrogen mustard	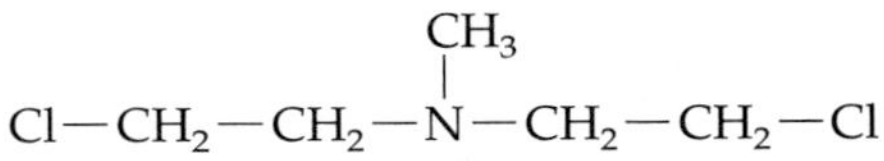 $Cl—CH_2—CH_2—N(CH_3)—CH_2—CH_2—Cl$
Ethyl methane sulfonate	EMS	$CH_3—CH_2—O—SO_2—CH_3$
Ethyl ethane sulfonate	EES	$CH_3—CH_2—O—SO_2—CH_2—CH_3$
N-Methyl-*N*′-nitro-*N*-nitrosoguanidine	NTG	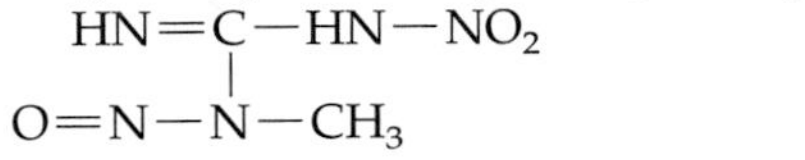 $HN{=}C—HN—NO_2$ $O{=}N—N—CH_3$
II. Base analogs		
5-Bromouracil	5-BU	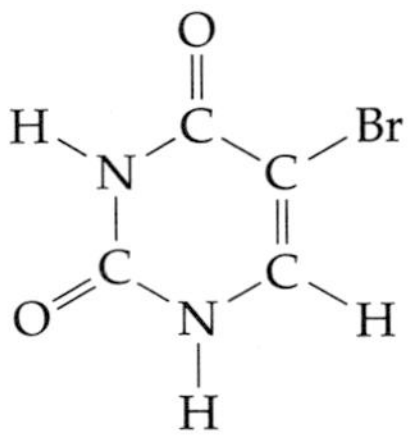
2-Aminopurine	2-AP	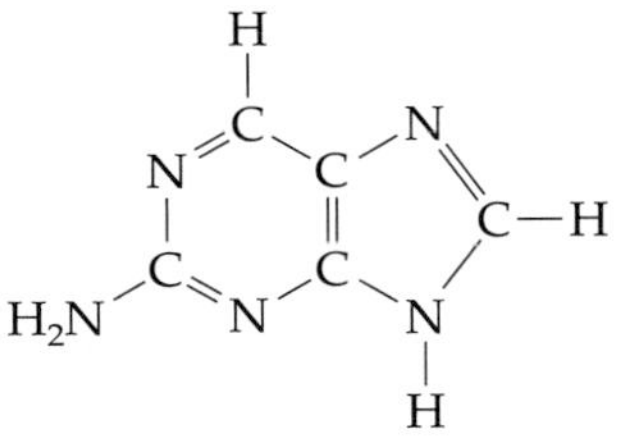
III. Acridines		
2,8-Diamino acridine	Proflavin	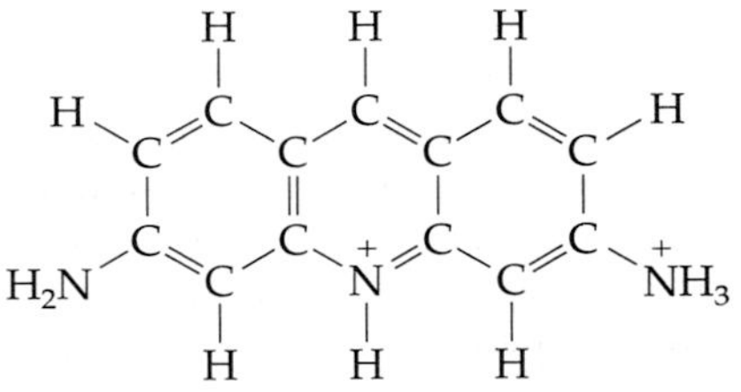
IV. Deaminating agents		
Nitrous acid	—	HNO_2
V. Miscellaneous		
Hydroxylamine	HA	NH_2OH

in thymine. However, the bromine at this position changes the charge distribution and increases the frequency of tautomeric shifts (see Figure 14.13). In its more stable keto form, 5-bromouracil pairs with adenine. After a tautomeric shift to its enol form, 5-bromouracil pairs with guanine (Figure 14.16). The mutagenic effect of 5-bromouracil is the same as that predicated for tautomeric shifts in normal bases (see Figure 14.14*b*), namely, transitions.

If 5-bromouracil is present in its less frequent enol form as a nucleoside triphosphate at the time of its incorporation into a nascent strand of DNA, it will be incorporated opposite guanine in the template strand and cause a G:C → A:T transition (Figure 14.17*a*). If,

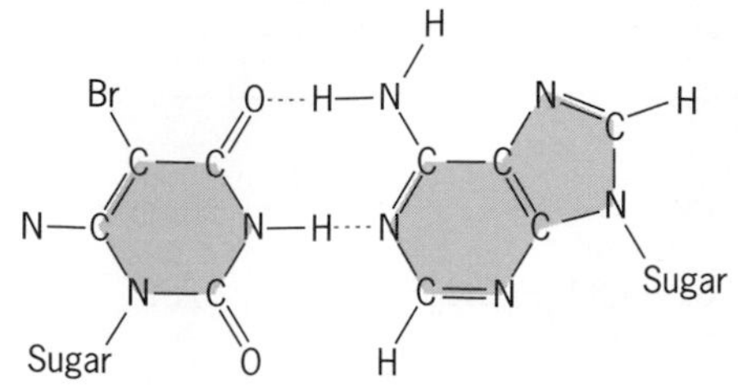

5-Bromouracil (keto form) Adenine

(*a*) 5-Bromouracil : adenine base pair.

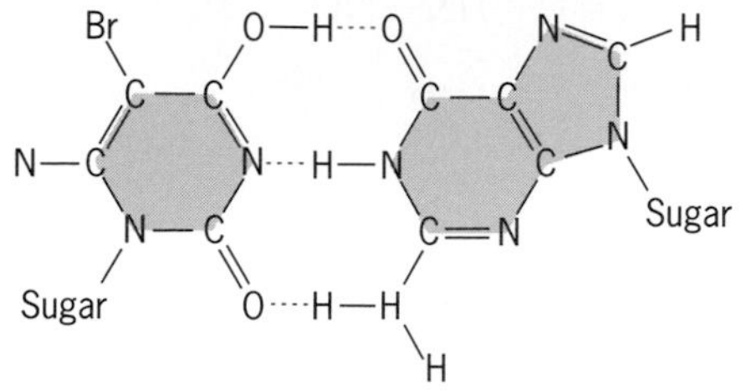

5-Bromouracil (enol form) Guanine

(*b*) 5-Bromouracil : guanine base pair.

Figure 14.16 Base-pairing between 5-bromouracil and (*a*) adenine or (*b*) guanine.

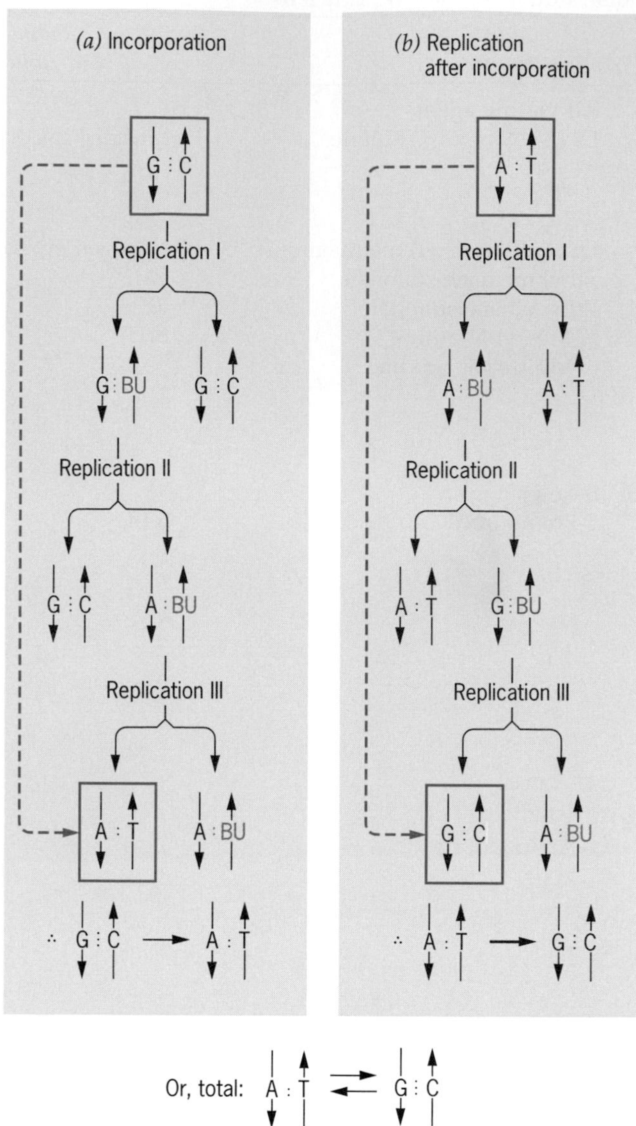

Figure 14.17 The mutagenic effects of 5-bromouracil. (*a*) When 5-bromouracil (BU) is present in its less frequent enol form (orange) at the time of incorporation into DNA, it induces G:C → A:T transitions. (*b*) When 5-bromouracil is incorporated into DNA in its more common keto form (blue) and shifts to its enol form during a subsequent replication, it induces A:T → G:C transitions. Thus, 5-bromouracil can induce transitions in both directions, A:T ⟷ G:C.

however, 5-bromouracil is incorporated in its more frequent keto form opposite adenine (in place of thymine) and undergoes a tautomeric shift to its enol form during a subsequent replication, it will cause an A:T → G:C transition (Figure 14.17*b*). Thus, 5-bromouracil induces transitions in both directions, A:T ⟷ G:C. An important consequence of the bidirectionality of 5-bromouracil-induced transitions is that mutations originally induced with this thymine analog can also be induced to mutate back to the wild-type with 5-bromouracil. 2-Aminopurine acts in a similar manner but is incorporated in place of adenine or guanine.

Nitrous acid (HNO_2) is a potent mutagen that acts on either replicating or nonreplicating DNA. Nitrous acid causes oxidative deamination of the amino groups in adenine, guanine, and cytosine. This reaction converts the amino groups to keto groups and changes the hydrogen-bonding potential of the modified bases (Figure 14.18). Adenine is deaminated to hypoxanthine, which base-pairs with cytosine rather than thymine. Cytosine is converted to uracil, which base-pairs with adenine instead of guanine. Deamination of guanine produces xanthine, but xanthine—just like guanine—base-pairs with cytosine. Thus, the deamination of guanine is not mutagenic. Because the deamination of adenine results in A:T → G:C transitions, and the deamination of cytosine produces G:C → A:T transitions, nitrous acid induces transitions in both directions, A:T ⟷ G:C. As a result, nitrous acid-induced mutations also are induced to mutate back to wild-type by nitrous acid.

The **acridine dyes** such as proflavin (Table 14.1), acridine orange, and a whole series of related compounds are potent mutagens that induce frameshift mutations (see Figure 14.15*b*). The positively charged

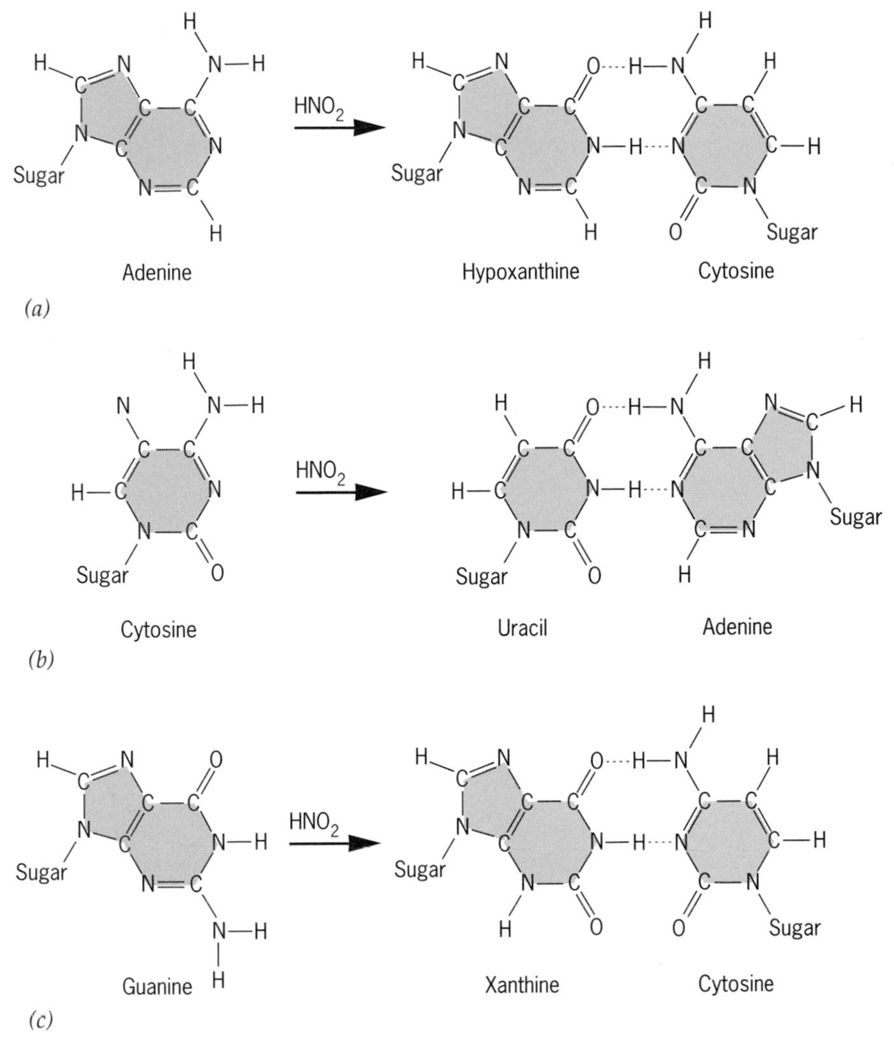

(a) Adenine $\xrightarrow{HNO_2}$ Hypoxanthine · Cytosine

(b) Cytosine $\xrightarrow{HNO_2}$ Uracil · Adenine

(c) Guanine $\xrightarrow{HNO_2}$ Xanthine · Cytosine

Figure 14.18 Nitrous acid induces mutations by oxidative deamination of the bases in DNA. Nitrous acid converts (*a*) adenine to hypoxanthine, causing A:T → G:C transitions; (*b*) cytosine to uracil, causing G:C → A:T transitions; and (*c*) guanine to xanthine, which is not mutagenic. Together, the effects of nitrous acid on adenine and cytosine explain its ability to induce transitions in both directions, A:T ↔ G:C.

acridines intercalate, or sandwich themselves, between the stacked base pairs in DNA (Figure 14.19). In so doing, they increase the rigidity and alter the conformation of the double helix, causing slight bends or kinks in the molecule. When DNA molecules containing intercalated acridines replicate, additions and deletions of one to a few base pairs occur. As we might expect, these small additions and deletions, usually of a single base pair, result in altered reading frames for the portion of the gene distal to the mutation (see Figure 14.15*b*). Thus, acridine-induced mutations usually result in nonfunctional gene products.

Alkylating agents are chemicals that donate alkyl groups to other molecules. They include the nitrogen and sulfur mustards, methyl and ethyl methane sulfonate (MMS and EMS), and nitrosoguanidine (NTG; Table 14.1)—chemicals that have multiple effects on DNA. One mechanism of mutagenesis by alkylating agents involves the transfer of methyl or ethyl groups to the bases, resulting in altered base-pairing potentials. For example, EMS causes ethylation of the bases in DNA at the 7-*N* and the 6-*O* positions. When 7-ethylguanine is produced, it base-pairs with thymine

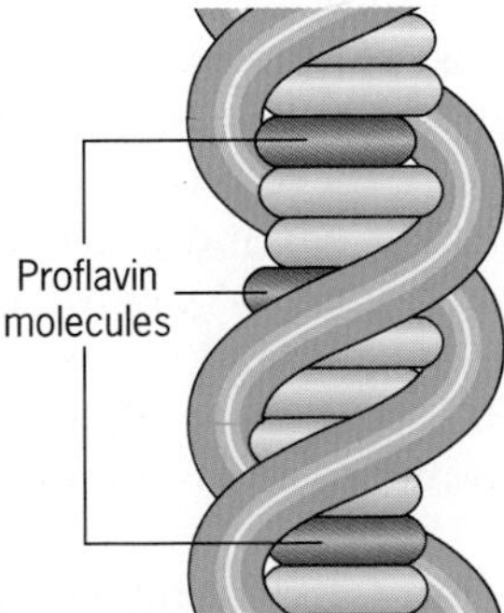

Figure 14.19 Intercalation of proflavin into the DNA double helix. X-ray diffraction studies have shown that these positively charged acridine dyes become sandwiched between the stacked base pairs.

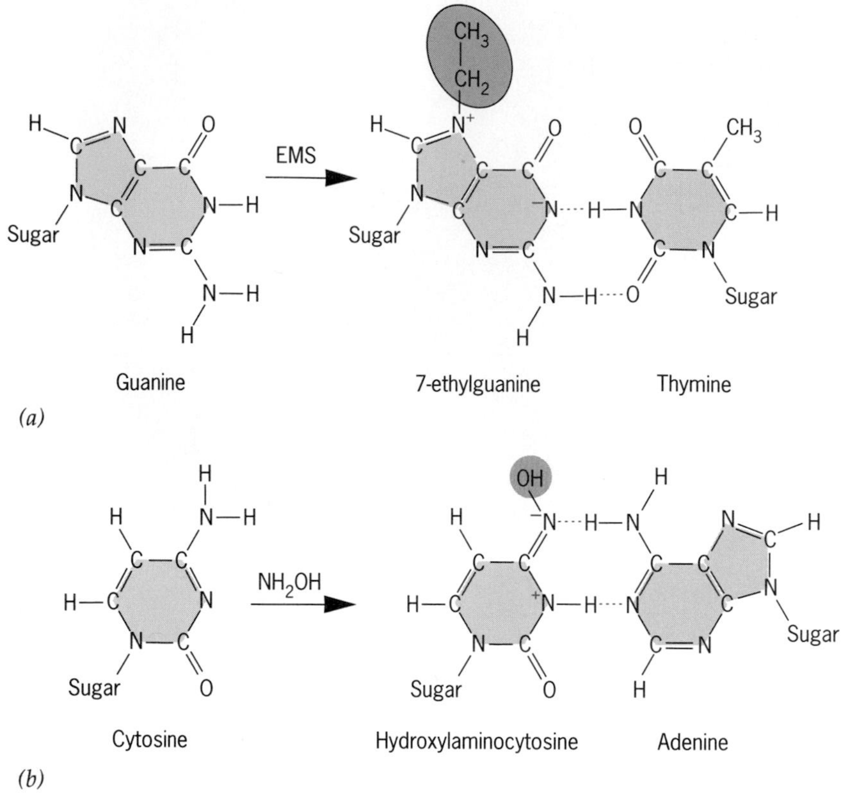

Figure 14.20 Proposed mutagenic effect of (*a*) the alkylating agent ethyl methane sulfonate (EMS) and (*b*) the hydroxylating agent hydroxylamine (NH_2OH). Alkylating agents such as EMS also induce mutations by other mechanisms.

(Figure 14.20*a*) to cause G:C → A:T transitions. Other base alkylation products activate error-prone DNA repair processes that introduce transitions, transversions, and frameshift mutations during the repair process. Some alkylating agents, particularly difunctional alkylating agents (those with two reactive alkyl groups), cross-link DNA strands or molecules and induce chromosome breaks, which result in various kinds of chromosomal aberrations (Chapter 7). Alkylating agents as a class therefore exhibit less specific mutagenic effects than do base analogs, nitrous acid, or acridines. Alkylating agents induce all types of mutations, including transitions, transversions, frameshifts, and even chromosome aberrations, with relative frequencies that depend on the reactivity of the agent involved.

Nitrosoguanidine (NTG), one of the most potent chemical mutagens known, induces clusters of closely linked mutations in segments of chromosomes that are replicated during the mutagenic treatment. Mutants isolated after NTG treatment often carry multiple, closely linked mutations, making them less useful for genetic studies.

In contrast to most alkylating agents, the **hydroxylating agent** hydroxylamine, (NH_2OH), has a specific mutagenic effect. It induces only G:C → A:T transitions. When DNA is treated with hydroxylamine, the amino group of cytosine is hydroxylated (Figure 14.20*b*). The resulting hydroxylaminocytosine base-pairs with adenine, leading to G:C → A:T transitions. Because of its specificity, hydroxylamine has been very useful in classifying transition mutations. Mutations that are induced to revert by nitrous acid or base analogs, and therefore were caused by transitions, can be divided into two classes on the basis of their revertibility with hydroxylamine. (1) Those with an A:T base pair at the mutant site will not be induced to revert by hydroxylamine. (2) Those with a G:C base pair at the mutant site will be induced to revert by hydroxylamine. Thus, hydroxylamine can be used to determine whether a particular mutation was an A:T → G:C or a G:C → A:T transition.

Mutations Induced by Radiation

The portion of the electromagnetic spectrum (Figure 14.21) with wavelengths shorter and of higher energy than visible light is subdivided into **ionizing radiation** (X rays, gamma rays, and cosmic rays) and **nonionizing radiation** (ultraviolet light). Ionizing radiations are of high energy and are useful for medical diagnosis because they penetrate living tissues for substantial distances. In the process, these high-energy rays collide with atoms and cause the release of electrons, leaving

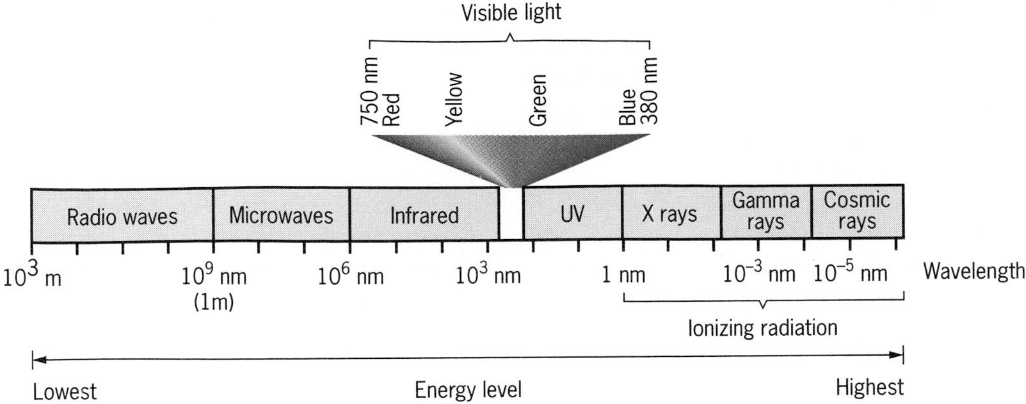

Figure 14.21 The electromagnetic spectrum.

positively charged free radicals or ions. The ions, in turn, collide with other molecules and cause the release of additional electrons. The result is that a cone of ions is formed along the track of each high-energy ray as it passes through living tissues. This process of ionization is induced by machine-produced X rays, protons, and neutrons, as well as by the alpha, beta, and gamma rays released by radioactive isotopes such as ^{32}P, ^{35}S, and the uranium 238 used in nuclear reactors. Ultraviolet rays, having lower energy than ionizing radiations, penetrate only the surface layer of cells in higher plants and animals and do not cause ionizations. Ultraviolet rays dissipate their energy to atoms that they encounter, raising the electrons in the outer orbitals to higher energy levels, a state referred to as **excitation.** Molecules containing atoms in either ionic forms or excited states are chemically more reactive than those containing atoms in their normal stable states. The increased reactivity of atoms present in DNA molecules is responsible for the mutagenicity of ionizing radiation and ultraviolet light.

X rays and other forms of ionizing radiation are quantitated in **roentgen (r)** units, which are measures of the number of ionizations per unit volume under a standard set of conditions. Specifically, one roentgen unit is a quantity of ionizing radiation that produces 2.083×10^9 ion pairs in one cubic centimeter of air at 0° C and a pressure of 760 mm of mercury. Note that the dosage of irradiation in roentgen units does not involve a time scale. The same dosage may be obtained by a low intensity of irradiation over a long period of time or a high intensity of irradiation for a short period of time. This point is important because in most studies the frequency of induced point mutations is directly proportional to the dosage of irradiation (Figure 14.22). For example, X-irradiation of *Drosophila* sperm causes an approximately 3 percent increase in mutation rate for each 1000 r increase in irradiation dosage. This linear relationship shows that the induction of mutations by X rays exhibits single-hit kinetics, which means that each mutation results from a single ionization event. That is, every ionization has a fixed probability of inducing a mutation under a standard set of conditions.

What is a safe level of irradiation? The development and use of the atomic bomb and the more recent accidents at nuclear power plants have generated concern about exposure to ionizing radiations. The linear relationship between mutation rate and radiation dosage indicates that there is no safe level of irradiation. Rather, the results indicate that the higher the dosage of irradiation, the higher the mutation rate, and the lower the dosage, the lower the mutation frequency. Even very low levels of irradiation have certain low, but real, probabilities of inducing mutations. However, the probabilities are not the same for all species.

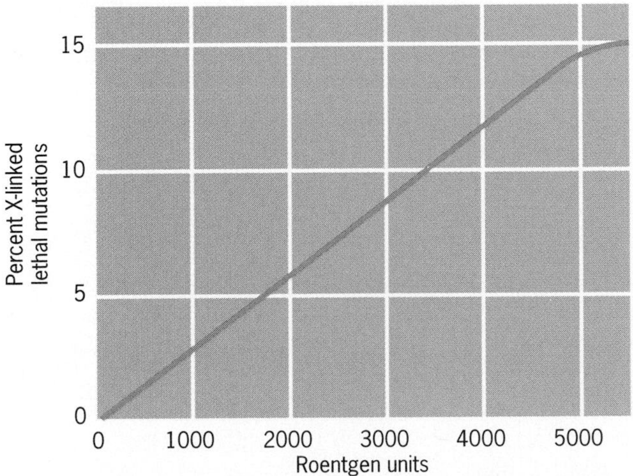

Figure 14.22 Relationship between irradiation dosage and mutation frequency in *Drosophila*.

In *Drosophila* sperm, chronic irradiation (low levels of irradiation over long periods of time) is as effective in inducing mutations as acute irradiation (the same total dosage of irradiation administered at high intensity for short periods of time). However, in mice, chronic irradiation results in fewer mutations than the same dosage of acute irradiation. Moreover, when mice are treated with intermittent doses of irradiation, the mutation frequency is slightly lower than when they are treated with the same total amount of irradiation in a continuous dose. The differential response of fruit flies and mammals to chronic irradiation is thought to result from differences in the efficiency with which these species repair irradiation-induced damage in DNA. Repair mechanisms may exist in the spermatogonia and oocytes of mammals that do not function in *Drosophila* sperm. Nevertheless, we should emphasize that all of these irradiation treatments are mutagenic, albeit to different degrees, in both *Drosophila* and mammals.

Ionizing radiation also induces gross changes in chromosome structure, including deletions, duplications, inversions, and translocations (Chapter 7). These chromosome aberrations result from radiation-induced breaks in chromosomes. Because these aberrations require two chromosomal breaks, they exhibit two-hit kinetics rather than the single-hit kinetics observed for point mutations.

Ultraviolet (UV) radiation does not possess sufficient energy to induce ionizations. However, it is readily absorbed by many organic molecules such as the purines and pyrimidines in DNA, which then enter a more reactive or excited state. UV rays penetrate tissue only slightly. Thus, in multicellular organisms, only the epidermal layer of cells usually is exposed to the effects of UV. However, ultraviolet light is a potent mutagen for unicellular organisms. The maximum absorption of UV by DNA is at a wavelength of 254 nm. Maximum mutagenicity also occurs at 254 nm, suggesting that the UV-induced mutation process is mediated directly by the absorption of UV by purines and pyrimidines. *In vitro* studies show that the pyrimidines absorb strongly at 254 nm and, as a result, become very reactive. Two major products of UV absorption by pyrimidines are pyrimidine hydrates and pyrimidine dimers (Figure 14.23). Several lines of evidence indicate that thymine dimerization is probably the major mutagenic effect of UV. Thymine dimers cause mutations in two ways. (1) Dimers perturb the structure of DNA double helices and interfere with accurate DNA replication. (2) Errors occur during the cellular processes that repair defects in DNA, such as UV-induced thymine dimers (see the section DNA Repair Mechanisms later in this chapter).

Mutations Induced by Transposable Genetic Elements

Living organisms contain remarkable DNA elements that can jump from one site in the genome to another site. These **transposons,** or transposable genetic elements, are the subject of Chapter 18. The insertion of one of these transposons into a gene will often render the gene nonfunctional (Figure 14.24). If the gene encodes an important product, a mutant phenotype is

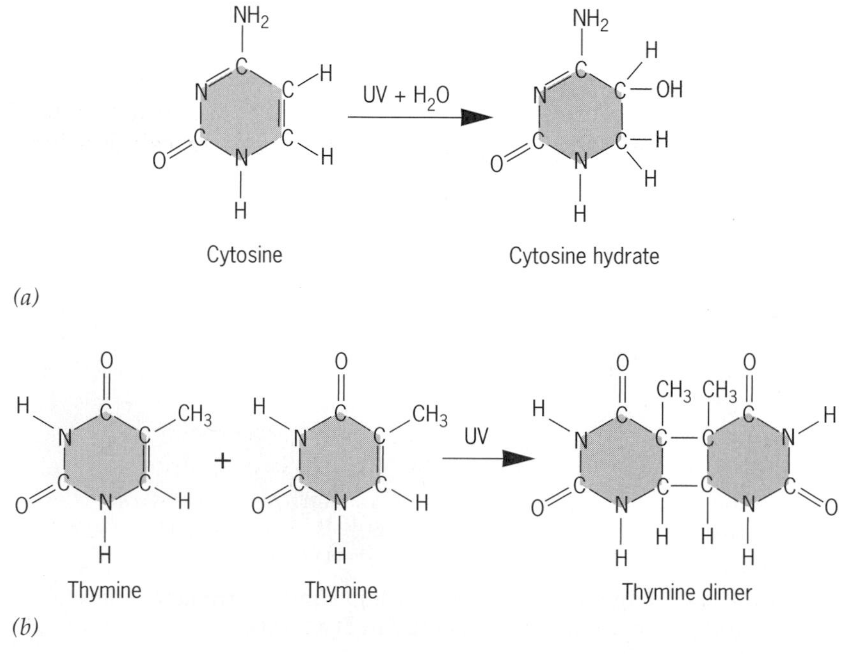

Cytosine

UV + H_2O

Cytosine hydrate

(a)

Thymine + Thymine

UV

Thymine dimer

(b)

Figure 14.23 Pyrimidine photoproducts of UV irradiation. (*a*) Hydrolysis of cytosine to a hydrate form that may cause mispairing of bases during replication. (*b*) Cross-linking of adjacent thymine molecules to form thymine dimers, which block DNA replication.

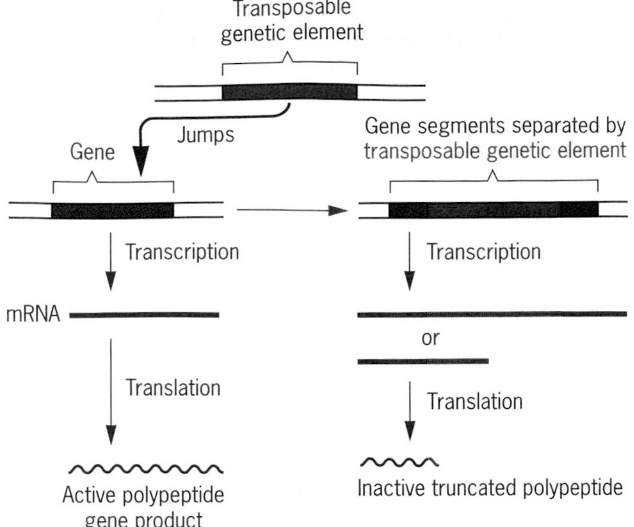

Figure 14.24 Mechanism of transposon-induced mutation. The insertion of a transposable genetic element (red) into a wild-type gene (left) will usually render the gene nonfunctional (right). A truncated gene product usually results from transcription- or translation-termination signals, or both, located within the transposon.

likely to result. Indeed, geneticists now know that many of the classical mutants of maize, *Drosophila, E. coli,* and other organisms were caused by the insertion of transposable genetic elements into important genes (see Figure 18.17). See Chapter 18 for additional details about the mechanisms by which transposons move and, in the process, produce mutations.

Expanding Trinucleotide Repeats and Inherited Human Diseases

All of the types of mutations discussed in the preceding sections of this chapter occur in humans. In addition, another type of mutation occurs that was detected first in humans. Tandemly repeated sequences of one to six nucleotide pairs are known as **simple tandem repeats.** Such repeats are dispersed throughout the human genome. Repeats of three nucleotide pairs, **trinucleotide repeats,** can increase in copy number and cause inherited diseases in humans. Several trinucleotides have been shown to undergo such increases in copy number. Expanded CGG trinucleotide repeats on the X chromosome are responsible for fragile X syndrome, the most common form of inherited mental retardation in humans (Chapter 6). Normal X chromosomes contain from 6 to about 50 copies of the CGG repeat at the *FRAXA* site. Mutant X chromosomes contain up to 1000 copies of the tandem CGG repeat at this site.

CAG and CTG trinucleotide repeats are involved in several inherited neurological diseases, including Huntington disease (discussed in detail in Chapter 22), myotonic dystrophy, Kennedy disease, dentatorubral pallidoluysian atrophy, Machado Joseph disease, and spinocerebellar ataxia. In all of these neurological disorders, the severity of the disease is correlated with trinucleotide copy number—the higher the copy number, the more severe the disease symptoms. In addition, the expanded trinucleotides associated with these diseases are unstable in somatic cells and between generations. This instability gives rise to the phenomenon of **anticipation,** which is the increasing severity of the disease or earlier age of onset that occurs in successive generations as the trinucleotide copy number increases. The mechanism of trinucleotide expansion is unknown.

Key Points: **Mutations are induced by chemicals, ionizing irradiation, ultraviolet light, and endogenous transposable genetic elements. Point mutations are of three types: (1) transitions, purine for purine and pyrimidine for pyrimidine substitutions; (2) transversions, purine for pyrimidine and pyrimidine for purine substitutions; and (3) frameshift mutations, additions or deletions of one or two nucleotide pairs, which alter the reading frame of the gene distal to the site of the mutation. Several inherited human diseases are caused by expanded trinucleotide repeats.**

SCREENING CHEMICALS FOR MUTAGENICITY: THE AMES TEST

Mutagenic agents are also **carcinogens;** that is, they induce cancers. The one characteristic that the hundreds of types of cancer have in common is that the malignant cells continue to divide after cell division would have stopped in normal cells. Of course, cell division, like all other biological processes, is under genetic control. Specific genes encode products that regulate cell division in response to intracellular, intercellular, and environmental signals. When these genes mutate to nonfunctional states, uncontrolled cell division sometimes results. Clearly, we wish to avoid being exposed to mutagenic and carcinogenic agents. However, our technological society often depends on the extensive use of chemicals in both industry and agriculture. Hundreds of new chemicals are being produced each year, and the mutagenicity and carcinogenicity of these chemicals need to be evaluated before their use becomes widespread.

Traditionally, the carcinogenicity of chemicals has been tested on rodents, usually newborn mice. These

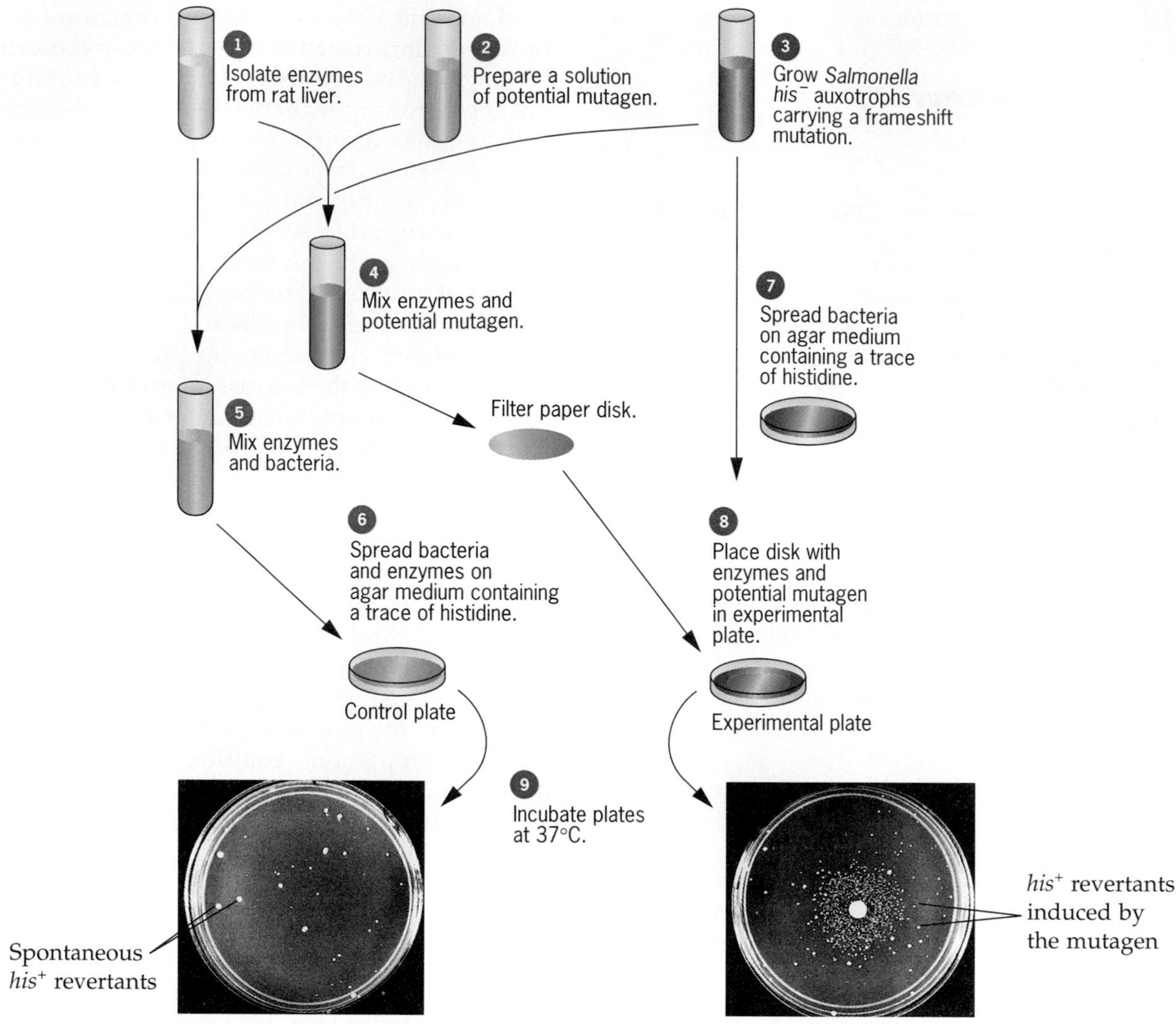

Figure 14.25 The Ames test for mutagenicity. The medium in each petri dish contains a trace of histidine and a known number of *his*⁻ cells of a specific *Salmonella typhimurium* "tester strain" harboring a frameshift mutation. The control plate shown on the left provides an estimate of the frequency of spontaneous reversion of this particular tester strain. The experimental plate on the right shows the frequency of reversion induced by the potential mutagen, in this case, the carcinogen 2-aminofluorene.

studies involved feeding or injecting the substance being tested and subsequently examining the animals for tumors. Mutagenicity tests have been done in a similar fashion. However, because mutation is a low-frequency event and because maintaining large populations of mice is an expensive undertaking, the tests have been relatively insensitive; that is, low levels of mutagenicity would not be detected.

Bruce Ames and his associates developed sensitive techniques that allow the mutagenicity of large numbers of chemicals to be tested quickly at relatively low cost. Ames and coworkers constructed auxotrophic strains of the bacterium *Salmonella typhimurium* carrying various types of mutations—transitions, transversions, and frameshifts—in genes required for the biosynthesis of the amino acid histidine. They monitored the reversion of these auxotrophic mutants to prototrophy by placing a known number of mutant bacteria on medium lacking histidine and scoring the number of colonies produced by prototrophic revertants. Because some chemicals are mutagenic only to replicating DNA, they added a small amount of histidine—enough to allow a few cell divisions, but not the formation of visible colonies—to the medium. They measured the mutagenicity of a chemical by comparing the frequency of reversion in its presence with the spontaneous reversion frequency (Figure 14.25). They assessed its ability to induce different types of mutations by using a set of tester strains that carry different types of mutations—one strain with a transition, one with a frameshift mutation, and so forth.

Over a period of several years during which they tested hundreds of different chemicals, Ames and his colleagues observed a greater than 90 percent correlation between the mutagenicity and the carcinogenicity of the substances tested. Initially, they found several potent carcinogens to be nonmutagenic to the tester strains. Subsequently, they discovered that many of these carcinogens are metabolized to strongly mutagenic derivatives in eukaryotic cells. Thus, Ames and his associates added a rat liver microsomal fraction to their assay systems in an attempt to detect the mutagenicity of metabolic derivatives of the substances being tested. Coupling of the rat liver microsomal activation system to the microbial mutagenicity tests expanded the utility of the system considerably. For example, nitrates are not themselves mutagenic or carcinogenic. However, in eukaryotic cells, nitrates are converted to nitrosamines, which are highly mutagenic and carcinogenic. Ames's mutagenicity tests demonstrated the presence of frameshift mutagens in several components of chemically fractionated cigarette smoke condensates. In some cases, activation by the liver microsomal preparation was required for mutagenicity; in other cases, activation was not required. The Ames test provides a rapid, inexpensive, and sensitive procedure for testing the mutagenicity of chemicals. Since mutagenic chemicals are also carcinogens, the Ames test can be used to prevent the widespread use of mutagenic and carcinogenic chemicals.

Key Point: **By using *Salmonella* tester strains that carry various types of mutations in genes encoding histidine biosynthetic enzymes, Bruce Ames and coworkers developed an inexpensive and sensitive method for detecting the mutagenicity of chemicals.**

DNA REPAIR MECHANISMS

The multiplicity of repair mechanisms that have evolved in organisms ranging from bacteria to humans emphatically documents the importance of keeping mutation, both somatic and germ-line, at a tolerable level. For example, *E. coli* cells possess at least five distinct mechanisms for the repair of defects in DNA: (1) light-dependent repair or photoreactivation, (2) excision repair, (3) mismatch repair, (4) postreplication repair, and (5) the error-prone repair system. Moreover, there are at least two different types of excision repair, and the excision repair pathways can be initiated by several different enzymes, each acting on a specific kind of damage in DNA. Mammals seem to possess all of the repair mechanisms found in *E. coli* except photoreactivation. Because most mammalian cells do not have access to light, photoreactivation would be of relatively little value to them. Humans and other mammals undoubtedly possess some DNA repair mechanisms that are not present in bacteria, but many details of these repair processes require further documentation.

The importance of DNA repair pathways to human health is clear. Inherited disorders such as xeroderma pigmentosum, which was discussed at the beginning of this chapter, vividly document the serious consequences of defects in DNA repair. We discuss these inherited disorders in a subsequent section of this chapter.

Light-Dependent Repair

Light-dependent repair or **photoreactivation** of DNA in bacteria is carried out by a light-activated enzyme called **DNA photolyase.** When DNA is exposed to ultraviolet light, thymine dimers are produced by covalent cross-linkages between adjacent thymine residues (see Figure 14.23). DNA photolyase binds to thymine dimers in DNA and uses light energy to cleave the covalent cross-links (Figure 14.26). Photolyase will bind to thymine dimers in DNA in the dark, but it cannot catalyze cleavage of the bonds joining the thymine moieties without energy derived from visible light, specifically light within the blue region of the spectrum. Photolyase also splits cytosine dimers and cytosine-thymine dimers. Thus, when ultraviolet light is used to induce mutations in bacteria, the irradiated cells are grown in the dark for a few generations to maximize the mutation frequency.

Excision Repair

Excision repair of damaged DNA involves at least three steps. In step 1, a DNA repair endonuclease or endonuclease-containing enzyme complex recognizes, binds to, and excises the damaged base or bases in DNA. In step 2, a DNA polymerase fills in the gap by using the undamaged complementary strand of DNA as template. In step 3, the enzyme DNA ligase seals the break left by DNA polymerase to complete the repair process. There are two major types of excision repair: **base excision repair** systems remove abnormal or chemically modified bases from DNA, whereas **nucleotide excision repair** pathways remove larger defects like thymine dimers. Both excision pathways are operative in the dark, and both occur by very similar mechanisms in *E. coli* and humans.

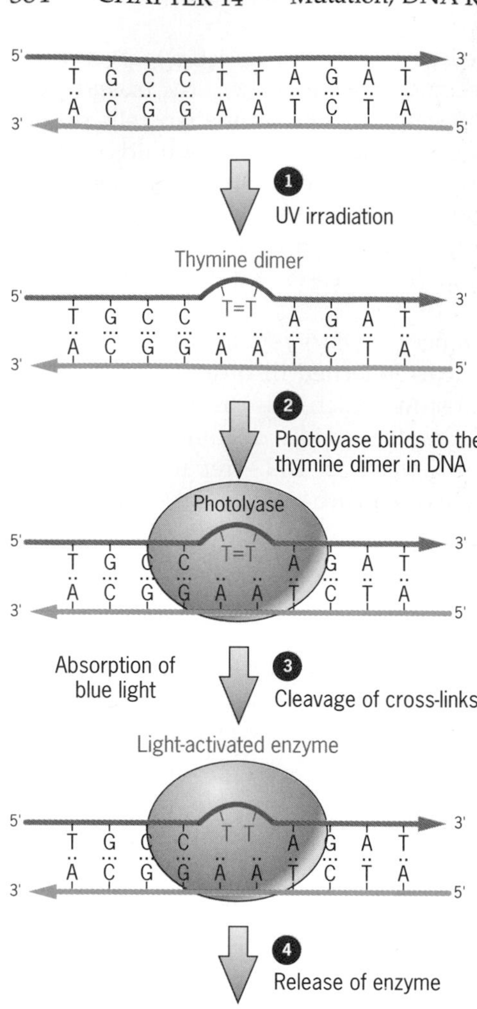

Figure 14.26 Cleavage of thymine dimer cross-links by light-activated photolyase. The arrows indicate the opposite polarity of the complementary strands of DNA.

Base excision repair (Figure 14.27) can be initiated by any of a group of enzymes called DNA glycosylases that recognize abnormal bases in DNA. Each glycosylase recognizes a specific type of altered base, such as deaminated bases, oxidized bases, and so on. The glycosylases cleave the glycosidic bond between the abnormal base and 2-deoxyribose, creating apurinic or apyrimidinic sites (AP sites) with missing bases. These AP sites are recognized by AP endonucleases, which act together with phosphodiesterases to excise the sugar-phosphate groups at sites where no base is present. DNA polymerase then replaces the missing nucleotide according to the specifications of the complementary strand, and DNA ligase seals the nick.

Nucleotide excision repair removes larger lesions like thymine dimers and bases with bulky side-groups from DNA. In nucleotide excision repair, a unique excision nuclease activity produces cuts on either side of the damaged nucleotide(s) and excises an oligonucleotide containing the damaged base(s). This nuclease is called an **excinuclease** to distinguish it from the endonucleases and exonucleases that play other roles in DNA metabolism.

The *E. coli* nucleotide excision repair pathway is shown in Figure 14.28. In *E. coli*, excinuclease activity requires the products of three genes, *uvrA*, *uvrB*, and

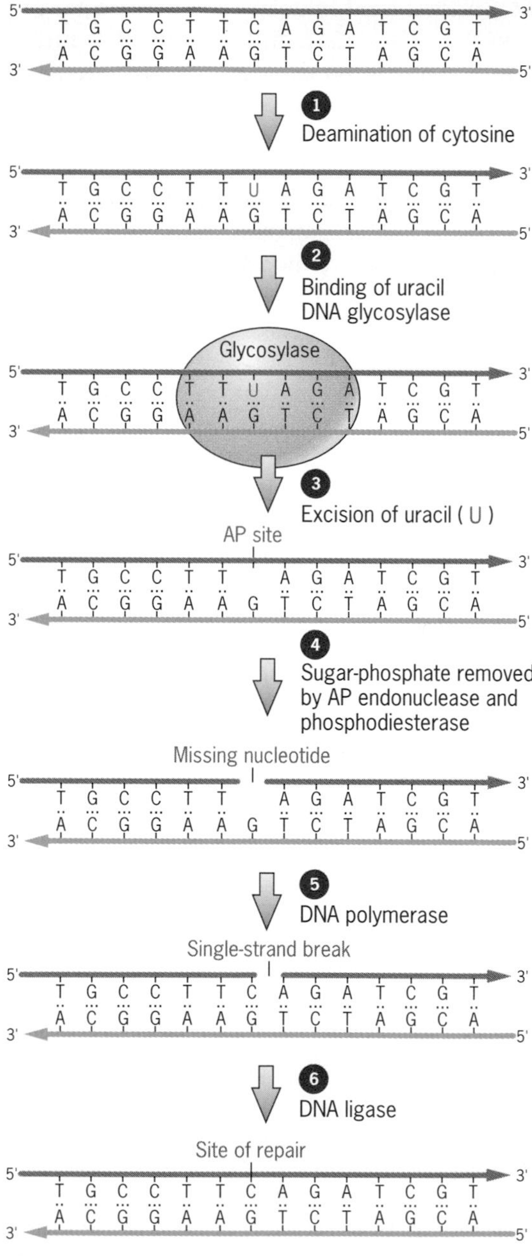

Figure 14.27 Repair of DNA by the base excision pathway. Base excision repair may be initiated by any one of several different DNA glycosylases. In the example shown, uracil DNA glycosylase starts the repair process.

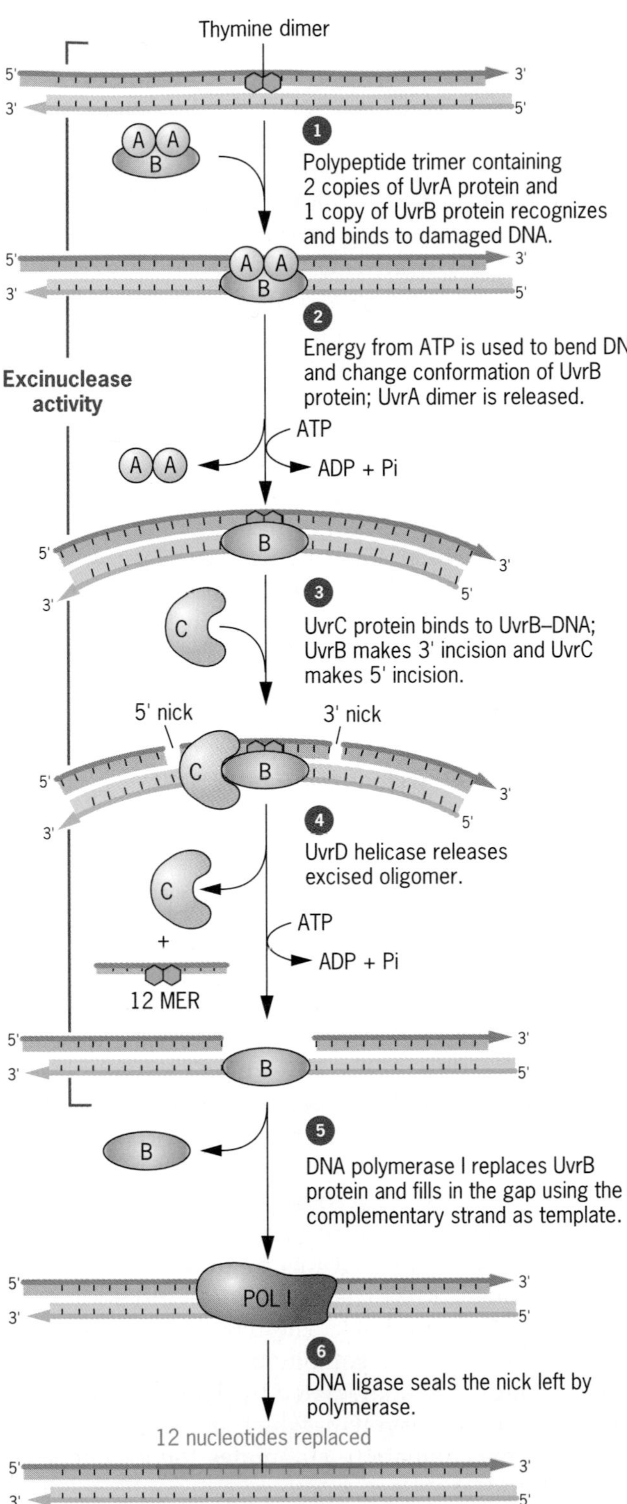

Figure 14.28 Repair of DNA by the nucleotide excision pathway in *E. coli*. The excinuclease (excision nuclease) activity requires the products of three genes—*UvrA, UvrB,* and *UvrC*. Nucleotide excision occurs by a similar pathway in humans, except that many more proteins are involved and a 29-nucleotide-long oligomer is excised.

uvrC (designated *uvr* for *UV* repair). A trimeric protein containing two UvrA polypeptides and one UvrB polypeptide recognizes the defect in DNA, binds to it, and uses energy from ATP to bend the DNA at the damaged site. The UvrA dimer is then released, and the UvrC protein binds to the UvrB-DNA complex. The UvrB protein cleaves the fifth phosphodiester bond from the damaged nucleotide(s) on the 3′ side, and the UvrC protein hydrolyzes the eighth phosphodiester linkage from the damage on the 5′ side. The *uvrD* gene product, DNA helicase II, releases the excised dodecamer. In the last two steps of the pathway, DNA polymerase I fills in the gap, and DNA ligase seals the remaining nick in the DNA molecule.

Nucleotide excision repair in humans occurs through a pathway similar to the one in *E. coli*, but it involves about four times as many proteins. In humans, the excinuclease activity requires at least 17 polypeptides. Protein XPA (for *x*eroderma *p*igmentosum protein *A*) recognizes and binds to the damaged nucleotide(s) in DNA. It then recruits the other proteins required for excinuclease activity. In humans, the excised oligomer is 29 nucleotides long rather than the 12-mer removed in *E. coli*. The gap is filled in by either DNA polymerase δ or ϵ in humans, and DNA ligase completes the job.

Mismatch Repair

In Chapter 11, we examined the mechanism by which the $3' \rightarrow 5'$ exonuclease activity built into DNA polymerases proofreads DNA strands during their synthesis, removing any mismatched nucleotides at the 3′ termini of growing strands. The **mismatch repair** pathway provides a backup to this replicative proofreading by correcting mismatched nucleotides remaining in DNA after replication. Mismatches often involve the normal four bases in DNA. For example, a T may be mispaired with a G. Because both T and G are normal components of DNA, mismatch repair systems need some way to determine whether the T or the G is the correct base at a given site. The repair system makes this distinction by identifying the template strand, which contains the original nucleotide sequence, and the newly synthesized strand, which contains the misincorporated base (the error). This distinction can be made based on the pattern of methylation in newly replicated DNA. In *E. coli*, the A in GATC sequences is methylated subsequent to its synthesis. Thus, an interval occurs during which the template strand is methylated and the newly synthesized strand is unmethylated. The mismatch repair system uses this difference in methylation state to excise the mismatched nucleotide in the nascent strand

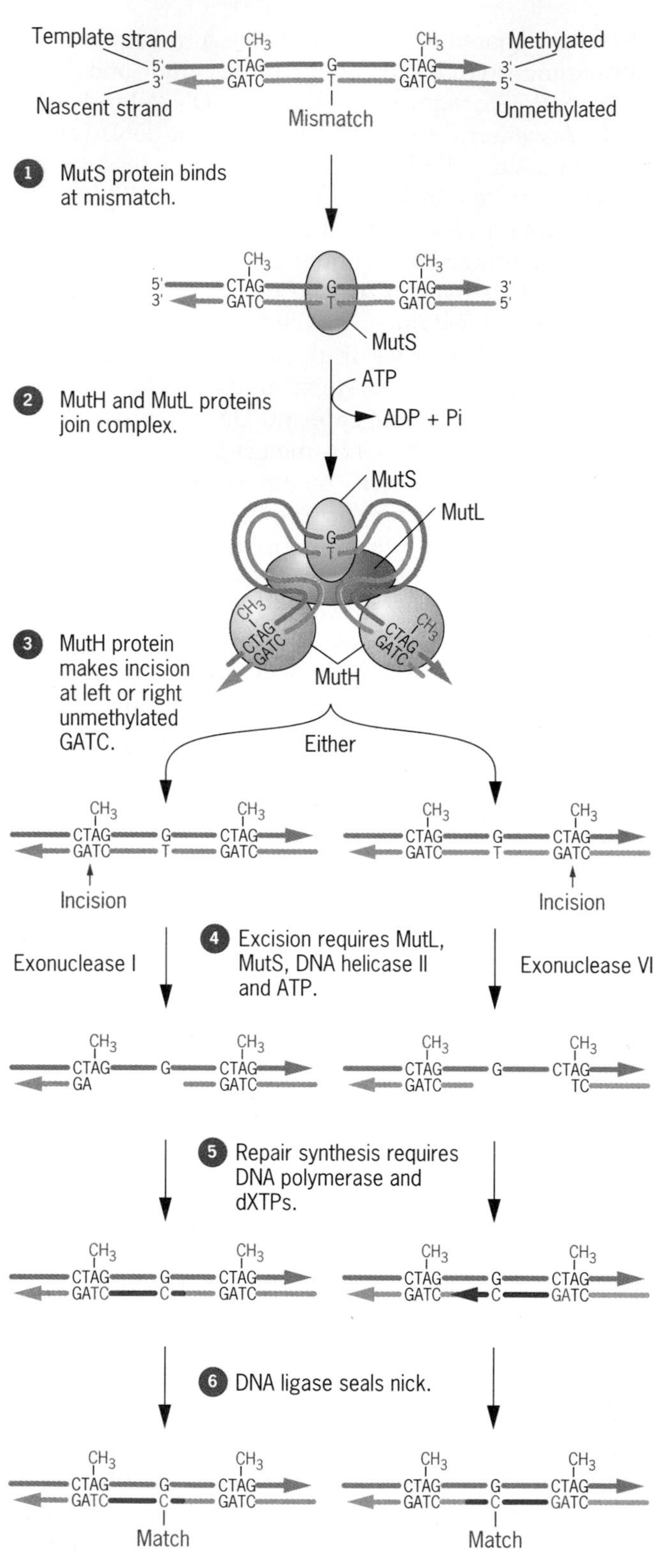

Figure 14.29 Mismatch repair of DNA in *E. coli*. The double helix has just replicated; the template strand is methylated, but the newly synthesized strand is still unmethylated. The MutH protein will cleave only unmethylated GATC sites, assuring that the original (template strand) sequence is restored.

and replace it with the correct nucleotide by using the methylated parental strand of DNA as template (Figure 14.29).

Mismatch repair of DNA in *E. coli* requires the products of four genes, *mutH*, *mutL*, *mutS*, and *mutU* (= *uvrD*). The MutS protein recognizes mismatches and binds to them to initiate the repair process. MutH and MutL proteins then join the complex. MutH contains a **GATC-specific endonuclease activity** that cleaves the unmethylated strand at hemimethylated (that is, half methylated) GATC sites either 5′ or 3′ to the mismatch. The incision sites may be 1000 nucleotide pairs or more from the mismatch. The subsequent excision process requires MutS, MutL, DNA helicase II (MutU), and an appropriate exonuclease. After the excision process has removed the mismatched nucleotide from the unmethylated strand, DNA polymerase fills in the gap, and DNA ligase seals the nick.

Homologs of the *E. coli* MutS and MutL proteins have been identified in both *Saccharomyces cerevisiae* and humans—an indication that similar mismatch repair pathways occur in eukaryotes. In fact, mismatch excision has been demonstrated *in vitro* with nuclear extracts prepared from human cells. Thus, mismatch repair is probably a universal or nearly universal mechanism for safeguarding the integrity of genetic information stored in double-stranded DNA.

Postreplication Repair

In *E. coli*, light-dependent repair, excision repair, and mismatch repair can be eliminated by mutations in the *phr* (*ph*otoreactivation), *uvr*, and *mut* genes, respectively. In multiple mutants deficient in these repair mechanisms, still another DNA repair system, called **postreplication repair,** is operative. When DNA polymerase III encounters a thymine dimer in a template strand, its progress is blocked. DNA polymerase restarts DNA synthesis at some position past the dimer, leaving a gap in the nascent strand opposite the dimer in the template strand. At this point, the original nucleotide sequence has been lost from both strands of this progeny double helix. The damaged DNA molecule is repaired by a recombination-dependent repair process (Figure 14.30) mediated by the *E. coli recA* gene product. The RecA protein, which is required for homologous recombination, stimulates the exchange of single strands between homologous double helices. During postreplication repair, the RecA protein binds to the single strand of DNA at the gap and mediates pairing with the homologous segment of the sister double helix. The gap opposite the dimer is filled with the homologous DNA strand from the sister DNA molecule. The resulting gap in the sister double helix is

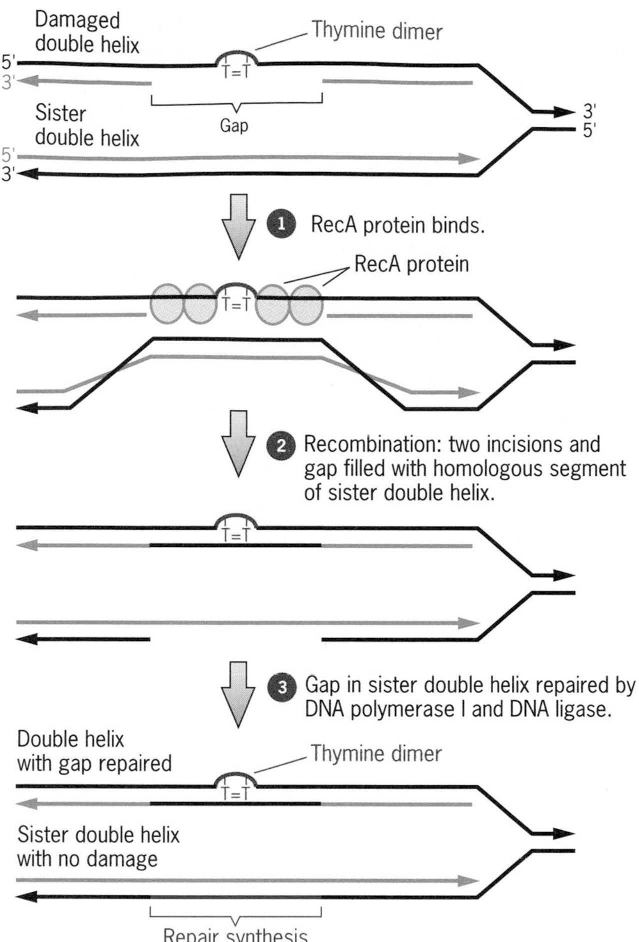

Figure 14.30 Postreplication repair of DNA in *E. coli*.

filled in by DNA polymerase, and the nick is sealed by DNA ligase. The thymine dimer remains in the template strand of the original progeny DNA molecule, but the complementary strand is now intact. If the thymine dimer is not removed by the nucleotide excision repair system, this postreplication repair must be repeated after each round of DNA replication.

The Error-Prone Repair System

The DNA repair systems described so far are quite accurate. However, when the DNA of *E. coli* cells is heavily damaged by mutagenic agents such as UV light, the cells take some drastic steps in their attempt to survive. They go through a so-called **SOS response,** during which a whole battery of DNA repair, recombination, and replication proteins are synthesized. Two of these proteins, encoded by the *umuC* and *umuD* (*UV mu*table) genes, encode proteins that allow DNA replication to proceed across damaged segments of template strands, even though the nucleotide sequences in the damaged region can't be accurately replicated. This **error-prone repair** system eliminates gaps in the newly synthesized strands opposite damaged nucleotides in the template strands but, in so doing, increases the frequency of replication errors. The SOS response appears to be a desperate attempt to escape the lethal effects of heavily damaged DNA. When the error-prone repair system is operative, mutation rates increase sharply.

Key Points: **Multiple DNA repair systems have evolved to safeguard the integrity of genetic information in living organisms. Each repair pathway corrects a specific type of damage in DNA.**

INHERITED HUMAN DISEASES WITH DEFECTS IN DNA REPAIR

As we discussed at the beginning of this chapter, individuals with xeroderma pigmentosum (XP) are extremely sensitive to sunlight. Exposure to sunlight results in a high frequency of skin cancer in XP patients. The cells of individuals with XP are deficient in the repair of UV-induced damage to DNA, such as thymine dimers. The XP syndrome can result from defects in any of at least nine different genes. The products of six of these genes, *XPA*, *XPB*, *XPC*, *XPD*, *XPF*, and *XPG*, are required for nucleotide excison repair. They have been purified and shown to be essential for excinuclease activity. Since excinuclease activity in humans requires at least 17 polypeptides, the list of *XP* genes will probably expand in the future. Two other human disorders, Cockayne syndrome and trichothiodystrophy, also result from defects in nucleotide excision repair. Individuals with Cockayne syndrome exhibit retarded growth and mental skills, but not increased rates of skin cancer. Patients with trichothiodystrophy have short stature, brittle hair, and scaly skin; they also have underdeveloped mental abilities. Individuals with either Cockayne syndrome or trichothiodystrophy are defective in a type of excision repair that is coupled to transcription. However, details of this transcription-coupled repair process are still being worked out.

In addition to the damage to skin cells, some individuals with XP develop neurological abnormalities, which appear to result from the premature death of nerve cells. This effect on the very long-lived nerve cells may have interesting implications with respect to the causes of aging. One theory is that aging results from the accumulation of somatic mutations. If so, a defective repair system would be expected to speed up the aging process, and this appears to be the case with the nerve cells of XP patients. However, at present, there is little evidence linking somatic mutation to senescence.

Ataxia-telangiectasia, Fanconi anemia, and Bloom syndrome are three other inherited diseases in humans associated with known defects in DNA metabolism. All three disorders exhibit autosomal recessive patterns of inheritance, and all result in a high risk of malignancy, especially leukemia in the case of ataxia-telangiectasia and Fanconi anemia. Cells of patients with ataxia-telangiectasia exhibit an abnormal sensitivity to ionizing radiation, suggesting a defect in the repair of radiation-induced DNA damage. Cells of individuals with Fanconi anemia are impaired in the removal of DNA interstrand cross-links, such as those formed by the antibiotic mitomycin C. Individuals with Bloom syndrome exhibit a high frequency of chromosome breaks that results in chromosome aberrations (Chapter 7) and sister chromatid exchanges. Although all three of these malignant-prone, inherited diseases are caused by defects in DNA metabolism, probably repair processes, the primary lesions are still unknown. The recent demonstration that the onset of hereditary nonpolyposis colorectal cancer is associated with a defect in the repair of mismatched base pairs in DNA hints that similar defects in other DNA repair pathways may be involved in the development of specific types of human cancer (Chapter 28).

Key Points: **The importance of DNA repair pathways is documented convincingly by inherited human disorders that result from defects in DNA repair. Recent evidence indicates that the onset of certain types of cancer may be associated with defects in specific DNA repair pathways.**

DNA RECOMBINATION MECHANISMS

In Chapter 8, we discussed the main features of recombination between homologous chromosomes, but we did not consider the molecular details of the process. Because many of the gene products involved in the repair of damaged DNA also are required for recombination between homologous chromosomes or crossing over, we will now examine some of the molecular aspects of this important process. Moreover, recombination usually, perhaps always, involves some DNA repair synthesis. Thus, much of the information discussed in the preceding sections is relevant to the process of recombination.

In eukaryotes, crossing over is associated with the formation of the synaptinemal complex, which forms during prophase of the first meiotic division. This structure is composed primarily of proteins and RNA. For unknown reasons, crossing over occurs only rarely in male *Drosophila*. (Crossing over does occur in both sexes of most species; the near absence of crossing over in the heterogametic sex is unique to *Drosophila* and a few other species.) Of interest here is the fact that no synaptinemal complex is present during the first meiotic division of spermatogenesis in *Drosophila*. In addition, mutations in the *c3G* gene of *Drosophila* eliminate both crossing over and the formation of the synaptinemal complex in females. Thus, crossing over and the formation of the synaptinemal complex appear to be linked. A small amount of DNA synthesis occurs during the formation of the synaptinemal complex, and this DNA synthesis may be involved in synapsis or crossing over. Unfortunately, geneticists still know little about the functions of the components of the synaptinemal complex. Presumably, it plays a role in synapsis or crossing over, or both.

Recombination: Cleavage and Rejoining of DNA Molecules

In Chapter 8, we discussed the experiment of Creighton and McClintock that indicated that crossing over occurs by breakage of parental chromosomes and rejoining of the parts in new combinations. Evidence demonstrating that recombination occurs by breakage and rejoining has also been obtained by autoradiography and other techniques. Indeed, the main features of the process of recombination are now well established, even though specific details remain to be elucidated.

Much of what we know about the molecular details of crossing over is based on the study of **recombination-deficient mutants** of *E. coli* and *S. cerevisiae*. Biochemical studies of these mutants have shown that they are deficient in various enzymes and other proteins required for recombination. Together, the results of genetic and biochemical studies have provided a fairly complete picture of recombination at the molecular level. Nevertheless, geneticists still do not understand all the details of crossing over in any species.

Many of the currently popular models of crossing over were derived from a model proposed by Robin Holliday in 1964. Holliday's model was one of the first that explained most of the genetic data available at the time by a mechanism involving the breakage, reunion, and repair of DNA molecules. An updated version of the Holliday model is shown in Figure 14.31. This mechanism, like many others that have been invoked, begins when an endonuclease cleaves single strands of each of the two parental DNA molecules (breakage). Segments of the single strands on one side of each cut are then displaced from their complementary strands

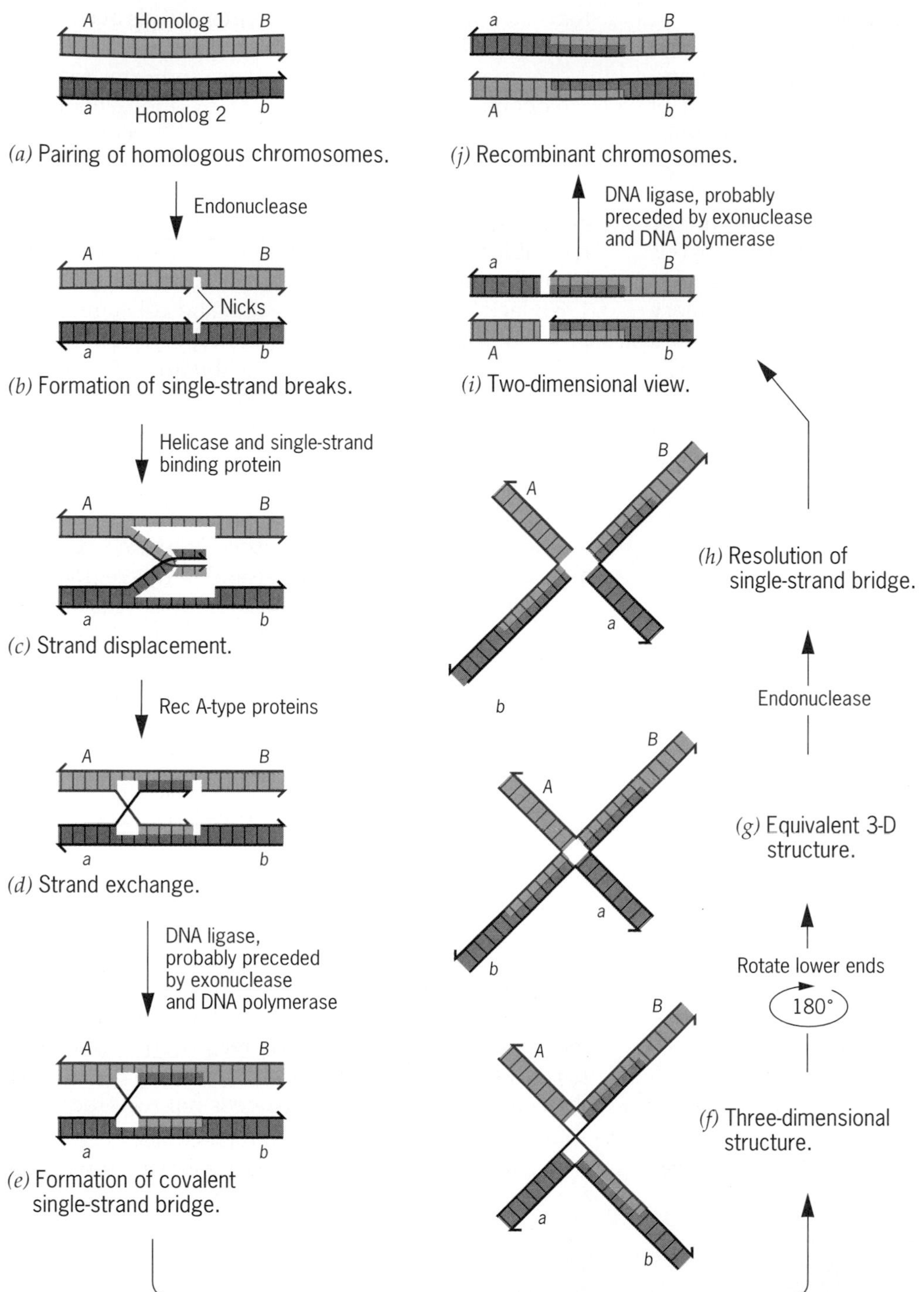

Figure 14.31 A mechanism for recombination between homologous DNA molecules. The pathway shown is based on the model originally proposed by Robin Holliday in 1964.

with the aid of DNA helicases and single-strand binding proteins. The helicases unwind the two strands of DNA in the region adjacent to single-strand incisions. In *E. coli*, the **RecBCD complex** contains both an endonuclease activity that makes single-strand breaks in DNA and a DNA helicase activity that unwinds the complementary strands of DNA in the region adjacent to each nick.

The displaced single strands then exchange pairing partners, base-pairing with the intact complementary strands of the homologous chromosomes. This process is stimulated by proteins like the *E. coli* **RecA**

protein. RecA-type proteins have been characterized in many species, both prokaryotic and eukaryotic. RecA protein and its homologs stimulate **single-strand assimilation,** a process by which a single strand of DNA displaces its homolog in a DNA double helix. RecA-type proteins promote reciprocal exchanges of DNA single strands between two DNA double helices in two steps. In the first step, a single strand of one double helix is assimilated by a second, homologous double helix, displacing the identical or homologous strand and base-pairing with the complementary strand. In the second step, the displaced single strand is similarly assimilated by the first double helix. The RecA protein mediates these exchanges by binding to the unpaired strand of DNA, aiding in the search for a homologous DNA sequence, and, once a homologous double helix is found, promoting the replacement of one strand with the unpaired strand. If complementary sequences already exist as single strands, the presence of RecA protein increases the rate of renaturation by over 50-fold.

The cleaved strands are then covalently joined in new combinations (reunion) by DNA ligase. If the original breaks in the two strands do not occur at exactly the same site in the two homologs, some tailoring will be required before DNA ligase can catalyze the reunion step. This tailoring usually involves the excision of nucleotides by an exonuclease and repair synthesis by a DNA polymerase. The sequence of events described so far will produce X-shaped recombination intermediates called **chi forms,** which have been observed by electron microscopy in several species (Figure 14.32). The chi forms are resolved by enzyme-catalyzed breakage and rejoining of the complementary DNA strands to produce two recombinant DNA molecules. In *E. coli*, chi structures can be resolved by the product of either gene *recG* or gene *ruvC* (*r*epair of *UV*-induced damage). Each gene encodes an endonuclease that catalyzes the cleavage of single strands at chi junctions (see Figures 14.31 and 14.33).

A substantial body of evidence indicates that homologous recombination occurs by more than one mechanism—probably by several different mechanisms. In *S. cerevisiae*, the ends of DNA molecules produced by double-strand breaks are highly recombinogenic. This fact and other evidence suggest that recombination in yeast often involves a double-strand break in one of the parental double helices. Thus, in 1983, Jack Szostak, Franklin Stahl, and colleagues proposed a **double-strand break model** of crossing over. According to their model, recombination involves a double-strand break in one of the parental double helices, not just single-strand breaks as in the Holliday model. The initial breaks are then enlarged to gaps in both strands. The two single-stranded termini produced at the double-stranded gap of the broken double helix invade the intact double helix and displace segments of the homologous strand in this region. The gaps are then filled in by repair synthesis. This process yields two homologous chromosomes joined by two single-strand bridges. The bridges are resolved by endonucleolytic cleavage, just as in the Holliday model. Both the double-strand-break model and the Holliday model nicely explain the production of chromosomes that are recombinant for genetic markers flanking the region in which the crossover occurs.

Gene Conversion: DNA Repair Synthesis Associated with Recombination

Up to this point, we have discussed only recombination events that can be explained by breakage of homologous chromatids and the reciprocal exchange of parts. However, tetrad analysis in Ascomycetes reveals that sometimes genetic exchange is not reciprocal. For

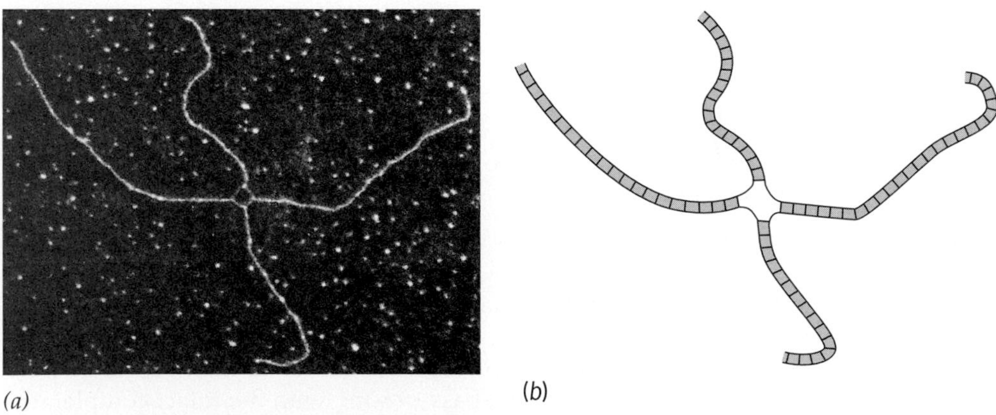

Figure 14.32 Electron micrograph (*a*) and diagram (*b*) of an X-shaped recombination intermediate or chi structure.

example, if crosses are performed between two closely linked mutations in *Neurospora*, and asci containing wild-type recombinants are analyzed, these asci frequently do not contain the reciprocal, double-mutant recombinant.

Consider a cross involving two closely linked mutations, m_1 and m_2. In a cross of $m_1\ m_2^+$ with $m_1^+\ m_2$, asci of the following type are sometimes observed:

Spore pair 1: $m_1^+\ m_2$
Spore pair 2: $m_1^+\ m_2^+$
Spore pair 3: $m_1\ \ m_2^+$
Spore pair 4: $m_1\ \ m_2^+$.

Wild-type $m_1^+\ m_2^+$ spores are present, but the $m_1\ m_2$ double-mutant spores are not present in the ascus. Reciprocal recombination would produce an $m_1\ m_2$ chromosome whenever an $m_1^+\ m_2^+$ chromosome was produced. In this ascus, the m_2^+: m_2 ratio is 3 : 1 rather than 2 : 2 as expected. One of the m_2 alleles appears to have been "converted" to the m_2^+ allelic form. Thus, this type of nonreciprocal recombination is called **gene conversion.** We might assume that gene conversion results from mutation, except that it occurs at a higher frequency than the corresponding mutation events, always produces the allele present on the homologous chromosome, not a new allele, and is correlated about 50 percent of the time with reciprocal recombination of flanking markers. The last observation strongly suggests that gene conversion results from events that occur during crossing over. Indeed, gene conversion is now believed to result from DNA repair synthesis associated with the breakage, excision, and reunion events of crossing over.

With closely linked markers, gene conversion occurs more frequently than reciprocal recombination. In one study of the *his1* gene of yeast, 980 of 1081 asci containing his^+ recombinants exhibited gene conversion, whereas only 101 showed classical reciprocal recombination.

The most striking feature of gene conversion is that the input 1:1 allele ratio is not maintained. This can be explained easily if short segments of parental DNA are degraded and then resynthesized with template strands provided by DNA carrying the other allele. Given the mechanisms of excision repair discussed earlier in this chapter, the Holliday model of crossing over explains gene conversion for genetic markers located in the immediate vicinity of the crossover. In Figure 14.31 *d-i*, there is a segment of DNA between the *A* and *B* loci where complementary strands of DNA from the two homologous chromosomes are base-paired. If a third pair of alleles located within this segment were segregating in the cross, mismatches in the two double helices would be present. DNA molecules containing such mismatches, or different alleles in the two complementary strands of a double helix, are called **heteroduplexes.** Such heteroduplex molecules occur as intermediates in the process of recombination.

If Figure 14.31*e* were modified to include a third pair of alleles, and the other two chromatids were added, the tetrad would have the following composition:

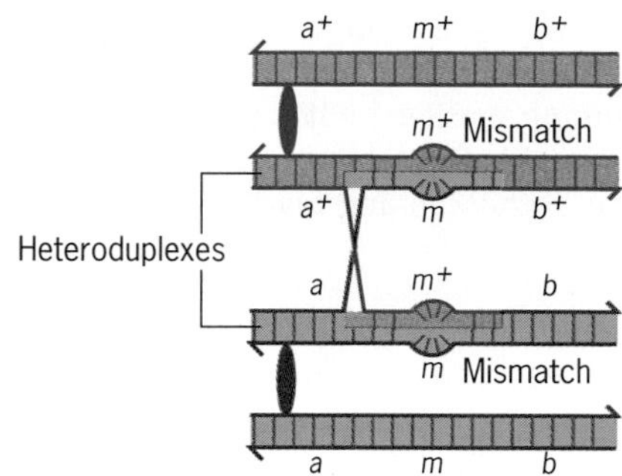

If the mismatches are resolved by excision repair (see Figure 14.28), in which the *m* strands are excised and resynthesized with the complementary m^+ strands as templates, the following tetrad will result:

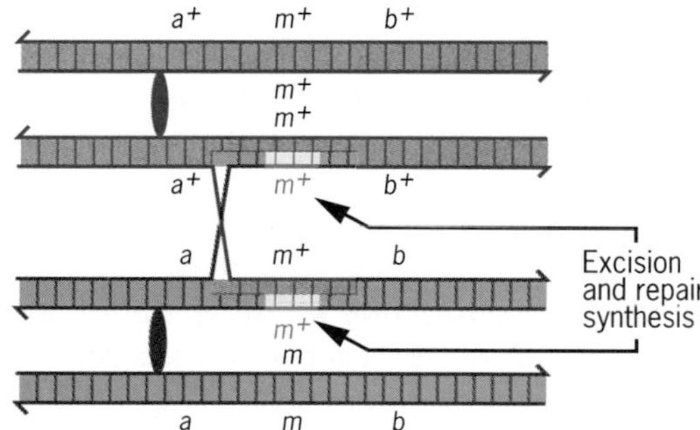

As a result of semiconservative DNA replication during the subsequent mitotic division, this tetrad will yield an ascus containing six m^+ ascospores and two *m* ascospores, the 3:1 gene conversion ratio.

Suppose that only one of the two mismatches in the tetrad just described is repaired prior to the mitotic division. In this case, the semiconservative replication of the remaining heteroduplex will yield one m^+ homoduplex and one *m* homoduplex, and the resulting ascus will contain a 5 m^+:3 *m* ratio of ascospores. Such 5:3 gene conversion ratios do occur. They result from postmeiotic (mitotic) segregation of unrepaired heteroduplexes.

Gene conversion is associated with the reciprocal recombination of flanking markers approximately 50 percent of the time. This correlation is nicely explained by the Holliday model of recombination presented in Figure 14.31. If the two recombinant chromatids of the tetrad just diagrammed are drawn in a form equivalent to that shown in Figure 14.31*g*, the association of gene conversion with reciprocal recombination of flanking markers can easily be explained (Figure 14.33). The single-strand bridge connecting the two chromatids must be resolved by endonucleolytic cleavage to complete

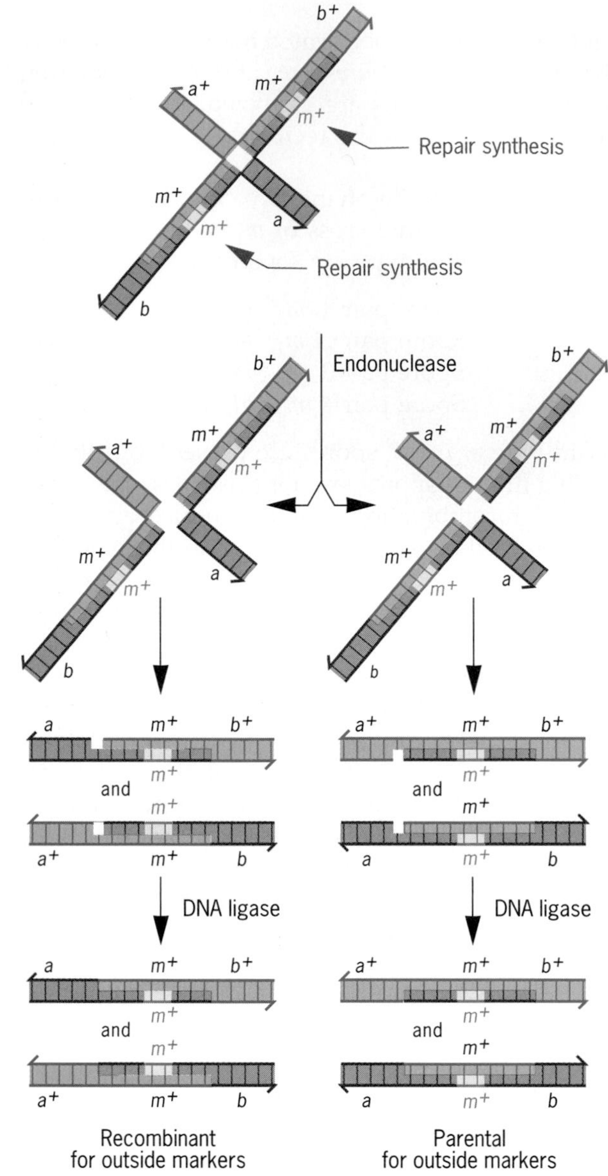

Figure 14.33 Formation of either the recombinant (bottom left) or parental combinations (bottom right) of flanking markers in association with gene conversion. The recombination intermediate at the top is the equivalent to that illustrated in Figure 14.31*g*, but shows the mismatch-repaired chromatids of the tetrad diagrammed in the text. This tetrad produces a 3 m^+ to 1 m gene conversion ascus pattern. Cleavage of the single-strand bridge in the vertical plane (left) produces the parental (a^+ b^+ and a b) arrangement of flanking markers, whereas cleavage in the horizontal plane yields the recombinant (a^+ b and a b^+) arrangement of the flanking markers.

the recombination process. This cleavage may occur either horizontally or vertically on the chi form drawn in Figure 14.33. Vertical cleavage will yield an ascus showing both gene conversion and reciprocal recombination of flanking markers. Horizontal cleavage will yield an ascus showing gene conversion and the parental combination of flanking markers. Thus, if cleavage occurs in the vertical plane half of the time and in the horizontal plane half of the time, gene conversion will be associated with reciprocal recombination of flanking markers about 50 percent of the time, as observed.

Key Points: **Crossing over involves the breakage of homologous DNA molecules and the rejoining of parts in new combinations. When genetic markers are closely linked, the four products of meiosis often contain three copies of one marker and one copy of the other marker. This gene conversion results from repair synthesis that occurs during the recombination process.**

TESTING YOUR KNOWLEDGE

1. Charles Yanofsky isolated a large number of auxotrophic mutants of *E. coli* that could grow only on medium containing the amino acid tryptophan. How can such mutants be identified? If a specific tryptophan auxotroph resulted from a nitrous acid-induced mutation, could it be induced to revert back to prototrophy by treatment with 5-bromouracil (5–BU)?

ANSWER

The culture of mutagenized bacteria must be grown in medium containing tryptophan so that the desired mutants can survive and reproduce. The bacteria should then be diluted, plated on agar medium containing tryptophan, and incubated until visible colonies are produced. The colonies are next replicated onto plates lacking tryptophan by the replica-plating technique developed by the Lederbergs (see Figure 14.3). The desired tryptophan auxotrophs will grow on the plates containing tryptophan, but not on the replica plates lacking tryptophan. Because nitrous acid and 5–BU produce transition mutations in both directions, A:T ⟷ G:C, any mutation induced with nitrous acid should be induced to back-mutate with 5–BU.

2. Assume that you recently discovered a new species of bacteria and named it *Escherichia mutaphilium*. During the last year, you have been studying the *mutA* gene and its polypeptide product, the enzyme trinucleotide mutagenase, in this bacterium. *E. mutaphilium* has been shown to use the established, nearly universal genetic code and to behave like *Escherichia coli* in all other respects relevant to molecular genetics.

The sixth amino acid from the amino terminus of the wild-type trinucleotide mutagenase is histidine, and the wild-type *mutA* gene has the triplet nucleotide-pair sequence

3′-G T A-5′
5′-C A T-3′

at the position corresponding to the sixth amino acid of the gene product. Seven independently isolated mutants with single nucleotide-pair substitutions within this triplet have also been characterized. Moreover, the mutant trinucleotide mutagenases have all been purified and sequenced. All seven are different; they contain glutamine, tyrosine, asparagine, aspartic acid, arginine, proline, and leucine as the sixth amino acid from the amino terminus.

Mutants *mutA1*, *mutA2*, and *mutA3* will not recombine with each other, but each will recombine with each of the other four mutants (*mutA4*, *mutA5*, *mutA6*, and *mutA7*) to yield true wild-type recombinants. Similarly, mutants *A4*, *A5*, and *A6* will not recombine with each other but will each yield true wild-type recombinants in crosses with each of the other four mutants. Finally, crosses between *mutA1* and *mutA7* yield about twice as many true wild-type recombinants as do crosses between *mutA6* and *mutA7*.

Mutants *A1* and *A6* are induced to back-mutate to wild-type by treatment with 5-bromouracil (5–BU), whereas mutants *A2*, *A3*, *A4*, *A5*, and *A7* are not induced to back-mutate by treatment with 5–BU. Mutants *A2* and *A4* grow slowly on minimal medium, whereas mutants *A3* and *A5* carry null mutations (producing completely inactive gene products) and are incapable of growth on minimal medium. This difference has been used to select for mutation events from genotypes *mutA3* and *mutA5* to genotypes *mutA2* and *mutA4*. Mutants *A3* and *A5* can be induced to mutate to *A2* and *A4*, respectively, by treatment with 5-bromouracil or hydroxylamine. However, mutant *A3* cannot be induced to mutate to *A4*, nor *A5* to *A2*, by treatment with either mutagen.

Use the information given above and the nature of the genetic code (Table 13.2) to deduce which mutant allele specifies the mutant polypeptide with each of the seven different amino acid substitutions at position 6 of trinucleotide mutagenase, and describe the rationale behind each of your deductions.

ANSWER

The following deductions can be made from the information given.

(1) The wild-type His codon must be CAU based on the nucleotide-pair sequence of the gene.

(2) The codons for the seven amino acids found at position 6 in the mutant polypeptides must be connected to CAU by a single-base change because the mutants were all derived from wild-type by a single nucleotide-pair substitution. Thus, the degeneracy of the genetic code is not a factor in deducing specific codon assignments.

(3) Because of the nature of the genetic code—specifically the degeneracy at the third (3′) position in each codon, there are three possible amino acid substitutions due to single-base substitutions at each of the first two positions (the 5′ base and the middle base), but only one possible amino acid change due to a single-base change at position 3 (the 3′ base in the codon). For ease of discussion, the three nucleotide-pair positions in the triplet under consideration will be referred to as position 1 (corresponding to the 5′ base in the codon), position 2 (the middle nucleotide pair), and position 3 (corresponding to the 3′ base in the mRNA codon).

(4) Since *A1*, *A2*, and *A3* do not recombine with each other, they must all result from base-pair substitutions at the same position in the triplet, at either position 1 or position 2. The same is true for *A4*, *A5*, and *A6*. Since *A7* recombines with each of the other six mutant alleles, it must result from the single base-pair substitution at position 3 that leads to an amino acid change.

(5) The only amino acid with a codon connected to the His codon CAU by a single-base change at position 3 is Gln (codon CAA). Thus, the *mutA7* polypeptide must have glutamine as the sixth amino acid.

(6) Since *mutA7* (the third position substitution) yields about twice as many wild-type recombinants in crosses with *mutA1* as in crosses with *mutA6*, the *A1* substitution must be at position 1 and the *mutA6* substitution must be at position 2. Combined with (4) above, this places the *A2* and *A3* substitutions at position 1 and the *A4* and *A5* substitutions at position 2.

(7) Since *mutA1* and *mutA6* are induced to revert to wild-type by 5-BU, they must be connected to the triplet of nucleotide pairs encoding His by transition mutations, that is,

$$(mut\ A1)\ \frac{\mathrm{ATA}}{\mathrm{TAT}} \xleftrightarrow{\text{5-BU}} \frac{\mathrm{GTA}}{\mathrm{CAT}} \xleftrightarrow{\text{5-BU}} \frac{\mathrm{GCA}}{\mathrm{CGT}}\ (mut\ A6)$$

(8) Since *mutA3* and *mutA5* are induced to mutate to *mutA2* and *mutA4*, respectively, by hydroxylamine, *A3* must be connected to *A2* and *A5* to *A4* specifically by G:C → A:T transitions, that is,

$$(mut\ A3)\ \frac{\mathrm{CTA}}{\mathrm{GAT}} \xrightarrow{\text{HA}} \frac{\mathrm{TTA}}{\mathrm{AAT}}\ (mut\ A2)$$

and

$$(mut\ A5)\ \frac{\mathrm{GGA}}{\mathrm{CCT}} \xrightarrow{\text{HA}} \frac{\mathrm{GAA}}{\mathrm{CTT}}\ (mut\ A4)$$

Collectively, these deductions establish that the following relationships between the amino acids, codons, and nucleotide-pair triplets are present at the position of interest in the trinucleotide mutagenase polypeptides, mRNAs, and genes in the seven different mutants.

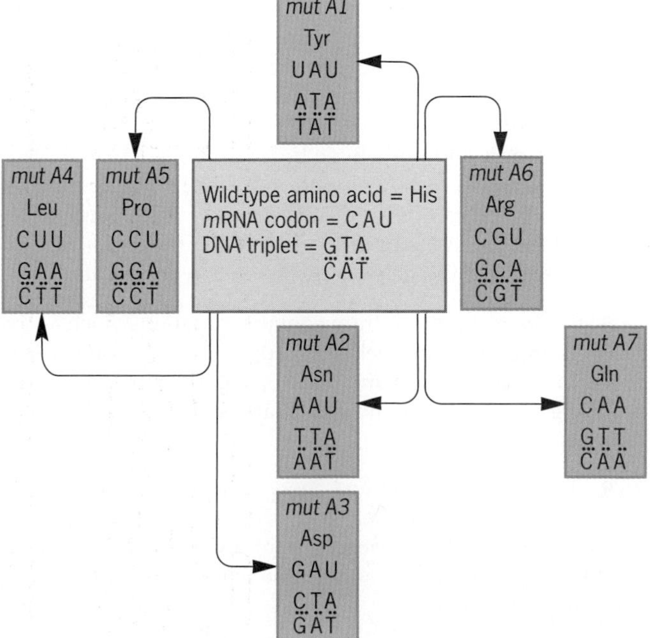

QUESTIONS AND PROBLEMS

14.1 Identify the following point mutations represented in DNA and in RNA as (1) transitions, (2) transversions, or (3) reading frameshifts. (a) A to G; (b) C to T; (c) C to G; (d) T to A; (e) UAU ACC UAU to UAU AAC CUA; (f) UUG CUA AUA to UUG CUG AUA.

14.2 Of all possible missense mutations that can occur in a segment of DNA encoding the amino acid tryptophan, what is the ratio of transversions to transitions if all single base-pair substitutions occur at the same frequency?

14.3 Both lethal and visible mutations are expected to occur in fruit flies that are subjected to irradiation. Outline a method for detecting (a) X-linked lethals and (b) X-linked visible mutations in irradiated *Drosophila*.

14.4 How can mutations in bacteria causing resistance to a particular drug be detected? How can it be determined whether a particular drug causes mutations or merely identifies mutations already present in the organisms under investigation?

14.5 Published spontaneous mutation rates for humans are generally higher than those for bacteria. Does this indicate that individual genes of humans mutate more frequently than those of bacteria? Explain.

14.6 A precancerous condition (intestinal polyposis) in a particular human family group is determined by a single dominant gene. Among the descendants of one woman who died with cancer of the colon, 10 people have died with the same type of cancer and 6 now have intestinal polyposis. All other branches of the large kindred have been carefully examined, and no cases have been found. Suggest an explanation for the origin of the defective gene.

14.7 Juvenile muscular dystrophy in humans is dependent on an X-linked recessive gene. In an intensive study, 33 cases were found in a population of some 800,000 people. The investigators were confident that they had found all cases that were well enough advanced to be detected at the time the study was made. The symptoms of the disease were expressed only in males. Most of those with the disease died at an early age, and none lived beyond 21 years of age. Usually, only one case was detected in a family, but sometimes two or three cases occurred in the same family. Suggest an explanation for the sporadic occurrence of the disease and the tendency for the gene to persist in the population.

14.8 Products resulting from somatic mutations, such as the navel orange and the Delicious apple, have become widespread in citrus groves and apple orchards. However, traits resulting from somatic mutations are seldom maintained in animals. Why?

14.9 If a single short-legged sheep should occur in a flock, suggest experiments to determine whether the short legs are the result of a mutation or an environmental effect. If due to a mutation, how can one determine whether the mutation is dominant or recessive?

14.10 How might enzymes such as DNA polymerase be involved in the mode of action of both mutator and antimutator genes (mutant genes that increase and decrease, respectively, mutation rates)?

14.11 How could spontaneous mutation rates be optimized by natural selection?

14.12 A mutator gene *Dt* in maize increases the rate at which the gene for colorless aleurone (*a*) mutates to the dominant allele (*A*), which yields colored aleurone. When reciprocal crosses were made (i.e., seed parent *dt/dt, a/a* × *Dt/Dt, a/a* and seed parent *Dt/Dt, a/a* × *dt/dt, a/a*), the cross with *Dt/Dt* seed parents produced three times as many dots per kernel as the reciprocal cross. Explain these results.

14.13 A single mutation blocks the conversion of phenylalanine to tyrosine. (a) Is the mutant gene expected to be pleiotropic? (b) Explain.

14.14 How can normal hemoglobin (hemoglobin A) and hemoglobin S be distinguished?

14.15 If CTT is a DNA triplet (transcribed strand of DNA) specifying glutamic acid, what DNA and mRNA base triplet alterations could account for valine and lysine in position 6 of the β-globin chain?

14.16 Why is sickle-cell anemia called a molecular disease?

14.17 Assuming that the β-globin chain and the α-globin chain shared a common ancestor, what mechanisms might explain the differences that now exist in these two chains? What changes in DNA and mRNA codons would account for the differences that have resulted in unlike amino acids at corresponding positions?

14.18 In a given strain of bacteria, all of the cells are usually killed when a specific concentration of streptomycin is present in the medium. Mutations occur that confer resistance to streptomycin. The streptomycin-resistant mutants are of two types: some can live with or without streptomycin; others cannot survive unless this drug is present in the medium. Given a streptomycin-sensitive strain of this species, outline an experimental procedure by which streptomycin-resistant strains of the two types could be established.

14.19 One stock of fruit flies was treated with 1000 roentgens (r) of X rays. The X-ray treatment increased the mutation rate of a particular gene by 2 percent. What percentage increases in the mutation rate of this gene would be expected if this stock of flies was treated with X-ray doses of 1500 r, 2000 r, and 3000 r?

14.20 Why does the frequency of chromosome breaks induced by X rays vary with the total dosage and not with the rate at which it is delivered?

14.21 A reactor overheats and produces radioactive tritium (H^3), radioactive iodine (I^{131}), and radioactive xenon (Xn^{133}). Why should we be more concerned about radioactive iodine than the other two radioactive isotopes?

14.22 One person was in an accident and received 50 roentgens (r) of X rays at one time. Another person received 5 r in each of 20 treatments. Assuming no intensity effect, what proportionate number of mutations would be expected in each person?

14.23 How does ultraviolet light produce revertible mutations?

14.24 How does nitrous acid induce mutations? What specific end results might be expected in DNA and mRNA from the treatment of viruses with nitrous acid?

14.25 Are mutational changes induced by nitrous acid more likely to be transitions or transversions?

14.26 You are screening three new pesticides for potential mutagenicity using the Ames test. Two *his*⁻ strains resulting from either a frameshift or a transition mutation were used and produced the following results:

	Transition Mutant Control (no chemical)	**Transition Mutant + Chemical**	**Transition Mutant + Chemical + Rat Liver Enzymes**
Pesticide #1	21	180	19
Pesticide #2	18	19	17
Pesticide #3	25	265	270

	Frameshift Mutant Control (no chemical)	**Frameshift Mutant + Chemical**	**Frameshift Mutant + Chemical + Rat Liver Enzymes**
Pesticide #1	5	4	5
Pesticide #2	7	5	93
Pesticide #3	6	9	7

What type of mutations, if any, do the three pesticides induce?

14.27 How does the action and mutagenic effect of 5-bromouracil differ from that of nitrous acid?

14.28 Sydney Brenner and A. O. W. Stretton found that nonsense mutations did not terminate polypeptide synthesis in the *rII* gene of the bacteriophage T4 when these mutations were located within a DNA sequence interval in which a single nucleotide insertion had been made on one end and a single nucleotide deletion had been made on the other. How can this finding be explained?

14.29 Seymour Benzer and Ernst Freese compared spontaneous and 5-bromouracil-induced mutants in the *rII* gene of the bacteriophage T4; the mutagen increased the mutation rate ($rII^+ \rightarrow rII$) several hundred times above the spontaneous mutation rate. Almost all (98 percent) of the 5-bromouracil-induced mutants could be induced to revert to wild-type ($rII \rightarrow rII^+$) by 5-bromouracil treatment, but only 14 percent of the spontaneous mutants could be induced to revert to wild-type by this treatment. Discuss the reason for this result.

14.30 How do acridine-induced changes in DNA result in inactive proteins?

Use the known codon-amino acid assignments given in Chapter 13 to work the following problems.

14.31 Mutations in the genes encoding the α and β subunits of hemoglobin lead to blood diseases such as thalassemias and sickle-cell anemia. You have found a family in China in which some members suffer from a new genetic form of anemia. The DNA sequences at the 5′ end of the nontemplate strand of the normal and mutant DNA encoding the α subunit of hemoglobin are as follows:

Normal 5′-ACGTTATGCCGTACTGCCAGCTAACTGC-TAAAGAACAATTA.......-3′

Mutant 5′-ACGTTATGCCCGTACTGCCAGCTAACTGC-TAAAGAACAATTA.......-3′

(a) What type of mutation is present in the mutant hemoglobin gene?

(b) What are the codons in the translated portion of the mRNA transcribed from the normal and mutant genes?

(c) What are the amino acid sequences of the normal and mutant polypeptides?

14.32 Bacteriophage MS2 carries its genetic information in RNA. Its chromosome is analogous to a polygenic molecule of mRNA in organisms that store their genetic information in DNA. The MS2 minichromosome encodes 4 polypeptides (i.e., it has four genes). One of these four genes encodes the MS2 coat protein, a polypeptide 129 amino acids long. The entire nucleotide sequence in the RNA of MS2 is known. Codon 112 of the coat protein gene is CUA, which specifies the amino acid leucine. If you were to treat a replicating population of bacteriophage MS2 with the mutagen 5-bromouracil, what amino acid substitutions would you expect to be induced at position 112 of the MS2 coat protein (i.e., Leu → other amino acid)? (*Note*: Bacteriophage MS2 RNA replicates using a complementary strand of RNA and base-pairing like DNA.)

14.33 Would the different amino acid substitutions induced by 5-bromouracil at position 112 of the coat polypeptide that you indicated in Problem 14.32 be expected to occur with equal frequency? If so, why? If not, why not? Which one(s), if any, would occur more frequently?

14.34 Would such mutations occur if a nonreplicating suspension of MS2 phage was treated with 5-bromouracil?

14.35 Recalling that nitrous acid deaminates adenine, cytosine, and guanine (adenine → hypoxanthine, which base-pairs with cytosine; cytosine → uracil, which base-pairs with adenine; and guanine → xanthine, which base-pairs with cytosine), would you expect nitrous acid to induce any mutations that result in the substitution of another amino acid for a glycine residue in a wild-type polypeptide (i. e., glycine → another amino acid) if the mutagenesis was carried out on a suspension of mature (nonreplicating) T4 bacteriophages. (*Note*: After the mutagenic treatment of the phage suspension, the nitrous acid is removed. The treated phage are then allowed to infect *E. coli* cells to express any induced mutations.) If so, by what mechanism? If not, why not?

14.36 Keeping in mind the known nature of the genetic code, the information given about phage MS2 in Problem 14.32, and the information you have learned about nitrous acid in Problem 14.35, would you expect nitrous acid to induce any mutations that would result in amino acid substitutions of the type glycine → another amino acid if the mutagenesis were carried out on a suspension of mature (nonreplicating) MS2 bacteriophage? If so, by what mechanism? If not, why not?

14.37 Would you expect nitrous acid to induce a higher frequency of Tyr → Ser or Tyr → Cys substitutions? Why?

14.38 Which of the following amino acid substitutions should you expect to be induced by 5-bromouracil with the highest frequency? (a) Met → Leu; (b) Met → Thr; (c) Lys → Thr; (d) Lys → Gln; (e) Pro → Arg; or (f) Pro → Gln? Why?

14.39 The wild-type sequence of part of a protein is

NH_2-Trp-Trp-Trp-Met-Arg-Glu-Trp-Thr-Met

Each mutant listed below differs from wild-type by a single point mutation. Using this information, determine the mRNA sequence coding for the wild-type polypeptide. If there is more than one possible nucleotide, list all possibilities.

Mutant	Amino Acid Sequence of Polypeptide
1	Trp-Trp-Trp Met
2	Trp-Trp-Trp-Met-Arg-Asp-Trp-Thr-Met
3	Trp-Trp-Trp-Met-Arg-Lys-Trp-Thr-Met
4	Trp-Trp-Trp-Met-Arg-Glu-Trp-Met-Met

14.40 Acridine dyes such as proflavin are known to induce primarily single base-pair additions and deletions. Suppose that the wild-type nucleotide sequence in the mRNA produced from a gene is

5′ -AUGCCCUUUGGGAAAGGGUUUCCCUAA-3′

Also, assume that a mutation is induced within this gene by proflavin and, subsequently, a revertant of this mutation is similarly induced with proflavin and shown to result from a second-site suppressor mutation within the same gene. If the amino acid sequence of the polypeptide encoded by this gene in the revertant (double mutant) strain is

NH_2-Met-Pro-Phe-Gly-Glu-Arg-Phe-Pro-COOH

what would be the most likely nucleotide sequence in the mRNA of this gene in the revertant (double mutant)?

BIBLIOGRAPHY

AMES, B. N., J. MCCANN, AND E. YAMASAKI. 1975. Methods for detecting carcinogens and mutagens with the *Salmonella*/mammalian-microsome mutagenicity test. *Mutation Research* 31:347–364.

DRAKE, J. W. 1970. *The Molecular Basis of Mutation*. Holden-Day, San Francisco.

DRAKE, J. W., B. CHARLESWORTH, D. CHARLESWORTH, AND J. F. CROW. 1998. Rates of spontaneous mutation. *Genetics* 148:1667–1686.

FRIEDBERG, E. C., G. C. WALKER, AND W. SIEDE. 1995. *DNA Repair and Mutagenesis*. ASM Press, Washington, DC.

HAWLEY, S., R. HAYNES, E. KUTTER, M. RESNICK, AND D. STADLER, eds. 1998. *Mutation and Repair*, John W. Drake Commemorative Issue. *Genetics* 148:1403–1687.

LEDERBERG, J., AND E. M. LEDERBERG. 1952. Replica plating and indirect selection of bacterial mutants. *J. Bacteriology* 63:399–406. (Reprinted in E. A. Adelberg, ed., 1966. *Papers on Bacterial Genetics*, 2nd ed., Little, Brown, Boston.)

MODRICH, P. 1994. Mismatch repair, genetic stability, and cancer. *Science* 266:1959–1960.

MULLER, H. J. 1927. Artificial transmutation of the gene. *Science* 66:84–87. (Reprinted in J. A. Peters, ed., 1959. *Classical Papers in Genetics*. Prentice-Hall, Englewood Cliffs, NJ).

SNYDER, L., AND W. CHAMPNESS. 1997. *Molecular Genetics of Bacteria*. ASM Press, Washington, DC.

STAHL, F. W. 1979. *Genetic Recombination*. Freeman, San Francisco.

WEST, S. C. 1997. Processing of recombination intermediates by the RuvABC proteins. *Annu. Rev. Genet.* 31:213–244.

An albino giraffe with three normally pigmented companions.

15

Definitions of the Gene

CHAPTER OUTLINE

Evolution of the Concept of the Gene: Summary
Evolution of the Concept of the Gene: Function
Evolution of the Concept of the Gene: Structure
A Genetic Definition of the Gene
Complex Gene–Protein Relationships

Sir Archibald Garrod and Human Inborn Errors of Metabolism

In 1902, just two years after the rediscovery of Mendel's work, Sir Archibald E. Garrod, a physician at the Hospital for Sick Children in London, published a paper entitled "The Incidence of Alkaptonuria: A Study in Chemical Individuality." Like Mendel's work, the concepts Garrod presented in this paper remained largely unknown by the scientists of the world for 40 years—until the concepts were independently formulated by George W. Beadle and Edward L. Tatum in 1941. Garrod's paper described the results of his studies on individuals with alkaptonuria, an innocuous, but easily detected, disorder. Because of the presence of a chemical called homogentisic acid (formerly called alkapton), the urine of affected individuals turns black when exposed to air. Garrod clearly recognized that alkaptonuria was inherited. In his 1902 paper, he wrote: "There are good reasons for thinking that alkaptonuria is not the manifestation of a disease but is rather the nature of an alternative course of metabolism, harmless and usually congenital and lifelong" (*Lancet* ii:1616).

In addition to alkaptonuria, Garrod studied albinism, cystinuria, pentosuria, and porphyrinurea in humans (see Human Genetics Sidelight: Human Inborn Errors of Metabolism). Garrod summarized the results of these studies in a treatise entitled "Inborn Errors of Metabolism," first presented as the Croonian Lectures to the Royal College of Physicians in London in 1908 and published in book form in 1909. In this book, Garrod clearly articulated the view that the metabolic defects observed in individuals with these disorders are caused by recessive mutant genes. Garrod probably did not develop his amazing insight into metabolic processes totally by himself. His father, Alfred B. Garrod, was also a physician and was the first to demonstrate the accumulation of the chemical uric acid in patients with gout, a painful inflammation of the joints in the hands and feet. In any case, Archibald E. Garrod was the first scientist to relate defects in genes to blocks in metabolic pathways. In this chapter, we focus on how Garrod's concept evolved into the current understanding of this basic unit of genetic information: the **GENE**.

The gene is to genetics what the atom is to chemistry. Thus, throughout this text we have focused our attention on the gene and the alternate forms of a gene or alleles. In the preceding chapters, we have examined the patterns of transmission of independently assorting and linked genes, the chromosomal location of genes, the chemical composition of genes and chromosomes, the mechanism of replication of genes, mutational events in genes, and the mechanisms by which genes exert their effects on the phenotype of the organism. What is this unit of genetic information that we call the gene? As we will see, the concept of a gene is not static; it has evolved through several phases since Wilhelm Johannsen introduced the term in 1909, and it will undoubtedly evolve through additional refinements in the future.

The gene has been defined as the unit of genetic information that controls a specific aspect of the phenotype. Such a description, though accurate, does not provide a precise, unambiguous definition that can be used to identify a gene at the molecular level. At a more fundamental level, the gene has been defined as the unit of genetic information that specifies the synthesis of one polypeptide. However, it is not a very good operational definition. An **operational definition** spells out an operation or experiment that can be carried out to define or delimit something. One gene specifying one polypeptide is a poor operational definition because experimentally relating all the segments of DNA that represent genes with all the polypeptides is not feasible in a complex organism. Moreover, it is preferable to define the gene by using genetic approaches rather than biochemical experiments. Thus, in this chapter, we focus on the **complementation test** as an operational definition of the gene. We will also consider the limitations of the complementation test, along with the unique structural features of selected genes.

EVOLUTION OF THE CONCEPT OF THE GENE: SUMMARY

Before discussing evidence supporting the various concepts, let's summarize the important stages in the evolution of the gene concept. The gene theory of inheritance began with the publication of Mendel's classic paper in 1866 but did not become an accepted part of scientific knowledge until after the rediscovery of Mendel's work in 1900. Mendel's gene (not so-named) was the "character" or "constant factor" that controlled one specific phenotypic trait such as flower color in peas. At the time of the rediscovery of Mendel's work, the English physician Sir Archibald E. Garrod was studying several inherited diseases in humans. Garrod first recognized that homozygosity for recessive mutant alleles can cause defects in the normal processes of metabolism. His concept of the gene is probably stated most accurately as one mutant gene–one metabolic block, which over 30 years later was refined to the one gene–one enzyme concept enunciated by George W. Beadle and Edward L. Tatum. Because many enzymes contain two or more different polypeptides, each encoded by a separate gene, the one gene–one enzyme concept subsequently was modified to one gene–one polypeptide.

Prior to 1940, genes were considered analogous to beads on a string; recombination occurred between, but not within, genes. The gene was both the basic functional unit, which controlled one phenotypic trait, and the elementary structural unit, which could not be subdivided by recombination or mutation. Clarence Oliver's 1940 report that recombination had occurred within the *lozenge* gene of *Drosophila* stimulated both excitement and much debate about its significance. When the debate ended, the nucleotide pair had replaced the gene as the basic unit of structure, the unit of genetic material not subdivisible by mutation or recombination.

In the early 1940s, Edward B. Lewis developed the complementation, or *cis-trans*, test for functional allelism in *Drosophila*. This test subsequently was exploited by Seymour Benzer to define experimentally the gene in bacteriophage T4. The complementation

HUMAN GENETICS SIDELIGHT

Human Inborn Errors of Metabolism

As discussed in the text, Garrod's concept of one mutant gene–one metabolic block was based on his studies of a few inherited human disorders. Alkaptonuria, on which much of Garrod's information was based, is described in the text. However, Garrod also studied familial cases of cystinuria, pentosuria, and porphyrinuria—inherited disorders characterized by elevated levels of the amino acid cystine, five-carbon sugars, and the iron-binding porphyrin component of hemoglobin, respectively, in urine. The most common disorder that Garrod studied was albinism (Figure 1), an autosomal recessive trait that occurs at a frequency of about 1 in 20,000 newborns in the United States. The red eyes and white skin and hair of albinos result from the absence of the black pigment melanin. The most common forms of albinism are caused by mutations that block the biosynthesis of melanin from the amino acid tyrosine. Of course, the pathway of melanin biosynthesis was unknown in 1909 when Garrod published his book *Inborn Errors of Metabolism*. Nevertheless, he clearly understood that mutations caused specific blocks in metabolic pathways.

Today, over 4000 inherited human disorders have been described, and the number of such hereditary abnormalities is constantly increasing. These disorders range from relatively innocuous disorders like alkaptonuria to those such as Tay-Sachs disease (characterized by rapid degeneration of the central nervous system) that are lethal in early childhood (see Human Genetics Sidelight: Tay-Sachs Disease in Chapter 14). Sickle-cell anemia and phenylketonuria (PKU) are perhaps the best known of the human inborn errors of metabolism. The severe anemia in individuals homozygous for the sickle-cell mutation is caused by a single amino acid substitution in their β-globin (Chapter 13). Children with PKU lack the enzyme phenylalanine hydroxylase. This enzyme deficiency results in the accumulation of phenylpyruvic acid, which is highly toxic to the central nervous system. If untreated, children with PKU develop severe mental retardation. However, if they are placed on a diet low in phenylalanine, they develop normal mental abilities (Chapter 26). Other inherited human diseases are discussed throughout the book.

In the United States alone, over 120,000 children with inherited defects are born each year. Long-term medical care of these birth defects is estimated to cost $10 to $20 billion. But the cost is even higher in terms of human suffering. For example, consider the tragic degeneration of the central nervous system that occurs in Huntington disease (discussed in Chapter 22) or the loss of body control resulting from progressive muscle degeneration associated with Duchenne muscular dystrophy or the consequences of the loss of memory that occurs in individuals with Alzheimer's disease (Chapter 26). What are the causes of these inherited diseases? What can humans do, if anything, to keep the frequency of these disorders as low as possible? How many different human inborn errors of metabolism are there? Can the deleterious symptoms of these inherited disorders be eliminated by treatment? Can any of these inherited diseases be cured so that they won't be transmitted from parents to their children?

Human inherited diseases are caused by all of the types of mutations discussed in Chapter 14 and by the changes in chromosome structure and number discussed in Chapter 7. Because mutations are required for evolution, they can never be completely eliminated. However, humans can minimize the frequency of new mutations by avoiding contact with highly mutagenic agents such as irradiation and chemicals that damage DNA. All genes can mutate to nonfunctional states. Therefore, the number of human inborn errors of metabolism is almost certainly equal to the number of genes required to develop and maintain a normal human phenotype. Several inherited disorders such as PKU have proven treatable once their molecular bases are known. Others should become treatable by somatic-cell gene therapy (Chapter 22). In the future, some inherited human diseases may be cured by germ-line gene therapy; but serious ethical issues must be addressed before germ-line gene therapy can be performed on humans. Given the rapidly increasing number of known inherited human diseases and the identification and characterization of the genes responsible for these disorders (Chapter 22), the tools of molecular genetics will play an increasingly important role in the practice of medicine in the future.

Figure 1 An individual (front) with albinism.

test provides an operational definition of the gene. This test allows a geneticist to determine whether two independent mutations occurred in the same gene or in two different genes. Moreover, the gene as defined by the complementation test is a perfect fit to the one gene–one polypeptide concept.

In the 1960s, elegant experiments by Charles Yanofsky, Sydney Brenner, and their collaborators showed that the gene and its polypeptide product were colinear structures, with a direct correlation between the sequence of nucleotide pairs in the gene and the sequence of amino acids in the polypeptide. However, this simple concept of the gene as a continuous sequence of nucleotide pairs specifying a colinear sequence of amino acids in the polypeptide gene product was short-lived. Overlapping genes and genes-within-genes were discovered in the late 1960s, and the coding sequences of eukaryotic genes were shown to be interrupted by noncoding intron sequences in the late 1970s. Moreover, some genes, for example, genes encoding immunoglobulins, were shown to be stored in germ-line chromosomes as short "gene segments," which are assembled into mature, functional genes during development.

Thus, the definition of the gene needs to remain somewhat pliable if it is to encompass all of the different structure/function relationships that occur in different organisms. Here, we define the gene as the unit of genetic information that controls the synthesis of one polypeptide or one structural RNA molecule. As just defined, the gene can be identified operationally by the complementation test. As such, the gene includes the 5′ and 3′ noncoding regions that are involved in regulating the transcription and translation of the gene and all noncoding sequences or introns within the gene (Figure 15.1). The structural gene refers

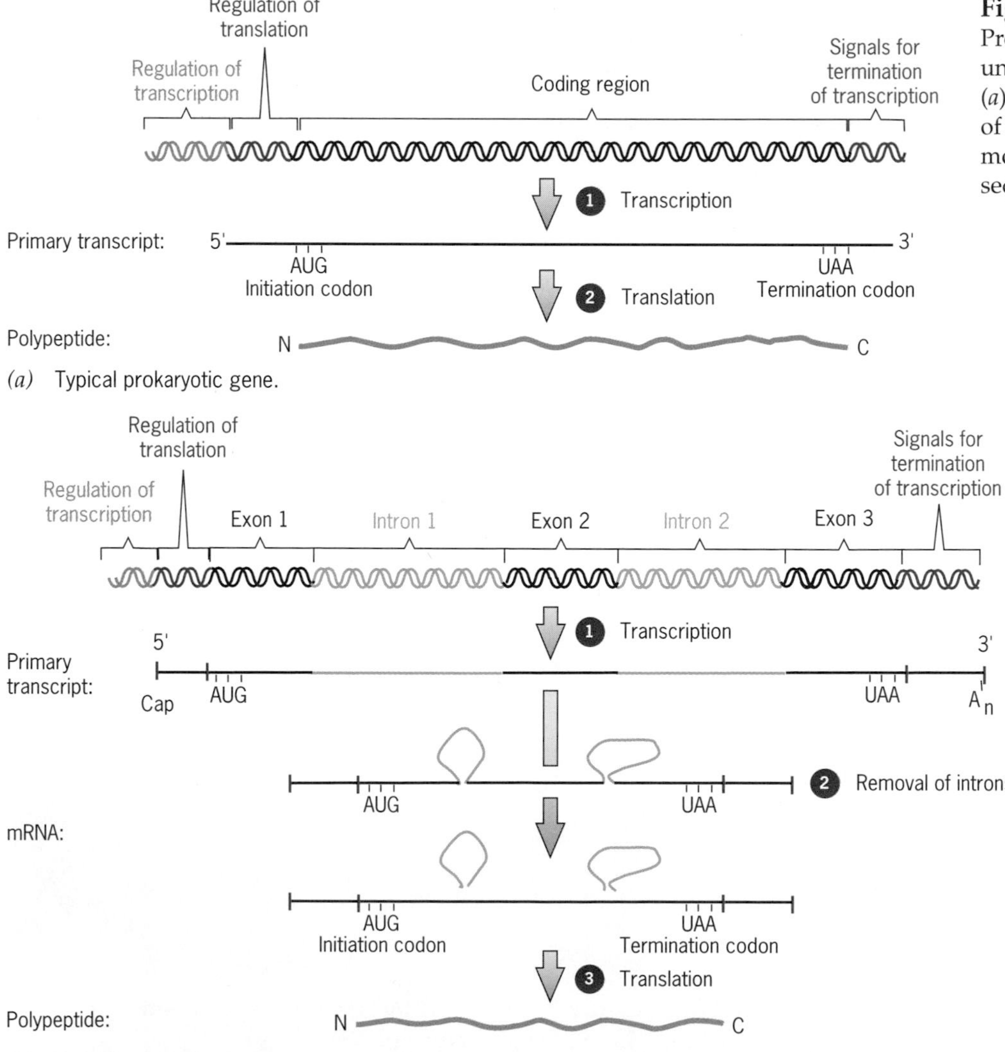

Figure 15.1 Gene structure. Prokaryotic genes usually contain uninterrupted coding sequences (*a*), whereas the coding sequences of eukaryotic genes are commonly interrupted by noncoding sequences (*b*).

to the portion that is transcribed to produce the RNA product. In the case of overlapping genes, this definition requires that some nucleotide-pair sequences be considered components of two or more genes. For those cases where exons are spliced together in various combinations to make related but different proteins, the gene may be defined as a DNA sequence that is a single unit of transcription and encodes a set of closely related polypeptides, sometimes called "protein isoforms." In germ-line chromosomes, the DNA sequences that encode segments of antibody chains probably should be called "gene segments," because this genetic information is not organized into units that fit any of the standard definitions of the gene (Chapter 27).

Key Points: **The concept of the gene has undergone many refinements since the rediscovery of Mendel's work in 1900. Most genes encode one polypeptide and can be operationally defined by the complementation test.**

EVOLUTION OF THE CONCEPT OF THE GENE: FUNCTION

In the preceding section, we briefly examined the evolution of the concept of the gene, the basic unit of genetic information. That summary included both functional and structural aspects of the gene. Now, let's take a closer look at the gene as a unit of function.

Mendel: Constant Factors Controlling Phenotypic Traits

The law of combination of different characters, which governs the development of the hybrids, finds therefore its foundation and explanation in the principle enunciated, that the hybrids produce egg cells and pollen cells which in equal numbers represent all constant forms which result from the combinations of the characters brought together in fertilisation. (Mendel, 1866; translation by William Bateson)

Mendel's characters or factors, which are now called genes, controlled specific phenotypic traits such as flower color, seed color, and seed shape. They were the basic units of function, the units of genetic information that governed one specific aspect of the phenotype. This definition of the gene as the basic unit of function is almost universally favored by present-day scientists. There has been no change in the concept of the gene as the basic unit of function since its discovery by Mendel in 1866. However, discovery of the chemical nature of the genetic material raised questions about the structure of the gene, and the concept of the molecular structure of this basic unit of function has undergone several changes and refinements since the rediscovery of Mendel's work in 1900.

If we examine what is known about how genes control phenotypic traits, the need for a more precise definition of the gene will be obvious. The pathway by which a gene exerts its effect on the phenotype of an organism is often very complex (Figure 15.2). Several genes may have similar effects on the same phenotypic

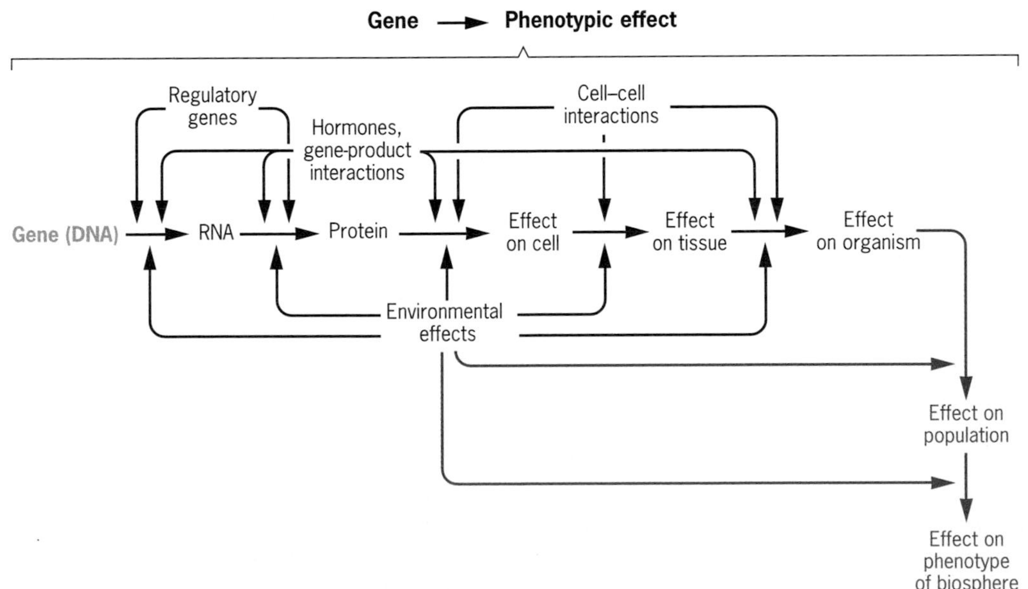

Figure 15.2 The complex pathway by which a gene exerts its effect on the phenotype of an organism, a population, or the biosphere.

trait, making it difficult to sort out the effects of individual genes. All the genes of an organism are located in the same nuclei, and they do not all function independently. The phenotype of an organism is the product of the action of all the genes acting within the restrictions imposed by the environment. Each gene also has an effect on the population to which the organism carrying the gene belongs. Ultimately, each gene has a potential effect, small though it may be, on the cumulative phenotype of the biosphere, for each gene may affect the ability of the organism, or the population, or the species to compete for an ecological niche in the biosphere (Chapter 29).

Garrod: One Mutant Gene–One Metabolic Block

At the time of the rediscovery of Mendel's work in 1900, Sir Archibald Garrod was studying several congenital metabolic diseases in humans. One of these was the inherited disease **alkaptonuria**, which is easily detected because of the blackening of the urine upon exposure to air. The substance responsible for this blackening is alkapton (or homogentisic acid), an intermediate in the degradation of the aromatic amino acids tyrosine and phenylalanine (Figure 15.3). Garrod believed that the presence of homogentisic acid in the urine was due to a block in the normal pathway of metabolism of this compound. Moreover, on the basis of the family pedigree studies, Garrod proposed that alkaptonuria was inherited as a single recessive gene. The results of Garrod's studies of alkaptonuria and a few other congenital diseases in humans, such as albinism, were presented in detail in his book *Inborn Errors of Metabolism*. Although the details of the biochemical pathway affected by the recessive mutations that cause alkaptonuria were not worked out until many years later, Garrod clearly understood the relationship between genes and metabolism. His concept might be best stated as **one mutant gene–one metabolic block.**

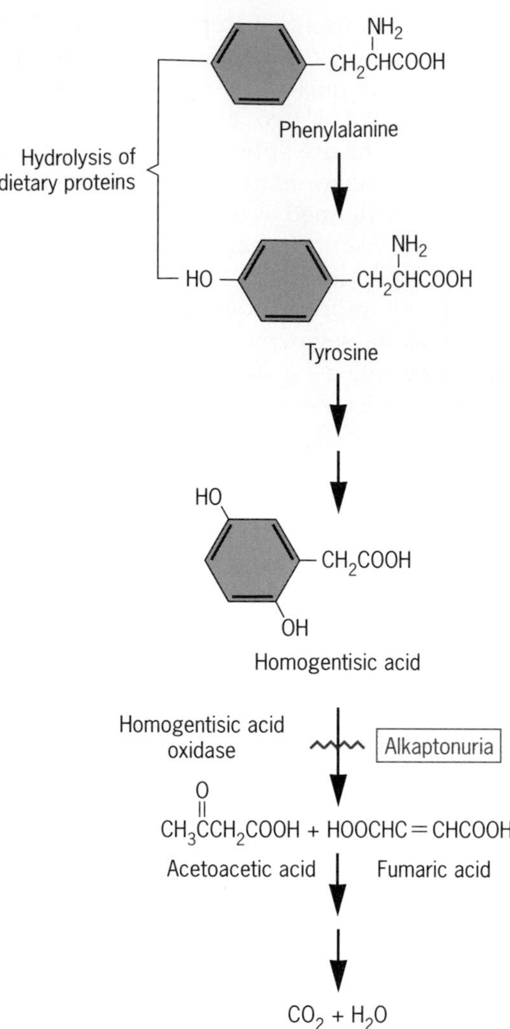

Figure 15.3 Alkaptonuria in humans results from a block in phenylalanine catabolism caused by a mutation in the gene encoding the enzyme homogentisic acid oxidase. When this enzyme is absent or inactive, its substrate, homogentisic acid, accumulates in tissues and in urine.

Early Evidence That Enzymes Are Controlled by Genes

In proposing that metabolic reactions were controlled by genes in 1902, Sir Archibald E. Garrod displayed great insight, because there was little direct evidence at the time to support his proposal. Some of the first evidence showing that genes control the enzyme-catalyzed reactions of metabolic pathways was obtained from studies of *Drosophila* eye color mutants. In 1935, George Beadle and Boris Ephrussi suggested that mutations in two genes controlling eye color resulted in blocks at two different steps in the biosynthesis of the brown eye pigment in *Drosophila*. Wild-type fruit flies have dark red eyes resulting from the presence of two eye pigments, one bright red and the other brown. The *vermilion* (v) and *cinnabar* (cn) mutant flies have bright red eyes owing to the absence of the brown pigment.

Like other insects, fruit flies undergo metamorphosis; fertilized eggs develop into larvae, then pupae, and finally adult flies (Chapter 2). The larval stages contain groups of embryonic cells called imaginal disks that will develop into the specific adult structures, such as eyes, wings, or legs, of the adult flies. Beadle and Ephrussi discovered that if they surgically transplanted an eye imaginal disk from one larva into the abdomen of another larva, the transplanted disk

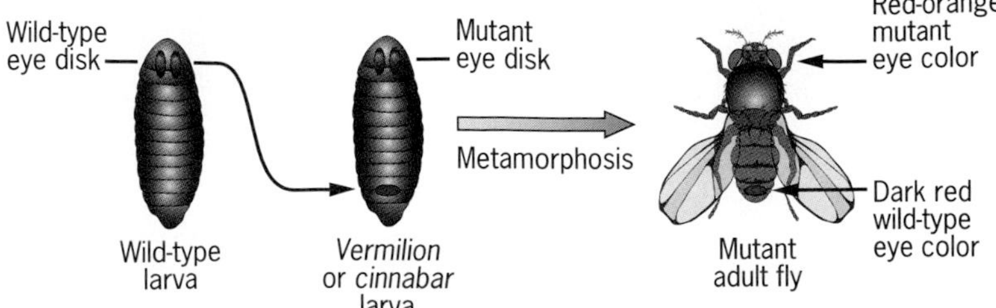

(*a*) Wild-type disks are transplanted into *vermilion* or *cinnabar* larvae.

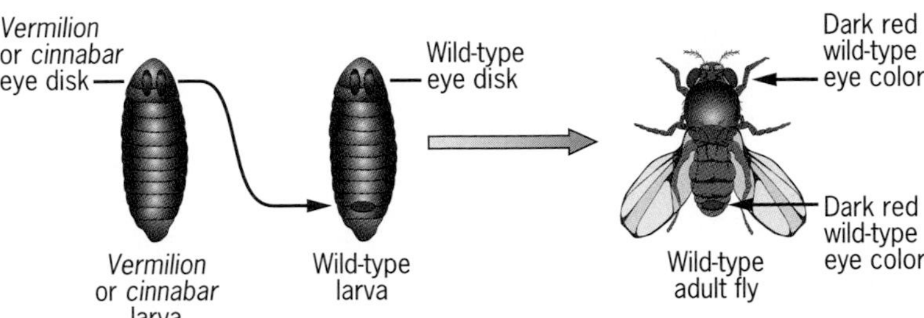

(*b*) *Vermilion* or *cinnabar* disks are transplanted into wild-type larvae.

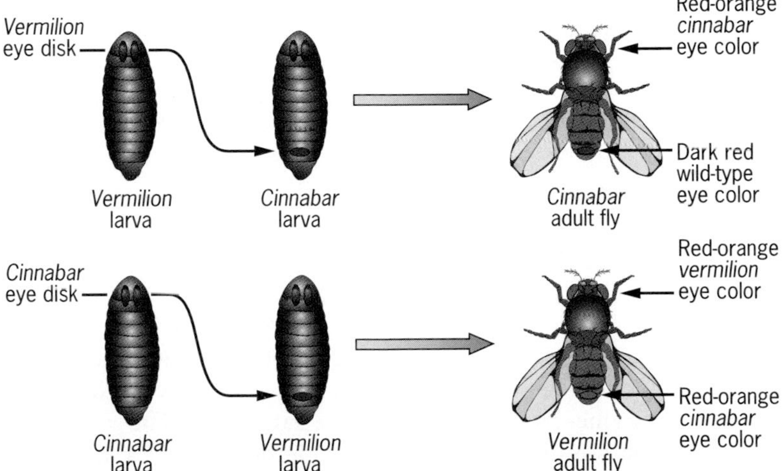

(*c*) *Vermilion* disks are transplanted into *cinnabar* larvae and *cinnabar* disks are transplanted into *vermilion* larvae.

Figure 15.4 Transplant experiments of *Drosophila* eye imaginal disks by Beadle and Ephrussi.

would develop into a third eye, albeit nonfunctional, in the adult fly (Figure 15.4). When they transplanted eye disks from wild-type larvae into *vermilion* or *cinnabar* larvae, the transplanted disks developed into dark red wild-type eyes (Figure 15.4*a*). In these two cases, the genotype of the transplanted eye disk controlled its own phenotype; the phenotype of the third eye was not influenced by the genotype of the larval host. However, this was not always the case.

When Beadle and Ephrussi transplanted eye disks from *vermilion* or *cinnabar* larvae into the abdomens of wild-type larvae, the disks developed into dark red wild-type eyes (Figure 15.4*b*). Beadle and Ephrussi reasoned that the wild-type larval hosts must have provided some diffusible substance to the transplanted mutant disks that allowed them to bypass the metabolic block and synthesize the brown eye pigment. In these two cases, development of the transplanted disks into adult eye structures was influenced by the genotype of the host organism.

Beadle and Ephrussi next transplanted eye disks from *vermilion* larvae into *cinnabar* hosts and eye disks from *cinnabar* larvae into *vermilion* hosts. The reciprocal transplants yielded quite different results! The *vermilion* disks transplanted into *cinnabar* larvae developed into dark red wild-type eyes, whereas the *cinnabar* disks transplanted into *vermilion* larvae developed into bright red *cinnabar* eyes (Figure 15.4*c*). To explain these results, Beadle and Ephrussi proposed that the *cinnabar* hosts supplied a diffusible substance

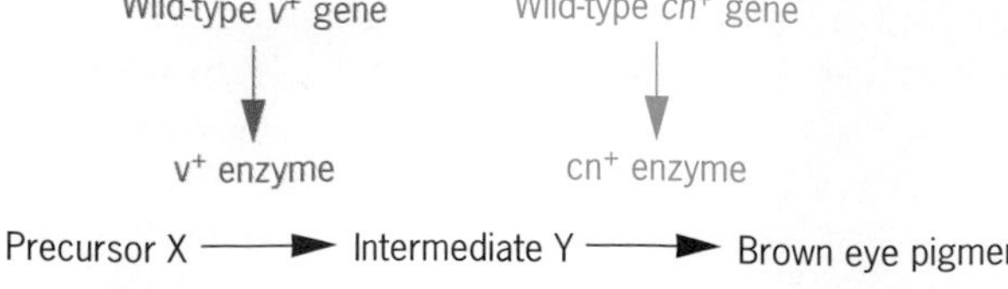

Figure 15.5 Beadle and Ephrussi's interpretation of the results of their reciprocal transplants of *vermilion* and *cinnabar* eye disks in *Drosophila*. They proposed that the v^+ and cn^+ genes control two different steps in the synthesis of the brown eye pigment and that the v^+ gene product acts before the cn^+ gene product.

that the implanted *vermilion* disks converted to the brown eye pigment. They further suggested that the *vermilion* hosts did not produce a metabolite that could be converted to the brown pigment by the *cinnabar* disks.

In fact, the results of Beadle and Ephrussi's reciprocal transplant experiments can be easily explained if the brown eye pigment is synthesized by a series of enzyme-catalyzed reactions and the v^+ and cn^+ genes control different steps in the biosynthetic pathway (Figure 15.5). Beadle and Ephrussi proposed that (1) a precursor molecule X is converted to a diffusible intermediate substance Y by the product of the v^+ gene, and (2) intermediate Y is converted to the brown eye pigment by the product of the cn^+ gene. If their proposal was correct, the *cinnabar* larval hosts, which were v^+, would provide diffusible substance Y to the transplanted *vermilion* disks. Because the *vermilion* disks were cn^+, they would convert substance Y to the brown eye pigment and would develop into dark red wild-type eyes. The *vermilion* larval hosts would not provide anything comparable to substance Y to the transplanted *cinnabar* disks. Since the *cinnabar* disks were v^+, they would convert precursor X to intermediate Y. However, in the absence of cn^+, intermediate Y could not be converted to the brown pigment. As a result, *cinnabar* disks transplanted into *vermilion* larvae developed into bright red mutant eyes.

During the 1940s, Beadle and Ephrussi's proposal was proven correct, albeit with the addition of a few steps to the pathway. The brown eye pigment, xanthommatin, was shown to be synthesized from the amino acid tryptophan (precursor X) by a sequence of enzyme-catalyzed reactions (Figure 15.6). The v^+ gene controls the synthesis of an enzyme called tryptophan pyrrolase, which converts tryptophan to *N*-formylkynurenine. Then, *N*-formylkynurenine is converted to kynurenine (intermediate Y) by the enzyme kynurenine formylase. The cn^+ gene controls the synthesis of the enzyme kynurenine hydroxylase, which converts kynurenine (substance Y) to 3-hydroxylkynurenine. Two additional enzyme-catalyzed reactions then convert 3-hydroxylkynurenine to the brown pigment xanthommatin.

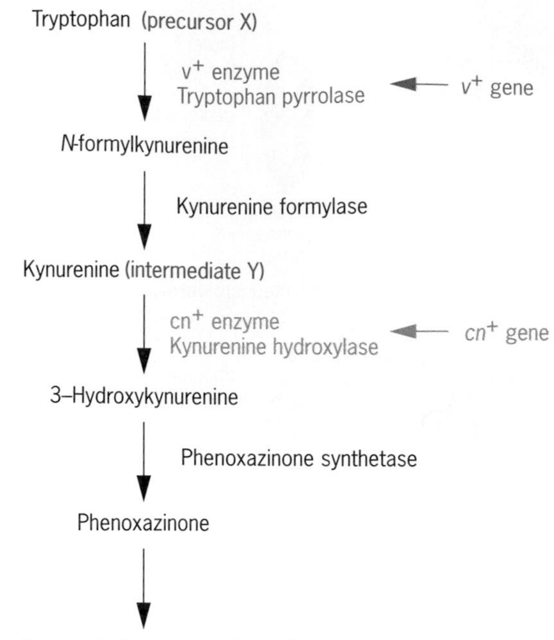

Figure 15.6 Biosynthesis of the brown eye pigment xanthommatin in *Drosophila*. The wild-type *vermilion* and *cinnabar* genes encode enzymes that catalyze two of the reactions in this pathway. Note the consistency of Beadle and Ephrussi's proposed sequence of reactions (Figure 15.5) with the actual biosynthetic pathway.

These early studies of *Drosophila* eye color mutants and the pathways of eye pigment biosynthesis provided direct evidence for the genetic control of enzyme-catalyzed metabolic pathways. The results of subsequent investigations demonstrated the genetic control of metabolism in all types of living organisms, from viruses to humans.

Beadle and Tatum: One Gene–One Enzyme

George Beadle's collaboration with Ephrussi in studies of the synthesis of the brown eye pigment in *Drosophila* led him to search for the ideal organism to use in extending this work. The pink bread mold *Neurospora crassa* can grow on medium containing only (1) inorganic salts, (2) a simple sugar, and (3) one vitamin, biotin. *Neurospora* growth medium containing only these components is called "minimal medium." George Beadle and Edward Tatum reasoned that *Neurospora* must be capable of synthesizing all the other essential metabolites, such as the purines, pyrimidines, amino acids, and other vitamins, *de novo*. Furthermore, they reasoned that the biosynthesis of these growth factors must be under genetic control. If so, mutations in genes whose products are involved in the biosynthesis of essential metabolites would be expected to produce mutant strains with additional growth-factor requirements.

Beadle and Tatum tested this prediction by irradiating asexual spores (conidia) of wild-type *Neurospora* with X rays or ultraviolet light, and screening the clones produced by the mutagenized spores for new growth-factor requirements (Figure 15.7). In order to select strains with a mutation in only one gene, they studied only mutant strains that yielded a 1:1 mutant to wild-type progeny ratio when crossed with wild-type. They identified mutants that grew on medium supplemented with all the amino acids, purines,

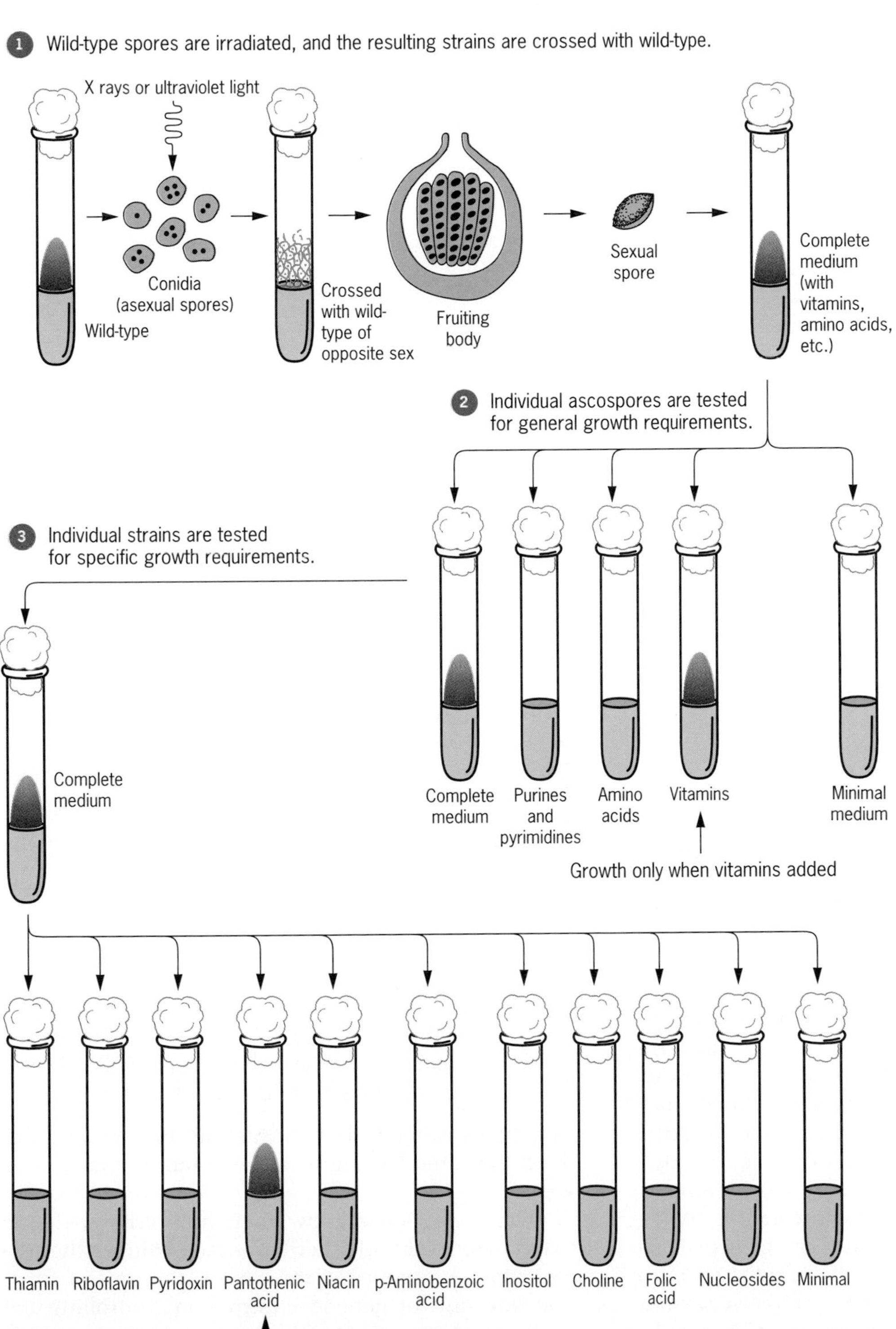

Figure 15.7 Diagram of Beadle and Tatum's experiment with *Neurospora* that led to the one gene–one enzyme hypothesis.

pyrimidines, and vitamins (called "complete medium"), but could not grow on minimal medium. They analyzed the ability of these mutants to grow on medium supplemented with just amino acids, or just vitamins, and so on (Figure 15.7, step 2). For example, Beadle and Tatum identified mutant strains that grew in the presence of vitamins but could not grow in medium supplemented with amino acids or other growth factors. They next investigated the ability of these vitamin-requiring strains to grow on media supplemented with each of the vitamins separately (Figure 15.7, step 3).

In this way, Beadle and Tatum demonstrated that each mutation resulted in a requirement for one growth factor. By correlating their genetic analyses with biochemical studies of the mutant strains, they demonstrated in several cases that one mutation resulted in the loss of one enzyme activity. This work, for which Beadle and Tatum received a Nobel Prize in 1958, was soon verified by similar studies of many other organisms in many laboratories. The **one gene–one enzyme** concept thus became a central tenet of molecular genetics.

Appropriately, in his Nobel Prize acceptance speech, Beadle stated:

> *In this long, roundabout way, we had discovered what Garrod had seen so clearly so many years before. By now we knew of his work and were aware that we had added little if anything new in principle. . . . Thus we were able to demonstrate that what Garrod had shown for a few genes and a few chemical reactions in man was true for many genes and many reactions in* Neurospora.

One Gene–One Polypeptide

Subsequent to the work of Beadle and Tatum, many enzymes and structural proteins were shown to be heteromultimeric, that is, to contain two or more different polypeptide chains, with each polypeptide encoded by a separate gene. For example, in *E. coli*, the enzyme tryptophan synthetase is a heterotetramer composed of two α polypeptides encoded by the *trpA* gene and two β polypeptides encoded by the *trpB* gene. Similarly, the hemoglobins, which transport oxygen from our lungs to all other tissues of our bodies, are tetrameric proteins that contain two α-globin chains and two β-globin chains, as well as four oxygen-binding heme groups (see Figure 13.3). In humans, the major form of adult hemoglobin contains two α-globin polypeptides encoded by the Hb^{A}_{α} gene on chromosome 16 and two β-globin polypeptides encoded by the Hb^{A}_{β} gene on chromosome 11. Other enzymes, for example, *E. coli* DNA polymerase III (Chapter 11) and RNA polymerase II (Chapter 12), contain many different polypeptide subunits, each encoded by a separate gene. Thus, the one gene–one enzyme concept was modified to **one gene–one polypeptide.**

Key Points: **The existence of a basic genetic element, the gene, that controlled a specific phenotypic trait was established by Mendel's work in 1866. Since the rediscovery of Mendel's results in 1900, the concept of the gene has evolved from the unit that can mutate to cause a specific block in metabolism, to the unit specifying one enzyme, to the sequence of nucleotide pairs in DNA encoding one polypeptide chain.**

EVOLUTION OF THE CONCEPT OF THE GENE: STRUCTURE

In the preceding section, we examined the evolution of the concept of the gene as the basic functional component of the genetic material. In this section, we examine the gene from a structural perspective. What is the structure of the gene? Do all genes have the same structure?

The Pre–1940 Beads-on-a-String Concept

Prior to 1940, the genes in a chromosome were considered analogous to beads on a string. Recombination was believed to occur only between the beads or genes, not within genes. The gene was believed to be indivisible. According to this beads-on-a-string concept, the gene was the basic unit of genetic information defined by three criteria: (1) function, (2) recombination, and (3) mutation. More specifically, the gene was

1. *The unit of function*, the unit of genetic material that controlled the inheritance of one "character" or one attribute of phenotype.
2. *The unit of structure*, operationally defined in two ways:
   a. *By recombination*: as the unit of genetic information not subdivisible by recombination.
   b. *By mutation*: as the smallest unit of genetic material capable of independent mutation.

Geneticists initially thought that all three criteria defined the same basic unit of inheritance, namely, the gene.

Geneticists now know that these criteria define two different units of inheritance. According to the current molecular concept, the gene is the unit of function, the unit of genetic information controlling the synthesis of one polypeptide chain or, in some cases, one RNA molecule. The unit of structure is simply the structural unit in DNA, the nucleotide pair. Because it does not make sense to call each nucleotide pair a gene, geneticists have focused on the original definition of the gene as the unit of function and have discarded the beads-on-a-string view that the gene is not subdivisible by recombination or mutation. This is clearly appropriate since the emphasis in Mendel's work was on the *Merkmal* (or gene, as it is now called) controlling one phenotypic characteristic.

Discovery of Recombination Within the Gene

In 1940, Clarence P. Oliver published the first evidence indicating that recombination could occur within a gene. Oliver was studying mutations at the *lozenge* locus on the X chromosome of *Drosophila melanogaster*. Two mutations, lz^s ("spectacle" eye) and lz^g ("glassy" eye), were thought to be alleles, that is, different forms of the same gene. The data available prior to 1940 indicated that they mapped at the same position on the X chromosome. They had similar effects on the phenotype of the eye, and heterozygous lz^s/lz^g females had lozenge rather than wild-type eyes. However, when lz^s/lz^g females were crossed with either lz^s or lz^g males and large numbers of progeny were examined, wild-type progeny occurred with a frequency of about 0.2 percent.

These rare wild-type progeny could be explained by reversion of either the lz^s or the lz^g mutation. But there were two strong arguments against the reversion explanation. (1) The frequency of reversion of lz^s or lz^g to wild-type in hemizygous lozenge males was much lower than 0.2 percent. (2) When the lz^s/lz^g heterozygotes carried genetic markers bracketing the *lozenge* locus, the rare progeny with wild-type eyes always carried an X chromosome with lz^+ that was flanked by recombinant outside markers. Moreover, the same combination of outside markers always occurred, as though the sites of lz^s and lz^g were fixed relative to each other and crossing over was occurring between them. Different sets of outside markers were used and yielded the same result. If the lz^s/lz^g heterozygous female carried X chromosomes of the type

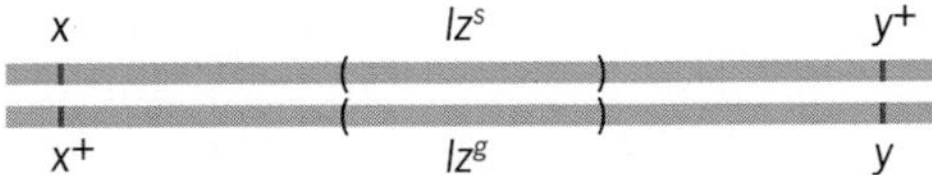

the rare progeny with wild-type eyes all (with one exception) contained an X chromosome with the following composition:

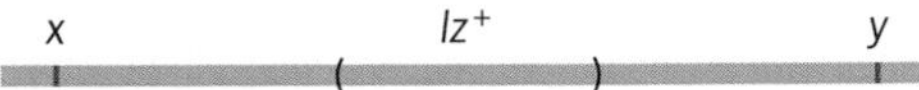

Among progeny of these matings, the reciprocal combination of outside markers (x^+-y^+) never appeared in combination with lz^+. This result strongly suggested that the lz^s and lz^g mutations were located at distinct sites in the *lozenge* locus, and that the lz^+ chromosome was produced by crossing over between the two sites, as shown in the following diagram.

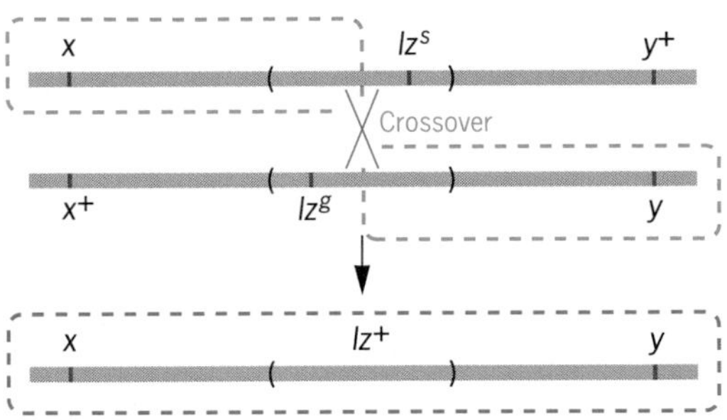

Definitive evidence for the involvement of recombination required the recovery and identification of the lz^s-lz^g double mutant with the reciprocal combination of outside markers—that is,

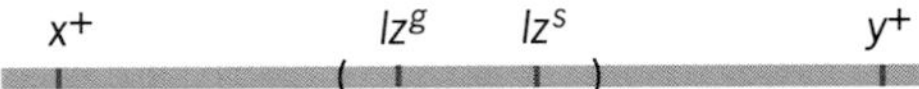

Oliver was not able to identify this double mutant because of the inability to distinguish it from the parental single-mutant phenotypes. The identification of both products, the wild-type and double mutant chromosomes, produced by crossing over within the *lozenge* gene, was first accomplished by Melvin M. Green, one of Oliver's students.

The results of these pioneering studies first indicated that the gene was more complex than a bead on a string. They showed that the gene was divisible, containing sites that were separable by crossing over. Oliver's and Green's results were the first step toward the present concept of the gene as a long sequence of nucleotide pairs, capable of mutating and recombining at many different sites along its length.

Recombination Between Adjacent Nucleotide Pairs

The results obtained by Oliver and Green in their studies of *Drosophila* genes all indicated that mutable sites that are separable by recombination can exist within a single gene. Seymour Benzer extended this picture of the gene by demonstrating the existence of 199 distinct sites of mutation that were separable by recombination within the *rIIA* gene of bacteriophage T4 (Chapter 16). Benzer's picture of the gene as a sequence of nucleotide pairs capable of mutating at many distinct sites was soon verified by the results of many researchers investigating gene structure in several different organisms, both prokaryotes and eukaryotes. Given this information about the structure of genes and the known structure of DNA, it followed that the smallest unit of genetic material capable of mutation might be the single nucleotide pair and that recombination might occur between adjacent nucleotide pairs, whether between or within genes. Recombination between adjacent nucleotide pairs of a gene was first demonstrated by Charles Yanofsky in his studies of the *trpA* gene encoding the α polypeptide of tryptophan synthetase in *E. coli*. This enzyme, a tetramer containing two α polypeptides and two β polypeptides, catalyzes the final step in the biosynthesis of the amino acid tryptophan.

Yanofsky and his colleagues isolated and characterized a large number of tryptophan auxotrophs with mutations in the *trpA* gene. The wild-type *trpA* gene encodes an α polypeptide that is 268 amino acids long. Yanofsky and associates used the laborious techniques of protein sequencing to determine the complete amino acid sequence of the wild-type α polypeptide. They also determined the amino acid substitutions that had occurred in several mutant forms of the tryptophan synthetase α polypeptide. They mapped the mutations within the *trpA* gene by two- and three-factor crosses, and compared the map positions with the locations of the amino acid substitutions in the mutant polypeptides.

Yanofsky's correlated genetic and biochemical data for the *trpA* gene and the tryptophan synthetase α polypeptide showed that recombination can occur between mutations that alter the same amino acid. Mutations *trpA23* and *trpA46* both result in the substitution of another amino acid (arginine in the case of *A23*, glutamic acid in the case of *A46*) for the glycine present at position 211 of the wild-type tryptophan synthetase α polypeptide. However, these two mutations occur at different mutable sites; that is, the *A23* and *A46* sites are separable by recombination. Yanofsky and colleagues determined the amino acids present at position 211 of the α polypeptide in other mutants as well as in revertants and partial revertants of the *trpA23* and *trpA46* mutants. By using this information and the known codon assignments, they were able to determine which of the glycine, arginine, and glutamic acid codons were present in the *trpA* mRNA at the position encoding amino acid residue 211 of the α polypeptides present in trp^+, *trpA23*, and *trpA46* cells, respectively (Figure 15.8).

Once the specific codons in mRNA are known, the corresponding base-pair sequences in the structural gene from which the mRNA is transcribed are also known. One strand of DNA will be complementary to the mRNA, and the second strand of DNA will be complementary to the first strand. Therefore, Yanofsky's data demonstrated that the mutational events that produced the *trpA23* and *trpA46* alleles were G:C to A:T transitions at adjacent nucleotide pairs. The trp^+ cells produced by recombination between chromosomes carrying mutations *A23* and *A46* demonstrated that recombination had occurred between adjacent nucleotide pairs in the *trpA* gene, as shown in Figure 15.9. Yanofsky's results showed that the unit of genetic material not divisible by recombination is the single nucleotide pair.

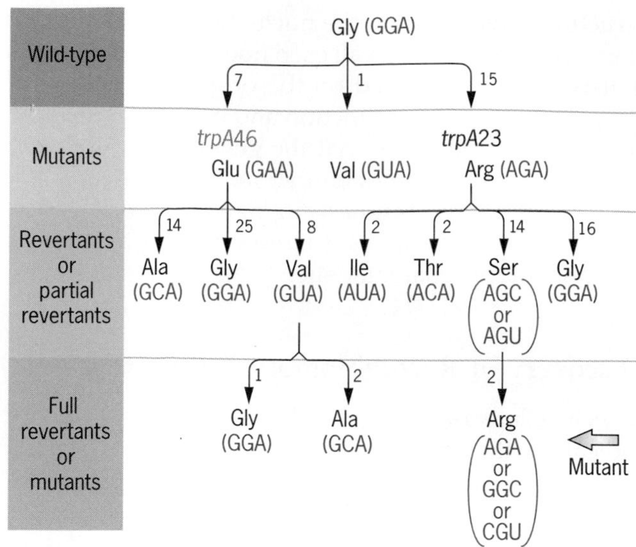

Figure 15.8 Pedigree of amino acid residue 211 (from the NH_2 terminus) of the α polypeptide of tryptophan synthetase of *E. coli*. Amino acid 211 is altered in *trpA23* and *trpA46* mutants (see Figure 15.11). The triplet codons shown in parentheses are the only codons specific to the indicated amino acids that will permit all of the observed amino acid replacements to occur by single base-pair substitutions. The number beside each arrow indicates the number of times that particular substitution was observed. These results indicate that the arginine and glutamic acid codons encoding amino acid 211 of the α polypeptides in *trpA23* and *trpA46* mutant cells are AGA and GAA, respectively. Thus, these two mutations alter adjacent nucleotide pairs in the *trpA* gene.

The pre–1940 concept of the gene as (1) the smallest unit of genetic material that could undergo mutation and (2) the unit of genetic information that could not be subdivided by recombination was clearly wrong. Each nucleotide pair of a gene can change or mutate independently, and recombination can occur between adjacent nucleotide pairs. The results of Oliver, Lewis, Green, Benzer, Yanofsky, and many others have compelled geneticists to focus on the gene as the unit of function, the sequence of nucleotide pairs controlling the synthesis and structure of one polypeptide or one RNA molecule.

Colinearity Between the Coding Sequence of a Gene and Its Polypeptide Product

The genetic information is stored in linear sequences of nucleotide pairs in DNA (or nucleotides in RNA, in

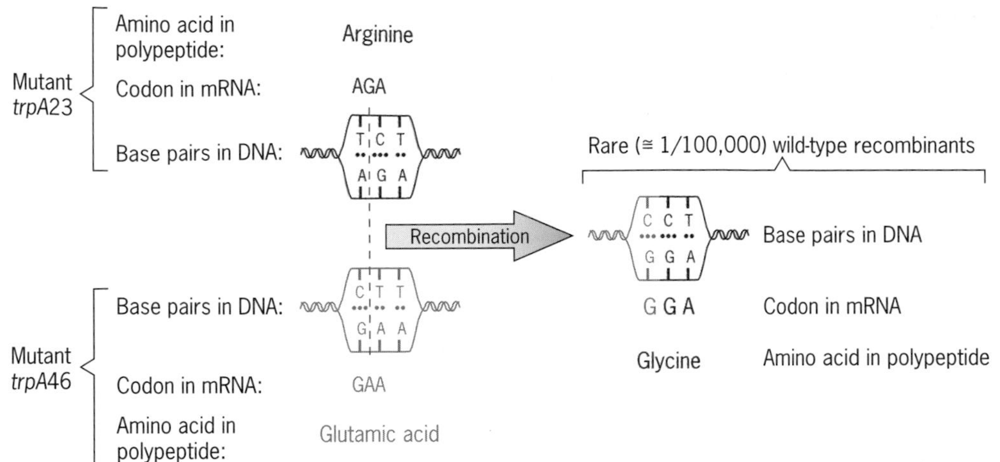

Figure 15.9 Recombination between mutations at adjacent nucleotide pairs in the *trpA* gene of *E. coli*. Mutations *A23* and *A46* both result in an amino acid substitution at position 211 of the tryptophan synthetase α polypeptide. Wild-type *E. coli* has a glycine residue at this position of the α polypeptide. *A23* causes a glycine to arginine substitution; *A46* causes a glycine to glutamic acid substitution (see Figure 15.8).

some cases). Transcription and translation convert this genetic information into linear sequences of amino acids in polypeptides, which function as the key intermediaries in the genetic control of the phenotype.

It is now known that the nucleotide-pair sequences of the coding regions of the structural genes and the amino acid sequences of the polypeptides that they encode are **colinear:** the first three base pairs of the coding sequence of a gene specify the first amino acid of the polypeptide, the next three base pairs (four to six) specify the second amino acid, and so on, in a colinear fashion (Figure 15.10*a*). It is also known that the coding regions of most of the genes in higher eukaryotes are interrupted by noncoding introns (Chapter 12). However, the presence of introns in genes does not invalidate the concept of colinearity. The presence of

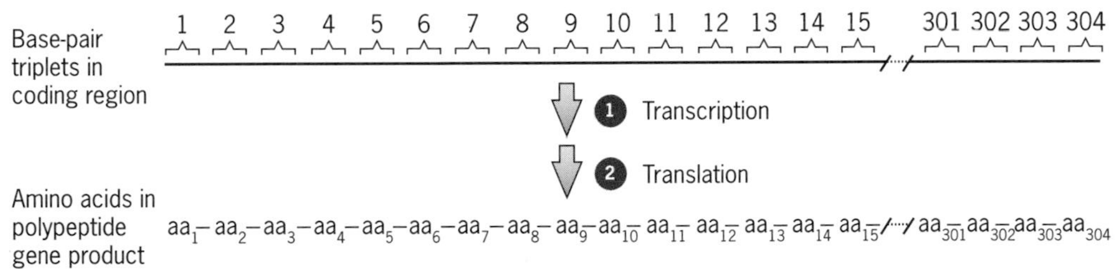

(*a*) Coding region of typical uninterrupted prokaryotic gene.

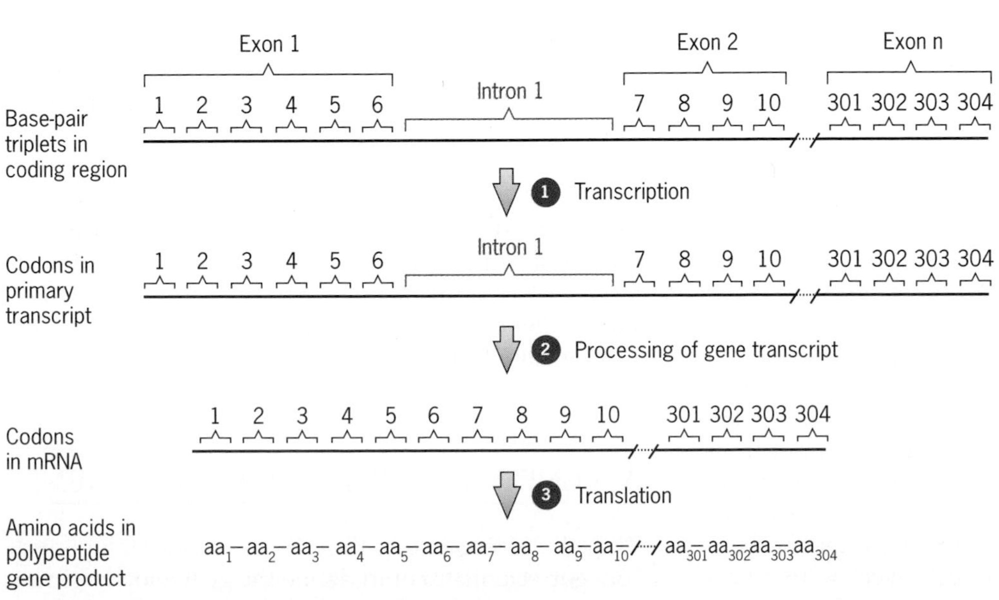

(*b*) Coding region of typical interrupted eukaryotic gene.

Figure 15.10 Colinearity between the coding regions of genes and their polypeptide products.

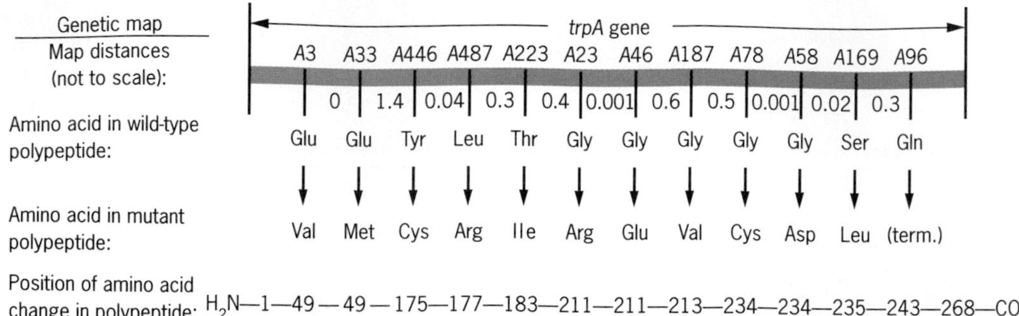

Figure 15.11 Colinearity between the *E. coli trpA* gene and its polypeptide product, the α polypeptide of tryptophan synthetase. The map positions of mutations in the *trpA* gene are shown at the top, and the locations of the amino acid substitutions produced by these mutations are shown below the map.

introns in genes simply means that there is no direct correlation in physical distances between the positions of base-pair coding triplets in a gene and the positions of amino acids in the polypeptide specified by that gene (Figure 15.10*b*).

The first strong evidence for colinearity between a gene and its polypeptide product resulted from studies by Charles Yanofsky and colleagues on the *E. coli* gene that encodes the α subunit of the enzyme tryptophan synthetase. As mentioned earlier, this enzyme contains two α polypeptides encoded by the *trpA* gene and two β polypeptides encoded by the *trpB* gene. Yanofsky and coworkers performed a detailed genetic analysis of mutations in the *trpA* gene and correlated the genetic data with biochemical data on the sequences of the wild-type and mutant tryptophan synthetase α polypeptides. They demonstrated that there was a direct correlation between the map positions of mutations in the *trpA* gene and the positions of the resultant amino acid substitutions in the tryptophan synthetase α polypeptide (Figure 15.11).

About the same time, Sydney Brenner and associates demonstrated a similar colinearity between the positions of mutations in the gene of bacteriophage T4 that encodes the major structural protein of the phage head and the positions in the polypeptide affected by these mutations. Brenner and colleagues studied *amber* (UAG chain-termination) mutations and demonstrated a direct correlation between the length of the polypeptide fragment produced and the position of the mutation within the gene.

In the yeast *S. cerevisiae*, Fred Sherman and colleagues demonstrated an uninterrupted colinear relationship between the map positions of mutations in the *CYCI* gene and the amino acid substitutions in the mutant forms of iso-1-cytochrome c, the polypeptide specified by the *CYC1* gene. More recently, colinearity has been documented for many interrupted genes in eukaryotes. The only difference observed with interrupted genes is that the linear sequence of nucleotide pairs encoding a colinear polypeptide is not one continuous sequence of nucleotide pairs. Instead, noncoding sequences (introns) intervene between coding sequences (exons).

Definitive evidence for colinearity has been provided by direct comparisons of the nucleotide sequences of genes and the amino acid sequences of their polypeptide products. One of the first cases where the amino acid sequence of a polypeptide and the nucleotide sequence of the gene encoding it were both determined experimentally involved the coat protein of bacteriophage MS2 and the gene that encodes it. This small virus has an RNA genome that encodes only four proteins, one being the coat protein that encapsulates the RNA. In 1972, Walter Fiers and coworkers used the genetic code to compare the nucleotide sequence of the coat protein gene with the amino acid sequence of the coat polypeptide and found that the sequences exhibited perfect colinearity (Figure 15.12). Since then, similar results have established colinearity between many genes and their protein products in organisms ranging from viruses to humans. Thus, colinearity between the coding regions of genes and the amino acid sequences of their polypeptide products is a universal or nearly universal feature of gene-protein relationships.

Key Points: **The concept of the gene has evolved from a bead on a string, not divisible by recombination or mutation, to a sequence of nucleotide pairs in DNA encoding one polypeptide chain. The unit of genetic material not divisible by recombination or mutation is the single nucleotide pair.**

A GENETIC DEFINITION OF THE GENE

With the emergence of the one gene–one polypeptide concept, scientists could define the gene biochemically, but they had no genetic tool to use in determining

Coat protein initiation codon

– AGC – AUG–

GCU – UCU – AAC – UUU – ACU – CAG – UUC – GUU – CUC – GUC – GAC – AAU – GGC – GGA – ACU – GGC – GAC – GUG – ACU – GUC – GCC – CCA – AGC – AAC – UUC –
Ala Ser Asn Phe Thr Gln Phe Val Leu Val Asp Asn Gly Gly Thr Gly Asp Val Thr Val Ala Pro Ser Asn Phe
1 5 10 15 20 25

GCU – AAC – GGG – GUC – GCU – GAA – UGG – AUC – AGC – UCU – AAC – UCG – CGU – UCA – CAG – GCU – UAC – AAA – GUA – ACC – UGU – AGC – GUU – CGU – CAG –
Ala Asn Gly Val Ala Glu Trp Ile Ser Ser Asn Ser Arg Ser Gln Ala Tyr Lys Val Thr Cys Ser Val Arg Gln
30 35 40 45 50

AGC – UCU – GCG – CAG – AAU – CGC – AAA – UAC – ACC – AUC – AAA – GUC – GAG – GUG – CCU – AAA – GUG – GCA – ACC – CAG – ACU – GUU – GGU – GGU – GUA –
Ser Ser Ala Gln Asn Arg Lys Try Thr Ile Lys Val Glu Val Pro Lys Val Ala Thr Gln Thr Val Gly Gly Val
55 60 65 70 75

GAG – CUU – CCU – GUA – GCC – GCA – UGG – CGU – UCG – UAC – UUA – AAU – AUG – GAA – CUA – ACC – AUU – CCA – AUU – UUC – GCU – ACG – AAU – UCC – GAC –
Glu Leu Pro Val Ala Ala Trp Arg Ser Tyr Leu Asn Met Glu Leu Thr Ile Pro Ile Phe Ala Thr Asn Ser Asp
80 85 90 95 100

UGC – GAG – CUU – AUU – GUU – AAG – GCA – AUG – CAA – GGU – CUC – CUA – AAA – GUA – GGA – AAC – CCG – AUU – CCC – UCA – GCA – AUC – GCA – GCA – AAC –
Cys Glu Leu Ile Val Lys Ala Met Gln Gly Leu Leu Lys Asp Gly Asn Pro Ile Pro Ser Ala Ile Ala Ala Asn
105 110 115 120 125

UCC – GGC – AUC – UAC – UAA – UAG – ACG – CCG – GCC – AUU – CAA – ACA – UGA – GGA –
Ser Gly Ile Tyr
129

Tandem *ochre* and *amber* termination codons

Opal termination codon

Figure 15.12 Colinearity between the nucleotide sequence of the bacteriophage MS2 coat protein gene and the amino acid sequence of the coat polypeptide that it encodes. Note that the amino acid sequence of this protein is precisely that predicted from the nucleotide sequence based on the genetic code. In addition, note that all three termination codons are present between the coat gene and the gene downstream from it on the MS2 chromosome.

whether two mutations were in the same or different genes. This deficiency was resolved when Edward Lewis developed the complementation test for functional allelism in 1942. Lewis was studying the *Star-asteroid* locus (named after two mutations that result in small, rough-surfaced eyes) in *Drosophila* and observed that flies of genotype $S\ ast^+/S^+\ ast$ had a more extreme mutant eye phenotype than flies of genotype $S\ ast/S^+\ ast^+$. His results were complicated by the partial dominance of the *Star* (*S*) mutation. Lewis subsequently performed similar experiments with two eye-color mutations, *white* (*w*) and *apricot* (*apr*), and obtained results that were easier to interpret.

The Complementation Test as an Operational Definition of the Gene

Fruit flies that are homozygous for the X-linked mutations *apr* (now w^a) and *w* have apricot-colored eyes and white eyes, respectively, in contrast to the red eyes of wild-type *Drosophila*.

Lewis reported that heterozygous *apr*/*w* females had light apricot-colored eyes and produced rare red-eyed progeny carrying recombinant apr^+w^+ chromosomes. In addition, Lewis was able to identify progeny flies that carried X chromosomes with the reciprocal *apr w* recombinant genotype. The observed frequency of recombination between the *apr* and *w* mutations was 0.03 percent. Clearly, the *apr* and *w* mutations were separable by recombination, and *apr* and *w* were not alleles according to the pre–1940 concept of the gene. The two mutations appeared to be in the same unit of function but in two different units of structure. When Lewis produced flies of genotype $apr\ w/apr^+\ w^+$, they had red eyes just like those of wild-type flies. When he constructed flies of genotype $apr\ w^+/apr^+\ w$, they had light apricot-colored eyes. Both genotypes contain the same mutant and wild-type genetic information, but in different arrangements. The presence of different phenotypes in organisms that contain the same genetic markers, but with the markers present in different arrangements, is called a **position effect,** and the type of position effect observed by Lewis is referred to as a ***cis-trans*** **position effect.**

Before we analyze Lewis's results in greater detail, we need to define some terms. A double heterozygote, which carries two mutations and their wild-type alleles, that is, m_1 and m_1^+ plus m_2 and m_2^+, can exist in either of two arrangements (Figure 15.13). When the

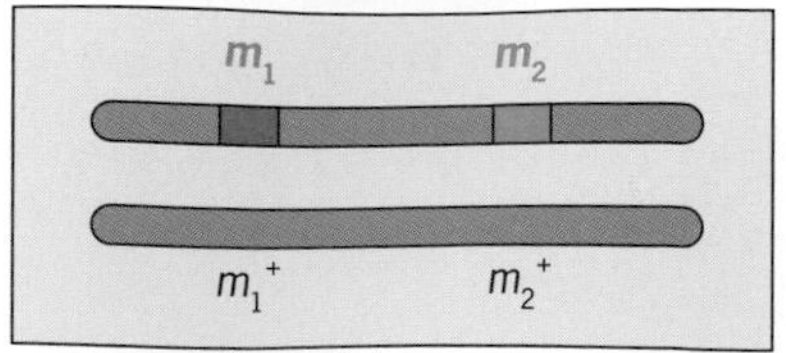

(a) *cis* heterozygote.

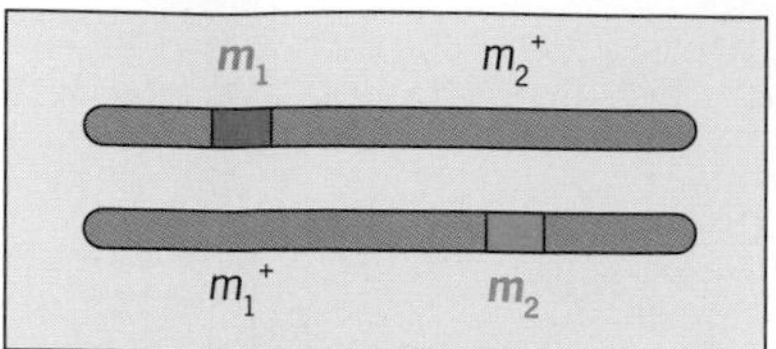

(b) *trans* heterozygote.

Figure 15.13 The arrangement of genetic markers in *cis* and *trans* heterozygotes.

two mutations are on the same chromosome, the arrangement is called the **coupling** or ***cis* configuration**; a heterozygote with this genotype is called a ***cis* heterozygote** (Figure 15.13*a*). When the two mutations are on different chromosomes, the arrangement is called the **repulsion** or ***trans* configuration.** An organism with this genotype is a ***trans* heterozygote** (Figure 15.13*b*).

Recall that *apr w*/*apr*$^+$ *w*$^+$ *cis* heterozygotes have wild-type red eyes, whereas *apr w*$^+$/*apr*$^+$ *w trans* heterozygotes have light apricot eyes. This is precisely the result that would be expected if *apr* and *w* are mutations at different sites in the same gene, the unit of genetic information encoding a single polypeptide (Figure 15.14). Thus, *apr* and *w* are considered alleles of the same gene, and *apr* is now designated *w*^{a} in recognition of this relationship. If *apr* and *w* had been mutations in two different units of function, two different genes, both the *cis* and the *trans* heterozygotes should have expressed the wild-type phenotype, namely, red eyes. In the *trans* heterozygote, the *apr*$^+$ gene product would be produced by the *apr*$^+$ gene on the chromosome carrying the *w* mutation, and the *w*$^+$ gene product would be specified by the *w*$^+$ gene on the chromosome harboring the *apr* mutation.

Lewis's discovery of the *cis-trans* position effects with *Star-asteroid* and *w*^{a}-*w* led to the development of the **complementation test** or ***trans* test** for functional allelism. The complementation test allows geneticists to determine whether mutations that produce the same or similar phenotypes are in the same gene or in different genes. They must test mutations pairwise by determining the phenotypes of *trans* heterozygotes. That is, they must construct *trans* heterozygotes with each pair of mutations to be analyzed and determine

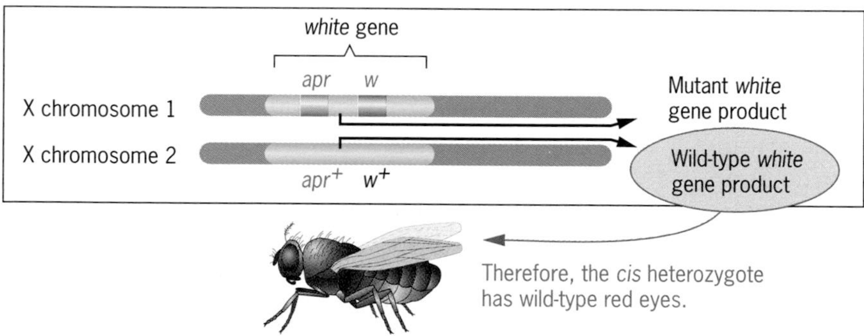

(a) *cis* heterozygote.

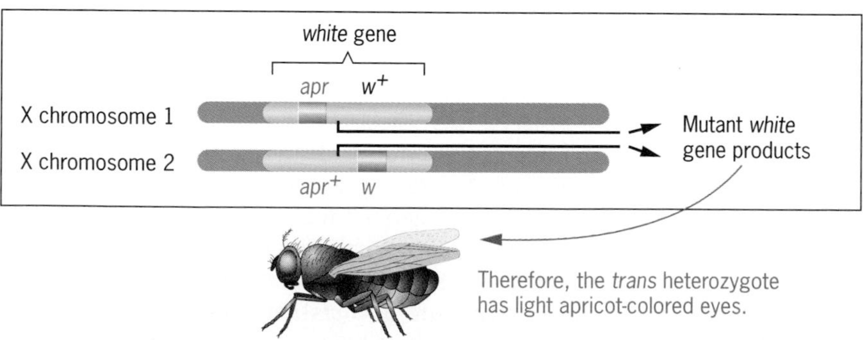

(b) *trans* heterozygote.

Figure 15.14 The *cis-trans* position effect observed by Edward Lewis with the *apr* and *w* mutations of *Drosophila*.

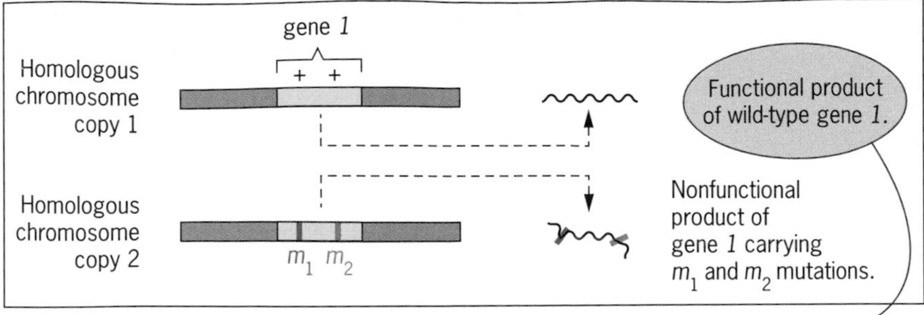

Therefore, the *cis* heterozygote will have the wild-type phenotype.

(a) *cis* heterozygote: mutations in one gene.

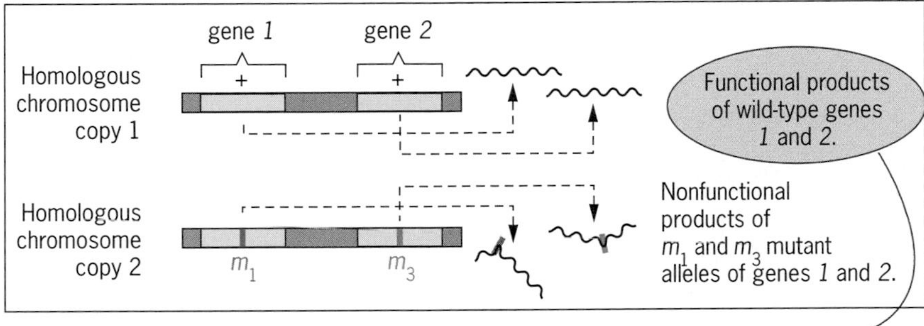

Therefore, the *cis* heterozygote will have the wild-type phenotype.

(b) *cis* heterozygote: mutations in two different genes.

Figure 15.15 The *cis* test. The *cis* heterozygote should have the wild-type phenotype whether the mutations are in the same gene (*a*) or in two different genes (*b*).

whether these heterozygotes have mutant or wild-type phenotypes.

Ideally, the complementation or *trans* test should be done in conjunction with the ***cis* test**—a control that is often omitted. *Cis* tests are performed by constructing *cis* heterozygotes for each pair of mutations to be analyzed and determining whether they have mutant or wild-type phenotypes. Together, the complementation or *trans* test and the *cis* test are referred to as the ***cis-trans* test.** Each *cis* heterozygote, which contains one wild-type chromosome, should have the wild-type phenotype whether the mutations are in the same gene or in two different genes (Figure 15.15). Indeed, the *cis* heterozygote must have the wild-type phenotype for the results of the *trans* test to be valid. If the *cis* heterozygote has the mutant phenotype, the *trans* test cannot be used to determine whether the two mutations are in the same gene. Thus, the *trans* test cannot be used with dominant mutations. Because *cis* heterozygotes contain one chromosome with wild-type copies of all relevant genes, these genes should specify functional products and a wild-type phenotype.

Whether two mutations are in the same gene or two different genes is determined by the results of the complementation or *trans* test. With diploid organisms, the *trans* heterozygote is produced simply by crossing organisms that are homozygous for each of the mutations of interest. With viruses, *trans* heterozygotes are produced by simultaneously infecting host cells with two different mutants. Regardless of how the two mutations are placed in a common protoplasm in the *trans* configuration, the results of the *trans* or complementation test provide the same information.

1. If the *trans* heterozygote has the mutant phenotype (the phenotype of organisms or cells homozygous for either one of the two mutations), then the two mutations are in the same unit of function, the same gene (Figure 15.16*a*).
2. If the *trans* heterozygote has the wild-type phenotype, then the two mutations are in two different units of function, two different genes (Figure 15.16*b*).

When the two mutations present in a *trans* heterozygote are both in the same gene, as shown for mutations m_1 and m_2 in Figure 15.16*a*, both chromosomes will carry defective copies of that gene. As a result, the *trans* heterozygote will contain only nonfunctional products of the gene involved and will have a mutant phenotype.

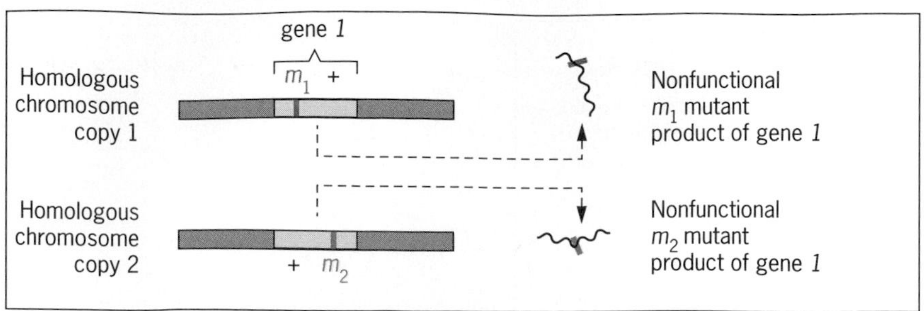

No functional gene 1 product is synthesized in the *trans* heterozygote; therefore, it will have a mutant phenotype.

(*a*) *trans* heterozygote: mutations in one gene.

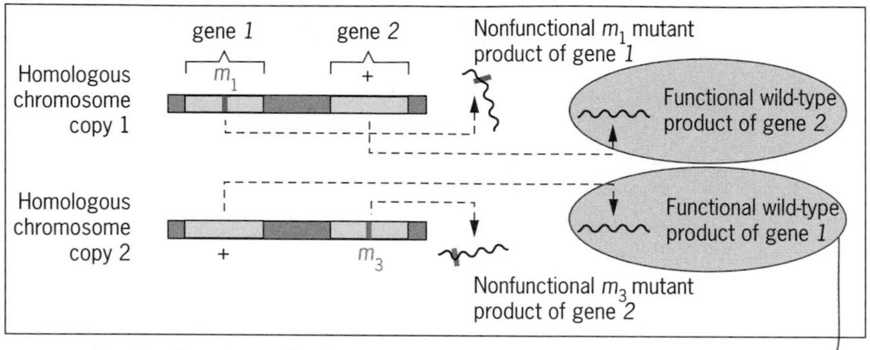

Functional products of both genes are synthesized in the *trans* heterozygote; therefore, it will have the wild-type phenotype.

(*b*) *trans* heterozygote: mutations in two different genes.

Figure 15.16 The *trans* test. The *trans* heterozygote should have (*a*) the mutant phenotype if the two mutations are in the same gene, and (*b*) the wild-type phenotype if the mutations are in two different genes.

When a *trans* heterozygote has the wild-type phenotype, the two mutations are said to exhibit complementation or to complement each other and are located in different genes. In the example illustrated in Figure 15.16*b*, the chromosome carrying mutation m_1 in gene *1* has a wild-type copy of gene *2*, which specifies functional gene *2* product, and the chromosome carrying mutation m_3 in gene *2* has a wild-type copy of gene *1*, which encodes functional gene *1* product. Thus, the *trans* heterozygote shown contains functional products of both genes and has a wild-type phenotype.

Only the complementation test, or the *trans* part of the *cis-trans* test, is included in most genetic analyses. Constructing a chromosome that carries both mutations for the *cis* test is often difficult, especially with eukaryotes. Moreover, the *cis* heterozygotes almost always have wild-type phenotypes if the mutations being analyzed are recessive. Thus, in most instances, the results of complementation or *trans* tests can be interpreted correctly without carrying out the laborious *cis* tests.

Seymour Benzer introduced the term **cistron** to refer to the unit of function operationally defined by the *cis-trans* test. However, today, most geneticists consider the terms *gene* and *cistron* to be synonyms. Thus, we will use gene, rather than cistron, throughout this text.

The gene is operationally defined as the unit of function by the complementation or *trans* test, which is used to determine whether mutations are in the same gene or different genes. The complementation test is one of the three basic tools of genetics. The other two genetic tools are recombination (Chapter 8) and mutation (Chapter 14).

The information provided by complementation tests is totally distinct from that obtained from recombination analyses.

The results of complementation tests indicate whether mutations are allelic, whereas the results of recombination analyses indicate whether mutations are linked and, if so, provide estimates of how far apart they are on a chromosome. Nevertheless, students sometimes confuse complementation and recombination. Thus, we will contrast complementation and recombination by illustrating both phenomena with the same three mutations in bacteriophage T4. Phage T4 is very similar to phage T2, which we discussed in Chapter 11. Because of the simple structure of the virus and the direct relationship between specific gene products and phenotypes, the phage T4 system provides an excellent visualization of the difference between complementation and recombination.

We discussed the *amber* mutations of bacteriophage T4 in Chapter 13 (see Figure 13.23) and the pathway of morphogenesis of phage T4 in Chapter 14 (see Figure 14.11) *Amber* mutations produce translation–termination triplets within the coding regions of genes. As a result, the products of the mutant genes are truncated polypeptides, which are almost always totally nonfunctional. Therefore, complementation tests performed with *amber* mutations are usually unambiguous. When *amber* mutations occur in essential genes, the mutant phenotype is lethality—that is, no progeny are produced when a restrictive host cell is infected; and the wild-type phenotype is a normal yield (about 300 phage per cell) of progeny phage in each infected restrictive host cell. In Chapter 14, we called such mutations conditional lethals and emphasized their utility in genetic analysis. With conditional lethals, the distinction between the mutant and wild-type phenotypes is maximal: lethality versus normal growth.

Two of the three *amber* mutations that we will consider (*am*B17 and *am*H32) are in gene *23,* which encodes the major structural protein of the phage head; the other mutation (*am*E18) is in gene *18,* which specifies the major structural protein of the phage tail. We can see from Figure 15.17 why complementation occurs between mutations *am*B17 (head gene) and *am*E18 (tail gene) and why complementation does not occur between mutations *am*B17 and *am*H32 (both in head gene). Complementation is the result of the functionality of the gene *products* specified by chromosomes carrying two different mutations when they are present in a common protoplasm. Complementation does not depend on recombination of the two chromosomes or involve any direct interaction between the chromosomes. *Complementation, or the lack of it, is assessed by the phenotype (wild-type or mutant) of each* trans *heterozygote.*

Figure 15.18 illustrates the occurrence of recombination between the *amber* mutations used to illustrate complementation in Figure 15.17. Although we haven't discussed viral genetics in detail (Chapter 16), we have covered the essential concepts of recombination (Chapter 8). Recombination of phage genes occurs by a process analogous to crossing over in eukaryotes, with linkage distances measured in map units, just as in eukaryotes. Recombination frequencies are measured by infecting permissive host cells with two mutants so that the mutant chromosomes can replicate and participate in crossing over. Then, the progeny are screened for wild-type recombinants by plating them on lawns of restrictive host cells (*E. coli* cells in which only the wild-type phage can grow). In the example shown in Figure 15.18, recombination is observed in both crosses: (1) *am*B17 × *am*E18, mutations in two different genes, and (2) *am*B17 × *am*H32, mutations in the same gene. The only difference is that more recombinants are produced in cross 1, which involves two *amber* mutations that are relatively far apart on the phage T4 chromosome, than in cross 2, which involves two mutations located near one another in the same gene. *Recombination involves direct interactions between the chromosomes carrying the mutations, the actual breakage of chromosomes, and reunion of parts to produce wild-type and double-mutant chromosomes.*

Complementation should occur in every *trans* heterozygote containing mutations in two different genes. Recombination is detected by examining the genotypes of the progeny of heterozygotes, not the phenotypes of the *trans* heterozygotes themselves. Moreover, a *trans* heterozygote will never produce more than 25 percent gametes (or progeny for haploids) with wild-type chromosomes. If the mutations are closely linked, like the *amber* mutations shown in Figure 15.18*b*, the frequency of wild-type recombinant chromosomes will be much lower than 25 percent.

Structural allelism is the occurrence of two or more different mutations at the same site and is determined by the recombination test. Two mutations that do not recombine are structurally allelic; the mutations either occur at the same site or overlap a common site. Functional allelism is determined by the complementation test as just described; two mutations that do not complement are in the same unit of function, the same gene. Mutations that are both structurally and functionally allelic are called **homoalleles;** they do not complement or recombine with each other. Mutant homoalleles have defects at the same site or overlap a common site in the same gene. Mutations that are functionally allelic, but structurally nonallelic, are called **heteroalleles;** they recombine with each other but do not complement one another. Mutant heteroalleles occur at different sites but within the same gene.

Intragenic Complementation

The results of complementation tests are usually unambiguous when mutations that result in the synthesis of no gene product, partial gene products, or totally defective gene products are used—for example, deletions of segments of genes, frameshift mutations, or polypeptide chain-terminating mutations. Of course, the mutations must be recessive. When mutations causing amino acid substitutions are used, the results of complementation tests are sometimes ambiguous

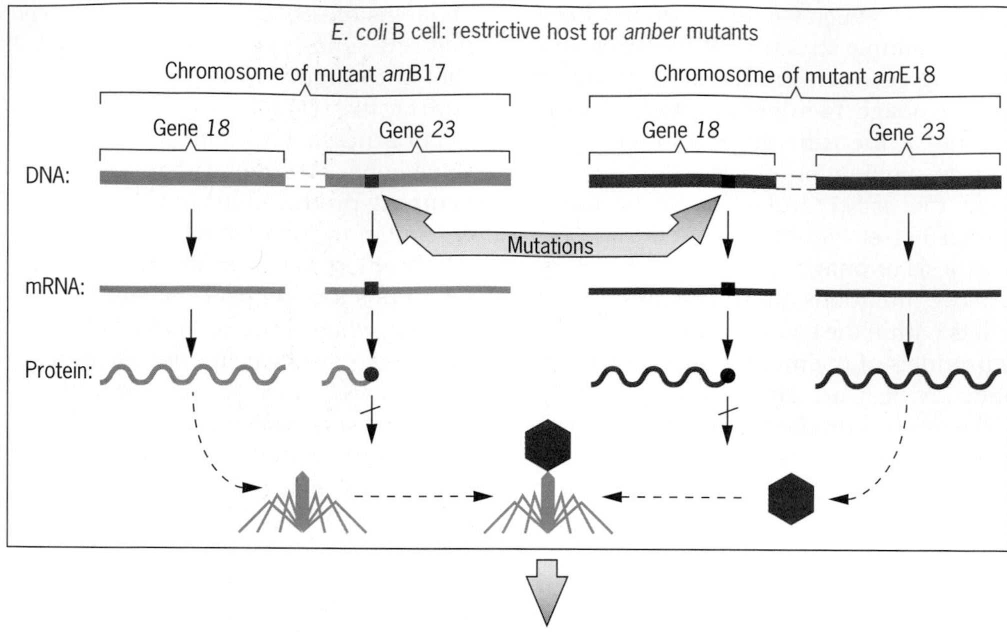

(*a*) Complementation between mutations *am*B17 and *am*E18.

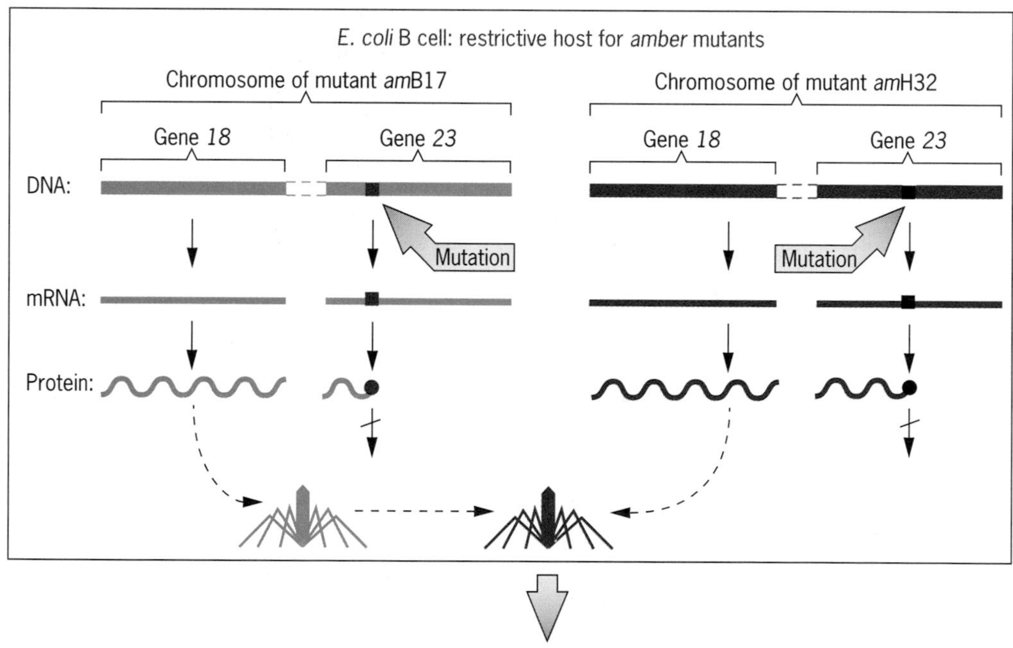

(*b*) Lack of complementation between mutations *am*B17 and *am*H32.

Figure 15.17 Complementation and noncomplementation in *trans* heterozygotes. (*a*) Complementation between mutation *am*B17 in gene *23*, which encodes the major structural protein of the phage T4 head, and mutation *am*E18 in gene *18*, which encodes the major structural protein of the phage tail. Both heads and tails are synthesized in the cell, with the result that infective progeny phage are produced. (*b*) When the *trans* heterozygote contains two mutations (*am*B17 and *am*H32) in gene *23*, no heads are produced, and no infective progeny phage can be assembled. Compare with Figure 15.18.

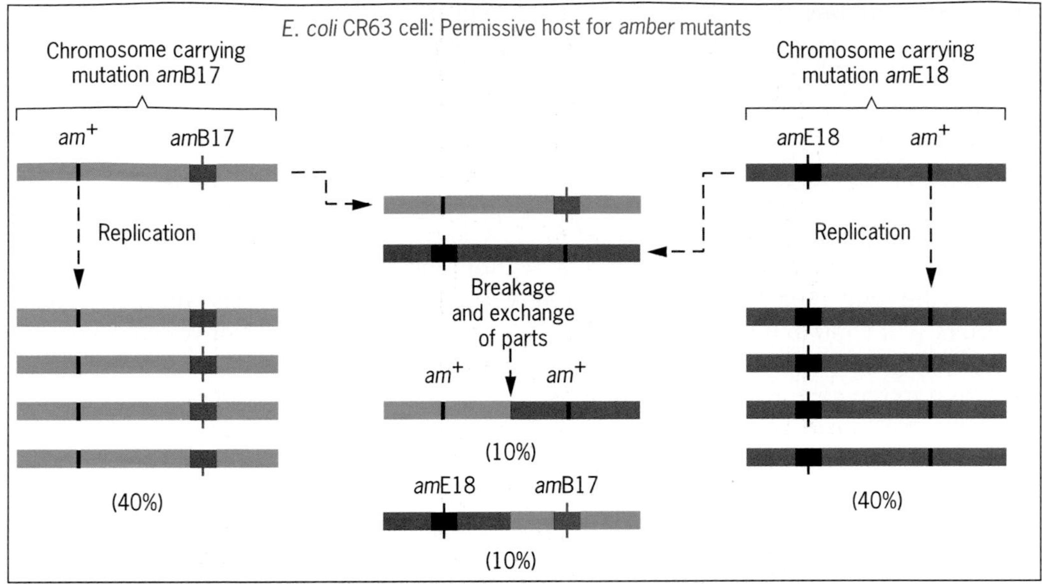

Lysis yields progeny phage of four genotypes:
Parental genotypes : ≃ 40% *am*B17 and ≃ 40% *am*E18;
Recombinant genotypes: ≃ 10% wild-type (am^+) and ≃ 10% double mutant (*am*E18-*am*B17).

(*a*) Recombination between phage T4 chromosomes carrying mutations *am*B17 and *am*E18.

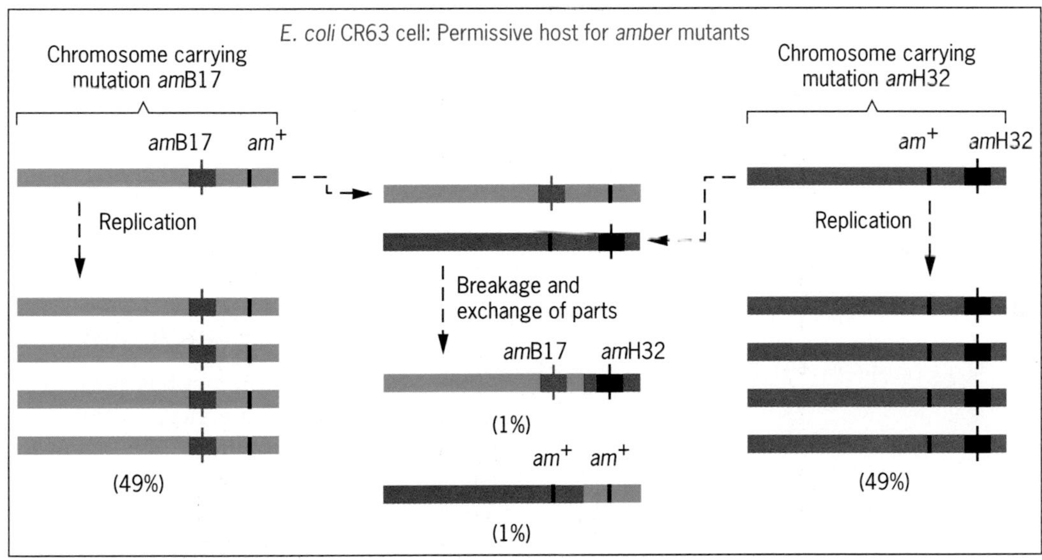

Lysis yields progeny phage of four genotypes:
Parental genotypes : ≃ 49% *am*B17 and ≃ 49% *am*H32;
Recombinant genotypes: ≃ 1% wild-type (am^+) and ≃ 1% double mutant (*am*B17-*am*H32).

(*b*) Recombination between phage T4 chromosomes carrying mutations *am*B17 and *am*H32.

Figure 15.18 Recombination between (*a*) the complementing mutations *am*B17 (gene *23*) and *am*E18 (gene *18*), and (*b*) the noncomplementing mutations *am*B17 and *am*H32 (both in gene *23*). Recombination occurs in both cases; however, fewer recombinants are produced in cells infected with *am*B17 and *am*H32 because the two mutations are located closer together on the phage T4 chromosome. Compare with Fig. 15.17.

because of the occurrence of a phenomenon called **intragenic complementation.**

The functional forms of some proteins are dimers or higher multimers consisting of two or more polypeptides. These polypeptides may be either homologous, the products of a single gene, or nonhomologous, the products of two or more distinct genes. When the active form of the protein contains two or more homologous polypeptides (it may or may not also contain nonhomologous polypeptides), intragenic complementation may occur. *Inter*genic complementation (discussed in the preceding section) and *intra*genic complementation (described below) are two distinct phenomena.

Let us consider an enzyme that functions as a homodimer, that is, a protein containing two copies of a specific gene product (Figure 15.19). In organisms that are homozygous for the wild-type allele of the gene, all the protein dimers will contain identical wild-type polypeptides. Similarly, organisms that are homozygous for any mutation in the gene will contain dimers with two mutant polypeptides. An organism that is heterozygous for two different mutations in the gene will produce some dimers that contain the two different mutant polypeptides. We call these heterodimers. Such heterodimers may have partial or complete (wild-type) function. If they do have partial or complete function, intragenic complementation has occurred, and the *trans* heterozygote has a wild-type phenotype or a phenotype intermediate between mutant and wild-type (Figure 15.19, bottom). In the case of noncomplementing mutations in a gene encoding a multimeric protein, the heteromultimers are nonfunctional, just like the mutant homomultimers (protein multimers composed of two or more identical mutant polypeptides).

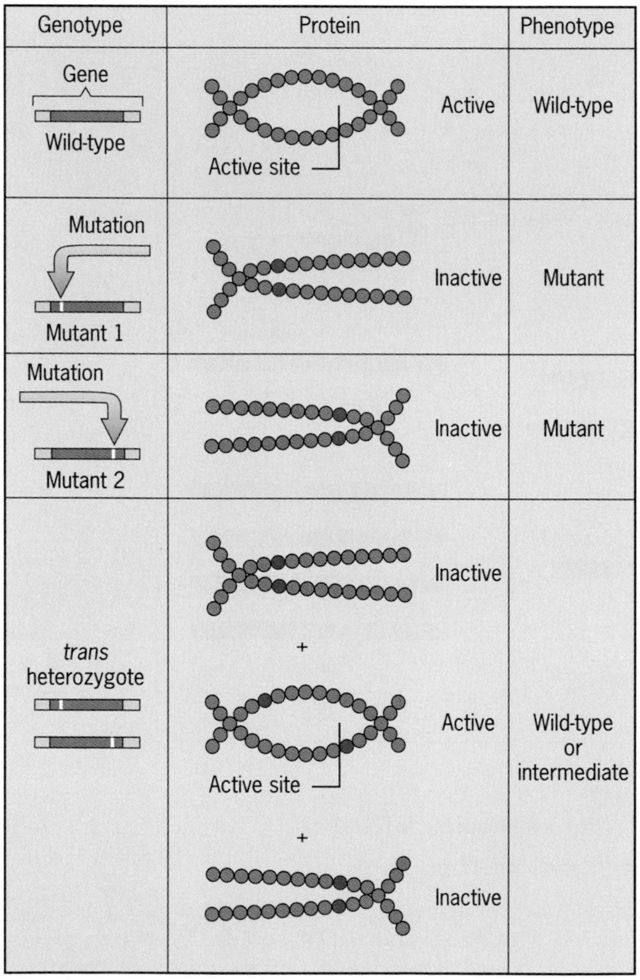

Genotype	Protein	Phenotype
Gene Wild-type	Active site — Active	Wild-type
Mutation Mutant 1	Inactive	Mutant
Mutation Mutant 2	Inactive	Mutant
trans heterozygote	Inactive + Active site — Active + Inactive	Wild-type or intermediate

Figure 15.19 *Intra*genic complementation sometimes occurs when the active form of an enzyme or structural protein is a multimer that contains at least two copies of any one gene product. Here, the functional form of the enzyme is a dimer composed of two polypeptides encoded by one gene. The amino acids altered by the mutations are shown as red circles in the polypeptide chains.

In several known cases of intragenic complementation, the active form of the protein in the heterozygote has been purified and shown to be a heterodimer or heterotetramer containing two different mutant polypeptides. Why such heteromultimers should be active when the two corresponding homomultimers are inactive is not clear. Apparently, the wild-type sequence of amino acids in the nonmutant segment of one mutant polypeptide somehow compensates for the mutant segment of the polypeptide encoded by the second mutant allele, and vice versa (Figure 15.19). However, most proteins have complex three-dimensional structures, and until the exact structures of a wild-type homomultimer, two mutant homomultimers, and an active heteromultimer composed of the two mutant polypeptides have been determined, the molecular basis of intragenic complementation will continue to be subject to speculation.

Limitations on the Use of the Complementation Test

The complementation test has been very useful in operationally delimiting genes. Usually, two or more mutations that produce the same phenotype can be assigned to one or more genes based on the results of complementation tests. However, in some cases, the results of complementation tests cannot be used to delimit genes. As previously mentioned, *complementation tests are not informative in studies of dominant or codominant mutations or in cases where intragenic complementation occurs.* In addition, complementation tests are sometimes uninformative because of epistatic interactions between the mutant gene products. If the *cis* test is done, such interactions are readily detected because the *cis* heterozygotes will have mutant phenotypes rather than the required wild-type phenotype.

Another limitation of the complementation test is encountered in working with so-called polar mutations. A **polar mutation** is a mutation that not only results in a defective product of the gene in which it is located but also interferes with the expression of one

Trans heterozygote: *E. coli* partial diploid.

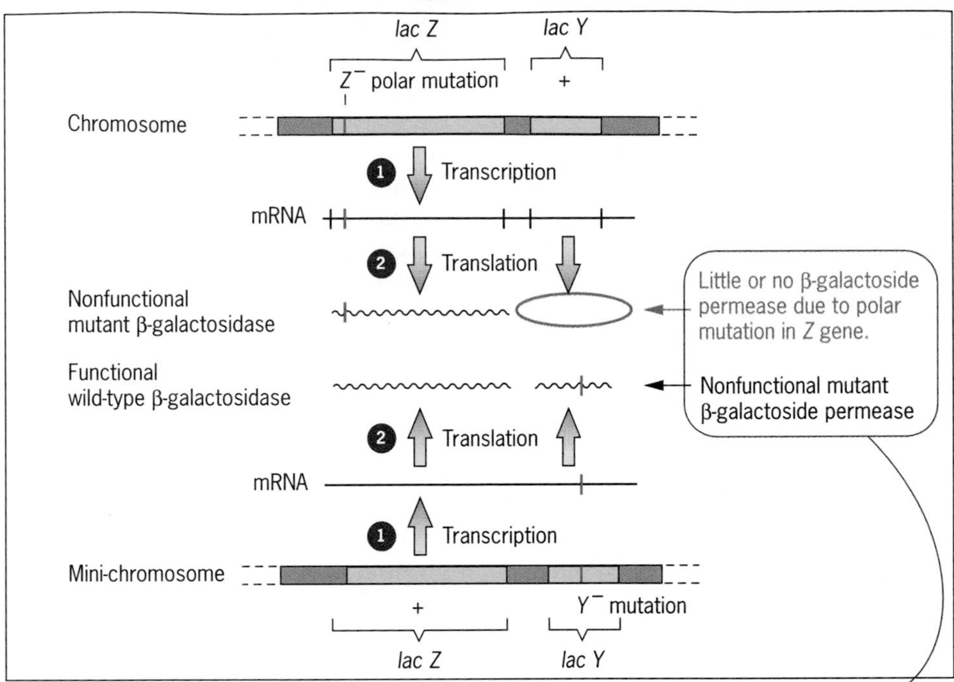

Little or no functional β-galactoside permease is produced. Therefore, the *trans* heterozygote has a mutant phenotype — that is, it is unable to utilize lactose as an energy source.

Figure 15.20 Lack of complementation between a polar mutation and a mutation in a downstream gene in the same transcription unit. The ability of *E. coli* cells to utilize lactose as an energy source depends on the products of two co-transcribed genes: *lacY*, which encodes β-galactoside permease, and *lacZ*, which encodes β-galactosidase. Transcription of the *lacY* and *lacZ* genes produces a multigenic mRNA, which is translated to provide the two proteins. A translation–termination mutation near the translation–start site in *lacZ* has a polar effect on translation of the *lacY* gene, reducing its translation efficiency to 1–2 percent of the normal level. Thus, the polar mutation in *lacZ* will not complement a null mutation in *lacY*, even though the two point mutations are in two different genes.

or more adjacent genes. The adjacent genes are always located on one side of the gene carrying the mutation (thus the term *polar mutation*). Such polar mutations are frequently observed in prokaryotes in coordinately regulated sets of genes called operons (Chapter 23). They usually are mutations resulting in polypeptide chain-termination signals (nucleotide-pair triplets yielding UAA, UAG, and UGA codons in mRNA) within genes. These polar mutations interfere with the expression of genes located downstream (relative to the direction of transcription) of the mutant gene. As a result, polar mutations fail to complement mutations in genes subject to the polar effect (Figure 15.20). Thus, the results of complementation tests performed with polar mutations are often ambiguous.

Key Points: **The complementation or *trans* test provides an operational definition of the gene; it is used to determine whether mutations are in the same gene or different genes. Intragenic complementation may occur when a protein is a multimer containing at least two copies of one gene product.**

COMPLEX GENE–PROTEIN RELATIONSHIPS

Most prokaryotic genes consist of continuous sequences of nucleotide pairs, which specify colinear sequences of amino acids in the polypeptide gene products. As we discussed in Chapter 12, most eukaryotic genes are split into coding sequences (exons) and noncoding sequences (introns). However, because the spliceosomes usually excise introns from primary transcripts by *cis*-splicing mechanisms (processes that join exons from the same RNA molecule), the presence of introns in genes does not invalidate the complementation test as an operational definition of the gene. Nevertheless, in some cases, transcripts of split genes may undergo several different types of splicing, making the relationships between genes and proteins more complex than the usual one gene–one polypeptide. In other cases, expressed genes are assembled from "gene pieces" during the development of the specialized cells in which they are expressed.

Alternate Pathways of Transcript Splicing: Protein Isoforms

Many interrupted eukaryotic genes, such as the mammalian hemoglobin genes and the chicken ovalbumin and 1α2 collagen genes discussed in Chapter 12, each encode a single polypeptide chain with a specific function. In these cases, the mRNA produced from a given gene contains all the exons of the gene joined together in the same order as they occur in the gene. However, the transcripts of some interrupted genes undergo alternate pathways of transcript splicing; that is, different exons of a gene may be joined to produce a related set of mRNAs encoding a small family of closely related polypeptides called **protein isoforms** (Figure 15.21). Note, however, that even in cases where alternate splicing occurs, colinearity between the genes and their polypeptide products is maintained.

The alternate splicing pathways are often tissue-specific, producing related proteins that carry out similar, but not necessarily identical, functions in different types of cells. The mammalian tropomyosin genes provide striking examples of genes which each produce a family of protein isoforms. Tropomyosins are proteins involved in the regulation of muscle contraction in animals. Because the various organs of an animal contain different muscle types, all of which need to be regulated, the availability of a family of related tropomyosins might be beneficial. In any case, one mouse tropomyosin gene is known to produce at least 10 different tropomyosin polypeptides as a result of alternate pathways of transcript splicing. Genes of this type obviously do not fit the one gene–one polypeptide concept very well. For such genes, where alternate splicing pathways give rise to two or more different polypeptides, the gene can be defined as a DNA sequence that is a single unit of transcription and encodes a set of protein isoforms.

Assembly of Genes During Development: Human Antibody Chains

Genetic information is not always organized into genes of the type described in the preceding sections of this chapter. In rare cases, genes are assembled from a storehouse of **gene segments** during the development of an organism. The immune system of vertebrate animals depends on the synthesis of proteins called **antibodies** to provide protection against infections by viruses, bacteria, toxins, and other foreign substances. Each antibody contains four polypeptides, two identi-

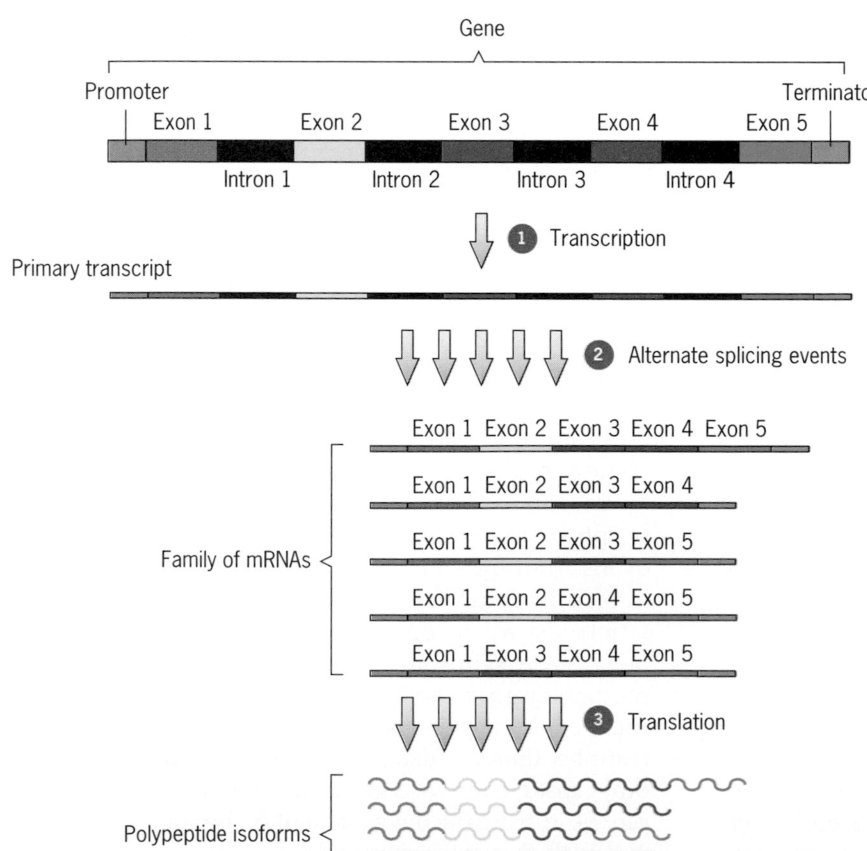

Figure 15.21 A single gene may produce a family of closely related polypeptides by using alternate pathways of exon splicing.

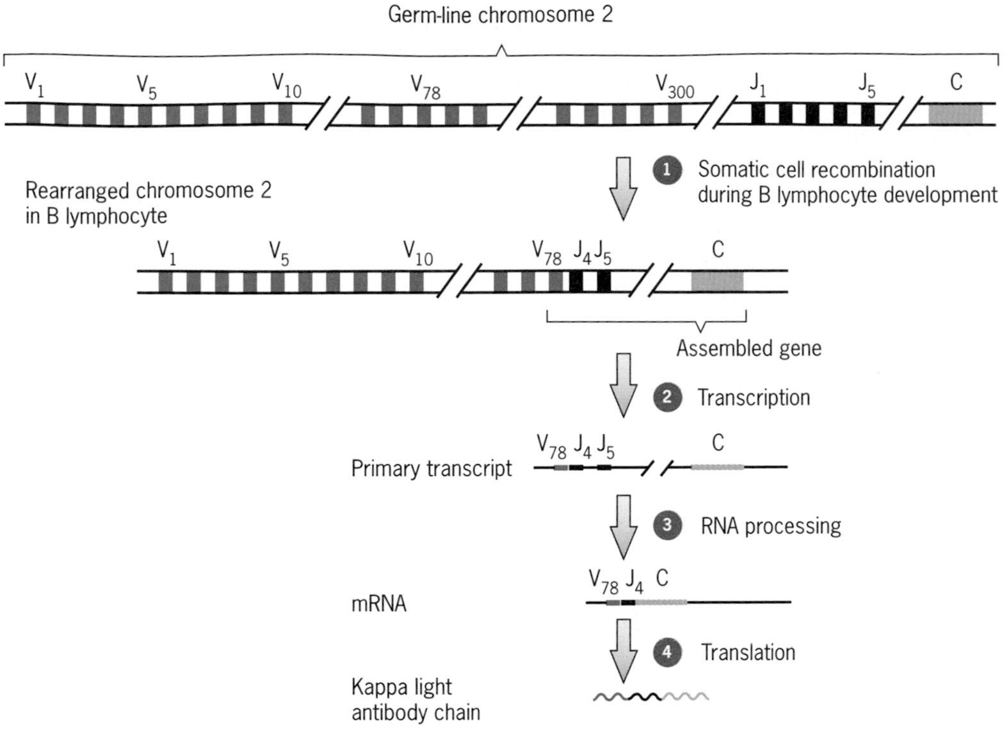

Figure 15.22 Assembly of a gene encoding an antibody kappa light chain from gene segments during the development of a B lymphocyte in humans.

cal heavy chains and two identical light chains. The light chains are of two types: kappa and lambda. Each antibody chain contains a variable region, which exhibits extensive diversity from antibody to antibody, and a constant region, which is largely the same in all antibodies. In germ-line chromosomes, the DNA sequences encoding these antibody chains are present in gene segments, and the gene segments are joined together to produce genes during the differentiation of the antibody-producing cells (B lymphocytes) from progenitor cells.

To illustrate this process of gene assembly during development, let us briefly consider the DNA sequences encoding kappa light chains in humans. (We discuss this topic in detail in Chapter 27.) A kappa light chain gene is assembled from three gene segments: V_k (**V** for **v**ariable region), J_k (**J** for **j**oining segment), and C_k (**C** for **c**onstant region), during B lymphocyte development. Together, the V_k and J_k gene segments encode the variable region of the kappa light chain, whereas the C_k gene segment encodes the constant region. No functional V_k-J_k-C_k kappa light chain gene is present in any human germ-line chromosome. Instead, human chromosome 2 contains a cluster of about 300 V_k gene segments, another cluster of five J_k gene segments, and a single C_k gene segment (Figure 15.22). During the differentiation of each B lymphocyte, recombination joins one of the V_k gene segments to one of the J_k gene segments. Any J_k segments remaining between the newly formed V_k-J_k exon and the C_k gene segment become part of an intron that is removed during the processing of the primary transcript. Similar somatic recombination events are responsible for the assembly of the genes encoding antibody heavy chains, lambda light chains, and T-lymphocyte receptor proteins (Chapter 27).

Key Points: **The transcripts of some genes undergo alternate pathways of splicing to produce mRNAs with different exons joined together. Translation of these mRNAs produces closely related polypeptides called protein isoforms. Other genes, such as those encoding antibody chains, are assembled from gene segments during development by regulated processes of somatic recombination.**

TESTING YOUR KNOWLEDGE

1. In *Drosophila, white, white cherry,* and *vermilion* are all sex-linked mutations affecting eye color. All three mutations are recessive to their wild-type allele(s) for red eyes. A white-eyed female crossed with a vermilion-eyed male produces white-eyed male offspring and red-eyed (wild-type) female offspring. A white-eyed female crossed with a white cherry-eyed male produces white-eyed sons and light cherry-eyed daughters. Do these results indicate whether or not any of the three mutations affecting eye color are located in the same gene? If so, which mutations?

ANSWER

The complementation test for allelism involves placing mutations pairwise in a common protoplasm in the *trans* configuration and determining whether the resulting *trans* heterozygotes have mutant or wild-type phenotypes. If the two mutations are in the same gene, both copies of the gene in the *trans* heterozygote will produce defective gene products, resulting in a mutant phenotype (see Figure 15.16*a*). However, if the two mutations are in different genes, the two mutations will complement each other, because the wild-type copies of each gene will produce functional gene products (see Figure 15.16*b*). When complementation occurs, the *trans* heterozygote will have the wild-type phenotype. Thus, the complementation test allows one to determine whether any two recessive mutations are located in the same gene or in different genes.

If the *trans* heterozygote has the mutant phenotype, the two mutations are in the same gene. If the *trans* heterozygote has the wild-type phenotype, the two mutations are in two different genes. Because the mutations of interest are sex-linked, all the male progeny will have the same phenotype as the female parent. They are hemizygous, with one X chromosome obtained from their mother. In contrast, the female progeny are *trans* heterozygotes. In the cross between the white-eyed female and the vermilion-eyed male, the female progeny have red eyes, the wild-type phenotype. Thus, the *white* and *vermilion* mutations are in different genes, as illustrated in the following diagram:

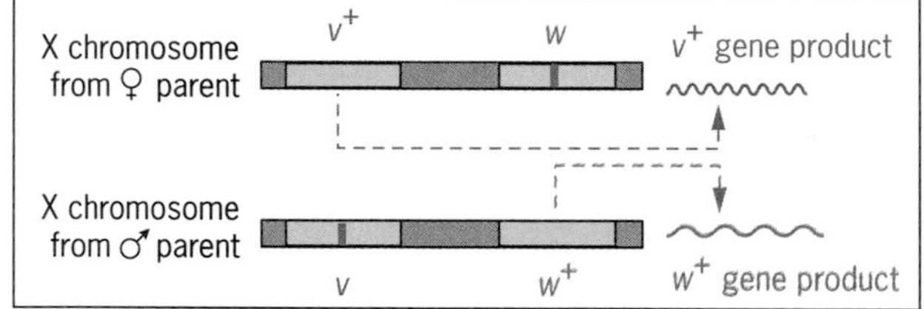

Complementation yields wild-type phenotype; both v^+ and w^+ gene products are produced in the *trans* heterozygote.

In the cross between a white-eyed female and a white cherry-eyed male, the female progeny have light cherry-colored eyes (a mutant phenotype), not wild-type red eyes as in the first cross. Since the *trans* heterozygote has a mutant phenotype, the two mutations, *white* and *white cherry*, are in the same gene:

trans heterozygote

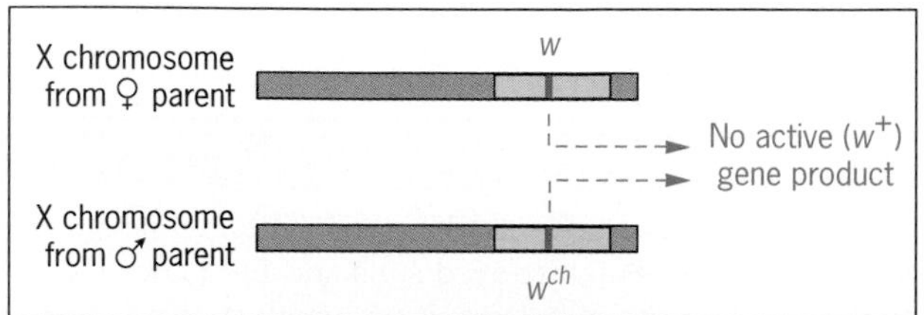

No w^+ gene product; therefore, mutant phenotype.

2. Suppressor-sensitive (*sus*) mutants of bacteriophage φ29 can grow on *Bacillus subtilis* strain L15 but cannot grow (that is, are lethal) on *B. subtilis* strain 12A. Wild-type (sus^+) φ29 phage can reproduce on both strains, L15 and 12A. Thus, the φ29 *sus* mutants are conditional lethal mutants like the *amber* mutants of bacteriophage T4 (see Figure 13.23). Seven different *sus* mutants of phage φ29 were analyzed for complementation by simultaneously infecting the restrictive host (*B. subtilis* strain 12A) with each possible pair of mutants. Single infections with each of the mutants and with wild-type φ29 were also done as controls. The results of these complementation or *trans* tests and the controls are given as progeny phage per infected cell in the accompanying table. Several infections performed with wild-type φ29 phage yielded 300 to 400 progeny phage per infected cell. The results of the *cis* controls are not given, but assume that all of the *cis* heterozygote controls yielded over 300 progeny phage per infected cell. Also assume that no intragenic complementation occurs between any of the *sus* mutants studied.

Phage φ29 Progeny per Infected Bacterium

Mutant:	*1*	*2*	*3*	*4*	*5*	*6*	*7*
7	365	384	344	371	347	333	0.01
6	341	301	351	369	329	0.1	
5	386	326	322	0.04	<0.01		
4	327	398	374	0.06			
3	354	387	<0.01				
2	0.01	<0.01					
1	0.02						

(a) Based on these data, how many genes are identified by the seven *sus* mutants? (b) Which *sus* mutations are located in the same gene(s)?

ANSWER

The seven *sus* mutants yielded <0.01 to 0.1 progeny phage per infected *B. subtilis* strain 12A cell; those data define the mutant phenotype (basically no progeny). Infections of strain 12A cells with wild-type φ29 produced 300 to 400 progeny phage per infected cell, defining the wild-type phenotype. We then examine the phenotypes of the *trans* heterozygotes to determine whether any of the mutations are located in the same gene(s). In each case, we must ask whether the *trans* heterozygote has the mutant or the wild-type phenotype. If a *trans* heterozygote has the mutant phenotype, the two *sus* mutations are in the same gene. If it has the wild-type phenotype, the two *sus* mutations are in different genes. If you are unsure of why this is true, review Figure 15.16. Of the 21 *trans* heterozygotes examined, 19 exhibited the wild-type phenotype, indicating that in each case the two mutations are in different genes. Two *trans* heterozygotes, (1) *sus*1 on one chromosome and *sus*2 on a second chromosome and (2) *sus*4 on one chromosome and *sus*5 on the another, had the mutant phenotype. Therefore, (*a*) the seven *sus* mutations are located in five different genes, with (*b*) mutations *sus*1 and *sus*2 in one gene and mutations *sus*4 and *sus*5 in another gene.

QUESTIONS AND PROBLEMS

15.1 In what ways does our present concept of the gene differ from the pre–1940 or classical concept of the gene?

15.2 What was the first evidence that indicated that the unit of function and the unit of structure of genetic material were not the same?

15.3 What is the currently accepted operational definition of the gene?

15.4 Based on our current concept of the gene, (a) what is the smallest unit of genetic material that can be changed by mutation, and (b) what is the smallest region of genetic material in which recombination can occur?

15.5 Four mutant strains of *Neurospora* require one or more of the supplemented metabolites A through D to grow. From the data below (where "+" = growth; "0" = no growth), draw the metabolic pathway for the synthesis of these metabolites and show where the mutant strains (1, 2, 3, and 4) are blocked.

	Metabolite			
Mutant	A	B	C	D
1	0	0	0	+
2	+	+	+	+
3	0	0	+	+
4	+	0	+	+

15.6 Seven mutants of *Neurospora* are unable to grow on minimal medium unless it is supplemented with one or more of the metabolites A through G. On the basis of the data given below (where "+" indicates growth and "0" indicates no growth), draw a biochemical pathway for the synthesis of these seven substances and show where the mutants are blocked in the pathway.

	Growth in the Presence of Metabolite(s)									
Mutant	A	B	C	D	E	F	G	B+C	D+G	B+C+E
1	+	0	0	0	0	0	0	0	+	+
2	0	+	0	0	0	0	0	+	0	+
3	0	0	+	0	0	0	0	+	0	+
4	0	0	0	+	+	0	0	0	+	+
5	0	0	0	0	+	0	0	0	0	+
6	+	0	0	0	0	+	0	0	+	+
7	0	0	0	0	0	0	+	+	+	+

15.7 Of what value are conditional lethal mutations for genetic fine structure analysis?

15.8 Eight independently isolated mutants of *E. coli*, all of which are unable to grow in the absence of histidine (his^-), were examined in all possible *cis* and *trans* heterozygotes (partial diploids). All of the *cis* heterozygotes were able to grow in the absence of histidine. The *trans* heterozygotes yielded two different responses: some of them grew in the absence of histidine; others did not. The experimental results, using "+" to indicate growth and "0" to indicate no growth, are given in the accompanying table. How many genes are defined by these eight mutations? Which mutant strains carry mutations in the same gene(s)?

Growth of *Trans* Heterozygotes (without Histidine)

Mutant:	1	2	3	4	5	6	7	8
8	0	0	0	0	0	0	+	0
7	+	+	+	+	+	+	0	
6	0	0	0	0	0	0		
5	0	0	0	0	0			
4	0	0	0	0				
3	0	0	0					
2	0	0						
1	0							

15.9 Assume that the mutants described in Problem 15.8 yielded the following results. How many genes would they have defined? Which mutations would have been in the same gene(s)?

Growth of *Trans* Heterozygotes (without Histidine)

Mutant:	1	2	3	4	5	6	7	8
8	+	+	+	+	+	+	0	0
7	+	+	+	+	+	+	0	
6	+	+	+	+	0	0		
5	+	+	+	+	0			
4	+	+	0	0				
3	+	+	0					
2	0	0						
1	0							

15.10 Propose a genetic map that is consistent with the complementation data provided below, where *m1*-*m7* are different point mutations and *Dfa*-*Dfe* are deletions. A "+" indicates that complementation occurs, whereas a "0" indicates that complementation does not occur.

	m1	*m2*	*m3*	*m4*	*m5*	*m6*	*m7*
Dfa	0	0	0	+	0	0	0
Dfb	0	+	+	0	+	0	0
Dfc	0	+	+	+	0	0	+
Dfd	0	0	+	0	0	0	0
Dfe	0	+	+	0	+	+	0

15.11 A researcher is interested in delineating the pathway for the production of blue colony color in yeast. She finds that it is possible to mate two mutant strains of yeast with gray colonies and produce diploid yeast cells that produce blue colonies. She then proceeds to mate nine different yeast mutants (*A* through *I*) with gray colonies in all possible pairwise combinations and inspects the diploid progeny for colony color. The results of her complementation analysis are presented in the following table, where a "+" = blue colonies and a "−" = gray colonies. (a) Place the mutations in the nine strains into complementation groups. (b) What is the minimum number of genes necessary for the production of blue color in yeast? (c) Can you estimate the maximum number of genes necessary for this trait, and if so, what is the maximum number?

	A	*B*	*C*	*D*	*E*	*F*	*G*	*H*	*I*
A	−	−	+	+	+	−	−	−	−
B	−	−	+	+	+	−	−	−	−
C	+	+	−	−	+	+	+	+	+
D	+	+	−	−	+	+	+	+	+
E	+	+	+	+	−	+	+	+	+
F	−	−	+	+	+	−	−	−	−
G	−	−	+	+	+	−	−	−	−
H	−	−	+	+	+	−	−	−	−
I	−	−	+	+	+	−	−	−	−

15.12 What determines the maximum number of different alleles that can exist for a given gene?

15.13 What is the difference between a pair of homoalleles and a pair of heteroalleles?

15.14 Two different inbred varieties of a particular plant species have white flowers. All other varieties of this species have red flowers. What experiments might be done to obtain evidence to determine whether the difference in flower color in these varieties is the result of different alleles of a single gene or the result of genetic variation in two or more genes?

15.15 The *amber* mutants of phage T4 are conditional lethal mutants. They grow on *E. coli* strain CR63 but are lethal on *E. coli* strain B. An *amber* mutant almost never exhibits *intragenic* complementation with any other *amber* mutant; for this problem, assume that no *intragenic* complementation occurs between any of the mutants involved. The following results were obtained when eight *amber* mutants were analyzed for complementation by infecting the restrictive host (*E. coli* strain B) with each possible pair of mutants. The results of mixed infections by pairs of mutants are shown as "0" if no progeny are produced and as "+" if progeny phage resulted from the infection with that particular pair of mutants.

Mutant:	1	2	3	4	5	6	7	8
8	+	+	+	+	+	+	0	0
7	+	+	+	+	+	+	0	
6	+	+	+	+	+	0		
5	0	+	0	+	0			
4	+	+	+	0				
3	0	+	0					
2	+	0						
1	0							

(a) These data indicate that the eight *amber* mutations are located in how many different genes?

(b) Which mutations are located in the same gene or genes?

15.16 Considering only base-pair substitutions, how many different mutant homoalleles can occur at one site in a gene?

15.17 Are the following statements concerning the genetic element referred to as the gene true or false?

(a) The classical (pre–1940) conception of the gene was that it was (1) a unit of physiological function or expression, (2) the smallest unit that could undergo mutation, and (3) a unit not subdivisible by recombination.

(b) In bacteria, the *cis-trans* test provides an operational definition by which we usually can identify a gene as the unit that specifies one mRNA molecule.

(c) Our present knowledge of the structure of the gene indicates that the units defined by criteria (2) and (3) in statement (a) above are both equivalent to a single nucleotide pair.

(d) Studies in the 1940s demonstrated the existence of heteroalleles, clearly indicating that many mutations that were allelic by the functional criterion could be separated by recombination, and thereby indicating that the units of function, mutation, and recombination are not equivalent.

(e) Homoalleles are functionally and structurally allelic; heteroalleles are functionally allelic but structurally nonallelic.

15.18 The *rosy* (*ry*) gene of *Drosophila* encodes the enzyme xanthine dehydrogenase; the active form of xanthine dehydrogenase is a dimer containing two copies of the *rosy* gene product. Mutations ry^2 and ry^{42} are both located within the region of the *rosy* gene that encodes the *rosy* polypeptide gene product. However, ry^2/ry^{42} *trans* heterozygotes have wild-type eye color. How can the observed complementation between ry^2 and ry^{42} be explained given that these two mutations are located in the same gene?

15.19 Both *temperature-sensitive* (*ts*) mutant alleles and *amber* (*am*) mutant alleles have been identified and studied for many of the genes of bacteriophage T4. Different *ts* mutations within the same gene are frequently found to complement each other, whereas different *am* mutations within the same gene practically never complement one another. Why is this difference to be expected?

15.20 Suppressor-sensitive (*sus*) mutants of bacteriophage lambda can grow on *E. coli* strain C600 but cannot grow (that is, are lethal) on *E. coli* strain W3350. In other words, *sus* mutants are conditional-lethal mutants. Seven *sus* mutants were analyzed for complementation by simultaneously infecting the restrictive host (*E. coli* strain W3350) with each possible pair of mutants. Single infections with each mutant and with wild-type lambda were also done as controls. The results of these complementation or *trans* tests and the controls are given as progeny per infected cell in the accompanying table. Several infections with wild-type lambda yielded 120 to 150 progeny phage per infected cell. The results of the *cis* heterozygote controls are not given, but assume that all of the *cis* heterozygotes yielded over 100 progeny phage per infected cell. Also assume that no intragenic complementation occurs between any of these *sus* mutants.

Lambda Progeny per Infected Cell

Mutant:	1	2	3	4	5	6	7
7	0.01	133	0.01	146	134	128	0.01
6	131	142	161	0.06	0.1	0.1	
5	120	126	134	0.05	<0.01		
4	147	129	134	0.06			
3	<0.01	147	<0.01				
2	170	<0.01					
1	0.02						

(a) Based on the above data, how many genes are defined by the seven *sus* mutants?

(b) Which *sus* mutations are located in the same gene(s)?

15.21 Is the number of potential alleles of a gene directly related to the number of nucleotide pairs in the gene? Is such a relationship more likely to occur in prokaryotes or in eukaryotes? Why?

15.22 In *Drosophila*, *white*, *eosin*, and *carnation* are all sex-linked recessive mutations affecting eye color. A white-eyed female crossed with a carnation-eyed male produced white-eyed male progeny and red-eyed (wild-type) female offspring. A white-eyed female crossed with an eosin-eyed male produced white-eyed sons and light eosin-eyed daughters. Based on these data, which of the three mutations (*white*, *eosin*, and *carnation*), if any, are located in the same gene(s)?

15.23 Suppressor-sensitive (*sus*) mutants of bacteriophage φ29 can grow on *Bacillus subtilis* strain L15 but cannot grow (that is, are lethal) on *B. subtilis* strain 12A. Wild-type (sus^+) φ29 phage can reproduce on both strains, L15 and 12A. Thus, the φ29 *sus* mutants are conditional lethal mutants like the *amber* mutants of bacteriophage T4. Seven different *sus* mutants of phage φ29 were analyzed for complementation by simultaneously infecting the restrictive host (*B. subtilis* strain 12A) with each possible pair of mutants. Single infections with each of the mutants and with wild-type φ29 were also done as controls. The results of these complementation or *trans* tests and the controls are given as progeny phage per infected cell in the accompanying table. Several infections performed with wild-type φ29 phage yielded 300 to 400 progeny phage per infected cell. The results of the *cis* controls are not given, but assume that all of the *cis* heterozygote controls yielded over 300 progeny phage per infected cell. Also assume that no intragenic complementation occurs between any of the *sus* mutants studied.

Phage φ29 Progeny per Infected Bacterium

Mutant	1	2	3	4	5	6	7
7	0.01	384	0.01	371	347	333	0.01
6	341	301	351	0.06	329	0.1	
5	386	326	322	367	<0.01		
4	327	398	374	0.06			
3	<0.01	387	<0.01				
2	354	<0.01					
1	0.02						

(a) Based on these data, how many genes are identified by the seven *sus* mutants?

(b) Which *sus* mutations are located in the same gene(s)?

15.24 Assume that the mutants described in Problem 15.23 had yielded the following results.

Phage φ29 Progeny per Infected Bacterium

Mutant	1	2	3	4	5	6	7
7	0.01	0.01	0.01	0.03	<0.01	0.01	0.01
6	0.08	0.09	0.05	0.06	0.1	0.1	
5	0.02	<0.01	<0.01	0.04	<0.01		
4	0.05	0.06	0.03	0.06			
3	<0.01	<0.01	<0.01				
2	0.01	<0.01					
1	0.02						

How many genes would they have defined? Which mutations would have been in the same gene(s)?

15.25 The recessive mutations *bl* (*black*) and *e* (*ebony*) in *Drosophila* both produce flies with black bodies rather than gray bodies like wild-type flies. Mapping studies showed that *bl* is located on chromosome 2, whereas *e* is on chromosome 3. When homozygous *bl*/*bl* flies are crossed with homozygous *e*/*e* flies, the heterozygous *bl*/*e* progeny have gray bodies. The observed complementation indicates that the two mutations are in two different genes. Was it necessary to perform a complementation test to conclude that the *bl* and *e* mutations were located in two different genes? If so, why? If not, why not?

15.26 Why was it necessary to modify Beadle and Tatum's one gene–one enzyme concept of the gene to one gene–one polypeptide?

15.27 In their analysis of gene function, Beadle and Tatum used *Neurospora* as an experimental organism, whereas Garrod had studied gene function in humans. What advantages does *Neurospora* have over humans for such studies?

15.28 Based on the information provided in Figure 15.11, (a) are mutations *trpA*3 and *trpA*33 heteroalleles or homoalleles? (b) Are mutations *trpA*78 and *trpA*58 heteroalleles or homoalleles?

15.29 Based on the information given in Figure 15.11, what is the maximum number of nucleotide pairs separating mutations *trpA*78 and *trpA*58?

15.30 Arthur Chovnick and colleagues have mapped a large number of recessive mutations that produce fruit flies with rose-colored eyes in the homozygous state. They also have performed complementation tests on these *ry* (*rosy*) mutations. Heterozygotes that carried mutations ry^{42} and ry^{406} in the *trans* configuration had wild-type eyes, whereas *trans* heterozygotes that harbored ry^5 and ry^{41} had rose-colored eyes. The results of two- and three-factor crosses unambiguously demonstrated that mutations ry^{42} and ry^{406} both map between mutations ry^5 and ry^{41}. How can these results be explained?

15.31 The sequences of nucleotide pairs that encode human antibody chains are usually referred to as gene segments rather than genes. Why?

15.32 Tropomyosins are proteins that mediate the interactions between actin and troponin and regulate muscle contractions. In *Drosophila*, six different tropomyosins that have some amino acid sequences in common, but differ in other sequences, are encoded by two tropomyosin genes (*TmI* and *TmII*). How can two genes encode six different polypeptides?

15.33 In *Drosophila*, *car* (*carnation*) and *g* (*garnet*) are X-linked mutations that produce brown eyes, in contrast to the dark red eyes of wild-type flies. The *g* and *car* mutations map at positions 44.4 and 62.5, respectively, on the linkage map of the X chromosome. Is a complementation test needed to determine whether these two mutations are in the same gene or two different genes? If so, why? If not, why not?

15.34 The *loz* (*l*ethal *o*n *Z*) mutants of bacteriophage X are conditional lethal mutants that can grow on *E. coli* strain Y but cannot grow on *E. coli* strain Z. The results shown in the following table were obtained when seven *loz* mutants were analyzed for complementation by infecting *E. coli* strain Z with each possible pair of mutants. A "+" indicates that progeny phage were produced in the infected cells, and a "0" indicates that no progeny phage were produced. All possible *cis* tests were also done, and all *cis* heterozygotes produced wild-type yields of progeny phage.

Mutant	*1*	*2*	*3*	*4*	*5*	*6*	*7*
7	+	+	0	+	0	0	0
6	+	+	+	+	+	0	
5	+	+	0	+	0		
4	0	0	+	0			
3	+	+	0				
2	0	0					
1	0						

Given that intragenic complementation does not occur between any of the seven *loz* mutants analyzed here, (a) propose three plausible explanations for the apparently anomalous complementation behavior of *loz* mutant number 7. (b) What simple genetic experiments can be used to distinguish between the three possible explanations? (c) Explain why specific outcomes of the proposed experiments will distinguish between the three possible explanations.

BIBLIOGRAPHY

Beadle, G. W., and E. L. Tatum. 1942. Genetic control of biochemical reactions in *Neurospora*. *Proc. Natl. Acad. Sci. USA* 27:499–506.

Carlson, E. A. 1966. *The Gene: A Critical History*. W. B. Saunders, Philadelphia.

Fincham, J. R. S. 1966. *Genetic Complementation*. Benjamin, Menlo Park, CA.

Garrod, A. E. 1909. *Inborn Errors of Metabolism*. Oxford University Press, New York. (Reprinted in H. Harris, 1963, *Garrod's Inborn Errors of Metabolism*, Oxford Monographs on Medical Genetics, Oxford University Press, London.)

Hawley, R. S., and C. A. Mori. 1999. *The Human Genome, A User's Guide*. Academic Press, New York.

Leder, P. 1982. The genetics of antibody diversity. *Sci. Amer.* 246(5):102–115.

Oliver, C. P. 1940. A reversion to wild type associated with crossing over in *Drosophila melanogaster*. *Proc. Natl. Acad. Sci. USA* 26:452–454.

Sturtevant, A. H. 1965. *A History of Genetics*. Harper & Row, New York.

Yanofsky, C., and V. Horn. 1972. Tryptophan synthetase α chain positions affected by mutations near the ends of the genetic map of *trpA* of *Escherichia coli*. *J. Biol. Chem.* 247:4494–4498.

A CONVERSATION WITH

GISELA MOSIG

When did you first become interested in genetics?

I grew up on a farm in a small village of Saxony, Germany, where we children had to help when help was needed. Thus, we had many opportunities to observe very early and first hand the obvious differences in crops and animals of different varieties. For example, my father had registered cattle, and I helped keep the books in which records and pedigrees were listed.

What specific event during your undergraduate years had a major effect on your future career as a geneticist?

It is difficult to describe this event to someone who has not experienced as a child the lies and then the collapse of the Nazi ideology, followed by the political climate and intellectual isolation of Eastern Europe after World War II. It actually happened during my last year in high school, although the last years in German high school curriculum correspond roughly to the first years in college in the United States. As we studied genetics, we learned about Mendel's rules, which finally brought some order and explanations into my varied experiences on the farm. Suddenly, one day, our biology teacher told us that these were all wrong: Genes were not transmitted according to Mendel's rules; instead, someone named Lysenko had definitely shown that one could change genes by changing the environment. By a certain treatment he had permanently changed winter wheat, which had to be seeded in the fall, to summer wheat, which had to be seeded in the spring.

Gisela Mosig is a leader in molecular genetics whose early work with T4 "petite" phage has greatly contributed to the understanding of the role of recombination in DNA replication of this virus. Recombination-dependent DNA replication is now known in many other organisms. She fled the stifling academic atmosphere of the German Democratic Republic after receiving a Ph.D. from the University of Cologne, and emigrated to the United States. Among her professional experiences was a position as Research Associate at Cold Spring Harbor, where she worked with A. D. Hershey. Dr. Mosig has been Professor of Molecular Biology at Vanderbilt University since 1971. She has served on the editorial boards and as a reviewer for numerous scientific journals and is a Fellow of the American Academy of Microbiologists.

From then on, this same teacher taught us the opposite of what he had taught before. Similar shifts in what was taught had happened two years earlier in history and in literature. Then old teachers were replaced by new teachers. I knew at this moment that I wanted to find out for myself. Fortunately, I had almost completed high school. When I was not admitted to the University, I became an apprentice in a plant breeding institute. Its library held many interesting books published in Eastern Europe (but no contemporary Western literature!). Among them was a detailed report translated into German about a meeting of Russian geneticists at which Lysenko and his friends had "condemned the reactionary and bourgeois Mendelists" and succeeded in having them deported to camps in Siberia for "reeducation." I also found Darwin's *Origin of Species.* I organized a discussion group to consider these and other books. We had no political agenda—we just wanted to find out—but this activity was considered subversive. I realized then that I had to get out of East Germany if I wanted to find out for myself, and I left with no money and no clear idea of how to go about my search. From today's perspective, with television and the World Wide Web connecting people in all parts of the world, this makes little sense. But at that time not even television reached across the borders. Western radio stations were blocked, and no Western newspapers or scientific journals were permitted.

When you were launching your career, did you have a role model?

I never thought about my launching a career. I simply wanted to know how things worked. Much later, at the university, we discussed Barbara McClintock's impressive work. Perhaps this helped me to neglect all the advice from different people that women don't belong in science, but I did not consider Barbara a role model. However, I was thrilled when I got to meet and know her in Cold Spring Harbor.

Your early work with T4 "petite" phage—T4 phage containing only two-thirds of the normal chromosome—made large contributions to our understanding of DNA replication and recombination in this virus. How did you discover these phage with incomplete genomes?

It was due both to serendipity and to Alfred Hershey's immense tolerance of letting me, as a postdoctoral student, work on an idea and a project that appeared to be inconsistent with the results of his recent experiments. Hershey and his coworkers had just shown that T4 phage have a single chromosome and that the shorter DNA pieces that appeared after certain extraction procedures were shearing artifacts. I attempted to test a then-popular hypothesis that T4 phage recombination involves reassortment of several chromosomes (partial replicas) that constitute the T4 genome. These partial replicas were thought to be connected by overlapping identical sequence segments. I had found T4 phages that banded in CsCl density gradients at lighter than average density and that appeared to recombine more frequently than average. I assumed that they were lighter because they contained less DNA in their overlaps, and I went to Cold Spring Harbor to use the "Hershey column" to look at the sizes of these DNA molecules. They were indeed smaller, but surprisingly, they were of rather uniform size—two-thirds of the normal size. Ultimately all contradictions were resolved because these short DNA molecules were packaged into small heads with a shape of non-elongated icosahedra. Because the small chromosomes corresponding to incomplete genomes are circular permutations of the T4 genome (consistent with a hypothesis proposed by George Streisinger), several of them regenerate—by recombination and recombination-dependent replication—complete chromosomes that are packaged into progeny viruses.

Can you single out any one research result as the most exciting of your career?

It was not a single result. The pattern of recombination in crosses involving genetically marked small T4 particles first suggested that DNA replication and recombination are intimately related, and how they might be related. When my coworkers and I used more markers, the results appeared more and more complex, causing us much frustration (and little sleep!). When I suddenly "saw" a simple way to explain them, I felt as elated as after listening to a beautiful piece of music. Of course, it took several years of more direct experiments by my students and myself to convince ourselves and others of the validity of the idea that DNA replication can be initiated from intermediates of recombination. Because the results of the later experiments were consistent with the original idea, they were simply fun.

Do you find an "artistic" element in science?

Very much so, and at different levels. Both in the arts and in science we try to understand the relationships of things that surround us and events that we encounter. In science, the efforts to design experiments and to construct models or working hypotheses from experimental results require a creativity similar to that of the arts, with the additional constraint of being subject to reality checks. Konrad Lorenz's essay on intuition in science beautifully describes this aspect, and Escher's graphics and their impact on structural biology give a visual illustration. To explain complex relationships in terms of simpler processes and models is simply aesthetically pleasing.

What are your current research projects?

(1) Testing our recent model on how homologous recombination, heteroduplex repair, and recombination-dependent DNA replication can contribute to mutagenesis and to evolution—specifically, we are investigating the roles of these processes in horizontal gene transfer, evolution and exclusion by and of T4-related bacteriophages. (2) Analyzing the coordination of redundant pathways of DNA replication and recombination with other aspects of DNA metabolism, such as repair, transcription, and packaging. Such coordination is important to allow optimal phage growth under different physiological conditions. We are intrigued by the redundancies because maintaining redundant mechanisms is thought to be one of the prerequisites for evolution of novel biological processes. This requires that different redundant mechanisms are subject to different kinds of selective pressure. The phage T4 system is an excellent model to test the validity of this premise.

How have you maintained your great enthusiasm throughout your long, productive career?

I have been so privileged to work on and teach something that interests me most. It far exceeded any expectation that I had when I grew up. Is it surprising that I am enthusiastic about it?

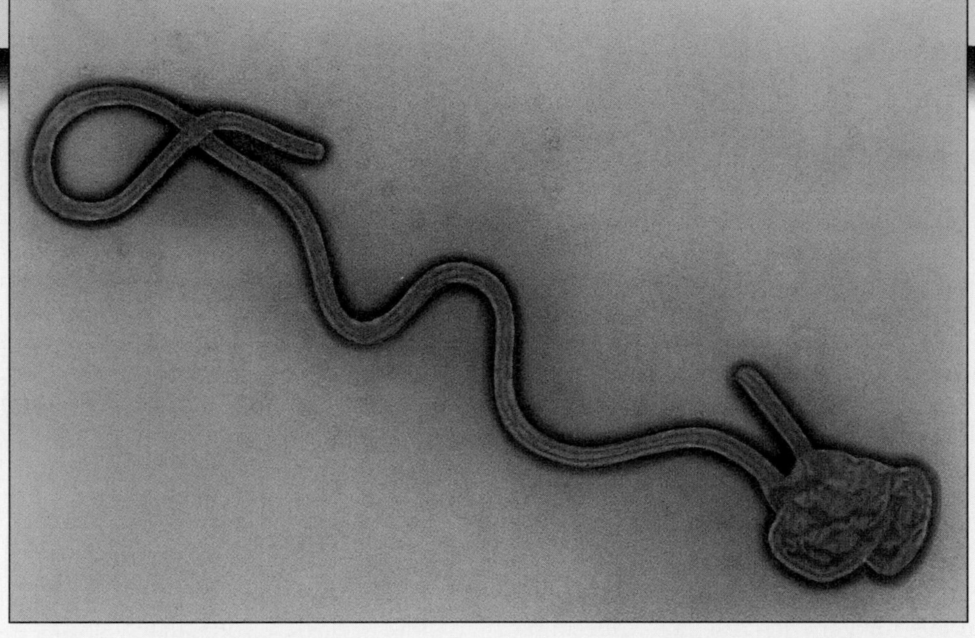

The Ebola virus pictured here (×19,000) is one of a group of viruses that causes deadly hemorrhagic fevers. Wild vertebrates such as rats serve as reservoir hosts.

16

The Genetics of Viruses

CHAPTER OUTLINE

The Discovery of Viruses
The Origin of Viruses
The Structure and Life Cycle of a Bacterial Virus
Mapping the Bacteriophage Genome
T4: A Circular Genetic Map But a Linear Chromosome
Genes-Within-Genes: Bacteriophage ΦX174
HIV: A Eukaryotic Virus

A Killer Unleashed

In the spring of 1995, Joseph Kinfumu, a laboratory technician working in Kikwit, Zaire (now the Democratic Republic of the Congo), came into the emergency room of the local hospital complaining of dysentery and a fever. By themselves, these symptoms were not alarming, but the situation changed rapidly. He soon began to bleed from his eyes, nose, ears, mouth, and anus. Within four days, he was dead. An autopsy revealed that his internal organs appeared to liquify. But Mr. Kinfumu was not going to be the only person to die like this. Soon others were diagnosed with the same disease, and they too died. The disease spread through Kikwit and neighboring villages. Blood samples from the victims were rushed to the Centers for Disease Control in Atlanta, and a terrifying diagnosis was made: the cause was the deadly Ebola virus, the same virus that had killed hundreds in Zaire and western Sudan in 1976 and 1986. Over 90 percent of people infected by Ebola die within nine days. This is a truly frightening disease.

The Ebola virus, like all viruses, requires a host cell in order to reproduce. Viral genetic material, which can be DNA or RNA (Ebola is RNA), is packaged inside a shell of proteins. Upon entering a host cell, the viral genetic material commandeers the cell's protein-synthesizing machinery to make copies of itself and in the process make proteins that, in the case of the Ebola virus infecting humans, make people so sick they usually die. The Ebola proteins cause a profound capillary leakage syndrome and severe blood loss. Progeny viruses are released by infected cells, they infect other cells, and the disease spreads rapidly throughout the body. Although the Ebola virus is more deadly than most viruses, many of these minute parasites cause serious diseases in plants, humans, and other animals.

Viruses straddle the line between the living and the nonliving. For example, the tobacco mosaic virus (TMV), which causes tobacco mosaic disease and often creates a major economic problem for tobacco farmers, can be crystallized and stored on a shelf for years. In this state, it exhibits none of the properties normally associated with living systems: it does not reproduce; it does not grow or develop; it does not utilize energy; and it does not respond to environmental stimuli. However, if TMV is rubbed onto the leaf of a tobacco plant, the viruses infect the cells, reproduce, utilize energy supplied by the plant cell, and are responsive to cellular signals. They exhibit the properties of living systems.

Indeed, it is the simplicity of viruses that has made them ideal materials for genetics research. Questions difficult to answer using more complicated eukaryote systems have been addressed using viruses. In Chapter 10, we discussed experiments that used viruses to demonstrate that genetic information is stored in DNA and RNA. In Chapters 11, 12, and 13, we discussed experiments that used viruses to elucidate the mechanisms of DNA replication, transcription, and translation. In this chapter, we will focus on the life cycles of viruses, the organization of viral genomes, the use of bacterial viruses to investigate gene structure, and our ongoing battle with the human immunodeficiency virus (HIV) that causes acquired immune deficiency syndrome (AIDS).

THE DISCOVERY OF VIRUSES

Today there is general agreement that viruses are ultramicroscopic, obligate, intracellular parasites capable of autonomous replication inside a host cell. However, viruses were not recognized as a distinct class of organisms until well into the twentieth century. They were defined simply as microbes that could pass through a filter specifically designed to trap bacteria. They could not be detected with a microscope, and they did not grow on any medium that supported the growth of bacteria. Perhaps they were just minute bacteria that could pass through a filter.

Viruses were established as distinct from bacteria by the work of Frederick Twort and Felix D'Hérelle during the period from 1910 to 1917. Each observed that the bacterial colonies that he was studying seemed to disappear when inoculated with fluids containing viruses. In a 1915 paper, Twort suggested as one possible explanation that a virus had infected and killed the bacteria. Two years later, Felix D'Hérelle reported similar observations and suggested even more forcefully that this bacterial killing was due to viral infection. D'Hérelle called these bacteria-destroying viruses **bacteriophage,** or "eater of bacteria."

Initially, the motivation for continuing research on bacteriophages was medical, not genetic. D'Hérelle and others believed that these bacterial viruses could be used therapeutically to attack and destroy bacteria that cause such devastating diseases as cholera, typhus, and scarlet fever (see D'Hérelle's Dream of Treating Dysentery in Humans by Phage Therapy in Chapter 23). Unfortunately, they soon discovered that when these bacterial viruses are introduced into the body, they are destroyed by the body's digestive and immune systems. However, phage therapy has reappeared in the 1990s. Some bacterial infections can be treated effectively with "cocktails" that contain several different viruses, all capable of infecting the same bacterial species.

Researchers began using bacteriophages, or simply phages, extensively as experimental organisms in the 1930s. They were attracted to bacteriophages primarily because of their molecular simplicity. Among these early investigators was Max Delbrück, who is often referred to as the father of modern phage genetics. Though trained as a physicist, Delbrück subsequently began asking biological questions. He became especially interested in genetics and in 1935 speculated as to the physical nature of the gene. To him, bacteriophages, because of their simplicity, were the ideal systems with which to explore the structure of the gene. In 1937 Delbrück and his colleague Emory Ellis described the key events in the life cycle of a bacteriophage. They demonstrated that within about 30 minutes after a phage infects a cell, the cell ruptures and releases over 100 progeny phage. Their results raised the question of how a bacteriophage can enter a cell, replicate, and produce so many progeny in such a short period of time.

The pioneering work of Twort, D'Hérelle, Delbrück, Ellis, and others on bacteriophages set the stage for the discovery of a vast array of bacterial, plant, and animal viruses (Figure 16.1), and new viruses are constantly being discovered. The deadly human immunodeficiency virus (HIV) was not characterized until the 1980s. Indeed, given the explosive growth rates and the large population sizes of viruses, scientists can predict that new viral "species" will continue to emerge.

Key Points: **Initially, viruses were thought to be small bacteria, but key observations by Twort and D'Hérelle indicated that viruses are a distinct class of organisms that can infect and destroy bacteria. Delbrück and Ellis recognized the genetic significance of viral simplicity and launched studies into the life cycles of viruses.**

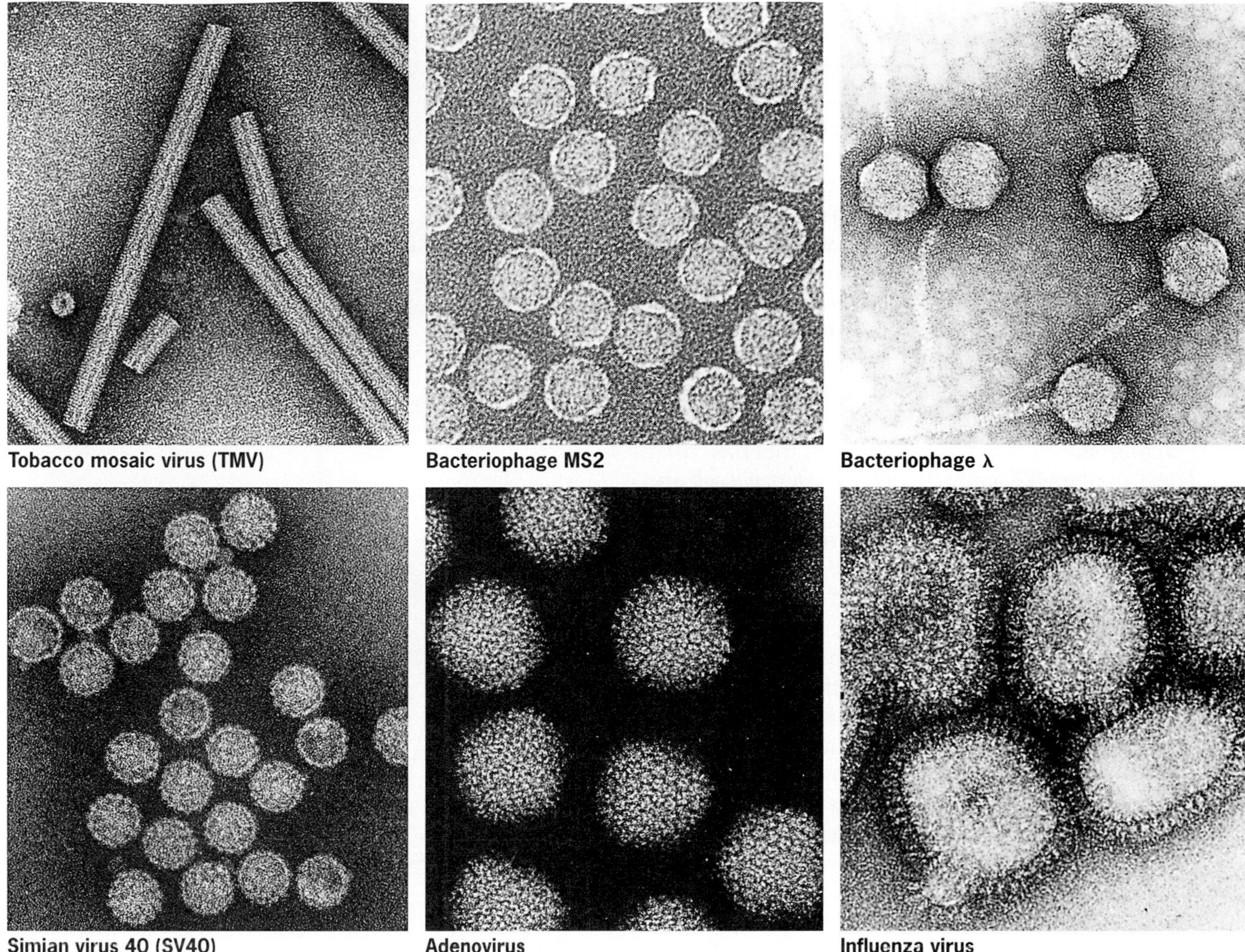

Figure 16.1 Electron micrographs showing the morphologies of a plant virus, two bacterial viruses, and three animal viruses (bottom). TMV, MS2, and influenza virus contain single-stranded RNA genomes; bacteriophage λ, SV40, and adenovirus store their genetic information in double-stranded DNA.

THE ORIGIN OF VIRUSES

All viruses, even the most complex ones, depend on the host cell, whether bacterial or eukaryotic, for their functions and their basic existence. Viruses do not encode ribosome components or the enzymes required for energy production. They are totally dependent on the metabolic machinery of the host cell. This dependence strongly suggests that viruses evolved *after* cells appeared. They probably evolved from small segments of host cell DNA or RNA that acquired the capacity to replicate independently of the genome of the host cell. If such nucleic acid segments acquired coding sequences that specified an encapsulating protein enabling the nucleic acid segment to exit the cell and to infect another cell, then this encapsulated nucleic acid molecule could well represent a primitive viral form.

If viral genomes evolved from segments of host cell genomes, they had regions of sequence homology with this cellular DNA. This homology probably served as a basis for increased recombination between viral and host cell genomes and created opportunities for increasing the complexity of the viral genome. Interacting selection strategies between the viruses and the cells they infect have led to the abundance of viruses that parasitize cells today.

Key Points: **Viruses probably evolved from segments of cellular DNA that acquired the ability to replicate independently of the rest of the cellular DNA and to encode encapsulating proteins.**

THE STRUCTURE AND LIFE CYCLE OF A BACTERIAL VIRUS

Structurally, viruses are the simplest of all organisms. Bacterial viruses, for example, are composed of two classes of macromolecules—proteins and nucleic acids. These viruses usually contain a single molecule of DNA or RNA packaged in a protein coat, which may be simple or complex. Animal viruses are often more complex in that the nucleoprotein complexes are encased in lipid membranes derived from the host cell. Viruses are classified according to three basic properties: (1) whether they contain DNA or RNA; (2) whether the nucleic acid is single- or double-stranded; and (3) whether or not the virus contains a membranous envelope.

The T phages, T1 through T7, of *Escherichia coli* are perhaps the most intensely studied group of bacterial viruses. Although some structural variation occurs between different T phages, they are all tadpole-shaped (Figure 16.2). The **head** is composed of various proteins and is shaped like an icosahedron, a polyhedron with 20 faces. Inside the head is a single molecule of double-stranded DNA. Attached to the head is a **tail** composed of two coaxial hollow tubes—an inner **needle** or **core** and an outer **sheath,** which is contractile. The **tail fibers,** which the virus uses to find a bacterium to infect, are attached to the **baseplate** at the end of the tail. The **tail pins** on the baseplate anchor the virus to the host cell during infection. After attachment, the tail sheath contracts like a small muscle pushing the tail core through the cell membrane. The phage DNA passes through the tail core into the host cell, where it is replicated and subsequently packaged into progeny virus particles. The protein coat remains outside the cell, as demonstrated in the classic Hershey and Chase experiment (Chapter 10).

The life cycle of bacteriophage T4 provides a good illustration of the growth of a bacterial virus. At 37° C (*E. coli* grows at normal human body temperature), the T4 life cycle spans about 22 minutes (Figure 16.3). The phage adsorbs to the bacterial cell wall and injects its DNA into the cell. Shortly after the phage DNA has entered the cell, the synthesis of bacterial DNA, RNA, and protein is shut down by **early phage proteins** (synthesized *early* after infection) that are translated from phage mRNAs synthesized within minutes after infection. Some of these early proteins are nucleases that degrade the host DNA to mononucleotides. Other early proteins are involved in the replication of phage DNA. About halfway through the life cycle, the **late proteins** that form the head and tail structures are synthesized. About 15 minutes after infection, complete progeny phage accumulate inside the cell. At 22 minutes after infection, a phage-encoded enzyme called

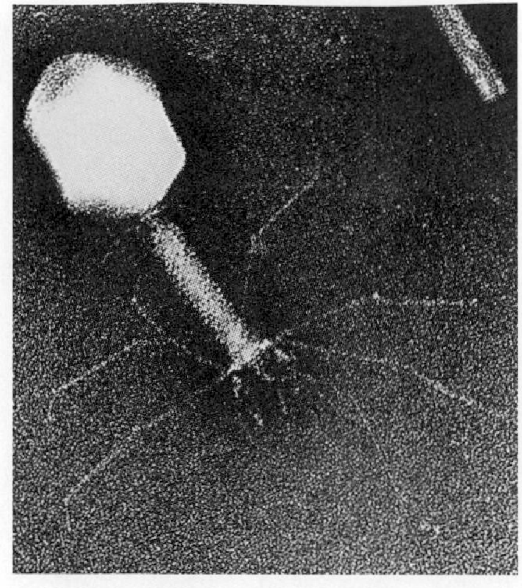

(*a*)

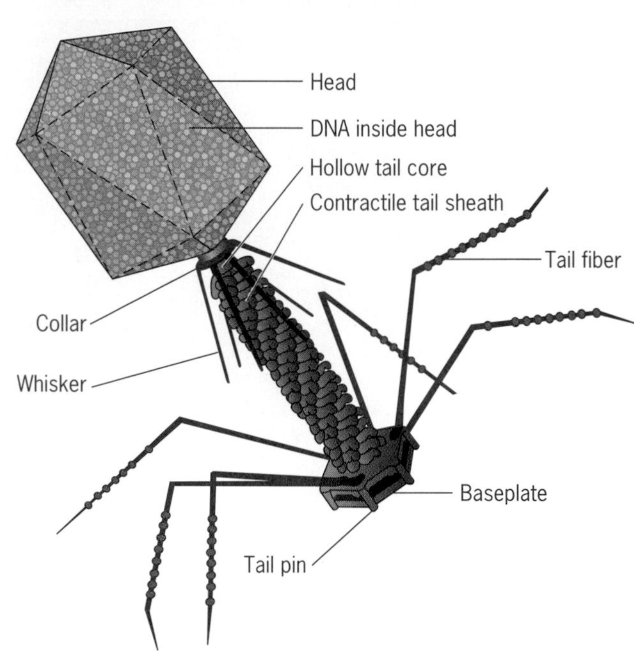

(*b*)

Figure 16.2 Electron micrograph (*a*) and diagram (*b*) showing the structure of bacteriophage T4.

lysozyme acts in concert with other proteins to lyse or rupture the host bacterium, and from 100 to 300 progeny phage are released (Figure 16.3).

Now, let's examine the T4 life cycle in slightly more detail. Within minutes after infection, the RNA polymerase of the host cell is modified by complexing with early phage proteins so that it no longer binds to the promoters of bacterial genes. Thus, this modification effectively shuts off the expression of the host genes. The modified RNA polymerase does, however, recognize the promoters of early phage genes and catalyze their transcription. The bacterial RNA polymerase is then further modified by complexing with other phage

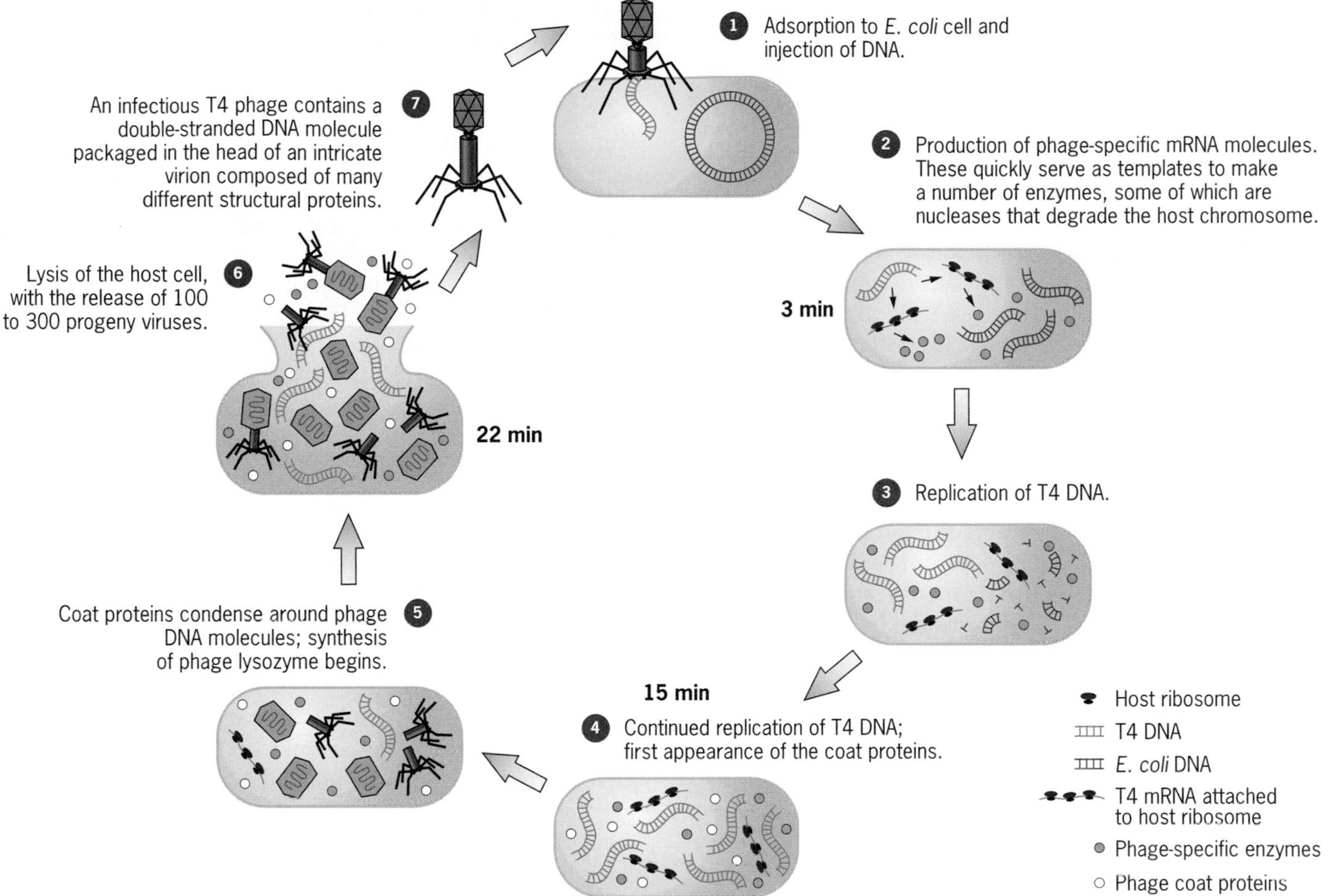

Figure 16.3 The life cycle of bacteriophage T4.

proteins so that it recognizes promoters of phage genes that are expressed during later stages of the life cycle. The sequential modification of the promoter specificity of RNA polymerase ensures a carefully orchestrated sequence of gene expression, with specific genes being switched on and others being switched off at progressive stages in the life cycle.

Another important aspect of the T4 life cycle is the composition of the phage DNA. T4 DNA is unusual in that it contains 5-hydroxymethylcytosine (HMC; cytosine with a —CH_2OH group attached to the 5 carbon) instead of cytosine. In addition, glucose derivatives are attached to the HMC. How is this glucosylated HMC synthesized, and what is its function in T4 DNA?

Recall, as mentioned above, that soon after the phage DNA enters the cell, the C-containing bacterial DNA is degraded. This bacterial DNA degradation is carried out by early phage gene products that cleave cytosine-containing host DNA into mononucleotides (dAMP, dCMP, dTMP, and dGMP). These mononucleotides, in addition to newly synthesized nucleotides, are used in the synthesis of progeny phage DNA. A phage-encoded enzyme converts C into HMC. Another phage-encoded enzyme prevents the incorporation of C into phage DNA. This exclusion is important because if phage DNA contained C instead of HMC, it would be degraded by the same enzymes that degrade host cell DNA. Therefore, a T4 phage with C-containing DNA would commit suicide every time it infected an *E. coli* cell. The addition of hydroxymethyl groups to cytosines protects the phage DNA from degradation by phage-encoded nucleases that degrade the DNA of the host bacterium.

Bacteria also encode endonucleases called **restriction enzymes** that help protect them from invasion by foreign DNAs, including infection by bacterial viruses. These restriction endonucleases degrade both C- and HMC-containing DNAs. Over time, in the competition for control of intracellular niches, viruses have evolved a mechanism to circumvent this bacterial defense. Certain viral genes encode enzymes that add glucose derivatives to HMC residues, and DNA that contains glucosylated HMC is not degraded by the *E. coli* restriction enzymes. Thus, T4 phage DNA must contain HMC and be glucosylated in order for the phage to grow normally in wild-type *E. coli* cells.

The life cycle of bacteriophage T4 typifies the coevolution of bacteria and the viruses that infect them. It is a fascinating story about the mechanisms that have evolved in bacteria to provide resistance to viral infection and the strategies that have evolved in viruses to counter these mechanisms. In Chapter 23, we examine the complex life cycle of bacteriophage lambda, another bewitching virus that infects *E. coli*. These viral-bacterial interactions illustrate the kinds of competition that occur between different genomes for control of intracellular niches, just as different species compete for environmental niches.

Key Points: **Bacteriophage require a bacterial host to complete their life cycle. The phage attach to the host cell and inject their DNA. The phage then take over the protein-synthesizing machinery of the host cell. Phage DNA directs the synthesis of progeny phage that are released from the cell and infect other cells. T4 viruses have glucosylated hydroxymethylcytosine instead of cytosine in their DNA. This protects their DNA from degradation by phage-encoded nucleases and bacterial restriction enzymes.**

MAPPING THE BACTERIOPHAGE GENOME

Constructing a genetic map in an organism requires crossing organisms that carry different alleles of a gene. These alleles cause such phenotypic differences as red eyes versus white eyes or wrinkled seeds versus smooth seeds. However, bacteriophage do not provide such obvious phenotypes. In fact, the virus itself can be seen only with an electron microscope. The phenotypes used to study bacteriophage genetics are usually based on the interaction between phage and host cell, or on variables in the phage life cycle.

Phage Phenotypes

Plaque morphology is a phenotypic trait that is often used in phage genetics research. A plaque is a clear area in a confluent lawn of bacterial cells that results from the lysis or killing of contiguous cells by several cycles of bacteriophage growth. Plaques may be large or small, with sharp or fuzzy edges. Plaque morphology is a function of two factors: (1) the physiological interaction between phage and bacterium; and (2) the phage genotype. The most intensely studied plaque morphology mutants are the **rapid lysis** (*r*) mutants of the T-even phages. Wild-type T phage (r^+) produce a plaque that is small with fuzzy edges, whereas *r* mutants produce large plaques with sharp edges (Figure 16.4).

Host range is another frequently studied phenotypic trait in bacteriophage. Host range mutants are able to infect cells of some bacterial strains but not those of other strains. For example, *E. coli* strain B cells can be infected by all wild-type T-even bacteriophages. However, *E. coli* strain B/2, a mutant derivative of *E. coli* B, is resistant to infection by phage T2. *E. coli* B/2 cells harbor a mutation that alters the phage T2 receptor on the bacterial surface so that T2 phage cannot attach to them. However, a mutant strain of T2, T2*h*, carries a host range mutation (*h*) that allows the virus to infect both *E. coli* B and B/2 cells. This alternation of resistance and susceptibility may continue with bacterial mutations that make *E. coli* resistant to *both* T2 and T2*h*, and with viral mutations that allow infection of the new resistant strains. Similar host range mutants occur in all other bacterial viruses that have been studied.

T2 and T2*h* bacteriophage have the same plaque morphology, but the mutants can be distinguished

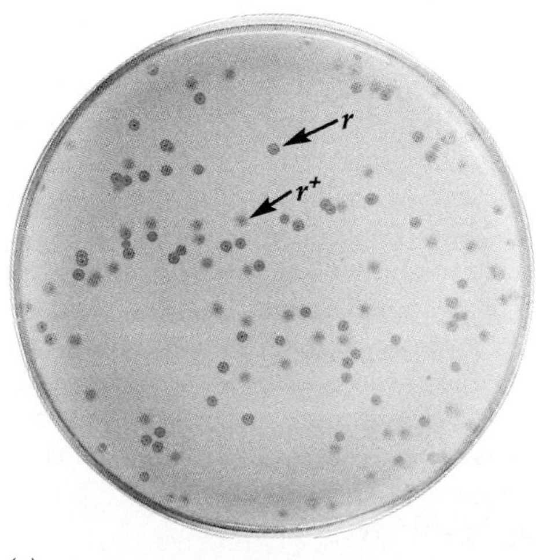

(*a*)

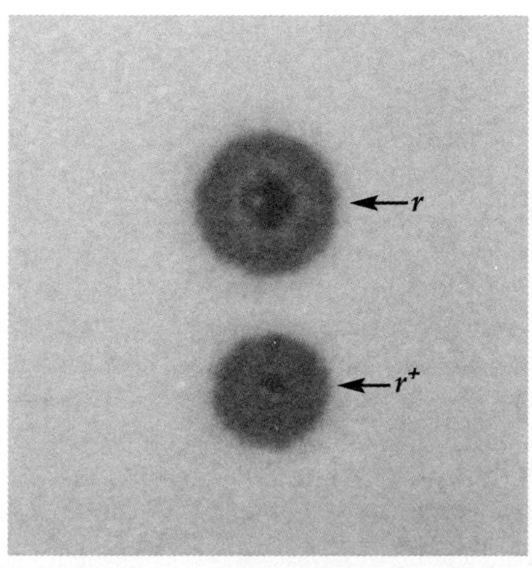

(*b*)

Figure 16.4 (*a*) Wild-type (r^+) and rapid lysis (*r*) mutant plaques of bacteriophage T4 growing on a confluent lawn of *E. coli* strain B cells. (*b*) Higher magnification of individual r^+ and *r* plaques.

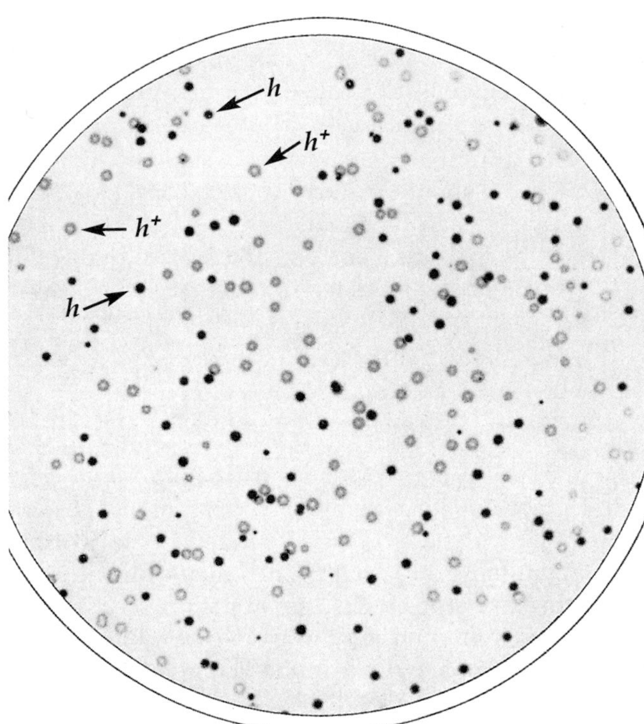

Figure 16.5 Plaques formed by phage T2 wild-type (h^+) and host range (h) mutants when grown on a mixed lawn of *E. coli* B and B/2 cells.

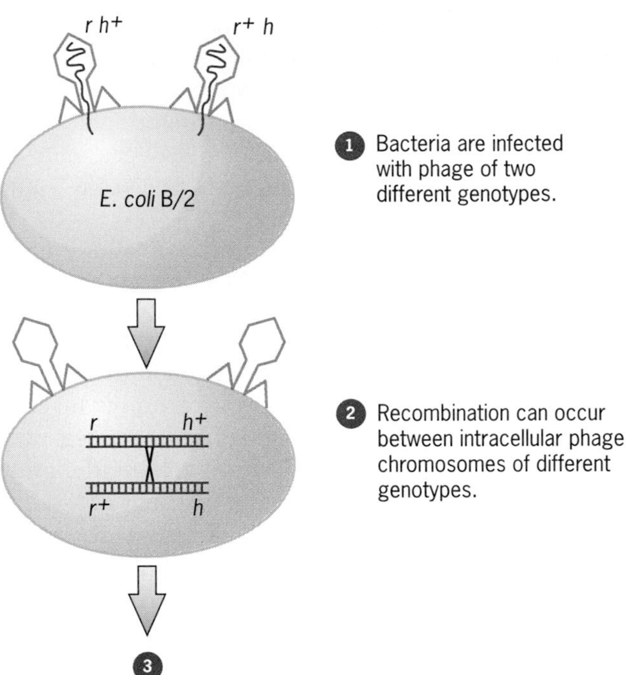

Figure 16.6 A cross between $r\,h^+$ and r^+h bacteriophage. Phage crosses are performed by simultaneously infecting bacteria with viruses of two different genotypes.

from each other if the viruses are grown on a mixed lawn of *E. coli* B and *E. coli* B/2 cells (Figure 16.5). The *h* mutants, which infect both B and B/2 cells, produce clear plaques because all the bacteria in the plaque area are lysed. T2h^+ viruses infect and destroy only the B cells, not the B/2 cells, and thus produce turbid plaques. The turbidity results from growth of the phage-resistant B/2 cells within the plaque.

Genetic Recombination in Phage

In 1946 Max Delbrück and Alfred Hershey independently announced the discovery of genetic recombination in phage. In the wake of this discovery, Hershey and his colleague Raquel Rotman carried out the first detailed study of genetic recombination in phage. They simultaneously infected *E. coli* cells with two different strains of bacteriophage T2—one of genotype h^+r and the other of genotype $h\,r^+$ (Figure 16.6). They collected the progeny phage from this mixed infection—the equivalent of a cross in viral research—and plated them on a mixture of *E. coli* B and B/2 cells. Each of the four possible genotypes produced a distinct plaque phenotype when it was grown on these mixed lawns (Figure 16.7). Wild-type (h^+r^+) plaques were small and turbid; $h\,r$ plaques were large and clear; h^+r plaques were large and turbid; and $h\,r^+$ plaques were small and

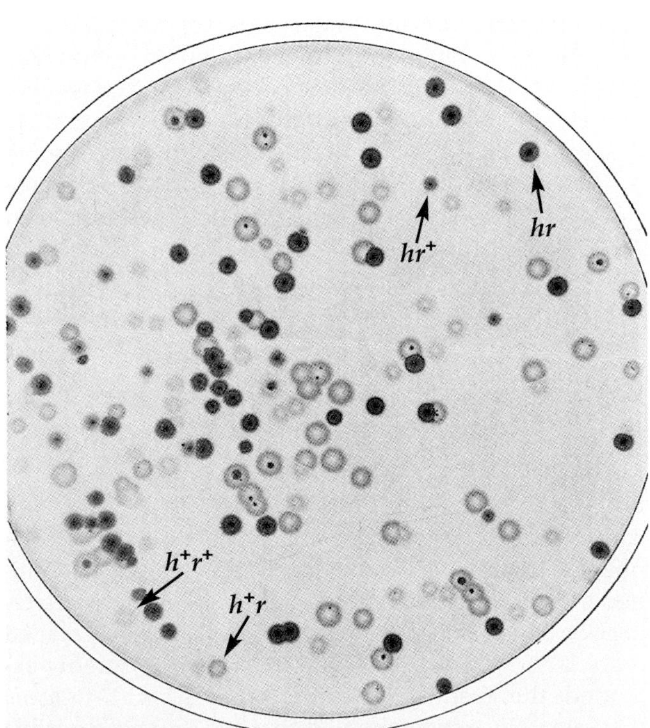

Figure 16.7 Photograph showing the types of plaques formed by T2 $r\,h$, $r^+\,h^+$, $r^+\,h$, and $r\,h^+$ phage on a mixed lawn of *E. coli* B and B/2 cells.

clear. Two of these genotypes are parental (h^+r and $h\ r^+$), and the other two are recombinant (h^+r^+ and $h\ r$). In one cross, Hershey and Rotman found that about 2 percent of the progeny were recombinant and 98 percent were parental. This exciting discovery allowed them to estimate that the distance between the *h* and *r* genes was about 2 map units.

Two important points emerged from these early phage mapping studies.

1. Recombination appeared to occur by a reciprocal process, as it frequently does during meiosis in eukaryotic organisms. Thus, if one recombinant class was present among the progeny produced by a population of doubly infected bacteria, the other recombinant class was usually present at approximately the same frequency. In contrast, lysates from single cells did not contain equal numbers of the reciprocal recombinants, which initially suggested that recombination between phage chromosomes is not reciprocal. However, this apparent lack of reciprocal recombination results in part from the peculiarities of the phage life cycle. Not all phage chromosomes are packaged into mature phage particles. When cells lyse, mature phage are released along with unpackaged DNA and other viral components. Clearly, if both members of a reciprocal event are not packaged into mature phage, the results will imply that recombination is not reciprocal.
2. Recombination in phage does not resemble meiotic recombination. In meiosis, which occurs only in sex cells, chromosomes synapse and cross over at a specific stage of cell division. In viruses, there is no process as complicated as meiosis. Recombination can occur at any time during the phage life cycle as long as the DNA has not been packaged into protein capsids. Multiple rounds of genetic exchange can and do occur.

Genetic Fine Structure

Gene mapping took on an entirely new dimension in the mid-1950s when Seymour Benzer mapped 2400 *rII* mutants to 304 sites separable by recombination in two contiguous genes in a small region of the bacteriophage T4 genome. Four additional sites were identified in another study, bringing the total number of sites to 308. Benzer's objective was to analyze the detailed organization of genes. His exquisite experiments extended the results of earlier studies on *Drosophila* (Chapter 15), which showed that the gene is divisible by both mutation and recombination. He studied the *rII* genes because of the unique properties of mutant phage with defects in these genes.

1. Phage T4 *rII* mutants produce large plaques with clear margins that are easily distinguished from the smaller, fuzzy-edged plaques produced by wild-type T4 phage (see Figure 16.4).
2. New *rII* mutants occur with a relatively high frequency—about 1 in every 100,000 r^+ progeny. This frequency can be enhanced by orders of magnitude by treating wild-type phage with mutagenic agents (Chapter 14).
3. Phage T4 *rII* mutants are unable to grow (are lethal) on a strain of *E. coli* called K12(λ), whereas wild-type T4 does grow on K12(λ).
4. Wild-type T4 phage and T4 *rII* mutants are both able to grow on *E. coli* strain B cells.

The system Benzer employed in his analysis of the *rII* region involved crossing two independently occurring *rII* mutants on *E. coli* B cells, then collecting and analyzing the progeny. If recombination occurred between the two mutants, a wild-type chromosome would result along with a double mutant (Figure 16.8). These recombinant chromosomes should be rare because the distance between the two mutant sites must be quite short. Indeed, a major advantage of the *rII* system is that recombination frequencies as low as 10^{-6} could easily have been detected in Benzer's experiments. The *rII* mutants are conditional lethal mutants like the *amber* mutants discussed in Chapter 13. Whereas wild-type T4 phage can grow on both *E. coli* strains B and K12(λ), *rII* mutants can grow only on *E. coli* strain B.

If progeny phage produced from a mixed infection of *E. coli* B cells are plated on a lawn of K12(λ) cells, only wild-type T4 phage can form plaques. The parental *rII* mutants and recombinant double mutants cannot grow on K12(λ). Thus, a defined volume of progeny phage lysate produced from a mixed infection of *E. coli* B cells is plated on a lawn of *E. coli* K12(λ). If 20 plaques form on the K12(λ) lawn, then 20 wild-type recombinant phage are present in the volume of lysate tested. However, if recombination is a reciprocal process, one *double mutant* recombinant will be present for each wild-type produced by recombination. The 20 wild-type plaques indicate that 40 recombinant phage are present. Thus, recombination frequencies and map distances are calculated as follows.

Recombination frequency =

$$\frac{2 \times \text{number of wild-type recombinants}}{\text{total number of progeny}}$$

and

Map distance (% recombinants) =

$$\frac{2 \times \text{number of plaques on K12}(\lambda)}{\text{number of plaques on B}} \times 100$$

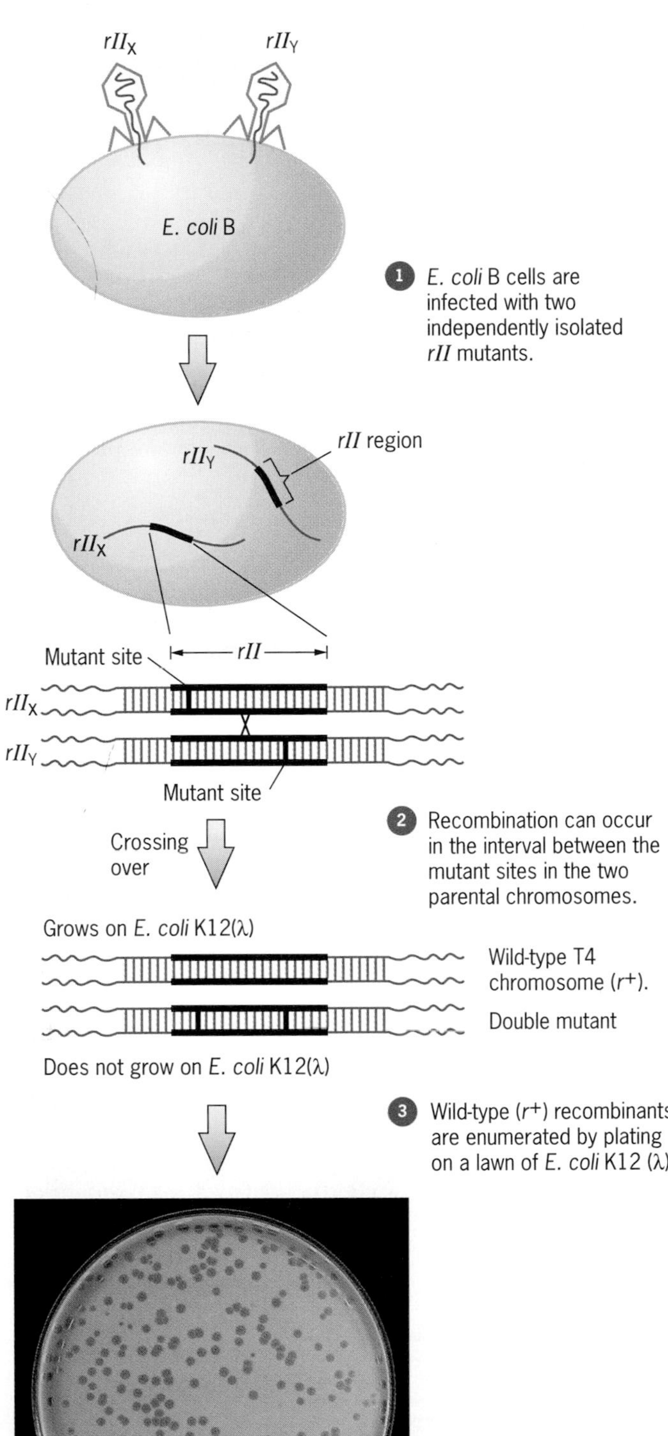

Figure 16.8 A genetic cross in which *E. coli* B cells are infected with two independently isolated phage T4 *rII* mutants. The recombinant progeny are wild-type (r^+) and double mutants.

To estimate the frequency of recombination, the total number of progeny phage produced per unit volume of mixed infection lysate must be determined. The total number of progeny phage produced is determined by diluting the lysate appropriately—perhaps a millionfold—and plating a sample of the diluted lysate on a lawn of *E. coli* B cells. All of the progeny phage (parental and recombinant) produced in a cross between two *rII* mutants can grow on *E. coli* B cells. If 100 plaques are counted in a sample diluted 10^6-fold, then the original sample contained 10^8 phage per unit volume (100×10^6). Thus, if 40 recombinant phage (wild-type and double mutant) are present among 10^8 total progeny, the recombination frequency between the two mutant sites will be $40/10^8 = 0.0000004$.

These calculations overlook one fact. An *rII* mutant may undergo spontaneous mutation back to wild-type (r^+). The frequency of reverse mutation can be estimated for the individual *rII* mutants, and the recombination frequencies can be corrected for these reversion events if necessary. However, except for very short distances or mutations with high reversion rates, such corrections are usually negligible.

All of Benzer's *rII* mutants contained mutations in one of two contiguous genes, designated *rIIA* and *rIIB*. He determined this fact by performing **complementation tests** (see Chapter 15) between each new *rII* mutant and reference mutants known to have mutations in the *rIIA* and *rIIB* genes, respectively. Benzer infected *E. coli* K12(λ) cells with pairs of *rII* mutants. If the two *rII* mutants contained mutations in the same gene (that is, both were *rIIA* mutants or both were *rIIB* mutants), they failed to grow on *E. coli* K12(λ); if they were mutant in different genes (one was *rIIA* and the other was *rIIB*), they grew on *E. coli* K12(λ).

Deletion Mapping

Benzer characterized a number of the *rII* mutants by the procedures described above. Although these procedures worked well, they were labor-intensive. After he had mapped about 60 mutations by two-factor crosses, he devised a new, more efficient procedure and used it to map the rest of the 2400 *rII* mutants in his collection.

Benzer's short-cut mapping procedure is called **deletion mapping.** He demonstrated that some of his *rII* mutants contained deletions of all or part of the *rII* region. A phage that carries a deletion cannot mutate back to wild-type, nor can it recombine with another phage that has a mutation in the region removed by the deletion. A deletion mutant cannot produce wild-type chromosomes by recombination with a point mutant that carries a defect in the interval that is missing in the deletion mutant. Neither phage will have the

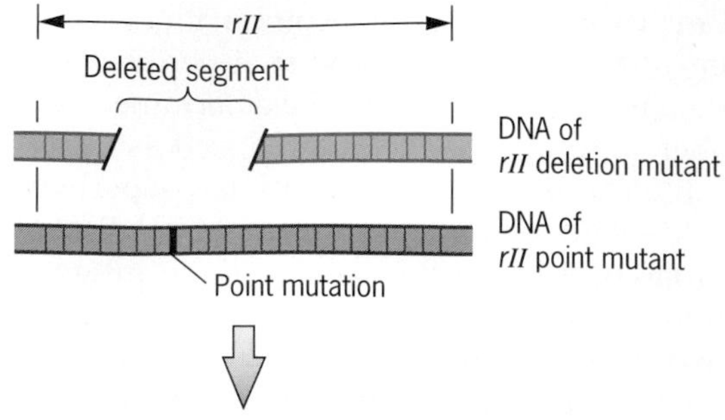

Figure 16.9 A cross between an *rII* point mutant and an *rII* deletion mutant.

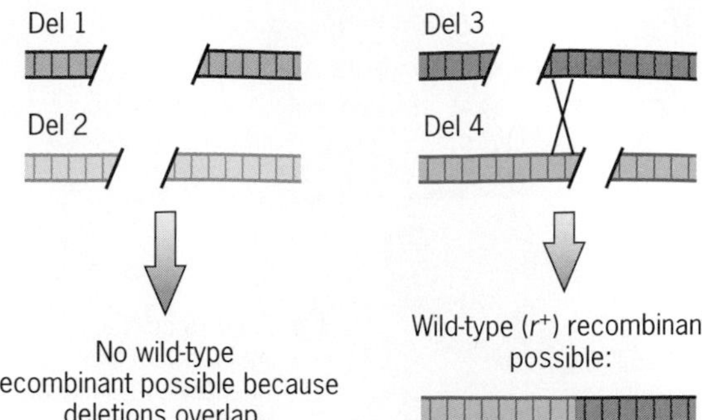

Figure 16.10 Crosses between pairs of deletion mutants. If the deletions overlap, no wild-type recombinants are produced; if they do not overlap, r^+ progeny are produced.

wild-type nucleotide sequence in this region of the gene (Figure 16.9). Any *rII* mutant that produces wild-type recombinants when crossed with a deletion mutant must have a mutation that maps *outside* the deleted segment, and any mutant that fails to generate wild-type recombinants must carry a mutation that maps *within* the area defined by the deletion.

Before he could use deletions in his short-cut mapping procedure, Benzer first had to determine the sizes of the deleted regions. He accomplished this by crossing different deletions with each other (Figure 16.10). If the two deletions overlapped, no wild-type recombinants formed because both were missing the same segment of the *rII* region. The appearance of wild-type recombinant progeny indicated that the deletions did

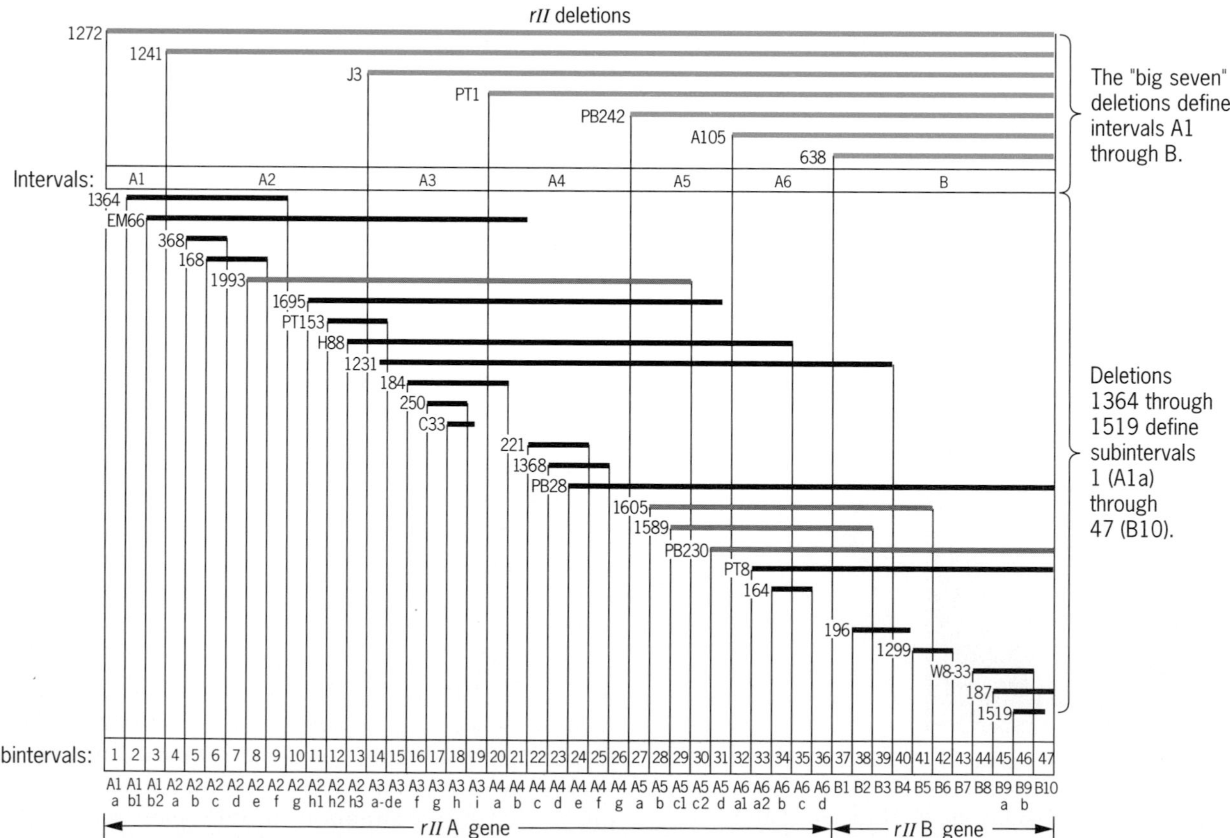

Figure 16.11 The deletion mutants used by Benzer to construct his fine structure map of the *rII* locus of bacteriophage T4. The four deletions shown in orange define the four subintervals of interval A5 (see Figure 16.12).

not overlap. Using this basic approach, Benzer estimated the extent of the deletions. He further refined the limits of each deletion by crossing them with more than 50 mutants that he had collected and mapped by the more tedious earlier methods.

Benzer identified seven large deletions (called the "big seven") that were missing overlapping segments of the *rII* locus, a short segment of the T4 chromosome containing the contiguous *rIIA* and *rIIB* genes (Figure 16.11). These seven deletions divided the *rII* region into seven intervals. By crossing a newly isolated point mutant with each of these seven large deletions, Benzer could map the point mutation to one of the seven intervals.

Benzer also isolated and characterized many smaller deletions that defined 47 short intervals within the *rII* region (Figure 16.11). After he had mapped a point mutation to one of the large intervals by crossing it with each of the "big seven" deletion mutants, he could map it to one of the 47 small intervals by crossing it with each of the deletions that had endpoints within the large interval (Figure 16.12). He then determined the precise location of the point mutation within the short interval by crossing it with each of the other point mutations that mapped to the same small interval. He determined the definitive order of the point mutations within each small interval by performing three-factor crosses as described in Chapter 8.

Let's consider an example. When point mutation *r*548 was crossed with the "big seven"—*r*1272, *r*1241, *r*J3, *r*PT1, *r*PB242, *r*A105, and *r*638—it recombined with *r*A105 and *r*638, but not with the other five large dele-

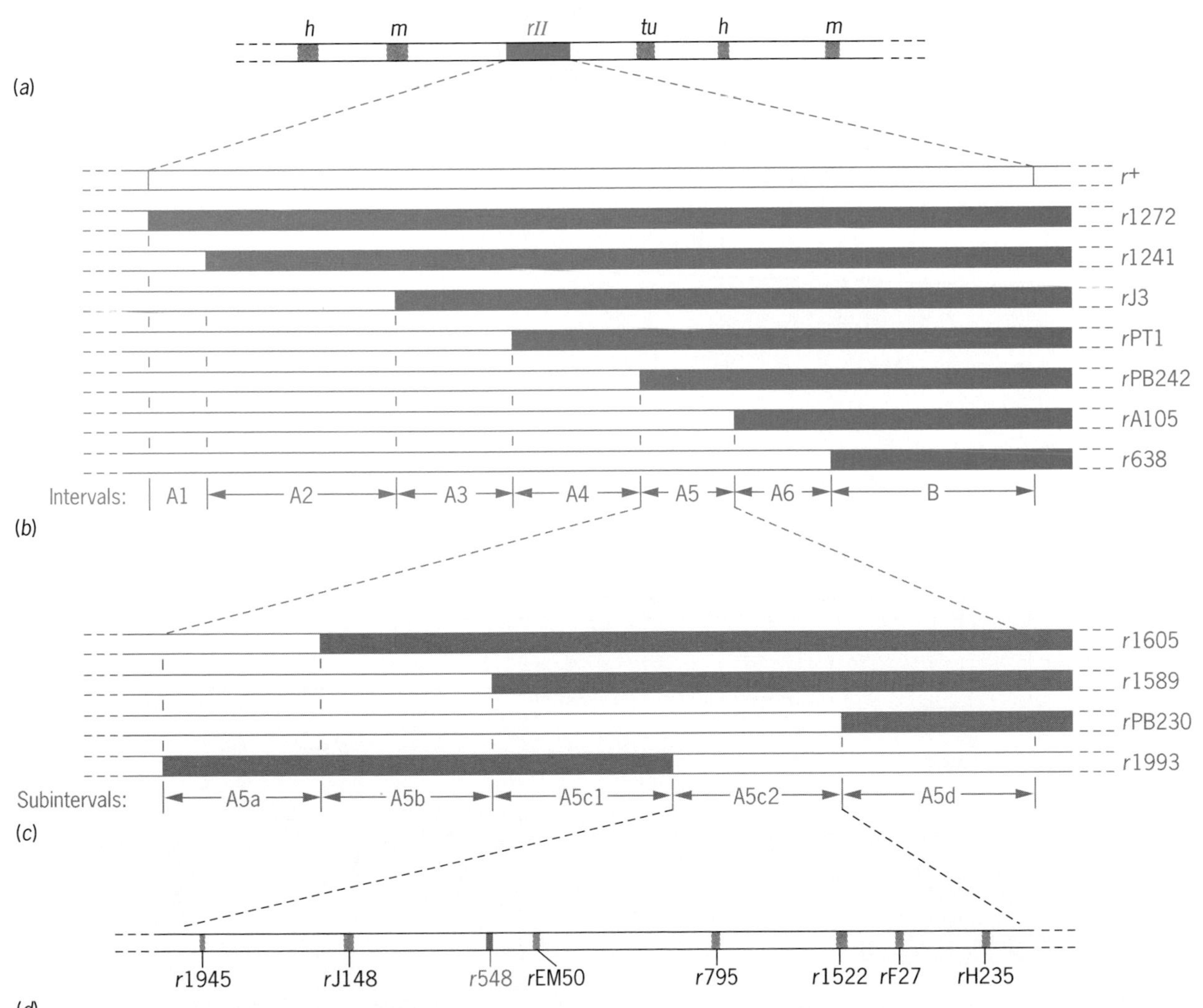

Figure 16.12 Benzer's short-cut deletion mapping procedure. Mutant *r*548 is used to illustrate how an *rII* point mutation is mapped by this procedure (see text). (*a*) Map of a short region of the phage T4 chromosome spanning the *rII* locus. (*b*) The "big seven" deletions were used to map *r*548 to interval A5. (*c*) Mutant *r*548 was next mapped to the A5c2 subinterval based on the results of crosses to deletions with endpoints in the A5 interval. (*d*) The location of *r*548 relative to other point mutations in the A5c2 subinterval was determined by standard two- and three-factor crosses with these mutations.

tions. These results showed that the mutation in *r*548 overlapped with deletions *r*1272, *r*1241, *r*J3, *r*PT1, and *r*PB242, but not with deletions *r*A105 and *r*638, and demonstrated that *r*548 is located in interval A5 (Figure 16.12*b*). Point mutation *r*548 was then crossed with deletions *r*1605, *r*1589, *r*PB230, and *r*1993—deletions with endpoints within the A5 interval. It recombined with *r*PB230 and *r*1993, but not with *r*1605 and *r*1589. These results demonstrated that *r*548 is located in subinterval A5c2, the segment of interval A5 missing in deletions *r*1605 and *r*1589 but present in deletions *r*PB230 and *r*1993 (Figure 16.12*c*). Benzer then only had to cross *r*548 with the other mutations that mapped in interval A5 to complete the fine structure map of this segment of the *rIIA* gene (Figure 16.12*d*).

Benzer used this procedure to map 2400 *rII* mutants within the *rIIA* and *rIIB* genes and to produce by far the most detailed genetic map available at the time (Figure 16.13).

Four important conclusions emerged from Benzer's analysis of the *rII* region.

1. The 2400 *rII* mutants mapped to 304 different sites that were separable by recombination. However, the mutants were not randomly distributed over the 304 sites. Some *rII* sites, called mutational **hot spots**, mutated more often than others; indeed, over 500 spontaneous mutations occurred at one hot spot in the *rIIB* gene (Figure 16.13).
2. All 2400 *rII* mutants mapped to either the *rIIA* gene or the *rIIB* gene or both genes. Benzer suggested that the two genes encoded two different polypeptides, although at the time he had no evidence to support this idea. He has since been proven correct.
3. The smallest recombination frequency that Benzer observed was 0.02 percent. Because the T4 map is about 1500 map units in length, this frequency represents 0.02/1500 = 0.00133 percent of the map. If recombinational events are equally likely over the entire T4 map, given that the T4 genome is about 173,000 base pairs in length, the smallest interval in which recombination can occur is (0.0000133) × (173,000), or about 2.3 base pairs. Thus, Benzer's results suggested, and later experiments proved, that recombination can occur between any pair of adjacent nucleotides.
4. Benzer's work extended the results of earlier experiments with *Drosophila* (Chapter 15) that dem-

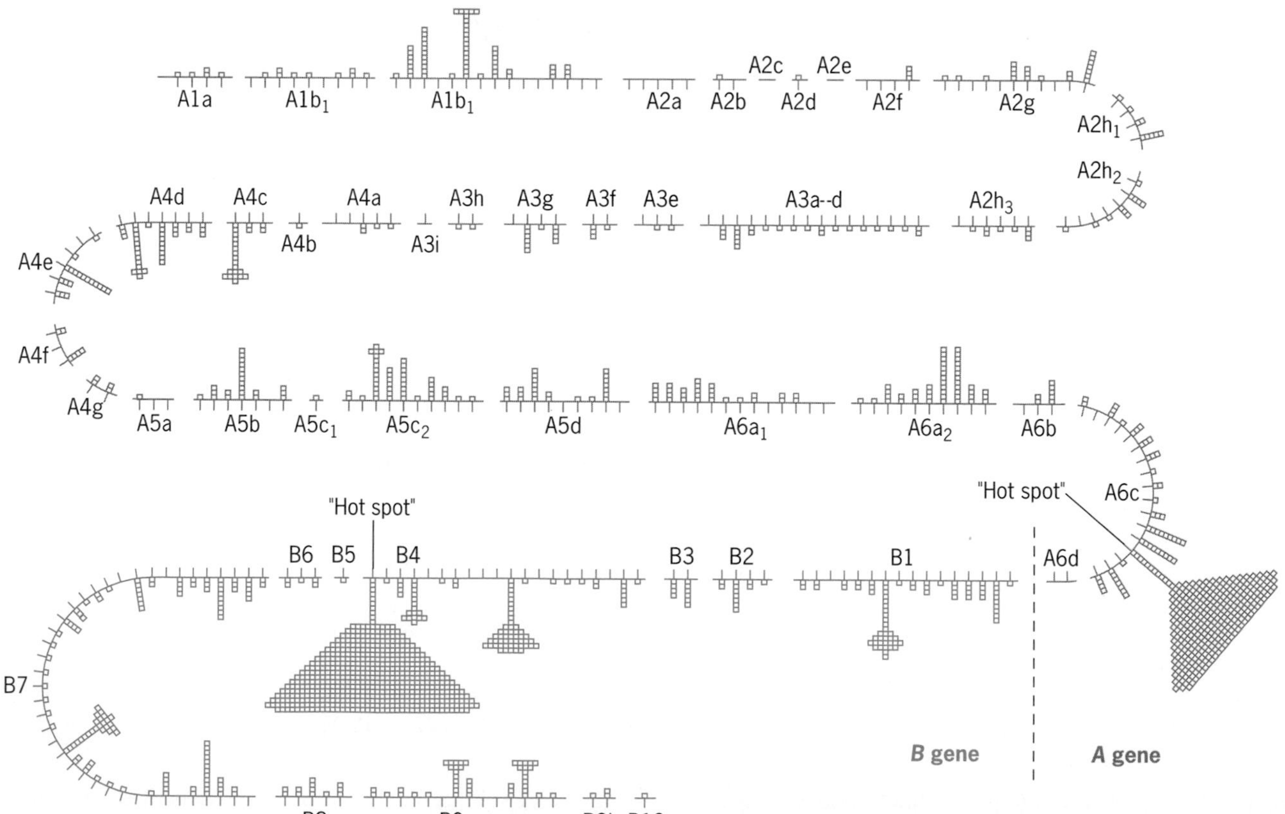

Figure 16.13 The genetic map of the *rII* locus of phage T4. The *rII* locus contains two genes, *rIIA* and *rIIB*. Each square represents the independent occurrence of a spontaneous mutation at the indicated site. Vertical lines at which no squares occur represent positions defined by mutations induced by mutagenic agents. Some sites, called hot spots, mutate at very high frequencies.

onstrated that the idea that the gene is not divisible by mutation and recombination is incorrect.

The existence of a detailed map of the *rII* locus of phage T4 provided researchers with opportunities to explore other important questions in genetics. For example, the *rII* system was used to elucidate mechanisms by which various chemicals induce mutations (Chapter 14). The first evidence that the genetic code was a triplet code resulted from studies of single base-pair addition and deletion mutants (Chapter 13). In addition, crosses involving *rII* mutants provided researchers with important insights about recombination mechanisms. Indeed, Benzer's analysis of the T4 *rII* locus stands today as one of the great achievements in modern genetics.

Key Points: **Plaque morphology and host range phenotypes were utilized in early phage mapping studies. Recombination is different in phage than in sexually reproducing organisms. Benzer mapped over 2400 mutations in the *rII* region of T4 and demonstrated that the base pair was the smallest unit of mutation and that recombination probably occurred between adjacent base pairs. His results solidified the concept of the gene as a unit of function that is divisible by mutation and recombination.**

T4: A CIRCULAR GENETIC MAP BUT A LINEAR CHROMOSOME

Bacteriophage T4 has been used for many important genetic studies. The results of early genetic mapping experiments indicated that the genes of phage T4 might be distributed over as many as seven different linkage groups. However, physical studies showed that the T4 virion contained a single linear molecule of DNA. Thus, all of the genes of T4 had to be located on the same DNA molecule.

The results of genetic crosses with T4 mutants and their wild-type alleles revealed some intriguing inconsistencies. For example, the results of mapping experiments established the following linear gene order: *h*42—*ac*41—*r*67. But when these three genes were used in three-point crosses, the rarest recombinant classes—presumably the double crossovers—indicated that *h*42, not *ac*41, was located in the middle. These inconsistencies between the results of different crosses could not be resolved by repeating the experiments.

However, the dilemma could be resolved by making the T4 genetic map circular (Figure 16.14). With a circular map, *h*42 maps to the left *and* to the right of both *ac*41 and *r*67, depending on the direction taken around the circle. Other crosses confirmed this circular map. But the next problem was that physical studies showed that the T4 DNA molecule is linear, not circular.

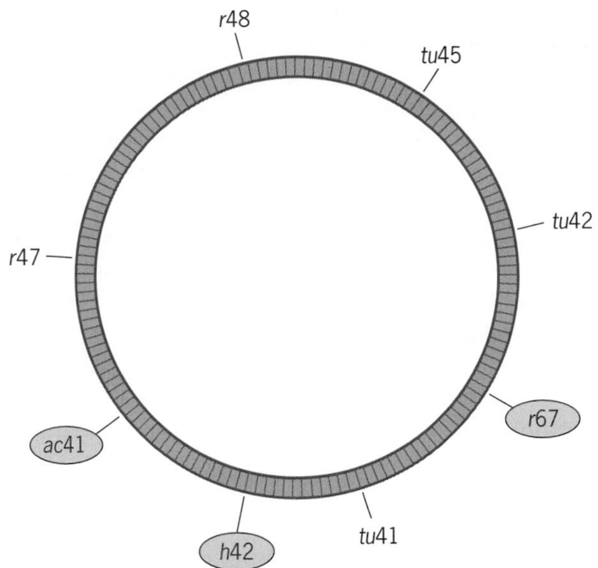

Figure 16.14 The circular genetic map of bacteriophage T4.

How can a circular genetic map be generated from a linear chromosome? In the mid-1960s, George Streisinger proposed that the T4 chromosome was both **terminally redundant** and **circularly permuted.** Initially, both geneticists and virologists were skeptical of this proposal. But both components of the hypothesis were soon proven correct.

Terminally redundant DNA molecules contain the same nucleotide sequence at both ends of a linear molecule, as in the sequence AAGGCCTTGACTA .... TACGTAAGGCCTT. This DNA strand is terminally redundant; the sequence AAGGCCTT is present at both ends.

Circularly permuted sequences are obtained by breaking a circular structure containing a linear sequence (with the ends joined) of markers at random in order to obtain a collection of linear structures, as shown below:

A, 1, B, 2, C, 3, D, 4

Cut at 1:	B	C	D	A	Circular permutations
Cut at 2:	C	D	A	B	
Cut at 3:	D	A	B	C	
Cut at 4:	A	B	C	D	

The ends of a T4 chromosome contain the same genes in the same order; that is, they are terminally redundant (*abcdefg........wxyzabc*). Note that the long, terminally redundant nucleotide sequences of the phage chromosome are denoted *abc* here for brevity. However, the endpoints for each chromosome are different; that is, one chromosome may have the sequence *abcdef .... xyzabc*, whereas another will have the sequence *defghi .... xyzabcdef*, and a third will have the sequence *fghijk .... xyzabcdefgh*. Thus, a population of T4 chromosomes consists of a set of circularly permutated DNA molecules that are also terminally redundant. Sometimes a particular segment is within the terminally redundant ends of the molecule, and sometimes it is in the middle of the DNA molecule.

The combination of these two features—circular permutations and terminal redundancy—results in a circular genetic map from a population of linear chromosomes. But how can a population of circularly permuted, terminally redundant progeny phage chromosomes be generated in a cell infected with a single T4 virus particle? That is, how can a parental phage that is redundant for *abc* at its ends produce progeny that are redundant for *cde*, *def*, *efg*, and so forth? The key to the answer is the way that T4 DNA replicates and is packaged into phage heads. During replication of T4 DNA molecules, recombination occurs between the terminally redundant ends of these molecules and generates long DNA molecules called **concatamers**

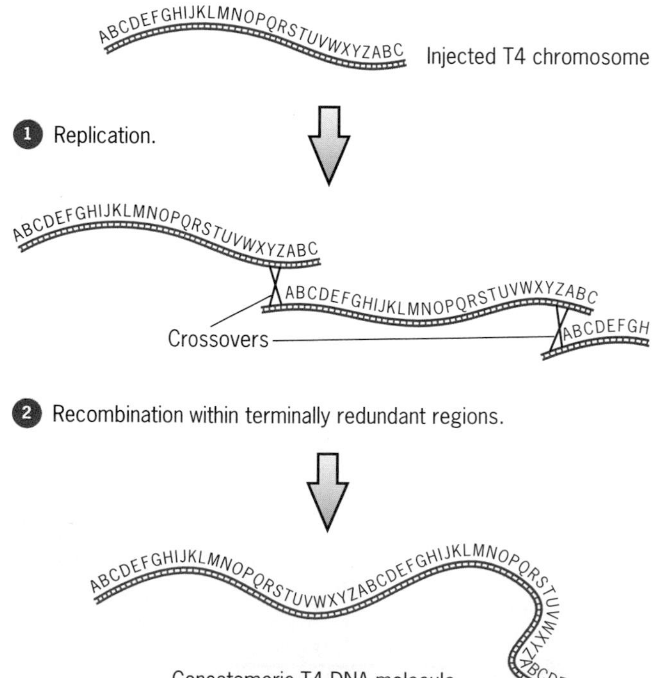

Figure 16.15 The formation of concatameric replicative DNA molecules by "head-to-tail" recombination within the terminally redundant regions of T4 chromosomes.

(Figure 16.15). Each concatamer contains many copies of the phage T4 genome joined end to end.

During the formation of progeny viruses, the T4 head proteins condense around these concatameric DNA molecules until the head is full. The amount of DNA required to fill a head is *slightly more* than one complete set of T4 genes, usually symbolized *abc .... xyz*. Because there is still room left in the head after one complete set of genes (*abc.........xyz*) has entered, a few additional genes are added, yielding a phage head containing gene sequence *abc.........xyzabc*. The head is now full and the DNA is cut. This particular virus is redundant for genes *a*, *b*, and *c*. The next virus particle begins packaging DNA at *d*, gets to *c*, and then keeps going to *d*, *e*, and *f*. This virus particle is redundant for genes *d*, *e*, and *f*. The next virus begins packaging at *g* and is redundant for *g*, *h*, and *i*. This mode of DNA packaging, called the **headful mechanism,** produces T4 progeny chromosomes that are both terminally redundant and circularly permuted (Figure 16.16). All progeny viruses contain terminally redundant DNA molecules, and the molecules are redundant for different genes.

The terminally redundant, circularly permuted nature of T4 chromosomes is supported by definitive physical evidence. Terminal redundancy can be demonstrated by treating T4 chromosomes with an exonuclease that digests DNA only from the 5′ end of each strand. This process creates single-stranded ends that are complementary to each other. Under appropriate conditions, the treated molecules fold back onto themselves, and the complementary single-stranded ends hybridize with each other, forming circular molecules that can be observed by electron microscopy. If the ends were not terminally redundant, there would be no circular molecules because the ends would not be complementary.

Evidence supporting circularly permuted chromosomes was obtained by denaturing a population of T4 chromosomes and then allowing them to renature. During renaturation, single DNA strands from different molecules randomly come together and hybridize. These strands, however, have different redundancies; thus, redundant ends of both strands cannot hybridize. These unhybridized single strands appear as tails on otherwise circular DNA molecules. If all the ends on all the DNA molecules were redundant for the same sequences, these tails would not appear.

The circularly permuted chromosomes of phage T4 will yield a circular genetic map, even though each phage contains a linear molecule of DNA. Map distances are based on average recombination frequencies, which are population parameters. Although two closely linked genes may be located at opposite (terminally redundant) ends of one phage chromosome, they will be present in contiguous segments of DNA within the nonredundant regions of the majority of chromosomes in the T4 progeny population.

Key Points: **Phage T4 has a circular genetic map but a linear chromosome. This apparent paradox is explained by the results of experiments showing that T4 chromosomes are terminally redundant and circularly permuted.**

GENES-WITHIN-GENES: BACTERIOPHAGE ΦX174

One of the most fascinating discoveries about gene organization in viruses has come from studies of the bacteriophage ΦX174. This phage contains a circular, single-stranded DNA molecule, 5386 nucleotides in length. The complete nucleotide sequence of this DNA molecule has been determined. Early studies suggested that there might be six or seven genes, but later studies showed that the ΦX174 genome encodes 11 different proteins comprising more than 2300 amino acids, and therein lies the enigma. The genetic code is a triplet code; that is, a sequence of three nucleotides specifies a single amino acid (Chapter 13). If all 5386 nucleotides of ΦX174 encoded amino acids, the maximum number of amino acids that it could specify would be 5386/3 or 1795, more than 500 amino acid residues fewer than observed. How can this discrepancy be explained?

This enigma was solved by the discovery that the ΦX174 genome contains **overlapping genes** and **genes-within-genes.** Genes overlap when different reading frames of the same DNA sequence encode different proteins. Some segments of the ΦX174 DNA molecule specify two or three different amino acid sequences by being translated in two or all three reading frames (Figure 16.17). For example, protein K is encoded by a DNA sequence shared with the *A* and *A** gene sequence. (*A* and *A** are read in the same frame, but they start at different points.) However, the reading frames for *A* (*A**) and *K* are different. Thus, even though the same nucleotides are used, the amino acid sequences for A and A* are different from K. In addition, the *B* gene is located entirely within the *A* gene, and the *E* gene is located entirely within the *D* gene. Use of the two different reading frames of a single DNA sequence to encode two different proteins is illustrated for the *D* and *E* genes in Figure 16.18. The key feature of all the overlapping sequences is the presence of the initiation codon, AUG (TAC in the DNA molecule), within the sequence of another gene. This initiation codon, which specifies N-formylmethionine

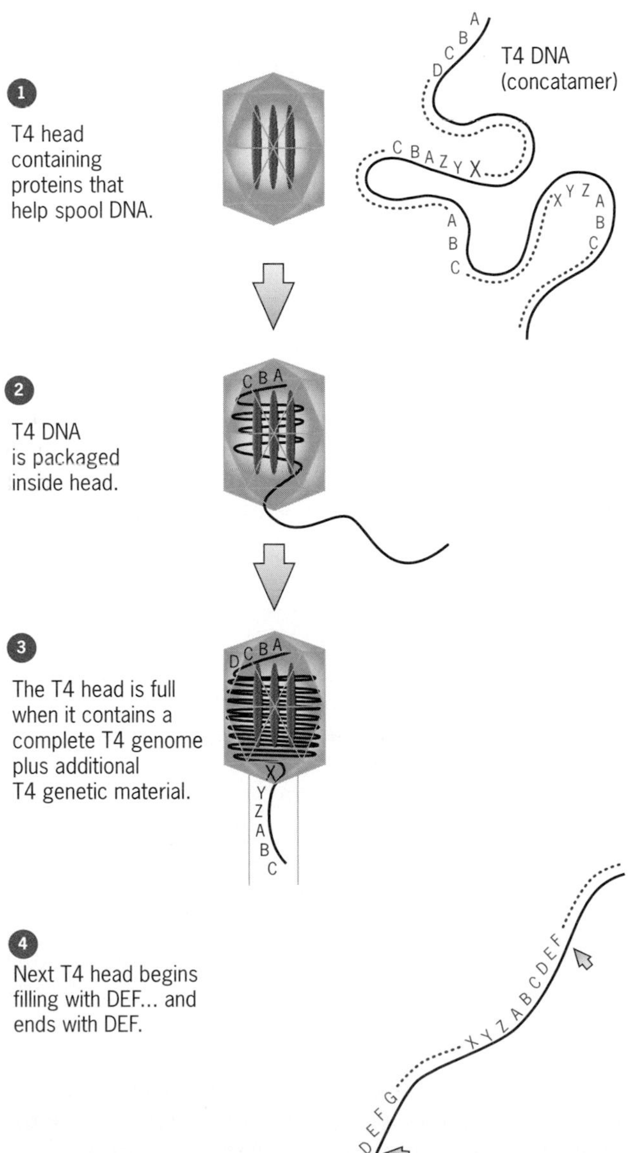

Figure 16.16 Generation of circularly permuted, terminally redundant T4 chromosomes by the headful mechanism.

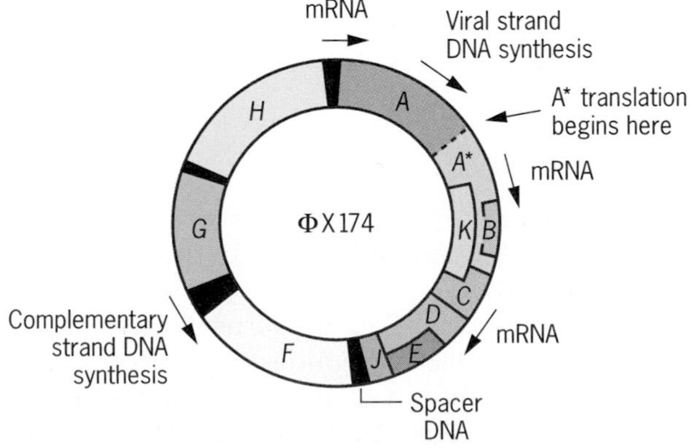

Figure 16.17 Physical map of the phage ΦX174 genome showing the overlapping genes and start and stop points for transcription. The solid black regions are nontranslated spacers.

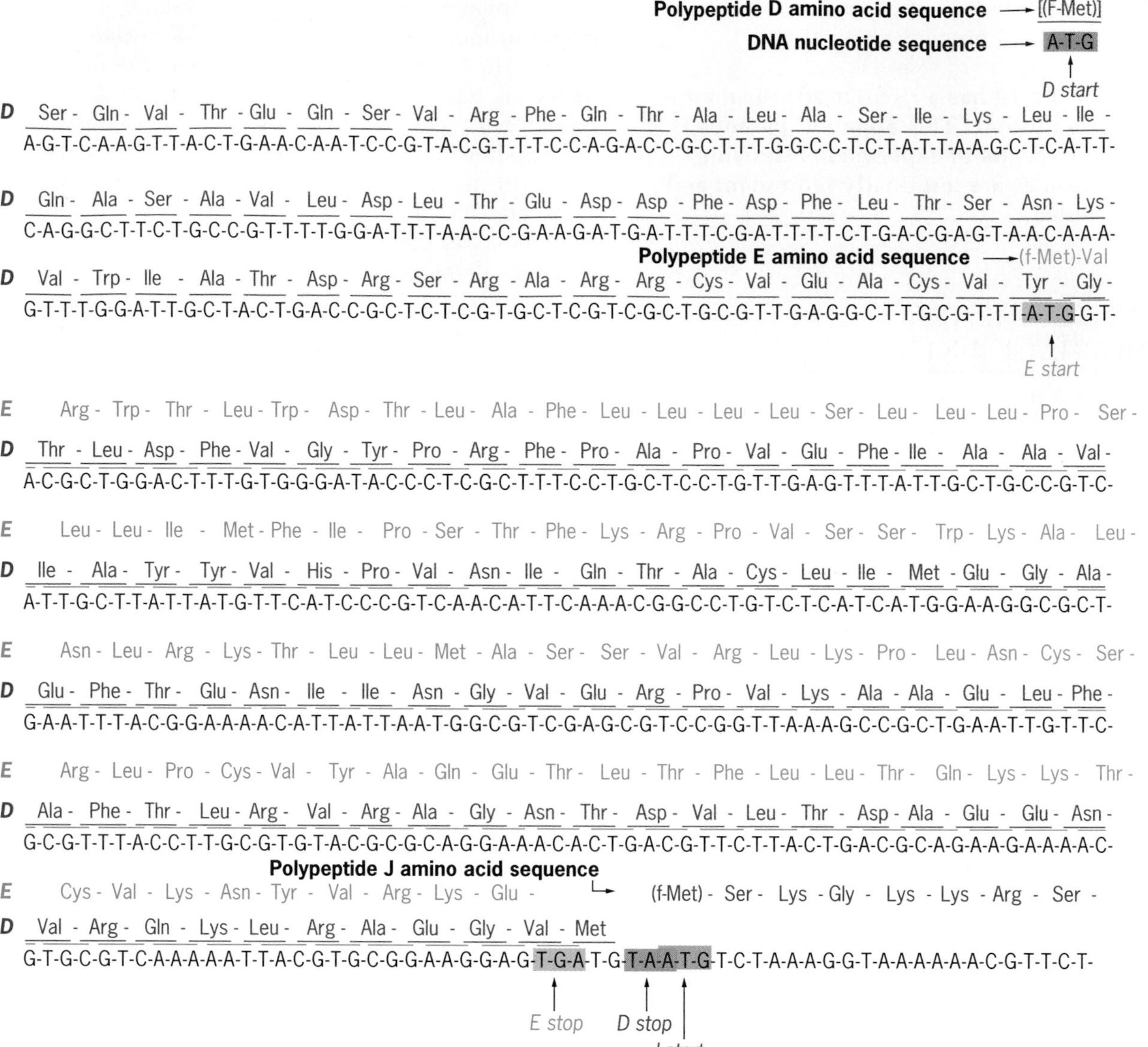

Figure 16.18 Overlapping genes in phage ΦX174. The results of Clyde Hutchison III and co-workers demonstrated that the *E* gene is located entirely within the *D* gene, but the two genes are translated using different reading frames. The reading frame of the *D* gene and the correlated amino acid sequence of the D polypeptide are shown in red above the nucleotide sequence. The reading frame of the *E* gene and the correlated amino acid sequence of the E polypeptide are shown in green. Note that the initiation triplet (ATG) of the *J* gene also overlaps the termination triplet (TAA) of the *D* gene.

(F-met), signals the first amino acid in nearly all proteins (Chapter 13).

Overlapping genes allow the virus to make maximum usage of its 5386 nucleotides. A greater variety of proteins results from a relatively small amount of DNA. However, there is a trade-off. A single mutation will affect the structure of more than one protein, thus slowing the evolutionary process, because it is unlikely that a single mutational event will have a positive or neutral impact on two or three proteins. In spite of this constraint, overlapping genes have evolved in ΦX174 and other viruses.

Key Points: **The single-stranded DNA chromosome of bacteriophage ΦX174 maximizes the use of its small genome by using overlapping genes—DNA sequences translated in different reading frames—to encode 11 different proteins.**

HIV: A EUKARYOTIC VIRUS

In the summer of 1981, the Centers for Disease Control (CDC) reported the unexplained occurrence of *Pneumocyctis carinii* pneumonia in five previously healthy homosexual men in Los Angeles. The same year, the CDC reported a rare skin cancer called Kaposi's sarcoma in 26 previously healthy homosexual men in New York and Los Angeles. These two reports from the CDC represent the first time that **acquired immune deficiency syndrome (AIDS)** was recognized in the United States. By 1985, researchers were referring to the 1980s as the decade of the AIDS pandemic. Many researchers thought that AIDS would be conquered before the 1990s began. They were wrong, for the 1990s marked the second decade of the AIDS pandemic. Today, 30 million people in the world are infected with the **human immunodeficiency virus (HIV),** the agent that causes this disease. In some African countries, one-fourth of the adults are infected with HIV. Whereas the majority of those infected with HIV are not yet showing symptoms of the disease, almost all eventually will unless a cure can be found.

What Is AIDS?

The universal symptom of AIDS is the slow and steady decline of the immune system. This deterioration is primarily the result of the decimation of one population of white blood cells. Because the immune system is so compromised, people with AIDS are vulnerable to a host of diseases that are rarely found in healthy, non-HIV-infected individuals. These opportunistic infections occur because organisms normally unable to infect the body are able to do so. For example, *Pneumocyctis carinii,* a parasite that causes a rare type of pneumonia, is normally controlled by a healthy immune system. In an immune-compromised person, the parasite is usually fatal.

The Structure of HIV

A virus of any type can be viewed as a bit of bad news wrapped in a protein coat. Certainly, bacterial viruses are bad news for bacteria, and eukaryotic viruses are bad news for eukaryotes. The basic structure of HIV has much in common with other eukaryotic viruses. Externally, HIV looks like a twenty-sided soccer ball (Figure 16.19). This shape is determined by the proteins that make up the outer protein coat, or **capsid.** Overlying the capsid is a **lipid membrane** derived from the host cell that HIV infected. Embedded in this membrane are various lollipop-shaped **glycoproteins,**

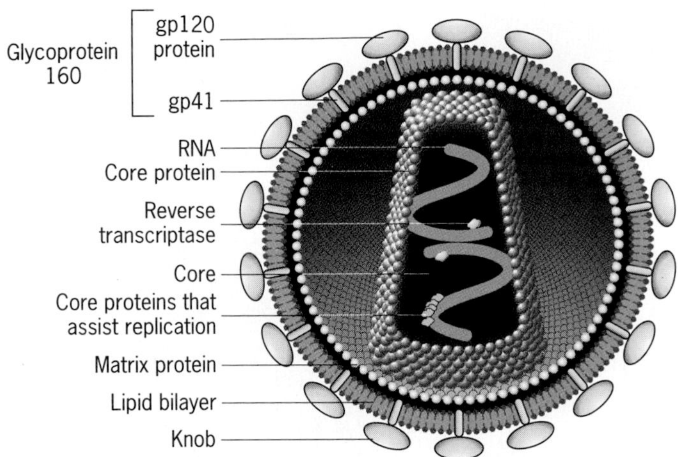

Figure 16.19 Structure of the human immunodeficiency virus (HIV). The core of the virion contains the genetic material and proteins required for replication. The protein capsid is surrounded by a lipid bilayer membrane derived from the host cell. Viral proteins are embedded in the membrane.

proteins complexed with sugars. The stem of the "lollipop" has a molecular weight of 41 kilodaltons (kd) and so is called gp41; the top part of the lollipop has a molecular weight of 120 kd and is referred to as gp120. Inside the capsid is a **protein core** made of various protein molecules.

The genome of HIV is not like that of the T phages discussed earlier. HIV has RNA, not DNA, as its genetic material. Two identical molecules of this RNA are packaged inside the protein core, along with an enzyme called **reverse transcriptase,** which catalyzes the synthesis of DNA copies of the genomic RNA after infection.

The HIV Life Cycle

Figure 16.20 presents an overview of the life cycle of HIV. In brief, HIV enters the host cell by interacting with specific receptor proteins on the cell surface. Upon entering the cell, the RNA genome of HIV is converted to double-stranded DNA by reverse transcriptase, and the HIV DNA is integrated into the chromosomal DNA of the host cell. The integrated HIV DNA is transcribed by the host cell's RNA polymerase, producing both HIV mRNA, which is translated into proteins, and genomic RNA, which is packaged in progeny virus particles.

Now, let's take a closer look at some key steps in the life cycle of HIV. An HIV infection begins with the binding of gp120 to the **CD4 receptor** on the membrane of a type of immune cell called the **helper T cell** (Chapter 27). The CD4 receptor is also found on an-

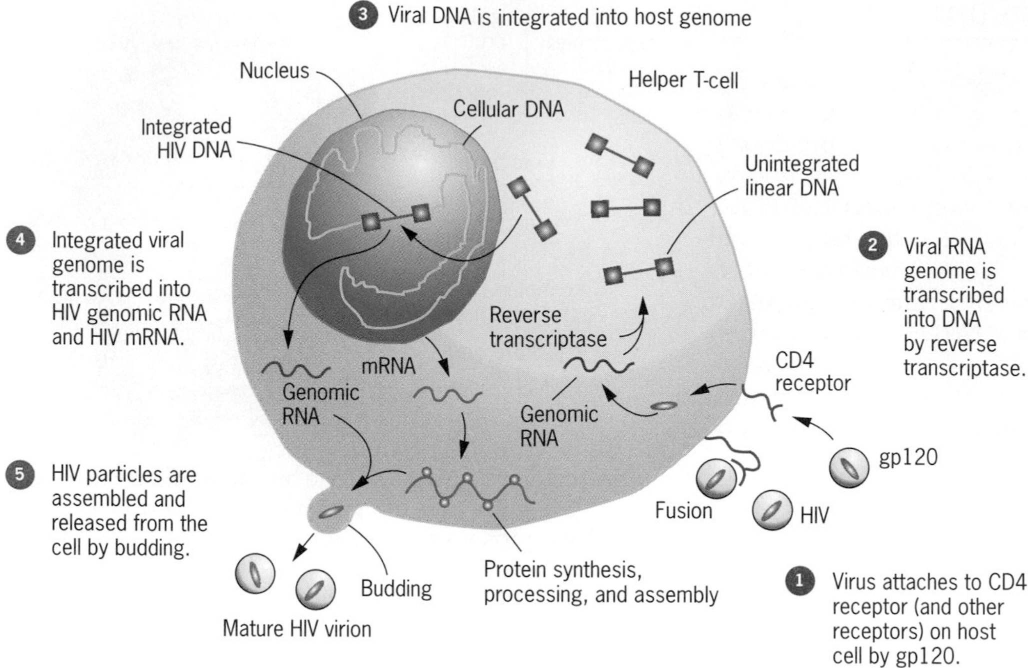

Figure 16.20 Overview of the life cycle of HIV.

other type of immune cell called a **macrophage.** Other HIV-binding receptors, such as fusin and CC-CKR-5, are essential for HIV entry into cells. Once HIV has bound to the host cell, the HIV membrane fuses with the host cell membrane. Fusion is mediated in part by viral protein gp41. Once the virus has entered the cell, the surrounding lipid bilayer and the protein capsids are removed, leaving uncoated viral RNA.

The hallmark of the HIV life cycle is the synthesis of double-stranded DNA from genomic RNA. This process, called reverse transcription (RNA → DNA, rather than DNA → RNA as in transcription), is relatively complex and is catalyzed by a viral enzyme called reverse transcriptase (Figure 16.21). The synthesis of the first DNA strand is primed by a tRNA that contains a sequence complementary to a sequence called PBS (*p*rimer *b*inding *s*ite) in the HIV RNA (step 1). The tRNA is packaged already hybridized to PBS in the HIV core. After reverse transcriptase catalyzes the synthesis of the 3′ end of the viral DNA, ribonuclease H (RNase H) degrades the genomic RNA in the RNA-DNA duplex (step 2). This degradation leaves the repeated (R) sequence of the nascent DNA strand free to hybridize with the R sequence at the 3′ end of the HIV RNA. The net result of this process is that the R region of the nascent DNA strand "jumps" from the 5′ end of the HIV RNA to the 3′ end of the HIV RNA (step 3). Reverse transcriptase next extends the DNA copy by using the 5′ region of the HIV RNA as template (step 4).

In step 5, RNase H degrades all the RNA in the RNA-DNA duplex except a small region, the polypurine tract, which is composed mostly of the purines adenine and guanine. This polypurine tract is used to prime second-strand synthesis of the 3′ half of the HIV genome (step 6). After the tRNA and the genomic RNA present in RNA-DNA duplexes are removed (step 7), a second DNA "jump" occurs during which the PBS sequence at the 3′ end of the second DNA strand hybridizes with the complementary PBS sequence at the 3′ end of the first DNA strand (step 8). The 3′-hydroxyl termini of the two DNA strands are then used to prime DNA synthesis to complete the synthesis of double-stranded HIV DNA (step 9). Note that the conversion of the viral RNA to viral DNA produces the *l*ong *t*erminal *r*epeats (LTRs) that are required for integration (covalent insertion) of the viral genome into the DNA of the host cell (Figure 16.22).

Integration of the double-stranded HIV DNA into the chromosomal DNA of the host cell is catalyzed by a viral enzyme called **integrase** (Figure 16.22). The endonuclease activity of integrase first produces recessed 3′ ends in the HIV DNA by making single-strand cuts near the ends of both LTRs (step 1). These recessed ends are next used for integrase-catalyzed attacks of phosphodiester bonds in the target sequence in the DNA of the host cell. This process results in the formation of new phosphodiester linkages between the 3′ ends of the HIV DNA and 5′ phosphates in the host DNA (step 2). In the final stage of integration, DNA

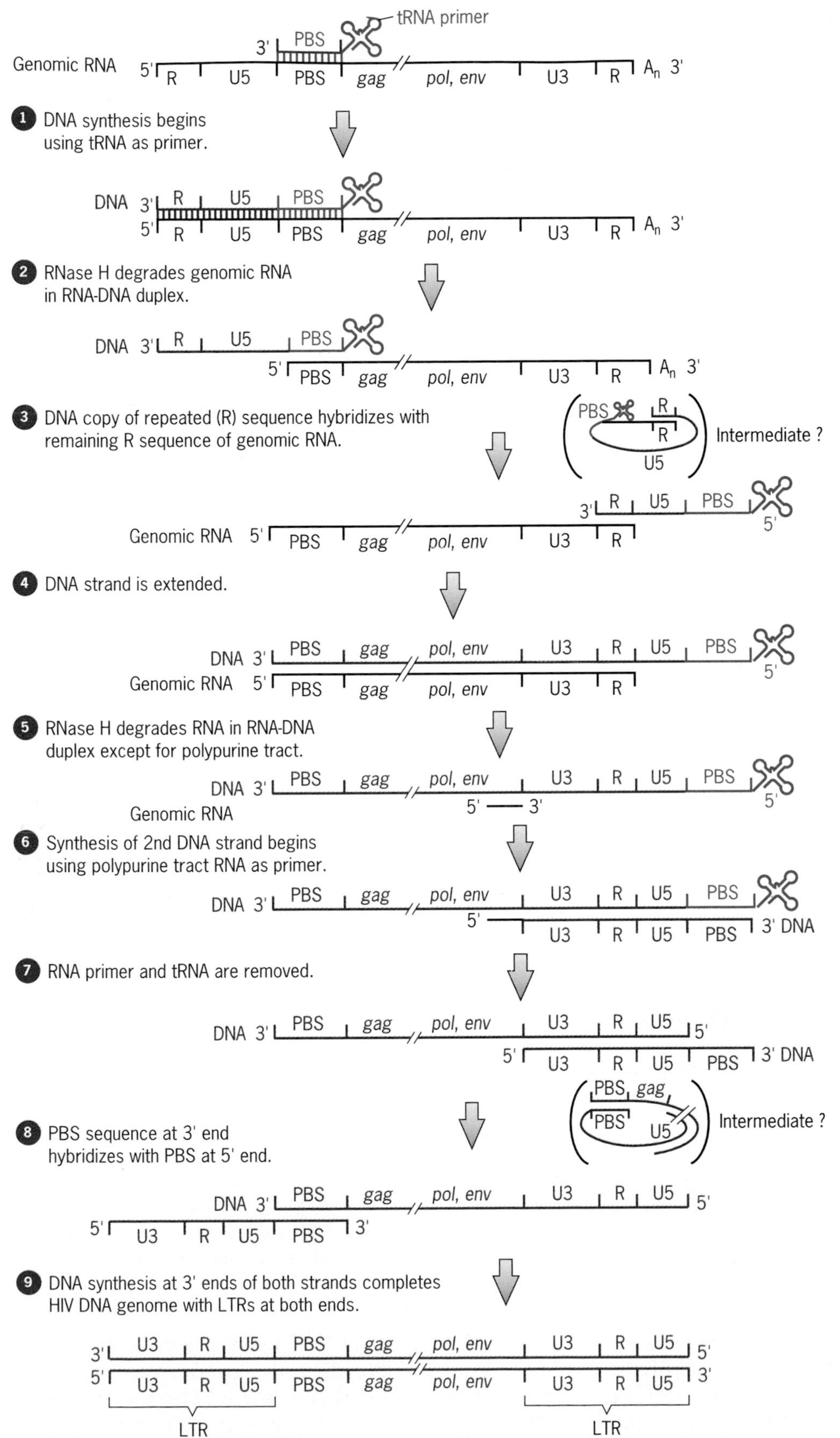

Figure 16.21 Conversion of the single-stranded RNA genome of HIV packaged in the virion to the double-stranded DNA form that is integrated into the DNA of the host cell. The viral reverse transcriptase uses the genomic RNA as template to synthesize a complementary strand of DNA. Reverse transcriptase and the tRNA primer are both packaged with the HIV RNA in the core of the virion. R, *re*peated sequence; U5, *u*nique sequence near *5'* terminus; U3, *u*nique sequence near *3'* terminus; PBS, *p*rimer *b*inding *s*ite; A_n, poly(*A*) tail; *gag*, *pol*, and *env*, sequences encoding HIV proteins (see Figure 16.23); polypurine tract, nucleotide sequence rich in the purines adenine and guanine; LTR, *l*ong *t*erminal *r*epeat.

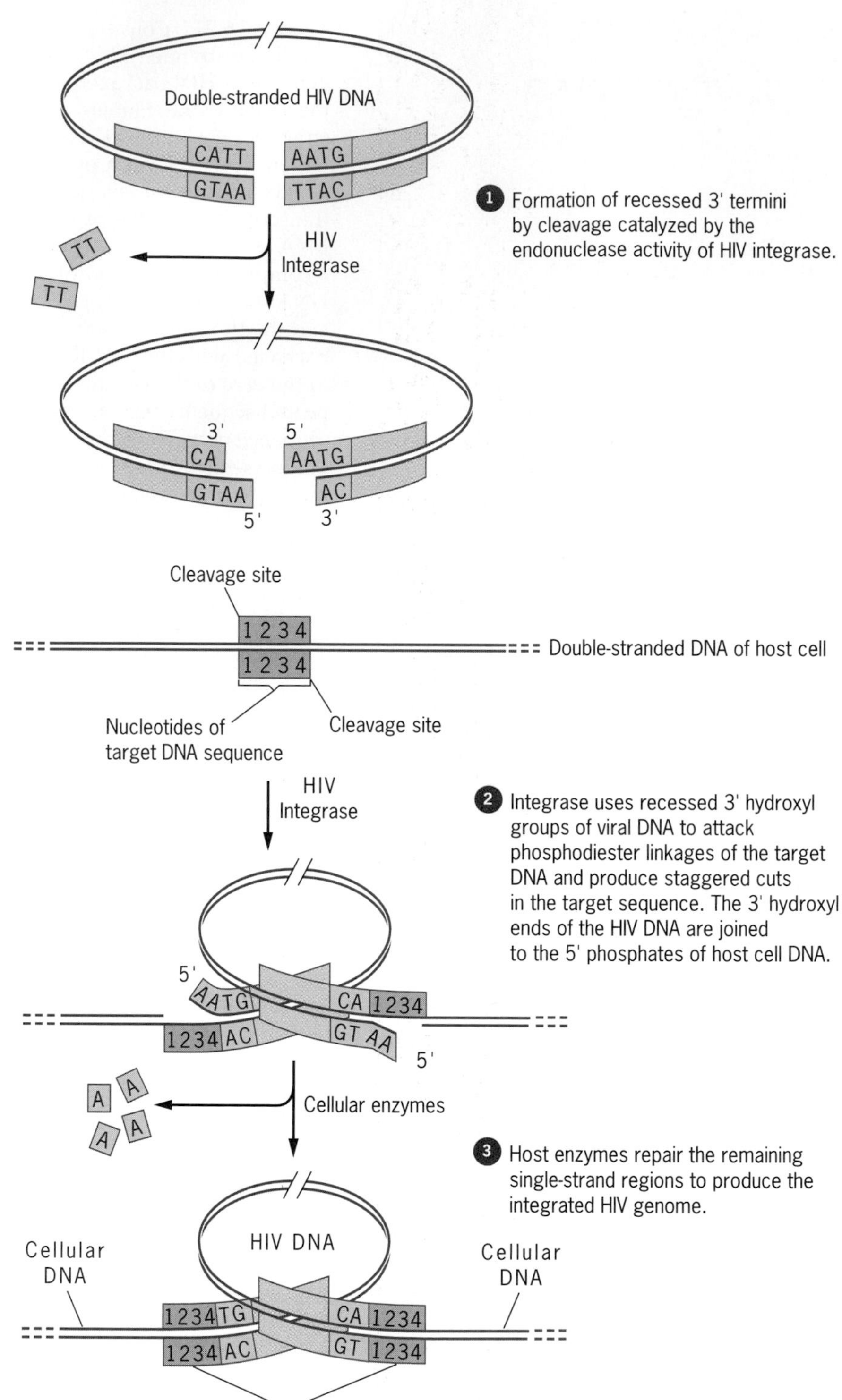

Figure 16.22 Integration of the HIV double-stranded DNA into the chromosomal DNA of the host cell.

repair enzymes of the host cell fill in the single-strand gaps to produce an HIV DNA genome covalently inserted into the chromosomal DNA of the host cell (step 3). The integrated HIV genome is thereafter a permanent part of the host cell genome, replicating just like any other segment of host DNA.

The integrated HIV genome may be inactive, or it may be transcriptionally active, producing progeny viral genomes. Activation of the HIV genome requires virus-encoded and host cell-encoded proteins. Following transcription, HIV mRNA is translated into viral proteins. The viral core is formed at the cell membrane. The core is composed of proteins surrounding viral genomic RNA and enzymes. Budding of progeny viruses occurs through the host cell membrane, where the virion acquires its external envelope.

Because HIV stores its genetic information in RNA and converts it into DNA before integration, it belongs to a class of viruses called **retroviruses** (*retro* viruses because genetic information is transferred RNA to DNA, rather than DNA to RNA as in other types of viruses).

The HIV Genome

The HIV genome is similar to that of other retroviruses (Figure 16.23). The *gag* gene encodes the proteins that form the core; the *pol* gene encodes reverse transcriptase and integrase; and the *env* gene encodes the envelope glycoproteins. These are the basic genes of all retroviruses, but HIV is more complex. At least six other genes (*tat, rev, nef, vif, vpr,* and *vpu*) encode proteins that regulate HIV expression. Flanking these genes are **long terminal repeats (LTRs),** which contain regulatory elements controlling gene expression. HIV contains genes-within-genes, a phenomenon first observed in bacteriophage ΦX174 (see the preceding section). For example, *vif, pol,* and *vpr* have DNA sequences in common. Also, two HIV proteins, Tat and Rev, are produced by alternate RNA splicing (Chapter 15).

The Course of HIV Infection

During the first two months following HIV infection, called the acute phase, the patient can experience flulike symptoms: fatigue, fever, headache, and muscle aches. Large numbers of free HIV particles can be detected in the bloodstream during this time. However, toward the end of the acute phase, the number of free HIV particles in the blood declines, and the virus becomes localized in the lymphatic system. During all stages of the infection, the immune system is actively fighting the virus. HIV-specific antibodies and killer T cells are deployed in this battle (Chapter 27).

Following the acute phase, the flulike symptoms disappear, and the infected person becomes asymptomatic. This phase lasts on average 8 to 10 years. During this period, there is a gradual decline in the number of helper T cells. Some of the infected cells continue to produce HIV particles, but the immune system still remains capable of managing the free viruses and the infected cells. Thus, the individual remains healthy.

Eventually, the immune system wears down. The population of helper T cells diminishes, and the load of HIV particles is not cleared from the system. Disease symptoms begin to appear. The function of helper T cells is to activate other cells of the immune system. When the helper T cells are destroyed, the immune system fails and opportunistic infections occur.

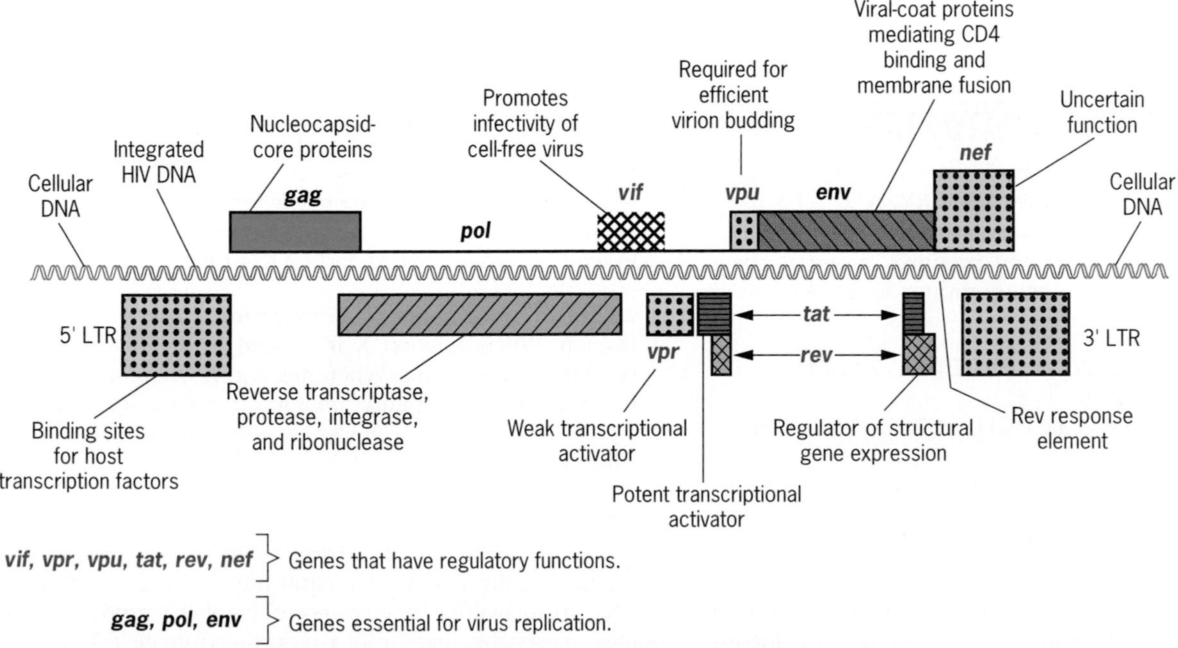

Figure 16.23 Map of the integrated HIV genome showing the location of regulatory genes and genes encoding important viral proteins.

The mechanism by which HIV kills helper T cells is not understood. An uninfected individual normally has about 800 helper T cells per microliter of blood. One diagnosis of AIDS is that the helper T cell count falls below 200 cells per microliter of blood. This means that 75 percent of the helper T cells have been lost. However, when these remaining helper T cells are examined, only about 1 percent of them are HIV infected. Why so many cells die when so few are infected is unknown.

Some researchers have suggested that HIV-infected cells have an indirect effect on uninfected cells, producing a product that kills them or a product that targets them for killing. One suggestion is that HIV-infected cells secrete gp120 fragments from the HIV "lollipop" into the blood. These gp120 proteins are picked up by helper T cells and displayed on their surface, causing them to look as if they are infected. The gp120 fragments are targets for destruction by killer T cells, even though the helper T cells are not HIV-infected.

Some researchers are focusing on other HIV proteins. They suggest that Tat and Nef may be secreted by infected cells to regulate or even destroy the activity of infected but latent and uninfected helper T cells. One strategy used by researchers is to target HIV genes that regulate the infection process and to look for ways to disrupt the viral life cycle. Unfortunately, HIV has evolved ways not only to evade the immune system but also to destroy it. As the virus replicates within the host cell, variants emerge that are more efficient killers, more effective at infecting cells than their progenitors.

Although there is no cure for AIDS, early treatment of HIV-infected individuals can prevent, or at least delay, the onset of AIDS. When HIV-positive individuals are treated daily with three drugs—Epivir®, Retrovir®, and Crixivan®—the growth of HIV is strongly inhibited. However, these drugs must be taken two to three times a day, and they have undesirable side effects such as nausea, weakness, and insomnia. Moreover, treatment costs about $15,000 per year. Two of the drugs—Epivir® and Retrovir®—inhibit reverse transcriptase, and Crixivan® is a protease inhibitor, which interferes with the cleavage of the primary translation product of HIV into individual viral proteins. All three drugs are required to contain the growth of HIV in the body. When any one of the three drugs was eliminated from an HIV-infected individual's treatment regime, the concentration of HIV increased rapidly. Although by no means ideal, this three-drug treatment strategy has dramatically improved the prognosis for individuals infected with HIV.

Key Points: **The human immunodeficiency virus (HIV) is a retrovirus that causes acquired immune deficiency syndrome (AIDS). HIV stores its genetic information in RNA rather than DNA. In infected host cells—usually immune system cells called helper T cells—this RNA genome is converted to double-stranded DNA by the viral enzyme reverse transcriptase. The HIV DNA is integrated into the chromosomal DNA of the host cell. In its integrated state, the HIV genome is transcribed into viral RNAs, which function as both mRNAs and progeny genomes. Eventually, HIV destroys the immune system of the infected individual, resulting in fatal infections by opportunistic microorganisms.**

TESTING YOUR KNOWLEDGE

1. A T4 bacteriophage has the terminally redundant chromosome: ABCDEF .... QRSTUVWXYZABC. This phage experiences a deletion that removes EF. This deletion mutant then infects an *E. coli* cell and produces progeny phage. Which of the following progeny groups would not be expected:

(a) ABCDG....WXYZABC, BCDGHIJ....XYZABCD, GHIJK....ABCDGHI
(b) ABCDG...ZABCDG, BCDGHI...YZABCDGH, GHIJK... BCDGHIJK
(c) ABCDGH...WXYZA, BCDGHI...WXYZAB, GHIJK... YZABCDG

ANSWER

The answer to this question depends on knowledge of the headful mechanism. Progeny group (b) is the only one that is consistent with the deletion of two genes from the T4 chromosome. Groups (a) and (c) are not expected. The redundant chromosome has 29 letters (26 + 3 extras). By deleting 2 letters, it means that the base of 26 letters + 3 extras now is 24 letters + 5 extras. Therefore, the redundancy will be 5 letters in length; (a) has 3 extras and (c) has 1 extra.

2. Three T4 *rII* deletion mutants are intercrossed in pairwise fashion. When deletion X is crossed to deletion Y, no recombinants are found. When deletion X is crossed to deletion Z, no recombinants are found. When deletion Y is crossed to deletion Z, wild-type recombinants are found. What is the map order of these deletions?

ANSWER

The results of the various crosses tell us that X overlaps both Y and Z because no recombinants are found. Y and Z, however, do not overlap because recombinants are produced. The most reasonable map order would therefore be Y X Z:

```
              ___________ X
Y ___________    ___________ Z
```

QUESTIONS AND PROBLEMS

16.1 A series of crosses is made using T4:

(a) $m^+ \; n \; r \times m \; n^+ \; r^+$

(b) $m \; n^+ \; r \times m^+ \; n \; r^+$

(c) $m \; n \; r^+ \times m^+ \; n^+ \; r$

Wild-type recombinants are much less frequent in cross (a) than in crosses (b) or (c). Using this information, determine the sequence of the genes.

16.2 Discuss the similarities and differences in recombination as it occurs in viruses on the one hand and eukaryotes on the other.

16.3 How has the study of bacteriophage genetics contributed to the chromosome theory of inheritance?

16.4 Two strains of T4 infect *E. coli* cells: strain 1 is *a b c*, and strain 2 is $a^+ \; b^+ \; c^+$. The recombination data are as follows:

Class	Frequency
$a \; b \; c$	0.34
$a^+b^+c^+$	0.36
$a \; b^+c^+$	0.05
$a^+b \; c$	0.05
$a \; b \; c^+$	0.08
$a^+b \; c$	0.09
$a \; b^+c$	0.02
$a^+b \; c^+$	0.02

Do these data support the hypothesis that the three genes are linked? If so, determine the sequence of the three genes and calculate the distances separating them. What is the coefficient of coincidence, and what does it mean?

16.5 Hershey and Chase performed mixed infections using three different T2 strains carrying alleles for three different genes (*h*, *m*, and *ri*) and obtained the following results:

	$h^+m^+ri^+$	h^+m^+ri	$h\;m^+ri^+$	$h^+m\;ri^+$	$h\;m^+ri$	$h\;m\;ri^+$	$h^+m\;ri$	$h\;m\;ri$
$h\;m^+\;ri^+ \times$ $h^+\;m\;ri^+ \times$ $h^+\;m^+\;ri$	25	17	18	16	7	8	6	3
$h\;m\;ri^+ \times$ $h^+\;m\;ri \times$ $h\;m^+ri$	3	4	7	10	16	22	15	23

What conclusions can you draw from this experiment pertaining to recombination?

16.6 The following T4 crosses are performed:

Cross	Progeny			
	$r\;h^+$	$r^+\;h$	$r^+\;h^+$	$r\;h$
$r_x\;h^+ \times r^+\;h$	0.38	0.44	0.09	0.09
$r_y\;h^+ \times r^+\;h$	0.43	0.48	0.04	0.05
$r_z\;h^+ \times r^+\;h$	0.47	0.51	0.01	0.01

where *x*, *y*, and *z* refer to different *r* genes.

(a) Construct a map for each cross.

(b) Construct the possible genetic map(s) for all four genes.

(c) The cross $r_y^+\;r_z \times r_y\;r_z^+$ yields 13 percent recombinants. What does this tell you about the h–r_x–r_z order?

(d) From what you know about the T4 genetic map, can you resolve the remaining gene maps so that they are all compatible?

16.7 Study the following phage map:

x — 1.2 — m — 0.9 — n

From 80,000 progeny scored in the cross $x^+ \; m^+ \; n^+ \times x \; m \; n$, 76 are $x^+ \; m \; n^+$ or $x \; m^+ \; n$. What is the interference in this cross? Refer to Chapter 8 to review interference if necessary.

16.8 Three independently arising *rIIB* mutants are crossed in T4, and the following results are obtained:

	Percent Recombinants
$rIIB_1 \times rIIB_2$	0
$rIIB_1 \times rIIB_3$	0
$rIIB_2 \times rIIB_3$	0.9

Interpret these results.

16.9 How do you determine whether a particular mutation in phage is a deletion?

16.10 *E. coli* B cells are infected by two different T4 *rII* mutants (*p* and *q*). The progeny collected from this mating produce *r* plaques on a lawn of B cells. These same two mutants do not grow on *E. coli* K12(λ) cells. If, however, the progeny from an *rIIp* × *rIIq* mating on *E. coli* B cells are plated on a lawn of *E. coli* K12(λ) cells, a few wild-type plaques appear. Interpret these observations.

16.11 How can a physically linear chromosome produce a circular genetic map?

16.12 The phage lambda chromosome is linear except when it is replicating inside a host bacterium. Then it is circular. Yet even though it is circular and recombines as a circular molecule, it produces a linear genetic map. How would you explain this?

16.13 A newly discovered *rIIA* point mutant, when mapped against several deletion mutants, produced wild-type recombinants only when crossed with deletion 638 of the "big seven" deletions. It also produced wild-type recombinants with deletion H88 but not with 1231. Using Figure 16.12, indicate the position of this new mutant.

16.14 T4 DNA is cleaved into fragments about one-fourth the size of the intact T4 chromosome. These fragments are then treated with a 3′ exonuclease to create single-stranded ends. Would you expect these fragments to circularize? Explain.

16.15 The following deletion map of the *rII* locus in phage T4 shows the extent of deletions present in T4 strains d1, d2, d3, d4, d5, and d6. Map the *rII* point mutations *r*41, *r*42, *r*43, *r*44, *r*45, and *r*46 relative to the deletions.

d1
d2
d3
d4
d5
d6

Segments: | A1 | A2 | A3 | A4 | A5 | A6 | A7

	Deletion Mutants					
Point Mutants	**d1**	**d2**	**d3**	**d4**	**d5**	**d6**
*r*41	0	0	+	+	+	0
*r*42	0	0	0	0	+	0
*r*43	0	0	0	+	0	+
*r*44	0	0	0	0	0	+
*r*45	0	0	+	0	+	0
*r*46	0	+	0	+	0	+

("+" = r^+ recombinants produced; "0" = no r^+ recombinants produced)

16.16 The horizontal lines in the topological map below represent the relative positions and extents of five deletions (R, S, T, U, and V) in the *rII* region of T4 phage.

Deletion |←----- gene *A* ----→|←------- gene *B* -------→|
R
S
T
U
V

Segments: | A1 | A2 | A3 | A4 | B1 | B2 | B3 | B4 | B5 |

(a) In which gene does a mutation occur if it complements with U but does not complement with S? (b) Where is a mutation located if it does not complement with S or V? (c) Where is a mutation located if it complements with T? (d) If a mutation is in gene *B*, will it complement with S or V, both, or neither? (e) Can a mutation complement with V but not with U? (f) Give the most probable position of a point mutation if it recombines with deletions S and V but not with the other three deletions. (g) Give the most probable position of a point mutation if it does not complement V, does not recombine with R, but does recombine with the other four deletions.

16.17 In bacteriophage ΨY271, one gene is responsible for cell lysis and is expressed late in the infection process. A DNA fragment containing the template strand for this gene was mixed, under re-annealing conditions, with RNA that was collected at various times after the initial infection. The results are summarized below.

RNA Collection Time (after infection)	DNA-RNA Duplexes Detected	Cell-lysis
3 minutes	no	no
6 minutes	no	no
9 minutes	yes	no
12 minutes	yes	no
15 minutes	no	no
18 minutes	no	no
21 minutes	yes	yes

Provide an explanation for these results.

16.18 Why does the infection of a human with the HIV retrovirus nearly always lead to acquired immune deficiency syndrome?

16.19 The enzyme reverse transcriptase lacks the ability to proofread and correct errors during the synthesis of DNA from RNA. How do you think this might complicate the medical treatment of individuals infected with HIV and the development of a vaccine that protects humans against this retrovirus?

BIBLIOGRAPHY

Benzer, S. 1959. On the topology of the genetic fine structure. *Proc. Natl. Acad. Sci. USA* 45:1607–1620.

Benzer, S. 1961. On the topography of the genetic fine structure. *Proc. Natl. Acad. Sci. USA* 47:403–415.

Cairns, J., G. S. Stent, and J. D. Watson (eds.). 1966. *Phage and the Origins of Molecular Biology*. Cold Spring Harbor Laboratory Press, Cold Spring Harbor, NY.

Goff, S. P. 1997. Retroviruses: Infectious genetic elements. In *Exploring Genetic Mechanisms*, M. Singer and P. Berg, eds. University Science Books, Sausalito, CA.

Hayes, W. 1968. *The Genetics of Bacteria and Their Viruses*, 2nd ed. John Wiley and Sons, New York.

Karam, J. D., ed.-in-chief. 1994. *Molecular Biology of Bacteriophage T4*. ASM Press, Washington, DC.

Levine, A. J. 1992. *Viruses*. Scientific American Library, New York.

Levy, J. A. 1998. *HIV and the Pathogenesis of AIDS*, 2nd ed. ASM Press, Washington, DC.

Stent, G. S. 1963. *The Molecular Biology of Bacterial Viruses*. W. H. Freeman, San Francisco.

Stent, G. S. 1966. *Papers on Bacterial Viruses*, 2nd ed. Little, Brown, Boston.

Stent, G. S., and R. Calendar. 1978. *Molecular Genetics: An Introductory Narrative*, 2nd ed. W. H. Freeman, San Francisco.

An elongated F^+ *Escherichia coli* cell (left), covered with thin hairs called fimbrae, is connected to a smaller F^- cell (right) by a single sex pilus during the process of conjugation.

17

The Genetics of Bacteria

CHAPTER OUTLINE

Genetic Exchange in Bacteria: An Overview
Mutant Phenotypes in Bacteria
Basic Test for Transformation, Conjugation, and Transduction
Transformation
Conjugation
Transduction
The Evolutionary Significance of Sexuality in Bacteria

Drug Resistance and Gene Transfer in Bacteria

Thomas Walker was a middle-aged male with a heart murmur. He entered the hospital for surgery to replace a defective heart valve. This is a serious operation with occasional complications, but the complications are seldom fatal. The surgery was successful; however, three days later he developed a high temperature. Chest X rays showed fluid in his lungs; he had pneumonia. Walker was treated with two powerful antibiotics, but they failed to stop the infection. Despite the addition of a third antibiotic, he developed respiratory failure and died five days after surgery. An autopsy showed that Walker's death was caused by an antibiotic-resistant bacterium in the genus *Pseudomonas*. The resistant bacteria that killed Walker were introduced into his surgical wounds by hospital personnel who did not properly wash their hands. Once in the wounds, the bacteria multiplied, unchecked by once lethal doses of antibiotics.

Why did drugs that in the past had been so successful in controlling these bacteria fail? The *Pseudomonas* strain that infected Walker had acquired antibiotic-resistance genes—genes that encode enzymes that modify or degrade antibiotics, rendering them ineffective. Many of these antibiotic- and drug-resistance genes are present on small, circular DNA molecules called plasmids, which exist separate from the main bacterial chromosome. Plasmids that carry antibiotic- and drug-resistance genes are called R plasmids (*r*esistance plasmids). What makes R plasmids so hazardous to human health is that (1) a single R plasmid may carry genes for resistance to several antibiotics and drugs; and (2) R plasmids can spread rapidly through bacterial populations, even different species, by natural processes of genetic exchange.

Drug-resistant strains of pathogenic bacteria are creating disturbing scenarios in hospitals throughout the world. Physicians are running out of options in selecting antibiotics to use in treating bacterial infections. Of 125 new antibiotics that were undergoing research and development in 1998, 28 were being tested specifically for use on antibiotic-resistant microorganisms. Nevertheless, new antibiotics are not being developed as rapidly as pathogenic bacteria are becoming resistant to those in current use. Many medical researchers fear that in the near future some bacterial pathogens will be resistant to all available antibiotics and drugs.

Bacteria reproduce asexually; that is, one cell divides by fission into two cells, two cells divide into four, and so on. Although we do not usually think of bacteria as having any kind of a sexual mode of reproduction, they do. In fact, genetic information is transferred from one bacterium to another by three totally distinct processes—transformation, conjugation, and transduction. Indeed, it is one type of sexual exchange—the conjugative transfer of R plasmids between bacteria—that has created serious problems in hospitals and clinics throughout the world. Today, the spread of drug-resistant pathogenic bacteria is a serious threat to the successful treatment of microbial disease. In this chapter, we discuss the three modes of genetic exchange in bacteria, the structures and functions of plasmids, the conjugative transfer of transmissible plasmids, and the evolution and importance of sexual reproduction in bacteria. With the complete nucleotide sequences of 18 species of bacteria now available, scientists can compare the genomes of various bacterial species and document the genetic changes that have allowed these species to adapt to various ecological niches. We discuss these comparisons in Chapter 21.

GENETIC EXCHANGE IN BACTERIA: AN OVERVIEW

The development of plasmid-mediated antibiotic resistance points out a mechanism for genetic exchange that has evolved in bacteria. It is not true sexual exchange in which meiotically produced male and female gametes unite in fertilization, such as that observed in peas, fruit flies, and humans. Even though the process may differ, the end result is similar: The transmission of plasmid DNA from one organism to another creates new genotypes, in this case, drug-resistant genotypes.

Plasmid genes can be mapped, and certain types of plasmids can be used to map genes in the main bacterial chromosome. To map genes in bacteria, geneticists apply the same basic experimental strategy that they use for other organisms. They cross genetically different strains and analyze the progeny for recombination events. They then estimate map distances between two genes by the frequency of exchange events between them. But there are important differences between crossing over in prokaryotes and crossing over in eukaryotes. Crossing over in eukaryotes occurs primarily between paired chromosomes during the early stages of meiosis I and results in an exchange of genetic material between two homologous chromosomes. Furthermore, both products of a single exchange can sometimes be recovered. The situation in prokaryotes is different. There is no meiotic process in prokaryotes; recombination is not necessarily reciprocal; and the products of a genetic exchange are not always recoverable.

A bacterium contains a single main chromosome—usually circular—carrying thousands of genes, all involved in various aspects of the bacterial life cycle. A bacterium may also contain one or more extrachromosomal elements called **plasmids.** Genetic exchanges in bacteria rarely involve the interaction between intact main chromosomes from different bacterial strains. Instead, recombination usually occurs between the main chromosome of one strain and a DNA fragment from another strain. The DNA fragment is contributed by a **donor strain** to the **recipient strain,** which carries an intact main chromosome. Recombination occurs in the recipient strain. The transfer of genetic material is thus **unidirectional.**

Because the recipient chromosome is circular and the donor DNA fragment linear, recombination requires an even number of exchange events (Figure 17.1). Only one of the crossover products (the main chromosome) is usually recovered in this process. Recall that in eukaryotes, there can be both odd and even numbers of exchanges. Sometimes, a genetic exchange occurs between the main bacterial chromosome and a circular DNA molecule (such as a plasmid) that entered the recipient cell. In these cases, a single exchange results in the integration of the entire circular molecule into the main chromosome (Figure 17.2).

We discuss three main types of genetic transfer in this chapter. All three are **parasexual** processes because recombination is achieved by mechanisms other than meiosis:

Transformation is the process by which a donor DNA molecule is taken up from the external environment and incorporated into the genome of a recipient cell.

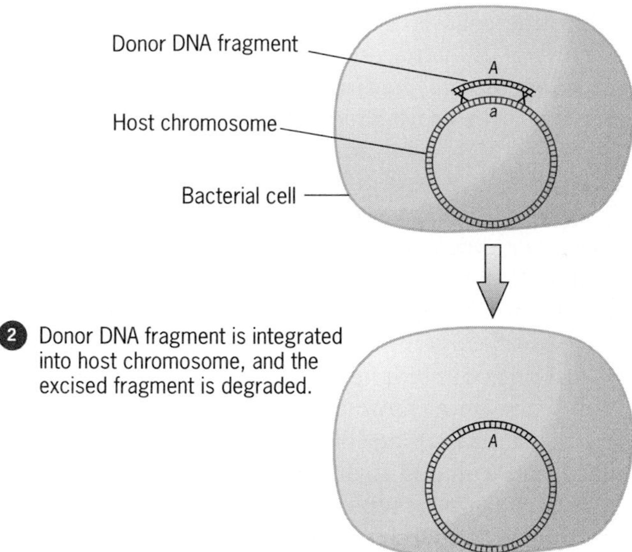

Figure 17.1 In order for successful recombination to occur, an even number of exchanges, two in this case, is required between the circular main chromosome and the linear fragment.

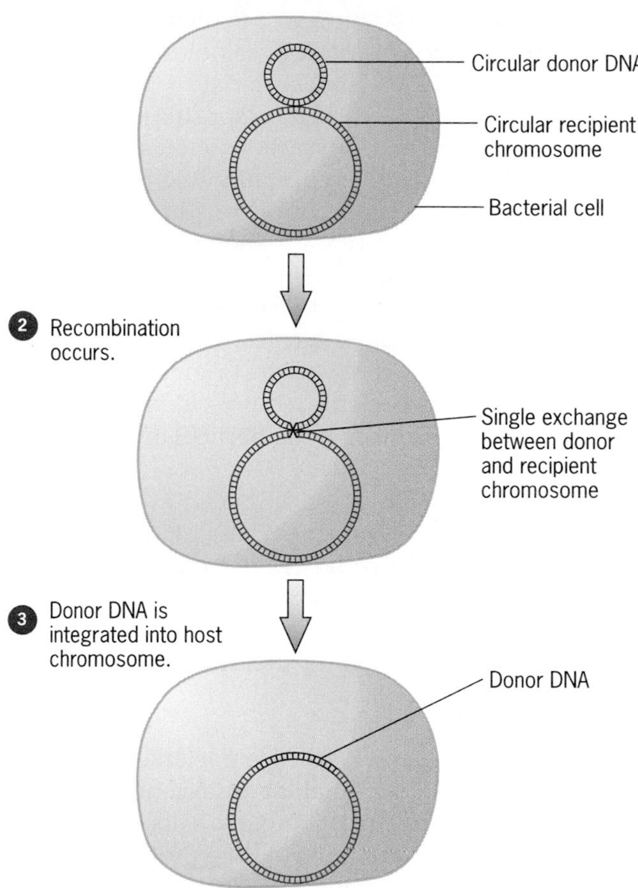

Figure 17.2 The integration of a circular donor DNA molecule into a circular recipient chromosome requires a single crossover event.

Conjugation is the process by which bacterial cells make effective contact with each other and DNA is transferred from one cell, the donor, to the other, the recipient cell.

Transduction is the process by which DNA is transferred from one bacterial cell to another by a bacterial virus.

Recombinational events that occur as a consequence of these processes can be used to construct genetic maps of bacterial chromosomes. These maps are generally quite different from the type of genetic map produced by using data from eukaryote crosses. For example, maps produced by conjugation analysis are measured in units of time, not frequencies of crossing over.

Key Points: **Three main mechanisms of genetic exchange in bacteria are transformation, conjugation, and transduction.**

MUTANT PHENOTYPES IN BACTERIA

In the laboratory, bacteria grow in a liquid medium or on the surface of a medium that has been solidified by agar. A single bacterium placed on an agar surface divides over and over again until it produces a visible cluster of cells called a **colony** (Figure 17.3). The number of bacteria growing in a liquid culture is estimated

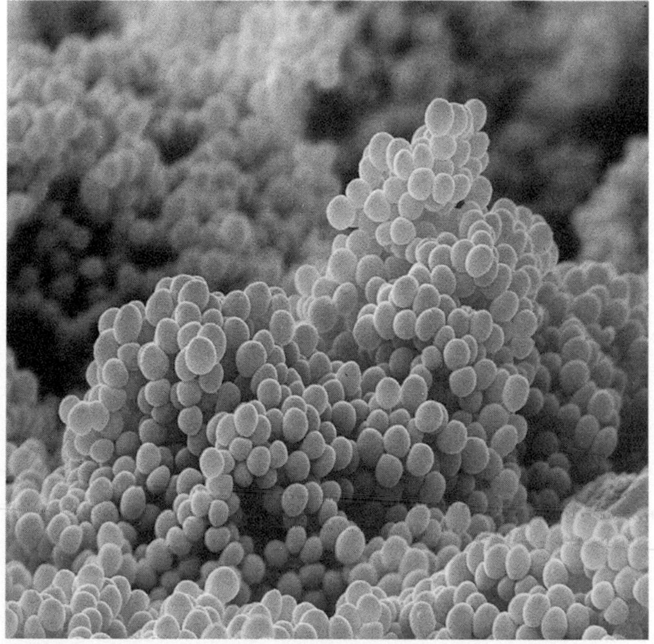

Figure 17.3 Colonies of *Staphylococcus aureus*.

by diluting the bacterial suspension, spreading a known volume of the final dilution on an agar surface, and then counting the number of colonies that form. For example, if we observe 100 colonies on a plate—after we diluted the original bacterial suspension by a factor of 1 million (10^6), and then spread 0.1 ml of this final dilution on an agar surface—we estimate that the bacterial cell concentration per ml in the original suspension was: 100 cells on the plate $\times$ 10 (the number of 0.1 ml aliquots per ml) $\times$ 1 million (the dilution factor) = 10^9 cells per ml.

The ability or inability to form colonies on a special medium can often be used to identify the genotype of the cell that produced the original colony. The genetic analysis of bacteria requires identifiable mutants. Of those mutants that are particularly useful, let us discuss three:

Antibiotic-resistant mutants. These mutants are able to grow and produce colonies on a medium containing an antibiotic such as streptomycin (Str) or tetracycline (Tet). For example, Str-sensitive cells (Str^s) fail to grow on medium containing streptomycin, but Str-resistant cells (Str^r) do grow.

Nutritional mutants. Wild-type bacteria are able to grow on a simple or **minimal medium** containing only a carbon source and some inorganic salts. From these materials, the cells synthesize all the proteins, nucleic acids, and lipids they require. Wild-type cells are said to be nutritionally independent, or **prototrophic.** Mutations may occur that result in the loss of enzyme function. Some of these mutants are no longer able to grow on minimal medium because they have lost the ability to synthesize an essential nutrient. They grow only if that nutrient is supplied in the medium. These mutants are called **auxotrophic** for the nutrient they require as a medium supplement. For example, if a bacterium has a mutation preventing it from synthesizing the amino acid tryptophan, it is unable to grow on the minimal medium containing only a carbon source and some inorganic salts. However, the tryptophan auxotroph is able to grow on a minimal medium supplemented with tryptophan.

Carbon-source mutants. Some mutants are unable to use certain substances as sources of energy or carbon atoms. For example, lactose mutants are unable to grow and produce colonies on a minimal medium containing the sugar lactose as the sole carbon source.

Other bacterial mutant phenotypes that have also been useful in genetic analysis include colony morphology, phage resistance, and temperature-sensitive mutants. These last mutants are especially useful for looking at mutations in essential genes, such as genes that encode enzymes involved in DNA replication.

Bacterial genotypes can be determined by first plating the cells on a nutritionally complete medium containing all the necessary materials for growth. This is a **nonselective medium** because both mutant and wild-type cells grow. Samples of cells from the colonies on the complete medium are placed on various **selective media** in which one or more nutrients are missing or antibiotics have been added. Failure to grow and produce colonies on a selective medium missing a nutrient indicates that the colony from which the cell came on the complete medium was mutant. For example, a selective medium containing only lactose as the carbon source identifies lactose mutants. The ability to grow on a minimal medium containing an antibiotic means that the colony from which the cell originated had mutated to antibiotic-resistance.

Bacterial phenotypes are designated by three letters, the first of which is capitalized, followed by a + or − superscript to denote the presence or absence of the designated character, and r or s to denote resistance or sensitivity to a compound or agent (e.g., phage). Bacterial genotypes use the same three letters, all lower case and italicized, followed by +, −, *r*, or *s* superscripts. Thus, a mutant requiring arginine as a supplement is Arg^- phenotypically and arg^- genotypically.

Key Points: **Three main types of bacterial mutants commonly used for genetic analysis are antibiotic-resistant mutants, nutritional mutants, and carbon-source mutants.**

BASIC TEST FOR TRANSFORMATION, CONJUGATION, AND TRANSDUCTION

The three parasexual processes of transformation, conjugation, and transduction can be distinguished by two simple criteria (Table 17.1): (1) sensitivity to deoxyribonuclease (DNase); and (2) dependence on cell–cell contact.

TABLE 17.1
Criteria for the Three Parasexual Processes

	Criterion	
Recombination Process	*Cell Contact Required?*	*Sensitive to DNase?*
Transformation	no	yes
Conjugation	yes	no
Transduction	no	no

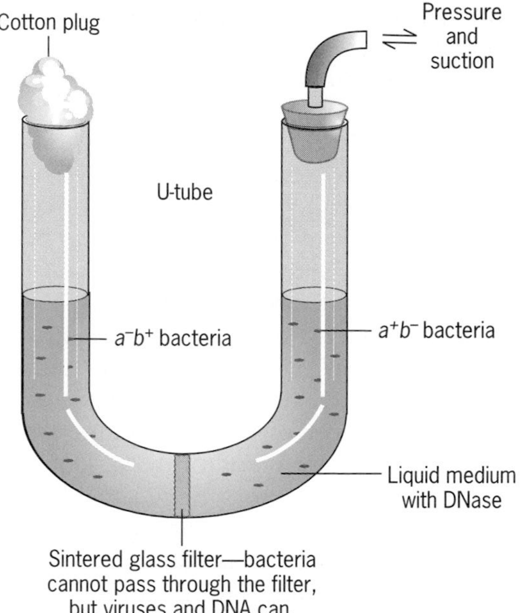

Figure 17.4 The U-tube experiment. Bacteria of different genotypes are in different arms of the tube, separated by a sintered glass filter that prevents cell–cell contact. This experiment tests for conjugation because cell–cell contact is required for the transfer of genetic material. DNase in the culture medium degrades free DNA, providing a test for transformation. If recombination occurs, it is likely taking place by means of transduction.

These two criteria are easily tested. The first criterion is tested by simply adding DNase to the medium containing the bacterial strains involved in recombination. If recombination occurs in the absence but not in the presence of DNase, transformation must be occurring because the DNA, which is free in the medium, is degraded by the enzyme.

In conjugation and transduction, DNA is not free in the medium, so it is not subject to degradation by DNase. Conjugation, which requires cell–cell contact, can be identified by a **U-tube experiment** in which bacteria of two different genotypes are placed in opposite arms of a U-shaped culture tube (Figure 17.4). The arms of the U-tube, and hence the two bacterial cultures, are separated by a sintered glass filter that allows DNA molecules and bacterial viruses but not bacterial cells to pass through. Conjugation cannot occur under these conditions because the filter prevents cell–cell contact. If DNase is present in the culture medium of the U-tube and if recombination occurs, it is likely taking place by the process of transduction.

Key Points: **To determine which of the three parasexual processes may be operating in a particular bacterial species, it is necessary to determine whether cell–cell contact is required and whether the process can be disrupted by the presence of DNase in the medium.**

TRANSFORMATION

Transformation was the first of the three mechanisms of genetic exchange in bacteria to be described. It was first discovered in 1928 in pathogenic strains of *Streptococcus pneumoniae* (also known as *Diplococcus pneumoniae*) by Frederick Griffith, although he knew nothing of the actual nature of the process (Chapter 10). Griffith's recorded observations were crucial to the eventual identification of DNA as the genetic material. In 1944 Avery, MacLeod, and McCarty demonstrated convincingly that the "transforming principle" described by Griffith was DNA. Some details of the mechanism by which transformation occurs remain unclear, 70 years after Griffith's discovery. However, a reasonably complete picture of the overall process of transformation has been established.

The Process of Transformation

The movement of donor DNA molecules across the cell membrane and into the cytoplasm of recipient bacteria is an active, energy-requiring process. It does not involve the passive diffusion of DNA molecules through permeable cell walls and membranes. Transformation is not a naturally occurring process in all species of bacteria; rather, it takes place only in those species that possess the protein and enzymatic machinery required to bind free DNA molecules in the medium and transport them to the cytoplasm. Most transformation studies have used three species of bacteria, *S. pneumoniae*, *Bacillus subtilis*, and *Haemophilus influenzae*. Even in these species, all cells in a given population are not capable of being transformed. Only **competent cells,** which secrete a **competence factor,** a small protein that induces the synthesis of 8 to 10 new proteins required for transformation, are capable of serving as recipients in transformation. The proportion of bacteria in a culture that are physiologically competent to be transformed depends on the growth conditions. In most bacterial species, cells that are likely to be transformed are dividing at their maximal rate. These populations of cells are growing exponentially and are fast approaching the plateau phase where nutrients in the medium become a limiting factor in the continued growth of the population (Figure 17.5).

B. subtilis cells become competent only in the presence of high concentrations of **competence phero-**

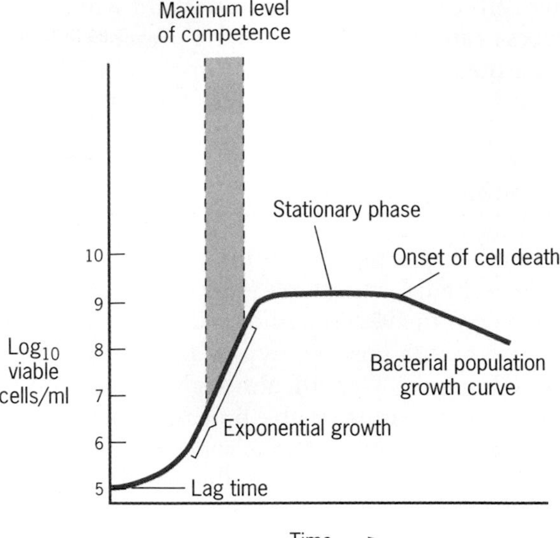

Figure 17.5 A bacterial growth curve. The maximum level of competence for transformation in most bacterial species occurs near the end of the exponential growth phase.

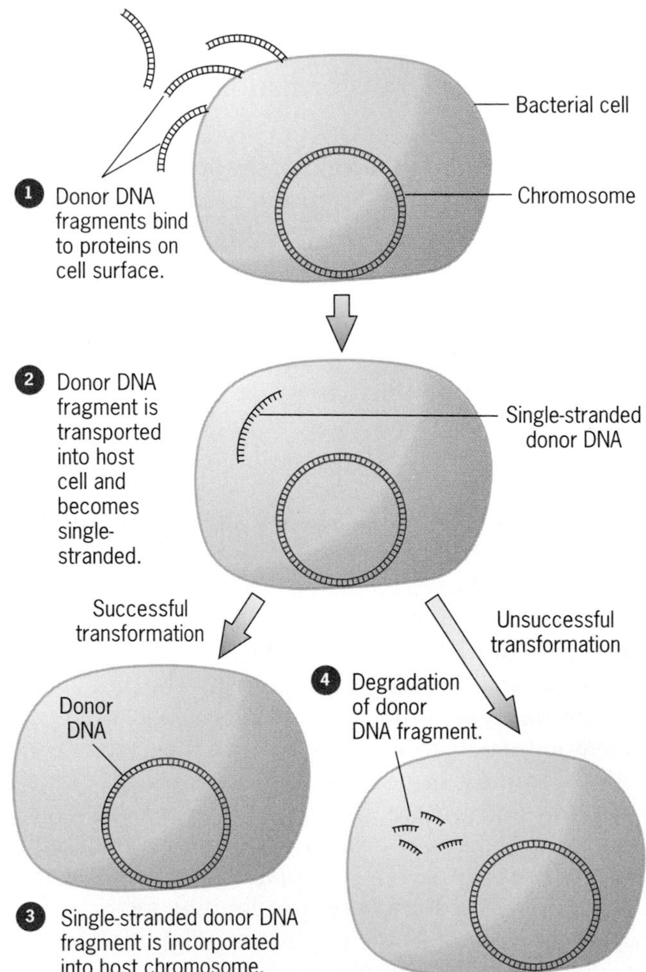

Figure 17.6 Bacterial transformation.

mones, small peptides that are excreted by growing bacteria. These pheromones reach high enough concentrations to induce competence only at high cell densities. Therefore, the bacteria are programmed to bind and internalize DNA only in the presence of large numbers of *B. subtilis* cells—some of which may lyse and release DNA to be taken up by the competent cells.

Since the mechanism of transformation has been studied extensively in *Streptococcus,* we will use that species as the basis for our discussion (Figure 17.6). A competent cell binds a large double-stranded fragment of DNA (at least 750 base pairs in length) at specific receptor sites on the surface of the bacterium. Because there are a limited number of these receptors, DNA molecules compete with each other for these sites. Following binding, DNA moves across the cell membrane. This is an active, energy-requiring process. One of the DNA strands is hydrolyzed by a membrane-bound exonuclease, providing energy to help fuel DNA transport across the membrane. The undegraded DNA strand associates with membrane proteins that were induced by the competence factor and is transported across the membrane. The single-stranded DNA fragment is coated with DNA-binding protein, and recombination proteins search for homologous regions in the chromosome of the recipient cell (Chapter 14). Other recombination enzymes catalyze the integration of the single strand of donor DNA, replacing the equivalent strand of the chromosome of the recipient cell (Figure 17.7). This type of recombination is unidirectional and **nonreciprocal** because only the recipient cell is genetically altered.

In *Haemophilis,* the process of transformation is different. This bacterial species does not produce a competence factor. While *Streptococcus* can take up DNA from a variety of sources, *Haemophilus* can take up DNA only from closely related species. Double-stranded DNA is not converted to single-stranded DNA but rather is complexed with proteins and then taken into the cell by membrane vesicles. The specificity of *Haemophilis* transformation is due to a special 11 base-pair sequence (5′–AAGTGCGGGTCA–3′) that is repeated about 600 times in the *Haemophilis* genome. For DNA to be bound by the membrane receptor and transported across the membrane, it must have this nucleotide sequence. This sequence-specific interaction between the DNA and the receptor protein prevents *Haemophilis* from taking up DNA from other species.

Transformation and Gene Mapping

Transformation is an important tool for use in gene mapping. When DNA is extracted or released from a cell, it fragments. DNA molecules extracted from *Strep-*

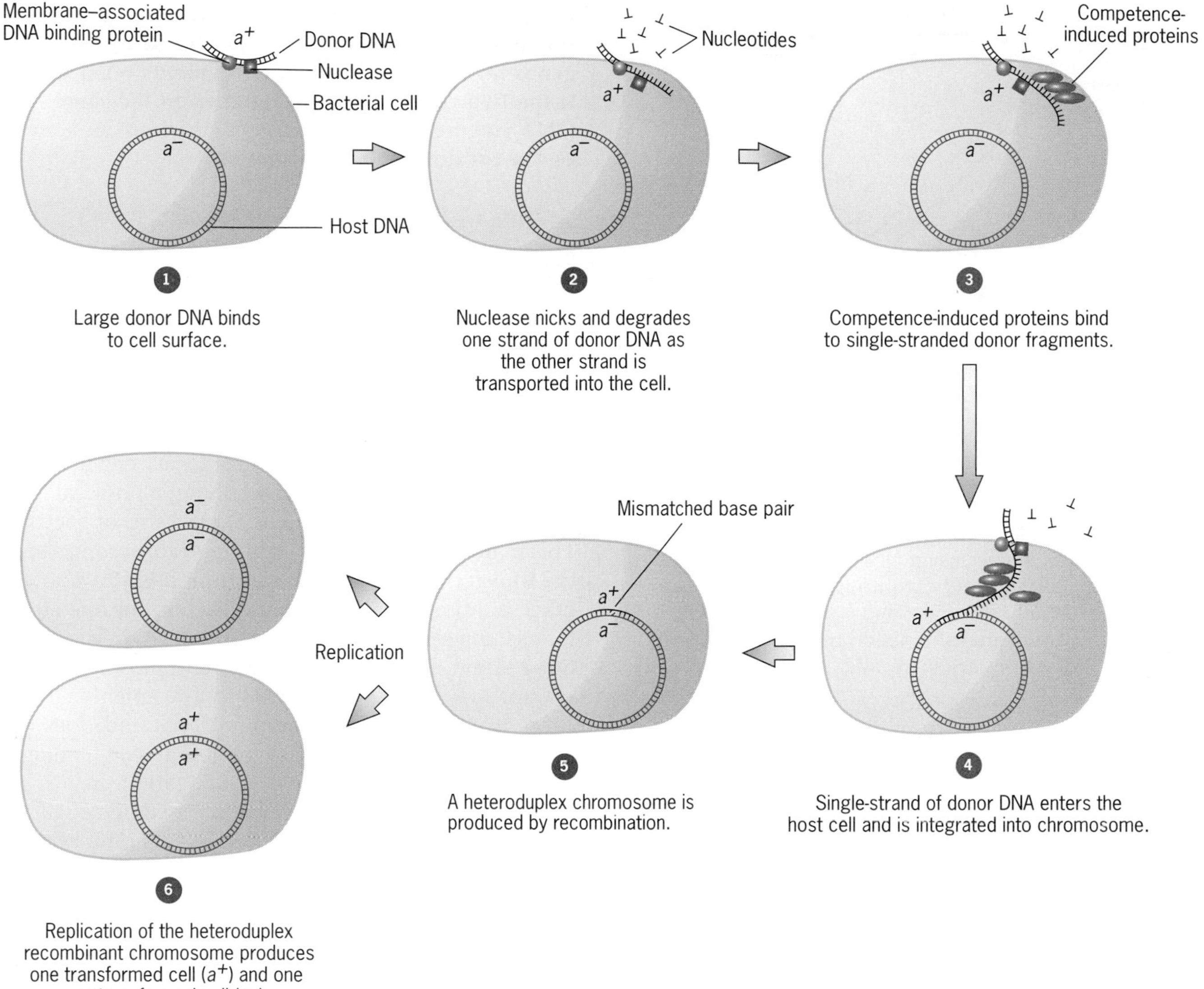

Figure 17.7 The mechanism of transformation.

tococcus bacteria typically break into about 50 fragments per genome depending on the extraction procedure; thus, each DNA fragment represents about 2 percent of the genome. In mapping experiments using transformation, two genes are said to be closely linked if they are frequently found on the same DNA fragment. This use of the term *linkage* is quite different from that in eukaryote gene mapping, where all genes on the same chromosome are said to be linked. Because bacteria have only one chromosome, all their genes are linked. If two genes are very close to each other, they will almost always be on the same DNA fragment and thus cells have a high probability of becoming **cotransformed** with these genes; if they are far apart, they will never be on the same fragment. The farther apart two genes are, the greater the chances that they will not be found on the same fragment. This is the basis of gene mapping experiments using transformation.

To determine whether two genes are closely linked, three different genetic classes of donor DNA molecules are used: a^+b^-, a^-b^+, and a^+b^+. The recipient cells are a^-b^-. Steadily decreasing concentrations of a^+b^- and a^-b^+ donor DNA are used in separate experiments to measure the single-transformation frequencies of a^-b^- to a^+b^- and a^-b^+. Theoretically, the frequency of single-transformants should decrease linearly as the concentration of donor DNA decreases (Figure 17.8*a*). Next, the double-transformation frequency of a^-b^- to a^+b^+ using a^+b^+ donor DNA is determined as a function of decreasing DNA concentration. If the two genes are closely linked—that is, they are frequently found on the same DNA fragment—a

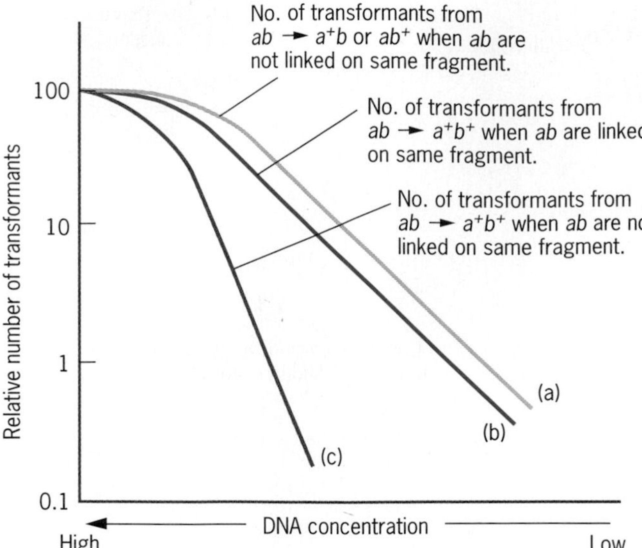

Figure 17.8 Single- and double-transformation frequencies as a function of decreasing DNA concentration. Double-transformation frequency requiring two events (c) decreases exponentially as a function of decreasing DNA concentration; double-transformation frequency requiring one event (b) decreases arithmetically.

curve with the same slope as single transformation is expected because double transformation, like single transformation, requires uptake and incorporation of a single DNA fragment (Figure 17.8*b*). If the two genes are far apart, double transformation can occur only if two different fragments are taken up by the recipient cells and two separate recombination events occur. Thus, if the DNA concentration decreases tenfold, the double-transformation frequency drops a hundredfold: For example, when the DNA concentration decreases from 1.0 to 0.1 units per ml, single-transformation frequency decreases tenfold from 1.0 to 0.1 cell per unit volume; the double-transformation frequency for two genes on two different fragments decreases to $0.1 \times 0.1 = 0.01$ cell per unit volume, a hundredfold decrease. This exponential decline generates a curve with a much steeper slope (Figure 17.8*c*). Thus, by analyzing single- and double-transformation frequencies solely as a function of decreasing DNA concentration, we are able unambiguously to ascertain linkage.

Key Points: **In transformation of some bacterial species, a fragment of double-stranded DNA is transported across the cell membrane of competent cells, is converted into a single-stranded DNA, and then is physically integrated into the recipient cell's genome, replacing its homologous sequence. Linkage in transformation experiments means that two genes are frequently found on the same DNA fragment. For a cell to be doubly transformed by genes far apart, two separate fragments must insert. If the two genes are close together on the same DNA fragment, only one fragment insertion is necessary for double transformation.**

CONJUGATION

Escherichia coli is the bacterial species most widely used in genetic research, but it does not normally undergo transformation. Nevertheless, other processes have evolved in this species to generate recombinant genotypes. In 1946, Joshua Lederberg and Edward L. Tatum established for the first time that *E. coli* engages in a form of sexual reproduction involving a physical mating between sexually differentiated strains of bacteria. This mating process, called **conjugation**, accomplishes the transfer of genetic material from one strain to another unidirectionally, enabling bacteria of one strain to exchange genetic information with bacteria of another strain. During conjugation, chromosomal DNA may or may not be transferred to a recipient cell. Because genetic transfer is unidirectional and does not involve gamete formation or gamete fusion, conjugation is a parasexual process. Conjugation creates recombinant genotypes that have provided investigators the opportunity to analyze the *E. coli* genome in great detail.

The Discovery of Conjugation

The discovery of a sexual process in bacteria by Lederberg and Tatum grew out of an effort to study the genetic control of metabolism. They analyzed various nutritional mutants in bacteria and in 1946 published a paper in the British journal *Nature* that began: "Analysis of mixed cultures of nutritional mutants has revealed the presence of new types which strongly suggest the occurrence of a sexual process in the bacterium *Escherichia coli*." They had mixed together two auxotrophic strains of *E. coli*, each requiring two different nutritional supplements in the medium. One of the auxotrophic strains (A) required methionine and biotin for growth (Met^- Bio^-); a second strain (B) required threonine and leucine for growth (Thr^- Leu^-). After plating the mixture of cells on a minimal medium that would not support the growth of either strain, they discovered prototrophic colonies that could grow. These results suggested that genetic recombination had occurred, but what was the mechanism?

When Lederberg and Tatum mixed strain A ($met^-bio^-thr^+leu^+$) cells with strain B ($met^+bio^+thr^-leu^-$)

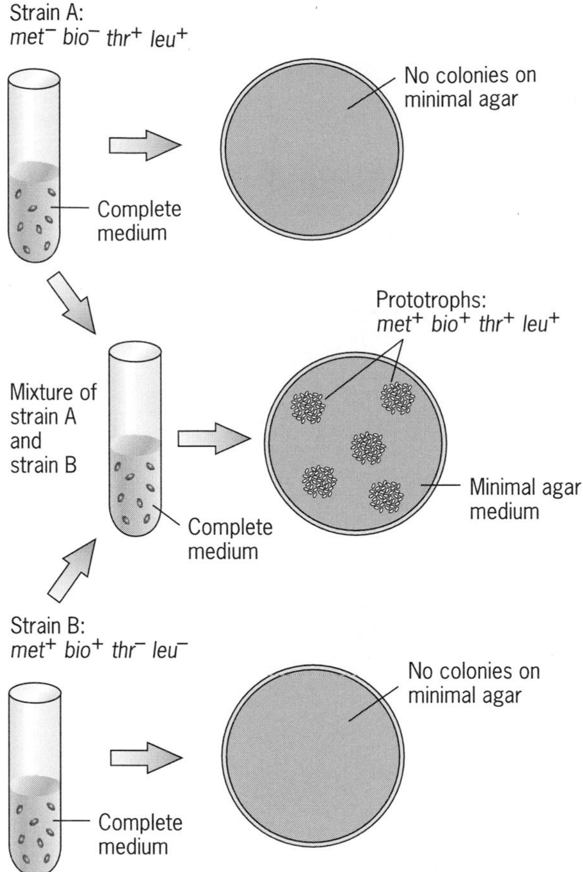

Figure 17.9 A summary of the classic Lederberg and Tatum experiment in which two auxotrophic strains of *E. coli* were mixed together, producing prototrophic genetic recombinants. Strains A and B alone did not produce prototrophic cells.

cells on a nutrient medium for several hours and then plated cells on a minimal medium, genetically stable $met^+bio^+thr^+leu^+$ prototrophs appeared (Figure 17.9). Strains A and B alone did not yield prototropic colonies. Considering the relatively high frequency of prototrophic colonies from the mixed strains, they concluded that the prototrophs could not have arisen by reverse mutation ($- \rightarrow +$) in the auxotrophic strains because two separate reversion events in each strain would be necessary to produce a prototroph. They suggested that either transformation was occurring or the two mutant strains were somehow mating and exchanging genetic material.

Lederberg and Tatum eliminated transformation as a possibility by showing that DNA extracted from either auxotrophic strain could not transform the other strain. Transformation does not normally occur in *E. coli*. They did not prove that physical contact was required. Proof that cell–cell contact was required for gene exchange came in 1950 when Bernard Davis developed the U-tube (see Figure 17.4). When strains A and B were separated by a sintered glass filter, gene transfer could not take place.

$F^+ \times F^-$ Mating

Two years after Davis proved that cell–cell contact was required for conjugation, William Hayes discovered that gene transfer was unidirectional from a donor (F^+) strain to a recipient (F^-) strain. Hayes further discovered that in $F^+ \times F^-$ matings, genes on the main bacterial chromosome were rarely transferred but that F^- cells usually became F^+.

Donor or F^+ cells carry a plasmid called the fertility or **F factor,** a circular extrachromosomal molecule containing about 94,000 base pairs. Genes on the F factor are responsible for facilitating cell–cell contact and F-factor transfer from donor to recipient cell. Many F-factor genes direct the synthesis of **sex pili** (singular, pilus), filamentous appendages on the bacterial surface that attach the F^+ cell to an F^- cell.

The F factor and the main bacterial chromosome have **insertion sequences (IS)** that enable the F factor to insert into the chromosome by homologous recombination. Insertion sequences are short, transposable DNA sequences that contain a gene coding for an enzyme called transposase which facilitates the movement of the insertion sequence around the genome (Chapter 18). Because the F factor can exist independently of the bacterial chromosome or be integrated into it, it is a special type of plasmid called an **episome.**

During $F^+ \times F^-$ conjugation, the F factor replicates by the rolling circle mechanism (Chapter 11), and a single-stranded copy of the F factor is transferred into the F^- cell (Figure 17.10). It is not clear whether the DNA passes through the hollow sex pilus or through another conjugation channel. The transferred single-stranded DNA is replicated in the recipient cell, producing a double-stranded molecule. Because genes on the main bacterial chromosome are rarely transferred into the recipient with the independent F factor, the recombination frequency is low. It can occur only if the F factor integrates into the chromosome.

Hfr Conjugation

Some strains that carry F factors transfer bacterial genes from the main chromosome of the donor cell into recipient cells with a high frequency, but they do not usually transfer the F factor. This result is quite different from what is observed with F^+ cells. In these special donor strains, the F factor does not exist independently of the main bacterial chromosome; rather, it

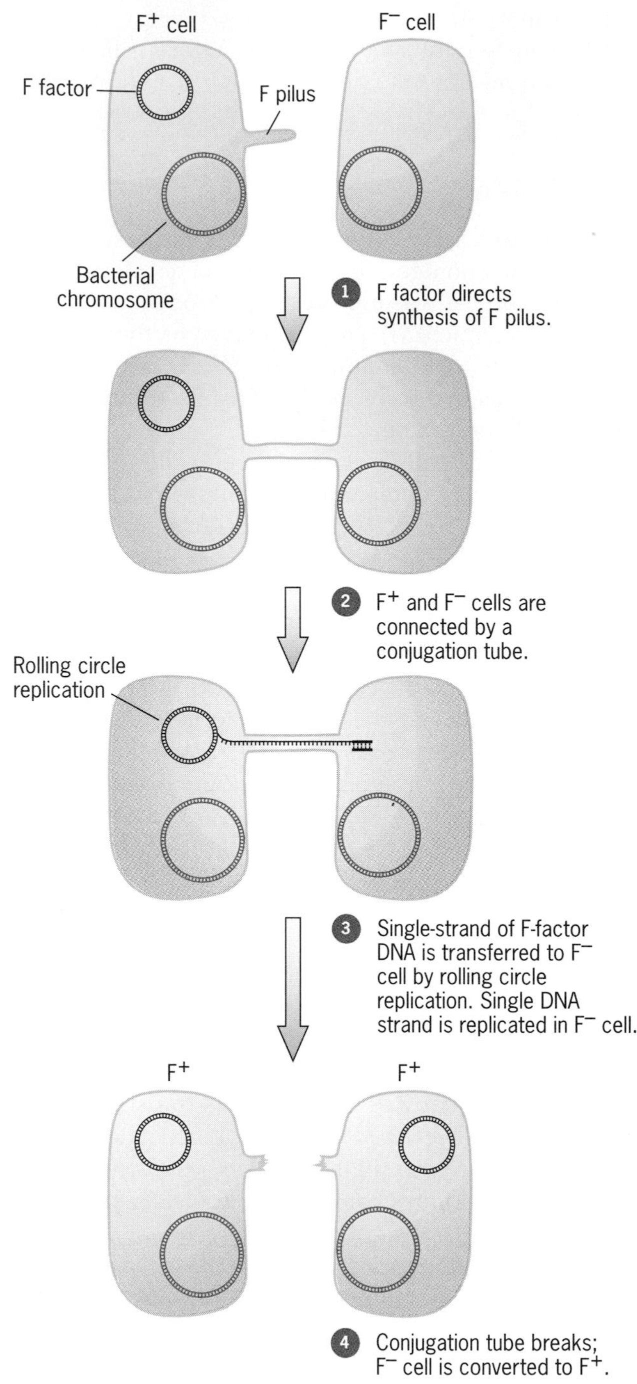

Figure 17.10 $F^+ \times F^-$ conjugation.

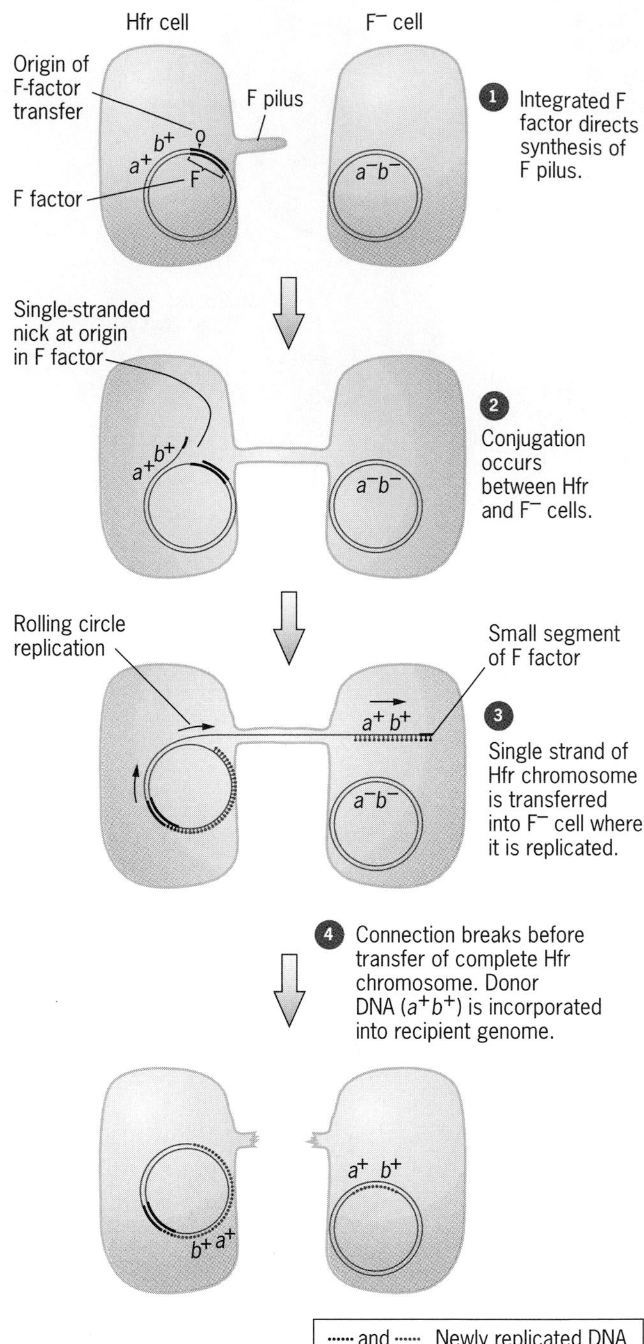

Figure 17.11 Hfr $\times F^-$ conjugation.

is integrated into it. The F factor can integrate into the bacterial chromosome at several locations by recombination between homologous insertion sequences present on both the F factor and the chromosome. The integrated F factor directs the synthesis of sex pili, undergoes rolling circle replication, and transfers genetic material to an F^- cell. This type of donor cell is called an **Hfr cell** (*h*igh-*f*requency *r*ecombination) because of the high frequency of chromosomal gene transfer into F^- cells.

DNA transfer is initiated at the origin of replication on the integrated F factor (Figure 17.11). Rolling circle replication transfers the donor strand from the main chromosome into the recipient cell where it becomes double-stranded. Because only a small part of the F factor is transferred to the recipient, the recipi-

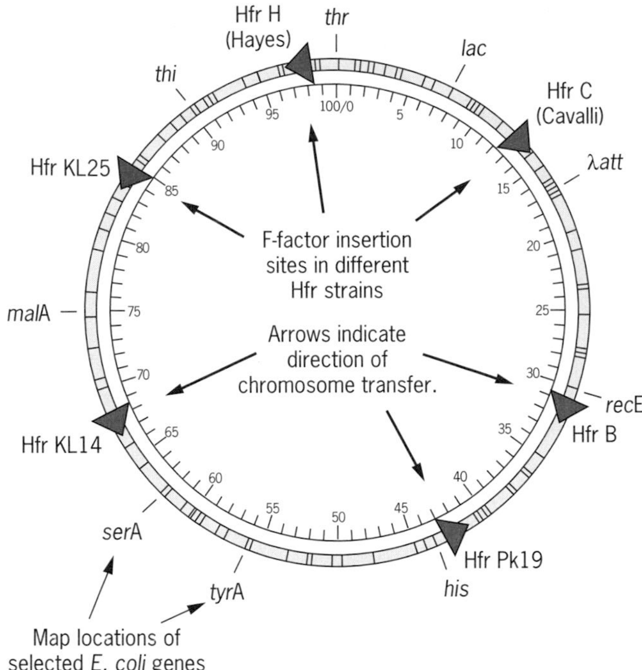

Figure 17.12 Some of the sites of F-factor integration in the *E. coli* chromosome. The arrows indicate the direction of transfer, with the genes immediately behind the arrowheads being transferred first. A few gene loci are shown as reference points.

ent remains F^- unless the entire bacterial chromosome is transferred. It takes about 100 minutes for the entire bacterial chromosome to be transferred, but the connection between conjugating cells is usually severed before the process is finished. Thus, the recipient usually remains F^-.

Hfr strains vary in the chromosomal location of F-factor integration and in the orientation of the F factor, which determines whether the direction of gene transfer is clockwise or counterclockwise around the circular chromosome (Figure 17.12). When the donor DNA enters the F^- cell, it may be degraded or it may recombine with the recipient chromosome. Hfr conjugation is the most efficient natural mechanism of gene transfer in bacteria.

F′ Conjugation or Sexduction

The integrated F factor in an Hfr cell may exit the bacterial chromosome by reversing the steps that resulted in its integration (Figure 17.13). Sometimes there is an error in the excision of the F factor from the chromosome. This error results in the formation of an **F′ factor**—an F factor containing a segment of the bacterial chromosome. The mating of an F′ cell with an F^-

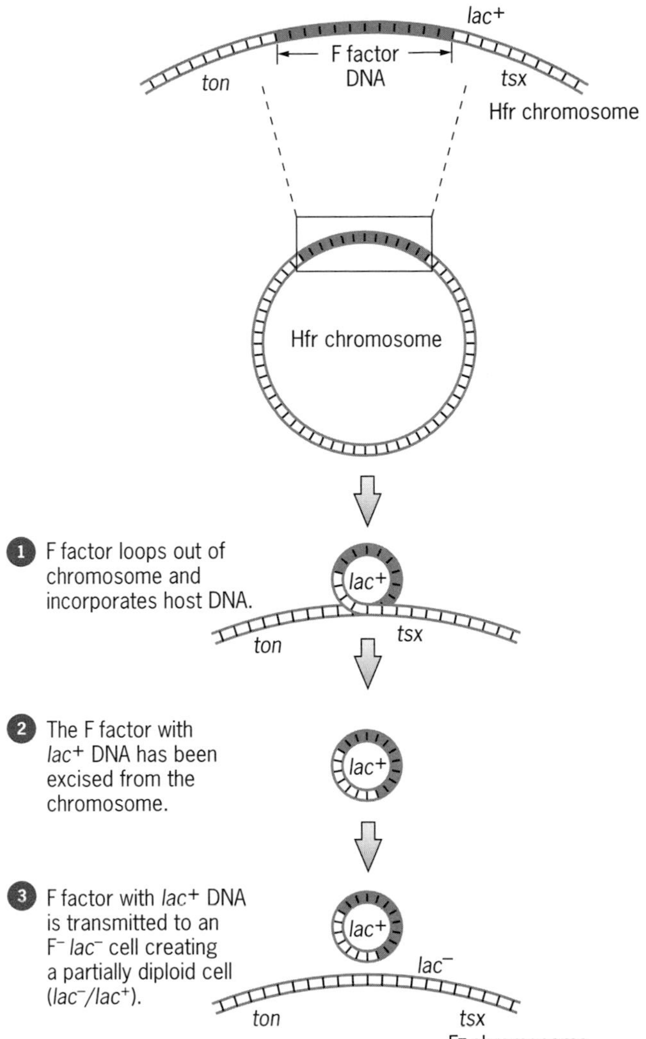

Figure 17.13 The formation of an F′ factor.

cell is basically the same as an $F^+ \times F^-$ mating, with one important difference: bacterial genes incorporated in the F′ plasmid are transferred with very high frequency to recipient cells. These genes do not have to be integrated into the recipient cell's chromosome to be expressed. The recipient acquires the F′ factor and is partially diploid for the genes that it carries. The transfer of bacterial genes by F′ factors is called **sexduction.** Sexduction results in the rapid spread of bacterial genes from the main chromosome through a population.

Sexduction is a valuable tool for microbial geneticists. Because of the cell's partially diploid condition, it allows investigators to answer questions about dominance and recessiveness for alleles carried by the F′ factor. In addition, F′ cells allow researchers to perform complementation tests in bacteria, which normally carry a single copy of each gene.

Conjugation and Gene Mapping

The discovery of Hfr cells opened the door to mapping genes in *E. coli*. The use of Hfr cells in gene mapping takes advantage of the fact that during conjugation, chromosomal DNA moves from donor to recipient at a constant rate. In an **interrupted mating experiment,** conjugating cells are placed in a blender and broken apart. The Hfr × F⁻ mating is stopped at various intervals after the start (Figure 17.14). The order and timing of gene transfer can be determined because the first gene to enter the recipient is closest to the origin of transfer, and genes farther from the origin enter the recipient later (Figure 17.15). In one Hfr × F⁻ conjugation experiment performed by Elie Wollman and François Jacob in the mid-1950s, an Hfr strain that carried the genetic markers azi^s ton^s lac^+ gal^+ was mated to an F⁻ strain that was azi^r ton^r lac^- gal^-. (The genes determine sodium azide resistance or sensitivity, phage T1 resistance or sensitivity, and the ability to use lactose or galactose, respectively, as a carbon source.) The *azi* gene was the first to enter the recipient, about nine minutes after the initiation of conjugation. The *gal* gene was the last to enter the recipient, about 25 minutes after conjugation began. The *ton* and *lac* genes entered about 10 and 17 minutes, respectively, after the initiation of conjugation.

The map generated by interrupted conjugation is circular and is marked off in units of time rather than crossover frequency. The technique can rather precisely locate genes that are three or more minutes apart. The genes that are farther from the origin enter later and have a lower plateau because interruption of mating occurs spontaneously, and later markers are less likely to be transferred.

Because of the large size of the *E. coli* genome, it is not possible to map all the genes using a single Hfr strain. Maps are generated by using different Hfr

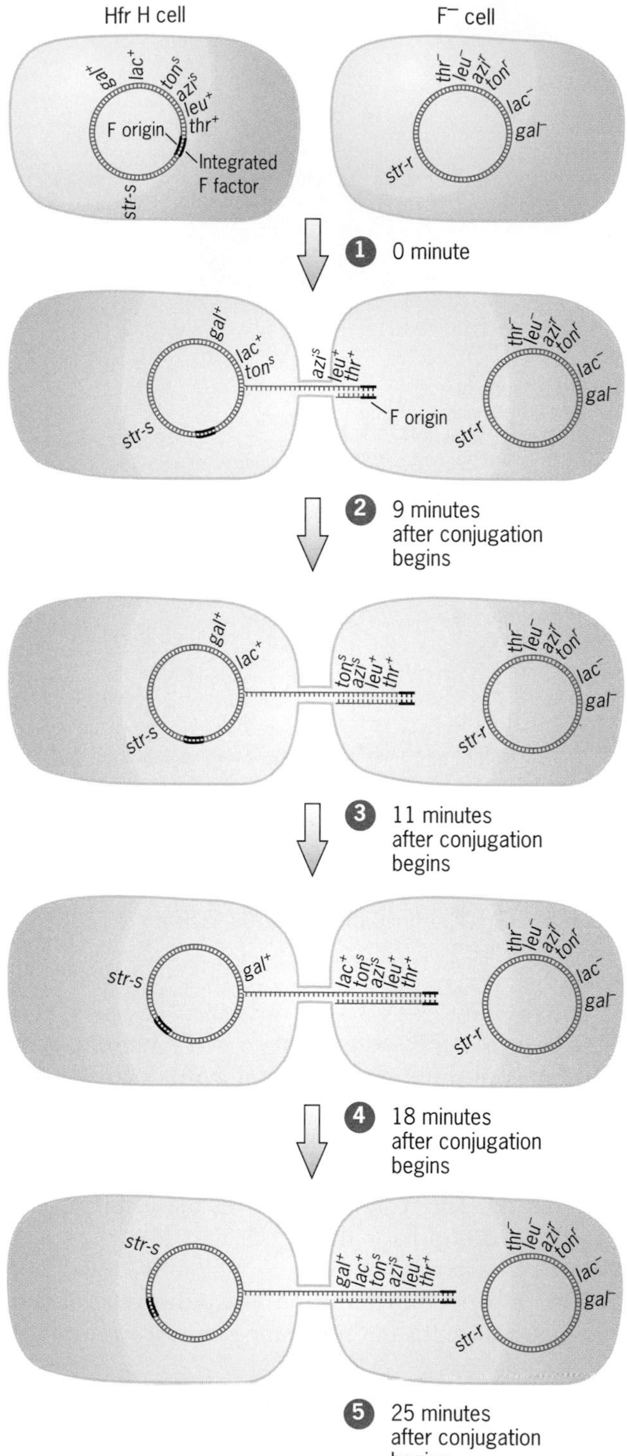

Figure 17.14 The interrupted mating experiment of Wollman and Jacob. The transfer of the Hfr chromosome to the F⁻ cell is linear, beginning at the origin point on the F factor.

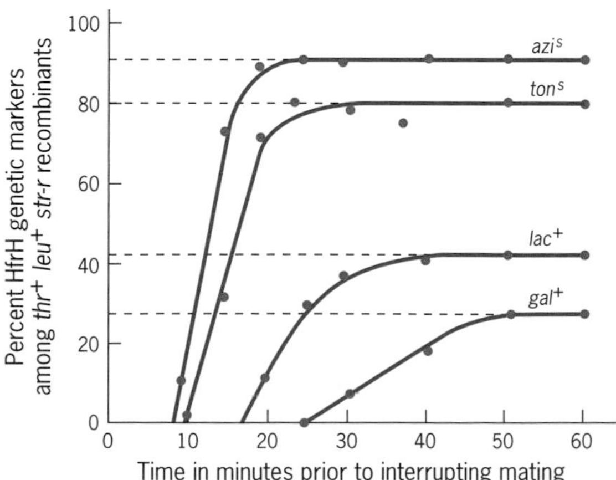

Figure 17.15 Data from the interrupted mating experiment. The frequencies of the unselected donor markers in thr^+ leu^+ str^r recombinants are shown on the ordinate and time of entry on the abscissa. The dashed lines indicate the plateau frequencies observed for the various donor markers.

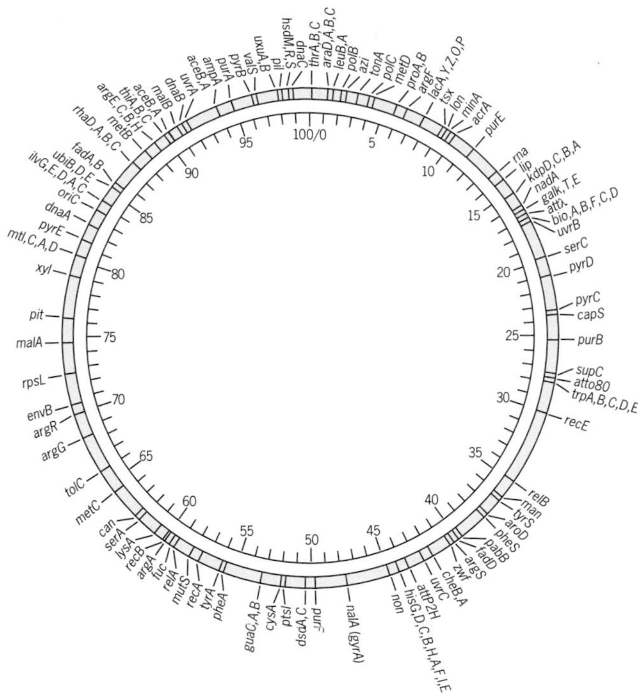

Figure 17.16 A map of the *E. coli* genome, showing selected genes.

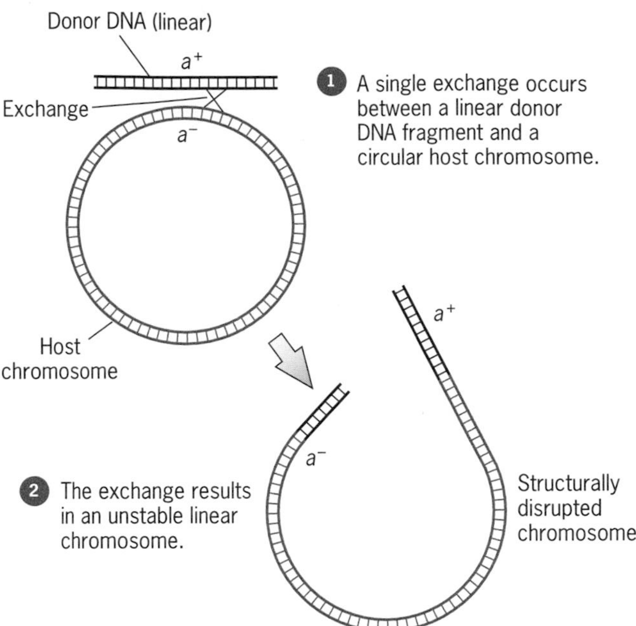

Figure 17.17 A single exchange between a linear donor DNA molecule and a circular recipient chromosome does not yield a structurally intact recombinant chromosome.

strains with different F-factor integration sites and different orientations. These maps are then combined to produce a final map of the *E. coli* chromosome (Figure 17.16). The map is divided into 100 minutes, the time required to transfer a complete chromosome from donor to recipient. The 0 point is set at the *thr* locus because this is the gene closest to the point of insertion of the F factor first described by Hayes.

Mapping Closely Linked Genes

Interrupted mating experiments provide approximate map distances for genes that are several map units apart, but they are usually uninformative for genes that are close together, say, one- or two-minute units apart. As in eukaryotes, mapping closely linked genes in bacteria usually necessitates three-point crosses. The rationale behind three-point crosses in bacteria is essentially the same as in eukaryotes: The double crossover class is the rarest, and this class identifies the middle gene.

Recombination in conjugating bacteria requires an interaction between the donor DNA fragment and the recipient cell's chromosome. The donor DNA segment must be incorporated into the recipient chromosome. Such an incorporation requires two crossovers or recombination events (see Figure 17.1). A single exchange between donor and recipient DNA does not yield a structurally intact chromosome (Figure 17.17).

In three-point mapping experiments with eukaryotes, double crossover events are much less frequent than single crossover events. The rarity of double crossover events allows geneticists to determine the order of three genes (Chapter 8). Three-point crosses for closely linked bacterial genes employ the same basic principle, with one minor variation: recombination involving two exchanges is more frequent than recombination involving four exchanges.

To use conjugation to map closely linked genes, each three-point cross is performed in two ways: (1) a single mutant donor × a double mutant recipient; and (2) a double mutant donor × a single mutant recipient. Suppose, for example, that we want to order two closely linked genes (*b* and *c*) relative to a third (*a*) that is nearby. Because the *a* gene is known from previous crosses to map at one end, it is called an **outside marker.** The two reciprocal crosses are:

	Donor		Recipient
(1)	$a^+\ b^+\ c^-$	×	$a^-\ b^-\ c^+$
(2)	$a^-\ b^-\ c^+$	×	$a^+\ b^+\ c^-$

Because the chromosome is circular, there are only two possible orders for these three genes: $a - b - c$ or $a - c - b$.

By comparing the frequency of $a^+\ b^+\ c^+$ recombinants in the reciprocal crosses, we can discriminate between these two orders. If the order of the genes is

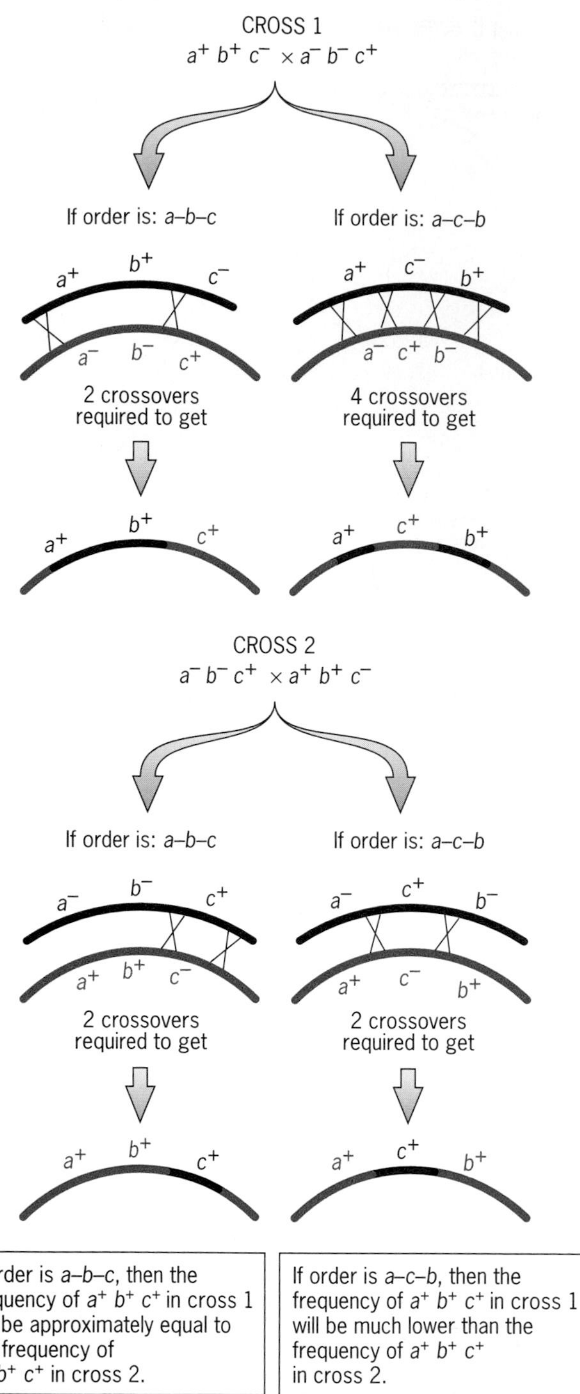

Figure 17.18 The rationale for ordering closely linked markers in a three-factor cross in bacteria. Reciprocal crosses predict different $a^+\ b^+\ c^+$ frequencies, depending on the gene order.

$a - b - c$, then the $a^+\ b^+\ c^+$ recombinant frequencies in the reciprocal crosses will be about the same because two crossover events are required in both cases (Figure 17.18). If the order is $a - c - b$, the $a^+\ b^+\ c^+$ recombinant frequencies will differ dramatically in the reciprocal crosses because in one cross a double exchange will produce the $a^+\ b^+\ c^+$ recombinant, whereas in the reciprocal cross four exchanges are necessary to produce the $a^+\ b^+\ c^+$ recombinant. Thus, the three-point cross can be used to establish the order of closely linked genes.

On the Origin of Plasmids

Ever since their discovery, investigators have pondered the significance and origin of extrachromosomal elements, such as the F factor. Their origin remains an intriguing mystery. Many view them as descendants of phage DNA that have entered into symbiotic or mutually advantageous relationships with bacterial hosts. In 1963, P. Fredericq, referring to the origin of the F factor, stated: "A phage is a particle that can establish effective contact with a cell and inject into it genetic material. . . . What else is an F^+ cell?"

It is not at all unreasonable to suggest that plasmids and episomes are remnants of bacterial viruses. These remnants, instead of encoding viral proteins that package phage DNA, encode proteins that are expressed on the bacterial surface. These cell-surface viral proteins function like the protein coat of the bacteriophage, enabling one bacterial cell to physically attach to another and inject plasmid or episomal DNA.

Numerous studies have shown that plasmids are dynamic genetic elements: new plasmids constantly evolve and old ones are lost. In the 1980s, Larry McKay and colleagues demonstrated that the number of plasmids in strains of *Streptococcus lactis* is highly variable, with some strains containing as many as 12 different plasmids. Moreover, some of the *S. lactis* plasmids contain transposable genetic elements that mediate their recombination with each other and their integration into the main bacterial chromosome (Chapter 18). Also, as we mentioned at the beginning of this chapter, new antibiotic- and drug-resistance plasmids are constantly evolving in pathogenic bacteria and are of major concern in the field of medicine.

Key Points: **Conjugation is a process in which bacteria make contact with each other and genetic material is transferred unidirectionally from donor to recipient cells. The F factor directs conjugation. Cells carrying an F factor are F^+, and those without it are F^-. A cell with the F factor integrated into the chromosome is called an Hfr. An F factor that has incorporated some genes from the main chromosome is called an F′ factor. Interrupted conjugation experiments showed that the Hfr chromosome is transferred unidirectionally and that the first genes to be transferred are located close to the point of F-**

factor integration. The *E. coli* chromosome map is calibrated in minutes.

TRANSDUCTION

Transduction, the third mechanism of gene transfer in bacteria, differs significantly from transformation and conjugation in that it is mediated by bacterial viruses. Bacterial genes are incorporated into a phage capsid because of errors that occur during the virus life cycle. The virus then injects these genes into another bacterium, transfering genetic material from one bacterium to another. Two different types of transduction have been identified: (1) generalized transduction, in which the phage can transfer any segment of the bacterial genome to another bacterium; and (2) specialized transduction, in which only restricted segments of the bacterial genome are transferred. Transduction is a common mechanism for gene exchange and recombination in bacteria and, like transformation and conjugation, has been extremely important in genetic studies of bacteria.

The Discovery of Transduction

Generalized transduction was discovered in 1952 by Joshua Lederberg and his student, Norton Zinder. They were attempting to show that conjugation, which Lederberg and Tatum discovered in *E. coli* in 1946, could occur in other bacterial species. While studying *Salmonella typhimurium*, they found that when two multiply auxotrophic strains were incubated together ($Phe^- Trp^- Tyr^-$ and $Met^- His^-$) on an amino-acid-free minimal medium, prototrophs appeared with a frequency of about 1 per 100,000 cells. Because this frequency was too high to be accounted for by reverse mutation, they concluded that the bacteria were undergoing a process of genetic exchange. To test for transformation and conjugation, they carried out a U-tube experiment (see Figure 17.4). The two strains were placed in the two arms of the U-tube, separated by a sintered glass filter, in a medium containing DNase. Prototrophs still appeared but in only one of the U-tube arms. These results could not be explained by transformation because DNase destroys free DNA. Conjugation could not explain the results because the filter prevented cell–cell contact. Lederberg and Zinder had discovered a new mechanism of bacterial gene transfer. Further analysis of this gene exchange process in *Salmonella* revealed that one of the bacterial strains harbored a virus called P22. When cells carrying these P22 viruses lysed, transducing particles were released. These particles passed through the filter and infected cells in the other arm of the U-tube, carrying with them genes from the virus-infected strain. These genes were subsequently incorporated into the genomes of the newly infected cells.

Generalized Transduction

In **generalized transduction,** a phage may carry any bacterial gene from one cell to another. Toward the end of the phage life cycle, when viral DNA is being packaged inside viral protein coats, fragments of the bacterial chromosome are mistakenly packaged into the phage head (Figure 17.19). This packaging error produces a phage particle that carries bacterial DNA and no phage DNA. The size of the phage head determines the amount of bacterial DNA that can be packaged. The size of the bacterial DNA fragment packaged into the phage varies, but it is not unusual for it to consist of 1 or 2 percent of the bacterial genome. Thus, for a bacterium with 3000 genes, 30 to 60 genes may be packaged into a transducing virus particle.

A phage particle that carries donor bacterial DNA is released upon lysis of the cell. The phage protein coat functions as a vehicle for the transfer of bacterial DNA from one cell to another. The transducing particle attaches to a recipient cell and injects donor bacterial DNA into that cell. This transducing particle, which carries no phage DNA, is incapable of going through a life cycle. Once the DNA is injected, it must be incorporated into the host genome in order to be preserved.

The donor DNA is not always incorporated into the recipient's genome. Occasionally, the fragment of donor DNA survives and is expressed in the recipient's cytoplasm. This cell is partially diploid for the introduced genes. The donor DNA is not replicated because it does not have the proper recognition site for DNA polymerase. Thus, when the bacterium divides, only one of the two daughter cells receives the donor fragment and expresses the donor genotype. The other cell, having lost the donor fragment, reverts back to its original phenotype. This phenomenon is called **abortive transduction;** it results in the formation of partially diploid bacteria containing nonintegrated, transduced DNA.

Specialized Transduction

In **specialized transduction**, the transducing particle carries only specific portions of the bacterial genome. In order to understand the nature of specialized transduction, it is first necessary to understand the life cy-

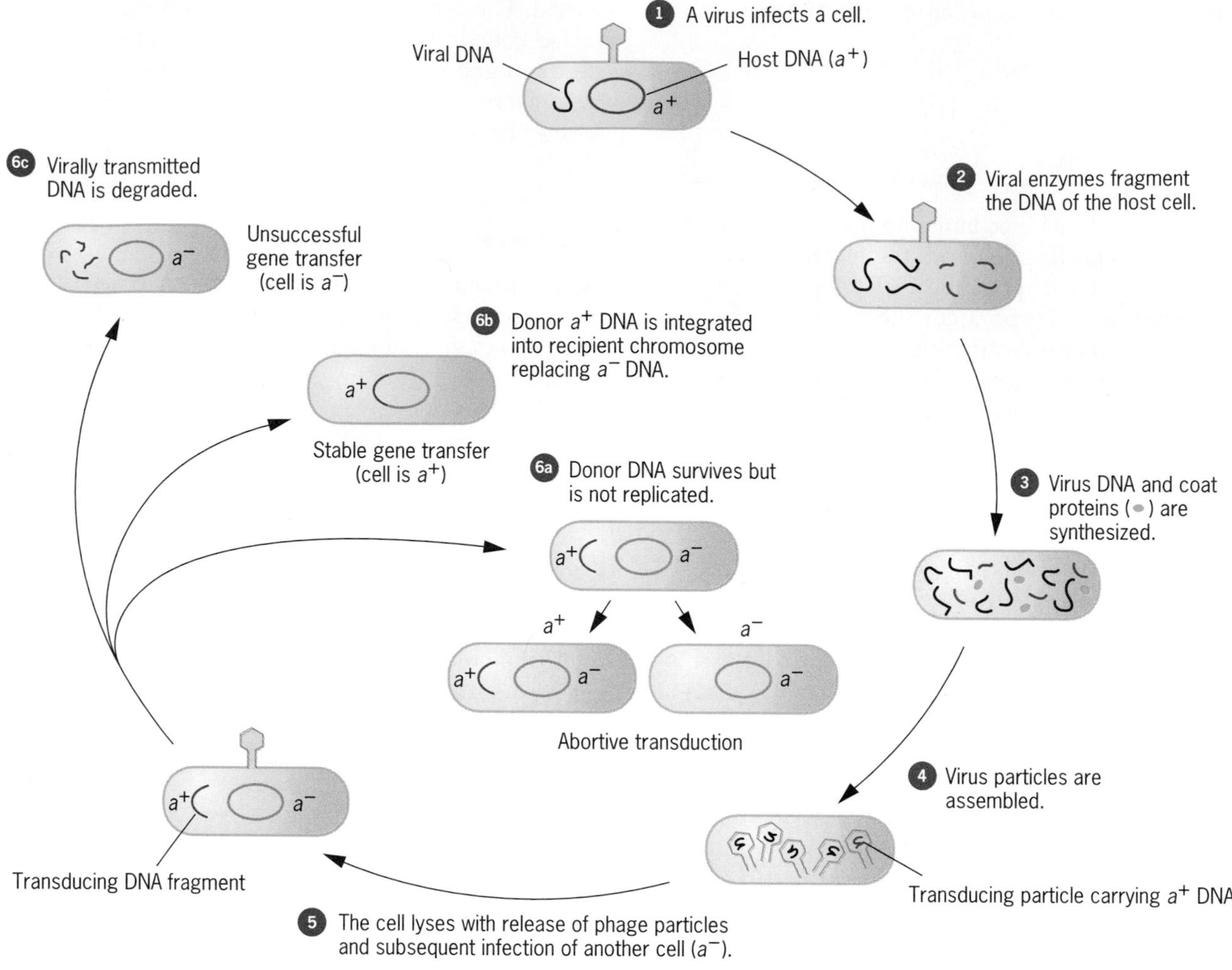

Figure 17.19 A model for the formation of generalized transducing viruses and the subsequent transduction of recipient bacteria.

cle of one of the most extensively studied and fascinating specialized transducing viruses, bacteriophage lambda (Figure 17.20). Lambda (λ), unlike the T phages, is a **temperate phage** because it has two options when it infects a cell. Lambda can proceed through the lytic cycle, like the T phages, and destroy the host cell; or it can enter into a **lysogenic relationship** with the host, a relationship in which the λ chromosome circularizes and then integrates into the host chromosome by recombination at a specific site. In the integrated state, the λ genome is genetically inactive, except for a single gene, encoding a repressor, that maintains all other λ genes in a transcriptionally inactive state. However, if the repression breaks down, the λ chromosome can excise from the bacterial chromosome and commence a lytic infection. Thus, an integrated λ is still able to cause cell lysis at some time in the future. Details of the regulatory mechanisms that determine whether λ enters the lytic or lysogenic state are discussed in Chapter 23. In the lysogenic state, the λ DNA is replicated along with the host chromosome during the cell's replication cycle.

The integrated λ genome, called a **prophage,** can be induced by various means to exit the host chromosome. Excision of the λ prophage occurs when the λ DNA forms a loop and a site-specific recombination event mediated by λ-encoded enzymes releases a circular phage chromosome from the larger bacterial chromosome. Sometimes this recombination occurs aberrantly, releasing a phage genome that has had some of its own genetic material replaced by bacterial genetic material (Figure 17.21) This process is similar to the formation of an F′ factor during conjugation. When the prophage excises improperly, it can pick up only those bacterial genes that bracket its insertion site in the bacterial chromosome. A λ chromosome that is produced in this way is usually defective because it has left behind some of its own genetic material. When the improperly excised λ chromosome is packaged into a phage coat, it forms a specialized transducing particle.

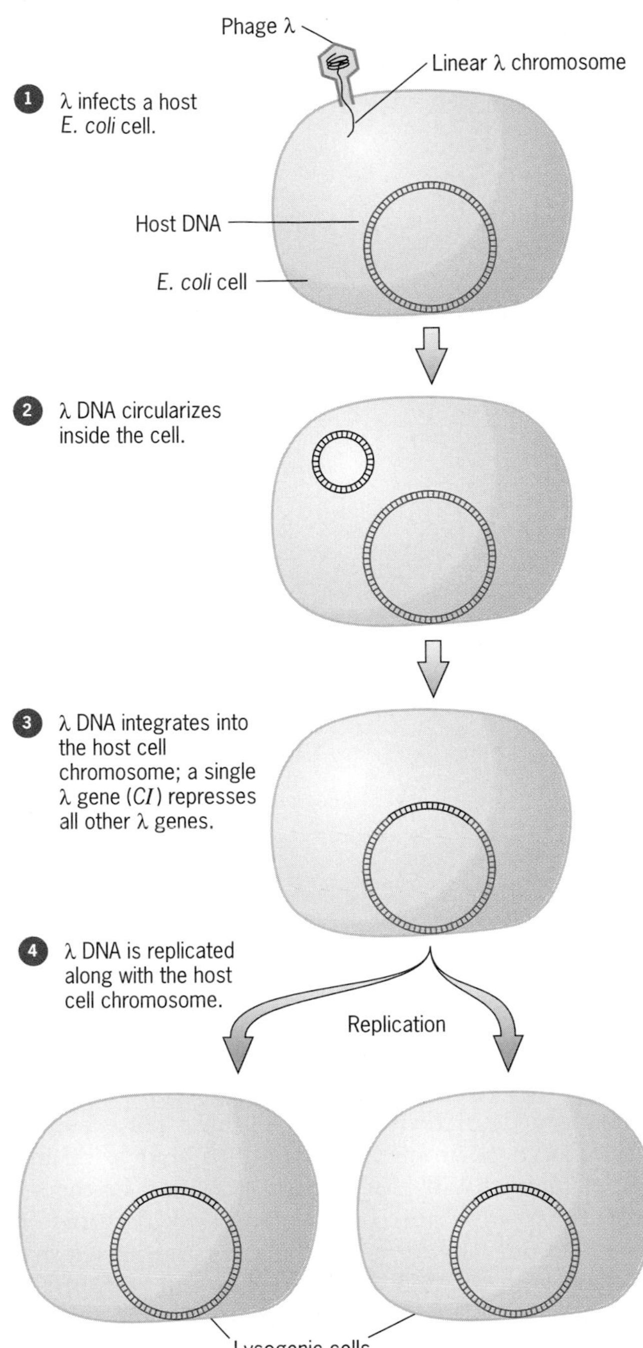

Figure 17.20 The establishment of lysogeny.

Unlike the generalized transducing particle discussed earlier (which carries only bacterial DNA), a specialized transducing particle carries a hybrid chromosome containing both bacterial and viral genetic material. Following lysis of the host cell, the specialized transducing particle injects its chromosome into another bacterium, thus transmitting bacterial genes from one cell to another. However, the defective chromosome of this particle is unable to direct the synthesis of progeny viruses. Instead, the bacterial genes carried by the transducing particle recombine with the genes of the newly infected cell and become incorporated into the recipient cell's genome.

Let's look more closely at λ's insertion site and how it becomes a transducing particle. Lambda inserts at a specific region of the *E. coli* chromosome called λ *att*, for *att*achment site (Figure 17.22). The *att* sites on the phage and bacterial chromosome are homologous. Pairing occurs, and crossing over integrates the λ chromosome into the bacterial chromosome. The λ *att* site on the bacterial chromosome lies between the *gal* and *bio* genes. When the λ prophage is improperly excised from the bacterial chromosome, it carries either galactose-utilization genes (*gal*) or biotin biosynthetic genes (*bio*). Lambda phage carrying *gal* genetic material are called lambda *dgal* because they are defective (*d*), having lost some of their own genes. Lambda *dbio* are defective and carry *bio* DNA.

When lysogenic cells lyse, most of the progeny are normal λ phage because excision is carried out properly. There are very few defective transducing particles in the lysate (about 1 in 10^5 or 10^6 particles); thus, these lysates are called **low-frequency transduction lysates (LFT lysates).** Normal λ phage have complete attachment sites, but defective transducing particles have a nonfunctional hybrid attachment site that is part bacterial and part phage in origin. Thus, defective phage chromosomes do not readily integrate into bacterial DNA.

Defective λ*d* phage can integrate into the recipient chromosome if there is a normal λ in the same cell. The normal λ integrates, producing two bacterial/phage hybrid *att* sites where the defective λ can now insert (Figure 17.22*c*). Furthermore, the normal λ supplies genes that the defective λ is missing. This normal λ is called a **helper phage** because it allows the defective λ to integrate into the host chromosome and reproduce. The bacterial chromosome carrying the helper λ and the λ*d* is an **unstable transductant** for two reasons:

1. The λ*d* prophage can exit the chromosome and be lost, thus causing the cell to revert to its original phenotype.
2. The transductant can be induced to enter the lytic cycle by UV irradiation or some other means. The helper λ and λ*d* are both excised from the chromosome and enter the lytic cycle. The lysate produced by this induction contains about 50 percent λ*d* and 50 percent normal λ. Because it has such a high proportion of transducing particles, this lysate is called a **high-frequency transduction lysate (HFT lysate)**.

A **stable transductant** may form when there is a crossover between the λ*d* chromosome and the bacte-

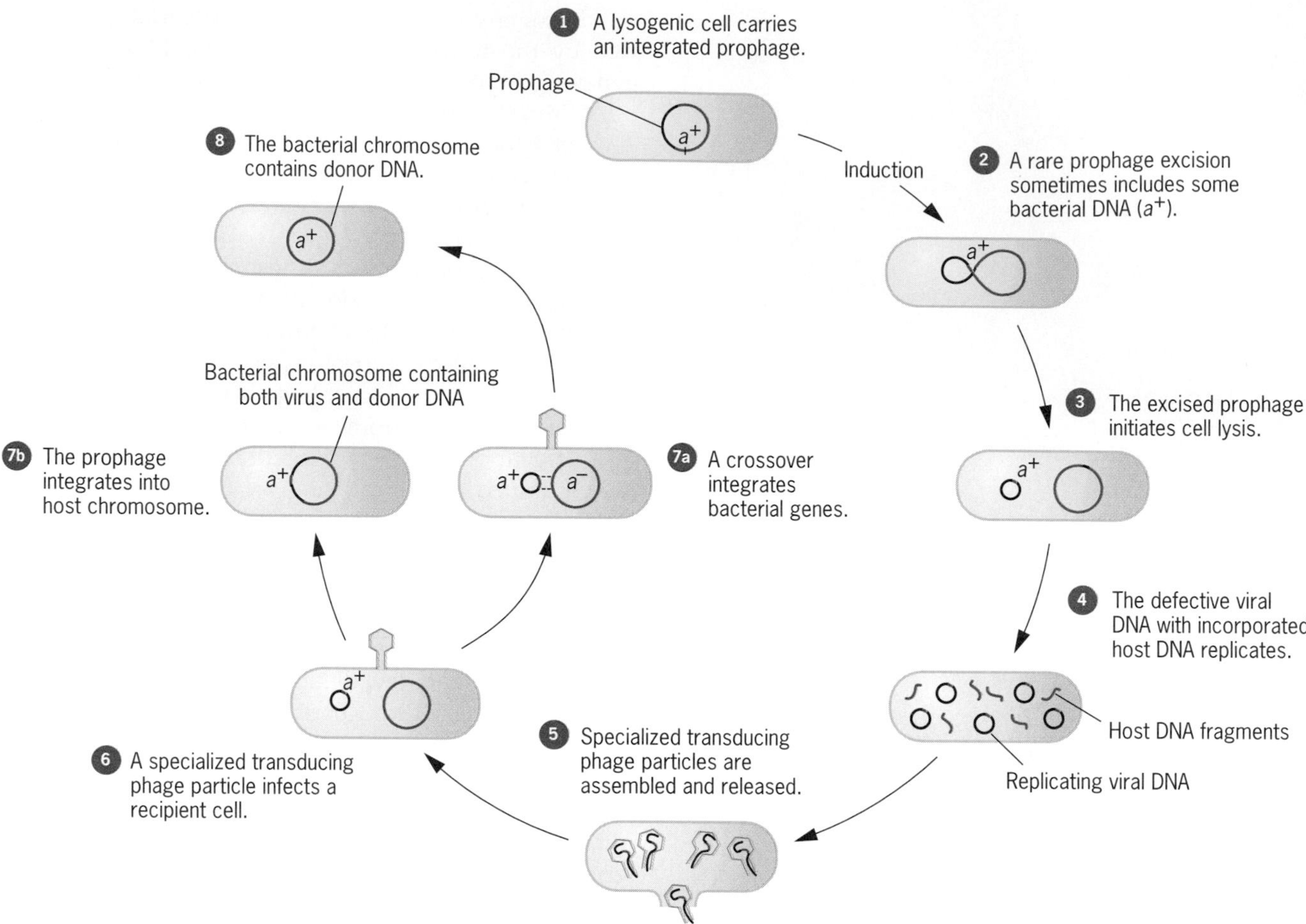

Figure 17.21 Specialized transduction.

rial chromosome. The crossover involves an exchange on both sides of the transduced DNA (Figure 17.22*d*) and the replacement of one allele by another. In this case, the λ genome is not inserted into the host chromosome, so the host does not become lysogenic.

Transduction and Gene Mapping

Generalized transduction provides linkage information in much the same way as transformation. Linkages are usually expressed as cotransduction frequencies: the closer two genes are, the greater the chance they will be on a single DNA fragment packaged together in the same transducing particle. The *E. coli* phage P1 forms generalized transducing particles and is useful for mapping bacterial genes. In one mapping study, P1-infected bacteria carried wild-type alleles of three genes: leu^+, thr^+, and azi^r (*leu* = leucine, *thr* = threonine, and azi^r = sodium azide resistant). Progeny phage from this infection were collected, and special techniques based on density differences were employed to separate transducing P1 phage from the regular P1 phage. The transducing phage were used to infect leu^- thr^- azi^s bacteria. The newly infected cells were placed on a medium that selected for one or two of the genetic markers but not all three. For example, cells placed on minimal medium without azide but supplemented with threonine, leu^+ would be the selected marker because only leu^+ cells would grow on this medium. The *thr* and *azi* markers are unselected because both thr^+ and thr^- genotypes will grow on the threonine-supplemented medium, and in the absence of azide, both azi^r and azi^s cells will grow.

For each of the three experiments with the selected marker, the frequency of the unselected marker(s) was

TABLE 17.2
Data from a Generalized Transduction Experiment Using Selected and Unselected Markers to Determine Gene Order

Experiment	*Selected Marker(s)*	*Unselected Marker(s)*	
1	leu^+	50% azi^r	2% thr^+
2	thr^+	3% leu^+	0% azi^r
3	leu^+, thr^+	0% azi^r	

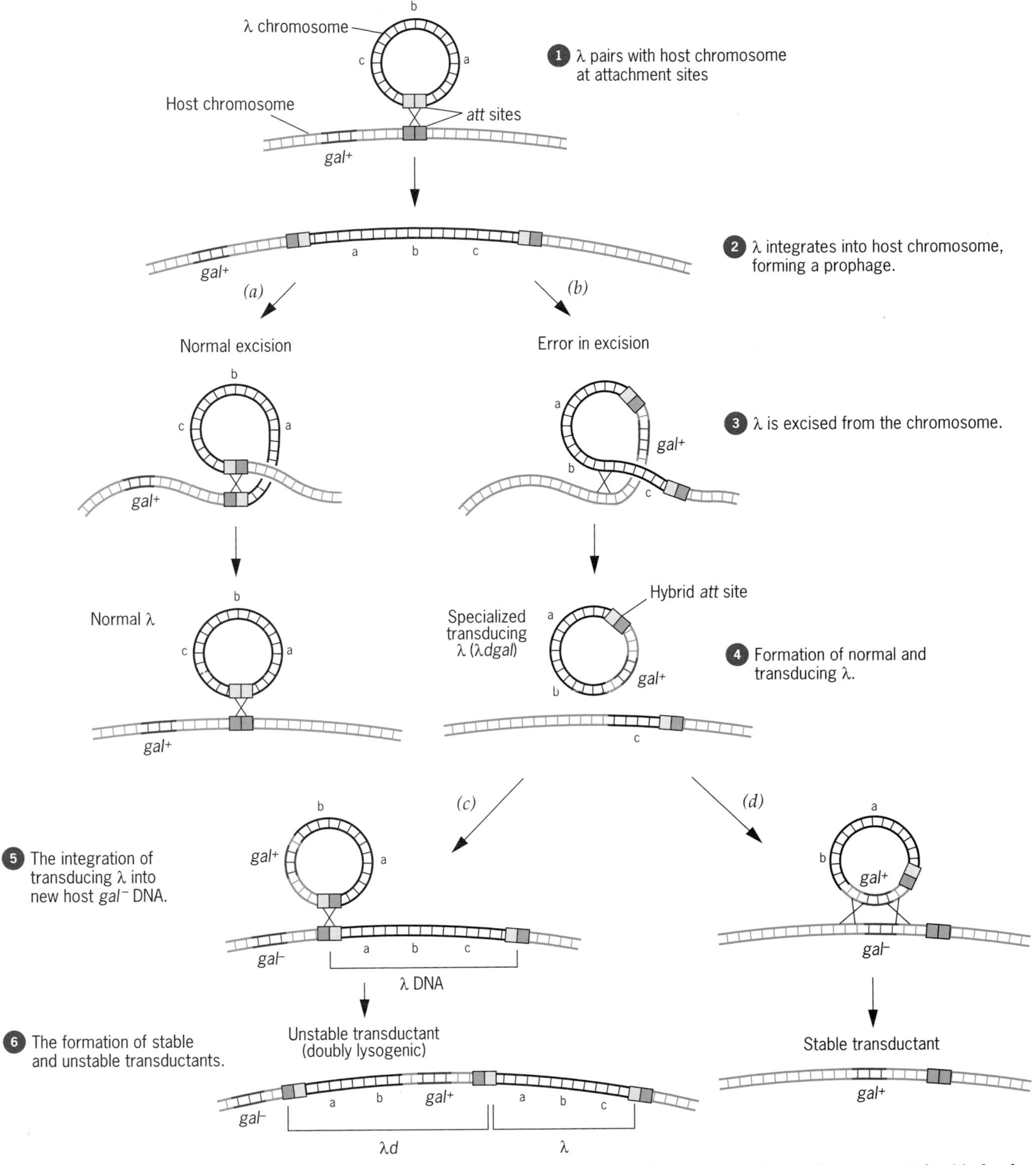

Figure 17.22 The mechanism of specialized transduction in λ bacteriophage. (*a*) Normal excision of λ from the *E. coli* chromosome; (*b*) anomalous excision leading to the formation of a specialized transducing particle; (*c*) the formation of an unstable transductant; (*d*) the formation of a stable transductant.

determined (Table 17.2). For example, those cells found to be leu^+ in the first experiment were further tested to see if they were thr^+ or thr^- and azi^r or azi^s.

Results from the first experiment indicated that the *leu* and *azi* loci were close together and that *leu* and *thr* were not close. The data showed that *leu* and *azi* were cotransduced 50 percent of the time, but in only 2 percent of the cases were *thr* and *leu* cotransduced. Thus there are two possible gene orders:

leu — *azi* ———————————— *thr*

or

azi — *leu* ———————————— *thr*

The second experiment indicated that because *leu* and *thr* were cotransduced 3 percent of the time, and because *thr* and *azi* were never cotransduced, *leu* must be closer to *thr* than is *azi*. Thus, the suggested gene order is

azi — *leu* ———————————— *thr*

The data from the third experiment were consistent with this arrangement because the *thr leu* transducing fragment never carried the *azi* marker.

This study also revealed some interesting information about the maximum size of the transducing DNA fragment carried by P1 phage. The length of DNA between *thr* and *leu* must be close to the maximum size limit because they are cotransduced only 2 to 3 percent of the time. The *azi* locus, which is close to *leu*, is never cotransduced with *thr*. Thus, the maximum size of the transducing fragment that P1 can carry extends from the *thr* locus to a point between *leu* and *azi*.

Specialized transduction can be used to map phage attachment sites and to analyze the genes that lie close to these sites. However, the attachment sites are usually studied using conjugation mapping. The genes closely linked to the phage attachment sites can be analyzed in detail by using specialized transducing particles that have incorporated segments of these genes. In general, however, specialized transducing particles are not as useful in gene mapping experiments as generalized transducing particles.

Key Points: **Transduction is the transport of bacterial genetic material from one bacterium to another by a phage. A generalized transducing phage can carry any segment of the bacterial genome. If the transduced DNA is not incorporated into the recipient genome, abortive transduction may result. The DNA molecules of specialized transducing phage insert into specific sites on the bacterial chromosome. Specialized transducing phage are able to incorporate and transfer only those bacterial genes located around the insertion site. Generalized transducing particles are useful for mapping bacterial genes. The closer together two genes are, the greater the likelihood that they will be cotransduced. Specialized transducing particles can be used to map phage attachment sites and to analyze the genes closely linked to these sites.**

THE EVOLUTIONARY SIGNIFICANCE OF SEXUALITY IN BACTERIA

Sexual reproduction maximizes the diversity of genotypes among progeny, which is highly adaptive under unpredictable conditions. Bacteria do not reproduce sexually, but parasexual mechanisms have evolved in these organisms. Genetic recombination is unquestionably as important to the evolution of bacteria as it is to the evolution of eukaryotes. The ability to generate new genotypes in rapidly changing environments allows members of a bacterial population the opportunity to avoid potentially catastrophic environmental changes, such as the presence of new toxins, bacterial viruses, and antibiotics, and to exploit new environments. Without parasexual processes, the only way that genetic variants would accumulate in bacterial populations would be through the relatively slow process of mutation.

Whereas parasexual processes are beneficial to bacteria, they create serious problems for humans who are attempting to combat bacterial diseases. These parasexual processes are partially responsible for the rapid evolution and spread of plasmids that confer antibiotic and drug resistance to bacteria. As is discussed in Chapter 18, the extensive use of antibiotics in agriculture and medicine has resulted in the evolution of plasmids that carry a whole battery of antibiotic-resistance genes. Indeed, the resurgence of tuberculosis in New York City in the 1990s has resulted in large part from the evolution of a strain of *Mycobacterium tuberculosis* that is resistant to seven different antibiotics. This strain has spread rapidly through hospitals and prisons in New York, causing over 400 cases of tuberculosis. Recent studies have shown that this strain is related to the predominant strain of *M. tuberculosis* in China. Thus, what is beneficial to pathogenic bacteria is often harmful to humans.

Key Points: Parasexual recombination mechanisms produce new combinations of genes in bacteria. These parasexual mechanisms enhance the ability of bacteria to adapt to changes in the environment. They also make it more difficult for humans to control pathogenic bacteria.

TESTING YOUR KNOWLEDGE

1. By means of the interrupted mating technique, five different Hfr strains (**A, B, C, D, E**) were analyzed for the sequence in which they transferred ten different genes (0, 1, 2, 3, 4, 5, 6, 7, 8, 9) to an F^- recipient. Each strain was found to transmit its genes in a unique sequence. The results of the experiment are presented in the following table (only the first six genes transferred are recorded):

		Hfr Strain				
		A	B	C	D	E
Order	First	5	0	6	3	5
of	Second	7	2	7	4	8
Transmission	Third	6	1	5	6	9
	Fourth	4	3	8	7	0
	Fifth	3	4	9	5	2
	Sixth	1	6	0	8	1

What is the gene sequence in the original strain from which these five strains were derived?

ANSWER

In the **A** strain, the sequence must be: origin 576431. Strain **B** has a different F insertion point, and the sequence is origin 021346. The 1346 sequence is common to both strains but in reverse order. The 02 genes are to the left of 1, and 75 are to the right, giving us a sequence of 02134675. In strain **C** the sequence is origin 675890, telling us that 890 map to the right of 5, giving us a sequence of 02134675890, a complete circle:

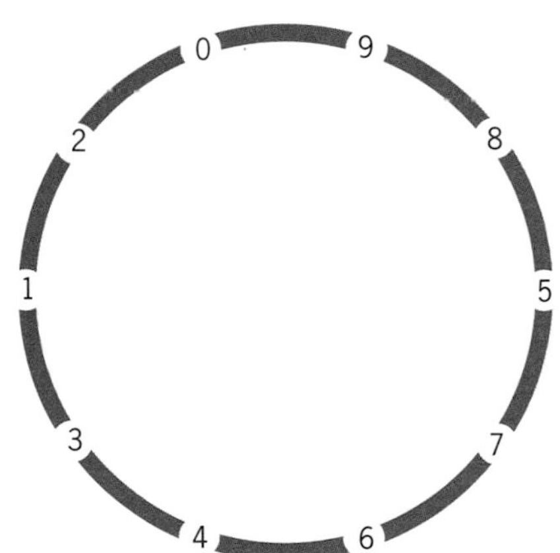

Strain **D** is 346758, which is consistent with this sequence; and strain **E** is 589021, which is also consistent. The F factors are inserted as follows: In **A** F is between 5 and 8, and the transmission is counterclockwise; in **B** it is between 0 and 9, and the transmission is clockwise; in **C** it is between 4 and 6, and the transmission is clockwise; in **D** it is between 1 and 3, and the transmission is clockwise; and in **E** it is between 5 and 7, and the transmission is clockwise.

2. In a transformation experiment, a number of crosses were made between strains that differed in two genes:

Donor	Recipient	Frequency of Wild-Type (++) Transformants
$a^- b^+$	$a^+ b^-$	0.450
$c^- b^+$	$c^+ b^-$	0.190
$d^- b^+$	$d^+ b^-$	0.260

Assuming that the b gene is at one end, what is the sequence of these four genes?

ANSWER

In each of these crosses, two exchanges are required to insert the wild-type allele into the recipient. These are diagrammed as follows:

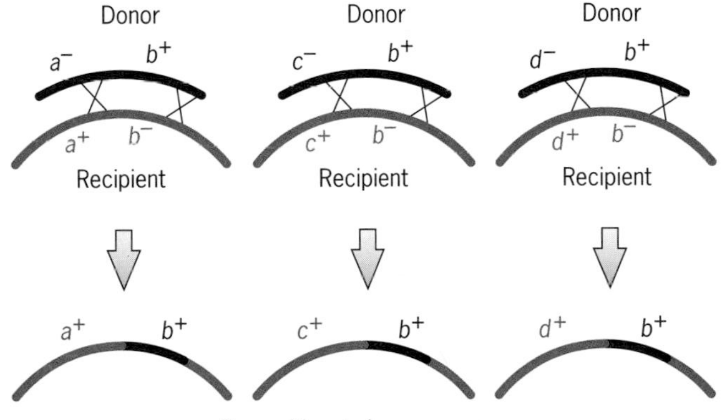

More exchanges occur between a and b than between any of the other genes and b, so a must be farthest from b. The c gene is closer to b than d because it has a lower recombination frequency. The b and c genes exhibit 19 percent recombination, b and d exhibit 26 percent recombination, and a and b exhibit 45 percent recombination; consequently the map is:

a------19------d---7---c------19------b

QUESTIONS AND PROBLEMS

17.1 A nutritionally defective *E. coli* strain grows only on a medium containing thymine, whereas another nutritionally defective strain grows only on medium containing leucine. When these two strains are grown together, a few progeny are able to grow on a minimal medium with neither thymine nor leucine. How can this result be explained?

17.2 Assume that you have just demonstrated genetic recombination in a previously undescribed species of bacteria (e.g., when a strain of genotype $a\ b^+$ is present with a strain of genotype $a^+\ b$, some recombinant genotypes, $a^+\ b^+$ and $a\ b$, are formed). Describe how you would go about (operationally) determining whether the observed recombination was the result of a process similar to transformation, transduction, or conjugation.

17.3 Compare, in table form, similarities and differences of the mechanisms through which (a) transduction, (b) sexduction, and (c) transformation may occur.

17.4 (a) What are the genotypic differences between F^- cells, F^+ cells, and Hfr cells? (b) What are the phenotypic differences? (c) By what mechanism are F^- cells converted to F^+ cells? F^+ cells to Hfr cells?

17.5 (a) Of what use are F′ factors in genetic analysis? (b) How are F′ factors formed? (c) By what mechanism does sexduction occur?

17.6 What are the basic differences between generalized transduction and specialized transduction?

17.7 What is the difference between a phage and a prophage?

17.8 How can genes be mapped by interrupted mating conjugation experiments?

17.9 What does the term *cotransduction* mean? How can cotransduction frequencies be used to map genetic markers?

17.10 An F^+ strain, marked at 10 loci, gives rise spontaneously to Hfr progeny whenever the F factor becomes incorporated into the chromosome of the F^+ strain. Because the F factor can integrate into the circular chromosome at many points, various Hfr segregants will transfer the genetic markers in different sequences. For any Hfr strain, the order of markers entering early can be determined by interrupted mating experiments. From the following data for several Hfr strains derived from the same F^+, determine the order of markers in the F^+ strain.

Hfr Strain	Markers Donated in Order
1	—Z-H-E-R→
2	—O-K-S-R→
3	—K-O-W-I→
4	—Z-T-I-W→
5	—H-Z-T-I→

17.11 The following table shows the time of entry of several genes transferred into F^- recipient strains of *E. coli* during conjugation with three different Hfr strains.

Hfr 1 Markers	*arg*	*pro*	*met*	*thr*	*his*
Time of entry (min.)	15	21	32	48	54
Hfr 2 Markers	*mal*	*met*	*thi*	*thr*	*arg*
Time of entry (min.)	10	17	22	33	57
Hfr 3 Markers	*his*	*phe*	*arg*	*mal*	*met*
Time of entry (min.)	17	23	35	45	52

(a) Construct a genetic map that shows the location of all eight markers. Place the *arg* gene at time 0 on your map. Show the position (in minutes) of all the genes relative to the *arg* gene. (b) How large is this map (in minutes)?

17.12 In *E. coli*, the ability to utilize lactose as a carbon source requires the presence of the enzymes β-galactosidase and β-galactoside permease. These enzymes are coded for by two closely linked genes, *lacZ* and *lacY*, respectively. Another gene, *proC*, controls, in part, the ability of *E. coli* cells to synthesize the amino acid proline. The alleles str^r and str^s control resistance and sensitivity, respectively, to streptomycin. HfrH is known to transfer the two *lac* genes, *proC*, and *str*, in that order, during conjugation.

A cross was made between an HfrH strain of genotype $lacZ^-$ $lacY^+$ $proC^+$ str^s and an F^- strain of genotype $lacZ^+$ $lacY^-$ $proC^-$ str^r. After about two hours of mating, the mixture was diluted and plated out on medium containing streptomycin but no proline. When the resulting $proC^+$ str^r recombinant colonies were checked for their ability to grow on medium containing lactose as the sole carbon source, very few of them were capable of fermenting lactose. When the cross HfrH $lacZ^+$ $lacY^-$ $proC^+$ str^s × F^- $lacZ^-$ $lacY^+$ $proC^-$ str^r was done, a large number of the $proC^+$ str^r recombinants were able to grow on medium containing lactose as the sole carbon source. What is the order of the *lacZ* and *lacY* genes relative to *proC*?

17.13 From the following transduction data, indicate the order of the genes.

Genes	Percent Cotransduction
a–b	6
a–c	0
a–d	15
b–c	3
c–d	1

17.14 Donor DNA from bacteria of genotype $a^+b^+c^+$ was used to transform bacteria of genotype $a^-b^-c^-$. Which transformant classes will occur with the lowest frequency if the three genes are closely linked in the order *a-b-c*?

17.15 A strain of wild-type bacteria was used as a donor to transduce recipient cells with the following genotypes, and the results given here were obtained.

Recipient Genotype	No. of Wild-Type Colonies
arg^+ ser^- glu^- his^+	125
arg^- ser^+ glu^- his^+	210
arg^+ ser^+ glu^- his^+	390
arg^- ser^- glu^+ his^+	42
arg^+ ser^- glu^+ his^+	312
arg^+ ser^- glu^+ his^-	0
arg^- ser^+ glu^+ his^-	256
arg^+ ser^+ glu^- his^-	67
arg^+ ser^+ glu^+ his^-	320
arg^+ ser^- glu^+ his^+	480

What is the order of the *arg*, *ser*, *glu*, and *his* genes on the bacterial chromosome?

17.16 Two reciprocal crosses (cross 1 = $a^+b^+c^-$ donor × $a^-b^-c^+$ recipient; cross 2 = $a^-b^-c^+$ donor × $a^+b^+c^-$ recipient) were used to determine the order of markers *b* and *c*, relative to *a*, by transduction. There are two possible orders relative to the outside marker, *a*: order I = *a-b-c* and order II = *a-c-b*. How many crossing over events would be required to produce $a^+b^+c^+$ recombinants in each of the two crosses, (a) if order I is the correct order, (b) if order II is correct? (c) How would one determine the correct order based on the information in parts (a) and (b)?

17.17 The data in the following table were obtained from three-point transduction tests made to determine the order of mutant sites in the *A* gene for tryptophan synthetase in *E. coli*. *Anth* is a linked, unselected marker. In each cross, trp^+ recombinants were selected and then scored for the *anth* marker ($anth^+$ or $anth^-$). What is the linear order of *anth* and the three mutant alleles of the *A* gene indicated by the data in the table?

Cross	Donor Markers	Recipient Markers	*anth* Allele in trp^+ Recombinants	% $anth^+$
1	$anth^+$ A34	$anth^-$ A223	72 $anth^+$:332 $anth^-$	18
2	$anth^+$ A46	$anth^-$ A223	196 $anth^+$:180 $anth^-$	52
3	$anth^+$ A223	$anth^-$ A34	380 $anth^+$:379 $anth^-$	50
4	$anth^+$ A223	$anth^-$ A46	60 $anth^+$:280 $anth^-$	20

17.18 Two additional mutations in the *trp A* gene of *E. coli*, *trp A58* and *trp A487*, were ordered relative to *trp A223* and the outside marker *anth* by three-factor transduction crosses as described in Problem 17.17. The results of these crosses are summarized in the following table. What is the linear order of *anth* and the three mutant sites in the *trp A* gene?

Cross	Donor Markers	Recipient Markers	*anth* Allele in trp^+ Recombinants	% $anth^-$
1	$anth^+$ A487	$anth^-$ A223	72 $anth^+$:332 $anth^-$	82
2	$anth^+$ A58	$anth^-$ A223	196 $anth^+$:180 $anth^-$	48
3	$anth^+$ A223	$anth^-$ A487	380 $anth^+$:379 $anth^-$	50
4	$anth^+$ A223	$anth^-$ A58	60 $anth^+$:280 $anth^-$	80

17.19 Four genes are present on the chromosome of *E. coli* in the order *abcd*. In an $a^+\ b^+\ c^-\ d^+ \times a^-\ b^-\ c^+\ d^-$ cross using conjugating *E. coli* cells, the $b^+\ c^+$ alleles were selected for among the recombinant offspring. The *a* and *d* alleles were not selected. When the $b^+\ c^+$ recombinants were examined, most were $a^-\ d^-$. Which strain was the donor (Hfr)?

17.20 Examine the results of two transformation experiments given in the following table.

Donor DNA	Recipient Cell Genotype	Percentage of Transformants a^+b	ab^+	a^+b^+
a^+b^+	a^-b^-	2.7	1.1	0.03
a^+b^- and a^-b^+	a^-b^-	1.9	1.2	0.02

Are the genes close to each other? Why or why not?

17.21 Donor DNA from bacteria of genotype $a^+b^+c^+$ was used to transform mutant bacteria carrying the linked genes $a^-b^-c^-$. The distribution of transformants was as follows:

$a^+b^+c^+$	1400	$a^-b^+c^+$	300
$a^+b^-c^-$	350	$a^-b^-c^+$	100
$a^+b^+c^-$	150	$a^+b^-c^+$	10
$a^-b^+c^-$	60		

Construct a linkage map showing the order of the genes on the chromosome and the distances between adjacent genes.

17.22 Two *Bacillus* genes, *a* and *b*, are studied in a transformation experiment. The results of the experiment are presented in the following table:

Experiment	Donor	Recipient	Transformant Classes *a*	*b*	Number
A	$a^+\ b^-$	$a^-\ b^-$	+	–	232
	and		–	+	341
	$a^-\ b^+$		+	+	7
B	a^+b^+	a^-b^-	+	–	130
			–	+	96
			+	+	247

(a) Are the two genes closely linked, according to the criteria for linkage in transformation experiments?

(b) If they are linked, what is the recombination frequency between them?

17.23 In a conjugation experiment, you find that there are two genes that lie so close together you are unable to determine a gene order by time-interval mapping. There are two possible gene orders to consider:

a ——— *b-c* or *a* ——— *c-b*

You set up reciprocal crosses so that the Hfr donor is + + *c* and the F^- recipient is *a b* + in one experiment (A) and the Hfr donor is *a b* + and the F^- recipient is + + *c* in another experiment (B). When you screen for the + + + recombinants and compare their frequencies in both experiments, you find that experiment A produces about 1/25th as many as experiment B. What does this suggest about the order of the three genes?

BIBLIOGRAPHY

ADELBERG, E. A. (ED). 1966. *Papers on Bacterial Genetics*, 2nd ed. Little, Brown, Boston.

BROCK, T. D. 1990. *The Emergence of Bacterial Genetics*. Cold Spring Harbor Laboratory Press, Cold Spring Harbor, NY.

DALE, J. W. 1994. *Molecular Genetics of Bacteria*, 2nd ed. John Wiley and Sons, New York.

DAVIS, R. W., D. BOTSTEIN, AND J. R. ROTH. 1980. *Advanced Bacterial Genetics*. Cold Spring Harbor Laboratory Press, Cold Spring Harbor, NY.

HAYES, W. 1968. *The Genetics of Bacteria and Their Viruses*, 2nd ed. John Wiley and Sons, New York.

HOLLOWAY, B. W. 1993. Genetics for all bacteria. *Annu. Rev. Microbio.* 47:659–684.

MALOY, S. R., E. J. CRONAN, AND D. FREIFELDER. 1994. *Microbial Genetics*, 2nd ed. Jones and Bartlett, Boston.

MILLER, R. V. 1998. Bacterial gene swapping in nature. *Sci. Amer.* 278(1):66–71.

SNYDER, L., AND W. CHAMPNESS. 1997. *Molecular Genetics of Bacteria*. ASM Press, Washington, DC.

STREIPS, U. N., AND R. E. YASBIN. 1991. *Modern Microbial Genetics*. Wiley-Liss, New York.

18

Color variation among kernels of maize. Studies of the genetic basis of this variation led to the discovery of transposable elements.

Transposable Genetic Elements

CHAPTER OUTLINE

Transposable Elements in Bacteria
Transposable Elements in Eukaryotes
Retrotransposons
The Genetic and Evolutionary Significance of Transposable Elements

Maize: A Staple Crop with a Cultural Heritage

Maize is one of the world's most important crop plants. In the United States alone over 79 million acres are devoted to maize cultivation. In 1994 this crop yielded an average of 138.6 bushels of grain per acre, enough to feed tens of millions of farm animals and to produce a large assortment of foods for human consumption, including cereal, sugar, corn oil, and corn syrup. The cultivation of maize began at least 5000 years ago in Central America. By the time Christopher Columbus arrived in the New World, maize cultivation had spread north to Canada and south to Argentina.

The native peoples of North and South America developed many different varieties of maize, each adapted to particular conditions. Maize was cultivated in lowland clearings, in upland terraces, in the rain forests of Amazonia, and in the deserts of Arizona and New Mexico. This diversification of maize varieties occurred through selective breeding. Ancient farmers cultivated plants with desirable characteristics—size, shape of the ears, number, texture and color of the kernels, dietary value, hardiness, and so on. After many generations of selective breeding, each locality developed a distinctive variety. This genetic manipulation was so extensive that today maize cannot grow without human assistance. It is a completely domesticated plant.

Maize cultivation has also had a cultural significance among these native peoples. They developed varieties that had colorful kernels—red, blue, yellow, white, and purple—and associated each color with a special aesthetic or religious value. To the peoples of the American Southwest, for example, blue maize is considered sacred, and each of the four cardinal directions of the compass is represented by a particular maize color. Some groups consider kernels with stripes and spots to be signs of strength and vigor.

The colorful patterns that we see on maize ears also have an important scientific significance. Modern research has shown that the stripes and spots on maize kernels are the result of a genetic phenomenon called *transposition.* Within the maize genome—indeed, within the genomes of most organisms—geneticists have found DNA sequences that can move from one position to another. These *transposable elements*—or more simply, *transposons*—constitute an appreciable fraction of the genome. In maize, for example, they account for more than half of all the DNA. When transposable elements move from one location to another, they may break chromosomes or mutate genes. Thus, these elements have a profound genetic significance.

The research that revealed the existence of transposable elements was performed in the 1940s and 1950s by Barbara McClintock (see Historical Sidelight: Barbara McClintock, the Discoverer of Transposable Elements), an American scientist who later received the Nobel Prize in Physiology or Medicine for her work. McClintock studied the genetic basis of kernel striping and spotting in maize and found that it was due to the insertion and excision of transposable elements. For many years, McClintock's work had no parallel, but in the 1960s and 1970s, as new techniques emerged to analyze genes at the molecular level, transposable elements were discovered in the genomes of assorted bacteria, fungi, plants, and animals. Then, in the 1980s, they were found in human beings. Extensive research since McClintock's time has indicated that these elements are present in the genomes of many—perhaps most—organisms, and suggests that they have played an important role in the evolution of many species.

In this chapter, we investigate the structure and behavior of different types of transposable elements. We begin with a discussion of the transposable elements that are found in bacteria because these were the first transposons to be characterized at the molecular level. Then we discuss some of the transposable elements found in eukaryotes such as maize, *Drosophila,* yeast, and human beings. Finally, we discuss the genetic and evolutionary significance of mobile genetic elements.

TRANSPOSABLE ELEMENTS IN BACTERIA

Bacterial transposons were the first to be studied at the molecular level. There are three main types: the Insertion Sequences, or IS elements, the composite transposons, and the Tn*3* elements. These three types of transposons differ in size and structure. The IS elements are the simplest, containing only genes that encode proteins involved in transposition. The composite transposons and Tn*3* elements are more complex, containing some genes that encode products unrelated to the transposition process.

IS Elements

The simplest bacterial transposons are the **Insertion Sequences,** or **IS elements,** so named because they can insert at many different sites in bacterial chromosomes and plasmids. IS elements were first detected in certain *lac*$^-$ mutations of *E. coli.* These mutations had the unusual property of reverting to wild-type at a high rate. Molecular analyses eventually revealed that these unstable mutations possessed extra DNA in or near the *lac* genes. When DNA from the wild-type revertants of these mutations was compared with that from the mutations themselves, it was found that the extra DNA had been lost. Thus, these genetically unstable mutations were caused by DNA sequences that had inserted into *E. coli* genes, and reversion to wild-type was caused by excision of these sequences. Similar insertion sequences have been found in many other bacterial species.

IS elements are compactly organized. Typically, they consist of less than 2500 nucleotide pairs and contain only genes whose products are involved in promoting or regulating transposition. Many distinct types of IS elements have been identified. The smallest, IS*1*, is 768 nucleotide pairs long. Each type of IS element is demarcated by short identical, or nearly identical, sequences at its ends (Figure 18.1). Because these terminal sequences are always in inverted orientation with

HISTORICAL SIDELIGHT

Barbara McClintock, the Discoverer of Transposable Elements

Figure 1 Barbara McClintock.

Scientific advances can come through the unflagging persistence of a single individual, someone who focuses on a question or problem and researches it for many years. Barbara McClintock was such an individual (Figure 1). Her long life was devoted to the study of maize genetics. Born in New England in 1902, she grew up in New York City and attended Cornell University, first as an undergraduate, then as a graduate student. After receiving her Ph.D., McClintock remained at Cornell for several years, collaborating with an illustrious group of maize geneticists, including George Beadle, R. A. Emerson, Charles Burnham, Marcus Rhoades, and Lowell Randolph. Together, these researchers developed the materials and methods of maize genetics into a rich intellectual discipline. McClintock played an important role in this work. Because of her skill in cytological analysis, she succeeded in identifying each of the 10 maize chromosomes and was able to connect them with linkage groups. She also did pioneering work on the mechanism of crossing over and on the origin of the nucleolus. However, McClintock's most notable achievement was to elucidate the genetic properties of transposable elements. Most of her research on this subject was done during the many years she spent in the Genetics Department of the Carnegie Institution at Cold Spring Harbor, New York. There she found the freedom to pursue her studies of transposon-mediated mutation and chromosome breakage.

McClintock published her first report about transposable elements in 1948. Several other reports followed, including a major paper in the 1951 *Cold Spring Harbor Symposium on Quantitative Biology.* For many reasons, the ideas that she espoused in these papers were not especially well received. The concept of transposition contradicted the established view that genes occupied fixed positions on chromosomes. McClintock's data were complex, and she had difficulty communicating them to her colleagues. In addition, transposition did not seem to be a general phenomenon. Although there was little doubt that it occurred in maize, no one had seen it in other organisms. This situation changed in the 1960s and 1970s, when transposition was discovered in bacteria and *Drosophila.* At last the scientific world awoke to the broad significance of McClintock's ideas.

McClintock was highly respected by her colleagues. In 1944, at the relatively young age of 42, she was admitted to the prestigious National Academy of Sciences of the United States. In 1945 she was elected president of the Genetics Society of America, and in 1970 she was awarded the National Medal of Science. Her Nobel Prize came in 1983, 35 years after her first publication on transposable elements. The day the prize was announced, she was out in the woods collecting mushrooms. During her life, McClintock was a model of "adamant individuality and self-containment."[1] She died in 1992, a few months after her ninetieth birthday.

[1]Federoff, N. V., 1994. Barbara McClintock (June 16, 1902–September 2, 1992). *Genetics* 136:1–10.

respect to each other, they are called **inverted terminal repeats.** Their lengths range from 9 to 40 nucleotide pairs. Inverted terminal repeats are characteristic of most—but not all—types of transposons. When nucleotides in these repeats are mutated, the transposon usually loses its ability to move. These mutations therefore demonstrate that inverted terminal repeats play an important role in the transposition process.

At least some IS elements encode a protein that is needed for transposition. This protein, called **transposase,** seems to bind at or near the ends of the element, where it cuts both strands of the DNA. Cleavage of the DNA at these sites excises the element from the chromosome or plasmid, making it free to insert at a new position in the same or a different DNA molecule. When IS elements insert into chromosomes or plasmids, they create a duplication of part of the DNA sequence at the site of the insertion. One copy of the duplication is located on each side of the element. These short (2 to 13 nucleotide pairs), directly repeated

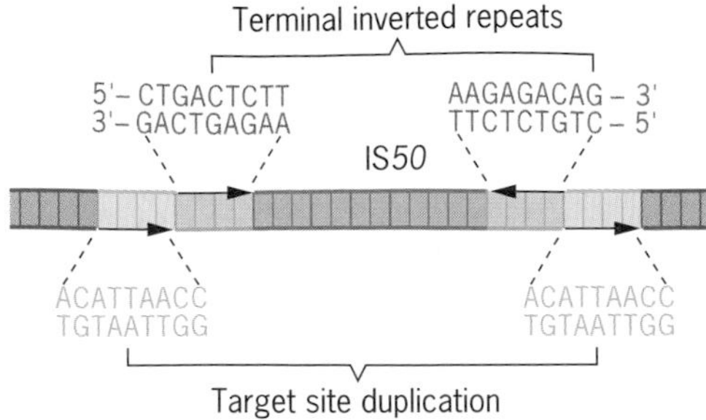

Figure 18.1 Structure of an inserted IS*50* element showing its terminal inverted repeats and target site duplication. The terminal inverted repeats are imperfect because the fourth nucleotide pair from each end is different.

sequences, called **target site duplications,** are thought to arise from staggered cleavage of the double-stranded DNA molecule (Figure 18.2).

A bacterial chromosome may contain several copies of a particular type of IS element. For example, 6 to 10 copies of IS*1* are found in the *E. coli* chromosome. Plasmids may also contain IS elements. The F plasmid, for example, typically has at least two different IS elements, IS*2* and IS*3*. When a particular IS element resides in both a plasmid and a chromosome, it creates the opportunity for homologous recombination between different DNA molecules. Such recombination appears to be responsible for the integration of the F plasmid into the *E. coli* chromosome (Chapter 17).

Both the *E. coli* chromosome and the F plasmid are circular DNA molecules. When the plasmid and the chromosome recombine in a region of homology—for example, in an IS element common to both of them—the smaller plasmid is integrated into the larger chromosome, creating a single circular molecule. Such integration events produce Hfr strains capable of transferring their chromosomes during conjugation. These strains vary in the integration site of the F plasmid because the IS elements that mediate recombination occupy different chromosomal positions in different *E. coli* strains—a result of their ability to transpose. Because of their role in forming Hfr strains, IS elements potentiate the exchange of chromosomal genes between different strains of bacteria. This exchange creates genetic variability, which, along with mutation, is the basis for evolutionary change in bacterial populations.

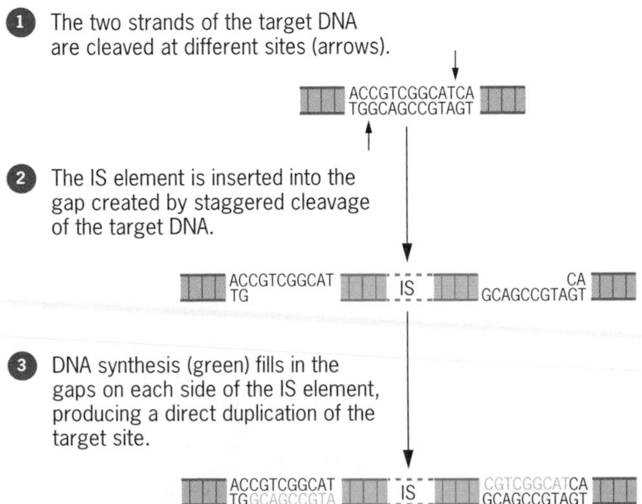

Figure 18.2 Production of target site duplications by the insertion of an IS element.

Composite Transposons

Composite transposons, denoted by the symbol Tn, are created when two IS elements insert near each other. The region between them can then be transposed by the joint action of the flanking elements. In effect, two IS elements "capture" a DNA sequence that is otherwise immobile and endow it with the ability to move. Figure 18.3 gives three examples. In Tn*9*, the flanking IS elements are in the same orientation with respect to each other, whereas in Tn*5* and Tn*10*, the orientation is inverted. The region between the IS elements in each

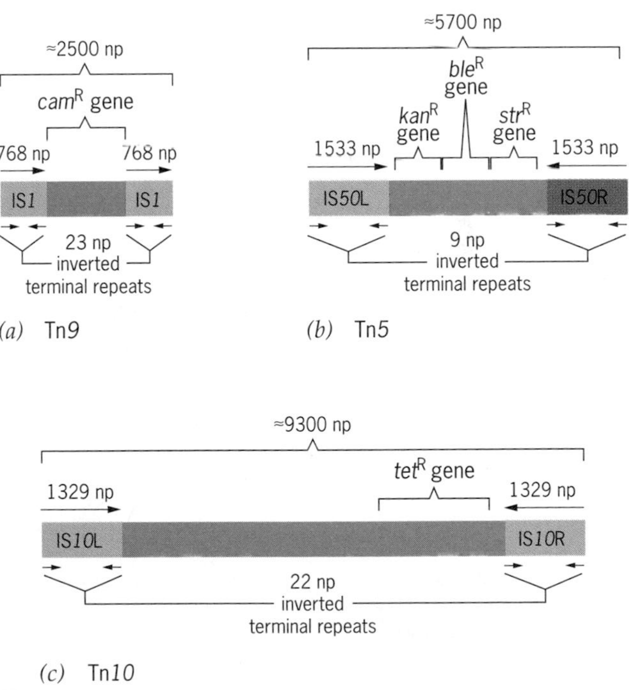

Figure 18.3 Genetic organization of composite transposons. The orientation and length (in nucleotide pairs, np) of the constituent sequences are indicated. (*a*) Tn*9* consists of two IS*1* elements flanking a gene for chloramphenicol resistance. (*b*) Tn*5* consists of two IS*50* elements flanking genes for kanamycin, bleomycin, and streptomycin resistance. (*c*) Tn*10* consists of two IS*10* elements flanking a gene for tetracycline resistance.

of these transposons contains genes that have nothing to do with transposition. In fact, in all three transposons, the genes between the flanking IS elements confer resistance to antibiotics—chloramphenicol resistance in Tn*9;* kanamycin, bleomycin, and streptomycin resistance in Tn*5;* and tetracycline resistance in Tn*10*.

Sometimes the flanking IS elements in a composite transposon are not quite identical. For instance, in Tn*5*, the element on the right, called IS*50*R, is capable of producing a transposase to stimulate transposition, but the element on the left, called IS*50*L, is not. This difference is due to a change in a single nucleotide pair that prevents IS*50*L from specifying the active transposase.

Tn*5* also illustrates another feature of the composite transposons: their movement is regulated (Figure 18.4). When a bacterial cell is infected with a nonlytic bacteriophage that carries Tn*5* on its chromosome, the frequency of Tn*5* transposition is dramatically reduced if the infected cell already carries a copy of Tn*5*. This reduction implies that the resident transposon inhibits the transposition of an incoming transposon, possibly by synthesizing a repressor. Analyses by Michael Syvanen, William Reznikoff, and their colleagues have shown that this hypothesis is correct. The IS*50*R element of Tn*5* actually produces two proteins. One, the transposase, catalyzes transposition, whereas the other, a truncated transposase created by translation from a start codon within the transposase gene, prevents transposition. Because the shorter protein is the more abundant, Tn*5* transposition tends to be repressed.

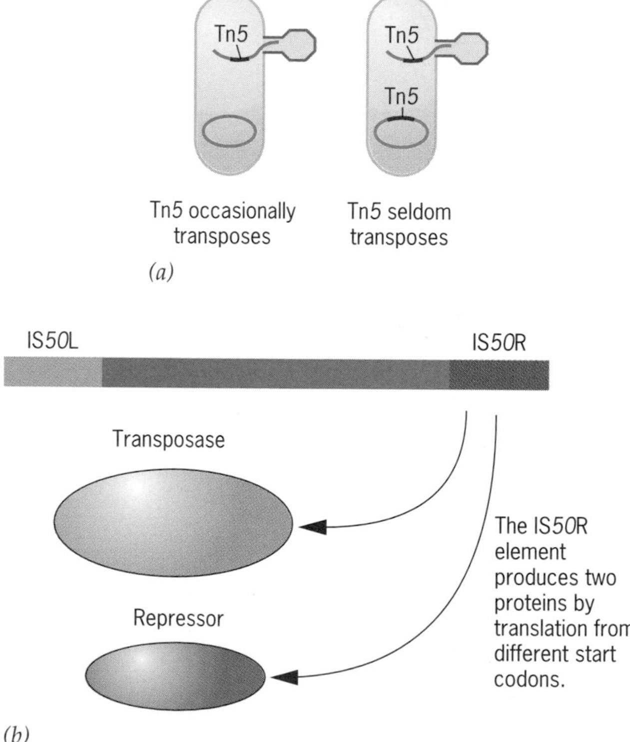

Figure 18.4 Regulation of Tn*5*. (*a*) Infection of *E. coli* cells with bacteriophages carrying Tn*5*. Cells that already possess a copy of Tn*5* repress transposition. (*b*) Genetic basis of Tn*5* regulation. One of the proteins produced by IS*50*R is a transposase that catalyzes transposition, but the other is a repressor that inhibits transposition. The effect of the repressor usually prevails.

Tn*3* Elements

The elements in this group of transposons are larger than the IS elements and usually contain genes that are not necessary for transposition—a feature that is also characteristic of the composite transposons. However, unlike the composite transposons, the Tn*3* elements do not have IS elements at each of their ends. Instead, the Tn*3* elements have simple inverted repeats 38 to 40 nucleotide pairs long at their termini. The Tn*3* elements also produce target site duplications when they insert into DNA.

The genetic organization of the element known specifically as Tn*3* is shown in Figure 18.5. There are three genes, *tnpA, tnpR,* and *bla,* encoding, respectively, a transposase, a resolvase/repressor, and an enzyme called beta lactamase. The beta lactamase confers resistance to the antibiotic ampicillin, and the other two proteins play important roles in transposition.

The transposition of Tn*3* occurs in two stages (Figure 18.6). First, the transposase mediates the fusion of two molecules, forming a structure called a **cointegrate.** During this process, the transposon is replicated, and one copy is inserted at each junction in the cointegrate; the two Tn*3* elements in the cointegrate are oriented in the same direction. In the second stage of transposition, the *tnpR*-encoded resolvase mediates a site-specific recombination event between the two Tn*3* elements. This event occurs at a sequence in Tn*3* called *res*, the *resolution site,* generating two molecules, each with a copy of the transposon.

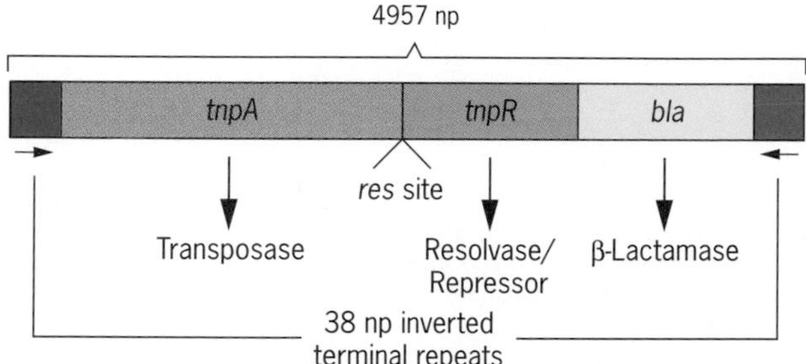

Figure 18.5 Genetic organization of Tn*3*. Lengths of DNA sequences are given in nucleotide pairs (np).

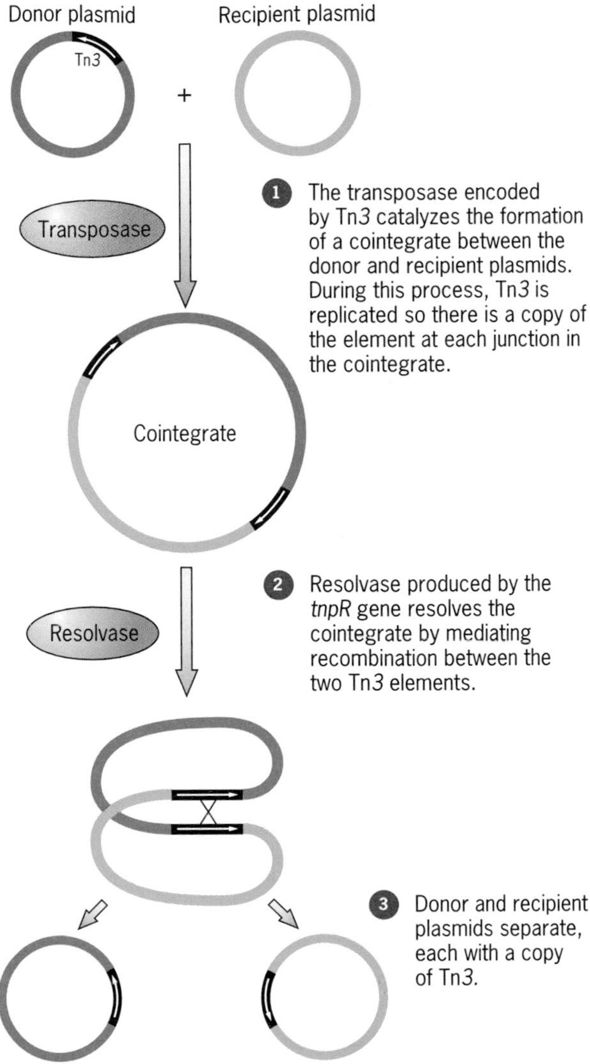

Figure 18.6 Transposition of Tn3 via the formation of a cointegrate.

The *tnpR* gene product has yet another function—to repress the synthesis of both the transposase and resolvase proteins. Repression occurs because the *res* site is located between the *tnpA* and *tnpR* genes. By binding to this site, the *tnpR* protein interferes with the transcription of both genes, leaving their products in chronic short supply. As a result, the Tn3 element tends to remain immobile.

The Medical Significance of Bacterial Transposons

Many bacterial transposons carry genes for antibiotic resistance. Consequently, it is a relatively simple matter for these genes to move from one DNA molecule to another—for instance, from a chromosome to a plasmid. This genetic flux has a profound medical significance because many of the DNA molecules that acquire resistance genes can be passed on to other cells. Resistance to a particular antibiotic can therefore be spread horizontally as well as vertically in a bacterial population. Eventually, all or nearly all the bacterial cells become resistant, and the antibiotic is no longer useful in combating the organism.

This process has occurred in several species pathogenic to humans, including strains of *Staphlyococcus, Enterococcus, Neisseria, Shigella,* and *Salmonella.* Today many bacterial infections causing diseases such as dysentery, tuberculosis, and gonorrhea are difficult to treat. In some cases, the pathogen has acquired resistance to several different antibiotics, making it particularly deadly.

The spread of multiple drug resistance in bacterial populations has been accelerated by the evolution of **conjugative R plasmids** that carry the resistance genes (Chapter 17). These plasmids have two components. One, called the **resistance transfer factor,** or **RTF,** contains the genes needed for conjugative transfer between cells; the other, called the **R-determinant,** contains the genes for antibiotic resistance. Often these resistance genes are carried by a transposon, or a set of transposons, that has been inserted into the plasmid (Figure 18.7). Conjugative R plasmids can be transferred rapidly between cells in a bacterial population. Surprisingly, this transfer is not limited to cells of the same species. Conjugative plasmids can, in fact, be passed from one species to another, even between quite dissimilar cell types—for example, between a *Coccus* and a *Bacillus.* Thus, once multiple drug resistance has evolved in a part of the microbial kingdom, it can spread to other parts with relative ease.

Key Points: **Bacterial transposons—including the IS elements, composite transposons made from them, and Tn3—have short, inverted repeats at their termini. Upon insertion into a chromosome or plasmid, they create a short, direct duplication of DNA at the insertion site. Transposition is catalyzed by a protein, the transposase, produced by a gene within the transposon. Other transposon-encoded proteins may repress transposition. Composite transposons and Tn3 elements carry genes that confer resistance to antibiotics.**

TRANSPOSABLE ELEMENTS IN EUKARYOTES

Geneticists have found many different types of transposons in eukaryotes. These elements vary in size,

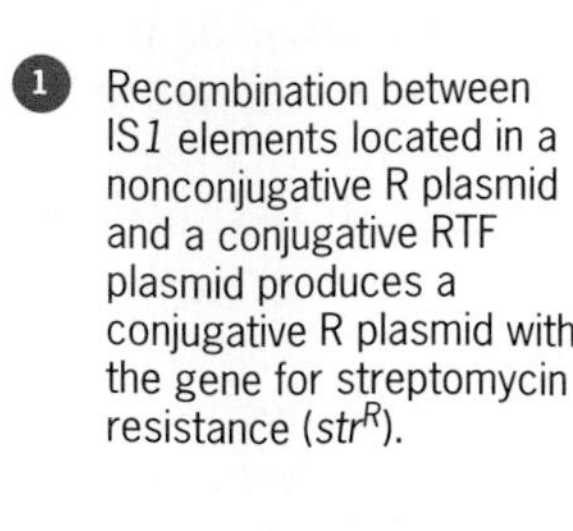

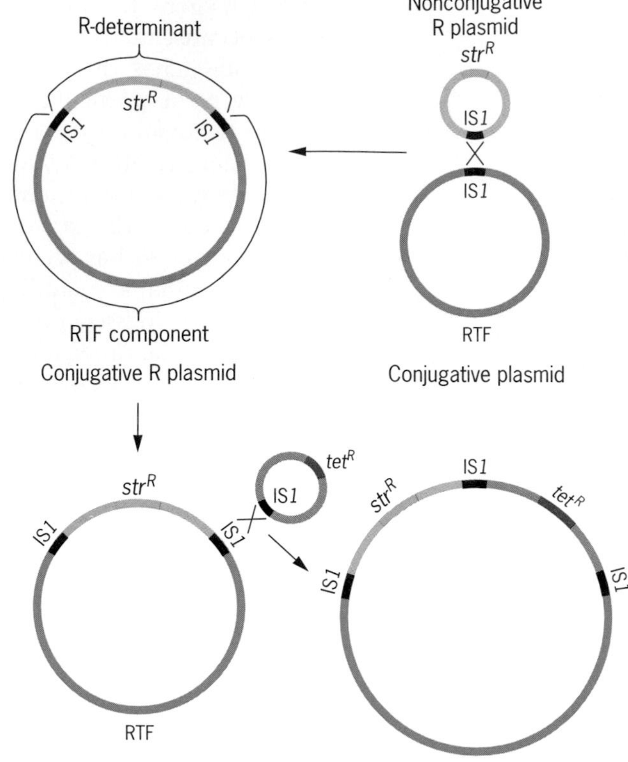

Figure 18.7 Evolution of conjugative plasmids carrying genes for antibiotic resistance.

structure, and behavior. Some are abundant in the genome, others rare. In the following sections, we discuss a few of the eukaryotic transposons that have been studied intensively. All these elements have inverted repeats at their termini and create target site duplications when they insert into DNA molecules. Some encode a transposase that catalyzes the movement of the element from one position to another.

Ac and *Ds* Elements in Maize

The *Ac* and *Ds* elements in maize were discovered through the pioneering work of Barbara McClintock. Through genetic analysis, McClintock showed that the activities of these elements are responsible for the striping and spotting of maize kernels. Many years later, Nina Federoff, Joachim Messing, Peter Starlinger, Heinz Saedler, Susan Wessler and their colleagues isolated the elements and determined their molecular structure.

McClintock discovered the *Ac* and *Ds* elements by studying chromosome breakage. She used genetic markers that controlled the color of maize kernels to detect the breakage events. When a particular marker was lost, McClintock inferred that the chromosome segment on which it was located had also been lost, an indication that a breakage event had occurred. The loss of a marker was detected by a change in the color of the aleurone, the outermost layer of the triploid endosperm of maize kernels.

In one set of experiments, the genetic marker that McClintock followed was an allele of the *C* locus on the short arm of chromosome 9. Because this allele, C^I, is a dominant inhibitor of aleurone coloration, any kernel possessing it is colorless. McClintock fertilized *CC* ears with pollen from C^IC^I tassels, producing kernels in which the endosperm was C^ICC. (Recall from Chapter 2 that the triploid endosperm receives two alleles from the female parent and one from the male parent.) Although McClintock found that most of these kernels were colorless, as expected, some showed patches of brownish-purple pigment (Figure 18.8). McClintock guessed that in such mosaics, the inhibitory C^I allele had been lost sometime during endosperm development, leading to a clone of tissue that was able to make pigment. The genotype in such a clone would be -*CC*, where the dash indicates the missing C^I allele.

Figure 18.8 Maize kernel showing loss of the C^I allele for the inhibition of pigmentation in the aleurone. The pigmented patches are -*CC*, whereas the colorless patches are C^ICC.

The mechanism that McClintock proposed to explain the loss of the C^I allele is diagrammed in Figure 18.9. A break at the site labeled by the arrow detaches a segment of the chromosome from its centromere, creating an acentric fragment. Such a fragment tends to be lost during cell division; thus, all the descendants of this cell will lack part of the paternally derived chromosome. Because the lost fragment carries the C^I allele, none of the cells in this clone is inhibited from forming pigment. If any of them produces a part of the aleurone, a patch of color will appear, creating a mosaic kernel similar to the one shown in Figure 18.8.

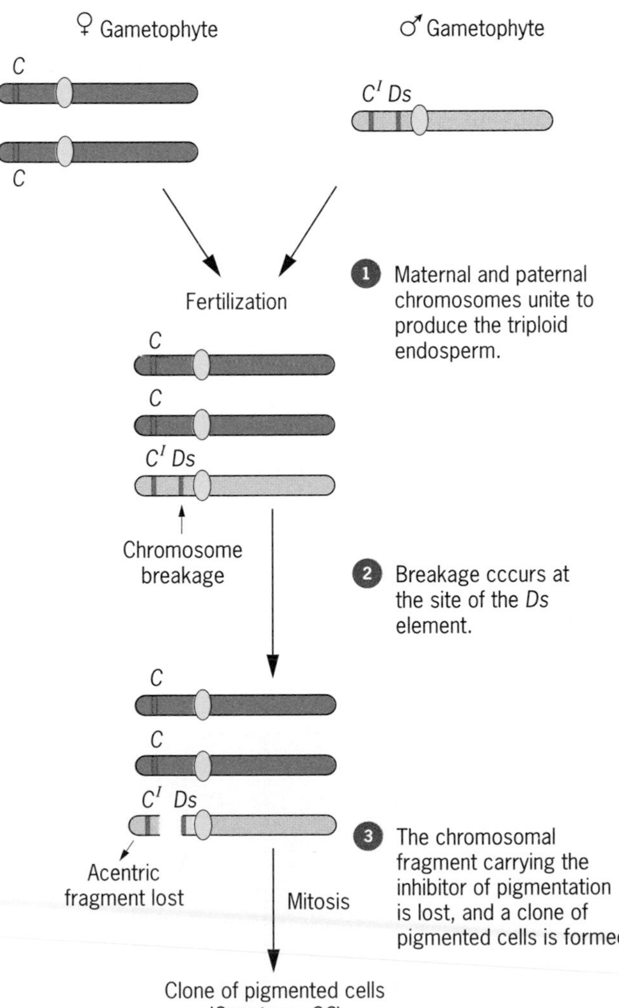

Figure 18.9 Chromosome breakage caused by the transposable element *Ds* in maize. The allele *C* on the short arm of chromosome 9 produces normal pigmentation in the aleurone; the allele C^I inhibits this pigmentation.

McClintock found that the breakage responsible for these mosaic kernels occurred at a particular site on chromosome 9. She named the factor that produced these breaks ***Ds***, for **Dissociation.** However, by itself, this factor was unable to induce chromosome breakage. In fact, McClintock found that *Ds* had to be stimulated by another factor, called ***Ac***, for **Activator.** The *Ac* factor was present in some maize stocks but absent in others. When different stocks were crossed, *Ac* could be combined with *Ds* to create the condition that led to chromosome breakage.

This two-factor *Ac/Ds* system provided an explanation for the genetic instability that McClintock had observed on chromosome 9. Additional experiments demonstrated that this was only one of many instabilities present in the maize genome. McClintock found other instances of breakage at different sites on chromosome 9 and also on other chromosomes. Because breakage at these sites depended on activation by *Ac*, she concluded that *Ds* factors were also involved. To explain all these observations, McClintock proposed that *Ds* could exist at many different sites in the genome and that it could move from one site to another.

This explanation has been borne out by subsequent analyses. The *Ac* and *Ds* elements belong to a family of transposons. These elements are structurally related to each other and can insert at many different sites on the chromosomes. Multiple copies of the *Ac* and *Ds* elements are often present in the maize genome. Through genetic analysis, McClintock demonstrated that both *Ac* and *Ds* can move. When one of these elements inserts in or near a gene, McClintock found that the gene's function is altered—sometimes completely abolished. Thus, *Ac* and *Ds* can induce mutations by inserting into genes. To emphasize this effect on gene expression, McClintock called the *Ac* and *Ds* transposons **controlling elements.**

DNA sequencing has shown that *Ac* elements consist of 4563 nucleotide pairs bounded by inverted repeats that are 11 nucleotide pairs long (Figure 18.10*a*); these inverted terminal repeats are essential for transposition. Each *Ac* element is also flanked by direct repeats 8 nucleotide pairs long. Because the direct repeats are created at the time the element is inserted into the chromosome, they are target site duplications, not integral parts of the element.

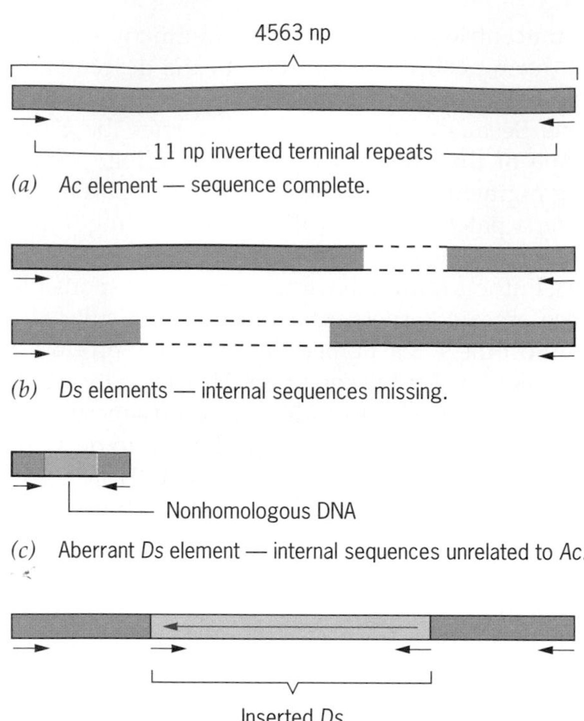

Figure 18.10 Structural organization of the members of the *Ac/Ds* family of transposable elements in maize. The terminal inverted repeats (short arrows underneath) and DNA sequence lengths (in nucleotide pairs, np) are indicated.

Unlike *Ac, Ds* elements are structurally heterogeneous. They all possess the same inverted terminal repeats as *Ac* elements, demonstrating that they belong to the same transposon family, but their internal sequences vary. Some *Ds* elements appear to have been derived from *Ac* elements by the loss of internal sequences (Figure 18.10*b*). The deletions in these elements may have been caused by incomplete DNA synthesis during replication or transposition. Other *Ds* elements contain non-*Ac* DNA between their inverted terminal repeats (Figure 18.10*c*). These unusual members of the *Ac/Ds* family are called *aberrant Ds* elements. A third class of *Ds* elements is characterized by a peculiar piggybacking arrangement (Figure 18.10*d*); one *Ds* element is inserted into another but in an inverted orientation. These so-called *double Ds* elements are apparently responsible for the chromosome breakage that McClintock observed in her experiments.

The activities of the *Ac/Ds* elements—excision and transposition, and all their genetic correlates, including mutation and chromosome breakage—are caused by a transposase encoded by the *Ac* elements. The *Ac* transposase apparently interacts with sequences at or near the ends of *Ac* and *Ds* elements, catalyzing their movement. Deletions or mutations in the gene that encodes the transposase abolish this catalytic function. Thus *Ds* elements, which have such lesions, cannot activate themselves. However, they can be activated if a transposase-producing *Ac* element is present somewhere in the genome. The transposase made by this element can diffuse through the nucleus, bind to a *Ds* element, and activate it. The *Ac* transposase is, therefore, a *trans*-acting protein.

P Elements and Hybrid Dysgenesis in *Drosophila*

Some of the most extensive research on transposable elements has focused on the *P* elements of *Drosophila melanogaster.* These transposons were identified through the cooperation of geneticists working in several different laboratories. In 1977 Margaret and James Kidwell, working in Rhode Island, and John Sved, working in Australia, discovered that crosses between certain strains of *Drosophila* produce hybrids with an assortment of aberrant traits, including frequent mutation, chromosome breakage, and sterility. The term **hybrid dysgenesis,** derived from Greek roots meaning "a deterioration in quality," was used to denote this syndrome of abnormalities.

Kidwell and her colleagues found that they could classify *Drosophila* strains into two main types based on whether or not they produce dysgenic hybrids in testcrosses. The two types of strains are denoted M and P. Only crosses between M and P strains produce dysgenic hybrids, and they do so only if the male in the cross is from the P strain. Crosses between two different P strains, or between two different M strains, produce hybrids that are normal. We can summarize the phenotypes of the hybrid offspring from these different crosses in a simple table:

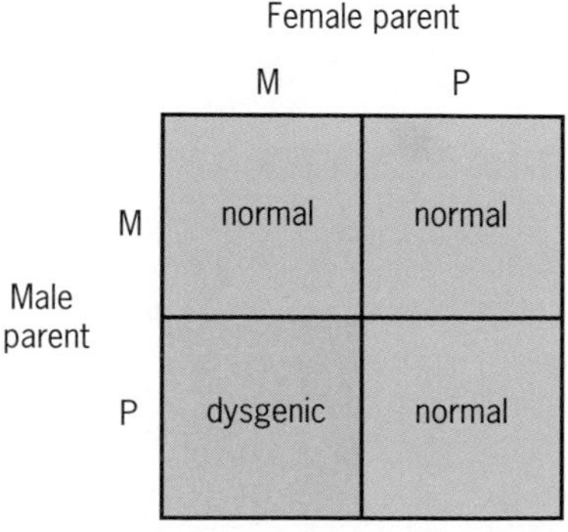

Male parent \ Female parent	M	P
M	normal	normal
P	dysgenic	normal

To Kidwell and her colleagues, these findings suggested that the chromosomes of P strains carry genetic factors that are activated when they enter eggs made by M females, and that once activated, these factors induce mutations and chromosome breakage. Inspired by this work, William Engels, a graduate student at the University of Wisconsin, began to study mutations induced in dysgenic hybrids. In 1979 Engels found a particular mutation that reverted to wild-type at a high rate. This instability, which is reminiscent of the behavior of IS-induced mutations in *E. coli,* strongly suggested that a transposable element was involved.

The discovery by Michael Simmons and Johng Lim of dysgenesis-induced mutations in the *white* gene allowed the transposon hypothesis to be tested. In 1980, Simmons and Lim, working in Minnesota and Wisconsin, respectively, sent the newly discovered *white* mutations to Paul Bingham, a geneticist in North Carolina. Bingham and his collaborator, Gerald Rubin, a geneticist in Maryland, had just finished isolating DNA from the *white* gene. Using this DNA as a probe, Bingham and Rubin were able to isolate DNA from the mutant *white* alleles and compare it to the wild-type *white* DNA. In each mutation, they found that a small element had been inserted into the coding region of the *white* gene. Additional experiments demonstrated that these elements are present in multiple copies and at different locations in the genomes of P strains; however, they are completely absent from the genomes of M strains. Geneticists therefore began calling these P-strain-specific transposons ***P* elements.**

DNA sequence analysis has shown that *P* elements vary in size. The largest elements are 2907 nucleotide pairs long, including terminal inverted repeats of 31 nucleotide pairs. These *complete P* elements carry a gene that encodes a transposase. When the P transposase binds near the ends of a complete *P* element, it can move that element to a new location in the genome. *Incomplete P* elements (Figure 18.11) lack the ability to produce the transposase because some of their internal sequences are deleted; however, they do possess the terminal and subterminal sequences that bind the transposase. Consequently, these elements can be mobilized if a transposase-producing complete element is present somewhere in the genome.

Surveys of natural populations of *Drosophila* conducted by Dominique Anxolabéhère and his colleagues in France have demonstrated that there is considerable variation in the number of *P* elements in the chromosomes. Some flies have as many as 50, whereas others have only a few. Perhaps the most surprising discovery is that flies derived from strains captured before 1950 have no *P* elements at all. Margaret Kidwell has suggested that these "empty" strains represent the primitive condition and that *P* elements have invaded natural populations of *Drosophila* during recent times. Curiously, the closest relatives of *D. melanogaster* have preserved the "empty" condition, but other, more distantly related species have acquired *P* elements. It is not possible to say how these species acquired their *P* elements, but one possibility is that the elements were carried into the genome by viruses that naturally infect *Drosophila.* Such a process would be analogous to the transduction of *E. coli* cells by a bacteriophage that carried an IS element.

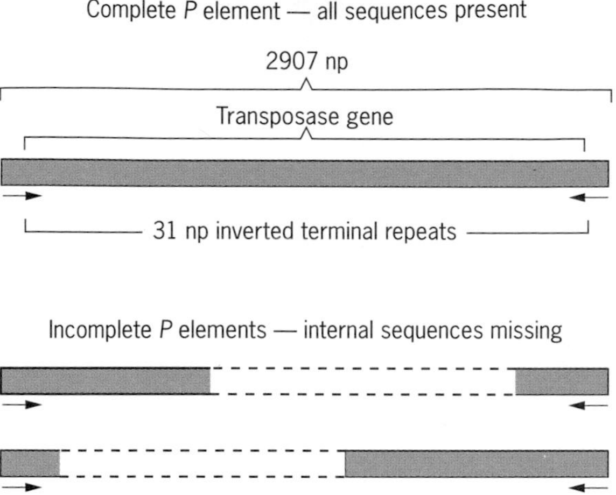

Figure 18.11 Structure of *P* elements in *Drosophila* showing orientations and lengths (in nucleotide pairs, np) of DNA sequences.

Populations of *Drosophila* that possess *P* elements have evolved mechanisms to regulate their movement. In some strains, this regulation depends on **cytotype,** a cellular condition that is transmitted maternally through the egg cytoplasm. The P cytotype represses *P* element movement, and the M cytotype permits it. P cytotype is characteristic of strains that have *P* elements on their chromosomes, whereas M cytotype is characteristic of strains that lack *P* elements. When *P* elements are combined with the M cytotype by making appropriate crosses, they are induced to transpose. Although the regulatory role of cytotype has been studied by numerous researchers, its cellular nature is still unknown. One possibility is that the P cytotype is produced by polypeptides encoded by the *P* elements themselves and that the M cytotype is simply the absence of these *P*-encoded polypeptides.

The maternal transmission of cytotype can be seen in the offspring of reciprocal crosses between P and M cytotype strains (Figure 18.12). There are two crosses: (1) P cytotype female × M cytotype male, and (2) P cytotype male × M cytotype female. The offspring from both crosses inherit *P* elements on their chromosomes and are, in fact, genotypically identical; however, only those from the second cross allow *P* movement. This difference between the reciprocal crosses indicates that the condition that permits or represses *P* movement is transmitted only by the female parent, presumably in

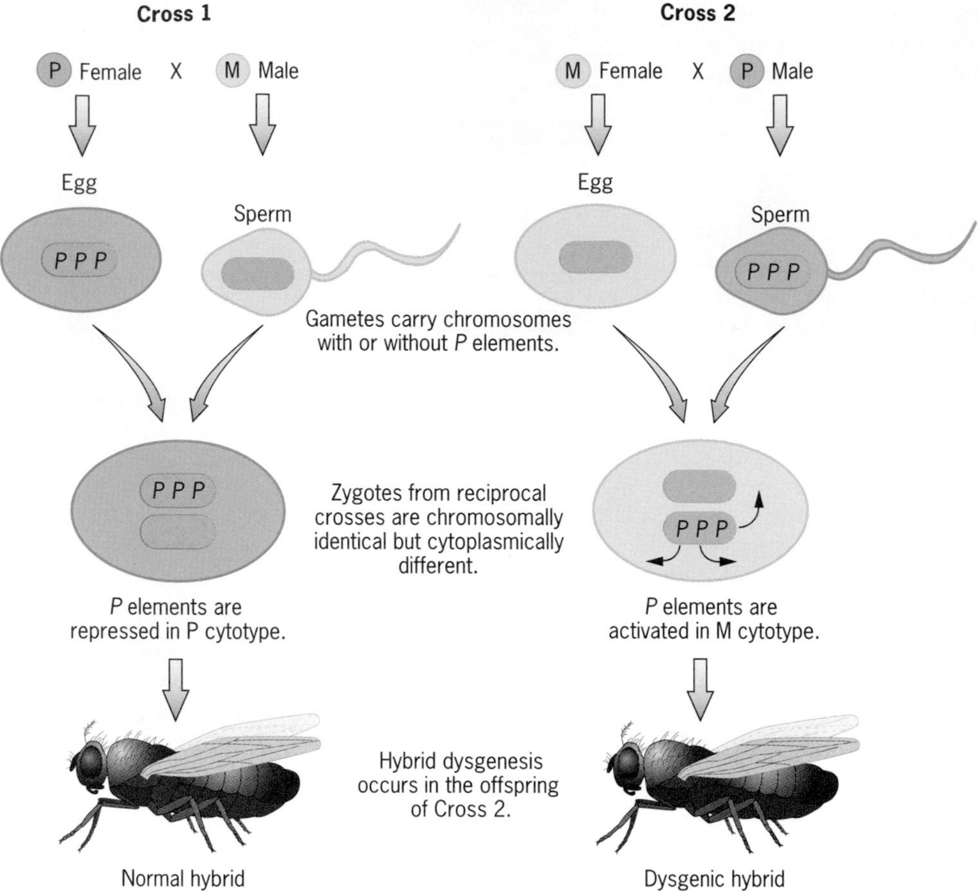

Figure 18.12 *P* element-mediated hybrid dysgenesis in *Drosophila.* Cytotype is inherited maternally. The P cytotype represses *P* element movement, whereas the M cytotype permits it. Dysgenesis occurs only in the M cytotype hybrids of Cross 2.

the egg cytoplasm. In Cross 1, the females transmit the P cytotype to their progeny, which then represses *P* movement, whereas in Cross 2, the females transmit the M cytotype, which allows *P* movement.

Given the damage that can be caused by extensive *P* element movement, it may seem surprising that P male × M female crosses produce any viable progeny at all. Donald Rio, Frank Laski, and Gerald Rubin have shown that these progeny are healthy because *P* elements move only in the germ line. In the somatic tissues, where *P* element movement would cause serious problems, there is little, if any, transposition because the P transposase is not synthesized there. The metabolic block occurs at the level of RNA splicing; one of the introns remains in the transcript of the transposase gene, creating a stop codon that prematurely terminates translation. Thus, instead of making the transposase, somatic cells synthesize a shorter protein that does not have the transposase's catalytic function. Without this function, *P* elements are unable to move, and the fly is protected from massive damage in its somatic tissues.

Geneticists routinely use hybrid dysgenesis to obtain *P*-insertion mutations in the laboratory. Dysgenic hybrids produced by crossing P males with M females are mated to recover mutations that have occurred in their germ lines. These mutations are detected by appropriate techniques, such as Muller's *ClB* crossing scheme to identify X-linked lethals (Chapter 14). Hybrid dysgenesis has an advantage over traditional methods of inducing mutations because a gene that has been mutated by the insertion of a transposable element is "tagged" with a known DNA sequence. The transposon tag may subsequently be used to isolate the gene from a large, heterogeneous mixture of DNA. Mutagenesis by **transposon tagging** is therefore a standard technique in molecular genetics. The Technical Sidelight: Genetic Transformation of *Drosophila* with *P* Elements discusses another important use of *P* elements in genetics research.

mariner, an Ancient and Widespread Transposon

Two of the closest relatives of *Drosophila melanogaster, D. simulans* and *D. mauritiana,* possess a small transposon called *mariner* (1286 nucleotide pairs long with 28-nucleotide-pair inverted terminal repeats). Although *mariner* is not present in *D. melanogaster,* similar transposons are present in many other insects, in nematodes, fungi, and even humans. This widespread distribution suggests that *mariner* elements are ancient, dating from the earliest evolutionary times.

Sequence analysis of *mariners* from many different organisms has also suggested that these elements are occasionally transferred horizontally between species. Hugh Robertson has shown that *mariner*-like elements are present in several different orders of insects. Curiously, the *mariners* in distantly related species are sometimes more similar to each other than the *mariners* in closely related species. For example, the *mariner* elements in the Mediterranean fruit fly are very similar to those in the honeybee; yet these two species are separated by hundreds of millions of years of evolutionary time. A plausible explanation is that the *mariners* in these two species are actually recent invaders from some outside source, perhaps another insect. It is not known how these elements could have been transferred between species, but one possibility is that they hitchhiked in the genome of a virus with a wide host range. During infection in one host species, the virus may have acquired the transposon; then during infection in another species, the transposon may have jumped from the virus into the host genome.

DNA sequencing has also identified another group of transposons related to the *mariner* elements. These are slightly larger, 1.6 to 1.7 kilobases long, with inverted repeats at their termini. Their repeat sequences typically range from 54 to 234 nucleotide pairs. The best studied transposon in this group, called *Tc1,* is found in the nematode *Caenorhabditis elegans.* Other *Tc1*-like elements have been found in vertebrates. The *Tc1* and *mariner* elements therefore belong to a transposon superfamily of ancient origin.

Key Points: **The maize transposable element *Ds,* discovered through its ability to break chromosomes, is activated by another transposable element, *Ac,* which encodes a transposase. In *Drosophila,* transposable *P* elements are responsible for hybrid dysgenesis, a syndrome of germ-line abnormalities that occurs in the offspring of crosses between P and M strains. The ancient *mariner* transposable elements are present in many distantly related organisms and probably have spread horizontally between species during the course of evolution.**

RETROTRANSPOSONS

In addition to transposons such as *Ac, P,* and *mariner,* eukaryotic genomes contain transposable elements whose movement depends on the reverse transcription of RNA into DNA. This reversal in the flow of genetic information has led geneticists to call these elements **retrotransposons,** from a Latin prefix meaning "backward." There are two main classes of retrotransposons: the retroviruslike elements and the retroposons. The members of the first class resemble the chromosomes of a group of viruses that depend on reverse transcription for their propagation, and the members of the second class have a structure reminiscent of polyadenylated RNA.

Retroviruslike Elements

Retroviruslike elements are found in many different organisms, including yeast, plants, and animals. Despite differences in size and nucleotide sequence, they all have the same basic structure: a central coding region flanked by **long terminal repeats,** or **LTRs,** which are oriented in the same direction. The repeated sequences are typically a few hundred nucleotide pairs long. Each LTR is, in turn, usually bounded by short, inverted repeats like those found in other types of transposons.

The coding region of a retroviruslike element contains a small number of genes, usually only two. These are homologous to the *gag* and *pol* genes found in retroviruses (see Chapter 16); *gag* encodes a structural protein of the virus capsule, and *pol* encodes a reverse transcriptase/integrase protein. The retroviruses have a third gene, *env,* which encodes a protein component of the virus envelope. In the retroviruslike elements, the gag and pol proteins play important roles in the transposition process.

One of the best-studied retroviruslike elements is the **Ty*1* transposon** from the yeast *Saccharomyces cerevisiae* (Figure 18.13*a*). This element is about 5.9 kilobase pairs long; its LTRs are about 340 base pairs long, and it creates a 5-bp target site duplication upon insertion into a chromosome. Most yeast strains have about 35 copies of the Ty*1* element; sometimes they also contain LTRs that have been detached from Ty*1* elements. These solo LTRs, or delta sequences as they are sometimes called, are apparently formed by recombination between the LTRs of complete Ty*1* elements (Figure 18.13*b*). The recombination event puts the central coding region and a portion of each LTR onto a circular molecule. When the circle leaves the chromosome, the remaining portions of the LTRs fuse, creating the solo delta sequence.

TECHNICAL SIDELIGHT

Genetic Transformation of *Drosophila* with *P* Elements

In the 1940s, Oswald Avery and colleagues discovered that bacteria could be genetically altered by treating them with isolated DNA. In this process, a DNA fragment enters a cell and is physically recombined into the chromosome. The recombinant cell can then be cultured to produce a strain of genetically transformed organisms. Avery's discovery raised the hope that someday it would be possible to alter the genomes

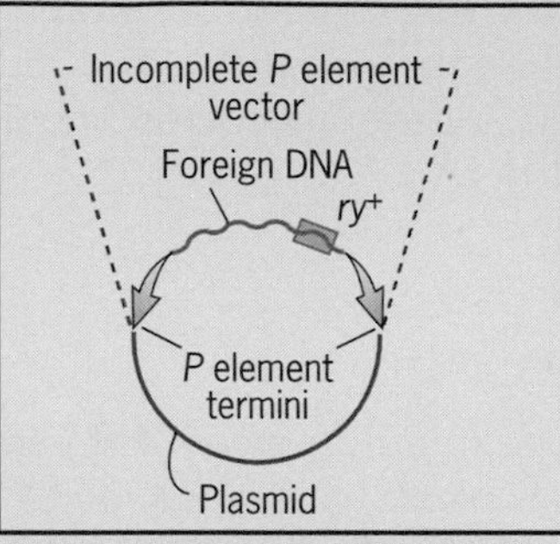

1 Foreign DNA is inserted into an incomplete *P* element in a plasmid. The insert also contains an eye color gene (ry^+) as a marker. In flies, this gene produces red eyes.

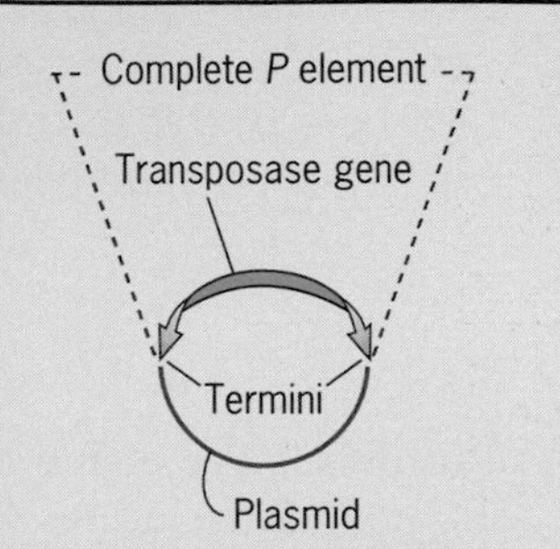

2 A complete *P* element is inserted into a different plasmid.

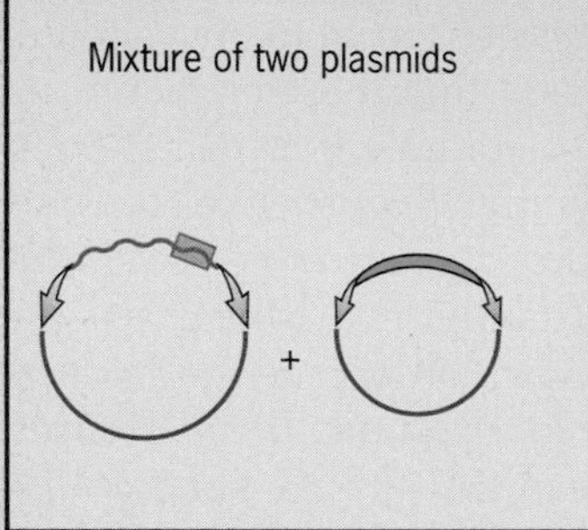

3 The two plasmids are mixed in solution.

Figure 1 Genetic transformation of *Drosophila* using *P* element vectors. Foreign DNA inserted between *P* element termini is integrated into the genome through the action of a transposase encoded by the complete *P* element. Flies with this DNA in their genomes can be propagated in laboratory cultures.

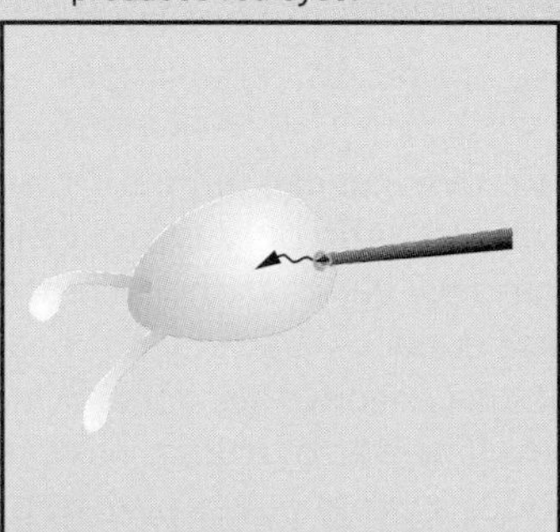

4 The plasmid mixture is microinjected into a ry^- mutant *Drosophilia* embryo. Flies with the ry^- mutation have brown (rosy) eyes.

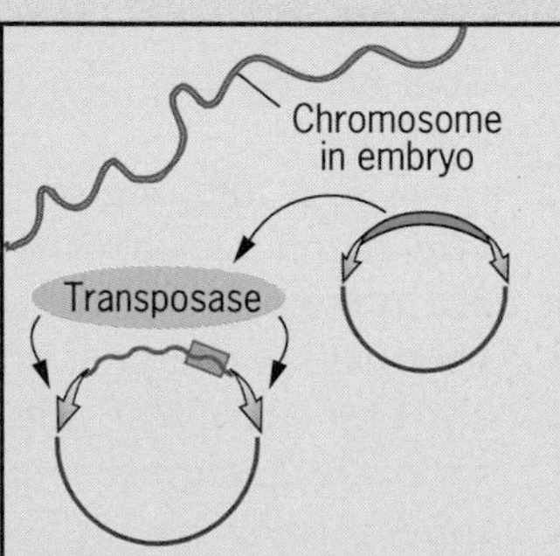

5 In the embryo's germ line, transposase from the complete *P* element catalyzes the incomplete *P* element to excise from its plasmid.

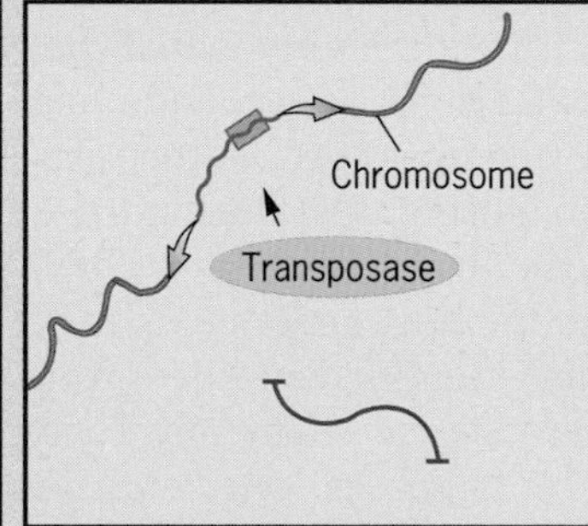

6 The excised *P* element is inserted into a chromosome in the embryo's germ line.

7 The embryo becomes an adult.

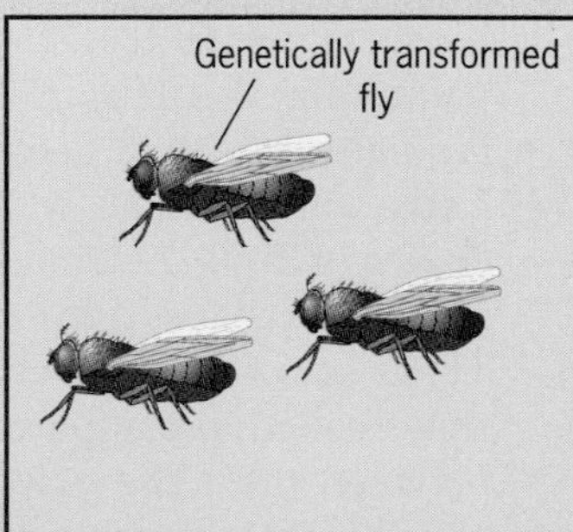

8 When the adult fly reproduces, the progeny are screened for incorporation of the incomplete *P* element with its ry^+ marker gene.

9 Genetically transformed progeny, recognized by their red eyes, are propagated.

of eukaryotic organisms by inserting specific DNA fragments into them. When applied to human beings, such a procedure might provide a way of correcting genetic diseases.

For nearly 40 years, the experimental production of genetic transformants was limited to microorganisms. Many researchers attempted to transform higher eukaryotes, but none succeeded. This string of failures was broken in 1982, when Allan Spradling and Gerald Rubin produced the first genetically transformed *Drosophila.*

Spradling and Rubin used transposable *P* elements to insert purified DNA into living *Drosophila* embryos. First, they constructed two bacterial plasmids that contained *P* elements. One plasmid contained a complete *P* element capable of producing the *P* transposase *in vivo.* The other contained an incomplete *P* element into which a gene for wild-type eye color had been inserted. Next, Spradling and Rubin injected a mixture of the two plasmids into *Drosophila* embryos that were homozygous for a recessive mutation of the eye color gene. They hoped that transposase produced by the complete *P* element would catalyze the incomplete *P* element to jump from its plasmid into the chromosomes of the injected animals, carrying with it the eye color gene. When these animals matured, Spradling and Rubin mated them to flies homozygous for the eye color mutation and looked for progeny that had wild-type eyes. They found many, indicating that the wild-type gene carried by the incomplete *P* element had been successfully incorporated into the genomes of some of the injected embryos. In effect, Spradling and Rubin had corrected the mutant eye color phenotype by inserting a copy of the wild-type gene into the fly genome.

The technique that Spradling and Rubin developed is now routinely used to transform *Drosophila* with isolated DNA (Figure 1). An incomplete *P* element serves as the **transformation vector,** and a complete *P* element serves as the source of the transposase that is needed to insert the vector into the chromosomes of an injected embryo. The term *vector* comes from the Latin word for "carrier"; it is used in this context because the incomplete *P* element *carries* a fragment of DNA into the genome. Practically any DNA sequence can be placed into the vector and ultimately inserted into the animal. Genes from organisms as diverse as bacteria and humans have successfully been incorporated into *Drosophila* chromosomes. It remains to be seen whether similar techniques will be developed for the genetic transformation of other organisms, including our own species.

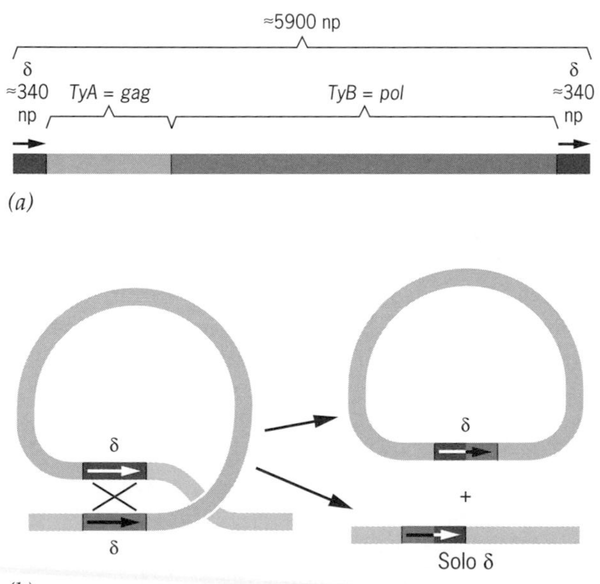

Figure 18.13 (*a*) Genetic organization of the yeast Ty*1* element, showing the long terminal repeat sequences (LTRs, denoted by the Greek letter delta) and the two genes (*TyA* and *TyB*). Lengths of sequences are in nucleotide pairs (np). (*b*) Formation of a solo delta sequence by homologous recombination between the delta sequences at the ends of the element.

Ty*1* elements have only two genes, *TyA* and *TyB,* which are homologous to the *gag* and *pol* genes of the retroviruses. Biochemical studies have shown that the products of these two genes can form viruslike particles inside yeast cells. However, it is not known whether these particles are genuinely infectious. The transposition of Ty*1* elements involves reverse transcription of RNA (Figure 18.14). After the RNA is synthesized from Ty*1* DNA, a reverse transcriptase encoded by the *TyB* gene uses it as a template to make double-stranded DNA. Then the newly synthesized DNA is inserted somewhere in the genome, creating a new Ty*1* element.

Retroviruslike elements have also been found in *Drosophila.* One of the first that was identified is called *copia,* so named because it produces copious amounts of RNA. The *copia* element is structurally similar to the Ty*1* element of yeast. The *gypsy* element, another *Drosophila* retrotransposon, is larger than the *copia* element because it contains a gene similar to the *env* gene of retroviruses. Both the *copia* and *gypsy* elements form viruslike particles inside *Drosophila* cells; however, only the particles that contain *gypsy* RNA can move across cell membranes, possibly because they also contain *gypsy*'s *env* gene product. The *gypsy* element therefore appears to be a genuine retrovirus. Many other fami-

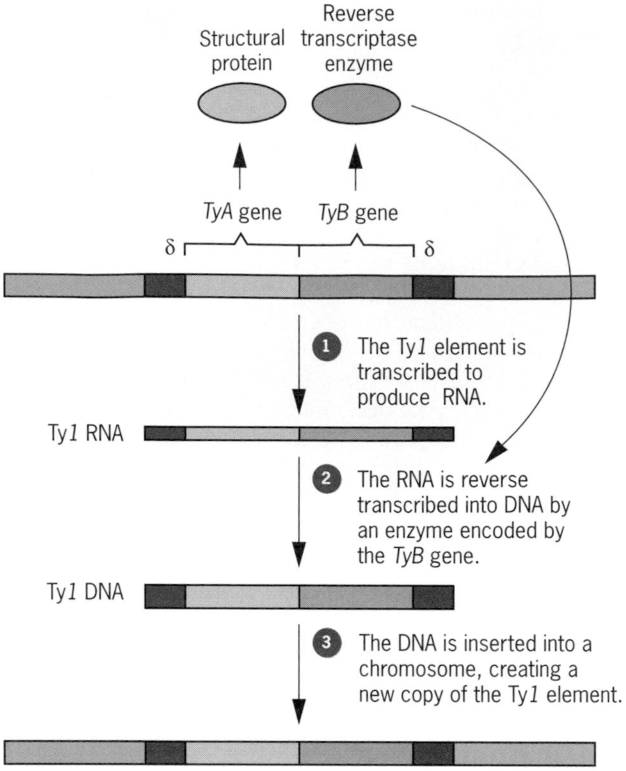

Figure 18.14 Transposition of the yeast Ty*1* element.

lies of retroviruslike transposons have been found in *Drosophila,* but their activities are poorly understood.

Retroposons

The **retroposons,** or non-LTR retrotransposons, are a large and widely distributed class of retrotransposons, including the *F, G,* and *I* elements of *Drosophila* and the **Long Interspersed Nuclear Elements (LINEs)** of mammals. These elements move through an RNA molecule that is reverse transcribed into DNA, probably by a protein encoded by the elements themselves. Although they create a target site duplication when they insert into a chromosome, they do not have inverted or direct repeats as integral parts of their termini. Instead, they are distinguished by a homogeneous sequence of A:T base pairs at one end. This sequence is derived from reverse transcription of the poly (A) tail that is added near the 3′ end of the retroposon RNA during its maturation. Integrated retroposons therefore exhibit a vestige of their origin as reverse transcripts of polyadenylated RNAs.

L1, a Human Retroposon The ***LINE-1* retroposon,** also known as ***L1,*** is the only transposable element known to be active in the human genome. (There may be others, but to date none has been found.) The first documented *L1* transposition event, reported by Haig Kazazian, Jr., and colleagues in 1988, involved an insertion into the X-linked gene for protein factor VIII, one of the clotting factors in the blood. The resulting mutation caused hemophilia, a failure of the blood to clot. The inserted *L1* element was apparently produced by reverse transcription of an RNA made by an *L1* element on chromosome 22. Curiously, the gorilla homolog of chromosome 22 also carries this *L1* element. Thus, it must have been present before the human and gorilla lineages diverged—about 6 million years ago.

L1 elements are heterogeneous in size, and most are truncated at the 5′ end—an indication of incomplete synthesis of the DNA by reverse transcription from an RNA template. These elements are present in 50,000 to 100,000 copies in the human genome and account for about 5 percent of the DNA. Similar elements are also found in the genomes of other mammals.

Telomere-associated Retroposons: HeT-A *and* TART In *Drosophila,* retroposons are found at the ends (telomeres) of chromosomes, where they perform the critical function of replenishing DNA that is lost by incomplete chromosome replication. With each round of DNA replication, a chromosome becomes shorter. Shortening takes place because the DNA polymerase can only move in one direction, adding nucleotides to the 3′ end of a primer (Chapter 11). Usually, the primer is RNA, and when it is removed, a single-stranded region is left at the end of the DNA duplex. In the next round of replication, the deficient strand produces a duplex that is shorter than the original. As this process continues, cycle after cycle, the chromosome loses material from its end.

To counterbalance this loss, *Drosophila* has evolved a curious mechanism involving at least two different retroposons, one called ***HeT-A*** and another called ***TART*** (for *T*elomere *A*ssociated *R*etro*t*ransposon). Mary Lou Pardue, Robert Levis, Harald Biessmann, James Mason, and their colleagues have shown that these two elements transpose preferentially to the ends of chromosomes, extending them by several kilobases. Eventually, the transposed sequences are lost by incomplete DNA replication, but then a new transposition occurs to restore them. The *HeT-A* and *TART* retroposons therefore perform the important function of regenerating lost chromosome ends.

Key Points: **The movement of retroviruslike elements and retroposons depends on reverse transcription of RNA into DNA. Retroviruslike elements have Long Terminal Repeat (LTR) sequences at both ends and resemble the integrated chromosomes of retroviruses. Retroposons have a sequence**

of A:T base pairs at one end. The retroposon *L1* is the only transposable element known to be active in the human genome. The *HeT-A* and *TART* retroposons are associated with the ends of *Drosophila* chromosomes.

THE GENETIC AND EVOLUTIONARY SIGNIFICANCE OF TRANSPOSABLE ELEMENTS

Transposable elements are widespread and probably ancient components of the genome. In some species, they constitute an appreciable fraction of the total DNA—in *Drosophila,* for example, 12 to 15 percent. Mobile DNA is therefore an important component of the genome. It also contributes significantly to the total mutation rate. In *Drosophila,* for example, perhaps half the mutations that occur spontaneously are caused by transposable element insertions. This ability to inflict mutational damage raises questions about the evolutionary status of transposable elements. Do they perform any useful function, or are they merely genetic parasites, causing mutations as they wander about the genome? Where did transposable elements come from? What mechanisms have evolved to control and limit their movement?

Transposons and Genome Organization

Certain chromosome regions are especially rich in transposon sequences. In maize, transposable elements are concentrated in the DNA between genes; collectively, these elements account for more than half the DNA in the maize genome. In *Drosophila,* transposons are concentrated in the centric heterochromatin and in the heterochromatin abutting the euchromatin of each chromosome arm. However, many of these transposons have mutated to the point where they cannot be mobilized; genetically, they are the equivalent of "dead." Heterochromatin therefore seems to be a kind of graveyard filled with degenerate transposable elements.

Transposable elements are also found in the euchromatin, where they are dispersed at many different sites. In *Drosophila,* approximately 40 distinct families of transposable elements have been identified. The number of members in each family ranges from a few to one or two hundred, but most families seem to have between 20 and 80 members.

There is some evidence, especially from cytological studies of *Drosophila* by Johng Lim, that transposable elements play a role in the evolution of chromosome structure. Several *Drosophila* transposons have been implicated in the formation of chromosome rearrangements, and a few seem to rearrange chromosomes at high frequencies. One possible mechanism is crossing over between homologous transposons located at different positions in a chromosome (Figure 18.15). If two transposons in opposite orientation pair

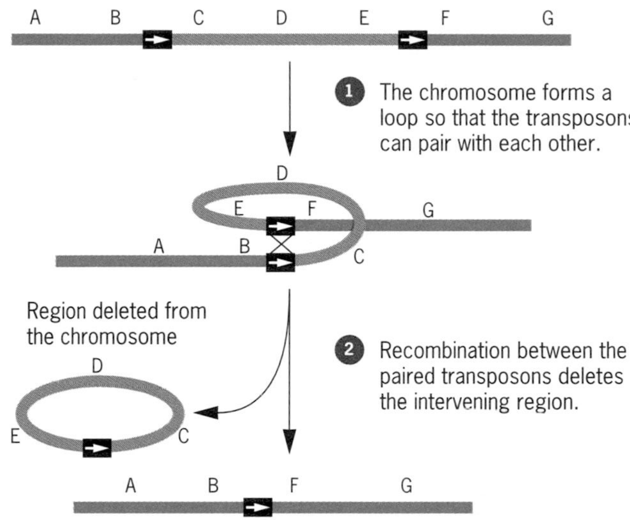

(*a*) Chromosome with two transposons (black rectangles) oriented in the same direction.

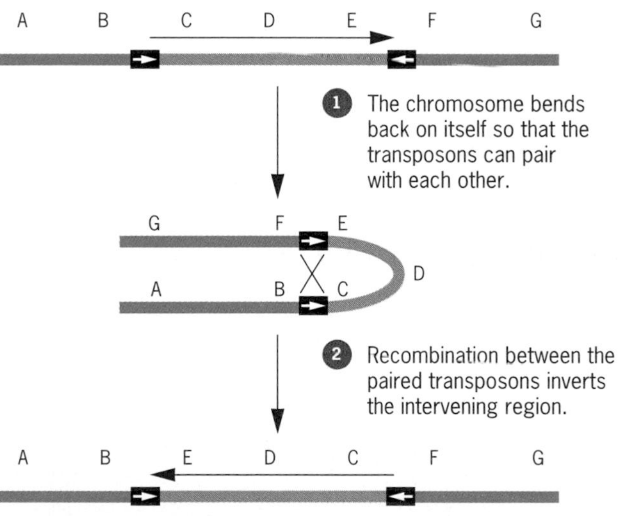

(*b*) Chromosome with two transposons (black rectangles) oriented in opposite directions.

Figure 18.15 Transposon-mediated chromosome rearrangements. (*a*) Formation of a deletion by intrachromosomal recombination between two transposons in the same orientation. (*b*) Formation of an inversion by intrachromosomal recombination between two transposons in opposite orientations.

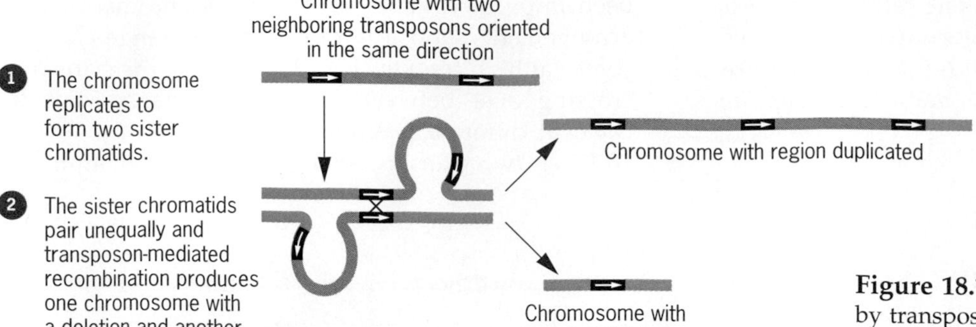

Figure 18.16 Origin of a gene duplication by transposon-mediated unequal crossing over between sister chromatids.

and cross over, the segment between them will be inverted. If two transposons in the same orientation pair and cross over, the segment between them will be deleted. These events are examples of **ectopic intrachromosomal exchanges**—that is, exchanges between sequences at different sites within a single chromosome. **Ectopic interchromosomal exchanges** are also possible. In these types of events, sequences in two different chromosomes (either homologous or nonhomologous) pair and cross over, generating novel products. Figure 18.16 gives an example of an ectopic exchange between transposable elements in two sister chromatids. Notice that one product from this exchange lacks the segment between the two transposons, whereas the other contains a duplication of this segment. Transposable elements can therefore change chromosome structure by mediating ectopic exchanges.

Another example of ectopic exchange is the crossover event that inserts the F plasmid into the *E. coli* chromosome. Such events are mediated by IS elements located in both of these circular molecules.

Transposons and Mutation

Transposable elements are responsible for mutations in a wide variety of organisms. Most transposons have been discovered as insertions in various mutant genes. Figure 18.17 shows some of the transposon insertions that have been found in different mutant alleles of the *Drosophila white* gene. These include several types of elements—*P*, retroviruslike elements, and retroposons. Some of these elements are inserted in exons, others in introns, and still others in regulatory DNA upstream of the actual gene. The very first mutant allele of *white*, w^1 discovered by T. H. Morgan, resulted from a transposon insertion.

Although transposon insertions are common in stocks of mutant organisms, the occurrence of new insertion mutations is a rare event. This suggests that the movement of many transposon families is regulated. When this regulation is upset, a burst of transposition may occur, causing many mutations simultaneously. This is apparently what happens when *P* elements are mobilized in dysgenic hybrids of *Drosophila*.

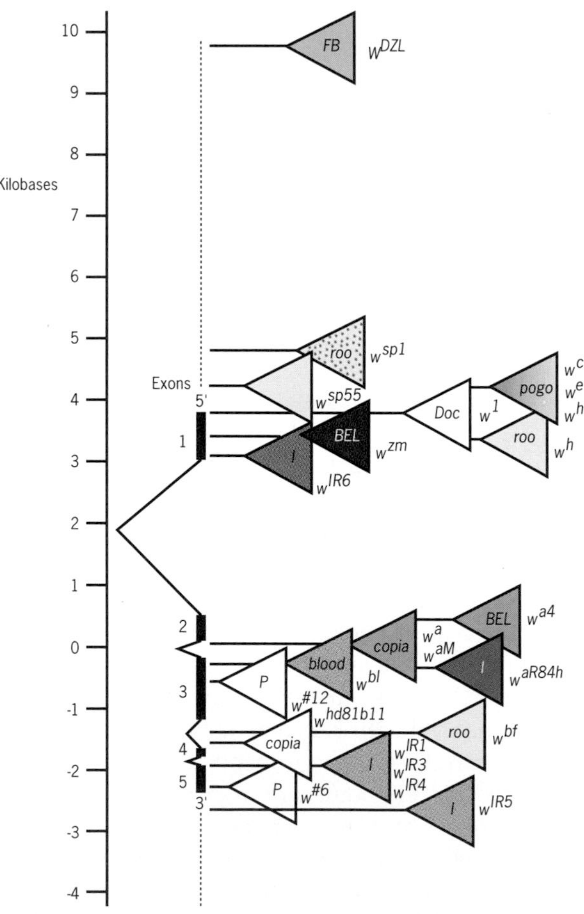

Figure 18.17 Transposon-insertion mutations in the *white* gene of *Drosophila*. Distance along the molecular map is given in kilobase pairs, with the zero coordinate arbitrarily positioned in the second intron of the gene at the site of the *copia* insertion in the w^a mutation. Each triangle represents a different transposon insertion in the *white* gene. Colors indicate the mutant eye colors.

Evolutionary Issues Concerning Transposable Elements

The widespread distribution of transposable elements suggests that they have played a role in evolution. One hypothesis is that these elements are nature's tools for genetic engineering. Their ability to copy, transpose, and rearrange other DNA sequences, such as genes for antibiotic resistance, can be construed as a benefit for the organisms that carry them. Thus, transposable elements may have spread because they confer a selective advantage. Another hypothesis is that transposable elements have spread simply because they have an ability to multiply independently of the normal replication machinery. According to this view, transposable elements are little more than genomic parasites—segments of DNA that replicate selfishly, possibly even to the detriment of their hosts.

How might the first transposons have evolved? Nancy Kleckner has suggested that a primordial transposon might arise by the modification of a gene that encodes an enzyme for the creation and repair of DNA breaks. All that would be needed is that the enzyme develop a modest degree of specificity, perhaps by recognizing a particular DNA sequence of six or eight nucleotide pairs. Such a sequence might occur by chance in inverted orientation on either side of the gene, creating a situation in which the gene's product could interact with each of these flanking sequences. By "cutting and pasting" the DNA, this modified enzyme could then transpose the entire unit to a new position in the genome. Such a unit would behave as a primitive transposon.

Other questions concern the evolutionary relationship between retroviruslike elements, such as Ty*1* in yeast, and full-fledged retroviruses. Collectively, these entities have been referred to as **retroelements.** Alan Kingsman and Susan Kingsman have proposed that retroviruses have developed from the simpler retrotransposons by the addition of a gene (called *env*) that synthesizes a membrane protein. With this addition, the retroelement could produce a particle capable of escaping from one cell and entering another one. Such a particle would be infectious and would therefore provide the retroelement with the opportunity to transpose between genomes as well as within them. Of course, the situation could be reversed—a retrovirus could lose its ability to escape from a cell and become trapped inside. Such a mutant virus would be reduced to the status of a retrotransposon, capable of moving within cells but not between them. Using a musical metaphor, the Nobel Prize-winning virologist Howard Temin once described these contrasting scenarios as ascending and descending evolutionary scales. Retrotransposons can rise to the level of retroviruses, and retroviruses can fall to the level of retrotransposons.

Key Points: **Transposons constitute a significant fraction of the DNA in the genomes of some organisms. In nature, they may be a major cause of mutations and chromosome rearrangements. Some transposons may confer a selective advantage on their carriers; others may simply be genetic parasites.**

TESTING YOUR KNOWLEDGE

1. A copy of the wild-type *white* gene (w^+) from *Drosophila* was inserted in the middle of an incomplete *P* element contained within a plasmid. The plasmid was mixed with another plasmid that contained a complete *P* element, and the mixture was carefully injected into *Drosophila* embryos homozygous for a null mutation (w^-) of the *white* gene. The adults that developed from these injected embryos all had white eyes, but when they were mated to uninjected white flies, some of their progeny had red eyes. Explain the origin of these red-eyed progeny.

ANSWER

The complete *P* element in one of the plasmids would produce the *P* transposase, the enzyme that catalyzes *P* element transposition, in the germ lines of the injected embryos. The incomplete *P* element in the other plasmid would be a target for this transposase. If this incomplete *P* element were mobilized by the transposase to jump from its plasmid into the chromosomes of the injected embryo, the fly that developed from this embryo would carry a copy of the wild-type *white* gene in its germ line. (*P* element movement is limited to the germ line; therefore, the incomplete *P* element would not jump into the chromosomes of the somatic cells, such as those that eventually form the eye.) Such a genetically transformed fly would, in effect, have the germ-line genotype w^-/w^-; $P(w^+)$ or w^-/Y; $P(w^+)$, where $P(w^+)$ denotes the incomplete *P* element that contains the w^+ gene. This element could be inserted on any of the chromosomes. If the transformed fly were mated to an uninjected white fly, some of its offspring would inherit the $P(w^+)$ insertion, which, because it carries a wild-type white gene, would cause red eyes to develop. The red-eyed progeny are therefore the result of genetic transformation of a mutant white fly by the w^+ gene within the incomplete *P* element.

QUESTIONS AND PROBLEMS

18.1 Which of the following pairs of DNA sequences could qualify as the terminal repeats of a bacterial IS element: (a) 5′-GAATCCGCA-3′ and 5′-ACGCCTAAG-3′, (b) 5′-GAATCCGCA-3′ and 5′-CTTAGGCGT-3′, (c) 5′-GAATCCGCA-3′ and 5′-GAATCCGCA-3′, (d) 5′-GAATCCGCA-3′ and 5′-TGCGGATTC-3′. Explain.

18.2 Which of the following pairs of DNA sequences could qualify as target site duplications at the point of an IS*50* insertion? (a) 5′-AATTCGCGT-3′ and 5′-AATTCGCGT-3′, (b) 5′-AATTCGCGT-3′ and 5′-TGCGCTTAA-3′, (c) 5′-AATT-CGCGT-3′ and 5′-TTAAGCGCA-3′, (d) 5′-AATTCGCGT-3′ and 5′-ACGCGAATT-3′. Explain.

18.3 One strain of *E. coli* is resistant to the antibiotic streptomycin, and another strain is resistant to the antibiotic ampicillin. The two strains were cultured together and then plated on selective medium containing streptomycin and ampicillin. Several colonies appeared, indicating that cells had acquired resistance to both antibiotics. Suggest a mechanism to explain the acquisition of double resistance.

18.4 What distinguishes IS and Tn*3* elements in bacteria?

18.5 The circular order of genes on the *E. coli* chromosome is **A B C D E F G H* *, with the * indicating that the ends of the chromosome are attached to each other. Two copies of an IS element are located in this chromosome, one between genes *C* and *D*, and the other between genes *D* and *E*. A single copy of this element is also present in the F plasmid. Two Hfr strains were obtained by selecting for integration of the F plasmid into the chromosome. During conjugation, one strain transfers the chromosomal genes in the order *D E F G H A B C*, whereas the other transfers them in the order *D C B A H G F E*. Explain the origin of these two Hfr strains. Why do they transfer genes in different orders? Does the order of transfer reveal anything about the orientation of the IS elements in the *E. coli* chromosome?

18.6 By chance, an IS*1* element has inserted near an IS*2* element in the *E. coli* chromosome. The gene between them, *sug*, confers the ability to metabolize certain sugars. Will the unit IS*1 sug* IS*2* behave as a composite transposon? Explain.

18.7 The composite transposon Tn*5* consists of two IS*50* elements, one on either side of a group of three genes for antibotic resistance. The entire unit IS*50*L *kan ble str* IS*50*R can transpose to a new location in the *E. coli* chromosome. However, of the two IS*50* elements in this transposon, only IS*50*R produces the catalytically active transposase. Would you expect IS*50*R to be able to excise from the Tn*5* composite transposon and insert elsewhere in the chromosome? Would you expect IS*50*L to be able to do this?

18.8 A researcher has found a new Tn*5* element with the structure IS*50*L *str ble kan* IS*50*L. What is the most likely origin of this element?

18.9 Would a Tn*3* element with a frameshift mutation early in the *tnpA* gene be able to form a cointegrate? Would a Tn*3* element with a frameshift mutation early in the *tnpR* gene be able to form a cointegrate?

18.10 What enzymes are necessary for replicative transposition of Tn*3*? What are their respective functions?

18.11 What is the medical significance of bacterial transposons?

18.12 Describe the structure of the *Ac* transposon in maize. In what ways do the *Ds* transposons differ structurally and functionally from the *Ac* transposon?

18.13 In homozygous condition, a deletion mutation of the *c* locus, c^n, produces colorless (white) kernels in maize; the dominant wild-type allele, *C*, causes the kernels to be purple. A newly identified recessive mutation of the *c* locus, c^m, has the same phenotype as the deletion mutation (white kernels), but when c^mc^m and c^nc^n plants are crossed, they produce white kernels with purple stripes. If it is known that the c^nc^n plants harbor *Ac* elements, what is the most likely explanation for the c^m mutation?

18.14 In maize, the *O2* gene, located on chromosome 7, controls the texture of the endosperm, and the *C* gene, located on chromosome 9, controls its color. The gene on chromosome 7 has two alleles, a recessive, *o2*, which causes the endosperm to be soft, and a dominant, *O2*, which causes it to be hard. The gene on chromosome 9 also has two alleles, a recessive, *c*, which allows the endosperm to be colored, and a dominant, C^I, which inhibits coloration. In one homozygous C^I strain, a *Ds* element is inserted on chromosome 9 between the *C* gene and the centromere. This element can be activated by introducing an *Ac* element by appropriate crosses. Activation of *Ds* causes the C^I allele to be lost by chromosome breakage. In $C^I/c/c$ kernels, such loss produces patches of colored tissue in an otherwise colorless background (see Figures 18.8 and 18.9). A geneticist crosses a strain with the genotype *o2/o2*; C^I *Ds*/C^I *Ds* to a strain with the genotype *O2/o2*; *c/c*. The latter strain also carries an *Ac* element somewhere in the genome. Among the offspring, only those with hard endosperm show patches of colored tissue. What does this tell you about the location of the *Ac* element in the *O2/o2*; *c/c* strain?

18.15 In maize, the recessive allele *bz* produces a lighter color in the aleurone than does the dominant allele, *Bz*. Ears on a homozygous *bz/bz* plant were fertilized by pollen from a homozygous *Bz/Bz* plant. The resulting cobs contained kernels that were uniformly dark except for a few on which light spots occurred. Suggest an explanation.

18.16 The X-linked *singed* locus is one of several in *Drosophila* that controls the formation of bristles on the adult cuticle. Males that are hemizygous for a mutant *singed* allele have bent, twisted bristles that are often much reduced in size. Several *P* element insertion mutations of the *singed* locus have been characterized, and some have been shown to revert to the wild-type allele by excision of the inserted element. What conditions must be present to allow such reversions to occur?

18.17 Dysgenic hybrids in *Drosophila* have elevated mutation rates as a result of *P* element transposition. How could you take advantage of this situation to obtain *P* element-insertion mutations on the X chromosome?

18.18 If DNA from a *P* element-insertion mutation of the *Drosophila white* gene and DNA from a wild-type *white* gene were purified, denatured, mixed with each other, renatured, and then viewed with an electron microscope, what would the hybrid DNA molecules look like?

18.19 When complete *P* elements are injected into embryos from an M strain, they transpose into the chromosomes of the germ line, and progeny reared from these embryos can be used to establish new P strains. However, when complete *P* elements are injected into embryos from insects that lack these elements, such as mosquitoes, they do not transpose into the chromosomes of the germ line. What does this failure to insert in the chromosomes of other insects indicate about the nature of *P* element transposition?

18.20 (a) What are retrotransposons? (b) Give examples of retrotransposons in yeast and *Drosophila.* (c) Describe how retrotransposons transpose. (d) After a retrotransposon has been inserted into a chromosome, is it ever expected to be excised?

18.21 Sometimes solo copies of the LTR of a retrotransposon called *gypsy* are found in *Drosophila* chromosomes. How might these solo LTRs originate?

18.22 Would you ever expect the genes in a retrotransposon to possess introns? Explain.

18.23 Suggest a method to determine whether the *TART* retroposon is situated at the telomeres of each of the chromosomes in the *Drosophila* genome.

18.24 Describe how two copies of a transposon can promote the formation of chromosomal rearrangements such as deletions and inversions.

18.25 It has been proposed that the *hobo* transposable elements in *Drosophila* mediate intrachromosomal recombination—that is, two *hobo* elements on the same chromosome pair and recombine with each other. What would such a recombination event produce if the *hobo* elements were oriented in the same direction on the chromosome? What if they were oriented in opposite directions?

18.26 What evidence suggests that some transposable elements are not simply genetic parasites?

18.27 Approximately half of all spontaneous mutations in *Drosophila* are caused by transposable element insertions. In human beings, however, the accumulated evidence suggests that the vast majority of spontaneous mutations are *not* caused by transposon insertions. Propose a hypothesis to explain this difference.

BIBLIOGRAPHY

DOMBROSKI, B. A., S. L. MATHIAS, E. NANTHAKUMAR, A. F. SCOTT, AND H. H. KAZAZIAN, JR. 1991. Isolation of an active human transposable element. *Science* 254:1805–1810.

ENGELS, W. R. 1989. *P* elements in *Drosophila.* In D. E. Berg and M. M. Howe (eds.), *Mobile DNA.* American Society for Microbiology.

FEDEROFF, N. V. 1989. About maize transposable elements and development. *Cell* 56:181–191.

GRINDLEY, N.D.F. 1983. Transposition of Tn3 and related transposons. *Cell* 32:3–5.

KIDWELL, M. G. 1994. The evolutionary history of the P family of transposable elements. *J. Heredity* 85:339–346.

KINGSMAN, A. J., AND S. M. KINGSMAN. 1988. Ty: A retroelement moving forward. *Cell* 53:333–335.

LIM, J. K., AND M. J. SIMMONS. 1994. Gross chromosome rearrangements mediated by transposable elements in *Drosophila melanogaster. BioEssays* 16:269–275.

MCCLINTOCK, B. 1956. Controlling elements and the gene. *Cold Spring Harbor Symp. Quant. Biol.* 21:197–216.

REZNIKOFF, W. S. 1982. Tn5 transposition and its regulation. *Cell* 31:307–308.

SHAPIRO, J. A., ED. 1983. *Mobile genetic elements.* Academic Press, New York.

19

A rain forest is a dramatic manifestation of the energy conversions that occur in chloroplasts and mitochondria.

The Genetics of Mitochondria and Chloroplasts

CHAPTER OUTLINE

The Classical Genetics of Organelles
The Molecular Genetics of Mitochondria
Mitochondrial DNA and Human Disease
The Molecular Genetics of Chloroplasts
The Origin and Evolution of Mitochondria and Chloroplasts

Mitochondria, Chloroplasts, and the Biological Energy Wheel

The primitive earth was a very different world from the one we know today. The land was barren, devoid of any living organisms, and the air consisted primarily of methane, ammonia, and water vapor. The earth was probably much hotter than it is today, radiating great quantities of heat from its molten core, and there were violent volcanic eruptions and intense electrical storms. Lethal ultraviolet radiation drenched the surface of the planet, energizing chemical reactions in the air and in shallow pools of water. In this hostile environment, life began, probably as a concretion of organic molecules that arose from the fusion of simpler compounds. Later, cellular organization evolved and diversified, leading eventually to a group of organisms that could think, speak, and conduct scientific research. Motivated by an intense curiosity, these organisms have attempted to look back over the long course of history and reconstruct the shadowy events of their own origins.

This research has revealed that the first true organisms were single cells that lived in water and fed on organic materials in the environment. These organisms sustained themselves by fermenting simple sugars, such as glucose, and by generating carbon dioxide (CO_2) gas as a waste product. The buildup of carbon dioxide provided a new resource on the primitive earth, and eventually organisms arose that could exploit it. These organisms evolved a mechanism to synthesize simple sugars from carbon dioxide and water by capturing energy from sunlight—

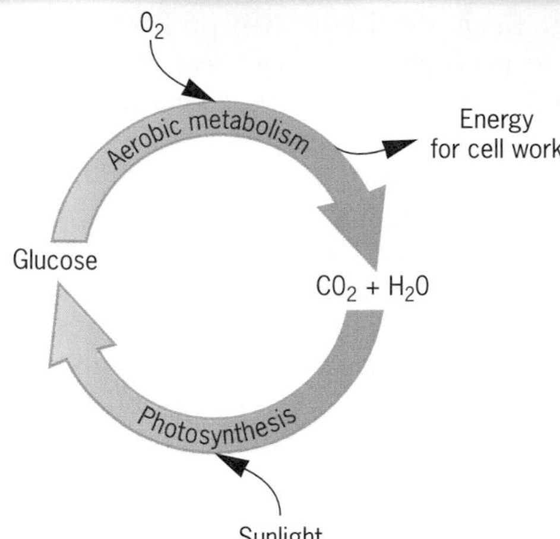

Figure 19.1 The biological energy wheel.

the process called photosynthesis; the sugars created by photosynthesis were then fermented to provide energy for life. As photosynthesis evolved and spread, it produced a new waste gas, oxygen (O_2), which then permitted a new group of organisms to evolve. These organisms were able to use the oxygen produced by photosynthesis to extract additional energy from sugars through a chemical process called oxidative, or aerobic, metabolism. Aerobic metabolism superseded the simpler anaerobic process of fermentation and led to the evolution of more complex, more active, and more diverse organisms.

Today, photosynthesis and aerobic metabolism are the complementary parts of a great energy cycle that encompasses the entire earth (Figure 19.1). Through photosynthesis, carbon dioxide and water are used to synthesize organic materials, and through oxidative metabolism, organic materials are decomposed into carbon dioxide and water. In eukaryotes, these processes are carried out in specialized subcellular organelles—photosynthesis in the **chloroplasts** and aerobic metabolism in the **mitochondria.** Both of these organelles appear to have arisen from simple, prokaryotic organisms that became established inside eukaryotic cells—a phenomenon called **endosymbiosis**—more than a billion years ago. Because these organisms brought along their genomes as well as their photosynthetic and oxidative powers, today all eukaryotic chloroplasts and mitochondria contain DNA. In this chapter, we explore the genetics and molecular biology of these important organelles.

THE CLASSICAL GENETICS OF ORGANELLES

The discovery of chloroplast and mitochondrial DNA was anticipated by studies that pointed to the existence of hereditary factors outside the nucleus. These studies date from the beginning of the twentieth century. In fact, the first of them was carried out by Carl Correns, one of the three botanists who rediscovered Mendel's principles.

Most of the early studies were performed with plants, which possess both mitochondria and chloroplasts. Consequently, it was not always possible to determine which of the two types of organelles was responsible for nonnuclear heredity. Later studies used yeast, for which the involvement of chloroplasts could be excluded. In the following sections, we review some of these classic studies. The common precept shared by all of them is that organelle heredity does not follow simple Mendelian rules. Instead, it is characterized by unequal contributions of the two parents and by an irregular segregation of alleles. Organelle heredity is therefore **non-Mendelian heredity.**

Leaf Variegation in Plants

The leaves in plants sometimes exhibit a striking pattern of **color variegation** (Figure 19.2). Some sectors are green, others pure white. Intermediate shades of pale or yellow green may also exist. These mosaic patterns are highly prized for their ornamental qualities, which partly explains why so many researchers have been interested in studying them.

Figure 19.2 Leaf variegation caused by the segregation of different types of chloroplasts.

Many cases of leaf variegation can be explained by the sorting of different types of chloroplasts during cell division (Figure 19.3). Cells may contain some chloroplasts that are capable of making the green pigment chlorophyll and others that are not. This mixture of

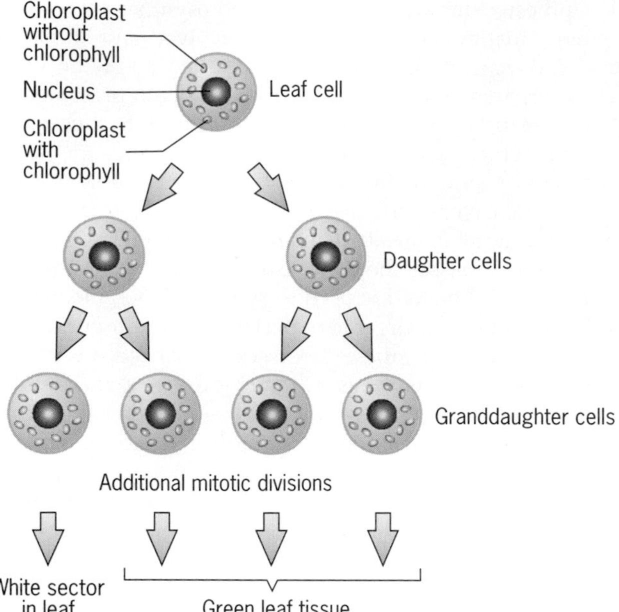

Figure 19.3 Chloroplast sorting during mitosis. The irregular distribution of chloroplasts during cell division may produce a cell that lacks chloroplasts capable of making chlorophyll. Through further divisions, such a cell will produce a white sector of tissue in an otherwise green leaf.

two types of organelles within a cell is called **heteroplasmy,** from Greek words meaning "two different bodies." The presence of a single type of organelle within a cell is called **homoplasmy.** When a heteroplasmic cell divides, the two types of chloroplasts are distributed to the daughter cells in an irregular fashion. Over several divisions, this irregular distribution may produce a cell that totally lacks pigment-producing chloroplasts. Such a cell will then proliferate into a white sector in an otherwise green leaf. The white sectors produced by this process are irregular in size, shape, and position. However, in some plants, white sectors are formed in definite, predictable patterns in the leaves. These sectors are not due to chloroplast sorting. Rather, they are created by physiological systems that regulate the synthesis of pigment in a spatial and temporal framework within the leaves.

The inheritance of leaf variegation was first studied by the German botanists Carl Correns and Erwin Baur. Correns worked with variegating strains of the four-o'clock, *Mirabilis jalapa,* a popular garden plant. He systematically made crosses between green and variegating strains. In these crosses it was possible to use flowers from a white sector that had developed on a variegating plant. Correns observed that the offspring of such crosses were always phenotypically identical to the tissue that produced the female gametes (Figure 19.4). Thus, a cross made by fertilizing ovules from a green plant with pollen from a white sector on a variegating plant produced only green offspring. However, a cross made with ovules from a white sector on a variegating plant and pollen from a green plant produced only pure white offspring. This strict **maternal inheritance** could be explained if plant color was controlled by factors that were transmitted through the ovules but not through the pollen. Chloroplasts were the obvious candidates to contain these factors.

In *Mirabilis,* the chloroplasts are transmitted to the offspring through the female reproductive cells but are largely, if not entirely, excluded from the male reproductive cells. When a plant inherits a mixture of pigmented and unpigmented chloroplasts from the fe-

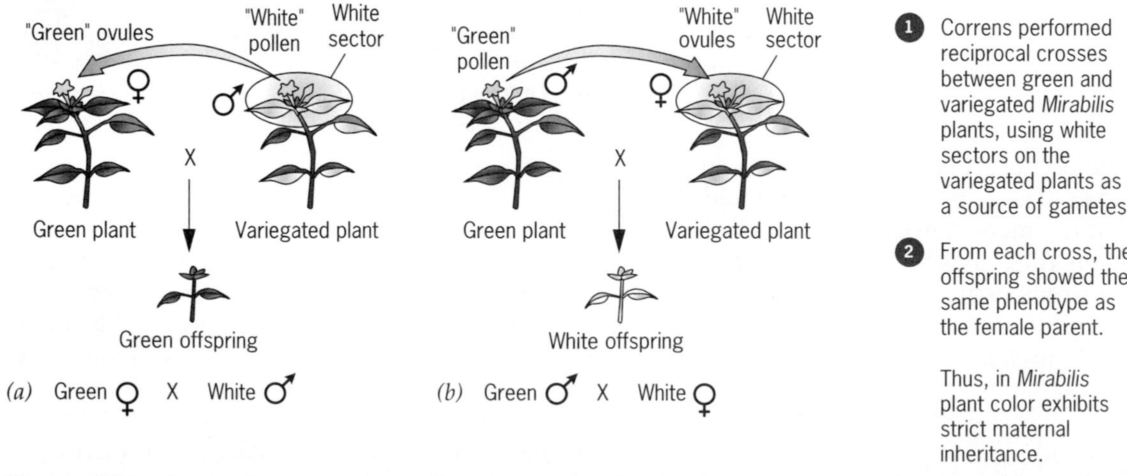

Figure 19.4 Correns' experiments on the inheritance of leaf variegation in *Mirabilis.* (*a*) Green female × white male (on a variegating plant). (*b*) Green male × white female (on a variegating plant).

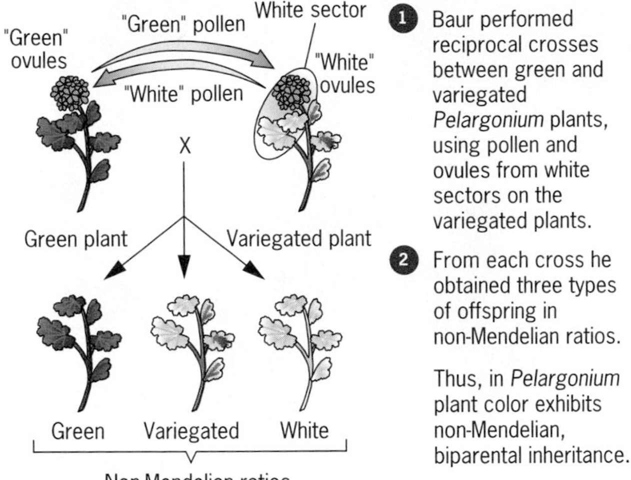

Figure 19.5 Baur's experiments on the inheritance of leaf variegation in *Pelargonium.* Both reciprocal crosses produced three types of progeny in non-Mendelian ratios.

male parent, its tissues may variegate because the two types of chloroplasts sort themselves out during development.

Baur studied leaf variegation in another ornamental species, *Pelargonium zonale* (Figure 19.5). In this plant, crosses between green and variegating strains produce a mixture of offspring, some green, some variegating, and some pure white. However, these different phenotypes are not recovered in Mendelian proportions. This **non-Mendelian, biparental inheritance** indicates that the color of *Pelargonium* leaves is determined by a mixture of maternal and paternal factors located outside the nucleus, presumably in the chloroplasts. It therefore seems that in *Pelargonium* the chloroplasts are transmitted through the pollen as well as through the ovule.

The experiments of Correns and Baur demonstrated that leaf variegation is inherited as a non-Mendelian trait, one that is almost certainly controlled by factors located in the chloroplasts. The discovery that chloroplasts contain DNA has made this hypothesis very attractive. Yet, in spite of the work of many scientists, the molecular basis of leaf variegation is still unknown. It probably involves differences in the DNA molecules that are present within normal and pigment-deficient chloroplasts. More work is needed to determine the cause of this interesting botanical phenomenon.

Cytoplasmic Male Sterility in Maize

In maize, the male and female reproductive organs are produced on different parts of the plant. The female organs are in the ears, which develop at various positions along the stalk, and the male organs are in the tassels, which develop only at the top of the stalk. In some strains of maize the tassels fail to develop; thus, these strains are male-sterile. However, because the ears develop and produce functional ovules, they can be crossed to other, male-fertile strains.

In the 1930s, Marcus Rhoades demonstrated that male sterility in maize can be caused by a maternally inherited factor (Figure 19.6). Rhoades pollinated the ears of a male-sterile strain with pollen from a male-fertile strain. All the offspring of this cross were male-sterile. Rhoades then backcrossed the male-sterile offspring to the male-fertile variety to see if male sterility was caused by a dominant allele inherited from the male-sterile strain. Because all the offspring were sterile, the dominant allele hypothesis was disproved. Rhoades continued to backcross the male-sterile offspring to male-fertile plants for several generations. At each step in the process, he found that the male-sterile phenotype persisted. This showed that male

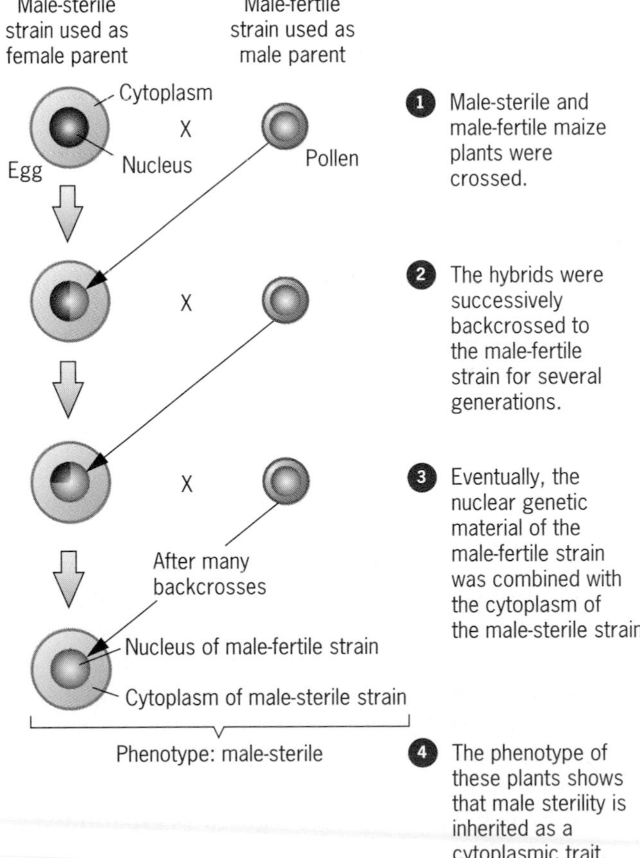

Figure 19.6 Cytoplasmic inheritance of male sterility in maize. Successive backcrosses place the nucleus of a male-fertile strain into the cytoplasm of a male-sterile strain. Because the end result is male-sterile, the trait must be inherited through the cytoplasm.

sterility was transmitted by a maternally inherited factor, evidently something in the cytoplasm of the female reproductive cells. As the research progressed, Rhoades discovered that dominant mutations in several nuclear genes could suppress the effect of this cytoplasmic factor. A plant that had the cytoplasmic factor in addition to one of these dominant mutations could make functional pollen. The male-sterile phenotype therefore depended on an interaction between the cytoplasmic factor and the products of several different nuclear genes. With one combination of nuclear genes, the cytoplasmic factor could cause sterility, but with another combination, it could not.

Maize breeders have taken advantage of cytoplasmic male sterility to streamline the production of hybrid maize seed. However, the method has not been a panacea. In the early 1970s, overreliance on the sterilizing cytoplasm of a particular strain of maize had a disastrous effect on the United States' corn crop. Plants that carried this cytoplasm proved to be especially susceptible to infection by a mutant race of *Helminthosporium maydis,* a fungus that causes southern leaf blight. In some parts of the United States nearly 50 percent of the corn crop was destroyed. Fortunately, maize varieties that did not have the blight-susceptible cytoplasm had been preserved for use in breeding programs. These varieties were pressed into service, and the U.S. corn crop quickly rebounded.

What is responsible for the non-Mendelian inheritance of male sterility in maize? Maize plants possess both mitochondria and chloroplasts, so either organelle could be involved. A partial understanding of the basis of cytoplasmic male-sterility emerged in the 1970s and 1980s. During this time, the procedures for isolating and manipulating organelle DNA were developed. The application of these techniques to male-sterile maize revealed that some varieties contain two small accessory DNA molecules, called S1 and S2, in their mitochondria. These molecules appear to be involved in the male sterility phenomenon, but the mechanism by which they cause it is not known. In other strains of maize, male sterility seems to be caused by a mutation in the mitochondrial DNA itself. Strains with this mutation produce a small polypeptide that may disrupt the function of proteins on the inner mitochondrial membrane. The strains that make this polypeptide are also the ones that are susceptible to southern leaf blight. Thus, it is possible that blight susceptibility is mediated by the mutant polypeptide.

Antibiotic Resistance in *Chlamydomonas*

Photosynthetic green algae are an important part of the biosphere. They are abundant in the oceans, where they form the foundation of the food chain, but they are also found on the land. One terrestrial species, *Chlamydomonas reinhardtii* (Figure 19.7), has been used extensively in genetic research. This unicellular, haploid organism exists in two different mating types, denoted plus and minus. Cells of opposite mating type can fuse to produce a diploid zygote. The zygote then undergoes meiosis to produce four haploid offspring, which divide mitotically to produce clones of vegetative cells. Analysis of these clones indicates that mating type is controlled by a nuclear gene with two alleles, mt^+ (specifying the plus mating type) and mt^- (specifying the minus mating type). Two of the four offspring from a *Chlamydomonas* zygote are therefore mt^+ and two are mt^-. Many other nuclear genes with a Mendelian pattern of segregation have been identified in *Chlamydomonas.*

Figure 19.7 Cells of the unicellular alga *Chlamydomonas reinhardtii.*

Each *Chlamydomonas* cell contains a single large chloroplast and several mitochondria. When mt^+ and mt^- cells fuse, their organelles are combined in the same cell. It is not altogether clear what happens to these organelles inside the zygote, but when the four haploid offspring are produced, each has one chloroplast and several mitochondria. The basic number of organelles is therefore preserved.

In 1954 Ruth Sager, an American geneticist, discovered that antibiotic resistance in *Chlamydomonas* is a non-Mendelian trait (Figure 19.8). Sager mated a mutant mt^+ strain that was resistant to the antibiotic streptomycin to an mt^- strain that was sensitive to this antibiotic. All the progeny from this cross were streptomycin resistant, even though the mating type alleles segregated in a 1:1 fashion. Thus, resistance to streptomycin in *Chlamydomonas* is controlled by a non-Mendelian factor. Sager then mated some of the resistant mt^- cells to sensitive mt^+ cells, and this time she found that all the progeny were sensitive. Resistance or sensitivity to streptomycin was therefore always inherited through the cytoplasm of the mt^+ cell. Sager soon found other traits that followed this same pattern of inheritance.

Sager and her associate Zenta Ramanis later discovered a way to alter the uniparental inheritance of these traits. This involved irradiating the mt^+ cells with UV light just prior to mating. When Sager and Ramanis mated irradiated mt^+ cells to unirradiated mt^- cells, they found that some of the offspring inherited the cytoplasmic trait of the mt^- parent instead of that of the mt^+ parent.

These phenomena were finally explained when DNA was discovered in the *Chlamydomonas* chloroplast. Both mt^+ and mt^- cells possess this DNA, but when mating occurs, the DNA contributed by the mt^- cell is degraded, probably by a nuclease that specifically recognizes it. The offspring from a cross therefore retain only the DNA inherited from the mt^+ chloroplast. As a result, the traits encoded by this DNA—and streptomycin resistance is one of them—are always inherited through the mt^+ parent. When mt^+ cells are irradiated prior to mating, the system that destroys the mt^- chloroplast DNA is partially incapacitated. Consequently, some of the mt^- chloroplast DNA survives and is transmitted to the offspring. This DNA may actually recombine with the mt^+ chloroplast DNA, generating offspring that inherit chloroplast genes from both parents. Sager and Ramanis used the frequency of such recombination to construct a map of

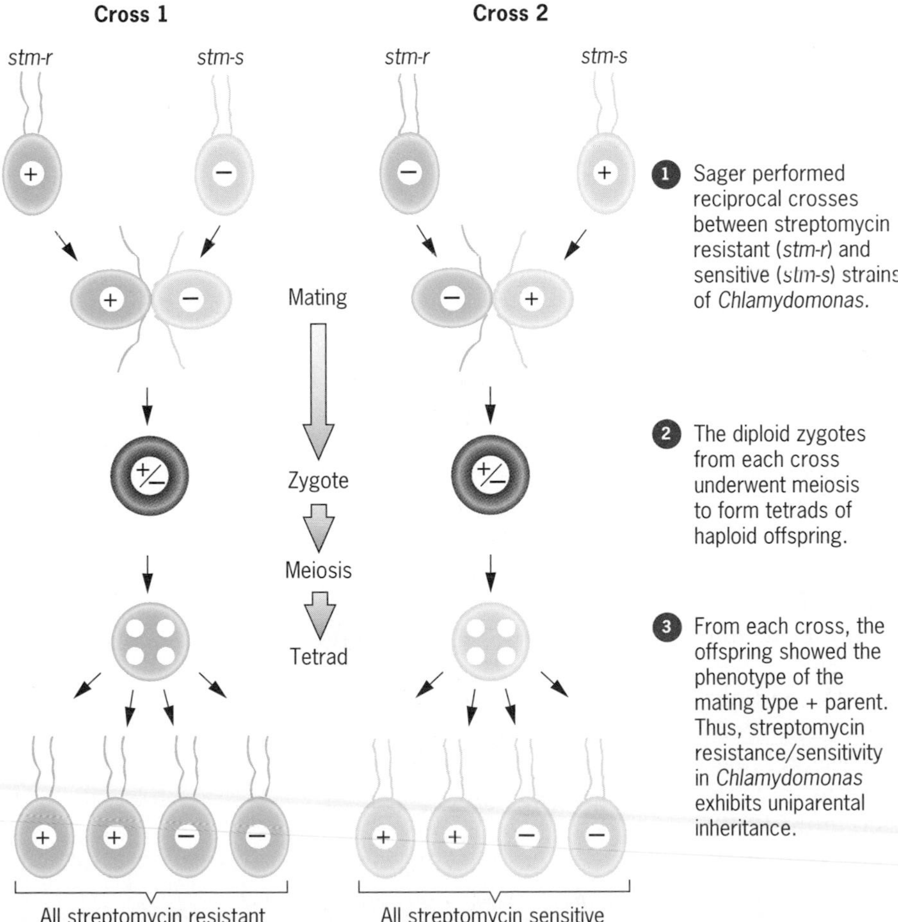

Figure 19.8 Sager's experiments showing uniparental inheritance of streptomycin resistance (*stm-r*) and sensitivity (*stm-s*) in *Chlamydomonas*. All the offspring of a cross have the same phenotype as the mt^+ parent.

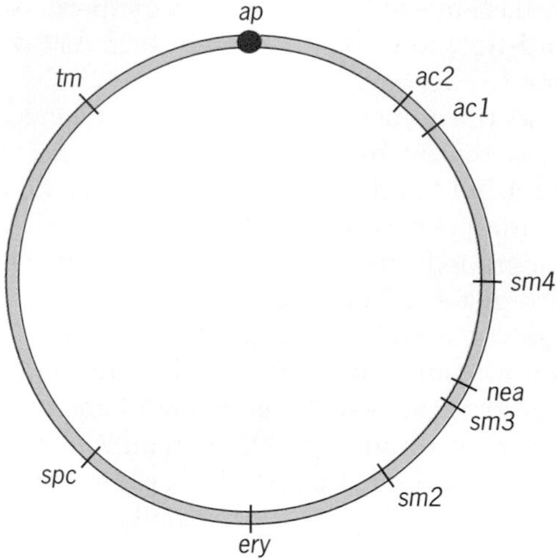

Figure 19.9 A genetic map of chloroplast DNA in *Chlamydomonas,* based on the work of Sager and Ramanis. Symbols: *ap,* attachment point; *ac1, ac2,* acetate requirement; *sm4,* streptomycin dependence; *nea,* neamine resistance; *sm3,* low-level streptomycin resistance; *sm2,* high-level streptomycin resistance; *ery,* erythromycin resistance; *spc,* spectinomycin resistance; *tm,* temperature sensitivity.

the chloroplast "chromosome" (Figure 19.9). This map was the first detailed picture of how genes are organized on a nonnuclear DNA molecule.

Metabolic Defects in Yeast

Some mutant strains of yeast form tiny colonies when grown on a rich, glucose-containing medium. These strains are called **petite mutants,** from the French word for "tiny." Wild-type yeast strains form large, or **grande,** colonies on glucose medium. Additional tests with different food sources suggest that petite mutants suffer from a defect in glucose metabolism. This defect is, in fact, traceable to the mitochondria of the petite cells, which are ill-formed and lack many of the macromolecules found in wild-type mitochondria. These mitochondrial defects prevent petite cells from carrying out aerobic metabolism.

The first genetic analysis of the petite condition was carried out by the French researcher Boris Ephrussi in the 1940s and 1950s. His analysis revealed that there are two general classes of petite mutants, neutral petites and suppressive petites (Figure 19.10). The **neutral petite mutants** are characterized by an inability to transmit the petite phenotype to the offspring of crosses with wild-type strains. When such crosses are performed, all four haploid offspring from a pair of mated cells grow into grande colonies—suggesting that the petite mutation has been lost. By contrast, **suppressive petite mutants** are, under appropriate conditions, able to transmit the petite phenotype to all their progeny—suggesting that the wild-type condition has been lost. With both types of mutants, reciprocal crosses between the mutants and wild-type strains give identical results; thus, the yeast mating types are not involved in the inheritance of the petite trait.

The molecular basis of the two types of petite mutants was determined in the 1970s and 1980s. Neutral petites lack any mitochondrial DNA, and suppressive petites have grossly mutated mitochondrial DNA (Figure 19.11). A cross between a neutral petite mutant and wild-type yields zygotes that inherit mitochondria from both parents; however, only the mitochondria from the wild-type parent contain any DNA. When these zygotes go through meiosis and form spores, the wild-type mitochondrial DNA is distributed to each of the four ascospores. As these grow and divide, healthy, functional mitochondria develop, allowing the cells to carry out aerobic metabolism. Thus, when colonies form, all of them have the grande phenotype.

A cross between a suppressive petite mutant and a wild-type strain yields zygotes with mitochondrial DNA from both parents. If these zygotes are sporu-

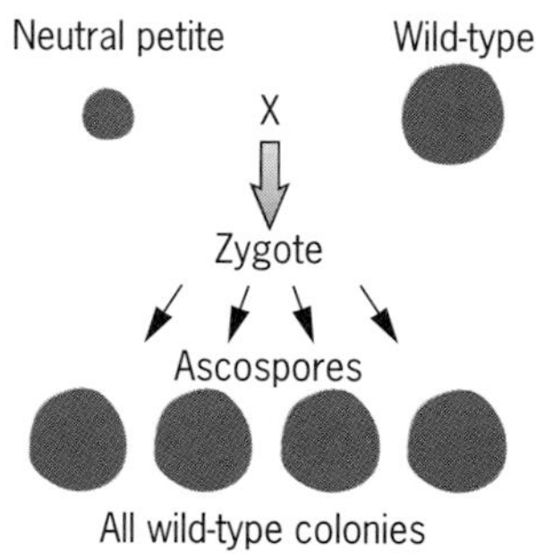

(*a*) A cross between a neutral petite strain and a wild-type strain produces only wild-type progeny.

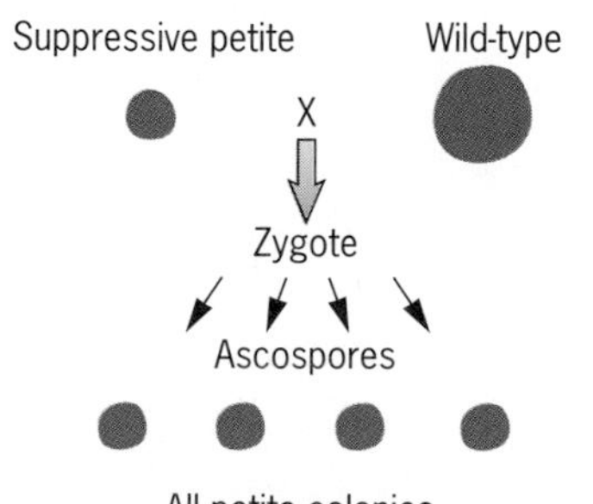

(*b*) A cross between a suppressive petite strain and a wild-type strain produces only petite progeny.

Figure 19.10 Non-Mendelian segregation of (*a*) neutral and (*b*) suppressive petite mutations in yeast.

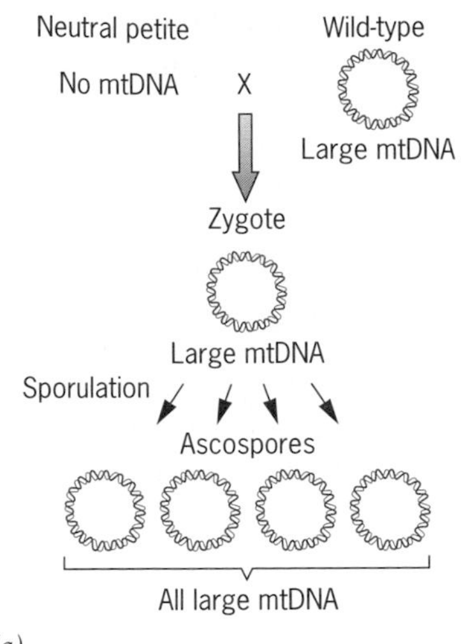

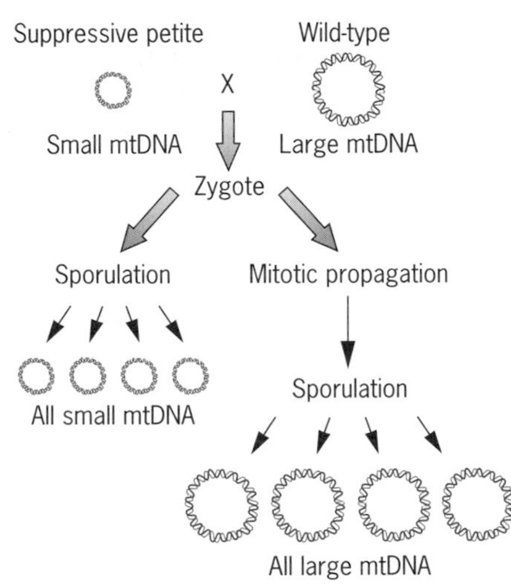

Figure 19.11 Inheritance of mitochondrial (mt) DNA in crosses between petite and wild-type strains of yeast. (*a*) Cross between neutral petite and wild-type strains. (*b*) Cross between suppressive petite and wild-type strains.

lated immediately, the resulting ascospores inherit only the mutant mitochondrial DNA and grow into petite colonies. However, if the zygotes are first propagated mitotically in liquid culture and then sporulated, each ascospore inherits wild-type mitochondrial DNA and grows into a grande colony. These results suggest that the mutant mitochondrial DNA initially has some sort of advantage—it "suppresses" the wild-type DNA, perhaps because the wild-type DNA cannot replicate as quickly. However, during mitotic propagation, the replication of the wild-type DNA catches up with that of the mutant DNA and eventually surpasses it.

What is the nature of the mitochondrial DNA in suppressive petite mutants? Physical and chemical analysis has shown that it is a smaller DNA molecule than wild-type mitochondrial DNA and that it is also very A:T-rich. These two features may endow the suppressive petite mitochondrial DNA with a replicative advantage over wild-type. However, a full understanding of why the suppressive petite DNA is so effectively transmitted through crosses will require more research.

Key Points: **Organelle heredity is characterized by unequal contributions of the two parents and by irregular segregation of phenotypes. These non-Mendelian phenomena are due to the preferential transmission of chloroplasts or mitochondria through the gametes of one sex, which in higher eukaryotes is usually the female. Traits due to mitochondrial or chloroplast genes are therefore often inherited in a uniparental fashion.**

THE MOLECULAR GENETICS OF MITOCHONDRIA

Mitochondrial genetic systems consist of DNA and the molecular machinery needed to replicate and express the genes contained in this DNA. This machinery includes the macromolecules needed for transcription and translation. (As remarkable as it may seem, mitochondria even possess their own ribosomes!) Many of these macromolecules are encoded by mitochondrial genes, but some are encoded by nuclear genes and are therefore imported from the cytosol. In the following sections, we explore the organization and function of the mitochondrial genetic system and the ways in which it interacts with the nuclear genetic system.

Mitochondrial DNA

Mitochondrial DNA, or **mtDNA** as it is sometimes abbreviated, was discovered in the 1960s, initially through electron micrographs that revealed DNA-like fibers within the mitochondria. Later, these fibers were extracted and characterized by physical and chemical procedures. The advent of recombinant DNA techniques made it possible to analyze mtDNA in great detail. In fact, the complete nucleotide sequences of mtDNA molecules from several different species have now been determined.

Mitochondrial DNA molecules vary enormously in size, from about 16 to 17 kb in vertebrate animals to 2500 kb in some of the flowering plants. Each mitochondrion appears to contain several copies of the DNA, and because each cell usually has many mito-

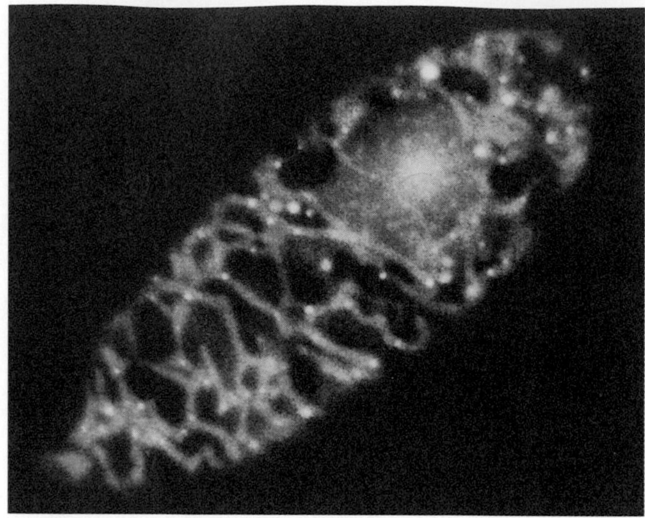

Figure 19.12 Mitochondrial DNA (yellow) in the unicellular organism *Euglena gracilis.* The nuclear DNA (red) is also visible.

chondria, the number of mtDNA molecules per cell can be very large (Figure 19.12). In a vertebrate oocyte, for example, it has been estimated that as many as 10^8 copies of the mtDNA are present. Somatic cells, however, have fewer copies, perhaps less than 1000.

Most mtDNA molecules are circular, but in some species, such as the alga *Chlamydomonas reinhardtii* and the ciliate *Paramecium aurelia,* they are linear. The circular mtDNA molecules, which have been studied the most thoroughly, appear to be organized in many different ways. The simplest arrangement is that seen in the vertebrates, where 37 distinct genes are packed into a 16- to 17-kb circle leaving little or no space between genes. The most complex arrangements exist in some of the flowering plants, where an unknown number of genes are dispersed over a very large circular DNA molecule hundreds or thousands of kilobases in circumference. In these plants the mitochondrial genes may become separated onto different circular molecules by a process of intramolecular recombination (Figure 19.13). This recombination is mediated by repetitive sequences located in the mtDNA. An exchange between two of the repetitive sequences can partition the "master" mtDNA circle into two smaller circles, a process that superficially resembles the excision of a lambda prophage from the *E. coli* chromosome. In some species, several DNA circles of different sizes are formed by recombination between pairs of repetitive sequences located at different positions around the master DNA circle. These molecules are difficult to study, and more research is needed to elucidate the mechanism that produces them.

The fine details of mtDNA organization have been studied by DNA sequencing. Animal mtDNA is small and compact (Figure 19.14). In human beings, for example, the mtDNA is 16,659 base pairs long and contains 37 genes, including two that encode ribosomal RNAs, 22 that encode transfer RNAs, and 13 that encode polypeptides involved in oxidative phosphorylation. In mice, cattle, and frogs, the mtDNA is similar to that of human beings—an indication of a basic conservation of structure within the vertebrate subphylum. Invertebrate mtDNA is about the same size as vertebrate mtDNA, but it has a somewhat different genetic organization. These differences seem to have been caused by structural rearrangements of the genes within the circular mtDNA molecule.

In fungi, the mtDNA is considerably larger than it is in animals. Yeast, for example, possesses circular mtDNA molecules 78 kb long. These molecules contain at least 33 genes, including two that encode ribosomal RNAs, 23 to 25 that encode transfer RNAs, one that encodes a ribosomal protein, and seven that encode different polypeptides involved in oxidative phosphorylation. The yeast mtDNA is larger than animal mtDNA because several of its genes contain introns and there are long noncoding sequences between

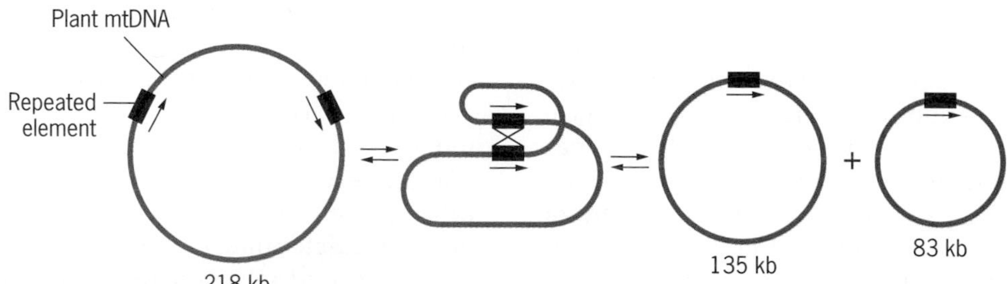

Figure 19.13 Intramolecular recombination in the mtDNA of the Chinese cabbage, *Brassica campestris.* Recombination between the repeated elements in the large circular DNA molecule partitions this molecule into two smaller ones. Alternatively, the repeated elements in the two small molecules may recombine with each other, thereby joining the molecules into a single large molecule.

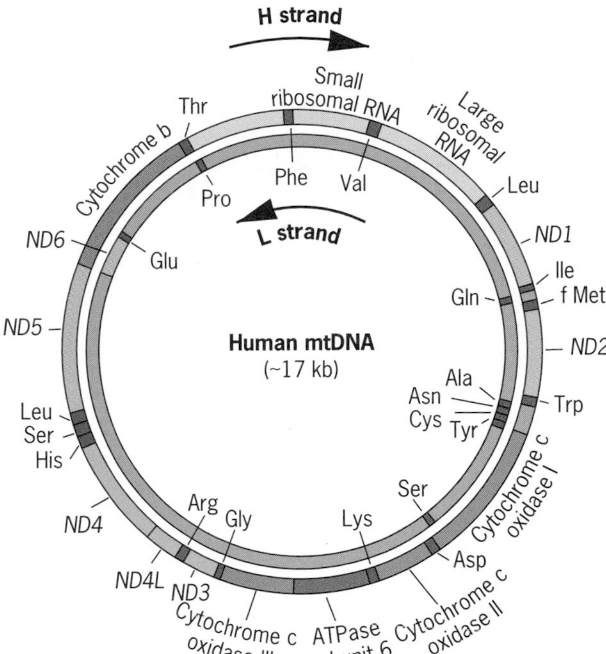

Figure 19.14 Map of human mtDNA showing the pattern of transcription. Genes on the inner circle are transcribed from the L strand of the DNA, whereas genes on the outer circle are transcribed from the H strand of the DNA. Arrows show the direction of transcription. *ND1–6* are genes encoding subunits of the enzyme NADH reductase; the tRNA genes in the mtDNA are indicated by abbreviations for the amino acids.

some of the genes. Animal mtDNA does not contain introns.

Plant mtDNA is much larger than the mtDNA of other organisms. It is also more variable in structure. These conclusions come from crude physical and chemical analyses and from limited DNA sequencing. One of the first plant mtDNAs to be sequenced is from the liverwort, *Marchantia polymorpha.* The mtDNA from this primitive, nonvascular plant is a 186-kb circular molecule with 94 substantial open reading frames (ORFs), some corresponding to known genes and others having still unassigned genetic functions. These latter ORFs are therefore called **URFs,** for *u*nassigned *r*eading *f*rames. Thirty-two distinct introns have been found in the *Marchantia* mtDNA, accounting for about 20 percent of the entire molecule. In vascular plants, the mtDNA is larger than it is in *Marchantia;* for example, it is a 570-kb circle in maize and a 300-kb circle in the watermelon. Higher plant mtDNA molecules contain many noncoding sequences, including some that are duplicated. The actual number of genes is unknown, but it is probably greater than in the mtDNA of other organisms. Physical mapping of some of these genes has shown that they are located in different positions on the mtDNA circles of different species, even when the species are fairly closely related. This implies that the mtDNA of higher plants has undergone many genetic rearrangements during its evolution.

Expression of Mitochondrial Genes

The simple mtDNAs of vertebrates are organized into two large transcription units, each encoding the information of several genes. As an example, consider the human mtDNA shown in Figure 19.14. When the two strands of human mtDNA are separated by centrifugation in an alkaline solution, one proves to be denser than the other. This strand is referred to as the H (for heavy) strand, and its complement is referred to as the L (for light) strand. Both strands are transcribed in human mitochondria. The promoters for the H and L transcription units are situated just upstream of the phenylalanine tRNA gene. The transcripts that initiate at these points are extended in opposite directions around the entire circumference of the mtDNA molecule. The transcript from the H strand encodes the two ribosomal RNAs, 14 tRNAs, and 12 polypeptides, and the transcript from the L strand encodes 8 tRNAs and one polypeptide. Each transcript is cleaved to separate the tRNAs from the ribosomal and mRNAs, and the mRNAs are polyadenylated. Each mRNA is then translated into polypeptides using the mitochondrial ribosomes and a combination of nuclear and ribosomal tRNAs.

Translation in the mitochondria proceeds much as it does on the ribosomes of the cytosol, except that some of the codons have a different meaning. AGA and AGG are termination codons in mammalian mitochondria, whereas in the cytosol they specify the incorporation of arginine into a polypeptide; UGA, which is a termination codon in the cytosol, is a tryptophan codon in the mitochondria, and AUA, which encodes isoleucine in the cytosol, is the methionine initiation codon in the mitochondria. These and other mitochondrial codon variations indicate that the genetic code is not completely universal. Evidently, mitochondria evolved their own variation of the genetic code—a consequence of their evolution from free-living organisms that were incorporated into eukaryotic cells more than a billion years ago. We shall return to the subject of mitochondrial evolution at the end of this chapter.

In fungi and plants, the mtDNA is organized into many separate transcription units, some containing the information for more than one gene. Little is known about the details of transcription, but in yeast, the mitochondrial RNA polymerase is a single polypeptide encoded by a nuclear gene. Thus, it is smaller than the RNA polymerase of *E. coli.* RNA processing separates plant mitochondrial transcripts into their constituent parts and also removes the introns, which are present in several plant mitochondrial genes. At present, the mechanics of these events are poorly understood.

Another peculiarity of plant mitochondrial gene expression is that many of the mtRNA transcripts undergo **editing;** that is, some of the nucleotides are changed after the transcript has been synthesized. The most frequent change is C to U, but occasionally, U is changed to C. Thus, RNA editing alters the composition of codons in plant mitochondrial transcripts, including some that would otherwise signal the end of polypeptide synthesis. Editing corrects the information that is actually encoded in the mtDNA and allows functional polypeptides to be synthesized. Curiously, RNA editing is not found in the nonvascular plants (mosses and algae), even though all groups of higher plants (ferns, gymnosperms, and angiosperms) seem to have it. Thus, the editing mechanism probably evolved sometime after plants had become established on the land. RNA editing also occurs in the mitochondria of protozoans, including the trypanosomes, where the mechanism has been studied in some detail. In these organisms, small RNA molecules that are partially complementary to the mtDNA transcripts serve as guides for the editing process (see Figure 12.22). They are, therefore, called guide RNAs (gRNAs). A similar guiding mechanism may operate in plants, but the details are currently unknown.

Yet a third peculiarity of plant mitochondrial gene expression is that some mitochondrial messenger RNAs are formed by the process of ***trans*-splicing.** *Trans*-splicing occurs when segments of a gene are scattered over the mtDNA molecule. Each gene segment is transcribed independently, and the exons of the different transcripts are spliced together by interactions between the introns that flank them. For example, in wheat, the *nad1* gene, which encodes a subunit of NADH reductase, a protein of oxidative phosphorylation, is partitioned into four segments in the mtDNA. Each of these segments is separately transcribed, and the resulting transcripts are then spliced together to form the mRNA (Figure 19.15). This process requires a single *cis*-splicing reaction and three *trans*-splicing reactions.

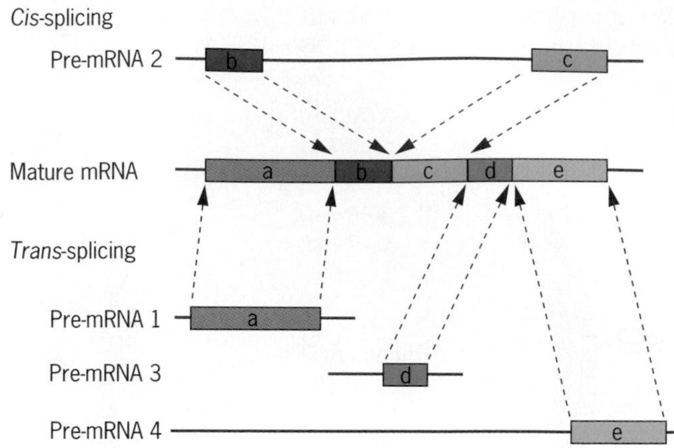

Figure 19.15 *Trans*-splicing in wheat mitochondria. Four different RNAs contribute to the final mRNA encoding a polypeptide of the enzyme NADH reductase.

Interplay Between Mitochondrial and Nuclear Gene Products

Most—perhaps all—mitochondrial gene products function solely within the mitochondrion. However, they do not function alone. Many nuclear gene products are imported to augment or facilitate their function. Mitochondrial ribosomes, for example, are constructed with ribosomal RNA transcribed from mitochondrial genes and with ribosomal proteins encoded by nuclear genes. The ribosomal proteins are synthesized in the cytosol and imported into the mitochondria for assembly into ribosomes.

Many of the polypeptides needed for aerobic metabolism are also synthesized in the cytosol. These include subunits of several proteins involved in oxidative phosphorylation, for example, the ATPase that is responsible for binding the energy of aerobic metabolism into ATP. However, because some of the subunits of this protein are synthesized in the mitochondria, the complete protein is actually a mixture of nuclear and mitochondrial gene products. This dual composition suggests that the nuclear and mitochondrial genetic systems are coordinated in some way so that equivalent amounts of their products are made; possible molecular mechanisms for this coordination are currently under investigation.

Key Points: **Mitochondrial DNA (mtDNA) molecules range from 16 kb to 2500 kb in size, and most appear to be circular. These molecules contain genes for some of the ribosomal RNAs, transfer RNAs, and polypeptides used within the mitochondrion. The structure, organization, and expression of these genes vary among species. In some groups of organisms, the transcripts of mitochondrial genes are edited after they are synthesized. Both mitochondrial and nuclear gene products are needed for proper mitochondrial function.**

MITOCHONDRIAL DNA AND HUMAN DISEASE

In yeast, petite mutants are associated with alterations in the structure of the mtDNA and, in some cases, with the complete loss of this DNA. Are any phenotypes in higher eukaryotes traceable to mtDNA mutations?

Recent research has demonstrated that several human diseases are caused by mitochondrial defects, and in some cases, these defects are due to mutations in the mtDNA. One such disease is **Leber's hereditary optic neuropathy (LHON),** a condition characterized by the sudden onset of blindness in adults. At a physiological level, this disease is associated with the death of the optic nerve, and at a molecular level, it is associated with mutations in any of several different mitochondrial genes. Each of these mutations changes an amino acid in one of the mitochondrial proteins, thereby reducing the efficiency of oxidative phosphorylation. The reduction is great enough to destroy the function of the optic nerve and cause total blindness. It is not known why this lethal effect is limited to the optic nerve. Perhaps the nerve cells are especially sensitive to disruptions in aerobic metabolism. As would be expected for a mitochondrial mutation, LHON is inherited strictly through the maternal line; there is never any transmission through males.

Another disorder caused by a mutation in the mtDNA is **Pearson marrow-pancreas syndrome.** This disease, characterized by a loss of bone-marrow cells during childhood, is frequently fatal. It is caused by a fairly large deletion in the mtDNA. People with the Pearson syndrome almost never have affected parents; thus, the causative deletion probably occurs spontaneously during development in the child or during oogenesis in the mother. Individuals with the Pearson syndrome actually have a mixture of deleted and normal mtDNA—an example of mitochondrial heteroplasmy. Individuals who are homoplasmic for the deleted mtDNA of Pearson's syndrome have never been observed, probably because they die very early in development.

Key Point: **Mutations in mtDNA can cause human diseases such as Leber's hereditary optic neuropathy and Pearson marrow-pancreas syndrome.**

THE MOLECULAR GENETICS OF CHLOROPLASTS

Chloroplasts are specialized forms of a general class of plant organelles called **plastids.** Botanists distinguish among several kinds of plastids, including chromoplasts (plastids containing pigments), amyloplasts (plastids containing starch), and elaioplasts (plastids containing oil or lipid). All three types seem to develop from small membrane-bounded organelles called proplastids, and, within a particular plant species, all seem to contain the same DNA. This DNA is generally referred to as **chloroplast DNA,** abbreviated simply as **cpDNA.**

Chloroplast DNA

In higher plants, cpDNAs typically range from 120 to 160 kb in size, and in algae, from 85 to 292 kb. In a few species of green algae in the genus *Acetabularia,* the cpDNA is much larger, about 2000 kb. Among the 200 or so species of plants whose chloroplast DNA has been at least partially characterized, the cpDNA seems to be organized as a covalently closed circular molecule. However, in some species, especially those with large cpDNAs, a linear arrangement cannot be ruled out.

The number of cpDNA molecules in a cell depends on two factors: the number of chloroplasts and the number of cpDNA molecules within each chloroplast. For example, in the unicellular alga *Chlamydomonas reinhardtii,* there is only one chloroplast per cell, and it contains about 100 copies of the cpDNA. In *Euglena gracilis,* another unicellular organism, there are about 15 chloroplasts per cell, and each contains about 40 copies of the cpDNA.

All cpDNA molecules carry basically the same set of genes, but in different species of plants these genes are arranged in different ways. The basic gene set includes genes for ribosomal RNAs, transfer RNAs, some ribosomal proteins, various polypeptide components of the photosystems that are involved in capturing solar energy, the catalytically active subunit of the enzyme ribulose 1,5-bisphosphate carboxylase, and four subunits of a chloroplast-specific RNA polymerase. A few cpDNA molecules have been sequenced in their entirety, including one from the liverwort, *Marchantia polymorpha* (Figure 19.16), and another from the tobacco plant, *Nicotiana tobacum.* The tobacco cpDNA is larger (155,844 bp) and probably contains about 150 genes. The best estimate for the gene number in the liverwort cpDNA (121,024 bp) is 136. Most cpDNAs have a pair of large inverted repeats that contain the genes for the ribosomal RNAs. These repeats range anywhere from 10 to 76 kb in length and are variously located in different cpDNA molecules.

Chloroplast Biogenesis

As mentioned earlier, all plastids develop from proplastids. In the case of chloroplasts, this development is stimulated by light and involves the transcription of many genes, including some located in the nucleus. Illumination triggers a profound change in each proplastid. The organelle increases in size, and the inner of its two surrounding membranes expands to bud off

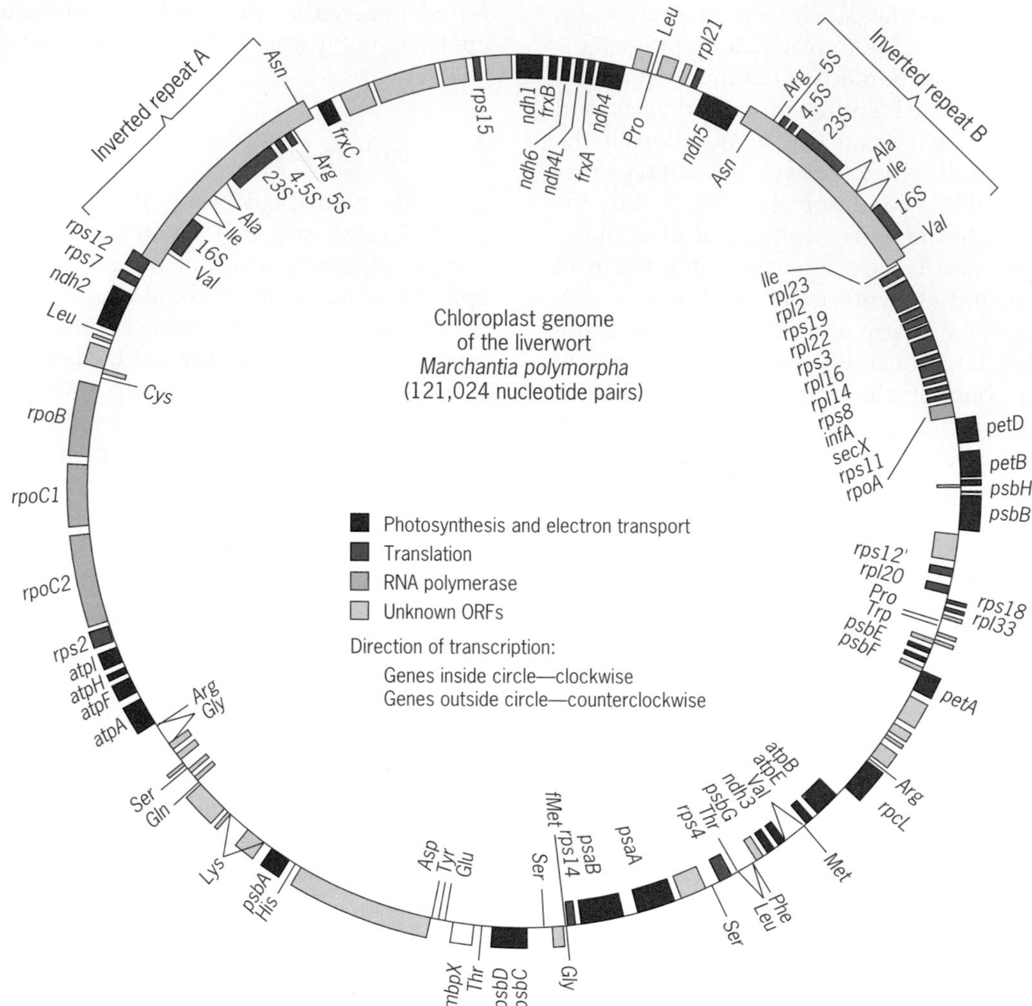

Figure 19.16 Genetic organization of the chloroplast DNA in the liverwort *Marchantia polymorpha.* Symbols: rpo, RNA polymerase; rps, ribosomal proteins of small subunit; rpl and secX, ribosomal proteins of large subunit; 4.5S, 5S, 16S, 23S, rRNAs of the indicated size; rbs, ribulose bisphosphate carboxylase; psa, photosystem I; psb, photosystem II; pet, cytochrome b/f complex; atp, ATP synthesis; infA, initiation factor A; frx, iron-sulfur proteins; ndh, putative NADH reductase; mpb, chloroplast permease (?); tRNA genes are indicated by abbreviations for the amino acids.

vesicles that eventually arrange themselves in stacks forming structures called grana (Figure 19.17). All the proteins and chlorophyll pigments needed for photosynthesis are made and targeted to their appropriate locations within the emerging chloroplast. Some of these molecules are positioned in the thylakoid membranes that make up the grana, and others are localized in the stroma, which is the protoplasmic space surrounding the grana.

The formation of functional chloroplasts is a process referred to as **biogenesis.** Only some of the details are known. However, it is clear that light plays an important role. For example, the nuclear gene that encodes the small subunit of ribulose 1,5-bisphosphate carboxylase is vigorously transcribed when light is provided. A special class of pigmented proteins called **phytochromes** seems to mediate this and other responses to light. By absorbing light energy, these proteins acquire the ability to trigger other proteins to stimulate the transcription of genes involved in chloroplast biogenesis.

The formation of chloroplasts and the maintenance of their structure and function during the life of a plant depend on the coordinated expression of nuclear and chloroplast genes. Each genome produces a distinct set of mRNAs, which are translated in the cytosol and the chloroplast, respectively. Chloroplast biogenesis therefore involves considerable interplay between nuclear and chloroplast gene products.

Key Points: **Circular chloroplast DNA (cpDNA) molecules are typically 120 to 292 kb in size and**

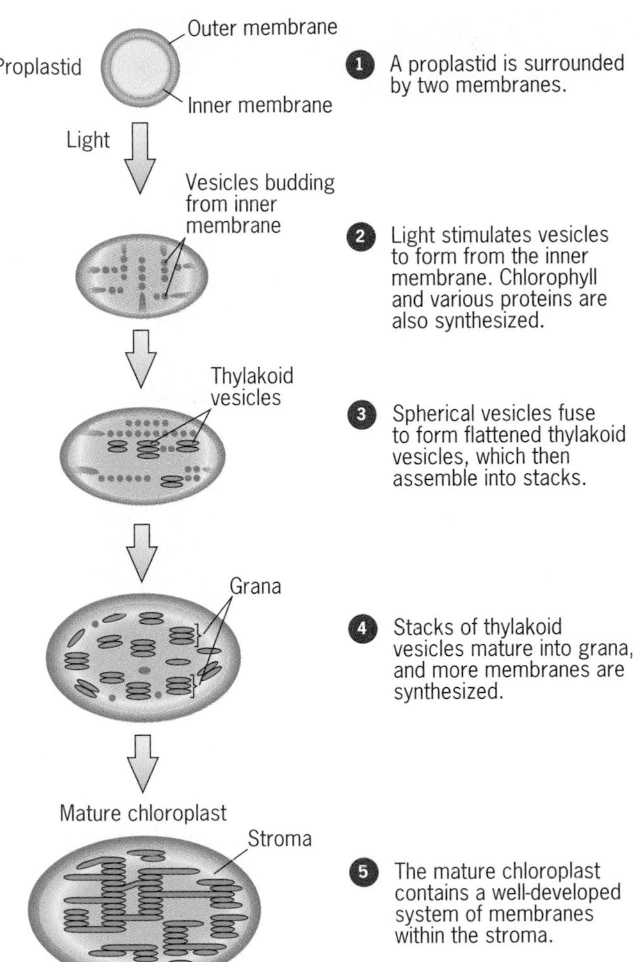

Figure 19.17 Chloroplast biogenesis. A mature chloroplast containing stacks of thylakoid membranes (grana) within its protoplasmic matrix (stroma) develops from a proplastid after exposure to light.

may contain 100 or more genes. The organization of these genes varies among species of plants and algae. Other plastids also contain cpDNA molecules. Light induces chloroplasts to develop from unpigmented proplastids. This process, called biogenesis, involves the interplay of chloroplast and nuclear gene products.

THE ORIGIN AND EVOLUTION OF MITOCHONDRIA AND CHLOROPLASTS

Life on earth originated more than 3 billion years ago, probably in shallow pools of nutrient-rich water. The first cellular organisms were prokaryotic; they lacked true nuclei and cytoplasmic organelles. Furthermore, these cells were heterotrophic; that is, they fed on nutrients present in the environment. Through the fermentation of nutrients, these primordial organisms recruited the energy they needed to survive. Later, as CO_2 accumulated in the atmosphere, photosynthetic organisms evolved, probably much like the cyanobacteria that are found in stagnant pools today. These organisms were also prokaryotic; their photosynthetic machinery was localized in membranes strewn within the cell—not organized within a special subcellular organelle. Photosynthesis generated substantial quantities of O_2, and as this gas built up in the atmosphere, aerobic metabolism evolved, probably 1 to 2 billion years ago. The first aerobic organisms were also prokaryotic.

Eukaryotic organisms evolved 1 to 1.5 billion years ago. The first known eukaryotes were filamentous green algae, with nuclei and an elaborate intracellular organization, including subcellular organelles. These organisms were capable of photosynthesis and aerobic metabolism, and probably contained both chloroplasts and mitochondria. Where did these cytoplasmic organelles come from?

In the 1970s the American scientist Lynn Margulis forcefully argued that mitochondria and chloroplasts were once free-living organisms that were incorporated into primitive eukaryotic cells. Today, the hypothesis that chloroplasts and mitochondria originated as separate organisms is widely accepted. These organisms evidently established a symbiotic relationship with the primitive eukaryotic cells that internalized them.

Eukaryotic Organelles as Endosymbionts

Symbiosis is a condition in which two or more organisms live together in close association for their mutual benefit. Lichens are a classic example. These widely distributed organisms are composed of an alga and a fungus that live together in an intimate, mutually beneficial association. **Endosymbiosis** is a special case in which one of the symbiotic partners lives inside the other. For example, cells of *Chlorella,* a unicellular alga, sometimes live inside the single-celled protozoan *Paramecium bursaria.* The chlorophyll in these endosymbiotic algae makes the *Paramecium* appear green.

Evidence from biochemical, genetic, and molecular studies suggests that mitochondria and chloroplasts originated as bacterial endosymbionts. Mitochondria and chloroplasts are both about the same size as some bacteria, and both contain circular DNA molecules much like circular bacterial chromosomes. Both mitochondria and chloroplasts contain their own distinctive ribosomes, and both utilize a distinctive RNA

HUMAN GENETICS SIDELIGHT

Using Mitochondrial DNA to Study Human Evolution

In biology few subjects are more fascinating than that of human evolution. Who are we? Where did we come from? Where are we going? Before the advent of molecular biology, the study of human evolution depended on the analysis of rare fossils—fragments of bone, a few teeth, an occasional weapon or tool. Today, human evolution can be studied by comparing DNA sequences. Each DNA sequence is descended from a sequence that was present in an ancestral organism. Thus, the DNA sequences that we find today are, in effect, living fossils—records of ancient DNA that has been transmitted through many generations to organisms currently alive. Because mutations may have occurred during this time, a modern DNA sequence is not likely to be an exact replica of its ancestor. However, by comparing modern DNA sequences, we can sometimes reconstruct features of the evolutionary process that produced them.

Some of the most insightful studies of human evolution have involved the analysis of mitochondrial DNA. There are two reasons why mtDNA is so useful: (1) it evolves faster than nuclear DNA, and (2) it is transmitted exclusively through the female. The rapidity of mtDNA evolution allows a scientist to detect significant genetic changes over a relatively short period of time (in evolutionary terms), and the strict maternal transmission of mtDNA allows a researcher to trace modern DNA sequences back to a common female ancestor.

Pioneering studies of human mtDNA were carried out in the 1980s by Allan Wilson, Rebecca Cann, Mark Stoneking, and their colleagues. These studies established that there is relatively little variation in the mtDNA from different human populations and that the greatest variation is found in the mtDNA from populations in Africa. Given the rate at which mtDNA is known to evolve, these discoveries suggested that modern human beings originated rather recently, probably within the last 200,000 years, and probably in Africa. Although these conclusions were initially controversial, later work has reinforced them.[1] Wilson's laboratory collected mtDNA samples from more than 200 individuals representing many different racial and ethnic groups. The mtDNA sequences in this collection were determined biochemically and then analyzed by a computer program that arranges the sequences in a phylogenetic, or evolutionary, tree. Wilson's conclusion was startling. The mtDNA in all modern groups of humans is descended from an mtDNA molecule that existed in a single woman who lived in Africa about 200,000 years ago. Applying a biblical metaphor, the popular press nicknamed this woman "Mitochondrial Eve."

By focusing on the evolution of mtDNA, Wilson's laboratory traced human ancestry back to a point where the maternal lineages of all modern mtDNA sequences coalesce in a single common ancestor—the mitochondrial mother of us all. However, these researchers never meant to imply that a single woman alone gave rise to all modern human beings. The mass of human nuclear DNA, which is inherited equally from males and females, and which varies among the members of a breeding population, cannot be traced to a single individual.

The work of Wilson and his colleagues strongly argues that all modern humans evolved from individuals who lived in Africa less than 200,000 years ago, and possibly as recently as 120,000 years ago. Migrants from this original African population presumably founded the archaic human populations of Europe and Asia, which, in turn, founded the early human populations of Australia, Oceana, and the Americas. This evolutionary scenario has been called the "Out of Africa" hypothesis. Another hypothesis proposes that humans evolved simultaneously in many regions of the world, from groups that were long established in those regions, perhaps for many hundreds of thousands of years, and that these groups probably interbred with other archaic populations such as the Neanderthals of Europe and western Asia.

The Neanderthals have always been an enigmatic group for students of human evolution. Fossil remains indicate that they were quite different from modern humans; thicker bones, greater musculature, and different body proportions clearly set them apart. Were the Neanderthals ancestral to modern humans? Did they interbreed with the populations that ultimately produced modern humans, or were they a separate and distinct species altogether?

In 1997 Matthias Krings, Anne Stone, Ralf Schmitz, Heike Krainitzki, Mark Stoneking, and Svante Pääbo published the sequence of 379 base pairs of mtDNA extracted from a fossilized Neanderthal arm bone.[2] This particular fossil, discovered in 1856 near Dusseldorf, Germany, has been the subject of many intensive studies. After lengthy negotiations, the fossil's custodians granted Krings and co-workers permission to remove a 3.5-g piece of bone from the right humerus. Small fragments from this piece were pulverized, and the DNA remnants within them were carefully extracted. Because of the fossil's age (between 30,000 and 100,000 years), most of the DNA was expected to be degraded. However, because mtDNA is much more abundant than any particular sequence of nuclear DNA, Krings and co-workers hoped that some of it had survived. Their first step was to use a technique called the polymerase chain reaction (PCR, see Chapter 20) to amplify small segments of surviving mtDNA molecules. PCR allows a researcher to generate millions of identical DNA molecules from just a few molecules by *in vitro* replication with a bacterial DNA polymerase. The sequence of the amplified DNA can then be determined biochemically.

In carefully controlled experiments, Krings and co-workers succeeded in amplifying mtDNA remnants extracted from the fossil. Biochemical analysis of this ampli-

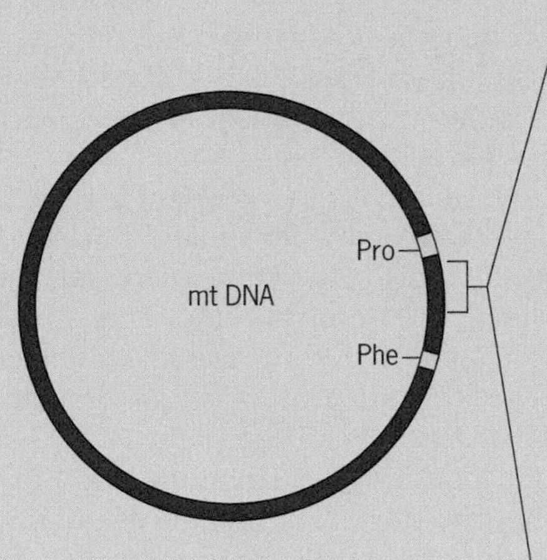

Figure 1. Nucleotide differences within a 379-bp noncoding region of the mtDNA of a Neanderthal fossil and that of a modern human being. The sequenced region lies between the genes for the phenylalanine (Phe) and proline (Pro) tRNAs. For each nucleotide difference (highlighted), the upper nucleotide is found in modern human mtDNA and the lower one is found in the Neanderthal mtDNA.

fied material showed that Neanderthal mtDNA differs from modern human mtDNA in 28 of the 379 nucleotides that were analyzed (Figure 1). The mtDNA isolated from different modern humans typically shows only 8 nucleotide substitutions in this region. Thus, Neanderthal mtDNA is significantly unlike that of modern humans. Computer analysis of the DNA sequences suggested that the human and Neanderthal mtDNA lineages began to evolve separately between 550,000 and 690,000 years ago, and that modern human mtDNAs originated between 120,000 and 150,000 years ago, apparently in Africa. Thus, Neanderthals were almost certainly not ancestral to modern humans. Rather, they evolved separately and, in the end, became extinct.

In the discussion section of their paper, Krings and coauthors concluded that "The Neanderthal mtDNA sequence thus supports a scenario in which modern humans arose recently in Africa as a distinct species and replaced Neanderthals with little or no interbreeding." They also added a caveat: "It must be emphasized that the above conclusions are based on a single individual sequence; the retrieval and analysis of mtDNA sequences from additional Neanderthal specimens is obviously desirable."[3] Of course, obtaining mtDNA sequences from other Neanderthals will entail the destruction of rare fossil material. Thus, the decision to collect such data should not be taken lightly. The benefit of collecting data from several individuals may not outweigh the cost of sacrificing so many valuable fossils. However, obtaining the sequence from at least one more Neanderthal does seem worthwhile, since this sequence could reinforce or invalidate the inferences that have to be made from the single sequence now available. We will have to wait and see if another Neanderthal fossil suitable for DNA analysis can be found. If it can, then the issue will be whether or not to allow part of that fossil to be destroyed to obtain a few molecules of mtDNA.

[1]Wilson, A. C., and R. L. Cann. 1992. The recent African genesis of humans. *Sci. Amer.*, 266(4):68–73.

[2]Krings, M., A. Stone, R. W. Schmitz, H. Krainitzki, M. Stoneking, and S. Pääbo. 1997. Neandertal DNA sequences and the origin of modern humans. *Cell* 90:19–30.

[3]ibid., p. 27.

polymerase. In chloroplasts, this RNA polymerase is similar to the one found in *E. coli.* DNA sequencing studies have demonstrated that some genes in the mitochondrial and chloroplast DNAs are similar to bacterial genes. For example, the mitochondrial genes for ribosomal RNA are most closely related to the genes for ribosomal RNA in a group of purple photosynthetic bacteria, and the chloroplast genes for ribosomal proteins seem to be related to bacterial genes for these proteins. All these facts strongly suggest that mitochondria and chloroplasts originated when bacteria were incorporated into primitive eukaryotic cells, probably more than a billion years ago. The endosymbiotic bacteria benefited from a stable biological environment that provided food and an ability to reproduce, and the eukaryotic cells benefited from the oxidative and photosynthetic energy-recruiting powers of the endosymbiotic bacteria.

Since the formation of these relationships, each of the symbiotic partners has undergone significant changes, and considerable genetic shuffling has taken place among the mitochondrial, chloroplast, and nuclear DNAs. For example, the mtDNA in some higher plants contains a transfer RNA gene from cpDNA, and in one plant species, the mtDNA contains part of the chloroplast gene for the large subunit of ribulose 1,5-bisphosphate carboxylase. In most plant species, the small subunit of ribulose 1,5-bisphosphate carboxylase is encoded by a nuclear gene, but in some species of algae, it is encoded by a chloroplast gene. These differences indicate that some of the genes that were originally present in the DNA of the endosymbiotic bacteria have been moved to other cellular compartments, including many that have been transferred into the nucleus. Because of this gene shuffling, neither mitochondria nor chloroplasts are able to sustain themselves without materials specified by the nucleus. Most of the materials that are needed for the assembly and function of these organelles are encoded by nuclear genes. Even major components of the genetic systems of the mitochondria and chloroplasts are derived from nuclear gene products. These include the DNA and RNA polymerases, some transfer RNAs, and many, if not all, of the ribosomal proteins. Over evolutionary time, the symbiotic partners have exchanged so many genes that they are now completely interdependent.

The Evolution of Mitochondria and Chloroplasts

Variation among the DNAs of mitochondria and chloroplasts indicates that the genomes of these organelles have undergone significant changes. It also suggests that these genomes may trace back to several different symbiotic partnerships that were established independently very long ago. Among organelle genomes that are clearly related, there are two kinds of variation.

First, there is structural variation, or variation in the order of genes within the DNA. For example, human and *Drosophila* mtDNAs contain the same set of genes, but these genes are arranged differently in their respective mtDNAs. The different arrangements suggest that regions of the mtDNA have been relocated since the vertebrate and invertebrate lineages diverged from a common ancestor 400 to 600 million years ago.

Second, there is variation in the nucleotide sequences of particular genes. For example, when we compare the genes for cytochrome b in human and *Drosophila* mtDNAs, we find many nucleotide differences, some of which alter the sequence of amino acids in the cytochrome b polypeptide. This kind of variation indicates that the sequences of these genes have evolved since they diverged from a common ancestor. Considerable effort has been expended to study the rate of mitochondrial gene evolution and to compare it with the rate of nuclear gene evolution. The surprising finding is that in some groups of organisms such as vertebrate animals mitochondrial genes evolve faster, perhaps 5 to 10 times faster, than nuclear genes. The reasons for this rapid evolution are not clear. Perhaps mitochondrial gene products are more flexible and are therefore able to tolerate more amino acid changes than are nuclear gene products, or perhaps the mutation rate is simply greater in the mitochondria than it is in the nucleus.

Mitochondria and chloroplasts are almost always transmitted through the female. There is some evidence for paternal transmission (see, for example, the earlier discussion of Baur's experiments on chloroplast inheritance in *Pelargonium*), but in many organisms it seems to be infrequent. Thus, the genomes of these organelles evolve mainly by an asexual process. However, because recombination apparently does take place between organelle genomes within particular cells, genetic variants on two different molecules can, in principle, be combined. The evolutionary significance of this possibility has not been thoroughly assessed.

Organellar DNA, especially that of mitochondria, has been very useful in studying the evolution of many groups of organisms. The Human Genetics Sidelight: Using Mitochondrial DNA to Study Human Evolution describes how the analysis of mtDNA has helped to elucidate the evolution of our own species.

Key Points: **Mitochondria and chloroplasts probably originated as bacteria that were incorporated into eukaryotic cells about a billion years ago. The mutually beneficial relationship that has existed between these cells and their internalized bacteria has evolved into a relationship of complete interdependence. In the course of this endosymbiosis, genes have been shuffled among the mitochondrial, chloroplast, and nuclear DNAs. In some groups of organisms, mitochondrial genes evolve 5 to 10 times faster than nuclear genes.**

TESTING YOUR KNOWLEDGE

1. A mutant mt^+ strain of *Chlamydomonas* that was resistant to the antibiotic streptomycin (*stm-r*) and the herbicide *benomyl* (*ben-r*) was crossed to a wild-type mt^- strain that was sensitive to these two chemicals (*stm-s* and *ben-s*). Twenty tetrads of progeny were analyzed for mating type and resistance to streptomycin and benomyl. When the analysis was completed, the tetrads were classified into three types:

	Type 1	Type 2	Type 3
	mt^+ *stm-r ben-r*	mt^+ *stm-r ben-s*	mt^+ *stm-r ben-r*
	mt^+ *stm-r ben-r*	mt^+ *stm-r ben-s*	mt^+ *stm-r ben-s*
	mt^- *stm-r ben-s*	mt^- *stm-r ben-r*	mt^- *stm-r ben-r*
	mt^- *stm-r ben-s*	mt^- *stm-r ben-r*	mt^- *stm-r ben-s*
Number	8	9	3

Which of these traits seems to be due to a nonnuclear gene?

ANSWER

Streptomycin resistance/sensitivity. All the progeny from the cross are streptomycin-resistant, just like the mt^+ parent. This uniparental inheritance is diagnostic of a chloroplast gene. The other two traits, mating type and benomyl resistance/sensitivity, segregate 1:1 in the progeny, indicating that they are determined by nuclear genes. In fact, from the numbers of the three types of tetrads, we can infer that the *mt* and *ben* loci are on different chromosomes. This inference follows because the number of parental ditype tetrads (Type 1) is equal to the number of nonparental ditype tetrads (Type 2); see Chapter 9 for a full explanation of tetrad analysis for nuclear genes.

QUESTIONS AND PROBLEMS

19.1 When Correns studied the inheritance of leaf variegation in *Mirabilis,* he made crosses between green and variegating plants. Why didn't he make crosses between green and white plants?

19.2 Leaf coloration in variegating strains is inherited maternally in *Mirabilis* and biparentally in *Pelargonium.* Furthermore, in *Pelargonium* the pattern of inheritance is non-Mendelian. What do these facts suggest about the inheritance of chloroplasts in these species?

19.3 Explain how plants such as *Mirabilis* could have green, pale green, and white sectors in their leaves. If such sectors reached sexual maturity, what color characteristics would each type be expected to transmit through male and female gametes?

19.4 Reciprocal crosses with experimental animals or plants sometimes give different results in the F_1. Two possible explanations are (a) sex-linked inheritance and (b) organelle inheritance. How could an investigator determine which explanation is correct?

19.5 Oscar Renner performed reciprocal crosses between two types of evening primroses, *Oenothera hookeri* and *O. muricata,* known to have the same chromosome constitution. When the female parent was *O. hookeri,* the leaves of the progeny were yellow, but when the female parent was *O. muricata,* the leaves of the progeny were green. How might this difference in the results of reciprocal crosses be explained?

19.6 What is the practical significance of cytoplasmic male sterility in maize?

19.7 A male-sterile maize plant was crossed with a male-fertile plant that carried male-sterilizing cytoplasm and one copy of a dominant nuclear allele, *Rf,* that restores male fertility. What fraction of the offspring will be male-sterile? If the male-fertile offspring of this cross are intercrossed with each other, what fraction of their offspring will be male-sterile?

19.8 A male-sterile variety of maize was crossed with a male-fertile variety and all the offspring were male-sterile. These offspring were then crossed to a male-fertile variety that had normal (male-fertile) cytoplasm. What fraction of the offspring of this second cross are expected to be male-sterile?

19.9 An mt^+ strain of *Chlamydomonas* that was resistant to the antibiotic spectinomycin was crossed with a sensitive mt^- strain. Half the progeny were mt^+ and half were mt^-, but all were resistant to spectinomycin. When the mt^- spectinomycin-resistant cells were crossed to mt^+ spectinomycin-sensitive cells, all the progeny were spectinomycin-sensitive. Explain the inheritance of spectinomycin resistance/sensitivity.

19.10 An mt^+ strain of *Chlamydomonas* that was resistant to the antibiotic spectinomycin was crossed with an mt^- strain that was resistant to the antibiotic streptomycin. The genes for resistance to these two antibiotics are known to reside in cpDNA. Just prior to mating, the mt^+ cells were irradiated with ultraviolet light. Each mated pair of cells produced four haploid progeny. In some of these tetrads, all the progeny were resistant to spectinomycin and sensitive to streptomycin. In others, all the progeny were resistant to streptomycin and sensitive to spectinomycin. In still others, at least one of the progeny was resistant to both antibiotics. Explain these different results.

19.11 If chloroplast DNA were inherited exclusively through the female gamete, would it ever be possible to observe recombination between chloroplast genes?

19.12 A mutant strain of yeast that produced tiny colonies when grown on glucose-rich medium was crossed to a wild-type strain. All the progeny were wild-type. What kind of mutation was responsible for the tiny colonies?

19.13 Another mutant yeast strain that produced tiny colonies was crossed to a wild-type strain. In this cross, however, all the progeny formed tiny colonies. What kind of mutation was responsible?

19.14 What are the main differences between plant and animal mtDNA molecules?

19.15 If a plant mitochondrial gene for NADH reductase were inserted into a plasmid and then placed into *E. coli* cells, would it produce catalytically active NADH reductase? Explain.

19.16 A mammalian mitochondrial gene has been fused with a bacterial gene by using recombinant DNA techniques. The fusion has been made so that the 3′ half of the mitochondrial gene (encoding 278 amino acids) is attached in frame to the 5′ half of the bacterial gene (encoding 200 amino acids). If this fusion gene is introduced into bacterial cells by transformation, will it produce a fusion polypeptide 478 amino acids long? Explain.

19.17 What implication does *trans*-splicing of mitochondrial RNA have for the one gene–one polypeptide concept?

19.18 Can both men and women develop Leber's hereditary optic neuropathy (LHON)? Explain.

19.19 The expression of hereditary cardiomyopathy in humans may be from mild to severe, with severe expression often resulting in premature death. The following pedigree shows the occurrence and severity of cardiomyopathy in a family. The severity of the disease is indicated by the letters m (mild), i (intermediate), and s (severe). (a) What type of inheritance is involved in cardiomyopathy? (b) What might explain the variable expressivity of cardiomyopathy in this pedigree?

19.20 During the study of an insect pest, a geneticist observes that four forms of an enzyme (A, B, C, and D) can be detected in the insects by biochemical analysis. However, individual insects show at most two different forms of this enzyme, and when two enzyme forms are present, they are equally abundant. The geneticist made a series of crosses between insects showing single forms of the enzyme. Then the F_1 progeny from these crosses were intercrossed to produce F_2 progeny. From the data that were collected from these crosses, what is the genetic basis for the different forms of the enzyme?

Cross	Mother	Father	F_1	F_2 (freq. in parentheses)
1	A	B	A, B	(1/4) A; (1/2) A, B; (1/4) B
2	B	A	A, B	(1/4) A; (1/2) A, B; (1/4) B
3	A	C	A	A
4	C	A	C	C
5	A	D	A, B	(1/4) A; (1/2) A, B; (1/4) B
6	D	A	C, D	(1/4) C; (1/2) C, D; (1/4) D
7	C	D	C, D	(1/4) C; (1/2) C, D; (1/4) D
8	D	C	C, D	(1/4) C; (1/2) C, D; (1/4) D
9	B	C	A, B	(1/4) A; (1/2) A, B; (1/4) B
10	C	B	C, D	(1/4) C; (1/2) C, D; (1/4) D
11	B	D	B	B
12	D	B	D	D

19.21 In plants, one subunit of the enzyme ribulose 1,5-bisphosphate carboxylase is encoded by a nuclear gene, and another is encoded by a chloroplast gene. Both subunits are needed for the enzyme to function in photosynthesis. What special problem does the separation of genes in different cellular compartments pose for the synthesis of this important protein?

19.22 What facts suggest that chloroplasts and mitochondria are derived from bacterial cells?

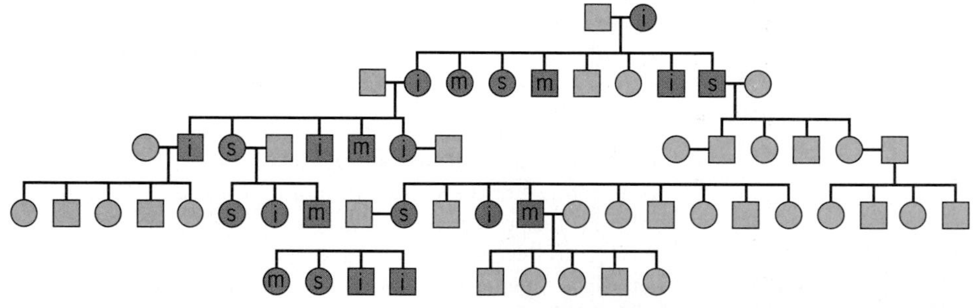

BIBLIOGRAPHY

AVISE, J. C. 1991. Ten unorthodox perspectives on evolution prompted by comparative population genetic findings on mitochondrial DNA. *Ann. Rev. Genet.* 25:45–69.

MARGULIS, L. 1971. Symbiosis and evolution. *Sci. Amer.* 225(2):49–57.

PALMER, J. D. 1985. Comparative organization of chloroplast genomes. *Ann. Rev. Genet.* 19:325–354.

SAGER, R. 1972. *Cytoplasmic Genes and Organelles.* Academic Press, New York.

SCHUSTER, W., AND A. BRENNICKE. 1994. The plant mitochondrial genome: Physical structure, information content, RNA editing and gene migration to the nucleus. *Ann. Rev. Plant Physiol. Plant Mol. Biol.* 45:61–78.

WALBOT, V. 1991. RNA editing fixes problems in plant mitochondrial transcripts. *Trends in Genetics.* 7:37–39.

WALLACE, D. C. 1993. Mitochondrial diseases: Genotype versus phenotype. *Trends in Genetics.* 9:128–133.

WISSINGER, B., A. BRENNICKE, AND W. SCHUSTER. 1992. Regenerating good sense: RNA editing and *trans* splicing in plant mitochondria. *Trends in Genetics.* 8:322–328.

A CONVERSATION WITH

NANCY WEXLER

Dr. Wexler, we appreciate this opportunity to discuss the personal and scientific dimensions of Huntington's disease with you. What inspired you to enter the field of genetics, and specifically neurogenetics?

Well, I was sort of yanked into the field by my hair. I know that everybody has a choice, but it really surprised me that I would become as intimately involved in this field as I am. My mother actually was a geneticist and was one of the early workers in *Drosophila* genetics. She worked in Thomas Hunt Morgan's famous fly room at Columbia. She came from a very unusual family. Her father came to the United States from a small village in Russia in the early 1800s. He was a garment salesman but eventually had to stop working because he developed Huntington's disease. Mother's three older brothers put her through college and graduate school. She got a Master's degree in genetics. So, when I was growing up, she would try to educate my sister Alice and me in genetics, but you don't usually listen to your mother at that age. But then she brought me into genetics in a different way, a way that I didn't anticipate. We discovered that she had Huntington's disease, just like her father.

Would you mind describing the circumstances of this shocking discovery for us?

Well, I knew all the time that I was growing up that there was something wrong with her, but I really didn't understand what it was. She was frequently depressed and became increasingly withdrawn. She had been, so I was told, a very lively, energetic person. But she became quiet and timid. We had moved from New York to California by this time, and she decided to go back to teaching high school. She needed to get certified, so she went to UCLA. For a while, she and I were hanging out at UCLA at the same time, she as a college student and I as a high school student participating in a special college program.

Things got worse all the time I was in college. I was taking a year off, I had a Fulbright Fellowship, and it was during this time she was diagnosed with Huntington's. But it was a very long odyssey up to that point. I was 21 at the time; she was 53. She died ten years later. We later learned that all of her brothers died of Huntington's disease. All three of them were diagnosed with Huntington's within the same week. Imagine, within a one-week period, my mother learned that all three of her brothers were dying, that she had a 50–50 risk of getting Huntington's, and that her two daughters were at risk. It was a quintuple whammy in her life.

How do you cope with the knowledge that you and your sister Alice could have inherited the Huntington's disease allele?

Our first reaction upon learning that we were at risk for inheriting Huntington's disease was one of horror. It's not anything you would ever expect. It feels like a total invasion of everything that you are. It feels like it's taking away everything that you wanted to be in the future. When you are in high school or college, the thought of getting se-

Nancy Wexler has made it her life's goal to find a treatment and cure for Huntington's disease (HD), a degenerative neurological disorder. She devotes her research to human genetics, specifically HD. Her personal experiences coupled with scientific discoveries have made Dr. Wexler the voice behind the Huntington's disease research. In the 1980s, her collaborative effort with other scientists led to the localization of the HD gene. One of her most phenomenal achievements is her work with several villages in Venezuela, where she and her team have put together a family tree of over 16,000 people—most living and under the age of 40—and *all* descendants of a woman appropriately named "Concepción." There are over 4653 adults and 73 children with symptoms of the disease and another 2600 children who are presymptomatic but will eventually die of the disease unless a cure is found in time. Currently, she is President of the Hereditary Disease Foundation and Higgins Professor of Neuropsychology at Columbia University.

riously ill or of dying . . . well, that's not on your radar. It's not what you think about. Suddenly, you feel as if you have been turned into a very old person. At this point, my sister and I decided not to have children.

Even though we both knew that this was a disease that tended to come on in your thirties and forties, we suddenly became extremely aware of every single little movement that we would make. I was at a Senate hearing on Huntington's disease, and I was really nervous. As I started the program, I spilled a glass of water all over everything, everybody. This was it. I was not only mortified and felt stupid, but in the back of my mind I was saying that now I have Huntington's and this is the first symptom and everyone knows it and is looking at me. So I become hypervigilant and very self-conscious. A person goes through a process of mourning. I know I did. I was depressed and irritable. But it was important to go through this phase. People just can't pretend that nothing has happened.

When people have a crisis in their life, if they don't allow themselves to mourn, they don't allow themselves to really recognize what has happened to them. It lasts much longer and it creeps out in other ways. For me, the turning-around point was when I started to fight it, started getting involved with it. At this point in time, there was not an awful lot going on in the science of Huntington's disease. I started meeting other families with Huntington's, and my father, who is a psychologist and really my role model in life, started the Hereditary Disease Foundation. So right away we started interacting with scientists. The science was so enthralling, so fascinating, so intriguing; and the people I met were and still are my closest friends; but this personal and scientific involvment in Huntington's disease was a life saver for me. It turned everything around.

What was your father's role in establishing the foundation?

My father came up with a brilliant idea that has been adopted by many other groups looking for cures for other diseases. The idea was an interdisciplinary workshop. At most scientific meetings, people stand up and present slides and everybody goes to sleep. You fall asleep because you spent the previous evening sitting at a bar, talking to people about the really interesting stuff in science. My father said, why not bring the bar to the workshop and forget the slide show. That is exactly what we did. We brought in 15 to 18 young people who were right in the midst of various research projects, graduate students, undergrads, post-docs, young instructors, and just got them thinking about the problem. It worked spectacularly. It has been an impetus for a tremendous amount of research on Huntington's disease, including the isolation of the gene.

The paper announcing the discovery of the Huntington's disease gene was authored by the Huntington's Collaborative Research Group. Were you pleased about this cooperative venture?

This was really a spectacular experience. If I had to choose whether I'd rather be at risk for Huntington's disease and know those people and have this experience or not, I'm not sure that I could make that choice. But the fact that you can think about it as choice is really wonderful. The experience I've gotten and the fun that I've had, it's been incredibly joyous and fantastically interesting. There have been frustrations, of course, but that goes with the terrritory. This was incredible! I wish the story could get across more. So many people think that science is so competitive and cut-throat, but more scientists are realizing that collaborations are essential because they cannot do it all by themselves and it's more fun to work together.

How do you feel about the future of science and your role in this future?

In a way, I wish I was starting all over again. This is such an exciting time. The intersection of genetics and medicine and biology is the future. What will we learn about human nature, how will this affect morality, what kind of world will open up to us?

We need to consider the challenges we face. What kind of people are we, and what kind of a future do we want to create for ourselves? This is really an extraordinary time, a unique epoch all made possible by discoveries of DNA's structure and function. It's a growing time. It's a good time to go to the bar and learn more.

The Little People of America at an athletic competition in California. Their short stature often results from a human growth hormone deficiency.

20

The Techniques of Molecular Genetics

CHAPTER OUTLINE

Basic Techniques Used to Clone Genes
Construction and Screening of DNA Libraries
The Manipulation of Cloned DNA Sequences *In Vitro*
The Molecular Analysis of DNA, RNA, and Protein
The Molecular Analysis of Genes and Chromosomes

Treatment of Pituitary Dwarfism with Human Growth Hormone

Kathy was a typical child in most respects—happy, playful, a bit mischievous, and intelligent. Indeed, the only thing unusual about Kathy was her small stature. She was born with pituitary dwarfism, which results from a deficiency of human growth hormone (HGH). Kathy seemed destined to remain abnormally small throughout her life. Then, at age 10, Kathy began receiving treatments of HGH synthesized in bacteria. She grew five inches during her first year of treatments. By continuing to receive HGH during maturation, Kathy reached the short end of the normal height distribution for adults. Without these treatments, she would have remained abnormally small in stature.

The HGH that allowed Kathy to grow to near-normal size was one of the first products of genetic engineering, the use of designed or modified genes to synthesize desired products. HGH was initially produced in *E. coli* cells harboring a modified gene composed of the coding sequence for HGH fused to synthetic bacterial regulatory elements. This chimeric gene was constructed *in vitro* and introduced into *E. coli* by transformation. In 1985, HGH produced in *E. coli* became the second pharmaceutical product of genetic engineering to be approved for use in humans by the U.S. Food and Drug Administration. Human insulin, which was the first such product, was approved in 1982 (see Human Genetics Sidelight: Treatment of Diabetes Mellitus with Human Insulin Produced in Bacteria).

How do scientists construct a gene that will produce HGH or human insulin in *E. coli?* If they do engineer this gene *in vitro,* how can they introduce it into living cells? In this chapter, we focus on the powerful tools of molecular genetics that allow researchers to construct and express such genes.

Much of what is known about the structure of genes has been obtained by molecular analyses through the application of **recombinant DNA technologies.** Recombinant DNA approaches begin with the **cloning** of the gene(s) of interest. The cloning of a gene involves its isolation and insertion into a small self-replicating genetic element such as a plasmid or viral chromosome. This small self-replicating genetic element is referred to as the **cloning vector.**

Once a gene has been cloned, it can be subjected to a whole array of manipulations that allow investigations of gene structure–function relationships. Usually, a cloned gene is sequenced; that is, the nucleotide-pair sequence of the gene is determined. If the function of the gene is unknown, its nucleotide sequence can be compared with thousands of gene sequences stored in three large computer gene banks, one in Germany, a second in Japan, and a third in the United States. Sometimes the function of a gene can be deduced based on its similarity to other genes whose functions are known. Given the nucleotide sequence of a gene and knowledge of the genetic code, the amino acid sequence of the polypeptide encoded by the gene can be predicted. The predicted amino acid sequence of the polypeptide can then be searched for functional domains that may provide clues as to its function. Nucleic acid and protein sequence databases have become important resources for research in molecular genetics, and they will become increasingly important in both basic biological research and diverse applications of this research (Chapters 21 and 22).

BASIC TECHNIQUES USED TO CLONE GENES

The haploid genome of a mammal contains about 3×10^9 nucleotide pairs. If the average gene is 3000 nucleotide pairs long (many are larger), each gene will represent one of a million such sequences in the genome. Thus, isolating any one gene is like searching for the proverbial needle in a haystack. Most techniques used in the analysis of genes and other DNA sequences require that the sequence be available in significant quantities in pure or essentially pure form. How can one identify the segment of a DNA molecule that carries a single gene and isolate enough of this sequence in pure form to permit molecular analyses of its structure and function?

The development of recombinant DNA and gene-cloning technologies has provided molecular geneticists with methods by which genes or other segments of large chromosomes can be isolated, replicated, and studied by nucleic acid sequencing techniques, electron microscopy, and other analytical techniques. The **gene-cloning** procedure involves two essential steps: (1) the incorporation of the gene of interest into a small self-replicating chromosome, and (2) the amplification of the recombinant minichromosome by its replication in an appropriate host cell. Step 1 involves the joining of two or more different DNA molecules *in vitro* to produce **recombinant DNA molecules,** for example, a human gene fused to an *E. coli* origin of replication. Step 2 is really the gene-cloning event in which the recombinant DNA molecule is replicated or "cloned" to produce many identical copies for subsequent biochemical analysis. Thus, although the entire procedure is often referred to as the recombinant DNA or gene-cloning technique, these terms actually refer to two separate steps in the overall process.

The Discovery of Restriction Endonucleases

The ability to clone and sequence essentially any gene or other DNA sequence of interest from any species depends on a special class of enzymes called **restriction endonucleases** (from the Greek term *éndon* meaning "within"; endonucleases make internal cuts in DNA molecules). Many endonucleases make random cuts in DNA, but the restriction endonucleases are site-specific. They cleave DNA molecules only at specific nucleotide sequences called **restriction sites.** Different restriction endonucleases are produced by different microorganisms and recognize different nucleotide sequences in DNA (Table 20.1). The restriction endonucleases are named by using the first letter of the genus and the first two letters of the species that produces the enzyme. If an enzyme is produced only by a specific strain, a letter designating the strain is appended to the name. The first restriction enzyme identified from a bacterial strain is designated I, the second II, and so on. Thus, restriction endonuclease ***Eco*RI** is produced by *Escherichia coli* strain RY13. About 400 different restriction enzymes have been characterized and purified; thus, restriction endonucleases that cleave DNA molecules at many different DNA sequences are available.

Restriction endonucleases were discovered in 1970 by Hamilton Smith and Daniel Nathans. Smith and Nathans shared the 1986 Nobel Prize in Physiology or Medicine with Werner Arber, who carried out pioneering research that led to the discovery of restriction

TABLE 20.1
Recognition Sequences and Cleavage Sites of Representative Restriction Endonucleases

Enzyme	*Source*	*Recognition Sequence*[a] *and Cleavage Sites*[b]	*Number of Recognition Sequences per Chromosome of*	
			Phage λ	*SV40 Virus*
*Eco*RI	*Escherichia coli strain* RY13	G↓AA·TTC CTT·AA↑G	5	1
*Hinc*II	*Haemophilus influenzae* strain R_c	GTPy↓·PuAC CAPu·↑PyTG	34	7
*Hind*III	*Haemophilus influenzae* strain R_d	A↓AG·CTT TTC·GA↑A	6	6
*Pst*I	*Providencia stuartii*	CTG·CA↓G G↑AC·GTC	28	2
*Hpa*II	*Haemophilus parainfluenzae*	C↓C·GG GG·C↑C	750	1
*Alu*I	*Arthobacter luteus*	AG↓·CT TC·↑GA	143	34

[a]The axis of symmetry in each palindromic recognition sequence is indicated by the red dot. Pu indicates that either purine (adenine or guanine) may be present at this position; Py indicates that either pyrimidine (thymine or cytosine) may be present.

[b]The position of each bond cleaved is indicated by an arrow. Note that the cuts are staggered (at different positions in the two complementary strands) with some restriction endonucleases.

enzymes. In the 1960s, Arber and colleagues demonstrated that bacteriophage DNA is often restricted from replicating in one strain of *E. coli* if the DNA was synthesized and packaged in phage particles in a different strain of *E. coli*. It was later shown that the phage DNA is restricted in the new host because it is degraded by a restriction endonuclease produced by the bacterial cell. The biological function of restriction endonucleases is to protect the genetic material of bacteria from "invasion" by foreign DNAs, such as viral DNAs. As a result, restriction endonucleases are sometimes referred to as the immune systems of prokaryotes.

All cleavage sites in the DNA of an organism must be protected from cleavage by the organism's own restriction endonucleases; otherwise the organism would commit suicide by degrading its own DNA. In many cases, this protection of endogenous cleavage sites is accomplished by **methylation** of one or more nucleotides in each nucleotide sequence that is recognized by the organism's own restriction endonuclease. Methylation occurs rapidly after replication, catalyzed by site-specific methylases produced by the organism. Each restriction endonuclease will cleave a foreign DNA molecule (a DNA molecule from another species) into a fixed number of fragments, the number depending on the number of restriction sites in the particular DNA molecule (Table 20.1).

An interesting feature of restriction endonucleases is that they commonly recognize DNA sequences that are **palindromes**—that is, nucleotide-pair sequences that read the same forward or backward from a central axis of symmetry, as in the nonsense phrase

⟵ ⟶
AND MADAM DNA

In addition, a useful feature of many restriction nucleases is that they make staggered cuts; that is, they cleave the two strands of a double helix at different points. (Other restriction endonucleases cut both strands at the

HUMAN GENETICS SIDELIGHT

Treatment of Diabetes Mellitus with Human Insulin Produced in Bacteria

Diabetes mellitus is the third leading cause of death in the United States; only heart disease and cancer cause more deaths. Over 2 million individuals have severe cases of this disease in the United States alone. In diabetics, the polypeptide hormone insulin is either not produced in adequate amounts or, if produced, does not stimulate the appropriate responses in muscle, liver, and adipose (fat storage) cells. The genetics of diabetes is complex; many genes, most of which appear to be autosomal recessive, contribute to a predisposition to the disease.

About 2 million diabetics in the United States require daily injections of insulin. Prior to 1982, this insulin was isolated from the pancreatic glands of cattle or pigs obtained from meat-packing plants. However, about 5 percent of humans with diabetes are allergic to these insulins, which differ slightly from human insulin. These individuals required injections of insulin isolated from other animals or from human cadavers—expensive treatments in either case. In 1982, human insulin produced in genetically engineered bacteria was approved for use in humans. Diabetics who were allergic to insulin from cattle and pigs were thus freed from their dependence upon the limited supply of human insulin obtained from cadavers.

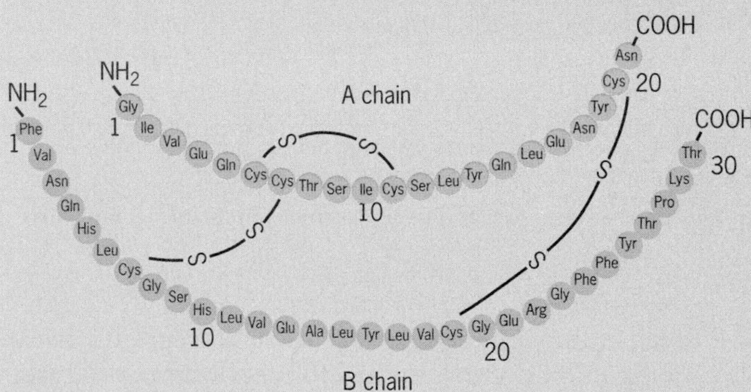

Figure 1 Structure of human insulin.

Insulin is a small peptide hormone composed of two chains joined by disulfide bonds (Figure 1). The A and B chains are 21 and 30 amino acids long, respectively. These two chains are synthesized as part of a larger gene product called preproinsulin. An amino-terminal signal sequence is excised during the secretion of preproinsulin from β cells, and a central region (the C chain) is subsequently removed by proteolytic cleavage. The primary secretion product, proinsulin, is converted to insulin by this proteolytic excision of the C chain and the formation of a third disulfide bond as shown in Figure 2.

Human insulin was initially produced in *E. coli* cells by recombinant DNA techniques in 1979. The DNA sequences encoding the A and B chains of human insulin were chemically synthesized and joined to the *E. coli* gene encoding β-galactosidase. The chimeric genes were inserted into plasmid pBR322 (Figure 20.4) and introduced into *E. coli* cells by transformation. The fusion proteins specified by these chimeric genes contained a methionine residue between the bacterial β-galactosidase and the human insulin portions. This facilitated the removal of the insulin chain from the fusion protein by treatment with cyanogen bromide (which cleaves peptide chains on the carboxyl sides of methionine residues). This procedure is illustrated in Figure 3.

The β-galactosidase/human insulin fusion proteins encoded by the chimeric genes were extracted, purified, and cleaved with cyanogen bromide. The A and B chains of human insulin were then purified, mixed, reduced, and reoxidized to form the disulfide bonds present in native insulin. The human insulin produced by this procedure was identical to human insulin produced *in vivo*. Today, human insulin produced by recombinant DNA techniques is still widely used by diabetics, but this insulin is produced in animal cells (for example, Chinese hamster cells) growing in culture.

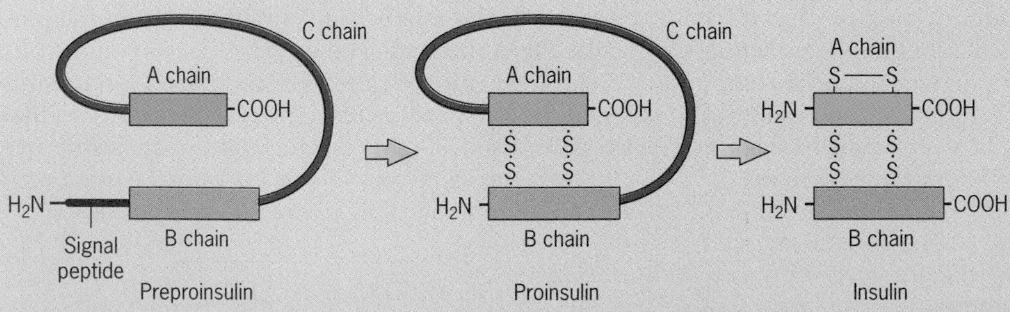

Figure 2 Conversion of preproinsulin to insulin.

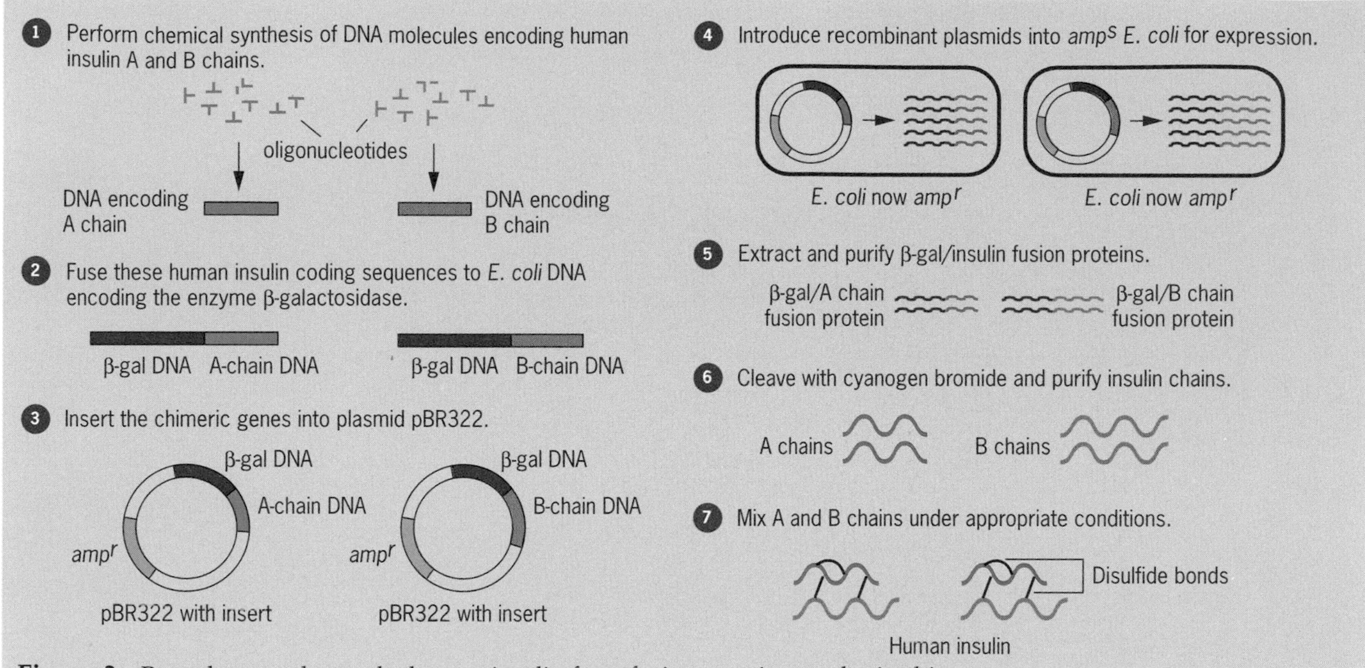

Figure 3 Procedure used to make human insulin from fusion proteins synthesized in *E. coli.*

same place and produce blunt-ended fragments.) Because of the palindromic nature of the restriction sites, the staggered cuts produce segments of DNA with complementary single-stranded ends. For example, cleaving a DNA molecule of the following type:

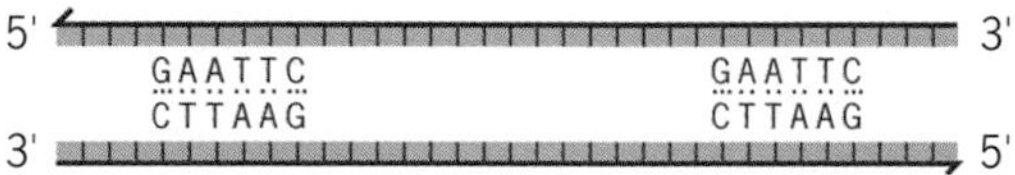

with the restriction endonuclease *Eco*RI will yield

Because all the resulting DNA fragments will have complementary single-stranded termini, they can be rejoined under the appropriate renaturation conditions by using the enzyme **DNA ligase** to reform the missing phosphodiester linkage in each strand. Thus, DNA molecules can be cut into pieces, called **restriction fragments,** and the pieces can be joined together again with DNA ligase, almost at will.

The Production of Recombinant DNA Molecules *In Vitro*

A restriction endonuclease catalyzes the cleavage of a specific nucleotide-pair sequence regardless of the source of the DNA. It will cleave phage DNA, *E. coli* DNA, corn DNA, human DNA, or any other DNA, as long as the DNA contains the nucleotide sequence that it recognizes. Thus, restriction endonuclease *Eco*RI will produce fragments with the same complementary single-stranded ends, AATT, regardless of the source of DNA, and two *Eco*RI fragments can be covalently fused regardless of their origin; that is, an *Eco*RI fragment from human DNA can be joined to an *Eco*RI fragment from *E. coli* DNA just as easily as two *Eco*RI fragments from *E. coli* DNA or two *Eco*RI fragments from human DNA can be joined. A DNA molecule of the type shown in Figure 20.1, containing DNA fragments from two different sources, is referred to as a recombinant DNA molecule. The ability of geneticists to construct such recombinant DNA molecules at will is the basis of the recombinant DNA technology that has revolutionized molecular biology in the last three decades.

The first recombinant DNA molecules were produced in Paul Berg's laboratory at Stanford University in 1972. Berg's research team constructed recombinant DNA molecules that contained phage lambda genes inserted into the small circular DNA molecule of simian virus 40 (SV40). In 1980, Berg was a co-recipient of the Nobel Prize in Chemistry as a result of this accomplishment. Shortly thereafter, Stanley Cohen and colleagues, also at Stanford, inserted an *Eco*RI restriction fragment from one DNA molecule into the

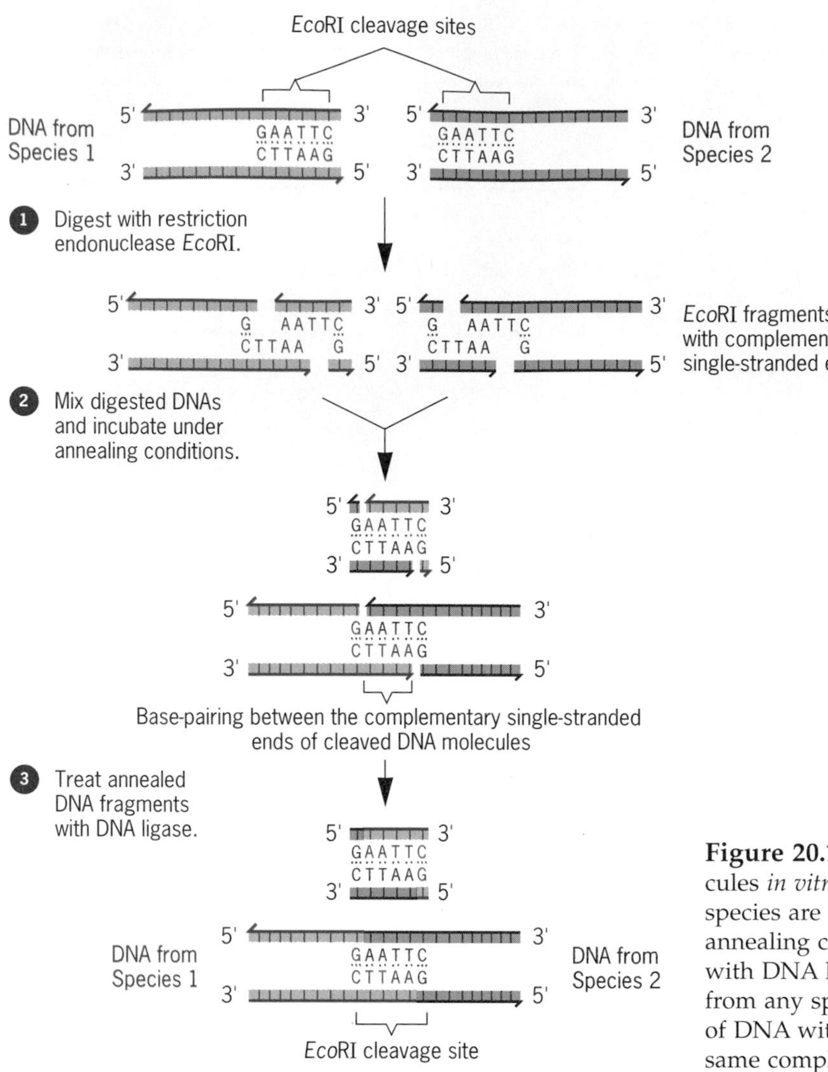

Figure 20.1 The construction of recombinant DNA molecules *in vitro.* DNA molecules isolated from two different species are cleaved with a restriction enzyme, mixed under annealing conditions, and covalently joined by treatment with DNA ligase. The DNA molecules can be obtained from any species—animal, plant, or microbe. The digestion of DNA with the restriction enzyme *Eco*RI produces the same complementary single-stranded AATT ends regardless of the source of the DNA.

cleaved, unique *Eco*RI restriction site of a circular, self-replicating plasmid. When this recombinant plasmid was introduced into *E. coli* cells by transformation, it exhibited normal autonomous replication, just like the original plasmid.

Amplification of Recombinant DNA Molecules in Cloning Vectors

The various applications of recombinant DNA techniques require not only the construction of recombinant DNA molecules, as shown in Figure 20.1, but also the **amplification** of these recombinant molecules, that is, the production of many copies or **clones** of these molecules. This is accomplished by making sure that one of the parental DNAs incorporated into the recombinant DNA molecule is capable of self-replication. In practice, the gene or DNA sequence of interest is inserted into a specially chosen cloning vector. Most of the commonly used cloning vectors have been derived from viral chromosomes (Chapter 16) or plasmids (Chapter 17).

A cloning vector has three essential components: (1) an origin of replication, (2) a **dominant selectable marker gene,** usually a gene that confers drug resistance to the host cell, and (3) at least one **unique restriction endonuclease cleavage site**—a cleavage site that is present only once in the DNA of the vector (Figure 20.2). For maximum usefulness, a cloning vector should contain unique cleavage sites for several different restriction enzymes. The currently used cloning vectors contain a cluster of unique restriction sites called a **polylinker** or **polycloning site** (Figure 20.3).

Plasmid vectors: Plasmids are extrachromosomal, double-stranded circular molecules of DNA present in microorganisms, especially bacteria (Chapter 17). They

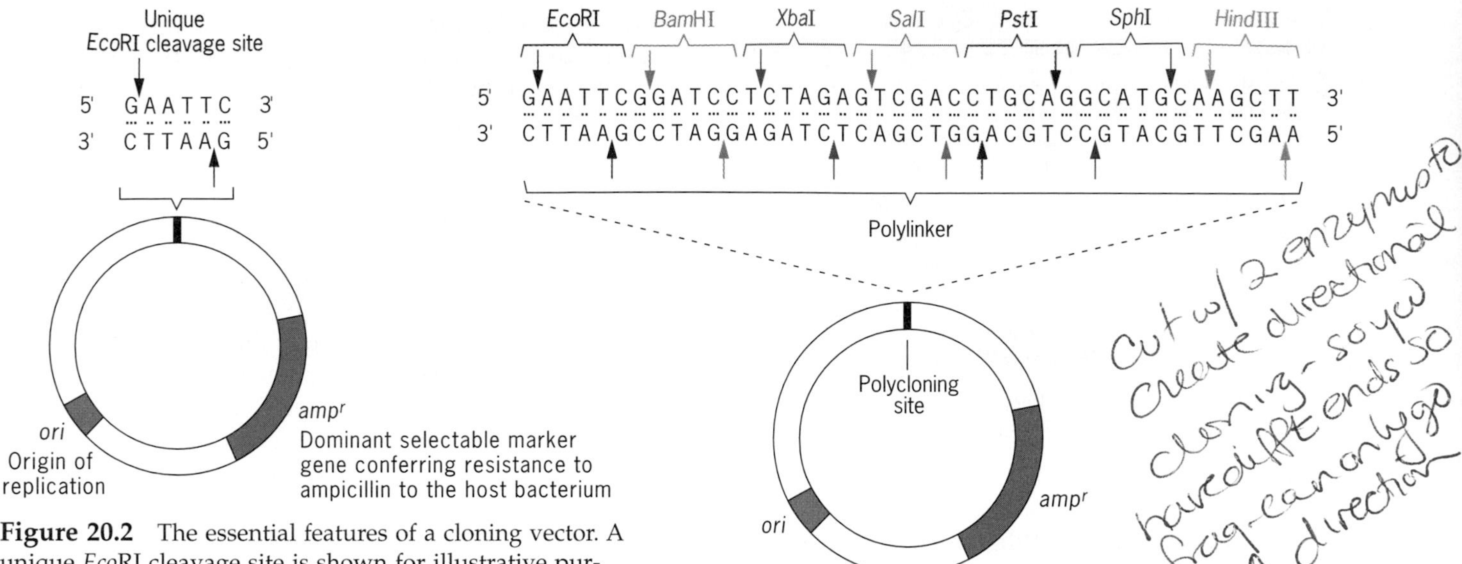

Figure 20.2 The essential features of a cloning vector. A unique *Eco*RI cleavage site is shown for illustrative purposes, but a unique cleavage site for any of a large number of different restriction endonucleases is just as appropriate.

Figure 20.3 Structure of a polycloning site in a cloning vector. The phosphodiester bonds cleaved by the restriction endonucleases are indicated by arrows. The restriction endonucleases shown are produced by the following microorganisms: *Eco*RI, *Escherichia coli* strain RY13; *Bam*HI, *Bacillus amyloliquefaciens* strain H; *Xba*I, *Xanthomonas badrii; Sal*I, *Streptomyces albus; Pst*I, *Providencia stuartii; Sph*I, *Streptomyces phaeochromogenes;* and *Hin*dIII, *Haemophilus influenzae* strain R_d.

range from about 1 kb (1 kilobase = 1000 base pairs) to over 200 kb in size and replicate autonomously. Long before the recombinant DNA revolution began, some bacterial plasmids were known to carry antibiotic-resistance genes, which are ideal dominant selectable markers. Thus, their use as cloning vectors was predictable. The first plasmid to be used as a cloning vector was *E. coli* plasmid pSC101, which contains a single *Eco*RI cleavage site and a tetracycline-resistance (*tet*^{r}) gene. However, more versatile plasmid vectors were soon developed. Plasmid pBR322 was one of the first widely used cloning vectors; it contains both ampicillin- and tetracycline-resistance genes and a number of unique restriction enzyme cleavage sites (Figure 20.4). Many of the second- and third-generation cloning vectors in use today were derived, at least in part, from plasmid pBR322.

Bacteriophage vectors: Most bacteriophage cloning vectors have been constructed from the bacteriophage λ chromosome. The complete 48,502 nucleotide-pair sequence of the wild-type lambda genome is known, along with the functions of all of its genes. The central one-third of the λ chromosome contains genes that are required for lysogeny (Chapter 17), but not for lytic growth. Thus, the central part (about 15 kb in length) of the λ chromosome can be excised with restriction enzymes and replaced with foreign DNA (Figure 20.5). The resulting recombinant DNA molecules can be packaged in phage heads *in vitro*. The phage particles can inject the recombinant DNA molecules into *E. coli* cells, where they will replicate and produce clones of the recombinant DNA molecules. DNA molecules that are too large or too small cannot be packaged in the lambda head; only molecules 45 to 50 kb in size are packaged. As a result, the lambda cloning vectors can only accommodate inserts of 10 to 15 kb.

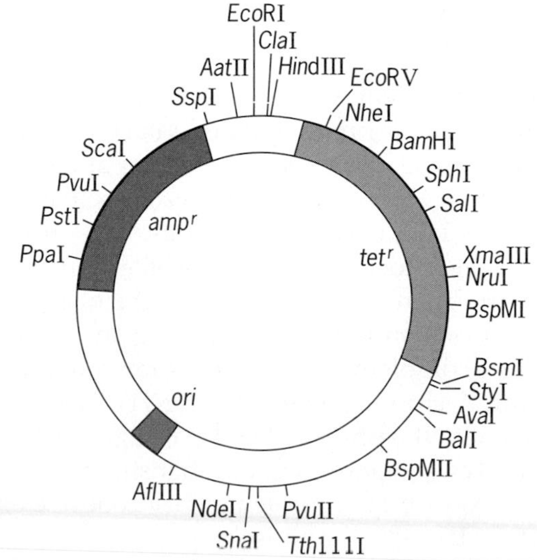

Figure 20.4 Structure of *E. coli* plasmid cloning vector pBR322, a circular DNA molecule 4.36 kb in size. The locations of unique restriction enzyme cleavage sites, the origin (*ori*) of replication, and the genes that confer resistance to the antibiotics ampicillin (*amp*^{r}) and tetracycline (*tet*^{r}) are shown on the map of the plasmid DNA molecule.

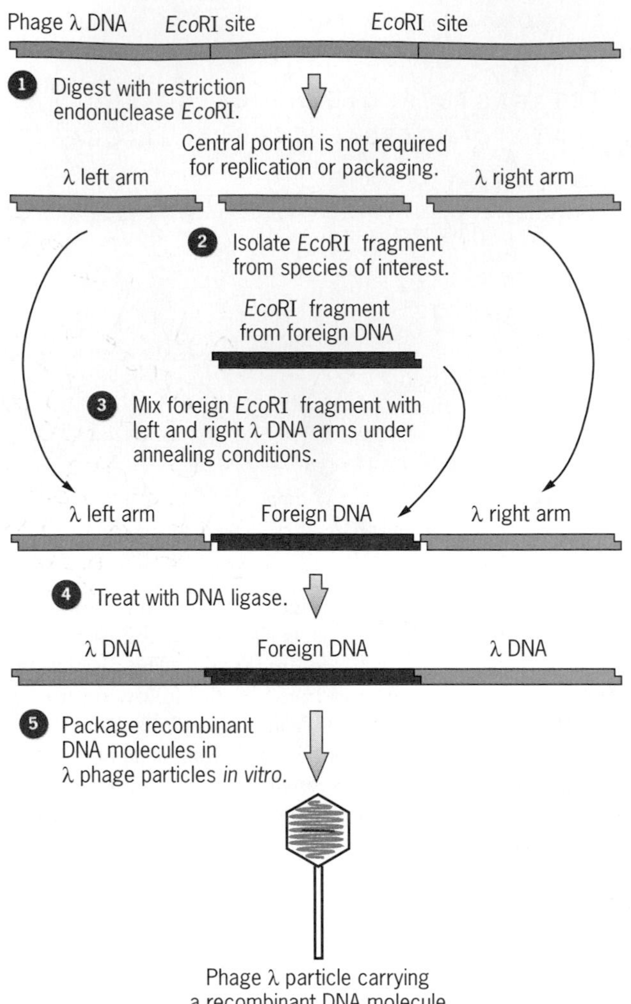

Figure 20.5 Strategy employed in using phage λ as a cloning vector.

Cosmid vectors: Some eukaryotic genes are larger than 15 kb in size and cannot be cloned intact in either plasmid or lambda cloning vectors. For this and other reasons, scientists have developed vectors that can accommodate larger DNA insertions. The first such vectors, called **cosmids** (for λ *cos* site and plas*mid*), were hybrids between plasmids and the phage λ chromosome. *Cos* stands for *co*hesive *s*ite, in reference to the 12-base complementary single-stranded termini on the mature λ chromosome (Chapter 11). The *cos* site is recognized by the phage λ DNA packaging apparatus, which makes staggered cuts at this site during packaging to produce the complementary cohesive ends of the mature lambda chromosome.

Cosmids combine the key advantages of plasmid and λ phage vectors; they possess (1) the plasmid's ability to replicate autonomously in *E. coli* cells and (2) the *in vitro* packaging capacity of the λ chromosome. The cosmid vectors (Figure 20.6) carry the origin of replication and one of the antibiotic-resistance genes of the parental plasmid and the λ *cos* site, which is needed to package DNA in λ heads, from the phage λ chromosome. The key advantage of cosmid vectors is their ability to accept foreign DNA inserts of 35 to 45 kb.

Eukaryotic and shuttle vectors: The plasmid, phage λ, and cosmid cloning vectors all replicate in *E. coli* cells. Because various taxonomic groups utilize distinct origins of replication and regulatory signals, different cloning vectors must be used in different species. Thus, special cloning vectors have been developed to introduce recombinant DNA molecules into other prokaryotes and into eukaryotic organisms. Many unique cloning vectors are available for use in *S. cerevisiae, D. melanogaster,* mammals, plants, and other taxonomic groups.

E. coli is the host cell of choice for many of the manipulations performed on cloned DNAs. Thus, some of the most useful cloning vectors are **shuttle vectors** that can replicate in both *E. coli* and another species, such as a eukaryotic organism. Shuttle vectors designed for use in the yeast *S. cerevisiae* contain both *E. coli* and *S. cerevisiae* origins of replication and selectable marker genes, along with a polycloning site (Figure 20.7). Such shuttle vectors are extremely useful for genetic dissections. A yeast gene can be cloned in a shuttle vector, subjected to site-specific mutagenesis in an *E. coli* vector, and then moved back to yeast to examine the effects of the induced modifications in native host cells. Similar shuttle vectors are available for use in *E. coli* and various animal systems.

Artificial chromosomes: Some eukaryotic genes are very large. For example, the human dystrophin (an elongated protein that links filaments to membranes in muscle cells) gene is over 2000 kb in length. With the goal of cloning large segments of chromosomes, researchers have worked on the development of vectors that accept DNA sequences longer than the 35 to 45 kb inserts of cosmid vectors. This research led to the de-

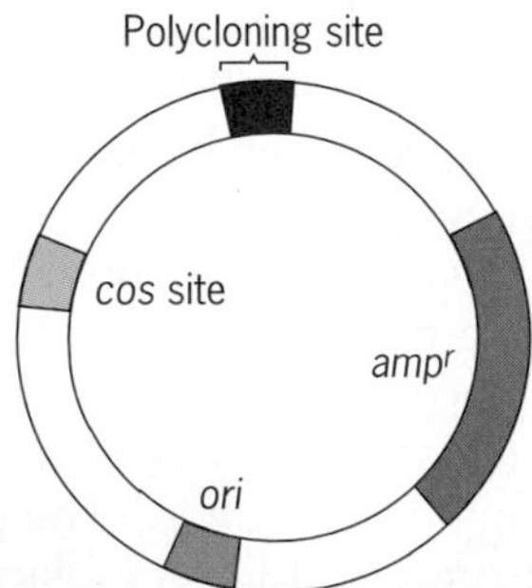

Figure 20.6 Structure of a typical cosmid cloning vector. Cosmids combine important features of phage λ and plasmid cloning vectors (see text).

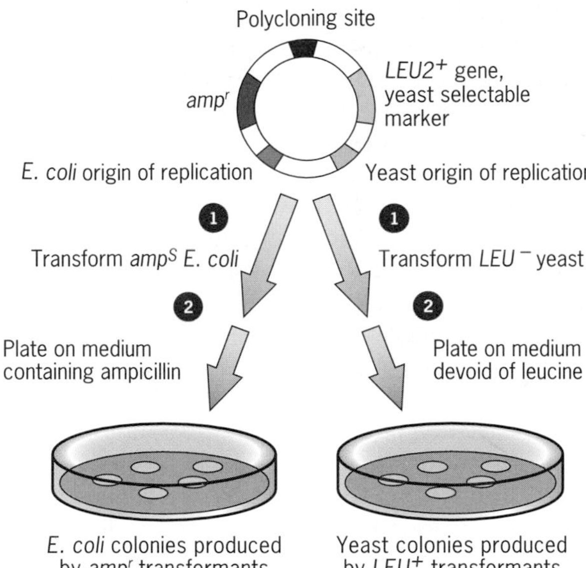

Figure 20.7 Basic structure and utility of an *E. coli*-yeast shuttle vector, which can replicate in both *E. coli* and *S. cerevisiae*. Shuttle vectors allow investigators to move genes back and forth between two organisms and study the structure or function of genes of interest in either host or in both hosts.

velopment of **yeast artificial chromosomes (YACs),** which can accommodate foreign DNA inserts of 200 to 500 kb. YAC vectors are genetically engineered yeast minichromosomes. They contain (1) a yeast origin of replication, (2) a yeast centromere, (3) two yeast telomeres, one at each end, (4) a selectable marker, and (5) a polycloning site (Figure 20.8). In yeast, origins of replication are called ***ARS*** (for *A*utonomously *R*eplicating *S*equence) elements. The selectable marker usually is a wild-type gene that confers prototrophy on the host cell. For example, $URA3^+$ can be used as the selectable marker in transformations of $URA3^-$ auxotrophs growing on medium lacking uracil.

YAC cloning vectors are especially valuable in investigations such as the Human Genome Project, which is designed to map, clone, and sequence large eukaryotic genomes (Chapter 21). The large DNA inserts make it much easier to identify and characterize sets of clones that collectively span an entire genome. If the average insert in cosmid vectors is 40 kb and the average insert in YAC vectors is 200 kb, five cosmid clones are needed to cover the region represented by one YAC clone. Because genome mapping projects require the isolation of overlapping clones, this difference in insert size translates into more than a fivefold difference in the effort required to obtain a complete physical map.

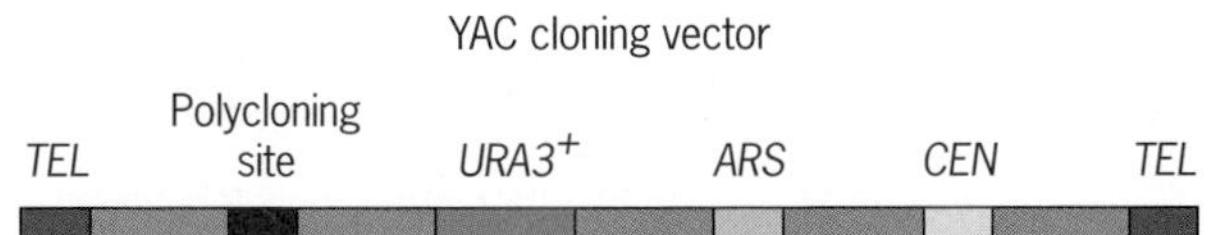

Figure 20.8 Structure of a YAC cloning vector. The components are: (1) *ARS*, autonomously replicating sequence (a yeast origin of replication), (2) *CEN*, a yeast centromere, (3) *TEL*, a yeast telomere, (4) $URA3^+$, a wild-type gene required for the biosynthesis of uracil, and (5) the polycloning site, containing the recognition and cleavage sites for a set of restriction endonucleases.

Bacterial artificial chromosomes (BACs), which have many of the advantages of YACs, have been constructed from bacterial fertility (F) factors. Like YACs, BACs accept large inserts. However, BACs are less complex and, thus, easier to construct. In addition, BACs replicate in *E. coli* like plasmid, lambda, and cosmid vectors. BAC vectors are therefore beginning to replace YAC vectors in studies where large inserts are needed, such as in the construction of physical maps of entire chromosomes.

Key Points: **Recombinant DNA and gene-cloning techniques allow scientists to isolate and characterize virtually any gene or other DNA sequence from any organism. These techniques became possible with the discovery of restriction endonucleases, enzymes that recognize and cleave DNA in a sequence-specific manner. DNA sequences of interest are inserted *in vitro* into small, self-replicating DNA molecules called cloning vectors, and the resulting recombinant DNA molecules are amplified by replication *in vivo* after being introduced into cells, usually bacteria, by transformation. A variety of cloning vectors have been constructed, each with advantages for certain research purposes.**

CONSTRUCTION AND SCREENING OF DNA LIBRARIES

The first step in cloning a gene from an organism usually involves the construction of a **genomic DNA library**—a set of DNA clones collectively containing the entire genome. Sometimes, individual chromosomes of an organism are isolated by a procedure that sorts chromosomes based on size and DNA content. The DNAs from the isolated chromosomes are then used to construct **chromosome-specific DNA libraries.** The availability of chromosome-specific DNA libraries facilitates the search for a gene that is known to reside on a particular chromosome, especially for organisms like humans with large genomes.

An alternative approach to gene cloning restricts the search for a gene to DNA sequences that are transcribed into mRNA copies. The RNA retroviruses

(Chapter 16) encode an enzyme called reverse transcriptase, which catalyzes the synthesis of DNA molecules complementary to single-stranded RNA templates. The DNA molecules that are synthesized from RNA templates are called **complementary DNAs (cDNAs).** They can be converted to double-stranded cDNA molecules with DNA polymerases (Chapter 11), and the double-stranded cDNAs can be cloned in plasmid or phage λ vectors. By starting with mRNA, geneticists are able to construct a **cDNA library** that contains only the coding regions of the expressed genes of an organism.

Construction of Genomic Libraries

Genomic DNA libraries are usually prepared by isolating total DNA from an organism, digesting the DNA with a restriction endonuclease, and inserting the restriction fragments into an appropriate cloning vector. Two different procedures are used to insert the DNA fragments into the cloning vector. If the restriction enzyme that is used makes staggered cuts in DNA, producing complementary single-stranded ends, the restriction fragments can be ligated directly into vector DNA molecules cut with the same enzyme (Figure 20.9). An advantage of this procedure is that the foreign DNA inserts can be precisely excised from the vector DNA by cleavage with the restriction endonuclease used to prepare the genomic DNA fragments for cloning.

If the restriction enzyme cuts both strands of DNA at the same position, producing blunt ends, or if fragments are produced by shearing DNA molecules, complementary single-stranded ends must be added to the DNA fragments *in vitro*. This is accomplished by us-

1 Isolate *E.coli* cosmid pJB8 DNA and mouse genomic DNA.

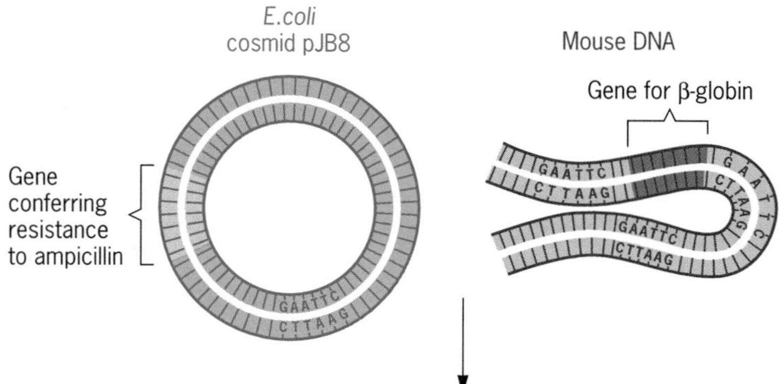

2 Cleave cosmid and mouse DNAs with restriction endonuclease *Eco*RI.

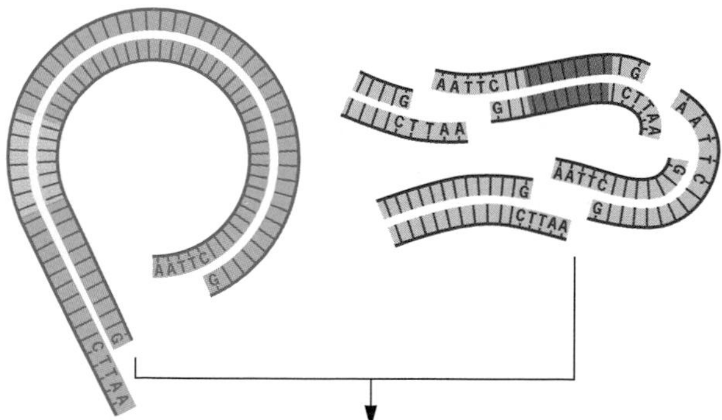

3 Mix cosmid and mouse DNAs under annealing conditions and treat with DNA ligase.

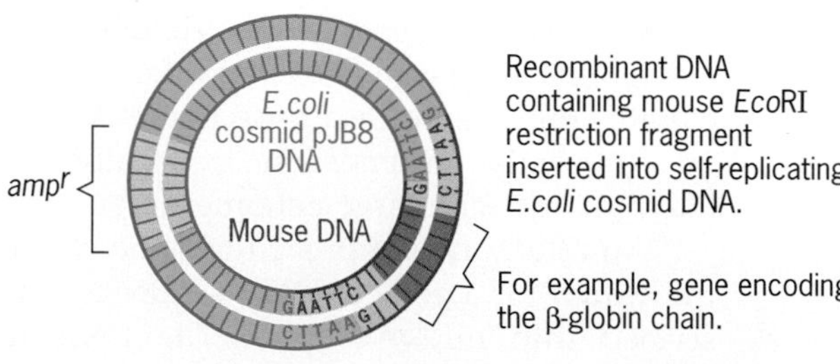

Figure 20.9 Procedure used to clone DNA restriction fragments with complementary single-stranded ends.

ing the enzyme **terminal transferase** to add nucleotides to the 3′ termini of the DNA strands after the 5′ ends are cut back with phage λ exonuclease. Usually, poly(A) tails are added to the cleaved vector DNA, and poly(T) tails are added to the genomic DNA fragments (Figure 20.10), or vice versa. Then, the T-tailed genomic DNA fragments are inserted into the A-tailed vector DNA molecules with DNA ligase. Since the T and A tails will not always be the same length, the *E. coli* enzymes exonuclease III and DNA polymerase I are used to cut back overhangs and fill in gaps, respectively. DNA ligase will only seal nicks between adjacent nucleotides; it will not add nucleotides if gaps are present.

Once the genomic DNA fragments are ligated into vector DNA, the recombinant DNA molecules must be introduced into host cells, usually *E. coli* cells, for amplification by replication *in vivo*. This step usually involves transforming antibiotic-sensitive recipient cells (Chapter 17). When *E. coli* is used, the bacteria must

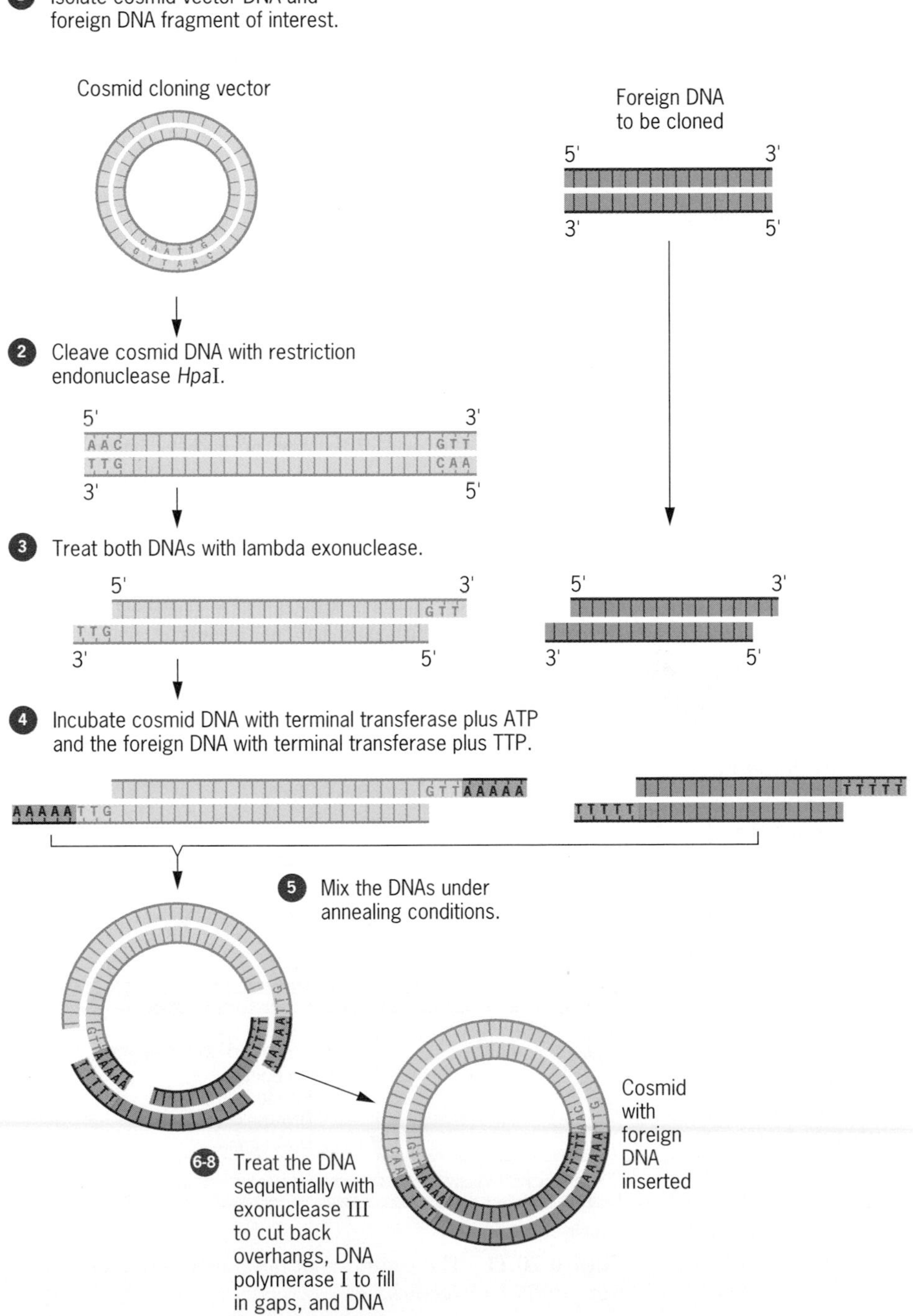

Figure 20.10 Terminal transferase procedure for cloning DNA fragments with blunt ends.

first be made permeable to DNA by treatment with a calcium salt or other chemicals. Transformed cells that harbor vector DNA are then selected by growing the cells under conditions where the selectable marker gene of the vector is essential for growth.

A good genomic DNA library contains essentially all of the DNA sequences in the genome of interest. For large genomes, complete libraries contain thousands to hundreds of thousands of different recombinant clones. If we assume that all DNA sequences have an equal chance of being inserted into a cloning vector, the number of clones required to produce a library that has a certain probability (P) of containing a given DNA sequence is given by the formula

$$N = \frac{ln\,(1-P)}{ln\,(1-f)}$$

where N is the required number of recombinant clones and f is the fraction of the genome present in a single clone of average size.

For example, suppose that you are constructing a cosmid library of the *Drosophila* genome, which is about 10^5 kb in size. If the average insert size is 40 kb, how many clones will be required to be 99 percent sure that a specific sequence is present in the library? The required number (N) of clones is 11,513, calculated by the equation

$$N = \frac{\ln\,[1-0.99]}{\ln\,[1-(40/100{,}000)]}$$

Construction of cDNA Libraries

Most of the DNA sequences present in the large genomes of higher animals and plants are not expressed. Thus, expressed DNA sequences can be identified more easily by working with complementary DNA (cDNA) libraries. Because most mRNA molecules contain 3′ poly(A) tails, poly(T) oligomers can be used to prime the synthesis of complementary DNA strands by reverse transcriptase (Figure 20.11). Then, the RNA-DNA duplexes are converted to double-stranded DNA molecules by the combined activities of ribonuclease H, DNA polymerase I, and DNA ligase. Ribonuclease H degrades the RNA template strand, and short RNA fragments produced during degradation serve as primers for DNA synthesis. DNA polymerase I catalyzes the synthesis of the second DNA strand and replaces RNA primers with DNA strands, and DNA ligase seals the remaining single-strand breaks in the double-stranded DNA molecules. These double-stranded cDNAs can be inserted into plasmid or phage λ cloning vectors by the terminal transferase procedure shown in Figure 20.10.

Screening DNA Libraries for Genes of Interest

The genomes of higher plants and animals are very large. For example, the human genome contains 3×10^9 nucleotide pairs. Thus, searching genomic DNA or cDNA libraries of multicellular eukaryotes for a specific gene or other DNA sequence of interest requires the identification of a single DNA sequence in a library that contains a million or more different sequences. The most powerful screening procedure is genetic selection: searching for a DNA sequence in the library that can restore the wild-type phenotype to a mutant organism. When genetic selection cannot be employed, more laborious molecular screens must be carried out. Molecular screens usually involve the use of DNA or RNA sequences as hybridization probes or the use of antibodies to identify gene products synthesized by cDNA clones. When genetic selection is not available, the identification of the desired clone often is more difficult than the construction of the genomic or cDNA library.

Genetic selection: The simplest procedure for identifying a clone of interest is the use of **genetic selection.** For example, the *Salmonella typhimurium* gene that confers resistance to penicillin can be easily cloned. A genomic library is constructed from the DNA of a *pen*^{r} strain of *S. typhimurium*. Penicillin-sensitive *E. coli* cells are transformed with the recombinant DNA clones in the library and are plated on medium containing penicillin. Only the transformed cells harboring the *pen*^{r} gene will be able to grow in the presence of penicillin.

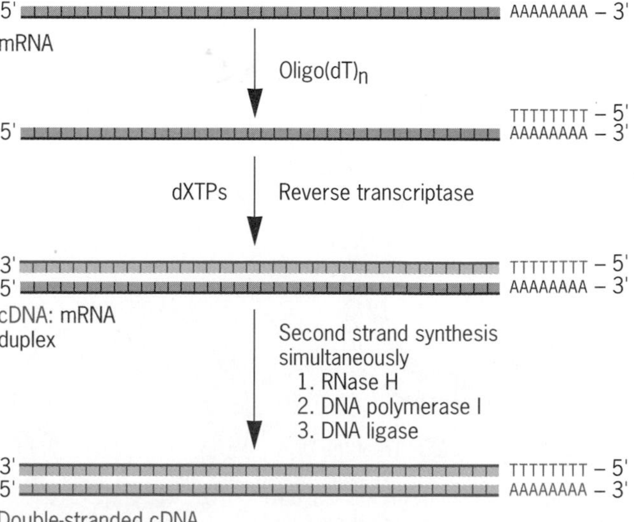

Figure 20.11 The synthesis of double-stranded cDNAs from mRNA molecules.

When mutations are available in the gene of interest, genetic selection can be based on the ability of the wild-type allele of a gene to restore the normal phenotype to a mutant organism. Although this type of selection is called **complementation screening,** it really depends on the dominance of wild-type alleles over mutant alleles that encode inactive products. For example, the genes of *S. cerevisiae* that encode histidine-biosynthetic enzymes were cloned by transforming *E. coli* histidine auxotrophs with yeast cDNA clones and selecting transformed cells that could grow on histidine-free medium. Indeed, many plant and animal genes have been identified based on their ability to complement mutations in *E. coli* or yeast.

One of the first genes of higher plants to be cloned by complementation screening encodes the enzyme glutamine synthetase in maize. Glutamine synthetase catalyzes the final step in the biosynthesis of the amino acid glutamine. Because glutamine is one of the 20 amino acids specified by the genetic code, it is essential for growth in all cellular organisms. Thus, an *E. coli* cell requires glutamine for growth, but it does not matter whether this glutamine is synthesized by a corn enzyme, a human enzyme, or an *E. coli* enzyme. Initially, glutamine synthetase cDNA clones from corn were identified on the basis of their ability to rescue glutamine synthetase-deficient *E. coli* mutants growing on medium lacking glutamine. The maize glutamine synthetase cDNA clone was then used to identify the corresponding genomic clone (the gene) from a genomic DNA library by the procedure outlined in the next section.

Complementation screening has some limitations. Eukaryotic genes contain noncoding introns, which must be spliced out of gene transcripts prior to their translation. Because *E. coli* cells do not possess the machinery required to excise introns from eukaryotic genes, complementation screening of eukaryotic clones in *E. coli* is restricted to cDNAs, from which the intron sequences have already been excised. In addition, the complementation screening procedure depends on the correct transcription of the cloned gene in the new host. Eukaryotes have signals that regulate gene expression that are different from those in prokaryotes; therefore, the complementation approach is more likely to work with prokaryotic genes in prokaryotic organisms, and eukaryotic genes in eukaryotic organisms. For this reason, researchers often use *S. cerevisiae* to screen eukaryotic DNA libraries by the complementation procedure.

Molecular hybridization: The first eukaryotic DNA sequences to be cloned were genes that are highly expressed in specialized cells. These genes included the mammalian α- and β-globin genes and the chicken ovalbumin gene. Red blood cells are highly specialized for the synthesis and storage of hemoglobin. Over 90 percent of the protein molecules sythesized in red blood cells during their period of maximal biosynthetic activity are globin chains. Similarly, ovalbumin is a major product of chicken oviduct cells. As a result, RNA transcripts of the globin and ovalbumin genes can be easily isolated from reticulocytes and oviduct cells, respectively. These RNA transcripts can be employed to synthesize radioactive cDNAs, which, in turn, can be used to screen genomic DNA libraries by ***in situ* colony** or **plaque hybridization** (Figure 20.12). Colony hybridization is used with libraries constructed in plasmid and cosmid vectors; plaque hybridization is used with libraries in phage lambda vectors. We shall focus on *in situ* colony hybridization here, but the two procedures are virtually identical.

The colony hybridization screening procedure involves replica plating the colonies formed by transformed cells onto nitrocellulose membranes, hybridization with a radioactive DNA or RNA probe, and autoradiography (Figure 20.12). If RNA transcripts of the gene of interest can be isolated from specialized cells, the RNAs can be used as templates for the *in vitro* synthesis of radioactive cDNAs by reverse transcriptase. Alternatively, if a cDNA clone can be isolated by genetic selection, as was done for the maize glutamine synthetase gene, the cDNA can be used to screen a genomic DNA library by colony hybridization. The labeled cDNA is employed as a probe for hybridization to denatured DNA from colonies grown on the nitrocellulose membranes. The DNA from the lysed cells is bound to the nitrocellulose membrane before hybridization so that it won't come off during subsequent steps in the procedure. After time is allowed for hybridization between complementary strands of DNA, the membranes are washed with buffered salt solutions to remove nonhybridized cDNA and are exposed to X-ray film to detect the presence of radioactivity on the membrane. Only colonies that contain DNA sequences complementary to the radioactive cDNA will yield radioactive spots on the autoradiographs (Figure 20.12). The locations of the radioactive spots are used to identify colonies that contain the desired sequence on the original replicated plates. These colonies are used to purify DNA clones harboring the gene or DNA sequence of interest.

Biological and Physical Containment of Recombinant DNA Molecules

As the methods for constructing recombinant DNA molecules and cloning were being developed, molecular biologists recognized that this technology would

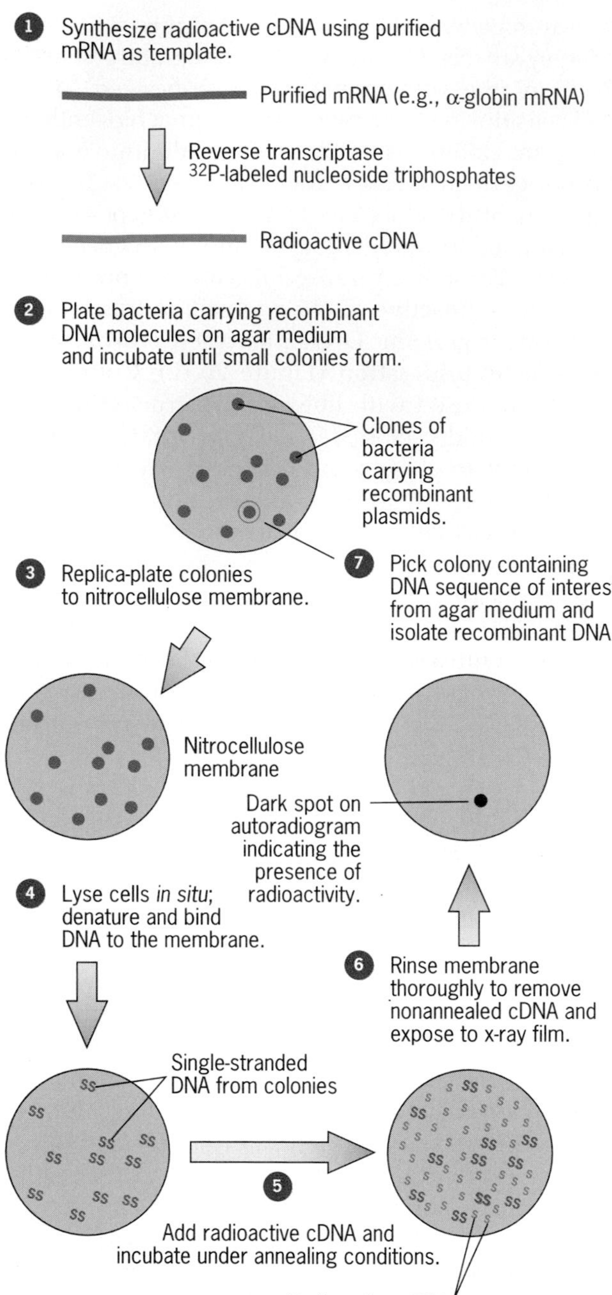

Figure 20.12 Screening DNA libraries by colony hybridization. Here, a radioactive cDNA is employed as a hybridization probe. Frequently, a purified restriction fragment is used as the hybridization probe. The colony hybridization procedure allows researchers to identify bacterial colonies harboring DNA sequences that hybridize to the radioactive DNA probe. These colonies are then isolated and used to prepare large quantities of the recombinant DNA molecules of interest.

provide a powerful tool with which to study the structure and function of genetic material. They also wondered whether it might have potential hazards. For example, some retroviruses carry genes that participate in the transformation of animal cells to the malignant state. If these genes were cloned in *E. coli* plasmids, might the *E. coli* strains carrying the viral genes serve as vectors for the transmission of cancer-causing genes? With such possibilities in mind, several prominent molecular biologists called for a voluntary moratorium on research of this type. Numerous conferences were held to discuss whether potential hazards existed, and, if so, what precautions should be taken to minimize the possibility of creating pathogenic recombinant organisms. A major concern of opponents of recombinant DNA work was that a pathogenic or otherwise harmful new organism might be introduced into the ecosystem. Proponents argued against the likelihood of this hazard and emphasized the benefits of recombinant DNA research.

After extensive discussions, the National Institutes of Health (NIH) of the U.S. Department of Health, Education, and Welfare (now the Department of Health and Human Services) established specific guidelines under which recombinant DNA research of various types would be done. These guidelines emphasized both physical and biological containment of the recombinant molecules constructed. **Physical containment** includes the use of sterile techniques, containment hoods, and specially designed laboratories to prevent vectors containing recombinant DNA molecules from escaping into natural ecosystems. **Biological containment** involves the use of organisms with specially constructed, weakened genotypes as vectors in cloning experiments. Ideally, these organisms should be unable to survive under conditions existing in any natural ecosystem. However, scientists cannot exclude the possibility that a newly constructed recombinant organism might escape from the laboratory and have a detrimental effect on the environment.

As more work with recombinant DNA was performed, it became evident that bacteria and viruses that carry foreign genes are not very healthy. They have been found not to survive in competition with wild-type organisms under natural ecosystem conditions. Thus, the NIH guidelines have been relaxed for the more routine types of gene-cloning experiments. Other kinds of experiments still require approval by an NIH panel and by a local institutional biosafety committee prior to their initiation.

Key Points: **DNA libraries can be constructed that contain complete sets of genomic DNA sequences or DNA copies (cDNAs) of mRNAs in an organism. Specific genes or other DNA sequences can be isolated from these libraries by genetic complementation and by hybridization to labeled nucleic acid probes containing DNA sequences of known function.**

THE MANIPULATION OF CLONED DNA SEQUENCES *IN VITRO*

Recombinant DNA technology has made huge contributions to biological and medical research. New information about the structures of eukaryotic genes and chromosomes has accumulated at an unprecedented rate since the development of these techniques. Practical applications also have been realized; human insulin and human growth hormone are now produced in genetically engineered bacteria. For diabetics who are allergic to nonhuman insulins or for individuals with defects in the synthesis of human growth hormone, these early achievements of scientists using recombinant DNA methodologies are of major significance.

As would be expected following the discovery of a major new research tool, technical advances have accumulated rapidly. Researchers can now clone and characterize any gene or other DNA sequence of interest from virtually any organism. These cloned genes can be modified *in vitro* and reintroduced into the same or different species for further study. A few of the most important methods for manipulating DNA are discussed in the following sections.

Phagemids: The Biological Purification of DNA Single Strands

Single-stranded DNA provides the optimal substrate for DNA sequencing experiments and is required for certain *in vitro* mutagenesis protocols. Several vectors have been constructed that provide a simple biological mechanism for purifying large amounts of single strands of cloned genes. The most useful of these are hybrid vectors called **phagemids,** which contain components from both *phage* chromosomes and plas*mids*. These vectors replicate in *E. coli* as normal double-stranded plasmids until a **helper phage** is provided. After addition of the helper phage, they switch to the phage mode of replication and package single strands of DNA in phage particles. The helper phage is a mutant that replicates its own DNA inefficiently, but provides viral replication enzymes and structural proteins for the production of phagemid DNA molecules that are packaged in phage coats.

Before discussing the phagemid vectors further, let's examine the life cycle of the filamentous phages that contain single-stranded DNA. The best-known of these phages are **M13, fl,** and **fd,** all of which have long threadlike morphologies and reproduce in *E. coli* cells. Their single-stranded DNA genomes replicate by the rolling-circle mechanism (Chapter 11). The filamentous single-stranded DNA phages infect cells by absorbing into and entering through F pili; thus, these phages only infect F^+ or Hfr cells, not F^- cells. Furthermore, these phages do not lyse the host cells like phage T4 (Chapter 16). Instead, the progeny viruses are extruded through the cell membrane and cell wall without killing the host cell. Infected cells continue to grow and extrude thousands of progeny virus particles, each containing a single-stranded genome, into the medium. Because the virus particles are much smaller than the host cells, the bacteria can be removed by low-speed centrifugation. The virus particles can then be collected from the supernatant suspension by high-speed centrifugation, and their single-stranded DNA molecules can be isolated by simple phenol-chloroform extractions. The same DNA strand of the virus is always packaged; it is called the + strand, and its complement is the − strand. The packaged + strand has the same sense as mRNA; its nucleotide triplets correspond to the mRNA codons, but with T in place of U.

The major features of the M13 life cycle are shown in Figure 20.13. Note that *the packaging of single strands of phage DNA in progeny phage provides a neat biological*

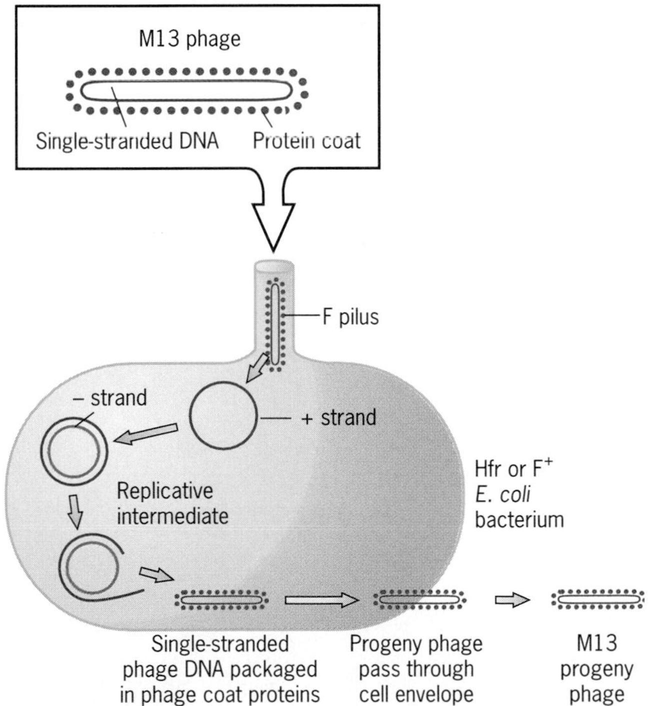

Figure 20.13 The life cycle of bacteriophage M13. The genetic information of the virus is stored in single-stranded DNA packaged within a long threadlike virion. The virus enters the cell through an F pilus. After the coat proteins are removed from the DNA, it replicates by the rolling-circle mechanism. Progeny single strands of DNA are packaged in new coats, and the progeny virions are extruded through the cell envelope without killing the host cell.

purification of single-stranded DNA. This is the case for a foreign gene cloned in the viral chromosome, just as it is for the phage genes themselves. The phagemid cloning vectors take advantage of this unique aspect of M13 reproduction.

The phagemid vectors **pUC118** and **pUC119** are virtually identical, but they contain the polycloning region in opposite orientations (turned end-for-end) relative to the rest of the genes of the vector (Figure 20.14). Thus, if a foreign DNA is inserted into a specific restriction site in both vectors, one vector will package one strand of the foreign DNA, and the other vector will package the complementary strand. Both strands of the cloned DNA can therefore be isolated, sequenced, subjected to site-specific mutagenesis, and so on. These vectors were designated **pUC** for *p*lasmid and *U*niversity of *C*alifornia, where they were constructed. Vectors pUC118 and pUC119 differ from earlier vectors in the pUC series by the addition of the origin of replication from phage M13. This distinction permits pUC118 and pUC119 to replicate either (1) as a double-stranded plasmid in the absence of helper phage or (2) as a single-stranded DNA, which is packaged in M13 phage coats and extruded from the cell, in the presence of helper phage. In the absence of helper phage, replication is controlled by the plasmid

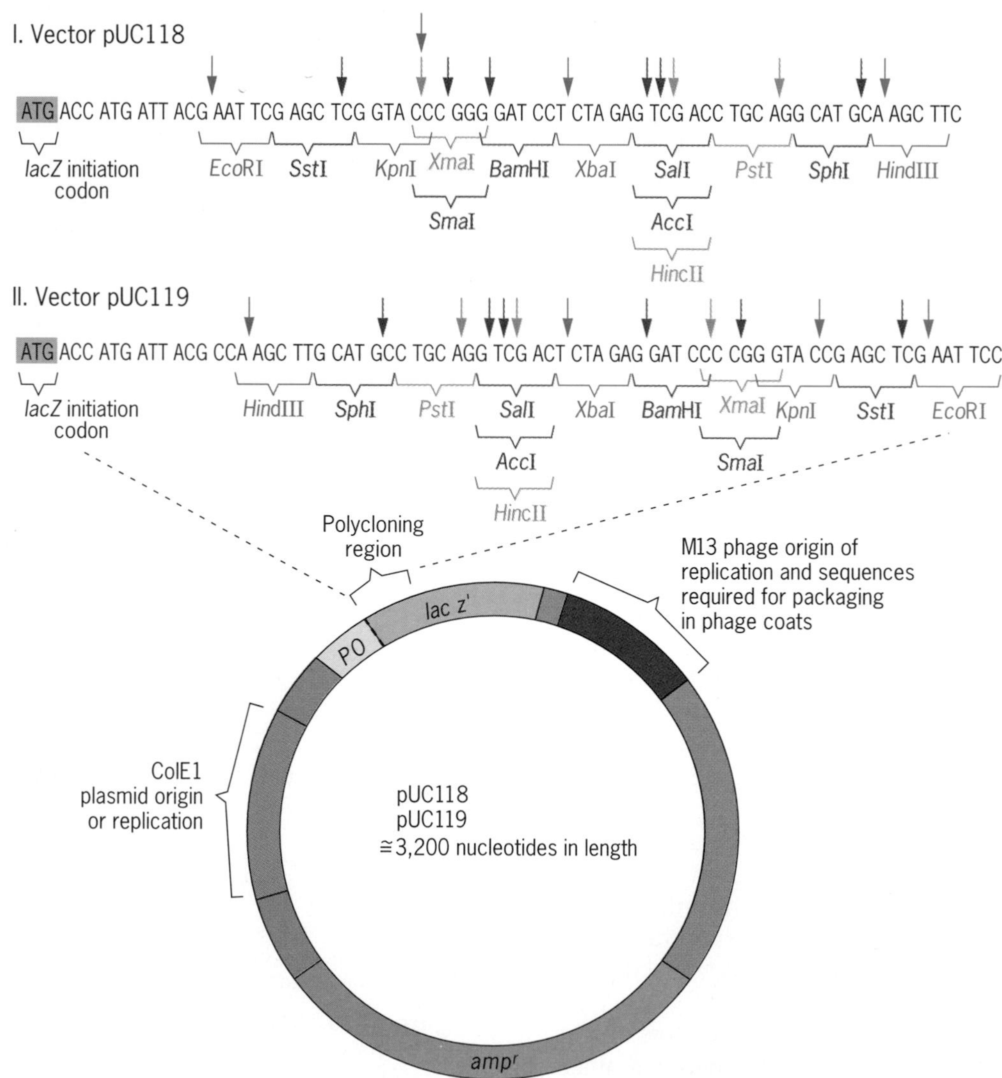

Figure 20.14 Important components of the phagemid vectors pUC118 and pUC119. *P* and *O* are the promoter (RNA polymerase binding site) and operator (repressor binding site) regions that regulate the transcription of the lactose biosynthetic enzymes (Chapter 23). Gene segment *lacZ'* encodes the amino terminal 147 amino acids of β-galactosidase. The polycloning site lies within *lacZ'* and contains a cluster of restriction enzyme cleavage sites (top). The ColE1 plasmid origin controls replication in the absence of helper phage, whereas the M13 origin controls replication in the presence of helper phage.

origin of replication. In the presence of helper phage, replication is directed by the phage M13 origin of replication.

The utility of the pUC vectors is greatly enhanced by a simple color test that allows cells harboring plasmids with foreign DNA inserts to be distinguished from those harboring plasmids with no insert. The basis of this color indicator test is the functional inactivation of the 5′ segment of the *lacZ* gene, which is present in the vector, by the insertion of foreign DNA into the polycloning region.

The *E. coli lacZ* gene encodes β-galactosidase, the enzyme that cleaves lactose into glucose and galactose. This is the first step in the catabolism of lactose in *E. coli* (Chapter 23). The presence of β-galactosidase in cells can be monitored on the basis of its ability to cleave the substrate 5-bromo-4-chloro-3-indolyl-β-D-galactoside (usually called Xgal) to galactose and 5-bromo-4-chloroindigo. Xgal is colorless; 5-bromo-4-chloroindigo is blue. Thus, cells containing active β-galactosidase produce blue colonies on agar medium containing Xgal. Cells lacking β-galactosidase activity produce white colonies on Xgal plates (Figure 20.15).

The molecular basis of the β-galactosidase activity that provides the color indicator test for pUC vectors is somewhat more complex. The *lacZ* gene of *E. coli* is over 3000 nucleotide pairs long, and placing the entire gene in the plasmid would make the vector larger than desired. The pUC vectors contain only a small part of the *lacZ* gene. This *lacZ* gene segment encodes only the amino terminal portion of β-galactosidase. However, the presence of a functional copy of the *lacZ* gene segment can be detected because of a unique type of intragenic complementation. When a functional copy of the *lacZ* gene segment on the pUC plasmid is present in a cell that contains a particular *lacZ* mutant allele on the chromosome or on an F′ plasmid, the two defective *lacZ* sequences yield polypeptide products that complement each other and produce β-galactosidase activity. The mutant allele, designated *lacZΔM15*, synthesizes a *lac* protein that lacks amino acids 11 through 14 from the amino terminus. The absence of these amino acids prevents the mutant polypeptides from interacting to produce the active tetrameric form of the enzyme.

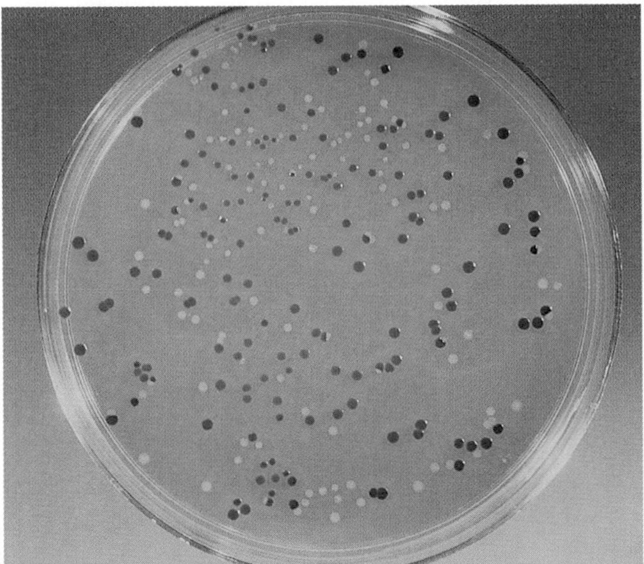

Figure 20.15 Photograph illustrating the use of X-gal to identify *E. coli* colonies containing (blue) or lacking (white) β-galactosidase activity. In this case, the cells in the white colonies harbor pUC119 plasmids with foreign DNA fragments inserted into the polycloning site, and the cells in the blue colonies contain pUC119 plasmids with no insert.

By some mechanism that is still not completely understood, the presence of the amino terminal fragment (the first 147 amino acids) of the *lacZ* polypeptide encoded by the *lacZ* gene fragment on pUC plasmids facilitates tetramer formation by the ΔM15 deletion polypeptides. This yields active β-galactosidase, which can be detected by the Xgal color test. The unique type of intragenic complementation observed between the amino terminal fragment of β-galactosidase and the defective ΔM15 polypeptide is called **α complementation.** It conveniently permits the Xgal color test to be utilized without placing the entire *lacZ* gene in the pUC vectors. However, for this color test to work with these vectors, the host cell must contain a mutant *lacZΔM15* gene on the chromosome, or, more commonly, on an F′ plasmid in a cell with a deletion of the chromosomal copy of the *lacZ* gene.

The standard protocol for cloning foreign DNAs into pUC vectors involves (1) ligation of the foreign DNA of interest into one of the restriction enzyme cleavage sites in the polycloning region of the vector, (2) transformation of cells of an appropriate amp^s *E. coli* strain carrying the *lacZΔM15* gene with the ligation products, and (3) plating the transformed cells in plates with a nutrient agar medium containing Xgal and ampicillin. Only transformed cells that contain a pUC plasmid will be able to grow in the presence of ampicillin. Colonies produced by cells harboring pUC plasmids without foreign DNA inserts will be blue (Figure 20.15). Colonies produced by bacteria harboring pUC vectors with foreign DNA inserts will be white because of the functional inactivation of the *lacZ* gene segment by the inserted DNA. Thus, white colonies are picked and used for further characterization of foreign DNAs.

The ability to perform **forced cloning,** the insertion of a foreign DNA into the polycloning region in a predetermined orientation, is another important advantage of the pUC118-119 vector system. Consider a

DNA sequence that has a *Sst*I site at one end and a *Pst*I site at the other end. If this DNA is cleaved with both enzymes, the resulting *Sst*I-*Pst*I fragment can be inserted into either pUC118 DNA or pUC119 DNA that has been cut with *Sst*I and *Pst*I (Figure 20.14). This *Sst*I-*Pst*I fragment will be present in pUC118 and pUC119 in opposite orientations. Thus, if a gene or other DNA fragment is force cloned in both pUC118 and pUC119, one vector will package one strand of DNA and the other vector will package the complementary strand of DNA after the infection of host cells with helper phage (Figure 20.16). By using both vectors, each of the two complementary single strands of a cloned gene or other DNA sequence of interest can therefore be isolated in large quantities.

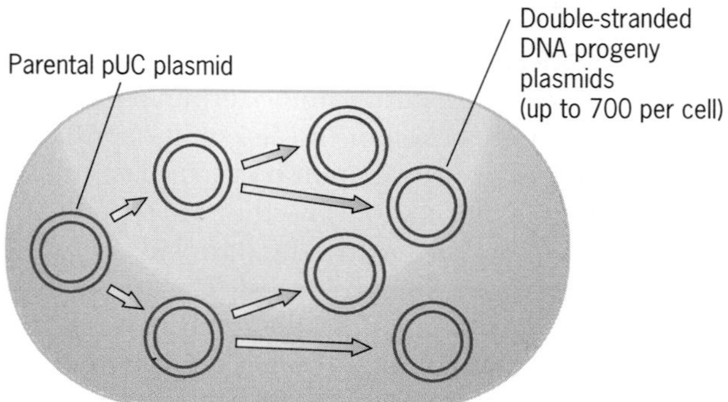

(*a*) Replication of pUC118 and pUC119 as double-stranded plasmid DNAs in the absence of helper phage.

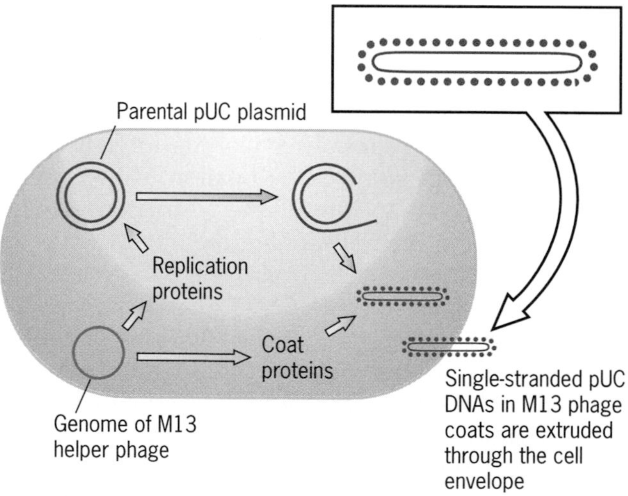

(*b*) Replication of pUC118 and pUC119 as single-stranded phage DNAs in the presence of helper phage.

Figure 20.16 The plasmid and phage modes of replication of the pUC118 and pUC119 phagemid vectors. (*a*) In the absence of M13 helper phage, replication is controlled by an origin of replication derived from plasmid ColE1. (*b*) In bacteria infected with an M13 helper phage, rolling-circle replication is controlled by the M13 origin of replication. The chromosome of the helper phage contains a modified origin of replication that results in inefficient replication; thus, most of the extruded phage particles contain single-stranded pUC DNA molecules.

In summary, the pUC118 and pUC119 cloning vectors (Figure 20.14) have several important features.

1. They consist of *small, supercoiled, covalently closed circular DNA;* thus, they are easily isolated and manipulated *in vitro.*
2. They carry *the* amp^r *gene as a selectable marker;* thus, only bacteria harboring a plasmid will grow on medium containing ampicillin.
3. They are present in *high copy number,* up to 700 copies per bacterium; thus, small cell cultures will yield large amounts of DNA.
4. They contain a *polycloning region* with several restriction enzyme cleavage sites; thus, many different types of restriction fragments can be inserted.
5. *The polycloning region interrupts the coding region of the 5′ end of the* E. coli lacZ *gene,* which encodes β-galactosidase; thus, colonies harboring plasmids with foreign DNA inserts can be distinguished from those carrying plasmids with no insert by a simple color test.
6. *The* lacZ *gene is under the control of the* lac *promoter;* thus, genes inserted in frame (with codons in the proper reading frame) can be expressed to produce β-galactosidase-foreign protein fusion products.
7. They carry a *plasmid origin of replication;* thus, replication in the absence of helper phage produces large amounts of double-stranded plasmid DNA.
8. They carry a *phage M13 origin of replication;* thus, single-stranded DNA is produced and packaged in phage coats when helper phage is present.
9. *The polycloning regions are present in opposite orientations in pUC118 and pUC119;* thus, if pUC118 packages one strand of a cloned gene, pUC119 will package the complementary strand.

In Vitro Site-Specific Mutagenesis

Recombinant DNA and gene-cloning technologies allow scientists to isolate and characterize any given gene from any living organism. In addition, these methodologies permit researchers to modify these genes *in vitro* in order to produce new restriction enzyme cleavage sites or to change a particular codon to a different codon or even a whole set of different codons. In short, scientists can now dissect genes at the nucleotide level by changing one nucleotide at a time and examining the effect of each change on the function of the gene or its product. Such nucleotide changes are accomplished by a procedure called **site-specific mutagenesis,** or, more precisely, **oligonucleotide-directed site-specific mutagenesis.**

Site-specific mutagenesis (Figure 20.17) can be carried out on genes cloned in vectors like pUC118 and pUC119 that provide for the isolation of DNA single strands (Figure 20.16). An oligonucleotide 12 to 15 nucleotides in length, which is largely complementary to the purified single-stranded copy of the gene in the region of interest, but which contains one or more noncomplementary or "mismatched" bases, is synthesized. The mismatched bases will provide the desired mutant sequences. The synthetic oligonucleotide is annealed to the single strand of the gene in the vector, and a complementary strand is synthesized *in vitro* with a DNA polymerase. DNA ligase is then used to seal the single-strand break left by polymerase, and the double-stranded circular DNA products are introduced into *E. coli* cells by transformation. Semiconservative replication of the DNA molecules in *E. coli* will produce a population of progeny DNA molecules, half of which carry the original DNA sequence and half of which carry the new or mutant DNA sequence. The two types of molecules are separated by a second cycle of transformation, and the desired mutant genes are identified by hybridization to radioactively labeled synthetic oligonucleotides under conditions of high stringency (conditions that require perfect complementarity of the strands for duplexes to form) or by sequencing the modified region of the cloned gene.

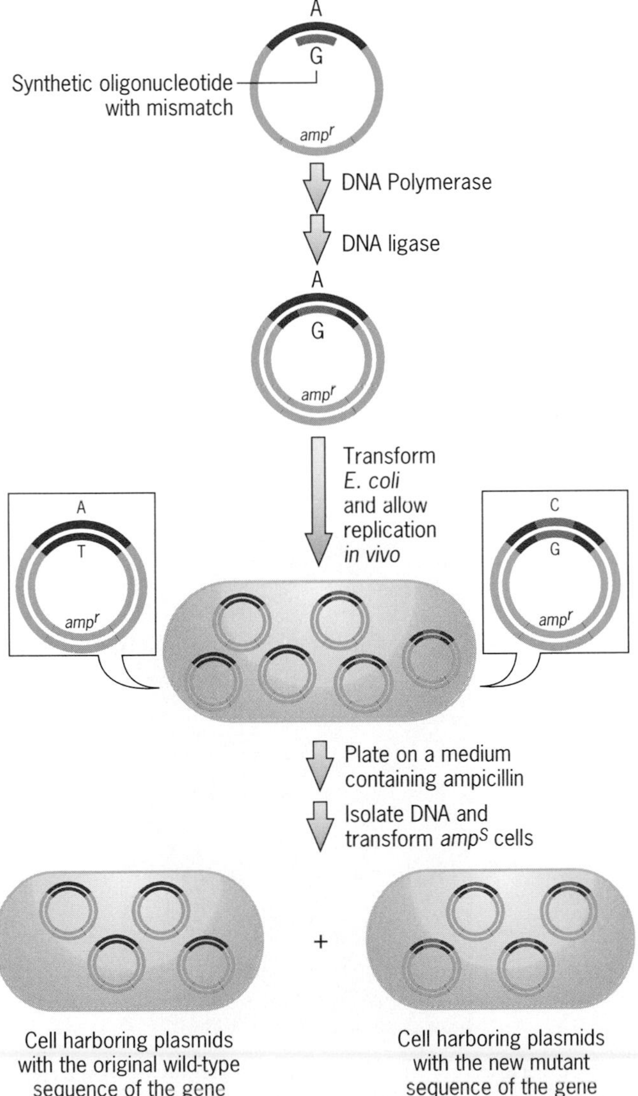

Figure 20.17 Oligonucleotide-directed site-specific mutagenesis. Single-stranded pUC119 DNA carrying the gene of interest is purified and annealed to a synthetic oligonucleotide that contains the desired change in sequence. Here, an A:T base pair is converted to a C:G base pair.

Key Points: **Cloned genes and other DNA sequences can be manipulated almost at will in the test tube. Phagemids provide a powerful tool for purifying the single strands of a cloned DNA sequence. The nucleotide sequences of cloned genes can be changed as desired by *in vitro* site-specific mutagenesis.**

THE MOLECULAR ANALYSIS OF DNA, RNA, AND PROTEIN

The development of recombinant DNA techniques has spawned a whole array of new approaches to the analysis of genes and gene products. Many questions that were totally unapproachable just 20 years ago can now be investigated with relative ease. Geneticists can isolate and characterize essentially any gene from any organism; however, the isolation of genes from large eukaryotic genomes is sometimes a long and laborious process (Chapter 22). Once a gene has been cloned, its expression can be investigated in even the most complex organisms such as humans.

Is a particular gene expressed in the kidney, the liver, bone cells, hair follicles, erythrocytes, or lymphocytes? Is this gene expressed throughout the development of the organism or only during certain stages of development? Is a mutant allele of this gene similarly expressed, spatially and temporally, during development? Or does the mutant allele have an altered pattern of expression? If the latter, is this altered pattern of expression responsible for an inherited syndrome or disease? These questions and many others can now be routinely investigated using well-established methodologies.

Clearly, a comprehensive discussion of all the techniques used to investigate gene structure and function is far beyond the scope of this text. However, let's consider some of the most important methods used to investigate the structure of genes (DNA), their transcripts (RNA), and their final products (usually proteins).

Analysis of DNAs by Southern Blot Hybridizations

Gel electrophoresis provides a powerful tool for the separation of macromolecules with different sizes and charges. DNA molecules have an essentially constant charge per unit mass; thus, they separate in agarose and acrylamide gels almost entirely on the basis of size or conformation. Agarose or acrylamide gels act as molecular sieves, retarding the passage of large molecules more than small molecules. Agarose gels are better sieves for large molecules (larger than a few hundred nucleotides); acrylamide gels are better for separating small DNA molecules. Figure 20.18 illustrates the separation of DNA restriction fragments by agarose gel electrophoresis. The procedures used to separate RNA

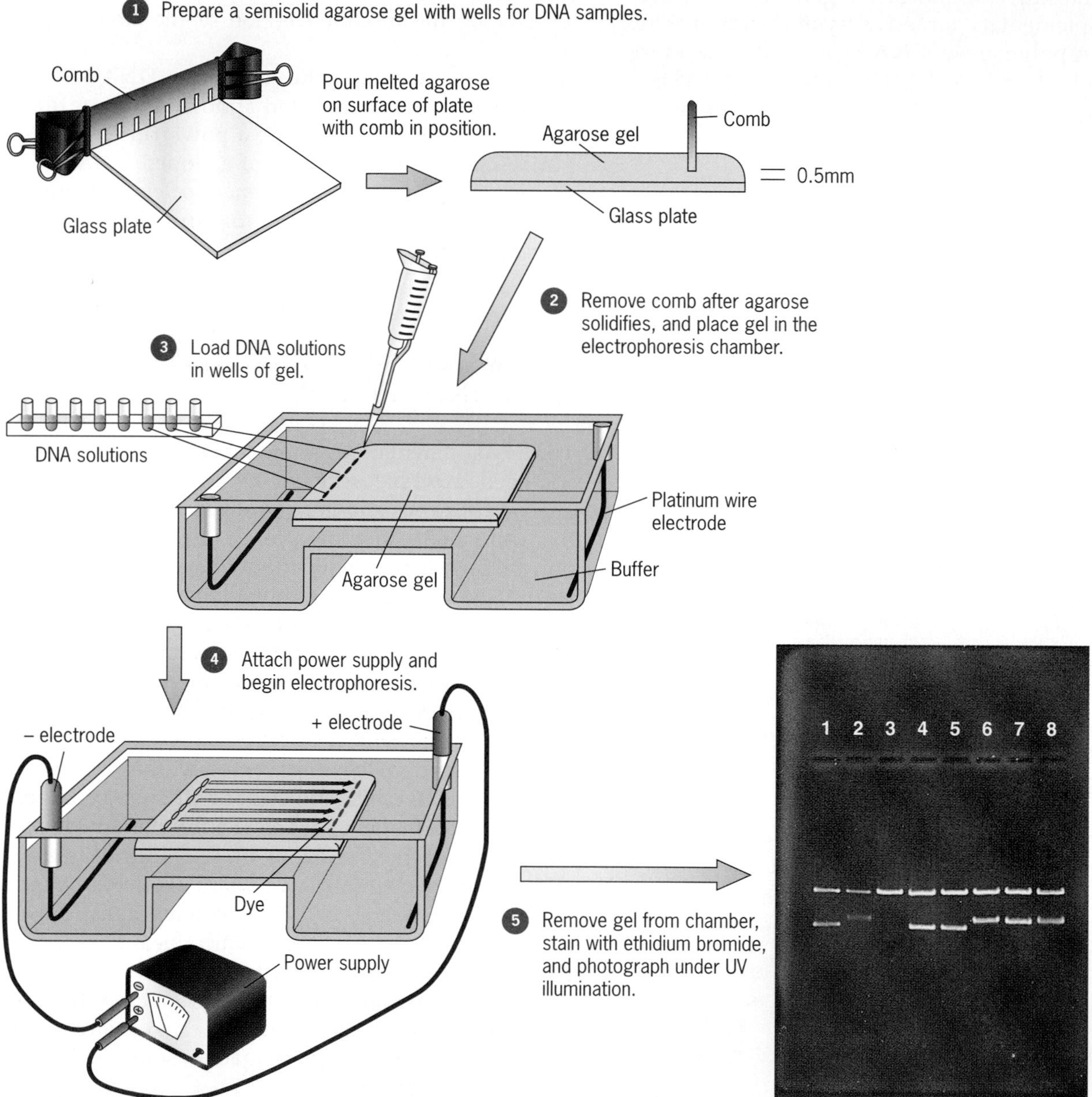

Figure 20.18 The separation of DNA molecules by agarose gel electrophoresis. The DNAs are dissolved in loading buffer with density greater than that of the electrophoresis buffer so that DNA samples settle to the bottoms of the wells, rather than diffusing into the electrophoresis buffer. The loading buffer also contains a dye to monitor the rate of migration of molecules through the gel. Ethidum bromide binds to DNA and fluoresces when illuminated with ultraviolet light. In the photograph shown, the third lane from the left contained *Eco*RI-cut pUC119 DNA; the other lanes contained *Eco*RI-cut pUC119 DNAs carrying maize glutamine synthetase cDNA inserts.

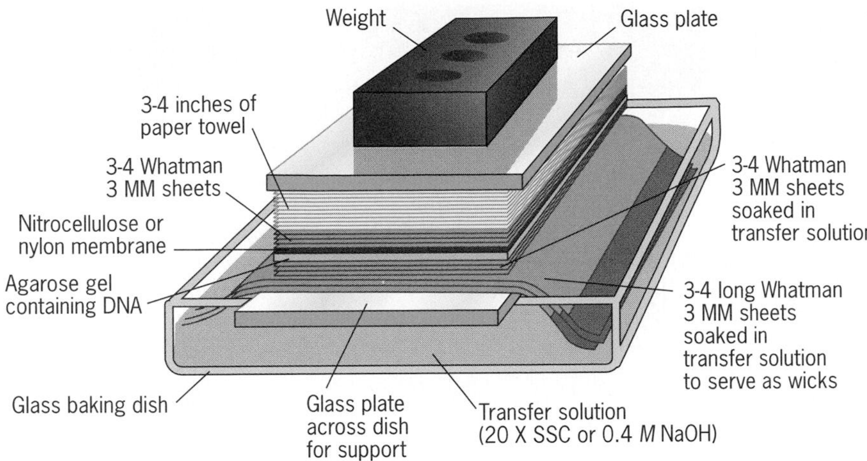

Figure 20.19 Procedure used to transfer DNAs separated by gel electrophoresis to nitrocellulose or nylon membranes. The transfer solution carries the DNA from the gel to the membrane as the dry paper towels on top draw the salt solution from the reservoir through the gel to the towels. The DNA binds to the membrane on contact. The membrane with the DNA bound to it is dried and baked under vacuum to affix the DNA firmly prior to hybridization.

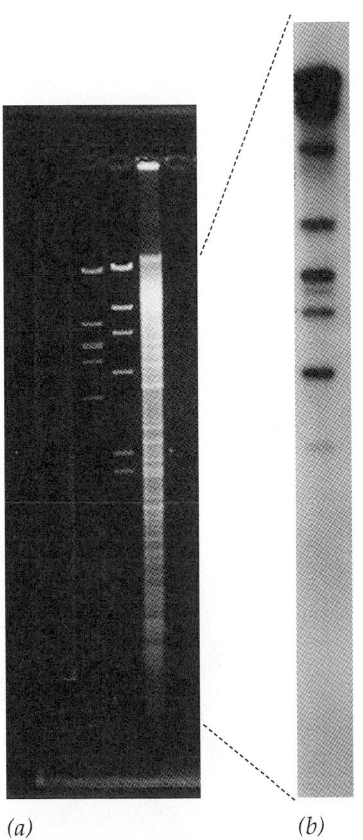

Figure 20.20 Identification of genomic restriction fragments harboring specific DNA sequences by the Southern blot hybridization procedure. (*a*) Photograph of an ethidium bromide-stained agarose gel containing phage λ DNA digested with *Eco*RI (left lane), phage λ DNA digested with *Hin*dIII (center lane), and *Arabidopsis thaliana* DNA digested with *Eco*RI (right lane). The λ DNA digests provide size markers. The *A. thaliana* DNA digest was transferred to a nylon membrane by the Southern procedure (Figure 20.19) and hybridized to a radioactive DNA fragment of a cloned β-tubulin gene. The resulting Southern blot is shown in (*b*); nine different *Eco*RI fragments hybridized with the β-tubulin probe.

and protein molecules are largely the same in principle but involve slightly different techniques because of the unique properties of each class of macromolecule.

In 1975, E. M. Southern published an important new procedure that allowed investigators to identify the locations of genes and other DNA sequences on restriction fragments separated by gel electrophoresis. The essential feature of this technique is the transfer of the DNA molecules that have been separated by gel electrophoresis onto a nitrocellulose or nylon membrane (Figure 20.19). Such transfers of DNA to membranes are called Southern blots after the scientist who developed the technique. The DNA is denatured either prior to or during transfer by placing the gel in an alkaline solution. After transfer is complete, the DNA is immobilized on the membrane by drying or UV irradiation. A radioactive DNA probe containing the sequence of interest is then hybridized or annealed (Chapter 10) with the immobilized DNA on the membrane. The probe will anneal only with DNA molecules on the membrane that contain a nucleotide sequence complementary to the sequence of the probe. Nonannealed probe is then washed off the membrane, and the washed membrane is exposed to X-ray film to detect the presence of the radioactivity in the bound probe. After the film is developed, the dark bands show the positions of DNA sequences that have hybridized with the probe (Figure 20.20).

The ability to transfer DNA restriction fragments or other DNA molecules that have been separated by gel electrophoresis to nitrocellulose or nylon membranes for hybridization studies and other types of analyses has proven to be extremely useful (see Technical Sidelight: Detection of a Mutant Gene Causing Cystic Fibrosis by Southern Blot Analysis). Radioactive probes for use in Southern blot hybridization experiments are prepared by a variety of procedures including (1) producing radioactive RNA probes by *in vitro*

TECHNICAL SIDELIGHT

Detection of a Mutant Gene Causing Cystic Fibrosis by Southern Blot Analysis

Southern blots are DNA gel blots prepared by the transfer of DNAs separated by gel electrophoresis to membranes and hybridized to labeled DNA or RNA probes (Figures 20.19 and 20.20). The Southern blotting procedure is one of the most important tools of molecular genetics. Southern blots are used to identify DNA molecules or restriction fragments that contain specific genes or DNA sequences (Figure 20.20). In some cases, Southern blots can be used to identify mutant alleles that are responsible for inherited diseases. Here, we examine how the Southern blotting procedure is used to identify the mutant allele responsible for the majority of the cases of cystic fibrosis (CF). This devastating disease, which is characterized by the accumulation of mucus in the lungs, pancreas, and liver, and the subsequent malfunction of these organs, is the most common inherited disease in humans of northern European descent. In Chapter 22, we discuss cystic fibrosis and the identification and characterization of the gene that causes it.

Approximately 70 percent of the cases of CF result from a specific mutant allele of the *CF* gene. This mutant allele, *CFΔF508,* contains a three-base deletion that eliminates a phenylalanine residue at position 508 in the polypeptide product. Because the nucleotide sequence of the *CF* gene is known and since the *CFΔF508* allele differs from the wild-type allele by three base pairs, it was possible to design oligonucleotide probes that hybridize specifically with the wild-type *CF* allele or the *ΔF508* allele under the appropriate conditions.

The wild-type *CF* gene and gene product have the following nucleotide and amino acid sequences in the region altered by the *ΔF508* mutation (top, right):

bases in the coding strand: 5′-AAA GAA AAT ATC ATC TTT GGT GTT-3′ (TTT: deleted in *ΔF508*)

amino acids in product: NH_2-Leu Glu Asn Ile Ile Phe Gly Val-COOH, (Phe: amino acid 508)

whereas the *ΔF508* allele and product have the sequences:

bases in the coding strand: 5′-AAA GAA AAT ATC AT. . .T GGT GTT-3′ (deletion)

amino acids in product: NH_2-Leu Glu Asn Ile Ile — Gly Val-COOH. (Phe absent)

Based on these nucleotide sequences, Lap-Chee Tsui and colleagues synthesized oligonucleotides spanning this region of the mutant and wild-type alleles of the *CF* gene and tested their specificity. They demonstrated that at 37° C under a standard set of conditions, one oligonucleotide probe (oligo-N: 3′-CTTTTATAGTAGAAACCAC-5′) hybridized only with the wild-type allele, whereas another oligonucleotide (oligo-ΔF: 3′-TTCTTTTATAGTA. . .ACCACAA-5′) hybridized only with the *ΔF508* allele. Their results indicated that the oligo-ΔF probe could be used to detect the *ΔF508* allele in either the homozygous or heterozygous state by Southern blot hybridization experiments. When Tsui and coworkers used these allele-specific oligonucleotide probes to analyze CF patients and their parents for the presence of the *ΔF508* mutation, they found that many of the patients were homozygous for this mutation, whereas most of their parents were heterozygous as is expected. Some of their results are as shown at the bottom of this page.

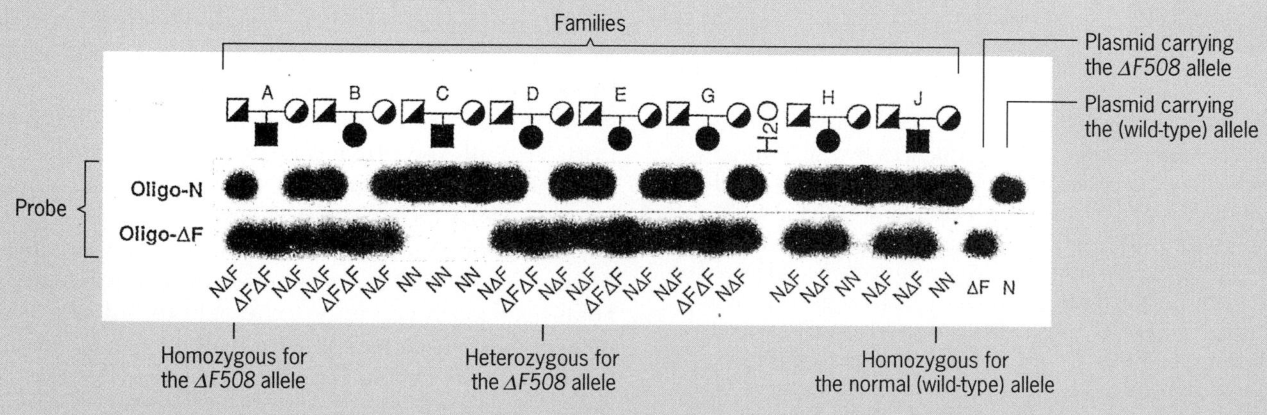

The Southern blot hybridization test for the presence of the Δ*F508* mutation can be carried out with small amounts of DNA if this segment of the *CF* gene is first amplified by the polymerase chain reaction (Figure 20.24). Indeed, the Southern blot test for the Δ*F508* mutation can be carried out with DNA obtained from fetal cells by amniocentesis or even with DNA from a single cell of an eight-cell pre-embryo produced by *in vitro* fertilization (Chapter 22).

transcription of cloned sequences using at least one radioactive ribonucleoside triphosphate precursor, (2) transferring ^{32}P groups to the ends of DNA molecules from labeled nucleotide triphosphates by the action of polynucleotide kinases, and (3) **nick-translation** of double-stranded DNA molecules. Nick-translation is performed by making single-strand cuts (nicks) in double helices with endonucleases, and then "translating" the nick along the DNA molecule by the concerted action of the 5′ ⟶ 3′ exonuclease activity and the 5′ ⟶ 3′ polymerase activity of *E. coli* DNA polymerase I (Chapter 11) in the presence of a labeled deoxyribonucleotide triphosphate precursor.

Analysis of RNAs by Northern Blot Hybridizations

If DNA molecules can be transferred from agarose gels to nitrocellulose or nylon membranes for hybridization studies, we might expect that RNA molecules separated by agarose gel electrophoresis could be similarly transferred and analyzed. Indeed, such RNA transfers are used routinely in molecular genetics laboratories. These RNA blots are called **northern blots** in recognition of the fact that the procedure is opposite to the Southern blotting technique. As we will discuss in the next section, this somewhat unusual terminology has been extended to the transfer of proteins from gels to membranes, a procedure called western blotting.

The northern blot procedure is essentially identical to that used for Southern blot transfers (Figure 20.19). However, RNA molecules are very sensitive to degradation by RNases. Thus, care must be taken to prevent contamination of materials with these extremely stable enzymes. Furthermore, most RNA molecules contain considerable secondary structure. RNA molecules must therefore be kept denatured during electrophoresis in order to separate them on the basis of size. Denaturation is accomplished by adding formaldehyde or some other chemical denaturant to the buffer used for electrophoresis. After transfer to an appropriate membrane, the RNA blot is hybridized to either RNA or DNA probes just as with a Southern blot.

Northern blot hybridizations (Figure 20.21) are extremely helpful in studies of gene expression. They can

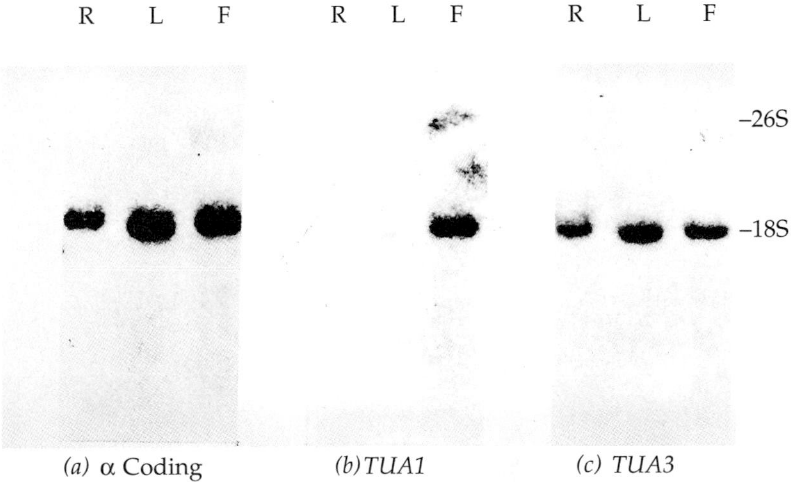

Figure 20.21 Typical northern blot hybridization data. Total RNAs were isolated from roots (R), leaves (L), and flowers (F) of *A. thaliana* plants, separated by agarose gel electrophoresis, and then transferred to nylon membranes. The autoradiogram shown in (*a*) is of a blot that was hybridized to a radioactive probe containing an α-tubulin coding sequence. This probe hybridizes to the transcripts of all six α-tubulin genes in *A. thaliana*. The autoradiograms shown in (*b*) and (*c*) are of RNA blots that were hybridized to DNA probes specific for the α1- and α3-tubulin genes, respectively. The results show that the α3-tubulin transcript is present in all organs analyzed, whereas the α1-tubulin transcript is present only in flowers. The 18S and 26S ribosomal RNAs provide size markers.

be used to determine whether a particular gene is transcribed in all tissues of an organism or only in certain tissues. They can also be used to study the temporal pattern of expression of individual genes during growth and development. However, we must remember that northern blot hybridizations measure only the accumulation of RNA transcripts. They provide no information about why the observed accumulation has occurred. Changes in transcript levels may be due to changes in the rate of transcription or changes in the rate of transcript turnover. More sophisticated procedures must be used to distinguish between these possibilities.

Analysis of Proteins by Western Blot Techniques

Polyacrylamide gel electrophoresis is an important tool for the separation and characterization of proteins. Because many functional proteins are composed of two or more subunits, individual polypeptides are separated by electrophoresis in the presence of the detergent sodium dodecyl sulfate (SDS), which denatures the proteins. After electrophoresis, the proteins are detected by staining with Coomassie blue or silver stain. However, the separated polypeptides also can be transferred from the gel to a nitrocellulose membrane, and individual proteins can be detected with antibodies. This transfer of proteins from acrylamide gels to nitrocellulose membranes, called **western blotting,** is performed by using an electric current to move the proteins from the gel to the surface of the membrane (Figure 20.22). After transfer, a specific protein of interest is identified by placing the membrane with the immobilized proteins in a solution containing an antibody to the protein. Nonbound antibodies are then

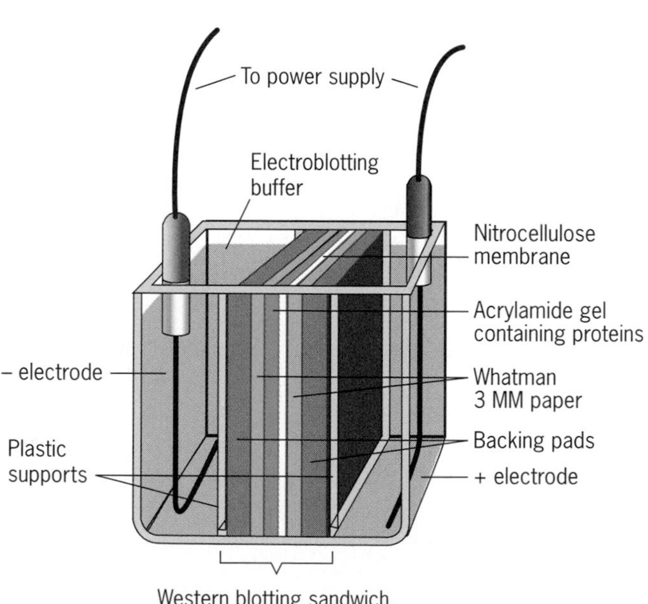

Figure 20.22 A typical western blotting or electroblotting apparatus. An electric current is used to transfer the proteins from a gel to a nitrocellulose membrane placed next to it in the blotting sandwich. All other components of the sandwich function to provide gentle but firm support; tight contact between the gel and the membrane is essential for good transfer.

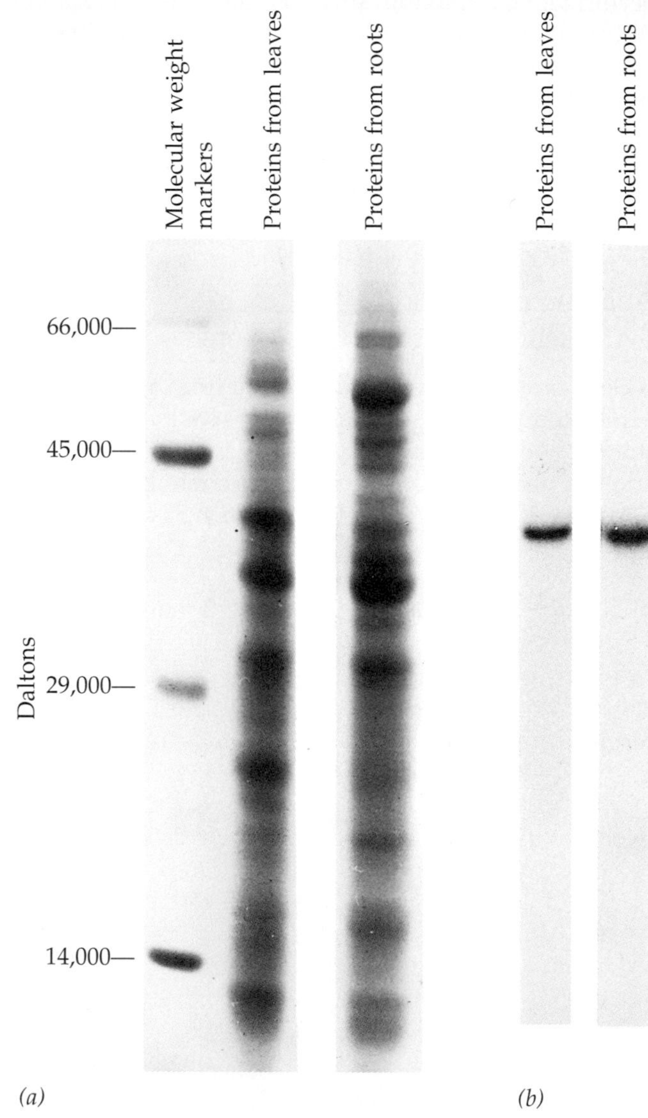

Figure 20.23 The use of western blots to identify individual proteins separated by polyacrylamide gel electrophoresis. (*a*) Proteins isolated from roots or leaves of maize were separated by polyacrylamide gel electrophoresis and stained with Coomassie blue. (*b*) The chloroplastic form of glutamine synthetase was identified by western blot analysis of the gel shown in (*a*).

washed off the membrane, and the presence of the initial (primary) antibody is detected by placing the membrane in a solution containing a secondary antibody. This secondary antibody reacts with immunoglobulins (the group of proteins comprising all antibodies) in general (Chapter 27). The secondary antibody is conjugated to either a radioactive isotope (permitting autoradiography) or an enzyme that produces a visible product when the proper substrate is added. Figure 20.23 shows the use of a western blot to detect a single protein in an acrylamide gel containing total cellular proteins from maize roots and leaves. Obviously, the western blot procedure is a powerful tool for identifying and characterizing specific gene products.

Key Points: **DNA restriction fragments and other small DNA molecules can be separated by agarose and acrylamide gel electrophoresis and transferred to nitrocellulose or nylon membranes to produce DNA gel blots called Southern blots. The DNAs on these blots can be hybridized to labeled DNA probes to detect sequences of interest by autoradiography. The same procedure can be applied to RNA molecules separated by gel electrophoresis to produce RNA gel blots called northern blots. When proteins are transferred from gels to membranes and detected with antibodies, the products are called western blots.**

THE MOLECULAR ANALYSIS OF GENES AND CHROMOSOMES

Recombinant DNA techniques allow geneticists to determine the structure of genes, chromosomes, and entire genomes. Indeed, molecular geneticists are constructing detailed genetic and physical maps of the genomes of many organisms (Chapter 21).

The ultimate physical map of a genetic element is its nucleotide sequence, and the complete nucleotide sequences of the genomes of many viruses, mitochondria, chloroplasts, 18 species of bacteria, and the yeast *S. cerevisiae,* have already been determined. As is discussed in Chapter 21, the goal of the Human Genome Project is to determine the complete nucleotide sequence of the human genome by 2003, and the cofounders of a new company are claiming that they will complete the sequence of the majority of the human genome by 2001. In the following sections, we discuss the amplification of specific segments of DNA, the construction of restriction enzyme cleavage site maps of genes and chromosomes, and the determination of DNA sequences.

Amplification of DNAs by the Polymerase Chain Reaction (PCR)

The **polymerase chain reaction,** usually referred to as **PCR,** is an extremely powerful procedure that allows the amplification of a selected DNA sequence in a genome a millionfold or more. The PCR procedure was devised by Kary Mullis, who received a Nobel Prize in 1993 for this work. PCR can be used to clone a given DNA sequence *in vitro*—without the use of living cells during the cloning process. However, the procedure can be applied only when the nucleotide sequence of at least one short DNA segment on each side of the region of interest is known. The PCR procedure involves using synthetic oligonucleotides complementary to these known sequences to prime enzymatic amplification of the intervening segment of DNA in the test tube.

The PCR procedure involves three steps, each repeated many times (Figure 20.24). In step 1, the genomic DNA containing the sequence to be amplified is denatured by heating to 95–97°C for 15–30 seconds. In step 2, the denatured DNA is annealed to an excess of the synthetic oligonucleotide primers by incubating them together at 50–60°C for 30 seconds. The ideal annealing temperature depends on the base composition of the primer. In step 3, DNA polymerase is used to replicate the DNA segment between the sites complementary to the oligonucleotide primers. The primer provides the free 3′-OH required for covalent extension, and the denatured genomic DNA provides the required template function (Chapter 11). Polymerization is usually carried out at 70–72°C for 1.5 minutes. The products of the first cycle of replication are then denatured, annealed to oligonucleotide primers, and replicated again with DNA polymerase. The procedure is repeated many times until the desired level of amplication is achieved. Note that *amplification occurs exponentially*. One DNA double helix will yield 2 double helices after one cycle of replication, 4 after two cycles, 8 after three cycles, 16 after four cycles, 1024 after ten cycles, and so on.

Initially, PCR was performed with DNA polymerase I of *E. coli* as the replicase. Because this enzyme is heat-inactivated during the denaturation step, fresh enzyme had to be added at step 3 of each cycle. A major improvement in PCR amplification of DNA came with the discovery of a heat-stable DNA polymerase in the thermophilic bacterium, *Thermus aquaticus*. This polymerase, called ***Taq* polymerase** (*T. aquaticus* polymerase), remains active during the heat denaturation step. As a result, polymerase does not have to be added after each cycle of denaturation. Instead, excess *Taq* polymerase and oligonucleotide primers can be added at the start of the PCR process, and amplification cycles can be carried out by sequential alterations in temperature. PCR machines or thermal cyclers change the

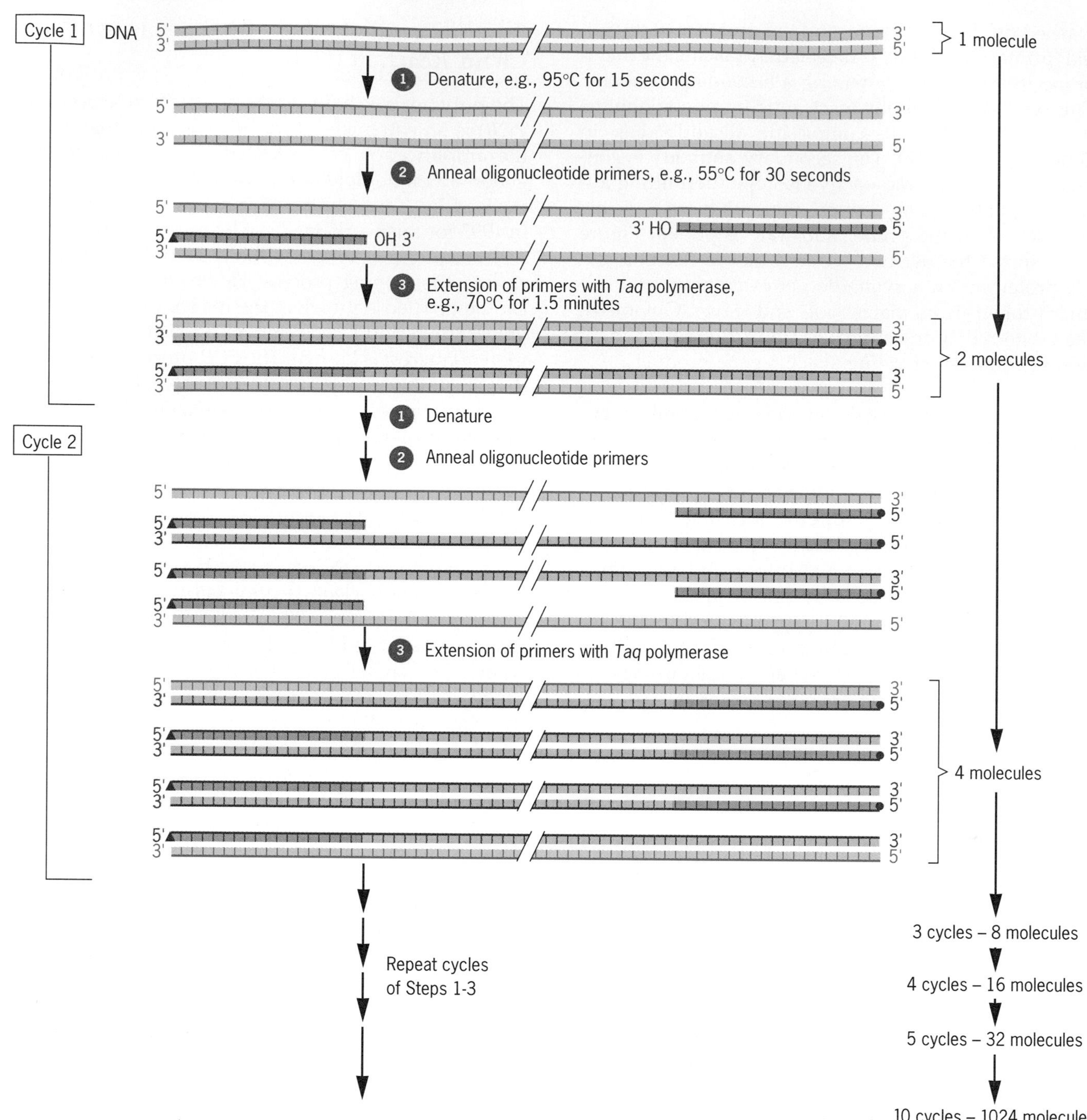

Figure 20.24 The use of PCR to amplify DNA molecules *in vitro*. Each cycle of amplification involves three steps: (1) denaturation of the genomic DNA being analyzed, (2) annealing of the denatured DNA to chemically synthesized oligonucleotide primers with sequences complementary to sites on opposite sides of the DNA region of interest, and (3) enzymatic replication of the region of interest by *Taq* polymerase.

temperature automatically and hold large numbers of samples, making PCR amplification of specific DNA sequences a simple task.

One disadvantage of PCR is that errors are introduced into the amplified DNA copies at low but significant frequencies. Unlike most DNA polymerases, *Taq* polymerase does not contain a built-in $3' \rightarrow 5'$ proofreading activity and, consequently, it produces a higher than normal frequency of replication errors. If an incorrect nucleotide is incorporated during an early

PCR cycle, it will be amplified just like any other nucleotide in the DNA sequence.

PCR technologies provide shortcuts for many cloning and sequencing applications. These procedures permit scientists to obtain definitive structural data on genes and DNA sequences when very small amounts of DNA are available. One important application occurs in the diagnosis of inherited human diseases, especially in cases of prenatal diagnosis, where limited amounts of fetal DNA are available. A second major application occurs in forensic cases involving the identification of individuals by using DNA isolated from very small tissue samples. Few criteria can provide more definitive evidence of identity than DNA sequences. By using PCR amplification, DNA sequences can be obtained from minute amounts of DNA isolated from small blood samples, semen, or even individual human hairs. Thus, **PCR DNA fingerprinting** experiments play important roles in legal cases involving uncertain identity. (See Technical Sidelight: DNA Tests and the Mystery of the Duchess Anastasia in Chapter 22.) These applications of PCR are discussed in Chapter 22.

Physical Maps of DNA Molecules Based on Restriction Enzyme Cleavage Sites

Most restriction endonucleases cleave DNA molecules in a site-specific manner (see Table 20.1). As a result, they can be used to generate **physical maps** of chromosomes that are of tremendous value in assisting researchers in isolating segments of DNA carrying genes or other DNA sequences of interest. The sizes of the restriction fragments can be determined by polyacrylamide or agarose gel electrophoresis (Figure 20.18). Because of the nucleotide subunit structure of DNA, with one phosphate group per nucleotide, DNA has an essentially constant charge per unit of mass. Thus, the rates of migration of DNA fragments during electrophoresis provide accurate estimates of their lengths; the rate of migration is inversely proportional to length.

A diagram of the procedure that is used to map the restriction enzyme cleavage sites on a DNA molecule is illustrated in Figure 20.25. The sizes of DNA restriction fragments are estimated by using a set of DNA markers of known size. In Figure 20.25, a set of DNA molecules that differ in length by 1000 nucleotide pairs are used as size markers. Consider a DNA molecule approximately 6000 nucleotide pairs (6 kb) in length. When the 6-kb DNA molecule is cut with *Eco*RI, two fragments of sizes 4000 and 2000 nucleotide pairs are produced. The possible positions of the single *Eco*RI cleavage site in the molecule are shown in Figure 20.25*b*. When the same DNA molecule is cleaved with *Hin*dIII, two fragments of sizes 5000 and 1000 nucleotide pairs result.

The possible locations of the single *Hin*dIII cleavage site are shown in Figure 20.25*c*. Note that at this stage of the analysis no deductions can be made about the relative positions of the *Eco*RI and *Hin*dIII cleavage sites. The *Hin*dIII cleavage site may be located in either of the two *Eco*RI restriction fragments. The molecule is then simultaneously digested with both *Eco*RI and *Hin*dIII, and three fragments of sizes 3000, 2000, and 1000 nucleotide pairs are produced. This result establishes the positions of the two cleavage sites relative to one another on the molecule. Since the 2000 nucleotide-pair *Eco*RI restriction fragment is still present (not cut by *Hin*dIII), the *Hin*dIII cleavage site must be at the opposite end of the molecule from the *Eco*RI cleavage site (Figure 20.25*d*). By extending this type of analysis to include the use of several different restriction enzymes, more extensive maps of restriction sites can be constructed. When large numbers of restriction enzymes are employed, detailed maps of entire chromosomes can be constructed. An important aspect of restriction maps is that, unlike genetic maps (Chapter 8), restriction maps reflect true physical distances along the DNA molecule.

During their early work on the characterization of restriction endonucleases, Hamilton Smith and Daniel Nathans used the DNA chromosome of simian virus 40 (SV40) for their studies. SV40 is an animal virus that can transform cells to the cancerous state. Its circular chromosome is only 5226 nucleotide pairs in length, making it well suited for restriction enzyme studies. As a result of their work, Smith and Nathans were able to construct a **restriction enzyme cleavage site map,** or more simply **restriction map,** of the SV40 chromosome.

By combining computer-assisted restriction mapping with other molecular techniques, it is possible to construct physical maps of entire genomes. The first multicellular eukaryote for which this was accomplished is *Caenorhabditis elegans,* a worm that is important for studies of the genetic control of development (Chapter 25). Moreover, the physical map of the *C. elegans* genome has been correlated with its genetic map. Thus, when an interesting new mutation is identified in *C. elegans,* its position on the genetic map often can be used to obtain clones of the wild-type gene from a large international *C. elegans* clone bank.

Nucleotide Sequences: The Ultimate Fine Structure Maps

The ultimate fine structure map of a specific gene or chromosome is its nucleotide-pair sequence, complete

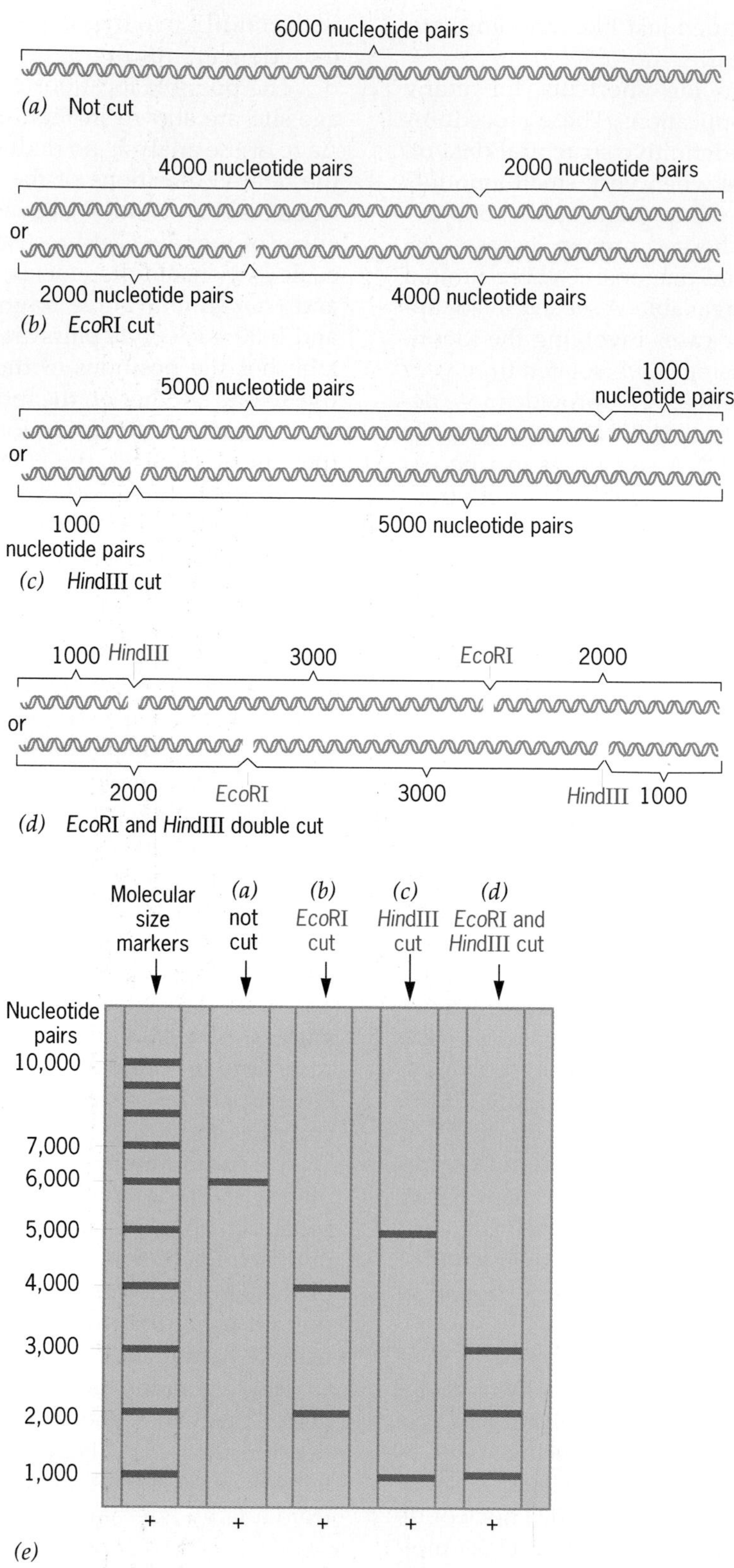

Figure 20.25 Procedure used to map restriction enzyme cleavage sites in DNA molecules. (*a-d*) Structures of the DNA molecule or of restriction fragments of the molecule either (*a*) uncut or cut with (*b*) *Eco*RI, (*c*) *Hin*dIII, or (*d*) *Eco*RI and *Hin*dIII. (*e*) The separation of these DNA molecules and fragments by agarose gel electrophoresis. The left lane on the gel contains a set of molecular size markers, a set of DNA molecules of size 1000 nucleotide pairs and multiples thereof.

with a chart of all nucleotide-pair changes that alter the function of that gene or chromosome. Prior to 1975, the thought of trying to sequence entire chromosomes was barely conceivable—at best, a laborious task requiring years of work. By late 1976, however, the entire 5386 nucleotide-long chromosome of phage ϕX174 had been sequenced. Today, sequencing is a routine laboratory procedure. The complete genomes of many viruses, 18 species of bacteria, the yeast *Saccharomyces cerevisiae*, mitochondria from several species, and chloroplasts from four species have already been determined. The sequence of the genome of the worm *Caenorhabditis elegans* will probably be reported before this book is in print, and that of the plant *Arabidopsis thaliana* will be completed shortly thereafter. Many other genome sequencing projects are in progress, and, as previously mentioned, one goal of the Human Genome Project is to complete the sequence of the human genome by 2003.

The present ability to sequence essentially any DNA molecule is the result of four major developments. The most important breakthrough was the discovery of restriction enzymes and their use in preparing homogeneous samples of specific segments of chromosomes. Another major advance was the improvement of gel electrophoresis procedures to the point where DNA fragments that differ in length by a single nucleotide could be resolved. Gene-cloning techniques to facilitate the preparation of large quantities of a particular DNA molecule were also developed. Finally, researchers invented two different procedures by which the nucleotide sequences of DNA molecules can be determined.

Both DNA sequencing procedures depend on the generation of a population of DNA fragments that all have one end in common (all end at exactly the same nucleotide) and terminate at all possible positions (every consecutive nucleotide) at the other end. These fragments are then separated on the basis of chain length by polyacrylamide gel electrophoresis. In both cases, four separate biochemical reactions are carried out simultaneously, each of which generates a set of fragments terminating at one of the four bases (A, G, C, or T) in DNA.

The first procedure, called the Maxam and Gilbert procedure after Allan Maxam and Walter Gilbert who invented it, uses four different chemical reactions to cleave DNA chains specifically at As, Gs, Cs, or Cs + Ts. The second procedure, developed by Fred Sanger and colleagues, uses *in vitro* DNA systhesis in the presence of radioactive nucleotides and specific chain-terminators to generate four populations of radioactively labeled fragments that end at As, Gs, Cs, and Ts, respectively. We will discuss the Sanger procedure.

2′,3′-Dideoxyribonucleoside triphosphates (Figure 20.26) are the chain-terminators most frequently used in the Sanger sequencing procedure. Recall that DNA polymerases have an absolute requirement for a free 3′-OH on the DNA primer strand (Chapter 11). If a 2′,3′-dideoxynucleotide is added to the end of a chain, it will block subsequent extension of that chain since the 2′, 3′-dideoxynucleotides have no 3′-OH. By using (1) 2′,3′-dideoxythymidine triphosphate (ddTTP), (2) 2′,3′-dideoxycytidine triphosphate (ddCTP), (3) 2′,3′-dideoxyadenosine triphosphate (ddATP), and (4) 2′,3′-dideoxyguanosine triphosphate (ddGTP) as chain-terminators in four separate DNA synthesis reactions, four populations of fragments can be generated, and each population will contain chains that all terminate with the same base (T, C, A, or G) (Figure 20.27).

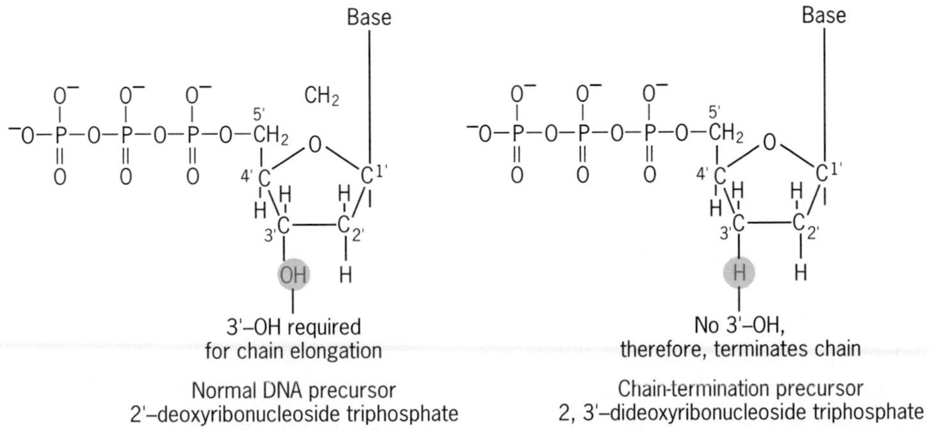

Figure 20.26 Comparison of the structures of the normal DNA precursor 2′-deoxyribonucleoside triphosphate and the chain-terminator 2′,3′-dideoxyribonucleoside triphosphate used in DNA sequencing reactions.

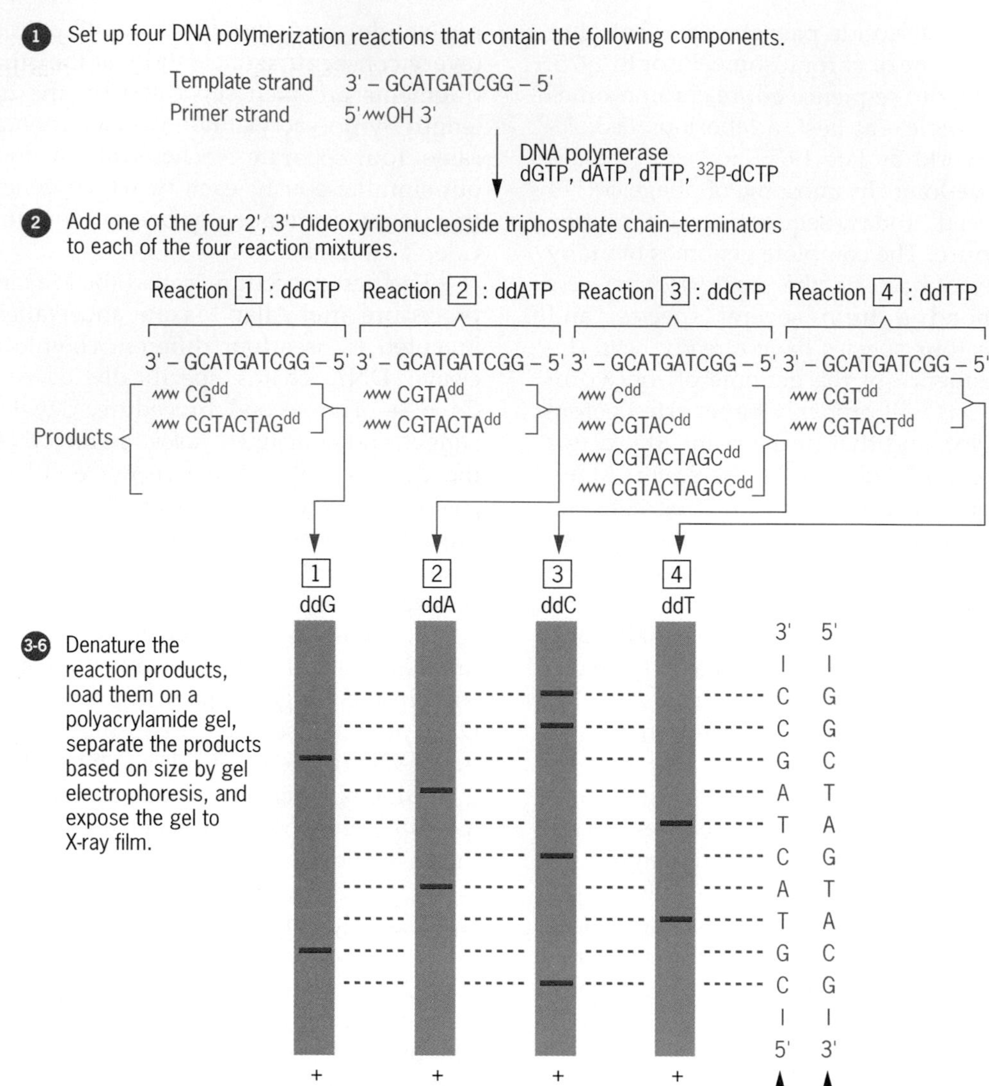

Figure 20.27 Sequencing DNA by the 2′,3′-dideoxynucleoside triphosphate chain-termination procedure. Four reactions are carried out in parallel, each of which contains one of the four 2′,3′-dideoxy chain-terminators: ddGTP, ddATP, ddCTP, and ddTTP. All four reaction mixtures contain the components required for DNA synthesis *in vitro*. One radioactive DNA precursor (^{32}P-dCTP here) is present in each reaction so that the products can be detected by autoradiography. The products of the four reactions are separated by polyacrylamide gel electrophoresis, and the positions of the nascent DNA chains in the gel are determined by autoradiography. Because the shortest chain migrates the greatest distance, the nucleotide sequence of the nascent chains (shown in red at the right of the autoradiogram) is obtained by reading the gel from the bottom (anode) to the top (cathode).

In a given reaction, the ratio of dXTP:ddXTP (where X can be any one of the four bases) is kept at approximately 100:1, so that the probability of termination at a given X in the nascent chain is about 1>100. This yields a population of fragments terminating at all potential (X) termination sites within a distance of a few hundred nucleotides from the original primer terminus.

After the DNA fragments generated in the four parallel reactions are released from the template strands by denaturation, they can be separated by polyacrylamide gel electrophoresis and their positions in the gel can be detected by autoradiography. The bands on the autoradiograms will correspond to radioactive chains of different lengths; they will produce a "ladder" defining the nucleotide sequence of the

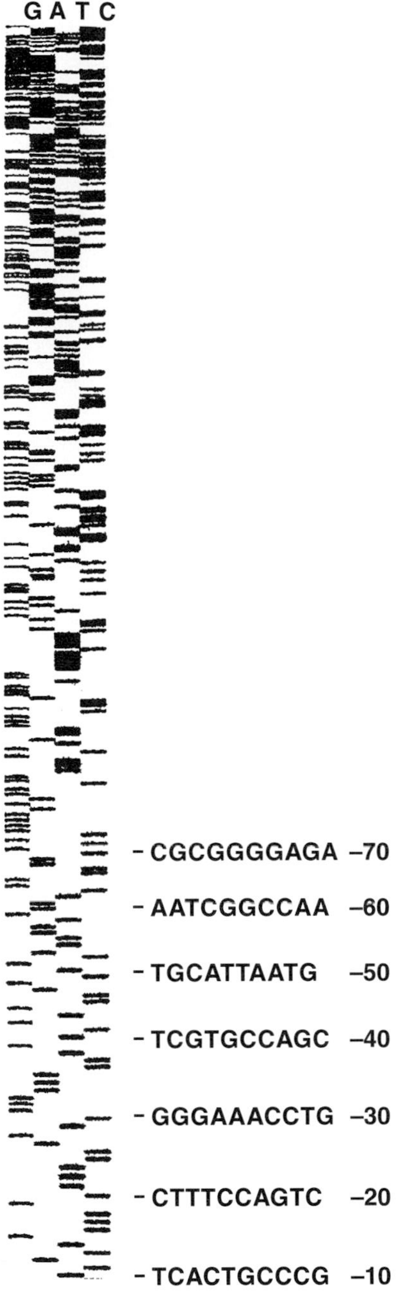

Figure 20.28 Photograph of an autoradiograph of a 2′,3′-dideoxynucleotide chain-terminator sequencing gel. The sequence defined by the lower portion of the gel is shown on the right.

longest chain that has been synthesized (Figures 20.27 and 20.28).

The shortest fragment will migrate the greatest distance and give rise to the band nearest the anode (the positive electrode). Each successive band will contain chains that are one nucleotide longer than the chains in the preceding band of the ladder. The 3′-terminal nucleotide of the chain in each band will be the dideoxynucleotide chain-terminator present in the reaction mixture (1, 2, 3, or 4) in which that specific chain was produced (Figure 20.27). By reading the ladder produced by autoradiography of the polyacrylamide gels used to separate the fragments generated in each of the four parallel reactions, the complete nucleotide sequence of a DNA chain can be determined. This is illustrated in Figure 20.27 for a hypothetical nucleotide sequence. An autoradiogram of an actual dideoxynucleotide chain-terminator sequencing gel is shown in Figure 20.28. Under optimal conditions, long sequences of several hundred nucleotides can be determined from a single sequencing gel.

Today, all large-scale DNA sequencing is done with automated DNA sequencing machines that use the dideoxy chain-termination procedure described above, but with modifications. Different DNA sequencing machines use slightly different protocols. However, the major differences between the slab-gel sequencing procedure illustrated in Figure 20.27 and automated DNA sequencing are (1) the use of fluorescent dyes, rather than radioactive isotopes, to detect DNA chains; (2) the separation of the products of all four dideoxy chain-terminator reactions by electrophoresis through a single gel or capillary tube; (3) the use of photocells to detect the fluorescence of the dyes as they pass through the gel or capillary tube; and (4) the direct transfer of the output of the photocell to a computer, which automatically analyzes, records, and prints out the results.

A different fluorescent dye is used to label the products of each of the four dideoxy chain-terminator sequencing reactions. As a result, the products of the four reactions can be distinguished by their fluorescence as they pass through a gel or capillary tube. The fluorescent dyes can be coupled to the primers used in the sequencing reactions or coupled directly to the dideoxynucleoside triphosphate chain-terminators. Figure 20.29 compares the automated DNA sequencing procedure with the slab-gel procedure described in Figure 20.27 and shows a computer printout of the results of automated sequencing of a short segment of DNA.

The newest automated DNA sequencing machines can perform 96 capillary electrophoresis separations simultaneously, and sample loading, electrophoresis, data collection, and data analysis are all fully automated. If operated continuously, a single 96-capillary machine can crank out over 100,000 nucleotides of sequence per day. Although that seems like a lot, remember that the human genome contains three billion nucleotide pairs!

Key Points: **The polymerase chain reaction can be used to amplify specific DNA sequences *in vitro* a millionfold or more. Detailed physical maps of DNA molecules can be prepared by identifying the**

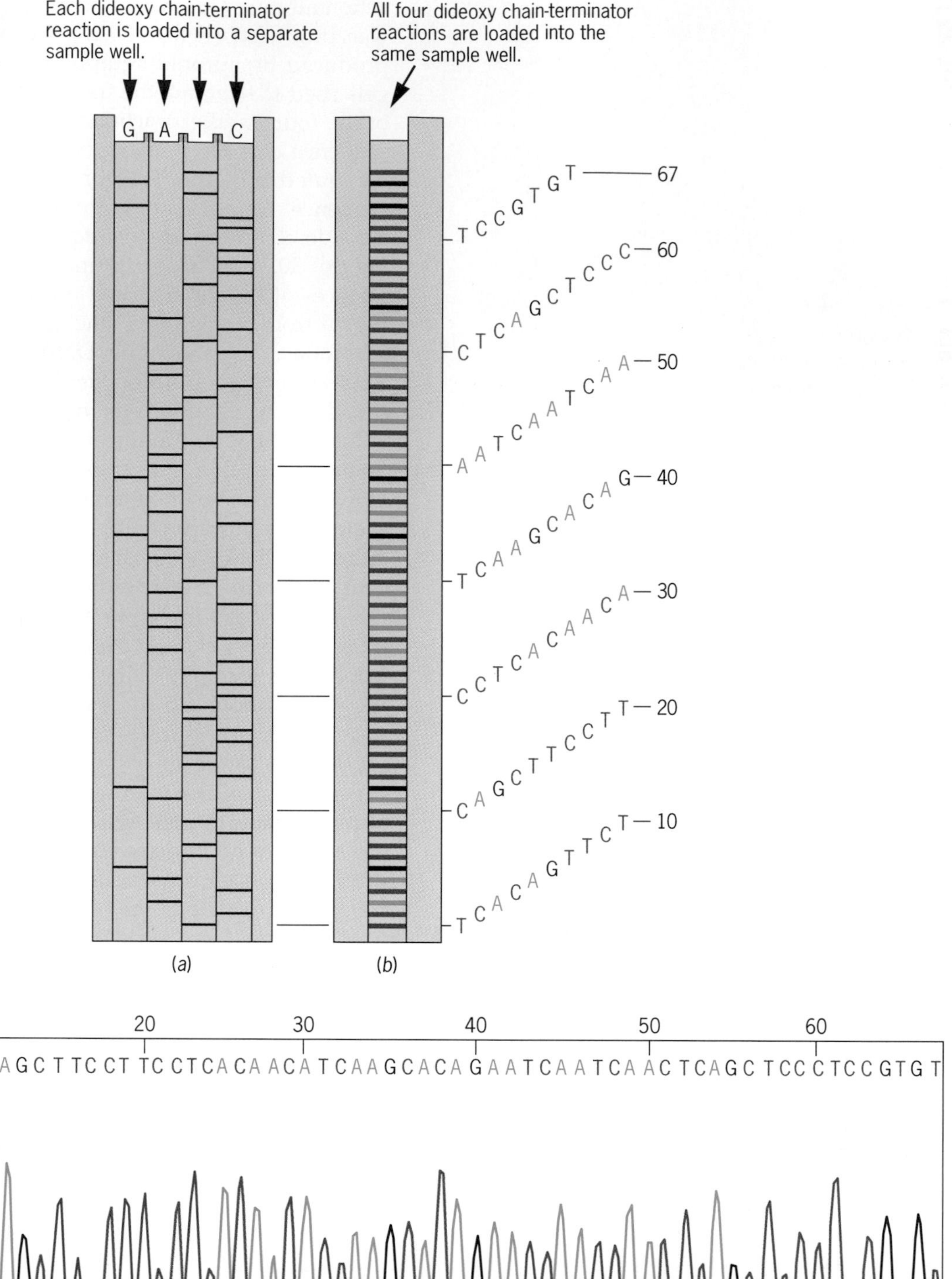

Figure 20.29 Comparison of the standard slab-gel (*a*) and automated (*b*) DNA sequencing procedures. (*c*) Computer printout of the results of an automated DNA sequencing run, showing the nucleotide sequence of a short segment of DNA.

sites that are cleaved by various restriction endonucleases. The nucleotide sequences of DNA molecules can be determined by the Maxam and Gilbert or Sanger procedures; these nucleotide sequences provide the ultimate physical maps of genes and chromosomes.

TESTING YOUR KNOWLEDGE

1. The human genome (haploid) contains about 3×10^9 nucleotide pairs of DNA. If you digest a preparation of human DNA with *Not*I, a restriction endonuclease that recognizes and cleaves the octameric sequence 5′-GCGGCCGC-3′, how many different restriction fragments would you expect to produce? Assume that the four bases (G, C, A, and T) are equally prevalent and randomly distributed in the human genome.

ANSWER

Assuming that the four bases are present in equal amounts and are randomly distributed, the chance of a specific nucleotide occurring at a given site is 1/4. The chance of a specific dinucleotide sequence (e.g., AG) occurring is $1/4 \times 1/4 = (1/4)^2$, and the probability of a specific octanucleotide sequence is $(1/4)^8$ or 1/65,536. Therefore, *Not*I will cleave such DNA molecules an average of once in every 65,536 nucleotide pairs. If a DNA molecule is cleaved at n sites, $n + 1$ fragments will result. A genome of 3×10^9 nucleotide pairs should contain about 45,776 ($3 \times 10^9/65{,}536$) *Not*I cleavage sites. If the entire human genome consisted of a single molecule of DNA, *Not*I would cleave it into 45,776 + 1 fragments. Given that these cleavage sites are distributed on 23 different chromosomes, complete digestion of the human genome with *Not*I should yield about 45,776 + 23 restriction fragments.

2. The maize gene (*gln2*) that encodes the chloroplastic form of the enzyme glutamine synthetase is known to contain a single cleavage site for *Hin*dIII, but no cleavage site for *Eco*RI. You are given an *E. coli* plasmid cloning vector that contains a unique *Hin*dIII cleavage site within the gene (amp^r) that confers resistance to the antibiotic ampicillin on the host cell and a unique *Eco*RI cleavage site within a second plasmid gene (tet^r) that makes the host cell resistant to the antibiotic tetracycline. You are also given an *E. coli* strain that is sensitive to both ampicillin and tetracycline (amp^s tet^s). How would you go about constructing a maize genomic DNA library that includes clones carrying a complete *gln2* gene?

ANSWER

Maize genomic DNA should be purified and digested with *Eco*RI. Vector DNA should be similarly purified and digested with *Eco*RI. The maize *Eco*RI restriction fragments and the *Eco*RI-cut plasmid DNA molecules will now have complementary single-stranded ends (5′-AATT-3′). The maize restriction fragments should next be mixed with the *Eco*RI-cut plasmid molecules and covalently inserted into the linearized vector molecules in an ATP-dependent reaction catalyzed by DNA ligase. The ligation reaction will produce circular recombinant plasmids, some of which will contain maize *Eco*RI fragment inserts. Insertion of maize DNA fragments into the *Eco*RI site of the plasmid disrupts the tet^r gene so that the resulting recombinant plasmids will no longer confer tetracycline resistance to host cells. Amp^s tet^s *E. coli* cells should then be transformed with the recombinant plasmid DNAs, and the cells should be plated on medium containing ampicillin to select for transformed cells harboring plasmids. The majority of the cells will not be transformed and, thus, will not grow in the presence of ampicillin. The cells that grow on ampicillin-containing medium should be pooled and frozen at -80°C in 20 percent glycerol. This collection of cells harboring different *Eco*RI fragments of the maize genome represents a clone library that should contain clones with an intact *gln2* gene since this gene contains no *Eco*RI cleavage site. Note that the *Hin*dIII site of the vector could be used to construct a similar maize genomic *Hin*dIII fragment library, but such a library would not contain intact *gln2* genes because of the *Hin*dIII cleavage site in *gln2*.

QUESTIONS AND PROBLEMS

20.1 (a) In what ways is the introduction of recombinant DNA molecules into host cells similar to mutation? (b) In what ways is it different?

20.2 Listed below are four different single strands of DNA. Which of these, in their double-stranded form, would you expect to be cleaved by a restriction endonuclease?

(a) ACTCCAGAATTCACTCCG (b) GCCTCATTCGAAGCCTGA
(c) CTCGCCAATTGACTCGTC (d) ACTCCACTCCCGACTCCA

20.3 If the sequence of base pairs along a DNA molecule occurs strictly at random, what is the expected frequency of a specific restriction enzyme recognition sequence of length (a) four and (b) six base pairs?

20.4 In what way(s) do restriction endonucleases differ from other endonucleases?

20.5 Of what value are recombinant DNA and gene-cloning technologies to geneticists?

20.6 What determines the sites at which DNA molecules will be cleaved by a restriction endonuclease?

20.7 Restriction endonucleases are invaluable tools for biologists. However, genes encoding restriction enzymes obviously did not evolve to provide tools for scientists. Of what possible value are restriction endonucleases to the microorganisms that produce them?

20.8 Why is the DNA of a microorganism not degraded by a restriction endonuclease that it produces, even though its DNA contains recognition sequences normally cleaved by the endonuclease?

20.9 One of the procedures for cloning foreign DNA segments takes advantage of restriction endonucleases like *Hin*dIII (see Table 20.1) that produce complementary single-stranded ends. These enzymes produce identical complementary ends on cleaved foreign DNAs and on the vector DNAs into which the foreign DNAs are inserted. What major advantage does this cloning strategy have over procedures that use terminal transferase to synthesize complementary single-stranded ends on foreign DNAs and vector DNAs *in vitro?*

20.10 You are working as part of a research team studying the structure and function of a particular gene. Your job is to clone the gene. A restriction map is available for the region of the chromosome in which the gene is located; the map is as follows.

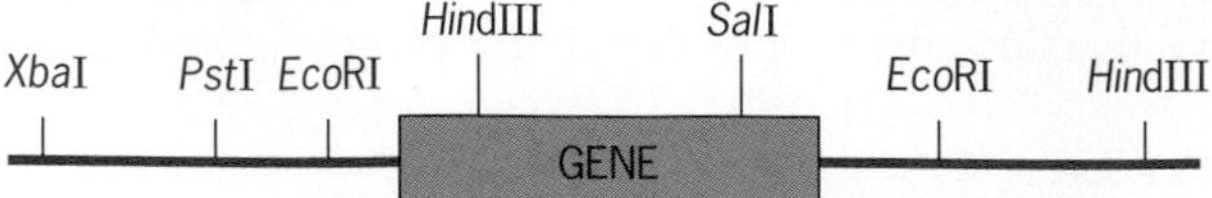

Your first task is to prepare a genomic DNA library that contains clones carrying the entire gene. Describe how you would prepare such a library in plasmid vector pUC118 (see Figure 20.14), indicating which restriction enzymes, media, and host cells you would use.

20.11 Compare the nucleotide-pair sequences of genomic DNA clones and cDNA clones of specific genes of higher plants and animals. What is the most frequent difference that you would observe?

20.12 Most of the genes of higher plants and animals that were cloned soon after the development of recombinant DNA and gene-cloning technologies were genes encoding products that are synthesized in large quantities in specialized cells or tissues. For example, about 90 percent of the protein synthesized in mature red blood cells of mammals consists of α- and β-globin chains, and the globin genes were among the first mammalian genes cloned. Why were genes of this type so prevalent among the first higher eukaryotic genes that were cloned?

20.13 Genomic clones of the chloroplastic glutamine synthetase gene (*gln2*) of maize are cleaved into two fragments by digestion with restriction endonuclease *Hin*dIII, whereas full-length maize *gln2* cDNA clones are not cut by *Hin*dIII. Explain these results.

20.14 In the following illustration, the upper line shows a gene composed of segments A-D. The lower circle shows a mutant version of this gene, consisting of two fused pieces (A′-B′, C′-D′), carried on a plasmid. You attempt a targeted mutagenesis of a diploid cell by transforming cells with the cloned mutant gene. The diagram below shows the desired pairing of the plasmid and chromosome just prior to recombination.

A x B C probe x D
A′ x B′ C′ x D′
plasmid

You prepare DNA from the cells, digest it with an enzyme that cuts at x, and hybridize the cleaved DNA with the probe shown above. The diagram below shows a Southern blot of possible results. (a) Which lane shows fragments produced from DNA in the cell before transformation? (b) Which lane shows fragments produced from DNA in the cell in which the anticipated targeted mutagenesis occurred? (c) Which of these blot patterns might be expected if two crossovers occurred, one between A and B, and the other between C and D?

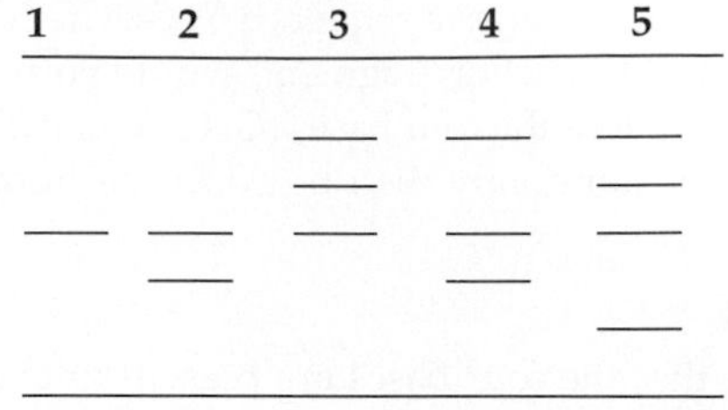

20.15 You are studying a gene in *E. coli* that is expressed at 45° C but not at 37° C. You have shown that the regulation of this gene occurs at the level of transcription. In addition, you have isolated a protein required for induction of transcription of the gene at high temperature and have demonstrated that it binds to a specific octameric nucleotide-pair sequence upstream (5′) from the gene. You have cloned the complete gene plus the upstream regulatory sequence in a plasmid, and you have available an *E. coli* mutant carrying a deletion of the entire gene. You now want to determine which of the nucleotide pairs in the 5′ octameric protein binding site are involved in the interaction with (binding of) the regulatory protein. How could you determine this experimentally?

20.16 What major advantage does the polymerase chain reaction (PCR) technique have over all other methods for analyzing nucleic acid structure and function?

20.17 Certain types of molecular analyses are facilitated by the availability of large quantities of pure single-stranded DNA from a normally double-stranded DNA molecule. However, the separation and purification of the two single strands of a given double helix are usually very difficult to accomplish by standard biochemical techniques. How have molecular biologists taken advantage of a natural biological mechanism for the purification of DNA single strands?

20.18 Almost all the sophisticated cloning vectors—phages, plasmids, or phage-plasmid hybrids—in use today have one component (in addition to some type of origin of replication) in common. What is this component, and what is its function?

20.19 The drawing below shows a restriction map of a segment of a DNA molecule. Eco refers to locations where the restriction endonuclease *Eco*RI cuts the DNA, and Pst refers to locations where the restriction enzyme *Pst*I cuts the DNA. Potential restriction site locations are numbered 1–6. Distances between restriction sites are shown on the bottom scale in base pairs (bp). The thick line represents the part of the molecule that has homology with a probe.

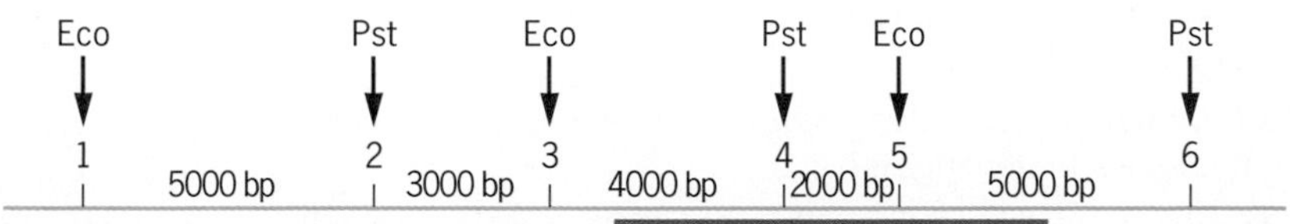

(a) Assume that individual 1 has restriction sites 1 through 6. If DNA is digested with *Pst*I, what are the expected sizes of the DNA fragments that will hybridize with the probe? (b) Assume that individual 2 has a mutation that eliminates site 4. If DNA is digested with *Pst*I, what are the expected sizes of the DNA fragments that will hybridize with the probe? (c) Assume that individual 3 has a mutation that eliminates site 5. If the DNA is digested with *Pst*I, what are the expected sizes of the DNA fragments that will hybridize with the probe? (d) If the DNA of individual 1 is digested with both *Pst*I and *Eco*RI, what are the expected sizes of the DNA fragments that will hybridize with the probe? (e) If the DNA of individual 3 is digested with both *Pst*I and *Eco*RI, what are the expected sizes of the DNA fragments that will hybridize with the probe?

20.20 An important tool of molecular biologists involves the transfer of proteins that have been separated by gel electrophoresis to nitrocellulose membranes and the detection of specific proteins on the membranes by using antibodies and coupled radioactive labels or coupled enzymatic reactions. When this procedure is used, the resulting display of the visualized protein bands is called a western blot. What is the significance of the name "western blot"?

20.21 The cystic fibrosis (*CF*) gene (location: chromosome 7, region q31) has been cloned and sequenced, and studies of CF patients have shown that about 70 percent of them are homozygous for a mutant *CF* allele that has a specific three nucleotide-pair deletion (equivalent to one codon). This deletion results in the loss of a phenylalanine residue at position 508 in the predicted CF gene product. Assume that you are a genetic counselor responsible for advising families with CF in their pedigrees regarding the risk of CF among their offspring. How might you screen putative CF patients and their parents and relatives for the presence of the *CFΔF508* mutant gene? What would the detection of this mutant gene in a family allow you to say about the chances that CF will occur again in the family?

20.22 Cereal grains are major food sources for humans and other animals in many regions of the world. However, most cereal grains contain inadequate supplies of certain of the amino acids that are essential for monogastric animals such as humans. For example, corn contains insufficient amounts of lysine, tryptophan, and threonine. Thus, a major goal of plant geneticists is to produce corn varieties with increased kernel lysine content. As a prerequisite to the engineering of high-lysine corn, molecular biologists need more basic information about the regulation of the biosynthesis and the activity of the enzymes involved in the synthesis of lysine. The first step in the anabolic pathway unique to the biosynthesis of lysine is catalyzed by the enzyme dihydrodipicolinate synthase. Assume (1) that you have recently been hired by a major U.S. plant research institute and (2) that you have been asked to isolate a clone of the nucleic acid sequence encoding dihydrodipicolinate synthase in maize. Briefly describe four different approaches you might take in attempting to isolate such a clone and include at least one genetic approach.

20.23 (a) What common experimental procedure is carried out in Southern, northern, and western blot analyses? (b) What is the major difference between Southern, northern, and western blot analyses?

20.24 You have isolated a cDNA clone encoding a protein of interest in a higher eukaryote. This cDNA clone is *not* cleaved by restriction endonuclease *Eco*RI. When this cDNA is used as a radioactive probe for blot hybridization analysis of *Eco*RI-digested genomic DNA, three radioactive bands are seen on the resulting Southern blot. Does this result indicate that the genome of the eukaryote in question contains three copies of the gene encoding the protein of interest?

20.25 A DNA molecule is subjected to single and double digestions with restriction endonucleases, and the following results are obtained.

Enzymes	Fragments (in kb)
*Eco*RI	2.9, 4.5, 7.4, 8.0
*Hin*dIII	3.9, 6.0, 12.9
*Eco*RI and *Hin*dIII	1.0, 2.0, 2.9, 3.5, 6.0, 7.4

Draw the restriction map defined by these data.

20.26 A DNA molecule is subjected to single and double digestions with restriction enzymes, and the products are separated by gel electrophoresis. The results are given below (fragment sizes are in kb):

*Eco*RI	*Eco*RI and *Hin*dIII	*Hin*dIII	*Bam*HI	*Eco*RI and *Bam*HI	*Hin*dIII and *Bam*HI
8	5	12	6	6	6
4	4		6	4	5
	3			2	1

Draw the restriction map of this DNA molecule.

20.27 You are studying a circular plasmid DNA molecule of size 10.5 kilobase pairs (kb). When you digest this plasmid with restriction endonucleases *Bam*HI, *Eco*RI, and *Hin*dIII, singly and in all possible combinations, you obtain linear restriction fragments of the following sizes.

Enzyme(s)	Fragment Size(s) in kb
*Bam*HI	7.3, 3.2
*Eco*RI	10.5
*Hin*dIII	5.1, 3.4, 2.0
*Bam*HI + *Eco*RI	6.7, 3.2, 0.6
*Bam*HI + *Hin*dIII	4.6, 2.7, 2.0, 0.7, 0.5
*Eco*RI + *Hin*dIII	4.0, 3.4, 2.0, 1.1
*Bam*HI + *Eco*RI + *Hin*dIII	4.0, 2.7, 2.0, 0.7, 0.6, 0.5

Draw a restriction map (map of the restriction enzyme cleavage sites) for the plasmid that fits your data.

20.28 The following is a segment of an autoradiogram of a dideoxy sequencing gel. The origin is at the top. What is the sequence (including polarity) of the single-stranded DNA that served as a template for generating this pattern? (Use all of the bands on the gel.)

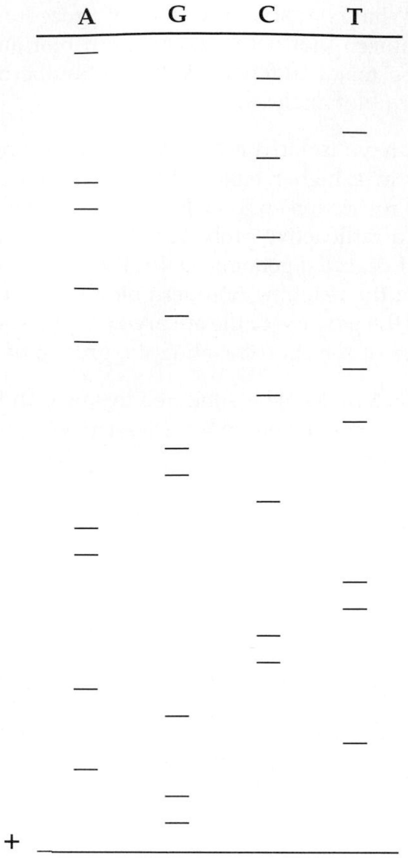

A G C T

+

20.29 Ten micrograms of a decanucleotide-pair *Hpa*I restriction fragment were isolated from the double-stranded DNA chromosome of a small virus. Octanucleotide poly-A tails were then added to the 3′-ends of both strands using terminal transferase and dATP, that is,

5′-X X X X X X X X X X-3′
3′-X′X′X′X′X′X′X′X′X′X′-5′
↓ terminal transferase, dATP
5′-X X X X X X X X X X A A A A A A A A-3′
3′-A A A A A A A A X′X′X′X′X′X′X′X′X′X′-5′

where X and X′ can be any of the four standard nucleotides, but X′ is always complementary to X.

The two complementary strands (Watson strand and Crick strand) were then separated and sequenced by the 2′, 3′-dideoxyribonucleoside triphosphate chain-termination method. The reactions were all primed using a radioactive (^{32}P-labeled) synthetic poly(T) octamer, that is,

Watson strand

3′-A A A A A A A A X′X′X′X′X′X′X′X′X′X′-5′
5′-^{32}P-T T T T T T T T OH

Crick strand

5′-X X X X X X X X X X A A A A A A A A-3′
HO T T T T T T T T-^{32}P-5′

The usual four parallel reactions: (1) ddTTP, (2) ddCTP, (3) ddATP, and (4) ddGTP (plus DNA polymerase and all other substrates and required components) were carried out for both strands. Each reaction mixture was applied to a lane in a polyacrylamide gel, fractionated by electrophoresis, and autoradiographed. The autoradiogram of the sequencing gel for one of the strands is shown on the left of the following diagram. Draw the banding pattern that would be expected on the autoradiogram of the gel for the complementary strand on the right in the diagram.

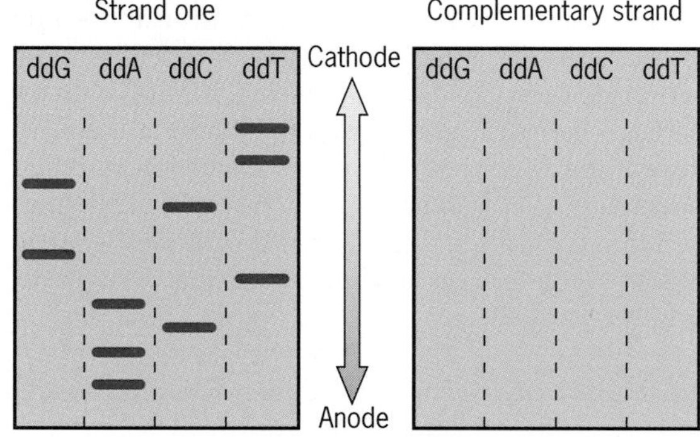

BIBLIOGRAPHY

Cohen, S. N. 1975. The manipulation of genes. *Sci. Amer.* 233(1):24–33.

Drlica, K. 1997. *Understanding DNA and Gene Cloning, a Guide for the Curious,* 3rd ed. John Wiley & Sons, New York.

Hawley, R. S., and C. A. Mori. 1999. *The Human Genome, a User's Guide.* Academic Press, New York.

Mullis, K. B. 1990. The unusual origin of the polymerase chain reaction. *Sci. Amer.* 262(4):56–65.

Rowen, L., G. Mahairas, and L. Hood. 1997. Sequencing the human genome. *Science* 278:605–607.

Sanger, F., S. Nicklen, and A. R. Coulson. 1977. DNA sequencing with chain-terminating inhibitors. *Proc. Natl. Acad. Sci., U.S.* 74:5463–5467.

Southern, E. M. 1975. Detection of specific sequences among DNA fragments separated by gel electrophoresis. *J. Mol. Biol.* 98: 503–517.

Vieira, J., and J. Messing. 1987. Production of single-stranded plasmid DNA. *Methods in Enzymology* 153: 3–11.

Watson, J. D., M. Gilman, J. Witkowski, and M. Zoller. 1992. *Recombinant DNA,* 2nd ed. Scientific American Books, New York.

Watson, J. D., and J. Tooze. 1981. *The DNA Story.* W. H. Freeman, San Francisco.

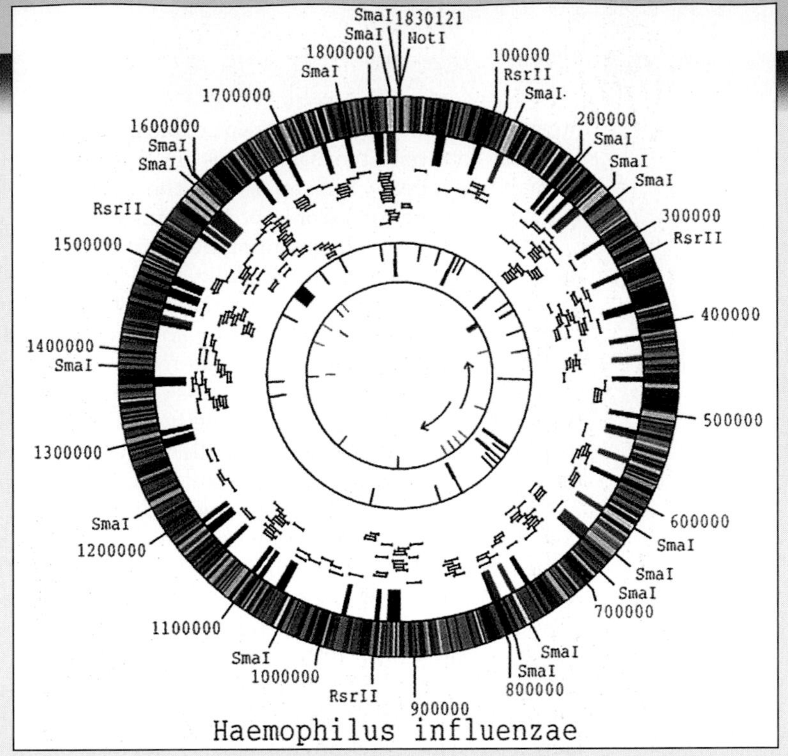

Map of the chromosome of Haemophilus influenzae. *This bacterium was the first free-living organism to have its entire genome—all 1,830,137 nucleotide pairs—sequenced. Distances in nucleotide pairs and the locations of* NotI, RsrII, *and* SmaI *restriction sites are given outside the circular map. The outer concentric circle shows genes color-coded according to function. For example, genes shown in green, yellow, and pink encode products that are involved in energy metabolism, DNA replication, and translation, respectively.*

21

Genomics

CHAPTER OUTLINE

Genomics: An Overview
Correlated Genetic, Cytological, and Physical Maps of Chromosomes
Map Position-Based Cloning of Genes
The Human Genome Project
RNA and Protein Assays of Genome Function
Evolution of Genomes

Human Gene Prospecting in Iceland

"Changes that will have effects comparable to the Industrial Revolution and the computer-based revolution are now beginning. The next great era, a genomics revolution, is in an early phase." (Philip H. Abelson, Science *279: 2019, March 27, 1998).*

The availability of detailed maps and nucleotide sequences of entire genomes of many species allows scientists to perform computer searches for sequences that encode enzymes with desired activities, to isolate these sequences, and to introduce them into the genomes of other species. Scientists have already engineered transgenic plants that are herbicide and insect resistant, as well as plants that synthesize antibodies, drugs, enhanced levels of vitamins, and even plastics.

Yes, the "genomics revolution" has begun, and the race to sequence the human genome and to identify genes that encode valuable products is underway. Although many scientists object to the patenting of living organisms and DNA sequences isolated from them, the courts have ruled that both genes and transgenic organisms are subject to patent protection. Indeed, patents that cover the nucleotide sequences of over a thousand genes and a large number of transgenic plants and animals have already been approved. The race to complete the sequence of the human genome and to patent important human genes is obviously on!

An interesting component of the effort to identify human genes with important pharmacological value is taking place in a somewhat unexpected location—the remote island of Iceland, located in the North Atlantic between Greenland and Scandinavia. Because of their history and geographical isolation, the 270,000 people of Iceland provide a unique resource for genetic studies. They are the genetically quite homogeneous descendants of Vikings who settled on this island more than 1100 years ago. This homogeneity has been enhanced by two "genetic bottlenecks" during which the population of Iceland was sharply reduced. During the fifteenth century, the population plummeted from about 70,000 to around 25,000 when bubonic plague ravished the island. During the 1700s, the population dropped below 50,000 on three occasions because of famine and disease caused in part by the eruption of the volcano Hekla. Thus, the human gene pool of Iceland is much more homogeneous than the gene pools of most other human populations. In addition, Iceland's national health service has kept superb family medical records since 1915.

In 1997, Kari Stefansson, a Harvard geneticist, recognized the uniqueness of Iceland's human gene pool and family records. He returned to his homeland to launch a private company, deCODE Genetics, with the goal of identifying human genes that would lead to the development of new pharmaceutical drugs and diagnostic tests. The company's first success was the identification of the *familial essential tremor* gene—a gene associated with shakiness in the elderly. In addition, deCODE scientists have made rapid progress in their studies of several other human disorders.

Based on these results, deCODE Genetics negotiated a $200 million contract with the Swiss pharmaceutical giant Hoffmann-LaRoche, which will give the Swiss firm exclusive rights to any drugs or diagnostic products resulting from the work of deCODE scientists. However, the contract specifies that Hoffmann-LaRoche must provide free of charge all drugs, diagnostic tests, and other products resulting from this research to the people of Iceland. Therefore, at least in this one case, the people who are providing the genetic data and the DNA samples for analysis will personally benefit from the results of research by a private company.

Whereas Mendel studied the effects of the alleles of seven genes (but no more than three in any one cross), today's geneticists will soon be able to study the expression of all the genes—the entire genome—of an organism in a single experiment. As of December 1998, complete nucleotide sequences are available for the genomes of eighteen species of bacteria, mitochondria from several species, chloroplasts from four species, and one eukaryote, the yeast *Saccharomyces cerevisiae.* The nucleotide sequence of the genome of the worm *Caenorhabditis elegans* is 99.9 percent complete, and the genomic sequences of the fruit fly *Drosophila melanogaster* and the plant *Arabidopsis thaliana* will soon follow. One of the original goals of the Human Genome Project was to determine the complete nucleotide sequence of the human genome by the year 2005. Now, the co-founders of a new company claim that they will sequence the human genome by the year 2001 (see Human Genetics Sidelight: The Race to Sequence the Human Genome). Once the complete sequence of the genome of an organism is known, the new "DNA chip" technologies will allow scientists to study the expression of all the genes in the organism simultaneously (see RNA and Protein Assays of Genome Function later in this chapter).

In this chapter, we will examine some of the tools and techniques that are used to study the structure and function of genomes, especially the human genome, and we will look at what has been learned to date by the analysis of complete genomes. In the following chapter, we will discuss additional uses of these tools and will see how they have been used to identify the defective human genes that are responsible for Huntington's disease and cystic fibrosis.

GENOMICS: AN OVERVIEW

Geneticists have used the term *genome* for over seven decades to refer to one complete copy of the genetic information or one complete set of chromosomes (monoploid or haploid) of an organism. In contrast, **genomics** is a relatively new term. The word *genomics* appears to have been coined by Thomas Roderick in 1986 to refer to the genetics subdiscipline of mapping, sequencing, and analyzing the functions of entire genomes, and to serve as the name of a new journal—*Genomics*—dedicated to the communication of new information in this subdiscipline.

As more detailed maps and sequences of genomes became available, the genomics subdiscipline was divided into **structural genomics**—the study of the structure of genomes—and **functional genomics**—the study of the function of genomes. (However, be forewarned! These terms are being used in many different ways by nongeneticists. To many protein chemists, "structural genomics" means the determination of the three-dimensional structures of all the proteins of an organism.)

Whereas structural genomics is quite advanced—given the availability of complete nucleotide sequences for numerous bacteria and a budding yeast, and rapid progress in sequencing the genomes of several multicellular eukaryotes including humans—functional genomics is just entering what promises to be an explosive phase of growth. The new array hybridization and gene-chip technologies allow researchers to monitor the expression of entire genomes—all the genes in an organism—at various stages of growth and development or in response to environmental changes. These powerful new tools promise to provide a wealth of information about genes and how they interact with each other and with the environment.

Key Point: **Genomics is the subdiscipline of genetics devoted to the mapping, sequencing, and functional analysis of genomes.**

CORRELATED GENETIC, CYTOLOGICAL, AND PHYSICAL MAPS OF CHROMOSOMES

The ability of scientists to identify and isolate genes based on information about their location in the genome was one of the first major contributions of genomics research. In principle, this approach, called *positional cloning,* can be used to identify and clone any gene with a known phenotypic effect in any species. Positional cloning has been used extensively in many species, including humans. Indeed, in Chapter 22, we will consider the use of positional cloning to identify the human genes responsible for Huntington's disease and cystic fibrosis.

Because the utility of positional cloning depends on the availability of detailed maps of the regions of the chromosomes where the genes of interest reside, major efforts have focused on the development of detailed maps of the human genome and the genomes of important model organisms such as *D. melanogaster, C. elegans,* and *A. thaliana.* The goal of this research is to construct correlated genetic and physical maps with markers distributed at relatively short intervals throughout the genome. In the case of the human and *Drosophila* genomes, the genetic and physical maps can also be correlated with cytological maps (banding patterns) of the chromosomes (Figure 21.1). We will discuss the construction of these maps in the following sections of this chapter.

Recall that genetic maps (Figure 21.1, left) are constructed from recombination frequencies, with 1 centiMorgan (cM) equal to the distance that yields an average frequency of recombination of 1 percent (Chapter 8). Genetic maps with markers spaced at short intervals—high-density genetic maps—are often constructed by using molecular markers such as restriction fragments of different lengths (*restriction fragment-length polymorphisms* or *RFLPs*). Cytological maps (Figure 21.1, center) are based on the banding patterns of chromosomes observed with the microscope after treatment with various stains (Chapter 7). **Physical maps** (Figure 21.1, right), such as the restriction maps (Figure 21.1, top right) discussed in Chapter 20, are based on the molecular distances—base pairs (bp), kilobases (kb, 1000 bp), and megabases (mb, 1 million bp)—separating sites on the giant DNA molecules present in chromosomes. Physical maps often contain the locations of overlapping genomic clones or *contigs* (Figure 21.1, right center) and unique nucleotide sequences called *sequence-tagged sites* or *STSs* (Figure 21.1, bottom right).

Physical maps of a chromosome can be correlated with the genetic and cytological maps in several ways. Genes that have been cloned can be positioned on the cytological map by *in situ* hybridization (Chapter 10). Correlations between the genetic and physical maps can be established by locating clones of genetically mapped genes or RFLPs on the physical map. Markers that are mapped both genetically and physically are called **anchor markers;** they anchor the physical map to the genetic map and vice versa. Physical maps

HUMAN GENETICS SIDELIGHT

The Race to Sequence the Human Genome

The Human Genome Project was launched in 1990 with the goal of sequencing the entire human genome—all 3 billion nucleotide pairs—by 2005 at an estimated cost of $3 billion. During 1997 and 1998, eight research teams in the United States and the team at the Sanger Centre in the United Kingdom were gearing up to begin high-output sequencing of their designated portions of the genome. During May 1998, J. Craig Venter (Figure 1), whose Institute for Genomic Research in Rockville, Maryland, has sequenced several bacterial genomes, stunned the genomics community by announcing that he had teamed up with Perkin-Elmer Corporation of Norwalk, Connecticut, and formed a new company, Celera Inc., that will sequence the human genome in just three years and at a cost of only $300 million. Shortly thereafter, another company, Incyte Pharmaceuticals Inc. in Palo Alto, California, announced plans to spend $200 million during the next two years to sequence the protein-coding regions (expressed genes) of the human genome. Obviously, the commercial race to identify, sequence, and patent important human genes is underway.

Venter's proposal is based on two key developments: (1) a sequencing strategy called **whole-genome shotgun sequencing** and (2) the development of faster, fully automated sequencing machines by Perkin-Elmer's Applied Biosystems division. Instead of marching down a chromosome sequencing carefully mapped genomic clones, whole-genome sequencing involves chopping the entire genome into small fragments, sequencing just the ends of the fragments, and using supercomputers to find overlapping sequences and put the complete sequence together. This procedure has worked well for the much smaller bacterial genomes, but some sequencing experts are skeptical about its utility with the large genomes of eukaryotes. Whereas bacterial genomes contain little repetitive DNA, eukaryotic genomes contain large amounts of highly repetitive DNA. These highly repetitive sequences will act as roadblocks to the assembly process and result in gaps in the computer-assembled sequence.

Venter admits that his sequence of the human genome will contain perhaps 2000 to 5000 gaps. Francis Collins (Figure 1), current director of the Human Genome Project, estimates that there will be tens of thousands of gaps in Venter's sequence, and other critics of the whole-genome shotgun sequencing strategy argue that there will be over 100,000 gaps, which will have to be filled in by the more methodical approaches being used in the other laboratories.

Venter's success will depend in large part on the performance of the new DNA sequencing machine developed by Perkin-Elmer. The sequencer can simultaneously separate the products of 96 dideoxy chain-terminator reactions (Chapter 20) and reportedly can operate unattended for 24 hours.

Francis Collins

J. Craig Ventner

Figure 1. Two of the major players in the race to sequence the human genome: Francis Collins, director of the Human Genome Project, and J. Craig Venter, president of Celera Inc., a new company that plans to sequence the human genome in just three years.

All functions, including gel loading, sample loading, electrophoresis, and data collection and analysis, are fully automated. The new sequencing machines are capable of performing about 10 sets of fully automated sequencing runs a day. According to Venter, the new company's sequencing center will house 230 of the new sequencers, which will crank out 100 million nucleotide pairs of sequence per day—a dramatic increase in productivity compared to sequencing in the past (Figure 2). It remains to be seen whether the new sequencers will perform up to expectations. Some skeptics doubt that all the bugs have been worked out of the new machines.

To test the new machines and the effectiveness of the whole-genome shotgun sequencing approach when applied to eukaryotic genomes, Venter's new company first plans to sequence the *Drosophila melanogaster* genome. Then, with the *Drosophila* sequence in hand, it will focus all of its effort on sequencing the human genome.

Some human geneticists wonder about the motivation behind this private sequencing venture. What is in it for Perkin-Elmer Corporation? Why would it invest $300 million to sequence the human genome? Will it release the data promptly to the public databases? Will it patent the sequences of important genes? Venter insists that sequence data will be deposited in the public databases every three months and that only selected genes with major pharmaceutical applications will be patented. Whereas definitive answers to these questions are not yet available, one thing is certain: Venter's proposal has stimulated investigators involved in the public human genome sequencing project to reevaluate their proposed timetable. In October 1998, Fran-

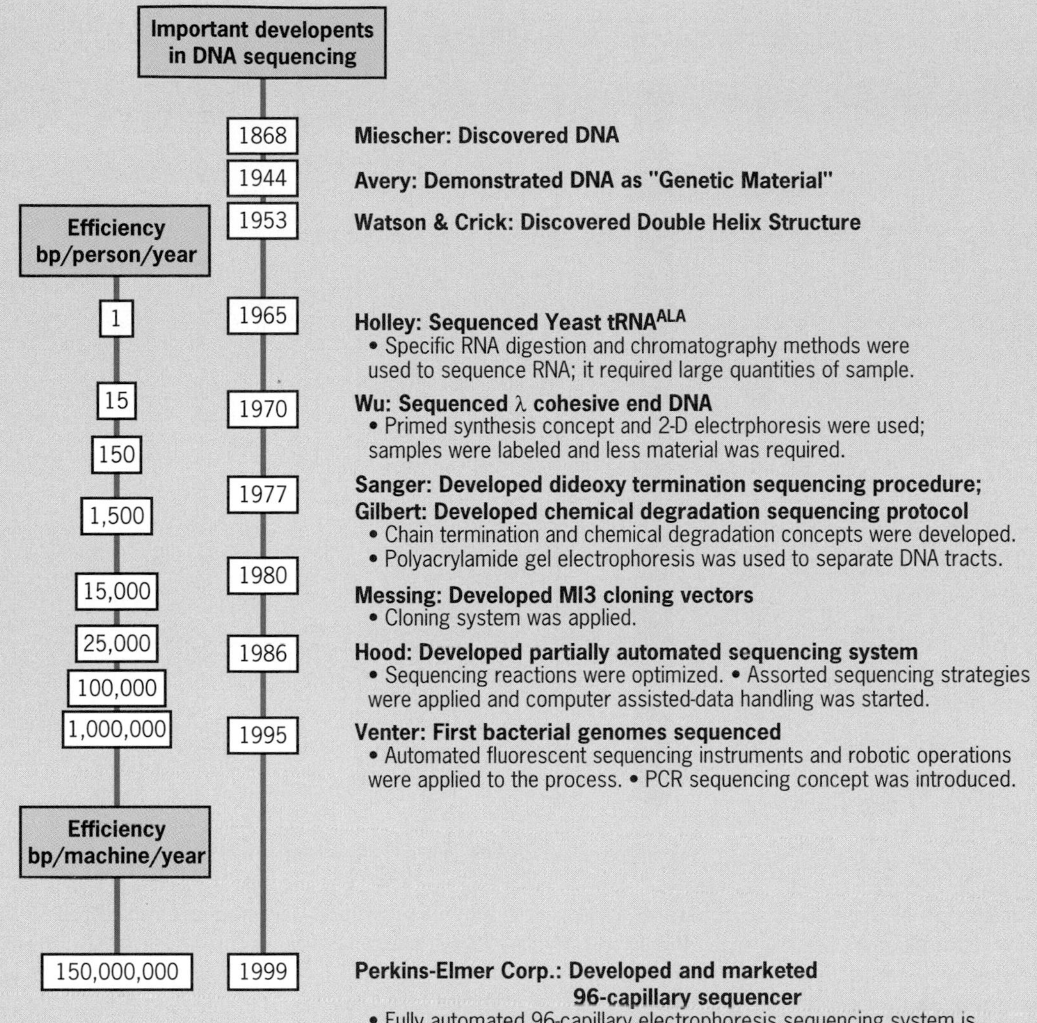

Figure 2. Advances in DNA sequencing efficiency and some of the technological developments that enhanced the productivity of sequencers. Initially, all the steps in DNA sequencing were performed manually, making it a very labor-intensive process. However, fully automated sequencing machines have now largely replaced human sequencers, greatly increasing efficiency.

cis Collins and colleagues announced that they would complete the sequence of the human genome by 2003—two years earlier than originally proposed. Moreover, they propose to have a "working draft" of the human genome—at least 90 percent sequenced with 99 percent accuracy—by 2001. Maybe a little competition will enhance the progress of both the private and public human genome projects. In any case, the race is on!

of chromosomes can also be correlated with genetic and cytological maps by using (1) PCR (see Figure 20.24) to amplify short—usually 200 to 500 bp—unique genomic DNA sequences, (2) Southern blots to relate these sequences to overlapping clones on physical maps, and (3) *in situ* hybridization to determine their chromosomal locations (cytological map positions). These short, unique anchor sequences are called **sequence-tagged sites** (**STSs**). Another approach uses short cDNA sequences (DNA copies of mRNAs) or **expressed-sequence tags** (**ESTs**) as hybridization probes to anchor physical maps to RFLP maps (genetic maps) and cytological maps.

Physical distances do not correlate directly with genetic map distances because recombination frequencies are not always proportional to molecular dis-

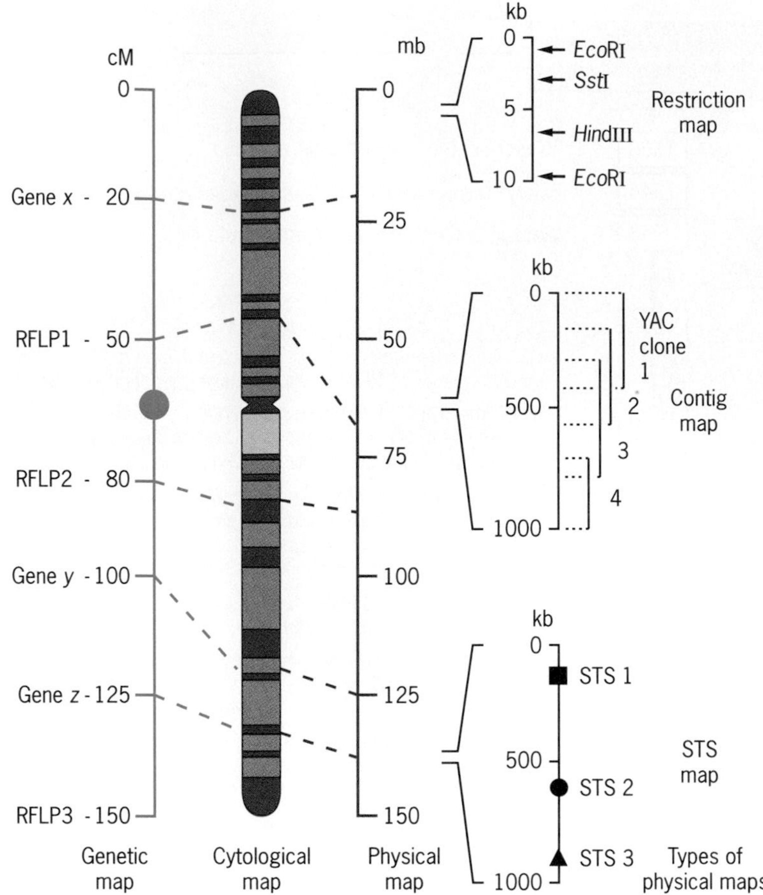

Figure 21.1 Correlation of the genetic, cytological, and physical maps of a chromosome. Genetic map distances are based on crossing over frequencies and are measured in percentage recombination or centiMorgans (cM), whereas physical distances are measured in kilobase pairs (kb) or megabase pairs (mb). Restriction maps, contig maps, and STS (sequence-tagged site) maps are described in the text.

tances. However, the two are often reasonably well correlated in euchromatic regions of chromosomes. In humans, 1 cM is equivalent, on average, to about 1 mb of DNA.

Restriction Fragment-Length Polymorphism (RFLP) and Microsatellite Maps

When mutations change the nucleotide sequences in restriction enzyme cleavage sites, the enzymes no longer recognize them (Figure 21.2*a*). Other mutations may create new restriction sites. These mutations result in variations in the lengths of the DNA fragments produced by digestion with various restriction enzymes (Figure 21.2*b*). Such restriction fragment-length polymorphisms, or RFLPs, have proven invaluable in constructing detailed genetic maps for use in positional cloning. The RFLPs are mapped just like other genetic markers; they segregate in crosses as though they are codominant alleles.

The DNAs of different geographical isolates, different ecotypes (strains adapted to different environmental conditions), and different inbred lines of a species contain many RFLPs that can be used to construct detailed genetic maps. Indeed, the DNAs of different individuals—even relatives—often exhibit RFLPs. Some RFLPs can be visualized directly when the fragments in DNA digests are separated by agarose gel electrophoresis, stained with ethidium bromide, and viewed under ultraviolet light. Other RFLPs can be detected only by using specific cDNA or genomic clones as radioactive hybridization probes on genomic Southern blots (Figure 21.2*b*). We discussed these techniques in Chapter 20. The RFLPs themselves are the phenotypes used to classify the progeny of crosses as parental or recombinant. Figure 21.3 illustrates the use of RFLPs as genetic markers in mapping experiments

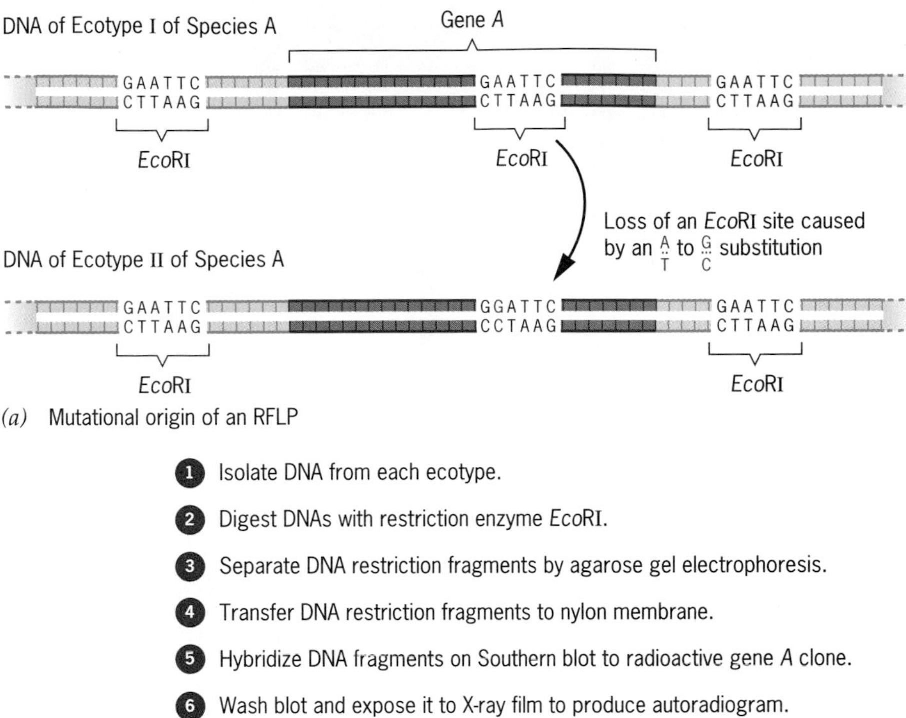

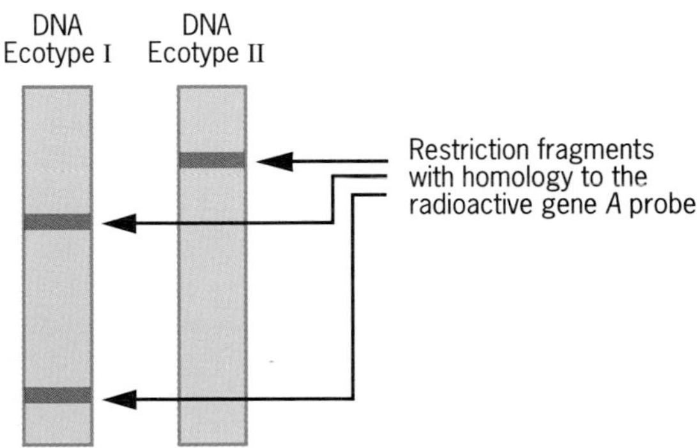

(*b*) Detection of an RFLP

Figure 21.2 The mutational origin (*a*) and detection (*b*) of RFLPs in different ecotypes of a species. In the example shown, an A:T → G:C base-pair substitution results in the loss of the central *Eco*RI recognition sequence present in gene *A* of the DNA of ecotype I. This mutation could have occurred in a common progenitor of the two ecotypes or in an ecotype II ancestor during the early stages of its divergence from ecotype I. In ecotype I, gene *A* sequences are present on two *Eco*RI restriction fragments, whereas in ecotype II, all gene *A* sequences are present on one large *Eco*RI restriction fragment.

in the model plant *Arabidopsis*. An RFLP map of chromosome 1 of *Arabidopsis* is shown in Figure 21.4. Because its genome is small (about 10^8 nucleotide pairs), *Arabidopsis* is an excellent system in which to clone genes based on their locations.

RFLP markers have proven especially valuable in mapping the chromosomes of humans, where researchers must rely on the segregation of spontaneously occurring mutant alleles in families to estimate map distances. Pedigree-based mapping of this type is done by comparing the probabilities that the genetic markers segregating in the pedigree are unlinked or linked by various map distances. In 1992, geneticists used this procedure to construct an early map of about

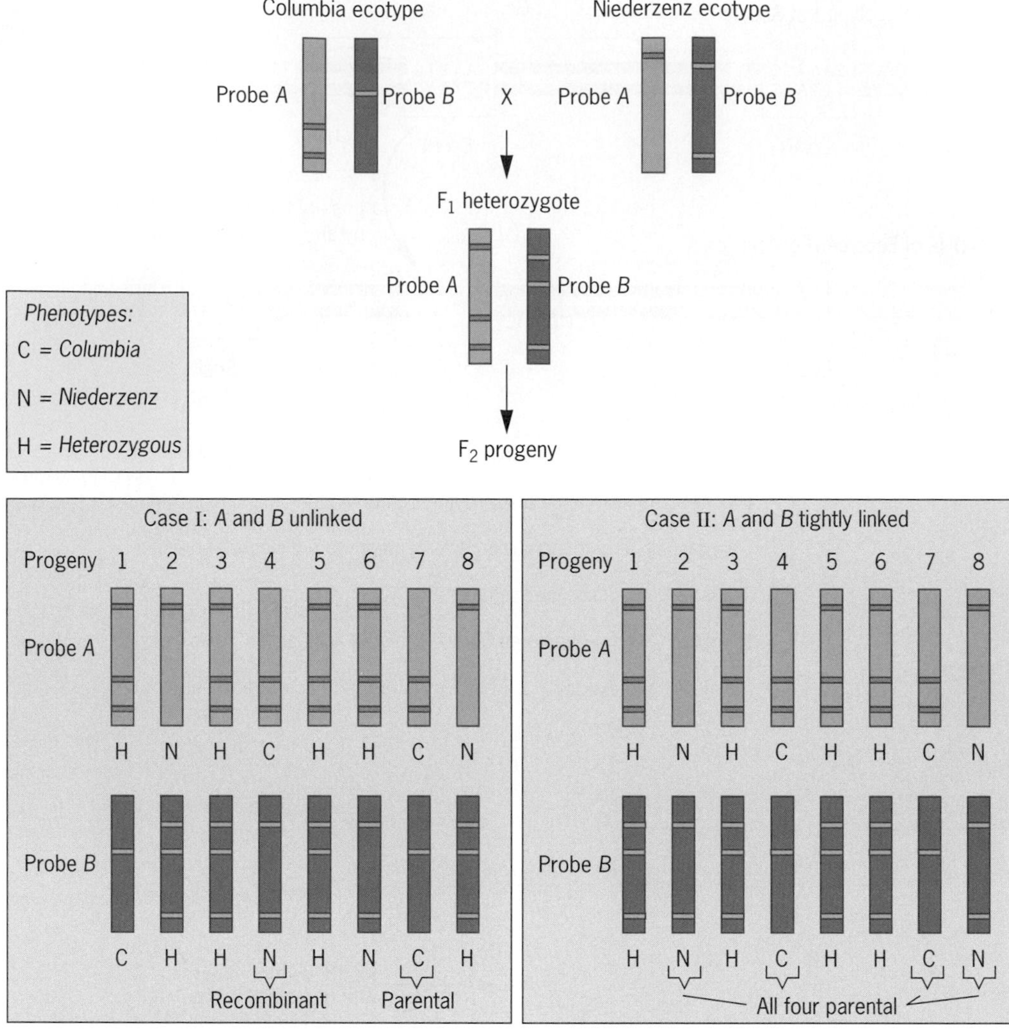

Figure 21.3 Mapping RFLPs in *Arabidopsis*. All genomic DNAs are digested with restriction endonuclease *Eco*RI. The DNA fragments in all lanes (rectangular boxes) of the Southern blots are hybridized sequentially to probe *A* (blue) and probe *B* (green). The F_1 progeny are heterozygous, containing the restriction fragments of both parents as expected. When such F_1 plants are self-pollinated, a 1:2:1 (C homozygote : heterozygote : N homozygote) segregation ratio is expected for each RFLP. If the two RFLPs assort independently (*left*), parental and recombinant combinations should occur with equal frequency among the F_2 progeny. If they are tightly linked (*right*), recombinants should be rare. The frequency of recombinants can be used to calculate the map distance between RFLPs, just as for other genetic markers.

2000 RFLPs on the 24 human chromosomes. Figure 21.5 shows an RFLP map of human chromosome 1 and its correlation with the cytological map of this chromosome.

In humans, the most useful RFLPs involve short sequences that are present as tandem repeats. The number of copies of each sequence present at a given site on a chromosome is highly variable. These sites, called **variable number tandem repeats,** or **VNTR**s, are therefore highly polymorphic. VNTRs vary in fragment length not because of differences in the positions of restriction enzyme cleavage sites, but because of differences in the number of copies of the repeated sequence between the restriction sites.

Microsatellites are another class of polymorphisms that have proven extremely valuable in constructing high-density maps of eukaryotic chromosomes. Microsatellites are polymorphic tandem repeats of sequences only two to five nucleotide pairs long. They are called microsatellites because they are a subset of the satellite sequences present in the highly repetitive DNA of eukaryotes (Chapter 10). In humans, microsatellite sequences composed of polymorphic tandem repeats of the dinucleotide sequence AC/TG (AC in one strand; TG in the complementary strand) provide especially useful markers. In 1996, a group of French and Canadian researchers published a comprehensive map of 5264 AC/TG microsatellites in the

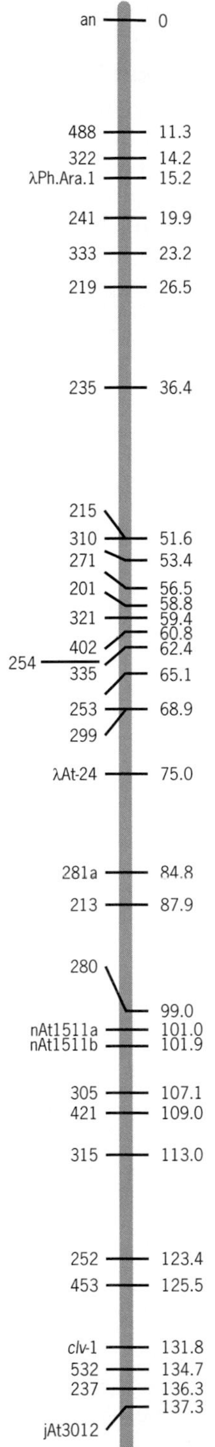

Figure 21.4 RFLP map of chromosome 1 in *Arabidopsis*. Clone designations are shown on the left; map positions are given in centiMorgans on the right, with the uppermost marker arbitrarily assigned position 0. Clone nAt1511 hybridized to sequences at two sites, designated a and b on the map.

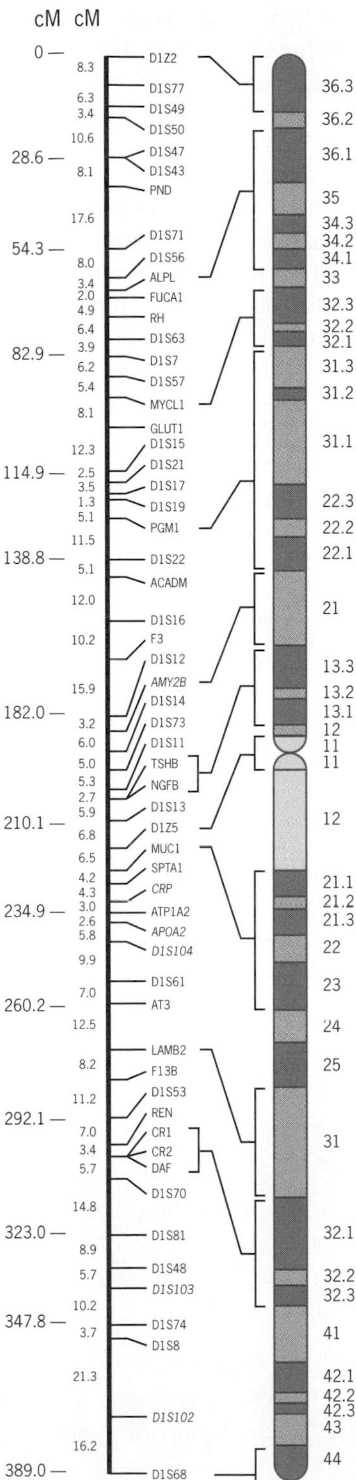

Figure 21.5 Correlation of the RFLP map (left) and the cytological map (right) of human chromosome 1. Distances are in centiMorgans (cM), with the uppermost marker set at position 0 on the left and distances between adjacent markers shown in the center.

human genome. These microsatellites defined 2335 sites with an average distance of 1.6 cM or about 1.6 mb between adjacent markers.

By 1997, a large international consortium had used RFLPs to map over 16,000 human genes (ESTs and cloned genes) and had integrated their map with the physical map of the human genome. In this collaborative study, over 20,000 STSs were mapped—many by researchers in two or more laboratories—to 16,354 distinct loci. These genetic maps composed primarily of

RFLP markers have made it possible to identify and characterize mutant genes that are responsible for several human diseases (see Figure 21.8 and Chapter 22).

Cytogenetic Maps

In some species, genes and clones can be positioned on the cytological maps of the chromosomes by *in situ* hybridization (Chapter 10). For example, in *Drosophila,* the banding patterns of the giant polytene chromosomes in the salivary glands (Chapter 7) provide high-resolution maps of the chromosomes (Figure 21.6*a*). Thus, a clone of unknown genetic content can be positioned on the cytological map with considerable precision. In mammals, including humans, fluorescent *in situ* hybridization (FISH; see Figure 10.35) can be used to position clones on chromosomes stained by any of several chromosome banding protocols. Figure 21.6 (*b* and *c*) illustrates the use of FISH to position clones on the cytological map of human chromosome 11. If RFLPs can be identified that overlap with these clones, they can be used as STS sites that anchor the genetic map of chromosome 11 to the cytological map producing a cytogenetic map. If the clones can be positioned on the physical map by Southern blot hybridization experiments (Chapter 20), they can also be used to tie the physical map to both the genetic and cytological maps of the chromosome.

Physical Maps and Clone Banks

The RFLP mapping procedure has been used to construct detailed genetic maps of chromosomes, which,

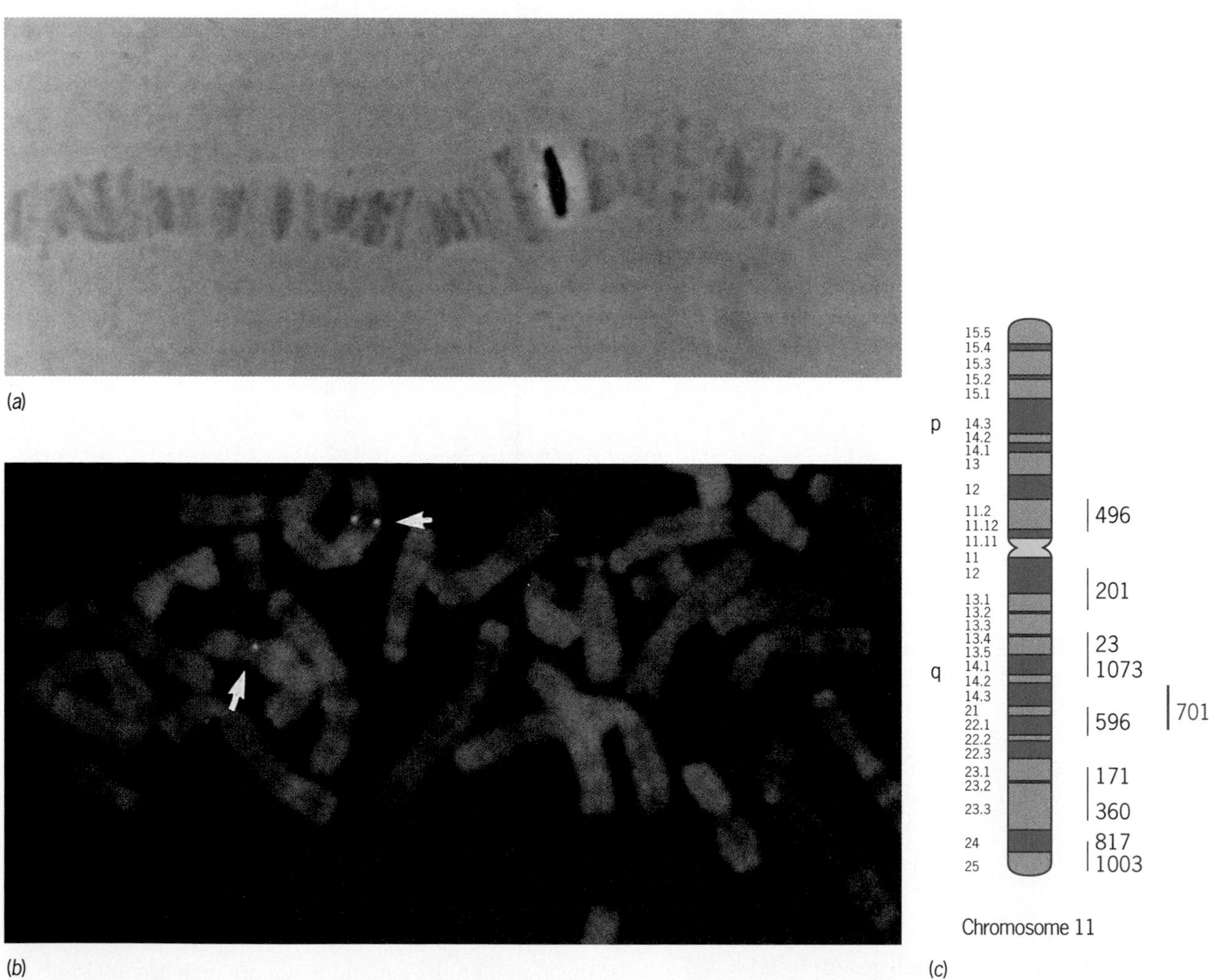

Figure 21.6 Determining the locations of genes and clones on chromosomes by *in situ* hybridization using radioactive (*a*) or fluorescent (*b*) probes. (*a*) *In situ* hybridization mapping of the *white* gene of *D. melanogaster* to region 3C2 of the X chromosome. (*b*) Fluorescent *in situ* hybridization (FISH) mapping of clone D11RP-701 to interval q14.3–22.1 of human chromosome 11. (*c*) Cytological map of chromosome 11 showing the locations of the clone (red bar) mapped by FISH in (*b*) and nine other clones (black bars) mapped to chromosome 11 in the same study.

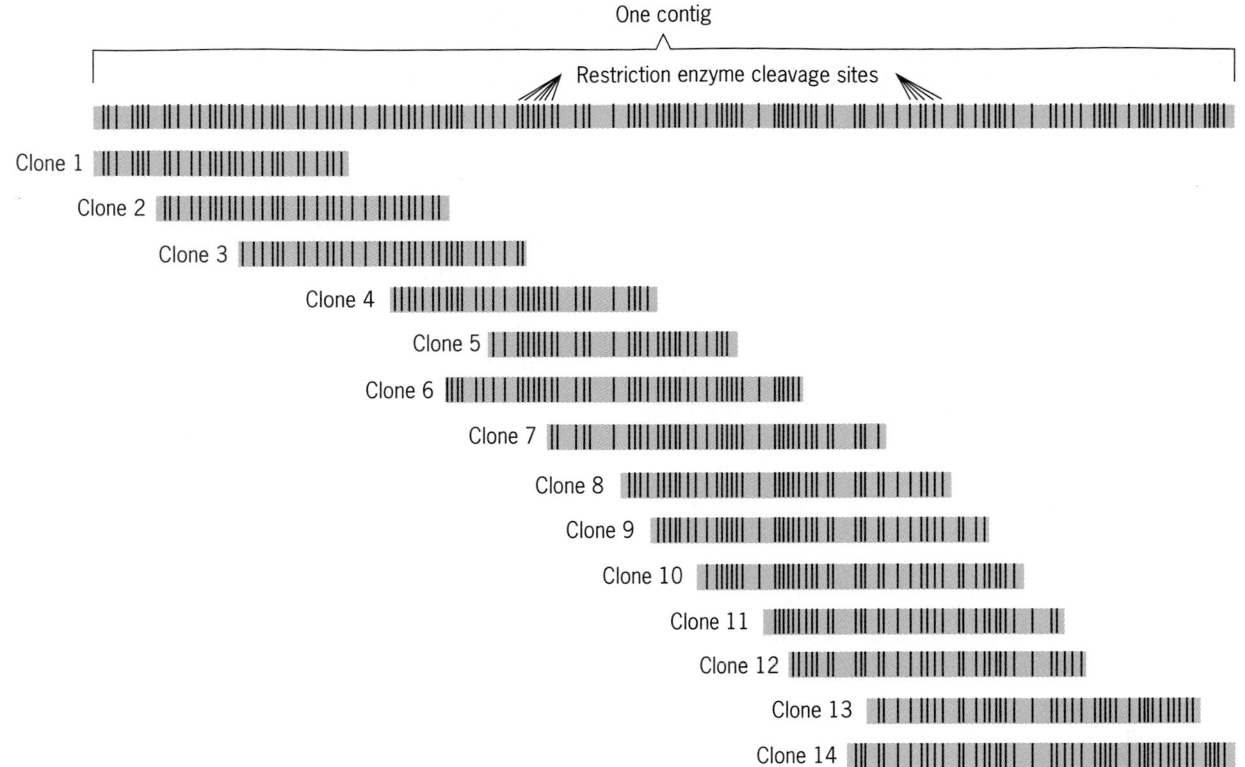

Figure 21.7 A contig map produced from overlapping genomic clones. Large—200 to 500 kb—genomic clones, such as those present in YAC or BAC vectors (Chapter 20), are used to construct contig maps. Restriction maps of individual clones are prepared and searched by computer for overlaps. Overlapping clones are then organized into contig maps like the one shown here. When the physical map of a genome is complete, each chromosome will be represented by a single contig map.

in turn, have made positional cloning feasible. These genetic maps have been supplemented with physical maps of chromosomes. By isolating and preparing restriction maps of large numbers of genomic clones, overlapping clones can be identified and used to construct physical maps of entire chromosomes and even entire genomes. In principle, this procedure is simple (Figure 21.7). However, in practice, it is a formidable task, especially for large genomes. The restriction maps of large genomic clones in YAC or BAC vectors (Chapter 20) are analyzed by computer and organized in overlapping sets of clones called **contigs.** As more data are added, adjacent contigs are joined; when the physical map of a genome is complete, each chromosome is represented by a single contig map.

The construction of physical maps of entire genomes requires that vast amounts of data be searched for overlaps. Nevertheless, detailed physical maps are available for several genomes, including the human, *C. elegans,* and *A. thaliana* genomes. These physical maps are being used to prepare clone banks that contain catalogued clones collectively spanning entire chromosomes. Thus, if a researcher needs a clone of a particular gene or segment of a chromosome, that clone may already have been catalogued in the clone bank and be available on request. Obviously, the availability of such clone banks and the correlated physical maps of entire genomes are dramatically accelerating genetic research. Indeed, searching for a specific gene with and without the aid of a physical map is like searching for a book in a huge library with and without a computer index giving the locations of the books in the library.

Key Points: **Genetic maps of chromosomes are based on recombination frequencies between markers. Cytogenetic maps are based on the location of markers within cytological features such as chromosome banding patterns observed by microscopy. Physical maps of chromosomes are determined by the molecular distances in base pairs, kilobase pairs, or megabase pairs separating markers. High-density maps that integrate the genetic, cytological, and physical maps of chromosomes have been constructed for all of the human chromosomes and many chromosomes of other model genetic organisms.**

MAP POSITION-BASED CLONING OF GENES

The first eukaryotic genes to be cloned were genes that are expressed at very high levels in specialized tissues or cells. For example, about 90 percent of the protein synthesized in mammalian reticulocytes is hemoglobin. Thus, α- and β-globin mRNAs could be easily isolated from reticulocytes and used to prepare radioactive cDNA probes for genomic library screens. However, since most genes are not expressed at such high levels in specialized cells, how are genes that are expressed at moderate or low levels cloned? One important approach is to map the gene precisely and to search for a clone of the gene by using procedures that depend on its location in the genome. This approach, called **positional cloning,** can be used to identify any gene, given an adequate map of the region of the chromosome in which it is located.

The steps in positional cloning are illustrated in Figure 21.8. The gene is first mapped to a specific region of a given chromosome by genetic crosses or, in the case of humans, by pedigree analysis, which usually requires large families. The gene is next localized on the physical map of this region of the chromosome by analysis of data involving STSs. Candidate genes in the segment of the chromosome identified by physical mapping are then isolated from mutant and wild-type organisms or individuals and sequenced to identify mutations that would result in a loss of gene function. As we shall discuss in Chapter 22, the human genes responsible for inherited disorders such as Huntington's disease and cystic fibrosis have been identified by using the positional cloning approach. In species where transformation is possible, copies of the wild-type alleles of candidate genes are introduced into mutant organisms to determine whether the wild-type genes will restore the wild phenotype. Restoration of the wild phenotype to a mutant organism provides strong evidence that the introduced wild-type gene is the gene of interest.

Chromosome Walks

Positional cloning is accomplished by mapping the gene of interest, identifying an RFLP or other molecular marker near the gene, and then "walking" or

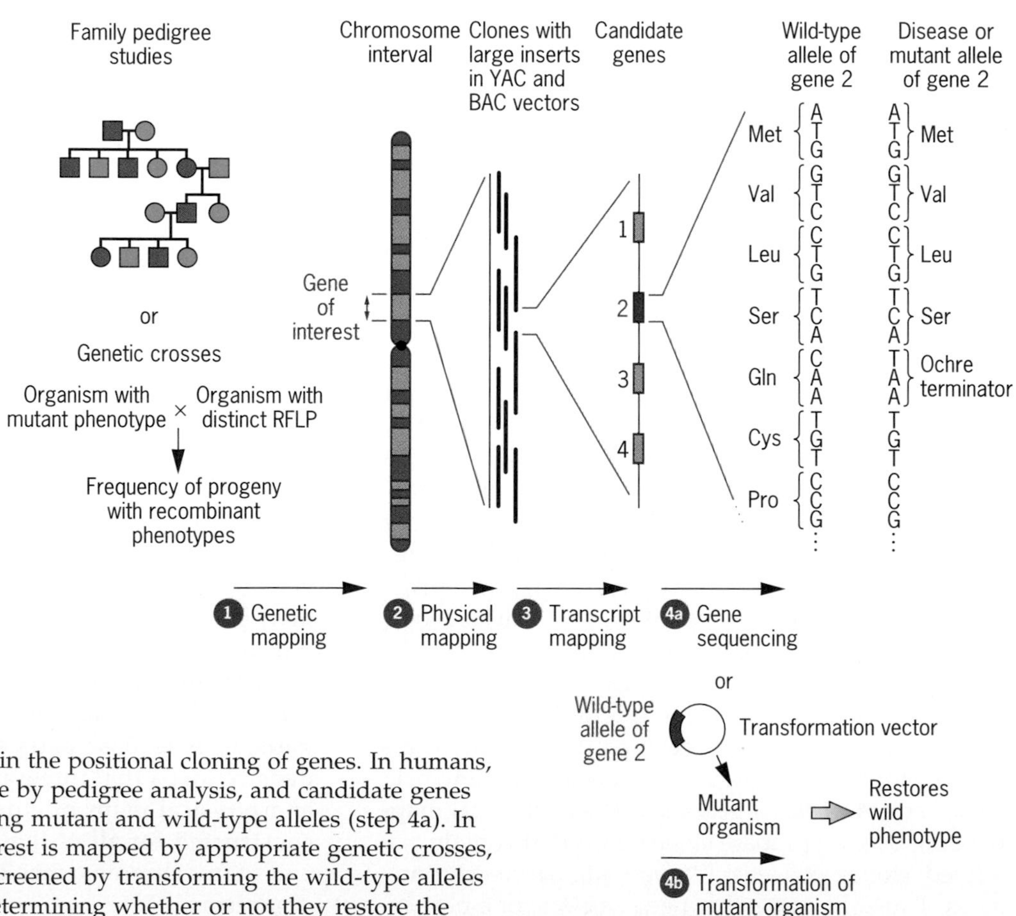

Figure 21.8 Steps involved in the positional cloning of genes. In humans, genetic mapping must be done by pedigree analysis, and candidate genes must be screened by sequencing mutant and wild-type alleles (step 4a). In other species, the gene of interest is mapped by appropriate genetic crosses, and the candidate genes are screened by transforming the wild-type alleles into mutant organisms and determining whether or not they restore the wild-type phenotype (step 4b).

"jumping" along the chromosome until the gene is reached. **Chromosome walking** is very difficult in species with large genomes (the walk is usually too far) and an abundance of dispersed repetitive DNA (each repeated sequence is a potential roadblock). Chromosome walking is easier in organisms such as *A. thaliana* and *C. elegans,* which have small genomes and little-dispersed repetitive DNA.

Chromosome walks are initiated by the selection of a molecular marker (RFLP or known gene clone) close to the gene of interest and the use of this clone as a hybridization probe to screen a genomic library for overlapping sequences. Restriction maps are constructed for the overlapping clones identified in the library screen, and the restriction fragment farthest from the original probe is used to screen a second genomic library constructed by using a different restriction enzyme or to rescreen a library prepared from a partial digest of genomic DNA. Repeating this procedure several times and isolating a series of overlapping genomic clones allow a researcher to walk the required distance down the chromosome to the gene of interest (Figure 21.9). Without information about the orientation of the starting clone on the linkage map, the initial walk will have to proceed in both directions until another RFLP is identified and it is determined whether the new RFLP is closer to or farther away from the gene of interest than is the starting RFLP.

Verification that a clone of the gene of interest has been isolated is accomplished in various ways. In experimental organisms such as *Drosophila* and *Arabidopsis,* verification is achieved by introducing the wild-type allele of the gene into a mutant organism and showing that it restores the wild-type phenotype. In *Drosophila,* this is done by transforming mutant embryos with transposable element vectors (see Technical Sidelight in Chapter 18) that carry the wild-type allele of the gene. In humans, verification usually involves determining the nucleotide sequences of the wild-type gene and several mutant alleles and showing that the

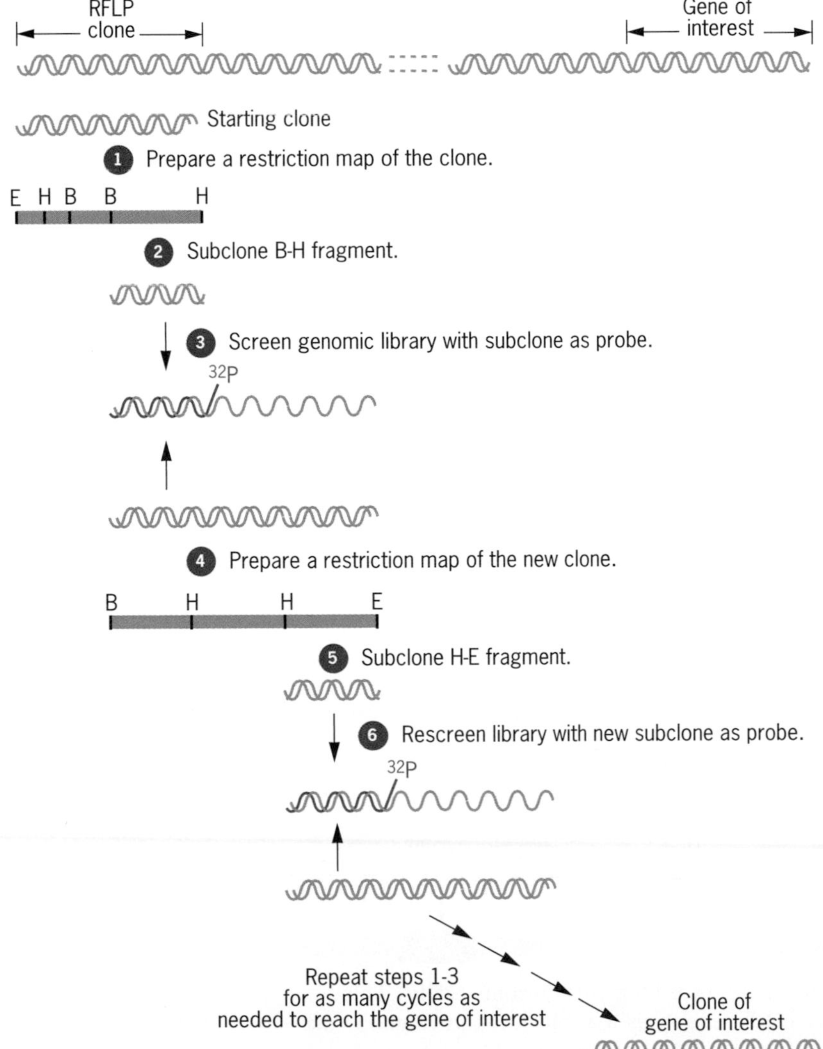

Figure 21.9 Positional cloning of a gene by chromosome walking. A chromosome walk starts with the identification of a molecular marker—such as the RFLP shown (top)—close to the gene of interest and proceeds by repeating steps 1 through 3 as many times as is required to reach the gene of interest (bottom).

coding sequences of the mutant genes are defective and unable to produce functional gene products.

Chromosome Jumps

When the distance from the closest molecular marker to the gene of interest is large, a technique called **chromosome jumping** can be used to speed up an otherwise long walk. Each jump can cover a distance of 100 kb or more. The chromosome jumping procedure is illustrated in Figure 21.10. Like a walk, a jump is initiated by using a molecular probe such as an RFLP as a starting point. However, with chromosome jumps, large DNA fragments are prepared by partial digestion of genomic DNA with a restriction endonuclease. The large genomic fragments are then circularized with DNA ligase. A second restriction endonuclease is used

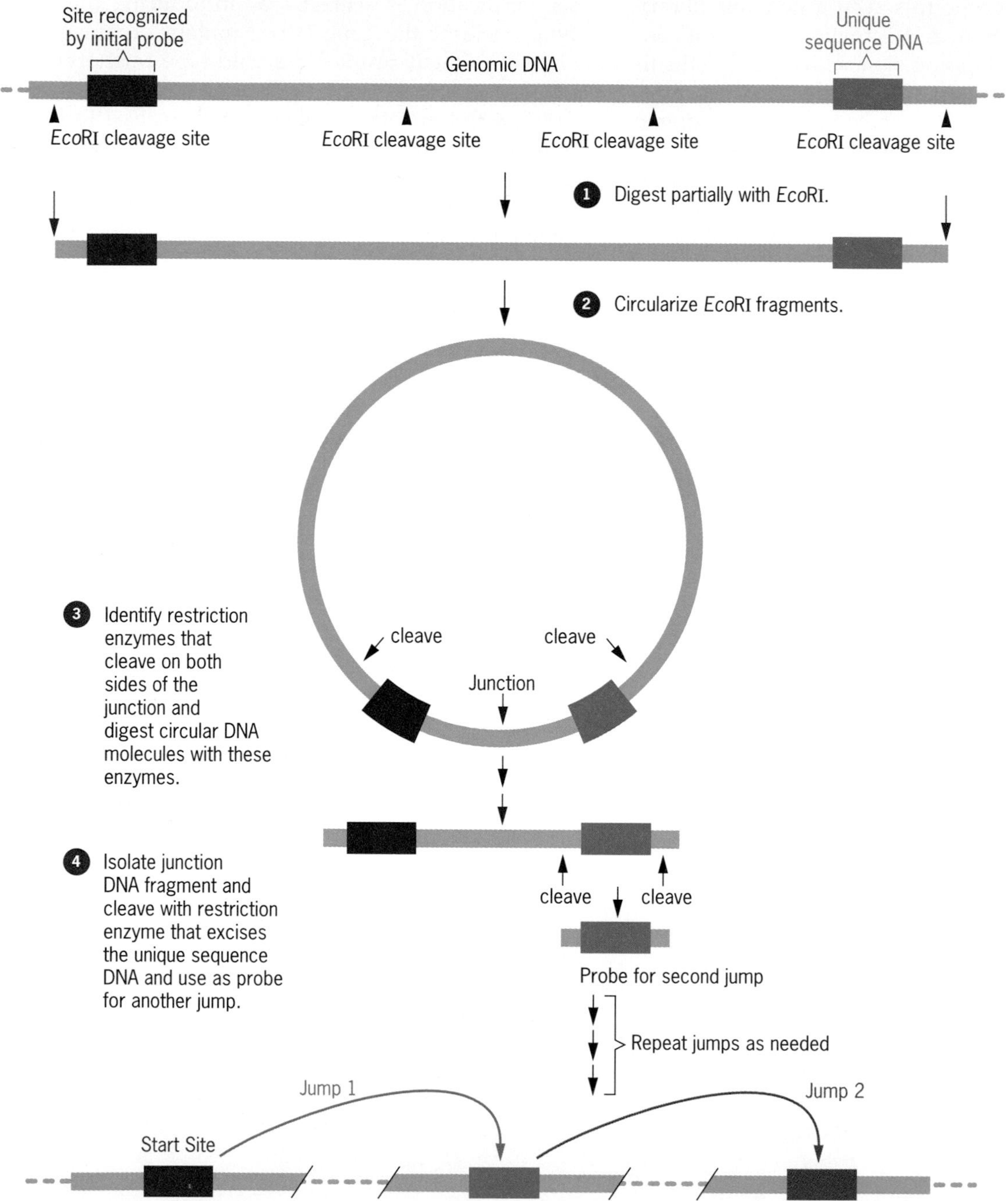

Figure 21.10 Chromosome jumping as a shortcut method for long chromosome walks. This procedure can also be used to jump over repetitive DNA sequences that block chromosome walks.

to excise the junction fragment from the circular molecule. This junction fragment will contain both ends of the long fragment; it can be identified by hybridizing the DNA fragments on Southern blots to the initial molecular probe. A restriction map of the junction fragment is prepared, and a restriction fragment that corresponds to the distal end of the long genomic fragment is cloned and used to initiate a chromosome walk or a second chromosome jump. Chromosome jumping has proven especially useful in work with large genomes such as the human genome. Chromosome jumps played a key role in identifying the human cystic fibrosis gene (Chapter 22).

Key Points: **The availability of detailed genetic, cytogenetic, and physical maps of chromosomes permit researchers to clone genes based on their location in the genome. If a molecular marker such as a restriction fragment-length polymorphism (RFLP) maps close to a gene, that gene can usually be cloned by chromosome walking or chromosome jumping.**

THE HUMAN GENOME PROJECT

As the recombinant DNA, gene cloning, and DNA sequencing technologies improved in the 1970s and early 1980s, scientists began discussing the possibility of sequencing all 3×10^9 nucleotide pairs in the human genome. These discussions led to the launching of the **Human Genome Project** in 1990. The initial goals of the Human Genome Project were (1) to map all of the 70,000 to 100,000 human genes, (2) to construct a detailed physical map of the entire human genome, and (3) to determine the nucleotide sequences of all 24 human chromosomes by the year 2005. Scientists soon realized that this huge undertaking should be a worldwide effort. Therefore, an international **Human Genome Organization (HUGO)** was organized to coordinate the efforts of human geneticists around the world.

James Watson, who, with Francis Crick, discovered the double-helix structure of DNA, was the first director of this ambitious project, which was expected to take nearly two decades to complete and to cost in excess of $3 billion. In 1993, Francis Collins, who, with Lap-Chee Tsui, led the research teams that identified the cystic fibrosis gene, replaced Watson as director of the Human Genome Project. In addition to work on the human genome, the Human Genome Project has served as an umbrella for similar mapping and sequencing projects on the genomes of several other organisms, including the bacterium *E. coli,* the yeast *S. cerevisiae,* the fruit fly *D. melanogaster,* the plant *A. thaliana,* and the worm *C. elegans.*

Bacterial Genomes

In 1995, *Haemophilus influenzae* became the first bacterium to have its genome sequenced in its entirety (see Chapter 21 frontispiece). By late 1998, the complete sequences of 18 bacterial genomes were available in the public databases. The genomes range in size from 580,070 bp for *Mycoplasma genitalium,* which is thought to have the smallest genome of any self-replicating organism, to 4,411,529 bp for *Mycobacterium tuberculosis,* which causes more human deaths than any other infectious bacterium, to 4,639,221 bp for *Escherichia coli,* the best-known cellular microorganism. The genome of *M. genitalium* is of special interest because it may approximate the "minimal gene set" for a self-replicating organism—the smallest set of genes that will allow an organism to reproduce itself. Of course, the genome of *M. tuberculosis* is of great interest because of the pathogenicity of this organism and the hope that a complete understanding of its metabolism will suggest ways to prevent tuberculosis in humans. The need for new ways to combat this pathogen has been enhanced by the recent evolution of antibiotic-resistant strains of *M. tuberculosis* (Chapter 17).

Of the bacterial genomes sequenced to date, the sequence of the *E. coli* genome (Figure 21.11) has undoubtedly caused the most excitement among geneticists. *E. coli* is the most studied and best understood cellular organism on our planet. Geneticists, biochemists, and molecular biologists have utilized *E. coli* as the preferred model organism for decades. Most of what is known about bacterial genetics was learned from research on *E. coli.* Thus, publication of the complete sequence of the *E. coli* genome in 1997 was a significant milestone in the history of genetics.

The *E. coli* genome contains 4288 putative protein-coding sequences or genes. Of these putative genes, about one-third are well-studied genes encoding known products, whereas 38 percent are of unknown function. Figure 21.12 shows a classification of the genes according to gene-product function. Putative protein-coding sequences that are not known to encode proteins are called **open reading frames** or **ORFs.** An ORF is a nucleotide sequence that begins with a translation-initiation codon (usually AUG), continues with a sequence of triplet codons specifying amino acids, and ends with one of the three translation-termination codons.

The average distance between genes (size of intergenic regions) in the *E. coli* genome is 118 bp. Known and putative genes specifying proteins and stable RNAs make up 87.8 percent and 0.8 percent of the genome, respectively, and noncoding repetitive elements account for 0.7 percent of the genome. Thus, 10.7 percent of the genome must involve regulatory se-

Key: ➡ tRNA genes; ➡ rRNA genes; — origin and terminus of replication

Figure 21.11 Sequence-based map of the chromosome of *Escherichia coli.* The blue arrows mark the halves of the chromosome traversed by the two replication forks. The outer concentric circle gives the positions of genes encoding proteins similar to bacteriophage proteins. The second concentric circle shows the location of genes that are transcribed clockwise (gold) from one strand or counterclockwise (yellow) from the complementary strand. The sunburst in the center is a histogram in which the length of each ray is proportional to the randomness of codon usage within each coding sequence.

quences, sequences with other unknown functions, and perhaps some nonfunctional nucleotide sequences. Six new tRNA genes were discovered during the analysis of the sequence, and similarity comparisons with sequenced regions of other genomes revealed two sets of genes encoding enzymes involved in the degradation of aromatic organic compounds and 12 new genes involved in the synthesis and assembly of flagella.

Once the complete sequence of a bacterial genome is available, computer searches can be performed to identify sequence similarities with other sequenced genomes. Such sequence comparisons can often be used to gain inferences about gene function. Because so much is known about gene function in *E. coli,* comparisons of other sequenced bacterial genomes with the *E. coli* genome are often very informative. For example, a comparison of the genomes of *Treponema pallidum,* the parasitic spirochete that causes syphilis, and *E. coli,* shows that *T. pallidum* contains the genes that encode proteins involved in DNA replication and repair, transcription, and translation, but carries few genes encoding biosynthetic enzymes and transport proteins. Thus, by comparing the coding potentials of entire genomes, geneticists can make important inferences about all aspects of metabolism, growth, and interactions with the environment in various bacteria. Such information should prove invaluable in devel-

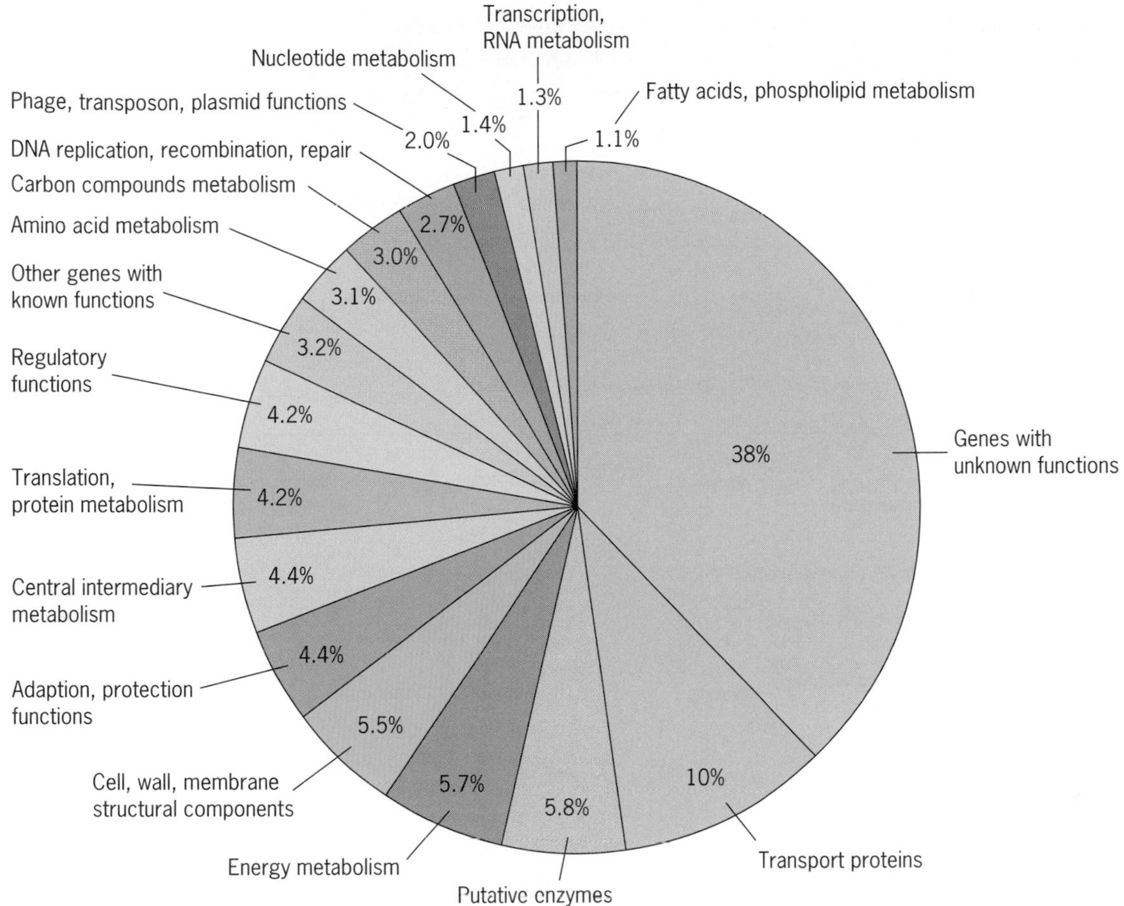

Figure 21.12 Classification of the 4,288 known and putative genes of *E. coli* according to function. The largest class of known genes encodes proteins involved in transport. However, note that 38 percent of the genes encode proteins of unknown function.

oping strategies to protect humans from bacterial pathogens.

The *Saccharomyces cerevisiae* (Yeast) Genome

Budding yeast, *S. cerevisiae,* was the first eukaryotic organism to have its entire genome sequenced. The complete 12,068-kb sequence of the *S. cerevisiae* genome was assembled in 1996 through an international collaboration of about 600 scientists working in Europe, North America, and Japan. However, the sequences of some of the yeast chromosomes were reported earlier. For example, the complete sequence of chromosome III (Figure 21.13)—the first yeast chromosome to be sequenced—was published in 1992. The yeast genome contains 5885 potential protein-coding genes, about 140 genes specifying ribosomal RNAs, 40 genes for small nuclear RNA molecules, and 275 tRNA genes.

A major difference between the yeast genome and the bacterial genomes that have been sequenced is the presence of considerable genetic redundancy—duplicate or multiple copies of very similar nucleotide sequences and genes encoding families of nearly identical proteins. The existence of two or more genes that encode identical or very similar proteins can provide the raw material for the evolution of new genes if one copy is sufficient for normal growth and reproduction. Nucleotide changes that alter one copy of a duplicated gene need not be detrimental to the survival of the organism. Thus, one copy of the gene is free to undergo sequential nucleotide-pair changes leading to the evolution of a new gene, a gene encoding a product with a function distinct from that of the original gene product. Other redundant genes may encode products that are required in large amounts or very similar, but functionally distinct, proteins that are important at different stages of the life cycle or under different environmental conditions.

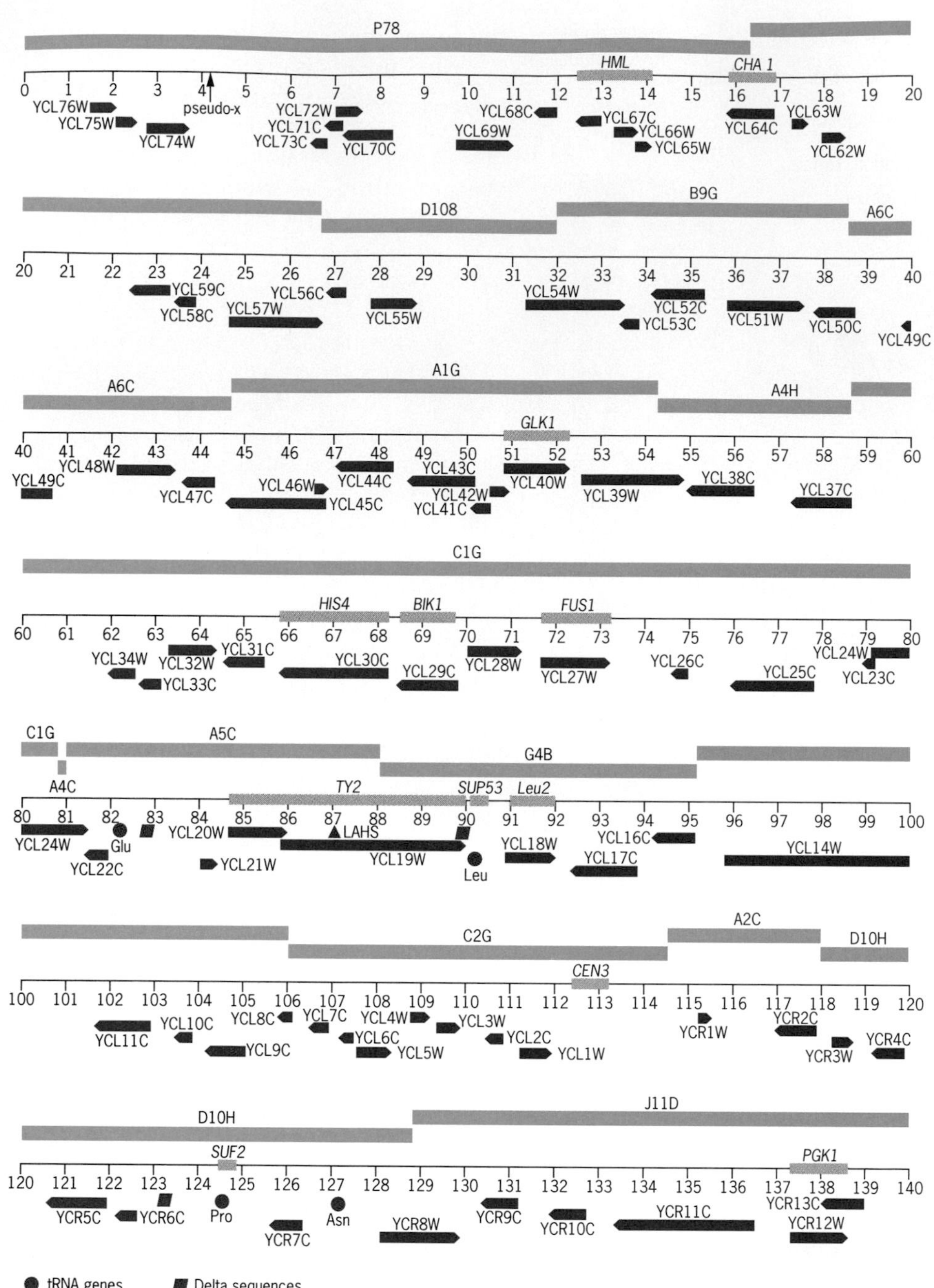

Figure 21.13 Sequence-based map of a 140-kb segment of chromosome III of the yeast *Saccharomyces cerevisiae*. This was the first chromosome of a free-living organism to be sequenced in its entirety. The positions of open reading frames (ORFs) and known genes are indicated by the royal-blue arrow bars and yellow boxes, respectively. The light blue and turquoise bars show the positions and extents of the clones used to sequence the various segments of the chromosome. Map positions are in kb.

Another example of this redundancy is the presence of 19 copies of a telomere-associated ORF whose predicted product would appear to be an RNA helicase. The function, if any, of the 19 copies of this ORF is unknown. However, speculation is that the 19 copies may be remnants of an ancient transposable element that integrated into subtelomeric sequences.

The presence of genetic redundancy is a common feature of eukaryotic genomes and one that distinguishes them from the genomes of prokaryotes. Genetic redundancy is even more prevalent, and copy number is usually higher, in the genomes of multicellular eukaryotes. Further research is needed to determine why the degree of genetic redundancy is often correlated with the developmental complexity of the species. Because of its many advantages for genetic analysis, *S. cerevisiae* is an ideal system with which to address this and other questions about genome function in eukaryotes.

The Human Genome

Progress in mapping the human genome has been quite spectacular. Complete physical maps of chromosomes Y and 21 and detailed RFLP maps of the X chromosome and all 22 autosomes were published in 1992. By 1995, the genetic map contained markers separated by, on average, 200 kb. A detailed microsatellite map of the human genome was published in 1996, and a comprehensive map of 16,354 distinct loci was released in 1997. All of these maps have proven invaluable to researchers cloning genes by their locations in the genome.

Unfortunately, the resolution of genetic mapping in humans is quite low—in the range of 1–10 mb. The resolution of fluorescent *in situ* hybridization (FISH) is also approximately 1 mb. Higher resolution mapping (down to 50 kb) can be achieved by **radiation hybrid mapping,** which is a modification of the somatic-cell hybridization mapping procedure discussed in Chapter 9. Standard somatic-cell hybridization involves the fusion of human cells and rodent cells growing in culture and the correlation of human gene products with human chromosomes retained in the hybrid cells.

Radiation hybrid mapping is done by fragmenting the chromosomes of the human cells with heavy irradiation prior to cell fusion (Figure 21.14). The irradiated human cells are then fused with Chinese hamster (or other rodent) cells growing in culture, usually in the presence of a chemical such as polyethylene glycol to increase the efficiency of cell fusion. The human-Chinese hamster somatic-cell hybrids are then identi-

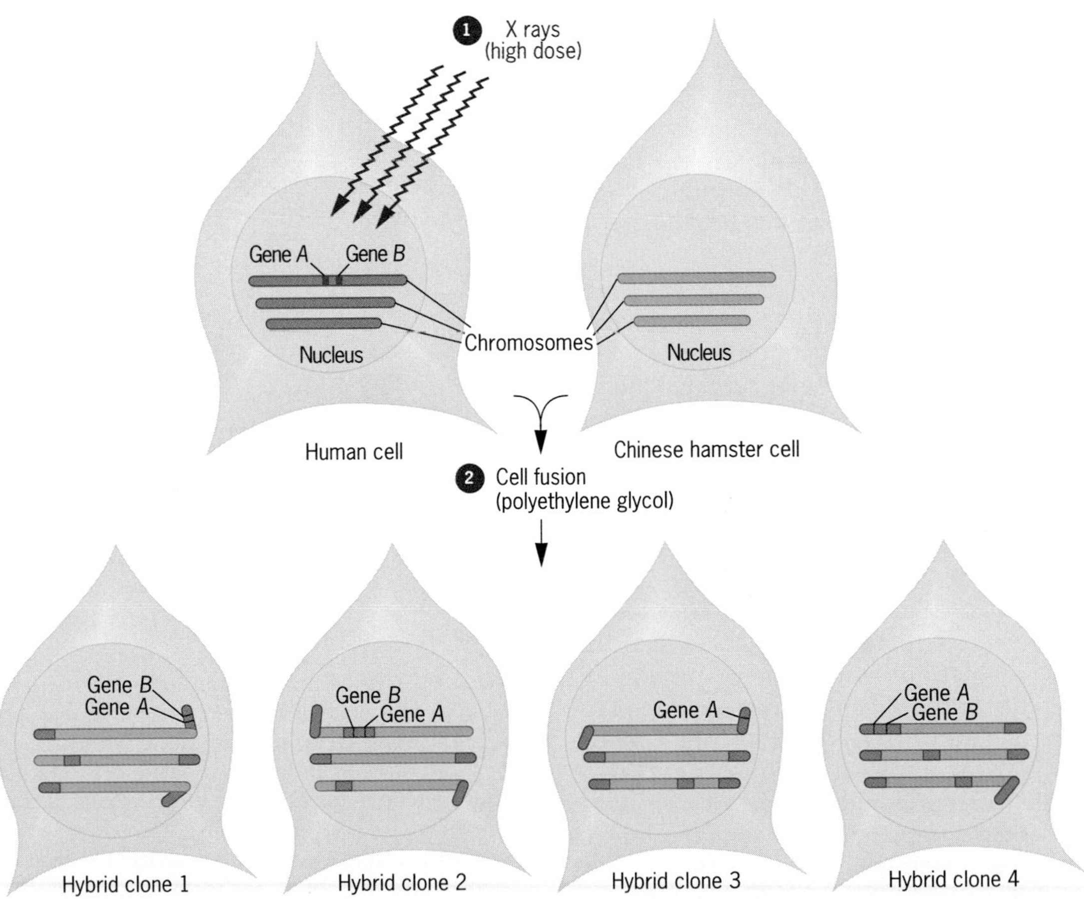

Figure 21.14 The use of radiation somatic cell hybrids for high-resolution mapping of the human genome. The rationale behind radiation hybrid maps is that the probability of an X ray-induced break between genes *A* and *B* is directly proportional to the distance between them on the DNA molecule. Note that genes *A* and *B* have remained together in hybrid clones 1, 2, and 4, but were separated by an X ray-induced break during the formation of hybrid clone 3.

fied by growth in an appropriate selection medium (see Chapter 9).

Many of the human chromosome fragments become integrated into the Chinese hamster chromosomes during this process and are transmitted to progeny cells just like the normal genes in the Chinese hamster chromosomes. The polymerase chain reaction (PCR; Chapter 20) is then used to screen a large panel of the selected hybrid cells for the presence of human genetic markers. Chromosome maps are constructed based on the assumption that the probability of an X ray-induced break between two markers is directly proportional to the distances separating them in chromosomal DNA.

Several groups have used the radiation hybrid mapping procedure to construct high-density maps of the human genome. In 1997, Elizabeth Stewart and coworkers published a map of 10,478 STSs based on radiation hybrid mapping; their map of human chromosome 1 is shown in Figure 21.15.

Whereas the gene mapping work has advanced quickly, progress toward sequencing the human genome has lagged behind schedule. However, its status is expected to change rapidly. In July 1998, the National Human Genome Research Institute of the National Institutes of Health awarded grants totaling $60.5 million to seven sequencing centers in the United States with the combined goal of sequencing 117 million base pairs during the next year—twice the output of all the U.S. sequencing centers to date.

However, the entire sequencing strategy of Human Genome Project laboratories has just been reevaluated because of new breakthroughs in automated DNA sequencing (Chapter 20) and new private human genome sequencing initiatives. J. Craig Venter, past president of the Institute for Genomic Research where several bacterial genomes have been sequenced, has joined forces with Perkin-Elmer Corporation and formed a private company, Celera Inc., with the goal of sequencing the human genome in just three years—by 2001 (see Human Genetics Sidelight: The Race to Sequence the Human Genome). Perhaps at least in part because of the competition with Venter's private initiative, the leaders of the seven Human Genome Project sequencing laboratories have revised their schedule. They now plan to complete the sequence by 2003—two years earlier than originally proposed.

In any case, the complete sequence of the human genome will almost certainly be known within the next few years, and its availability will raise a whole new set of questions about the proper use of this new knowledge. Many of these questions focus on an individual's right to privacy. For example, if a mutation that causes a late-onset disorder such as Huntington's disease (Chapter 22) is discovered in a family, who should have access to this information? If such information were available to the public, widespread discrimination might occur. Employers might not hire members of the family, and medical schools might not admit talented young scholars to their M.D. programs

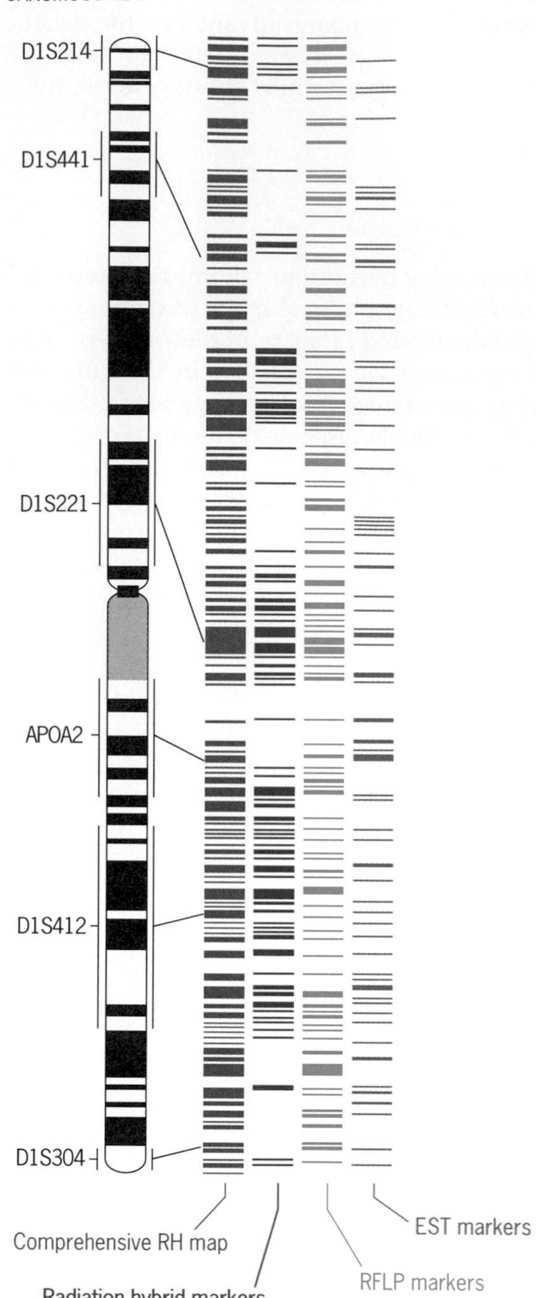

Figure 21.15 A high-resolution radiation hybrid map of human chromosome 1. The cytogenetic map of chromosome 1 is shown on the left, along with the locations of six anchor markers. To the right of the cytogenetic map are four genetic maps showing the locations of the ESTs (purple lines), RFLPs (green lines), high-confidence radiation hybrid markers (blue lines), and the comprehensive radiation hybrid map of all the markers (red lines).

(see A Conversation with Nancy Wexler, which precedes Chapter 20).

Key Points: **Important goals of the Human Genome Project are to map all 70,000 to 100,000 human genes and to determine the complete nucleotide sequence of all 24 human chromosomes. Additional goals associated with the Human Genome Project are to map and sequence the genomes of several important model organisms. The complete sequences of the genomes of 18 species of bacteria, mitochondria and chloroplasts from several species, and the yeast *S. cerevisiae* are now available; 99.9 percent of the sequence of the genome of the worm *C. elegans* is known; and the sequences of the genomes of the fruit fly *D. melanogaster* and the plant *A. thaliana* will soon be available.**

RNA AND PROTEIN ASSAYS OF GENOME FUNCTION

Knowing the complete sequence of the human genome will help identify genes responsible for human diseases and might lead to successful gene therapies for these diseases. However, it will not tell us what these genes do or how they control biological processes. Indeed, by itself, the nucleotide sequence of a gene, a chromosome, or an entire genome is uninformative. Only when supplemented with information about their functions do sequences become truly meaningful. Thus, information about the functions of nucleotide sequences must still be obtained by traditional genetic studies and by molecular analyses. If geneticists want to understand the genetic control of the growth and development of a mature human from a single fertilized egg (Chapter 25), they will need to know much more than the sequence of the human genome. But the availability of the ultimate map of the human genome, its nucleotide sequence, will certainly accelerate progress toward understanding the programs of gene expression that control morphogenesis. Indeed, the development of new technologies such as "gene chip" hybridizations is designed to take advantage of the availability of the sequences of complete genomes (see the later subsection, Array Hybridizations and Gene Chips).

Expressed Sequences

In large eukaryotic genomes, only a small proportion of the DNA encodes proteins. In the yeast *S. cerevisiae*, almost 70 percent of the genome encodes proteins, and there is 1 gene for every 2 kb of sequence. In humans, less than 5 percent of the genome encodes amino acid sequences, and there is about 1 gene for every 30 kb of sequence. Thus, in order to focus on the protein-coding content of genomes, many scientists have analyzed cDNA clones (DNAs complementary to RNA molecules; Chapter 20) or ESTs rather than genomic clones. By 1996, the public databases contained more than 600,000 cDNA sequences, about 450,000 of which were human cDNAs. The number of human cDNA sequences nearly doubled to about 800,000 by late 1997. However, many of these cDNA sequences are derived from the same gene transcripts. Multiple cDNAs can be obtained from different segments of a single gene transcript or from alternative splicing of a gene transcript. For example, the human gene that encodes serum albumin is represented by more than 1300 EST sequences in public databases.

The transcripts of different genes can usually be recognized by distinct 3′ untranslated regions—different nucleotide sequences in the region between the 3′ translation-termination codon and the 3′ terminus of the transcript. When sequence comparisons were made between the regions of cDNAs corresponding to the 3′ untranslated regions of the transcripts, the cDNAs were grouped into 49,625 clusters (with 97 percent sequence identity within clusters). The number of clusters is considered a good estimate of the number of distinct genes identified by EST sequencing to date. Therefore, cDNA sequencing projects have probably identified 50 to 70 percent of the genes in the human genome. Of the sequence clusters, 4563 correspond to known human genes. One problem with the EST approach to genome analysis is that genes that are expressed at very low levels or only in a few specialized cells will be missing or very rare in cDNA libraries.

Similar EST sequencing projects at private companies have identified over 90 percent of the estimated 70,000 genes in the corn genome. However, these sequences have not been released to the public databases because of the competitive advantage they give the companies involved.

Array Hybridizations and Gene Chips

Given the sequence of an entire genome, geneticists can immediately begin to study the expression of every gene in the organism. Oligonucleotide hybridization probes can be synthesized that are complementary to segments of the transcripts of every ORF, or PCR can be used to make millions of copies of each gene in a genome. Thus, scientists can monitor changes in total genome expression over time, throughout development, or in response to changes in the environment.

Such knowledge should prove invaluable in understanding human diseases such as cancer and, perhaps, even the aging process.

In Chapter 20, we discussed the use of northern blot hybridizations—RNAs separated by gel electrophoresis and transferred to membranes for hybridization to radioactive probes—to study gene expression. To investigate the expression of entire genomes or large sets of genes, scientists basically reverse the northern blot procedure and perform **dot blot hybridizations.** In dot blot hybridizations, the gene-specific nucleotide sequences or probes are applied and bound to membranes in specific patterns or arrays (Figure 21.16), and these probe arrays or dot blots are hybridized to radioactive or fluorescent RNA or cDNA preparations (Figure 21.17*a*). The amounts of RNA or cDNA hybridized to each probe on a dot blot or hybridization array can be measured by scanning the blot with an imaging system that measures the amount of radioactivity or fluorescence and analyzing the results with computer programs that compare the signals with those produced by known control probes and RNAs or cDNAs. Figure 21.17*b* shows the results of scanning and computer analysis of the autoradiograms shown in Figure 21.17*a*.

The hybridization arrays shown in Figure 21.17 contain 588 human EST probes. However, new technologies now allow scientists to produce microarrays that contain thousands of hybridization probes on a single membrane or other solid support. The gene chips shown in Figure 21.18 can array over 10,000 oligonucleotide probes on a silicon wafer of only a few square centimeters. Microarrays of oligonucleotide or cDNA probes are produced in several ways: (1) microsynthesis of oligonucleotides *in situ* (on the chip), (2) spotting prefabricated oligonucleotide probes on solid supports, and (3) spotting DNA fragments or cDNAs on the supports. The probes on the microarrays are then hybridized to labeled (usually with a fluorescent tag) RNA or cDNA samples. The amount of RNA or cDNA hybridized to each probe in a microarray is quantified by using sophisticated scanners with micrometer resolution and appropriate computer software.

The genome sequencing projects and the microarray hybridization technologies have spawned the new subdiscipline of functional genomics, which focuses on the expression of entire genomes. However, some geneticists have argued that this has been the goal of the science of genetics since its inception. As knowledge in the field has advanced, geneticists have been able to study the expression of more and more genes. Now, for the first time, they can study the expression of all the genes of an organism simultaneously. Probe microarrays are already being produced that will allow geneticists to analyze the expression of the nearly 6000 genes of budding yeast, and chips that permit scientists to study the expression of the 70,000 or so genes in the human genome will undoubtedly become available soon after the sequence of the human genome is known. The ability to analyze the expression of entire genomes should enhance the current rapid progress in biology and eventually lead to an understanding of the normal process of human development and the causes of at least some human diseases.

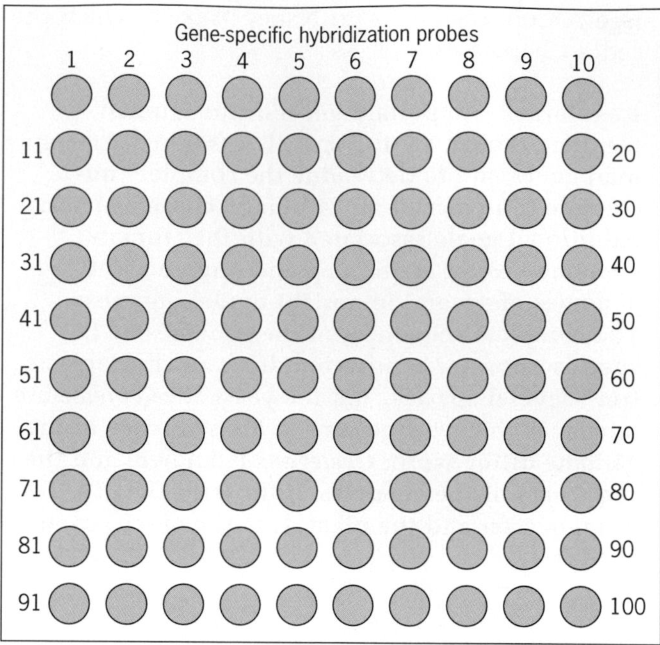

Figure 21.16 Dot blot or array hybridization analysis of gene expression. Gene-specific hybridization probes are applied to a membrane with the aid of a vacuum filtration apparatus. The membrane is then placed in hybridization solution containing RNAs or cDNAs labeled with a radioactive isotope or a fluorescent dye. After hybridization, the labeled RNA bound to each probe is either visualized by autoradiography (see Figure 21.17) or quantitated by densitometry or by scanners that measure the amount of radioactivity or fluorescence.

Use of Green Fluorescent Protein as a Reporter of Protein Synthesis

Array hybridizations and gene chips can be used to determine whether genes are transcribed, but they provide no information about the translation of the gene transcripts. Thus, biologists often use antibodies to detect the protein products of genes of interest. Western blots are used to detect proteins separated by

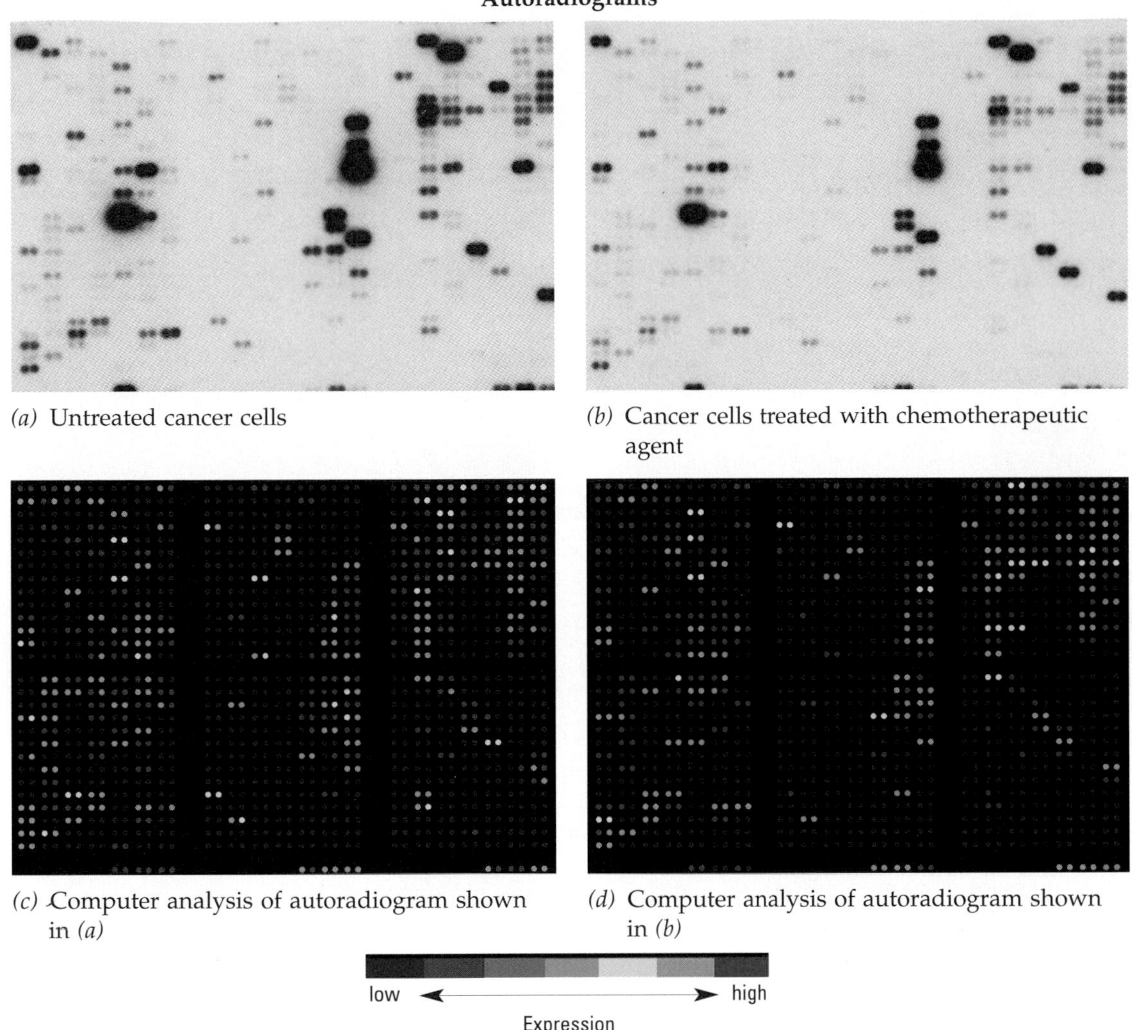

Figure 21.17 The use of dot blot or array hybridizations to investigate changes in levels of gene expression. The autoradiograms (top) show the results of hybridizations of arrays of 588 human ESTs to ^{32}P-labeled cDNAs prepared from poly(A$^+$) RNAs isolated from human cancer cells growing in culture, either (*a*) untreated or (*b*) treated with a chemotherapeutic agent. The photographs in (*c*) and (*d*) show the results of scanning and computer analysis of the autoradiograms in (*a*) and (*b*), respectively. Changes in levels of gene expression that are induced by the chemotherapeutic agent can be determined by comparing the two arrays.

electrophoresis (Chapter 20), and antibodies coupled to fluorescent compounds are used to detect the location of proteins *in vivo.* However, both of these approaches provide only a single time-point assay of a protein in a cell, tissue, or organism.

The discovery of a naturally occurring fluorescent protein, the **green fluorescent protein** (**GFP**) of the jellyfish *Aequorea victoria,* has provided a powerful new tool that can be used to study gene expression at the protein level. During the last few years, GFP has been used to monitor the synthesis and localization of specific proteins in a wide variety of living cells. These studies entail constructing fusion genes that contain the nucleotide sequence encoding GFP, coupled in frame to the nucleotide sequence encoding the protein of interest; introducing the chimeric gene into cells by transformation; and studying the fluorescence of the fusion protein in transgenic cells exposed to blue or UV light (Figure 21.19). Because GFP is a small protein, it can often be coupled to proteins without interfering with their activity or interaction with other cellular components.

As the name implies, GFP fluoresces bright green when exposed to blue or ultraviolet light. The chromophore of GFP is produced by the posttranslational cyclization and oxidation of an encoded serine-tyrosine-glycine tripeptide. This chromophore is largely protected from ion and solvent effects by encasement in a barrel-like fold of the mature protein. Unlike other bioluminescent proteins, GFP does not require the ad-

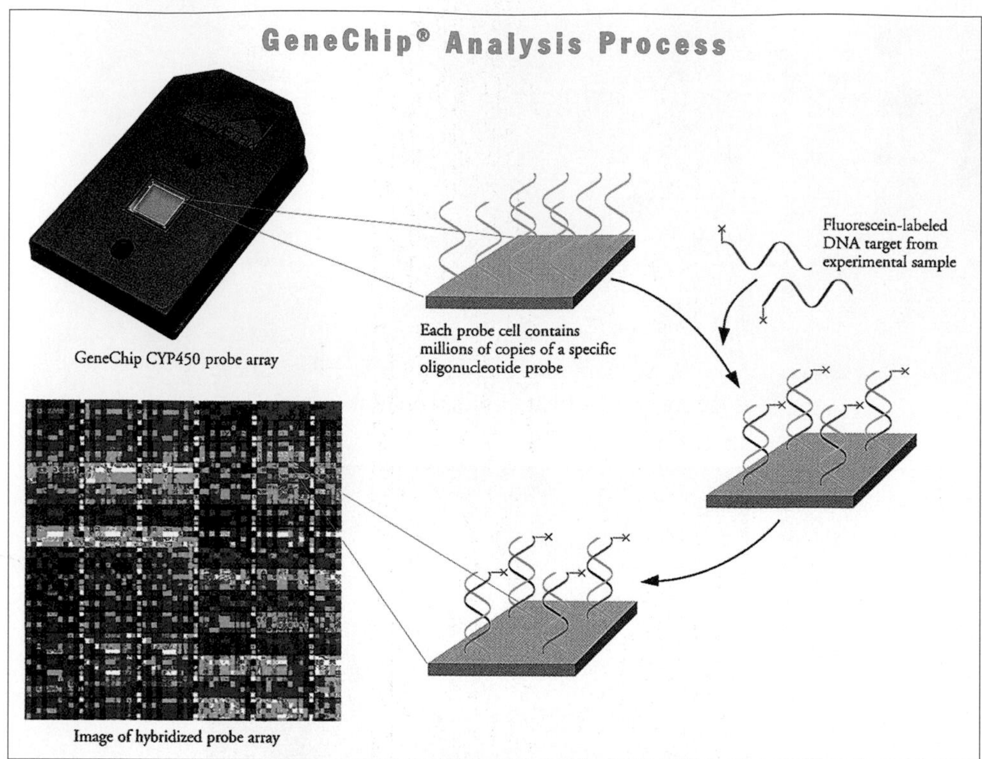

Figure 21.18 Photograph of a gene chip (top left) and an illustration of the use of the gene chip to detect mutations in the human *p53* tumor-suppressor gene. Mutations in this gene are associated with several types of human cancer (see Chapter 28). Other gene chips permit researchers to detect the transcripts of thousands of genes in one experiment.

dition of substrates, cofactors, or any other substances to fluoresce—only exposure to blue or UV light. Thus, GFP can be used to study gene expression in living cells and to study protein localization and movement in cells over time. By mutagenizing the GFP gene, molecular biologists have produced variant forms of GFP that emit blue or yellow light, variants that fluoresce up to 35 times more intensely than the wild-type GFP, and variants whose fluorescence depends on the pH of its microenvironment. These GFP variants can be used to study the synthesis and intracellular localization of two or more proteins simultaneously (Figure 21.19*d*).

Some geneticists are using GFP fusions to study changes in the expression of all the genes encoding proteins that are involved in a particular metabolic pathway in response to treatment of cells or tissues with a specific drug or potentially therapeutic agent. They construct an entire set of GFP-other protein fusion genes, introduce them into host cells, and monitor their expression by quantifying the fluorescence of fusion proteins separated by electrophoresis or other techniques. Technologies are being developed that will allow scientists to observe changes in the levels of large arrays of GFP fusion proteins by capillary electrophoresis (electrophoresis performed in small capillary tubes), monitored by sensitive microphotodetectors and sophisticated computer software. Indeed, in the not-too-distant future, functional genomics may involve the use of DNA chips to detect gene transcripts and "protein chips" to detect the polypeptides encoded by these transcripts.

Key Points: **Once the complete sequence of a genome has been determined, scientists can study the temporal and spatial patterns of expression of all the genes of the organism. One way to study expressed genes is to characterize ESTs (cDNAs) produced from gene transcripts present in a specific cell type. Almost 50,000 distinct human ESTs have been characterized to date. Microarrays of gene-specific hybridization probes on gene chips allow researchers to study the transcription of thousands of genes simultaneously. Chimeric genes that contain the coding region of the green fluorescent protein of the jellyfish fused with the coding regions of the genes of experimental organisms can be used to study the localization of proteins in living cells.**

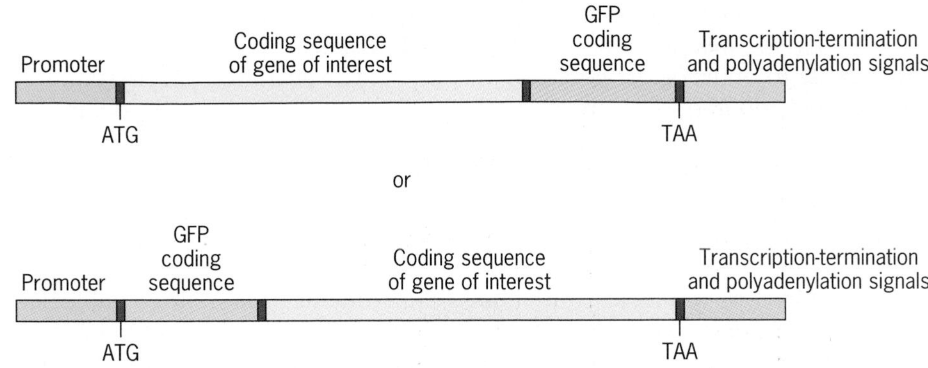

(*a*) Structure of GFP fusion genes.

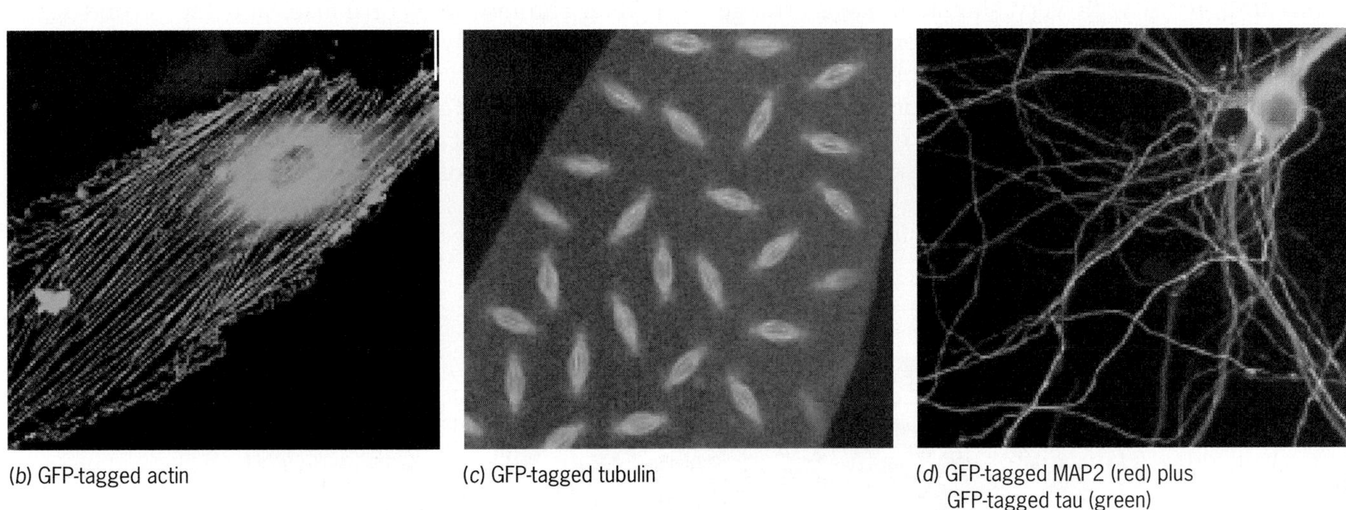

(*b*) GFP-tagged actin (*c*) GFP-tagged tubulin (*d*) GFP-tagged MAP2 (red) plus GFP-tagged tau (green)

Figure 21.19 Use of the green fluorescent protein (GFP) of the jellyfish to study protein localization in living cells. (*a*) Structure of GFP fusion genes. The GFP coding sequence may be placed at either end of the gene of interest or at internal positions. (*b-d*) Immunofluorescence localization of GFP-tagged proteins: (*b*) smooth-muscle actin in a fibroblast cell; (*c*) the microtubule structural protein tubulin in Chinese hamster ovary cells; and (*d*) double labeling of two microtubule-binding proteins, MAP2 with blue light-emitting GFP and tau with green light-emitting GFP, in a rat neuron. With the light filters used for microscopy, MAP2 and tau appear red and green, respectively.

EVOLUTION OF GENOMES

Genome mapping and sequencing projects have made major contributions to our understanding of genome evolution. The results of mapping and sequencing projects have demonstrated striking similarities in the genomes of diverse species as well as some unexpected differences. In bacteria, for example, the genes that participate in processes that are essential in all cellular organisms—DNA replication, transcription, translation, energy metabolism, and the like—are highly conserved, whereas genes that control processes that allow a species to adapt to a specific environment are often unique to that species. In eukaryotes, the striking conservation of genome structure can be illustrated by examining the genomes of the cereal grasses and mammals.

Genome Evolution in the Cereal Grasses

The cereal crops provide much of the food for humans and their domestic livestock, which provide additional human foodstuff. Therefore, enhanced cereal productivity is an important component of the effort to feed the constantly expanding human population of our planet. Increased knowledge of the genomes of these agronomic species is key to achieving this objective. Recent comparative analyses of the genomes of several cereal species indicate that much of the information obtained by mapping and sequencing the smallest of these genomes—the 400-mb genome of rice—may be directly applicable to other cereal grasses because of the conservation of genome structure in these species.

When Graham Moore and colleagues compared the high-density genetic maps of the chromosomes of several cereal grasses, they discovered that despite large differences in genome size and chromosome number, the linkage relationships of blocks of unique DNA sequences and known genes were remarkably conserved. In contrast, the quantities and locations of repetitive DNA sequences were highly variable. Moore and coworkers detected large chromosome segments containing genes with very similar linkage arrangements in all of the cereal grass genomes studied. By simply rearranging the rice chromosome segments, Moore and colleagues were able to construct the chromosomes of the other cereals. Based on their comparative analysis of the genomes of these cereal species, Moore and colleagues were also able to infer a putative genome for the common ancestor of the cereal grasses. This genome is quite similar to the rice genome.

The striking conservation of genome structure in the cereal grasses is illustrated most clearly by drawing the rice genome as a circular array and aligning the conserved blocks of genes in the other species with the rice genome (Figure 21.20). The circular display of the cereal genomes does not imply any circularity of ancestral chromosomes; it simply permits maximal alignment of homologous blocks of genes. The alignment also emphasizes the presence of duplicate copies of each block of genes in the maize genome, indicating that maize has evolved from a tetraploid ancestor. Interestingly, one set of genes is present largely in the small chromosomes of maize, and the second set is present primarily in the large chromosomes.

The conserved structures of the cereal grass genomes should assist plant breeders in their attempts to produce varieties with increased yield, pest resistance, drought tolerance, and other desired traits. Because of the conserved genome structure, information obtained by sequencing the relatively small rice

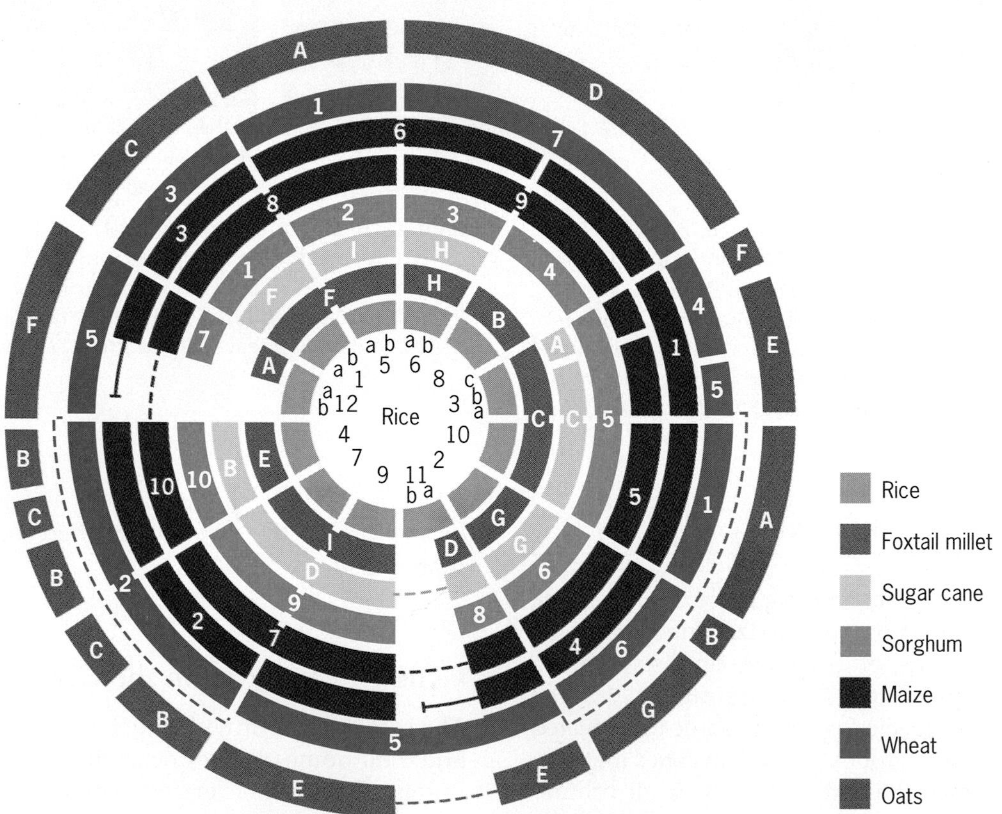

Figure 21.20 Simplified comparative map of the genomes of seven cereal grasses. Chromosomes and segments of chromosomes of the various cereal grasses are aligned with the chromosomes of rice, the grass species with the smallest genome (center). The circular shape of the comparative map allows for maximum alignment of the chromosomes of the different species; it is not meant to imply the existence of a circular ancestral genome. The maize genome has two similar copies of each block of genes and thus occupies two rings of the circle. The dashed lines connect adjacent segments of wheat chromosomes. Similar segments of chromosomes in the oats genome are not connected by dashed lines for the sake of simplicity.

genome should be more easily applied to breeding and genetic engineering projects on the other cereal crop species.

Genome Evolution in Mammals

Mammalian genomes exhibit conservation of chromosome structure similar to that observed in the cereal grasses. Although genetic mapping is being done in over 200 mammalian species, high-density maps are currently available for only human, mouse, rat, and a few agriculturally important farm animals such as swine and cattle. These detailed chromosome maps can be used to demonstrate the conserved linkage relationships of genes in species where such maps are available. For other species, a procedure called **chromosome painting** has been used for comparative genome analyses. Chromosome painting is a variation of fluorescent *in situ* hybridization (FISH; see Figure 10.35) in which chromosomes are "painted" different colors by using DNA hybridization probes labeled with fluorescent dyes that emit light of different wavelengths.

In comparative genomic studies, DNA sequences from one species are used to paint the chromosomes of a related species (Figure 21.21). Such cross-species chromosome painting experiments (also called Zoo-FISH experiments) are usually done under conditions of reduced hybridization stringency (for example, lower temperature). Since the nucleotide sequences of genes of even closely related species diverge somewhat during their evolution from a common ancestor, little or no interspecific hybridization would occur if the experiments were done using high-stringency conditions (conditions that require perfect matches for hybridization to occur). Even the nucleotide sequences of the closely related human and ape genomes, for example, differ by 1.6 percent or an average of 4 base-pair changes in every 250 base pairs. In more distantly related genomes, nucleotide sequences are more divergent, and low-stringency conditions must be used to detect cross-hybridization between the partially complementary strands of homologous genes.

Some of the most interesting chromosome painting studies have used the sequences in a chromosome-specific genomic library (Chapter 20) from one species to paint the chromosomes of related species. Indeed, different fluorescent dyes can be used to label sequences from two or more different chromosomes of one species, and these fluorescence-tagged sequences can be used to "paint" the chromosomes of related species. Chromosome-specific libraries are available for all 24 human chromosomes, and sequences from

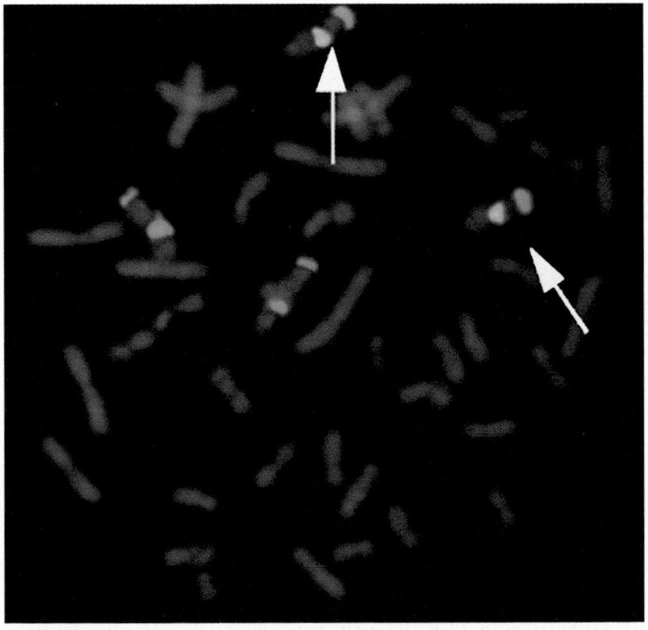

(*a*)

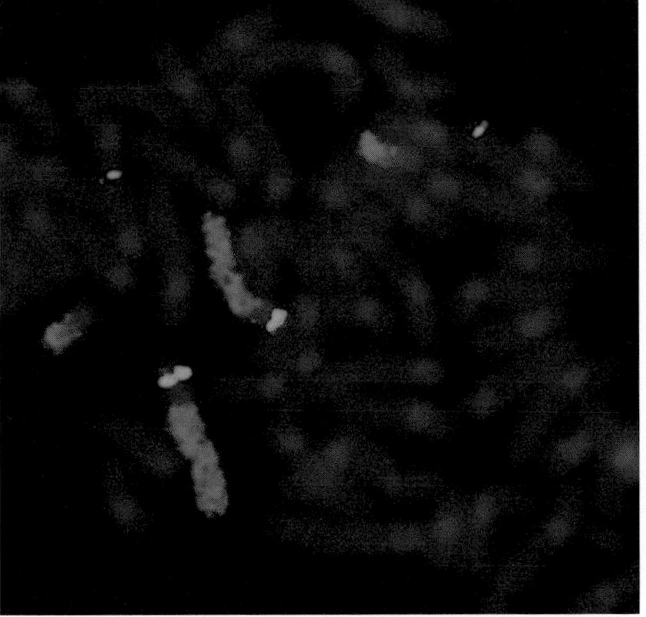

(*b*)

Figure 21.21 Illustration of the use of chromosome painting (see text) to study evolutionary relationships between chromosomes. (*a*) Fluorescent *in situ* hybridization performed on chromosomes of the gibbon *Hylobates lar* using probes prepared from chromosome-specific clone libraries of human chromosomes 5 (orange), 16 (green), 17 (pink), and 22 (yellow). The arrows point to gibbon chromosome 8, which contains sequences that hybridize to fluorescent probes derived from all four of these human chromosomes. (*b*) African green monkey chromosomes painted with fluorescent probes from human chromosome 4 (orange) and a YAC clone specific to human chromosome segment 4q.1 (green).

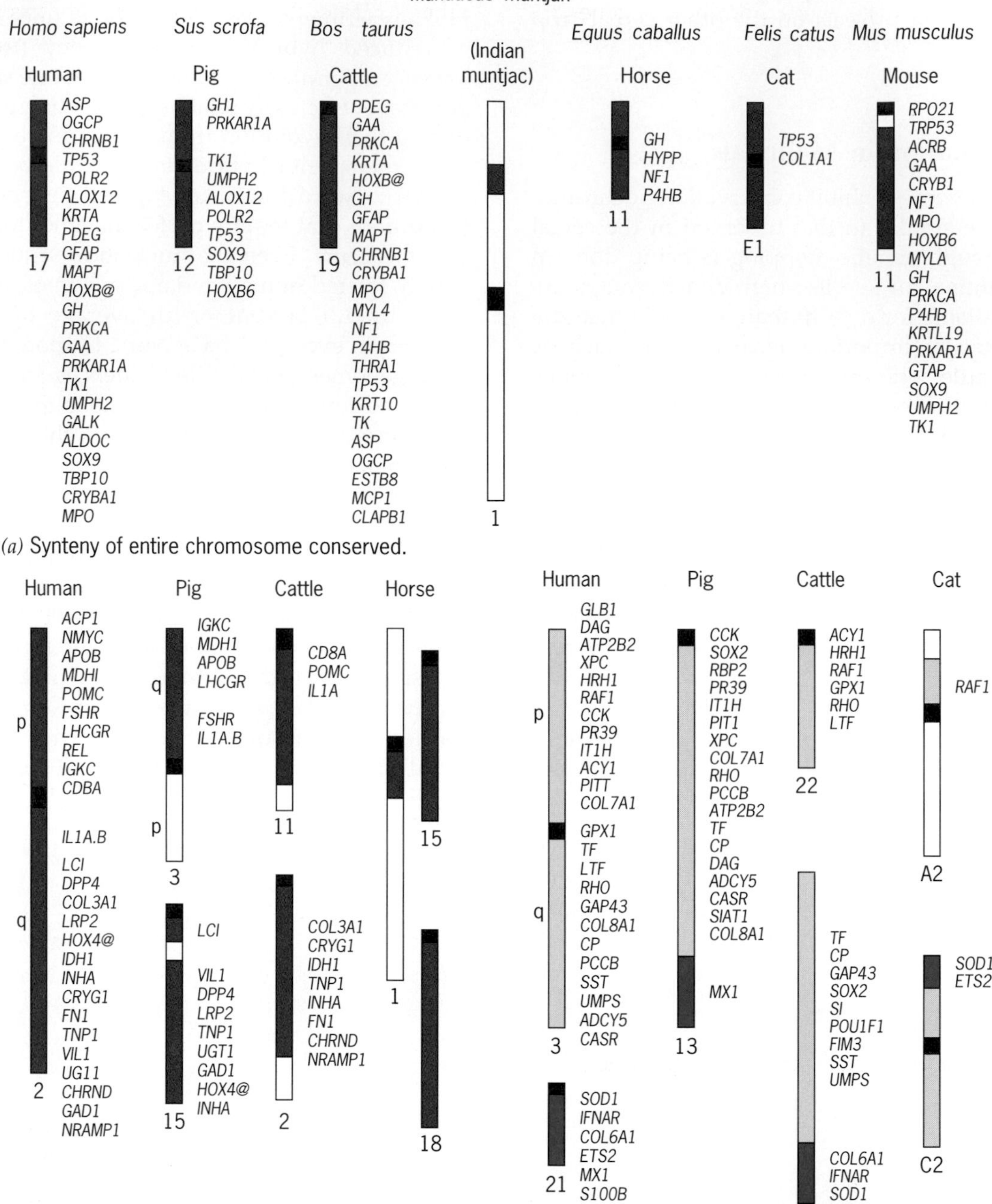

Figure 21.22 Chromosome evolution in mammals. Examples of three classes of conserved synteny (conserved blocks of linked genes) are illustrated. (*a*) Human chromosome 17 is an example of the conservation of entire chromosomes. (*b*) Human chromosome 2 provides an example of the conservation of large chromosomal segments. (*c*) Human chromosomes 3 and 21 illustrate the formation of new synteny by chromosome fusion. The chromosomal locations of a few genes are shown to the right of the chromosomes. Species-specific chromosome designators are given below the chromosomes.

these libraries have been used to "paint" the chromosomes of several related species, including most of the primates and a few more distantly related mammals.

Upon reviewing comparative linkage and chromosome painting data, Bhanu Chowdhary and colleagues concluded that the evolution of mammalian chromosomes involved three classes of conserved synteny. (Recall that synteny is the presence of genes on the same chromosome). The three classes are (1) conservation of entire chromosomes, (2) conservation of large segments of chromosomes, and (3) the joining of segments of different chromosomes to produce new synteny. Each type of chromosome evolution is illustrated in Figure 21.22.

The synteny of genes on human chromosome 17 is conserved on chromosome 12 of pig and chromosome 19 of cattle, is largely conserved on chromosomes 11 and E1 of horse and cat, respectively, and is partially conserved on chromosomes 1 and 11 of Indian muntjac and mouse, respectively (Figure 21.22*a*). Similar patterns of conservation of human chromosomes 13 and 20 are observed in these species. Note that although we are discussing the conservation of synteny in mammals by using the human chromosomes as references because they are the most thoroughly mapped, the conservation is actually the result of the evolution of these species from a common ancestor.

Human chromosome 2 provides an example of the second class of conserved synteny, the conservation of large segments of chromosomes (Figure 21.22*b*). Major portions of the long and short arms of human chromosome 2 are conserved on two chromosomes of pig, cattle, horse, and cat. The genes on human chromosome 2 are more dispersed on the chromosomes of Indian muntjac and mouse. Human chromosomes 4, 5, 6, 9, and 11 also have large chromosome segments conserved in the other mammals studied.

The third pattern of conserved synteny—the joining of segments of chromosomes to produce new synteny—is illustrated by human chromosomes 3 and 21 (Figure 21.22*c*). Chromosome 13 of pig appears to contain the genes of both human chromosomes arranged as if they had simply fused together. In cattle, one block of genes present on human chromosome 3 is present on a separate chromosome (22), whereas most of the rest of the genes from human chromosome 3 are on a chromosome (1) with the genes from human chromosome 21. More complex rearrangements are observed in the other species; however, even in the other mammals, blocks of genes from human chromosomes 3 and 21 are present together on the same chromosome. Other examples of this pattern of conserved synteny involve human chromosomes 12 and 22, 14 and 15, and 16 and 19.

Despite the extensive conservation of blocks of genes illustrated by the examples discussed above, more detailed comparisons of chromosome structure in mammals clearly show that numerous chromosome rearrangements have occurred during the evolution of even closely related species. A comparison of human chromosomes with the chromosomes of the white-cheeked gibbon, *Hylobates concolor,* revealed that at least 21 translocations have occurred during the evolution of these two species from their common ancestor. Inversions and other intrachromosomal rearrangements are especially prevalent in the genomes of closely related species. The availability of the complete sequence of the human genome will greatly facilitate comparative genomic analyses, which will, in turn, provide a much more complete picture of our evolutionary relationship with primate relatives and other mammalian species.

Key Points: **Comparative genomic studies have revealed a remarkable conservation of synteny in related species. This conservation of blocks of linked genes is especially notable in the cereal grasses. Conserved synteny is also prevalent in evolutionarily divergent mammalian species.**

TESTING YOUR KNOWLEDGE

1. Best disease is a form of blindness in humans that develops gradually in adults. It is caused by an autosomal dominant mutation on chromosome 11. Nine RFLPs, designated 1 through 9, map on chromosome 11 in numerical order. The polymorphisms at each site are designated by superscripts 0 through N, where N + 1 is the number of polymorphisms present at a site in the family represented by the pedigree below. DNA was obtained from each member of the family, digested with the appropriate restriction enzyme, subjected to gel electrophoresis, transferred to a nylon membrane by Southern blotting, denatured, and hybridized to radioactive probes that detect all the RFLPs. After hybridization, the membranes were exposed to X-ray film, and the autoradiograms were used to determine which RFLP was present in each member of the family. The results are shown in the pedigree below. Circles represent females; squares represent males; solid symbols indicate individuals with Best disease.

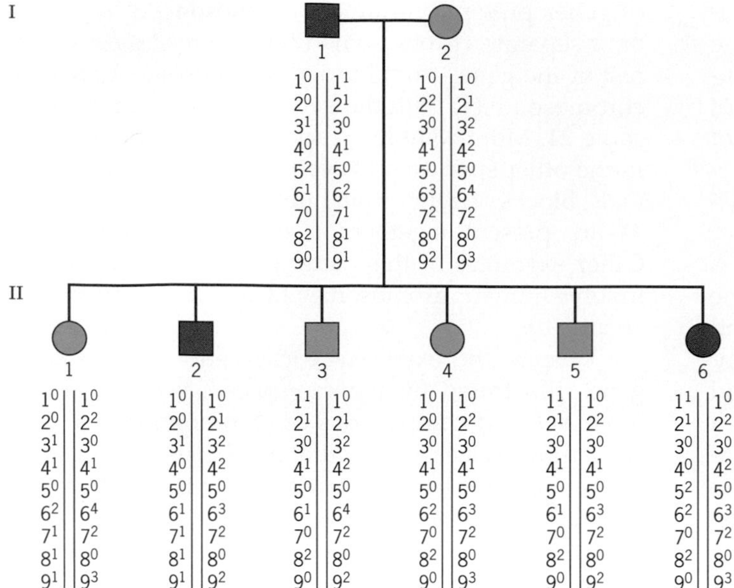

Which RFLP site is closest to the mutation that causes Best disease? Which allele of this RFLP is present on the chromosome that carries the Best disease mutation?

ANSWER

RFLP site 4 is closest to the Best disease mutation, which is present on the copy of chromosome 11 carrying the 4^0 allele of the polymorphism. Of the polymorphisms on chromosome 11, only the 4^0 allele is present in all three family members with Best disease and absent from all five members with normal vision.

2. Eleven genomic clones, each containing DNA from chromosome 4 of *Drosophila melanogaster,* were tested for cross-hybridization in all pairwise combinations. The clones are designated A through K, and the results of the hybridizations are shown in the table below. A plus sign indicates that hybridization occurred; a minus sign indicates that no cross-hybridization was observed.

	A	B	C	D	E	F	G	H	I	J	K
K:	–	–	–	–	+	–	–	–	–	–	+
J:	–	–	+	+	–	+	–	+	–	+	
I:	–	+	–	–	+	–	–	–	+		
H:	–	–	–	–	–	+	–	+			
G:	+	–	+	–	–	–	+				
F:	–	–	–	+	–	+					
E:	–	–	–	–	+						
D:	–	–	+	+							
C:	–	–	+								
B:	–	+									
A:	+										

Based on the hybridization results shown in the table, how many contigs do these clones define? Draw the contig map(s) defined by these data.

ANSWER

Maps of the 2 contigs defined by the 11 mutations are as follows:

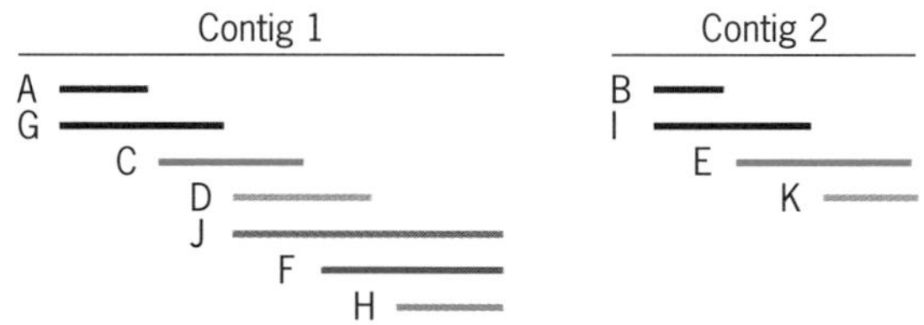

QUESTIONS AND PROBLEMS

21.1 Distinguish between a genetic map, a cytogenetic map, and a physical map. How can each of these types of maps be used to identify a gene by positional cloning?

21.2 What is the difference between chromosome walking and chromosome jumping? Why must chromosome jumps, rather than just chromosome walks, sometimes be used to identify a gene of interest?

21.3 What is a contig? an RFLP? a VNTR? an STS? an EST? How are each of these used in the construction of chromosome maps?

21.4 The following is a Southern blot of *Eco*RI-digested DNA of rye plants from two different inbred lines, A and B. Developed autoradiogram I below shows the bands resulting from probing the blot with ^{32}P-labeled cDNA1. Autoradiogram II below shows the same Southern blot after it was stripped of probe and reprobed with ^{32}P-labeled cDNA2.

	I: A	I: B		II: A	II: B
a1	—	—	b1	—	
			b2	—	
a2	—				
a3	—		b3		—
a4	—				

(a) Which bands would you expect to see in the autoradiogram of a similarly probed Southern blot prepared using *Eco*RI-digested DNA from F_1 hybrid plants produced by crossing the two inbred lines? (b) What can you conclude about the gene(s) represented by band a1 on blot I in the two inbreds? (c) The F_1 plants were crossed to plants possessing only bands a1, a4, and b3. DNA was isolated from several individual progeny and digested with *Eco*RI. The resulting DNA fragments were separated by gel electrophoresis, transferred to a nylon membrane, and hybridized with radioactive cDNA1 and cDNA2 probes. The following table summarizes the bands present in autoradiograms obtained using DNA from individual progeny. Interpret these data. Do the data provide evidence for RFLPs? At how many loci? Are any of the RFLPs linked? If so, what are the linkage distances defined by the data?

	Bands Present						
Plant No.	**a1**	**a2**	**a3**	**a4**	**b1**	**b2**	**b3**
1	+	+	+	+			+
2	+	+	+	+			+
3	+	+	+	+			+
4	+	+	+	+			+
5	+	+	+	+	+	+	+
6	+			+	+	+	+
7	+			+	+	+	+
8	+			+	+	+	+
9	+			+	+	+	+
10	+			+			+

21.5 As part of the Human Genome Mapping Project, you are trying to clone a gene involved in colon cancer. Your first step is to localize the gene using RFLP markers. In the following table, RFLP loci are defined by STS (sequence tagged site) number, for example, STS1, and the gene for colon cancer is designated *C*.

Loci	% recombination	Loci	% recombination
C, STS1	50	STS1, STS5	10
C, STS2	15	STS2, STS3	30
C, STS3	15	STS2, STS4	14
C, STS4	1	STS2, STS5	50
C, STS5	40	STS3, STS4	16
STS1, STS2	50	STS3, STS5	25
STS1, STS3	35	STS4, STS5	41
STS1, STS4	50		

(a) Given the percentage recombination between different RFLP loci and the gene for colon cancer shown in the table, draw a genetic map showing the order and genetic distances between adjacent RFLP markers and the gene for colon cancer. (b) Given that the human genome contains approximately 3.3×10^9 base pairs of DNA and that the human genetic map contains approximately 3300 centiMorgans, approximately how many base pairs of DNA are located along the stretch of chromosome defined by this RFLP map? (*Hint:* First figure how many base pairs of DNA are present per cM in the human genome.) (c) How many base pairs of DNA are present in the region between the colon cancer gene and the nearest STS?

21.6 What are microsatellites? Why are they called microsatellites?

21.7 You have cloned a previously unknown human gene. What procedure will allow you to position this gene on the cytological map of the human genome without performing any pedigree analyses? Describe how you would carry out this procedure.

21.8 You have identified a previously undescribed human EST. What must be done before this new EST can be called an STS?

21.9 VNTRs and microsatellites are specific classes of polymorphisms. What is the difference between a VNTR and a microsatellite? Why would it be inappropriate to call most VNTRs microsatellites?

21.10 An RFLP and a mutant allele that causes albinism in humans cannot be shown to be separated by recombination based on pedigree analysis or by radiation hybrid mapping. Do these observations mean that the RFLP occurs within or overlaps the gene harboring the mutation that causes albinism? If so, why? If not, why not?

21.11 An RFLP and a mutation that causes deafness in humans both map to the same location on the same chromosome. How can you determine whether or not the RFLP overlaps with the gene containing the deafness mutation?

21.12 Why is the resolution of radiation hybrid mapping of human genes higher than the resolution of standard somatic-cell hybrid mapping?

21.13 What are the goals of the Human Genome Project? What impact will achieving these goals have on the practice of medicine in the future? What are some of the possible misuses of human genome data?

21.14 To be useful as a genetic marker for positional cloning of a mutant allele that causes an inherited abnormality in humans, an RFLP must be present on the same homolog as the mutation. Why?

21.15 Which type of molecular marker, RFLP or EST, is most likely to mark a disease-causing mutant gene in humans? Why?

21.16 Bacteriophage ΦX174 contains 11 genes in a genome of 5386 bp; *E. coli* has a predicted 4288 genes in a genome of about 4.639 kb; *S. cerevisiae* has about 6000 genes in a genome of size 12.1 mb; *C. elegans* has about 15,000 genes present in a genome of about 100 mb; and *H. sapiens* has an estimated 70,000 genes in its 3000-mb genome. Which genome has the highest gene density? the lowest gene density? Does there appear to be any correlation between gene density and developmental complexity? If so, describe the correlation.

21.17 A contig map of one segment of chromosome 3 of *Arabidopsis* is shown below.

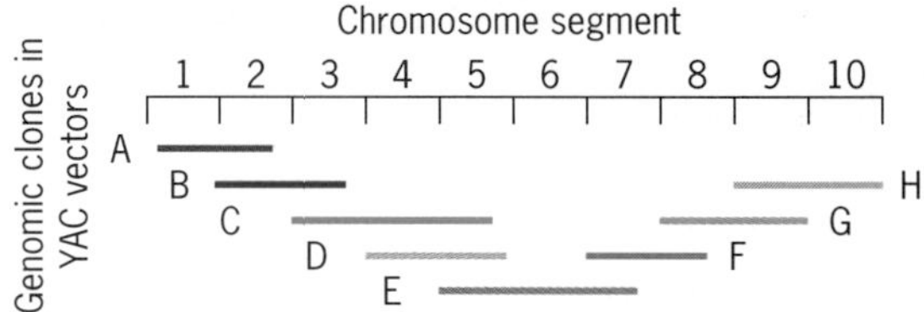

(a) If an EST hybridizes with genomic clones C, D, and E, but not with the other clones, in which segment of chromosome 3 is the EST located? (b) If a clone of gene *ARA* hybridizes only with genomic clones C and D, in which chromosome segment is the gene located? (c) If a restriction fragment hybridizes with only one of the genomic clones shown above, in which chromosome segment(s) could the fragment be located?

21.18 Eight human-Chinese hamster radiation hybrids were tested for the presence of six human ESTs designated A through F. The results are shown in the table below, where a plus indicates that a marker was present and a minus indicates that it was absent.

	Radiation hybrid							
Marker	1	2	3	4	5	6	7	8
A	−	+	−	−	+	+	−	+
B	+	−	+	−	−	+	−	+
C	−	+	+	+	−	−	−	+
D	+	−	+	+	−	−	+	−
E	+	−	+	+	−	−	+	−
F	−	+	−	+	+	+	+	−

Based on these data, do any of the ESTs appear to be closely linked? Which ones? What would be needed for you to be more certain of your answer?

21.19 What advantages do gene chips have over traditional dot blot hybridization techniques?

21.20 What major advantage does the green fluorescent protein of the jellyfish have over other methods for studying protein synthesis and localization?

21.21 You are given chromosome-specific cDNA libraries for all 24 human chromosomes. How might these libraries be used to study chromosome evolution in primates?

21.22 Of the cereal grass species, only maize contains two copies of each block of linked genes. What does this duplication of sets of maize genes indicate about the origin of this agronomically important species?

21.23 Five human genomic DNA clones present in YAC vectors were tested by hybridization for the presence of six sequence-tagged sites designated STS1 through STS6. The results are given in the following table; a plus indicates the presence of the STS and a minus indicates the absence of the STS.

	STS					
YAC clone	1	2	3	4	5	6
A	+	−	+	+	−	−
B	+	−	−	−	+	−
C	−	−	+	+	−	+
D	−	+	−	−	+	−
E	−	−	+	−	−	+

(a) What is the order of the STS sites on the chromosome? (b) Draw the contig map defined by these data.

BIBLIOGRAPHY

CHOWDHARY, B. P., T. RAUDSEPP, L. FRÖNICKE, AND H. SCHERTHAN. 1998. Emerging patterns of comparative genome organization in some mammalian species as revealed by Zoo-FISH. *Genome Research* 8:577–589.

DOOLITTLE, R. F. 1998. Microbial genomes opened up. *Nature* 392:339–342.

GALE, M. D., AND K. M. DEVOS. 1998. Comparative genetics of grasses. *Proc. Natl. Acad. Sci., U.S.* 95:1971–1974.

MARSHALL, A., AND J. HODGSON. 1998. DNA chips: an array of possibilities. *Nature Biotechnology* 16:27–31.

MARSHALL, E., AND E. PENNISI. 1998. Hubris and the human genome. *Science* 280:994–995.

ROWEN, L., G. MAHAIRAS, AND L. HOOD. 1997. Sequencing the human genome. *Science* 278:605–607.

SCHULER, G. D., AND 100 CO-AUTHORS. 1996. A gene map of the human genome. *Science* 274:540–546.

STEWART, E. A., AND 27 CO-AUTHORS. 1988. An STS-based radiation hybrid map of the human genome. *Genome Research* 7:422–433.

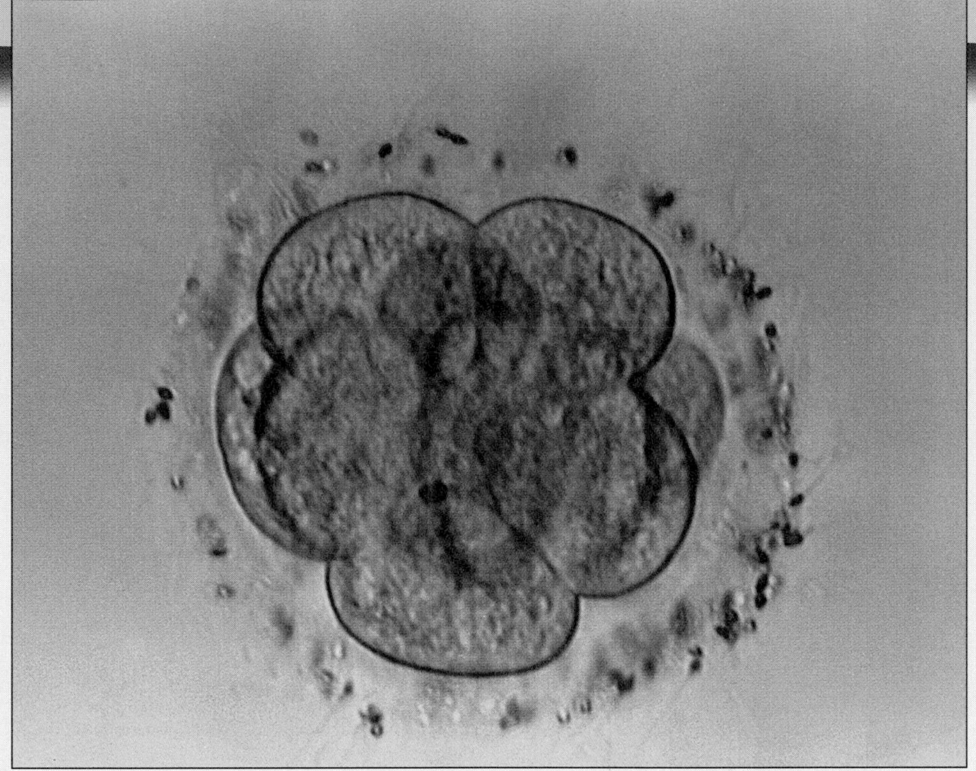

An eight-cell human pre-embryo.

22

Molecular Analysis of Genes and Gene Products

CHAPTER OUTLINE

Use of Recombinant DNA Technology to Identify Human Genes
Molecular Diagnosis of Human Diseases
DNA Fingerprints
Human Gene Therapy
Production of Eukaryotic Proteins in Bacteria
Transgenic Plants and Animals
Blocking Gene Expression with Antisense RNA

Detection of the Tay-Sachs Mutation in Eight-Cell Pre-embryos

Brittany was conceived *in vitro* from ova and sperm obtained from her mother and father. Three days after conception, one of the eight cells of the pre-embryo from which Brittany developed was used to conduct DNA tests for a mutant gene that causes Tay-Sachs disease. The results of the tests were negative, and the pre-embryo was implanted in the womb of Brittany's mother. Nine months later, Brittany was delivered with no concern that she might someday die from Tay-Sachs disease as her older sister had. The removal of one cell from an eight-cell pre-embryo does not affect its subsequent development.

Tay-Sachs disease (see Human Genetics Sidelight in Chapter 14) is a lethal autosomal recessive disorder. Infants with Tay-Sachs are normal at birth, but undergo rapid neurological degeneration leading to blindness, paralysis, mental retardation, and death at three to four years of age. The degeneration results from the absence of an enzyme called hexosaminidase A, which catalyzes the first step in the breakdown of a complex lipid named ganglioside G_{M2}. In the absence of hexosaminidase A, ganglioside G_{M2} accumulates in neurons and causes progressive degeneration of the central nervous system.

Both of Brittany's parents are heterozygous for the mutant gene that causes Tay-Sachs disease. They knew that if they conceived another child, it would have a 1/4th chance of suffering from Tay-Sachs disease, as did their first daughter. Although they longed for a child, Brittany's parents were not willing to risk having another infant with Tay-Sachs disease. Tay-Sachs disease can be detected prenatally; however, because of religious convictions, the couple had ruled out the abortion of an affected fetus. *In vitro* fertilization and a DNA test for the mutant hexosaminidase A gene that both parents carried allowed confirmation that Brittany was not homozygous for the mutant gene before implantation at the preembryo stage. To her parents, who suffered through the degeneration and death of their first child, Brittany is a priceless treasure.

The DNA tests for Tay-Sachs and other human disorders became possible with the discovery of the genetic defects that cause these diseases. These defects have been identified by searching the human genome using sophisticated chromosome mapping and recombinant DNA techniques. In this chapter, we will consider a few of the many applications of the recombinant DNA, gene-cloning, and genomic technologies described in the preceding two chapters. These technologies are, without doubt, the most powerful tools ever developed in the field of biology. The ability to carry out carefully designed genetic engineering of living cells has many potential applications. Several of these applications have already reached fruition. If used wisely, genetic engineering promises to enhance the quality of human life. On the other hand, if used carelessly or in an unethical manner, genetic engineering could have a negative impact on our quality of life. As with any powerful tool, the impact of recombinant DNA techniques on our lives will depend on the wisdom of the humans who use them.

Just as geneticists now know the complete pathway of morphogenesis of bacteriophage T4 (Chapter 14), in the future they will know the complete pathway of morphogenesis of a yeast cell, a fruit fly, an *Arabidopsis* plant, or, indeed, even a human being. Moreover, at some point, biologists will understand the molecular basis of learning and memory and will know what molecular events underlie the aging process. Most important, they will understand the complex mechanisms that regulate cell division in humans and should be able to use this knowledge to prevent or cure at least some types of human cancer and life-threatening viral infections. Clearly, much important work remains to be done by the next generation of biologists, and the powerful tools of molecular genetics will play central roles in expanding our understanding of the world we live in. In this chapter, we discuss some of the applications of these tools.

USE OF RECOMBINANT DNA TECHNOLOGY TO IDENTIFY HUMAN GENES

Recombinant DNA techniques have revolutionalized the search for defective genes that cause human diseases. Indeed, several major "disease genes" have already been identified by positional cloning (Chapter 21). In addition, the mutations responsible for the diseases have been determined by comparing the nucleotide sequences of wild-type and mutant alleles of the genes. The coding sequences of the wild-type alleles have been translated by computer to predict the amino acid sequences of the gene products. Oligopeptides have been synthesized based on the predicted amino acid sequences and used to produce antibodies, which, in turn, have been used to localize the gene products and to investigate their functions *in vivo.* In some cases, the resulting information about gene function has suggested possible approaches to treating the diseases. The results of these studies will allow future treatment of some of these diseases by gene therapy.

Huntington's Disease

Huntington's disease (HD) is an insidious disorder caused by an autosomal dominant mutation, which occurs in about one of 10,000 individuals of European descent. Individuals with HD undergo progressive degeneration of the central nervous system, usually beginning at age 30 to 50 years and terminating in death 10 to 15 years later. To date, HD is untreatable. However, identification of the gene and the mutational defect responsible for HD has kindled hope for an effective treatment in the future. Because of the late age of onset of the disease, most HD patients already have children before the disease symptoms appear. Since the disorder is caused by a dominant mutation, each child of a heterozygous HD patient has a 50 percent chance of being afflicted with the disease. These children observe the degeneration and death of their HD parent, knowing that they have a 50:50 chance of suffering the same fate.

The HD gene was one of the first human genes shown to be tightly linked to an RFLP. In 1983, James Gusella, Nancy Wexler, and coworkers demonstrated that the HD gene cosegregated with an RFLP that mapped near the end of the short arm of chromosome 4. They based their findings largely on data from studies of two large families, one in Venezuela and one in the United States. Subsequent research showed that the linkage was about 96 percent complete; 4 percent of the offspring of HD heterozygotes were recombinant for the RFLP and the HD allele. Given this early localization of the HD gene to a relatively short segment of chromosome 4, some geneticists predicted that the HD gene would soon be cloned and characterized. However, the task was more difficult than anticipated and took a full 10 years to accomplish. Unlike many other inherited disorders, no case of Huntington's disease is caused by a chromosome rearrangement with a breakpoint that could be used to pinpoint the location of the gene. Therefore, even though the RFLP cosegregation data focused the search on a 500-kb region of chromosome 4, researchers still had to identify all of the genes in this region and determine which one was responsible for HD. Detailed restriction and contig maps were prepared of the 500-kb region (Figure 22.1). Then, a procedure called **exon amplification** was used to identify coding sequences of genes within the region of interest.

Exon amplification (Figure 22.2) is used to identify coding sequences in a DNA cassette that is flanked by 5′ and 3′ intron splice sites. In principle, the procedure is simple. Genomic DNA fragments are inserted into a cloning site within an intron from a mammalian virus. The viral intron contains 5′ and 3′ splice sites flanking the cloning site. Therefore, if an exon flanked by 3′ and 5′ intron splice sites is inserted into the cloning site of the vector, the exon will lie between two introns that should be spliced out *in vivo.* This intron-cloning site is present in a viral cloning vector, which can be introduced into host cells by transfection. The vector contains a viral origin of replication and a strong promoter adjacent to the cassette so that the entire region is transcribed efficiently. After transfection, the mammalian culture cells are grown for a few days to allow transcription and RNA splicing to occur. Then, cytoplasmic RNAs are isolated, cDNAs are synthesized with reverse transcriptase (Figure 20.11), and the exon is amplified by PCR (Figure 20.24) using primers complementary to flanking vector DNA sequences. The amplified exon sequences are then used as probes to screen cDNA libraries by colony or plaque hybridization (Figure 20.12). The cDNA clones, in turn, are used to identify complementary genomic DNA clones. The nucleotide sequences of the cDNAs and genomic DNAs are compared in order to define the introns and

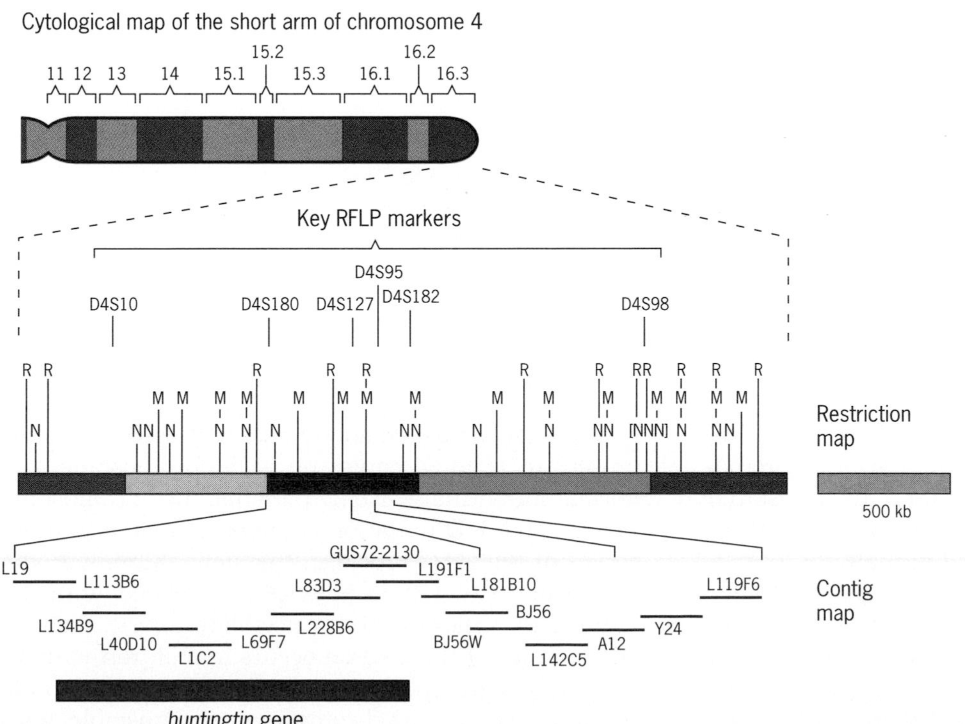

Figure 22.1 Identification of the gene responsible for Huntington's disease by positional cloning. The cytological map of the short arm of chromosome 4 is shown at the top. The RFLP markers, restriction map, and contig map used to locate the *huntingtin* gene are shown below the cytological map. M, N, and R represent *Mlu*I, *Not*I, and *Nru*I restriction sites, respectively.

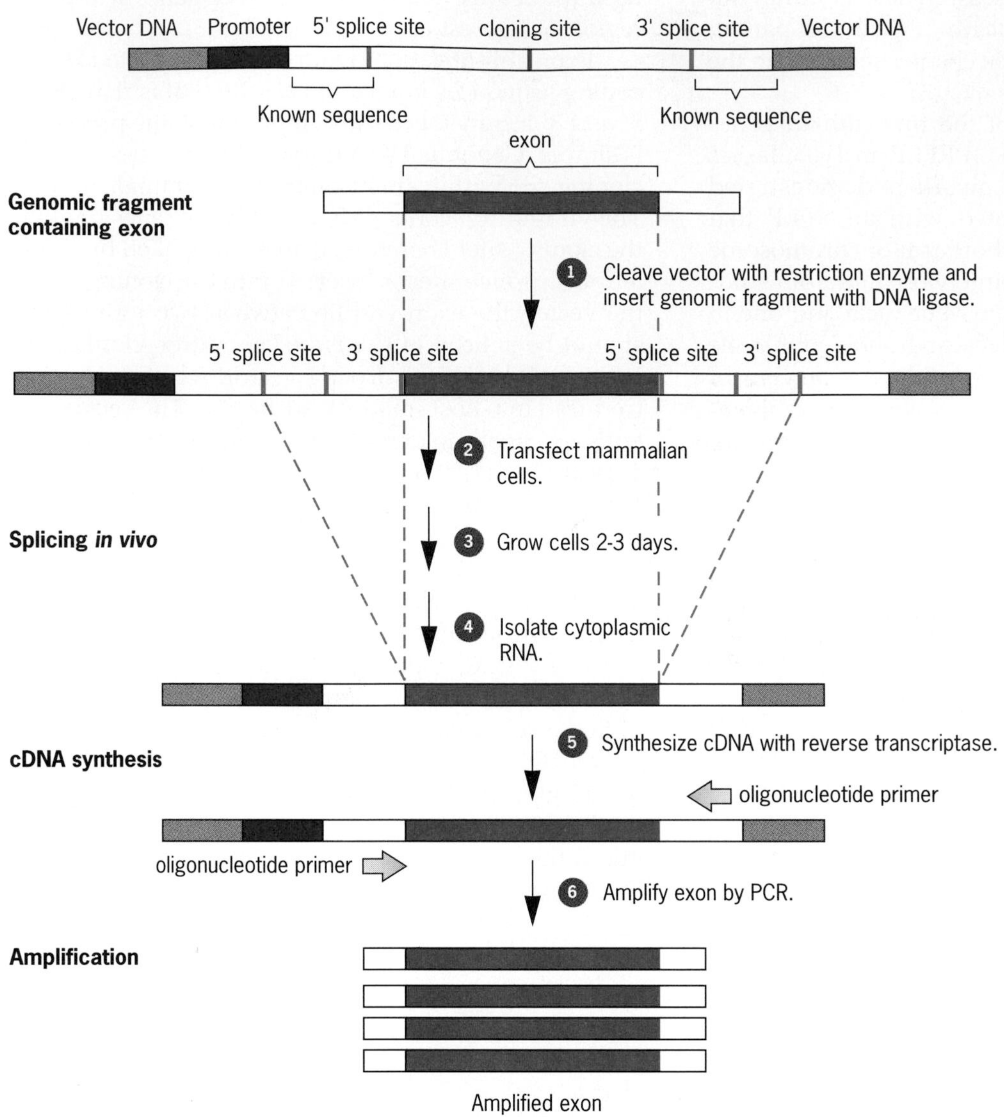

Figure 22.2 Identification of exons by the exon amplification procedure. If a genomic DNA fragment contains an exon with intact 3′ and 5′ splice sites, the adjoining intron sequences will be spliced out *in vivo* after its insertion into a viral splicing vector. The splicing events will move the oligonucleotide primer sites adjacent to the exon so that it is amplified during the subsequent PCR step. If 3′ and 5′ splice sites are not present in the fragment, splicing will not occur, and the exon will not be amplified by this procedure.

exons in the genes. These procedures allow identification of all genes within a given region of a chromosome. Then, the gene responsible for a given disorder must be identified by comparing the sequences of the alleles in affected and normal individuals.

Using all of the above procedures, Gusella, Wexler, and coworkers identified a gene, first called IT15 (for *I*nteresting *T*ranscript number *15*) and subsequently named *huntingtin,* that spans about 210 kb near the end of the short arm of chromosome 4 (Figure 22.1). This gene contains a trinucleotide repeat, $(CAG)_n$, which is present in 11 to 34 copies on each chromosome 4 of normal individuals. In HD patients, the chromosome carrying the HD mutation contains 42 to 100 or more copies of the CAG repeat in this gene. Moreover, the age of onset of HD is inversely correlated with the number of copies of the trinucleotide repeat. Rare juvenile onset of the disease occurs in children with an unusually high repeat copy number. The trinucleotide repeat regions of HD chromosomes are unstable, with repeat number often expanding and sometimes contracting between generations. Gusella, Wexler, and collaborators detected expanded CAG repeat regions in chromosomes from 72 different families with HD, leaving little doubt that they had identified the correct gene.

The *huntingtin* gene is expressed in many different cell types, producing a large 10- to 11-kb mRNA. The coding region of the *huntingtin* mRNA predicts a protein 3144 amino acids in length. Unfortunately, the predicted amino acid sequence of the huntingtin protein has provided no hint as to its function. The amino acid sequence is not closely related to that of any known protein. The dominance of the HD mutation indicates that the mutant protein has some new function that causes the disease, but what this function might be is the focus of ongoing research.

HD was the fourth human disease to be associated with an unstable trinucleotide repeat. Fragile X syndrome (the most common form of mental retardation in humans), as well as myotonic dystrophy and spinobulbar muscular atrophy (both diseases associated with loss of muscle control), had previously been shown to result from expanded trinucleotide repeats. More recently, two other neurodegenerative diseases, spinocerebellar ataxia type 1 and dentatorubro-pallidoluysian atrophy, have been shown to result from similar enlarged trinucleotide repeat regions. These results indicate that the expansion of such repeat regions may be a common mutational event in humans.

Although the identification of the genetic defect, the expanded trinucleotide repeat in the *huntingtin* gene, has contributed little insight into the molecular and cellular basis of the neurological degeneration that occurs in HD patients, it has provided a simple and accurate DNA test for the HD mutation (Figure 22.3). Once the nucleotide sequences of the *huntingtin* gene on either side of the trinucleotide repeat region were known, oligonucleotide primers could be synthesized and used to amplify the region by PCR, and the num-

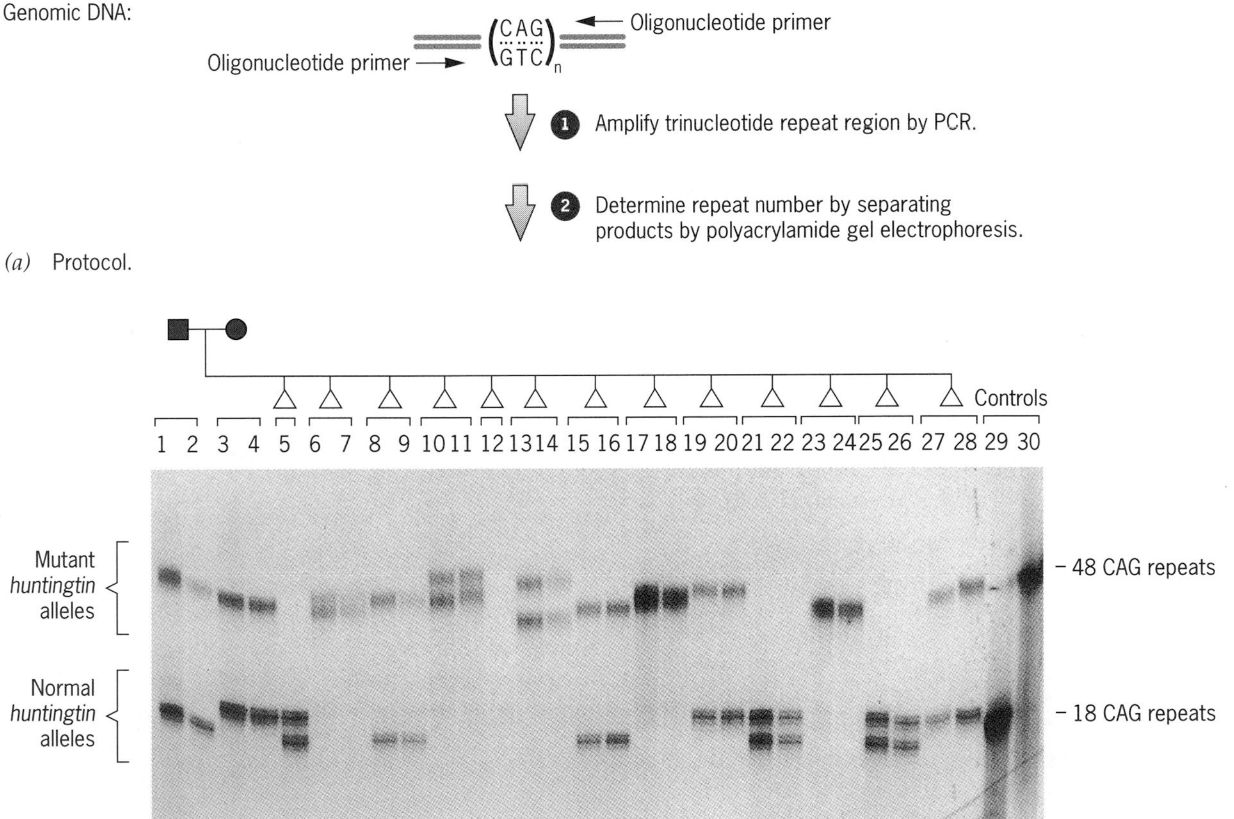

Figure 22.3 Testing for the expanded trinucleotide repeat regions in the *huntingtin* gene that are responsible for Huntington's disease by PCR. The results shown are from a Venezuelan family in which the parents are heterozygous for the same mutant *huntingtin* allele. The order of birth of the children has been changed, and their sex is not given to assure anonymity. All of the family members but two (DNA samples 5 and 12) were tested twice using DNA from cells obtained at least one full year apart.

ber of CAG repeats could be determined by polyacrylamide gel electrophoresis. Thus individuals at risk of carrying the mutant *huntingtin* gene can easily be tested for its presence. Because the PCR procedure requires very little DNA, the test for HD also can be performed prenatally on fetal cells obtained by amniocentesis or chorionic biopsy (see Human Genetics Sidelight in Chapter 7).

Given the availability of the DNA test for the HD mutation, individuals who are at risk of transmitting the defective gene to their children can determine whether they carry it before starting a family. However, the potential psychological effects of a positive identification of the HD mutation can be traumatic. Given a 50:50 chance of a diagnosis of HD:no HD, would you want to know early that you carry the mutation, or, instead, hold on to the hope for as long as possible that you do not carry it? For those individuals who test negative for the HD mutation, a heavy burden is lifted. For those who test positive, will knowing help them prepare emotionally and financially for the onset of the disease? Clearly, the DNA tests for HD must be performed with great care and only in conjunction with proper counseling. Still, for the young men and women at risk for HD and not wanting to transmit the mutation to their children, the DNA test offers hope. Each person with a heterozygous parent has a 50 percent chance of not carrying the defective gene. If the test is negative, she or he can begin a family with no concern about transmitting the mutation.

If the test is positive, the fetus can be tested prenatally, or the couple can consider *in vitro* fertilization, as did the parents we discussed at the beginning of this chapter. If the eight-cell pre-embryo tests negative for the HD mutation, it can be implanted in the mother's uterus with the knowledge that it carries two normal copies of the *huntingtin* gene. If used conscientiously, the DNA test for the HD mutation should diminish human suffering from this dreaded disease.

Cystic Fibrosis

Cystic fibrosis (CF) is one of the most common inherited diseases in humans, affecting 1 in 2000 newborns of northern European heritage. CF is inherited as an autosomal recessive mutation, and the frequency of heterozygotes is estimated to be about 1 in 25 in Caucasian populations. In the United States alone, over 30,000 people suffer from this devastating disease. One easily diagnosed symptom of CF is excessively salty sweat, a largely benign effect of the mutant gene. Other symptoms are anything but benign. The lungs, pancreas, and liver become clogged with a thick mucus, which results in chronic infections and the eventual malfunction of these vital organs. In addition, mucus often builds up in the digestive tract, causing individuals to be malnourished no matter how much they eat. Lung infections are recurrent, and patients often die from pneumonia or related infections of the respiratory system. In 1940, the average life expectancy for a newborn with CF was less than two years. With improved methods of treatment, life expectancy has gradually increased. Today, the life expectancy for someone with CF is about 30 years, but the quality of life is poor.

Identification of the *CF* gene is one of the major successes of positional cloning (Chapter 21). Biochemical analyses of cells from CF patients had failed to identify any specific metabolic defect or mutant gene product. Then, in 1989, Francis Collins and Lap-Chee Tsui and their coworkers identified the *CF* gene and characterized some of the mutations that cause this tragic disease. The cloning and sequencing of the *CF* gene quickly led to the identification of its product, which, in turn, has suggested approaches to clinical treatment of the disease and hope for successful gene therapy for CF patients in the future.

The *CF* gene was first mapped to the long arm of chromosome 7 by its cosegregation with specific RFLPs. Further RFLP mapping indicated that the gene was located within a 500-kb region of chromosome 7. The two RFLP markers closest to the *CF* gene were then used to initiate chromosome walks (see Figure 21.9) and jumps (see Figure 21.10) and to begin construction of a detailed physical map of the region (Figure 22.4). Three kinds of information were used to narrow the search for the *CF* gene.

1. Human genes are often preceded by clusters of cytosines and guanines called **CpG islands** (Chapter 24). Three such clusters are present just upstream from the *CF* gene (Figure 22.4).
2. Important coding sequences usually are conserved in related species. When exon sequences from the *CF* gene were used to probe Southern blots containing restriction fragments from human, mouse, hamster, and bovine genomic DNAs (often called **zoo blots**), the exons were found to be highly conserved.
3. As previously mentioned, CF is known to be associated with abnormal mucus in the lungs, pancreas, and sweat glands. A cDNA library was prepared from mRNA isolated from sweat gland cells growing in culture and screened by colony hybridization using exon probes from the *CF* gene (candidate *CF* gene at the time).

Use of the sweat gland cDNA library proved to be critical in identifying the *CF* gene, because northern blot experiments subsequently showed that this gene

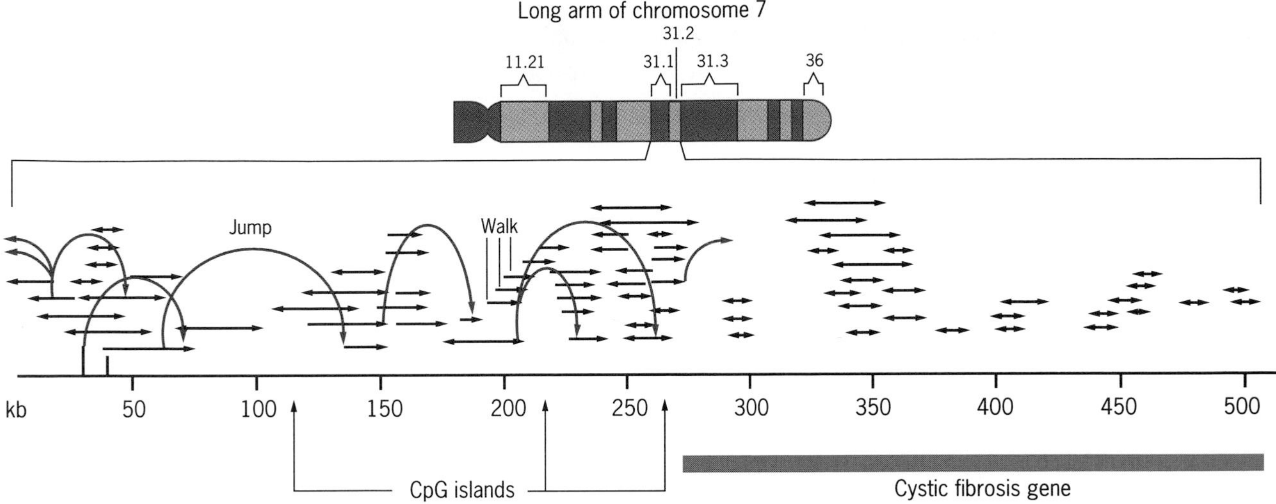

Figure 22.4 The sequence of chromosome walks and jumps used to locate and characterize the cystic fibrosis gene. The positions of CpG islands used as landmarks in locating the 5′ end of the gene are also shown.

is expressed only in epithelial cells of the lungs, pancreas, salivary glands, sweat glands, intestine, and reproductive tract. Thus, cDNA clones of the *CF* gene would not have been identified using cDNA libraries prepared from other tissues and organs. The northern blot results also showed that the putative *CF* gene is expressed in the appropriate tissues.

Identification of a candidate gene as a disease gene hinges on comparisons of normal and mutant alleles from several different families. CF is unusual in that 70 percent of the mutant alleles contain the same three-base deletion, *ΔF508,* which eliminates the phenylalanine residue at position 508 in the *CF* gene product. Unlike the *huntingtin* gene, the nucleotide sequence of the *CF* gene proved very informative. The gene is huge, spanning 250 kb and containing 24 exons (Figure 22.5). The *CF* mRNA is about 6.5 kb in length and encodes a protein of 1480 amino acids. A computer search of the protein data banks quickly showed that the *CF* gene product is similar to several ion channel proteins, which form pores between cells through which ions pass. The *CF* gene product, called the **cystic fibrosis transmembrane conductance regulator,** or **CFTR protein,** forms ion channels (Figure 22.5) through the membranes of cells that line the respiratory tract, pancreas, sweat glands, intestine, and other organs and regulates the flow of salts in and out of these cells. A large body of evidence indicates that the CFTR ion channel is activated by a cyclic ATP-dependent protein kinase, although the exact mechanism of activation remains to be worked out. Because the mutant CFTR protein does not function properly in CF patients, salt accumulates in epithelial cells and mucus builds up on the surfaces of these cells.

Although 70 percent of the cases of CF are due to the *ΔF508* trinucleotide deletion, over 170 different *CF* mutations have been identified (Figure 22.6). About 20 of these mutations are quite common, others are rare, and many have been identified in only one individual. Several of these mutations can be detected by DNA screens such as the test for the *ΔF508* deletion illustrated in the Technical Sidelight in Chapter 20. These tests can be performed on fetal cells obtained by amniocentesis or chorionic biopsy. They have also been done successfully on eight-cell pre-implantation embryos produced by *in vitro* fertilization.

The rapid accumulation of new information about the CFTR protein has dramatically increased prospects for the successful treatment of CF by gene therapy. The ion channel defect has been corrected in CF culture cells by the introduction of a functional *CF* gene, and many strategies for gene therapy treatment of the disease are currently being evaluated. One possibility involves the delivery of the wild-type *CF* gene to lung epithelial cells with one of the viruses that causes the common cold. Given the wealth of new information about the function of the CFTR protein at the molecular and cellular level, we can at least be optimistic that CF, once a deadly childhood killer, may be successfully treated in the not-too-distant future.

Key Points: **The human genes that cause Huntington's disease, cystic fibrosis, and several other disorders have been identified by positional cloning. The nucleotide sequences of these genes were used to predict the amino acid sequences of their polypeptide products and to obtain valuable information about the functions of the gene products.**

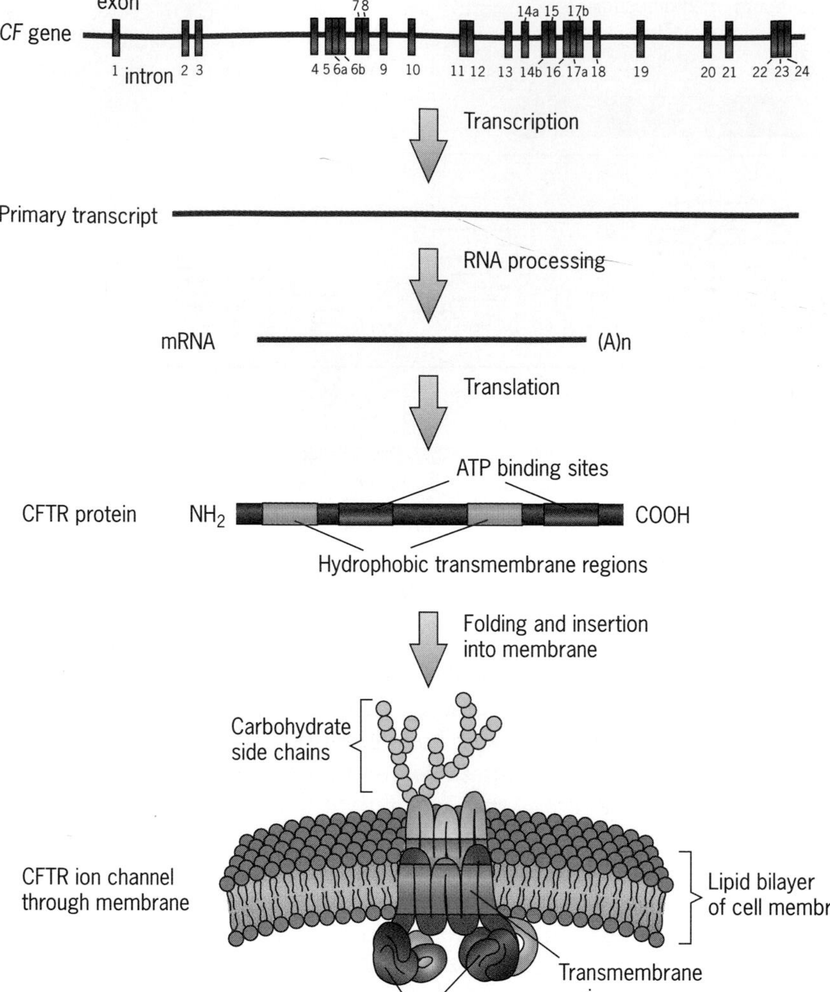

Figure 22.5 The structure of the *CF* gene and its product, the CFTR protein. The CFTR protein forms ion channels through the membranes of epithelial cells of the lungs, intestine, pancreas, sweat glands, and some other organs.

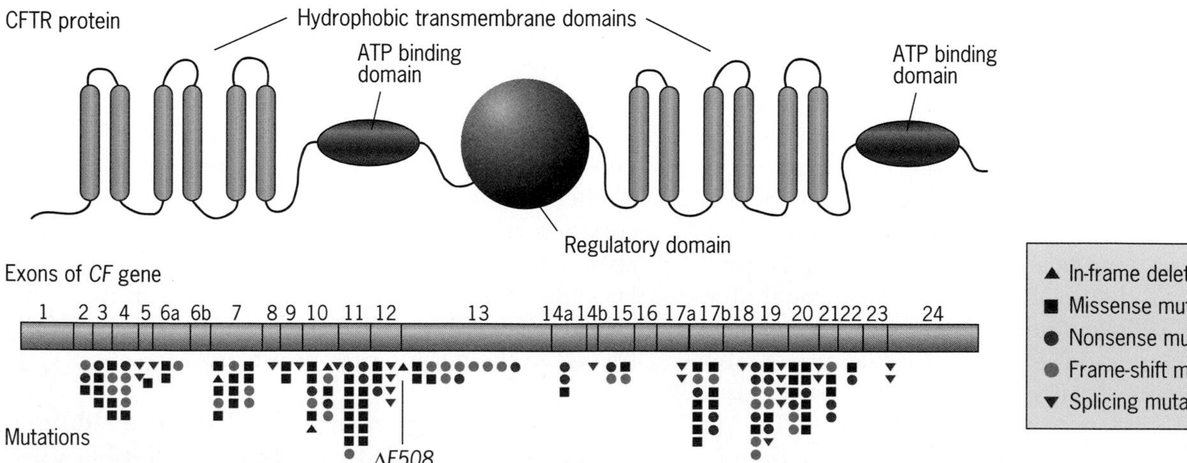

Figure 22.6 The distribution and classification of the mutations that cause cystic fibrosis are shown below the exons of the *CF* gene. A schematic diagram of the CFTR protein is shown above the exon map to illustrate the domains of the protein that are altered by the mutations. About 70 percent of all cases of CF result from mutation *ΔF508*, which deletes the phenylalanine present at position 508 of the normal CFTR protein.

Identification of these genes has led directly to new methods of treating the diseases, possible approaches to gene therapy, and DNA tests for the mutations that cause the diseases.

MOLECULAR DIAGNOSIS OF HUMAN DISEASES

Once the gene responsible for a human disease has been cloned and sequenced and the nature of the mutations that cause the disorder is known, a molecular test for the mutant alleles usually can be designed. These tests can be performed on small amounts of DNA by using PCR to amplify the DNA segment of interest (Figure 20.24). Thus, they can be performed prenatally on fetal cells obtained by amniocentesis or chorionic biopsy, or even on a single cell from a pre-embryo produced by *in vitro* fertilization.

Some molecular diagnoses involve simply testing for the presence or absence of a specific restriction enzyme cleavage site in DNA. For example, the mutation that causes sickle-cell anemia (Chapter 14) removes a cleavage site for the restriction enzyme *Mst*II (Figure 22.7). The Hb_{β}^{S} (sickle-cell) allele can be distinguished from the normal β-globin allele (Hb_{β}^{A}) by amplifying part of the β-globin gene by PCR, cutting the amplified DNA with *Mst*II, separating the resulting restriction fragments by agarose gel electrophoresis, preparing a Southern blot of the separated fragments, and hybridizing the DNA on the blot to a probe spanning the site of the mutation. The probe will hybridize with two small fragments from the normal β-globin gene, but with only one fragment from the sickle-cell β-globin gene (Figure 22.7). Thus, this test allows the detection of heterozygotes as well as homozygotes.

For inherited disorders such as Huntington's disease and fragile X syndrome, which result from expanded trinucleotide repeat regions in genes, PCR and Southern blots can be used to detect the mutant alleles. The DNA test for the *huntingtin* gene is illustrated in Figure 22.3. Other types of mutations can be detected by using allele-specific oligonucleotides to probe genomic Southern blots. This procedure is illustrated for the *ΔF508* mutation in the *CF* gene—the most frequent cause of cystic fibrosis—in the Technical Sidelight in Chapter 20. Indeed, once the mutations responsible for a disease have been characterized, the development of DNA tests to detect the most common ones is usually routine. Clearly, the availability of diagnostic tests for mutations that cause human diseases

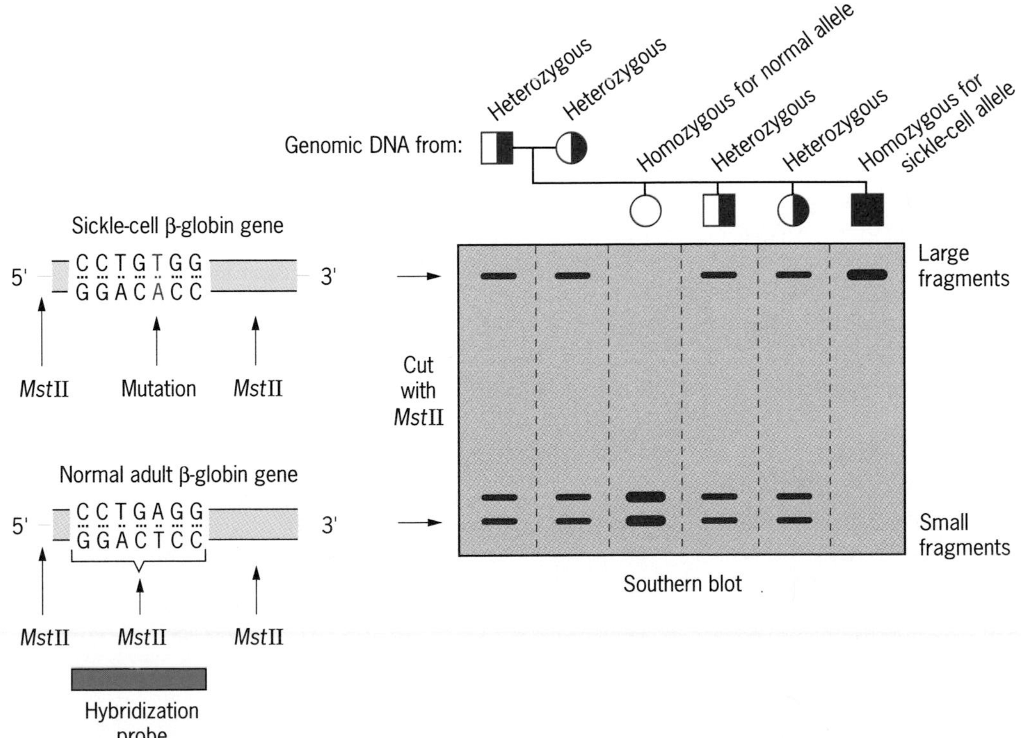

Figure 22.7 Detection of the sickle-cell hemoglobin mutation by Southern blot analysis of genomic DNAs cut with restriction enzyme *Mst*II.

has contributed greatly to the field of genetic counseling, providing families in which the genetic defects occur with invaluable information.

Key Points: **The mutant genes responsible for several inherited human diseases can be accurately diagnosed by screening genomic DNAs for the genetic defect. The results of these tests provide information that allows genetic counselors to inform families of the risk of an affected child.**

DNA FINGERPRINTS

Fingerprints have played a central role in human identity cases for decades. Indeed, fingerprints often provide the key evidence that places a suspect at a crime scene. The use of fingerprints in forensic cases is based on the premise that no two individuals will have identical prints. Similarly, no two individuals, except for identical twins, will have genomes with the same nucleotide sequences. The human genome contains 3×10^9 nucleotide pairs; each site is occupied by one of the four base pairs in DNA. Many base-pair substitutions are silent; they occur in unessential noncoding sequences or at genomic positions corresponding to the third base of codons and do not change the amino acid sequences of the gene products because of the degeneracy in the code. Therefore, such nucleotide-pair substitutions accumulate in genomes during the course of evolution. In addition, duplications and deletions of DNA sequences and other genome rearrangements contribute to the evolutionary divergence of genomes. Indeed, recent evidence has demonstrated that the human genome contains large families of DNA polymorphisms of many different types, polymorphisms that can provide valuable evidence in cases of uncertain human identity. These polymorphisms can be used to produce **DNA fingerprints,** specific banding patterns on Southern blots of genomic DNA cleaved with a specific restriction enzyme and hybridized to an appropriate DNA probe.

The power and utility of the DNA fingerprinting procedure in personal identity cases are obvious to anyone familiar with molecular genetics and the techniques utilized in the production of DNA prints. The controversies regarding the use of DNA fingerprints in forensic cases relate to the competency of the research laboratories involved, the probability of human error in producing prints, and the methods for calculating the probability that two individuals have identical fingerprints. To make accurate estimates of the likelihood of identical prints, researchers must have reliable information about the frequency of the polymorphisms in the population in question. For example, if inbreeding (matings between related individuals) is common in the population, the probability of identical fingerprints will increase. Thus, accurate estimates of the probability that two individuals will have matching fingerprints requires reliable information about the frequencies of the polymorphisms in the relevant population. Data obtained from one population should never be extrapolated to another population, because different polymorphism frequencies may be present in different populations.

DNA fingerprinting provides an extremely powerful forensic tool if used properly. The DNA prints can be prepared from extremely minute amounts of blood, semen, hair bulbs, or other cells of the body. The DNA is extracted from these cells, amplified by PCR, and analyzed with carefully chosen DNA probes by the Southern blot procedure (Chapter 20). Indeed, fingerprinting can sometimes be done with DNA isolated from preserved tissues of individuals long after their deaths (Technical Sidelight: DNA Tests and the Mystery of the Duchess Anastasia). As previously mentioned, the human genome contains numerous short DNA sequences that are present as tandem repeats of varied lengths at several chromosomal locations. These variable number tandem repeats, or VNTRs, are important components of DNA fingerprints (Figure 22.8).

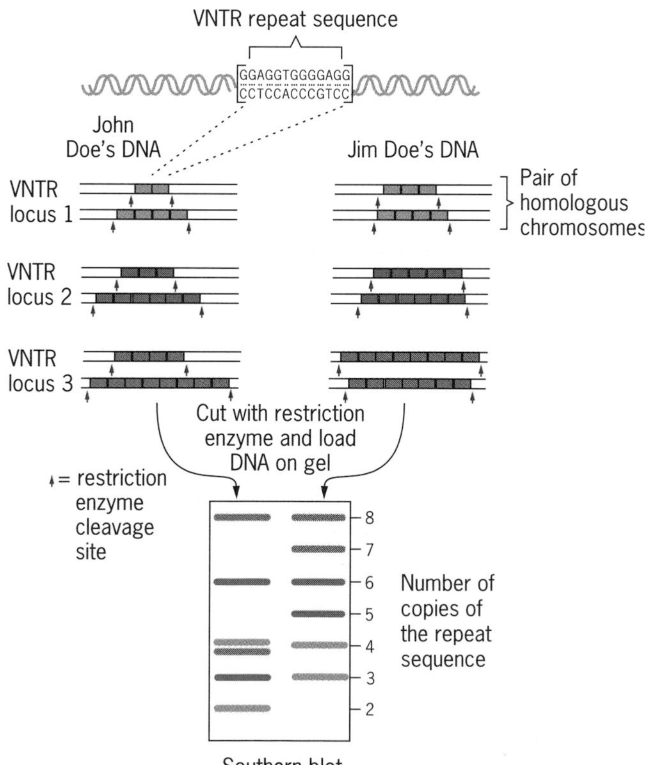

Figure 22.8 Simplified diagram of the use of variable number tandem repeats in preparing DNA fingerprints.

Although DNA prints are applicable in all cases of questionable identity, they have proven especially useful in paternity and forensic cases.

Paternity Tests

In the past, cases of uncertain paternity often have been decided by comparing the blood types of the child, the mother, and possible fathers. Blood-type data can be used to prove that men with particular blood types could not have fathered the child. Unfortunately, these blood-type comparisons contribute little toward a positive identification of the father. In contrast, DNA fingerprints not only exclude misidentified fathers, but also come close to providing a positive identification of the true father. DNA samples are obtained from cells of the child, the mother, and possible fathers, and DNA fingerprints are prepared as described in Figure 22.8. When the fingerprints are compared, all the bands in the child's DNA print should be present in the DNA prints of the parents. For each pair of homologous chromosomes, the child will have received one from each parent. Thus, approximately half of the bands in the child's DNA print will result from DNA sequences inherited from the mother, and the other half from DNA sequences inherited from the father.

Figure 22.9 shows the DNA fingerprints of a child, the mother, and two men suspected of being the child's father. In this case, the DNA prints indicate that the second father candidate is probably the child's biological father. The accuracy of DNA fingerprints in identifying child-parent relationships can be enhanced by increasing the number of hybridization probes used in the analysis. With the use of more probes, more polymorphisms can be surveyed and a larger proportion of the genomes of the child and parents can be compared; the result is a more reliable identification.

Forensic Applications

DNA fingerprints were first used as evidence in a criminal case in 1988. In 1987, a Florida judge denied the prosecutor's request to present statistical interpretations of DNA evidence against an accused rapist. After a mistrial, the suspect was released. Three months later, he was again in court, accused of another rape. This time the judge allowed the prosecutor to present a statistical analysis of the data, based on appropriate population surveys. The analysis showed that the DNA fingerprint prepared from semen recovered from the victim had a probability of about one in 10 billion of matching the DNA fingerprint of the suspect purely by chance. This time the suspect was convicted. There can be no question about the value of DNA prints in forensic cases of this type when good tissue or cell samples are obtained from the scene of the crime. If performed carefully by trained scientists and interpreted conservatively using valid population-based data on the distributions of the polymorphisms involved, DNA fingerprints can provide a much needed and powerful tool in the ongoing fight against crime.

Figure 22.10 illustrates one type of DNA fingerprints used in forensic cases, namely, VNTR prints. Even a novice can see that the DNA fingerprint prepared from the bloodstain at the crime scene matches the DNA print from suspect 1, but not the prints from the other two suspects. Of course, these matching DNA prints by themselves do not prove that suspect 1 com-

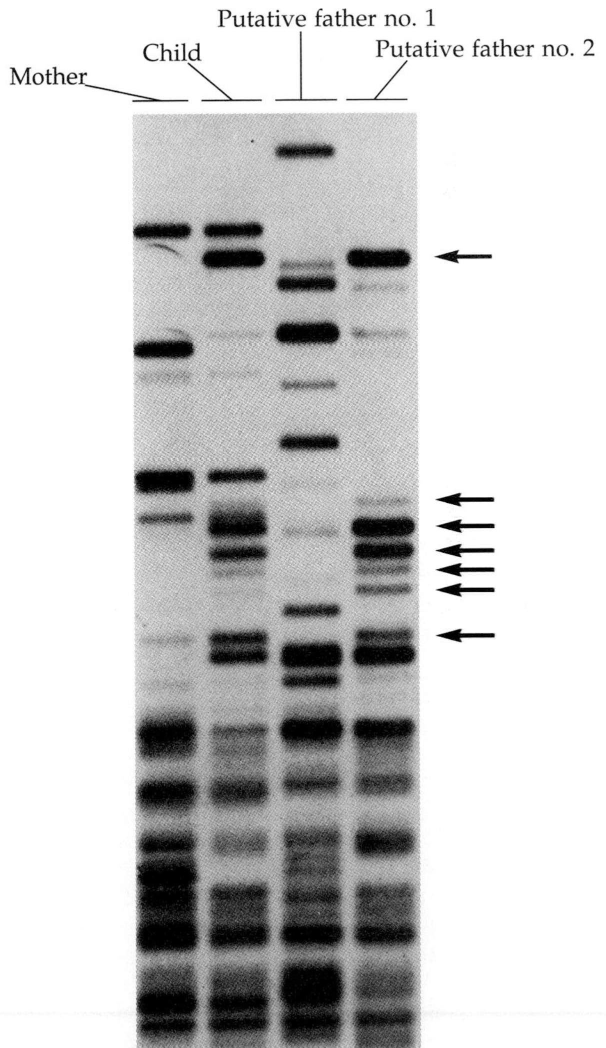

Figure 22.9 DNA fingerprints of a mother, her child, and two men, each of whom claimed to be the child's father. Arrows mark bands that identify male no. 2 as the biological father.

TECHNICAL SIDELIGHT

DNA Tests and the Mystery of the Duchess Anastasia

According to historical records, the Russian royal family—Tsar Nicholas II, Tsarina Alexandra, and their five children: Alexis, Olga, Tatiana, Marie, and Anastasia (Figure 1)—were executed on July 16, 1918, by a revolutionary Bolshevik firing squad and then were buried in a single grave in the Ural Mountains. However, in 1920, an unknown woman, "Fraulein Unbekannt," who was pulled from a canal in Berlin in a state of hypothermia, claimed that she was the Duchess Anastasia. Although she did not speak Russian, Fraulein Unbekannt, or Anna Anderson Manahan, as she was subsequently known, was amazingly well informed about details of life in the imperial Russian court. Her claim to be Anastasia was vigorously rejected by the surviving relatives of the Russian royal family. The Grand Duke of Hesse even hired a private detective to investigate Anna's heritage. The detective concluded that Anna was really Franzisca Schanzkowska, but the dispute continued. Although little is known about Franzisca, she was born in the northern part of Germany, lived in Berlin during World War I, and was severely injured by an explosion while working in a munitions factory. She was subsequently admitted to two mental hospitals for treatment. She disappeared in 1920, about the same time that Anna Anderson Manahan was rescued from the Berlin canal and claimed to be Anastasia.

When Princess Irene of Prussia, Anastasia's aunt, was persuaded to meet with the woman who claimed to be her niece, Anna ran and hid in her room. Anna's bizarre behavior made her claim to be Anastasia difficult to evaluate, and the controversy over the identity of Anna Anderson Manahan continued for over 70 years. Was Anna really Anastasia? Her supporters were steadfast in their belief that she was indeed the Duchess. Disbelievers were equally adamant that she was not Anastasia.

In 1979, a Russian geologist discovered a shallow grave believed to contain the remains of the royal family. Because of the political climate in the Soviet Union at the time, the geologist reburied the bodies. Twelve years later, when the political climate was more favorable, the bodies were exhumed, and their authenticity was established by comparing DNA from the skeletons with DNA from surviving relatives. However, the controversy about the identity of Anna was rekindled by the absence of two bodies, those of Anastasia and her brother Alexis. Had they somehow escaped execution? Although there is still no definitive answer to this question, the results of recent DNA tests indicate that Anna Anderson Manahan was not the Duchess Anastasia.

Anna Anderson Manahan died in 1984 at the age of 83. However, during surgery performed in 1979 at the Martha Jefferson Hospital in Charlottesville, Virginia, intestinal tissues were removed, fixed in formaldehyde, and preserved

Figure 1 The children of Tsar Nicholas II: (left to right) Marie, Tatiana, Anastasia, Olga, and Alexis.

in paraffin blocks. In addition, a few of Anna's hair follicles were preserved. DNA tests—VNTR (variable number tandem repeat) prints and nucleotide sequences of specific noncoding regions of mitochondrial DNA—were performed on these preserved tissues and on relatives of Franzisca Schanzkowska and of the royal family. These tests were performed independently in three different laboratories: (1) the Armed Forces DNA Identification Laboratory in the United States, (2) the Forensic Science Service in the United Kingdom, and (3) the Department of Anthropology at Pennsylvania State University. The results obtained by the three laboratories all indicate that Anna Anderson Manahan was not Anastasia. Indeed, the results strongly suggest that Anna was Franzisca Schanzkowska.

Of five different VNTRs examined, four were inconsistent with the possibility that Anna was the daughter of Tsar Nicholas II and Tsarina Alexandra. DNA sequence comparisons also argued that Anna was not related to the royal family. Instead, the nucleotide sequence data indicated that Anna was Ms. Schanzkowska. At the six variable positions shown below, Anna's mitochondrial DNA contained the same nucleotides as the DNA from Carl Maucher, Franzisca Schanzkowska's grand nephew, and differed from those in the DNA of the Duke of Edinburgh, the grand nephew of Tsarina Alexandra.

	Variable Nucleotides in Mitochondrial DNA					
Position:	**1**	**2**	**3**	**4**	**5**	**6**
Anna Anderson Manahan	C	C	T	T	C	T
Carl Maucher (Grand nephew of Franzisca)	C	C	T	T	C	T
Duke of Edinburgh (Grand nephew of Alexandra)	T	T	C	C	T	C

When these mitochondrial DNA sequences were compared with over 300 published and unpublished DNA sequences in data banks, no matches to Carl Maucher and Anna Anderson Manahan's sequence were found. These results indicate that the Maucher-Manahan sequence is rare. If the database sequences are representative of Caucasians of northern European origin, the chance of matching DNA profiles is less than 1 in 300 if Carl and Anna are unrelated. The results clearly support the hypothesis that Anna was Franzisca Schanzkowska rather than the Duchess Anastasia, and finally put to rest the controversy about the identity of Anna Anderson Manahan.

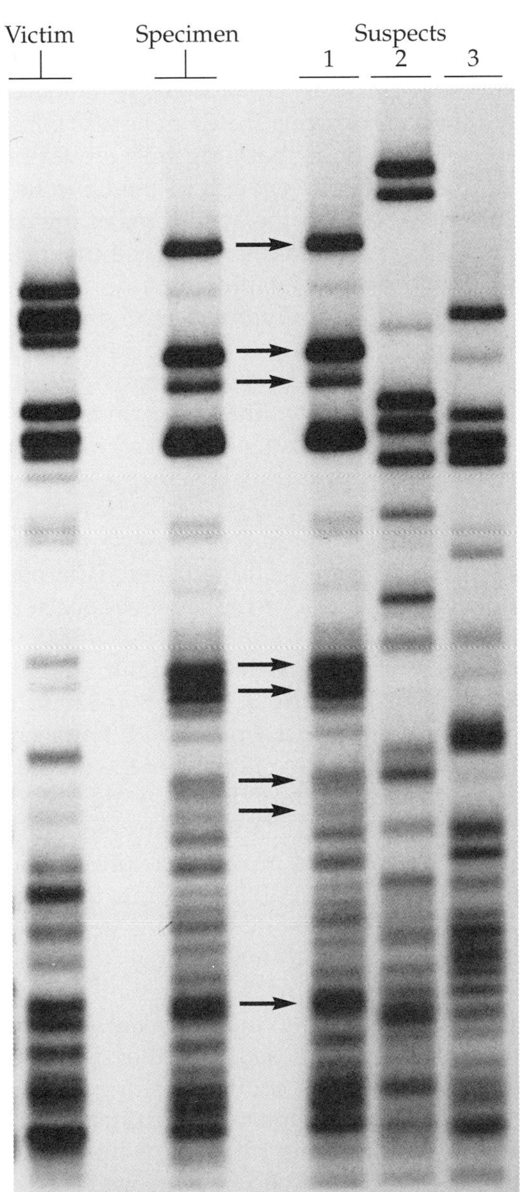

Figure 22.10 DNA fingerprints prepared from DNA isolated from a bloodstain at the site of a crime and from blood obtained from three individuals suspected of committing the crime. The arrows denote DNA fragments from suspect 1—not present in suspects 2 or 3—that match those from the bloodstain at the crime scene.

mitted the crime, but, if combined with additional DNA prints and supporting evidence, they can provide strong indications that suspect 1 was at the scene of the crime. In addition, these prints clearly show that the blood cells in the stain were not from either of the other two suspects.

By combining VNTR fingerprints with prints produced using other types of DNA probes, the possibility that DNA fingerprints from two individuals will match just by chance can virtually be eliminated. The rationale behind using DNA fingerprints to identify individuals is that each person's DNA has a unique nucleotide sequence. Despite the human population explosion, there are far more possible combinations of the four base pairs in a human genome of 3×10^9 base pairs than there are humans on planet Earth. Thus, except for identical twins, no two humans should have identical genomes. DNA fingerprints provide a tool by which these differences can be detected and recorded, just as fingerprints have been recorded for decades.

Key Points: **DNA fingerprints detect and record polymorphisms in the genomes of individuals. DNA prints provide strong evidence of individual identity, evidence that may be extremely valuable in paternity and forensic cases.**

HUMAN GENE THERAPY

Of the over 4000 inherited human diseases catalogued to date, only a few are currently treatable. For many of these diseases, the missing or defective gene product cannot be supplied exogenously, as insulin is supplied to diabetics. Most enzymes are unstable and cannot be delivered in functional form to their sites of action in the body, at least not in a form that provides for long-term activity. Cell membranes are impermeable to large macromolecules such as proteins; thus, enzymes must be synthesized in the cells where they are needed. Therefore, treatment of inherited diseases is largely restricted to those cases where the missing metabolite is a small molecule that can be distributed to the appropriate tissues of the body through the circulatory system. For many other inherited diseases, **gene therapy** offers the most promising approach to successful treatment. Gene therapy involves adding a normal (wild-type) copy of a gene to the genome of an individual carrying defective copies of the gene. A gene that has been introduced into a cell or organism is called a **transgene** (for *trans*ferred *gene*) to distinguish it from endogenous genes, and the organism carrying the introduced gene is said to be **transgenic.** If gene therapy is successful, the transgene will synthesize the missing gene product and restore the normal phenotype.

Before considering specific examples, we need to discuss two types of gene therapy: **somatic-cell** or **nonheritable gene therapy,** and **germ-line** or **heritable gene therapy.** In higher animals such as humans, the reproductive or germ-line cells are produced by a cell lineage separate from all somatic-cell lineages. Thus, somatic-cell gene therapy will treat the disease symptoms of the individual but will not cure the disease. That is, the defective gene(s) will still be present in the germ-line cells of the patient after somatic-cell gene therapy and may be transmitted to his or her children. All of the gene-therapy treatments of human diseases that we will discuss here are somatic-cell gene therapies. Germ-line gene therapy is being performed on mice and other animals, but not on humans.

The distinction between somatic-cell and germ-line gene therapy is important when we discuss humans. The frequently expressed concerns about humankind's "tinkering with nature" or "playing God" apply to germ-line gene transfers, not to somatic-cell gene therapy. Major moral and ethical considerations are involved in any decision to perform germ-line modifications of human genes. In contrast, somatic-cell gene therapy is no different from enzyme (gene-product) therapy or cell, tissue, and organ transplants. In transplants, entire organs, with all the foreign genes present in the genome of every cell in the organ, are implanted in patients. In current somatic-cell gene therapies, some of the patient's own cells are removed, repaired, and reimplanted in the patient. Thus, somatic-cell gene therapy is less complex and less life-threatening for an individual than an organ transplant.

To perform somatic-cell gene therapy, wild-type genes must be introduced into and expressed in cells homozygous or hemizygous for a mutant allele of the gene. In principle, the wild-type gene could be delivered to the mutant cells by any of several different procedures. Most commonly, viruses are used to carry the wild-type gene into cells. In the case of retroviral vectors, the wild-type transgene is integrated—along with the retroviral DNA—into the DNA of the host cell. Thus, when retroviral vectors are used, the transgene is transmitted to all progeny cells in the affected cell lineage.

With other viral vectors, such as those derived from adenoviruses, the transgenes are present only transiently in host cells, because the genomes of these viruses replicate autonomously and persist only until the immune system eliminates the viruses from the infected cells. The advantage of these vectors over retroviral vectors is that no potentially harmful mutations are induced during the integration step (Chapter 14). However, they have two major disadvantages: (1) transgene expression is transient, lasting only as long as the viral infection persists, and (2) most humans exhibit strong immune responses to these viruses, presumably because of prior exposure to the same or closely related viruses. For example, in early attempts to treat cystic fibrosis by somatic-cell gene therapy, an adenoviral vector carrying the *CF* gene was inhaled by patients, with the hope that lung cells would become infected and synthesize enough of the *CF* gene product to alleviate some of the symptoms of the disease. Unfortunately, these treatments proved ineffective, at least in part because of rapid immune responses to these viruses in the individuals receiving the treatments.

With diseases such as cystic fibrosis, where effective gene therapy will require long-term transgene expression, the standard adenovirus vectors probably will not work. Because transgene expression is transient, the treatments will need to be repeated periodically. However, given that secondary immune responses are very rapid and efficient (Chapter 27), subsequent treatments with the same viral vector probably will be ineffective.

Transgenes can also be delivered into somatic cells by coating them with lipids, which mediate fusion with cell membranes. Such lipid-coated transgenes are called **liposomes.** They avoid most of the problems associated with viral vectors, but they do not deliver transgenes to somatic cells efficiently enough to be very effective. Moreover, transgene expression is transient in the few cells that do incorporate the liposome. Thus, at present, most somatic-cell gene therapy research is being done with retroviral vectors.

Human gene therapy is performed under strict guidelines developed by the National Institutes of Health (NIH). Each proposed gene-therapy procedure is scrutinized by review committees at both the local (institution or medical center) and national (NIH) levels. Several requirements must be fulfilled before a gene-therapy procedure will be approved:

1. The gene must be cloned and well characterized; that is, it must be available in pure form.
2. An effective method must be available for delivering the gene into the desired tissue(s) or cells.
3. The risks of gene therapy to the patient must have been carefully evaluated and shown to be minimal.

4. The disease must not be treatable by other strategies.
5. Data must be available from preliminary experiments with animal models or human cells and must indicate that the proposed gene therapy should be effective.

A gene-therapy proposal will not be approved by the local and national review committees until they are convinced that all of the above conditions have been fulfilled.

The first use of gene therapy in humans occurred in 1990, when a four-year-old girl with **adenosine deaminase-deficient severe combined immunodeficiency disease (ADA⁻SCID)** received her first transgene treatment. Several other ADA⁻SCID patients have subsequently been treated, along with a few individuals with other inherited disorders. SCID is a rare autosomal disease of the immune system. Individuals with SCID have essentially no immune system, so that even minor infections can become serious and often fatal. Some SCID patients lack an enzyme called adenosine deaminase (ADA). In the absence of this enzyme, toxic levels of the phosphorylated form of its substrate, deoxyadenosine, accumulate in T lymphocytes (white blood cells essential to an immune response) and kill them. T lymphocytes stimulate cells called B lymphocytes to develop into antibody-producing plasma cells. Thus, in the absence of T lymphocytes, no immune response is possible, and newborns with ADA⁻SCID seldom live more than a few years.

Four factors made ADA⁻SCID a good candidate for somatic-cell gene therapy. First, the *ADA* gene was one of the first human disease genes to be cloned and characterized. Second, white blood cells can easily be obtained from ADA⁻SCID patients and reintroduced after functional copies of the *ADA* gene are added. Third, even a small amount of functional ADA will restore partial immune function. Fourth, the overproduction of ADA does not appear to have toxic effects on the patient.

As a prerequisite to gene therapy, a retroviral gene-transfer vector was constructed with the human *ADA* gene under the control of a strong viral promoter (Figure 22.11), and expression of the chimeric gene was demonstrated in white blood cells infected with the retroviral vector. The actual treatment of ADA⁻SCID by gene therapy involves four sequential steps: (1) isolation of white blood cells from the patient, (2) introduction of functional copies of the *ADA* gene into these cells, (3) demonstration of transgene expression in these cells, and (4) infusion of the transgenic cells back into the patient (Figure 22.12).

Although the number of ADA⁻SCID patients who have been treated by gene therapy is still small and these patients have continued to receive treatments with purified ADA, the therapy has resulted in improved immune function. Most are attending normal schools, and some have survived infectious diseases such as chicken pox. It was anticipated that the effects of gene therapy using white blood cells would be short-lived because of the limited lifespan of these cells. Thus, repeated infusions of white blood cells car-

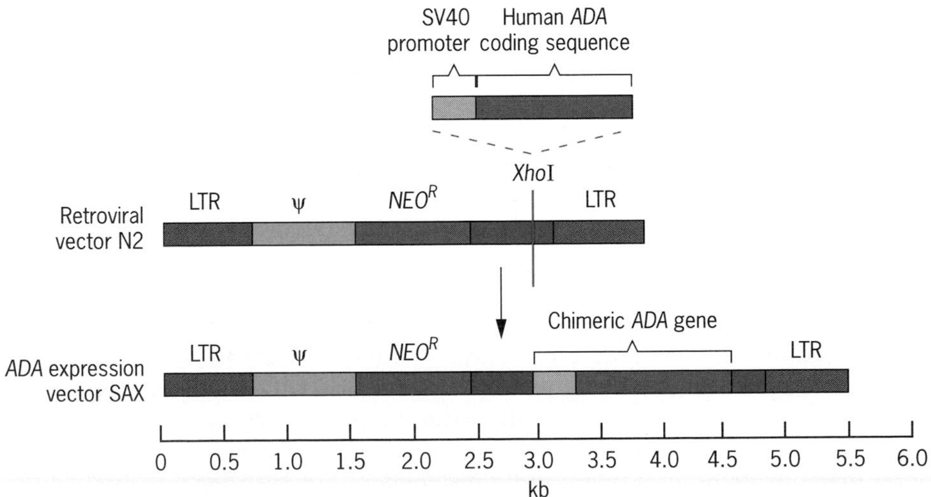

Figure 22.11 Construction of the human adenosine deaminase (*ADA*) gene-transfer vector SAX (for simian virus 40, *ADA*, *Xho*I site). In this vector, expression of the human *ADA* coding sequence is controlled by a strong promoter obtained from simian virus 40. The NEO^R gene makes cells resistant to the antibiotics neomycin, kanamycin, and G418. LTR refers to the long terminal repeats from the Moloney murine leukemia virus; Ψ is a DNA sequence required to package vector DNA in virions.

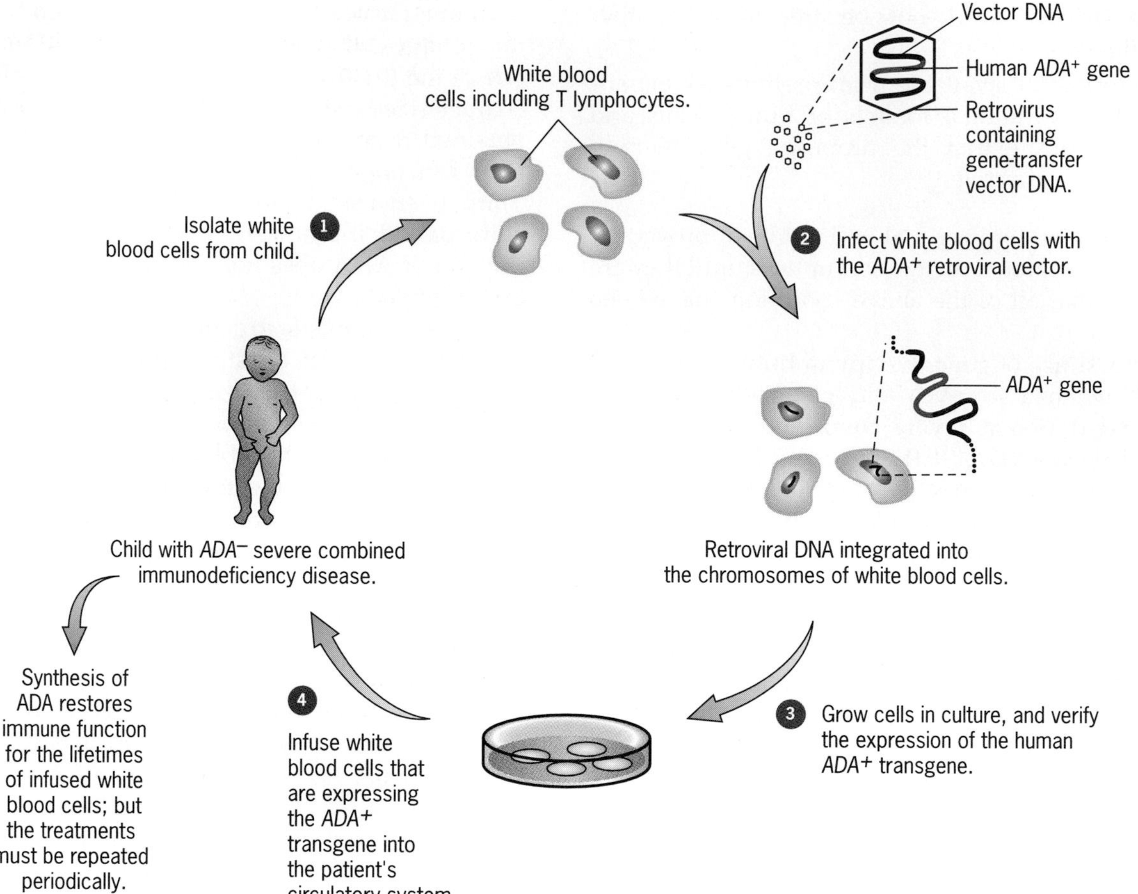

Figure 22.12 Treatment of adenosine deaminase-deficient severe combined immunodeficiency disease (ADA^-SCID) by somatic-cell gene therapy. ADA^-SCID results from the loss or lack of activity of the enzyme adenosine deaminase (ADA). Gene therapy is performed by isolating white blood cells from the patient, introducing a wild-type copy of the *ADA* gene into these cells with a retroviral vector, verifying the expression of the transgene in cultured cells, and infusing the transformed white blood cells back into the patient.

rying functional *ADA* genes have been performed on the patients. The major problem encountered in attempts to treat ADA^-SCID patients by gene therapy has not been the short lifespan of the white blood cells, but the transient expression of the introduced *ADA* transgenes. Within a few weeks after treatment, transcription of the *ADA* transgene has subsided. The activity of the viral promoters that control this transcription has apparently been arrested by an unknown mechanism. These results indicate that additional research is needed to understand why transgenes are often silenced (expression turned off) in their new hosts.

To avoid the limitations resulting from the short lifespan of white blood cells, the bone marrow stem cells that give rise to white blood cells (Chapter 27) could be used to treat immune disorders such as ADA^- SCID. The modified stem cells should continually produce T lymphocytes with the *ADA* transgene and could provide a permanent or long-term treatment of the disease. Indeed, stem-cell gene therapy was first used to treat two infants with ADA^-SCID in 1993, and this procedure will undoubtedly be the method of choice in the future.

Although the initial somatic-cell gene therapy experiments have been less effective than scientists and physicians had hoped, this approach to the treatment of human diseases holds promise for the future. With the recent identification and cloning of several genes that cause serious human diseases, gene therapy is clearly on the horizon for individuals suffering from several of these disorders. We previously discussed the possible treatment of cystic fibrosis by gene therapy. Other candidates for somatic-cell gene therapy include phenylketonuria (Chapter 26), the thalassemias (hemoglobin deficiencies), sickle-cell anemia (Chapter 14), citrullinemia (a life-threatening hyperammonemia caused by a deficiency of the urea-cycle enzyme argininosuccinate synthetase), Lesch-Nyhan syndrome (caused by a defect in purine metabolism; see Chapter 26), and Hurler syndrome (caused by a defect in polysaccharide metabolism).

Current somatic-cell gene therapy protocols are **gene-addition** procedures; they simply add functional copies of the gene that is defective in the patient to the genomes of recipient cells. They do not replace the defective gene with a functional gene. In fact, the introduced genes are inserted at random or nearly random sites in the chromosomes of the host cells. The ideal gene-therapy protocol would replace the defective gene with a functional gene. **Gene replacements** would be mediated by homologous recombination and would place the introduced gene at its normal location in the host genome. In humans, gene replacements are usually referred to as targeted gene transfers. Oliver Smithies and coworkers first used homologous recombination to target DNA sequences to the β-globin locus of human tissue-culture cells in 1985. However, the frequency of the targeted gene transfer was very low (about 10^{-5}). Since then, Smithies, Mario Capecchi, and others have developed improved gene targeting vectors and selection strategies. As a result, more efficient targeted gene replacements are possible, and cells with the desired gene replacement can be identified more easily. In the future, targeted gene replacements will probably become the method of choice for somatic-cell gene therapy treatment of human diseases.

Key Points: **Gene therapy involves the addition of a normal (wild-type) copy of a gene to the genome of an individual who carries defective copies of the gene. Physicians are currently testing the effectiveness of somatic-cell gene therapy in the treatment of patients with adenosine deaminase-deficient severe combined immunodeficiency disease. Although many technical details must still be worked out, somatic-cell gene therapy holds promise for the treatment of many inherited human diseases. In the future, somatic-cell gene therapy will probably involve gene replacements, rather than gene additions.**

PRODUCTION OF EUKARYOTIC PROTEINS IN BACTERIA

For decades, microorganisms have been used to produce important products for humans. We are all aware of the impact of antibiotics on human health; fewer of us are aware of their economic importance. The wholesale market value of antibiotics in the United States is over $2 billion annually. Microbes also play important roles in the production of many other materials, for example, antifungal drugs, amino acids, and vitamins. Today, because of genetic engineering, bacteria are being used in the production of important eukaryotic proteins such as human insulin, human growth hormone, and the entire family of human interferons. In addition, genetically engineered microbes are being used to synthesize valuable enzymes and other organic molecules and to provide metabolic machinery for the detoxification of pollutants and the conversion of biomass to combustible compounds.

Human Growth Hormone

In 1982, human insulin became the first commercial success of the new recombinant DNA technologies in the field of pharmaceuticals (see Human Genetics Sidelight in Chapter 20). Since then, several other human proteins with medicinal value have been synthesized in bacteria. Some of the first human proteins to be produced in microorganisms were blood-clotting factor VIII (lacking in individuals with one type of hemophilia), plasminogen activator (a protein that disperses blood clots), and human growth hormone (a protein deficient in certain types of dwarfism). As an example, let's examine the synthesis of **human growth hormone (HGH)** in *E. coli.* HGH, which is required for normal growth, is a single polypeptide chain 191 amino acids in length. In contrast to insulin, porcine and bovine pituitary growth hormones do not work in humans. Only growth hormones from humans or from closely related primates function effectively in humans. Thus, prior to 1985, the major source of growth hormone suitable for treatment of humans was from human cadavers.

To obtain expression in *E. coli,* the HGH coding sequence must be placed under the control of *E. coli* regulatory elements. Therefore, the HGH coding sequence was joined to the promoter and ribosome-binding sequences of the *E. coli lac* operon (a set of genes encoding proteins required for growth on the sugar lactose; see Chapter 23). To accomplish this, a *Hae*III cleavage site in the nucleotide-pair triplet specifying codon 24 of HGH was used to fuse a synthetic DNA sequence encoding amino acids 1–23 to a partial cDNA sequence encoding amino acids 24–191. This unit was then inserted into a plasmid carrying the *lac* regulatory signals and introduced into *E. coli* by transformation. The structure of the first plasmid used to produce HGH in *E. coli* is shown in Figure 22.13.

The HGH produced in *E. coli* in these first experiments contained methionine at the amino terminus (the methionine specified by the ATG initiator codon). Native HGH has an amino-terminal phenylalanine; the initial methionine is enzymatically removed after syn-

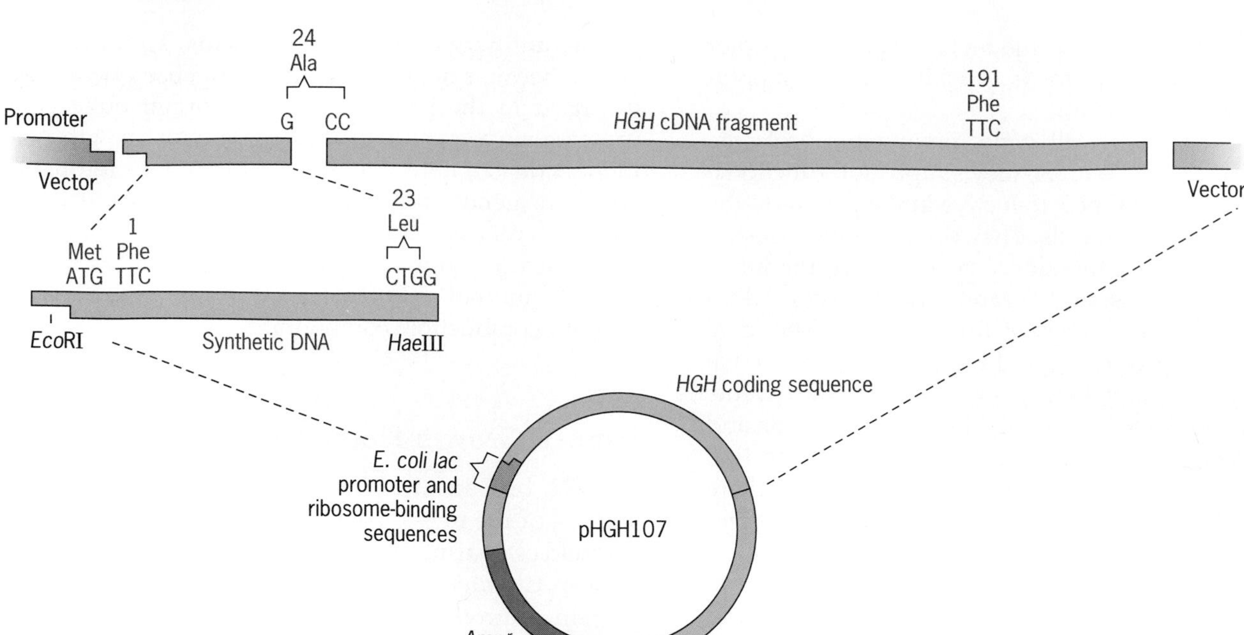

Figure 22.13 Structure of the first vector used to produce human growth hormone (HGH) in *E. coli.* The *Amp*^{r} gene provides resistance to ampicillin; *Ori* is the plasmid's origin of replication.

thesis. *E. coli* also removes many amino-terminal methionine residues posttranslationally. However, the excision of the terminal methionine is sequence dependent, and *E. coli* cells do not excise the amino-terminal methionine residue from HGH. Nevertheless, the HGH synthesized in *E. coli* was found to be fully active in humans despite the presence of the extra amino acid. More recently, a DNA sequence encoding a signal peptide (the amino acid sequence required for transport of proteins across membranes) has been added to an *HGH* gene construct similar to the one shown in Figure 22.13. With the signal sequence added, HGH is both secreted and correctly processed. That is, the methionine residue is removed with the rest of the signal peptide during the transport of the primary translation product across the membrane. This product is identical to native HGH. In 1985, HGH became the second genetically engineered pharmaceutical to be approved for use in humans by the U.S. Food and Drug Administration. Human insulin produced in *E. coli* was approved for use by diabetics in 1982.

Proteins with Industrial Applications

Some enzymes with important industrial applications have been manufactured for many years by using microorganisms to carry out their synthesis. For example, proteases have been produced from *Bacillus licheniformis* and other bacteria. These proteases have been employed extensively as cleaning aids in detergents and in smaller amounts as meat tenderizers and as digestive aids in animal feeds. Amylases have been widely used to break down complex carbohydrates such as starch to glucose. The glucose is then converted to fructose with the enzyme glucose isomerase, and this fructose is used as a food sweetener. The amylases and glucose isomerase are all manufactured by microbiological processes.

The protein rennin is used in making cheeses. Prior to the advent of genetic engineering, rennin was extracted from the fourth stomach of cattle. Genetically engineered bacteria are now used for the commercial production of rennin. These examples are all proteins that have had important industrial applications for some time. In the future, we can expect many additional enzymes to be manufactured and used in industrial applications because of the ease of producing these proteins by means of recombinant microorganisms (or by transgenic plants and animals; see the next section).

In fact, the major tools of molecular biology today are themselves enzymes—namely, the vast array of restriction endonucleases, ligases, polymerases, and reverse transcriptases that are used to manipulate DNA molecules *in vitro.* These enzymes are now commonly produced by cloning the respective genes and expressing these genes on high-copy-number plasmids in bacteria. Thus, the development of the recombinant DNA and gene-cloning technologies has spawned a whole new industry, a group of companies that manufacture and sell the enzymes, vectors, and so on that are used by genetic engineers. New applications in this field are inevitable, as it is now possible to produce essentially any enzyme or other protein of interest by cloning the gene that encodes this protein and expressing that gene at very high levels in microorganisms.

Key Points: **Valuable proteins that could be isolated from eukaryotes only in small amounts and at great expense can now be produced in large quantities in genetically engineered bacteria. Proteins such as human insulin and human growth hormone are valuable pharmaceuticals used to treat diabetes and pituitary dwarfism, respectively. Other proteins have important industrial applications.**

TRANSGENIC PLANTS AND ANIMALS

Although the first recombinant DNA molecules were constructed and expressed in microorganisms, methods are now available that allow foreign or synthetic genes to be introduced into and expressed in higher plants and animals. Although a complete discussion of the methods used to produce transgenic plants and animals is beyond the scope of this book, let us discuss a few commonly used procedures and some of the initial applications of recombinant DNA technologies in plant and animal breeding.

Transgenic Animals: Microinjection of DNA into Fertilized Eggs

Two methods are predominantly used to produce transgenic animals: (1) microinjection of DNA into the pronuclei of fertilized eggs, and (2) infection of preimplantation embryos with retroviral vectors. We will only describe the microinjection procedure here. See Figure 22.12 for an illustration of the retroviral procedure.

Most of the transgenic animals studied to date were produced by microinjection of DNA into fertilized eggs. Prior to microinjection, the eggs are surgically removed from the female parent and are fertilized *in vitro.* The DNA is then microinjected into the male pronucleus (the haploid nucleus contributed by the sperm, prior to nuclear fusion) of the fertilized egg through a very fine-tipped glass needle (Figure 22.14). The injected DNA can be either linear or circular. However, linear DNA integrates with a higher (about fivefold) efficiency than circular DNA. Although the reason for this difference is not known, it probably results from the high frequency of recombination at the ends of DNA molecules owing to their ability to invade homologous double helices. Usually, several hundred to several thousand copies of the gene of interest are injected into each egg, and multiple integrations often occur. Surprisingly, when multiple copies do integrate into the genome, they usually do so as tandem, head-to-tail arrays at a single chromosomal site. These DNA molecules carrying multiple tandem copies of the transgene are believed to form prior to their integration into the host genome, by recombination events between injected DNA molecules. The integration of injected DNA molecules appears to occur at random sites in the genome.

Because the DNA is injected so early during development, even before fusion of the haploid nuclei to form the diploid nucleus of the zygote, integration of the injected DNA molecules usually occurs early during embryonic development. As a result, some germ-line cells usually carry the transgene. As would be expected, the animals that develop from the injected eggs—called the **G_0 generation**—are almost always genetic mosaics, with some somatic cells carrying the transgene and others not carrying it. The initial (G_0) transgenic animals must be mated and G_1 progeny produced to obtain animals in which all cells carry the transgene. In most of the cases where progeny studies have been performed, the transgenes were transmitted to progeny. In a few cases, transgenes have undergone rearrangements in progeny generations. In mammals, injected DNA molecules are not maintained by autonomous replication as they are in frogs, sea urchins, and the soil nematode *Caenorhabditis elegans.*

Transgenic mice are produced routinely in laboratories throughout the world. They provide valuable tools for the study of gene expression in mammals and an excellent model system with which to test various gene-transfer vectors and methodologies for possible use in humans. Hundreds of different transgenes have been studied in mice, and, in most cases, the transgenes show normal patterns of inheritance, indicating that they have been integrated into the host genome.

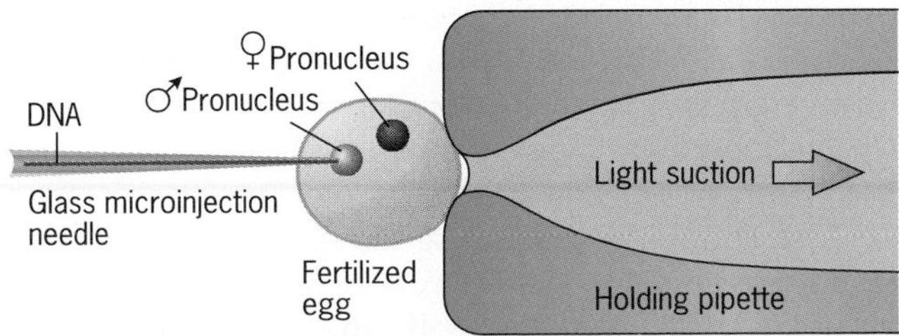

Figure 22.14 The production of transgenic animals by microinjection of DNA into fertilized eggs.

Figure 22.15 The transgenic mouse on the left, which carries a chimeric human growth hormone gene, is about twice the size of the control mouse on the right.

We discuss some of the applications of transgenic mice in studies of development in Chapter 25.

One of the early dramatic effects of a transgene was the increased growth rate that occurred when rat, bovine, or human growth hormone genes were expressed in mice (Figure 22.15). This prompted animal breeders to ask whether the introduction of either (1) extra copies of the homologous (same-species) growth hormone gene or (2) copies of heterologous growth hormone genes from related species might result in domestic animals with enhanced growth rates. Transgenic pigs were produced with the hope that enhanced growth hormone levels might result in leaner pigs with improved meat quality and with faster growth. Other scientists introduced growth hormone transgenes into fish and chickens with similar objectives. In several studies, transgenic fish that were expressing high levels of growth hormone exhibited striking increases in growth rate (Figure 22.16).

The experiments with transgenic pigs indicated that growth rates were increased not on standard diets but on high-protein diets. However, the transgenic pigs were found to be leaner than the controls, as was expected, because growth hormone favors the synthesis of proteins instead of fats.

Unfortunately, the transgenic pigs also exhibited several undesirable side effects of the higher growth hormone levels. Most notably, the female transgenic pigs were sterile. In addition, transgenic animals of both sexes were lethargic with weak muscles and were highly susceptible to arthritis and ulcers. Although scientists are hopeful that ways can be found to overcome these side effects, the initial results suggest that attempts to improve growth rates and enhance meat quality in domestic animals by increasing growth hormone levels with transgenes may not be very effective.

Clearly, the genetic control of the native levels of growth hormone represents the end product of many generations of natural selection for optimal fitness in these animals, and any changes in hormone levels induced by transgenes or other manipulations are likely to decrease fitness by upsetting the normal metabolic balance. Thus, deleterious side effects are to be expected. These do not mean that increased growth rates or improved meat quality cannot be achieved without unacceptable decreases in fitness of the transgenic animals. Moderate decreases in vigor may well be offset by highly desirable changes in meat quality (for ex-

Figure 22.16 Comparison of transgenic coho salmon fingerlings that are overexpressing the growth hormone gene from sockeye salmon (top) with normal siblings (bottom). The average weight of the transgenic fish was 11 times the mean weight of their normal siblings; the range was from no increase in weight to 37 times the average weight of the normal siblings.

ample, lower cholesterol levels) or by increases in the efficiency of conversion of feed into meat products.

Other transgenic animals are being tested for resistance to viral infections. Avian leukosis virus (ALV) is a major viral pathogen of chickens, causing losses to the poultry industry of $50 million to $100 million per year. Obviously, the availability of an ALV-resistant strain of chickens would be of major commercial value. Therefore, researchers have produced transgenic chickens that carry a defective ALV genome. These chickens produce viral RNA and the viral envelope protein, but no progeny viruses. Most importantly, they are resistant to infection by ALV. The synthesis of large amounts of the retroviral envelope protein somehow blocks the reproductive cycle of intact, pathogenic ALV viruses. That ALV resistance has been transmitted to several generations of progeny indicates that the trait is stable. These encouraging results suggest that the introduction of defective viral genomes into domestic animals may be a useful tool in the production of virus-resistant genotypes.

Another potentially important use of transgenic animals is for the production and secretion of valuable proteins in milk. Many native human proteins contain carbohydrate or lipid side groups that are added post-translationally. Bacteria do not contain the enzymes that catalyze the addition of these moieties to nascent proteins. In such cases, recombinant bacteria cannot be used to synthesize the final product; they will synthesize the polypeptide only in its unmodified form. For this reason, some researchers have begun to explore alternative methods for producing valuable human proteins, especially glycoproteins and lipoproteins.

One promising approach involves the use of transgenic animals that secrete human proteins in milk. In some of the first experiments of this kind, transgenic sheep were produced that secrete either of two human proteins, the blood-clotting protein factor IX or the elastase inhibitor α1-antitrypsin, in their milk. A deficiency of factor IX occurs in one type of hemophilia in humans, and a lack of α1-antitrypsin sometimes contributes to the development of emphysema (degenerative lung disease) in humans. Thus, both proteins have important medicinal applications. Researchers inserted the coding sequences for these proteins into the β-lactoglobulin (a milk protein) gene of sheep and microinjected the chimeric constructs into fertilized eggs. The injected eggs were then implanted into foster mothers to produce transgenic lambs. Although the concentrations of human factor IX and human α1-antitrypsin were low in the milk of these sheep, researchers believe that the concentrations can be increased by appropriately manipulating the regulatory sequences of the chimeric genes. Importantly, the transgenic sheep exhibited no apparent side effects from the production of either human protein in their milk. Nevertheless, further research must be done before the potential applicability of this approach can be adequately evaluated.

Transgenic Plants: The Ti Plasmid of *Agrobacterium tumefaciens*

Plants have been genetically manipulated by plant breeders for decades. Today, however, plant breeders can directly modify the DNA of plants and can quickly add genes from other species to plant genomes by recombinant DNA techniques. Indeed, transgenic plants can be produced by several different procedures. For example, one widely used procedure, called **microprojectile bombardment,** involves shooting DNA-coated tungsten or gold particles into plant cells. Another procedure, called **electroporation,** uses a short burst of electricity to get the DNA into the cell. However, more transgenic plants are produced by ***Agrobacterium tumefaciens*-mediated transformation** than by the other procedures. *Agrobacterium tumefaciens* is a soil bacterium that has evolved a natural genetic engineering system; it contains a segment of DNA that is transferred from the bacterium to plant cells.

An important feature of plant cells is their **totipotency**—that is, the ability of a single cell to produce all the differentiated cells of the mature plant. Many differentiated plant cells are able to dedifferentiate to the embryonic state and subsequently to redifferentiate to new cell types. Thus, there is no separation of germ-line cells from somatic cells as in higher animals. When excised tissues from mature plants are placed in the appropriate sterile tissue culture conditions (notably in the presence of the plant growth regulator 2,4-dichlorophenoxyacetic acid, 2,4-D), cells in these tissue explants will dedifferentiate and grow into highly unorganized cell masses called **calli** (or calluses; singular = callus). If these undifferentiated callus cell-clumps are subsequently transferred to growth medium favoring differentiation (medium lacking 2,4-D, but containing growth regulators such as kinetin), plantlets will often regenerate. This totipotency of plant cells is a major advantage for genetic engineering because it permits the regeneration of entire plants from individual modified somatic cells.

Agrobacterium tumefaciens is the causative agent of crown gall disease of dicotyledonous plants. The name refers to the galls or tumors that often form at the crown (junction between the root and the stem) of infected plants (Figure 22.17*a*). Because the crown of the plant is usually located at the soil surface, it is here that a plant is most likely to be wounded (for example, from a soil abrasion as it blows in a strong wind)

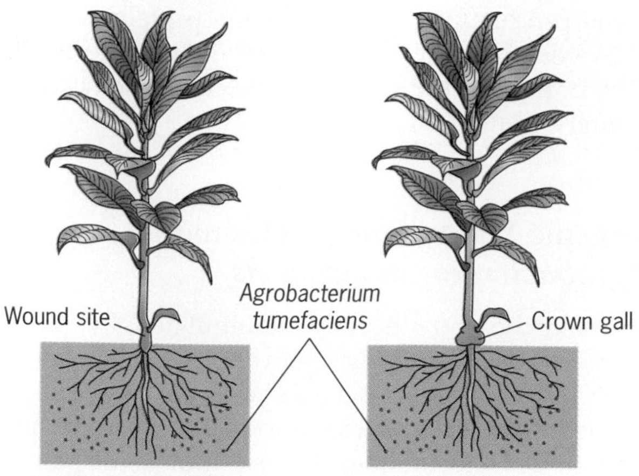

(*a*) Formation of a crown gall at the soil surface.

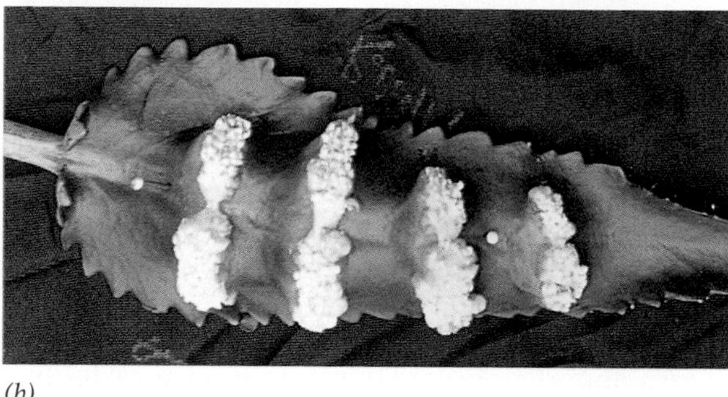

(*b*)

Figure 22.17 The formation of tumors or "galls" on plants at wound sites infected by *Agrobacterium tumefaciens.*

and infected by a soil bacterium such as *A. tumefaciens.* However, *A. tumefaciens* can infect a plant and induce a tumor at any wound site (Figure 22.17*b*). After the infection of a wound site by *A. tumefaciens,* two key events occur: (1) the plant cells begin to proliferate and form tumors, and (2) they begin to synthesize an arginine derivative called an opine. The opine synthesized is usually either nopaline or octopine depending on the strain of *A. tumefaciens.* These opines are catabolized and used as energy sources by the infecting bacteria. *Agrobacterium tumefaciens* strains that induce the synthesis of nopaline can grow on nopaline, but not on octopine, and vice versa. Clearly, an interesting interrelationship has evolved between *A. tumefaciens* strains and their plant hosts. *Agrobacterium tumefaciens* is able to divert the metabolic resources of the host plant to the synthesis of opines, which are of no apparent benefit to the plant but which provide sustenance to the bacterium.

The ability of *A. tumefaciens* to induce crown galls in plants is controlled by genetic information carried on a large (about 200,000 nucleotide pairs) plasmid called the **Ti plasmid** for its *t*umor-*i*nducing capacity. Two components of the Ti plasmid, the **T-DNA** and the ***vir* region,** are essential for the transformation of plant cells. During the transformation process, the T-DNA (for *T*ransferred *DNA*) is excised from the Ti plasmid, transferred to a plant cell, and integrated (covalently inserted) into the DNA of the plant cell. The available data indicate that integration of the T-DNA occurs at random chromosomal sites; moreover, in some cases, multiple T-DNA integration events occur in the same cell. In nopaline-type Ti plasmids, the T-DNA is a 23,000-nucleotide-pair segment that carries 13 known genes. In octopine-type Ti plasmids, there are two separate T-DNA segments. For the sake of brevity, we consider only nopaline-type Ti plasmids in the subsequent discussion.

The structure of a typical nopaline Ti plasmid is shown in Figure 22.18. Some of the genes on the T-DNA segment of the Ti plasmid encode enzymes that catalyze the synthesis of phytohormones (the auxin indoleacetic acid and the cytokinin isopentenyl adenosine). These phytohormones are responsible for the tumorous growth of cells in crown galls. The T-DNA region is bordered by 25-nucleotide-pair imperfect repeats, one of which must be present in *cis* for T-DNA excision and transfer. The deletion of the right border sequence completely blocks the transfer of T-DNA to plant cells.

The *vir* (for virulence) region of the Ti plasmid contains the genes required for the T-DNA transfer process. These genes encode the DNA processing enzymes required for excision, transfer, and integration of the T-DNA segment during the transformation process. The *vir* genes can supply the functions needed for T-DNA transfer when located either *cis* or *trans* to the T-DNA. They are expressed at very low levels in *A. tumefaciens* cells growing in soil. However, exposure of the bacteria to wounded plant cells or exudates from plant cells induces enhanced levels of expression of the *vir* genes. Surprisingly, this induction process is very slow for bacteria, taking 10 to 15 hours to reach maximum levels of expression. Phenolic compounds such as acetosyringone act as inducers of the *vir* genes, and transformation rates can often be increased by adding acetosyringone to plant cells inoculated with *Agrobacterium.* The transformation of plant cells by the Ti plasmid of *A. tumefaciens* occurs as illustrated in Figure 22.19.

Once it had been established that the T-DNA region of the Ti plasmid of *A. tumefaciens* is transferred to plant cells and becomes integrated in plant chromosomes, the potential use of *Agrobacterium* in plant genetic engineering was obvious. Foreign genes could be inserted into the T-DNA with the hope that these genes would be transferred to the plant with the rest of the T-DNA segment. This procedure works very

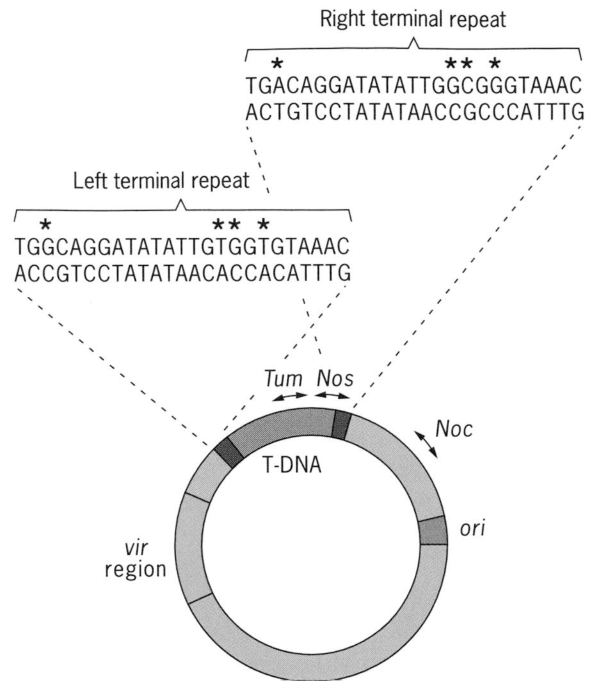

Figure 22.18 Structure of the nopaline Ti plasmid pTi C58, showing only components required for DNA transfer from *Agrobacterium* to plant cells. The Ti plasmid is 210 kb in size. Symbols used are: *ori*, origin of replication; *Tum*, genes responsible for tumor formation; *Nos*, genes involved in nopaline biosynthesis; *Noc*, genes involved in the catabolism of nopaline; *vir*, virulence genes required for T-DNA transfer. The nucleotide-pair sequences of the left and right terminal repeats are shown at the top; the asterisks mark the four base pairs that differ in the two border sequences.

well. The problem is that the transformed plant cells carrying wild-type T-DNAs lose their normal control of cell division and form tumors. This feature of T-DNA renders wild-type Ti plasmids incompatible with the goals of most gene-transfer experiments. Fortunately, the solution to this problem came early with the identification of the genes in the T-DNA that were responsible for tumor formation (Figure 22.18). The deletion of one or more of these genes produces a disarmed Ti plasmid. Unfortunately, the deletion of the tumor-causing genes also makes it extremely difficult to identify plant cells that have received the disarmed T-DNA. With wild-type Ti plasmids, the recipient plant cells form tumors and are easily identified by the tumor phenotype; with disarmed Ti plasmids, the recipient plant cells continue to grow just like their neighbors that do not harbor the T-DNA. Thus, a way to identify plant cells transformed with disarmed Ti plasmids—ideally, a good selectable marker gene located within the T-DNA region of the disarmed Ti plasmid—is needed.

A good selectable marker gene is one that will provide resistance to a drug, an antibiotic, or another agent that arrests the growth of normal plant cells. The selective agent should inhibit the growth of plant cells or kill them slowly. Agents that kill cells rapidly result in the release of phenolic compounds and other substances that are toxic to the growth of the remaining, otherwise resistant cells. To date, three selectable marker genes have been used extensively in plant systems; they provide resistance to the antibiotics (1) kanamycin and the related aminoglycoside G418

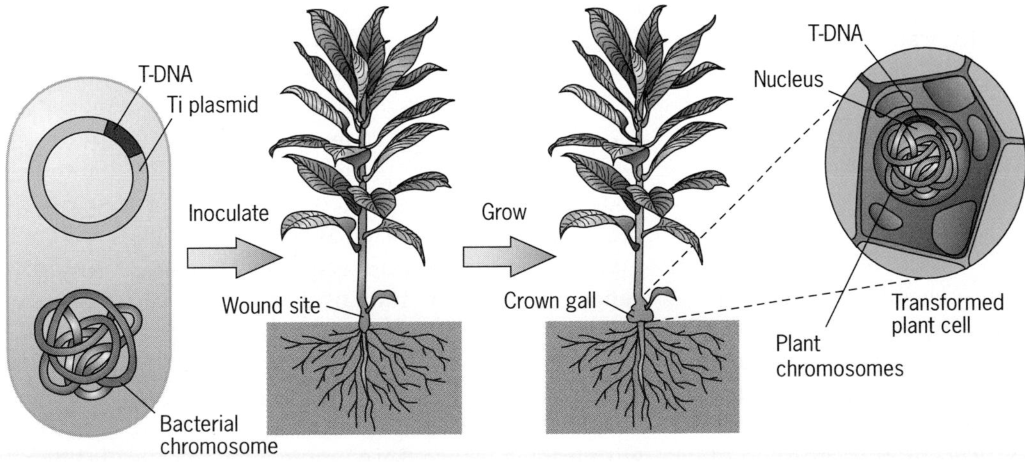

Figure 22.19 Transformation of plant cells by *Agrobacterium tumefaciens* harboring a wild-type Ti plasmid. Plant cells in the tumor contain the T-DNA segment of the Ti plasmid integrated into chromosomal DNA.

and (2) hygromycin, and to the drug (3) methotrexate. Of these, kanamycin has been the most widely used selective agent in plants.

The *Kan*^{r} gene from the *E. coli* transposon Tn*5* has been used extensively as a selectable marker in plants; it encodes an enzyme called neomycin phosphotransferase type II (NPTII). NPTII is one of several prokaryotic enzymes that detoxify the kanamycin family of aminoglycoside antibiotics by phosphorylating them. Because the promoter sequences and transcription-termination signals are different in bacteria and plants, the native Tn*5 Kan*^{r} gene cannot be used in plants. Instead, the NPTII coding sequence must be provided with a plant promoter (5′ to the coding sequence) and plant termination and polyadenylation signals (3′ to the coding sequence). Such constructions with prokaryotic coding sequences flanked by eukaryotic regulatory sequences are called **chimeric selectable marker genes.**

Regulatory sequences from several different plant genes have been used to construct chimeric marker genes. The two most frequently used regulatory sequences are (1) the promoter of the nopaline synthase (*nos*) gene of the Ti plasmid and (2) the promoter that controls the synthesis of the 35S transcript of cauliflower mosaic virus (CaMV). The most frequently used termination and polyadenylation sequence is that from the nopaline synthase gene of the Ti plasmid. One widely used chimeric selectable marker gene has the structure CaMV 35S promoter/NPTII coding sequence/Ti *nos* termination sequence; this chimeric gene is usually symbolized 35S/NPTII/*nos*. The Ti vectors used to transfer genes into plants have the tumor-inducing genes of the plasmid replaced with a chimeric selectable marker gene such as 35S/NPTII/*nos*. A large number of sophisticated Ti plasmid gene-transfer vectors are now used routinely to transfer genes into plants.

BLOCKING GENE EXPRESSION WITH ANTISENSE RNA

Another new tool that became possible with the advent of recombinant DNA and gene-cloning technologies is the use of **antisense RNA** to regulate gene expression. The antisense RNA method involves the synthesis of RNA molecules that are complementary to the mRNA molecules produced by transcription of a given gene. The normal mRNA of a gene is said to be sense because it carries the codons that are read during translation to produce the specified sequence of amino acids in the polypeptide gene product. Normally, the complement to the mRNA sense strand will not contain a sequence of codons that can be translated to produce a functional protein; thus, this complementary strand is called antisense RNA. In addition, an antisense RNA usually will not contain the regulatory sequences required for translation.

When antisense RNA molecules are present in the same cytosol with sense (mRNA) molecules from a gene, the antisense RNA and mRNA molecules will anneal to form duplex RNA molecules. These duplex RNA molecules cannot be translated; thus the presence of antisense RNA will block translation of the mRNA of the affected gene.

The antisense RNA approach has proven useful in dissecting pathways of gene expression in plants, animals, and microorganisms. It has also been a commercial success. The first genetically engineered plant product approved for human consumption, the FlavrSavr™ tomato, was introduced in supermarkets in California and Illinois in 1994. The FlavrSavr™ tomato, which remains firm longer during the ripening process, was produced by using antisense RNA to decrease the rate of expression of an endogenous gene encoding an enzyme called polygalacturonase to 10 percent of the normal level. Polygalacturonase is an enzyme that breaks down cell walls and causes softening in tomatoes as they ripen.

As most of us know from first-hand experience, vine-ripened tomatoes are much more flavorful than those that are picked green and allowed to ripen en route to the marketplace. However, vine-ripened tomatoes are too soft to survive the handling required during transport; they bruise too easily. The FlavrSavr™ tomato has partially solved this problem. FlavrSavr™ tomatoes remain firm further into the ripening process, allowing them to remain on the vines longer before they are picked and shipped to market.

The simplest way to produce antisense RNA of a gene within a cell or an organism is to (1) clone the gene of interest, (2) separate the coding sequence of the gene from its promoter by cleavage of the DNA with an appropriate restriction enzyme, (3) ligate the coding sequence to its promoter in the inverse orientation (flipped end-for-end) and (4) introduce the promoter/inverted coding sequence construct or **antisense gene** into the host cell or organism by transformation. This procedure is illustrated in Figure 22.20. The net result of this procedure is that transcription of the antisense gene will produce antisense RNA transcripts. These antisense RNAs will hybridize with mRNA (sense RNA) molecules in cells and block translation of the mRNAs. As a result, little or no protein gene product will be synthesized in these cells.

Another potential application of the antisense RNA technology is the treatment or prevention of viral diseases. In principle, this approach is simple. A bi-

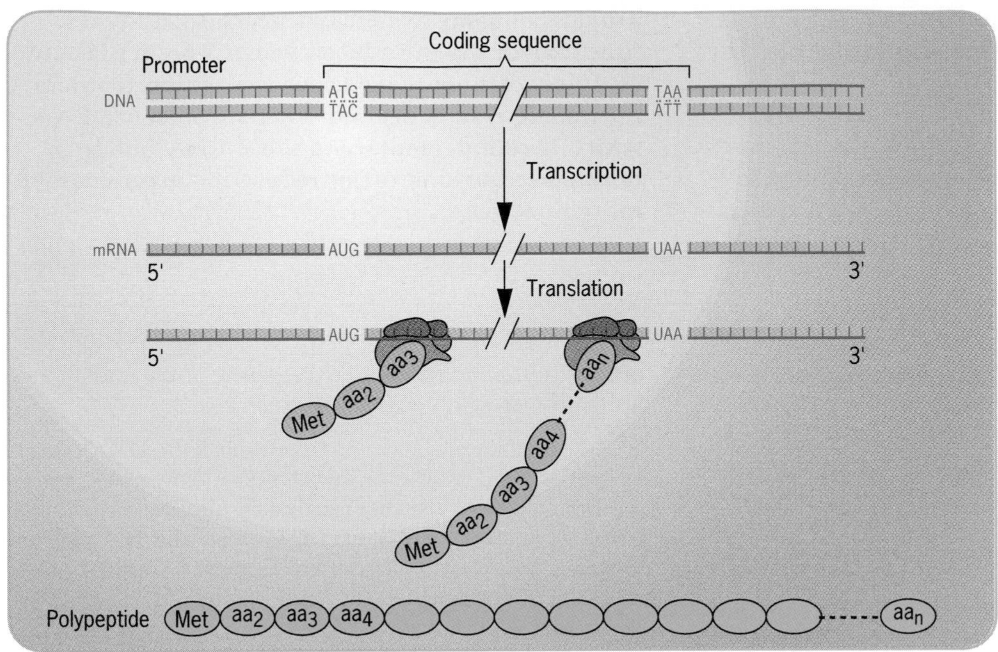

(a) Normal cell: only "sense RNA" (=mRNA).

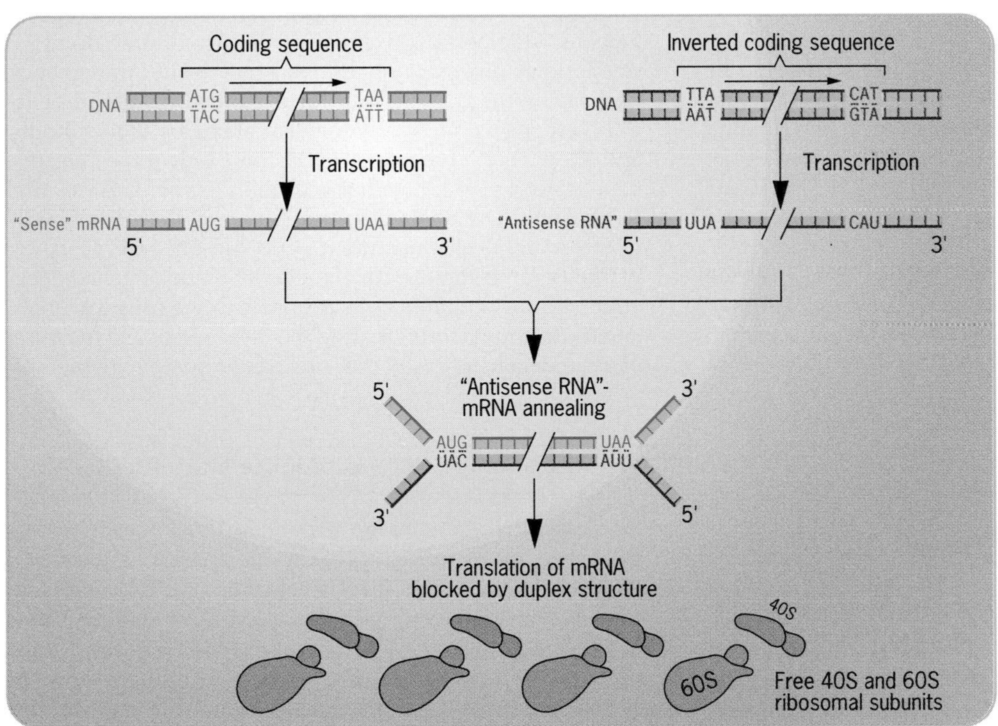

(b) Recombinant cell: "sense RNA" + "antisense RNA".

Figure 22.20 The antisense RNA procedure for blocking or reducing the level of expression of a specific gene.

ologist identifies a gene that is completely essential for the reproduction of a virus and blocks its expression with antisense RNA. The difficult step in this approach is delivering the antisense gene to the right cells in the desired tissue(s) in order to intercept the reproduction of the target virus. Nevertheless, some scientists believe that the antisense RNA strategy may be the best hope for treating and/or preventing AIDS in the future. Whether antisense RNA will prove useful in combating pathogenic viruses remains uncertain. Currently, geneticists know that antisense RNAs shut off some genes and do not shut off other genes, but they do not know why.

Key Points: **Synthetic, modified, or other foreign genes can now be introduced into most plant and animal species. The resulting transgenic organisms provide valuable systems in which to study the functions of the introduced genes. The Ti plasmid of *Agrobacterium tumefaciens* is an important tool for transferring genes into plants. Antisense RNA, which is complementary to sense RNA (mRNA), can be used to shut off or reduce the expression of individual genes.**

TESTING YOUR KNOWLEDGE

1. Spinocerebellar ataxia (type 1) is a progressive neurological disease with onset typically occurring between ages 30 and 50. The neurodegeneration results from the selective loss of specific neurons. Although it is not understood why selective neuronal death occurs, it is known that the disease is caused by the expansion of a CAG trinucleotide repeat, with normal alleles containing about 28 copies and mutant alleles harboring 43 to 81 copies of the trinucleotide. Given the nucleotide sequences on either side of the repeat region, how would you test for the presence of the expanded trinucleotide repeat region responsible for type 1 spinocerebellar ataxia?

ANSWER

The DNA test for spinocerebellar ataxia (type 1) would be similar to the test for the *huntingtin* allele described in Figure 22.3. You would first make PCR primers corresponding to DNA sequences on either side of the CAG repeat region. These primers would be used to amplify the desired CAG repeat region from genomic DNA of the individual being tested by PCR. Then, the sizes of the trinucleotide repeat regions would be determined by measuring the sizes of the PCR products by gel electrophoresis. Any gene with less than 30 copies of the CAG repeat would be considered a normal allele, whereas the presence of a gene with 40 or more copies of the trinucleotide would be diagnostic of the mutant alleles that cause spinocerebellar ataxia.

2. Assume that you have just performed the DNA test for spinocerebellar ataxia on a 25-year-old woman whose mother died from the disease. The results came back positive for the ataxia mutation. The woman and her husband long for their own biological children, but do not want to risk transmitting the defective gene to any of these children. What are their options?

ANSWER

Their options will depend on their religious and moral convictions. One possibility involves the use of amniocentesis or chorionic biopsy to obtain fetal cells early in pregnancy, performing the DNA test for the expanded trinucleotide region responsible for spinocerebellar ataxia on the fetal cells, and allowing the pregnancy to continue only if the defective gene is not present. Another possibility is the use of *in vitro* fertilization. The ataxia DNA test is then performed on a cell from an eight-cell pre-embryo, and the pre-embryo is implanted only if the test for the defective ataxia gene is negative. A third option may be available in the future, namely, an effective method of treating the disease prior to the onset of neurodegeneration, perhaps by gene-replacement therapy.

QUESTIONS AND PROBLEMS

22.1 What are CpG islands? Of what value are CpG islands in positional cloning of human genes?

22.2 Why is the mutant gene that causes Huntington's disease called *huntingtin?* Why might this gene be renamed in the future?

22.3 How was the nucleotide sequence of the *CF* gene used to obtain information about the structure and function of its gene product?

22.4 How might the characterization of the *CF* gene and its product lead to the treatment of cystic fibrosis by somatic-cell gene therapy? What obstacles must be overcome before cystic fibrosis can be treated successfully by gene therapy?

22.5 Myotonic dystrophy (MD), occurring in about 1 of 8000 individuals, is the most common form of muscular dystrophy in adults. The disease, which is characterized by progressive muscle degeneration, is caused by a dominant mutant gene that contains an expanded CAG repeat region. Wild-type alleles of the *MD* gene contain 5 to 30 copies of the trinucleotide. Mutant *MD* alleles contain 50 to over 2000 copies of the CAG repeat. The complete nucleotide sequence of the *MD* gene is available. Design a diagnostic test for the mutant gene responsible for myotonic dystrophy that can be carried out using genomic DNA from newborns, fetal cells obtained by amniocentesis, and single cells from eight-cell pre-embryos produced by *in vitro* fertilization.

22.6 In humans, the absence of an enzyme called purine nucleoside phosphorylase (PNP) results in a severe T-cell immunodeficiency similar to that of severe combined immunodeficiency disease (SCID). PNP deficiency exhibits an autosomal recessive pattern of inheritance, and the gene encoding human PNP has been cloned and sequenced. Would PNP deficiency be a good candidate for treatment by gene therapy? Design a procedure for the treatment of PNP deficiency by somatic-cell gene therapy.

22.7 Human proteins can now be produced in bacteria such as *E. coli*. However, one cannot simply introduce a human gene into *E. coli* and expect it to be expressed. What steps must be taken to construct an *E. coli* strain that will produce a mammalian protein such as human growth hormone?

22.8 You have constructed a synthetic gene that encodes an enzyme that degrades the herbicide glyphosate. You wish to introduce your synthetic gene into *Arabidopsis* plants and test the transgenic plants for resistance to glyphosate. How could you produce a transgenic *Arabidopsis* plant harboring your synthetic gene by *A. tumefaciens*-mediated transformation?

22.9 A human VNTR locus contains a tandem repeat $(TAA)_n$, where n may be between 5 and 15. How many alleles of this locus would you expect to find in the human population?

22.10 DNA fingerprints have played central roles in many recent rape and murder trials. What is a DNA fingerprint? What roles do DNA prints play in these forensic cases? In some cases, geneticists have been concerned that DNA fingerprint data were being used improperly. What were some of their concerns, and how can these concerns be properly addressed?

22.11 Most forensic experts agree that fingerprinting of DNA from blood samples obtained at crime scenes and on personal items can provide convincing evidence for murder convictions. However, the defense attorneys sometimes argue successfully that sloppiness in handling blood samples results in contamination of the samples. What problems would contamination of blood samples present in the interpretation of DNA fingerprints? Would you expect such errors to lead to the conviction of an innocent person or the acquittal of a guilty person?

22.12 A group of bodies are found buried in a woods. The police suspect that they may include the missing Jones family (two parents and two children). They extract DNA from bones and examine (using PCR) genes *A* and *B*, which are known to contain tandem triplet repeats of variable length. They also analyze DNA from two other men. The results are shown below where the numbers indicate the number of copies of a tandem repeat in a particular allele; for example, male 1 has one allele with 8 and another allele with 9 copies of tandem repeat in gene *A*.

	Gene *A*	Gene *B*
male 1	8/9	5/7
male 2	6/8	5/5
male 3	7/10	7/7
woman	8/8	3/5
child 1	7/8	5/7
child 2	8/8	3/7

Could the woman have been the mother of both children? Why or why not? Which man, if any, could have been the father of child 1?

22.13 The DNA fingerprints shown below were prepared using genomic DNA from blood cells obtained from a woman, her daughter, and three men who all claim to be the girl's father. Based on the DNA prints, what can be determined about paternity in this case?

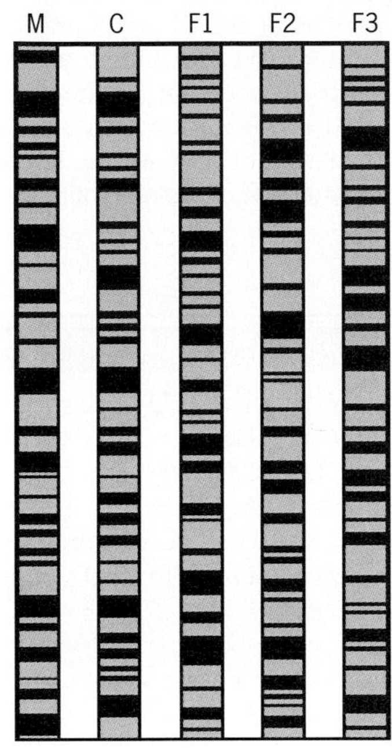

22.14 The Ti plasmid contains a region referred to as T-DNA. Why is this region called T-DNA, and what is its significance?

22.15 The generation of transgenic plants using *A. tumefaciens*-mediated transformation often results in multiple sites of insertion. These sites frequently vary in the level of transgene expression. What approaches could you use to determine whether or not transgenic plants carry more than one transgene and, if so, where the transgenes are inserted into chromosomes?

22.16 Disarmed retroviral vectors can be used to introduce genes into higher animals, including humans. What advantages do retroviral vectors have over other kinds of gene-transfer vectors? What disadvantages?

22.17 Transgenic mice are now routinely produced and studied in research laboratories throughout the world. How are transgenic mice produced? What kinds of information can be obtained from studies performed on transgenic mice? Does this information have any importance to the practice of medicine? If so, what?

22.18 What is antisense RNA? Of what use is antisense RNA in genetic research? in agriculture? How can a cell or an organism that produces a specific antisense RNA be produced?

22.19 Many valuable human proteins contain carbohydrate or lipid components that are added posttranslationally. Bacteria do not contain the enzymes needed to add these components to primary translation products. How might these proteins be produced using transgenic animals?

22.20 Richard Meagher and coworkers have cloned a family of 10 genes that encode actins (a major component of the cytoskeleton) in *Arabidopsis thaliana*. The 10 actin gene products are similar, often differing by just a few amino acids.

Thus, the coding sequences of the 10 genes are also very similar, so that the coding region of one gene will cross-hybridize with the coding regions of the other nine genes. In contrast, the noncoding regions of the 10 genes are quite divergent. Meagher has hypothesized that the 10 actin genes exhibit quite different temporal and spatial patterns of expression. You have been hired by Meagher to test this hypothesis. Design experiments that will allow you to determine the temporal and spatial pattern of expression of each of the 10 actin genes in *Arabidopsis.*

22.21 The first transgenic mice resulted from microinjection of vector DNA similar to that diagrammed in Figure 22.13 (except that it contained a promoter for the mammalian metallothienin gene linked to the *HGH* gene) into mice eggs. The resulting transgenic mice showed elevated levels of HGH in tissues of organs other than the pituitary gland, for example, in heart, lung, and liver, and the pituitary gland underwent atrophy. How might the production of HGH in transgenic animals be better regulated, with expression restricted to the pituitary gland?

BIBLIOGRAPHY

Blaese, R. M., K. W. Culver, A. D. Miller, C. S. Carter, T. Fleisher, M. Clerici, G. Shearer, L. Chang, Y. Chiang, P. Tolstoshev, J. J. Greenblatt, S. A. Rosenberg, H. Klein, M. Berger, C. A. Mullen, W. J. Ramsey, L. Muul, R. A. Morgan, and W. F. Anderson. 1995. T lymphocyte-directed gene therapy for ADA^-SCID: Initial trial results after 4 years. *Science* 270:475–480.

Bordignon, C., L. D. Notarangelo, N. Nobili, G. Ferrari, G. Casorati, P. Panina, E. Mazzolari, D. Maggioni, C. Rossi, P. Servida, A. G. Ugazio, and F. Mavilio. 1995. Gene therapy in peripheral blood lymphocytes and bone marrow for ADA^- immunodeficient patients. *Science* 270:470–475.

Huntington's Disease Collaborative Research Group. 1993. A novel gene containing a trinucleotide repeat that is expanded and unstable on Huntington's disease chromosomes. *Cell* 72:971–983.

Kristjansson, K., S. S. Chong, I. B. van den Veyver, S. Subramarian, M. C. Snabes, and M. R. Hughes. 1994. Preimplantation single cell analyses of dystrophin gene deletions using whole genome amplification. *Nature Genetics* 6:19–23.

Mol, J.N.M., R. van Blokland, P. de Lange, M. Stam, and J. M. Kooter. 1994. Post-translational inhibition of gene expression: sense and antisense genes. In *Homologous Recombination and Gene Silencing in Plants,* pp. 309–334, J. Paszkowski, ed. Kluwer, Dordrecht, The Netherlands.

Rommens, J. M., M. C. Iannuzzi, B. Kerem, M. L. Drumm, G. Melmer, M. Dean, R. Rozmahel, J. L. Cole, D. Kennedy, N. Hidaka, M. Zsiga, M. Buchwald, J. R. Riordan, L-C. Tsui, and F. S. Collins. 1989. Identification of the cystic fibrosis gene: chromosome walking and jumping. *Science* 245:1059–1065.

Stewart, T. 1997. Genetic modification of animals. In *Exploring Genetic Mechanisms,* M. Singer and P. Berg, eds., University Science Books, Sausalito, CA.

Weintraub, H. M. 1990. Antisense RNA and DNA. *Sci. Amer.* 262(1):40–46.

Willems, R. J. 1994. Dynamic mutations hit double figures. *Nature Genetics* 8:213–215.

Zambryski, P. 1997. Genetic manipulation of plants. In *Exploring Genetic Mechanisms,* M. Singer and P. Berg, eds., University Science Books, Sausalito, CA.

A CONVERSATION WITH

MITZI KURODA

What was your first experience with biology? How did your interests develop?

I didn't focus on biology until I was a college student. The emergence of recombinant DNA technology really made the difference for me. The possibility of understanding the molecular mechanisms of gene regulation captured my imagination and sent me straight to graduate school at Stanford University. I owe my scientific education in a large part to my Ph.D. mentor, Dr. Charley Yanofsky. He has been an outstanding role model who pursues science for its pure, intrinsic interest. I am the first to admit that my career development since then has involved a tremendous amount of good luck. I am one of many molecular biologists who have profited from the seminal genetic studies of others—in my case, the work of Drs. John Lucchesi and Kugao Oishi and colleagues. Choosing to work on intriguing, unusual biological problems in both bacteria and fruit flies has been a very rewarding path for me.

You concentrate most of your research on the study of dosage compensation in *Drosophila*. What is the thrust of this research?

In the fruit fly, most genes on the X chromosome are expressed at a twofold higher level per gene dose in males (XY) than in females (XX); the result is equivalent levels of X-linked products in the two sexes. It is fascinating to me that a whole chromosome can be regulated without disturbing the individual control mechanisms of the many diverse genes on that chromosome. A set of proteins is required for this process, and we are studying how they specifically recognize the male X chromosome and alter transcription levels. We are very excited about recent evidence that noncoding RNAs co-localize with the protein complex on the male X chromosome, indicating a role in dosage compensation. An intriguing parallel is that a noncoding RNA called Xist is essential for dosage compensation in mammals and coats the inactivated X chromosome. We would like to know how these RNAs function and are excited by the possibility that noncoding RNAs may play general roles in chromatin structure and gene regulation.

Mitzi Kuroda is the daughter of immigrants. Her father came from Japan and her mother from the Netherlands. She is representative of a large group of talented young researchers who entered the field of molecular genetics just as recombinant DNA revolutionized biological research. She was introduced to the beauty of the *Drosophila* polytene chromosomes as an undergraduate student at Tulane University. She then obtained her Ph.D. degree from Stanford University in 1987, studying transcriptional attenuation in bacteria. Dr. Kuroda has been a leader in the study of dosage compensation in *Drosophila* since she was a postdoctoral fellow at Stanford. Soon after starting her own lab, she was named an NSF Presidential Young Investigator and a Searle Scholar. Currently, Dr. Kuroda is an Associate Professor in the Departments of Cell Biology and Molecular and Human Genetics at Baylor College of Medicine and an Associate Investigator of the Howard Hughes Medical Institute.

Can you single out any one research result as the most exciting of your career? If so, what was it, and what were your feelings at the time?

This is an easy question. My most exciting result was seeing that the Maleless dosage compensation protein was binding along the length of the male X chromosome. I had raised an antibody to the Maleless protein and immunostained the giant polytene chromosomes of *Drosophila* larvae. Seeing the X chromosome binding was a dramatic event for me. First of all, selective binding of a protein along the length of a specific chromosome had not, to my knowledge, ever been seen. Second, X localization fulfilled the genetic predictions for a key player in dosage compensation, indicating that I should be able to pursue the mechanism directly by studying the Maleless protein. Third, it was visually stunning. To this day I continue to be amazed by the result, even though X chromosome staining is a daily occurrence in the laboratory.

In your career as a scientist, what was one of your greatest challenges, and how did you overcome it?

Starting my own lab and gaining the respect of other scientists have been tremendous challenges. Science is a very competitive enterprise, and respect does not come automatically. I have established relationships with many of my scientific colleagues by exchanging reagents and ideas, communicating results before publication, and engaging in collaborative research. However, it is a continuing challenge to balance open communication with the fundamentally competitive nature of scientific research.

Please describe your typical day.

I spend my day talking to people, reading, and working at my computer. As the head of a lab, my time is essentially spent in determining strategy, summarizing results, editing, and critiquing the work of others. On the one hand, it is fun to solve problems, hear about successful experiments, and develop the big picture. On the other hand, there is nothing quite so satisfying as doing an informative experiment with your own hands.

With ever increasing technologies, do you believe that bioethical issues are being thoroughly explored?

Technology is moving very fast, responding to the ever-present urgency to alleviate human suffering. I believe that the development of technology is overwhelmingly positive, but I would like to see a general public that is much more informed with regard to possible developments in the future. Several major science organizations have been making progress in the educational arena in recent years, but we all need to contribute. As the human genome project unleashes vast amounts of information, there will be a great need for scientists to play a larger role in public policy.

What skill sets do you think are mandatory for students starting out in biology?

You must have an intense interest in the subject matter, coupled with the determination to excel. You also need to derive satisfaction from the day-to-day experimental work, as the ultimate goals often require years to attain. Written and verbal communication skills are critical for most jobs. Finally, having broad interests in biology can prepare you to make the insightful connections that we all hope to attain with our specific research.

Do you have a favorite hobby or activity unrelated to science?

My husband and I are raising two young children, Daniel and Susanna. Parenthood is an intense experience, fun and fascinating. These days my activities outside work are geared almost exclusively to family life. As a family, one of our favorite activities is to visit the zoo and talk about all the interesting animals. During quiet times, I enjoy my lifelong love for reading, especially novels.

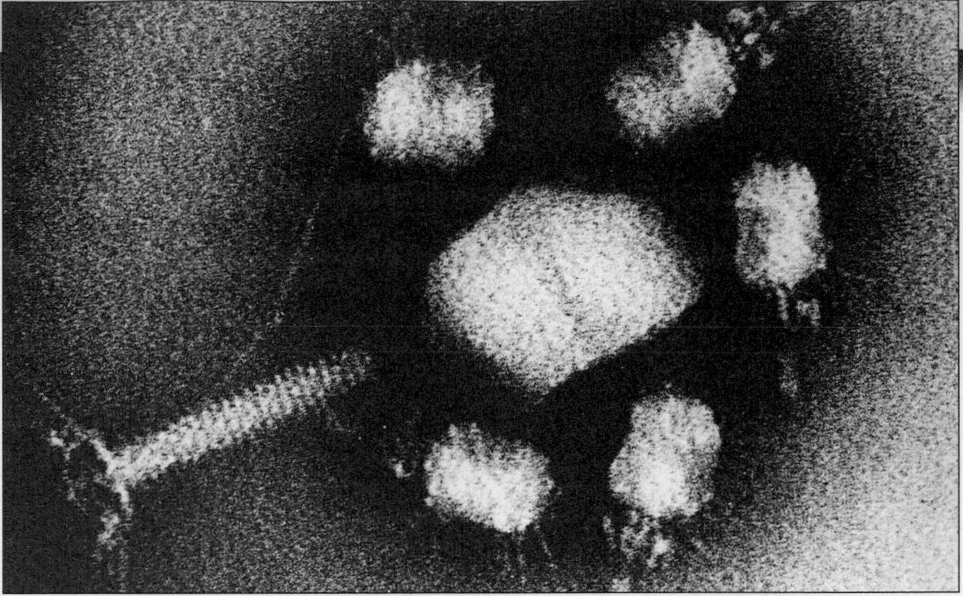

A quintet of Bacillus subtilis *bacteriophage* ϕ*29 particles surrounding an* Escherichia coli *T2 bacteriophage.*

23

Regulation of Gene Expression in Prokaryotes and Their Viruses

CHAPTER OUTLINE

Constitutive, Inducible, and Repressible Gene Expression
Positive and Negative Control of Gene Expression
Operons: Coordinately Regulated Units of Gene Expression
The Lactose Operon in *E. coli:* Induction and Catabolite Repression
The Tryptophan Operon in *E. coli:* Repression and Attenuation
Bacteriophage Lambda: Lysogeny or Lysis
Temporal Sequences of Gene Expression During Phage Infection
Translational Control of Gene Expression
Posttranslational Regulatory Mechanisms

D'Hérelle's Dream of Treating Dysentery in Humans by Phage Therapy

In 1910, the French-Canadian microbiologist Felix d'Hérelle was in Mexico investigating a bacterial disease that was killing entire populations of locusts. The infected locusts developed severe diarrhea, excreting almost pure suspensions of bacilli prior to death. When he isolated and studied the bacteria in the feces of the locusts, d'Hérelle observed circular clear spots in the bacterial cultures grown on agar. However, when he examined the material in the clear spots microscopically, he could not see anything. In 1915, d'Hérelle returned to the Pasteur Institute in Paris, where he studied an epidemic of bacterial dysentery that was raging through army units stationed in France. He once again observed clear spots in lawns of bacteria. In addition, he demonstrated that whatever was killing the *Shigella*—the bacterium that causes dysentery in humans—could pass through a porcelain filter that retained all known bacteria. In 1917, d'Hérelle published his results and named the submicroscopic bacteriocidal agents bacteriophages (from the Greek word for "bacteria-devouring"). About the same time, an English medical bacteriologist, Frederick W. Twort, discovered a similar submicroscopic agent that killed micrococci. Twort's results were published in 1915 but remained largely unknown until after the publication of d'Hérelle's work. Unfortunately, Twort's research was soon interrupted by a call to serve in the Royal Army Medical Corps in World War I.

Meanwhile, d'Hérelle continued to study the submicroscopic agents that killed *Shigella* and the bacteria responsible for diarrhea in locusts. He provided the following first-hand account of one of his experiments: "in a flash I had understood: what caused my clear spots was in fact an invisible microbe, a filtrable virus, but a virus parasitic on bacteria. . . . 'If this is true, the same thing has probably occurred during the night in the sick man, who yesterday was in serious condition. In his intestine, as in my test-tube, the dysentery baccilli will have dissolved away under the action of their parasite. He should now be cured.' I dashed to the hospital. In fact, during the night, his condition had greatly improved and convalescence was beginning" (d'Hérelle, 1949, "The Bacteriophage," *Science News*, No. 14, Penguin, Harmondsworth).

Indeed, d'Hérelle became obsessed with his belief that human diseases caused by bacteria could be treated, perhaps even eradicated, by bacteriophage therapy. Unfortunately, it was soon proven that bacteriophage therapy is not effective in treating bacterial infections because, too frequently, the bacteria mutate to phage-resistant forms. Nevertheless, d'Hérelle's work set the stage for research that would eventually produce a whole new field—microbial genetics—and yield insights into the mechanisms by which gene expression is regulated. In this chapter, we examine some of these mechanisms.

Microorganisms exhibit remarkable capacities to adapt to diverse environmental conditions. This adaptability depends in part on their ability to turn on and turn off the expression of specific sets of genes in response to changes in the environment. Most microorganisms exhibit a striking ability to regulate the expression of specific genes in response to environmental signals. The expression of particular genes is turned on when the products of these genes are needed for growth. Their expression is turned off when the gene products are no longer needed. Clearly, the ability of an organism to regulate gene expression in this way increases its ability to reproduce under a variety of environmental conditions. The synthesis of gene transcripts and translation products requires the expenditure of considerable energy. By turning off the expression of genes when their products are not needed, an organism can save energy and can utilize the conserved energy to synthesize products that maximize growth rate. What, then, are the mechanisms by which microorganisms regulate gene expression in response to changes in the environment? Is there a single mechanism by which the expression of different genes is regulated? Or are different genes controlled by different mechanisms?

Gene expression in prokaryotes is regulated at several different levels: transcription, mRNA processing, mRNA turnover, translation, and posttranslation (Figure 23.1). However, the regulatory mechanisms with the largest effects on phenotype act at the level of transcription.

Based on what is known about the regulation of transcription, the various regulatory mechanisms seem to fit into two general categories:

1. *Mechanisms that involve the rapid turn-on and turn-off of gene expression in response to environmental changes.* Regulatory mechanisms of this type are important in microorganisms because of the frequent exposure of these organisms to sudden changes in environment. They provide microorganisms with considerable "plasticity," an ability

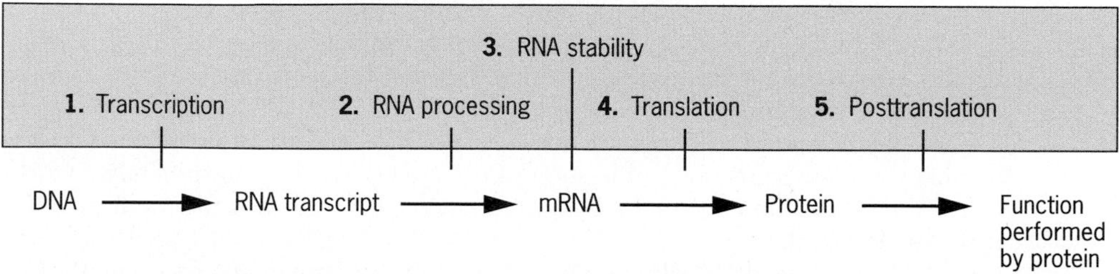

Figure 23.1 An abbreviated pathway of gene expression, showing five important levels of regulation in prokaryotes.

to adjust their metabolic processes rapidly in order to achieve maximal growth and reproduction under a wide range of environmental conditions. These rapidly responding on-off switches seem to be less important in multicellular eukaryotes, perhaps because the circulatory systems of higher animals buffer cells against many of the sudden environmental changes that microorganisms encounter.

2. *Mechanisms referred to as preprogrammed circuits or cascades of gene expression.* In these cases, some event triggers the expression of one set of genes. The product (or products) of one (or more) of these genes functions by turning off the transcription of the first set of genes or turning on the transcription of a second set of genes. Then, one or more of the products of the second set acts by turning on a third set, and so on. In these cases, the sequential expression of genes is genetically preprogrammed, and the genes cannot usually be turned on out of sequence. Such preprogrammed sequences of gene expression are well documented in both prokaryotes and eukaryotes. When a lytic bacteriophage infects a bacterium, the viral genes are expressed in a predetermined sequence, and this sequence is directly correlated with the temporal sequence of gene-product involvement in the reproduction and morphogenesis of the virus. In most of the known examples of preprogrammed gene expression, the circuitry is cyclical. For example, during viral infections, some event associated with the packaging of the viral DNA or RNA in the protein coat resets the genetic program so that the proper sequence of gene expression occurs once again when a progeny virus infects a new host cell.

CONSTITUTIVE, INDUCIBLE, AND REPRESSIBLE GENE EXPRESSION

Certain gene products—such as tRNA molecules, rRNA molecules, ribosomal proteins, RNA polymerase subunits, and enzymes catalyzing metabolic processes that are frequently referred to as cellular "housekeeping" functions—are essential components of almost all living cells. Genes that specify products of this type are continually being expressed in most cells. Such genes are said to be expressed constitutively and are referred to as **constitutive genes.**

Other gene products are needed for cell growth only under certain environmental conditions. Constitutive synthesis of such gene products would be wasteful, using energy that could otherwise be utilized for more rapid growth and reproduction. The evolution of regulatory mechanisms that provide for the synthesis of such gene products only when and where they are needed would clearly endow the organisms that possess these regulatory mechanisms with a selective advantage over organisms that lack them. This undoubtedly explains why currently existing organisms, including the "primitive" bacteria and viruses, exhibit highly developed and efficient mechanisms for the control of gene expression.

Escherichia coli and most other bacteria are capable of growth using any one of several carbohydrates, for example, glucose, sucrose, galactose, arabinose, and lactose, as an energy source. If glucose is present in the environment, it will be preferentially metabolized by *E. coli* cells. However, in the absence of glucose, *E. coli* cells can grow very well on other carbohydrates. Cells growing in medium containing the sugar lactose, for example, as the sole carbon source synthesize two enzymes, β-galactosidase and β-galactoside permease, which are uniquely required for the catabolism of lactose. A third enzyme, β-galactoside transacetylase, is also synthesized, but its function is unknown. β-Galactoside permease pumps lactose into the cell, where β-galactosidase cleaves it into glucose and galactose. Neither of these enzymes is of any use to *E. coli* cells if no lactose is available to them. The synthesis of these two enzymes requires the utilization of considerable energy (in the form of ATP and GTP; see Chapters 12 and 13). Thus, *E. coli* cells have evolved a regulatory mechanism by which the synthesis of these lactose-catabolizing enzymes is turned on in the presence of lactose and turned off in its absence.

In natural environments (intestinal tracts and sewers), *E. coli* cells probably encounter an absence of glucose and the presence of lactose relatively infrequently. Therefore, the *E. coli* genes encoding the enzymes involved in lactose utilization are probably turned off most of the time. If cells growing on a carbohydrate other than lactose are transferred to medium containing lactose as the only carbon source, they quickly begin to synthesize the enzymes required for lactose utilization (Figure 23.2*a*). This process of turning on the expression of genes in response to a substance in the environment is called **induction.** Genes whose expression is regulated in this manner are called **inducible genes;** their products, if enzymes, are called **inducible enzymes.**

Enzymes that are involved in **catabolic** (degradative) **pathways,** such as in lactose, galactose, or arabinose utilization, are characteristically inducible. As we discuss below, induction occurs at the level of transcription. Induction alters the rate of synthesis of enzymes, not the activity of existing enzyme molecules. Induction should not be confused with enzyme acti-

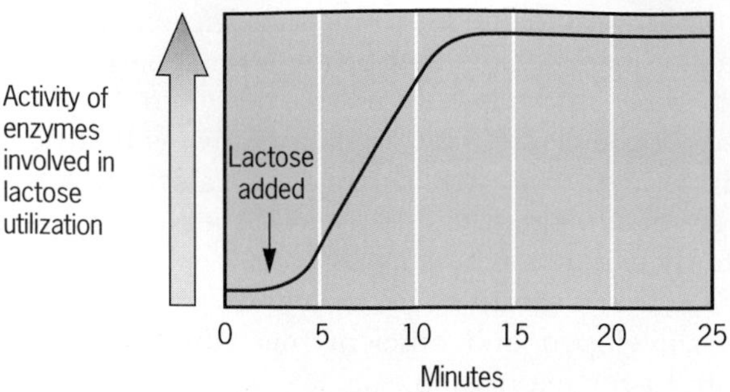

(*a*) Induction of enzyme synthesis.

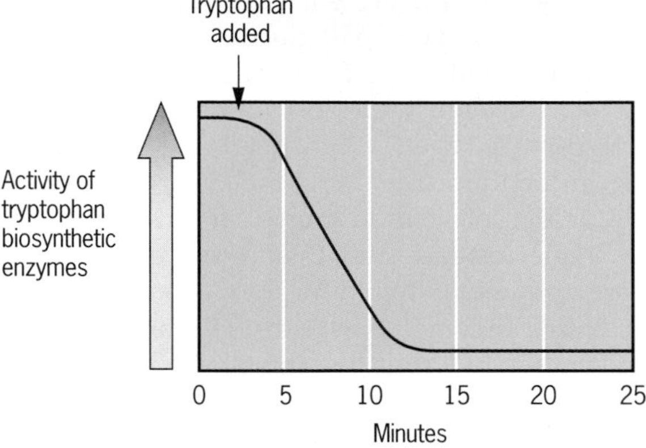

(*b*) Repression of enzyme synthesis.

Figure 23.2 (*a*) Induction of the synthesis of enzymes required for the utilization of lactose as an energy source and (*b*) repression of the synthesis of the enzymes required for the biosynthesis of tryptophan, both in *E. coli*. Note that low levels of enzyme synthesis occur whether the metabolites are present or absent.

vation, which occurs when the binding of a small molecule to an enzyme increases the activity of the enzyme, but does not affect its rate of synthesis.

Bacteria can synthesize most of the organic molecules, such as amino acids, purines, pyrimidines, and vitamins, required for growth. For example, the *E. coli* genome contains five genes encoding enzymes that catalyze steps in the biosynthesis of tryptophan. These five genes must be expressed in *E. coli* cells growing in an environment devoid of tryptophan in order to provide adequate amounts of this amino acid for ongoing protein synthesis.

When *E. coli* cells are present in an environment containing concentrations of tryptophan sufficient to support optimal growth, the continued synthesis of the tryptophan biosynthetic enzymes would be a waste of energy, because these bacteria can accumulate and utilize tryptophan from the environment. Thus, a regulatory mechanism has evolved in *E. coli* that turns off the synthesis of the tryptophan biosynthetic enzymes when external tryptophan is available (Fig. 23.2*b*). A gene whose expression has been turned off in this way is said to be repressed; the process is called **repression.** When the expression of this gene is turned on, it is said to be derepressed; such a response is called **derepression.**

Enzymes that are components of **anabolic** (biosynthetic) **pathways** often are repressible. Repression, like induction, occurs at the level of transcription. Repression should not be confused with feedback inhibition, which occurs when the product of a biosynthetic pathway binds to and inhibits the activity of the first enzyme in the pathway, but does not affect the synthesis of the enzyme.

Key Points: **In prokaryotes and viruses, genes that specify housekeeping functions such as rRNAs, tRNAs, ribosomal proteins, and DNA and RNA polymerase subunits are expressed constitutively. Other genes usually are expressed only when their products are needed. Genes that encode enzymes involved in catabolic pathways are often expressed only in the presence of the substrates of the enzymes; their expression is inducible. Genes that encode enzymes involved in anabolic pathways usually are turned off in the presence of the end product of the pathway; their expression is repressible. Although gene expression can be regulated at many levels, transcriptional regulation is the most common.**

POSITIVE AND NEGATIVE CONTROL OF GENE EXPRESSION

The regulation of gene expression—induction or turning genes on, and repression or turning genes off—can be accomplished by both positive control mechanisms and negative control mechanisms. Both mechanisms involve the participation of **regulator genes**—genes encoding products that regulate the expression of other genes. In **positive control mechanisms,** the product of the regulator gene is required to turn on the expression of one or more structural genes (genes specifying the amino acid sequences of enzymes or structural proteins), whereas in **negative control mechanisms,** the product of the regulator gene is necessary to shut off the expression of structural genes. Positive and negative regulation are illustrated for both inducible and repressible systems in Figure 23.3.

Recall that a gene is expressed when RNA polymerase binds to its promoter and synthesizes an RNA transcript that contains the coding region of the gene (Chapter 12). The product of the regulator gene acts by binding to a site called the regulator protein binding site (*RBS*) adjacent to the promoter of the structural

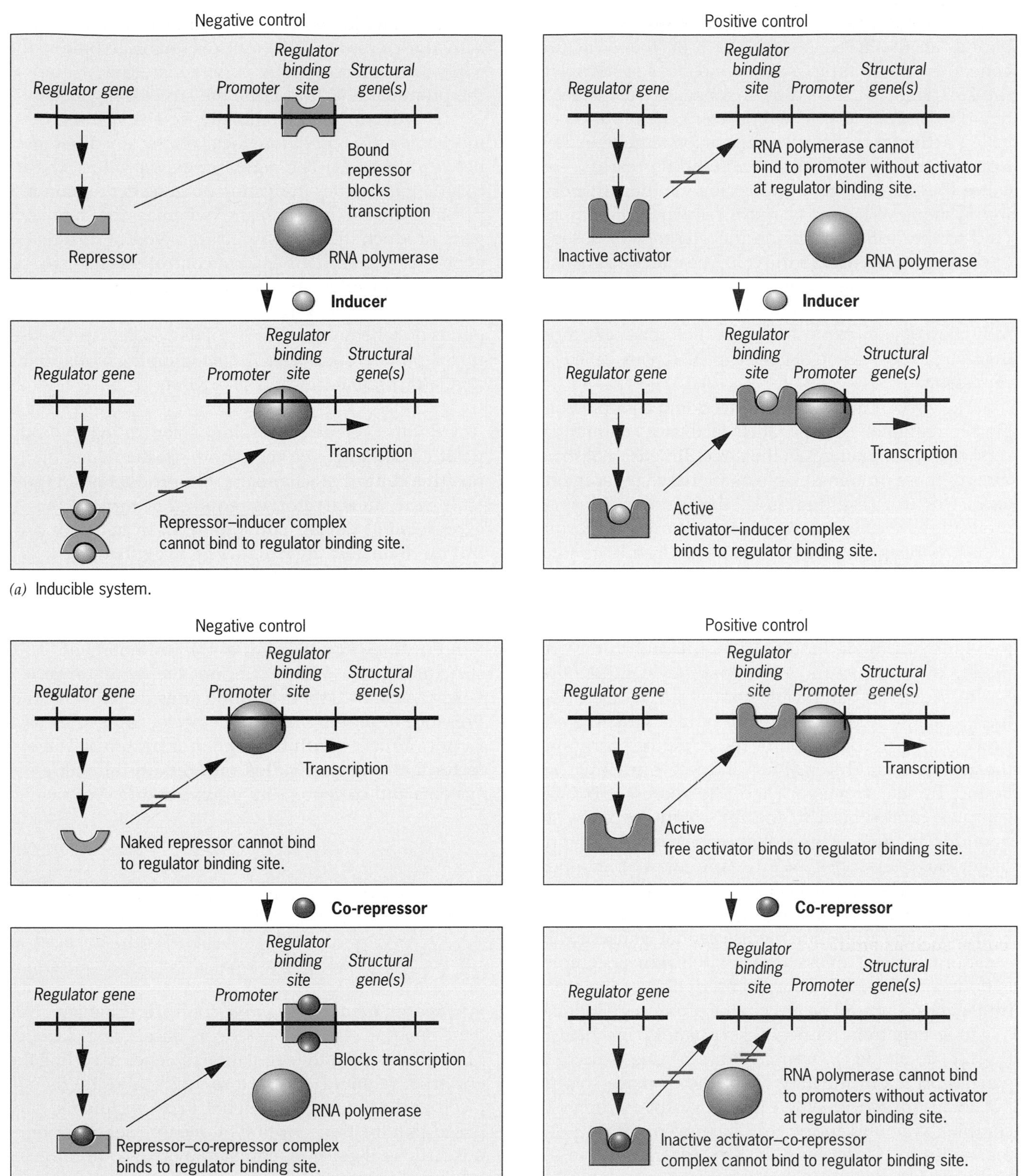

Figure 23.3 Negative and positive control of inducible and repressible gene expression. The regulator gene product is required to turn on gene expression in positive control circuits and to turn off gene expression in negative control systems.

gene(s). When the product of the regulator gene is bound at the *RBS*, transcription of the structural gene(s) is turned on in a positive control system (Figure 23.3, right) or turned off in a negative control system (Figure 23.3, left). The regulator gene products are called **activators**—because they activate gene expression—in positive control systems and **repressors**—because they repress gene expression—in negative control systems. Whether or not a regulator protein can bind to the *RBS* depends on the presence or absence of **effector molecules** in the cell. The effectors are usually small molecules such as amino acids, sugars, and similar metabolites. The effector molecules involved in induction of gene expression are called **inducers;** those involved in repression of gene expression are called **co-repressors.**

The effector molecules (inducers and co-repressors) bind to regulator gene products (activators and repressors) and cause changes in the three-dimensional structures of these proteins. Conformational changes in protein structure resulting from the binding of small molecules are called **allosteric transitions.** Conformational changes in proteins frequently result in alterations in their activity. In the case of activators and repressors, allosteric transitions caused by the binding of effector molecules usually alter their ability to bind to regulator protein binding sites adjacent to the promoters of the structural genes that they control.

In a negative, inducible control mechanism (Figure 23.3*a*, left), the free repressor binds to the *RBS* and prevents the transcription of the structural gene(s) in the absence of inducer. When inducer is present, it is bound by the repressor, and the repressor-inducer complex cannot bind to the *RBS*. With no repressor bound to the *RBS*, RNA polymerase binds to the promoter and transcribes the structural gene(s). In a positive, inducible control mechanism (Figure 23.3*a*, right), the activator cannot bind to the *RBS* unless the inducer is present, and RNA polymerase cannot transcribe the structural gene(s) unless the activator-inducer complex is bound to the *RBS*. Thus, transcription of the structural genes is turned on only in the presence of inducer.

In a negative, repressible regulatory mechanism (Figure 23.3*b*, left), transcription of the structural gene(s) occurs in the absence of the co-repressor, but not in its presence. When the repressor-co-repressor complex is bound to the *RBS*, it prevents RNA polymerase from transcribing the structural genes. In the absence of co-repressor, free repressor cannot bind to the *RBS*; thus, RNA polymerase can bind to the promoter and transcribe the structural genes. In a positive, repressible control mechanism (Figure 23.3*b*, right), the product of the regulator gene, the activator, must be bound to the *RBS* in order for RNA polymerase to bind to the promoter and transcribe the structural gene(s). When co-repressor is present, it forms a complex with the activator protein, and this activator-co-repressor complex is unable to bind to the *RBS*; consequently, RNA polymerase cannot bind to the promoter and transcribe the structural gene(s).

In order to understand the details of these four mechanisms of regulation, focus on the key differences between them. (1) The regulator gene product, the activator, participates in turning on gene expression in a positive control mechanism, whereas the regulator gene product, the repressor, is involved in turning off gene expression in a negative control mechanism. (2) With both positive and negative control mechanisms, whether gene expression is inducible or repressible depends on whether the free regulator protein or the regulator protein-effector molecule complex binds to the regulator protein binding site (*RBS*).

Key Points: **Gene expression is controlled by both positive and negative regulatory mechanisms. In positive control mechanisms, the product of a regulator gene, an activator, is required to turn on the expression of the structural gene(s). In negative control mechanisms, the product of a regulator gene, a repressor, is necessary to turn off the expression of the structural gene(s). Activators and repressors regulate gene expression by binding to regulator protein binding sites adjacent to the promoters of structural genes. Whether or not the regulator proteins can bind to their binding sites depends on the presence or absence of small effector molecules that form complexes with the regulator proteins. The effector molecules are called inducers in inducible systems and co-repressors in repressible systems.**

OPERONS: COORDINATELY REGULATED UNITS OF GENE EXPRESSION

The **operon model**, a negative control mechanism, was developed in 1961 by François Jacob and Jacques Monod to explain the regulation of genes encoding the enzymes required for lactose utilization in *E. coli.* Jacob and Monod, 1965 Nobel Prize recipients, proposed that the transcription of one or a set of contiguous structural genes is regulated by two controlling elements (Figure 23.4*a*). One of these elements, the repressor gene, encodes a repressor, which, under the appropriate conditions, binds to the second element, the **operator.** The operator is always contiguous with the structural gene or genes whose expression it regulates.

When the repressor is bound to the operator, transcription of the structural genes cannot occur. Tran-

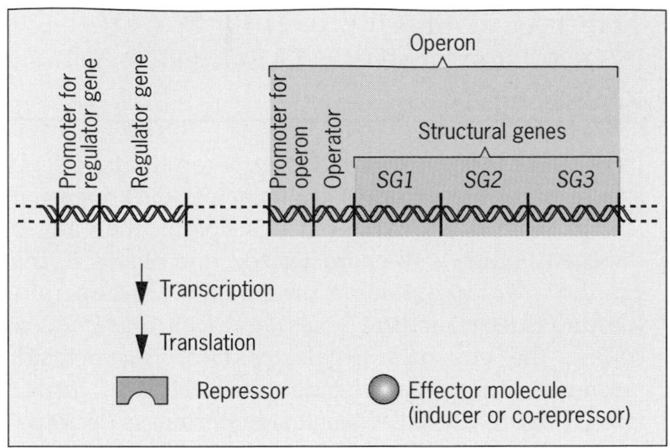

(*a*) The operon: components.

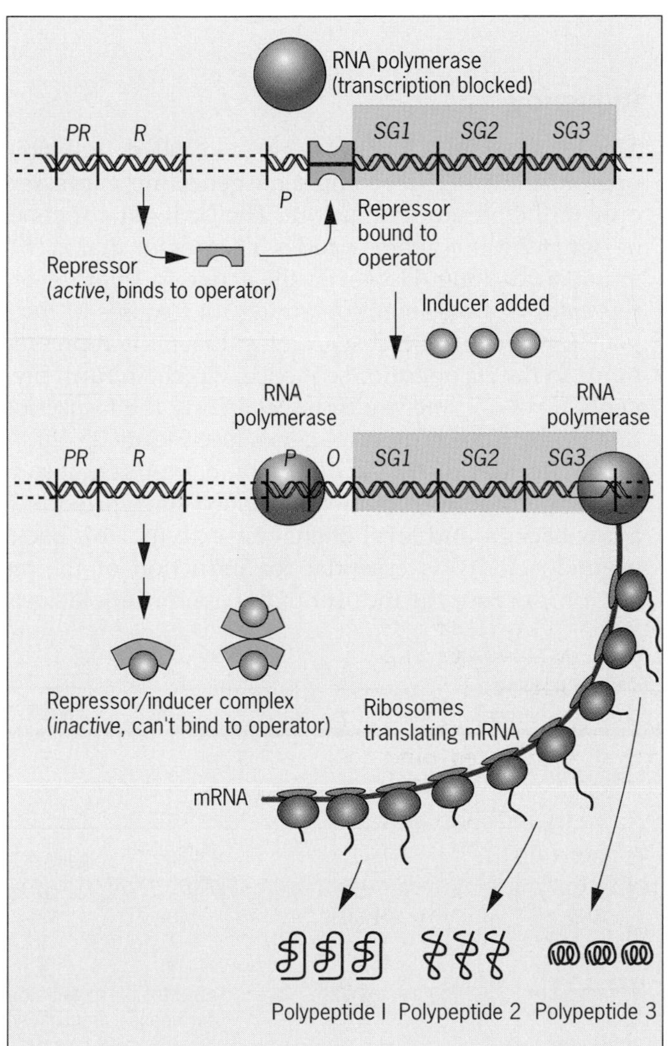

(*b*) The operon: induction.

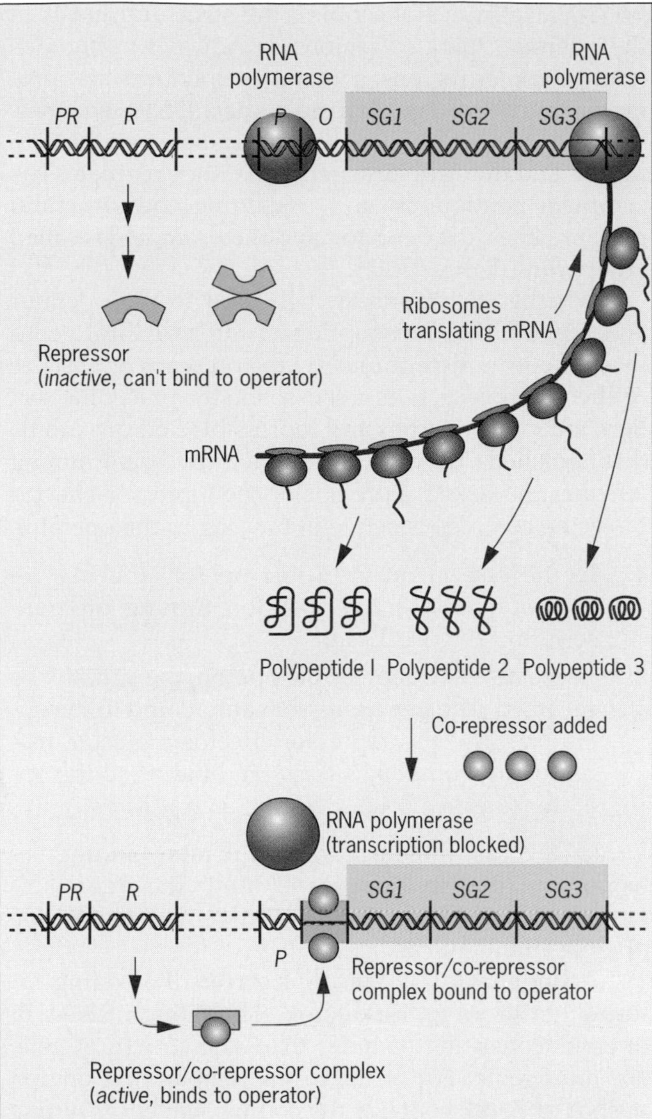

(*c*) The operon: repression.

Figure 23.4 Regulation of gene expression by the operon mechanism. (*a*) Essential components of an operon: one or more structural genes (three, *SG1, SG2,* and *SG3,* are shown) and the adjoining operator (*O*) and promoter (*P*) sequences. The transcription of the regulator gene (*R*) is initiated by RNA polymerase, which binds to its promoter (*PR*). When the repressor is bound to the operator, it sterically prevents RNA polymerase from initiating transcription of the structural genes. The difference between an inducible operon (*b*) and a repressible operon (*c*) is that free repressor binds to the operator of an inducible operon, whereas the repressor-effector molecule complex binds to the operator of a repressible operon. Thus an inducible operon is turned off in the absence of the effector (inducer) molecule, and a repressible operon is turned on in the absence of the effector (co-repressor) molecule.

scription is initiated at promoters located just upstream (5′) from the coding regions of structural genes. Promoters were unknown at the time of Jacob and Monod's proposal, but they have since been shown to be essential components of operons. When repressor is bound to the operator, it sterically prevents RNA polymerase from transcribing the structural genes in the operon. Operator regions are always contiguous with promoter regions; sometimes operators and promoters even overlap, sharing a short DNA sequence. Operator regions are usually located between the promoters and the structural genes that they regulate. The complete contiguous unit, including the structural gene or genes, the operator, and the promoter, is called an **operon** (Figure 23.4*a*).

Whether the repressor will bind to the operator and turn off the transcription of the structural genes in an operon is determined by the presence or absence of effector molecules as discussed in the preceding section. Inducible operons and repressible operons can be distinguished from one another by determining whether the naked repressor or the repressor-effector molecule complex is active in binding to the operator.

1. In the case of an inducible operon, the free repressor binds to the operator, turning off transcription (Figure 23.4*b*).
2. In the case of a repressible operon, the situation is reversed. The free repressor cannot bind to the operator. Only the repressor-effector molecule (corepressor) complex is active in binding to the operator (Figure 23.4*c*).

Except for this difference in the operator-binding behavior of the free repressor and the repressor-effector molecule complex, inducible and repressible operons are identical.

A single mRNA transcript carries the coding information of an entire operon. Thus, the mRNAs of operons consisting of more than one structural gene are multigenic. For example, the tryptophan operon mRNA of *E. coli* contains the coding sequences of five different genes. Because they are co-transcribed, all structural genes in an operon are coordinately expressed. Whereas the molar quantities of the different gene products need not be the same (because of different efficiencies of initiation of translation), the relative amounts of the different polypeptides specified by genes in an operon remain the same, regardless of the state of induction or repression.

Key Points: **In bacteria, genes with related functions frequently occur in coordinately regulated units called operons. Each operon contains a set of contiguous structural genes, a promoter (the binding site for RNA polymerase) and an operator (the binding site for a regulatory protein called a repressor). When the repressor is bound to the operator, RNA polymerase cannot transcribe the structural genes in the operon. When the operator is free of repressor, RNA polymerase can transcribe the operon.**

THE LACTOSE OPERON IN *E. COLI*: INDUCTION AND CATABOLITE REPRESSION

Jacob and Monod proposed the operon model largely as a result of their studies of the lactose (*lac*) operon in *E. coli*. More is known about this operon than any of the many other well-characterized operons in *E. coli*. The *lac* operon contains a promoter (*P*), an operator (*O*), and three structural genes, *lacZ*, *lacY*, and *lacA*, encoding the enzymes β-galactosidase, β-galactoside permease, and β-galactoside transacetylase, respectively (Figure 23.5). β-Galactoside permease "pumps" lactose into the cell, where β-galactosidase cleaves it into glucose and galactose (Figure 23.6). The biological role of the transacetylase is not yet known.

Induction

The lac operon is a negatively controlled inducible operon; the *lacZ*, *lacY*, and *lacA* genes are expressed only in the presence of lactose. The *lac* regulator gene, designated the *I* gene, encodes a repressor that is 360 amino acids long. However, the active form of the *lac* repressor is a tetramer containing four copies of the *I* gene product. In the absence of inducer, the repressor binds to the *lac* operator sequence, which, in turn, prevents RNA polymerase from catalyzing the transcription of the three structural genes (see Figure 23.4*b*). A few molecules of the *lacZ*, *lacY*, and *lacA* gene products are synthesized in the uninduced state, providing a low background level of enzyme activity. This background activity is essential for induction of the *lac* operon, because the inducer of the operon, allolactose,

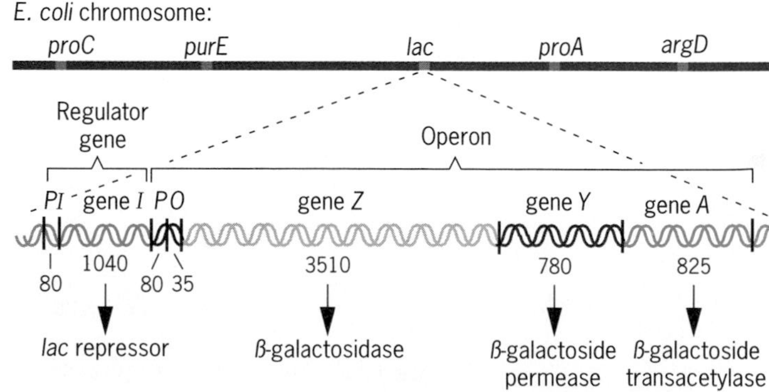

Figure 23.5 The *lac* operon of *E. coli*. The *lac* operon consists of three structural genes, *Z*, *Y*, and *A*, plus the promoter (*P*) and operator (*O*) regions adjoining the *Z* gene. The regulator gene (*I*) is contiguous with the operon in the case of *lac*. The regulator gene has its own promoter (*PI*). The numbers below the various genetic elements indicate their sizes in nucleotide pairs.

is derived from lactose in a reaction catalyzed by β-galactosidase (Figure 23.6). Once formed, allolactose is bound by the repressor, causing the release of the repressor from the operator. In this way, allolactose induces the transcription of the *lacZ*, *lacY*, and *lacA* structural genes (see Figure 23.4*b*).

The *lacI* gene, the *lac* operator, and the *lac* promoter were all initially identified genetically by the isolation of mutant strains that exhibited altered expression of the *lac* operon genes. Mutations in the *I* gene and the operator frequently result in constitutive synthesis of the *lac* gene products. These mutations are designated I^- and O^c, respectively. The I^- and O^c constitutive mutations can be distinguished not only by map position, but also by their behavior in partial diploids (Chapter 17) in which they are located in *cis-* and *trans-*configurations relative to mutations in *lac* structural genes.

Like monoploid wild-type ($I^+P^+O^+Z^+Y^+A^+$) cells, partial diploids of genotype F′ $I^+P^+O^+Z^+A^+/I^+P^+O^+Z^-Y^-A^-$ or of genotype F′ $I^+P^+O^+Z^-Y^-A^-/I^+P^+O^+Z^+Y^+A^+$ are inducible for the utilization of lactose as a carbon source. The wild-type alleles (Z^+, Y^+, and A^+) of the three structural genes are dominant to their mutant alleles (Z^-, Y^-, and A^-). This dominance is to be expected because the wild-type alleles produce functional enzymes, whereas the mutant alleles produce no enzymes or defective (inactive) enzymes. Partial diploids of genotype $I^+P^+O^+Z^+Y^+A^+/I^-P^+O^+Z^+Y^+A^+$ are also inducible for the synthesis of the three enzymes specified by the *lac* operon. Thus, I^+ is dominant to I^- as expected, since I^+ encodes a functional repressor molecule and its I^- allele specifies an inactive repressor. The dominance of I^+ over I^- also indicates that the repressor is diffusible, because the repressor produced by the $lacI^+$ allele on one chromosome can turn off the *lac* structural genes on both operons in the cell (Figure 23.7*a*).

Figure 23.6 Two physiologically important reactions catalyzed by β-galactosidase: (1) conversion of lactose to the *lac* operon inducer allolactose, and (2) cleavage of lactose to produce the monosaccharides glucose and galactose.

E. coli partial diploid

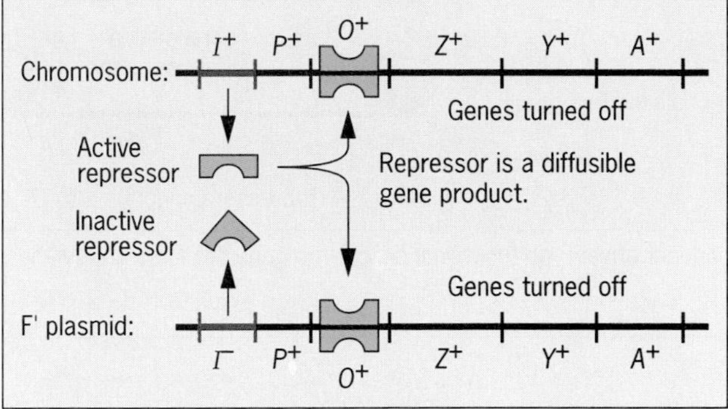

Inducible synthesis of *lac* operon gene products because the wild-type ($lacI^+$) repressor binds to the *lac* operators on both chromosomes

(*a*) Dominance of $lacI^+$ over $lacI^-$.

E. coli partial diploid

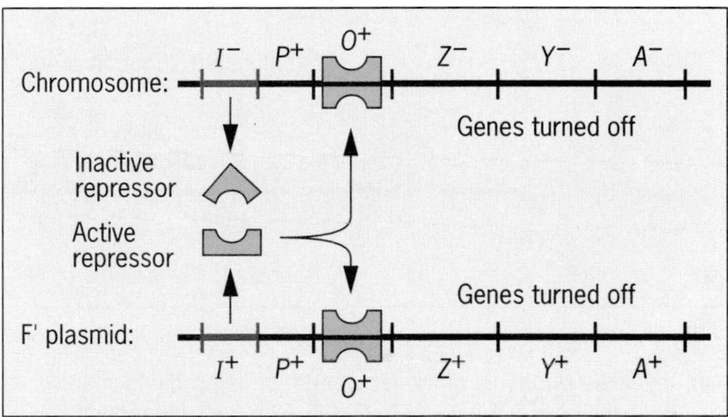

Inducible synthesis of the *lac* operon gene products

(*b*) *Cis*–dominance of $lacI^+$: I^+ located *cis* to Z^+, Y^+ and A^+.

E. coli partial diploid

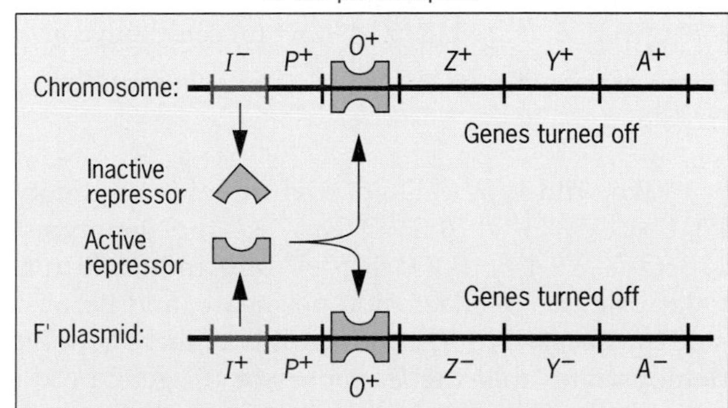

Inducible synthesis of the *lac* operon gene products

(*c*) *Trans*–dominance of $lacI^+$: I^+ located *trans* to Z^+, Y^+ and A^+.

Figure 23.7 The $lacI^+$ gene is dominant to $lacI^-$ alleles (*a*) and controls *lac* operators located either *cis* (*b*) or *trans* (*c*) to itself. These effects demonstrate that the *lacI* gene product is diffusible.

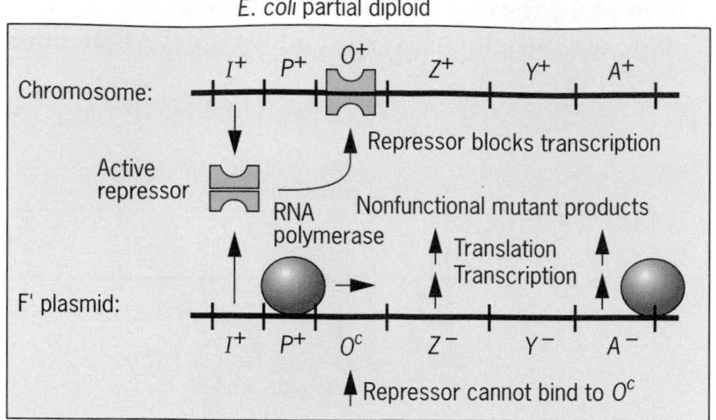

Inducer absent; no functional *lac* operon gene products are synthesized

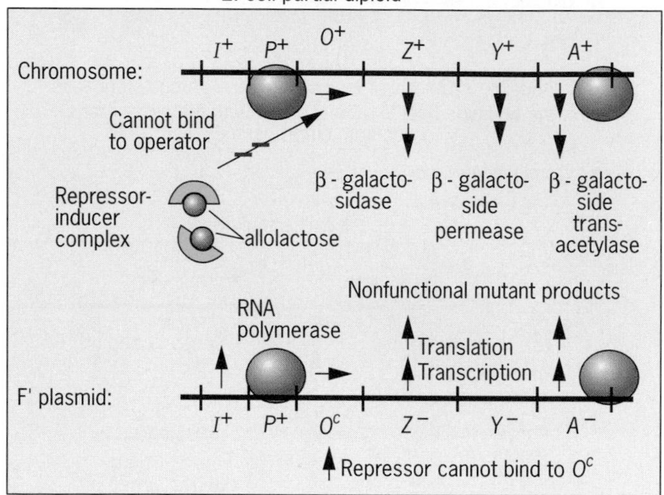

Inducer present; functional *lac* operon gene products are synthesized

(*a*) Inducible synthesis of the *lac* operon gene products in an F′ $I^+ P^+ O^c Z^- Y^- A^- / I^+ P^+ O^+ Z^+ Y^+ A^+$ bacterium.

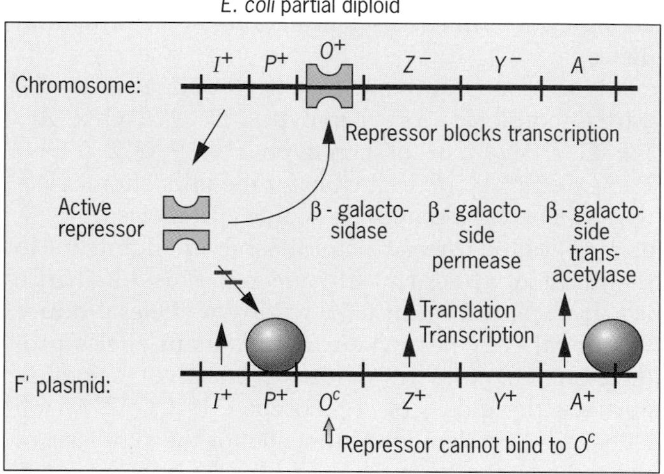

Inducer absent; functional *lac* operon gene products are synthesized

(*b*) Constitutive synthesis of the *lac* operon gene products in an F′ $I^+ P^+ O^c Z^+ Y^+ A^+ / I^+ P^+ O^+ Z^- Y^- A^-$ bacterium.

Figure 23.8 The operator acts only in the *cis* configuration. The synthesis of functional β-galactosidase, β-galactoside permease, and β-galactoside transacetylase is (*a*) inducible in an *E. coli* partial diploid of genotype F′ $I^+P^+O^cZ^-Y^-A^-/I^+P^+O^+Z^+Y^+A^+$ and (b) constitutive in an *E. coli* partial diploid of genotype F′ $I^+P^+O^cZ^+Y^+A^+/I^+P^+O^+Z^-Y^-A^-$. These results demonstrate that the operator (*O*) is *cis*-acting; that is, it regulates only those structural genes located on the same chromosome.

Like wild-type cells, partial diploids of genotype *F′* $I^+P^+O^+Z^+Y^+A^+/I^-P^+O^+Z^-Y^-A^-$ or genotype F′ $I^+P^+O^+Z^-Y^-A^-/I^-P^+O^+Z^+Y^+A^+$ are inducible for β-galactosidase, β-galactoside permease, and β-galactoside transacetylase. The inducibility of these genotypes demonstrates that the *lac* repressor (I^+ gene product) controls the expression of structural genes located either *cis* (Figure 23.7*b*) or *trans* (Figure 23.7*c*) to the $lacI^+$ allele.

The operator constitutive (O^c) mutations act only in ***cis***. That is, O^c mutations affect the expression of only those structural genes located on the same chromosome. The *cis*-acting nature of *O* mutations is logical given the function of the operator. *O* mutations should not act in *trans* if the operator is the binding site for the repressor; as such, the operator does not encode any product, diffusible or otherwise. A regulator gene should act in *trans* only if it specifies a diffusible product. Therefore, a partial diploid of genotype *F′* $I^+P^+O^cZ^-Y^-A^-/I^+P^+O^+Z^+Y^+A^+$ is inducible for the three enzymes specified by the structural genes of the *lac* operon (Figure 23.8*a*), whereas a partial diploid of genotype *F′* $I^+P^+O^cZ^+Y^+A^+/I^+P^+O^+Z^-Y^-A^-$ synthesizes these enzymes constitutively (Figure 23.8*b*).

Some of the *I* gene mutations, those designated I^{-d}, are dominant to the wild-type allele (I^+). This dominance results from the inability of heteromultimers (proteins composed of two or more different forms of a polypeptide; recall that the *lac* repressor functions as a tetramer) that contain both wild-type and mutant polypeptides to bind to the operator. Other *I* gene mutations, those designated I^s (*s* for *s*uperrepressed), cause the *lac* operon to be uninducible. In strains carrying these I^s mutations, the *lac* structural genes can usually be induced to some degree with high concentrations of inducer, but they are not induced at normal concentrations of inducer. When studied *in vitro,* the mutant I^s polypeptides form tetramers that bind to *lac* operator DNA. However, they either do not bind inducer or exhibit a low affinity for inducer. Thus, the I^s mutations alter the inducer binding site of the *lac* repressor.

Promoter mutations do not change the inducibility of the *lac* operon. Instead, they modify the levels of gene expression in the induced and uninduced state by changing the frequency of initiation of *lac* operon transcription—that is, the efficiency of RNA polymerase binding.

The *lac* promoter actually contains two separate components: (1) the RNA polymerase binding site, and (2) a binding site for another protein called *Catabolite Activator Protein* (abbreviated *CAP*) that prevents the *lac* operon from being induced in the presence of glucose. This second control circuit, which we consider next, assures the preferential utilization of glucose as an energy source when it is available.

Catabolite Repression

Jacob and Monod proposed the operon model to explain the induction of the biosynthesis of the enzymes involved in lactose utilization when this sugar is added to the medium in which *E. coli* cells are growing. However, the presence of glucose has long been known to prevent the induction of the *lac* operon, as well as other operons controlling enzymes involved in carbohydrate catabolism. This phenomenon, called **catabolite repression** (or the **glucose effect**), assures that glucose is metabolized when present, in preference to other, less efficient, energy sources.

The catabolite repression of the *lac* operon and several other operons is mediated by a regulatory protein called **CAP** (for **Catabolite** *A***ctivator** *P***rotein**) and a small effector molecule called **cyclic AMP** (adenosine-3′, 5′-monophosphate; abbreviated cAMP) (Figure 23.9). Because CAP binds cAMP when this mononucleotide is present at sufficient concentrations, it is sometimes called the cyclic AMP receptor protein.

We know that the *lac* promoter contains two separate binding sites: (1) one for RNA polymerase and (2) one for the CAP-cAMP complex (Figure 23.10). The CAP-cAMP complex must be present in its binding site in the *lac* promoter in order for the operon to be induced. The CAP-cAMP complex thus exerts positive control over the transcription of the *lac* operon. It has an effect exactly opposite to that of repressor binding to an operator. Although the precise mechanism by which CAP-cAMP stimulates RNA polymerase binding to the promoter is still uncertain, its positive control of *lac* operon transcription is firmly established by the results of both *in vivo* and *in vitro* experiments. CAP is known to function as a dimer; thus, like the *lac* repressor, it is multimeric in its functional state.

Only the CAP-cAMP complex binds to the *lac* promoter; in the absence of cAMP, CAP does not bind. Thus, cAMP acts as the effector molecule, determining the effect of CAP on *lac* operon transcription. The intracellular cAMP concentration is sensitive to the presence or absence of glucose. High concentrations of glucose cause sharp decreases in the intracellular concentration of cAMP. Glucose prevents the activation of adenylcyclase, the enzyme that catalyzes the formation of cAMP from ATP. Thus, the presence of glucose results in a decrease in the intracellular concentration of

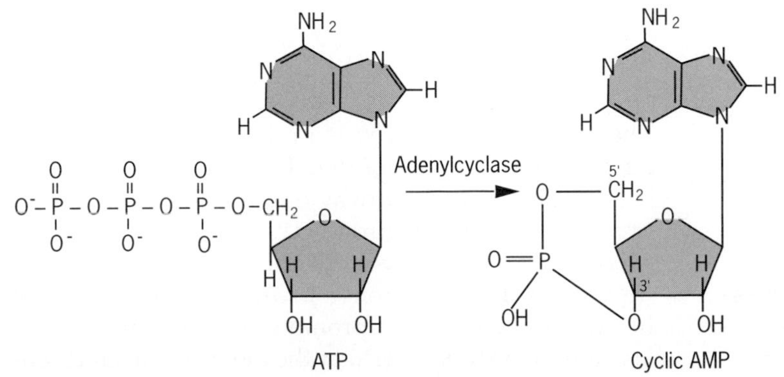

Figure 23.9 The adenylcyclase-catalyzed synthesis of cyclic AMP (cAMP) from ATP.

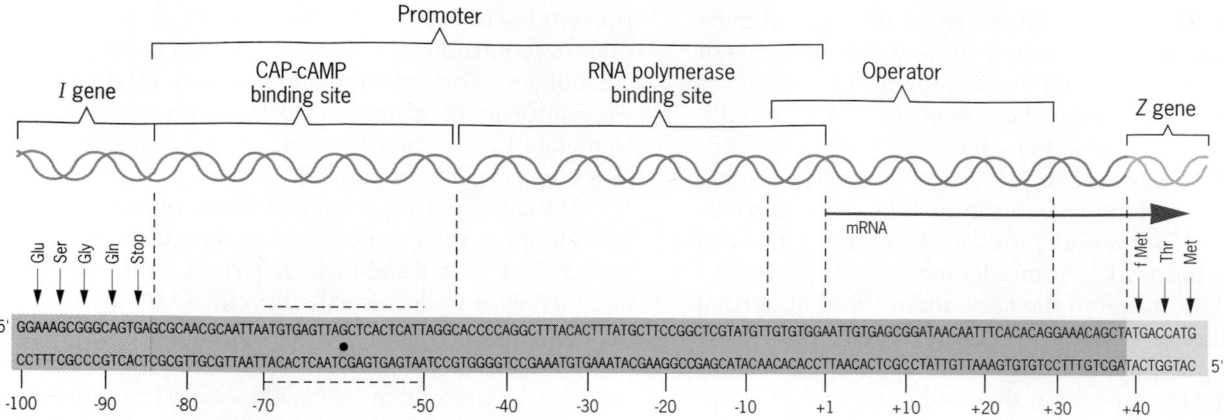

Figure 23.10 Organization of the promoter-operator region of the *lac* operon. The promoter consists of two components: (1) the site that binds the CAP-cAMP complex and (2) the RNA polymerase binding site. The adjacent segments of the *lacI* (repressor) and *lacZ* (β-galactosidase) structural genes are also shown. The horizontal line labeled mRNA shows the position at which transcription of the operon begins (the 5′-end of the *lac* mRNA). The numbers at the bottom give distances in nucleotide pairs from the site of transcript initiation (position +1). The dot between the two nucleotide strands indicates the center of symmetry of an imperfect palindrome.

cAMP. In the presence of a low concentration of cAMP, CAP cannot bind to the *lac* operon promoter. In turn, RNA polymerase cannot bind efficiently to the *lac* promoter in the absence of bound CAP-cAMP. The overall result of the positive control of transcription of the *lac* operon by the CAP-cAMP complex is that, in the presence of glucose, *lac* operon transcription never exceeds 2 percent of the induced rate observed in the absence of glucose. By similar mechanisms, CAP and cAMP keep the arabinose (*ara*) and galactose (*gal*) operons of *E. coli* from being induced in the presence of glucose.

The complete nucleotide-pair sequence of the *lac* operon regulatory region is shown in Figure 23.10. Comparative nucleotide-sequence studies of mutant and wild-type promoters and operators, in addition to *in vitro* CAP-cAMP, RNA polymerase, and repressor binding studies, are providing detailed information about the nature of these important **sequence-specific protein-nucleic acid interactions.** The results of these studies provide an excellent example of the advantages of integrating genetic and biochemical approaches in attempting to understand biological phenomena.

Key Points: **The *E. coli lac* operon is a negative inducible and catabolite repressible system; the three structural genes in the *lac* operon are transcribed only in the presence of lactose. In the absence of lactose, the *lac* repressor binds to the *lac* operator and prevents RNA polymerase from initiating transcription of the operon. Catabolite repression keeps operons such as *lac* that encode enzymes involved in carbohydrate catabolism from being induced in the presence of glucose, the preferred energy source.**

THE TRYPTOPHAN OPERON IN *E. COLI*: REPRESSION AND ATTENUATION

The *trp* operon of *E. coli* controls the synthesis of the enzymes that catalyze the biosynthesis of the amino acid tryptophan. The functions of the five structural genes and the adjacent regulatory sequences of the *trp* operon have been analyzed in detail by Charles Yanofsky and colleagues. The five structural genes encode enzymes that convert chorismic acid to tryptophan. The expression of the *trp* operon is regulated at two levels: repression, which controls the initiation of transcription, and attenuation, which governs the frequency of premature transcript termination. We will discuss these regulatory mechanisms in the following two sections.

Repression

The *trp* operon of *E. coli* is probably the best known negative repressible operon. The organization of the *trp* operon and the pathway of biosynthesis of tryptophan are shown in Figure 23.11. The *trpR* gene, which encodes the *trp* repressor, is not closely linked to the *trp* operon. The operator (*O*) region of the *trp* operon lies within the primary promoter (P_1) region. There is also a weak promoter (P_2) at the operator-distal end of

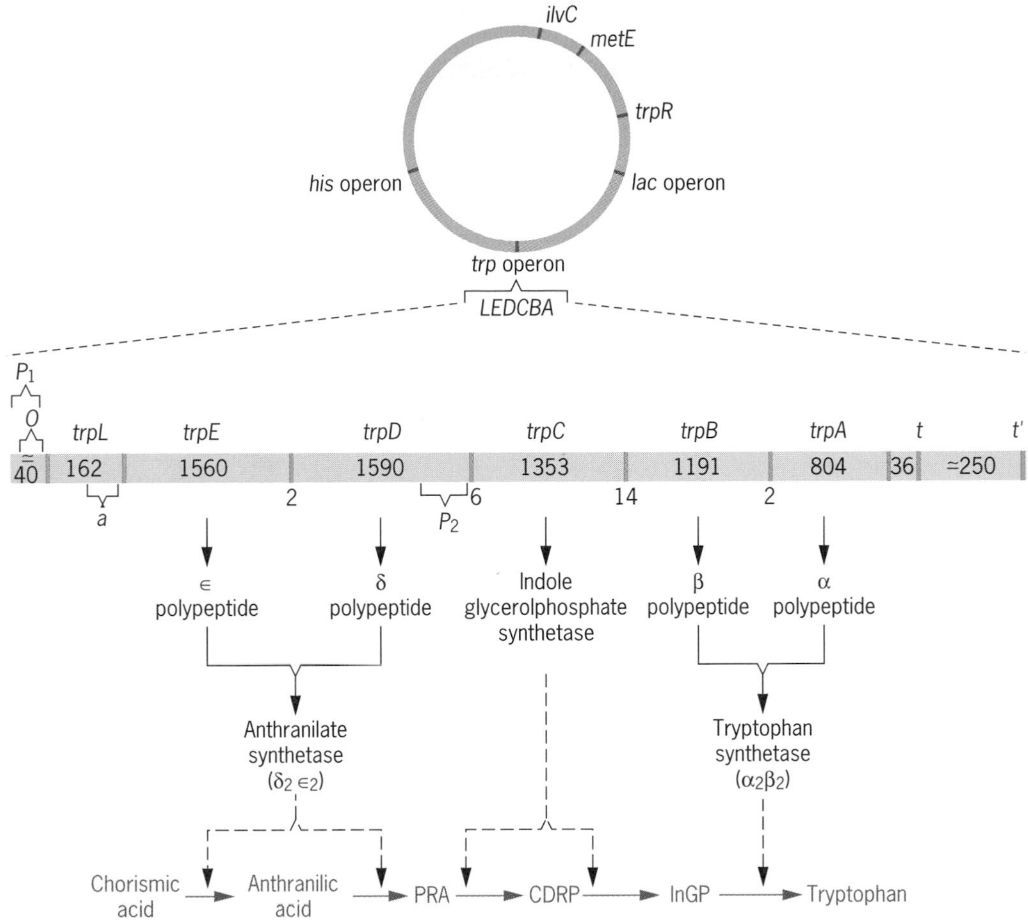

Figure 23.11 Organization of the *trp* (tryptophan) operon in *E. coli*. The *trp* operon contains five structural genes that encode enzymes involved in the biosynthesis of tryptophan, as shown at the bottom. The length of each gene or region is given in nucleotide pairs; the intergenic distances are shown below the gene sequence. Key: PRA, phosphoribosyl anthranilate; CDRP, carboxyphenylamino-deoxyribulose phosphate; InGP, indole-glycerol phosphate.

the *trpD* gene. The P_2 promoter results in a somewhat increased basal level of transcription of the *trpC*, *trpB*, and *trpA* genes. Two transcription termination sequences *(t* and *t')* are located downstream from *trpA*. The *trpL* region specifies a 162-nucleotide-long mRNA leader sequence.

The regulation of transcription of the *trp* operon is diagrammed in Figure 23.4*c*. In the absence of tryptophan (the co-repressor), RNA polymerase binds to the promoter region and transcribes the structural genes of the operon. In the presence of tryptophan, the co-repressor/repressor complex binds to the operator region and prevents RNA polymerase from initiating transcription of the genes in the operon.

The rate of transcription of the *trp* operon in the derepressed state (absence of tryptophan) is 70 times the rate that occurs in the repressed state (presence of tryptophan). In *trpR* mutants that lack functional repressor, the rate of synthesis of the tryptophan biosynthetic enzymes (the products of the structural genes of the *trp* operon) is still reduced about tenfold by the addition of tryptophan to the medium. This additional reduction in *trp* operon expression is caused by attenuation, which is discussed next.

Attenuation

Deletions that remove part of the *trpL* region (Figure 23.11) result in increased rates of expression of the *trp* operon. However, these deletions have no effect on the repressibility of the *trp* operon; that is, repression and derepression occur just as in $trpL^+$ strains. These results indicate that the synthesis of the tryptophan biosynthetic enzymes is regulated at a second level by a mechanism that is independent of the state of repression/derepression of the *trp* operon and requires nucleotide sequences present in the *trpL* region of the *trp* operon.

This second level of regulation of the *trp* operon is called **attenuation,** and the sequence within *trpL* that

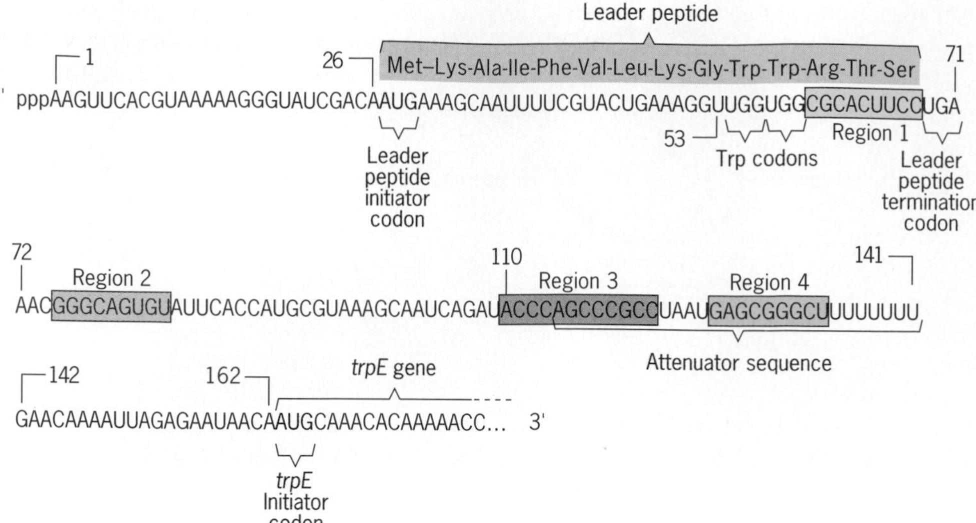

(*a*) Regulatory components of the *trpL* region.

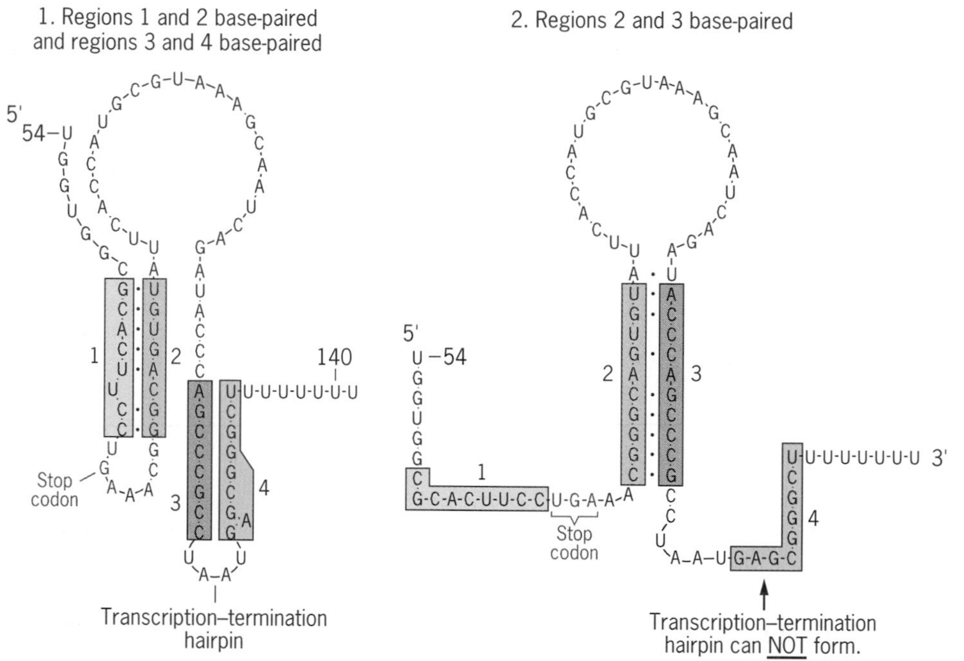

(*b*) Alternate secondary structures formed by the *trpL* transcript.

Figure 23.12 Sequences in the leader region of the *trp* mRNA responsible for attenuation. (*a*) The *trpL* sequence, highlighting the sequence encoding the leader peptide, the two tandem tryptophan codons responsible for the control of attenuation by tryptophan, and the four regions (shaded) that form the stem-and-loop or hairpin structures shown in (*b*). (*b*) Alternate secondary structures formed by the *trpL* mRNA—either (1) region 1 will pair with region 2 and region 3 with region 4, forming a transcription-termination hairpin, or (2) region 2 will base-pair with region 3, preventing region 3 from pairing with region 4. The presence or absence of tryptophan determines which of these structures will form during the transcription of the *trp* operon.

controls this phenomenon is called the **attenuator** (Figure 23.12*a*). Attenuation occurs by control of the termination of transcription at a site near the end of the mRNA leader sequence. This "premature" termination of *trp* operon transcription occurs only in the presence of tryptophan-charged $tRNA^{Trp}$. When this premature termination or attenuation occurs, a truncated (140 nucleotides) *trp* transcript is produced.

The attenuator region has a nucleotide-pair sequence essentially identical to the **transcription-**

termination signals found at the ends of most bacterial operons (including the *trp* operon). These termination signals contain a G:C-rich palindrome followed by several A:T base pairs. Transcription of these termination signals yields a nascent RNA with the potential to form a hydrogen-bonded hairpin structure followed by several uracils. When a nascent transcript forms this hairpin structure, it is believed to cause a conformational change in the associated RNA polymerase, resulting in termination of transcription within the following, more weakly hydrogen-bonded $[(A:U)_n]$ region of DNA-RNA base-pairing.

The nucleotide sequence of the attenuator therefore explains its ability to terminate *trp* operon transcription prematurely. But how can this be regulated by the presence or absence of tryptophan?

First, recall that transcription and translation are coupled in prokaryotes; that is, ribosomes begin translating mRNAs while they are still being produced by transcription. Thus, events that occur during translation may also affect transcription.

Second, note that the 162-nucleotide-long leader sequence of the *trp* operon mRNA contains sequences that can base-pair to form alternate stem-and-loop or hairpin structures (Figure 23.12*b*). The four leader regions that can base-pair to form these structures are: (1) nucleotides 60–68, (2) nucleotides 75–83, (3) nucleotides 110–121, and (4) nucleotides 126–134. The actual lengths of these regions involved in base-pairing vary somewhat depending on which regions pair. The nucleotide sequences of these four regions are such that region 1 can base-pair with region 2, region 2 can pair with region 3, and region 3 can base-pair with region 4. Region 2 can base-pair with either region 1 or region 3, but, obviously, it can pair with only one of these regions at any given time. Thus, there are two possible secondary structures for the *trp* leader sequence: (1) region 1 paired with region 2 and region 3 paired with region 4 or (2) region 2 paired with region 3, leaving regions 1 and 4 unpaired. The pairing of regions 3 and 4 produces the previously mentioned transcription-termination hairpin. If region 3 is base-paired with region 2, it cannot pair with region 4, and the transcription-termination hairpin cannot form. As you have probably guessed by now, the presence or absence of tryptophan determines which of these alternative structures will form.

Third, note that the leader sequence contains an AUG translation-initiation codon, followed by 13 codons for amino acids, followed, in turn, by a UGA translation-termination codon (Figure 23.12*a*). In addition, the *trp* leader sequence contains an efficient ribosome-binding site located in the appropriate position for the initiation of translation at the leader AUG initiation codon. All the available evidence indicates that a 14 amino acid "leader peptide" is synthesized as diagrammed in Figure 23.12*a*. The synthesis of this putative leader peptide *in vivo* has not been verified directly. However, because short peptides of this type are very unstable in cells, the inability to detect it is not surprising.

The formation of the normal *trp* operon transcription-termination hairpin is shown in Figure 23.13*a*, and the proposed mechanism of attenuation of *trp* operon transcription is diagrammed in Figure 23.13*b* and *c*. The leader peptide contains two contiguous tryptophan residues. The two Trp codons are positioned such that in the absence of tryptophan (and thus the absence of $Trp\text{-}tRNA^{Trp}$), the ribosome will stall before it encounters the base-paired structure formed by leader regions 2 and 3 (Figure 23.13*b*). Because the pairing of regions 2 and 3 precludes the formation of the transcription-termination hairpin by the base-pairing of regions 3 and 4, transcription will continue past the attenuator into the *trpE* gene in the absence of tryptophan.

In the presence of tryptophan, the ribosome can translate past the Trp codons to the leader-peptide termination codon. In the process, it will disrupt the base-pairing between leader regions 2 and 3. This disruption leaves region 3 free to pair with region 4, forming the transcription-termination hairpin (Figure 23.13*c*). Thus, in the presence of tryptophan, transcription frequently (about 90 percent of the time) terminates at the attenuator, reducing the amount of mRNA for the *trp* structural genes.

The transcription of the *trp* operon can be regulated over a range of almost 700-fold by the combined effects of repression (up to 70-fold) and attenuation (up to 10-fold).

Regulation of transcription by attenuation is not unique to the *trp* operon. Five other operons (*thr, ilv, leu, phe,* and *his*) are known to be regulated by attenuation. Of these six operons, *trp* and possibly *phe* are also regulated by repression. The *his* operon, which was thought to be repressible for many years, is now believed to be regulated entirely by attenuation. Although minor details of the attenuation mechanism vary from operon to operon, the main features of attenuation are the same for all six operons.

Key Points: **The *E. coli trp* operon is a negative repressible system; transcription of the five structural genes in the *trp* operon is repressed in the presence of tryptophan. Operons such as *trp* that encode enzymes involved in amino acid biosynthetic pathways often are controlled by a second regulatory mechanism called attenuation. The level of expression of these operons is reduced by the premature termination of transcription at an attenuator site located in the mRNA leader sequence when the amino acid produced by the pathway is present.**

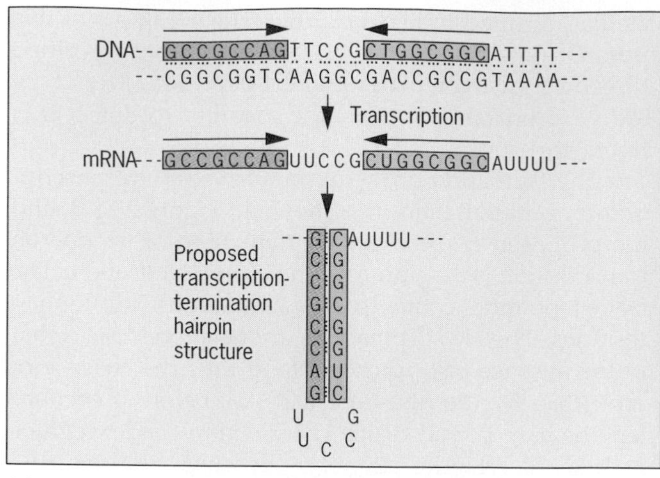

(a) Structure of *trp* operon transcription-termination sequence *t* and formation of the transcription-termination hairpin.

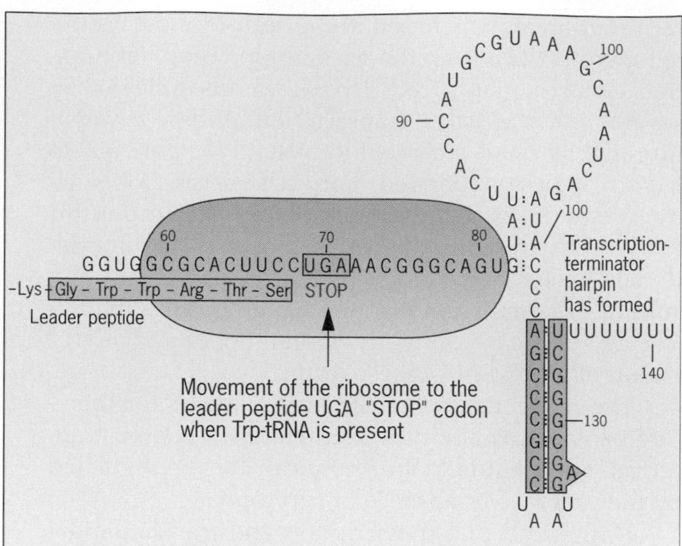

(c) In the presence of tryptophan, translation proceeds past the Trp codons to the termination codon and disrupts the base-pairing between leader regions 2 and 3. This process leaves region 3 free to pair with region 4 to form the transcription-termination hairpin, which stops transcription at the attenuator sequence.

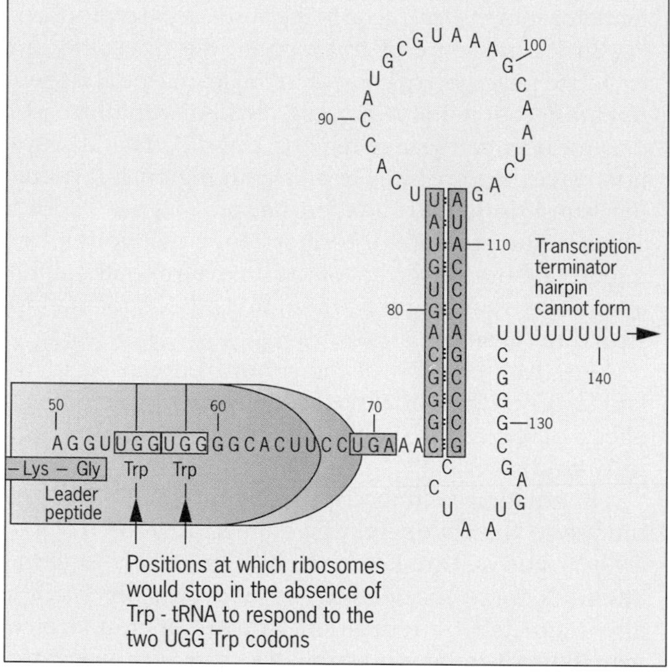

(b) In the absence of tryptophan, translation of the leader sequence stalls at one of the Trp codons. This stalling allows leader regions 2 and 3 to pair, which prevents region 3 from pairing with region 4 to form the transcription-termination hairpin. Thus transcription proceeds through the entire *trp* operon.

Figure 23.13 Control of the *trp* operon by attenuation. (*a*) The transcription-termination signal in *E. coli* contains a region of dyad symmetry (arrows) that results in mRNA sequences that can form hairpin structures. (*b*) In the absence of tryptophan, transcription proceeds past the attenuator sequence through the entire *trp* operon. (c) In the presence of tryptophan, transcription frequently terminates at the attenuator sequence.

BACTERIOPHAGE LAMBDA: LYSOGENY OR LYSIS

When a temperate bacteriophage such as lambda (λ) infects a bacterium, it can follow either of two pathways of development (Figure 23.14). A temperate phage can either (1) enter the lytic cycle, during which it reproduces and lyses the host cell, just like a virulent phage (Chapter 16), or (2) enter the lysogenic pathway (Chapter 17), during which its chromosome is inserted into the chromosome of the host and thereafter is replicated like any other segment of that chromosome. When integrated in the chromosome of the host cell, the temperate phage chromosome is called a prophage. In a lysogenic bacterium, the genes of the prophage that encode products involved in the lytic pathway must not be expressed. The prophage genes specifying enzymes involved in the replication of

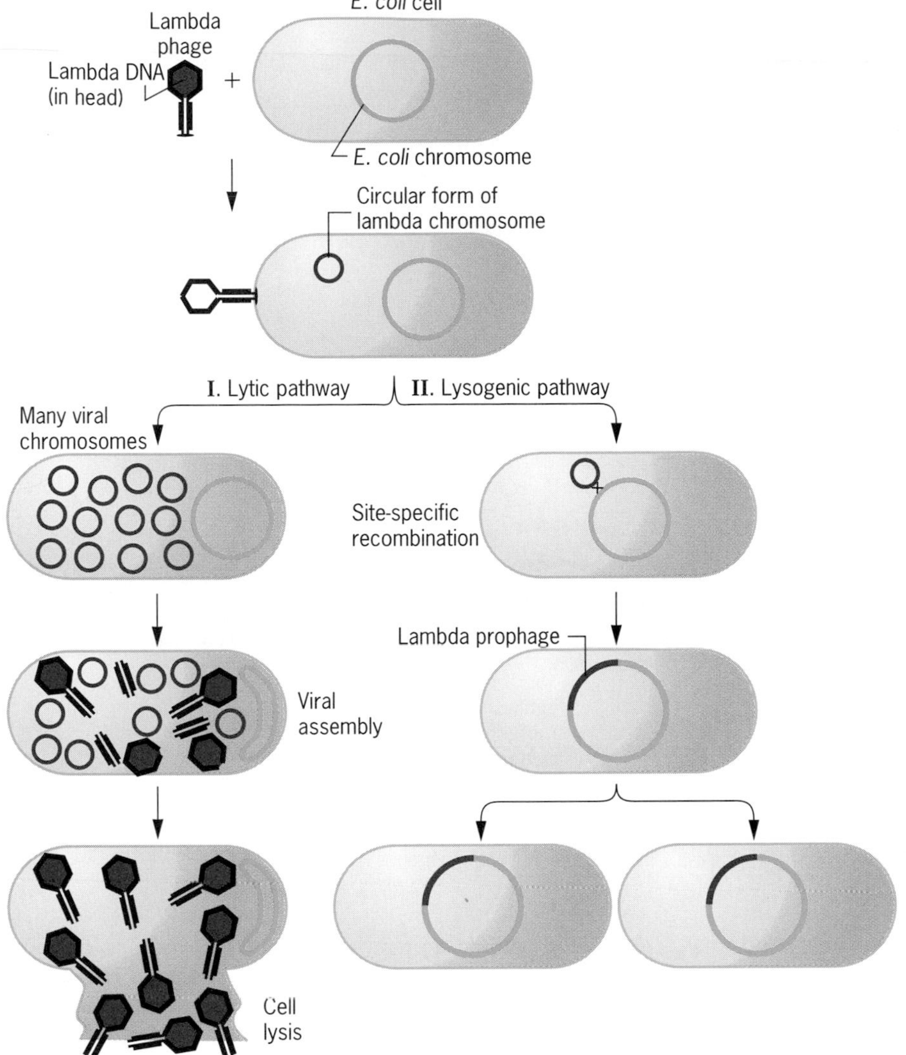

Figure 23.14 The two intracellular states of bacteriophage lambda: lytic growth and lysogeny.

phage DNA, structural proteins required for phage morphogenesis, and the lysozyme that catalyzes cell lysis must be kept turned off to maintain a stable lysogenic state. Indeed, André Lwoff shared the 1965 Nobel Prize in Physiology or Medicine with Jacob and Monod for his discovery and characterization of the prophage state of temperate bacteriophages.

The lytic pathway genes of the prophage are kept turned off in lysogenic cells by a simple repressor-operator-promoter circuit, much like the negative regulatory circuits of bacterial operons. However, the really interesting aspect of phage lambda is the mechanism that determines whether λ enters the lytic pathway or the lysogenic pathway. This determination involves an elegant genetic switch that sends the infecting phage chromosome down one of the two pathways: (1) the lytic regulatory cascade or (2) the autogenously maintained repression circuit of lysogeny. The interplay between these two regulatory networks is one of the most fascinating stories in the field of molecular genetics. In the end, if one protein—the λ repressor—occupies key regulatory sites, lysogeny occurs. If another λ protein—called Cro—occupies these sites, lytic development occurs.

Repression of Lambda Lytic Pathway Genes During Lysogeny

The phage lambda genes that encode functions involved in lytic development are maintained in the off state by a negative regulatory mechanism analogous to those controlling the *lac* and *trp* operons of *E. coli*. The C_I gene of phage λ encodes a repressor, a well-

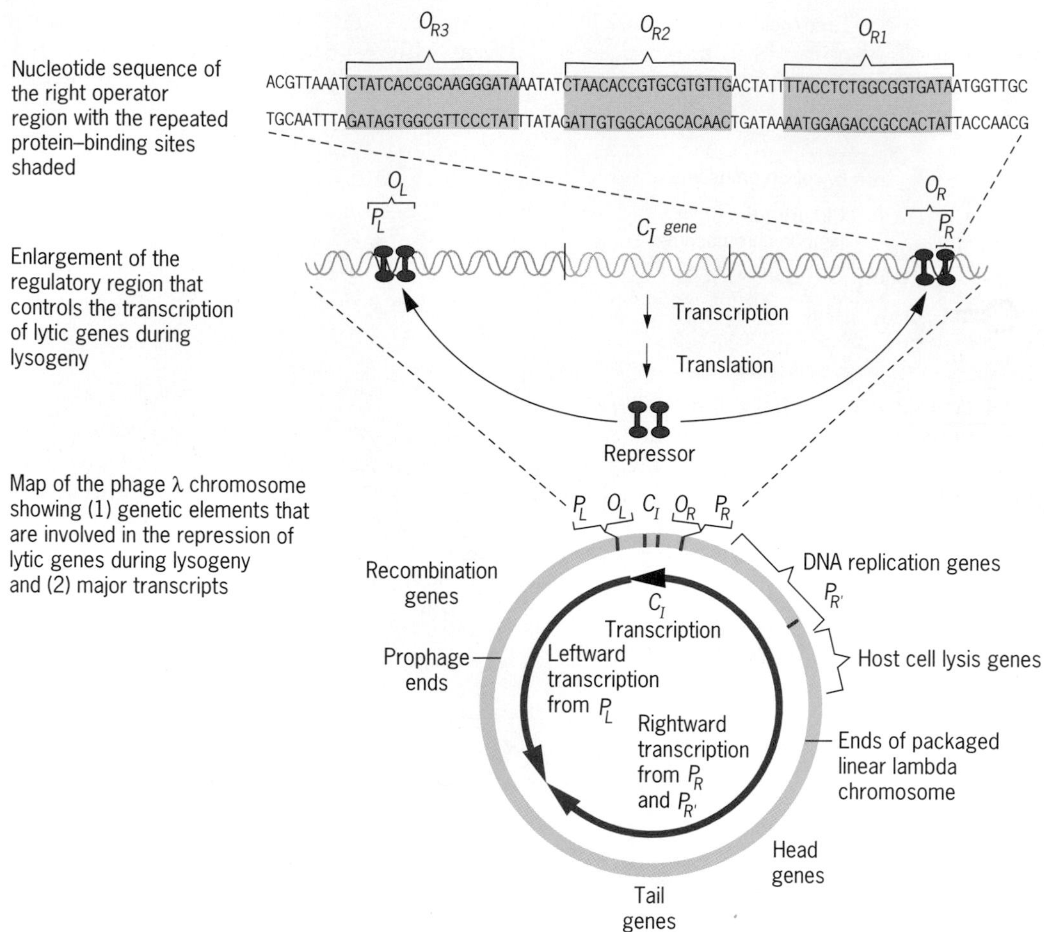

Figure 23.15 Repression of the lambda lytic genes in a lysogenic *E. coli* cell. Transcription of the lytic genes is repressed by the binding of the lambda repressor to two operator sequences (O_L and O_R), which regulate leftward and rightward transcription of the lambda chromosome (bottom). The lambda repressor is encoded by the C_I gene (center) and represses the synthesis of the major transcripts by binding to triplicate protein binding sites in O_L and O_R. The arrows show the relative sizes and directions of synthesis of the major lambda transcripts.

characterized protein with a weight of 27,000 daltons. The C_I repressor, as a dimer, binds to two operator regions that control transcription of the λ genes involved in lytic growth (Figure 23.15). These two operator regions, designated O_L (for transcription in the *l*eftward direction) and O_R (for transcription in the *r*ightward direction), overlap with promoter sequences at which RNA polymerase binds to initiate transcription of the genes controlling lytic development. With repressor bound to the two operators, RNA polymerase cannot bind to the two promoters and cannot initiate transcription. In this way, the lambda lytic genes continue to be repressed, allowing the dormant prophage to be transmitted from parental host cells to progeny cells generation after generation.

When the phage lambda operator and promoter regions were sequenced, each operator was found to contain three repressor binding sites, each 17 nucleotide pairs long, with similar, but not identical, nucleotide sequences. Each repressor binding site has partial twofold symmetry around the central base pair. This symmetry plays an important role in the interaction of repressor dimers with these operator binding sites (Figure 23.16).

The lambda repressor has been studied in great detail. The repressor monomer contains three regions: (1) a **DNA-binding domain** at the amino terminus, (2) a **dimerization domain** at the carboxyl terminus, and (3) a central **connector region.** The three-dimensional structure is known only for the DNA-binding domain. This domain, 92 amino acids in length, contains five regions that form α-helical structures; the α-helices are numbered 1 through 5, starting from the NH_3 terminus. Two of the α-helical regions, numbers 2 and 3, are largely responsible for the DNA-binding specificity of the repressor. When a repressor dimer makes contact with its DNA binding site (Figure 23.16), the two region 3 helices lie almost entirely within adjacent major grooves on one face of the λ operator DNA. Each monomer contacts one side (half-site) of the palindromic nucleotide-pair sequence in each repressor binding site. The two region 2 helices lie across the major grooves occupied by the region 3 helices.

Because the lambda repressor has the highest affinity for sites O_{L1} and O_{R1}, the first repressor dimers usually bind at these two sites. The λ repressor exhibits cooperative binding to (1) sites O_{L1} and O_{L2}, and (2) sites O_{R1} and O_{R2}. Thus, the presence of a dimer at site O_{L1} or O_{R1} increases the affinity of repressor for sites O_{L2} or O_{R2}, respectively. This cooperativity does not extend to O_{L3} or O_{R3}; at normal intracellular concentrations of repressor, only sites O_{L1}, O_{L2}, O_{R1}, and O_{R2} are occupied by repressor. When repressor dimers are bound to O_{R1}-O_{R2} and O_{L1}-O_{L2}, RNA polymerase cannot bind to P_R and P_L and, therefore, cannot initiate transcription. Thus, the λ genes encoding functions involved in lytic development are maintained in a repressed state.

Lambda lysogeny is quite stable. The lambda lytic pathway genes are tightly repressed in a lysogenic cell; thus, lytic-function gene products are not produced by transcription of prophage genes. In addition, spontaneous switches from lysogeny to lytic development are rare under normal growth conditions. Populations of lysogenic cells can be induced to enter the lytic pathway only by drastic treatments, such as by irradiation with ultraviolet light (UV). UV irradiation of lambda lysogens activates a host cell protease that cleaves the connector region of the λ repressor and renders it nonfunctional.

A major reason for the stability of the λ lysogenic state is that the synthesis of the repressor is autogenously regulated; that is, the presence of λ repressor stimulates the synthesis of more repressor. By what mechanism is the synthesis of the λ repressor self-regulated?

The λ C_I gene, which encodes the repressor, is transcribed from a promoter designated P_{RM} (for *p*romoter for *r*epressor *m*aintenance), located between the C_I gene and O_{R3} (Figure 23.17). Promoter P_{RM} overlaps operator O_{R3}, much as P_R overlaps O_{R1} (Figure 23.15). However, transcription initiated at P_{RM} occurs in the opposite direction (to the left as drawn in Figure 23.15) from transcription initiated at P_R. The key factor in the maintenance of the lysogenic state is that repressor bound at O_{R2} acts as a positive regulator of transcription from P_{RM}. In the absence of repressor at O_{R2} (lytic growth), RNA polymerase cannot bind at P_{RM} to initi-

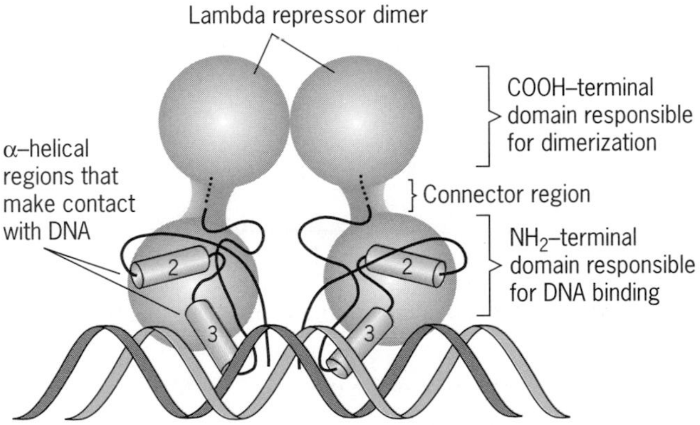

Figure 23.16 The lambda repressor dimer contacts its binding sites in the lambda operator (O_L and O_R) regions primarily through two α-helical regions (green cylinders) in the amino-terminal DNA-binding domain. Although the exact conformation of the carboxyl-terminal domain is unknown, the three-dimensional structure (black line and green cylinders) of the amino-terminal domain has been determined. This DNA-binding domain contains five α-helical regions, two of which (labeled 2 and 3 on the diagram) are primarily responsible for the specific binding of the repressor to the lambda operator regions. (*Based on Benjamin Lewin,* Genes V, *Cell Press, 1994. Used by permission.*)

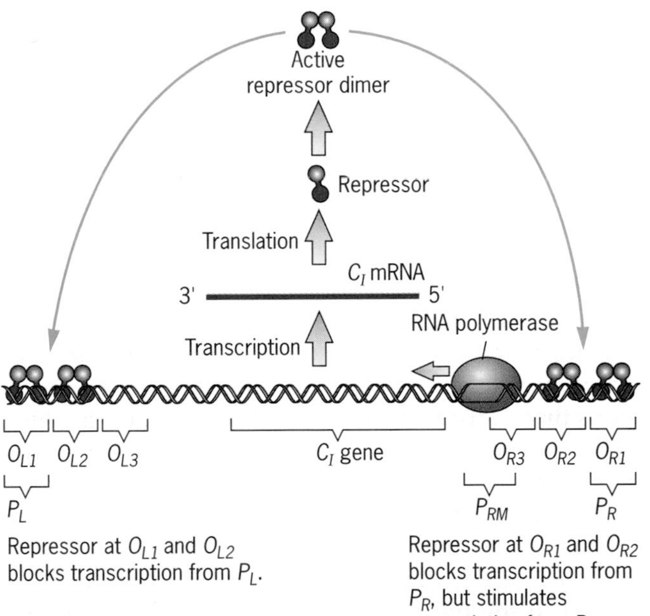

Figure 23.17 Autogenous regulation of phage lambda repressor synthesis. The presence of λ repressor stimulates the synthesis of more repressor by acting as a positive regulator of transcription of the C_I gene from promoter P_{RM}. At the same time, the repressor functions as a negative regulator of genes involved in lytic development. Repressor dimers bound at O_{L1}–O_{L2} and O_{R1}–O_{R2} prevent transcription of lytic function genes from promoters P_L and P_R, respectively.

ate transcription of the C_I gene. In the presence of repressor, the O_{R1} and O_{R2} sites are occupied by repressor dimers (Figure 23.17). Repressor bound at O_{R2} stimulates the binding of RNA polymerase at P_{RM}, leading to the synthesis of more repressor (positive self-regulated synthesis).

Although the repressor has lower affinity for O_{R3} than for O_{R2} or O_{R1}, it will bind at O_{R3} and block transcription from P_{RM} if present at high concentration. This block will lead to a lower concentration of repressor, which, in turn, will yield an open O_{R3} and renewed transcription from P_{RM}. The control of repressor synthesis by this autogenous regulatory circuit enhances the stability of the lambda lysogenic state. It also helps explain why repressor is not synthesized during the lytic growth of this virus. If no repressor is present in the cell, the O_{R2} site will be free, and repressor will not be synthesized from transcripts initiated at P_{RM}. As a result, once lambda has started down the lytic pathway, P_{RM} will not be utilized to produce repressor, and the absence of repressor will allow lytic development to proceed unabated.

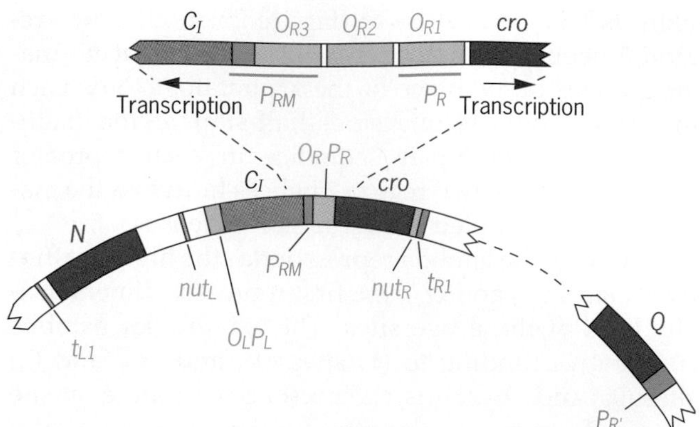

Figure 23.18 Genes and recognition sites involved in the lambda lytic regulatory cascade. Genes *cro*, *N*, and *Q* encode regulatory proteins required for lytic development. Operators O_L and O_R control leftward and rightward transcription from promoters P_L and P_R, respectively. O_{R1}, O_{R2}, and O_{R3} are the three repressor binding sites within O_R. Sequences nut_L and nut_R are the sites at which the *N* gene product must act to prevent transcription termination at sites t_L and t_{R1}, respectively.

The Lambda Lytic Regulatory Cascade

The lambda lytic regulatory cascade is straightforward, once a commitment has been made to enter the lytic pathway. The λ genes can be placed in three groups based on when they are expressed during the lytic cycle: (1) immediate early genes, (2) delayed early genes, and (3) late genes. Lambda has only two immediate early genes, *cro* and *N*, both encoding regulatory proteins required for lytic development. The λ chromosome contains a dozen delayed early genes, which specify products required for DNA replication, recombination, and additional regulation. The remaining 23 genes of phage lambda encode late-function proteins involved in head and tail morphogenesis and in lysis of the host cell. One of the immediate early gene products, **N protein,** is required to express the delayed early genes. In turn, one of the delayed early gene products, **Q protein,** must be present to express the λ late genes. Let's look at the functions of the key regulatory genes, *cro*, *N*, and *Q*, involved in the lytic growth of phage λ (Figure 23.18).

In the absence of the lambda repressor, the C_I gene product described earlier, RNA polymerase initiates transcription at promoters P_L and P_R. The first gene to be transcribed starting at P_R is *cro* (for *c*ontrol of *r*epressor and *o*ther gene products). **Cro protein** is a repressor with a DNA-binding region very similar to that of the λ repressor. Cro binds to the same DNA sites in O_L and O_R as the λ repressor; but Cro has a higher affinity for O_{R3} than for O_{R1} and O_{R2}. Thus, when Cro is first synthesized, it binds to O_{R3} and keeps the synthesis of repressor turned off by blocking the initiation of transcription at P_{RM} (Figure 23.19*a*). Later, when more Cro protein has accumulated in the cell, it also binds to O_{R1}, O_{R2}, O_{L1}, and O_{L2}, which suppresses transcription from P_R and P_L, decreasing the rate of synthesis of early gene products (Figure 23.19*b*).

Synthesis of the delayed early gene products requires the product of the second immediate early gene,

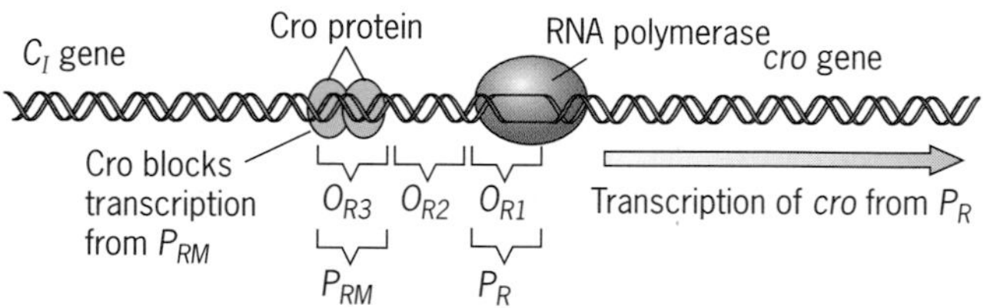

(*a*) Early: low concentration of Cro protein.

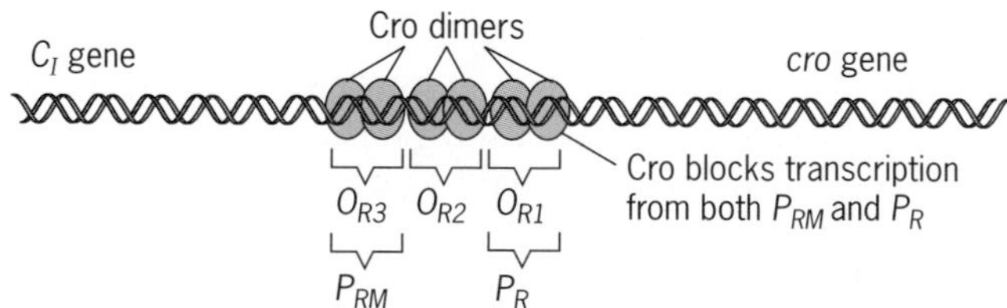

(*b*) Late: high concentration of Cro protein.

Figure 23.19 Functions of the *cro* gene product at early (*a*) and late (*b*) stages of lambda lytic growth. Cro protein has higher affinity for O_{R3} than O_{R1} or O_{R2}, resulting in its presence only at O_{R3} at low concentrations and at all three O_R sites (and O_L sites) at high concentrations.

N, which is the first gene transcribed from promoter P_L. The gene *N* product functions to prevent the termination of transcription at two sites, one (t_{R1}) located adjacent to *cro* and the other (t_{L1}) just downstream from gene *N* itself (Figure 23.18). Thus, N protein acts as a **transcriptional antiterminator;** it permits RNA polymerase to continue transcription past termination signals t_{L1} and t_{R1} and on through the adjacent delayed early genes. N protein carries out its antitermination function only in the presence of DNA sequences called *nut* (for *N utilization*) sites located upstream from the termination sites (Figure 23.20). N protein binds to the *nut* sites and, with the aid of *E. coli* proteins called Nus (for *N utilization substance*) factors and ribosomal protein S10, modifies the specificity of RNA polymerase as it passes the *nut* site so that termination does not occur when the RNA polymerase complex reaches termination signals t_{L1} and t_{R1}.

The antiterminator role of N protein is illustrated in Figure 23.20. In the absence of functional N protein, transcription terminates at t_{R1} and t_{L1}. In the presence of active N protein, it binds to the *nut* sites, and, along with proteins NusA, NusB, and S10, modifies RNA polymerase so that termination of transcription does not occur at t_{R1} and t_{L1}. As a result, transcription continues into the adjacent delayed early genes.

One of the delayed early genes, *Q*, encodes another transcription antiterminator that is required for late gene expression. Q protein functions much like N protein but recognizes a different DNA sequence, called *qut* (for *Q utilization*). Q protein prevents the arrest of transcription initiated at the constitutive promoter $P_{R'}$ at termination site t_{R3}, allowing transcription to continue through the late genes.

Nondefective lambda prophages in lysogenic *E. coli* cells can be induced to enter the lytic pathway by exposure to ultraviolet light (UV). UV irradiation of cells causes damage to DNA, which results in a number of physiological changes collectively called the SOS response (Chapter 14). One component of the SOS response in *E. coli* is the conversion of the RecA protein (an important recombination protein; see Chapter 14) to a protease that cleaves the connector region of the λ repressor. This cleavage event prevents the dimerization of the DNA-binding domain of the repressor. In the absence of active repressor dimers, Cro and N protein are synthesized. Cro binds to O_{R3} and prevents the synthesis of repressor. With Cro and N protein, but no repressor, present in the cell, lytic development is almost inevitable.

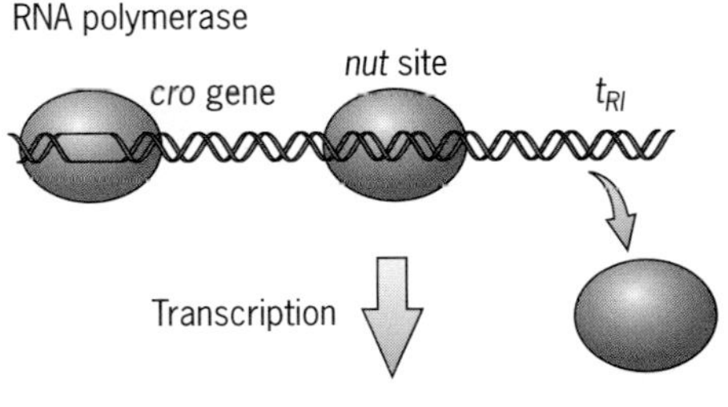

(*a*) Transcription is terminated at t_{R1} in the absence of N protein.

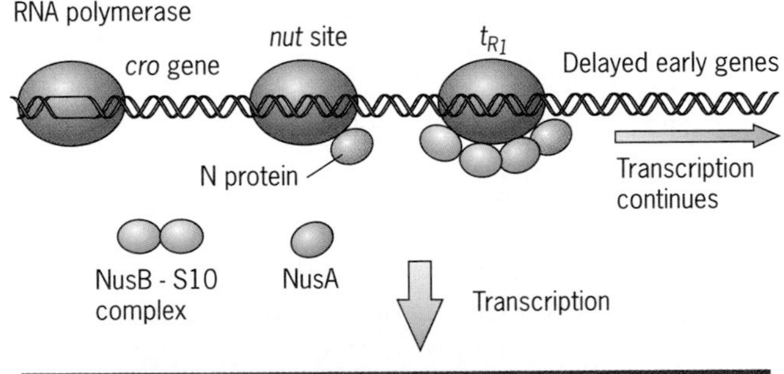

(*b*) Transcription continues through t_{R1} in the presence of N protein.

Figure 23.20 The phage lambda N protein functions as an antiterminator of transcription at sites t_{R1} and t_{L1} (not shown). (*a*) In the absence of N protein, transcription of the early gene *cro*, which is initiated at P_R, terminates at t_{R1}. (*b*) In the presence of N protein, transcription initiated at P_R continues through t_{R1} into the delayed early genes.

The Lambda Switch: Lytic Development or Lysogeny

When phage lambda injects its DNA into an *E. coli* cell, that cell contains neither repressor nor Cro protein. What then determines whether the injected lambda chromosome will (1) enter the lysogenic pathway and insert itself into the host chromosome, or (2) enter the lytic pathway and produce progeny phage at the expense of the host cell? Having examined the lambda regulatory circuits involved in the lysogenic and lytic pathways, let's consider the switch that controls whether lysogeny or lysis follows an infection of *E. coli* by a lambda phage. At the outset, we should realize that no one can accurately predict which response will occur in a specific cell that has been infected with a wild-type λ phage.

The decision between lysogeny and lytic growth hinges on a delicate balance between the λ repressor and Cro protein. Which protein will occupy the three binding sites in operators O_L and O_R and, therefore, control transcription initiation at P_L, P_R, and P_{RM}? The decision is finalized when one of these proteins gains occupancy of the O_L and O_R sites. If λ repressor occu-

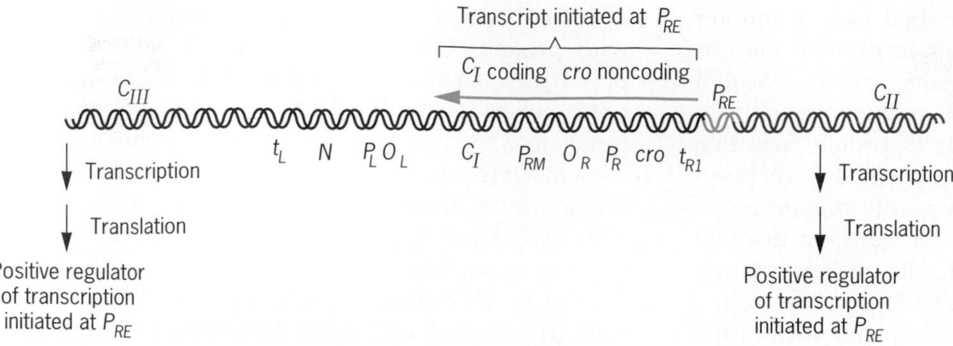

Figure 23.21 Regulatory elements involved in setting the lambda genetic switch to lysogeny or to lytic growth. The C_{II} and C_{III} genes encode proteins that are required for the initiation of transcription at promoter P_{RE}.

pies these operators, lysogeny will result. If Cro protein occupies O_L and O_R, lytic growth will occur.

The genetic switch that sends lambda down either the lysogenic or the lytic pathway involves two key regulatory genes, C_{II} and C_{III}, and one promoter, P_{RE} (*p*romoter for *r*epressor *e*stablishment). P_{RE} is located on the opposite side of the *cro* gene from P_{RM} (Figure 23.21). The transcript initiated at P_{RE} contains the noncoding or antisense sequence of the *cro* gene and the normal mRNA coding or sense sequence of the C_I gene. Both C_{II} and C_{III} are delayed early genes; therefore, their expression is dependent on N protein. The C_{II} protein is required for the initiation of transcription at P_{RE}; RNA polymerase can initiate transcription at P_{RE} only when C_{II} protein is present (Figure 23.22). However, C_{II} protein is very unstable; it is rapidly degraded by a host cell protease. The function of the C_{III} protein is to stabilize the C_{II} protein, allowing RNA polymerase to initiate transcription at P_{RE}. Actually, the effect of C_{III} on the stability of C_{II} is a function of the growth conditions. Under poor growth conditions, C_{III} protects C_{II} from degradation by proteases more completely, favoring lysogeny, whereas under optimal growth conditions, C_{III} is less active and, thus, C_{II} is less stable, favoring lytic growth.

The transcript initiated at P_{RE} favors establishment of the lysogenic pathway in two ways: (1) Lambda repressor is produced by translation of the C_I coding sequence. (2) The 5′ portion of this transcript contains the noncoding or antisense sequence of the *cro* gene. This *cro* antisense sequence is complementary to the *cro* coding or sense sequence in authentic *cro* mRNA synthesized by transcription initiated at P_R. Thus, these *cro* antisense and sense RNAs hybridize (form double helices) with each other, which blocks translation of the *cro* mRNA.

Once repressor has been synthesized by use of the C_{II}-C_{III}-P_{RE} circuit, it binds at O_{R1} and O_{R2}. The presence of repressor at O_{R1} and O_{R2} inhibits the expression of

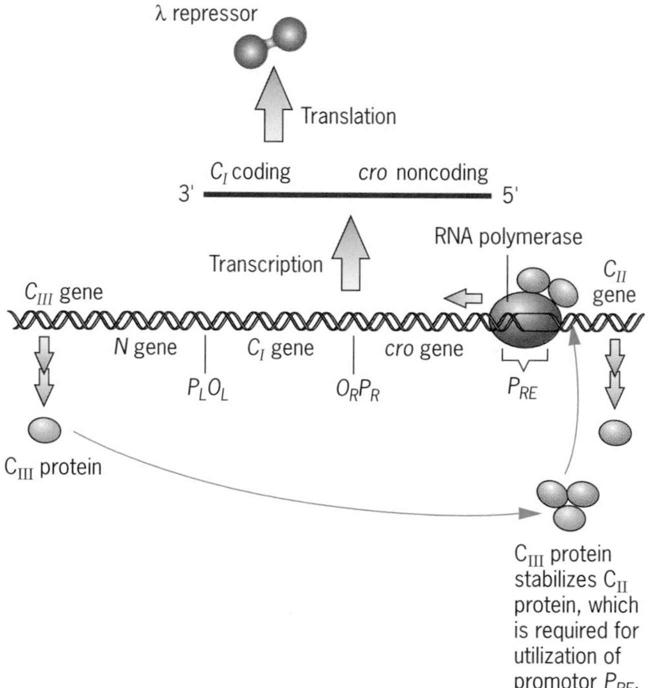

Figure 23.22 Diagram of the phage λ C_{II}-C_{III}-P_{RE} regulatory circuit involved in determining whether lytic and lysogenic development will occur.

lytic function genes by blocking the initiation of transcription at P_R and establishes the autogenous repressor maintenance circuit mediated by P_{RM}. Figure 23.23 summarizes the regulatory genes and *cis*-acting elements involved in setting the lysogeny-lytic switch and in the lysogenic and lytic pathways of development after the switch has been set.

In summary, in each infected cell, the genetic switch for either lysogeny or lytic development is set by the race between Cro protein and λ repressor for occupancy of the O_L and O_R sites. If Cro protein occupies these operators, lytic development ensues. If re-

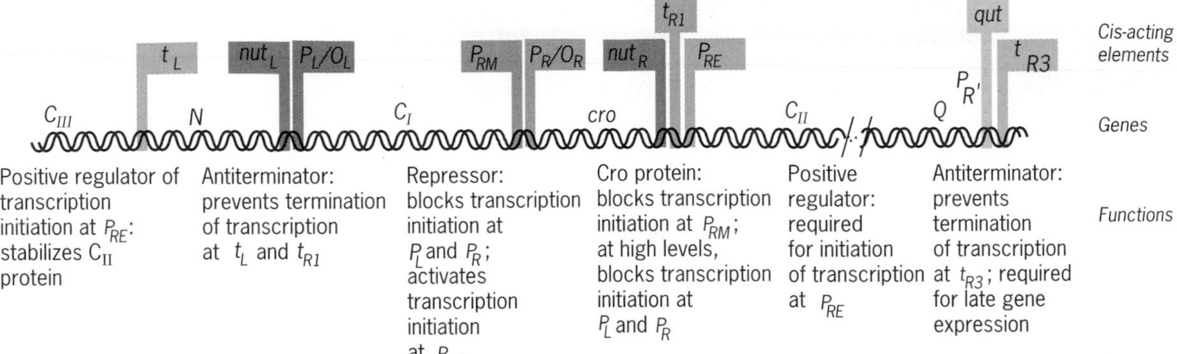

Figure 23.23 Summary of the lambda regulatory elements that control lysogenic and lytic development. *Cis*-acting elements are shown at the top, genes encoding *trans*-acting factors are given directly above the DNA molecule, and the functions of the *trans*-acting regulators are described at the bottom. (*Based on Benjamin Lewin,* Genes V, *Cell Press, 1994. Used by permission.*)

pressor occupies these operators, lysogeny is established. The C_{II} and C_{III} regulatory proteins play important roles in deciding the outcome of this race between Cro protein and repressor. Environmental influences on the outcome of the competition between Cro and repressor act through C_{III} and its effect on the stability of C_{II} protein. Increased stability of C_{II} favors lysogeny; decreased stability of C_{II} favors lytic growth.

Key Points: **Temperate bacteriophages can follow either of two pathways: (1) lytic growth, during which they reproduce and kill the host cells, and (2) lysogeny, during which their chromosomes exist as a dormant prophage covalently inserted in the chromosomes of the host bacteria. During lysogeny, the lytic genes of the prophage are kept turned off by a repressor-operator-promoter circuit similar to those of bacterial operons. Whether a lambda phage will enter the lysogenic state or undergo lytic development is determined by which of two regulatory proteins, λ repressor or Cro protein, occupies key operator sites that control transcription of the λ genome. The lysogenic state is maintained by the autogenous control of λ repressor, whereas lytic development is controlled by a regulatory cascade in which transcriptional antiterminators play a major role.**

TEMPORAL SEQUENCES OF GENE EXPRESSION DURING PHAGE INFECTION

Regulation of gene expression during the life cycles of virulent bacteriophages is quite different from the reversible on-off switches characteristic of bacterial operons. In phage-infected bacteria, the viral genes are expressed in genetically preprogrammed sequences or cascades. Although different bacterial viruses exhibit variations of the specific mechanisms involved, a common picture emerges. One set of phage genes, usually called early genes, is expressed immediately after infection. The product(s) of one or more of the early genes is (are) responsible for turning on the expression of the next set of genes and turning off the expression of the early genes, and so on. Two to four sets of genes are usually involved, depending on the virus. In all cases studied so far, the regulation of sequential gene expression during phage infection occurs primarily at the level of transcription.

In *Bacillus subtilis* phage SP01, sequential gene expression is controlled by modifying the specificity of RNA polymerase for different promoter sequences through phage-induced alterations of the host cell's RNA polymerase.

Bacillus subtilis phage SP01 contains three sets of genes, called early, middle, and late genes, in reference to their time of expression during the phage reproductive cycle (Figure 23.24). The phage SP01 early genes are transcribed by the *B. subtilis* RNA polymerase. One of the early gene products is a polypeptide that binds to the host cell's RNA polymerase, changing its specificity so that the modified RNA polymerase transcribes the middle genes of SP01. Two of the products of the middle genes are, in turn, polypeptides that associate with the *B. subtilis* RNA polymerase, further changing its specificity so that it then transcribes the SP01 late genes.

Key Points: **Preprogrammed temporal sequences of viral gene expression occur in bacteriophage-infected cells. The first viral genes expressed in an infected cell are transcribed by unmodified bacterial RNA polymerase. Subsequent sets of expressed**

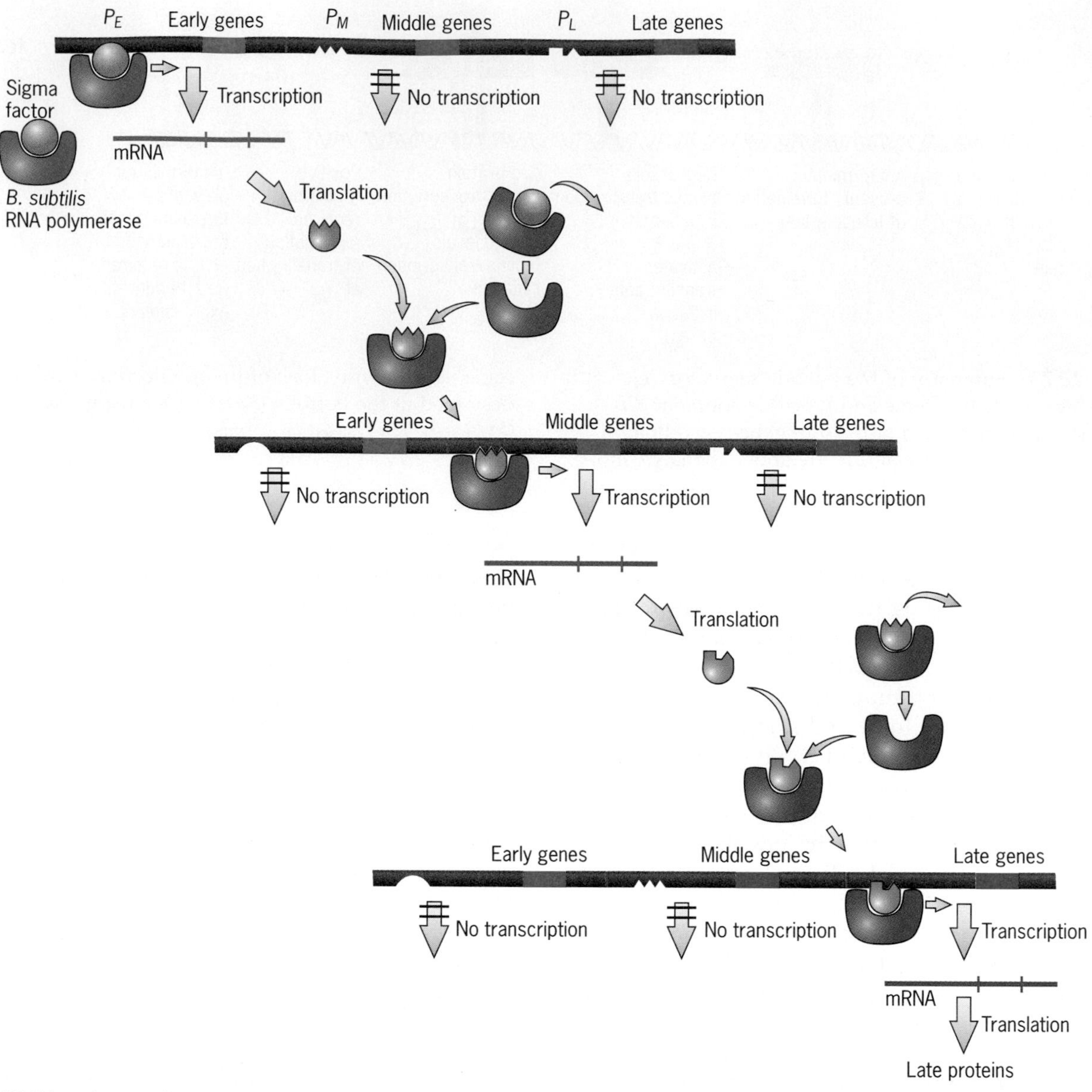

Figure 23.24 The *Bacillus subtilis* phage SP01 regulatory cascade. Early genes are transcribed by the RNA polymerase of the host cell (top). Middle and late genes are transcribed by RNA polymerases in which the host σ factor has been replaced by one or two SP01 proteins, respectively (middle and bottom). P_E, P_M, and P_L are the promoters for early, middle, and late transcripts, respectively.

viral genes are transcribed either by an RNA polymerase encoded by the phage genome or by bacterial RNA polymerase modified by the addition of viral protein(s).

TRANSLATIONAL CONTROL OF GENE EXPRESSION

In the preceding sections of this chapter, we have focused on regulatory mechanisms that act at the level of transcription. Transcriptional regulation of gene expression is paramount in prokaryotes. Nevertheless, gene expression is often fine-tuned by regulatory mechanisms that act at the translation level.

In prokaryotes, mRNA molecules are frequently multigenic, carrying the coding sequences of several genes. For example, the *E. coli lac* operon mRNA harbors nucleotide sequences encoding β-galactosidase, β-galactoside permease, and β-galactoside transacetylase. Thus, the three genes encoding these proteins must be turned on and turned off together at the transcription level, because the genes are co-transcribed. Nevertheless, the three gene products are not synthesized in equal amounts. An *E. coli* cell that is growing on rich medium with lactose as the sole carbon source contains about 3000 molecules of β-galactosidase, 1500 molecules of β-galactoside permease, and 600 molecules of β-galactoside transacetylase. Clearly, the different molar quantities of these proteins per cell must be controlled posttranscriptionally.

Remember that transcription, translation, and mRNA degradation are coupled in prokaryotes; an mRNA molecule usually is involved in all three processes at any given time. Thus, gene products may be produced in different amounts from the same transcript by several mechanisms.

1. **Unequal efficiencies of translational initiation** are known to occur at the ATG start codons of different genes.
2. **Altered efficiencies of ribosome movement** through intergenic regions of a transcript are quite common. Decreased translation rates often result from hairpins or other forms of secondary structure that impede ribosome migration along the mRNA molecule.
3. **Differential rates of degradation** of specific regions of mRNA molecules also occur.

The synthesis of the ribosomal proteins of *E. coli* provides several well-documented examples of translational regulation of gene expression. *E. coli* cells that are growing rapidly under optimal conditions need more ribosomes for protein synthesis than those that are growing slowly under adverse conditions. Thus, under optimal growth conditions, *E. coli* cells devote a larger share of their energy to the production of ribosomes than do cells growing under less favorable conditions. Recall (Chapter 13) that the *E. coli* ribosome contains three RNA molecules and 52 polypeptides. The syntheses of these structural components must be coordinated to assure proper stoichiometry for ribosome assembly. The synthesis of one or more components in excess would be a waste of energy. Thus, regulatory mechanisms have evolved that assure that ribosomal RNAs and proteins are synthesized in the appropriate amounts, and some of these regulatory mechanisms act at the level of translation.

Most of the *E. coli* genes encoding ribosomal proteins are located in clusters, and the genes in each cluster are co-transcribed. All but one of the ribosomal proteins are utilized in equimolar amounts during ribosome assembly. Moreover, the synthesis of the ribosomal proteins must be coordinated with the synthesis of the three ribosomal RNAs. This coordination between ribosomal protein synthesis and ribosomal RNA synthesis occurs at the level of translation.

The ribosomal protein S10 (for *s*mall subunit protein number *10*) gene cluster provides a good illustration of how this regulation works. The *S10* transcriptional unit contains 11 coordinately regulated genes, all encoding ribosomal proteins (Figure 23.25*a*). In the case of the *S10* transcriptional unit and at least five other such units containing ribosomal protein genes, the product of one of the genes inhibits the translation of the multigenic transcript. Thus, the gene encoding the regulatory protein is itself one of the regulated genes, so that the regulatory gene is negatively self-regulated. In the *S10* transcriptional unit, the regulatory gene is *rplD* (for *r*ibosomal *p*rotein *l*arge subunit gene *D*), which encodes protein L4 (*l*arge subunit protein number *4*). When free rRNAs are present in the cell, nascent L4 protein binds to the RNA (Figure 23.25*b*) and is assembled into ribosomes. In the absence of rRNA, the L4 protein binds to the 5′ end of the *S10* transcriptional unit mRNA and inhibits its translation (Figure 23.25*c*). These events prevent the synthesis of ribosomal proteins that cannot be used by the cell.

The inhibition of translation of an mRNA molecule by one of the products that it encodes is common in both prokaryotes and eukaryotes. This mechanism is referred to as **negative self-regulation** or **negative autogenous regulation.** When the gene product is a structural component of the cell or of some organelle within the cell, the autoregulation often is carried out by the free monomers present in the cell.

Key Points: **Regulatory fine-tuning often occurs at the level of translation by modulation of the rate of either polypeptide chain initiation or chain elongation. Sometimes regulation occurs by the differential degradation of specific regions of polygenic mRNAs. The inhibition of translation of a specific mRNA by a protein that it encodes is called negative autogenous regulation.**

POSTTRANSLATIONAL REGULATORY MECHANISMS

Earlier in this chapter, we discussed the mechanism by which the transcription of bacterial genes encoding enzymes in a biosynthetic pathway is repressed when the product of the pathway is present in the medium in which the cells are growing. A second, and more rapid, regulatory fine-tuning of metabolism often occurs at the level of enzyme activity. The presence of a sufficient concentration of the end product of a biosynthetic pathway frequently results in the inhibition of the first enzyme in the pathway (Figure 23.26). This phenomenon is called **feedback inhibition** or **end-product inhibition.** Feedback inhibition results in an almost instantaneous arrest of the synthesis of the end product when it is added to the medium.

Feedback inhibition-sensitive enzymes contain an end-product binding site (or sites) in addition to the substrate binding site (or sites). In the case of multi-

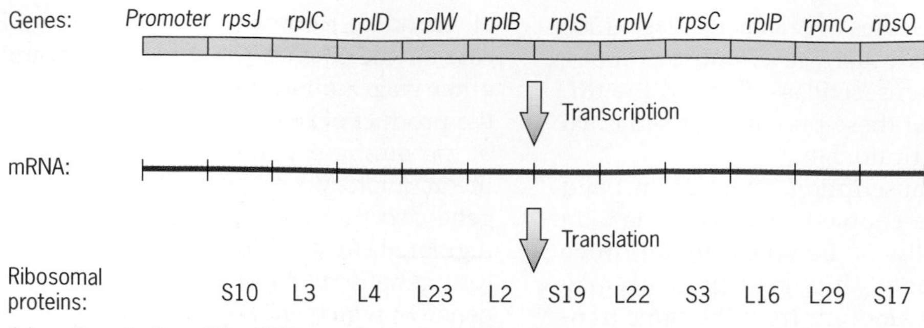

(*a*) Organization of the *S10* transcriptional unit of *E. coli*.

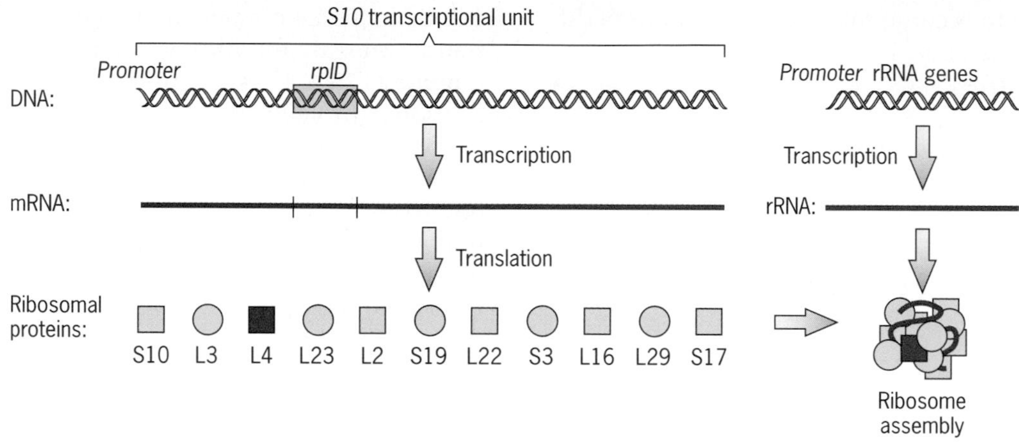

(*b*) The *S10* mRNA is translated when free ribosomal RNA is present.

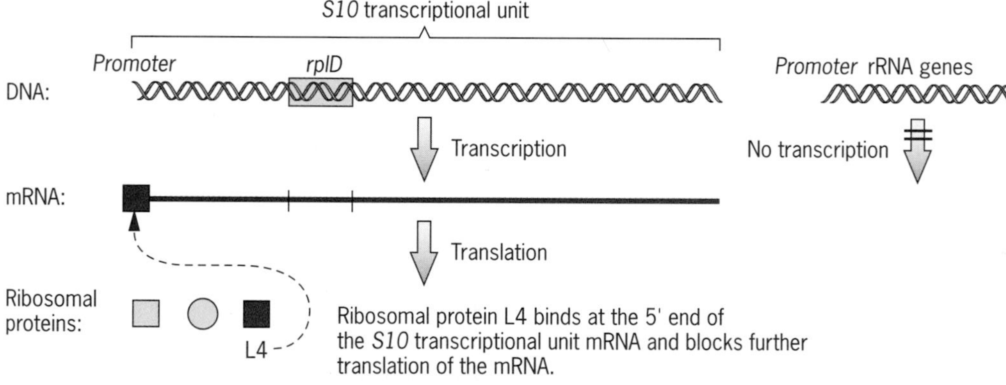

(*c*) Protein L4 blocks translation of the *S10* mRNA when no free ribosomal RNA is present.

Figure 23.25 Organization (*a*) and translational regulation (*b* and *c*) of the *E. coli S10* transcriptional unit, which contains 11 genes encoding ribosomal proteins. Translation of the mRNA of the *S10* transcriptional unit is regulated by ribosomal protein L4, which binds to a nucleotide sequence near the 5′ end of the *S10* transcript. (*b*) In the presence of ribosomal RNA, the ribosomal proteins interact with the rRNA in the assembly of ribosomes. (*c*) In the absence of ribosomal RNA, ribosomal protein L4 binds near the 5′ end of the *S10* transcript and blocks its translation.

meric enzymes, the **end-product or regulatory binding site** often is on a subunit (polypeptide) different from that of the substrate site. Upon binding the end product, such enzymes undergo allosteric transitions that reduce their affinity for the substrates. Proteins that undergo such conformational changes are referred to as allosteric proteins. Many, perhaps most, enzymes undergo allosteric transitions of some kind.

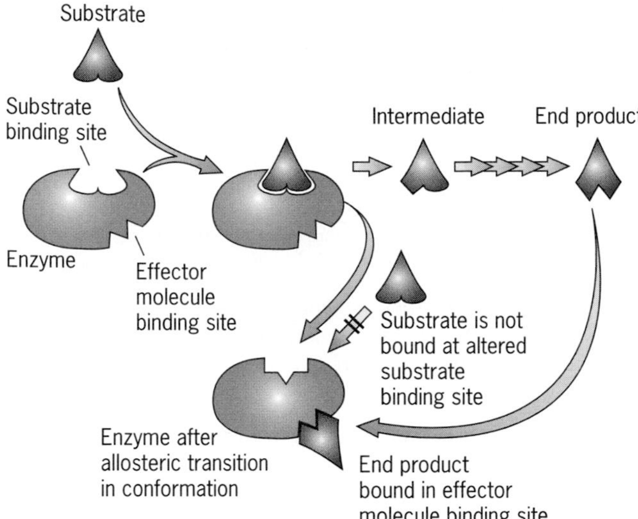

Figure 23.26 Feedback inhibition of gene-product activity. The end product of a biosynthetic pathway often binds to and arrests the activity of the first enzyme in the pathway, quickly blocking the synthesis of the end product.

Allosteric transitions also appear to be responsible for enzyme activation, which often occurs when an enzyme binds one or more of its substrates or some other small molecule. Some enzymes exhibit a broad spectrum of activation and inhibition by many different effector molecules. The enzyme glutamine synthetase catalyzes the final step in the biosynthesis of the amino acid glutamine. Glutamine synthetase is a complex multimeric enzyme in both prokaryotes and eukaryotes. The glutamine synthetase of *E. coli* has been shown to respond, either by activation or inhibition, to 16 different metabolites, presumably through allosteric transitions.

Key Points: **The end product of a biosynthetic pathway often inhibits the activity of the first enzyme in the pathway, rapidly shutting off the synthesis of the product. This regulatory mechanism is called feedback inhibition. Enzyme activation occurs when a substrate or other effector molecule enhances the activity of an enzyme.**

TESTING YOUR KNOWLEDGE

1. The operon model for the regulation of enzyme synthesis concerned in lactose utilization by *E. coli* includes a regulator gene (I), an operator region (O), a structural gene (Z) for the enzyme β-galactosidase, and another structural gene (Y) for β-galactoside permease. β-Galactoside permease transports lactose into the bacterium, where β-galactosidase cleaves it into galactose and glucose. Mutations in the *lac* operon have the following effects: Z^- and Y^- mutant strains are unable to make functional β-galactosidase and β-galactoside permease, respectively, whereas I^- and O^c mutant strains synthesize the *lac* operon gene products constitutively. The following figure shows a partially diploid strain of *E. coli* that carries two copies of the *lac* operon. On the diagram, fill in a genotype that will result in the constitutive synthesis of β-galactosidase and the inducible synthesis of β-galactoside permease by this partial diploid.

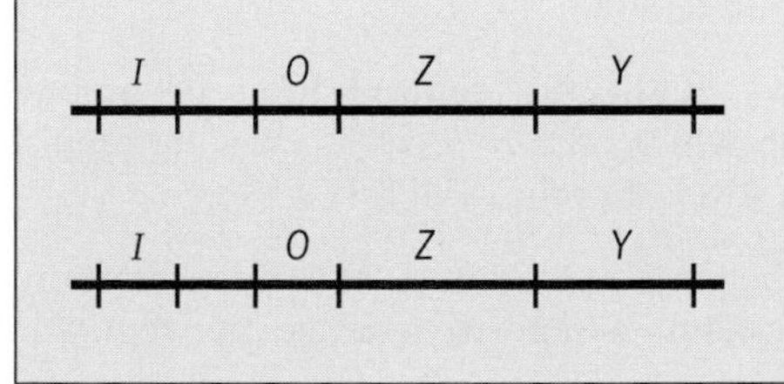

ANSWER

Several different genotypes will produce β-galactosidase constitutively and β-galactoside permease inducibly. They must meet two key requirements: (1) the cell must contain at least one copy of the I^+ gene, which encodes the repressor, and (2) the Z^+ gene and an O^c mutation must be on the same chromosome, because the operator acts only in *cis*; that is, it only affects the expression of genes on the same chromosome. In contrast, the cell can be either homozygous or heterozygous for the I^+ gene, and, if heterozygous, I^+ may be on either chromosome, because I^+ is dominant to I^- and I^+ acts in both the *cis* and *trans* arrangement. One possible genotype is given in the following diagram. How many other genotypes can you devise that will synthesize β-galactosidase constitutively and β-galactoside permease inducibly?

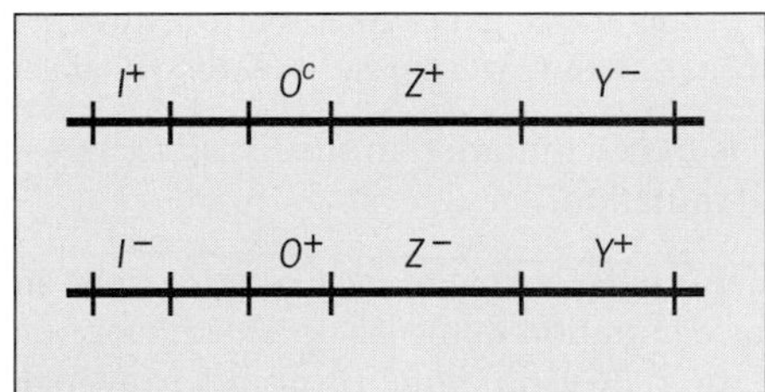

2. When phage λ infects an *E. coli* cell, it can either undergo lytic cell growth and lyse the host cell or enter the lysogenic pathway. (a) What determines which response will occur in any given infected cell? (b) Will lytic or lysogenic development occur in a cell infected with a λ phage with a deletion of the C_I gene? (c) Which will occur in a cell infected with a λ phage with a deletion of the *Cro* gene?

ANSWER

(a) Whether a λ phage will undergo lytic growth or enter the lysogenic pathway when it infects an *E. coli* cell is determined by a genetic switch controlled by two regulatory proteins—λ repressor and Cro protein—that bind to the λ operators and govern transcription of the λ genome. If λ repressor occupies these sites, lysogeny results (see Figure 23.14, right).

If Cro protein occupies these sites, lytic development occurs (see Figure 23.14, left). (b) The C_I gene encodes the λ repressor. Thus, lytic development occurs in cells infected with λ phage that carry a deletion of the C_I gene. Cro protein will occupy the λ operator sites, because no λ repressor will be present if the C_I gene has been deleted. (c) In contrast, lysogeny will occur in bacteria infected with λ phage that harbor a deletion of the *cro* gene. Repressor will be synthesized by expression of the C_I gene, and in the absence of Cro protein, it will bind to the λ operator sites and block the expression of the λ lytic genes.

QUESTIONS AND PROBLEMS

23.1 How can inducible and repressible enzymes of microorganisms be distinguished?

23.2 Distinguish between (a) repression and (b) feedback inhibition caused by the end product of a biosynthetic pathway. How do these two regulatory phenomena complement each other to provide for the efficient regulation of metabolism?

23.3 In the lactose operon of *E. coli,* what is the function of each of the following genes or sites: (a) regulator, (b) operator, (c) promoter, (d) structural gene *Z* and (e) structural gene *Y*?

23.4 What would be the result of inactivation by mutation of the following genes or sites in the *E. coli* lactose operon: (a) regulator, (b) operator, (c) promoter, (d) structural gene *Z*, and (e) structural gene *Y*?

23.5 Groups of alleles associated with the lactose operon are as follows (in order of dominance for each allelic series): repressor, I^s (superrepressor), I^+ (inducible), and I^- (constitutive); operator, O^c (constitutive, *cis*-dominant) and O^+ (inducible, *cis*-dominant); structural, Z^+ and Y^+. *(a)* Which of the following genotypes will produce β-galactosidase and β-galactoside permease if lactose is present: (1) $I^+O^+Z^+Y^+$, (2) $I^-O^cZ^+Y^+$, (3) $I^sO^cZ^+Y^+$, (4) $I^sO^+Z^+Y^+$, and (5) $I^-O^+Z^+Y^+$? (b) Which of the above genotypes will produce β-galactosidase and β-galactoside permease if lactose is absent? Why?

23.6 Why is the O^c mutation in the *E. coli lac* operon epistatic to the I^s mutation?

23.7 Assume that you have discovered a new strain of *E. coli* that has a mutation in the *lac* operator region that causes the wild-type repressor protein to bind irreversibly to the operator. You have named this operator mutant O^{sb} for "super-binding" operator. (a) What phenotype would a partial diploid of genotype $I^+O^{sb}Z^-Y^+/I^+O^+Z^+Y^-$ have with respect to the synthesis of the enzymes β-galactosidase and β-galactoside permease? (b) Does your new O^{sb} mutation exhibit *cis* or *trans* dominance in its effects on the regulation of the *lac* operon?

23.8 For each of the following partial diploids indicate whether enzyme synthesis is constitutive or inducible (see Problem 23.5 for dominance relationships): (a) $I^+O^+Z^+Y^+/I^+O^+Z^+Y^+$, (b) $I^+O^+Z^+Y^+/I^+O^cZ^+Y^+$, (c) $I^+O^cZ^+Y^+/I^+O^cZ^+Y^+$, (d) $I^+O^+Z^+Y^+/I^-O^+Z^+Y^+$, (e) $I^-O^+Z^+Y^+/I^-O^+Z^+Y^+$. Why?

23.9 Write the partial diploid genotype for a strain that will (a) produce β-galactosidase constitutively and permease inducibly and (b) produce β-galactosidase constitutively but not permease either constitutively or inducibly, even though a Y^+ gene is known to be present.

23.10 Constitutive mutations produce elevated enzyme levels at all times; they may be of two types: O^c or I^-. Assume that all other DNA present is wild-type. Outline how the two constitutive mutants can be distinguished with respect to (a) map position, (b) regulation of enzyme levels in O^c/O^+ versus I^-/I^+ partial diploids, and (c) the position of the structural genes affected by an O^c mutation versus the genes affected by an I^- mutation in a partial diploid.

23.11 As a genetics historian, you are repeating some of the classic experiments conducted by Jacob and Monod with the lactose operon in *E. coli*. You use an F′ plasmid to construct several *E. coli* strains that are partially diploid for the *lac* operon. You construct strains with the following genotypes: (1) $I^+O^cZ^+Y^-/I^+O^+Z^-Y^+$, (2) $I^+O^cZ^-Y^+/I^+O^+Z^+Y^-$, (3) $I^-O^+Z^+Y^+/I^+O^+Z^-Y^-$, (4) $I^sO^+Z^-Y^-/I^+O^+Z^+Y^+$, and (5) $I^+O^cZ^+Y^+/I^sO^+Z^-Y^+$. (a) Which of these strains will produce functional β-galactosidase in both the presence and absence of lactose? (b) Which of these strains will exhibit constitutive synthesis of functional β-galactoside permease? (c) Which of these strains will express both gene *Z* and gene *Y* constitutively and will produce functional products (β-galactosidase and β-galactoside permease) of both genes? (d) Which of these strains will show *cis* dominance of *lac* operon regulatory elements? (e) Which of these strains will exhibit *trans* dominance of *lac* operon regulatory elements?

23.12 How could the tryptophan operon in *S. typhimurium* have developed and been maintained by evolution?

23.13 Of what biological significance is the phenomenon of catabolite repression?

23.14 How might the concentration of glucose in the medium in which an *E. coli* cell is growing regulate the intracellular level of cyclic AMP?

23.15 Is the CAP-cAMP effect on the transcription of the *lac* operon an example of positive or negative regulation? Why?

23.16 Would it be possible to isolate *E. coli* mutants in which the transcription of the *lac* operon is not sensitive to catabolite repression? If so, in what genes might the mutations be located?

23.17 Using examples, distinguish between negative regulatory mechanisms and positive regulatory mechanisms.

23.18 The following table gives the relative activities of the enzymes β-galactosidase and β-galactoside permease in cells with different genotypes at the *lac* locus in *E. coli*. The level of activity of each enzyme in wild-type *E. coli* not carrying F′s was arbitrarily set at 100; all other values are relative to the observed levels of activity in these wild-type bacteria. Based on the data given in the table for genotypes 1 through 4, fill in the levels of enzyme activity that would be expected for the fifth genotype.

	β-galactosidase		β-galactoside permease	
Genotype	**−Inducer**	**+Inducer**	**−Inducer**	**+Inducer**
1. $I^+O^+Z^+Y^+$	0.1	100	0.1	100
2. $I^-O^+Z^+Y^+$	100	100	100	100
3. $I^+O^cZ^+Y^+$	25	100	25	100
4. $I^-O^+Z^+Y^-$/F′ $I^-O^+Z^+Y^+$	200	200	100	100
5. $I^-O^cZ^-Y^+$/F′ $I^+O^+Z^+Y^+$	—	—	—	—

23.19 The rate of transcription of the *trp* operon in *E. coli* is controlled by both (1) repression/derepression and (2) attenuation. By what mechanisms do these two regulatory processes modulate *trp* operon transcript levels?

23.20 What effect will deletion of the *trpL* region of the *trp* operon have on the rates of synthesis of the enzymes encoded by the five genes in the *trp* operon in *E. coli* cells growing in the presence of tryptophan?

23.21 By what mechanism does the presence of tryptophan in the medium in which *E. coli* cells are growing result in premature termination or attenuation of transcription of the *trp* operon?

23.22 Suppose that you used site-specific mutagenesis to modify the *trpL* sequence such that the two UGG Trp codons at positions 54–56 and 57–60 (see Figure 23.12) in the mRNA leader sequence were changed to GGG Gly codons. Will attenuation of the *trp* operon still be regulated by the presence or absence of tryptophan in the medium in which the *E. coli* cells are growing?

23.23 What is a prophage? In what ways does the phage λ prophage differ from the λ chromosome present during lytic infections?

23.24 In what ways does the phage λ lytic regulatory cascade differ from the autogenously maintained repression of viral gene expression that occurs during lysogeny?

23.25 What is the major structural difference between the operator region in the *lac* operon of *E. coli* and the phage λ operators O_L and O_R?

23.26 The λ lysogenic state is very stable, in part because the synthesis of the λ repressor is autogenously regulated. What does this mean? How does autogenous regulation of repressor enhance the stability of lysogeny?

23.27 In λ, the C_I gene produces a repressor of the lytic pathway. C_I^- mutations cause clear plaques, in which no lysogenic cells are present. A plasmid carries this C_I^- mutant allele under the control of its normal (wild-type) promoter. If this plasmid is introduced into a lysogenic cell that contains an integrated wild-type λ genome (with a normal C_I^+ gene), what will happen and why?

23.28 During lytic growth, the phage λ genes exhibit a rather precise pattern of temporal expression. The immediate early genes are expressed first, then the delayed early genes, and lastly the late genes. How does this regulatory cascade work? Can the late genes be expressed without the delayed early genes being expressed first? If not, why not? What role does transcriptional termination play in the λ lytic regulatory cascade?

23.29 The N protein of phage λ is a transcriptional antiterminator. What role does it play in the lambda regulatory cascade?

23.30 A lambda phage has a UAG chain-termination mutation in the middle of the C_I gene. What phenotype will result from this mutation?

23.31 Would attenuation of the type that regulates the level of *trp* transcripts in *E. coli* be likely to occur in eukaryotic organisms?

BIBLIOGRAPHY

Beckwith, J. R., and D. Zipser. 1970. *The Lactose Operon*. Cold Spring Harbor Laboratory Press, Cold Spring Harbor, New York.

Henkin, T. M. 1996. Control of transcription termination in prokaryotes. *Ann. Rev. Genetics* 30:35–57.

Jacob, F., and J. Monod. 1961. Genetic regulatory mechanisms in the synthesis of proteins. *J. Mol. Biol.* 3:318–356.

Kolb, A., S. Busby, H. Buc, S. Garges, and S. Adhya. 1993. Transcriptional regulation by cAMP and its receptor. *Ann. Rev. Biochem.* 62:749–795.

Neidhardt, F. C., R. Curtiss III, J. L. Ingraham, E.C.C. Lin, K. B. Low, B. Magasanik, W. S. Reznikoff, M. Riley, M. Schaechter, and H. E. Umbarger. (eds.) 1996. Escherichia coli *and* Salmonella typhimurium, *Cellular and Molecular Biology*, 2nd ed. American Society for Microbiology, Washington, DC.

Ptashne, M. 1987. *A Genetic Switch: Gene Control and Phage λ*. Cell Press, Cambridge, MA.

Ptashne, M. 1989. How gene activators work. *Sci. Amer.* 260(1):40–47.

Ptashne, M., A. D. Johnson, and C. O. Pabo. 1982. A genetic switch in a bacterial virus. *Sci. Amer.* 247(5):128–140.

Snyder, L., and W. Champness. 1997. *Molecular Genetics of Bacteria*. ASM Press, Washington, DC.

Streips, U. N., and R. E. Yasbin. 1991. *Modern Microbial Genetics*. Wiley-Liss, New York.

Yanofsky, C. 1981. Attenuation in the control of expression of bacterial operons. *Nature* 289:751–758.

24

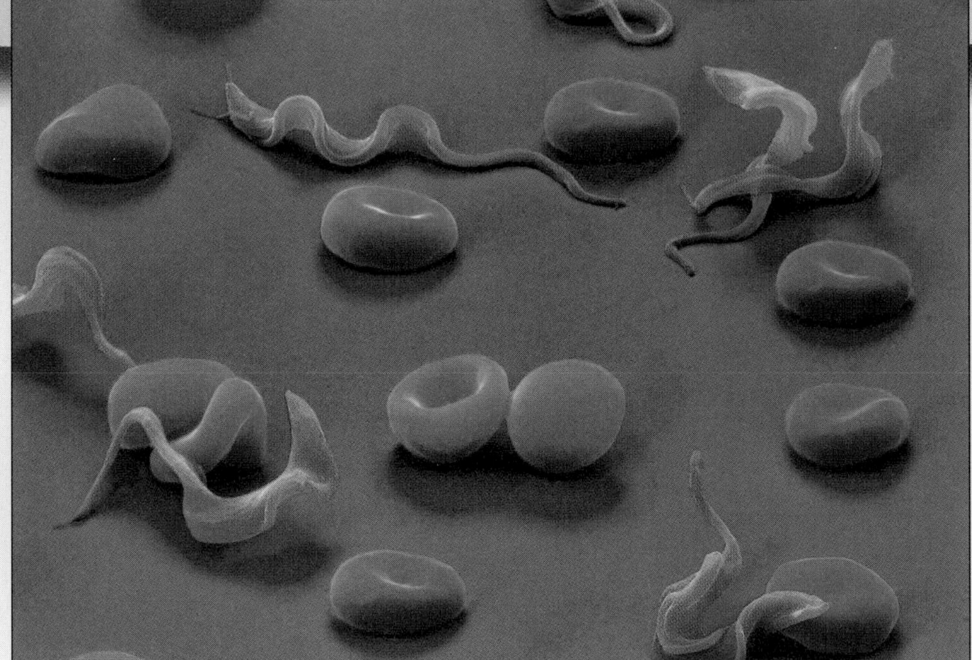

Trypanosomes among red blood cells.

Regulation of Gene Expression in Eukaryotes

CHAPTER OUTLINE

Spatial and Temporal Regulation of Eukaryotic Genes
Ways of Regulating Eukaryotic Gene Expression
Induction of Transcriptional Activity by Environmental and Biological Factors
Molecular Control of Transcription in Eukaryotes
Gene Expression and Chromosome Organization
Activation and Inactivation of Whole Chromosomes

African Trypanosomes: A Wardrobe of Molecular Disguises

Near the end of the nineteenth century, David Bruce, a surgeon in the British Medical Service, summarized his observations and experiments on a disease of wild and domesticated animals in southern Africa. The disease, called nagana from a Zulu word meaning "loss of spirit," is characterized by fever, swelling, lethargy, and emaciation. Bruce recognized that nagana is transmitted by the tsetse, a type of biting fly common in the open spaces of the African scrub plain. Furthermore, his examination of diseased animals led him to conclude that the causative agent is a flagellated, unicellular protozoan that is injected into the animal's blood when the tsetse bites. This blood parasite, a type of trypanosome, is now called *Trypanosoma brucei* in Bruce's honor. Humans can also be infected with tsetse-borne trypanosomes, whereupon they develop the debilitating illness known as African sleeping sickness.

In both animals and humans, trypanosome infections last a long time. This is remarkable because, in the blood, trypanosomes are subjected to repeated attacks by the immune system. With each immune attack, most of the trypanosomes are destroyed; however, a few always survive to repopulate the blood and maintain the infection.The key to this resurgence is the trypanosome's ability to change the protein that coats its surface. Each trypansome is covered with about 10 million molecules of a single glycoprotein. When the immune system recognizes this protein coat, the infecting trypanosome is in trouble; immune cells will trap and destroy it. However, before all the trypanosomes in the animal are completely wiped out, a few manage to change their surface glycoprotein to one that is not immediately recognized by the immune system. These altered trypanosomes escape destruction and proliferate. Eventually, the immune system will learn to recognize them too, but in the meantime another group of altered trypanosomes arises to keep the infection going. The seemingly endless supply of molecular disguises available to trypanosomes is due to a large array of genes that encode the variant surface glycoproteins (VSGs) coating these organisms. At any one time, only one of these genes is expressed; all the others are silent. However, during the course of an infection the identity of the expressed gene changes. With each change, the trypanosomes acquire a new surface protein and manage to stay one step ahead of the animal's immune defenses. Thus, the infection is maintained for weeks or even months until, through exhaustion, the animal dies.

In this chapter, we consider the mechanisms that regulate gene expression in eukaryotes. These mechanisms are more elaborate than those in prokaryotes because many eukaryotic organisms are multicellular, and within each such organism there may be many different types of cells. The formation of these cell types requires that groups of genes be expressed in precise spatial and temporal patterns within the organism. We will see that this spatial and temporal regulation of gene expression is accomplished by a diverse array of molecular mechanisms.

SPATIAL AND TEMPORAL REGULATION OF EUKARYOTIC GENES

The thousands of genes that are present in the genomes of multicellular eukaryotes are normally expressed in a controlled fashion. One dimension of this control is spatial. Not every gene product is needed in every tissue. Some genes are expressed in nerve cells, others in blood cells, and still others in reproductive cells. In fact, the complexity of multicellular eukaryotes is partly due to the tissue-specific expression of many different genes. A second dimension of eukaryotic gene regulation is temporal. Different genes are expressed at different times, some in response to biological signals such as hormones, and others in response to environmental stimuli. Temporal specificity is most dramatically seen during development, when a fertilized egg grows into a multicellular organism. As this organism forms, batteries of genes are expressed in an orderly sequence to direct the formation of tissues and organs. The temporal and spatial regulation of genes is therefore an important aspect of eukaryotic biology.

Spatial Regulation of Tubulin Genes in Plants

The genes for the tubulin polypeptides provide a dramatic example of expression that is spatially regulated. These polypeptides are the building blocks of microtubules (Chapter 2). There are two general types of tubulin polypeptides, α and β, and one molecule of each type aggregates to form a dimer. These dimers then assemble in parallel rows to form hollow, cylindrical microtubules about 24 nm in diameter. Several microtubules may aggregate with each other to create more complex structures, such as cilia and flagella. Within cells, microtubules are found in many places—in the cytoplasm just beneath the plasma membrane, around the nuclear membrane, and in a special region called the microtubule-organizing center. Microtubules play an important role in cell movement. In cilia and flagella their wavelike bending helps to move a cell from one position to another, and inside cells they are responsible for moving chromosomes during mitosis.

The α and β tubulins are encoded by distinct sets of genes. Typically, several different genes for each type of tubulin are present in the genome. For example, the weedy plant *Arabidopisis thaliana* has six α tubulin genes and nine β tubulin genes. Each of these genes has been isolated, and the expression of some of them has been studied in whole plants. The results of this work show that each gene is expressed in a specific spatial pattern. For example, the *TUA1* gene, which encodes a variant of the α tubulin polypeptide, is expressed preferentially in pollen grains (Figure 24.1); there is also some expression of *TUA1* in young anthers, but little or no expression of it in leaves, stems, or roots. In contrast, *TUB1*, a β tubulin gene, is ex-

Figure 24.1 Flowers of *Arabidopsis thaliana* stained to detect expression from the promoter of the *TUA1* alpha tubulin gene. The *TUA1* promoter has been fused to the coding sequence of the β-glucouronidase gene from *Escherichia coli,* which acts as a reporter of expression (blue color). Expression of *TUA1* is limited to the anthers.

pressed only in a section of the root immediately above the rapidly dividing root tip. Another β tubulin gene, *TUB8*, is preferentially expressed in the vascular tissues of the plant. The spatially specific patterns of expression of these and other tubulin genes suggest that slightly different types of microtubules are needed in different parts of *Arabidopsis* plants.

Temporal Regulation of Globin Genes in Animals

Perhaps one of the most dramatic examples of temporally regulated gene expression comes from the study of hemoglobin, the protein that is responsible for transporting oxygen in the blood of vertebrate animals. In higher vertebrates, this protein is a tetramer of polypeptides called globins; in each tetramer there are two α and two β globin chains. Molecules of an iron-containing compound called heme are loosely joined to each of these polypeptides, forming pockets that can bind molecular oxygen. In human beings, multiple genes for the α and β globins are located in two separate sites in the genome. The α globin genes occupy a 28-kb region on chromosome 16, and the β globin genes occupy a 45-kb region on chromosome 11 (Figure 24.2). Because the genes within each cluster are duplicates of an ancestral globin gene, they form a small multigene family. Over evolutionary time, the members of these families have diverged from each other by random mutation so that today, each one encodes a slightly different polypeptide. In some of these duplicate genes, a frameshifting or chain-terminating mutation has abolished the ability to make a polypeptide; such noncoding genes are called **pseudogenes** and are denoted by the Greek letter psi (ψ).

A remarkable feature of both the α and β gene clusters is that their members are expressed at different times during development. The genes on one side in the clusters are expressed only in the embryo, those in the middle are expressed only in the developing fetus, and those on the other side are expressed only after birth. This sequential activation of genes from one side to the other in a cluster is apparently related to the need to produce slightly different kinds of hemoglobin during the course of human development. Embryo, fetus, and infant have different oxygen requirements, different circulatory systems, and different physical environments. The temporal switching in globin gene expression is apparently an adaptation to this changing array of conditions.

Key Points: **In eukaryotes, gene expression is spatially and temporally regulated. Some genes, such as those encoding the α and β tubulins in *Arabidopsis*, are expressed in a tissue-specific manner. Other genes, such as those encoding the α and β globins in vertebrates, are expressed in a specific temporal pattern during development.**

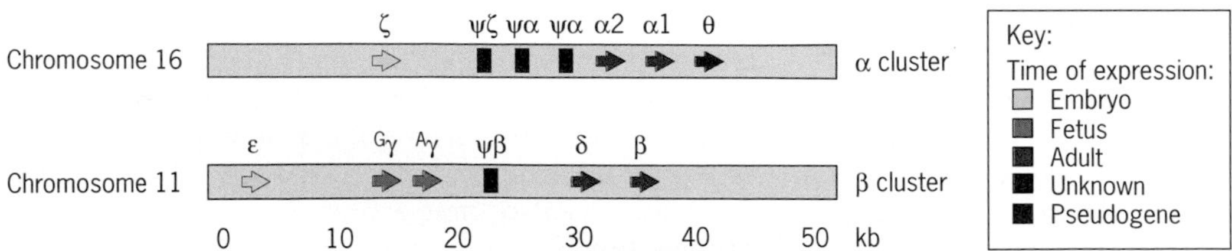

Figure 24.2 Organization and temporally specific expression of the human α and β globin genes.

WAYS OF REGULATING EUKARYOTIC GENE EXPRESSION

As in prokaryotes, the expression of genes in eukaryotes involves the transcription of DNA into RNA and the subsequent translation of that RNA into polypeptides. However, prior to translation, most eukaryotic RNA is "processed." During processing, the RNA is capped at its 5′ end, polyadenylated at its 3′ end, and altered internally by losing its noncoding intron sequences (see Chapter 12). Prokaryotic RNAs typically do not undergo these terminal and internal modifications.

Gene expression in eukaryotes is more complicated than it is in prokaryotes because eukaryotic cells are compartmentalized by an elaborate system of membranes. This compartmentalization subdivides the cells into separate organelles, the most conspicuous one being the nucleus; eukaryotic cells also possess mitochondria, chloroplasts (if they are plant cells), and an endoplasmic reticulum. Each of these organelles performs a different function. The nucleus stores the genetic material, the mitochondria and chloroplasts recruit energy, and the reticulum transports materials within the cell.

The subdivision of eukaryotic cells into organelles physically separates the events of gene expression. The primary event, transcription of DNA into RNA, occurs in the nucleus. RNA transcripts are also modified in the nucleus by capping, polyadenylation, and the removal of introns. The resulting messenger RNAs are then exported to the cytoplasm where they become associated with ribosomes, many of which are located on the membranes of the endoplasmic reticulum. Once associated with ribosomes, these mRNAs are translated into polypeptides. This physical separation of the events of gene expression makes it possible for regulation to occur in different places (Figure 24.3). Regulation can occur in the nucleus at either the DNA or RNA level, or in the cytoplasm at either the RNA or polypeptide level.

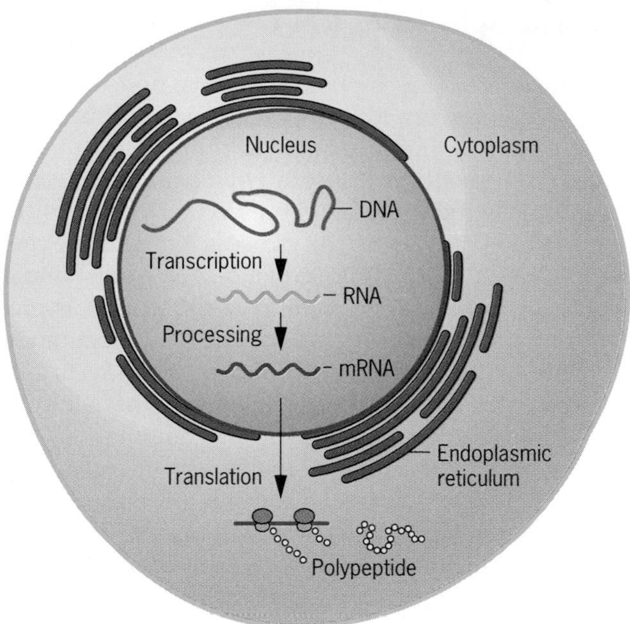

Figure 24.3 Eukaryotic gene expression showing the stages at which expression can be regulated: transcription, processing, and translation.

Controlled Transcription of DNA

In prokaryotes, gene expression is regulated mainly by controlling the transcription of DNA into RNA. A gene that is not transcribed is simply not expressed. Transcription occurs in prokaryotes when negative regulatory molecules such as the *lac* repressor protein have been removed from the vicinity of a gene and positive regulatory molecules such as the cyclic AMP-catabolite activator protein (CAP) complex have bound to it (Chapter 23). These protein-DNA interactions control whether or not a gene is accessible to RNA polymerase. Furthermore, the mechanisms that have evolved to control transcription in these organisms respond quickly to environmental changes. As we discussed in Chapter 23, this hair-trigger control is an efficient strategy for prokaryotic survival.

The control of transcription is more complex in eukaryotes than it is in prokaryotes. One reason is that genes are sequestered in the nucleus. Before environmental signals can have any effect on the level of transcription, they must be transmitted from the cell surface, where they are usually received, through the cytoplasm and the nuclear membrane, and onto the chromosomes. Eukaryotic cells therefore need internal signaling systems to control the transcription of DNA. Another complicating factor is that many eukaryotes are multicellular. Environmental cues may have to pass through layers of cells in order to have an impact on the transcription of genes in a particular tissue. Intercellular communication is therefore an important aspect of eukaryotic transcriptional regulation.

As in prokaryotes, eukaryotic transcriptional regulation is mediated by protein-DNA interactions. Positive and negative regulator proteins bind to specific regions of the DNA and stimulate or inhibit transcription. As a group, these proteins are called **transcription factors.** Many different types have been identified, and most seem to have characteristic domains that allow them to interact with DNA. The structure of these proteins, and the nature of their interactions with DNA, will be discussed in a later section.

Alternate Splicing of RNA

Most eukaryotic genes possess introns, noncoding regions that interrupt the sequence that specifies the amino acids of a polypeptide. Each intron must be removed from the RNA transcript of a gene in order for the coding sequence to be expressed properly. As we discussed in Chapter 12, this process involves the precise joining of the coding sequences, or exons, into a messenger RNA. The formation of the mRNA is mediated by tiny nuclear organelles called spliceosomes.

Genes with multiple introns present a curious problem to the RNA splicing machinery. These introns can be removed separately or in combination, depending on how the splicing machinery interacts with the RNA. If two successive introns are removed together, the exon between them will also be removed. Thus, the splicing machinery has the opportunity to modify the coding sequence of an RNA by deleting some of its exons. This phenomenon of splicing an RNA transcript in different ways is apparently a way of economizing on genetic information. Instead of duplicating genes, or pieces of genes, the **alternate splicing** of transcripts makes it possible for a single gene to encode different polypeptides.

One example of alternate splicing occurs during the expression of the gene for troponin T, a protein found in the skeletal muscles of vertebrates; the size of this protein ranges from about 150 to 250 amino acids. In the rat, the troponin T gene is more than 16 kb long and contains 18 different exons (Figure 24.4). Transcripts of this gene are spliced in different ways to create a large array of mRNAs. When these are translated, many different troponin T polypeptides are produced. All these polypeptides share amino acids from exons 1–3, 9–15, and 18. However, the regions encoded by exons 4–8 may be present or absent, depending on the splicing pattern, and apparently in any combination. Additional variation is provided by the presence or absence of regions encoded by exons 16 and 17; if 16 is present, 17 is not, and vice versa. These different forms of troponin T presumably function in slightly different ways within the muscles, contributing to the variability of muscle cell action.

Another example of alternate splicing occurs during the expression of genes involved in sex determination in *Drosophila*. The master regulator of the sex-determination process is the X-linked *Sex-lethal (Sxl)* gene. In chromosomal females, the transcript of this gene is spliced to produce an mRNA that encodes a regulatory protein. In chromosomal males, the *Sxl* transcript is alternately spliced to include an exon with a stop codon; thus, when this RNA is translated, it generates a short polypeptide without regulatory function (Figure 24.5). In XX embryos, where the regulatory *Sxl* protein is present, a particular set of genes is expressed that causes these embryos to develop as females. In XY embryos, where the regulatory *Sxl* protein is absent, a different set of genes is expressed that causes these embryos to develop as males. Female-specific or male-specific splicing is ultimately controlled by the ratio of X chromosomes to autosomes. If the X:A ratio is 1.0, the female-specific pattern of splicing occurs; if the X:A ratio is 0.5, the male-specific pattern occurs. Thus, alternate splicing of the *Sxl* RNA is responsible for sexual differention in *Drosophila*.

Exons in rat troponin T gene

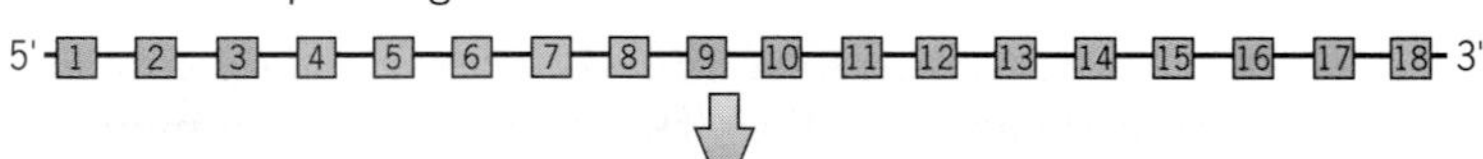

Alternate splicing of exons produces 64 different mRNAs.

Exons 1–3, 9–15, and 18 present in all mRNAs.

Exons 4–8 present in various combinations in mRNAs.

Exons 16 or 17, but not both, present in all mRNAs.

Examples of mRNAs

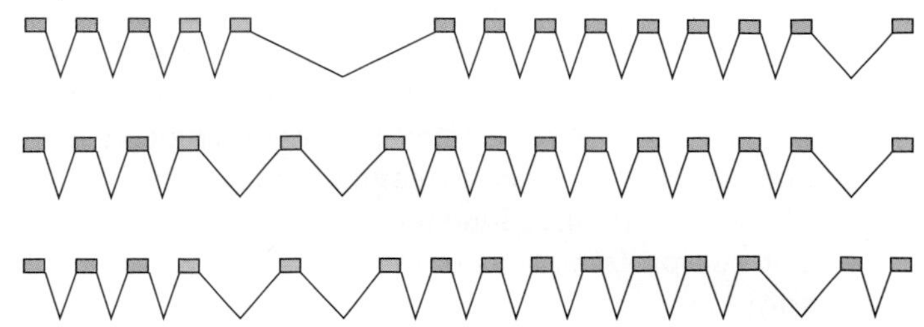

Figure 24.4 Alternate splicing of transcripts from the rat troponin T gene. Only 3 of the possible 64 different mRNAs are shown.

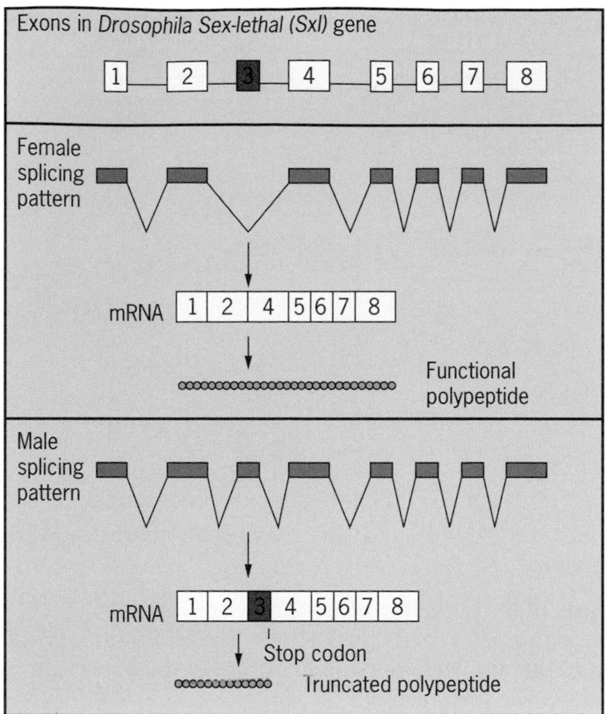

Figure 24.5 Alternate splicing of transcripts from the *Sex-lethal* gene in male and female *Drosophila*.

Cytoplasmic Control of Messenger RNA Stability

Messenger RNAs are exported from the nucleus to the cytoplasm where they serve as templates for polypeptide synthesis. Once in the cytoplasm, a particular mRNA can be translated by several ribosomes that move along it in sequential fashion. This translational assembly line continues until the mRNA is degraded. Messenger RNA degradation is therefore another control point in the overall process of gene expression. Long-lived mRNAs can support multiple rounds of polypeptide synthesis, whereas short-lived mRNAs cannot.

An mRNA that is rapidly degraded must be replenished by additional transcription; otherwise, the polypeptide it encodes will cease to be synthesized. This cessation of polypeptide synthesis may, of course, be part of a developmental program. Once the polypeptide has had its effect, it may no longer be needed; in fact, its continued synthesis may be harmful. In such cases, rapid degradation of the mRNA would be a reasonable way of preventing undesired polypeptide synthesis.

Although the longevity of mRNAs has been difficult to study, a few insights have been obtained by monitoring radioactively labeled RNAs over time. The results of these labeling experiments indicate that mRNA longevity can be influenced by several factors, including the length of the poly-A tail, the structure of the 3′ untranslated region preceding the tail, and the metabolic state of the cell.

Poly-A tails seem to stabilize mRNAs. Histone mRNAs, for example, lack poly-A tails and are extremely short-lived. The sequence of the 3′ untranslated region (3′UTR) also seems to affect mRNA stability. Several short-lived mRNAs have the sequence AUUUA repeated several times in their 3′ untranslated regions. When this sequence is artificially transferred to the 3′ untranslated region of more stable mRNAs, they too become unstable. Other studies have suggested that the stability of mRNAs can be influenced by chemical factors such as hormones. In the toad *Xenopus laevis*, the vitellogenin gene is transcriptionally activitated by the steroid hormone estrogen. However, in addition to inducing transcription of this gene, estrogen also increases the longevity of its mRNA. All these findings demonstrate that by controlling the rate of mRNA degradation, cells have additional ways of modulating the overall process of gene expression.

Key Points: **Eukaryotic gene expression can be regulated at the transcriptional, processing, or translational levels. Transcriptional regulation involves protein-DNA interactions, processing regulation involves the alternate splicing of primary gene transcripts, and translational regulation involves mRNA stability.**

INDUCTION OF TRANSCRIPTIONAL ACTIVITY BY ENVIRONMENTAL AND BIOLOGICAL FACTORS

In their study of the *lactose* operon in *E. coli*, Jacob and Monod discovered that the genes for lactose metabolism were specifically transcribed when lactose was given to the cells. Thus, they demonstrated that lactose was an **inducer** of gene transcription. Following in the footsteps of Jacob and Monod, many researchers have attempted to identify specific inducers of eukaryotic gene transcription. Although these efforts have met with considerable success, the overall extent to which eukaryotic genes are induced by environmental and nutritional factors seems to be less than it is in prokaryotes. Here we will consider three examples of inducible gene expression in eukaryotes. The first two involve induction by environmental factors—temperature and light—and the third involves induction by that special group of signaling molecules called hormones.

Temperature: The Heat-shock Genes

When organisms are subjected to the stress of high temperature, they respond by synthesizing a group of proteins that help to stabilize the internal cellular environment. These **heat-shock proteins,** found in both prokaryotes and eukaryotes, are among the most conserved polypeptides known. Comparisons of the amino acid sequences of heat-shock proteins from organisms as diverse as *E. coli* and *Drosophila* show that they are 40 to 50 percent identical—a remarkable finding considering the length of evolutionary time separating these organisms.

The expression of the heat-shock proteins is regulated at the transcriptional level; that is, heat stress specifically induces the transcription of the genes encoding these proteins (Figure 24.6). In *Drosophila,* for example, one of the heat-shock proteins called **HSP70** (or *h*eat-shock *p*rotein, molecular weight *70* kilodaltons) is encoded by a family of genes located in two nearby clusters on one of the autosomes. Altogether, there are five to six copies of these *hsp70* genes in the two clusters. When the temperature exceeds 33°C, as it does on hot summer days, each of the genes is transcribed into RNA, which is then processed and translated to produce HSP70 polypeptides. This heat-induced transcription of the *hsp70* genes is mediated by a polypeptide called the **Heat-Shock Transcription Factor,** or **HSTF,** which is present in the nuclei of *Drosophila* cells. When *Drosophila* are heat stressed, the HSTF is chemically altered by phosphorylation. In this altered state, it binds specifically to nucleotide sequences upstream of the *hsp70* genes and makes the genes more accessible to RNA polymerase II, the enzyme that transcribes most protein-encoding genes. The transcription of the *hsp70* genes is then vigorously stimulated. The sequences to which the phosphorylated HSTF binds are called **heat-shock response elements (HSEs).**

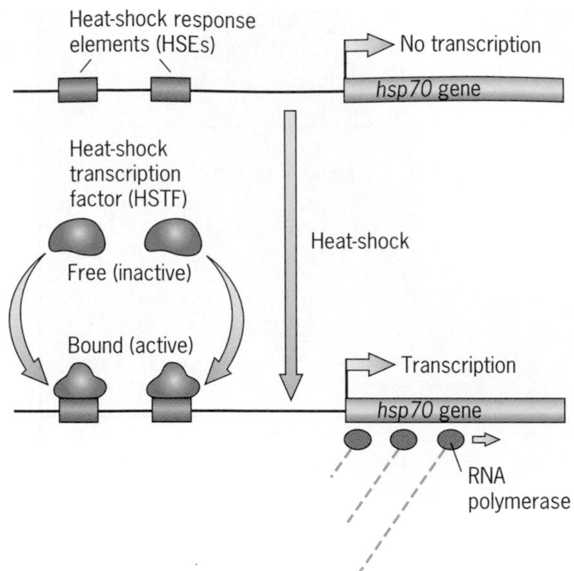

Figure 24.6 Induction of transcription from the *Drosophila hsp70* gene by heat shock. The HSEs are located between 40 and 90 base pairs upstream of the transcription initiation site (bent arrow).

Light: The Ribulose 1,5-bisphosphate Carboxylase Genes in Plants

The most abundant protein on earth is ribulose 1,5-bisphosphate carboxylase (RBC), an enzyme that plays a critical role in photosynthesis in green plants. Through the work of this enzyme, carbon dioxide is incorporated into molecules of sugar, which are then metabolized to provide energy for cells. This process ultimately depends on the ability of plants to absorb light energy. Without light, the entire process comes to a halt, and there is no need for enzymes such as RBC. It is therefore no surprise that the production of RBC is specifically induced when plants are exposed to light.

RBC is a complex enzyme consisting of large and small subunits, each encoded by different genes. In some plant species, both genes are located in the DNA of the chloroplast (Chapter 19), but in others, the gene for the small subunit is located in the nuclear DNA and the gene for the large subunit is located in the chloroplast DNA. The expression of the nuclear gene for the small subunit (denoted *rbcS)* has been analyzed in several different plant species. These studies have shown that *rbcS* is vigorously transcribed after plants have been exposed to light (Figure 24.7).

The light-inducible transcription of *rbcS* is only partly understood. One important event is the ab-

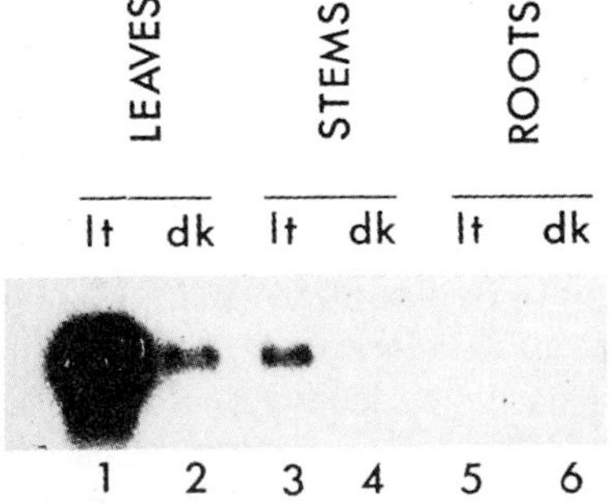

Figure 24.7 Light-induced transcription of the gene for the small subunit of ribulose 1,5-bisphosphate carboxylase. Plants were either exposed to light (lt) or kept in the dark (dk), and RNA was extracted from different plant tissues. The extracted RNA was analyzed by northern blotting using a radioactively labeled probe for the *rbcS* gene. The resulting autoradiogram shows that *rbcS* RNA is produced in leaves and stems, especially after exposure to light.

sorbtion of light by a cytoplasmic protein called phytochrome (Chapter 19). A light-absorbing molecule called a chromophore is attached to each molecule of phytochrome. The absorption of light by the chromophore apparently causes a conformational change in the phytochrome polypeptide, which then triggers changes in other proteins. Although the details of subsequent events are sketchy, it seems that eventually some of these proteins bind to regions upstream of the *rbcS* gene and stimulate its transcription. This response is quick and vigorous; a substantial amount of *rbcS* RNA is generated after exposure to light, providing numerous templates for the synthesis of the small subunit of RBC. A similar process may be involved in the production of the large subunit. Exposure to light therefore induces the production of one of the key enzymes needed for photosynthesis.

Signal Molecules: Genes That Respond to Hormones

In multicellular eukaryotes, one type of cell can signal another by secreting a **hormone.** Hormones circulate through the body, make contact with their target cells, and then initiate a series of events that regulate the expression of particular genes. In animals there are two general classes of hormones. The first class, the **steroid hormones,** are small, lipid-soluble molecules derived from cholesterol. Because of their lipid nature, they have little or no trouble passing through cell membranes. Examples are estrogen and progesterone, which play important roles in female reproductive cycles, testosterone, a hormone of male differentiation and behavior, the glucocorticoids, which are involved in regulating blood sugar levels, and ecdysone, a hormone that controls developmental events in insects. Once these hormones have entered a cell, they interact with cytoplasmic proteins called **hormone receptors.** The receptor-hormone complex that is formed then enters the nucleus where it acts as a transcription factor to regulate the expression of certain genes (Figure 24.8).

The second class of hormones, the **peptide hormones,** are linear chains of amino acids. Like all other polypeptides, these molecules are encoded by genes. Examples are insulin, which regulates blood sugar levels, somatotropin, which is a growth hormone, and prolactin, which targets tissues in the breasts of female mammals. Because peptide hormones are typically too

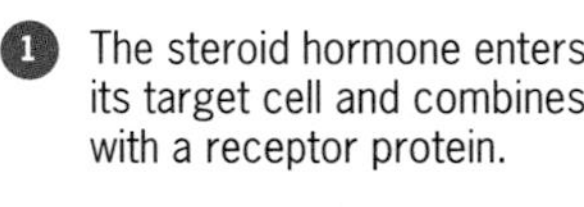

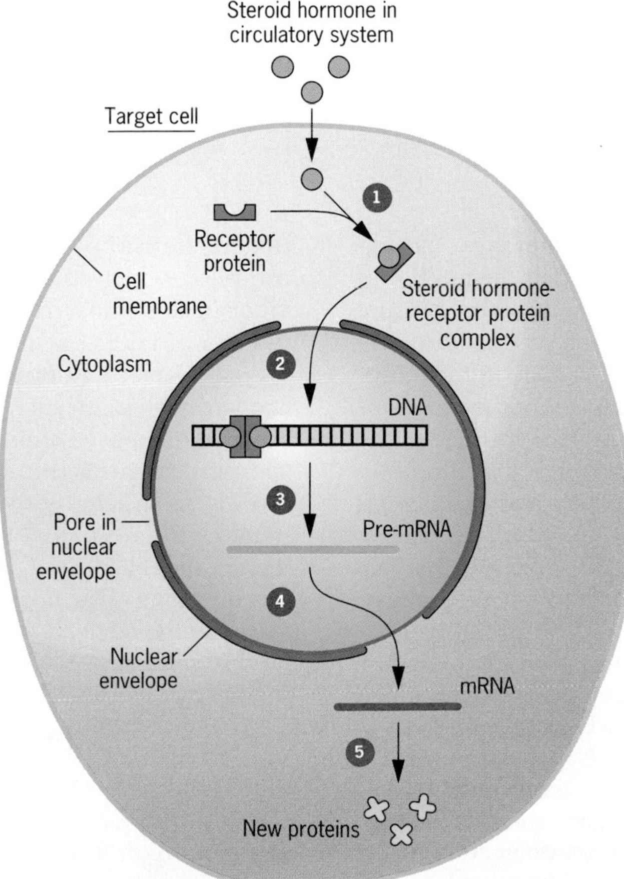

Figure 24.8 Regulation of gene expression by steroid hormones. The hormone interacts with a receptor inside its target cell, and the resulting complex moves into the nucleus where it activates the transcription of particular genes.

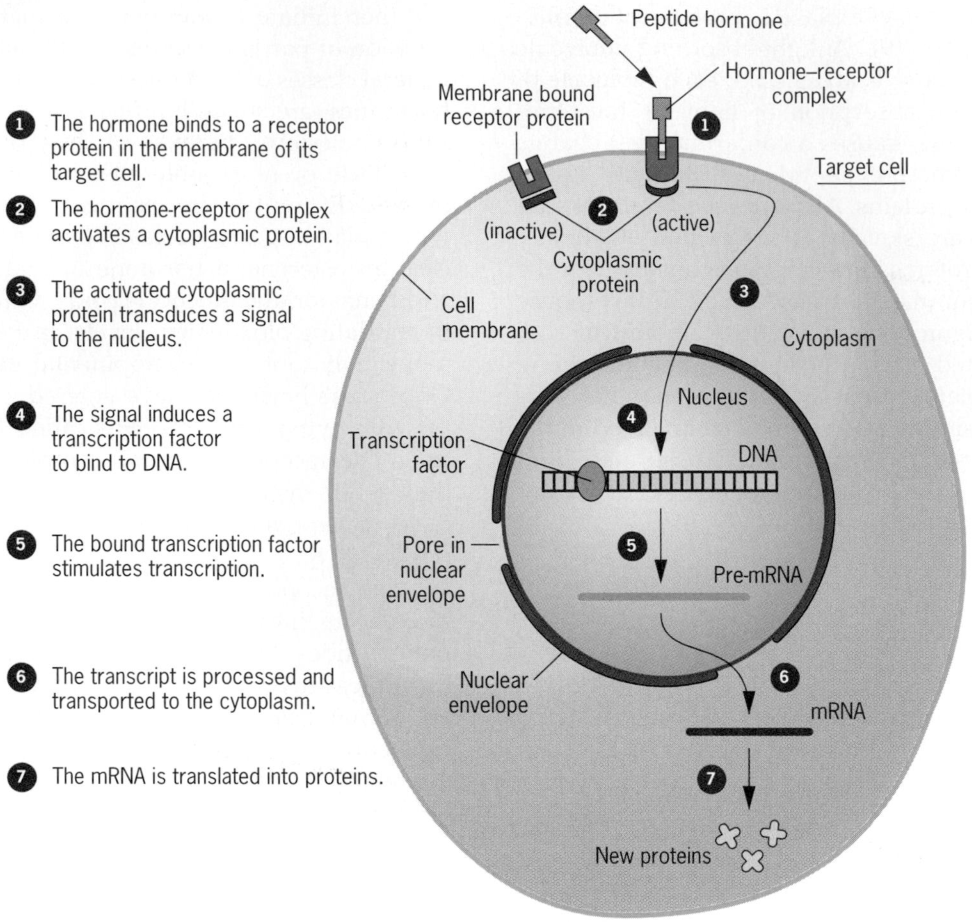

1. The hormone binds to a receptor protein in the membrane of its target cell.
2. The hormone-receptor complex activates a cytoplasmic protein.
3. The activated cytoplasmic protein transduces a signal to the nucleus.
4. The signal induces a transcription factor to bind to DNA.
5. The bound transcription factor stimulates transcription.
6. The transcript is processed and transported to the cytoplasm.
7. The mRNA is translated into proteins.

Peptide hormone
Hormone–receptor complex
Membrane bound receptor protein
Target cell
(inactive)
(active)
Cytoplasmic protein
Cell membrane
Cytoplasm
Nucleus
Transcription factor
DNA
Pore in nuclear envelope
Pre-mRNA
Nuclear envelope
mRNA
New proteins

Figure 24.9 Regulation of gene expression by peptide hormones. The hormone (an extracellular signal) interacts with a receptor in the membrane of its target cell. The resulting hormone-receptor complex activates a cytoplasmic protein that triggers a cascade of intracellular changes. These changes transmit the signal into the nucleus, where a transcription factor stimulates the expression of particular genes.

large to pass freely through cell membranes, the signals they convey must be transmitted to the interior of cells by **membrane-bound receptor proteins** (Figure 24.9). When a peptide hormone interacts with its receptor, it causes a conformational change in the receptor that eventually leads to changes in other proteins inside the cell. Through a cascade of such changes, the hormonal signal is transmitted through the cytoplasm of the cell and into the nucleus, where it ultimately has the effect of regulating the expression of specific genes. This process of transmitting the hormonal signal through the cell and into the nucleus is called **signal transduction.**

Hormone-induced gene expression is mediated by specific sequences in the DNA. These sequences, called **hormone response elements (HREs),** are analogous to the heat-shock response elements discussed earlier. They are situated near the genes they regulate and serve to bind specific proteins, which then act as transcription factors. With steroid hormones such as estrogen, the HREs are bound by the hormone-receptor complex, which then stimulates transcription. The vigor of this transcriptional response depends on the number of HREs present. When there are multiple response elements, hormone-receptor complexes bind cooperatively with each other, significantly increasing the rate of transcription; that is, a gene with two response elements is transcribed more than twice as vigorously as a gene with only one. With peptide hormones, the receptor usually remains in the cell membrane, even after it has formed a complex with the hormone. The hormonal signal is therefore conveyed to the nucleus by other proteins, some of which bind to sequences near the genes that are regulated by the hormone. These proteins then act as transcription factors to control the expression of the genes.

Transcriptional activity can be induced by many other kinds of proteins that are not hormones in the classical sense, that is, not produced by a particular gland or organ. These include a variety of secreted, circulating molecules such as nerve growth factor, epidermal growth factor, and platelet-derived growth factor, and other noncirculating molecules associated with cell surfaces or with the matrix between cells. Although each of these proteins has its own peculiarities, the general mechanism whereby they induce transcription resembles that of the peptide hormones. An interaction between the signaling protein and a membrane-bound receptor initiates a chain of events inside the cell that ultimately results in specific transcription factors binding to particular genes, which are then transcribed.

Key Points: **Eukaryotic gene expression can be induced by environmental factors such as heat and light, and by signaling molecules such as hormones and growth factors. Hormone-induced gene expression is mediated by proteins that interact with the hormones. Some of these hormone receptors act directly as transcription factors by binding to DNA sequences in the vicinity of a gene; others control transcription indirectly through a signal transduction pathway that targets transcription factors to a gene.**

MOLECULAR CONTROL OF TRANSCRIPTION IN EUKARYOTES

Much of the current research on eukaryotic gene expression focuses on the factors that control transcription. This heavy emphasis on transcriptional control is partly due to the development of experimental techniques that have permitted this aspect of gene regulation to be analyzed in great detail. However, it is also due to the appeal of ideas that emerged from the study of prokaryotic genes. In both prokaryotes and eukaryotes, transcription is the primary event in gene expression; it is therefore the most fundamental level at which gene expression can be controlled.

DNA Sequences Involved in the Control of Transcription

Transcription is initiated in the promoter of a gene, the region recognized by the RNA polymerase. However, as we discussed in Chapter 12, the accurate initiation of transcription from eukaryotic gene promoters requires several accessory proteins, or **basal transcription factors.** Each of these proteins binds to a sequence within the promoter to facilitate the proper alignment of the RNA polymerase on the template strand of the DNA.

The transcription of eukaryotic genes is also controlled by a variety of **special transcription factors,** such as those involved in the regulation of the heat, light, and hormone inducible genes we have discussed. These factors bind to response elements, or, more generally, to sequences called **enhancers** located in the vicinity of a gene. The special transcription factors that bind to these enhancers may interact with the basal transcription factors and the RNA polymerase, which bind to the promoter of a gene. The interactions that take place among the special transcription factors, the basal transcription factors, and the RNA polymerase regulate the transcriptional activity of a gene.

Enhancers exhibit three fairly general properties: (1) they act over relatively large distances—up to several thousand base pairs from their regulated gene(s); (2) their influence on gene expression is independent of orientation—they function equally well in either the normal or inverted orientation within the DNA; and (3) their effects are independent of position—they can be located upstream, downstream, or within an intron of a gene and still have profound effects on the gene's expression. These three characteristics distinguish enhancers from promoters, which are typically located immediately upstream of the gene and which function only in one orientation.

Enhancers can be relatively large, up to several hundred base pairs long. They sometimes contain repeated sequences that have partial regulatory activity by themselves. Most enhancers function in a tissue-specific manner; that is, they stimulate transcription only in certain tissues. In other tissues they are simply ignored. A clear example of this tissue specificity comes from the study of the *yellow* gene in *Drosophila* (Figure 24.10). This gene is responsible for pigmentation in many parts of the body—in the wings, legs, thorax, and abdomen. Wild-type flies show a dark brownish-black pigment in all these structures, whereas mutant flies show a lighter yellowish-brown pigment. However, in some mutants, there is a mosaic pattern of pigmentation, brownish-black in some tissues and yellowish-brown in others. These mosaic patterns are due to mutations that alter the transcription of the *yellow* gene in some tissues but not in others. Pamela Geyer and Victor Corces have shown that the *yellow* gene is regulated by several enhancers, some of which are located within an intron, and that each enhancer activates transcription in a different tissue. If, for example, the enhancer for expression in the wing is mutated, the bristles on the wings are yellowish-brown instead of brownish-black. The battery of enhancers associated with the *yellow* gene allows its expression to be controlled in a tissue-specific way.

One of the first enhancers to be studied extensively is located in the chromosome of the eukaryotic virus

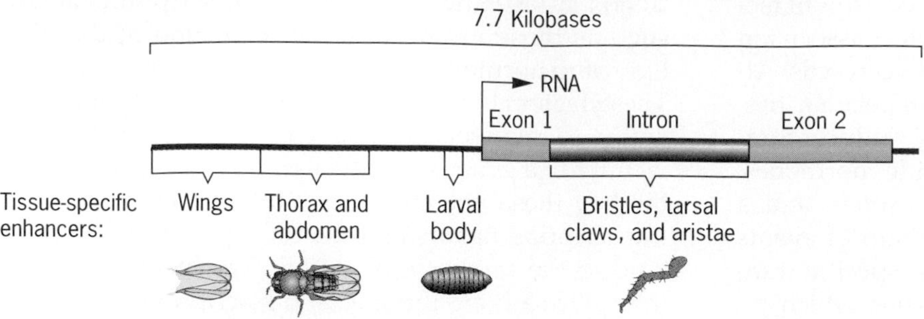

Figure 24.10 The tissue-specific enhancers of the *Drosophila yellow* gene.

SV40. This virus infects monkey cells and has been widely used in biological research. Its 5.2-kb circular chromosome contains a single prominent enhancer about 220 base pairs long (Figure 24.11*a*, *b*). The enhancer regulates the transcription of two groups of genes on the virus chromosome. One group, situated to the right of the enhancer, is transcribed early during infection, and the other group, situated to the left, is transcribed later. The SV40 enhancer contains two 72-bp direct repeats, either of which is sufficient for en-

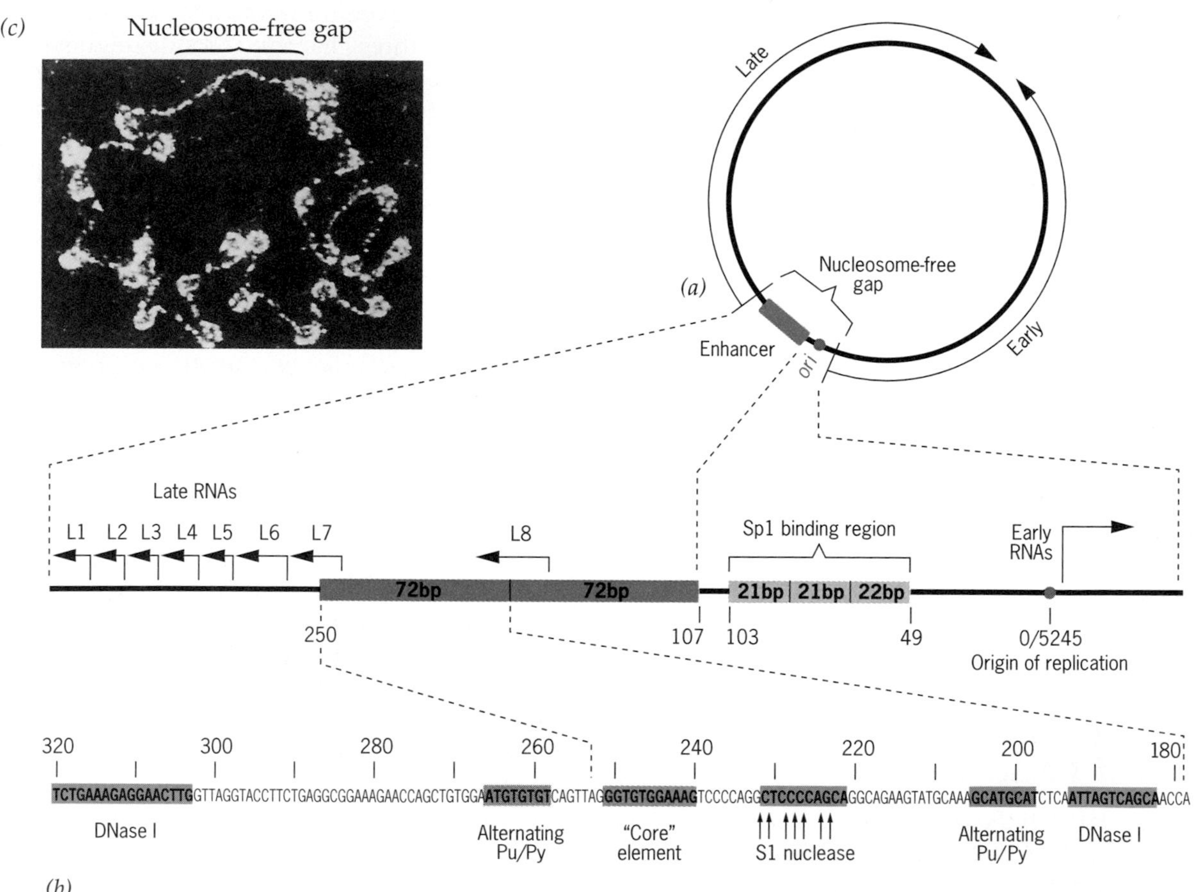

Figure 24.11 Structure of the enhancer of the simian virus 40 (SV40). (*a*) Diagram of the SV40 chromosome showing the location of the enhancer and the origin of replication (*ori*). (*b*) Diagram showing the components of the enhancer. Each of the two 72-base-pair repeats contains regions that are sensitive to DNase I and S1 nuclease digestion, regions with alternating purines (Pu) and pyrimidines (Py), and a core-enhancer element. A 64-base-pair element that binds the transcription factor SP1 is adjacent to one of the repeats. The origin or replication is at coordinate 0/5245 on the circular chromosome. (*c*) Electron micrograph of an SV40 chromosome showing nucleosomes everywhere except in the vicinity of the enhancer.

hancer function. It can be inverted or moved to different locations on the SV40 chromosome and still retains its regulatory ability. Furthermore, if it is inserted upstream or downstream from other eukaryotic genes, it stimulates their transcription. These effects are presumably mediated by proteins that bind to the enhancer. Curiously, when SV40 chromosomes are examined with the electron microscope, the enhancer region is not wrapped around nucleosomes (Figure 24.11*c*). A plausible interpretation is that enhancer-binding transcription factors prevent nucleosome formation.

How do enhancers influence the transcription of genes? Although a detailed answer is not known, it appears that the proteins that bind to enhancers influence the activity of the proteins that bind to promoters, including the basal transcription factors and the RNA polymerase. This finding suggests that the two groups of proteins are in physical contact, perhaps because the enhancer and promoter elements are brought together by some sort of DNA folding.

Proteins Involved in the Control of Transcription: Transcription Factors

Recent research has identified a large number of eukaryotic proteins that stimulate transcription. Many of these proteins appear to have at least two important chemical domains: a DNA-binding domain and a transcriptional activation domain. These domains may occupy separate parts of the molecule, or they may be overlapping. In the GAL4 transcription factor from yeast (see Technical Sidelight: GAL4, a Transcription Factor That Regulates the Genes Involved in Galactose Metabolism in Yeast), for example, the DNA-binding domain is situated near the amino terminus of the polypeptide; two transcriptional activation domains are present in this polypeptide, one more or less in the middle and one near the carboxy terminus. In the steroid hormone receptor proteins, which are transcription factors in animals, the DNA-binding domain is centrally located and seems to overlap a transcriptional activation domain that extends toward the amino terminus; steroid hormone receptors also have a third domain that specifically binds the steroid hormone.

Although the detailed mechanism of transcriptional activation is not fully understood, it appears to involve physical interactions between proteins. A transcription factor that has bound to an enhancer may make contact with one or more proteins at other enhancers, or it may interact directly with proteins that have bound in the promoter region. Through these contacts and interactions, the transcriptional activation domain of the factor may then induce conformational changes in the assembled proteins, paving the way for the RNA polymerase to initiate transcription.

Many eukaryotic transcription factors have characteristic structural motifs that result from associations between amino acids within their polypeptide chains. One of these motifs is the **zinc finger,** a short peptide loop that forms when two cysteines in one part of the polypeptide and two histidines in another part nearby jointly bind a zinc ion; the peptide segment between the two pairs of amino acids then juts out from the main body of the protein as a kind of finger (Figure 24.12*a*). Mutational analysis has demonstrated that these fingers play important roles in DNA binding.

A second motif in many transcription factors is the **helix-turn-helix,** a stretch of three short helices of amino acids separated from each other by turns (Figure 24.12*b*). Genetic and biochemical analyses have shown that the helical segment closest to the carboxy

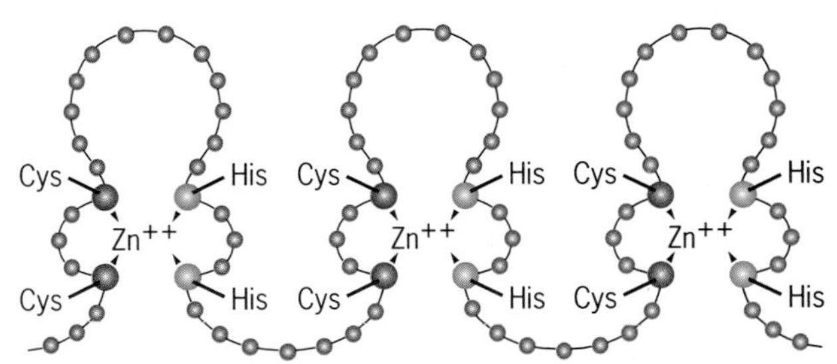

(*a*) Zinc finger motif.

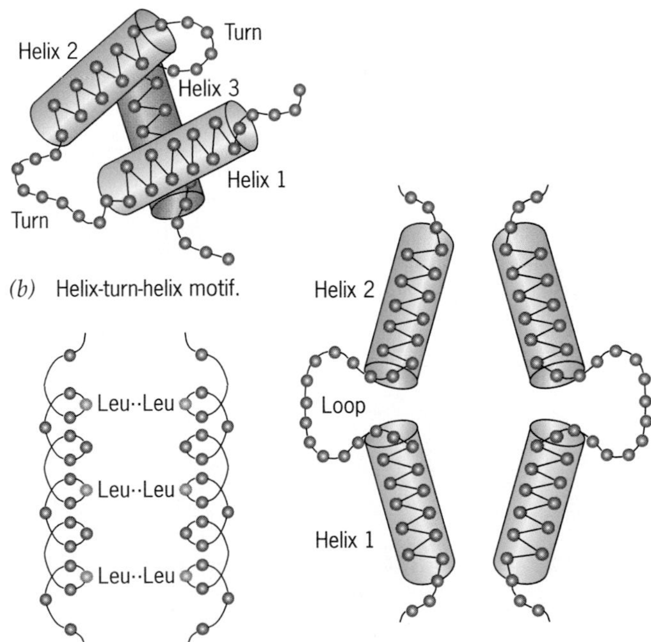

(*b*) Helix-turn-helix motif.

(*c*) Leucine zipper motif. (*d*) Helix-loop-helix motif.

Figure 24.12 Structural motifs within different types of transcription factors. (*a*) Zinc-finger motifs in the mammalian transcription factor SP1. (*b*) Helix-turn-helix motif in a homeodomain transcription factor. (*c*) A leucine zipper motif that allows two polypeptides to dimerize and then bind to DNA. (*d*) A helix-loop-helix motif that allows two polypeptides to dimerize and then bind to DNA.

TECHNICAL SIDELIGHT

GAL4, a Transcription Factor That Regulates the Genes Involved in Galactose Metabolism in Yeast

Yeast cells growing in a medium that contains galactose but not glucose specifically express several genes whose products are involved in galactose metabolism. Two of these genes, *GAL1* and *GAL10*, are less than a kilobase apart on one of the yeast chromosomes (Figure 1). The transcription of these genes is controlled by a sequence located between them. This sequence, called the Upstream Activating Sequence, or UAS, is the binding site for a protein that activates transcription from the *GAL1* and *GAL10* promoters. Because these promoters are oriented in opposite directions, the *GAL1* and *GAL10* genes are divergently transcribed.

The protein that binds to the UAS between *GAL1* and *GAL10* is encoded by an unlinked gene, *GAL4*. This gene is constitutively transcribed in yeast cells. In the presence of galactose, the GAL4 protein is able to activate the transcription of *GAL1* and *GAL10*, but, in the absence of galactose, the GAL4 protein cannot activate the transcription of these genes. Thus galactose can be considered an inducer of *GAL1* and *GAL10* gene expression—through an effect of the GAL4 protein.

The GAL4 protein is 881 amino acids long. Experiments with truncated GAL4 proteins have shown that the first 73 amino acids are sufficient for binding to the UAS. This portion of the protein contains a zinc finger motif, a structural feature frequently associated with the ability to bind DNA. Once the GAL4 protein has bound to DNA, it can activate the transcription of nearby genes. Studies with portions of the GAL4 protein have shown that this activating property resides in two separate domains. One spans amino acids 148–196, and the other spans amino acids 768–881. Transcriptional activation is thought to involve contact between these parts of the GAL4 protein and other proteins that have bound to the gene's promoter, including, possibly, RNA polymerase II.

The UAS located between the *GAL1* and *GAL10* genes resembles the enhancers found in higher eukaryotes. It functions in either orientation and at some distance from the genes it regulates; however, unlike an enhancer, it cannot function downstream of a gene's transcription initiation site. Molecular analysis has revealed that the UAS contains four GAL4 binding sites. Each is an imperfect palindrome of eight base pairs centered on an A:T base pair.

When the UAS is artificially inserted upstream of a gene, that gene comes under the control of the GAL4 protein. This phenomenon was first demonstrated by Hitoshi Kakidani and Mark Ptashne in experiments with Chinese hamster cells that were transfected with two plasmids, one carrying a version of the *GAL4* gene and the other carrying a "reporter" gene located downstream of the UAS. The reporter gene encoded an enzyme (chloramphenicol acetyltransferase, CAT) for which a simple biochemical assay was available. Kakidani and Ptashne set out to determine whether the GAL4 protein expressed from one of the plasmids could stimulate the expression of the *CAT* reporter gene on the other plasmid by binding to the UAS near it. Their results (Figure 2) showed that plasmids carrying an intact *GAL4* gene, or a *GAL4* gene encoding the UAS-binding domain and at least one of the transcriptional activation domains of the protein, were able to stimulate expression of the *CAT* reporter gene. However, a plasmid with a *GAL4* gene encoding only the UAS-binding domain could not stimulate *CAT* gene expression. Thus, both the UAS-binding and transcriptional activation domains of the GAL4 protein are necessary for biological function.

The discovery that the GAL4 protein can regulate the expression of a gene near a UAS has become the basis of a procedure to study the effects of eukaryotic gene expression *in vivo*. In this procedure, a UAS is inserted upstream of a gene's transcription initiation site, and the resulting fusion

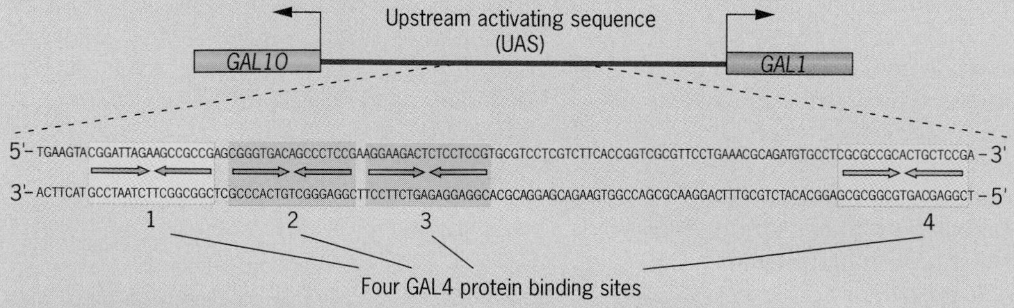

Figure 1 The *GAL1* and *GAL10* genes in yeast. The GAL4 protein coordinately regulates these divergently transcribed genes by binding to four sites within the Upstream Activating Sequence (UAS) between them.

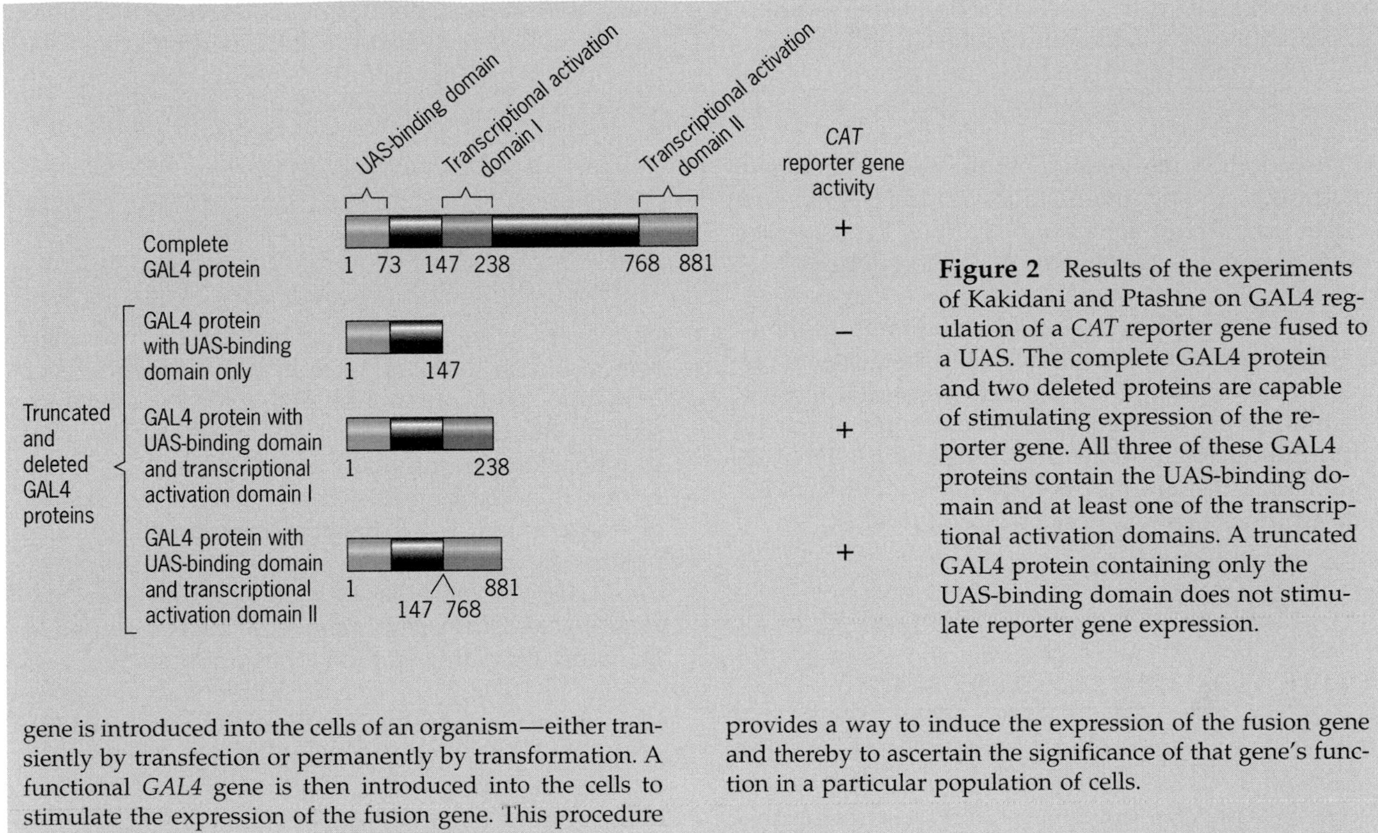

Figure 2 Results of the experiments of Kakidani and Ptashne on GAL4 regulation of a *CAT* reporter gene fused to a UAS. The complete GAL4 protein and two deleted proteins are capable of stimulating expression of the reporter gene. All three of these GAL4 proteins contain the UAS-binding domain and at least one of the transcriptional activation domains. A truncated GAL4 protein containing only the UAS-binding domain does not stimulate reporter gene expression.

gene is introduced into the cells of an organism—either transiently by transfection or permanently by transformation. A functional *GAL4* gene is then introduced into the cells to stimulate the expression of the fusion gene. This procedure provides a way to induce the expression of the fusion gene and thereby to ascertain the significance of that gene's function in a particular population of cells.

terminus is required for DNA binding; the other helices seem to be involved in the formation of protein dimers. In many transcription factors, the helix-turn-helix motif coincides with a highly conserved region of approximately 60 amino acids called the **homeodomain,** so named because it occurs in proteins encoded by the homeotic genes of *Drosophila*. Classical analyses have demonstrated that mutations in these genes alter the developmental fates of groups of cells (Chapter 25). Thus, for example, mutations in the *Antennapedia* gene can cause antennae to develop as legs. This bizarre phenotype is an example of a homeotic transformation—the substitution of one body part for another during the developmental process. Molecular analyses of the homeotic genes in *Drosophila* have demonstrated that each encodes a protein with a homeodomain and that these proteins can bind to DNA. The homeodomain proteins stimulate the transcription of particular genes in a spatially and temporally specific manner during development. Homeodomain proteins have also been identified in other organisms, including human beings, where they may play important roles as transcription factors.

A third structural motif found in transcription factors is the **leucine zipper,** a stretch of amino acids with a leucine at every seventh position (Figure 24.12*c*). Polypeptides with this feature can form dimers by interactions between the leucines in each of their zipper regions. Usually, the zipper sequence is adjacent to a positively charged stretch of amino acids. When two zippers interact, these charged regions splay out in opposite directions, forming a surface that can bind to negatively charged DNA.

A fourth structural motif found in some transcription factors is the **helix-loop-helix,** a stretch of two helical regions of amino acids separated by a nonhelical loop (Figure 24.12*d*). The helical regions permit dimerization between two polypeptides. Sometimes the helix-loop-helix motif is adjacent to a stretch of basic (positively charged) amino acids, so that when dimerization occurs, these amino acids can bind to negatively charged DNA. Proteins with this feature are denoted basic HLH, or bHLH, proteins.

Transcription factors with dimerization motifs such as the leucine zipper or the helix-loop-helix could, in principle, combine with polypeptides like themselves to form homodimers, or they could combine with different polypeptides to form heterodimers. This second possibility suggests a way in which complex patterns of gene expression can be achieved. The transcription of a gene in a particular tissue might depend on activation by a heterodimer, which could form only if its constituent polypeptides were synthesized in that tissue. Moreover, these two polypeptides would have to be present in the correct amounts to favor the formation of the heterodimer over the corresponding homodimers. Subtle modulations in gene expression might therefore be achieved by shifting the concentrations of the two components of a heterodimer.

Key Points: **Enhancers act in an orientation-independent manner over considerable distances to regulate transcription from a gene's promoter. This regulation is mediated by different types of transcription factor proteins that bind to these DNA sequences. Different motifs have been identified in the amino acid sequences of these proteins.**

GENE EXPRESSION AND CHROMOSOME ORGANIZATION

The primary event in gene expression is transcription of DNA into RNA. For this to occur, the DNA must be accessible to RNA polymerase and an assortment of other proteins that help to initiate transcription. If the DNA is tightly bound by histones or other kinds of "packaging" proteins, it may be too condensed to allow transcription. Chromosomal position may also affect the transcription of genes because genes that have been transposed to different sites within a chromosome often show altered expression. These alterations may be caused by enhancers at the new chromosomal site or by other aspects of chromosome structure. All these phenomena indicate that gene expression is influenced by chromosome organization.

Transcription in Lampbrush Chromosome Loops

Few if any of the genes in highly condensed chromosomes, such as those found at metaphase of mitosis or in mature sperm, are transcribed. In order to be transcribed, the chromatin must be "open" to the transcriptional apparatus. Some of the first evidence supporting this idea came from the cytological localization of transcription in the meiotic chromosomes of amphibian oocytes. These very large (400 to 800 μm long), duplicated chromosomes—called lampbrush chromosomes—consist of a highly condensed axis surrounded by numerous pairs of lateral loops (Chapter 10).

The lateral loops in lampbrush chromosomes are regions of intense transcriptional activity (Figure 24.13). Researchers have demonstrated this activity by pulse-labeling amphibian oocytes with ^{3}H-uridine and then examining the chromosomes by autoradiography. The ^{3}H-uridine is incorporated into newly synthesized RNA. Autoradiography of pulse-labeled oocytes reveals that the radioactive uridine is localized around the lateral loops of the lampbrush chromosomes rather than around the condensed axes; thus, the more loosely organized loops are actively involved in RNA synthesis.

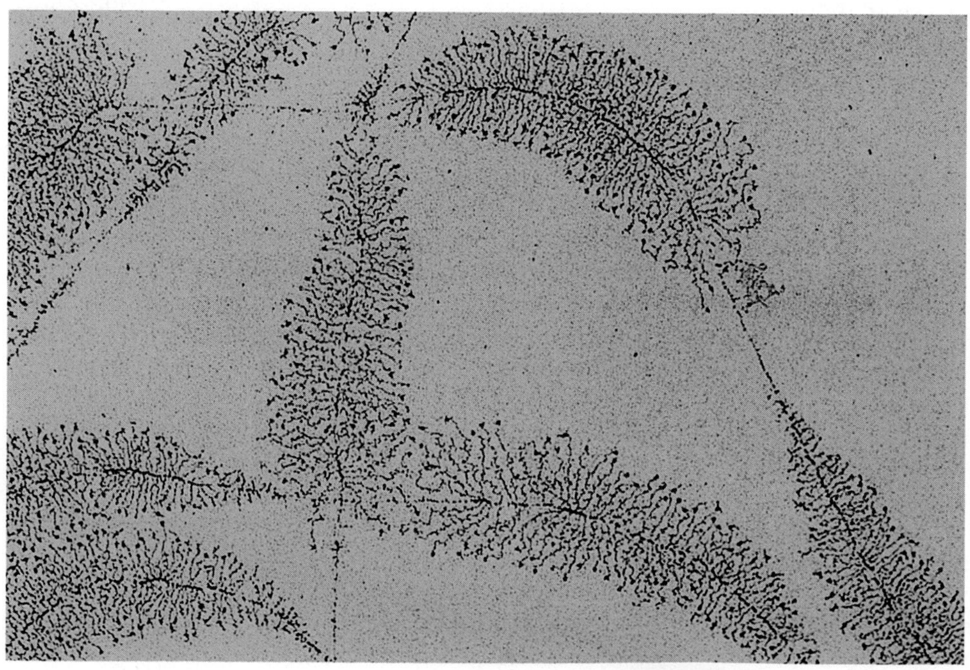

Figure 24.13 Photomicrograph showing transcription from the DNA loops of amphibian lampbrush chromosomes.

Transcription in Polytene Chromosome Puffs

Additional cytological evidence for transcription in "open" chromosome regions comes from the study of polytene chromosomes in *Drosophila* and other Dipteran insects. As discussed in Chapter 7, these chromosomes consist of hundreds of sister chromatids aligned side by side, creating a large, thick cable that is longitudinaly differentiated into light and dark bands. For many years, the bands, called **chromomeres**, have been thought to have a functional significance. One reason for this speculation is that during the course of development, particular bands expand into diffuse, less densely staining structures called **puffs** (Figure 24.14). *In situ* hybridization experiments have shown that puffs contain genes that are actively transcribed. In these experiments, a radioactively labeled RNA or DNA that contains a specific sequence is denatured and hybridized with denatured RNA or DNA in tissues that have been prepared for cytological analysis. The labeled RNA or DNA probe binds to its complementary sequence in the tissues, that is, it binds *in situ*; the location of the bound probe is then revealed by autoradiography. Application of this technique shows that puffs in polytene chromosomes contain DNA sequences that are actively transcribed into RNA. Over time, particular puffs regress and new ones form at different sites in the chromosomes. This temporal sequence of puffing is controlled by **ecdysone,** a steroid hormone that appears in pulses at different times during development. A plausible interpretation is that ecdysone induces transcription in puffed bands. Puffing may also be induced by other factors, for example, heat shock. When *Drosophila* larvae are incubated at temperatures greater than 33°C, several new puffs appear in the polytene chromosomes and old ones regress. The new puffs correspond to the loci for the heat-shock genes. These cytological studies with *Drosophila* therefore show that diffuse, expanded chromosomal regions are sites of intense RNA synthesis.

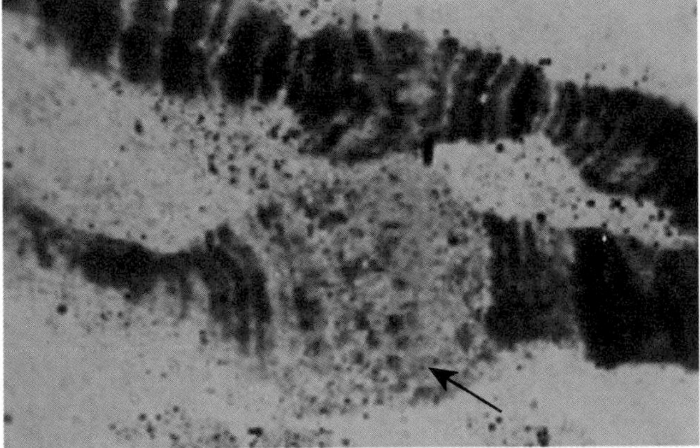

Figure 24.14 A puff (arrow) in the polytene chromosomes of *Drosophila*.

Molecular Organization of Transcriptionally Active DNA

The findings with lampbrush and polytene chromosomes raise questions about the molecular organization of transcriptionally active DNA. Is this DNA packaged in nucleosomes? If it is, what structural changes occur in the nucleosomes during transcription? Are the nucleosomes "opened" and "closed" as the RNA polymerase passes along the DNA template? Are there sites in the DNA that define transcriptionally active domains such as lampbrush chromosome loops and polytene chromosome puffs?

Various techniques have been used to answer these questions, including electron microscopy, enzymatic digestion of DNA, and biochemical analysis of chromosomal proteins. The application of these techniques has shown that DNA that is being transcribed is indeed packaged into nucleosomes, and that it has the same frequency and spacing of core particles as nontranscribed DNA. However, the fact that transcribed DNA is considerably more sensitive to degradation by enzymes such as pancreatic deoxyribonuclease I (DNase I) suggests that it is more "open" than nontranscribed DNA.

This increased sensitivity to DNase was demonstrated in 1976 by Mark Groudine and Harold Weintraub in a comparative study of transcribed and nontranscribed genes in chicken red blood cells. Groudine and Weintraub extracted chromatin from these cells and partially digested it with DNase I. They then probed the residual material for sequences of two genes, β-globin, which is actively transcribed in red blood cells, and ovalbumin, which is not. They found that over 50 percent of the β-globin DNA had been digested by the DNase I enzyme, compared with only 10 percent of the ovalbumin DNA. The results strongly implied that the actively transcribed gene was more "open" to nuclease attack. Subsequent research has shown that the nuclease sensitivity of transcriptionally active genes depends on at least two small nonhistone proteins, HMG14 and HMG17 (HMG for *h*igh *m*obility *g*roup, because they have high mobility during gel electrophoresis). When these proteins are removed from active chromatin, nuclease sensitivity is lost; when they are again added, it is restored.

The treatment of isolated chromatin with a very low concentration of DNase I causes the DNA to be cleaved at a few specific sites, appropriately called **DNase I hypersensitive sites.** Some of these sites have been shown to lie upstream of transcriptionally active genes, in either promoter or enhancer regions. The functional significance of these hypersensitive sites is still unclear, but some evidence suggests that they may mark regions in which the DNA is locally unwound, perhaps because transcription has begun.

In the case of the human β-globin genes, several DNase I hypersensitive sites are located in a 15 kb-long **locus control region (LCR)** upstream of the genes themselves. This LCR contains binding sites for transcription factors that preactivate the β-globin genes for transcription. Preactivation is detected by an increase in the sensitivity of the DNA within the LCR to digestion with low concentrations of DNase I. The actual transcription of the β-globin genes appears to require this preactivation and is stimulated by transcription factors that bind to specific enhancers in the β-globin gene complex. Thus, expression of the β-globin genes depends on changes in the LCR that alter the structure of the adjacent chromatin, perhaps by making it more accessible to the action of particular transcription factors.

Are chromosomes organized into functional domains by special sites in the DNA? Some evidence from the study of *Drosophila* heat-shock genes suggests that the answer is yes. Paul Schedl, Rebecca Kellum, and colleagues have identified short elements in the DNA on either side of a pair of *hsp70* genes situated in polytene chromosome band 87A7 (Figure 24.15). These elements, called **specialized chromatin structures** (denoted **scs** and **scs'**), are themselves relatively impervious to DNase I digestion, but each is flanked by DNase I hypersensitive sites. The scs and scs' elements apparently insulate the *hsp70* genes within the 87A7 chromomere from the surrounding chromatin. This insulating property was demonstrated when these elements were artificially placed on either side of another gene, and the resulting scs–gene–scs' complex was tested for function in different genomic positions. As hypothesized, the "sandwiched" gene was expressed uniformly in a wide variety of chromosomal sites; thus, it could function independently of the surrounding chromatin. However, when either of the bordering elements was eliminated, the gene's ability to function was influenced by chromosomal position. These results therefore suggest that chromosomes may be organized into functional domains by short DNA elements.

Euchromatin and Heterochromatin

Variation in the density of chromatin within the nuclei of cells leads to differential staining of sections of chromosomes. The deeply staining material is called **heterochromatin,** and its lightly staining counterpart is called **euchromatin.** What, if any, is the functional significance of these different types of chromatin?

A combination of genetic and molecular analyses has shown that the vast majority of eukaryotic genes are located in euchromatin. Moreover, when euchromatic genes are artificially transposed to a heterochromatic environment, they tend to function abnormally, and, in some cases, not to function at all. This impaired ability to function can create a mixture of normal and mutant characteristics in the same individual, a condition referred to as **position-effect variegation.** This term is used because the variability in the phenotype is caused by changing the position of the euchromatic gene, specifically by relocating it to the heterochromatin. Many examples of position–effect variegation have been discovered in *Drosophila*, usually in association with inversions or translocations that move a euchromatic gene into the heterochromatin. The *white* mottled allele, w^{m4}, discussed in Chapter 7 is a good example. In this case, a wild-type allele of the *white* gene has been relocated by an inversion, with one break near the *white* locus and the other in the basal heterochromatin of the X chromosome. This rearrangement interferes with the normal expression

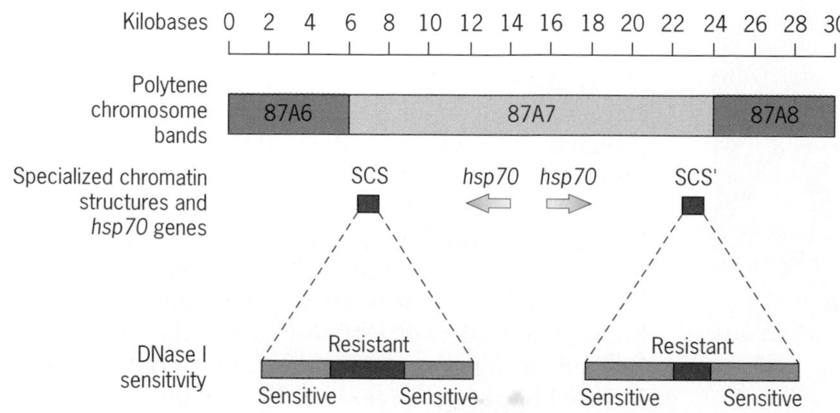

Figure 24.15 Location of the specialized chromatin structures (scs and scs') in the vicinity of the *hsp70* genes in band 87A7 of chromosome 3 in *Drosophila*. The regions of DNase I sensitivity that flank the structures are indicated.

of the *white* gene and causes a mottled-eye phenotype. Apparently, the euchromatic *white* gene cannot function well in a heterochromatic environment. This and other examples have led to the view that heterochromatin represses gene function, perhaps because it is condensed into a form that is not accessible to the transcriptional machinery. Ongoing research is attempting to identify the proteins that might be involved in this condensation process.

One research strategy has been to identify genes that, when mutant, suppress or enhance position-effect variegation. Several screens for such mutations have used the variegating w^{m4} allele of the *white* locus in *Drosophila* as a convenient assay. When w^{m4} is combined with a *su*ppressor of position-effect *var*iegation, that is, with a *su(var)* mutation, the eye color of the fly will be wild type, or nearly so; when w^{m4} is combined with an *e*nhancer of position-effect *var*iegation, that is, with an *e(var)* mutation, the eye color will be more mottled than it is with w^{m4} alone. Figure 24.16 shows the effect of a dominant suppressor of position-effect variegation, *Su(var)205*^{4}, on the w^{m4} phenotype. Molecular analysis has demonstrated that this suppressor mutation abolishes the production of a nonhistone chromosomal protein that associates specifically with the heterochromatin. Flies heterozygous for the *Su(var)205*^{4} mutation suppress the w^{m4} phenotype; however, flies that are homozygous for it die early in development. Thus, the protein encoded by the wild-type *Su(var)205* gene, called Heterochromatin Protein 1, or simply HP1, is essential for life. Perhaps a mutation in one copy of the *Su(var)205* gene reduces the amount of HP1 enough to relax the heterochromatin but not enough to unravel it. In this relaxed state, a euchromatic gene such as w^{m4} that has come under the repressive influence of the heterochromatin is expressed at normal levels; consequently, the phenotype that it controls ceases to variegate.

Approximately 30 percent of *Drosophila* DNA is heterochromatic. Most of this DNA is located around the centromeres of the chromosomes. However, some of it is interspersed in the euchromatin, and some is located at the ends, or telomeres, of the chromosomes. Thus, in *Drosophila*, three types of heterochromatin are present: *centric* heterochromatin (around the centromeres), *intercalary* heterochromatin (scattered in the euchromatin), and *telomeric* heterochromatin (at the ends of the chromosomes).

Most of the heterochromatic regions in the *Drosophila* genome appear to be rich in repetitive DNA sequences. Both highly repetitive and moderately repetitive sequences can be found in these regions, and the moderately repetitive sequences comprise many types of transposable elements, especially inactive retrotransposons (Chapter 18). The heterochromatic regions of the *Drosophila* genome also contain some expressed genes. Several hundred genes for ribosomal RNA are located in the centric heterochromatin of the X and Y chromosomes, and about 20 protein-encoding genes are known to reside in other heterochromatic regions of the *Drosophila* genome. Thus, heterochromatin should not be associated with complete genetic inactivity. Curiously, when a heterochromatic gene is moved into a euchromatic environment by a chromosome rearrangement, its expression tends to variegate. Furthermore, this variegation tends to be enhanced by the *su(var)* mutations that suppress the variegation of a euchromatic gene that has been placed in a heterochromatic environment. The fact that heterochromatic genes exhibit position effects opposite to those of euchromatic genes indicates that they are specially adapted to function in a heterochromatic environment. However, the nature of this adaptation is currently unknown.

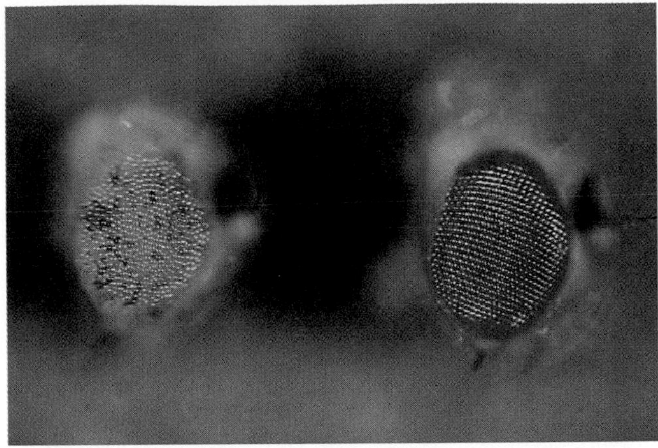

Figure 24.16 Photographs showing the eye phenotype of a fly carrying the w^{m4} allele (left) and a fly carrying this allele along with a dominant suppressor of variegation, *Su(var)205*^{4} (right).

Gene Silencing

Although heterochromatin has the ability to repress euchromatic genes that have been placed near it by a chromosome rearrangement, this feature of chromatin is not the only one that prevents gene expression in eukaryotes. Studies with many different organisms indicate that other aspects of chromatin organization, some possibly related to the protein components of heterochromatin, control gene activity. We have already seen that transcription factors, acting in concert with particular enhancer and promoter sequences, turn genes on in a spatially and temporally specific manner. However, at a particular time in a given eukaryotic cell, most genes are not being transcribed. Thus, the rule, rather than the exception, is that genes are transcriptionally silent. What features of chromatin organization bring about this widespread gene silencing?

In *Drosophila*, studies of genes that control the formation of particular body parts have identified proteins that are involved in gene silencing. These proteins are encoded by a special class of genes known as the *Polycomb Group*, abbreviated *PcG*. This class is named for *Polycomb*, a gene defined by a dominant mutation that causes extra bristles to form in clusters on the legs. These clusters, known as sex combs, are normally found on the forelegs of male *Drosophila*; however, in males that carry a *Polycomb* mutation, they are found on the middle and hind legs as well. *Polycomb* mutations also cause other phenotypic anomalies, including the transformation of antennae into rudimentary legs.

Polycomb and the other *PcG* genes encode proteins that regulate the expression of transcription factors involved in *Drosophila* development. These regulatory proteins appear to form large multiprotein complexes that associate with specific regions of the genome—in particular, with the loci that encode these developmentally important transcription factors. When and how they do this remain a mystery. However, several facts are known. First, the PcG proteins associate with chromosomal regions that contain special DNA sequences known as Polycomb Response Elements (PREs). Individually, most of the PcG proteins are not able to bind to the PRE DNA itself; however, they may be able to bind as a complex, or they may interact with other types of proteins that can bind to the PRE DNA. Second, the PcG proteins repress the genes located near the PREs. Whether or not the PcG proteins associate with a particular PRE and repress the genes in its neighborhood depends on factors present in the animal very early in its embryonic development. Third, once established, the association between the PcG proteins and a PRE is maintained faithfully throughout development. Thus, the repressed state of the genes near a PRE is transmitted from a cell to its daughters through each mitotic division. This form of gene silencing therefore assures that clones of cells develop consistently into particular body parts. When the PcG mechanism of gene silencing is compromised by mutations in any of the *PcG* genes, some body parts develop inappropriately, as for example, with the Polycomb mutant itself, where antennae may develop as legs, and middle and hind legs may develop as forelegs.

The phenomenon of gene silencing has also been studied in yeast where one of the most informative phenotypes is mating type. The haploid yeast *Saccharomyces cerevisiae* has two possible mating types, "a" and α. The type that is expressed by a particular cell is controlled by the mating type locus (*MAT*) situated on chromosome 3. Genetic information in this locus specifies the mating type as either "a" or α. Thus, the *MAT* locus has two alleles, *MAT**a*** and *MAT*α. Colonies grown from single cells all express the same mating type information. However, at a frequency that depends on the particular yeast strain, the mating type can switch from "a" to α, or from α to "a."

The basis for this switching is now well understood. At some distance on either side of the *MAT* locus there are DNA segments called *silent mating type cassettes* that contain extra genetic information for mating type (Figure 24.17). The cassette on the left, denoted *HML*α, contains information for the α mating type and the cassette on the right, denoted *HMR**a***, contains information for the "a" mating type. However, a gene silencing mechanism prevents either of these cassettes from being expressed. Thus, the mating type of a cell is exclusively controlled by the genetic information at the *MAT* locus. When a yeast cell expresses an enzyme called the HO endonuclease, DNA in the *MAT* locus can be replaced by DNA from one of the flanking cassettes. If the *MAT* locus carries the ***a*** allele, the silent α allele located in the cassette to its left can be copied into the *MAT* locus and subsequently expressed. This DNA replacement will cause the mating type of the cell to switch from "a" to α. Conversely, if the *MAT* locus carries the α allele, the silent ***a*** allele in the cassette to its right can be copied into the *MAT* locus and then expressed, switching the mating type from α to "a."

An important feature of the entire yeast mating type system is that a particular yeast cell expresses the genetic information for only one of the two possible mating types. Thus, the cassettes that flank the expressed *MAT* locus must be silenced. Recent research has demonstrated that the silencing of these cassettes involves several proteins, including one called the *re*pressor/*a*ctivator *p*rotein, abbreviated Rap1p, several *s*ilent *i*nformation *r*egulator *p*roteins, abbreviated Sir(n)p, where the numeral n identifies the specific protein, and several proteins in the *o*rigin *r*ecognition *c*omplex, abbreviated ORC—a complex that is also involved in DNA replication. Experimental evidence indicates that histone H4 is involved in silencing the left and right mating type cassettes. Although the detailed mechanism remains to be elucidated, it appears that the ORC proteins and Rap1p interact with specific DNA sequences situated alongside the mating type loci in the left and right cassettes. These bound proteins then facilitate the formation of a more elaborate protein complex that includes the Sir proteins, which interact with Rap1p, histones, and each other to si-

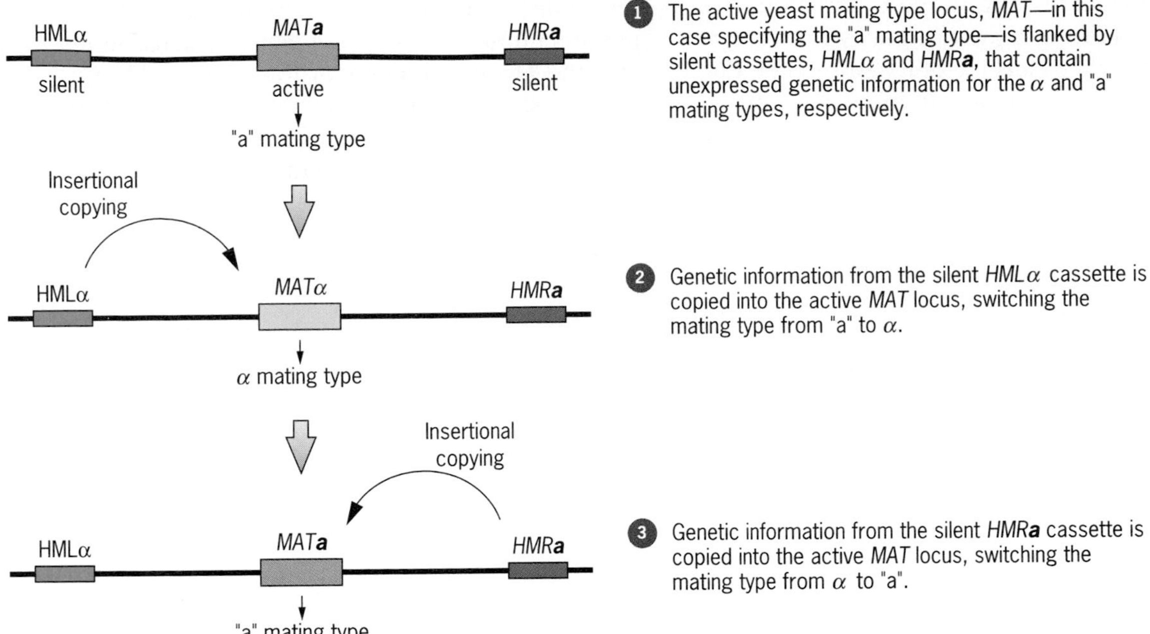

Figure 24.17 The process of mating type switching in yeast.

lence the genes in the left and right mating type cassettes. The DNA sequences that mediate this process are therefore appropriately called **silencers.** When this large protein complex has assembled around the silencer sequences, it makes the left and right mating type cassettes inaccessible to the enzymes that catalyze transcription.

Silencing has also been observed at the telomeres of yeast chromosomes, where it may control the activity of certain types of transposable elements. In the trypanosomes that cause nagana and African sleeping sickness, a form of silencing controls the expression of the variant surface gylcoprotein (*vsg*) genes mentioned at the beginning of this chapter. These genes are located at numerous sites in the trypanosome genome, most of them being near the ends of chromosomes. Altogether there may be as many as 1000 different *vsg* genes. However, in the blood of an infected animal, trypanosomes express only one of the *vsg* genes, and it is always located near a chromosome end. All the other *vsg* genes, including all the *vsg* genes that are located near telomeres, are silenced. The single expressed *vsg* gene resides in a multigenic unit controlled by a promoter about 40–50 kb away (Figure 24.18); other unrelated genes lie between the promoter and the *vsg* gene. This entire multigenic unit is called an **Expression Site (ES).**

Expression Sites appear to exist at the ends of each of the trypanosome's 24 large chromosomes; however, at any one time, only one of them is transcribed. An as-yet-unknown mechanism silences the other potential expression sites, thereby restricting the infecting trypanosomes to the production of a single type of variant surface glycoprotein (VSG). During the course of an infection, the ES that is transcribed may change. The active ES may be silenced and a previously inac-

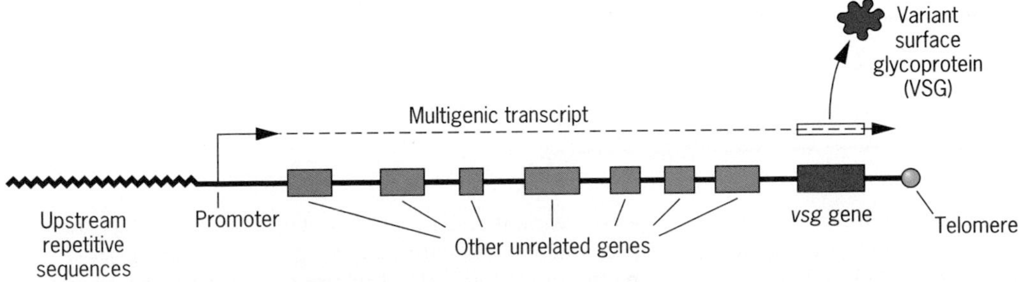

Figure 24.18 Structure of an Expression Site (ES) containing an active *vsg* gene near the end of one of the chromosomes in *Trypanosoma brucei.*

tive ES become active, leading to the production of a different VSG. Such changes provide a way for the trypanosomes to alter their surface coats and evade destruction by the infected animal's immune system.

The trypanosome's surface coat can be changed in two other ways. In addition to the *vsg* genes located at the telomeres of the large chromosomes, trypanosomes possess numerous other *vsg* genes situated at internal loci and at the telomeres of more than 100 minichromosomes; these extra *vsg* genes are all transcriptionally silent. However, as with the silent yeast mating type cassettes, one of the silent *vsg* genes can be copied

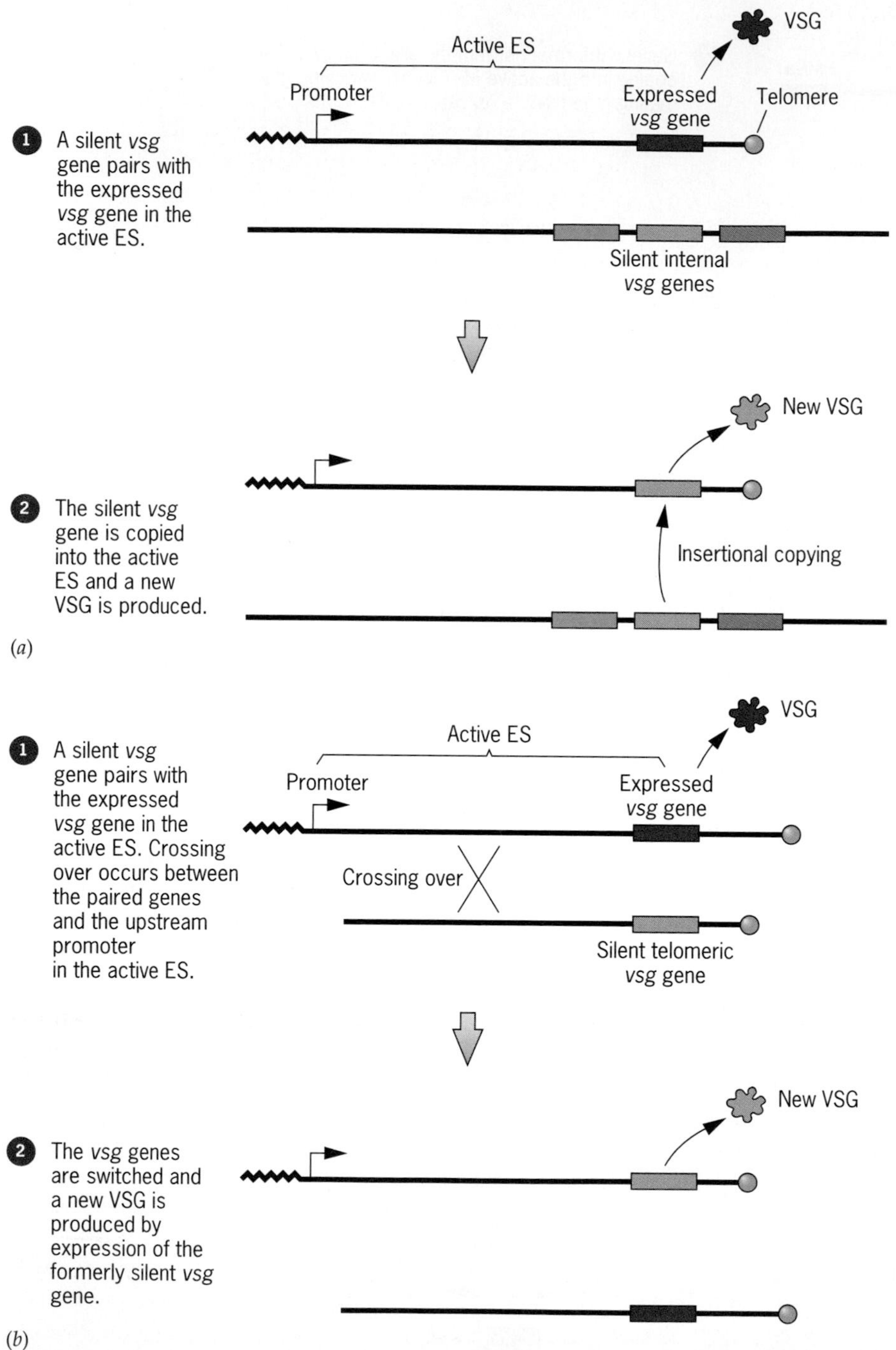

Figure 24.19 (*a*) Replacement of the telomeric *vsg* gene in an active trypanosome ES by a silent internal *vsg* gene encoding a different variant surface glycoprotein (VSG). The replacement mechanism excises the currently active *vsg* gene and inserts a copy of the silent *vsg* gene into the vacant site within the active ES. At the end of this process, a new variant surface glycoprotein is produced. (*b*) Replacement of the telomeric *vsg* gene in an active trypanosome ES by a silent telomeric *vsg* gene encoding a different VSG. The replacement mechanism involves crossing over between the paired *vsg* genes and the promoter of the ES.

into the ES that is currently transcribed, thereby causing the organism to produce a different variant surface glycoprotein (Figure 24.19*a*). The nature of the VSG protein can also be altered by a crossover that exchanges the active *vsg* gene for one of the many inactive telomeric *vsg* genes (Figure 24.19*b*); after the inactive gene becomes linked to the promoter of the transcribed ES, it becomes active. Thus, either by a nonreciprocal copying mechanism or by reciprocal crossing over, the *vsg* gene in the active ES can be replaced by a different *vsg* gene, with the result that a different VSG is produced.

The *vsg* genes outside the telomeric ESs on the trypanosome's large chromosomes are inactive because each of these genes lacks a promoter. However, the absence of a promoter cannot explain why all but one of the *vsg* genes located within the telomeric ESs are inactive. Recently, researchers have discovered that inactive ESs contain β-glucosyl-hydroxymethyluracil, a modified nucleotide, and that the single active ES does not. Thus, modification of nucleotide sequences within the ESs may play a role in telomeric *vsg* gene silencing.

DNA Methylation and Imprinting

The chemical modification of nucleotides also appears to be important for the regulation of genes in other eukaryotes, especially mammals. Of the approximately 3 billion base pairs in a typical mammalian genome, about 40 percent are G:C base pairs and about 2–7 percent of these are modified by the addition of a methyl group to the cytosine (Figure 24.20). Most of the methylated cytosines are found in base-pair doublets with the structure

$$\begin{array}{l} 5'\ \mathrm{mC\ p\ G}\ \ 3' \\ 3'\ \ \ \mathrm{G\ p\ Cm}\ 5' \end{array}$$

where mC denotes methylcytosine and the p between C and G denotes the phosphodiester bond between adjacent nucleotides in each DNA strand. This structure is often simply abbreviated by giving the composition of one strand; thus, mCpG. Methylated CpG dinucleotides can be detected by digesting DNA with restriction enzymes that are sensitive to chemical modifications of their recognition sites. For example, the enzyme *Hpa*II recognizes and cleaves the sequence CCGG; however, when the second cytosine in this sequence is methylated, *Hpa*II cannot cleave the sequence. Thus, methylated and unmethylated DNAs give different patterns of restriction fragments when they are digested with this enzyme.

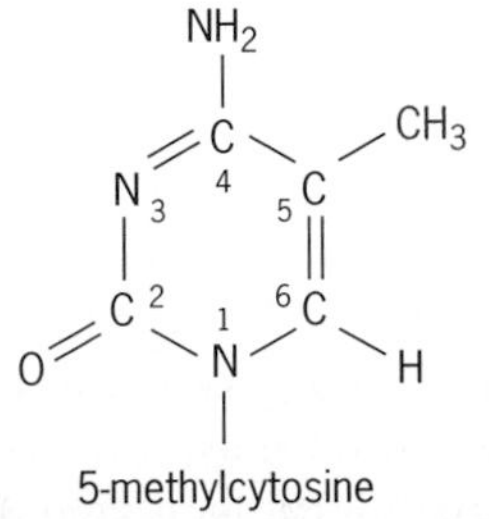

5-methylcytosine

Figure 24.20 The structure of 5-methylcytosine.

CpG dinucleotides occur less often than expected in mammalian genomes, probably because they have been mutated into TpG dinucleotides over the course of evolution. Moreover, the distribution of CpG dinucleotides is uneven, with numerous short segments of DNA having a much higher density of CpG dinucleotides than other regions of the genome. These CpG-rich segments, usually about 1–2 kb long, are called **CpG islands.** In the human genome, there are about 45,000 such islands, most being situated near transcription start sites. Molecular analysis has demonstrated that the cytosines in these islands are rarely, if ever, methylated, and that this un- or undermethylated state is conducive to transcription. Thus, DNA in the vicinity of a CpG island is hypersensitive to digestion with DNase I, and its nucleosomes are usually somewhat different than nucleosomes elsewhere in the genome—typically there is less histone H1, and some of the core histones are acetylated.

Where methylated DNA is found, it is associated with transcriptional repression. This is most dramatically seen in female mammals where the inactive X chromosome is extensively methylated. Regions of the mammalian genome that contain repetitive sequences, including those regions that are rich in transposable elements, are also methylated, perhaps as a way of protecting the organism against the deleterious effects of transposon expression and movement. The mechanisms that cause methylated DNA to be transcriptionally silent are not thoroughly understood; however, at least two proteins that repress transcription are known to bind to methylated DNA, and one of these, denoted MeCP2, has been shown to cause changes in chromatin structure. Thus, it is possible that methylated CpG dinucleotides bind specific proteins and that these proteins form a complex that prevents the transcription of neighboring genes. In this respect, methylated DNA in mammals may act like the Polycomb Response Elements in *Drosophila*, which, incidentally, does not have any methylated DNA. As discussed earlier, these elements bind protein complexes that silence nearby genes.

DNA methylation in mammals is also responsible for unusual cases in which the expression of a gene is controlled by its parental origin. For example, in mice, the *Igf2* gene, which encodes an *i*nsulin-like *g*rowth *f*actor, is expressed when it is inherited from the father but not from the mother. By contrast, a gene known as *H19* is expressed when it is inherited from the mother but not from the father. Whenever the expression of a gene is conditioned by its parental origin, geneticists say that the gene has been **imprinted**—a term intended to convey the idea that the gene has been marked in some way so that it "remembers" which parent it came from.

Recent molecular analysis has demonstrated that the mark that conditions the expression of a gene is methylation of one or more CpG dinucleotides in the gene's vicinity. These methylated dinucleotides are initially formed in the parental germ line (Figure 24.21). Thus, for example, the *Igf2* gene is methylated in the female germ line but not in the male germ line. At fertilization, a methylated, maternally contributed *Igf2* gene is combined with an unmethylated, paternally contributed *Igf2* gene. During embryogenesis, the methylated and unmethylated states are preserved each time the genes replicate. Because the methylated gene is silent, only the paternally contributed *Igf2* gene is expressed in the developing animal. Exactly the opposite happens with the *H19* gene, which is methylated in the male germ line but not in the female germ line. More than 20 different imprinted genes have been identified in mice and humans. For each, the methylation imprint is established in the parental germ line. However, a methylated gene that was inherited from one sex can be unmethylated when it passes through an offspring of the opposite sex. Thus, the methylation imprints are re-set each generation, depending on the sex of the animal. The fact that some genes are methylated in one sex but not in the other implies that there are sex-specific factors that control the methylation machinery.

Gene Amplification

Sometimes, gene expression is facilitated by an increase in gene number. This process, called **gene amplification,** is designed to increase the number of DNA templates for RNA synthesis. Perhaps the most dramatic example of expression-related gene amplification involves the ribosomal RNA genes in amphibian oocytes. These genes are needed to produce structural components of the ribosomes. In eukaryotes, there are three main types of rRNA genes: the 5S gene, which encodes a 120-base rRNA; the 18S gene, which encodes a 1.8-kb rRNA; and the 28 S gene, which encodes a 4.7-kb rRNA. In the genome of the toad *Xenopus laevis,* thousands of 5S genes are scattered over all the chromosomes, but there are far fewer copies of the 18S and 28S genes, perhaps only 800 to 1000 per diploid cell. These latter genes are all concentrated at a single site called the **nucleolar organizer,** so named because it forms the nucleolus. Molecular mapping has shown that the 18S and 28S rRNA genes belong to a single transcriptional unit that is tandemly repeated within the nucleolar organizer.

All three types of ribosomal RNA genes must be transcribed to produce RNAs for *Xenopus* ribosomes. In oocytes, this transcription must be especially vigorous because the egg needs to accumulate enough ribosomes to support the early development of the embryo after fertilization. Perhaps as many as 10^{12} ribosomes must be synthesized in each oocyte. Because each ribosome contains exactly one molecule of each type of rRNA, 10^{12} molecules of each of the 5S, 18S, and 28S rRNAs must be synthesized. This requirement is fulfilled in two ways. First, *Xenopus* has about 24,000 5S genes, including many that are specifically activated in oocytes. This enormous set of genes is therefore able to generate the 5S rRNAs that are needed for ribosome production in the egg. Second, the smaller number of 18S and 28S genes is specifically amplified in oocytes by the creation of extrachromosomal copies of these genes. Small, covalently closed circular DNAs carrying the 18S and 28S genes are formed. These replicate by a rolling-circle mechanism (Chapter 11) to produce many copies, which form supernumerary nucleoli within the oocyte. Transcription from these circular DNAs provides much of the 18S and 28S rRNA that is needed for ribosome assembly in the egg. The mechanism that generates these extrachromosomal DNA molecules is not known. However, it might involve some sort of intrachromosomal recombination between repetitive units within the nucleolar organizer.

Researchers have identified other cases of selective gene amplification in eukaryotes. During oogenesis in *Drosophila,* for example, a group of X-linked genes encoding eggshell proteins is amplified within the follicle cells of the ovary by localized replication of a small part of the X chromosome. This process forms a branching network of DNA duplexes within the single DNA duplex of the chromosome. Transcription from each of the replicated molecules increases the overall production of mRNA from this group of genes, which, in turn, leads to vigorous synthesis of the eggshell proteins. Large quantities of these proteins are needed to form protective shells around *Drosophila* eggs.

In mammals, the gene for dihydrofolate reductase (DHFR), an enzyme required for DNA synthesis, is sometimes amplified when cells are exposed to methotrexate, a chemotherapeutic drug that binds to dihydrofolate reductase and inhibits its activity. This amplification is accomplished either by multiplying the gene within the chromosome to form a tandem array or by creating numerous extrachromosomal copies. The first mechanism visibly increases the size of the

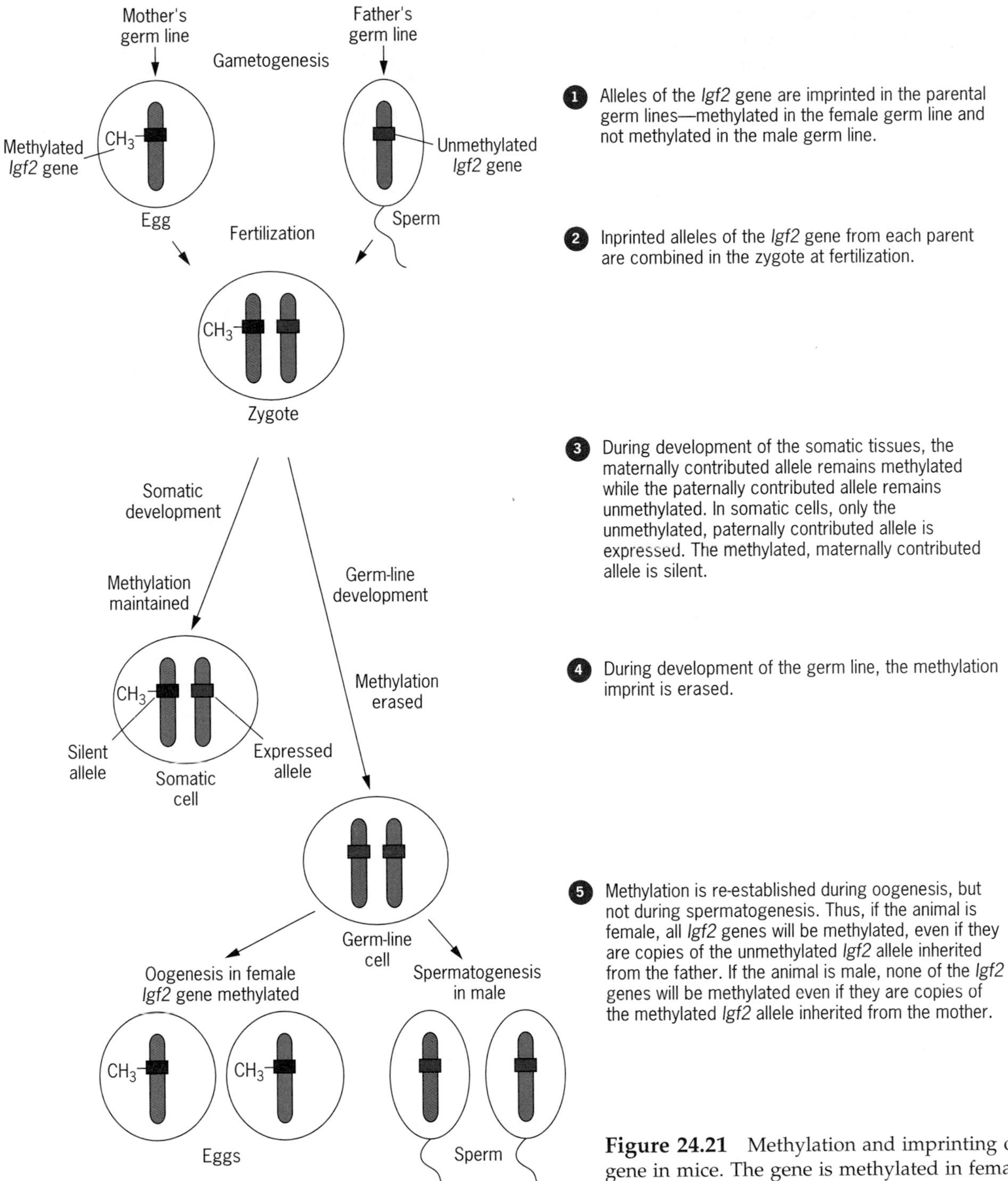

Figure 24.21 Methylation and imprinting of the *Igf2* gene in mice. The gene is methylated in females but not in males.

chromosome, forming a region within the chromosome that stains uniformly during mitotic metaphase—a **homogeneously staining region (HSR).** The second mechanism generates many tiny chromatin fragments that segregate irregularly during cell division—dotlike bodies called **double minutes (DMs).** Amplification of the *DHFR* gene by either of these mechanisms causes dihydrofolate reductase to be overproduced. Because some of the excess enzyme remains unbound by methotrexate, cells in which the *DHFR* gene has been amplified are able to replicate their DNA and divide. Thus, amplification of the *DHFR* gene can thwart a physician's attempts to inhibit the spread of a cancer by using methotrexate as a chemotherapeutic drug.

Key Points: Transcription occurs preferentially in loosely organized chromosome regions exemplified by the loops of lampbrush chromosomes and the puffs of polytene chromosomes. Transcriptionally active DNA tends to be more sensitive to digestion with DNase I and may be delimited by specialized chromatin structures that insulate it from neighboring DNA. Heterochromatin is associated with the repression of transcription. Protein complexes are responsible for gene silencing in *Drosophila* and yeast. In trypanosomes, only one telomeric *vsg* gene is expressed; all the other *vsg* genes, including those at telomeres, are silent. Methylation of DNA is associated with gene silencing in mammals. Increased gene expression may be achieved by amplifying the DNA, either within chromosomes or on extrachromosomal DNA molecules.

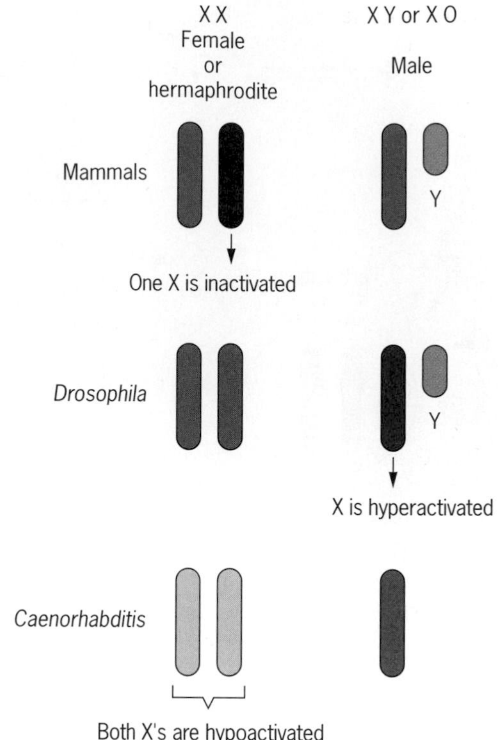

Figure 24.22 Three mechanisms of dosage compensation for X-linked genes: inactivation, hyperactivation, and hypoactivation.

ACTIVATION AND INACTIVATION OF WHOLE CHROMOSOMES

Organisms with an XX/XY or XX/XO sex-determination system face the problem of equalizing the activity of X-linked genes in the two sexes. In mammals, this problem is solved by randomly inactivating one of the two X chromosomes in females; each female therefore has the same number of transcriptionally active X-linked genes as a male. In *Drosophila*, neither of the two X chromosomes in a female is inactivated; instead, the genes on the single X chromosome in a male are transcribed more vigorously to bring their output in line with that of the genes on the two X chromosomes in a female. Still another solution to the problem of unequal numbers of X-linked genes has been found in the nematode *Caenorhabiditis elegans*. In this organism, XX individuals are hermaphrodites (they function as both male and female) and XO individuals are males. X-linked transcriptional activity is equalized in these two genotypes by partial repression of the genes on both of the X chromosomes in the hermaphrodites. Therefore, mammals, flies, and worms have solved the problem of X-linked gene dosage in different ways (Figure 24.22). In mammals, one of the X chromosomes in females is inactivated; in *Drosophila*, the single X chromosome in males is hyperactivated; and in *C. elegans*, both of the X chromosomes in hermaphrodites are hypoactivated.

These three different mechanisms of **dosage compensation—inactivation, hyperactivation**, and **hypoactivation**—have an important feature in common: many different genes are coordinately regulated because they are on the same chromosome. This chromosomewide regulation is superimposed on all other regulatory mechanisms involved in the spatial and temporal expression of these genes. What might be responsible for such a global regulatory system? For decades, geneticists have been trying to elucidate the molecular basis of dosage compensation. The working hypothesis has been that some factor or factors bind specifically to the X chromosome and alter its transcriptional activities. Recent discoveries indicate that this idea is correct.

Inactivation of X Chromosomes in Mammals

In mammals, X chromosome inactivation begins at a particular site called the **X inactivation center (XIC)** and then spreads in opposite directions toward the ends of the chromosome (Chapter 6). Curiously, not all genes on an inactivated X chromosome are transcriptionally silent. One that remains active is called *XIST* (for *X inactive specific transcript*); this gene is located within the XIC (Figure 24.23). Recent research has shown that, in human beings, the *XIST* gene encodes a 17-kb transcript devoid of any significant open reading frames. It therefore seems unlikely that the *XIST* gene codes for a protein. Instead, the RNA itself is

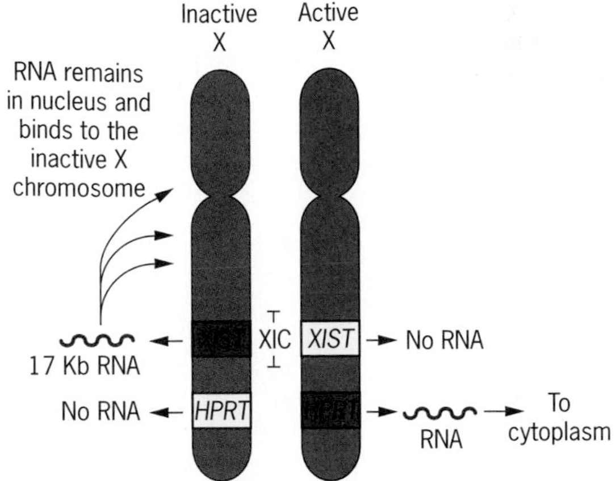

Figure 24.23 Expression of the *XIST* gene in the inactive X chromosome of human females. For comparison, the expression of the *HPRT* gene on the active X chromosome is shown. This gene encodes hypoxanthine phosphoribosyl transferase, an enzyme that plays a role in the metabolism of purines.

probably the functional product of the *XIST* gene. Though polyadenylated, this RNA is restricted to the nucleus and is specifically localized to inactivated X chromosomes; it does not appear to be associated with active X chromosomes in either males or females.

In mice, where fairly detailed experimental analysis has been possible, researchers have found that the homolog of the human *XIST* gene is transcribed during the early stages of embryonic development at a low level from both of the X chromosomes that are present in females. The transcripts from each of a female mouse's *Xist* genes are unstable and remain closely associated with their respective genes. As development proceeds, the transcripts from one of the genes stabilize and eventually envelope the entire X chromosome on which that gene is located; the transcripts from the other *Xist* gene disintegrate, and further transcription from that gene is repressed by methylation of nucleotides in the gene's promoter. Thus, in the female mouse, one X chromosome—the one whose *Xist* gene continues to be transcribed—becomes coated with *Xist* RNA and the other does not. The choice of the chromosome that becomes coated is apparently random. Although the coating mechanism is not yet understood, the consequence of coating is clear: most of the genes on the coated chromosome are repressed, and that chromosome becomes the inactive X chromosome. In the mammalian dosage compensation system, therefore, the X chromosome that remains active is, paradoxically, the one that represses its *Xist* gene.

Inactive X chromosomes are readily identified in mammalian cells. During interphase, they condense into a darkly staining mass associated with the nuclear membrane. This mass, the Barr body, decondenses during S phase to allow the inactive X chromosome to be replicated. However, because this process takes some time, the inactive X replicates later than the rest of the chromosomes. Inactive X chromosomes must therefore have a very different chromatin structure than that of other chromosomes. Recent evidence suggests that this difference is partly determined by the kinds of histones associated with the DNA. One of the four core histones, H4, can be chemically modified by the addition of acetyl groups to any of several lysines in the polypeptide chain. Acetylated H4 is associated with all the chromosomes in the human genome. However, on the inactive X it seems to be restricted to three fairly narrow bands, each corresponding to a region that contains some active genes. Acetylated H4 is also depleted in areas of heterochromatin on the other chromosomes. These findings suggest that acetylated H4 is involved in the maintenance of transcriptional activity.

Hyperactivation of X Chromosomes in *Drosophila*

In *Drosophila*, dosage compensation requires the protein products of at least four different genes. Null mutations in these genes result in male-specific lethality because the single X chromosome in males is not hyperactivated. Mutant males usually die during the late larval or early pupal stages. These dosage compensation genes are therefore called male-specific lethal (*msl*) loci, and their products are called the MSL proteins. Antibodies prepared against these proteins have been used as probes to localize the proteins inside cells. The remarkable finding is that each of the MSL proteins binds specifically to the X chromosome in males (Figure 24.24). One of these proteins, the product of the *msl* gene called *maleless* (*mle*), is homologous to DNA and RNA helicases, which are enzymes capable of unwinding nucleic acids. The putative helicase function of the maleless protein is consistent with the idea that the X chromosome in a *Drosophila* male must be "opened" for vigorous transcription. It is also consistent with the observation that polytene X chromosomes in males appear bloated and have a diffuse banding pattern, perhaps because they are hyperactivated.

The four MSL proteins apparently form a complex that binds to the single X chromosome in *Drosophila* males. Recent studies indicate that this MSL protein complex also contains specific types of RNA. Two different RNAs, *roX1* and *roX2* (for *R*NA *o*n the *X* chromosome), co-localize on the male's X chromosome with the MSL proteins. Both of these RNAs are en-

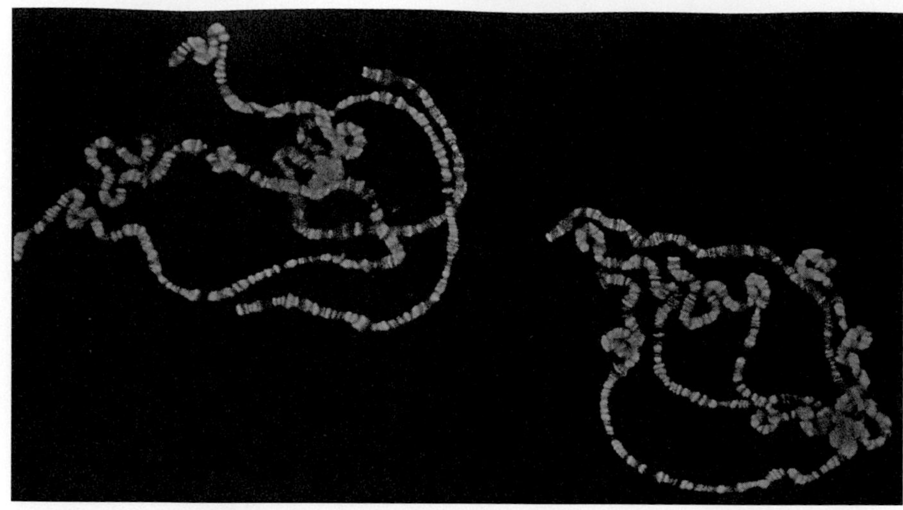

Figure 24.24 Binding of the protein product of the *Drosophila mle* gene to the single X chromosome in males.

coded by X-linked genes, and both are spliced and polyadenylated. However, they do not contain long open reading frames and are therefore probably not translated into proteins. Like the *Xist* RNAs in mammals, the *roX* RNAs in *Drosophila* seem to be the functional end-products of the genes that encode them. Curiously, the expression of the *roX* genes depends on the proper functioning of the *msl* genes. Furthermore, this expression is turned off when *Sxl,* the master regulatory gene of the *Drosophila* dosage compensation and sex determination systems (Chapter 6), is turned on, as it naturally is in *Drosophila* females. The *roX* RNAs are therefore produced only in male *Drosophila.*

Much remains to be learned about the association between the MSL proteins and the *roX* RNAs, and about their roles in hyperactivating transcription of X-linked genes in male *Drosophila.* Recent research has indicated that a particular acetylated version of histone H4 is exclusively associated with the X chromosome in males. It is possible that the MSL/roX protein-RNA complex facilitates this chromosome-specific acetylation and that the acetylated histone brings about an alteration in chromatin structure that doubles the amount of transcriptional activity on the male's single X chromosome.

Hypoactivation of X Chromosomes in *Caenorhabditis*

In *C. elegans,* dosage compensation involves the partial repression of X-linked genes in the somatic cells of hermaphrodites. The mechanism is not fully understood, but the products of several genes are involved. Like the MSL proteins in *Drosophila,* the proteins encoded by these genes bind specifically to the X chromosome. However, unlike the situation in *Drosophila,* they bind only when two X chromosomes are present. The proteins apparently do not bind to the single X chormosome in males, nor do they bind to any of the autosomes in either males or hermaphrodites. Dosage compensation in *C. elegans* therefore seems to involve a mechanism exactly opposite to the one in *Drosophila.* A protein complex binds to the X chromosomes and represses rather than enhances transcription.

Key Points: **In mammals, dosage compensation is achieved by inactivating one of the X chromosomes in XX females, in *Drosophila,* by hyperactivating the single X chromosome in XY males, and in *Caenorhabditis,* by hypoactivating the two X chromosomes in XX hermaphrodites.**

TESTING YOUR KNOWLEDGE

The bacterial *lacZ* gene for β-galactosidase was inserted into a transposable *P* element from *Drosophila* (Chapter 18) so that it could be transcribed from the *P* element promoter. This fusion gene was then injected into the germ line of a *Drosophila* embryo along with an enzyme that catalyzes the transposition of *P* elements. During development, the modified *P* element became inserted into the chromosomes of some of the germ-line cells. Progeny from this injected animal were then individually mated to flies from a standard laboratory stock to establish strains that carried the *P/lacZ* fusion gene in their genomes. Three of these strains were analyzed for *lacZ* expression by staining dissected tissues from adult flies with X-gal, a chromogenic substrate that turns blue in the presence of β-galactosidase. In the first strain, only the eyes stained blue, in the second, only the intestines stained blue, and in the third, all the tissues stained blue. How do you explain these results?

ANSWER

The three strains evidently carried different insertions of the *P/lacZ* fusion gene. In each strain, the expression of the *P/lacZ* fusion gene must have come under the influence of a different regulatory sequence, or enhancer, capable of interacting with the *P* promoter and initiating transcription into the *lacZ* gene. In the first strain, the modified *P* element must have inserted near an eye-specific enhancer, which would drive transcription only in eye tissue. In the second strain, it must have inserted near an enhancer that drives transcription in the intestinal cells, and in the third strain, it must have inserted near an enhancer that drives transcription in all, or nearly all, cells, regardless of tissue affiliation. Presumably each of these different enhancers lies near a gene that would normally be expressed under its control. For example, the eye-specific enhancer would be near a gene needed for some aspect of eye function or development. These results show that random insertions of the *P/lacZ* fusion gene can be used to identify different types of enhancers and, through them, the genes they control. These fusion gene insertions are therefore often called enhancer traps.

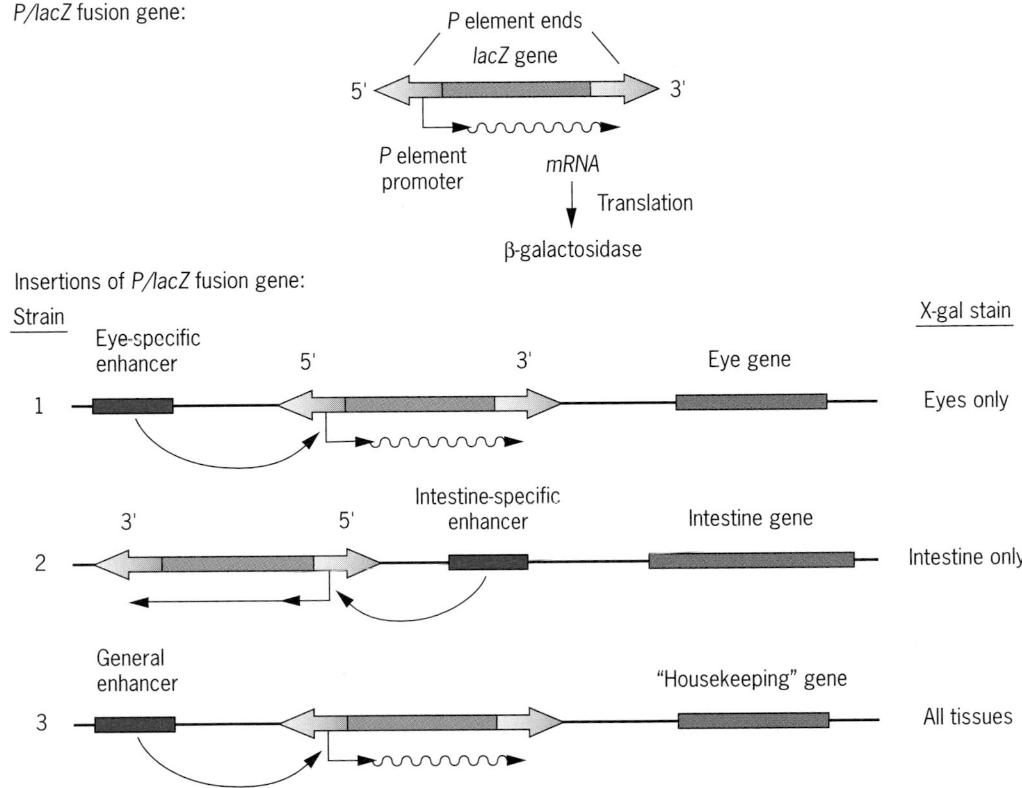

QUESTIONS AND PROBLEMS

24.1 Operons are common in bacteria but not in eukaryotes. Suggest a reason why.

24.2 Using examples, describe the temporal and spatial dimensions of eukaryotic gene regulation.

24.3 What factors influence the concentration and longevity of mRNA molecules?

24.4 Why do steroid hormones interact with receptors inside the cell, whereas peptide hormones interact with receptors on the cell surface?

24.5 How can you use the polytene chromosomes of Dipteran insects to study the regulation of transcription?

24.6 How would you distinguish between an enhancer and a promoter?

24.7 Tropomyosins are proteins that mediate the interaction of actin and troponin, two proteins involved in muscle contractions. In higher animals, tropomyosins exist as a family of closely related proteins that share some amino acid sequences but differ in others. Explain how these proteins could be created from the transcript of a single gene.

24.8 In *Arabidopsis*, one β tubulin gene (*TUB1*) is expressed in the roots and another β tubulin gene (*TUB8*) is expressed in the vascular tissues. Suggest a mechanism to account for this tissue-specific expression.

24.9 What techniques could be used to show that a plant gene is transcribed when the plant is illuminated with light?

24.10 Does a pseudogene specify an RNA? Does it specify a functional polypeptide?

24.11 Using the techniques of genetic engineering, a researcher has constructed a fusion gene containing the heat-shock response elements from a *Drosophila hsp70* gene and the coding region of a jellyfish gene for green fluorescent protein (*gfp*). This fusion gene has been inserted into the chromosomes of living *Drosophila* by the technique of transposon-mediated transformation (Chapter 18). Under what conditions will the green fluorescent protein be synthesized in these genetically transformed flies? Explain.

24.12 Suppose that the segment of the *hsp70* gene that was used to make the *hsp70/gfp* fusion in the preceding problem had mutations in each of its heat-shock response elements. Would the green fluorescent protein encoded by this fusion gene be synthesized in genetically transformed flies?

24.13 Why does it make sense for the photosynthetic enzyme ribulose1,5-bisphophate carboxylase (RBC) to be synthesized specifically when plants are exposed to light?

24.14 Distinguish between the zinc finger and leucine zipper motifs found in some eukaryotic transcription factors.

24.15 The protein product of the *Drosophila Sex-lethal* gene is needed in females but not in males. Predict the phenotype of XY *Drosophila* that are hemizygous for a null mutation in the X-linked *Sex-lethal* gene.

24.16 The alternately spliced forms of the RNA from the *Drosophila doublesex* gene encode proteins that are needed to block the development of one or the other set of sexual characteristics. The protein that is made in female animals blocks the development of male characteristics, and the protein that is made in male animals blocks the development of female characteristics. Predict the phenotype of XX and XY animals homozygous for a null mutation in the *doublesex* gene.

24.17 In *Drosophila*, expression of the *yellow* gene is needed for the formation of dark pigment in many different tissues; without this expression, a tissue appears yellow in color. In the wings, the expression of the *yellow* gene is controlled by an enhancer located upstream of the gene's transcription initiation site. In the tarsal claws, expression is controlled by an enhancer located within the gene's only intron. Suppose that by genetic engineering, the wing enhancer is placed within the intron and the claw enhancer is placed upstream of the transcription initiation site. Would a fly that carried this modified *yellow* gene in place of its natural *yellow* gene have darkly pigmented wings and claws? Explain.

24.18 Describe the cytological evidence that RNA is preferentially transcribed from "loosely" organized segments of chromosomes.

24.19 Would a *Drosophila* gene that was flanked by specialized chromatin structures be susceptible to position-effect variegation?

24.20 In *Drosophila* larvae, the single X chromosome in males appears diffuse and bloated in the polytene cells of the salivary gland. Is this observation compatible with the idea that X-linked genes are hyperactivated in *Drosophila* males?

24.21 What chemical modification of histone H4 is associated with transcriptionally active genes?

24.22 Why do null mutations in the *mle* gene in *Drosophila* have no effect in females?

24.23 What are the processes involved in X chromosome inactivation in mammals?

BIBLIOGRAPHY

AMREIN, H., AND R. AXEL. 1997. Genes expressed in neurons of adult male *Drosophila*. *Cell* 88:459–469.

BAKER, B. S. 1989. Sex in flies: The splice of life. *Nature* 340: 521–524.

BLACKWOOD, E. M., AND J. T. KADONAGA. 1998. Going the distance: A current view of enhancer action. *Science* 281: 60–63.

BROWN, C. J., B. D. HENDRICH, J. L. RUPERT, R. G. LAFRENIERE, Y. XING, J. LAWRENCE, AND H. F. WILLARD. 1992. The human *XIST* gene: Analysis of a 17 kb inactive X-specific RNA that contains conserved repeats and is highly localized within the nucleus. *Cell* 71:527–542.

EVANS, R. M. 1988. The steroid and thyroid hormone receptor superfamily. *Science* 240:889–895.

GORMAN, M., AND B. S. BAKER. 1994. How flies make one equal two: Dosage compensation in *Drosophila*. *Trends in Genet.* 10:376–380.

KASS, S. U., D. PRUSS, AND A. P. WOLFFE. 1997. How does DNA methylation repress transcription? *Trends in Genet.* 13: 444–449.

MELLER, V. H., K. H. WU, G. ROMAN, M. I. KURODA, AND R. L. DAVIS. 1997. *roX1* RNA paints the X chromosome of male *Drosophila* and is regulated by the dosage compensation system. *Cell* 88:445–457.

PANNING, B., AND R. JAENISCH. 1998. RNA and the epigenetic regulation of X chromosome inactivation. *Cell* 93:305–308.

PARKHURST, S. M., AND P. M. MENEELY. 1994. Sex determination and dosage compensation: Lessons from flies and worms. *Science* 264:924–932.

PIRROTTA, V. 1998. Polycombing the genome: PcG, trxG, and chromatin silencing. *Cell* 93:333–336.

RUDENKO, G., M. CROSS, AND P. BORST. 1998. Changing the end: antigenic variation orchestrated at the telomeres of African trypanosomes. *Trends in Micro.* 6:113–117.

SURANI, M. A. 1998. Imprinting and the initiation of gene silencing in the germ line. *Cell* 93:309–312.

WAKIMOTO, B. 1998. Beyond the nucleosome: Epigenetic aspects of position-effect variegation in *Drosophila*. *Cell* 93: 321–324.

A CONVERSATION WITH

ERIC WIESCHAUS

What was your first experience with biology? How did your interests in biology develop?

Two chance events: After my junior year in high school, I attended a summer science camp at the University of Kansas. I had never been away from home and it was a great experience, being with a lot of other kids my age who were smart and interested in science. It was also my first chance to work in a lab and do real experiments. All this made science seem less abstract, made it something that I could see myself doing. Even after that experience, I wasn't sure that I wanted to be a professional scientist. I have a strong visual side to my character, and I thought I wanted to be an artist.

A second pivotal event in my life occurred when I was a junior in college. As a part of the embryology lab, we watched cleaving frog embryos. For me, it was the first time I had seen actual embryos. The cells are big, and the process is rapid and amazing. It had the same visual impact I had felt in looking at great works of art, but it was happening right in front of me! Moreover, it seemed to me that there was a chance I might be able to understand at least some part of it.

What led you to focus your research on the process of development?

It was looking at embryos that got me into science, and that was what interested me most. Besides, the early stages of embryonic development offered advantages to get at specific genes, since morphological events might correlate with the earliest times when particular genes first become active.

Eric Wieschaus was the 1995 recipient, with Christiane Nüsslein-Volhard and Edward Lewis, of the Nobel Prize in Physiology or Medicine. Their research on the development of *Drosophila* has provided a basis for studies on the genetic control of development in all multicellular organisms. After receiving a B.S. in Biology with high honors from the University of Notre Dame in 1969 and a Ph.D. in Biology from Yale University in 1974, Dr. Wieschaus pursued postdoctoral work—primarily focused on *Drosophila* embryology—in laboratories first in Basel, then in Zurich. He began his close collaboration with Dr. Nüsslein-Volhard in Heidelberg. In 1981, Dr. Wieschaus moved from Heidelberg to Princeton University; since then he has taught genetics and development courses at both the graduate and undergraduate level and currently is a professor in the Department of Molecular Biology. The author of numerous publications, he is also an Investigator of the Howard Hughes Medical Institute, a Fellow of the American Academy of Arts and Sciences, and a member of the National Academy of Sciences. Dr. Wieschaus is married to Dr. Trudi Schüpbach, with whom he has collaborated in the study of cell lineages of the embryonic epidermis. They have three daughters, Ingrid, Eleanor, and Laura.

How did your collaboration with Christiane Nüsslein-Volhard begin? How did you partition the project?

The collaboration was a natural extension of our earlier conversations. We met in Basel, Switzerland, just after finishing our graduate work. It was clear that we had many common interests in development. The mutagenesis experiments were something we had talked about for many years, long before we both arrived in Heidelberg and began working on the project. As to our work itself, we both did everything in parallel and scored the phenotypes of all the mutants using a discussion microscope with two sets of eye pieces so that we would both be looking at the same embryos at the same time.

How long did it take to complete the work, and what were your major accomplishments?

Although it took eight months to identify the mutations, it took four years to complete the characterization of the loci and publish the results. The original idea of the screen was to survey genes that had to be transcribed in the fruit fly embryo for normal development. We looked at enough mutant lines to be pretty sure that we had obtained at least one mutation in all the relevant genes. Over the eight-month period, we tested 27,000 lines, probably 18,000 different mutations. From these lines we identified 580 lines where the mutations had drastic effects on embryonic patterning and differentiation. When we did complementation tests among these lines, we found that the 580 mutations could be assigned to 139 genes and that we had hit most genes more than once. We decided not to score any more lines, since we would just get more mutations in the same genes. Therefore, the 139 represent most of the identifiable genes in *Drosophila* whose transcription is required for normal morphology of the embryo. These turned out to be many of the same genes that control development of all multicellular organisms.

How has your research helped to make *Drosophila* one of the premier organisms for the study of embryological development?

The possibility of identifying mutations with discrete effect on embryonic development means that in *Drosophila,* extrapolation from genes to mechanisms can be very direct. What we didn't know when we started our experiments, but what helped us a lot, was the relative simplicity of embryonic development in *Drosophila.* Most processes occur in sheets of cells that are only one cell layer thick; often cell fates are controlled by concentration gradients of single molecules within these sheets.

Some people have remarked that the study of gene expression in higher organisms has become the study of many special cases, each with its own peculiarities. Are there unifying themes among the many special cases?

That subtle differences in concentration (sometimes less than twofold) matter. That once cell fate decisions are made, even though the triggering differences may not be great, organisms have mechanisms to stabilize and amplify those decisions.

Has winning the Nobel Prize changed your professional activities?

Not really—I still spend time at the bench sorting flies. I still teach. It does make a somewhat more public person out of you. Rightly or wrongly, people value your opinions on things.

What are some of your current research projects?

I am interested in how cells change shape, how they reorganize their internal contents. A lot of this is still done on embryos, but the questions have a broader cell biology impact.

How has genetics helped biologists to study development?

Embryos are packed with gene products. Biochemical and molecular techniques don't really allow you to distinguish which RNAs or proteins are important for what processes. Because genetics allows you to remove single gene products, you can study a single function in isolation.

What advice do you have for biology students today?

Find a problem you love, where the big picture fascinates you. Don't worry about learning too many facts. Facts are essential, but learning them will follow naturally if you are really interested in the problem.

These modifications of insect wings are the result of differences in the way that sets of genes control development.

25

The Genetic Control of Animal Development

CHAPTER OUTLINE

The Process of Development in Animals
Genetics Analysis of Development in Model Organisms
Genetic Analysis of Developmental Pathways
The Use of Genetic Mosaics to Study Development
Molecular Analysis of Genes Involved in Development
Maternal Gene Activity in Development
Zygotic Gene Activity in Development
Genetic Analysis of Development in Vertebrates

Insect Wings: Evidence for the Genetic Control of Development

No one knows how many different species of animals are currently on the earth, but the vast majority of them are insects. These small creatures originated about 400 million years ago when life spread from the seas onto the land. Many different groups of insects were established in those ancient times; today, their descendants are the most abundant eukaryotes on the planet. All insects have segmented bodies and three pairs of legs, but beyond that, their features are amazingly diverse. Consider, for example, insect wings. Some groups of insects, such as butterflies, moths, and dragonflies, have two pairs of membranous wings, but others, such as grasshoppers, beetles, and flies, have only one. The wings of butterflies and moths are delicate and colorful, whereas those of dragonflies are sturdy and plain. In grasshoppers what would have been an anterior pair of wings has been modified into protective sheaths, and in beetles this modification has been extended even further, forming tough, shell-like structures called elytra. In flies what would have been a posterior pair of wings has been transformed into short, knoblike structures called halteres, which help the animal to balance during flight.

Figure 25.1 A fruit fly with four wings instead of the normal two. This unusual phenotype is caused by a trio of tightly linked autosomal mutations.

All these differences demonstrate that one of the basic features of the insect body plan—two pairs of flight-related appendages—has been extensively modified during evolution. These modifications are, of course, due to genetic differences among the various groups of insects. Why does a fly have halteres instead of a second pair of wings? Because its genes make it that way. Geneticists know this because mutations in certain fly genes cause wings to develop instead of halteres (Figure 25.1). This remarkable transformation produces a fly with two pairs of wings, like a dragonfly, and convincingly shows that the development of specific body parts is under genetic control.

In this chapter, we explore the genetic basis of animal development—how genes shape the basic body plan and all its anatomical details. This is one of the most exciting areas in modern biology, one that is growing at a rapid pace because of intensive research, much of it focused on a few carefully chosen organisms. The goal of all this effort is to obtain a general understanding of how the body of a multicellular animal—particularly that of humans—develops from a fertilized egg.

THE PROCESS OF DEVELOPMENT IN ANIMALS

Animal development actually begins before fertilization of the egg. Critical substances are synthesized by the female and sequestered in the egg to nourish and control the development of the zygote into a multicellular organism. After fertilization, a rapid series of mitotic divisions increases the number of cells in the developing animal, usually without much growth in overall size (Figure 25.2). The resulting mass of cells, the **blastula,** is relatively nondescript, but even at this stage, some of its cells have been earmarked to form specific tissues in the animal's body. Such cells have been fated, or determined, to develop in a certain way. As development continues, the blastula is reorganized by a series of cell migrations, collectively called **gastrulation,** to produce an embryo with three distinct cell layers—**ectoderm** (on the outside), **mesoderm** (in the middle), and **endoderm** (on the inside). These three layers give rise to specific tissues, which are then assembled to form organs. From gastrulation onward, cells become different from each other, a process called **differentiation.** This is a key feature of animal development, providing the many specialized functions that are needed for life.

Oogenesis and Fertilization

All the developmental potential of an animal is contained in a fertilized egg. When sperm and egg unite, they bring together the paternal and maternal sets of

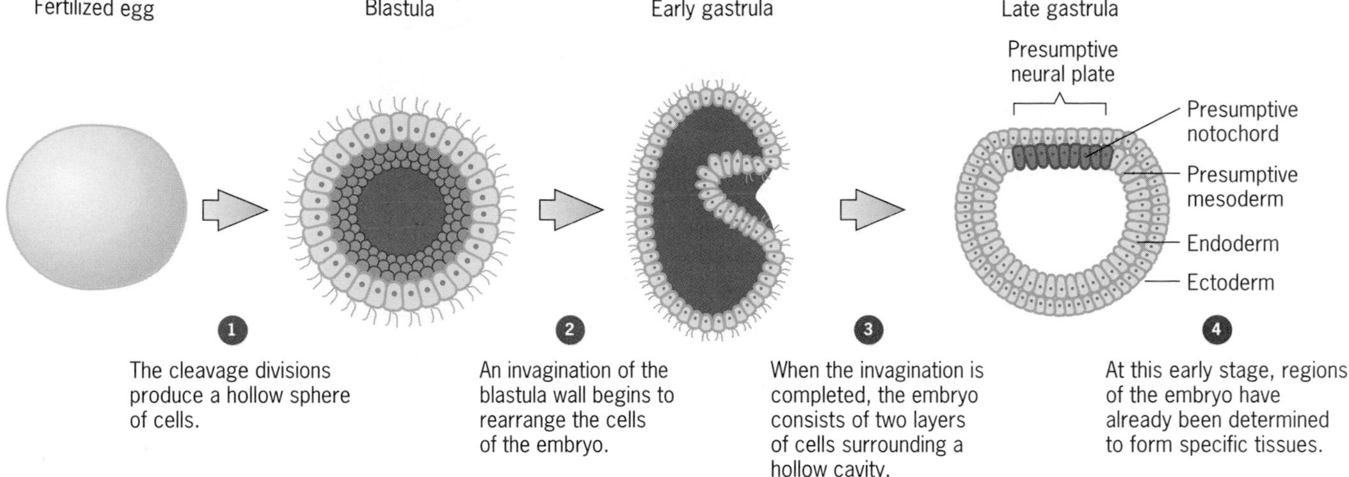

(*a*) Early stages of development in *Amphioxus*.

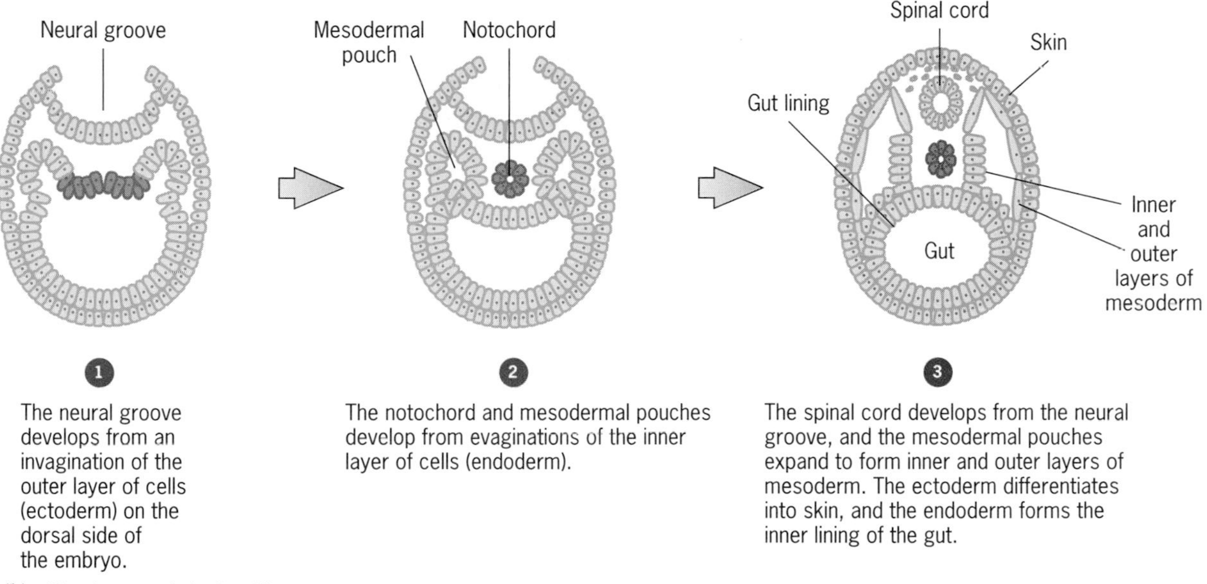

(*b*) Morphogenesis in *Amphioxus*.

Figure 25.2 Embryonic development of the marine animal *Amphioxus,* a protochordate. (*a*) Early stages of embryonic development. (*b*) Morphogenesis.

genes, which provide the detailed instructions for development. However, development is also profoundly affected by cytoplasmic materials present in the egg. Some of these materials, such as yolk, are nutritive and serve as food for the developing animal. Other materials are determinative in that they help to lay out the basic pattern of development itself. Both the nutritive and determinative components of the egg cytoplasm are manufactured during oogenesis, the process of egg formation. This involves a complex series of events, including the two meiotic divisions, which reduce the chromosome number from diploid to haploid. Curiously, the second meiotic division is sometimes not completed until after the egg has been fertilized.

In female animals, the meiotic divisions produce cells of unequal size (Chapter 2). The first division separates a large cell, the secondary oocyte, from a small cell, the first polar body. The secondary oocyte goes on to develop into the egg, whereas the polar body degenerates. This unequal division of cells is clearly a way of mustering materials into the egg cytoplasm. Very little cytoplasm is allotted to the polar body; instead, most of it is reserved for the oocyte. The second meiotic division is also unequal in that most of the cytoplasm is reserved for the egg; only a tiny bit is pinched off to form the second polar body.

Neighboring cells also enrich the cytoplasm of the egg. In many organisms, these cells synthesize materials that are transported into the oocyte, performing a nourishing or nursing function. Surrounding cells may also provide materials to envelop the oocyte. In birds, for example, a hard, calcium-rich shell is formed by secretions in the oviduct, a part of the female reproductive tract through which the egg passes on its way out the body.

Fertilization occurs when an egg is penetrated by a sperm. This is a complex process, involving interactions between different proteins on the surfaces of the two gametes. When the sperm makes contact with the egg, it triggers a response in the egg that causes the cytoplasm near the surface of the egg to mound up around the sperm and internalize it. After the sperm has entered the egg, the egg completes meiosis (if it has not already done so) and the sperm and egg nuclei fuse to form the nucleus of the zygote. In some organisms, for example, *Drosophila,* the sperm and egg nuclei actually divide once mitotically *before* fusing.

The Embryonic Cleavage Divisions and Blastula Formation

A newly formed zygote quickly divides into two cells, each of which divides into two more, and so on, until many cells are present in the developing embryo. These initial divisions are called **cleavage divisions.** Sometimes they occur asymmetrically so that the daughter cells receive different amounts of cytoplasm (Figure 25.3). However, except for a few curious cases, each cell seems to get the same genetic material. Thus, their subsequent differentiation into specific cell types is not due to the segregation of genes or chromosomes, but rather to the activation or repression of specific sets of genes.

The differentiation of cell types is preceded by **determination,** a process in which cells in the blastula are assigned developmental fates. This has been confirmed by experiments in which a piece of the blastula is surgically excised and moved to a different position, or even to a different organism. In some cases, these transplants survive and differentiate just as they would have if they had not been moved. Thus, even before there is any visible evidence of specialization, some of the cells in the blastula are determined to develop in a specific way. By the time of blastula formation, the basic developmental plan of the animal has already been laid down.

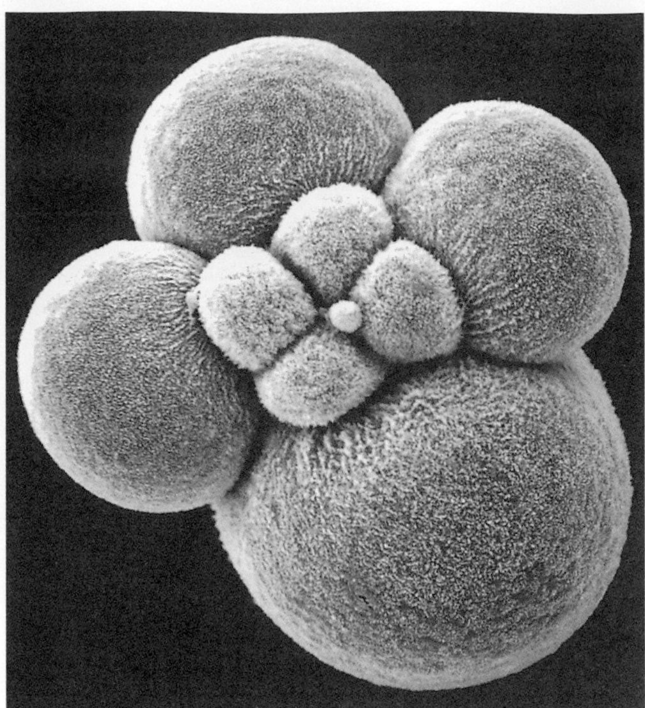

Figure 25.3 Scanning electron micrograph of an embryo of the marine snail *Ilyanassa obsoleta* at the eight-cell stage. The tiny cell in the center is a polar body.

Gastrulation and Morphogenesis

The cells of the blastula are reorganized by gastrulation, a series of movements in which groups of cells migrate to new positions in the embryo. This converts the nondescript blastula, which is essentially a sphere of cells, into an organism with distinct cell layers, often called the **primary germ layers** (not to be confused with the germ cells). These layers form the primitive tissues of the animal—ectoderm, mesoderm, and endoderm—and eventually differentiate into specific tissues—nerve, epithelium, muscle, blood, bone, hair, glands, viscera, and so forth. It is therefore through a combination of cell movement and differentiation that the body of an animal takes shape. In the course of this process, often called **morphogenesis,** individual organs such as brain, heart, intestine, and liver are formed.

Key Points: **The development of a fertilized egg into a multicellular animal involves cell division, the assignment of fates to individual cells (determination), and the subsequent realization of those fates (differentiation). Key materials for the early phases of this process are transferred into the egg during oogenesis. After fertilization, the egg proceeds through several cleavage divisions to form a sphere of cells (the blastula), which is subsequently reorganized by cell movements (gastrulation) to form the primary germ layers (ectoderm, endoderm, and mesoderm). These primitive tissues eventually produce mature tissues and organs.**

GENETIC ANALYSIS OF DEVELOPMENT IN MODEL ORGANISMS

Modern developmental biology depends heavily on genetic analysis, which must begin with the identification of relevant genes. This is usually accomplished by collecting mutations with diagnostic phenotypes; for example, if we wish to study the development of an insect's wings, we would begin by collecting mutations that alter or prevent wing development. We would then map these mutations and test them for allelism with each other to define and position the relevant genetic loci. Once we have identified these loci, we would combine representative mutations from each in pairwise fashion to determine whether some are epistatic over others. Such epistasis testing can sometimes provide valuable insights into how different genes contribute to a developmental process (Chapter 4). Further analysis might involve studying the nature of gene expression by creating animals that are composed of a mixture of genotypes; these so-called genetic mosaics permit us to study the cell lineages in which gene expression is required. Finally, to investigate the molecular basis of gene action, we would have to clone individual genes involved in the developmental process and study them with the full panoply of techniques now available, including DNA sequencing.

Not many animals are suitable for such an aggressive research program. Consequently, developmental geneticists have concentrated their efforts on a few model species in which it is possible to do intensive genetic analysis *and* high-resolution molecular biology. Currently, the best species are the fruit fly *Drosophila melanogaster* and the roundworm *Caenorhabditis elegans*.

Drosophila as a Model Organism

Drosophila was introduced to the world of genetics in 1909 through the laboratory of T. H. Morgan. Since then, intensive research has made it one of the best understood eukaryotes. Adult *Drosophila* develop from ellipsoidal eggs about 1 mm long and 0.5 mm wide at their maximum diameter (Figure 25.4*a*). Each egg is surrounded by a **chorion,** a tough shell-like structure that is made of materials synthesized by somatic cells in the ovary. The anterior end of the egg is distinguished by two filaments that help to bring oxygen into the interior. Sperm enter the egg through another anterior structure, the **micropyle.** The divisions that follow fertilization are rapid—so rapid that there is no time for membranes to form between daughter cells. Consequently, the early *Drosophila* embryo is actually a single cell with many identical nuclei; such a cell is called a **syncytium** (Figure 25.4*b*). After division cycle 9 within the syncytium, the 512 nuclei that have been created migrate to the cytoplasmic membrane on the periphery of the embryo, where they continue to divide four more times. In addition, a few of the nuclei migrate to the posterior pole of the embyro. At division cycle 13, all the nuclei in the syncytium become separated from each other by cell membranes, creating a single layer of cells on the embryo's surface. This single layer, called the **cellular blastoderm,** is equivalent to the blastula of other organisms; it will give rise to all the somatic tissues of the animal. Cellularization of the nuclei at the posterior pole creates the **pole cells,** which give rise to the adult germ line. Thus, at this very early stage of development, the somatic and germ-cell lineages of the future adult have already been separated.

It takes about a day for the *Drosophila* embryo to develop into a wormlike **larva.** This larva hatches by chewing its way through the egg shell and begins feeding voraciously. It sheds its skin twice to accommodate increases in size and then, about five days after hatching, becomes immobile and hardens its skin to form a **pupa.** During the next four days, many of the larval tissues are destroyed, and flat packets of cells that were sequestered during the larval stages expand and differentiate into adult structures such as antennae, eyes, wings, and legs. Because an adult insect is called an **imago,** these packets are referred to as **imaginal discs** (Figure 25.5). When this anatomical reorganization is completed, a radically different animal emerges from the pupal casing—an animal that can fly and reproduce!

Caenorhabditis as a Model Organism

Caenorhabditis elegans (Figure 25.6) became a part of the genetics scene in the 1960s, long after *Drosophila* had been firmly established as an experimental organism. Sydney Brenner, who initially worked with bacteriophages, chose this small, free-living soil nematode as a model for studying animal development. Adult *C. elegans* are about 1 mm long, roughly the same size as a *Drosophila* egg. They reproduce quickly and prolifically and can be reared easily on agar plates that have been seeded with *E. coli* bacteria as food. Under optimal conditions, the life cycle is completed in a mere three days.

C. elegans is a hermaphroditic species, that is, one in which individual organisms are capable of producing both sperm and eggs. Because there is no system of self-incompatibility, sperm can fertilize eggs from the same animal. To the researcher, this self-fertilization is a convenient way of making recessive mutations

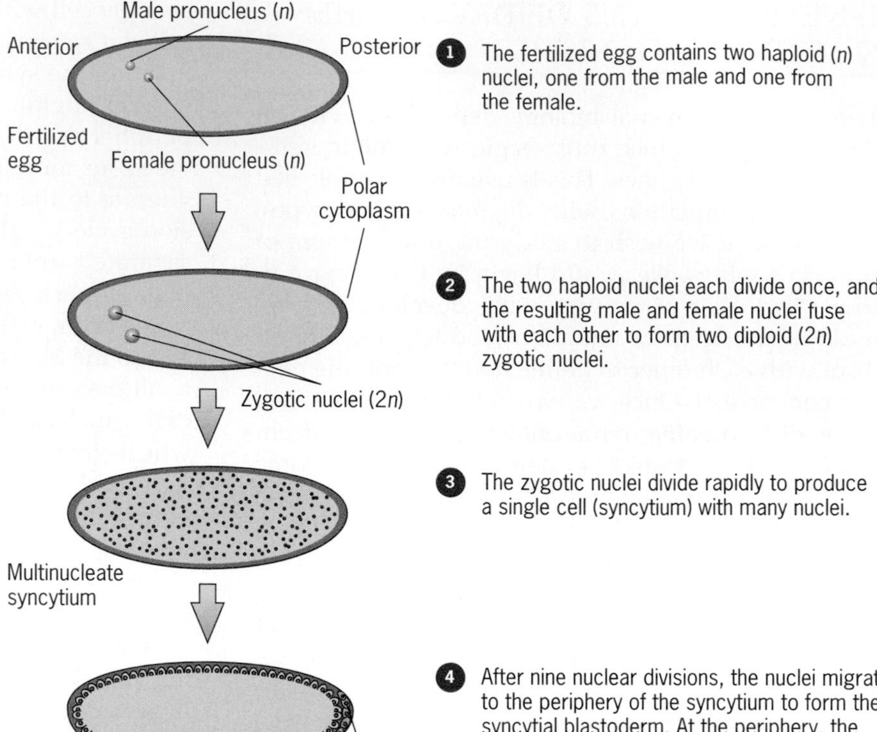

Figure 25.4 (*a*) Photograph of *Drosophila* eggs, with (top) and without (bottom) the surrounding chorion. (*b*) Early embryonic development in *Drosophila*.

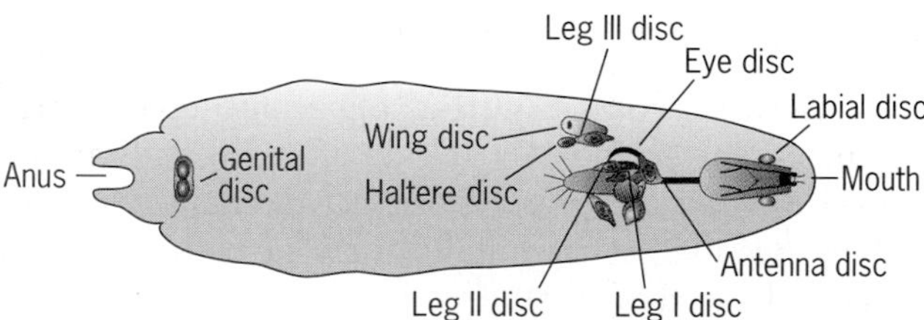

Figure 25.5 The location of the imaginal discs in *Drosophila* larvae.

homozygous. Hermaphrodites have two X chromosomes and five pairs of autosomes. Occasionally, animals with a single X chromosome and five pairs of autosomes are produced by meiotic nondisjunction; these animals are males, capable of making sperm but not eggs. XO males can be crossed with XX hermaphrodites to carry out standard genetic procedures such as recombination mapping and complementation testing.

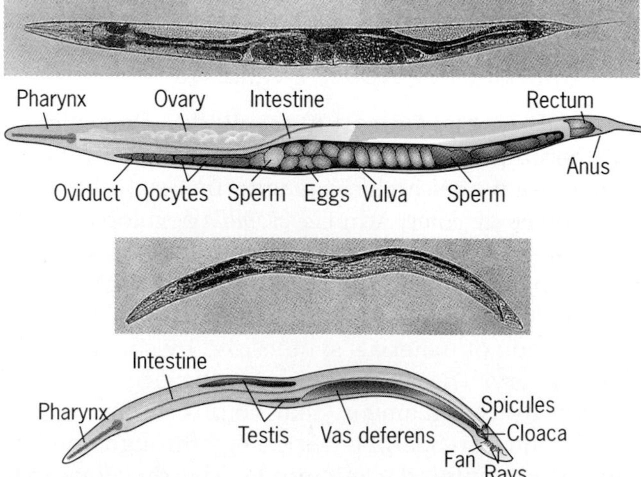

Figure 25.6 Photographs and diagrams showing the phenotypes of a *Caenorhabditis* hermaphrodite (top) and male (bottom).

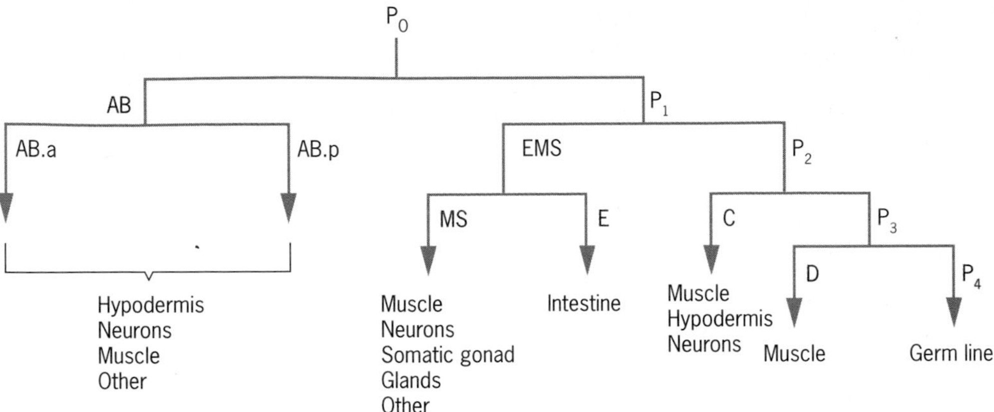

Figure 25.7 A portion of the cell lineage of the *C. elegans* hermaphrodite. P_0, the fertilized egg, divides to produce two cells denoted AB and P_1. Each of these then divides to produce two pairs of cells, one pair denoted AB.a and AB.p and the other denoted EMS and P_2. Further divisions of these cells and their descendants produce all the cells of the adult hermaphrodite. The six founder cells are indicated in red.

C. elegans is a transparent animal, affording researchers the opportunity to observe each cell in the course of development. John Sulston and coworkers have been able to trace the lineage of all the cells in the adult worm from the single-celled zygote (Figure 25.7). This work has shown that the events of *C. elegans* development are essentially invariant. Each adult hermaphrodite consists of 959 somatic nuclei (some cells are multinucleate) as well as an indeterminant number of germ cells. A *C. elegans* zygote goes through a series of asymmetric cleavage divisions to produce six "founder" cells. One of these eventually gives rise to the entire germ line, one to the intestine, and one to the body wall muscles. The other three founder cells produce mainly nerve and muscle cells. The **invariant cell lineages** that form the tissues of the adult make *C. elegans* an excellent organism for the study of development.

Key Points: **The fruit fly *Drosophila melanogaster* and the hermaphroditic nematode *Caenorhabditis elegans* are model organisms with features that make them ideal for studies of the genetic control of development.**

GENETIC ANALYSIS OF DEVELOPMENTAL PATHWAYS

The notion that phenotypes result from a series of steps in a pathway and that genes control each of these steps emerged in the 1930s and 1940s from a combination of genetic and biochemical investigations. By that time, it had already been established that cells carry out a multitude of biochemical reactions, each catalyzed by an enzyme, and that several reactions can be linked to form a pathway. Genetics provided the key insight that each enzyme in a pathway is encoded by a gene. A mutation in one of these genes could therefore inactivate an essential enzyme, blocking the entire pathway and causing a mutant phenotype. Geneticists quickly realized that by studying such mutations, they could identify each step in the pathway. Furthermore, by analyzing pairs of mutations, they could sometimes arrange the steps in the correct temporal order. This genetic dissection of biochemical pathways was soon extended to other biological processes, including development.

A developmental pathway consists of the events involved in the differentiation of tissues and organs. Different gene products participate in these events, including some that are signal molecules, some that are signal receptors, others that are signal transducers, and still others that are transcription factors. Other kinds of regulatory proteins may also be involved. Ultimately, a pathway generates the components that form the structures of particular tissues and organs. It therefore creates a phenotype.

The components of a pathway are causally ordered by the way in which they bring about the phenotype. Gene *A* may produce a secreted protein that acts as a signal to stimulate the transcription of gene *X*, whose product is a component of a differentiated cell. This stimulation may be mediated by other gene products, including, for example, a membrane-bound receptor for the A protein, intracellular proteins that are activated by this receptor, and one or more transcription factors that respond to this activation by binding to enhancers near gene *X* to induce its expression. The overall structure of the pathway is

Gene *A* →	Gene *R* →	Gene *C* →	Gene *T* →	Gene *X*
↓	↓	↓	↓	↓
secreted signal protein A	membrane-bound receptor protein	cytoplasmic protein	transcription factor	protein X in a differentiated cell

In this developmental pathway, the different proteins are ordered by a causal chain that ultimately produces a protein characteristic of a particular differentiated cell. The arrows between the genes in the pathway indicate the order in which the gene products act to bring about a differentiated state.

In *Drosophila* and *Caenorhabditis,* the developmental pathways that have been among the most thoroughly studied are responsible for the phenotypes of the two sexes. These pathways control the differentiation of the somatic tissues into male or female structures. Other, less well understood pathways govern the development of the tissues in the male and female germ lines. The *Drosophila* and *Caenorhabditis* somatic sex-determination pathways involve entirely different molecular mechanisms. In *Drosophila,* the key genes encode proteins that regulate RNA splicing, whereas in *Caenorhabditis,* they encode transcription factors, signaling molecules and their receptors. In both animals, however, the sex-determinaton pathway responds to the same fundamental signal, which is the ratio of X chromosomes to autosomes. When this ratio is 1.0 or greater, the pathway produces a female phenotype, whereas when it is 0.5 or less, it produces a male phenotype.

Sex Determination in *Drosophila*

The sex-determination pathway in *Drosophila* has three components: (1) a system to ascertain the X:A ratio very early in the embryo; (2) a system to convert this ratio

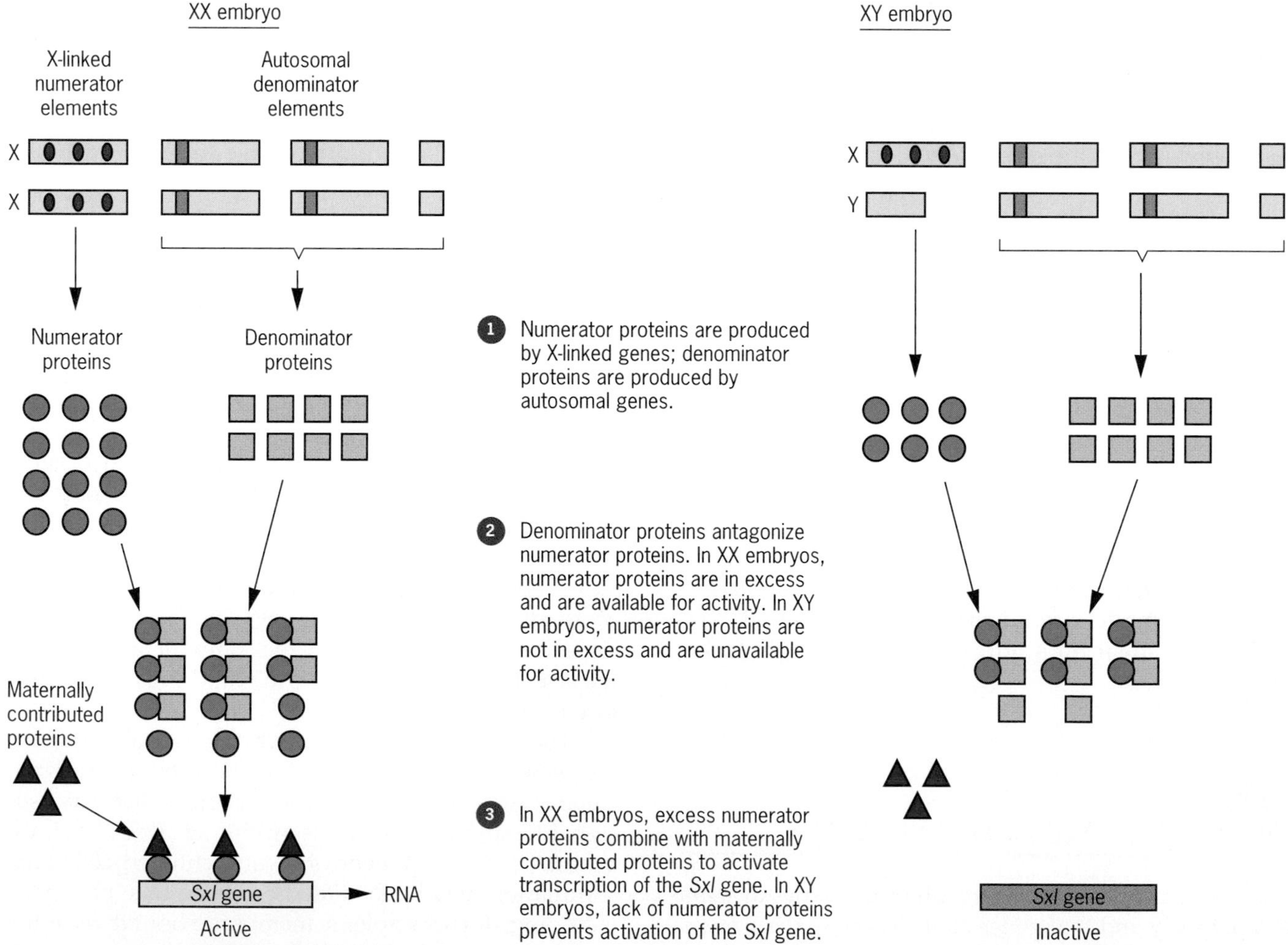

Figure 25.8 Ascertainment of the X:A ratio by numerator and denominator elements in *Drosophila*. The ratio of X chromosomes to sets of autosomes is ascertained by interactions between the protein products of these genes.

into a developmental signal; and (3) a system to respond to this signal by producing either male or female structures. Intensive research has led to a detailed understanding of the first two components; however, the last component is still something of a mystery.

The system to ascertain the X:A ratio involves interactions between maternally synthesized proteins that have been deposited in the egg cytoplasm, and embryonically synthesized proteins that are encoded by several X-linked genes (Figure 25.8). These latter proteins are twice as abundant in XX embryos as in XY embryos, and therefore provide a means of "counting" the number of X chromosomes present. Because the genes that encode these proteins affect the numerator of the X:A ratio, they are called **numerator elements.** Other genes located on the autosomes affect the denominator of the X:A ratio. These so-called **denominator elements** encode proteins that antagonize the products of the numerator elements. Consequently, as the dosage of denominator elements is increased, the "perceived" dosage of numerator elements is decreased, and the number of X chromosomes present in the genotype is underestimated. This process occurs in *Drosophila* with two X chromosomes and three pairs of autosomes (genotype XX; AAA); such flies develop as intersexes rather than as females. The system for ascertaining the X:A ratio in *Drosophila* is therefore based on antagonism between X-linked (numerator) and autosomal (denominator) gene products.

Once the X:A ratio has been ascertained, it is converted into a molecular signal that controls the expression of the X-linked *Sex-lethal* (*Sxl*) gene, the master regulator of the sex-determination pathway (Figure 25.9). Early in development, this signal activates transcription of the *Sxl* gene from P_E, the gene's "early" promoter, but only in XX embryos. The early transcripts from this promoter are processed and translated to produce functional Sex-lethal protein, denoted SXL. After only a few cell division cycles, transcription from the P_E promoter is replaced by transcription from another promoter, P_M, the so-called maintenance promoter of the *Sxl* gene. Curiously, transcription from the P_M promoter is also initiated in XY embryos. However, the transcripts from P_M are correctly processed only if SXL protein is present. Consequently, in XY embryos, where this protein has not yet been made, the *Sxl* transcripts are alternately spliced to include an exon with a stop codon, and when these alternately spliced transcripts are translated, they generate a short polypeptide without regulatory function. Thus, alternate splicing of the *Sxl* transcripts in XY embryos does not lead to the production of functional SXL protein, and in the limits the synthesis of the SXL protein to XX embryos, which develop as females. The absence of SXL protein in XY embryos causes them to develop as males.

1 Transcription:

In XX embryos, a molecular signal based on the X:A ratio initiates transcription of the *Sxl* gene from promoter P_E.

Later, transcription is initiated at promoter P_M in both XX and XY embryos.

2 Splicing:

In XX embryos, the *Sxl* transcripts are spliced to contain all the exons except exon 3.

In XY embryos, the *Sxl* transcripts are spliced to contain all the exons, including exon 3.

3 Translation:

In XX embryos, the *Sxl* mRNA is translated into a polypeptide (SXL) that regulates splicing, including the splicing of *Sxl* transcripts.

In XY embryos, a stop codon in exon 3 prevents the *Sxl* mRNA from being translated into a functional polypeptide.

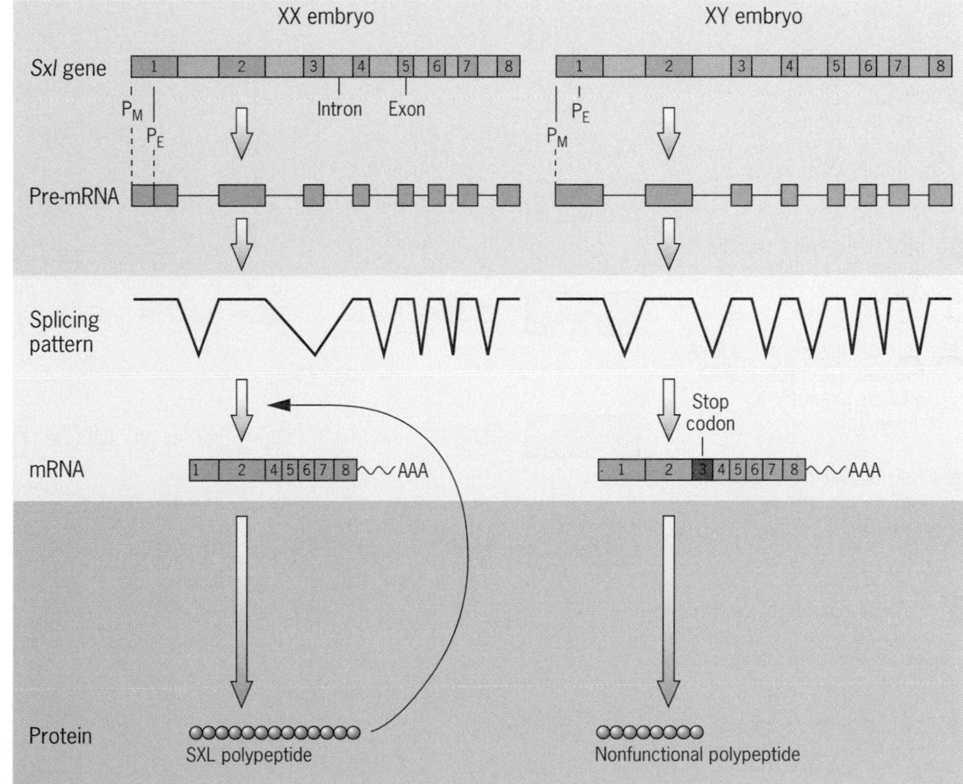

Figure 25.9 Sex-specific expression of the *Sex-lethal* (*Sxl*) gene in *Drosophila*. Although this gene is transcribed in both XX and XY embryos, alternate splicing of its RNA

absence of this protein, these embryos develop as males. In XX embryos, where SXL protein was initially made in response to the X:A signal, *Sxl* transcripts from the P_M promoter are spliced to produce more SXL protein. This expression pattern is maintained because the SXL protein regulates the splicing of transcripts from the *Sxl* gene in such a way that they encode the full-length SXL protein. In XX embryos, this protein is therefore a positive regulator of its own synthesis—a curious feedback mechanism that maintains the expression of the SXL protein in XX embryos and prevents its expression in XY embryos.

The SXL protein also regulates the splicing of transcripts from another gene in the sex-determination pathway, *transformer* (*tra*) (Figure 25.10) These transcripts can be processed in two different ways. In chromosomal males, where the SXL protein is absent, the splicing apparatus always leaves a stop codon in the second exon of the *tra* RNA. Thus, when spliced *tra* RNA is translated, it generates a truncated (and nonfunctional) polypeptide. In females, where the SXL protein is present, this premature stop codon is removed by alternate splicing in at least some of the transcripts. Thus, when they are translated, some functional *transformer* protein (denoted TRA) is produced. The SXL protein therefore allows the synthesis of functional TRA protein in XX embryos but not in XY embryos.

The TRA protein also turns out to be a regulator of RNA processing. Along with TRA2, a protein encoded by the *transformer2* (*tra2*) gene, it controls the expression of *doublesex* (*dsx*), an autosomal gene that can produce two different proteins through alternate splicing of its RNA. In XX embryos, where the TRA protein is present, *dsx* transcripts are processed to encode a DSX protein that represses the genes required for male development. Therefore, such embryos develop into females. In XY embryos, where the TRA protein is absent, *dsx* transcripts are processed to encode a DSX protein that represses the genes required for female development. Consequently, such embryos develop into males. The *dsx* gene is therefore the switch point at which a male or female developmental pathway is explicitly chosen. From this point, different sets of genes are specifically expressed in males and females to bring about sexual differentiation.

Mutations have been obtained in each of the genes of the *Drosophila* sex-determination pathway (Table 25.1). Loss-of-function mutations in *Sxl* prevent SXL protein from being made in females. Homozygous mutants would therefore develop into males; however, they die as embryos. This embryonic death is not due to the incipient sexual transformation but rather to an abnormality in the dosage compensation system (Chapter 24). In addition to regulating the sex-determination pathway, *Sxl* also regulates dosage com-

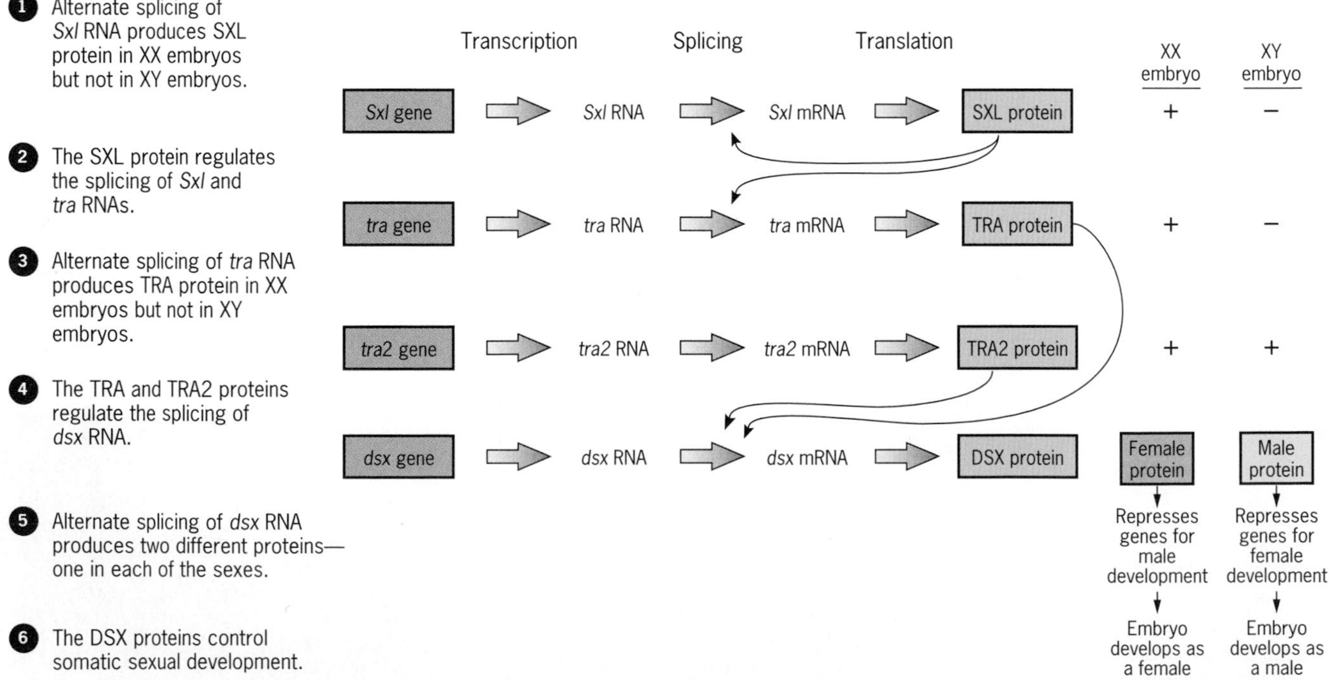

Figure 25.10 Regulation of sex determination in *Drosophila* by the *Sex-lethal* (*Sxl*) gene. The *Sxl* gene regulates the expression of the *transformer* (*tra*) gene, which, in turn, regulates the expression of the *doublesex* (*dsx*) gene; the *transformer2* (*tra2*) gene also participates in the regulation of *dsx*.

TABLE 25.1
Phenotypes of Loss-of-Function Mutations in Sex-Determination Genes in *Drosophila melanogaster* and *Caenorhabditis elegans*

Gene	*XX mutant phenotype*	*XY (or XO) mutant phenotype*
Drosophila melanogaster		
Numerator gene	lethal	no effect
Denominator gene	no effect	reduced viability
Sxl	lethal	no effect
tra	male	no effect
tra2	male	sterile male
dsx	sterile intersex	sterile intersex
Caenorhabditis elegans		
xol-1	no effect	lethal
sdc-1	masculinized	no effect
sdc-2	masculinized	no effect
sdc-3	no sex-determination effect	no effect
her-1	no effect	fertile hermaphrodite
tra-2	male	no effect
tra-3	male	no effect
fem-1	female	female
fem-2	female	female
fem-3	female	female
tra-1	male	minor effects in gonad

Source: S. M. Parkhurst and P. M. Meneely, *Science* 264:924–932, 1994.

pensation. Although the mechanism is not fully understood, it apparently prevents the hyperactivation of X-linked genes in XX animals. When this hyperactivation occurs, as it does in homozygous *Sxl* mutants, XX embryos die because there is too much X-linked gene expression. However, in XY animals, loss-of-function mutations in the *Sxl* gene have no effect, which is consistent with the fact that the SXL protein is normally not made in males.

Loss-of-function mutations in *transformer* and *transformer2* have the same phenotype: both XX and XY animals develop into males. The sexual transformation in XX animals demonstrates that both the tra^+ and $tra2^+$ genes are needed for female development; however, they are perfectly dispensable for male development. Loss-of-function mutations in the *dsx* gene cause both XX and XY embryos to develop into intersexes. The intersexual phenotype appears because *both* the male and the female developmental pathways are activated in *dsx* mutants.

Sex Determination in *Caenorhabditis*

The somatic sex-determination pathway in *Caenorhabditis* involves at least 10 different genes (Figure 25.11). As in *Drosophila,* mutations in these genes alter sexual development (Table 25.1). For example, loss-of-function mutations in the two *transformer* genes *tra-1* and *tra-2* (not to be confused with the *tra* and *tra2* genes of *Drosophila*) cause XX animals to develop into males, and loss-of-function mutations in the *hermaphrodite her-1* gene cause XO animals to develop into hermaphrodites. These phenotypic alterations show that the *tra-1* and *tra-2* gene products are needed for normal hermaphrodite development and that the *her-1* gene product is needed for normal male development. Loss-of-function mutations in three other genes, *fem-1, fem-2,* and *fem-3* (for feminization), cause XO animals to develop into spermless females. This feminized phenotype shows that the *fem* gene products are needed for normal male development.

The phenotypes of double mutant combinations have established an epistasis hierarchy for these six sex-determination genes:

tra-1 >> *fem-1, fem-2, fem-3* >> *tra-2* >> *her-1*

For example, a *tra-1, her-1* double mutant develops as a male, whether it is XO or XX. The masculinizing *tra-1* mutation is therefore epistatic over the feminizing *her-1* mutation. Using these epistatic relationships and additional information about how the genes might function, researchers have been able to arrange these six genes in a developmental pathway.

The first gene, *her-1,* encodes a secreted protein that is likely to be a signaling molecule. The next gene, *tra-2,* encodes a membrane-bound protein, which may function as a receptor for the *her-1* signaling protein.

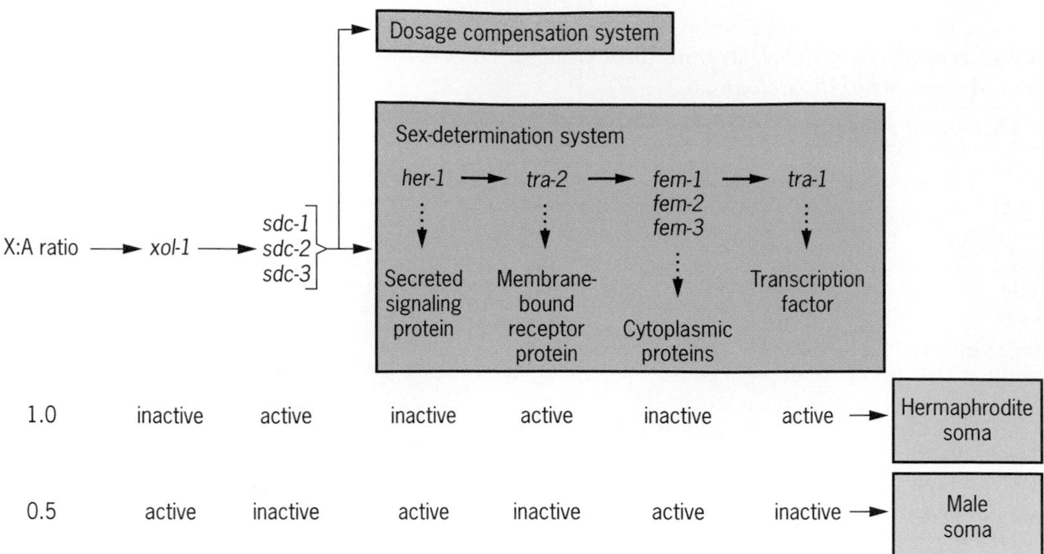

Figure 25.11 Developmental pathway for sex determination in *Caenorhabditis*. Differentation of the animal as a male or as a hermaphrodite depends on a cascade of gene expression, which, in turn, depends on the X:A ratio. Dosage compensation also depends on the X:A ratio.

The products of the *fem* genes are cytoplasmic proteins that may transduce the *her-1* signal, and the last gene in the pathway, *tra-1,* apparently encodes a zinc-finger type transcription factor, which may regulate the genes involved in sexual differentiation.

In *Caenorhabditis,* the sex-determination pathway seems to involve a series of negative regulators of gene expression. In XO animals, the secreted *her-1* gene product apparently interacts with the *tra-2* gene product, causing it to become inactive. This interaction allows the three *fem* gene products to be activated, and they collectively inactivate the *tra-1* gene product, which is a positive regulator of female differentiation. Because the animal cannot develop as a hermaphrodite without active *tra-1* protein, it develops as a male. In XX animals, the *her-1* protein is not made; thus, its putative receptor, the *tra-2* protein, remains active. Active *tra-2* protein causes the *fem* gene products to be inactivated, which in turn allows the *tra-1* protein to stimulate female differentiation. The animal therefore develops as a hermaphrodite.

Sexual development in *Caenorhabditis* fundamentally depends on the X:A ratio, just as it does in *Drosophila*. There appears to be a system of numerator and denominator elements, but to date only numerator elements have been identified. The X:A ratio is somehow converted into a molecular signal that controls sexual differentiation. This same signal also controls the phenomenon of dosage compensation, which in *Caenorhabditis* involves hypoactivation of the two X chromosomes in the hermaphrodite (Chapter 24). The signal from the X:A ratio is funneled into the sex-determination and dosage-compensation pathways through a short pathway involving at least four genes. One of these, *xol-1,* is required in males but not in hermaphrodites. Loss-of-function mutations in *xol-1* cause XO animals to die, hence the name of the gene, *XO-lethal*. Three other genes, *sdc-1, sdc-2,* and *sdc-3* (for sex determination and dosage compensation), are negatively regulated by *xol-1*. Loss-of-function mutations in these genes either kill XX animals or turn them into males. Thus, the *sdc* genes are needed in hermaphrodites but not in males.

Key Points: **The genetic dissection of a developmental pathway consists of analyzing genes whose products are involved in the differentiation of specific tissues and organs. In *Drosophila,* for example, the pathway that controls sexual differentiation into male or female involves some genes that ascertain the X:A ratio, some that convert this ratio into a developmental signal, and others that respond to the signal by producing either male or female structures. In *Caenorhabditis,* the sexual-differentiation pathway involves genes that encode signaling proteins, their receptors, signal transducers, and transcription factors.**

THE USE OF GENETIC MOSAICS TO STUDY DEVELOPMENT

Because all the cells in a multicellular organism are derived mitotically from a single fertilized egg, we would expect every one of them to have the same genotype.

For the most part, this seems to be the case. The phenotypic differences that emerge among cells during development are caused by differential gene expression rather than by differences in gene content. However, abnormalities in mitosis sometimes produce daughter cells that have different genotypes. Replication of these daughter cells within the developing animal can give rise to genetically different clones. An animal with such clones is called a **genetic mosaic.** The analysis of genetic mosaics has allowed researchers to trace cell lineages during development as well as to ascertain whether genes control intracellular or intercellular processes.

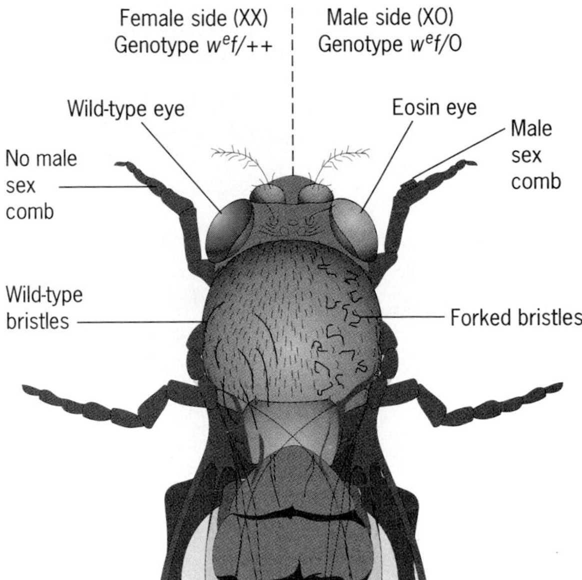

Figure 25.12 A bilateral *Drosophila* gynandromorph. The left side is XX and female, and the right side is XO and male. Note the sex comb on the right foreleg, a male characteristic. This gynandromorph arose from an embryo heterozygous for two recessive, X-linked mutations, w^e (*white-eosin* eye) and *f* (*forked* bristles). The X chromosome carrying the wild-type alleles of these mutations was lost early in development in the part of the embryo that produced the right side of the fly. Thus, this side of the fly shows the two mutant phenotypes.

X Chromosome Loss in *Drosophila:* Gynandromorphs

The first genetic mosaics to be studied in *Drosophila* arose from loss of an X chromosome very early in development (Figure 25.12). These mosaics occur spontaneously at a low rate and are composed of two types of cells, one containing two X chromosomes and the other containing only one. Such XX/XO mosaics develop into flies that are part female and part male, called gynandromorphs (Chapter 7). The XX tissues in a gynandromorph differentiate into female structures, and the XO tissues differentiate into male structures. The perfect coincidence between the sexual phenotype of a tissue and its chromosomal genotype demonstrates that in *Drosophila,* sex determination occurs wholly within cells. Communication between cells—for example, by hormones—does not seem to play a role. Instead, every cell develops its sexual phenotype independently of every other cell. Thus, sexual development in *Drosophila* is said to be a **cell-autonomous process.**

Other aspects of *Drosophila* development are also cell-autonomous—for example, the formation and pigmentation of the sensory bristles and hairs that cover much of the adult body. Two genes controlling these processes are located on the X chromosome. The *singed* (*sn*) gene controls bristle formation and the *yellow* (*y*) gene controls bristle pigmentation. Loss-of-function mutations in the *singed* gene cause the bristles to be gnarled and bent, instead of straight; loss-of-function mutations in the *yellow* gene cause the bristles to be brownish-yellow instead of black. If a zygote that is heterozygous for these recessive mutations (genotype *y sn*/+ +) loses the wild-type X chromosome during one of the early cleavage divisions, it will develop into a gynandromorph with brownish-yellow and singed bristles in the male tissue and wild-type bristles in the female tissue. The male tissue will be *y sn*/O and the female tissue will be *y sn*/+ +. Within the male tissue, the bristles will manifest both of the mutant phenotypes, even if they are adjacent to female tissue on the adult body. Careful examination shows that each bristle develops from a single cell. Thus, formation and pigmentation of bristles in *Drosophila* are cell-autonomous processes.

Somatic Recombination in *Drosophila:* Twin Spots

Genetic mosaics can also be produced by somatic recombination between homologous chromosomes. This is a rare event in *Drosophila,* even though homologous chromosomes pair in the somatic cells. However, its frequency can be increased by X-irradiation during development. Radiation-induced breaks in the DNA apparently initiate recombination events between paired chromosomes. As an example, let's consider an animal that is a repulsion heterozygote for mutations in the *yellow* and *singed* genes (genotype *y* +/+ *sn*). Both of these genes are far away from the centromere of the X chromosome. Consequently, it is possible to induce a recombination event between the centromere and the segment of the X that carries the two genes.

Figure 25.13 shows such a recombination event in a tetrad of chromatids produced by chromosome replication. When these chromatids disjoin during the fol-

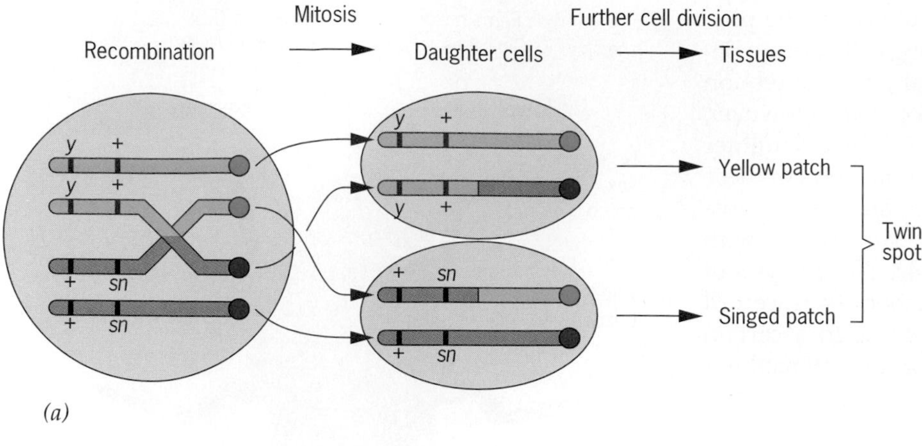

(a)

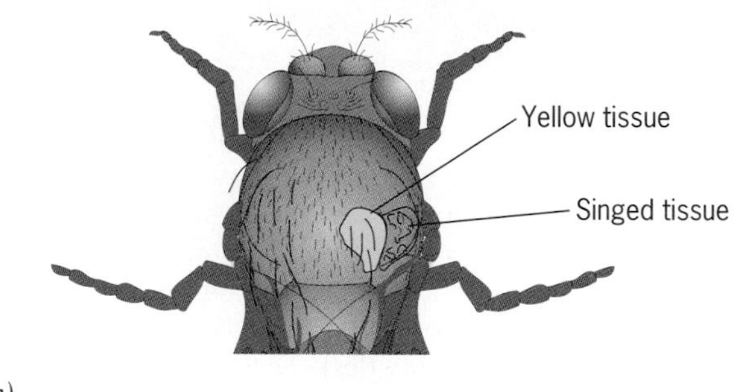

(b)

Figure 25.13 Production of twin spots in *Drosophila* by somatic recombination. (*a*) Recombination between the genes *yellow* (*y*) and *singed* (*sn*) and the centromere produces genetically different daughter cells, one homozygous for the *yellow* mutation and the other homozygous for the *singed* mutation. (*b*) A yellow/singed twin spot on the thorax of a fly.

lowing mitosis, each daughter cell may wind up homozygous for both of the loci. One cell will be homozygous for the mutant allele of *yellow* and the wild-type allele of *singed* (genotype $y\ +/y\ +$), and the other cell will have the reciprocal genotype ($+\ sn/+\ sn$). Of course, this is not the only possible outcome. The chromatids may disjoin so that both cells remain heterozygous for both of the loci. However, if disjunction produces cells that are homozygous for the different mutants, and if these cells generate clones in the bristle-forming tissues, the adult will show a patch of yellow bristles adjacent to a patch of singed bristles in an otherwise wild-type background. This condition is referred to as a **twin spot.**

Somatic recombination can be induced at any time during development. When the recombination is induced early, the resulting mosaic patches are large; when it is induced late, they are small. The size of a mosaic patch therefore indicates how early in development the causative recombination event occurred. Because all the cells in a patch are descendants of a common ancestor, the shape of a patch can be used to trace cell lineages during development.

Cell-autonomous genes such as *yellow* and *singed* can be used as markers to determine whether other genes linked to them are also cell autonomous (Figure 25.14). Let's suppose, for example, that we are interested in a gene located between *yellow* and *singed* on the X chromosome that is essential for life. Loss-of-function mutations in this gene would be recessive lethals (denoted by *l*). To determine whether the gene is cell-autonomous, we could induce somatic recombination in $y\ l\ +/+\ +\ sn$ heterozygotes. We would expect that exchanges between the *singed* locus and the centromere would produce twin spots on the adult body, as we discussed above. However, if the lethal mutation, which is on the same X chromosome as the *yellow* mutation, functions autonomously, the homozygous $y\ l\ +/y\ l\ +$ cells produced by such exchanges will die, leaving their $+\ +\ sn/+\ +\ sn$ sister cells to grow into solitary patches of singed tissue on the adult. The appearance of these solitary singed patches would therefore be an indication that the lethal mutation is cell-autonomous. If it were nonautonomous, homozygous $y\ l\ +/y\ l\ +$ cells could be rescued by the surrounding heterozygous tissue, and yellow/singed twin spots would appear. This technique therefore allows us to determine whether a gene controls intracellular (autonomous) or intercellular (nonautonomous) functions.

Key Points: **Genetic mosaics caused by mitotic chromosome loss or somatic recombination can be used to trace cell lineages during development and to ascertain whether a gene functions autonomously within cells. In *Drosophila,* such mosaics can be created by X chromosome loss or by radiation-induced somatic recombination.**

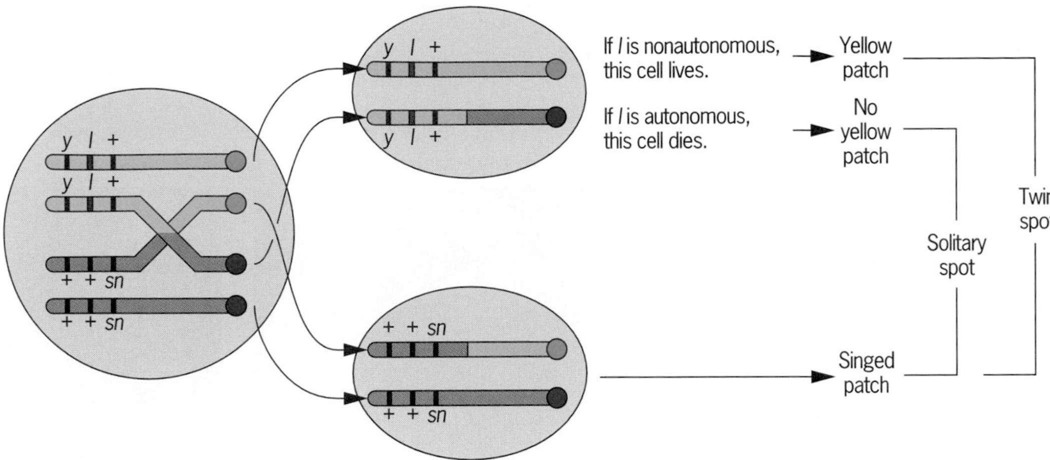

Figure 25.14 Somatic recombination in *Drosophila* with markers to determine if a lethal mutation (*l*) is autonomous or nonautonomous.

MOLECULAR ANALYSIS OF GENES INVOLVED IN DEVELOPMENT

The detailed analysis of developmentally significant genes requires that they be cloned and sequenced. For both *Drosophila* and *Caenorhabditis,* this is a fairly straighforward process involving the techniques we discussed in Chapters 20 and 21. Genomic or cDNA libraries can be constructed from either species and screened for particular genes. Screening is facilitated by the relatively small sizes of the genomes (170,000 kb in *Drosophila* and 80,000 kb in *Caenorhabditis*) and by the wealth of genetic, cytological, and molecular data already collected from these organisms. In *Drosophila,* for example, cloned DNA sequences can be localized cytologically by *in situ* hybridization to polytene chromosome squashes (Chapter 10). In *Caenorhabditis,* clones can be localized on a detailed physical map of the genome by hybridization with a comprehensive array of tester clones fixed to nylon membranes. Transposon tagging, a procedure whereby genes are mutated by the insertion of mobile genetic elements (Chapter 18), is also possible in these organisms. Once tagged, a mutant gene can be cloned by using a transposon probe to screen a recombinant DNA library made from mutant animals.

Cloning a gene makes it possible to conduct experiments that can help scientists understand the gene's developmental significance. Cloned sequences can be used to make probes to detect transcripts from the gene in different tissues and at different times during development. After a gene has been cloned, it is also possible to generate antibodies to the protein encoded by the gene. Once antibodies have been made, they can be used to detect the protein in specific tissues at different developmental stages. All these procedures help researchers understand where and when a gene is expressed in a developing animal.

For both *Drosophila* and *Caenorhabditis,* a cloned gene can also be reintroduced into the animal, perhaps after altering it *in vitro,* to determine how the gene functions *in vivo*. Researchers first developed the techniques for genetic transformation with cloned DNA by using transposable elements as vectors to insert the cloned DNA into the *Drosophila* genome (Chapter 18). These techniques involve injecting the DNA into the primordial germ line of *Drosophila* embryos prior to blastoderm formation. At this stage, it is possible for the injected DNA to be integrated into the chromosomes of the future germ line. If integration occurs and the injected animal survives to become an adult, the inserted DNA may be passed on to the animal's offspring. This procedure of germ-line transformation can therefore create a stock of flies in which cloned DNA has been inserted into the genome. The inserted DNA is called a *transgene*, and the flies are called transgenic animals. Although the most common procedure involves transforming *Drosophila* with *Drosophila* DNA, it is also possible to transform them with DNA from other organisms—bacteria, other invertebrates, even vertebrates. In this way, *Drosophila* can be used as a living laboratory to investigate the functions of genes from many different species, including human beings.

Cloned DNA can also be introduced into *Caenorhabditis* by injecting it into a syncytial region in the gonad of a hermaphrodite. In this case, however, the DNA usually does not integrate into the chromosomes but rather, remains unincorporated, probably as a tandem array of linked molecules. By chance, some of the DNA may be packaged into sperm that are formed in this part of the gonad and then may be subsequently transmitted to a zygote. During development, this extrachromosomal DNA segregates irregularly during cell division, creating mosaic animals that may or may not pass the DNA on to their offspring.

Key Points: **Cloning a gene is fundamental to determining its role in development. Genetically transformed, or transgenic, animals can provide information on how a cloned gene functions *in vivo*.**

MATERNAL GENE ACTIVITY IN DEVELOPMENT

Important events occur in animal development even before an egg is fertilized. At this time, nutritive and determinative materials are transported into the egg from surrounding cells, laying up food stores and organizing the egg for its subsequent development—the molecular equivalent of a mother's love. These materials are generated by the expression of genes in the female reproductive system, some being expressed in somatic reproductive tissues and others only in germline tissues. Collectively, these genes help to form eggs that can develop into embryos after fertilization. In some species, these maternal gene products lay out the basic body plan of the embryo, distinguishing head from tail and back from belly. These maternally supplied materials therefore establish a molecular coordinate system to guide an embryo's development.

Maternal-Effect Genes

Mutations in genes that contribute to the formation of healthy eggs may have no effect on the viability or appearance of the female making those eggs. Instead, their effects may be seen only in the next generation. Such mutations are called **maternal-effect mutations** because the mutant phenotype in the offspring is caused by a mutant genotype in its mother.

Genes identified by such mutations are called **maternal-effect genes.** The *dorsal* (*dl*) gene in *Drosophila* is a good example (Figure 25.15). Matings between flies homozygous for recessive mutations in this gene produce inviable progeny. This lethal effect is strictly maternal. A cross between homozygous mutant females and homozygous wild-type males produces inviable progeny, but the reciprocal cross (homozygous mutant males × homozygous wild-type females) produces viable progeny. The lethal effect of the *dorsal* mutation is therefore manifested only if females are homozygous for it. The male genotype is irrelevant.

Molecular characterization of the *dorsal* gene has revealed the basis for this maternal effect. The *dorsal* gene encodes a transcription factor that is produced

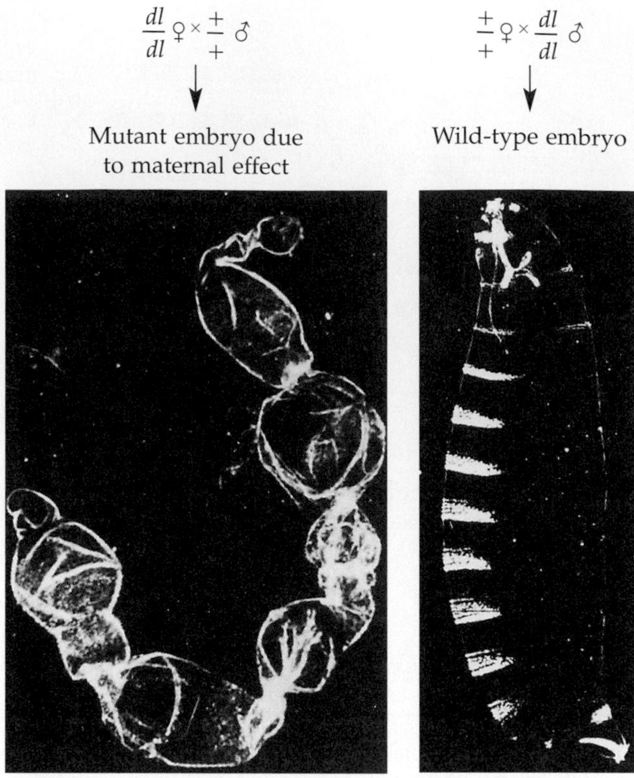

Figure 25.15 The maternal effect of a mutation in the *dorsal* (*dl*) gene of *Drosophila*. The mutant phenotype is an embryo that lacks ventral tissues; that is, it is dorsalized.

during oogenesis and stored in the egg. Early in development, this transcription factor plays an important role in the differentiation of the dorsal and ventral parts of the embryo. When it is missing, the ventral parts incorrectly differentiate as if they were on the dorsal side, creating an embryo with two dorsal surfaces. This lethal condition cannot be prevented by a wild-type *dorsal* allele inherited from the father because *dorsal* is not transcribed in the embryo. Rather, because its expression is limited to the female germ line, mutations in *dorsal* are strict maternal-effect lethals.

Determination of the Dorsal-Ventral and Anterior-Posterior Axes in *Drosophila* Embryos

Animals with bilateral symmetry have two primary body axes, one distinguishing back from belly (dorsal from ventral) and the other distinguishing head from tail (anterior from posterior). Both of these axes are established very early in development, in some species even before fertilization. In *Drosophila,* the processes of axis formation have been dissected genetically by collecting mutations that affect early embryonic development.

In the 1970s and 1980s, massive searches for such mutations were carried out by Christiane Nüsslein-

6 Dorsal cells differentiate into epidermis.

5 The genes *twist* and *snail* are repressed.
The genes *zerknüllt* and *decapentaplegic* are induced.

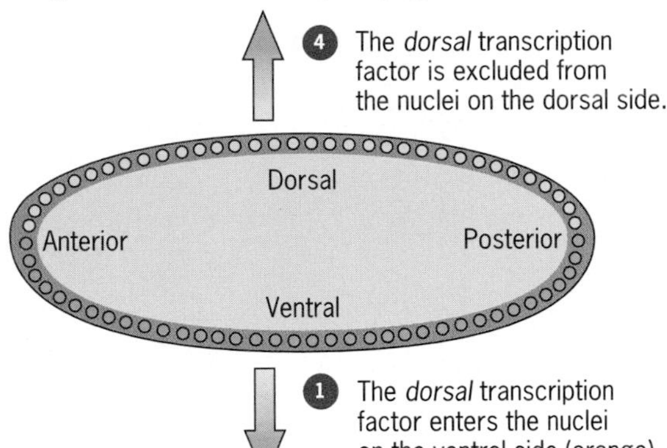

2 The genes *twist* and *snail* are induced.
The genes *zerknüllt* and *decapentaplegic* are repressed.

3 Ventral cells differentiate into mesoderm.

Figure 25.16 Determination of the dorsal-ventral axis in *Drosophila* by the *dorsal* protein, a transcription factor that acts only in the nuclei on the ventral side of the embryo. The genes *twist, snail, zerknüllt,* and *decapentaplegic* are regulated by *dorsal* protein.

Volhard, Eric Weischaus, Trudi Schüpbach, Gerd Jurgens, and others. These researchers used chemical mutagens to induce mutations in each of the *Drosophila* chromosomes. Many mutations were identified, including maternal-effect lethals in genes such as *dorsal*. Molecular and genetic analyses of these mutations have provided considerable insight into the events of early *Drosophila* development.

In *Drosophila* differentiation of the embryo along the dorsal-ventral axis hinges on the action of the transcription factor encoded by the *dorsal* gene (Figure 25.16). This protein is synthesized maternally and stored in the cytoplasm of the egg. At the time of blastoderm formation, the *dorsal* protein enters the nuclei on the ventral side of the embryo, inducing the transcription of two genes called *twist* and *snail* (whimsically named for their mutant phenotypes). In these same nuclei, it represses the genes *zerknüllt* (from the German for "crumpled") and *decapentaplegic* (from the Greek words for "15" and "stroke"). The selective induction and repression of genes cause the ventral cells to differentiate into embryonic mesoderm. On the opposite side of the embryo, where the *dorsal* protein is excluded from the nuclei, *twist* and *snail* are not induced and *zerknüllt* and *decapentaplegic* are not repressed. Consequently, these cells differentiate into embryonic epidermis. The entrance of the *dorsal* transcription factor into the ventral nuclei and its excusion from the dorsal nuclei therefore initiate differentiation along the dorsal-ventral axis.

But what triggers the *dorsal* protein to move into the nuclei on only one side of the embryo? The answer is an interaction between two proteins on the ventral surface of the developing embryo (Figure 25.17). One protein, the product of the *Toll* gene (from the German word for "tuft"), is distributed uniformly over the em-

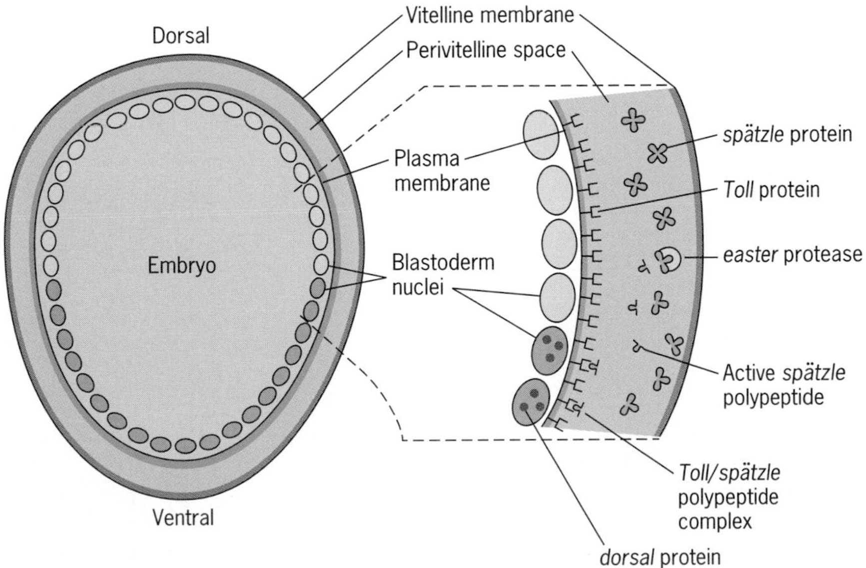

Figure 25.17 Cross section through a *Drosophila* embryo showing the interaction between the membrane-bound *Toll* receptor protein and a polypeptide from the *spätzle* protein that induces differentiation along the dorsal-ventral axis. Formation of the interacting *spätzle* polypeptide occurs in the space between the plasma membrane and the vitelline membrane on the ventral side of the embryo.

bryo's surface; this protein is embedded in the plasma membrane that surrounds the embryo. The other protein, the product of the *spätzle* gene (from the German word for "little dumpling"), is found in the perivitelline space, a fluid-filled cavity between the plasma membrane and the external vitelline membrane. Through the action of a protease encoded by a gene called *easter* (because it was discovered on Easter Sunday), the *spätzle* protein is cleaved to produce a polypeptide that interacts with the *Toll* protein. However, cleavage of the *spätzle* protein occurs only in the perivetilline space on the ventral side of the embryo. When the *Toll* protein interacts with the ventrally generated *spätzle* polypeptide, it initiates a cascade of

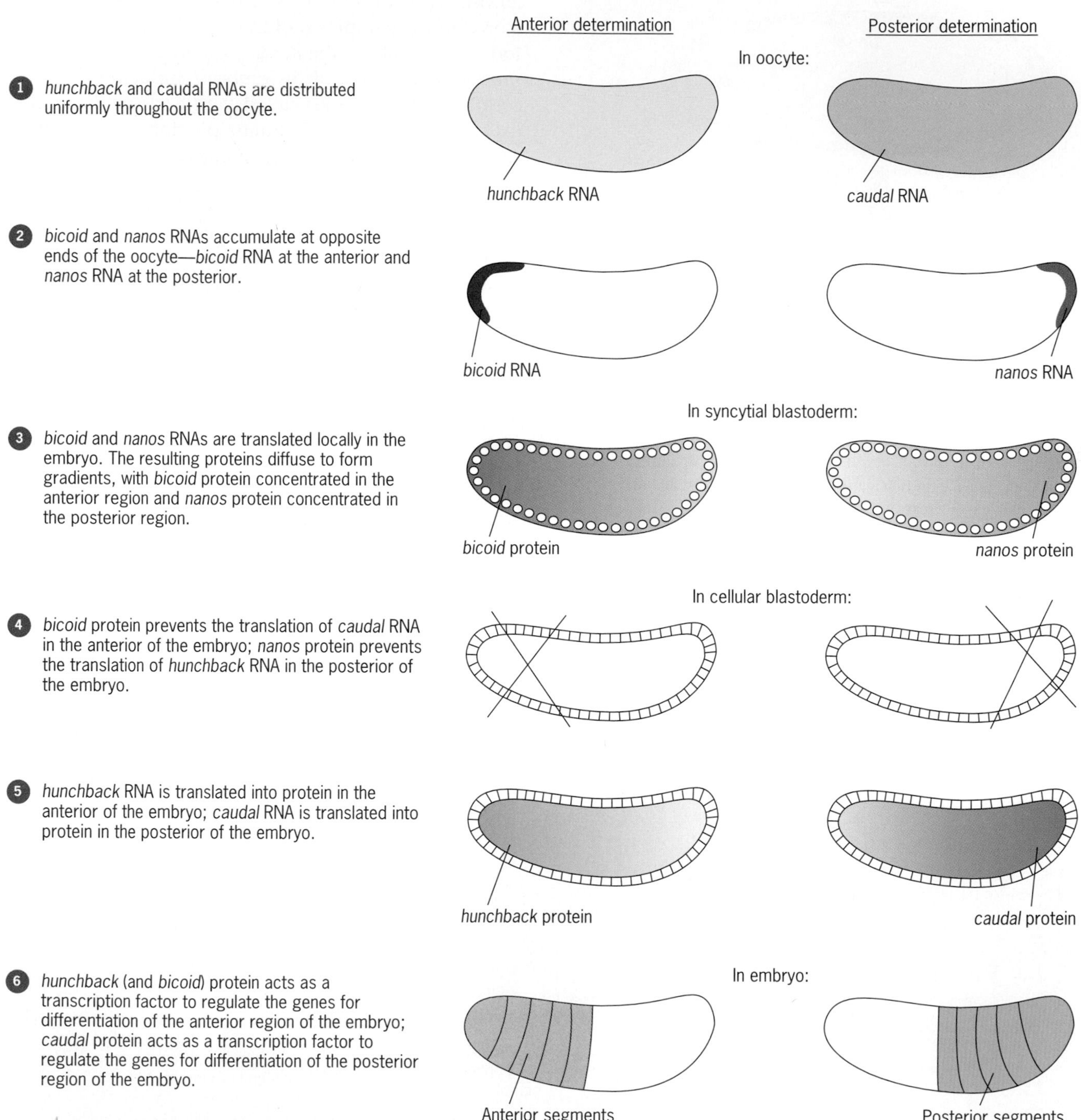

Figure 25.18 Determination of the anterior-posterior axis in *Drosophila* by maternally supplied RNAs from the *hunchback, caudal, bicoid,* and *nanos* genes. For each oocyte or embryo, anterior is at the left and posterior is at the right.

events within the embryo that ultimately sends the *dorsal* protein into the embryonic nuclei. There the *dorsal* protein functions as a transcription factor to regulate the expression of the genes *twist, snail, decapentaplegic,* and *zerknüllt*. Thus, the membrane-bound *Toll* protein acts as a receptor for the determinative *spätzle* polypeptide, and the physical interaction between these two molecules acts as a signal to trigger a genetic program for the differentiation of the embryo along its dorsal-ventral axis.

The anterior-posterior axis in *Drosophila* is created by the regional synthesis of transcription factors encoded by the *hunchback* and *caudal* genes (Figure 25.18). These two genes are transcribed in the nurse cells of the maternal germ line—that is, the cells that support the growth and development of the oocyte.The maternal transcripts of the *hunchback* and *caudal* genes are then carried into the oocyte where they become uniformly distributed in the cytoplasm. However, both types of transcripts are translated in different parts of the embryo. The *hunchback* RNA is translated only in the anterior part, and the *caudal* RNA is translated only in the posterior part. This differential translation produces concentration gradients of the proteins encoded by these two genes; *hunchback* protein is concentrated in the anterior part of the embryo, and *caudal* protein is concentrated in the posterior part. These two proteins then function to activate or repress transcription of the genes whose products are involved in the differentiation of the embryo along its anterior-posterior axis.

What limits the translation of *hunchback* RNA to the anterior part of the embryo and *caudal* RNA to the *posterior* part? It turns out that two maternally supplied RNAs are involved, one transcribed from the *bicoid* gene and the other from the *nanos* gene. Both of these RNAs are synthesized in the nurse cells of the maternal germ line and are then transported into the oocyte. The *bicoid* RNA becomes anchored at the anterior end of the developing oocyte, and the *nanos* RNA becomes anchored at the posterior end. After fertilization, each type of RNA is translated locally, and the resulting protein products diffuse through the embryo to form concentration gradients; *bicoid* protein is concentrated at the anterior end, and *nanos* protein is concentrated at the posterior end.

The *bicoid* protein has two functions. First, it acts as a transcription factor to stimulate the synthesis of RNAs from several genes, including *hunchback*. These RNAs are then translated into proteins that control the formation of the anterior structures of the embryo. Second, *bicoid* protein prevents the translation of *caudal* RNA by binding to sequences in the 3′ untranslated region of that RNA. Thus, wherever *bicoid* protein is abundant (that is, in the anterior of the embryo), *caudal* RNA is not translated into protein. Conversely, wherever *bicoid* protein is scarce (that is, in the posterior of the embryo), *caudal* RNA is translated into protein. The translational regulation of *caudal* RNA by *bicoid* protein is therefore responsible for the gradient of *caudal* protein that forms in the embryo. Because *caudal* protein is a specific activator of genes that control posterior differentation, the part of the embryo that has the highest concentration of *caudal* protein develops posterior structures.

Unlike *bicoid* protein, *nanos* protein does not function as a transcription factor. However, like *bicoid* protein, it does function as a translational regulator. *Nanos* protein is concentrated in the posterior of the embryo, and there it binds to the 3′ untranslated region of *hunchback* RNA and causes the degradation of that RNA. Consequently, *hunchback* protein is not synthesized in the posterior of the embryo. Instead, its synthesis is restricted to the anterior of the embryo where it acts as a transcription factor to regulate the expression of genes involved in anterior-posterior differentiation. Wherever *hunchback* protein is synthesized, the embryo develops anterior structures.

As the preceding discussion has indicated, differentiation along the dorsal-ventral and anterior-posterior axes in *Drosophila* embryos depends on the products of several maternal-effect genes: *bicoid, nanos, dorsal, hunchback, Toll, spätzle, easter,* and many other genes whose roles are beyond the scope of this textbook. The products of these genes are synthesized or activated in specific regions of the embryo in response to cues present during oogenesis. Recent research has identified how these cues are established. When the oocyte first forms, it is nestled at the posterior end of an egg chamber (Figure 25.19). The nurse cells that support its growth and development are located at the anterior end, and the entire chamber is surrounded by a layer of somatic follicle cells. Nutritive and determinative molecules, including, for example, the *bicoid* and *nanos* RNAs, are synthesized in the nurse cells and transported into the oocyte by a system of microtubules. As we have already discussed, the asymmetric distribution of *bicoid* and *nanos* RNAs in the oocyte ultimately determines the anterior-posterior axis of the embryo that will develop from it after fertilization.

About midway through oogenesis, the nucleus of the oocyte communicates with the undifferentiated follicle cells at the posterior end of the egg chamber. This communication is mediated by a signaling molecule encoded by a maternal-effect gene called *gurken* (from the German for "pickle"). The protein product of the *gurken* gene is homologous to Transforming Growth Factor alpha (TGFα), a type of growth factor first identified in mammalian cells. The *gurken* protein is secreted from the oocyte and interacts with a receptor

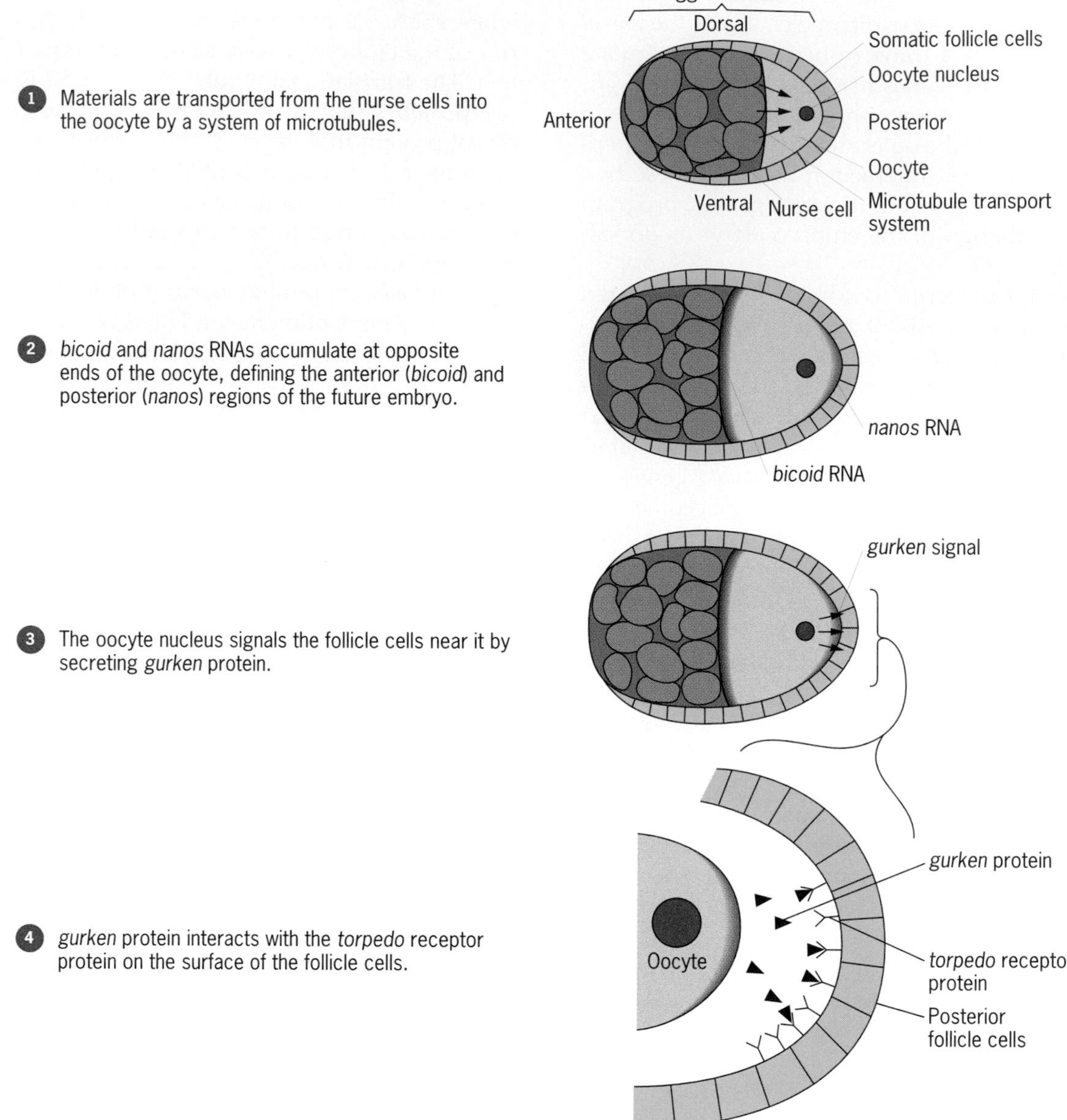

Figure 25.19 Establishment of anterior-posterior and dorsal-ventral polarity in the *Drosophila* egg chamber by the interaction between a secreted signaling protein encoded by the *gurken* gene and a membrane-bound receptor protein encoded by the *torpedo* gene.

protein on the surface of the adjacent somatic follicle cells. This receptor, encoded by a gene called *torpedo,* is then activated to send a signal into the nuclei of the follicle cells, inducing them to respond by sending a signal back to the oocyte. The molecular nature of this return signal is not yet known. However, it causes the microtubules within the oocyte to reorient themselves so that the oocyte nucleus migrates from its posterior position to the anterior-dorsal corner of the egg chamber. There, the oocyte nucleus sends another *gurken* signal to the follicle cells, activating their *torpedo* receptor. This time, however, the activated *torpedo* receptor induces the follicle cells to acquire a specific dorsal identity, thereby distinguishing them from the follicle cells on the other side of the egg chamber, which acquire a ventral identity. This dorsal/ventral differentiation then leads to the cleavage of the *spätzle* protein in the perivitelline space on the ventral side of the embryo, and this cleavage event plays a key role in the development of the embryo along its dorsal-ventral axis.

Key Points: **Materials transported into the egg during oogenesis play a major role in embryonic**

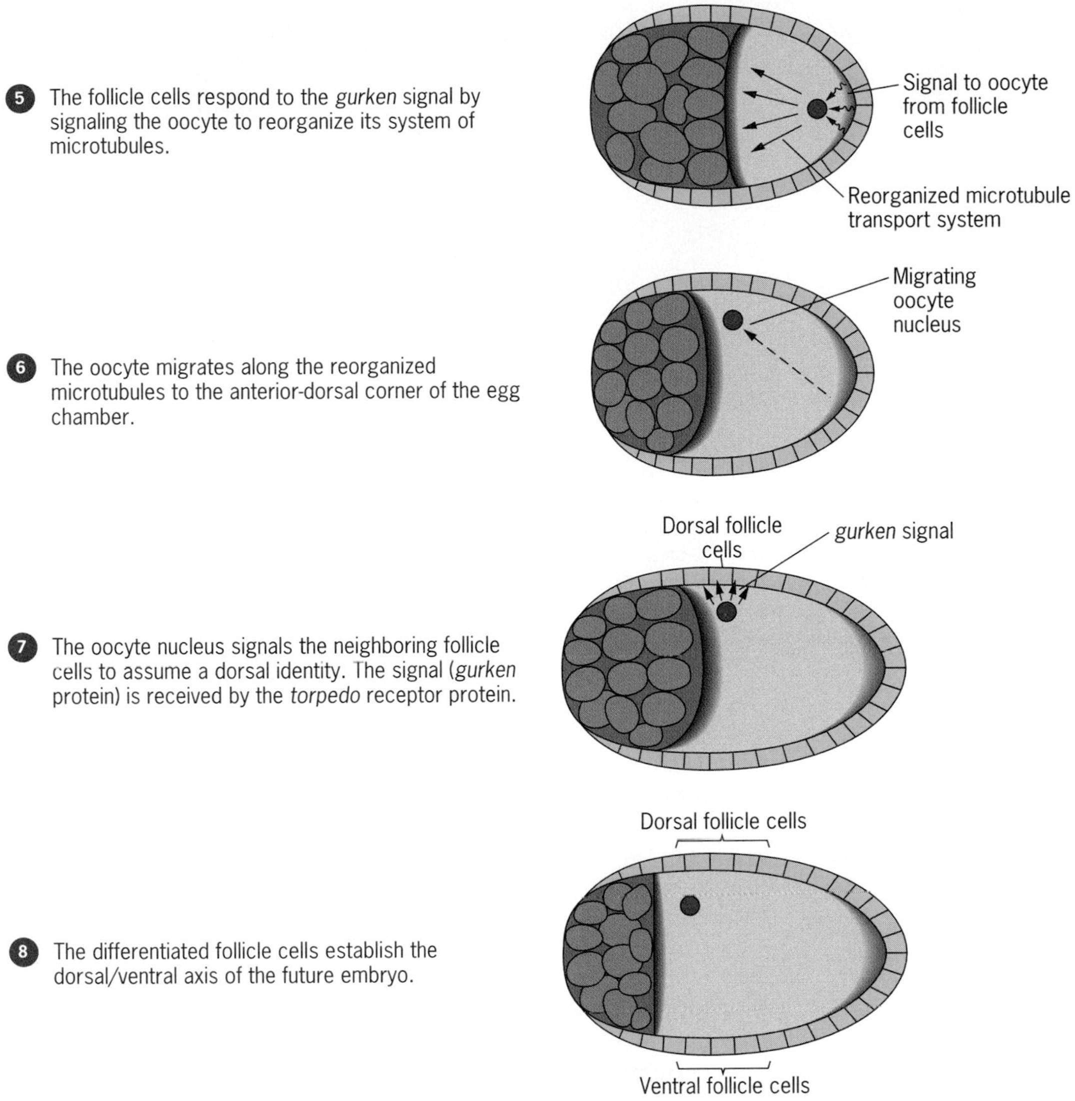

5 The follicle cells respond to the *gurken* signal by signaling the oocyte to reorganize its system of microtubules.
Signal to oocyte from follicle cells
Reorganized microtubule transport system
6 The oocyte migrates along the reorganized microtubules to the anterior-dorsal corner of the egg chamber.
Migrating oocyte nucleus
7 The oocyte nucleus signals the neighboring follicle cells to assume a dorsal identity. The signal (*gurken* protein) is received by the *torpedo* receptor protein.
Dorsal follicle cells
gurken signal
8 The differentiated follicle cells establish the dorsal/ventral axis of the future embryo.
Dorsal follicle cells
Ventral follicle cells

Figure 25.19 *(Continued)*

development. In *Drosophila,* these include proteins and RNAs encoded by maternal-effect genes such as *dorsal, hunchback, bicoid,* and *nanos.* The products of these genes are involved in the determination of the dorsal-ventral and anterior-posterior embryonic axes. Recessive mutations in maternal-effect genes are expressed only in embryos produced by females homozygous for these mutations.

ZYGOTIC GENE ACTIVITY IN DEVELOPMENT

The earliest events in animal development are controlled by maternally synthesized factors. However, at some point, the genes in the embryo are selectively activated, and new materials are made. This process is referred to as **zygotic gene expression,** because it occurs after the egg has been fertilized. The initial wave of zygotic gene expression is a response to maternally synthesized factors. In *Drosophila,* for example, the maternally supplied *dorsal* transcription factor activates the zygotic genes *twist* and *snail.* As development proceeds, the activation of other zygotic genes leads to complex cascades of gene expression. We shall now examine how these zygotic genes carry the process of development forward.

Body Segmentation

In many invertebrates the body consists of an array of adjoining units called **segments.** An adult *Drosophila,* for example, has a head, three distinct thoracic segments, and eight abdominal segments. Within the tho-

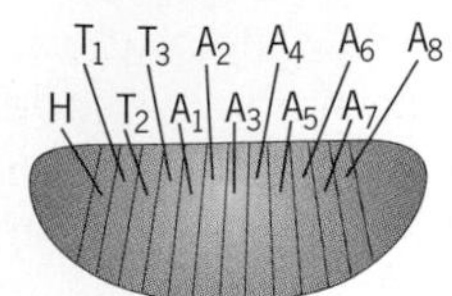

(*a*) Blastoderm

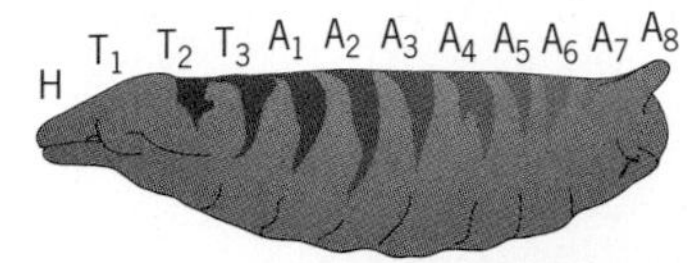

(*b*) Larva

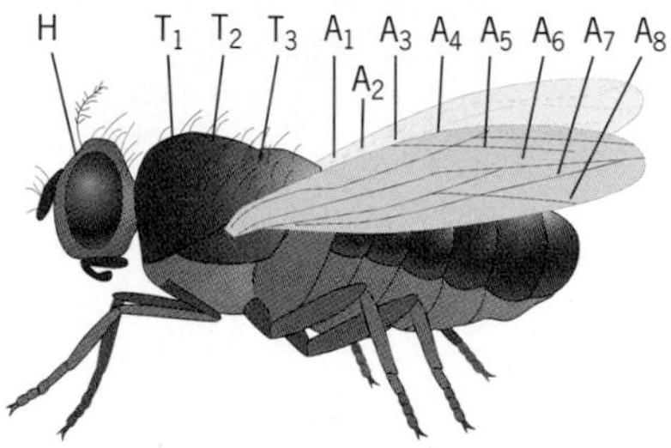

(*c*) Adult

Figure 25.20 Segmentation in *Drosophila* at the (*a*) blastoderm, (*b*) larval, and (*c*) adult stages of development. Although segments are not visible in the blastoderm, its cells are already committed to form segments as shown; H, head segment; T, thoracic segment; A, abdominal segment.

Figure 25.21 The phenotype of a *bithorax* mutation in *Drosophila*.

rax and abdomen, each segment can be identified by coloration, bristle pattern, and the kinds of appendages attached to it. These segments can also be identified in the embryo and the larva (Figure 25.20). In vertebrates, a segmental pattern is not so evident in the adult, but it can be recognized in the embryo from the way that nerve fibers grow from the central nervous system, from the formation of branchial arches in the head, and from the organization of muscle masses along the anterior-posterior axis. Later in development, these features are modified, and the original segmental pattern becomes obscured. Nonetheless, in both vertebrates and many invertebrates, segmentation is a key aspect of the overall body plan.

Homeotic Genes Interest in the genetic control of segmentation began with the discovery of mutations that transform one segment into another. The first such mutation was found in *Drosophila* in 1915 by Calvin Bridges. He named it *bithorax* (*bx*) because it affected two thoracic segments. In this mutant, the third thoracic segment was transformed, albeit weakly, into the second, creating a fly with a small pair of rudimentary wings in place of the halteres (Figure 25.21). Later, other segment-transforming mutations were found in *Drosophila,* for example, *Antennapedia* (*Antp*), a mutant that partially transforms the antennae on the head into legs, which characteristically grow from the thorax. These mutations have come to be called **homeotic mutations** because they cause one body part to look like another. The word homeotic comes from William Bateson, who coined the term **homeosis** to refer to cases in which "something has been changed into the likeness of something else." Like so many other words Bateson coined, this one has become a standard term in the modern genetics vocabulary.

The bithorax and Antennapedia phenotypes result from mutations in homeotic genes. Several such genes have now been identified in *Drosophila,* where they form two large clusters on one of the autosomes (Figure 25.22). The **Bithorax Complex,** usually denoted **BX-C,** consists of three genes, *Ultrabithorax* (*Ubx*), *abdominal-A* (*abd-A*), and *Abdominal-B* (*Abd-B*); the **Antennapedia Complex,** denoted **ANT-C,** consists of five genes, *labial (lab), proboscipedia (pb), Deformed (Dfd), Sex combs reduced (Scr),* and *Antennapedia (Antp).* Molecular analysis of these genes has demonstrated that they all encode helix-turn-helix transcription factors with a conserved region of 60 amino acids. This region, called the homeodomain, is involved in DNA binding.

The BX-C was the first of the two homeotic gene complexes to be dissected genetically. Analysis of this complex began in the late 1940s with the work of Edward Lewis. By studying mutations in the BX-C, Lewis showed that the wild-type function of each part of the complex is restricted to a specific region in the developing animal. Molecular analyses later reinforced and refined this conclusion. Study of the ANT-C began in the 1970s, principally through the work of Thomas Kaufman, Matthew Scott, and their collaborators. Through a combination of genetic and molecular analyses, these investigators showed that the genes of

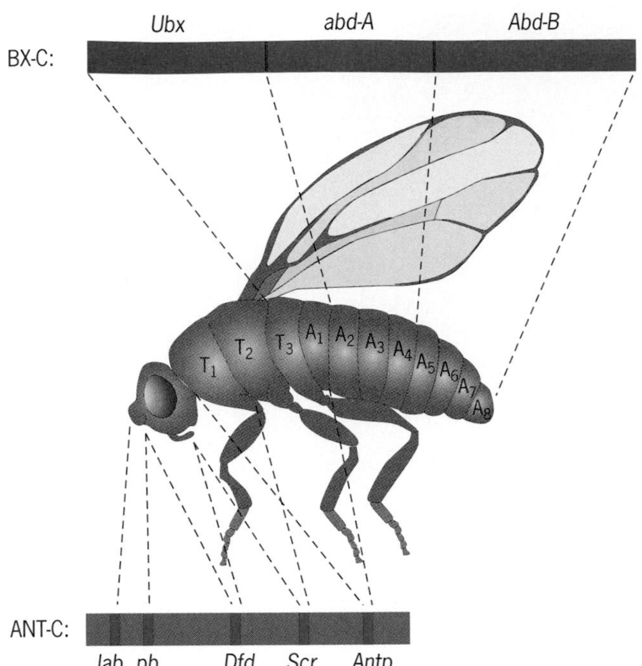

Figure 25.22 The homeotic genes in the Bithorax Complex (BX-C) and Antennapedia Complex (ANT-C) of *Drosophila*. The body regions in which each gene is expressed are indicated.

the ANT-C are also expressed in a regionally specific fashion. However, the ANT-C genes are expressed more anteriorly than the BX-C genes. Curiously, the pattern of expression of the ANT-C and BX-C genes along the anterior-posterior axis corresponds exactly to the order of the genes along the chromosome (Figure 25.22); it is not yet clear why this is so. The analysis of genetic mosaics has shown that these homeotic genes control the identities of individual segments in a cell-autonomous way. In fact, the developmental pathway that each cell takes seems to depend simply on the set of homeotic genes that are expressed within it. Because the homeotic genes play such a key role in selecting the segmental identities of individual cells, they are often called **selector genes.**

The proteins encoded by the homeotic genes are homeodomain transcription factors. These proteins bind to regulatory sequences in the DNA, including some within the Bithorax and Antennapedia Complexes themselves. For example, the UBX and ANTP proteins bind to a sequence within the promoter of the *Ubx* gene—a suggestion that the homeotic genes can regulate themselves and each other. Other gene targets of the homeodomain transcription factors have been identified, including some that encode other types of transcription factors. The homeotic genes therefore seem to control a regulatory cascade of target genes, which in turn act to determine the segmental identities of individual cells. However, the homeotic genes do not stand at the top of this regulatory cascade. Their activities are controlled by another group of genes expressed earlier in development—the segmentation genes.

Segmentation Genes Most of the homeotic genes were identified by mutations that alter the phenotype of the adult fly. However, these same mutations also have phenotypic effects in the embryonic and larval stages. This finding suggested that other genes involved in segmentation might be discovered by screening for mutations that cause embryonic and larval defects. In the 1970s and 1980s, Christiane Nüsslein-Volhard and Eric Wieschaus carried out such screens. They found a whole new set of genes required for segmentation along the anterior-posterior axis. Nüsslein-Volhard and Wieschaus classified these **segmentation genes** into three groups based on embryonic mutant phenotypes.

1. ***Gap Genes.*** These genes define segmental regions in the embryo. Mutations in the gap genes cause an entire set of contiguous body segments to be missing; that is, they create an anatomical gap along the anterior-posterior axis. Four gap genes have been well characterized: *Krüppel* (from the German for "cripple"), *giant, hunchback,* and *knirps* (from the German for "dwarf"). Each is expressed in characteristic regions in the early embryo under the control of the maternal effect genes *bicoid* and *nanos*. The gap genes encode transcription factors.
2. ***Pair-rule Genes.*** These genes define a pattern of segments within the embryo. The pair-rule genes are regulated by the gap genes and are expressed in seven alternating bands, or stripes, along the anterior-posterior axis, in effect dividing the embryo into 14 distinct zones, or **parasegments** (Figure 25.23). Mutations in each of the several pair-rule genes produce embryos with only half as many parasegments as wild-type. In each mutant, every other parasegment is missing, although the missing parasegments are not the same in different pair-rule mutants. Examples of pair-rule genes are *fushi tarazu* (from the Japanese for "something missing") and *even-skipped*. In *fushi tarazu* mutants, each of the odd-numbered parasegments is missing; in *even-skipped* mutants, each of the even-numbered parasegments is missing. Most of the pair-rule genes encode transcription factors.
3. ***Segment-Polarity Genes.*** These genes define the anterior and posterior compartments of individual segments along the anterior-posterior axis. Mutations in segment-polarity genes cause part of

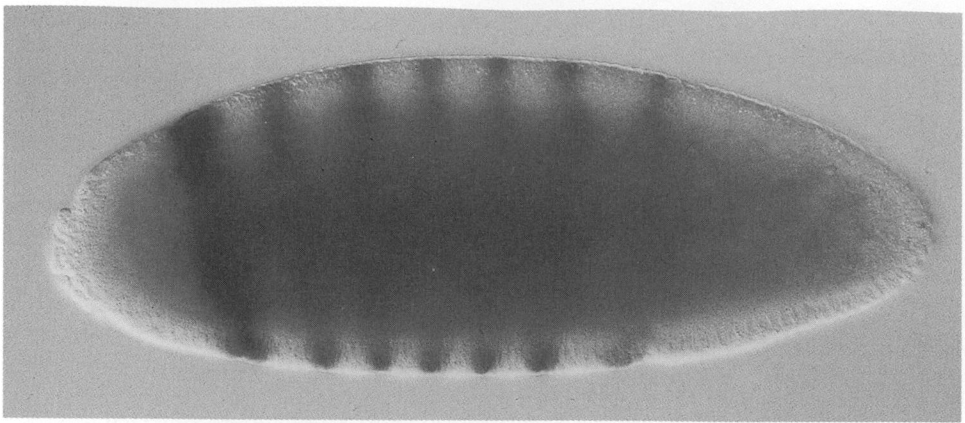

Figure 25.23 The seven-stripe pattern of RNA expression of the pair-rule gene *fushi tarazu* (*ftz*) in a *Drosophila* blastoderm embryo. The RNA was detected by *in situ* hybridization with a *ftz*-specific probe. Anterior is at the left; dorsal is at the top.

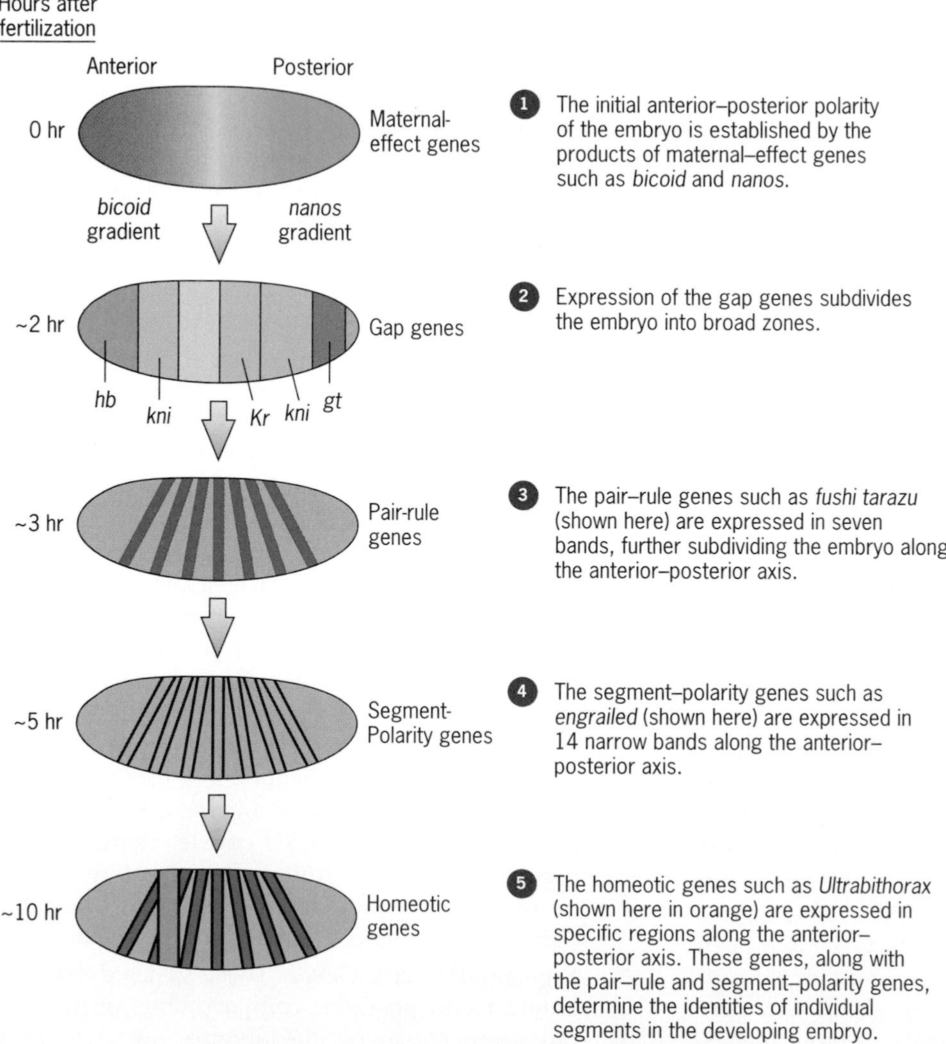

Figure 25.24 Cascade of gene expression to produce segmentation in *Drosophila* embryos.

each segment to be replaced by a mirror-image copy of an adjoining half-segment. For example, mutations in the segment-polarity gene *gooseberry* cause the posterior half of each segment to be replaced by a mirror-image copy of the adjacent anterior half-segment. Many of the segment-polarity genes are expressed in 14 narrow bands along the anterior-posterior axis. Thus, they refine the segmental pattern established by the pair-rule genes. Two of the best-studied segment-polarity genes are *engrailed* and *wingless*; *engrailed* encodes a transcription factor, and *wingless* encodes a signaling molecule.

These three groups of genes form a regulatory hierarchy (Figure 25.24). The gap genes, which are regionally activated by the maternal-effect genes, regulate the expression of the pair-rule genes, which in turn, regulate the expression of the segment-polarity genes. Concurrent with this process, the homeotic genes are activated under the control of the gap and pair-rule genes to give unique identities to the segments that form along the anterior-posterior axis. Interactions among the products of all these genes then refine and stabilize the segmental boundaries. In this way, the *Drosophila* embryo is progressively subdivided into smaller and smaller developmental units. For an examination of developmental events in another group of insects, see the Technical Sidelight: The Development of Butterfly Wings: Insights from *Drosophila* Genetics.

Specification of Cell Types

The genetic control of segmentation shows how sets of genes work in a regulatory cascade to determine the identities of groups of cells in different regions of the embryo. However, within these regions, what causes differentiation into specific cell types? What causes some cells to become neurons and others to become neuronal support cells? The answer to this kind of question is only now coming into scientific grasp, largely through the study of very simple situations involving a few distinct cell types. One such situation occurs in the development of the *Drosophila* eye (Figure 25.25).

Each of the large compound eyes in *Drosophila* originates as a flat sheet of cells in one of the imaginal discs. Initially, all the cells in this epithelial sheet look the same, but late during the larval stage, a furrow forms near the posterior margin of the disc. As this furrow moves in the anterior direction across the disc, it triggers a wave of cell divisions in its wake. The newly divided cells then differentiate into specific cell types to form the 800 individual facets of the adult eye. Each facet consists of 20 cells. Eight are photoreceptor neurons designed to absorb light; four are cone cells that secrete a lens to focus light into the photoreceptors; six are sheath cells to provide insulation and support; and the two remaining cells form sensory hairs on the eye's surface. Thus, a highly patterned array of intricately differentiated facets develops from what had been a flat sheet of identical cells. What brings this array about?

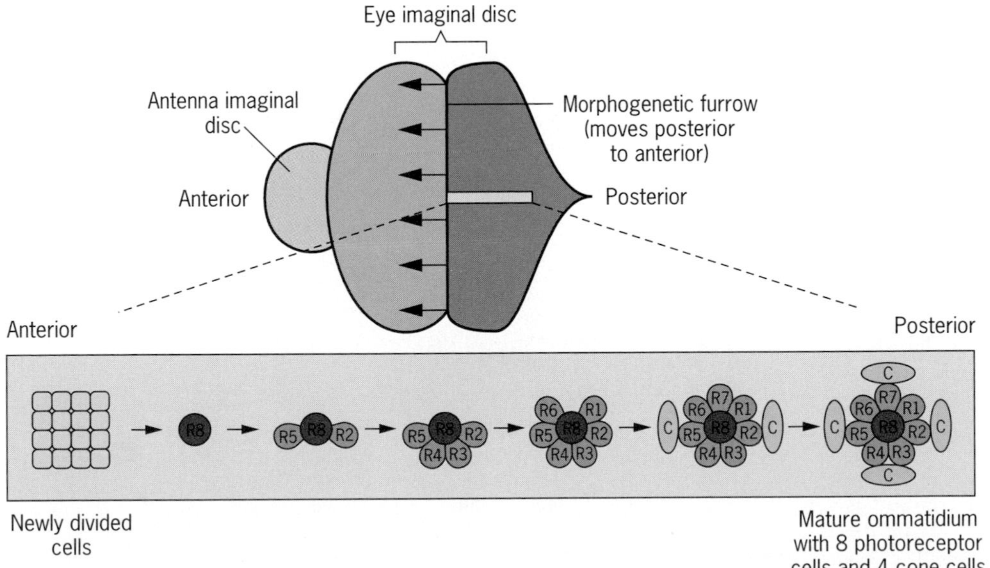

Figure 25.25 Development of the *Drosophila* eye. As the morphogenetic furrow moves toward the anterior of the eye-antenna imaginal disc, a wave of cell divisions follows in its wake. The newly divided cells then begin to differentiate into specific types. The insert shows the differentiation of the photoreceptor (R1-R8) and cone (C) cells that form each ommatidium of the compound eye.

TECHNICAL SIDELIGHT

The Development of Butterfly Wings: Insights from *Drosophila* Genetics

More than 17,000 species of butterflies have been identified, including many with beautiful patterns on their wings. These insects have long been the object of collectors and recently have caught the eyes of geneticists. What genes are involved in the formation of butterfly wings? What molecular events are responsible for producing their colorful patterns? How might the system for wing development have evolved in different groups of insects?

Sean Carroll, a molecular geneticist at the University of Wisconsin, is attempting to answer these questions by applying knowledge gained from the study of *Drosophila* wing development. The wings of both butterflies and fruit flies develop from imaginal discs, which are packets of undifferentiated cells sequestered in the larva. Initially, each disc is organized as a relatively small number of cells within the thousands of cells that are found in a particular body segment. During larval growth, the disc cells proliferate, and regions of the disc become determined to produce characteristic adult structures. For example, a pair of discs that form in one of the thoracic segments of *Drosophila* larvae will eventually differentiate into membranous wings that consist of two cell layers—dorsal and ventral; these discs also produce the thoracic tissue that anchors the wings to the adult's body. Carroll's research has shown that the developmental events that produce a butterfly's wing are similar to those that occur in *Drosophila*.

A wing disc in third instar *Drosophila* larvae consists of a single layer of cells (Figure 1). In one portion of the disc, the cells are determined to become the dorsal side of the adult wing; in another portion, they are determined to be-

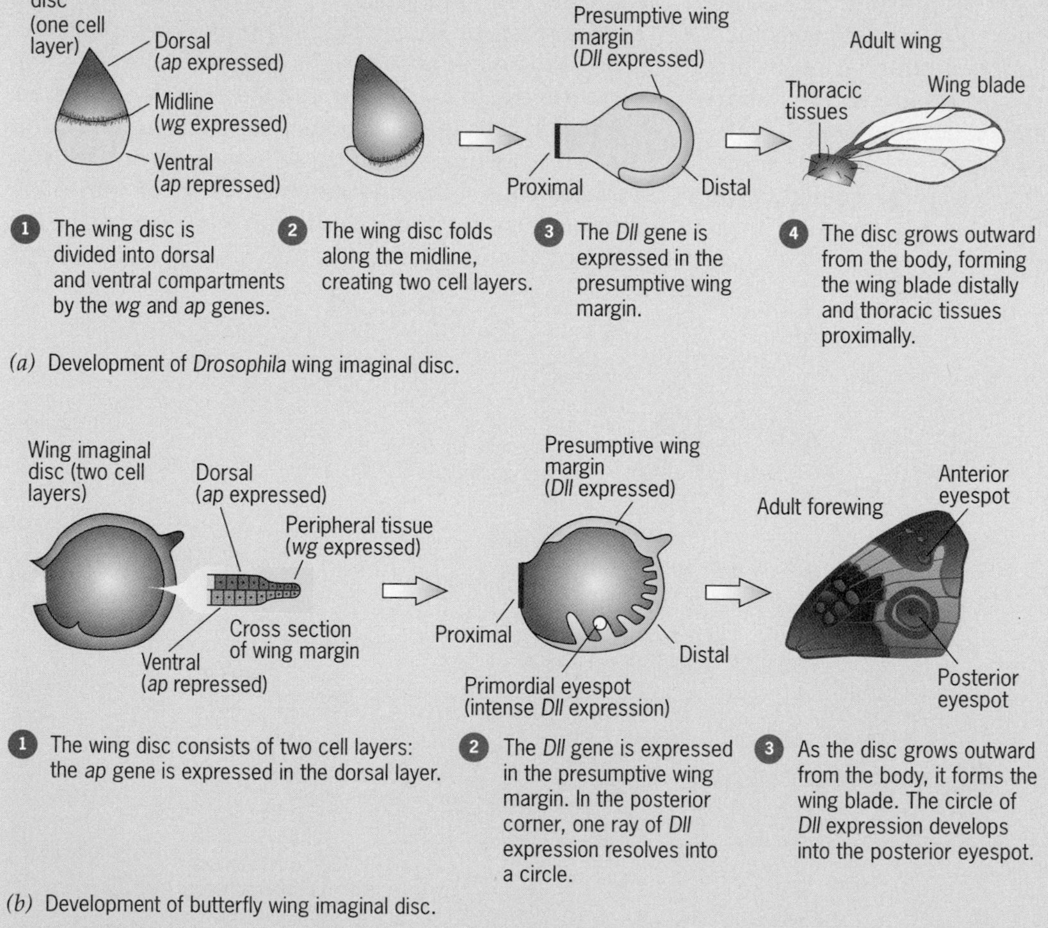

Figure 1 Comparison of development in *Drosophila* and butterfly wings. (*a*) Development of the *Drosophila* wing imaginal disc. (*b*) Development of a butterfly wing imaginal disc.

come the ventral side. This determination is associated with the expression of a gene called *apterous* (*ap*), which encodes a transcription factor. The *ap* gene is expressed in the presumptive dorsal portion of the disc; in the presumptive ventral portion, it is repressed. Another gene, *wingless* (*wg*), is expressed along the midline that divides the dorsal and ventral portions of the disc. This gene encodes a signaling molecule that organizes the cells destined to form the distal margin of the adult wing. During metamorphosis, the wing disc folds along its midline, creating dorsal and ventral wing surfaces. After folding, another gene, *Distal-less* (*Dll*), produces a transcription factor in the wing margin. As the wing develops, it grows outward—that is, distal—from the body. The expression of the *Dll* gene therefore helps to define the proximal-distal axis for wing development.

In butterflies, the wing discs consist of two layers of cells, one destined to form the dorsal surface of the adult wing and the other the ventral surface (Figure 1). Carroll studied the development of these two layers by monitoring the expression of the butterfly homologs of *Drosophila* genes. As in *Drosophila,* the presumptive dorsal layer is marked by the expression of a gene that is homologous to *ap,* and the boundary between the presumptive dorsal and ventral layers is marked by the expression of a gene that is homologous to *wg*. Later in development, a homolog of the *Dll* gene is expressed in the presumptive margin of the butterfly wing. Thus, these three genes—*ap, wg* and *Dll*—are expressed in similar patterns in the wing discs of both types of insects. However, in the developing butterfly wing, the expression of *Dll* has a feature not seen in *Drosophila*. In butterflies, the zone of *Dll* expression extends inward from the wing margin in a series of rays. One of these rays eventually resolves itself into a circle of intense expression. Carroll and his colleagues discovered that this circle ultimately forms one of the pigmented eyespots on the adult butterfly wing. Thus, the *Dll* transcription factor appears to be responsible for producing a striking element in the pattern of a butterfly's wings—one that probably has a significant adaptive value.

Carroll's research shows that homologous genes in butterflies and fruit flies have preserved some of their basic functions, even though these organisms diverged from a common ancestor about 200 million years ago. However, these genetic functions are not absolutely conserved, as the difference in *Dll* expression indicates. Additional comparative studies of the type that Carroll has initiated may eventually reveal how variations in the expression of particular genes—such as *Dll*—alter the form and function of wings in many different types of insects. Thus, these studies may provide an understanding of how insect wings evolved.

Gerald Rubin and his collaborators have attempted to answer this question by collecting mutations that disrupt eye development. Their research has led to the idea that the specification of cell types within each facet depends on a series of cell–cell interactions. This is illustrated in the differentiation of the eight photoreceptor cells, denoted R1, R2, . . . R8 (Figure 25.26). In a fully formed facet, six of the photoreceptors (R1-R6) are arranged in a circle around the other two (R7, R8). One of the central cells, R8, is the first to differentiate in the developing facet. Its appearance is followed by the differentiation of the peripheral cells R2 and R5, then by R3 and R4, and R1 and R6; finally, the second central cell, R7, differentiates into a photoreceptor.

This last event has been studied in great detail. Rubin and his colleagues have shown that the differentiation of the R7 cell depends on reception of a signal

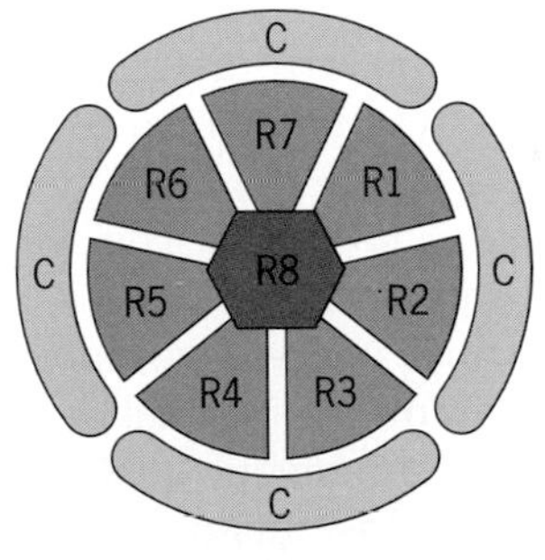

(*a*)

(*b*)

Figure 25.26 Determination of the R7 photoreceptor of an ommatidium in the *Drosophila* compound eye. (*a*) Arrangement of the eight photoreceptors (1–8) and four cone cells (C) in an ommatidium. (*b*) Signaling between the differentiated R8 cell and the presumptive R7 cell. The *bride of sevenless* protein (BOSS) on the R8 cell is the ligand for the *sevenless* receptor protein (SEV) on the surface of the R7 cell. Activation of this receptor initiates a signaling cascade within the R7 cell that induces it to differentiate.

from the already differentiated R8 cell. To receive this signal, the R7 cell must synthesize a specific receptor, a membrane-bound protein encoded by a gene called *sevenless* (*sev*). Mutations in this gene abolish the function of the receptor and prevent the R7 cell from differentiating as a neuron; instead, it differentiates as a cone cell. The signal for the R7 receptor is produced by a gene called *bride of sevenless* (*boss*) and is specifically expressed on the surface of the R8 cell. Contact between the differentiated R8 cell and the undifferentiated R7 cell allows the R8 signal, or **ligand** as it is technically called, to interact with the R7 receptor and activate it. This activation induces a cascade of changes within the R7 cell that ultimately prompt it to differentiate as a light-receiving neuron. This differentiation is presumably mediated by one or more transcription factors acting on genes within the R7 nucleus. Thus, the signal from the R8 cell is "transduced" into the R7 nucleus, where it alters the pattern of gene expression. The analysis of eye development in *Drosophila* therefore shows that **induction,** the process of determining the fate of an undifferentiated cell by a signal from a differentiated cell, can play an important role in the specification of cell types.

Organ Formation

When many different types of cells are organized for a specific purpose, they form an organ. The heart, stomach, kidney, liver, and eye are all examples of organs. One of the remarkable features of an organ is that it forms in a specific part of the body. The development of a heart in the head or an eye in the thorax would be extremely abnormal, and we would wonder what had gone wrong. Anatomically correct organ formation is obviously under tight genetic control.

Geneticists have obtained insights into the nature of this control from the study of another gene in *Drosophila*. This gene is called *eyeless* after the phenotype of flies that are mutant for it (Figure 25.27). The wild-type *eyeless* gene encodes a homeodomain transcription factor whose action switches on a developmental pathway that involves several thousand genes. Initially, several subordinate regulatory genes are activated. Their products then trigger a cascade of events that create specific cell types within the developing eye.

The role of the *eyeless* gene has been demonstrated by expressing it in tissues that normally do not form eyes (Figure 25.28). Walter Gehring and colleagues did this by creating transgenic flies in which the *eyeless* gene was fused to a promoter that could be activated in specific tissues. Activation of this promoter caused transcription of the *eyeless* gene outside its normal domain of expression. This, in turn, caused eyes to form in unorthodox places such as wings, legs, and antennae. These extra (or ectopic) eyes were anatomically well developed and functional; in fact, their photoreceptors responded to light.

An even more remarkable finding is that a mammalian homolog of the *eyeless* gene also produces these extra eyes when it is inserted into *Drosophila* chromosomes. Gehring and coworkers used the mouse homolog of *eyeless* to transform *Drosophila* and got the same result as they did with the *eyeless* gene itself. This showed that the mouse gene, which also encodes a homeodomain protein, is functionally equivalent to the *Drosophila* gene; that is, it regulates the pathway for eye development. However, when the mouse gene

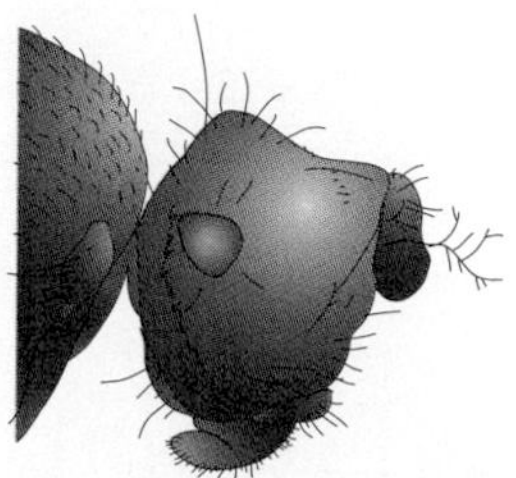

Figure 25.27 The phenotype of an *eyeless* mutant in *Drosophila*.

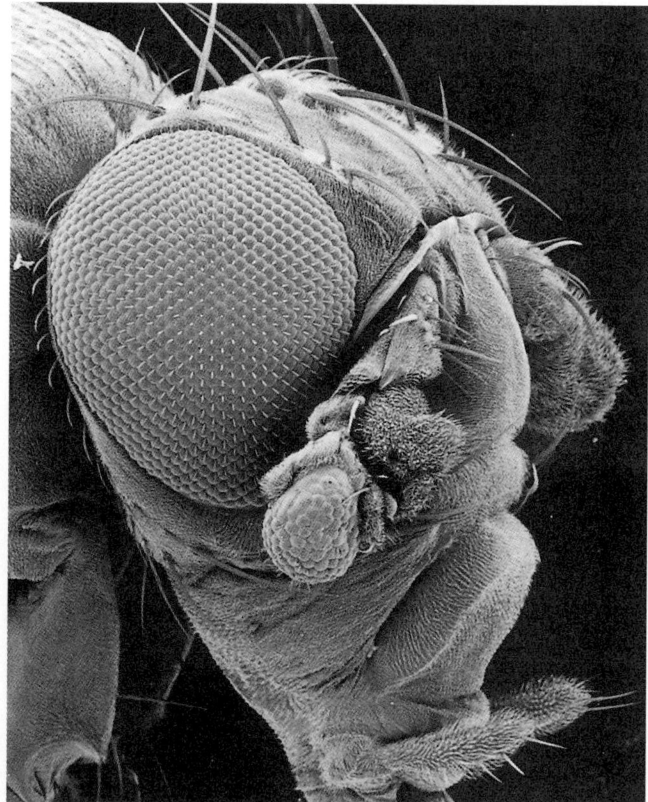

Figure 25.28 An extra eye produced by expressing the wild-type *Drosophila eyeless* gene in the antenna of a fly.

is put into *Drosophila,* it produces *Drosophila* eyes, not mouse eyes. *Drosophila* eyes develop because the genes that respond to the regulatory command of the inserted mouse gene are normal *Drosophila* genes, which must, of course, specify the formation of a *Drosophila* eye. In mice, mutations in the homolog of the *eyeless* gene reduce the size of the eyes. For that reason, the gene is called *Small eye*. A homolog of *eyeless* and *Small eye* has also been found in human beings. Mutations in this gene cause a syndrome of eye defects called *aniridia* (also the name of the gene) in which the iris is reduced or missing.

The discovery of homologous genes that control eye development in different organisms has profound evolutionary implications. It suggests that the function of these genes is very ancient, dating back to the common ancestor of flies and mammals. Perhaps the eyes in this ancestral organism were nothing more than a cluster of light-sensitive cells organized through the regulatory effects of a primitive *eyeless* gene. Over evolutionary time, this gene continued to regulate the increasingly more complicated process of eye development, so that today, eyes as different as those in insects and mammals are still formed under its control.

Key Points: **The zygotic genes are activated after fertilization in response to maternal gene products. In *Drosophila,* a cascade of zygotic gene activity brings about the differentiation of the dorsal-ventral and anterior-posterior dimensions of the embyro. Ultimately, this activity subdivides the embryo into a series of segments. The identity of each segment is determined by the homeotic genes of the Bithorax and Antennapedia Complexes, acting in response to a regulatory hierarchy of segmentation genes. Once segmental identities have been established, specific cell types differentiate. The formation of an entire organ may depend on the product of a master regulatory gene, such as *eyeless* in *Drosophila*.**

GENETIC ANALYSIS OF DEVELOPMENT IN VERTEBRATES

Much of the knowledge about the genetic control of development comes from the study of *Drosophila* and *Caenorhabditis,* two model invertebrates. Geneticists would like to apply and extend this knowledge to other groups of animals, in particular, to vertebrates. The ultimate goal would be to learn about the genetic control of development in our own species. One strategy for achieving this goal is to use invertebrate genes as probes to isolate and clone their vertebrate homologs. Another is to study model vertebrate species with techniques similar to those that are being used in *Drosophila* and *Caenorhabditis*.

Vertebrate Homologs of Invertebrate Genes

Once a gene has been cloned, it can be used as a probe to clone homologous genes from other organisms. If the gene's sequences have been reasonably well conserved over evolutionary time, this procedure works even for distantly related species. Thus, it has been possible to clone genes from various vertebrate species by using probes made from *Drosophila* and *Caenorhabditis* genes. The cloning of a vertebrate gene then makes many kinds of experimental analyses possible, including assays for the gene's expression at both the RNA and protein levels.

One of the most dramatic applications of this approach has shown that vertebrates contain homologs of the homeotic genes of *Drosophila*. These so-called *Hox* genes were initially identified by probing Southern blots of mouse and human genomic DNA with segments of the *Drosophila* homeotic genes. Subsequently, the cross-hybridizing DNA fragments were cloned, mapped with restriction enzymes, and sequenced. The results of all these analyses have established that mice, humans, and all other vertebrates so far examined have 38 *Hox* genes in their genomes (Figure 25.29). These genes are organized in four clusters (a, b, c, and d), each about 120 kb long; in mice and humans each cluster is located on a different chromosome. It seems that the four *Hox* gene clusters were created by the quadruplication of a primordial cluster very early in the evolution of the vertebrates, probably 500 to 600 million years ago.

The molecular analysis of the vertebrate *Hox* genes has revealed striking structural and functional similarities to the homeotic genes of the Bithorax and Antennapedia Complexes in *Drosophila*. These latter genes actually seem to have been members of a larger cluster, called HOM-C, which was split by a chromosome rearrangement during the evolution of the flies. In more primitive insects such as the flour beetle, *Tribolium castaneum,* the Bithorax and Antennapedia complexes are united in a single cluster. Comparison between vertebrates and invertebrates shows that the basic organization of the HOM/*Hox* genes has been preserved during evolution. The structural homologs of the ANT-C genes are at one end of each vertebrate *Hox* gene cluster and the structural homologs of the *BX-C* genes are at the other end. Moreover, within each cluster, the physical order of the *Hox* genes corresponds to the position of their expression along the anterior-posterior axis of the embryo, just as it does for

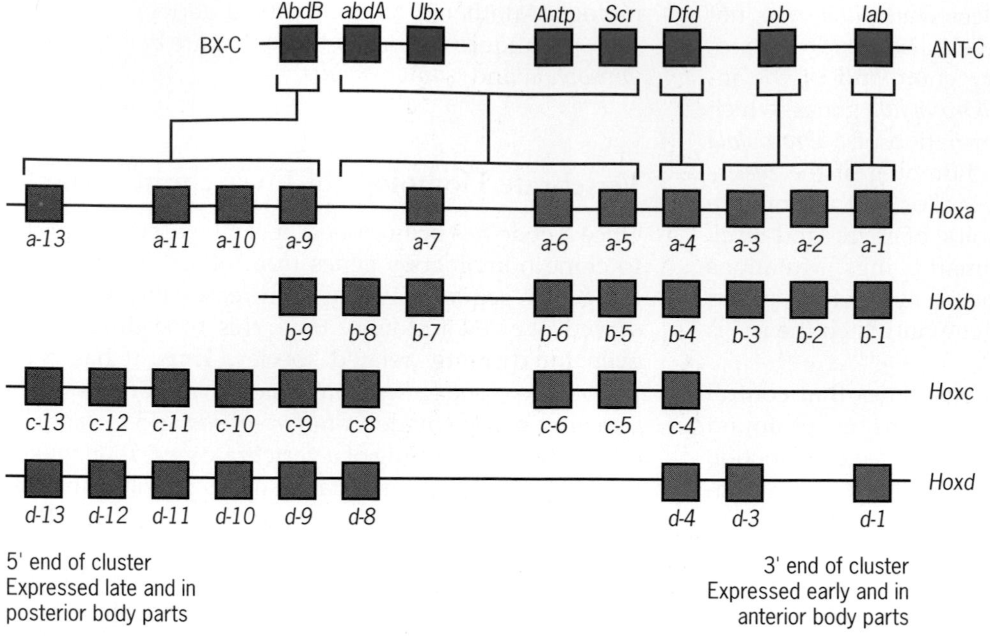

Figure 25.29 Organization and expression of the mammalian *Hox* genes homologous to the *Drosophila* genes in the BX-C and ANT-C. Homologies are indicated by brackets. All the genes except *Dfd* are transcribed from left to right. The time and anatomical location of expression are indicated.

the homeotic genes in *Drosophila* (Figure 25.30). With one exception (the *Deformed* gene in *Drosophila*), all the HOM and *Hox* genes are transcribed in the same direction, with expression proceeding from the 3′ end of a cluster to the 5′ end, both spatially (anterior to posterior in the embryo) and temporally (early to late in development). This phenomenon, called **colinearity,** suggests that the HOM and *Hox* genes provide a common molecular mechanism for establishing the identities of specific regions in many different types of embryos. However, even in *Drosophila,* researchers have only begun to investigate how this mechanism operates.

The Mouse: Insertional Mutagenesis, Transgenics, and Knockout Mutations

The genetic control of development cannot be studied in vertebrates with the same thoroughness as it can in invertebrates such as *Drosophila*. Obviously, there are technical and logistical constraints. Vertebrates have comparatively long life cycles, their husbandry is expensive, and typically few mutant strains, especially those with a developmental significance, are available. In spite of these shortcomings, geneticists have been able to make some headway in the genetic analysis of development in one complex vertebrate—the mouse—and efforts are currently underway to analyze development in a simpler vertebrate—the zebrafish. Over 500 loci responsible for genetic diseases have been identified in the mouse, and some are involved in developmental processes. Most of these loci were discovered through ongoing projects to collect spontaneous mutations. Such work requires that very large numbers of mice—hundreds of thousands, if not millions—be reared and examined for phenotypic differences, and that whatever differences are found be tested for genetic transmission. This is painstaking, costly work that can be supported only at a few facilities in the entire world—one in England and two in the United States. Once a mutation is detected, it can be mapped and tested for allelism and epistatic interactions with other mutations. However, cloning the mutant gene may prove to be as difficult and costly as finding the original mutant mouse. Fortunately, the discovery of new techniques has expedited this process.

These techniques are based on the introduction of particular DNA molecules into the mouse genome. This DNA may cause mutations by inserting into genes in the mouse chromosomes. Such insertional mutations are much easier to clone than spontaneous mutations because they have been tagged by the inserted DNA (Figure 25.31). A probe made from this DNA can be used to isolate the mutated gene from a recombinant DNA library constructed from the mutant animal. Insertional mutagenesis in the mouse is therefore

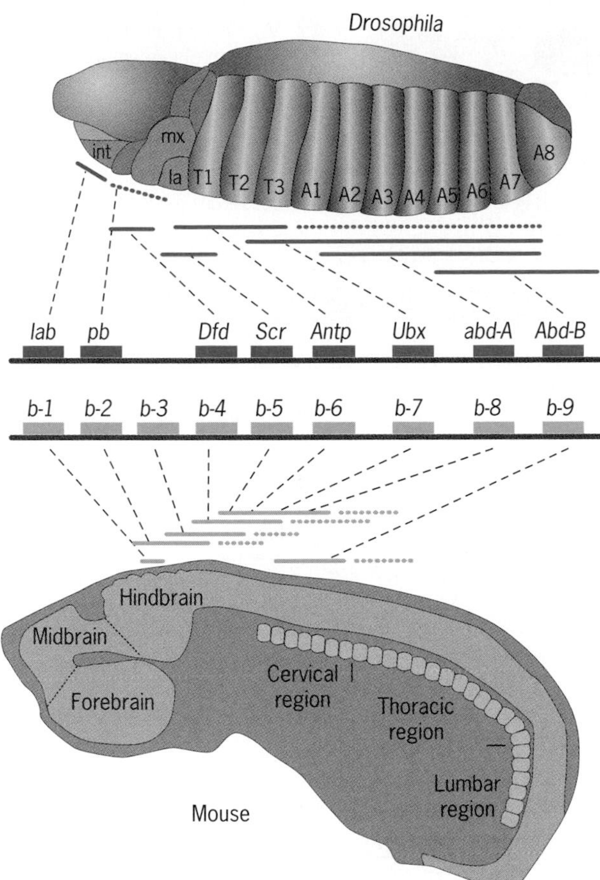

Figure 25.30 Expression patterns of *Drosophila* and mouse homeotic genes. Top: A 10-hour-old *Drosophila* embryo showing the approximate expression of the homeotic genes in the epidermis. In the head, int, mx, and la designate intercalary, maxillary, and labial segments, respectively. Thoracic and abdominal segments are indicated as T1-T3, and A1-A8. Bottom: A 12-day-old mouse embryo showing the approximate expression of the *Hoxb* genes in the central nervous system. The dotted lines indicate extension of the expression toward the posterior.

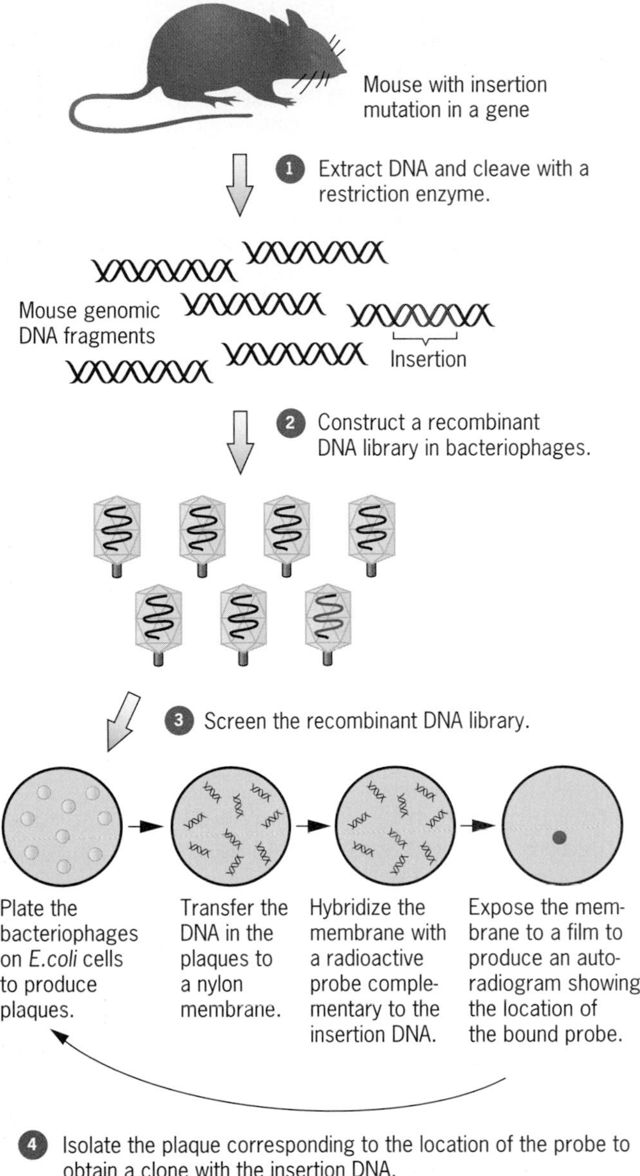

Figure 25.31 Cloning a gene containing an insertion mutation from a mouse.

analogous to transposon tagging in *Drosophila* and *Caenorhabditis*. However, the DNA used for insertional mutagenesis in the mouse is not a *bona fide* transposon but some other sequence capable of integrating more or less randomly into the chromosomes. For example, replication-defective retroviruses have been used for this purpose.

As we discussed in an earlier section, a "foreign" DNA sequence that has been inserted into the genome is called a transgene. There are two general methods of introducing transgenes into mouse chromosomes. One relies on the injection of DNA into fertilized eggs or embryos (Figure 25.32), and the other relies on the injection or transfection of DNA into large populations of cultured cells that were derived from very young mouse embryos (Figure 25.33). These **embryonic stem cells** (or **ES cells**) come from the inner cell mass, a group of cells found in the blastula stage of mouse embryos. Such cells can be cultured *in vitro,* transfected or injected with DNA, and then introduced into other developing mouse embryos. By chance, some of the introduced ES cells may contribute to the formation of adult tissues, so that when the mouse is born, it may consist of a mixture of two types of cells, its own and those derived from the cultured (and potentially mutagenized) ES cells. Such mice are called **chimeras.** If the ES cells happened to contribute to the chimera's germ line, a mutation that was induced in them by the

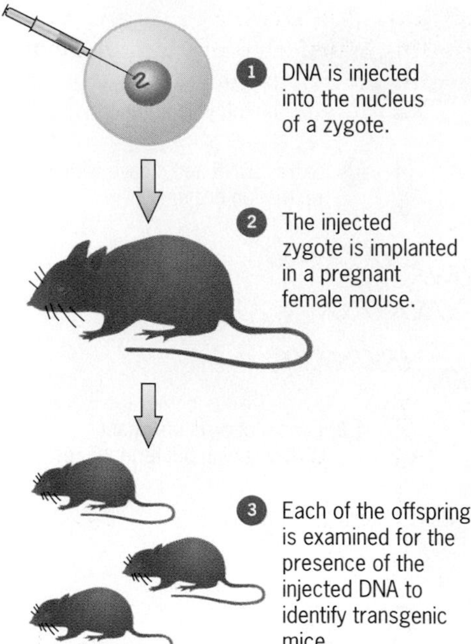

Figure 25.32 Producing transgenic mice by injecting eggs and implanting them into females to complete their development.

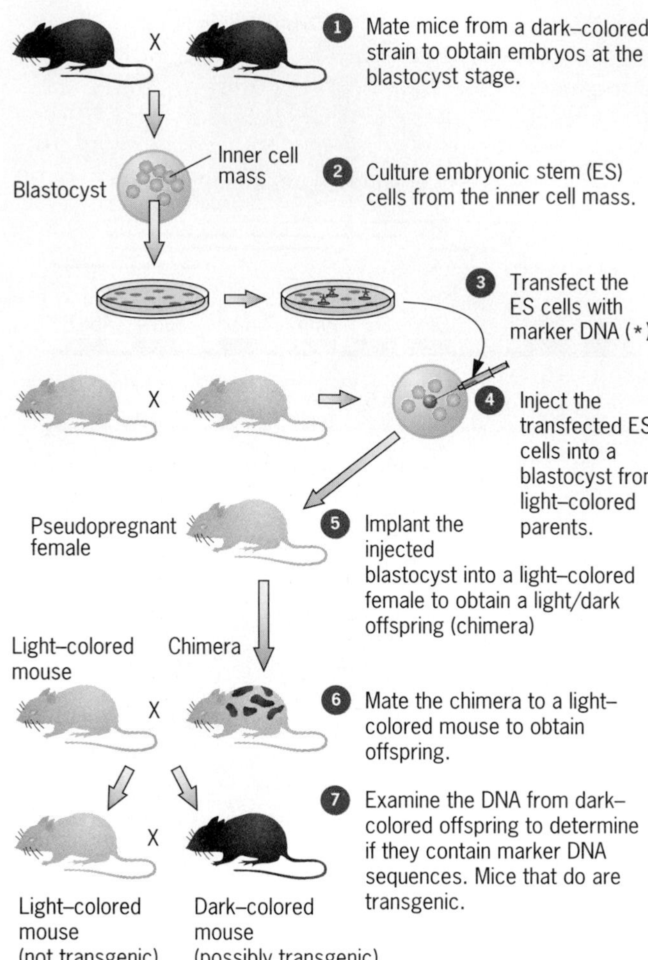

Figure 25.33 Producing transgenic mice by embryonic stem (ES) cell technology.

insertion of foreign DNA has a chance of being transmitted to the next generation. Breeding a chimeric mouse may therefore establish a mutant transgenic strain. Transgenic strains can also be produced by breeding mice derived from embryos that were directly injected with DNA. To date, by using either of these methods, thousands of transgenic strains have been created, and among them, many developmentally significant mutations have been detected.

A special application of transgenic mutagenesis is possible with genes that have already been cloned. For such genes, it is possible to create mutations that knock out the gene's function. A **"knockout" mutation** can help a reseacher determine what role the normal gene plays during development. To create a knock-out mutation, the sequence of the cloned gene must be altered *in vitro* and then introduced into cultured ES cells. At a low frequency, the mutated transgene will replace its normal allele on the chromosome by homologous recombination, a process called **targeting.** ES cells that contain a targeted knockout mutation can be used to create chimeras, which can then be bred to produce transgenic strains that carry the knockout mutation. It is then a simple matter of intercrossing two heterozygotes to make the knockout mutation homozygous and determine what effect it has on development. As an example, let's consider the effect of a knockout mutation in the *Hoxc-8* gene. Mice that are homozygous for this mutation develop an extra pair of ribs posterior to the normal sets of ribs; they also have clenched toes on their forepaws. The extra-rib phenotype in these mutant mice is reminiscent of the segmental transformations that are seen with homeotic mutations in *Drosophila*.

The genetic analysis of development in mice is providing clues about development in our own species. For example, mutations in at least two different mouse genes mimic the development of abnormal left-right asymmetries in human beings. Normally, humans, mice, and other vertebrates exhibit structures that are asymmetric along the left-right body axis. The heart tube always loops to the right, and the liver, stomach, and other viscera are shifted either to the left or right away from the body's midline. In mutant individuals, these characteristic asymmetries are not seen, perhaps because of a defect in the mechanisms that establish the basic body plan. Studying these types of mutants in the mouse may therefore help to elucidate how the organs are positioned in human beings.

Key Points: **Many vertebrate genes have been identified by homology to genes isolated from**

model organisms such as *Drosophila* and *Caenorhabditis*. For example, the mammalian *Hox* genes are structurally and functionally similar to the homeotic genes of *Drosophila*. Among vertebrates, the mouse provides opportunities to study mutations that affect development. Transgenic mice can be created by injecting DNA into eggs or embryos, or by inserting transfected embryonic stem cells into developing embryos. These mice can be a source of insertional mutations, including knockout mutations, in genes that have a developmental significance.

TESTING YOUR KNOWLEDGE

1. The protein product of the *dorsal* (*dl*) gene in *Drosophila* has been called a ventral morphogen—that is, a substance that brings about the formation of ventral structures in the embryo by virtue of its high concentration in the nuclei on the ventral side of the blastoderm. However, the *dorsal* protein can enter these ventral nuclei only if a receptor on the embryo's ventral surface has been activated. This receptor is encoded by the *Toll* (*Tl*) gene. The extracellular ligand for the *Toll* receptor is encoded by the *spätzle* (*spz*) gene. This ligand can exist in two states, "native" and "modified," and the modified state is needed for the activation of the *Toll* receptor. The products of three genes, *snake* (*snk*), *easter* (*ea*) and *gastrulation defective* (*gd*), are required to convert the native ligand into the modified ligand. All three of these gene products are serine proteases, proteins capable of cleaving other proteins at certain serines in the polypeptide chain. Using these facts, diagram the developmental pathway that ultimately causes the *dorsal* protein to induce the formation of ventral structures in the *Drosophila* embryo.

ANSWER

Here is one representation.

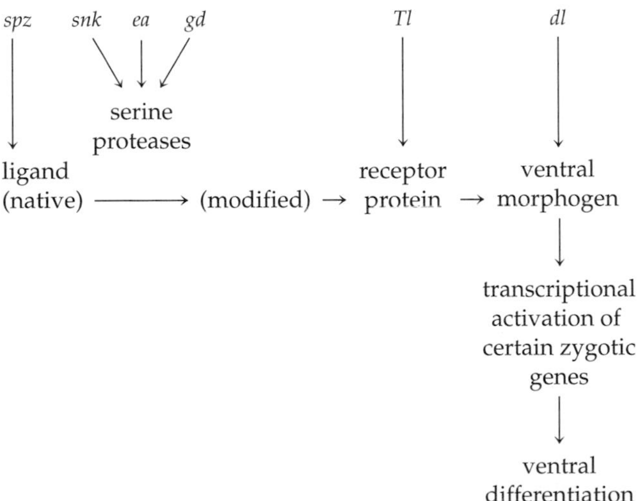

The protein product of the *spz* gene is modified by the serine proteases made by the *snk*, *ea*, and *gd* genes. In its modified form, this ligand is able to activate the *Toll* receptor protein, but the activation is restricted to the ventral side of the embryo. When the *Toll* receptor has been activated (presumably, by binding the modified *spätzle* ligand), it transduces a signal into the cytoplasm of the embryo. This signal ultimately causes the *dorsal* protein to move into the nuclei on the ventral side of the embryo, where it acts as a transcription factor to regulate the expression of the zygotic genes involved in the differentiation of ventral fates.

2. Considering the pathway described above, what would be the phenotypes of recessive loss-of-function mutations in the *spz* and *Tl* genes?

ANSWER

For reference, we should note that loss-of-function mutations in *dl* are maternal effect lethals; that is, embryos from *dl/dl* mothers die during development. When these dying embryos are examined, they are found to lack ventral structures. Geneticists say that they are "dorsalized." This peculiar phenotype is due to the failure of the *dorsal* transcription factor to induce appropriate development in the ventral nuclei of the embryo. In the absence of this induction, the ventral cells differentiate as if they were on the dorsal side of the embryo. Mutations in *spz* and *Tl* might be expected to have the same phenotypic effect because they would block steps in the pathway that ultimately causes the *dorsal* protein to induce ventral differentiation. Recessive mutations in *spz* and *Tl* are therefore maternal effect lethals. Females homozygous for these mutations produce dorsalized embryos that die during development.

QUESTIONS AND PROBLEMS

25.1 What events form the blastula during animal development? What events form the gastrula?

25.2 During oogenesis, what mechanisms enrich the cytoplasm of animal eggs with nutritive and determinative materials?

25.3 Distinguish between "determination" and "differentiation."

25.4 Are the "primary germ layers" components of the germ line? Explain.

25.5 Outline the main steps in the genetic analysis of development in a model organism such as *Drosophila* or *Caenorhabditis*.

25.6 Why is the early *Drosophila* embryo a syncytium?

25.7 In *Drosophila,* what larval tissues produce the external organs of the adult?

25.8 In *Caenorhabditis,* homozygosity for a *dumpy* (*dpy*) mutation causes the animal to be shorter than wild-type. If a hermaphrodite heterozygous for such a mutation were self-fertilized, what fraction of its progeny would be dumpy?

25.9 A researcher wants to perform a complementation test between two independently isolated *dumpy* mutations in *Caenorhabditis*. Both mutations are autosomal recessives. In addition to being homozygous for the *dumpy* mutation, one strain is homozygous for an X-linked recessive mutation (*unc*) that causes the worms to be uncoordinated. The other strain is homozygous for the *dumpy* mutation alone. How should the complementation test be performed?

25.10 Loss-of-function mutations in the *Drosophila Sxl* gene cause females to die but have no effect in males. What phenotype would a gain-of-function mutation most likely have?

25.11 A *Drosophila* researcher has discovered an autosomal recessive mutation that causes XX animals to develop into sterile males. Propose a scheme to find out if the mutation is an allele of either the *tra* or *tra2* genes.

25.12 Diagram the sex-determination pathway in *Drosophila.*

25.13 What phenotype would you expect in an XX *Drosophila* that is homozygous for loss-of-function mutations in both the *tra* and *dsx* genes? Explain.

25.14 Triploid *Drosophila* with two X chromosomes develop as intersexes. Suggest a mechanism to explain this phenotype.

25.15 Predict the phenotype of XO and XX *Caenorhabiditis* that are homozygous for loss-of-function mutations in both the *tra-1* and *fem-1* genes.

25.16 Loss-of-function mutations in the *Caenorhabditis xol-1* gene cause males to die, whereas loss-of-function mutations in the *Drosophila Sxl* gene cause females to die. What is the reason for these opposite effects?

25.17 A *Drosophila* researcher has observed a peculiar fly. The forelegs have sex combs, which are characteristic of males, and the abdomen has an ovipositor, which is characteristic of females. Furthermore, the bristles on the forelegs and part of the thorax are yellow whereas those on the rest of the body are black. Suggest an explanation for the unusual phenotype of this fly.

25.18 *Drosophila* larvae that were heterozygous for a null mutation of the X-linked *white* gene (*w*) and its wild-type allele (w^+) were irradiated with X rays and then reared to adulthood. When the adults emerged from the pupal cases, a few had white patches in their otherwise red eyes. What caused these patches to develop? Do the patches indicate whether the *white* gene functions autonomously or nonautonomously? Explain.

25.19 A *Drosophila* developmental geneticist has identified a recessive lethal mutation (*l*) on an X chromosome that is marked with a recessive allele of the *singed* gene (*sn*). Through recombination analysis, this lethal has been mapped between *sn* and *yellow* (*y*), which is at the tip of the X chromosome far away from the centromere (denoted by a dot). The geneticist has produced *y* + + . / + *l sn* . heterozygous larvae and has irradiated them with X rays to induce recombination between *sn* and the centromere. Some of the adults that have developed from these larvae have solitary patches of yellow tissue on an otherwise black body; however, none of the tissue (either yellow or black) is singed. What does this pattern of pigmentation reveal about the function of the lethal mutation?

25.20 A researcher is trying to clone the *dpy*-3 gene from *Caenorhabditis*. Recessive mutations in this gene cause worms to be shorter than normal, a phenotype referred to as "dumpy." The researcher has mapped the *dpy*-3 gene relative to other genes on one of the *Caenorhabditis* chromosomes and, using the map data, has obtained cosmid clones with wild-type DNA from the vicinity of the *dpy*-3 gene. To determine which clone contains the *dyp*-3 gene, the researcher has injected each of them into the gonads of hermaphrodites homozygous for a recessive mutation in *dpy*-3. She has then scored the transgenic progeny derived from these injected hermaphrodites for the dumpy phenotype. From the results shown below, where on the molecular map should the researcher place the *dpy*-3 gene?

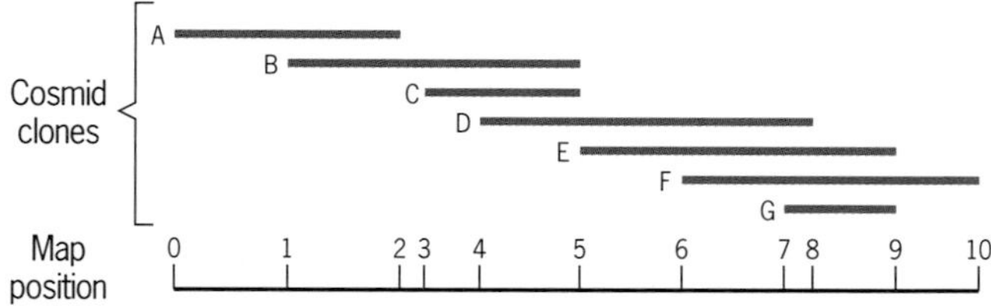

Cosmid Clone	Phenotype of Transgenic Progeny
A	dumpy
B	dumpy
C	dumpy
D	wild type
E	wild type
F	dumpy
G	dumpy

25.21 Like *dorsal, bicoid* is a strict maternal-effect gene in *Drosophila*; that is, it has no zygotic expression. Recessive mutations in *bicoid* (*bcd*) cause embryonic death by preventing the formation of anterior structures. Predict the phenotypes of (a) *bcd/bcd* animals produced by mating heterozygous males and females; (b) *bcd/bcd* animals produced by mating *bcd/bcd* females with *bcd*/+ males; (c) *bcd*/+ animals produced by mating *bcd/bcd* females with *bcd*/+ males; (d) *bcd/bcd* animals produced by mating *bcd*/+ females with *bcd/bcd* males; (e) *bcd*/+ animals produced by mating *bcd*/+ females with *bcd/bcd* males.

25.22 In *Drosophila,* recessive mutations in the dorsal-ventral axis gene *dorsal* (*dl*) cause a dorsalized phenotype in embryos produced by *dl/dl* mothers; that is, no ventral structures develop. Predict the phenotype of embryos produced by females homozygous for a recessive mutation in the anterior-posterior axis gene *nanos.*

25.23 A researcher is planning to collect mutations in maternal-effect genes that control the earliest events in *Drosophila* development. What phenotype should the researcher look for in this search for maternal-effect mutations?

25.24 A researcher is planning to collect mutations in the gap genes, which control the first steps in the segmentation of *Drosophila* embryos. What phenotype should the researcher look for in this search for gap gene mutations?

25.25 Why do the children of women, but not men, who are homozygous for the mutant allele that causes phenylketonuria produce children that are physically and mentally retarded?

25.26 Diagram a pathway that shows the contributions of the *sevenless* (*sev*) and *bride of sevenless* (*boss*) genes to the differentiation of the R7 photoreceptor in the ommatidia of *Drosophila* eyes. Where would *eyeless* (*ey*) fit in this pathway?

25.27 When the mouse gene *Small eye,* which is homologous to the *Drosophila* gene *eyeless,* is expressed in *Drosophila,* it produces extra compound eyes with ommatidia, just like normal *Drosophila* eyes. If the *Drosophila eyeless* gene were introduced into mice and expressed there, what effect would you expect? Explain.

25.28 How might you show that two mouse *Hox* genes are expressed in different tissues and at different times during development?

BIBLIOGRAPHY

ANDERSON, K. V. 1995. One signal, two body axes. *Science* 269:489–490.

ANDERSON, K. V. 1998. Pinning down positional information: Dorsal-ventral polarity in the *Drosophila* embryo. *Cell* 95:439–442.

BEDDINGTON, R. 1992. Transgenic mutagenesis in the mouse. *Trends in Genet.* 8:345–347.

CLINE, T. W., AND B. J. MEYER. 1996. Vive la différence: Males vs females in flies and worms. *Ann. Rev. Genet.* 30:637–702.

GREENWALD, I., AND G. M. RUBIN. 1992. Making a difference: the role of cell-cell interactions in establishing separate identities for equivalent cells. *Cell* 68:271–281.

HALDER, G., P. CALLAERTS, AND W. J. GEHRING. 1995. Induction of ectopic eyes by targeted expression of the *eyeless* gene in *Drosophila. Science* 267:1788–1792.

KENYON, C. 1988. The nematode *Caenorhabditis elegans. Science* 240:1448–1453.

LEWIS, E. B. 1994. Homeosis: The first 100 years. *Trends in Genet.* 10:341–343.

MCGINNIS, W., AND R. KRUMLAUF. 1992. Homeobox genes and axial patterning. *Cell* 68:283–302.

MORISATO, D., AND K. V. ANDERSON. 1995. Signaling pathways that establish the dorsal-ventral pattern of the *Drosophila* embryo. *Ann. Rev. Genet.* 29:371–399.

PARKHURST, S. M., AND P. M. MENEELY. 1994. Sex determination and dosage compensation: Lessons from flies and worms. *Science* 264:924–932.

RIVERA-POMAR, R., AND H. JACKLE. 1996. From gradients to stripes in *Drosophila* embryogenesis: filling in the gaps. *Trends in Genet.* 12:478–483.

RUBIN, G. M. 1988. *Drosophila melanogaster* as an experimental organism. *Science* 240:1453–1459.

26

A pair of monozygotic twin sisters.

The Genetic Control of Behavior

CHAPTER OUTLINE

Genetic Analysis of Behavior in Experimental Animals
Chromosome Abnormalities and Insights into Human Behavior
Single-Gene Mutations and Human Behavior
Complex Human Behaviors

Twins and Personality

Oskar Stohr and Jack Yufe are monozygotic, or identical, twins who were separated shortly after their birth. Oskar was taken by his mother to Germany where he was raised by his grandmother as a Christian and a Nazi. Jack was raised by his father in the Caribbean as a Jew, spending part of his youth on an Israeli kibbutz. At the age of 47, these two brothers were reunited. The similarities in their personalities and behavior were remarkable. Both men had mustaches, wore two-pocket shirts with epaulets, and used wire-rimmed glasses. Each loved spicy foods and sweet liqueurs and tended to fall asleep in front of the television. Each thought it amusing to sneeze in a room full of strangers. Both flushed the toilet before using it, stored rubber bands on their wrists, read magazines from back to front, and dipped buttered toast in their morning coffee. They were raised very differently, yet it was clear to all observers that they shared many mannerisms.

Oskar and Jack are only one of many pairs of identical twins being studied by Thomas Bouchard and his colleagues at the University of Minnesota. Bouchard's research team has been identifying twins who were separated shortly after birth and then raised in different environments in order to determine the extent to which genetic variation influences complex human behavioral traits such as intelligence and personality. Their work has led them to the conclusion that these traits are, indeed, influenced by genetic variation. However, twins are not the only subjects used to investigate the genetic control of behavior. Sometimes chromosome abnormalities and gene mutations provide important insights, and, of course, much can be learned from experimental analysis with animals.

The term *behavior* refers to a coordinated neuromuscular response to changes or signals in the external or internal environment. This response requires the integration of sensory, neural, and hormonal factors. Many gene products play a role, directly or indirectly, in the structure and function of the nerves and muscles. If one of these gene products is altered, a behavior may be changed. In this chapter we explore some of the approaches to study the genetic control of behavior. First, we discuss efforts to investigate the genetic basis of behavior in experimental animals; then we consider the genetic analysis of behavior in human beings.

GENETIC ANALYSIS OF BEHAVIOR IN EXPERIMENTAL ANIMALS

Animals exhibit a wide range of behaviors associated with feeding, courtship, reproduction, and a host of other activities. The genetic determinants of these behaviors are only now beginning to be identified through experimental work. In the following sections, we discuss the genetic analysis of particular behaviors in honeybees, *Drosophila*, and mice.

Nest-Cleaning Behavior in Honeybees

The social insects—ants, bees, wasps, and termites—exhibit a fascinating array of behaviors. These species live in colonies, sometimes with millions of members, and their lifestyles are cooperative. The activities within a colony are divided among different types of individuals—workers and soldiers. In some cases, the behaviors are truly remarkable. In certain tropical species of ants, for example, the workers cut leaves in the forest and carry them into their colonies where they chew the leaves and use them as a medium to cultivate fungi, which they eat. Unfortunately, little is known about the genetic basis of this and most other social insect behaviors.

One behavior that has been analyzed is a type of nest cleaning found in some strains of honeybees. These strains routinely clean their hives to maintain a "hygienic" environment. The cleaning behavior includes the removal of larvae that have been killed by a bacterial disease called American foulbrood (Figure 26.1*a*). Removal of the dead larvae requires two genetically controlled behaviors. First, worker bees must uncap the cells that contain the dead larvae, and second, they must remove these larvae. Bee strains that do not exhibit these behaviors are said to be "nonhygienic."

In 1964, Walter Rothenbuhler reported that nest-cleaning behavior in honeybees is controlled by two independently assorting genes. When Rothenbuhler crossed nonhygienic, haploid drones to hygienic, diploid queens, all the diploid worker offspring were nonhygienic. Thus, nonhygienic behavior is dominant. When haploid drones derived from the unfertilized eggs of F_1 hybrid queens (these drones are genetically equivalent to the gametes of the F_1 hybrid queens) were backcrossed to hygienic, diploid queens, Rothenbuhler obtained four classes of F_2 colonies:

1. Colonies with hygienic behavior.
2. Colonies in which the workers uncapped cells with dead larvae but did not remove the larvae.
3. Colonies in which the workers did not uncap cells with dead larvae, but if an experimenter uncapped the cells for them, the workers removed the larvae.
4. Colonies in which the workers neither uncapped cells with dead larvae nor removed those larvae.

Because the four types of colonies were equally frequent, Rothenbuhler's results suggested that two genes, each with two alleles, control nest-cleaning behavior. One gene, with alleles u and u^+, governs the uncapping behavior, and another gene, with alleles r and r^+, governs the behavior of removing dead larvae (Figure 26.1*b*). Workers homozygous for the u allele uncap cells with dead larvae, and workers homozygous for the r allele remove dead larvae. The alternate alleles, u^+ and r^+, lead to nonhygienic behaviors, even in the heterozygous condition.

The gene products responsible for nest-cleaning behavior in honeybees are unknown. Perhaps a chemical signal given off by cells with dead larvae attracts the hygienic workers but not the unhygienic ones. Once a cap is removed, the chemical signal may become strong enough to trigger the expression of the gene that determines the larval removal behavior. For workers that uncap cells but do not remove dead larvae, the initial signal may be sufficient to trigger the uncapping behavior, but, after the cell is uncapped, it may become so strong that it inhibits the larval removal behavior. Whatever the mechanism, hygienic behavior in honeybees is clearly under genetic control.

Biological Rhythms in *Drosophila* and Mice

Anyone who has traveled by airplane across several time zones has recognized the importance of biological rhythms. Human beings and many other organisms exhibit a daily cycle of sleeping and waking that is difficult to change. Biological rhythms are pervasive; they allow organisms to adapt to daily and seasonal cycles that are generated by the earth's rotation and by its revolution around the sun. Through careful study, researchers have discovered that rhythmic behavior patterns are under genetic control. The most detailed analyses have been performed with *Drosophila*, which exhibit both 24-hour (circadian) and short-term (ultradian) rhythmic behaviors.

In *Drosophila*, locomotor activity follows a 24-hour cycle. This activity can be measured by placing a fly in a darkened glass tube and shining a narrow beam of red light across the tube (Figure 26.2*a*); the fly cannot see red light. Every time the fly moves through the light beam, it disrupts that beam and causes an electronic photoreceptor cell to record the event. The more active a fly is, the more it will interrupt the light beam. A typical record of wild-type *Drosophila* activity is shown in Figure 26.2*b*. Clearly, the pattern has a 24-hour periodicity.

The time when adult *Drosophila* emerge from their pupal cases is also synchronized to a 24-hour cycle. Wild-type flies tend to eclose from the pupal case in the morning (Figure 26.2*c*). Some flies eclose at other times of the day, and a few also eclose at night. However, in most *Drosophila* stocks, eclosion occurs primarily in the morning. This rhythm persists even if cultures that have been reared in standard day/night cycles are transferred to complete darkness. Under these so-called free-running conditions, the fly's circa-

(*a*)

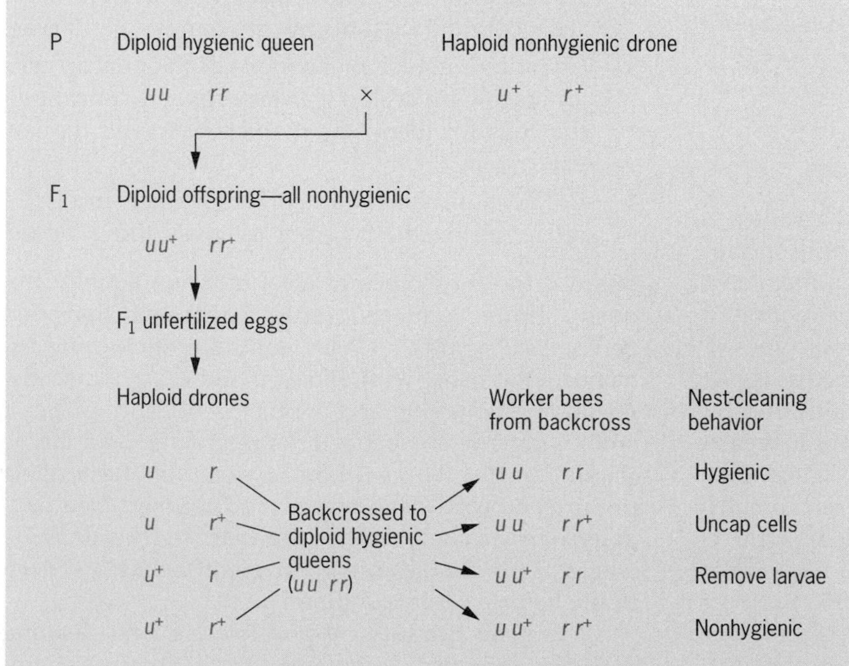

(*b*)

Figure 26.1 (*a*) A honeybee comb. The empty cells have been uncapped, and the dead larvae have been removed by hygienic bees. (*b*) Hygienic behavior in honeybees follows a classic dihybrid inheritance pattern. Two independently assorting genes, each with two alleles (u and u^+; r and r^+), control uncapping and larval removal.

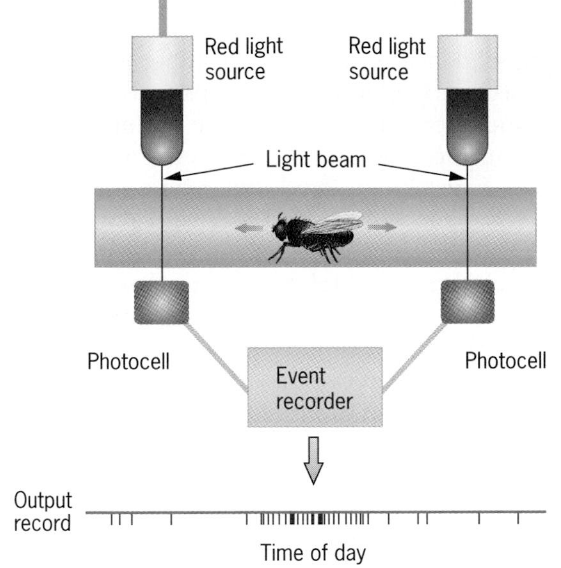

(a)

24 hours

12 18 24 6 12 18 24 6

Time of day (24-hour clock)

(b)

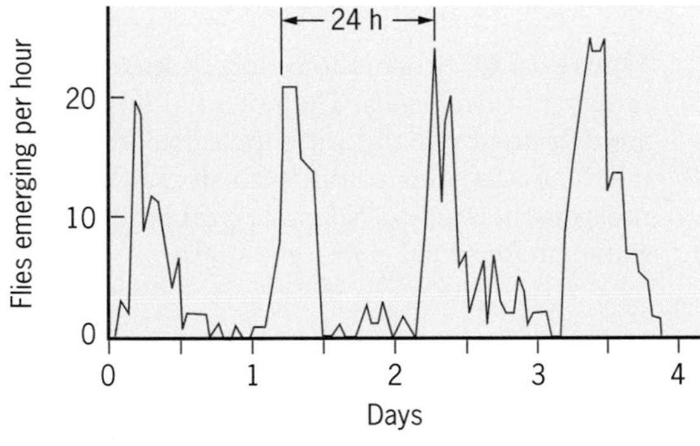

(c)

Figure 26.2 (*a*) Procedure to monitor locomotor activity in *Drosophila*. The fly is placed in a darkened glass tube that is illuminated with a red light at each end. When the fly moves through the light beam, the event is detected by a photocell and registered on an output record. Each vertical line on the output record represents one activity event. (*b*) Locomotor activity pattern of a wild-type fly. The activity is maximal where the vertical lines are the densest. The pattern has a 24-hour periodicity. (*c*) Eclosion rhythm of wild-type flies under free-running conditions (total darkness after pupation). On each free-running day, eclosion is maximal during the hours corresponding to morning.

dian clock triggers eclosion at what would be morning in a 24-hour cycle.

In addition to these two circadian rhythms, *Drosophila* exhibit short-term, or ultradian, rhythms that are associated with the courtship ritual between males and females. Each *Drosophila* species has its own distinctive courtship behavior. In general, however, the male approaches the female from the front. He extends his wing at a 90° angle and vibrates it, generating a song that makes the female receptive to his advances. Then the male moves to the rear of the female and touches her genitalia with his proboscis. After this encounter, the male attempts to copulate with the female. If the female accepts him, mating is completed; if she does not, the male will repeat the entire courtship ritual again.

The song generated by the male's rapid wing vibrations is a species characteristic. A typical *Drosophila* song pattern, as recorded by a special device, consists of a "hum" followed by a series of pulses (Figure 26.3*a*). The intervals between the pulses vary from 15 to 100 milliseconds. The characteristic interval pattern usually starts short, becomes longer, and then becomes short again (Figure 26.3*b*). Thus, over a span of about 45 to 60 seconds, the frequency of pulses in the courtship song shows a rhythmic pattern.

In 1971 Ronald Konopka and Seymour Benzer discovered recessive X-linked mutations that alter these rhythmic behaviors. Konopka and Benzer mutagenized wild-type males with the chemical ethylmethane sulfonate (EMS). The mutagenized males were mated

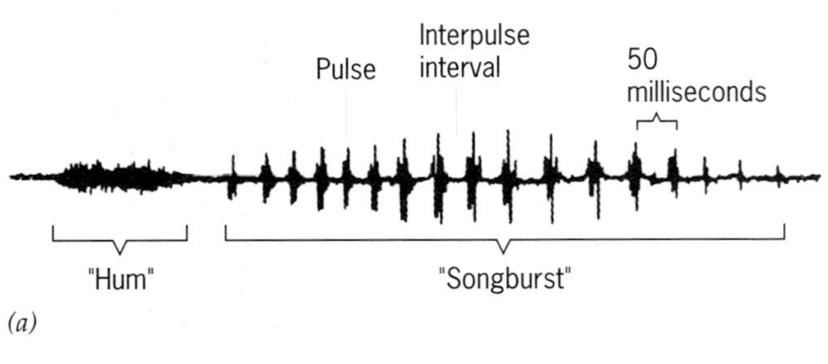

(a)

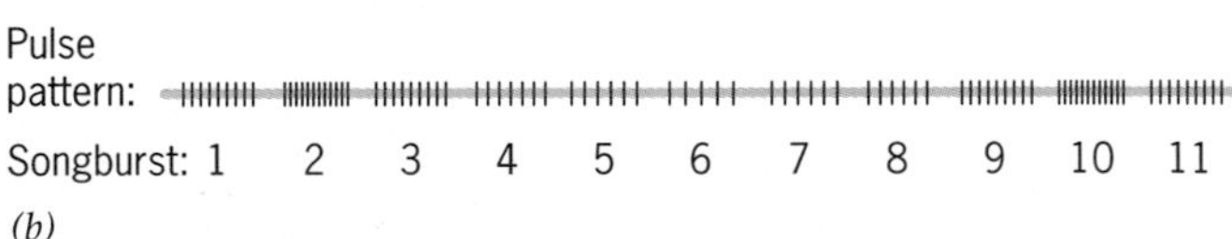

(b)

Figure 26.3 (*a*) Typical songburst pattern of *Drosophila* during courtship. The burst in wing vibrations is preceded by a "hum," followed by a series of discrete pulses. The time between two pulses is an interpulse interval. A cluster of pulses constitutes the songburst. (*b*) A pattern of 11 separate songbursts over a 60-second period. Variation in the length of the interval between pulses within songbursts shows a rhythmic pattern. Beginning with burst 2, the interpulse intervals within the burst are short, then long, then short again.

to females with attached-X chromosomes. In this type of mating, the mutagenized X chromosome is transmitted from the father to his sons. Consequently, Konopka and Benzer were able to establish a large number of *Drosophila* stocks in which mutagenized X chromosomes were carried in males but not in females (Figure 26.4). The males in each of these stocks were then tested for abnormalities in the circadian eclosion rhythm under free-running conditions.

Among 2000 stocks that were screened, Konopka and Benzer found three in which the eclosion rhythm was abnormal. In one stock, the male flies emerged at all times in the 24-hour cycle—that is, the flies were apparently arrhythmic; in another stock, the males emerged on a 19-hour cycle; and in a third stock, they emerged on a 28-hour cycle. Konopka and Benzer mapped the mutations responsible for these abnormal eclosion rhythms to a small interval at the left end of the X chromosome. In addition, through complementation analysis, they demonstrated that all three mutations were alleles of the same gene. Subsequently, this gene was given the name *period* (*per*).

Mutations in the *per* gene affect other behaviors in *Drosophila* as well as the eclosion rhythm. The allele that causes arrhythmic eclosion, for example, also causes arrhythmic locomotor activity and arrhythmic courtship songs; *per* is therefore a central component of *Drosophila*'s biological clock. In 1984, the *per* gene was cloned. This achievement made it possible to elucidate the molecular nature of the mutations that Konopka and Benzer had discovered. The arrhythmic mutation, per^0, is due to a premature stop codon in the coding sequence of the gene. The other two mutations, per^S (short period) and per^L (long period), are both missense mutations that cause single amino acid substitutions in the PER protein. Analyses of *per* gene expression have shown that in wild-type flies, the levels of *per* RNA and PER protein oscillate on a 24-hour cycle. PER protein accumulates late in the day and persists through the night; however, by dawn, it is nearly gone. Thus, PER protein is a molecular oscillator in *Drosophila's* biological time-keeping system.

Recently, researchers have identified several other proteins involved in this biological clock. They include the products of four autosomal genes: *timeless* (*tim*), *clock* (*clk*), *cycle* (*cyc*), and *double-time* (*dbt*). All four of these genes were initially identified through genetic screens for rhythm-altering mutations. Later they were cloned, and their expression patterns were characterized. TIM, the protein product of the *timeless* gene, cycles along with PER and interacts with PER to form a PER/TIM protein complex. CLK and CYC, the respective protein products of the *clock* and *cycle* genes, are basic helix-loop-helix (bHLH) transcription factors that positively regulate the expression of the *per* and *tim* genes. However, this positive regulation is suppressed by the PER and TIM proteins, which act as negative regulators of their own synthesis. Thus, the PER, TIM, CLK, and CYC proteins collectively control the oscillatory expression of the *per* and *tim* genes. DBT, the protein product of the *double-time* gene, seems to influence the stability of PER protein by altering its phosphorylation state. The molecular mechanisms that govern the operation of this biological clock are not

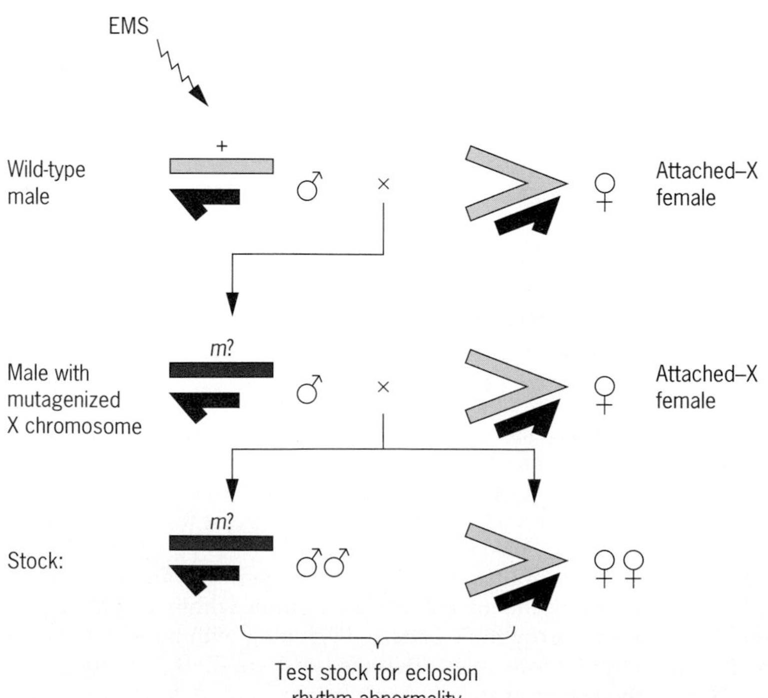

Figure 26.4 Scheme to induce X-linked rhythm mutants in *Drosophila*. The potential mutations (*m*?) are detected by monitoring the eclosion patterns of males and females within each stock. Compared to nonmutant attached-X females, mutant males will show an abnormal eclosion pattern.

1 During the day, the CLK and CYC transcription factors may bind as a heterodimer to an enhancer upstream of the promoter in both the *per* and *tim* genes. Binding of this heterodimer may stimulate transcription of these genes.

2 The *per* and *tim* RNAs are translated into the PER and TIM proteins. Initially these proteins exist as monomers in the cytoplasm.

3 As night falls, the PER and TIM proteins form a heterodimer that migrates from the cytoplasm into the nucleus.

4 During the night, the PER/TIM heterodimers are destabilized. Either PER or TIM (here PER), or both proteins reassociate to form heterodimers with the CLK and/or CYC proteins. Because CLK and CYC cannot dimerize with each other, the *per* and *tim* genes cannot be transcriptionally activated.

5 At dawn, TIM, which is light-sensitive, is degraded. PER protein is also degraded. The CLK and CYC proteins may then once again form heterodimers that transcriptionally activate the *per* and *tim* genes.

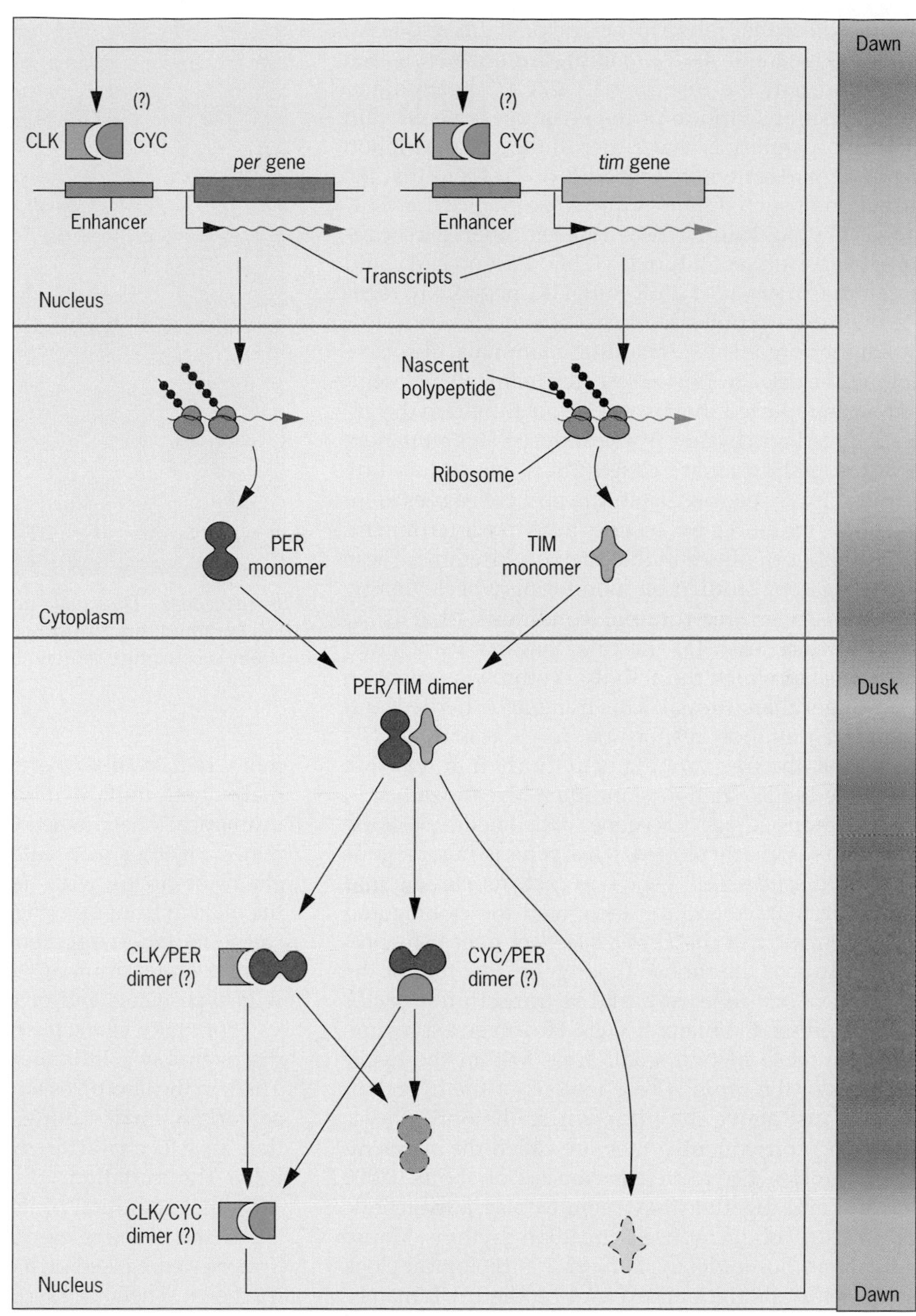

Figure 26.5 Possible molecular mechanisms governing the biological clock in *Drosophila* during one 24-hour cycle. Hypothetical protein dimers are denoted by a question mark.

fully understood. Figure 26.5 presents a plausible model.

Protein-protein interactions appear to be a key feature of the mechanism of PER and TIM oscillation. Three of the five proteins involved in this molecular clock—PER, CLK, and CYC—contain a stretch of amino acids known as the PAS domain, named for its occurrence in *P*ER, in the human proteins *A*RNT and *A*HR, and in another *Drosophila* protein called *S*IM. Studies with other PAS-domain proteins have indicated that the PAS domain facilitates dimerization between polypeptides. Thus, the PAS domain in PER might allow it to form a dimer with TIM (which, by the way, does not contain a PAS domain). Another possibility is that the transcription factors CLK and CYC, which both contain PAS domains, dimerize with each other, and that the resulting CLK/CYC heterodimer stimulates transcription of the *per* and *tim* genes. Still another possibility is that either of these transcription factors dimerizes with either PER or TIM and that the formation of such dimers reduces the chance that CLK and CYC would dimerize with each other. Dimerization between either CLK or CYC and PER or TIM could therefore explain how PER and TIM negatively regulate their own synthesis.

Higher organisms, including mammals, also have biological clocks. In 1994, a research team led by Joseph Takahashi reported the discovery of a mutation that alters a circadian rhythm in the mouse. This mutation, an autosomal dominant called *Clock*, was induced by chemical mutagenesis. Takahashi and colleagues identified the mutation by testing heterozygous mutant mice for abnormalities in their activity rhythms. These rhythms can be studied by monitoring wheel-running behavior under free-running conditions (that is, in complete darkness). Takahashi's reasearch team found one mouse in which the activity rhythm was about an hour longer than normal. This trait was inherited, and when the mutation responsible for it was made homozygous, the free-running activity rhythm was first lengthened to 27–28 hours and then lost altogether. In 1997 the mouse *Clock* (*Clk*) gene was cloned, and its expression was characterized. This gene is homologous to the *Drosophila clock* gene—in fact, its cloning and molecular characterization facilitated the cloning and characterization of the *Drosophila clock* gene. The protein product of the mouse *Clock* gene, like that of the *Drosophila clock* gene, is a bHLH transcription factor that is expressed in many tissues. However, expression is most intense in two small areas within the hypothalamus of the brain. These regions of the hypothalamus lie just above the place where the optic nerves cross over; consequently, they are called the suprachiasmatic nuclei. Each nucleus consists of about 10,000 neurons, and together they seem to play a major role in the control of the mouse's circadian rhythms. Mouse and human homologs of *Drosophila*'s biological clock are currently being isolated and studied. Ultimately, analysis of these proteins will help us to understand the molecular mechanisms that govern rhythmic behavior in our own species.

Sexual Preference in *Drosophila*

In 1963 at the annual meeting of the American Society of Zoologists, Kulbir Gill reported the discovery of a new male-sterile mutation in *Drosophila*. Males homozygous for this mutation did not attempt to mate with females even though they courted them. Even more remarkably, the mutant males courted other males, both mutant and wild-type. This behavior was dramatically seen when Gill placed a group of mutant males together in a culture. The males formed long chains or circles, with each male in the chain courting the male in front of it (Figure 26.6). However, the male-sterile mutation apparently had no effects in females. Homozygous mutant females mated readily with wild-type males and produced offspring.

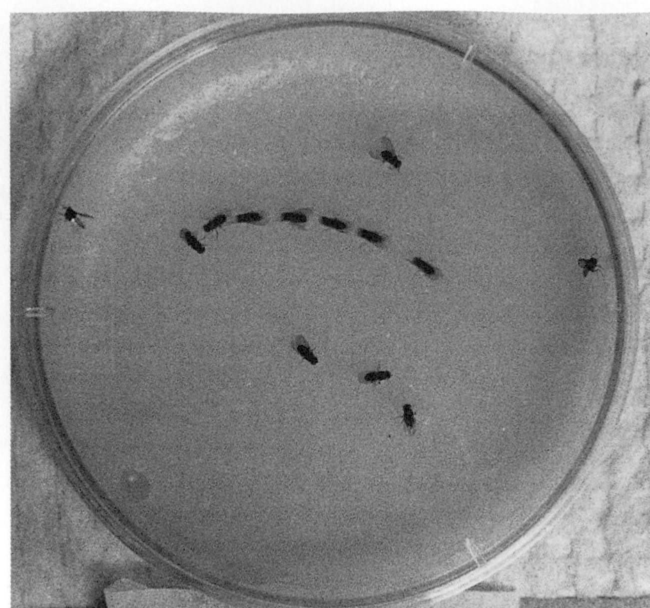

Figure 26.6 *Drosophila* males homozygous for the *fruitless* (*fru*) mutation courting each other. Each male in the chain is courting the male in front of it.

For many years, the male-sterile mutation that Gill discovered was little more than a laboratory curiosity. Then, in the late 1970s and early 1980s, Jeffrey Hall and coworkers carried out experiments to map the mutation and to characterize its phenotypic effects more fully. The mutation was also given a name, *fruitless* (*fru*), in recognition of its male-sterile effect.

In the 1990s, researchers from several different laboratories embarked on a cooperative effort to clone the *fru* gene. Their work, published in 1996 with Lisa Ryner as the principal author, established that *fru* encodes a zinc-finger type transcription factor that regulates the genes for male sexual behavior—indeed, male sexual preference—in *Drosophila*.

Earlier genetic experiments by Barbara Taylor had suggested that *fru* might control a branch in the *Drosophila* sex-determination pathway (Figure 26.7). As discussed in Chapter 25, sex determination in *Drosophila* is controlled by the X-linked *Sex-lethal* (*Sxl*) gene in response to the X:autosome ratio. The product of *Sxl* is a protein, SXL, that in females regulates the splicing of transcripts from *Sxl* and another sex-deter-

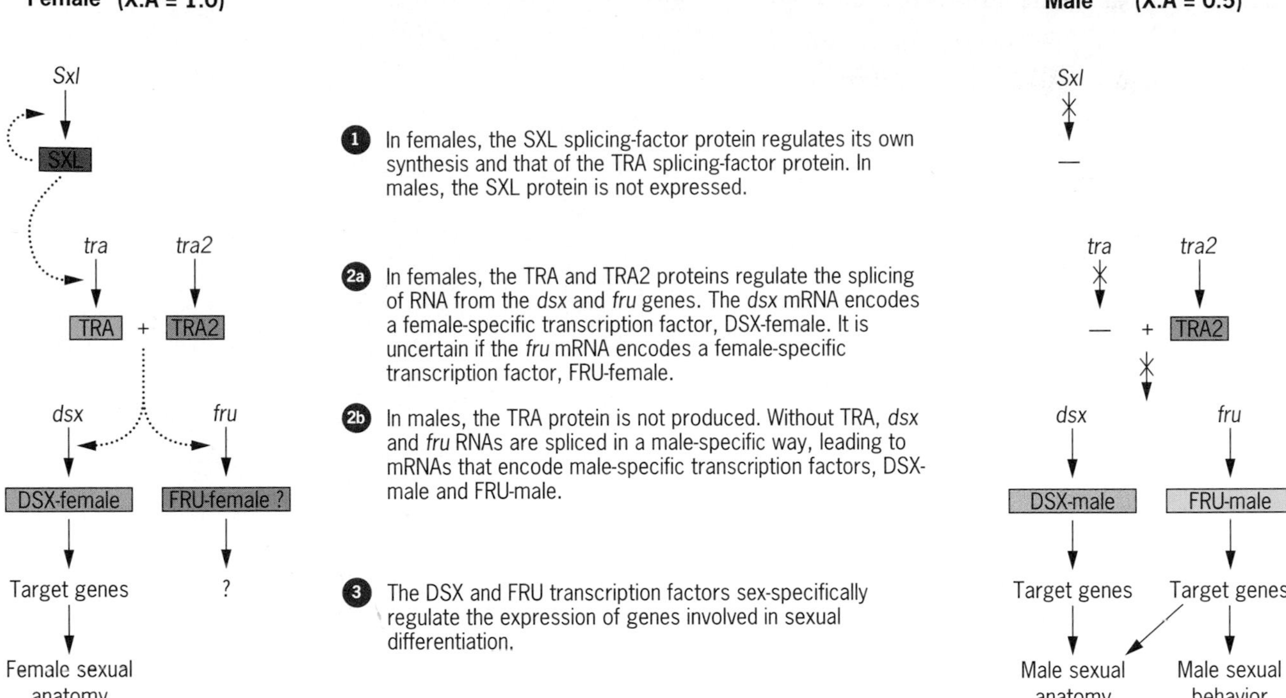

Figure 26.7 Role of the *fru* gene in sex determination in *Drosophila*. Genes are denoted by lower case letters, their protein products by upper case letters. Dotted-lines indicate regulation of RNA splicing. Sex-specific expression of *fru* is regulated jointly by the splicing-factor proteins TRA and TRA2. In males, where TRA is not produced, *fru* RNA is spliced to produce an mRNA that encodes the FRU-male protein, a transcription factor that regulates the genes for male sexual behavior and at least one aspect of male sexual anatomy (an abdominal muscle called the Muscle of Lawrence). In females, the TRA/TRA2 complex splices *fru* RNA to produce an alternate mRNA. The nature of the protein encoded by this mRNA is currently unknown. However, because null alleles of *fru* are lethal in both males and females, the putative FRU-female protein must perform an essential function.

mination gene called *transformer* (*tra*). The protein product of *tra*, denoted TRA, is also a regulator of splicing, and, together with the constitutively produced protein of the *tra2* gene, denoted TRA2, it controls the splicing of transcripts from the *doublesex* (*dxs*) gene. Alternate splicing of *dsx* transcripts in males and females produces different forms of a transcription factor (DSX-female and DSX-male) that sex-specifically regulates the genes involved in sexual differentiation. However, Taylor had observed that the formation of a particular abdominal muscle in males was not controlled by the *dsx* gene. This finding implied that the sex-determination pathway had a branch, regulated not by *dsx* but by another as yet unidentifed gene, and that this unknown gene might, like *dsx*, encode sex-specific transcription factors. Furthermore, it seemed likely that the unknown gene would itself be regulated posttranscriptionally by the splicing-factor proteins TRA and TRA2.

The TRA and TRA2 proteins bind to an RNA sequence 13 nucleotides long. This sequence is repeated multiple times in transcripts from the *dsx* gene. Thus, when Ryner and her colleagues began their efforts to clone *fru*, they searched for other genes that possessed the 13-nucleotide TRA/TRA2 RNA-binding sites. Their search led them to a locus on chromosome 3 where the *fru* gene had been mapped. Cloning and analysis of the DNA at this locus revealed the molecular structure of the *fru* gene. It contains three copies of the 13-nucleotide TRA/TRA2 RNA-binding sequence, and, as expected, its expression is regulated at the splicing level by the TRA and TRA2 proteins.

The pattern of *fru* expression is complex; there are multiple promoters and the transcripts from the gene are alternately spliced in males and females. *In situ* RNA hybridization experiments indicate that *fru* expression is limited to a relatively small number of neurons in the fly's central nervous system, some of which have been implicated in the control of male sexual behavior.

The discovery of *Drosophila*'s *fru* gene raises the possibility that sexual orientation is under genetic control in other species, including our own. Indeed, Dean Hamer and his associates have obtained evidence from pedigree analysis that a gene in the long arm of the human X chromosome contributes to male homosexuality (see the Human Genetics Sidelight: A Genetic

HUMAN GENETICS SIDELIGHT

A Genetic Basis for Male Homosexuality?

Recent studies have suggested that sexual orientation is influenced by biological factors and that some of these factors are genetic. In 1991 Simon LeVay reported that a morphological difference exists between the brains of homosexual and heterosexual men. LeVay examined a small region of the hypothalamus and found that in homosexual men and in heterosexual women this region is about half the size that it is in heterosexual men. Also in 1991, J. Bailey and R. Pillard investigated the incidence of homosexuality among identical and fraternal twins and among adopted brothers. They studied 161 homosexual men and found that 52 percent of the identical twins (who share 100 percent of their genes) and 22 percent of the fraternal twins (who share 50 percent of their genes) were homosexual, but that only 11 percent of the adopted brothers of homosexuals (who are, of course, genetically unrelated) were also homosexual. Thus, Bailey and Pillard's study suggested that a homosexual orientation is at least partly determined by the genotype.

In 1993, researchers at the National Cancer Institute identified a candidate region of the X chromosome that appears to influence male sexual orientation.[1] These researchers, led by Dean Hamer, analyzed the pedigrees of a set of homosexual men whom they had identified. The subjects of the study completed an interview or a questionnaire covering childhood gender identification, childhood and adolescent sexual development, and adult sexual behavior. Hamer and his colleagues found that 13.5 percent of the homosexual men's brothers were also homosexual, compared to 2 percent of men in the general population. In addition, Hamer and his colleagues noticed that there were more homosexual male relatives on the maternal side in the pedigrees they studied than on the paternal side (Figure 1). Thus, their analysis suggested that homosexuality is preferentially transmitted through females—in other words, that it seems to follow an X-linked inheritance pattern.

Hamer and his colleagues looked for regions of the X chromosome that might carry a gene influencing sexual orientation. Their approach was a simple one. A pair of brothers have about half their X chromosome DNA in common. If they are both homosexual, then the putative genetic factor causing a predisposition to homosexuality should be present in the X-linked DNA they share. Hamer's research

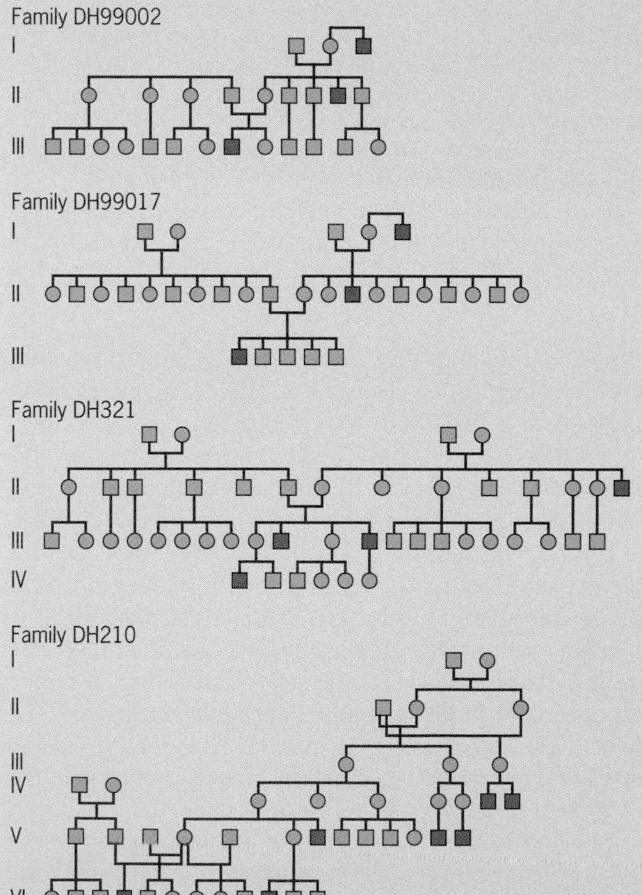

Figure 1 Pedigrees showing apparent X-linked inheritance of male homosexuality. Red symbol: homosexual; blue symbol: nonhomosexual.

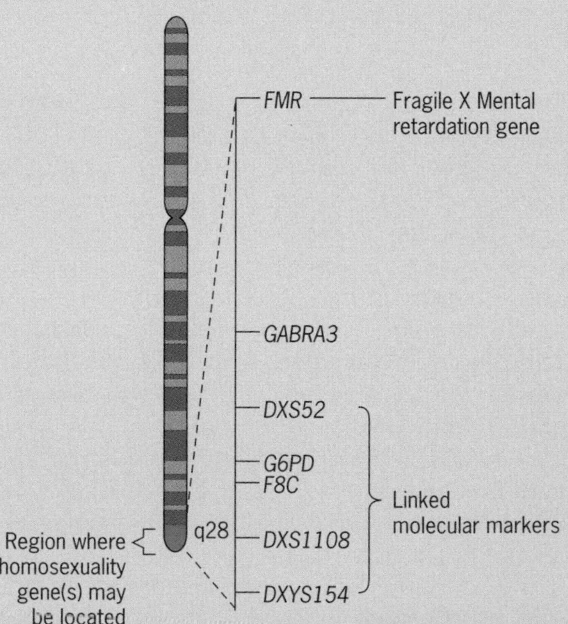

Figure 2 The human X chromosome showing molecular markers that appear to be linked to a gene that causes a predisposition to male homosexuality. The markers and the putative homosexuality gene are located in region Xq28.

team screened 40 pairs of homosexual brothers for molecular markers on the X chromosome. They found that 33 pairs shared five molecular markers, all located near the tip of the long arm of the chromosome in a region denoted as Xq28 (Figure 2). Thus, the molecular analysis suggested that a gene for male homosexuality is located in region q28 on the X chromosome. A followup study published in 1995 reinforced this conclusion and also addressed whether or not female homosexuality is influenced by an X-linked gene.[2] There was no evidence for X-linkage of female homosexuality. Efforts are now underway to identify candidate genes for male homosexualtity in the q28 region of the X chromosome. Other genetic factors that might influence sexual orientation are also being sought.

[1]Hamer, D. H., S. Hu, V. L. Magnuson, N. Hu, and A. M. L. Pattatucci. 1993. A linkage between DNA markers on the X chromosome and male sexual orientation. *Science* 261:321–327.

[2]Hu, S., A. M. L. Pattatucci, C. Patterson, L. Li, D. W. Fulker, S. S. Cherney, L. Kruglyak, and D. H. Hamer. 1995. Linkage between sexual orientation and chromosome Xq28 in males but not in females. *Nature Genet.* 11:248–256.

Basis for Male Homosexuality?). However, the nature of this gene and its protein product are currently unknown.

Nurturing Behavior in Mice

The *per* and *fru* genes were identified in *Drosophila* by inducing mutations that alter behavior. In higher animals such as the mouse, this approach is very expensive. However, new techniques have made it possible to mutate particular mouse genes *in vitro* and then determine if the artificially created mutations affect mouse behavior *in vivo*. Jennifer Brown and coworkers took this approach to determine if the mouse *fosB* gene, which encodes a transcription factor known to be expressed in the brain, exerts any control over behavior.

Brown and colleagues worked with a cloned copy of the *fosB* gene. Through standard recombinant DNA techniques, they inserted a gene for resistance to the antibiotic G418 into the second exon of the *fosB* gene, thereby disrupting *fosB*'s functional integrity. The disrupted *fosB* gene was then transfected into mouse embryonic stem (ES) cells (see Chapter 25), and colonies of cells that incorporated it were selected by virtue of their ability to survive in the presence of G418. DNA extracted from each surviving colony was then examined by Southern blotting to determine if the disrupted gene had replaced one of the normal *fosB* genes present in the diploid ES cells. When colonies with such mutant gene replacements were identified, cells from them were introduced into normal mouse embryos. Some of these embryos developed into adults that had tissues derived from the heterozygous mutant ES cells. These chimeric mice were allowed to reproduce, and in a few cases, they transmitted the mutant *fosB* gene replacement to their offspring. From these offspring, Brown and coworkers succeeded in establishing strains of mice heterozygous for the *fosB* mutant gene replacement. When these mice were intercrossed, they produced homozygous mutant offspring, which, though normal in appearance, consistently failed to nurture their young (Figure 26.8). Normal mice create a nest, clean their pups regularly, crouch over them to warm and nurse them, and bring them back to the nest if they wander off. However, mice homozygous for the disrupted *fosB* gene do not exhibit any of these behaviors.

Certain regions of the mouse brain are important for nurturing behavior. Merely showing mouse pups to a normal adult female will elicit this behavior, even if the female is not the pups' mother. Brown and

(*a*)

(*b*)

Figure 26.8 (*a*) Typical nursing posture of a wild-type female mouse shortly after giving birth. (*b*) Typical behavior of a *fosB* mutant female shortly after giving birth. The mutant female's pups are scattered and neglected and die shortly after birth.

coworkers therefore looked for the expression of FosB protein in the brains of normal female mice that had been shown pups for a brief period of time. They found that FosB expression was specifically induced in the preoptic area of the hypothalamus within six hours of pup presentation. This induction of the FosB transcription factor presumably results from a signaling cascade within the cells of the hypothalamus. Once induced, the FosB protein apparently orchestrates the components of nurturing behavior by activating the transcription of specific target genes.

Key Points: **Nest-cleaning behavior in honeybees is controlled by two independently assorting genes. Genes that regulate biological rhythms have been identified in *Drosophila* and mice. In *Drosophila*, the expression of two such genes, *period* and *timeless*, follows a 24-hour cycle. Transcription of these genes is regulated by bHLH proteins encoded by the genes *clock* and *cycle*. The *fruitless* gene in *Drosophila* encodes a transcription factor that controls male sexual behavior. Another transcription factor encoded by the *fosB* gene regulates nurturing behavior in mice.**

CHROMOSOME ABNORMALITIES AND INSIGHTS INTO HUMAN BEHAVIOR

In Chapter 7 we discussed the phenotypes associated with chromosome abnormalities in humans. One of these abnormalities, trisomy for chromosome 21 (Down syndrome), is a major cause of mental retardation. Individuals with this condition are usually diagnosed at birth or shortly thereafter. When Down syndrome children are tested for intelligence, they are typically found to have intelligence quotients (IQs) between 25 and 50; among karyotypically normal individuals, the mean IQ is 100, and the standard deviation is 15. Thus, the IQs of people with Down syndrome are very much below average—an indication that genes and gene dosage influence human intelligence.

In addition to the obvious effect on intelligence, trisomy 21 also influences behavior. If people with Down syndrome survive beyond age 30—and today, with improved medical care, many of them do—they invariably develop **Alzheimer's disease**, a form of dementia that also occurs in karyotypically normal individuals. About 5 percent of karyotypically normal individuals over age 65 develop Alzheimer's disease, and nearly 25 percent of those over age 80 do so. People with Alzheimer's disease gradually, but inexorably, lose their memories and intellectual functions; they become progressively more forgetful and disoriented, and need to be monitored constantly to prevent their hurting themselves or others.

The brains of Alzheimer's patients show a marked loss of neurons. They also show an accumulation of *senile plaques*, which are thickened nerve cell processes (axons and dendrites) surrounding a deposit of a particular type of polypeptide called *amyloid* β *protein*. These plaques are found in the extracellular spaces outside the cell bodies. Within the cell bodies, there are lesions called *neurofibrillary tangles*—masses of twisted filaments composed of cytoskeletal proteins. As the neurons accumulate these neurofibrillary tangles, they die.

The occurrence of Alzheimer's disease in people with Down syndrome suggests that a gene or genes on chromosome 21 is involved. Genetic mapping has, in fact, demonstrated that the gene for amyloid β protein is located on chromosome 21. This gene encodes an amyloid precursor protein (APP) that is enzymatically cleaved to produce amyloid β proteins. APP is a membrane-bound protein, and approximately two-thirds of it protrudes from the membrane into the extracellular space. The protruding portion of APP is susceptible to cleavage by different proteases, which have yet to be identified. However, the sites at which they cleave APP are known.

Cleavage of APP occurs in either of two pathways (Figure 26.9). In the α pathway, a protease called the α-secretase cleaves APP between amino acids 687 and 688. In the β pathway, APP is internalized in the cell and transported to the lysosomes—small intracellular vesicles that contain proteolytic enzymes (see Chapter 2)—where it is sequentially cleaved by two different proteases. The first of these, called the β-secretase, cleaves APP between amino acids 671 and 672, and the second, called the γ-secretase, cleaves the resulting C-terminal polypeptide fragment, either between amino acids 712 and 713, or between amino acids 714 and 716. Thus, within the β pathway, alternate cleavage by the γ-secretase leads to two different APP polypeptide fragments. One, called the short amyloid β protein (short Aβ) is 40 amino acids long; the other, called the long amyloid protein β (long Aβ), is 42 or 43 amino acids long. The short Aβ protein is more abundant than the long Aβ protein. However, in Alzheimer's patients, with or without trisomy 21, excessive amounts of the long Aβ proteins are produced, and ultimately these molecules are deposited in the extracellular spaces, where they form senile plaques. It is still unclear whether plaque formation is a cause or an effect of the disease, although most researchers tend to believe that the plaques are a cause of Alzheimer's disease.

Mutations in the *APP* gene can predispose individuals to develop Alzheimer's disease. Approximately 20 families that segregate these mutations have been identified, and in each, the disease is inherited as

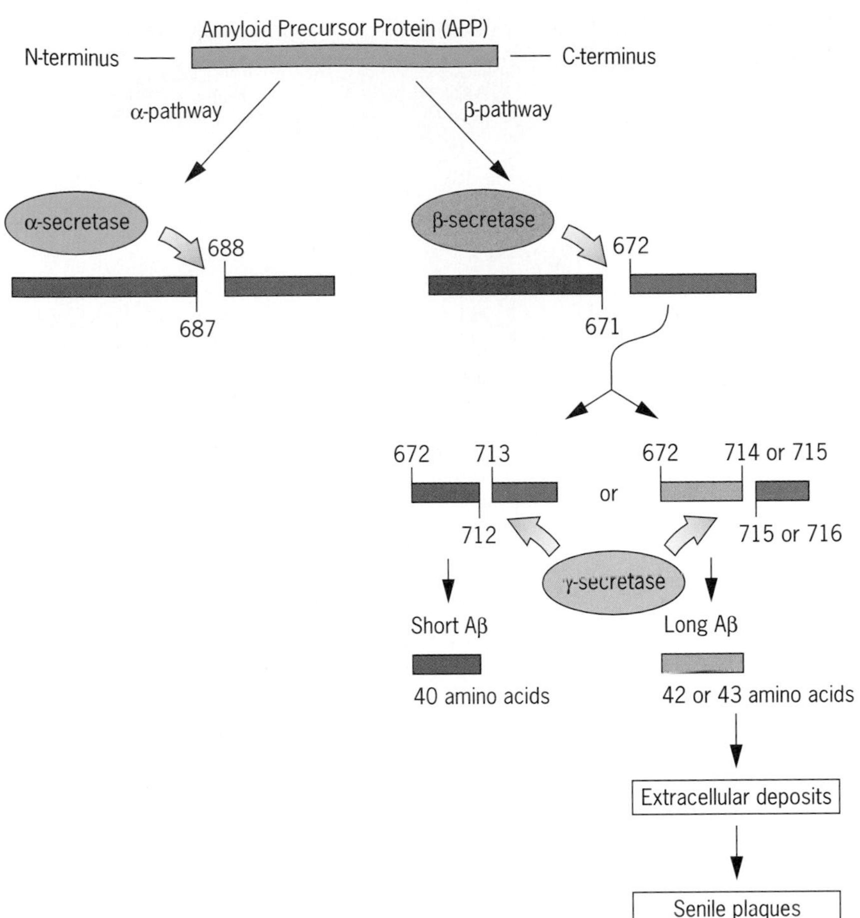

Figure 26.9 Alternate pathways for cleavage of amyloid precursor protein (APP) to form Aβ polypeptides. Only the long Aβ polypeptide has been implicated in the formation of extracellular plaques in the brains of patients with Alzheimer's disease.

an autosomal dominant. In these cases, the symptoms appear before age 65—much earlier than in the general population. Thus, this inherited form of Alzheimer's disease is characterized by early onset.

Other early-onset cases of Alzheimer's have been traced to dominant mutations in genes on two other chromosomes. Both genes encode transmembrane proteins known as the *presenilins* (*PS*s). The *PS1* gene is located on chromosome 14, and the *PS2* gene is located on chromosome 1. The mechanism by which mutations in these genes cause excessive production of the long Aβ is not known. However, because the PS proteins are located in neural membranes, they may mediate signaling processes that influence how APP is transported and proteolytically cleaved within cells. Further research is needed to clarify the role of the presenilins in both the metabolism of APP and the etiology of Alzheimer's disease.

Key Points: **Trisomy for chromosome 21 is associated with mental retardation and the development of Alzheimer's disease, a condition involving extracellular deposits of polypeptides derived from the amyloid precursor protein.**

SINGLE-GENE MUTATIONS AND HUMAN BEHAVIOR

Many single-gene mutations have a dramatic effect on human behavior. Some of these mutations have been identified because they produce recessive disorders, and others because they produce dominant disorders. The list of mutations associated with behavioral phenotypes is expected to grow as the Human Genome Project advances toward its goal of identifying and characterizing all the genes in the human genome.

Phenylketonuria

Phenylketonuria (PKU), a disorder of amino acid metabolism, affects about 1 in 10,000 Caucasians and about 1 in 16,500 Asians. PKU is an autosomal recessive trait that results from a defective liver enzyme, phenylalanine hydroxylase (PAH). This enzyme converts the amino acid phenylalanine into tyrosine. If the enzyme is defective, this conversion does not occur and phenylalanine accumulates in the blood plasma. The excessive phenylalanine is then metabolized by alternative biochemical pathways, generating an array of substances which, like phenylalanine itself, build up in the blood. These phenylalanine-derived metabolites cause abnormal myelination of the nerve cells and impair brain development. Mental retardation usually ensues. In addition, people with PKU usually fail to make adequate amounts of tyrosine. This tyrosine deficiency limits the synthesis of melanin pigment, which causes the hair and skin of affected individuals to be very lightly pigmented.

The range of PKU phenotypes is broad—from mildly elevated levels of plasma phenylalanine and nearly normal mental functions to very high levels of plasma phenylalanine and profound mental retardation. This phenotypic heterogeneity is probably due to variation among the mutant alleles of the *PAH* gene. Null alleles produce the most severe form of the disease, and hypomorphic alleles produce less severe forms.

PKU is now successfully treated in infants and children by restricting the dietary intake of phenylalanine. This therapy keeps the phenylalanine level at an appropriate level, thereby preventing harmful byproducts from accumulating in the blood plasma and, ultimately, in the nervous system.

Lesch-Nyhan Syndrome

Like PKU, the Lesch-Nyhan syndrome (LNS) is an inborn error of metabolism. This X-linked trait is caused by mutations in the gene encoding the enzyme hypoxanthine-guanine phosphoribosyltransferase (HPRT). The symptoms of the disease are disquieting. Affected individuals exhibit a ceaseless and involuntary writhing movement, and they suffer from an uncontrollable urge to bite themselves, starting with their lips and the insides of their mouths, and progressing to their fingers and hands (Figure 26.10). LNS patients are protected from this destructive behavior by being physically restrained and sometimes by having their teeth removed.

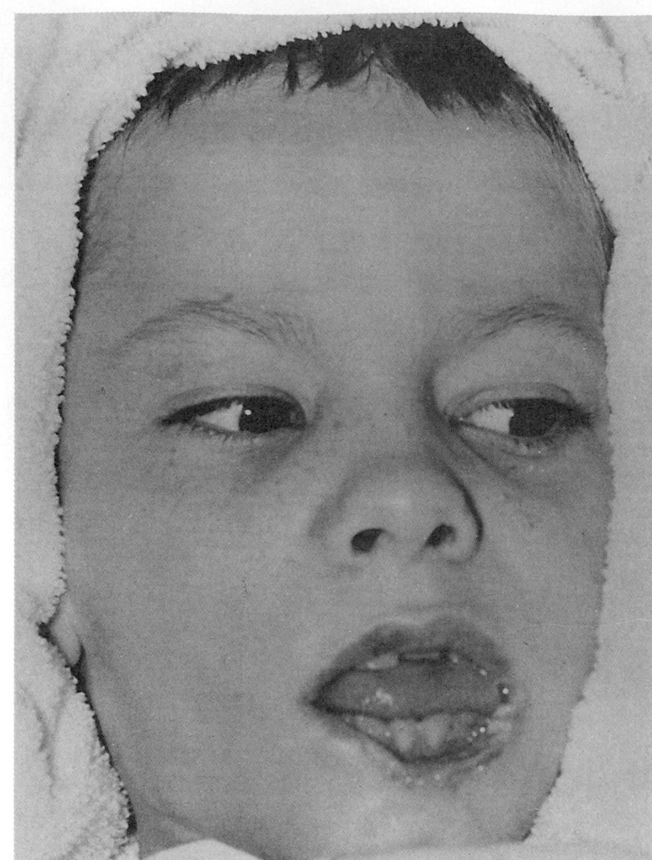

Figure 26.10 The consequences of the self-mutilation behavior in Lesch-Nyhan syndrome.

HPRT is an enzyme of purine metabolism—it catalyzes a step in one of the biochemical pathways that synthesizes purines. Normally, this enzyme is expressed at higher levels in the brain than in other tissues of the body. The known mutations of the *HPRT* gene reduce or eliminate HPRT enzyme activity. Thus, the behavioral disorders associated with LNS may result from abnormally low purine levels in some tissues of the brain.

Huntington's Disease

Huntington's disease (HD) is a fatal neurodegenerative disorder that follows an autosomal dominant pattern of inheritance. People who inherit the HD mutation suffer a gradual loss of neurons, mainly in the caudate nucleus and putamen located in the central region of the brain (Figure 26.11). Both of these areas are important for the coordination of body movement. Other nerve cells in the brain are also affected by the HD mu-

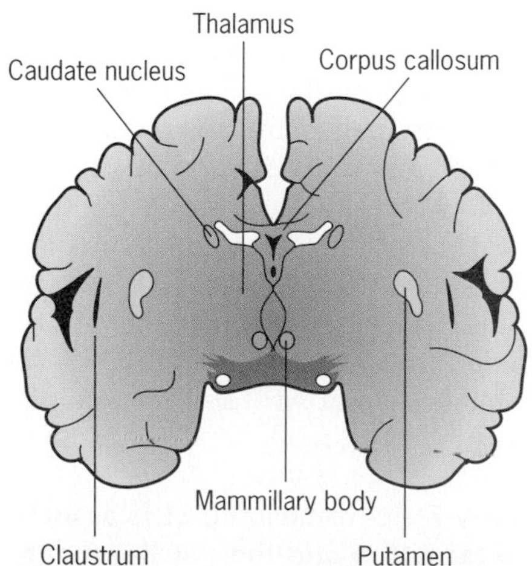

Figure 26.11 Drawing of a frontal section of the brain showing the caudate nucleus and putamen, areas of the brain severely affected by Huntington's disease.

tation, but it is unclear whether their loss is a primary or secondary effect of the disease.

The HD phenotype is manifested in adults, usually after age 30, and it worsens with time. HD patients are unable to control their movements, their speech is disturbed, they sometimes become depressed, and they may even become psychotic. Most patients undergo significant mental decline as the disease progresses; they have difficulty acquiring and recalling information, and they lose their ability to organize information.

The *HD* gene was isolated in 1993, and the protein that it encodes—huntingtin—was identified shortly thereafter (see Chapter 22). Despite intensive work, little is known about how huntingtin functions within cells. It is a large protein, and it is known to interact physically with other proteins, although the significance of these interactions is not clear. Mutant forms of huntingtin contain extra glutamine residues in a polyglutamine segment of the polypeptide chain. The extra glutamines arise from expansion of a repeating CAG nucleotide triplet in the coding sequence of the *HD* gene. Normal individuals have fewer than 37 copies of this triplet, whereas affected individuals typically have more than 38 copies. Some researchers have speculated that the extra glutamines found in the mutant forms of huntingtin alter the way the protein functions, perhaps by influencing its ability to interact with other proteins. Another hypothesis is that the excessively large polyglutamine tract within mutant huntingtin is toxic to certain types of neurons in the brain. More work is needed to elucidate the normal role for huntingtin within cells and to determine what goes wrong in individuals who carry the HD mutation.

Key Points: **Phenylketonuria, Lesch-Nyhan syndrome, and Huntington's disease have led to the identification of particular genes that influence behavior.**

COMPLEX HUMAN BEHAVIORS

Traits such as PKU, LNS, and HD are the equivalent of behavioral mutants in *Drosophila*. Although they allow us to identify individual genes influencing human behavior, they do not offer much insight into the nature of the behavioral differences that we see in the general population. Does genetic variation account for some of these differences, and if it does, what proportion of the overall variability is due to genetic factors? These provocative questions fall within the purview of quantitative genetics—the branch of genetics that uses statistical techniques to study genetic variation in complex traits. In the following sections, we discuss key features of this statistical methodology, and then we apply it to the analysis of two important aspects of human behavior, intelligence and personality.

Quantitative Human Genetics: The Twin Correlation Method

It is not ethical to subject human beings to breeding experiments. However, nature provides the equivalent of a genetics experiment when a fertilized human egg splits and forms two genetically identical zygotes. The individuals who develop from these zygotes are referred to as one-egg, or **monozygotic** (**MZ**), twins. MZ twins are relatively rare in the population—less than 0.3 percent of live births—and when they occur, they usually remain together during their childhood and adolescence. Such twins are termed MZ twins reared together (MZT). In some cases, MZ twins are separated early in life and are subsequently reared in different homes. Such twins are referred to as MZ twins reared apart (MZA). Geneticists have long recognized that MZ twins, especially MZA twins, provide a unique opportunity to study the genetic component of complex behavioral phenotypes.

Twins that originate from two different fertilized eggs are also found in human populations. Two-egg, or **dizygotic** (**DZ**), twins are genetically as related as ordinary siblings; consequently, they share one-half their genes. As with MZ twins, DZ twins may be reared together (DZT) or apart (DZA).

The use of twins to study the genetic control of complex human phenotypes originated with the British scientist Francis Galton in the late nineteenth century. Galton developed a statistical procedure to quantify the phenotypic similarity between twins. The degree of similarity is estimated by the correlation coefficient. This statistic, symbolized by the letter r, takes values between 0 and 1. A value of 0 means that the twins in a sample are phenotypically unrelated to each other, and a value of 1 means that they are phenotypically identical to each other. Thus, the closer the correlation coefficient is to 1, the more similar are the members of each twin pair.

The genetic significance of the correlation coefficient is presented in Figure 26.12, where we consider five kinds of relationships: MZT, MZA, DZT, DZA, and unrelated individuals reared together (URT). The underlying hypothesis is that the complex phenotype being studied results from the combined effects of genetic and environmental factors. Letting T represent a quantitative measure of the trait under study, we have the simple expression

$$T = \mu + g + e,$$

where μ is the mean value of the trait in the population,

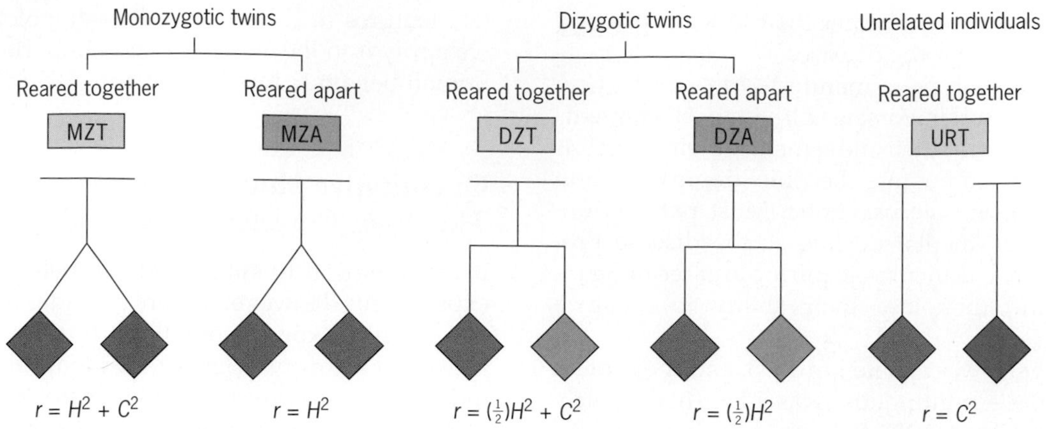

Monozygotic twins | Dizygotic twins | Unrelated individuals

Reared together MZT | Reared apart MZA | Reared together DZT | Reared apart DZA | Reared together URT

$r = H^2 + C^2$ | $r = H^2$ | $r = (\frac{1}{2})H^2 + C^2$ | $r = (\frac{1}{2})H^2$ | $r = C^2$

Figure 26.12 The genetic significance of the correlation coefficient (r) in five different relationships. H^2 is the broad-sense heritability, and C^2 is the environmentability. It is assumed that all the genetic variance in the trait is additive. Any amount of nonadditive genetic variance would reduce the value of H^2 for DZ twins to a value closer to the narrow-sense heritability, h^2 (see Chapter 5).

g represents the deviation from the mean that is due to genetic factors, and e represents the deviation from the mean that is due to environmental factors. To keep the analysis simple, we assume that the genetic factors encompassed by the term g influence the phenotype in a strictly additive fashion, and that the genetic and environmental factors affecting the phenotype are independent of each other. Under this hypothesis, the correlation coefficient r for each relationship in Figure 26.12 represents the proportion of the total variation in the trait (as measured by the statistic called the variance—see Chapter 5) that is due to the genetic and environmental factors shared by the individuals in the relationship.

For MZT, there are two sources of shared factors: those in the genotypes of the MZ twins, which are identical, and those in the common environment of the twins. Thus, for MZ twins reared together, the correlation coefficient r estimates the proportion of the total variance in the trait that is due to genetic factors (H^2) plus the proportion that is due to shared environmental factors (C^2). H^2 is known as the **broad-sense heritability** (see Chapter 5), and C^2 is known as the **environmentality**. We write each of these quantities with the exponent 2 as a reminder that they represent proportions of the total phenotypic variance, which is calculated from the squared differences between each phenotypic measure and the overall population mean.

For MZA, individual twins are genetically identical, but they do not share a common environment. Thus, r reflects only the effects of the identical genotypes; that is, it estimates H^2, the broad-sense heritability. For DZ twins, the genotypes of indivdual twins are only 50 percent identical. Thus, r estimates $(1/2)H^2$ plus, in the case of DZT, the effect of a common environment, C^2. For URT, there is no genetic correlation between individuals; thus, r simply estimates C^2, the proportion of the total variance due to common environmental effects.

Human geneticists have obtained correlation coefficients for each relationship in Figure 26.12, and with these statistics they have estimated the extent to which genetic factors influence aspects of behavior in the general population. It is important to recognize that these estimates depend on several simplifying assumptions and that they tell us only about aggregate genetic variation in the population studied, not about the genetic determinants of individual behavioral differences. We now consider applications of the twin correlation method to estimate the broad-sense heritability for two complex human traits, intelligence and personality.

Intelligence

Intelligence is a complex trait influenced by both genetic and environmental factors. It is a constellation of mental abilities including the skills needed to (1) define and understand words, (2) think of words rapidly, (3) analyze mathematical relationships, (4) analyze spatial relationships, (5) memorize and recall information, (6) perceive similarities and differences among objects, and (7) formulate rules, principles, and concepts for solving problems and understanding situations.

For more than a century, psychologists have tried to characterize and quantify these abilities by administering intelligence tests. The tests—and many different ones have been used—attempt to measure general reasoning ability. The score that an individual makes on one of these tests is converted into an *intelligence quotient*, or *IQ*, which is considered to be an indication of the person's intelligence. The IQ is calculated by dividing the person's mental age (as determined by the test score) by his or her chronological age and then multiplying by 100. Thus, a 10-year-old whose test score is equivalent to the test score of an average 12-year-old has an IQ of $(12/10) \times 100 = 120$. IQ scores are scaled so that the mean of the population is 100 and the standard deviation is 15. Whether IQ scores are valid indicators of intelligence is a topic of much debate, especially when the scores are used to compare individuals from different social, economic, and cultural backgrounds. Whatever their shortcomings, researchers have frequently used them to study the extent to which genetic factors influence intelligence.

For IQ scores, the correlation coefficients of MZ twins, reared together or apart, are very high—in the range of 0.7–0.8 (Table 26.1). By comparison, the correlation coefficients of DZ twins tend to be lower—presumably because, like ordinary siblings, they share only 50 percent of their genes; the correlation coefficients for unrelated individuals reared together are essentially zero. Such analyses strongly suggest that whatever an IQ test measures, it has a large genetic component. This conclusion is supported by other correlation analyses. For example, the IQs of adopted children are more strongly correlated with the IQs of their biological parents than with those of their adoptive parents. Thus, in the determination of IQ, the biological (that is, genetic) link between parents and children seems to be more influential than the environmental one.

What fraction of the variation among IQ scores is attributable to genetic differences among people? The most direct estimate comes from the correlation coefficient for MZ twins reared apart. Observed values of this correlation coefficient are around 0.7; thus, as much as 70 percent of the variation in IQ scores is attributable to genetic variability in the population. This estimate of the broad-sense heritability implies that, with respect to intelligence (as measured by an IQ test), people differ one from another more because of genetic factors than because of environmental factors.

TABLE 26.1

Correlation Coefficients for IQ Test Scores for MZ and DZ Twins, Reared Together or Apart[a]

Study	*MZT*	*MZA*	*DZT*	*DZA*
Newman et al. 1937		0.71		
Juel-Nielsen 1980		0.69		
Shields 1962		0.75		
Bouchard et al. 1990	0.83	0.75		
Pedersen et al. 1992	0.80	0.78	0.22	0.32
Newman et al. 1998				0.47
Average	0.82	0.75	0.22	0.38

[a]Data and references from Bouchard, T. J. 1998. Genetic and environmental influences on adult intelligence and special mental abilities. *Human Biol.* 70:257–279. By permission of the Wayne State University Press.

Personality

The similarities between the MZ twins Oskar Stohr and Jack Yufe who were introduced at the beginning of this chapter raise questions about the genetic determination of personality. Oskar and Jack share many mannerisms even though they were reared in very different environments. Are such characteristics and other, more important aspects of personality influenced by genetic factors?

Personality traits, like intelligence, can be assessed by testing. Psychologists make use of many different tests, some to measure personality characteristics, and others to measure vocational and social interests. The results of these tests tend to be less reliable than those of IQ tests because the scores are less reproducible. Nevertheless, they quantify aspects of human personality in ways that allow them to be analyzed for genetic influences.

Perhaps the most thorough genetic analysis of personality in the general population has come from the Minnesota Study of Twins Reared Apart, a long-term research project directed by Thomas Bouchard. In this project, Bouchard and coworkers have identified more than 50 pairs of MZA twins from all over the world. Each twin is subjected to more than 50 hours of medical and psychological evaluation, including testing for intelligence and personality traits. The results of Bouchard's research suggest that genetic differences explain a significant fraction of the overall variation in human personality, perhaps as much as 50 percent (Table 26.2). The correlation coefficient for the personality and psychological interest test scores of MZA twins ranges from 0.34 to 0.50. Thus, the broad-sense heritability for these traits is reasonably high.

TABLE 26.2

Mean Correlation Coefficients for MZ Twins Reared Together or Apart Who Were Evaluated for Personality Traits, Psychological Interests, and Social Attitudes as Part of the Minnesota Study of Twins Reared Apart.[a]

Test Instrument	*MZT*	*MZA*
Personality traits		
Multidimensional Personality Questionnaire	0.49	0.50
California Psychological Inventory	0.49	0.48
Psychological interests		
Strong Campbell Interest Inventory	0.48	0.39
Jackson Vocational Interest Survey	NA	0.43
Minnesota Occupational Interest Scales	0.49	0.40
Social attitudes		
Religiosity Scales	0.51	0.49
Nonreligious Social Attitude Items	0.28	0.34
MPQ Traditionalism Scale	0.50	0.53

[a]Abstracted with permission from Bouchard et al. 1990. *Science* 250:223–228. Copyright 1990 American Association for the Advancement of Science.

Additional insight into the genetic control of personality has come from study of conditions such as manic depression, schizophrenia, and alcoholism. The occurrence of these traits in the members of MZ and DZ twin pairs has been estimated, and the general finding is that MZ twins are more similar than DZ twins. Thus, for example, among MZ twin pairs with at least one member identified as alcoholic, the co-twin is alcoholic 55 percent of the time. By contrast, a DZ co-twin is alcoholic only 28 percent of the time. The greater concordance for alcoholism between MZ twins suggests that this trait is influenced by genetic factors.

Key Points: **Study of monozygotic and dizygotic twins, reared together or apart, has been useful in assessing the extent to which genes influence behavior in the general population. The broad-sense heritability for intelligence, as measured by IQ tests, is estimated to be 70 percent, and the broad-sense heritability for personality traits is estimated to be between 34 and 50 percent.**

TESTING YOUR KNOWLEDGE

1. You discover a *Drosophila* mutant that exhibits abnormal behavior. Not only does the locomotor activity run on a cycle that is longer than 24 hours, but also the male's courtship song is different from that of wild-type. The new mutation is recessive and X-linked. How would you determine if it is an allele of the *period* gene?

ANSWER

Males carrying the new mutation, *m*, should be crossed to homozygous per^0 females, and the locomotor activity of their m/per^0 daughters should be examined. If this activity is abnormal—arrhythmic or rhythmic with a period longer than 24 hours—the new mutation is an allele of *per*. If locomotor activity is wild-type, the new mutation is not an allele of *per*.

2. In a study of MZ and DZ twins, reared together and apart, a group of Swedish researchers obtained the following correlation coefficients for IQ: MZT, 0.80; MZA, 0.78; DZT, 0.22; DZA, 0.32. What do these correlations suggest about the extent to which variation in IQ scores is attributable to genetic variation? Are the results internally consistent?

ANSWER

The correlation for MZ twins reared apart, 0.78, implies that 78 percent of the population's variability in IQ is due to genetic variation; that is, the broad-sense heritability is 0.78. The slightly higher correlation for MZ twins reared together reinforces this conclusion and suggests that the effect of a common environment on the correlation for IQ is negligible. Thus, common environmental influences seem to account for a very small percentage of the overall variation in IQ within the population. The correlations for the DZ twins are generally in agreement with this view, but there is one inconsistency: the correlation for DZ twins reared together is less than that for DZ twins reared apart. We might have expected the correlation for DZ twins reared together to be as great or greater than the correlation for DZ twins reared apart. This inconsistency is probably due to sampling error. If we accept the correlation for DZ twins reared apart at face value, then doubling it should estimate the broad-sense heritability of IQ, under the assumption that all the underlying genetic variation in this trait is additive; $2 \times 0.32 = 0.64$. However, this result is less than the broad-sense heritability estimated from the correlation between MZ twins reared apart (0.78). This inequality suggests (albeit not too strongly given all the statistical uncertainties associated with these data) that some of the genetic variation in IQ is due to nonadditive genetic factors such as dominance and epistasis (see Chapter 5).

QUESTIONS AND PROBLEMS

26.1 Can two nonhygienic bees produce hygienic offspring? Explain.

26.2 Two mice expressing a behavioral aberrancy called zigzag were crossed. They produced 84 zigzag offspring and 58 normal offspring. What can you conclude about the inheritance of zigzag from these data? Explain.

26.3 You discover a new *Drosophila* mutant that is extremely sensitive to loud noises. When you bang on the table sharply the fly stops, quivers for a while, and then falls over. It recovers after a few minutes. You call this mutant "sensitive" (*sen*). In order to investigate the genetic basis for the trait, you cross the mutant to a wild-type fly from the same population from which *sen* originated. Half of the offspring are wild-type and half express *sen*. When the *sen* mutants are crossed among themselves, all the offspring are *sen*; when wild-type flies from the small population that produced *sen* are intercrossed, some produce all wild-type offspring, but some of the wild-type × wild-type matings produce 25 percent *sen* offspring. On the basis of this information, is *sen* due to a dominant or a recessive mutation? Explain.

26.4 If honeybee queens from a colony in which the workers uncap cells but do not remove dead larvae are crossed with drones derived from the unfertilized eggs of hygienic queen bees, what types of worker offspring would be expected, and in what proportions?

26.5 Suppose that EMS-induced mutations in the *period* gene are associated with a recessive lethal phenotype. Why would Konopka and Benzer have failed to detect such mutations in their screen? Suggest a way in which they might have circumvented the problem of recessive lethality.

26.6 Suppose that a X-linked gene in *Drosophila* controls the courtship behavior of males. Mutations in this gene abolish the courtship song, causing the males to be sterile. Suggest a way in which a researcher could propagate mutations in this gene so that they are available for genetic analysis.

26.7 A geneticist has carried out an experiment to map a dominant autosomal mutation called *Awry* that abolishes *Drosophila*'s locomotor activity rhythm. Males homozygous for *Awry* were crossed to females homozygous for three linked autosomal markers: *cinnabar* (*cn*) eyes, *vestigial* (*vg*) wings, and *brown* (*bw*) eyes. (Note that *cn bw*/*cn bw* double homozygotes have white eyes.) The F_1 daughters were then backcrossed to males homozygous for the three recessive markers, and the F_2 offspring were classified for wing and eye phenotype. Next they were tested for locomotor activity, yielding the following data:

cinnabar, vestigial, brown	25; none awry
wild-type	28; all awry
cinnabar, vestigial	17; 7 awry, 10 nonawry
brown	19; 10 awry, 9 nonawry
cinnabar	4; all awry
vestigial, brown	5; none awry
cinnabar, brown	1; nonawry
vestigial	1; nonawry

If the known map positions of *cn*, *vg*, and *bw* are 57, 67, and 104, respectively, where on the map would you place the *Awry* mutation?

26.8 Null mutations in the X-linked *per* gene and the autosomal *tim* gene both cause *Drosophila* to be arrhythmic. Predict the phenotype of a *per*; *tim* double mutant.

26.9 Design an experiment to demonstrate that the protein products of the *Drosophila clock* and *cycle* genes bind to DNA sequences upstream of the *per* and *tim* transcription start sites.

26.10 The *cycle* and *clock* mutations in *Drosophila* have similar effects on locomotor and eclosion rhythms, and both map to an autosome, chromosome 3. How would you demonstrate that *cycle* and *clock* are mutations in different genes?

26.11 Design an experiment to show that the protein product of the *Drosophila clock* gene positively regulates expression of the *period* gene.

26.12 The behavioral significance of the mouse *Clock* gene was identified by using a potent chemical mutagen to induce a dominant mutation in the gene. By contrast, the behavioral significance of the mouse *fosB* gene was identified by using a cloned copy of the gene to knock out the wild-type gene and replace it with a disrupted copy of the cloned gene. Analysis of the *Clock* gene exemplifies the so-called forward genetic approach, and analysis of the *fosB* gene exemplifies the so-called reverse genetic approach. Discuss the advantages and disadvantages of these two approaches to study the genetic control of behavior in the mouse.

26.13 Predict the sexual phenotype (anatomical and behavioral) of the following mutant genotypes in *Drosophila*: (a) XY; *tra fru*/*tra fru*; (b) XX; *tra fru*/*tra fru*; (c) XY; *dsx*/*dsx*; *fru*/*fru*; (d) XX; *dsx*/*dsx*; *fru*/*fru*.

26.14 Design an experiment to show that the expression of the *fru* gene in *Drosophila* is controlled by the products of the *tra* and *tra2* genes but not by the product of the *dsx* gene.

26.15 In what ways are phenylketonuria and Lesch-Nyhan syndrome pleiotropic disorders?

26.16 In *Drosophila*, the PER protein does not appear to bind to DNA, yet it is implicated in the regulation of its own synthesis. How would you explain this?

26.17 What are the similarities and differences between PKU and LNS on the one hand and Huntington's disease on the other?

26.18 What is the connection between amyloid precursor protein and Alzheimer's disease?

26.19 What is the significance of the observation that patients with Down syndrome invariably develop Alzheimer's disease?

26.20 How could you determine if the dominant, early-onset Alzheimer's disease that occurs in families with mutations in the *PS1* or *PS2* genes is due to a loss of protein function or a gain of protein function?

26.21 On the basis of the observed correlations for personality traits shown in Table 26.2, predict the correlations for the same traits for DZ twins reared apart. What assumptions are necessary to make these predictions?

BIBLIOGRAPHY

Bouchard, Jr., T. J. 1998. Genetic and environmental influences on adult intelligence and special mental abilities. *Human Biol.* 70:257–279.

Bouchard, Jr., T. J., D. T. Lykken, M. McGue, N. L. Segal, and A. Tellegen. 1990. Sources of human psychological differences: The Minnesota Study of Twins Reared Apart. *Science* 250:223–228.

Brown, J. R., H. Yee, R. T. Bronson, P. Dikkes, and M. E. Greenberg. 1996. A defect in nurturing in mice lacking the immediate early gene *fosB*. *Cell* 86:297–309.

Hall, J. C. 1994. The mating of a fly. Science 264:1702–1714.

Hardy, J. 1997. Amyloid, the presenilins, and Alzheimer's disease. *Trends in Neurosci.* 20:154–159.

Konopka, R., and S. Benzer. 1971. Clock mutants of *Drosophila melanogaster*. *Proc. Natl. Acad. Sci. USA* 68:2112–2116.

Reppert, S. M., and D. R. Weaver. 1997. Forward genetic approach strikes gold: cloning of a mammalian *Clock* gene. *Cell* 89:487–490.

Ryner, L. C., S. F. Goodwin, D. H. Castrillon, A. Anand, A. Villella, B. S. Baker, J. C. Hall, B. J. Taylor, and S. A. Wasserman. 1996. Control of male sexual behavior and sexual orientation in Drosophila by the *fruitless* gene. *Cell* 87:1079–1089.

27

A human T lymphocyte (green) that is infected with human immunodeficiency viruses (red). Individuals infected with HIV usually develop acquired immune deficiency syndrome (AIDS), the plague of the twentieth century.

The Genetic Control of the Vertebrate Immune System

CHAPTER OUTLINE

The Immune Response: An Overview
Components of the Mammalian Immune System
The Antibody-Mediated (Humoral) Immune Response
The T Cell-Mediated (Cellular) Immune Response
Immunological Memory
Genome Rearrangements During B Lymphocyte Differentiation
Antibody Class Switching
Assembly of T Cell Receptor Genes by Somatic Recombination
Regulation of Immunoglobin Gene Expression

AIDS Emphasizes the Vital Role of the Immune System

In 1971, when Brandon was just a toddler, he fell on the front step of his house and cut his knee. When the bleeding would not stop, his mother took him to the emergency room of the local hospital. Blood tests revealed that Brandon had hemophilia A, an inherited deficiency of blood-clotting factor VIII (Chapter 6). Individuals with this X-linked recessive disorder exhibit excessive bleeding from even small cuts and lacerations. However, the bleeding can be arrested by treatment with the missing clotting factor. In 1982, when Brandon was injured in a car accident, he received blood transfusions and treatment with clotting factor VIII. He recovered from his injuries and continued to live a normal, active life. Eight years later, Brandon began to experience fatigue, recurrent fever, and weight loss. He seemed to become infected with every virus that he encountered. Shortly thereafter, the frightening diagnosis was made: Brandon had *a*cquired *i*mmune *d*eficiency *s*yndrome (AIDS), the fatal disease caused by the *h*uman *i*mmunodeficiency *v*irus (HIV). A few months later, Brandon contracted a type of pneumonia caused by *Pneumocystis carinii.* This protozoan is an opportunistic pathogen that infects individuals, such as AIDS patients, with weakened immune systems. HIV destroys the immune system of its hosts by killing a type of white blood cell that is required for a normal immune response (Chapter 16). Brandon died less than a year later.

Brandon became infected with HIV in contaminated blood used in the transfusion after his car accident. Prior to 1985, blood serum used in transfusions was not tested for HIV, which was identified in 1983. Today, all donated blood is tested for HIV before being deposited in blood banks. Thus, the risk of contracting AIDS from a blood transfusion is extremely low—approximately one case in 10^5 transfusions.

AIDS emphasizes the vital role that a functional immune system plays in our survival. Many of us have watched the health of a relative, friend, or neighbor with AIDS deteriorate before our eyes after HIV destroyed essential cells of his or her immune system. If not, we have observed the horrible effects of AIDS on the human body through the news media. The pain and suffering that occurs in terminal AIDS patients illustrate what happens when the immune system can no longer do its job. As a result, most of us are more aware of the important roles that our immune systems play in our lives. On a nearly daily basis, our immune systems fight off a vast array of pathogenic microorganisms.

The environment in which we live contains an enormous number of different viruses, bacteria, fungi, and other agents. These organisms—many of which are infectious—are constantly evolving and producing new strains. Viruses, in particular, accumulate new mutations and evolve very rapidly (Chapter 16). How, then, can our immune systems protect us from all of these different pathogens, some of which are newly evolved and are confronting cells of the immune system for the first time? In this chapter, we will discuss the answer to this question. We will examine the components of the human immune system and the processes by which immune responses protect us from infectious agents. In doing so, we will encounter some fascinating genes that are assembled from gene pieces during the differentiation of the cells that produce them. We will see how this assembly process facilitates the production of an army of proteins with the specificity required to combat a seemingly unlimited variety of infectious agents.

All vertebrate animals are protected by immune systems. However, for the sake of brevity and focus, we will restrict our discussion in this chapter to the immune systems of mammals, with special emphasis on humans.

THE IMMUNE RESPONSE: AN OVERVIEW

When a mammal is infected by a pathogenic virus, bacterium, fungus, or other invading cell, the body mounts a complex imune response. Each immune response involves three essential steps:

1. *Recognition* of the foreign entity.
2. *Communication* of this recognition to the appropriate cells.
3. *Elimination* of the invading entity.

Some components of the immune response are directed against infectious agents in general and are called **nonspecific immune responses.** They include inflammation of the infected tissues resulting in increased blood flow and the recruitment of a special type of blood cells called *phagocytes* (Figure 27.1) that ingest and destroy viruses, bacteria, and fungi in a nonspecific manner. However, the most important responses are specific to the foreign substance and thus are called **specific immune responses.** For the rest of this chapter, we focus on the two major specific immune responses: the synthesis of special proteins and the production of cells that protect the mammal from infectious agents. We should emphasize that these two responses do not occur independently; rather, the cells involved in the two responses must communicate with each other for either response to be effective. Indeed, the immune responses to foreign agents are complex processes, and we should familiarize ourselves with the cells and proteins that are involved before discussing these processes in more detail.

Key Points: **The immune response in mammals involves three steps: (1) recognition of the foreign substance, (2) communication of this recognition to the responding cells, and (3) elimination of the in-**

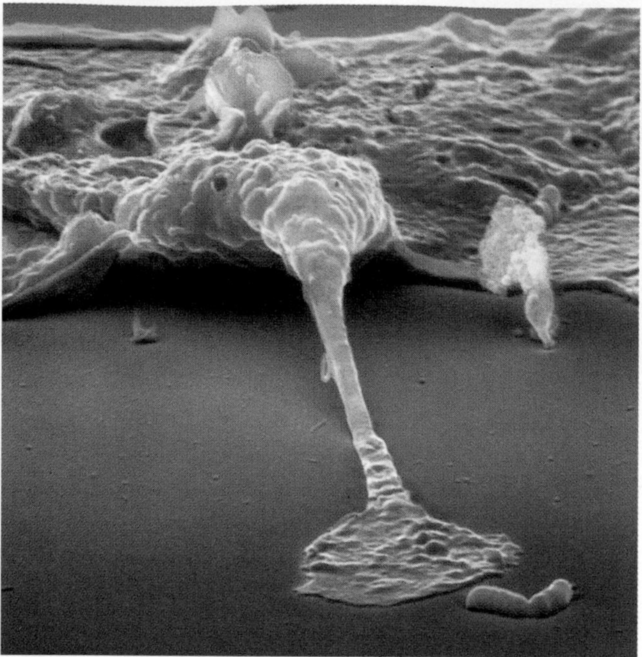

Figure 27.1 A phagocyte stalking a rod-shaped bacterium.

vading agent. Nonspecific responses include the recruitment of phagocytes to ingest and destroy viruses and microorganisms. Specific responses include the synthesis of special proteins and cells that work together to remove the foreign entity from the body.

COMPONENTS OF THE MAMMALIAN IMMUNE SYSTEM

The most striking feature of the immune system is its **specificity.** This specificity is provided by (1) a team of specialized cells, each performing its job in a highly coordinated fashion; (2) two armies of proteins, *antibodies* and *T cell receptors,* each with seemingly unlimited ability to recognize foreign substances; and (3) a specialized set of proteins, the *histocompatibility antigens,* that facilitate communication between cells and allow cells of the immune system to recognize foreign substances and to distinguish an individual's own cells (self) from foreign cells (nonself). The most important of these immune system cells and proteins are listed in Table 27.1. In total, the immune system of an adult human consists of something like a trillion (10^{12}) white blood cells and 10^{15} or more special protein molecules, all acting in concert to eliminate foreign substances.

When a foreign agent such as a virus or bacterium invades the human body, it is recognized by cells of the immune system and triggers the production of antibodies and T cell receptors that specifically recognize and bind to the foreign substance. A substance that is bound by an antibody or T cell receptor is called an **antigen** (*anti*body-*gen*erating substance), and an antigen that elicits an immune response is called an immunogen. Virtually all biological macromolecules, including polypeptides, polysaccharides, and nucleic acids, can act as antigens, and most can function as immunogens.

The mammalian immune response has two major components: the antibody-mediated or humoral response (based on the Latin term *humor* meaning "fluid"—thus, body fluids) and the T cell-mediated or cellular response. The humoral immune response involves the production and secretion of antibodies into the circulatory system, where these antibodies bind to and sequester their respective antigens. The antibody-antigen complexes are then ingested and destroyed by a special class of white blood cells. The humoral immune system is primarily reponsible for protecting mammals from foreign cells (for example, bacteria and fungi) and from free viruses before they infect cells of the host organism. The cellular immune response involves the production of T cell receptors that coat the surfaces of special white blood cells, and allow these cells to recognize and kill infected cells of the host organism. Thus, the cellular immune system is largely reponsible for protecting mammals from viral infections. The cells and macromolecules that participate in the antibody-mediated and T cell-mediated immune responses of mammals are described in the following sections.

Specialized Cells Mediate the Immune Response

The immune response to a foreign substance relies on the coordinated activities of a variety of highly specialized cells. All of these cells are derived from undifferentiated precursor cells, called **stem cells,** present in bone marrow. One major component of the immune response is carried out primarily by **B lymphocytes,** called **B cells** because, in birds, they mature in the *B*ursa of Fabricius, a saclike structure that performs some of the same functions as the *b*one marrow in mammals. The other major component of the immune response is performed largely by **T lymphocytes,** called **T cells** because they undergo key differentiation events in the *t*hymus gland. The Greek roots of the word *lymphocyte* mean colorless cell, an appropriate name for these small white blood cells. The nearly colorless phenotype of these lymphocytes is a striking contrast to the highly pigmented phenotype

TABLE 27.1
Components of the Mammalian Immune System

Component	*Description and Function*
I. Proteins	
Antigens	Substances that trigger an immune response; *anti*body-*gen*erating substances
Antibodies	Proteins produced by the immune system that bind to antigens and assist in their destruction
T cell receptors	Proteins produced by the immune system in response to antigens; they become anchored on the surface of killer T cells and bind antigens with the help of major histocompatibility antigens
Major histocompatibility complex (MHC) antigens	Cell surface proteins that (1) allow immune system cells to distinguish foreign substances from self and (2) facilitate communication between cells
II. Cells	
Stem cells	Undifferentiated bone marrow cells that give rise to the various cells of the immune system
Phagocytes	Large cells that capture, ingest, and destroy invading foreign agents such as viruses, bacteria, and fungi
Macrophages	Phagocytic cells that ingest antigens and display them on the cell surface for interactions with other cells of the immune system
B lymphocytes	Cells that differentiate (in the *b*one marrow of mammals) into antibody-producing plasma cells and memory B cells
Plasma cells	Antibody-producing white blood cells derived from B lymphocytes
Memory B cells	B cells that facilitate the rapid production of a given antibody in response to a second and subsequent encounters with an antigen
T lymphocytes	Cells that undergo important differentiation events in the *t*hymus gland and further differentiate into the various types of T cells
Helper T cells	T cells that respond to the display of an antigen by a macrophage by stimulating B lymphocytes to produce antibodies and T lymphocytes to produce T cell receptors
Suppressor T cells	T cells that suppress the production of antibodies and T cell receptors by B cells and T cells, respectively
Cytotoxic or killer T cells	T cells that carry T cell receptors and kill cells displaying the recognized antigens
Memory T cells	T cells that facilitate the rapid production of a given T cell receptor in second and subsequent encounters with an antigen

of their hemoglobin-containing neighbors, the red blood cells. Another group of immune system cells are the **phagocytes** (from the Greek for "eating cells"). One type of phagocyte, the **macrophage** (from the Greek for "big eater"), plays a central role in initiating both components of the immune response.

Upon receiving the appropriate signals, B lymphocytes differentiate into **plasma cells,** which produce antigen-binding proteins called antibodies, and **memory B cells,** which facilitate a more rapid production of antibodies during a later exposure to the same antigen. The T lymphocytes differentiate into four important types of T cells.

1. **Cytotoxic** or **killer T cells** carry antigen-binding proteins called T cell receptors on their cell surfaces and, as the name implies, kill cells displaying the appropriate antigens.
2. **Helper T cells** stimulate B and T lymphocytes to differentiate into antibody-producing plasma cells and killer T cells, respectively.
3. **Suppressor T cells** assist in "suppressing" or down-regulating the activity of plasma cells and killer T cells.
4. **Memory T cells** "remember" the antigen and provide for a rapid production of killer T cells during subsequent encounters with the same antigen.

Figure 27.2 summarizes the origins and functions of the cells involved in important aspects of the immune response of mammals.

Some individuals have inherited or acquired defects in cells of the immune system. The most extreme of these disorders are **severe combined immunodeficiency syndrome** (**SCID;** inherited) and *acquired immune deficiency syndrome* (*AIDS;* discussed in Chapter 16). Individuals with SCID are unable to mount an immune response of any kind, either antibody-mediated or T cell-mediated. These individuals have no functional B or T lymphocytes; in most cases, no B or T cells are even present. There are multiple forms of SCID, with about half of the cases caused by X-linked recessive mutations and the other half by autosomal recessive defects. Among the autosomally inherited cases of SCID, about two-thirds are due to autosomal recessive mutations in a gene or genes of unknown function. The other one-third are caused by mutant alleles of the gene encoding an enzyme called adenosine

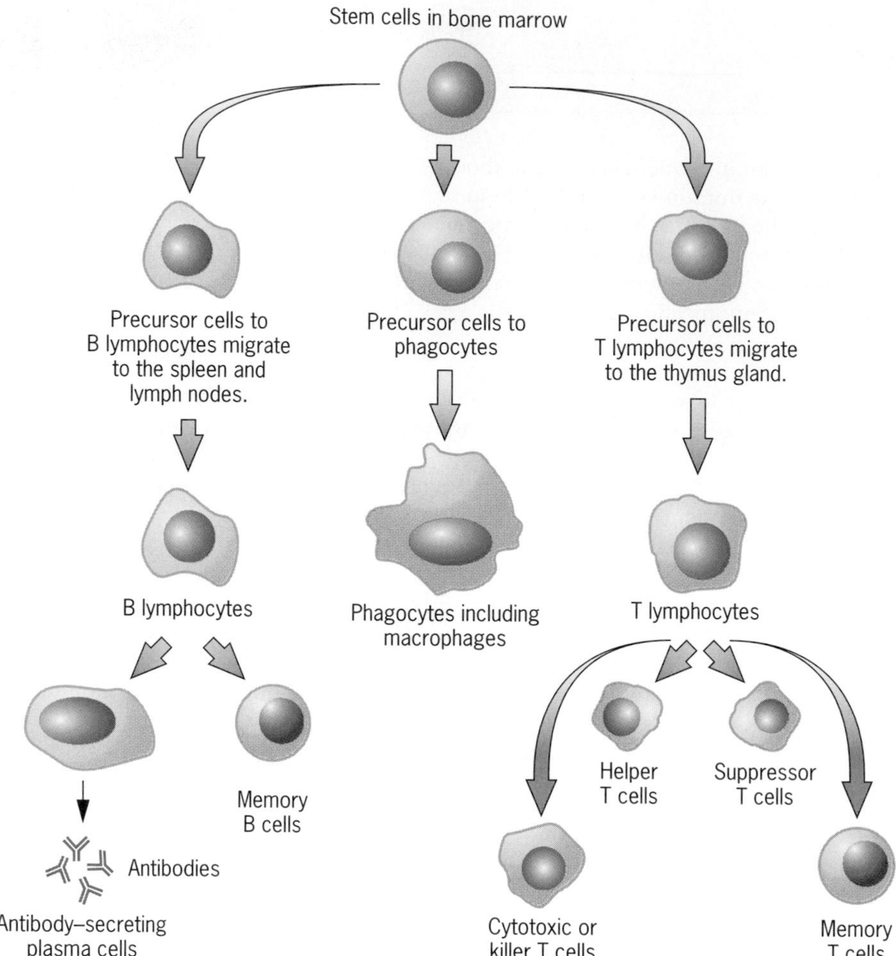

Figure 27.2 The most important cells of the immune system are derived from bone marrow stem cells by three separate cell lineages. They give rise to (1) B cells and antibody-producing plasma cells, (2) the various types of T cells, and (3) the phagocytes, including macrophages.

deaminase (ADA). If untreated, autosomally inherited ADA-deficient SCID usually results in death from severe infections early in childhood. We discussed the treatment of ADA-deficient SCID patients by infusions of normal white blood cells and somatic-cell gene therapy in Chapter 22.

Specialized Proteins Provide Immunological Specificity

Immunological specificity is provided by two classes of proteins, antibodies and T cell receptors, which mediate the two major components—the antibody-mediated or humoral response and the T cell-mediated or cellular response—of the mammalian immune system, respectively. Each of these classes of proteins displays a seemingly unlimited variety of antigen-binding sites, and the mechanisms by which this variability is generated are most fascinating.

Antibodies belong to the class of proteins called **immunoglobulins.** Each antibody is a tetramer composed of four polypeptides, two identical **light chains** and two identical **heavy chains,** joined by disulfide bonds (Figure 27.3*a*). The light chains are about 220 amino acids long, and the heavy chains are about 440 to 450 amino acids long. Every chain, heavy and light, has an amino-terminal **variable region,** within which the amino acid sequence varies among antibodies specific for different antigens, and a carboxyl-terminal **constant region,** within which the amino acid sequence is the same for all antibodies of a given immunoglobulin (Ig) class, regardless of antigen-binding specificity. The variable regions of all antibody chains are about 110 amino acids long.

Regions of proteins that carry out particular functions are called **domains.** Each antibody has two **antigen-binding sites** or domains, each of which is formed by the variable regions of one light chain and one heavy chain (Figure 27.3*b*). These variable region domains form binding sites that interact with the antigen in a lock-and-key fashion (Figure 27.4). In addition, the constant regions of the two heavy chains interact to form a third domain, called the **effector function do-**

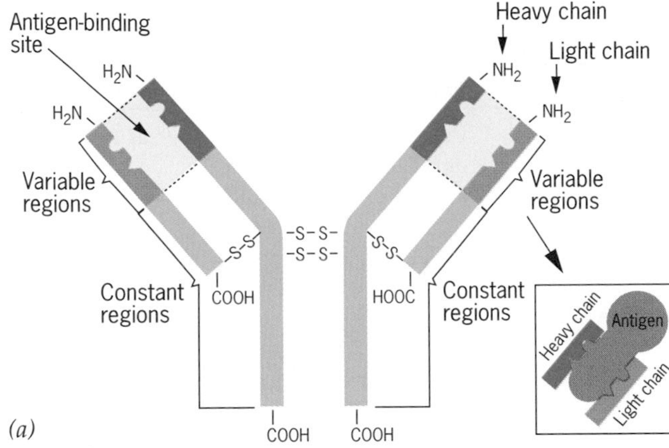

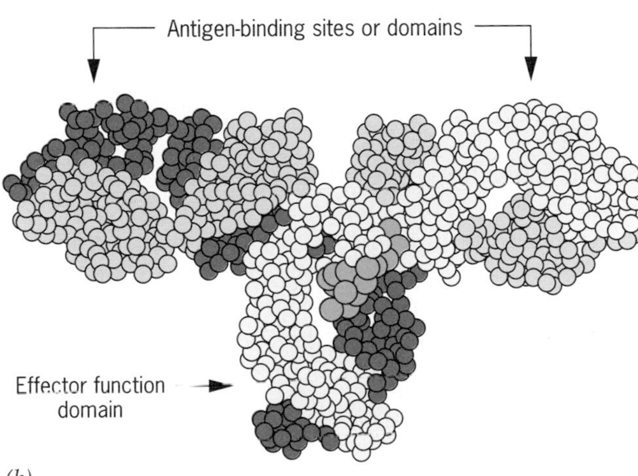

Figure 27.3 *(a)* Diagram and *(b)* space-filling model of antibody structure. Each antibody is a tetramer composed of four polypeptide chains: two identical light chains and two identical heavy chains. *(a)* Each chain consists of a variable region (blue) and a constant region (green), and each antibody has two antigen-binding sites, formed by heavy and light chain variable regions. The inset shows the "lock-and-key" interaction between an antigen and an antibody. *(b)* The two heavy chains are shown in white and blue, respectively. The two light chains are shown in yellow. The green-colored components are the associated carbohydrate moieties; their location and structure will vary depending on the immunoglobulin (Ig) class. The structure shown is for a human IgG molecule.

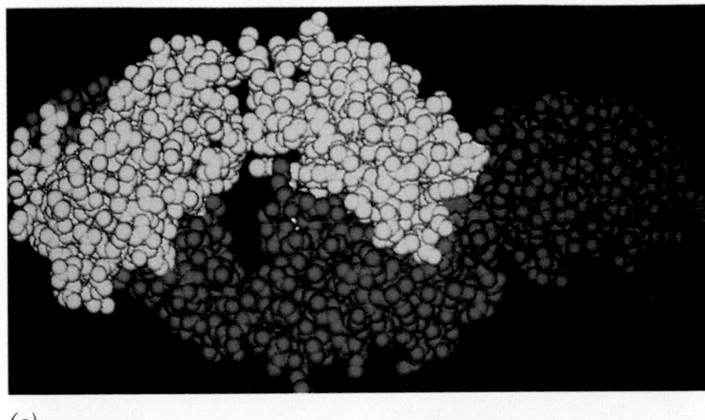
(a)

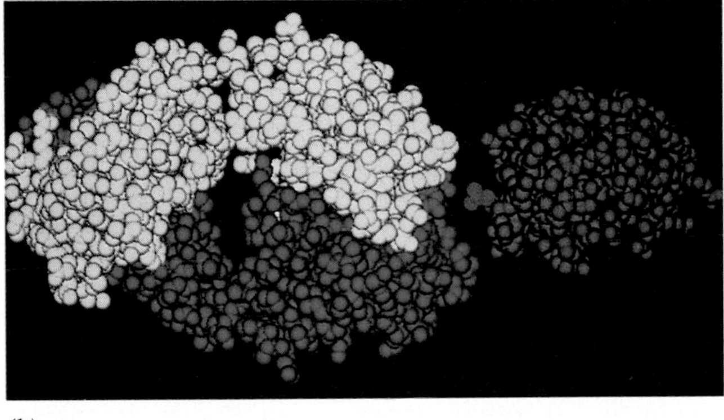
(b)

Figure 27.4 Three-dimensional structure of an antigen-antibody complex. Only one of the two antigen-binding sites of a typical antibody is shown. The antigen (green) is the enzyme lysozyme. The antigen-binding site of the antibody is formed by the amino-terminal portions of a light chain (yellow) and a heavy chain (blue). A glutamine residue that protrudes from lysozyme at the site where the antibody binds is shown in red. The structures shown are based on X-ray diffraction data. *(a)* The fully engaged antigen-antibody complex. *(b)* The antigen and antibody have been pulled apart to show their complementary lock-and-key structures.

main, which is responsible for the proper interaction of the antibody with other components of the immune system.

The light chains of antibodies are of two types, **kappa chains** and **lambda chains,** with type being determined by the structure of the constant region. As we will see, antibodies may have the same antigen-binding specificity, as determined by the variable regions of the four chains, but different immunological functions, as determined by the constant regions of the two heavy chains.

There are five classes of antibodies: **IgA, IgD, IgE, IgG,** and **IgM** (Table 27.2). The class to which an antibody belongs, and thus the function that it carries out, is determined by the structure of its heavy chain constant region (i.e., the structure of its effector function domain). IgA, IgD, IgE, IgG, and IgM antibodies have α, δ, ϵ, γ, and μ heavy chains, respectively. In humans, there are two subclasses of IgA antibodies and four subclasses of IgG antibodies; each subclass has different, but closely related, heavy chains. The first class of antibody produced in a developing B cell is always IgM. The initial IgM antibodies are anchored in the membrane of the cell, providing it with cell-surface antigen receptors. In a primary immune response (the

TABLE 27.2
Classes of Immunoglobulins

Type	*Heavy Chain*	*Structure*[a]	*Proportion*	*Location*	*Function*
IgA	α	α_2L_2	14%	Secretions: milk, saliva, tears	Protection against microbes at potential entry sites
IgD	δ	δ_2L_2	1%	Blood; B cells	Uncertain; may stimulate B cells to make other classes of antibodies
IgE	ϵ	ϵ_2L_2	<1%	Tissues; mast cells[b]	Receptors for antigens that lead to the secretion of histamines by mast cells
IgG	γ	γ_2L_2	80%	Blood; macrophages, plasma cells	Activation of complement during a secondary immune response[c]
IgM	μ	μ_2L_2	5%	Blood; B cells	Activation of complement during a primary immune response[c]

[a]L represents an antibody *l*ight chain, either kappa or lambda.
[b]Cells that are widely distributed in various tissues and release histamines during the nonspecific inflammatory response to an infection.
[c]Complement is a system of approximately 20 soluble proteins that circulate in the bloodstream and form large protein complexes in response to antibody-antigen complexes. The complement proteins kill cells by a variety of mechanisms, such as punching holes in the cell membrane. This system of proteins is called "complement" because it complements the activity of antibodies in fighting pathogenic foreign cells.

first encounter with a given antigen), a B cell subsequently will begin producing **secreted IgM antibodies,** which are pentameric molecules with ten antigen-binding sites. Later, the B cell may switch to the production of other classes of antibodies, a process called **class switching.**

IgG antibodies account for about 80 percent of the immunoglobulin molecules in human blood. They are produced in large amounts during the second and subsequent encounters with a given antigen, the secondary immune response. IgA molecules are the predominant immunoglobulins in secretions such as saliva, tears, and milk, whereas IgE molecules are found as membrane-bound antigen receptors on special cells called mast cells in various tissues. The binding of antigens by these mast cell surface receptors triggers the secretion of histamines and other amines, which are largely responsible for the allergic reactions observed in individuals with asthma and hay fever.

If we were to examine the structure of antibodies, we would see that their diversity resides almost entirely within the variable regions of the molecules. If these polypeptides were synthesized from colinear nucleotide-pair sequences of genes, one gene per polypeptide chain, the genome would have to contain a vast array of genes with highly variable sequences at one end and essentially identical sequences at the other end. However, this is not the case. Recombinant DNA techniques have made it possible to isolate and sequence many of the segments of chromosomal DNA of mice and humans encoding antibody chains. The results of these studies have shown that antibody genes are stored as gene segments in germ-line cells and that these gene segments are spliced together into genes during the differentiation of antibody-producing plasma cells. We will examine this process in subsequent sections of this chapter.

The cellular immune response is highly antigen-specific, just like the antibody-mediated immune response, and is equally important in defending mammals against infections. In the cell-mediated immune response, the antigen specificity is provided by **T cell receptors.** Each T cell receptor is a dimer composed of an α polypeptide and a β polypeptide (Figure 27.5).

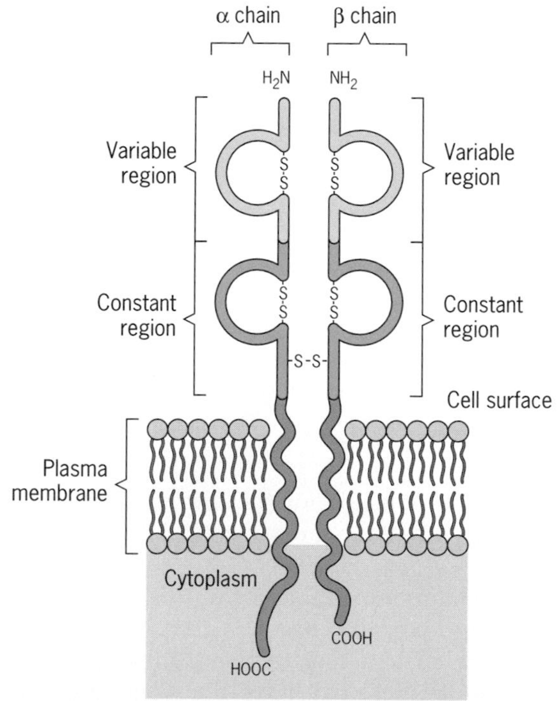

Figure 27.5 Structure of a T cell receptor anchored in the cell membrane.

Both polypeptides have amino-terminal variable regions and carboxyl-terminal constant regions, like antibody chains. However, T cell receptors have only one antigen-binding site, rather than two, as in antibodies. As the name implies, these receptors are present on the surfaces of T cells, anchored in the cell membrane by their constant regions. T cell receptors are present on both cytotoxic or killer T cells and on helper T cells, where, in collaboration with the major histocompatibility antigens, they bind antigens and participate in the activities of these important immune cells.

Major Histocompatibility Antigens: Distinguishing Self from Nonself

Organ transplants and skin grafts between unrelated individuals usually result in immune rejections of the transplanted tissues. In contrast, organ transplants between identical twins usually are accepted. Also, skin can be transplanted from one location to another on the same individual without immune rejection. These results indicate that the immune system can distinguish foreign or "nonself" cells from "self" cells. Carefully controlled experiments have shown that the observed rejections of nonself tissue grafts are controlled by a complex set of cell-surface macromolecules called **histocompatibility antigens.** The most important of these histocompatibility antigens are the **major histocompatibility complex (MHC)** proteins. Many of the proteins encoded by the MHC are deposited on cell surfaces where they are highly antigenic to other mammals as well as "nonidentical" individuals of the same species. In humans, the MHC proteins are called the **HLA antigens** (*h*uman-*l*eukocyte-*a*ssociated antigens) because they were discovered on the surfaces of leukocytes (white blood cells). The HLA antigens are encoded by a large cluster of genes, the ***HLA* locus,** on chromosome 6 (Figure 27.6). The *HLA* gene cluster is over 2×10^6 nucleotide pairs in length.

The genes of the *HLA* locus are highly polymorphic, with some genes having up to 100 or more different alleles. Thus, the probability that two unrelated individuals will carry the same alleles of all the *HLA* genes is very low. As a result, finding organ donors with HLA antigens that precisely match those of a recipient is very difficult.

The *HLA* genes encode three different classes of MHC proteins, with each class of molecules involved in a different aspect of the immune response. The ***HLA* class I genes** encode proteins that are called the **transplantation antigens** because of their role in tissue rejections during organ and tissue transplants. The class I proteins exist as glycoproteins anchored as integral membrane proteins with the antigenic determinants exposed on the outside of cells. They are present on virtually all cells of an organism and provide killer T

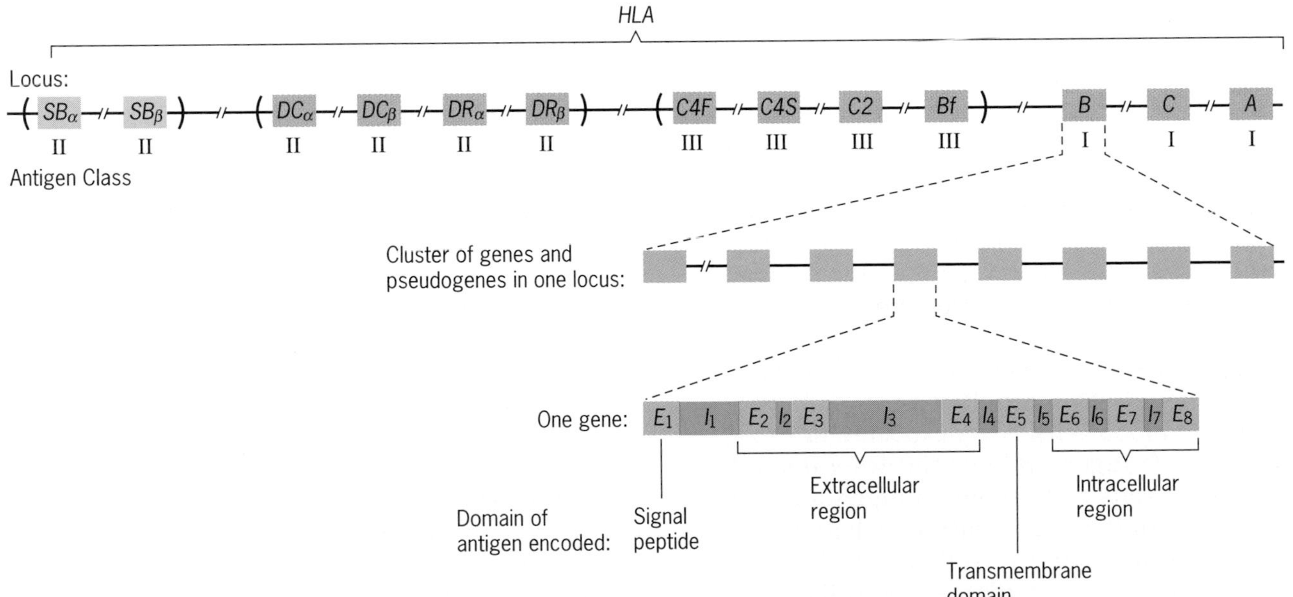

Figure 27.6 Organization of the major histocompatibility complex (*HLA*) on human chromosome 6. The relative positions of mapped loci within this huge gene complex are shown at the top. The entire *HLA* complex is over 2×10^6 nucleotide pairs in length. The class of histocompatibility antigen encoded by genes in each locus is indicated below the map. Note that each of the loci within the *HLA* complex is itself a complex locus containing several genes and pseudogenes (center). The structure of a typical class I gene is shown at the bottom. Note that different exons encode different functional domains of the polypeptide gene product. Key: E = exon; I = intron.

cells with a mechanism for distinguishing "foreign" from "self." Each class I MHC protein can bind a rather broad spectrum of antigens and present them to killer T cells.

The ***HLA* class II genes** encode polypeptides that are located primarily on the surfaces of B lymphocytes and macrophages, cells that interact with helper T cells. The class II MHC proteins contain antigen-binding grooves that bind a wide range of related peptides and displays them to helper T cells. Finally, the ***HLA* class III genes** encode **complement proteins** that interact with antibody-antigen complexes and help destroy them by proteolysis. Complement proteins also kill infected cells by disrupting the integrity of cell membranes.

Sometimes the mechanisms by which the body distinguishes "self" from "foreign" malfunction, and antibodies are produced against "self" antigens—antigens in the individual producing the antibodies. When this happens, the individual suffers from an **autoimmune disease.** These antibodies may cause damage to specific tissues or organs or general damage throughout the body. In most autoimmune diseases, this damage results in severe pain to the affected individuals. Crohn's disease, juvenile diabetes, hemolytic anemia, myasthenia gravis, scleroderma, and systemic lupus erythematosus are examples of autoimmune diseases in which individuals produce antibodies that attack self-antigens on cells of the small intestine, pancreatic beta cells, red blood cells, skeletal muscles, connective tissue cells, and neurons plus blood cells, respectively. In juvenile diabetes, these antibodies attack and destroy the insulin-producing beta cells in the pancreas, which results in the dependence of diabetics on insulin obtained from animals or human insulin produced in bacteria or cultured animal cells by recombinant DNA techniques (see Human Genetics Sidelight in Chapter 20). Without diagnosis and treatment with insulin, juvenile diabetes results in eventual blindness, kidney malfunction, heart disease, and premature death.

Key Points: **The immune response of mammals involves the coordinated activities of several specialized white blood cells. After exposure to a foreign substance (antigen), B lymphocytes differentiate into plasma cells that produce antigen-binding proteins called antibodies. Similarly, T lymphocytes develop into killer T cells that carry T cell antigen receptors on their surfaces and kill cells displaying the recognized antigens. Individuals with severe combined immunodeficiency disease (SCID) are unable to mount any immune response. Highly polymorphic major histocompatibility antigens on cells allow killer T cells to distinguish foreign cells from "self" cells, the cells of the individual mounting the immune response. Individuals with autoimmune diseases produce antibodies against "self" antigens.**

THE ANTIBODY-MEDIATED (HUMORAL) IMMUNE RESPONSE

Now that we are familiar with the most important components of the immune system of mammals, let's examine the immune responses in more detail.

When a foreign antigen, for example, a glycoprotein on the surface of a bacterium, enters the bloodstream of a mammal, it triggers a defense mechanism, the **antibody-mediated** or **humoral immune response,** which results in the synthesis of antibodies. These antibodies bind to the antigen with exceptional specificity, thus facilitating the removal of the invading bacterium from the circulatory system. The antibody-mediated immune response is primarily responsible for protecting mammals from bacterial and fungal infections, and for removing viruses before they infect host cells. It involves several different classes of immune cells and leads to the production of antibodies specific to the invading antigen by plasma cells derived from B lymphocytes (Figure 27.7, *left*).

The antibody-mediated immune response is initiated when a macrophage encounters a foreign substance such as a bacterium and engulfs and destroys it. Antigens from the partially degraded bacterium are anchored on the surface of the macrophage, where they are recognized by helper T cells. Once presented with antigens by macrophages, the activated helper T cells stimulate B lymphocytes to differentiate into plasma cells, which synthesize and secrete large quantities of antibodies, and memory B cells, which facilitate a faster immune response during subsequent encounters with the same antigen.

In AIDS patients, helper T cells are infected and killed by the human immunodeficiency virus (Chapter 16), leaving the patients with weakened immune systems and, therefore, susceptible to infection by a variety of opportunistic pathogens.

One specific structural feature of an antigen, called an **epitope,** stimulates the production of antibodies, and the resulting antibodies specifically recognize and bind to that particular structural component of the antigen. An antigen may contain several epitopes. For proteins, an epitope may consist of only a short peptide, sometimes containing as few as six amino acids.

Antibodies bind specifically to the antigens eliciting the immune response and help inactivate them or remove them from the body in several ways. Antibody-antigen complexes are eliminated by two major path-

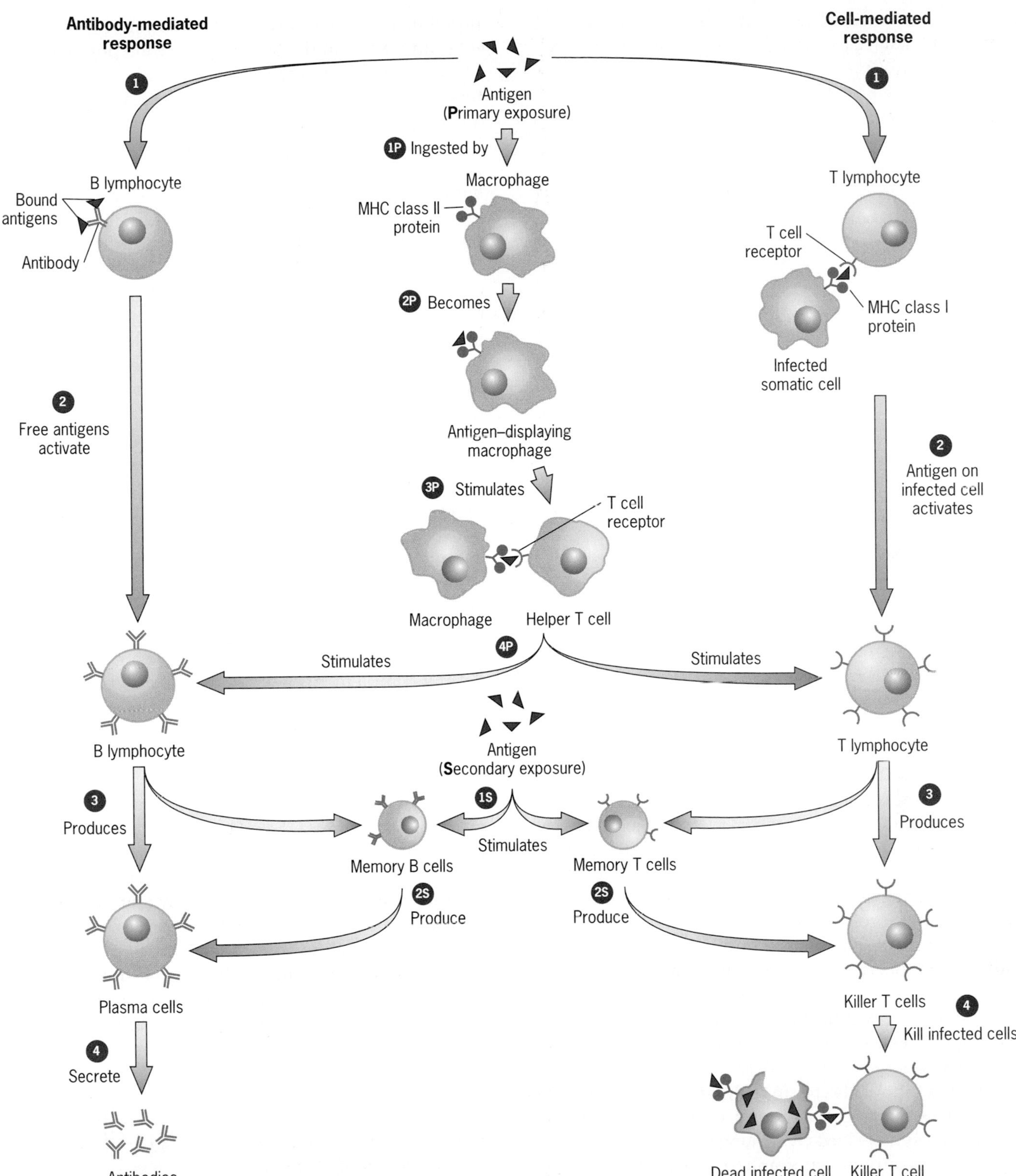

Figure 27.7 Summary of the immune response in mammals. The antibody-mediated (humoral) immune response (left), which is activated by free antigens, is carried out by B lymphocytes, and the cell-mediated (cellular) immune response (right), which is activated by antigens displayed on infected cells, is performed by T lymphocytes. Both pathways are stimulated by activated helper T cells produced in response to the display of antigens by macrophages (center; steps 1P through 4P). The rapid immune responses that occur after a second or subsequent exposure to an antigen are mediated by memory B and T lymphocytes that rapidly produce antibody-producing plasma cells and killer T cells, respectively (center; steps 1S and 2S).

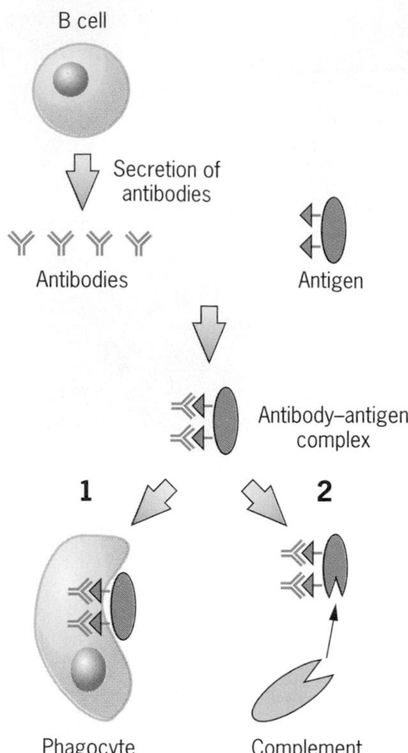

Figure 27.8 Antibody-antigen complexes are eliminated by two major pathways: (1) ingestion and degradation by phagocytes and (2) the proteolytic activity of complement proteins, which degrade the protein coats of viruses and lyse cellular pathogens.

ways (Figure 27.8). (1) Many antibody-antigen complexes are ingested and degraded by phagocytes. (2) IgG and IgM antibodies recruit the complement proteins, which degrade viruses and other proteinaceous substances by proteolysis and kill bacteria and fungi by cell lysis. In other instances, the binding of the antibody to the foreign substance may inactivate it. For example, viruses contain specific surface proteins that are used to attach to and infect host cells. When antibodies bind to these attachment proteins, the viruses can no longer infect host cells. Similarly, toxins and related agents usually are inactivated when agglutinated by antibodies.

As B lymphocytes begin to differentiate, each B cell synthesizes one specific type of antibody. That is, the antibodies synthesized by any one B cell are all identical, with exactly the same antigen-binding specificity. How, then, does an individual produce large amounts of a particular antibody in order to combat an infection by an invading pathogen? Remember that the circulatory system of a mammal contains millions of B cells. Thus the entire population of B cells in an individual will be producing millions of different antibodies. If a B cell is producing an antibody and is exposed to an antigen bound by that antibody, the B cell is stimulated to multiply and produce a population of plasma cells all producing the same antibody. Because this process results in a **clone** of identical plasma cells, all synthesizing a given antibody, it is called **clonal selection.** Because a mature plasma cell can synthesize and secrete 2,000 to 20,000 antibodies per second, an entire clone of such cells can unleash an enormous molecular army to combat an invading pathogen. Once the foreign antigen has been eliminated from the body, the suppressor T cells signal the antibody-producing cells to turn off the production of that particular antibody, at least until a subsequent encounter with the offending antigen.

Some individuals have inherited defects in the antibody-mediated immune response. **X-linked agammaglobulinemia** is among the most severe of these disorders. People with this disease are unable to synthesize antibodies of any kind. B lymphocytes and plasma cells are usually absent in affected individuals. In some cases, immature B lymphocytes are present, but they never develop into functional antibody-producing plasma cells. Because the genetic defect is X-linked, the vast majority of individuals with this form of agammaglobulinemia are hemizygous males. As expected, these individuals are highly susceptible to bacterial and fungal infections. However, they have a normal complement of T cells and exhibit a normal T cell-mediated immune response (see the next section). Thus, they are usually able to combat viral infections effectively.

Key Points: **When presented with antigens by macrophages, helper T cells stimulate B lymphocytes to differentiate into antibody-producing plasma cells. The synthesis of large amounts of antibodies specific for a particular antigen results from clonal selection, the stimulation of a B cell producing an antibody that recognizes the antigen to multiply and produce a population of plasma cells all producing the same antibody. Individuals with X-linked agammaglobulinemia are unable to synthesize antibodies.**

THE T CELL-MEDIATED (CELLULAR) IMMUNE RESPONSE

In addition to the antibody-mediated immune response, mammals possess a second antigen-specific defense mechanism called the **T cell-mediated** or **cellular immune response** (Figure 27.7, *right*). As the name implies, this defense mechanism is mediated directly by cells, specifically by cytotoxic or killer T cells,

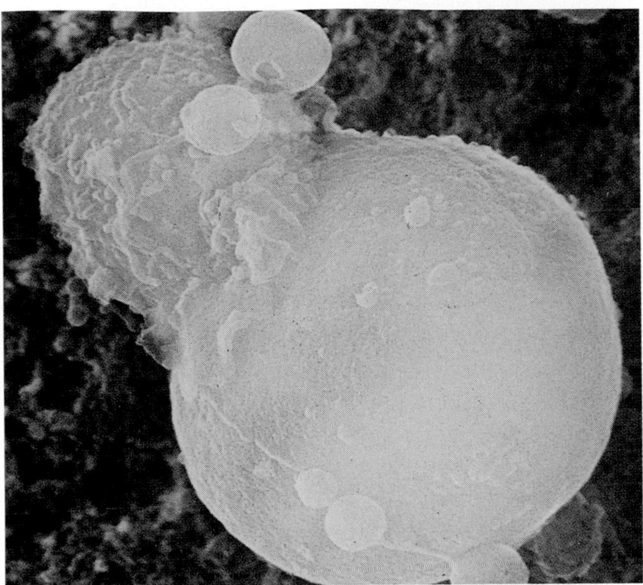

Figure 27.9 Lysis of a target cell displaying a foreign antigen by a killer T cell (upper left).

which attack and kill cells displaying foreign antigens on their surfaces. Killer T cells seek out and destroy cells infected with viruses, bacteria, and other pathogens, as well as some types of cancer cells (Figure 27.9). The killer T cells have antibody-like T cell receptors on their surfaces and use these receptors to seek out infected cells carrying the appropriate antigen. The antigens that are recognized by T cell receptors are usually fragments of partially degraded viral or bacterial proteins; these antigens are displayed on the surfaces of infected cells by proteins encoded by the major histocompatibility complex (see Figure 27.7, *right,* steps 1 and 2).

The T cell-mediated immune response begins just like the antibody-mediated response when a macrophage ingests and partially degrades a foreign agent. Antigens produced by partial degradation of the foreign substance are then displayed on the surface of the macrophage by the major histocompatibility complex proteins. The displayed antigens are recognized by developing killer T cells and by helper T cells. Having been activated by antigen-bearing macrophages, helper T cells stimulate the development of immature B and T lymphocytes by secreting classes of signal molecules collectively called cytokines, lymphokines, and interleukins. Once activated, the immature killer T cells develop into mature killer T cells and memory T cells. As with B lymphocytes, a population of killer T cells produces an enormous number of different antigen-specific T cell receptors. Moreover, effective populations of killer T cells armed with receptors for a given antigen are produced by clonal selection, just as they are for B cells.

Mature killer T cells attach to target cells displaying the appropriate antigens and kill them by a special form of cell lysis. The killer cells secrete a protein called perforin, which inserts itself into the membranes of target cells, producing pores. The cytoplasm of the target cells leaks out through these pores, causing cell death.

Key Points: **When presented with antigens by macrophages, helper T cells stimulate T lymphocytes to differentiate into cytotoxic or killer T cells that recognize and lyse cells displaying these antigens on the cell surfaces.**

IMMUNOLOGICAL MEMORY

When a human is first exposed to a particular foreign antigen, it takes a week to 10 days for the body to produce a significant concentration of antibodies specific to the antigen and two to three weeks to reach maximal antibody production. The first encounter with a foreign antigen results in a relatively slow immune response, called the **primary immune response.** A subsequent encounter with the same antigen produces a more rapid response, called the **secondary immune response.** Not only is a secondary immune response faster, but it also results in a higher concentration of antibodies. Similar slow primary responses and rapid secondary responses are observed for T cell receptors. Thus, the immune system has memory; it can remember previously encountered antigens, which is why vaccinations against pathogens are often effective.

The cellular basis of immunological memory is known, but the molecular mechanisms responsible for this memory remain to be determined. B and T lymphocytes that have not encountered a foreign antigen are referred to as **virgin** or **naive cells.** After exposure to an antigen, the virgin cells differentiate into two types of cells: **memory cells** and so-called **activated B or T cells** (Figure 27.7, step 3). The activated cells will multiply and differentiate into antibody-producing plasma cells or receptor-producing T cells, respectively.

In contrast to the short lifespan (usually days to a week) of plasma cells and killer T cells, memory cells are long-lived, with life spans of months or even years. Moreover, an individual's memory B and T cells are already in an activated state. They can, therefore, develop into antibody-producing plasma cells and receptor-producing killer T cells, respectively, more rapidly than can virgin cells. During a second or subsequent exposure to an antigen, the memory B cells and memory T cells are stimulated to multiply and differentiate into plasma cells and killer T cells. Therefore, secondary immune responses are faster and result in larger populations of plasma cells and killer T cells than primary immune responses.

Key Points: **Long-lived memory cells facilitate a faster immune response during a second encounter with a foreign substance.**

GENOME REARRANGEMENTS DURING B LYMPHOCYTE DIFFERENTIATION

Genetically, the most remarkable aspect of the immune response is the seemingly infinite variety of antibodies that can be synthesized in response to antigens that an animal has not previously encountered. How can the human genome store enough genetic information to encode all these antibodies? The answer to this question is simple if we overlook the molecular details.

The genetic information encoding antibody chains is stored in bits and pieces, and these bits and pieces are put together in the appropriate sequences by genome rearrangements occurring during the development of the antibody-producing cells of the body. Each antibody chain is synthesized using information stored as several different gene segments in germ-line chromosomes. The antibody genes that are expressed in B lymphocytes and plasma cells are spliced together from these gene segments by recombination events that occur during the differentiation of these cells from stem cells in bone marrow. During these genomic rearrangements, the appropriate gene segments are joined by sequence-specific recombination processes that delete the regions of DNA between these gene segments.

Lambda and kappa light chain genes and heavy chain genes are spliced together by very similar mechanisms. The only major difference is that lambda light chain genes, kappa light chain genes, and heavy chain genes are assembled from two, three, and four different gene segments, respectively. However, note that the gene segments encoding lambda, kappa, and heavy chains are located on three different human chromosomes (chromosomes 22, 2, and 14, respectively). Thus, the assembly of human lambda, kappa, and heavy chain genes involves three distinct somatic recombination processes.

Lambda Light Chain Genes Assembled from Two Gene Segments

In germ-line chromosomes, the DNA sequences that encode lambda light chains are present in two distinct types of gene segments (Figure 27.10). One type of gene segment, designated $L_\lambda V_\lambda$ (for *l*eader peptide and *v*ariable region), encodes (1) the amino-terminal hydrophobic leader peptide that is excised from the antibody chain after it directs the nascent polypeptide through the membrane of the rough endoplasmic reticulum and (2) the NH_2-terminal 97 amino acids of the **variable region** of the lambda light chain. The other type of gene segment, called $J_\lambda C_\lambda$ (for *j*oining segment and *c*onstant region), contains the so-called **joining sequence** and the region encoding the COOH-terminal **constant region** of the lambda light chain. The joining sequence encodes the last 13 to 15 amino acids of the variable region of lambda light chains. A complete lambda light chain gene is assembled by joining an $L_\lambda V_\lambda$ gene segment with a $J_\lambda C_\lambda$ gene segment, with the deletion of the intervening DNA, during B lymphocyte differentiation (Figure 27.10). The gene assembly process is controlled by sequence-specific recombination events; the sequences that regulate this process are discussed in a subsequent section of this chapter.

In humans, there are about 300 different $L_\lambda V_\lambda$ gene segments clustered near the centromere on chromosome 22 and nine different $J_\lambda C_\lambda$ gene segments—at least six of which encode functional gene products—located somewhat farther from the centromere on chromosome 22. During the differentiation of B lymphocytes, one $L_\lambda V_\lambda$ gene segment is joined to one $J_\lambda C_\lambda$ gene segment, with the deletion of the intervening DNA (Figure 27.10). Any $L_\lambda V_\lambda$ gene segment may be joined to any $J_\lambda C_\lambda$ gene segment. The newly assembled $L_\lambda V_\lambda J_\lambda C_\lambda$ gene contains two introns, one 93 nucleotide pairs long between the L_λ and V_λ exons, and one about 1200 nucleotide pairs long between the J_λ and C_λ exons. These two introns are spliced out of the primary transcript during RNA processing just like the introns of other genes.

In contrast to the over 300 $L_\lambda V_\lambda$ gene segments in the human genome, the mouse genome contains only two $L_\lambda V_\lambda$ gene segments. Thus, it is not surprising that only 5 percent of the antibodies of mice are of the lambda type, whereas 40 percent of the antibodies of humans have lambda light chains.

Kappa Light Chain Genes Assembled from Three Gene Segments

Synthesis of kappa light chains is controlled by three different gene segments: (1) an $L_K V_K$ gene segment, which encodes the leader peptide and the NH_2-terminal 95 amino acids of the variable region, (2) a J_K gene segment, which encodes the last 13 amino acids of the variable region, and (3) a C_K gene segment, which encodes the COOH-terminal constant region. In humans,

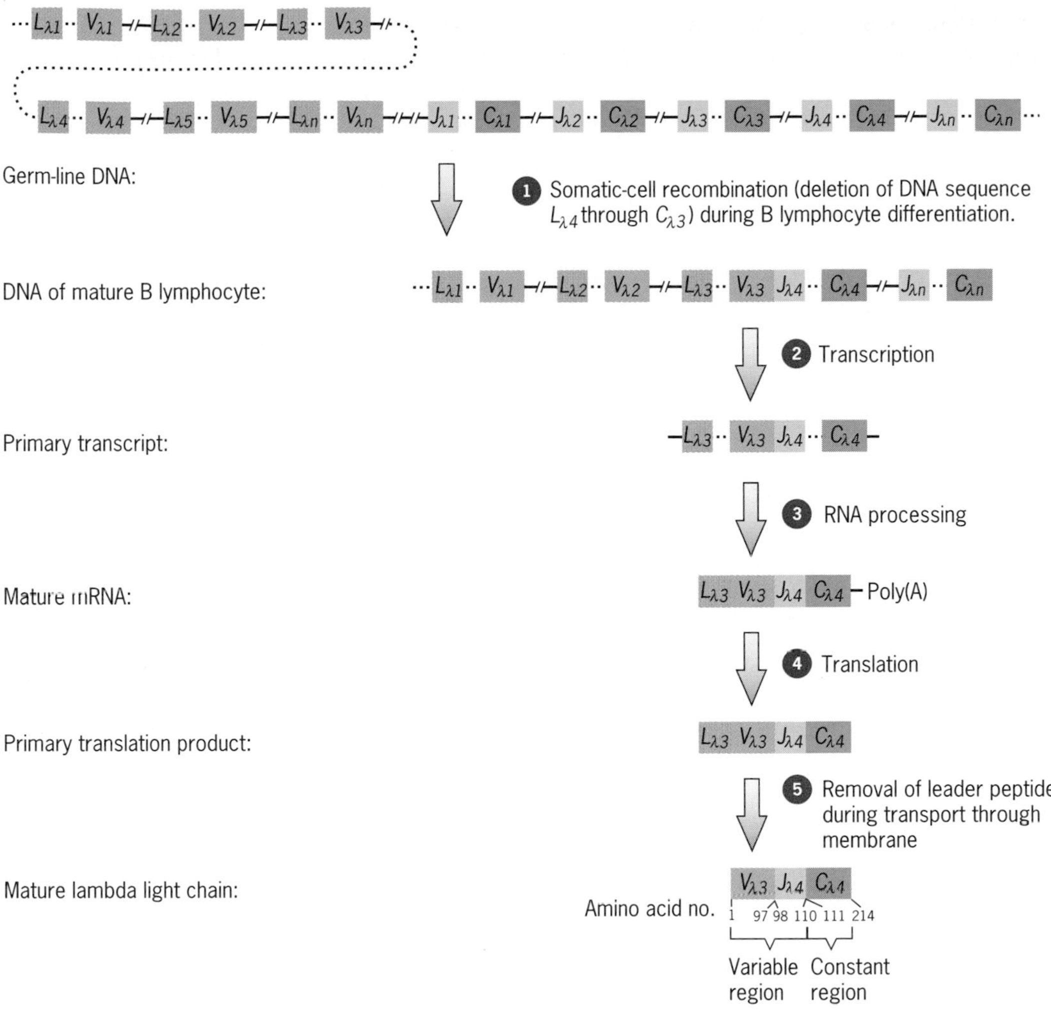

Figure 27.10 The genetic control of human antibody lambda light chains. Each lambda light chain gene is assembled from two pools of gene segments present on germ-line chromosome 22 during B lymphocyte differentiation.

chromosome 2 contains about 300 $L_K V_K$ gene segments, five J_K gene segments, and a single C_K gene segment. The J_K gene segments are located between the $L_K V_K$ gene segments and the C_K gene segment.

In germ-line cells, the five J_K segments are separated from the $L_K V_K$ segments by a long noncoding sequence and from the C_K segment by an approximately 2000 nucleotide-pair-long noncoding sequence. During the development of a B lymphocyte, the particular kappa light chain gene that will be expressed in that cell is assembled from one $L_K V_K$ segment, one J_K segment, and the single C_K segment by somatic recombination. During this process, recombination can join any one of the approximately 300 $L_K V_K$ gene segments with any one of the five J_K segments, with the deletion of all intervening DNA. It produces an $L_K V_K J_K$ fusion that encodes the entire variable region of the kappa light chain. The noncoding sequence between the cluster of J_K gene segments and the C_K gene segment along with any nondeleted C_K-proximal J_K gene segments are retained (between the new $L_K V_K J_K$ fusion and the C_K segment) in the DNA of plasma cells. This entire DNA sequence ($L_K V_K J_K$-noncoding-C_K) is transcribed, and the noncoding sequences are removed during RNA processing, just as are the introns of any other eukaryotic gene. The resulting mRNA is translated, and the leader peptide is excised from the nascent kappa polypeptide during its transport through the membrane of the rough endoplasmic reticulum.

Heavy Chain Genes Assembled from Four Gene Segments

The genetic information encoding antibody heavy chains is organized into $L_H V_H$, J_H, and C_H gene seg-

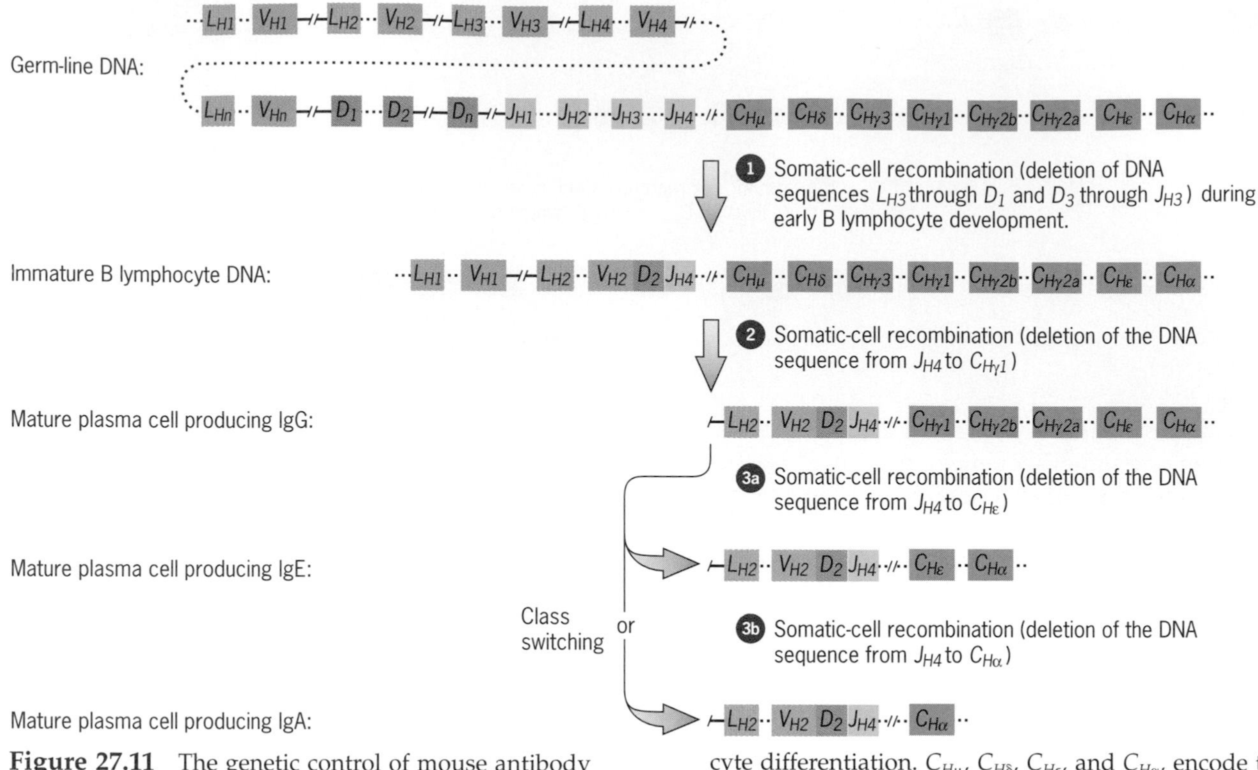

Figure 27.11 The genetic control of mouse antibody heavy chains. Each heavy chain gene is assembled from four different gene segments located on chromosome 12 by somatic recombination events that occur during B lymphocyte differentiation. $C_{H\mu}$, $C_{H\delta}$, $C_{H\epsilon}$, and $C_{H\alpha}$, encode the heavy chain constant regions of IgM, IgD, IgE, and IgA, respectively. $C_{H\gamma3}$, $C_{H\gamma1}$, $C_{H\gamma2b}$, and $C_{H\gamma2a}$ encode heavy chain constant regions of four closely related forms of IgG.

ments analogous to those for kappa light chains; but an additional gene segment, called ***D*** for *d*iversity, encodes 2 to 13 amino acids of the variable region. The variable region of the heavy chain is thus specified by three separate gene segments, L_HV_H, J_H, and D, which must be joined during B-lymphocyte development. In addition, there are one to four C_H gene segments for each Ig class.

In the mouse, there are a total of eight C_H gene segments, all functional, arranged on chromosome 12 in the sequence $C_{H\mu}$, $C_{H\delta}$, $C_{H\gamma3}$, $C_{H\gamma1}$, $C_{H\gamma2b}$, $C_{H\gamma2a}$, $C_{H\epsilon}$, $C_{H\alpha}$, (Figure 27.11). $C_{H\mu}$, $C_{H\delta}$, $C_{H\epsilon}$ and $C_{H\alpha}$ encode the heavy chain constant regions of IgM, IgD, IgE, and IgA, respectively. Four gene segments, $C_{H\gamma3}$, $C_{H\gamma1}$, $C_{H\gamma2b}$, and $C_{H\gamma2a}$, encode IgG heavy chain constant regions. In humans, there are 9 or 10 functional C_H gene segments: $C_{H\mu}$, $C_{H\delta}$, $C_{H\gamma1}$, $C_{H\gamma2}$, $C_{H\gamma3}$, $C_{H\gamma4}$, $C_{H\epsilon1}$, probably $C_{H\epsilon2}$, $C_{H\alpha1}$, and $C_{H\alpha2}$. The human C_H gene cluster also contains two pseudogenes—sequences that are very similar to genes, but that contain in-frame stop codons and are thus nonfunctional.

In mouse germ-line cells, there are about 300 L_HV_H gene segments, from 10 to 50 D gene segments, 4 J_H gene segments, and 8 C_H gene segments, arranged on chromosome 12 in the order given above (Figure 27.11). During B lymphocyte differentiation, somatic recombination events join one L_HV_H gene segment with one D gene segment and one J_H gene segment, deleting the two intervening sequences of DNA, to form one continuous DNA sequence, $L_HV_HDJ_H$, which encodes the entire heavy chain variable region (Figures 27.11, step 1).

Signal Sequences Control Somatic Recombination Events

How are the genome rearrangements that occur during B lymphocyte development regulated? What controls the somatic recombination events such that a *V* gene segment is joined to a *J* segment and not to another *V* segment or directly to a *C* segment? Several long segments of chromosomal DNA carrying clusters of *V* gene segments, *D* gene segments, and *J* gene segments of both mice and humans have been sequenced, and the resulting nucleotide-pair sequences suggest the presence of specific *V-J*, *V-D*, and *D-J* joining signals. The same signal sequences are found adjacent to all *V* gene segments. Similarly, all *J* gene segments have identical signal sequences located adjacent to their coding sequences; however, their signal sequence is different from that adjacent to *V* gene segments. *D* and *C* gene segments also have their own signal sequences.

The signal sequences controlling *V-J*, *V-D*, and *D-J* joining contain 7-base-pair-long (heptamer) and 9-base-pair-long (nonamer) sequences separated by spacers of different, but specific, lengths. For L_kV_k-J_k joining, the spacer in the L_kV_K signal sequence is 12 nucleotide pairs long, whereas that in the J_K signal sequence is 23 nucleotide pairs long. The heptamer and nonamer sequences located 3′ to or "after" the L_kV_k gene segments are complementary (with the exception of one base pair) to those 5′ to or "preceding" the J_k gene segments. These signal sequences have the potential to form "stem-and-loop" structures (Figure 27.12), thus bringing the L_kV_k and J_K gene segments into juxtaposition for joining. Apparently, joining will occur only when one signal sequence contains a 12-base-pair spacer and the other contains a 23-base-pair spacer. This requirement presumably is enforced by the RAG1 (*r*ecombination-*a*ctivating *g*ene *1*) and RAG2 proteins that mediate the DNA splicing processes. Very similar signal sequences appear to control L_HV_H-*D* and *D*-J_H joining.

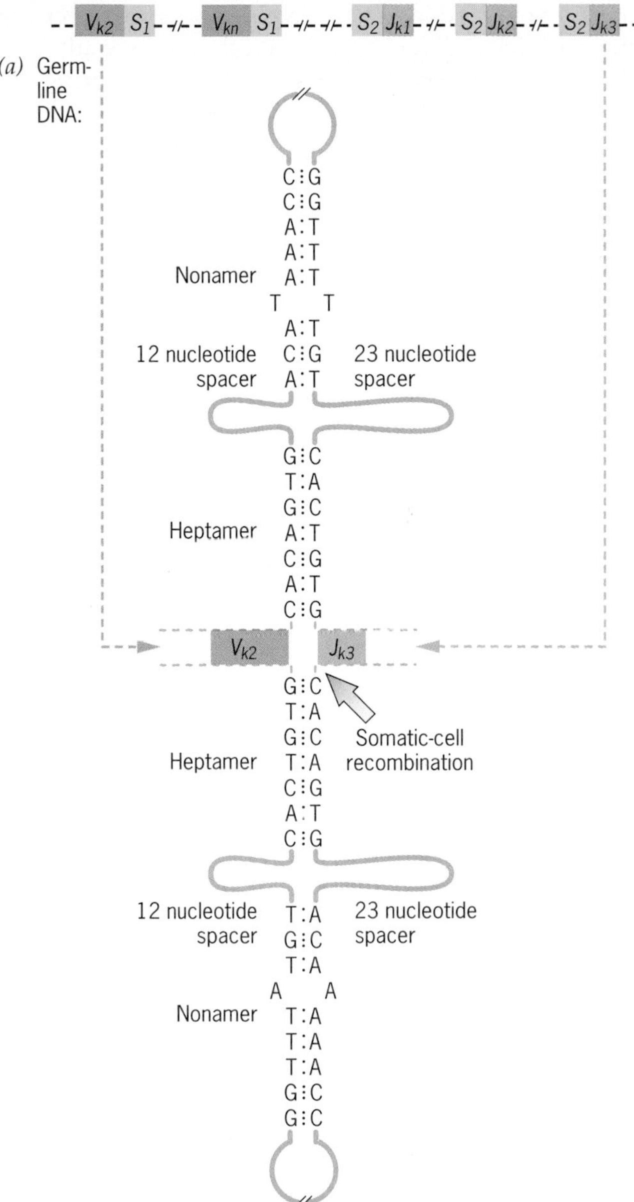

Figure 27.12 Signal sequences direct V_K–J_k joining during the assembly of kappa light chain genes. The germ-line and plasma cell DNA arrangements are shown in *(a)* and *(c)*, respectively. The heptameric and nonameric signal sequences and the spacer regions that control the DNA splicing events are shown in *(b)*. The "stem-and-loop" structure shown in *(b)* is a hypothetical intermediate.

Additional Diversity: Variable Joining Sites and Somatic Hypermutation

A vast amount of antibody diversity is generated by the combinatorial joining of antibody gene segments as previously described. For example, consider the number of different kappa light chains possible in humans. If there are 300 L_KV_K gene segments and 5 J_k segments, then 1500 fused $L_kV_kJ_k$ gene segments can be produced by joining the L_kV_k and J_k gene segments in all possible combinations. Similarly, given 300 $L_\lambda V_\lambda$ and 6 $J_\lambda C_\lambda$ gene segments, 1800 different lambda light chains can be produced by combinatorial joining. The heavy chain variable region provides even greater diversity because of the multiple *D* gene segments. If there are 300 V_H gene segments, 25 *D* gene segments, and 6 J_H gene segments in human germ-line cells, 45,000 different heavy chain variable regions can be assembled. If there are 3300 (1500 + 1800) different light chains and 45,000 different heavy chains, then 3300 × 45,000 or 148,500,000 different antibodies can be produced by combinatorial joining alone. Clearly, the combinatorial joining of antibody gene segments produces a vast amount of antigen-binding diversity. However, molecular analyses of antibodies and antibody genes have shown that additional diversity is generated by other mechanisms.

A comparison of the diversity of amino acid sequences present in antibody molecules with that predicted from the sequences of gene segments that encode these antibodies reveals that there is more variation in amino acid sequences at the *V-J* junctions than is predicted by the nucleotide sequences. Some of this additional diversity has been shown to result from the use of alternate sites of recombination during the joining events that are involved in the assembly of mature antibody genes. An example of the use of alternate sites of joining of V_K and J_K gene segments in the mouse is illustrated in Figure 27.13. During the joining of gene segments V_{K41} and J_5, recombination events have been shown to occur between four adjacent nucleotide positions at the junction sites. As shown in Figure 27.13*d*, these recombination events produce four different nucleotide sequences that encode three distinct amino acids at position 96 in the mouse kappa light chain. Since amino acid 96 occurs in a region of the antibody chain that is involved in antigen binding, alternate *V-J* joining events of this type contribute significantly to the great diversity of antibody specificity that is observed in mammals. Similar alternate joining events have been documented for V_λ-J_λ and V_H-D-J_H joining reactions.

Despite the vast array of antibody diversity produced by (1) the joining of large families of *V*, *D*, and *J* gene segments, and (2) the use of alternate positions of recombination during the joining reactions, considerable data demonstrate that still another mechanism must be involved in the generation of antibody diversity. This has been established by comparing (1) the nucleotide-pair sequences of expressed genes with the sequences of germ-line gene segments and (2) the actual amino acid sequences of antibody chains with the amino acid sequences predicted from the nucleotide sequences of the genes. When the actual amino acid sequences of different mouse λ_1 chains were compared with the amino acid sequences predicted from the nucleotide-pair sequences of λ_1 gene segments, differences were found at sites in the variable regions other than the junction sites. Similar observations have been made in studies of heavy chain variable regions. In essentially all cases, the changes have resulted from single nucleotide-pair substitutions. Such substitutions may represent up to 2 percent of the nucleotide pairs of the gene segments encoding the variable regions of antibodies. These nucleotide-pair substitutions must occur by some mechanism of **somatic mutation** that is restricted to the DNA sequences encoding the variable regions of antibody chains. Because these changes in the variable segments of antibody genes occur at such a high frequency, the process by which they occur is called **somatic hypermutation.** The mechanism by which somatic hypermutation occurs is unknown.

Somatic hypermutation of regions of antibody genes that encode antigen-binding sites may be of great value to the organism. Without this mechanism for generating antibody diversity, the range of available antibody specificity would be fixed in terms of the sequences present in the genome at birth and the combinations that could be produced by the various levels of gene-segment joining reactions. Viruses and other pathogens are constantly evolving and producing new variants with new antigenic determinants. To provide an adequate defense against the changing antigenic composition of these viruses and other components of the environment, the immune system must also be capable of responding rapidly to these changes. What better way to provide this safeguard than to endow antibody genes with their own mechanism—so-

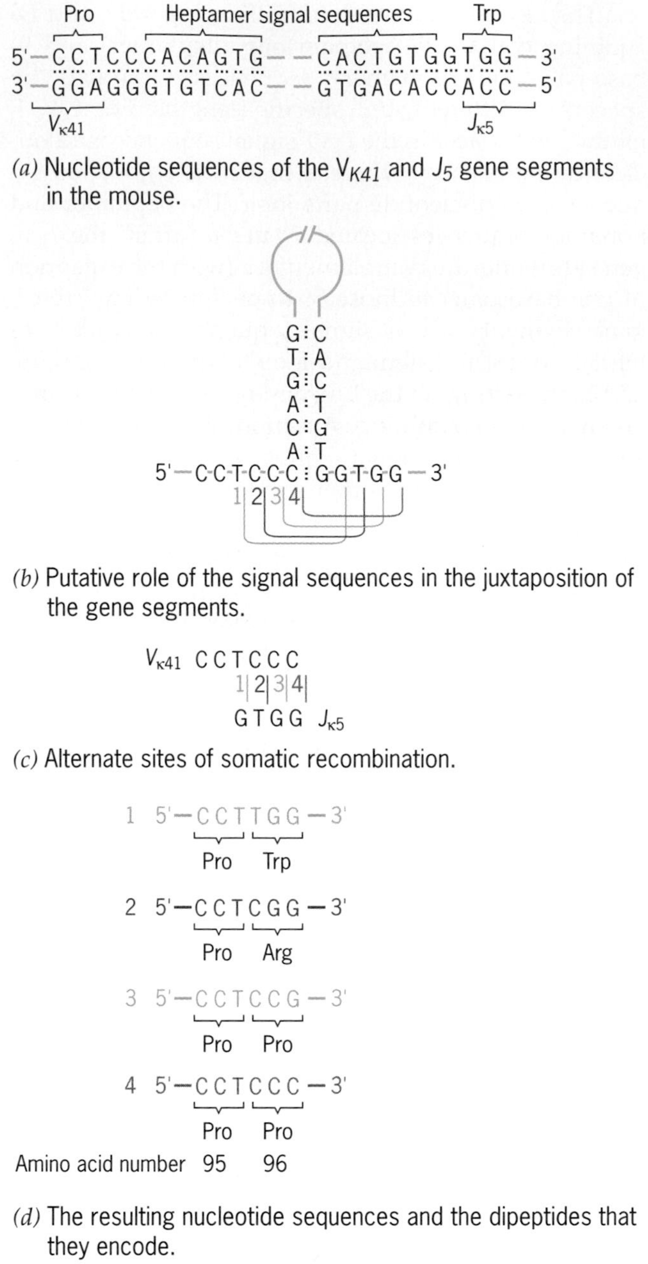

Pro Heptamer signal sequences Trp
5'—CCTCCCACAGTG— —CACTGTGGTGG—3'
3'—GGAGGGTGTCAC— —GTGACACCACC—5'
$V_{\kappa 41}$ $J_{\kappa 5}$

(*a*) Nucleotide sequences of the V_{K41} and J_5 gene segments in the mouse.

G:C
T:A
G:C
A:T
C:G
A:T
5'—CCTCCC:GGTGG—3'
1 2 3 4

(*b*) Putative role of the signal sequences in the juxtaposition of the gene segments.

$V_{\kappa 41}$ CCTCCC
1 2 3 4
GTGG $J_{\kappa 5}$

(*c*) Alternate sites of somatic recombination.

1 5'—CCTTGG—3'
Pro Trp

2 5'—CCTCGG—3'
Pro Arg

3 5'—CCTCCG—3'
Pro Pro

4 5'—CCTCCC—3'
Pro Pro

Amino acid number 95 96

(*d*) The resulting nucleotide sequences and the dipeptides that they encode.

Figure 27.13 Antibody diversity at the V_k–J_k junction is produced by variation in the exact position of the joining reaction.

matic hypermutation—for rapid adaptation to new antigens that might evolve in the future.

Key Points: **The enormous variety of antibodies produced by mammals results from the combinatorial joining of antibody gene segments by somatic recombination during the differentiation of B lymphocytes into antibody-producing plasma cells. Signal sequences flanking antibody gene segments control these somatic recombination events. Additional antibody diversity is produced by variability in the sites of gene-segment joining and by a high frequency of mutation—somatic hypermutation—in the DNA sequences that encode antigen-binding sites.**

ANTIBODY CLASS SWITCHING

At the time that antibody synthesis begins in a developing B lymphocyte, all the C_H gene segments are still present, separated from the newly formed $L_HV_HDJ_H$ gene fusion by a short noncoding sequence (Figure 27.11). At this stage, all antibodies synthesized have IgM heavy chains ($C_{H\mu}$ gene products). However, some plasma cells may soon begin to produce other classes of antibodies (with IgA, IgD, IgE, or IgG heavy chains). This process is called **class switching.** Remember, as we discussed earlier in this chapter, different classes of antibodies perform different functions in the body.

The first type of class switching occurs when certain plasma cells begin to produce both IgM and IgD antibodies. These antibodies differ only in their effector function domains; they have identical antigen-binding domains, specified by the same fused V_kJ_k or $V_\lambda J_\lambda$ and V_HDJ_H gene segments. In these cells, a primary transcript that extends through both the $C_{H\mu}$ and $C_{H\delta}$ gene segments is synthesized. During processing, the V_HDJ_H transcript sequence may be spliced to either the $C_{H\mu}$ sequence or the $C_{H\delta}$ sequence. As a result, both types of heavy chain are synthesized in the same cell (Figure 27.14).

Other B cells will switch from producing antibodies of class IgM to producing IgA, IgD, IgE, or IgG antibodies. This type of class switching involves further genome rearrangements during which the C_H gene segments closest to the previously joined $L_HV_HDJ_H$ gene segments are deleted (Figure 27.11, step 3). The class of antibodies produced after class switching depends on which C_H coding region remains juxtaposed to the previously fused $L_HV_HDJ_H$ coding sequence.

A further complexity observed in the antibody-mediated immune response is the sequential production of membrane-bound and secreted forms of a given antibody. The first antibodies to appear in developing B lymphocytes are membrane-bound IgM molecules. Subsequently, these cells switch to the production of a secreted form of IgM. These two forms of IgM differ only in the COOH-terminal portions of their heavy chains. The heavy chain of the membrane-bound form is 21 amino acids longer than that of the secreted form. The membrane-bound heavy chain has a 41 amino acid-long hydrophobic sequence at the carboxyl terminus that anchors it to the cell surface. This hydrophobic sequence is replaced by a 20-amino-acid-long hydrophilic sequence in the secreted form. The distinct heavy chains of the membrane-bound and secreted forms of IgM are produced by alternate pathways of transcript splicing.

Key Points: **Developing B lymphocytes can switch from the production of IgM antibodies to other classes of antibodies by genome rearrangements or by alternate pathways of transcript splicing.**

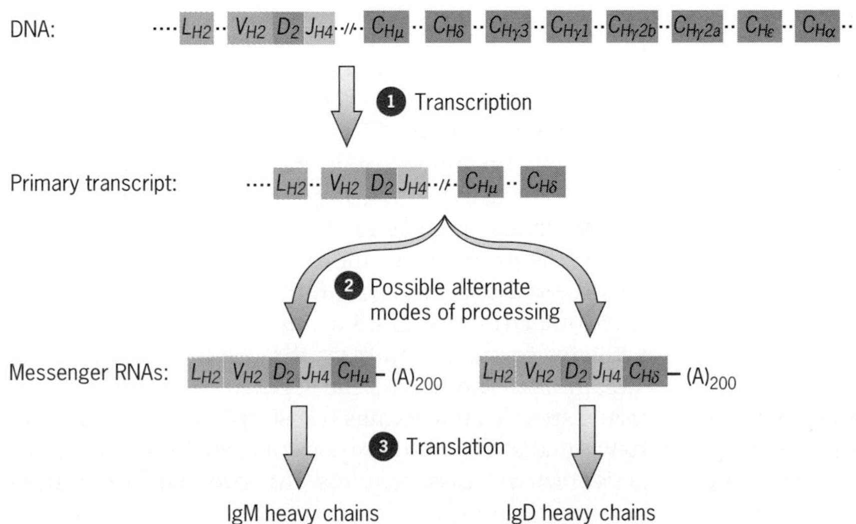

Figure 27.14 Antibody class switching may occur by alternate pathways of transcript splicing. The two splicing pathways shown result in the production of both IgM and IgD heavy chains by the same plasma cell.

ASSEMBLY OF T CELL RECEPTOR GENES BY SOMATIC RECOMBINATION

Like antibody heavy chains, the two polypeptide chains of T cell receptors are encoded by *L-V, D, J,* and *C* gene segments. The variable regions of the T cell receptor proteins are encoded by multiple *L-V, D,* and *J* gene segments; the constant regions are encoded by a small number of *C* gene segments. The T cell receptor genes are assembled by genomic rearrangements that occur during the differentiation of T lymphocytes from stem cells just as in the case of antibody genes in developing B lymphocytes. The α and β receptor proteins are encoded by gene segments that are arranged in clusters much like those of gene segments encoding antibody chains. In humans, the α and β gene-segment clusters are located on chromosomes 14 and 7, respectively.

Key Points: **T cell receptor genes are assembled from gene segments during T lymphocyte differentiation by genome rearrangements analogous to those involved in antibody production.**

REGULATION OF IMMUNOGLOBIN GENE EXPRESSION

Antibody gene segments are not transcribed or are transcribed at very low levels in germ-line cells. In plasma cells, 10 to 20 percent of the mRNA molecules are antibody gene transcripts. What activates the transcription of antibody genes during or immediately after their assembly? Why is only one type of antibody produced by a given plasma cell? What prevents the synthesis of more than one type of antibody by a plasma cell?

Allelic Exclusion: Only One Functional Rearrangement per Cell

Each plasma cell produces only one type of antibody. Why? Mammalian cells are diploid; they carry two sets of genetic information encoding each of the antibody chains. However, only one productive genome rearrangement of light chain coding sequences and one productive genome rearrangement of heavy chain coding sequences occur during the differentiation of each B lymphocyte. This phenomenon is called **allelic exclusion** because one of the "alleles" is excluded from being expressed. How? Why? At present, geneticists don't know. They do know that both "alleles" are sometimes rearranged in the same cell, but, in these cases, only one "allele" has undergone a productive rearrangement to a functional gene. The other "allele" is always only partially assembled or contains deletions of essential coding sequences.

Clearly, there must be some type of feedback mechanism that arrests the antibody gene-assembly process after one functional gene has been constructed and the cell has started to synthesize antibodies. The simplest mechanism would involve inhibition of the assembly process by the antibody itself. However, further work will be required to establish the mechanism by which allelic exclusion is enforced. It is known that the allelic exclusion mechanism works independently for the heavy chain gene segments and the light chain gene segments. Only one heavy chain and one light chain are synthesized in a given plasma cell. Regardless of the mechanism, allelic exclusion, which assures the production of a single type of antibody by each plasma cell, has one invaluable practical consequence: it permits researchers to obtain large amounts of an antibody that binds a single epitope, a so-called **monoclonal antibody** (Technical Sidelight: Monoclonal Antibodies: Powerful Probes for Detecting Proteins and Other Antigens), without using laborious protein purification methodologies.

Heavy Chain Gene Transcription: A Tissue-Specific Enhancer

By what mechanism are antibody genes activated during their assembly from gene segments? In the case of heavy chain genes, the answer is that the assembly process brings the promoters located upstream of L_HV_H gene segments into the range of influence of a strong enhancer element located in the intron between the J_H gene segments and the $C_{H\mu}$ gene segment (Figure 27.15). Each L_HV_H gene segment contains an upstream promoter. Prior to the genomic rearrangement events that lead to heavy chain synthesis, the closest L_HV_H promoter is over 100,000 nucleotide pairs away from the enhancer (Figure 27.15, *top*), and enhancers have little effect on transcription initiated at promoters located that far away. During B cell differentiation, the heavy chain gene-assembly process (Figure 27.11) brings the promoter of the closest L_HV_H gene segment and the enhancer to within 2000 nucleotide pairs of each other (Figure 27.15, *bottom*). The enhancer now activates transcription from the promoter located upstream from the L_HV_H gene segment. This enhancer is tissue specific; it activates transcription only in B cells, having no effect in other types of cells. Presumably, the activation process requires the presence of a tran-

TECHNICAL SIDELIGHT

Monoclonal Antibodies: Powerful Probes for Detecting Proteins and Other Antigens

In addition to defending the body against viruses, bacteria, fungi, and other foreign substances, antibodies provide scientists with powerful research tools that can be used to identify specific proteins or other antigenic molecules. For many years, the problem with using antibodies as molecular probes was the difficult task of purifying those specific to the antigen of interest. In 1975, Cesar Milstein and Georges Köhler developed a procedure that solved this problem. They discovered how to produce immortal, antibody-secreting cell culture clones. They fused B lymphocytes with cancer cells called myelomas to produce hybrid somatic cells. The resulting hybrid cells, called **hybridomas,** were immortal; moreover, they could be cloned and screened for the production of the antibodies of interest. Once identified, the desired hybridoma cell lines could be used to produce essentially unlimited quantities of pure **monoclonal antibodies,** antibodies all recognizing the same antigenic epitope.

The procedure used to produce monoclonal antibodies is shown in Figure 1. The first step is to inject a mouse with the antigen of interest. This will stimulate the immune system of the mouse to begin producing antibodies specific to the antigen. The second step is to remove the spleen from the injected mouse and to place the spleen cells, many of which are B lymphocytes, in cell culture medium. The third step involves fusion of the B cells with rapidly growing myeloma cells to produce hybridoma cells. This is the key step in the production of monoclonal antibodies. Each B cell will produce a single type of antibody due to allelic exclusion. However, one cannot simply clone B cells because they will live for only a few days whether *in vitro* or *in vivo.* Cancer cells, such as the myelomas, are immortal in culture; they divide rapidly and continuously, exhibiting totally uncontrolled growth and division. But myeloma cells do not produce antibodies. Fortunately, the hybrid cells acquire the immortality of the parental myeloma cells and the capacity to produce antibodies from the parental B lymphocytes. Individual hybrid cells are separated and cultured to produce clones of antibody-producing hybridomas. Each clone is assayed for the production of the antibody of interest. Once identified, the hybridomas that are producing the desired antibodies can be grown continuously in culture or stored frozen in liquid nitrogen for later use. Thus, the monoclonal antibody procedure allows one to produce an essentially unlimited quantity of an important antibody and to store the antibody-producing cells for use many years later.

Because of their phenomenal specificity, monoclonal antibodies are invaluable research tools. They can be used to identify individual polypeptides that have been separated by gel electrophoresis (Chapter 20) or to determine the location of macromolecules in tissues and cells by combining microscopy with immunolocalization techniques. Monoclonal antibodies also can be used in the purification of proteins by immunoprecipitation. In recent years, monoclonal antibodies have begun to be used in the diagnosis of inherited and acquired diseases. Although monoclonal antibodies have not provided the hoped-for cure of cancer, they have proven to be invaluable tools in biological research and medical diagnosis.

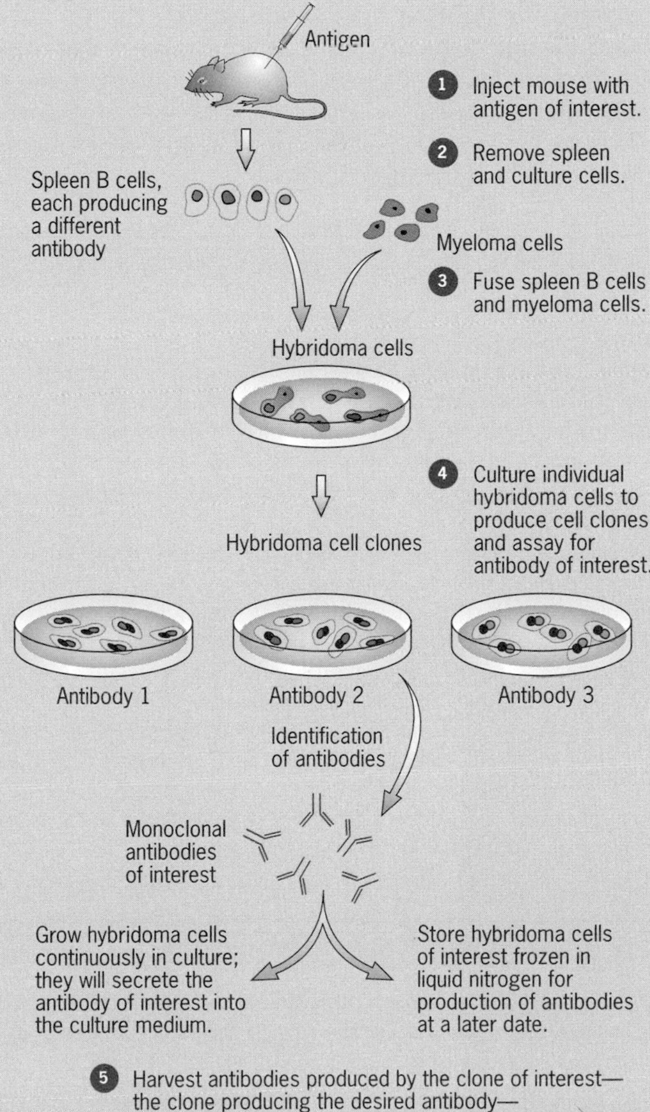

Figure 1 The production of monoclonal antibodies by hybridoma cell clones.

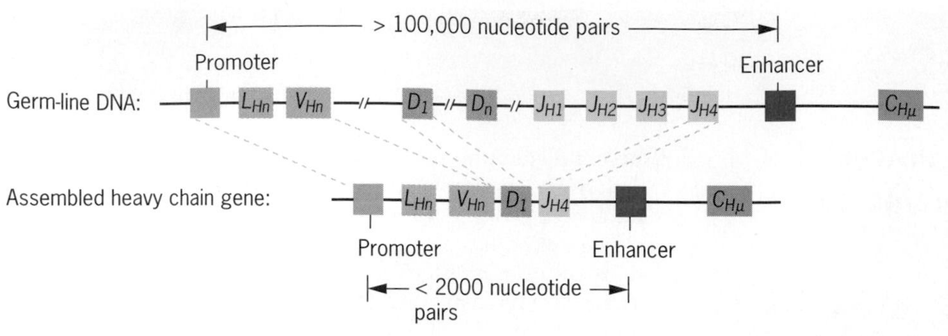

Figure 27.15 Activation of heavy chain gene transcription by movement of a heavy chain gene promoter (orange box) into proximity with a tissue-specific enhancer (red box) during gene assembly in B lymphocytes.

scription factor that is present in B lymphocytes but not in other types of cells.

A similar enhancer element is present in the intron between the light chain J_K gene segment cluster and the C_k coding sequence. Thus, it seems likely that the movement of antibody gene promoters into the range of influence of tissue-specific enhancers may be a general mechanism of activation of antibody genes during B lymphocyte differentiation.

Key Points: **Antibody gene segments are not transcribed at significant levels in stem cells. Their transcription is activated by the genome rearrangements that occur during the differentiation of B lymphocytes. Only one productive rearrangement occurs per cell. Transcription of assembled heavy chain genes is activated by transfer of their promoters to positions where they are controlled by a strong tissue-specific enhancer.**

TESTING YOUR KNOWLEDGE

1. If the genome of a particular primate contains 300 V_k, 5 J_k, 1 C_k, 150 V_λ, 5 J_λ–C_λ, 300 V_H, 4 J_H, 50 D, and 10 C_H gene segments, and these can be joined in all possible appropriate combinations during the genome arrangements that occur in the differentiation of B lymphocytes, how many different antibodies could this primate produce?

ANSWER

Over a billion different antibodies (1.35 billion) could be assembled by combinatorial joining of the various gene segments and the different light and heavy antibody chains. There would be 2250 different light chains, consisting of 1500 kappa light chains (300 V_k gene segments × 5 J_k segments × 1 C_k segment) and 750 lambda light chains (150 V_λ gene segments × 5 J_λ–C_λ segments), and 600,000 heavy chains (300 V_H gene segments × 4 J_H segments × 50 D segments × 10 C_H gene segments). The number of combinations of light and heavy chains would be 2250 light chains × 600,000 heavy chains, or 1,350,000,000 different antibodies.

2. In humans, the absence of an enzyme called purine nucleoside phosphorylase (PNP) results in a severe deficiency of T cells. Deoxyguanosine, the PNP substrate, accumulates in the absence of the enzyme and is especially toxic to T lymphocytes. What effect would you expect this PNP deficiency to have on the cellular and humoral immune systems of individuals with the disorder?

ANSWER

Helper T cells are required to stimulate the development of both antibody-producing plasma cells and killer T cells (see Figure 27.7). Given that deoxyguanosine—the substrate of purine nucleoside phosphorylase—is toxic to T lymphocytes, the absence of PNP activity would be expected to block both the humoral and cellular immune responses in an individual with this disorder.

QUESTIONS AND PROBLEMS

27.1 Is the genetic information specifying antibody chains stored in germ-line cells in the same format as that specifying most other polypeptides?

27.2 How many polypeptide chains are present in each antibody molecule? How many antigen-binding sites are present per antibody? How many different antibodies are produced in each plasma cell?

27.3 What are three different sources of antibody variability?

27.4 Antibodies are of five different classes. What determines the class of an antibody?

27.5 Which class of antibody is always produced first in a developing B cell?

27.6 Which class of immunoglobin is the most common in human blood?

27.7 Which class of immunoglobin binds to receptors on mast cells and stimulates the secretion of histamines and other amines that are involved in an allergic reaction?

27.8 Considering the size of the mammalian genome, why is it not possible to have a separate gene coding for each antibody polypeptide?

27.9 Does class switching during B lymphocyte differentiation occur at the DNA level or the RNA level? By what mechanism does it occur?

27.10 (a) Name three different types of white blood cells that play important roles in the immune response of vertebrates. (b) What function(s) does each of these types of cells perform in the immune response?

27.11 Of what importance is the *MHC* locus of humans in medical practice?

27.12 The genes of the *MHC* locus of humans are said to be highly polymorphic. (a) What does this mean? (b) Of what significance is this in the field of medicine?

27.13 What are the two major pathways by which antibody-antigen complexes are eliminated?

27.14 What is an epitope, and what is its role in triggering an immune response?

27.15 Where are molecules such as cytokines, lymphokines, and interleukens produced, and what is their role in the immune response?

27.16 Plasma cells make up a population of cells that collectively produce a vast array of different antibodies. Yet, each plasma cell produces only a single type of antibody (sometimes slightly different antibodies, but all specific for a single antigen). After exposure of a human or other vertebrate to a foreign antigen such as the coat protein of a virus, the individual rapidly initiates the production of large quantities of antibodies specific to the foreign antigen to which she or he was exposed. By what mechanism does this response occur?

27.17 Antibodies are encoded by families of gene segments that are assembled into the genes that direct their synthesis during the development of the plasma cells that produce them. Are any other vertebrate proteins encoded by gene segments that are assembled in a similar manner during cell differentiation?

27.18 In the differentiation of B lymphocytes, what controls somatic recombination events such that a *V* gene segment is joined to a *J* segment and not another *V* segment or directly to a *C* segment?

27.19 In what ways are the structures of antibodies and T cell antigen receptors similar?

27.20 Why are secondary immune responses faster and stronger than primary immune responses?

27.21 Microbial infections of mammals trigger several different responses, some specific to the invading agent, others nonspecific. What are the major nonspecific immune responses? What are the major specific immune responses?

27.22 In what way do memory cells differ from most other cells of the immune system?

27.23 Monoclonal antibodies have proven to be very powerful tools for studies of biological processes. (a) How are monoclonal antibodies produced? (b) How are they used in biological research?

27.24 How many different antibody heavy chains are usually produced by a given plasma cell? What role does allelic exclusion play in antibody production by plasma cells?

27.25 Helper T cells are required to mount both antibody-mediated and T cell-mediated immune responses. Why?

27.26 What are the roles of phagocytes and complement proteins, respectively, in the immune responses of mammals? In what ways are the activities of phagocytes and complement proteins dependent on the production of antibodies?

27.27 What are autoimmune diseases, and why are they so-named?

27.28 If the genome of a particular mammal contains 200 V_k, 4 J_k, 1 C_k, 150 V_λ, 6 J_λ–C_λ, 200 V_H, 4 J_H, 100 D, and 8 C_H gene segments, and these can be joined in all possible appropriate combinations during the genome rearrangements that occur in the differentiation of B lymphocytes, how many different antibodies could this mammal produce?

27.29 Propose a mechanism by which an autoimmune disease such as juvenile diabetes occurs?

BIBLIOGRAPHY

AGRAWAL, A., Q. M. EASTMAN, AND D. G. SCHATZ. 1998. Transposition mediated by RAG1 and RAG2 and its implications for the evolution of the immune system. *Nature* 394:744–751.

ALBERTS, B., D. BRAY, J. LEWIS, M. RAFF, K. ROBERTS, AND J. D. WATSON. 1994. *Molecular biology of the cell.* 3rd ed. (Chapter 23, "The immune system"). Garland Publishing, New York.

CHEN, J., AND F. ALT. 1993. Gene rearrangements and B-cell development. *Curr. Opin. Immunol.* 5:194–200.

CRABTREE, G. R., AND N. A. CLIPSTONE. 1994. Signal transmission between the plasma membrane and the nucleus of T lymphocytes. *Annu. Rev. Biochem.* 63:1045–1083.

GOLDE, D. 1991. The stem cell. *Sci. Amer.* 265(12):86–93.

LEDER, P. 1982. The genetics of antibody diversity. *Sci. Amer.* 246(5):102–115.

LEVY, J.A. 1998. *HIV and the Pathogenesis of AIDS.* 2nd ed. ASM Press, Washington, DC.

LEWIS, R. 1999. *Human Genetics, Concepts and Applications.* 3rd ed. WCB/McGraw-Hill, Boston.

LIEBER, M. R. 1992. The mechanism of V(D)J recombination: A balance of diversity, specificity, and stability. *Cell* 70:873–876.

TONEGAWA, S. 1985. The molecules of the immune system. *Sci. Amer.* 253(10):122–131.

YOUNG, J. D-E., AND Z. A. COHN. 1988. How killer cells kill. *Sci. Amer.* 258(1):38–44.

28

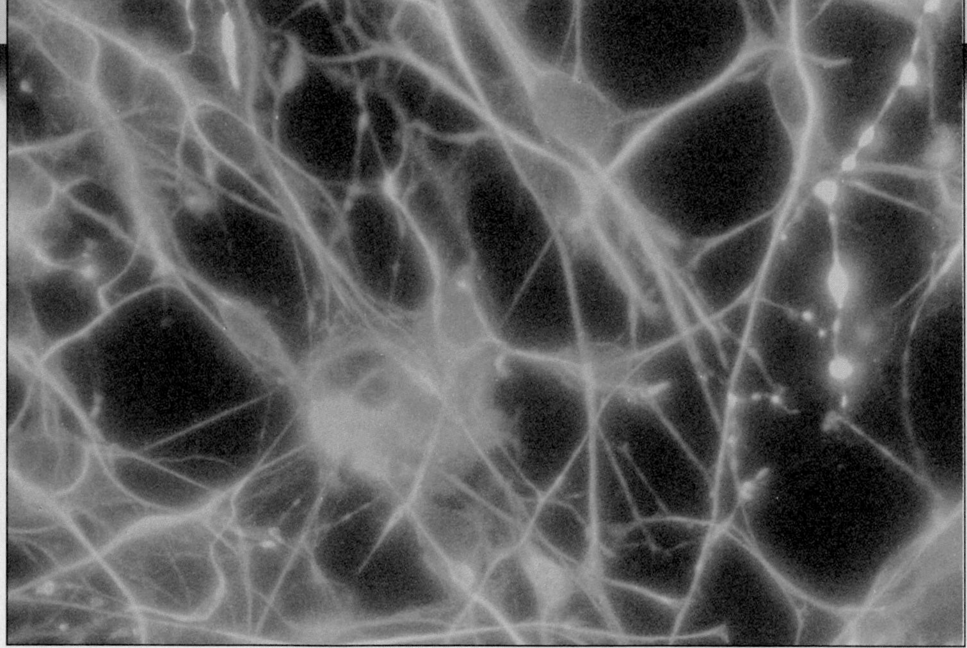

Fluorescense light micrograph of a culture of human brain tumor cells. The cells are interconnected by fibrous axons and dendrites.

The Genetic Basis of Cancer

CHAPTER OUTLINE

Cancer: A Genetic Disease
Oncogenes
Tumor Suppressor Genes
Genetic Pathways to Cancer

A Common Killer

Lawrence Taylor, a biology professor at a large American university, had trouble finishing one of his lectures. His thoughts were labored and unfocused, and he felt tired and disoriented. After dismissing his class, Lawrence sought medical attention. The preliminary diagnosis was that Lawrence's mental and motor abilities had been impaired by a slight stroke, the result of a rupture in a tiny blood vessel in the brain. However, further examination revealed a more serious problem. Lawrence's brain had developed a tumor, probably derived from another tumor located on one of his kidneys. These two masses of cancerous tissue had grown to the point where they were beginning to interfere with important biological functions. Lawrence had surgery to remove the kidney tumor, and he was given radiation therapy to kill the brain tumor. Unfortunately, the growth of the brain tumor could not be stopped, and 18 months after teaching his last class, Lawrence died.

Cancerous tumors kill several hundred thousand Americans every year. What causes tumors to form, and what causes some of them to spread? Why do some types of tumors tend to be found in families? Is the tendency to develop cancer inherited? Do environmental factors contribute to the development of cancer? In recent years, these and other questions have stimulated an enormous amount of research on the basic biology of cancer. Although many details are still unclear, the fundamental finding is that cancers result from genetic malfunctions. In some instances, these malfunctions may be triggered or exacerbated by environmental factors such as diet, excessive exposure to sunlight, or chemical pollutants. Cancers arise when critical genes are mutated. These mutations can cause biochemical processes to go awry and lead to the unregulated proliferation of cells. In this chapter we consider the nature of cancer and the genetic mechanisms that cause it.

CANCER: A GENETIC DISEASE

Cancer is a collection of diseases in which cell growth and division are unregulated. Without regulation, cells divide ceaselessly, piling up on top of each other to form tumors. When cells detach from a tumor and invade the surrounding tissues, the tumor is malignant. When the cells do not invade the surrounding tissues, the tumor is benign. Malignant tumors may spread to other locations in the body, forming secondary tumors. This process is called **metastasis,** from Greek words meaning to "change state." In both benign and malignant tumors, something has gone wrong with the systems that control cell division. Two decades of research have now firmly established that this loss of control is due to underlying genetic defects.

The Many Forms of Cancer

In the human population, cancer is responsible for no small amount of misery. Approximately one-third of all people in the United States will eventually die of some form of cancer. Cancer is not a single disease, but rather a group of diseases characterized by uncontrolled cell growth. Cancers can orginate in many different tissues of the body. Some grow aggressively, others more slowly. Figure 28.1 shows the frequencies of new cases of different types of cancer in the United States, as well as the number of fatalities attributed to each type. The most common type of cancer involves the prostate gland in males. Breast cancer in females and lung cancer in both sexes are also common. In some cases, a predisposition to develop cancer is inherited. However, most cases of cancer are due to the accumulation of spontaneous mutations in the somatic tissues.

Although the death rate from cancer is high, during the last two decades enormous progress has been made in detecting and treating different types of cancer. The techniques of molecular genetics have enabled scientists to characterize cancers in ways that were not previously possible, and they have allowed them to devise new strategies for cancer therapy. There is little doubt that the large investment in basic cancer research is paying off.

Cancer cells can be obtained for experimental study by removing tissue from a tumor and dissociating it into its constituent cells. With appropriate nutri-

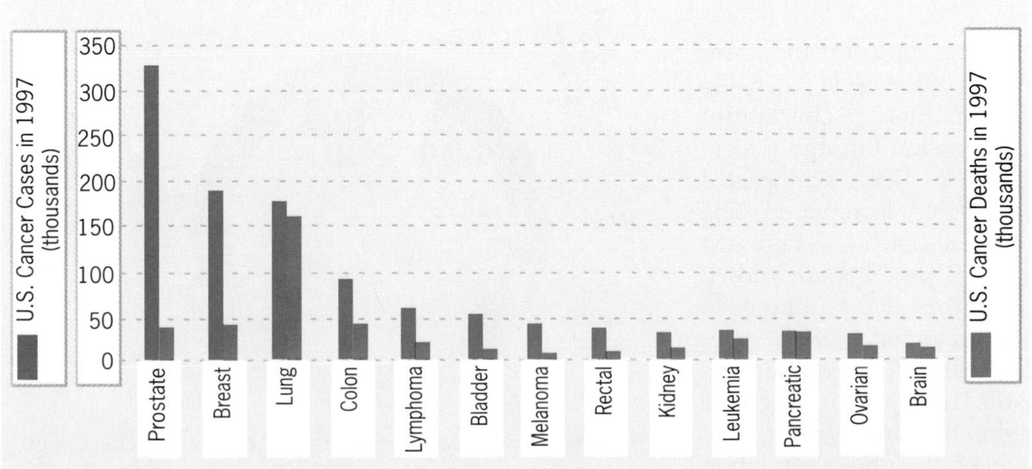

Figure 28.1 Estimated number of new cases and deaths from specific types of cancer in the United States in 1997.

ents, these dissociated tumor cells can be cultured *in vitro,* sometimes indefinitely. Cancer cells can also be derived from cultures of normal cells by treating the cells with agents that induce the cancerous state. Radiation, mutagenic chemicals, and certain types of viruses can irreversibly transform normal cells into cancerous cells. The agents that cause this type of transformation are called **carcinogens.**

The abiding characteristic of all cancer cells is that their growth is unregulated. When normal cells are cultured *in vitro,* they form a single cell layer—a monolayer—on the surface of the culture medium. Cancer cells, by contrast, overgrow each other, piling up on the surface of the culture medium to form masses. This unregulated pile-up occurs because cancer cells do not respond to the chemical signals that inhibit cell division and because they cannot form stable associations with their neighbors.

The external abnormalities that are apparent in a culture of cancer cells are correlated with profound intracellular abnormalities. Cancer cells often have a disorganized cytoskeleton, they may synthesize unusual proteins and display them on their surfaces, and they frequently have abnormal chromosome numbers—that is, they are aneuploid.

Cancer and the Cell Cycle

The cell cycle consists of periods of growth, DNA synthesis, and division. The length of this cycle and the duration of each of its components are controlled by external and internal chemical signals. The transition from each phase of the cycle, such as from G_1 to S or from G_2 to M, requires the integration of specific chemical signals and precise responses to these signals. If the signals are incorrectly sensed or if the cell is not properly prepared to respond, the cell could become cancerous.

The current view of cell-cycle control is that transitions between different phases of the cycle (G_1, S, G_2, and M) are regulated at "checkpoints." A **checkpoint** is a mechanism that halts progression through the cycle until a critical process such as DNA synthesis is completed, or until damaged DNA is repaired. The molecular machinery that operates each checkpoint is complex. However, two types of proteins are known to play especially critical roles: the *cyclins* and the *cyclin-dependent kinases,* often abbreviated CDKs.

The CDKs are the catalytically active components of the cell-cycling mechanism. These proteins regulate the activities of other proteins by transferring phosphate groups to them. However, the phosphorylation activity of the CDKs depends on the presence of the cyclins. The cyclins enable the CDKs to carry out their function by forming cyclin-CKD complexes. When the cyclins are absent, these complexes cannot form, and the CDKs are inactive. Progression through the cell cycle—that is, passage through each of its checkpoints—therefore requires the alternate formation and degradation of cyclin-CKD complexes.

One of the most important cell-cycle checkpoints, called *START,* is in mid-G_1 (Figure 28.2). The cell receives both external and internal signals at this checkpoint to determine when it is appropriate to move into the S phase. This checkpoint is regulated by D-type cyclins in conjunction with CDK4. If a cell is driven past the *START* checkpoint by the cyclin D-CDK4 complex, it becomes committed to another round of DNA replication. Inhibitory proteins with the capability of sensing problems in the late G_1 phase, such as low levels of nutrients or DNA damage, can put a brake on the cyclin-CDK motor and prevent the cell from entering the S phase. In the absence of such problems, the cyclin D-CDK4 motor drives the cell through the end of the G_1 phase and into the S phase, thereby initiating the DNA replication that is a prelude to cell division.

In tumor cells, checkpoints in the cell cycle are often deregulated. This deregulation is due to genetic defects in the machinery that alternately raise and lower the abundance of the cyclin-CDK complexes. For example, the genes encoding the cyclins or the CDKs may be mutated, or the genes encoding the proteins that respond to specific cyclin-CDK complexes or that regulate the abundance of these complexes may be mutated. Many different types of genetic defects can deregulate the cell cycle, with the ultimate consequence that the cells may become cancerous.

Cells in which the *START* checkpoint is dysfunctional are especially prone to become cancerous. The

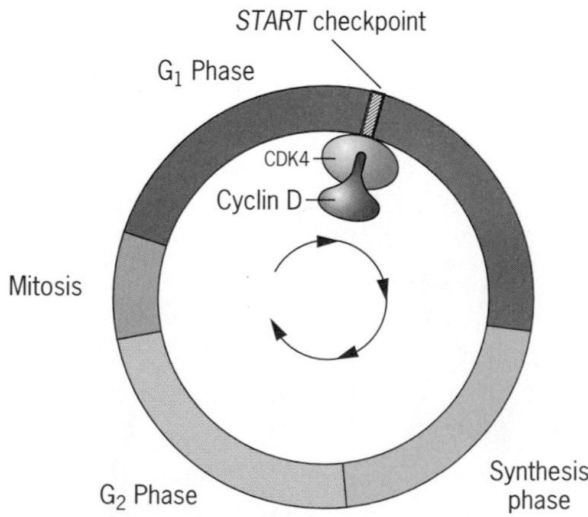

Figure 28.2 A schematic view of the *START* checkpoint in the mammalian cell cycle. Passage through the checkpoint depends on the activity of the cyclin D-CDK4 protein complex.

START checkpoint controls entry into the S phase of the cell cycle. If DNA within a cell has been damaged, it is important that entry into the S phase be delayed to allow for the damaged DNA to be repaired. Otherwise, the damaged DNA will be replicated and transmitted to all the cell's descendants. Normal cells are programmed to pause at the *START* checkpoint to ensure that repair is completed before DNA replication commences. By contrast, cells in which the *START* checkpoint is dysfunctional move into S phase without repairing their damaged DNA. Over a series of cell cycles, mutations that result from the replication of unrepaired DNA may accumulate and cause further deregulation of the cell cycle. A clone of cells with a dysfunctional *START* checkpoint may therefore become aggressively cancerous.

A Genetic Basis for Cancer

The recent great advances in understanding cancer have come through application of molecular genetic techniques. However, before these techniques were available to researchers, there was strong evidence that the underlying causes of cancer are genetic. First, it was known that the cancerous state is clonally inherited. When cancer cells are grown in culture, their descendants are all cancerous. The cancerous condition is therefore transmitted from each cell to its daughters at the time of division—a phenomenon indicating that cancer has a genetic basis. Second, it was known that certain types of viruses can induce the formation of tumors in experimental animals. The induction of cancer by viruses implies that the proteins encoded by viral genes are involved in the production of the cancerous state. Third, it was known that cancer can be induced by agents capable of causing mutations. Mutagenic chemicals and ionizing radiation had been shown to induce tumors in experimental animals. In addition, a wealth of epidemiological data had implicated these agents as the causes of cancer in human beings. Fourth, it was known that certain types of cancer tend to run in families. In particular, susceptibility to retinoblastoma, a rare cancer of the eye, and susceptibility to some forms of colon cancer appeared to be inherited as simple dominant conditions, albeit with incomplete penetrance and variable expressivity. Because susceptibility to these special types of cancer is inherited, it seemed plausible that all cancers might have their basis in genetic defects—either inherited mutations or mutations acquired somatically during a person's lifetime. Finally, it was known that certain types of white blood cell cancers (leukemias and lymphomas) are associated with particular chromosomal aberrations. Collectively, these diverse observations strongly suggested that cancer is due to genetic malfunctions.

In the 1980s, when molecular genetic techniques were first used to study cancer cells, researchers discovered that the cancerous state is, indeed, traceable to specific genetic defects. Typically, however, not one but several such defects are required to convert a normal cell into a cancerous cell. Cancer researchers have identified two broad classes of genes that, when mutated, can contribute to the development of a cancerous state. In one of these classes, mutant genes actively promote cell division; in the other class, mutant genes fail to repress cell division. Genes in the first class are called oncogenes, from the Greek word for "tumor." Genes in the second class are called tumor suppressor genes. In the sections that follow, we discuss the discovery, characteristics, and significance of each of these classes of cancer-related genes.

Key Points: **Cancer is an array of diseases in which cell growth and division are unregulated. This lack of regulation is due to mutations in genes whose protein products are involved in the control of the cell cycle.**

ONCOGENES

Oncogenes comprise a diverse group of genes whose products play important roles in the regulation of biochemical activities within cells, including those activities related to cell division. These genes were first discovered in the genomes of RNA viruses that are capable of inducing tumors in vertebrate hosts. Later, cellular counterparts of these viral oncogenes were discovered in many different organisms, ranging from *Drosophila* to humans.

Tumor-Inducing Retroviruses and Viral Oncogenes

Fundamental insights into the genetic basis of cancer have come from the study of tumor-inducing viruses. Many of these viruses have a genome composed of RNA instead of DNA. After entering a cell, the viral RNA is used as a template to synthesize complementary DNA, which is then inserted at one or more positions in the cell's chromosomes. The synthesis of DNA from RNA is catalyzed by the viral enzyme reverse transcriptase. This reversal of the normal flow of genetic information from DNA to RNA has prompted biologists to call these pathogens **retroviruses** (see Chapter 16).

The first tumor-inducing virus was discovered in 1910 by Peyton Rous; it caused a special kind of tumor, or sarcoma, in the connective tissue of chickens and has since been called the Rous sarcoma virus. Modern research has shown that the RNA genome of this retrovirus contains exactly four genes: *gag,* which encodes the capsid protein of the virion; *pol,* which encodes the reverse transcriptase; *env,* which encodes a protein of the viral envelope; and *v-src,* which encodes a protein kinase that attaches to the plasma membranes of infected cells. The distinguishing feature of a kinase is that it can phosphorylate other proteins. Of these four genes, only one, the *v-src* gene, is responsible for the virus's ability to form tumors. A virus in which the *v-src* gene has been deleted is infectious but unable to induce tumors. Genes such as *v-src* that cause cancer are called **oncogenes.**

Studies with other tumor-inducing retroviruses have uncovered at least 20 different viral oncogenes, usually denoted *v-onc* (Table 28.1). Some of these are related to cellular genes that encode growth factors. For example, *v-sis,* an oncogene from the simian sarcoma virus, codes for a version of the platelet-derived growth factor (PDGF). PDGF is normally produced by platelet cells to promote wound healing, which it does by stimulating the growth of cells at the wound site. Simian sarcoma viruses carrying the *v-sis* gene induce tumors in monkeys. They also transform cultured cells to a cancerous state, presumably by producing large amounts of the *v-sis* version of PDGF, which then causes uncontrolled cell growth.

Other viral oncogenes encode proteins that are similar to growth-factor and hormone receptors. For example, the *v-erbB* gene from the avian erythroblastosis virus encodes a protein much like the cellular receptor for epidermal growth factor (EGF), and the *v-fms* gene from the feline sarcoma virus encodes a protein much like the receptor for the cellular CSF-1 (colony stimulating factor 1) growth factor. Both of these growth-factor receptors are transmembrane proteins with a growth-factor-binding domain on the outside of the cell and a protein kinase domain on the inside. This domain on the inside enables the protein to phosphorylate certain amino acids, usually tyrosine, in other proteins that interact with it.

Many viral oncogenes, including *v-src,* encode tyrosine kinases that do not span the plasma membrane. Instead, these proteins are situated on the inner face of the plasma membrane, where they perform their phosphorylating function. Some of the best-studied of the membrane-associated tyrosine kinases are encoded by the *v-ras* oncogenes. These proteins bind GTP and exhibit GTPase activity, much like the cellular G proteins, which play an important role in regulating the level of cyclic AMP.

Another group of viral oncogenes encode proteins that apparently function as transcription factors. These include the *v-jun, v-fos, v-erbA,* and *v-myc* genes, each

TABLE 28.1
Retroviral Oncogenes

Oncogene	*Virus*	*Host Species*	*Function of Gene Product*
abl	Abelson murine leukemia virus	Mouse	Tyrosine-specific protein kinase
erbA	Avian erythroblastosis virus	Chicken	Analog of thyroid hormone receptor
erbB	Avian erythroblastosis virus	Chicken	Truncated version of epidermal growth factor (EGF) receptor
fes	ST feline sarcoma virus	Cat	Tyrosine-specific protein kinase
fgr	Gardner-Rasheed feline sarcoma virus	Cat	Tyrosine-specific protein kinase
fms	McDonough feline sarcoma virus	Cat	Analog of colony stimulating growth factor (CSF-1) receptor
fos	FJB osteosarcoma virus	Mouse	Transcriptional activator protein
fps	Fuginami sarcoma virus	Chicken	Tyrosine-specific protein kinase
jun	Avian sarcoma virus 17	Chicken	Transcriptional activator protein
mil (mht)	MH2 virus	Chicken	Serine/threonine protein kinase
mos	Moloney sarcoma virus	Mouse	Serine/threonine protein kinase
myb	Avian myeloblastosis virus	Chicken	Transcription factor
myc	MC29 myelocytomatosis virus	Chicken	Transcription factor
raf	3611 murine sarcoma virus	Mouse	Serine/threonine protein kinase
H-ras	Harvey murine sarcoma virus	Rat	GTP-binding protein
K-ras	Kirsten murine sarcoma virus	Rat	GTP-binding protein
rel	Reticuloendotheliosis virus	Turkey	Transcription factor
ros	URII avian sarcoma virus	Chicken	Tyrosine-specific protein kinase
sis	Simian sarcoma virus	Monkey	Analog of platelet-derived growth factor (PDGF)
src	Rous sarcoma virus	Chicken	Tyrosine-specific protein kinase
yes	Y73 sarcoma virus	Chicken	Tyrosine-specific protein kinase

carried by a different retrovirus. These proteins are homologous to cellular proteins that bind to DNA and regulate transcription.

Each type of viral oncogene therefore appears to encode a protein that could theoretically play a role in regulating the expression of cellular genes, including those involved in the processes of growth and division. Some of these proteins may act as signals to stimulate certain types of cellular activity; others may act as receptors to pick up these signals or as intracellular agents to convey them from the plasma membrane to the nucleus; yet another category of viral oncogene proteins may act as transcription factors to stimulate gene expression.

Cellular Homologs of Viral Oncogenes: The Proto-Oncogenes

The proteins encoded by viral oncogenes are similar to cellular proteins with important regulatory functions. Many of these cellular proteins were actually identified by isolating the cellular homolog of the viral oncogene. For example, the cellular homolog of the *v-src* gene was obtained by screening a genomic DNA library made from uninfected chicken cells. For this screening, the *v-src* gene was used as a hybridization probe to detect recombinant DNA clones that could base-pair with it. Analysis of these clones established that chicken cells contain a gene that is similar to *v-src*—indeed, that is related to it in an evolutionary sense. However, this gene is not associated with an integrated sarcoma provirus, and it differs from the *v-src* gene in a very important respect: it contains introns (Figure 28.3). There are, in fact, 11 introns in the chicken homolog of *v-src,* compared to zero in the *v-src* gene itself. This startling discovery suggested that perhaps *v-src* had evolved from a normal cellular gene, and that, concomitantly, it had lost its introns.

The cellular homologs of viral oncogenes are called **proto-oncogenes,** or sometimes, **normal cellular oncogenes,** denoted *c-onc*. The cellular homolog of *v-src* is therefore *c-src*. The coding sequences of these two genes are very similar, differing only in 18 nucleotides; *v-src* encodes a protein of 526 amino acids, and *c-src* encodes a protein of 533 amino acids. By using *v-onc* genes as probes, other *c-onc* genes have been isolated from many different organisms, including human be-

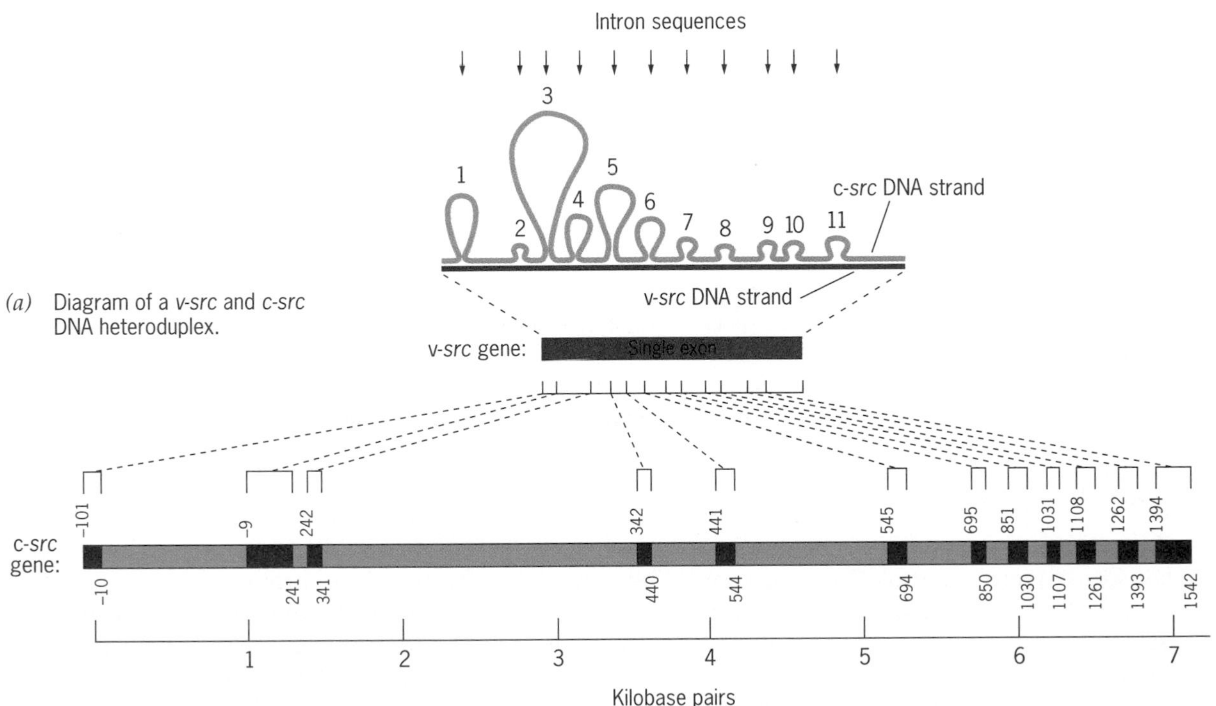

(*a*) Diagram of a *v-src* and *c-src* DNA heteroduplex.

(*b*) Comparison of the structure of the *v-src* and *c-src* genes.

Figure 28.3 Structures of the *v-src* and *c-src* genes. (*a*) Diagram of the predicted DNA heteroduplex formed by hybridization of a strand from the *c-src* gene (top) and a partially complementary strand from the *v-src* gene (bottom). The introns (numbered 1–11) form single-stranded loops. (*b*) Schematic comparison of these two genes, with exons shown in black. The coordinate system for the exons in the *c-src* gene is based on the first nucleotide in the coding sequence (position 1). The first exon (position -101 to -10) is in the 5′ leader of the gene.

ings. As a rule, these cellular oncogenes show considerable conservation in structure. *Drosophila,* for example, carries very similar homologs of the vertebrate cellular oncogenes *c-abl, c-erbB, c-fps, c-raf, c-ras,* and *c-myb.* The similarity of oncogenes from different species strongly suggests that the proteins they encode are involved in important cellular functions.

Why do *c-onc*s have introns whereas *v-onc*s do not? The most plausible answer is that *v-onc*s were derived from *c-onc*s by the insertion of a fully processed *c-onc* mRNA into the genome of a retrovirus. A virion that packaged such a recombinant molecule would then be able to transduce the *c-onc* gene whenever it infected another cell. During infection, the recombinant RNA would be reverse-transcribed into DNA and then integrated into the cell's chromosomes. What could be of greater value to a virus than to have a new gene that stimulates increased growth of its host, while its integrated genome goes along for the ride?

In many cases, the acquisition of an oncogene by a retrovirus has been accompanied by the loss of some viral genetic material. Because this lost material is needed for viral replication, these oncogenic viruses are able to reproduce only if a helper virus is present. In this respect, they resemble the defective transducing bacteriophages we discussed in Chapter 17.

Why do *v-onc*s induce tumors, whereas normal *c-onc*s do not? In some cases it appears that the viral oncogene produces much more protein than its cellular counterpart, perhaps because it has been transcriptionally activated by enhancers embedded in the viral genome. In chicken tumor cells, for example, the *v-src* gene produces 100 times as much tyrosine kinase as the *c-src* gene. This vast oversupply of the kinase evidently upsets the delicate signaling mechanisms that control cell division, causing unregulated growth. Other *v-onc* genes may induce tumors by expressing their proteins at inappropriate times, or by expressing altered—that is, mutant—forms of these proteins.

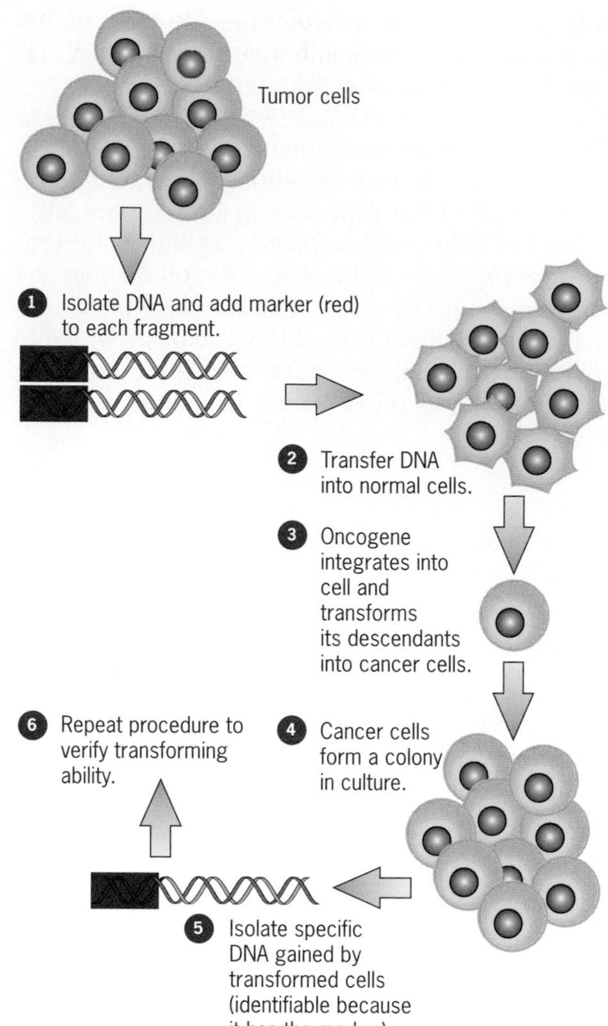

Figure 28.4 The transfection test to identify DNA sequences capable of transforming normal cells into cancer cells.

Mutant Cellular Oncogenes and Cancer

The products of the *c-onc*s play key roles in regulating cellular activities. Consequently, a mutation in one of these genes can upset the biochemical balance within a cell and put it on the track to becoming cancerous. Studies of many different types of human cancer have demonstrated that mutant cellular oncogenes are associated with the development of a cancerous state.

The first evidence linking cancer to a mutant *c-onc* came from the study of a human bladder cancer. The mutation responsible for this bladder cancer was isolated by Robert Weinberg and colleagues using a transfection test (Figure 28.4). DNA was extracted from the cancerous tissue and fragmented into small pieces; then each of these pieces was joined to a segment of bacterial DNA, which served as a molecular marker. The marked DNA fragments were then introduced, or transfected, into cells growing in culture to determine if any of them could transform the cells into a cancerous state. This state could be recognized by the tendency of the cancer cells to form small clumps, or foci, when grown on soft agar plates. The DNA from such cells was extracted and screened to see if it carried the molecular marker that was linked to the original transfecting fragments. If it did, this DNA was retested for its ability to induce the cancerous state. After several tests, Weinberg's research team identified a DNA fragment from the original bladder cancer that reproducibly transformed cultured cells into cancer cells. This fragment carried an allele of the *c-H-ras* oncogene, a homolog of an oncogene in the Harvey strain of the rat sarcoma virus. DNA sequence analysis subsequently showed that a nucleotide in codon 12 of this allele had been mutated, with a substitution of a valine for the glycine normally found at this position in the *c-H-ras* protein.

Geneticists now have some understanding of how this mutation causes cells to become cancerous. Unlike viral oncogenes, the mutant *c-H-ras* gene does not synthesize abnormally large amounts of protein. Instead, the valine-for-glycine substitution at position 12 impairs the ability of the mutant *c-H-ras* protein to hydrolyze one of its substrates, GTP. Because of this impairment, the mutant protein is kept in an active signaling mode, transmitting information that ultimately stimulates the cells to divide in an uncontrolled way (Figure 28.5).

Mutant versions of the *c-ras* oncogenes have now been found in a large number of different human tumors, including lung, colon, mammary, prostate, and

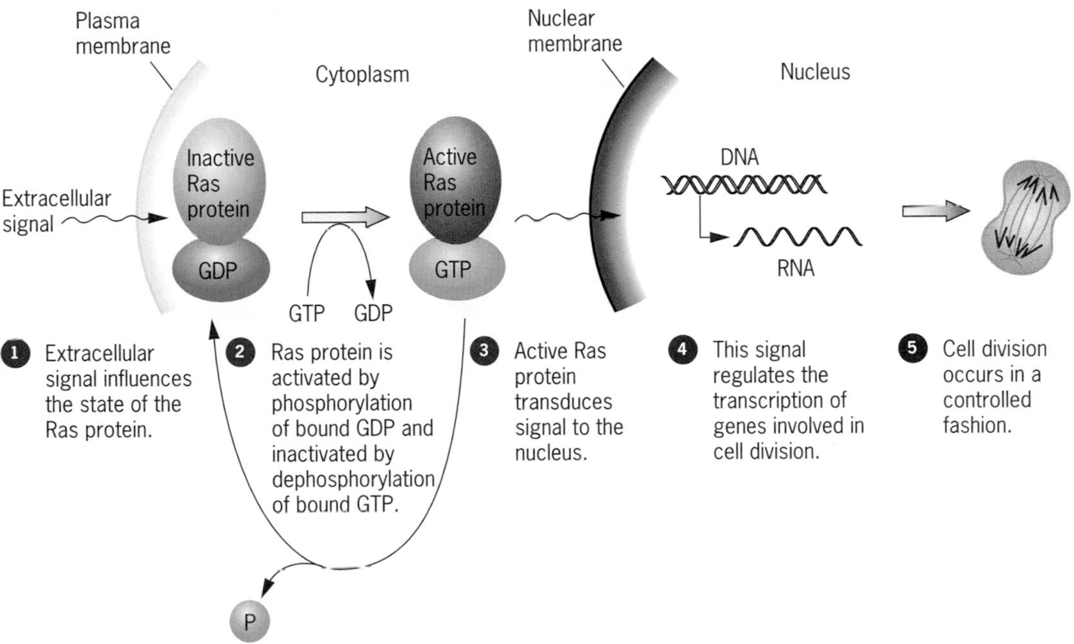

(*a*) Normal Ras protein signaling.

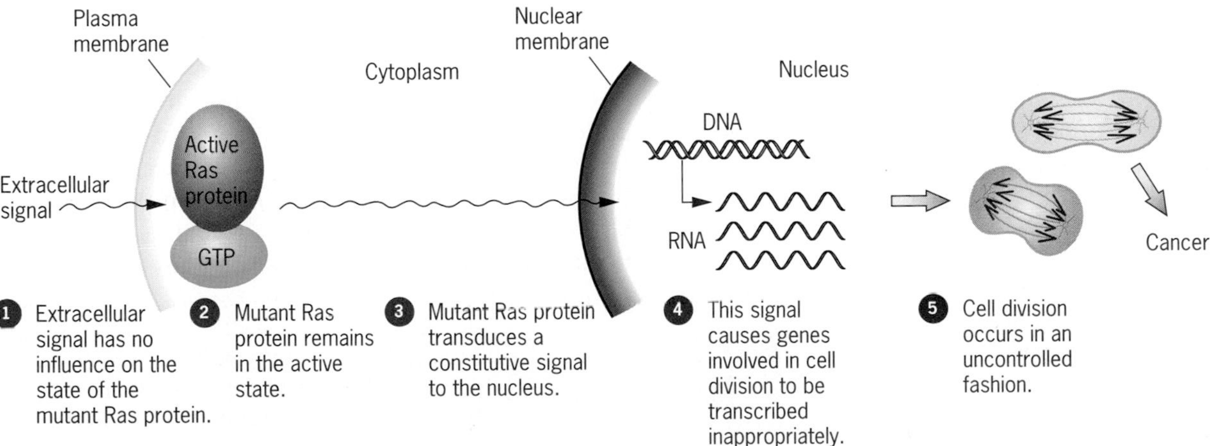

(*b*) Oncogenic Ras protein signaling.

Figure 28.5 Ras protein signaling and cancer. (*a*) The normal protein product of the *ras* gene alternates between inactive and active states, depending on whether it is bound to GDP or GTP. Extracellular signals such as growth factors stimulate the conversion of inactive Ras to active Ras. Through active Ras, these signals are transmitted to other proteins and eventually to the nucleus, where they induce the expression of genes involved in cell division. Because this signaling is intermittent and regulated, cell division occurs in a controlled manner. (*b*) Mutant Ras proteins exist mainly in the active state. These proteins transmit their signals more or less constantly, leading to uncontrolled cell division, the hallmark of cancer.

bladder tumors, as well as neuroblastomas (nerve cell cancers), fibrosarcomas (cancers of the connective tissues), and teratocarcinomas (cancers that contain different embryonic cell types). In all cases, the mutations involve amino acid changes in one of three positions—12, 59, or 61. Each of these amino acid changes impairs the ability of the mutant Ras protein to switch out of its active signaling mode. These types of mutations are therefore *dominant activators* of uncontrolled cell growth.

Mutations in *c-ras* and other cellular oncogenes that contribute to the formation of tumors are seldom inherited through the germ line; rather, the vast majority of them occur spontaneously in the soma during the course of cell division. Because the number of cell divisions in a human life is very large—more than 10^{16}— thousands of potentially oncogenic mutations are bound to occur, and if each one functioned as a dominant activator of uncontrolled cell growth, the development of a tumor would be inevitable. However, many people lead long lives without developing tumors. The explanation for this paradox is that each individual oncogene mutation is, by itself, seldom able to induce a cancerous state. In human beings, cells apparently have ways to adjust for the deleterious effects of a single mutant oncogene. However, when several different growth-regulating genes within a cell have been altered by mutations, the cell cannot compensate for their separate effects, its growth becomes unregulated, and cancer ensues. In many tumors, at least one of these deleterious mutations is in a cellular oncogene. Thus, this group of genes plays an important role in the etiology of human cancer.

Chromosome Rearrangements and Cancer

Certain types of human cancer are associated with chromosome rearrangements. For example, chronic myelogenous leukemia (CML) is associated with an aberration of chromosome 22. This abnormal chromosome was originally discovered in the city of Philadelphia and thus is called the *Philadelphia chromosome.* Initially it was thought to have a simple deletion in its long arm; however, subsequent analysis using molecular techniques has shown that the Philadelphia chromosome is actually the result of a reciprocal translocation between chromosomes 9 and 22. (For a general discussion of translocations, see Chapter 7.) In the Philadelphia translocation, the tip of the long arm of chromosome 9 has been joined to the body of chromosome 22, and the distal portion of the long arm of chromosome 22 has been joined to the body of chromosome 9 (Figure 28.6). The translocation breakpoint on chromosome 9 is in the *c-abl* oncogene, which encodes a tyrosine kinase, and the breakpoint on chromosome 22 is in a gene called *bcr*. Through the translocation, the *bcr* and *c-abl* genes have been physically joined, creating a fusion gene whose polypeptide product has the amino terminus of the *bcr* protein and the carboxy terminus of the *abl* protein. Although it is not understood precisely why, this fusion polypeptide causes white blood cells to become cancerous.

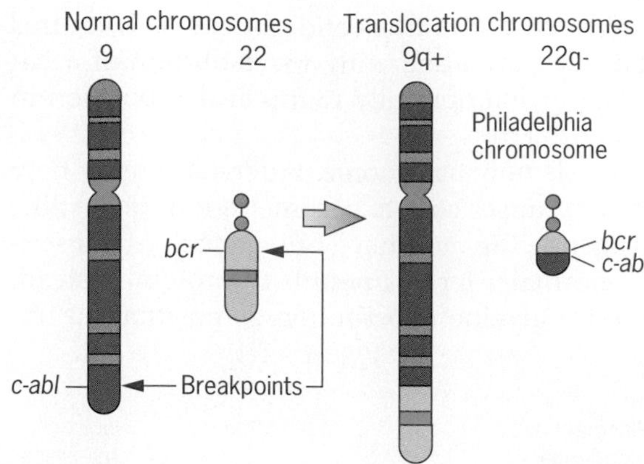

Figure 28.6 The reciprocal translocation involved in the Philadelphia chromosome associated with chronic myelogenous leukemia.

Burkitt's lymphoma is another example of a white blood cell cancer associated with reciprocal translocations. These translocations invariably involve chromo-

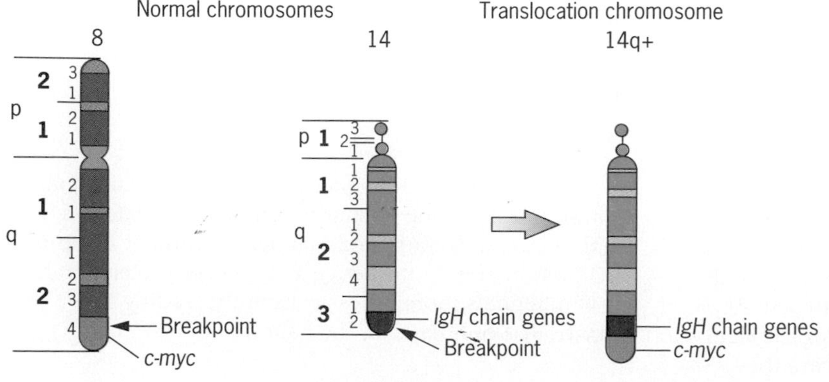

Figure 28.7 A reciprocal translocation involved in Burkitt's lymphoma. Only the translocation chromosome (14q+) that carries both the *c-myc* oncogene and the immunoglobulin heavy chain genes (*IgH*) is shown.

some 8 and one of the three chromosomes (2, 14, and 22) that carry genes encoding the polypeptides that form immunoglobulins (also known as antibodies; see Chapter 27). Translocations involving chromosomes 8 and 14 are the most common (Figure 28.7). In these translocations, the *c-myc* oncogene on chromosome 8 is juxtaposed to the genes for the immunoglobulin heavy chains on chromosome 14. This rearrangement results in the overexpression of the *c-myc* oncogene in cells that produce immunoglobulin heavy chains—that is, in B lymphocytes—and this overexpression causes those cells to become cancerous.

Key Points: **Some viruses carry genes that can induce the formation of tumors in animals. These viral oncogenes have cellular homologs called proto-oncogenes, which can induce tumors when they are overexpressed or when they are mutated to produce an activated protein. Mutations in proto-oncogenes actively promote cell proliferation. Some cancers are associated with chromosome rearrangements that enhance the expression of proto-oncogenes or that alter the nature of their protein products.**

TUMOR SUPPRESSOR GENES

The normal alleles of genes such as *c-ras* and *c-myc* produce proteins that regulate the cell cycle. When these genes are overexpressed, or when they produce proteins that function as dominant activators, the cell is predisposed to become cancerous. However, the full development of a cancerous state usually requires additional mutations, and typically these mutations affect genes that are normally involved in the restraint of cell growth. These mutations therefore define a second class of cancer-related genes—the antioncogenes, or, as they are more often called, the **tumor suppressor genes.**

Inherited Cancers and Knudson's Two-Hit Hypothesis

Many of the tumor suppressor genes were initially discovered through the analysis of rare cancers in which a predisposition to develop the cancer follows a dominant pattern of inheritance. This predisposition is due to heterozygosity for an inherited, loss-of-function mutation in the tumor suppressor gene. A cancer develops only if a second mutation occurs in the somatic cells and if this mutation knocks out the function of the wild-type allele of the tumor suppressor gene. Thus, development of the cancer requires two loss-of-function mutations—that is, two inactivating "hits," one in each of the two copies of the tumor suppressor gene.

In 1971 Alfred Knudson proposed this explanation for the occurrence of *retinoblastoma,* a rare, childhood cancer of the eye. In most human populations, the incidence of retinoblastoma is about 5 in 100,000 children. Pedigree analysis indicates that approximately 40 percent of the cases involve an inherited mutation that predisposes the individual to develop the cancer. The other 60 percent of the cases cannot be traced to a specific inherited mutation. These noninherited cases are said to be *sporadic.* On the basis of statistical analyses, Knudson proposed that both the inherited and sporadic cases of retinoblastoma occur because the two copies of a particular gene have been inactivated (Figure 28.8). In the inherited cases, one of the inactivating mutations has

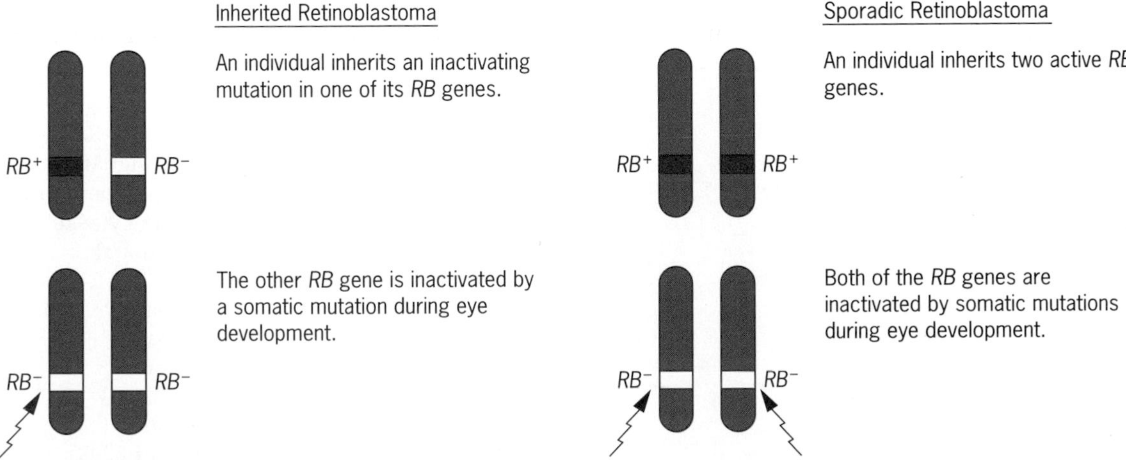

Figure 28.8 Knudson's two-hit hypothesis to explain the occurrence of inherited and sporadic cases of retinoblastoma. Two inactivating mutations are required to eliminate the function of the *RB* gene.

been transmitted through the germ line, and the other occurs during the development of the somatic tissues of the eye. In the sporadic cases, both of the inactivating mutations occur during eye development. Thus, in either type of retinoblastoma, two mutational "hits" are required to knock out a gene that normally functions to suppress tumor formation in the eye.

Subsequent research findings have verified the correctness of Knudson's two-hit hypothesis. First, several cases of retinoblastoma were found to be associated with a small deletion in the long arm of chromosome 13. The gene that normally prevents retinoblastoma—symbolized *RB*—must therefore be located in the region defined by this deletion. More refined cytogenetic mapping subsequently placed the *RB* gene in locus 13q14.2. Second, positional cloning techniques were used to isolate a candidate *RB* gene. Once isolated, the gene's structure, sequence, and expression patterns were determined. Third, the structure of the candidate gene was examined in cells taken from tumorous eye tissue. As predicted by Knudson's two-hit hypothesis, both copies of this gene were inactivated in retinoblastoma cells. Thus, the candidate gene appeared to be the authentic *RB* gene. Finally, cell culture experiments demonstrated that a cDNA from the wild-type allele of the candidate gene could revert the cancerous properties of cultured tumor cells. These cancer reversion experiments proved beyond a doubt that the candidate gene was the authentic *RB* tumor suppressor gene. The protein product of this gene—denoted pRB—was subsequently found to be a ubiquitously expressed protein that interacts with a family of transcription factors involved in regulating the cell cycle.

Knudson's two-hit hypothesis has since been applied to other inherited cancers, including Wilms tumor, Li-Faumeni Syndrome, neurofibromatosis, von Hippel-Lindau Syndrome, and certain types of colon and breast cancer (Table 28.2). In each case a different tumor suppressor gene is involved. For example, in Wilms tumor, a cancer of the urogenital system, the relevant tumor suppressor gene is the *WT1* gene located in the short arm of chromosome 11; in neurofibromatosis, a disease characterized by benign tumors and skin lesions, it is the *NF1* gene located in the long arm of chromosome 17; and in familial adenomatous polyposis, a condition characterized by the occurrence of numerous tumors in the colon, it is the *APC* gene located in the long arm of chromosome 5. Like retinoblastoma, these three diseases are rare, and only a fraction of the observed cases involve an inherited mutation in the relevant tumor suppressor gene. The

TABLE 28.2
Inherited Cancer Syndromes

Syndrome	*Primary Tumor*	*Gene*	*Chromosomal Location*	*Proposed Protein Function*
Familial retinoblastoma	Retinoblastoma	*RB*	13q14.3	Cell cycle and transcriptional regulation
Li-Fraumeni Syndrome	Sarcomas, breast cancer	*TP53*	17p13.1	Transcription factor
Familial adenomatous polyposis (FAP)	Colorectal cancer	*APC*	5q21	Regulation of β-catenin
Hereditary nonpolyposis colorectal cancer (HNPCC)	Colorectal cancer	*MSH2* *MLH1* *PMS1* *PMS1*	2p16 3p21 2q32 7p22	DNA mismatch repair
Neurofibromatosis type 1	Neurofibromas	*NF1*	17q11.2	Regulation of Ras-mediated signaling
Neurofibromatosis type 2	Acoustic neuromas, meningiomas	*NF2*	22q12.2	Linkage of membrane proteins to cytoskeleton
Wilms tumor	Wilms tumor	*WT1*	11p13	Transcriptional repressor
Familial breast cancer 1	Breast cancer	*BRCA1*	17q21	DNA repair?
Familial breast cancer 2	Breast cancer	*BRCA2*	13q12	DNA repair?
von Hippel-Lindau Syndrome	Renal cancer	*VHL*	3p25	Regulation of transcriptional elongation
Familial melanoma	Melanoma	*p16*	9p21	Inhibitor of CDKs
Ataxia telangiectasia	Lymphoma	*ATM*	11q22	DNA repair
Bloom's syndrome	Solid tumors	*BLM*	15q26.1	DNA helicase?

From E. R. Fearon. 1997. Human cancer syndromes: clues to the origin and nature of cancer. *Science* 278:1043–1050.

other cases are caused either by two independent somatic mutations in that gene or by mutations in other, as-yet-unidentified tumor suppressor genes.

Cellular Roles of Tumor Suppressor Proteins

Only about 1 percent of all cancers are hereditary. However, more than 20 different inherited cancer syndromes have been identified, and in nearly all of them the underlying defect is in a tumor suppressor gene rather than in an oncogene. The proteins encoded by these tumor suppressor genes function in a diverse array of cellular processes, including division, differentiation, programmed cell death, and DNA repair. In the following sections, we discuss some of the tumor suppressor proteins that have been studied intensively.

pRB Recent research has revealed that the RB tumor suppressor protein plays a key role in regulation of the cell cycle. Although the *RB* gene was discovered through its association with retinoblastoma, mutations in this gene are also associated with other types of cancer, including small cell lung carcinomas, osteosarcomas, and bladder, cervical, and prostate carcinomas. Furthermore, mice that are homozygous for an *RB* knock-out mutation die during embryonic development. Thus, the RB gene product is essential for life.

The RB gene product is a 105 kD nuclear protein that is involved in cell-cycle regulation. Two genes homologous to *RB* have been found in mammalian genomes, and their protein products, p107 and p130 (each named for its mass in kilodaltons), may also play key roles in cell-cycle regulation. No human tumors are known to have inactivating mutations in either of these two genes, and mice homozygous for a knock-out mutation in either of them do not show abnormal phenotypes. However, mice that are homozygous for knock-out mutations in both of these genes die shortly after birth. Thus, the p107 and p130 members of the RB family of proteins are involved in important cellular processes.

Molecular and biochemical analyses have elucidated the role of pRB in cell-cycle regulation (Figure 28.9). Early in the G_1 phase of the cell cycle, pRB binds to the E2F proteins, a family of transcription factors that control the expression of several genes whose products move the cell through its cycle. When E2F transcription factors are bound to pRB, they cannot bind to specific enhancer sequences in their target genes. Consequently, the cell-cycle factors encoded by these genes are not produced, and the machinery for DNA synthesis and cell division remains quiescent. Later in G_1, pRB is phosphorylated through the action of cyclin-dependent kinases. In this changed state, pRB releases the E2F transcription factors that have bound to it. These released transcription factors are then free to activate their target genes, which encode proteins that induce the cell to progress through S phase and into mitosis. After mitosis, pRB is dephosphorylated, and each of the daughter cells enters the quiescent phase of a new cell cycle.

This orderly and rhythmic progression through the cell cycle is disrupted in cancer cells. In many types of cancer—not just retinoblastoma—both copies of the *RB* gene have been inactivated, either by deletions or by mutations that impair or abolish the ability of the RB protein to bind E2F transcription factors. The inability of pRB to bind to these transcription factors leaves them free to activate their target genes, thereby setting in motion the machinery for DNA synthesis and cell division. In effect, one of the natural brakes on the process of cell division has been released. In the absence of this brake, cells have a tendency to move through their cycle quickly. If other cell-cycle brakes fail, the cells divide ceaselessly to form tumors.

p53 The 53-kilodalton tumor suppressor protein p53 was discovered through its role in the induction of cancers by certain DNA viruses. This protein is encoded by a tumor suppressor gene called *TP53*. Inherited mutations in *TP53* are associated with the Li-Fraumeni Syndrome, a rare, dominant condition in which any of several different types of cancer may develop. Somatic mutations that inactivate both copies of the *TP53* gene are also associated with a variety of cancers. In fact, such mutations are found in a majority of all human tumors. Loss of p53 function is therefore a key step in carcinogenesis.

The p53 protein is a 393 amino acid-long transcription factor that consists of three distinct domains: an N-terminal transcription activation domain (TAD), a central DNA-binding core domain (DBD), and a C-terminal homo-oligomerization domain (OD) (Figure 28.10*a*). Most of the mutations that inactivate p53 are located in the DBD. These mutations evidently impair or abolish the ability of p53 to bind to specific DNA sequences that are embedded in its target genes, thereby preventing the transcriptional activation of these genes. Thus, mutations in the DBD are typically recessive, loss-of-function mutations. Other types of mutations are found in the OD portion of the polypeptide. Molecules of p53 with these types of mutations dimerize with wild-type p53 polypeptides and prevent the wild-type polypeptides from functioning as transcriptional activators. Thus, mutations in the OD have a *dominant negative* effect on p53 function.

The p53 protein plays a key role in cellular responses to stress (Figure 28.10*b*). In normal cells the level of p53 is low, but when the cells are treated with a DNA-damaging agent such as radiation, the level of p53 increases dramatically. This response to DNA dam-

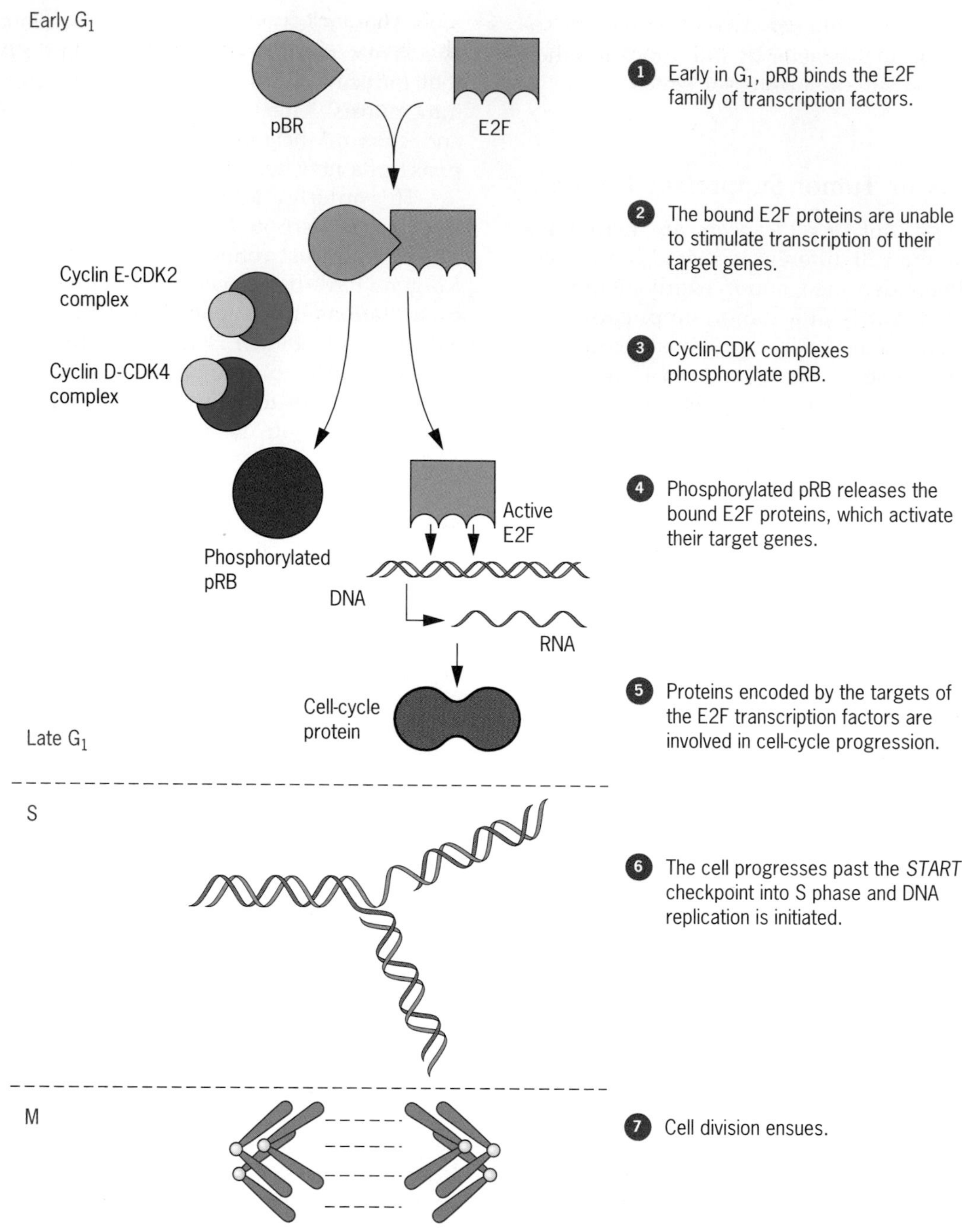

Figure 28.9 Role of pRB in progression of the cell cycle. Through its negative interaction with E2F transcription factors, pRB stalls the cell cycle in the G_1 phase. Phosphorylation of pRB by the cyclin-CDK complexes frees E2F proteins to activate their target genes, which encode proteins that are instrumental in moving the cell past the *START* checkpoint into the S phase.

age is mediated by an unknown signal transduction pathway that increases the transcription of the *TP53* gene. This pathway also causes p53 protein to be converted into a stable and active form, perhaps by phosphorylation. Once activated, p53 either stimulates the transcription of genes whose products arrest the cell cycle, thereby allowing the damaged DNA to be repaired, or it activates another set of genes whose products ultimately cause the damaged cell to die—a process called **apoptosis** (from Greek roots meaning "falling away").

One prominent factor in the response that arrests the cell cycle is p21, a protein encoded by a gene that is activated by the p53 transcription factor. The p21 protein is an inhibitor of cyclin-CDK protein complexes. When p21 is synthesized in response to cell stress, the cyclin-CDK complexes are inactivated and the cell cycle is arrested. During this timeout, the cell's damaged DNA can be repaired. Thus, p53 is responsible for activating a brake on the cell cycle, and this brake allows the cell to maintain its genetic integrity. Cells that lack functional p53 have difficulty applying this brake. If these cells progress through the cell cycle and proceed into subsequent divisions, additional mutations that cause them to be unregulated may accumulate. Mutational inactivation of p53 is therefore often a key step in the pathway to cancer.

The p53 protein can also mediate another response to cell stress. Instead of orchestrating efforts to repair damage within a cell, p53 may trigger a suicidal response in which the damaged cell is programmed for destruction. The way in which p53 programs cell death

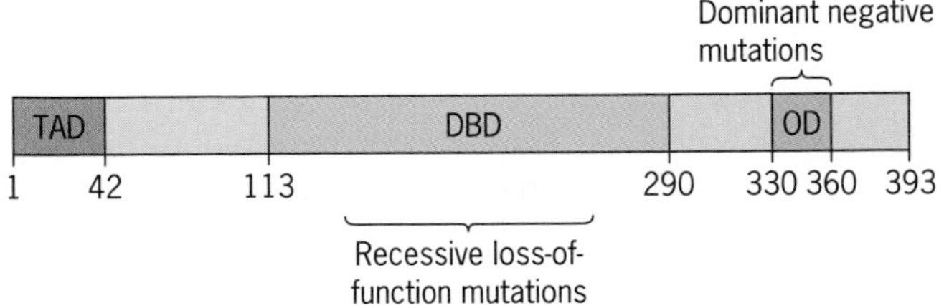

Dominant negative mutations
TAD
DBD
OD
1
42
113
290
330
360
393
Recessive loss-of-function mutations
(*a*)

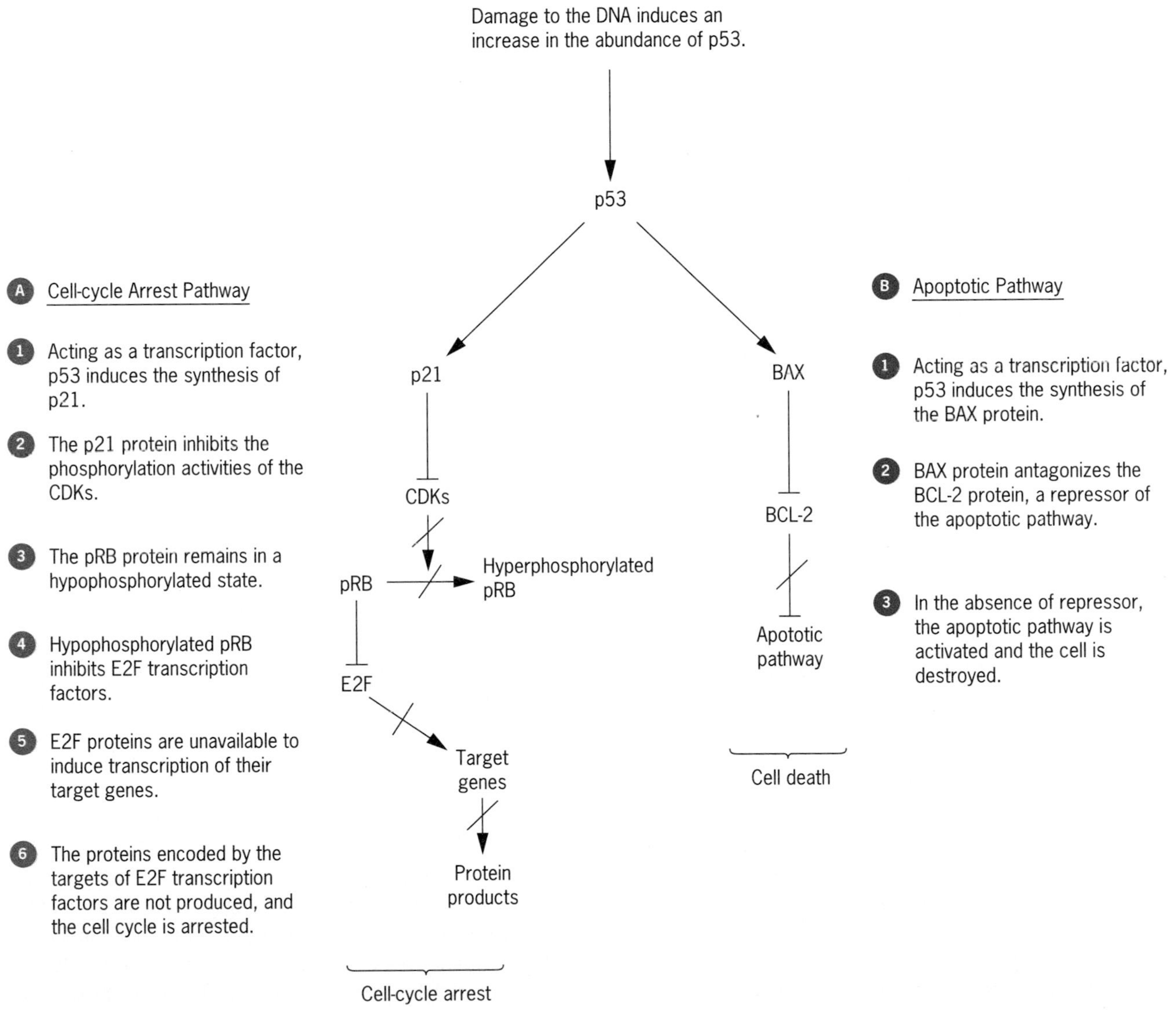

Damage to the DNA induces an increase in the abundance of p53.
p53
A Cell-cycle Arrest Pathway
1 Acting as a transcription factor, p53 induces the synthesis of p21.
2 The p21 protein inhibits the phosphorylation activities of the CDKs.
3 The pRB protein remains in a hypophosphorylated state.
4 Hypophosphorylated pRB inhibits E2F transcription factors.
5 E2F proteins are unavailable to induce transcription of their target genes.
6 The proteins encoded by the targets of E2F transcription factors are not produced, and the cell cycle is arrested.
p21
CDKs
pRB
Hyperphosphorylated pRB
E2F
Target genes
Protein products
Cell-cycle arrest
B Apoptotic Pathway
1 Acting as a transcription factor, p53 induces the synthesis of the BAX protein.
2 BAX protein antagonizes the BCL-2 protein, a repressor of the apoptotic pathway.
3 In the absence of repressor, the apoptotic pathway is activated and the cell is destroyed.
BAX
BCL-2
Apototic pathway
Cell death
(*b*)

Figure 28.10 (*a*) Principal domains within p53. TAD = transcription activation domain; DBD = DNA-binding domain; OD = oligomerization domain. The numbers refer to amino acid positions in the polypeptide. (*b*) Role of p53 in the cellular response to DNA damage. Two response pathways have been identified. Within each pathway, a pointed arrow indicates a positive influence or a directional change (e.g., a protein is synthesized or phosphorylated, a protein catalyzes a reaction, or a gene is expressed), and a blunted arrow indicates a negative influence (e.g., repression of protein synthesis or protein activity, or repression of a pathway). A slash through an arrow indicates that the influence—positive or negative—is blocked.

is not well understood. One mechanism seems to involve the protein product of the *BAX* gene. The BAX protein is an antagonist of another protein called BCL-2, which normally suppresses the apoptotic, or cell death, pathway. When the *BAX* gene is activated by p53, its protein product releases the BCL-2 protein from its suppressing mode. This release then opens the apoptotic pathway, and the cell proceeds to its own destruction.

Programmed cell death is a drastic response to stress. However, this response makes sense in so far as it eliminates cells that have been so seriously damaged that their continued existence could jeopardize the well-being of the organism. Programmed cell death also plays an important role in unstressed cells. During embryogenesis many different types of cells are programmed to die after they have fulfilled their functions. This process is necessary for normal development. For example, during the development of the hands and feet of many vertebrate animals, the cells that lie between the developing digits must die. If they do not, the digits remain fused.

Curiously, the p53 protein does not seem to play a significant role in the programmed cell death that occurs during embryogenesis. Mice that are homozygous for knock-out mutations in *TP53* develop normally, although they have a tendency to develop tumors as they age. Thus, despite its pivotal role in regulating cellular responses to stress, p53 does not seem to influence the course of embryonic development.

pAPC The 310 kilodalton pAPC protein was discovered through the study of *adenomatous polyposis coli,* an inherited condition that often leads to colorectal cancer. This large protein, 2843 amino acids long, plays a key role in regulating the renewal of cells in the epithelium of the large intestine. Although the mechanisms that regulate this process are not fully understood, current information suggests that pAPC controls the delicate balance between cell death and cell proliferation. When pAPC function is lost, this balance is upset and cell proliferation exceeds cell death. The resulting increase in the number of cells causes many small, benign tumors to form in the intestinal epithelium. These tumors are called *polyps* or *adenomas,* and the predisposition to form them is inherited as a rare, autosomal dominant condition called *familial adenomatous polyposis (FAP).* In Western countries, its population frequency is about 1 in 7000.

Patients with FAP develop multiple adenomas during their teens and early twenties. Although the adenomas are initially benign, there is a high probability that at least one of them will become a malignant tumor. Thus, at a relatively early age—in the United States, the median is 42—carriers of an FAP mutation develop full-fledged colorectal cancer.

Multiple adenomas develop in the intestines of people who are heterozygous for an FAP mutation because the wild-type *APC* allele they carry mutates multiple times during the natural regeneration of the intestinal epithelium. When such mutations occur, the cells lose their ability to synthesize functional pAPC protein. The absence of this protein releases an important brake on cell proliferation, and cell division proceeds unchecked. Thus, the formation of numerous benign tumors in the intestines of FAP heterozygotes results from the independent occurrence of second mutational "hits" in the cells of the intestinal epithelium. Individuals who do not carry an FAP mutation seldom form multiple adenomas. However, they may produce one or a few adenomas if by chance both of their *APC* genes are inactivated by somatic mutations.

The pAPC protein appears to regulate the balance between cell proliferation and cell death through its ability to bind β-catenin, a protein that interacts with several other proteins inside cells. The pAPC polypeptide contains two centrally located β-catenin binding domains (Figure 28.11*a*). Cells with mutations that truncate pAPC tend to have high levels of monomeric β-catenin, which is then free to bind to other proteins, including certain transcription factors that promote cell proliferation (Figure 28.11*b*). In normal cells, pAPC suppresses cell proliferation by sequestering β-catenin. It also appears to promote apoptosis. Thus, pAPC regulates the opposing forces of cell proliferation and cell death within the intestinal epithelium.

phMSH2 The phMSH2 protein is the human homolog of a DNA repair protein called MutS found in bacteria and yeast. Its involvement in human cancer was elucidated through the study of *hereditary nonpolyposis colorectal cancer (HNPCC),* a dominant autosomal condition with a population frequency of about 1 in 500. Unlike FAP, HNPCC is characterized by the occurrence of a small number of adenomas, one of which eventually progresses to a cancerous condition. In the United States, the median age at which the cancer occurs is 42, the same age at which malignant cancer occurs in FAP patients.

The *hMSH2* gene was implicated in the inheritance of HNPCC after researchers found that cells in HNPCC tumors suffer from a general genetic instability. In these cells, di- and trinucleotide microsatellite repeat sequences (see Chapter 21) throughout the genome exhibit frequent changes in length. This instability is reminiscent of the types of DNA sequence changes observed in bacteria with mutations in the genes that control DNA mismatch repair (see Chapter 14). The human homolog of one of these bacterial genes maps to the short arm of chromosome 2, a chromosome that had previously been implicated in HNPCC by linkage analysis. Sequence analysis of this gene—

denoted *hMSH2*—indicated that it was inactivated in tumors removed from some HNPCC patients. Thus, loss of *hMSH2* function was causally connected to the genome-wide instability observed in HNPCC tumors. Subsequent analysis has demonstrated that germ-line mutations in *hMSH2,* or in three other human homologs of bacterial mismatch repair genes, account for the inherited cases of HNPCC.

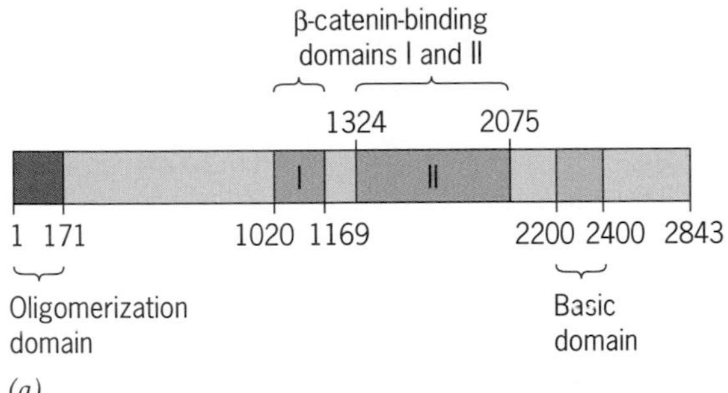

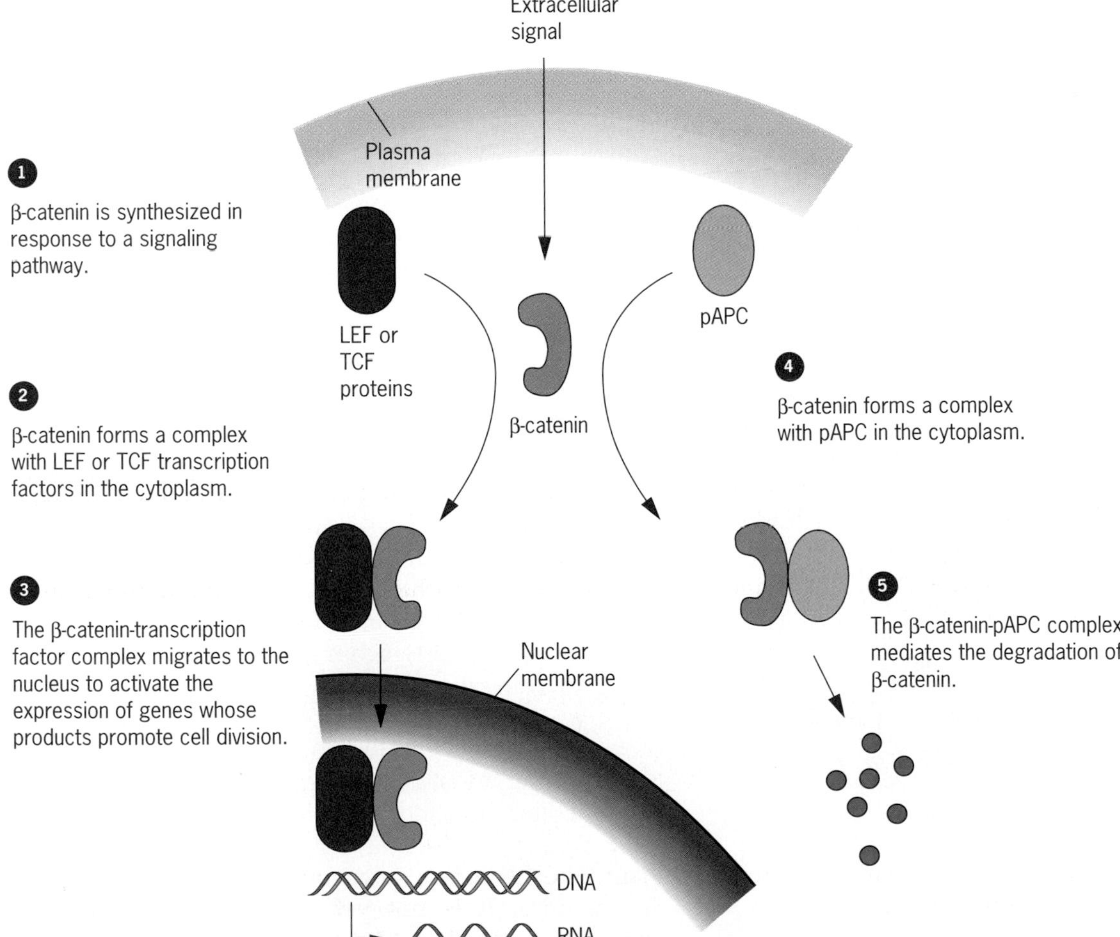

Figure 28.11 (*a*) Principal domains within pAPC. The numbers refer to amino acid positions in the polypeptide. (*b*) Role of pAPC in cell-cycle control. The pAPC protein influences progression through the cell cycle by interacting with β-catenin, a protein that can activate LEF or TCF transcription factors. Binding of β-catenin to pAPC reduces the chance that β-catenin will activate these transcription factors and stimulate cell division.

pBRCA1 and pBRCA2 Mutant versions of the tumor suppressor genes *BRCA1* and *BRCA2* genes have been implicated in hereditary breast and ovarian cancer. *BRCA1* was mapped to chromosome 17 in 1990 and isolated in 1994, and *BRCA2* was mapped to chromosome 13 in 1994 and isolated in 1995. Both genes encode large proteins; pBRCA1 is a 220-kilodalton polypeptide and pBRCA2 is a 384-kilodalton polypeptide. Cellular and biochemical studies have shown that each of these proteins is located within the nuclei of normal cells and that each contains a putative transcriptional activation domain. However, it is not known if either of these proteins functions as a bona fide transcription factor. The pBRCA1 and pBRCA2 proteins also contain a domain that allows them to interact physically with other proteins, in particular with pRAD51, a eukaryotic homolog of the bacterial DNA-repair protein known as RecA. Thus, pBRCA1 and

HUMAN GENETICS SIDELIGHT

Cancer and Genetic Counseling[1]

The identification of inherited mutations in tumor suppressor genes has opened a new era in genetics counseling. The carriers of such mutations are often at high risk to develop potentially life-threatening tumors, sometimes at a relatively early age. If molecular tests reveal that an individual carries a mutant tumor suppressor gene, medical treatment can be given to reduce the chance that he or she will develop a lethal cancer. For example, a child who carries a mutation in the *APC* gene could be checked periodically by endoscopy and suspicious lesions in the intestine could be removed, or a woman who carries a mutation in either of the *BRCA* genes could undergo a prophylactic mastectomy (removal of the breasts) or oophorectomy (removal of the ovaries).

A negative result from a test for a mutant tumor suppressor gene would, of course, be a cause for celebration—at least to the extent that the test can be trusted. For a large gene with many different mutant alleles segregating in the population, it is difficult to design a cost-effective test to detect mutations located anywhere in the gene. Typically these tests are based on the polymerase chain reaction, and most of them are designed to detect specific mutant alleles. An individual who is at risk to carry a mutant tumor suppressor gene can be tested for the known mutations—at least the most frequent ones. However, a negative result is not definitive because that individual could carry a "private" mutation—that is, one that has not previously been identified in the population.

The existence of private alleles makes counseling for inherited cancers difficult. For example, over 300 different mutations have been identified in the *BRCA1* gene, and about 50 percent of them are private. If an individual with a family history of breast cancer comes to a genetic counselor for evaluation, which mutations should the counselor look for? Sometimes data from other family members or information collected from the individual's ethnic group can provide clues. If other individuals in the family have been found to carry a particular mutant allele, then the counselor should test for that allele. If certain mutant alleles are characteristic of the individual's ethnic group, then the counselor should test for them. In Ashkenazi Jewish populations, for example, some *BRCA1* and *BRAC2* mutant alleles have frequencies as high as 2.5 percent. By comparison, the combined frequency of all mutant alleles in non-Jewish Caucasian populations is only 0.1 percent. Thus, an Ashkenazi Jew at risk for inherited breast or ovarian cancer should be tested for the mutant alleles that are likely to be segregating in Ashkenazi Jewish families.

Genetic testing for mutant tumor suppressor genes raises a host of psychological issues. In cases where therapeutic medical treatment is not available, an individual might choose not to be tested because the psychological burden of living with the knowledge that one carries a potentially lethal mutant gene could be overwhelming. Knowledge that one is a carrier would be expected to influence career plans and decisions about marriage and child-bearing. The prospect of an early death might dissuade an individual from seeking permanent commitments—to a spouse, to children, or to a vocation—and the chance of transmitting a mutant allele to children might deter the individual from reproducing. Knowledge that one is a carrier can also influence other people—family members, friends, and coworkers. A young daughter whose mother has tested positively for a *BRCA1* mutation must herself begin to grapple with the prospect of being a carrier, and a husband whose wife carries a *BRCA1* mutation must share in the decision of whether or not she should undergo a prophylactic oophorectomy and preclude the couple from ever having children of their own.

Testing for mutant tumor suppressor genes raises many ethical issues. To whom should the test results be revealed? The patient? The patient's family? Parents? Children? Employer? Landlord? Insurance agent? What measures should society take to safeguard the privacy of genetic test results? What policies should governments adopt to protect individuals from discrimination on the basis of their genotypes? How should insurance and employment policies be modified? Should the reproductive rights of individuals who carry harmful mutations be limited? As with any technological advance, the ability to detect mutations in tumor suppressor genes leaves us with many questions about how we should proceed. Currently, the answers to these questions are far from clear.

[1]Ponder, Bruce. 1997. Genetic testing for cancer risk. *Science* 278:1050–1054.

pBRCA2 may be involved in one of the many systems that repair damaged DNA in human cells.

Both pBRCA1 and pBRCA2 carry out important functions within cells. Mice that are homozygous for a knock-out mutation in the *BRCA1* gene die early during embryogenesis, and mice that are homozygous for a 3′ truncation of the *BRCA2* gene are small, short-lived, and prone to develop thymic lymphomas. Mutant *BRCA2* mice also accumulate chromosomal abnormalities. It is not yet clear what roles mutant pBRCA1 and pBRCA2 proteins play in human carcinogenesis. Possibly they compromise a cell's ability to detect or repair damaged DNA.

Mutations in the *BRCA1* and *BRCA2* genes account for about 7 percent of all cases of breast cancer and about 10 percent of all cases of ovarian cancer in the

United States. For each gene, the predisposition to develop these cancers is inherited as a dominant allele with high penetrance. Carriers have a 10- to 25-fold greater risk than noncarriers of developing breast or ovarian cancer, and in some families, the risk of developing colon or prostate cancer is also increased. Because many different inactivating mutations in *BRCA1* and *BRCA2* are found in the human population, genetics counseling for families that are segregating these mutations can be difficult (see the Human Genetic Sidelight: Cancer and Genetic Counseling).

Key Points: **Tumor suppressor genes were discovered through their association with rare, inherited cancers such as retinoblastoma. Mutational inactivation of various tumor suppressor genes is characteristic of most forms of cancer. Two mutational hits are required to eliminate both functional copies of a tumor suppressor gene within a cell. The proteins encoded by tumor supressor genes play key roles in regulating the cell cycle.**

GENETIC PATHWAYS TO CANCER

In most cancer cases, the formation of a malignant tumor is not attributable to the uncontrolled activation of a single proto-oncogene or to the inactivation of a single tumor suppressor gene. Rather, tumor formation, growth, and metastasis usually depend on the accumulation of mutations in several different genes. Thus, the genetic pathways to cancer are diverse and complex.

We can see this diversity and complexity in the formation and development of different types of tumors. For example, benign tumors of the large intestine develop in individuals with inactivating mutations in the *APC* gene. However, the progression of these tumors to potentially lethal cancers requires mutations in several other genes. This mutational pathway is summarized in Figure 28.12*a*. Inactivating mutations in the *APC* gene initiate the process of tumor formation by causing the development of abnormal tissues within the intestinal epithelium. These abnormal tissues contain dysplastic cells—cells with unusual shapes and enlarged nuclei—that may grow into early stage adenomas. If the *K-ras* proto-oncogene is activated in one of these adenomas, that adenoma may grow and develop more fully. Inactivating mutations in any of several tumor suppressor genes located in the long arm of chromosome 18 may then induce the adenoma to progress further, and inactivating mutations in the *TP53* tumor suppressor gene on chromosome 17 may transform it into a vigorously growing carcinoma. Additional tumor suppressor gene mutations may allow carcinoma cells to break away and invade other tissues. Thus, no less than seven independent mutations (two inactivating hits in the *APC* gene, one activating mutation in the *K-ras* gene, two inactivating hits in a tumor suppressor gene on chromosome 18, and two inactivating hits in the *TP53* gene) are required for the development of an intestinal carcinoma, and still more mutations are probably required for the metastasis of that carcinoma to other parts of the body.

The genetic pathways to prostate and brain cancer have also been elucidated (Figure 28.12*b* and Figure 28.12*c*). Mutations in *HPC1*, a gene for hereditary prostate cancer located in the long arm of chromosome 1, have been implicated in the origin of prostate tumors. Mutations in other tumor suppressor genes located in chromosomes 13, 16, 17, and 18 can transform prostate tumors into metastatic cancers, and overexpression of the *BCL-2* proto-oncogene gene can make these cancers immune to androgen deprivation therapy, a standard technique for the treatment of prostate cancer. The steroid hormone androgen is required for the proliferation of cells in the prostate epithelium. In the absence of androgen, these cells are programmed to die. However, prostate tumor cells may acquire the ability to survive in the absence of androgen, probably because an excess of the *BCL-2* gene product represses the programmed cell death pathway. Prostate cancers that have progressed to the stage of androgen independence are almost always fatal.

Glioblastomas are tumors that develop from glial cells in the brain. Primary glioblastomas usually occur in older individuals; they grow rapidly and are almost always fatal. Secondary glioblastomas usually occur in children and young adults; they develop from preexisting, low-grade tumors called astrocytomas and grow slowly—a feature that makes them more treatable than primary glioblastomas. Two proto-oncogenes have been implicated in the formation of glioblastomas. The gene for the epidermal growth factor receptor (EGFR) is frequently overexpressed in primary glioblastomas, and the gene for the platelet-derived growth factor (PDGF) protein is often overexpressed in secondary glioblastomas. Mutations in a variety of tumor suppressor genes, including *TP53*, *RB*, and *NF2*, have been implicated in the formation of glioblastomas, mainly in the development of secondary glioblastomas from low-grade astrocytomas. Many other tumor suppressor genes appear to be involved in the formation of glioblastomas, but they have not yet been identified.

Studies on the development and metastasis of different types of tumors have demonstrated that somatic mutations are almost always involved. Some of these mutations are specific for particular types of tumors, whereas others, like those that inactivate *TP53*, are found in almost all tumor types. In the progressive de-

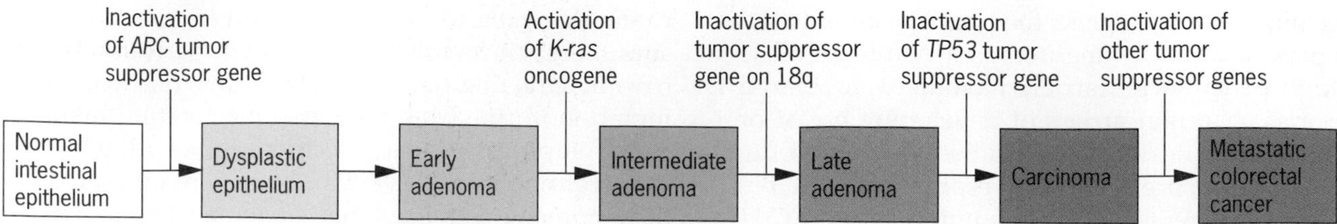

(*a*) Pathway to metastatic colorectal cancer.

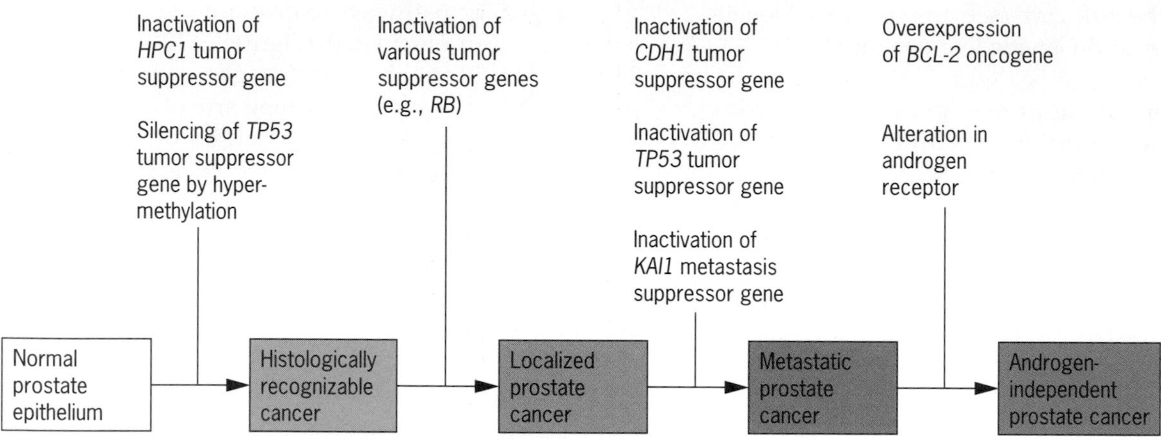

(*b*) Pathway to androgen-independent prostate cancer.

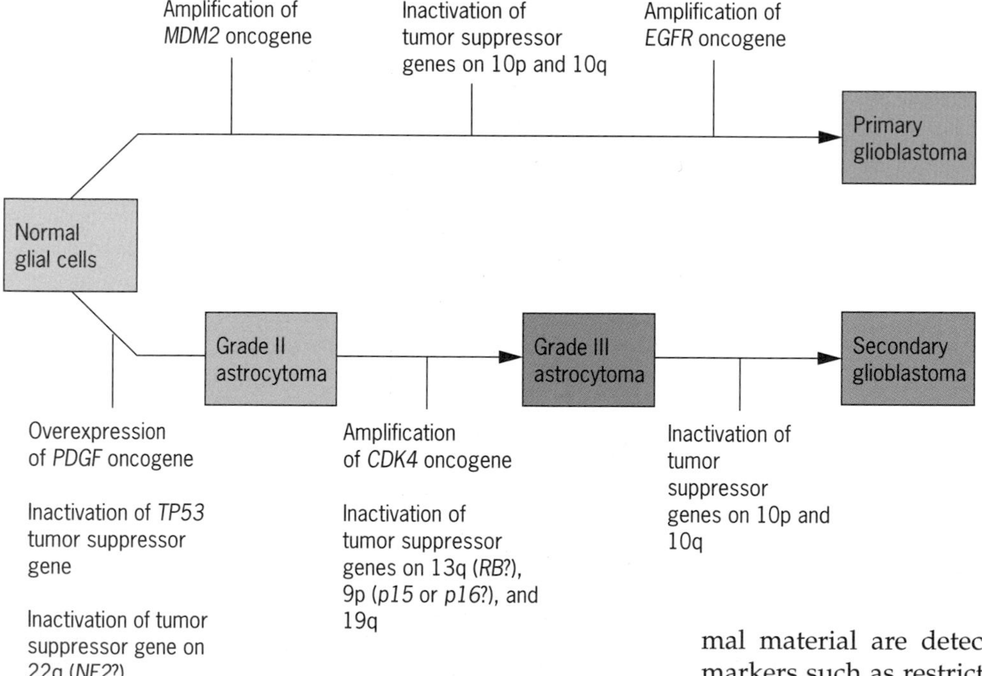

(*c*) Pathways to primary and secondary glioblastomas.

Figure 28.12 Genetic pathways to cancer.

velopment of a cancer, tumor suppressor genes may be inactivated when whole chromosomes or parts of chromosomes are lost. These deletions of chromosomal material are detected by the loss of molecular markers such as restriction fragment-length polymorphisms or microsatellite sequences that are linked to the tumor suppressor gene. Often it is possible to detect the loss of a segment within a specific member of a chromosome pair because the markers on that segment are heterozygous. Thus, even when a tumor suppressor gene has not been identified precisely, its role in the etiology of a cancer can be inferred by the consistent loss of genetically linked markers in different tumor specimens. Cancer researchers refer to this way of detecting a tumor suppressor gene's role as the *loss of heterozygosity*.

Because of the importance of somatic mutations in the etiology of cancer, factors that increase the mutation rate are bound to increase the incidence of cancer. Today many countries maintain research programs to identify mutagenic and carcinogenic agents (see Chapter 14 for a discussion of the Ames Test to identify chemical mutagens). When such agents are identified, public health authorities devise policies to minimize human exposure to them. However, no environment is carcinogen-free, and human behaviors that contribute to the risk of cancer such as smoking, excessive exposure to sunlight, and consumption of fatty foods that contain little fiber are difficult to change. Understanding of the processes that cause cancer has advanced significantly during the last decade. In the next decade, we can expect this understanding to lead to more effective strategies for cancer prevention and treatment.

Key Points: **Cancers develop through an accumulation of somatic mutations in proto-oncogenes and tumor suppressor genes. Different types of cancer are associated with mutations in different genes.**

TESTING YOUR KNOWLEDGE

An oncogene within the genome of a retrovirus has a high probability of causing cancer, but an oncogene in its normal chromosomal position does not. If these two oncogenes encode exactly the same polypeptide, how can we explain their different properties?

ANSWER

There are at least three possibilities. One (*a*) is that the virus simply adds extra copies of the oncogene to the cell and that collectively these produce too much of the polypeptide. An excess of polypeptide might cause uncontrolled cell division, that is, cancer. Another possibility (*b*) is that the viral oncogene is expressed inappropriately under the control of enhancers in the viral DNA. These enhancers might trigger the oncogene to be expressed at the wrong time or to be overexpressed constitutively. In either case, the polypeptide would be inappropriately produced and might thereby upset the normal controls on cell division. A third possibility (*c*) is that integration of the virus into the chromosomes of the infected cell might put the viral oncogene in the vicinity of an enhancer in the chromosomal DNA and that this enhancer might elicit inappropriate expression. All three explanations stress the idea that the expression of an oncogene must be correctly regulated. Misexpression or overexpression could lead to uncontrolled cell division.

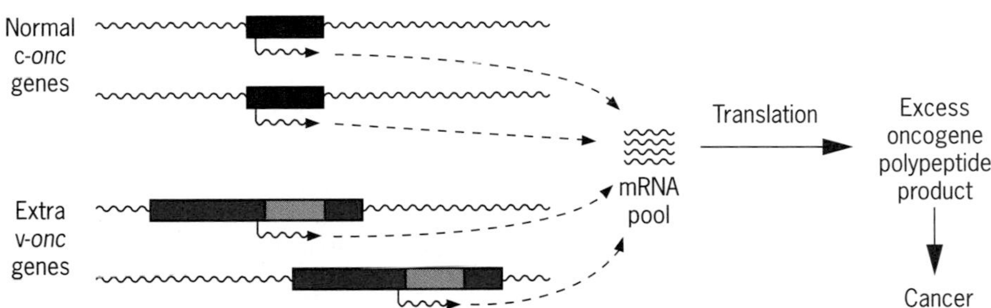

(*a*) The virus adds extra copies of the oncogene to the cell.

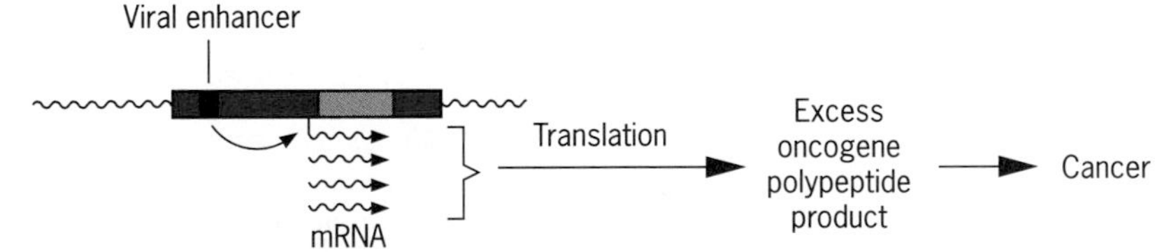

(*b*) The viral oncogene is expressed inappropriately under the control of a viral enhancer.

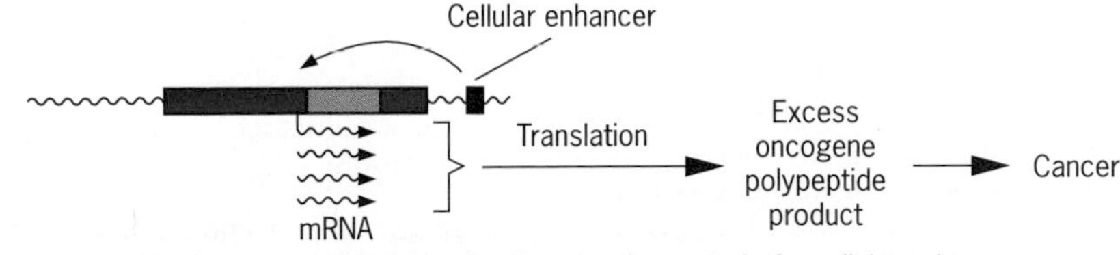

(*c*) The viral oncogene is expressed inappropriately under the control of a cellular enhancer.

QUESTIONS AND PROBLEMS

28.1 Many cancers seem to involve environmental factors. Why, then, is cancer called a genetic disease?

28.2 Both embryonic cells and cancer cells divide quickly. How can these two types of cells be distinguished from each other?

28.3 Most cancer cells are aneuploid. Suggest how aneuploidy might contribute to deregulation of the cell cycle.

28.4 Would you ever expect to find a tumor-inducing retrovirus that carried a processed cellular tumor suppressor gene in its genome?

28.5 How do we know that normal cellular oncogenes are not simply integrated retroviral oncogenes that have acquired the proper regulation?

28.6 How might the absence of introns in a retroviral oncogene explain that gene's overexpression in the tissues of an infected animal?

28.7 When cellular oncogenes are isolated from different animals and compared, the amino acid sequences of the polypeptides they encode are found to be very similar. What does this suggest about the functions of these polypeptides?

28.8 The majority of the *c-ras* oncogenes obtained from cancerous tissues have mutations in codon 12, 59, or 61 in the coding sequence. Suggest an explanation.

28.9 When a mutant *c-H-ras* oncogene with a valine for glycine substitution in codon 12 is transfected into cultured NIH 3T3 cells, it transforms those cells into cancer cells. When the same mutant oncogene is transfected into cultured embryonic cells, it does not transform them. Why?

28.10 A mutation in the *ras* cellular oncogene can cause cancer when it is in heterozygous condition, but a mutation in the *RB* tumor suppressor gene can cause cancer only when it is in homozygous condition. What does this difference between dominant and recessive mutations imply about the roles that the *ras* and *RB* gene products play in normal cellular activities?

28.11 Explain why individuals who develop nonhereditary retinoblastoma usually have tumors in only one eye, whereas individuals with hereditary retinoblastoma usually develop tumors in both eyes.

28.12 Inherited cancers like retinoblastoma show a dominant pattern of inheritance. However, the underlying genetic defect is a recessive loss-of-function mutation—often the result of a deletion. How can the dominant pattern of inheritance be reconciled with the recessive nature of the mutation?

28.13 Approximately 5 percent of the individuals who inherit an inactivated *RB* gene do not develop retinoblastoma. Use this statistic to estimate the number of cell divisions that form the retinal tissues of the eye. Assume that the rate at which somatic mutations inactivate the *RB* gene is 1 mutation per 10^6 cell divisions.

28.14 The following pedigree shows the inheritance of familial ovarian cancer caused by a mutation in the *BRCA1* gene. Should II-1 be tested for the presence of the predisposing mutation? Discuss the advantages and disadvantages of testing.

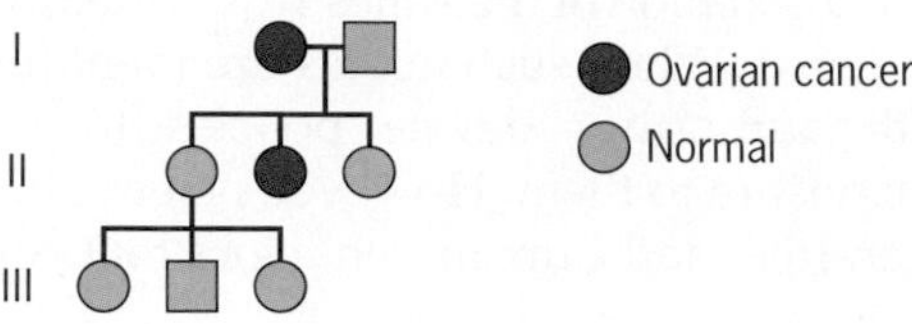

28.15 In what sense is pRB a negative regulator of E2F transcription factors?

28.16 A particular E2F transcription factor recognizes the sequence TTTCGCGC in the promoter of its target gene. A temperature-sensitive mutation in the gene encoding this E2F transcription factor alters the ability of its protein product to activate transcription; at 25°C the mutant protein activates transcription normally, but at 35°C, it fails to activate transcription at all. However, the ability of the protein to recognize its target DNA sequence is not impaired at either temperature. Would cells heterozygous for this temperature-sensitive mutation be expected to divide normally at 25°C? At 35°C? Would your answers change if the E2F protein functions as a homodimer?

28.17 During the cell cycle, the p16 protein is an inhibitor of cyclin-CDK activity. Predict the phenotype of cells homozygous for a loss-of-function mutation in the gene that encodes p16. Would this gene be classified as a proto-oncogene or as a tumor suppressor gene?

28.18 The *BCL-2* gene encodes a protein that represses the pathway for programmed cell death. Predict the phenotype of cells heterozygous for a dominant activating mutation in this gene. Would the *BCL-2* gene be classified as a proto-oncogene or as a tumor suppressor gene?

28.19 The protein product of the *BAX* gene negatively regulates the protein product of the *BCL-2* gene—that is, BAX protein interferes with the function of the BCL-2 protein. Predict the phenotype of cells homozygous for a loss-of-function mutation in the *BAX* gene. Would this gene be classified as a proto-oncogene or as a tumor suppressor gene?

28.20 Cancer cells frequently are homozygous for loss-of-function mutations in the *TP53* gene, and many of these mutations map in the portion of *TP53* that encodes the DNA-binding domain of p53. Explain how these mutations contribute to the cancerous phenotype of the cells.

28.21 Suppose that a cell is heterozygous for a mutation that caused p53 to bind tightly and constitutively to the DNA of its target genes. How would this mutation affect the cell cycle? Would such a cell be expected to be more or less sensitive to the effects of ionizing radiation?

28.22 Mice homozygous for a knock-out mutation of the

TP53 gene are viable. Would they be expected to be more or less sensitive to the killing effects of ionizing radiation?

28.23 Would cancer-causing mutations of the *APC* gene be expected to increase or decrease the ability of pAPC to bind β-catenin?

28.24 It has been demonstrated that individuals with diets poor in fiber and rich in fatty foods have an increased risk to develop colorectal cancer. Fiber-poor, fat-rich diets may irritate the epithelial lining of the large intestine. How could such irritation contribute to the increased risk for colorectal cancer?

28.25 Mice that are heterozygous for a knock-out mutation in the *RB* gene develop pituitary and thyroid tumors. Mice that are homozygous for this mutation die during embryonic development. Mice that are homozygous for a knock-out mutation in the gene encoding the p130 homolog of *RB* and heterozygous for a knock-out mutation in the gene encoding the p107 homolog of *RB* do not have a tendency to develop tumors. However, homozygotes for knock-out mutations in both of these genes die during embryonic development. What do these findings suggest about the roles of the *RB*, *p139*, and *p107* genes in embryos and adults?

28.26 Messenger RNA from the *KAI1* gene is strongly expressed in normal prostate tissues but weakly expressed in cell lines derived from metastatic prostate cancers. What does this finding suggest about the role of the *KAI1* gene product in the etiology of prostate cancer?

28.27 The p21 protein is strongly expressed in cells that have been irradiated. Researchers have thought that this strong expression is elicited by transcriptional activation of the *p21* gene by the p53 protein acting as a transcription factor. Does this hypothesis fit with the observation that p21 expression is induced by radiation treatment in mice homozygous for a knock-out mutation in the *TP53* gene? Explain.

BIBLIOGRAPHY

BLACKWOOD, M. A., AND B. L. WEBER. 1998. BRCA1 and BRCA2: From molecular genetics to clinical medicine. *J. of Clinical Oncology* 16:1969–1977.

FEARON, E. R. 1997. Human cancer syndromes: Clues to the origin and nature of cancer. *Science* 278:1043–1050.

KINZLER, K. W., AND B. VOGELSTEIN. 1996. Lessons from hereditary colorectal cancer. *Cell* 87:159–170.

KNUDSON, A. G. 1971. Mutation and cancer: Statistical study of retinoblastoma. *Proc. Natl. Acad. Sci. USA* 68:820–823.

KNUDSON, A. G. 1993. Antioncogenes and human cancer. *Proc. Natl. Acad. Sci. USA* 90:10914–10921.

LATIL, A. AND R. LIDEREAU. 1998. Genetic aspects of prostate cancer. *Virchow's Archiv* 432:389–406.

LOUIS, D. N., AND J. F. GUSELLA. 1995. A tiger behind many doors: Multiple genetic pathways to malignant glioma. *Trends in Genetics* 11:412–415.

MANSUR, C. P. 1997. The regulation and function of the p53 tumor suppressor. *Adv. in Dermatology* 13:121–166.

MULLIGAN, G., AND T. JACKS. 1998. The retinoblastoma gene family: Cousins with overlapping interests. *Trends in Genetics* 14:223–229.

NG, H. K., AND P. V. P. LAM. 1998. The molecular genetics of central nervous system tumors. *Pathology* 30:196–202.

ZHANG, H., G. TOMBLINE, AND B. L. WEBER. 1998. BRCA1, BRCA2 and DNA damage response: Collision or collusion? *Cell* 92:433–436.

A CONVERSATION WITH

DEBORAH AND BRIAN CHARLESWORTH

How did you become interested in genetics and specifically in evolutionary genetics?

D.C.: Actually, I have been interested in genetics since I was a teenager, though I was interested in biology much earlier than that. When I was in high school, my biology teacher suggested that I should listen to some programs on the radio which were talks by Julian Huxley, a renowned evolutionary biologist and a member of the famous Huxley family of biologists. I found the talks fascinating and later managed to get hold of Huxley's book, *Evolution: The Modern Synthesis.* From that point on, I was hooked. It was an exciting book to read. It really introduced me to genetics, which was not taught in our high school, and it introduced me to evolutionary genetics. So I got both of them right away at the beginning. It was an eye opener to me that biology could have such generality and be so beautiful. Brian, I believe you had a similar type of experience.

B.C.: Well, I can be brief because my experience was indeed highly similar, except that it was not a particular book that ignited my interest. As a child, I was always very interested in natural history. In high school I had a very good biology teacher who taught us some genetics. I got excited about biology and genetics, and how genetics connected with evolution; it just went on from there.

Brian Charlesworth and **Deborah Charlesworth** are an amazing team; they collaborate as scientists and have authored many of their scientific papers together. They were born in England and received their Ph.D. degrees in genetics from the University of Cambridge. In 1985, they emigrated to the United States. Their work in population genetics and evolutionary biology has won them international recognition. Until 1997, they were at the University of Chicago, where Brian Charlesworth was George Wells Beadle Distinguished Service Professor and Deborah Charlesworth was Professor of Ecology and Evolution.

In 1997, they returned to the UK, where they work at the University of Edinburgh. Brian Charlesworth is a Royal Society Research Professor, and Deborah Charlesworth is a Senior Research Fellow.

You both have published extensively on a wide variety of topics such as transposable genetic elements, mating systems, mutational load, and chromosome evolution. How do you decide which research topics to pursue?

B.C.: Well, I don't think I actually plan ahead very carefully. Basically, you think about one thing, and it occurs to you that it might relate to something else, and I think you go on from day to day building on ideas and making connections. What do you think, Deborah?

D.C.: Yes, I think one thing leads to another, quite unexpectedly sometimes. For example, our work on the evolution of separate sexes was, oddly enough, the result of previous research we were doing on mimicry in butterflies and on the close linkage of genes involved in controlling the different mimetic patterns. So we got interested in looking at other evolutionary or genetic situations where you knew that several genes were involved but they were so closely linked they behaved as a block. This block of tightly linked genes looked like a single locus with different alleles. The control of sex by the sex chromosomes is a classic case of the same phenomenon, and that led me further into plant mating systems and breeding systems in general.

This line of investigation led us later into studies of deleterious mutations because we were looking at how inbreeding leads to a loss of vigor. We were thinking about possible reasons why organisms might have systems to ensure outbreeding. Our interest in deleterious mutations has recently led us into studying the effect of deleterious mutations in shaping molecular evolution. It's quite strange really how things that at first might seem quite disparate do relate to one another. Our research is actually a very nice interconnected system.

I think that this is one of the wonderful things about genetics. It's trite to say that it integrates so much of biology, but that is one of the attractions of studying and working in genetics. You learn about all different types of biological systems, and you don't have to specialize too narrowly on a single system. From reading the literature you find useful information from a different organism or a different system that relates to your own system, and this is always fascinating.

Let's explore Darwinism for a moment. There continues to be resistance to Darwinism, the most recent example being the attempt by the Tennessee State Legislature to pass a law requiring teachers to refer to evolution as a theory and not fact. Why do you think that there is this continuing resistance to evolution and what can we do as biologists to counter it?

B.C.: To some extent, this is more of a problem in the United States than in Europe, and I think ultimately the answer to why people are hostile to Darwinism is religious prejudice. Opponents of evolution feel that the idea that humans and other organisms are the product of natural forces extending over billions of years overthrows their fundamental beliefs about the nature of their relation to the universe. It is not something that most professional biologists have a problem with, because we are obviously trained to think of the world in terms of scientific evidence, not something which is written down in a book and taken for granted. I think that it is the same kind of problem that Galileo encountered with the Catholic Church 300 years ago. People do not want to see themselves as cosmically insignificant, so to speak.

What we can do about it is a more difficult question. I think there should certainly be a push for much better science education in schools at all levels, and that people should be introduced to science in a way which is not just *here are the facts, learn them.* Students need to understand the ongoing process of exploring nature. They should be taught how to relate ideas and concepts to data and to learn how to test ideas against facts very early on. One should be absolutely hard-nosed about the fact that the evidence of evolution is based on the same kind of scientific evidence as any other theory in science. If you begin to question evolution, you might as well give up as far as the whole of science is concerned. I think it is not really widely appreciated that the creationist attitude is fundamentally irrational and is essentially a threat to our entire enterprise of rational investigation of the natural world.

We should be concerned about efforts by the fundamentalists to interject religion into science. They are politically astute in getting control of school boards and this kind of thing, and they will undoubtedly use that power as a vehicle for propagating creationist views. Anti-evolutionary views could be a problem at the national level if they affect the funding for basic research, although I don't see any evidence for this at the moment.

D.C.: I think Brian put it very clearly, and I agree with everything he says. I do think that the main thing we can do against this type of irrational thinking is educational. The idea of evolution is really simple in outline, and I think that many people do not accept it because they haven't been taught it in a clear way. The idea that we are animals and the product of evolution along with other animals comes very naturally of course, once you see the evidence and follow the reasoning.

I certainly think that one of the defects of a lot of science education is that people are just told to learn the facts. Science teaching ought to include more of a debate component. People need to understand that the scientific process involves a lot of debate. I remember as a high school student arguing with teachers. Often I was being ridiculous, of course, and I was trying to defend indefensible positions, but in doing that sort of debate you get an idea of what the support really is for ideas, and where it is lacking. You quickly realize the weakness of your position when you are arguing for something that is unsupported by data. Presenting only what is accepted in textbooks is a difficulty for young people studying science. They don't want to swallow something merely because it's written in a book. Science does not work that way.

"Where do we come from? What are we? Where are we going?" An 1897 painting by the French artist Paul Gauguin.

29

Population and Evolutionary Genetics

CHAPTER OUTLINE

The Emergence of Evolutionary Theory and Population Genetics
The Theory of Allele Frequencies
Natural Selection
Random Genetic Drift
Populations in Genetic Equilibrium
Genetic Variation in Natural Populations
Speciation
Human Evolution

D'ou venons nous? Que sommes nous? Ou allons nous?

In 1897 in Tahiti, the French artist Paul Gauguin created an enormous painting with a provocative title: "Where do we come from? What are we? Where are we going?" The painting, now on display in the Boston Museum of Fine Arts, shows a group of Polynesian people, both young and old, reclining, sitting, walking, and eating in a strangely colored landscape. The figures are forlorn and abstracted, and a few of them seem to stare interrogatively at the viewer, posing, as it were, those three haunting questions that Gauguin inscribed in the painting's margin. This melancholy canvas, created near the end of Gauguin's life, seems to depict the artist's personal search for answers to some of life's deep questions. However, it is more than the statement of an individual who sought inspiration, freedom, and fulfillment in the South Seas. Gauguin's painting reflects a universal quest for what it means to be human. During the nineteenth century, people began to see this issue in a new light, especially with the emergence of evolutionary theory. Charles Darwin's *Origin of Species,* first published in 1859, advanced the ideas that species were not fixed and that populations of organisms changed over time. In a later book, *The Descent of Man,* Darwin proposed that the human species was also subject to evolutionary forces. Darwin's ideas troubled many people.

In this chapter, we consider the genetic basis of evolutionary change—how populations of organisms are modified over time, and how they split into different populations to form new species. We begin by discussing the emergence of evolutionary theory and its incorporation of the principles of Mendelian genetics. Then we discuss the basic ideas of population genetics, their applications in different situations, and the ways in which evolutionary forces change a population's genetic composition over time. In the final sections of this chapter, we consider the process of speciation and the evolution of the human species.

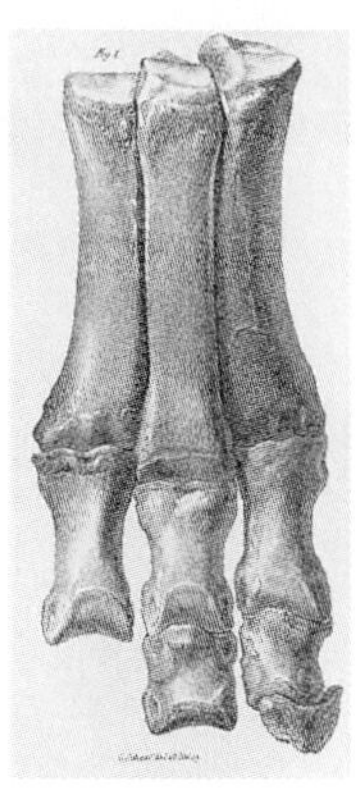

(*a*)

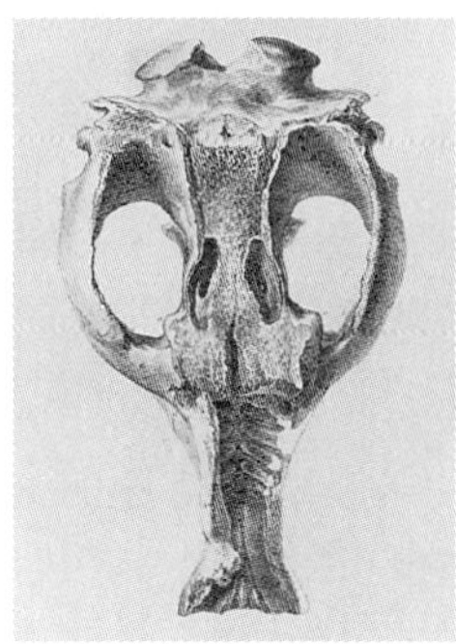

(*b*)

Figure 29.1 (*a*) The Patagonian guanaco is similar to the extinct Macrauchenia. (*b*) The cabybara is similar to the extinct Toxodon.

THE EMERGENCE OF EVOLUTIONARY THEORY AND POPULATION GENETICS

Darwin's theory of evolution precipitated one of the most complex and profound intellectual revolutions in human history. This revolution affected philosophy, religion, economics, politics, and social theory, as well as biology. The idea of evolution—that species change in appearance and behavior over time—had existed for more than 100 years before Darwin. However, it was not until Darwin published *The Origin of Species* that this idea attracted much attention.

Darwin's Theory of Evolution

Darwin proposed that a species changes as a result of competition between variants within the species. Variants that favor survival and reproduction tend to be preserved in the species, and variants that impair survival and reproduction tend to be lost. Over a long period of time, this process, which Darwin called *natural selection,* changes the characteristics of the species—that is, the species evolves.

What led Darwin to his theory of evolution? Probably the most important influence was a five-year voyage on the British survey ship *H.M.S. Beagle.* On the recommendation of one of his university teachers, Darwin was appointed as the ship's naturalist. The *Beagle* departed from England in 1831, traveled to South America, and returned to England in 1836. The lengthy sojourn along the coast of South America afforded Darwin many opportunities to observe plants, animals, and geological formations. From these observations, he was led to the view that species were not static; rather, he inferred that they change slowly over time and that some even become extinct.

Darwin amassed voluminous data on his trip around South America. For example, in Patagonia he observed fossils that bore a marked similarity to living animals (Figure 29.1). Did this similarity imply that the fossils were the ancestors of the current inhabitants of the region? Had these species changed subtly over time through the inheritance of anatomical modifications? In Chile, Darwin observed the effects of a recent earthquake. Land had been moved and rearranged. Could such upheavals explain why fossil sea creatures were found in rocks in the Andes Mountains more than 4 kilometers above the current sea level? In the Galapagos Islands, Darwin observed 13 species of finches that all resembled birds on the South American mainland (Figure 29.2). However, each finch species was distinct in size, beak shape, and feeding behavior. Darwin hypothesized that these island species were descended from migrants from the mainland and that over time they had adapted to different ecological niches.

Darwin spent more than 20 years analyzing and interpreting the data that he collected on the voyage of the *Beagle.* In addition, at his country estate in Kent,

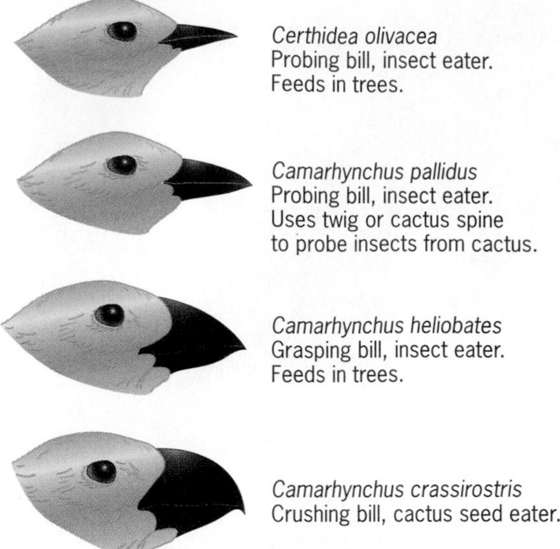

Figure 29.2 Examples of beak shape variation in different species of Galapagos finches, as correlated with feeding.

England, he performed experiments with a variety of plants and domesticated animals. The observations that he made in this experimental work, along with the data collected on the *Beagle*'s voyage, gave Darwin the insights that eventually led to the publication of *The Origin of Species*. However, as odd as it might seem, Darwin was actually forced into the publication of this important book. Another indefatigable naturalist, Alfred Wallace, had come to the same conclusions as Darwin through his study of plants and animals in South America and the East Indies. In 1858, Wallace sent a manuscript explaining his ideas to Darwin and asked that Darwin transmit the manuscript to the Linnean Society in London for publication. What could Darwin do but comply—and transmit his own preliminary manuscripts as well? The following year, 1859, Darwin published a full account of his ideas on evolution—the book entitled *The Origin of Species by Means of Natural Selection or the Preservation of Favored Races in the Struggle for Life.*

Population Genetics

Darwin's theory of evolution had one major gap. It offered no explanation for the origin of variation, and it could not explain how particular variants are inherited. Eventually, Darwin did propose a theory of inheritance based on the transmission of acquired characteristics. However, his theory was flawed. Biologists who were attracted to Darwin's ideas on evolution struggled, as he did, to explain how the variants that natural selection favors are transmitted from parents to offspring. In 1900 the rediscovery of Mendel's principles provided the long-sought-after explanation: Traits are determined by genes, which segregate different alleles, and genes are transmitted to the offspring in gametes produced by their parents. The analysis of genetic transmission in experimental crosses and pedigrees quickly gave rise to a new type of analysis that involved whole populations. The discipline of population genetics was born, and by 1930, especially through the contributions of Sewall Wright, R. A. Fisher, and J. B. S. Haldane (Figure 29.3), it had

Figure 29.3 (*a*) Sewall Wright. (*b*) R. A. Fisher. (*c*) J. B. S. Haldane.

become the foundation for Darwin's evolutionary theory.

Population genetics studies the origin of variation, the transmission of variants from parents to offspring generation after generation, and the temporal changes that occur in a population because of systematic and random evolutionary forces. Genetic variation originates when one allele mutates to another. Mutation is, therefore, the ultimate source of all genetic variability. Over time, mutational variation is sifted and shaped by evolutionary forces. Through natural selection, mutants with impaired survival and reproductive abilities decrease in frequency, and mutants with superior survival and reproductive abilities increase in frequency. In addition to the systematic changes caused by natural selection, random changes sometimes occur in populations. For example, if a particular mutant is neither better nor worse than any of the other genetic variants in the population, natural selection does not affect it—we say that the mutant is selectively neutral. The frequency of such a mutant may rise or fall in a population purely by accident. Population genetics studies both the systematic and random forces of evolution and in so doing, attempts to elucidate the fundamental mechanisms that cause genetic changes in a population over time.

Key Points: **Evolution is a process in which genetic variation in a population changes over time. Population genetics is the study of how this variation changes under the influence of systematic and random evolutionary forces such as mutation and natural selection.**

THE THEORY OF ALLELE FREQUENCIES

The theory of population genetics is a theory of allele frequencies. Each gene in the genome exists in different allelic states, and, if we focus on a particular gene, a diploid individual is either a homozygote or a heterozygote. Within a population of individuals, we can calculate the frequencies of the different types of homozygotes and heterozygotes of a gene, and from these frequencies we can estimate the frequency of each of the gene's alleles. These calculations are the foundation for population genetics theory.

Estimating Allele Frequencies

Because an entire population is usually too large to study, we resort to analyzing a representative sample of individuals from it. Table 29.1 presents data from a sample of people who were tested for the M-N blood types. These blood types are determined by two alleles of a gene on chromosome 4: L^M, which produces the M blood type, and L^N, which produces the N blood type (see Chapter 4). People who are L^ML^N heterozygotes have the MN blood type.

TABLE 29.1
Frequency of the M-N Blood Types in a Sample of 6129 Individuals

Blood Type	*Genotype*	*Number of Individuals*
M	L^ML^M	1787
MN	L^ML^N	3039
N	L^NL^N	1303

To estimate the frequencies of the L^M and L^N alleles, we simply calculate the incidence of each allele among all the alleles sampled:

1. The total number of alleles in the sample is two times the sample size: $2 \times 6129 = 12{,}258$.
2. The frequency of the L^M allele is two times the number of L^ML^M homozygotes plus the number of L^ML^N heterozygotes, all divided by the total number of alleles sampled: $[(2 \times 1787) + 3039]/12{,}258 = 0.5395$.
3. The frequency of the L^N allele is two times the number of L^NL^N homozygotes plus the number of L^ML^N heterozygotes, all divided by the total number of alleles sampled: $[(2 \times 1303) + 3039]/12{,}258 = 0.4605$.

Thus, letting p represent the frequency of the L^M allele and letting q represent the frequency of the L^M allele, we estimate that in the population from which the sample was taken, $p = 0.5395$ and $q = 0.4605$. Furthermore, because L^M and L^N are the only two alleles of this particular gene, $p + q = 1$.

When directly counting the number of alleles in a sample is not possible because one of the alleles is phenotypically dominant, we cannot use this method to estimate the allele frequencies. However, another method, discussed in a section below, does provide these estimates.

When the gene under study is X-linked, we only need to count the different alleles in males. For example, in a sample of 200 men, 24 have X-linked color blindness and all the others have normal color vision. Assuming that each color blind man is hemizygous for the same mutant allele, we estimate the frequency of that allele to be $24/200 = 0.12$ and the frequency of the normal allele to be $1 - 0.12 = 0.88$.

In these examples, each of the alleles has a reasonably high frequency—one that can be estimated reliably with a sample of moderate size. However, some alleles have frequencies of 0.01 or less, and estimating their frequencies, or even detecting them, requires a large sample. Whenever the second most frequent allele of a gene has a frequency greater than 0.01, we refer to the situation as a genetic **polymorphism.** Later in this chapter we shall discuss the evolutionary forces that maintain genetic polymorphisms in nature.

Relating Genotype Frequencies to Allele Frequencies: The Hardy–Weinberg Principle

Do the estimated allele frequencies have any predictive power? Can we use them to predict the frequencies of genotypes? In the first decade of the twentieth century, these questions were posed independently by G. H. Hardy, a British mathematician, and by Wilhelm Weinberg, a German physician. In 1908 Hardy and Weinberg each published papers describing a mathematical relationship between allele frequencies and genotype frequencies. This relationship, now called the **Hardy–Weinberg Principle,** allows us to predict a population's genotype frequencies from its allele frequencies.

Let's suppose that in a population a particular gene is segregating two alleles, *A* and *a*, and that the frequency of *A* is p and that of *a* is q. If we assume that the members of the population mate randomly, then the diploid genotypes of the next generation will be formed by the random union of haploid eggs and haploid sperm (Figure 29.4). The probability that an egg (or sperm) carries *A* is p, and the probability that it carries *a* is q. Thus, the probability of producing an *AA* homozygote in the population is simply $p \times p = p^2$, and the probability of producing an *aa* homozygote is $q \times q = q^2$. For the *Aa* heterozygotes, there are two possibilities: An *A* sperm can unite with an *a* egg, or an *a* sperm can unite with an *A* egg. Each of these events occurs with probability $p \times q$, and because they are equally likely, the total probability of forming an *Aa* zygote is $2pq$. Thus, on the assumption of random mating, the predicted frequencies of the three genotypes in the population are:

Genotype	Frequency
AA	p^2
Aa	$2pq$
aa	q^2

These predicted frequencies are the terms in the expansion of the binomial expression $(p + q)^2 = p^2 + 2pq + q^2$. Population geneticists refer to them as the Hardy–Weinberg genotype frequencies.

The key assumption underlying the Hardy–Weinberg Principle is that the members of the population mate at random with respect to the gene under study.

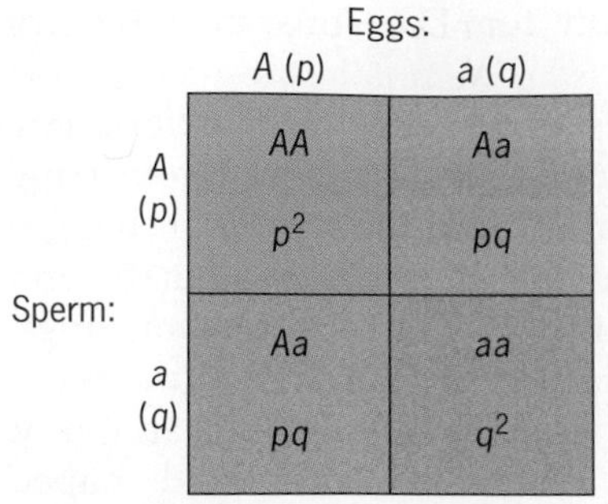

Figure 29.4 Punnett Square showing the Hardy–Weinberg Principle.

This assumption means that the adults of the population essentially form a pool of gametes that, at fertilization, combine randomly to produce the zygotes of the next generation. If these zygotes have equal chances of surviving to the adult stage, then the genotype frequencies created at the time of fertilization will be preserved, and when the next generation reproduces, these frequencies will once again appear in the offspring. Thus, with random mating and no differential survival or reproduction among the members of the population, the Hardy–Weinberg genotype frequencies—and, of course, the underlying allele frequencies—persist generation after generation. This condition is referred to as the **Hardy–Weinberg equilibrium.** Later in this chapter we shall consider forces that upset this equilibrium by altering allele frequencies; these forces—mutation, migration, natural selection, and random genetic drift—play key roles in the evolutionary process.

Applications of the Hardy–Weinberg Principle

Let's return to the M-N blood type example to see how the Hardy–Weinberg Principle applies to a real population. From the sample data given in Table 29.1, the frequency of the L^M allele was estimated to be $p = 0.5395$ and the frequency of the L^N allele was estimated to be $q = 0.4605$. With the Hardy–Weinberg Principle, we can now use these frequencies to predict the genotype frequencies of the M-N blood type gene:

Genotype	Hardy–Weinberg Frequency
L^ML^M	$p^2 = (0.5395)^2 = 0.2911$
L^ML^N	$2pq = 2\,(0.5395)\,(0.4605) = 0.4968$
L^NL^N	$q^2 = (0.4605)^2 = 0.2121$

Do these predictions fit with the original data from which the two allele frequencies were estimated? To answer this question, we must compare the observed genotype numbers with numbers predicted by the Hardy–Weinberg Principle. We obtain these predicted numbers by multiplying the Hardy–Weinberg frequencies by the size of the sample taken from the population. Thus,

Genotype	Predicted Number
$L^M L^M$	$0.2911 \times 6129 = 1784.2$
$L^M L^N$	$0.4968 \times 6129 = 3044.8$
$L^N L^N$	$0.2121 \times 6129 = 1300.0$

The results are extraordinarily close to the original sample data presented in Table 29.1.

We can check for agreement between the observed and predicted numbers by calculating a chi-square statistic (see Chapter 3):

$$\chi^2 = \frac{(1787 - 1784.2)^2}{1784.2} + \frac{(3039 - 3044.8)^2}{3044.8} + \frac{(1303 - 1300.0)^2}{1300.0} = 0.0223$$

This chi-square statistic has $3 - 2 = 1$ degree of freedom because (1) the sum of the three predicted numbers is fixed by the sample size, and because (2) the allele frequency p was estimated directly from the sample data. (The frequency q can be estimated indirectly as $1 - p$ and therefore does not reduce the degrees of freedom any further.) The critical value for a chi-square statistic with one degree of freedom is 3.841, which is much greater than the observed value. Consequently, we conclude that the predicted genotype frequencies are in agreement with the observed frequencies in the sample, and furthermore, we infer that in the population from which the sample was obtained, the M-N genotypes are in Hardy–Weinberg proportions—a finding that is not too surprising given that marriage is usually not based on blood type.

The preceding analysis indicates how we can use the Hardy–Weinberg Principle to predict genotype frequencies from allele frequencies. Can we turn the Hardy–Weinberg Principle around and use it to predict allele frequencies from genotype frequencies? For example, in the United States, the incidence of the recessive metabolic disorder phenylketonuria (PKU) is about 0.0001. Does this statistic allow us to calculate the frequency of the mutant allele that causes PKU?

We cannot proceed as before by counting the different types of alleles, mutant and normal, that are present in the population because heterozygotes and normal homozygotes are phenotypically indistinguishable. Instead, we must proceed by applying the Hardy–Weinberg Principle in reverse to estimate the mutant allele frequency. The incidence of PKU, 0.0001, represents the frequency of mutant homozygotes in the population. Under the assumption of random mating, these individuals should occur with a frequency equal to the square of the mutant allele frequency. Denoting this allele frequency by q, we have

$$q^2 = 0.0001$$
$$q = \sqrt{0.0001} = 0.01$$

Thus, 1 percent of the alleles in the population are estimated to be mutant. Using the Hardy–Weinberg Principle in the usual way, we can then predict the frequency of people in the population who are heterozygous carriers of the mutant allele:

$$\text{Carrier frequency} = 2pq = 2\,(0.99)\,(0.01) = 0.019$$

Thus, approximately 2 percent of the population are predicted to be carriers.

The Hardy–Weinberg Principle also applies to X-linked genes and to genes with multiple alleles. For an X-linked gene such as the one that controls color vision, the allele frequencies are estimated from the frequencies of the genotypes in males, and the frequencies of the genotypes in females are obtained by applying the Hardy–Weinberg Principle to these estimated allele frequencies. (We assume, of course, that the allele frequencies are the same in the two sexes.) If the frequency of the allele for normal color vision (C) is $p = 0.88$ (taken from our earlier example) and the frequency of the allele for color blindness (c) is $q = 0.12$, then, under the assumptions of random mating and equal allele frequencies in the two sexes, we have:

Sex	Genotype	Frequency	Phenotype
Males	C	$p = 0.88$	Normal vision
	c	$q = 0.12$	Color blind
Females	CC	$p^2 = 0.77$	Normal vision
	Cc	$2pq = 0.21$	Normal vision
	cc	$q^2 = 0.02$	Color blind

For genes with multiple alleles, the Hardy–Weinberg genotype proportions are obtained by expanding a multinomial expression. For example, the A-B-O blood types are determined by three alleles I^A, I^B, and I^O. If the frequencies of these are p, q, and r, respectively, then the frequencies of the six different genotypes in the A-B-O blood-typing system are obtained by expanding the trinomial $(p + q + r)^2 = p^2 + q^2 + r^2 + 2pq + 2qr + 2pr$:

Blood type	Genotype	Frequency
A	$I^A I^A$	p^2
	$I^A I^O$	$2pr$
B	$I^B I^B$	q^2
	$I^B I^O$	$2qr$
AB	$I^A I^B$	$2pq$
O	$I^O I^O$	r^2

Exceptions to the Hardy–Weinberg Principle

There are many reasons why the Hardy–Weinberg Principle might not apply to a particular population. Mating might not be random, the members of the population carrying different alleles might not have equal chances of surviving and reproducing, the population might be subdivided into partially isolated units, or it might be an amalgam of different populations that have come together recently by migration. We now briefly consider each of these exceptions to the Hardy–Weinberg Principle.

1. *Nonrandom mating.* Random mating is the key assumption underlying the Hardy–Weinberg Principle. If mating is not random, the simple relationship between allele frequencies and genotype frequencies breaks down. There are two ways in which the members of a population might mate nonrandomly. First, they might mate with each other because they are genetically related—for example, because they are siblings or first-cousins. We refer to this type of nonrandom mating as *consanguineous mating* (see Chapter 5). Second, the members of the population might mate with each other because they are phenotypically similar—for example, because the mates have the same stature or skin color. We refer to this type of nonrandom mating as *assortative mating.*

Consanguineous mating and assortative mating have the same qualitative effect; they reduce the frequency of heterozygotes and increase the frequency of homozygotes compared to the Hardy–Weinberg genotype frequencies. For the case of consanguineous mating, we can quantify this effect by using the inbreeding coefficient, F (see Chapter 5). Let's suppose that a gene has two alleles, A and a, with respective frequencies p and q, and that the population in which the gene is segregating has reached a level of inbreeding measured by F. (Recall from Chapter 5 that the range of F is between 0 and 1, with 0 corresponding to no inbreeding and 1 corresponding to complete inbreeding.) The genotype frequencies in this population are given by the following formulae:

Genotype	Frequency with Consanguineous Mating
AA	$p^2 + pqF$
Aa	$2pq - 2pqF$
aa	$q^2 + pqF$

From these formulae, it is clear that the frequencies of the two homozygotes have increased compared to the Hardy–Weinberg frequencies and that the frequency of the heterozygotes has decreased compared to the Hardy–Weinberg frequency. For each genotype, the change in frequency is directly proportional to the inbreeding coefficient. For a population that is completely inbred, $F = 1$, and the genotype frequencies become:

Genotype	Frequency with $F = 1$
AA	p
Aa	0
aa	q

With assortative mating, mathematical expressions for the genotype frequencies are more complicated than those for consanguineous mating and are beyond the scope of this book. However, assortative mating has the same general effect as consanguineous mating; it increases the frequency of homozygotes and decreases the frequency of heterozygotes. These changes occur because phenotypically similar individuals tend to have similar genotypes. Thus, when such individuals mate, they tend to produce more homozygous offspring than do randomly mated individuals.

2. *Unequal survival.* If zygotes produced by random mating have different survival rates, we would not expect the genotype frequencies of the individuals that develop from these zygotes to conform to the Hardy–Weinberg predictions. For example, let's consider data from a study of polytene chromosome structure in the larvae of *Drosophila tropicalis,* a relative of *D. melanogaster.* There are two polytene banding sequences, *A* and *D*, that can be identified in one of the autosomes of the *D. tropicalis* genome. Because these sequences do not recombine with each other, they can be treated as if they are alleles:

Genotype	Observed Number	Hardy–Weinberg Prediction
AA	3	35
AD	134	70
DD	3	35

The Hardy–Weinberg predictions were calculated by estimating the frequencies of each of the "alleles" (frequency of A is $p = 0.5$; frequency of D is $q = 0.5$) and then using these frequencies to predict the genotype frequencies under the assumptions of random mating and equal survival of the three genotypes. Obviously, the predictions are not in agreement with the observed data, which show an excess of heterozygotes and a dearth of both homozygotes. (Here the disagreement is so obvious that a chi-square calculation is unnecessary.) The explanation for these discrepancies probably lies with differential survival among the larvae; the heterozygotes have a much better chance of surviving to the late larval stage (when the chromosome banding patterns are analyzed) than do either of the two homozygotes. Thus, unequal survival rates can lead to genotype frequencies that deviate from the Hardy–Weinberg predictions.

3. *Population subdivision.* When a population is a single interbreeding unit, it is said to be **panmictic.** Panmixis (the noun) implies that any member of the population is able to mate with any other member—that is, there are no geographical or ecological barriers to mating in the population. In nature, however, populations are often subdivided. We can think of fish living in a group of lakes that are intermittently connected by rivers, or of birds living on a chain of islands in an archipelago. Such populations are structured by geographical and ecological features that might be correlated with genetic differences. For example, the fish in one lake might have a high frequency of allele *A*, while those in another lake might have a low frequency of this allele. Although the genotype frequencies might conform to Hardy–Weinberg predictions within each lake, across the entire range of the fish population, they will not. Geographical subdivision makes the population genetically inhomogeneous, and such inhomogeneity violates a tacit assumption of the Hardy–Weinberg Principle: that allele frequencies are uniform throughout the population.

4. *Migration.* When individuals move from one territory to another, they carry their genes with them. The introduction of genes by recent migrants can alter allele and genotype frequencies within a population and disrupt the state of Hardy–Weinberg equilibrium. As an example, let's consider the situation in Figure 29.5. Two populations of equal size are separated by a geographical barrier. In population I the frequencies of *A* and *a* are both 0.5, whereas in population II the frequency of *A* is 0.8 and that of *a* is 0.2. With random mating within each population, the Hardy–Weinberg Principle predicts that the two populations will have different genotype frequencies (see Figure 29.5).

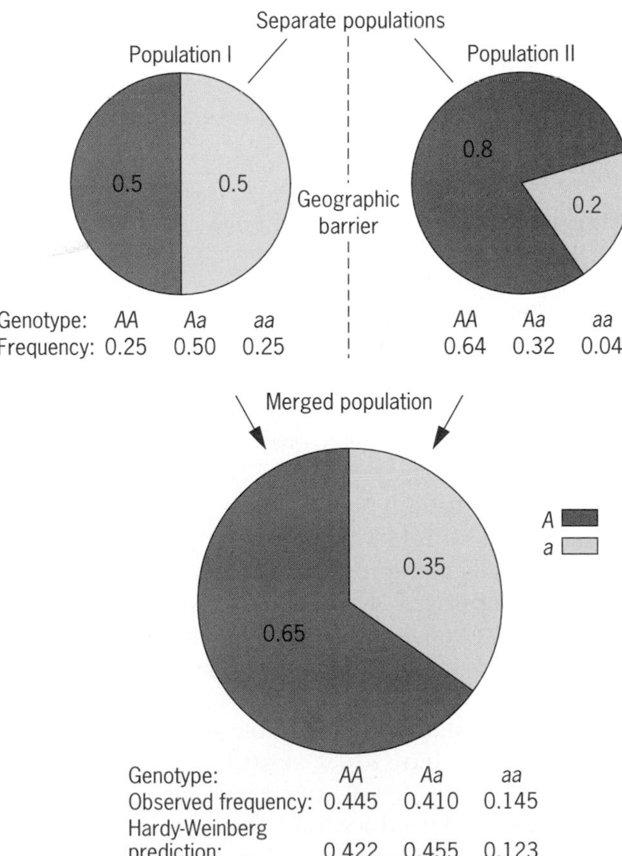

Figure 29.5 Effects of population merger on allele and genotype frequencies.

Let's suppose that the geographic barrier between the populations breaks down and that the two populations merge completely. In the merged population, the allele frequencies will be the simple averages of the frequencies of the separate populations; the frequency of *A* will be $(0.5 + 0.8)/2 = 0.65$, and the frequency of *a* will be $(0.5 + 0.2)/2 = 0.35$. Moreover, the genotype frequencies in the merged population will be the simple averages of the genotype frequencies in the separate populations; the frequency of *AA* will be $(0.25 + 0.64)/2 = 0.445$, that of *Aa* will be $(0.50 + 0.32)/2 = 0.410$, and that of *aa* will be $(0.25 + 0.04)/2 = 0.145$. Notice, however, that these observed genotype frequencies are not equal to the frequencies predicted by the Hardy–Weinberg Principle: $(0.65)^2 = 0.422$ for *AA*, $2(0.65)(0.35) = 0.455$ for *Aa*, and $(0.35)^2 = 0.123$ for *aa*. The reason for this discrepancy is that the observed genotype frequencies were not created by random mating within the entire merged population. Rather, they were created by amalgamating genotype frequencies from separate randomly mating populations. Thus, the merger of two randomly mating populations does not produce a population with Hardy–Weinberg genotype frequencies. However, if the merged population mates randomly for just one generation, Hardy–Weinberg genotype frequencies will be established, and the allele frequencies of the merged population will allow prediction of these genotype frequencies.

This example demonstrates that merging randomly mating populations temporarily upsets Hardy–Weinberg equilibrium. The migration of individuals from one population to another also causes a temporary upset in Hardy–Weinberg equilibrium. However, if a population that has received migrants mates randomly for just one generation, Hardy–Weinberg equilibrium will be restored.

Using Allele Frequencies in Genetic Counseling

Knowledge of allele frequencies in specific populations is important in genetic counseling. Consider, for example, a family with a history of the autosomal reces-

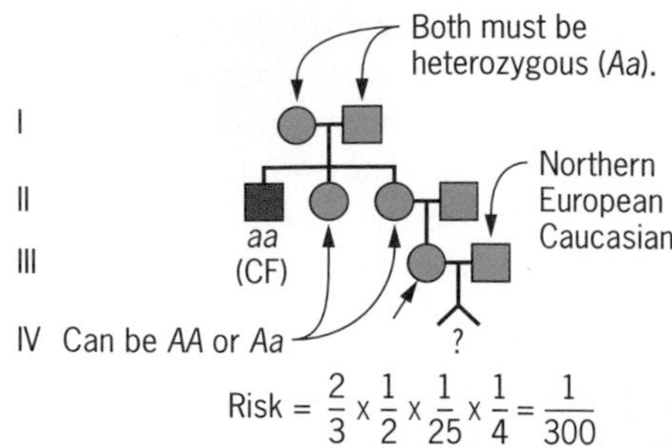

Figure 29.6 Calculation of the risk of cystic fibrosis appearing in the child of a couple of northern European ancestry. The frequency of carriers of cystic fibrosis in northern European populations is about 4 percent.

sive disorder cystic fibrosis, CF (Figure 29.6). A woman (III-1) whose uncle (II-1) has cystic fibrosis wants to know the chance that she will have an affected baby. First, we must determine the probability that the woman is a carrier of the deleterious recessive allele, *a*. Both of her grandparents (I-1 and I-2) must be carriers because they have had an affected child. The woman's mother (II-3) could also be a carrier because both of her parents were carriers. To determine the probability that II-3 carries the recessive allele, we note that one of her possible genotypes (*aa*) is excluded because she does not have CF; there are two remaining possibilities, *AA* and *Aa*, with probabilities 1/3 and 2/3, respectively. Thus, the chance that II-3 is a carrier is 2/3, and the chance that her daughter, III-1, is a carrier is $2/3 \times 1/2 = 1/3$.

Second, we must determine the probability that III-1's husband carries the recessive allele. This man is of northern European ancestry, and the incidence of CF in that population is about 0.0004. Using the Hardy–Weinberg Principle in reverse, we estimate the frequency of the deleterious allele to be $\sqrt{0.0004} = 0.02$; then, using the Hardy–Weinberg Principle in the usual way, we estimate the frequency of carriers in the northern European population to be $2 \times 0.98 \times 0.02$, or approximately 1/25.

Finally, we combine these two probabilities to obtain the risk that the couple will have an affected child. The chance that the husband and wife are both carriers is $(1/3) \times (1/25) = 1/75$, and, if they are, the chance that their child will have CF is 1/4. Thus, the overall risk for the child is $(1/75) \times (1/4) = 1/300$. If III-1's husband came from a population in which the frequency of the CF allele was very low, then the chance that the child would have CF would be essentially 0. Other genetic disorders, such as Tay-Sachs disease and sickle-cell anemia, have strong ethnic associations. Thus, knowledge of allele frequencies in specific ethnic groups can help genetic counselors in the process of risk assessment.

Key Points: **Allele frequencies can be estimated by enumerating the genotypes in a sample from a population. Under the assumption of random mating, the Hardy–Weinberg Principle allows genotype frequencies for autosomal and X-linked genes to be predicted from allele frequencies in a population. The Hardy–Weinberg Principle does not apply to populations with consanguineous or assortative mating, unequal survival among genotypes, population subdivision, or migration. The Hardy–Weinberg Principle is useful in genetic counseling.**

NATURAL SELECTION

Darwin described the key force that drives evolutionary change in populations. He argued that organisms produce more offspring than the environment can support and that a struggle for survival ensues. In the face of this competition, the organisms that survive and reproduce transmit to their offspring traits that favor survival and reproduction. After many generations of such competition, traits associated with strong competitive ability become prevalent in the population, and traits associated with weak competitive ability disappear. Selection for survival and reproduction in the face of competition is therefore the mechanism that changes the physical and behavioral characteristics of a species. Darwin called this process **natural selection.**

Natural Selection at the Level of the Gene

To put the mechanism of natural selection into a genetic context, we must recognize that the ability to survive and reproduce is a phenotype—arguably the most important phenotype of all—and that it is determined, at least partly, by genes. Geneticists refer to this ability to survive and reproduce as **fitness,** a quantitative variable they usually symbolize by the letter w. Each member of a population has its own fitness value: 0 if it dies or fails to reproduce, 1 if it survives and produces 1 offspring, 2 if it survives and produces 2 offspring, and so forth. The average of all these values is the average fitness of the population, usually symbolized $\overline{w}$.

For a population with a stable size, the average fitness is 1; each individual in such a population produces, on average, one offspring. Of course, some individuals will produce more than one offspring, and some will not produce any offspring at all. However, when the population size is not changing, the average number of offspring (that is, the average fitness) is 1. In a declining population, the average number of off-

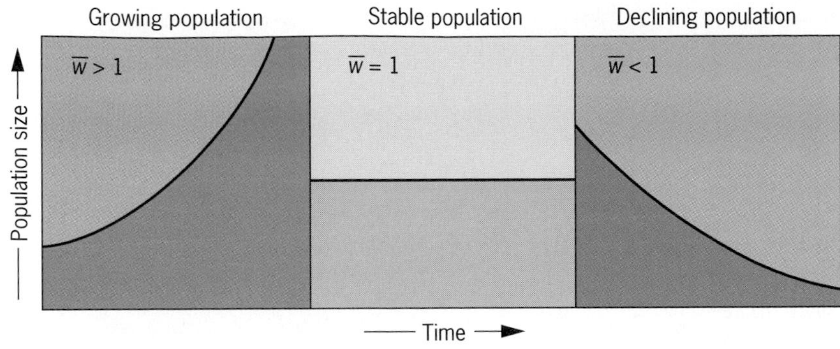

Figure 29.7 Significance of average fitness ($\overline{w}$) for population size as a function of time. Population size grows, is stable, or declines depending on the value of the average fitness.

spring is less than 1, and in a growing population it is greater than 1 (Figure 29.7).

To see how fitness differences among individuals lead to change in the characteristics of a population, let's assume that fitness is determined by a single gene segregating two alleles, *A* and *a*, in a particular species of insect. Furthermore, let's assume that allele *A* causes the insects to be dark in color, that allele *a* causes them to be light in color, and that *A* is completely dominant to *a*. In a forest habitat, where plant growth is luxuriant, the dark form of the insect survives better than the light form. Consequently, the fitnesses of genotypes *AA* and *Aa* are greater than the fitness of genotype *aa*. By contrast, in open fields, where plant growth is scarce, the light form of the insect survives better than the dark form, and the fitness relationships are reversed.

We can express these relationships mathematically by applying the concept of **relative fitness.** In each of the two environments we arbitrarily define the fitness of the competitively superior genotype(s) to be equal to 1 and express the fitness of the inferior genotype(s) as a deviation from 1. This fitness deviation, symbolized by the letter *s*, is called the **selection coefficient;** it measures the intensity of natural selection acting on the genotypes in the population. We can summarize the fitness relationships among the three insect genotypes in each of the two habitats in the following table:

Genotype:	*AA*	*Aa*	*aa*
Phenotype:	dark	dark	light
Relative fitness in forest habitat:	1	1	$1 - s_1$
Relative fitness in field habitat:	$1 - s_2$	$1 - s_2$	1

These relative fitnesses tell us nothing about the absolute reproductive abilities of the different genotypes in the two habitats. However, they do tell us how well each genotype competes with the other genotypes within a particular environment. Thus, for example, we know that *aa* is a weaker competitor than either *AA* or *Aa* in the forest habitat. How much weaker depends, of course, on the actual value of the selection coefficient, s_1. If $s_1 = 1$, then *aa* is effectively a lethal genotype (its relative fitness is 0), and we would expect natural selection to cull the *a* allele from the population. If s_1 were much smaller, say only 0.01, natural selection would still cull the *a* allele, but it would do so very slowly.

To see the effect of natural selection on allele frequencies, let's focus on an insect population in the forest habitat. We shall assume that initially the frequency of *A* is $p = 0.5$, that the frequency of *a* is $q = 0.5$, and that $s_1 = 0.1$. Furthermore, let's assume that the population mates randomly and that the genotypes are present in Hardy–Weinberg frequencies at fertilization each generation. (Differential survival among the genotypes will change these frequencies as the insects mature.) Under these assumptions, the initial genetic composition of the population is:

Genotype:	*AA*	*Aa*	*aa*
Relative fitness:	1	1	$1 - 0.1 = 0.9$
Frequency: (at fertilization)	$p^2 = 0.25$	$2pq = 0.50$	$q^2 = 0.25$

In forming the next generation, each genotype will contribute gametes in proportion to its frequency and relative fitness. Thus, the relative contributions of the three genotypes will be:

Genotype:	*AA*	*Aa*	*aa*
Relative contribution to next generation:	(0.25) × 1 = 0.25	(0.50) × 1 = 0.50	(0.25) × (0.9) = 0.225

If we divide each of these relative contributions by their sum, we obtain the proportional contributions of each of the genotypes to the next generation:

Genotype:	*AA*	*Aa*	*aa*
Proportional contribution to next generation:	0.256	0.513	0.231

From these numbers we can calculate the frequency of the *a* allele after one generation of selection simply by noting that all the genes transmitted by the *aa* homozygotes are a and that half the genes transmitted by the Aa heterozygotes are a. In the next generation, the frequency of a, symbolized q', will be

$$q' = 0.231 + (1/2)\ (0.513) = 0.487$$

which is slightly less than the starting frequency of 0.5. Thus, in the forest habitat, natural selection, acting through the lower fitness of the aa homozygotes, has decreased the frequency of a from 0.5 to 0.487. In every subsequent generation, the frequency of a will be reduced slightly because of selection against the aa homozygotes, and eventually, this allele will be eliminated from the population altogether. Figure 29.8*a* shows how natural selection will drive the a allele to extinction.

In the field habitat, aa homozygotes are selectively superior to the other two genotypes. Thus, starting with $q = 0.5$, Hardy–Weinberg genotype frequencies, and the selection coefficient $s_2 = 0.1$, we have:

Genotype:	*AA*	*Aa*	*aa*
Relative fitness:	$1 - 0.1 = 0.9$	$1 - 0.1 = 0.9$	1
Frequency:	0.25	0.50	0.25

After one generation of selection in the field habitat, the frequency of a will be $q' = 0.513$, which is slightly greater than the starting frequency. Every generation afterward, the frequency of a will rise, and eventually it will equal 1, at which point it is said that the allele has been fixed in the population. Figure 29.8*b* shows the selection-driven path toward fixation of a.

These two scenarios illustrate selection for or against a recessive allele. In the forest habitat, the recessive allele a is deleterious in homozygous condition and selection acts against it. In the field habitat, a is selectively favored over the dominant allele A, which is deleterious in both homozygous and heterozygous condition.

Studies of the moth *Biston betularia*, an inhabitant of wooded areas in Great Britain, have shown that selection of the type we have been discussing does operate to change allele frequencies in nature. This species, commonly known as the peppered moth, exists in two color forms, light and dark (Figure 29.9); the light form is homozygous for a recessive allele c, and the dark form carries a dominant allele C. Since 1850, the frequency of the dark form has increased in

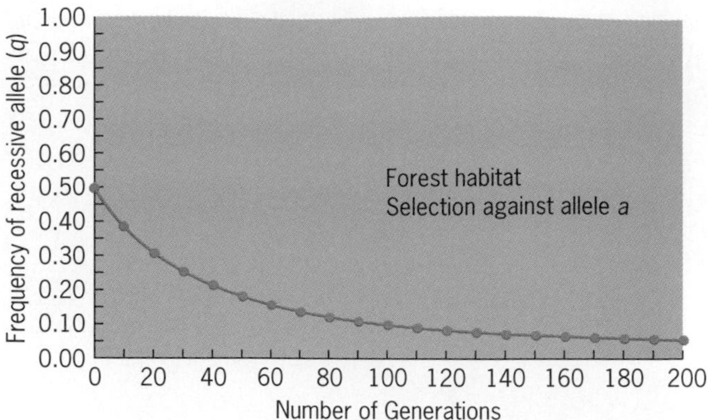

(*a*)

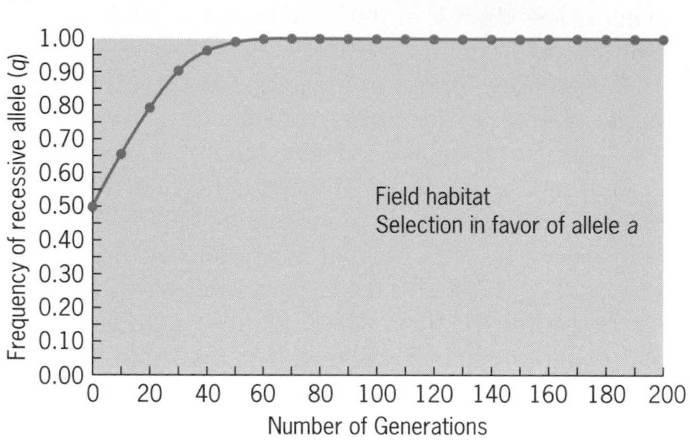

(*b*)

Figure 29.8 (*a*) Selection against the recessive allele a in the forest habitat. (*b*) Selection in favor of the recessive allele a in the field habitat.

(*a*)

(*b*)

Figure 29.9 (*a*) Dark form of the moth *Biston betularia* resting on a lichen-covered tree trunk. (*b*) Light form of the same moth resting on a soot-covered tree trunk.

certain areas of England, particularly in the industrialized Midlands section of the country. Around the heavily industrialized cities of Manchester and Birmingham, for example, the frequency of the dark form has increased from 1 to 90 percent. This dramatic increase has been attributed to selection against the light form in the soot-polluted landscapes of industrialized areas.

Peppered moths are active at night. During the day, they rest on tree trunks in the woodlands. Because birds may prey upon the moths during this resting period, camouflage is the moth's only defense. On white or gray tree bark, the light moths are protectively colored, especially if the bark is overgrown with lichens—the normal condition in unpolluted woodlands. However, in industrialized areas, pollution has killed the lichens and soot has blackened the tree bark. Under these conditions, the light moths are not camouflaged, and they are readily consumed by predatory birds. Experiments by H. B. D. Kettlewell have shown that the dark moths are protectively colored in polluted landscapes. Birds apparently have difficulty seeing them on the lichen-bare, sooty tree trunks. Protection against predation by birds is therefore the reason why the dark form of *Biston betularia* has become prevalent in the industrialized areas of England.

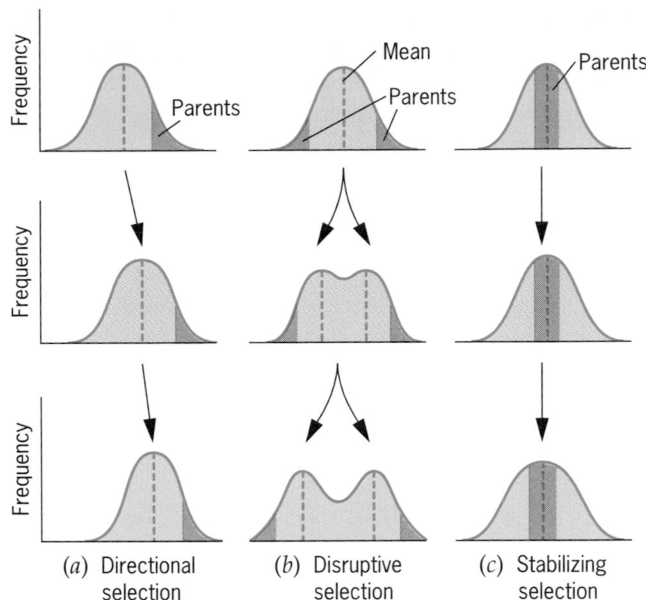

Figure 29.10 Three basic modes of selection and their effect on the mean (dashed line) and variation (width of distribution) in a quantitative trait.

Natural Selection at the Level of the Phenotype

Although the example of *Biston betularia* shows that fitness can be dramatically influenced by different alleles of a single gene, in most circumstances, it is influenced by the alleles of many genes. Typically, fitness depends on sets of genes that control quantitative traits such as body size, disease susceptibility, and fecundity. Thus, we would expect natural selection to affect the statistical distributions of these kinds of traits within a population. We now consider three ways in which selection can affect the distribution of a quantitative trait (Figure 29.10).

1. *Directional selection.* Selection that favors values of a trait at one end of its distribution is *directional selection.* This type of situation commonly occurs in agriculture where plant and animal breeders practice artificial selection to improve traits such as crop yield, nutritional content, and egg production (see Chapter 5). In nature, directional selection may occur when a deterioriating environment steadily challenges the population to adapt. R. A. Fisher, who studied this situation theoretically, concluded that directional selection increases the average fitness of a population at a rate that is proportional to the additive genetic variance for fitness—a principle that he immodestly called the Fundamental Theorem of Natural Selection. Although a discussion of Fisher's Theorem is beyond the scope of this textbook, it should be noted that the Theorem parallels the principle that the response to artificial selection depends on the proportion of variance in a trait that is additive genetic variance—that is, it depends on the narrow-sense heritability (see Chapter 5). Thus, the rate at which natural selection can change fitness in a population is a function of the narrow-sense heritability for fitness.

Evolutionary biologists have discovered many examples of directional selection. The increase in body size of the horse during the last 40 million years, the development of extravagant body ornaments such as antlers in deer and feathers in birds, and the increase in brain size in our own species all probably involved directional selection.

2. *Disruptive selection.* Selection that favors extreme values of a trait at the expense of intermediate values is *disruptive selection.* It is, in effect, directional selection that works simultaneously to increase and decrease the values of the trait in the population. For this type of selection to be effective, either there must be strong assortative mating for the trait—that is, matings preferentially take place between individuals with the same extreme values of the trait—or the population must become subdivided by geographical or ecological barriers.

Disruptive selection in a subdivided population seems to have played a role in the evolution of the elephants that inhabited large sections of Europe and Asia during the Pleistocene period of geologic time. On both of these continents, selection favored the evolution of mammoths that stood as much as 5 meters tall. However, on certain islands in the Arctic Ocean, selection worked in the opposite direction, producing mammoths that were barely the size of a pony. Although some would regard a pony-sized mammoth as a contradiction in terms, the fact that such animals existed at the same time as their truly mammoth relatives shows that disruptive selection has the power to change a trait in different directions.

3. *Stabilizing selection.* Selection may also operate to conserve the distribution of a quantitative trait by favoring intermediate values. Such a process is called *stabilizing selection.* This process occurs when intermediate values of the trait are associated with high fitness and extreme values are associated with low fitness. An example is selection for birth weight in human babies. The optimum birth weight is around 8 pounds; babies that weigh more than 9 pounds or less than 7 pounds have reduced chances for survival.

Key Points: **Natural selection occurs when genotypes differ in the ability to survive and reproduce (fitness). The intensity of natural selection is quantified by the selection coefficient. At the level of the gene, natural selection changes the frequencies of alleles in populations. At the level of the phenotype, it influences the distributions of quantitative traits. Natural selection may be directional, disruptive, or stabilizing.**

RANDOM GENETIC DRIFT

In *The Origin of Species,* Darwin emphasized the role of natural selection as a systematic force in evolution. However, he also recognized that evolution is affected by random processes. New mutants appear unpredictably in populations. Thus, mutation, the ultimate source of all genetic variability, is a random process that profoundly affects evolution; without mutation, evolution could not occur. Darwin also recognized that inheritance (which, of course, he did not understand) is unpredictable. Traits are inherited, but offspring are not exact replicas of their parents; there is always some unpredictability in the transmission of a trait from one generation to the next. In the twentieth century, after Mendel's principles were re-discovered, the evolutionary implications of this unpredictability were investigated by Sewall Wright, R. A. Fisher, and Motoo Kimura. From their theoretical analyses, it is clear that the randomness associated with the Mendelian mechanism profoundly affects the evolutionary process. In the following sections, we explore how the uncertainties of genetic transmission can lead to random changes in allele frequencies—a phenomenon called **random genetic drift.**

Random Changes in Allele Frequencies

To investigate how the uncertainties associated with the Mendelian mechanism can lead to random changes in allele frequencies, let's consider a mating between two heterozygotes, *Cc* × *Cc*, that produces two offspring, which is the number expected if each individual in the population replaces itself (Figure 29.11). We can enumerate the possible genotypes of the two offspring and compute the probability associated with each of the possible combinations by using the methods discussed in Chapter 3. For example, the probability that the first offspring is *CC* is 1/4 and the probability that the second offspring is *CC* is also 1/4; thus, the probability that both offspring are *CC* is (1/4) × (1/4) = 1/16. The probability that one of the offspring is *CC* and the other is *Cc* is (1/4) × (1/2) × 2 (because there are two possible birth orders: *CC* then *Cc* or *Cc* then *CC*); thus, the probability of observing the genotypic combination *CC* and *Cc* in the two offspring is 1/4. The entire probability distribution for the various genotypic combinations of offspring is given in Figure 29.11. This figure also gives the frequency of the *c* allele associated with each combination.

Among the parents, the frequency of *c* is 0.5. This frequency is the most probable frequency for *c* among the two offspring; in fact, the probability that the frequency of *c* will not change between parents and off-

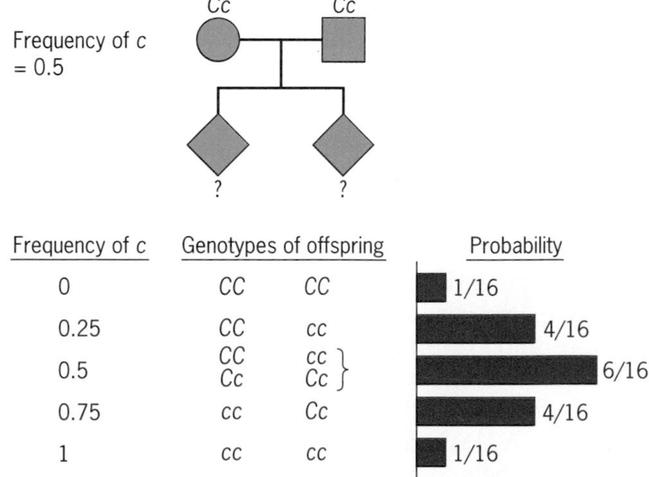

Figure 29.11 Probabilities associated with possible frequencies of the allele *c* among the two children of heterozygous parents.

spring is 6/16. However, there is an appreciable chance that the frequency of *c* will increase or decrease among the offspring simply because of the uncertainties associated with the Mendelian mechanism.The chance that the frequency of *c* will increase is 5/16, and the chance that it will decrease is also 5/16. Thus, the chance that the frequency of *c* will change in one direction or the other, $5/16 + 5/16 = 10/16$, is actually greater than the chance that it will remain the same.

This situation illustrates the phenomenon of random genetic drift. For every pair of parents in the population that is segregating different alleles of a gene, there is a chance that the Mendelian mechanism will lead to changes in the frequencies of those alleles. When these random changes are summed over all pairs of parents, there may be aggregate changes in the allele frequencies. Thus, the genetic composition of the population can change even without the force of natural selection.

Random genetic drift is essentially the result of a gene sampling process that occurs when organisms reproduce. This sampling process has two components. First, the alleles of segregating genes are randomly incorporated into gametes. The offspring produced by a heterozygote with genotype *Cc* inherit either allele *C* or allele *c*, each with probability 1/2. Thus, in segregating individuals, there is always uncertainty as to which allele a given offspring will receive. Second, there is random variation in the number of offspring that a parent produces. Some parents produce many offspring, some produce a few, and some produce none. Although part of this variation may be due to intrinsic fitness differences among the members of the population, part of it may be due to purely random factors—accidental deaths, bad weather, environmental catastrophes. This purely random variation in reproductive output compounds the randomness associated with Mendelian segregation, and the net result is random change in allele frequencies.

The Effects of Population Size

A population's susceptibility to random genetic drift depends on its size. In large populations, the effect of genetic drift is minimal, whereas in small ones, it may be the primary evolutionary force. Geneticists gauge the effect of population size by monitoring the frequency of heterozygotes over time. Let's focus, once again, on alleles *C* and *c*, with respective frequencies *p* and *q*, and let's assume that neither allele has any effects on fitness; that is, *C* and *c* are *selectively neutral.* Further, let's assume that the population mates randomly and that in any given generation, the genotypes are present in Hardy–Weinberg proportions.

In a very large population—essentially infinite in size—the frequencies of *C* and *c* will be constant and the frequency of the heterozygotes that carry these two alleles will be $2pq$. In a small population of finite size N, the allele frequencies will change randomly as a result of genetic drift. Because of these changes, the frequency of heterozygotes, often called the **heterozygosity,** will also change. To express the magnitude of this change over one generation, let's define the current frequency of heterozygotes as H, and the frequency of heterozygotes in the next generation as H'. Then the mathematical relationship between H' and H is

$$H' = \left(1 - \frac{1}{2N}\right) H$$

This equation tells us that in one generation, random genetic drift causes the heterozygosity to decline by a factor of $1/2N$. In a total of t generations, we would expect the heterozygosity to decline to a level given by the equation

$$H_t = \left(1 - \frac{1}{2N}\right)^t H$$

This equation enables us to see the cumulative effect of random genetic drift over many generations. In each generation, the heterozygosity is expected to decline by a factor of $\frac{1}{2N}$, over many generations, the heterozygosity will eventually be reduced to 0, at which point all genetic variability in the population will be lost. At this point the population will possess only one allele of the gene, and either $p = 1$ and $q = 0$, or $p = 0$ and $q = 1$. Thus, through random changes in allele frequencies, drift steadily erodes the genetic variability of a population, ultimately leading to the fixation and loss of alleles. It is important to recognize that this process depends critically on the population size (Figure 29.12). Small populations are the most sensitive to

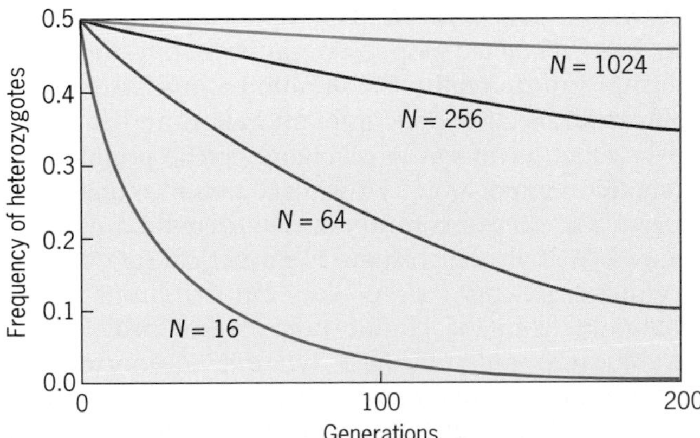

Figure 29.12 Decline in the frequency of heterozygotes due to random genetic drift in populations of different size N. The populations begin with $p = q = 0.5$.

the variability-reducing effects of drift. Large populations are less sensitive.

If selectively neutral alleles of the sort we have been discussing are ultimately destined for fixation or loss, can we determine the probabilities that are associated with these two ultimate outcomes? Let's suppose that at the current time, the frequency of C is p and that of c is q. Then, as long as the alleles are selectively neutral and the population mates randomly, the probability that a particular allele will ultimately be fixed in the population is its current frequency—p for allele C and q for allele c—and the probability that the allele will ultimately be lost from the population is 1 minus its current frequency; that is, $1 - p$ for allele C and $1 - q$ for allele c. Thus, when random genetic drift is the driving force in evolution, we can assign specific probabilities to the possible evolutionary outcomes and, remarkably, these probabilities are independent of population size. For more discussion of evolution by random genetic drift, see the Technical Sidelight: Molecular Evolution and the Neutral Theory.

Key Points: **Genetic drift is the process of random allele frequency change in populations. This process is due to intrinsic uncertainties in the mechanism of inheritance and to random variation in the reproductive ability of individuals in the population. Drift erodes genetic variability in populations. Small populations are more susceptible to drift than large populations.**

POPULATIONS IN GENETIC EQUILIBRIUM

In a randomly mating population without selection or drift to change allele frequencies, and without migration or mutation to introduce new alleles, the Hardy–Weinberg genotype frequencies persist indefinitely. Such an idealized population is in a state of genetic equilibrium. In reality, the situation is much more complicated; selection and drift, migration and mutation are almost always at work changing the population's genetic composition. However, these evolutionary forces may act in contrary ways to create a dynamic equilibrium in which there is no net change in allele frequencies. This type of equilibrium differs fundamentally from the equilibrium of the ideal Hardy–Weinberg population. In a dynamic equilibrium, the population simultaneously tends to change in opposite directions, but these opposing tendencies cancel each other and bring the population to a point of balance. In the ideal Hardy–Weinberg equilibrium, the population does not change because there are no evolutionary forces at work. We now explore how opposing evolutionary forces can create a dynamic equilibrium within a population.

Balancing Selection

One type of dynamic equilibrium arises when selection favors the heterozygotes at the expense of each type of homozygote in the population. In this situation, called *balancing selection* or *heterozygote advantage*, we can assign the relative fitness of the heterozygotes to be 1 and the relative fitnesses of the two types of homozygotes to be less than 1:

Genotype:	AA	Aa	aa
Relative fitness:	$1 - s$	1	$1 - t$

In this formulation, the terms $1 - s$ and $1 - t$ contain selection coefficients that are assumed to lie between 0 and 1. Thus, each of the homozygotes has a lower fitness than the heterozygotes. The superiority of the heterozygotes is sometimes referred to as *overdominance.*

In cases of heterozygote advantage, selection tends to eliminate both the A and a alleles through its effects on the homozygotes, but it also preserves these alleles through its effects on the heterozygotes. At some point these opposing tendencies balance each other, and a dynamic equilibrium is established. To determine the frequencies of the two alleles at the point of equilibrium, we must derive an equation that describes the process of selection, and then solve this equation for the allele frequencies when the opposing selective forces are in balance—that is, when the allele frequencies are no longer changing (Table 29.2). At the balance point, the frequency of A is $p = t/(s + t)$, and the frequency of a is $q = s/(s + t)$.

As an example, let's suppose that the AA homozygotes are lethal ($s = 1$) and that the aa homozygotes are 50 percent as fit as the heterozygotes ($t = 0.5$). Under these assumptions, the population will establish a dynamic equilibrium when $p = 0.5/(0.5 + 1) = 1/3$ and $q = 1/(0.5 + 1) = 2/3$. Both alleles will be maintained at appreciable frequencies by selection in favor of the heterozygotes—a condition known as a **balanced polymorphism.**

In humans, the disease sickle-cell anemia is associated with a balanced polymorphism. Individuals with this disease are homozygous for a mutant allele of the β-globin gene, denoted Hb^S, and they suffer from a severe form of anemia in which the hemoglobin molecules crystallize in the blood. This crystallization causes the red blood cells to assume a characteristic sickle shape. Because sickle-cell anemia is usually fatal early in life, the fitness of Hb^SHb^S homozygotes is

TABLE 29.2
Calculating Equilibrium Allele Frequencies with Balancing Selection

Genotypes:	AA	Aa	aa
Relative fitnesses:	$1 - s$	1	$1 - t$
Frequencies:	p^2	$2pq$	q^2
Average relative fitness:	$\bar{w} = p^2 \times (1 - s) + 2pq \times 1 + q^2 \times (1 - t)$		
Frequency of A in the next generation after selection:	$p' = [p^2(1 - s) + (1/2)\,2pq]/\bar{w} = p(1 - sp)/\bar{w}$		
Change in frequency of A due to selection:	$\Delta p = p' - p = pq(tq - sp)/\bar{w}$		
At equilibrium, $\Delta p = 0$:	$p = t/(s + t)$ and $q = s/(s + t)$		

essentially 0. However, in some parts of the world the frequency of the Hb^S allele is as high as 0.2 (Figure 29.13*a*). With such harmful effects, why does the Hb^S allele remain in the population at all?

The answer is that there is moderate selection against homozygotes that carry the wild-type allele Hb^A. These homozygotes are less fit than the Hb^SHb^A heterozygotes because they are more susceptible to infection by the parasites that cause malaria, a fitness-reducing disease that is widespread in regions where the frequency of the Hb^S allele is high (Figure 29.13*b*). We can schematize this situation by assigning relative fitnesses to each of the genotypes of the β-globin gene:

Genotype:	Hb^SHb^S	Hb^SHb^A	Hb^AHb^A
Relative fitness:	$1 - s$	1	$1 - t$

If we assume that the equilibrium frequency of Hb^S is $p = 0.1$—a typical value in west Africa—and if we note that $s = 1$ because the Hb^SHb^S homozygotes are lethal, we can estimate the intensity of selection against the Hb^AHb^A homozygotes because of their greater susceptibility to malaria:

$$p = t/(s + t)$$
$$0.1 = t/(1 + t)$$
$$t = (0.1)/(0.9) = 0.11$$

This result tells us that the Hb^AHb^A homozygotes are about 11 percent less fit than the Hb^SHb^A heterozygotes. Thus, the selective inferiority of the Hb^SHb^S and Hb^AHb^A homozygotes compared to the heterozygotes creates a balanced polymorphism in which both alleles of the β-globin gene are maintained in the population.

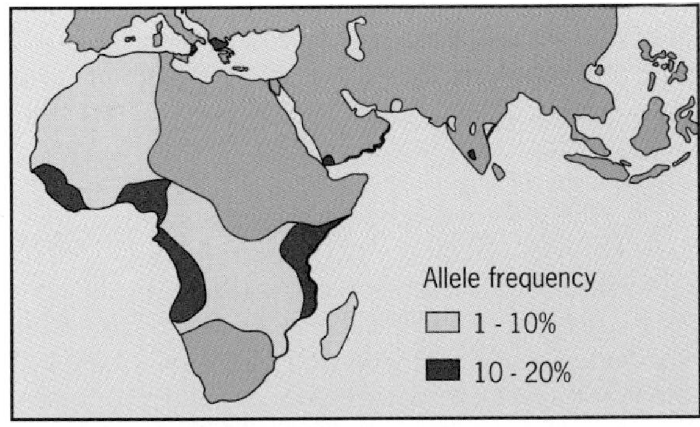

(a) Distribution of sickle-cell anemia allele (Hb^S).

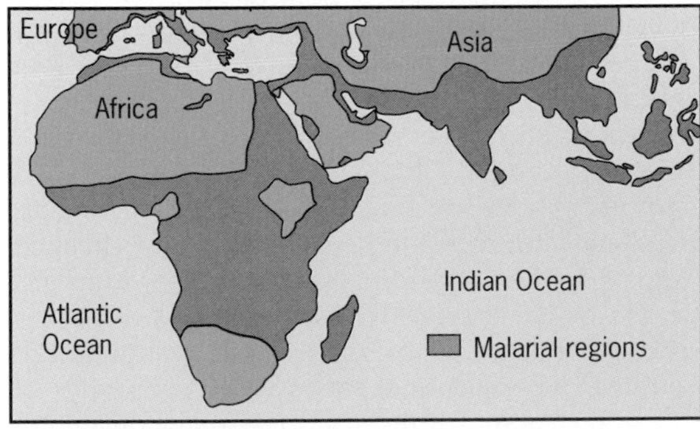

(b) Distribution of Falciparum malaria.

Figure 29.13 Distribution of the sickle-cell anemia allele Hb^S (*a*) and malaria (*b*) in Africa, the Mediterranean basin, and southern Asia.

Mutation-Selection Balance

Another type of dynamic equilibrium is created when selection eliminates deleterious alleles that are produced by recurrent mutation. For example, let's consider the case of a deleterious recessive allele a that is produced by mutation of the wild-type allele A at rate u. A typical value for u is 3×10^{-6} mutations per generation. Even though this rate is very low, over time, the mutant allele will accumulate in the population, and, because it is recessive, it can be carried in heterozygous condition without having any harmful effects. At some point, however, the mutant allele will become frequent enough for aa homozygotes to appear

TECHNICAL SIDELIGHT

Molecular Evolution and the Neutral Theory

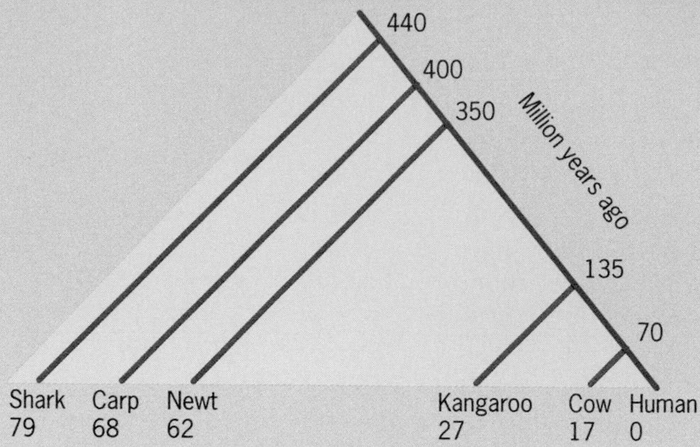

Figure 1 The evolution of α-globin in several vertebrate groups.

At a fundamental level, evolution is change. DNA sequences change by base-pair substitution, insertions, or deletions. If these mutational events occur in a protein-encoding region of DNA, an amino acid substitution may result. Given enough time, such changes can accumulate, leading to a molecule that is very different from its progenitor. Recent advances in molecular biology have made it possible to determine the nucleotide sequences of DNA and the amino acid sequences of polypeptides. By comparing related sequences, the molecular details of evolution can be analyzed.

Amino acid sequence analysis when combined with fossil data provides considerable insight into evolutionary events. For example, the analysis of the α-globin protein in various vertebrates allows researchers to estimate the evolutionary rate of the α-globin gene (Figure 1). All of the organisms in Figure 1 are derived from an evolutionary line that appeared about 440 million years ago. Since that time, this line has split into many different branches, each giving rise to a different vertebrate class—the cartilaginous fishes, the bony fishes, the amphibians, the marsupials, and the placental mammals. Figure 1 shows the number of amino acids that are different when human α-globin is compared to each of the other α-globin molecules. Altogether, there are 141 amino acids in the α-globin polypeptide. The shark α-globin is the most unlike the human α-globin, followed by carp, newt, kangaroo, and cow α-globins, in that order. The degree of difference depends on the time that has elapsed since the evolutionary lines diverged from a common ancestor: the greater the time, the greater the amino acid differences.

The relationship between the degree of difference (D) and the evolutionary time (T) can be used to estimate the rate at which α-globin has evolved. Consider, for example, the human and carp evolutionary lines. The common ancestor of these two species existed about 400 million years ago, so the total time that has elapsed since the two lines diverged is 800 million years (400 million years for each line). During this time, human and carp α-globins have accumulated differences in 68 amino acid sites. The degree of difference is therefore $68/141 = 0.482$ difference per amino acid site. Thus, we can estimate that α-globin has evolved at a rate of $D/T = 0.482$ amino acid substitution per site over 800 million years, or 0.6×10^{-9} substitution at each site per year.

Actually, this value underestimates the true evolutionary rate because some amino acid sites may have changed more than once and there is no way of knowing which ones, or how many times they may have changed. Fortunately, it is possible to make adjustments for these unseen changes and to obtain a more realistic value for the evolutionary rate of α-globin. The corrected rate is about one substitution per amino acid site every 1 billion years.

Similar analyses have been performed with other polypeptides, and the results indicate that each polypeptide has its own evolutionary rate. For instance, the fibrinopeptides, which are components of blood clots, are evolving about eight times faster than α-globin, whereas the histones, which bind to eukaryotic DNA, are evolving about 1000 times slower.

How can these observations be explained? Motoo Kimura and Tomoko Ohta proposed that the great majority of these molecular changes are caused by the random fixation of selectively neutral, and nearly neutral, mutations. In this neutral theory of molecular evolution, mutation and random genetic drift play the critical roles. The probability of random fixation of a selectively neutral mutation is equal to the mutation rate. Thus, in the neutral theory, the rate of evolution is completely determined by the force of mutation. Because mutation pressure is more or less constant, this would explain why polypeptides like α-globin evolve at a fairly uniform rate in different taxonomic groups. However, it does not explain why there are different rates for different polypeptides.

To account for such differences, it is necessary to invoke natural selection. Kimura, Ohta, and others have proposed that some polypeptides are functionally constrained. This means that some sites in a polypeptide cannot be changed without impairing the polypeptide's function. For example, the active site of an enzyme would not easily tolerate a change while other areas of the enzyme molecule might. The more constrained a polypeptide, the lower the probability of a neutral amino acid substitution and the lower the evolutionary rate. In this view, polypeptides such as the histones are highly constrained, whereas those such as the fibrinopeptides are not.

The theory of molecular evolution has been controversial because it places so much emphasis on random processes. Although the neutral theory admits a role for selection as a purifying agent—that is, as a force that eliminates harmful mutations—it has little or no place for selection in the positive Darwinian sense. According to the neutral theory, species do not get better by fixing beneficial mutations; they simply do not get worse by fixing only neutral or nearly neutral mutations. For this reason, the neutral theory is sometimes called the non-Darwinian theory of evolution. Most evolutionary geneticists would agree that it has been an important influence on our thinking about the evolutionary process.

in the population, and these will be subject to the force of selection in proportion to their frequency and the value of the selection coefficient s. Selection against these homozygotes will counteract the force of mutation, which introduces the mutant allele into the population.

If we assume that the population mates randomly, and if we denote the frequency of A as p and that of a as q, then we can summarize the situation as follows:

Mutation: produces a		Selection: eliminates a		
$A \rightarrow a$	Genotype:	AA	Aa	aa
rate $= u$	Relative fitness:	1	1	$1 - s$
	Frequency:	p^2	$2pq$	q^2

Mutation introduces mutant alleles into the population at rate u, and selection eliminates them at rate sq^2 (Figure 29.14). When these two processes are in balance, a dynamic equilibrium will be established. We can calculate the frequency of the mutant allele at the equilibrium created by mutation-selection balance by equating the rate of mutation to the rate of elimination by selection:

$$u = sq^2$$

Thus, after solving for q, we obtain

$$q = \sqrt{u/s}.$$

For a mutant allele that is lethal in homozygous condition, $s = 1$, and the equilibrium frequency of the mutant allele is simply the square root of the mutation rate. If we use the value for u that was given above, then for a recessive lethal allele the equilibrium frequency is $q = 0.0017$. If the mutant allele is not completely lethal in homozygous condition, then the equilibrium frequency will be higher than 0.0017 by a factor that depends on $1/\sqrt{s}$. For example, if s is 0.1, then at equilibrium the frequency of this slightly deleterious allele will be $q = 0.0055$—3.2 times greater than the equilibrium frequency of a recessive lethal allele.

Studies with natural populations of *Drosophila* have indicated that lethal alleles are less frequent than the preceding calculations predict. The discrepancy between the observed and predicted frequencies has been attributed to partial dominance of the mutant alleles—that is, these alleles are not completely recessive. Natural selection appears to act against deleterious alleles in heterozygous condition as well as in homozygous condition. Thus, the equilibrium frequencies of these alleles are lower than we would otherwise predict. Selection that acts against mutant alleles in homozygous or heterozygous condition is sometimes called *purifying selection.*

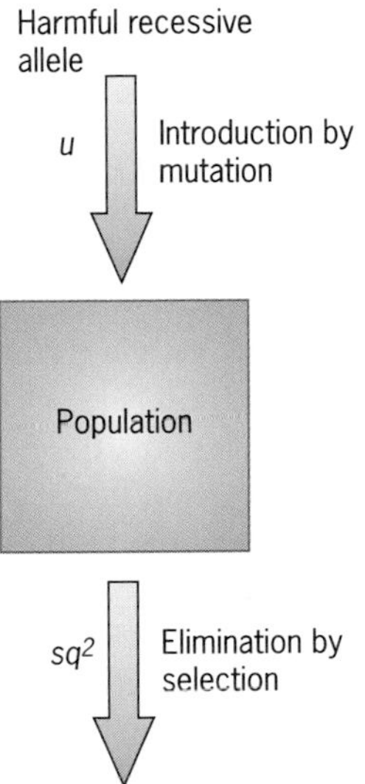

Figure 29.14 Mutation-selection balance for a deleterious recessive allele with frequency q. Genetic equilibrium is reached when the introduction of the allele into the population by mutation at rate u is balanced by the elimination of the allele by selection with intensity s against the recessive homozygotes.

Mutation-Drift Balance

We have already seen that random genetic drift eliminates variability from a population. Without any counteracting force, this process would eventually make all populations completely homozygous. However, mutation replenishes the variability that is lost by drift. At some point, the opposing forces of mutation and genetic drift come into balance and a dynamic equilibrium is established.

Previously we saw that genetic variability can be quantified by calculating the frequency of heterozygotes in a population—a statistic called the heterozygosity that is symbolized by the letter H. The frequency of homozygotes in a population—often called the *homozygosity*—is equal to $1 - H$. Over time, genetic drift decreases H and increases $1 - H$, and mutation does just the opposite (Figure 29.15). Let's assume that each new mutation is selectively neutral. In a randomly mating population of size N, the rate at which drift decreases H is $(\frac{1}{2N})H$ (see p. 787). The rate at which mutation increases H is proportional to the frequency of the homozygotes in the population $(1 - H)$ and the probability that one of the two alleles in a particular homozygote mutates to a different allele, thereby converting that homozygote into a heterozygote. This probability is simply the mutation rate u for each of the two alleles in the homozygote; thus, the total probability of mutation converting a particular homozygote into a heterozygote is $2u$. The rate at which mutation increases H in a population is therefore equal to $2u(1 - H)$.

When the opposing forces of mutation and drift come into balance, the population will achieve an equilibrium level of variability denoted by $\hat{H}$. We can calculate this equilibrium value of H by equating the rate at which mutation increases H to the rate at which drift decreases it:

$$2u(1 - H) = \left(\frac{1}{2N}\right)H$$

By solving for H, we obtain the equilibrium heterozygosity at the point of mutation-drift balance:

$$\hat{H} = 4Nu/(4Nu + 1)$$

Thus, the equilibrium level of variability (as measured by the heterozygosity) is a function of the population size and the mutation rate.

If we assume that the mutation rate is $u = 1 \times 10^{-6}$, we can plot $\hat{H}$ for different values of N (Figure 29.16). For $N < 10{,}000$, the equilibrium frequency of heterozygotes in the population is quite low; thus, drift dominates over mutation in small populations. For N equal to $1/u$, the reciprocal of the mutation rate, the equilibrium frequency of heterozygotes is 0.8, and for even greater values of N, the frequency of heterozygotes increases asymptotically toward 1. Thus, in large populations, mutation dominates over drift; every mutational event creates a new allele, and each new allele contributes to the heterozygosity because the large size of the population protects it from being lost by random genetic drift.

Values of $\hat{H}$ in natural populations vary among species. In the African cheetah, for example, $\hat{H}$ is 1

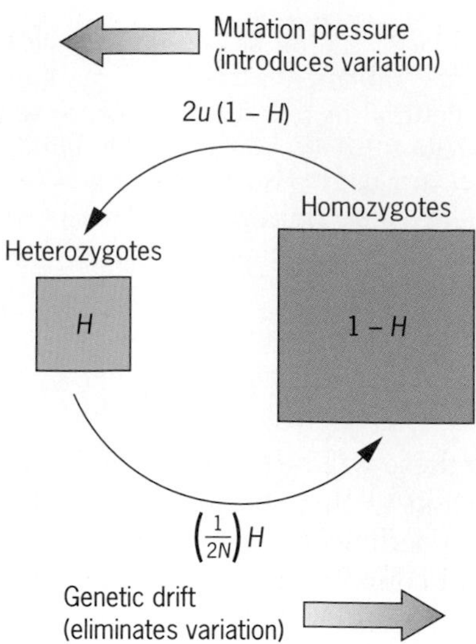

Figure 29.15 Mutation-drift balance for variability as measured by the frequency of heterozygotes H in a population of size N. An equilibrium frequency of heterozygotes is reached when the introduction of variability by mutation at rate u is balanced by the elimination of variability by genetic drift at rate $\frac{1}{2N}$.

percent or less, suggesting that over evolutionary time, population size in this species has been small. In humans, $\hat{H}$ is about 12 percent, suggesting that over evolutionary time population size has averaged about 30–40 thousand individuals. Estimates of population size that are derived from heterozygosity data are typically much smaller than estimates obtained from census data. The reason for this discrepancy is that the estimates based on heterozygosity data are *genetically effective* population sizes—sizes that take into account restrictions on mating and reproduction, as well as temporal fluctuations in the number of mating individuals. The ge-

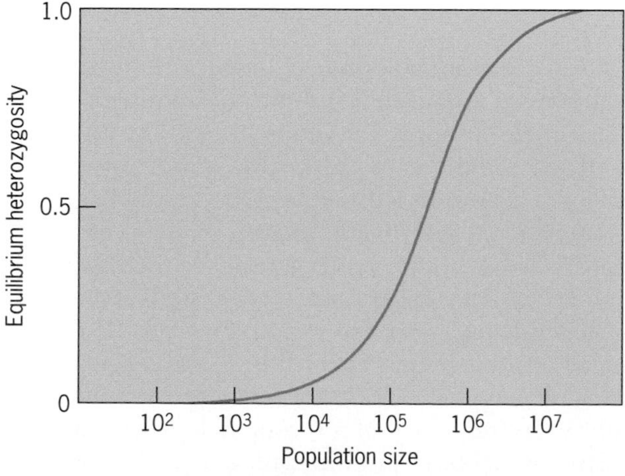

Figure 29.16 Equilibrium frequency of heterozygotes (heterozygosity) under mutation-drift balance as a function of genetically effective population size.

netically effective size of a population is almost always less than the census size of a population.

Key Points: **A dynamic equilibrium may be established by balancing selection (heterozygote advantage), by mutation-selection balance, or by mutation-drift balance.**

GENETIC VARIATION IN NATURAL POPULATIONS

Darwin's *Origin of Species* begins with a discussion of variation. Without variation, populations cannot evolve. Soon after Mendel's Principles were rediscovered, geneticists began to document genetic variation in natural populations. Initially these efforts focused on conspicuous features of the phenotype—pigmentation, size, and so forth. Later, they emphasized characteristics that are more directly related to chromosomes and genes. In the following sections, we discuss variation at the phenotypic, chromosomal, and molecular levels.

(a)

(b)

Figure 29.17 (*a*) Red and orange color in the salmonberry, and (*b*) blue and white plumage in the snow goose are polymorphisms controlled by single genes with two alleles.

Phenotypic Variation

Individuals in natural populations often exhibit morphological variation that is due to genetic differences. This type of variation can be continuous or discontinuous. Discontinuous variation is seen in the salmonberry (*Rubus spectabilis*) and in the snow goose (*Chen hyperborea*). In the salmonberry, the fruits are either red or orange (Figure 29.17*a*), and in the snow goose, the plumage is either white or blue (Figure 29.17*b*). In both of these cases, the difference between the two color forms is controlled by a single gene with two alleles. Thus, they are examples of genetic polymorphisms.

Continuous phenotypic variation is also seen in natural populations. For example, the Asiatic ladybeetle (*Harmonia axyridis*) exhibits variation in its color pattern, and this variation has a genetic basis (Figure 29.18*a*); however, it is not traceable to a single gene. Color variation in the starfish (*Pisaster ochraceus*) is also genetically determined. In this organism, body pig-

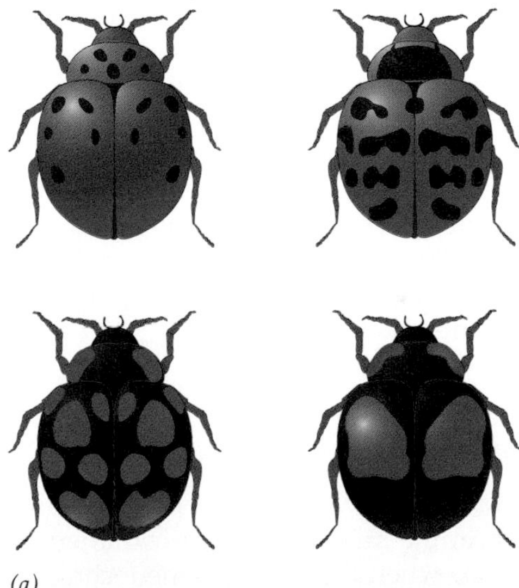

(a)

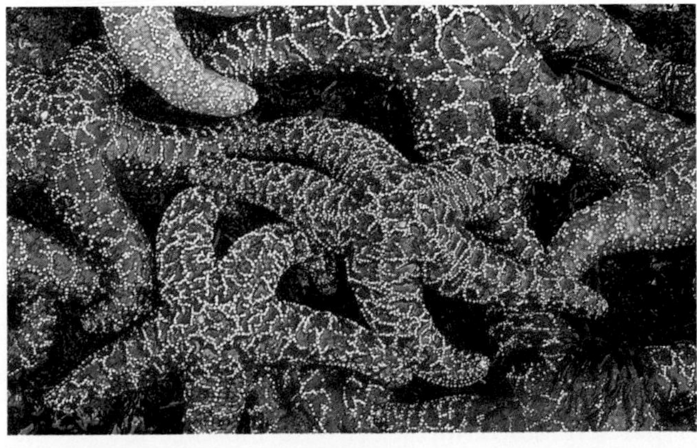

(b)

Figure 29.18 (*a*) Color patterns in the Asiatic ladybeetle, and (*b*) color variants in a starfish.

mentation varies from purple to orange (Figure 29.18*b*).

Polymorphism of Chromosome Structure

As a general rule, each species has a distinctive karyotype; that is, all members of a species have the same chromosome number and the same basic chromosome structure. However, the number or structure of the chromosomes sometimes varies. For example, in populations of *Drosophila pseudoobscura* in the Mount San Jacinto area of California, Theodosius Dobzhansky and colleagues identified three inversion patterns in one of the autosomes. These patterns have been designated as Standard (ST), Arrowhead (AR), and Chiricahua (CH). Each of these patterns occurs at appreciable frequencies in the Mount San Jacinto population; furthermore, the frequencies change during the year. From March to June, the frequency of ST decreases and that of CH increases, and from June to August, the frequency of ST increases and that of CH decreases (Figure 29.19). Although these changes suggest that the inversion patterns are associated with fitness differences, it is not known why CH is favored during the spring and ST is favored during the summer. However, laboratory experiments have shown that the ST/CH heterozygotes are selectively superior to both the ST/ST and CH/CH homozygotes. Thus, variation among inversion patterns in the Mount San Jacinto population of *D. pseudoobscura* appears to be a balanced polymorphism at the chromosomal level.

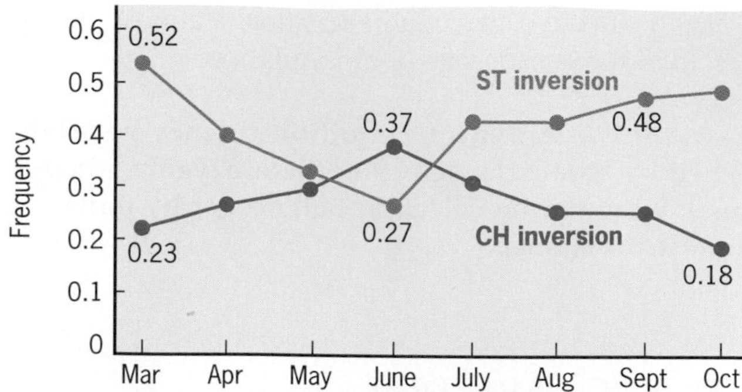

Figure 29.19 Seasonal changes in the frequencies of the chromosome inversions ST and CH in a population of *Drosophila pseudoobscura.*

Genetic Variation at the Molecular Level

In 1966 R. C. Lewontin, J. L. Hubby, and H. Harris initiated a new era in the study of genetic variation in natural populations when they applied the technique of gel electrophoresis to detect amino acid differences in proteins. In this procedure, tissues are homogenized to release proteins; then samples of the homogenates are placed in a gel and subjected to an electric current (Figure 29.20). The proteins in these samples migrate in the gel toward the positive or the negative pole in the electric field, and the rate of migration is a function of a protein's size and charge. After sufficient time has elapsed to separate different forms of a protein, the gel is stained to locate each form.

As an example, consider the staining pattern shown in Figure 29.21. Eleven *D. pseudoobscura* flies were analyzed for molecular variation in the structure of phosphoglucomutase, an enzyme encoded by the gene *Pgm-1*. This enzyme has two electrophoretically detectable forms, 100 and 104, each denoted by a meas-

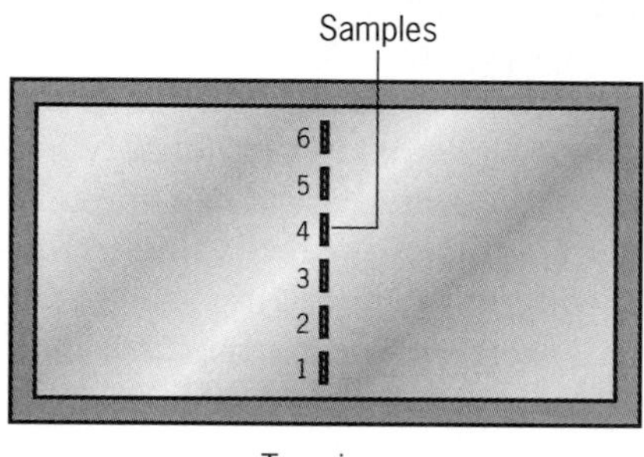

1 Tissue homogenates are placed in wells in a gel.

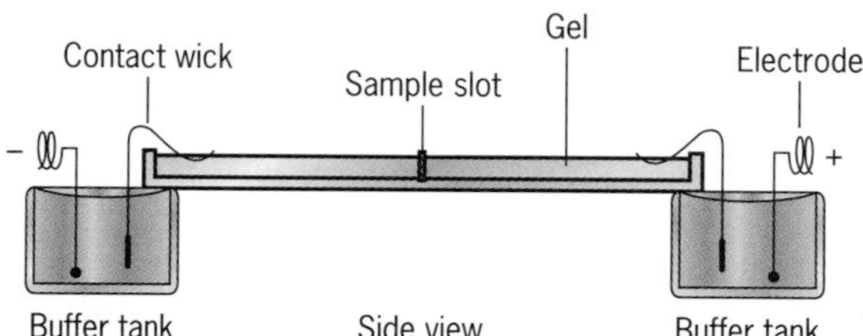

2 The gel with the samples in it is placed in an electric field for several hours.

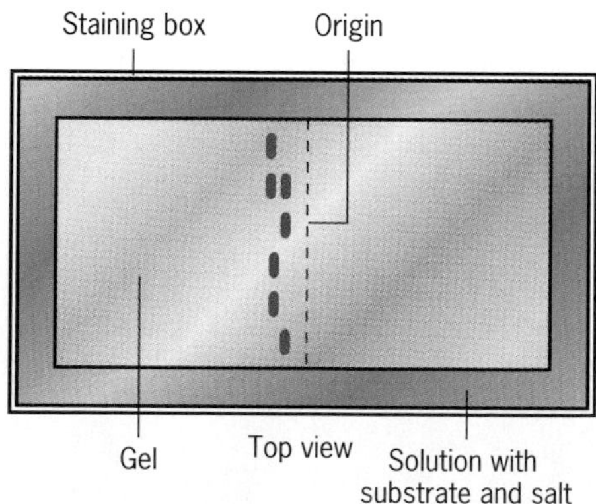

3 The gel is placed in a solution with an appropriate substrate for the enzyme to be assayed and a salt that produces a colored band when it reacts with the product of the enzyme's activity.

Figure 29.20 A schematic representation of the technique of gel electrophoresis.

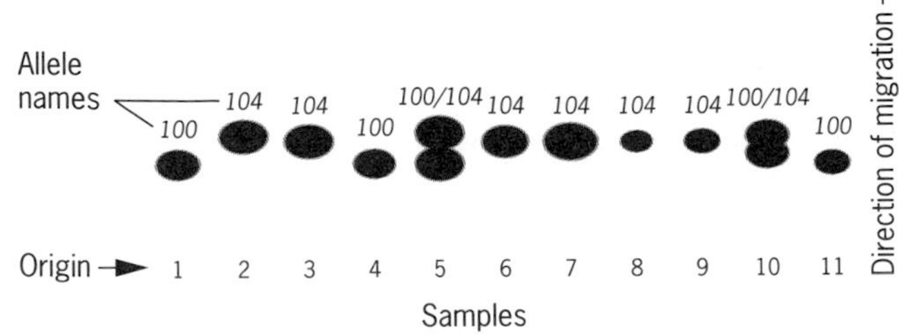

Figure 29.21 Electrophoretic variation in the *Pgm-1* gene, which encodes the enzyme phosphoglucomutase, in a sample of 11 *D. pseudoobscura* females from a natural population. Homozygotes produce only one band of protein in the gel, whereas heterozygotes produce two bands.

ure of the distance that the protein migrates in the gel. Different forms of an enzyme are called *allozymes*, and each allozyme is encoded by a different allele of the gene. From the data in Figure 29.21, it is clear that *Pgm-1* has two alleles and that individuals within the *D. pseudoobscura* population can be homozygous for either allele (for example, individuals 1 and 2), or they can be heterozygous (individuals 5 and 10).

Gel electrophoresis allows a researcher to detect variation at the molecular level. Although some proteins do not exhibit electrophoretic variation—we say they are *monomorphic*—others do. Table 29.3 summarizes the incidence of electrophoretic polymorphisms among different types of human enzymes. Overall, one-third of these enzymes are polymorphic.

Protein gel electrophoresis provided the first estimates of how much variation exists at the molecular level. In many species, approximately one-third of all genes that encode soluble proteins exhibit electrophoretic polymorphisms, and about 12 to 15 percent of individuals within a population are heterozygotes. These two statistics—the proportion of loci that are polymorphic and the proportion of individuals that are heterozygous—are simple and convenient measures of the amount of genetic variation within a population. Other techniques are now used to detect genetic variability at the molecular level. These include Southern blotting to identify restriction fragment-length polymorphisms (RFLPs) in genomic DNA and a plethora of PCR-based procedures to uncover DNA sequence variation (see Chapters 20–22). The application of all these techniques has reinforced the conclusion that most populations are rich in genetic variation. Some of this variation is probably maintained by balancing selection, and some is probably due to the recurrent production of neutral or slightly deleterious mutations.

Key Points: **Genetic variation within a population can be documented at the phenotypic, chromosomal, and molecular levels.**

SPECIATION

The evolutionary forces of mutation, genetic drift, migration, and natural selection create and destroy genetic variability in a population. Over time, a population may split into two or more subpopulations that eventually are unable to interbreed; that is, they become reproductively isolated from each other. The attainment of reproductive isolation between subpopulations is the key event in the formation of species.

Definition of Species

In classical taxonomy, a species is defined on the basis of phenotypic characteristics. If the characteristics of two groups of organisms are sufficiently different, the groups are considered to be different species. In evolutionary biology, a species is defined on the basis of a common gene pool. A group of interbreeding, or potentially interbreeding populations, that is reproduc-

TABLE 29.3
Incidence of Electrophoretic Polymorphisms in Different Types of Human Enzymes

Enzyme Type	*No. of Loci Studied*	*Polymorphic Loci*	*Percent Polymorphic Loci*
1. Oxidoreductases	24	7	29
2. Transferases	29	10	34
3. Hydrolases	38	13	34
4. Lyases	10	3	30
5. Isomerases	3	—	—
Totals	104	33	

TABLE 29.4
A Classification of Prezygotic and Postzygotic Isolating Mechanisms

1. *PREZYGOTIC MECHANISMS prevent the formulation of hybrid zygotes.*
(a) **Ecological Isolation.** Populations occupy different habitats in the same general region.
(b) **Temporal Isolation.** Mating or flowering times occur at different seasons.
(c) **Behavioral Isolation.** Mutual attraction between the sexes of different species is weak.
(d) **Mechanical Isolation.** Physical noncorrespondence of the genitalia or flower parts prevents copulation or pollen transfer.
(e) **Isolation by Different Pollinators.** Related flowering plant species may be specialized to attract different insects as pollinators.
(f) **Gametic Isolation.** Gametes may not attract each other, or gametes may be inviable in the reproductive tract of the opposite sex.
2. *POSTZYGOTIC MECHANISMS reduce the viability or fertility of hybrid zygotes.*
(g) **Hybrid Inviability.** Hybrid zygotes are inviable or have reduced viability.
(h) **Hybrid Sterility.** Hybrids of one or both sexes fail to produce functional gametes.
(i) **Hybrid Breakdown.** The F_2 or backcross hybrids have reduced fertility or viability.

tively isolated from other such groups constitutes a species.

These two ways of defining species are not always in agreement. Two populations may be reproductively isolated, but they may not have evolved distinct, easily recognized phenotypic characteristics. In the tradition of classical taxonomy, such populations would be regarded as a single species, whereas in the tradition of evolutionary biology they would be regarded as two separate species. Conversely, two populations may have evolved distinct phenotypic characteristics, but they may not be reproductively isolated. A classical taxonomist would regard such populations as two separate species, whereas an evolutionary biologist would regard them as a single species. When it is possible to determine if two populations are reproductively isolated, we can apply the evolutionary definition of species. However, when such determinations are not possible—as, for example, with fossilized organisms—we are limited to the taxonomic definition of species.

Reproductive isolation is the key to the evolutionary definition of species. Different types of mechanisms can cause populations to be reproductively isolated (Table 29.4). These mechanisms fall into two general categories: (1) **prezygotic isolating mechanisms,** which prevent the formation of hybrid offspring from individuals that belong to different species, and (2) **postzygotic isolating mechanisms,** which reduce the viability or fertility of hybrid offspring produced by an interspecific mating. Prezygotic and/or postzygotic mechanisms may operate together to reinforce reproductive isolation between species. For example, two closely related species of toads (*Bufo fowleri* and *Bufo americanus*) are reproductively isolated because they have different breeding seasons (temporal isolation) and different mating calls (behavioral isolation).

Modes of Speciation

The splitting of a population into separate species requires that subdivisions of the population become reproductively isolated from each other. The most straightforward way for this to happen is for the subpopulations to become geographically separated (Figure 29.22). During the period of separation, reproductive isolating mechanisms can evolve. Then, when the subpopulations are reunited by the disappearance of geographic barriers, they are unable to exchange genes. Speciation that involves the geographic separation of subpopulations is called **allopatric speciation** (from Greek roots meaning "other villages").

The evolution of two populations of tuft-eared squirrels (*Sciurus aberti*) on the north and south rims of the Grand Canyon in the American Southwest is an example of allopatric speciation. Phenotypic and molecular characterization of these squirrels suggests that they evolved from the same parent population. The formidable canyon that separates them has reduced interpopulational gene flow to near zero. Consequently, the two populations have developed their own identities. The northern population is darker and has whiter tails than the southern population (Figure 29.23). Until recently, many mammologists regarded the two populations as different species; today, however, they are increasingly considered to belong to the same species: *S. aberti aberti* on the south rim and *S. aberti kaibabensis* on the north rim. The Grand Canyon is an effective physical barrier to gene flow between the two populations, but it is not perfect. Infrequently, a few squirrels may cross the canyon, mate with members of the other population, and produce viable hybrids. Reproductive isolation—and therefore speciation—is not quite complete.

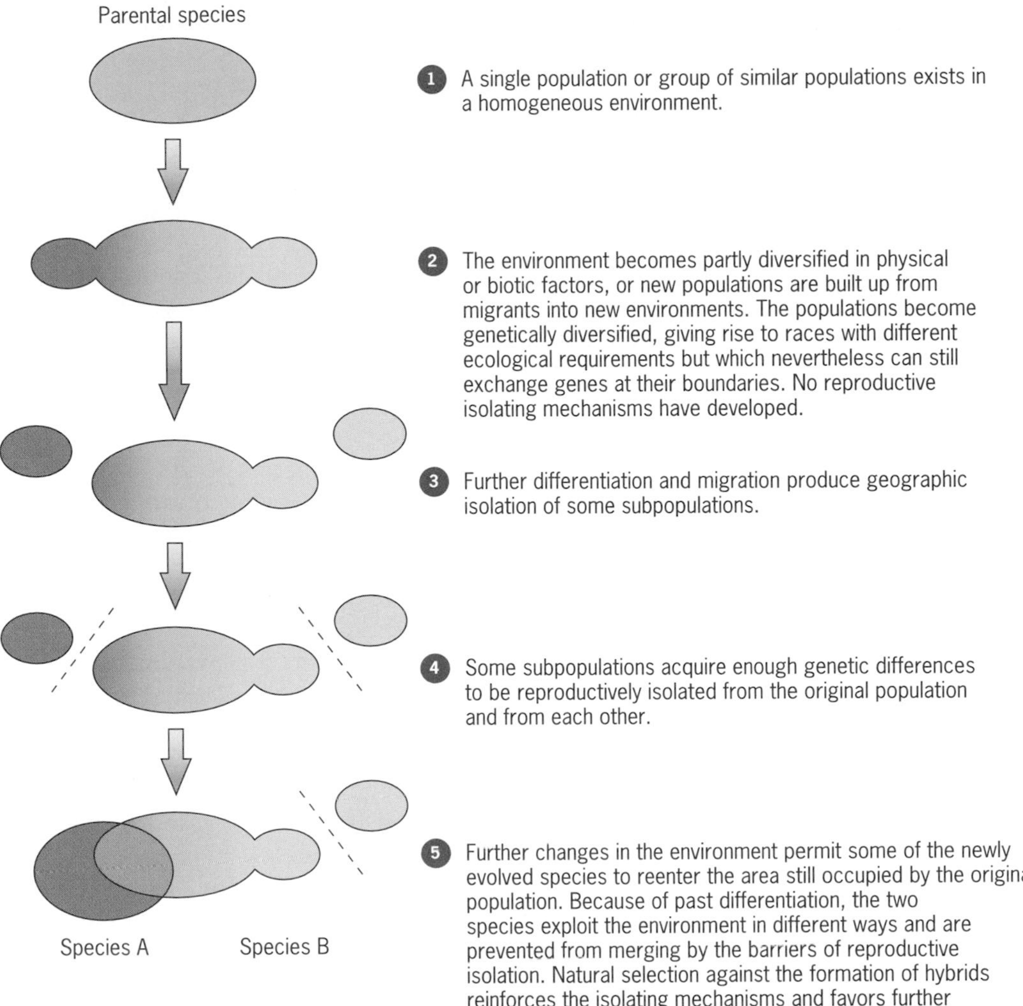

Figure 29.22 A schematic model of allopatric speciation.

Figure 29.23 The Grand Canyon looking to the East. The North Rim, home to the Kaibab population of squirrels, is to the left and the South Rim, home to the Abert population of squirrels, is to the right.

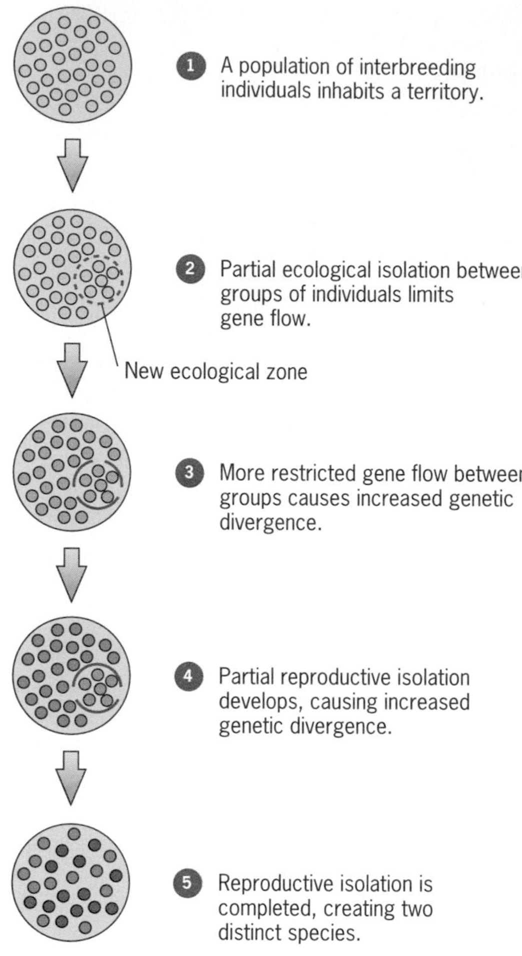

Figure 29.24 A schematic model of sympatric speciation.

Speciation might also occur in the absence of any geographical barrier (Figure 29.24). This process, called **sympatric speciation** (from Greek roots meaning "same village"), requires that subpopulations develop reproductive isolation while they occupy the same territory. Perhaps the subpopulations become separated ecologically so that they evolve more or less independently, and during the period of independent evolution, reproductive isolating mechanisms develop.

The evolution of two species of green lacewing flies, *Chrysoperla carnea* and *C. downesi* (Figure 29.25), is considered to be an example of sympatric speciation. These two species, analyzed by Catherine and Maurice Tauber, occupy the same territory in North America. Their subtle coloration differences are associated with habitat preferences. *C. carnea* is light green in spring and early summer and changes to brown in the fall. This species lives in grassy areas in the early part of the year and moves to deciduous trees in the fall. Its coloration is adapted to these habitats. *C. downesi* is dark green all year long and lives on evergreens that do not change color during the year. No evidence exists that the two species were ever physically separated from each other. However, they are reproductively isolated by habitat preference and by temporal mechanisms. *C. carnea* breeds in the winter and summer, and *C. downesi* breeds only in the spring. Breeding time in these insects is controlled by day length. If the two species are brought into the laboratory and the day length is manipulated, they can be induced to interbreed. Furthermore, the hybrid offspring are fertile.

A single gene with two alleles controls color in these insects. The genotype G_1G_1 produces light green pigment that changes to brown upon aging, G_2G_2 produces dark green pigment, and G_1G_2 produces an intermediate pigment that is never observed in nature. The fact that the intermediate is not observed in nature indicates that the two species do not normally interbreed. The breeding pattern is controlled by two genes. A dominant allele at either of the two genes causes a winter/summer breeding pattern, whereas homozygous recessive alleles at both of the genes causes a spring breeding pattern.

The Taubers hypothesize that polymorphism of the color gene initially led to restricted gene flow between two populations of lacewings. One homozygote was better adapted to the grass/deciduous tree habitat; the other homozygote was better adapted to the evergreen habitat, and the heterozygote was not well adapted to either. Disruptive selection for coloration favored the development of assortative mating for this trait. Eventually, different breeding cycles evolved to keep the light and dark populations of lacewings reproductively isolated.

Although this study and others suggest that sympatric speciation may occur, the idea remains controversial. Usually it is difficult to exclude geographic separation as a factor in the speciation process.

Key Points: **In evolutionary biology, a species is a group of populations that share a common gene**

Figure 29.25 The lacewings *Chrysoperla carnea* (*left*) and *C. downesi* (*right*).

pool. The development of reproductive isolation between populations is a key event in the speciation process. This process may occur when the populations are geographically separated (allopatric) or when they are together in the same territory (sympatric).

HUMAN EVOLUTION

When Darwin proposed his theory of evolution in 1859, and later, when he suggested that human beings had evolved from more primitive organisms, he provoked a great controversy. The idea that organisms evolve, and more specifically, that humans evolve, troubled many people. In the ensuing 140 years, much has been learned about the course of human evolution. Paleontologists have analyzed the fossilized remains of organisms that are likely to have been the ancestors of modern humans, and geneticists have analyzed DNA sequence data in order to study the relationships among humans and their closest nonhuman relatives, the great apes. In the following sections, we discuss some of these analyses.

Humans and the Great Apes

Several morphological features distinguish human beings from chimpanzees and gorillas. The apes have larger canine and incisor teeth than modern humans, and their jaws are larger and heavier. Ape brains are smaller than human brains, and the point where the ape brain attaches to the spinal cord is placed farther to the back of the skull than it is in humans. The shape and proportions of an ape's body are also different from those of a human. In an ape, the body's trunk widens toward the base, whereas in a human, it tends to have the same width from the shoulders to the waist. The legs of an ape are proportionately shorter than those of a human, and the pelvis is not constructed to accommodate a regular upright stance. Although apes can walk upright on two legs, they cannot do so for long periods of time. By contrast, humans are exclusively bipedal—except, of course, in early childhood. The hands and feet of apes also differ from those of humans. Apes do not have apposable thumbs, and their feet do not provide the support that is needed for bipedal locomotion.

Despite all these morphological differences, the DNA of apes and humans is remarkably similar. When the genomes of chimpanzees and humans are compared by denaturation/renaturation techniques, they are found to be more than 99 percent identical. This high degree of identity implies that chimpanzees and humans are quite closely related, and suggests that they diverged from a common ancestor rather recently in evolutionary time, perhaps 5–9 million years ago. The other great ape species, the gorilla, appears to be less closely related to humans than the chimpanzee is.

Human Evolution in the Fossil Record

Though rare, fossils have provided important information about human evolution (Figure 29.26). The oldest fossils that appear to be in the human evolutionary line come from East Africa where they were formed 4–5 million years ago. These first human-like—that is, *hominid*—creatures have been given the name *Ardipithecus ramidus*. Later in the fossil record, 3–4 million years ago, another hominid creature appeared. This or-

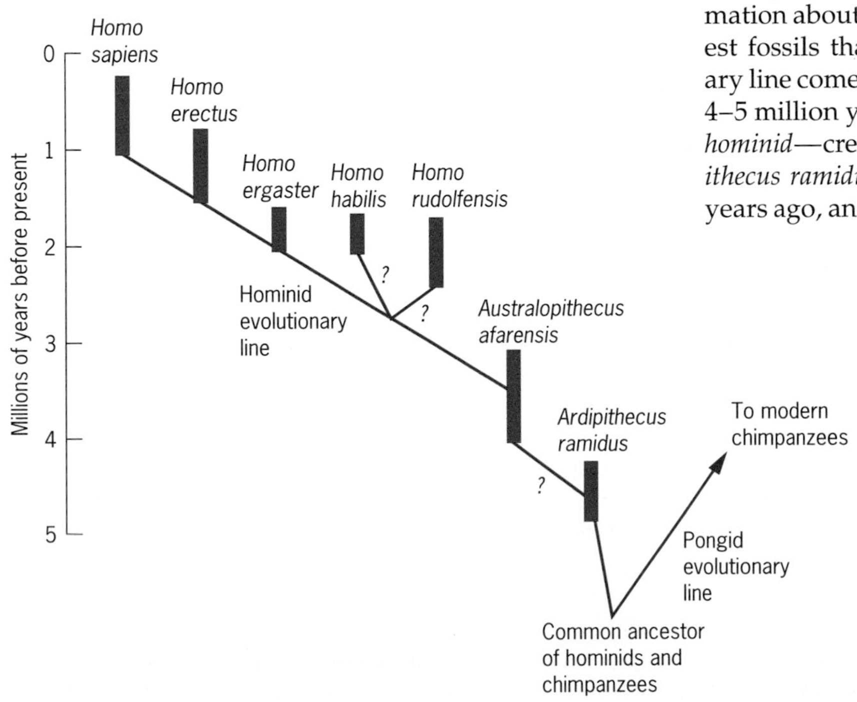

Figure 29.26 Ancestors of human beings that have been discovered through fossil evidence. The hominid evolutionary line leads from the common ancestor of humans and chimpanzees to modern humans (*Homo sapiens*). The pongid evolutionary line leads from this common ancestor to modern chimpanzees. Uncertainties in the hominid evolutionary line are indicated by question marks.

ganism, known as *Australopithecus afarensis,* probably stood 1–1.5 m tall and walked upright, at least for short distances. The fossil known as Lucy is a specimen of *Australopithecus afarensis.*

The first organisms to be classified in the same genus as *Homo sapiens* appeared 2–2.5 million years ago. Two species have been named, *H. rudolfensis* and *H. habilis.* Both of these "early *Homo*" species have many ape-like features; however, compared to *Australopithecus,* the opening for the spinal cord is closer to the middle of the skull, and the skull itself is reduced in length and increased in width—all hominid characteristics. Nevertheless, many paleontologists have questioned the inclusion of these two species within the genus *Homo,* and there is some sentiment to reclassify them in the genus *Australopithecus.*

Between 1.9 and 1.5 million years ago, a more advanced hominid appeared in the fossil record. This creature, called *Homo ergaster,* had a body shape and limb proportions like those of modern humans, and its teeth and jaws were also human in structure. Thus, *H. ergaster* is the first hominid that can confidently be placed within the genus *Homo.*

All the early hominid fossils come from Africa. The first hominid species to produce fossils outside of Africa was *Homo erectus.* These fossils, formed about 1 million years ago, have been found in China and Indonesia. Thus, *H. erectus* was widespread and probably gave rise to archaic populations of humans in Europe, Asia, and Africa. The best known of the archaic humans were the Neanderthals, a species that evolved in Europe and the Near East several hundred thousand years ago. Analysis of DNA from a fossilized Neanderthal bone suggests that the Neanderthals did not contribute genetic material to the modern human species, *H. sapiens.* (See the Technical Sidelight: Using Mitochondrial DNA to Study Human Evolution in Chapter 19.) In fact, they may have been competitors with the populations that evolved into modern humans, and eventually lost out in the competition and became extinct.

Modern humans may have evolved simultaneously in Europe, Asia, and Africa from the archaic human populations that existed on each of those continents, or they may have evolved on one continent—probably Africa—and subsequently spread to the others. The first of these ideas about the origin of modern humans is called the Multiregional Hypothesis, and the second is called the Out-of-Africa Hypothesis. According to the Multiregional Hypothesis, the evolution of modern humans involved archaic populations in different regions of the Old World. According to the Out-of-Africa Hypothesis, it involved only populations in Africa, which then migrated to each of the other continents and replaced the archaic human populations that were living there. Fossil evidence cannot discriminate between these two hypotheses. However, genetic evidence obtained by studying DNA sequences in living human beings has provided ways of testing them.

DNA Sequence Variation and Human Origins

Genetic data allow researchers to study human evolution by investigating the relationships among extant human populations. Populations that are closely related share genetic properties that distantly related populations do not. Thus, by analyzing variation in genes, gene products, and DNA sequences, it is possible to determine the relatedness of different racial and ethnic groups, and to arrange them in an evolutionary—or phylogenetic—tree. Genetic analysis also permits researchers to decipher key events in human evolutionary history.

Many types of genetic variation have been used to study human evolution: blood group and allozyme polymorphisms, restriction fragment-length polymorphisms, variation in the length of tandemly repeated DNA sequences, and variation in the composition of DNA sequences themselves. Both nuclear and mitochondrial genetic variation has been investigated. The nuclear genome contains the preponderance of human polymorphisms, but the mitochondrial genome has the unique property of being transmitted exclusively through females. Variation in mitochondrial DNA therefore provides a way of tracing maternal lineages in human evolutionary history.

Compared to other species, the human species is genetically rather uniform. At the nucleotide level, humans have about one-fourth the genetic variation of chimpanzees and about one-tenth that of *Drosophila.* Furthermore, most of the genetic variation in the human species—perhaps 85 to 95 percent of it—is within rather than between populations.

The relative absence of genetic variation in human populations implies that during its evolutionary history, the genetically effective size of the human population was small—between 10,000 and 100,000 individuals. The census size may have been larger (today, it certainly is), but the mating system, various constraints on reproduction, and bottlenecks in size caused by famine, disease, or weather-related catastrophes apparently conspired to keep the effective population size under 100,000. In such a population, random genetic drift dominates over mutation to determine the equilibrium level of variability for selectively neutral alleles (see p. 792).

When different human populations are analyzed for genetic variation, those in Africa are found to have more variation than those in other continents. The greater accumulation of genetic variation in African

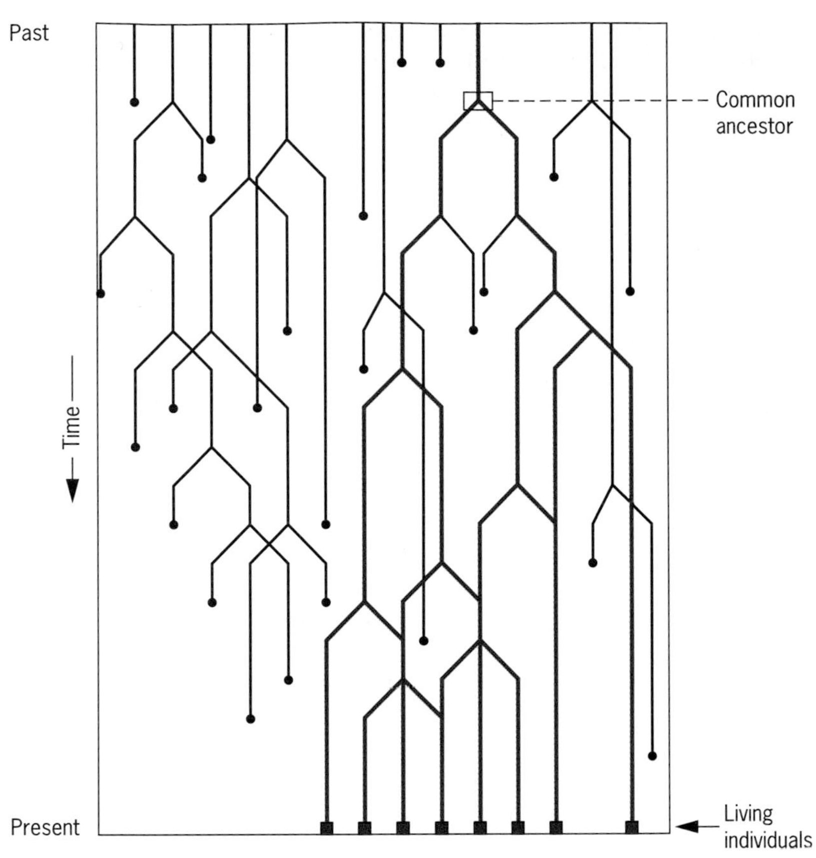

Figure 29.27 A coalescence process. If the lineages of the DNA sequences found in living individuals are traced back into the past, they coalesce in a common ancestor. These lineages are highlighted in red in the timeline. Other DNA sequences from the past are not represented in living individuals; the time each became extinct is indicated by a dot.

populations suggests that these populations are the oldest—an idea that is consistent with the Out-of-Africa Hypothesis of human origins. Fairly strong evidence for this hypothesis has come from studies of mitochondrial DNA sequences from different human populations. By analyzing sequences from living individuals, it is possible to work back to the ancestral sequence from which all the existing sequences could have sprung. This ancestral sequence represents the point at which the lineages of the living individuals coalesce into one individual, the common ancestor of them all (Figure 29.27). Then, by counting the number of mutations that occurred between the ancestral DNA sequence and the current sequences, and by dividing this number by the known mutation rate, it is possible to calculate the time that has elapsed since the common ancestor existed.

When this type of analysis is performed on mitochondrial DNA sequences, the elapsed time between the present and the time when the common ancestor lived is estimated to be 100–200 thousand years. Analyses of DNA sequences on the Y chromosome, which is transmitted exclusively through males, yield a similar estimate. Thus, the coalescent principle suggests that all modern humans are descended from maternal and paternal common ancestors who lived between 100,000 and 200,000 years ago. This result does not imply, however, that these common ancestors were the only two people alive at that remote time. Almost certainly many others were alive too. Their genetic lineages—mitochondrial in the case of females and Y chromosome in the case of males—simply became extinct. With the coalescent method, current DNA sequences can be traced back to the individuals whose mitochondrial or Y chromosomal lineages were lucky enough to survive and spread through the species, modified, of course, by the random process of mutation.

If these analyses can be trusted—and there are many uncertainties in both the methods and the data—they suggest that the modern human species is relatively young and that it probably originated in the archaic human populations of Africa. From Africa, humans migrated to Asia and Europe, and later, to Australia and the Americas, ultimately becoming the dominant species on the Earth.

Key Points: **Fossil evidence indicates that the remote ancestors of human beings evolved in Africa, beginning about 4–5 million years ago. Genetic evidence indicates that modern human populations may have emerged from Africa about 100–200 thousand years ago and subsequently spread to other continents.**

TESTING YOUR KNOWLEDGE

1. The following blood-type data were collected in a sample of 700 Native Americans from South Dakota:

Blood Type	Number of Individuals
A	326
B	20
AB	16
O	338

Estimate the frequencies of the I^A, I^B, and I^O alleles from the sample data.

ANSWER

Let's symbolize the frequencies of the *A*, *B*, and *O* alleles of the *I* gene as *p*, *q*, and *r*, respectively, and let's assume that the genotypes of the *I* gene are in Hardy–Weinberg proportions. We begin by estimating *r*, the frequency of the I^O allele. To obtain this estimate, we note that the frequency of the O blood type, which is $338/700 = 0.483$ in the sample data, should correspond to the Hardy–Weinberg frequency of the I^OI^O genotype, r^2. Thus, if we use the Hardy–Weinberg Principle in reverse, we can estimate the frequency of the I^O allele as $r = \sqrt{0.483} = 0.695$.

To estimate *p*, the frequency of the I^A allele, we note that $(p + r)^2 = p^2 + 2pr + r^2$ corresponds to the combined frequencies of the A ($p^2 + 2pr$) and O (r^2) blood types. From the sample data, these combined frequencies are estimated to be $(326 + 338)/700 = 0.949$. If we set $(p + r)^2 = 0.949$ and take the square root, we obtain $p + r = 0.974$; then, by subtraction, we can estimate the frequency of the I^A allele as $p = 0.974 - r = 0.974 - 0.695 = 0.279$.

To estimate *q*, the frequency of the I^B allele, we note $p + q + r = 1$. Thus, $q = 1 - p - r = 1 - 0.279 - 0.695 = 0.026$.

2. Pattern baldness in humans is an autosomal, sex-influenced trait; it is dominant in males and recessive in females. A male with at least one copy of the pattern baldness allele (*PB*) is bald; however, two copies of this allele must be present for a female to be bald. The frequency of the *PB* allele in a particular population has been estimated to be 0.3. Assuming random mating, what is the expected frequency of pattern bald males and pattern bald females in the population?

ANSWER

If the frequency of *PB* is 0.3, the frequency of PB^+ is 0.7. With random mating, the genotype frequencies are:

Genotype:	PB/PB	PB/PB^+	PB^+/PB^+
Frequency:	$(0.3)^2 = 0.09$	$2(0.3)(0.7) = 0.42$	$(0.7)^2 = 0.49$

Based on these frequencies, we expect $(0.09 + 0.42) = 0.51$ of the males in the population to be bald and 0.09 of the females to be bald.

3. The Hb^S allele responsible for sickle-cell anemia is maintained in many human populations because in heterozygous condition it confers some resistance to infection by malaria parasites; however, in homozygous condition, this allele is essentially lethal. Thus, as malaria is eradicated we might expect the Hb^S allele to disappear from human populations. If the normal allele Hb^A mutates to Hb^S at a rate of 10^{-8} per generation, what ultimate frequency would you predict for the Hb^S allele in a malaria-free world?

ANSWER

In a malaria-free world, the advantage of maintaining the Hb^S allele in a balanced polymorphism would disappear. Hb^SHb^A heterozygotes would have the same fitness as Hb^AHb^A homozygotes, and Hb^SHb^S homozygotes would continue to have very low fitness—essentially zero compared to the other two genotypes. Under these circumstances, the frequency of the Hb^S allele (q) would be determined by a balance between selection against it in homozygous condition (selection coefficient $s = 1$) and introduction into the population by mutation at rate $u = 10^{-8}$ per generation. The equilibrium frequency of the Hb^S allele would be $q = \sqrt{u/s} = 0.0001$, a thousandfold less than its current frequency in malaria-infested regions of the world.

QUESTIONS AND PROBLEMS

29.1 The following data for the M-N blood types were obtained from Indian villages in North and Central America:

Group	Sample Size	M	MN	N
Central American	86	53	29	4
North American	278	78	61	139

Calculate the frequencies of the L^M and L^N alleles for the two groups.

29.2 The frequency of an allele in a large randomly mating population is 0.2. What is the frequency of heterozygous carriers?

29.3 The incidence of recessive albinism is 0.0004 in a human population. If mating for this trait is random in the population, what is the frequency of the recessive allele?

29.4 In a sample from an African population, the frequencies of the L^M and L^N alleles were 0.78 and 0.22, respectively. If the population mates randomly with respect to the M-N blood types, what are the expected frequencies of the M, MN, and N phenotypes?

29.5 A gene has three alleles, A_1, A_2, and A_3, with frequencies 0.6, 0.3, and 0.1, respectively. If mating is random, predict the combined frequency of all the heterozygotes in the population.

29.6 Human beings carrying the dominant allele *T* can taste the substance phenylthiocarbamide (PTC). In a population in which the frequency of this allele is 0.4, what is the probability that a particular taster is homozygous?

29.7 Hemophilia is caused by an X-linked recessive allele. In a particular population, the frequency of males with hemophilia is 1/4000. What is the expected frequency of females with hemophilia?

29.8 A trait determined by an X-linked dominant allele shows 100 percent penetrance and is expressed in 36 percent of the females in a population. Assuming that the population is in Hardy–Weinberg equilibrium, what proportion of the males in this population express the trait?

29.9 In *Drosophila* the ruby eye phenotype is caused by a recessive, X-linked mutant allele. The wild-type eye color is red. A laboratory population of *Drosophila* is started with 25 percent ruby-eyed females, 25 percent homozygous red-eyed females, 5 percent ruby-eyed males, and 45 percent red-eyed males. (a) If this population mates randomly for one generation, what is the expected frequency of ruby-eyed males and females? (b) What is the frequency of the recessive allele in each of the sexes?

29.10 A population in Hardy–Weinberg equilibrium is segregating two alleles, *A* and *a*. The frequency of heterozygotes in the population is 0.5. What are the frequencies of each of the alleles?

29.11 What frequencies of alleles *A* and *a* in a randomly mating population maximize the frequency of heterozygotes?

29.12 In a class of 25 students, 14 were unable to taste phenylthiocarbamide (genotype *tt*). The other 11 were able to taste this substance (genotype *TT* or *Tt*). Among the 11 tasters, how many were heterozygous?

29.13 T. Dobzhansky and O. Pavlovsky studied two autosomal inversions (*A* and *D*) in *Drosophila tropicalis*. Individuals can be either *AA*, *AD*, or *DD* for the two inversions. In a sample of 200 fly larvae, they found 20 *AA*, 150 *AD*, and 30 *DD*. Are these genotypes in Hardy–Weinberg proportions? Explain your answer.

29.14 A population of Hawaiian *Drosophila* is segregating two alleles, P^1 and P^2, of the phosphoglucose isomerase (*PGI*) gene. In a sample of 100 flies from this population, 30 were P^1P^1 homozygotes, 60 were P^1P^2 heterozygotes, and 10 were P^2P^2 homozygotes. (a) What are the frequencies of the P^1 and P^2 alleles in this sample? (b) Perform a chi-square test to determine if the genotypes in the sample are in Hardy–Weinberg proportions. (c) Assuming that the sample is representative of the population, how many generations of random mating would be required to establish Hardy–Weinberg proportions in the population?

29.15 The frequencies of the alleles *A* and *a* are 0.6 and 0.4, respectively, in a particular plant population. After many generations of random mating, the population goes through one cycle of self-fertilization. What is the expected frequency of heterozygotes in the progeny of the self-fertilized plants?

29.16 In a large population that reproduces by random mating, the frequencies of the genotypes *GG*, *Gg*, and *gg* are 0.04, 0.32, and 0.64, respectively. Assume that a change in the climate induces the population to reproduce exclusively by self-fertilization. Predict the frequencies of the genotypes in this population after many generations of self-fertilization.

29.17 Each of two isolated populations is in Hardy–Weinberg equilibrium with the genotype frequencies shown below:

Genotype:	*AA*	*Aa*	*aa*
Frequency in Population 1:	0.04	0.32	0.64
Frequency in Population 2	0.64	0.32	0.04

(a) If the populations are equal in size and they merge to form a single large population, predict the allele and genotype frequencies in the large population immediately after merger. (b) If the merged population reproduces by random mating, predict the genotype frequencies in the next generation. (c) If the merged population continues to reproduce by random mating, will these genotype frequencies remain constant?

29.18 A population consists of 25 percent tall individuals (genotype *TT*), 25 percent short individuals (genotype *tt*), and 50 percent individuals of intermediate height (genotype *Tt*). Predict the ultimate phenotypic and genotypic composition of the population if generation after generation, mating is strictly assortative (that is, tall individuals mate with tall individuals, short individuals mate with short individuals, and intermediate individuals mate with intermediate individuals).

29.19 A population is composed of equal numbers of the genotypes *AA*, *Aa*, and *aa*. The offspring produced by these genotypes number 996, 996, and 224, respectively. What are the relative fitnesses of the three genotypes?

29.20 In a large randomly mating population, 0.84 of the individuals express the phenotype of the dominant allele *A* and 0.16 express the phenotype of the recessive allele *a*. (a) What is the frequency of the dominant allele? (b) If the *aa* homozygotes are 5 percent less fit than the other two genotypes, what will the frequency of *A* be in the next generation?

29.21 Because individuals with cystic fibrosis die before they can reproduce, the coefficient of selection against them is $s = 1$. Assume that heterozygous carriers of the recessive mutant allele responsible for this disease are as fit as wild-type homozygotes and that the population frequency of the mutant allele is 0.02. (a) Predict the incidence of cystic fibrosis in the population after one generation of selection. (b) Explain why the incidence of cystic fibrosis hardly changes even with $s = 1$.

29.22 A study of 108 achondroplastic dwarfs showed that they produced 27 offspring. These dwarfs had 457 nondwarf siblings who produced 582 children. What are the relative fitnesses of dwarfs and nondwarfs?

29.23 The frequency of newborn infants homozygous for a recessive lethal allele is about 1 in 25,000. What is the expected frequency of carriers of this allele in the population?

29.24 A population of size 50 reproduces in such a way that the population size remains constant. If mating is random, how rapidly will genetic variability, as measured by the frequency of heterozygotes, be lost from this population?

29.25 A population is segregating three alleles, A_1, A_2, and A_3, with frequencies 0.2, 0.5, and 0.3, respectively. If these alleles are selectively neutral, what is the probability that A_2 will ultimately be fixed by genetic drift? What is the probability that A_3 will ultimately be lost by genetic drift?

29.26 A small island population of mice consists of roughly equal numbers of males and females. The Y chromosome in one-fourth of the males is twice as long as the Y chromosome in the other males because of an expansion of heterochromatin. If mice with the large Y chromosome have the same fitness as mice with the small Y chromosome, what is the probability that the large Y chromosome will ultimately be fixed in the mouse population?

29.27 In some regions of west Africa, the frequency of the Hb^S allele is 0.2. If this frequency is the result of a dynamic equilibrium due to the superior fitness of Hb^SHb^A heterozygotes, and if Hb^SHb^S homozygotes are essentially lethal, what is the intensity of selection against the Hb^AHb^A homozygotes?

29.28 Mice with the genotype *Hh* are twice as fit as either of the homozygotes *HH* and *hh*. With random mating, what is the expected frequency of the *h* allele when the mouse population reaches a dynamic equilibrium because of balancing selection?

29.29 A completely recessive allele *g* is lethal in homozygous condition. If the dominant allele *G* mutates to *g* at a rate of 10^{-6} per generation, what is the expected frequency of the lethal allele when the population reaches mutation-selection equilibrium?

29.30 Individuals with the genotype *bb* are 20 percent less fit than individuals with the genotypes *BB* or *Bb*. If *B* mutates to *b* at a rate of 10^{-6} per generation, what is the expected frequency of the allele *b* when the population reaches mutation-selection equilibrium?

29.31 You discover two geographically separate populations of *Drosophila*. How would you ascertain whether the two populations are members of the same species or of different species?

29.32 The human species is genetically diverse, yet we still consider it a single species. Explain.

29.33 On a trip to a zoo, your friend sees a gorilla and remarks, "you mean we descended from that animal?" How would you respond to this remark about the relationship between gorilla and *Homo sapiens*?

BIBLIOGRAPHY

Crow, J. F. 1986. *Basic Concepts in Population, Quantitative and Evolutionary Genetics.* W. H. Freeman, New York.

Dobzhansky, T. 1970. *Genetics of the Evolutionary Process.* Columbia University Press, New York.

Hartl, D. L. 1988. *A Primer of Population Genetics.* 2nd ed. Sinauer, Sunderland, MA.

Jorde, L. B., M. Bamshad, and A. R. Rogers. 1998. Using mitochondrial and nuclear DNA markers to reconstruct human evolution. *BioEssays* 20:126–136.

Kimura, M. 1983. *The Neutral Theory of Molecular Evolution.* Cambridge University Press, Cambridge.

Lewontin, R. C. 1974. *The Genetic Basis of Evolutionary Change.* Columbia University Press, New York.

Price, P. W. 1996. *Biological Evolution.* Saunders, Philadelphia.

Ridley, M. 1993. *Evolution.* Blackwell Scientific Publications, Boston.

Tauber, C. A., and M. J. Tauber. 1989. Sympatric speciation in insects: Perception and perspective. In: D. Otte and J. A. Endler, Eds. *Speciation and Its Consequences.* Sinauer, Sunderland, MA.

Wood, B. 1996. Human evolution. *BioEssays* 18:945–954.

A CONVERSATION WITH
JAMES F. CROW

When did you first become interested in genetics, and what stimulated that interest?

I was fascinated by Mendelian ratios in my first college biology course and enjoyed the problems. Chemistry was also appealing, and the outcome was a biology-chemistry double major. I applied for graduate study in both fields and accepted an assistantship at the University of Texas to study genetics. Later, an offer came in biochemistry. If these had come in the other order, I would be a biochemist.

You have been a role model to many geneticists. Did you have a role model?

I never had a role model in the sense of wanting to mimic such a person or of feeling the need for one. But I certainly did have heroes, and still do.

During graduate school days at the University of Texas, the person whose work I admired was Sewall Wright. I found his papers interesting, but bordering on the occult, and hoped for a postdoctoral fellowship to study with him. Since I graduated in 1941, and war was imminent, a fellowship was out of the question. But I spent a great deal of my spare time reading his papers.

My other hero at the time was H. J. Muller. During World War II my job was teaching naval trainees at Dartmouth College. Muller was then at Amherst, and I visited him several times. After the first visit, I was so excited by new ideas that I didn't sleep at all that night—the only time this has ever happened.

Later, I became well acquainted with both men, not as teachers, but as colleagues. Since then I've added R. A. Fisher, J. B. S. Haldane, and A. H. Sturtevant to the list of heroes that I was fortunate in knowing personally.

Dr. James F. Crow, Professor Emeritus at the University of Wisconsin, Madison, has the credentials to write about genetics yesterday, today, and tomorrow. He received a Ph.D. in 1941 from the University of Texas, taught at Dartmouth College for a while, and then joined the faculty of the University of Wisconsin, where he served as Chairman of the Genetics Department and Acting Dean of the Medical School. Dr. Crow's expertise is in population and *Drosophila* genetics. He is the author of many scholarly articles and three textbooks, and has been active in numerous scientific organizations, including the Genetics Society of America and the National Academy of Sciences. For many years he has edited the Perspective series of articles in *Genetics* and has written many of the articles himself. Dr. Crow's contributions to science and education have been widely recognized. He received Distinguished Teaching and Service Awards from the University of Wisconsin, and an honorary doctorate from the University of Chicago. His avocation is music, and for more than forty years he was a violist in the Madison Symphony Orchestra. Though officially "retired," Dr. Crow still gives lectures in genetics courses at the University of Wisconsin.

What advice would you give to undergraduate students interested in the study of genetics?

First, get a good background in the physical sciences and math; it's probably unnecessary to mention computer literacy. I am sure that geneticists in the future will be increasingly dependent on such knowledge. There is a risk that the beautifully neat, clean results of molecular genetics will produce a generation of super-technicians. If you are to be an innovator, not a recipe-follower, you need the underlying principles.

Let me urge you also to get a broad general education. I hope you will continue to learn throughout your lives, but this is the last time in which learning is your main business. Knowledge of literature and the arts may not make you a better geneticist, but it will make you a better person and your life a more satisfying one.

In addition to your career as a geneticist, you are known for your avocation as a musician. How did you find the time to pursue this interest?

As a child, I inveigled my parents into giving me violin lessons, and I started out to be a musician. I gradually came to realize that I lacked the talent and drive to perform professionally. The final nail in my musician's coffin was harmony, which didn't come easily. I was recently asked by a group of aspiring high-school students if I had ever had a teacher who inspired me to be a scientist. I answered: Yes, my harmony teacher. I have not given up music, however, and have continued to play viola with undiminished enthusiasm. I managed to squeeze out time to play in the Madison Symphony for 47 years and to see it grow from a loose coterie of light-hearted amateurs into a band of serious professionals, now setting a standard too high for my arthritic fingers.

You have known most of the famous geneticists of the twentieth century. What is your favorite story about one of them?

My favorite story about Wright—everybody's favorite story—is his absent-mindedly erasing the blackbroad with a guinea pig; but it is probably apocryphal. Here is a true story. In the early 1950s Newton Morton and I took a train to Chicago to ask Wright about some ideas on effective population number. To each of our questions, he simply said that he didn't know the answer, and the conference was over in a few minutes. On the way back, Morton, who then as now was not given to understating his views, said that if Wright was the world's greatest population geneticist, the subject was in bad shape. Then, a few days later I got a 15-page letter from Wright with every question answered. Later, I had many confirmations of this Wrightian trait; he would not guess or speculate, but worked out each idea carefully before speaking about it.

During your many years of teaching genetics, do any classroom incidents stand out?

One of the most satisfying teaching experiences is eliciting an unexpectedly deep insight from a student. After one lecture, a student corrected an erroneous formula for predicting yield of some corn crosses. I had unthinkingly accepted a wrong formula in a technical paper. I tried to seduce this student into genetics, but he was committed to organic chemistry.

Is there an "artistic" element in science?

There is certainly an aesthetic element. Some of my greatest intellectual satisfacton has come from neat, clever, mathematical derivations, often of old results previously arrived at in a clumsy way. R. A. Fisher's genetical and statistical work epitomizes elegance, as does Malecot's formulation of the inbreeding coefficient. Some of the work that Motoo Kimura and I did together has given me aesthetic pleasure not unlike that of an artfully crafted Mozart String Quartet.

Can you single out any one research result as the most exciting of your career?

Most of my experimental research has involved long-time experiments planned well in advance, so the results were more solid than exciting. One purely accidental discovery, however, brought me genuine excitement. In 1957, a student Yuichiro Hiraizumi, discovered a gene in a natural *Drosophila* population that violated Mendel's laws by segregating, not in the regulation 1:1 ratio, but more like 98:1. At first I didn't believe the result and thought there was a mistake somewhere. But when it became clear, I realized that we now had, in a species where a thorough genetic analysis was possible, a beautiful example of meiotic drive. The late Larry Sandler was joining me the next year as a postdoctoral fellow, and I was pleased to tell him that we had something great for him to work on—an instance of the phenomenon that he had predicted and called meiotic drive. The study of Segregation Distortion is still going on, with recent new breakthroughs from molecular technology.

In your long life as a scientist you have spent a great deal of time teaching and on administrative and committee work, both local and national. How do you feel about it now?

I have indeed spent a great deal of time teaching, and I don't regret a minute of it. Teaching was fun, but I thought it important and took it very seriously. As for committee and administrative work, this was less satisfying, but on the whole I am glad to have done it—to have had some influence on University and national policy. Of course, if I had not done all this, I would have published more papers. But I doubt that my scientific contribution would have amounted to any more; all the ideas that I regard as important have found their way into print.

Epilogue
Genetics Yesterday, Today, and Tomorrow
A Personal View

James F. Crow

Because genetics has such a short past, the authors of this book thought that a history of the subject, as seen by one who has lived through a large part of that history, might be appropriate. I am delighted to oblige. This review is spotty and personal, but I hope it conveys some of the excitement that came with the rapid succession of often-surprising discoveries. I also hope that it conveys some of the uncertainties I feel about the future.

You have now finished the book and have perforce acquired a great deal of genetic knowledge. I trust you have found the experience interesting, even exciting. I would hope that you are just a bit curious as to how all this came about. So, I invite you to view the subject through the eyes of one who was around when much of what you have learned was being discovered.

Genetics is a new science, a twentieth-century phenomenon. The subject is unusual in that its origin can be pinpointed to a single event, the publication of Gregor Mendel's paper in 1866. Or perhaps I could better say it started in 1900 when Mendel's work, ignored for 34 years, was finally rediscovered.

I find it clarifying to divide genetics into three periods. The first period dates from the rediscovery of Mendel's laws in 1900 to the beginning of World War II. During this period the subject was dominated by breeding experiments and cytology. The main emphasis, and the main success, was transmission genetics. The second period starts at the end of the war and extends to the present. It was dominated by the genetics of microorganisms and molecular techniques. The emphasis was on molecular understanding, and it is still going on. Thanks to new tools, development and neurobiology have come to the front. The third period, overlapping the second, crept in gradually just one or two decades ago. Still in its early stages, this period marks the beginning of a major impact of molecular genetics on agriculture and medicine.

I started college in 1933 and completed my graduate work in 1941, just at the end of the first period. It was a different time. What did we know then? And, perhaps more interesting, what did we *not* know?

THE FIRST PERIOD, 1900–1941

The year 1900 brought a genetic spectacular. Of course, the main event was the rediscovery of Mendel's laws by three men, working in three different countries. But there were significant sideshows. The year also saw Karl Pearson's invention of the χ^2 test and Karl Landsteiner's discovery of blood groups, which gave genetics its first useful human polymorphism and marked the beginning of immunogenetics.

In the years before 1900, a number of discoveries prepared the way for the instant acceptance of Mendel's work. By the turn of the century, the details of mitosis and meiosis had been worked out; thus, once Mendel's laws were announced, it was immediately obvious that his particles were carried by the chromosomes. The key papers pointing this out were written by T. H. Montgomery and W. S. Sutton in the United States and T. Boveri in Germany. Undoubtedly, the idea occurred to many others, but they didn't write about it. Sutton, incidentally, was still a student and later went into medicine. The other two are well-known scientists.

Another nineteenth-century discovery turned out to be premature. In the 1870s F. Miescher discovered a mixture of proteins and nucleic acids in pus and sperm cells and called the substance nuclein. In 1895 the cytologist E. B. Wilson wrote:

> *Now, chromatin is well known to be similar to, if not identical with, a substance known as nuclein. . . . And thus we reach the remarkable conclusion that inheri-*

tance may, perhaps, be effected by the physical transmission of a particular chemical compound from parent to offspring.

A year later Wilson suggested that nucleic acid was the component responsible for genetic transmission. He was a half-century ahead of his time.

I'll mention one other nineteenth-century finding: the realization that a single egg and sperm (or pollen grain) unite to produce a zygote. Without knowing this fact, Mendel could hardly have arrived at his pathbreaking laws.

With Mendel's discovery of particulate inheritance, the identification of chromosomes as the vehicles of inheritance, and the suggestion that nucleoproteins constitute the chemical basis of inheritance, the stage was set for the subject to take off. And take off it did. But the successes and the emphasis were mainly on the rules of transmission, not on the chemical basis of heredity. The time wasn't ripe for this; chemical techniques were not up to the job.

Before 1900 the understanding of heredity was vague, and the subject seemed hopelessly complicated. There were no general rules, only countless anecdotes. Then Mendel's elegant rules appeared, bringing simplicity and order out of chaos. Thus came the realization that the elements of heredity—Mendel's *Merkmale,* or in contemporary English, genes—are particulate and are carried by the chromosomes. What could be neater?

The X and Y chromosomes were soon discovered, bringing a solution to the mystery of sex determination. And along with X chromosomes came X-linked genes, the classical example being the white-eyed *Drosophila* mutant in T. H. Morgan's Columbia University laboratory. Actually, the first sex-linked trait had been discovered earlier, in *Abraxis,* a moth. The genetic data did not fit the cytological observations, however, and you can imagine the confusion before it was realized that in moths the female is XY and the male is XX.

The next big step was the use of linkage to map the position of genes on the chromosome. The idea that the frequency of crossing over between pairs of genes could be used as a measure of their distance apart occurred to A. H. Sturtevant, an undergraduate student in Morgan's laboratory. He took the data home one evening, and the next morning, having neglected his regular homework, produced the first chromosome map (see Chapter 8). It is remarkably similar to the contemporary one shown in Figure 8.1.

Mutation as a phenomenon was realized earlier in the century. It is interesting to note that the original "mutations" found by Hugo deVries in the evening primrose, *Oenothera,* were in fact not mutations at all, but segregants from complex translocation heterozygotes. But the idea was on target; this was not the only time in science that a good hypothesis has come from misinterepreted data. The kinetics of mutation (for example, its independent occurrence in different genes and its temperature dependence) were laboriously worked out, mainly by the *Drosophila* group, who developed techniques for measuring mutation rates accurately. The climax came in 1927 when H. J. Muller, a former Morgan student, discovered that the mutation rate could be greatly enhanced by ionizing radiation.

Population genetics had its beginning with the Hardy–Weinberg rule. G. H. Hardy, who was England's most distinguished mathematician, regarded this simple application of the binomial theorem as utterly trivial. Wilhelm Weinberg was a busy physician responsible for delivering hundreds of babies; yet he found time to invent all sorts of clever ways to bypass the difficulties caused by the fact that humans have small families and do not choose to mate so as to facilitate genetic analysis. Subsequent developments came almost exclusively from the great trinity, J. B. S. Haldane and R. A. Fisher in Britain and Sewall Wright in the United States. Wright invented the inbreeding coefficient in 1922. By the end of the first period, population genetics was solidly established, and a mathematical theory of evolution had emerged. It still stands.

Meanwhile, much of the pleasing simplicity disappeared. The particulate nature of the gene and its chromosomal location remained intact, but all sorts of complications were tacked on. Various exceptions to Mendel's rules turned up, at first puzzling, then interesting, and eventually revealing and sometimes useful. For example, many plants reproduce asexually. The dandelion is one. Although the plant produces pollen and goes through the motions of sex, meiosis is bypassed and the seed is a genetic replica of the mother. As an instance of the occasional perversity of nature, the hawkweed reproduces this way. This explains the great disappointment that Mendel experienced when he tried to duplicate his pea results in this species. Genetics might have gotten off to an earlier start and its history might have been quite different had the hawkweed behaved in a more orthodox way.

Sex determination was not always found to be standard XX-XY. Sometimes the Y was missing. Sometimes there were multiple X's. As mentioned earlier, the system is reversed in moths and butterflies. And the mechanism in bees and wasps turned out to depend on haploidy versus diploidy. In still other organisms, sex determination was not chromosomal at all but environmental. Alligators and some turtles develop into males or females according to the temperature at a crucial period. (Curiously, this discovery, which might have been made at any time, didn't come until later.)

Many plants and some animals turned out to be polyploid. At first, the inheritance pattern in these species seemed utterly bewildering, but it was immediately clarified when the chromosomal basis was straightened out. Plant breeders quickly put polyploidy to use, for example, using allopolyploidy to produce useful, fertile hybrids. The process was eased by the discovery that colchicine induces polyploidy. Breeders also discovered that conveniently seedless fruits, like bananas, had odd chromosome numbers and hence could not have a normal meiosis. Hitoshi Kihara in Japan used triploidy (produced by crossing a diploid with a tetraploid) to engineer seedless watermelons.

There were many studies of developmental and physiological genetics. The creative use of mutant genes permitted a fine dissection of developmental processes, much finer than the surgical techniques of the time permitted. Early in the century, Garrod discovered that mutant genes, usually recessive, led to metabolic diseases and that the diseased state was the consequence of enzymatic failure. So biochemical genetics had its origins long ago. The great mathematical theorists, J. B. S. Haldane and Sewall Wright, applied the theory of enzyme kinetics to genetic problems with considerable success. But there was no deep knowledge of the chemical details such as we have today. And embryology lacked the tools for major advances.

The theory of quantitative genetics developed rapidly during this first period. R. A. Fisher, while teaching high school in 1915, wrote a paper giving the basis for statistical analysis of quantitative traits. The paper was rejected by the referees and was not published until 1918, and then with the help of a subsidy from Charles Darwin's son Leonard. Even the greatest can have trouble getting their work published. This must surely be one of the deepest analyses ever made by a schoolteacher, working in his spare time. Our whole theory of efficient breeding for performance in plants and livestock had its origins in this paper. Fisher, along with Sewall Wright in the United States, made us realize that what an individual inherits from its parents is best described as a deviation from the population average, and the very useful concepts of additive genetic variance and heritability emerged.

So, to view this period with the advantage of your much deeper knowledge, what did we know and what did we not know at the beginning of World War II? In particular, how much did we know about the deepest genetic question, the nature of the gene? So, let's look at what was understood in 1941, through the hindsight of your vastly superior knowledge, through a retrospectroscope, say. What *did* geneticists know?

We knew a great deal about the gene. We knew from Mendel that it is particulate. We knew from cytology and chromosome mapping that the genes are arranged in linear order along the chromosome. We knew from developmental and chemical studies that the genes often exert their influence by producing proteins, enzymes in particular. We knew a great deal about the kinetics of mutation. The chief idea-man, Muller, had clearly delineated what the gene had to do: (1) It had to carry information, enormous amounts of it; (2) it had to replicate itself every time the cell divides; (3) it had to mutate, meaning that it had to copy errors; and (4) it had to control developmental and physiological processes.

But the most interesting point is what we *didn't* know. We didn't know *what* a gene is. Needless to say, geneticists speculated endlessly about this, *the* central question. To quote a reviewer of the 1932 International Genetics Congress: "Oceans of words were spilled in formal and informal gatherings to discuss the vital question: 'What is the gene?' but that important entity is still elusive." My fellow graduate students in the late 1930s, usually over glasses of beer on Saturday night, opined—parroting the views of our teachers—that we would not live long enough to see the day when we would know what a gene is. So we had better work on other things. How wrong we were! Although the advice to work on other things might have been good.

INTERLUDE: WORLD WAR II

In most genetics laboratories, especially in Europe and Japan, the lights were out. All the stocks of *Drosophila ananasse,* a species extensively studied in Japan, were lost. Research went on in other parts of the world but at a considerably reduced pace. Yet some key findings emerged during this dark period. One was the discovery of biochemical mutations in *Neurospora* by George Beadle and Edward L. Tatum. Not only was this the beginning of modern biochemical genetics, but also it was the first step toward applying genetic methods to microorganisms.

A second great discovery during this period was Oswald Avery and colleagues' demonstration that the transforming principle in *Pneumococcus* was DNA. This came as a big surprise to most geneticists, who were betting on protein. But recall that Wilson had made the right guess in the 1890s.

The third big breakthrough during the war was the discovery of chemical mutagens by Charlotte Auerbach in Edinburgh. This discovery led to increasingly sophisticated mutagenesis studies and ultimately to a rational chemical study of the mutation process. That mustard gas might cause mutations was suggested by the similarity of mustard and X-ray injuries, as Auerbach painfully discovered for herself. Although her discoveries were made during the war, they were not published until afterward. Because mustard was a war gas, all studies of it were military secrets and her results could not be disclosed. (Actually they were not completely secret, for during the war Muller told me that Auerbach had found something very exciting. He clearly deduced what she had done and would like to have discussed it with someone else privy to the information; but since I was not, he was a good citizen and didn't tell me. I had to wait until the war was over.)

All three of these discoveries were forerunners of the great molecular advances soon to follow. In 1945, Muller wrote about "the coming chemical attack on the gene," and the attack came almost immediately. Within a few years studies of ever finer resolution by geneticists and of ever larger molecules by chemists had converged.

THE SECOND PERIOD

Shortly after the discovery of *Neurospora* mutants with specific vitamin or amino acid requirements, Tatum extended these methods to *E. coli.* This opened the door to a major breakthrough. Joshua Lederberg was a medical student at Columbia University who had reason (based on the permutational patterns of antigen types) to think there might be a scrambling process in bacteria. He also thought that such a sexual process would be rare, so some selective mechanism would be needed to isolate recombinants. The idea seemed promising enough that he took leave from his medical studies to work with Tatum at Yale. There they mixed strains with different nutritional requirements chosen so that only recombinants would grow in minimal medium. They were lucky; the first experiment worked, and a medical career ended. At last, the stage was set for the genetic analysis of a prokaryote. At about the same time, virologists, led by Max Delbrück, found recombination in bacteriophage. Much of genetic progress, such as the study of mutations and fine-structure mapping, had already depended on devising tricks to study very rare events. Microorganisms extended this power by several orders of magnitude. Genetic resolution was carried to its logical extreme by Seymour Benzer, with recombination mapping down to the nucleotide level in phage T4.

Avery's *Pneumococcus* transforming experiments argued very convincingly for DNA as the genetic material, but curiously the less definitive experiments of Al Hershey and Martha Chase in phage had a larger impact on the genetics community. By this time, 1952, DNA was becoming increasingly accepted as the key substance. A year later James Watson and Francis Crick proposed their model of DNA, and genetics was forever changed.

Seldom in science does a single finding bring immediate order out of confusion, but it's wonderful when it does. The Watson–Crick model was unexpectedly simple and surprisingly elegant. The very structure itself shouted the ways, quickly confirmed, in which Muller's questions could be answered. The information lay in the base sequences, a simple copying mechanism insured the continuity of the sequence, and mutation was explained as the copying of errors by blindly trusting enzymes of replication. Furthermore, the linearity of the information suggested an answer to Muller's fourth point: some sort of translating mechanism transforms the linear DNA information into one-dimensional polypeptides, which fold into three-dimensional proteins.

In this period, genetics was again beautifully simple, as it had been in the teens. Messenger and transfer RNA were soon discovered. The simple three-letter code was solved. The *lac* operon provided a particularly neat means of developmental control. It appeared that biology would have some of the elegant simplicity of physics. Just as the period after 1900 brought a great generality and simplication to genetics, so did the DNA revolution and the beginnings of molecular biology.

Then the complications started, complications of which you, as the reader of this book, are well aware, perhaps painfully so. The deepest generalities remained, but questions of replication, transcription, translation, and regulation became more and more complex and, still more distressing, different from one species to another. The eukaryotic genome turned out to be surprisingly complicated. Genetics since the 1960s has been very much like the genetics of the 1930s. The basic principles, particulate inheritance in the early period and the structure of DNA in the later, are intact. But there are innumerable detailed ways in which the system operates, and, as before, much of the interest in biology lies in the diverse ways that different species do a job, often the same job. But for students, this means learning a large number of facts rather than deducing them from a few general principles.

One of the most remarkable of the new findings is error correction. DNA replication is much too accurate to be operating on the simple principles of physical chemistry. We now know about such things as excision repair and proofreading that collectively bring DNA replication to its astonishingly low error rate. And be thankful, for the last thing we need is a higher mutation rate.

Who would have expected Okazaki fragments, leading and trailing strands, and the need for an RNA primer in DNA replication? Everyone thought that there would be both 5′ and 3′ polymerases. An engineer would surely have arranged it that way. Many other surprises and complications I will simply mention. They are all in the book, along with many others. Some of these are overlapping genes, introns, reverse transcriptase, Z-DNA, antisense RNA, junk DNA, lines, sines, different codes in mitiochondria and some microorganisms, imprinting, pseudogenes, attenuation, microsatellites, chaperones, and transposons.

One of the most striking developments of recent years has been the rapid growth of gene mapping in higher plants and animals, including humans. The discovery of an almost unlimited number of molecular markers has totally changed human gene mapping. A few years ago only half a dozen linkages were known. Cytological studies had localized a few genes to specific chromosomes, or chromosome regions, but the process was slow. Molecular polymorphisms, most of which have no known function, can be used as markers. Through positional cloning and other methods, genes can be identified and then mapped by standard methods. The human linkage map has jumped from a dozen or so to thousands of genes in less than a decade. And now the program is underway to determine the entire DNA sequence. At this moment, the largest organism whose total sequence is known is yeast, but it won't be long until the list is extended to nematodes, to various plants, to flies, and to mice and men.

Some of the most striking advances have been in the genetic control of development. Again, the details differ from species to species. At first, it looked as if the *lac*-operon would be *the* method of gene regulation. Now it turns out to be only one of many.

DNA chemistry has broken a major barrier in evolution studies. Previously, these studies were confined to animals or plants that could be crossed, and this usually meant only closely related species. There was even doubt as to whether differences between more remotely related forms were due to genes. Now, it is commonplace to compare genes throughout the animal and plant kingdoms. You have to have lived through an earlier period to realize how remarkable this is.

Molecular biology has created the totally new field of molecular evolution. This has permitted an accurate record of evolutionary changes at the DNA level. Although nothing in molecular studies has undermined the centrality of Darwin's theory of evolution by natural selection, Motoo Kimura in Japan has made us conscious of the large role that random changes play in molecular evolution.

One byproduct of molecular evolution studies has been more accurate phylogenetic trees. It is remarkable that some genes, such as homeobox genes, have been maintained through a whole evolutionary history. Molecular studies have also shown how species can become more complicated by adding new genes. But the new genes are not really new, they are adaptations of old genes recently duplicated and redundant, and therefore ready to take on a new function while the old one is retained. By gene duplication, evolution really *can* have its cake and eat it.

Now here's a curious question. Why does genetics, and especially developmental genetics, have to be so complicated—so needlessly complicated, it seems? And why do many processes that might be done in a uniform way differ so greatly from species to species? Every species seems to want to do things its own way. As I have said, the most basic rules are general. But detailed implementation of these rules is anything but general; it is remarkably disorderly. The key to understanding this, I think, is history. Remember that each contemporary species is at the end of a long evolutionary tree. As changes occurred, they altered something that was already there, patching up and tacking on. As a result, development does not proceed in the orderly way that it might if it were being designed from scratch using engineering principles of simplicity and efficiency. Rather, it is a series of changes grafted on other changes, and often the process would make sense only if we could reconstruct the history. Stated another way, this seeming disorder reflects the opportunism of evolution. Evolution incorporates whatever change is best at the moment, and it may not fit into any rational long-range plan. So, biology is complicated and varied, and often illogical; history has made it so. I know you find it bewildering, but I hope you also find it challenging and interesting.

Let's pause a moment before going into the next section. What, among the many unknowns in genetics, would you most like to understand? When I was a student, almost all geneticists would have given the same answer: What is the gene? What is it made of? What is its structure? How does it carry information? How does it replicate? How does it mutate? How does it control development and physiology?

Now things are different. There doesn't seem to be a single, central question. Would you opt for an understanding of aging? Or how life originated, on earth or elsewhere? Or what is mind? Or might you prefer less profound questions, such as what is the use of introns? Or are transposons useful to the organism, or simply DNA's way of doing what it does best, replicate? Let me suggest that, before reading on, you pause a moment and ask yourself this question. You might even write down your answer, and stash it away. You will enjoy looking at it 10, 25, or 50 years hence, and perhaps your question will have been answered.

THE THIRD PHASE: THE FUTURE IS NOW

Deep new scientific discoveries usually bring important new applications. The profound and revolutionary understanding that came with the Watson–Crick model and the development of molecular genetics ought to have practical consequences. Yet, these were slow in coming. The earliest applications involved the use of microorganisms: for example, developing strains to produce specific chemicals.

The third phase of genetics, the practical application of molecular genetics to agriculture and medicine, began more recently. The beginning was gradual, but the acceleration phase is here. About 15 years ago, it first became feasible to clone human genes and their mouse homologs. The ability to detect genes at the DNA level means that geneticists can now begin to perform in plants and animals, including humans, the kinds of tricks previously possibly only in microbes.

In agriculture, molecular methods have made it feasible to introduce useful genes into farm crops. An example is a tomato that carries an introduced gene delaying the normal softening process, so that the tomatoes can be shipped with less spoilage. Similarly, genes for disease resistance can be introduced, as they have recently been in the important food crop, rice. The old process of introgressing useful genes from the wild relatives of cultivars involved several generations of backcrossing to remove unwanted hitchhikers that came in with the desired gene. Molecular methods can introduce only the genes that are wanted, without unwelcome linked interlopers.

Improvement of quantitative traits still proceeds by selection, and the methods used by animal and plant breeders are as efficient as sophisticated statistical methods can make them. But it would be better if some of the individual gene components of performance could be identified. To some extent this has happened as quantitative trait loci (QTLs) have been discovered. These discoveries can partially, but surely not totally, replace the slower process of crossing and selection to get improved performance.

The greatest successes in human genetics have been aids to diagnosis and genetic prognosis. These advances can be valuable in warning future parents of a possible diseased child. But improved diagnosis is often of little comfort if there is no treatment. Except for a few very rare diseases, human gene therapy has so far been a disappointment. But the successes, at least partial, in mammals argue that there will be successes in humans as well. But how soon? We shall have to wait to see.

WHAT OF THE FUTURE?

The human genome project has an intermediate goal of physically mapping 30,000 sites by 1998. This is one marker sequence every 100,000 bases. As I write this epilogue, the intermediate goal has been passed. The ultimate goal is the complete nucleotide sequence by 2003. Whether this is too optimistic remains to be seen, but surely a substantial part will be completed and will include virtually all the functional regions.

Then what? Knowing the complete sequence and finding all the genes won't tell us a lot of things we should like to know about development. There are problems of regulation, and who knows how complicated the interactions will be? But the sequence will provide an enormous technical boost as the process of development, complex as it is, is further dissected, with both intellectual satisfaction and practical uses.

We can look forward to a time in which the great bulk of severe monogenic diseases will be preventable. The main exception will be new mutations, since it is not likely that screening all fetuses for a large number of diseases will be feasible for some time to come. But we can count on better, cheaper, and less invasive methods of detecting fetal abnormalities, especially morphological ones. Almost all parents will be assured of a "perfect" baby—that is, one with no overt abnormalities. We can also count on increasing ethical dilemmas (e.g., privacy) as increasingly detailed knowledge of ourselves, and our neighbors, becomes available.

We can safely predict that chemical understanding, partly from genetics and partly from other sources, will bring mental disease into the realm of chemical intervention. We can expect simpler and better treatments, with untold reduction in human despair. The methods of discovery will be more rational and less empirical. Future advances in genetics, if intelligently applied, can surely decrease human suffering and anguish. But, as with any revolutionary change, there will be problems.

Gene therapy, as it develops, will almost certainly not be perfect, especially at first. And it is likely to be expensive. So we can expect an increase in medical costs. Diseases, genetic and nongenetic, that once caused early death will be treated. But everyone dies eventually, and a frequent consequence of better treatment will be trading an early, quick death for a protracted, expensive one later one. For this, and many other reasons, we can expect an increasingly aged and infirm population. Natural selection has no interest in people past the reproductive and child-caring ages and has made no provisions for the welfare of those euphemistically referred to as "seniors" or "mature" (postmature would be better). Are we really interested in prolonging life of low quality? Who wants to spend years dying? Will society soon condone assisted suicide for the terminally ill? Or perhaps go farther and institute some sort of euthanasia, as some societies that we like to call "primitive" have long done? The ethical dilemmas are clear: balancing societal needs and personal desires against cherished traditions and deeply held beliefs.

One problem that is not likely to have an easy solution is mutation. Molecular methods can deal with major mutations. But if *Drosophila* tells us anything, it is that the great bulk of mutations are those that cause small, nonspecific, harmful effects. This can only mean a gradual weakening of the population, unless this is compensated by environmental improvements or opposed by selection. But natural selection has been and we hope will continue to be less and less effective as each zygote has a high probability of surviving and being able to reproduce. Will future societies try to do something about this? Fortunately, this is a problem that will arise gradually over dozens of generations. We have time to learn more before acting precipitously.

We can be sure that human evolution will not stop, but can we predict the course of future genetic changes? Some things, like mutation accumulation, we can foresee. But for changes in frequencies of genes for health, intelligence, special talents, emotional stability, and other matters that we care about, there is no predicting. The reason is obvious: the genetic future depends on who produces more children than someone else. And that depends on human decisions, and human decisions are notoriously unpredictable. So you will get no predictions from me, only a feeling of uneasiness.

This raises the question of eugenics. Will human society be content to let its genetic future be determined by individual human decisions? How far can we count on molecular intervention and improved environments to substitute for selection? In Chapter 1 you learned of the crudity and naiveté of some of the early eugenicists in this country. Yet the idea, as formulated by Galton and many followers, had a large element of idealism and respect for individual liberties. The subject was dealt a death blow by Nazism. But genetic problems won't go away. Can we hope for a society that is sufficiently sophisticated, far-sighted, and compassionate to use its power wisely? Or will our successors decide not to step in where angels fear to tread?

And there are other problems. Try global warming. But to stay with biology, our greatest biological problem is not genetic; it is demographic. Unless the birthrate declines, and declines rapidly, this already overcrowded planet will become unlivable for most of humanity. If we don't solve the numbers problem, we won't have the luxury of worrying about mutation accumulation, eugenics, and ethical dilemmas arising from genetic knowledge. Yours may be the last generation in which reproduction is regarded as a right rather than a rationed privilege; this is already happening in parts of China. It is hard to be optimistic about a political solution to the world population problem. Maybe there will be a technical fix (a cheap, candy-coated, long-lasting contraceptive pill, perhaps?). The alternative to population control is an ever-widening extension of the poverty, disease, and starvation that we already see in too many parts of the globe.

Well, you have read one person's mixture of optimism and pessimism, of excitement and apprehension. One thing you can be sure of: the years ahead won't be dull. The road is full of potholes and land mines. But it promises the thrill of new insights, and the hope for a better world. I would like to be traveling it with you, but I'm afraid that is not in the cards.

Glossary

This glossary provides an introduction to some basic and recurring terms in the text. Names of chemical compounds, definitions of specialized terms, and variants of basic names have been omitted from the glossary but are given in the index. Please locate terms that are not in the glossary by referring to the index.

Abscissa. The horizontal scale on a graph.

Acentric chromosome. Chromosome fragment lacking a centromere.

Acridine dyes. A class of positively charged polycyclic molecules that intercalate into DNA and induce frameshift mutations.

Acrocentric. A modifying term for a chromosome or chromatid that has its centromere near the end.

Acrosome. An apical organelle in the head of the sperm.

Adaptation. Adjustment of an organism or a population to an environment.

Additive allelic effects. Genetic factors that raise or lower the value of a phenotype on a linear scale of measurement.

Additive genetic variance. The portion of the total phenotypic variance in a quantitive trait that is due to the additive effects of alleles.

Adenine. A purine base found in RNA and DNA.

A-DNA. A right-handed DNA double helix that has 11 base pairs per turn. DNA exists in this form when partially dehydrated.

***Agrobacterium tumefaciens*-mediated transformation**. A naturally occurring process of DNA transfer from the bacterium *A. tumefaciens* to plants.

AIDS (acquired immunodeficiency syndrome). The usually fatal human disease in which the immune system is destroyed by the human immunodeficiency virus (HIV).

Albinism. Absence of pigment in skin, hair, and eyes of an animal. Absence of chlorophyll in plants.

Aleurone. The outermost layer of the endosperm in a seed.

Alkaptonuria. An inherited metabolic disorder. Alkaptonurics excrete excessive amounts of homogentisic acid (alkapton) in the urine.

Alkylating agents. Chemicals that transfer alkyl (methyl, ethyl, and so on) groups to the bases in DNA.

Allelic exclusion. A phenomenon whereby only one functional allele of an antibody gene can be assembled in a given B lymphocyte. The "allele" on the other homologous chromosome in a diploid mammalian cell cannot undergo a functional rearrangement, which would result in the production of two different antibodies by a single plasma cell.

Allele (allelomorph; *adj.*, allelic, allelomorphic). One of a pair, or series, of alternative forms of a gene that occur at a given locus in a chromosome. Alleles are symbolized with the same basic symbol (for example, *D* for tall peas and *d* for dwarf). (See **Multiple alleles.**)

Allele frequency. The proportion of one allele relative to all alleles at a locus in a population.

Allopatric speciation. Speciation occurring at least in part because of geographic isolation.

Allopolyploid. A polyploid having chromosome sets from different species; a polyploid containing genetically different chromosome sets derived from two or more species.

Allosteric transition. A reversible interaction of a small molecule with a protein molecule that causes a change in the shape of the protein and a consequent alteration of the interaction of that protein with a third molecule.

Allotetraploid. An organism with four genomes derived from hybridization of different species. Usually, in forms that become established, two of the four genomes are from one species and two are from another species.

Amino acid. Any one of a class of organic compounds containing an amino (NH_2) group and a carboxyl (COOH) group. Amino acids are the building blocks of proteins. Alanine, proline, threonine, histidine, lysine, glutamine, phenylalanine, tryptophan, valine, arginine, tyrosine, and leucine are among the common amino acids.

Aminoacyl (*A*) site. The ribosome binding site that contains the incoming aminoacyl-tRNA.

Aminoacyl-tRNA synthetases. Enzymes that catalyze the formation of high energy bonds between amino acids and tRNA molecules.

Amniocentesis. A procedure for obtaining amniotic fluid from a pregnant woman. Chemical contents of the fluid are studied directly for the diagnosis of some diseases. Cells are cultured, and metaphase chromosomes are examined for irregularities (for example, trisomy).

Amnion. The thin membrane that lines the fluid-filled sac in which the embryo develops in higher vertebrates.

Amniotic fluid. Liquid contents of the amniotic sac of higher vertebrates containing cells of the embryo (not of the mother). Both fluid and cells are used for diagnosis of genetic abnormalities of the embryo or fetus.

Amorph. A mutation that obliterates gene function; a null mutation.

Amphidiploid. A species or type of plant derived from doubling the chromosomes in the F_1 hybrid of two species; an allopolyploid. In an amphidiploid the two species are known, whereas in other allopolyploids they may not be known.

Anabolic pathway. A pathway by which a metabolite is synthesized; a biosynthetic pathway.

Anaphase. The stage of mitosis or meiosis during which the daughter chromosomes pass from the equatorial plate to opposite poles of the cell (toward the ends of the spindle). Anaphase follows metaphase and precedes telophase.

Anchor gene. A gene that has been positioned on both the physical map and the genetic map of a chromosome.

Androgen. A male hormone that controls sexual activity in vertebrate animals.

Anemia. Abnormal condition characterized by pallor, weakness, and breathlessness, resulting from a deficiency of hemoglobin or a reduced number of red blood cells.

Aneuploid. An organism or cell

having a chromosome number that is not an exact multiple of the monoploid (n) with one genome, that is, hyperploid, higher (for example, $2n + 1$); or hypoploid, lower (for example, $2n - 1$). Also applied to cases where part of a chromosome is duplicated or deficient.

Antibody. Substance in a tissue or fluid of the body that acts in antagonism to a foreign substance (antigen).

Antibody class. In mammals, there are five classes of antibodies—IgA, IgD, IgE, IgG, and IgM; the class to which an antibody belongs depends on the type of heavy chain present.

Antibody-mediated (humoral) immune response. The synthesis of antibodies by plasma cells in response to an encounter of the cells of the immune system with a foreign immunogen.

Anticodon. Three bases in a transfer RNA molecule that are complementary to the three bases of a specific codon in messenger RNA.

Antigen. A substance, usually a protein, that is bound by an antibody or a T cell receptor when introduced into a vertebrate organism (cf. **Immunogen**).

Antisense gene. A gene that produces a transcript that is complementary to pre-mRNA or mRNA of a normal gene (usually constructed by inverting the coding region relative to promoter).

Antisense RNA. RNA that is complementary to the pre-mRNA or mRNA produced from a gene.

Apoptosis. A phenomenon in which eukaryotic cells die because of genetically programmed events within those cells.

Artificial selection. The practice of choosing individuals from a population for reproduction, usually because these individuals possess one or more desirable traits.

Ascospore. One of the spores contained in the ascus of certain fungi such as *Neurospora.*

Ascus (*pl.,* **asci**). Reproductive sac in the sexual stage of a type of fungi (Ascomycetes) in which ascospores are produced.

Asexual reproduction. Any process of reproduction that does not involve the formation and union of gametes from the different sexes or mating types.

Assortative mating. Mating in which the partners are chosen because they are phenotypically similar.

Asynapsis. The failure or partial failure in the pairing of homologous chromosomes during the meiotic prophase.

ATP. Adenosine triphosphate: an energy-rich compound that promotes certain activities in the cell.

Attenuation. A mechanism for controlling gene expression in prokaryotes that involves premature termination of transcription.

Attenuator. A nucleotide sequence in the 5′ region of a prokaryotic gene (or in its RNA) that causes premature termination of transcription, possibly by forming a secondary structure.

Autocatalytic reaction. A reaction catalyzed by a substrate without the involvement of any other catalytic agent.

Autoimmune diseases. Disorders in which the immune systems of affected individuals produce antibodies against self antigens—antigens synthesized in their own cells.

Autonomous. A term applied to any biological unit that can function on its own, that is, without the help of another unit. For example, a transposable element that encodes an enzyme for its own transposition (cf. **Nonautonomous**).

Autopolyploid. A polyploid that has multiple and identical or nearly identical sets of chromosomes (genomes). A polyploid species with genomes derived from the same original species.

Autoradiograph. A record or photograph prepared by labeling a substance such as DNA with a radioactive material such as tritiated thymidine and allowing the image produced by radioactive decay to develop on a film over a period of time.

Autosome. Any chromosome that is not a sex chromosome.

Auxotroph. A mutant microorganism (for example, bacterium or yeast) that will not grow on a minimal medium but that requires the addition of some compound such as an amino acid or a vitamin.

Backcross. The cross of an F_1 hybrid to one of the parental types. The offspring of such a cross are referred to as the backcross generation or backcross progeny. (See **Testcross.**)

Back mutation. A second mutation at the same site in a gene as the original mutation, which restores the wild-type nucleotide sequence.

BACs (bacterial artificial chromosomes). Cloning vectors constructed from bacterial fertility (F) factors; like YAC vectors, they accept large inserts of size 200 to 500 kb.

Bacteriophage. A virus that attacks bacteria. Such viruses are called bacteriophages because they destroy their bacterial hosts.

Balanced lethal. Lethal mutations in different genes on the same pair of chromosomes that remain in repulsion because of close linkage or crossover suppression. In a closed population, only the trans-heterozygotes ($l_1 +/+ l_2$) for the lethal mutations survive.

Balanced polymorphism. Two or more types of individuals maintained in the same breeding population by a selection mechanism.

Barr body. A condensed mass of chromatin found in the nuclei of placental mammals that contains one or more X chromosomes; named for its discoverer, Murray Barr.

Basal body. Small granule to which a cilium or flagellum is attached.

Base analogs. Unnatural purine or pyrimidine bases that differ slightly from the normal bases and that can be incorporated into nucleic acids. They are often mutagenic.

Base substitution. A single base change in a DNA molecule. (See **Transition; Transversion.**)

Binomial expansion. Exponential multiplication of an expression consisting of two terms connected by a plus (+) or minus (−) sign, such as $(a + b)^n$.

Biometry. Application of statistical methods to the study of biological problems.

Bivalent. A pair of synapsed or associated homologous chromosomes that have undergone the duplication process to form a group of four chromatids.

Blastomere. Any one of the cells formed from the first few cleavages in animal development.

Blastula. In animals, an early embryo form that follows the morula stage; typically, a single-layered sheet or ball of cells.

B lymphocytes (B cells). An important class of cells that mature in bone marrow and are largely responsible for the antibody-mediated or humoral immune response; they give rise to the antibody-producing plasma cells and some other cells of the immune system.

Breeding value. In quantitative genetics, the part of the deviation of an individual phenotype from the population mean that is due to the additive effects of alleles.

Broad-sense heritability. In quantitative genetics, the proportion of the total phenotypic variance that is due to genetic factors.

CAAT box. A conserved nucleotide sequence in eukaryotic promoters involved in the initiation of transcription.

Carcinogen. An agent capable of inducing cancer in an organism.

Carrier. An individual who carries a recessive allele that is not expressed (that is, is obscured by a dominant allele).

Catabolic pathway. A pathway by which an organic molecule is degraded in order to obtain energy for growth and other cellular processes; degradative pathway.

Catabolite repression. Glucose-mediated reduction in the rates of transcription of operons that specify enzymes involved in catabolic pathways (such as the *lac* operon).

cDNA (complementary DNA). A DNA molecule synthesized *in vitro* from an RNA template.

cDNA library. A collection of cDNA clones containing copies of the RNAs isolated from an organism or a specific tissue or cell type of an organism.

Cell cycle. The cyclical events that occur during the divisions of mitotic cells. The cell cycle oscillates between mitosis and the interphase, which is divided into G_1, S, and G_2.

Cellular immune response. See **T cell-mediated immune response.**

Centimorgan. See **Crossover unit.**

Centriole. An organelle in many animal cells that appears to be involved in the formation of the spindle during mitosis.

Centromere. Spindle-fiber attachment region of a chromosome.

Character (*contraction of the word* **characteristic).** One of the many details of structure, form, substance, or function that make up an individual organism.

Checkpoint. A mechanism that halts progression through the eukaryotic cell cycle.

Chemotaxis. Attraction or repulsion of organisms by a diffusing substance.

Chiasma (*pl.,* **Chiasmata**). A visible change of partners in two of a group of four chromatids during the first meiotic prophase. In the diplotene stage of meiosis, the four chromatids of a bivalent are associated in pairs, but in such a way that one part of two chromatids is exchanged. This point of "change of partner" is the chiasma.

Chimera (animal). Individual derived from two embryos by experimental intervention.

Chimera (plant). Part of a plant with a genetically different constitution as compared with other parts of the same plant. It may result from different zygotes that grow together or from artificial fusion (grafting); it may either be pernical, with parallel layers of genetically different tissues, or sectorial.

Chi-square. A statistic used to test the goodness of fit of data to the predictions of an hypothesis.

Chloroplast. A green organelle in the cytoplasm of plants that contains chlorophyll and in which starch is synthesized. A mode of cytoplasmic inheritance, independent of nuclear genes, has been associated with these cytoplasmic organelles.

Chromatid. In mitosis or meiosis, one of the two identical strands resulting from self-duplication of a chromosome.

Chromatin. The complex of DNA and proteins in eukaryotic chromosomes; originally named because of the readiness with which it stains with certain dyes.

Chromatin fibers. A basic organizational unit of eukaryotic chromosomes that consists of DNA and associated proteins assembled into strands of average diameter 30 nm.

Chromatography. A method for separating and identifying the components from mixtures of molecules having similar chemical and physical properties.

Chromocenter. Body produced by fusion of the heterochromatic regions of the chromosomes in the polytene tissues (for example, the salivary glands) of certain *Diptera.*

Chromomeres. Small bodies that are identified by their characteristic size and linear arrangement along a chromosome.

Chromonema (*pl.,* **chromonemata**). An optically single thread forming an axial structure within each chromosome.

Chromosome aberration. Abnormal structure or number of chromosomes; includes deficiency, duplication, inversion, translocation, aneuploidy, polyploidy, or any other change from the normal pattern.

Chromosome banding. Staining of chromosomes in such a way that light and dark areas occur along the length of the chromosomes. Lateral comparisons identify pairs. Each human chromosome can be identified by its banding pattern.

Chromosome jumping. A procedure that uses large DNA fragments to move discontinuously along a chromosome from one site to another site. (See **Positional cloning.**)

Chromosome painting. The study of the organization and evolution of chromosomes by *in situ* hybridization using DNA probes labeled with fluoresent dyes that emit light at different wavelengths.

Chromosomes. Darkly staining nucleoprotein bodies that are observed in cells during division. Each chromosome carries a linear array of genes.

Chromosome theory of inheritance. The theory that chromosomes carry the genetic information and that their behavior during meiosis provides the physical basis for the segregation and independent assortment of genes.

Chromosome walking. A procedure that uses overlapping clones to move sequentially down a chromosome from one site to another site. (See **Positional cloning.**)

Cilium (*pl.,* **cilia;** *adj.,* **ciliate**). Hair-like locomotor structure on certain cells; a locomotor structure on a ciliate protozoan.

***cis*-acting sequence.** A nucleotide sequence that only affects the expression of genes located on the same chromosome, that is, *cis* to itself.

***cis* configuration.** See **Coupling.**

***cis* heterozygote.** A heterozygote that contains two mutations arranged in the *cis* configuration—for example, $a^+ \, b^+ \; / \; a \, b$)

Cistron. See **Gene.**

Class switching. The process during which a plasma cell stops producing antibodies of one class and begins producing antibodies of another class.

Clonal selection. The production of a population of plasma cells all producing the same antibody in reponse to the interaction between a B lymphocyte producing that specific antibody and the antigen bound by that antibody.

Clone. All the individuals derived by vegetative propagation from a single original individual. In molecular biology, a population of identical DNA molecules all carrying a particular DNA sequence from another organism.

Cloning vector. A small, self-replicating DNA molecule—usually a plasmid or viral chromosome—into which foreign DNAs are inserted in the process of cloning genes or other DNA sequences of interest.

Codominant alleles. Alleles that produce independent effects when heterozygous.

Codon. A set of three adjacent nucleotides in an mRNA molecule that specifies the incorporation of an amino acid into a polypeptide chain or that signals the end of polypeptide synthesis. Codons with the latter function are called termination codons.

Coefficient. A number expressing the amount of some change or effect under certain conditions (for example, the coefficient of inbreeding).

Coenzyme. A substance necessary for the activity of an enzyme.

Coincidence. The ratio of the observed frequency of double crossovers to the expected frequency, where the expected frequency is calculated by assuming that the two crossover events occur independently of each other.

Cointegrate. A DNA molecule formed by the fusion of two different DNA molecules, usually mediated by a transposable element.

Colchicine. An alkaloid derived from the autumn crocus that is used as an agent to arrest spindle formation and interrupt mitosis.

Colinearity. A relationship in which the units in one molecule occur in the same sequence as the units in another molecule which they specify; for example, the nucleotides in a gene are colinear with the amino acids in the polypeptide encoded by that gene.

Competence. Ability of a bacterial cell to incorporate DNA and become genetically transformed.

Complementarity. The relationship between the two strands of a double helix of DNA. Thymine in one strand pairs with adenine in the other strand, and cytosine in one strand pairs with guanine in the other strand.

Complementation test (*trans* test). Introduction of two recessive mutations into the same cell to determine whether they are alleles of the same gene, that is, whether they affect the same genetic function. If the mutations are allelic, the genotype $m_1\ +/+\ m_2$ will exhibit a mutant phenotype, whereas if they are nonallelic, it will exhibit the wild phenotype.

Complement proteins. Proteins (encoded by the *HLA* class III genes in humans) that bind to antibody-antigen complexes and help degrade the complexes by proteolysis.

Composite transposon. A transposable element formed when two identical or nearly identical transposons insert on either side of a nontransposable segment of DNA—for example, the bacterial transposon Tn*5*.

Compound chromosome. A chromosome formed by the union of two separate chromosomes, as in attached-X chromosomes or attached X-Y chromosomes.

Concordance. Identity of matched pairs or groups for a given trait—for example, identical twins both expressing the same genetic syndrome.

Conditional lethal mutation. A mutation that is lethal under one set of environmental conditions—the restrictive conditions—but is viable under another set of environmental conditions—the permissive conditions.

Conidium (*pl.,* conidia). An asexual spore produced by a specialized hypha in certain fungi.

Conjugation. Union of sex cells (gametes) or unicellular organisms during fertilization; in *Escherichia coli,* a one-way transfer of genetic material from a donor ("male" cell) to a recipient ("female" cell).

Consanguineous mating. A mating between relatives.

Consanguinity. Related by descent from a common ancestor.

Consensus sequence. The nucleotide sequence that is present in the majority of genetic signals or elements that perform a specific function.

Constitutive enzyme. An enzyme that is synthesized continually regardless of growth conditions (cf. **Inducible enzyme** and **Repressible enzyme**).

Constitutive gene. A gene that is continually expressed in all cells of an organism.

Contig. A set of overlapping clones that provide a physical map of a portion of a chromosome.

Continuous variation. Variation not represented by distinct classes. Individuals grade into each other, and measurement data are required for analysis (cf. **Discontinuous variation**). Multiple genes are usually responsible for this type of variation.

Controlling element. In maize, a transposable element such as *Ac* or *Ds* that is capable of influencing the expression of a nearby gene.

Coordinate repression. Correlated regulation of the structural genes in an operon by a molecule that interacts with the operator sequence.

Copolymers. Mixtures consisting of more than one monomer; for example, polymers of two kinds of organic bases such as uracil and cytosine (poly-UC) have been combined for studies of the genetic code.

Co-repressor. An effector molecule that forms a complex with a repressor and turns off the expression of a gene or set of genes.

Correlation. A statistical association between variables.

Cosmids. Cloning vectors that are hybrids between phage λ chromosomes and plasmids; they contain λ ***cos*** sites and plasmid origins of replication.

Coupling (*cis* configuration). The condition in which a double heterozygote has received two linked mutations from one parent and their wild-type alleles from the other parent (for example, $a\ b/a\ b \times +\ +/+\ +$ produces $a\ b/+\ +$ (cf. **Repulsion**).

Covalent bond. A bond in which an electron pair is equally shared by protons in two adjacent atoms.

Covariance. A measure of the statistical association between variables.

cpDNA. The DNA of plant plastids, including chloroplasts.

CpG islands. Clusters of cytosines and guanines that often occur upstream of human genes.

Crossbreeding. Mating between members of different races or species.

Crossing over. A process in which chromosomes exchange material through the breakage and reunion of their DNA molecules. (See **Recombination.**)

Crossover unit. A measure of distance on genetic maps that is based on the average number of crossing-over events that take place during meiosis. A map interval that is one crossover unit in length (sometimes called a centimorgan) implies that only one in every hundred chromatids recovered from meiosis will have undergone a crossing-over event in this interval.

Cytogenetics. Area of biology concerned with chromosomes and their implications in genetics.

Cytokinesis. Cytoplasmic division and other changes exclusive of nuclear division that are a part of mitosis or meiosis.

Cytology. The study of the structure and function of cells.

Cytoplasm. The protoplasm of a cell outside the nucleus in which cell organelles (mitochondria, plastids, and the like) reside; all living parts of the cell except the nucleus.

Cytoplasmic inheritance. Hereditary transmission dependent on the cytoplasm or structures in the cytoplasm rather than the nuclear genes; extrachromosomal inheritance. Example: Plastid characteristics in plants may be inherited by a mechanism independent of nuclear genes.

Cytosine. A pyrimidine base found in RNA and DNA.

Cytotype. A maternally inherited cellular condition in *Drosophila* that regulates the activity of transposable P elements.

Dalton. The mass of a hydrogen atom.

Deficiency (deletion). Absence of a segment of a chromosome, reducing the number of loci.

Degeneracy (of the genetic code). The specification of an amino acid by more than one codon.

Deme. A local population of organisms.

Denaturation. Loss of native configuration of a macromolecule, usually accompanied by loss of biological activity. Denatured proteins often unfold their polypeptide chains and express changed properties of solubility.

***de novo*.** Arising anew, afresh, once more.

Deoxyribonuclease (DNase). Any enzyme that hydrolyzes DNA.

Deoxyribonucleic acid. See **DNA.**

Derepression. The process of turning on the expression of a gene or set of genes whose expression has been repressed (turned off).

Determination. Process by which undifferentiated cells in an embryo become committed to develop into specific cell types, such as neuron, fibroblast, and muscle cell.

Deviation. As used in statistics, a departure from an expected value.

Diakinesis. A stage of meiosis just before metaphase I in which the bivalents are shortened and thickened.

Dicentric chromosome. One chromosome having two centromeres.

Dicot. A plant with two cotyledons, or seed leaves.

Differentiation. A process in which unspecialized cells develop characteristic structures and functions.

Dihybrid. An individual that is heterozygous for two pairs of alleles; the progeny of a cross between homozygous parents differing in two respects.

Dimer. A compound having the same percentage composition as another but twice the molecular weight; one formed by polymerization.

Dimorphism. Two different forms in a group as determined by such characteristics as sex, size, or coloration.

Diploid. An organism or cell with two sets of chromosomes ($2n$) or two genomes. Somatic tissues of higher plants and animals are ordinarily diploid in chromosome constitution in contrast with the haploid (monoploid) gametes.

Diplonema (*adj.*, **diplotene**). That stage in prophase of meiosis I following the pachytene stage, but preceding diakinesis, in which the chromosomes of bivalents separate from each other at and around their centromeres.

Discontinuous variation. Phenotypic variability involving distinct classes such as red versus white, tall versus dwarf (cf. **Continuous variation**).

Discordant. Members of a pair showing different, rather than similar, characteristics.

Disjunction. Separation of homologous chromosomes during anaphase of mitotic or meiotic divisions. (See **Nondisjunction.**)

Disome. See **Monosomic.**

Ditype. In fungi, a tetrad that contains two kinds of meiotic products (spores) (for example, 2 *AB* and 2 *ab*).

Dizygotic twins. Two-egg or fraternal twins.

DNA. Deoxyribonucleic acid; the information-carrying genetic material that comprises the genes. DNA is a macromolecule composed of a long chain of deoxyribonucleotides joined by phosphodiester linkages. Each deoxyribonucleotide contains a phosphate group, the five-carbon sugar 2-deoxribose, and a nitrogen-containing base.

DNA fingerprint. The banding pattern on a Southern blot of an individual's genomic DNA that has been cleaved with a restriction enzyme(s) and hybridized to an appropriate nucleic acid probe(s).

DNA gyrase. An enzyme in bacteria that catalyzes the formation of negative supercoils in DNA.

DNA helicase. An enzyme that catalyzes the unwinding of the complementary strands of a DNA double helix.

DNA ligase. An enzyme that catalyzes covalent closure of nicks in DNA double helices.

DNA polymerase. An enzyme that catalyzes the synthesis of DNA.

DNA primase. An enzyme that catalyzes the synthesis of short strands of RNA that initiate the synthesis of DNA strands.

DNA repair enzymes. Enzymes that catalyze the repair of damaged DNA.

DNA topoisomerase. An enzyme that catalyzes the introduction or removal of supercoils from DNA.

Dominance. A condition in which one member of an allele pair is manifested to the exclusion of the other.

Dominant selectable marker gene. A gene that allows the host cell to survive under conditions where it would otherwise die.

Dosage compensation. A phenomenon in which the activity of a gene is increased or decreased according to the number of copies of that gene in the cell.

Dot blot hydridization. Hybridization of a labeled DNA or RNA probe to multiple DNA sequences fixed to a membrane in a specific pattern.

Drift. See **Random genetic drift.**

Duplication. The occurrence of a segment more than once in the same chromosome or genome; also, the multiplication of cells.

Ecdysone. A hormone that influences development in insects.

Eclosion. Emergence of an adult insect from the pupal stage.

Ecotype. A population or strain of organisms that is adapted to a particular habitat.

Effector molecule. A molecule that influences the behavior of a regulatory molecule, such as a repressor protein, thereby influencing gene expression.

Egg (ovum). A germ cell produced by a female organism.

Electrophoresis. The migration of suspended particles in an electric field.

Electroporation. A process whereby cell membranes are made permeable to DNA by applying an intense electric current.

Elongation factors. Soluble proteins that are required for polypeptide chain elongation.

Embryo. An organism in the early stages of development; in humans, the first two months in the uterus.

Embryo sac. A large thin-walled space within the ovule of the seed plant in which the egg and, after fertilization, the embryo develop; the mature female gametophyte in higher plants.

Endogenote. The part of the bacterial chromosome that is homologous to a genome fragment (exogenote) transferred from the donor to the recipient cell in the formation of a merozygote.

Endomitosis. Duplication of chromosomes without division of the nucleus, resulting in increased chromosome number within a cell. Chromosomes strands separate, but the cell does not divide.

Endonuclease. An enzyme that breaks strands of DNA at internal positions; some are involved in recombination of DNA.

Endoplasmic reticulum. Network of membranes in the cytoplasm to which ribosomes adhere.

Endopolyploidy. A state in which the cells of a diploid organism contain multiples of the diploid chromosome number (that is, $4n$, $8n$, and so on).

Endosperm. Nutritive tissue that develops in the embryo sac of most angiosperms. It usually forms after the fertilization of the two fused primary endosperm nuclei of the embryo sac with one of the two male gamete nuclei. In most diploid plants, the endosperm is triploid ($3n$).

End-product inhibition. See **Feedback inhibition.**

Enhancer. A substance or object that increases a chemical activity or a physiological process; a major or modifier gene that increases a physiological process; a DNA sequence that influences transcription of a nearby gene.

Environment. The aggregate of all the external conditions and influences affecting the life and development of an organism.

Environmentality. The porportion of the total phenotypic variance in a quantitative trait that is due to the effects of a shared environment.

Enzyme. A protein that accelerates a specific chemical reaction in a living system.

Epigenetic. A term referring to the nongenetic causes of a phenotype.

Episome. A genetic element that may be present or absent in different cells and that may be inserted in a chromosome or independent in the cytoplasm (for example, the fertility factor (F) in *Escherichia coli*).

Epistasis. Interactions between products of nonallelic genes. Genes suppressed are said to be hypostatic. Dominance is associated with members of allelic pairs, whereas epistasis is interaction among products of nonalleles.

Epitope. A specific structural feature of an antigen that stimulates the production of antibodies.

Equational division. Mitotic-type division that is usually the second division in the meiotic sequence; somatic mitosis and the nonreductional division of meiosis.

Equatorial plate. The figure formed by the chromosomes in the center (equatorial plane) of the spindle in mitosis.

Equilibrium. A state of dynamical systems in which there is no net change.

Equilibrium density gradient centrifugation. A procedure used to separate macromolecules based on their density (mass per unit volume).

Estrogen. Female hormone or estrus-producing compound.

ESTs (expressed sequence tags). Short cDNA sequences that are used to link physical maps and genetic (RFLP) maps.

Euchromatin. Genetic material that is not stained so intensely by certain dyes during interphase and that comprises many different kinds of genes (cf. **Heterochromatin**).

Eugenics. The application of the principles of genetics to the improvement of humankind.

Eukaryote. A member of the large group of organisms that have nuclei enclosed by a membrane within their cells (cf. **Prokaryote**).

Euploid. An organism or cell having a chromosome number that is an exact multiple of the monoploid (n) or haploid number. Terms used to identify different levels in an euploid series are diploid, triploid, tetraploid, and so on (cf. **Aneuploid**).

Excinuclease. The endonuclease-containing protein complex that excises a segment of damaged DNA during excision repair.

Excision repair. DNA repair processes that involve the removal of the damaged segment of DNA and its replacement by the synthesis of a new strand using the complementary strand of DNA as template.

Exit (*E*) site. The ribosome binding site that contains the free tRNA prior to its release.

Exogenote. Chromosomal fragment homologous to an endogenote and donated to a merozygote.

Exon amplification. A procedure that is used to identify coding regions (exons) that are flanked by 5′ and 3′ intron splice sites.

Exons. The segments of a eukaryotic gene that correspond to the sequences in the final processed RNA transcript of that gene.

Exonuclease. An enzyme that digests DNA or RNA, beginning at the ends of strands.

Expressivity. Degree of expression of a trait controlled by a gene. A particular gene may produce different degrees of expression in different individuals.

Extrachromosomal. Structures that are not part of the chromosomes; DNA units in the cytoplasm that control cytoplasmic inheritance.

F_1. The first filial generation; the first generation of descent from a given mating.

F_2. The second filial generation produced by crossing *inter se* or by self-pollinating the F_1. The inbred "grandchildren" of a given mating, but in controlled genetic experimentation, self-fertilization of the F_1 (or equivalent) is implied.

F factor. A bacterial episome that confers the ability to function as a genetic donor ("male") in conjugation; the fertility factor in bacteria.

Feedback inhibition. The accumulated end product of a biochemical pathway stops synthesis of that product. A late metabolite of a synthetic pathway regulates synthesis at an earlier step of the pathway.

Fertilization. The fusion of a male gamete (sperm) with a female gamete (egg) to form a zygote.

Fetus. Prenatal stage of a viviparous animal between the embryonic stage and the time of birth; in humans, the final seven months before birth.

Filial. See F_1 and F_2.

Fitness. The number of offspring left by an individual, often compared with the average of the population or with some other standard, such as the number left by a particular genotype.

Fixation. An event that occurs when all the alleles at a locus except one are eliminated from a population. The remaining allele, with frequency 100 percent, is said to have been fixed.

Flagellum (*pl.*, **flagella;** *adj.* **flagellate**). A whiplike organelle of locomotion in certain cells; locomotor structures in flagellate protozoa.

Fluorescence *in situ* hybridization (FISH). *In situ* hybridization performed using a DNA or RNA probe coupled to a fluorescent dye.

Folded genome. The condensed intracellular state of the DNA in the nucleoid of a bacterium. The DNA is segregated into domains, and each domain is independently negatively supercoiled.

Forced cloning. The insertion of a foreign DNA into a cloning vector in a predetermined orientation.

Founder principle. The possibility that a new, small, isolated population may diverge genetically because the founding individuals are a random sample from a large, main population.

Frameshift mutation. A mutation that changes the reading frame of an mRNA, either by inserting or deleting nucleotides.

Frequency distribution. A graph showing either the relative or absolute incidence of classes in a population. The classes may be defined by either a discrete or a continuous variable; in the latter case, each class represents a different interval on the scale of measurement.

Fusion protein. A polypeptide made from a recombinant gene that contains portions of two or more different genes. The different genes are joined so that their coding sequences are in the same reading frame.

Gall. A tumorous growth in plants.

Gamete. A mature male or female reproductive cell (sperm or egg).

Gametogenesis. The formation of gametes.

Gametophyte. That phase of the plant life cycle that bears the gametes; the cells have *n* chromosomes.

Gametophytic incompatibility. A botanical phenomenon controlled by the complex *S* locus in which a pollen grain cannot fertilize an ovule produced by a plant that carries the same *S* allele as the pollen grain. For example, S_1 pollen cannot fertilize an ovule made by an S_1/S_2 plant.

Gap gene. A gene that controls the formation of adjacent segments in the body of *Drosophila*.

Gastrula. An early animal embryo consisting of two layers of cells; an embryological stage following the blastula.

Gene. A hereditary determinant of a specific biological function; a unit of inheritance (DNA) located in a fixed position on a chromosome; a segment of DNA encoding one polypeptide and defined operationally by the *cis-trans* or complementation test.

Gene addition. The addition of a functional copy of a gene to the genome of an organism.

Gene cloning. The incorporation of a gene of interest into a self-replicating DNA molecule and the amplification of the resulting recombinant DNA molecule in an appropriate host cell.

Gene conversion. A process, often associated with recombination, during which one allele is replicated at the expense of another, leading to non-Mendelian segregation ratios. In whole tetrads, for example, the ratio may be 6:2 or 5:3 instead of the expected 4:4.

Gene expression. The process by which genes produce RNAs and proteins and exert their effects on the phenotype of an organism.

Gene flow. The spread of genes from one breeding population to another by migration, possibly leading to allele frequency changes.

Gene pool. The sum total of all different alleles in the breeding members of a population at a given time.

Gene replacement. The incorporation of a transgene into a chromosome at its normal location by homologous recombination, thus replacing the copy of the gene originally present at the locus.

Gene therapy. The treatment of inherited diseases by introducing wild-type copies of the defective gene causing the disorder into the cells of affected individuals. If reproductive cells are modified, the procedure is called *germ-line* or *heritable gene therapy*. If cells other than reproductive cells are modified, the procedure is called *somatic-cell* or *noninheritable gene therapy*.

Genetic code. The set of 64 nucleotide triplets that specify the 20 amino acids and polypeptide chain initiation and termination.

Genetic drift. See **Random genetic drift.**

Genetic equilibrium. Condition in a group of interbreeding organisms in which the allele frequencies remain constant over time.

Genetics. The science of heredity and variation.

Genome. A complete set (*n*) of chromosomes (hence, of genes) inherited as a unit from one parent.

Genomic DNA library. A collection of clones containing the genomic DNA sequences of an organism.

Genomics. The study of the structure and function of entire genomes.

Genotype. The genetic constitution (gene makeup) of an organism (cf. **Phenotype**).

Germ cell. A reproductive cell capable when mature of being fertilized and reproducing an entire organism (cf. **Somatic cell**).

Germ plasm. The hereditary material transmitted to the offspring through the germ cells.

Globulins. Common proteins in the blood that are insoluble in water and soluble in salt solutions. Alpha, beta, and gamma globulins can be distinguished in human blood serum. Gamma globulins are important in developing immunity to diseases.

Glucocorticoid. A steroid hormone that regulates gene expression in higher animals.

Gonad. A sexual organ (that is, ovary or testis) that produces gametes.

Green fluorescent protein (GFP). A naturally occurring fluorescent protein synthesized by the jellyfish *Aequorea victoria*.

Guanine. A purine base found in DNA and RNA.

Guide RNAs. RNA molecules that contain sequences that function as templates during RNA editing.

Gynandromorph. An individual in which one part of the body is female and another part is male; a sex mosaic.

Haploid (monoploid). An organism or cell having only one complete set (*n*) of chromosomes or one genome.

Haptoglobin. A serum protein, alpha globulin, in the blood.

Hardy-Weinberg Principle. Mathematical relationship that allows the frequencies of genotypes in a population to be predicted from their constituent allele frequencies; a consequence of random mating.

Helix. Any structure with a spiral shape. The Watson and Crick model of DNA is in the form of a double helix.

Helper T cells. T cells that respond to an antigen displayed by a macrophage by stimulating B and T lymphocytes to develop into antibody-producing plasma cells and killer T cells, respectively.

Hemizygous. The condition in which only one allele of a pair is present, as in sex linkage or as a result of deletion.

Hemoglobin. Conjugated protein compound containing iron, located in erythrocytes of vertebrates; important in the transportation of oxygen to the cells of the body.

Hemolymph. The mixture of blood and other fluids in the body cavity of an invertebrate.

Hemophilia. A bleeder's disease; tendency to bleed freely from even a slight wound; hereditary condition dependent on a sex-linked recessive gene.

Heredity. Resemblance among individuals related by descent; transmission of traits from parents to offspring.

Heritability. Degree to which a given trait is controlled by inheritance. (See **Broad-sense heritability** and **Narrow-sense heritability**.)

Hermaphrodite. An individual with both male and female reproductive organs.

Heteroalleles. Mutations that are functionally allelic but structurally nonallelic; mutations at different sites in a gene.

Heterochromatin. Chromatin staining darkly even during interphase, often containing repetitive DNA with few genes.

Heteroduplex. A double-stranded nucleic acid containing one or more mismatched (noncomplementary) base pairs.

Heterogametic sex. Producing unlike gametes with regard to the sex chromosomes. In humans, the XY male is heterogametic, and the XX female is homogametic.

Heterogeneous nuclear RNA (hnRNA). The population of primary transcripts in the nucleus of a eukaryotic cell.

Heterokaryon. A cell containing two or more different nuclei.

Heteroplasmy. A cellular condition in which two genetically different types of an organelle are present (cf. **Homoplasmy**).

Heteropyknosis (*adj.,* **heteropyknotic**). Property of certain chromosomes, or of their parts, to remain more dense and to stain more intensely than other chromosomes or parts during the cell cycle.

Heterosis. Superiority of heterozygous genotypes in respect to one or more traits in comparison with corresponding homozygotes.

Heterozygosity. The proportion of heterozyous individuals in a population; used as a measure of genetic variability.

Heterozygote (*adj.,* **heterozygous**). An organism with unlike members of any given pair or series of alleles that consequently produces unlike gametes.

Hfr. High-frequency recombination strain of *Escherichia coli;* in such strains, the F episome is integrated into the bacterial chromosome.

Histones. Group of proteins rich in basic amino acids. They function in the coiling of DNA in chromosomes and in the regulation of gene activity.

HIV (human immunodeficiency virus). The retrovirus that causes AIDS in humans.

HLA antigens. See **Major histocompatibility antigens.**

HLA **locus.** See **Major histocompatibility complex.**

Holoenzyme. The form of a multimeric enzyme in which all of the component polypeptides are present.

Homeobox. A DNA sequence found in several genes that are involved in the specification of organs in different body parts in aminals; characteristic of genes that influence segmentation in animals. The homeobox corresponds to an amino acid sequence in the polypeptide encoded by these genes; this sequence is called the homeodomain.

Homeotic mutation. A mutation that causes a body part to develop in an inappropriate position in an organism; for example, a mutation in *Drosophila* that causes legs to develop on the head in the place of antennae.

Homoalleles. Mutations that are both functionally and structurally allelic; mutations at the same site in the same gene.

Homogametic sex. Producing like gametes with regard to the sex chromosomes (cf. **Heterogametic sex**).

Homologous chromosomes. Chromosomes that occur in pairs and are generally similar in size and shape, one having come from the male parent and the other from the female parent. Such chromosomes contain the same array of genes.

Homoplasmy. A cellular condition in which all copies of an organelle are genetically identical (cf. **Heteroplasmy**).

Homozygote (*adj.,* homozygous). An individual in which the two copies of a gene are the same allele.

Hormone. An organic product of cells of one part of the body that is transported by the body fluids to another part where it influences activity or serves as a coordinating agent.

Human Genome Project. A huge international effort to map and sequence the entire human genome.

Humoral immune response. See **Antibody-mediated immune response.**

Hybrid. An offspring of homozygous parents differing in one or more genes; more generally, an offspring of a cross between unrelated strains.

Hybrid dysgenesis. In *Drosophila,* a syndrome of abnormal germ-line traits, including mutation, chromosome breakage, and sterility, which results from transposable element activity.

Hybridization. Interbreeding of species, races, varieties, and so on, among plants or animals; a process of forming a hybrid by cross pollination of plants or by mating animals of different types.

Hybrid vigor (heterosis). Unusual growth, strength, and health of heterozygous hybrids from two less vigorous homozygous parents.

Hydrogen bonds. Weak interactions between electronegative atoms and hydrogen atoms (electropositive) that are linked to other electronegative atoms.

Hydrophobic interactions. Association of nonpolar groups with each other when present in aqueous solutions because of their insolubility in water.

Hyperploid. A genetic condition in which a chromosome or a segment of a chromosome is overrepresented in the genotype (cf. **Hypoploid**).

Hypersensitive sites. Regions in the DNA that are highly susceptible to digestion with endonucleases.

Hypomorph. A mutation that reduces but does not completely abolish gene expression.
Hypoploid. A genetic condition in which a chromosome or segment of a chromosome is underrepresented in the genotype (cf. **Hyperploid**).
Hypostasis. See **Epistasis.**

Imaginal disc. A mass of cells in the larvae of *Drosophila* and other holometabolous insects that gives rise to particular adult organs such as antennae, eyes, and wings.
Immunogen. A substance, usually a protein, that elicits an immune response when introduced into a vertebrate.
Immunoglobulin. See **Globulin.**
Inbreeding. Matings between related individuals.
Inbreeding coefficient. The probability that two alleles in an individual are identical to each other by descent from a common ancestor.
Incomplete dominance. Expression of two alleles in a heterozygote that allows the heterozygote to be distinguished from either of its homozygous parents.
Independent assortment. The random distribution of alleles to the gametes that occurs when genes are located in different chromosomes. The distribution of one pair of alleles is independent of other genes located in nonhomologous chromosomes.
Inducer. A substance of low molecular weight that is bound by a repressor to produce a complex that can no longer bind to the operator; thus, the presence of the inducer turns on the expression of the gene(s) controlled by the operator.
Inducible enzyme. An enzyme that is synthesized only in the presence of the substrate that acts as an inducer.
Inducible gene. A gene that is expressed only in the presence of a specific metabolite, the inducer.
Induction. The process of turning on the expression of a gene or set of genes by an inducer.
Inhibitor. Any substance or object that retards a chemical reaction; a major or modifier gene that interferes with a reaction.
Initiation factors. Soluble proteins required for the initiation of translation.
In situ. From the Latin, meaning in the natural place; refers to experimental treatments performed on cells or tissue rather than on extracts from them.
***In situ* colony or plaque hybridization.** A procedure for screening colonies or plaques growing on plates or membranes for the presence of specific DNA sequences by the hybridization of nucleic acid probes to the DNA molecules present in these colonies or plaques.
***In situ* hybridization.** A method for determining the location of specific DNA sequences in chromosomes by hybridizing labeled DNA or RNA to denatured DNA in chromosome preparations and visualizing the hybridized probe by autoradiography or fluorescence microscopy.
Interaction. In statistics, an effect that cannot be explained by the additive action of contributing factors; a departure from strict additivity.
Intercalating agent. A chemical capable of inserting between adjacent base pairs in a DNA molecule.
Interference. Crossing over at one point that reduces the chance of another crossover nearby; detected by studying the pattern of crossing over with three or more linked genes.
Interphase. The stage in the cell cycle when the cell is not dividing; the metabolic stage during which DNA replication occurs; the stage following telophase of one division and extending to the beginning of prophase in the next division.
Intersex. An organism displaying secondary sexual characters intermediate between male and female; a type that shows some phenotypic characteristics of both males and females.
Intragenic complementation. Complementation that occurs between two mutant alleles of a gene; common only when the product of the gene functions as a homomultimer.
Introns. Intervening sequences of DNA bases within eukaryotic genes that are not represented in the mature RNA transcript because they are spliced out of the primary RNA transcript.
Invariant. Constant, unchanging, usually referring to the portion of a molecule that is the same across species.
Inversion. A rearrangement that reverses the order of a linear array of genes in a chromosome.
Inverted repeat. A sequence present twice in a DNA molecule but in reverse orientation.
in vitro. From the Latin meaning "within glass"; biological processes made to occur experimentally outside the organism in a test tube or other container.
in vivo. From the Latin meaning "within the living organism."
Ionic bonds. Attractions between oppositely charged chemical groups.
Ionizing radiation. The portion of the electromagnetic spectrum that results in the production of positive and negative charges (ion pairs) in molecules. X rays and gamma rays are examples of ionizing radiation.
IS element (insertion sequence). A short (800–1400 nucleotide pairs) DNA sequence found in bacteria that is capable of transposing to a new genomic location; other DNA sequences that are bounded by IS elements may also be transposed.
Isoalleles. Different forms of a gene that produce the same phenotype or very similar phenotypes.
Isochromosome. A chromosome with two identical arms and identical genes. The arms are mirror images of each other.
Isoform. A member of a family of closely related proteins—proteins that have some amino acid sequences in common and some different.
Isogenic stocks. Strains of organisms that are genetically uniform; completely homozygous.

Kappa chain. One of two classes of antibody light chains (cf. **Lambda chain**).
Karyotype. The chromosome constitution of a cell or an individual; chromosomes arranged in order of length and according to position of centromere; also, the abbreviated formula for the chromosome constitution, such as 47, XX + 21 for human trisomy-21.
Killer T cells (cytotoxic T cells). T cells that carry T-cell receptors and kill cells displaying the recognized antigens.
Kinetics. A dynamic process involving motion.
Kinetosome. Granular body at the base of a flagellum or a cilium.

Lagging strand. The strand of DNA that is synthesized discontinuously during replication.
Lambda chain. One of two classes of antibody light chains (cf. **Kappa chain**).
Lamella. A double-membrane structure, plate, or vesicle that is formed

by two membranes lying parallel to each other.

Lampbrush chromosomes. Large diplotene chromosomes present in oocyte nuclei, particularly conspicuous in amphibians. These chromosomes have extended regions called loops, which are active sites of transcription.

Leader sequence. The segment of an mRNA molecule from the 5′ terminus to the translation initiation codon.

Leading strand. The strand of DNA that is synthesized continuously during replication.

Leptonema (*adj.*, leptotene). Stage in meiosis immediately preceding synapsis in which the chromosomes appear as single, fine, threadlike structures (but they are really double because DNA replication has already taken place).

Lethal mutation. A mutation that renders an organism or a cell possessing it inviable.

Ligand. A molecule that can bind to another molecule in or on cells.

Ligase. An enzyme that joins the ends of two strands of nucleic acid.

Ligation. The joining of two or more DNA molecules by covalent bonds.

LINEs (long interspersed nuclear elements). Families of long (average length = 6500 bp) moderately repetitive transposable elements in eukaryotes.

Linkage. A relationship among genes in the same chromosome. Such genes tend to be inherited together.

Linkage equilibrium. A state in which the alleles of linked loci are randomized with respect to each other on the chromosomes of a population.

Linkage map. A linear or circular diagram that shows the relative positions of genes on a chromosome as determined by genetic analysis.

Locus (*pl.*, loci). A fixed position on a chromosome that is occupied by a given gene or one of its alleles.

Lod score. The logarithm of the ratio of odds for pedigree data, where the odds are calculated under the assumption of linkage with a specified frequency of recombination and under the assumption of no linkage, that is, 50 percent recombination.

LTR (Long terminal repeat). A DNA sequence present at each end of a retrotransposon.

Lymphocyte. A general class of white blood cells that are important components of the immune system of vertebrate animals.

Lysis. Bursting of a cell by the destruction of the cell membrane following infection by a virus.

Lysogenic bacteria. Those harboring temperate bacteriophages.

Macromolecule. A large molecule; term used to identify molecules of proteins and nucleic acids.

Macrophages. Large, white blood cells that ingest foreign substances and display antigens produced from them on their surfaces to be recognized by other cells of the immune system.

Major histocompatibility antigens. A complex set of cell-surface macromolecules that allow the immune system to distinguish foreign or "nonself" cells from "self" cells. These are the antigens that must be matched between donors and recipients during organ and tissue transplants to prevent rejections. Called **HLA (human-leukocyte-associated) antigens** in humans.

Major histocompatibility complex. The large cluster of genes that encode the major histocompatibility antigens in mamals. Called the ***HLA* locus** in humans.

Map unit. See **Crossover unit.**

Mass selection. As practiced in plant and animal breeding, the choosing of individuals for reproduction from the entire population on the basis of the individual's phenotypes rather than on the phenotypes of their relatives.

Maternal effect. Trait controlled by a gene of the mother but expressed in the progeny.

Maternal inheritance. Inheritance controlled by extrachromosomal (that is, cytoplasmic) factors that are transmitted through the egg.

Maturation. The formation of gametes or spores.

Mean. The arithmetic average; the sum of all measurements or values in a sample divided by the sample size.

Median. In a set of measurements, the central value above and below which there are an equal number of measurements.

Meiosis. The process by which the chromosome number of a reproductive cell becomes reduced to half the diploid ($2n$) or somatic number; results in the formation of gametes in animals or of spores in plants; important source of variability through recombination.

Meiotic drive. Any mechanism that causes alleles to be recovered unequally in the gametes of a heterozygote.

Melanin. Brown or black pigment.

Memory cells. Long-lived B and T cells that mediate rapid secondary immune responses to a previously encountered antigen.

Mendelian population. A natural interbreeding unit of sexually reproducing plants or animals sharing a common gene pool.

Merozygote. Partial zygote produced by a process of partial genetic exchange, such as transformation in bacteria. An exogenote may be introduced into a bacterial cell in the formation of a merozygote.

Mesoderm. The middle germ layer that forms in the early animal embryo and gives rise to such parts as bone and connective tissue.

Messenger RNA (mRNA). RNA that carries information necessary for protein synthesis from the DNA to the ribosomes.

Metabolic cell. A cell that is not dividing.

Metabolism. Sum total of all chemical process in living cells by which energy is provided and used.

Metacentric chromosome. A chromosome with the centromere near the middle and two arms of about equal length.

Metafemale (superfemale). In *Drosophila*, abnormal female, usually sterile, with an excess of X chromosomes compared with sets of autosomes (for example, XXX; AA).

Metaphase. That stage of cell division in which the chromosomes are most discrete and arranged in an equatorial plate; stage following prophase and preceding anaphase.

Metastasis. The spread of cancer cells to previously unaffected organs.

Microprojectile bombardment. A procedure for transforming plant cells by shooting DNA-coated tungsten or gold particles into the cells.

Microsatellites. DNA sequences composed of polymorphic tandem repeats of two to five nucleotide pairs.

Microtubules. Hollow filaments in the cytoplasm making up a part of the locomotor apparatus of a motile cell; component of the mitotic spindle.

Midparent value. In quantitative genetics, the average of the phenotypes of two mates.

Mismatch repair. DNA repair processes that correct base pairs that are not properly hydrogen-bonded.
Missense mutation. A mutation that changes a codon specifying an amino acid to a codon specifying a different amino acid.
Mitochondria. Organelles in the cytoplasm of plant and animal cells where oxidative phosphorylation takes place to produce ATP.
Mitosis. Disjunction of duplicated chromosomes and division of the cytoplasm to produce two genetically identical daughter cells.
Modal class. In a frequency distribution, the class having the greatest frequency.
Model. A mathematical description of a biological phenomenon.
Modifier (modifying gene). A gene that affects the expression of some other gene.
Monoclonal antibody. An antibody produced by a population (clone) of identical hybridoma cells growing in culture.
Monohybrid. An offspring of two homozygous parents that differ from one another by the alleles present at only one gene locus.
Monohybrid cross. A cross between parents differing in only one trait or in which only one trait is being considered.
Monomer. A single molecular entity that may combine with others to form more complex structures.
Monoploid. Organism or cell having a single set of chromosomes or one genome (chromosome number n).
Monosomic. A diploid organism lacking one chromosome of its proper complement ($2n - 1$); an aneuploid. Monosome refers to the single chromosome, disome to two chromosomes of a kind, and trisome to three chromosomes of a kind.
Monozygotic twins. One-egg or identical twins.
Morphogen. A substance that stimulates the development of form or structure in an organism.
Morphology. Study of the form of an organism; developmental history of visible structures and the comparative relation of similar structures in different organisms.
Mosaic. An organism or part of an organism that is composed of cells of different genotypes.
mtDNA. The DNA of mitochondria.
Multifactorial trait. A trait determined by a combination of several genetic and environmental factors.
Multigene family. A group of genes that are similar in nucleotide sequence or that produce polypeptides with similar amino acid sequences.
Multiple alleles. A condition in which a particular gene occurs in three or more allelic forms in a population of organisms.
Mutable genes. Genes with an unusually high mutation rate.
Mutagen. An environmental agent, either physical or chemical, that is capable of inducing mutations.
Mutant. A cell or individual organism that shows a change brought about by a mutation; a changed gene.
Mutation. A change in the DNA at a particular locus in an organism. The term is used loosely to include point mutations involving a single gene change as well as a chromosomal change.
Mutation pressure. A constant mutation rate that adds mutant genes to a population; repeated occurrences of mutations in a population.
Mycelium (*pl.,* mycelia). Threadlike filament making up the vegetative portion of thallus fungi.

Narrow-sense heritability. In quantitative genetics, the proportion of the phenotypic variance that is due to the additive effects of alleles.
Natural selection. Differential survival and reproduction in nature that favors individuals that are better adapted to their environment; elimination of less fit organisms.
Negative autogenous regulation (negative self-regulation). Inhibition of the expression of a gene or set of coordinately regulated genes by the product of the gene or the product of one of the genes.
Negative control system. A mechanism in which the regulatory protein(s) is required to turn off gene expression.
Neutral mutation. A mutation that changes the nucleotide sequence of a gene but has no effect on the fitness of the organism.
Neutral theory. The theory that the evolution of traits with little or no effect on fitness is a random process involving mutation and genetic drift.
Nick-translation. A procedure for labeling DNA by nicking it with an endonuclease and "translating" the nick along the DNA molecule in the presence of labeled deoxyribonucleoside triphosphates by the concerted action of the 5′ —> 3′ exonuclease and 5′ —> 3′ polymerase activities of *E. coli* DNA polymerase I.
Nonautonomous. A term referring to biological units that cannot function by themselves; such units require the assistance of another unit, or "helper" (cf. **Autonomous**).
Nondisjunction. Failure of disjunction or separation of homologous chromosomes in mitosis or meiosis, resulting in too many chromosomes in some daughter cells and too few in others. Examples: In meiosis, both members of a pair of chromosomes go to one pole so that the other pole does not receive either of them; in mitosis, both sister chromatids go to the same pole.
Nonhistone chromosomal proteins. All of the proteins in chromosomes except the histones.
Nonsense mutation. A mutation that changes a codon specifying an amino acid to a termination codon.
Nontemplate strand. In transcription, the nontranscribed strand of DNA. It will have the same sequence as the RNA transcript, except that T is present at positions where U is present in the RNA transcript.
Northern blot. The transfer of RNA molecules from an electrophoretic gel to a cellulose or nylon membrane by capillary action.
Nuclease. An enzyme that catalyzes the degradation of nucleic acids.
Nucleic acid. A macromolecule composed of phosphoric acid, pentose sugar, and organic bases; DNA and RNA.
Nucleolar Organizer (NO). A chromosomal segment containing genes that control the synthesis of ribosomal RNA, located at the secondary constriction of some chromosomes.
Nucleolus. An RNA-rich, spherical sack in the nucleus of metabolic cells; associated with the nucleolar organizer; storage place for ribosomes and ribosome precursors.
Nucleoprotein. Conjugated protein composed of nucleic acid and protein; the material of which the chromosomes are made.
Nucleoside. An organic compound consisting of a base covalently linked to ribose or deoxyribose.
Nucleosome. Spherical subunits of eukaryotic chromatin that are composed of a core particle consisting of an octamer of histones and 146 nucleotide pairs.

Nucleotide. A subunit of DNA and RNA molecules containing a phosphate group, a sugar, and a nitrogen-containing organic base.

Nucleus. The part of a eukaryotic cell that contains the chromosomes; separated from the cytoplasm by a membrane.

Nullisomic. An otherwise diploid cell or organism lacking both members of a chromosome pair (chromosome formula $2n - 2$).

Null mutation. A mutation that abolishes the expression of a gene (See **Amorph**).

Octoploid. Cell or organism with eight genomes or sets of chromosomes (chromosome number $8n$).

Oligonucleotide-directed site-specific mutagenesis. A procedure by which a specific nucleotide sequence can be changed to another predetermined sequence.

Oncogene. A gene that can cause cancerous transformation in animal cells growing in culture and tumor formation in animals themselves; a gene that promotes cell division.

Oocyte. The egg-mother cell; the cell that undergoes two meiotic divisions (oogenesis) to form the egg cell. Primary oocyte—before completion of the first meiotic division; secondary oocyte—after completion of the first meiotic division.

Oogenesis. The formation of the egg or ovum in animals.

Oogonium (*pl.*, **oogonia**). A germ cell of the female animal before meiosis begins.

Open Reading Frame (ORF). A DNA segment containing the sequences required to encode a polypeptide. The RNA transcript of an ORF begins with a translation start codon, followed by a sequence of codons specifying amino acids, and ending with a translation stop codon. An ORF is presumed, but not known, to encode a polypeptide.

Operational definition. An operation or procedure that can be carried out to define or delimit something.

Operator. A part of an operon that controls the expression of one or more structural genes by serving as the binding site for one or more regulatory proteins.

Operon. A group of genes making up a regulatory or control unit. The unit includes an operator, a promoter, and structural genes.

Ordinate. The vertical axis in a graph.

Organelle. Specialized part of a cell with a particular function or functions (for example, the cilium of a protozoan).

Organizer. An inductor; a chemical substance in a living system that determines the fate in development of certain cells or groups of cells.

Outbreeding. Mating of unrelated individuals.

Ovary. The swollen part of the pistil of a plant flower that contains the ovules; the female reproductive organ or gonad in animals.

Overdominance. A condition in which heterozygotes are superior (on some scale of measurement) to either of the associated homozygotes.

Ovule. The macrosporangium of a flowering plant that becomes the seed. It includes the nucellus and the integuments.

P. Symbol for the parental generation or parents of a given individual.

Pachynema (*adj.*, **pachytene**). A midprophase stage in meiosis immediately following zygonema and preceding diplonema. In favorable microscopic preparations, the chromosomes are visible as long, paired threads. Rarely, four chromatids are detectable.

Pair-rule gene. A gene that influences the formation of body segements in *Drosophila.*

Palindrome. A segment of DNA in which the base-pair sequence reads the same in both directions from a central point of symmetry.

Panmictic population. A population in which mating occurs at random.

Panmixis. Random mating in a population.

Paracentric inversion. An inversion that is entirely within one arm of a chromosome and does not include the centromere.

Parameter. A value or constant based on an entire population (cf. **Statistic**).

Parthenogenesis. The development of a new individual from an egg without fertilization.

Paternal. Pertaining to the father.

Pathogen. An organism that causes a disease.

PCR. See **Polymerase chain reaction.**

Pedigree. A table, chart, or diagram representing the ancestry of an individual.

Penetrance. The percentage of individuals that show a particular phenotype among those capable of showing it.

Peptide. A compound containing amino acids; a breakdown or buildup unit in protein metabolism.

Peptide bond. A chemical bond holding amino acid subunits together in proteins.

Peptidyl (*P*) site. The ribosome binding site that contains the tRNA to which the growing polypeptide chain is attached.

Peptidyl transferase. An enzyme activity—built into the large subunit of the ribosome—that catalyzes the formation of peptide bonds between amino acids during translation.

Pericentric inversion. An inversion including the centromere, hence involving both arms of a chromosome.

Petite mutant. A respiration-deficient yeast mutant that produces small colonies when grown on glucose-containing medium.

Phage. See **Bacteriophage.**

Phagemids. Cloning vectors that contain components derived from both phage chromosomes and plasmids.

Phagocytes. Immune system cells that ingest and destroy viruses, bacteria, fungi, and other foreign substances or cells.

Phenocopy. An organism whose phenotype (but not genotype) has been changed by the environment to resemble the phenotype of a different (mutant) organism.

Phenotype. The observable characteristics of an organism.

Phenylalanine. See **Amino acid.**

Phenylketonuria. Metabolic disorder resulting in mental retardation; transmitted as a Mendelian recessive and treated in early childhood by special diet.

Photoreactivation. A DNA repair process that is light-dependent.

Phylogeny. Evolutionary history of populations of related organisms.

Plaque. Clear area on an otherwise opaque culture plate of bacteria where the bacteria have been killed by a virus.

Plasma cells. Antibody-producing white blood cells devived from B lymphocytes.

Plasmid. An extrachromosomal hereditary determinant that exists in an autonomous state and is transferred independently of chromosomes.

Plastid. A cytoplasmic body found

in the cells of plants and some protozoa. Chloroplasts, for example, produce chlorophyll that is involved in photosynthesis.

Pleiotropy (*adj.*, **Pleiotropic**). Condition in which a single gene influences more than one trait.

Point mutations. Changes that occur at specific sites in genes. They include nucleotide-pair substitutions and the insertion or deletion of one or a few nucleotide pairs.

Polar bodies. In female animals, the smaller cells produced at meiosis that do not develop into egg cells. The first polar body is produced at division I and may not go through division II. The second polar body is produced at division II.

Polar mutation. A mutation that influences the functioning of genes that are downstream in the same transcription unit.

Pole cells. A group of cells in the posterior of *Drosophila* embryos that are precursors to the adult germ line.

Polyadenylation. The addition of poly(A) tails to eukaryotic gene transcripts (RNAs).

Poly(A) polymerase. An enzyme that adds the poly(A) tails to the 3′ termini of eukaryotic gene transcripts (RNAs).

Polycloning site. See **Polylinker.**

Polydactyly. The occurrence of more than the usual number of fingers or toes.

Polygene (*adj.*, **polygenic**). One of many genes involved in quantitative inheritance.

Polylinker (polycloning site). A segment of DNA that contains a set of unique restriction enzyme cleavage sites.

Polymer. A compound composed of many smaller subunits; results from the process of polymerization.

Polymerase. An enzyme that catalyzes the formation of DNA or RNA.

Polymerase chain reaction (PCR). A procedure involving multiple cycles of denaturation, hybridization to oligonucleotide primers, and polynucleotide synthesis that amplifies a particular DNA sequence.

Polymerization. Chemical union of two or more molecules of the same kind to form a new compound having the same elements in the same proportions but a higher molecular weight and different physical properties.

Polymorphism. Two or more kinds of individuals present in a breeding population.

Polynucleotide. A linear sequence of joined nucleotides in DNA or RNA.

Polypeptide. A linear molecule with two or more amino acids and one or more peptide groups. They are called dipeptides, tripeptides, and so on, according to the number of amino acids present.

Polyploid. An organism with more than two sets of chromosomes ($2n$ diploid) or genomes—for example, triploid ($3n$), tetraploid ($4n$), pentaploid ($5n$), hexaploid ($6n$), heptaploid ($7n$), octoploid ($8n$).

Polysaccharide capsules. Carbohydrate coverings with antigenic specificity that are present on some types of bacteria.

Polytene chromosomes. Giant chromosomes produced by interphase replication without division and consisting of many identical chromatids arranged side by side in a cablelike pattern.

Population. Entire group of organisms of one kind; an interbreeding group of plants or animals; the extensive group from which a sample might be taken.

Population (effective). Breeding members of the population.

Population genetics. The branch of genetics that deals with frequencies of alleles and genotypes in breeding populations.

Positional cloning. The isolation of a clone of a gene or other DNA sequence based on its map position in the genome.

Position effect. A difference in phenotype that is dependent on the position of a gene or group of genes, often caused by heterochromatin that is nearby.

Positive control system. A mechanism in which the regulatory protein(s) is required to turn on gene expression.

Postreplication repair. A recombination-dependent mechanism for repairing damaged DNA.

Primary immune response. The immune response that occurs during the first encounter of a mammal with a given antigen (cf. **Secondary immune response**).

Primary transcript. The RNA molecule produced by transcription prior to any post-transcriptional modifications; also called a pre-mRNA in eukaryotes.

Primer. A short nucleotide sequence with a reactive 3′ OH that can initiate DNA synthesis along a template.

Primosome. A protein replication complex that catalyzes the initiation of Okazaki fragments during discontinuous synthesis. It contains DNA primase and DNA helicase activities.

Probability. The frequency of occurrence of an event.

Proband. The individual in a family in whom an inherited trait is first identified.

Progeny testing. The practice of ascertaining the genotype of an individual by mating it to an individual of known genotype and examining the progeny.

Progerias. Inherited diseases characterized by premature aging.

Prokaryote. A member of a large group of organisms (including bacteria and bluegreen algae) that lack true nuclei in their cells and that do not undergo meiosis.

Promoter. A nucleotide sequence to which RNA polymerase binds and initiates transcription; also, a chemical substance that enhances the transformation of benign cells into cancerous cells.

Proofreading. The enzymatic scanning of DNA for structural defects such as mismatched base pairs.

Prophage (provirus). The genome of a temperate bacteriophage integrated into the chromosome of a lysogenic bacterium and replicated along with the host chromosome.

Prophase. The stage of mitosis between interphase and metaphase. During this phase, the centriole divides and the two daughter centrioles move apart. Each sister DNA strand from interphase replication becomes coiled, and the chromosome is longitudinally double except in the region of the centromere. Each partially separated chromosome is called a chromatid. The two chromatids of a chromosome are sister chromatids.

Protamines. Small basic proteins that replace the histones in the chromosomes of some sperm cells.

Protease. Any enzyme that hydrolyzes proteins.

Protein. A macromolecule composed of one to several polypeptides. Each polypeptide consists of a chain of amino acids linked together by peptide bonds.

Proto-oncogene. A normal cellular gene that can be changed to an oncogene by mutation.

Protoplast. A plant or bacterial cell

from which the wall has been removed.

Prototroph. An organism such as a bacterium that will grow on a minimal medium.

Provirus. A viral chromosome that has integrated into a host—either prokaryotic or eukaryotic—genome (cf. **Prophage**).

Pseudoautosomal gene. A gene located on both the X and Y chromosomes.

Pseudogene. An inactive but stable component of a genome resembling a gene; apparently derived from active genes by mutation.

Pulsed-field gel electrophoresis. A procedure used to separate very large DNA molecules by alternating the direction of electric currents across a semisolid gel in a pulsed manner.

Punctuated equilibrium. The occurrence of speciation events in bursts separated by long intervals of species stability.

Purine. A double-ring nitrogen-containing base present in nucleic acids; adenine and guanine are the two purines present in most DNA and RNA molecules.

Pyrimidine. A single-ring nitrogen-containing base present in nucleic acids; cytosine and thymine are commonly present in DNA, whereas uracil usually replaces thymine in RNA.

Quantitative inheritance. Inheritance of measurable traits (height, weight, color intensity) that depend on the cumulative action of many genes, each producing a small effect on the phenotype.

Quantitative trait loci (QTL). Two or more genes that affect a single quantitative trait.

Quantum speciation. The formation of a new species in one or a few generations by selection and genetic drift.

Race. A distinguishable group of organisms of a particular species.

Radiation hybrid mapping. The use of human-rodent hybrid cells containing fragments of human chromosomes (produced by irradiation) fused to rodent chromosomes to determine the linkage relationships of human genes.

Radioactive isotope. An unstable isotope (form of an atom) that emits ionizing radiation.

Random genetic drift. Changes in allele frequency in small breeding populations due to chance fluctuations.

Reading frame. The series of nucleotide triplets that are sequentially positioned in the *A* site of the ribosome during translation of an mRNA; also, the sequence of nucleotide-pair triplets in DNA that correspond to these codons in mRNA.

Receptor. A molecule that can accept the binding of a ligand.

Recessive. A term applied to one member of an allelic pair lacking the ability to manifest itself when the other or dominant member is present.

Reciprocal crosses. Crosses between different strains with the sexes reversed; for example, female A × male B and male A × female B are reciprocal crosses.

Recombinant DNA molecule. A DNA molecule constructed *in vitro* by joining all or parts of two different DNA molecules.

Recombination. The production of gene combinations not found in the parents by the assortment of nonhomologous chromosomes and crossing over between homologous chromosomes during meiosis. For linked genes, the frequency of recombination can be used to estimate the genetic map distance; however, high frequencies (approaching 50%) do not yield accurate estimates.

Reduction division. Phase of meiosis in which the maternal and paternal chromosomes of the bivalent separate (cf. **Equational division**).

Regulator gene. A gene that controls the rate of expression of another gene or genes. Example: The *lacI* gene produces a protein that controls the expression of the structural genes of the *lac* operon in *Escherichia coli.*

Release factors. Soluble proteins that recognize termination codons in mRNAs and terminate translation in response to these codons.

Renaturation. The restoration of a molecule to its native form. In nucleic acid biochemistry, this term usually refers to the formation of a double-stranded helix from complementary single-stranded molecules.

Repetitive DNA. DNA sequences that are present in a genome in multiple copies—sometimes a million times or more.

Replica plating. A procedure for duplicating the bacterial colonies growing on agar medium in one petri plate to agar medium in another petri plate.

Replication. A duplication process that is accomplished by copying from a template (for example, reproduction at the level of DNA).

Replicon. A unit of replication. In bacteria, replicons are associated with segments of the cell membrane that control replication and coordinate it with cell division.

Replisome. The complete replication apparatus—present at a replication fork—that carries out the semiconservative replication of DNA.

Repressible enzyme. An enzyme whose synthesis is diminished by a regulatory molecule.

Repression. The process of turning off the expression of a gene or set of genes in response to an environmental signal.

Repressor. A protein that binds to DNA and turns off gene expression.

Repressor gene. A gene that encodes a repressor.

Repulsion (*trans* configuration). The condition in which a double heterozygote has received a mutant and a wild-type allele from each parent; for example, $a\,+/a\,+ \times +\,b/+\,b$ produces $a\,+/+\,b$ (cf. **Coupling.**)

Resistance factor A plasmid that confers antibiotic resistance to a bacterium.

Restitution nucleus. A nucleus with unreduced or doubled chromosome number that results from the failure of a meiotic or mitotic division.

Restriction enzyme. An endonuclease that recognizes a specific short sequence in DNA and cleaves the DNA molecule at or near that site.

Restriction fragment. A fragment of DNA produced by cleaving a DNA molecule with one or more restriction endonucleases.

Restriction map. A linear or circular physical map of a DNA molecule showing the sites that are cleaved by different restriction enzymes.

Restriction site. A DNA sequence that is cleaved by a restriction enzyme.

Reticulocyte. A young red blood cell.

Retroelement. Any of the integrated retroviruses or the transposable elements that resemble them.

Retroposon. A transposable element that moves via reverse transcription of RNA into DNA but lacks the long terminal repeat sequences.

Retrotransposon. A transposable element that resembles the integrated form of a retrovirus.

Retrovirus. A virus that stores its genetic information in RNA and replicates by using reverse transcriptase to synthesize a DNA copy of its RNA genome.

Reverse transcriptase. An enzyme that catalyzes the synthesis of DNA using an RNA template.

Reversion (reverse mutation). Restitution of a mutant gene to the wild-type condition, or at least to a form that gives the wild phenotype; more generally, the appearance of a trait expressed by a remote ancestor.

RFLP (Restriction Fragment Length Polymorphism). A genetic difference among individuals that is detected by comparing DNA fragments released by digestion with one or more restriction enzymes.

Rh factor. Antigen in the red blood corpuscles of certain people. A pregnant Rh negative woman carrying an Rh positive child may develop antibodies, causing the child to develop a hemolytic disease.

Ribonuclease. Any enzyme that hydrolyzes RNA.

Ribonucleic acid. See **RNA.**

Ribosomal RNAs (rRNAs). The RNA molecules that are structural components of ribosomes.

Ribosome. Cytoplasmic organelle on which proteins are synthesized.

R-loops. Single-stranded DNA regions in RNA-DNA hybrids formed *in vitro* under conditions where RNA-DNA duplexes are more stable than DNA-DNA duplexes.

RNA. Ribonucleic acid; the information-carrying material in some viruses; more generally, a molecule derived from DNA by transcription that may carry information (messenger or mRNA), provide subcellular structure (ribosomal or rRNA), transport amino acids (transfer or tRNA) or facilitate the biochemical modification of itself or other RNA molecules.

RNA editing. Post-trancriptional processes that alter the information encoded in gene transcripts (RNAs).

RNA polymerase. An enzyme that catalyzes the synthesis of RNA.

Roentgen (r). Unit of ionizing radiation.

Satellite DNA. A component of the genome that can be isolated from the rest of the DNA by density gradient centrifugation. Usually, it consists of short, highly repetitious sequences.

Scaffold. The central core structure of condensed chromosomes. The scaffold is composed of nonhistone chromosomal proteins.

SCID (severe combined immunodeficiency syndrome). A group of diseases characterized by the inability to mount an immune response, either humoral or cellular.

Secondary immune response. The rapid immune response that occurs during the second and subsequent encounters of the immune system of a mammal with a specific antigen (cf. **Primary immune response**).

Secondary oocyte. See **Oocyte.**

Secondary spermatocyte. See **Spermatocyte.**

Segment-polarity gene. A gene that functions to define the anterior and posterior components of body segments in *Drosophila.*

Segregation. The separation of paternal and maternal chromosomes from each other at meiosis; the separation of alleles from each other in heterozygotes; the occurrence of different phenotypes among offspring, resulting from chromosome or allele separation in their heterozygous parents; Mendel's first principle of inheritance.

Selection. Differential survival and reproduction among genotypes; the most important of the factors that change allele frequencies in large populations.

Selection coefficient. A number that measures the fitness of a genotype relative to a standard.

Selection differential. In plant and animal breeding, the difference between the mean of the individuals selected to be parents and the mean of the overall population.

Selection pressure. Effectiveness of differential survival and reproduction in changing the frequency of alleles in a population.

Selection response. In plant and animal breeding, the difference between the mean of the individuals selected to be parents and the mean of their offspring.

Selector gene. A gene that influences the development of specific body segments in *Drosophila;* a homeotic gene.

Self-fertilization. The process by which pollen of a given plant fertilizes the ovules of the same plant. Plants fertilized in this way are said to have been selfed. An analogous process occurs in some animals, such as nematodes and molluscs.

Semisterility. A condition of only partial fertility in plant zygotes (for example, maize); usually associated with translocations.

Sense RNA. A primary transcript or mRNA that contains a coding region (contiguous sequence of codons) that is translated to produce a polypeptide.

Serology. (*adj.,* **serological**). The study of interactions between antigens and antibodies.

Sex chromosomes. Chromosomes that are connected with the determination of sex.

Sexduction. The incorporation of bacterial genes into F factors and their subsequent transfer by conjugation to a recipient cell.

Sex factor. A bacterial episome (for example, the F plasmid in *E. coli*) that enables the cell to be a donor of genetic material. The sex factor may be propagated in the cytoplasm, or it may be integrated into the bacterial chromosome.

Sex-influenced dominance. A dominant expression that depends on the sex of the individual. For example, horns in some breeds of sheep are dominant in males and recessive in females.

Sex-limited. Expression of a trait in only one sex. Examples: milk production in mammals; horns in Rambouillet sheep; egg production in chickens.

Sex linkage. Association or linkage of a hereditary trait with sex; the gene is in a sex chromosome, usually the X; often used synonymously with X-linkage.

Sex mosaic. See **Gynandromorph.**

Sexual reproduction. Reproduction involving the formation of mature germ cells (that is, eggs and sperm).

Shine-Dalgarno sequence. A conserved sequence in prokaryotic mRNAs that is complementary to a sequence near the 5′ terminus of the 16S ribosomal RNA and is involved in the initiation of translation.

Shuttle vector. A plasmid capable of replicating in two different organisms, such as yeast and *E. coli.*

Sib-mating (crossing of siblings). Matings involving two individuals of the same parentage; brother-sister matings.

Sigma factor. The subunit of prokaryotic RNA polymerases that is responsible for the initiation of transcription at specific initiation sequences.

Silencer. A DNA sequence that helps to reduce or shut off the expression of a nearby gene.

SINEs (short interspersed nuclear elements). Families of short (150 to

300 bp), moderately repetitive transposable elements of eukaryotes. The best known SINE family is the Alu family in humans.

Single-strand DNA-binding protein. A protein that coats DNA single strands, keeping them in an extended state.

Site-specific mutagenesis. See **Oligonucleotide-directed site-specific mutagenesis.**

Small nuclear ribonucleoproteins (snRNPs). RNA-protein complexes that are components of spliceosomes.

Small nuclear RNAs (snRNAs). Small RNA molecules that are located in the nuclei of eukaryotic cells; most snRNAs are components of the spliceosomes that excise introns from pre-mRNAs.

Somatic cell. A cell that is a component of the body, in contrast with a germ cell that is capable, when fertilized, of reproducing the organism.

Somatic hypermutation. A high frequency of mutation that occurs in the gene segments encoding the variable regions of antibodies during the differentiation of B lymphocytes into antibody-producing plasma cells.

SOS response. The synthesis of a whole set of DNA repair, recombination, and replication proteins in bacteria containing severely damaged DNA (for example, following exposure to UV light).

Southern blot. The transfer of DNA fragments from an electrophoretic gel to a cellulose or nylon membrane by capillary action.

Species. Interbreeding, natural populations that are reproductively isolated from other such groups.

Sperm (abbreviation of spermatozoon, *pl.,* **spermatozoa).** A mature male germ cell.

Spermatids. The four cells formed by the meiotic divisions in spermatogenesis. Spermatids become mature spermatozoa or sperm.

Spermatocyte (sperm mother cell). The cell that undergoes two meiotic divisions (spermatogenesis) to form four spermatids; the *primary* spermatocyte before completion of the first meiotic division; the *secondary* spermatocyte after completion of the first meiotic division.

Spermatogenesis. The process by which maturation of the gametes (sperm) of the male takes place.

Spermatogonium (*pl.,* **spermatogonia).** Primordial male germ cell that may divide by mitosis to produce more spermatogonia. A spermatogonium may enter a growth phase and give rise to a primary spermatocyte.

Spermiogenesis. Formation of sperm from spermatids; the part of spermatogenesis that follows the meiotic divisions of spermatocytes.

Spheroplast. A plant or bacterial cell from which the wall has been removed. See **Protoplast.**

Splicing. The process that covalently joins exon sequences of RNA and eliminates the intervening intron sequences.

Sporophyte. The diploid generation in the life cycle of a plant that produces haploid spores by meiosis.

Standard deviation. A measure of variability in a set of data; the square root of the variance.

Standard error. A measure of variation among a population of means.

Statistic. A value based on a sample or samples of a population from which estimates of a population value or parameter may be obtained.

Stem cell. A mitotically active somatic cell from which other cells arise by differentiation.

Sterility. Inability to produce offspring.

Structural gene. A gene that specifies the synthesis of a polypeptide.

STSs (sequence-tagged sites). Short, unique DNA sequences (usually 200 to 500 bp) that are amplified by PCR and used to link physical maps and genetic maps.

Subspecies. One of two or more morphologically or geographically distinct but interbreeding populations of a species.

Supercoil. A DNA molecule that contains extra twists as a result of overwinding (positive supercoils) or underwinding (negative supercoils).

Suppressor mutation. A mutation that partially or completely cancels the phenotypic effect of another mutation.

Suppressor-sensitive mutant. An organism that can grow when a second genetic factor—a suppressor—is present, but not in the absence of this factor.

Symbiont. An organism living in intimate association with another, dissimilar organism.

Sympatric speciation. The formation of new species by populations that inhabit the same or overlapping geographic regions.

Synapsis. The pairing of homologous chromosomes in the meiotic prophase.

Synaptinemal complex. A ribbon-like structure formed between synapsed homologs at the end of the first meiotic prophase, binding the chromatids along their length and facilitating chromatid exchange.

Syndrome. A group of symptoms that occur together and represent a particular disease.

Synkaryon. A nucleus formed by the fusion of nuclei from two different somatic cells during somatic-cell hybridization.

Synteny. The occurrence of two loci on the same chromosome, without regard to the distance between them.

Taq polymerase. A heat-stable DNA polymerase isolated from the thermophilic bacterium *Thermus aquaticus.*

Target site duplication. A sequence of DNA that is duplicated when a transposable element inserts; usually found at each end of the insertion.

TATA box. A conserved promoter sequence that determines the transcription start site.

Tautomeric shift. The transfer of a hydrogen atom from one position in an organic molecule to another position.

T cell-mediated (cellular) immune response. The synthesis of antigen-specific T cell receptors and the development of killer T cells in response to an encounter of immune system cells with a foreign immunogen.

T cell receptor. An antigen-binding protein that is located on the surfaces of killer T cells and mediates the cellular immune response of mammals. The genes that encode T cell antigens are assembled from gene segments by somatic recombination processes that occur during T lymphocyte differentiation.

T-DNA. The segment of DNA in the Ti plasmid of *Agrobacterium tumefaciens* that is transferred to plant cells and inserted into the chromosomes of the plant.

Telomerase. An enzyme that adds telomere sequences to the ends of eukaryotic chromosomes.

Telomere. The unique structure found at the end of eukaryotic chromosomes.

Telophase. The last stage in each mitotic or meiotic division in which the chromosomes are assembled at the poles of the division spindle.

Temperate phage. A phage (virus)

that invades but may not destroy (lyse) the host (bacterial cell) (cf. **Virulent phage**). However, it may subsequently enter the lytic cycle.

Temperature-sensitive mutant. An organism that can grow at one temperature but not at another.

Template. A pattern or mold. DNA stores coded information and acts as a model or template from which information is copied into complementary strands of DNA or transcribed into messenger RNA.

Template strand. In transcription, the DNA strand that is copied to produce a complementary strand of RNA.

Terminalization. Repelling movement of the centromeres of bivalents in the diplotene stages of the meiotic prophase that tends to move the visible chiasmata toward the ends of the bivalents.

Terminal transferase. An enzyme that adds nucleotides to the 3′ termini of DNA molecules.

Termination signal. In transcription, a nucleotide sequence that specifies RNA chain termination.

Testcross. Backcross to the recessive parental type, or a cross between genetically unknown individuals with a fully recessive tester to determine whether an individual in question is heterozygous or homozygous for a certain allele. It is also used as a test for linkage.

Tetrad. The four cells arising from the second meiotic division in plants (pollen tetrads) or fungi (ascospores). The term is also used to identify the quadruple group of chromatids that is formed by the association of duplicated homologous chromosomes during meiosis.

Tetraploid. An organism whose cells contain four haploid ($4n$) sets of chromosomes or genomes.

Tetrasomic (*noun,* **tetrasome).** Pertaining to a nucleus or an organism with four members of one of its chromosomes whereas the remainder of its chromosome complement is diploid. (Chromosome formula: $2n + 2$).

Tetratype. In fungi, a tetrad of spores that contains four different types; for example, *AB, aB, Ab* and *ab.*

Threshold trait. A trait that is manifested discontinuously but that is a function of underlying continuous genetic and environmental variation.

Thymine. A pyrimidine base found in DNA. The other three organic bases—adenine, cytosine, and guanine—are found in both RNA and DNA, but in RNA, thymine is replaced by uracil.

Ti plasmid. The large plasmid in *Agrobacterium tumefaciens.* It is responsible for the induction of tumors in plants with crown gall disease and is an important vector for transferring genes into plants, especially dicots.

T lymphocytes (T cells). Cells that differentiate in the thymus gland and are primarily responsible for the T cell-mediated or cellular immune response.

Topoisomerase. An enzyme that introduces or removes supercoils from DNA.

Totipotent cell (or **nucleus).** An undifferentiated cell (or nucleus) such as a blastomere that when isolated or suitably transplanted can develop into a complete embryo.

***trans*-acting.** A term describing substances that are diffusable and that can affect spatially separated entities within cells.

***trans* configuration.** See **Repulsion.**

Transcription. Process through which RNA is formed along a DNA template. The enzyme RNA polymerase catalyzes the formation of RNA from ribonucleoside triphosphates.

Transcriptional antiterminator. A protein that prevents RNA polymerase from terminating transcription at specfic transcription-termination sequences.

Transcription factor. A protein that regulates the transcription of genes.

Transcription unit. A segment of DNA that contains transcription initiation and termination signals and is transcribed into one RNA molecule.

Transduction (t). Genetic recombination in bacteria mediated by bacteriophage. Abortive t: Bacterial DNA is injected by a phage into a bacterium, but it does not replicate. Generalized t: Any bacterial gene may be transferred by a phage to a recipient bacterium. Restricted t: Transfer of bacterial DNA by a temperate phage is restricted to only one site on the bacterial chromosome.

Transfection. The uptake of DNA by a eukaryotic cell, followed by the incorporation of genetic markers present in the DNA into the cell's genome.

Transfer RNAs (tRNAs). RNAs that transport amino acids to the ribosomes, where the amino acids are assembled into proteins.

Transformation (genetic). Genetic alteration of an organism brought about by the incorporation of foreign DNA into cells.

Transformation (cancerous). The conversion of eukaryotic cells growing in culture to a state of uncontrolled cell growth (similar to tumor cell growth).

Transgene. A foreign or modified gene that has been introduced into an organism.

Transgenic. A term applied to organisms that have been altered by introducing DNA molecules into them.

Transgressive variation. The appearance in the F_2 (or later) generation of individuals showing more extreme development of a trait than either of the original parents.

trans heterozygote. A heterozygote that contains two mutations arranged in the *trans* configuration—for example, $a\ b^+ / a^+\ b$).

Transition. A mutation caused by the substitution of one purine by another purine or one pyrimidine by another pyrimidine in DNA or RNA.

Translation. Protein (polypeptide) synthesis directed by a specific messenger RNA; occurs on ribosomes.

Translocation. Change in position of a segment of a chromosome to another part of the same chromosome or to a different chromosome.

Transposable genetic element. A DNA element that can move from one location in the genome to another.

Transposase. An enzyme that catalyzes the movement of a DNA sequence to a different site in a DNA molecule.

Transposons. DNA elements that can move ("transpose") from one position in a DNA molecule to another.

Transposon tagging. The insertion of a transposable element into or nearby a gene, thereby marking that gene with a known DNA sequence.

Transversion. A mutation caused by the substitution of a purine for a pyrimidine or a pyrimidine for a purine in DNA or RNA.

Trihybrid. The offspring from homozygous parents differing in three pairs of genes.

Trinucleotide repeats. Tandem repeats of three nucleotides that are present in many human genes. In several cases, these trinucleotide repeats have undergone expansions in copy number that have resulted in inherited diseases.

Trisomic. An otherwise diploid cell or organism that has an extra chromosome of one pair (chromosome formula: $2n + 1$).

Tubulin. The major protein component of the microtubules of eukaryotic cells.

Tumor suppressor gene. A gene whose product is involved in the repression of cell division.

Ultraviolet (UV) radiation. The portion of the electromagnetic spectrum—wavelengths from about 1 to 350 nm—between ionizing radiation and visible light. UV is absorbed by DNA and is highly mutagenic to unicellular organisms and to the epidermal cells of multicellular organisms.

Unequal crossing over. Crossing over between repeated DNA sequences that have paired out of register, creating duplicated and deficient products.

Univalent. An unpaired chromosome at meiosis.

Universality (of the genetic code). The codons have the same meaning, with minor exceptions, in virtually all species.

Uracil. A pyrimidine base found in RNA but not in DNA. In DNA, uracil is replaced by thymine.

Van der Waals interactions. Weak attractions between atoms placed in close proximity.

Variance. A measure of variation in a population; the square of the standard deviation.

Variation. In biology, the occurrence of differences among individuals.

Vector. A plasmid or viral chromosome that may be used to construct recombinant DNA molecules for introduction into living cells.

Velocity density gradient centrifugation. A procedure used to separate macromolecules based on their rate of movement through a density gradient.

Viability. The capability to live and develop normally.

Virulent phage. A phage (virus) that destroys the host (bacterial) cell (cf. **Temperate phage**).

Viscoelastometry. A method to study the physical properties of molecules in solution.

VNTR (variable number tandem repeat). A short DNA sequence that is present as tandem repeats and in highly variable copy number.

Western blot. The transfer of proteins from an electrophoretic gel to a cellulose or nylon membrane by means of an electric force.

Wild type. The customary phenotype or standard for comparison.

Wobble hypothesis. Hypothesis to explain how one tRNA may recognize two codons. The first two bases of the mRNA codon and anticodon pair properly, but the third base in the anticodon has some play (or wobble) that permits it to pair with more than one base.

X chromosome. A chromosome associated with sex determination. In most animals, the female has two, and the male has one X chromosome.

YACs (yeast artificial chromosomes). Linear cloning vectors constructed from essential elements of yeast chromosomes. They can accommodate foreign DNA inserts of 200 to 500 kb in size.

Y chromosome. The partner of the X chromosome in the male of many animal species.

Z-DNA. A left-handed double helix that forms in GC-rich DNA molecules. The Z refers to the zig-zagged paths of the sugar-phosphate backbones in this form of DNA.

Zygonema (*adj.,* zygotene). Stage in meiosis during which synapsis occurs; after the leptotene stage and before the pachytene stage in the meiotic prophase.

Zygote. The cell produced by the union of two mature sex cells (gametes) in reproduction; also used in genetics to designate the individual developing from such a cell.

Answers to Odd-Numbered Questions and Problems

CHAPTER 2

2.1 (a) 0; (b) 0; (c) 0; (d) 0; (e) 0; (f) +; (g) +.
2.3 (a) 200; (b) 50
2.5 Parents: $M//m$ female × $m//m$ male
Gametes: $\frac{1}{2}$ M and $\frac{1}{2}$ m; all m
Offspring: $\frac{1}{2}$ $M//m$ and $\frac{1}{2}$ $m//m$; $\frac{1}{2}$ of the progeny will have myopia.
2.7 Recombination occurring during prophase I results in gametes having different combinations of the alleles of various genes. During anaphase I, the maternal and paternal chromosomes of each pair of homologs segregate at random into each of the daughter nuclei. During anaphase II, the sister chromatids of each chromosome separate at random into each of the two daughter cells. The sister chromatids are not identical due to crossing over.
2.9 Meiosis includes a pairing (synapsis) of homologous maternal and paternal chromosomes. In the cell division that follows, the chromosomes that have previously paired separate. This results in a reduction of chromosome number from $2n$ (diploid) to n (haploid).
2.11 (a) Many plants have male and female parts on the same plant or in the same flower. Unlike animals, plants have a gametophyte stage that consists (in higher plants) of a few cell divisions. (b) The chromosome mechanism is essentially the same during the formation of **gametes in plants and animals.**
2.13 The father is *Aa hh* and the mother is *aa Hh*; There are two classes of sperm: *A h* and *a h*; there are two classes of eggs: *a H* and *a h*. An *a H* egg fertilized by an *A h* sperm will produce an *Aa Hh* zygote.
2.15 See Figure 2.14 (spermatogenesis). The primary spermatocyte is *Aa hh*; following meiosis I and II, half of the mature sperm will carry the *A* and *h* chromosomes.
2.17 Bisexual organisms. Asexual reproduction provides for no genetic variation, except that of rare mutations. Self-fertilization tends toward homozygosity or pure lines. Bisexual reproduction in higher organisms is associated with great hereditary variation through segregation, independent assortment, and crossing over between maternal and paternal chromosomes.
2.19 For each pair of chromosomes, there are 2 different classes of gametes possible. Therefore, the number of different classes of gametes is $2 \times 2 \times 2 \times 2 = 2^4 = 16$.
2.21 Four.
2.23 Haploid cells differentiate into gametes that fuse to produce a diploid zygote. The diploid zygote then undergoes meiosis to produce four haploid cells, each of which divides mitotically to produce the haploid individual. A haploid individual cannot undergo meiosis directly because its chromosomes do not exist as homologous pairs.

CHAPTER 3

3.1 (a) All tall; (b) 3/4 tall, 1/4 dwarf; (c) all tall; (d) 1/2 tall, 1/2 dwarf.
3.3 The data suggest that coat color is controlled by a single gene with two alleles, *C* (gray) and *c* (albino), and that *C* is dominant over *c*. On this hypothesis, the crosses are: gray (*CC*) × albino (*cc*) → F_1 gray (*Cc*); $F_1 \times F_1$ → 3/4 gray (2 *CC*: 1 *Cc*), 1/4 albino (*cc*). The expected results in the F_2 are 203 gray, 67 albino. To compare the observed and expected results, compute χ^2 with one degree of freedom: $(198 - 203)^2/203 + (67 - 72)^2/72 = 0.470$, which is not significant at the 5% level. Thus, the results are consistent with the hypothesis.
3.5 (a) Checkered, red (*CC BB*) × plain, brown (*cc bb*) → F_1 all checkered, red (*Cc Bb*); (b) F_2 progeny: 9/16 checkered, red (*C- B-*), 3/16 plain, red (*cc B-*), 3/16 checkered, brown (*C- bb*), 1/16 plain, brown (*cc bb*).
3.7 Among the F_2 progeny with long, black fur, the genotypic ratio is 1 *BB RR*: 2 *BB Rr*: 2 *Bb RR*: 4 *Bb Rr*; thus, 1/9 of the rabbits with long, black fur are homozygous for both genes.
3.9

	F_1 gametes	F_2 genotypes	F_2 phenotypes
(a)	2	3	2
(b)	$2 \times 2 = 4$	$3 \times 3 = 9$	$2 \times 2 = 4$
(c)	$2 \times 2 \times 2 = 8$	$3 \times 3 \times 3 = 27$	$2 \times 2 \times 2 = 8$
(d)	2^n	3^n	2^n, where n is the number of genes

3.11 On the hypothesis, the expected number in each class is 27.5; χ^2 with three degrees of freedom is calculated as $(31 - 27.5)^2/27.5 + (26 - 27.5)^2/27.5 + (27 - 27.5)^2/27.5 + (26 - 27.5)^2/27.5 = 0.618$, which is not significant at the 5% level. Thus, the results are consistent with the hypothesis of two independently assorting genes, each segregating two alleles.
3.13 $\chi^2 = (30 - 25)^2/25 + (20 - 25)^2/25 = 2$, which is less than 3.84, the 5 percent critical value for a chi-square statistic with one degree of freedom; consequently, the observed segregation ratio is consistent with the expected ratio of 1:1.
3.15 Half the children from *Aa* × *aa* matings would be albino. In a family of three children, the chance that one will be normal and two albino is $3 \times (1/2)^1 \times (1/2)^2 = 3/8$.
3.17 Man (*Cc ff*) × woman (*cc Ff*). (a) *cc ff*, $(1/2) \times (1/2) = 1/4$; (b) *Cc ff*, $(1/2) \times (1/2) = 1/4$; (c) *cc Ff*, $(1/2) \times (1/2) = 1/4$; (d) *Cc Ff*, $(1/2) \times (1/2) = 1/4$.
3.19 $(1/2)^3 = 1/8$
3.21 $(20/64) + (15/64) + (6/64) + (1/64) = 42/64$
3.23 (a) $(1/2) \times (1/4) = 1/8$; (b) $(1/2) \times (1/2) \times (1/4) = 1/16$; (c) $(2/3) \times (1/4) = 1/6$; (d) $(2/3) \times (1/2) \times (1/2) \times (1/4) = 1/24$
3.25 For III-1 × III-2, the chance of an affected child is 1/2. For IV-2 × IV-3, the chance is zero.
3.27 (a) $(1/4)^3 = 1/64$; (b) $(1/2)^3 = 1/8$; (c) $3 \times (1/2)^1 \times (1/2)^2 = 3/8$; (d) 1 − probability that the offspring is not homozygous for the recessive allele of any gene = $1 - (3/4)^3 = 37/64$.

CHAPTER 4

4.1 M and MN.
4.3 One gene with four alleles. (a) purple: c^pc^p, c^pc^b, c^pc^t, c^pc^w; (b) blue: c^bc^b, c^bc^t, c^bc^w; (c) turquoise: c^tc^t, c^tc^w; (d) light-blue: c^tc^w; (e) white: c^wc^w.
4.5 (a) S^1S^2, S^1S^3, S^2S^3; (b) S^1S^3, S^1S^4, S^2S^3, S^2S^4; (c) none; (d) S^3S^5, S^3S^6, S^4S^5, S^4S^6.
4.7 No. The woman must be I^OI^O; if her mate is I^AI^B; they could not have an I^OI^O child.
4.9 The woman is I^OI^O L^ML^M; the man is I^AI^B L^ML^N; the blood types of the children will be A and M, A and MN, B and M, and B and MN, all equally likely.
4.11 The individuals III-4 and III-5 must be homozygous for recessive mutations in different genes; that is, one is *aa BB* and the other is *AA bb*; none of their children is deaf because all of them are heterozygous for both genes (*Aa Bb*).
4.13 No. The test for allelism cannot be performed with dominant mutations.
4.15 The mother is *Bb* and the father is *bb*. The chance that a daughter is *Bb* is 1/2.

(a) The chance that the daughter will have a bald son is $(1/2) \times (1/2) = 1/4$. (b) The chance that the daughter will have a bald daughter is zero.

4.17 (a) 3/4 walnut, 1/4 rose; (b) 1/2 walnut, 1/2 pea; (c) 3/8 walnut, 3/8 rose, 1/8 pea, 1/8 single; (d) 1/2 rose, 1/2 single.

4.19 12/16 white, 3/16 yellow, 1/16 green.

4.21 9/16 dark red (wild-type), 3/16 brownish-purple, 3/16 bright red, 1/16 white.

4.23 9 black: 39 gray: 16 white.

4.25 (a) purple × red; (b) proportion white (*aa*) = 1/4; (c) proportion red (*A- B- C- dd*) = (3/4)(3/4)(3/4)(1/2) = 27/128, proportion white (*aa*) = 1/4 = 32/128, proportion blue (*A- B- cc Dd*) = (3/4)(3/4)(1/4)(1/2) = 9/128.

4.27 (a) Because the F_2 segregation is approximately 9 red: 7 white, flower color is due to epistasis between two independently assorting genes: red = *A-B-* and white = *aa B-*, *A-bb*, or *aa bb*. (b) colorless precursor —*A*→ colorless product —*B*→ red pigment.

CHAPTER 5

5.1 The phenotypes are determined by the number of primed alleles in the genotypes. This is a classic dihybrid cross with no dominance; the phenotypic frequencies in the F_2 are: 1/16 white (0 primed alleles), 4/16 light red (1 primed allele), 6/16 intermediate red (2 primed alleles), 4/16 darker red (3 primed alleles), 1/16 darkest red (4 primed alleles).

5.3 Three genes are segregating in the series of crosses. The parental cross was *AA BB CC* (white) × *A'A' B'B' C'C'* (red), and the F_1 was *AA' BB' CC'*. In the F_2, 1/64 of the progeny are expected to be true-breeding white plants (genotype *AA BB CC*). Thus, in a sample of 78 F_2 plants, $(1/64) \times 78 = 1.2$ plants would be expected to be in this class—a number small enough to have been missed by chance.

5.5 (a) 4; (b) 3 inches; (c) frequency of 1/4.

5.7 Because $\Sigma(X_i - \text{mean}) = 0$

5.9 $3.17/6.08 = 0.52$

5.11 V_e is estimated by the average of the variances of the inbreds: 9.4 cm². V_g is estimated by the difference between the variances of the randomly pollinated population and the inbreds: $(26.4 - 9.4) = 7.4$ cm². The broad-sense heritability is $H^2 = V_g/V_t = 7.4/26.4 = 0.64$.

5.13 Broad-sense heritability must be greater than narrow-sense heritability because $H^2 = V_g/V_t > V_a/V_t = h^2$.

5.15 $(15 - 12)(0.3) + 12 = 12.9$ bristles.

5.17 $h^2 = R/S = (12.5 - 10)/(15 - 10) = 0.5$; selection for increased growth rate should be effective.

5.19 $(40)(0.3)(10) + 2000 = 2120$ μg

5.21 $F_I = (1/2)^3(1 + F_C) = 0.25$; thus, $F_C = 1$

5.23 From the pedigree below, the inbreeding coefficient is $(1/2)^3(1 + F_C) + (1/2)^4(1 + F_B) = 3/8$ because $F_B = F_C = 1$.

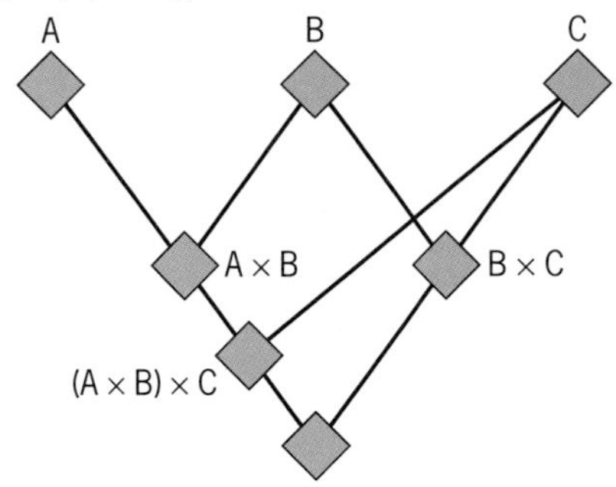

CHAPTER 6

6.1 The male-determining sperm carries a Y chromosome; the female-determining sperm carries an X chromosome.

6.3 All the daughters will be green and all the sons will be rosy.

6.5 XX is female, XY is male, XXY is female, XXX is female (but barely viable), XO is male (but sterile).

6.7 No. Defective color vision is caused by an X-linked mutation. The son's X chromosome came from his mother, not his father.

6.9 The risk for the child is P(mother is *C/c*) × P(mother transmits *c*) × P(child is male) = $(1/2) \times (1/2) \times (1/2) = 1/8$; if the couple has already had a child with color blindness, P(mother is *C/c*) = 1, and the risk for each subsequent child is 1/4.

6.11 Each of the rare vermilion daughters must have resulted from the union of an X(*v*) X(*v*) egg with a Y-bearing sperm. The diplo-X eggs must have originated through nondisjunction of the X chromosomes during oogenesis in the mother. However, we cannot determine if the nondisjunction occurred in the first or the second meiotic division.

6.13 1/8

6.15 Female.

6.17 Male.

6.19 (a) 1/2 X^{bb} X^{bb} females, 1/2 X^{bb} Y^+ wild-type males; (b) 1/2 X^+ X^{bb} wild-type females, 1/2 X^{bb} Y^{bb} bobbed males; (c) 1/4 X^+ X^+ wild-type females, 1/4 X^+ X^{bb} wild-type females, 1/4 X^+ Y^{bb} wild-type males, 1/4 X^{bb} Y^{bb} bobbed males; (d) 1/4 X^+ X^{bb} wild-type females, 1/4 X^{bb} X^{bb} bobbed females, 1/4 X^+ Y^+ wild-type males, 1/4 X^{bb} Y^+ wild-type males

6.21 *Drosophila* does not achieve dosage compensation by inactivating one of the X chromosomes in females.

6.23

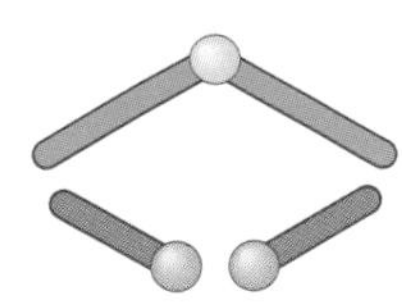

CHAPTER 7

7.1 Use one of the banding techniques.

7.3 In allotetraploids, each member of the different sets of chromosomes can pair with a homologous partner during prophase I and then disjoin during anaphase I. In triploids, disjunction is irregular because homologous chromosomes associate during prophase I by forming either bivalents and univalents or by forming trivalents.

7.5 The fertile plant is an allotetraploid with 7 pairs of chromosomes from species A and 9 pairs of chromosomes from species B; the total number of chromosomes is $(2 \times 7) + (2 \times 9) = 32$.

7.7 XX is female, XY is male, XO is female (but sterile), XXX is female, XXY is male (but sterile), XYY is male.

7.9 The fly is a gynandromorph, that is, a sexual mosaic. The yellow tissue is X(*y*)/O and the gray tissue is X(*y*)/X(+). This mosaicism must have arisen through loss of the X chromosome that carried the wild-type allele, presumably during one of the early embryonic cleavage divisions.

7.11 The daughter with Turner and Hunter syndrome in family A must have received her single X chromosome from her mother, who is heterozygous for the mutation causing Hunter syndrome. The daughter did not receive a sex chromosome from her father because sex chromosome nondisjunction must have occurred during meiosis in his germline. The son with Klinefelter syndrome in family B is karyotypically XXY, and both of his X chromosomes carry the mutant allele for Hunter syndrome. This individual must have received two mutant X chromosomes from his heterozygous mother due to X chromosome nondisjunction during the second meiotic division in her germline. The daughter with Hunter syndrome in family C is karyotypically XX, and both of her X chromosomes carry the mutant allele for Hunter syndrome. This individual received the two mutant X chromosomes from her heterozygous mother through nondisjunction during the second meiotic division in the mother's germline. Furthermore, because the daughter did not receive a sex chromosome from her father, sex chromosome nondisjunction must have occurred during meiosis in his germline too.

7.13 XYY men would produce more children with sex chromosome abnormalities because their three sex chromosomes will disjoin irregularly during meiosis. This irregular disjunction will produce a variety of aneuploid gametes, including the XY, YY, XYY, and nullo sex chromosome constitutions.

7.15 (a) Deletion:

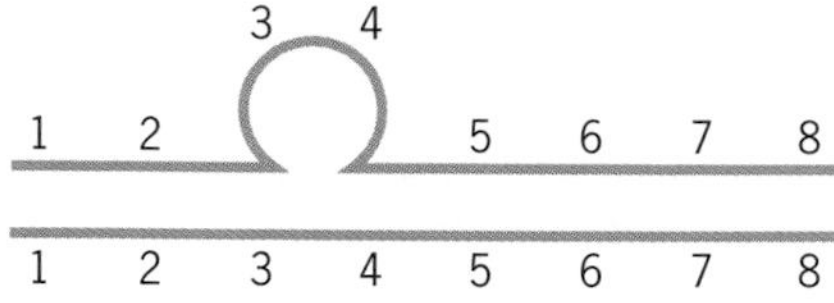

(b) Duplication:

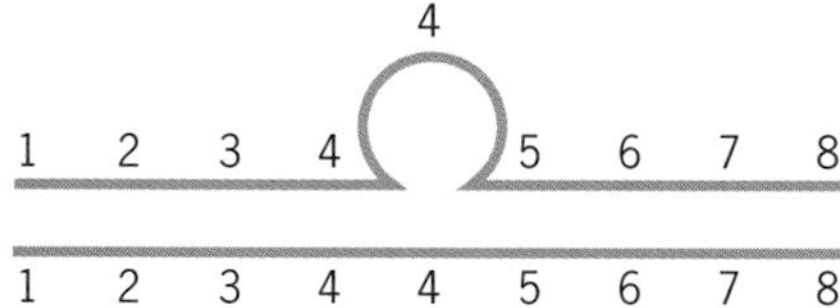

(c) A terminal inversion:

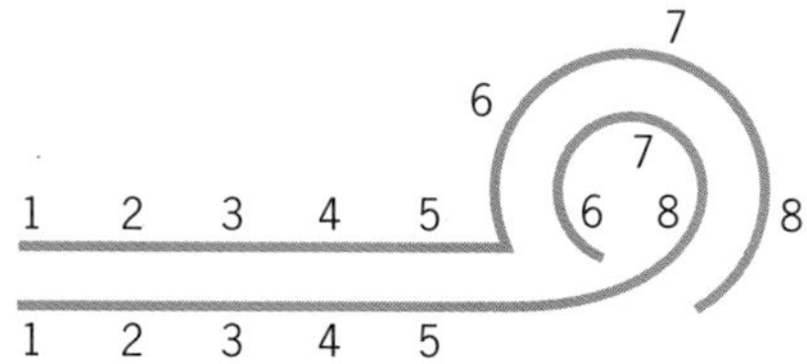

7.17

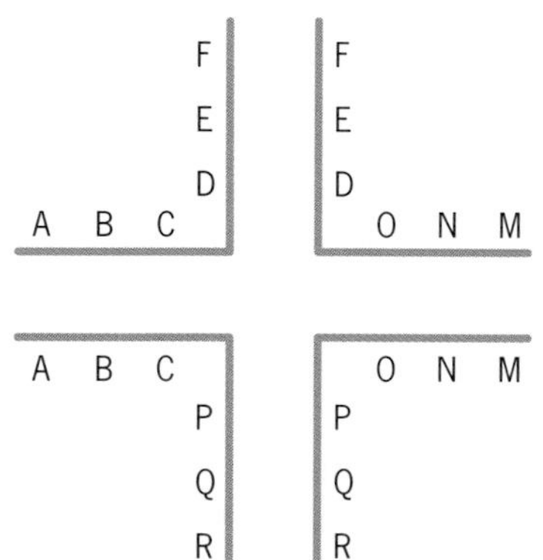

7.19 The boy carries a translocation between chromosome 21 and another chromosome, say chromosome 14. He also carries a normal chromosome 21 and a normal chromosome 14. The boy's sister carries the translocation, one normal chromosome 14, and two normal copies of chromosome 21.

7.21 All the daughters will be yellow-bodied and all the sons will be white-eyed.

7.23 The three populations are related by a series of inversions:

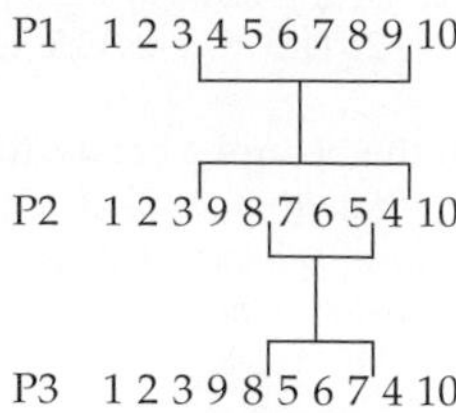

7.25 The mother is heterozygous for a reciprocal translocation between the long arms of the large and small chromosomes; a piece from the long arm of the large chromosome has been broken off and attached to the long arm of the short chromosome. The child has inherited the rearranged large chromosome and the normal small chromosome from the mother. Thus, because the rearranged large chromosome is deficient for some of its genes, the child is hypoploid.

7.27 The sons will have bright red eyes because they will inherit the Y chromosome with the bw^+ allele from their father. The daughters will have white eyes because they will inherit an X chromosome from their father.

7.29 XX zygotes will develop into males because one of their X chromosomes carries the *SRY* gene that was translocated from the Y chromosome. XY zygotes will develop into females because their Y chromosome has lost the *SRY* gene.

CHAPTER 8

8.1 If Mendel had known of the existence of chromosomes, he would have realized that the number of factors determining traits exceeds the number of chromosomes, and he would have concluded that some factors must be linked on the same chromosome. Thus, Mendel would have revised the Principle of Independent Assortment to say that factors on different chromosomes (or far apart on the same chromosome) are inherited independently.

8.3 No. The genes *a* and *d* could be very far apart on the same chromosome—so far apart that they recombine freely, that is, 50 percent of the time.

8.5 $(1/19)^2 = 0.0027$

8.7 (a) Cross: $a^+ b^+/a^+ b^+ \times a\ b/a\ b$
Gametes: $a^+ b^+$ from one parent, $a\ b$ from the other
F_1: $a^+ b^+/a\ b$

(b) 40% $a^+ b^+$, 40% $a\ b$, 10% $a^+ b$, 10% $a\ b^+$

(c) F_2 from testcross: 40% $a^+ b^+/a\ b$, 40% $a\ b/a\ b$, 10% $a^+ b/a\ b$, 10% $a\ b^+/a\ b$

(d) Coupling linkage phase

(e) F_2 from intercross:

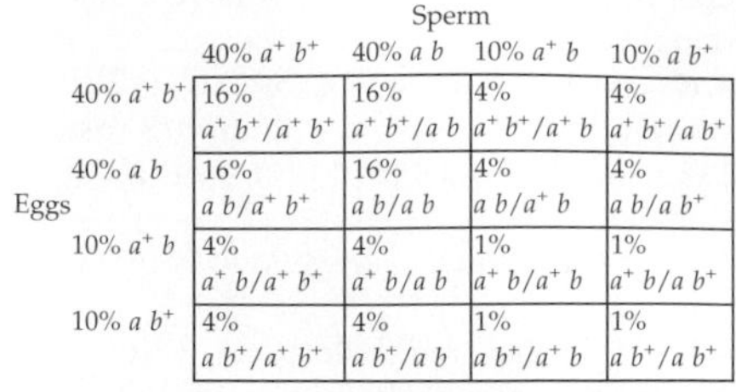

Eggs \ Sperm	40% $a^+ b^+$	40% $a\ b$	10% $a^+ b$	10% $a\ b^+$
40% $a^+ b^+$	16% $a^+ b^+/a^+ b^+$	16% $a^+ b^+/a\ b$	4% $a^+ b^+/a^+ b$	4% $a^+ b^+/a\ b^+$
40% $a\ b$	16% $a\ b/a^+ b^+$	16% $a\ b/a\ b$	4% $a\ b/a^+ b$	4% $a\ b/a\ b^+$
10% $a^+ b$	4% $a^+ b/a^+ b^+$	4% $a^+ b/a\ b$	1% $a^+ b/a^+ b$	1% $a^+ b/a\ b^+$
10% $a\ b^+$	4% $a\ b^+/a^+ b^+$	4% $a\ b^+/a\ b$	1% $a\ b^+/a^+ b$	1% $a\ b^+/a\ b^+$

Summary of phenotypes:

a^+ and b^+	66%
a^+ and b	9%
a and b^+	9%
a and b	16%

8.9 Coupling heterozygotes $a^+ b^+/a\ b$ would produce the following gametes: 30% $a^+ b^+$, 30% $a\ b$, 20% $a^+ b$, 20% $a\ b^+$; repulsion heterozygotes $a^+ b/a\ b^+$ would produce the following gametes: 30% $a^+ b$, 30% $a\ b^+$, 20% $a^+ b^+$, 20% $a\ b$. In each case, the frequencies of the testcross progeny would correspond to the frequencies of the gametes.

8.11 Yes. Recombination frequency = $(24 + 26)/(126 + 24 + 26 + 124) = 0.167$. Cross:

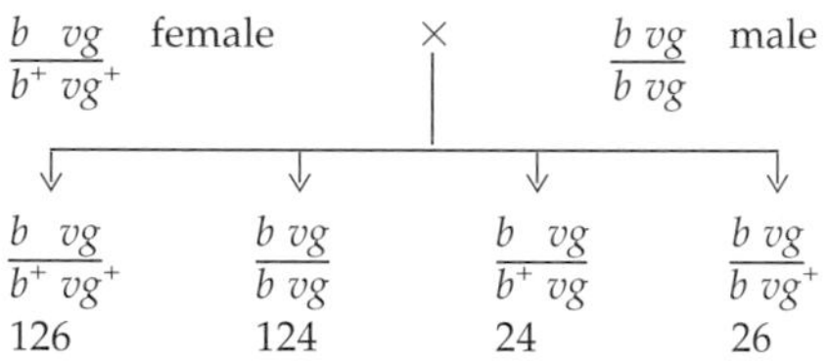

8.13 Yes. Recombination frequency is estimated by the frequency of black offspring among the colored offspring: $34/(66 + 34) = 0.34$. Cross:

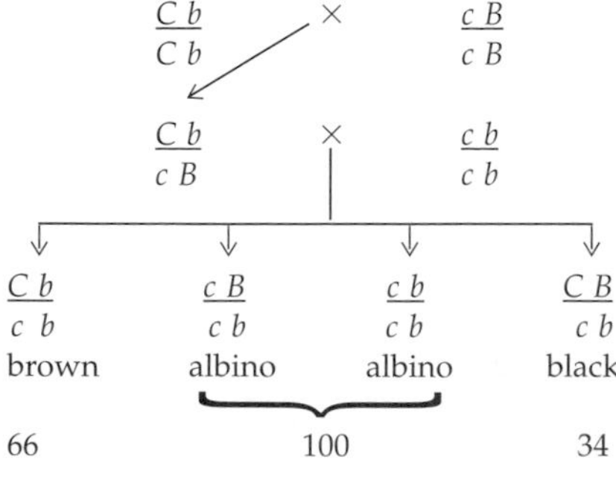

8.15 Because the two chromosomes assort independently, the genetic makeup of the gametes (and, therefore, of the backcross progeny) can be obtained from the following table.

		Chromosome 3 in gametes			
		$c\ d$ 0.425	$c^+\ d^+$ 0.425	$c\ d^+$ 0.075	$c^+\ d$ 0.075
Chromosome 2 in gametes	$a\ b$ 0.40	$a\ b\ c\ d$ 0.17	$a\ b\ c^+\ d^+$ 0.17	$a\ b\ c\ d^+$ 0.03	$a\ b\ c^+\ d$ 0.03
	$a^+\ b^+$ 0.40	$a^+\ b^+\ c\ d$ 0.17	$a^+\ b^+\ c^+\ d^+$ 0.17	$a^+\ b^+\ c\ d^+$ 0.03	$a^+\ b^+\ c^+\ d$ 0.03
	$a^+\ b$ 0.10	$a^+\ b\ c\ d$ 0.0425	$a^+\ b\ c^+\ d^+$ 0.0425	$a^+\ b\ c\ d^+$ 0.0075	$a^+\ b\ c^+\ d$ 0.0075
	$a\ b^+$ 0.10	$a\ b^+\ c\ d$ 0.0425	$a\ b^+\ c^+\ d^+$ 0.0425	$a\ b^+\ c\ d^+$ 0.0075	$a\ b^+\ c^+\ d$ 0.0075

8.17 (a) The F_1 females, which are *cn* vg^+/cn^+ *vg*, produce four types of gametes: 45% *cn* vg^+, 45% cn^+ *vg*, 5% cn^+ vg^+, 5% *cn vg*. (b) 45% cinnabar eyes, normal wings; 45% reddish-brown eyes, vestigial wings; 5% reddish-brown eyes, normal wings; 5% cinnabar eyes, vestigial wings.

8.19 In the enumeration below, classes 1 and 2 are parental types, classes 3 and 4 result from a single crossover between *Pl* and *Sm*, classes 5 and 6 result from a single crossover between *Sm* and *Py*, and classes 7 and 8 result from a double crossover, with one of the exchanges between *Pl* and *Sm* and the other between *Sm* and *Py*.

Class	Phenotypes	(a) Frequency with no interference	(b) Frequency with complete interference
1	purple, salmon, pigmy	0.405	0.40
2	green, yellow, normal	0.405	0.40
3	purple, yellow, normal	0.045	0.05
4	green, salmon, pigmy	0.045	0.05
5	purple, salmon, normal	0.045	0.05
6	green, yellow, pigmy	0.045	0.05
7	purple, yellow, pigmy	0.005	0
8	green, salmon, normal	0.005	0

8.21 The double crossover classes, which are the two that were not observed, establish that the gene order is *y*—*w*—*ec*. Thus, the F_1 females had the genotype *y w ec*/+ + +. The distance between *y* and *w* is estimated by the frequency of recombination between these two genes: (8 + 7)/1000 = 0.015; similarly, the distance between *w* and *ec* is (18 + 23)/1000 = 0.041. Thus, the genetic map for this segment of the X chromosome is *y*—1.5 cM—*w*—4.1 cM—*ec*.

8.23 (a) Two of the classes (the parental types) vastly outnumber the other six classes (recombinant types); (b) *st* + +/+ *ss e*; (c) *st*—*ss*—*e*; (d) [(145 + 122) × 1 + (18) × 2]/1000 = 30.3 cM; (e) (122 + 18)/1000 = 14.0 cM; (f) (0.018)/(0.163 × 0.140) = 0.789. (g) *st* + +/+ *ss e* females × *st ss e*/*st ss e* males → 2 parental classes and 6 recombinant classes.

8.25 Ignore the female progeny and base the map on the male progeny. The parental types are + + *z* and *x y* +. The two missing classes (+ *y* + and *x* + *z*) must represent double crossovers; thus, the gene order is *y*—*x*—*z*. The distance between *y* and *x* is (32 + 27)/1000 = 5.9 cM and that between *x* and *z* is (31 + 39)/1000 = 7.0 cM. Thus, the map is *y*—5.9 cM—*x*—7.0 cM—*z*. The coefficient of coincidence is zero.

8.27 $(P/2)^2$

8.29 From the parental classes, + + *c* and *a b* +, the heterozygous females must have had the genotype + + *c*/*a b* +. The missing classes, + *b* + and a + *c*, which would represent double crossovers, establish that the gene order is *b*—*a*—*c*. The distance between *b* and *a* is (96 + 110)/1000 = 20.6 cM and that between *a* and *c* is (65 + 75)/1000 = 14.0 cM. Thus, the genetic map is *b*—20.6 cM—*a*—14.0 cM—*c*.

8.31 5.4 chiasmata.

8.33 M2 carries an inversion that suppresses recombination in the chromosome.

CHAPTER 9

9.1 PD >> NPD, so the genes are linked; the distance between the genes is estimated as [(1/2) × 23 + 3]/48 = 30 cM.

9.3 The distance is half the frequency of second division segregation asci: (1/2) × (84/200) = 21 cM.

9.5 Genes *b* and *c* are linked and are 17 cM apart. Gene *a* is located on a different chromosome.

9.7 (a) 2 crossovers; (b) one exchange between *x* and *y*, the other between *y* and *z*; (c) 4-strand double crossover.

9.9 (a) 8 cM; (b) 21.5 cM; (c) the *arg* and *w* loci are located in different arms of a chromosome.

9.11 Band 7.

9.13 *C h*/*c H*.

9.15 *A B c*/*a b C*.

9.17 *G6PD*, *HPRT*, and *PGK* are in Xq distal to the translocation breakpoint; *NP* is in 14q proximal to the translocation breakpoint.

9.19 The fragile site and the *HPA* gene are separated by 8 cM.

9.21 *A*—10 cM—*B*—6 cM—*C*.

9.23 The gene for E1 is on chromosome 8, that for E2 is on the short arm of chromosome 2, that for E3 is on the short arm of chromosome 5, and that for E4 is on the long arm of chromosome 17.

CHAPTER 10

10.1 (a) Griffith's *in vivo* experiments demonstrated the occurrence of transformation in pneumococcus. They provided no indication as to the molecular basis of the transformation phenomenon. Avery and colleagues carried out *in vitro* experiments, employing biochemical analyses to demonstrate that transformation was mediated by DNA. (b) Griffith showed that a transforming substance existed; Avery et al. defined it as DNA. (c) Griffith's experiments did not include any attempt to characterize the substance responsible for transformation. Avery et al. isolated DNA in "pure" form and demonstrated that it could mediate transformation.

10.3 Purified DNA from Type III cells was shown to be sufficient to transform Type II cells. This occurred in the absence of any dead Type III cells.

10.5 DNA contains phosphorus (normally ^{31}P) but no sulfur; it can be labeled with ^{32}P. Proteins contain sulfur (normally ^{32}S) but usually no phosphorus; they can be labeled with ^{35}S.

10.7 (a) The ladderlike pattern was known from X-ray diffraction studies. Chemical analyses had shown that a 1:1 relationship existed between the organic bases adenine and thymine and between cytosine and guanine. Physical data concerning the length of each spiral and the stacking of bases were also available. (b) Watson and Crick developed the model of a double helix, with the rigid strands of sugar and phosphorus forming spirals around an axis, and hydrogen bonds connecting the complementary bases in nucleotide pairs.

10.9 (a) 400,000; (b) 20,000; (c) 400,000; (d) 68,000 nm.

10.11 DNA has one atom less of oxygen than RNA in the sugar part of the molecule. In DNA, thymine replaces the uracil that is present in RNA. (In certain bacteriophages, DNA-containing uracil is present.) DNA is most frequently double-stranded, but bacteriophages such as ϕX174 contain single-stranded DNA. RNA is most frequently single-stranded. Some viruses, such as the Reoviruses, however, contain double-stranded RNA chromosomes.

10.13 No. TMV RNA is single-stranded. Thus, the base-pair stoichiometry of DNA does not apply.

10.15 (a) double-stranded DNA; (b) single-stranded DNA; (c) single-stranded RNA.

10.17 The value of T_m increases with the GC content because GC base pairs, connected by three hydrogen bonds, are stronger than AT base pairs connected by two hydrogen bonds.

10.19 (1) The nucleosome level; the core containing an octamer of histones plus 146 nucleotide pairs of DNA arranged as $1\frac{3}{4}$ turns of a supercoil (see Figure 10.26), yielding an approximately 10 nm diameter spherical body; or juxtaposed, a roughly 10 nm diameter fiber. (2) The 30 nm fiber observed in condensed mitotic and meiotic chromosomes; it appears to be formed by coiling or folding the 10-nm nucleosome fiber. (3) The highly condensed mitotic and meiotic chromosomes (for example, metaphase chromosomes); the tight folding or coiling maintained by a "scaffold" composed of nonhistone chromosomal proteins (see Figure 10.29).

10.21 In the diploid nucleus of *D. melanogaster*, 10^6 nucleosomes would be present; these would contain 2×10^6 molecules of each histone, H2a, H2b, H3, and H4.

10.23 It indicates that highly repetitive DNA sequences do not contain structural genes specifying RNA and polypeptide gene products.

10.25 Interphase. Chromosomes are for the most part metabolically inactive (very little transcription) during the various stages of condensation in mitosis and meiosis.

10.27 Viscoelastometry is a procedure used to measure the viscosity of molecules in solution. In addition, viscoelastometric methods can be used to estimate the sizes of the largest DNA molecules present in aqueous solutions. By using viscoelastometry, scientists have obtained evidence that indicates all of the DNA present in chromosomes of eukaryotes exists as giant, "chromosome-size" DNA molecules (one huge DNA molecule per chromosome). These data eliminated early models of chromosome structure with multiple DNA molecules joined end-to-end by protein or RNA "linkers."

10.29 Nonhistone chromosomal proteins. The "scaffold" structures of metaphase chromosomes can be observed by light microscopy after removal of the histones by differential extraction procedures.

10.31 The rate of DNA renaturation is proportional to the square of the concentration of single strands (C^2), because renaturation requires a collision between and a reassociation of two complementary molecules (single strands) that are present in the reaction mixture in *identical concentrations*. These single strands were originally present in the same double helix. Thus, the reaction depends on the concentrations of both strands (with both concentrations equal, therefore, $C \times C = C^2$), whereas the rates of most other bimolecular reactions are a function of the concentrations (often different, therefore, $C_1 \times C_2$) of both reacting substances. The fact that DNA renaturation rates are proportional to C^2 is thus a direct consequence of the double-helix structure of DNA with each double helix composed of two complementary single strands.

10.33 (a) One μg of human DNA will contain, on average, 3.04×10^5 copies of the genome. Using an average molecular weight per nucleotide pair of 660, the "molecular" weight of the entire human genome is 1.98×10^{12} ($3 \times 10^9 \times 660$). Thus, 1.98×10^{12} grams (1 "mole" = number of grams equivalent to the "molecular" weight) of human DNA will contain, on average, 6.02×10^{23} molecules [Avogadro's number = number of molecules (here, copies of the genome) present in one "mole" of a substance]. One gram will contain, on average, $3.04 \times 10^{11} = \frac{6.02 \times 10^{23}}{1.98 \times 10^{12}}$ copies of the genome; thus, 1 μg will contain, on average, 3.04×10^5 copies of the human genome. (b) One copy of the human genome weighs approximately 3.3×10^{-12} g = $\frac{1.98 \times 10^{12} \text{ g per "mole"}}{6.02 \times 10^{23} \text{ molecules per "mole"}}$ or 3.3×10^{-6} μg. (c) By analogous calculations, 1 μg of *Arabidopsis thaliana* DNA contains, on average, 1.18×10^7 copies of the genome. (d) Similarly, one copy of the *A. thaliana* genome weighs approximately 8.4×10^{-8} μg. (e) In carrying out molecular analyses of the structures of genomes, geneticists frequently need to know how many copies of a genome are present, on average, in a given quantity of DNA.

10.35 Number of copies is inversely proportional to $C_ot_{1/2}$ value. (a) $10^3/(2 \times 10^{-1}) = 5 \times 10^3$; (b) $10^3/(5 \times 10^{-3}) = 2 \times 10^5$

CHAPTER 11

11.1 (a) (i) One-half of the DNA molecules with ^{15}N in both strands and $\frac{1}{2}$ with ^{14}N in both strands; (ii) all DNA molecules with one strand containing ^{15}N and the complementary strand containing ^{14}N; (iii) all DNA molecules with both strands containing roughly equal amounts of ^{15}N and ^{14}N. (b) (i) $\frac{1}{4}$ of the DNA molecules with ^{15}N in both strands and $\frac{3}{4}$ with ^{14}N in both strands; (ii) $\frac{1}{2}$ of the DNA molecules with one strand containing ^{15}N and the complementary strand containing ^{14}N and the other $\frac{1}{2}$ with ^{14}N in both strands; (iii) all DNA molecules with both strands containing about $\frac{1}{4}$ ^{15}N and $\frac{3}{4}$ ^{14}N. See Figure 11.2.

11.3 (a) Both $3' \longrightarrow 5'$ and $5' \longrightarrow 3'$ exonuclease activities. (b) The $3' \longrightarrow 5'$ exonuclease "proofreads" the nascent DNA strand during its synthesis. If a mismatched base pair occurs at the 3′-OH end of the primer, the $3' \longrightarrow 5'$ exonuclease removes the incorrect terminal nucleotide before polymerization proceeds again. The $5' \longrightarrow 3'$ exonuclease is responsible for the removal of RNA primers during DNA replication and functions in pathways involved in the repair of damaged DNA (see Chapter 14). (c) Yes, both exonuclease activities appear to be very important. Without the $3' \longrightarrow 5'$ proofreading activity during replication, an intolerable mutation frequency would occur. The $5' \longrightarrow 3'$ exonuclease activity is essential to the survival of the cell. Conditional mutations that alter the $5' \longrightarrow 3'$ exonuclease activity of DNA polymerase I are lethal to the cell under conditions where the exonuclease is nonfunctional.

11.5 If nascent DNA is labeled by exposure to ^3H-thymidine for very short periods of time, continuous replication predicts that the label would be incorporated into chromosome-sized DNA molecules, whereas discontinuous replication predicts that the label would first appear in small pieces of nascent DNA (prior to covalent joining, catalyzed by polynucleotide ligase).

11.7

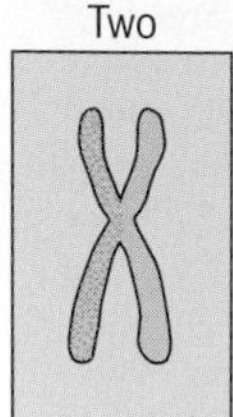

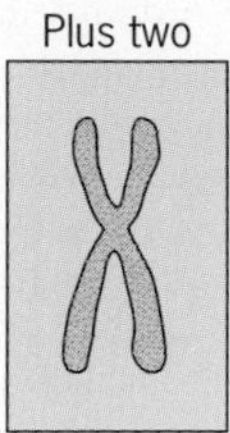

For both the large and small chromosomes

11.9 That DNA replication was unidirectional rather than bidirectional. As the intracellular pools of radioactive ^3H-thymidine are gradually diluted after transfer to nonradioactive medium, less and less ^3H-thymidine will be incorporated into DNA at each replicating fork. This will produce autoradiograms with tails of decreasing grain density at each growing point. Since such tails appear at only one end of each track, replication must be unidirectional. Bidirectional replication would produce such tails at both ends of an autoradiographic track (see Figure 11.30).

11.11 DNA polymerases α, β, δ, and ϵ are located in the nuclei of cells; polymerase γ is located in mitochondria and chloroplasts. Current evidence suggests that polymerases α and δ are both required for the replication of nuclear DNA. Polymerase δ is thought to catalyze the continuous synthesis of the leading strand, and polymerase α is believed to catalyze discontinuous synthesis of the lagging strand because it contains the primase activity required for the repeated initiation of "Okazaki fragments." Polymerase γ presumably catalyzes replication of organellar chromosomes. Polymerases β and ϵ function in DNA repair pathways, like DNA polymerase I of *E. coli*.

11.13 Sucrose velocity density gradient centrifugation is the standard technique for separating DNA molecules in this size range. Pulsed-field gel electrophoresis (Chapter 10) could also be used.

11.15 (a) DNA gyrase; (b) primase; (c) the $5' \longrightarrow 3'$ exonuclease activity of DNA polymerase I; (d) the $5' \longrightarrow 3'$ polymerase activity of DNA polymerase III; (e) the $3' \longrightarrow 5'$ exonuclease activity of DNA polymerase III.

11.17 In eukaryotes, the rate of DNA synthesis at each replication fork is about 2,500 to 3,000 nucleotide pairs per minute. Large eukaryotic chromosomes often contain 10^7 to 10^8 nucleotide pairs. A single replication fork could not replicate the giant DNA in one of these large chromosomes fast enough to permit the observed cell generation times.

11.19 No. *E. coli* strains carrying *polA* mutations that eliminate the 3′ ⟶ 5′ exonuclease activity of DNA polymerase I will exhibit unusually high mutation rates.

11.21 Rolling-circle replication begins when an endonuclease cleaves one strand of a circular DNA double helix. This cleavage produces a free 3′-OH on one end of the cut strand, allowing it to function as a primer. The discontinuous synthesis of the lagging strand requires the *de novo* initiation of each Okazaki fragment, which requires DNA primase activity.

11.23 DNA helicase unwinds the DNA double helix, and single-strand DNA-binding protein coats the unwound strands, keeping them in an extended state. DNA gyrase catalyzes the formation of negative supercoiling in *E. coli* DNA, and this negative supercoiling behind the replication forks is thought to drive the unwinding process because superhelical tension is reduced by unwinding the complementary strands.

11.25 DnaA protein initiates the formation of the replication bubble by binding to the 9-bp repeats of *OriC*. DnaA protein is known to be required for the initiation process because bacteria with temperature-sensitive mutations in the *dnaA* gene cannot initiate DNA replication at restrictive temperatures.

11.27 Nucleosomes and replisomes are both large macromolecular structures, and the packaging of eukaryotic DNA into nucleosomes raises the question of how a replisome can move past a nucleosome and replicate the DNA in the nucleosome in the process. The most obvious solution to this problem would be to completely or partially disassemble the nucleosome to allow the replisome to pass. The nucleosome would then reassemble after the replisome had passed. One popular model has the nucleosome disassembling into two half-nucleosomes, allowing the replisome to move past it (see Figure 11.33*b*).

11.29 The product of the first gene is required for DNA chain extension, whereas the product of the second gene is only required for the initiation of DNA synthesis.

11.31 (a) 2,000 to 4,000 Okazaki fragments. (b) 300,000 to 600,000 Okazaki fragments.

11.33 Without telomerase, the 5′ end of the newly replicated strand will be missing some bases. The exact number of missing bases will vary:

5′-GATTCCCCGGGAAGCTTGGGGGGC-
3′-CTAAGGGGCCCTTCGAACCCCCCG-

CCATCT T CGT ACGTCTTTGCA-3′
CGGGTAGAAGCAT G-5′

5′-AAGCTT GGGGGG-
3′-CTAAGGGGCCCTTCGAA CCCCCC-

CCCATCT T CGT ACGTCT T T G CA-3′
GGGTAGAAGCATG CAGAAACG T-5′

CHAPTER 12

12.1 (a) RNA contains the sugar ribose, which has an hydroxyl (OH) group on the 2-carbon; DNA contains the sugar 2-deoxyribose, with only hydrogens on the 2-carbon. RNA usually contains the base uracil at positions where thymine is present in DNA. However, some DNAs contain uracil, and some RNAs contain thymine. DNA exists most frequently as a double helix (double-stranded molecule); RNA exists more frequently as a single-stranded molecule; but some DNAs are single-stranded and some RNAs are double-stranded. (b) The main function of DNA is to store genetic information and to transmit that information from cell to cell and from generation to generation. RNA stores and transmits genetic information in some viruses that contain no DNA. In cells with both DNA and RNA: (1) mRNA acts as in intermediary in protein synthesis, carrying the information from DNA in the chromosomes to the ribosomes (sites at which proteins are synthesized). (2) tRNAs carry amino acids to the ribosomes and function in codon recognition during the synthesis of polypeptides. (3) rRNA molecules are essential components of the ribosomes. (c) DNA is located primarily in the chromosomes (with some in cytoplasmic organelles, such as mitochondria and chloroplasts), whereas RNA is located throughout cells.

12.3 3′—GACTA—5′

12.5 Protein synthesis occurs on ribosomes. In eukaryotes, most of the ribosomes are located in the cytoplasm and are attached to the extensive membranous network of endoplasmic reticulum. Some protein synthesis also occurs in cytoplasmic organelles such as chloroplasts and mitochondria.

12.7 Both prokaryotic and eukaryotic organisms contain messenger RNAs, transfer RNAs, and ribosomal RNAs. In addition, eukaryotes contain small nuclear RNAs. Messenger RNA molecules carry genetic information from the chromosomes (where the information is stored) to the ribosomes in the cytoplasm (where the information is expressed during protein synthesis). The linear sequence of triplet codons in an mRNA molecule specifies the linear sequence of amino acids in the polypeptides produced during translation of that mRNA. Transfer RNA molecules are small (about 80 nucleotides long) molecules that carry amino acids to the ribosomes and provide the codon-recognition specificity during translation. Ribosomal RNA molecules provide part of the structure and function of ribosomes; they represent an important part of the machinery required for the synthesis of polypeptides. Small nuclear RNAs are structural components of spliceosomes, which excise noncoding intron sequences from nuclear gene transcripts.

12.9 "Self-splicing" of RNA precursors demonstrates that RNA molecules can also contain catalytic sites; this property is not restricted to proteins.

12.11 The introns of protein-encoding nuclear genes of higher eukaryotes almost invariably begin (5′) with GT and end (3′) with AG. In addition, the 3′ subterminal A in the "TACTAAC box" is completely conserved; this A is involved in bond formation during intron excision.

12.13 (a) Sequence 5. It contains the conserved intron sequences: a 5′ GU, a 3′ AG, and a UACUAAC internal sequence providing a potential bonding site for intron excision. Sequence 4 has a 5′ GU and a 3′ AG, but contains no internal A for the bonding site during intron excision. (b) 5′—UAGUCUCAA—3′; the putative intron from the 5′ GU through the 3′ AG has been removed.

12.15

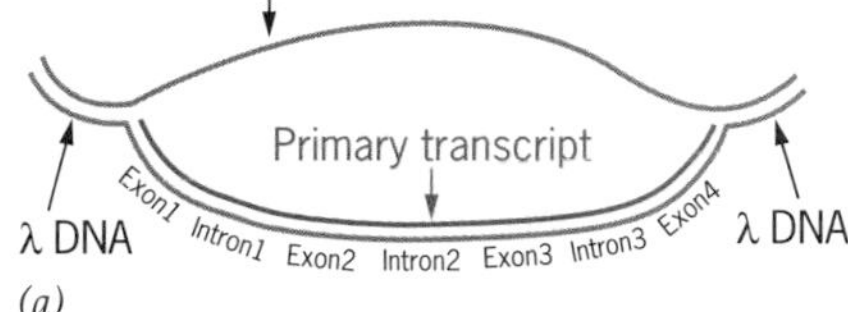

(*a*)

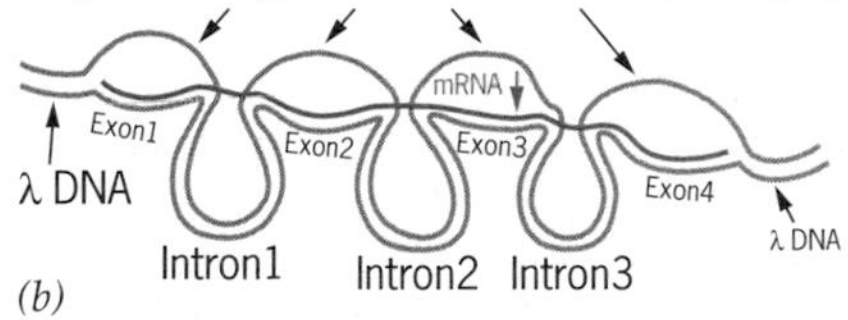

(*b*)

12.17 Assuming that there is a -35 sequence upstream from the consensus -10 sequence in this segment of the DNA molecule, the nucleotide sequence of the transcript will be 5′-ACCCGACAUAGCUACGAUGACGAUAAGCGACAUAGC-3′.

12.19 Assuming that there is a CAAT box located upstream from the TATA box shown in this segment of DNA, the nucleotide sequence of the transcript will be 5′-ACCCGACAUAGCUACGAUGACGAUA-3′.

12.21 Although in theory it would be possible to produce six different enzyme activities by alternative splicing of one gene transcript or by combining the polypeptide products of two or three genes into different combinations with distinct enzyme activities, in reality these possibilities are unlikely. In living organisms, six enzymes are usually specified by at least six genes—with each gene encoding a single polypeptide. However, many enzymes are composed of two or more distinct polypeptides. Thus the synthesis of six enzymes may require more than six genes.

12.23 DNA, RNA, and protein synthesis all involve the synthesis of long chains of repeating subunits. All three processes can be divided into three stages: chain initiation, chain elongation, and chain termination.

12.25 The primary transcripts of eukaryotes undergo more extensive post-transcriptional processing than those of prokaryotes. Thus the largest differences between mRNAs and primary transcripts occur in eukaryotes. Transcript processing is usually restricted to the excision of terminal sequences in prokaryotes. In contrast, eukaryotic transcripts are usually modified by: (1) the excision of intron sequences; (2) the addition of 7-methyl guanosine caps to the 5′ termini; (3) the addition of poly(A) tails to the 3′ termini. In addition, the sequences of some eukaryotic transcripts are modified by RNA editing processes.

12.27 The gene that contains a single exon will be transcribed in the least time. The rate of RNA chain extension is the same—about 30 nucleotides per second—for both intron and exon sequences. Thus the synthesis of the long intron sequence will take about 23 minutes. In contrast, the time required to translate the two mRNAs will be the same because intron sequences are spliced out of transcripts prior to their translation.

12.29 A simple pulse-and pulse/chase-labeling experiment will demonstrate that RNA is synthesized in the nucleus and is subsequently transported to the cytoplasm. This experiment has two parts: (1) Pulse label eukaryotic culture cells by growing them in ^{3}H-uridine for a few minutes, and localize the incorporated radioactivity by autoradiography. (2) Repeat the experiment, but this time add a large excess of nonradioactive uridine to the medium in which the cells are growing after the labeling period, and allow the cells to grow in the nonradioactive medium for about an hour. Then localize the incorporated radioactivity by autoradiography. The expected results are shown in Figure 12.6.

12.31 During DNA synthesis, (1) both strands of DNA are used as templates for the synthesis of complementary nascent strands, (2) the precursors are deoxribonucleoside triphosphates, (3) new chains are initiated with RNA primers, (4) the parental DNA strands are completely separated, (5) chain extension occurs at a rate of about 500 nucleotides per second (prokaryotes). During RNA synthesis, (1) only one strand of DNA is used as a template for the synthesis of a complementary nascent strand, (2) the precursors are ribonucleoside triphosphates, (3) new chains are initiated *de novo*, (4) synthesis occurs within a localized region of strand separation, (5) chain extension occurs at a rate of 40 to 50 nucleotides per second (prokaryotes).

12.33 The first preparation of RNA polymerase is probably lacking the sigma subunit and, as a result, initiates the synthesis of RNA chains at random sites along both strands of the *argH* DNA. The second preparation probably contains the sigma subunit and initiates RNA chains only at the site used *in vivo*, which is governed by the position of the -10 and -35 sequences of the promoter.

12.35 The simplest procedure for determining which of the three RNA polymerases catalyzes the transcription of the gene is to measure the sensitivity of its transcription to α-amanitin. Since the gene is expressed in cells growing in culture, you can simply add α-amanitin to the culture medium. If the gene is transcribed by RNA polymerase I, α-amanitin will have no effect on its transcription. If the gene is transcribed by RNA polymerase II, its transcription will be completely blocked by the presence of α-amanitin. If the gene is transcribed by RNA polymerase III, the rate of transcription will be reduced, but not blocked, by α-amanitin.

12.37 (1) Intron sequences are spliced out of gene transcripts to provide contiguous coding sequences for translation. (2) The 7-methyl guanosine caps added to the 5′ termini of most eukaryotic mRNAs help protect them from degradation by nucleases and are recognized by proteins involved in the initiation of translation. (3) The poly(A) tails at the 3′ termini of mRNAs play an important role in their transport from the nucleus to the cytoplasm and enhance their stability.

12.39 The introns of tRNA precursors, *Tetrahymena* rRNA precursors, and nuclear pre-mRNAs are excised by completely different mechanisms. (1) Introns in tRNA precursors are excised by cleavage and joining events catalyzed by splicing nucleases and ligases, respectively. (2) Introns in *Tetrahymena* rRNA precursors are excised autocatalytically. (3) Introns of nuclear pre-mRNAs are excised by spliceosomes. snRNAs are involved in nuclear pre-mRNA splicing as structural components of spliceosomes. In addition, snRNA U1 is required for the cleavage events at the 5′ termini of introns; U1 is thought to base-pair with a partially complementary consensus sequence at this position in pre-mRNAs.

12.41 This zygote will probably be inviable because the gene product is essential and the elimination of the 5′ splice site will almost certainly result in the production of a nonfunctional gene product.

CHAPTER 13

13.1 Proteins are long chainlike molecules made up of amino acids linked together by peptide bonds. Proteins are composed of carbon, hydrogen, nitrogen, oxygen, and usually sulfur. They provide the enzymatic capacity and much of the structure of living organisms. DNA is composed of phosphate, the pentose sugar 2-deoxyribose, and four nitrogen-containing organic bases (adenine, cytosine, guanine, and thymine). DNA stores and transmits the genetic information in most living organisms. Protein synthesis is of particular interest to geneticists because proteins are the primary gene products—the key intermediates through which genes control the phenotypes of living organisms.

13.3 Ribosomes are from 10 to 20 nm in diameter. They are located primarily in the cytoplasm of cells. In bacteria, they are largely free in the cytoplasm. In eukaryotes, many of the ribosomes are attached to the endoplasmic reticulum. Ribosomes are complex structures composed of over 50 different polypeptides and three to five different RNA molecules.

13.5 Messenger RNA molecules carry genetic information from the chromosomes (where the information is stored) to the ribosomes in the cytoplasm (where the information is expressed during protein synthesis). The linear sequence of triplet codons in an mRNA molecule specifies the linear sequence of amino acids in the polypeptide(s) produced during translation of that mRNA. Transfer RNA molecules are small (about 80 nucleotides long) molecules that carry amino acids to the ribosomes and provide the codon-recognition specificity during translation. Ribosomal RNA molecules provide part of the structure and function of ribosomes; they represent an important part of the machinery required for the synthesis of polypeptides.

13.7 A specific aminoacyl-tRNA synthetase catalyzes the formation of an amino acid-AMP complex from the appropriate amino acid and ATP (with the release of pyrophosphate). The same enzyme then catalyzes the formation of the aminoacyl-tRNA complex, with the release of AMP. The amino acid-AMP and aminoacyl-tRNA linkages are both high-energy phosphate bonds.

13.9 (a) The genetic code is degenerate in that all but 2 of the 20 amino acids are

specified by two or more codons. Some amino acids are specified by six different codons. The degeneracy occurs largely at the third or 3′ base of the codons. "Partial degeneracy" occurs where the third base of the codon may be either of the two purines or either of the two pyrimidines and the codon still specifies the same amino acid. "Complete degeneracy" occurs where the third base of the codon may be any one of the four bases and the codon still specifies the same amino acid. (b) The code is ordered in the sense that related codons (codons that differ by a single base change) specify chemically similar amino acids. For example, the codons CUU, AUU, and GUU specify the sructurally related amino acids, leucine, isoleucine, and valine, respectively. (c) The code appears to be almost completely universal. Known exceptions to universality include strains carrying suppressor mutations that alter the reading of certain codons (with low efficiencies in most cases) and the use of UGA as a tryptophan codon in yeast and human mitochondria.
13.11 (a) Met ⟶ Val. This substitution occurs as a result of a transition. All other amino acid substitutions listed would require transversions.
13.13 (a) By a complex reaction involving mRNA, ribosomes, initiation factors (IF-1, IF-2, and IF-3), GTP, the initiator codon AUG, and a special initiator tRNA ($tRNA_f^{Met}$). It also appears to involve a base-pairing interaction between a base sequence near the 3′-end of the 16S rRNA and a base sequence in the "leader sequence" of the mRNA. (b) By recognition of one or more of the chain-termination codons (UAG, UAA, and UGA) by the appropriate protein release factor (RF-1 or RF-2).
13.15 Crick's wobble hypothesis explains how the anticodon of a given tRNA can base-pair with two or three different mRNA codons. Crick proposed that the base-pairing between the 5′ base of the anticodon in tRNA and the 3′ base of the codon in mRNA was less stringent than normal and thus allowed some "wobble" at this site. As a result, a single tRNA often recognizes two or three of the related codons specifying a given amino acid (see Table 13.3).
13.17 Because of wobble (Table 13.3), one tRNA can recognize both UUA and UUG codons for leucine. However, it takes two more tRNAs to recognize CUU, CUC, CUA, and CUG. Therefore, a minimum of three tRNAs are required to recognize the six codons for leucine.
13.19 (a) Singlet and doublet codes provide a maximum of 4 and $(4)^2$ or 16 codons, respectively. Thus neither code would be able to specify all 20 amino acids. (b) 20. (c) $(20)^{146}$.
13.21 (a) Attachment of an amino acid to the correct tRNA. (b) Recognition of termination codons UAA and UAG and release of the nascent polypeptide from the tRNA in the *P* site of the ribosome. (c) Formation of a peptide bond between the amino group of the aminoacyl-tRNA in the *A* site and the carboxyl group of the growing polypeptide on the tRNA in the *P* site. (d) Formation of the initiation complex required for translation; all steps leading up to peptide bond formation. (e) Translocation of the peptidyl-tRNA from the *A* site on the ribosome to the *P* site.
13.23 (a) A nonsense mutation changes a codon specifying an amino acid to a chain-termination codon, whereas a missense mutation changes a codon specifying one amino acid to a codon specifying a different amino acid. (b) Missense mutations are more frequent. (c) Of the 64 codons, only three specify chain termination. Thus the number of possible missense mutations is much larger than the number of possible nonsense mutations. Moreover, nonsense mutations almost always produce nonfunctional gene products. As a result, nonsense mutations in essential genes are usually lethal in the homozygous state.
13.25 The incoming aminoacyl-tRNA enters the *A* site of the ribosome, the nascent polypeptide-tRNA occupies the *P* site, and the uncharged exiting tRNA occupies the *E* site.
13.27 (a) The Shine-Dalgarno sequence is a conserved polypurine tract, consensus AGGAGG, that is located about seven nucleotides upstream from the AUG initiation codon in mRNAs of prokaryotes. It is complementary to, and is believed to base-pair with, a sequence near the 5′ terminus of the 16S ribosomal RNA. (b) Prokaryotic mRNAs with the Shine-Dalgarno sequence deleted are either not translated or are translated inefficiently.
13.29 Incorporation of alanine into polypeptide chains.
13.31 NH_2-Met-Ala-Ile-Cys-Leu-Phe-Gln-Ser-Leu-Ala-Ala-Gln-Asp-Arg-Pro-Gly-COOH.
13.33 (a) 5′-CAAUCAUGGACUGC-CAUGCUUCAUAUGAAUAGUUGACAU-3′. (b) NH_2 -fMet-Asp-Cys-His-Ala-Ser-Tyr-Glu-COOH. (c) NH_2-fMet-Asp-Cys-Met-Leu-His-Met-Asn-Ser-COOH.

CHAPTER 14

14.1 (a) Transition, (b) transition, (c) transversion, (d) transversion, (e) frameshift, (f) transition.
14.3 (a) *ClB* method, (b) attached-X method (see Chapter 7).
14.5 Probably not. A human is larger than a bacterium, with more cells and a longer life span. If mutation frequencies are calculated in terms of cell generations, the rates for human cells and bacterial cells are similar.
14.7 The X-linked gene is carried by mothers, and the disease is expressed in half of their sons. Such a disease is difficult to follow in pedigree studies because of the recessive nature of the gene, the tendency for the expression to skip generations in a family line, and the loss of the males who carry the gene. One explanation for the sporadic occurrence and tendency for the gene to persist is that, by mutation, new defective genes are constantly being added to the load already present in the population.
14.9 The sheep with short legs could be mated to unrelated animals with long legs. If the trait is expressed in the first generation, it could be presumed to be inherited and to depend on a dominant gene. On the other hand, if it does not appear in the first generation, F_1 sheep could be crossed back to the short-legged parent. If the trait is expressed in one-half of the backcross progeny, it might be presumed to be inherited as a simple recessive. If two short-legged sheep of different sex could be obtained, they could be mated repeatedly to test the hypothesis of dominance. In the event that the trait is not transmitted to the progeny that result from these matings, it might be considered to be environmental or dependent on some complex genetic mechanism that could not be identified by the simple test used in the experiments.
14.11 If both mutators and antimutators operate in the same living system, an optimum mutation rate for a particular organism in a given environment may result from natural selection.
14.13 (a) Yes. (b) A block would result in the accumulation of phenylalanine and a decrease in the amount of tyrosine, which would be expected to result in several different phenotypic expressions.
14.15

Amino Acid	mRNA	DNA
Glumatic acid	—GAA→	—GAA→ ⋯⋯ ←CTT—←Transcribed strand ↓ Mutation
Valine	—GUA→	—GTA→ ⋯⋯ ←CAT— Mutation
Lysine	—AAA→	—AAA→ ⋯⋯ ←TTT—

14.17 Mutations: transitions, transversions, and frameshifts.
14.19 3%; 4%; 6%.
14.21 Radioactive iodine is concentrated by living organisms and food chains.
14.23 During the replicating process, untraviolet light produces mispairing alterations mostly in pyrimidines (for example, cytosine to thymine transitions). Thymine

may be altered to cytosine (or a modified pyrimidine with the base-pairing potential of cytosine), which pairs with guanine. A reverse mutation may occur when cytosine is changed to thymine (or a derivative of cytosine with the hydrogen-bonding potential of thymine), which pairs with adenine. The T-A base pair may thus be changed to a C-G, and the reverse mutation may occur from C-G to T-A.

14.25 Transitions.

14.27 Nitrous acid acts as a mutagen on either replicating or nonreplicating DNA and produces transitions from A to G or C to T, whereas 5-bromouracil does not affect nonreplicating DNA but acts during the replication process causing GC ⟺ AT transitions. 5-Bromouracil must be incorporated into DNA during the replication process in order to induce mispairing of bases and thus mutations.

14.29 5-BU causes GC ⟺ AT transitions. 5-BU can, therefore, revert almost all of the mutations that it induces by enhancing the transition event that is the reverse of the one that produced the mutation. In contrast, the spontaneous mutations will include transversions, frameshifts, deletions, and other types of mutations, including transitions. Only the spontaneous transitions will show enhanced reversion after treatment with 5-BU.

14.31 (a) Frameshift due to the insertion of C at the 9th, 10th, or 11th nucleotide from the 5′ end; (b) normal: 5′-AUGCCGUACUGCCAGCUAACUGCUAAAGAACAAUUA-3′; mutant: 5′-AUGCCCGUACUGCCAGCUAACUGCUAAAGAACAAUUA-3′; (c) normal: NH_2-fMet-Pro-Tyr-Cys-Gln-Leu-Thr-Ala-Lys-Glu-Gln-Leu; mutant: NH_2-fMet-Pro-Val-Leu-Pro-Ala-Asn-Cys.

14.33 No. Leucine ⟶ proline would occur more frequently. Leu (C**U**A) $\xrightarrow{5\text{-BU}}$ Pro (C**G**A) occurs by a single base-pair transition, whereas Leu (**CU**A) $\xrightarrow{5\text{-BU}}$ Ser (**UC**A) requires two base-pair transitions. Recall that 5-bromouracil (5-BU) induces only transitions (see Figure 14.17).

14.35 Yes:

DNA:	←GGX— / —CCX′→	$\xrightarrow{HNO_2}$	←GGX— / —UCX′→
	↓		↓
mRNA:	GGX		AGX
	↓		↓
Polypeptide:	Gly		Ser or Arg (depending on X)

or

DNA:	←GGX— / —CCX′→	$\xrightarrow{HNO_2}$	←GGX— / —CUX′→
	↓		↓
mRNA:	GGX		GAX
	↓		↓
Polypeptide:	Gly		Asp or Glu (depending on X)

or

DNA:	←GGX— / —CCX′→	$\xrightarrow{HNO_2}$	←GGX— / —UUX′→
	↓		↓
mRNA:	GGX		AAX
	↓		↓
Polypeptide:	Gly		Asn or Lys (depending on X)

Note: The X at the third position in each codon in mRNA and in each triplet of base pairs in DNA refers to the fact that there is complete degeneracy at the third base in the glycine codon. Any base may be present in the codon, and it will still specify glycine.

14.37 Tyr ⟶ Cys substitutions; Tyr to Cys requires a transition, which is induced by nitrous acid. Tyr to Ser would require a transversion, and nitrous acid is not expected to induce transversions.

14.39 UGG UGG UGG AUG CGA GAG UGG ACG AUG AGA GAA

CHAPTER 15

15.1 Prior to 1940, the gene was considered a "bead-on-a-string," not subdivisible by recombination or mutation. Today, the gene is considered to be the unit of genetic material that codes for one polypeptide. The unit of structure, not subdivisible by recombination or mutation, is known to be the single nucleotide pair.

15.3 The *cis-trans* test, which defines the unit of genetic material specifying the amino acid sequence of one polypeptide.

15.5 Precursor ⟶ B ⟶ A ⟶ C ⟶ D
2 4 3 1

15.7 They provide powerful selective sieves for identifying rare recombinants. This is accomplished by using the restrictive environmental conditions to select wild-type recombinant progency from crosses between pairs of conditional lethal mutants.

15.9 Four genes; mutations 1 and 2 in one gene; mutations 3 and 4 in a second gene; mutations 5 and 6 in a third gene; mutations 7 and 8 in a fourth gene.

15.11 (a) (A, B, F, G, H, I) and (C,D) and (E); (b) Blue colony requires at least three genes; (c) The maximum number of genes for blue color cannot be estimated from the above data.

15.13 Homoalleles are structurally and functionally allelic; they are not separable by recombination. Heteroalleles are functionally allelic (based on *cis-trans* tests) but are structurally nonallelic (based on recombination tests). Heteroalleles thus result from mutations occurring at different sites within a gene.

15.15 (a) Five genes. (b) Mutations 1, 3, and 5 are in one gene; mutations 7 and 8 are in a second gene; mutations 2, 4, and 6 identify genes 3, 4, and 5, respectively.

15.17 (a) True. (b) False. (c) True. (d) True. (e) True.

15.19 *am* mutations result in UAG chain-termination codons within the coding sequence of the mRNA product of a gene; they thus produce truncated polypeptide gene-products. Since all *am* mutant alleles of a gene will produce polypeptides lacking the COOH-terminus, they would not be expected to exhibit intragenic complementation except in very rare cases. In contrast, most *ts* mutations are caused by missense mutations that change the amino acid sequence of the polypeptide gene-product making it more heat-labile. However, most *ts* mutant alleles produce a complete, although altered, gene-product. As a result, *ts* mutant alleles often exhibit intragenic complementation when the active form of the protein gene-product is a homomultimer. For this reason, *cis-trans* tests carried out with *am* mutants, not *ts* mutants, have been used whenever possible to define the genes of phage T4 operationally.

15.21 It depends on how you define alleles. If every variation in nucleotide sequence is considered to be a different allele, even if the gene product and the phenotype of the organism carrying the mutation are unchanged, then the number of alleles will be directly related to gene size. However, if the nucleotide sequence change must produce an altered gene product or phenotype before it is considered a distinct allele, then there will be a positive correlation, but not a direct relationship, between the number of alleles of a gene and its size in nucleotide pairs. The relationship is more likely to occur in prokaryotes where most genes lack introns. In eukaryotic genes, nucleotide sequence changes within introns usually are neutral; that is, they do not affect the activity of the gene product or the phenotype of the organism. Thus, in the case of eukaryotic genes with introns, there may be no correlation between gene size and number of alleles producing altered phenotypes.

15.23 (a) Four genes. (b) Mutations 1, 3, and 7 are in one gene; mutations 4 and 6 are in a second gene; mutation 2 is in a third gene; and mutation 5 is in a fourth gene.

15.25 No. Because the two mutations map to different chromosomes, they could not be located within the same gene, at least, based on our current concept of the gene. However, if two transcripts are spliced in *trans*, it is possible for two parts of a gene—a nucleotide sequence encoding one polypeptide—to be located on two different chromosomes. Remember that our concept of the gene has evolved considerably since it was introduced by Mendel in 1865, and it will undoubtedly continue to evolve in the future.

15.27 *Neurospora* has many advantages over humans as an experimental organism.

The most important advantages are the ability to grow organisms under carefully controlled conditions, to enhance mutation frequency by treatment with mutagenic agents, and to perform controlled crosses for genetic analysis.

15.29 One. The *trpA*58 and *trpA*78 mutations alter the same codon. If they alter the 5′ and 3′ nucleotide pairs of the triplet, they will be separated by the middle nucleotide pair of the triplet specifying one mRNA codon.

15.31 They are usually referred to as gene segments because they are joined together by somatic recombination to produce a sequence of nucleotide pairs that encodes a single polypeptide.

15.33 No. These mutations are located far apart on the X chromosome. They are separated by millions of nucleotide pairs and many other genes, and it is virtually impossible for mutations located that far apart to be part of the same gene. For these mutations to be located in the same gene, all of the intervening genes would have to be part of a huge intron or some type of *trans* splicing would have to occur.

CHAPTER 16

16.1 Because wild-type recombinants are rarest in cross (a), this must be the double crossover class, and *m* must be the middle gene. In crosses (b) and (c), wild-type recombinants can be generated by single exchanges.

16.3 The chromosome theory of inheritance is the theory that genes are located on chromosomes. The gene is DNA, and Hershey and Chase's studies on bacterial viruses (Chapter 10) demonstrated that DNA is the genetic material. Studies on the mechanism of recombination have correlated genetic recombination with the physical exchange of DNA molecule segments.

16.5 These data show that recombination occurred between the chromosomes of all three phage strains in each cross to produce recombinants carrying genes from all three parental strains. This result demonstrates that phage chromosomes participate in multiple recombination events during the infection of a single bacterium.

16.7 We expect 9 double crossovers (0.012 × 0.009 × 80,000), and we observe 76. The coefficient of coincidence = 76/9 = 8.44; the interference is 1 − 8.44 = −7.44 (negative interference).

16.9 The most reliable way to determine if mutation is a deletion is to perform a reverse mutation analysis. Deletions do not revert to wild type, but point mutations do.

16.11 Linear chromosomes that are terminally redundant and circularly permuted produce a circular genetic map.

16.13 The new mutant maps in the A6c-A6d region.

16.15

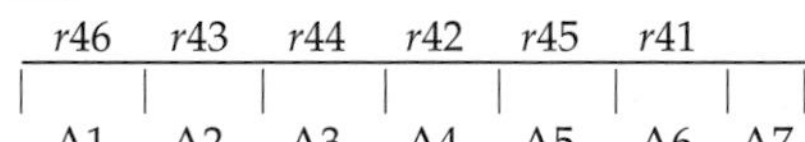

A1 A2 A3 A4 A5 A6 A7

16.17 These results suggest that the same stretch of DNA is transcribed in different ways to produce two mRNAs, one for cell lysis and the other one for some other function. Presumably, transcription does not begin at the same point but does overlap for the two genes and does use the same DNA strand as a template.

16.19 Since reverse transcriptase cannot detect and repair mutations occurring in the production of DNA from RNA, variability is generated by new mutations. This variability contributes to the rapid divergence and the evolution of new HIV strains. The highly variable nature of HIV means that drug treatments of AIDS will also not remain the same and makes the development of vaccines difficult.

CHAPTER 17

17.1 Recombination has occurred between the two strains, producing wild-type (prototrophic) bacteria.

17.3

Recombination Process	Agent Mediating DNA Transfer	Size of DNA Units Transferred	State of Donor DNA in Recombinant Cell
Transformation	Active uptake of free DNA	Small (about 1/200 to 1/100 of a chromosome)	Single-stranded; integrated
Transduction	Bacteriophage	Small (about 1/100 to 1/50 of a chromosome)	Double-stranded; integrated into host chromosome (except in abortive transduction)
Sexduction	F factor	Variable	Initially added to the host cell as separate plasmid; may undergo recombination with host chromosome to yield stable transductant

17.5 (a) F′ factors are useful for genetic mapping, for complementation analysis (see Chapter 15), and for studies of dominance relationships. (b) F′ factors are formed by errors during excision of F factors from Hfr chromosomes. (c) Sexduction occurs by the conjugative transfer of an F′ factor from a donor to a recipient (F^-) cell.

17.7 A prophage is a phage chromosome that has become integrated into the host chromosome. The prophage is dormant in the sense that the phage genes involved in lytic development (replication and maturation) are repressed. The prophage is replicated during host chromosome replication just like any other segment of that chromosome.

17.9 Cotransduction refers to the simultaneous transduction of two different genetic markers into a single recipient cell. Since bacteriophage particles can package a maximum of 1 to 2 percent of the total bacterial chromosome, only markers that are relatively closely linked can be cotransduced. The frequency of cotransduction of any two markers will be an inverse function of the distance between them on the chromosome. As such, this frequency can be used as an estimate of the linkage distance. Specific cotransduction-linkage relationships must be prepared for each phage-host system studied.

17.11 (a)

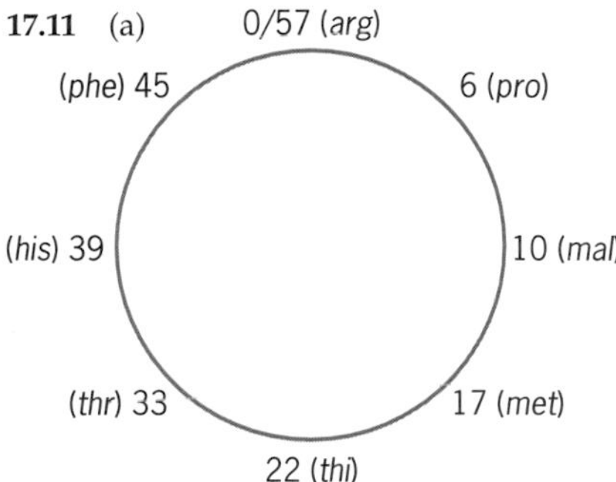

(b) 57

17.13 The order is *a-d-b-c* because the closer the genes, the higher the cotransduction percentage.

7.15 Since loci that are close together will show a higher frequency of cotransduction than those that are farther apart, the correct gene order must be *ser-glu-arg-his*.

17.17 *anth* − *A34* − *A223* − *A46*.

17.19 The donor strain (Hfr) was $a^-b^+c^-d^+$ and the recipient (F^-) was $a^-b^-c^+d^-$. We conclude this because with $a^+b^+c^-d^+$ as the donor, 2 exchanges are required to produce an $a^-b^+c^+d^-$ recombinant; 4 exchanges are necessary to produce the same recombinant in the reciprocal type of cross, a much rarer event.

7.21 The distance $(a - b)$ = single transformants / (single transformants + double transformants) = (350 + 60 + 300 + 10)/[(350 + 60 + 300 + 10) + (1400 + 150)] = 720/2270 = 0.317; similarly, distance $(a - c)$ = 900/(900 + 1410) = 0.390; distance $(b - c)$ = 320/(320 + 1700) = 0.158. Therefore, *a* 0.317 *b* 0.158 *c*.

17.23 If the sequence is *a-b-c*, the reciprocal crosses (A and B) should give the same frequency of + + + recombinants because in both crosses, two exchanges are required:

```
(A)   Hfr      +    +    c
F⁻ →         X       X
             a    b    +

(B)   Hfr      a    b    +
F⁻ →                X    X
               +    +    c
```

If the sequence is *a-c-b*, the reciprocal crosses give different frequencies of + + + recombinants because four exchanges are required in (A) and would be much less frequent than (B), where two are required:

```
(A)   Hfr      +    c    +
F⁻ →         X    X    X    X
               a    +    b

(B)   Hfr      a    +    b
F⁻ →              X    X
               +    c    +
```

Thus, the sequence is: *a-c-b*.

CHAPTER 18

18.1 The pair in (d) are inverted repeats and could therefore qualify.
18.3 Resistance for the second antibiotic was acquired by conjugative gene transfer between the two types of cells.
18.5 In the first strain, the F factor integrated into the chromosome by recombination with the IS element between genes *C* and *D*. In the second strain, it integrated by recombination with the IS element between genes *D* and *E*. The two strains transfer their genes in different orders because the two chromosomal IS elements are in opposite orientation.
18.7 Both IS*50* elements should be able to excise from the transposon and insert elsewhere in the chromosome.
18.9 The *tnpA* mutation: no; the *tnpR* mutation: yes.
18.11 Many bacterial transposons carry genes for antibiotic resistance and it is relatively simple for these genes to move from one DNA molecule to another. DNA molecules that acquire resistance genes can be passed to other cells in a bacterial population, both vertically (by descent) and horizontally (by conjugative transfer). Over time, continued exposure to an antibiotic will select for cells that have acquired a gene for resistance to that antibiotic. The antibiotic will therefore no longer be useful in combating these bacteria.
18.13 The c^m mutation is due to a *Ds* or an *Ac* insertion.
18.15 The paternally inherited *Bz* allele was inactivated by a transposable element insertion.
18.17 Cross dysgenic (highly mutable) males carrying a wild type X chromosome to females homozygous for a balancer X chromosome; then cross the heterozygous F_1 daughters individually to their brothers and screen the F_2 males that lack the balancer chromosome for mutant phenotypes, including failure to survive (lethality). Mutations identified in this screen are probably due to *P* element insertions in X-linked genes.
18.19 Factors made by the host's genome are required.
18.21 Through crossing over between the LTRs of a *gypsy* element.
18.23 *In situ* hybridization to polytene chromosomes using a *TART* probe (see Chapter 10).
18.25 Same orientation: a deletion; opposite orientation: an inversion.
18.27 The transposition rate in humans may be very much less than it is in *Drosophila*.

CHAPTER 19

19.1 The white phenotype is lethal in whole plants.
19.3 Color depends on the proportion of wild-type and mutant chloroplasts in the tissue. Female gametes from green sectors should transmit the green color, female gametes from white sectors should transmit the white color, and female gametes from pale green sectors should transmit all three colors. Male gametes should have no effect on the trait.
19.5 If *O. hookeri* had yellow plastids and *O. muricata* had green plastids.
19.7 Half the offspring will be male-sterile. After the intercross, one-fourth of the offspring will be male-sterile.
19.9 Spectinomycin resistance is encoded by a chloroplast gene.
19.11 Very rarely. Mutations would have to occur in two different cpDNA molecules within the same cell, and then these would have to recombine to produce a molecule that carried both mutations—an extremely unlikely series of events.
19.13 Suppressive petite.
19.15 No. The genetic code for mitochondrial genes is not the same as that for *E. coli* genes.
19.17 More than one gene encodes a polypeptide.
19.19 (a) Cardiomyopathy is transmitted from mother to both male and female offspring, but not from father to offspring. This pattern of maternal inheritance suggests that the trait is due to a mutation in the mitochondrial genome. (b) The varying degrees of severity of cardiomyopathy may be due to varying numbers of mutant mitochondria that result from cytoplasmic segregation during cell division. Other factors that could influence the severity of the condition are (1) interactions with nuclear genes, and (2) environmental variables.
19.21 The expression of genes in two different genetic systems must be coordinated.

CHAPTER 20

20.1 (a) Both introduce new genetic variability into the cell. In both cases, only one gene or a small segment of DNA representing a small fraction of the total genome is changed or added to the genome. The vast majority of the genes of the organism remain the same. (b) The introduction of recombinant DNA molecules, if they come from a very different species, is more likely to result in a novel, functional gene product in the cell, if the introduced gene (or genes) is capable of being expressed in the foreign protoplasm. The introduction of recombinant DNA molecules is more analogous to duplication mutations (see Chapter 7) than to other types of mutations.
20.3 (a) $(1/4)^4 = 1/256$; (b) $(1/4)^6 = 1/4096$
20.5 Recombinant DNA and gene-cloning techniques allow geneticists to isolate essentially any gene or DNA sequence of interest and to characterize it structurally and functionally. Large quantities of a given gene can be obtained in pure form, which permits one to determine its nucleotide-pair sequence (to "sequence it" in common lab jargon). From the nucleotide sequence and our knowledge of the genetic code, geneticists can predict the amino acid sequence of any polypeptide encoded by the gene. By using an appropriate subclone of the gene as a hybridization probe in northern blot analyses, geneticists can identify the tissues in which the gene is expressed. Based on the predicted amino acid sequence of a polypeptide encoded by a gene, geneticists can synthesize oligopeptides and use these to raise antibodies that, in turn, can be used to identify the actual product of the gene and localize it within cells or tissues of the organism. Thus, recombinant DNA and gene-cloning technologies provide very powerful tools with which to study the genetic control of essentially all biological processes. These tools have played major roles in the explosive progress in the field of biology during the last two decades.
20.7 Restriction endonucleases are believed to provide a kind of primitive immune system to the microorganisms that produce them—protecting their genetic material from "invasion" by foreign DNAs from viruses or other pathogens or just DNA in the environment that might be taken up by the microorganism. Obviously, these microorganisms do not have a sophisticated immune system like that of higher animals (Chapter 27).
20.9 A foreign DNA cloned using an enzyme that produces single-stranded com-

plementary ends can always be excised from the cloning vector by cleavage with the same restriction enzyme that was originally used to clone it. For example, if a *Hin*dIII fragment from the human genome is cloned into *Hin*dIII-cleaved pUC119, the human *Hin*dIII fragment can be excised from a plasmid DNA preparation of this clone by cleavage with restriction endonuclease *Hin*dIII. The human *Hin*dIII fragment will be flanked in the recombinant plasmid DNA clone by *Hin*dIII cleavage sites. When terminal transferase is used to add complementary single-stranded ends during cloning, the original restriction endonuclease cleavage sites are destroyed. Thus, the restriction enzyme used to generate the fragment for cloning cannot be used to excise the original fragments from the cloning vector.

20.11 Most genes of higher plants and animals contain noncoding intron sequences. These intron sequences will be present in genomic clones, but not in cDNA clones, because cDNAs are synthesized using mRNA templates and intron sequences are removed during the processing of the primary transcripts to produce mature mRNAs.

20.13 The maize *gln2* gene contains many introns, and one of the introns contains a *Hin*dIII cleavage site. The intron sequences (and thus the *Hin*dIII cleavage site) are not present in mRNA sequences and thus are also not present in full-length *gln2* cDNA clones.

20.15 By oligonucleotide-directed site-specific mutagenesis, you can change each of the eight nucleotide pairs in the protein binding site to each of the other three possible nucleotide pairs. You can then examine the binding of the regulatory protein to each of the mutant sequences (24 single nucleotide-pair changes are possible at the eight positions), and you can examine the ability of each of the mutant sequences to mediate induction of transcription of the gene at 45°C *in vivo*. You could study the binding of the protein using synthetic oligonucleotide sequences more easily, but that approach would not let you study induction of transcription of the gene *in vivo*. The site-specific mutagenesis approach will allow you to carry out a saturation mutational analysis of the octameric regulatory protein binding sequence.

20.17 Some bacterial viruses package single-stranded DNA rather than double-stranded DNA. These bacteriophages employ a double-stranded DNA intermediate for replication, but then switch to the production of single-stranded DNAs for packaging during phage maturation. When a foreign DNA segment is cloned into the replicative form of the chromosome of one of these phages, only one of the two strands gets packaged during the subsequent maturation processes. In the case of the filamentous single-stranded DNA phages such as M13 and f2, the progeny phage particles are simply extruded through the cell membrane and wall without lysing the host bacterium. Because the phage are so small relative to the size of the bacteria, the phage particles can be separated from the bacteria by a simple low-speed centrifugation step. The bacteria form a pellet at the bottom of the centrifuge tube; the phage remain in the supernatant suspension. The bacterial pellets are discarded, and the phage particles can be collected by high-speed centrifugation or by precipitation with polyethylene glycol and low-speed centrifugation. Pure single-stranded DNA can then be separated from the phage proteins by phenol-chloroform extractions. Many cloning vectors like pUC118 and pUC119 (see Figure 20.14) are phage-plasmid hybrids. They have both plasmid and M13 phage origins of replication as well as M13 phage packaging signals so that they can replicate either as plasmids or as phages and can package single-stranded phage DNA when in the phage mode of replication. The switch from the plasmid mode of replication to the phage mode is accomplished by superinfecting the host bacteria with a mutant "helper" phage.

20.19 (a) 7000 bp + 7000 bp; (b) 14,000 bp; (c) 7000 bp + 7000 bp; (d) 4000 bp + 2000 bp + 5000 bp; (e) 4000 bp + 7000 bp.

20.21 Because the nucleotide-pair sequences of both the normal *CF* gene and the *CF*Δ*508* mutant gene are known, labeled oligonucleotides can be synthesized and used as hybridization probes to detect the presence of each allele (normal and Δ*508*). Under high-stringency hybridization conditions, each probe will hybridize only with the *CF* allele that exhibits perfect complementarity to itself. Since the sequences of the *CF* gene flanking the Δ*508* site are known, oligonucleotide PCR primers can be synthesized and used to amplify this segment of the DNA obtained from small tissue explants of putative CF patients and their relatives by PCR. The amplified DNAs can then be separated by agarose gel electrophoresis, transferred to nylon membranes, and hybridized to the respective labeled oligonucleotide probes, and the presence of each *CF* allele can be detected by autoradiography. For a demonstration of the utility of this procedure, see B. Kerem et al. "Identification of the cystic fibrosis gene: Genetic analysis." *Science 245*: 1073–1080, 1989. Kerem and coworkers used two synthetic oligonucleotide probes (oligo-N = 3′-CTTTTATAGTAGAAACCAC-5′ and oligo-ΔF = 3′-TTCTTTTATAGTA—ACCACAA-5′; the dash indicates the deleted nucleotides in the *CF*Δ*508* mutant allele) to analyze the DNA of CF patients and their parents. For confirmed CF families, the results of these Southern blot hybridizations with the oligo-N (normal) and oligo-ΔF (*CF*Δ*508*) labeled probes were often as follows:

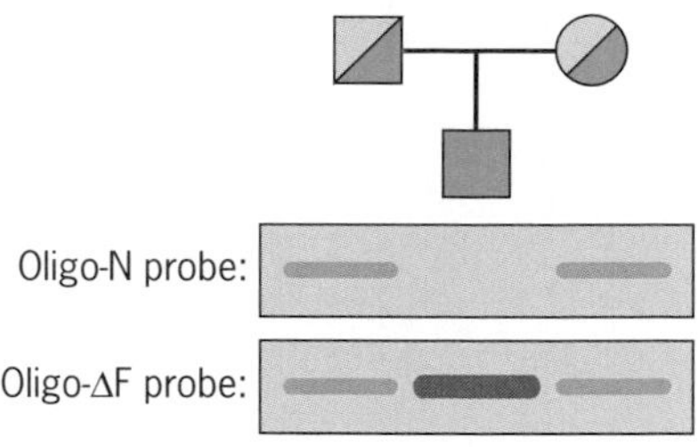

Both parents were heterozygous for the normal *CF* allele and the mutant *CF*Δ*508* allele as would be expected for a rare recessive trait, and the CF patient was homozygous for the *CF*Δ*508* allele. In such families, one-fourth of the children would be expected to be homozygous for the Δ*508* mutant allele and exhibit the symptoms of CF, whereas three-fourths would be normal (not have CF). However, two-thirds of these normal children would be expected to be heterozygous and transmit the allele to their children. Only one-fourth of the children of this family would be homozygous for the normal *CF* allele and have no chance of transmitting the mutant *CF* gene to their offspring. Note that the screening procedure described here can be used to determine which of the normal children are carriers of the *CF*Δ*508* allele: that is, the mutant gene can be detected in heterozygotes as well as homozygotes.

20.23 (a) Southern, northern, and western blot procedures all share one common step, namely, the transfer of macromolecules (DNAs, RNAs, and proteins, respectively) that have been separated by gel electrophoresis to a solid support—usually a nitrocellulose or nylon membrane—for further analysis. (b) The major difference between these techniques is the class of macromolecules that are separated during the electrophoresis step: DNA for Southern blots, RNA for northern blots, and protein for western blots.

20.25

H H
6.0 2.0 7.4 3.5 1.0 2.9
E E E

20.27 There are two possible retriction maps for these data as shown below:

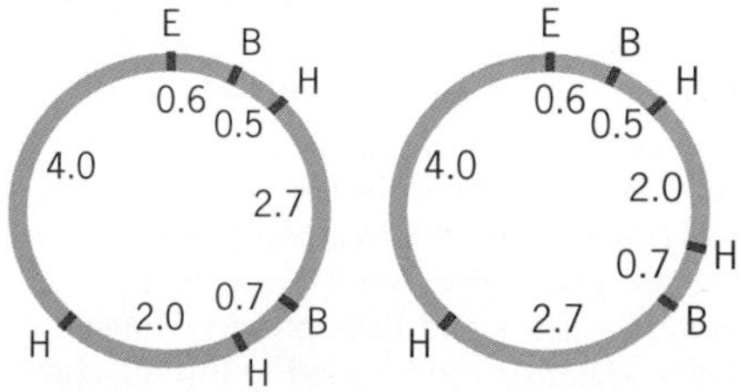

Restriction enzyme cleavage sites for *Bam*HI, *Eco*RI, and *Hin*dIII are denoted by

B, E, and H, respectively. The numbers give distances in kilobase pairs.

20.29

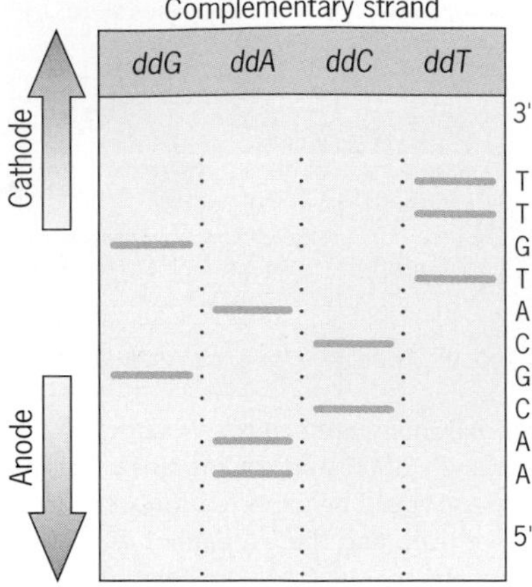

CHAPTER 21

21.1 Genetic map distances are determined by crossover frequencies. Cytogenetic maps are based on chromosome morphology or physical features of chromosomes. Physical maps are based on actual physical distances—the number of nucleotide pairs (0.34 nm per bp)—separating genetic markers. If a gene or other DNA sequence of interest is shown to be located near a mutant gene, a specific knob on a chromosome, or a particular DNA restriction fragment, that genetic or physical marker (mutation, knob, or restriction fragment) can be used to initiate a chromosome walk (see Figure 21.9).

21.3 A contig (*contig*uous clones) is a physical map of a chromosome or part of a chromosome prepared from a set of overlapping genomic DNA clones. An RFLP (*r*estriction *f*ragment *l*ength *p*olymorphism) is a variation in the length of a specific restriction fragment excised from a chromosome by digestion with one or more restriction endonucleases. A VNTR (*v*ariable *n*umber *t*andem *r*epeat) is a short DNA sequence that is present in the genome as tandem repeats and in highly variable copy number. An STS (*s*equence *t*agged *s*ite) is a unique DNA sequence that has been mapped to a specific site on a chromosome. An EST (*e*xpressed *s*equence *t*ag) is a cDNA sequence—a genomic sequence that is transcribed. Contig maps permit researchers to obtain clones harboring genes of interest directly from DNA Stock Centers—to "clone by phone." RFLPs are used to construct the high density genetic maps that are needed for positional cloning. VNTRs are especially valuable RFLPs that are used to identify multiple sites in genomes. STSs and ESTs provide molecular probes that can be used to initiate chromosome walks to nearby genes of interest.

21.5 (a)

| 10 cM | 25 cM | 15 cM | 1 cM | 14 cM |

STS1 STS5 STS3 C STS4 STS2

(b) 3.3×10^9 bp / 3.3×10^3 cM = 1×10^6 bp/cM. The total map length is 65 cM, which equates to about 65×10^6 or 65 million bp. (c) The cancer gene (C) and STS4 are separated by 1 cM or about one million base pairs.

21.7 With a clone of the gene available, fluorescent *in situ* hybridization (FISH) can be used to determine which human chromosome carries the gene and to localize the gene on the chromosome. Single-stranded copies of the clone are coupled to a fluorescent probe and hybridized to denatured DNA in chromosomes spread on a slide (see Figure 10.35). After hybridization and removal on nonhybridized probe, autoradiography is performed to determine the location of the fluorescent probe (see Figure 21.6*b*).

21.9 VNTRs and microsatellites are both polymorphic tandem repeats of relatively short nucleotide sequences. Microsatellites, however, are repeats of sequences only 2–5 nucleotide pairs in length and are a subset of the satellite sequences (sequences that separate from the main band of DNA and form satellite bands in density gradients) in the highly repetitive DNAs of eukaryotes. Most VNTRs are present in mainband DNA—not satellite bands—when eukaryotic DNAs are analyzed by equilibrium density gradient centrifugation (see Chapter 11).

21.11 You can start a chromosome walk using the hybridization probe that detects the RFLP (see Figure 21.9). However, if a physical map of this region of the chromosome already exists (see Figure 21.7), a chromosome walk might not be necessary. cDNAs can be used to locate candidate genes in the region covered by the chromosome walk, and the sequences of genes in individuals with this form of inherited deafness can be compared with the sequences of homologous genes of individuals with normal hearing, looking for changes that would be expected to cause a loss of gene function. The overall process is illustrated in Figure 21.8.

21.13 The goals of the Human Genome Project are to prepare genetic and physical maps showing the locations of all the genes in the human genome and to determine the nucleotide sequences of all 24 chromosomes in the human genome. These maps and nucleotide sequences of the human chromosomes will help scientists identify mutant genes that result in inherited diseases and, hopefully, will lead to successful treatments, including gene therapies, for at least some of these diseases. Potential misuses of these data include invasions of privacy by governments and businesses—especially employment agencies and insurance companies. Individuals must not be denied educational opportunities, employment, or insurance because of inherited diseases or mutant genes that result in a predisposition to mental or physical abnormalities.

21.15 An EST is more likely than an RFLP to occur in a disease-causing human gene. ESTs all correspond to expressed sequences in a genome. RFLPs occur throughout a genome, in both expressed and unexpressed sequences. Because less than 5 percent of the human genome encodes proteins, most RFLPs occur in noncoding DNA.

21.17 (a) Segment 5; (b) segment 4; (c) segment 1, 6, or 10.

21.19 The advantage that gene chips have over traditional dot blot hybridization methods is that a single gene chip can be used to quantify thousands of distinct nucleotide sequences simultaneously. The hybridization probes are arranged on gene chips in microarrays on a solid surface, whereas traditional dot blots were macroarrays of a few to, at most, a few hundred sequences. The gene-chip technology allows researchers to investigate the levels of expression of large numbers of genes more efficiently than was possible using traditional dot blot techniques.

21.21 The DNA sequences in human chromosome-specific cDNA libraries can be coupled to fluorescent dyes and hybridized *in situ* to the chromosomes of other primates. The hybridization patterns can be used to detect changes in genome structure that have occurred during the evolution of the various species of primates from common ancestors (see Figure 21.21). Such comparisons are especially effective in detecting new linkage relationships resulting from translocations and centric fusions.

21.23

(*a*) Order of STS sites: 2-5-1-4-3-6.

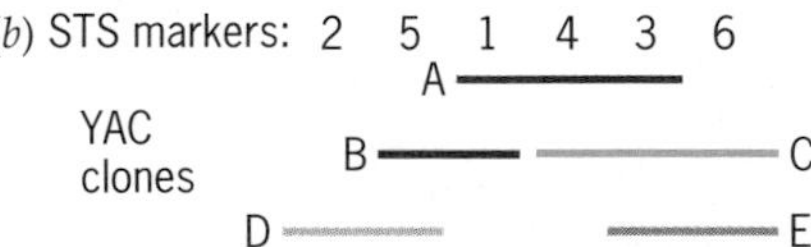

CHAPTER 22

22.1 CpG islands are clusters of cytosines and guanines that are often located just upstream (5′) from the coding regions of human genes. Their presence in nucleotide sequences can provide hints as to the location of genes in human chromosomes.

22.3 The *CF* gene was identified by map position-based cloning, and the nucleotide sequences of *CF* cDNAs were used to predict the amino acid sequence of the *CF* gene product. A computer search of the protein data banks revealed that the *CF* gene product was similar to several ion channel proteins. This result focused the attention of scientists studying cystic fibrosis on proteins involved in the transport of salts between cells and led to the discovery that the *CF* gene product was a transmembrane conductance regulator—now called the CFTR protein.

22.5 Oligonucleotide primers complementary to DNA sequences on both sides (upstream and downstream) of the CAG repeat region in the *MD* gene can be synthesized and used to amplify the repeat region by PCR. One primer must be complementary to an upstream region of the template strand, and the other primer must be complementary to a downstream region of the nontemplate strand. After amplification, the size(s) of the CAG repeat regions can be determined by gel electrophoresis (see Figure 22.3). Trinucleotide repeat lengths can be measured by including repeat regions of known length on the gel. If less than 30 copies of the trinucleotide repeat are present on each chromosome, the newborn, fetus, or pre-embryo is homozygous for a wild-type *MD* allele or heterozygous for two different wild-type *MD* alleles. If more than 50 copies of the repeat are present on each of the homologous chromosomes, the individual, fetus, or cell is homozygous for a dominant mutant *MD* allele or heterozygous for two different mutant alleles. If one chromosome contains less than 30 copies of the CAG repeat and the homologous chromosome contains more than 50 copies, the newborn, fetus, or pre-embryo is heterozygous, carrying one wild-type *MD* allele and one mutant *MD* allele.

22.7 The transcription initiation and termination and translation initiation signals or eukaryotes differ from those of prokaryotes such as *E. coli*. Therefore, to produce a human protein in *E. coli,* the coding sequence of the human gene must be joined to appropriate *E. coli* regulatory signals—promoter, transcription terminator, and translation initiator sequences. Moreover, if the gene contains introns, they must be removed or the coding sequence of a cDNA must be used, because *E. coli* does not possess the spliceosomes required for the excision of introns from nuclear gene transcripts. In addition, many eukaryotic proteins undergo post-translational processing events that are not carried out in prokaryotic cells. Such proteins are more easily produced in transgenic eukaryotic cells growing in culture.

22.9 Eleven, ranging in multiples of 3, from 15 to 45 nucleotides long.

22.11 Contamination of blood samples would introduce more variability into the fingerprint. This would lead to a lack of allelic matching of fingerprints obtained from the blood samples and obtained from the defendant. Mixing errors would be expected only to lead to the acquittal of a guilty person and not to the conviction of an innocent person. Only the mislabeling of samples could implicate someone who is innocent.

22.13 Neither F1 nor F2 could be the girl's biological father; only individual F3 could be the child's father.

22.15 Probing Southern blots of restriction enzyme-digested DNA of the transgenic plants with ^{32}P-labeled transgene may provide evidence of multiple insertions, but would not reveal the genomic location of the inserts. Fluorescence *in situ* hybridization (FISH) is a powerful procedure for determining the genomic location of gene inserts. FISH is used to visualize the location of transgenes in chromosomes.

22.17 Transgenic mice are usually produced by microinjecting the genes of interest into pronuclei of fertilized eggs or by infecting pre-implantation embryos with retroviral vectors containing the genes of interest. Transgenic mice provide invaluable tools for studies of gene expression, mammalian development, and the immune system of mammals. Transgenic mice are of major importance in medicine; they provide the model system most closely related to humans. They have been, and undoubtedly will continue to be, of great value in developing the tools and technology that will be used for human gene therapy in the future.

22.19 Posttranslationally modified proteins can be produced in transgenic eukaryotic cells growing in culture or in transgenic plants and animals. Indeed, transgenic sheep have been produced that secrete human blood-clotting factor IX and α1-antitrypsin in their milk. These sheep were produced by fusing the coding sequences of the respective genes to a DNA sequence that encodes the signal peptide required for secretion and introducing this chimeric gene into fertilized eggs that were then implanted and allowed to develop into transgenic animals. In principle, this approach could be used to produce any protein of interest.

22.21 The vector described contains the HGH gene; however, it does not contain a mammalian HGH-promoter that will regulate the expression of the transgene in the appropriate tissues. Construction of vectors containing a properly positioned mammalian HGH-promoter sequence should result in transgenic mice in which HGH synthesis is restricted to the pituitary gland.

CHAPTER 23

23.1 By studying the synthesis or lack of synthesis of the enzyme in cells grown on chemically defined media. If the enzyme is synthesized only in the presence of a certain metabolite or a particular set of metabolites, it is probably inducible. If it is synthesized in the absence but not in the presence of a particular metabolite or group of metabolites, it is probably repressible.

23.3

Gene or Regulatory Element	Function
(a) Regulator gene	Codes for repressor
(b) Operator	Binding site of repressor
(c) Promoter	Binding site of RNA polymerase and CAP-cAMP complex
(d) Structural gene *Z*	Encodes β-galactosidase
(e) Structural gene *Y*	Encodes β-galactoside permease

23.5 (a) 1, 2, 3*, and 5; (b) 2, 3*, and 5; *3 may be either noninducible or constitutive, depending on whether the specific O^c mutation eliminates binding of the I^s "super-repressor."

23.7 (a) β-Galactosidase will be produced only when lactose is present. Permease will not be produced at all. (b) *Cis* dominance

23.9 (a) $\frac{I^+O^c\,Z^+Y^-}{I^+O^+Z^-Y^+}$; (b) $\frac{I^sO^c\,Z^+Y^-}{I^sO^+Z^-Y^+}$

23.11 (a) 1, 2, 3, 5; (b) 2, 5; (c) 5; (d) 1, 2, 5; (e) 3, 4.

23.13 Catabolite repression has apparently evolved to assure the use of glucose as a carbon source when this carbohydrate is available, rather than less efficient energy sources.

23.15 Positive regulation; the CAP-cAMP complex has a positive effect on the expression of the *lac* operon. It functions in turning on the transcription of the structural genes in the operon.

23.17 Negative regulatory mechanisms such as that involving the repressor in the lactose operon block the transcription of the structural genes of the operon, whereas positive mechanisms such as the CAP-cAMP complex in the *lac* operon promote the transcription of the structural genes of the operon.

23.19 Repression/derepression of the *trp* operon occurs at the level of transcription initiation, modulating the frequency at which RNA polymerase initiates transcription from the *trp* operon promoters. Attenuation modulates *trp* transcript levels by altering the frequency of termination of transcription within the *trp* operon leader region (*trpL*).

23.21 First, remember that transcription and translation are coupled in prokaryotes. When tryptophan is present in cells, tryptophan-charged tRNA^{Trp} is produced. This allows translation of the *trp* leader sequence through the two UGG Trp codons to the *trp* leader sequence UGA termination codon. This translation of the *trp* leader region prevents base-pairing between the partially complementary mRNA leader sequences 75–83 and 110–121 (see Figure 23.13*b*), which in turn permits formation of the transcription-termination "hairpin" involving leader sequences 110–121 and 126–134 (see Figure 23.13*c*).

23.23 A prophage is a chromosome of a bacteriophage after it has become integrated into the chromosome of the host bacterium. The lambda prophage is present as a transcriptionally inactive linear structure in the *E. coli* chromosome. During lytic infections, the lambda chromosome begins its reproductive cycle by replicating as a transcriptionally active circular DNA molecule.

23.25 The operator region of the *lac* operon in *E. coli* contains a single repressor binding site, whereas the lambda operators O_L and O_R each contain three repressor binding sites that are differentially occupied under different conditions. The differential affinity of lambda repressor for the three binding sites in these operators plays an important role in the maintenance of the lysogenic state of a lambda prophage in *E. coli*.

23.27 Lysogeny will continue, because C_I^+ acts in *trans* and is dominant.

23.29 The N protein, the product of the first gene transcribed from promoter P_L, prevents the termination of transcription at two sites (t_{R1} and t_{L1}), which results in the transcription of the delayed early genes located downstream from these sites. N protein binds to *nut* sites just upstream from the termination sites, and, together with the Nus protein and ribosomal protein S10, modifies the specificity of RNA polymerase so that termination does not occur at sites t_{R1} and t_{L1} (see Figure 23.20).

23.31 No. Since transcription (nucleus) and translation (cytoplasm) are not coupled in eurkaryotes, attenuation of the type occurring in prokaryotes would not be possible.

CHAPTER 24

24.1 In multicellular eukaryotes, the environment of an individual cell is relatively stable. There is no need to respond quickly to changes in the external environment. In addition, the development of a multicellular organism involves complex regulatory hierarchies composed of hundreds of different genes. The expression of these genes is regulated spatially and temporally, often through intricate intercellular signaling processes.

24.3 The concentration of mRNA is determined by the rate of transcription relative to the rate of degradation. The longevity of an mRNA is influenced by the length of its poly(A) tail, the structure of its 3′ untranslated region, and the metabolic state of the cell.

24.5 By monitoring puffs in response to environmental signals, such as heat shock, or to hormonal signals.

24.7 By alternate splicing of the transcript.

24.9 Northern blotting of RNA extracted from plants grown with and without light, or PCR amplification of cDNA made by reverse transcribing these same RNA extracts.

24.11 The green fluorescent protein will be made after the flies are heat shocked.

24.13 This enzyme plays an important role in photosynthesis, a light-dependent process. Thus, it makes sense that its production should be triggered by exposure to light.

24.15 The flies would be phenotypically normal males.

24.17 Yes. Enhancers are able to function in different positions in and around a gene.

24.19 No.

24.21 H4 is acetylated.

24.23 X chromosome inactivation begins at the X inactivation center (XIC) which contains the *XIST* gene. During early embryogenesis in XX mammals, the *XIST* genes produce unstable 17-kb-long RNAs that remain closely associated with the XIC region. Transcripts from one of the two *XIST* genes stabilize and eventually coat the entire X chromosome containing that gene; transcripts from the other *XIST* gene disintegrate and further transcription from that gene is repressed by methylation of nucleotides in the gene's promoter. The chromosome that becomes coated with *XIST* RNA becomes the inactive X chromosome; it condenses and forms a Barr body.

CHAPTER 25

25.1 Cell division forms the blastula; cell movement forms the gastrula.

25.3 Determination establishes cell fates; differentiation realizes those fates.

25.5 Collect mutations with diagnostic phenotypes; map the mutations and test them for allelism with one another; perform epistasis tests with mutations in different genes; clone individual genes and analyze their function at the molecular level.

25.7 Imaginal discs.

25.9 Mate homozygous *unc* hermaphrodites that carry one of the *dumpy* mutations with males that carry the other *dumpy* mutation and score the non-uncoordinated (*unc*/+) F_1 hermaphrodites for the dumpy phenotype. If these worms are dumpy, the two mutations are allelic; if they are wild-type, the two mutations are not allelic.

25.11 Cross heterozygous carriers separately with *tra*/+ and *tra2*/+ flies. If any of the XX progeny of these matings are transformed into sterile males, the new mutation is an allele of the *tra* or *tra2* tester.

25.13 Intersex. The *dsx* mutation is epistatic to the *tra* mutation.

25.15 XX, male phenotype; XO, female phenotype.

25.17 The yellow phenotype is due to the expression of a recessive X-linked mutation (*y*) and the fly is a gynandromorph. It probably arose through the loss of a wild-type X chromosome early in the development of a *y*/+ zygote.

25.19 The lethal is autonomous.

25.21 (a) Wild-type; (b) embryonic lethal; (c) embryonic lethal; (d) wild-type; (e) wild-type.

25.23 Female sterility.

25.25 In homozygous condition, the mutation that causes phenylketonuria has a maternal effect. Women homozygous for this mutation influence the development of their children *in utero*.

25.27 Extra mouse eyes or eye primorida.

CHAPTER 26

26.1 Yes; in a mating of $u^+u\ r^+r$ x $u^+u\ r^+r$, 1/16 of the offspring will be *uu rr*.

26.3 *Sensitive* is probably a recessive mutation. When *sen* was crossed to a wild-type from the same population, the wild-type was apparently heterozygous; *sen* × *sen* crosses would be expected to produce all *sen* offspring if the mutation was recessive; and in wild-type × wild-type crosses, we would expect some of the crosses to involve two heterozygotes and thus produce 1/4 *sensitive* offspring.

26.5 Konopka and Benzer would have failed to have detected lethal alleles of the *period* gene because they maintained the mutagenized X chromosomes in males by mating the males to females with attached-X chromosomes. To circumvent this problem, Konopka and Benzer could have used a balancer X chromosome to maintain the mutagenized X chromosomes in heterozygous condition in females: balancer-X/mutagenized-X female × balancer-X male. Of course, recessive lethal mutations could not be tested for a behavioral phenotype such as an abnormal eclosion rhythm. To demonstrate that a recessive lethal mutation is associated with such a phenotype, Konopka and Benzer would have had to have studied conditional lethal mutations, perhaps mutations that are lethal at 27° C but viable at 21° C. The lethal mutations could have been identified at the restrictive temperature (27°) and subsequently tested for rhythm abnormalities at the permissive temperature (21°).

26.7 The *Awry* mutation is located between the *vg* and *bw* loci. Of 36 crossovers between *vg* and *bw* observed in the offspring, 16 are between *vg* and *Awry* and 20 are between *Awry* and *bw*. Thus, the *Awry* mutation is located to the right of *vg*, $(16/36) \times 100 = 44$ percent of the 37 cM distance between *vg* and *bw*. The map is *vg* (67)—*Awry* (83.3)—*bw* (104).
26.9 One approach is to test for protein/DNA binding *in vitro*. The *clk* and *cyc* genes have been cloned. With recombinant DNA techniques, cDNAs made from these genes could be fused to a bacterial promoter sequence in a plasmid vector, introduced separately into *E. coli* cells by transformation, and then induced to synthesize mRNA. The transformed cells would be expected to translate this mRNA into CLK and CYC proteins, which could then be extracted from the cells and combined *in vitro* with purified *per* and *tim* DNA. If the CLK and CYC proteins bind to the *per* and *tim* DNA, the electrophoretic mobility of that DNA will be retarded. Thus, by agarose gel electrophoresis of protein/DNA mixtures, it might be possible to determine if CLK and CYC bind to *per* and *tim* DNA. Several different protein/DNA mixtures would have to be tested: each protein separately mixed with each DNA sequence, or with portions of that sequence (for example, only the sequences upstream of the transcription start site); both proteins mixed together with each DNA, and with each portion of that DNA, because CLK and CYC may bind to the *per* and *tim* genes only as a heterodimer. It would also be necessary to test proteins known not to bind to *per* and *tim* DNA as negative controls for the experiments.
26.11 One approach is to use northern blots to assay for synthesis of *per* RNA in homozygous *clk* mutants and in homozygous wild-type flies. If the CLK protein positively regulates the expression of *per*, the homozygous *clk* mutants should synthesize much less *per* RNA than the homozygous wild-type flies. Another approach is to transform *Drosophila* with a reporter gene such as the green fluorescent protein gene from jellyfish under the control of the *per* promoter. If the CLK protein positively regulates *per* transcription, this reporter gene should not be expressed (or should be minimally expressed) in homozygous *clk* mutants, whereas it should be vigorously expressed in homozygous wild-type flies.
26.13 (a) Male with abnormal Muscle of Lawrence (MOL) anatomy and abnormal courting behavior; (b) Male with abnormal MOL anatomy and abnormal courting behavior; (c) intersex, probably with abnormal MOL anatomy and abnormal courting behavior; (d) intersex, probably with abnormal MOL anatomy and abnormal courting behavior.
26.15 Both are pleiotropic because they affect several systems. PKU affects neurological functions, pigment formation, head diameter, phenylalanine metabolism, and other traits. Lesch-Nyhan syndrome affects behavior, neurological functions, purine metabolism, and viability.
26.17 All three disorders affect behavior. However, both PKU and Lesch-Nyhan syndrome are recessive conditions and HD is a dominant condition. PKU can be managed successfully by dietary control. Neither Lesch-Nyhan nor HD can be so treated. PKU and Lesch-Nyhan syndrome are metabolic disorders caused by the failure of an enzyme to function normally. The abnormalities in HD arise from a polyglutamine tract in the huntingtin polypeptide; this polyglutamine tract may predispose huntingtin to form insoluble aggregates within neurons in a specific part of the brain. The mutations that lead to PKU and Lesch-Nyhan syndrome are typically point mutations in the coding region of the gene; the mutations that lead to HD are expansions in a trinucleotide repeat region within the gene's coding region (the region encoding the polyglutamine tract).
26.19 Down syndrome is caused by an extra copy of chromosome 21, the chromosome that carries the gene for amyloid precursor protein (APP). The extra copy of the *APP* gene in Down syndrome patients apparently causes them to develop Alzheimer's disease.
26.21 For DZ twins reared apart, the observed correlations should be half the observed correlations for MZ twins reared apart.

CHAPTER 27

27.1 The genetic information specifying antibody chains is stored in sequences of nucleotide pairs encoding sequences of amino acids, just like the genetic information specifying other polypeptides. However, the information specifying an antibody chain is stored in bits and pieces that are assembled into functional genes encoding antibody chains by genome rearrangements (somatic-cell recombination events) occurring during the development of the B lymphocytes (the antibody-producing cells). See Figures 27.10 and 27.11.
27.3 (1) The joining of different *V, D,* and *J* gene segments by somatic recombination during B lymphocyte development (see Figures 27.10 and 27.11); (2) variability in the exact location of the joining reaction during *V-D-J* joining events (see Figure 27.13); (3) somatic mutation.
27.5 IgM is the initial antibody produced by a B cell. Later, B cells may switch to the production of other classes of antibodies.
27.7 IgE.
27.9 At the DNA level; class switching occurs by somatic recombination during B lymphocyte differentiation (see Figure 27.11, steps 3a and 3b).
27.11 The *MHC* locus encodes the transplantation antigens that play a major role in the rejection of foreign tissues after organ and tissue transplant operations.
27.13 Antibody-antigen complexes are eliminated by (1) ingestion and degradation by phagocytes or (2) the proteolytic activity of complement proteins, which degrades the protein coats of viruses and lyses cellular pathogens.
27.15 Cytokines, lymphokines, and interleukens are secreted by helper T cells. These molecules stimulate the development of immature B and T lymphocytes. The immature T lymphocytes develop into killer T cells and memory T cells.
27.17 Yes, the T lymphocyte antigen receptor proteins.
27.19 Both contain polypeptides with variable regions that form the antigen-binding domains and constant regions that facilitate interactions with cell membranes and other components of the immune response.
27.21 Nonspecific immune responses include inflammation of infected tissues resulting in increased blood flow to the tissues and recruitment of phagocytes to ingest and destroy the infecting agent(s). The two major specific immune responses are the production of antigen-specific antibodies and killer T cells.
27.23 (a) Monoclonal antibodies are produced by fusing B lymphocytes from mammals that have been exposed to a specific antigen with cancer cells called myelomas to produce immortal hybrid cells called hybridomas. Each hybridoma clone produces a single antibody. The most difficult step in monoclonal antibody production is screening the resulting population of hybridoma cells for the clone producing the desired antibody. Once identified, the antibody-producing cells can be cultured indefinitely as a result of their cancerous heritage. (b) Monoclonal antibodies are invaluable for use in identifying individual polypeptides separated by gel electrophoresis (see Figure 20.23) and in determining the location of specific macromolecules in tissues and cells by immunolocalization techniques. They are also used to immunoprecipitate proteins during purification processes and to diagnose human diseases.
27.25 When helper T cells encounter antigens on the surfaces of macrophages, they secrete cytokines, lymphokines, and interleukins that stimulate B and T lympho-

cytes to differentiate into antibody-producing plasma cells and killer T cells, respectively. Thus helper T cells are involved in both the antibody-mediated and T cell-mediated immune responses.

27.27 Autoimmune diseases are disorders in which the immune system of an individual attacks specific tissues or cells of her or his own body ("self" cells). The combining term *auto* means self; they are called autoimmune diseases because of the immune system's attack on the patient's own cells.

27.29 One hypothesis regarding the cause of autoimmune diseases is that an immune response is induced against a foreign antigen—for example, one found on an invading pathogen—that is similar to an epitope of a human protein. If the immune response is not suppressed after the antigen has been eliminated, then the antibodies may attack the epitope-containing protein of the individual producing them and cause an autoimmune disease.

CHAPTER 28

28.1 Cancer has been called a genetic disease because it results from mutations of genes that regulate cell growth and division. Nonhereditary forms of cancer result from mutations in somatic cells. These mutations, however, can be induced by environmental factors including tobacco smoke, chemical pollutants, ionizing radiation, and UV light. Hereditary forms of cancer also frequently involve the occurrence of environmentally-induced somatic mutations.

28.3 Aneuploidy might involve the loss of functional copies of tumor suppressor genes, or it might involve the inappropriate duplication of proto-oncogenes. Loss of tumor suppressor genes would remove natural brakes on cell division, and duplication of proto-oncogenes would increase the abundance of factors that promote cell division.

28.5 They possess introns.

28.7 The products of these genes play important roles in cell activities.

28.9 The cultured NIH 3T3 cells probably carry other mutations that predispose them to become cancerous; transfection of such cells with a mutant *c-H-ras* oncogene may be the last step in the process of transforming the cells into cancer cells. Cultured embryonic cells probably do not carry the predisposing mutations needed for them to become cancerous; thus, when they are transfected with the mutant *c-H-ras* oncogene, they continue to divide normally.

28.11 Retinoblastoma results from homozygosity for a loss-of-function (recessive) allele. The sporadic occurrence of retinoblastoma requires two mutations of this gene in the same cell or cell lineage. Therefore retinoblastoma is rare among individuals who, at conception, are homozygous for the wild-type allele of the *RB* gene. For such individuals, we would expect the frequency of tumors in both eyes to be the square of the frequency of tumors in one eye. Individuals who are heterozygous for a mutant *RB* allele require only one somatic mutation to occur for them to develop retinoblastoma. Because there are millions of cells in each retina, there is a high probability that this somatic mutation will occur in at least one cell in each eye, causing both eyes to develop tumors.

28.13 The probability that a carrier does not develop retinoblastoma is 0.05, which is equal to the probability that the wild-type *RB* allele is not mutationally inactivated during the cell divisions that form the retinas of the eyes. If the rate of mutational inactivation is $u = 10^{-6}$ per cell division, and n is the number of cell divisions, then $0.05 = (1 - u)^n$. If we take logarithms of both sides, $\log(0.05) = n \log(1 - u)$, which to a good approximation, is equal to $n \log(u)$. After substituting 10^{-6} for u and solving, we find that $n = 1.3 \times 10^6$.

28.15 By binding to E2F transcription factors, pRB prevents those transcription factors from activating their target genes—which encode proteins involved in progression of the cell cycle; pRB is therefore a negative regulator of transcription factors that stimulate cell division.

28.17 Cells homozygous for a loss-of-function mutation in the *p16* gene might be expected to divide in an uncontrolled fashion because the p16 protein would not be able to inhibit cyclin-CDK activity during the cell cycle. The *p16* gene would therefore be classified as a tumor suppressor gene.

28.19 Cells homozygous for a loss-of-function mutation in the *BAX* gene would be unable to prevent repression of the programmed cell death pathway by the *BCL-2* gene product. Consequently, these cells would be unable to execute that pathway in repsonse to DNA damage induced by radiation treatment. Such cells would continue to divide and accumulate mutations; ultimately, they would have a good chance of becoming cancerous. The *BAX* gene would therefore be classified as a tumor suppressor gene.

28.21 If a cell were heterozygous for a mutation that caused p53 to bind tightly and constitutively to the DNA of its target genes, its growth and division might be retarded, or it might be induced to undergo apoptosis. Such a cell would be expected to be more sensitive to the effects of ionizing radiation because radiation increases the expression of p53, and in this case, the p53 would be predisposed to activate its target genes, causing the cell to respond vigorously to the radiation treatment.

28.23 They would probably decrease the ability of pAPC to bind β-catenin.

28.25 All three genes (*RB*, *p130*, and *p107*) are essential for embryonic development, although by themselves, *p130* and *p107* are dispensable, possibly because their products are functionally redundant. (Both gene products must be inactivated before any deleterious effect is seen.) In adults, only pRB appears to play a role in suppressing tumor formation.

28.27 No. Apparently there is another pathway—one not mediated by p53—that leads to the activation of the *p21* gene.

CHAPTER 29

29.1 Frequency of L^M in Central American population: $p = (2 \times 53 + 29)/(2 \times 86) = 0.78$; $q = 0.22$. Frequency of L^M in North American population: $p = (2 \times 78 + 61)/(2 \times 278) = 0.39$; $q = 0.61$.

29.3 $q^2 = 0.0004$; $q = 0.02$.

29.5 Frequency of heterozygotes combined $= 2[(0.6)(0.3) + (0.3)(0.1) + ((0.6)(0.1)] = 0.54$.

29.7 $(0.00025)^2 = 6.25 \times 10^{-8}$.

29.9 (a) Half the males will be ruby-eyed and 5 percent ($0.50 \times 0.10 \times 100$ percent) of the females will be ruby-eyed. (b) Among males, the frequency of the recessive allele will be 0.5, which was its frequency among females in the previous generation; among females, the frequency of the recessive allele will be $(0.5 + 0.1)/2 = 0.3$, which is the average of the frequencies of this allele in males and females in the previous generation.

29.11 Frequency of heterozygotes $= H = 2pq = 2p(1 - p)$. Using calculus, take the derivative of H and set the result to zero to solve for the value of p that maximizes H: $dH/dp = 2 - 4p = 0$ implies that $p = 2/4 = 0.5$.

29.13

Genotype	*AA*	*AD*	*DD*
Observed No.	20	150	30

From these data, the frequency of A is $p = (2 \times 20 + 150)/(2 \times 200) = 0.475$, and the frequency of D is $q = 1 - p = 0.525$. Using these frequencies and the Hardy–Weinberg Principle, we predict the genotype frequencies to be 0.225 (*AA*), 0.500 (*AD*), and 0.275 (*DD*), and multiplying these by the sample size (200), we predict the genotype numbers to be 45 (*AA*), 100 (*AD*), and 55 (*DD*). To compare observed numbers with predicted numbers, we use a chi-square statistic with 1 degree of freedom:

$\chi^2 = (20 - 45)^2/45 + (150 - 100)^2/100 + (30 - 55)^2/55 = 50.3$

The result strongly warrants rejection of the hypothesis that the sample data are in Hardy–Weinberg proportions.

29.15 $2pq(1 - F) = 2(0.6)(0.4)(1 - 0.5) = 0.24$.

29.17 (a) Frequency of A in merged population is 0.5, and that of a is also 0.5; (b) 0.25 (AA), 0.50 (Aa), and 0.25 (aa); (c) Frequencies in (b) will persist.
29.19 Divide each measure of reproductive ability by the largest number (996) to obtain the relative fitnesses 1 (AA), 1 (Aa), and 0.224 (aa).
29.21 Use the following scheme:

Genotype	CC	Cc	cc
Hardy–Weinberg frequency	$(0.98)^2$ = 0.9604	2(0.98)(0.02) = 0.0392	$(0.02)^2$ = 0.0004
Relative fitness	1	1	0
Relative contribution to next generation	(0.9604) × 1	(0.0392) × 1	0
Proportional contribution	0.9604/0.9996 = 0.9608	0.0392/0.9996 = 0.0392	

New frequency of the allele for cystic fibrosis = (0.5)(0.0392) = 0.0196; thus, the incidence of the disease will be $(0.0196)^2$ = 0.00038, which is very slightly less than the incidence in the previous generation.
29.23 $q^2 = 4 \times 10^{-5}$; thus $q = 6.3 \times 10^{-3}$ and $2pq = 0.0126$.
29.25 Probability of ultimate fixation of A_2 is 0.5; probability of ultimate loss of A_3 is $1 - 0.3 = 0.7$.
29.27 $p = 0.2$; at equilibrium, $p = t/(s + t)$. Because $s = 1$, we can solve for t; $t = 0.25$.
29.29 At mutation-selection equilibrium $q = \sqrt{u/s} = \sqrt{10^{-6}/1} = 0.001$.
29.31 At one level, morphological similarities should be evaluated. The populations should also be tested for reproductive isolation by trying to interbreed them. If they do interbreed, are the offspring viable and fertile?
29.33 The human and the gorilla share a common ancestor, but each species has undergone extensive evolutionary divergence since that ancestor existed.

Photo Credits

Chapter 1. Opener: Courtesy of the American Museum of Natural History. Figures 1.1a and e: Reprinted with permission from *The Scientist*, ©1998 American Association for the Advancement of Science. Figure 1.1b: Reprinted with permission from *St. Paul Pioneer Press*. Figure 1.1c: Reprinted with permission. ©1998 The New York Times Company. Figure 1.1d: Reprinted with permission. ©1998 *The Washington Post.* Figure 1.2: Courtesy National Portrait Gallery. Figure 1.4: Courtesy Affymetrix. Figure 1.5: From *Breeding Field Crops,* 3e, by John Milton Poehlman. Avi Publishing Company, Inc., 1987. Figure 1.6: Courtesy International Maize and Wheat Improvement Center, Mexico. Figure 1.7: D. Cavagnaro/Visuals Unlimited. Figure 1.8: Courtesy Poultry World. Figure 1.9a: Courtesy American Angus Association. Figure 1.9b: Courtesy Beefmaster Breeders United. Figure 1.9c: Courtesy American Simmental Association. Figure 1.9d: Courtesy American International Charolais Association. Figure 1.9e, f & h: Courtesy Sally Anne Thompson. Figure 1.9g: Courtesy R. Willbie. Figure 1.10: Courtesy Mycogen. Figure 1.14: Stan Flegler/ Visuals Unlimited. ***Historical Sidelight:*** Figure 1: The Granger Collection. Figure 2: Dennis Kunkel/Phototake.

Chapter 2. Opener: Courtesy Evelyn Schrock, Stan du Manoir, Thomas Reid. Figure 2.1: CNRI/Photo Researchers. Figure 2.3a: K.R. Porter/Photo Researchers. Figure 2.3b: Newcomb & Wergin/Tony Stone Images. Figure 2.4: Gunther F. Bahr/Tony Stone Images. Figure 2.7: Courtesy Andrew S. Bajer. Figure 2.9: David M. Phillips/Visuals Unlimited. Figure 2.11: From J. McLeish and B. Snoad, *Looking at Chromosomes,* ©1958, St. Martin's, Macmillan. Figure 2.12a: From M. Westergaard and D. von Wettstein, *Annu. Rev. Genetics,* vol. 6; copyright 1972 by Annual Reviews Inc. Original photography courtesy of D. von Wettstein. Figure 2.13: From Bernard John, *Meiosis,* Cambridge, 1990. Figure 2.18: J. Forsdyke & Gene Cox/SPL/ Photo Researchers. Figure 2.21: Carolina Biological Supply/Phototake. Figure 2.22: Lennart Nilsson, *A Child Is Born,* Dell Publishing Company. ***Technical Sidelight:*** Figure 1a: I. Wilmut, A. E. Schnieke, J. McWhir, A. J. Kind, & K. H. S. Campbell. Reproduced with permission from *Nature.* From the cover of *Nature,* 385:6619, 1997. Figure 1b: Wakayama, et al. Reproduced with permission from *Nature.* From *Nature,* 394:370, 1998. Figure 2: James King-Holmes/SPL/Photo Researchers.

Chapter 3. Opener: Courtesy of the New York Public Library and the American Museum of Natural History. Figure 3.10: Larry Lefever/Grant Heilman Photography. Figure 3.15: UPI/Corbis-Bettmann. Figure 3.16: Courtesy of McKay Kunz and Charles Kunz.

Chapter 4. Opener: Blair Seitz/Photo Researchers. Figure 4.10: Verjik Martin/Amstock. Figure 4.11: Courtesy Lester V. Bergman & Associates, Inc. Figure 4.13: Courtesy Ralph Somes. Figure 4.16a: Courtesy New York Public Library. ***Historical Sidelight:*** Figure 1a & b: The Granger Collection. Figure 1c: Courtesy Isabella Stewart Gardner Museum, Boston. Figure 1d: Roger Viollet/Gamma Liaison.

Chapter 5. Opener: Science Photo Library/Photo Researchers. ***Technical Sidelight:*** Figure 1: Courtesy Frank D. Enfield, University of Minnesota.

Chapter 6. Opener: Courtesy Stanley J.P. Iyadurai. Figure 6.9: Corbis-Bettmann. Figure 6.11: From Richards and Sutherland, *Trends in Genetics,* 8(7):249, 1992. Photo Courtesy Grant Sutherland. Figure 6.20: Grant Heilman/Grant Heilman Photography. Figure 6.22: From N. Ason/DeHaan, *Biological World,* Fig. 9.8, 1973. ***Historical Sidelight:*** Figure 1: Columbia University, Columbiana Collection.

Chapter 7. Opener: Kunio Owaki/The Stock Market. Figure 7.2: From S. E. Schlarbaum and T. Tsuchiya, "Chromosomes of Incense Cedar," *Journal of Heredity,* 66:41–42, 1975. Figure 7.3: From C. G. Vosa, "The Quinacrine-Flourescent Patterns of the Chromosomes of *Albium carinatum,*" *Chromosoma,* 33:382–385, 1971. Figure 7.4a: From R. M. Patterson & J. C. Petricciani, "A Comparison of Prophase & Metaphase G-bands in the Muntjak," *Journal of Heredity,* 64(2):80–82, 1973. Figure 7.4b: From R. S. Verma, H. Dosik, and H. A. Lubs, Jr., "Demonstration of Color and Size Polymorphisms in Human Acrocentric Chromosomes," *Journal of Heredity,* 68:262–263, 1977. Figure 7.4c: From A. N. Bruere, H. M. Chapman, P. M. Jaine & R. M. Morris, "Origin and Significance of Centric Fusions in Domestic Sheep," *Journal of Heredity,* 67:149–154, 1976. Figure 7.5: From "The Normal Human Karyotype," originally reproduced from "An International System for Human Cytogenetic Nomenclature," (1978) in *Birth Defects, Original Article Series,* XIV: 8, 1978, The National Foundation, New York, and *Cytogenetics and Cell Genetics,* 21:6, 1978, S. Karger. AG, Basel, Switzerland. Figure 7.6 (left): C. G. Maxwell/National Audubon Society/Photo Researchers. Figure 7.6 (middle top): Eric Crichton/Bruce Coleman, Inc. Figure 7.6 (right): M. Timothy O'Keefe/Bruce Coleman, Inc. Figure 7.6 (middle bottom): Ken Wagner/Phototake. Figure 7.12: Peter J. Bryant/Tony Stone Images. Figure 7.15a: Steve Dunwell/The Image Bank. Figure 7.15b: Courtesy R. M. Fineman. Figure 7.18a: Courtesy Irene A. Uchida, Department of Pediatrics, McMaster University. Figure 7.18b: Courtesy University of Utah Cytogenetics Laboratory. Figures 7.19 & 7.23b: Courtesy J. K. Lim. ***Historical Sidelight:*** Figure 1: IMS Creative/Custom Medical Stock. ***Technical Sidelight:*** Figure 1: David Jackson/A-Z Botanical Collection Ltd.

Chapter 8. Opener: B. John/Cabisco/Visuals Unlimited. Figure 8.2: UPI/Corbis-Bettmann. Figure 8.10: Courtesy Bernard John.

Chapter 9. Opener: Biophoto Associates/Photo Researchers. Figure 9.6: From *Introduction to Genetic Analysis,* 5e., by Griffiths, et al. ©1993. W. H. Freeman and Company. Reproduced with permission. Figure 9.8: Science Source/Photo Researchers. Figure 9.21: Courtesy Beverly Emanuel, Children's Hospital, Philadelphia, PA. ***Technical Sidelight:*** Figure 1a: From B. H. Judd, M. W. Shen and T. C. Kaufman, *Genetics,* 71:139–152, 1972.

Chapter 10. Opener: Biology Media/Science Source/Photo Researchers. Figure 10.1: From O. T. Avery, C. M. MacLeod, and M. McCarty, *Journal of Experimental Medicine,* 79:153, 1944; by copyright permission of Rockefeller University Press. Figure 10.9: Ted Thai/Time Magazine. Figure 10.10: Courtesy M. H. F. Wilkins, Biophysics Dept., King's College, London. Figure 10.15: Courtesy Svend Freytag. Figure 10.21: Omikron/Science Source/Photo Researchers. Figure 10.22: From Vollrath & Davis, *Nucleic Acid Res.* 15:7876, 1987. Oxford University Press. Photo courtesy Douglas Vollrath. Figure 10.23: From R. Kavenoff, L. C. Klotz, & B. H. Zimm, *Cold Spring Harbor Symposium of Quantitative Biology,* 38:1–8, 1973. Copyright © 1973 by Cold Spring Harbor Laboratory. Photo courtesy R. Kavenoff. Figure 10.24: Courtesy F. Thoma. Figure 10.26b & c: Reprinted with permission from Karolyn Luger, et al. *Nature* 389:251, 1997. Photo courtesy Timothy J. Richmond. Figure 10.26d: Courtesy D. Rhodes. ©1997 Macmillan Magazines Limited. Figure 10.27: From E. DuPraw, *DNA & Chromosomes,* Holt, Rinehart, & Winston, New York, 1970. Photo courtesy E. DuPraw. Figure 10.29: From J. R. Paulson & U. K. Laemmli, *Cell,* 12:817–828, 1977. ©1977 MIT; published by MIT Press. Photo courtesy U. K. Laemmli. Figure 10.35b: Courtesy Mary Lou Pardue, MIT. Figure 10.35d: From J. Meyne, in R. P. Wagner, *Chromosomes: A Synthesis,* Copyright ©1993. Reproduced with permission of Wiley-Liss, Inc., a subsidiary of John Wiley & Sons, Inc.

Chapter 11. Opener: Courtesy Karl Fredga, Department of Genetics, Uppsala University, Sweden. Figure 11.4: From J. H. Taylor, "The Replication and Organization of DNA in Chromosomes," *Molecular Genetics,* Part I, J. H. Taylor (ed.), Academic Press, New York, 1963. Figure 11.5: From J. Cairns, Imperial Cancer Research Fund, London, England. Reproduced with permission from *Cold Spring Harbor Symposium of Quantitative Biology,* 28:43, 1963. Photo courtesy John Cairns. Figure 11.8: Photo reproduced with permission from M. Schnös and R. B. Inman, *Journal of Molecular Biology,* 51:61–73, 1970. Copyright ©1970 Academic Press, Inc, London. Figure 11.10: From J. Wolfson, D. Dressler, and M. Magazin, *Proceedings of the National Academy of Sciences,* U.S.A., 69:499, 1972. Micrographs courtesy D. Dressler. Figure 11.14: From S. Doublié, S. Tabor, A. M. Long, C. C. Richardson, and T. Ellenberger, *Nature,* 391:251–258. Figure 11.15: From X. P. Kong, et al., *Cell,* 69:425–437, 1992. ©1992 Cell Press. Photograph courtesy John Kuriyan, Howard Hughes Medical Institute, Rockefeller University. Figure 11.29 (inset): From K. Koths and D. Dressler, Harvard Medical School. Figure 11.30: Reproduced with permission from J. A. Huberman and A. D. Riggs, *Journal of Molecular Biology,* 32:327–341, 1968. Copyright ©1968 by Academic Press. Photographs courtesy J. A. Huberman. Figure 11.31: From D. R. Wolstenholme, *Chromosoma,* 43:1, 1973. Photograph courtesy D. R. Wolstenholme. Figure 11.33: Courtesy Steven L. McKnight and Oscar L. Miller, Jr. Figure 11.35: AP/Wide World Photos.

Chapter 12. Opener: Reproduced with permission from R. Reed, J. Griffith, and T. Maniatis, *Cell,* 53:949–961, 1988. © 1988 by Cell Press. Micrograph courtesy J. D. Griffith. Figure 12.6: From D. Prescott, "Cellular Sites of RNA Synthesis," *Prog. Nucleic Acid Res. Mol. Biol.,* 3:33–57, 1964. Figure 12.14: From O. L. Miller, Jr., B. A. Hamkalo, and C. A. Thomas, Jr., *Science,* 169:392–395, 1970. ©1970 American Association for the Advancement of Science. Micrograph courtesy O. L. Miller, Jr. Figure 12.18: Courtesy Seth Darst. Figure 12.25: From Tilghman et al., *Proceedings of the National Academy of Sciences,* U.S.A., 75:1309–1313. Figure 12.27: Courtesy Pierre Chambon. Figure 12.31: Courtesy Jack Griffith, Lineberger Comprehensive Cancer Center, University of North Carolina at Chapel Hill.

Chapter 13. Opener: Stanley Flegler/Visuals Unlimited. Figure 13.8: Courtesy E. G. Jordon, Molecular Biology, King's College, London. Figure 13.9: O. L. Miller, B. R. Beatty, D. W. Fawcett /Visuals Unlimited. Figure 13.12a: From S. H. Kim, F. L. Suddath, G. J. Quiqley, A. McPherson, J. L. Sussman, A. H. J. Wang, N. C. Seeman, and A. Rich, *Science,* 185:435–440, 1974. ©1974 American Association for the Advancement of Science. Photo courtesy S. H. Kim. Figure 13.14: Courtesy Joachim Frank. From Frank, et al., *Biochemistry and Cell Biology,* 73:357, 1995. Figure 13.18: From S. L. McKnight, N. L. Sullivan, and O. L. Miller, Jr., *Prog. Nucleic Acid Res. Mol. Biol.* 19:313–318, 1976. Micrograph courtesy S. L. McKnight and O. L. Miller, Jr., University of Virginia.

Chapter 14. Opener: David Scharf/Peter Arnold. Figure 14.1: Photograph by P. E. Polani, Guy's Hospital, London. Reprinted with permission of Macmillan Publishing Co., Inc., from E. Novitski, *Human Genetics,* ©1977 by Edward Novitski. Figure 14.2a: Frank P. Rossotto/The Stock Market. Figure 14.2b: Kenneth W. Fink/Bruce Coleman, Inc. Figure 14.25: From B. N. Ames, J. McCann, and E. Yamasaki, *Mutat. Res.* 31:347, 1975. Photograph courtesy B. N. Ames. Figure 14.32: Courtesy H. Potter and D. Dressler, Harvard Medical School.

Chapter 15. Opener: Reinhard Kunkel/Peter Arnold, Inc. ***Historical Sidelight:*** Figure 1: Nancy M. Hamilton/Photo Researchers.

Chapter 16. Opener: Barry Dowsett/SPL/Photo Researchers. Figure 16.1: Courtesy Robley Williams, University of California, Berkeley and Harold Fisher, University of Rhode Island. Figure 16.2: Courtesy John Finch, Cambridge University. Figure 16.4: Photographs by R. Wagner and D. P. Snustad. Figures 16.5 & 16.7: From G. S. Stent, *Molecular Biology of Bacterial Viruses.* © W. H. Freeman and Company, 1963. Figure 16.8: ©Bruce Iverson.

Chapter 17. Opener: Courtesy Charles C. Brinton, Jr., and Judith Carnahan. Figure 17.3: Lennart Nilsson/Bonnier Alba AB, © Boehringer Ingelheim International GmbH.

Chapter 18. Opener: Tom Bean/Tony Stone Images. ***Historical Sidelight:*** Figure 1: Reproduced with permission of Marjorie M. Bhavnani. Photo courtesy Cold Spring Harbor Laboratory Archives.

Chapter 19. Opener: Frans Lanting/Minden. Figure 19.2: Courtesy Scott Poethig, University of Pennsylvania. Figure

19.7: R. Kessel & C. Y. Shih/Visuals Unlimited. Figure 19.12: From Y. Hayashi and K. Ueda, *Journal of Cell Science,* 93:565, 1989, Company of Biologists, Ltd. Photograph courtesy K. Ueda.

Chapter 20. Opener: Grantpix/Photo Researchers. Figure 20.15: Courtesy S. Kopczak and D. P. Snustad, University of Minnesota. Figure 20.18: Courtesy D. P. Snustad. Figure 20.20 Courtesy S. R. Ludwig and D. P. Snustad, University of Minnesota. Figure 20.21: Courtesy S. R. Ludwig and D. P. Snustad, University of Minnesota. Figure 20.23: Courtesy M. G. Li and D. P. Snustad, University of Minnesota. Figure 20.28: From *DNA Sequencing,* 2e., Amersham Life Science. ***Technical Sidelight:*** Figure 1: From Kerem, et al., *Science,* 245:1073–1080, 1989.

Chapter 21. Opener: From Fleischman, et. al, *Science,* 269:496–512, 1995. Figure 21.6a: Courtesy Johng Lim. Figure 21.6b: From Y-S. Fan, L. M. Davis, and T. B. Shows, *Proceedings of the National Academy of Sciences,* USA, 87:6223–6227, 1990. Figure 21.11: Courtesy Frederick Blattner. Figure 21.17: Courtesy CLONTECH Laboratories, Inc. Figure 21.18: Courtesy Affymetrix. Figure 21.19b, c, & d: From Ludin & Matus, *Trends in Cell Biology,* 8:72, 1998. Figure 21.21a & b: Weinberg & Stanyon, *Current Opinions in Genetics & Development,* 5:792–797, 1995. ***Human Genetics Sidelight:*** Figure 1 (left): Courtesy National Institutes of Health. Figure 1 (right): Courtesy Celera Genomics.

Chapter 22. Opener: Courtesy Susan Lanzendorf, Jones Institute for Reproductive Medicine, Eastern Virginia Medical School. Figure 22.3: From *Cell,* 72:971–983, 1993. ©1993 Cell Press. Figures 22.9 & 22.10: Courtesy Cellmark Diagnostics, Germantown, Maryland. Figure 22.15: Courtesy R. L. Brinster, School of Veterinary Medicine, University of Pennsylvania. Figure 22.16: From Devlin, et al., *Nature,* 371:209–210, 1994. Figure 22.17: Photograph courtesy G. Panzour and A. Das, University of Minnesota. ***Technical Sidelight:*** Figure 1: Hulton Deutsch Collection/Corbis.

Chapter 23. Opener: Courtesy Dwight Anderson.

Chapter 24. Opener: Oliver Meckes/Photo Researchers. Figure 24.1: Courtesy D. P. Snustad. Figure 24.7: From G. Coruzzi, R. Broglie, C. Edwards and N. H. Chua, "Tissue-specific and light-regulated expression of a pea nuclear gene encoding the small subunit of ribulose-1, 5-bisphosphate carboxylase," *EMBO J.* 3:1671–1679, 1984. Figure 24.11: Reproduced with permission from E. Serfling, M. Jasin, and W. Schaffner, *Trends in Genetics,* 1:224–230, 1985. Figure 24.13: B. R. Beatty, and D. W. Fawcett/Visuals Unlimited. Figure 24.14: Jack M. Bostrack/Visuals Unlimited. Figure 24.16: Courtesy Joseph Fong, University of Minnesota. Figure 24.24: From M. I. Kuroda et al., *Cell,* 66:935–947, 1991. Photograph courtesy Mitzi Kuroda.

Chapter 25. Opener: Butterfly: Ross Hamilton/Tony Stone Images. Dragonfly: Rod Planck/Tony Stone Images. Beetle: James H. Carmichael, Jr./The Image Bank. Drone fly: Stephen Dalton/Photo Researchers. Figure 25.1: Courtesy Edward B. Lewis, California Institute of Technology. Figure 25.3: Courtesy Michael M. Craig, Southwest Missouri State University. Figure 25.4: Courtesy Kevin Haley, University of Minnesota. Figure 25.6: From J. E. Sulston and H. R. Horvitz, *Developmental Biology,* 56:110–156, 1977. Photograph courtesy John Sulston. Figure 25.15: Daniel St. Johnston and Christiane Nüsslein-Volhard, *Cell,* 68:201–219, 1992. Photograph courtesy Christiane Nüsslein-Volhard. Figure 25.23: Courtesy Matthew Scott, Howard Hughes Medical Institute. Figure 25.28: Courtesy Walter Gehring, Universität Basel, Switzerland.

Chapter 26. Opener: Richard Wiess/Peter Arnold, Inc. Figure 26.1: Treat Davidson/Photo Researchers. Figure 26.6: Courtesy Jeffrey Hall. Figure 26.8: From J. R. Brown et al., "Nurturing and Not Nurturing in Mice," *Cell,* 86:297–309, 1996. ©1996 Cell Press. Photo courtesy Jennifer R. Brown, Children's Hospital Neurobiology Lab, Harvard Medical School. Figure 26.10: Courtesy William L. Nyhan, Department of Pediatrics, University of California, San Diego.

Chapter 27. Opener: NIBSC/SPL/Photo Researchers. Figure 27.1: Lennart Nilsson/Bonnier Alba AB, © Boehringer Ingelheim International GmbH. Figure 27.4: From Amit et al., *Science,* 233:747, 1986. ©1986 American Association for the Advancement of Science. Photograph courtesy R. J. Poljak. Figure 27.9: A. Liepins/Photo Researchers.

Chapter 28. Opener: Dr. Neal Scolding/Photo Researchers.

Chapter 29. Opener: Painting by Paul Gauguin, "Where Do We Come From? What Are We? Where Are We Going?," 1897. Museum of Fine Arts, Boston. Figure 29.1a: Courtesy Victory Ranch Alpacas. Mora, New Mexico. Figure 29.1b: Courtesy New York Public Library. Figures 29.3a & c: SPL/Photo Researchers. Figure 29.3b: Corbis. Figure 29.9: Courtesy Lawrence Cook, University of Manchester. Figure 29.17a (left): Alexander Lowry/Photo Researchers. Figure 29.17a (right): Ron Sanford/Photo Researchers. Figure 29.17b: Jim Brandenburg/Minden. Figure 29.18b: Darrell Gulin/Tony Stone Images. Figure 29.23 (center): Carr Clifton. Figure 29.23 (left): Pat & Tom Leeson/Photo Researchers. Figure 29.23 (right) C. K. Lorenz/Photo Researchers. Figure 29.25: Courtesy Maurice J. Tauber and Catherine A. Tauber, Cornell University.

Conversations: Lewis: Courtesy Edward B. Lewis. Soyfer: Courtesy Valery N. Soyfer. Mosig: Courtesy Gisela Mosig. Wexler: Courtesy Nancy Wexler. Kuroda: Courtesy Mitzi Kuroda. Wieschaus: Courtesy Eric Wieschaus. Charlesworth: Courtesy Deborah and Brian Charlesworth. Crow: Courtesy James F. Crow.

Illustration Credits

Chapter 4. Figure 4.11b: From *Principles of Human Genetics,* 3/e by Curt Stern, © 1973 by W. H. Freeman and Company; © renewal 1988 by W. H. Freeman and Company. Used with permission.

Chapter 5. Figure 5.9: From Nance, W. E., Jackson, C. E., and Witkop, C. J., Jr. 1970. *American Journal of Human Genetics* 22:579–586. Used with permission of the University of Chicago Press. Figure 5.12: Data from Hallauer, A. R., and Sears, J. H., 1973. *Crop Science* 13:327–330.

Chapter 7. ***Technical Sidelight,*** Figure 2: From Burnham, C. 1962. *Discussions in Cytogenics.* Reprinted by permission of the author's estate.

Chapter 9. Figure 9.14: Detera-Wadleigh, S. D., et al. 1997. *American Journal of Medical Genetics* 74:254–262. Copyright © 1997 John Wiley and Sons. Reprinted by permission of Wiley-Liss, a division of John Wiley and Sons, Inc. ***Technical Sidelight,*** Figure 1b: Judd, B. H., Shen, M. W., and Kaufman, T. C. 1972. *Genetics* 71:139. Reprinted by permission of the Genetics Society of America.

Chapter 11. Figure 11.26: Adapted from *DNA Replication* by Kornberg and Baker © 1992 by W. H. Freeman and Company. Used with permission.

Chapter 12. Figure 12.30: Reprinted with permission of *Nature* from Zang, N. J., Grabowski, P. J., and Cech, T. R. 1983. *Nature* 301:578–583. Copyright 1983 Macmillan Magazines Limited. Figure 12.32: Reprinted with permission of *Nature* from Brow, D. A., and Guthrie, C. 1989. *Nature* 337:14–15. Copyright 1989 Macmillan Magazines Limited.

Chapter 13. Figures 13.3, 13.4: From Alberts, B., et al., *Molecular Biology of the Cell, 3/e,* page 114. New York: Garland Publishing Inc., 1994. Figure 13.5: Figure from *Biology,* Second Edition by Claude A. Villee, Eldra Pearl Solomon, Charles E. Martin, Diana W. Martin, Linda R. Berg, and P. William Davis, copyright © 1989 by Saunders College Publishing, reproduced by permission of Harcourt Brace & Company. Figure 13.11: Reprinted with permission from Holley, R. W., et al. 1965. *Science* 147:1462–1465. Copyright 1965 American Association for the Advancement of Science.

Chapter 16. Figures 16.11, 16.12, 16.13: From Benzer, S. 1961. *Proceedings of the National Academy of Sciences* 47:403–415. Figure 16.18: Reprinted with permission of *Nature* based on the data of Barrell, B. G., Air, G. M., and Hutchinson, C. A., III. 1976. *Nature* 264:34. Copyright 1976 Macmillan Magazines Limited. Figure 16.20: From Fauci, A. S. 1988. *Science* 239:617. Figure 16.23: From Greene, W. C. 1991. *New England Journal of Medicine* 324:308. Copyright © 1991 Massachusetts Medical Society. All rights reserved. Adapted with permission.

Chapter 18. Figure 18.15: From Lim, J. K., and Simmons, M. J. 1994. *BioEssays* 16:269–275. © ICSN Press. Figure 18.17: From Lindsley, D. L., and Zimm, G. *The Genome of* Drosophila Melanogaster. Orlando: Academic Press, Inc., 1992.

Chapter 19. Figures 19.8, 19.9: After Sager, R. *Cytoplasmic Genes and Organelles,* Orlando: Academic Press, Inc., 1972. Reprinted by permission. Figures 19.13, 19.17: From *Molecular Cell Biology* by Darnell, Lodish, and Baltimore © 1995, 1990, 1986 by Scientific American Books. Used with permission by W. H. Freeman and Company. Figure 19.15: From Wissinger, B., Brennicke, A., and Schuster, W. 1992. *Trends in Genetics* 8:322–328.

Chapter 21. Figure 21.5: Based on data from NIH/CEPH Collaboration Mapping Group, 1992. *Science* 258:67–86. Figure 21.12: Data used with permission from Blattner, F. R., et al. 1997. *Science* 277:1453, Table 4. Copyright 1997 American Association for the Advancement of Science. Figure 21.15: From Stewart, E. A., et al. 1997. *Genome Research* 7:422–433. Figure 21.20: After Moore, et al. 1995. *Current Opinion Genetics & Development* 5:537 and Gale, M. D., and Devos, K. M. 1998. *Proceedings of the National Academy of Sciences* 95:1971–1973. Figure 21.22: From Chowdhary B. P., et al., 1998. *Genome Research* 8:577–584. ***Human Genetics Sidelight,*** Figure 2: After Messing, J. W., and Llaca, V. 1998. *Proceedings of the National Academy of Sciences* 95:2017. Copyright 1998 National Academy of Sciences, U.S.A.

Chapter 22. Figure 22.1: From Huntington's Disease Collaborative Research Group. 1993. *Cell* 72:971–983. Copyright Cell Press. Figure 22.4: From Marx, J. L. 1989. *Science* 245:923–925. Reprinted by permission of Dr. Lap-Chee Tsui, The Hospital for Sick Children, Toronto, Canada. Figures 22.5, 22.6: Reprinted by permission from Collins, F. S. 1992. *Science* 256:774–779. Copyright 1992 American Association for the Advancement of Science.

Chapter 23. Figure 23.16, 23.23. From Lewin, B. 1994. *Genes V.* Copyright Cell Press. Figure 23.25: From Jinks-Robertson, S., and Nomura, M. 1982. *Escherichia coli* and *Salmonella typhimurium, Cellular and Molecular Biology* 2: 1358–1385.

Chapter 24. Figure 24.6: From *Molecular Cell Biology* by Darnell, Lodish, and Baltimore. © 1996, 1990, 1985 by Scientific American Books. Used with Permission by W. H. Freeman and Company.

Chapter 25. Figure 25.29: Reprinted from *Trends in Genetics* 10, Duboule and Morata, pages 358–364, copyright 1994, with permission from Elsevier Science. Figure 25.30: From McGinnis, W., and Krumlauf, R. 1992. *Cell* 68:283–302. Copyright Cell Press.

Chapter 26. Figure 26.2a: From Hartl, D. *Genetics* third edition, © 1994 Sudbury, MA: Jones and Bartlett Publishers, www.jbpub.com. Reprinted with permission. Figure 26.2b: Reprinted with permission from *Nature* from Citri, Y., et al., 1987. *Nature* 346:82. Copyright 1987 Macmillan Magazines Limited. Figure 26.2c: From Konpka, R. J., and Benzer, S. 1971. *Proceedings of the National Academy of Sciences* 68:2112. Figure 26.3a, b: Kryiacou, C. P., et al., in *Molecular Genetics of Biological Rhythms,* edited by M. W. Young. New York: Marcel Dekker, Inc., 1992, p. 171. Reprinted by courtesy of Marcel Dekker, Inc.

Chapter 27. Figure 27.3b: From Silverton, E. W., Navia, M. A., and Davies, D. R. 1977. *Proceedings of the National Academy of Sciences* 75:5140–5144.

Chapter 28. Figure 28.1: From Karp, G., *Cell and Molecular Biology, 2/e,* Copyright © 1997 John Wiley & Sons, Inc. Reprinted by permission of John Wiley & Sons, Inc. Figure 28.10a: From Oren, M. 1997. *Cell* 90:829–832. Copyright 1990 Cell Press. Figure 28.11a, 28.12a: From Kinzler, K. W., and Vogelstein, B. 1996. *Cell* 87:159–170. Copyright Cell Press. Figure 28.11b: Reprinted from *Trends in Genetics* 13, Moon, R. T., and Miller, J. R. The APC tumor suppressor protein in development and cancer, pages 256–258, copyright 1997, with permission from Elsevier Science. Figure 28.12b: Based on Latil, A., and Lidereau, R. 1998. *Virchow's Archiv* 432:389–406. Figure 28.12c: From Ng, H. K., and Lam, P. V. P. 1998. *Pathology* 30:196–202. Carfax Publishing, 11 New Fetter Lane, London, EC4P 4EE.

Chapter 29. Figure 29.19: From *Genetics of the Evolutionary Process,* by Theodosius Grigorievich Dobzhansky. Copyright © 1972 Columbia University Press. Reprinted with permission of the publisher. Figures 29.20, 29.22: From *Evolution* by Dobzhansky et al. © 1977 by W. H. Freeman and Company. Used with permission. Figure 29.26: Based on Wood, B. 1996. *BioEssays* 18:945–954. Copyright © 1996 John Wiley & Sons, Inc. Reprinted by permission of Wiley-Liss, Inc., a division of John Wiley & Sons, Inc.

Index

NOTE: An *f* after a page number denotes a figure; *t* denotes a table; *HG* denotes a Human Genetics Sidelight; *TS* denotes a Technical Sidelight, and *HS* denotes a Historical Sidelight.

A

A-DNA, 231, 231t
Abortive transduction, 467, 468f
Acquired immune deficiency syndrome, *see* AIDS; HIV infection
Activators, of prokaryote gene expression, 619f, 620
Additive genetic variance, 102
Adenine, 226f
  base-pairing in DNA, 228, 230f
Adenosine deaminase-deficient severe combined immunodeficiency syndrome (ADA⁻SCID), 599–600, 600f, 732
Adenosine monophosphate, cyclic, *see* Cyclic AMP
Adjacent disjunction, 158, 158f
Aerobic (oxidative) metabolism, 497
Aging
  and abnormal chromosome number, 36, 37, 151
  and mutations, 387–388
  and nondisjunction, 151
  and telomere length, 285–287
Agricultural genetics, 11–15
  see also Selective breeding; Selection, artificial
AIDS, 445, 450, 728–729, 736
Albinism, 67–68, 68f, 368f, 369, 399HG
Alcoholism, 726
Alkaptonuria, 7
Allele frequencies, 777–782
  see also Population genetics
  dynamic equilibrium, 788
  effect of natural selection, 783–785, 784f
  how to calculate, 777–778
  random changes, 786–787
  relation to genotype frequencies, 778, 779
  use in genetic counseling, 781–782
Alleles, 55
  exchange between chromosomes, *see* Crossing over; Recombination
  gain-of-function, 79–80, 79f
  heteroalleles vs. homoalleles, 415
  hypomorphic, 75
  isoalleles, 363
  loss-of-function, 79–80, 79f
    and cancer, 759, 759f
  Mendel's principles, 56
  multiple, 74–75
    see also Polymorphisms, genetic
    Hardy-Weinberg analysis, 779
  mutant, 74
  neomorphic, 79
  nonfunctional (null, amorphic), 75, 363
  segregation, 56
  symbols for, 55–56, 56f, 64–65, 64f
  wild-type, 74
Allelic exclusion, 746
Allelic segregation, *see* Segregation
Allelic series, 75, 75f
Allelic variation, 73–80
  symbols for, 73f, 74, 74f, 171
Allelism, structural vs. functional, 415
Allelomorph, 78HS
Allopolyploids, 146
Allosteric proteins, 640
Allosteric transitions, 620, 640–641, 641f
Alternate splicing, *see under* Splicing
Alu sequences, 251
Alzheimer's disease, 720–721
Amber mutants, 350t, 353, 353f, 354, 415, 351HS
Ames test for mutagens, 381–383, 382f
Amino acids, 18, 293
  abbreviations, 328f
  attachment to tRNAs, 332f, 335
  creation, 294–295, 297f
  in histones, 235, 235f
  inteins, 331
  peptide bonds, 327, 328f, 329f
  in polypeptides, 19f, 327–329, 328f
    colinearity with nucleotide pairs, 400, 409, 409f, 410f, 411f
  predicting from gene sequence, 518
  R groups (side chains), 327, 328f
  sequencing, in evolution studies, 790TS-791TS
  structure, 327, 328f
Aminoacyl-tRNA synthetase, 335
Aminoacyl-tRNAs, 332f, 335, 336–337, 337f
Amniocentesis, 154HG
Amphidiploids, 147
Amplification
  see also Repetitive DNA; Genetic redundancy
  of exons, 587–588, 588f
  of genes, 666–668
  by polymerase chain reaction (PCR), 541–543, 542f
  of recombinant DNA molecules, 522–525
Analysis of variance, 100
Anchor markers, anchor sequences, 555, 557, 558f
Aneuploidy, 142, 149–155
  detection in human fetus, 154HG
  hypo- vs. hyperploidy, 150
  from nondisjunction, in humans, 150–151, 150f, 151f, 152t
  in translocation heterozygotes, 158–159, 158f
  vs. polyploidy, 142–143
Animal cell, 24f
Antennapedia complex (ANT-C), 696–697, 697f
  vertebrate homologs, 703–704, 704f
Antibiotic resistance, 361–362, 453–454, 472, 500–502, 501f
  and transposons, 480, 481, 482f
Antibodies, 730, 731t, 732–734, 732f
  allelic exclusion, 746
  class switching, 734, 745, 745f
  classes, 733–734, 734t
  clonal selection, 738
  complexes with antigen, 730, 733f, 736, 738, 738f
  diversity, 734, 740–745
    from somatic hypermutation, 744–745
    from variable gene segments, 740–743
    from variable joining sites, 743–745, 744f
  domains, 732–733
  gene assembly, 420–421, 421f, 734, 740–745, 741f, 742f
    for heavy chain genes, 741–742, 742f
    for light chain genes, 740–741, 741f
    signal sequences, 742–743, 743f
    somatic recombination, 740, 741, 741f
    tissue-specific enhancers, 746, 748, 748f
  in immune response, 736–738, 737f
  membrane-bound vs. secreted, 745
  monoclonal, 747TS
  regulation of gene expression, 746, 748
  structure, 732–733, 733f
  use in protein analysis, 540–541
Antibody-mediated (humoral) immune response, 730, 736–738, 737f
Anticodons, 296, 335, 336, 336f, 352
Antigen-antibody complexes, 730, 733f, 736, 738, 738f
Antigen-binding sites
  of antibodies, 732, 733f
  of T cell receptors, 735
Antigens, 730, 731t
  in immune responses, 736, 737f, 739
  in transplantation, 735
Antioncogenes, 759
Antisense and sense strands, 300
Antisense RNA, 608, 609f
  research uses, 608–610
APC gene, APC gene product, 764, 767
Apes, 799
Apomixis, 145
Apoptosis, 762, 763f
  see also Cell death
Array (dot blot) hybridization, 574, 574f, 575f

ARS elements (autonomously replicating sequences), 263, 525, 525f
Artificial chromosomes, 524–525, 525f
Artificial selection, *see* Selection, artificial; Selective breeding
Ascomycetes, *see Saccharomyces*
Asexual reproduction, in plants, 145
Assortment, *see* Independent assortment
Attached-X chromosomes, 159, 161, 161f
Attenuation, of *trp* operon, 627–629, 628f, 630f
Autoantibodies, 321
Autocatalytic splicing, 319–320, 320f
Autogenous (self-) regulation, 633–634, 633f, 639
Autoimmune diseases, 736
Autonomous genes, processes, *see* Cell-autonomous genes
Autonomously replicating sequences (ARS elements), 263, 525, 525f
Autopolyploids, 146
Autoradiography, 238, 239f
Autosomes, 119
  human
    in karyotype, 141, 142f-143f
    linkage, 201
    mapping, 203–204, 204–211, 204f
    translocation with X chromosome, 206–208, 208f
Auxotrophic mutants (auxotrophs), 369, 456
Average, statistical, 97
Axes of body
  determination of, 690–695, 691f, 692f, 694f-695f
  left-right asymmetries, 706

B

B cells (B lymphocytes), 730, 731t, 732f
  activated, 737f, 739–740
  in antibody-mediated immune response, 736, 737f
  genome rearrangements, 740–745, 741f, 742f
  memory B cells, 731, 731t, 732f, 737f, 739–740
Bacteria, 23–24, 23f
  see also Prokaryotes; *E. coli*
  as ancestors of organelles, 509, 511–512
  chromosomes, 233–234, 234f
  genetic exchange, 454–455, 455f
    see also Conjugation; Transduction; Transformation
    evolutionary significance, 472
  genetics, 453–475
  Hfr strains, 461–463, 462f, 479
  mapping of genes, 454, 567–569
    closely linked genes, 465–466, 466f
    and conjugation, 464–465
    and transduction, 470–472
    and transformation, 458–460
  mutant phenotypes, 455–456
  as protein factories
    for eukaryotic proteins, 601–603
      human insulin production, 520HG-521HG
    for industrial proteins, 602–603
  recombination, evolutionary significance, 472
  resistance to antibiotics, 361–362, 453–454, 472
    and transposons, 480, 481, 482f
  sexual exchange, 454
    see also Conjugation; Transduction; Transformation
    evolutionary significance, 472
  symbols for phenotypes and genotypes, 456
  transposable genetic elements, 477–481
  viruses, *see* Bacteriophage *entries*
Bacterial artificial chromosomes (BACs), 525
Bacterial plaques, 434, 434f, 435f
Bacteriophage lambda
  as cloning vector, 523, 524f
  lambda repressor, 631–633, 632f, 633f
    autogenous (self-) regulation, 633–634, 633f
    and lambda switch, 635–637
  lysis vs. lysogeny, 468, 630–638, 631f, 637f
    lambda switch, 635–637, 636f
  lysogeny (lysogenic state), 468, 469f, 630–631, 631f, 633
  lytic pathway, 630–631, 631f
    repression, 631–634, 632f
  lytic regulatory cascade, 634–635, 634f
  prophage, 468
  and specialized transduction, 468–470, 471f
  transcription antiterminators, 635, 635f, 637f
Bacteriophage T2, 222–224, 224f
Bacteriophage T4
  genetic map, 439f, 440f, 441–443, 441f
  life cycle, 432–434, 433f
  linear chromosome, 441–443, 442f, 443f
  lysis by, 432, 433f
  morphogenesis, 370, 370f
  structure, 432f
Bacteriophages, 430
  cascades (preprogrammed circuits) of gene expression, 617, 634, 637–638, 638f
  as cloning vectors, 523, 524f
  discovery, 615–616
  filamentous, 531
  genome mapping, 434–441
    by recombination techniques, 436–437
    deletion mapping, 437–441, 438f, 439f, 440f
  helper phage, 531, 534f
  life cycle, 432–434
  plaque morphology, 434, 434f, 435f
  recombination, 435–437, 435f, 437f
  regulation of gene expression, 615–643
  role in transduction, 467
  structure, 432
  temperate, 468, 630–631
  therapeutic use, 615–616
  transducing particles, 467, 468–469, 468f, 470f, 471f
  virulent, 637–638
Baker's yeast, *see Saccharomyces*
Balance
  mutation-drift balance, 791–793, 792f, 800
  mutation-selection balance, 789, 791, 791f
Balancers (balancer chromosomes), 185
  use in gene mapping, 194–196, 195f
Balancing selection, 788–789, 789t
Baldness, 81, 81f
Banding of chromosomes, 140–141, 140f, 141f
  of *Drosophila* polytene chromosomes, 148–149, 148f, 149f
  relation to genes, 198, 200, 202TS-203TS
Barr bodies, 134, 134f, 669, 151, 153
Base excision repair, 383–384, 384f
Base-pair substitutions, 20–21, 20f, 374f
  see also Bases, mismatched; Mutations
  and antibody diversity, 744–745
  and tautomeric shifts, 372, 372f, 373f
Base-pairing, 17, 17f
  in DNA, 228, 230, 230f
    Chargaff's rules, 225–226
    and semiconservative replication, 256, 257f
  in RNA synthesis, 18, 19f
Bases, in nucleic acids, 17
  see also Nucleotides; Nucleotide pairs
  in DNA vs. RNA, 225, 226f
  mismatched
    see also Base-pair substitutions
    and heteroduplexes, 391, 391f
    repair, 385–386, 386f, 391, 392f
    and site-specific mutagenesis, 535, 535f
  ratios in DNA, 225–226, 228t
  tautomeric forms, 372, 372f
Basic genetic concepts, overview, 16–21
B-DNA, 230, 231, 231f, 231t
Bees
  nest-cleaning behavior, 711–712, 712f
  sex determination, 131–132, 131f, 132f
Behavior, 5, 710–727
  chromosome abnormalities, 720–721
  Drosophila biological rhythms, 712–716, 714f, 715f
  in experimental animals, 711–720
  nest-cleaning in bees, 711–712, 712f
  nurturing in mice, 719–720, 719f
  sexual preference (orientation)
    in *Drosophila*, 716–717, 716f, 719
    in humans, 717, 718HG-719HG
  single-gene mutations, 721–723
  statistical approach, 723–726
Bell-shaped (normal) distribution, 87, 87f, 98, 98f, 99
Binomial probabilities, 66, 67TS
Biological clock, 714–716, 715f
  see also Circadian rhythms
Biological pathways, 681
Biological rhythms, 712–716
Birds and bees, sex determination, 131–132, 131f, 132f
Bithorax complex (BX-C), 696–697, 697f
  vertebrate homologs, 703–704, 704f
Bivalents, 35

Blood coagulation, 11HGf
Blood types
  allelic variation, 73–74, 73f, 74–75, 75t
  linkage, 201, 201f, 203
Blotting techniques, 536–541, 537f, 538TS-539TS, 539f, 540f
  dot blot (array) hybridization, 574, 574f, 575f
  northern blot hybridization, 539–540, 539f
  probes for blotting techniques, 537–539, 537f, 538TS, 574, 574f
  Southern blot hybridization, 536–537, 537f, 538TS-539TS, 539
    in testing for genetic diseases, 593, 593f
  western blotting, 540–541, 540f
Body axes
  determination of, 690–695, 691f, 692f, 694f-695f
  left-right asymmetries, 706
Body organs
  formation, 702–703
  left-right asymmetries, 706
  transplants, 735
Body segmentation, 695–699
  regulatory cascade, 698f, 699
Bonds
  in DNA structure, 228, 229t, 230
  in protein structure, 327, 328f, 329f, 330, 330f, 331f
Bone marrow stem cells, for gene therapy, 600
Boxes, in eukaryotic transcription, 307–308, 308f, 309f
BRCA1, BRCA2 genes, 8, 9, 765–767, 766HG
Bread mold, *see Neurospora crassa*
Breakpoints, *see under* Chromosomes
Breast cancer genes, 8, 9, 765–767, 766HG
Breeding, *see* Mating; Inbreeding; Selective breeding; Selection, artificial
Broad-sense heritability, 101, 724, 724f
Butterflies, wing development, 700TS-701TS

C

CAAT box, 308, 308f
Caenorhabditis (C. elegans), 680f, 681f
  as development model, 679–681
  sex determination, 685–686, 685t, 686f
cAMP, *see* Cyclic AMP
Cancer, 750–771
  see also Tumors
  cell characteristics, 752
  and cell cycle, 752–753
  and cell division regulation, 8, 751, 757, 757f
  and chromosome rearrangements, 758–759, 758f
  and DNA repair defects, 388
  genetic basis, 753
  genetic counseling, 766HG
  as genetic disease, 8, 751–753
  genetic pathways, 767–769, 768f
  hereditary syndromes, 759–761, 760t, 766HG
  and oncogenes, 756–758
  and tumor-suppressor genes, 759–767
Caps, on mRNA, 306–307, 306f, 308–310, 309f
Carcinogens, 381–383, 382f
Cardiovascular disease, 91–92, 105, 107t
Cascades (preprogrammed circuits, regulatory hierarchies) of gene expression, 617, 637–638, 638f
  in development, 695, 698f, 699
  in phage lambda lytic pathway, 634–635, 634f
Cassettes, *see* DNA cassettes
Catabolite activator protein (CAP; cyclic AMP receptor protein), 625–626
Catabolite repression, 625–626
Celera, Inc., 556HG-557HG, 572
Cell-autonomous genes, 687, 688, 689f
Cell-cell interactions, *see* Cells, communication systems
Cell cycle, 28, 28f
  and cancer, 752–753
  checkpoints, 752–753, 752f
  and DNA replication, 281
  and pAPC protein, 765f
  and p53 protein, 763, 763f
  regulation, 752–753, 752f
    and RB gene product, 761, 761f
Cell death
  see also Apoptosis
  and pAPC protein, 764
  programmed, 763 764, 763f
Cell division
  see also Mitosis; Meiosis; Cell cycle
  regulation, and cancer, 8, 751, 757, 757f
Cell hybridization (cell fusion), 205–206, 205f, 206f, 207f, 207t
Cell-mediated (T cell-mediated, cellular) immune response, 730, 734–735, 737f, 738–739
Cell proliferation
  see also Mitosis; Meiosis; Cell cycle; Cell division
  and cancer, 752
  and pAPC protein, 764
  uncontrolled, dominant activators, 758
Cells
  as basic unit of life, 23–27
  communication (signaling) systems, 647
    and cell differentiation, 701–702
  cytological analysis, 139–143, 139f
  differentiation, 699–703
  eukaryotic, 24–27
    structure, 24f
  immortal, 285
  meristematic, 146–147, 146f
  plant vs. animal, 24f
  prokaryotic, 23, 23f, 27
Cellular oncogenes, normal (*c-onc;* proto-oncogenes), 755–756, 755f
  and cancer, 756–758
Cellular (T cell-mediated) immune response, 730, 734–735, 737f, 738–739
Central dogma of molecular biology, 294–297, 295f
Centrioles, 29f, 31, 31f
Centromeres, 27, 27f
  and chromosome inversions, 156, 156f
  mapping, 194, 196–198
  structure, 243–244, 243f, 244f
Centrosomes, 29f, 31, 31f
Cereals
  see also Maize
  genome evolution, 577–579, 578f
Cesium chloride equilibrium density-gradient centrifugation, 260TS-261TS
CFTR protein (cystic fibrosis gene product), 9, 591
Chain-termination (stop) codons, 343–344, 343f, 350, 350t
  in RNA editing, 310–311, 311f
  and suppressor mutations, 353, 353f
Chargaff's rules, 225–226
Checkpoints, in cell cycle, 752–753, 752f
Chemical bonds, *see* Bonds
Chemical mutagens, 373–378, 375t
  testing for, 381–383, 382f
Chi-square test, 62–64, 63f, 63t
Chi structure, 390, 390f
Chiasmata, 36, 36f, 167f, 168, 173–174, 173f, 174f
  see also Crossing over
  and genetic map distance, 182, 182f
  what they are, 174
Chimeras, 705–706
Chimeric genes, 520HG-521HG
Chimeric selectable marker genes, 608
Chimpanzees, 799, 799f
Chinese hamster-human cell hybrids, 571–572, 571f
Chlamydomonas
  antibiotic resistance, 500–502, 501f
  map of chloroplast DNA, 501–502, 502f
Chloroplast DNA (cpDNA), 507
  map, 501–502, 502f, 508f
Chloroplasts
  biogenesis, 507–508, 509f
  as endosymbionts, 509, 511–512
  evolution, 512
  gene shuffling, 512
  genetics, 496–514
  and leaf variegation, 497–499, 498f
  molecular genetics, 507–509
  origin and evolution, 509–512
  RNA processing, 318, 319
Cholesterol blood levels, and repetitive DNA, 251
Chorionic biopsy, 154HG
c-H-ras, see ras genes
Chromatids, 27f
  crossing over, 171–172, 171f, 172f, 173f
Chromatin, 117, 235
  composition, 235–236, 235f, 239–241
  nucleosome substructure, 239–240, 240f, 241, 241f, 242f
  specialized structures (scs, scs'), 660
Chromatin fibers, 241, 242f
Chromosomal basis of Mendelism, 116–137
Chromosomal position effect, 162, 162f

Chromosome jumps, jumping, 566–567, 566f, 591f
Chromosome number, 117, 118t, 142–155
  see also Ploidy; Polyploidy; Aneuploidy; Trisomies
  abnormalities, 36, 37
  chromosome doubling, 146–147, 146f
  in humans, 141
  and meiosis, 32
  nomenclature, 142–143
  and parental age, 151
  polyteny, 148–149
Chromosome painting, 579, 579f
Chromosome rearrangements, 143, 156–162
  see also Inversions; Translocations; Deletions; Duplications
  and cancer, 758–759, 758f
  during evolution, 580f, 581
  and gene mapping, 206–211
  phenotypic effects, 162
  and recombination frequency, 184–185, 184f
  seasonal changes, 794, 794f
  and transposable genetic elements, 491, 491f, 492f
Chromosome rings, 159, 160TS-161TS
Chromosome theory of heredity, 119–126
  nondisjunction as proof, 121, 123
Chromosome walks, walking, 564–566, 565f, 591f
Chromosomes, 27–28, 117–119
  see also Sex chromosomes; Autosomes
  activation and inactivation, 668–670
  arms, 141
  as arrays of genes, 120–121, 168
  artificial, 524–525, 525f
  in bacteria, 233–234, 234f
  banding, 140–141, 140f, 141f, 148–149, 148f, 149f
    relation to genes, 198, 200, 202TS-203TS
  breakpoints, use in mapping, 208, 208f, 209, 209f, 210–211, 210f
  chemical composition, 220, 235–236
  circular permutations, 441–442, 441f
  compound, 159, 161–162
  condensation, 241–242
  cytological analysis, 139–143
  deletions, *see* Deletions
  duplications, *see* Duplications
  DNA packaging, 239–242
  ectopic exchanges, 492
  exchange of alleles, *see* Crossing over; Recombination
  functional domains, 660
  homologous, 32, 33f, 36, 36f
  in interphase, *see* Chromatin
  karyotype, 141, 142f-143f
  lampbrush, 236, 237f, 658, 658f
  mammalian
    conserved sequences, 580f, 581
    evolutionary relationships, 579–581, 579f, 580f
  mapping, *see* Maps and mapping; Human Genome Project; Sequencing
  in meiosis, 32, 33f, 34–37, 34f
  at metaphase, 241–242, 242f, 243f
  in mitosis, 29f, 30f, 31
  molecular analysis, techniques, 541–548
  organization, and transcription control, 658–668
  pairing (synapsis), 33f, 35
    in inversions, 157, 157f
    in polyteny, 148
    in translocations, 157, 158f
  Philadelphia chromosome, 758, 758f
  polytene, *see* Polytene chromosomes
  prenatal inspection, 154HG
  recombinant, 171f
  replication in eukaryotes, 280–287
  segment deletions, *see* Deletions
  segment duplications, *see* Duplications
  sequencing, *see* Sequencing; Human Genome Project; Maps and mapping
  shortening, 490
  staining, 117, 140–141, 140f, 141f
  structural aberrations, 153, 153, 155
    see also Chromosome rearrangements
    from radiation, 380
    from transposable genetic elements, 491–492, 491f, 492f
  structural polymorphism, 794, 794f
  structure, 27f, 141
    in eukaryotes, 234–246
    in prokaryotes, 233–234
    unineme vs. multineme, 236
  univalent and trivalent, 144, 144f
Circadian (24-hour) rhythms, 712, 716
  see also Biological clock
Cis configuration, 412
Cis heterozygote, 412, 412f
Cis-splicing, 419, 506, 506f
Cis test, 413
  see also Complementation test
Cis-trans position effect, 411, 412f
Cis-trans test, 413, 414
  see also Complementation test
Class switching of antibodies, 734
ClB technique, 371, 371f
Cleavage sites, *see* Restriction enzyme cleavage sites
clock genes, 714, 716
Clonal selection, 738
Clone banks, 562–563
Clones
  Dolly the sheep, 42TS-43TS
  overlapping (contigs), 555, 558f, 563, 563f
  of plasma cells, 738
Cloning
  of genes, 518
    and DNA library construction, 525
    forced cloning, 533–534
    positional (map position-based), 555, 564–567, 564f
      chromosome jumps, 566–567, 566f
      chromosome walks, 564–566, 565f
      for identifying disease genes, 587f
    in studies of development, 689–690
    techniques used, 518–525
  of mammals, 42TS-43TS
  nuclear transfer techniques, 43TS
Cloning vectors, *see under* Vectors
Coagulation pathway, 11HGf
Coalescent principle, 801, 801f
Codominance, 73–74
Codon reading frame, 20
Codons, 294, 344–346
  code assignments, 350t
  deciphering them, 346–349, 348TS-349TS
  interactions with tRNAs, 336–337, 337f, 352–354
  in mitochondria, 351, 505, 506
  in transcription/translation, 19, 19f
Coefficient
  inbreeding coefficient, 110–111, 111f, 112f
Coefficient of coincidence, 180
Coefficient of relationship, 112–113
Cohesive (sticky) ends of DNA, 263, 264f
Colinearity, 704
  between gene and polypeptide, 400, 408–410, 409f, 410f, 411f
Colony hybridization, 529, 530f
Color-blindness, 127, 127f
  linkage with hemophilia, 178HG-179HG
Communication between cells, *see* Cells, communication systems
Competence pheromones, 457–458, 457–458
Complement proteins, 734t, 738, 738f
Complementary DNAs (cDNAs), 526
  libraries, 526, 528, 590–591
  as mapping anchors, 557
  synthesis, 526, 528, 528f
  use in finding genes, 573, 575f, 587, 588f, 590–591
Complementation
  intergenic, 418
  intragenic, 415–418, 418f
  vs. recombination, 414–415
Complementation screening, 529
Complementation test, 76, 398, 400, 411–415
  limitations, 418–419, 419f
Complex inheritance patterns, *see under* Inheritance
Complex traits, 91–115
  see also Multifactorial traits; Quantitative traits; Inheritance, complex patterns; Inheritance, polygenic; Variation, continuous phenotypic behavioral, statistical analysis, 723–726
Compound chromosomes, 159, 161–162
Computer chips, biological, 9, 9f, 11, 574, 576, 576f
c-onc genes, *see* Oncogenes, normal cellular
Concatamers, 442, 442f
Concordance rate, 95, 95t
Conjugation, 455, 460–466
  $F^+ \times F^-$ mating, 461, 462f
  and gene mapping, 464–465
  $Hfr \times F^-$ mating, 461–463, 462f
  interrupted mating experiment, 464, 464f
  sexduction, 463
  testing for, 456–457, 456t, 457f
Conjugative plasmids, *see* R plasmids
Consensus sequences, 303

Conserved genes, 577
in cereals, 578, 578f
in mammals, 579, 580f, 581
Constant regions
of antibody chains, 732, 733f
of T cell receptors, 735, 735f
Constitutive genes, 617
Contig maps, *see under* Maps and mapping, map types
Contigs (overlapping clones), 555, 558f, 563, 563f
Continuous phenotypic variation, *see under* Variation
Co-repressors
in operon model, 621f, 622
in prokaryote gene expression, 619f, 620
in *trp* operon, 627
Corn, *see* Maize
Coronary heart disease, *see* Diseases, cardiovascular
Correlation coefficient, 723–724, 724f
Cosmids, 524, 524f
Counseling, 66–68
in cancer, 766HG
and multifactorial traits, 96HG
predicting phenotypes, 103–104, 103f
using allele frequencies, 781–782
Coupling (*cis*) configuration, 412
CpG islands, 590, 591f, 665
c-ras, *see ras* genes
Crick, Francis H. C., photo, 228f
Crimefighting, *see* DNA fingerprinting
Crop plants, *see* Food crops
Cross-fertilization, 54, 54f
Crosses
of bacteriophage, 435–437, 435f, 437f
dihybrid, 56–58, 57f
intercrosses, 58, 59f
in Mendel's studies, 54, 54f, 55t
monohybrid, 54–56, 54f, 55t
definition, 55
predicting outcomes, 58–61
symbolic representation, 55–56, 56f
testcrosses, 59, 59f
trihybrid, 59
Crossing over, 35–36, 36f, 167f, 168
see also Recombination
as basis for recombination, 171–173, 171f
and chromosome rearrangements, 184–185, 184f
coefficient of coincidence, 180
double-strand break model, 390
and evolution, 183
and genetic mapping, 174–175, 175f, 177, 177f, 179–180
Holliday (single-strand break) model, 388–389, 389f
intentional suppression, 185
interference vs. independence, 180
likely regions, 183
molecular mechanisms, 388–392
multiple exchanges, 171–172, 171f, 181–182, 181f
in yeast, 192–194, 193f
in prokaryotes vs. eukaryotes, 454
when it occurs, 171, 173–174, 174f
Crossovers, *see* Crossing over
Cruciform pairing, in translocations, 157, 158f
Culture media, 456
Cyclic AMP
and *lac* operon, 625–626
synthesis from ATP, 625f
Cyclic AMP receptor protein (catabolite activator protein; CAP), 625–626
Cyclin-dependent kinases (CDKs), 752, 752f
Cyclins, 752, 752f
Cystic fibrosis (CF), 590
finding the gene, 538TS-539TS, 590–591, 591f, 592f
gene therapy, 9, 591
molecular basis, 591, 592f
prenatal testing, 539TS, 591
social issues, 15
Cystic fibrosis gene product (CFTR protein), 9, 591
Cytogenetic mapping, 198–200, 202TS-203TS
Cytogenetic variation, 142–143
Cytogenetics, 139
Cytokinesis, 28, 29f, 31, 32f, 33f
Cytological maps, *see under* Maps and mapping, map types
Cytological techniques, 139–143
Cytoplasmic inheritance, 499–500, 499f
Cytosine, 226f
base-pairing in DNA, 228, 230f
methylation, 665, 665f
Cytotoxic (killer) T cells, 731, 731t, 732f, 737f, 738–739, 739f

D
Darwin, Charles, photo, 3f
Darwin's evolutionary theory, 775–776, 777, 782
Databases
FlyBase electronic database, 122HS
of amino acid sequences, 518
of nucleic acid sequences, 518
Death of cell
see also Apoptosis
and pAPC protein, 764
programmed, 763–764, 763f
Deletions (deficiency) of chromosome segments, 153, 153f, 155
in cystic fibrosis, 591, 592f
phenotypic effects, 162
symbols used, 153, 153f, 155, 199
from transposons, 491f, 492, 492f
use in gene mapping, 198–200, 199f, 202TS-203TS, 208–209, 209f, 437–441
Denominator elements, proteins, genes, 682f, 683, 685t
Density-gradient centrifugation, 260TS-261TS
Deoxyribonuclease (DNase) hypersensitive sites, 659–660
Deoxyribonucleic acid, *see* DNA
Deoxyribonucleotides, 227f
2-Deoxyribose, 226f
Derepression, of gene expression, 618, 627
Determination (developmental fate), 676, 678
Development, 675–709
in *Caenorhabditis*, 681, 681f
control of, 18–20
see also Gene expression
in *Drosophila*, 679, 680f
of body axes, 690–695, 691f, 692f, 694f-695f
cell-autonomous genes, 687, 688, 689f
of eye, 699, 699f, 701–702, 701f, 702–703
mammalian gene homologs, 702–703
embryonic, 676–678, 677f, 680f
programmed cell death, 764
finding and analyzing genes, 689–690
gene silencing, 662
homeotic genes, 657, 696–697, 697f, 698f, 699
vertebrate homologs, 703–704, 704f, 705f
how it's studied, 679
imaginal discs, 402–404, 679, 680f, 700TS-701TS
of insect wings, 675–676, 675f, 676f, 700TS-701TS
in mice, 704–706
in model organisms, 679–681
morphogenesis, 18–20, 677f, 678
mutational analysis, 370
in mosaics, 686–688
organ formation, 702–703
overview, 676–678
role of maternal genes, 690–695
temporal regulation of gene expression, 645, 646
in vertebrates, 703–707
zygotic gene activity, 695–703
Developmental fate (determination), 676, 678
Developmental pathways, 681–682
cascades (preprogrammed circuits, regulatory hierarchies) of gene expression, 695, 698f, 699
Diabetes mellitus, 736
genetically engineered insulin, 520HG-521HG
Dideoxy chain termination (Sanger) sequencing procedure, 545–547, 546f, 547f, 548f
Differentiation, 676, 678, 699–703
of eye, in *Drosophila*, 699, 699f, 701–702, 701f, 702–703
organ formation, 702–703
Dihybrid crosses, 56–58, 57f
Dimerization, *see* Thymine dimers
Diplo-X (XX) egg, 121, 123, 123f
Diploid (2n) cells, organisms, 28, 117, 142
amphidiploids, 147
Diploid genes, 55
Discontinuous traits, 95
Diseases
see also Cancer; Gene therapy
autoimmune, 736

Diseases, *(contiued)*
cardiovascular, 91–92, 105, 107t
genetic, 7–9
DNA testing, 593–594
and expanded trinucleotide repeats, 128, 128f, 381, 589
finding the genes, 586–593
prenatal testing, 154HG
for cystic fibrosis, 539TS, 591
for Huntington's disease, 590
for Tay-Sachs disease, 585–586
inherited, 5, 7–9, 64t
with DNA repair defects, 387–388
metabolic, 7, 397, 399HG
from mitochondrial DNA defects, 506–507
recessive, and inbreeding, 108, 108f, 111–112
Disjunction, 121, 124, 124f, 125f
adjacent vs. alternate, 158, 158f
in translocations, 158, 158f
Distances between genes, *see under* Maps and mapping
Distribution, normal (bell-shaped), 87, 87f, 98, 98f, 99
DNA
amplification, *see* Amplification
bacterial exchange, 454, 455f
base composition, 225–226, 228t
vs. RNA, 225, 226f
base pairs, *see* Base-pairing; Nucleotide pairs
breaks, 275–276, 279
closure, 273, 273f
double-strand, 276, 278f
nicks vs. gaps, 273
single-strand, 273f, 276, 277f
Chargaff's rules, 225–226
in chloroplasts (cpDNA), 507, 508f
in chromosomes, 27, 27f, 236–239, 237f, 239f
circular permutations, 441–442, 441f
cleavage, *see* Restriction enzyme cleavage sites
cohesive (sticky) ends, 263, 264f
complementarity, 17, 17f, 18f, 230
complementary (cDNAs), *see* Complementary DNAs
condensation, 239, 241–242
damage
see also DNA, repair mechanisms
and cell cycle checkpoints, 752, 753
denaturation, 246, 264, 265f
of eukaryotes vs. prokaryotes, 234
in HIV, 445, 446–449, 447f, 448f
insertion into vector, 526–527, 526f, 527f
intercalations, 377, 377f
linear, 264, 266f
mapping, *see* Maps and mapping
methylation, 665–666
in mitochondria (mtDNA), *see* Mitochondrial DNA
molecular analysis, 517–552, 585–612
see also Molecular genetics; Maps and mapping; Sequencing
in Human Genome Project, 211
parental strands, 17, 18f
in prokaryotes, 233–234, 234f
recombinant, *see* Recombinant DNA *entries*
relation to histones, 240, 240f, 241f, 242f
renaturation, 246
renaturation kinetics, 246–249, 248TS, 249f
repair mechanisms, 383–387
base vs. nucleotide excision repair, 383
and cancer, 388
and cell cycle checkpoints, 752, 753
defective, and inherited diseases, 387–388
error-prone, 387
excision repair, 383–385, 384f, 385f
and inherited diseases, 387
and gene conversion, 390–392
light-dependent repair, 383
mismatch repair, 385–386, 386f
and p53 protein, 761–763
postreplication repair, 386–387, 387f
repeated sequences, *see* Repetitive DNA
replication, *see* DNA replication
restriction fragments, *see* Restriction fragments
satellites, 246, 249–251, 250f
sense and antisense strands, 300
sequencing, *see* Sequencing; Human Genome Project; Maps and mapping
single-copy, 247, 249
as site of genetic information, 220–225
structure, 17f, 225–233, 226f, 227f, 229f, 231f
alternate forms, 230–231, 231f, 231t
chemical bonds, 229t
double helix, 17, 225–230, 229f, 230f, 231f
Watson-Crick model, 225, 228, 229f
hairpins, 275, 275f
hydrogen bonding, 228, 229t, 230, 230f
nucleotide pairs per turn, 230, 231, 231t
polarity, 227f, 230, 230f
polynucleotide chain, 227f
right- vs. left-handed, 231t
R-loops, 313, 313f
sugar-phosphate backbone, 227f, 230, 231f
supercoils, 232–233, 232f, 275–277, 276f, 277f, 278f, 279
superhelix, 240, 241f
X-ray diffraction patterns, 228, 229f
structure in chromosomes, 236–239, 237f
synthesis, *see* DNA replication
template and nontemplate strands, 300, 300f
terminal redundancy, *see under* Repetitive DNA
transcription, *see* Transcription
transcription-controlling sequences, 653–655
transcriptionally active, 659–660
unwinding
in DNA replication, 259, 275–277, 275f, 276f, 277f, 278, 278f, 279, 280f
in RNA synthesis, 302f, 303, 304f, 307
DNA cassettes, 662–663, 663f
DNA chips, 576
DNA fingerprinting, 543, 594–597
DNA gyrase, 233, 276, 278f, 279
DNA helicases, 275–277, 275f, 278f, 279, 279f, 280f, 283, 284f
DNA libraries, 525–530, 590–591
construction, 526–528
genomic vs. chromosome-specific, 525
screening for a sequence, 528–529, 530f
DNA ligases, 279
in DNA repair, 384, 384f, 385, 385f
in DNA replication, 272f, 273, 273f, 274, 274f, 278f
DNA markers, *see* Marker genes
DNA packaging, 236, 239–242
DNA polymerases, 264–271
in DNA repair, 384, 384f, 385, 385f
in eukaryotes, 268, 283–284, 284f
exonuclease actions, 267, 270, 271f, 274, 274f
how they work, 269, 269f
in polymerase chain reaction (PCR), 541–543, 542f
polymerase I, 266–267, 267f, 268f, 274, 274f, 278f, 279
polymerase II, 268
polymerase III, 268, 269–270, 274, 274f, 279, 280f
catalytic core, 269–270, 270f, 279, 280f
holoenzyme, 269–270, 270f, 278f, 279
polymerases α, β, γ, δ, ϵ, 268, 271
proofreading actions, 270–271, 271f
properties, 268–269
repair actions, 267
DNA primases, 273, 273f, 274f, 278f, 279, 280f, 283, 284f
DNA repeats, *see* Repetitive DNA
DNA replicase, 267, 269–270
DNA replication (DNA synthesis), 17, 18f, 255–291
accuracy, 270, 279
basic features, 256–264, 277–279, 278f, 280f
bidirectionality, 261–262, 262f, 263–264, 265f, 282f, 283
and cell cycle, 281
and cell cycle checkpoints, 752, 752f, 753
continuous vs. discontinuous, 272, 272f
direction of synthesis, 267, 271, 272, 272f
in eukaryotes, 280–287, 284f
in eukaryotes vs. prokaryotes, 281
how it works, 271–280, 274f
initiation, 273–274, 273f, 274f
initiation complex, 279, 279f
leading and lagging strands, 272, 272f, 278f, 279, 280f, 284f
in linear DNA, 264, 266f
Okazaki fragments, 272, 273–274, 279, 280f
origin (initiation site), 261, 262–263, 263f, 278–279

multiplicity in eukaryotes, 281–283, 282f
permuted sequences, 441–442, 441f
prepriming proteins, 278, 279, 279f
primer strands, 274f
primers, 266–267, 266f, 268, 269, 269f, 273–274, 273f, 274f, 278f, 279, 280f
protein complexes, 278–279, 278f, 279f
replication bubble, 262–263, 278–279, 279f
replication forks, 259, 262, 280f
direction of synthesis, 272, 272f
in eukaryotes, 284, 284f, 285f
formation, 279
what happens there, 278f
rolling-circle replication, 280, 281f
semiconservative nature, 256–259, 257f, 258f, 259f
template, 256, 266–267, 266f, 269, 269f, 273–274, 273f, 275f
unwinding of double helix (strand separation), 259, 275–277, 275f, 276f, 277f, 278, 278f, 279, 280f
when it occurs, 174, 281
DNA-RNA hybrids, duplexes, 313, 313f
DNA sequence variants, use in gene mapping, 204
DNA synthesis, *see* DNA replication
DNA testing
see also DNA fingerprinting
in genetic diseases, 593–594
for cystic fibrosis, 539TS, 591
for Huntington's disease, 589–590, 589f
for Tay-Sachs disease, 585–586
DNA topoisomerases, 233, 275–276, 278f, 279
DNase (deoxyribonuclease) hypersensitive sites, 659–660
Dogma, central, of molecular biology, 294–297, 295f
Dolly the cloned sheep, 42, 42f
Domains, 732
of antibodies, 732–733
Dominance, 73
incomplete (partial), 73–74, 73f, 93f
Mendel's principle, 56
Dominance hierarchy, 75, 75f
Dominance variance, 102
Dominant activators of uncontrolled cell growth, 758
Dominant mutations, *see under* Mutations
Dominant traits, 55, 56
in pedigrees, 64f, 65
Dosage, *see* Gene dosage; Radiation, dosage
Dot blot hybridization, 574, 574f, 575f
Double helix of DNA, *see under* DNA, structure
Down syndrome (trisomy 21), 22–23, 22f, 150–151, 150f, 151f, 152t
and Alzheimer's disease, 720
and maternal age, 37, 151
Downstream direction, 302
Drosophila
behavioral studies
biological rhythms, 712–716, 714f, 715f
sexual preference (sexual orientation), 716–717, 716f, 719
cell-autonomous genes, 687
chromosome bands, 148–149, 148f, 149f
relation to genes, 198, 200, 202TS-203TS
and chromosome theory of heredity, 119–121, 119f, 120f
development
of body axes, 690–695, 691f, 692f, 694f-695f
of body segmentation, 695–696, 696f, 698f
of eye, 699, 699f, 701–702, 701f, 702–703
mammalian homologs, 702–703
of wing, 700TS-701TS
as development model, 679, 680f
dosage compensation, of X-linked genes, 132, 132f, 133, 668, 669–670, 683–684
FlyBase electronic database, 122HS
"Fly Room," 122HS
genome size, 200
gynandromorphs, 687, 687f
homeotic genes, 696–697, 697f
vertebrate homologs, 703–704, 704f, 705f
imaginal discs, 679, 680f
life cycle, 45–46, 46f
mapping of chromosomes, 121f, 149, 149f, 167, 167f, 180f, 183f, 198–200, 199f, 200f, 202TS-203TS
P elements, 484–487, 485f, 488TS-489TS, 492
polytene chromosomes, 148–149, 148f
bands, 148–149, 148f, 149f, 202TS-203TS
deletions and duplications, 155, 155f
maps and mapping, 149, 149f, 183f, 198–200, 202TS-203TS
puffs, 659, 659f
research use, 122HS
retroposons, 490
retrovirus-like transposable elements, 489–490
sex determination, 130–131, 131f, 131t, 682–685, 682f, 683f, 684f, 685t, 687
and sexual preference (orientation), 716–717, 717f
and XX/XO mosaics, 153
somatic recombination, 687–688, 688f, 689f
transposable genetic elements, 484–487, 485f, 488TS-489TS, 489–490, 491, 492, 492f
whole-genome shotgun sequencing, 556HG-557HG
Duchenne muscular dystrophy, 208, 209, 209f
Duplications of chromosome segments, 155
phenotypic effects, 162
from transposons, 492, 492f
use in gene mapping, 198–200, 200f, 202TS-203TS, 209–211, 210f
Dwarfism, 517
probability calculations, 66, 66f
Dysgenesis, 484–486, 492

E

E. coli
see also Bacteria; Prokaryotes
chromosome map, 568f
chromosome structure, 233–234, 234f
DNA polymerases, 266–270, 273–274, 274f
DNA replication, 259, 261–263, 262f, 277–279, 278f, 279f
gene mapping, 464–465, 465f, 567–568, 568f
genes, 567–568, 569f
genome, 465f, 567–568
human growth hormone production, 601–602, 602f
human insulin production, 520HG-521HG
operons, *see lac* operon; *trp* operon
viruses, *see* Bacteriophage *entries*
Ebola virus, 429, 429f
*Eco*RI restriction endonuclease, 518, 519t, 521
Effector molecule (end-product, regulatory) binding site, 639–640, 641f
Effector molecules
of operons, 621f, 622, 626
in prokaryote gene expression, 620
Eggs
formation, *see* Oogenesis
preparation for development, 690
Electromagnetic spectrum, 379f
Electrophoresis, *see* Gel electrophoresis
Embryo, *see* Development
Embryonic stem cells (ES cells), 705, 706f
End-product (feedback) inhibition, 639–641, 641f
End-product (regulatory, effector molecule) binding site, 639–640, 641f
Endomitosis, 148
Endonucleases, 267, 518
see also Restriction enzymes
in intron excision, 318–319, 319f
in RNA chain termination, 310, 310f
in rolling-circle replication, 280, 281f
Endosymbiosis, 509
Enhancers of transcription, 653–655, 654f
tissue-specific, in antibody gene assembly, 746, 748, 748f
Environment
and complex inheritance patterns, 93–97, 94f
and complex traits, 723–724
effects on genes, 80–81
Environmental variance, 100–101
Enzymes, 293, 294f, 295f
see also Restriction enzymes *and specific enzymes, e.g.* DNA polymerases
activation vs. induction, 617–618
allosteric transitions, 640–641, 641f
binding sites, 639–640, 641f

Enzymes, *(contiued)*
catalytic core, 269
control by genes, 402–404
feedback (end-product) inhibition, 639–641, 641f
heteromultimeric, 406
inducible, 617–618, 618f
for industry, from bacteria, 602–603
multimeric, 269
repressible, 618, 618f
repression vs. feedback inhibition, 618
Episomes, 461
Epistasis, 84–86, 85f
Epistatic variance, 102
Epitopes, 736
Equilibrium density gradients, 260TS-261TS
Equilibrium level ($\hat{H}$), of genetic variability, 792, 800
Erythrocytes, *see* Red blood cells
Escherichia coli, see E. coli
Ethical questions, *see* Social issues
Euchromatin, 117, 660–661
Eugenics, 16
Eukaryotes, 24–27
cell structure, 24–27, 24f
chromosome replication, 280–287
chromosome structure, 234–246
cloning vectors for, 524
DNA polymerases, 268, 283–284, 284f
DNA replication (DNA synthesis), 280–287, 284f
gene structure, 400f
genome, vs. prokaryotes, 234
primary transcripts, 295, 296f, 306–307, 306f, 308–310, 309f, 310f, 649
regulation of gene expression, *see under* Gene expression
RNA synthesis and processing, 305–312
transcription, 305–312, 309f
regulation, 647, 653–658
translation, 340–341, 342, 649
transposable genetic elements, 481–487
Euploidy, 142
"Eve, Mitochondrial," 510HG
Evolution, 774–804
see also Natural selection; Selection
Darwin's theory, 775–776, 777, 782
of chromosomes, in mammals, 579–581, 580f
of genomes, 577–581
in cereals, 577–579, 578f
in mammals, 579–581
and homologous genes, in eye development, 703
of humans, 799–801
fossil record, 799–800, 799f
mitochondrial DNA evidence, 510HG-511HG
and meiosis, 37, 183
of mitochondria and chloroplasts, 512
molecular, and neutral theory, 790TS-791TS
mutation theory, 160TS
and mutations, 359–360, 777, 786, 790TS
random forces affecting, 786
and random genetic drift, 788
rate of, 790TS
and recombination, 183–185
and transposable genetic elements, 493
Exchange between chromosomes, *see* Crossing over; Recombination
Excision repair, 383–385, 384f, 385f
and inherited diseases, 387
Exon amplification procedure, 587–588, 588f
Exon-intron junctions, 316, 317–318
Exons, 295, 296f, 312–313, 315f
and definition of gene, 401
Exonucleases, 267
DNA polymerase actions, 267, 270, 271f, 274, 274f
Expressed-sequence tags (ESTs), 557, 573, 575f
Expression sites, 663–665, 663f, 664f
Expressivity, 82, 82f, 83HG
Eye development
in *Drosophila*, 699, 699f, 701–702, 701f, 702–703
mammalian gene homologs, 702–703

F

F factor (F plasmid), 461, 462f, 463f, 466, 479, 492
F′ conjugation, 463
F′ factor, 463, 463f
$F^+ \times F^-$ conjugation, 461, 462f
Feedback (end-product) inhibition of enzymes, 639–641, 641f
Females, sex determination, 129–130, 129f
Fertile polyploids, 145–146, 145f
Fertilization, 39–40, 676, 678
Filamentous phages, 531
Finger motif, 655, 655f
Fingerprinting, *see* DNA fingerprinting
Fitness (w), 782–783, 783f, 785
5′ caps, on mRNA, 306–307, 306f, 308–310, 309f
5′→3′ (5′ to 3′) activities, of DNA polymerase I, 267, 268f
5′→3′ (5′ to 3′) polarity, 227f, 230, 230f
Flanking (outside) markers, 390, 465
reciprocal recombination, 391–392, 392f, 407
FlavrSavr™ tomato, 608
Fluorescent *in situ* hybridization (FISH), 562, 562f, 579, 579f
FlyBase, *Drosophila* electronic database, 122HS
"Fly Room," 122HS
Food crops, food products
genetic improvements, 11–12, 12f, 13
early practices, 14–15
genetically engineered, 5
FlavrSavr™ tomato, 608
genome evolution in cereals, 577–579, 578f
hybridization and polyploidy, 145–146, 145f, 147–148
Foreign molecules, distinguishing from self, 735–736
Forensic medicine, *see* DNA fingerprinting
Forked-line method, for predicting outcome of crosses, 59, 59f
Fossil record, and human evolution, 799–800, 799f
Fragile X syndrome, 127–128, 128f, 381
Frameshift mutations, 20, 372, 374f, 377
Frequency distributions, 97–98, 97f, 98f
Frequency of recombination, *see* Recombination frequency
Fruit fly, *see Drosophila*
Functional allelism, 415
Functional genomics, 574, 576
Fungi, 41, 41f, 43
see also Saccharomyces
Fusion proteins, 520HG-521HG
and cancer, 758

G

Gain-of-function alleles, mutations, 79–80, 79f
GAL4 transcription factor, of yeast, 656TS-657TS
Gametes, 37–40
Gametogenesis, 38–40, 40f
Gap genes, 697, 698f, 699
Garrod, Sir Archibald, photo, 7f
GC box, 308, 308f
Gel electrophoresis, 536–537, 536f, 540, 543, 544f, 794–795, 794f, 795f
pulsed-field, 236–237, 237f
Gene banks, 518
Gene chips, 9, 9f, 11, 574, 576f
Gene conversion, 390–392, 392f
Gene dosage
dosage compensation
see also Barr bodies
in translocation heterozygotes, 159, 159f
of X-linked genes, 132–134, 132f, 668–670, 668f, 683–684, 686
and Duchenne muscular dystrophy, 208, 209f
and IQ, 720
Gene expression, 18–20, 19f, 295f, 297f, 298–302
see also Transcription
and chromosome rearrangements, 162
constitutive, 617
derepression, 618, 627
environmental effects, 81
as function of genetic material, 220
inducible, 617
in eukaryotes, 649–653
negative and positive control, 619f, 620
and DNA methylation, 665–666
mitochondrial, 505–506
and parental source, 665–666
pathways, 19
preprogrammed circuits (cascades, regulatory hierarchies), 617, 637–638, 638f
in development, 695, 698f, 699
prevention, 661–665
using antisense RNA, 608–610, 609f
regulation in bacteriophages, 630–637, 637–638

regulation in eukaryotes, 644–672
in antibody synthesis, 746
chromosome activation and inactivation, 668–670
chromosome organization, 658–668
gene amplification, 666–668
by hormones, 651–653, 651f, 652f
mechanisms, 647–649
messenger RNA stability, 649
methylation, 665–666
nucleotide modification, 665–666
and repetitive DNA, 249
spatial regulation, 645–646
temporal regulation, 645–646
transcriptional control mechanisms, 653–658
where it occurs, 647, 647f
regulation in prokaryotes, 615–643
levels, 616, 616f
mechanisms, 616–617
negative vs. positive control, 618–620, 619f
operons, 620–622, 621f
post-translational control, 639–641
translational control, 638–639, 640f
repressible, repressed, 618
negative and positive control, 619f, 620
sex-specific, 683f
study techniques
dot blot (array hybridization) analysis, 574, 574f, 575f
northern blot hybridization, 539–540
RNA and protein assays, 573–574
use of green fluorescent protein, 574–576
temporal specificity, 645
tissue specificity, 645
total genome expression, 573–574
and viral oncogenes, 755
zygotic, 695
Gene products, 293–294
see also Proteins
mitochondrial, 506
molecular analysis, 585–612
Gene rearrangements, *see* Genome rearrangements
Gene replacements (targeted gene transfers, gene targeting), 601
Gene segments
in antibody gene assembly, 420–421, 421f, 740–745, 741f, 742f
for heavy chains, 741–742, 742f
joining signal sequences, 742–743, 743f
for light chains, 740–741, 741f
and definition of gene, 400, 401
in T cell receptor gene assembly, 746
Gene segregation, *see* Segregation
Gene silencing, 661–665
Gene targeting, *see* Gene replacements
Gene therapy, 8–11, 10HG-11HG, 597–601, 600f
candidate diseases, 600–601
for cystic fibrosis, 9, 591
gene additions vs. replacements, 601
germ-line (heritable) vs. somatic-cell (nonheritable), 598
guidelines, 598–599
vectors for (gene-transfer vectors), 598, 599f, 600f
Gene transcripts, *see under* Transcripts
Gene transfer, in bacteria, 453–454, 454–455
see also Conjugation; Transduction; Transformation
Genes
see also Alleles *and specific gene types, e.g., ras* genes; Breast cancer genes
amplification, 666–668
see also Repetitive DNA; Genetic redundancy
assembly, from segments, *see* Gene segments, *above*
as basic functional unit, 398, 401–406, 406
as basic structural unit, 398, 406
cancer-related types, 753
cell-autonomous, 687, 688, 689f
chimeric, 520HG-521HG
in chloroplasts, 507, 511–512
in chromatin vs. heterochromatin, 660–661
cloning, *see* Cloning, of genes
conserved, 577, 578, 578f, 579, 580f, 581
constitutive, 617
definitions (changing concepts), 397–426
definition used in this book, 400
functional aspects, 401–406
genetic definition, 410–419
Mendel's concept, 398, 401–402
one gene—one enzyme concept, 398, 404–406, 405f
one gene—one polypeptide concept, 398, 400, 406
one mutant gene—one metabolic block concept, 398, 402
operational definition, 398, 400, 411–415
and overlapping genes, 400, 401
overview, 398–401
present concept, 407
structural aspects, 406–410
developmental
see also Homeotic genes
maternal genes, 690–695
study techniques, 689–690
zygotic, 695–703
diploid, 55
environmental influences, 80–81
in enzyme control, 402–404
eukaryotic vs. prokaryotic, 400f
exchange in bacteria, 454–455
see also Conjugation; Transduction; Transformation
exchange between chromosomes, *see* Crossing over; Recombination
finding their function, 518
gap genes, 697, 698f, 699
haploid, 55
haplo- and triplo-lethal, 162
heat-shock, 650, 650f
homologous, 703–704, 704f, 705f
in eye development, 702–703
homozygous vs. heterozygous, 55, 56f
how they evoke phenotype, 80–87, 401–402, 401f
see also Gene expression
identifying them, 528–529, 573, 586–593, 689–690
see also Maps and mapping; Sequencing
imprinted, 666, 667f
inducible, 617
interactions, 82, 82f, 84, 84f
jumping, *see* Transposable genetic elements
length, average, in mammals, 518
linear array on chromosomes, 120–121, 168
linked, *see* Linkage; Synteny; X-linked genes; Sex-linked genes
locus (loci) on chromosomes, 105, 121, 121f
maps, *see* Maps and mapping; Human Genome Project
as markers, *see* Marker genes
maternal-effect genes, 690, 698f, 699
mitochondrial, 504–505, 511–512
molecular analysis, 517–552
see also Molecular genetics; Maps and mapping; Genetic engineering
techniques, 535, 541–548, 585–612
multiple copies, *see* Genetic redundancy
mutant, 7, 20, 20f
noncoding sequences, *see* Introns
number in humans, 567
oncogenes, 753–759
one gene—multiple products, 317
origin of term, 55
overlapping, 400, 401, 443–444, 443f, 444f
pair-rule, 697, 698f, 699
patenting, 556HG
and personality, 5, 710–711, 725–726, 725t
polymorphic, 75
pseudoautosomal, 128
pseudogenes, 646
recombinant, *see* Recombinant *entries*; Recombination *entries*
regulator, 618–620, 619f
of *lac* operon (*lacI*), 622, 622f, 623–625, 623f, 624f, 626f
in operon model, 621f
relation to chromosome bands, 198, 200, 202TS-203TS
relation to polypeptides, 77–79, 77f, 400, 408–410, 409f, 410f, 411f
relation to proteins, 419–421
repressed and derepressed, 618
screening for, in DNA libraries, 528–529, 530f
segmentation genes, 697–699, 698f
segment-polarity genes, 697, 698f, 699
sequencing, *see* Sequencing; Human Genome Project; Maps and mapping
and sex determination, 116–117
sex-linked, *see* X-linked genes; Sex-linked genes

Genes, *(continued)*
split, 295, 312–317
see also Introns; Splicing
structural, 249, 400–401, 618
control of expression, 618–620, 619f
in operons, 620–622, 621f, 622f, 624f
structure, 400f
beads on a string concept, 398, 406–407
tumor-suppressing, *see* Tumor suppressor genes
wild-type vs. mutant, 7
X-linked, *see* X-linked genes; Sex-linked genes
Y-linked, 128
zygotic, 695–703
Genes-within-genes, 400, 443–444, 444f, 449
Genetic code, 294, 344–352
codon meanings, 350t
cracking it, 346–349, 348TS-349TS
degeneracy, 350–351
in mitochondria, 505
near-universality, 351
properties, 344
and translation, 19, 19f
triplet nature, 344–346, 345f
Genetic concepts, overview, 16–21
Genetic counseling, *see* Counseling
Genetic diseases, *see under* Diseases
Genetic drift, 786–788, 790TS-791TS
balance with mutation, 791–793, 792f
in humans, 800
Genetic engineering
of food products, 5, 11–12, 12f, 13
early practices, 14–15
FlavrSavr™ tomato, 608
guidelines, 530
of human growth hormone, 517
of human insulin production, 520HG-521HG
industrial applications, 602–603
of pharmaceuticals, 602
of plants, 606–608
potential hazards, 529–530
tools and techniques, 517–552, 585–612
Genetic engineering industry, 603
Genetic equilibrium, 788–793
Genetic exchange
in bacteria, 454–455
see also Conjugation; Transduction; Transformation
between chromosomes, *see* Crossing over; Recombination
Genetic information
basic unit, 406
DNA as site, 220–225
transfer (flow), from DNA to protein, 294–297, 295f
transmission to progeny, 17, 294
Genetic instability, *see* Transposable genetic elements
Genetic maps, *see under* Maps and mapping, map types
Genetic markers, *see* Marker genes; Markers
Genetic material
functions, 220
in RNA viruses, 224–225, 225f
Genetic mosaics, *see* Mosaicism
Genetic polymorphism, 75, 778, 793
Genetic redundancy, 334, 569–571
see also Repetitive DNA
Genetic selection, to screen for genes, 528–529
Genetic transformation, *see* Transformation
Genetic transmission, 17, 294
Mendel's rules, 56, 58
Genetic variance, *see under* Variance
Genetic variation, variability, *see under* Variation; Variability
Genetics
birth of, 52–53
classical vs. molecular, 4–5
misuse of, 15–16
modern, 4
origin of term, 53
Genome rearrangements
in antibody gene assembly, 421, 421f, 740–745, 741f, 742f
DNA signal sequences, 742–743, 743f
in T cell receptor gene assembly, 746
Genomes
chromosome number, 117, 118t
comparative sizes, 234f
definition, 17
of eukaryotes vs. prokaryotes, 234
evolution, 577–581
in cereals, 577–579, 578f
in mammals, 579–581
finding protein-coding parts, 573
monitoring total expression, 573–574
of prokaryotes, 233, 234f, 567–569
sequences now known, 545
sequencing, *see* Sequencing; Maps and mapping; Human Genome Project
types of variations, 512
Genomics, 553–584
functional, 574, 576
origin of term, 553
structural vs. functional, 553
Genotype frequencies, 778, 779
Genotypes, 55, 56, 56f
relation to phenotype and environment, 80–81
Germ cell line, 38
Germ cell precursors, 679
Globin chains, in hemoglobin, 329f, 365
Globin genes, 18, 19f
evolution, 790TS-791TS
introns and exons, 313–314, 314f, 315f
mutations, 364–366, 366f
temporal expression, 646
Glucose effect, 625
gp120 (gene product 120) of HIV, 445, 446f
Green algae, antibiotic resistance, 500–502, 501f
Green fluorescent protein (GFP), 574–576, 577f
Growth, *see* Development; Cell division
Growth factor receptors, and oncogenes, 754
Growth factors
and oncogenes, 754
as transcription inducers, 653
Growth hormone
human (HGH), 517
from bacteria, 601–602, 602f
transgenes, 604–605, 604f
Guanine, 226f
base-pairing in DNA, 228, 230f
Guide RNAs (gRNAs), 311f, 312, 312f, 506
Gynandromorphs, 153, 687, 687f
Gyrases, *see* DNA gyrase

H

Hairpin structures, 275, 275f
in RNA synthesis, 304, 305f
in *trp* operon, 628f, 629, 630f
Haplo-lethal gene, 162
Haploid (n) cells, 28, 33f, 34f, 36–37, 117
Haploid genes, vs. diploid, 55
Haploid genome, 117, 118t
Hardy-Weinberg equilibrium, 778
Hardy-Weinberg principle, 778–779, 778f
exceptions, 780–781
use in genetic counseling, 781–782
Health insurance issues, 15
Heart disease, 91–92, 105, 107t
Heat-induced transcription, 650, 650f
Heat-shock proteins, genes, 650, 650f
Heavy chains of antibodies, 732, 733f
in class switching, 745, 745f
gene assembly, 741–742, 742f, 743, 746, 748, 748f
Helicases, *see* DNA helicases
Helix, of DNA, *see under* DNA, structure
Helix-loop-helix motif, 655f, 657
Helix-turn-helix motif, 655, 655f, 657
Helper phage, 531, 534f
Helper T cells, 449, 731, 731t, 732f
in antibody-mediated immune response, 736, 737f
in HIV infection, 449–450
in T cell-mediated immune response, 737f, 739
Hemizygote, 120
Hemoglobin, 326–327, 329f, 365
mutations, 364–366, 366f
in sickle-cell anemia, 326–327, 365, 366f
Hemophilia, 8, 10HG-11HG, 126–127, 126f
linkage with color blindness, 178HG-179HG
and transposable elements, 490
Heredity
see also Inheritance; Mendel's laws
chromosome theory, 119–126
Mendel's studies, 52–53, 53–58
Heritability, 100–103
broad-sense, 101, 724, 724f
narrow-sense, 102–105, 102t, 103f, 104f
of psychological traits, 723–726
Heritable (germ-line) gene therapy, 598
Hermaphroditism, 679–680, 680f, 681f, 686, 686f
Heteroalleles, 415

Heterochromatin, 117, 660–661
  and chromosome rearrangements, 162
Heteroduplexes, 391, 391f
Heterogeneous nuclear RNA (hnRNA), 307
Heteromultimeric proteins, 406
Heterosis, 109
Heterozygosity, 787, 792, 792f
  loss of, 769
Heterozygote advantage, 788
Heterozygotes, 55, 56, 56f
  cis heterozygotes, 412, 412f
  effects of chromosome rearrangements, 162
  and inbreeding, 109, 109f
  inversion heterozygotes, 157, 157f
    and recombination frequency, 184–185, 184f
  Mendel's principles, 56
  trans heterozygotes, 412–414, 412f
  translocation heterozygotes, 157–159, 158f, 159f
Hfr bacterial strains, 461–463, 462f, 479
Histocompatibility antigens, 735–736
Histones, 235, 284
  relation to DNA, 240, 240f, 241f, 242f
HIV (human immunodeficiency virus), 445–450, 728–729
  activation in host, 448–449
  genome, 449, 449f
  integration into host DNA, 446, 448, 448f, 449f
  life cycle, 445–449, 446f
  structure, 445, 445f
HIV infection, 449–450
  see also AIDS
HLA (human-leukocyte-associated) antigens, genes, 735–736, 735f
HLA locus, 735, 735f
hMSH2 gene, 764–765
Homeodomain, 657, 696–697
Homeodomain transcription factors, 657, 696, 697
Homeosis, 696
Homeotic genes, 657, 696–697, 697f
  regulation by segmentation genes, 697, 698f, 699
  vertebrate homologs, 703–704, 704f, 705f
Hominids, 799–800, 799f
Homoalleles, 415
Homoduplexes, 391
Homologous chromosomes, 32, 33f
  crossing over, 36, 36f
Homologous genes, 703–704, 704f, 705f
  in eye development, 702–703
Homosexuality
  in *Drosophila*, 716, 716f
  in humans, 717, 718HG-719HG
Homozygosity, 792
Homozygotes, 55, 56, 56f
  effects of chromosome rearrangements, 162
  and inbreeding, 109, 109f
Honeybees, *see* Bees
Hormone receptors, 651, 651f, 652, 652f
Hormone response elements (HREs), 652
Hormones, in gene expression, 651–653, 651f
Hox genes, 703–704, 704f
  knockout mutations, 706
Human Genome Project, 15, 211, 567–573
  competitors, 556HG-557HG, 572
  other genomes sequenced, 567
Human growth hormone, *see under* Growth hormone
Human immunodeficiency virus, *see* HIV
Humans
  aging, and telomere length, 285–287
  cloning of, 42–43
  diseases, *see* Diseases; Cancer
  evolution, 799–801
    finding our common ancestor, 801, 801f
    fossil record, 799–800, 799f
    mitochondrial DNA evidence, 510HG-511HG
  and great apes, 799
  homosexuality, and X chromosome, 717, 718HG-719HG
  karyotype, 141, 142f-143f
  life cycle, 46–47
  linkage detection, 178HG-179HG
  mapping of genes, of genome, 201, 202–211, 571–573
    see also Human Genome Project
    of autosomes, 203–204, 204–211, 204f
    cell fusion (hybridization) techniques, 205–206, 205f, 206f, 207f, 207t
    chromosomal rearrangements, 206–211
    use of DNA sequence variants, 204
    use of RFLPs, 559–561, 561f, 587f
  Mendel's principles applied, 64–68
  mitochondrial DNA (mtDNA), 504, 505f, 510HG
    vs. Neanderthal mtDNA, 511HG
  monosomies, 152–153, 152f
  mutations and behavior, 721–723
  nondisjunction, 150–151, 151f, 152t
  number of genes, 567
  number of nucleotide pairs, 567
  sex determination, 129–130, 129f
  sex-linked genes, 126–128
  size of largest chromosome, 239
  transposable genetic elements, 490
  trisomies, 150–152, 150f
Humoral (antibody-mediated) immune response, 730, 736–738, 737f
Huntington's disease (HD), 8–9, 190–191, 586–587, 722–723, 722f
  and expanded trinucleotide repeats, 588–589, 589f
  finding the gene, 586–590, 587f
  prenatal testing, 590
Hybrid dysgenesis, 484–486, 492
Hybridization array, 574, 574f, 575f
Hybridization techniques
  see also Blotting techniques
  array (dot blot) hybridization, 574, 574f, 575f
  colony hybridization, 529, 530f
  fluorescent *in situ* hybridization (FISH), 562, 562f, 579, 579f
  in situ hybridization, 249–251, 250f, 529, 562, 562f
  in mapping, 562, 562f
    radiation hybrid mapping, 571–572, 571f, 572f
  molecular, in screening for genes, 529, 530f
  plaque hybridization, 529
  probes for, 529, 530f, 537–539, 537f, 538TS, 557, 574, 574f
  R-loop hybridization, 313, 313f
  somatic-cell hybridization (cell fusion), 205–206, 205f, 206f, 207f, 207t
Hybridomas, 747TS
Hybrids
  between species, 205, 205f, 206f, 207f
    Chinese hamster-human cell hybrids, 571–572, 571f
  DNA-RNA hybrids, duplexes, 313, 313f
  in food crops, and polyploidy, 145–146, 145f, 147–148, 147f
  in Mendel's crosses, 54–56, 54f, 55t, 56–58, 57f, 59
  radiation hybrid mapping, 571–572, 571f, 572f
Hyperploidy, 150, 155
Hypoploidy, 150, 155
Hypotheses, formulating and testing, 61–64

I

Identity, proving with DNA fingerprints, 594–597, 596TS-597TS
Identity by descent, 109–110
Imaginal discs, 402, 679, 680f
  and *Drosophila* eye color, 402–404
  for insect wings, 700TS-701TS
Immortal cells, 285
Immune response, 729, 730
  cell-mediated (T cell-mediated, cellular), 730, 738–739
  humoral (antibody-mediated), 730, 736–738, 737f
  primary, 733–734, 737f, 739–740
  secondary, 734, 737f, 739–740
Immune system, 728–749
  in AIDS, 736
  components, 730–736, 731t, 732f
  defects, 731, 738
  gene assembly, 420–421, 421f, 740–745, 741f, 742f, 746
  in HIV infection, 449–450
  specificity, 730, 732–735, 734
Immunodeficiency disease, severe combined (SCID), 599–600, 600f, 731–732
Immunogens, 730
Immunoglobulins (Igs), 732
  see also Antibodies
  classes, 733–734, 734t
  class switching, 745, 745f
Immunological memory, 739–740
Imprinted genes, 666, 667f
In situ hybridization, 249–251, 250f, 529, 562, 562f, 579, 579f

Inactivation
of X chromosome, 133–134, 133f, 134f, 668–669, 668f
of X-linked genes, 133–134
and Duchenne muscular dystrophy, 208, 209f
Inborn errors of metabolism, 397, 399HG
Inbreeding, 107–113, 780
effects, 107–109, 108f, 111–112
genetic analysis, 109–112
Inbreeding coefficient, 110–111, 111f, 112f
Inbreeding depression, 108, 112, 112f
Incomplete (partial) dominance, 73–74, 73f, 93f
Incomplete penetrance, 81–82, 83HG
Independent assortment, 56–58, 57f, 58f
chromosomal basis, 124–126, 125f
and complex inheritance patterns, 93–95, 93f, 94f
and frequency of recombination, 169
and meiosis, 37
Mendel's principle, 58
exceptions, 168–169, 169f
Inducers
of eukaryotic gene transcription, 649–653
of *lac* operon, 622–623, 624f
of prokaryotic gene expression, 619f, 620
Induction
and cell differentiation, 702
of gene expression, 617
negative and positive control, 619f, 620
Information, genetic, 17, 294–297, 295f, 406
DNA as site, 220–225
Inheritance
see also Heredity; Mendel's laws
basic units, 406–407
complex patterns, 92–97
see also Inheritance, polygenic; Multifactorial traits; Complex traits; Quantitative traits; Variation, continuous phenotypic
and Mendel's principles, 93–95, 93f, 94f
cytoplasmic, 499–500, 499f
extranuclear, 497–503
fundamental laws, 4
mitochondrial, 498, 499–500, 507
non-Mendelian, 497–503
polygenic, 96HG, 100
Inherited diseases, *see under* Diseases
Initiation codons, 338f, 339, 341, 349–350, 350t
and overlapping genes, 443–444, 444f
Initiation complexes, 308, 309f, 338f, 339
Initiation factors (IFs), 338f, 339
Initiation site (origin), 261, 262–263, 263f, 278–279
in eukaryotes, 281–283, 282f
Insecticides, 13
Insertion sequences (IS elements), 477–480
Insertional mutagenesis, 704–706
Insulin, genetically engineered, 520HG-521HG
Insurance, and genetic issues, 15
Inteins, 331
Intelligence, 724–725, 725t
Interactions between cells, *see* Cells, communication systems
Interactions between genes, 82, 82f, 84, 84f
Intercalations in DNA, 377, 377f
Intercrosses, 58, 59f
Interference between crossovers, 180
Introns, 295, 296f, 306f, 312–316, 315f
biological roles, 316–317
and definition of gene, 400
discovery, 313–315
excision, 313, 314f, 317–322, 317f, 419
see also Splicing
mechanisms, 318
and gene-polypeptide colinearity, 409–410
junctional consensus sequences, 316, 317
in mitochondrial DNA, 504–505
in normal cellular vs. viral oncogenes, 755, 755f, 756
and screening for genes, 529
Inversion heterozygotes, 157, 157f
and recombination frequency, 184–185, 184f
Inversions of chromosome segments, 156–157, 156f, 157f
and evolution, 581
paracentric; pericentric, 156–157, 156f
phenotypic effects, 162, 162f
and recombination frequency, 184–185, 184f
seasonal changes, 794, 794f
from transposons, 491f, 492
use in gene mapping, 194–196
Inverted terminal repeats, *see under* Repetitive DNA
Isoalleles, 363
Isochromosome, 159
Isolation, 795–796, 796t

J

Joining (*J*) segments
in antibody gene assembly, 740–745, 741f, 742f, 743f, 744f
in T cell receptor assembly, 746
Joining sites, and antibody diversity, 743–745, 744f
Joining signal sequences, in antibody gene assembly, 742–743, 743f
Jumping, *see* Transposable genetic elements; Chromosome jumps

K

Kappa light chains, of antibody molecule, 733
gene assembly, 421, 421f, 740–741, 743–744, 743f, 744f, 748
Karyotype, 141, 142f–143f
in Down syndrome, 150f
prenatal determination, 154HG
Killer (cytotoxic) T cells, 731, 731t, 732f, 737f, 738–739, 739f
Kinases, 318, 754, 756
and cell cycle regulation, 752, 752f
Kinetochores, 27–28, 27f, 29f, 33f
"Knockout" mutations, 706
Kornberg's enzyme, 266

L

lac (lactose) operon, 622–626, 622f
catabolite repression, 625–626
components, 622, 622f
induction, 622–625
mutational changes, 623–625, 624f
and test for vectors, 533, 534
lacI gene, 622, 622f, 623–625, 623f, 624f, 626f
Lagging strand, in DNA replication, 272, 272f, 274f, 278f, 279, 280f, 284f
Lambda light chains, of antibody molecule, 421, 733
gene assembly, 740, 741f, 743–744
Lambda phage, *see* Bacteriophage lambda
Lambda repressor, *see under* Bacteriophage lambda
Lampbrush chromosomes, 236, 237f, 658, 658f
Laws of inheritance, 4
Leading strand, in DNA replication, 272, 272f, 278f, 279, 280f, 284f
Legal questions, *see* Social issues
Lethal chromosomal rearrangements, 162
Lethal mutations, 77, 77t, 363–364, 364f, 369–370, 370–371, 371f
maternal-effect, 690
and mutation-selection balance, 791
Leucine zipper motif, 655f, 657
Libraries, *see* DNA libraries
Ligases
see also DNA ligases
in intron excision, 318–319, 319f
Light chains of antibodies, 732, 733, 733f
gene assembly, 421, 421f, 740–741, 741f, 743–744, 743f, 744f, 748
Light-dependent DNA repair (photoreactivation), 383
Light-induced transcription, 650–651, 650f
LINE transposable elements (long interspersed nuclear elements), 251, 490
Linkage, 168–171
see also Synteny; X-linked genes; Sex-linked genes
detecting it, 191–196
hemophilia and color blindness, 178HG-179HG
and independent assortment, 125–126, 168–169, 170f
measuring intensity, 169–171
Linkage analysis, 190–216
see also Maps and mapping
in humans, 201, 203–211
in bacterial gene mapping, 459–460, 460f, 465
Linkage maps, 211
Linkage phase, 170–171, 171f
Linked genes, *see* Linkage; Synteny; X-linked genes; Sex-linked genes
Linkers, in chromatin, 239, 240, 240f, 242f
Livestock, selective breeding, 12–13, 14f
Locus control regions (LCRs), 660
Locus (loci) of genes on chromosomes, 105, 121, 121f

Long interspersed nuclear elements (LINE transposable elements), 251, 490
Loss-of-function alleles, mutations, 79–80, 79f
  and cancer, 759, 759f
Loss of heterozygosity, 769
Lucy, 800
Lymphocytes, *see* B cells; T cells
Lysis, Lysogeny, *see under* Bacteriophage lambda; Bacteriophage T4

M

Maize (corn, *Zea mays*)
  color variation, 476–477, 476f, 482–483, 483f
  genetic alterations, 11, 12f, 13, 14f
  genome evolution, 578, 578f
  life cycle, 44–45, 45f
  male sterility, 499–500, 499f
  transposable genetic elements, 482–484, 483f, 484f
Major histocompatibility complex (MHC), 735
  MHC antigens (MHC proteins), 731t, 735–736, 737f, 739
  structure, 735f
Males, sex determination, 129–130, 129f
Map position-based cloning, *see* Cloning, positional
Map unit, 176
Maps and mapping, 174–183, 190–216, 555–567
  see also Human Genome Project; Sequencing; Linkage analysis
  distances between genes, 174–175
    calculation, 177, 177f, 179–180, 180f
    and chiasma frequency, 182, 182f
    on physical vs. genetic maps, 182–183, 183f, 557–558, 558f
    and recombination frequency, 180–182, 181f, 182f
    tetrad analysis, 193–194, 193f
    true genetic distance, 181, 181f
  first map ever, 167–168, 167f
  gene order, 176
  genomes now known, 545
  map types
    contig maps, 558f, 563, 563f
      use in finding genes, 587f
    cytogenetic maps, mapping, 198–200, 202TS-203TS, 563
    cytological maps, 555, 558f
      correlation with physical and genetic maps, 555–563, 558f
    use in finding genes, 587f
    genetic maps, 211, 555, 558f
      correlation with physical and cytological maps, 555–563, 558f
      high-density, 555
    linkage maps, 211
    physical maps, 211, 555, 558f
      and clone banks, 562–563
      correlation with genetic and cytological maps, 555–563, 558f
      use of restriction enzyme cleavage sites, 543, 544f
    restriction maps, 543, 555, 558f
      use in finding genes, 587f
    sequence-tagged-site (STS) maps, 558f
    what they're based on, 555
  of bacterial chromosomes, 454, 455
    closely linked genes, 465–466, 466f
    and conjugation, 464–465
    and transduction, 470–472
    and transformation, 458–460
  of bacteriophage genome, 434–441
    by recombination techniques, 436–437
    deletion mapping, 437–441, 438f, 439f, 440f
  of centromeres, 194, 196–198
  of chloroplast DNA, 501–502, 502f, 508f
  of *Drosophila* chromosomes, 121f, 149, 149f, 167, 167f, 180f, 183f, 198–200, 199f, 200f, 202TS-203TS
  of *E. coli* genes, 464–465, 465f, 567–568, 568f
  of human genes, 201, 202–211, 571–573
    on autosomes, 203–204, 204–211, 204f
    levels of resolution, 211
    use of DNA sequence variants, 204
  of human mitochondrial DNA, 505f
  of nucleotide sequences, 543, 544–548
  of polytene chromosomes, 149, 149f, 198–200, 202TS-203TS
  techniques
    cell fusion (hybridization) techniques, 205–206, 205f, 206f, 207f, 207t
    chromosome jumps, 566–567, 566f
    chromosome walks, 564–566, 565f
    deletion mapping, 198–200, 202TS-203TS
      of human genes, 208–209, 209f
      of phage genome, 437–441, 438f, 439f, 440f
    duplication mapping, 200, 200f, 209–211, 210f
    high-resolution radiation hybrid mapping, 571–572, 571f, 572f
    hybridization techniques, 562, 562f
    recombination mapping, 175–180, 436–437
    somatic-cell genetics, 204–211
    specialized techniques, 196–200
    use of chromosomal rearrangements, 206–211
    use of crossovers, 174–175, 177, 177f, 179–180
    use of recombinants, 175–177, 179–182
    use of restriction fragment-length polymorphisms, 558–562, 560f, 561f, 587f
mariner transposon, 487
Marker genes (marker DNA), 180
  chimeric selectable, 608
  in cloning vectors, 522
  DNA sequence variants, 204
  and Human Genome Project, 211
  selectable, 607–608
  for transgenic plants, 607–608
  use in gene mapping, 194–196, 195f
  use in linkage maps, 204, 204f
Markers
  see also Marker genes; Restriction fragment-length polymorphisms
  anchor markers, 555, 557, 558f
  and gene conversion, 391–392, 392f
  microsatellites, 560–562
  outside (flanking) markers, 390, 465
    reciprocal recombination, 391–392, 392f, 407
  and position effect, 411, 412f
  in recombination studies, 173
  "reporter" genes, 656TS
  use in agriculture, 13–14
  use in mapping
    recombinations, 174, 175, 180–181
    signs of crossovers, 174–175, 180–181
    types used, 555
Maternal-effect genes, 690, 698f, 699
Maternal-effect mutations, 690
Maternal inheritance, mitochondrial, 498, 499–500, 507
Mating
  see also Selective breeding; Inbreeding
  assortative, 780
  in bacteria, *see* Conjugation
  consanguineous, 107–108, 780
  panmixis, 781
Mean (average), statistical, 97
Medical genetics, 7–11
Meiosis, 32–37, 33f, 34f
  abnormalities, 36, 37, 151
  chromosome inversion loops, 157, 157f
  crossing over, 171, 171f, 173–174, 173f, 174f
  evolutionary significance, 37, 183
  and Mendel's principles, 123–126, 124f, 125f
  stages, 33f, 34–37, 34f
  vs. mitosis, 28
Memory, immunological, 739–740
Memory B and T cells, 731, 731t, 732f, 736, 737f, 739–740
Mendel, Gregor Johann, 4, 52f
  studies of heredity, 52–53, 53–58
Mendel's laws, principles, 4, 56, 58
  see also Dominance; Independent assortment; Segregation
  applications, 58–61, 64–68
  chromosomal basis, 116–137, 123–126
  and complex inheritance patterns, 93–95, 93f, 94f
  fusion with Darwinian theory, 776
  and meiosis, 37
  and random genetic drift, 786–787
Meristematic cells, 146–147, 146f
Messenger RNAs (mRNAs), 295, 297f
  see also pre-mRNAs
  editing, 310–312, 311f, 312f
  as intermediary in protein synthesis, 298, 298f, 299TS-300TS
  sense vs. antisense, 608
  splicing in mitochondria, 506, 506f
  stability, and control of gene expression, 649
  synthesis, 18–19, 19f
Metabolic disorders, 7, 397, 399HG

Metabolic pathways, 293, 294f, 295f
  effects of mutation, 364, 364f, 366–369
  mutational analysis, 369–370
Metabolism, 293
  genetic control, 293–294, 295f
  inborn errors, 397, 399HG
  oxidative (aerobic), 497
Metaphase chromosomes, 241–242, 242f, 243f
Metastasis, 751
Methionyl-tRNAs, 339
Methylation
  of DNA strands, and mismatch repair, 385–386, 386f
  of nucleotides, and gene expression, 665–666
  and restriction endonucleases, 519
5-Methylcytosine, 665f
7-Methyl guanosine caps, on mRNA, 306–307, 306f, 308–310, 309f
MHC, *see* Major histocompatibility complex
Mice
  biological rhythms, 712, 716
  chimeric, 705–706
  cloned, 42, 42f
  developmental studies, 704–706
  insertional mutations, 704–706
  nurturing behavior, 719–720, 719f
  as research tool, 704–706
  transgenic, 603–604, 604f, 705–706, 706f
Microarrays, 574
Microsatellites, 560–562
Microtubule organizing centers (MTOCs), 31
Microtubules, 33f
  in mitosis, 29f, 31, 31f
  and spatially regulated gene expression, 645–646
Migration, and Hardy-Weinberg equilibrium, 781
Mismatch repair, 385–386, 386f, 391, 392f
Mismatched bases
  see also Base-pair substitutions
  and heteroduplexes, 391, 391f
  and site-specific mutagenesis, 535, 535f
Missense mutations, vs. nonsense mutations, 353
Misuse of genetics, 15–16
Mitochondria, 25–26, 26f, 496–514
  as endosymbionts, 509, 511–512
  gene expression, 505–506
  gene shuffling, 512
  genetic code differences, 351
  and male-sterile maize, 500
  maternal inheritance, 498, 499–500, 507
  molecular genetics, 503–506
  origin and evolution, 509–512
  RNA processing, 311–312, 311f, 312f, 317–318, 319
Mitochondrial DNA (mtDNA), 26, 503–505, 504f
  in humans, 504, 505f
    and disease, 506–507
    use in evolution studies, 510HG-511HG, 800, 801
    vs. Neanderthals, 511HG
  and yeast petite mutants, 502–503, 503f
"Mitochondrial Eve," 510HG
Mitosis, 28–32, 29f, 30f
  phases, 28f, 29f, 30
Modal class (mode), statistical, 97
Modern genetics, 4
Molds, 41, 41f
Molecular analysis, *see* Molecular genetics; Maps and mapping; Sequencing
Molecular diagnosis, *see* DNA testing
Molecular evolution, and neutral theory, 790TS-791TS
Molecular genetics
  see also Maps and mapping
  in human evolution studies, 800–801
  potential hazards, 529–530
  tools and techniques, 517–552, 585–612
Molecular markers, *see* Markers
Monoclonal antibodies, 747TS
Monohybrid crosses, 54–56, 54f, 55t
Monosomies, 152–153, 152f, 152t
Moral questions, *see* Social issues
Morphogenesis, *see under* Development
Mosaicism, 133–134, 133f, 152–153
  how it develops, 687, 688
  in transgenic animals, 603
  use in development studies, 686–688
Motifs, of transcription factors, 655, 655f, 657
mRNA, *see* Messenger RNAs; pre-mRNAs
MTOCs, *see* Microtubule organizing centers
Multifactorial traits, 93–97, 93f, 94f
  see also Complex traits; Quantitative traits; Inheritance, complex patterns; Inheritance, polygenic; Variation, continuous phenotypic
  genetic counseling, 96HG
Multiple Factor Hypothesis, 99–100, 100f
Mutagenesis
  insertional, 704–706
site-specific (oligonucleotide-directed site-specific), 534–535, 535f
Mutagens, 361
  and cancer, 753
  chemicals, 373–378, 375t
    testing for, 381–383, 382f
  radiation, 370–371, 371f, 378–380
Mutant alleles, 74
Mutant genes, 7, 20, 20f
Mutant phenotype, 7
Mutation-drift balance, 791–793, 792f, 800
Mutation-selection balance, 789, 791, 791f
Mutation theory of evolution, 160TS
Mutations, 8, 20–21, 20f, 75–77, 79–80, 359–383, 365f
  and aging, 387–388
  back mutations, 363, 363f
  in bacterial cultures, 455–456
  basic features, 360–363, 365f
  and behavior, 721–723
  and cancer, 753, 756–758, 759–761, 766HG, 767–769
    genetic pathways, 767–769, 768f
    two-hit hypothesis, 759–761, 759f
  in cellular oncogenes, 756–758
  conditional lethal, 369–370
  dominant, 79–80, 79f
    testing for allelism, 76
    vs. recessive, 76, 79–80, 363–364
  dominant lethal, 77
  dysgenesis-induced, 485, 492
  effects on metabolic pathways, 364, 364f, 366–369
  effects on phenotype, 363–370
  effects on polypeptides, 79–80, 79f
  and evolution, 359–360, 777, 786
  evolutionary rate, 790TS
  forward mutations, 363
  frameshift mutations, 372, 374f
  as function of genetic material, 220
  gain-of-function, 79–80, 79f
  germinal, 360
  homeotic, 696–697
  homoalleles vs. heteroalleles, 415
  "hot spots," 440, 440f
  in human globin gene, 364–366, 366f
  induced, 360–361
  insertional, 704–706
  "knockout," 706
  lethal, 77, 77t, 363–364, 364f, 369–370, 370–371, 371f, 690
    and mutation-selection balance, 791
  loss-of-function (inactivating), 79–80, 79f
    and cancer, 759, 759f
  maternal-effect, 690
  missense, 353
  mitochondrial, 507
  molecular basis, 371–381
  neutral, 364
  nonsense, 353–354
  point (single-site) mutations, 20, 20f, 359, 372, 374f
  polar, 418–419, 419f
  from radiation, 370–371, 371f, 378–380
  randomness, 361–362
  recessive, 79–80, 79f
    testing for allelism, 75–76
    vs. dominant, 76, 79–80, 363–364
  recessive lethal, 77, 77f, 363–364, 364f, 370–371, 371f
  and recombination, 183
  restrictive vs. permissive conditions, 369
  reverse mutations, 363, 363f
  reversion to wild-type, 362–363, 363f
  somatic, 360
    and pathways to cancer, 767–769
  somatic hypermutation, 744–745
  spontaneous, 360–361
    causes, 372
  sterile, 77
  structural vs. functional allelism, 415
  suppressor mutations, 345, 352–354, 363, 363f
  suppressor-sensitive, 369
  tautomeric shifts, 372, 372f, 373f
  temperature-sensitive, 81, 369
  testing for allelism, 75–76, 76f
  and transposable genetic elements (transposons), 380–381, 381f, 491, 492, 492f

in tumor suppressor genes, 759–761, 765–767, 767–769
testing for, 766HG
use in research, 369–370
transposon tagging, 486
visible, 76
X-linked recessive, 364, 364f, 370–371, 371f

N

Narrow-sense heritability, 102–105, 102t, 103f, 104f
Natural populations, variation in, 793–795
Natural selection, 92, 775, 777, 782–786
see also Selection
Darwin's principle, 782
effect on allele frequencies, 783–785, 784f
at level of gene, 782–785
at level of phenotype, 785–786, 785f
misapplication, 15–16
Neanderthals, 800
mitochondrial DNA (mtDNA), 510HG-511HG
Negative autogenous (self-) regulation, 639
Neomorphic alleles, 79
Neurospora crassa (bread mold)
life cycle, 41, 41f, 43
mapping of centromeres, 196–198
Neutral theory of molecular evolution, 790TS-791TS
Nomenclature, 78HS
see also Symbols
for allelic variations, 74
amino acid abbreviations, 328f
for chromosome number (ploidy), 142–143
of chromosomes, 141, 142f–143f
for deoxyribonucleotides, 227f
for DNA, RNA, and components, 225
in karyotyping, 141, 142f-143f
in polytene chromosome maps, 149, 149f
in twin studies, 723–724
Noncoding sequences, *see* Introns
Nondisjunction, 37, 121, 123, 123f
and aneuploidy, 150–151, 151f, 152t
and Down syndrome, 150–151, 151f
Nonheritable (somatic-cell) gene therapy, 598
Nonhistone chromosomal proteins, 235, 236
Nonself, vs. self, 730, 735–736
Nontranscribed (nontemplate) strand, of DNA, 300, 300f
Normal (bell-shaped) distribution, 87, 87f, 98, 98f, 99
Northern blot hybridization, 539–540, 539f
Nuclease, 267
Nucleic acids, 17
density-gradient analyses, 260TS-261TS
discovery, 219–220
as site of genetic information, 220
structure, 225, 226f
types, 220
Nuclein, 219–220
Nucleoids, of prokaryotes, 233
Nucleosomes, 235, 239–240, 240f, 241, 241f, 659
nucleosome core, 239–240, 240f, 241f, 242f
nucleosome fibers, 241, 242f
at replication forks, 284, 285f
Nucleotide excision repair, 383–384, 385f
vs. base excision repair, 383
and inherited diseases, 387
Nucleotide pairs (base pairs), 17, 17f
as basic unit of genetic material, 398, 407
changes in, *see* Mutations; Base-pair substitutions
colinearity with polypeptide, 400, 408–410, 409f, 411f
number in humans, 528, 567
number per turn, 230, 231, 231t
recombination between, 407–408, 409f
Nucleotides
see also Bases
chemical modification, 665
in codons, 344–346
in DNA, number per turn, 230, 231, 231t
in DNA vs. RNA, 225, 226f
purines vs. pyrimidines, 225, 226f, 227f
Nullo-X egg, 121, 123, 123f
Numerator elements, proteins, genes, 682f, 683, 685t

O

Octamer box, 308, 308f
Octaploid (8n) cells, 117
Okazaki fragments, 272, 273–274, 279, 280f, 284f
Oligonucleotide-directed site-specific mutagenesis, 534–535, 535f
Oncogenes, 753–759
normal cellular (*c-oncs*; proto-oncogenes), 755–756, 755f
and cancer, 756–758, 767
viral (*v-oncs*), 753–755, 754t
cellular homologs, 755–756, 755f
Oogenesis, 38–39, 38f, 676–678
and Barr bodies, 134
Open reading frames (ORFs), 567, 570, 570f
Operators
of *lac* operon, 622, 622f, 623, 623f, 624, 624f, 626f
in operon model, 620–622, 621f
of phage lambda lytic pathway, 632, 632f
Operons, 620–622, 621f
see also lac operon; *trp* operon
inducible vs. repressible, 621f, 622
Ordered tetrads, 194, 196–198, 196f
Organelles, 497–503
Organs, *see* Body organs
OriC, 262–263, 263f, 278–279, 279f
Origin, in DNA replication, 261, 262–263, 263f, 278–279
in eukaryotes, 281–283, 282f
Origin recognition complex (ORC), 662
Outside (flanking) markers, 390, 465
reciprocal recombination, 391–392, 392f, 407
Ova, *see* Eggs; Oogenesis
Overhangs, 244
Overlapping clones (contigs), in mapping, 555, 558f, 563, 563f
Overlapping genes, 443–444, 443f, 444f
and definition of gene, 400, 401
Overviews
base-pairing, 17, 17f
basic genetic concepts, 16–21
chromosome rearrangements, 156–162
chromosomes, 27–28
cytogenetic variation, 142–143
development, in animals, 18–20, 676–678
DNA replication, synthesis, 17, 18f, 277–279, 278f, 280f
DNA structure, 17, 17f
gene expression, 18–20, 19f, 295f
genes, changing concepts, 398–401
genetic code, 19, 344
genetic control of metabolism, 293–294, 295f
genetic exchange in bacteria, 454–455
genomics, 553
immune response, 729, 730
mutations, 20–21, 20f
nucleic acids, 17
protein synthesis, 18, 19f, 296f, 297f, 331–332, 332f
transcription, 18–19, 19f, 294–295, 295f, 297f, 300–302, 301f, 302f
translation, 18–19, 19f, 294–295, 295f, 297f, 331–332, 332f
Oxidative (aerobic) metabolism, 497

P

p (short) arm of chromosome, 141
P elements, 484–487, 485f, 492
as research tool, 486, 488TS-489TS
p21 protein, 762–763, 763f
p53 protein (*TP53* gene product), 761–764, 763f
Pair-rule genes, 697, 698f, 699
Pairing, *see* Base pairing; Nucleotide pairs; Chromosomes, pairing
Palindromes, 519, 519t
Panmixis, 781
pAPC protein, 764, 765f
Parental age, and nondisjunction, 151
Parental DNA molecule, 17, 18f
Paternity tests, by DNA fingerprinting, 595, 595f
Pathways
biological, 681
to cancer, 767–769, 768f
of coagulation, 11HGf
of development, 681–682
cascades (preprogrammed circuits, regulatory hierarchies), 695, 698f, 699
of gene expression, 19
preprogrammed circuits (regulatory hierarchies, cascades), 617, 637–638, 638f, 695, 698f, 699
lytic, of phage lambda, 630–631, 631f
operators, 632, 632f
regulatory cascade, 634–635, 634f
repression, 631–634, 632f
metabolic, 293, 294f, 295f

Pathways, *(continued)*
effects of mutation, 364, 364f, 366–369
mutational analysis, 369–370
pBRCA1, pBRCA2 proteins, 765–767
Pedigrees, 64, 64f
analysis for gene mapping, 201, 203–204, 559, 564f
analysis for linkage, 178HG-179HG
in inbreeding, 109
symbols used, 64–65, 64f, 109
Penetrance, incomplete, 81–82, 83HG
Pentoses, 225, 226f
Peptides, 327
Permuted sequences, 441–442, 441f
Personality, 5, 710–711, 725–726, 725t
Pesticides, 13
Petite mutants in yeast, 502–503, 502f, 503f
Phage, *see* Bacteriophage
Phagemids, 531–534, 532f, 534f
Phagocytes, 730f, 731, 731t, 732f
Pharmaceuticals, genetically engineered, 602
Phenotypes, 56, 56f
effects of chromosome rearrangements, 162
effects of mutations, 363–370
how gene evokes, 80–87, 401–402, 401f
how they develop, 681
mutant, 7
predicting, 103–104, 103f
Phenotypic variance, 100
partitioning for heritability, 100–101
Phenotypic variation, continuous, 86, 92, 793–794
see also Inheritance, complex patterns; Inheritance, polygenic; Multifactorial traits; Complex traits; Quantitative traits
Phenylalanine, mutations, 366–369, 368f
Phenylketonuria (PKU), 7, 81, 86, 367–368, 368f, 721–722
Pheromones, bacterial, 457–458
Philadelphia chromosome, 758, 758f
phMSH2 protein, 764–765
Phosphate group, in DNA and RNA, 226f
Phosphodiester linkages, 227f
Phosphorylation, *see* Kinases
Photoreactivation DNA repair, 383
Photosynthesis, 497
Physical maps, *see under* Maps and mapping, map types
Plant cell, 24f
Plants
see also Food crops; Chloroplasts
aneuploidy, 149–150, 150f
asexual reproduction, 145
cell totipotency, 605
gamete formation, 40, 40f
inbreeding, 108–109
leaf variegation, 497–499, 497f, 498f, 499f
mitochondrial DNA, 505
polyploidy, 143–144, 144f, 147–148, 147f
and chromosome doubling, 146–147, 146f
and reproduction, 145–146, 145f
reproductive structures, 44, 44f
transgenic, 605–608
how they're made, 605
trisomies, 150f
tubulin genes, spatial regulation, 645–646
tumors, 605–606, 606f
Plaque hybridization, in screening for genes, 529
Plaque morphology, in bacteriophage, 434, 434f, 435f
Plasma cells, 731, 731t, 732f
in immune response, 736, 737f
Plasmids, 466
and bacterial drug resistance, 453, 454, 472, 481, 482f
and bacterial genetic exchange, 454
F factor (F plasmid), 461, 462f
IS elements, 479
R plasmids, 453, 454, 481, 482f
Ti plasmid, 606–608, 607f
transfer between species, 481, 488TS-489TS
as vectors
for cloning, 522–523, 523f
for HGH production in bacteria, 601, 602f
Plastids, 507
Pleiotropy, 86
Ploidy, 117, 142–143
nomenclature, 142–143
Pneumococci, transformation, 221–222, 221f, 221t, 222f, 223f
Point (single-site) mutations, 20, 20f, 359, 372, 374f
Polar bodies, 38, 38f, 39
Polar mutations, 418–419, 419f
Polarity
of DNA, 230, 230f
of polynucleotide chain, 227f
Poly(A) tails
on mRNA, 306–307, 306f, 310, 310f, 649
on retroposons, 490
Polyacrylamide gel electrophoresis, 540
Polyadenylation, 310, 310f
and retroposons, 490
Polycloning (polylinker) sites, in cloning vectors, 522, 523f, 534
Polycomb group (PcG) genes, proteins, 662
Polycomb response elements (PREs), 662
Polydactyly, 81–82, 81f
Polygenic inheritance, traits, 96HG, 100
see also Multifactorial traits; Complex traits; Quantitative traits; Inheritance, complex patterns; Variation, continuous phenotypic
Polylinker (polycloning) sites, in cloning vectors, 522, 523f, 534
Polymerase chain reaction (PCR), 5, 541–543, 542f
Polymerases, *see* DNA polymerases; RNA polymerases; *Taq* polymerase
Polymorphic genes, 75
Polymorphisms
see also Restriction fragment-length polymorphisms; Microsatellites
balanced, 788–789
of chromosome structure, 794, 794f
and DNA fingerprinting, 594
electrophoretic, 795, 795f, 795t
genetic, 75, 778, 793
Polynucleotides, 225, 227f
Polypeptides, 18, 78, 327
amino acid subunits, 327–329
cellular roles, 293
chain formation (synthesis), 19f, 297f
see also Translation; Proteins, synthesis
effects of mutations, 79–80, 79f
finding their function, 518
levels of organization, 329, 329f
relation to genes, 77–79, 77f, 400, 408–410, 409f, 410f, 411f
separation by gel electrophoresis, 540
Polyploidy, 143–149
allo- and autopolyploids, 146
fertile polyploids, 145–146
nomenclature, 142–143
in plants, 143–144, 144f, 147–148, 147f
origin, 146–147, 146f
and reproduction, 145–146, 145f
sterile polyploids, 144–145
tissue-specific, 148–149
vs. aneuploidy, 142–143
Polytene chromosomes, 148–149, 148f
bands, 148–149, 148f, 149f
relation to genes, 202TS-203TS
deletions and duplications, 155, 155f
maps and mapping, 149, 149f, 183f, 198–200, 199f, 202TS-203TS
puffs, 659, 659f
transcription, 659
Polyteny, 148–149
Pongids, 799f
Population genetics, 774–804
see also Allele frequencies
emergence, 776–777
Population size
genetically effective size vs. census size, 792–793
and heterozygosity, 792, 792f
and random genetic drift, 787–788, 787f
Populations
in dynamic equilibrium, 788
in genetic equilibrium, 788–793
natural, genetic variation, 793–795
inhomogeneous, and Hardy-Weinberg principle, 781
Position effect, 162, 162f, 411, 412f
Position-effect variegation, 660–661
Positional cloning, *see under* Cloning
pRB protein, 760, 761, 761f
Pre-mRNAs, 295, 296f, 306–307, 306f, 308–310, 309f
see also Primary transcripts; Heterogeneous nuclear RNA; Messenger RNAs
editing, 310–312, 311f, 312f
splicing, 320–321, 321f
Predicting phenotypes, 103–104, 103f
see also Counseling
Prenatal testing, 154HG, 539TS, 585–586, 590, 591, 593

Prepriming proteins, in DNA replication, 278, 279, 279f
Preprogrammed circuits of gene expression, *see* Cascades
Primary transcripts, 295, 297f
see also pre-mRNAs; Heterogeneous nuclear RNA; Splicing
in eukaryotes vs. prokaryotes, 295
modification in eukaryotes, 295, 296f, 306–307, 306f
5′ caps on mRNA, 306–307, 306f, 308–310, 309f
3′ poly(A) tails on mRNA, 306–307, 306f, 310, 310f, 649
Primases, *see* DNA primases
Primers, for DNA synthesis, 266–267, 266f, 268, 269, 269f, 278f
RNA primers, 273–274, 273f, 274f, 279, 280f
Primosome, 278f, 279, 280f
Principles of genetics, 16–21
see also Mendel's laws, principles
Probability
binomial, 66, 67TS
rules, principles, 61TS
Probability distribution, 65f, 67TS
Probability method, 59–60, 60f, 65–66
Probes, for hybridization and blotting, 529, 530f, 537–539, 537f, 538TS, 557, 574, 574f
use as mapping anchors, 557
Prokaryotes
see also Bacteria; *E. coli*
cells, 23–24, 23f, 27
chromosome structure, 233–234
genes, 400f
genomes, 233, 234f, 567–569
vs. eukaryotes, 234
regulation of gene expression, 615 643
posttranslational control, 639–641
translational control, 638–639, 640f
transcription, 302–305, 304f, 305f
translation, 338f, 339–340, 339f-340f, 341–344, 343f
Promoters of transcription, 301
in antibody gene assembly, 746, 748, 748f
in eukaryotes, 307, 308f, 309f
in operons, 621–622, 621f, 622f, 623, 625, 626f
in prokaryotes, 303, 303f
regulation, 618–620, 619f
Proofreading, by DNA polymerases, 270–271, 271f
Prophage, 468, 470f, 630–631
Proplastids, 507, 509f
Prostate cancer, genetic pathway, 768f
Protein isoforms, 401, 420, 420f
Protein kinases, 754
Proteins, 18
in chromatin, 235, 236
complex relations with genes, 419–421
as computer chips, 576
folding, 329
fusion, from chimeric genes, 520HG-521HG
human, production by bacteria, 601–602
for industry, production by bacteria, 602–603
localization in cells, 575–576, 577f
sequence databases, 518
structure, 327–331, 329f, 330f, 331f
allosteric transitions, 620, 640–641, 641f
levels of organization, 329–331, 329f, 330f
study techniques
separation by gel electrophoresis, 540
use of green fluorescent protein, 574–576
western blotting, 540–541, 540f
synthesis, 296f, 297f, 331–344
see also Transcription; Translation
components required, 332–337
mRNA as intermediary, 298, 298f, 299TS–300TS
overview, 18, 19f, 331–332, 332f
Proto-oncogenes, *see* Oncogenes, normal cellular
Prototrophic organisms, 369, 456
Pseudoautosomal genes, 128
Pseudogenes, 646
Psychological traits, and heritability, 723–726
pUC vectors, 532–534, 532f, 534f
Pulse-chase, pulse-labeling experiments, 272
Pulsed-field gel electrophoresis, 236–237, 237f
Punnett Square method, 57f, 58
Purines and pyrimidines, 225, 226f, 227f
base-pairing in DNA, 228, 230f
ratios in DNA, 226, 228t
tautomeric shifts, 372, 372f
transitions and transversions, 372, 374f
Pyrimidine dimers, *see* Thymine dimers

Q

q (long) arm of chromosome, 141
Quantitative genetics, 97–99, 723–724
see also Statistics
Quantitative trait (QT) loci, 14, 105, 107t
Quantitative traits, 92–95
see also Multifactorial traits; Complex traits; Inheritance, complex patterns; Inheritance, polygenic; Variation, continuous phenotypic
and natural selection, 785–786, 785f
normal distribution, 99
statistical analysis, 99–107

R

R groups (side chains) of amino acids, 327, 328f
R-loops, of DNA, 313, 313f
R plasmids, 453, 454, 481, 482f
Radiation
dosage, and mutation rate, 379–380, 379f
electromagnetic spectrum, 379f
ionizing vs. nonionizing, 378
mutations from, 370–371, 371f, 378–380
ultraviolet (UV), as mutagen, 379, 380, 380f
Radiation hybrid mapping, 571–572, 571f, 572f
Random genetic drift, *see* Genetic drift
ras genes, Ras proteins, 754
and cancer, 754t, 757–758, 757f, 767, 768f
Ratios, phenotypic and genotypic, 54, 56, 56f, 57f
RB gene product (pRB), 760, 761, 761f
Reading frames, 20, 345, 345f
see also Frameshift mutations
open (ORFs) and unassigned (URFs), 505
and overlapping genes, 443, 444f
RecA proteins
in DNA repair, 386, 387f
and pBRCA proteins, 765
in recombination, 389–390, 389f
Receptors
for hormones, 651, 651f, 652, 652f
and oncogenes, 754, 755
for T cells, *see* T cell receptors
Recessive disorders, and inbreeding, 108, 108f, 111–112
Recessive mutations, *see under* Mutations
Recessive traits, 55, 56
in pedigrees, 64f, 65
Reciprocal exchanges, in recombination, 390–391
Recombinant chromosomes, 171f
Recombinant DNA molecules, 518, 521
amplification in vectors, 522–525
manipulation *in vitro*, 531–535
potential hazards, 529–530
how they're made, 521–522, 522f, 526–527, 526f, 527f
Recombinant DNA technology, 518
see also Molecular genetics; Genetic engineering
identifying disease genes, 586–593
potential hazards, 529–530
research guidelines, 530
vs. gene cloning, 518
Recombinants, 169, 169f, 170f
use in mapping, 175–177, 179–182
Recombination, 168–174, 388–392
see also Crossing over
in bacteria, 454–455, 455f, 458, 472
see also Conjugation; Transduction; Transformation
in bacteriophage, 435–437, 435f, 437f
between nucleotide pairs, 407–408, 409f
crossing over as basis, 171–173, 171f
and evolution, 183–185, 472
gene conversion, 390–392, 392f
genetic control, 185
and independent assortment, 169–170
intentional suppression, 185
and IS elements, 479
and markers, in mapping, 174, 175, 180–181
as measure of linkage, 169–171
in mitochondrial DNA, 504, 504f
molecular mechanisms, 388–392
nonreciprocal, 391, 436, 458
reciprocal, 390–391
of outside (flanking) markers, 391–392, 392f, 407

Recombination, *(continued)*
in antibody gene assembly, 421, 421f, 740, 741, 741f
DNA signal sequences, 742–743, 743f
in *Drosophila*, 687–688, 688f, 689f
in T cell receptor gene assembly, 746
vs. complementation, 414–415
within genes, 398, 407
Recombination frequency
and chromosome rearrangements, 184–185, 184f
and true map distance, 176, 180–182, 181f, 182f
use in mapping, 176, 179–182
Recombination intermediates, 390, 390f, 391, 391f, 392f
Recombination mapping, 175–180, 436–437
Red blood cells, in sickle cell anemia, 20–21, 20f, 326–327
Redundancy, *see* Repetitive DNA; Genetic redundancy
Regulator genes, 618–620, 619f
of *lac* operon (*lacI*), 622, 622f, 623–625, 623f, 624f, 626f
in operon model, 621f
Regulator protein binding site (RBS), 618–620, 619f
Regulatory (end-product, effector molecule) binding site, for enzymes, 639–640, 641f
Regulatory hierarchy, of gene expression, *see* Cascades
Relatives
correlations between, 112–113
intermarriage, *see* Mating, consanguineous; Inbreeding
sharing of alleles, 65, 109
Release factors (RFs, eRF), 343f, 344, 350
Repair, of DNA, *see* DNA, repair mechanisms
Repeats, repeat sequences, *see* Repetitive DNA
Repetitive DNA, 246–251
see also Genetic redundancy; Amplification
detection, 246–249, 248TS
expanded repeats, and disease, 128, 128f, 381
finding in chromosomes, 249–251, 250f
in fragile X syndrome, 127–128, 128f, 381
functions, 249
highly vs. moderately repetitive, 247, 249
microsatellites, 560–562
in mitochondria, 504
renaturation kinetics, 249f, 248TS
and restriction fragment-length polymorphisms, 560
tandem repeats (tandem duplications), 155, 155f
simple, 381
target site duplications, 479, 479f
telomere-associated sequences, 244
terminal redundancy, 441–442
inverted terminal (terminal inverted) repeats, 477–478, 479f, 483, 484f, 485, 485f
long terminal repeats (LTRs), 449, 487, 489f
in transposable genetic elements (transposons), 249, 477–479, 479f, 483, 484f, 485, 485f
trinucleotide repeats, 381, 588–589, 589f
variable number tandem repeats (VNTRs), 560
and DNA fingerprinting, 594, 594f, 596TS-597TS
Replication
see also DNA replication
as function of genetic material, 220
Replication factors, 283, 284f
Replication forks, *see under* DNA replication
Replicons, 262, 281–283, 282f, 283f
Replisomes, 279, 280f, 284, 285f
"Reporter" genes, 656TS
Repressors
see also lambda repressor, *under* Bacteriophage lambda
of *lac* operon, 622–623, 622f, 623f, 624, 624f, 626f
in operon model, 620–622, 621f
of prokaryote gene expression, 619f, 620
Reproductive isolation, 795–796, 796t
Repulsion (*trans*) configuration, 412
Response elements
for heat-shock genes (HSEs), 650, 650f
for hormones (HREs), 652
Restriction endonucleases, *see* Restriction enzymes
Restriction enzyme cleavage sites, 519, 519t, 522f
in cloning vectors, 522, 523f
and site-specific mutagenesis, 534
use in mapping, 543, 544f
Restriction enzymes (restriction endonucleases), 433, 518–519, 519t, 521, 522f, 523f
nomenclature, 518
use in mapping, 543, 544f
what they "restrict," 519
Restriction fragment-length polymorphisms (RFLPs)
how they're created and detected, 558, 559f
use in mapping, 555, 558–562, 558f, 560f, 561f, 587f
Restriction fragments, 521, 522f
separation by gel electrophoresis, 536
Southern blotting, 537, 537f
use in mapping, 543, 544f
Restriction maps, *see under* Maps and mapping, map types
Restriction sites, *see* Restriction enzyme cleavage sites
Retroposons, 487, 490
Retrotransposons, 487, 489–491, 493
Retrovirus-like transposable elements, 487, 489–490, 489f, 493
Retroviruses, 449
see also RNA viruses
and retrotransposons, 487, 493
tumor-inducing, 753–755, 754t
Reverse transcriptase, 445, 446, 446f
use in DNA library construction, 526
Reverse transcription, 295f, 446, 447f
and retroposons, 490
and retrotransposons, 487, 490f
Rho-dependent and -independent transcription terminators, 304, 305f
Rhythms, in behavior patterns, 712–716
circadian (24-hour), 712, 716
see also Biological clock
Ribonucleic acid, *see* RNA
Ribonucleoproteins, small nuclear (snRNPs), 320–321, 321f
Ribose, 226f
Ribosomal RNAs (rRNAs), 296, 297f
and gene amplification, 666
genetic redundancy, 334
synthesis, 333–334, 335f, 639
Ribosomes, 332–334, 332f, 338f, 340f–341f, 343f
mitochondrial, 506
in prokaryotes vs. eukaryotes, 332–333, 333f
sizes, 261TS, 333
structure, 332–333, 333f
subunits, 261TS, 333, 333f, 338f, 343f
tRNA binding sites, 336–337, 337f
RNA, 17
see also specific types, e.g., Messenger RNAs
alternate splicing, 648, 648f
antisense strands, 300, 608, 609f
uses, 608–610
base composition, vs. RNA, 225, 226f
editing, 310–312, 311f, 312f, 506
as genetic material, in RNA viruses, 224–225, 225f
guide RNAs (gRNAs), 311f, 312, 312f, 506
in mitochondria, 505–506
northern blot hybridization, 539–540, 539f
processing, 308–312, 647
see also Primary transcripts
sense strands, 300
structure, 225, 226f
synthesis, *see* Transcription
types, 295–297
RNA polymerases, 301, 301f, 302f
in DNA synthesis, 273
in eukaryotes, 307, 307t, 309f
how they work, 302f, 303–304, 304f
phage-induced changes, 432–433
regulation, 618–620, 619f
by *lac* operon, 622–623, 624f
in operon model, 621–622, 621f
structure, 302–303, 309f
RNA primers, for DNA synthesis, 273–274, 273f, 274f, 279, 280f, 284f
RNA splicing, *see* Splicing
RNA transcripts, *see* Transcripts; Primary transcripts; Pre-mRNAs; Heterogeneous nuclear RNA
RNA viruses, 224–225, 225f, 449
see also Retroviruses
Rolling-circle replication, 280, 281f

Rous sarcoma virus, 754
"Royal hemophilia," 10, 10f, 126–127, 126f
Royal prognathism, 83HG
Russian royal family
and DNA fingerprinting, 596TS-597TS
and hemophilia, 126–127, 126f

S

S (Svedberg) units, 261TS
of ribosomes, 333, 333f
s (sedimentation coefficient), 261TS
Saccharomyces cerevisiae (baker's yeast)
see also Yeast
chromosome map, 570f
genome, 569–571
life cycle, 43–44, 44f
linkage studies, 171, 171f, 191–194, 192f, 193f
Sample, statistical, 97
Sanger (dideoxy chain termination) sequencing procedure, 545–547, 546f, 547f, 548f
Satellite bands, 246
Satellite DNA, 246, 249–251, 250f
Screening for genes, in DNA libraries, 528–529, 530f
Sedimentation coefficient (s), 261TS
Segment-polarity genes, 697, 698f, 699
Segmentation genes, 697–699, 698f
Segments of body, 695–696, 696f
development, 695–699, 698f
Segregation, 56, 65–66
chromosomal basis, 123–124, 124f
and complex inheritance patterns, 93–95, 94f
and meiosis, 37
Mendel's principle, 56
in *Neurospora*, 196, 197, 197f
in X-linked inheritance, 119–120
Selection
see also Natural selection; Selective breeding
artificial, 104–105, 104f, 106TS–107TS, 785
balancing, 788–789, 789t
clonal, 738
directional, 785, 785f
disruptive, 785–786, 785f
genetic, to screen for genes, 528–529
purifying, 791
stabilizing, 785f, 786
Selection coefficient (*s*), 783
Selection-mutation balance, 789, 791, 791f
Selection pressure, *see* Natural selection
Selective breeding
see also Selection, artificial
eugenics movement, 16
of livestock, 12–13, 14f
predicting outcomes, 104–105, 104f
Self, vs. nonself, 730, 735–736
Self-fertilization, 108–109, 109f
Self-(autogenous) regulation, 633–634, 633f, 639
Self-splicing, 319–320, 320f
Semidominance, 73, 94, 95
Sense strands, of DNA and RNA, 300
Sequence-tagged sites (STSs), 555, 557, 558f, 561
Sequencing
see also Human Genome Project; Maps and mapping; DNA fingerprinting
of bacterial genomes, 567–569
databases, 518
of DNA, automated, 547, 548f, 556HG-557HG, 572
in evolution studies, 510HG, 790TS-791TS, 800–801
of genes, 518
genomes now known, 545
of nucleotide pairs, 543, 544–548
procedures, 545–547, 546f, 547f, 548f
history, 557HG
of proteins/amino acids, 518, 790TS-791TS
sequencing machines, 556HG–557HG
social issues, 15
of whole genome, 15
shotgun approach, 556HG–557HG
Sex chromosomes, 116–117, 117–119, 118f
see also X chromosome; Y chromosome
in human karyotype, 141, 143f
and sex determination, 116–117, 128–132, 129f
Sex determination, 116–117, 128–132
in bees, 131, 132f
in birds, 131f
in *Caenorhabditis*, 685–686, 685t, 686f
in *Drosophila*, 130–131, 131f, 131t, 682–685, 682f, 683f, 684f, 685t, 687
and sexual preference (orientation), 716–717, 717f
and XX/XO mosaics, 153
gynandromorphs, 153
in humans, 129–130, 129f
and sex chromosomes, 116–117, 128–132, 129f
Sex-lethal gene, protein, *see Sxl* gene
Sex-linked genes, 126–128
see also X-linked genes; Synteny
on Y chromosome, 128
Sexduction, 463
Sexual preference (orientation)
in *Drosophila*, 716–717, 716f, 719
in humans, 717, 718HG–719HG
Short interspersed nuclear elements (SINE elements), 251
Shotgun sequencing, 556HG–557HG
Shuttle vectors, 524, 525f
Siblings, *see* Relatives; Inbreeding; Mating, consanguineous
Sickle cell anemia, 20–21, 20f, 326–327
balanced polymorphism, 788–789
DNA testing, 593, 593f
Sickle-cell hemoglobin, 326–327, 365, 366f
Side (R) groups (side chains) of amino acids, 327, 328f
Sigma (Σ), statistical meaning, 98
Signal transduction, 652
Signaling, signal sequences
in antibody gene assembly, 742–743, 743f
and cancer, 757, 757f
in cell differentiation, 701–702
in developmental pathways, 681–682
and hormones, 651–653
in transcription, 19, 647
Silencing of genes, 661–665
Silent mating type cassettes, 662–663, 663f
Single-strand DNA-binding (SSB) proteins, 275, 275f, 278f, 279, 280f
Site-specific (oligonucleotide-directed site-specific) mutagenesis, 534–535, 535f
Small nuclear ribonucleoproteins (snRNPs), 320–321, 321f
Small nuclear RNAs (snRNAs), 296, 297f, 320–321, 321f
Social issues, 15–16
breast cancer genes, 9
cloning, 42–43
DNA fingerprinting, 594
gene therapy, 598–599
genetic engineering, 529–530
guidelines, 530
genome sequencing, 15
Human Genome Project, 572–573
vs. Celera, Inc., 556HG–557HG
insurance issues, 15
patenting genes, 556HG
testing for Huntington's disease, 590
testing for mutant tumor suppressor genes, 766HG
Solenoid model of chromatin structure, 241, 242f
Somatic cell precursors, 679
Somatic-cell genetics, and gene mapping, 204–211
Somatic-cell hybridization, 205–206, 205f, 206f, 207f, 207t
radiation hybrid mapping, 571–572, 571f, 572f
Somatic-cell (nonheritable) gene therapy, 598
Somatic hypermutation, 744–745
Somatic mutations, 360
and genetic pathways to cancer, 767–769
Somatic recombination, *see under* Recombination
SOS response, 387, 635
Southern blot hybridization, 536–537, 537f, 538TS–539TS, 539
in testing for genetic diseases, 593, 593f
Speciation, 795–799
allopatric (geographic), 796, 797f
how it happens, 796–799
sympatric, 798, 798f
Species
hybrids between, *see under* Hybrids
taxonomic vs. evolutionary definitions, 795–796
Spermatogenesis, 39–40, 39f
Spindles, in cell division, 29f, 31, 31f, 33f, 243f
Spliceosomes, 292f, 295, 296, 297f, 320–321, 320f, 321f
Splicing
alternate splicing, 317, 420, 420f, 648, 648f, 649f

Splicing, *(continued)*
in *Drosophila* sex determination, 648, 649f, 683–684, 683f, 684f
in antibody gene assembly, 734, 740, 745, 745f
autocatalytic, 319–320, 320f
cis-splicing, 419
of mitochondrial mRNA, 506, 506f
of pre-mRNAs, 320–321, 321f
of primary transcripts, 295, 296f, 306f, 313, 314f, 317–322, 317f
mechanisms, 318–322
of transfer RNA (tRNA) precursor, 318–319, 319f
Split genes, 295, 312–317
see also Introns; Splicing
Standard deviation, 98f, 99
Statistical analysis, 97–105, 106TS-107TS
in human behavioral studies, 723–726
of quantitative traits, 99–105, 106TS-107TS
Statistics
see also Quantitative genetics
analysis of variance, 100
basic concepts, 97–99
bell-shaped distribution, 87, 87f, 98, 98f
chi-square test, 62–64
correlation coefficient, 723–724, 724f
frequency distributions, 97–98, 97f, 98f
mean (average), 98
modal class (mode), 97
normal distribution, 98, 98f, 99
sigma (Σ), 98
standard deviation, 98f, 99
variance, 98–99
Stem-and-loop structures, 628f, 629, 743, 743f
Stem cells
of bone marrow, 730, 731t, 732f
for gene therapy, 600
embryonic (ES cells), 705, 706f
Sterile polyploids, 144–145
Sterilization, for eugenics, 16
Steroid hormones, 651, 651f
Sticky (cohesive) ends of DNA, 263, 264f
Stop codons, *see* Termination codons
Streptococcus pneumoniae, *see* Pneumococci
Stress, and p53 protein, 761–764, 763f
Structural genes, 249, 400–401, 618
control of expression, 618–620, 619f
in operons, 620–622, 621f, 622f, 624f
STSs (sequence-tagged sites), 555, 557, 558f, 561
Substrate binding site, on enzymes, 639–640, 641f
Sugars, in DNA vs. RNA, 225, 226f
Supercoils, in DNA, 232–233, 232f, 234f, 275–277, 276f, 277f, 278f, 279
Superhelix, 240, 241f
Suppressor mutations, 345, 352–354, 363, 363f
Suppressor-sensitive mutants, mutations, 369
Suppressor T cells, 731, 731t, 732f, 738
Suppressor tRNAs, 353f, 354
Survivability, as phenotype, 782
Survival rate, and Hardy-Weinberg principle, 780
Sxl (Sex-lethal) gene, protein, in *Drosophila*, 131, 131f, 132f, 133, 683–685, 683f, 684f, 685t
Symbiosis, 509
Symbols, 77, 78HS
see also Nomenclature
for alleles, 55–56, 56f, 64–65, 64f, 73f, 74, 74f, 171
for bacterial phenotypes and genotypes, 456
for chromosome segment deletions, 153, 153f, 155, 199
in crosses, 55, 56f, 57, 57f
for dinucleotides, 665
in pedigrees, 64–65, 64f, 109
in population genetics, 778, 780, 782, 783, 787, 792
for protein gene products, 761
for pseudogenes, 646, 646f
for restriction endonucleases, 518
in trisomy, 150
in twin studies, 723–724
Synapsis, 33f, 35
Synaptinemal complex, 35, 35f, 36, 388
Synteny
see also Linkage; X-linked genes; Sex-linked genes
conserved, in mammals, 580f, 581

T

T cell-mediated (cellular) immune response, 730, 734–735, 737f, 738–739
T cell receptors, 730, 731t, 734–735, 739
gene assembly, 746
in immune response, 737f
structure, 734–735, 735f
T cells (T lymphocytes), 730, 731, 731t, 732f
activated, 737f, 739–740
in immune response, 737f
memory T cells, 731, 731t, 732f, 737f, 739–740
types, 731
Tails, *see* Poly(A) tails
Tandem repeats, *see under* Repetitive DNA
Taq polymerase, 541, 542–543, 542f
Target site duplications, 479, 479f
Targeted gene transfers (gene replacements; gene targeting), 601
Targeting, in knockout mutations, 706
TATA box, 307, 308f, 309f
Tautomeric shifts, 372, 372f, 373f, 375–376
Tay-Sachs disease, 367HG
detecting in pre-embryos, 585–586
Telomerase, 284–285, 286f
Telomeres, 27f, 28, 244, 250f
functions, 244
length, and aging, 285–287
replication, 284–285, 286f
and retroposons, 490
structure, 244–245
Template (transcribed) strand of DNA, 300, 300f, 301f, 302f
Templates
for DNA synthesis, 17, 18f, 256, 266–267, 266f, 269, 269f, 273, 273f, 275f
for RNA synthesis, 18, 19f, 294, 300, 300f, 301f, 302f
Terminal repeats (terminally redundant DNA), *see under* Repetitive DNA
Termination (stop) codons, 343–344, 343f, 350, 350t
in RNA editing, 310–311, 311f
and suppressor mutations, 353, 353f
Terminology, *see* Symbols; Nomenclature
Testcrosses, 59, 59f
Testis-determining factor (TDF), 129, 129f, 130f
Tetrad analysis, for gene mapping, 191–194, 196–198
Tetrads, ordered, 194, 196–198, 196f
Tetraploid (4n) cells, 117
Tetraploids, 142, 145, 145f, 146f, 147
3′ overhangs, 244
3′ poly(A) tails, *see* Poly(A) tails
3′→5′ (3′ to 5′) activity, of DNA polymerase I, 267, 268f
3′→5′ (3′ to 5′) polarity, 230, 230f
3′ untranslated regions, 573, 649
Threshold traits, 95–96, 95f, 95t, 96t
Thymine, 226f
base-pairing in DNA, 228, 230f
Thymine dimers, 380, 380f, 383
light-dependent repair, 383, 384f, 385f
postreplication repair, 386–387, 387f
Ti plasmid, 606–608, 607f
Tissue-specific enhancers, in antibody gene assembly, 746, 748, 748f
Tissue-specific gene expression, 645
Tn elements (composite transposons), 479–481, 479f, 480f
Tomatoes, genetic alterations, 5, 12, 12f, 608
Topoisomerases, *see* DNA topoisomerases
TP53 gene product (p53 protein), 761–764, 763f
Trans (repulsion) configuration, 412
Trans heterozygote, 412–414, 412f
Trans-splicing, of mitochondrial mRNA, 506, 506f
Trans test, 413–414, 414f
see also Complementation test
Transcriptase, *see* Reverse transcriptase
Transcription, 18 – 19, 19f, 294 – 295, 295f, 297f, 300 – 302, 301f, 302f
see also Proteins, synthesis; Gene expression
and acetylated histone H4, 669, 670
activation mechanisms, 655
see also Enhancers; Promoters; Transcription factors
antiterminators, 635, 635f, 637f
chain elongation, 301, 301f, 302f
in eukaryotes, 308–310, 309f
in prokaryotes, 303–304, 304f
chain initiation, 302f
in eukaryotes, 307–308, 309f
in prokaryotes, 303
chain termination, 302f
in eukaryotes, 310

- in prokaryotes, 304
- and chromosomal organization, 658–668
- DNA template strand, 300, 300f, 301f, 302f
- DNA unwinding and rewinding, 302f, 303, 304f, 307
- in eukaryotes, 305–312
- in eukaryotes vs. prokaryotes, 305–306
- heat-induced, 650, 650f
- inducers, 649–653
- initiation complex, 308, 309f
- light-induced, 650–651, 650f
- in mitochondria, 505–506
- prevention, 661–665
  - and methylated DNA, 665
- in prokaryotes, 302–305
- regulation, 618–620, 619f
  - in antibody synthesis, 746
  - by attenuation, 627–629, 630f
  - in eukaryotes, 647
    - molecular mechanisms, 653–658
  - in *lac* operon, 622–623, 624f, 625–626
  - in operon model, 620–622, 621f
  - in prokaryotes vs. eukaryotes, 647
  - in *trp* operon, 626–629
- reverse transcription, 295f, 446, 447f
  - and retroposons, 490
  - and retrotransposons, 487, 490f
- sense and antisense strands, 300
- stages, 302f
- termination signals, 628–629, 630f
- terminators, rho-dependent and -independent, 304, 305f

Transcription bubble, 301–302, 302f, 303–304
Transcription factors, 653, 655–658, 656TS–657TS
- of homeodomain, 657, 696, 697
- and oncogenes, 754–755
- p53 protein, 761–764
- regulation, 662
- for RNA polymerase II (TFIIs), 308, 309f
- structural motifs, 655, 655f, 657
- of yeast, 656TS–657TS

Transcription unit, 302
Transcripts
- gene transcripts, 294, 307
  - finding protein-coding parts, 573
- primary, *see* Primary transcripts

Transducing particles, 467, 468–469, 468f, 470f, 471f
Transductants, 469–470, 471f
Transduction, 455, 467–472
- abortive, 467, 468f
- and gene mapping, 470–472, 470t
- generalized, 467, 468f
- specialized, 467–470, 470f, 471f
- testing for, 456–457, 456t, 457f

Transfection experiments, tests, 224, 756, 756f
Transfer RNAs (tRNAs), 296, 297f, 332f, 334–337
- interactions with codons, 336–337, 337f, 352–354
- precursor processing, 318–319, 319f, 335
- structure, 335–336, 336f
- suppressor mutations, 352–354, 353f
- in translational initiation, 338f, 339, 349

Transformation, 220–222, 222f, 223f, 454, 457–460, 458f, 459f
- and gene mapping, 458–460
- homeotic, 657
- testing for, 456–457, 456t, 457f
- vectors, 488TS–489TS

Transgenes, 598, 600f, 689, 705
Transgenic organisms, 598, 603–608
- animals, 603–605, 689
  - how they're made, 600f, 603, 603f
  - mice, 603–604, 604f, 705–706, 706f
- plants, 605–608
  - how they're made, 605

Transitions, 372, 374f, 375–376, 378
Translation, 18–19, 19f, 294–295, 295f, 331–332, 332f, 337–344
- *see also* Proteins, synthesis
- chain elongation, 340f-341f, 341–343, 342f
- chain termination, 343–344, 343f
- codon-tRNA interactions, 352–354
- and control of gene expression
  - in eukaryotes, 649
  - in prokaryotes, 638–639, 640f
- green fluorescent protein studies, 574–576
- initiation, 338–341, 338f
  - in eukaryotes, 340–341
  - in prokaryotes, 338–340
- in mitochondria, 505–506

Translocations, 157–159
- during evolution, 580f, 581
- and gene mapping, 206–208, 208f, 210–211
- phenotypic effects, 162
- reciprocal, 157, 158f
  - and cancer, 758–759, 758f
- Robertsonian, 161–162, 161f
- serial, 159, 160TS-161TS
- vs. compound chromosomes, 159

Transmission, genetic, 17, 56, 58, 294
Transplantation antigens, 735
Transposable genetic elements (transposons), 249, 476–495
- in bacteria, 477–481
- and chromosome structure, 491–492, 491f, 492f
- cointegrates, 480, 481f
- composite (Tn), 479–481, 479f, 480f
- discovery, 476, 478HS
- DNA repeats, 477–479, 479f
- in *Drosophila*, 484–487, 485f, 488TS–489TS, 489–490, 491, 492, 492f
- in eukaryotes, 481–487
- and evolution, 493
- genetic significance, 491–493
- in humans, 251, 490
- LINE transposable elements, 251, 490
- in maize, 482–484, 483f, 484f
- *mariner*, 487
- medical significance, 481
- and mutations, 380–381, 381f, 491, 492, 492f
- origins, 493
- *P* elements, 484–487, 485f, 492
- retroposons, 487, 490
- retrotransposons, 487, 489–491
- retrovirus-like elements, 487, 489–490, 489f, 493
- transfer between species, 481, 487, 488TS–489TS
- transposon tagging, 486

Transposition, 477, 478HS
Transposons, *see* Transposable genetic elements
Transversions, 372, 374f
Trihybrid crosses, 59
Triplet code, codons, 19, 19f, 344
Triplo-lethal gene, 162
Triplo-X (XXX) karyotype, 151, 152t
Triploids, 142
Trisomies, 150
- in humans, 150–152, 150f, 152t
- in plants, 149–150, 150f
- trisomy 21, *see* Down syndrome

trp (tryptophan) operon, 626–629, 627f
- attenuation, 627–629, 628f, 630f
- repression, 626–627
- transcription termination, 629, 630f

True-breeding strains, 54
Trypanosomes, 644–645, 663–665
Tryptophan operon, *see trp* operon
Tumor-inducing viruses, 753–755, 754t
Tumor suppressor genes, proteins, 753, 759–767, 760t
- cellular roles, 761–767
- mutations
  - and genetic pathways to cancer, 767–769
  - testing for, 766HG
- nomenclature, 761

Tumors
- *see also* Cancer
- malignant vs. benign, 751
- in plants, 605–606, 606f

24-hour (circadian) rhythms, 712
Twin correlation method, 723–724
Twins, 92, 95, 95t, 255–256, 255f
- concordance rates, 95, 95t
- and personality, 710–711, 725–726, 725t
- in studies of complex traits, 723–724

Two-hit hypothesis, 759–761, 759f
Tyrosine, mutations, 366–369, 368f
Tyrosine kinases, 754, 756

U

Ultraviolet repair proteins, 385, 385f
Ultraviolet (UV) radiation, as mutagen, 379, 380, 380f
Untranslated regions, 573
Upstream direction, 302
Upstream activating sequences (UASs), 656TS–657TS
Uracil, 226f

V

Variability
- *see also* Variation
- genetic equilibrium level ($\hat{H}$), 792, 800
  - and heterozygosity, 787, 792

Variable number tandem repeats (VNTRs), *see under* Repetitive DNA
Variable regions
of antibody chains, 732, 733f
of T cell receptors, 735, 735f
Variance, 98–99
and analysis of quantitative traits, 99–107
components, 102
genetic, 102
vs. environmental, 100–101
partitioning for heritability, 100–101
phenotypic, 100
Variant surface glycoprotein genes (*vsg* genes), proteins (VSGs), 645, 663–665, 663f, 664f
Variation
see also Variability
allelic, 73–80
symbols for, 73f, 74, 74f
continuous, vs. discontinuous, 793
continuous phenotypic, 86, 92, 793–794
see also Inheritance, complex patterns; Inheritance, polygenic; Multifactorial traits; Complex traits; Quantitative traits;
cytogenetic, 142–143
genetic
and human evolution, 800–801
and meiosis, 37
at molecular level, 794–795, 795f, 795t
and mutation, 359–360
in natural populations, 793–795
and population genetics, 777
Variegation, position-effect, 660–661
Vectors, 489TS
cloning vectors, 518
artificial chromosomes (YACs, BACs), 524–525, 525f
bacteriophages, 523, 524f
cosmids, 524, 524f
DNA amplification in, 522–525
features, 522, 523f
inserting DNA, 526–527, 526f, 527f
phagemids, 531–534, 532f, 534f
plasmids, 522–523, 523f
pUC118, pUC119, 532–534, 532f, 534f
shuttle vectors, 524, 525f
use in finding genes, 587, 588f
for gene therapy (gene-transfer vectors), 598, 599f, 600f
plasmids
as cloning vectors, 522–523, 523f
for human growth hormone production in bacteria, 601, 602f
for transformation, 488TS-489TS
Velocity density gradients, 260TS-261TS
Vertebrates
body segmentation, 696
development, 703–707
homologs of invertebrate genes, 703–704, 704f, 705f
immune system, 728–749
mitochondrial DNA, 503–504
Viral diseases, genetically engineered resistance, 605, 608–610
Viral oncogenes, *see under* Oncogenes
Virion, of HIV, 445f
Viruses, 222–223
see also RNA viruses; Retroviruses; Bacteriophage *entries*; HIV
and cancer, 753
chromosome structure, 233–234
genetics, 429–452
how classified, 432
origin, 431
of prokaryotes, *see* Bacteriophage *entries*
tumor-inducing, 753–755, 754t
as vectors for gene therapy, 598, 599f, 600f
Viscoelastometry, 238
v-onc genes, *see* Oncogenes, viral
v-ras, *see ras* genes

W

Watson, James D., photo, 228f
Western blotting, 540–541, 540f
Wild-type genes, alleles, 7, 74
Wing development, in insects, 675–676, 675f, 676f, 700TS-701TS
Wobble hypothesis, 352, 352f, 352t

X

X chromosomes, 117–119, 118f
see also Sex chromosomes; Barr bodies
absence, *see* XO state
of *Drosophila*
loss of, 687, 687f
map, 167, 167f
polytene, maps and mapping, 183f, 198–200, 199f, 200f, 202TS–203TS
fused, 159, 161, 161f
genes on, *see* X-linked genes
and homosexuality, 717, 718HG–719HG
hyperactivation, 668, 668f, 669–670, 685
hypoactivation, 668, 668f, 670, 686
inactivation, 133–134, 133f, 134f, 665, 668–669, 668f
and Duchenne muscular dystrophy, 208, 209f
multiple copies, 134, 151, 152t
and nondisjunction, 121, 123, 123f
translocation with autosome, and gene mapping, 206–208, 208f
X:A (X:autosome) ratio
in *Caenorhabditis*, 686
in *Drosophila*, 131, 131t, 682–683, 682f
X-gal color test, 533, 533f
X-linked genes, 119–121
see also Sex-linked genes; Synteny
dosage compensation, 132–134, 132f, 133f, 153, 668–670, 668f, 685, 686
and Duchenne muscular dystrophy, 208, 209f
in *Drosophila*, 119–121, 119f, 120f, 121f, 131
map, 167, 167f, 180f
in humans, 126–128, 126f, 127f
dominant disorders, 127–128
recessive disorders, 126
calculating risk, 127, 127f
X-linked recessive mutations, 126, 364, 364f, 370–371, 371f
X rays, 370–371, 371f, 378–379, 379f
XO state, 117–118, 129, 152–153
in humans (45,X), 152–153, 152f
from nondisjunction, 121, 123, 123f
XX males, 129–130, 130f
XX state, 117–118, 129
XXX state, 151, 152t
from nondisjunction, 121, 123, 123f
XXY state, 129, 151, 152t
from nondisjunction, 121, 123f
XY females, 129–130, 130f
XY state, 118, 129
XYY state, 151, 152t

Y

Y chromosome, 118, 118f
see also Sex chromosomes
genes on, 128
multiple copies, 151
Y-linked genes, in humans, 128
Yeast
see also Saccharomyces
mating type switching, 662–663, 663f
metabolic defects, 502–503, 502f, 503f
mitochondrial DNA, 504
petite mutants, 502–503, 502f, 503f
transcription factors, 656TS–657TS
Yeast artificial chromosomes (YACs), 525, 525f
YO state, from nondisjunction, 123, 123f

Z

Z-DNA, 231, 231f, 231t
Zea mays, *see* Maize
Zinc finger motif, 655, 655f
Zipper motif, 655f, 657
Zoo-FISH experiments, 579
Zygote, 39, 40
Zygotic genes, role in development, 695–703